C0-AKG-524

Neurobiology of Cerebrospinal Fluid

1

Neurobiology of Cerebrospinal Fluid

1

Edited by

JAMES H. WOOD, M.D.

Division of Neurosurgery
University of Pennsylvania School of Medicine
Philadelphia, Pennsylvania

PLENUM PRESS NEW YORK AND LONDON

Library of Congress Cataloging in Publication Data

Main entry under title:

Neurobiology of cerebrospinal fluid.

Includes index.

1. Cerebrospinal fluid. 2. Central nervous system—Diseases—Diagnosis. 3. Neurotransmitters. I. Wood, James H., 1948- [DNLM: 1. Cerebrospinal fluid—Physiology. WL203 N494]
RB55.N48 616.8'04'0756 79-21731
ISBN 0-306-40369-2

A Division of Plenum Publishing Corporation
227 West 17th Street, New York, N.Y. 10011

Printed in the United States of America

Contributors

George S. Allen, M.D., Ph.D.
Associate Professor
Department of Neurosurgery
The Johns Hopkins University School of Medicine and Hospital
Baltimore, Maryland 21205

James C. Ballenger, M.D.
Director of Research
Associate Professor
Department of Behavioral Medicine and Psychiatry
University of Virginia School of Medicine
Charlottesville, Virginia 22908

Bruce J. Biller, M.D.
Instructor
Department of Medicine
Massachusetts Institute of Technology
Cambridge, Massachusetts 02139
Division of Endocrinology
Department of Medicine
Tufts–New England Medical Center Hospital
Boston, Massachusetts 02111

W. Archie Bleyer, M.D.
Associate Professor
Division of Hematology–Oncology
Department of Pediatrics
Children's Orthopedic Hospital Medical Center
Seattle, Washington 98105

Floyd E. Bloom, M.D.
Director
Arthur Vining Davis Center for Behavioral Neurobiology
The Salk Institute
Professor
Department of Psychiatry
University of California School of Medicine
San Diego, California 92112

Malcolm B. Bowers, Jr., M.D.
Professor
Department of Psychiatry
Yale University School of Medicine
New Haven, Connecticut 06510

Benjamin Rix Brooks, M.D.
Assistant Professor
Department of Neurology
The Johns Hopkins University School of Medicine
Baltimore, Maryland 21205
Consultant, Medical Neurology Branch
National Institute of Neurological and Communicative Disorders and Stroke
National Institutes of Health
Bethesda, Maryland 20205
Consultant, Neuropharmacologic Drug Advisory Council
Food and Drug Administration
Rockville, Maryland

Derek A. Bruce, M.B., Ch.B.
Associate Neurosurgeon
Children's Hospital of Philadelphia
Associate Professor
Division of Neurosurgery
University of Pennsylvania School of Medicine
Philadelphia, Pennsylvania 19104

Saul W. Brusilow, M.D.
Professor
Department of Pediatrics
The Johns Hopkins University School of Medicine
Baltimore, Maryland 21205

Dennis E. Bullard, M.D.
Division of Neurosurgery
Fellow
Department of Microbiology and Immunology
Duke University Medical Center
Durham, North Carolina 27710

Barry B. Burns, Ph.D.
Associate Professor
Department of Environmental Medicine
The Johns Hopkins University School of Hygiene and Public Health
Baltimore, Maryland 21205

Ian J. Butler, M.D., F.R.A.C.P.
Associate Professor
Departments of Neurology, Pediatrics, and Neurobiology and Anatomy
The University of Texas Medical School at Houston
Houston, Texas 77025

Thomas N. Chase, M.D.
Director
Intramural Research
National Institute of Neurological and Communicative Disorders and Stroke
National Institutes of Health
Bethesda, Maryland 20205

Donald J. Cohen, M.D.
Professor
Departments of Pediatrics, Psychiatry, and Psychology
Child Study Center
Yale University School of Medicine
New Haven, Connecticut 06510

Ronald J. Cohen, M.D.
Department of Neurosurgery
The Johns Hopkins University School of Medicine and Hospital
Baltimore, Maryland 21205

Steven R. Cohen, Ph.D.
Assistant Professor
Neurochemistry Laboratory
Department of Neurology
The Johns Hopkins University School of Medicine
Baltimore, Maryland 21205

Jay D. Cook, M.D.
Chief, Neuromuscular Section
Assistant Professor
Department of Neurology
University of Texas Southwestern Medical School
Chief, Muscle Disease Section
Neurology Service
Veterans Administration Hospital
Chief of Child Neurology
Scottish Rite Hospital for Crippled Children
Dallas, Texas 75216

Francisco Correa-Paz, M.D.
Assistant Professor
Neuroradiology Section
Department of Radiology and Radiological Sciences
Vanderbilt University School of Medicine
Nashville, Tennessee 37232

Rex W. Cowdry, M.D.
Clinical Associate
Clinical Psychobiology Branch
National Institute of Mental Health
National Institutes of Health
Bethesda, Maryland 20205

Robert W. P. Cutler, M.D.
Professor
Department of Neurology
Stanford University Medical Center
Stanford, California 94305

Carol Czerwinski, Ph.D.
Assistant Professor
Department of Obstetrics and Gynecology
Uniformed Services University of Health Sciences
School of Medicine
Bethesda, Maryland 20014

Maria Diaz, B.S.
Research Chemist
Endocrinology Division
Department of Medicine
National Naval Medical Center
Bethesda, Maryland 20014

Burton P. Drayer, M.D.
Associate Professor
Division of Neuroradiology
Department of Radiology
Duke University Medical Center
Durham, North Carolina 27710

Michael H. Ebert, M.D.
Chief
Experimental Therapeutics Section
Laboratory of Clinical Science
National Institute of Mental Health
National Institutes of Health
Bethesda, Maryland 20205

W. King Engel, M.D.
Chief
Medical Neurology Branch
National Institute of Neurological and Communicative Disorders and Stroke
National Institutes of Health
Bethesda, Maryland 20205

S. J. Enna, Ph.D.
Associate Professor
Departments of Pharmacology and Neurobiology and Anatomy
The University of Texas Medical School at Houston
Houston, Texas 77025

Mel H. Epstein, M.D.
Associate Professor
Director of Pediatric Neurosurgery
Department of Neurosurgery
The Johns Hopkins University School of Medicine
Baltimore, Maryland 21205

Arthur M. Feldman, Ph.D
Clinical Instructor
Department of Otolaryngology
Louisiana State University School of Medicine in Shreveport
Shreveport, Louisiana 71130

Alan S. Fleischer, M.D.
Chief
Division of Neurosurgery
Grady Memorial Hospital
Atlanta, Georgia 30303
Associate Professor
Division of Neurosurgery
Emory University School of Medicine
Atlanta, Georgia 30322

William J. Flor, Ph.D.
Chief
Department of Experimental Pathology
Armed Forces Radiobiological Research Institute
National Naval Medical Center
Bethesda, Maryland 20014

Leon P. Georges, M.D.
Professor and Chairman
Endocrinology Division
Department of Medicine
National Naval Medical Center
Bethesda, Maryland 20014

Bruce S. Glaeser, Ph.D.
Research Associate
Department of Nutrition and Food Science
Massachusetts Institute of Technology
Cambridge, Massachusetts 02139

Frederick K. Goodwin, M.D.
Chief
Clinical Psychobiology Branch
National Institute of Mental Health
National Institutes of Health
Bethesda, Maryland 20205

Robert G. Grossman, M.D.
Professor and Chairman
Division of Neurosurgery
Baylor University School of Medicine
Houston, Texas 77030

Bernard Haber, Ph.D.
Chief, Neurochemistry Section
The Marine Biomedical Institute
Galveston, Texas 77550
Associate Professor
Departments of Neurology and Biochemistry
University of Texas Medical Branch
Galveston, Texas 77550

Hajime Handa, M.D.
Professor and Chairman
Department of Neurosurgery
Kyoto University School of Medicine
Kyoto, Japan

Theodore A. Hare, Ph.D.
Associate Professor
Department of Pharmacology
Jefferson Medical College of Thomas Jefferson University
Philadelphia, Pennysylvania 19107

Robert M. Herndon, M.D.
Professor
Department of Neurology
Director
Center for Brain Research
University of Rochester School of Medicine and Dentistry
Rochester, New York 14642

Suellen A. Hill, R.N.
Research Associate
Division of Neurosurgery
The University of Texas Health Science Center at Dallas
Dallas, Texas 75235

Marc E. Horowitz, M.D.
Clinical Associate
Pediatric Oncology Branch
National Cancer Institute
National Institutes of Health
Bethesda, Maryland 20205

Ivor M. D. Jackson, M.D.
Associate Professor
Division of Endocrinology
Department of Medicine
Tufts–New England Medical Center Hospital
Boston, Massachusetts 02111

A. Everette James, Jr., Sc.M., J.D., M.D.
Professor and Chairman
Department of Radiology and Radiological Sciences
Vanderbilt University School of Medicine
Nashville, Tennessee 37232

David C. Jimerson, M.D.
Staff Psychiatrist
Laboratory of Clinical Science
Biological Psychiatry Branch
National Institute of Mental Health
National Institutes of Health
Bethesda, Maryland 20205

Burk Jubelt, M.D.
Fellow
Neurovirology Laboratory
Department of Neurology
The Johns Hopkins University School of Medicine
Baltimore, Maryland 21205

Ronald Kartzinel, M.D., Ph.D.
Director
Division of Neuropharmacology
Food and Drug Administration
Department of Health, Education and Welfare
Rockville, Maryland

David C. Klein, Ph.D.
Chief
Section on Neuroendocrinology
Laboratory of Developmental Neurobiology
National Institute of Child Health and Human Development
National Institutes of Health
Bethesda, Maryland 20205

Enrique L. Labadie, M.D.
Chief
Neurology Service
Tucson Veterans Administration Medical Center
Tucson, Arizona 85723
Assistant Professor
Department of Neurology
University of Arizona School of Medicine
Tucson, Arizona 85724

C. Raymond Lake, M.D., Ph.D.
Associate Professor
Departments of Psychiatry and Pharmacology
Uniformed Services University of Health Sciences
School of Medicine
Bethesda, Maryland 20014

John P. Laurent, M.D.
Assistant Professor
Division of Neurosurgery
St. Luke's–Texas Children's Hospital
Baylor University School of Medicine
Houston, Texas 77030

Hans H. Loeschcke, M.D.
Professor
Institut für Physiologie
Ruhr-Universität
Bochum, Germany

D. Lynn Loriaux, M.D., Ph.D.
Head, Endocrine Service Unit
Developmental Endocrinology Section
Endocrinology and Reproduction Research Branch
National Institute of Child Health and Human Development
National Institutes of Health
Bethesda, Maryland 20205

Thomas G. Luerssen, M.D.
Department of Neurosurgery
Indiana University School of Medicine
Indianapolis, Indiana 46223

N. V. Bala Manyam, M.D.
Neurology Service
Veterans Administration Hospital
Wilmington, Delaware 19805
Instructor
Department of Neurology
Jefferson Medical College of Thomas Jefferson University
Philadelphia, Pennsylvania 19107

Samuel P. Marynick, M.D.
Assistant Professor
Division of Endocrinology
Department of Internal Medicine
Baylor University Medical Center
Dallas, Texas 75246
Assistant Clinical Professor
Department of Internal Medicine
University of Texas Southwestern Medical School
Dallas, Texas 75235

Guy M. McKhann, M.D.
Professor and Chairman
Department of Neurology
The Johns Hopkins University School of Medicine
Baltimore, Maryland 21205

John Stirling Meyer, M.D.
Professor
Department of Neurology
Baylor College of Medicine
Chief
Cerebral Blood Flow Laboratory
Veterans Administration Medical Center
Houston, Texas 77211

Edward A. Neuwelt, M.D.
Assistant Professor
Division of Neurosurgery
The University of Texas Health Science Center at Dallas
Dallas, Texas 75235

Ragnar Norrby, M.D., Ph.D.
Associate Professor
Department of Infectious Diseases
East Hospital
University of Gothenburg
S-41685 Gothenburg, Sweden

Gary R. Novak, L.A.T.
Associate Director for Research
Department of Radiology and Radiological Sciences
Vanderbilt University School of Medicine
Nashville, Tennessee 37232

Shinichiro Okamoto, M.D.
Assistant Professor
Department of Neurosurgery
Kyoto University School of Medicine
Kyoto, Japan

William H. Oldendorf, M.D.
Professor and Lasker Award Recipient
Department of Neurology
Center for Health Sciences
University of California at Los Angeles School of Medicine
Los Angeles, California 90024
VA Brentwood Medical Center
Los Angeles, California 90073

Kunihiko Osaka, M.D.
Instructor
Department of Neurosurgery
Kyoto University School of Medicine
Kyoto, Japan

Mark J. Perlow, M.D.
Staff Neurologist
Laboratory of Clinical Psychopharmacology
National Institute of Mental Health
National Institutes of Health
St. Elizabeth's Hospital
Washington, D.C. 20032

John D. Pickard, M.D.
Assistant Professor
Division of Neurosurgery
Institute of Neurological Sciences
Southern General Hospital
GB Glasgow, G51 4TF, Scotland

David G. Poplack, M.D.
Senior Investigator
Pediatric Oncology Branch
National Cancer Institute
National Institutes of Health
Bethesda, Maryland 20205

Kalmon D. Post, M.D.
Assistant Professor
Department of Neurosurgery
Tufts–New England Medical Center Hospital
Boston, Massachusetts 02111

Robert M. Post, M.D.
Chief
Section on Psychobiology
Biological Psychiatry Branch
National Institute of Mental Health
National Institutes of Health
Bethesda, Maryland 20205

Steven M. Reppert, M.D.
Clinical Associate
Section on Neuroendocrinology
Laboratory of Developmental Neurobiology
National Institute of Child Health and Human Development
National Institutes of Health
Bethesda, Maryland 20205

Gary L. Robertson, M.D.
Professor
Division of Endocrinology
Department of Medicine
University of Chicago
Pritzker School of Medicine
Chicago, Illinois 60637

Arthur E. Rosenbaum, M.D.
Chief
Division of Neuroradiology
Professor
Department of Radiology
University of Pittsburgh Health Center
Pittsburgh, Pennsylvania 15261

S. Clifford Schold, M.D.
Assistant Professor
Division of Neurology
Duke University Medical Center
Durham, North Carolina 27710

David S. Segal, Ph.D.
Professor
Department of Psychiatry
University of California School of Medicine
San Diego, California 92112

William E. Seifert, Ph.D.
Assistant Professor
Department of Biochemistry and Molecular Biology
The University of Texas Medical School at Houston
Houston, Texas 77025

Bennett A. Shaywitz, M.D.
Associate Professor
Departments of Pediatrics and Neurology
Yale University School of Medicine
New Haven, Connecticut 06510

Ira Shoulson, M.D.
Assistant Professor
Division of Neurology
University of Rochester School of Medicine
Rochester, New York 14642

Frederick A. Simeone, M.D.
Chief
Division of Neurosurgery
Pennsylvania Hospital
Philadelphia, Pennsylvania 19107
Associate Professor
Division of Neurosurgery
University of Pennsylvania School of Medicine
Philadelphia, Pennsylvania 19104

Frederick H. Sklar, M.D.
Associate Professor
Division of Neurological Surgery
University of Texas Health Science Center at Dallas
Dallas, Texas 75235

Jonas Sode, M.D.
Chief of Medicine
Medical Service
Carbondale Veterans Administration Hospital
Carbondale, Illinois 62901
Professor
Department of Internal Medicine
Southern Illinois University School of Medicine
Carbondale, Illinois 62901

David E. Sternberg, M.D.
Clinical Associate
Section on Neuropsychopharmacology
Biological Psychiatry Branch
National Institute of Mental Health
National Institutes of Health
Bethesda, Maryland 20205
Assistant Professor
Department of Psychiatry
Georgetown University School of Medicine
Washington, D.C. 20007

Ernst-Peter Strecker, M.D.
Director of Laboratory for Research
Institut für Röntgendiagnostik der Universität Freiburg
Freiburg, West Germany

Paul F. Teychenne, M.D.
Associate Professor
Department of Neurology
George Washington University Medical School
Washington, D.C. 20037

George T. Tindall, M.D.
Professor and Chairman
Division of Neurosurgery
Emory University School of Medicine
Atlanta, Georgia 30322

John L. Trotter, M.D.
Assistant Professor
Department of Neurology and Neurosurgery (Neurology)
Barnes Hospital Plaza
Washington University School of Medicine
St. Louis, Missouri 63110

Daniël P. van Kammen, M.D., Ph.D.
Unit Chief
Section in Neuropsychopharmacology
Biological Psychiatry Branch
National Institute of Mental Health
National Institutes of Health
Bethesda, Maryland 20205

Phillip E. Vinall, B.S.
Research Associate
Division of Neurosurgery
Pennsylvania Hospital
Philadelphia, Pennsylvania 19107

K. M. A. Welch, M.B. Ch.B., M.R.C.P. (U.K.)
Associate Professor
Department of Neurology
Baylor College of Medicine
Neurosensory Center
Houston, Texas 77030

James H. Wood, M.D.
Division of Neurosurgery
University of Pennsylvania School of Medicine and Hospital
Philadelphia, Pennsylvania 19104
Consultant, Pediatric Oncology Branch
National Cancer Institute
National Institutes of Health
Bethesda, Maryland 20205

J. Gerald Young, M.D.
Assistant Professor
Departments of Pediatrics and Psychiatry
Child Study Center
Yale University School of Medicine
New Haven, Connecticut 06510

Michael G. Ziegler, M.D.
Department of Medicine
University of California
San Diego, California 92103

Preface

Physiologic compartmentalization effectively isolates the central nervous system from the rest of the body. This isolation not only provides protection of its delicate function from aberrant peripheral influences but also impedes its diagnostic evaluation. Cerebrospinal fluid (CSF) bathes the brain and spinal cord, is in dynamic equilibrium with its extracellular fluid, and tends to reflect the state of health and activity of the central nervous system. CSF examination is the most direct and popular method of assessing the central chemical and cellular environment in the living patient or mammal.

The purpose of this multidisciplined reference text is to provide the sophisticated knowledge of CSF physiology and pathology necessary for the meaningful interpretation of data obtained by various types of CSF analysis. The methodology for reliable CSF collection, storage, preparation, and analysis is discussed with respect to individual, somatotropic, chronologic, endorcinologic, pharmacologic, and possible artifactual variations in CSF composition. These essential aspects, which ensure the validity of CSF data, are presented to aid the investigator in clinical and experimental protocol formulation and in elimination of possible sources of error.

This volume has been written by international experts in each field of CSF investigation and covers degenerative, convulsive, cerebrovascular, traumatic, immunologic, and demyelinating disorders with respect to historical as well as recent advances in diagnosis, evaluation, and therapy. Cellular and chemical data of patients with infections of the central nervous system are categorized to aid the investigator or clinician in his differential diagnosis and choice of antimicrobial therapy. This reference text also examines the use of tumor markers and discusses antineoplastic drug pharmacology in order to provide insight into the latest advances in the evaluation and treatment of patients with cancer of the central nervous system. The significance of CSF neurotransmitters, metabolites, hormones, peptides, and endorphins is discussed in detail with respect to neurologic and psychiatric disorders.

More specifically, each chapter is extensively referenced and contains ample charts of original data, summary charts, and anatomical diagrams. Detailed illustrations of experimental and clinical techniques have been included to facilitate their practical application.

Intimate experience with each aspect of the broad field of CSF pathophysiology is required for the preparation of a reference text as comprehensive as *Neurobiology of Cerebrospinal Fluid.* Recognition of the impossibility of a single individual's or small group of participants' possessing this first-hand knowledge necessitated the formation of an extensive international contributorship. In addition, a network of readily available sources of consultation enabled each chapter to be critically reviewed by two experts in the respective field.

As editor, I wish to acknowledge the profound contributions of Dr. Benjamin Rix Brooks of the Department of Neurology at The Johns Hopkins University School of Medicine. His excellent editorial criticisms, based upon his broad, intense knowledge of neuroscience, and his superb chapter contributions ensured the success of this project. In addition, appreciation must be expressed to my mentors, Dr. A. Everette James of the Vanderbilt University School of Medicine and Dr. Robert M. Post of the National Institute of Mental Health, for their chapter contributions and for introducing me to the physiology and neurochemistry of CSF. Dr. S. J. Enna of the University of Texas Medical Center at Houston is to be thanked for his consultation involving the contractual arrangements.

Appreciation is to be extended to Drs. Theodore A. Hare, Robert D. Myers, Joseph D. Fenstermacher, Victor A. Levin, Michael Pollay, Michael H. Ebert, T. J. Malkinson, Hymie Anisman, and a British neurologist for the preparation of some illustrations in this volume. I am most indebted to the many contributing authors and reviewers for their preparation of such excellent chapters and their punctual observance of submission deadlines. The editorial and production staff at Plenum Publishing Corporation is to be commended for their superb execution of this first volume of *Neurobiology of Cerebrospinal Fluid*.

I am, of course, grateful to my wife, Mary, and my parents, Naomi and Harold, for their encouragement and support.

James H. Wood, M.D.

Philadelphia

Introduction

Cerebrospinal fluid (CSF), the bathing solution of the central nervous system, has great clinical and basic scientific importance. Because of its availability by relatively innocuous lumbar puncture for modern biochemical analysis, CSF offers a valuable, although indirect, vehicle for defining brain malfunctions. Inferences from CSF data, however, require a clear understanding of the relationship between CSF and brain.

Much evidence exists to support the concept that CSF is an extension of the brain and spinal extracellular fluid (ECF) serving to meet certain of the unique mechanical and chemical needs of the very specialized and fragile central nervous system. Tracer substances introduced into CSF readily diffuse without impediment into the interstitial ECF of brain and spinal cord. This observation implies that the CSF is in continuity with the interstitial ECF and is supported by histological studies.

The CSF could thus be expected to reflect the composition of the brain interstitial fluid within the limits imposed by rates of diffusion across the ependymal or pial surfaces. The generous convolutions of the cerebral and cerebellar cortices considerably reduce this diffusion distance. Almost all of adult human gray matter is within a few millimeters of a CSF interface, excepting a few regions in the basal ganglia and thalamus. The farther a portion of ECF is from the CSF compartment, the more its composition is expected to be dissimilar from that of CSF.

Solutes in the ECF of all tissues must be returned to the blood plasma by some nonspecific mechanism that is independent of their relative concentrations in CSF and plasma. This required transfer of solute passing into plasma may occasionally proceed against a concentration gradient and is accomplished systemically by the bulk flow of lymph into the venous system. In nonneural tissues, a considerable net flux outward is possible because the small solutes can move with the water in response to hydrostatic gradients. This lymph can be formed from blood plasma and travel through lymph channels coalescing ultimately to the thoracic duct, which "injects" the lymph into venous blood. Such bulk-flow injections allow ECF solutes of any concentration to return to the blood.

However, no recognizable lymphatic system such as found systemically exists in the central nervous system to react to foreign constituents entering tissue ECF. Accordingly, the brain, encased in bone and having singularly impermeable capillaries, has little exposure to foreign protein. The blood–brain barrier (BBB) is quite impermeable to the osmotically active solutes; therefore, a large net flux of water from plasma into brain ECF is unlikely. Thus, the brain might have evolved a mechanism for turnover of ECF solutes that is considerably different from the mechanisms by which other organs do so.

The return of brain ECF solutes to blood appears to be a more indirect mechanism. In the absence of a large net flux of water moving through brain ECF in an "irrigating" fashion, the brain appears to generate most of its ECF-cleansing fluid (the CSF) from the choroid plexus. The CSF then circulates about the ventricular and subarachnoid spaces. In addition to the mixing due to its formation, flow, and resorption, the major homogenizing force acting on CSF is the to-and-fro surging from changes in intracranial blood volume as a result of heart action, respiration, swallowing, and other changes in intrathoracic pressure. The CSF acts as a "sink" for any ECF solute that cannot otherwise be disposed of through the BBB. These solutes, after dispersion throughout the CSF system, are injected through several hundred "thoracic ducts of the brain," the arachnoid villi. The forward bulk flow into the dural sinus needs only to be faster than the solutes can diffuse from blood to CSF to prevent their central accumulation.

The rate of CSF formation by the choroid plexus in healthy, intact mammals is controversial. Various experimental arrangements have demonstrated the existence of a substantial potential for extrachoroidal formation of CSF that conceivably might originate from the brain capillary wall. The uniquely large mitochondrial content of brain capillary endothelial cells implies a metabolic capability greater than that of systemic capillaries that could be employed to pump electrolytes along with water into the brain. Whether or not these mechanisms are operant *in vivo* is speculative.

Obliteration of the mixed layer of CSF in the cerebral cortical subarachnoid space by enlargement of the brain would be expected to impair the ability of CSF to act as a sink for cortical metabolic end products unable to pass through the BBB into blood. The dementia of normal pressure hydrocephalus conceivably could be an autointoxication due to a failure of this CSF sink function. The occasional dramatic clearing of the dementia coincident with restoration of cortical subarachnoid circulation after ventricular shunting is compatible with this hypothesis.

A teleology of CSF would not be complete without mention of its function of mechanically supporting the brain. The subarachnoid and ventricular spaces filled with CSF are actually "lakes of ECF" in continuity with interstitial ECF. The subarachnoid CSF (perhaps a total of 100–200 ml) floats the brain and distributes the shear forces that might otherwise injure the brain during rapid head movement. The evolution of this flotation system and of the internal brain accelerators (falx cerebri and tentorium cerebelli) have made possible the parallel evolution of huge cerebral hemispheres in man, while allowing a highly mobile neck. The ability of the brain to tolerate high rotary accelerations despite its low intrinsic stiffness and great sensitivity to shearing forces is dependent upon this CSF flotation effect, which, in essence, frees the brain from the influence of gravity.

Just as the whale could evolve into the largest mammal because it was freed from gravity in its water environment, so the human brain and that of other large mammals could evolve to their present sizes. This degree of physical brain development is far beyond that attainable if it were subjected to the mechanical forces experienced by systemic viscera not floating in a "perfect" fluid environment surrounded by a rigid bony encasement of just the optimal size.

William H. Oldendorf, M.D.
Department of Neurology
Center for Health Sciences
University of California at Los Angeles School of Medicine
Los Angeles, California 90024

VA Brentwood Medical Center
Los Angeles, California 90073

Contents

CHAPTER 7: Technical Aspects of Clinical and Experimental Cerebrospinal Fluid Investigations

James H. Wood

CHAPTER 8: Cerebrospinal Fluid Amine Metabolites and Probenecid Test

Michael H. Ebert, Ronald Kartzinel, Rex W. Cowdry, and Frederick K. Goodwin

CHAPTER 9: Extracellular Cyclic Nucleotide Metabolism in Human Central Nervous System

Benjamin Rix Brooks, James H. Wood, Maria Diaz, Carol Czerwinski, Leon P. Georges, Jonas Sode, Michael H. Ebert, and W. King Engel

CHAPTER 14: **Cerebrospinal Fluid γ-Aminobutyric Acid: Correlation with Cerebrospinal Fluid and Blood Constituents and Alterations in Neurological Disorders**

S. J. Enna, Michael G. Ziegler, C. Raymond Lake, James H. Wood, Benjamin Rix Brooks, and Ian J. Butler

CHAPTER 15: **Cerebrospinal Fluid Studies in Parkinson's Disease: Norepinephrine and γ-Aminobutyric Acid Concentrations**

Paul F. Teychenne, C. Raymond Lake, and Michael G. Ziegler

CHAPTER 16: **Neurochemical Alterations in Parkinson's Disease**

Thomas N. Chase

CHAPTER 17: **Cerebrospinal Fluid Monoamine Metabolites in Neurological Disorders of Childhood**

Bennett A. Shaywitz, Donald J. Cohen, and Malcolm B. Bowers, Jr.

CHAPTER 22: **Response of Extraparenchymal Cerebral Arteries to Biochemical Environment of Cerebrospinal Fluid**
Frederick A. Simeone, Phillip E. Vinall, and John D. Pickard

CHAPTER 23: **Subarachnoid Erythrocytes and Their Contribution to Cerebral Vasospasm**
Kunihiko Osaka, Hajime Handa, and Shinichiro Okamoto

CHAPTER 24: **Neurochemical Alterations in Cerebrospinal Fluid in Cerebral Ischemia and Stroke**
K. M. A. Welch and John Stirling Meyer

CHAPTER 25: **Cerebrospinal Fluid Cyclic Nucleotide Alterations in Traumatic Coma**
Alan S. Fleischer and George T. Tindall

CHAPTER 26: **Acetylcholine Metabolism in Intracranial and Lumbar Cerebrospinal Fluid and in Blood**
Bernard Haber and Robert G. Grossman

CHAPTER 27: Cerebrospinal Fluid Pressure Dynamics and Brain Metabolism

Derek A. Bruce

CHAPTER 28: Non-Steady-State Measurements of Cerebrospinal Fluid Dynamics: Laboratory and Clinical Applications

Frederick H. Sklar

CHAPTER 29: Experimental Studies Relating to Diagnostic Imaging in Disorders of Cerebrospinal Fluid Circulation

A. Everette James, Jr., Gary R. Novak, Ernst-Peter Strecker, Barry B. Burns, Francisco Correa-Paz, and William J. Flor

CHAPTER 30: Dynamics of the Cerebrospinal Fluid System as Defined by Cranial Computed Tomography

Burton P. Drayer and Arthur E. Rosenbaum

CHAPTER 31: Cerebrospinal Fluid Alterations Associated with Central Nervous System Infections

Enrique L. Labadie

CHAPTER 32: Penetration of Antimicrobial Agents into Cerebrospinal Fluid: Pharmacokinetic and Clinical Aspects

Ragnar Norrby

CHAPTER 33: Pathophysiology of Cerebrospinal Fluid Immunoglobulins

John L. Trotter and Benjamin Rix Brooks

CHAPTER 34: Myelin Basic Protein in Cerebrospinal Fluid: Index of Active Demyelination

Steven R. Cohen, Benjamin Rix Brooks, Burk Jubelt, Robert M. Herndon, and Guy M. McKhann

CHAPTER 35: Electron-Microscopic Studies on Cerebrospinal Fluid Sediment

Robert M. Herndon

CHAPTER 36: Lymphocyte Subpopulations in Human Cerebrospinal Fluid: Effects of Various Disease States and Immunosuppressive Drugs

Jay D. Cook and Benjamin Rix Brooks

CHAPTER 37: Intrathecal Lymphocyte Infusions: Clinical and Animal Toxicity Studies

Edward A. Neuwelt and Suellen A. Hill

CHAPTER 38: Cerebrospinal Fluid Analysis in Central Nervous System Cancer

S. Clifford Schold and Dennis E. Bullard

CHAPTER 39: Pharmacology of Antineoplastic Agents in Cerebrospinal Fluid

David G. Poplack, W. Archie Bleyer, and Marc E. Horowitz

CHAPTER 40: Cerebrospinal Fluid Melatonin

Steven M. Reppert, Mark J. Perlow, and David C. Klein

CHAPTER 41: Cerebrospinal Fluid Pituitary Hormone Concentrations in Patients with Pituitary Tumors

Kalmon D. Post, Bruce J. Biller, and Ivor M. D. Jackson

CHAPTER 42: Cerebrospinal Fluid Steroid Hormones

Samuel P. Marynick, James H. Wood, and D. Lynn Loriaux

CHAPTER 43: Cerebrospinal Fluid Vasopressin and Vasotocin in Health and Disease

Thomas G. Luerssen and Gary L. Robertson

CHAPTER 44: Significance and Function of Neuropeptides in Cerebrospinal Fluid

Ivor M. D. Jackson

CHAPTER 45: Endorphins in Cerebrospinal Fluid

Floyd E. Bloom and David S. Segal

CHAPTER 46: **Cerebrospinal Fluid Monoamine Metabolites in Neuropsychiatric Disorders of Childhood**
Donald J. Cohen, Bennett A. Shaywitz, J. Gerald Young, and Malcolm B. Bowers, Jr.

CHAPTER 47: **Cerebrospinal Fluid Studies of Neurotransmitter Function in Manic and Depressive Illness**
Robert M. Post, James C. Ballenger, and Frederick K. Goodwin

CHAPTER 48: **Cerebrospinal Fluid Studies in Schizophrenia**
Daniël P. van Kammen and David E. Sternberg

CHAPTER 49: **Cerebrospinal Fluid Calcium: Clinical Correlates in Psychiatric and Seizure Disorders**
David C. Jimerson, James H. Wood, and Robert M. Post

CHAPTER 1

Physiology, Pharmacology, and Dynamics of Cerebrospinal Fluid

James H. Wood

1. Introduction

Galen described cerebrospinal fluid (CSF) as a vaporous humor produced in the cerebral ventricles that provides energy and motion to the entire body. In 1760, von Haller suggested that this vapor condenses after death to form water, which then fills the spaces surrounding the brain and spinal cord. However, four years later, Cotugno proposed that the subarachnoid space is filled with water prior to death. In 1825, Magendie[55] confirmed the continuity between the ventricular system and the subarachnoid spaces, performed the first cisternal puncture, and attempted to analyze CSF. Luschka identified the lateral connections of the fourth ventricle to the cisterna magna in 1855. Quincke[78a] performed the first lumbar puncture *in vivo* in 1891.

Goldmann[34,36] demonstrated the presence of a blood–brain barrier to trypan blue dye in 1909, and Dandy and Blackfan[15] in 1914 produced experimental ventricular distension by obstructing the aqueduct with a cotton plug. These latter two reports prompted renewed interest in CSF physiology that has endured through the present. The purpose of this chapter is to discuss more recent contributions to the understanding of the formation, absorption, flow, and function of CSF in health and disease. Several of these topics shall be assessed in greater detail in subsequent chapters.

James H. Wood, M.D. • Division of Neurosurgery, University of Pennsylvania School of Medicine and Hospital, Philadelphia, Pennsylvania 19104.

2. Embryology and Anatomy

The formation of the cerebral ventricles in man begins at about 4 weeks of gestation with dorsal closure of the primitive neural groove, followed by segmentation of the neural tube. The choroid plexuses arise from invaginations of the roof plates at 6–8 weeks.[44] The choroid plexus of the fourth ventricle develops initially, followed by that of the lateral ventricle and finally the third ventricle. Although choroid plexus secretion has not been observed prior to 24 weeks of gestation,[44] Milhorat[66] has reported congenital hydrocephalus in human fetuses several weeks prior to that time.

The ventricular volume in an adult man is approximately 23 ml, whereas the total CSF volume is approximately 140 ml.[20] The ventricles are lined by a single layer of ependymal cells. The choroid plexuses are suspended within the ventricles on vascular invaginations of the pia mater called the *tela choroidea* and enlarge in the trigone of the lateral ventricles to form the glomus. The blood vessels supplying the choroid plexuses in the body of the lateral ventricles, third ventricles, and temporal horns include the posterior choroidal, anterior choroidal, and superior cerebellar arteries, respectively. The choroid plexus of the fourth ventricle is supplied by the posterior inferior cerebellar arteries.[67] The nervous supply to the choroid plexus includes branches of the vagus, glossopharyngeal, and sympathetic nerves.

Light-microscopic examination of the choroid plexus reveals tightly packed villous folds consisting of cuboidal epithelium overlying a central core of highly vascularized stroma (Fig. 1). The ventricular surface of the epithelial cells contains a brush border and, in infants, cilia.[66] Electron-microscopic studies[68] have demonstrated fenestrations in the capillaries of the choroid plexus stroma that are not present in brain capillaries. Contrarily, the choroidal epithelial cells have tight apical intercellular junctions.

3. CSF Production

Pappenheimer *et al.*[73] developed steady-state ventriculocisternal perfusion techniques that were modified by Rubin *et al.*[84] and Cutler *et al.*[13] to measure CSF formation in man. The mean rate of CSF production has been calculated to be approximately 0.3–0.4 ml/min or 500–600 ml/day. Approximately 0.25% of the total CSF volume in adults is replaced by freshly secreted fluid every minute; thus, total CSF volume is renewed every 5–7 hr.[12,20] The choroid plexus accounts for approximately 70% of CSF production, whereas the remaining CSF is derived from the capillary bed of the brain and metabolic water production.[74,86,87]

3.1. Choroid Plexus

In 1919, Dandy[16] observed that if the choroid plexus of one lateral ventricle was removed, and if the foramina of Monro of both lateral ventricles were obstructed, the ventricle containing choroid plexus would dilate and the ventricle lacking choroid plexus would collapse. This finding suggested that the choroid plexus was a source of CSF.

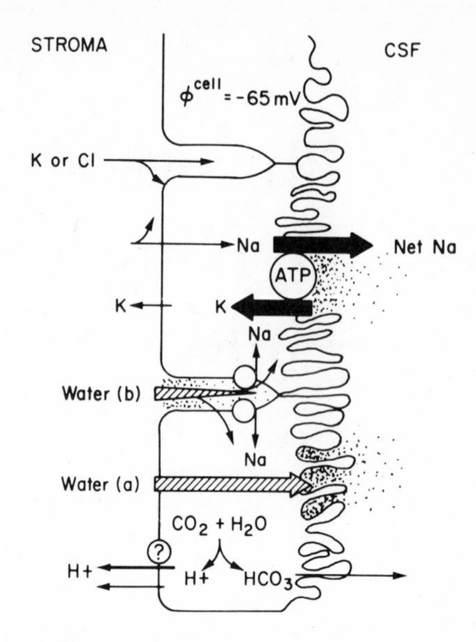

Figure 2. Diagram illustrating ionic and water movement at choroid plexus. Sodium (Na)–potassium (K) pump at apical surface of choroidal epithelium actively drives Na into CSF and K into cell. This pump may also operate in lateral intercellular spaces. Bicarbonate (HCO_3) and chloride (Cl) move into CSF, while hydrogen ions (H^+) and K migrate in opposite direction. Carbonic anhydrase regulates hydration of carbon dioxide (CO_2) to supply H^+ and HCO_3. Ions move either through or between epithelial cells. Water flows into CSF along osmotic gradient produced by Na pump (a,b). Adapted from Woodbury[113] and Wright[115] and reprinted from Rapoport,[79] with permission.

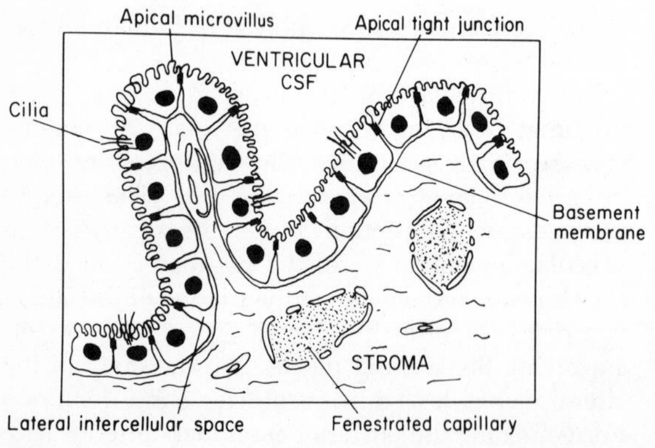

Figure 1. Choroid plexus villus extending from choroidal stroma into ventricle. Villus is covered by single-layered cuboidal epithelium with apical microvilli along ventricular surface and basal basement membrane. Tight junctions connect apical regions of epithelial cell membranes. Lateral intercellular spaces enlarge and shrink as a function of choroidal secretory activity. Stroma capillaries have fenestrated endothelium. Adapted from Millen and Woollam[60] and Wright[115] and reprinted from Rapoport,[79] with permission.

Several reviews of choroid plexus physiology have appeared in the recent literature.[12,28,46,68,76,79,116] In brief, the fenestrated endothelium of choroid plexus capillaries (Fig. 1) is permeable to most solutes; however, the apical tight junctions of the epithelial cells restrict passive solute exchange.[9] Protein-tagged Evans blue dye stains the choroidal stroma, but does not enter the CSF.[79] Following passage through the leaky choroidal capillaries, this ultrafiltrate enters the intercellular clefts by bulk flow.

The sodium–potassium pump modulated by sodium-potassium-adenosine triphosphatase (Na-K-ATPase) operates within the choroidal epithelial cells.[79] This energy-dependent pump moves sodium ions toward the ventricular surface and potassium ions in the direction of the stroma (Fig. 2) and is inhibited by ouabain. The chloride and bicarbonate anions may move passively into CSF;[99] however, Maren[57a] suggests that all of the bicarbonate reaching the CSF is formed from carbon dioxide rather than by transfer from the plasma. The passage of protein into CSF may occur by pinocytosis[76,80] and through pores.[80] Water appears to migrate across the choroidal epithelium into the ventricles via the osmotic gradient established by the active sodium secretion.[79] Vates *et al.*[95] linked sodium transport with CSF production by simultaneously inhibiting choroidal Na-K-ATPase activity and CSF flow with ouabain (Fig. 3). Sodium enters the epithelial cell in exchange for hydrogen ions, and the intracellular supply of these protons depends on the availability of carbon dioxide procured through cell metabolism and inward diffusion. The importance of this intracellular hydrogen-ion pool is demonstrated by the reduction of CSF secretion by carbonic anhydrase inhibitors that impede the hydration of carbon dioxide.[72,84,99]

3.2. Extrachoroidal Secretion

The presence of extrachoroidal CSF secretion has been demonstrated by perfusing areas of ependyma or pia that have been isolated from the choroid plexus. Pollay and Curl[74] observed that approximately 30% of CSF is secreted by the ventricular ependyma in rabbits. Sato and Bering[87] reported that 40% of CSF in dogs originates from the subarachnoid space. Milhorat *et al.*[65] correlated the CSF concentration of intravenously administered labeled sodium with CSF production and noted this sodium to be concentrated in the gray matter surrounding the ventricles and subarachnoid spaces. These data suggest that a major fraction of CSF is formed within the cerebral parenchyma. This extrachoroidal CSF secretion may explain the failure of choroid plexectomy as a treatment for hydrocephalus.[63a,68]

3.3. Pathological and Pharmacological Aspects of CSF Formation

3.3.1. Choroid Plexus Papillomas. Milhorat *et al.*[64] documented severalfold CSF overproduction by ventriculolumbar perfusion techniques in a child with a large choroid plexus papilloma. Normal CSF formation rates and resolution of the hydrocephalus were achieved following surgical removal of this benign tumor. The CSF produced by these tumors is not different from that formed by normal choroid plexuses. The localization of the ouabain-sensitive, potassium-dependent phosphatase activity is similar to that found in normal choroid plexus epithelium.[63] The blood–CSF barrier of these tumors is intact.

The hydrocephalus associated with the presence of choroid plexus tumors may be secondary to observed reductions in CSF absorptive capacity or obstruction of CSF pathways by tumor bulk.[68] Contrarily, these tumors may be present in the absence of ventricular dilatation.

3.3.2. Choroid Plexectomy. Experimental studies by Milhorat *et al.*[62] have suggested that intraventricular CSF production is reduced by only one third after choroid plexectomy and that obstructed choroid-plexectomized ventricles enlarge progressively. The composition of CSF after plexectomy is

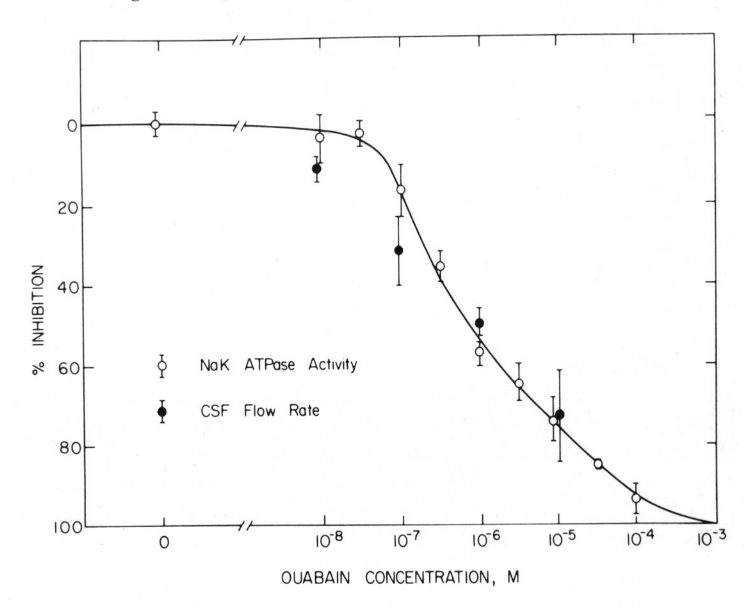

Figure 3. Ouabain inhibition of Na-K-ATPase in choroid plexus and of CSF formation *in vivo*, as measured with ventriculocisternal perfusion in cat. Vertical brackets indicate two standard errors of mean. Reprinted from Vates *et al.*,[95] with permission.

unaltered. These findings are consistent with clinical observations that hydrocephalus does not resolve after choroid plexectomy except in cases of removal of choroid plexus papillomas.

3.3.3. Inflammation. Commonly, increased CSF pressure is observed in patients with meningitis. Recent experimental studies by Dacey *et al.*[14] document 5-fold reductions in endogenous CSF production in the presence of meningitis. Thus, this increased intracranial pressure is due to markedly increased CSF outflow resistance that persists even after eradication of viable bacteria from the CSF by antibiotics. Contrarily, bacterial enterotoxins in the absence of their bacteria of origin have been shown to induce choroid plexus production of CSF.[27]

Cutler *et al.*[13] have reported normal CSF formation rates among patients with subacute sclerosing leukoencephalitis.

3.3.4. Alterations of Intracranial and Cerebral Perfusion Pressures. Heisey *et al.*[38] observed that hydrostatic pressure had no significant effect on CSF formation over the physiological pressure range. Although Lorenzo *et al.*[54] failed to demonstrate significant alterations in CSF production in hydrocephalic infants with chronic increased intracranial pressure, Hochwald and Sahar[39] noted reductions in CSF formation with increases in intraventricular pressure. More recently, Weiss and Wertman[108] demonstrated diminished CSF production when intracranial pressure is raised to the point where cerebral perfusion pressure (systemic arterial pressure minus intracranial pressure) falls below 55 torr. Factors which reduce cerebral perfusion pressure also decrease choroidal blood flow.[76a] Elevations in intracranial pressure do not appear to influence CSF formation rate if cerebral perfusion pressure is maintained above 70 torr.

3.3.5. Hypotension. According to Welch,[110] choroidal blood flow expressed in volume per tissue weight is 5 times that of whole brain or any nuclear mass. This choroidal perfusion may place an upper limit on the CSF secretory rate, since approximately 25% of this choroidal blood flow is secreted as CSF. Severe systemic hypotension or the combination of impaired cerebral autoregulation and hypotension may evoke reductions in CSF production by lowering choroidal blood flow.

3.3.6. Alterations of Serum or CSF Osmolarity. The volume flow of CSF into the cerebral ventricles is inhibited by either decreasing the osmolarity of the fluid perfusing the ventricles[102] or increasing the osmolarity of the serum.[40] Hochwald and associates[40,101,102] conclude that the increase in CSF production during ventricular perfusion with hypertonic fluids occurred at the choroid plexus, whereas the source of the augmented CSF formation observed during intravenous infusion of hypotonic fluids is the brain.

3.3.7. Drug-Induced Alterations in CSF Production

3.3.7.1. Metabolic Inhibitors. The formation of CSF is an active secretory process requiring the expenditure of energy. Drugs or environmental conditions that interfere with local metabolism reduce CSF production rates. Dinitrophenol uncouples oxidative phosphorylation and decreases CSF formation.[75] Hypothermia,[20] hypercapnia, and alkalosis[2] produce similar effects. Digoxin[92] and ouabain[95] inhibit the Na-K-ATPase of the sodium–potassium pump of the choroidal epithelial cell and blocks its secretory activity.

3.3.7.2. Diuretics. Acetazolamide (Diamox) inhibits carbonic anhydrase, the enzyme that catalyzes the hydration of carbon dioxide. This drug has been shown to reduce the CSF production rate by approximately 50%[19,57] in dogs, cats, and rabbits, but the exact mechanism is unknown. Vogh and Maren[99] suggest that acetazolamide has an indirect action on ion transport mediated by an action on bicarbonate. Tschirgi *et al.*[94] have suggested that acetazolamide inhibits CSF production from metabolically produced carbon dioxide. This reduction of carbonic acid would decrease the amount of hydrogen ions available for sodium exchange; thus, the CSF secretory rate would be reduced. Macri *et al.*[56] observed that acetazolamide evokes vasoconstriction of choroidal arteries. The resultant reduction in choroidal blood flow would decrease CSF production. Other carbonic anhydrase inhibitors such as methazolamide also decrease CSF production by about 50%.[72] The combination of intravenous acetazolamide and intraventricular ouabain has been reported to cause 95% reductions in CSF formation in dogs.[92]

According to Maren's review of carbonic anhydrase physiology,[57] intracranial pressure in normal monkeys and man undergoes only small reductions after acetazolamide, and this manifestation of lowering CSF secretion may be delayed for 25 hr. Oral acetazolamide (40–100 mg/kg) reduces elevated intracranial pressure in some slowly progressive hydrocephalic children, but requires 8–24 hr for its clinical effect.[41] Improvement without head enlargement has been maintained on chronic acetazolamide therapy, but was accompanied by sustained metabolic acidosis.

Ethacrynic acid[24,61,92] and amiloride[21] depress CSF formation by inhibiting sodium exchange and entry into cells at the transport site, respectively. Furosemide (Lasix) also inhibits CSF secretion, possibly by reducing chloride[24] or sodium[9a] transport.

3.3.7.3. Steroids. Garcia-Bengochea[33] in 1965 observed significant reductions in CSF formation during chronic administration of cortisone to cats. Weiss and Nulsen[107] noted 40 and 50% reductions in CSF production in hydrocephalic and normal dogs, respectively, within 30 min after the administration of 0.25 mg/kg of dexamethasone. These findings were later verified by Sato *et al.*,[88] who observed reductions of CSF formation rate in dogs within 10 min, the reductions reaching a maximum of 50% within 50 min following intravenous 0.15 mg/kg dexamethasone. Contrarily, Vela *et al.*[96] recently reported the lack of acute alterations and only a 9% decrease in CSF production in dogs at 4 hr following intravenous 0.4 mg/kg dexamethasone administration. The rapid uptake of radioactive hydrocortisone by the choroid plexus suggests a direct and specifically located action that reduces CSF formation.[89] Dexamethasone has been shown to inhibit Na-K-ATPase in the choroid plexus,[58] which impedes the secretory function of the epithelial cells.

3.3.7.4. Hormones, Neurotransmitters, and Cyclic Nucleotides. The inhibitory action of high doses of vasopressin on CSF transport may be secondary to vasoconstriction of choroidal blood vessels.[21] However, at physiological doses of vasopressin, this action has been disputed.[114,116] Although reductions in choroidal blood flow may contribute to the decreases in CSF production rate observed after the addition of norepinephrine to ventriculocisternal perfusates,[95] direct inhibition of choroid epithelium by norepinephrine may also slow CSF secretion.[52]

The choroid plexus is innervated by both cholinergic and adrenergic neurons.[26] Chemical stimulation of the cholinergic pathways to the choroid plexus increases CSF production.[37] Electrical stimulation of sympathetic nerve trunks bilaterally has been demonstrated to produce a reduction in the rate of bulk CSF formation.[52] Sympathectomy and catecholamine depletion with reserpine increases choroid plexus carbonic anhydrase activity and augments CSF production.[26,52] Theophylline, a phosphodiesterase inhibitor, elevates choroid plexus cyclic adenosine 3′,5′-monophosphate levels, thereby stimulating the activity of the sodium–potassium pump and CSF production.[116]

4. CSF Composition and Gradients

According to Rapoport,[79] the composition of CSF is determined by several factors: (1) metabolism, production, or uptake of solutes by cells of the central nervous system; (2) restriction of intercellular diffusion coupled with special transport mechanisms at both the blood–brain and blood–CSF barriers; (3) rates of CSF production and excretion by bulk flow. The concentrations of various solutes in plasma and lumbar CSF in man are listed in Table 1.

4.1. Ions and Glucose

In 1938, Flexner[30] compared the chemical composition of plasma with that of CSF and observed higher concentrations of sodium, chloride, and magnesium but lower concentrations of potassium, bicarbonate, calcium, phosphate, and glucose in CSF. Flexner suggested that CSF could not be formed entirely as an ultrafiltrate of plasma and that active processes must account for these concentration disparities. Davson[20] reported the ratio of CSF to plasma sodium concentrations (mEq/kg H_2O) in man to be 0.98.

The concentration of these constituents is dependent on sampling sites, since diffusionary exchanges do occur between brain extracellular fluid (ECF) and CSF during passage of CSF through the ventricles and subarachnoid spaces.[20] According to Bito and Davson,[7] the concentration of potassium ion in rabbits decreases steadily as CSF passes from the cerebral ventricles to the subarachnoid space. Ventricular CSF has a somewhat higher chloride concentration while the calcium level tends to rise on passing from the ventricle to cisterna magna and on to cortical and lumbar subarachnoid spaces. The sodium, bicarbonate, and glucose levels are similar throughout the CSF. Contrarily, the review of Cserr[12] states that during circulation from the choroid plexus of the lateral ventricle to the cisterna magna in cats, calcium and bicarbonate concentrations fall while chloride levels rise. Thus, species variations in CSF composition exist. Subarachnoid CSF appears to have a significantly higher osmolality than ventricular CSF.[7]

In general, the CSF glucose concentration should be two thirds that of a simultaneously drawn plasma sample.[59]

4.2. Amino Acids, Urea, and Protein

Flexner[30] originally reported lower levels of amino and uric acids and proteins in CSF than in plasma. Rapoport and Pettigrew[80] have postulated two pathways for protein transfer at the choroid plexus epithelium: 117-Å-radius pores that allow the transfer

Table 1. Concentrations of Solutes in Plasma and Lumbar CSF in Man[a]

Substance	Plasma	CSF	CSF/plasma
Sodium (mM)	150	147	0.98
(mEq/liter)*	140	144	1.03
Potassium (mM)	4.63	2.86	0.62
Magnesium (mM)	0.81	1.12	1.4
Calcium (mM)	2.35	1.14	0.49
Chloride (mM)	99	113	1.1
Bicarbonate (mM)	26.8	23.3	0.87
Inorganic phosphate (mg/100 cm^3)	4.70	3.40	0.73
Protein (mg/100 cm^3)	6800	28	0.004
Glucose (mg/100 cm^3)	110	50–80	0.6
Osmolality	0.289	0.289	1.0
pH	7.397	7.307	—
P_{CO_2} (mm Hg)	41.1	50.5	—

[a] Modified from Rapoport,[79] with permission. Data from Davson,[20] except CSF protein from Fremont-Smith *et al.*,[32] CSF glucose from Merritt and Fremont-Smith,[59] blood glucose from Linder and Blomstrand,[51] and plasma protein from Altman and Dittmer.[1] mM(H_2O) data for sodium assumes CSF water content is 99% CSF volume and plasma water is 92% plasma volume.[46] *mEq/liter data for sodium employs actual volumes of respective fluids.

of smaller proteins by diffusion or ultrafiltration, or both, and 250-Å-radius pinocytotic vesicles that account for the exchange of larger proteins. Although not described in the choroidal epithelium, the 117-Å-radius pores may represent a 0.08% defect in the normally continuous tight junctions that surround and closely connect choroidal epithelial cells. The total protein concentration of CSF is less than 0.5% of that of plasma.[116] CSF protein normally varies with the patient's age according to the equation $23.8 + 0.39 \times \text{age} \pm 15.0$ mg/dl.[69]

Regional variations in the concentrations of these CSF components have been described by Bito and Davson.[7] The concentration of urea in CSF rises significantly on passing from cisternal to cortical subarachnoid spaces.

In a recent communication, Weisner and Bernhardt[106] analyzed the protein fractions in ventricular CSF of patients with psychiatric or extrapyramidal disorders undergoing stereotaxic operations and in cisternal and lumbar CSF of patients lacking confirmation of suspected neurological disease. Lumbar CSF contains approximately 1.6 times more total protein than ventricular CSF, whereas cisternal CSF protein content is 1.2 times that of ventricular CSF. Generally, total protein concentrations in the ventricles are 6–15 mg/dl, whereas those in the cisterna magna and lumbar sac are 15–25 and 20–50 mg/dl, respectively. This higher level of protein in lumbar CSF is thought to result from the addition of plasma protein and not from dehydration.[29]

This mechanism does not appear to affect all protein fractions equally. Lumbar CSF albumin and immunoglobulin G (IgG) concentrations were 2.2 and 2.6 times higher, respectively, than their levels in ventricular CSF. CSF albumin and IgG concentrations correlate highly in all regions of the subarachnoid space despite the presence of a 10% IgG fraction that does not originate from the serum. Contrarily, the concentration of prealbumin is reduced by a factor of 0.7 in lumbar CSF in comparison to ventricular CSF levels. Weisner and Bernhardt[106] were not able to detect the trace levels of IgA in the CSF, despite its similarity to IgG in size and molecular weight. This disparity between IgA and IgG concentrations suggests the presence of selective functions of the blood–CSF barrier with respect to protein fractions. The normal lumbar CSF and serum protein composition as determined by cellulose acetate protein electrophoresis is summarized in Table 2.[70]

4.3. Organic Constituents and Drugs

According to the review of Fenstermacher and Rall,[28] the transport of lactate from blood to brain or CSF is slow. Alterations of brain and CSF lactate concentrations are independent of blood lactate changes; however, shifts in brain and CSF levels are correlated. Apparently, the CSF may serve as a route for the clearance of lactate from the central nervous system. Thus, the CSF is a better indicator

Table 2. Composition of Human CSF and Serum as Determined by Cellulose Acetate Protein Electrophoresis[a]

	Normal CSF (mg%)	Normal serum (g%)
Total protein	20–50	5–7
Protein Component	Normal range of CSF fraction	Normal range of serum fraction
Prealbumin	0.015–0.065	0–0.02
Albumin	0.45–0.75	0.43–0.55
Alpha₁ globulins	0.04–0.08	0.03–0.08
Alpha₂ globulins	0.05–0.09	0.09–0.14
Beta globulins	0.08–0.16	0.09–0.19
Gamma globulins	0.06–0.14	0.15–0.22

[a] Reprinted from Neuwelt and Clark,[70] with permission.

of central lactic acid levels and neural metabolism than cerebral venous blood.

CSF uptake of drugs that are undissociated at body pH is quick, and their respective permeability coefficients rise as their relative lipid solubilities increase.[28] The pH of CSF is slightly lower than that of plasma (Table 1); thus the distribution of basic drugs between the CSF and plasma would favor the accumulation of the ionized form of the drug in CSF. The CSF uptake of lipid-insoluble drugs is exceedingly slow, with the order of permeability decreasing with increasing molecular size.

5. CSF Absorption

5.1. Cranial and Arachnoid Villi

In 1875, Key and Retzius[47] infused dye under pressure into the subarachnoid space of cadavers and noted entrance of the dye into the venous sinuses via the pacchionian granulations. Weed[104] confirmed these findings *in vivo* to add support to the concept that CSF is absorbed at the arachnoid villi (Fig. 4). Davson *et al.*[17] studied the relationship of CSF flow rate to the interventricular pressure and observed that as the rate of flow increased, the resistance of the outflow system appeared to decrease. This increased intracranial pressure evokes a compensatory rise in CSF absorption (Fig. 5) without enlarging CSF drainage pathways.[41a]

The characteristics of cranial CSF absorption as reviewed by Pollay[76] include (1) unidirectional CSF flow across the villi; (2) nonlinear flow requiring a critical opening pressure of approximately 5 cm water; (3) particulate passage from CSF to blood, but not vice versa; (4) CSF absorption reductions as protein content increases; and (5) uniform absorption rates for different-sized molecules. This type of

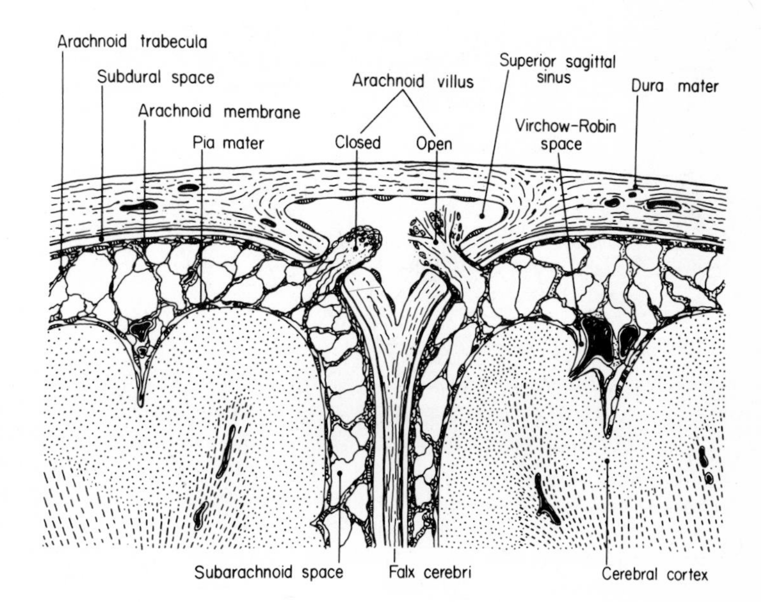

Figure 4. Coronal section of superior sagittal sinus, demonstrating meninges and subarachnoid spaces. Two arachnoid villi protrude into superior sagittal sinus. Closed arachnoid villus (*left*) has an overlapping surface layer of endothelial cells. Open arachnoid villus (*right*) allows entry of CSF into venous blood of sinus by bulk flow through tubular channels between endothelial cells. Adapted from Weed[105] by Rapoport[79]; reprinted with permission.

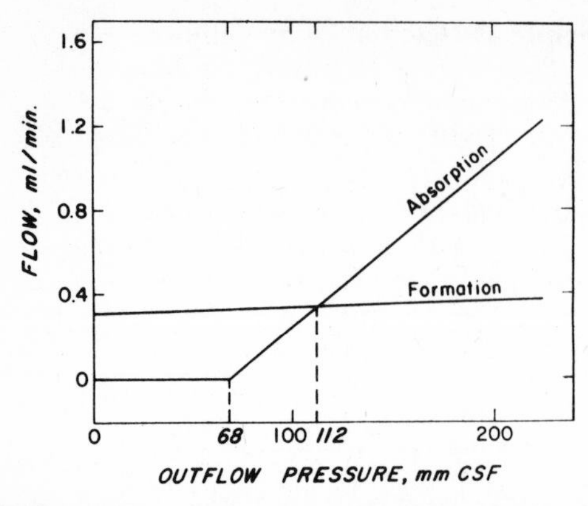

Figure 5. Superimposed regression lines for CSF formation and absorption as a function of outflow pressure. Intercept at 112 mm indicates CSF pressure at which formation and absorption are equal. Pressure at which absorption is zero is approximately 68 mm water. Reprinted from Cutler *et al.*,[13] with permission.

process is most consistent with a valvular system; however, Tripathi and Tripathi[93] have recently proposed that CSF passes by bulk flow through temporary transmesothelial channels into the venous sinuses (Fig. 6).

5.2. Spinal Arachnoid Villi

Dye injected into the spinal subarachnoid space also passes into the dural and epidural veins.[8] Thus, CSF is also absorbed through spinal arachnoid villi located on the dorsal roots projecting into the dural sinusoids (Fig. 7).[109]

6. CSF Circulation

The net formation of CSF at the choroid plexus and its absorption into the venous sinuses provide the forces necessary for the circulation of CSF through the ventriculospinal system. Obstruction to CSF flow or elevations in sagittal sinus pressure impede CSF absorption at the arachnoid villi.

6.1. Normal Patterns

CSF is formed at a hydrostatic pressure of 15 cm water, which provides force for CSF movement.[20] Additional momentum is supplied by the pulse pressure generated by the vascular choroid plexus.[6] Ependymal cilia produce currents that propel CSF toward the fourth ventricle and its foramina into the

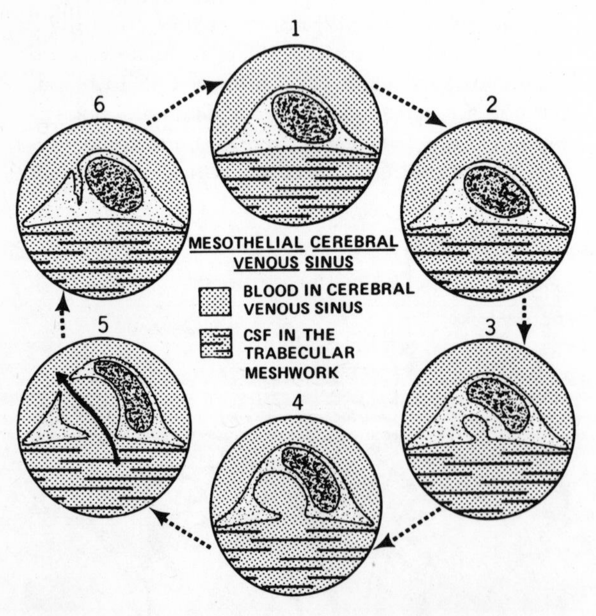

Figure 6. Diagram of endothelial vacuolation cycle. Temporary transcellular channels or pores (stage 5) allow CSF to flow from subarachnoid space to venous sinus down a pressure gradient. Arrows indicate bulk overflow of CSF. Modified from Tripathi and Tripathi[93] by Pollay[76]; reprinted with permission.

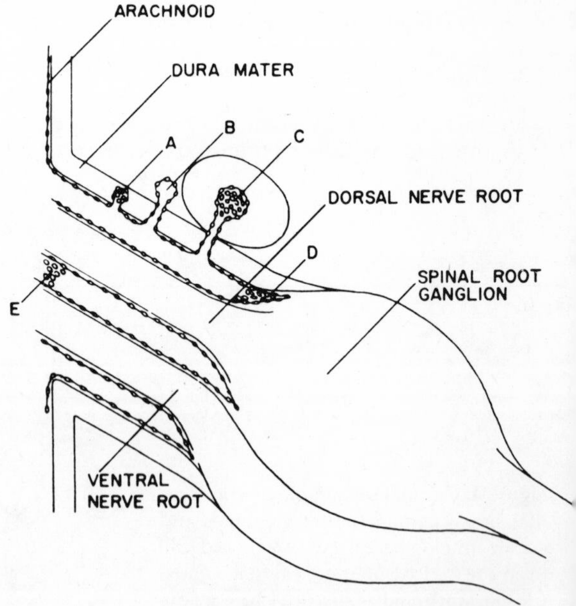

Figure 7. Diagram of spinal arachnoid villus in relation to nerve root: (A) within dura mater; (B) complete penetration of dura; and (C) penetration of villous tissue into spinal vein. (D,E) Arachnoid proliferation within subarachnoid space. Reprinted from Welch and Pollay,[109] with permission.

subarachnoid spaces.[10] In addition, the CSF–venous gradient at the sinuses in man is approximately 5–6 cm water and augments CSF circulation.[91]

DiChiro[23] observed the flow of CSF in man employing radioisotope-labeled albumin injected into the CSF (Fig. 8). When injected into the lumbar sac, this tracer is detected in the basal cisterns in approximately 1 hr. From the basal cisterns, the activity proceeds mainly through the anterior subarachnoid spaces and the Sylvian fissures, to the convexity of the brain. Approximately 20–33% of the isotope injected intrathecally reaches the intracranial cavity within 12 hr. At 12–24 hr, most of the activity is detectable along the superior longitudinal sinus. DiChiro[23] also noted that this circulatory pattern revealed by isotope cisternography was not altered by posture or ambulation, although physical activity is known to disturb CSF concentration gradients by promoting CSF mixing.[78]

DiChiro[23] also observed CSF flow patterns after injecting the radioisotope-labeled albumin intraventricularly in man. The radioactivity flowed into the basal cisterns within a few minutes and began to pass through the anterior subarachnoid spaces and Sylvian fissures at 2 hr to collect along the superior sagittal sinus area at 12–24 hr. These investigations employing isotope encephalography were expanded to assess the spinal descent of CSF in man.[22] Activity was observed to reach the low cervical–high thoracic area in 10–20 min. At 30–40 min, the entire thoracic area and thoracolumbar segment contained significant amounts of radioactivity. This flow of activity reached the lumbo-sacral *cul de sac* at 60–90 min. Thus, the descent of CSF in spinal subarachnoid space is faster than its ascent either thecocranially or over the cerebral convexities. Under certain conditions, such as coughing, the spinal CSF flow undergoes a complete reversal in the direction of the cisterna magna.[111]

6.2. Pathological Obstructions

Hydrocephalus is an imbalance of CSF formation and absorption of sufficient magnitude to produce a net accumulation of CSF within the cerebral ventricles.[67] This condition usually evokes an elevation in intracranial pressure; however, compensatory adjustments may occur in infants and young children that can reduce CSF pressure to normal levels. CSF pressure is normally less than 240 mm of water.[33a] More detailed physiological, radiological, and biochemical reviews of hydrocephalus are discussed in subsequent chapters.

In her 1949 review, Russell[85] concludes that almost every case of hydrocephalus has a pathological obstruction at some point along the pathways of CSF flow. Obstructions can usually be demonstrated in cases of choroid plexus papillomas. Mil-

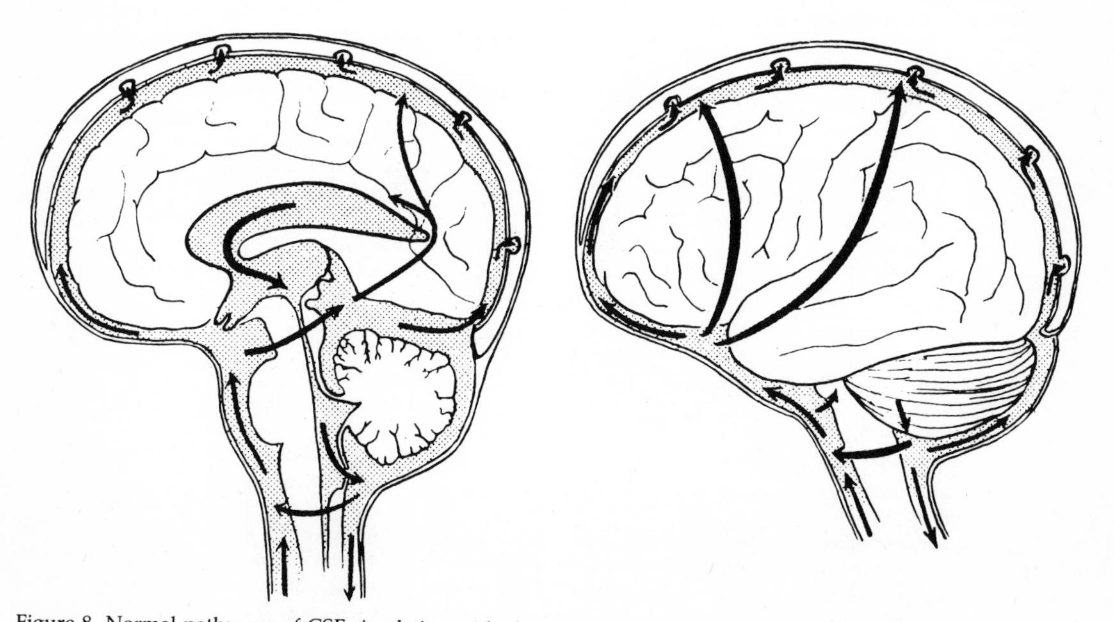

Figure 8. Normal pathways of CSF circulation with absorption at arachnoid villi in sagittal (*left*) and lateral (*right*) projections. Reprinted from Milhorat,[66] with permission.

horat[67] has divided these absorption defects into two types: (1) lesions that obstruct the ventricular system, preventing free circulation of CSF out of the ventricles (noncommunicating hydrocephalus); and (2) lesions that obstruct the subarachnoid space, preventing access of the subarachnoid CSF to the arachnoid villi (communicating hydrocephalus). Milhorat[67] classified the etiologies of these types of CSF flow obstructions as presented in Table 3.

Abnormal CSF circulation with flow into the ventricular system has been demonstrated in patients with communicating hydrocephalus (Fig. 9) employing radioisotopic CSF imaging techniques.[23,42,114] Recently, Drayer and Rosenbaum[25] described techniques of metrizamide computed tomographic cisternography and ventriculography. Normal, intermediate, delayed, and obstructive CSF circulation patterns are readily demonstrated by evaluation of ventricular stasis of metrizamide on serial computed tomograms.

6.2.1. Mechanisms of Compensation in Hydrocephalus. Although CSF production rates decrease when intracranial pressure significantly lowers cerebral arterial perfusion pressure,[39,108] other mechanisms of compensation for CSF circulatory obstructions and the resultant ventricular enlargement have been proposed.[112] The pathological establishment of alternative pathways of CSF drainage such as transependymal absorption[43] and spinal cord central canal dilatation[5] would explain the abnormal entry of radioactivity into the ventricles and central canal during isotope cisternography[23,42,114] and ventriculography,[35] respectively. However, the central canal is not a significant pathway of CSF absorption in persons with dilated ventricles who are older than 20 years of age.[45] Progressive head enlargement compensates for increases in intracranial pressure

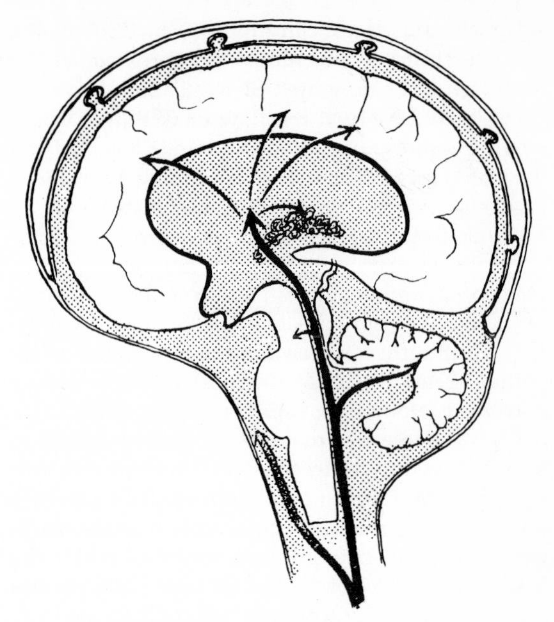

Figure 9. CSF circulation with ventricular entry and transependymal absorption in communicating hydrocephalus. Reprinted from Milhorat,[66] with permission.

Table 3. Classification of CSF Circulatory Obstructions[a]

Noncommunicating type	Communicating type
I. Congenital lesions	I. Congenital lesions
A. Aqueductal obstruction (stenosis)	A. Arnold–Chiari malformation
1. Gliosis	B. Encephalocele
2. Forking	C. Leptomeningeal inflammations
3. True narrowing	D. Lissencephaly
4. Septum	E. Congenital absence of arachnoidal granulations
B. Atresia of foramina of Luschka and Magendie (Dandy–Walker cyst)	II. Acquired lesions
C. Masses	A. Leptomeningeal inflammations
1. Benign intracranial cysts	1. Infections
2. Vascular malformations	2. Hemorrhage
3. Tumors	3. Particulate matter
II. Acquired lesions	B. Masses
A. Aqueductal stenosis (gliosis)	1. Tumors
B. Ventricular inflammation and scars	2. Nonneoplastic masses
C. Masses	C. Platybasia
1. Tumors	III. Oversecretion of CSF (choroid plexus papilloma)
2. Nonneoplastic masses	

[a] Reprinted from Milhorat,[67] with permission.

in infants and young children with hydrocephalus. CSF absorption is augmented at the arachnoid villi during periods of increased intracranial pressure. These topics are assessed in greater detail in subsequent chapters.

7. Functions of CSF

7.1. Protection of Brain and Spinal Cord

The CSF provides buoyancy and protection to the central nervous system. The difference in specific gravities of the brain (1.040) and the CSF (1.007) reduces the effective mass of a 1500-g brain to only 50 g.[12,53] According to Cserr,[12] this buoyancy prevents the brain's full weight from producing traction on emerging nerve roots, blood vessels, and delicate membranes. Severe pain is experienced when CSF is partially replaced by air during pneumoencephalography. The buoyancy afforded by the CSF also reduces the momentum and acceleration of the brain and tends to decrease the concussive damage to nervous tissue when the head is suddenly displaced.[71]

7.2. ECF–CSF Exchange

The maximum distance for diffusion between the CSF and any area of brain is 15 mm in the adult man and less in the human fetus.[12] Alterations in the composition of CSF profoundly influence central functions of the nervous system. As reviewed by Leusen[50] and in several chapters in this volume, manipulation of CSF calcium, potassium, and magnesium concentrations is known to alter heart rate, blood pressure, vasomotor and other autonomic reflexes, respiration, muscle tone, and emotional states. The acid–base characteristics of CSF influence respiration, cerebral blood flow autoregulation, and brain metabolism. Wald *et al.*[101] have demonstrated the movement of fluid, macromolecules, and ions from the brain extracellular fluid (ECF) into the CSF. These phenomena indicate a relatively free exchange between the brain ECF and CSF compartments. The brain ECF provides the chemical environment for the neurons and glia and equals approximately 18% of brain wet weight.[28]

7.2.1. Nutrition. Early investigators[34] suggested that CSF provides a major route of substrate transport from the blood into the brain. However, recent reviews[12] indicate a significant nutritive function for CSF only in embryos and lower vertebrates whose brains are poorly vascularized. Spector and Lorenzo[92a] have found evidence that the CSF may mediate the brain uptake of required vitamins (e.g., ascorbic acid).

7.2.2. Excretion. The existence of extracellular channels by which brain metabolites could be transported into the ventriculosubarachnoid space and the absence of lymphatic channels in the central nervous system suggest an excretory role for CSF.[12] In 1914, Weed[103] postulated that ECF formed from plasma drains into the CSF. Davson *et al.*[18] suggested that concentrations of metabolic products and unwanted drugs that enter the ECF slowly are rapidly diluted by the "sink" action of CSF, thereby preventing their accumulation in high levels in the brain.

Although two mechanisms, bulk flow of ECF and net diffusion, have been proposed by which substances may travel from the brain into CSF, direct evidence of net transport by diffusion is lacking.[12] Cserr *et al.*[11] report similar rates of transport of substrates of different molecular weights between the ECF and CSF and suggest that bulk flow accounts for ECF circulation. Reulen *et al.*[81] have demonstrated that vasogenic brain edema is cleared by bulk flow down a pressure gradient into CSF. Possibly, increasing this pressure gradient by reducing CSF pressure would result in an augmentation of brain ECF clearance into CSF. Contrarily, in other pathological conditions such as hydrocephalus and brain dehydration by intravenous administration of hypertonic solutions, there exists a pressure gradient between the CSF and brain capillaries that accelerates solute movements into the brain.[31,112]

Cserr *et al.*[11] have also noted that less than 20% of solute clearance from the ECF could be accounted for by the amount of solute drained into CSF; thus, some solute may be absorbed into the veins of the choroid plexus. Drugs such as penicillin and methotrexate and neurotransmitter acid metabolites are removed rapidly from the CSF by the choroid plexus employing a probenecid-sensitive mechanism.[4]

7.2.3. Control of the Central Chemical Environment. The unique permeability characteristics of the choroid plexus and the blood–brain barrier contribute to the homeostasis of the chemical composition of the ECF. Many potentially toxic polar molecules including drugs, humoral agents, and metabolites are almost totally excluded from the central nervous system. These mechanisms under physiological conditions minimize the disruptive effects of variations in plasma composition on the functional activity of the nervous system.

Pathological breakdown of the blood–brain barrier, interference with cell metabolism, or fluid and electrolyte disorders may not only alter the composition of ECF and CSF but also compromise brain function. Contrarily, these homeostatic mechanisms may impede the successful treatment of central chemical derangements by systemic means. Lactic acidosis of the brain and CSF following head injury may persist despite systemic acid–base correction and require intrathecal administration of bicarbonate.[90]

7.2.4. Intracerebral Transport of Hormones. Recently Knigge *et al.*[48] and Rodriguez[83] reviewed the topic of CSF as a pathway in neuroendocrine integration. In some areas, the ependyma of the ventricles is highly specialized for secretory functions. The CSF may act as a vehicle for intracerebral transport of neuroendocrine factors such as hypothalamic hormones[49] and the pineal hormone melatonin.[3]

Neurohormone-releasing factors are synthesized in the hypothalamus and released into the ECF and CSF by neurons that make axonal contact with specialized ependymal cells.[3,97,98] These factors are carried via the CSF to the median eminence, where stimulation of the dendritic processes of the "CSF-contacting neurons" occurs,[97] or are transferred to the hypophyseal portal system by specialized ependymal cells called "tanycytes."[77,79] This ependymal transport (Fig. 10) is known to be preserved after disconnection of the hypothalamus from the median eminence and modified by peripheral hormone levels (feedback).

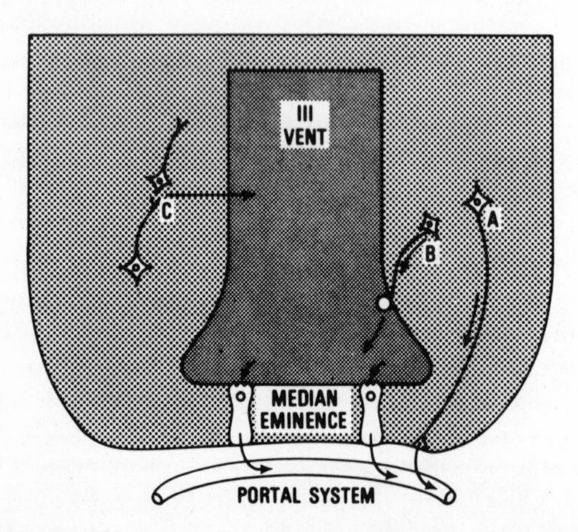

Figure 10. Schematic diagram of possible relationship between hypothalamic neurons containing releasing hormones and third ventricle (vent). (A) Neuron releasing hormones directly on portal capillaries; (B) neuron delivering hormones into ventricular CSF which transports hormones to median eminence tanycyte (specialized ependyma) through which hormones pass to portal system; (C) hormone or neurotransmitter released by neuron and reaching third ventricular CSF by diffusion or bulk flow between lining ependymal cells. Reprinted from Pollay,[75a] with permission.

Most adenohypophyseal hormones have been demonstrated in the CSF in quantities that, when compared to plasma concentrations, suggest that these hormones reach the CSF by a saturable active transport system rather than by simple diffusion.[83] Experimental conditions known to provoke a discharge of vasopressin from the neurohypophysis also evoke a rise in antidiuretic activity in CSF.[100] Release of antidiuretic hormone during perfusion of the cerebral ventricles with hypertonic saline suggests that CSF alterations may also affect central osmoreceptors.[82]

References

1. Altman, P. L., Dittmer, D. S. (eds.): *Blood and Other Body Fluids.* Washington, D.C., Federation of American Societies for Experimental Biology, 1961.
2. Ames, A., Higashi, K., Nesbett, F. B.: Effects of P_{CO_2}, acetazolamide and ouabain on volume and composition of choroid plexus fluid. *J. Physiol. (London)* **181**:516–524, 1965.
3. Anton-Tay, R., Wurtman, R. J.: Regional uptake of ^{3}H-melatonin from blood or cerebrospinal fluid by rat brain. *Nature (London)* **222**:474–475, 1969.
4. Baramy, E. H.: Inhibition by hippurate and probenecid of *in vitro* uptake of iodipamide and *O*-iodohippurate: A composite uptake system for iodipamide in choroid plexus, kidney cortex and anterior uvea of several species. *Acta Physiol. Scand.* **86**:12–27, 1972.
5. Becker, D. P., Wilson, J. A., Watson, G. W.: The spinal cord central canal: Response to experimental hydrocephalus and canal occlusion. *J. Neurosurg.* **36**:416–424, 1972.
6. Bering, E. A., Jr.: Pathophysiology of hydrocephalus. In Shulman, K. (ed.): *Workshop in Hydrocephalus.* Philadelphia, Children's Hospital of Philadelphia, 1965, pp. 9–19.
7. Bito, L. Z., Davson, H.: Local variations in cerebrospinal fluid composition and its relationship to the composition of the extracellular fluid of the cortex. *Exp. Neurol.* **14**:264–280, 1966.
8. Brierley, J. B., Field, E. J.: The connections of the spinal subarachnoid space with the lymphatic system. *J. Anat.* **82**:153–166, 1948.
9. Brightman, M. W., Reese, T. S., Feder, N.: Assessment with the electron microscope of the permeability to peroxidase of cerebral endothelium and epithelium in mice and sharks. In Crane, C., Lassen, N. A. (eds.): *Capillary Permeability.* New York, Academic Press, 1970, pp. 468–476.

9a. BUHRLEY, L. E., REED, D. J.: The effect of furosemide on sodium-22 uptake into cerebrospinal fluid and brain. *Exp. Brain Res.* **14**:503–510, 1972.
10. CATHCART, R. S., WORTHINGTON, W. C.: Ciliary movement in the rat cerebral ventricles, clearing action and directions of currents. *J. Neuropathol. Exp. Neurol.* **23**:609–618, 1964.
11. CSERR, H. F., COOPER, D. N., MILHORAT, T. H.: Flow of cerebral interstitial fluid as indicated by removal of extracellular markers from rat caudate nucleus. In Bito, L. Z., Davson, H., Fenstermacher, J. D. (eds.): *The Ocular and Cerebrospinal Fluids.* New York, Academic Press, 1977, pp. 461–473.
12. CSERR, H. F.: Physiology of the choroid plexus. *Physiol. Rev.* **51**:273–311, 1971.
13. CUTLER, R. W. P., PAGE, L., GALICICH, J., WATTERS, G. V.: Formation and absorption of cerebrospinal fluid in man. *Brain* **91**:707–720, 1968.
14. DACEY, R. G., WELSH, J. E., SCHELD, W. M., WINN, H. R., SANDE, M. A., JANE, J. A.: Alterations of cerebrospinal fluid outflow resistance in experimental bacterial meningitis. *Ann. Neurol.* **4**:173, 1978 (abstract).
15. DANDY, W. E., BLACKFAN, K. D.: Internal hydrocephalus, an experimental, clinical and pathological study. *Am. J. Dis. Child.* **8**:406–482, 1914.
16. DANDY, W. E.: Experimental hydrocephalus. *Ann. Surg.* **70**:129–142, 1919.
17. DAVSON, H., HOLLINGSWORTH, G., SEGAL, M. D.: The mechanism of drainage of the cerebrospinal fluid. *Brain* **93**:665–678, 1970.
18. DAVSON, H., KLEEMAN, C. R., LEVIN, E.: Quantitative studies of the passage of different substances out of the cerebrospinal fluid. *J. Physiol. (London)* **161**:126–142, 1962.
19. DAVSON, H., LUCK, C. P.: The effect of acetazoleamide on the chemical composition of the aqueous humour and cerebrospinal fluid of some mammalian species and on the role of turnover of ^{24}Na in these fluids. *J. Physiol. (London)* **137**:279–293, 1957.
20. DAVSON, H.: *Physiology of the Cerebrospinal Fluid.* London, Churchill, 1967.
21. DAVSON, H., SEGAL, M. B.: The effects of some inhibitors and accelerators of sodium transport on the turnover of ^{22}Na in the cerebrospinal fluid and the brain. *J. Physiol. (London)* **209**:131–153, 1970.
22. DICHIRO, G., HAMMOCK, M. K., BLEYER, W. A.: Spinal descent of cerebrospinal fluid in man. *Neurology (Minneapolis)* **26**:1–8, 1976.
23. DICHIRO, G.: Movement of the cerebrospinal fluid in human beings. *Nature (London)* **204**:290–291, 1964.
24. DOMER, F. R.: Effects of diuretics on cerebrospinal fluid formation and potassium movement. *Exp. Neurol.* **24**:54–64, 1969.
25. DRAYER, B. P., ROSENBAUM, A. E.: Studies in the third circulation, amipaque CT cisternography and ventriculography. *J. Neurosurg.* **48**:946–956, 1978.
26. EDVINSSON, L., NIELSEN, K. C., OWMAN, C.: Innervation of choroid plexus in rabbits and cats. *Brain Res.* **63**:500–503, 1973.
27. EPSTEIN, M. H., FELDMAN, A. M., BRUSILOW, S. W.: Cerebrospinal fluid production: Stimulation by cholera toxin. *Science* **196**:1012–1013, 1977.
28. FENSTERMACHER, J. D., RALL, D. P.: Physiology and pharmacology of cerebrospinal fluid. In Capri, A. (ed.): *Pharmacology of the Cerebral Circulation,* Vol. 1. Oxford, Pergamon Press, 1972, pp. 35–79.
29. FISHMAN, R. A., RANSOHOFF, A. J., OSSERMAN, E. F.: Factors influencing the concentration gradient of protein in cerebrospinal fluid. *J. Clin. Invest.* **37**:1419–1428, 1958.
30. FLEXNER, L. B.: Changes in the chemistry and nature of the cerebrospinal fluid during fetal life in the pig. *Am. J. Physiol.* **124**:131–135, 1938.
31. FOLEY, F. E. B.: Alterations in the currents and absorption of cerebrospinal fluid following salt administration. *Arch. Surg.* **6**:587–604, 1923.
32. FREMONT-SMITH, F., DAILEY, M. E., MERRITT, H. H., CARROLL, M. P., THOMAS, G. W.: The equilibrium between cerebrospinal fluid and blood plasma. I. The composition of the human cerebrospinal fluid and blood plasma. *Arch. Neurol. Psychiatry* **25**:1271–1289, 1931.
33. GARCIA-BENGOCHEA, F.: Cortisone and the cerebrospinal fluid of non-castrated cats. *Am. Surg.* **31**:123–125, 1965.
33a. GILLAND, O., TOURTELLOTTE, W. W., O'TAUMA, L., HENDERSON, W. G.: Normal cerebrospinal fluid pressure. *J. Neurosurg.* **40**:587–593, 1974.
34. GOLDMANN, E. E.: Die aussere and innere Sekretion des gesunden und kranken Organismus im Lichte der "vitalen Farbung." *Beitr. Klin. Chir.* **65**:192–265, 1909.
35. HALL, P. V., KALSBECK, J. E., WELLMAN, H. N., BALNITZKY, S., CAMPBELL, R. L., LEWIS, S.: Clinical radioisotope investigations in hydrosyringomyelia and myelodysplasia. *J. Neurosurg.* **45**:188–194, 1976.
36. HAYMAKER, W., SCHILLER, F.: *The Founders of Neurology.* Springfield, Illinois, Charles C. Thomas, 1970.
37. HAYWOOD, J. R., VOGH, B. P.: Some measurements of autonomic nervous system influence on production of cerebrospinal fluid. *J. Pharmacol. Exp. Ther.* **208**:341–346, 1979.
38. HEISEY, S. R., HELD, D., PAPPENHEIMER, J. R.: Bulk flow and diffusion in the cerebrospinal fluid system of the goat. *Am. J. Physiol.* **203**:775–781, 1962.
39. HOCHWALD, G. M., SAHAR, A.: The effect of spinal fluid pressure on cerebrospinal fluid formation. *Exp. Neurol.* **32**:30–40, 1971.
40. HOCHWALD, G. M., WALD, A., DIMATTIO, J., MALHAN, C.: The effects of serum osmolarity on cerebrospinal fluid volume flow. *Life Sci.* **15**:1309–1316, 1974.
41. HUTTENLOCKER, P. R.: Treatment of hydrocephalus with acetazolamide. *J. Pediatr.* **66**:1023–1030, 1965.
41a. JAMES, A. E., MCCOMB, J. G., CHRISTIAN, J., DAVSON, H.: The effect of cerebrospinal fluid pressure on the size of drainage pathways. *Neurology (Minneap.)* **26**:659–663, 1976.
42. JAMES, A. E., NEW, P. F. J., HEINZ, E. R., HODGES, F. J., DELAND, F. H.: A cisternographic classification

of hydrocephalus. *Am. J. Roentgenol. Radium Ther. Nucl. Med.* **115**:39–49, 1972.

43. JAMES, A. E., STRECKER, E. P., SPERBER, E., FLOR, W. J., MERZ, T., BURNS, B.: An alternative pathway of cerebrospinal fluid absorption in communicating hydrocephalus. *Radiology* **111**:143–146, 1974.
44. KAPPERS, J. A.: Cerebrospinal fluid; production, circulation and absorption. Summit, New Jersey, *Ciba Foundation Symposium*, 1957, pp. 3–31.
45. KASANTIKUL, V., NETSKY, M. G., JAMES, A. E.: Relation of age and cerebral ventricle size to central canal in man; morphological analysis. *J. Neurosurg.* **51**:85–93, 1979.
46. KATZMAN, R., PAPPIUS, H. M.: *Brain Electrolytes and Fluid Metabolism.* Baltimore, Williams & Wilkins, 1973, pp. 14–32.
47. KEY, G., RETZIUS, M.: *Anatomie des Nerven-systems und des Bindergewebes.* Stockholm, Samson and Wallin, 1875.
48. KNIGGE, K. M., SCOTT, D. E., KOBAYASHI, H., ISHII, S.: *Brain–Endocrine Interaction. II. The Ventricular System in Neuroendocrine Mechanisms.* Basel, S. Karger, 1975.
49. KUMAR, T. C. A., THOMAS, G. H.: Metabolites of ^{3}H-oestradiol-17β in the cerebrospinal fluid of the rhesus monkey. *Nature (London)* **219**:628–629, 1968.
50. LEUSEN, I.: Regulation of cerebrospinal fluid composition with reference to breathing. *Physiol. Rev.* **52**:1–56, 1972.
51. LINDER, E., BLOMSTRAND, R.: Technic for collection of thoracic duct lymph of man. *Proc. Soc. Exp. Biol.* **97**:653–657, 1958.
52. LINDVALL, M., EDVINSSON, L., OWMAN, C.: Sympathetic nervous control of cerebrospinal fluid production from the choroid plexus. *Science* **201**:176–178, 1978.
53. LIVINGSTON, R. B.: Cerebrospinal fluid. In Fulton, J. F. (ed.): *A Textbook of Physiology.* Philadelphia, W. B. Saunders, 1955, pp. 950–963.
54. LORENZO, A. V., PAGE, L. K., WATTERS, G. V.: Relationship between cerebrospinal fluid formation, absorption and presence in human hydrocephalus. *Brain* **93**:679–692, 1970.
55. MAGENDIE, F.: Memoire sur le liquide qui se trouvre dans le crane et l'épine de l'homme et des animaux vertebraux. *J. Physiol. Exp.* **5**:27–37, 1825.
56. MACRI, F. J., POLITOFF, A., RUBIN, R., DIXON, R., RALL, D.: Preferential vasoconstrictor properties of acetazolamide on the arteries of the choroid plexus. *Int. J. Neuropharmacol.* **5**:109–115, 1966.
57. MAREN, T. H.: Carbonic anhydrase: Chemistry, physiology and inhibition. *Physiol. Rev.* **47**:595–781, 1967.

57a. MAREN, T. H.: Effect of varying CO_2 equilibria on rates of HCO_3 formation on cerebrospinal fluid. *J. Appl. Physiol. Respirat. Environ. Exercise Physiol.* **47**:471–477, 1979.

58. MAYMAN, C. I.: Inhibitory effect of dexamethasone on sodium–potassium activated adenosine triphosphatase of choroid plexus in the cat and rabbit. *Fed. Proc. Fed. Am. Soc. Exp. Biol.* **31**:591, 1972.
59. MERRITT, H. H., FREMONT-SMITH, F.: *The Cerebrospinal Fluid.* Philadelphia, W. B. Saunders, 1937.
60. MILLEN, J. W., WOOLLAM, D. H. M.: *The Anatomy of the Cerebrospinal Fluid.* London, Oxford University Press, 1962.
61. MINER, L. C., REED, D. J.: The effect of ethacrynic acid on Na^+ uptake into the cerebrospinal fluid of the rat. *Arch. Intern. Pharmacodynam. Ther.* **190**:316–321, 1971.
62. MILHORAT, T. H.: Choroid plexus and cerebrospinal fluid production. *Science* **166**:1514, 1969.
63. MILHORAT, T. H., DAVIS, D. A., HAMMOCK, M. K.: Choroid plexus papilloma. II. Ultrastructure and ultracytochemical localization of Na-K-ATPase. *Child's Brain* **2**:290–303, 1976.

63a. MILHORAT, T. H., HAMMOCK, M. K., CHIEN, T., D'AVIS, D. A.: Normal rate of cerebrospinal fluid formation five years after bilateral choroid plexectomy. *J. Neurosurg.* **44**:735–739, 1976.

64. MILHORAT, T. H., HAMMOCK, M. K., DAVIS, D. A., FENSTERMACHER, J. D.: Choroid plexus papilloma. I. Proof of cerebrospinal fluid overproduction. *Child's Brain* **2**:273–289, 1976.
65. MILHORAT, T. H., HAMMOCK, M. K., FENSTERMACHER, J. D., RALL, D. P., LEVIN, V. A.: Cerebrospinal fluid production by the choroid plexus and brain. *Science* **173**:330–332, 1971.
66. MILHORAT, T. H.: *Hydrocephalus and the Cerebrospinal Fluid.* Baltimore, Williams & Wilkins, 1972.
67. MILHORAT, T. H.: Pediatric neurosurgery. In Plum, F., McDowell, F. H. (eds.): *Contemporary Neurology Series,* Vol. 16. Philadelphia, F. A. Davis, 1978, pp. 91–135.
68. MILHORAT, T. H.: Structures and function of the choroid plexus and other sites of cerebrospinal fluid formation. In Bourne, G. H., Danielli, J. F. (eds.): *International Review of Cytology,* Vol. 47. New York, Academic Press, 1976, pp. 225–288.
69. MILLER, O. H., JAWORSKI, A. A., SILVERMAN, A. C., ELWOOD, M. J.: The effect of age on the protein concentration of cerebrospinal fluid of "normal" individuals and patients with poliomyelitis and other diseases. *Am. J. Med. Sci.* **228**:510–519, 1954.
70. NEUWELT, E. A., CLARK, W. K.: *Clinical Aspects of Neuroimmunology.* Baltimore, Williams & Wilkins, 1978, pp. 39–72.
71. OMMAYA, A. K., CORRAO, P., LETCHER, F. S.: Head injury in the chimpanzee. Part I. Biodynamics of traumatic unconsciousness. *J. Neurosurg.* **39**:152–166, 1973.
72. OPPELT, W. W., PATLAK, C. S., RALL, D. P.: Effect of certain drugs on cerebrospinal fluid production in the dog. *Am. J. Physiol.* **206**:247–250, 1964.
73. PAPPENHEIMER, J. R., HEISEY, S. R., JORDAN, E. F., DOWNER, J. D. C.: Perfusion of the cerebral ventricular system in unanesthetized goats. *Am. J. Physiol.* **203**:763–774, 1962.

74. Pollay, M., Curl, F.: Secretion of cerebrospinal fluid by the ventricular ependyma of the rabbit. *Am. J. Physiol.* **213:**1031–1038, 1967.

75. Pollay, M., Davson, H.: The passage of certain substances out of the cerebrospinal fluid. *Brain* **86:**137–150, 1963.

75a. Pollay, M.: Recent developments in cerebrospinal fluid physiology. In Tindall, G. T., Long, D. M. (eds.): *Contemporary Neurosurgery 14.* Baltimore, Williams and Wilkins, 1979.

76. Pollay, M.: Review of spinal fluid physiology: Production and absorption in relation to pressure. *Clin. Neurosurg.* **24:**254–269, 1977.

76a. Pollay, M., Stevens, A. F., Roberts, P. A.: Alteration in choroid plexus blood flow and cerebrospinal fluid formation by increased ventricular pressure. In Wood, J. H. (ed.): *Neurobiology of Cerebrospinal Fluid II.* New York, Plenum Press, 1981 (in press).

77. Porter, J. C.: Neuroendocrine systems: The need for precise identification and rigorous description of their operation. *Prog. Brain Res.* **39:**1–6, 1973.

78. Post, R. M., Allen, F. H., Ommaya, A. K.: Cerebrospinal fluid flow and iodide^{131} transport in the spinal subarachnoid space. *Life Sci.* **14:**1885–1894, 1974.

78a. Quincke, H.: Die Lumbalpunction des Hydrocephalus. *Berl. Klin. Wschr.* **28:**929–933, 1891.

79. Rapoport, S. I.: *The Blood–Brain Barrier in Physiology and Medicine.* New York, Raven Press, 1975, pp. 43–86.

80. Rapoport, S. I., Pettigrew, K. D.: A heterogenous, pore-vesicle membrane model for protein transfer from blood to cerebrospinal fluid at the choroid plexus. *Microvasc. Res.* **18:**105–119, 1979.

81. Reulen, H. J., Graham, R., Spatz, M., Klatzo, I.: Role of pressure gradients and bulk flow in dynamics of vasogenic brain edema. *J. Neurosurg.* **46:**24–35, 1977.

82. Rodriguez, E. M., Baigorria, Z., Rodriguez, A., Ciocca, D.: Evidence for the periventricular localization of the hypothalamic osmoreceptors. In Knowles, F., Vollrath, L. (eds.): *Neurosecretion—The Final Neuroendocrine Pathway.* Berlin, Springer-Verlag, 1973, pp. 319–320.

83. Rodriguez, E. M.: The cerebrospinal fluid as a pathway in neuroendocrine integration. *J. Endocrinol.* **71:**407–443, 1976.

84. Rubin, R. C., Henderson, E. S., Ommaya, A. K., Walker, M. D., Rall, D. P.: The production of cerebrospinal fluid in man and its modification by acetazolamide. *J. Neurosurg.* **25:**430–436, 1966.

85. Russell, D. S.: *Observations on the Pathology of Hydrocephalus.* London, Her Majesty's Stationery Office, 1949.

86. Sahar, A.: Choroidal origin of cerebrospinal fluid. *Isr. J. Med. Sci.* **8:**594–596, 1972.

87. Sato, O., Bering, E. A.: Extra-ventricular formation of cerebrospinal fluid. *Brain Nerve* **19:**883–885, 1967.

88. Sato, O., Hara, M., Asai, T., Tsugane, R., Kageyama, N.: The effect of dexamethasone phosphate on the production rate of cerebrospinal fluid in the subarachnoid space of dogs. *J. Neurosurg.* **39:**480–484, 1973.

89. Schwartz, M. L., Tator, C. H., Hoffman, H. J.: The uptake of hydrocortisone in mouse brain and ependymoblastoma. *J. Neurosurg.* **36:**178–183, 1972.

90. Seitz, H. D., Ocker, K.: The prognostic and therapeutic importance of changes in the CSF during the acute stage of brain injury. *Acta Neurochir.* **38:**211–231, 1977.

91. Shulman, K., Yarnell, P., Ransohoff, J.: Dural sinus pressure in normal and hydrocephalic dogs. *Arch Neurol.* **10:**575–580, 1964.

92. Smith, R. V., Roberts, P. A., Fisher, R. G.: Alteration of cerebrospinal fluid production in the dog. *Surg. Neurol.* **2:**267–270, 1974.

92a. Spector, R., Lorenzo, A. V.: Specificity of ascorbic acid transport system of the central nervous system. *Am. J. Physiol.* **226:**1468–1473, 1974.

93. Tripathi, B. S., Tripathi, R. C.: Vacuolar transcellular channels as a drainage pathway for cerebrospinal fluid. *J. Physiol. (London)* **239:**195–206, 1974.

94. Tschirgi, R. C., Frost, R. W., Taylor, J. L.: Inhibition of cerebrospinal fluid formation by a carbonic anhydrase inhibitor, 2 acetylamino-1,3,4-thiadiazole-5-sulfonamide (Diamox). *Proc. Soc. Exp. Biol. Med.* **87:**373–376, 1954.

95. Vates, T. S., Bonting, S. J., Oppelt, W. W.: Na-K-activated adenosine triphosphatase formation of cerebrospinal fluid in the cat. *Am. J. Physiol.* **206:**1165–1172, 1964.

96. Vela, A. R., Carey, M. E., Thompson, B. M.: Further data on the acute effect of intravenous steroids in canine CSF secretion and absorption. *J. Neurosurg.* **50:**477–482, 1979.

97. Vigh, B., Vigh-Teichman, I.: Comparative ultrastructure of the cerebrospinal fluid contacting neurons. In Bourne, G. H., Danielli, J. F. (eds.): *International Review of Cytology,* Vol. 35. New York, Academic Press, 1973, pp. 189–251.

98. Vigh-Teichman, I., Vigh, B.: Structure and function of the liquor contacting neurosecretory system. In Bargmann, W., Scharrer, B. (eds.): *Aspects of Neuroendocrinology.* Berlin, Springer-Verlag, 1970, pp. 329–337.

99. Vogh, B. P., Maren, T. H.: Sodium, chloride and bicarbonate movement from plasma to cerebrospinal fluid in cats. *Am. J. Physiol.* **228:**673–683, 1975.

100. Vorherr, H., Bradbury, M. W. B., Hoghoughi, M., Kleeman, C. R.: Antidiuretic hormones in cerebrospinal fluid during endogenous and exogenous changes in its blood levels. *Endocrinology* **83:**246–250, 1968.

101. Wald, A., Hochwald, G. M., Gandhi, M.: Evidence for the movement of fluid, macromolecules and ions from the brain extracellular space to the CSF. *Brain Res.* **151:**283–290, 1978.

102. Wald, A., Hochwald, G. M., Malhan, C.: Rela-

tionship between osmolarity and formation of cerebrospinal fluid (CSF). *Fed. Proc. Fed. Am. Soc. Exp. Biol.* **33:**418, 1974 (abstract #1173).

103. WEED, L. H.: Studies on cerebrospinal fluid. IV. The dual source of cerebrospinal fluid. *J. Med. Res.* **26:**91–113, 1914.
104. WEED, L. H.: Studies on cerebrospinal fluid. III. The pathways of escape from the subarachnoid spaces with particular reference to the arachnoid villi. *J. Med. Res.* **31:**51–92, 1914.
105. WEED, L. H.: The absorption of cerebrospinal fluid into the venous system. *Am. J. Anat.* **31:**191–221, 1923.
106. WEISNER, B., BERNHARDT, W.: Protein fractions of lumbar, cisternal and ventricular cerebrospinal fluid, separate areas of reference. *J. Neurol. Sci.* **37:**205–214, 1978.
107. WEISS, M. H., NULSEN, F. E.: The effect of glucocorticoids on CSF flow in dogs. *J. Neurosurg.* **32:**452–458, 1970.
108. WEISS, M. H., WERTMAN, N.: Modulation of CSF production by alterations in cerebral perfusion pressure. *Arch. Neurol.* **35:**527–529, 1978.
109. WELCH, M. H., POLLAY, M.: The spinal arachnoid villi of the monkeys *Cercopithecus aethiops sabaeus* and *Macacairus*. *Anat. Rec.* **145:**43–48, 1963.
110. WELCH, K.: Secretion of cerebrospinal fluid by choroid plexus of the rabbit. *Am. J. Physiol.* **205:**617–624, 1963.
111. WILLIAMS, B.: Cerebrospinal fluid pressure changes in response to coughing. *Brain* **99:**331–346, 1976.
112. WISLOCKI, G. B., PUTNAM, T. J.: Absorption from the ventricles in experimentally produced internal hydrocephalus. *Am. J. Anat.* **29:**313–320, 1921.
113. WOODBURY, J. W.: An epilogue, an hypothetical model for CSF formation and blood-brain barrier function. In Siesjo, B. K., Sorenson, S. C. (eds.): *Ion Homeostasis of the Brain.* Copenhagen, Munksgaard, 1971, pp. 465–471.
114. WOOD, J. H., BARTELT, D., JAMES, A. E., UDVARHELYI, G. B.: Normal-pressure hydrocephalus: Diagnosis and patient selection for shunt surgery. *Neurology (Minneapolis)* **24:**517–526, 1974.
115. WRIGHT, E. M.: Mechanisms of ion transport across the choroid plexus. *J. Physiol. (London)* **226:**545–571, 1972.
116. WRIGHT, E. M.: Transport processes in the formation of cerebrospinal fluid. *Rev. Physiol. Biochem. Pharmacol.* **83:**1–34, 1978.

CHAPTER 2

Role of Cyclic AMP in Cerebrospinal Fluid Production

Arthur M. Feldman, Mel H. Epstein, and Saul W. Brusilow

1. Introduction: The Adenylate Cyclase–Cyclic AMP System and Cell Secretion

The cyclic nucleotide 3′,5′-adenosine monophosphate (cyclic AMP) was first described by Sutherland[58] and his co-workers and since that time has been exhaustively studied in a large number of mammalian as well as nonmammalian tissues.[29,51] That cyclic AMP acts as a second messenger in mediating the actions of numerous hormones has been conclusively demonstrated. Cyclic AMP mediates such diverse functions as thyroid-stimulating hormone (TSH) stimulation of the thyroid gland, the renal response to parathyroid hormone, the effects of epinephrine, glucagon, and TSH on lipolysis, and the release of insulin by the pancreas.[29] A significant body of literature has developed showing that cyclic AMP plays a major role in the movement of solute and water across cellular membranes. Vasopressin (ADH) initiates antidiuresis in the kidney by stimulating adenylate cyclase activity.[14,32,42] Other hormones control fluid and electrolyte movement through cyclic AMP mediation in the gut,[23] the toad bladder,[47] the inner ear,[19] the pancreas,[8] and the salivary glands.[15]

The reactions involved in the adenylate cyclase–cyclic AMP system have been studied in the epinephrine- and glucagon-induced regulation of glycogen breakdown and synthesis.[29] It is generally accepted that a specific hormone or other exogenous or endogenous substance binds to a specific receptor site located on the vascular side of the cell. The interaction of the hormone and the receptor initiates the stimulation of the membrane-bound adenylate cyclase with which it is associated. It appears that this activation might occur, not directly, but through the activation of an unknown compound called a "transducer." The adenylate cyclase then catalyzes the formation of cyclic AMP from adenosine triphosphate (ATP). Cyclic AMP can travel one of two pathways. Either it is rapidly metabolized by the enzyme phosphodiesterase to the inactive 5′-AMP or it binds to the inhibitory subunit of a specific protein kinase. The active unit of the protein kinase then catalyzes the phosphorylation of certain cytosolic or membrane-bound proteins, the active phosphate group being provided by ATP. The phosphorylation of a cellular enzyme can result in both inhibition and excitation, and it is now known that a variety of cellular components such as electrolytes and prostaglandins (PG) can influence steps in the cyclic AMP pathway. Recent studies have also shown that various cells contain many adenylate cyclases, each of which may be associated with a specific receptor site. It has been proposed that each of these adenylate cyclases might be compartmentalized with its site of action.[26] In the toad bladder,

Arthur M. Feldman, Ph.D. • Department of Otolaryngology, Louisiana State University School of Medicine in Shreveport, Shreveport, Louisiana 71130. **Mel H. Epstein, M.D.** • Department of Neurosurgery, The Johns Hopkins University School of Medicine, Baltimore, Maryland 21205. **Saul W. Brusilow, M.D.** • Department of Pediatrics, The Johns Hopkins University School of Medicine, Baltimore, Maryland 21205.

PGE_1 stimulates sodium transport by stimulating adenylate cyclase, while also inhibiting ADH-induced water flow by inhibiting adenylate cyclase.[41] Since both these functions occurred in the same cell type, it is virtually axiomatic that some type of compartmentalization must be present. In addition, Corbin *et al.*[10] have demonstrated compartmentalization of cyclic AMP and cyclic-AMP-dependent protein kinase in heart tissue. Although it is possible that microtubules function in compartmentalization,[13] the exact nature of cellular compartmentalization is unclear.

Sutherland[58] and his co-workers established several criteria for establishing the role of cyclic AMP as a mediator of hormone action in a particular organ: (1) if a given response to a particular hormone is mediated by the cyclic AMP system, then adenylate cyclase in that system should be activated by the hormone; (2) the levels of cyclic AMP in the tissues should change in response to the hormone; (3) agents that inhibit phosphodiesterase, such as theophylline, should act synergistically with hormones that stimulate adenylate cyclase; and (4) the addition of exogenous cyclic AMP should cause a cellular reaction that mimics the addition of the hormone. While these have served as general guidelines for the study of the role of cyclic AMP in various organ systems, several problems are inherent in this approach. Cyclic AMP, like most organic phosphate compounds, does not easily penetrate cellular membranes, and therefore the addition of exogenous cyclic AMP or of cyclic AMP derivatives is difficult to rationalize. Furthermore, the addition of theophylline systemically causes inhibition of phosphodiesterase throughout the body.

Any attempt to study the cyclic AMP system in the choroid plexus and its effect on cerebrospinal fluid (CSF) production was hindered by the fact that the adenylate cyclase enzyme has a receptor site with great specificity, and no hormone or endogenous agent had been found that significantly stimulated the rate of CSF production. However, studies in the gut,[56] the kidney, and the inner ear[19] had shown that cholera toxin, a purified enterotoxin isolated from the *Vibrio cholerae* bacterium, could serve as an adenylate cyclase probe. Cholera toxin is an effective probe for several reasons: (1) It stimulates adenylate cyclase in the small intestinal mucosa,[57] liver,[31] thyroid,[44] leukocyte,[6] and kidney.[39] (2) It induces urinary electrolyte excretion in the kidney[27] and isotonic fluid secretion in the gut.[7] (3) It is a soluble, high-molecular-weight protein that could be easily injected into the ventricles. (4) It can produce its effect from the luminal side of epithelial cells. (5) Its activity is not associated with receptor-site binding, but it binds directly to the adenylate cyclase molecule and continues to stimulate the production of cyclic AMP until the particular cell in which it is located dies.[11,55]

To determine the effects of cholera toxin and other agents on CSF production as well as on the choroid plexus adenylate cyclase–cyclic AMP system, we developed *in vivo* and *in vitro* experimental models.

2. *In Vivo* Model

Initial *in vivo* experiments were performed on mongrel dogs weighing 20–25 kg, anesthetized with pentobarbital, and maintained on positive-pressure ventilation. Arterial pressure, heart rate, intracranial pressure, body temperature, PO_2, and PCO_2 were constantly monitored and stabilized. CSF production rate was measured using a modification of the technique described by Pappenheimer *et al.*[48] Both lateral ventricles were cannulated and perfused with Elliott's B artificial CSF containing [^{14}C]carboxyinulin. The outflow catheter was placed at zero pressure with respect to the ear, so that bulk collection of fluid represented CSF production plus the volume of perfusate less the amount absorbed, and CSF production rate was calculated by the indicator dilution method.[18] Considerable variation was seen within the population of dogs, and therefore each animal had to serve as its own control. After the dog was perfused for 100 min, and a control production rate was established, 250 μl purified cholera toxin, heat-inactivated cholera toxin, or a bovine serum albumin control was injected into each lateral ventricle. The perfusion was again started after a 2-hr incubation period, and the CSF formation was again measured. While significant results were obtained using this technique, it had the major deficiency that each animal had to serve as its own control and therefore it required an experimental time of 7–8 hr. To alleviate this problem, we began to use cats as the experimental animals and found that population differences were much smaller and that we could therefore compare populations of controls and populations of treated.

3. *In Vitro* Model

The *in vitro* experiments were performed on choroid plexuses isolated from male Sprague–Dawley rats weighing 150–250 g. After the animals were anesthetized with pentobarbital and the cerebral hemispheres were exposed, the rats were perfused with isotonic saline containing heparin through a needle placed in the left ventricle to clear the choroid plexuses of blood. Both choroid plexuses were then removed and placed in cold Krebs–Ringer bicarbonate buffer that contained 0.3 M glucose and that was constantly gassed. The choroid plexuses were then placed in individual incubation vials (2–4/vial) and placed in a shaker bath maintained at 38°C and constantly agitated and gassed with 95% O_2, 5% CO_2. After a 15-min equilibration period, cholera toxin, heat-inactivated cholera toxin, other pharmacological agents, or appropriate controls were added to the incubation vessel. The incubation period was stopped by the addition of cold 7% trichloroacetic acid and immediately homogenized. Following centrifugation, the supernatant was removed, ether-washed, acetylated, and assayed for cyclic AMP activity using a modification of the radioimmunoassay developed by Harper and Brooker.[34] The assay involved the competition of succinylated cyclic AMP and an acetylated labeled sample for binding sites on a cyclic AMP antibody. The assay was linear between 10 and 100 fmol and was therefore appropriate for use with the small tissue samples obtained from the choroid plexus. Additionally, the presence of bovine serum albumin in the assay reaction reduced the effect of buffer salts on cyclic AMP binding and thereby obviated the purification of the samples by column chromatography that was necessary with many earlier assay techniques. By modifying the basic protocol and using the cyclic AMP assay method, adenylate cyclase activity could be measured using the technique of Kurokawa *et al.*[39]

4. Effects of Cholera Toxin on CSF Production and Choroid Plexus Cyclic AMP

The addition of purified cholera toxin to both lateral ventricles resulted in a greater than twofold increase in CSF production in dogs when compared with their own control values (Fig. 1).[18] No difference was observed between bovine serum albumin, heat-inactivated cholera toxin, and saline controls. Additionally, an identical increase was seen when CSF production was measured using the inulin dilution technique as well as when production was measured by bulk outflow from the cisterna magna

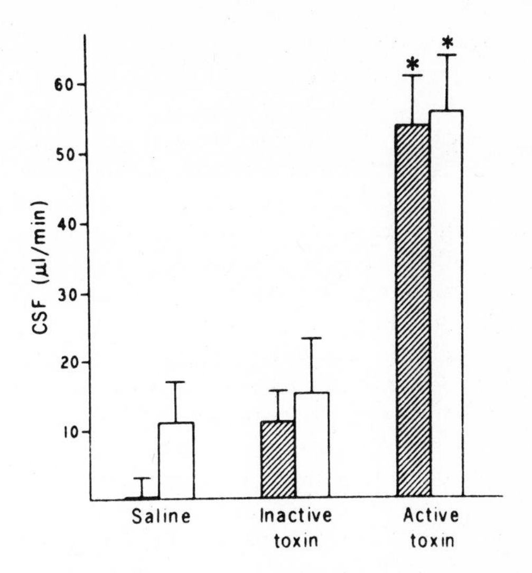

Figure 1. Differences between baseline and experimental CSF production in the dog after intraventricular administration of saline, heat-inactivated cholera toxin, and active cholera toxin. (■) Change in CSF formation calculated according to the inulin dilution technique; (□) change in bulk collection at the cisterna magna outflow catheter. Brackets represent the standard errors of the mean; $n = 5$ in all groups. (*) Indicates a significant difference ($P < 0.001$). From Epstein *et al.*[18]

catheter. Cholera-toxin challenge had a similar effect in the cat, there being a 2-fold difference between CSF production rates in the control and the experimental population.[22] The standard errors within the control as well as within the treated population were very small, making comparisons of populations appropriate.

It has been demonstrated that the intraventricular administration of cholera toxin in the unanesthetized cat causes a hyperthermia of almost 2°C.[9] However, the time period required to see an effect with dosages of cholera toxin used in our anesthetized cats was almost 5 hr. Therefore, any temperature effect would have occurred after our experiments were concluded, and we observed no hypo- or hyperthermia in our animals. When we switched from using the dog to the cat as the experimental model, we encountered an interesting problem that points up one of the major pitfalls in *in vivo* measurement of CSF production and particularly in its relationship to cyclic AMP generation. In our initial experiments with the cat, we used the anesthetic ketamine. Following the normal experimental protocol, there was only a 25% increase in CSF production after challenge with cholera toxin—considerably smaller than that seen in the dog. However, when we anesthetized the cats with pentobarbital, we got over a 2-fold increase in CSF production and an increase that was very similar to that seen in dogs. Since the experimental protocol in both these situations required the administration of either cholera toxin or a control solution to the ventricles and an incubation period of 2 hr prior to measurement of the CSF production rate, we were interested in seeing what would happen if CSF production rate was measured immediately after the administration of the anesthetic. As can be seen in Fig. 2, the CSF production rate in cats that had been anesthetized with ketamine was twice the rate in cats that had been anesthetized with pentobarbital. It is also interesting to note that in both the ketamine cholera-toxin-challenged animals and the pentobarbital cholera-toxin-challenged animals, the CSF production rate after challenge was almost identical. This would indicate that the effects of ketamine and cholera toxin on increasing CSF production may not be additive. However, the possibility that pentobarbital lowers normal CSF production at the same time that ketamine raises CSF production must also be considered. The important point to note is that anesthetics do have a significant effect on CSF pro-

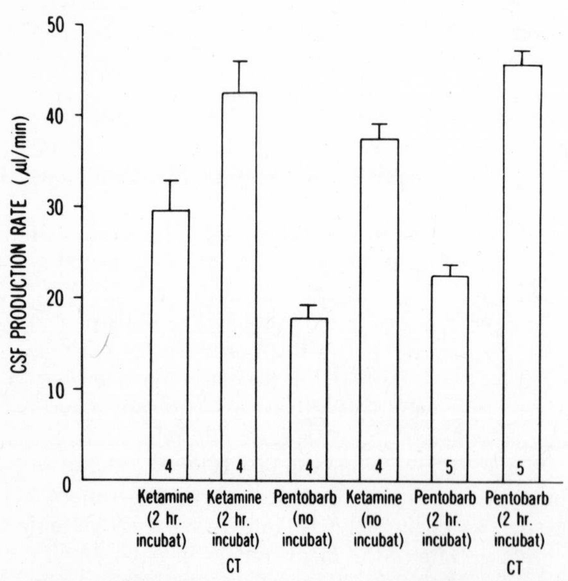

Figure 2. Effects of pentobarbital and ketamine on CSF production and cholera toxin (CT)-induced CSF secretion. In some of the groups, secretion was measured after the intraventricular administration of CT (100 μg) or a control solution and a 2-hr incubation period. In the other experiments, CSF production was measured immediately after preparation of nonchallenged animals (unpublished results).

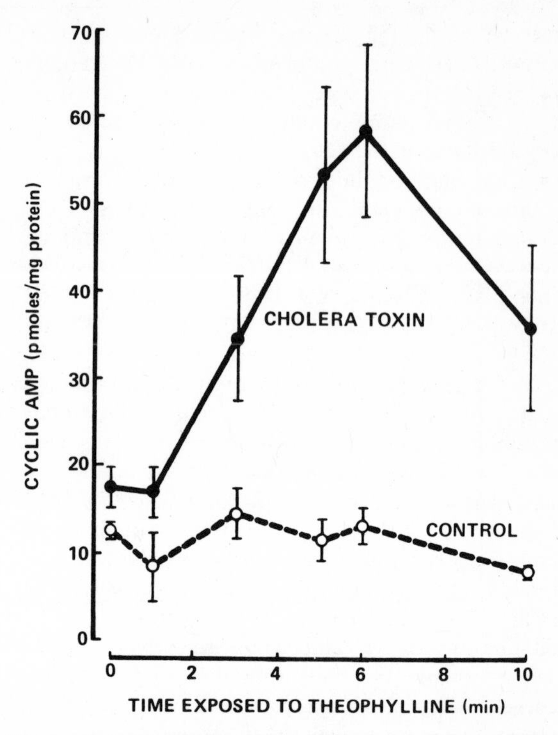

Figure 3. Cyclic AMP accumulation in choroid plexus treated with cholera toxin (100 μg/ml) or inactivated cholera toxin (control) and exposed to theophylline for varying periods of time. Tissues were incubated at 38°C for 60 min with cholera toxin prior to the addition of the theophylline. Points represent the mean ±S.E. of four experiments. From Feldman *et al.*[20]

duction rate and can dramatically alter experimental results in CSF studies.

The brains of cats used in the *in vivo* experiments showed no significant pathology in either the choroid plexus or the ventricular lining when observed with the light microscope, and both control and treated specimens appeared to be identical. This was consistent with earlier pathological studies in the gut after cholera-toxin challenge.[17]

That the increase in CSF production after cholera-toxin challenge was due to an increase in cyclic AMP production was demonstrated in a series of *in vitro* studies (Figs. 3 and 4).[20] Cholera toxin caused a 5-fold increase in cyclic AMP accumulation in choroid plexuses that had been exposed to the toxin for 60 min. This increase was dose-responsive, with maximum activity occurring at a cholera toxin concentration of 10 μg/ml, and was seen only when the phosphodiesterase inhibitor theophylline was present in the incubation medium 5 min prior to the

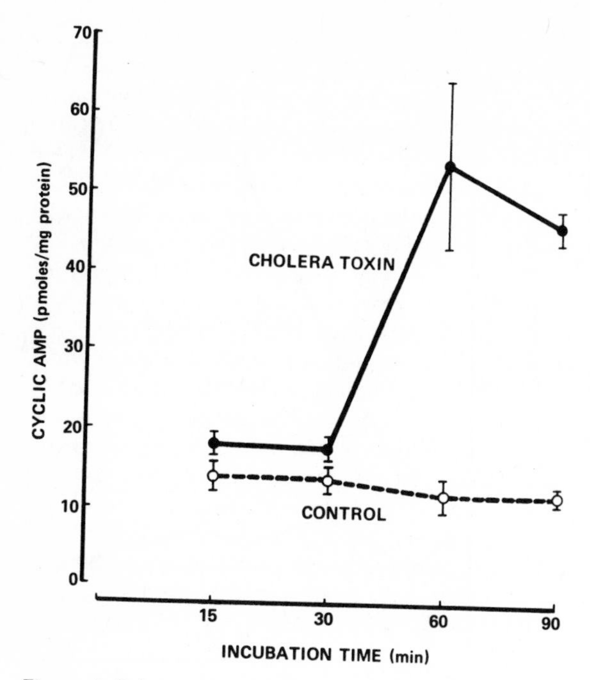

Figure 4. Relationship between time of cholera toxin (10 μg/ml) incubation (at 38°C) and cyclic AMP accumulation in the choroid plexus. Heat-inactivated cholera toxin served as a control. Theophylline (10 mM) was added 5 min prior to termination of the incubation. Each point represents the mean ±S.E. of three incubations. From Feldman *et al.*[20]

termination of the experiment. The theophylline requirement was also seen in studies in the kidney[4,39] and in the toad bladder.[46] It is believed that without theophylline, the cyclic AMP that is produced and that does not bind to a protein kinase is quickly metabolized by phosphodiesterase and is therefore not recovered in the assay procedure. As was demonstrated in both the kidney[39] and the gut,[37,49] cholera toxin produced an increase in cyclic AMP accumulation after a lag time of approximately 60 min.[20] However, when choroid plexuses were exposed to cholera toxin for only 5 min and then washed several times and allowed to incubate in cholera-toxin-free buffer for the remainder of the 60-min incubation period, a 5-fold increase in cyclic AMP accumulation still occurred. Therefore, the toxin binds rapidly and irreversibly to the choroid plexus, and the required lag period is not a consequence of membrane binding. That the increase in cyclic AMP accumulation was due to stimulation of the adenylate cyclase enzyme and not to inhibition of phosphodiesterase was shown by the fact that cholera toxin caused over a 2-fold increase in *in vitro* adenylate cyclase activity in the choroid plexus.[20]

If the increased CSF production seen after administration of cholera toxin was in fact due to increased production of cyclic AMP, we would expect that the administration of high doses of cyclic AMP to the ventricles or to the choroid plexus should result in an increase in CSF production. When the lateral ventricles were perfused with 1 mM dibutyryl cyclic AMP for 1 hr and then perfused with labeled inulin and cyclic AMP, there was a significant 30% increase in CSF production (unpublished results). The fact that this increase was less than that seen after cholera toxin administration and occurred only with high concentrations of cyclic AMP could well be due to the fact that only a small portion of the cyclic AMP perfused was able to cross the membrane into the cells of the choroid plexus or that a large portion of that cyclic AMP that was taken up by the cells of the choroid plexus was metabolized by phosphodiesterase within the cell before it was able to bind effectively to a protein kinase.

The molecular mechanism of the cholera-toxin-induced CSF production has not been established. However, studies in the gut have shown that the massive secretion of fluid associated with cholera toxin challenge was due to inhibition of Na^+ transport and stimulation of an outwardly directed Cl^- pump.[23,24]

5. Prostaglandins and CSF Production

Prostaglandins, naturally occurring fatty acids that are derivatives of prostanoic acid, have been shown to influence fluid and electrolyte movement in a number of secretory epithelia including the toad bladder,[26,41] the gut,[1] and the kidney,[2,33] These actions were mediated through either a stimulatory or an inhibitory effect on cyclic AMP production.[26,39,41,57] Of the five naturally occurring prostaglandins that were studied (PGA_1, PGA_2, PGE_1, PGE_2, and PGF_2), only PGE_2 had a significant effect on cyclic AMP accumulation when measured at concentrations that had been shown to be effective in stimulating adenylate cyclase activity in other organ systems.[20] The effect of PGE_2 was does-responsive, with a maximal accumulation occurring at a concentration of 10 μg/ml after 1 min of incubation (Fig. 5). The choroid plexus appears to be highly organ-specific for PGE_2 in that all five prostaglandins resulted in a 2-fold increase in cyclic AMP accumu-

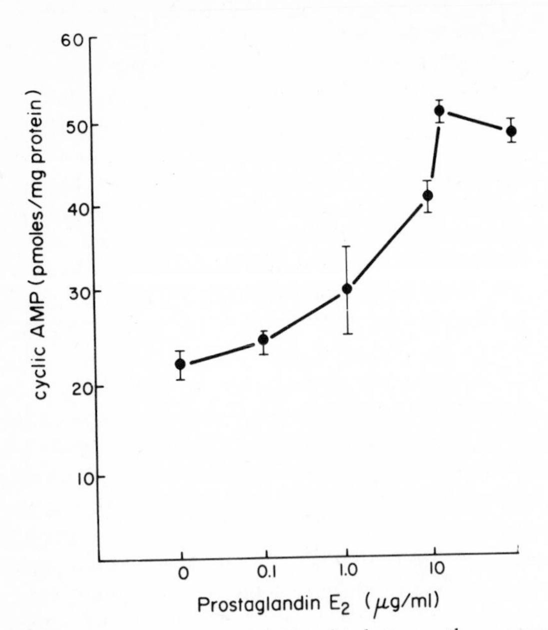

Figure 5. Dose–response relationship between the concentration of PGE_2 in the incubation medium and cyclic AMP accumulation in the choroid plexus. Plexuses were pretreated with theophylline (10 mM) for 10 min and then incubated with PGE_2 for 1 min at 38°C. Points represent the means ±S.E. of four experiments. From Feldman *et al.*[20]

lation in kidney tissue,[39] while PGE_1 stimulated accumulation in the toad bladder[41] and the intestines.[23]

That a 1- to 2-hr incubation period was required for cholera toxin to exert its secretory activity has been seen both in our *in vivo* and *in vitro* studies[18,20] and in studies in other organs. Two hypotheses have been proposed to explain the presence of this lag time. One is that protein synthesis occurred during the time period, and the other is that prostaglandin synthesis was required for cholera-toxin-induced activity to occur. Studies in other tissues have shown that the administration of agents that block protein synthesis had no effect on either cyclic AMP production or secretory processes.[36] However, blockage of prostaglandin synthesis with either of the nonsteroidal antiinflammatory agents indomethacin and aspirin inhibited cholera-toxin-induced intestinal secretion.[25,35] This effect was not accompanied by a decrease in cyclic AMP accumulation.[5]

When animals were pretreated with indomethacin prior to the intraventricular administration of cholera toxin, there was a significant 50% decrease in CSF production when compared with cholera-toxin-stimulated CSF production. Paradoxically, when indomethacin alone was administered to the animals, there was a 50% increase in CSF production (Fig. 6).[22] Similar results were seen in the gut, in which pretreatment with indomethacin decreased cholera-toxin-induced fluid secretion, but initiated net intestinal reabsorption when given alone.[60] Unlike studies in the intestine, when choroid plexuses were pretreated with indomethacin *in vitro*, the cholera-toxin-induced increase in cyclic AMP production was blunted.[22] One explanation of the paradoxical effect of indomethacin on CSF secretion is the presence of multiple secretory pathways, some of which are inhibited and others of which are stimulated by prostaglandins. When all prostaglandin synthesis is inhibited, as occurs after indomethacin treatment, secretion can occur via those pathways that are normally inoperative in the presence of a specific inhibitory prostaglandin. Although the exact role of the prostaglandins in the mechanism of CSF production is unclear, it is evident that prostaglandins are intimately linked with cyclic AMP and CSF secretion.

6. Adenylate Cyclase Agonists and CSF Production

To find other endogenous agents that might stimulate CSF production by acting as cyclic AMP agon-

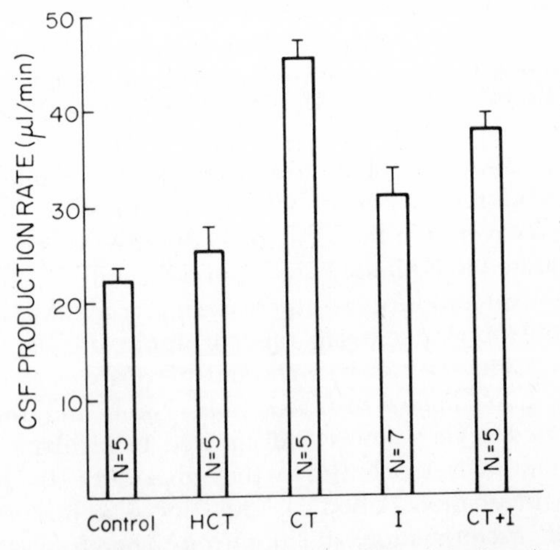

Figure 6. Effects of cholera toxin (CT), heat-inactivated cholera toxin (HCT), and indomethacin (I) on CSF production in the cat. Production rates were measured 2–3 hr after administration of 50 μg purified CT to each lateral ventricle. Indomethacin was given intravenously (30 mg/kg) and subcutaneously (30 mg/kg) at the time of CT challenge, and an intravenous booster dose (30 mg/kg) was given 2.5 hr later. From Feldman *et al.*[22]

Table 1. Choroid Plexus Adenylate Cyclase Agonists[a]

Agent	Minimal effective dose (μg/ml)	Species	Approximate increase after maximal dose
Isoproterenol	0.001	Rat	3×
Epinephrine	0.01	Rat	2×
Norepinephrine	10	Rat	2×
ACTH	0.1	Rabbit	10×
Serotonin	10	Rabbit	4×
PGE_2	20	rat	2×

[a] Adapted from Rudman *et al.*[53] and Feldman *et al.*[20]

ists, we studied the effects of various pharmacological agents that were known to stimulate the cyclic AMP system in other organs (Tables 1 and 2). The first of these agents that we looked at were the catecholamines, which are known adenylate cyclase agonists.[4,50,51] However, in low concentrations, neither epinephrine, isoproterenol, nor epinephrine with propranolol had any effect on cyclic AMP accumulation when compared with controls. Rudman *et al.*[53] recently found that isoproterenol, epinephrine, and norepinephrine, in slightly higher concentrations, stimulated cyclic AMP production in the rat choroid plexus. However, this effect was species-specific in that catecholamines failed to increase cyclic AMP accumulation in the rabbit choroid plexus. This is particularly significant in light of the recent work by Lindvall *et al.*[40] They found that in the rabbit, the choroid plexus receives a well-developed adrenergic nerve supply that originates almost entirely from the superior cervical sympathetic ganglia. The nerve terminals appear to innervate the secretory epithelia as well as the vascular beds. Additionally, sympathetic denervation resulted in a 33% increase in bulk CSF production, while stimulation of the nerve trunk beneath the superior cervical ganglia resulted in a 32% reduction in CSF production. Moreover, it was shown by Edvinsson *et al.*[16] that sympathetic denervation results in an increase in carbonic anhydrase activity, an enzyme believed to function in CSF formation. The obvious species diversity makes these data difficult to interpret, and it would be of great interest to learn

Table 2. Pharmacological Agents That Are Known Adenylate Cyclase Agonists or Have an Effect on CSF Secretion but Have No Effect on Choroid Plexus Cyclic AMP Accumulation[a]

Agent	Incubation time (min)	Maximal dose studied (μg/ml)	Species
Dopamine	2–30	1000	Rat and rabbit
Histamine	2–30	1000	Rat and rabbit
Serotonin	2–30	1000	Rat and rabbit
Arginine vasopressin	2–30	1000	Rat and rabbit
Lysine vasopressin	2–30	1000	Rat and rabbit
Oxytocin	2–30	1000	Rat and rabbit
Angiotensin	2–30	1000	Rat and rabbit
ACTH	2–30	1000	Rat and rabbit
β-Melanocyte-stimulating hormone	2–30	1000	Rat and rabbit
Choroid plexus peptide IIF	2–30	1000	Rat and rabbit
Carbachol	2	182	Rat[b]
Acetazolamide	10	23	Rat[b]
Vitamin A	10	20	Rat[b]
Prostaglandins E_1, A_1, A_2, F_{2a}	10	10	Rat[b]

[a] Adapted from Rudman *et al.*[53]
[b] From Feldman *et al.* (1978, unpublished results).

whether the epithelia of the choroid plexus in other species also receives direct sympathetic innervation. Also, it would be interesting to know whether sympathetic stimulation in the rabbit induces production of cyclic guanosine 3′,5-monophosphate (GMP), which could have an action antagonistic to that of cyclic AMP.[30]

The effects of several other substances that are known adenylate cyclase agonists in other organs were studied in the choroid plexus (unpublished results). Vasopressin, which controls water movement in the kidney,[32] as well as in the toad bladder,[46] through cyclic AMP mediation, had no effect on cyclic AMP accumulation in the choroid plexus in dosages ranging from 10^{-5} to 10^{-7} M and incubation times ranging from 1 to 10 min. Similar negative results were obtained with vitamin A, the massive ingestion of which has been associated with hydrocephalus.[38] Carbachol, serotonin, and acetazolamide, all of which can stimulate adenylate cyclase,[51,52] had no effect on the adenylate cyclase of the choroid plexus. Rudman *et al.*[53] found that rat choroid plexus cyclic AMP concentrations were not elevated by dopamine, histamine, serotonin, arginine and lysine vasopressins, oxytocin, angiotensin, choroid plexus peptides IIF, adrenocorticotropin (ACTH), or β-melanocyte-stimulating hormone (β-MSH). However, in the rabbit choroid plexus, ACTH and serotonin affected cyclic AMP content. These results were particularly interesting in that ACTH, serotonin, dopamine, and histamine all stimulate adenylate cyclase in other regions of the brain, while the intracisternal injection of ACTH, β-MSH, and IIF caused an increase in cyclic AMP concentration of the CSF within 30 min.[54] Unfortunately, the significance of the species specificity and of the effect of ACTH and serotonin on rabbit CSF production is unclear.

7. Carbonic Anhydrase, Cyclic AMP, and Cellular Compartmentalization

Considerable attention has been given to the role of carbonic anhydrase in choroid plexus secretion of the CSF. Maren and his co-workers[43,59] have proposed that carbonic anhydrase functions in anion, cation, and fluid secretion into the CSF by hydroxylation of CO_2 in the choroid plexus. This theory is supported by the fact that the carbonic anhydrase inhibitor acetazolamide caused a 60% reduction in CSF production.[12] Several attempts have been made to correlate cyclic AMP production and carbonic anhydrase activity. Narumi and Maki[45] demonstrated that gastric acid secretion in the gut was associated with potentiation of carbonic anhydrase by cyclic AMP. Beck *et al.*[3] demonstrated that parathyroid hormone, which initiates ion and water excretion by depressing proximal tubular reabsorption through cyclic AMP mediation, inhibited renal carbonic anhydrase activity *in vitro*. However, another investigator[28] was unable to verify these results. Studies in our laboratory have shown that when choroid plexuses were incubated with cholera toxin for 60 min, there was a significant 2-fold increase in carbonic anhydrase activity when compared with heat-inactivated cholera toxin controls.[21] Incubation with 10^{-3} M dibutyryl cyclic AMP caused a significant but small (19%) increase in choroid plexus carbonic anhydrase activity. However, PGE_2, which stimulated choroid plexus adenylate cyclase activity, had no effect on carbonic anhydrase activity. Because of

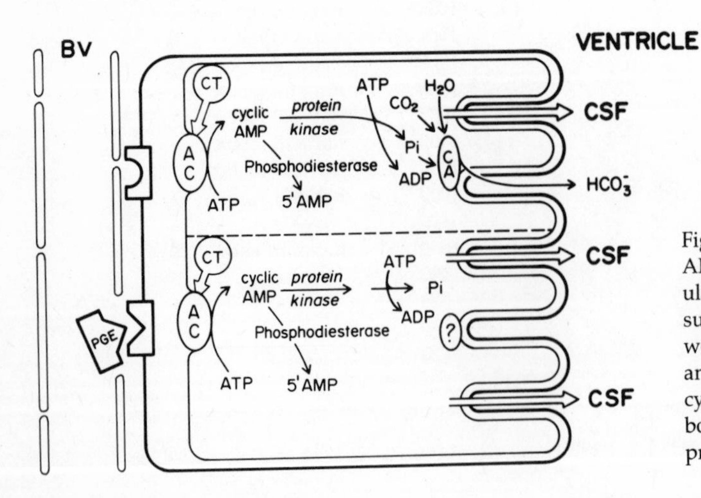

Figure 7. A hypothetical model for the role of cyclic AMP in CSF production. Cholera toxin (CT) stimulates all adenylate cyclase enzymes (AC) and results in an increased production of cyclic AMP as well as an increase in CSF secretion and carbonic anhydrase activity. PGE_2 initiates an increase in cyclic AMP accumulation, but has no effect on carbonic anhydrase. Unfortunately, its effect on CSF production is unknown.

the large concentration of carbonic anhydrase in the choroid plexus and the fact that carbonic anhydrase must be inhibited by 99% to inhibit CSF production, it is possible that the 2-fold increase in activity seen after challenge with cholera toxin is physiologically insignificant. However, if a small portion of the cellular carbonic anhydrase was compartmentalized with a specific adenylate cyclase and sequestered from the general cystolic pool, a 2-fold increase after cholera toxin challenge would be significant (Fig. 7). This theory is supported by the studies in the toad bladder that have already been mentioned as well as by the recent finding that denervation results in an increase in carbonic anhydrase activity as well as in bulk CSF production in the rabbit.[16] Compartmentalization would also explain why cholera toxin and PGE_2 stimulated choroid plexus adenylate cyclase, but only cholera toxin, which binds to all adenylate cyclases irrespective of their receptor sites, stimulated carbonic anhydrase activity.

8. Conclusions

In conclusion, it appears that cyclic AMP plays an important role in CSF production. However, specific hormones or endogenous agents that stimulate CSF secretion via cyclic AMP mediation have not yet been identified. Prostaglandins affect choroid plexus cyclic AMP concentrations and cause changes in CSF production rate; however, the molecular mechanisms of their actions are unclear.

The practical implications of understanding CSF production concern the modification of many pathological states involving abnormal accumulation of CSF. As these mechanisms are better understood, new pharmacological agents may be found for the medical treatment of these disorders.

ACKNOWLEDGMENTS. The authors gratefully acknowledge the invaluable assistance of Mr. Thomas Smith of the Department of Neurosurgery. This work was supported in part by USPHS Grants AMO7145, NS 11274, and HD 00091, the Kerr Foundation, and the National Foundation–March of Dimes.

References

1. Al-Awqati, Q., Greenough, W. B.: Prostaglandins inhibit intestinal sodium transport. *Nature (London) New Biol.* **238**:26–27, 1972.
2. Beck, N. P., Kaneko, T., Zor, U., Field, J. B., Davis, B. B.: Effects of vasopressin and prostaglandin E_1 on the adenyl cyclase cyclic 3′,5′-adenosine monophosphate system of the renal medulla of the rat. *J. Clin. Invest.* **50**:2461–2465, 1971.
3. Beck, N., Kim, K. S., Wolak, M., Davis, B. B.: Inhibition of carbonic anhydrase by parathyroid hormone and cyclic AMP in rat renal cortex *in vitro*. *J. Clin. Invest.* **55**:149–156, 1975.
4. Beck, N. P., Reed, S. W., Murdaugh, H. V., Davis, B. B.: Effects of catecholamines and their interaction with other hormones on cyclic 3′,5′-adenosine monophosphate of the kidney. *J. Clin. Invest.* **51**:939–944, 1972.
5. Bourne, H. R.: Cholera enterotoxin: Failure of anti-inflammatory agents to prevent cyclic AMP accumulation. *Nature (London)* **241**:399, 1973.
6. Bourne, H. R., Lehrer, R. I., Lichtenstein, L. M., Weissmann, G., Zurier, R.: Effects of cholera enterotoxin on adenosine 3′,5′-monophosphate and neutrophil function: Comparison with other compounds which stimulate leukocyte adenyl cyclase. *J. Clin. Invest.* **52**:698–708, 1973.
7. Carpenter, C. C. J., Sack, R. B., Feeley, J. C., Steenberg, R. W.: Site and characteristics of electrolyte loss and effect of intraluminal glucose in experimental canine cholera. *J. Clin. Invest.* **47**:1210–1220, 1968.
8. Case, R. M., Laundy, T. J., Scratcherd, T.: Adenosine 3′,5′-monosphophate (cyclic AMP) as the intracellular mediator of the action of secretin as the exocrine pancreas. *J. Physiol. (London)* **204**:45–47P, 1969.
9. Clark, W. G., Cumby, H. R., Davis, H. E., IV: The hyperthermic effect of intracerebroventricular cholera enterotoxin in the unanaesthetized cat. *J. Physiol. (London)* **240**:493–504, 1974.
10. Corbin, J. D., Sugden, P. H., Lincoln, T. M., Keely, S. L.: Compartmentalization of adenosine 3′5′-monophosphate and adenosine 3′5′-monophosphate-dependent protein kinase in heart tissue. *J. Biol. Chem.* **252**:3854–3861, 1977.
11. Cuatrecasas, P.: Cholera toxin–fat cell interaction and the mechanism of action of the lipolytic response. *Biochemistry* **12**:3567–3577, 1973.
12. Davson, H.: *Physiology of the Cerebrospinal Fluid.* London, Churchill, 1967, p. 134.
13. Dousa, T. P., Barnes, L. D.: Effects of colchicine and vinblastine on the cellular action of vasopressin in mammalian kidney. *J. Clin. Invest.* **54**:252–262, 1974.
14. Dousa, T. P., Walter, R., Schwartz, I. L., Sands, H., Hechter, O.: Role of cyclic AMP in the action of neurohypophyseal hormones on kidney. In Greengard, P., Robison, G. A. (eds.): *Advances in Cyclic Nucleotide Research*, Vol. 1. New York, Raven Press, 1972, pp. 121–135.
15. Durham, J. P., Butcher, F. R.: The effect of catecholamine analogues upon amylase secretion from the mouse parotid gland *in vivo*: Relationship to changes in cyclic AMP and cyclic GMP levels. *FEBS Lett.* **47**:218–221, 1974.

16. EDVINSSON, L., HAKANSON, R., LINDVALL, M., OWMAN, C., SVENSSON, K.-G.: Ultrastructural and biochemical evidence for a sympathetic neural influence on the choroid plexus. *Exp. Neurol.* **48**:241–251, 1975.
17. ELLIOTT, H. L., CARPENTER, C. C. J., SACK, R. B., YARDLEY, J. H.: Small bowel morphology in experimental canine cholera. *Lab. Invest.* **22**:112–120, 1970.
18. EPSTEIN, M. H., FELDMAN, A. M., BRUSILOW, S. W.: Cerebrospinal fluid production: Stimulation by cholera toxin. *Science* **196**:1012–1013, 1977.
19. FELDMAN, A. M., BRUSILOW, S. W.: Effects of cholera toxin on cochlear endolymph production: Model for endolymphatic hydrops. *Proc. Natl. Acad. Sci. U.S.A.* **73**:1761–1764, 1976.
20. FELDMAN, A. M., EPSTEIN, M. H., BRUSILOW, S. W.: Effects of cholera toxin and prostaglandins on the rat choroid plexus *in vitro*. *Brain Res.* **167**:119–128, 1979.
21. FELDMAN, A. M., EPSTEIN, M. H., MAYLACK, F., BRUSILOW, S. W.: Effects of cholera toxin on choroid plexus carbonic anhydrase activity *in vitro*. *Neurol. Res.* (in press).
22. FELDMAN, A. M., SMITH, T., EPSTEIN, M. H., BRUSILOW, S. W.: Effects of indomethacin on cholera toxin–induced cerebrospinal fluid production. *Brain Res.* **142**:379–383, 1978.
23. FIELD, M.: Ion transport in rabbit ileal mucosa. II. Effects of cyclic 3′,5′-AMP. *Am. J. Physiol.* **221**(4):992–997, 1971.
24. FIELD, M., FROMM, D., AL-AWQATI, Q., GREENOUGH, W. B., III: Effect of cholera enterotoxin on ion transport across isolated ileal mucosa. *J. Clin. Invest.* **51**:796–803, 1972.
25. FINCH, A. D., KATZ, R. L.: Prevention of cholera-induced intestinal secretion in the cat by aspirin. *Nature (London)* **238**:273–274, 1972.
26. FLORES, J., WITKUM, P. A., BECKMAN, B., SHARP, G. W. G.: Stimulation of osmotic water flow in toad bladder by prostaglandin E_1: Evidence for different compartments of cyclic AMP. *J. Clin. Invest.* **56**:256–262, 1975.
27. FRIEDLER, R. M., KUROKAWA, K., COBURN, J. W., MASSRY, S. G.: Renal action of cholera toxin. I. Effects on urinary excretion of electrolyte and cyclic AMP. *Kidney Int.* **7**:77–85, 1975.
28. GARG, L. C.: Effect of parathyroid hormone and adenosine 3′,5′-monophosphate on renal carbonic anhydrase. *Biochem. Pharmacol.* **24**:437–439, 1975.
29. GOLDBERG, N. D.: Cyclic nucleotides and cell function in cell membranes. In Weissman, G., Claiborne, R. (eds.): *Biochemistry, Cell Biology and Pathology*. New York, HP Publishing Co., 1975, pp. 185–202.
30. GOLDBERG, N. D., HADDOX, M. K., DUNHAM, E., LOPEZ, C., HADDEN, J. W.: The yin yang hypothesis of biological control: Opposing influences of cyclic GMP and cyclic AMP in the regulation of cell proliferation and other biological processes. In Clarkson, B., Baserga, R. (eds.): *The Cold Spring Harbor Symposium on the Regulation of Proliferation in Animal Cells*. New York, Cold Spring Harbor Laboratory, 1974, pp. 609–626.
31. GORMAN, R. E., BITENSKY, M. W.: Selective effects of cholera toxin on the adrenaline responsive component of hepatic adenyl cyclase. *Nature (London)* **235**:439–440, 1972.
32. GRANTHAM, J. J., BURG, M. B.: Effect of vasopressin and cyclic AMP on permeability of isolated collecting tubules. *Am. J. Physiol.* **211**:255–259, 1966.
33. GRANTHAM, J. J., ORLOFF, J.: Effect of prostaglandin E_1 on the permeability response of the isolated collecting tubule to vasopressin, adenosine 3′,5′-monophosphate, and theophylline. *J. Clin. Invest.* **47**:1154–1161, 1968.
34. HARPER, J. F., BROOKER, G.: Femtomole sensitive radioimmunoassay for cyclic AMP and cyclic GMP after 2′0 acetylation by acetic anhydride in aqueous solution. *J. Cyclic Nucleotide Res.* **1**:207–218, 1975.
35. JACOBY, H. J., MARSHAL, C. H.: Antagonism of cholera enterotoxin by anti-inflammatory agents in the rat. *Nature (London)* **235**:163–165, 1972.
36. KIMBERG, D. V., FIELD, M., GERSHON, E., SCHOOLEY, R. T., HENDERSON, A.: Effects of cycloheximide on the response of intestinal mucosa to cholera enterotoxin. *J. Clin. Invest.* **52**:1376–1383, 1973.
37. KIMBERG, D. V., FIELD, M., JOHNSON, J., HENDERSON, A., GERSHON, E.: Stimulation of intestinal mucosal adenyl cyclase by cholera toxin and prostaglandins. *J. Clin. Invest.* **50**:1218–1230, 1971.
38. KNUDSON, A. G., JR., ROTHMAN, P.: Hypervitaminosis A: Review with discussion of vitamin A. *Am. J. Dis. Child.* **85**:316–334, 1953.
39. KUROKAWA, K., FRIEDLER, R. M., MASSRY, S. G.: Renal action of cholera toxin. II, Effects on adenylate cyclase–cyclic AMP system. *Kidney Int.* **7**:137–144, 1975.
40. LINDVALL, M., EDVINSSON, L., OWMAN, C.: Sympathetic nervous control of cerebrospinal fluid production from the choroid plexus. *Science* **201**:176–178, 1978.
41. LIPSON, L. C., SHARP, G. W. G.: Effect of prostaglandin E_1 on sodium transport and osmotic water flow in the toad bladder. *Am. J. Physiol.* **220**:1046–1052, 1971.
42. LORENTZ, W. B., JR.: The effect of cyclic AMP and dibutyryl cyclic AMP on the permeability characteristics of the renal tubule. *J. Clin. Invest.* **53**:1250–1257, 1974.
43. MAREN, T. H.: Bicarbonate formation in the cerebrospinal fluid: Role in sodium transport and pH regulation. *Am. J. Physiol.* **222**:885–899, 1972.
44. MASHITER, K., MASHITER, G. D., HAUGER, R. L., FIELD, J. B.: Effects of cholera and *E. coli* enterotoxins on cyclic adenosine 3′,5′-monophosphate levels and intermediary metabolism in the thyroid. *Endocrinology* **92**:541–549, 1973.
45. NARUMI, S., MAKI, Y.: Possible role of cyclic AMP in gastric acid secretion in rat—activation of carbonic anhydrase. *Biochim. Biophys. Acta* **311**:90–97, 1973.
46. OMACHI, R. S., ROBBIE, D. E., HANDLER, J. S., ORLOFF, J.: Effects of ADH and other agents on cyclic AMP accumulation in toad bladder epithelium. *Am. J. Physiol.* **226**:1152–1157, 1974.
47. ORLOFF, J., HANDLER, J.: The role of adenosine 3′,5′-

phosphate in the action of anti-diuretic hormone. *Am. J. Med.* **42:**757–768, 1967.
48. PAPPENHEIMER, J. R., HEISEY, S. R., JORDAN, E. F., DOWNER, J.: Perfusion of the cerebral ventricular system in unanesthetized goats. *Am. J. Physiol.* **203:**763–774, 1962.
49. PIERCE, N. P., CARPENTER, C. C. J., GREENOUGH, W. B., III: Cholera toxin and its mode of action. *Bacteriol. Rev.* **35:**1–13, 1971.
50. ROBISON, B. A., BUTCHER, R. W., SUTHERLAND, E. W.: Adenyl cyclase as an adrenergic receptor. *Ann. N.Y. Acad. Sci.* **139:**703–723, 1967.
51. ROBISON, G. A., BUTCHER, R. W., SUTHERLAND, E. W.: *Cyclic AMP.* New York, Academic Press, 1971.
52. RODRIGUEZ, H. J., WALLS, J., YATES, J., KLAHR, S.: Effects of acetazolamide on the urinary excretion of cyclic AMP and on the activity of renal adenyl cyclase. *J. Clin. Invest.* **53:**122–130, 1974.
53. RUDMAN, D., HOLLINS, B. M., LEWIS, N. C., SCOTT, J. W.: Effects of hormones on 3′,5′-cyclic adenosine monophosphate in choroid plexus. *Am. J. Physiol.* **232:**E353–E357, 1977.
54. RUDMAN, D., ISAACS, J. W.: Effect of intrathecal injection of melanotropic–lipolytic peptides on the concentration of 3′,5′-cyclic adenosine monophosphate in cerebrospinal fluid. *Endocrinology* **97:**1476–1480, 1975.
55. SAHYOUN, N., CUATRECASAS, P.: Mechanism of activation of adenylate cyclase by cholera toxin. *Proc. Natl. Acad. Sci. U.S.A.* **72:**3438–3442, 1975.
56. SCHAFER, D. E., LUST, W. D., SIRCAR, B., GOLDBERG, N. D.: Elevated concentration of adenosine 3′,5′-cyclic monophosphate in intestinal mucosa after cholera toxin. *Proc. Natl. Acad. Sci. U.S.A.* **67:**851–856, 1970.
57. SHARP, G. W. G., HYNIE, S.: Stimulation of intestinal adenyl cyclase by cholera toxin. *Nature (London)* **229:**266–269, 1971.
58. SUTHERLAND, E. W.: Studies on the mechanism of hormone action. *Science* **177:**401–408, 1972.
59. VOGH, B. P., MAREN, T. H.: Sodium, chloride, and bicarbonate movement from plasma to cerebrospinal fluid in cats. *Am. J. Physiol.* **228:**673–683, 1975.
60. WALD, A., GOTTERER, G. S., RAJENDRA, G. R., TURJMAN, N. A., HENDRIX, T. R.: Effect of indomethacin on cholera-induced fluid movement, unidirectional sodium fluxes, and intestinal cAMP. *Gastroenterology* **72:**106–110, 1977.

CHAPTER 3

Chemical Alterations of Cerebrospinal Fluid Acting on Respiratory and Circulatory Control Systems

Hans H. Loeschcke

1. CSF and ECF in Central Nervous System

Electron microscopy has revealed that the extracellular spaces of the brain are open to the subarachnoid space.[18] Free exchange of substance has been demonstrated for nonelectrolytes such as inulin,[81] for ampholytes such as horseradish peroxidase,[18] and for charged particles such as potassium ions.[15,44] Molecular size of the particle and its charge do not prevent exchange between the cerebrospinal fluid (CSF) and the extracellular space. Thus, this exchange occurs by *diffusion*[74] and possibly by local *convection* currents. The movement of the head causes mechanical distortions of the brain and may "pump" fluid from regions of momentarily high pressure to areas of lower pressure.[107,108] The ciliary movements of the ependyma may also contribute to such exchange.[70,107]

The CSF and the brain extracellular fluid (ECF) are solutions that normally contain little protein and therefore in comparison with most other body fluids exert minimal if any colloidosmotic pressure. The entrance of these fluids into the venous portions of the capillaries and filtration of plasma from the arterial end may contribute to the local fluid circulation. Capillary permeability within the brain is highly variable. For instance, most capillaries have relatively tight endothelial junctions that account for the blood–brain barrier, whereas those in the area postrema are quite permeable.[16] Local circulation of fluids within the brain, however, has not yet been investigated quantitatively.

The bicarbonate (HCO_3^-) and chloride (Cl^-) fluxes between CSF and blood vary with manipulation of the CSF ion concentrations.[28,74] No net fluxes are observed if the ion concentrations are normal. Therefore, in the steady state, the composition of the CSF and ECF of the brain with respect to hydrogen ions (H^+), potassium (K^+), Cl^-, and HCO_3^- appears to be similar.[15] Contrary to the steady-state situation, HCO_3^- exchange between brain cells and ECF is quite rapid during respiratory acidosis or alkalosis and is somewhat less between blood and brain ECF during metabolic acidosis or alkalosis.[1,2,45,46,62,69,73] Normally, changes of the carbon dioxide tension (P_{CO_2}) and fixed acid occur all the time and thus evoke temporary gradients of HCO_3^- and Cl^-.

Furthermore, the production of metabolites in the brain yields a net output of acid into the ECF. During severe hypoxia[95] and respiratory alkalosis, lactic acid is produced and liberated into the brain ECF.[50,51,52,78] A gradient of acidity from the brain ECF to the CSF is negligible under normal conditions, but has appreciable magnitude during hypoxia. This gradient is maintained as the acid is cleared into the CSF. This is, however, a very slow

Hans H. Loeschcke, M.D. • Institut für Physiologie, Ruhr-Universität, Bochum, Germany.

process that should depend on the acidity of the freshly produced CSF. Hydrogen ions from the lactic acid produced in the brain react with HCO_3^- to form carbonic acid (H_2CO_3), which is then cleaved into CO_2 and water. Carbon dioxide will leave the brain by diffusion into the blood, but the HCO_3^- will be diminished and replaced stoichiometrically by lactate ions. The correction of this situation is not complete until the lactate ion has again been replaced by HCO_3^-.

In conclusion, the CSF accurately reflects brain ECF during steady-state situations. However, discrepancies occur during transient disturbances of brain metabolism.

Dissolved substances may enter the extracellular compartment of the brain from the CSF without impediment. If such a substance does not enter the capillaries and is not catabolized in the CSF, then its concentration will tend to equilibrate. If, however, this substance is cleared by permeating the capillaries, a stationary gradient from the surface into the depth of the brain tissue will be produced. A similar gradient may be produced by the degradation of the substance by enzymes fixed to cell membranes. For instance, brain acetylcholine is degraded by the abundant quantities of specific and nonspecific acetylcholinesterases.[47,63] The same fate is assumed for procaine, which is split by a nonspecific cholinesterase.[42] These processes preclude accurate estimations of the local composition of ventricularly administered substances that are destroyed or cleared by the brain tissue. The concentrations of these substances in the ECF deep within the brain may be much lower than brain-surface concentrations. When injected into the CSF, many substances such as procaine may not enter deeper brain layers.[92]

2. Physiological and Experimental Mediators of Respiratory and Circulatory Control Systems

As previously discussed, substances enter the extracellular space of the brain from the CSF if a concentration gradient exists. Whether the origin for these substances found in the CSF, such as endorphins, sleep substances, and other hormones, is the brain or the blood passing through the choroid plexus is controversial. Since CSF is almost free of protein, the latter mechanism cannot play a role for large protein hormones. Steroid hormones may enter the ECF more easily, but penetration varies with lipid solubility. Whether transmitter substances under physiological conditions enter the CSF and reach sufficient concentrations to exert their influence distant from the site of origin or entrance is unclear.

Acids and alkalis take part in physiological control processes and especially in the respiratory control system via feedback modulation. Increased acidity augments the discharge of chemosensitive neuronal structures in the ventral surface layer of the medulla oblongata. These neurons project to the respiratory centers and provide a substantial part of the normal respiratory drive to increase ventilation. Augmented ventilation diminishes arterial P_{CO_2} and also reduces brain tissue P_{CO_2}. The P_{CO_2} in the ECF surrounding the cells of these chemosensitive structures is equal to local tissue P_{CO_2}. This local P_{CO_2} together with the local HCO_3^- determine the local tissue pH. Increased ventilation will evoke an alkaline shift in the local extracellular pH, correcting an initial acidosis. This feedback mechanism protects the homeostasis of the extracellular pH in the brain. For quantitative understanding of this control system, the exchange of ions from ECF to the blood on one side and to the brain cells on the other side must be known.

When acid or alkali is *injected* into the CSF or mock CSF with altered acid–base balance is experimentally *perfused* through the CSF system, large deviations in CSF pH occur. Analysis of evoked alterations in respiration activity provide quantitative assessments of the nature and magnitude of this reaction. This experimental model may serve as a tool to investigate central respiratory chemosensitivity to H^+ and CO_2. It cannot be concluded, however, that the chemosensory apparatus normally responds only to changes of CSF pH. Since these reacting structures lie in the substance of the brainstem, the chemical stimulus when applied locally must alter the composition of the ECF surrounding these respiratory chemosensors. Thus, changes in CSF pH would be sensed by the receptor within the brain tissue only to the extent that these pH variations are reflected in the ECF surrounding the chemosensitive receptor. This mechanism of alteration of brain ECF environment by manipulating the CSF composition provides a marvelous tool for the investigation of the properties of the respiratory and circulatory control systems.

Similarly, a response of these chemosensitive structures to transmitter substances such as acetylcholine when applied to the bulbar surface indicates only that these structures react to the transmitter. This does not necessarily imply that under physiological conditions, acetylcholine flows into the ECF from the CSF. The presence of such a phenomenon would depend on the local concentrations and concentration gradients of normally occurring transmitters. However, the cholinergic nature of a given synapse might be established by studying the receptor reaction to cholinomimetic drugs and their antagonists applied to the bulbar surface.

In summary, the fact that substances enter the *extracellular* spaces of the brain freely from the ventricular or subarachnoid space may be used to test the responses of the respiratory or the circulatory system either to drugs or to physiological stimuli. These evoked reactions, when present, help to elucidate the properties of the physiological systems under investigation.

3. Central Chemosensitivity of Respiration

The respiratory muscles are driven by impulses originating in the anterior spinal medulla. The respiratory motoneurons of the anterior horn serving the respiratory muscles receive their dominant input from the respiratory centers in the medulla oblongata. These respiratory centers are controlled by afferent impulses that include those arriving from the peripheral chemoreceptors of the carotid and aortic regions and from the central chemosensory structures located in the ventral superficial layer of the medulla oblongata. These respiratory centers themselves are not chemosensitive to CO_2 or H^+ as has been discussed.[64,66,86] The site and properties of the medullary chemosensory structures have been previously reviewed.[54] The physiological responses to peripheral carotid and aortic nerve stimulation are eliminated by cutting these nerves. The central chemosensitive receptor activity is lost after electrocoagulation or reversibly blocked by bilateral cooling of small areas (the "intermediate" areas) close to the rostral end of the fanning hypoglossal rootlets. If both peripheral and central chemosensory afferents are eliminated, an anesthetized or decerebrated animal (cat) stops breathing. This respiratory arrest also occurs in animals inhaling high concentrations of CO_2.[87] In this situation, stimulation of the femoral nerve or the cornea is likewise ineffective in inducing respiratory activity.[83,91] Stimulation of the central end of the severed carotid sinus nerve or the thermoregulative zone of the hypothalamus reestablishes ventilatory drive.[93]

In cats surviving bilateral electrocoagulation of the "intermediate" area, the respiratory response to inhaled CO_2 was either strongly diminished or completely lost.[86] These cats apparently survived because the peripheral chemoreceptors provided enough respiratory drive to maintain a much-reduced ventilation. Accordingly, the arterial P_{CO_2} was elevated to higher levels than noted in animals that had undergone denervation of the peripheral chemosensitive afferents.[11,61,65]

4. Alteration of CSF as a Tool in Analysis of Site and Properties of the Central Chemosensitivity of Respiration

In addition to enabling the discovery of the medullary chemosensitive areas, the application of altered CSF to the surface of the brain has proved to be an important investigative tool. First, Leusen,[48,49] extending an experimental approach of Stern and Gautier,[96,97] perfused solutions of various P_{CO_2} or pH levels through the ventricles and subarachnoid spaces in dogs and described the effect of increased acidity on ventilation. Later, surgical techniques for enabling access to the dorsal or ventral medulla oblongata were developed and provided means to study brain-surface localization of physiological functions.

Two chemosensitive areas rostral and caudal to the aforementioned intermediate area have been identified by local application of acid buffers,[28,60,67,68] neurotropic drugs such as acetylcholine and nicotine,[19,20,24,67,68] or electrical stimulation.[55] The intermediate area seems to be a structure to which the chemosensory signals converge in order to enter the depth of the medulla oblongata.[19,54,90] The cells beneath this area constitute the nucleus paragigantocellularis[98] and receive fibers from the intermediate area. A separate but more superficial population of small nerve cells has been described in this anatomical region.[77,84] Spontaneous action potentials have been observed in all three areas that could be activated by increasing the H^+ concentration of solutions applied to the medullary surface.[79,80,88] Techniques to measure ECF pH rather than CSF pH have been developed to aid in the quantitative study of the dependence of ventilation on the extracellular pH and exogenous drugs. These methods involve placement of macroelectrodes on the brain surface[1,89] and introduction of microelectrodes into the tissue.[14] Action potentials have been recorded in thin slices of the ventral surface of the medulla that have survived in a perfusion chamber.[31,32] This neuronal activity was found to be dependent on the pH of the perfused buffer. Thus, experimental alteration of the acid–base balance of CSF as well as the direct application of drugs or electrical stimuli to the superficial structures of the medulla oblongata have contributed to the localization of the central chemosensitivity of respiration to areas distinct from the classic respiratory centers.

5. Effects on Circulatory Control System of Altering Acid–Base Balance of CSF or Application of Drugs via CSF

Alterations in the CSF acidity, which strongly influences ventilatory activity, appear ineffective in modifying arterial blood pressure.[60] However, direct electrical stimulation or veratridine application to the ventral surface of the medulla increases the arterial pressure. Contrarily, local application of procaine[58,59] to the ventral medullary surface evokes a depression of arterial pressure. Thus, the central chemosensitive structures that react to increased acidity and project to the respiratory centers do not innervate the vasomotor control system. These differential and separate circulatory responses to electrical stimulation and some drugs imply that this brainstem region must contain at least two independent chemosensitive cell populations, one projecting to the circulatory control system and the other acting on respiration. This conclusion is supported by the observation that locally applied veratridine increases both ventilation and arterial pressure, whereas acetylcholine increases ventilation but diminishes arterial pressure.

This interpretation is, however, complicated by the observation of Trzebski *et al.*[100] that increased CSF acidity evokes an increase in the impulse frequency of several sympathetic nerve branches in the *neck*. Such increased peripheral sympathetic tone would be expected to influence peripheral vasomotor tone and blood pressure. The mock CSF solutions employed in these experiments had a very low pH; thus, this discrepancy in the results may be due to difference in experimental technique. Meningeal receptors may possibly be influenced by high CSF acidity and may project to the sympathetic system. Recently, diminished sympathetic discharge has been observed on stimulation of the nucleus paragigantocellularis.[89]

Thus, moderate variations in the acid–base balance of the CSF appear to have no systematic effect on circulation. However, augmentation of the discharge frequency in sympathetic neurons is induced by markedly low CSF pH. The lack of response in the circulatory system to moderate CSF pH changes contrasts greatly with the marked responses evoked in the respiratory system. Drugs administered via the CSF do exert dramatic effects on the peripheral circulation.

6. Confirmation of Reaction Theory of Central Chemosensitivity by Alteration of CSF Acid–Base Parameters

In an effort to differentiate between the effects of H^+ and the molecular effects of CO_2, Loeschcke *et al.*[60] varied the mock CSF perfused through the subarachnoid space in such a way that pH was varied at constant P_{CO_2} or P_{CO_2} was altered at constant pH. The results of these experiments implied that the CO_2 effects are mediated by H^+ and that the chemosensitive structures react to the extracellular H^+ concentration. This "reaction theory" has been supported by Mitchell *et al.*[67,68] Increasing the P_{CO_2} in the CSF at constant H^+ concentration actually depressed ventilation slightly. This effect cannot be fully explained, although a depressant action of CO_2 on neuronal activity may have contributed to this discrepancy. The effect of CSF H^+ concentrations on respiration has also been studied in rats,[39] anesthetized dogs, and awake goats.[28,74,75] Fencl *et al.*[29] employed indirect methods to study the ventilatory response to H^+ variation in human subjects.

7. Ventilation in Respiratory and Nonrespiratory Acidosis

Fencl *et al.*[28] varied the HCO_3^- concentration in the CSF perfusate to study the HCO_3^- fluxes and to establish stationary HCO_3^- gradients from the perfusate to the blood in the brain capillaries. In these experiments, a unique response of ventilation to an alteration of pH was obtained when the pH was manipulated by variation of the HCO_3^- (metabolic acidosis) or P_{CO_2} (respiratory acidosis). This phenomenon, which had been postulated since 1965,[53,62] occurred at a point approximately one third of the pH gradient from the blood side. If the H^+ concentration gradient from the surface to the interior of the medulla is calculated,[5] this point of the unique reaction to H^+ appears to be located between the surface and a depth of approximately 300 μm. The influence of metabolic acidosis on respiration via the central chemosensitive structures implies exchange of either H^+ or HCO_3^- between the blood and brain ECF. This exchange has been confirmed[50,75] and occurs over a period of seconds to minutes.[1,2,45,46,62,69] Mitchell[65] suggests that the CSF pH may become alkaline during systemic metabolic acidosis, which would diminish the central chemical stimulus. Thus, the increase in respiratory activity noted during metabolic acidosis is secondary to the influence of low blood pH on peripheral chemoreceptors. However, Katsaros[43] demonstrated that ventilation is also augmented in the presence of metabolic acidosis despite denervation of the peripheral chemosensitive afferents. This apparent discrepancy might be explained by HCO_3^- exchange between the blood and ECF, which may thereby mediate the chemical signal.

The experiments reported above indicate that only one type of central receptor reacting to H^+ is necessary to explain the responses of ventilation to respiratory and to nonrespiratory acidosis. Central chemosensitivity alone could account for respiratory increases in both types of acidosis and does so even if peripheral chemosensitivity is eliminated. Peripheral chemosensitivity in the intact animal, however, comprises the main reaction in nonrespiratory acidosis.

8. Effects of Chemoreceptor Stimulants Applied via CSF

The peripheral and central chemoreceptors may be similar structures in that both react to H^+.[103] Winterstein and Gokhan[104] introduced ammonium chloride (NH_4Cl), an acid salt, into the CSF and evoked an increase of ventilatory activity. This initial augmentation of ventilation by NH_4Cl was, however, followed by a subsequent depression of respiration.[17,56] Winterstein and co-workers[105,106] injected lobeline and cyanide into the cisterna magna and the cisterna pontis and concluded that the increased ventilatory responses were similar to those effects evoked by the same drugs on the peripheral chemoreceptors. These intracisternal injections may be associated with increases of CSF pressure that may in turn alter local brainstem circulation. The solutions used in these experiments may have had different pH and P_{CO_2} than the CSF, which may have induced central chemosensitive receptor activity.

The results of Winterstein's group, however, were not supported by either previous[4] or subsequent[57,101,102,106] experiments. In perfusion experiments that avoided the shortcomings of suboccipital injection, lobeline depressed ventilation.[57,101,102,106] Ventricular perfusion with cyanide also evoked a depression of ventilation, but only at low concentrations.[106] High concentrations of cyanide in perfusates caused convulsions. Veratridine perfusions were associated with not only increased ventilation but also elevation of arterial pressure.[57] Thus, central and peripheral chemosensitive receptors appear to react differently. Accordingly, hypoxia has long been known to be the main stimulus to peripheral chemoreceptors, yet hypoxia depresses the ventilation following the destruction of these peripheral chemoreceptors.

9. Effects on Respiration of Transmitter Substances and Neurotropic Drugs Applied via CSF

The superficial location of respiratory chemosensitive structures in the ventral surface of the medulla oblongata allows investigation of the effects of neurotransmitters on the central chemosensitivity of respiration. Mitchell *et al.*[67,68] and Wiemer[102] applied acetylcholine to the ventral medullary surface and observed an increase of ventilation. The topical distribution of this response was identical with the distribution of pH-evoked and also electrical-stimulation-induced responses.[19] Local application of atropine to the chemosensitive areas strongly diminished the response of respiration to inhaled CO_2,[20] thus suggesting that acetylcholine may mediate this chemosensitive response. Furthermore, augmented discharge frequencies of some cells have been observed in isolated ventral medullary surface slices prepared by Fukuda's technique when acetylcholine was added to the perfusion fluid.[32] These same cells also reacted similarly to increases of perfusate H^+ concentration. This increase in ventral medullary activity evoked by acetylcholine and by H^+ is further enhanced by the addition of physostigmine to the perfusate. Contrarily, this discharge could be completely blocked by atropine and, in some cells, by hexamethonium. This result must be interpreted as meaning that muscarinic as well as nicotinic receptors are involved in the ventral medullary action of acetylcholine. Atropine also blocked the effect of H^+; thus, the action of H^+ is mediated by a cholinergic mechanism. Furthermore, the influence of acetylcholine and H^+ on the ventral medullary discharge could be reversed by increasing the magnesium and decreasing the calcium concentration in the surrounding perfusion fluid. Since this procedure is usually thought to block synaptic transmission, H^+ was felt to influence cholinergic synaptic transmission in the ventral medulla. This involvement of acetylcholine in the central respiratory drive supports previous concepts promoted by Gesell and co-workers.[33,34,35]

4-Aminopyridine is used to counteract the respiratory depression of relaxant drugs and acts as a releaser of acetylcholine in synapses such as the neuromuscular junction.[30] This drug acts on the ventral medullary structures by increasing the discharge of action potentials of H^+-activated cells.[94] This observation supports the theory of a cholinergic link in central respiratory chemosensitivity.

The effect of nicotine on respiration when applied via the CSF parallels the effects of acetylcholine. In addition, the topical distribution of ventral medullary responses to nicotine coincides with that of acetylcholine.[19,20] After having discussed the driving effects on respiration of H^+, acetylcholine, and nicotine and the blockade by atropine, it is not surprising that application of novocaine (procaine) to the ventral surface of the medulla inhibits respiration.[58,59] If the peripheral chemoreceptor nerves are transected beforehand, novocaine can lead to respiratory arrest.[6,19,91] Nicotine and procaine perfused from the third ventricle to the cisterna magna evoke biphasic responses in ventilatory activity.[8,82] The authors conclude that this type of perfusion inflow and outflow enables the procaine to enter the depth of the brain parenchyma. This hypothesis, however, contradicts the direct histochemical determinations[92] revealing procaine stasis in the superficial layers of the medulla oblongata. This finding supports the localization of the chemosensitive mechanism in the superficial layer of the ventral medulla. Respiratory neurons in the depth of the medulla (the centers) become rather quiet after procaine application on the surface, indicating that the central integrative mechanism relies on afferent information from the chemosensitive structures.[92]

Actions of acetylcholine and nicotine were also observed when these drugs were applied to the dorsal surface of the medulla oblongata.[67] The area postrema could be identified as the site of action. Both drugs applied to this rather small area inhibited respiration. The area postrema is one of the few regions in the central nervous system where no blood–brain barrier exists, and its close proximity to the tractus solitarius and nucleus of the tractus solitarius may contribute this apparent involvement in the central respiratory organization.

Respiratory alterations have also been elicited by application of cocaine, nicotine, and lobeline to the floor of the fourth ventricle.[71] These drugs may or may not have had access to the area postrema. All three drugs depressed ventilation but had different effects on respiratory frequency.

10. Effects on Circulation of Transmitter Substances and Neurotropic Drugs Applied via CSF

Electrical stimulation and topical drug application to the ventral surface of the medulla oblongata influence the circulatory as well as the respiratory control system. This double action cannot be the result of a common receptor projecting to both systems. Thus, the effects of this medullary region on respiration and circulation are independent. For example, acidification of the CSF and intrathecal administration of 4-aminopyridine[29] alter respiration, whereas some pyrimidine derivatives strongly influence circulation but minimally alter ventilation. Veratridine has similar actions on both ventilation and vasomotor tone,[57] whereas acetylcholine or carbachol increases ventilation but depresses the arterial pressure.[19,20,24]

The influence on systemic arterial pressure of topical application of acetylcholine and carbachol to the ventral medullary surface was enhanced by eserine but blocked by atropine.[20,24] The frequency of action potentials when recorded from the underlying superficial layer of the ventral medulla oblongata was increased following the application of acetylcholine.[84,89] Some of these cells responding to acetylcholine also reacted to elevations in the local H^+ concentration, while others did not. Similarly, some acetylcholine-activated neurons also increased their discharge following the topical application of an imidazoline derivative, BA 168 (Lofexidine). When administered intravenously or applied locally to the ventral medullary surface, this drug caused the arterial pressure to drop. Similarly, the diminutions in systemic arterial pressure were also observed following similar local application of clonidine[10] (a clinically used vasodepressor substance) to the ventral medullary surface.

These observations indicate the presence of neurons in the superficial layer of the ventral medulla that are excited by acetylcholine, BA 168, or clonidine, and that act on the cardiovascular control system in such a way that the vasomotor tone is diminished. These neurons are depressed by another imidazoline derivative, AN 180, which raises the arterial pressure when administered intravenously or applied locally to the ventral medullary surface. Similar depressions of the arterial pressure have been observed with pentobarbitone,[23] γ-aminobutyric acid (GABA), and glycine.[22,23] Feldberg and Guertzenstein[23] induced arterial pressure reductions with nicotine; however, Dev and Loeschcke[19,20] observed blood pressure elevations with small doses and blood pressure depressions with higher doses of nicotine.

In most cases, the augmentation or reduction of arterial pressure was associated with only minor alterations of heart rate. Thus, these circulatory reactions were predominantly due to alterations of vasomotor tone via the direct influence of the medullary neurons on the sympathetic structures in the mediolateral spinal medulla.[3] These blood pressure alterations may also be influenced by humoral mechanisms. The ventral medullary application of nicotine is associated with a discharge of vasopressin.[7] Angiotensin II may contribute to these blood pressure alterations, since stimulation of the (sympathetic) renal nerves releases renin.

The topographical distribution of these drug effects on the circulation is similar to that of those on respiration. Two medullary areas of maximal blood pressure response to acetylcholine have been demonstrated, although the localization is less precise than in the regions associated with the maximum respiratory responses. The area of maximal response to nicotine has been located in a relatively caudal position, just medial to the posterior root of the hypoglossal nerve.[24] This localization has been confirmed,[19,20] but an action of nicotine on a more rostral field has also been demonstrated. These observations suggest that nicotine acts on the same areas as acetylcholine.

Procaine added to a CSF perfusion or applied to the ventral medullary surface strongly depresses arterial pressure.[58,59] A similar vasomotor response is observed when procaine[58,59,87] or pentabarbitone[23] is applied to a small area which is well defined and appears to be located in the rostral portion of the intermediate area (Feldberg's[24] area A).

Electrical stimulation of the ventral medullary surface, especially on the rostral and caudal chemosensitive areas, raises the arterial blood pressure.[55] The question arises how this observation may be reconciled with the depressing action of acetylcholine. Since the discharge of the neurons that react to vasoactive drugs is augmented by acetylcholine, acetylcholine must be exciting rather than inhibiting the neurons that project to the cardiovascular system. The blood pressure reduction evoked by acetylcholine cannot be mediated by the same neuron that raises the arterial pressure when stimulated. One model that might explain these phenomena assumes that the acetylcholine-sensitive neuron, when activated, inhibits another neuron that is involved in vasomotor innervation. Electrical stimulation may act only on the latter neuron or on both neurons (Fig. 1). If both neurons are arranged in series, the effect of electrical stimulation will be on the second neuron. Also, the inhibitory transmitters GABA and glycine would act on the second neuron. The apparently paradoxical effect of nicotine, which raises the blood pressure in small doses and depresses arterial pressure in higher doses, may also be interpreted as nicotine-evoked stimulation of both neurons. The possibility also exists that nicotine stimulates one neuron in small doses but par-

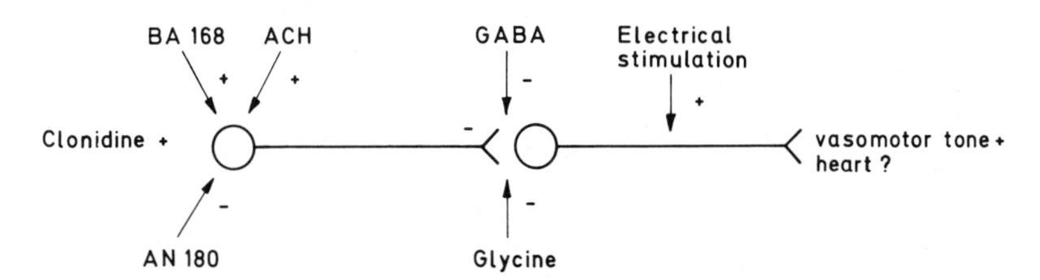

Figure 1. Two-neuron model of cardiovascular responses induced by direct application of drugs to ventral surface of medulla oblongata. The first neuron is excited by acetylcholine (ACH), BA 168 (Lofexidine), and clonidine, and is inhibited by AN 180. The second neuron projects to sympathetic areas of spinal cord and facilitates peripheral vasomotor tone and possibly sympathetic output to heart. Glycine and γ-aminobutyric acid (GABA) inhibit this second neuron, whereas nicotine may stimulate both neurons. Electrical stimulation of ventral medullary surface alters discharge of second neuron.

alyzes it in higher doses. This latter phenomenon has been observed following nicotine application to the sympathetic ganglia.

The outgoing neuron appears to project directly to the mediolateral column of the spinal medulla. Horseradish peroxidase injected into the mediolateral column is taken up by these spinal medullary neurons and transported in a retrograde direction to areas of the ventral medullary surface.[3] These neurons may well be part of the descending defense system of Hilton.[38] The experiments of Schlaefke and See[89] demonstrated a diminution of impulse frequency in the splanchnic nerves following electrical stimulation of the paragigantocellular nucleus. This nucleus may be the location of the first neuron emanating from the model discussed above.

Small elevations of arterial blood pressure can be induced by the application of only high doses of epinephrine and norepinephrine.[19,20] Contrarily, the similar application of serotonin depresses the arterial pressure.

In conclusion, alterations of the CSF and application of neurotropic drugs via the CSF elicit alterations in the function of the cardiovascular system. The organization of these cardiovascular structures differs from that of the respiratory structures with respect to the descending neuronal system. The cardiovascular control system does not react to alterations in the local CSF H^+ concentration. The efferent discharge from this ventral medullary center influences the sympathetic tone as well as the release of vasoactive substances like vasopressin and possibly renin-angiotensin.

11. Influences of Morphine, Apomorphine, Endorphins, Enkephalins, and Other Substances in CSF

As early as 1923, emesis was induced by intrathecal injections of morphine and apomorphine in the dog.[37] Later, the area postrema was identified as the trigger zone for vomiting.[9] Vomiting interferes with the respiratory regulation and also influences circulatory control.

Recently, "opiate receptors" have been demonstrated to be concentrated in the medial thalamus, limbic system, and spinal cord.[72] Although morphine accumulation is wide spread in the brain, the analgesic effect of morphine in cats seems to be located in superficial structures of the ventral medulla oblongata.[21] This action appears to be mediated by serotonergic nerve fibers that originate in the raphe nuclei. Interestingly, morphine depresses respiration as much as the reaction to painful stimuli. These two actions of morphine are usually inseparable.

Endorphins and enkephalins produced within the brain appear to bind to the opiate receptors. Direct investigations of the actions of these peptides in the CSF on respiration and circulation are still lacking. Preliminary studies indicate that these peptides occur naturally in the CSF and mimic morphine by blocking pain reception and depressing respiration. The actions of endorphins are antagonized by naloxone and by substance P.[13,36,40,99] There is no indication yet that endorphins and enkephalins need to be transported by the CSF to exert their action; however, their action when administered via the CSF shall be discussed in a subsequent chapter.[7a] These peptides may modulate the synaptic function of other transmitters. A peptide hormone-transmitter that depresses arterial pressure and increases vascular permeability, neurotensin, has already been identified.[12]

Many drugs, when introduced into the third or lateral ventricle, evoke pronounced alterations of behavior, temperature regulation, or motor control.[25–27] Major modifications of body temperature are accompanied by secondary alterations of respiration and circulation. The functions of substances that cause sleep may be more closely connected to the respiratory and circulatory control systems.[41,76]

References

1. AHMAD, H. R., BERNDT, J., LOESCHCKE, H. H.: Bicarbonate exchange between blood, brain extracellular fluid and brain cells at maintained pCO_2. In Loeschcke, H. H. (ed.): *Acid–Base Homeostasis of the Brain Extracellular Fluid and the Respiratory Control System*. Stuttgart, Georg Thieme, 1976, pp. 19–27.
2. AHMAD, H. R., LOESCHCKE, H. H., WOIDTKE, H. H.: Three compartments model for the bicarbonate exchange of the brain extracellular fluid with blood and cells. In Fitzgerald, S., Gautier, H., Lahiri, S. (eds.): *The Regulation of Respiration during Sleep and Anesthesia, Advances in Experimental Medicine and Biology*, Vol. 99, New York and London, Plenum Press, 1977, pp. 195–209.
3. AMENDT, K., CZACHURSKI, J., DEMBROVSKY, K., SELLER, H.: Neurones within the "chemosensitive area" on the ventral surface of the brainstem which project to the intermediolateral column. *Pfluegers Arch.* **375**:289–292, 1978.
4. BEKAERT, J., LEUSEN, I.: Au sujet de l'influence res-

piratoire de l'injection sousoccipitale de lobeline. *Schweiz. Med. Wochenschr.* **80:**1236, 1950.

5. Berndt, J., Berger, W., Berger, K., Schmidt, M.: Untersuchungen zum zentralen chemosensiblen Mechanismus der Atmung. II. Die Steuerung der Atmung durch das extrazellulare pH im Gewebe der Medulla oblongata, *Pfluegers Arch. Physiol.* **332:**146–170.
6. Berndt, J., Berger, W., Trouth, C. O.: Respiratory and circulatory effects of 100 meq/l potassium or 2% procaine in the cerebrospinal fluid of cats. *Pfluegers Arch.* **321:**346–363, 1970.
7. Bisset, G. W., Feldberg, W., Guertzenstein, P. G., Rocha e Silva, M., Jr.: Vasopressin release by nicotine: The site of action. *Br. J. Pharmacol.* **54:**463–474, 1975.

7a. Bloom, F. E., Segal, D. S.: Endorphins in cerebrospinal fluid. In Wood, J. H. (ed.): *Neurobiology of Cerebrospinal Fluid I.* New York, Plenum Press, 1980.

8. Borison, H. L., Haranath, P. S. R. K., McCarthy, L. E.: Respiratory responses to chemical pulses in the cerebrospinal fluid. *Br. J. Pharmacol.* **44:**605–616, 1972.
9. Borison, H. L., Wang, S. C.: Physiology and pharmacology of vomiting. *Pharmacol. Rev.* **5:**193–230, 1953.
10. Bousquet, P., Guertzenstein, P. G.: Localization of the central cardiovascular action of clonidine. *Br. J. Pharmacol.* **49:**573–579, 1973.
11. Bouverot, P., Flandrois, R., Puccinelli, R., Dejours, P.: Étude du role des chemorecepteurs arteriels dans la regulation de la respiration pulmonaire chez le chien éveillé. *Arch. Int. Pharmacodyn.* **157:**253–271, 1965.
12. Carraway, R., Leeman, G. E.: The isolation of a new hypotensive peptide, neurotensin, from bovine hypothalami, *J. Biol. Chem.* **248:**6854–6861, 1973.
13. Catt, K. J., Dufau, M. L.: Peptide hormone receptors. *Annu. Rev. Physiol.* **39:**529–557, 1977.
14. Cragg, P., Patterson, L., Purves, M. J.: The pH of brain extracellular fluid in the cat. *J. Physiol.* **272:**137–166, 1977.
15. Cserr, H.: Potassium exchange between cerebrospinal fluid, plasma and brain, *Am. J. Physiol.* **209:**1219–1226, 1965.
16. Davson, H.: *Physiology of the Cerebrospinal Fluid.* London, Churchill, 1967.
17. De Bersaques, J., Leusen, I. R.: The direct influence of ammonium chloride on the respiratory system, *Arch. Int. Pharmacodyn.* **97:**13–16, 1954.
18. Dermietzel, R.: Central chemosensitivity: Morphological studies. In Loeschcke, H. H. (ed.): *Acid–Base Homeostasis of the Brain Extracellular Fluid and the Respiratory Control System.* Stuttgart, Georg Thieme, 1976, pp. 52–65.
19. Dev, N. B., Loeschcke, H. H.: Topography of the respiratory and circulatory responses to acetylcholine and nicotine on the ventral surface of the medulla oblongata. *Pfluegers Arch. Physiol.* **379:**19–27, 1979.
20. Dev, N. B., Loeschcke, H. H.: A cholinergic mechanism involved in the respiratory chemosensitivity of the medulla oblongata in the cat. *Pfluegers Arch. Physiol.* **379:**29–36, 1979.
21. Dey, P. K., Feldberg, W.: Analgesia produced by morphine when acting from the liquor space. *Br. J. Pharmacol. Chemother.* **58:**383–394, 1976.
22. Feldberg, W.: The ventral surface of the brain stem: A scarcely explored region of pharmacological sensitivity. *Neuroscience* **1:**427–441, 1976.
23. Feldberg, W., Guertzenstein, P. G.: A vasedepressor effect of pentobarbitone sodium. *J. Physiol.* **224:**83–103, 1972.
24. Feldberg, W., Guertzenstein, P. G.: Vasedepressor effects obtained by drugs acting on the ventral surface of the brain stem. *J. Physiol.* **258:**337–355, 1976.
25. Feldberg, W., Hellong, R. F., Myers, R. D.: Effects on temperature of monoamines injected into the cerebral ventricles of anaesthetized dogs. *J. Physiol.* **186:**416–423, 1966.
26. Feldberg, W., Myers, R. D., Veale, W L.: Perfusion from cerebral ventricle to cisterna magna in the unanaesthetized cat: Effect of calcium on body temperature. *J. Physiol.* **207:**403–416, 1970.
27. Feldberg, W., Saxena, P. N.: Further studies on prostaglandin E_1 fever in cats. *J. Physiol.* **219:**739–745, 1971.
28. Fencl, V., Miller, T. B., Pappenheimer, J. R.: Studies on the respiratory response to disturbances of acid–base balance, with deductions concerning the ionic composition of cerebral interstitial fluid. *Am. J. Physiol.* **210:**459–472, 1966.
29. Fencl, V., Vale, J. R., Broch, J. A.: Respiration and cerebral blood flow in metabolic acidosis and alkalosis in humans. *J. Appl. Physiol.* **27:**67–76, 1969.
30. Folgering, H., Rutten, F., Agoston, S.: Stimulation of ventilation by an acetylcholine releasing drug: 4-aminopyridine, *Pfluegers Arch. Physiol.* **379:**181–185, 1979.
31. Fukuda, Y., Loeschcke, H. H.: Effect of H^+ on spontaneous neuronal activity in the surface layer of the rat medulla oblongata *in vitro. Pfluegers Arch.* **371:**125–134, 1977.
32. Fukuda, Y., Loeschcke, H. H.: A cholinergic mechanism involved in the neuronal excitation by H^+ in the respiratory chemosensitive structures of the ventral medulla oblongata of rats *in vivo. Pfluegers Arch.* **379:**125–135, 1979.
33. Gesell, R., Brassfield, C. R., Hamilton, M. A.: An acid-neurohumoral mechanism of nerve cell activation. *Am. J. Physiol.* **136:**604–608, 1942.
34. Gesell, R., Hansen, E. T.: Eserine, acetylcholine, atropine and nervous integration. *Am. J. Physiol.* **139:**371–385, 1943.
35. Gesell, R., Hansen, E. T.: Anticholinesterase activity as a biological instrument of nervous integration. *Am. J. Physiol.* **144:**126–163, 1945.

36. GUILLEMIN, R.: Peptides in the brain: The new endocrinology of the neuron. *Science* **202:**390–402, 1978.
37. HATCHER, R. A., WEISS, S.: Studies on vomiting, *J. Pharmacol. Exp. Ther.* **22:**139–193, 1923.
38. HILTON, S. M.: Ways of viewing the central nervous control of the circulation—old and new. *Brain Res.* **87:**213–219, 1975.
39. HORI, T., ROTH, G. I., YAMAMOTO, W. S.: Respiratory sensitivity of rat brain-stem surface to chemical stimuli. *J. Appl. Physiol.* **28:**721–724, 1970.
40. HUGHES, J.: Les morphines du cerveau. *Recherche* **93:**866–875, 1978.
41. JOUVET, M.: The role of monoamines and acetylcholine-containing neurons in the regulation of the sleep–waking cycle. *Ergeb. Physiol.* **64:**166–307, 1972.
42. KALOW, W.: Hydrolysis of local anaesthetics by human serum-cholinesterase. *J. Pharmacol. Exp. Ther.* **104:**122–134, 1952.
43. KATSAROS, B.: Die Rolle der Chemoreceptoren des Carotisgebiets der narkotisierten Katze für die Antwort der Atmung auf isolierte Änderung der Wasserstoffionen-Konzentration und des CO_2-Drucks des Blutes. *Pfluegers Arch.* **282:**157–178, 1965.
44. KATZMAN, R., GRAZIANI, L., GINSBURG, S.: Cation exchange in blood, brain and CSF. In Lajtha, A., Ford, D. H. (eds.): *Brain Barrier Systems, Progress in Brain Research.* Amsterdam, London, New York, Elsevier, 1968, pp. 283–293.
45. KAZEMI, H., JAVAHERI, S.: Interaction between P_{CO_2} and plasma HCO_3^- in regulation of CSF HCO_3^- in respiratory alkalosis and metabolic acidosis. In Fitzberald, R., Gautier, H., Lahiri, S. (eds.): *The Regulation of Respiration during Sleep and Anesthesia, Advances in Experimental Medicine and Biology,* Vol. 99, New York and London, Plenum Press, 1978, pp. 173–183.
46. KAZEMI, H., WEYNE, J., VAN LEUVEN, F., LEUSEN, I.: The CSF HCO_3^- increase in hypercapnia: Relationship to HCO_3^-, glutamate, glutamine and NH_3 in brain. *Respir. Physiol.* **28:**338–401, 1976.
47. KOELLE, G. B.: The histochemical localization of cholinesterase in the central nervous system of the cat. *J. Comp. Neurol.* **100:**211–235, 1954.
48. LEUSEN, I.: Chemosensitivity of the respiratory center: Influence of CO_2 in the cerebral ventricles on respiration. *Am. J. Physiol.* **176:**39–44, 1954.
49. LEUSEN, I.: Chemosensitivity of the respiratory center: Influence of changes in the H^+ and total buffer concentrations in the cerebral ventricles on respiration. *Am. J. Physiol.* **176:**45–51, 1954.
50. LEUSEN, I.: Regulation of cerebrospinal fluid composition with reference to breathing. *Physiol. Rev.* **52:**1–56, 1972.
51. LEUSEN, I., DEMEESTER, G.: Acid–base balance in cerebrospinal fluid during prolonged artificial hyperventilation. *Arch. Int. Physiol. Biochem.* **72:**721–724, 1964.
52. LEUSEN, I., LACROIS, E., DEMEESTER, G.: Lactate and pyruvate in the brain of rats during changes in acid–base balance. *Arch. Int. Physiol. Biochem.* **75:**310–324, 1967.
53. LOESCHCKE, H. H.: A concept of the role of intracranial chemosensitivity in respiratory control. In Brooks, McC., Kao, F. F., Lloyd, B. B. (eds.): *Cerebrospinal Fluid and the Regulation of Ventilation.* Oxford, Blackwell, 1965, pp. 183–207.
54. LOESCHCKE, H. H.: Central nervous chemoreceptors. In Guyton, A. C., Widdicombe, J. G. (eds.): *Medical and Technical Publishing Co. International Review of Science, Respiratory Physiology, Physiology Series I,* Vol. 2, University Park Press, Baltimore, 1974, pp. 167–196.
55. LOESCHCKE, H. H., DE LATTRE, J., SCHLAEFKE, M. E., TROUTH, C. O.: Effects on respiration and circulation of electrically stimulating the ventral surface of the medulla oblongata. *Respir. Physiol.* **10:**184–197, 1970.
56. LOESCHCKE, H. H., KATSAROS, B.: Die Wirkung von in den Liquor cerebrospinalis eingebrachtem Ammoniumchlorid auf Atmung und Vasomotorik. *Pfluegers Arch.* **270:**147–160, 1959.
57. LOESCHCKE, H. H., KOEPCHEN, H. P.: Über das Verhalten der Atmung und des arteriellen Drucks bei Einbringen von Veratridin, Lobelin und Cyanid in den Liquor cerebrospinalis. *Pfluegers Arch. Physiol.* **266:**586–610, 1958.
58. LOESCHCKE, H. H., KOEPCHEN, H. P.: Beeinflussung von Atmung und Vasomotorik durch Einbringen von Novocain in die Liquorraume. *Pfluegers Arch. Physiol.* **266:**611–627, 1958.
59. LOESCHCKE, H. H., KOEPCHEN, H. P.: Versuch zur Lokalisation des Angriffsortes der Atmungs- und Kreislaufwirkung von Novocain im Liquor cerebrospinalis, *Pfluegers Arch. Physiol.* **226:**628–641, 1958.
60. LOESCHCKE, H. H., KOEPCHEN, H. P., GERTZ, K. H.: Über den Einfluss von Wasserstoffionenkonzentration und CO_2-Druck im Liquor cerebrospinalis auf die Atmung, *Pfluegers Arch.* **266:**569–585, 1958.
61. LOESCHCKE, H. H., SCHLAEFKE, M. E.: Central chemosensitivity. In Paintal, A. S. (ed.): *Morphology and Mechanisms of Chemoreceptors.* Delhi, Vallabhbhai Patel Chest Institute, University of Delhi, 1977, pp. 282–296.
62. LOESCHCKE, H. H., SUGIOKA, K.: pH of cerebrospinal fluid in the cisterna magna and on the surface of the choroid plexus of the 4th ventricle and its effect on ventilation in experimental disturbances of acid base balance: Transients and steady states. *Pfluegers Arch.* **312:**161–188, 1969.
63. MAREN, T. H.: Carbonic anhydrase: Chemistry, physiology and inhibition. *Physiol. Rev.* **47:**595–781, 1967.
64. MARINO, P. L., LAMB, T. W.: Effects of CO_2 and extracellular H^+ iontophoresis on single cell activity in the cat brainstem. *J. Appl. Physiol.* **38:**688–695, 1975.
65. MITCHELL, R. A.: The regulation of respiration in metabolic acidosis and alkalosis. In Brooks, McC., Kao, F. F., Lloyd, B. B. (eds.): *Cerebrospinal Fluid and the Regulation of Ventilation,* Oxford, Blackwell, 1965, pp. 109–131.

66. MITCHELL, R. A., HERBERT, D. A.: The effect of carbon dioxide on the membrane potential of medullary respiratory neurons. *Brain Res.* **75:**345–349, 1974.
67. MITCHELL, R. A., LOESCHCKE, H. H., MASSION, W. H., SEVERINGHAUS, J. W.: Respiratory responses mediated through superficial chemosensitive areas on the medulla. *J. Appl. Physiol.* **18:**523–533, 1963.
68. MITCHELL, R. A., LOESCHCKE, H. H., SEVERINGHAUS, J. W., RICHARDSON, B. W., MASSION, W. H.: Regions of respiratory chemosensitivity on the surface of the medulla. *Ann. N. Y. Acad. Sci.* **109:**661–681, 1963.
69. NATTIE, E. E., ROMER, L.: The role of chloride and other anions in cerebrospinal fluid bicarbonate regulation. In Fitzgerald, S., Gautier, H., Lahiri, S. (eds.): *The Regulation of Respiration during Sleep and Anesthesia, Advances in Experimental Medicine and Biology,* Vol. 99, New York and London, Plenum Press, 1977, pp. 211–218.
70. NELSON, D. J., WRIGHT, E. M.: The distribution, activity and function of the cilia in the frog brain. *J. Gen. Physiol.* **243:**63–78, 1974.
71. NICHOLSON, H. C., SOBIN, S.: Respiratory effects from application of cocaine, nicotine and lobeline to the floor of the fourth ventricle. *Am. J. Physiol.* **123:**766–774, 1938.
72. NICOLL, R. A., SIGGINS, G., LING, N., BLOOM, F., GUILLEMIN, R.: Neuronal actions of endorphins and enkephalins among brain regions—comparative microiontophoretic study. *Proc. Natl. Acad. Sci. U.S.A.* **74:**2584–2588, 1977.
73. PANNIER, J. L., WEYNE, J., LEUSEN, I.: The CSF blood potential and the regulation of the bicarbonate concentration of CSF during acidosis in the cat. *Life Sci.* **10:**287–300, 1971.
74. PAPPENHEIMER, J. R.: The ionic composition of cerebral extracellular fluid and its relation to control of breathing. *Harvey Lect.* **61:**71–94, 1967.
75. PAPPENHEIMER, J. R., FENCL, V., HEISEY, S. R., HELD, D.: Role of cerebral fluids in control of respiration as studied in unanesthetized goats. *Am. J. Physiol.* **208:**436–450, 1965.
76. PAPPENHEIMER, J. R., MILLER, T. B., GOODRICH, C. A.: Sleep-promoting effects of cerebrospinal fluid from sleep-deprived goats. *Proc. Natl. Acad. Sci. U.S.A.* **58:**513–518, 1967.
77. PETROVICKY, P.: Über die Glia marginalis und oberflächliche Nervenzellen im Hirnstamm der Katze. *Z. Anat. Entwicklungsgesch.* **127:**221–231, 1968.
78. PLUM, F., POSNER, J. B.: Blood and cerebrospinal fluid lactate during hyperventilation. *Am. J. Physiol.* **212:**864–870, 1967.
79. POKORSKI, M.: Neurophysiological studies on central chemosensor in medullary ventrolateral areas. *Am. J. Physiol.* **230:**1288–1295, 1976.
80. PRILL, R.: Des Verhalten von Neuronen des caudalen chemosensiblen Felds in der Medulla oblongata der Katze gegenüber intravenosen Injektionen von $NaHCO_3$ und HCl. Thesis, Abteilung für Naturwissenschaftliche Medizin, Ruhr-Universitat Bochum, 1977.
81. RALL, D. P., OPPELT, W. W., PATLAK, C. S.: Extracellular space of brain as determined by diffusion of inulin from the ventricular system. *Life Sci.* **2:**43–48, 1962.
82. ROSENSTEIN, R., MCCARTHY, L. E., BORISON, H. L.: Respiratory effects of ethanol and procaine injected into the cerebrospinal fluid of the brainstem in cats. *J. Pharmacol. Exp. Ther.* **162:**174–181, 1968.
83. SCHLAEFKE, M. E.: "Specific" and "non-specific" stimuli in the drive of respiration. *Acta Neurobiol. Exp.* **33:**149–154, 1973.
84. SCHLAEFKE, M. E., SEE, W. R., KILLE, J. F., Origin and afferent modification of respiratory drive from ventral medullary areas. In von Euler, C., and Lagercrantz, H. (eds.): *Central Nervous Control Mechanisms in Breathing.* Oxford, Pergamon Press, 1979, pp. 25–34.
85. SCHLAEFKE, M. E., KILLE, J., FOLGERING, H., HERKER, A., SEE, W. R.: Breathing without central chemosensitivity. In Umbach, W., Koepchen, H. P. (eds.): *Central Rhythmic and Regulation.* Stuttgart, Hippokrates Verlag, 1974, pp. 97–104.
86. SCHLAEFKE, M. E., KILLE, J. F., LOESCHCKE, H. H.: Elimination of central chemosensitivity by coagulation of a bilateral area on the ventral medullary surface: Studies in awake cats. *Pfluegers Arch.* **378:**231–242, 1979.
87. SCHLAEFKE, M. E., LOESCHCKE, H. H.: Lokalisation eines an der Regulation von Atmung und Kreislauf beteiligten Gebietes an der ventralen Oberfläche der Medulla oblongata durch Kalteblockade. *Pfluegers Arch.* **297:**201–220, 1967.
88. SCHLAEFKE, M. E., POKORSKI, M., SEE, W. R., PRILL, R. K., LOESCHCKE, H. H.: Chemosensitive neurons on the ventral medullary surface. *Bull. Physiopath. Resp.* **11:**277–284, 1975.
89. SCHLAEFKE, M. E., SEE, W. R.: Ventral surface stimulus response in relation to ventilatory and cardiovascular effects. In Koepchen, H. P., Hilton, S. M., Trzebski, A. (eds.): *Central Interaction between Respiratory and Cardiovascular Control Systems.* Berlin, Heidelberg, and New York, Springer-Verlag, 1980 (in press).
90. SCHLAEFKE, M. E., SEE, W. R., LOESCHCKE, H. H.: Ventilatory response to alterations of H^+ ion concentration in small areas of the ventral medullary surface. *Respir. Physiol.* **10:**198–212, 1970.
91. SCHLAEFKE, M. E., SEE, W. R., MASSION, W. H., LOESCHCKE, H. H.: Die Rolle spezifischer und unspezifischer Afferenzen für den Antrieb der Atmung: Untersucht durch Reizung und Blockade von Afferenzen an der dezerebrierten Katze. *Pfluegers Arch.* **312:**189–205, 1969.
92. SCHWANGHARDT, F., SCHROTER, R., KLUSSENDORF, D., KOEPCHEN, H. P.: Influence of novocaine block

of superficial brain stem structures. In Umbach, W., Koepchen, H. P. (eds.): *Central Rhythmic and Regulation*, Stuttgart, Hippokrates Verlag, 1974, pp. 104–110.

93. SEE, W. R.: Respiratory drive in hyperthermia: Interaction with central chemosensitivity. In Loeschcke, H. H. (ed.): *Acid–Base Homeostasis of the Brain Extracellular Fluid and the Respiratory Control System*. Stuttgart, Georg Thieme, 1976, pp. 122–127.
94. SEE, W. R., FOLGERING, M., SCHLAEFKE, M. E.: Central respiratory and cardiovascular effects of the ACH releaser 4-aminopyridine (r-AP). *Pfluegers Arch.* **377**:R20, 1978.
95. SORENSEN, S. C.: The chemical control of ventilation. *Acta Physiol. Scand., Suppl.* 371, 1971.
96. STERN, L. S., GAUTIER, E.: L'emploi de l'injection intraventriculaire comme methode d'étude de l'action directe des substances sur les centres nerveux. *C. R. Soc. Biol.* **86**:648–649, 1922.
97. STERN, L. S., ROZSINE, J. A., CHUOLES, G. J.: Le mechanisme de l'action du K et du Ca injectés dans les ventricules cerebraux. *C. R. Soc. Biol.* **114**:674–677, 1933.
98. TABER, E.: The cytoarchitecture of the brain stem of the cat. *J. Comp. Neurol.* **116**:27, 1961.
99. TERENIUS, L.: Endogenous peptides and analgesia. *Annu. Rev. Pharmacol.* **18**:189–204, 1978.
100. TRZEBSKI, A., MAJHERCZYK, P., SZULCZYK, P., CHRUSCIELEWSKI, L.: Direct nervous mechanisms as the possible pathways of interaction of the central and peripheral chemosensitive areas. In Loeschcke, H. H. (ed.): *Acid–Base Homeostasis of the Brain Extracellular Fluid and the Respiratory Control System*. Stuttgart, Georg Thieme, 1945, pp. 130–145.
101. WIEMER, W.: Die Wirkung von Chemorezeptoren-Reizstoffen vom Liquor aus, *Z. Biol.* **111**:287–320, 1959.
102. WIEMER, W.: Zur Wirkung des Acetylcholins auf die Atmung. IV. Wirkung bei Injektion in die Cisterna pontis, Cisterna cerebellomedullaris, in den Seitenventrikel und den Liquor spinalis, *Pfluegers Arch. Physiol.* **276**:568–578, 1963.
103. WINTERSTEIN, H.: Die chemische Steurerung der Atmung. *Ergeb. Physiol.* **48**:328–528, 1955.
104. WINTERSTEIN, H., GÖKHAN, N.: Ammoniumchlorid-Acidose und Reaktionstheorie der Atmungsregulation. *Arch. Int. Pharmacodyn.* **93**:212–232, 1953.
105. WINTERSTEIN, H., GÖKHAN, N.: Chemoreceptoren Reizstoffe und Blut/Hirn-Schranke. *Naunyn-Schmiedebergs Arch. Exp. Pathol. Pharmakol.* **219**:192–196, 1953.
106. WINTERSTEIN, H., WIEMER, W.: Die Wirkung von Lobelin und Natriumcyanid bei suboccipitaler Injektion. *Naunyn-Schmiedebergs Arch. Exp. Pathol. Pharmakol.* **235**:235–242, 1959.
107. WRIGHT, E. M.: Factors influencing the composition of the cerebrospinal fluid. In Loeschcke, H. H. (ed.): *Acid–Base Homeostasis of the Brain Extracellular Fluid and the Respiratory Control System.* Stuttgart, Georg Thieme, 1976, pp. 2–7.
108. WRIGHT, E. M.: Transport processes in the formation of the cerebrospinal fluid. *Rev. Physiol. Biochem. Pharmacol.* **83**:1–34, 1978.

CHAPTER 4

Neurochemical Aspects of Blood–Brain–Cerebrospinal Fluid Barriers

Robert W. P. Cutler

1. Introduction

The blood–brain barrier and the blood–cerebrospinal fluid (CSF) barrier are terms that are deeply entrenched in the literature of neurobiology. The concept of a permeability barrier between the blood and the nervous system evolved from observations of Ehrlich that the brain did not become stained after intravenous injection of aniline dyes (e.g., trypan blue). Goldmann later showed that the brain became stained after injection of trypan blue into the CSF. For many years, the barrier was considered to be an anatomical one that absolutely restricted the passage of certain substances into the brain. When tracer methodology became available, it became clear that most substances were not absolutely restricted from the brain, but rather that their rates of entry or ultimate ability to accumulate were reduced with respect to other organs. In this way, the chemical environment of the brain could be carefully guarded and exquisitely regulated. This chapter provides a brief account of the mechanisms that contribute to the regulation of concentrations of various classes of solutes in the brain and CSF. It seems likely that no fundamental differences exist in the mechanisms of transport at the blood–brain and blood–CSF interfaces. Several monographs expand the subject for interested readers.[11,35,48,52,53,81]

Robert W. P. Cutler, M.D. • Department of Neurology, Stanford University Medical Center, Stanford, California 94305.

2. General Mechanisms of Blood–Brain–CSF Barriers

2.1. Brain Capillary Transport

Exchange of metabolites between the blood and the brain takes place across the capillary endothelial cell wall. The area of this membrane has been estimated to be 240 cm^2/g brain.[19] Capillaries in most regions of brain have distinctive structural properties in comparison with capillaries in other organs. These special properties include overlapping endothelial cells joined to one another by tight junctions, absence of fenestrations in the endothelial cell wall, a paucity of pinocytotic vesicles in the endothelial cell, and a complete investiture of the capillary by an astrocytic membrane. These morphological features are found in the capillaries of regions of brain that do not become stained after intravenous injection of trypan blue. Ultrastructural study of cerebral capillaries after injection of low-molecular-weight substances such as microperoxidase or lanthanum has disclosed that the intercellular tight junction is the site of impermeability for these substances.[14] When they are injected into the blood, their inward movement is stopped at the first tight junction on the luminal side of the endothelium. When they are injected into the brain, their outward movement is stopped at the first tight junction on the abluminal side; movement through the glial envelope and basement membrane is not restricted.

Other regions of brain, such as the median eminence, area postrema, pineal gland, and choroid plexus, contain capillaries that resemble those in other body organs such as muscle and kidney. These vessels are comprised of a discontinuous endothelial lining containing fenestrations, and they are not impervious to trypan blue. Movement of solutes through these regions of the nervous system appears to be influenced more by the tight junctions between parenchymal cells—for example, intravenously injected peroxidase enters the stroma of the choroid plexus, but its passage into the CSF is impeded by tight junctions between epithelial cells.[15]

The integrity of the capillary endothelial lining ensures that substances will not move indiscriminately into and out of the brain and gives to the endothelial cell membrane the responsibility of regulating solute movement by a transcellular route. Capillary membrane transport is probably the most important aspect of the blood–brain barrier, and one that has only recently been subjected to quantitative study. It is of interest that the brain capillaries that do not permit passive solute movement between endothelial cells have apparently developed a special energy source to assist in active solute movement across endothelial membranes. Thus, Oldendorf *et al.*[66] have determined stereologically the mitochondrial volume in electron micrographs of capillary endothelial cells from the brain and several other organs. In the regions of brain possessing a blood–brain barrier, mitochondria comprised 8–11% of the endothelial cell volume, whereas in regions of brain not possessing a blood–brain barrier, and in other body organs, mitochondria comprised only 2–5% of the endothelial cell volume.

Another relatively unique property of brain capillary endothelial cells is their high content of certain enzymes that may contribute barriers to the movement of neurotransmitters or their precursors from blood to brain. The best known example is that of L-DOPA decarboxylase, a mitochondrial enzyme that converts L-DOPA into dopamine, an impermeable solute.[6] γ-Aminobutyric acid (GABA) transaminase is also present in large amounts in the mitochondria of cerebral capillaries and is thought to participate in the exclusion of systemically administered GABA from the brain.[94]

Recently, it has proved possible to isolate metabolically active capillaries from rat cerebral cortex.[44] These capillaries accumulate glucose by a temperature-dependent, saturable process, and oxidize glucose to carbon dioxide. They are enriched in γ-glutamyl transpeptidase, an enzyme that catalyzes the transfer of the γ-glutamyl group of glutathione to amino acids or peptides.[44,68] The γ-glutamyl cycle has been proposed as an amino acid transport system in the kidney.[60] The presence of high concentrations of the transpeptidase in the choroid plexus and brain capillaries has raised the possibility that a similar mechanism for amino acid transport may be present in the brain.[68] It is known that isolated cerebral capillaries possess mechanisms for the stereospecific accumulation of α-methylaminoisobutyric acid and L-leucine.[10,83]

2.2. Choroid Plexus Transport

The choroid plexus has the morphological characteristics of a secretory organelle. Each microvillus of the extensively folded plexus has a core containing a capillary with fenestrated endothelial cells and a loose connective tissue. The core is surrounded by cuboidal epithelial cells, continuous with the ventricular ependyma, that have a brush border at their apical surfaces and extensive interdigitations with adjacent cells at their basilar surfaces.[38]

The choroid plexus is involved in the secretion of CSF; this aspect of its function will not be covered here. Equally important is its role in controlling the chemical composition of the CSF. This control is accomplished by membrane transport mechanisms for a large variety of substances including ions, organic acids, organic bases, amino acids, hexoses, purines, and vitamins.[22,56]

Several methods have been used to study the transport properties of the choroid plexus. These include analysis of the composition of fluid collected at the surface of the plexus,[37] extracorporeal perfusion of the choroid plexus,[21,77] ventriculocisternal perfusion,[72] and incubation of the choroid plexus *in vitro*.[57] The last two methods have been most widely used. The ventriculocisternal technique permits study of the kinetics of transfer of solutes into and out of the CSF, but does not permit precise identification of the transport site. Removal of solutes from the CSF could occur across the choroidal epithelium or arachnoid membrane into the blood, or across the ependymal epithelium and pial membrane into the brain, with subsequent transport across the brain capillary into the blood. There is evidence that net solute transport out of the CSF may occur at each of these sites.

The choroid plexus, when incubated *in vitro*, is capable of concentrative uptake of a variety of substances from the medium. For example, amino acids are accumulated by the choroid plexus by processes showing saturation kinetics, stereospecificity, amino acid group specificity, and energy dependence.[55,57]

On the basis of quite similar kinetic constants (binding affinity and maximum transport velocity) for amino acid clearance during ventriculocisternal perfusion and amino acid uptake by incubated choroid plexus, it has been suggested that the plexus is the principal site of transport of amino acids from the cerebral ventricles.[56] Further examples of the role of the choroid plexus in CSF homeostasis will be given below.

2.3. Arachnoidal Transport

It is clear that transport of amino acids and other solutes between blood and CSF may occur at sites that do not contain choroid plexus, as, for example, the spinal and cranial subarachnoid spaces.[58,61] In these studies, saturable transport of amino acids from the CSF during isolated perfusion of the subarachnoid space was found. Because the amino acids did not accumulate in the brain or spinal cord, it seems plausible that their transport occurred at the arachnoidal membrane; direct proof that the arachnoid membrane may actively transport glycine in an outward direction has been obtained by Wright[99] in the frog.

2.4. Bulk Flow

The rate of formation of CSF is quite rapid; the CSF turnover rate is approximately 25% per hour in several animals studied and in children.[29] The drainage of CSF occurs by bulk flow through membranes of the arachnoid villi that cannot discriminate among molecules of different sizes. Thus, by bulk removal alone, all substances in the CSF turn over at a minimum rate of 25% per hour. Provided there is restricted entry of the solute from blood to CSF, because of either slow diffusion or saturable transport, the rapid fractional exit rate will lead to a lower concentration in the CSF than in the blood. This mechanism accounts for the very low CSF concentration of slowly diffusing solutes such as proteins.[26]

2.5. Brain–Blood Transport

It is difficult to directly demonstrate brain–to–blood transport *in vivo* under normal conditions. However, strong evidence of passive and active transcapillary movement of solutes from brain to blood has been found when substances are introduced into the brain by perfusion through the cerebral ventricles. Fenstermacher and Patlak[40] analyzed tissue concentration profiles of various solutes in the caudate nucleus after ventriculocisternal perfusion. From this analysis, they could determine passive brain–blood transcapillary transfer coefficients. Davson and Hollingsworth[36] measured the accumulation of [^{131}I]iodide in the brain extracellular space during ventriculocisternal perfusion. When intravenous perchlorate was given to block transport processes, the brain iodide concentration increased in consequence of inhibition of brain-to-blood transcapillary transport. These indirect experiments provide evidence that outwardly oriented brain–blood capillary transport mechanisms exist, but they do not indicate the extent to which such transport contributes to the blood–brain barrier.

2.6. Sink Action of CSF

The influence of choroidal and arachnoidal transport mechanisms, and of rapid bulk turnover of CSF, on the composition of brain extracellular fluid (ECF) is of great importance. The configuration of the ventricular and arachnoid compartments is such that most regions of gray matter of the brain and spinal cord are within a few millimeters of CSF. It is known that solutes diffuse freely from CSF to brain.[40] Because of active transport and bulk flow, the concentrations of most metabolites are lower in the CSF than in the plasma. Davson[34] theorized that such plasma-to-CSF concentration gradients would influence the concentration of solute in the extracellular space of the brain through a sink action. The concentration in the brain extracellular space would be expected to lie somewhere between the concentrations in the plasma and CSF. It has been difficult to demonstrate this gradient directly because there is no way of sampling brain ECF. Bito *et al.*[12] approached the problem by implanting small dialysis sacs in the brain parenchyma to collect a dialysate of brain fluid. They found that the concentrations in the dialysate of several amino acids were intermediate between the concentrations in plasma and CSF. The influence of the sink is evident when considering the plasma, brain, and CSF distribution of iodide. The concentration of tracer amounts of radioactive iodide in the CSF is 1% of that in plasma; the volume of distribution of iodide in the brain is 1–2%. At first, this finding resulted in the conclusion that the extracellular space of the brain was very small. When the potential sink action of the CSF was eliminated, however, by maintaining equal concentrations of iodide in the plasma and CSF (by ventriculocisternal perfusion), the volume of distribution of iodide in the brain rose to 10–20% in regions that were well bathed by the perfusate.[28] Similar experiments have been performed for a variety of substances. Many investigators believe that the sink action of the spinal fluid represents an important route for the elimination of products of cerebral metabolism.

3. Movement of Solutes across Blood–Brain–CSF Barriers

3.1. General Aspects

The flux of nonelectrolytes through the cells of the blood–brain–CSF barriers may be regulated by passive diffusion, pinocytosis, or carrier-mediated transport. As in other cellular membranes, the rate of passive diffusion is determined by the transcellular concentration gradient, the lipid solubility, and, in the case of electrolytes, the degree of ionization. The rate of entry of solutes into brain and CSF is very slow if they move solely by diffusion. This fact forms the basis for the use of water-soluble solutes such as urea, fructose, mannitol, and glycerol in the treatment of brain edema or in the prevention of blood–brain osmotic dysequilibrium during hemodialysis. Because of their nearly complete restriction of entry into the brain, these solutes generate an osmotic pressure gradient that draws water from the brain.

Pinocytosis (and subsequent vesicular transport) is generally considered to be an unimportant mode of transport of solutes across the blood–brain barrier. Normally, brain capillaries contain very few membranous vesicles. There is evidence that vesicular transport may lead to protein exudation under conditions of experimental convulsions or trauma to the nervous system.[5,59]

By far the most important route of entry of solutes into the brain is by carrier-mediated transport. The concept of carrier transport is useful in explaining many of the phenomena of transmembranous solute flux, although the molecular mechanisms of carrier transport are not understood.[98] The essential features of the carrier hypothesis are: (1) solute binds to a mobile component of the cell membrane, generally described as a carrier protein, at the extracellular face of the membrane; (2) the solute–carrier protein complex moves freely across the membrane and releases the solute at the intracellular face. The process is called *facilitated diffusion* if the binding and membrane translocation phases do not require energy and result in the equilibration of intracellular and extracellular solute concentration. The process is called *active transport* if those steps utilize energy and result in the uphill movement of molecules against a concentration gradient. There is substantial evidence in bacterial membranes for the presence of carrier proteins that participate in the transport of amino acids, sugars, and ions.[70] Such evidence is not yet available for capillary endothelial or choroidal epithelial cells. Nonetheless, several aspects of the blood–brain–CSF barriers are explained best by a carrier-mediated transport model. These include the following:

1. The transport of many substances is saturable—that is, the transport velocity increases hyperbolically as the solute concentration is increased. The relationship between transport velocity and solute concentration conforms to Michaelis–Menton enzyme–substrate kinetics. Both maximum transport velocity (V_{max}) and the carrier–solute binding constant (K_m) may be calculated from the Michaelis–Menton equation.
2. The transport system is specific for certain classes of solutes, as, for example, neutral, acidic, or basic amino acids. A solute in one class has little or no affinity for binding to the carrier of solutes in another class.
3. The affinity of binding to a carrier is determined by the structural configuration of the solute. For example, the stereoisomers of a solute have markedly different binding affinities and rates of transport into the brain.
4. The movement of solute molecules in one direction across cell membranes may be accompanied by the countermovement of molecules in the other direction. This process is referred to as *exchange*. When the inward movement of a molecule is accompanied by the outward movement of a molecule of the same species, the process is called *homoexchange*. When the inward movement of a molecule is accompanied by the outward movement of a molecule of a different species, the process is called *heteroexchange*. This is an extremely important concept to consider when evaluating studies of the blood–brain barrier, particularly those studies employing tracer (isotope) methodology. Failure to consider exchange processes has resulted in many misconceptions about the blood–brain barrier.

3.2. Amino Acids

The principles of carrier-mediated transport and exchange are best illustrated by considering the blood-to-brain and blood-to-CSF transfer of amino acids. It has long been known that the concentration of amino acids in the brain and CSF changes very little in response to marked increases in concentration in the plasma. A "barrier" for entry of amino acids was thought to exist. When the brain uptake of isotopically labeled amino acids was studied, however, it was found that there was a rapid entry of labeled amino acids, equilibrium being reached generally in a few minutes. Thus, no barrier to entry exists. The labeled amino acids enters the brain rapidly, but then exchanges with its nonlabeled homologue, and no net uptake of amino acid occurs.

The concentration of amino acids in the brain ECF appears to be regulated in part by carrier-mediated transport mechanisms located in the cerebral capillary. These have been extensively characterized by Lajtha[51] and by Oldendorf[64,67,75] and Baños[2–4] and their respective co-workers. High influx rates are found for amino acids that are nutritionally essential for brain metabolism as well as for those that the brain cannot synthesize (e.g., tyrosine). There are three known transport mechanisms that mediate the influx of neutral, acidic, and basic amino acids. Neutral amino acid transport systems may be subdivided into the L-system (leucine-preferring) and the A-system (alanine-preferring) described by Oxender and Christensen[71] in ascites tumor cells. The L-system transports mainly large neutral amino acids, is predominantly an equilibrating system, and does not require the presence of sodium. The A-system transports mainly small amino acids, can maintain steep concentration gradients, and requires sodium for its operation. Unidirectional, carrier-mediated influx into brain has been shown only for amino acids carried by the L-system but not the A-system.[73,95,100] However, it has been shown that isolated cerebral capillaries contain an A-system for amino acid transport.[10] It has been proposed that this system serves to transport small amino acids *out of* the brain.[10,25] Thus, the polarity of the L-system is for inward transport, that of the A-system for outward transport. Whereas there is a large unidirectional influx of essential amino acids, there is little or no net flux when cerebral arterial–venous concentration differences are considered.[8] This fact provides the most convincing evidence for bidirectional amino acid transport across the cerebral capillary.

The plasma concentration of an amino acid is generally of the same order of magnitude as the binding constant for its transport into brain, indicating that the rate of transport will be very sensitive to changes in plasma concentration of either that amino acid or amino acids that compete with it for transport. Such an equivalence of plasma amino acid concentration and binding constant is not found in other organs. Pardridge and Oldendorf[76] have suggested that the apparent selective vulnerability of the brain in cases of phenylketonuria in which there is an elevated plasma concentration of phenylalanine is explained by this unique set point of the transport mechanism in the brain capillary. The sensitive relationship between influx and plasma concentration is also important in meeting the brain's variable need for amino acids in the processes of protein and neurotransmitter synthesis. There appears to be a good correlation between rates of amino acid influx into brain and amino acid incorporation into brain protein.[51] For example, in immature animals in which there is a high rate of protein synthesis, there is also a high rate of blood–brain amino acid transport.[4]

The CSF concentrations of amino acids are much lower than those in plasma, with the exception of glutamine, which has an equal concentration in the two compartments. The lower CSF concentration is maintained by transport mechanisms that actively transport amino acids from CSF to blood against the concentration gradient. Separate mechanisms for the carrier-mediated transport of large and small neutral amino acids,[84] dibasic amino acids,[23] and GABA[85] have been identified. Thus, both L-system and A-system neutral amino acid transport systems operate at the blood–CSF interface. Transport consists of a nonsaturable component, presumably representing simple diffusion, and a saturable component conforming to Michaelis–Menton kinetics.[27] Carrier-mediated transport with the same group specificity has been found in isolated choroid plexus,[57] and the similarities of kinetic constants from studies of ventricular clearance of amino acids *in vivo* and choroid plexus uptake *in vitro* have suggested that the plexus is the site of transport of amino acids from the cerebral ventricles to the blood.[55] Other sites undoubtedly exist, including the cranial[58] and spinal[39] subarachnoid spaces. The concentration of amino acids is not uniform throughout the spinal fluid; the differences are most likely accounted for by different rates of entry and exit of amino acids in various regions of the CSF compartment.[42]

In all likelihood, the low concentration of amino acids in the CSF plays a role in the maintenance of brain extracellular amino acid concentration through a sink action. Transmitter amino acids are known to be lost to the CSF under certain conditions after their release at synapses; tissue-to-fluid efflux of amino acids has been found in the ventricles,[62] as well as in the cranial[47] and spinal[24] subarachnoid spaces.

3.3. Glucose

The blood–brain transfer of glucose has received extensive study since the original demonstration of Crone[20] of facilitated transport of glucose into the dog brain *in vivo.* Evidence that the passage of glucose from blood to brain is accomplished by carrier-mediated transport includes saturability of transfer, stereospecificity of transfer, competitive inhibition by structural analogues, and exchange diffusion. The kinetic parameters of glucose transport have been studied most extensively in the isolated perfused dog brain and in the rat. These features have been discussed in recent reviews.[9,74,78]

The affinity constant for binding of D-glucose to its carrier in the cerebral capillary is approximately 6–10 mM. Therefore, the carrier is approximately half-saturated at normal blood glucose concentrations, and the influx transport velocity can change considerably as the blood glucose concentration is raised or lowered. In the normal adult animal, the influx rate greatly exceeds the rate of cerebral retention of glucose as measured by the cerebral arterial–venous concentration difference. The difference between influx and retention represents efflux of glucose from brain to blood. Efflux of glucose from the brain is mediated by the same carrier as influx, as demonstrated by counterflow experiments.[30] Pardridge and Oldendorf[74] have estimated that the fraction of 3-*O*-methyl-D-glucose transferred from brain to blood per unit time is the same as the fraction transferred from blood to brain. The bidirectional equivalence of fractional extraction of the nonmetabolized glucose analogue suggests that the mechanism for glucose transport across the blood–brain barrier is one of facilitated diffusion rather than active transport. This suggestion is supported by the finding that sodium does not seem to be required for blood–brain glucose transport and that ouabain does not inhibit glucose transport. However, as pointed out by Betz *et al.*,[9] the concentration of glucose in the brain compartment into which it is transported from blood is unknown, so that the possibility of uphill transport cannot be definitely excluded.

The rate of influx of glucose at a normal blood glucose concentration is approximately 1 μmol/min per g brain. The rate of metabolic utilization of glucose is approximately 0.3 μmol/min per g brain under normal conditions; thus, transport of glucose across the blood–brain barrier is not rate-limiting for brain metabolism. The excess influx of glucose normally provides a considerable safety margin for the brain. However, blood–brain transport becomes increasingly rate-limiting under conditions of hypoglycemia. It has been calculated that the velocity of glucose transport across the cerebral capillary in the dog will be insufficient to satisfy the metabolic needs for glucose when the blood concentration falls below 2.8 mM.[9]

Developmental studies in the rat have shown that the velocity of transport of glucose is a function of age, whereas the carrier–substrate binding constant does not change with age. The rate of influx of glucose doubles between the ages of 3 and 8 weeks, and then falls to an intermediate level in the adult.[31] These changes in glucose influx appear to parallel changes in metabolic requirement for glucose. The margin of safety (i.e., excess glucose influx) is greatest in the young adult. In the neonatal animal, increased metabolic demands for glucose, as, for example, during seizures, cannot be met by increased influx of glucose unless the blood glucose is raised above normal.[96]

Characteristics of the glucose transport system have been studied by Betz *et al.*[7] and Pardridge and Oldendorf.[74] The glucose carrier is shared by other hexoses such as mannose and galactose; however, their binding affinity is much less than that of glucose, and they probably have little influence on glucose transport under normal conditions. In pathological states, such as galactosemia, the elevated blood levels of galactose conceivably interfere with blood–brain glucose transport. From inhibition studies with a large number of analogues, Betz *et al.*[7] concluded that hydroxyl groups on carbons 1, 3, 4, and 6 were involved in the binding of glucose to its carrier. Phloretin and phlorizin, which inhibit glucose transport in the erythrocyte, also inhibit blood–brain transport of glucose.[7,74] Most studies have shown that the influx of glucose is not dependent on insulin and is not altered by fasting or starvation. It is possible, however, that hyperinsulinemia inhibits efflux of glucose from brain to blood.[78]

Regulation of the glucose concentration in the CSF is achieved by mechanisms that are, in all likelihood, indistinguishable from those operating at the blood–brain interface. The CSF glucose concentration is normally about two-thirds the plasma glucose concentration. Facilitated diffusion appears to be the mechanism responsible for maintaining this ratio; the fractional entry and exit of glucose in CSF is equal, as measured during perfusion of the cerebral ventricles.[13] It is likely that the choroid plexus is a major transport site mediating both influx and efflux. Welch and coworkers[82,97] have found that the concentration of glucose in newly secreted CSF is only 60% of that in plasma. The transfer of glucose from CSF to blood is not inhibited by 2,4-dinitrophenol, but is partially inhibited by ouabain.[17] Fishman[41] has clearly demonstrated the bidirectional nature of glucose transport between blood and CSF and has shown that under certain circumstances, glucose may be transported into CSF against a concentration gradient. When the CSF compartment was loaded with 2-deoxyglucose, the efflux of this hexose (mediated by the glucose carrier) caused a counter influx of glucose against a gradient.[41] Bidirectional transport of glucose across the blood–CSF barrier was shown to be partially inhibited in dogs with experimental bacterial meningitis, but it is unlikely that transport inhibition plays a major role in lowering the CSF glucose in this condition.[80]

3.4. Monocarboxylic Acids

3.4.1. Ketones. Glucose is the principal source of energy for the brain in the normal adult. However, during the neonatal period, there appears to be a substantial utilization of ketone bodies, principally acetoacetate and β-hydroxybutyrate, to support brain metabolism.[46,50,93] Increased oxidation of ketones by the brain also occurs after prolonged fasting in the adult.[69] Until recently, it was believed that ketones crossed the blood–brain barrier by a process of simple diffusion, the rate of entry being directly proportional to the plasma concentration of the ketones.[81] Recent studies have provided evidence, however, that these acids are transported into the brain by a carrier-mediated process. Using the brain uptake index method of Oldendorf,[63] Gjedde and Crone[43] found that the uptake of D-3-hydroxybutyrate conformed to Michaelis–Menton kinetics, with a K_m of approximately 3 mM. Cremer *et al.*[18] also found self-inhibition of D-3-hydroxybutyrate extraction, as well as cross-inhibition by L-lactate. The capacity of the transport system for these acids was 6-fold higher in suckling than in adult rats. These workers suggested that a common carrier might mediate the inward blood-to-brain transport of 3-hydroxybutyrate and the outward brain-to-blood transport of lactate.[18] Further evidence for carrier-mediated transport of ketones was found by Daniel *et al.*[32] They found that the influx of ketones into the brains of normal adult rats per millimole of ketone in the blood was 10 nmol/min per g brain, a rate 5 times higher than would be expected from passive diffusion alone. There was a higher transport rate for the D-isomer than for a racemic mixture of D,L-hydroxybutyrate. The rate of influx per unit of concentration in blood was much higher in suckling rats and fell rapidly at about the time of weaning. An increased influx of ketones was also found in rats that were fasted for several days, a result similar to that found by Gjedde and Crone.[43] These latter workers found an increase in both V_{max} and K_m for ketone transport, suggesting that there was an induction of the transport mechanism during starvation. A similar induction of blood–brain ketone transport by the relative hyperketonemia of infancy is thought to underlie the high rate of transport during this period of life.[32]

The increased influx and utilization of ketones for brain metabolic energy during infancy may have protective value. The influx of glucose into the brain and the stability of the blood glucose concentration are much lower in infancy than in adulthood. During early maturation, the brain cannot rely on the glucose transfer system to satisfy its metabolic needs; the ketone transfer system provides an increased margin of safety.

3.4.2. Lactate and Pyruvate. Brain capillary transport of lactate and pyruvate is governed by the same system as that for ketones.[65] It has been suggested that influx of 3-hydroxybutyrate is accompanied by efflux of lactate.[18] Saturability of transport of lactate and pyruvate may be demonstrated easily; the estimated K_m value for lactate is 1.9 mM and for pyruvate 0.4 mM.[65] These half-saturation values are close to the normal plasma concentrations of each metabolite, and transport occurs at a maximum velocity when the plasma concentrations are elevated about 4-fold. Saturation of influx at relatively low plasma concentrations serves as a mechanism to protect the brain in cases of lactic acidosis.

Normally, the direction of net transport of lactate and pyruvate is from brain to blood; net efflux is indicated by their slightly higher concentrations in cerebral venous blood than in cerebral arterial blood.[33] Outward transport serves to eliminate at least a fraction of the pool of lactate and pyruvate produced by glycolysis. When lactic acid accumulates in the brain under conditions of hypoxia, as a result of anaerobic glycolysis, the low-capacity transport system cannot clear the excess metabolite into the blood. Saturation of the brain–blood transport mechanism results in an increase in the concentration of lactate in brain ECF followed by diffusion down a concentration gradient into the CSF, where it may accumulate to a high level.[16] Whereas many brain metabolites are cleared rapidly from the CSF, this does not appear to be the case for lactate. Prockop[79] found that the rate of clearance of lactate from the CSF of the dog was approximately that expected from simple diffusion. Alternatively, lactate may be transported out of CSF by a carrier system of low capacity that is saturated at low concentrations.

3.5. Vitamins

The metabolically important solutes discussed thus far are transported from blood to brain across the brain capillary by an equilibrating process. Regulation of their concentration in the brain is governed principally by the favorable set point of the transport mechanism; that is, the binding constant (K_m) is approximately the same as the plasma concentration. When the plasma concentration falls, a greater fraction is transported into the brain. When the plasma concentration rises, a smaller fraction is transported into the brain.

The blood–brain–CSF transport of vitamins appears to be quite different. The mechanisms have been studied extensively by Spector and Lorenzo.[86–92] Their studies have disclosed relatively unique mechanisms governing the distribution of ascorbic acid, folates, and inositol among these three compartments in the rabbit. It is likely that similar principles apply to the distribution of vitamins in man.[87]

The concentration of each of these substances is 4–5 times *higher* in the CSF than in the plasma of the rabbit. Since each is water-soluble and has a fairly large molecular size (mol. wt. 175–465), entry from blood to CSF by diffusion would be very slow and removal by bulk flow would be sufficiently rapid to maintain a low CSF/plasma ratio. Thus, from consideration of the concentrations alone, specialized transport mechanisms would be anticipated.

Ascorbic acid has been studied most extensively. In many animals, including man, this vitamin cannot be synthesized and is supplied by the diet. It appears to be required in the brain as a cofactor for dopamine β-hydroxylase.[49] According to the results of autoradiographic[45] and physiological studies,[88,89] ascorbic acid is supplied to the brain by way of the CSF. Results of these studies suggest that ascorbic acid is actively accumulated by the choroid plexus and transported against a concentration gradient into the CSF, from which it is distributed to the brain. The transport system in the isolated choroid plexus of the rabbit is energy-dependent, stereospecific, and saturable. The half-saturation value for plexus uptake is 44 μM, a concentration similar to the plasma concentration and similar to the half-saturation concentration for transport into the CSF *in vivo*[88,89] The effect of this transport mechanism is illustrated by the fact that a 100-fold increase in plasma ascorbic acid results in only a 13-fold increase in CSF and a 2-fold increase in brain.[92] Evidence for entry of ascorbic acid into the brain via the CSF includes the autoradiographic pattern of distribution[45] and the higher initial concentration in CSF and cortical and periventricular tissues than in deep cerebral tissues after intravenous injection of [^{14}C]ascorbic acid.[89] If these results are confirmed, an important new function must be ascribed to the CSF, namely, that of a nutrient medium.

Unique features have also been found for the blood–brain–CSF regulation of folic acid (FA), another exclusively dietary vitamin. The principal form of this vitamin in the blood and CSF is 5-methyltetrahydrofolic acid (MeTHF), and its concentration is higher in the CSF than in the blood.[54] In addition, the concentration of the vitamin in the nervous system is carefully regulated; in vitamin-deficiency states, the concentration in the brain falls more slowly than that in the blood.[1] Again, transchoroidal transport of MeTHF appears to account for this homeostasis. An unusual feature of the transport mechanism is its unidirectional polarity. Thus, MeTHF appears to be transported by the choroid plexus only in the direction from blood to CSF, whereas FA is transported only in the direction from CSF to blood.[91]

As indicated above, the transport constants for these vitamins are close to their plasma concentrations, so that the transport V_{max} will be reached with only moderate increases in plasma concentration. Since blood-to-brain transfer by diffusion would not be likely to increase the concentration in brain substantially, Spector[87] has raised the question whether the megavitamin therapy advocated for the treatment of psychotic disorders could have as its therapeutic basis the elevation of brain vitamin concentration.

References

1. ALLEN, C. D., KLIPSTEIN, F. A.: Brain folate concentrations in rats receiving diphenylhydantoin. *Neurology* **20**:403, 1970.
2. BAÑOS, G., DANIEL, P. M., MOORHOUSE, S. R., PRATT,

O. E.: The influx of amino acids into the brain of the rat *in vivo:* The essential compared with some non-essential amino acids. *Proc. R. Soc. Lond. Ser. B* **183:**59–70, 1973.

3. BAÑOS, G., DANIEL, P. M, PRATT, O. E.: Saturation of a shared mechanism which transports L-arginine and L-lysine into the brain of the living rat. *J. Physiol.* **236:**29–41, 1974.
4. BAÑOS, G., DANIEL, P. M., PRATT, O. E.: The effect of age upon the entry of some amino acids into the brain, and their incorporation into cerebral protein. *Dev. Med. Child Neurol.* **20:**335–346, 1978.
5. BEGGS, I. L., WAGGENER, J. D.: Transendothelial vesicular transport of protein following compression injury to the spinal cord. *Lab. Invest.* **34:**428–439, 1976.
6. BERTLER, A., FALCK, B., OWMAN, C. H., *et al.:* The localization of mono-aminergic blood–brain barrier mechanisms. *Pharmacol. Rev.* **18:**369–385. 1966.
7. BETZ, A. L., DREWES, L. R., GILBOE, D. D.: Inhibition of glucose transport into brain by phlorizin, phloretin and glucose analogues. *Biochim. Biophys. Acta* **406:**505–515, 1975.
8. BETZ, A. L., GILBOE, D. D.: Effect of pentobarbital on amino acid and urea flux in the isolated dog brain. *Am. J. Physiol.* **244:**580–587, 1973.
9. BETZ, A. L., GILBOE, D. D., DREWES, L. R.: The characteristics of glucose transport across the blood brain barrier and its relation to cerebral glucose metabolism. In Levi, G., Battistin, L., Lajtha, A. (eds.): *Transport Phenomena in the Nervous System.* New York, Plenum Press, 1976, pp. 133–149.
10. BETZ, A. L., GOLDSTEIN, G. W.: Polarity of the blood–brain barrier: Neutral amino acid transport into isolated brain capillaries. *Science* **202:**225–227, 1978.
11. BITO, L. Z., DAVSON, H., FENSTERMACHER, J. D. (eds.): The ocular and cerebrospinal fluids, *Exp. Eye Res.* **25**(Suppl. 1-561), 1977.
12. BITO, L. Z., DAVSON, H., LEVIN, E., MURRAY, M., SNIDER, N.: The concentrations of free amino acids and other electrolytes in cerebrospinal fluid, *in vivo* dialysate of brain and blood plasma of the dog, *J. Neurochem.* **13:**1057–1067, 1966.
13. BRADBURY, M. W. B., DAVSON, H.: The transport of urea, creatinine and certain monosaccharides between the blood and fluid perfusing the cerebral ventricular system of rabbits. *J. Physiol.* **170:**195–211, 1964.
14. BRIGHTMAN, M. W., REESE, T. S.: Junctions between intimately apposed cell membranes in the vertebrate brain. *J. Cell Biol.* **40:**648–677, 1969.
15. BRIGHTMAN, M. W., REESE, T. S., FEDER, N.: Assessment with the electron-microscope of the permeability to peroxidase of cerebral endothelium and epithelium in mice and sharks. In Crone, C., Lassen, N. A., (eds.): *Capillary Permeability.* New York, Academic Press, 1970, pp. 463–476.
16. BRODERSEN, P., JØRGENSEN, E. O.: Cerebral blood flow and oxygen uptake, and cerebrospinal fluid biochemistry in severe coma. *J. Neurol. Neurosurg. Psychiatry* **37:**384–391, 1974.
17. BRØNDSTED, H. E.: Ouabain-sensitive carrier-mediated transport of glucose from the cerebral ventricles to surrounding tissues in the cat. *J. Physiol.* **208:**187–201, 1970.
18. CREMER, J. E., BRAUN, L. D., OLDENDORF, W. H.: Changes during development in transport processes of the blood–brain barrier. *Biochim. Biophys. Acta* **448:**633–637, 1976.
19. CRONE, C.: The permeability of capillaries in various organs as determined by use of the "indicator diffusion" method. *Acta. Physiol. Scand.* **58:**292–305, 1963.
20. CRONE, C.: Facilitated transfer of glucose from blood into brain tissue. *J. Physiol.* **181:**103–113, 1965.
21. CSÁKY, T. Z., RIGOR, B. M.: The choroid plexus as a glucose barrier. In Lajtha, A., Ford, D. H., (eds.): *Progress in Brain Research,* Vol. 29, *Brain Barrier Systems.* Amsterdam, Elsevier, 1968, pp. 147–154.
22. CSERR, H. F.: Physiology of the choroid plexus. *Physiol. Rev.* **51:**273–311, 1971.
23. CUTLER, R. W. P.: Transport of lysine from cerebrospinal fluid of the cat. *J. Neurochem.* **17:**1017–1027, 1970.
24. CUTLER, R. W. P.: Release of amino acids from the spinal cord *in vitro* in *in vivo.* In Levi, G., Battistin, L., Lajtha, A. (eds.): *Transport Phenomena in the Nervous System.* New York, Plenum Press, 1976, pp. 435–446.
25. CUTLER, R. W. P., COULL, B. M.: Amino acid transport in brain. In Barbeau, A., Huxtable, R. J. (eds.): *Taurine and Neurological Disorders.* New York, Raven Press, 1978, pp. 95–107.
26. CUTLER, R. W. P., DEUEL, R. K., BARLOW, C. F.: Albumin exchange between plasma and cerebrospinal fluid. *Arch. Neurol.* **17:**261–270, 1967.
27. CUTLER, R. W. P., LORENZO, A. V.: Transport of 1-aminocyclopentane-carboxylic acid from feline cerebrospinal fluid. *Science* **161:**1363–1364, 1968.
28. CUTLER, R. W. P., LORENZO, A. V., BARLOW, C. F.: Sulfate and iodide concentration in brain: The influence of cerebrospinal fluid. *Arch. Neurol.* **18:**316–323, 1968.
29. CUTLER, R. W. P., PAGE, L., GALICICH, J., WATTERS, G. V.: Formation and absorption of cerebrospinal fluid in man. *Brain* **91:**707–720, 1968.
30. CUTLER, R. W. P., SIPE, J. C.: Mediated transport of glucose between blood and brain in the cat. *Am. J. Physiol.* **220:**1182–1186, 1971.
31. DANIEL, P. M., LOVE, E. R., PRATT, O. E.: The effect of age upon the influx of glucose into the brain. *J. Physiol.* **274:**141–148, 1978.
32. DANIEL, P. M., LOVE, E. R., MOORHOUSE, S. R., PRATT, O. E.: The transport of ketone bodies into the brain of the rat (*in vivo*). *J. Neurol. Sci.* **34:**1–13, 1977.
33. DANIEL, P. M., LOVE, E. R., MOORHOUSE, S. R.,

PRATT, O. E., WILSON, P.: The movement of ketone bodies, glucose, pyruvate and lactate between the blood and the brain of rats. *J. Physiol.* **221:**22p–23p, 1972.
34. DAVSON, H.: The cerebrospinal fluid. *Ergeb. Physiol.* **52:**20–73, 1963.
35. DAVSON, H.: *Physiology of the Cerebrospinal Fluid.* Boston, Little, Brown, 1967.
36. DAVSON, H., HOLLINGSWORTH, J. R.: Active transport of ^{131}I across the blood–brain barrier. *J. Physiol.* **233:**327–347, 1973.
37. DE ROUGEMONT, J., AMES, A., III, NESBETT, F. B., HOFFMAN, H. F.: Fluid formed by choroid plexus: A technique for its collection and a comparison of its electrolyte composition with serum and cisternal fluid. *J. Neurophysiol.* **23:**485–495, 1960.
38. DOHRMANN, G. J.: The choroid plexus: A historical review, *Brain Res.* **18:**197–218, 1970.
39. DUDZINSKI, D. S., CUTLER, R. W. P.: Spinal subarachnoid perfusion in the rat: Glycine transport from spinal fluid. *J. Neurochem.* **22:**355–361, 1974.
40. FENSTERMACHER, J. D., PATLAK, C. S.: The exchange of material between cerebrospinal fluid and brain. In Cserr, H. F., Fenstermacher, J. D., Fencl, V. (eds.): *Fluid Environment of the Brain.* New York, Academic Press, 1975, pp. 201–214.
41. FISHMAN, R. A.: Carrier transport of glucose between blood and cerebrospinal fluid. *Am. J. Physiol.* **206:**836–844, 1964.
42. FRANKLIN, G. M., DUDZINSKI, D. S., CUTLER, R. W. P.: Amino acid transport into the cerebrospinal fluid of the rat. *J. Neurochem.* **24:**367–372, 1975.
43. GJEDDE, A., CRONE, C.: Induction processes in blood–brain transfer of ketone bodies during starvation. *Am. J. Physiol.* **229:**1165–1169, 1975.
44. GOLDSTEIN, G. W., WOLINSKY, J. S., CSEJTEY, J., DIAMOND, I.: Isolation of metabolically active capillaries from rat brain. *J. Neurochem.* **25:**715–717, 1975.
45. HAMMARSTRÖM, L.: Autoradiographic studies on the distribution of ^{14}C-labelled ascorbic acid and dehydroascorbic acid. *Acta Physiol. Scand. Suppl.* **289:**1–70, 1966.
46. HAWKINS, R. A., WILLIAMSON, D. H., KREBS, H. A.: Ketone-body utilization by adult and suckling rat brain *in vivo*. *Biochem. J.* **122:**13–18, 1971.
47. JASPER, H. H., KOYAMA, I.: Rate of release of amino acids from the cerebral cortex in the cat as affected by brainstem and thalamic stimulation. *Can. J. Physiol. Pharmacol.* **47:**889–905, 1969.
48. KATZMAN, R., PAPPIUS, H. M.: *Brain Electrolytes and Fluid Metabolism.* Baltimore, Williams and Wilkins, 1973.
49. KAUFMAN, S., FRIEDMAN, S.: Dopamine-β-hydroxylase. *Pharmacol. Rev.* **17:**71–100, 1965.
50. KRAUS, H., SCHLENKER, S., SCHWEDESKY, D.: Developmental changes of cerebral ketone body utilization in human infants. *Hoppe-Seyler's Z. Physiol. Chem.* **355:**164–170, 1974.
51. LAJTHA, A.: Amino acid transport in the brain *in vivo* and *in vitro*. In Wolstenholme, G. E. W., Fitzsimons, D. W. (eds.): *Aromatic Amino Acids in the Brain.* Amsterdam, Elsevier, 1974, pp. 25–41.
52. LAJTHA, A., FORD, D. H. (eds.): *Progress in Brain Research,* Vol. 29, *Brain Barrier Systems.* Amsterdam, Elsevier, 1968.
53. LEVI, G., BATTISTIN, L., LAJTHA, A. (eds.): *Advances in Experimental Medicine and Biology,* Vol. 69, *Transport Phenomena in the Nervous System.* New York, Plenum Press, 1976.
54. LEVITT, M., NIXON, P. F., PINCUS, J. H., BERTINO, J. R.: Transport characteristics of folates in cerebrospinal fluid: A study utilizing doubly labelled 5-methyltetrahydrofolate and 5-formyltetrahydrofolate. *J. Clin. Invest.* **50:**1301–1308, 1971.
55. LORENZO, A. V.: Amino acid transport mechanisms of the cerebrospinal fluid. *Fed. Proc. Fed. Am. Soc. Exp. Biol.* **33:**2079–2085, 1974.
56. LORENZO, A. V.: Factors governing the composition of the cerebrospinal fluid. *Exp. Eye Res. Suppl.* **25:**205–228, 1977.
57. LORENZO, A. V., CUTLER, R. W. P.: Amino acid transport by choroid plexus *in vitro*. *J. Neurochem.* **16:**577–585, 1969.
58. LORENZO, A. V., SNODGRASS, S. R.: Leucine transport from the ventricles and cranial subarachnoid space in the cat. *J. Neurochem.* **19:**1287–1298, 1972.
59. LORENZO, A. V., HEDLEY-WHITE, E. T., EISENBERG, H. M., HSU, D. W.: Increased penetration of horseradish peroxidase across the blood–brain barrier induced by metrazol seizures. *Brain Res.* **88:**136–140, 1975.
60. MEISTER, A.: On the enzymology of amino acid transport. *Science* **180:**33–39, 1973.
61. MURRAY, J. E., CUTLER, R. W. P.: Transport of glycine from the cerebrospinal fluid: Factors regulating amino acid concentrations in feline cerebrospinal fluid. *Arch. Neurol.* **23:**23–31, 1970.
62. OBATA, K., TAKEDA, K., Release of γ-aminobutyric acid into the fourth ventricle induced by stimulation of the cat's cerebellum. *J. Neurochem.* **16:**1043–1047, 1969.
63. OLDENDORF, W. H.: Measurement of brain uptake of radiolabeled substances using a tritiated water internal standard. *Brain Res.* **24:**372–376, 1970.
64. OLDENDORF, W. H.: Brain uptake of radiolabeled amino acids, amines and hexoses after arterial injection. *Am. J. Physiol.* **221:**1629–1639, 1971.
65. OLDENDORF, W. H.: Carrier-mediated blood–brain barrier transport of short-chain monocarboxylic organic acids. *Am. J. Physiol.* **224:**1450–1453, 1973.
66. OLDENDORF, W. H., CORNFORD, M. E., BROWN, W. J.: The large apparent work capability of the blood–brain barrier: A study of the mitochondrial content of capillary endothelial cells in brain and other tissues of the rat. *Ann. Neurol.* **1:**409–417, 1977.
67. OLDENDORF, W. H., SZABO, J.: Amino acid assign-

ment to one of three blood–brain barrier amino acid carriers. *Am. J. Physiol.* **230:**94–98, 1976.
68. Orlowski, M., Sessa, G., Green, J. P.: γ-Glutamyl transpeptidase in brain capillaries: Possible site of a blood–brain barrier for amino acids. *Science* **184:**66–68, 1974.
69. Owen, O. E., Morgan, A. P., Kemp, H. G., *et al.*: Brain metabolism during fasting. *J. Clin. Invest.* **46:**1589–1595, 1967.
70. Oxender, D. L.: Membrane transport. *Annu. Rev. Biochem.* **41:**777–814, 1972.
71. Oxender, D. L., Christensen, H. N.: Distinct mediating systems for the transport of neutral amino acids by the Ehrlich cell. *J. Biol. Chem.* **238:**3686–3699, 1963.
72. Pappenheimer, J. R., Heisey, S. R., Jordan, E. F.: Active transport of diodrast and phenolsulfonphthalein from cerebrospinal fluid to blood. *Am. J. Physiol.* **200:**1–10, 1961.
73. Pardridge, W. M.: Kinetics of competitive inhibition of neutral amino acid transport across the blood–brain barrier. *J. Neurochem.* **28:**103–108, 1977.
74. Pardridge, W. M., Oldendorf, W. H.: Kinetics of blood–brain barrier transport of hexoses. *Biochim. Biophys. Acta* **382:**377–392, 1975.
75. Pardridge, W. M., Oldendorf, W. H.: Kinetic analysis of blood–brain barrier transport of amino acids. *Biochim. Biophys. Acta* **401:**128–136, 1975.
76. Pardridge, W. M., Oldendorf, W. H.: Transport of metabolic substrates through the blood–brain barrier. *J. Neurochem.* **28:**5–12, 1977.
77. Pollay, M.: Transport mechanisms in the choroid plexus. *Fed. Proc. Fed. Am. Soc. Exp. Biol.* **33:**2064–2069, 1974.
78. Pratt, O. E.: The transport of metabolizable substances into the living brain. In Levi, G., Battistin, L., Lajtha, A. (eds.): *Advances in Experimental Medicine and Biology,* Vol. 69, *Transport Phenomena in the Nervous System.* New York, Plenum Press, 1976, pp. 55–75.
79. Prockop, L. D.: Cerebrospinal fluid lactic acid. *Neurology* **18:**189–196, 1968.
80. Prockop, L. D., Fishman, R. A.: Experimental pneumococcal meningitis: Permeability changes influencing the concentration of sugars and macro-molecules in cerebrospinal fluid. *Arch. Neurol.* **19:**449–463, 1968.
81. Rapoport, S. I.: *Blood–Brain Barrier in Physiology and Medicine.* New York, Raven Press, 1976.
82. Sadler, K., Welch, K.: Concentration of glucose in new choroidal cerebrospinal fluid of the rat. *Nature (London)* **215:**884–885, 1967.
83. Sershen, H., Lajtha, A.: Capillary transport of amino acids in the developing brain. *Exp. Neurol.* **53:**465–474, 1976.
84. Snodgrass, S. R., Cutler, R. W. P., Kang, E. S., Lorenzo, A. V.: Transport of neutral amino acids from feline cerebrospinal fluid. *Am. J. Physiol.* **217:**974–980, 1969.
85. Snodgrass, S. R., Lorenzo, A. V.: Transport of GABA from the perfused ventricular system of the cat. *J. Neurochem.* **20:**761–769, 1973.
86. Spector, R.: Thiamine transport in the central nervous system. *Am. J. Physiol.* **230:**1101–1107, 1976.
87. Spector, R.: Vitamin homeostasis in the central nervous system. *N. Engl. J. Med.* **296:**1393–1398, 1977.
88. Spector, R., Lorenzo, A. V.: Ascorbic acid homeostasis in the central nervous system. *Am. J. Physiol.* **225:**757–763, 1973.
89. Spector, R., Lorenzo, A. V.: Specificity of ascorbic acid transport system of the central nervous system. *Am. J. Physiol.* **226:**1468–1473, 1974.
90. Spector, R, Lorenzo, A. V.: Folate transport by the choroid plexus *in vitro. Science* **187:**540–542, 1975.
91. Spector, R., Lorenzo, A. V.: Folate transport in the central nervous system. *Am. J. Physiol.* **229:**777–782, 1975.
92. Spector, R., Spector, A. Z., Snodgrass, S. R.: Model for transport in the central nervous system. *Am. J. Physiol.* **232:**73–79, 1977.
93. Spitzer, J. J., Weng, J. T.: Removal and utilization of ketone bodies by the brain of newborn puppies. *J. Neurochem.* **19:**2169–2173, 1972.
94. Van Gelder, N. M.: A possible enzyme barrier for γ-aminobutyric acid in the central nervous system. In Lajtha, A., Ford, D. H., (eds.): *Progress in Brain Research,* Vol. 29, *Brain Barrier Systems.* Amsterdam, Elsevier, 1968, pp. 259–268.
95. Wade, L. A., Katzman, R.: Rat brain regional uptake and decarboxylation of L-dopa following carotid injection. *Am. J. Physiol.* **228:**352–359, 1975.
96. Wasterlain, C. G., Duffy, T. E.: Status epilepticus in immature rats: Protective effects of glucose on survival and brain development. *Arch. Neurol.* **33:**821–827, 1976.
97. Welch, K., Sadler, K., Hendee, R.: Cooperative phenomena in the permeation of sugars through the lining epithelium of the choroid plexus. *Brain Res.* **19:**465–482, 1970.
98. Wilbrandt, W., Rosenberg, T.: The concept of carrier transport and its corollaries in pharmacology. *Pharmacol. Rev.* **13:**109–183, 1961.
99. Wright, E. M.: Active transport of glycine across the frog arachnoid membrane. *Brain Res.* **76:**354–358, 1974.
100. Yudilevich, D. L., deRose, N., Sepúlveda, F. V.: Facilitated transport of amino acids through the blood–brain barrier of the dog studied in a single capillary circulation. *Brain Res.* **44:**569–578, 1972.

CHAPTER 5

Sites of Origin and Cerebrospinal Fluid Concentration Gradients

Neurotransmitters, Their Precursors and Metabolites, and Cyclic Nucleotides

James H. Wood

1. Introduction

Determination of neurotransmitter precursors, neurotransmitters, and their metabolites in cerebrospinal fluid (CSF) has recently become a popular method for studying pathological and drug-induced alterations in central nervous system metabolism of living patients. According to Moir *et al.*,[58] interpretation of this CSF analysis requires the absence of CSF contamination with peripheral neurotransmitters and their metabolites. The CSF should also reflect the chemical composition of adjacent areas of nervous tissue. Unfortunately, CSF circulation, regional selective neurochemical absorption, and pathological alterations in the blood–CSF barrier complicate data evaluation.

The purpose of this chapter is to discuss the blood–brain–CSF permeability and sites of origin of CSF neurochemical substances. In addition, the presence and magnitude of CSF neurochemical concentration gradients will be assessed.

James H. Wood, M. D. • Division of Neurosurgery, University of Pennsylvania School of Medicine and Hospital, Philadelphia, Pennsylvania 19104.

2. γ-Aminobutyric Acid

γ-Aminobutyric acid (GABA), a putative inhibitory neurotransmitter, induces hyperpolarization and reduces membrane resistance by increasing chloride ion permeability.[25,51] Although orally administered L-glutamine has been shown to increase CSF GABA in a patient with Huntington's disease,[11] the ability of GABA to penetrate the blood–brain–CSF barrier has been much disputed.

Pharmacological investigations of parenterally administered GABA have generally failed to demonstrate central GABA accumulations in normal adult animals.[48,64,69,71] Apparently, penetration of intravenously injected GABA does take place after experimental local breakdown of the blood–brain barrier.[69] Most therapeutic trials in patients with Huntington's disease have not documented clinical improvement despite high oral doses of GABA.[7,65] However, Tower[81] reported suppression of petit mal and grand mal seizure activity in patients after oral GABA administration.

The extrapyramidal areas, hypothalamus, base of the pons, and cerebellum are the brain regions containing the highest concentrations of GABA.[30,34] In addition, small interneurons in the cerebral cortex,[51] hippocampus,[78] and thalamus[26] release GABA. Levels of GABA in the spinal cord are low compared to those in the brain.[34,42] Accordingly, Enna *et al.*[31] observed an increasing GABA concentration gradient during continuous serial sampling of lumbar CSF (Fig. 1). CSF GABA gradients have been confirmed by Hare and Manyam,[47] who noted the 20th milliliter to have an approximately 30% higher GABA concentration than the first milliliter in six of their seven reported patients. The one patient in their study who lacked such a gradient also had Huntington's disease. The presence of CSF GABA gradients suggests that lumbar CSF GABA levels reflect brain GABA metabolism.[90,92] Accordingly, Bohlen *et al.*[10] have documented close dose-dependent correlations between cisternal CSF GABA levels and brain GABA concentrations in rats following intraperitoneal injections of some GABA-transaminase inhibitors. The association of reduced lumbar CSF GABA concentrations and spinal or lower bulbar degenerative disorders suggests that the spinal cord may contribute to the lumbar CSF GABA content.[97] In addition, GABA-containing peptides in CSF such as homocarnisine may contribute to CSF GABA concentrations.[43]

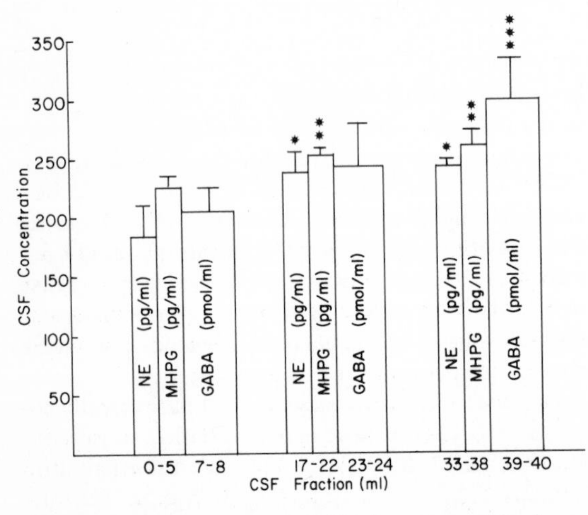

Figure 1. Norepinephrine (NE), 3-methoxy-4-hydroxyphenylethylene glycol (MHPG), and γ-aminobutyric acid (GABA) concentrations (S.E.M.) in serial samples of lumbar CSF in man (data taken with permission from Ziegler *et al.*[99] and Enna *et al.*[31] Mean incremental increases in *NE and **MHPG levels in the 17–22 ml and 33–38 ml CSF fractions over those respective levels in the first 5-ml CSF aliquot are significant ($p < 0.002$, two-tailed paired Student's *t* test). Mean incremental changes in NE and MHPG concentrations in 17–22 ml and 33–38 ml CSF fractions are not significantly different. ***Mean GABA level in the 39–40 ml CSF fraction is significantly higher ($p < 0.05$) than that in the 7–8 ml CSF aliquot.

3. Norepinephrine

Norepinephrine is the primary adrenergic neurotransmitter within the central nervous system. Efforts to increase endogenous brain norepinephrine by systemic administration have not been successful.

Increased oral intake of tyrosine, the dietary source of norepinephrine, is not practical because tyrosine hydroxylase, the rate-limiting step in the synthesis of norepinephrine, is saturated at physiological concentrations of tyrosine.[56] Intravenously injected norepinephrine is unable to cross the blood–brain barrier.[70,86] Although Ziegler *et al.*[98] observed a significant relationship between plasma and CSF levels of norepinephrine, acute norepinephrine elevations in the plasma were not reflected in the CSF. In addition, intravenous infusions of labeled norepinephrine in monkeys induce peak CSF levels of labeled norepinephrine that are equal to only 2% of the peak plasma concentrations. Thus, Ziegler *et al.*[98] conclude that the correlation between CSF and plasma norepinephrine cannot be explained by penetration of blood norepinephrine into CSF. Peripheral sympathetic activity may account for this blood–CSF correlation by making a minor contribution to the CSF norepinephrine content via innervation to the major subarachnoid blood vessels at the base of the brain (circle of Willis).[72] However, the majority of norepinephrine in the CSF is of central origin.

As reviewed by Smith and Sweet,[75] the central noradrenergic system in man includes the locus coeruleus and the lateral tegmental nucleus. The locus coeruleus lies in the floor of the aqueduct and rostral fourth ventricle and provides innervation, usually inhibitory, to the mesencephalic tegmentum, basal telencephalic areas representing the limbic system, cerebral cortex, cerebellum, and lateral column of the spinal cord. The noradrenergic locus coeruleus participates in the regulation of levels of consciousness and arousal, behavior patterns such as reinforcement and learning, cerebral blood flow, and permeability of cerebral microvasculature. The lateral tegmental system innervates the medial hypothalamus and acts to augment gonadotropic, thyrotropin-releasing, luteinizing, and growth hormone secretion as well as to probably inhibit adrenocorticotropic and antidiuretic hormone release.[55]

Ziegler *et al.*[99] demonstrated a concentration gradient for norepinephrine in lumbar CSF with a 30% increase in norepinephrine concentration from the first CSF sample to one taken after 17 ml of CSF had drained from the spinal needle (Fig. 1). Additional lumbar drainage of CSF was not associated with further elevation of CSF norepinephrine levels. The lower norepinephrine concentrations in the initial CSF samples were attributed to the minimal noradrenergic contribution of the cauda equina, whereas the higher concentrations observed in more rostral aliquots represented the activity of noradrenergic tracts descending from brainstem nuclei such as the locus coeruleus.[99]

3.1. 3-Methoxy-4-hydroxyphenylethylene Glycol

The primary mechanism for termination of the action of norepinephrine is synaptic reuptake. Catabolism of central norepinephrine proceeds via the monoamine oxidase system to form the major metabolite, 3-methoxy-4-hydroxyphenylethylene glycol (MHPG), and lesser quantities of vanillylmandelic acid (VMA).[20] Relatively little entry of intravenously infused labeled MHPG into lumbar CSF occurs; thus, MHPG in CSF arises largely from central rather than peripheral metabolism.[1,19] Ziegler *et al.*[99] demonstrated a significant correlation between lumbar CSF norepinephrine and MHPG in man and concluded that CSF MHPG reflects central noradrenergic metabolism.

The highest concentrations of norepinephrine and MHPG in man are present in structures adjacent to the third and fourth ventricles,[53] and accordingly, CSF MHPG levels are highest in the third and fourth ventricles in monkeys.[32] Adér *et al.*[11] have concluded that a considerable portion of CSF MHPG is dependent upon the activity of noradrenergic locus coeruleus neurons. Gordon and Oliver[40] and Chase *et al.*[19] noted similar MHPG levels in lateral ventricular and lumbar CSF in man, and Sjöström *et al.*[74] observed no craniocaudal MHPG concentration gradient in draining lumbar CSF. Patients with and without spinal canal blockage appear to have similar lumbar CSF MHPG levels.[68] These studies suggest that the spinal cord contributes to lumbar CSF MHPG levels in man. Post *et al.*[68] demonstrated lower-than-normal CSF MHPG levels in patients with spinal cord transections and concluded that the spinal source of MHPG in the lumbar CSF may be the descending noradrenergic tracts. Ziegler *et al.*[99] observed only a 12% increase in MHPG up to the 17th milliliter of CSF removed during lumbar puncture (Fig. 1) and suggested that the slower MHPG turnover and lack of control of physical activity might have obliterated the small spinal MHPG gradient in previous studies.

3.2. Vanillylmandelic Acid

Gordon *et al.*[41] demonstrated lateral ventricular CSF VMA concentrations to be 3-fold higher than those in lumber CSF, although the highest levels were noted in third and fourth ventricular CSF of monkeys. Jimerson *et al.*[49] found that VMA represented 8% of the total noradrenergic metabolites in lumbar CSF and extrapolated their data to suggest that VMA might represent as much as 20% of the major noradrenergic metabolites in lateral ventricular CSF. However, other investigators[74] have not confirmed a rostrocaudal gradient for CSF VMA.

4. Dopamine

Dopaminergic neurons in the rat connect the pars compacta of the substantia nigra to the striatum, and the nucleus interpeduncularis to the nucleus accumbens, olfactory tuberculum, and amygdala. Similarly, dopamine-containing neurons project from the arcuate and periventricular nuclei of the hypothalamus to the median eminence.[56] In man, the highest concentrations of dopamine are located in the nucleus accumbens and caudate nucleus; however, plentiful quantities have been demonstrated in the olfactory area, hypothalamus, and substantia nigra.[53] Dopamine is almost absent in the spinal cord.[54] The nigrostriatal pathway is primarily concerned with motor function, and the infundibular system inhibits the release of prolactin, thyrotropin-stimulating hormone, and possibly lutenizing hormone and growth hormone.[55,56] The exact function of the mesolimbic system is unknown. Dopamine is generally regarded as an inhibitory neurotransmitter and does not cross the blood–brain[36] or blood–CSF[46] barrier. At present, analytical methods for free or conjugated dopamine in CSF lack the sensitivity to measure this neurotransmitter without precursor stimulation.[83a]

4.1. L-3,4-Dihydroxyphenylalanine

Parentally administered L-3,4-dihydroxyphenylalanine (L-DOPA), a precursor of dopamine, crosses the blood–brain barrier.[33,36] Labeled brain L-DOPA and dopamine increase to maximal levels within ½ hr following intravenous injection of labeled L-DOPA.[33] The dopamine formed from the exogenously administered L-DOPA becomes part of the functional endogenous dopamine pool.

4.2. Homovanillic Acid

The principal metabolite of central dopamine is homovanillic acid (HVA). Intravenous administration of labeled L-DOPA induces a rapid elevation in both labeled brain and CSF HVA.[33,66] Peripherally injected dopamine or HVA does not increase CSF HVA in dogs[45] or cats.[8] Correlation of HVA concentrations in the caudate nucleus with CSF HVA levels[46] and the evoked release of HVA into lateral ventricular CSF during substantia nigral stimulation[67] provides more evidence that the origin of CSF HVA is the brain parenchyma; however, some contribution from brain capillary walls has been observed.[9] Sourkes[77] estimates that the amount of HVA entering the CSF represents only approximately 30% of the dopamine turnover of the caudate nucleus and may originate largely in the part adjacent to the ventricular surface. In patients given radioactive L-DOPA, labeled HVA reaches a maximum in the cisternal CSF at 2–4 hr, but does not attain its maximum concentration in lumbar CSF until 8 hr.[66] This lag in HVA accumulation in lumbar CSF corresponds to the rostrocaudal CSF circulation time.

The concentrations of HVA in man are highest in the substantia nigra, caudate nucleus, olfactory area, and especially the nucleus accumbens.[53] Marked rostrocaudal HVA concentration gradients have been documented in dog,[46,58] monkey,[41] and human[38,41,74] CSF (Fig. 2). The highest HVA levels are present in lateral ventricular CSF. Lumbar CSF HVA levels in patients with spinal canal blockage are greatly reduced or undetectable below the blocked site.[23,37,68,96] In addition, lumbar CSF HVA concentrations are greatly reduced in patients whose lateral ventricles have become isolated secondary to bilateral foramina of Monro obliteration.[37] Thus, most of the HVA in CSF originates from the lateral ventricles and their surrounding brain. Lumbar CSF concentrations of HVA, however, reflect only part of the total dopamine metabolism.[49a]

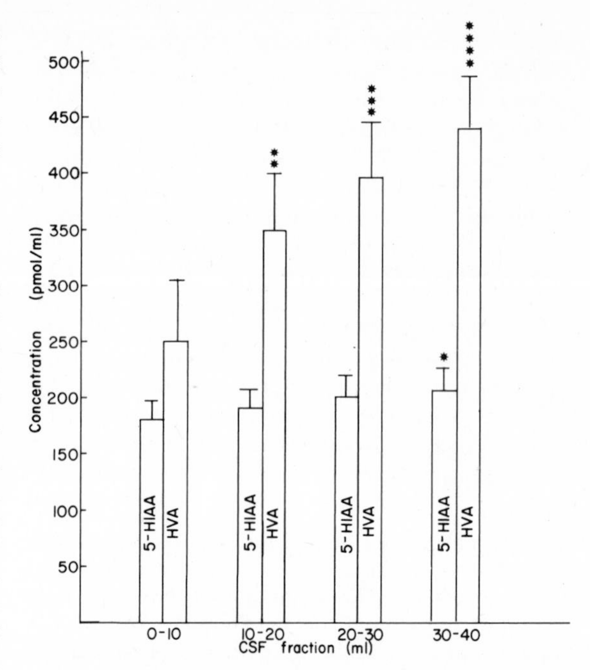

Figure 2. 5-Hydroxyindoleacetic acid (5-HIAA) and homovanillic acid (HVA) concentrations (S.E.M.) in serial samples of lumbar CSF in man (data taken with permission from Sjöstrom *et al.*[74]). *Mean level of 5-HIAA in the 30–40 ml CSF fraction is significantly higher ($p < 0.05$, Wilcoxon matched-pairs rank test) than that in the first 10-ml CSF aliquot. Mean levels of HVA in the **10–20 ml, ***20–30 ml, and ****30–40 ml CSF fractions were significantly higher ($p < 0.05$, 0.01, and 0.001, respectively; two-tailed paired Student's t test) than that in the first 10-ml CSF sample.

4.3. 3,4-Dihydroxyphenylacetic Acid

Another acid metabolite of dopamine, 3,4-dihydroxyphenylacetic acid (DOPAC), behaves similarly to HVA in CSF. Parentally administered L-DOPA is associated with increase in striatal levels of DOPAC as well as HVA.[3] Intravenous injection of L-DOPA induces earlier elevations in lateral ventricular CSF DOPAC concentrations than HVA levels.[46] The CSF level of DOPAC begins to decline while that of HVA is still rising. The inability of intravenous DOPAC to penetrate the blood–brain barrier[18] implies a central origin of the DOPAC located in CSF.[46]

DOPAC is most plentiful in the human substantia nigra, caudate nucleus, olfactory area, and nucleus accumbens.[53] The concentration of DOPAC in the dog caudate nucleus is approximately one tenth of that of HVA, and a similar ratio for these acids occurs in CSF.[4] The concentration of DOPAC in the caudate nucleus is about 6 times higher than in lateral ventricular CSF.[46] Thus, the level of DOPAC in ventricular CSF appears to be related to its concentration in the caudate nucleus. Rostrocaudal CSF DOPAC concentration gradients have been observed in monkeys[39] and man.[89] Wiesel[89] documented the ventricular CSF DOPAC level in man to be about 20-fold higher than that in lumbar CSF, while DOPAC concentrations in cisternal CSF were within an intermediate range. The ratio of DOPAC to HVA was roughly 3 times higher in ventricular and cristernal CSF than in lumbar CSF. The DOPAC content in lumbar CSF is only 1% of that of HVA; thus, HVA determinations remain the most useful method of studying dopamine metabolism in human lumbar CSF.

5. Serotonin

Serotonergic neurons of the dorsal and median raphe nuclei in the pons and mesencephalon send fibers rostrally to essentially all regions of the diencephalon and telencephalon. The descending serotonergic tracts in the spinal cord originate in the medulla oblongata.[56,76] In man, the highest concentration of serotonin is found in the midbrain, with similarly high levels in the substantia nigra and the pons. Serotonin levels in the hypothalamus, caudate nucleus, thalamus, and amygdala are approximately one third as high as those in the pons.[53] Depending on the particular region of origin, serotonergic stimulation evokes either excitation or depression.[76] Excitatory responses to iontophoretically applied serotonin are more common in areas of the brain and spinal cord in which the density of serotonergic terminals is low or scattered.[1a] Serotonergic activity has been associated with increases in body temperature,[60] sleep induction and maintenance,[13] decreased pain sensitivity,[2] vasoconstriction,[84] and increased vascular permeability.[80] In addition, serotonin may be implicated in the suppression of antidiuretic hormone release and the augmentation of prolactin, growth hormone, and corticotropin-releasing factor output.[55] Systemically administered serotonin does not readily cross the blood–brain barrier,[20,56] and a reliable assay for CSF serotonin is not yet available.

5.1. L-Tryptophan

Systemic administration of L-tryptophan, the essential dietary precursor of serotonin, increases brain serotonin.[6,35] Unlike tyrosine hydroxylase, tryptophan hydroxylase, the rate-limiting enzyme in the synthesis of serotonin, is not saturated at physiological levels of tryptophan.[27] Tryptophan crosses both the blood–brain and the blood–CSF barrier.[29,95] The tryptophan concentration in brain and CSF varies with the dietary content of tryptophan.[57] A concentration gradient for CSF tryptophan exists between the ventricle and spinal subarachnoid space.[95]Tryptophan levels are similar in cisternal and lumbar CSF among patients with blocked spinal canals.[96] In addition, CSF tryptophan concentrations are unaltered by the spinal CSF mixing that occurs with subarachnoid air injections during pneumoencephalography.[95] Thus, tryptophan in lumbar CSF may be derived almost totally from spinal sources.

5.2. L-5-Hydroxytryptophan

The immediate precursor of serotonin, L-5-hydroxytryptophan (L-5-HTP), appears to cross the blood–brain barrier and influence cerebral serotonin metabolism.[32,44,82,85] Oral administration of L-5-HTP to patients with movement, mood, or seizure disorders has been noted to elevate CSF serotonin metabolite concentrations.[44,82,85] Most of the brain aromatic L-amino-acid decarboxylase, the enzyme that converts L-5-HTP to serotonin, is located outside serotonergic neurons[52]; thus, much of the serotonin synthesized from exogenously administered L-5-HTP is formed extraneuronally, possible in glial cells.[21,59]

5.3. 5-Hydroxyindoleacetic Acid

The distribution of 5-hydroxyindoleacetic acid (5-HIAA) in the human brain parallels that of serotonin.[53] Systemically administered 5-HIAA does not cross the blood–brain[59] or the blood–CSF barrier.[5] Brainstem concentrations of 5-HIAA correlate with cisternal CSF levels.[29] L-Tryptophan-induced elevations in brain 5-HIAA concentrations[6,12,29] are reflected in corresponding changes in CSF 5-HIAA levels.[12,28,29,57] The accumulation of trypotophan in lumbar CSF occurs as early as 2 hr after tryptophan administration, whereas the accumulation of 5-HIAA is present at 6 hr.[28] Young *et al.*[94a] suggest that the cisternal CSF elevations in tryptophan and 5-HIAA induced in rats by tryptophan loading occur at 1 and 2 hr, respectively. No significant correlation between CSF tryptophan and 5-HIAA has been noted in untreated patients,[95] but significant correlations between these two substances have been observed in both the brains and CSF of tryptophan-treated rats.[57]

Moir *et al.*[58] demonstrated a rostrocaudal 5-HIAA concentration gradient in dogs and man; the ventricular/lumbar CSF 5-HIAA ratio in man was approximately 5:1. Sjöström *et al.*[74] observed this 5-HIAA gradient in human lumbar CSF to be less pronounced than that of lumbar HVA and suggested the possibility of some 5-HIAA contribution from the spinal cord (Fig. 2). Post *et al.*[68] reported that patients with apparent blockage of CSF flow in the spinal subarachnoid space had lumbar 5-HIAA concentrations similar to those in patients without evidence of CSF blockage. In addition, spinal cord transection did not alter lumbar CSF 5-HIAA levels. Garelis and Sourkes[37] estimated that this spinal cord contribution to the 5-HIAA content of lumbar CSF is of the order of 23–37%. Although Weir *et al.*[87] also estimated that 70% of the lumbar CSF 5-HIAA has its origin rostral to the foramen magnum, Bulat[17] maintained that almost all the 5-HIAA in lumbar CSF arises exclusively from spinal metabolism.

6. Cyclic Nucleotides

Extensive reviews of the cyclic nucleotides in central nervous system function have been prepared by Daly[24] and Nathanson.[62] In addition, a detailed discussion of CSF cyclic nucleotide metabolism is contained in this volume.[15]

6.1. Cyclic Adenosine 3′,5′-Monophosphate

Intravenously administered cyclic adenosine 3′,5′-monophosphate (AMP) does not penetrate into CSF.[22,73] Brooks *et al.*[14,15] demonstrated that lumbar CSF cyclic AMP concentrations are not acutely altered by systemic pharmacological manipulations that markedly increase plasma cyclic AMP levels. This demonstration of a blood–CSF barrier for cyclic AMP suggests a central origin for CSF cyclic AMP. Sebens and Korf[73] suggest that cisternal CSF cyclic AMP levels reflect brain cyclic AMP content.

Tsang *et al.*[83] reported higher cyclic AMP concentrations in ventricular CSF than in lumbar CSF. The elevations in lumbar CSF cyclic AMP occurring after cerebral infarction or ischemia with breakdown of the blood–brain barrier[50,88] and following epileptic discharges[61] suggest an intracranial contribution to the lumbar CSF content of this cyclic nucleotide. However, lumbar CSF cyclic AMP levels were not altered in patients with complete or partial spinal canal blockage. These investigators also noted cisternal cyclic AMP concentrations to be similar to that of lumbar CSF. Brooks *et al.*[15] did not observe lumbar CSF cyclic AMP elevations after serial sampling or cerebral cortical resections.[16] Lumbar levels of this cyclic nucleotide are not altered by electrical stimulation of the cerebellum[93,94] despite plentiful quantities of cyclic AMP in the cerebellum.[79] Reduced lumbar CSF cyclic AMP levels have been associated with degenerative disorders of the spinal cord.[16] Thus, cyclic AMP in lumbar CSF may not be totally dependent on the brain as the major source of this nucleotide, but may also reflect spinal nucleotide metabolism.[83]

6.2. Cyclic Guanosine 3′,5′-Monophosphate

High levels of guanylate cyclase are present in the cerebellum and cerebrum, but low concentrations are found in the pons, medulla, and spinal cord.[24] Brooks *et al.*[15,16] were not able to demonstrate cyclic guanosine 3′,5′-monophosphate (GMP) concentration gradients in lumbar CSF on serial sampling or alterations after cerebral resection. Wood *et al.*[91] did not observe lumbar CSF cyclic GMP alterations during electrical stimulation of the cerebellum in epileptic patients. Again, lumbar CSF cyclic GMP levels often may not accurately reflect intracranial cyclic nucleotide metabolism.

7. Conclusions

The blood–brain barrier makes significant contributions to central nervous system function. This barrier not only maintains the brain extracellular fluid constituents at concentrations that are optimal for neuronal transmission, but also excludes most polar toxic substances and peripherally synthesized neurotransmitters. Although their precursors readily diffuse from the blood to the brain, the respective neurotransmitters cannot cross this barrier. The blood–brain barrier promotes reuptake and reutilization of central neurotransmitters by somewhat restricting them to the immediate regions of their origin.[63] The CSF concentration gradients reflect not only variations in the distribution and metabolism of these constituents within the central nervous system but also their absorption mechanisms and CSF circulation patterns. Knowledge of these gradients is an absolute necessity for the interpretation of precursor, neurotransmitter, and metabolite concentrations in CSF.

ACKNOWLEDGMENTS. The author is grateful to Dr. Theodore A. Hare for the preparation of illustrations and thanks Drs. S. J. Enna, Michael G. Ziegler, C. Raymond Lake, Benjamin R. Brooks, Jonas Sode, Rolf Sjöström, Jan Ekstedt, Erik Änggård, W. King Engel, Solomon H. Snyder, and Irwin J. Kopin for their contributions to the data on which this chapter is based.

References

1. ADÉR, J.-P., AIZENSTEIN, M. L., POSTEMA, F., KORF, J.: Origin of free 3-methoxy-4-hydroxyphenylethylene glycol in rat cerebrospinal fluid. *J. Neural Trans.* **46**:279–290, 1979.

1a. AGHAJANIAN, G. K., HAIGLER, H. J. BENNETT, J. L.: Amine receptors in CNS. III. 5-Hydroxytryptamine in brain. In Iversen, L. L., Iversen, S. D., Snyder, S. H. (eds.): *Handbook of Psychopharmacology*, Vol. 6, *Biogenic Amine Receptors*. New York, Plenum Press, 1975, pp. 63–96.

2. AKIL, H., LIEBESKIND, J. C.: Monoaminergic mechanisms of stimulation-produced analgesia. *Brain Res.* **94**:279–296, 1975.

3. ANDEN, N.-E., ROOS, B.-E., WERDINIUS, B.: On the occurrence of homovanillic acid in brain and cerebrospinal fluid and its determination by a fluorimetric method. *Life Sci.* **2**:448–458, 1963.

4. ASHCROFT, G. W., CRAWFORD, T. B. B., DOW, R. C., GULDBERT, H. C.: Homovanillic acid, 3,4-dihydroxyphenylacetic acid and 5-hydroxyindol-3-ylacetic acid in serial samples of cerebrospinal fluid from the lateral ventricle of the dog. *Br. J. Pharmacol. Chemother.* **33**:441–456, 1968.

5. ASHCROFT, G. W., DOW, R. C., MOIR, A. T. B.: The active transport of 5-hydroxyindol-3-ylacetic acid and 3-methoxy-4-hydroxyphenylacetic acid from a recirculatory perfusion system of the cerebral ventricles of the unanesthetized dog. *J. Physiol. (London)* **199**:397–425, 1968.

6. ASHCROFT, G. W., ECCLESTON, D., CRAWFORD, T. B. B.: 5-Hydroxyindole metabolism in rat brain: A study of intermediate metabolism using the technique of tryptophan loading. I. Methods. *J. Neurochem.* **12**:483–492, 1965.

7. BARBEAU, A.: GABA and Huntington's chorea. *Lancet* **2**:1499–1500, 1973.

8. BARTHOLINI, G., PLETSCHER, A., TISSOT, R.: On the origin of homovanillic acid in the cerebrospinal fluid. *Experientia* **22**:609–610, 1966.

9. BARTHOLINI, G., PLETSCHER, A., TISSOT, R.: Brain capillaries as a source of homovanillic acid in the cerebrospinal fluid. *Brain Res.* **27**:163–168, 1971.

10. BOHLEN, P., HUOT, S., PALFREYMAN, M. G.: The relationship between GABA concentrations in brain and cerebrospinal fluid. *Brain Res.* **167**:297–305, 1979.

11. BERRY, H. C., STEINER, J. C.: L-Glutamine increases CSF GABA in patient with Huntington's disease. *Neurology (Minneapolis)* **29**:535, 1979 (abstract).

12. BOWERS, M. B.: 5-Hydroxyindoleacetic acid in the brain and cerebrospinal fluid of the rabbit following administration of drugs affecting 5-hydroxytryptamine. *J. Neurochem.* **17**:827–828, 1970.

13. BREMER, F.: Cerebral hypnogenic centers. *Ann. Neurol.* **2**:1–6, 1977.

14. BROOKS, B. R., ENGEL, W. K., SODE, J.: Blood-to-cerebrospinal fluid barrier for cyclic adenosine monophosphate in man. *Arch. Neurol.* **34**:468–469, 1977.

15. BROOKS, B. R., WOOD, J. H., DIAZ, M., CZERWINSKI, C., GEORGES, L. P., SODE, J., EBERT, M. H., ENGEL, W. K.: Extracellular cyclic nucleotide metabolism in human central nervous system. In Wood, J. H. (ed.): *Neurobiology of Cerebrospinal Fluid I*. New York, Plenum Press, 1980.

16. BROOKS, B. R., WOOD, J. H., SODE, J., ENGEL, W. K.: Cyclic nucleotide metabolism in neurological disease. *Trans. Am. Neurol. Assoc.* **101**:221–222, 1976.

17. BULAT, M.: On the cerebral origin of 5-hydroxyindoleacetic acid in the lumbar cerebrospinal fluid. *Brain Res.* **122**:388–391, 1977.

18. CARLSEN, A., HILLARP, N.-A.: Formation of phenolic acids in brain after administration of 3,4-dihydroxyphenylalanine. *Acta Physiol. Scand.* **55**:95–100, 1962.

19. CHASE, T. N., GORDON, E. K., NG, L. K. Y.: Norepinephrine metabolism in the central nervous system of man: Studies using 3-methoxy-4-hydroxyphenylethylene glycol levels in cerebrospinal fluid. *J. Neurochem.* **21**:581–587, 1973.

20. COOPER, J. R., BLOOM, F. E., ROTH, R. H.: *The Biochemical Basis of Neuropharmacology*. New York, Oxford University Press, 1978.

21. CORRODI, H., FUXE, K., HÖKFELT, T.: Replenishment of 5-hydroxytryptophan of the amine stores in the central 5-hydroxy-tryptamine neurons after depletion induced by reserpine or by an inhibitor of mono-amine synthesis. *J. Pharm. Pharmacol.* **19**:433–438, 1967.

22. CRAMER, H.: Cyclic 3′,5′ nucleotides in extracellular fluids of neural systems. *J. Neurosci. Res.* **3**:241–246, 1977.

23. CURZON, G., GUMPERT, E. J. W., SHARPE, D. M.: Amine metabolites in the lumbar cerebrospinal fluid of humans with restricted flow of cerebrospinal fluid. *Nature (London) New Biol.* **231**:189–191, 1971.

24. DALY, J. W.: The formation, degradation and function of cyclic nucleotides in the nervous system. In Smythies, J. R., Bradley, R. J. (eds.): *International Review of Neurobiology*, Vol. 20. New York, Academic Press, 1977, pp. 105–168.

25. DREIFUSS, J. J., KELLY, J. S., KRNJEVIC, K.: Cortical inhibition and gamma-aminobutyric acid. *Exp. Brain Res.* **9**:137–154, 1969.

26. DUGGAN, A. W., MCLENNAN, H.: Bicuculline and inhibition in the thalamus. *Brain Res.* **25**:188–191, 1971.

27. ECCLESTON, D., ASHCROFT, G. W., CRAWFORD, T. B. B.: 5-Hydroxyindole metabolism in rat brain: A study of intermediate metabolism using the technique of tryptophan loading. II. Applications and drug studies. *J. Neurochem.* **12**:493–503, 1965.
28. ECCLESTON, D., ASHCROFT, G. W., CRAWFORD, T. B. B., STANTON, J. B., WOOD, D., MCTURK, P. H.: Effect of tryptophan administration on 5-HIAA in cerebrospinal fluid in man. *J. Neurol. Neurosurg. Psychiatry* **33**:269–272, 1970.
29. ECCLESTON, D., ASHCROFT, G. W., MOIR, A. T. B., PARKER-RHODES, A., LUTZ, W., O'MAHONEY, D. P.: A comparison of 5-hydroxyindoles in various regions of dog brain and cerebrospinal fluid. *J. Neurochem.* **15**:947–957, 1968.
30. ENNA, S. J., BENNETT, J. P., BYLUND, D. B., CRESE, I., BURT, D. R., CHARNESS, M. E., YAMAMURA, H. I., SIMANTOV, R., SNYDER, S. H.: Neurotransmitter receptor binding: Regional distribution in human brain. *J. Neurochem.* **28**:233–236, 1977.
31. ENNA, S. J., WOOD, J. H., SNYDER, S. H.: γ-Aminobutyric acid (GABA) in human cerebrospinal fluid: Radioreceptor assay. *J. Neurochem.* **28**:1121–1124, 1977.
32. EVERETT, G. M.: Effect of 5-HTP on brain levels of dopamine, norepinephrine and serotonin in mice. In Costa, E. (ed.): *Advances in Biochemical Pharmacology*, Vol. 10. New York, Raven Press, 1974, pp. 261–262.
33. EXTEIN, I., ROTH, R. H., BOWERS, M. B.: Accumulation of ^{3}H-homovanillic acid in rabbit brain and cerebrospinal fluid following intravenous ^{3}H-L-DOPA. *Biol. Psychiatry* **9**:161–170, 1974.
34. FAHN, S., CÔTÉ, L. J.: Regional distribution of γ-aminobutyric acid (GABA) in brain of the rhesus monkey. *J. Neurochem.* **15**:209–213, 1968.
35. FERNSTROM, J. D., WURTMAN, R. J.: Brain serotonin content: Physiological dependence on plasma tryplophan levels. *Science* **173**:149–152, 1971.
36. FRIEDMAN, A. H., EVERETT, G. M.: Pharmacological aspects of parkinsonism. *Adv. Pharmacol.* **3**:83–127, 1964.
37. GARELIS, E., SOURKES, T. L.: Sites of origin in the central nervous system of monoamine metabolites measured in human cerebrospinal fluid. *J. Neurol. Neurosurg. Psychiatry* **36**:625–629, 1973.
38. GARELIS, E., SOURKES, T. L.: Use of cerebrospinal fluid drawn at pneumoencephalography in the study of monoamine metabolism in man. *J. Neurol. Neurosurg. Psychiatry* **37**:704–710, 1974.
39. GORDON, E. K., MARKEY, S. P., SHERMAN, R. L., KOPIN, I. J.: Conjugated 3,4-dihydroxy phenol acetic acid (DOPAC) in human and monkey cerebrospinal fluid and rat brain and the effects of probenecid treatment. *Life Sci.* **18**:1285–1292, 1976.
40. GORDON, E. K., OLIVER, J.: 3-Methoxy-4-hydroxyphenylethylene glycol in human cerebrospinal fluid. *Clin. Chem. Acta* **35**:145–150, 1971.
41. GORDON, E., PERLOW, M., OLIVER, J., EBERT, M., KOPIN, I.: Origins of catecholamine metabolites in monkey cerebrospinal fluid, *J. Neurochem.* **25**:347–349, 1975.
42. GRAHAM, L. T., JR., SHANK, R. P., WERMAN, R., APRISON, M. H.: Distribution of some synaptic transmitter suspects in cat spinal cord: Glutamic acid, aspartic acid, gamma-aminobutyric acid, glucine and glutamine. *J. Neurochem.* **14**:465–472, 1967.
43. GROSSMAN, M. H., HARE, T. A., MANYAM, N. V. B., GLAESER, B. S., WOOD, J. H.: Stability of GABA levels in CSF under various conditions of storage. *Brain Res.* **182**:99–106, 1980.
44. GUILLEMINAULT, C., THARP, B. R., COUSIN, D.: HVA and 5-HIAA CSF measurements and 5-HTP trials in some patients with involuntary movements. *J. Neurol. Sci.* **18**:435–441, 1973.
45. GULDBERT, H. C.: Changes in amine metabolite concentrations in cerebrospinal fluid as an index of turnover. In Hooper, G. (ed.): *Metabolism of Amines in the Brain*. London, Macmillan, 1969, pp. 55–64.
46. GULDBERT, H. C., YATES, C. M.: Some studies of the effects of chlorpromazine, reserpine and dihydroxyphenylalanine on the concentrations of homovanillic acid, 3,4-dihydroxyphenylacetic acid and 5-hydroxyindol-3-ylacetic acid in ventricular cerebrospinal fluid of the dog using the technique of serial sampling of the cerebrospinal fluid. *Br. J. Pharmacol.* **33**:457–471, 1968.
47. HARE, T. A., MANYAM, N. V. B.: Rapid and sensitive ion-exchange/fluorometric measurement of GABA in physiological fluids. *Anal. Biochem.* **101**:349–355, 1980.
48. HESPE, W., ROBERTS, E., PRINS, H.: Autoradiographic investigation of the distribution of C^{14}-GABA in tissues of normal and aminooxyacetic acid-treated mice. *Brain Res.* **14**:663–671, 1969.
49. JIMERSON, D. C., GORDON, E. K., POST, R. M., GOODWIN, F. K.: Central noradrenergic function in man: Vanillylmandelic acid in CSF. *Brain Res.* **99**:434–439, 1975.
49a. KESSLER, J. A., FENSTERMACHER, J. D., PATLAK, C. S.: Homovanillic acid transport by the spinal cord. *Neurology (Minneapolis)* **26**:434–440, 1976.
50. KOBAYASHI, M., LUST, W. D., PASSONNEAU, J. V.: Concentrations of energy metabolites and cyclic nucleotides during and after bilateral ischemia in the gerbil cerebral cortex. *J. Neurochem.* **29**:53–59, 1977.
51. KRNJEVIC, K., SCHWARTZ, S.: The action of γ-aminobutyric acid on cortical neurons. *Exp. Brain Res.* **3**:320–336, 1967.
52. KUHAR, M. J., ROTH, R. H., AGHAJANIAN, G. K.: Selective reduction of tryptophan hydroxylase activity in rat forebrain after midbrain raphe lesions. *Brain Res.* **35**:167–176, 1971.
53. MACKAY, A. V. P., YATES, C. M., WRIGHT, A., HAMILTON, P., DAVIES, P.: Regional distribution of monoamines and their metabolites in the human brain. *J. Neurochem.* **30**:841–848, 1978.
54. MAGNUSSON, T., ROSENGREN, E.: Catecholamines in the spinal cord normally and after transection. *Experientia* **19**:229–230, 1963.

55. MARTIN, J. B., REICHLIN, S., BROWN, G. M.: Clinical neuroendocrinology. In Plum, F., McDowell, F. H. (eds.): *Contempory Neurology Series*, Vol. 14. Philadelphia, F. A. Davis, 1977.
56. MAYNERT, E. W., MARCZYNSKI, T. J., BROWNING, R. A.: The role of the neurotransmitters in the epilepsies. In Friedlander, W. J. (ed.): *Current Reviews, Advances in Neurology*, Vol. 13. New York, Raven Press, 1975, pp. 103–114.
57. MODIGH, K.: The relationship between the concentrations of tryptophan and 5-hydroxy-indoleacetic acid in rat brain and cerebrospinal fluid. *J. Neurochem.* **25**:351–352, 1975.
58. MOIR, A. T. B., ASHCROFT, G. W., CRAWFORD, T. B. B., ECCLESTON, D., GULDBERT, H. C.: Cerebral metabolites in cerebrospinal fluid as a biochemical approach to the brain. *Brain* **93**:357–368, 1970.
59. MOIR, A. T. B., ECCLESTON, D.: The effects of precursor loading in the cerebral metabolism of 5-hydroxyindoles. *J. Neurochem.* **15**:1093–1108, 1968.
60. MYERS, R. D.: The role of hypothalamic serotonin in thermoregulation. In Barchas, J., Usdin, E. (eds.): *Serotonin and Behavior*. New York, Academic Press, 1973, pp. 293–302.
61. MYLLYLA, V. V., HEIKKINEN, E. R., VAPAATALO, H., HOKKANEN, E.: Cyclic AMP concentration and enzyme activities of cerebrospinal fluid in patients with epilepsy or central nervous system damage. *Eur. Neurol.* **13**:123–130, 1975.
62. NATHANSON, J. A.: Cyclic nucleotides and nervous system function. *Physiol. Rev.* **57**:157–256, 1977.
63. OLDENDORF, W. H.: The blood–brain barrier. In Bito, L. Z., Davson, H., Fenstermacher, J. D., (eds.): *The Ocular and Cerebrospinal Fluids*. London, Academic Press, 1977, pp. 177–190.
64. PERRY, T. L., HANSEN, S.: Sustained drug-induced elevation of brain GABA in the rat. *J. Neurochem.* **21**:1167–1175, 1973.
65. PERRY, T. L., HANSEN, S., URQUHART, N.: GABA in Huntington's chorea. *Lancet* **1**:995–996, 1974.
66. PLETSCHES, A., BARTHOLINI, G., TISSOT, R.: Metabolic fate of L-^{14}C-DOPA in cerebrospinal fluid and blood plasma of humans. *Brain Res.* **4**:106–109, 1967.
67. PORTIG, P. J., VOGT, M.: Release to the cerebral ventricles of substances with possible transmitter function in the caudate nucleus. *J. Physiol. (London)* **204**:687–715, 1969.
68. POST, R. M., GOODWIN, F. K., GORDON, E., WATKIN, D. M.: Amine metabolites in human cerebrospinal fluid: Effects of cord transection and spinal fluid block. *Science* **179**:897–899, 1973.
69. PURPURA, D. P., GIRADO, M., SMITH, T. G., GOMEZ, J. A.: Synaptic effects of systemic γ-aminobutyric acid in cortical regions of increased vascular permeability. *Proc. Soc. Exp. Biol.* **97**:348–353, 1958.
70. REIS, D. J., WURTMAN, R.: Diurnal changes in brain noradrenalin. *Life Sci.* **7**:91–98, 1968.
71. ROBERTS, E., LOWE, I. P., GUTH, L., JELINEK, B.: Distribution of γ-aminobutyric acid and other amino acids in nervous tissue of various species. *J. Exp. Zool.* **138**:313–328, 1958.
72. SATO, S., SUZUKI, J.: Anatomical mapping of the cerebral nervi vasorum in the human brain. *J. Neurosurg.* **43**:559–568, 1975.
73. SEBENS, J. B., KORF, J.: Cyclic AMP in cerebrospinal fluid: Accumulation following probenecid and biogenic amines. *Exp. Neurol.* **46**:333–344, 1975.
74. SJÖSTRÖM, R., EKSTEDT, J., ÄNGGÅRD, E.: Concentration gradients of monoamine metabolites in human cerebrospinal fluid. *J. Neurol. Neurosurg. Psychiatry* **38**:666–668, 1975.
75. SMITH, B. H., SWEET, W. H.: Monoaminergic regulation of central nervous system function. I. Noradrenergic systems. *Neurosurgery* **3**:109–119, 1978.
76. SMITH, B. H., SWEET, W. H.: Monoaminergic regulation of central nervous system function. II. Serotonergic systems. *Neurosurgery* **3**:257–272, 1978.
77. SOURKES, T. L.: On the origin of homovanillic acid (HVA) in the cerebrospinal fluid. *J. Neural Transm.* **34**:153–157, 1973.
78. STORM-MATHISEN, J., FONNUM, F.: Quantitative histochemistry of glutamate decarboxylase in the rat hippocampal region. *J. Neurochem.* **18**:1105–1111, 1971.
79. SUTHERLAND, E. W., RALL, T. W., MENON, T.: Adenyl cyclase. I. Distribution, preparation and properties. *J. Biol. Chem.* **237**:1220–1227, 1962.
80. SWANK, R. L., HISSON, W.: Influence of serotonin on cerebral circulation. *Arch. Neurol.* **10**:468–472, 1964.
81. TOWER, D. B.: The neurochemistry of convulsive states. In Folch-Pi, J. (ed.): *Chemical Pathology of the Nervous System*. Oxford/London, Pergamon Press, 1961, pp. 307–344.
82. TRIMBLE, M., CHADWICK, D., REYNOLDS, E. H., MARSDEN, C. D.: L-5-Hydroxytryptophan and mood. *Lancet* **1**:583, 1975.
83. TSANG, D., LAL, S., SOURKES, T. L., FORD, R. M., ARONOFF, A.: Studies on cyclic AMP in different compartments of cerebrospinal fluid. *J. Neurol. Neurosurg. Psychiatry* **39**:1186–1190, 1976.
83a. TYCE, G. M., SHARPLESS, N. S., KERR, F. W. L., MUENTER, M. D.: Dopamine conjugate in cerebrospinal fluid. *J. Neurochem.* **34**:210–212, 1980.
84. VON ESSEN, C.: Effects of dopamine, noradrenaline and 5-hydroxytryptamine on the cerebral blood flow in the dog. *J. Pharm. Pharmacol.* **24**:668, 1972.
85. VON WOERT, M. H., SETHY, V. H.: Therapy of intention mycolonus with L-5-hydroxytryptophan and a peripheral decarboxylase inhibitor, MK-486. *Neurology (Minneapolis)* **25**:135–140, 1975.
86. WEIL-MALHERBE, H., AXELROD, J., TOMCHICK, R.: Blood–brain barrier for adrenaline. *Science* **129**:1226–1227, 1959.
87. WEIR, R. L., CHASE, T. N., NG, L. K. Y., KOPIN, I. J.: 5-Hydroxyindoleacetic acid in spinal fluid: Relative

contribution from brain and spinal cord. *Brain Res.* **52**:409–412, 1973.

88. WELCH, K. M. A., MEYER, J. S., CHEE, A. N. C.: Evidence for disordered cyclic AMP metabolism in patients with cerebral infarction. *Eur. Neurol.* **13**:144–154, 1975.
89. WIESEL, F.-A.: Mass fragmentographic determination of acidic dopamine metabolites in human cerebrospinal fluid. *Neurosci. Lett.* **1**:219–224, 1975.
90. WOOD, J. H., GLAESER, B. S., ENNA, S. J., HARE, T. A.: Verification and quantification of GABA in human cerebrospinal fluid. *J. Neurochem.* **30**:291–293, 1978.
91. WOOD, J. H., GLAESER, B. S., HARE, T. A., SODE, J., BROOKS, B. R., VAN BUREN, J. M.: Cerebrospinal fluid GABA reductions in seizure patients evoked by cerebellar surface stimulation. *J. Neurosurg.* **47**:582–589, 1977.
92. WOOD, J. H., HARE, T. A., GLAESER, B. S., BALLENGER, J. C., POST, R. M.: Low cerebrospinal fluid γ-aminobutyric acid content in seizure patients. *Neurology (Minneapolis)* **29**:1203–1208, 1979.
93. WOOD, J. H., LAKE, C. R., ZIEGLER, M. G., SODE, J., BROOKS, B. R., VAN BUREN, J. M.: Cerebrospinal fluid norepinephrine alterations during electrical stimulation of cerebellar and cerebral surfaces in epileptic patients. *Neurology (Minneapolis)* **27**:716–724, 1977.
94. WOOD, J. H., ZIEGLER, M. G., LAKE, C. R., SODE, J., BROOKS, B. R., VAN BUREN, J. M.: Elevations in cerebrospinal fluid norepinephrine during unilateral and bilateral cerebellar stimulation in man. *Neurosurgery* **1**:260–265, 1977.

94a. YOUNG, S. N., ANDERSON, G. M., PURDY, W. C.: Indoleamine metabolism in rat brain studied through measurements of tryptophan, 5-hydroxyindoleacetic acid, and indoleacetic acid in cerebrospinal fluid. *J. Neurochem.* **34**:309–315, 1980.

95. YOUNG, S. N., GARELIS, E., LAL, S., MARTIN, J. B., MOLINA-NEGRO, P., SOURKES, T. L.: Tryptophan and 5-hydroxyindoleacetic acid in human cerebrospinal fluid. *J. Neurochem.* **22**:777–779, 1974.
96. YOUNG, S. N., LAL, S., MARTIN, J. B., FORD, R. M., SOURKES, T. L.: 5-Hydroxyindoleacetic acid, homovanillic acid and tryptophan levels in CSF above and below a complete block of CSF flow. *Psychiatr. Neurol. Neurochir.* **76**:439–444, 1973.
97. ZIEGLER, M. G., BROOKS, B. R., LAKE, C. R., WOOD, J. H., ENNA, S. J.: Norepinephrine and gamma-aminobutyric acid in amyotrophic lateral sclerosis. *Neurology (Minneapolis)* **30**:98–101, 1980.
98. ZIEGLER, M. G., LAKE, C. R., WOOD, J. H., BROOKS, B. R., EBERT, M. H.: Relationship between norepinephrine in blood and cerebrospinal fluid in the presence of a blood–cerebrospinal fluid barrier for norepinephrine. *J. Neurochem.* **28**:677–679, 1977.
99. ZIEGLER, M. G., WOOD, J. H., LAKE, C. R., KOPIN, I. J.: Norepinephrine and 3-methoxy-4-hydroxyphenyl glycol gradients in human cerebrospinal fluid. *Am. J. Psychiatry* **134**:565–568, 1977.

CHAPTER 6

Daily Fluctuations in Catecholamines, Monoamine Metabolites, Cyclic AMP, and γ-Aminobutyric Acid

Mark J. Perlow and C. Raymond Lake

1. Introduction

It seems that a very large number of physiological variables fluctuate in a rhythmic fashion with a period length of approximately 24 hr.[1-3,6,11,14,15,20,33,34] In man and other primates, this is manifested by many rhythmic alterations including sleeping–waking activity, the level of attention, pain threshold, urinary excretory rates, body temperature, and plasma hormone concentrations. The cause of these changes is a matter of considerable interest. Although nonnervous peripheral tissues may be capable of generating some rhythms, it is generally felt that most daily biological rhythms originate within the central nervous system.[21]

To understand the mechanism by which the central nervous system generates these rhythms, we decided to investigate the neurotransmitter changes that take place during the 24-hr day in the nonhuman primate.

Mark J. Perlow, M.D. • Laboratory of Clinical Psychopharmacology, National Institute of Mental Health, National Institutes of Health, St. Elizabeth's Hospital, Washington, D.C. 20032. **C. Raymond Lake, M.D., Ph.D.** • Departments of Psychiatry and Pharmacology, Uniformed Services University of Health Sciences, School of Medicine, Bethesda, Maryland 20014.

2. Materials and Methods

We studied 5- to 6.5-kg male rhesus monkeys. The animals were adapted to primate restraining chairs in isolation chambers with a 12:12 light–dark cycle (lights on from 06:00 to 18:00 hours, lights off from 18:00 to 06:00 hours). Using a peristaltic pump, lateral cerebral ventricular fluid (Fig. 1) was withdrawn continuously in 1½-, 2-, or 3-hr aliquots. Cerebrospinal fluid (CSF) remained at room temperature in the collecting tubing for approximately 2 hr until it could be refrigerated (4°C). For norepinephrine (NE) analyses, each collecting tube contained 10 mg ascorbic acid. Fractions from a 24-hr period were collected each morning and frozen (−20°C) until assayed. Homovanillic acid (3-methoxy-4-hydroxyphenylacetic acid) (HVA), vanillylmandelic acid (3-methoxy-4-hydroxymandelic acid) (VMA), and 3-methoxy-4-hydroxyphenylethylene glycol (MHPG) were then assayed by the gas chromatographic mass spectroscopic procedure of Gordon *et al.*,[9] NE by the enzymatic procedure of Lake *et al.*,[17] and γ-aminobutyric acid (GABA) by the receptor binding method of Enna *et al.*[5]

3. Norepinephrine and Its Metabolites

The concentration of NE was measured in lateral ventricular fluid obtained (1.0 ml/hr) every 3 hr for

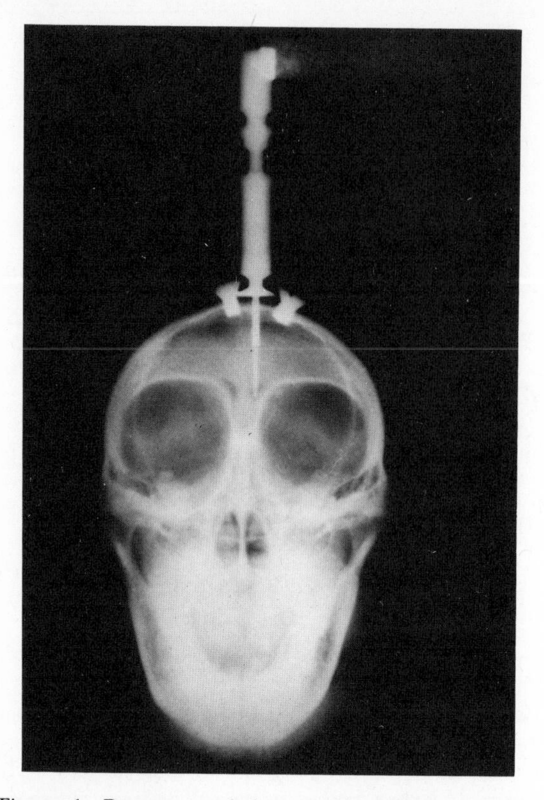

Figure 1. Pneunoencephalogram of a rhesus monkey showing a cannula in the lateral ventricle.

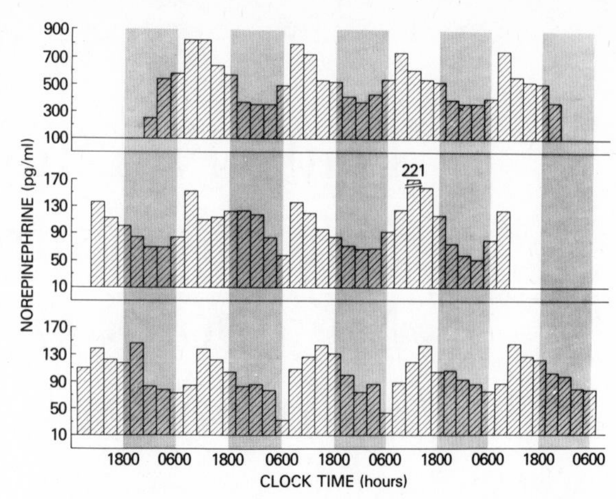

Figure 2. Circadian fluctuation of NE concentration in CSF obtained from the lateral ventricle of three conscious primates at a rate of 1.0 ml/hr. The height of each bar represents the mean concentration over a 3-hr interval. Shaded areas indicate times of day when lights were off, and unshaded areas indicate when lights were on. From Perlow *et al.*,[25] with permission.

four or five contiguous 24-hr periods in three animals. Although the mean concentration of NE varied in each animal over this period of time, there was a consistent day-to-day variation in concentration, with high concentrations occurring during the light hours and low concentrations during the dark hours (Figs. 2 and 3). The NE found in the CSF comes from the central nervous system rather than the circulation, since following intravenous administration of [^{14}C]-NE, the maximum concentration of [^{14}C]-NE in the ventricular fluid was less than 2% of its corresponding plasma concentration (Fig. 4).[25]

In contrast to NE released into extracellular spaces, NE released into the CSF by lateral and third periventricular tissues is not readily subject to reuptake or enzymatic degradation. Thus, it seems reasonable to suggest that the concentrations of NE in the CSF reflect the rate of release or turnover of NE in periventricular tissues. This fluctuation in concentration of NE indicates that there is a daily increase in turnover during the light hours and a daily decrease in turnover during the dark hours in at least a portion of the brain.

The concentrations of MHPG and VMA, major metabolites of NE, were determined in samples of lateral and fourth ventricular fluid obtained every 2 hr for three or four contiguous 24-hr periods in six animals. The results of a representative experiment on samples obtained from the lateral ventricle (0.4 ml/hr) are presented in Fig. 5. In individual animals, the patterns of VMA and MHPG were random and without a discernible rhythm. Dispite simultaneous analysis of these compounds, there was no correlation between the concentration of these two metabolites.[25]

The discrepancy in the pattern of concentration changes between NE and its metabolites is probably the result of a number of considerations: (1) while NE in the CSF is the product of cells close to the ventricular surface, VMA and MHPG are the products of cells both deep and close to the ventricular surface; (2) VMA and MHPG dissolve readily in the cells of the brain, forming large pools of exchangeable metabolites that easily buffer any fluctuations in metabolite synthesis rates; and (3) VMA and MHPG can be transported directly from the brain into the blood as well as from the blood into the brain and CSF.[25] Thus, while the concentrations of catecholamine metabolites in ventricular CSF may reflect the concentrations in neighboring nervous tissue, they do so only in gross terms, and only when averaged over a relatively long period of time.

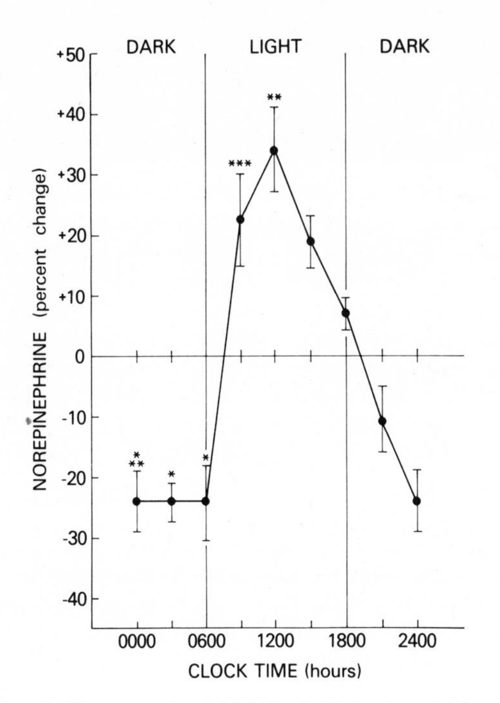

Figure 3. Concentration of NE in the lateral ventricle as a function of time of day. The curve represents the mean (± S.E.M) percentage of concentration change from the daily mean concentration for the 13 24-hr periods for the three primates illustrated in Fig. 2. Using Student's *t* test, mean NE concentrations at different times vary in the following manner: $^{*}P < 0.001$ from values at 09:00, 12:00, 15:00, and 18:00 hours; $^{**}P < 0.001$ from values at 18:00 and 21:00 hours; $^{***}P < 0.005$ from values at 21:00 hours. From Perlow *et al.*,[25] with permission.

4. Dopamine and Its Major Metabolite

The concentration of HVA, the principal metabolite of dopamine, was measured in samples of ventricular fluid obtained (0.4 ml/hr) every 2 hr from three or four contiguous 24-hr periods from four monkeys. When analyzed as to variation about a daily mean, a graph of HVA concentration described a daily pattern as illustrated in Fig. 6.[24]

The concentration changes appear to reflect fluctuation in the rate of HVA synthesis occurring in the caudate nucleus during the preceding 6 hr. The cause of this delayed response in the CSF is the fact that HVA is synthesized throughout the caudate nucleus, and appears in the CSF as a result of diffusional transport from brain tissue. Several studies suggest that an alteration in intraneuronal dopamine metabolism requires 4–6 hr before it is reflected by a change in the concentration of HVA in the CSF.[24] Although the change in HVA concentration is small, it may reflect a much larger change in HVA synthesis. The presence of a large exchangeable pool of HVA, as exists in the caudate, reduces the magnitude of any short-term change in HVA concentration, in addition to delaying its appearance in CSF.

Therefore, dopamine metabolism, as manifested by the concentration of HVA in the ventricular CSF, fluctuates in a daily pattern, with maximum turnover occurring during the light hours and minimum turnover occurring during the dark hours.

5. Cyclic Adenosine 3′,5′-Monophosphate

The concentration of cyclic adenosine 3′,5′-monophosphate (cAMP) was measured in samples of ventricular fluid obtained (0.4 ml/hr) every 2 hr for three or four contiguous 24-hr periods from four monkeys. When analyzed as to a variation about a daily mean, a graph of cAMP concentration described a daily pattern as illustrated in Fig. 7. This pattern parallels the ones described for dopamine and NE, and may reflect dopamine-stimulated postsynaptic activity in the caudate nucleus.[23]

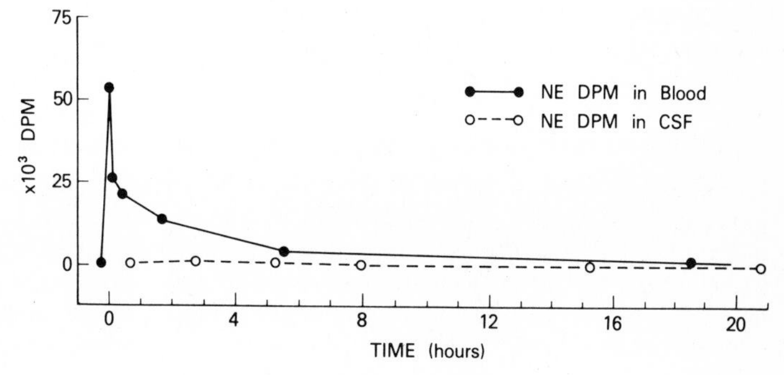

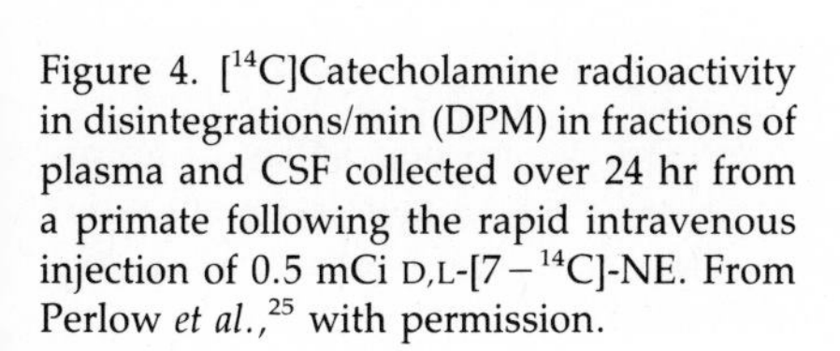
Figure 4. [^{14}C]Catecholamine radioactivity in disintegrations/min (DPM) in fractions of plasma and CSF collected over 24 hr from a primate following the rapid intravenous injection of 0.5 mCi D,L-[7 − ^{14}C]-NE. From Perlow *et al.*,[25] with permission.

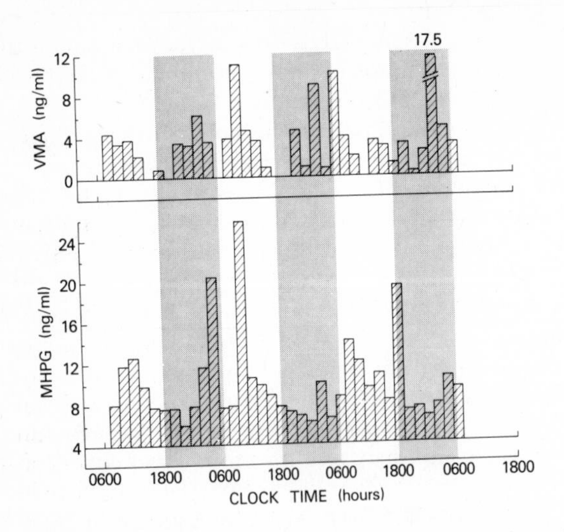

Figure 5. Representative profile of CSF concentration of MHPG and VMA obtained from the lateral ventricle of a conscious primate at a rate of 0.4 ml/hr. The height of each bar represents the mean concentration over a 2-hr interval. Shaded areas indicate the times of day when lights were off, and unshaded areas indicate when lights were on. From Perlow *et al.*,[25] with permission.

6. γ-Aminobutyric Acid

The concentration of GABA was determined in samples of ventricular CSF obtained (1.0 ml/hr) every 2 hr for five or six contiguous 24-hr periods in four monkeys. Profiles of CSF GABA concentrations are illustrated in Fig. 8. Despite wide and irregular fluctuations in concentrations over the time period in all animals, and marked variations among animals, there is a suggestion of a daily rhythm in GABA concentration, with highest concentrations occurring during the daytime (Fig. 9).[26] We feel that, like NE, GABA in the CSF reflects GABA metabolism near the surface rather than in deep periventricular tissues. The low concentrations of GABA in the lumbar CSF of patients with Huntington's disease supports this idea.[4,7]

Recent studies by Grossman *et al.*[9a] on human lumbar CSF indicate that GABA concentrations, as measured by the ion exchange–fluorometric procedure, more than double when exposed to room temperature and atmosphere for 2 hr. If these ob-

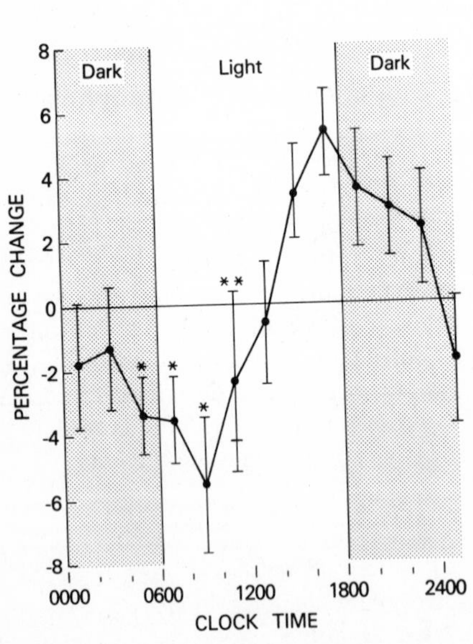

Figure 6. HVA concentration in the lateral ventricular fluid of rhesus monkeys. Variation from daily mean concentration during the 24-hr day. Each point represents the mean of 15 samples ± S.E.M. Using Student's *t* test: (*) indicates that $P < 0.005$ as compared with values at 19:00, 23:00 and 01:00 hours, $P < 0.001$ as compared with values at 21:00 hours, $P < 0.02$ as compared with values at 03:00 hours; (**) indicates that $P < 0.005$ as compared with values at 21:00 and 03:00 hours. From Perlow *et al.*,[24] with permission.

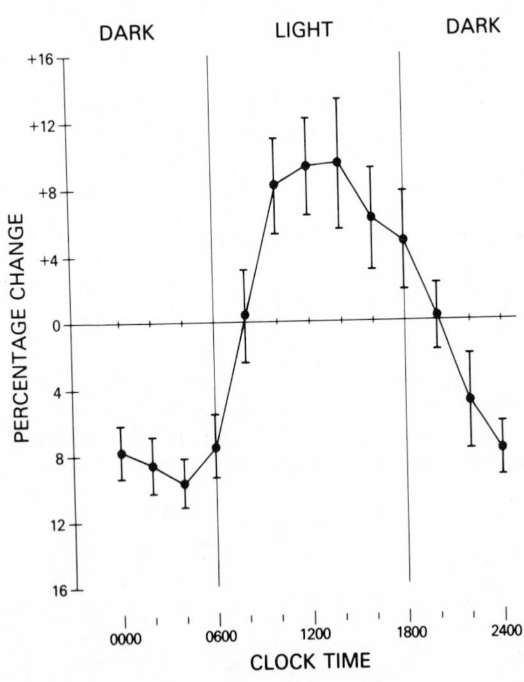

Figure 7. cAMP concentration in the lateral ventricular fluid of rhesus monkeys. Variation from daily mean concentration during the 24-hr day. Each point represents the mean value of 13 samples ± S.E.M. Repeated single-factor analysis of variance was performed on the data, and significance was found at the 0.001 level, $F^{11,33} = 9.017$. From Perlow *et al.*,[23] with permission.

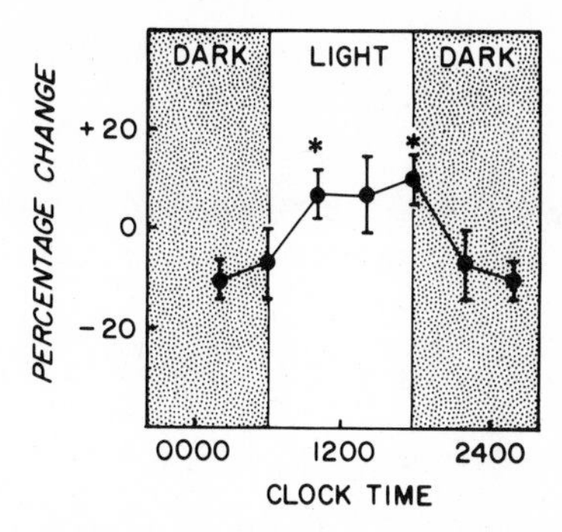

Figure 9. GABA concentration in the lateral ventricular fluid of rhesus monkeys. Variation from daily mean concentration during the 24-hr day. Each point represents the mean ± S.E.M. concentration of the means of each of four animals. Using Student's *t* test: (*) indicates that $P < 0.005$ as compared with values at 02:00 hours. From Perlow *et al.*,[26] with permission.

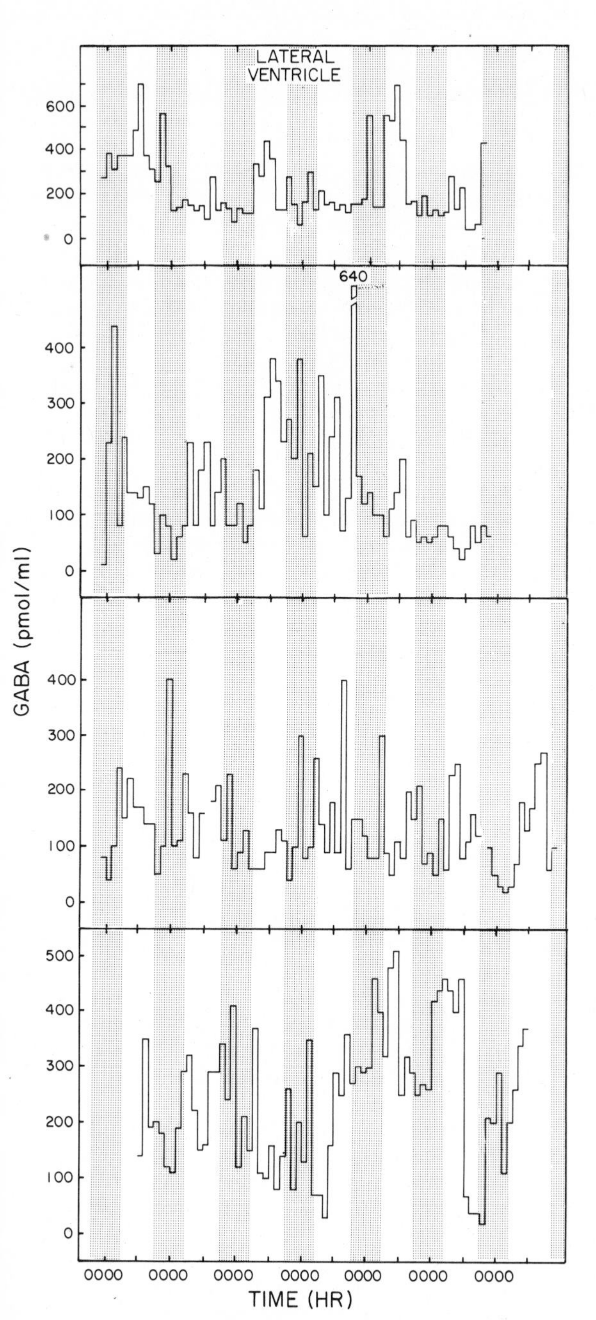

Figure 8. Profiles of CSF concentration of GABA obtained from the lateral ventricular fluid of conscious primates at a rate of 1.0 ml/hr. The height of each bar represents the mean concentration of a 2-hr collection period. Shaded areas indicate the times of day when lights were off, and unshaded areas indicate when lights were on. From Perlow *et al.*,[26] with permission.

servations can be confirmed by other laboratories, the results described above may not accurately reflect GABA concentrations in the CSF or in neural tissue, since our samples were exposed to body temperature (37°C) for unknown periods of time and to room temperature (20–24°C) for 2 hr prior to refrigeration.

7. Conclusions

In the rhesus monkey, the increase and decrease in catecholamine and GABA metabolism during the day and night, respectively, parallels wakefulness,[16,31,35] body activity,[10,33,36] and body temperature,[22,33,36] and is out of phase with increases in circulating melatonin,[32] cortisol,[12,13,19,27] prolactin,[29,30] and testosterone.[8,18] Similar daily phase relationships in humans are numerous.[6,11,14,20] Although there is a considerable amount of correlative evidence that the catecholamine and GABA neuronal systems are responsible for some daily behavioral and physiological changes, additional studies of other neurotransmitter systems are necessary before a causal relationship can be established. The extent to which these neuronal systems fluctuate independently or as part of a hierarchy of dependent oscillators remains to be determined.[1,21,28]

References

1. Ashoff, J.: Circadian rhythms in man. *Science* **148:**1427–1432, 1965.

2. Ashoff, J. (ed.): *Circadian Clock.* Amsterdam, North-Holland, 1966.
3. Bunning, E.: *The Physiological Clock.* Heidelberg, Springer-Verlag, 1964.
4. Enna, S. J., Stern, L. Z., Wastek, G. J., Yamamura, H. I.: Cerebrospinal fluid γ-aminobutyric acid variations in neurological disorders. *Arch. Neurol.* **34:**683–685, 1977.
5. Enna, S. J., Wood, J. H., Snyder, S. H.: Gamma-aminobutyric acid (GABA) in human cerebrospinal fluid: Radioreceptor assay. *J. Neurochem.* **28:**1121–1124, 1977.
6. Ferin, M., Halberg, F., Richart, R. M., VandeWiele, R. L. (eds.): *Biorhythms and Human Reproduction.* New York, John Wiley, 1974.
7. Glaeser, B. S., Hare, T. A., Vogel, W. H., Olewiler, D. B., Beasley, B. L.: Low GABA levels in CSF in Huntington's chorea. *N. Engl. J. Med.* **292:**1029–1030, 1975.
8. Goodman, R. L., Hotchkiss, J., Karsch, F., Knobil, E.: Diurnal variation in serum testosterone concentrations in the adult male rhesus monkey. *Biol. Reprod.* **11:**624–630, 1974.
9. Gordon, E. K., Oliver, J., Black, K., Kopin, I.: Simultaneous assay by mass fragmentography of vanillylmandelic acid, homovanillic acid, and 3-methoxy-4-hydroxy-phenylethylene glycol in cerebrospinal fluid and urine. *Biochem. Med.* **11:**32–40, 1974.
9a. Grossman, M. H., Hare, T. A., Manyam, N. V. B., Glaeser, B. S., Wood, J. H.: Stability of GABA levels in CSF under various conditions of storage. *Brain Res.* **182:**99–106, 1980.
10. Hawking, F.: Circadian rhythms in monkeys, dogs and other animals. *J. Interdiscipl. Cycle Res.* **2:**153–156, 1971.
11. Hedlund, L. W., Franz, J. M., Kenny, A. D. (eds.): *Biological rhythms and Endocrine Function* New York, Plenum Press, 1973.
12. Holaday, J. W., Mejerhoff, J. L., Natelson, B. H.: Cortisol secretion and clearance in the rhesus monkey. *Endocrinalogy* **100:**1178–1185, 1977.
13. Jacoby, J. H., Sassin, J. F., Greenstein, M., Weitzman, E. D.: Patterns of spontaneous cortisol and growth hormone secretion in rhesus monkeys during sleep–wake cycle. *Neuroendocrinology* **14:**165–173, 1974.
14. Kawakami, M. (ed.): *Biological Rhythms in Neuroendocrine Activity.* Tokyo, Igaku Shoin, 1974.
15. Kletiman, N.: *Sleep and Wakefulness.* Chicago, University of Chicago Press, 1963.
16. Kripke, D. F., Reite, M. L., Pegram, G. V., Stephens, L. M., Lewis, O. F.: Nocturnal sleep in rhesus monkeys. *Electroencephalogr. Clin. Neurophysiol.* **24:**582–586, 1968.
17. Lake, C. R., Ziegler, M. G., Kopin, I. J.: Use of plasma norepinephrine for evaluation of sympathetic neuronal function in man. *Life Sci.* **18:**1315–1326, 1976.
18. Michael, R. P., Setchell, K. D. R., Plaut, T. M.: Diurnal changes in plasma testerone and studies on plasma corticosteroid in non-anesthetized male rhesus monkeys (*Macaca mulatta*). *J. Endocrinol.* **63:**325–335, 1974.
19. Migeon, C. J., French, A. B., Samuels, L. T., Bowers, J. Z.: Plasma 17-hydroxycorticosteroid levels and leucocyte values in the rhesus monkeys, including normal variation and the effect of ACTH. *Am. J. Physiol.* **182:**462–468, 1955.
20. Mills, J. N. (ed.): *Biological Aspects of Circadian Rhythms.* New York, Plenum Press, 1973.
21. Moore-Ede, M. C., Shemelzer, W. S., Kass, D. A., Herd, J. A.: Internal organization of the circadian timing system in multicellular animals. *Fed. Proc. Fed. Am. Soc. Exp. Biol.* **35:**2333–2380, 1976.
22. Perlow, M., Dinarello, C. A., Wolff, S. M.: A primate model for the study of human fever. *J. Infect. Dis.* **132:**157–164, 1975.
23. Perlow, M. J., Festoff, B., Gordon, E. K., Ebert, M. H., Johnson, D. K., Chase, T. N.: Daily fluctuation in the concentrations of cAMP in the conscious primate brain. *Brain Res.* **126:**391–396, 1977.
24. Perlow, M. J., Gordon, E. K., Ebert, M. E., Hoffman, H. J., Chase, T. N.: The circadian variation in dopamine metabolism in the subhuman primate. *J. Neurochem.* **28:**1381–1383, 1977.
25. Perlow, M., Ebert, M. H., Gordon, E. K., Ziegler, M. G., Lake, C. R., Chase, T. N.: The circadian variation of catecholamine metabolism in the subhuman primate. *Brain Res.* **139:**101–113, 1978.
26. Perlow, M. J., Enna, S. J., O'Brien, B. J., Hoffman, H. J., Wyatt, R. J.: Cerebrospinal fluid gamma-aminobutyric acid: Daily pattern and response to haloperidol. *J. Neurochem.* **32:**265–268, 197.
27. Perlow, M. J., Reppert, S. M., Boyar, R. M., Wyatt, R. J., Klein, D. C.: Daily rhythms in cortisol and melatonin in primate CSF: Effects of constant light and dark (submitted).
28. Pittenrigh, C. S.: Circadian oscillators in cells and the circadian organization of multicellular systems. In Schmitt, F. O., Worden, F. G. (eds.) *The Neurosciences: Third Study Program.* Cambridge, MIT Press, 1974, pp. 437–458.
29. Quabbe, H.-J., Gregor, M., Bumke-Vogt, C., Eckhof, A., Bohlscheid, P., Schoppenhorst, M.: 24-Hour pattern of growth hormone, prolactin and cortisol in the rhesus monkey. *Acta. Endocrinol.* **87**(Suppl. 215):8–9, 1978.
30. Quadu, S. K., Sipes, H. G.: Cyclic and diurnal patterns of serum prolactin in the rhesus monkey. *Biol. Med.* **14:**495–501, 1976.
31. Reite, M. L., Rhodes, J. M., Kavan, E., Adey, W. R.: Normal sleep patterns in macaque monkey, *Arch. Neurol. (Chicago)* **12:**133–144, 1965.
32. Reppert, S. M., Perlow, M. J., Tarmarkin, L., Kelin, D. C.: A diurnal melatonin rhythm in the primate cerebrospinal fluid. *Endocrinology* **104:**295–301, 1979.

33. ROHLES, F. H. (ed.): *Circadian Rhythms in Non-Human Primates.* New York, S. Karger, 1969.
34. SCHEVING, L. E., HALBERG, F., PAULY, J. E. (eds.): *Chronobiology.* Tokyo, Igaku Shoin, 1974.
35. WEITZMAN, E. D., KIRPKE, D. F., POLLACK, C., DOMIGUEZ, J.: Cyclic activity in sleep of *Macaca mulatta. Arch. Neurol. (Chicago)* **12**:463–467, 1965.
36. WINGET, C. W.: Circadian rhythms of the rhesus monkey. In Bourne, G. H. (ed.): *The Rhesus Monkey,* Vol. 2. New York, Academic Press, 1969, pp. 277–302.

CHAPTER 7

Technical Aspects of Clinical and Experimental Cerebrospinal Fluid Investigations

James H. Wood

1. Introduction

In recent years, the physical and chemical examination of cerebrospinal fluid (CSF) has become increasingly important in diagnostic patient evaluations and in both clinical and animal research. The CSF bathes the brain and spinal cord, and thus tends to reflect the state of health and activity of the central nervous system.[109] Meaningful interpretation of CSF findings requires relatively sophisticated knowledge of CSF physiology and pathology. It is hoped that this information will be supplied by this multidisciplined volume. The purpose of this chapter is to discuss aspects of protocol formulation that ensure the validity of CSF data. In addition, various clinical and experimental techniques of obtaining ventricular, cisternal, and lumbar CSF will be assessed.

2. Clinical Protocols

The state of activity of nervous tissue is subject to individual, somatotopic, chronological, endocrinological, and pharmacological variations. Therefore, accurate clinical evaluations of CSF data require protocols that minimize these variations by appropriate subject selection and preparation as well as by employing reliable CSF sampling and storage techniques.

James H. Wood, M.D. • Division of Neurosurgery, University of Pennsylvania School of Medicine, Philadelphia, Pennsylvania 19104.

2.1. Patient Selection

The nonuniformity of neuronal metabolism among patient populations may result in part from individual variations. The composition of brain extracellular fluid and CSF would then reflect these variations.

Anderson and Roos[3] have noted reductions in CSF levels of 5-hydroxyindoleacetic acid (5-HIAA), the major metabolite of serotonin, with increasing age among pediatric patients under 10 years of age. Accordingly, Rogers and Dubowitz[91] have reported higher CSF 5-HIAA levels in very young patients as compared to older children and adults. Bowers[11] has noted a tendency for baseline concentrations of 5-HIAA and homovanillic acid (HVA), the major metabolite of dopamine, to increase in older age groups; however, this relationship does not hold for values obtained following probenecid administration. Cohen *et al.*, in Chapter 46, will discuss their previously reported finding of a significant negative correlation between age and dopamine turnover observed after probenecid loading in pediatric patients.[19] Similarly, Post *et al.*[81] have observed a negative correlation between age and accumulations of HVA following probenecid administration in depressed patients. Ziegler, Lake, and co-workers[56,113] have noted a tendency for CSF norepinephrine concentrations to rise with advancing age. Hare *et al.*[49,50] have demonstrated that the tendency of lumbar CSF levels of γ-aminobutyric acid (GABA) to decrease with age is more pronounced among normal females; however, Enna *et al.*[30] were not able to confirm this age-dependent phenomenon in sexually unmatched patients. As yet, no correlation between cyclic nucleotide[14] levels in CSF and age has been reported.

The CSF levels of 5-HIAA and HVA have been reported not to vary with the sex of the patient.[11] No correlations between sex and CSF norepinephrine[56,115] have been reported; however, Robinson *et al.*[90] have recently observed a positive correlation between age and brain catechol-*O*-methyltransferase, a degradative enzyme in catecholamine metabolism, as well as higher levels of this enzyme in women than in men. Hare *et al.*[49,50] have observed a tendency of CSF GABA concentrations to be lower in older normal females than in older normal males. Thus, age and sex appear to be determinants of amine and GABA metabolism in man.

Patients being considered for inclusion in clinical studies should be screened carefully for complicating conditions that might add variability to CSF data. Patients with degenerative disorders have pathologically low CSF levels of monoamine metabolites,[16,18] norepinephrine,[111] and GABA.[30,43,49] Hypertensive[113,115] or manic[85] patients tend to have elevated CSF concentrations of norepinephrine and its major central metabolite, 3-methoxy-4-hydroxyphenylethylene glycol (MHPG). Altered catecholamine metabolite, GABA, and cyclic nucleotide levels in CSF have been assocoated with seizure disorders. [103a] Acute cerebral infarctions may elevate CSF cyclic nucleotide content.[100] Patients with acute head or spine injuries, central nervous system tumors, and subarachnoid hemorrhage have CSF contamination from peripheral sources as a result of breakdown of the blood–brain or blood–CSF barrier. Obstructions to CSF circulation such as hydrocephalus[3] or spinal canal blockage[82] alter CSF composition and gradients. More specifically, cervical spondylosis may produce partial spinal canal obstruction in older age groups.

In addition, the selection of comparison groups requires considerable attention to minimize CSF variations not secondary to the variable being studied. Normal volunteers, when available, are preferred; however, sex- and age-matched patients with normal autonomic, neurological, and psychiatric examinations may serve as an appropriate control population.

2.2. Patient Preparation

Patients participating in clinical CSF investigations need not be placed on special diets prior to CSF sampling. Tyrosine hydroxylase, the rate-limiting enzyme in the synthesis of dopamine and norepinephrine, is saturated at physiological levels of tyrosine.[63] Thus, dietary increases in tyrosine content do not appear to increase CSF HVA, norepinephrine, or MHPG levels.[56] However, tyrptophan hydroxylase is normally unsaturated with its substrate[25]; thus, dietary increases in tryptophan content elevate CSF 5-HIAA concentrations.[64] Dietary alterations in tryptophan may also affect CSF GABA metabolism, since Enna *et al.*[30] have observed a positive correlation between CSF 5-HIAA and GABA levels in CSF. The recommended daily intake of tryptophan, an essential amino acid, is approximately 0.5 g/day, while the minimal daily requirement in adult males is 0.25 g/day.[15] Obviously, fad diets, overeating foods high in tryptophan (egg albumin, casein), or vitamin/amino acid supplements should be avoided several days prior to CSF sampling for 5-HIAA or GABA. Supplemental L-glutamine intake has been reported to elevate CSF GABA in man[7] and thus should be avoided.

Preferably, patients under consideration should be free of medications, especially those known to alter neurotransmitter metabolism. Antiemetic, hypnotic, sedative, antipsychotic, antiparkinsonian, neuroleptic, analgesic, and anesthetic medications are known to alter brain and CSF levels of neurotransmitters. Even aspirin should be avoided. Antihypertensive medications such as α-methyldopa may alter central catecholamine metabolism. Some GABA transaminase inhibitors have been reported to elevate brain and CSF GABA levels[10] and thus should be avoided. Anticonvulsant medications cannot ethically be withdrawn for the sake of research in seizure patients. Dosages of anticonvulsant drugs should be adjusted to yield blood levels within therapeutic range[57] and then should not be acutely altered during the preparation or CSF-sampling periods.

Patients involved in clinical CSF studies should be restricted to bedrest with a bedpan because physical activity affects CSF concentration gradients by promoting mixing[80] and alters CSF monoamine metabolite levels.[83] Fluid intake should likewise be restricted for several hours prior to CSF sampling because the intake of copious amounts of fluids alters serum osmolarity and may in turn modify CSF production.[51] Stress has been noted to elevate CSF levels of HVA, 5-HIAA,[46] and norepinephrine.[56,115] The initiation of the activity and oral-intake restrictions during the evening prior to CSF sampling will reduce the discomfort and stress of the protocol by allowing the patient to sleep through most of the immediate preparation period.

2.3. CSF Sampling Techniques

Monoamine metabolites, norepinephrine, GABA, cyclic nucleotide, and neurohormone concentrations in CSF exhibit circadian rhythms.[84,88,114] Thus, the sampling of CSF should be performed at the same time of day on all members of the test and control populations.

The central nervous system is relatively heterogeneous with respect to regional distribution of neurotransmitters and their metabolites. Concentrations of various CSF components reflect local neuronal metabolism, CSF circulation patterns, and absorption mechanisms. These factors contribute to the ventriculospinal CSF concentration gradients that have been demonstrated for neurotransmitters.[105,115,116] their metabolites[81,116] ions, and protein fractions.[106] This topic is discussed in detail by Wood[110] in a separate chapter. The presence of these CSF gradients restricts the valid comparison of CSF findings to similar CSF fractions or aliquots.

2.3.1. Lumbar Puncture and Chronic Drainage. The technique of lumbar puncture has been reviewed in detail by Fishman[38] and Patten.[77] The patient is usually placed in the lateral knee–chest position with the lumbar site of puncture at the same level as the external occipital proturberance. An 18- to 20-gauge spinal needle is preferred when manometric recordings are to be obtained. A 22-gauge needle should be employed to avoid large dural holes in patients suspected of having increased intracranial pressure. Tourtelotte *et al.*[96] suggest that this small-gauge needle decreases the incidence of post-lumbar-puncture headache. Spinal needles should have stylettes to ensure patency, to avoid the transfer of skin or subcutaneous material into the subarachnoid space, and to prevent iatrogenic intraspinal epidermoid tumors.[5]

The ideal puncture site in adults lies in the midline beneath the spinous process at the interspace between the third and fourth lumbar vertebrae (Fig. 1). This level is located on an imaginary line stretching between the superior iliac crests and avoids the conus medullaris, which may extend as low as the second lumbar vertebra. Accordingly, lumbar punctures should be performed at the interspace between the fourth and fifth lumbar or the fifth lumbar and first sacral vertebrae in children and infants so as

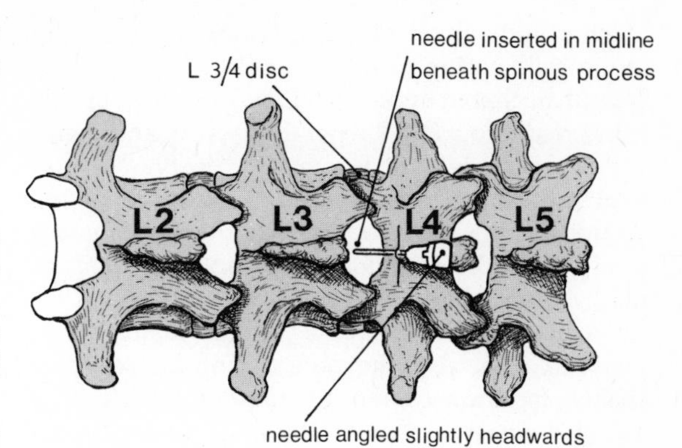

Figure 1. Ideal site for lumbar puncture in adult. Interspace between third and fourth lumbar vertebrae is located at midpoint of imaginary line between superior iliac crests.

to avoid the conus medullaris, which may lie as caudal as the interspace between the third and fourth lumbar vertebrae. Skin and subcutaneous tissues are then infiltrated with local anesthetic agents.

The bevel of the needle point should enter the dura parallel to the longitudinal fibers of the dura. This needle orientation will minimize the size of the dural hole by spreading rather than cutting the dural fibers. Occasionally, a tactile sensation may be experienced on puncturing the dura, and initiation of CSF flow may be prompted by performing a half-turn of the needle hub. The manometer should be attached to the needle hub immediately after CSF flow is established and an opening pressure in millimeters of water recorded. Needle patency is documented by abdominal-compression-evoked or respiration-induced fluctuations in CSF pressure. The presence of CSF pressure above 240 mm of water in a relaxed patient with legs partially extended precludes further CSF drainage.[41] The CSF within the manometer should be used for diagnostic testing.

Lumbar punctures may be difficult in obese patients, and the sitting position may be necessary for needle orientation. Immediately following the initiation of CSF flow, sterile plastic tubing should be attached to the spinal needle hub. The CSF pressure should normally elevate the CSF within the tubing only to the level of the foramen magnum.

The standard technique for the establishment of chronic spinal drainage has been plagued by difficulties in passing a fine epidural catheter through a Touhy needle (Becton-Dickinson Company) and maintaining patency. Recently, Post and Stein[79] have described a method for placing a No. 5 French

Stamey ureteral catheter (American Latex Corporation) with multiple side holes at the tip in the lumbar subarachnoid space. A 14-gauge Touhy needle is inserted into a low lumbar interspace, and thereafter the bevel is turned cephalad. The Stamey ureteral catheter is passed through the bore of the Touhy needle and advanced to the first lumbar level. The catheter stylus and needle are then removed, and a blunt-ended No. 20 needle with Intramedic Luer Stub Adapter (Clay Adams Company) is connected to the catheter. This Luer adapter enables the connection of the ureteral catheter to a closed drainage system.[101] Prophylactic antibodies are usually administered to patients undergoing prolonged spinal drainage.

2.3.1.1. Spinal Subarachnoid Blockage. Manual bilateral jugular vein compression (Queckenstedt's test) evokes cerebral venous engorgment and a rapid elevation of intracranial pressure that normally is transmitted to lumbar subarachnoid space.[86] A prompt rise and return to baseline CSF pressure on light and deep jugular vein compression indicates the absence of spinal canal obstruction (Fig. 2). A prompt rise but a slow fall in CSF pressure after deep jugular vein compression and a failure of CSF pressure to return to the baseline level after release suggests an incomplete subarachnoid block (Fig. 3). In the presence of complete spinal canal blockage, the rise in CSF pressure is greater after straining or abdominal compression than following deep jugular vein compression (Fig. 4). More accurately quantitated manometric data may be obtained employing cuff manometrics.[45] The presence of a complete spinal block precludes further removal of CSF and may require intrathecal injection of positive contrast medium and immediate myelography. The presence of an incomplete spinal obstruction may alter lumbar CSF concentrations of those neu-

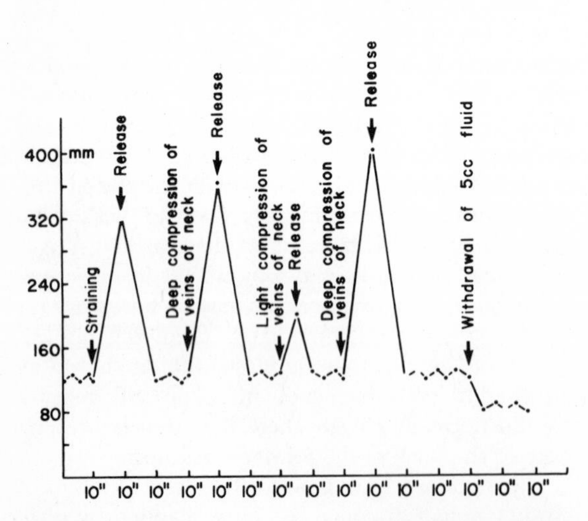

Figure 2. Lumbar CSF manometric chart in patient with normal spinal subarachnoid space. Note normal oscillations of respiration and pulse, rapid rise and fall of CSF pressure on straining, and prompt rise and fall of CSF pressure on superficial and deep jugular vein compression. Reprinted from Spurling,[94] with permission.

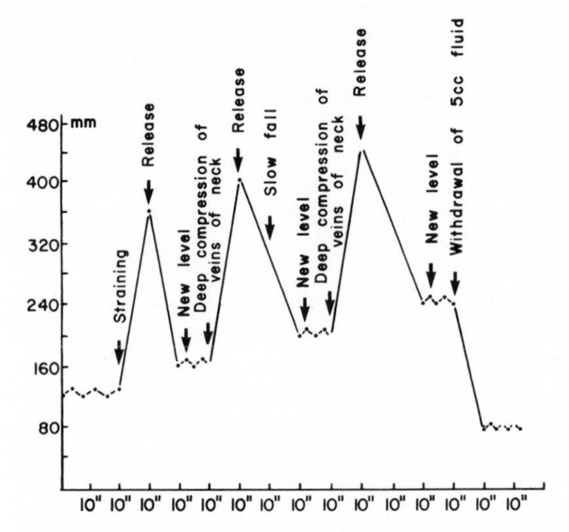

Figure 3. Lumbar CSF manometric chart in patient with incomplete spinal subarachnoid block. Note new CSF pressure level after each trial of straining or jugular compression. Also note prompt rise and slow fall of CSF pressure after deep compression of jugular veins. Reprinted from Spurling,[94] with permission.

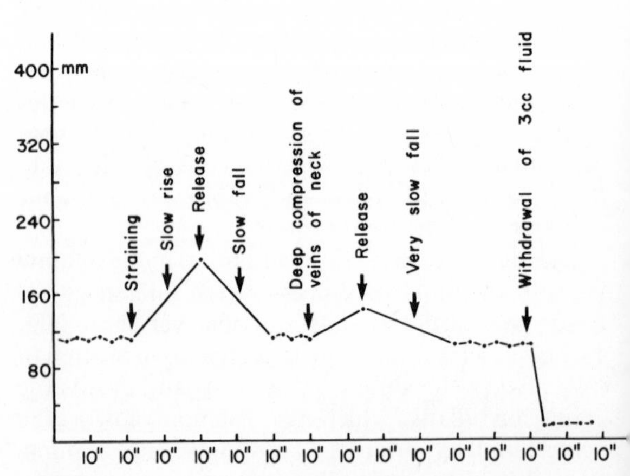

Figure 4. Lumbar CSF manometric chart in patient with complete spinal subarachnoid block. Note that rise in CSF pressure is greater on straining than after deep jugular vein compression. Reprinted from Spurling,[94] with permission.

rotransmitters and their metabolites that have intracranial origins.[82,110]

2.3.1.2. Contraindications to Lumbar Puncture. Lumbar punctures should be avoided if evidence of infection is present in the region of the puncture site.[38] Although Korein *et al.*[55] reported complications following lumbar punctures among only 1.2% of patients with papilledema, this procedure should be avoided in the presence of known or suspected increased intracranial pressure or posterior fossa mass lesions if CSF data will not alter treatment. Elective lumbar punctures are contraindicated in the presence of known spinal canal blockage in patients with neurological function below the block. This procedure should also be avoided in patients with blood dyscrasias or known spinal cord arteriovenous malformations and those receiving anticoagulant therapy.[26,103]

2.3.1.3. Complications of Lumbar Puncture. As reviewed by Dripps and Vandam,[21] lumbar punctures may be complicated by backache, damage to intervertebral disc, nerve root or spinal cord injury, and subarachnoid bleeding from epidural vein. Painful contact with sensory nerves of cauda equina occurs in 13% of procedures. According to Fishman,[38] post-lumbar-puncture headaches occur in approximately 20% of patients. These headaches usually remit after 2–5 days, but may persist for 8 weeks.[38] Ballenger *et al.*[4a] recently reported the performance of 1341 lumbar punctures with substantial CSF withdrawal (26–30 ml) without complication other than headache. The incidence of headache among their schizophrenic patients, affectively ill patients, and healthy subjects was 18%, 22%, and 35%, respectively. The mean duration of the post-lumbar-puncture headaches in these groups was approximately 1, 4, and 7 days, respectively. Maintenance of the prone position for several hours, bedrest, adequate hydration, and mild analgesics lessen the patient's discomfort. The incidence of this headache syndrome may be reduced by employing small-bore needles[96] and draining only the minimum amount of CSF required for study.

2.3.2. Cisternal Puncture. The technique of cisternal puncture has been described by Fishman.[38] The patient is placed in the lateral decubitus position with shoulders vertical and head flexed. A mark is placed 7.5 cm from the spinal needle point prior to needle insertion into the anesthetized scalp in the midline just above the spinous process of the axis. The needle is advanced in an upward direction toward the midpoint of an imaginary line through both external auditory meatuses. After striking the occiput, the needle is redirected slightly downward so as to slide under the occipital bone into the cisterna magna. The needle should not be advanced more than 7.5 cm, to avoid injury to the medulla. If no CSF is obtained, an alternative method such as lateral cervical puncture should be employed.

Cisternal puncture is generally safe for cooperative patients, but major complications such as large-vessel perforation, medullary injury with vomiting, or cessation of breathing and compromise of vertebral arterial blood supply in elderly patients have been reported.[60]

2.3.3. Lateral Cervical Puncture. The lateral cervical approach was originally employed for percutaneous cordotomy,[65] but later Zivin[117] described a technique that did not require fluoroscopic control. The patient is placed in the supine position, without a pillow and with neck straightened. The puncture site, located 1 cm caudal and 1 cm dorsal to the tip of the mastoid process (Fig. 5), is infiltrated with local anesthetic. The interspace between the first and second cervical vertebrae lacks bony overlap laterally at the atlantoaxial joint and offers a reasonably wide intervertebral space. The standard 20-gauge spinal needle is inserted perpendicular to the neck and parallel to the plane of the bed. Tactile sensations are experienced as a number of tissue planes are traversed; thus, the stylette must be removed frequently to check for CSF flow. Slow needle advancement prevents overpenetration of the subarachnoid space and injury to the spinal

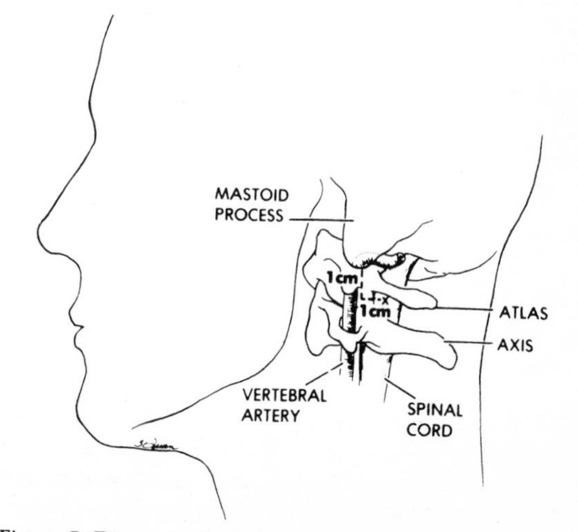

Figure 5. Diagram of relationships of major structures and landmarks in lateral cervical region. Lateral cervical puncture target lies 1 cm caudal and 1 cm dorsal to tip of mastoid process. Reprinted from Zivin,[117] with permission.

cord. If the vertebral artery is inadvertently pierced, then the needle should be withdrawn and local pressure applied. Occasionally, a nerve root may be irritated, causing local pain, or the procedure may be followed by headache. In general, the rate of traumatic taps employing this procedure has been less than 20%.[117]

2.3.4. Ventricular Puncture and Chronic Drainage. In infants, access to the lateral ventricles is provided by the insertion of an 18-gauge spinal needle through the anesthetized scalp and open coronal suture. The needle is aligned in the sagittal plane on a line with the pupil and advanced in the coronal plane on a line toward the inner canthus of the eye. CSF flow should normally be observed prior to a needle depth of 3.5 cm in infants and 6 cm in school-age children after removal of the stylette.[63a]

Robertson and Denton[89] have described techniques for ventriculostomy in older children and adults whose sutures have closed. Three standard points for the placement of trephines (burr holes) are illustrated in Fig. 6. The posterior aspect of the lateral ventricle is approached through a posterior parietal trephine placed 8 cm above the inion and 2.5–3 cm lateral to the midline. A cottonoid soaked in local anesthetic is applied to the underlying dura prior to dural perforation. The ventricular cannula is directed toward the inner aspect of the eye within the transverse plane of the supraorbital ridge. The trigone of the lateral ventricle is approached via a trephine at Keen's point located 2.5 cm above and 2.5 cm behind the pinnal helix of the ear. The cannula should be inserted along a path perpendicular to the cortex. Cannulation of the frontal horn of the lateral ventricle may be performed through a trephine opening placed at Kocher's point located 3 cm posterior to the normal hairline and 2.5 cm lateral to the midline. The ventricular cannula is then advanced perpendicularly into the cortex within a plane passing through the inner canthus of the ipsilateral eye. A blunt ventricular needle allows a keen sensation of increased and then suddenly decreased resistance as the ependymal wall of the ventricle is pierced. CSF should flow from the hub after withdrawal of the stylette.

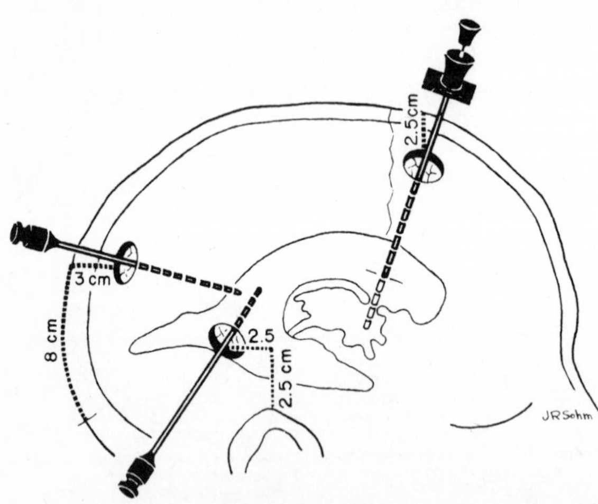

Figure 6. Three standard trephination sites used to gain access to lateral ventricles. Anterior site (Kocher's point) is 3 cm posterior to normal hairline and 2.5 cm lateral to midline. Keen's point trephine opening permits tapping of trigone of lateral ventricle and is located 2.5 cm behind and 2.5 cm above pinnal helix of ear. Posterior aspect of lateral ventricle is approached through posterior parietal trephine opening that is 8 cm above inion and 2.5–3 cm lateral to midline. Reprinted from Robertson and Denton,[89] with permission.

White *et al.*[101] have described a system of closed drainage of ventricular CSF that avoids repeated ventricular punctures and open-tube ventriculostomy. A flexible ventricular catheter connected to a Rickham reservoir is implanted in the lateral ventricle such that the reservoir overlies the trephine. The reservoir is connected in series to a medium- or low-pressure Spitz-Holter valve via silicone tubing that exists through a separate distant stab incision (catheter, reservoir, valve, and tubing are made by the Holter Company). The valve is then connected to a sterile calibrated bottle or bag. The subcutaneous reservoir enables direct sampling of ventricular CSF and intraventricular injection of drugs. This system provides continuous pressure-controlled CSF drainage as well as prevents retrograde migration of bacteria.

Complications of ventriculostomy include infection, development of porencephaly following multiple needle punctures in the same site, upward herniation of the cerebellum in the presence of mass lesions in the posterior fossa, and intracranial hemorrhage. Multiple needle passes through the occipital cortex may produce visual field deficits, which usually clear within a few days but may persist.

The connection of an Ommaya CSF reservoir (Heyer-Schulte Corporation) to the ventricular catheter offers a closed system for reliable chronic ventricular access with fewer complications than open-tube ventriculostomy.[73] The technical aspects of Ommaya reservoir placement in the right frontal region have recently been described for use in patients with meningeal cancer.[58]

2.4. CSF Storage and Analysis

Entire patient populations usually cannot undergo simultaneous CSF sampling; thus, CSF specimens collected over several days must be appropriately stored to ensure the stability of the CSF constituents under investigation. Failure to consider sample stability may result in artifactual variations of data and invalid conclusions.

CSF samples that are to be assayed for monoamine metabolites usually require the immediate addition of 2 mg/ml ascorbic acid for extended stability.[1,40,44] Aliquots intended for norepinephrine analysis should be collected in polypropylene tubes containing either abscorbic acid or reduced glutathione.[112] The currently available GABA[29,42,104] and cyclic nucleotide[92,95,100] assays do not require preservatives.

The polypropylene tubes containing CSF, and, if necessary, preservative, should be immediately placed on ice (4°C) or preferably dry ice (−57°C) and then into freezer storage at −70°C within 15–30 min after collection. Monoamine metabolites, norepinephrine,[112] and cyclic guanosine monophosphate (GMP)[14] in CSF degrade whereas CSF GABA levels increase[47,49] over time if the CSF aliquots are left at room temperature and not adequately frozen. If different assays are to be done on different days, thawing and refreezing of CSF samples should be avoided by aliquoting the CSF prior to freezing. CSF samples undergoing ultracold storage should be collected in polypropylene tubes, since glass tubes may fracture on thawing.

Simultaneous analysis of both the test and control CSF samples may eliminate major interassay variations.[109] Those CSF components that are unstable at room temperature should be analyzed at the time of the first thawing of the CSF samples. Care must be taken to employ verified assays with appropriate sensitivities capable of determining the CSF constituents under investigation. Many assay systems capable of making plasma determinations cannot be adapted for CSF analysis. Accurate determination of CSF neurotransmitter alterations employing radioreceptor assays may be precluded by the therapeutic administration of receptor agonist or antagonist medications. Overlapping radiation spectra of various radioisotopes may produce mutual interference that compromises multiple radioassays on the same unaliquoted CSF sample.

The absolute values of amine metabolite levels in CSF collected from patients who have received probenecid must be corrected for variations in probenecid blockage of metabolite egress. The calculation of the corrected CSF metabolite level requires determination of the CSF probenecid concentration at the time of CSF sampling.[24,99,109]

3. Animal Investigations

Meaningful physiological, pharmacological, endocrinological, and pathological investigations require animal systems that approximate the human situation. Unfortunately, most experimental systems require immobilization of the animal for long periods by physical restraints or general anesthesia.

The stress of immobilization alters CSF levels of neurotransmitters[56] and their metabolites[46]; thus, awake animals should be conditioned by spending several days in the experimental situation prior to CSF sampling. The lighting schedule of the laboratory may disturb normal circadian rhythms and produce artifactual variations in CSF neurotransmitters, their metabolites, cyclic nucleotides, and hormones.[84,88,114]

Animals requiring electrode implantations should undergo CSF analysis before and after surgery to document the return to baseline levels of the CSF component under investigation. General anesthesia usually evokes acute alterations in the neurohumoral composition of CSF; thus, a CSF steady state should be documented prior to pharmacological or physiological manipulation. Obviously, CSF sampling of unanesthetized, conditioned animals is preferred.

3.1. Model Selection

Most ventricular CSF sampling techniques require the passage of cannulae through the cerebrum. These methods are associated with tissue damage[33,34] edema,[27] infection, and catheter obstruction.[48,74] This acute tissue disruption is associated with a breakdown of the blood–brain and blood–CSF barriers, which would tend to increase the penetration of normally excluded or slowly crossing agents. Intravenous injection of vital dyes or albumin-bound radioisotopes may verify this alteration of the blood–brain barrier. Thus, intravenously administered drugs may have an artifactually augmented central action secondary to tissue damage during cannula insertion. In addition, the vasogenic edema associated with blood–brain barrier disruption, extravasated blood, and cell necrosis surrounding the cannula tract may acutely alter baseline CSF neurotransmitter and cyclic nucleotide levels, but should resolve after several days to a week.[66] Iatrogenic opening of the blood–brain barrier may occur transiently following intravenous administration of hypertonic solutions.[76] Reestablishment of steady-state CSF concentrations should be documented by serial CSF sampling and analysis prior to experimental manipulation. This documentation of baseline conditions is especially important in CSF investigations involving catecholamines, their metabolites, GABA, and cyclic nucleotides.[109,111]

In general, the ependymal lining of the ventricles offers little resistance to the penetration of the brain extracellular space by substances contained in CSF.[13,35,106] Pharmacological studies of agents that do not cross the blood–brain barrier may be facilitated by employing intraventricular administration. Animal models that require the constant drainage or perfusion of the ventriculosubarachnoid spaces may not be optimal for the investigation of intraventricularly injected drugs. The relatively rapid washout and reduction of CSF concentration of drugs injected as a bolus tend to decrease their central action. CSF drug concentrations may be maintained by administering the agent into the ventricles as a constant infusion rather than a bolus or may be prolonged by employing closed animal systems that do not require CSF drainage for catheter patency.

Animal models and sampling methods should be appropriate for the stability characteristics of the CSF components under investigation so as to avoid artifactual variations of data. Animal models employing automated serial sampling systems should be designed not to allow unpreserved CSF to flow through lengthy plastic tubing exposed to room temperature. The GABA content of CSF rapidly increases over time secondary to the breakdown of CSF homocarnosine into GABA if not frozen properly.[47] Contrarily, CSF cyclic GMP concentrations decrease with exposure to room temperature.[14] Preservative should be added immediately to CSF samples that are to undergo monoamine metabolite or norepinephrine analysis.[1,40,44]

Unfortunately, many currently available animal models have been plagued by technical problems and have not provided long-term access to ventricular CSF with any reliability. The following sections of this chapter will discuss the technical aspects of and indications for various experimental systems for CSF sampling.

3.2. Methods of Chronic Ventricular Access

Serial sampling of CSF permits the comparison of CSF levels of a substance under both control and experimental situations in the same animal. This methodology amplifies the statistical analysis of the data by enabling each animal to serve as its own control. The study of the cumulative effects of pharmacological agents on the brain and the secondary development of pathological or behavioral alterations requires chronic ventricular access. Isolated intraventricular drug injections or perfusions yield information concerning the action of chemical substrates on specific areas of cerebrum lining the ventricles. In addition, ventricular cannulation allows assessment of CSF pressure fluctuations. Most important, chronic catheter implantations permit the investigation of unanesthetized animals.

In general, aseptic techniques should be employed for all chronic preparations. Rats appear to be more resistent to infection than are cats or monkeys; however, all animals should receive prophylactic antibiotics.

3.2.1. Cannulation. Klee and Praestholm[53] have described a simple, reliable, free-hand method for cerebral ventricular puncture in the rat. Midline skin incisions are placed over the top of the head of the anesthetized rat, and a 26-gauge, $\frac{3}{8}$-inch short-beveled intradermal needle fitted with a rubber stop such that 2 mm protrudes is employed to bore holes through the skull 2 mm behind the coronal suture and 2 mm lateral to the sagittal suture. A second identical needle fitted with a stop at 4 mm, attached to a tuberculin syringe, is inserted through the hole, and the contents of the syringe are injected into the lateral ventricle. Thereafter, the skin incision is closed with a clip. This method allows multiple intraventricular drug injections; however, it is not appropriate for ventricular perfusions. The small amounts of CSF obtained may also be insufficient for analysis.

Most cannulation techniques require the stereotaxic insertion of guide tubes so as to locate portions of the ventricular system that maximize the inflow of drug solutions and outflow of CSF. Myers[69] has selected stereotaxic coordinates (Table 1) for cannula tip placement in the anterior horn of the lateral ventricle close to the foramen of Monro in monkeys and cats and 1 mm distant to the foramen in rats. External bony landmarks such as sutural lines should be replaced by the interaural line as the zero reference.

Table 1. Range of Optimal Stereotaxic Coordinates for Placement of Cannulae in Lateral, Third, or Fourth Ventricle of Rat, Cat, and Monkey[a]

Target	Projection	Coordinates (mm)		
		Rat[b]	Cat[c]	Monkey[d]
Lateral ventricle[e]	Anteroposterior	5.4 to 6.2	11.5 to 13.5	13.0 to 16.0
	Lateral	1.2 to 2.0	3.0 to 4.5[f]	3.0 to 5.0[f]
	Horizontal	+1.5 to +2.5	+6.5 to +8.5	+11.0 to +16.0
	"Dura"[g]	−2.8 to −3.8	−11.5 to −13.5	−14.0 to −19.0
Third ventricle	Anteroposterior	5.5 to 6.5	12.0 to 13.5	12.5 to 15.5
	Lateral	0.0	0.0	0.0
	Horizontal	−1.0 to −3.0	+3.0 to −4.0	+9.0 to −2.0
	"Dura"[g]	−7.0 to −9.0	−17.0 to −24.0	−21.0 to −32.0
Fourth ventricle	Anteroposterior	P 3.0 to 4.0[h]	P 0.5 to 2.0[h]	P 6.5 to 9.5[h]
	Lateral	0.0	0.0	0.0
	Horizontal	−4.0 to −5.0	+1.0 to +3.0	−5.5 to −8.5
	"Dura"[g]	−4.0 to −5.0[i]	−16.0 to −18.0[j]	−35.5 to −38.5

[a] Reprinted from Myers,[69] with permission.
[b] Atlas of DeGroot[20]; coordinates based on stereotaxic zero, not bregma.
[c] Atlas of Jasper and Ajmone-Marsan.[52]
[d] Atlases of Olszewski[72] and Winters *et al.*,[102] for macaques.
[e] Lateral angle of 5–6° off vertical is often employed for lateral ventricular cannulation in cat and monkey.
[f] Position of tip when cannula is implanted at 5–6° angle.
[g] "Dura" refers to distance (mm) that cannula tip should rest below surface of exposed dura mater.
[h] P signifies posterior (in mm caudal to stereotaxic zero).
[i] Below cerebellar meninges.
[j] The bony tentorium must be bypassed, necessitating angled placement.

Kokkinidis *et al.*[54] described a simple compact cannula system for the application of substances to the brain and ventricles of freely moving mice. The cannula (Fig. 7) is constructed from a flat-head stainless steel jeweler's screw through which a 0.508-mm hole has been drilled in the long axis. A 26-gauge stainless steel needle is inserted through the screw hole and soldered into place such that 2.7 mm of the stylette protrudes below the tip of the screw. The device is attached to the skull by four rotations of the screw following stereotaxic implantation. Acrylic dental cement provides additional bonding of the screw to the skull. An obturator constructed from 0.254-mm stainless steel wire and a polyethylene cap is inserted into the stylette barrel to keep the barrel free from fibrotic material between intracranial injections. This delicate system offers the opportunity of installing several of these cannulae into several regions of the brain of the same animal.

Ventricular cannulation for chronic infusion in the rat has been described by Myers.[69] This system (Fig. 8, left) incorporates a plastic base made from the head of a 1-ml disposable tuberculin syringe (Becton-Dickinson Company), the outer surface of which is threaded with a $\frac{3}{16}$-inch tap. A section of 22-gauge stainless steel needle tubing is mounted inside this plastic base to serve as a guide tube. The guide needle is implanted employing a stereotaxic head rest such that the tip is resting in one lateral ventricle (Fig. 8, right). The guide tube is held in place by cementing the plastic base to stainless steel anchor screws inserted into the rat's calvarium. A plastic syringe cap is internally threaded with a $\frac{3}{16}$-inch tap and screwed onto the base to protect the cannula. Polyethylene (PE-10 or PE-20) tubing is attached to a 28-gauge injector needle and filled with artificial CSF. This injector needle is then inserted through an undersized hole drilled in the top of the plastic syringe cap and advanced into the ventricle extending 1 mm beyond the tip of the guide tube. A 2.5-mm segment of larger flexible tubing is placed over the PE tubing and attached to the plastic cap to protect the junction of the PE tubing and injector needle. This apparatus weighs approximately 0.75 g and, when attached to a swivel, allows complete freedom of movement. The PE tubing may be suspended from a bar over the cage employing a pulley system with a 30-g counterweight.

Large animals such as cats or monkeys are preferred for chronic CSF sampling because their ven-

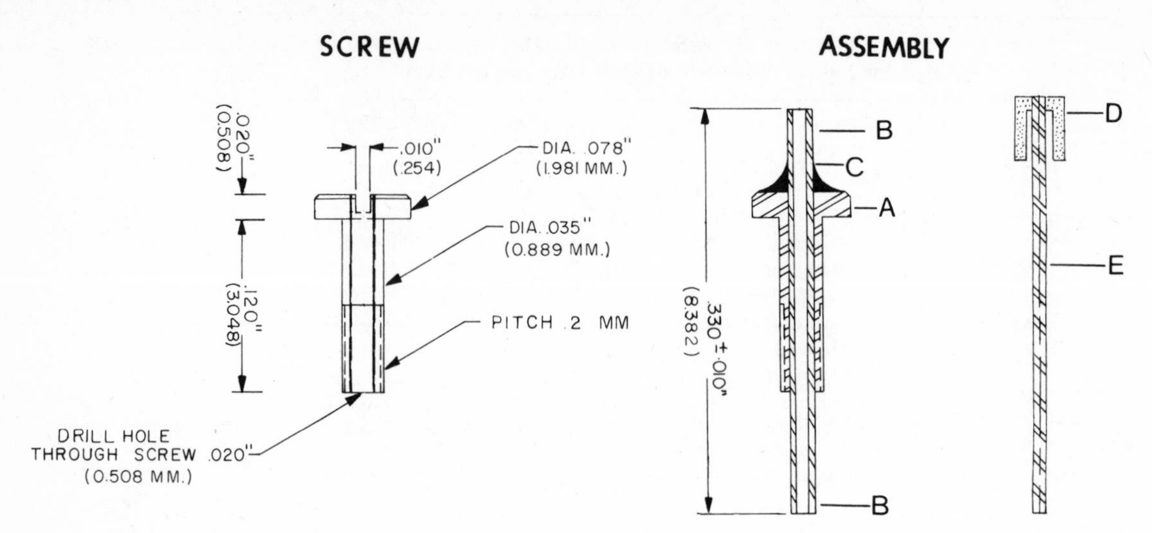

Figure 7. Schematic representation of cannula system for mice. (A) Screw; (B) stylette; (C) solder; (D) polyethylene cap; (E) wire. Reprinted from Kokkinidis *et al.*,[54] with permission.

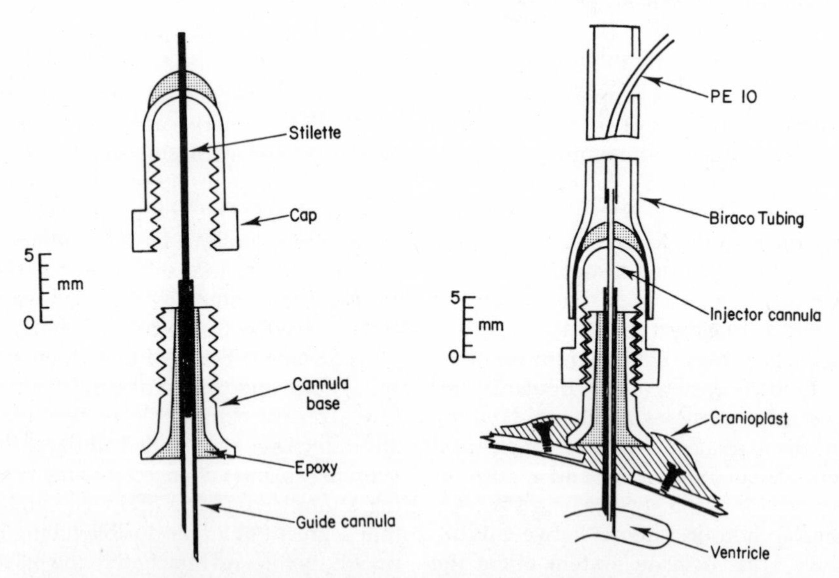

Figure 8. Simplified intracranial cannula for chemical stimulation or chronic intraventricular infusion into small animals. Stainless steel injector and guide cannula are fitted to plastic cap and Luer end of disposable tuberculin syringe. *Left*: Base with 22-gauge guide cannula cemented in place. Twenty-eight-gauge stylette that is affixed to cap before implantation is shown partially lowered within guide cannula. *Right*: Guide cannula–base assembly after implantation into lateral ventricle of rat. Tip of injector cannula rests in ventricle. Distal end of cannula is attached to polyethylene tubing (PE-10) that is threaded through and taped to protective Biraco tubing. Cannula base is fixed to cranium with miniature screws and cranioplast. Reprinted from Myers *et al.*,[68] with permission.

tricles contain larger volumes of CSF. Myers[69] has described a cannula system for these larger animals requiring the stereotaxic implantation of the guide cannula at a 6–10° angle to the parasagittal plane (Fig. 9). The guide cannula is constructed from 17- or 18-gauge stainless steel tubing and is placed 0.5–1.5 mm dorsal to the lateral ventricle. After flexible PE-60 tubing filled with mock CSF is attached, the tapping needle is inserted through the guide tube with the fluid-filled PE tubing held above the animal's head. On penetration of the ependymal wall of the ventricle, the mock CSF contained in the tubing spontaneously flows into the ventricle. The tapping needles are constructed from 20-gauge stainless steel tubing by sealing the tip with solder and cutting a hole in the side of the tube (Fig. 10, right). The sealed tip prevents occlusion of the tapping needle with cerebral tissue during insertion.

According to Myers,[69] repeated puncturing of the ventricle often produces a channel between the ventricle and the tip of the guide tube such that CSF flows spontaneously into the guide tube. At this point, a tapping needle without a sealed tip may be used to provide two lumens for entry of CSF (Fig. 10, left). A stylette attached to a sterile threaded syringe cap be employed to seal the guide tubing during periods between CSF sampling.

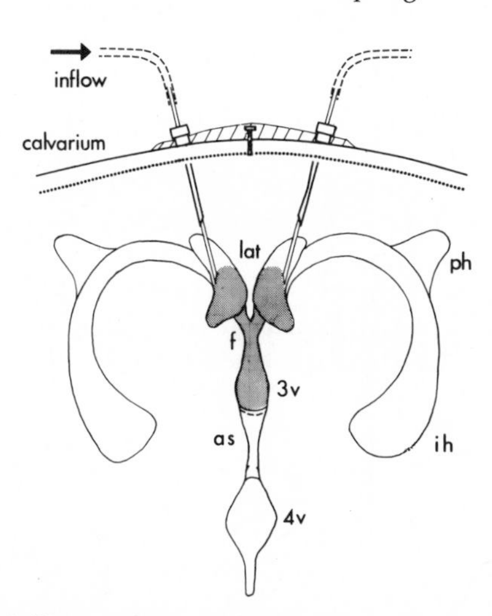

Figure 9. Diagrammatic representation in coronal plane of bilateral lateral ventricular cannulae of unanesthetized monkey. Shaded area indicates region immediately irrigated by infusion or perfusion. Cranioplast cement retains cannula hubs in position together with anchor screw placed on midline. (as) Aqueduct of Sylvius; (f) foramen of Monro; (ih) inferior horn of the lateral ventricle (lat) lateral ventricle; (3v) third ventricle; (4v) fourth ventricle. Reprinted from Myers *et al.*,[71] with permission.

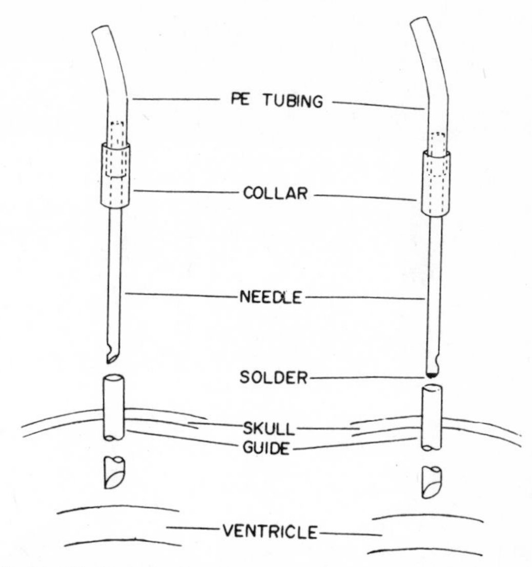

Figure 10. Ventricular tapping needles with sideopening at their base with tip (*left*) open to permit CSF outflow or (*right*) sealed in case of occlusion. Sealed needle (*right*) is employed when considerable tissue must be penetrated to reach ventricular lumen. Reprinted from Myers,[69] with permission.

A similar model has been developed by Ebert *et al.*[23] for serial collection of lateral ventricular CSF in conscious rhesus monkeys. This system consists of a base plate and guide tube (in one piece), a cannula, a stylette to block the guide when the cannula is not in place, and a protective cap (Fig. 11). CSF collection from the lateral ventricle cannula is begun immediately following installation. The cannula equipped with a standard Luer fitting is connected with PE tubing to a fraction collector housed in a refrigerator placed 2 feet below the animal's head. The CSF spontaneously flows down the siphon in the PE tubing into collection tubes for 20–40 sec, after which an automatic solenoid valve closes for 3–6 min. This sequence is repeated continuously so that 0.3–0.5 ml of CSF is removed per hour. This sampling system remains functional indefinitely as long as intermittent CSF flow is maintained. The cannula may be replaced with a stylette prior to capping of the base plate so as to allow the animal to return to the cage between experiments.

Malkinson *et al.*[61] have developed for larger animals (cats, rabbits) a cerebral cannula assembly that can be employed experimentally to sample CSF from or inject drugs into the cerebral ventricles, perfuse the ventricle, electrically record or stimulate the activity in specific brain areas, and monitor CSF pres-

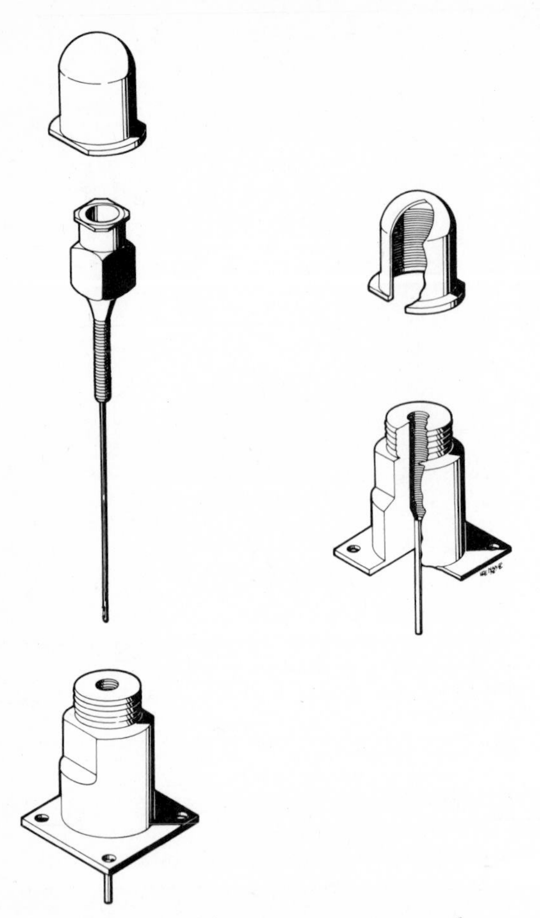

Figure 11. Cannulation system for conscious monkey, consisting of base plate and guide tube (manufactured in one piece), cannula stylette to block guide tube when cannula is not in place, and protective cap. Reprinted from Ebert *et al.*,[23] with permission.

sure. This system also incorporates a vascular catheter capable of recording arterial or venous pressures, infusing drugs into the bloodstream, and collecting blood samples.

In brief, a silastic vascular catheter (Dow Corning Corporation) containing heparinized (10 U/ml) saline is implanted into a distant vessel and brought subcutaneously to exit through the scalp. This catheter is passed through PE conduit tubing (Clay-Adams Company). Stainless steel guide needles with stylettes (Bectin-Dickinson Company) of appropriate gauge and length are stereotaxically implanted to a depth approximately 2–4 mm above the target site and secured to the skull with dental cement (L. P. Caulk Company). A pedestal (Fig. 12A) is formed from a portion of a PE bottle that has been fixed to the skull with four 0-80, ¼-inch stainless steel binding head machine screws (Small Parts, Inc.). This supporting bottle section mold, which surrounds the guide needles and protective conduit tube, is then filled with dental cement. The pedestal is capped for protection between sampling periods. The catheter is flushed once per week with heparininzed saline, and the stylettes are removed and cleaned with 70% ethanol every 4 days. Tubing may be attached to the guide needle hub via an adapter (Fig. 12B,C) constructed from a male "Luer-Lok" tip (Becton-Dickinson Company) by drilling a threaded side hole that will accept a 1-72, ⅜-inch stainless steel fillister machine screw (Small Parts, Inc.). This adapter is capped to secure a rubber diaphragm through which a brain probe may be passed. The security of this system allows complete freedom of movement of the caged animal without alteration of probe placement. According to its developers, this apparatus has been maintained in animals for periods of over 6 months.[61]

Recently, Faulhauer and Donauer[30a] have used a permanent ventriculostomy system employing a subcutaneous reservoir in cats (some of which had hydrocephalus) for durations up to 5 months. The distal 12- to 15-mm portion is removed from a "type C" atrial catheter (Holter Company) (Fig. 13, top) and the tip of the thin portion is sealed with silicone elastomer (Fig. 13, middle). A side opening is then placed above the closed tip and the shaft of a Salmon–Rickham low-profile reservoir (Holter Company) is reduced in length (Fig. 13, bottom). A trephine is drilled over the parietal bone approximately 6–7 mm from the midline and 14–15 mm in front of the external auditory meatus. The catheter, reenforced with a removable guide wire (Fig. 14, top), is then inserted at an inclined angle to the parasagittal plane into the lateral ventricle. Therefore, the catheter is removed, attached to the reservoir (Fig. 14, middle), and reinserted into the lateral ventricle. The reservoir is sutured to the temporal musculature prior to skin closure.

Wood *et al.*[107,108] have developed a primate model (Fig. 15) that does not require insertion of cannulae through cerebral tissue and that is based on an adaptation of the subcutaneous Ommaya CSF reservoir.[73] A midline posterior incision is made between the inion and the arch of the axis in anesthetized monkeys. After removal of the posterior margin of the foramen magnum (Fig. 16A), the dura mater and arachnoid are opened in the midline. The foramen of Magendie is dilated to avoid retraction of the cerebellum (Fig. 16B). A 2-mm outside-diameter silicone

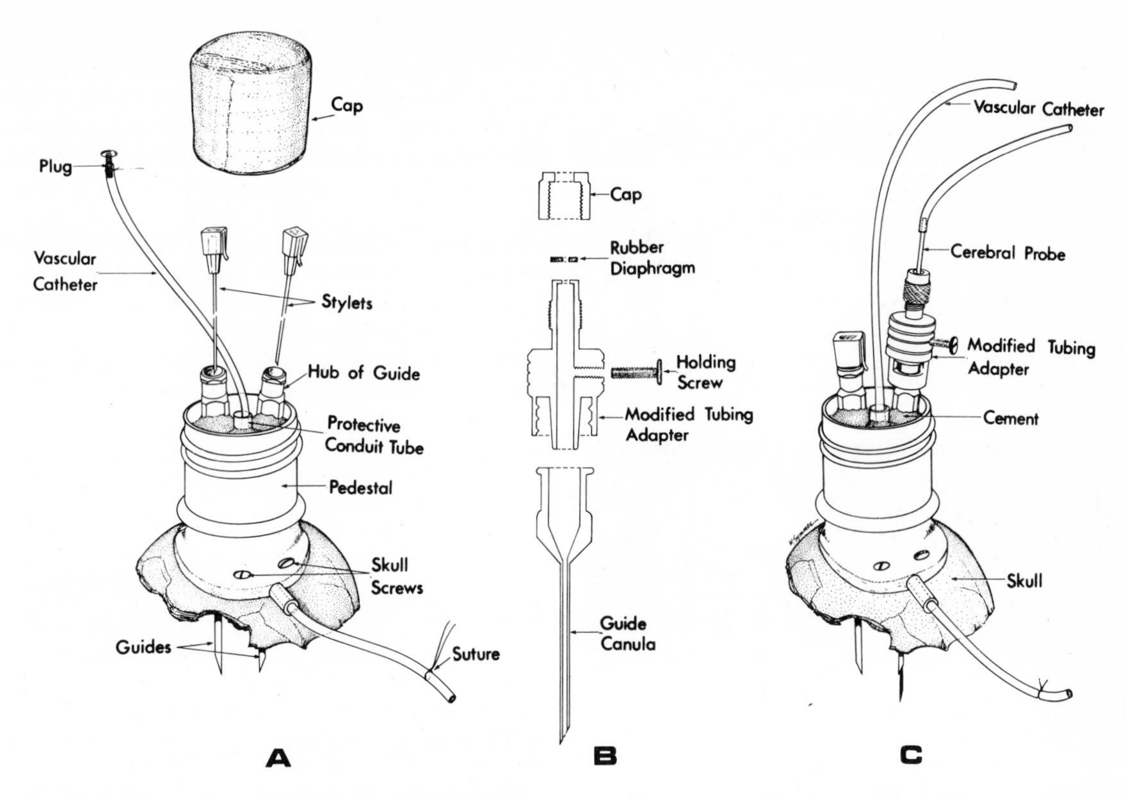

Figure 12. Diagram of multipurpose cannula system for access to brain or ventricles and/or systemic vascular system of unanesthetized animals. (A) Pedestal and its component parts; (B) modified tubing adapter and guide system; (C) system used in one experimental situation. Reprinted from Malkinson *et al.*,[61] with permission.

catheter is introduced into the fourth ventricle and advanced until the catheter tip makes contact with the anterior wall of the cavity (Fig. 16C). The catheter is then withdrawn 3 mm to prevent occlusion of the aqueduct. Following watertight dural closure (Fig. 16D) and placement of gelfoam over the dural suture line, the CSF within the catheter is noted to pulsate in response to respiratory activity. A 2.5-cm side-armed CSF reservoir is connected to the catheter using a right-angled connector and is secured subcutaneously to prevent displacement during pumping maneuvers (Fig. 16E) (Pudenz Catheter, Pudenz right-angled connector, and Ommaya CSF reservoir made by Heyer-Schulte Corporation). The skin is closed with two suture layers to avoid wound dehiscence in case the animal pulls out several skin sutures. All animals should receive prophylactic antibiotics during the first 4 postoperative days.

The CSF reservoir is depressed several times every 2 days to prevent catheter occlusion and to reduce animal anxiety during CSF sampling. The rapidity of reexpansion of the reservoir can be improved by flushing the catheter system (Fig. 17) with small amounts of sterile mock CSF or Elliott's B solution[28,73] (Tavenol Laboratories, Inc.). Four depressions and reexpansions of the reservoir yield ventricular distribution of drugs injected into the reservoir (Fig. 18). Sterile 3-ml CSF samples may be removed from the reservoir several times per day indefinitely without anesthesia employing 25-gauge scalp vein needles, provided sterile techniques are used during sampling. A 23-gauge scalp vein needle may be inserted percutaneously into the reservoir with the animal in the sitting position to record fluctuations in intraventricular pressure when connected to a physiological pressure transducer (Statham Instruments Division of Gould, Inc.) (Figs. 17 and 19). The animals may be returned to their cages between sampling periods.

In conclusion, this method of fourth ventricular cannulation using subcutaneous CSF reservoirs in rhesus monkeys (1) provides chronic access to sterile CSF without chronic immobilization; (2) enables mixing of injected drugs with lateral ventricular

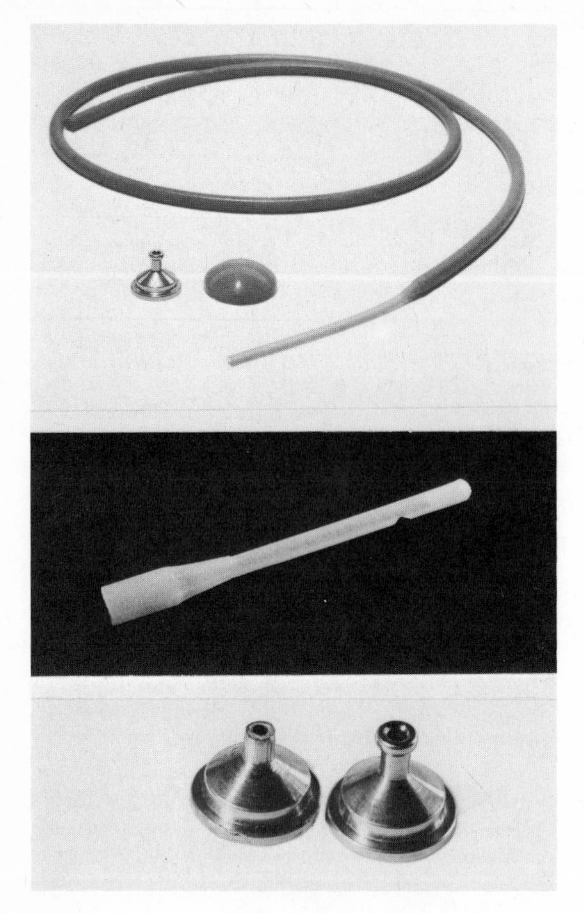

Figure 13. *Top*: "Type C" Holter atrial catheter and two components of Salmon–Rickham low profile reservoir. *Middle:* Ventricular catheter (1.3-mm outer-tip diameter, 0.8-mm inner diameter) with closed tip and lateral opening. *Bottom:* Reservoir bases before (right) and after (left) shortening of shaft. Reprinted from Faulhauer and Donauer,[30a] with permission.

CSF; (3) permits sensitive monitoring of intraventricular pressure; and (4) most important, does not produce tissue damage during cannula implantation or breakdown of the blood–brain barrier. This model has been validated in several neuropharmacological and neurotoxicological investigations.[78,107]

3.2.2. Intraventricular Drug Administration and Infusions. The animal models described above allow chronic intermittent injection or continuous infusion of drugs into the ventricular system. Normal (0.9%) saline should not be used as a vehicle for intrathecal drug injection or as a substitute for CSF because this solution has potent central actions on neuronal activity and physiological functions.[2,28,97] In addition, unbuffered salt solutions evoke marked dilatation of pial blood vessels.[28] Myers[69] has prepared artificial CSF and five-ion solutions that evoke no notable behavioral or physiological response in animals. Fenstermacher *et al.*[36,37] have also described a balanced salt solution suitable for use as artificial CSF. The use of Elliott's B solution[28] (Travenol Laboratories) as a diluent reduces the incidence of fever, headaches, and vomiting following intrathecal drug administration.[22,39]

According to Myers,[70] the ventricles of animals used for chronic infusion studies should be tapped immediately following cannula implantation, but should not be infused for a period of 2–3 days thereafter. Prolonged infusion of microliter quantities of

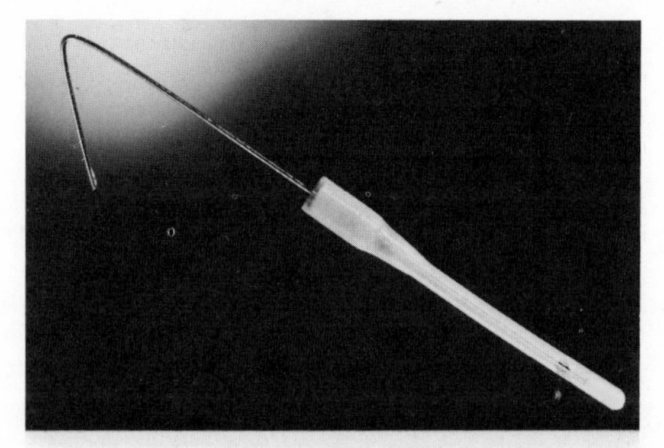

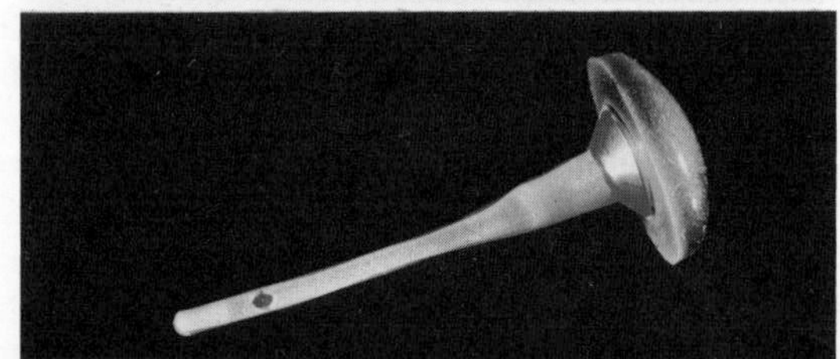

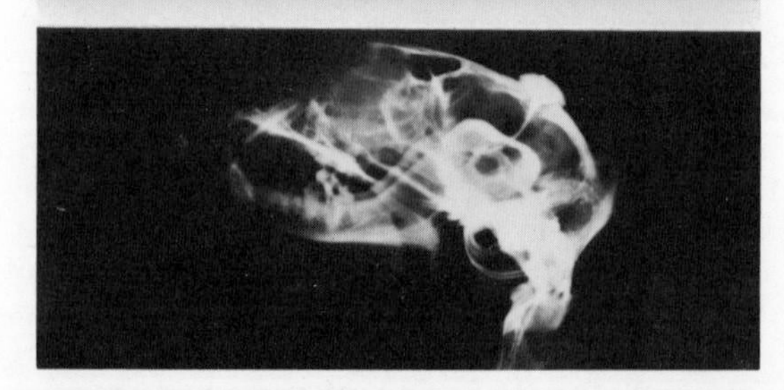

Figure 14. *Top:* Ventricular catheter with wire guide. *Middle:* Ventricular catheter attached to assembled reservoir. *Bottom:* Right lateral ventriculogram of cat demonstrating position of ventriculostomy. Reprinted from Faulhauer and Donauer,[30a] with permission.

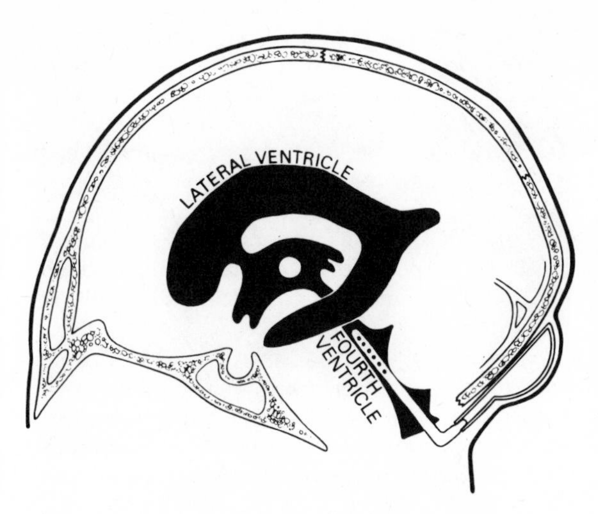

Figure 15. Catheter placed in fourth ventricle and connected to subcutaneous Ommaya CSF reservoir. Note absence of catheter penetration of brain. Reprinted from Wood *et al.*,[107] with permission. © 1977 by the American Association for the Advancement of Science.

solutions does not produce hydrocephalus, even in the rat. The rate of infusion may be determined by connecting the injection needle to a calibrated 10- to 50-μl syringe (Hamilton Company) and observing the filling of the syringe barrel following activation of the infusion pump. Air bubbles in the PE tubing are avoided by allowing the infusate to come to room temperature, adequate greasing of the infusion pump plunger, and positioning the infusion pump at the same level as the floor of the animal's cage. Polyvinyl chloride tubing should not be used in CSF studies because this material contains a highly active substance.[12] Care must also be taken to avoid decomposition of the drug to be infused prior to ventricular instillation.

3.2.2.1. Documentation of Infusate Distribution. Prior to sacrifice, cannulated animals should be infused with a vital dye at the same rate employed in the experiment to document the anatomical distribution of the infusate. According to Myers,[70] a 2- to 5-μl infusion of dye should be delivered to the ventricles of rats, whereas 100- to 200-μl dye infusions may be employed in monkeys. Following perfusion of the brain with formalin solutions, the ventricles should be examined for the extent of dye staining prior to prolonged formalin fixation.

3.2.2.2. Control Groups. As proposed by Myers,[70] experimental infusion protocols should contain a sham-operated or nonimplanted group of control animals. Another control group of animals undergoing artificial CSF infusion is required as a pharmacological control, and a third group undergoing subarachnoid-space rather than ventricular infusion is recommended as an anatomical control.

3.3. Experimental Perfusion Systems

The perfusion of the brain or ventricular system offers an *in vivo* method of documenting that an increase or decrease in the release of specific transmitter substances contributes to the activity of an individual efferent pathway that in turn elicits a physiological or behavioral response. The effects of various pharmacological agents on specific brain regions have been assessed using central perfusion techniques. Ventricular perfusions have also been employed to study the physiology of CSF production and absorption.

3.3.1. Perfusion of Ventricles. A technique for ventriculocisternal perfusion of the anesthetized rat has been described by Myers and Brophy.[67] A 22-gauge guide tube is fastened to a stereotaxic frame and inserted to the level of the lateral ventricle through a hole in the skull. The inflow cannula constructed from 28-gauge tubing is filled with artificial CSF and connected to the infusion pump by PE-20 tubing. This inflow cannula tip is beveled at 45° and advanced within the guide tube to a position 1 mm beyond the guide tube tip. Inflow cannula patency is verified by observing the mock CSF flow into the ventricle as the PE tubing is raised about 10 mm above the skull. Cisternal outflow is established by puncturing the atlanto-occipital membrane with a 22-gauge needle and inserting PE-50 tubing 2 mm beyond this needle tip. Care must be taken to document that the outflow rate matches the inflow rate of 25–50 μl/min and that the level of the outflow tubing is not more than 1 cm below the level of the lateral ventricle to prevent a siphoning effect. Unfortunately, this method is not appropriate for unanesthetized rats.

Ashcroft *et al.*[4] implanted similar guide tubes and needles into the lateral ventricle and cisternal magna of dogs (Fig. 20). This method of closed ventriculocisternal perfusion allows recirculation of the dog's own CSF, thereby minimizing electrolyte disequilibrium, and can be applied to conscious and unrestrained animals. Pappenheimer *et al.*[75] have described a commonly used open perfusion system for larger unanesthetized animals. Methods of perfusion are also reviewed Chapter 28 by Sklar.[93] These systems enable the determination of bulk for-

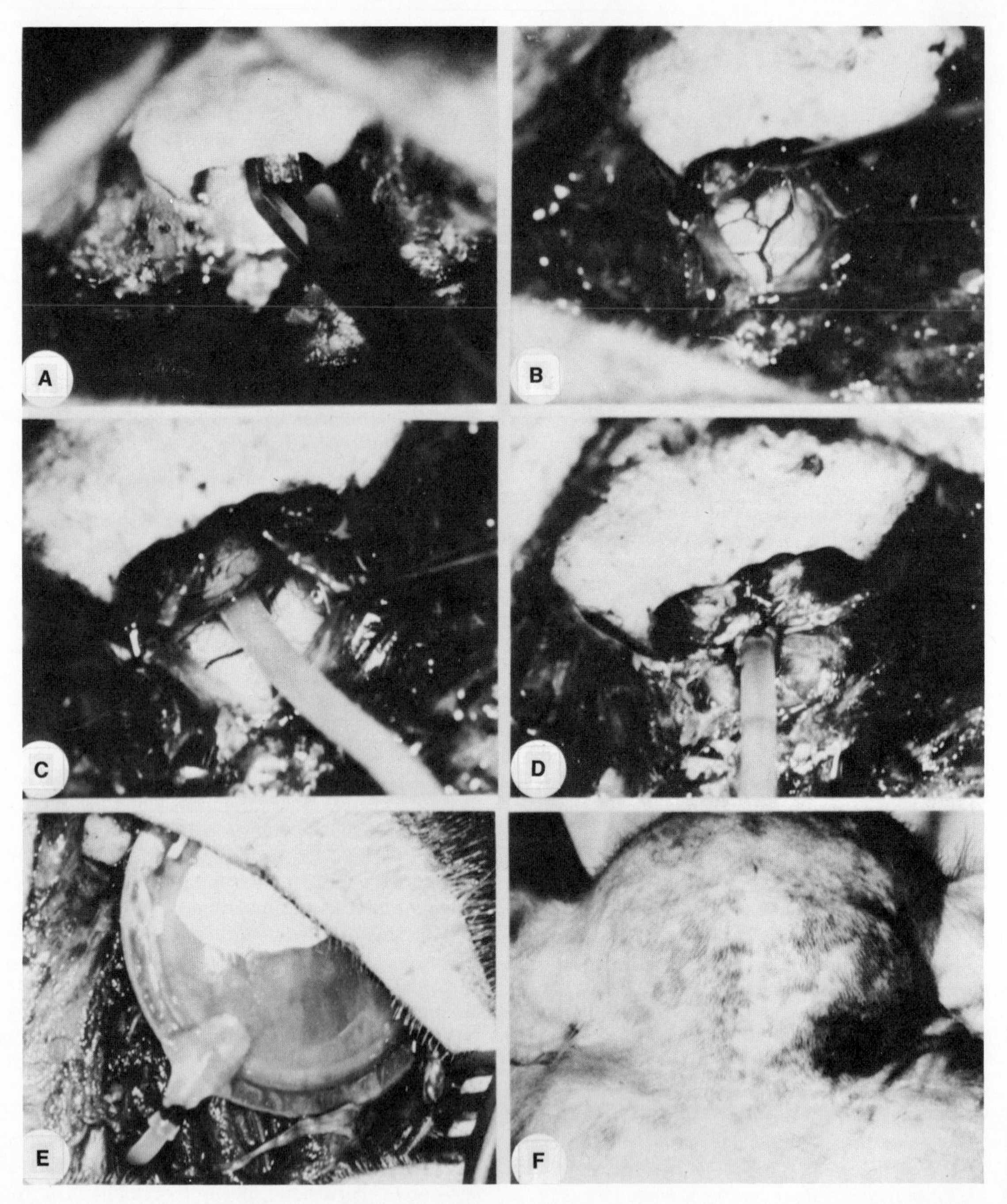

Figure 16. Operative technique for catheter and subcutaneous Ommaya CSF reservoir implantation. (A) Removal of posterior rim of foramen magnum; (B) midline incision of dura mater and arachnoid followed by dilation of foramen of Magendie; (C) catheter placement into fourth ventricle via foramen of Magendie (note withdrawal 3 mm from aqueduct exit to avoid obstruction); (D) watertight dural closure including purse-string suture around catheter; (E) subcutaneous side-armed CSF reservoir is connected to catheter using right-angled connector and sutured to occipital musculature; (F) external appearance of subcutaneous CSF reservoir after wound healing and suture removal. Reprinted from Wood *et al.*,[108] with permission.

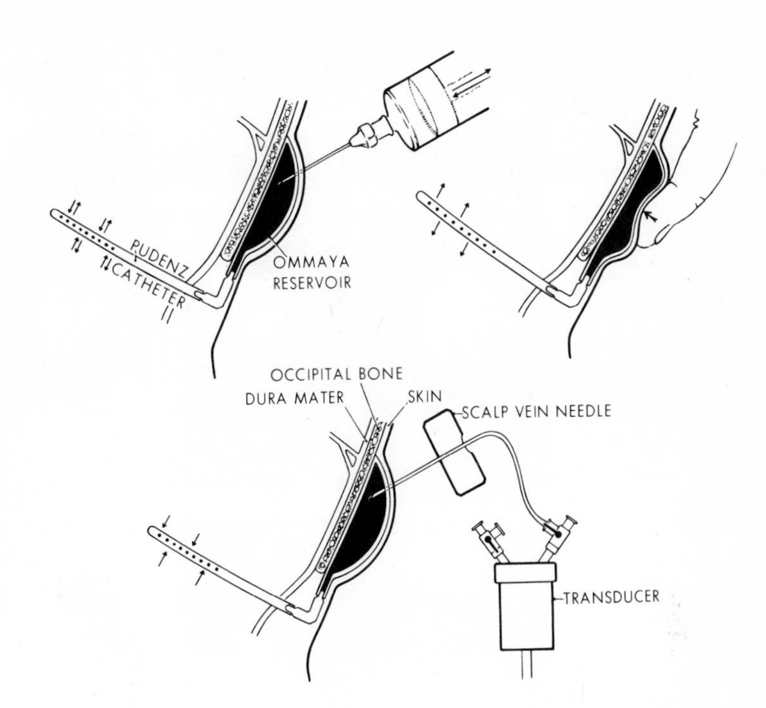

Figure 17. Diagram of applications of subcutaneous Ommaya CSF reservoir. *Top left*: Method for intraventricular injection of drugs or withdrawal of ventricular CSF by means of syringe with 25-gauge needle. *Top right*: Manual depression of reservoir (pumping) to expel contents into the fourth ventricle. *Bottom*: Intraventricular pressure being recorded with 23-gauge scalp vein needle attached to soft catheter that is connected to physiological pressure transducer. Reprinted from Wood *et al.*,[107] with permission. ©1977 by the American Association for the Advancement of Science.

mation or absorption of CSF and its constituents as well as the volume of the ventricular space perfused.

The ventricular cannulation method employed in these perfusion systems is similar to that of Myers,[69] which is described in detail in Section 3.2.1. In brief, Feldberg *et al.*[32] developed a reliable technique for the establishment of the cisternal outflow tract in cats and other larger animals (Fig. 21). The interparietal and supraoccipital bones are scraped clean, and two stainless steel screws are inserted into each of these bones on either side of the midline. The posterior lip of the foramen magnum is enlarged to expose the atlanto-occipital membrane. The 18-mm 20-gauge cisternal guide cannula is cemented to the supraoccipital bone and screws with acrylic dental cement such that the guide cannula tip rests approximately 1 mm above the atlanto-occipital membrane. The scalp wound incision is then closed around the cisternal guide tube, which has been capped with a diaphragm. All animals are given prophylactic antibiotics. Artificial CSF-filled PE tubing is connected to a 5-cm 25-gauge hypodermic needle from which the hub has been removed. The cisternal puncture is performed by introducing this 25-gauge needle into the guide tube and advancing it through the atlanto-occipital membrane until CSF drains from the PE tubing. Feldberg *et al.*[32] perfused cats at a constant rate of 0.1 ml/min, and cisternal outflow was approximately 3.5 ml in 30 min. An alternative method for placement of a permanent cisternal cannula in cats has been described by Radulovacki and Girgis.[87]

Perfusion systems have been modified by limiting the perfusate access to only portions of the ventricular system. Bhattacharya and Feldberg[9] isolated the ventricular system of anesthetized cats from the subarachnoid space by advancing the 2-mm-diameter outflow cannula into the aqueduct. Carmichael *et al.*[17] developed a method of multiple cannulation of the ventricular system to exclude drugs from or limit their perfusion to a given segment of the ventricle and its surrounding brain tissue. One cannula acts as an outflow tract, whereas all the remaining cannulae serve as inflow ports. Usually, only one of these inflow cannulae delivers the drug under study, whereas the other inflow cannulae deliver artificial CSF. In this manner, drug perfusion is limited to either the temporal or anterior horns of the lateral ventricles or the third ventricle. This method enables the individual study of the hippocampus and amygdala lining the temporal horn, the olfactory gray matter, septum and caudate nucleus bordering the anterior horn, or the hypothalamus and thalamus lining the third ventricle. Besson *et al.*[8] have developed a "cup technique" for specifically superfusing the region of ventricular surface be-

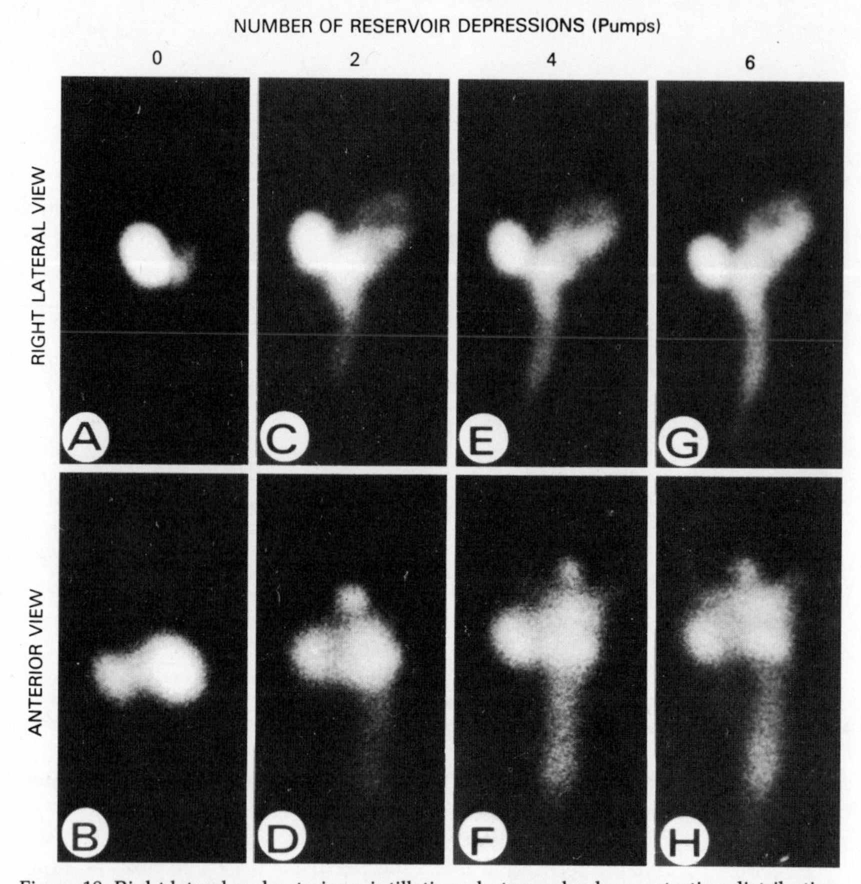

Figure 18. Right lateral and anterior scintillation photographs demonstrating distribution of radiopharmaceutical (technetium-99m-labeled human serum albumin) in CSF spaces of monkey following intrareservoir injection and each set of double pump sequences. (A,B) Activity outlining reservoir and catheter after isotope injection; (C,D) radionuclide activity in fourth ventricle, cisterna magna, and adjacent spinal CSF spaces following two reservoir depressions; (E,F) activity extending into third and lateral ventricles after four reservoir depressions; (G,H) minimal alteration of radiopharmaceutical migration after six pumps as compared to activity distribution following four reservoir depressions. Reprinted from Wood *et al.*,[108] with permission.

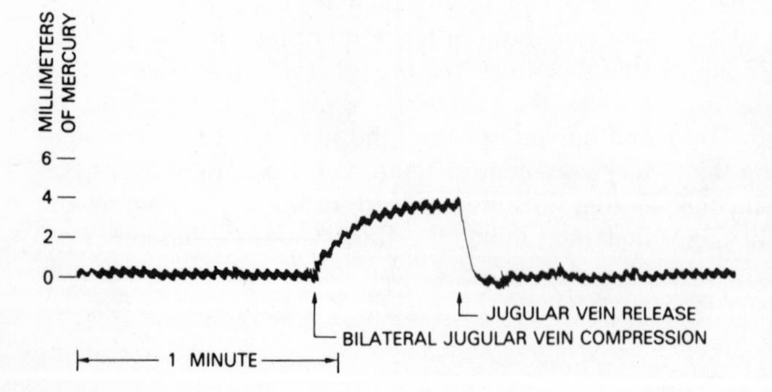

Figure 19. Fourth ventricular pressure tracing recorded from subcutaneous Ommaya CSF reservoir calibrated at level of foramen magnum. Note demonstration of CSF pressure fluctuations in response to both arterial pulsations and respiratory activity. Also note bilateral jugular vein compression resulting in rapid elevation of intraventricular pressure followed by prompt return of pressure to baseline level after jugular vein release. Reprinted from Wood *et al.*,[108] with permission.

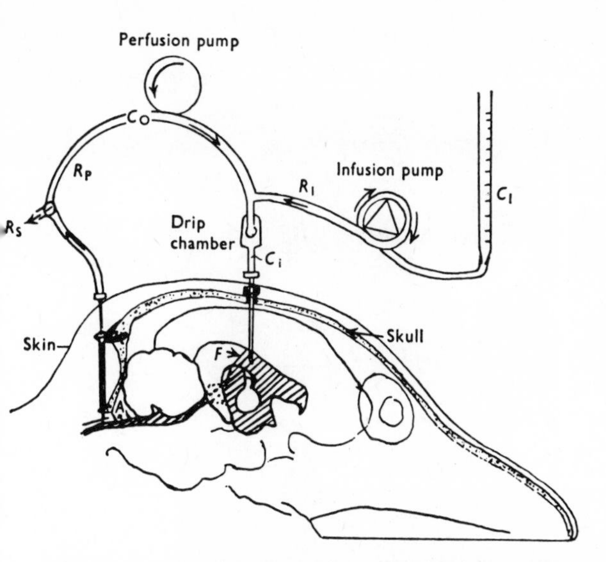

Figure 20. Diagram of ventriculocisternal perfusion in dog. CSF obtained from anesthetized dog by cisternal puncture is pumped into perfusion circuit for priming and is also employed as solvent for infusion solution. Infusate is drawn up into reservoir, with care taken to avoid air bubbles in system. The 26-gauge needle of infusion system is then inserted through wall of tubing of perfusion circuit just above drip chamber. (C_I) Infusion fluid concentration; (Co) cisternal outflow fluid concentration; (C_i) ventricular inflow fluid concentration; (R_I) volume infused/unit time; (R_p) rate of perfusion; (R_s) average rate of sampling (which is 0.0167 ml/min). Reprinted from Ashcroft *et al.*,[4] with permission.

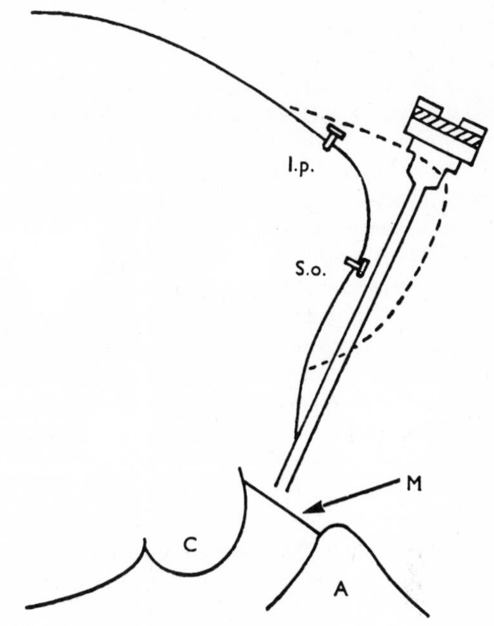

Figure 21. Diagram of cisternal guide cannula with its tip resting about 1 mm above atlanto-occipital membrane (M). (– – –) Outline of acrylic cement that permanently fixes cannula to back of skull and four small anchoring screws (two shown). (I.p.) Interparietal bone; (S.o.) supraoccipital bone; (C) occipital condyle; (A) atlas. Reprinted from Feldberg *et al.*,[32] with permission.

neath the cup; however, this method requires extensive operative manipulation of the surrounding brain.

Contrarily, Wood *et al.*[107] have described the installation in the spinal subarachnoid space of a T-shaped Hoffman catheter (Holter Company) that is attached to a subcutaneous CSF reservoir (Fig. 22). This system, when combined with the previously discussed fourth ventricular cannula and CSF reservoir, may provide a model for spino–fourth ventricular perfusion.

Myers's review[70] stressed several considerations in experimental perfusions. The recommended rates of infusion of perfusates in cats and monkeys are 0.05 or 0.1 ml/min and 0.2 ml/min, respectively. The rate of outflow from the cisternal catheter varies with the level at which the open end is held; thus, this end should be maintained at approximately the level of the foramen magnum to avoid the siphoning effect. Slowing of CSF outflow suggests partial occlusion, and a lack of CSF pulsation on elevation of

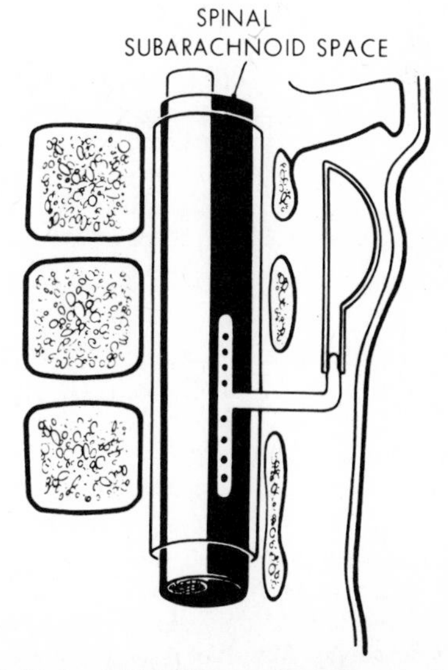

Figure 22. T-shaped Hoffman catheter in spinal subarachnoid space connected to subcutaneous Ommaya CSF reservoir for chronic intrathecal infusions or drug injections. Reprinted from Wood *et al.*,[107] with permission. ©1977 by the American Association for the Advancement of Science.

the outflow tubing implies cannula occlusion. Outflow catheter occlusion produces increased intracranial pressure to which the animal usually vocalizes, retches, becomes agitated, or convulses. If occlusion occurs, then the cannulae should be withdrawn and flushed with sterile mock CSF. The presence of bloody CSF flowing from the outflow cannula necessitates termination of the experiment.

3.3.2. Perfusion of Subarachnoid Spaces. Beleslin and Myers[6] developed a stainless steel reservoir to repeatedly superfuse small isolated areas of the exposed cortex of unanesthetized animals (Fig. 23). This reservoir is capped with a watertight rubber diaphragm that enables the insertion of 21-gauge inflow and outflow needles to a position of approximately 2 mm above the cortical surface. The rate of outflow should be identical to the 0.01–0.1 ml/min infusion rate. The cortex should be perfused for 20–30 min prior to fluid sampling to remove accumulated metabolites and to establish steady-state neurohumoral concentrations. Diffusion of substances from the cortex into the reservoir fluid may be augmented by perforation of the pia mater in areas void of blood vessels.

Levin *et al.*[59] have described a method for perfusing the nonisolated cerebral subarachnoid space of larger animals (rabbits, cats, dogs, monkeys). The

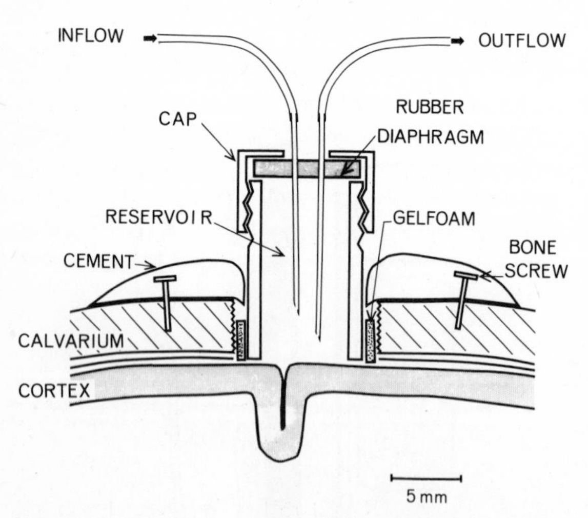

Figure 23. Diagram of stainless steel perfusion reservoir resting on surface of cerebral cortex. Dura mater is excised to edge of craniectomy. Note gelfoam placed between calvarial edge and external wall of reservoir to prevent cranioplast cement from making contact with bone or pia–arachnoid. Maintenance of equality between inflow and outflow rates enables cortical perfusion without compression. Reprinted from Beleslin and Myers,[6] with permission.

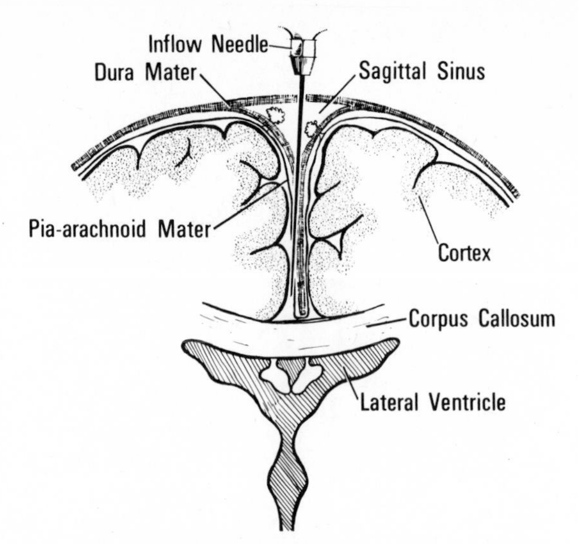

Figure 24. Diagram of approximate position of inflow needle in supracallosal subarachnoid space. Note that insertion of inflow needles through superior sagittal sinus causes no apparent intracranial bleeding. Reprinted from Levin *et al.*,[59] with permission.

animals are placed in stereotaxic head-holders, and trephines are made in the midline of the skull. A 20- to 22-gauge Pitkin spinal needle with stylette is then inserted through the sagittal sinus and positioned so that the tip is in the supracallosal subarachnoid space (Fig. 24). This method of inflow-needle placement is not associated with intracranial bleeding. A 19- to 20-gauge needle is percutaneously inserted into the cisterna magna for outflow sampling. A slightly positive outflow pressure is maintained by elevating the PE tubing attached to the outflow needle 10 mm above the level of the foramen magnum. The infusion rates for rabbits, cats, dogs, and monkeys are 0.08, 0.16, 0.32, and 0.35 ml/min, respectively.

Edvinsson *et al.*[27] reported alterations in both the intracranial pressure and the blood–brain barrier and production of brain edema after the implantation of brain cannulae in conscious animals. Malkinson *et al.*[62] recently modified the "Richmond screw"[98] to monitor intracranial pressure in the unanesthetized, minimally restrained rabbit. The subarachnoid screw is installed into a hole in the skull placed 6 mm posterior to the coronal suture and 4 mm lateral to the midline after perforation of the underlying dura mater (Fig. 25). The intracranial pressure is transmitted from the subarachnoid space within the cranium via a flexible saline-filled catheter to a fixed external physiological pressure trans-

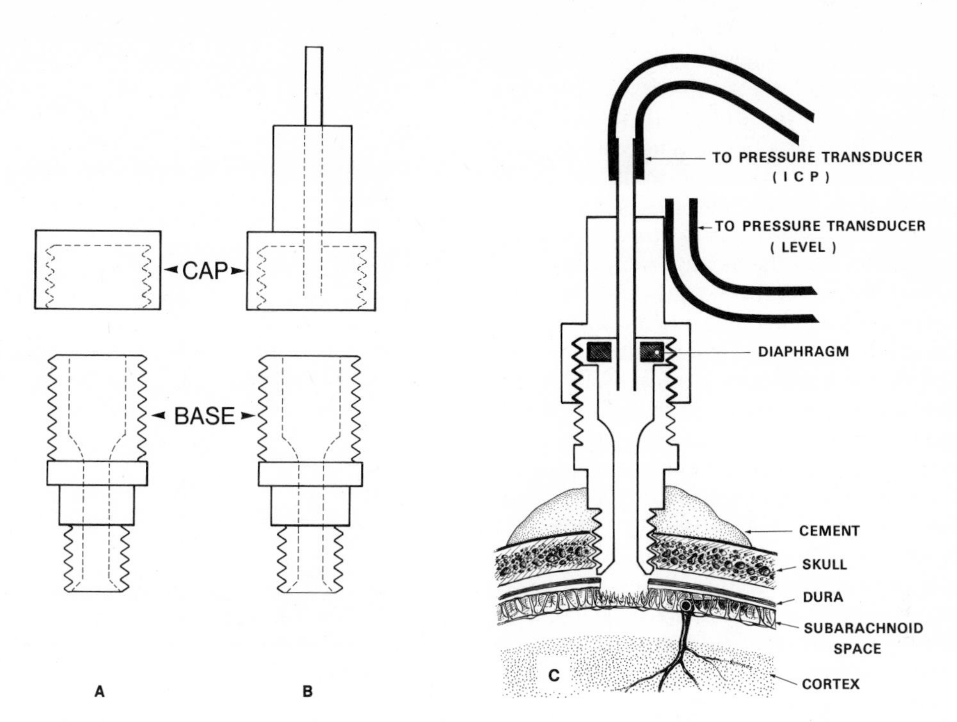

Figure 25. Diagram of subarachnoid screw system employed to monitor intracranial pressure in unanesthetized animals. (A) Base and protective cap; (B) base and experimental cap; (C) base and experimental cap installed in twist drill hole and communicating with subarachnoid space through dural perforation. Reference level of transducer is skull surface at point of attachment of base assembly. Reprinted from Malkinson *et al.*,[62] with permission.

ducer referenced with respect to the screw's position on the skull. Small amounts of subarachnoid fluid may be withdrawn from this screw for neurochemical analysis.

3.3.3. Documentation of Perfusate Distribution. Following regional ventricular or subarachnoid perfusion with both mock CSF and the substance under investigation, the precise anatomical distribution of the perfusate must be documented to determine which brain structures may have been exposed to the substance in the perfusate. Several investigators[31,70] recommend the substitution of 0.2% bromophenol blue dye for 15 min followed by replacement of the original perfusate in the inflow cannula. The rate of perfusion of this vital dye solution should be the same as that employed in the experiment. Bromophenol blue dye stains exposed neural tissue, but the color may bleach out if its exposure to formalin is prolonged. The extent of this vital dye staining should be recorded immediately following formalin intravascular perfusion and extraction of the brain.

3.3.4. Control Groups. In addition to the sham-operated and nonimplanted control animals, Myers[70] recommends the inclusion of both pharmacological and anatomical control groups in experimental protocols. Pharmacological control animals should be perfused with only mock CSF, whereas the anatomical control animals in ventricular-perfusion studies should undergo perfusion of the subarachnoid space. Contrarily, in subarachnoid-perfusion investigations, ventricular perfusion of a group of animals would likewise offer the opportunity for anatomical comparison.

4. Conclusions

This chapter has briefly reviewed the sources of CSF neurochemical variability that may be operant during periods of clinical CSF evaluation. Circadian rhythms, ventriculospinal concentration gradients, impediments of normal CSF circulation, physical activity, stress, medications, precursor intake, concomitant illness, age, and possibly sex may alter the baseline neurochemical composition of CSF. Differential probenecid blockade may confuse monoamine metabolite accumulation interpretations. In addition, decomposition of CSF constituents during collection, storage, and analysis may add artifactual variations to CSF data. Methodology has been presented that minimizes these physiological and pharmacological sources of CSF variability in an effort to aid investigators in protocol formulation.

Many of these sources of artifacts pertain to animal model systems. Anesthesia, stress of immobilization, breakdown of the blood–brain barrier during cannula implantation, and infection may also alter CSF composition. Insufficient volumes of available CSF and recurrent cannula obstruction may preclude CSF analysis in some animal systems. The experimental aspects of intraventricular or intrathecal drug injection, infusion, and perfusion have been discussed with respect to animal model and control group selection. In addition, the technical aspects of the implantation of reliable devices that allow chronic CSF access in conscious, unrestrained animals have been described in detail. It is hoped that this chapter has enabled the reader to choose the animal system that is optimal for his experimental needs and that will produce valid data mimicking the human situation.

ACKNOWLEDGMENTS. The author is grateful to Drs. Joseph D. Fenstermacher, Victor A. Levin, Robert D. Myers, Michael H. Ebert, K. Faulhauer, E. Donauer, T. J. Malkinson, and Hymie Anisman for submitting some of the illustrations in this chapter.

References

1. ANDEN, N. E., ROOS, B.-E., WERDINIUS, B.: On the occurrence of homovanillic acid in brain and cerebrospinal fluid and its determination by a fluorometric method. *Life Sci.* **2:**448–458, 1963.
2. ANDERSON, B., MCCANN, S. M.: A further study of polydipsia evoked by hypothalamic stimulation in the goat. *Acta Physiol. Scand.* **33:**333–346, 1955.
3. ANDERSON, H., ROOS, B. E.: 5-Hydroxyindoleacetic acid in cerebrospinal fluid of hydrocephalic children. *Acta Paediatr. Scand.* **58:**601–608, 1969.
4. ASHCROFT, G. W., DOW, R. C., MOIR, A. T. B.: The active transport of 5-hydroxyindol-3-ylacetic acid and 3-methoxy-4-hydroxyphenylacetic acid from a recirculatory perfusion system of the cerebral ventricles of the unanaesthetized dog. *J. Physiol. (London)* **199:**397–425, 1968.

4a. BALLENGER, J. C., POST, R. M., STERNBERG, D. E., VAN KAMMEN, D. P., COWDRY, R. W., GOODWIN, F. K.: Headaches after lumbar punctures and insensitivity to pain in psychiatric patients. *N. Engl. J. Med.* **301:**110, 1979.

5. BATNITZKY, S., KAUCHER, T. R., MEALEY, J., CAMPBELL, R. L.: Iatrogenic intraspinal epidermoid tumors. *J. Am. Med. Assoc.* **237:**148–150, 1977.
6. BELESLIN, D. B., MYERS, R. D.: A technique for repeated superfusion or withdrawal of fluid from the exposed cerebral cortex of a conscious animal. *Physiol. Behav.* **5:**1173–1175, 1970.
7. BERRY, H. C., STEINER, J. C.: L-Glutamine increases CSF GABA in patient with Huntington's disease. *Neurology (Minneapolis)* **29:**535, 1979.
8. BESSON, M. J., CHERAMY, A., FELTZ, P., GLOWINSKI, J.: Dopamine: Spontaneous and drug-induced release from the caudate nucleus in the cat. *Brain Res.* **32:**407–424, 1971.
9. BHATTACHARYA, B. K., FELDBERG, W.: Perfusion of the ventricular system of the brain in the anaesthetized cat. *J. Physiol. (London)* **135:**4P–5P, 1957.
10. BOHLEN, P., HUOT, S., PALFREYMAN, M. G.: The relationship between GABA concentrations in brain and cerebrospinal fluid. *Brain Res.* **167:**297–305, 1979.
11. BOWERS, M. B.: Clinical measurements of central dopamine and 5-hydroxytryptamine metabolism: Reliability and interpretation of cerebrospinal fluid acid monoamine metabolite measures. *Neuropharmacology* **11:**101–111, 1972.
12. BOWERY, N. G., LEWIS, G. P.: Pharmacological activity in polyvinyl chloride (PVC) tubing. *Br. J. Pharmacol.* **34:**207, 1968.
13. BRIGHTMAN, M. W.: The distribution within the brain of ferritin injected into the cerebrospinal fluid compartments. *J. Cell Biol.* **26:**99–123, 1965.
14. BROOKS, B. R., WOOD, J. H., DIAZ, M., CZERWINSKI, C., GEORGES, L. P., SODE, J., EBERT, M. H., ENGEL, W. K.: Extracellular cyclic nucleotide metabolism in the human central nervous system. In Wood, J. H. (ed.): *Neurobiology of Cerebrospinal Fluid I.* New York, Plenum Press, 1980.
15. BURTON, B. T., (ed.): *Human Nutrition,* 3rd ed. New York, McGraw-Hill, 1976, pp. 45–60.
16. CARACENI, T., CALDERINI, G., CONSOLAZIONE, A., RIVA, E., ALGERI, S., GIROTTI, F., SPREAFICO, R., BRANCIFORTI, A., DALL'OLIO, A., MORSELLI, P. L.: Biochemical aspects of Huntington's chorea. *J. Neurol. Neurosurg. Psychiatry* **40:**581–587, 1977.
17. CARMICHAEL, E. A., FELDBERG, W., FLEISCHHAUER, K.: Methods for perfusing different parts of the cat's cerebral ventricles with dogs. *J. Physiol. (London)* **173:**354–367, 1965.
18. CHASE, T. N., NG, L. K. Y.: Central monoamine metabolism in Parkinson's disease. *Arch. Neurol.* **27:**486–491, 1972.
19. COHEN, D. J., CAPARULO, B. K., SHAYWITZ, B. A., BOWERS, M. B.: Dopamine and serotonin metabolism in neuropsychiatrically disturbed children. *Arch. Gen. Psychiatry* **34:**545–550, 1977.
20. DEGROOT, J.: The rat forebrain in stereotaxic coordinates. *Proc. K. Ned. Akad. Wet. C* **52:**1–40, 1959.

21. DRIPPS, R. D., VANDAM, L. D.: Hazards of lumbar puncture. *J. Am. Med. Assoc.* **147**:1118–1121, 1951.
22. DUTTERA, M. J., GALLELLI, J. F., KLEINMAN, L. M., TANGREA, J. A., WITTGROVE, A. C.: Intrathecal methotrexate. *Lancet* **1**:540, 1972.
23. EBERT, M. H., HAVENS, W. W., MATTHYSSE, S. W., DENNING, R., SIEGEL, P.: A technique for serial collection of lateral ventricle cerebrospinal fluid of a conscious rhesus macaque. *J. Appl. Physiol.* (in press).
24. EBERT, M. J., KARTZINEL, R., COWDRY, R. W., GOODWIN, F. K.: Cerebrospinal fluid amine metabolites and the probenecid test. In Wood, J. H. (ed.): *Neurobiology of Cerebrospinal Fluid I.* New York, Plenum Press, 1980.
25. ECCLESTON, D., ASHCROFT, G. W., CRAWFORD, T. B. B.: 5-Hydroxyindole metabolism in rat brain; a study of intermediate metabolism using the technique of tryptophan loading. II. Applications and drug studies. *J. Neurochem.* **12**:493–503, 1965.
26. EDELSON, R. N., CHERNIK, N. L., POSNER, J. B.: Spinal subdural hematomas complicating lumbar punctures. *Arch. Neurol.* **31**:134–137, 1974.
27. EDVINSSON, L., NIELSEN, K. C., OWMAN, C. H., WEST, K. A.: Alterations in intracranial pressure, blood–brain barrier, and brain edema after sub-chronic implantation of a cannula into the brain of conscious animals. *Acta Physiol. Scand.* **82**:527–531, 1971.
28. ELLIOTT, K. A. C., JASPER, H. H.: Physiological salt solutions for brain surgery. *J. Neurosurg.* **6**:140–143, 1949.
29. ENNA, S. J., WOOD, J. H., SNYDER, S. H.: γ-Aminobutyric acid (GABA) in human cerebrospinal fluid: Radioreceptor assay. *J. Neurochem.* **28**:1121–1124, 1977.
30. ENNA, S. J., ZIEGLER, M. G., LAKE, C. R., WOOD, J. H., BROOKS, B. R., BUTLER, I. J.: Cerebrospinal fluid γ-aminobutyric acid: Correlation with cerebrospinal fluid and blood constituents and alterations in neurological disorders. In Wood, J. H. (ed.): *Neurobiology of Cerebrospinal Fluid I.* New York, Plenum Press, 1980.
30a. FAULHAUER, K., DONAUER, E.: Permanent ventriculostomy in cats, technical note. *Acta Neurochir.* **46**:169–172, 1979.
31. FELDBERG, W., FLEISCHHAUER, K.: Penetration of bromophenol blue from the perfused cerebral ventricles into the brain tissue. *J. Physiol.* **150**:451–462, 1960.
32. FELDBERG, W., MYERS, R. D., VEALE, W. L.: Perfusion from cerebral ventricle to cisterna magna in the unanesthetized cat; effect of calcium on body temperature. *J. Physiol. (London)* **207**:403–416, 1970.
33. FELDBERG, W., SHERWOOD, S. L.: A permanent cannula for intraventricular injections in cats. *J. Physiol. (London)* **120**:3P–5P, 1953.
34. FELDBERG, W., SHERWOOD, S. L.: Injections of drugs into the lateral ventricle of the cat. *J. Physiol. (London)* **123**:148–167, 1954.
35. FENCL, V.: Ion exchange between blood and interstitial fluid in the brain. In Loeschcke, H. H. (ed.): *Acid–Base Homeostasis of the Brain Extracellular Fluid and the Respiratory Control System.* Littleton, Massachusetts, Publishing Sciences Group, 1976, pp. 15–18.
36. FENSTERMACHER, J. D., LI, C.-L., LEVIN, V. A.: The extracellular space of the cerebral cortex of normothermic and hypothermic cats. *Exp. Neurol.* **27**:101–114, 1970.
37. FENSTERMACHER, J. D., RALL, D. P., PATLAK, C. S., LEVIN, V. A.: Ventriculocisternal perfusion as a technique for analysis of brain capillary permeability and extracellular transport. In Crone, C., Lassen, N. A. (eds.): *Capillary Permeability, Alfred Benzon Symposium II.* Copenhagen, Munksgaard, 1970, pp. 483–490.
38. FISHMAN, R. A.: Cerebrospinal fluid. In Baker, A. B., Baker, L. H. (eds.): *Clinical Neurology.* Hagerstown, Maryland, Harper & Row, 1977, Chapter 5, pp. 1–40.
39. GEISER, C. F., BISHOP, Y., JAFFE, N., FURMAN, L., TRAGGIS, D., FREI, E.: Adverse effects of intrathecal methotrexate in children with acute leukemia in remission. *Blood* **45**:189–195, 1975.
40. GERBODE, F. A., BOWERS, M. B.: Measurement of acid monoamine metabolites in human and animal cerebrospinal fluid. *J. Neurochem.* **15**:1053–1055, 1968.
41. GILLAND, O., TOURTELLOTTE, W. W., O'TAUMA, L., HENDERSON, W. G.: Normal cerebrospinal fluid pressure. *J. Neurosurg.* **40**:587–593, 1974.
42. GLAESER, B. S., HARE, T. A.: Measurement of GABA in human cerebrospinal fluid. *Biochem. Med.* **12**:274–282, 1975.
43. GLAESER, B. S., VOGEL, D. B., OLEWEILER, D. B., HARE, T. A.: GABA levels in cerebrospinal fluid of patients with Huntington's chorea: A preliminary report. *Biochem. Med.* **12**:380–385, 1975.
44. GORDON, E. K., OLIVER, J., BLACK, K., KOPIN, I. J.: Simultaneous assay by mass fragmentography of vanillyl mandelic acid, homovanillic acid and 3-methoxy-4-hydroxyphenylethylene glycol in cerebrospinal fluid and urine. *Biochem. Med.* **11**:32–40, 1974.
45. GRANT, W. T., CONE, W. V.: Graduated jugular compression in the lumbar manometric test for spinal subarachnoid block. *Arch. Neurol. Psychiatry* **32**:1194–1201, 1934.
46. GRIAUZDE, M., RADULOVACKI, M.: 5-Hydroxyindoleacetic acid and homovanillic acid in cisternal fluid of cats subjected to stress. *J. Neurochem.* **26**:1301–1306, 1976.
47. GROSSMAN, M. H., HARE, T. A., MANYAM, N. V. B., GLAESER, B. S., WOOD, J. H.: Stability of GABA levels in CSF under various conditions of storage. *Brain Res.* **182**:99–106, 1980.
48. HALEY, T. J., DICKINSON, R. W.: A note on an indwelling cannula for intraventricular injection of drugs in the unanesthetized dog. *J. Am. Pharm. Assoc.* **45**:432, 1956.
49. HARE, T. A., MANYAM, N. V. B., GLAESER, B. S.: Evaluation of cerebrospinal fluid γ-aminobutyric acid con-

tent in neurological and psychiatric disorders. In Wood, J. H. (ed.): *Neurobiology of Cerebrospinal Fluid I*. New York, Plenum Press, 1980.

50. Hare, T. A., Wood, J. H., Manyam, N. V. B., Ballenger, J. C., Post, R. M., Gerner, R. H.: Selection of control populations for clinical cerebrospinal fluid GABA investigations based on comparison with normal volunteers. *Brain Res. Bull.* (in press).
51. Hockwald, G. M., Wald, A., DiMattio, J., Malhan, C.: The effects of serum osmolarity on cerebrospinal fluid volume flow. *Life Sci.* **15**:1309–1316, 1974.
52. Jasper, H. H., Ajmone-Marsan, C.: Diencephalon of the cat. In Sheer, D. E. (ed.): *Electrical Stimulation of the Brain*. Austin, University of Texas Press, 1961, pp. 203–231.
53. Klee, J. G., Praestholm, J.: A simple, reliable freehand method for cerebral ventricular puncture in the rat. *Invest. Radiol.* **9**:109–110, 1974.
54. Kokkinidis, L., Raffler, L., Anisman, H.: Simple and compact cannula system for mice. *Pharmacol. Biochem. Behav.* **6**:595–597, 1977.
55. Korein, J., Cravioto, H., Leicach, M.: Re-evaluation of lumbar puncture, a study of 129 patients with papilledema or intracranial hypertension. *Neurology (Minneapolis)* **9**:290–297, 1959.
56. Lake, C. R., Ballenger, J. C., Ziegler, M. G., Post, R. M., van Kammen, D. P., Ebert, M. H.: Pitfalls in assessing brain monoamine metabolism in man. Presented at Annual Meeting of Society for Neuroscience, St. Louis, Missouri, November 5–9, 1978.
57. Leal, K. W., Troupin, A. S.: Clinical pharmacology of anti-epileptic drugs, a summary of current information. *Clin. Chem.* **23**:1964–1968, 1977.
58. Leavens, M. E., Aldama-Luebert, A.: Ommaya reservoir placement; technical note. *Neurosurgery* **5**:264–266, 1979.
59. Levin, V. A., Fenstermacher, J. D., Patlak, C. S.: Sucrose and inulin space measurement of cerebral cortex in four mammalian species. *Am. J. Physiol.* **219**:1528–1533, 1970.
60. Lups, S., Haan, A. M. F. H., Bailey, P.: *The Cerebrospinal Fluid*. New York, Elsevier, 1954, pp. 48–64.
61. Malkinson, T. J., Jackson-Middelkoop, L. M., Veale, W. L.: A simple multi-purpose cannula system for access to the brain and/or systemic vascular system of unanesthetized animals. *Brain Res. Bull.* **2**:57–59, 1977.
62. Malkinson, T. J., Veale, W. L., Cooper, K. E.: Measurement of intracranial pressure in the unanesthetized rabbit. *Brain Res. Bull.* **3**:635–638, 1978.
63. Maynert, E. W., Marczynski, T. J., Browning, R. A.: The role of the neurotransmitters in the epilepsies. In Friedlander, W. J. (ed.): *Advances in Neurology, Current Reviews*, Vol. 13. New York, Raven Press, 1975, pp. 79–147.

63a. Milhorat, T. H.: Pediatric neurosurgery. In Plum, F., McDowell, F. H. (eds.): *Contemporary Neurology Series*, Vol. 16. Philadelphia, F. A. Davis, 1978, pp. 8–12.

64. Modigh, K.: The relationship between the concentrations of tryptophan and 5-hydroxyindoleacetic acid in rat brain and cerebrospinal fluid. *J. Neurochem.* **25**:351–352, 1975.
65. Mullan, S., Harper, P. V., Hekmatpanah, J., Torres, H., Dobbin, G.: Percutaneous interruption of spinal-pain tracts by means of strontium-90 needle. *J. Neurosurg.* **20**:931–939, 1963.
66. Myers, R. D.: Blood–brain barrier: Technique for the intracerebral administration of drugs. In Iversen, L. L., Iversen, S. D., Snyder, S. H. (eds.): *Handbook of Psychopharmacology*, Vol. 2. New York, Plenum Press, 1975, pp. 1–28.
67. Myers, R. D., Brophy, P. D.: Temperature changes in the rat produced by altering the sodium–calcium ratio in the cerebral ventricles. *Neuropharmacology* **11**:351–361, 1972.
68. Myers, R. D., Casaday, G., Holman, R. B.: A simplified intracranial cannula for chemical stimulation or long-term infusion of the brain. *Physiol. Behav.* **2**:87–88, 1967.
69. Myers, R. D.: Chronic methods: Intraventricular infusion, cerebrospinal fluid sampling, and push–pull perfusion. In Myers, R. D. (ed.): *Methods in Psychobiology*, Vol. 3. New York, Academic Press, 1977, pp. 281–315.
70. Myers, R. D.: Methods of perfusing different structures of the brain. In Myers, R. D. (ed.): *Methods in Psychobiology*, Vol. 2. New York, Academic Press, 1972, pp. 69–211.
71. Myers, R. D., Veale, W. L., Yaksh, T. L.: Preference for ethanol in the rhesus monkey following chronic infusion of ethanol into the cerebral ventricles. *Physiol. Behav.* **8**:431–435, 1972.
72. Olszewski, J.: *The Thalamus of the Macaca mulatta; An Atlas for Use with the Stereotaxic Instrument*. Basel, Karger, 1952.
73. Ommaya, A. K.: Subcutaneous reservoir and pump for sterile access to ventricular cerebrospinal fluid. *Lancet* **2**:983–984, 1963.
74. Palmer, A. C.: Injection of drugs into the cerebral ventricles of sheep. *J. Physiol. (London)* **149**:209–214, 1959.
75. Pappenheimer, J. R., Heisey, S. R., Jordan, E. F., Downer, J. deC.: Perfusion of the cerebral ventricular system in unanesthetized goats. *Am. J. Physiol.* **203**:763–744, 1962.
76. Pappius, H. M., Savaki, H. E., Fieschi, C., Rapoport, S. I., Sokoloff, L.: Osmotic opening of the blood–brain barrier and local cerebral glucose utilization. *Ann. Neurol.* **5**:211–219, 1979.
77. Patten, J.: *Neurological Differential Diagnosis*. New York, Springer-Verlag, 1977, pp. 259–266.
78. Poplack, D. G., Bleyer, W. A., Wood, J. H., Kostolich, M., Savitch, J., Ommaya, A. K.: A primate

model for study of methotrexate pharmacokinetics in the central nervous system. *Cancer Res.* **37**:1982–1985, 1977.
79. Post, K. D., Stein, B. M.: Technique for spinal drainage. *Neurosurgery* **4**:255, 1979.
80. Post, R. M., Allen, F. H., Ommaya, A. K.: Cerebrospinal flow and iodide[131] transport in the spinal subarachnoid space. *Life Sci.* **14**:1885–1894, 1974.
81. Post, R. M., Ballenger, J. C., Goodwin, F. K.: Cerebrospinal fluid studies of neurotransmitter function in manic and depressive illness. In Wood, J. H. (ed.): *Neurobiology of Cerebrospinal Fluid I.* New York, Plenum Press, 1980.
82. Post, R. M., Goodwin, F. K., Gordon, E., Watkin, D. M.: Amine metabolites in human cerebrospinal fluid: Effects of cord transection and spinal fluid block. *Science* **179**:897–899, 1973.
83. Post, R. M., Kotin, J., Goodwin, F. K., Gordon, E. K.: Psychomotor activity and cerebrospinal fluid amine metabolites in affective illness. *Am. J. Psychiatry* **130**:67–72, 1973.
84. Perlow, M. J., Lake, C. R.: Daily fluctuations in cerebrospinal fluid concentrations of catecholamines, monoamine metabolites, cyclic AMP, and γ-aminobutyric acid in rhesus monkeys. In Wood, J. H. (ed.): *Neurobiology of Cerebrospinal Fluid I.* New York, Plenum Press, 1980.
85. Post, R. M., Lake, C. R., Jimerson, D. C., Bunney, W. E., Wood, J. H., Ziegler, M. G., Goodwin, F. K.: Cerebrospinal fluid norepinephrine in affective illness. *Am. J. Psychiatry* **135**:907–912, 1978.
86. Queckenstedt, H.: Zur Diagnose der Rückenmarkskompression. *Dtsch. Z. Nervenheilkd.* **55**:325–330, 1916.
87. Radulovacki, M., Girgis, M.: The permanent cannula to cisterna magna in cats. *Sudan Med. J.* **6**:170–173, 1968.
88. Reppert, S. M., Perlow, M. J., Klein, D. C.: Cerebrospinal fluid melatonin. In Wood, J. H. (ed.): *Neurobiology of Cerebrospinal Fluid I.* New York, Plenum Press, 1980.
89. Robertson, J. T., Denton, I. C.: Surgical considerations of ventriculography. In Youmans, J. R. (ed.): *Neurological Surgery.* Philadelphia, W. B. Saunders, 1973, pp. 220–234.
90. Robinson, D. S., Sourkes, T. L., Nies, A., Harris, L. S., Spector, S., Bartlett, D. L., Kaye, I. S.: Monoamine metabolism in human brain. *Arch. Gen. Psychiatry* **34**:89–92, 1977.
91. Rogers, J. J., Dubowitz, V.: 5-Hydroxyindoles in hydrocephalus; a comparative study of cerebrospinal fluid and blood levels. *Dev. Med. Child. Neurol.* **12**:461–466, 1970.
92. Rudman, D., O'Brien, M. S., McKinney, A. S., Hoffman, J. C., Patterson, J. H.: Observations of the cyclic nucleotide concentrations in human cerebrospinal fluid. *J. Clin. Endocrinol. Metab.* **42**:1088–1097, 1976.
93. Sklar, F. H.: Non-steady-state measurements of cerebrospinal fluid dynamics: laboratory and clinical applications. In Wood, J. H. (ed.): *Neurobiology of Cerebrospinal Fluid I.* New York, Plenum Press, 1980.
94. Spurling, R. G.: *Practical Neurological Diagnosis,* 6th ed. Springfield, Illinois, Charles C. Thomas, 1960, pp. 204–206.
95. Steiner, A. L., Pagliari, A. S., Chase, L. R., Kipnis, D. M.: Radioimmunoassay for cyclic nucleotides. II. Adenosine 3′,5′-monophosphate and guanosine 3′,5′-monophosphate in mammalian tissues and body fluids. *J. Biol. Chem.* **247**:1114–1120, 1972.
96. Tourtellotte, W. W., Haerer, A. F., Heller, G. L., Somers, J. E.: *Post-Lumbar Puncture Headaches.* Springfield, Illinois, Charles C. Thomas, 1964.
97. Veale, W. L., Myers, R. D.: Emotional behavior, arousal and sleep produced by sodium and calcium ions perfused within the hypothalamus of the cat. *Physiol. Behav.* **7**:601–607, 1971.
98. Vries, J. K., Becker, D. P., Young, H. F.: A subarachnoid screw for monitoring intracranial pressure. *J. Neurosurg.* **39**:416–419, 1973.
99. Watson, E., Wilk, S.: Determination of probenecid in small volume of cerebrospinal fluid. *J. Neurochem.* **21**:1569–1571, 1973.
100. Welch, K. M. A., Meyer, J. S., Chee, A. N. C.: Evidence for disordered cyclic AMP metabolism in patients with cerebral infarction. *Eur. Neurol.* **13**:144–154, 1975.
101. White, R. J., Dakters, J. G., Yashon, D., Albin, M. S.: Temporary control of cerebrospinal fluid volume and pressure by means of an externalized valve-drainage system. *J. Neurosurg.* **30**:264–269, 1969.
102. Winters, W. D., Kado, R. T., Adey, W. R.: *Stereotaxic Brain Atlas for Macaca nemistrina.* Berkeley, University of California Press, 1969.
103. Wolcott, G. J., Grunnet, M. L., Lahey, M. E.: Spinal subdural hematoma in a leukemic child. *J. Pediatr.* **77**:1062, 1970.
104. Wood, J. H., Glaeser, B. S., Enna, S. J., Hare, T. A.: Verification and quantification of GABA in human cerebrospinal fluid. *J. Neurochem.* **30**:291–293, 1978.
105. Wood, J. H., Hare, T. A., Enna, S. J., Manyam, N. V. B.: Sites of origin and rostrocaudal concentration gradients of GABA in cerebrospinal fluid. *Brain Res. Bull.* (in press).
106. Wood, J. H.: Physiology, pharmacology, and dynamics of CSF. In Wood, J. H. (ed.): *Neurobiology of Cerebrospinal Fluid I.* New York, Plenum Press, 1980.
107. Wood, J. H., Poplack, D. G., Bleyer, W. A., Ommaya, A. K.: Primate model for long-term study of intraventricularly and intrathecally administered drugs and intracranial pressure. *Science* **195**:499–501, 1977.
108. Wood, J. H., Poplack, D. G., Flor, W. J., Gunby, E. N., Ommaya, A. K.: Chronic ventricular cerebro-

spinal fluid sampling, drug injections and pressure monitoring using subcutaneous reservoirs in monkeys. *Neurosurgery* **1**:132–135, 1977.
109. WOOD, J. H.: Neurochemical analysis of cerebrospinal fluid. *Neurology (Minneapolis)* **30**:645–651, 1980.
110. WOOD, J. H.: Sites of origin and concentration gradients of CSF neurotransmitters, their precursors and metabolites, and CSF cyclic nucleotides. In Wood, J. H. (ed.): *Neurobiology of Cerebrospinal Fluid I.* New York, Plenum Press, 1980.
111. WOOD, J. H., ZIEGLER, M. G., LAKE, C. R., SHOULSON, I., BROOKS, B. R., VANBUREN, J. M.: Cerebrospinal fluid norepinephrine reductions in man after degeneration and electrical stimulation of the caudate nucleus. *Ann. Neurol.* **1**:94–99, 1977.
112. ZIEGLER, M. G., LAKE, C. R., FOPPEN, F. H., SHOULSON, I., KOPIN, I. J.: Norepinephrine in cerebrospinal fluid. *Brain Res.* **108**:436–440, 1976.
113. ZIEGLER, M. G., LAKE, C. R., WOOD, J. H., BROOKS, B. R.: Relationship between cerebrospinal fluid norepinephrine and blood pressure in neurologic patient. *Clin. Exp. Hypertension* (in press).
114. ZIEGLER, M. G., LAKE, C. R., WOOD, J. H., EBERT, M. H.: Circadian rhythm in cerebrospinal fluid noradrenaline of man and monkey. *Nature* **264**:656–658, 1976.
115. ZIEGLER, M. G., LAKE, C. R., WOOD, J. H., EBERT, M. H.: Norepinephrine in cerebrospinal fluid: Basic studies, effects of drugs and disease. In Wood, J. H. (ed.): *Neurobiology of Cerebrospinal Fluid I.* New York, Plenum Press, 1980.
116. ZIEGLER, M. G., WOOD, J. H., LAKE, C. R., KOPIN, I. J.: Norepinephrine and 3-methoxy-4-hydroxyphenyl glycol gradients in human cerebrospinal fluid. *Am. J. Psychiatry* **134**:565–568, 1977.
117. ZIVIN, J. A.: Lateral cervical puncture, an alternative to lumbar puncture. *Neurology (Minneapolis)* **28**:616–618, 1978.

CHAPTER 8

Cerebrospinal Fluid Amine Metabolites and the Probenecid Test

Michael H. Ebert, Ronald Kartzinel, Rex W. Cowdry, and Frederick K. Goodwin

1. Introduction

In the past decade, studies of cerebrospinal fluid (CSF) amine metabolites have become an important aspect of clinical metabolic studies of central nervous system disease. Investigators have pursued these studies with the hope of obtaining a direct measure of central nervous system metabolism to test hypotheses concerning putative neurotransmitter abnormalities in various neurological and psychiatric illnesses. These hypotheses usually arise indirectly from clinical trials of drugs with effects on amine biochemistry and physiology that are known from *in vivo* and *in vitro* animal studies. Such hypotheses as the catecholamine hypothesis of affective disorders or the dopamine hypothesis of schizophrenia have been powerful conceptual tools to bring diverse pharmacological data to bear on the pathophysiology of several neurological and psychiatric illnesses. It is clear, however, that direct examination, in humans, of the proposed metabolic abnormalities is important, not only to confirm or reject these hypotheses, but also to provide a means of differentiating subgroups of patients with similar clinical syndromes. This is particularly important in psychiatry and neurology, where there are few complete animal models of the major diseases under study, and because species differences are extremely important in the study of brain function.

However, studies of CSF monoamine biochemistry have been plagued by a number of difficulties, and the data deriving from them have developed slowly and uncertainly. First of all, the necessity of performing a lumbar puncture has created a serious impediment to the collection of data. The lumbar puncture places a serious restriction on the number of observations that can be made on a single patient, as well as limits the number of patients that can be studied or the number of research groups contributing data. Serial data points cannot be gathered in a given study in an individual patient. Satisfactory control subjects cannot be studied in a systematic way. Statistical techniques are available for defining abnormal values of a clinical biochemical measurement, (without determinations being made on a normal population) when large numbers of the determination are performed in a standardized manner on patients with a variety of diagnoses. It is much more difficult to do this when small numbers of patients are being studied by each of several research groups that are often using slightly different biochemical techniques.

Michael H. Ebert, M.D. • Laboratory of Clinical Science, National Institute of Mental Health, National Institutes of Health, Bethesda, Maryland 20205. **Ronald Kartzinel, M.D., Ph.D.** • Division of Neuropharmacology, Food and Drug Administration, Department of Health, Education and Welfare, Rockville, Maryland. **Rex W. Cowdry, M.D.**, and **Frederick K. Goodwin, M.D.** • Clinical Psychobiology Branch, National Institute of Mental Health, National Institutes of Health, Bethesda, Maryland 20205.

Second, until recently, some of the biochemical methods that were used to study lumbar CSF were operating at the lower limits of their sensitivity. Examples of these analytical problems exist in studies of homovanillic acid (HVA) and 5-hydroxyindoleacetic acid (5-HIAA) in lumbar CSF by spectrophotofluorometry when probenecid is not administered to the patient to elevate the levels of these acid metabolites. In the last several years, a number of improved assays have been published and widely adapted. Gas chromatography–mass spectrometry, or mass fragmentography, has been applied to the measurement of catecholamine metabolites[18,21,44] and indoleamine metabolites.[13] These assays are performed with deuterated internal standards and offer improved specificity and sensitivity and more accurate estimation of losses during the extraction and derivatization procedures. Deuterated internal standards are now commercially available for several of the catecholamine metabolites and for 5-HIAA. Improved electron capture–gas chromatography techniques have been reported.[51] High-pressure liquid chromatography has been applied to the measurement of amine metabolites.[41]

Finally, only recently have quantitative physiological studies begun to appear on the clearance of amine metabolites from the various regions of the CSF circulation. The relationship of amine metabolites measured in lumbar CSF to brain amine metabolism is complex. Additional animal studies need to be performed to interpret fully the physiological significance of amine metabolite changes observed in lumbar CSF of patients.

2. Brain Amine Metabolism Reflected in Cerebral Ventricular CSF

The distribution of catecholamines, indoleamines, and their synthetic and degradative enzymes is well defined in rats, subhuman primates, and other mammalian animals. Norepinephrine (NE)[1a,28a] and serotonin[5a] neuronal systems have a wide distribution in the central nervous system, in contrast to the more discretely localized dopamine (DA)-containing neuronal systems in the basal ganglia, mesolimbic system, and tuberoinfundibular system.[46] The largest concentration of dopaminergic nerve endings is found in the basal ganglia in the nigrostriatal system. The majority of these nerve endings are found in the caudate nucleus, which forms the lateral wall of the lateral ventricle.

The regional concentration of amines and amine metabolites in the CSF is in part a reflection of the concentration of catecholamines and indoleamines in the immediately adjacent neuronal parenchyma.[15,32] Thus, one would expect lateral ventricular CSF to reflect the metabolism of the caudate nucleus and the CSF of the third and fourth ventricles to possibly reflect cord metabolism. This is complicated by the circulation of CSF so that CSF downstream from a structure would continue to reflect the metabolism of that structure until the metabolites were cleared from the CSF by active transport or diffusion. The sites of origin and concentration gradients of various neurotransmitters and their metabolites in CSF have been reviewed in detail by Wood in Chapter 5.

The regional concentration of amine metabolites in the ventricular system of the conscious rhesus macaque has been studied by placing permanent cannulas in various regions of the ventricular system and slowly and constantly withdrawing CSF through a pump or siphon for weeks at a time. Such a sampling procedure obtains CSF from one region without draining other regions and minimizes acute disruptions in CSF production rates that might change the concentration of metabolites.

The major metabolites of NE are vanillylmandelic acid (VMA) and 3-methoxy-4-hydroxyphenyl glycol (MHPG). MHPG has been demonstrated to be the major metabolite of NE in the brain of several species including rat, man, and rhesus macaque. MHPG and VMA are detectable in nanogram amounts in all ventricles, cistern, and subarachnoid spaces (Table 1). The concentrations of MHPG and VMA are, respectively, in the range of 10–15 and 1–2 ng/ml in the lateral ventricle, cistern, and lumbar space. In the third and fourth ventricles, the concentration of these compounds increases 50–100%, and is significantly higher than in the lateral ventricle.[19] This increase may reflect the increased concentration of NE found in hypothalamic and brainstem structures. In contrast to the acidic metabolites of DA and serotonin, we have not been able to demonstrate a decreasing MHPG or VMA concentration gradient from the ventricles to the lumbar space. The relatively similar and low concentrations of MHPG and VMA in all parts of the CSF system give indirect evidence that they are cleared rapidly from the central nervous system by a diffusional mechanism.

HVA, the major metabolite of DA, is present in high concentrations in the lateral ventricle (Table 1) because of its proximity to the caudate nucleus.

Table 1. Catecholamine Metabolites in CSF in Various Regions of Rhesus Macaque Central Nervous System[a]

Area	HVA (ng/ml)[b]	MHPG (ng/ml)[b]	VMA (ng/ml)[b]
Lateral ventricle	1367 ± 206	15.8 ± 1.6	1.75 ± 0.23
Third ventricle	744	20.3	2.54
Fourth ventricle	601 ± 92[c]	24.8 ± 3.2[c]	3.46 ± 1.11
Cervical subarachnoid space	132	20.7	1.07
Lumbar subarachnoid space	26 ± 7.8[c,d]	21.8 ± 4.7	0.79 ± 0.21[c,d]

[a] Abbreviations: (HVA) Homovanillic acid; (MHPG) 3-methoxy-4-hydroxyphenyl glycol; (VMA) vanillylmandelic acid.
[b] Values are mean values ± S.E.M. for the total compound.
[c] Significantly different from the lateral ventricle ($P < 0.01$).
[d] Significantly different from the fourth ventricle ($P < 0.01$).

There is a 2-fold decrease as CSF flows into the fourth ventricle and an additional decrease as CSF is sampled from the cisterna magna. The lumbar concentration of HVA is 1/50th of that found in the lateral ventricle. The high ventricular concentrations and striking gradient in this acid metabolite suggest that it is actively transported out of the CSF, and also that the major portion of HVA probably enters the CSF in the lateral ventricle from the metabolism of DA released in the caudate nucleus. Another possible source of HVA in CSF is from brain capillaries,[6] but this probably does not account for a substantial proportion of CSF HVA.[12]

5-HIAA, the major metabolite of serotonin, also declines considerably in concentration as CSF moves from the ventricles to the lumbar space. The gradient is not as steep as for HVA. In the lateral ventricle, the concentration of 5-HIAA is 100–125 ng/ml. In primates with lateral- and fourth-ventricle cannulas, CSF obtained simultaneously from both areas demonstrates a 50–100% increase in concentration of 5-HIAA in the fourth ventricle. This probably again reflects the increased number of serotonin-containing nerves and nerve endings in the hypothalamus and brainstem.

A variety of evidence suggests that meaningful physiological information can be obtained about the activity and metabolism of DA and serotonin neuronal systems adjacent to the ventricular system from the levels of their acid metabolites in the ventricles. Portig and Vogt[38] demonstrated that electrical stimulation of the substantia nigra leads, after a short delay, to increased levels of HVA in the lateral ventricle. Administration of a neuroleptic, such as chlorpromazine, increases turnover of dopamine in the caudate and increases the level of HVA in the caudate and in lateral ventricle to a similar degree.[32] HVA is decreased to a similar extent in the caudate nucleus of patients with Parkinson's disease and in the lateral ventricular CSF of such patients.[32] There is a significant diurnal rhythm in HVA in the lateral ventricle, with highest levels during the day.[37] The concentration of cyclic adenosine monophosphate in lateral ventricle also fluctuates in a diurnal rhythm in phase with the cycle in catecholamine metabolites.[37]

3. Clearance of Amine Metabolites from Central Nervous System

Guldberg *et al.*[20] demonstrated the gradient between lateral ventricle and cisterna magna for HVA and 5-HIAA, and the reduction in the gradient following the administration of probenecid by elevation of cisternal levels of the acid metabolites. Ashcroft *et al.*[4] subsequently demonstrated by recirculatory perfusion of the ventricular system that the efflux of 5-HIAA from CSF had components due to diffusion, a saturable transport system, and bulk flow of CSF. They postulated that the transport system was located in the region of the fourth ventricle. Subsequently, it has become less clear that the fourth ventricle is the primary site of active transport. The choroid plexus has been shown *in vitro* to accumulate probenecid against a concentration gradient, and this process is blocked by probenecid.[14] The majority of 5-HIAA and HVA formed in the brain probably does not leave via the CSF through the choroid plexus or site of active transport in the ventricular system. Neff and co-workers[34,35] demonstrated in the rat that only about 10% of the

5-HIAA that is formed and eliminated from rat brain is eliminated via the CSF pathway.

Wolfson *et al.*[53] have recently carried out studies of 5-HIAA clearance from the CSF of cats by ventriculocisternal and cortical subarachnoid perfusions. Low 5-HIAA clearances were found during ventriculocisternal perfusions (0.009 ml/min), whereas perfusions of the cerebral subarachnoid space were 10-fold greater (0.107 ml/min). Probenecid pretreatment decreased these clearance rates 4-fold. The relatively high clearance rates from the cerebral subarachnoid space suggested that this was where the bulk of 5-HIAA clearance was occurring. The authors reasoned that the large surface area of the hemispheres and rich capillary beds made the cerebral subarachnoid space a logical place for active transport to occur, probably at the capillary wall. Regardless of the exact location of the active transport, it is clear that probenecid blocks the transport system and that the acid metabolites of DA and serotonin increase in concentration in the cistern and spinal subarachnoid space to reflect more closely the levels in the cerebral ventricles.

A different situation exists for MHPG, the neutral compound that is the major metabolite of NE in the brain of many species, including man. Wolfson and Escriva[54] have reported similar clearance studies for MHPG in the cat using perfusion techniques. Both ventriculocisternal and cerebral subarachnoid perfusions demonstrated relatively high clearance rates (0.07–0.08 ml/min) that were not diminished by probenecid. There was no evidence of a saturable clearance mechanism. The studies suggest a diffusional mechanism for clearance of MHPG from the central nervous system. This rapid, unblockable diffusional clearance suggests that MHPG levels in CSF may be a poor index of brain NE metabolism; however, this conclusion has been recently disputed by Adèr *et al.*[1] and Ziegler *et al.*[55]

4. Measurement of Amine Metabolites in Lumbar CSF

The concentration of biogenic amine metabolites in lumbar CSF, and their elevation by probenecid, is often used by clinical investigators to assess indirectly the rate of turnover of amines in the central nervous system of man. The HVA[7] and 5-HIAA[11] found in the CSF appear to be derived principally from the central nervous system and not from peripheral tissues. In contrast, MHPG and VMA in the CSF appear to be derived from both central nervous system and peripheral tissues.[37] The extent of the peripheral contribution remains to be determined.

Assuming the proposition that biogenic amine metabolites in CSF are primarily products of central nervous system metabolism, we are still left with the question of whether or not metabolites in lumbar CSF are a reflection of brain or spinal cord metabolism. There are two approaches that have been used to determine the origin of amine metabolites found in lumbar CSF. The first involves perfusion of the spinal column of an experimental animal with a radiolabeled metabolite. The spinal cord and the perfusate are then analyzed for distribution of the radioactivity. The second approach is a clinical one, involving multiple samples of lumbar CSF.

In studies of radiolabeled 5-HIAA transport by perfusion of the spinal subarachnoid space in the rhesus macaque, 5-HIAA was found to distribute in a tissue space of 50–55% and to exchange readily between tissue and blood across parenchymal capillaries. Probenecid had only a modest effect in this system on clearance rates of 5-HIAA.[23] Similar results were obtained in the same system for HVA, and the clearance rates were approximately the same.[24] The authors concluded that lumbar CSF concentrations of 5-HIAA reflect only a small proportion of total 5-HIAA production by the brain. This study contrasts with tracer perfusion studies of spinal transport of 5-HIAA in cats. The authors of one study[49] concluded that about 70% of the 5-HIAA measured in the lumbar space derived from structures rostral to the foramen magnum. A second study demonstrated that spinal subarachnoid clearance of 5-HIAA can be resolved into a saturable mediated transport mechanism and a diffusional component, and that the transport mechanism is effectively blocked by probenecid.[53] It appears that there are species differences in spinal subarachnoid transport mechanisms for 5-HIAA and HVA.

The clinical approach to the origin of lumbar amine metabolites has involved analysis of sequential lumbar CSF samples. Such studies show a progressive increase in 5-HIAA and HVA concentration,[43] suggesting that the metabolites enter the CSF primarily above the spinal cord, and are transported out of the CSF as they move caudally. Post *et al.*[40] attempted to make these observations more quantitative by studying a group of patients with spinal cord block. Because the presence of a spinal fluid block did not affect the concentration of 5-HIAA and MHPG, but did lower the concentration of HVA, they concluded that the spinal cord contributed significantly to lumbar CSF concentration of MHPG and 5-HIAA and very little to HVA.

The exact quantitative contributions of spinal cord and brain to the levels of amine metabolites in lumbar CSF in man are obviously not known. It is reasonable to generalize from the available data that the lumbar concentrations of 5-HIAA and MHPG reflect primarily brain metabolism. This is probably accounted for by the fact that the spinal cord contains serotonin and NE, but very little DA, rather than by a difference in spinal subarachnoid transport of 5-HIAA and HVA in man. Since there is a decreasing concentration gradient moving caudally and an active transport mechanism for HVA and 5-HIAA, the effect of probenecid is to increase the proportion of contribution from brain as well as the absolute level.

5. Development of Probenecid Test

The major clinical application to date of amine metabolites in CSF has been the probenecid test. The probenecid technique was developed in an attempt to block the active transport of acidic amine metabolites out of CSF. Probenecid is an organic acid that competitively inhibits the active transport of acid monoamine metabolites from brain and CSF to the bloodstream.[32,50] If egress of these acid metabolites from CSF were effectively blocked by probenecid, the rate of accumulation of 5-HIAA and HVA might then reflect primarily the rate of production of these metabolites from the parent amines, serotonin and DA, and thereby provide a measure of central neurotransmitter activity.

Neff and Costa[34] have shown that in rats treated with 200 mg/kg of probenecid, the rate at which 5-HIAA accumulates in brain tissue almost exactly equals the rate of formation of this acid from serotonin. Application of steady-state kinetics allowed the calculation of quantitative turnover rates after probenecid blockade.

Clinical investigators have tried to approach this experiment by giving large doses of probenecid to patients and measuring the accumulation of 5-HIAA and HVA in lumbar CSF as indirect indicators of turnover rates of serotonin and DA in brain. However, in human studies, it has been necessary to use lower doses spaced over a number of hours, to avoid the toxicity (primarily severe nausea and vomiting) that would occur at doses approaching 200 mg/kg. At present, the probenecid test is carried out in a variety of designs by different investigators, making comparison among research groups somewhat difficult. Such designs include: 1 g hourly for 4–5 hr (lumbar puncture at 8 hr), 40 mg/kg intravenously over 1 hr (lumbar puncture at 8 hr), 2 g every 2 hr for 3 doses (lumbar puncture at 8 hr), and 100 mg/kg divided over 18 hr (lumbar puncture at 18 hr). These various designs obviously result in different kinetics of probenecid in the central nervous system, and in general result in lower central nervous system probenecid levels than those achieved by a single 200 mg/kg dose.

Despite use of lower amounts of the drug, probenecid studies in humans have produced results that support the validity of this test as a measure of central monoamine turnover. Thus, probenecid-induced accumulation of HVA in lumbar CSF is low in patients with Parkinson's disease, a disease presumably involving decreased central DA activity.[10] Similarly, pharmacological manipulations with demonstrated effects on brain serotonin or DA turnover in animals influence the probenecid-induced rise in CSF monoamine metabolites in the appropriate direction. HVA accumulation in CSF increases with L-DOPA and decreases with α-methyl-*p*-tyrosine (an inhibitor of DA synthesis). Similarly, the rate of 5-HIAA accumulation increases with L-tryptophan (the amino acid precursor of serotonin) and decreases with *p*-chlorophenylalanine (an inhibitor of serotinin synthesis).[17] However, when the method has been used to explore hypothesized differences in central monoamine turnover in psychiatric disorders, such as schizophrenia and manic–depressive illness, conflicting results have been obtained.[39] The discrepancies have prompted researchers to explore unresolved methodological issues regarding the probenecid technique—in particular, whether variations in CSF probenecid levels achieved by these techniques are associated with varying degrees of transport inhibition. In that case, observed metabolite levels might reflect variable transport blockade as well as variable central monoamine turnover.

As an example of the application of the probenecid test in psychiatric research, Table 2 summarizes a group of studies utilizing the probenecid test in different dosage schedules to study patients with affective disorders. Several conclusions can be drawn. First, as might be expected, probenecid levels differ widely depending on both total dose and schedule of administration. Second, metabolite accumulations seem to be more strongly related to average probenecid levels in CSF rather than total time of blockade. Resolution of this issue would require

Table 2. Results of Probenecid Studies in Primary Affective Disorder

Reference	Dosage	Hours to LP[a]	CSF levels			Probenecid correlation with		Metabolite levels in PAD[a] different from controls?
			Probenecid	5-HIAA	HVA	5-HIAA	HVA	
Sjöström[45]	1 g b.i.d. × $2\frac{1}{2}$ days	60	1.7	38	53	0.63	0.32	Yes
Bowers[8]	1 g t.i.d. × $2\frac{1}{2}$ days	60	—	49	95	—	—	Yes
Korf and van Praag[25]	1 g i.v. + 1 g p.o. q.h. × 3	8	6.0	—	—	0.42	0.39	Yes
Van Praag[47]	1 g i.v. + 1 g p.o. q.h. × 4	8	9.2	77	122	—	—	Yes
Van Praag[47]	1 g i.v. + 1 g p.o. q.h. × 4	8	9.7	61	—	0.64	—	Yes
Goodwin *et al.*[17]	115 mg/kg in divided doses	9	—	90	90	—	—	Yes
Bowers[8]	100 μg/kg in divided doses		22.7	123	—	0.45	—	No
Ebert *et al.*[10b]	100 μg/kg in divided doses	18	22.8	140	218	0.52	0.45	No

[a] Abbreviations: (LP) lumbar puncture; (PAD) primary affective disorder; (5-HIAA) hydroxyindoleacetic acid; (HVA) homovanillic acid.

complex pharmacokinetic studies that have not been attempted. Third, CSF probenecid level is an important determinant of acid metabolite levels throughout the range of studies reported, accounting for 10–40% of the variance observed; this suggests that probenecid-induced transport blockade is incomplete in patients with affective disorders even at high CSF probenecid levels.

6. Significance of CSF Probenecid Levels in Probenecid Test

Several investigators have studied CSF probenecid concentrations in the course of performing probenecid tests. Sjöström[45] studied CSF probenecid concentrations in manic–depressive patients and controls under steady-state conditions of an oral dose of 2 g per day. The concentration of probenecid was low and varied considerably among subjects (0.2–5.0 μg/ml); there was a significant positive correlation in CSF between probenecid and both 5-HIAA (r = 0.63) and HVA (r = 0.32). Korf and Van Praag,[25] using 1 g i.v. followed by 1 g p.o. q.h. × 3, observed similar strong correlations between probenecid and both 5-HIAA (r = 0.42) and HVA (r = 0.39) at 8 hr after starting probenecid. Bowers,[8] using 100 mg/kg in divided doses over 18 hr, found a similar strong probenecid/5-HIAA correlation (r = 0.45). The only report failing to find significant correlations is that of Perel *et al.*[36] They observed CSF probenecid levels ranging from 16.4 to 35.7 μg/ml after 100 mg/kg probenecid administered in divided doses in an identical manner to the study of Bowers,[8] and reported that concentrations of CSF 5-HIAA and HVA in their study were not significantly associated with CSF probenecid concentration.

During the course of our research utilizing the 100 mg/kg oral probenecid test, we have studied the relationship between CSF probenecid levels and amine metabolite accumulation in lumbar CSF. A total of 54 patients with primary affective disorders were studied on metabolic research units at the National Institute of Mental Health specifically designed for the collection of behavioral and biochemical data on a longitudinal basis. Two groups of lumbar puncture protocols are discussed: First, all the initial lumbar punctures performed on each patient while the patient was "medication-free" (N = 54); second, all lumbar punctures performed on these patients, regardless of what medication they were taking concurrently (N = 166). "Drug-free" subjects were left off all medications for at least 2 weeks prior to the probenecid procedure. During the study, diet was limited in foods that might affect indoleamine or catecholamine metabolism. The lumbar puncture was performed at 3:00 P.M. following 15 hr of bedrest. During the 18 hr preceding the lumbar puncture, probenecid (100 mg/kg) was administered in divided doses (approximately 30 mg/kg at 9:00 P.M., and 23 mg/kg at 2:00 A.M., 7:00 A.M., and 12:00 noon). Total probenecid dose was calculated by the patient's weight at the time of the first lumbar puncture and was administered in multiples of 250 mg tablets.

CSF samples were collected in 20 mg ascorbic acid/10 ml CSF and stored at −50°C until analyzed. 5-HIAA and HVA were determined by fluorometric assay in the first 8 ml of CSF obtained according to modifications of the methods of Ashcroft and Sharman,[5] Gerbode and Bowers,[16] and Anden *et al.*[2] Probenecid levels were determined by electron capture–gas chromatography by the method of Watson and Wilk.[48] A 0.1-ml aliquot of CSF was diluted to 0.6 ml with distilled water, acidified with 0.025 ml hydrochloric acid, extracted once into 20 ml ethyl acetate and proprionic anhydride at 750°C for 5 min, and dried under nitrogen. The derivative was dissolved in ethyl acetate and chromatographed on a 3% OV-17 column at 180°C. The butyl derivative of probenecid was used as an internal standard.

Data are reported as Pearson's coefficient of correlation. This is a proper measure only for the 54 medication-free values, since they each represent an independent subject. In the larger group, there is more than one value for some subjects, and thus the correlation may be somewhat inflated due to autocorrelation; this effect is balanced, however, by the addition of a further source of variation in metabolite and probenecid levels, namely, medication.

At 18 hr after a total oral probenecid dose of 100 mg/kg (in divided doses as described above), there was a wide variability of CSF probenecid levels that ranged from 9.1 to 44.6 μg/ml in all probenecid tests. Mean CSF probenecid level (mean ± S.D.) was 20.4 ± 5.2 for all probenecid tests and 23.1 ± 8.1 for medication-free probenecid tests. The corresponding values for HVA were 222 ± 22 and 250 ± 14 ng/ml, and for 5-HIAA were 128 ± 9 and 148 ± 7 ng/ml.

Figure 1 illustrates the relationship between probenecid concentration and HVA concentration in lumbar CSF for medication-free probenecid tests. There is a significant positive correlation (r = 0.379, p = 0.01), with probenecid concentration accounting for 14% of the total variance. When all the avail-

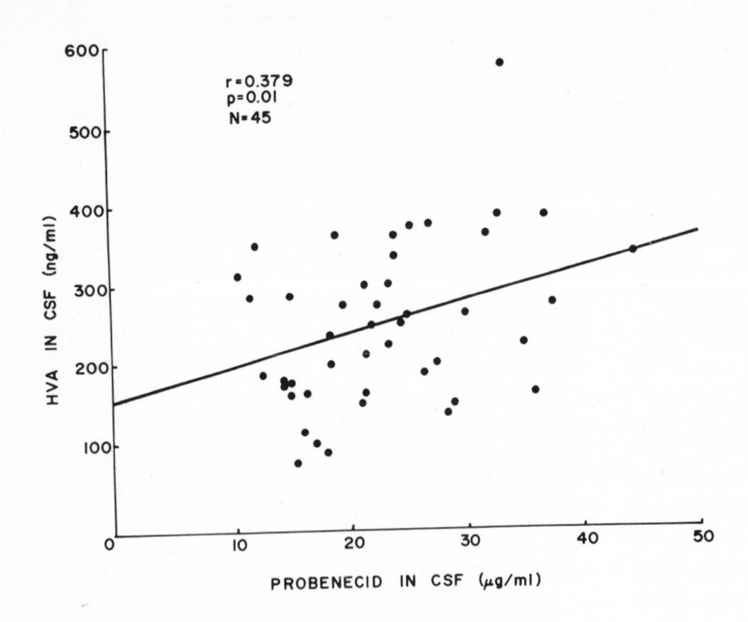

Figure 1. Relationship between homovanillic acid (HVA) and probenecid levels in CSF.

able probenecid tests are considered, a similar relationship exists ($r = 0.311$, $p = 0.0001$). When only lumbar punctures achieving CSF probenecid concentrations of 20 μg/ml or greater are considered a significant correlation with CSF HVA still exists ($r = 0.409$, $p = 0.03$).

Figure 2 illustrates the relationship between probenecid concentration and 5-HIAA concentration in lumbar CSF in medication-free probenecid tests. Again, there is a significant positive correlation ($r = 0.489$, $p = 0.0006$), and probenecid concentration accounts for 24% of the total variance. When all the available probenecid tests are considered, a similar relationship exists ($r = 0.351$, $p = 0.00001$). Again, if only lumbar punctures achieving CSF probenecid concentrations of 20 μg/ml or greater are considered, a significant correlation with CSF 5-HIAA still exists ($r = 0.55$, $p = 0.005$).

Side effects were examined in relation to CSF probenecid levels. In both the medication-free and unselected group of probenecid tests, the occurrence of vomiting was clearly related to higher CSF probenecid concentrations as shown in Table 3, which is based on 78 probenecid tests for which reliable ratings of side effects were available.

After 100 mg/kg of probenecid is given over 18 hr, 4-fold differences in CSF levels of probenecid are found. This range of concentrations may be due to differences in absorption, distribution (including such variables as transport and binding to plasma proteins), or metabolism. In this study, the coefficient of variation for CSF probenecid levels is 0.352. The observed variability in CSF probenecid is not likely a result of individual differences in protein binding, since plasma protein-binding sites are usually saturated at the plasma probenecid levels achieved with the 100 mg/kg probenecid test.[36] It is noteworthy that intravenous administration of probenecid in a single dose of 40 mg/kg produces a coefficient of variation of probenecid levels of only 0.121 at 8 hr.[22] There is a strong suggestion that the oral route of administration may involve a significant degree of variability due to individual differences in absorption.

This variability in CSF probenecid levels would be of no practical significance for the probenecid test unless probenecid levels are associated with metabolite accumulation in CSF. In fact, we observe that the CSF probenecid levels are significantly associated with acidic metabolite accumulation even at the highest CSF probenecid levels achieved in human studies (levels of 10–40 μg/ml). The observed correlations are similar to those reported by other authors reporting on small groups of patients. Since these reports taken as a group involve CSF probenecid levels ranging from 0.2 to 40.0 μg/ml, they suggest that CSF probenecid is associated with a significant effect on CSF metabolite accumulations throughout this range.

At least three interpretations of this phenomenon are possible. Since the blockade of acid transport is competitive, levels of 10–30 μg/ml may still achieve only a partial blockade in which the degree of blockade increases with increasing probenecid levels.

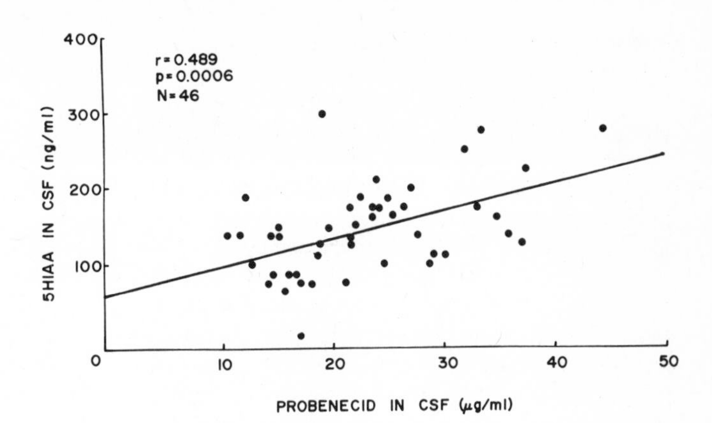

Figure 2. Relationship between 5-hydroxyindoleacetic acid (5-HIAA) and probenecid levels in CSF.

This would be consistent with *in vivo* studies in ox erythrocyte[33] and choroid plexus[31] that suggest that 50% of inhibition of organic acid transport occurs at probenecid concentrations of 15–30 μg/ml. The second possibility is that higher levels of probenecid induce increased central monoamine turnover; the finding in this study of a markedly increased incidence of vomiting associated with higher CSF levels (but, by definition, not with higher probenecid doses) provides some support for the existence of central nervous system effects of the drug. Finally, there may be no direct association between higher CSF probenecid levels and higher CSF metabolite levels; rather, high probenecid levels and high central monoamine turnover may both be related to some third variable influencing each independently. For example, both probenecid and metabolite levels in CSF are significantly higher in women than in men.

These interpretations present a dilemma regarding the significance of metabolite levels in CSF. If the increased metabolite levels are caused by higher probenecid levels, then the 14–24% of total variance in CSF metabolite levels that is associated with variations in CSF probenecid levels could be removed statistically through an analysis of covariance, provided the metabolite–probenecid interaction is similar across significant subgroups of the population. Removing this method-associated variance may increase the proportion of the overall variance associated with true differences in central monoamine turnover, and thus provide a more sensitive tool for exploring subgroup differences in central monoamine turnover. Since the causal connections involved are not clear, we routinely assay CSF probenecid levels in conjunction with CSF metabolite studies. Statistical analyses are then performed using both the observed metabolite values and metabolite values adjusted for variations in probenecid levels through an analysis of covariance.

Table 3. CSF Probenecid Levels in 100 mg/kg Oral Probenecid Test in Relation to Side Effects

Side Effect	Number of patients	CSF probenecid levels
No nausea	36	18.8 ± 0.8
Nausea	25	19.5 ± 1.2
Vomited	10	25.4 ± 1.8
Vomited several times	7	30.1 ± 3.4
TOTAL:	78	20.9 ± 0.8

$F = 10.04$; $P < 0.001$.

Several authors have recently reported CSF metabolite findings that attempt to control for varying probenecid levels by dividing the metabolite value by the probenecid level.[42,52] Based on the regression equations derived from our correlational studies, we feel that this approach to controlling for probenecid differences is statistically unwarranted and would overcorrect for probenecid-associated variations in metabolite levels.

Finally, since high doses of probenecid do not markedly decrease the strength of the probenecid–metabolite association, but do increase the incidence of side effects, and since oral administration may be associated with greater variability of CSF probenecid levels, possibly due to variable absorption, a lower-dose intravenous probenecid technique may have certain advantages over the 100 mg/kg oral study.

7. Sources of Variance in Probenecid Test

After variations in the amount of probenecid in the central nervous system are taken into account by measurement of CSF probenecid levels, the assumption underlying the clinical application of the test remains that the probenecid-induced accumulation of metabolites in CSF passively reflects the turnover rate of the parent amine without directly affecting its metabolism. Although the bulk of evidence suggests that this is true, several lines of evidence suggest other neuropharmacological effects of probenecid in addition to transport blockade. Probenecid induces alterations in sleep EEG patterns.[26] As discussed earlier, high CSF probenecid levels are associated with vomiting during the probenecid procedure, suggesting that this is an early sign of central nervous system toxicity of the drug. Probenecid administration in the 100 mg/kg multiple oral dose format is associated with significant increases in NE levels in CSF, suggesting that it may directly increase turnover of NE.[27] Probenecid in a dose of 1 g b.i.d. for 3 days produces a significant increase in free tryptophan in plasma, which raises the question of whether the synthesis rate of brain serotonin could be increased by the drug.[28]

Second, in studies in which the probenecid test is used to investigate the neurochemical responses of patients to drugs, the possibility of drug interactions between probenecid and other drugs must be kept in mind. This possibility can be monitored by comparing CSF probenecid levels during periods without and intervals with the administration of a given drug. Preliminary evidence from our group suggests that lithium produces a decrease in CSF probenecid levels.

Finally, when the probenecid test is used to compare different diseases or subcategories of a syndrome, the possibility that probenecid pharmacokinetics or transport systems differ among patient groups must be considered. These possibilities can be monitored by comparing CSF probenecid levels or probenecid–metabolite correlations among patient groups. The literature on amine metabolites in affective disorders suggests that there may be subgroup differences in probenecid–amine metabolite relationships. A study of van Praag,[47] in a reanalysis of data to explore possible group differences, indicates that probenecid/5-HIAA correlations are significantly higher in depressed patients ($r = 0.64$) than in a "nondepressed" control group ($r = 0.15$). Sjöström,[45] in a study achieving CSF probenecid levels of less than 5 μg/ml, also found diagnostic group differences; however, his study reports higher correlations ($r = 0.90$, $r = 0.83$) and steeper regression lines (slopes = 14.4 and 31) in his control group than in his manic–depressive group ($r = 0.63$, $r = 0.32$; slopes = 4.2 and 9). While these findings may be confounded by medications (most notably, 7 of 15 manic–depressive patients who were on lithium), they may also be interpreted as further indication of different probenecid–metabolite relationships in the two diagnostic groupings.

8. Intravenous Administration of Probenecid

In a majority of clinical studies of central monoamine metabolism, probenecid has been administered orally. Total drug quantities, dose intervals, and timing of lumbar punctures have varied considerably, making it difficult to compare CSF monoamine metabolite data among investigators. Nausea and vomiting frequently attend the oral administration of probenecid, further complicating the interpretation of results. Moreover, probenecid-induced changes in CSF monoamine metabolites may reflect individual differences in the rate and degree of gastrointestinal absorption of probenecid rather than in the central metabolism of DA or serotonin. Some of these difficulties should be mitigated through the use of intravenously administered probenecid. In previous studies, however, probenecid given intravenously was followed by multiple oral doses. In view of the marked interindividual differences in plasma probenecid concentrations after oral administration of this drug, we have developed a variation of the probenecid test in which the drug is given as an intravenous infusion.[22]

A total of 55 probenecid infusions were performed on 24 consenting patients (17 men and 7 women, ranging in age from 26 to 62 years). Neurologic diagnoses were as follows: Parkinson's disease, 12; Huntington's disease, 9; familial spinocerebellar degeneration, 2; schizophrenia, 1. Patients in whom CSF HVA and 5-HIAA levels were measured received no drugs for at least 1 week before the study, consumed a diet low in monoamine precursors during the preceding 2 days, and remained at bedrest from midnight before the test until its completion.

Infusions were done with probenecid doses of either 20 or 40 mg/kg. Probenecid ampules were added to 50 or 150 ml normal saline and infused over 60 min. All infusions were begun at 9:00 A.M.; CSF samples were obtained by lumbar puncture immediately before the infusion and after 4, 6, 8, 10, and 12 hr. A maximum of three lumbar punctures were performed in one day on a given patient. In patients receiving intravenous probenecid, 4 ml of venous blood was obtained 1, 2, 4, 6, 8, 10, and 12 hr after the infusion was started. During the oral test, blood for probenecid determination was obtained at the time of the second lumbar puncture.

At the end of the 1-hr infusion period, plasma probenecid levels averaged 139 ± 11 μg/ml after a probenecid dose of 20 mg/kg and 323 ± 18 μg/ml after a dose of 40 mg/kg (Fig. 3). Over the next 12 hr, these levels declined as a logarithmic function of time. Half-lives of 5.2 and 6.6 hr were found at the 20 and 40 mg/kg probenecid doses, respectively, which were not significantly different. Throughout this period, plasma probenecid levels after the higher doses were approximately twice those produced by the lower dose. Probenecid levels in CSF peaked between 6 and 8 hr after the 40 mg/kg infusion was started and had decreased significantly by 12 hr (Fig. 3A). Beginning 4 hr after the infusion was initiated, the ratio of CSF to plasma probenecid increased steadily, reaching a maximum value of 0.04 at 10 hr.

HVA levels in CSF were significantly increased above baseline values at 4 hr ($p < 0.05$) and peaked in samples obtained 8 hr after the infusion was started (Fig. 3B). The time course for 5-HIAA changes in CSF was essentially the same (Fig. 3C).

A significant positive correlation was found between CSF levels of probenecid and the increase in 5-HIAA, but not HVA. There was no apparent correlation between plasma and CSF probenecid concentration or between the increase in CSF content of either monoamine metabolite and the plasma concentration of probenecid when evaluated in each patient at the five time intervals studied. Since CSF levels of HVA and 5-HIAA peaked 8 hr after the probenecid infusion (40 mg/kg) was started, we have selected this time for future tests.

The advantages of performing the probenecid test by intravenous infusion are that side effects are rarely encountered, that the kinetics of probenecid are more uniform among patients, and that the test can be completed in a shorter period of time. On the other hand, the oral probenecid loading test (100 mg/kg) achieves higher CSF probenecid levels and larger amine metabolite accumulations. In either procedure, it seems imperative that CSF probenecid levels be measured, since in none of the clinical probenecid experiments is a complete blockade of transport achieved.

9. Kinetics of CSF Probenecid and Amine Metabolites

To further describe the kinetic aspects of the intravenous probenecid test, a dose-response study to 40, 80, and 160 mg/kg of intravenous probenecid was pursued in rhesus monkeys. A total of 27 probenecid infusions were performed on ten rhesus monkeys (5.5–6.0 kg) that were acclimated to re-

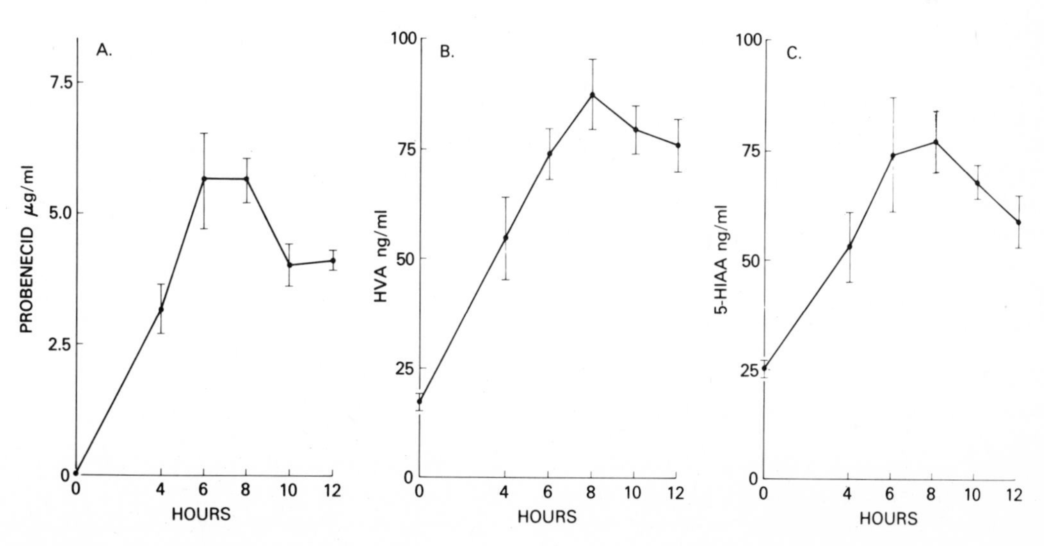

Figure 3. Time course for changes in probenecid (A), homovanillic acid (HVA) (B), and 5-hydroxyindoleacetic acid (5-HIAA) (C) in lumbar CSF after single 1-hr probenecid infusion at 40 mg/kg. Each point represents mean for 3–11 infusions. Vertical brackets indicate S.E.M.

straining chairs and fed a standard chow diet. Chronically implanted cannulas were placed in the anterior horn of the right lateral ventricle in six primates. Lumbar cannulas were placed in two of these monkeys and in four others using a 19-gauge angiocath needle with a 21-gauge tubing and wire stylet. CSF was collected continuously by gravity drainage from the ventricular or lumbar catheter; the flow rate was maintained at 1.0–1.5 ml every 2 hr by regulating the opening and closing times of an LKB solenoid valve, and samples were collected in a fraction collector at 4°C.

After complete recovery from the surgical procedure, probenecid was administered intravenously over 1 min at a dose of 40 or 80 mg/kg, or infused over 10 min with a Harvard pump at a dose of 160 mg/kg. Probenecid was determined by electron capture–gas chromatography, as previously described. The HVA and 5-HIAA content in the CSF samples were measured by a gas chromatography–mass spectrometry method.

There was a rapid increase in ventricular CSF levels of probenecid, with a maximum 3 hr after intravenous administration (Fig. 4). Thereafter, there was a linear decrease in the natural logarithm of probenecid concentration with time. Probenecid levels in ventricular CSF showed a significant increase ($p < 0.001$) with increasing dose. However, 2-fold and 4-fold increases in the dose of probenecid led to a 3-fold and 10-fold increase in ventricular CSF probenecid levels.

Similar results were found for probenecid concentrations in lumbar CSF (Fig. 4). Probenecid levels were lower, and maximum concentrations were reached 2 hr later, in lumbar than in ventricular CSF. The half-time for elimination of probenecid ranged from 2.8 to 4.4 hr and showed small but significant increases with increasing doses of probenecid. At a given dose, however, there was no significant difference in half-time between ventricular and lumbar CSF.

In ventricular CSF, maximum HVA concentrations were obtained 3–5 hr after the administration of probenecid (Fig. 5). Increases in HVA concentration at the 80 and 160 mg/kg dose of probenecid were almost identical, and significantly greater ($p < 0.01$) than at the 40 mg/kg dose. In lumbar CSF, there were significant increases ($p < 0.001$) in HVA concentrations with each dose of probenecid (Fig. 5), and the time at which the maximum probenecid-induced increase in HVA occurs appeared to be dose-related (5 hr at 40 mg/kg, 7 hr at 80 mg/kg, and 9 hr at 160 mg/kg). Ventricular HVA levels were significantly greater ($p < 0.001$) than lumbar levels

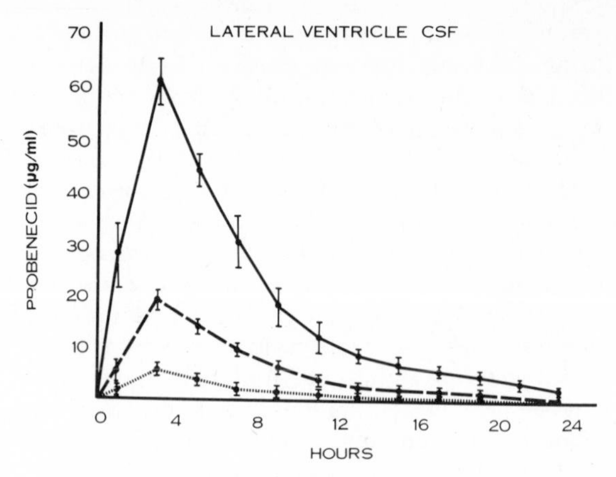

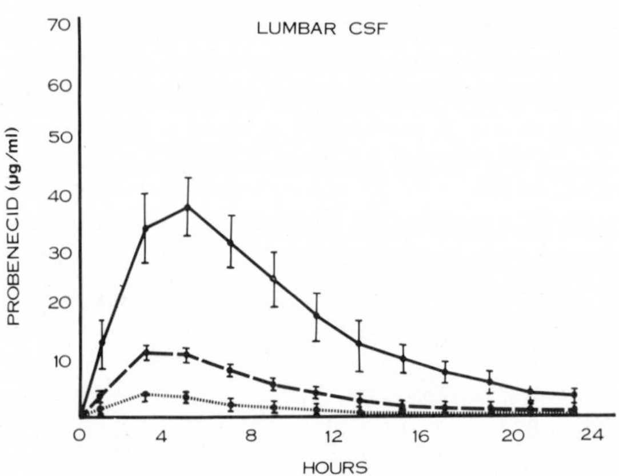

Figure 4. Probenecid concentration in lateral ventricular and lumbar CSF as function of time. Probenecid dose administered intravenously: (· · · · ·) 40 mg/kg; (– – –) 80 mg/kg; (——) 160 mg/kg.

at all probenecid doses as well as in the baseline period. In general, there were significant positive correlations between CSF levels of probenecid and HVA.

There was no significant difference in the probenecid-induced increase in ventricular 5-HIAA concentrations with the three doses of probenecid (Fig. 6). In lumbar CSF, in contrast, there was a significant increase ($p < 0.001$) in the probenecid-induced rise in 5-HIAA levels with increasing doses of probenecid (Fig. 6). Although baseline values for 5-HIAA in ventricular CSF were similar, significant correlations were obtained between CSF levels of 5-HIAA and probenecid.

Probenecid entered lumbar CSF within the first 2-hr sampling period at a time that cannot be entirely explained by bulk flow from the ventricle; direct transfer of probenecid from blood to lumbar CSF undoubtedly occurs. Since probenecid concentrations were lower in lumbar than in ventricular CSF, reabsorption may have occurred along the ventriculocisternal lumbar axis. A less likely possibility is that probenecid enters the lumbar space from the blood less readily than it enters the lateral ventricle. In addition, the time at which peak probenecid concentrations in lumbar CSF are reached was delayed at the 160 mg/kg dose. Thus, at a dose that produces the greatest blockade of acid transport from CSF, less probenecid may be able to leave the central nervous system and more probenecid is transferred by bulk flow to the lumbar space at a longer time interval.

There was an increase in the half-time for elimination of probenecid from CSF with increasing dose. This appears to be a direct reflection of changes in probenecid kinetics in the circulation, since preliminary studies in primates revealed the same prolon-

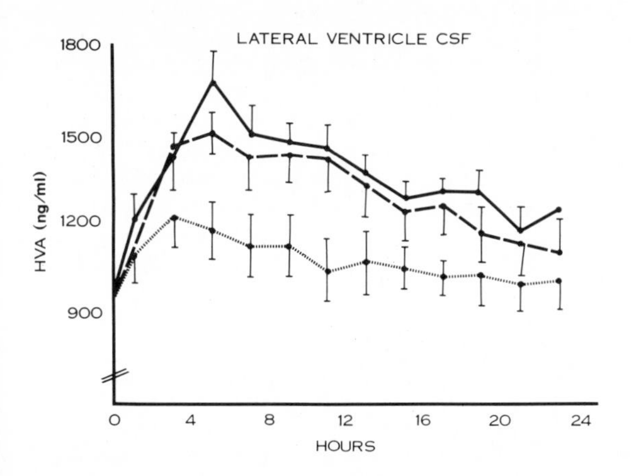

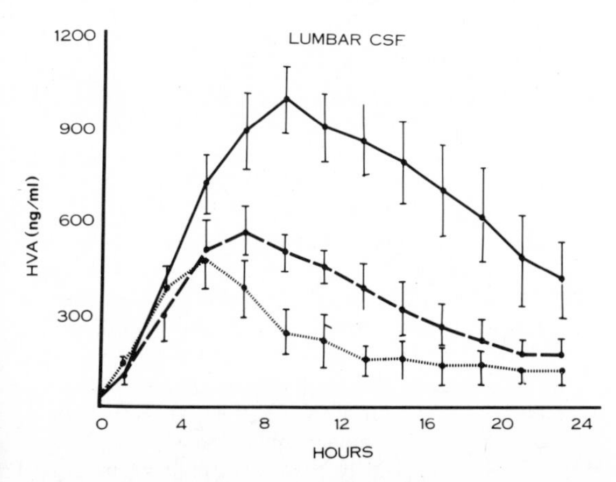

Figure 5. Homovanillic acid (HVA) concentration in lateral ventriclar and lumbar CSF as function of time following probenecid. Probenecid dose administered intravenously: (· · · · ·) 40 mg/kg; (– – –) 80 mg/kg; (——) 160 mg/kg.

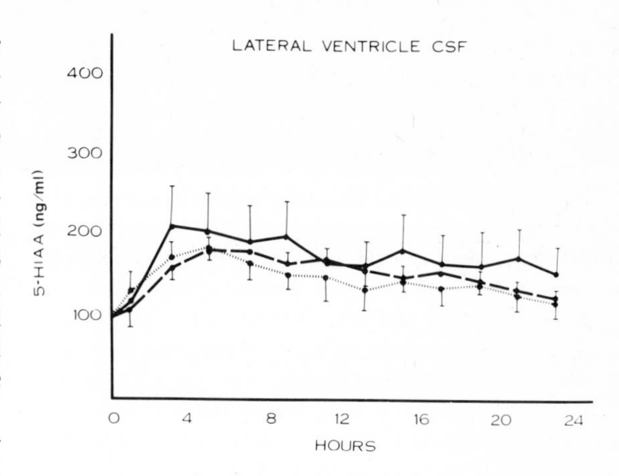

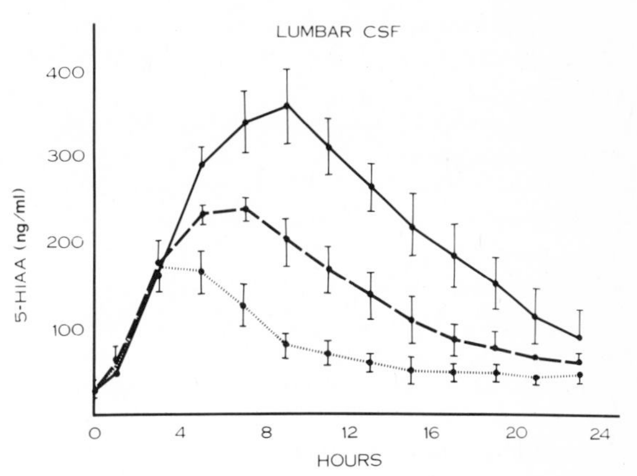

Figure 6. 5-Hydroxyindoleacetic acid (5-HIAA) concentration in lateral ventriclar and lumbar CSF as function of time following probenecid. Probenecid dose administered intravenously: (· · · · ·) 40 mg/kg; (– – –) 80 mg/kg; (——) 160 mg/kg.

gation of half-time in plasma with increasing doses of probenecid. Similar results were reported in an earlier study in dogs,[10a] and the same trend was noted in patients with Huntington's disease. Therefore, the clearance rate of probenecid decreased with increasing dose.

After the administration of probenecid, dose–response curves of a similar configuration are obtained for both 5-HIAA and HVA regardless of dose. The first portion of the curve represents accumulation of the monoamine metabolite in CSF during the period of transport blockage, whereas the downslope represents the excretion phase. Although there are relatively few points, the upgoing portion of all six curves appears parallel. Since the inflection points of both the 5-HIAA and the HVA curve occur at later times with increasing doses of probenecid, the blockade of acid transport appears to last longer with increasing dose. It may therefore be possible to extrapolate from the lower curves for the CSF monoamine metabolites obtained with smaller doses of probenecid to the highest curves at which complete block of transport may occur.

5-HIAA levels were higher in lumbar than in ventricular CSF at the 160 mg/kg dose, a reversal of the anticipated concentration gradient. However, this may reflect the fact that 5-HIAA levels are approximately twice as high in the fourth ventricle as in the third ventricle. Although the exact location of the transport system for HVA and 5-HIAA is uncertain, it must be distal to the lateral ventricle but proximal to the lumbar space, since there are smaller probenecid-induced changes in the former but a clear dose–response relationship in the latter location.

10. Conclusions

At least in the near future, studies of CSF biochemistry in man will continue to play an important role in clinical studies of brain amine metabolism. There remain many practical and theoretical problems in such studies. The probenecid test, although imperfect in many ways, has become one of the only widely utilized metabolic procedures to study brain monoamine function in man. In the last several years, new analytical methods have become available to more accurately measure amine metabolites in CSF. New approaches to CSF amine biochemistry are available, including the direct measurement of trace amounts of neurotransmitters and stable isotopic labeling of brain metabolism. Nevertheless, it is incumbent on clinical investigators in this field to work vigorously on alternative metabolic experiments that derive conclusions about brain metabolism of amines from serial observations in blood and urine. Only in this way can proper control subjects be studied, and data derived on large enough groups of patients and controls to draw consistent conclusions among research groups.

References

1. ADÈR, J.-P., AIZENSTEIN, M. L., POSTEMA, F., KORF, J.: Origin of free 3-methoxy-4-hydroxyphenylethylene glycol in rat cerebrospinal fluid. *J. Neurol. Trans.* **46**:279–290, 1979.

1a. ANDEN, N. E.: Distribution of monoamines and dihydroxyphenylalanine decarboxylase activity in the spinal cord. *Acta Physiol. Scand.* **64**:197–203, 1965.

2. ANDEN, N. E., ROOS, B. E., WERDINIUS, B.: The occurence of homovanillic acid in brain and cerebrospinal fluid and its determination by a fluorimetric method. *Life Sci.* **2**:448–458, 1963.

3. ANDERSON, E. G., HOSGERSON, L. O.: The distribution of 5-hydroxytrypatamine and norepinephrine in cat spinal cord. *J. Neurochem.* **13**:479–485, 1966.

4. ASHCROFT, G. W., DOW, R. C., MOIR, A. T. B.: The active transport of 5-hydroxyindole acetic acid and 3-methoxy-4-hydroxyphenylacetic acid from recirculatory perfusion system of the cerebral ventricles of the unanaesthetized dog. *J. Physiol.* **199**:397–425, 1968.

5. ASHCROFT, G. W., SHARMAN, D. F.: Measurement of acid monoamine metabolites in human CSF. *Br. J. Pharmacol.* **19**:153–160, 1962.

5a. AZMITIG, E. C.: The serotonin-producing neurons of the midbrain median and dorsal raphe nuclei. In Iversen, L. L., Iversen, S. D., and Snyder, S. H. (eds.): *Handbook of Psychopharmacology*, Vol. 9, *Chemical Pathways in the Brain.* New York, Plenum Press, 1978.

6. BARTHIOLINI, G., PLETSCHER, A., TISSOT, R.: On the origin of homovanillic acid in the cerebrospinal fluid. *Experientia* **22**:609–610, 1966.

7. BARTHOLINI, G., TISSOT, R., PLETSCHER, A.: Brain capillaries as a source of homovanillic acid in cerebrospinal fluid. *Brain Res.* **27**:163–168, 1971.

8. BOWERS, M. B., JR.: Fluorometric measurement of 5-hydroxyindoleacetic acid (5HIAA) and tryptophan in human CSF: Effect of high doses of probenecid. *Biol. Psychiatry* **9**:93–97, 1974.

9. CARLSSON, A., FALCK, B., FUXE, K., HILLARP, N. A.: Cellular localization of monoamines in the spinal cord. *Acta Physiol. Scand.* **60**:112–119, 1964.

10. CHASE, T. N., NG, L. K. Y.: Central monoamine metabolism in Parkinson's disease. *Arch. Neurol.* **27**:486–491, 1972.

10a. DAYTON, P. G., CUCINELL, S. A., WEISS, M., PEREL, J. M.: Dose-dependence of drug plasma level decline in dogs. *J. Pharmacol. Exp. Ther.* **158**:305–316, 1967.

10b. EBERT, M. H., COWDRY, R. W., POST, R. M., GOODWIN, F. K.: CSF probenecid levels in the oral probenecid test. *Psychopharmacology* (in press).

11. ECCLESTON, D., ASHCROFT, G. W., MOIR, A. T. B., PARKER-RHODES, A., LUTZ, W., O'MAHONEY, D. P.: A comparison of 5-hydroxyindoles in various regions of dog brain and cerebrospinal fluid. *J. Neurochem.* **15**:947–957, 1968.

12. EXTEIN, I., ROTH, R. H., BOWERS, M. B.: Accumulation of ^{3}H-homovanillic acid in rabbit brain and cerebrospinal fluid following intravenous ^{3}H-L-Dopa. *Biol. Psychiatry.* **9**(2):161–170, 1974.

13. FRI, C. G., WIESEL, F. A., SEDVALL, G.: Simultaneous quantification of homovanillic acid and 5-hydroxyindoleacetic acid in cerebrospinal fluid by mass fragmentography. *Life Sci.* **14**(12):2469–2480, 1974.

14. FORN, J.: Active transport of 5-hydroxyindoleacetic acid by the rabbit choroid plexus *in vitro*. *Biochem. Pharmacol.* **21**:619–624, 1972.

15. GARELIS, E., YOUNG, S. N., LAL, S., SOURKES, T. L.:

Monoamine metabolites in human clinical studies. *Brain Res.* **79**:1–8, 1974.

16. GERBODE, F., BOWERS, M. B.: Measurement of acid monoamine metabolites in human and animal cerebrospinal fluid. *J. Neurochem.* **15**:1053–1055, 1968.
17. GOODWIN, F. K., POST, R. M., DUNNER, D. L., GORDON, E. K.: Cerebrospinal fluid amine metabolites in affective illness: The probenecid technique. *Am. J. Psychiatry* **130**:73–79, 1973.
18. GORDON, E. K., OLIVER, J., BLACK, K., KOPIN, I. J.: Simultaneous assay by mass fragmentography of vanillyl mandelic acid, homovanillic acid, and 3-methoxy-4-hydroxy-phenylethylene glycol in cerebrospinal fluid and urine. *Biochem. Med.* **11**:32–40, 1974.
19. GORDON, E., PERLOW, M., OLIVER, J., EBERT, M., KOPIN, I.: Origins of catecholamine metabolites in monkey cerebrospinal fluid. *J. Neurochem.* **25**:347–349, 1975.
20. GULDBERG, H. C., ASHCROFT, G. W., CRAWFORD, T. B. B.: Concentration of 5-hydroxindolylacetic acid and homovanillic acid in the cerebrospinal fluid of the dog before and during treatment with probenecid. *Life Sci.* **5**:1571–1575, 1966.
21. KAROUM, F., GILLIN, J. G., WYATT, R. J., COSTA, E.: mass-fragmentography of nanogram quantities of biogenic amine metabolites in human cerebrospinal fluid and whole rat brain. *Biomed. Mass Spectrom.* **2**:183–189, 1975.
22. KARTZINEL, R., EBERT, M. H., CHASE, T. N.: Intravenous probenecid loading. *Neurology* **26**:992–996, 1976.
23. KESSLER, J. A., PATLAK, C. S., FENSTERMACHER, J. D.: Transport of 5-hydroxy-3-indoleacetic acid by spinal cord during subarachnoid perfusion. *Brain Res.* **116**:471–483, 1976.
24. KESSLER, J. A., FENSTERMACHER, J. D., PATLAK, C. S.: 3-Methoxy-4-hydroxyphenylethyleneglycol (MHPG) transport from the spinal cord during spinal subarachnoid perfusion. *Brain Res.* **102**:131–141, 1976.
25. KORF, J., VAN PRAAG, H. M.: Amine metabolism in the human brain: Further evaluation of the probenecid test. *Brain Res.* **35**:221–230, 1971.
26. KUPFER, D. J., BOWERS, M. B.: REM sleep and central monoamine oxidase inhibition. *Psychopharmacology* **27**:183, 1972.
27. LAKE, C. R., WOOD, J. H., ZIEGLER, M. G., EBERT, M. H., KOPIN, I. J.: Probenecid-induced norepinephrine elevations in plasma and CSF. *Arch. Gen. Psychiat.* **35**:237–240, 1978.
28. LEWANDER, T., SJÖSTRÖM, R.: Increase in the plasma concentration of free tryptophan caused by probenecid in humans. *Psychopharmacology* **33**:81–86, 1973.

28a. LINDVALL, O., BJÖRKLUND, A.: Organization of catecholamine neurons in the rat central nervous system. In Iversen, L. L., Iversen, S. D., and Snyder, S. H. (eds.): *Handbook of Psychopharmacology*, Vol. 9, *Chemical Pathways in the Brain*. New York, Plenum Press, 1978.

29. MCGEER, E. G., MCGEER, P. L.: Catecholamine content of spinal cord. *Can. J. Biochem. Physiol.* **40**:1141–1151, 1962.
30. MEEK, M. L., NEFF, N. H.: Is cerebrospinal fluid the major avenue for the removal of 5-hydroxindoleacetic acid from the brain? *Neuropharmacology* **12**:497–499, 1973.
31. MILLER, T. B., ROSS, C. R.: Transport of organic cations and anions by choroid plexus. *J. Pharmacol. Exp. Ther.* **196**:771–777, 1976.
32. MOIR, A. T. B., ASHCROFT, G. W., CRAWFORD, T. B. B., ECCLESTON, D., GUILDBERG, H. C.: Cerebral metabolites in cerebrospinal fluid as a biochemical approach to the brain. *Brain* **93**:357–368, 1970.
33. MOTAIS, R., COUSIN, J. L.: Inhibitor effect of probenecid and structural analogues on organic anions and chloride permeabilities in ox erythrocytes. *Biochim. Biophys. Acta* **419**:309–313, 1976.
34. NEFF, N. H., COSTA, E.: Application of steady-state kinetics to the study of catecholamine turnover after monoamine oxidase inhibition or reserpine administration. *J. Pharmacol. Exp. Ther.* **160**(1):40–47, 1968.
35. NEFF, N. H., TOZER, T. N., BRODIE, B. B.: Application of steady-state kinetics to studies of the transfer of 5-hydroxyindoleacetic acid from brain to plasma. *J. Pharmacol. Exp. Ther.* **158**:214–218, 1967.
36. PEREL, J. M., LEVITT, M., DUNNER, D. L.: Plasma and cerebrospinal fluid probenecid concentrations as related to accumulation of acidic biogenic amine metabolites in man. *Psychopharmacology* **35**:83–90, 1974.
37. PERLOW, M., EBERT, M., GORDON, E. K., ZIEGLER, M., LAKE, C. R., CHASE, T. N.: The circadian variation of catecholamine metabolism in the subhuman primate. *Brain Res.* **139**:101–113, 1978.
38. PORTIG, P. J., VOGT, M.: Release to the cerebral ventricles of substances with possible transmitter function in the caudate nucleus. *J. Physiol.* **204**:687, 1969.
39. POST, R. M., GOODWIN, F. K.: Estimation of brain amine metabolism in affective illness: Cerebrospinal fluid studies utilizing probenecid. *Psychother. Psychosom.* **23**:142–158, 1974.
40. POST, R. M., GOODWIN, F. K., GORDON, E., WATKIN, D. M.: Amine metabolites in human cerebrospinal fluid: Effects of cord transection and spinal fluid block. *Science* **179**:897–899, 1973.
41. REFSHAUGE, C., KISSINGER, P. T., DREILING, R., BLANK, L., FREEMAN, R., ADAMS, R. N.: New high performance liquid chromatographic analysis of brain catecholamines. *Life Sci.* **14**(2):311–322, 1974.
42. SHAYWITZ, B. A., COHEN, D. J., BOWERS, M. B., JR.: CSF monoamine metabolites in children with minimal brain dysfunction: Evidence for alteration of brain dopamine. A preliminary report. *J. Pediatr.* **90**(1):67–71, 1977.
43. SIEVER, L., KRAEMER, H., SACK, R., ANGWIN, P., BERGER, P., ZARCONE, V., BARCHAS, J., BRODIE, H. K. H.: Gradients of biogenic amine metabolites in cerebrospinal fluid. *Dis. Nerv. Syst.* **36**:13–16, 1975.
44. SJOQUIST, B.: Mass fragmentographic determination of 4-hydroxy-3-methoxymandelic acid in human urine,

cerebrospinal fluid, brain, and serum using a deuterium-labelled internal standard. *J. Neurochem.* **24**(1):199–201, 1975.

45. SJÖSTRÖM, R.: Steady-state labels of probenecid and their relation to acod monoamine metabolites in human cerebrospinal fluid. *Psychopharmacology* **25**:96–100, 1972.
46. UNGERSTEDT, U.: Stereotaxic mapping of the monoamine pathways in the rat brain. *Acta Physiol. Scand. Suppl.* **367**:1–122, 1971.
47. VAN PRAAG, H. M.: Significance of biochemical parameters in the diagnosis, treatment and prevention of depressive disorders. *Biol. Psychiatry* (in press).
48. WATSON, E., WILK, S.: Determination of probenecid in small volumes of cerebrospinal fluid. *J. Neurochem.* **21**:1569–1571, 1973.
49. WEIR, R. L., CHASE, T. N., NG, K. Y., KOPIN, I. J.: 5-Hydroxyindoleacetic acid in spinal fluid: Relative contribution from brain and spinal cord. *Brain Res.* **52**:409–412, 1973.
50. WERDINIUS, B. G.: Effects of probenecid on the levels of monoamine metabolites in the rat brain. *Acta Pharmacol. Toxicol.* **25**:18–23, 1967.
51. WILK, S., WATSON, B.: Evaluation of dopamine metabolism in rat striatum by a gas chromatographic technique. *Eur. J. Pharmacol.* **30**(2):238–243, 1975.
52. WINSBERG, B. G., HERWIC, M. H., PEREL, J.: Neurochemistry of withdrawal emergent symptoms in children. *Psychopharmacol. Bull.* **13**(3):38–40, 1977.
53. WOLFSON, L. I., KATZMAN, R., ESCRIVA, A.: Clearance of amine metabolites from the cerebrospinal fluid: The brain as a "sink." *Neurology* **24**:772–779, 1974.
54. WOLFSON, L. I., ESCRIVA, A.: Clearance of 3-methoxy-4-hydroxy-phenylglycol from the cerebrospinal fluid. *Neurology* **26**:781–784, 1976.
55. ZIEGLER, M. G., WOOD, J. H., LAKE, C. R., KOPIN, I. J.: Norepinephrine and 3-methoxy-4-hydroxyphenyl glycol gradients in human cerebrospinal fluid. *Am. J. Psychiatry* **134**:565–568, 1977.

CHAPTER 9

Extracellular Cyclic Nucleotide Metabolism in the Human Central Nervous System

Benjamin Rix Brooks, James H. Wood, Maria Diaz, Carol Czerwinski, Leon P. Georges, Jonas Sode, Michael H. Ebert, and W. King Engel

1. Introduction

The cyclic nucleotides—adenosine-3′,5′-cyclic monophosphate (cAMP) and guanosine-3′,5′-cyclic monophosphate (cGMP)—serve important metabolic functions in many mammalian tissues (Fig. 1). Glycogen metabolism is the best-studied example of the regulatory function of cAMP in the phosphorylation of phosphorylase b to phosphorylase a.[134] The proven and hypothetical roles of cyclic nucleotides in human nonneurological disease have been described in detail elsewhere.[7,12,13,69,84,97,149,151] The role of cyclic nucleotides in human neurological diseases is beginning to be explored.[24,52,53,158] The evidence defining the importance of cyclic nucleotides in many facets of central and peripheral nervous system functions is briefly reviewed below.[34,35,38,81,106,114]

Benjamin Rix Brooks, M.D. • Department of Neurology, Johns Hopkins University School of Medicine, Baltimore, Maryland 21205; Medical Neurology Branch, National Institute of Neurological and Communicative Disorders and Stroke, National Institutes of Health, Bethesda, Maryland 20205. **James H. Wood, M.D.** • Division of Neurosurgery, University of Pennsylvania School of Medicine and Hospital, Philadelphia, Pennsylvania 19104. **Maria Diaz, B.S.**, and **Leon P. Georges, M.D.** • Endocrinology Division, Department of Medicine, National Naval Medical Center, Bethesda, Maryland 20014. **Carol Czerwinski, Ph.D.** • Department of Obstetrics and Gynecology, Uniformed Services University of Health Sciences, School of Medicine, Bethesda, Maryland 20014. **Jonas Sode, M.D.** • Medical Service, Carbondale Veterans Administration Hospital, Carbondale, Illinois 62901; Department of Internal Medicine, Southern Illinois University School of Medicine, Carbondale, Illinois 62901. **Michael H. Ebert, M.D.** • Experimental Therapeutics Branch, National Institute of Mental Health, National Institutes of Health, Bethesda, Maryland 20205. **W. King Engel, M.D.** • Medical Neurology Branch, National Institute of Neurological and Communicative Disorders and Stroke, National Institutes of Health, Bethesda, Maryland 20205.

2. CNS Cyclic Nucleotide Metabolism

Central nervous system (CNS) tissues contain large amounts of cyclic nucleotides and the enzymes for their synthesis and destruction.[35,48] Immunocytochemical demonstration of cAMP in brainstem and cerebellum indicates higher cAMP content in neuronal soma and dendrites than in glia, endothelial cells, or pericytes.[9] No cAMP is demonstrable by this technique in white matter. Adenyl cyclase, which forms cAMP from ATP, is present in high amounts in the cortical gray matter of the cerebrum and cerebellum and the subcortical gray matter structures.[153,154]

Dopamine-responsive adenyl cyclase activity is found primarily in the caudate nucleus and peripheral sympathetic ganglia,[35] while norepinephrine-responsive adenyl cyclase is widespread, yet highest in cerebellum,[49] and also found in astrocytoma

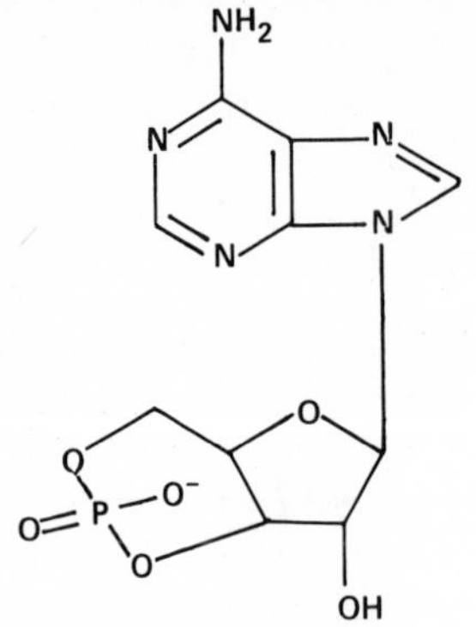

CYCLIC AMP (cAMP)
adenosine-3′,5′-cyclic monophosphate

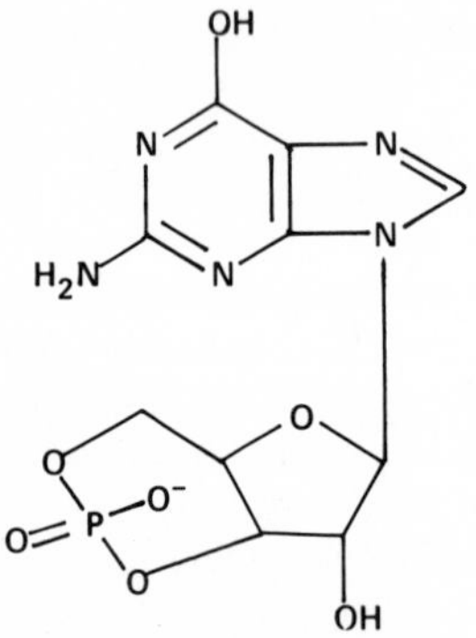

CYCLIC GMP (cGMP)
guanosine-3′,5′-cyclic monophosphate

Figure 1. Chemical structure of cyclic nucleotides.

cells in culture.[111] Histochemical localization of adenyl cyclase activity without lead-containing reagents indicate capillary and astrocytic membrane localization in addition to neuronal localization.[40,75,91] Adenyl cyclase is localized in particulate fractions of CNS tissue homogenates.[86,151] Enzyme activity is found in nuclear, microsomal, and synaptic subcellular membrane fractions.[153] A variety of physiological maneuvers[35] and pharmacological agents[114] affect adenyl cyclase and hence tissue cAMP concentrations *in vivo* and *in vitro*.[35] Regulation of tissue levels of cAMP is accomplished by cellular extrusion[119] and hydrolysis.[66]

Nucleotide-3′,5′-phosphodiesterases, which convert cyclic nucleotides to 5′-nucleotide monophosphates, are present at high concentration in CNS tissues of different species, including man.[42,152] Nucleotide 3′,5′-cyclic monophosphate diesterase activity is present in higher concentrations (100-fold) than adenyl cyclase in mammalian tissues.[152] The greatest activity is present in the gray matter of cerebral and cerebellar cortex, brainstem, and spinal cord. White matter structures (optic nerve, cerebellar white matter, and dorsal columns of spinal cord) contain smaller amounts of enzyme activity.[11,152] Enzyme activity is primarily localized to synaptic membranes and microsomal membranes, although more of this enzyme can be solubilized as compared with adenyl cyclase.[22] Cytochemical localization of nucleotide phosphodiesterase activity with the electron microscope indicates that the reaction product is located almost exclusively in postsynaptic dendritic process membranes.[54,55]

Biochemically, nucleotide phosphodiesterase activity is found in multiple forms in different amounts througout the CNS.[152] These different forms display unique biochemical properties.[76] The more soluble form preferentially reacts with cAMP only at high substrate concentration.[73]

Guanyl cyclase, which forms cGMP from GTP, is present in high amounts in gray matter throughout the CNS.[105] Both particulate and soluble forms of guanyl cyclase are present, although the soluble form predominates in chick neurons in culture.[65,105] Many pharmacological agents affect cerebellar cGMP concentrations *in vivo* but not *in vitro*, indicating that these effects are mediated indirectly.[48] At present, no receptor for guanyl cyclase activity, responsive at physiological concentrations of neurotransmitters, has been categorically demonstrated in the CNS.[123,124] Guanyl cyclase activity has not been demonstrated histochemically. Immunocytochemical demonstration of cGMP in the mouse cerebellum, contrasted with the diffuse intracellular localization of cAMP, indicates that the distribution of cGMP is primarily in proximity to cell membranes.[33] Quantitative analysis indicates that most of the cGMP is in the molecular layer, whereas cAMP is highest in the granular layer.[48] A variety of physiological maneuvers,[82] pharmacological agents,[49] and pathological states[17,96] affect cGMP concentrations *in vivo*. Agents or activities that excite or increase motor activity are associated with increased tissue cGMP concentrations particularly in the molecular layer of the cerebellum.[48] Nevertheless, the same activity or agent may have different effects on different parts of the CNS.[93–95]

Both particulate and soluble forms of cGMP phosphodiesterase activity are present in the CNS. The latter form is more commonly associated with astrocytes.[76]

The exact role of cyclic nucleotides in the CNS is not clearly delineated because the data, although numerous, are diverse, incomplete, sometimes species-specific, and sometimes contradictory.[35] In general, metabolic functions of nervous tissues are probably influenced by cyclic nucleotides in a similar fashion to metabolic changes seen in extraneural tissues.[114] In addition, strong evidence from animal studies implicates a role for cyclic nucleotides in electrochemical events within peripheral nervous system and CNS tissues.[8,82] Recent studies have indicated that the main effect of muscarinic cholinergic synaptic transmission was to elevate cGMP in the postsynaptic neurons, while the effect of dopaminergic synaptic transmission was to elevate cAMP in corresponding neurons.[83]

Electrophysiological and iontophoretic studies of the rat cerebellum indicate that norepinephrine-containing axons from the locus coeruleus have an inhibitory effect on Purkinje cells.[132] The spontaneous discharge rate of Purkinje cells was decreased by iontophoretically applied norepinephrine, cAMP, and dibutyryl cAMP. Pretreatment with theophylline, a phosphodiesterase inhibitor, intravenously, but not iontophoretically, markedly increased the response of Purkinje cells to norepinephrine. Iontophoretically applied acetylcholine and cGMP increased the firing rate of rat cortical pyramidal tract neurons.[142] Norepinephrine and cAMP had effects similar to those observed on Purkinje cells. All these studies, however, measured the possible extracellular role of cyclic nucleotides in the CNS. Iontophoresis of cAMP or cGMP into feline spinal α-motor neurons caused an acceleration of the action potential and a potentiation of the afterhyperpolarization, suggesting a more probable electrochemical role for cyclic nucleotides in synaptic facilitation.[88,89] Thus, evidence is accumulating for a role for cyclic nucleotides in postsynaptic events following electrical stimulation in nervous system tissue, but little evidence is found for a role of these nucleotides presynaptically.

In addition to modulation of short-term metabolic and electrochemical events, cyclic nucleotides may be important in differentiation and maintenance of neuron structures dependent on microtubule assembly.[39] Axon elongation from chick embryo dorsal root ganglion cultures was promoted by cAMP,[122] and cAMP can reverse the colchicine-induced block in axon elongation.[121] The physiological significance of these effects is uncertain.[56] Morphological and biochemical changes in mouse neuroblastoma cells in cell cultures are affected by cAMP. Neurites are formed, cell soma and nucleus increase in size, cell total RNA and protein increase, tyrosine hydroxylase, choline acetyltransferase, and acetylcholine esterase increase, sensitivity of adenyl cyclase to catecholamines increases, and the cells are less efficient at causing tumors.[118] Thus, important differentiated functions appear to be promoted directly or indirectly by cAMP in neuroblastoma cells in culture.

The role of cyclic nucleotides in CNS pathological states has only recently been explored. Cerebellar cGMP, but not cAMP, rises *in vivo* after administration of oxotremorine, glutamic acid, harmoline, or cold stress.[93-95] The mouse cerebellar mutant, "nervous" (nr), suffers selective loss of 90% of the cerebellar Purkinje cells during the second postnatal month.[90] These animals have decreased cerebellar cGMP, but not cAMP, compared to unaffected controls, and diazepam does not cause a further reduction in cerebellar cGMP as it does in control animals.[96] Studies with cerebellar slices *in vitro* show normal cGMP synthesis in response to kainic acid in both mutant and control mice.[128] Total guanyl cyclase activity is also identical in both groups of mice. At present, there is no definite explanation for the decreased tissue cGMP concentrations in the cerebellum of this mouse mutant. Another mouse mutant, "wobbler" (wr), manifests decreased spinal cord cGMP but not cAMP concentrations.[17] *In vitro* studies with this model of a disease affecting large motor neurons have not been accomplished. In contrast to the decrease in tissue concentrations of cGMP in the CNS parenchymal degenerations discussed above is the mouse retinal degeneration mutant, the retinal levels of cGMP of which accumulate due to a deficiency in the development of active cGMP phosphodiesterase activity as photoreceptor cells degenerate.[43]

3. Relationship of CNS Cyclic Nucleotide Metabolism to Extracellular Cyclic Nucleotide Metabolism in Animals

Although animal models of genetic or acquired disease permit evaluation of tissue concentrations of cyclic nucleotides, this capacity is not available for the study of human CNS cyclic nucleotide metabolism. Therefore, the relationship of cyclic nucleotide metabolism in CNS tissue to cyclic nucleotide metabolism in CNS extracellular fluid (CSF) *in vivo* must be delineated.

Electroconvulsive shock can raise cAMP in the brain by 300%.[63] By 1–3 hr following a single shock, cAMP will increase by 50% in the CSF of animals and be cleared by 3–6 hr.[23,100] Acute cerebral trauma will raise cAMP in the brain,[150] and within 1 hr following injury, CSF cAMP will increase by 50% and remain elevated through 14 days postinjury.[101]

The effects of pharmacological agents on CNS cyclic nucleotide metabolism and the reflection of changes in CNS cyclic nucleotide metabolism by the CSF have not yet been fully studied. A variety of animal investigations have related increases of CSF cAMP to particular pharmacological perturbations.[85,87] Increased tissue cAMP is seen in the caudate nucleus following L-DOPA.[57] Such tissue concentrations increase before increases are noted in the CSF following intraperitoneal administration of this drug.[85,87] In another investigation, a cataleptic dose of morphine was administered to ketamine-anesthetized rhesus monkeys. Decreased cerebellar cortical cGMP but not cerebral cortical cGMP was observed in one group that had biopsies taken 45 min following drug administration.[80] This decrease was mitigated by nalaxone administration. CSF cGMP rose between 15 and 60 min following morphine in another group of monkeys, and this increase could also be blocked by naloxone. Thus, under some circumstances, there may be a direct relationship between the content of CNS tissue cAMP and CSF cAMP, while in different situations there may be an inverse relationship between the content of CNS tissue cGMP and CSF cGMP.

Two studies in rats have suggested that multiple pharmacological systems may lead to increased CSF cAMP.[85,87] Norepinephrine, isoprenaline, dopamine, or adenosine when administered intraventricularly will increase CSF cAMP, but serotonin and histamine will not.[87] The dopamine effect is delayed compared to the norepinephrine effect. The dopamine effect is inhibited by haloperidal, but the norepinephrine effect is not. However, the norepinephrine effect is blocked by propranolol. In another species, rabbit, intracisternal histamine also did not raise CSF cAMP.[130]

Animal studies therefore show possible contrasting relationships between the tissue concentration of cyclic nucleotides and the observed concentration in the extracellular fluid (CSF). Further studies are required to define the exact kinetic relationship between the changes in cyclic nucleotide metabolism in varying regions of the CNS and the observed changes in CSF cyclic nucleotide metabolism for a variety of subprimate and primate species.[81,112,113]

4. Assay of Cyclic Nucleotides in Human CSF

A variety of assay methods have been used to measure cyclic nucleotide concentrations in human CSF. CSF cAMP (Table 1) ranges between 5 and 30 pmol/ml in a variety of neurologically stable patients whether measured by protein-binding assays,[25,27,28,58,101,115,133,148,157] radioimmunoassays,[18,19,62] or enzyme-coupling assays.[14,98,120] In only four studies has validity of the measured CSF cAMP been proven by chromatographic separation.[14,98,120,157] Similar values have been obtained with neet, perchloric-acid-treated, or trichloroacetic-acid-treated CSF.

Interfering substances most commonly plague protein-binding assays.[60,61] Brain extracts can show biphasic effects on protein-binding assays[159] by inhibiting binding at high concentrations of extract (apparent increased cAMP concentration when none is actually present) or promoting binding at low concentrations of extract (apparent decreased cAMP concentration when some is actually present). These effects are due to nondialyzable, presumably protein, constituents, since albumin will cause similar effects.[146] Similar studies with human CSF that contains little protein have not been reported. The effect of such interfering substances may be minimized by preparing standard curves in pooled activated-charcoal-treated CSF–buffer mixtures.[133] Nevertheless, unknown CSF samples may also contain sufficient concentration of other cyclic nucleotides (presumably cGMP) that may yield increased apparent "cAMP" concentrations due to cross-reactivity with the binding protein.[44]

Radioimmunoassay (RIA) is perhaps the easiest, most reproducible means to measure cyclic nucleotides in human CSF. It is less susceptible to interfering substances[45,138] and permits the use of neet or diluted CSF directly in the reaction mixture.[57,126,133] Both cAMP and cGMP can be measured by RIA.[138–141] CSF cGMP (Table 2) ranges between 0 and 15 pmol/ml in a variety of neurologically stable patients when measured by RIA[18,19,62] or enzyme coupling.[14] In only one study has validity of the measured CSF GMP been proven by chromatographic separation.[14,15] Possible interfering substances for cAMP and cGMP RIAs of human CSF have not been reported.[44,129,138–141]

4.1. Radioimmunoassay of Cyclic Nucleotides

Adenosine-3′-5′-cyclic monophosphate and guanosine-3′,5′-cyclic monophosphate were measured in human CSF employing commercially available RIA preparations (Schwartz-Mann). Antibody was raised in rabbits to the succinyl–cAMP– or cGMP–albumin conjugate.[138–141] To obtain standard curves, known amounts of cAMP or cGMP were

Table 1. cAMP in Human Lumbar CSF

Method and pretreatment	Patients	Number	cAMP (pmol/ml) Mean	Range	Ref. Nos.
Radioimmunoassay					
None	No CNS disease	59	10.4	1–24	18, 19
None	Lumbar disc disease	17	7.5	0–17	79
None	No neurological disease				
	Adults	14	21.0	5–38	126
	Children	9	21.0	0–38	
Perchloric acid	No CNS disease	16	26.3	10–42	104
Acetylation	Low back pain	14	8.7	0–25	62
Protein binding					
None	No CNS disease	27	17.0	7–35	133
None	No CNS disease	9	11.1	8–16	28, 115
	CNS disease	10	14.0	6–24	
None	Orthopedic disease	20	14.8	12–20	58
	Neurological disease	11	10.0	7–13	
None	No CNS disease	11	36.3	16–52	148
Boiling for 3 min	Neurological disease	—	14.1	5–28	10
Trichloroacetic acid	No CNS disease	13	24.0	16–32	101
Chromatographic separation	No CNS disease	8	17.0	10–30	157
Enzyme coupling					
Chromatographic separation	CNS disease	16	31.1	24–40	98
Chromatographic separation	CNS disease	6	—	5–22	14
Chromatographic separation	Neurological disease	—	16.2	—	120

Table 2. cGMP in Human Lumbar CSF

Method and pretreatment	Patients	Number	cGMP (pmol/ml) Mean	Range	Ref. Nos.
Radioimmunoassay					
None	No CNS disease	32	2.4	0–4	18, 19
None	No neurological disease				
	Adults	14	2.4	0–5	126
	Children	9	2.1	0–10	
	Low back pain	18	1.6	0–10	
None	CNS disease	10	2.5	0–11	79
	Low back pain	25	0.7	0–2	
None	No CNS disease	12	3.0	1–6	133
None	No neurological disease	9	3.6	2–7	41
None	No neurological disease	16	1.9	0–4	147
Acetylation	No CNS disease	14	3.4	0–12	62
Enzyme coupling					
Chromatographic separation	CNS disease	6	—	0–7	14

added to each tube in duplicate containing a standard amount of either anti-succinyl–cAMP or anti-succinyl–cGMP antibody, and either succinyl–cAMP– or succinyl–cGMP–tyrosine methyl ester labeled with ^{125}I in 0.05 M sodium acetate buffer, pH 6.2. Binding was allowed to proceed for 18–20 hr at 4°C, followed by precipitation of the antibody complex with 60% saturated ammonium sulfate. The tubes were centrifuged for 10 min at 600 *g* to pack the precipitate, and the supernatant was decanted. The tubes were then counted on a solid crystal scintillation counter. The ratio of the antibody-bound succinyl–cAMP– or succinyl–cGMP–[^{125}I]tyrosine methyl ester in the presence of added cyclic nucleotide (B) to antibody-bound labeled cyclic nucleotide in the absence of added cyclic nucleotide (B_0) was obtained for duplicate standards of known amounts of cyclic nucleotides added. Standard curves prepared in Elliott's B solution (artificial CSF) showed no shift of the binding equilibrium compared to standard cyclic nucleotides in buffer alone (Fig. 2).

CSF was obtained at the time of lumbar puncture and cooled at 4°C prior to being frozen at −70°C within 30–45 min. CSF was thawed just prior to assay. Samples of pre-probenecid CSF for cGMP assay were concentrated by evaporation for 1–2 hr at 50°C and taken up in 0.5 ml 0.05 M sodium acetate buffer, pH 6.2, just prior to assay. Samples of pre-probenecid CSF for cAMP assay were diluted directly in 0.05 M acetate buffer, pH 6.2, to a final dilution of 1:5 prior to assay in duplicate. CSF samples from lumbar punctures in patients receiving probenecid were diluted 1:10 and 1:20 prior to assay. The intraassay coefficient of variation ranged from 12 to 16% for cAMP and from 11 to 17% for cGMP in plasma, urine, or CSF. The interassay coefficient of variation ranged from 13 to 20% for cAMP and from 16 to 24% for cGMP in plasma, urine, or CSF. These ranges for the coefficient of variation are to be expected when ejection micropipettes are used. If microsyringes are used, the respective coefficients of variation may be significantly reduced.[62]

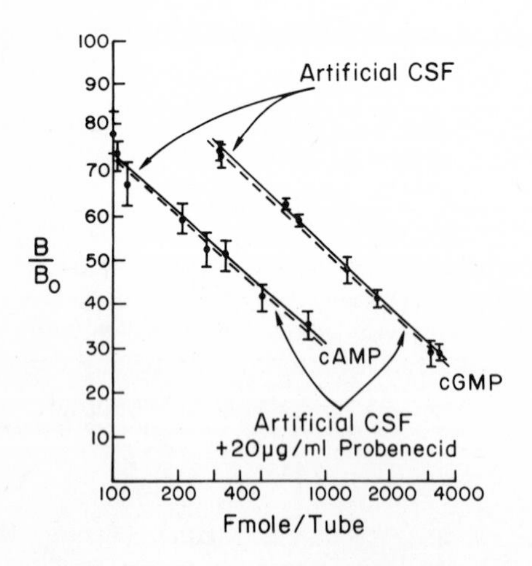

Figure 2. Radioimmunoassay (RIA) curves for cyclic nucleotides in artificial CSF. Standard solutions of weighed exogenous cyclic nucleotides (Sigma Biochemical) were prepared in Elliott's B solution with (- - -) and without (——) added probenecid (20 μg/ml). Samples in standard 0.05 M acetate buffer, pH 6.2, yielded identical results.[141] Each point (±1 S.D.) represents 4–8 separate assays performed in duplicate.

The diluted baseline CSF samples showed proportional dilutional behavior at 50, 100, and 200 μl added sample volume in the cAMP RIA. The baseline CSF samples for the cGMP RIA do not show proportional dilutional behavior above 100 μl added sample. With this latter assay, the binding curve is not changed significantly by changes in pH in the range from pH 4.0 to 8.0, or by changes in ionic strength. Standard curves prepared with exogenous cyclic nucleotide in artificial CSF (Elliott's B solution) show no significant difference in the presence or absence of 20 μg/ml probenecid (Fig. 2). Addition of exogenous cAMP or cGMP to nine human CSF samples prior to assay resulted in quantitative recoveries of 109 ± 2% for cAMP and 115 ± 3% for cGMP. Incubation of CSF samples with 1–3 mU of beef heart phosphodiesterase for 1 hr completely removed cyclic nucleotide activity measured by RIA (see Fig. 4).

Recent reports indicate that pretreatment of unknown samples by 2′-*O*-succinylation with triethylamine allows increased sensitivity of the RIA.[50,62,70] The assay reported here for cAMP is sensitive between 200 and 500 fmol/tube (see Fig. 2), whereas succinylation allows detection between 20 and 200 fmol/tube. The cGMP assay without succinylation is sensitive between 500 and 2000 fmol/tube. These new techniques for RIA of cyclic nucleotides allow greater dilution of the unknown samples in assay buffer and will allow for greater precision of measurement. In one recent study of a small number of lumbar CSF samples, the same range of cAMP and cGMP concentrations in the CSF were observed.[62]

4.2. Probenecid Protocols

Patients received no hypnotic medication for at least 3 days prior to lumbar puncture, but all patients on anticonvulsant medications remained on these medications throughout the investigation. All lumbar punctures in the baseline state were performed, after an overnight fast, between 7:00 A.M. and 9:00 A.M. following 8–12 hr of absolute bedrest unless otherwise indicated. After baseline lumbar puncture, patients received either oral probenecid (100 mg/kg) in four divided doses for 18 hr or intravenous probenecid (40 mg/kg) in normal saline for 70–90 min. The second lumbar puncture for the oral probenecid group occurred at 18 hr following the initiation of probenecid.[116] The second and third lumbar punctures for the intravenous group occurred at 6 and 10 hr following the initiation of the probenecid infusion.[78] CSF probenecid concentrations were measured by electron capture–gas chromatography.[78]

4.3. Properties of Cyclic Nucleotides in Human CSF

Fresh human CSF stored at 4–25°C for 24 hr lost 80–90% of the initial cAMP and 90–100% of the initial cGMP present as measured by RIA (Fig. 3). The nature of the reaction products has not been precisely defined. Frozen CSF stored at −70°C showed no decrement in cAMP or cGMP compared with freshly assayed human CSF samples. Freezing to −20°C and thawing twice significantly lowers cGMP in both baseline and probenecid-stimulated CSF samples measured in the same RIA (Table 3). No change is seen in CSF cAMP under these conditions.

Incubation of previously frozen human CSF samples collected in poly-carbonate tubes at 25°C or 37°C indicated that there may be persistent degradation of cAMP (Fig. 4). CSF samples collected in unsiliconized borosilicate vials before freezing showed no evidence of this activity. Similar degradation of

Table 3. Effect of Freeze–Thaw Cycles on Measurement of cGMP in the Human CSF RIA

Patients with neurological disease[a]	cGMP (pmol/ml)[b]	
	First thaw	Second thaw
Pre-probenecid	3.6 ± 1.1	0.9 ± 0.8
Post-probenecid	15.5 ± 2.0	5.5 ± 1.8

[a] Patients with seizure disorders (three patients) and movement disorder (one patient).
[b] Means ± S.E.M. All measurements were performed by RIA[139] in duplicate on the same day.

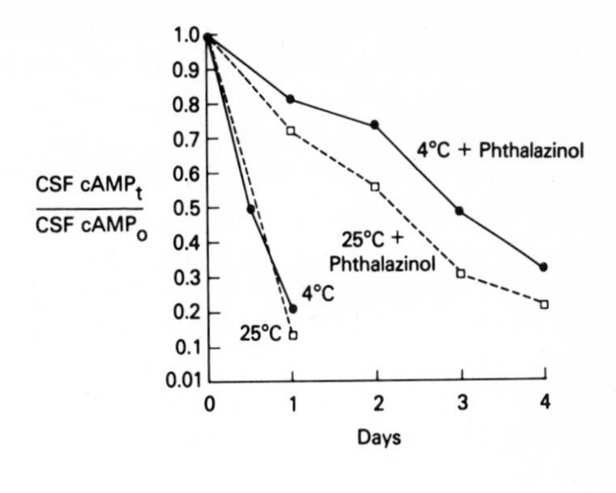

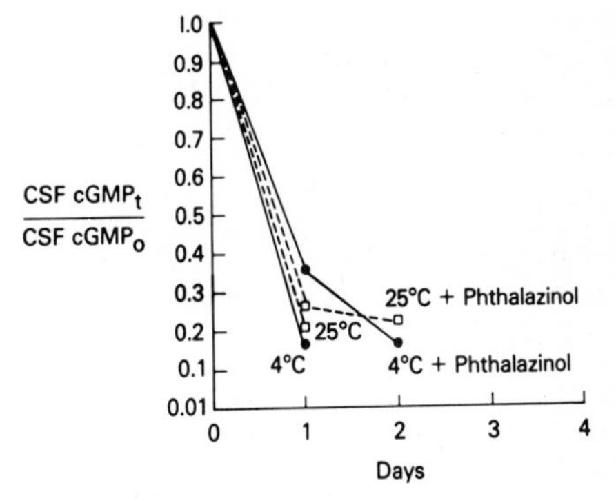

Figure 3. Degradation of cyclic nucleotides in freshly collected CSF. CSF was aliquoted into 1-ml volumes, and two aliquots were frozen at various times after incubation at 4°C and 25°C. Samples from two patients receiving the phosphodiesterase inhibitor phthalazinol[21] were compared with samples from two patients not receiving this drug. Initial cyclic nucleotide concentrations were measured on samples frozen immediately on dry ice (cAMP: 15.2, 10.3, 9.8, 12.1 pmol/ml; cGMP: 7.2, 6.3, 4.5, 4.0 pmol/ml).

cGMP in frozen samples of human CSF collected in either polycarbonate tubes or borosilicate glass vials did not occur. Cyclic nucleotide phosphodiesterase activity has been concentrated from the ventricular CSF of humans with nonmalignant brain tumors and increased intracranial pressure, but not from lumbar CSF of patients without tumors.[74] The concentrated phosphodiesterases in CSF from brain tumor patients behave kinetically similar to the cAMP and cGMP phosphodiesterases isolated from human cerebral cortex and are stable at −80°C. The reason for the difficulty in demonstrating cyclic nucleotide phosphodiesterase activity in frozen normal human CSF is unclear. At present, only degradation consistent with persistent cAMP phosphodiesterase activity can be demonstrated in previously frozen CSF.

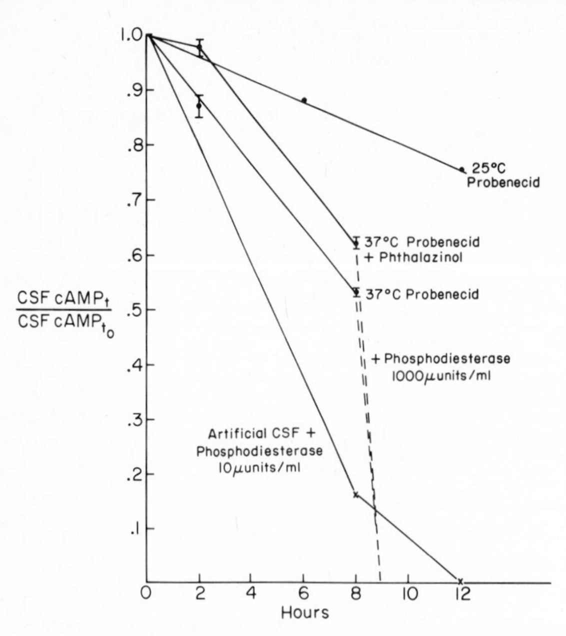

Figure 4. Degradation of cyclic nucleotides in frozen CSF. CSF was collected as described from five patients before and after phthalazinol administration.[21] CSF samples from patients at 10 hr following intravenous probenecid were frozen at −70°C. Samples were thawed and incubated at 25°C (two patients) and at 37°C (five patients) for various intervals before cAMP determination by radioimmunoassay.[139] Each point (±S.E.M.) represents two separate assays performed in duplicate. After 8-hr hydrolysis, 1 mU/ml beef heart phosphodiesterase (Sigma Biochemical) was added per tube, and hydrolysis was continued for 1 hr before cAMP determination. Initial cAMP concentrations ranged between 40 and 55 pmol/ml. Artificial CSF consisted of 20 pmol/ml cAMP in Elliott's B solution to which 10 μU/ml beef heart phosphodiesterase was added just prior to hydrolysis at 25°C.

5. CSF Cyclic Nucleotide Concentrations

Lumbar CSF cyclic nucleotide concentrations should be measured in patients free of CNS disease, other systemic disease, endocrinopathy, or exogenous medications to provide an adequate range of normal CSF cyclic nucleotide concentrations. Several studies (see Tables 1 and 2) have evaluated CSF cyclic nucleotide levels in orthopedic patients, lumbosacral disc patients, or neuromuscular disease patients with no CNS involvement.[18, 19, 25, 27–29, 58, 62, 71–73, 101–104, 115, 120, 126, 130, 156] In general, the baseline CSF cyclic nucleotide concentrations in control patients can be divided into high, medium, and low values, regardless of the assay method or pretreatment of the CSF samples (Table 1). For cAMP, values ranging between 5–20 pmol/ml[14,15,18,19,25,27,28,58,62,115,120,133,156] and 15–40 pmol/ml[98,103,104,126,148] have been observed. Chromatographic isolation of cAMP from CSF corrected for recoveries would favor the lower range.[14,15,120,156] For cGMP, values between 0–3 pmol/ml[18,79,126,147] and 1–7 pmol/ml[6,19,41,133] have been observed. Chromatographic isolation of cGMP from CSF corrected for recoveries would favor the higher range.[14,15] The observed instability of cGMP on repeated freezing and thawing could explain the number of studies that have reported CSF cGMP concentrations in the lower range.[18,147] If cAMP and cGMP concentrations are to be determined on the same CSF sample, one thawing and refreezing could lower the observed cGMP concentration when different assays are to be done on different days. Therefore, all CSF samples should be aliquoted prior to freezing and thawed only prior to assay.

In 22 myopathy patients free of CNS disease, endocrinopathy, or exogenous medication, the CSF cyclic nucleotide concentrations are approximately one half to two thirds of the plasma concentration [CSF cAMP = 2.2 + 0.6 (plasma cAMP) ± 3.2; $r = 0.5806$, $p < 0.01$; CSF cGMP = −1.2 + 0.5 (plasma cGMP) ± 0.8; $r = 0.8509$, $p < 0.01$]. There is no significant correlation of CSF cyclic nucleotide concentration with age,[18,102–104] sex,[19,102–104,126] weight, pulse rate, or blood pressure. Cyclic nucleotide concentrations in the CSF do not correlate with CSF cell count, glucose, protein, or immunoglobulin concentrations.[18,19] There is, however, a correlation between CSF cAMP and CSF cGMP ($r = 0.5100$, $p < 0.01$), as well as between CSF cAMP, but not plasma cAMP, and CSF γ-aminobutyric acid ($r = 0.40$, $p < 0.01$). A variety of weak correlations also exist between the CSF cyclic nucleotide concentrations and the concentrations of homovanillic acid (HVA) and 5-hydroxyindoleacetic acid (5-HIAA), but not norepinephrine or 3-methoxy-4-hydroxyphenylethylene glycol (MHPG). In systemic lupus erythematosus with CNS involvement, there was a weak correlation between the CSF leukocyte count and the CSF cGMP concentration that was not seen in control patients.[79]

5.1. Reproducibility of CSF Cyclic Nucleotide Determinations

Reproducibility of CSF cyclic nucleotide determinations under identical pharmacological and physiological conditions is a prerequisite for further study of cyclic nucleotide metabolism in the human

Table 4. Lack of Rostrocaudal Gradient for Cyclic Nucleotides in Human CSF

Patients with neurological disease[a]	Pre-probenecid aliquot			Post-probenecid aliquot		
	0–5 ml	17–22 ml	33–38 ml	0–5 ml	17–22 ml	33–38 ml
cAMP (pmol/ml)[b]	11.0 ± 1.5	10.7 ± 1.1	11.3 ± 1.0	34.9 ± 3.7	33.4 ± 3.1	39.2 ± 4.5
cGMP (pmol/ml)[b]	6.0 ± 0.1	7.2 ± 0.2	6.4 ± 0.7	15.0 ± 1.6	16.8 ± 2.0	17.0 ± 2.1

[a] There was a total of 13 patients with seizure disorders on anticonvulsant medication as well as patients with movement disorders, pain, and tumor.
[b] Means ± S.E.M.

CNS. In 71 duplicate determinations at different times in 26 patients, baseline and post-probenecid CSF cAMP concentrations showed excellent correlation ($r = 0.9369$, $p < 0.001$) with a slope approaching identity (Fig. 5A). Similar results were obtained for CSF cGMP concentrations under identical conditions ($r = 0.8664$, $p < 0.001$) for 56 duplicate determinations in 26 patients (Fig. 5B).

5.2. Apparent Lack of Rostrocaudal Gradient for CSF Cyclic Nucleotides

Several biological amine metabolites, including HVA and 5-HIAA, demonstrate rostrocaudal gradients in the CSF consistent with increased cerebral rather than spinal cord contributions to the observed metabolite concentration in the CSF.[117,164] Cyclic nucleotide concentrations were measured in the CSF of six seizure patients and five nonseizure patients after at least 10 hr of bedrest. Aliquots from three volumes were taken at 0–5, 17–22, and 33–38 ml of continuous CSF removal.[164] Baseline and post-probenecid concentrations of CSF cAMP and cGMP showed no concentration gradient (Table 4).

Two previous studies employing a protein-binding assay suggested differences between ventricular and lumbar CSF cAMP concentrations.[30,148] Two recent studies employing the RIA method could demonstrate no difference between lumbar and ventricular CSF cAMP in patients without increased intracranial pressure.[52,53,125] Our own previous studies with electrical stimulation of the caudate nucleus or cerebellum in humans demonstrated no evoked

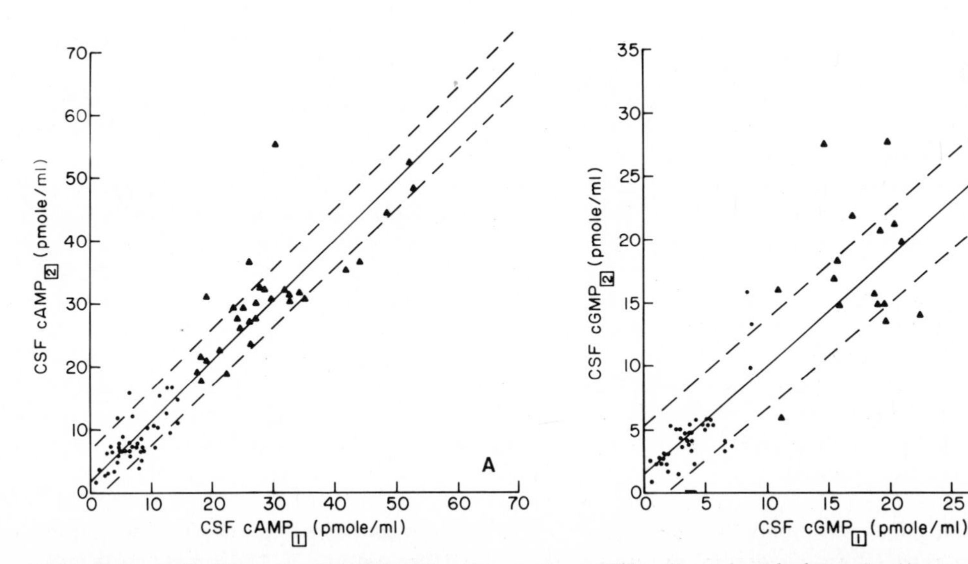

Figure 5. Reproducibility of cyclic nucleotide concentrations. CSF cyclic nucleotide determinations were performed in 26 patients under identical pharmacological and physiological conditions. Determinations were made on baseline CSF (●) and following probenecid (▲). (A) 71 duplicate determinations: $(cAMP)_2 = 1.7 \pm 0.95 \cdot (cAMP)_1 \pm 4.4$; $r = 0.9369$, $p < 0.001$. (B) 56 duplicate determinations: $(cGMP)_2 = 1.5 \pm 0.85 \cdot (cGMP)_1 \pm 3.5$; $r = 0.8664$, $p < 0.001$. (——) Mean; (– – –) ± one S.D. (standard deviation).

alteration in lumbar CSF cyclic nucleotide concentrations at a time when significant changes in CSF norepinephrine and GABA had occurred.[19,162,163] In addition, lumbar CSF cAMP and cGMP concentrations did not change chronically following unilateral frontal lobectomy.[19]

Increased clearance of cyclic nucleotides from the fourth ventricle could dampen observed changes in the lumbar CSF. The direct observation of ventricular CSF cyclic nucleotide concentrations, however, suggests that this is apparently not the case.[52,53,125,126]

6. CSF Cyclic Nucleotide Concentrations in Neurological and Psychiatric Disease

Baseline CSF cyclic nucleotide concentrations have been measured in patients with a variety of neurological and psychiatric diseases to see whether there is any obvious evidence for disordered cyclic nucleotide metabolism in these diseases. Increased plasma and CSF cAMP concentrations have been observed in patients within the first few days after cerebral infarction in man.[145,157,158] These changes are similar in magnitude to those observed in animals following ischemic injury.[100,101] Seizures in man are reported to produce elevations in CSF cAMP within the first 3 days postictally,[101] but this was not confirmed in two studies of depressed patients receiving electroconvulsive therapy.[58,115] Migraine,[155] aseptic meningitis,[72] and bacterial meningitis[73] are reported to raise the baseline CSF cAMP concentration. Acute cerebral trauma is reported to raise CSF cAMP,[98] but several head injury patients in coma had decreased cAMP that increased toward normal during the course of recovery.[52,53,125] With the protein-binding assay, CSF cAMP is reportedly elevated in hepatic coma,[148] but a recent report employing the RIA indicated a decreased CSF cAMP during hepatic coma that rose following charcoal hemoperfusion.[59] The baseline concentrations of CSF cAMP in Parkinson's disease, spinocerebellar degeneration, Huntington's chorea, dystonia, dyskinesia, neuropathy, and myopathy are not significantly different from normal.[25,27,28]

Abnormalities in urinary cAMP metabolism in manic–depressive patients compared to that in control patients sparked efforts by several groups to determine whether CNS cyclic nucleotide metabolism was altered in a variety of psychiatric diseases.[10,25,27,58,115,133] No significant differences in baseline CSF cAMP concentrations were found between patients with either orthopedic or neurological disease (control) and psychiatric patients (manic–depressive disease, character disorders, or schizophrenia). The stage of the disease process or pharmacological therapy had no effect on baseline CSF cAMP concentrations. Lithium, tricyclic antidepressants, Piribedil, Pimozide, tryptophan, and methadone did not effect baseline CSF cAMP. Neuroleptics[6] and ethanol[108] may decrease baseline CSF cAMP, but the measured concentrations of CSF cAMP in these reports were above the accepted values. In only one study of psychotic patients following oral probenecid administration has a significant reduction in CSF cAMP been demonstrated after chlorpromazine administration.[10] Cerebellar or caudate stimulation in man does not effect the lumbar CSF cAMP concentration.[162,163]

CSF cGMP concentrations have been less extensively studied. CSF cGMP levels are increased in patients with increased intracranial pressure,[52,63,125] brain tumors,[147] and systemic lupus erythematosus with CNS involvement.[79] One recent report suggested that CSF cGMP was decreased in Parkinson's disease.[5] No significant difference from controls in CSF cGMP concentrations was observed in psychiatric patients when compared to control patients.[5,133] The same group that reported that neuroleptics decreased CSF cAMP noted in a separate paper that these drugs increased CSF cGMP in treated schizophrenics.[41] Cerebellar or caudate stimulation in man does not affect the lumbar CSF cGMP concentration.[162,163]

Baseline CSF cyclic nucleotide concentrations were measured in 154 patients with a variety of neuromuscular and neurological disorders (Fig. 6). Patients with muscle disease[62] were considered to be "disease controls" having varying amounts of weakness but no dramatic involvement of the CNS. CSF cAMP [10.4 ± 0.5 (S.E.M.) pmol/ml] and CSF cGMP (2.4 ± 0.3 pmol/ml) in these weak patients were not significantly different from the reported values for nondisease control patients.[58,62,79] Similar values were obtained for patients[39] with peripheral nerve disease. In 40 patients with various forms of the amyotrophic lateral sclerosis complex (amyotrophic lateral sclerosis, spinal muscular atrophy, and progressive bulbar palsy), the mean CSF cAMP (5.9 ± 0.5) level was decreased 40–55% and the mean CSF cGMP concentration (1.6 ± 0.1) was decreased 33–50% as compared to those concentrations in muscle and nerve disease patients ($p < 0.01$, two-tailed Student's t test). In contrast, both mean CSF cyclic nucleotide concentrations were slightly in-

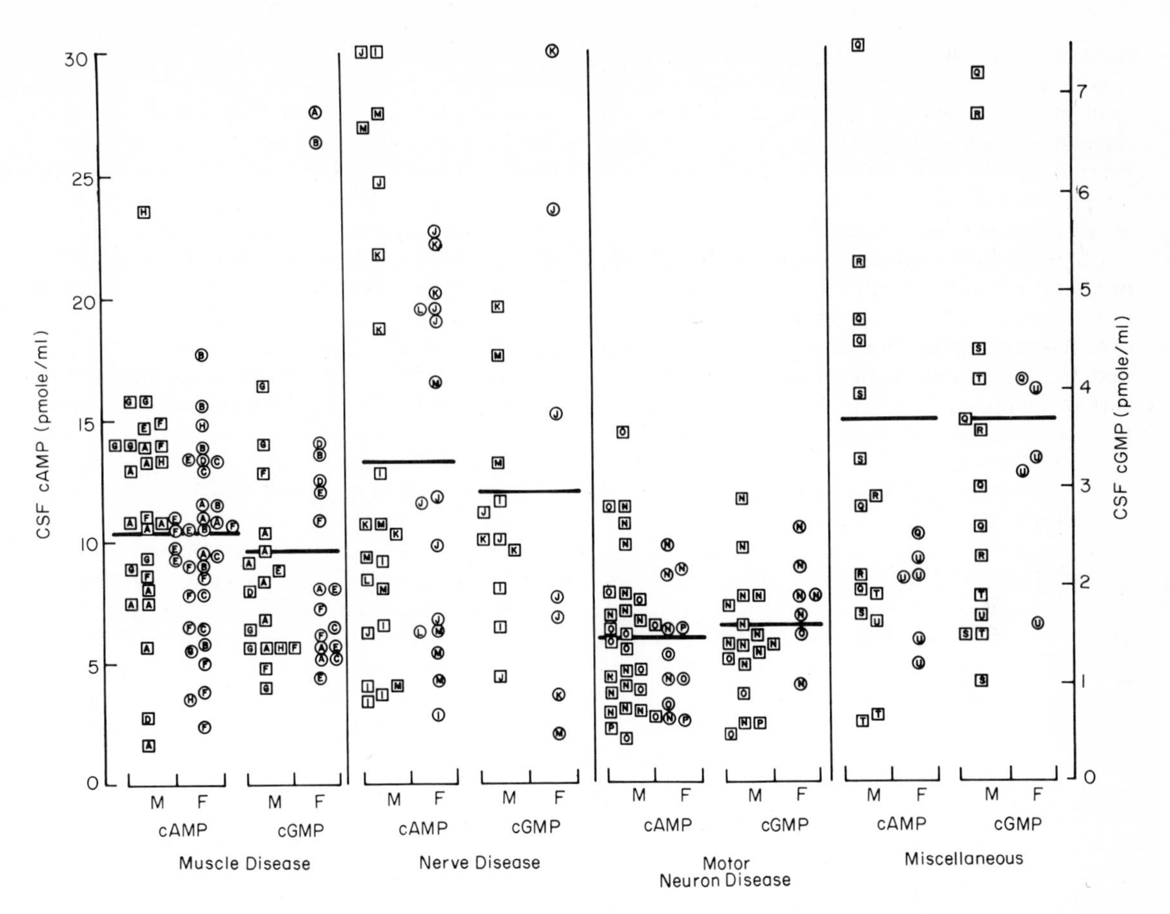

Figure 6. CSF cyclic nucleotide concentrations in neurological disease. CSF cyclic nucleotide concentrations were determined by radioimmunoassay[139] in males and females with a variety of neurological disorders: (A) myopathy; (B) type II fiber atrophy, no prednisone; (C) type II fiber atrophy, prednisone; (D) polymyositis, no prednisone; (E) polymyositis, prednisone; (F) myasthenia gravis; (G) myotonic atrophy; (H) oculocranial somatic disease; (I) motor neuropathy; (J) sensory neuropathy; (K) relapsing polyneuropathy, no prednisone; (L) relapsing polyneuropathy, prednisone; (M) peroneal muscular atrophy; (N) amyotrophic lateral sclerosis; (O) spinal muscular atrophy; (P) progressive bulbar palsy; (Q) demyelinating disease; (R) familial spastic paraparesis; (S) CNS vascular or degenerative lesions with spasticity; (T) arteriovenous malformation of the spinal cord; (U) cerebellar ataxia of diverse etiology.

creased in 24 patients with a variety of disorders associated with increased spasticity. It is of interest that the CSF cAMP/cGMP ratio is approximately 4:1, which is the inverse of the observed ratio for cAMP and cGMP phosphodiesterases in the CSF.[74]

7. Blood–CSF Barrier to Cyclic Nucleotides

Measurements of cyclic nucleotides in the CSF would provide information concerning certain intracellular events in the CNS tissues only in the presence of an effective blood-to-CSF barrier for these cyclic nucleotides. To determine whether such a barrier is present in man, we have measured CSF cAMP in 14 patients with various neuromuscular diseases after significantly elevating the endogenous plasma cAMP concentration by means of a glucagon infusion.[16] In patients free of CNS disease and systemic endocrinopathy, there is a correlation between the baseline plasma and CSF cyclic nucleotide concentrations. This correlation suggests that cyclic nucleotides might pass through the blood–brain barrier and be measurable in the CSF.[36] Glucagon is a potent stimulating agent that activates liver adenyl cyclase in man, leading to significant rises in plasma cAMP but not cGMP.[14,15] Lumbar punctures were aseptically performed in the lateral decubitus position under local anesthesia with a 20-gauge needle that remained in position for 180 min. Simultaneous plasma and CSF (12–15 ml) samples were obtained at 30 min, 15 min, and immediately prior to, as well as 30, 60, 90, 120, and 150 min following, the start of the glucagon (Eli Lilly) infusion at 100 ng/kg per min for 30 min.[16]

The intravenous infusion of glucagon for 30 min produced a 40-fold (mean) rise in the plasma cAMP concentration, which then declined exponentially over the ensuing 120 min (Fig. 7). During this period, no significant change in plasma cGMP occurred. Lumbar CSF cAMP and cGMP concentrations were not altered significantly over the entire period of observation. In contrast, plasma glucose rose approximately 2-fold (mean) and declined to slightly below preinfusion concentrations at 90 and 120 min following the infusion ($p < 0.05$; paired two-tailed Student's t test). Lumbar CSF glucose began to rise 60 min following the infusion and was significantly higher than baseline ($p < 0.02$; paired two-tailed Student's t test) by 120 min following infusion. During the period of observation, there was a significant rise in plasma insulin but not CSF insulin measured by RIA.[67]

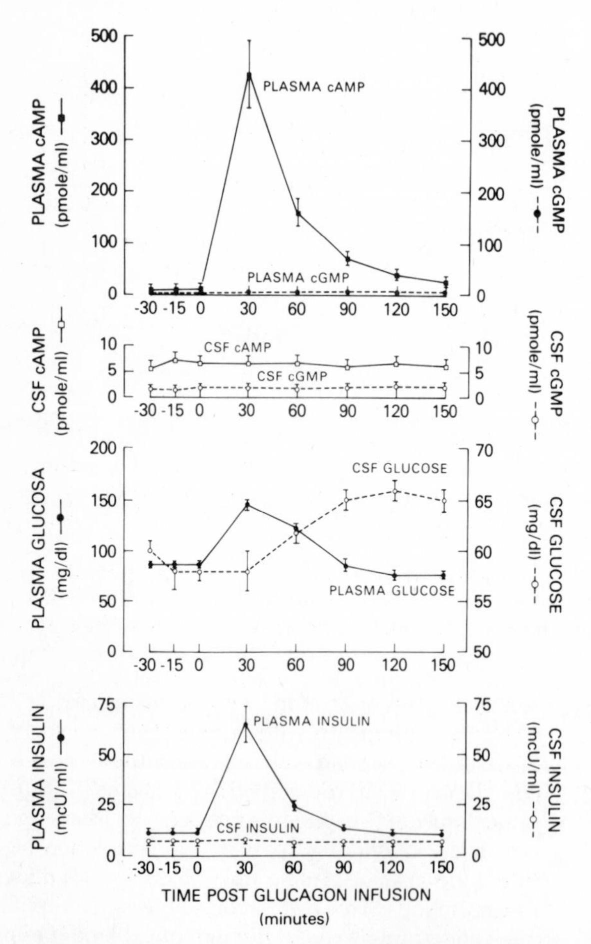

Figure 7. Effect of glucagon infusion on CSF cyclic nucleotide concentrations. Glucose, insulin, cAMP, and cGMP were measured in plasma and CSF of 14 patients following intravenous glucagon at 100 ng/kg per min for 30 min.[16] Each point (± S.E.M.) represents two separate assays done in duplicate.

These results indicate the presence of a significant blood-to-lumbar CSF barrier for cAMP in man under physiological conditions. Transfer of glucose into the lumbar CSF occurred during the period of observation, indicating that the lack of cAMP transfer was not due simply to a truncated sampling period if a facilitated transport mechanism were operational.[51] Similar results employing comparable doses of intravenously infused exogenous cAMP have been obtained in the rabbit.[130]

The maximal rise in lumbar CSF glucose in our patients occurred approximately 90 min following the establishment of the peak plasma glucose concentration. These results are in agreement with those reported by others who utilized oral or intravenous glucose loads to elevate plasma glucose concentrations.[3] In patients without meningitis, the cisternal CSF glucose concentration rises only slightly faster than the lumbar CSF glucose concentration, indicating that transfer of glucose into the CSF is more generalized and not simply dependent on diffusion from the cisterna magna.[131] Comparison of the kinetics of plasma and CSF concentration changes of cAMP and glucose in our patients indicates that glucose, but not cAMP, is preferentially transferred into the CSF. Glucose transfer into the CSF is mediated by a carrier transport system.[51] Our data would suggest that in man, no similar system exists for cAMP.

8. Carrier-Mediated Transport of Cyclic Nucleotides from CSF

Studies in single-cell systems indicate that one important mechanism for inactivation of cyclic nucleotides is extrusion from the cell.[119] Investigations of cyclic nucleotide metabolism in the human CNS are greatly facilitated by the presence of the CSF, which serves to remove active metabolites from the CNS.[31] As shown above, systemic pharmacological perturbations that change plasma cAMP will probably not affect CSF cAMP concentrations unless there are direct effects of pharmacological agents on CNS cAMP metabolism. It is possible, however, that clearance of cyclic nucleotides from the CSF occurs at such a rapid rate that changes in CSF concentrations of cyclic nucleotides may not be seen even with pharmacological agents that affect CNS cyclic nucleotide metabolism. Thus, an important step in determining the usefulness of the study of CSF cyclic nucleotide concentrations as an indicator of CNS cyclic nucleotide metabolism is a delineation of the clearance mechanisms of cyclic nucleotides from the CSF.

Clearance of monoamine metabolites from the CSF is determined by bulk absorption of CSF into venous blood and arachnoid villi throughout the CNS and by exchange of CSF components via transport processes at choroidal and extrachoroidal sites.[31] Mediated transport processes include a particular carrier system for weakly anionic organic molecules such as benzyl-penicillin, 5-HIAA, and *p*-aminohippuric acid.[135] Several reports in rats,[26,85] rabbits,[130] and humans[19,27,28] indicate that clearance of cAMP from the CSF is inhibited by probenecid. These results suggest that an important mechanism for cAMP (and possibly cGMP) clearance from the CSF is transport via a weakly anionic organic molecule carrier system. Knowledge of the nature of this transport system in man and its inhibition by exogenously administered probenecid will provide important data necessary to the study of CNS cyclic nucleotide metabolism.[107]

The proposed strategy for investigation of CNS cyclic nucleotide metabolism requires the following assumptions; first, the major means of inactivating cyclic nucleotides is extrusion from CNS cells into the CSF[110,119] and to a much lesser extent degradation either intra- or extracellularly by the majority of the CNS cells[66]; second, clearance of cyclic nucleotides from the CSF is primarily by a carrier-mediated transport process (which can be inhibited), and not by bulk absorption or diffusion back into the CNS tissues.[31] If clearance from the CSF can be totally prevented, then measurement of CSF cyclic nucleotide concentrations at two time points following inhibition will provide an approximation of the turnover of cyclic nucleotides in the CNS.

Methods are available that may be used to inhibit the weakly anionic organic molecule transport system in man.[135] Probenecid administered orally or intravenously effectively raises the CSF concentration of HVA and 5-HIAA, but usually not of MHPG.[78] *In vitro* experiments with choroid plexus from animals have suggested that probenecid inhibits 5-HIAA uptake competitively in this system,[32] while *in vivo* experiments showed little effect of probenecid directly on metabolite synthesis.[26] Similar experiments have shown that probenecid competitively blocks cAMP uptake by choroid plexus *in vitro*.[68]

8.1. Oral Probenecid Technique

The properties of the carrier-mediated transport processes for cyclic nucleotide clearance from the CSF in man have now been determined. The accumulation of cAMP and cGMP in the CSF after oral or intravenous probenecid administration revealed several differences. The extent of cAMP and cGMP accumulation with either oral or intravenous probenecid administration showed a different relationship to the CSF probenecid concentration achieved. Probenecid was administered orally at 100 mg/kg in four divided doses for 18 hr in 37 patients with seizure disorders or other CNS diseases.[19,161] Lumbar punctures were performed prior to and 18 hr after beginning the administration of probenecid. The concentration of cAMP in the CSF rose significantly ($p < 0.001$) following probenecid, but a wide range of CSF probenecid concentrations (4–34 μg/ml) accrued (Fig. 8A). Similar results were obtained for cGMP in the CSF (Fig. 8B). If the transport mechanism for cAMP is competitively inhibited by probenecid and other methods for cAMP clearance from the CSF are negligible, then the ratio of the final cAMP concentration in the CSF to the baseline cAMP concentration should be proportional to the probenecid concentration achieved by 18 hr. Transformation of the experimental data for cAMP in the CSF (Fig. 9A) did indeed show a clear linear relationship between the CSF cAMP accumulation ratio and the CSF probenecid concentration [(CSF cAMP) post/(CSF cAMP) pre $= 0.9 + 0.25 \cdot$(CSF probenecid) ± 0.9; $r = 0.8576$, $p < 0.01$). Similar results were obtained for cGMP (Fig. 9B) in these patients, but the slope of the least-squares regression line was altered [(CSF cGMP) post/(CSF cGMP) pre $= 0.6 \pm 0.17 \cdot$(CSF probenecid) ± 1.0; $r = 0.6622$, $p < 0.01$). Further study is required before a definite statement can be made as to whether or not the transport systems for cAMP and cGMP are identical or have different kinetics of inhibition.[24]

The results presented above indicate that over the 18-hr period under investigation, there is no apparent defined threshold CSF probenecid level that must be reached before the transport process is inhibited. In addition, cAMP and cGMP accumulation after probenecid relative to baseline rose 1.4- to 9-fold and was directly proportional to the CSF probenecid concentration. Therefore, absolute total blockade of the transport process does not occur. This finding indicates that any study by this technique of CNS cyclic nucleotide turnover, as measured in the CSF, will require measurement of the CSF probenecid concentration. Comparison of cyclic nucleotide metabolism in the same patient under different pharmacological conditions or among groups of patients under varying disease conditions will require normalization of all data. Pharmacological perturbations may affect the CSF probenecid concentration achieved, and observed differences in cyclic nucleotide accumulation will reflect, not differences in cyclic nucleotide turnover, but the extent of blockade of the clearance mechanism from the CSF. With appropriate controls, oral probenecid administration may be a powerful tool in our understanding of CNS cyclic nucleotide metabolism because of the profound amplification in CSF cyclic nucleotide concentrations obtained by even partial blockade of the clearance of these nucleotides from the CSF.

8.2. Intravenous Probenecid Technique

Because the accumulation of cyclic nucleotides in the CSF is proportional to the CSF probenecid concentration, the ideal situation for the study of CNS cyclic nucleotide metabolism would result from quickly achieving a known constant CSF probenecid concentration for a known period of time. Measurement of CSF cyclic nucleotide concentrations before and then at two points in time during the period of constant CSF probenecid concentration would then provide information on the synthesis and release into the CSF of cyclic nucleotides when transport of these cyclic nucleotides has been blocked at a constant level.

Administration of probenecid intravenously in prolonged infusions provides a short period of relatively constant CSF probenecid concentration.[78] Probenecid was given in 150-ml volumes of normal saline at 40 mg/kg over 70–100 min to 9 patients with

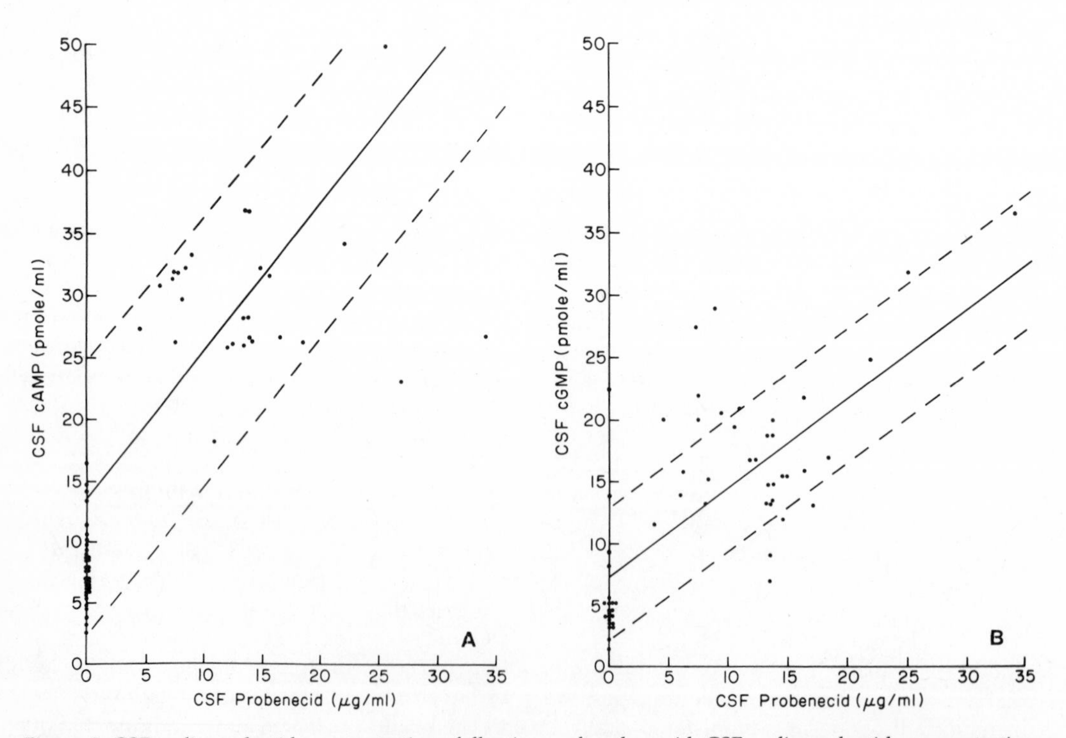

Figure 8. CSF cyclic nucleotide concentrations following oral probenecid. CSF cyclic nucleotide concentrations were measured by radioimmunoassay[139] at 18 hr and related to the CSF probenecid concentration.[78] (A) 37 patients; (cAMP) = 11.1 + 0.94 · (CSF probenecid) ± 8.8; $r = 0.6740$, $p < 0.01$. (B) 35 patients: (cGMP) = 7.2 + 0.74 · (CSF probenecid) ± 5.4; $r = 0.7182$, $p < 0.01$. (——) Mean; (– – –) ± one S.D. (standard deviation).

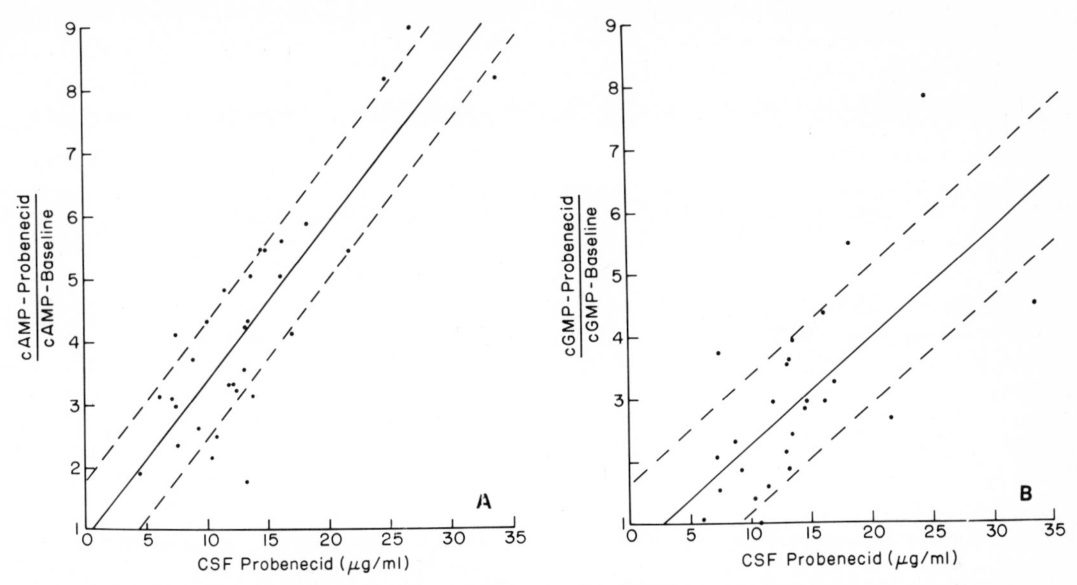

Figure 9. CSF cyclic nucleotide increase following oral probenecid. The ratio of the CSF cyclic nucleotide concentration following oral probenecid to the baseline (pre-probenecid) CSF cyclic nucleotide concentration was related to the CSF probenecid concentration. (A) 31 patients: (cAMP) post-probenecid/(cAMP) pre-probenecid $= 0.9 + 0.25 \cdot$ (CSF probenecid) $\pm$ 0.9; $r = 0.8576$, $p < 0.01$. (B) 35 patients: (cGMP) post-probenecid/(cGMP) pre-probenecid $= 0.6 + 0.17 \cdot$ (CSF probenecid) $\pm$ 1.0; $r = 0.6622$, $p < 0.01$. (——) Mean; (- - -) ± one S.D. (standard deviation).

amyotrophic lateral sclerosis (ALS) and 6 patients with peripheral nerve and muscle disease. Lumbar punctures were performed prior to and then at 6 and 10 hr following the beginning of the probenecid infusions. The concentration of cAMP in the CSF rose significantly ($p < 0.001$) following probenecid in all 15 patients, and a narrow range of CSF probenecid concentrations (3–8 μg/ml) accrued (Fig. 10A). The concentration of cGMP in the CSF of the patients without ALS showed similar results (Fig. 10B). Surprisingly, the CSF cGMP concentrations of patients with ALS clustered in a group below those for control patients in the same range of CSF probenecid concentrations. Transformation of the experimental data (as described above) for cAMP indicated that within a narrow range of CSF probenecid concentrations, the ratio of the CSF cAMP concentration following probenecid to the baseline CSF cAMP concentration ranged from 2 to 6.5 for both groups of patients (Fig. 11A). The ratio of the CSF cGMP concentration post-probenecid to the baseline CSF cGMP concentration ranged from 0.3 to 7 for ALS patients and from 2 to 6.5 for the control patients (Fig. 11B). This transformation of data corrects for the low baseline CSF cGMP concentrations in ALS patients and indicates that the carrier-mediated transport mechanism to remove cGMP from the CSF is equally susceptible to probenecid inhibition in both ALS and control patients. Thus, the decreased accumulation of CSF cGMP in ALS patients following intravenous probenecid administration was due to either decreased intracellular synthesis, decreased release into the CSF, or increased intra- or extracellular degradation of cGMP.[24]

9. Investigations of CNS Cyclic Nucleotide Metabolism in Amyotrophic Lateral Sclerosis

The intravenous probenecid (40 mg/kg) test produces reproducible CSF probenecid concentrations over an extended period of at least 4 hr. By providing a constant CSF probenecid concentration over this time period (Fig. 12), the inhibition of the carrier-mediated transport mechanism removing cyclic nucleotides from the CSF was kept at a constant level.[78] Cyclic nucleotide accumulation in the CSF before 6 hr is a function of synthesis, degradation, and release of cyclic nucleotides into the CSF as well as increasing blockade of transport of cyclic nucleotides from the CSF. Between 6 and 10 hr, this block-

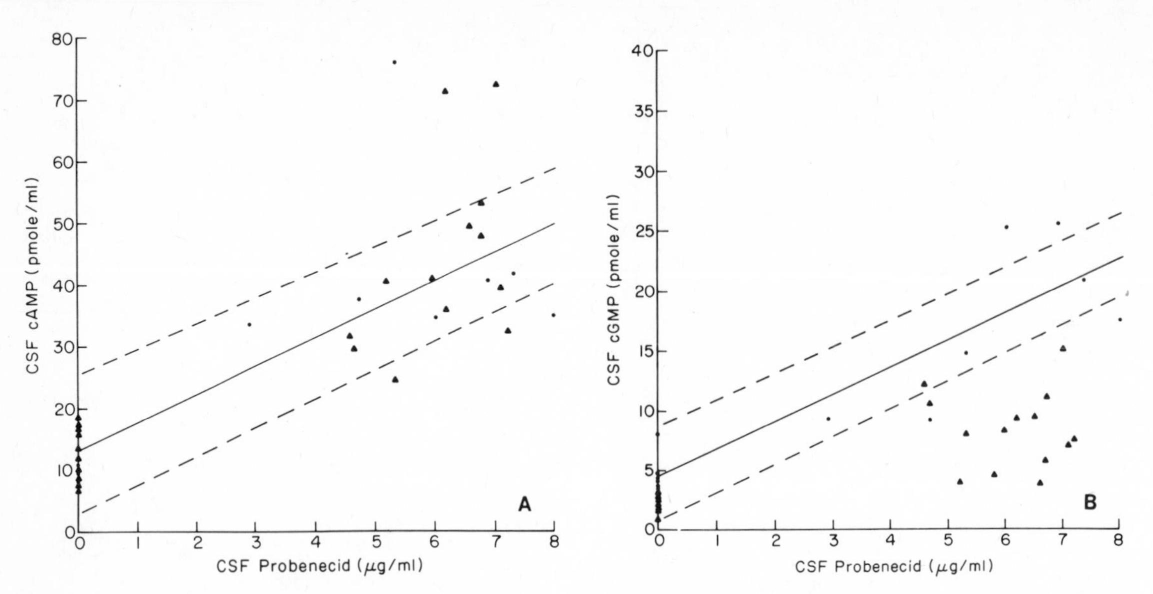

Figure 10. CSF cyclic nucleotide concentrations following intravenous probenecid. CSF cyclic nucleotide concentrations were measured by radioimmunoassay[139] at 6 and 10 hr following intravenous probenecid at 40 ng/kg for 70–100 min and related to the CSF probenecid concentration.[78] (A) 15 patients of whom 9 had amyotrophic lateral sclerosis (ALS): (cAMP) = 12.4 + 4.7 · (CSF probenecid) ± 12.0; $r = 0.7786$, $p < 0.01$. (B) 6 non-ALS patients: (cGMP) = 4.4 + 2.3 · (CSF probenecid) ± 3.6; $r = 0.8938$, $p < 0.01$ (——) Mean; (- - -) ± one S.D. (standard deviation). (●) Non-ALS; (▲) ALS.

ade of egress is relatively constant, as evidenced by constant CSF probenecid concentrations. Thus, any accumulation in CSF cyclic nucleotides during this 4-hr period represents the synthesis and degradation as well as the release of cyclic nucleotides into the CSF, and not changing blockade of egress from the CSF.

Several assumptions are inherent in this interpretation: First, probenecid administration does not directly affect cyclic nucleotide synthesis or degradation (intra- or extracellularly). In animal systems *in vivo* and cell systems *in vitro*, probenecid at concentrations approximately 10- to 100-fold higher than that achieved in human CSF did not affect cAMP synthesis.[26,110,119] No direct data are available from either *in vivo* or *in vitro* studies on the effect of probenecid on cyclic nucleotide degradation. Second, probenecid administration does not affect release of cyclic nucleotides into the CSF from those cells synthesizing cyclic nucleotides. *In vitro* studies of cAMP efflux from astrocytes indicate that probenecid can inhibit efflux of cAMP from these cells, but the concentration that inhibits cellular efflux is 10- to 100-fold higher than that achieved in human CSF.[37,110,119] Indeed, at the concentrations achieved in human CSF, no inhibition of cellular efflux of cAMP has been observed *in vitro*.[68,110] Third, administration of probenecid does not affect CSF production and thereby alter other methods for removal of CSF constituents by bulk flow or diffusion. A comprehensive study of the possible effects of probenecid on CSF production has not been attempted. Nevertheless, CSF pressure was not affected significantly at 6 and 10 hr after probenecid in our patients. Fourth, increased concentrations of cyclic nucleotides in the CSF after probenecid do not significantly affect cyclic nucleotide metabolism during the interval under investigation. Likewise, other metabolites that are increased by probenecid administration, such as HVA or 5-HIAA, are assumed not to significantly affect cyclic nucleotide metabolism. Little information is now available to shed light on the possible metabolic interactions. At the present level of understanding, some of these assumptions appear valid, but further clarification will be required before the assumptions can be fully substantiated.[68]

The observation of decreased baseline cyclic nucleotide concentrations in the CSF of patients with ALS suggested that an abnormality (or abnormalities, since the two nucleotides may function and be metabolized differently) in CNS cyclic nucleotide metabolism may be important in the pathogenesis of this tragic, crippling disease.[18] Indeed, one animal model of spinal muscular atrophy in mice is accom-

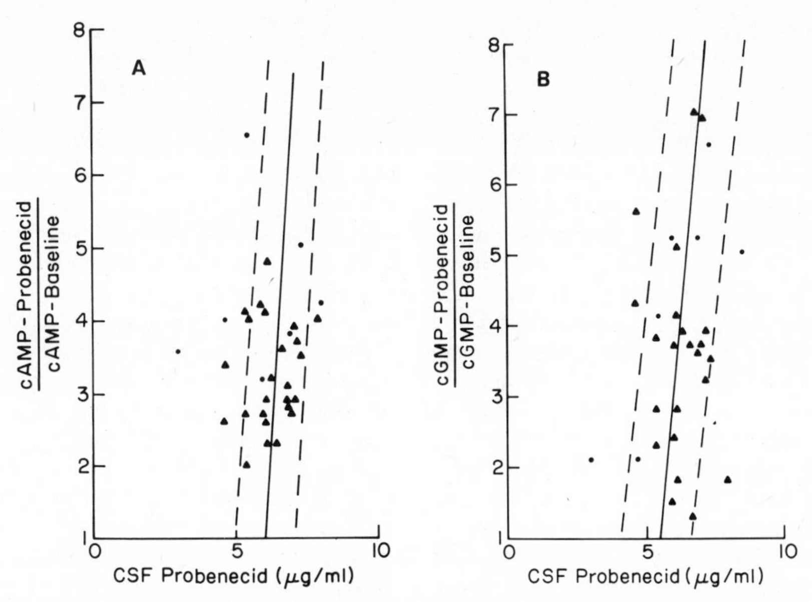

Figure 11. CSF cyclic nucleotide increase following intravenous probenecid. Ratio of CSF cyclic nucleotide concentration following intravenous probenecid to baseline (pre-probenecid) CSF cyclic nucleotide concentration was related to CSF probenecid concentration: 9 amyotrophic lateral sclerosis (ALS) (▲) and 6 non-ALS (●) patients. (A) (CSF probenecid) = 6.0 + 0.06 · [(cAMP) post-probenecid/(cAMP) pre-probenecid)] ± 1.0; $r = 0.0513$, $p < 0.05$. (B) (CSF probenecid) = 5.4 + 0.20 · [(cGMP) post-probenecid/(cGMP) pre-probenecid)] ± 1.0; $r = 0.2938$, $p < 0.05$. Over the narrow range of CSF probenecid concentrations achieved, a family of least-squares regression lines could describe observed data. Relative to oral probenecid technique, intravenous probenecid yielded constant but incomplete blockade of cyclic nucleotide egress over narrow reproducible range. (——) Mean; (– – –) ± one S.D. (standard deviation).

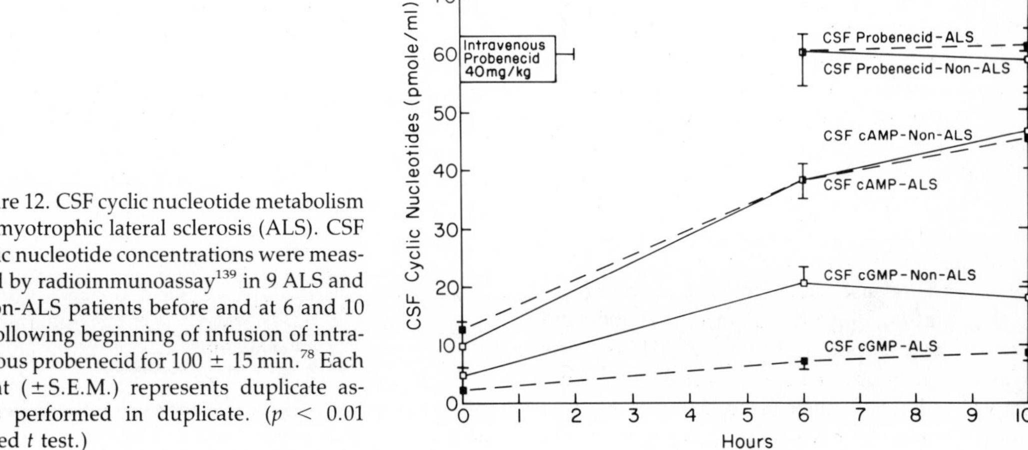

Figure 12. CSF cyclic nucleotide metabolism in amyotrophic lateral sclerosis (ALS). CSF cyclic nucleotide concentrations were measured by radioimmunoassay[139] in 9 ALS and 6 non-ALS patients before and at 6 and 10 hr following beginning of infusion of intravenous probenecid for 100 ± 15 min.[78] Each point (±S.E.M.) represents duplicate assays performed in duplicate. ($p < 0.01$ paired t test.)

panied by decreased spinal cord concentrations of cGMP in proportion to the loss of motor neurons.[17] To clarify the nature of the abnormality in cyclic nucleotide metabolism, 9 patients with ALS and 6 control patients with peripheral nerve and muscle disease were studied before and during intravenous probenecid administration. Accumulation of cAMP in the CSF of ALS and non-ALS patients following probenecid was identical at 6 and 10 hr following initiation of probenecid infusion (Fig. 12). Mean CSF probenecid concentrations were identical at 6 and 10 hr. When blockade of the cAMP transport mechanism was constant during this period, the net synthesis and release of cAMP into the CSF was 2.4 pmol/ml per hr more than the transport from the CSF in both patient groups. Before this steady state was achieved in the blockade of the cAMP transport system, cAMP accumulation averaged 4.3 pmole/ml per hr up to 6 hr. Accumulation of cGMP in the CSF of non-ALS patients during the first 6 hr averaged 2.7 pmole/ml per hr. During the steady-state blockade period between 6 and 10 hr following the initiation of the probenecid infusion, net synthesis and release of cGMP into the CSF apparently equaled the remaining capacity for transport of cGMP from the CSF in non-ALS patients (Fig. 12).

The degree of probenecid blockade of the transport systems for cAMP and cGMP from the CSF was similar (see Fig. 11A, B). Therefore, the observed differences in turnover between cAMP and cGMP could be due to differences in synthesis and release into the CSF or to differences in degradation intra- or extracellularly. After freezing, we observed no cGMP phosphodiesterase activity in human CSF and cAMP phosphodiesterase activity of approximately 2 pmol/ml per hr. CSF from patients with brain tumors contained cAMP phosphodiesterase activity approximating 5 pmole/ml per hr and cGMP phosphodiesterase activity approximating 18 pmole/ml per hr.[74] The latter estimates would be consistent with our observation on unfrozen CSF (see Fig. 3). Thus, analysis of net synthesis of cyclic nucleotides in human CSF will require correction for extracellular phosphodiesterase-associated degradation.

The markedly diminished absolute accumulation of cGMP at 6 and 10 hr in ALS patients as compared with controls could be due to either decreased cGMP synthesis and release or increased extracellular cGMP degradation, since the degree of probenecid blockade of the transport system for cGMP from the CSF was similar in both ALS and non-ALS patients (see Fig. 11A, B). Several techniques were used to determine whether increased degradation of cGMP in ALS CSF explained the observed results. First, exogenous cAMP or cGMP was added to CSF from 5 ALS patients and 4 non-ALS patients. Following 1 hr at 25°C, samples were frozen at −70°C prior to assay for the respective cyclic nucleotide. Recoveries were quantitative, and there was no difference in the recoveries for either cAMP or cGMP between ALS and non-ALS patients. Second, ALS patients were given an oral phosphodiesterase inhibitor, phthalazinol (Fig. 13), which does not block the adenosine receptor (as does theophylline) and which is 10-fold more potent against the high and low K_m mouse brain cAMP phosphodiesterases than theophylline, which inhibits both phosphodiesterases equally.[1,21] Baseline CSF cAMP, but not cGMP, increased significantly ($p < 0.01$; paired two-tailed Student's t test) in 8 ALS patients on phthalazinol and decreased following cessation (Fig. 14). Four ALS patients received intravenous probenecid tests before, and again 2 months following, 40 mg/kg per

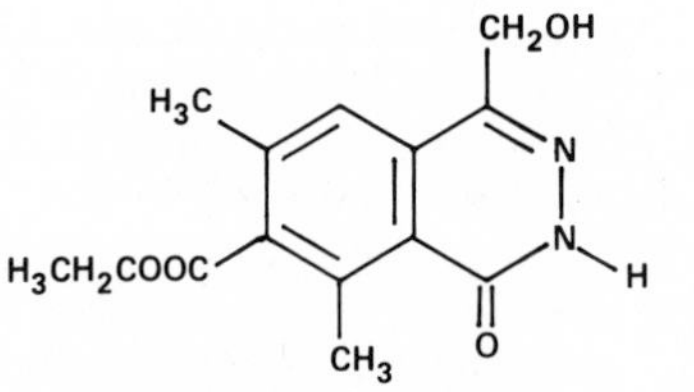

PHTHALAZINOL (EG-626)
1-oxo-2-hydro-4-hydroxymethyl-6,8-dimethyl-7-ethoxycarbamyl-phthalazine

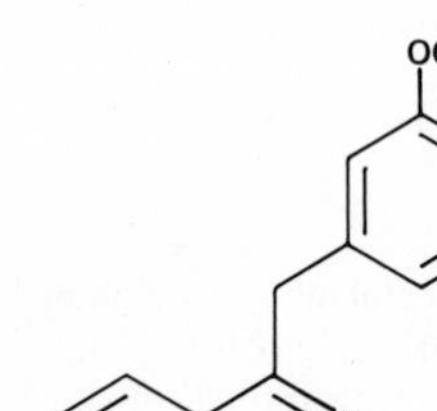

THEOPHYLLINE
1,3-dimethylxanthine

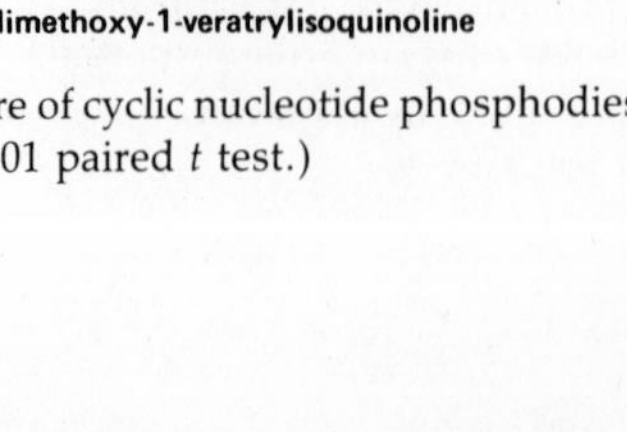

PAPAVERINE
6,7-dimethoxy-1-veratrylisoquinoline

Figure 13. Structure of cyclic nucleotide phosphodiesterase inhibitors. ($p < 0.01$ paired t test.)

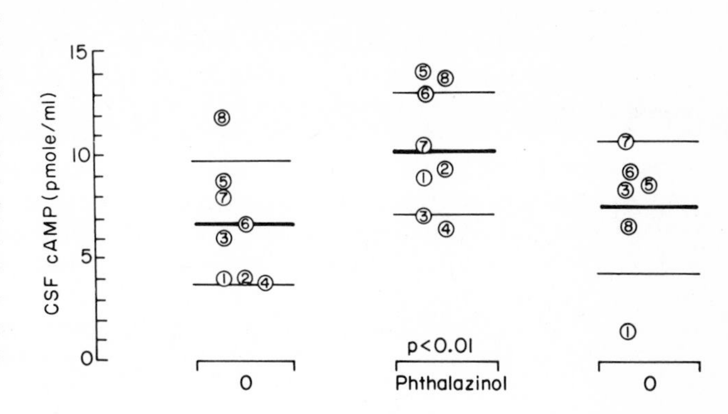

Figure 14. Effect of phthalazinol administration on baseline CSF cAMP concentrations. Eight patients received phthalazinol orally in doses ranging from 25 to 60 mg/kg per day for 2–12 months. CSF cAMP was determined by radioimmunoassay[139] before and while patients were on highest dose of phthalazinol. Six patients had repeat lumbar punctures over 1–8 months following cessation of phthalazinol. Each point is mean CSF cAMP concentration for each patient of 2–4 separate lumbar punctures under identical pharmacological conditions. CSF cAMP rose significantly ($p < 0.01$; paired two-tailed t test) on phthalazinol and decreased on cessation of drug.

day phthalazinol. CSF cAMP was increased significantly at 6 and 10 hr after probenecid following phthalazinol administration, but CSF cGMP was unchanged (Fig. 15). During the steady state between 6 and 10 hr following probenecid, CSF cAMP accumulation was increased above the pre-phthalazinol level by 2.4 pmol/ml per hr, or approximately 2-fold (Fig. 15). CSF cGMP accumulation during this period was unchanged. Comparison of *in vitro* degradation of cAMP at 37°C in previously frozen CSF samples indicated that degradation in CSF from phthalazinol-treated patients was slower than that noted in CSF from these patients prior to phthalazinol (see Fig. 4). Comparison studies of cGMP degradation in these samples was not possible because of inactivation of cGMP phosphodiesterase activity by freezing.

The data from intravenous probenecid tests in ALS patients and controls as well as in ALS patients before and after phosphodiesterase inhibitor administration indicated a unique abnormality in cGMP metabolism in patients with ALS. This abnormality is most likely a decrease in synthesis and release of cGMP, rather than increased degradation of cGMP in the CSF. Support for this interpretation derives from the following observations: (1) baseline CSF cGMP concentrations were decreased 50% in ALS patients; (2) CSF cGMP absolute accumulation following intravenous probenecid was abnormally low, although the ratio of CSF cGMP concentrations before probenecid to CSF cGMP concentration after probenecid was similar in ALS and non-ALS patients; and (3) phosphodiesterase inhibitor administration, which was effective *in vivo* in augmenting CSF cAMP accumulation, did not augment CSF cGMP accumulation. The present data did not rule

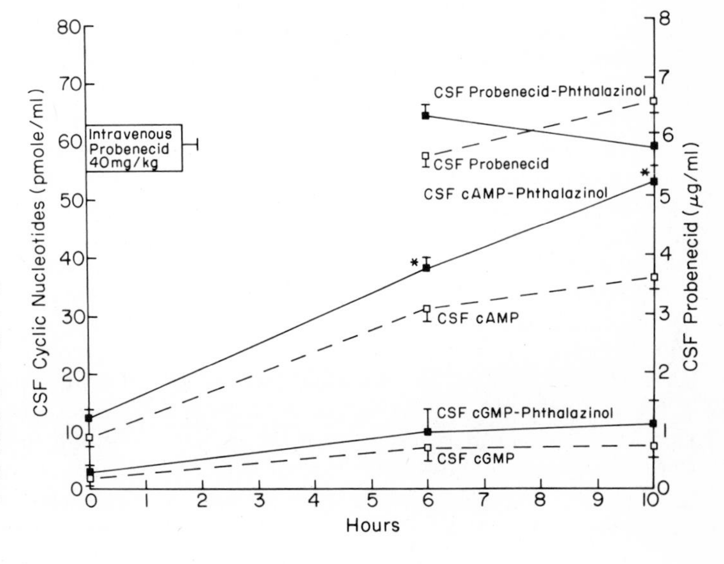

Figure 15. CSF cyclic nucleotide metabolism in amyotrophic lateral sclerosis (ALS) following phthalazinol administration. CSF cyclic nucleotide concentrations were measured by radioimmunoassay[139] in 4 ALS patients following intravenous probenecid (40 mg/kg) before and after receiving phthalazinol (42 ± 5 mg/kg) daily for 2 months. Each point (±S.E.M.) represents duplicate assays performed in duplicate.

out, however, the possibility that rather than decreased synthesis and release of cGMP into the CSF, there was increased intracellular degradation of cGMP in ALS.

10. Application of Probenecid Techniques to Study of Cyclic Nucleotide Metabolism in Neurological and Psychiatric Disease

To date, evaluations of CNS cyclic nucleotide metabolism in human disease have utilized primarily the oral probenecid technique.[10,19,23–25,27,28,115,133] Only two of these studies have determined the CSF probenecid concentrations in an attempt to correct for differences in the blockade of the carrier-mediated transport systems of cyclic nucleotides from the CSF.[10,19] Differences in the protocols significantly affect the observed ratio of the CSF cyclic nucleotide concentration after probenecid to the baseline CSF cyclic nucleotide concentration (Table 5). If the second lumbar puncture is performed following oral probenecid at 100 mg/kg for 9 hr, the ratio for CSF cAMP is 2.5 ± 0.3[26] but increases to 4.3 ± 1.8 at 18 hr following probenecid at the same dose schedule.[115] At this time, there has been no significant difference observed in the accumulations of cAMP following oral probenecid in the CSF of patients with neurologic disease (spinocerebellar degenerations, Parkinson's disease) or psychiatric disease (depression, mania, schizophrenia).[19,23–25,27,28,115,133] The discrepancy between two studies indicating either no change[133] or a decrease[10] in CSF cAMP in treated schizophrenics following oral probenecid administration will require further evaluation. The ratio method will give falsely elevated values for cGMP (as observed in psychiatric disease) if the baseline cGMP concentration is measured after thawing and refreezing.[133]

Methodological difficulties with the oral probenecid technique include (1) possible differences in the individual absorption of probenecid; (2) differences in the CSF probenecid concentration obtained; and (3) intercurrent side effects, particularly nausea and vomiting, due to probenecid administration.[78] The wide range of CSF probenecid concentrations attained decreases the reproducibility of the test in the same patient under differing pharmacological conditions.

The intravenous probenecid technique provides a narrower, more precisely defined range of CSF probenecid concentrations (see Fig. 11) with excellent reproducibility (see Figs. 12 and 15). To date, only two studies in man have employed the intravenous probenecid technique in the study of CNS cyclic nucleotide metabolism.[19,29] Despite differences in the probenecid dose administered, similar increases in cAMP and cGMP are noted in the CSF between 5 and 6 hr after initiation of the probenecid infusion. The ratio method, however, suggests that the transport mechanisms for cAMP and cGMP do have different properties (Table 6).

The intravenous probenecid technique has documented efficacy in studies of monoamine metabolite metabolism in man.[78] Its usefulness in animal studies is being carefully delineated. One study in normal dogs and a strain of dogs with hyperactivity described as "nervous" dogs has employed the intravenous probenecid technique. At 6 hr following intravenous probenecid at 50 mg/kg, cisternal CSF concentrations of cAMP and cGMP were significantly elevated in "nervous" dogs compared with normal dogs of the same species.[2] The CSF concentration of 5-HIAA was decreased. Amphetamine administration significantly elevated CSF cAMP but not cGMP in both control and "nervous" dogs. The amphetamine-induced CSF cAMP concentrations were more marked following probenecid administration.[38] In contrast, chlordiazepoxide reduced the accumulation of cGMP in the CSF of "nervous"

Table 5. Effect of Time of Lumbar Puncture on Oral-Probenecid-Induced Increase in Cyclic Nucleotides[a] in Human CSF

Patients	Time following 100 mg/kg probenecid orally	
	9 hr	18 hr
Neurological disease[b]		
cAMP (post-probenecid) / cAMP (pre-probenecid)	2.5 ± 0.3 (15)	4.3 ± 1.8 (34)
cGMP (post-probenecid) / cGMP (pre-probenecid)	—	2.9 ± 1.3 (31)
Psychiatric disease[c]		
cAMP (post-probenecid) / cAMP (pre-probenecid)	—	4.7 ± 0.5 (65)
cGMP (post-probenecid) / cGMP (pre-probenecid)	—	10.6 ± 3.9 (32)

[a] Means ± S.E.M. (number of observations in parentheses).
[b] Patients with Parkinson's disease,[27] spinocerebellar degeneration,[27] and seizure disorders.[19]
[c] Patients with manic–depressive disease, unipolar depression, and schizophrenia.[115,133]

Table 6. Effect of Dose of Probenecid on Intravenous-Probenecid-Induced Increase in Cyclic Nucleotides[a] in Human CSF

Patients with neurological disease[b]	Probenecid dose	
	40 mg/kg[18,19]	75 mg/kg[29]
cAMP (post-probenecid) / cAMP (pre-probenecid)	3.9 ± 0.2 (15)	1.8 ± 0.7 (59)
cGMP (post-probenecid) / cGMP (pre-probenecid)	4.1 ± 0.3 (15)	4.6 ± 0.5 (59)

[a] Means ± S.E.M. (number of observations in parentheses).
[b] Patients with neuromuscular diseases including ALS[18,19] and various CNS diseases.[29]

dogs, but not in normal dogs. This effect of chlordiazepoxide was apparent only following intravenous probenecid administration and was not reflected in baseline CSF cyclic nucleotide concentrations. The excellent reproducibility of CSF probenecid concentrations during these experiments allowed refined studies of the pharmacological effects of different doses of drugs on CNS cyclic nucleotide metabolism.[41,99] Thus, some pharmacological and possible physiological effects on CNS cyclic nucleotide metabolism may be seen only following probenecid administration.

11. Conclusion

A wealth of information from *in vivo* and *in vitro* studies highlights the putative pivotal roles of cyclic nucleotides in the metabolism, differentiation, and maintenance of the central and peripheral nervous systems.[34,35] In addition, an important role for cyclic nucleotides in the mediation of electrochemical events, principally in the peripheral nervous system, has received much support from the available evidence.[82] Different physiological and pathological states may alter CNS cyclic nucleotide metabolism.[12,47,136,156,158] These changes may be reflected in the CSF—either in the baseline cyclic nucleotide concentrations or the accumulation following probenecid administration. Further investigation of physiological, pathological, and pharmacological alterations in CNS cyclic nucleotide metabolism can now be conducted in animals and man in conjunction with an intravenous probenecid technique. This technique is very reproducible and easily performed, and may lead to greater understanding of CNS cyclic nucleotide metabolism.

ACKNOWLEDGMENTS. This work was supported in part by the Wallace Memorial Fund (B.R.B.) of the George Washington University School of Medicine and project CICC 2-06-322 from the Bureau of Medicine and Surgery, Navy Department, Washington, D.C. (M.D., C.C., L.P.G., J.S.). B.R.B. is the recipient of a Teacher-Investigator Development Award from the National Institute of Neurological and Communicative Disorders and Stroke (KO7-NS00385). The excellent assistance of the nursing staff of the National Institute of Neurological and Communicative Disorders and Stroke is gratefully acknowledged.

References

1. ADACHI, K., NUMANO, F.: Phosphodiesterase inhibitors: Their comparative effectiveness *in vitro* in various organs. *Jpn. J. Pharmacol.* **27**:97–103, 1977.
2. ANGEL, C., DELUCA, D. C., MURPHREE, O. D.: Probenecid-induced accumulation of cyclic nucleotides, 5-hydroxyindoleacetic acid, and homovanillic acid in cisternal spinal fluid of genetically nervous dogs. *Biol. Psychiatry* **11**:743–753, 1976.
3. ASTIN, K. J., WILDE, C. E., DAVIES-JONES, G. A. B.: Glucose metabolism and insulin response in the plasma and CSF in motor neuron disease. *J. Neurol. Sci.* **25**:205–210, 1975.
4. BALL, J. H., KAMINSKY, N. I., HARDMAN, J. G., BROADUS, A. E., SUTHERLAND, E. W., LIDDLE, G. W.: Effect of catecholamines and adrenergic blocking agents on plasma and urinary cyclic nucleotides in man. *J. Clin. Invest.* **51**:2124–2129, 1972.
5. BELMAKER, R. H., EBSTEIN, R. P., BIEDERMAN, J., STERN, J., BERMAN, M., VAN PRAAG, H. M.: The effect of L-DOPA and propranolol on human CSF cyclic nucleotides. *Psychopharmacology* **58**:307–310, 1978.
6. BIEDERMAN, J., RIMON, R., EBSTEIN, R., ZOHAR, J., BELMAKER R.: Neuroleptics reduce spinal fluid cyclic AMP in schizophrenic patients. *Neuropsychobiology* **2**:324–327, 1976.
7. BITENSKY, M. W., KEIRNS, J. J., FREEMAN, J.: Cyclic adenosine monophosphate and clinical medicine. Part I. Calcium and phosphate metabolism. *Am. J. Med. Sci.* **266**:320–347, 1973.
8. BLOOM, F. E.: The role of cyclic nucleotides in central synaptic function. *Rev. Physiol. Biochem. Pharmacol.* **74**:1–104, 1975.
9. BLOOM, F. E., HOFFER, B. J., BATTENBERG, E. F., SIGGINS, G. R., STEINER, A. L., PARKER, C. W., WEDNER, H. J.: Adenosine 3′,5′-monophosphate is localized in cerebellar neurons: Immunofluorescence evidence. *Science* **177**:436–438, 1972.
10. BOWERS, M. B., JR., STUDY, R. E.: Cerebrospinal fluid cyclic AMP and acid monoamine metabolites follow-

ing probenecid: Studies in psychiatric patients. *Psychopharmacology* **62**:17–22, 1979.

11. Breckenridge, B. McL. Johnson, R. E.: Cyclic 3′,5′-nucleotide phosphodiesterase in brain. *J. Histochem. Cytochem.* **17**:505–511, 1969.
12. Broadus, A. E.: Clinical cyclic nucleotide research. *Adv. Cyclic Nucleotide Res.* **8**:509–548, 1977.
13. Broadus, A. E., Hardman, J. G., Kaminsky, N. I., Ball, J. H., Sutherland, E. W., Liddle, G. W.: Extracellular cyclic nucleotides. *Ann. N.Y. Acad. Sci.* **185**:50–69, 1971.
14. Broadus, A. E., Kaminsky, N. I., Hardman, J. G., Sutherland, E. W., Liddle, G. W.: Kinetic parameters and renal clearances of plasma adenosine 3′,5′-monophosphate and guanosine 3′,5′-monophosphate in man. *J. Clin. Invest.* **49**:2222–2236, 1970.
15. Broadus, A. E., Kaminsky, N. I., Northcutt, R. C., Hardman, J. G., Sutherland, E. W., Liddle, G. W.: Effects of glucagon on adenosine 3′,5′-monophosphate and guanosine 3′,5′-monophosphate in human plasma and urine. *J. Clin. Invest.* **49**:2237–2245, 1970.
16. Brooks, B. R., Engel, W. K., Sode, J.: Blood-to-cerebrospinal fluid barrier for cyclic adenosine monophosphate in man. *Arch. Neurol.* **34**:468–469, 1977.
17. Brooks, B. R., Lust, W. D., Andrews, J. M., Engel, W. K.: Decreased spinal cord cGMP in murine (WOBBLER) spontaneous lower motor neuron degeneration. *Arch. Neurol.* **35**:590–591, 1978.
18. Brooks, B. R., Sode, J., Engel, W. K.: Cyclic nucleotide metabolism in neuromuscular disease. In Andrews, J. M., Johnson, R. T., Brazier, M. A. B. (eds.): *Amyotrophic Lateral Sclerosis: Recent Research Trends,* U.C.L.A. Forum in Medical Sciences, Vol. 19. New York, Academic Press, 1976, pp. 101–118.
19. Brooks, B. R., Wood, J., Sode, J., Engel, W. K.: Cyclic nucleotide metabolism in neurological disease. *Trans. Am. Neurol. Assoc.* **101**:221–222, 1976.
20. Butcher, R. W., Ho, R. J., Meng, H. C., Sutherland, E. W.: Adenosine 3′,5′-monophosphate in biological materials. II. The measurement of adenosine 3′,5′-monophosphate in tissues and the role of the cyclic nucleotide in the lipolytic response of fat to epinephrine. *J. Biol. Chem.* **240**:4515–4523, 1965.
21. Castaner, J., Hillier, K.: Phthalazinol: Cardiovascular agent and phosphodiesterase inhibitor. *Drugs Future* **3**:55–58, 1978.
22. Cheung, W. Y., Salganicoff, L.: Cyclic 3′,5′-nucleotide phosphodiesterase: Localization and latent activity in rat brain. *Nature (London)* **214**:90–91, 1967.
23. Clarenbach, P. A., Wenzel, D. C., Cramer, H.: Cyclic AMP in cerebrospinal fluid of rats: Effects of electroconvulsive shock. *Eur. Neurol.* **17**:83–86, 1978.
24. Cramer, H.: Cyclic 3′,5′-nucleotides in extracellular fluids of neural systems. *J. Neurosci. Res.* **3**:241–246, 1977.
25. Cramer, H., Goodwin, F. K., Post, R. M., Bunney, W. E., Jr.: Effects of probenecid and exercise on cerebrospinal fluid cyclic AMP in affective illness. *Lancet* **1**:1346–1347, 1972.
26. Cramer, H., Lindl, T.: Probenecid inhibits efflux of adenosine 3′,5′-monophosphate (cAMP) from cerebrospinal fluid (CSF) in the rat. *Psychopharmacology* **26**(Suppl.):49, 1972.
27. Cramer, H., Ng, L. K. Y., Chase, T. N.: Effect of probenecid on levels of cyclic AMP in human cerebrospinal fluid. *J. Neurochem.* **19**:1601–1602, 1972.
28. Cramer, H., Ng, L. K. Y., Chase, T. N.: Adenosine 3′,5′-monophosphate in cerebrospinal fluid: Effect of drugs and neurologic disease. *Arch. Neurol.* **29**:197–199, 1973.
29. Cramer, H., Renaud, B., Billiard, M., Mouret, J., Hammers, R.: Monoamine metabolites and cyclic nucleotides in the cerebrospinal fluid with bismuth or mercury poisoning. *Arch. Psychiatr. Nervenkr.* **226**:173–181, 1978.
30. Cramer, H., Renaud, B., Ortega-Suhr Kamp, E.: Concentration of cyclic AMP in lumbar, cisternal, and ventricular cerebrospinal fluid in neurological patients. *Z. Klin. Chem. Klin. Biochim.* **13**:245, 1975.
31. Cserr, H.: Physiology of the choroid plexus. *Physiol. Rev.* **51**:273–311, 1971.
32. Cserr, H., Van Dyke, D. H.: 5-Hydroxyindole-acetic acid accumulation in isolated choroid plexus. *Am. J. Physiol.* **220**:718–723, 1971.
33. Cumming, R., Eccleston, D., Steiner A.: Immunohistochemical localization of cyclic GMP in rat cerebellum. *J. Cyclic Nucleotide Res.* **3**:275–282, 1977.
34. Daly, J. W.: The formation, degradation, and function of cyclic nucleotides in the nervous system. *Int. Rev. Neurobiol.* **20**:105–168, 1977.
35. Daly, J. *Cyclic Nucleotides in the Nervous System.* New York, Plenum Press, 1977.
36. Dascombe, M. J., Milton, A. S.: Cyclic adenosine-3′,5′-monophosphate in cerebrospinal fluid. *Br. J. Pharmacol.* **54**:254P–255P, 1975.
37. Davoren, R. P., Sutherland, E. W.: The effect of L-epinephrine and other agents on the synthesis and release of adenosine 3′,5′-phosphate by whole pigeon erythrocytes. *J. Biol. Chem.* **238**:3009–3015, 1963.
38. DeLuca, D. C., Angel, C., Murphree, O. D.: Effects of amphetamine and chlordiazepoxide on probenecid-induced accumulation of acidic metabolites in the cerebrospinal fluid of the dog. *Biol. Psychiatry* **12**:577–582, 1977.
39. Drummond, G. I.: Metabolism and functions of cyclic AMP in nerve. *Prog. Neurobiol.* **2**:120–176, 1972.
40. Dubrovsky, A. L., Engel, W. K.: New histochemical technique for the demonstration of adenyl cyclase (AC) in nervous tissue and muscle. *Trans. Am. Neurol. Assoc.* **101**:83–84, 1976.
41. Ebstein, R. P., Biederman, J., Rimon, R., Zohar, J., Belmaker, R. H.: Cyclic CMP in the CSF of patients with schizophrenia before and after neuroleptic treatment. *Psychopharmacology* **51**:71–74, 1976.

42. EGRIE, J. C., CAMPBELL, J. A., FLANGAS, A. L., SIEGEL, F. L.: Regional, cellular, and subcellular distribution of calcium-activated cyclic nucleotide phosphodiesterase and calcium dependent regulator in porcine brain. *J. Neurochem.* **28**:1207–1213, 1977.
43. FARBER, D. B., LOLLEY, R. N.: Cyclic guanosine monophosphate: Elevation in degenerating photoreceptor cells of the C3H mouse retina. *Science* **186**:449–451, 1974.
44. FEINGLOS, M. N., DREZNER, M. K., LEBOVITZ, H. E.: Measurement of plasma adenosine 3′,5′-monophosphate. *J. Clin. Endocrinol. Metab.* **46**:824–829, 1978.
45. FELBER, J. P.: Radioimmunoassay in the clinical laboratory. *Adv. Clin. Chem.* **20,** 130–179, 1978.
46. FELDMAN, A. M., EPSTEIN, M. H., BRUSILOW, S. W.: Role of cyclic AMP in cerebrospinal fluid production. In Wood, J. H. (ed.) *Neurobiology of Cerebrospinal Fluid I.* New York, Plenum Press, 1980.
47. FERRENDELLI, J. A.: Role of cyclic GMP in the function of the central nervous system. In Weiss, B. (ed.) *Cyclic Nucleotides in Disease.* Baltimore, University Park Press, 1975, pp. 377–390.
48. FERRENDELLI, J. A.: Distribution and regulation of cyclic GMP in the central nervous system. *Adv. Cyclic Nucleotide Res.* **9**:453–464, 1978.
49. FERRENDELLI, J. A., KINSCHERF, D. A., CHANG, M. M.: Comparison of the effects of biogenic amines on cyclic GMP and cyclic AMP levels in mouse cerebellum *in vitro. Brain Res.* **84**:63–73, 1975.
50. FERRENDELLI, J. A., RUBIN, E. H., ORR, H. T., KINSCHERF, D. A., LOWRY, O. H.: Measurement of cyclic nucleotides in histologically defined samples of brain and retina. *Anal. Biochem.* **78**:252–259, 1977.
51. FISHMAN, R. A.: Carrier transport of glucose between blood and cerebrospinal fluid. *Am. J. Physiol.* **206**:836–844, 1964.
52. FLEISCHER, A. S., RUDMAN, D. R., FRESH, C. B., TINDALL, G. T.: Concentration of 3′,5′-cyclic adenosine monophosphate in ventricular CSF of patients following severe head trauma. *J. Neurosurg.* **47**:517–524, 1977.
53. FLEISCHER, A. S., TINDALL, G. T.: Cerebrospinal fluid cyclic nucleotide alterations in traumatic coma. In Wood, J. H. (ed.): *Neurobiology of Cerebrospinal Fluid I.* New York, Plenum Press, 1980.
54. FLORENDO, N. T., BARRNETT, R. J., GREENGARD, P.: Cyclic 3′,5′-nucleotide phosphodiesterase cytochemical localization in cerebral cortex. *Science* **173**:745–747, 1971.
55. FLORENDO, N. T., GREENGARD, P., BARRNETT, R. T.: Fine structural localization of cyclic 3′,5′-nucleotide phosphodiesterase in rat cerebral cortex. *J. Histochem. Cytochem.* **18**:682–685, 1970.
56. FRAZIER, W. A., OHLENDORF, C. E., BOYD, L. F., ALDE, L., JOHNSON, E. M., FERRENDELLI, J. A., BRADSHAW, R. A.: Mechanism of action of nerve growth factor and cyclic adenosine 3′,5′-monophosphate in neurite outgrowth in embryonic chick sensory ganglia: Demonstration of independent pathways of stimulation. *Proc. Natl. Acad. Sci. U.S.A.* **70**:2884–2452, 1973.
57. GARELIS, E., NEFF, N. H.: Cyclic adenosine monophosphate: Selective increase in caudate nucleus after administration of L-dopa. *Science* **183**:532–533, 1974.
58. GEISLER, A., BECH, P., JOHANNESEN, M., RAFAELSEN, O. J.: Cyclic AMP levels in cerebrospinal fluid in manic–melancholic patients. *Neuropsychobiology* **2**:211–220, 1976.
59. GELFAND, M., COHAN, S. L., WINCHESTER, J. F., KNEPSHIELD, J. H.: The treatment of hepatic coma by charcoal hemoperfusion. *Neurology* **29**:540, 1979.
60. GILMAN, A. G.: A protein binding assay for adenosine 3′,5′-cyclic monophosphate. *Proc. Natl. Acad. Sci. U.S.A.* **67**:305–312, 1970.
61. GILMAN, A. G.: Protein binding assays for cyclic nucleotides. *Adv. Cyclic Nucleotide Res.* **2**:9–24, 1972.
62. GOLDBERG, M. L.: Radioimmunoassay for adenosine 3′,5′-cyclic monophosphate and guanosine 3′,5′-cyclic monophosphate in human blood, urine, and cerebrospinal fluid. *Clin. Chem.* **23**:576–580, 1977.
63. GOLDBERG, N. D., LUST, W. D., O'DEA, R. F., WEI, S., O'TOOLE, A. G.: A role of cyclic nucleotides in brain metabolism. *Adv. Biochem. Psychopharmacol.* **3**:67–87, 1970.
64. GOLDBERG, N. D., O'TOOLE, A. G., HADDOX, M. K.: Analysis of cyclic AMP and cyclic GMP by enzymic cycling procedures. *Adv. Cyclic Nucleotide Res.* **2**:63–80, 1972.
65. GORIDIS, C., MASSARELLI, R., SENSENBRENNER, M., MANDEL, P.: Guanyl cyclase in chick embryo brain cell cultures: Evidence of neuronal localization. *J. Neurochem.* **23**:135–138, 1974.
66. GORIN, E., BRENNER, T.: Extracellular metabolism of cyclic AMP. *Biochim. Biophys. Acta* **451**:20–28, 1976.
67. HALES, C. N., RANDLE, P. J.: Immunoassay of insulin with insulin–antibody precipitate. *Biochem. J.* **88**:137–146, 1963.
68. HAMMERS, R., CLARENBACH, P., LINDL, T., CRAMER, H.: Uptake and metabolism of cyclic AMP in rabbit choroid plexus *in vitro. Neuropharmacology* **16**:135–141, 1977.
69. HARDMAN, J. G., ROBINSON, G. A., SUTHERLAND, E. W.: Cyclic nucleotides. *Annu. Rev. Physiol.* **33**:311–336, 1971.
70. HARPER, J. F., BROOKER, G.: Femtomole sensitive radioimmunoassay for cyclic AMP and cyclic GMP after 2′O acetylation by acetic anhydride in aqueous solution. *J. Cyclic Nucleotide Res.* **1**:207–218, 1975.
71. HEIKKINEN, E. R., MYLLYLA, V. V., HOKKANEN, E., VAPAATALO, H.: Cerebrospinal fluid concentration of cyclic AMP in cerebrovascular diseases. *Eur. Neurol.* **14**:129–137, 1976.
72. HEIKKINEN, E. R., MYLLYLA, V. V., VAPAATALO, H., KOKKANEN, E.: Urinary excretion and cerebrospinal

fluid concentration of cyclic adenosine-3′,5′-monophosphate in various neurological diseases. *Eur. Neurol.* **11**:270–280, 1974.

73. HEIKKINEN, E. R., SIMILA, S., MYLLYLA, V. V., HOKKANEN, E., VAPAATALO, H.: Cyclic adenosine-3′,5′-monophosphate concentration and enzyme activities of cerebrospinal fluid in meningitis of children. *Z. Kinderheilkd.* **120**:243–250, 1975.
74. HIDAKA, H., SHIBUYA, M., ASANO, T., HARA, F.: Cyclic nucleotide phosphodiesterase of human cerebrospinal fluid. *J. Neurochem.* **25**:49–53, 1975.
75. JOO, F., TOTH, I.: Brain adenylate cyclase: Its common occurrence in the capillaries and astrocytes. *Naturwissenschaften* **62**:397–398, 1975.
76. KAKIUCHI, S., YAMAZAKI, R., TESHIMA, Y., UENISHI, K., MIYAMOTO, E.: Multiple cyclic nucleotide phosphodiesterase activities from rat tissues and occurrence of a calcium-plus-magnesium-non-dependent phosphodiesterase and its protein activator. *Biochem. J.* **146**:109–120, 1975.
77. KAMINSKY, N. I., BROADUS, A. E., HARDMAN, J. G., JONES, D. J., JR., BALL, J. H., SUTHERLAND, E. W., LIDDLE, E. W.: Effects of parathyroid hormone on plasma and urinary adenosine 3′,5′-monophosphate in man. *J. Clin. Invest.* **49**:2387–2395, 1970.
78. KARTZINEL, R., EBERT, M. H., CHASE, T. N.: Intravenous probenecid loading: Effects on plasma and cerebrospinal fluid probenecid levels and on monoamine metabolites in cerebrospinal fluid. *Neurology* **26**:992–996, 1976.
79. KASSAN, S. S., KAGEN, L. J.: Elevated levels of cerebrospinal fluid guanosine 3′,5′-cyclic monophosphate (cGMP) in systemic lupus erythematosus. *Am. J. Med.* **64**:732–741, 1978.
80. KATZ, J. B., CATRAVAS, G. N., VALASES, C., WEIGHT, S. J., JR.: Morphine reduces cerebellar guanosine 3′,5′-cyclic monophosphate content and elevates cerebrospinal fluid guanosine-3′,5′-cyclic monophosphate content in rhesus monkeys. *Life Sci.* **22**:467–472, 1978.
81. KATZ, J. B., VALASES, C., CATRAVAS, G. N., WRIGHT, S. J.: Cerebrospinal fluid cyclic AMP levels in rhesus monkeys: Daily fluctuations. *Life Sci.* **22**:445–450, 1978.
82. KEBABIAN, J. W.: Biochemical regulation and physiological significance of cyclic nucleotides in the nervous system. *Adv. Cyclic Nucleotide Res.* **8**:421–508, 1977.
83. KEBABIAN, J. W., STEINER, A. L., GREENGARD, P.: Muscarinic cholinergic regulation of cyclic guanosine 3′,5′-monophosphate in autonomic ganglia: Possible role in synaptic transmission. *J. Pharmacol. Exp. Ther.* **193**:474–488, 1975.
84. KEIRNS, J. J., FREEMAN, J., BITENSKY, M. W.: Cyclic adenosine monophosphate and clinical medicine. Part II. Carbohydrate and lipid metabolism. *Am. J. Med. Sci.* **268**:62–92, 1974.
85. KIESSLING, M., LINDL, T., CRAMER, H.: Cyclic adenosine-monophosphate in cerebrospinal fluid: Effects of theophylline, L-dopa and a dopamine receptor stimulant in rats. *Arch. Psychiatr. Nervenkr.* **220**:325–333, 1975.
86. KODAMA, T., MATSUKADO, Y., SHIMIZU, H.: The cyclic AMP system of human brain. *Brain Res.* **50**:135–146, 1973.
87. KORF, J., BOER, P. H., FEKKES, D.: Release of cerebral cyclic AMP into push–pull perfusates in freely moving rats. *Brain Res.* **113**:551–561, 1976.
88. KRNJEVIC, K., PUIL, E., WERMAN, R.: Is cyclic guanosine monophosphate the internal "second messenger" for cholinergic actions on central neurons? *Can. J. Physiol. Pharmacol.* **54**:172–176, 1976.
89. KRNJEVIC, K., VAN METER, W. G.: Cyclic nucleotides in spinal cells. *Can. J. Physiol. Pharmacol.* **54**:416–421, 1976.
90. LANDIS, S. C.: Ultrastructural changes in mitochondria of cerebellar Purkinje cells of "nervous" mutant mice. *J. Cell. Biol.* **57**:782–797, 1973.
91. LEMAY, A., JARETT, L.: Pitfalls in the use of lead nitrate for the histochemical demonstration of adenylate cyclase activity. *J. Cell Biol.* **65**:39–50, 1975.
92. LINDL, T., CRAMER, H.: Formation, accumulation and release of adenosine 3′,5′-monophosphate induced by histamine in the superior cervical ganglion of the rat *in vitro*. *Biochim. Biophys. Acta* **343**:182–191, 1974.
93. MAO, C. C., GUIDOTTI, A., COSTA, E.: Interactions between gamma-aminobutyric acid and cyclic guanosine 3′,5′-monophosphate in rat cerebellum. *Mol. Pharmacol.* **10**:736–745, 1974.
94. MAO, C. C., GUIDOTTI, A., COSTA, E.: The regulation of cyclic GMP in rat cerebellum: Possible involvement of putative amino-acid neurotransmitters. *Brain Res.* **79**:510–514, 1974.
95. MAO, C. C., GUIDOTTI, A., COSTA, E.: Inhibition by diazepam of the tremor and the increase of cerebellar cGMP content elicited by harmoline. *Brain Res.* **83**:516–519, 1975.
96. MAO, C. C., GUIDOTTI, A., LANDIS, S. C.: Cyclic GMP: Reduction of cerebellar concentrations in "nervous" mutant mice. *Brain Res.* **90**:335–339, 1975.
97. MURAD, F.: Clinical studies and applications of cyclic nucleotides. *Adv. Cyclic Nucleotide Res.* **3**:356–383, 1973.
98. MURAD, F., RALL, T. W., VAUGHAN, M.: Conditions for the formation, partial purification and assay of an inhibitor of adenosine 3′,5′-monophosphate. *Biochim. Biophys. Acta* **192**:430–445, 1969.
99. MURPHREE, O. D., ANGEL, C., DELUCA, D. C., NEWTON, J. E. O.: Longitudinal studies of genetically nervous dogs. *Biol. Psychiatry* **12**:573–576, 1977.
100. MYLLYLA, V. V.: Effect of convulsions and anticonvulsive drugs on cerebrospinal fluid cyclic AMP in rabbits. *Eur. Neurol.* **14**:97–107, 1976.
101. MYLLYLA, V. V.: Effect of cerebral injury on cerebrospinal fluid cyclic AMP concentration. *Eur. Neurol.* **14**:413–425, 1976.

102. MYLLYLA, V. V., HEIKKINEN, E. R., SIMILA, S., HOKKANEN, E., VAPAATALO, H.: Cerebrospinal fluid concentration and urinary excretion of cyclic adenosine-3',5'-monophosphate in various diseases of children: A preliminary study. *Z. KinderHeilkd.* **118**:259–264, 1976.
103. MYLLYLA, V. V., KEIKKINEN, E. R., VAPAATALO, H., HOKKANEN, E.: Cyclic AMP concentration and enzyme activities of cerebrospinal fluid in patients with epilepsy or central nervous system damage. *Eur. Neurol.* **13**:123–130, 1975.
104. MYLLYLA, V. V., VAPAATALO, H., HOKKANEN, E., HEIKKINEN, E. R.: Cerebrospinal fluid concentration of cyclic adenosine 3',5'-monophosphate and pneumoencephalography. *Eur. Neurol.* **12**:28–32, 1974.
105. NAKAZAWA, K., SANO, M.: Studies on guanylate cyclase: A new assay method for guanylate cyclase and properties of the cyclase from rat brain. *J. Biol. Chem.* **249**:4207–4211, 1974.
106. NATHANSON, J. A.: Cyclic nucleotides and nervous system function. *Physiol. Rev.* **57**:157–256, 1977.
107. NEFF, N. H., TOZER, T. N., BRODIE, B. B.: Application of steady-state kinetics to studies of the transfer of 5-hydroxyindole acetic acid from brain to plasma. *J. Pharmacol. Exp. Ther.* **158**:214–218, 1967.
108. ORENBERG, E. K., ZARCONE, V. P., RENSON, J. F., BARCHAS, J. D.: The effects of ethanol ingestion on cyclic AMP, homovanillic acid and 5-hydroxyindole acetic acid in human cerebrospinal fluid. *Life Sci.* **19**:1669–1672, 1976.
109. ORTMANN, R., PERKINS, J. P.: Stimulation of adenosine 3',5'-monophosphate formation by prostaglandins in human astrocytoma cells. *J. Biol. Chem.* **252**:6018–6025, 1977.
110. PENIT, J., JARD, S., BENDA, P.: Probenecid sensitive 3',5'-cyclic AMP secretion by isoproterenol stimulated glial cells in culture. *FEBS Lett.* **41**:156–160, 1974.
111. PERKINS, J. P., MOORE, M. M., KALISKER, A., SU, Y. F.: Regulation of cyclic AMP content in normal and malignant brain cells. *Adv. Cyclic Nucleotide Res.* **5**:641–660, 1975.
112. PERLOW, M., GORDON, E., EBERT, M., FESTOFF, B., CHASE, T. N.: The circadian pattern of catecholamine metabolites and cAMP in the cerebrospinal fluid of subhuman primates. *Trans. Am. Neurol. Assoc.* **101**:279–280, 1976.
113. PERLOW, M. J., LAKE, C. R.: Daily fluctuations in cerebrospinal fluid concentrations of catecholamines, monoamine metabolites, and cyclic AMP, and γ-aminobutyric acid in rhesus monkeys. In Wood, J. H. (ed.): *Neurobiology of Cerebrospinal Fluid I.* New York, Plenum Press, 1980.
114. PHILLIS, J. W.: The role of cyclic nucleotides in the CNS. *Can. J. Neurol. Sci.* **4**:151–195, 1977.
115. POST, R. M., CRAMER, H., GOODWIN, F. K.: Cyclic AMP in cerebrospinal fluid in patients with affective illness: Effects of probenecid, activity, and psychotropic medications. In Usdin, E., Hamburg, D. A., Barchas, J. D. (eds.): *Neuroregulators and Psychiatric Disorders.* New York, Oxford University Press, 1977, pp. 464–469.
116. POST, R. M., GOODWIN, F. K.: Estimation of brain amine metabolism in affective illness: Cerebrospinal fluid studies utilizing probenecid. *Psycother. Psychosom.* **23**:142–158, 1974.
117. POST, R. M., GOODWIN, F. K., GORDON, E., WATKIN, D. M.: Amine metabolites in human cerebrospinal fluid: Effects of cord transection and spinal fluid block. *Science* **179**:897–899, 1973.
118. PRASAD, K. N., SAHU, S. K., KUMAR, S.: Relationship between cyclic AMP level and differentiation of neuroblastoma cells in culture. In Nakahara, W., Ono, T., Sugimura, T., Sugano, H. (eds.): *Differentiation and Control of Malignancy of Tumor Cells.* Baltimore, University Park Press, 1974, pp. 287–309.
119. RINDLER, M. J., BASHOR, M. M., SPITZER, N., SAIER, M. H., JR.: Regulation of adenosine 3',5'-monophosphate efflux from animal cells. *J. Biol. Chem.* **253**:5431–5436, 1978.
120. ROBISON, G. A., COPPEN, A. J., WHYBROW, P. C., PRANGE, A. J.: Cyclic AMP in affective disorders. *Lancet* **2**:1028–1029, 1970.
121. ROISEN, F. J., MURPHY, R. A., BRADEN, W. G.: Dibutyryl cyclic adenosine monophosphate stimulation of colcemide-inhibition axonal elongation. *Science* **177**:809–811, 1972.
122. ROISEN, F. J., MURPHY, R. A., PICHICHERO, M. E., BRADEN, W. G.: Cyclic adenosine monophosphate stimulation of axonal elongation. *Science* **177**:73–74, 1972.
123. RUBIN, E. H., FERRENDELLI, J. A.: Distribution and regulation of cyclic nucleotide levels in cerebellum *in vivo*. *J. Neurochem.* **29**:43–51, 1977.
124. RUDLAND, P. S., GOSPODAPROWICZ, D., SIEFFERT, W. E.: Activation of guanyl cyclase and intracellular cyclic GMP by fibroblast growth factor. *Nature (London)* **250**:741–744, 1974.
125. RUDMAN, D., FLEISCHER, A., KUTNER, M. H.: Concentration of 3',5'-cyclic adenosine monophosphate in ventricular cerebrospinal fluid of patients with prolonged coma after head trauma or intracranial hemorrhage. *N. Engl. J. Med.* **295**:635–638, 1976.
126. RUDMAN, D., O'BRIEN, M. S., MCKINNEY, A. S., HOFFMAN, J. C., JR., PATTERSON, J. H.: Observations of the cyclic nucleotide concentrations in human cerebrospinal fluid. *J.Clin.Endocrinol.Metab.* **42**:1088–1097, 1976.
127. SCHIMMER, B. P.: Effects of catecholamines and monovalent cations on adenylate cyclase activity in cultured glial tumor cells. *Biochim. Biophys. Acta* **252**:567–573, 1971.
128. SCHMIDT, M. J., NADI, N. S.: Cyclic nucleotide accumulation *in vitro* in the cerebellum of "nervous" neurologically mutant mice. *J. Neurochem.* **29**:87–90, 1977.
129. SCHWARTZEL, E. H., JR., BACHMAN, S., LEVINE, R. A.:

Cyclic nucleotide activity in gastrointestinal tissues and fluids. *Anal. Biochem.* **78**:395–405, 1977.

130. SEBENS, J. B., KORF, J.: Cyclic AMP in cerebrospinal fluid: Accumulation following probenecid and biogenic amines. *Exp. Neurol.* **46**:333–344, 1975.
131. SIFONTES, J. E., BROOKE-WILLIAMS, R. D., LINCOLN, E. M., CLEMONS, H.: Observations on the effect of induced hyperglycemia on the glucose content of the cerebrospinal fluid in patients with tuberculous meningitis. *Am. Rev. Tuberc.* **67**:732–754, 1953.
132. SIGGINS, G. R., HOFFER, B. J., BLOOM, F. E.: Studies on norepinephrine containing afferents to Purkinje cells of the rat cerebellum. II. Evidence for mediation of norepinephrine effects by cyclic 3′,5′-AMP. *Brain Res.* **25**:535–553, 1971.
133. SMITH, C. C., TALLMAN, J. F., POST, R. M., VAN-KAMMEN, D. F., JIMERSON, D. C., BROWN, G. L., BROOKS, B. R., BUNNEY, W. E., JR.: An examination of baseline and drug-induced levels of cyclic nucleotide in the cerebrospinal fluid of control and psychiatric patients. *Life Sci.* **19**:131–136, 1976.
134. SODERLING, T. R., PARK, C. R.: Recent advances in glycogen metabolism. *Adv. Cyclic Nucleotide Res.* **4**:284–333, 1974.
135. SPECTOR, R., LORENZO, A. V.: The effects of salicylate and probenecid on the cerebrospinal fluid transport of penicillin, amino salicyclic acid and iodide. *J. Pharmacol. Exp. Ther.* **188**:55–65, 1974.
136. STEINER, A. L., FERRENDELLI, J. A., KIPNIS, D. M.: Radioimmunoassay for cyclic nucleotides. III. Effect of ischemia, changes during development and regional distribution of adenosine 3′,5′-monophosphate and guanosine 3′,5′-monophosphate in mouse brain. *J. Biol. Chem.* **247**:1121–1124, 1972.
137. STEINER, A. L., ONG, S., WEDNER, H. J.: Cyclic nucleotide immunocytochemistry. *Adv. Cyclic Nucleotide Res.* **7**:116–155, 1976.
138. STEINER, A. L., PAGLIARA, A. S., CHASE, L. R., KIPNIS, D. M.: Radioimmunoassay for cyclic nucleotides. II. Adenosine 3′,5′-monophosphate and guanosine 3′,5′-monophosphate in mammalian tissues and body fluids. *J. Biol. Chem.* **247**:1114–1120, 1972.
139. STEINER, A. L., PARKER, C. W., KIPNIS, D. M.: The measurement of cyclic nucleotides by radioimmunoassay. *Adv. Biochem. Psychopharmacol.* **3**:89–111, 1970.
140. STEINER, A. L., PARKER, C. W., KIPNIS, D. M.: Radioimmunoassay for cyclic nucleotides. I. Preparation of antibodies and iodinated cyclic nucleotides. *J. Biol. Chem.* **247**:1106–1113, 1972.
141. STEINER, A. L., WEHMANN, R. E., PARKER, C. W., KIPNIS, D. M.: Radioimmunoassay for the measurement of cyclic nucleotides. *Adv. Cyclic Nucleotide Res.* **2**:51–61, 1972.
142. STONE, T. W., TAYLOR, D. A.: Microiontophoretic studies of the effects of cyclic nucleotides on excitability of neurons in the rat cerebral cortex. *J. Physiol. (London)* **266**:523–544, 1977.
143. SUTHERLAND, E. W., RALL, T. W.: Fractionation and characterization of a cyclic adenine ribonucleotide formed by tissue particles. *J. Biol. Chem.* **232**:1077–109, 1958.
144. TIHON, C., GOREN, M. B., SPITZ, E., RICKENBERG, H. V.: Convenient elimination of trichloroacetic acid prior to radioimmunoassay of cyclic nucleotides. *Anal. Biochem.* **80**:652–653, 1977.
145. TOMINAGA, S., MURAKAMI, M., KOJIMA, S., SUZUKI, T., NAKAMURA, T.: Venous plasma cyclic AMP in acute cerebrovascular disease. *Tohoku J. Exp. Med.* **120**:151–158, 1976.
146. TOVEY, K. C., OLDHAM, K. G., WHELAN, J. A. M.: A simple direct assay for cyclic AMP in plasma and other biological samples using an improved competitive protein binding technique. *Clin. Chem. Acta* **56**:221–234, 1974.
147. TRABUCCHI, M., CERRI, C., SPANO, P. F., KUMAKURA, K.: Guanosine 3′,5′-monophosphate in the CSF of neurological patients. *Arch. Neurol.* **34**:12–13, 1977.
148. TSANG, D., LAL, S., SOUKES, T. L., FORD, R. M., ARONOFF, A.: Studies on cyclic AMP in different compartments of cerebrospinal fluid. *J. Neurol. Neurosurg. Psychiatry* **39**:1186–1190, 1976.
149. VOLICER, L. (ed.): *Clinical Aspects of Cyclic Nucleotides.* Jamaica, New York, Spectrum, 1977.
150. WATANABE, H., PASSONEAU, J. V.: Cyclic adenosine monophosphate in cerebral cortex: Alterations following trauma. *Arch. Neurol.* **32**:181–184, 1975.
151. WEISS, B. (ed.): *Cyclic Nucleotides in Disease.* Baltimore, University Park Press, 1975.
152. WEISS, B.: Differential activation and inhibition of the multiple forms of cyclic nucleotide phosphodiesterase. *Adv. Cyclic Nucleotide Res.* **5**:195–211, 1975.
153. WEISS, B., COSTA, E.: Regional and subcellular distribution of adenyl cyclase and 3′,5′-cyclic nucleotide phosphodiesterase in brain and pineal gland. *Biochem. Pharmacol.* **17**:2107–2116, 1968.
154. WEISS, B., COSTA, E.: Selective stimulation of adenyl cyclase of rat pineal gland by pharmacologically active catecholamines. *J. Pharmacol. Exp. Ther.* **161**:310–319, 1968.
155. WELCH, K. M. A., CHABI, E., NELL, J. H., BARTOSH, K., CHEE, A. N. C., MATHEW, N. T., ACHAR, U. S.: Biochemical comparison of migraine and stroke. *Headache* **16**:160–167, 1976.
156. WELCH, K. M. A., MEYER, J. S.: Neurochemical alterations in cerebrospinal fluid in cerebral ischemia and stroke. In Wood, J. H. (ed.): *Neurobiology of Cerebrospinal Fluid I.* New York, Plenum Press, 1980.
157. WELCH, K. M. A., MEYER, J. S., CHEE, A. N. C.: Evidence for disordered cyclic AMP metabolism in patients with cerebral infarction. *Eur. Neurol.* **13**:144–154, 1975.
158. WELCH, K. M., A., NELL, J., CHABI, E.: The role of cyclic AMP in neurologic and affective disorders. In Volicer, L. (ed.): *Clinical Aspects of Cyclic Nucleotides.* Jamaica, New York, Spectrum, 1977, pp. 327–360.

159. WELLER, M., RODNIGHT, R., CARRERA, D.: Determination of adenosine 3′,5′-cyclic monophosphate in cerebral tissues by saturation analysis: Assessment of a method using a binding protein from ox muscle. *Biochem. J.* **129**:113–121, 1972.
160. WILLIAMS, R. H., LITTLE, S. A., BEUG, A. G., ENSINCK, J. W.: Cyclic nucleotide phosphodiesterase activity in man, monkey and rat. *Metabolism* **20**:743–748, 1971.
161. WILLIAMS, R. H., LITTLE, S. A., ENSINCK, J. W.: Adenyl cyclase and phosphodiesterase activities in brain areas of man, monkey and rat. *Am. J. Med. Sci.* **258**:190–202, 1969.
161a. WOOD, J. H., BROOKS, B. R.: Neurotransmitter, metabolite, and cyclic nucleotide alterations in cerebrospinal fluid of seizure patients. In Wood, J. H. (ed.): *Neurobiology of Cerebrospinal Fluid I.* New York, Plenum Press, 1980.
162. WOOD, J. H., GLAESER, B. S., HARE, T. A., SODE, J., BROOKS, B. R., VAN BUREN, J. M.: Cerebrospinal fluid GABA reductions in seizure patients evoked by cerebellar surface stimulation. *J. Neurosurg.* **47**:582–589, 1977.
163. WOOD, J. H., LAKE, C. R., ZIEGLER, M. G., SODE, J., BROOKS, B. R., VAN BUREN, J. M.: Cerebrospinal fluid norepinephrine alterations during electrical stimulation of cerebellar and cerebral surfaces in epileptic patients. *Neurology* **27**:716–724, 1977.
164. ZIEGLER, M. G., WOOD, J. H., LAKE, C. R., KOPIN, I. J.: Norepinephrine and 3-methoxy-4-hydroxyphenyl glyol gradients in human cerebrospinal fluid. *Am. J. Psychiatry* **134**:565–568, 1977.
165. ZIMMERMAN, T. P., WINSTON, M. S., CHU, L. C.: A more sensitive radioimmunoassay (RIA) for guanosine 3′,5′-cyclic monophosphate (cGMP) involving prior 2′-O-succinylation of samples. *Anal. Biochem.* **71**:79–95, 1976.

CHAPTER 10

Norepinephrine in Cerebrospinal Fluid

Basic Studies, Effects of Drugs and Disease

Michael G. Ziegler, C. Raymond Lake, James H. Wood, and Michael H. Ebert

1. Introduction

In peripheral tissues, noradrenergic nerves have a diffuse distribution, but in the brain, noradrenergic neurons arise in the brainstem and are inhomogeneously distributed. The locus coeruleus is the largest noradrenergic nucleus in rat brain, and it contains only about 1500 neurons. Although norepinephrine is present in the brain in small amounts, it has been extensively studied for many reasons. Central norepinephrine is involved in the regulation of blood pressure and emotions, and the two most common diseases in Western civilization are hypertension and mental illness. Many of the drugs used in the treatment of these and other diseases interact with norepinephrine. Sensitive fluorometric and radioenzymatic assays make analysis of the small amount of norepinephrine present in the brain possible so that we know a great deal about this chemical that occurs in such small amounts.

The concentration of norepinephrine in some areas of brain tissue is 1000 times as great as norepinephrine levels in cerebrospinal fluid (CSF). Norepinephrine levels in CSF reported by most investigators of about 100–300 pg/ml[9,11,35,36] are at the lower limits of detection of most fluorometric assays. These low levels can be reliably determined by gas chromatography–mass spectroscopy, though with considerable difficulty. Recently developed radioenzymatic techniques also have sufficient sensitivity to measure norepinephrine in CSF and can be performed more rapidly. The most successful of these methods utilizes the conversion of norepinephrine to [^{3}H]epinephrine by the transfer of a [^{3}H]methyl group from [^{3}H]methyl-*S*-adenosyl-methionine to the primary amine of norepinephrine.[76] In this assay method, 2 ml of CSF is shaken with alumina and the norepinephrine is adsorbed on the alumina at basic pH and then eluted from alumina with acid. This eluate is incubated with the enzyme phenylethanolamine-*N*-methyl-transferase (PNMT) and [^{3}H]methyl-*S*-adenosyl-methionine. The product of this reaction, [^{3}H]epinephrine, is then purified and counted by scintillation spectroscopy. The assay has a sensitivity of 14 pg norepinephrine/ml CSF.

Michael G. Ziegler, M.D. • Department of Medicine, University of California, San Diego, California 92103. **C. Raymond Lake, M.D., Ph. D.** • Departments of Psychiatry and Pharmacology, Uniformed Services University of Health Sciences, School of Medicine, Bethesda, Maryland 20014. **James H. Wood, M.D.** • Division of Neurosurgery, University of Pennsylvania School of Medicine, Philadelphia, Pennsylvania 19104. **Michael H. Ebert, M.D.** • Experimental Therapeutics Section, Laboratory of Clinical Science, National Institute of Mental Health, National Institutes of Health, Bethesda, Maryland 20205.

When this assay is compared with gas chromatographic–mass spectroscopic techniques, the mean difference between the two methods is 12% and correlation between the methods is r = 0.95.[76] Both techniques can reliably measure the small amounts of norepinephrine present in CSF, and the radioenzymatic technique is practical for determination of norepinephrine in a fairly large number of samples.

Norepinephrine in CSF can also be measured by a catechol-*O*-methyl-transferase (COMT)-based assay that converts norepinephrine to [^{3}H]normetanephrine. In this assay, 100–200 μl of CSF is incubated with COMT and [^{3}H]methyl-*S*-adenosyl-methionine. The *O*-methylated metabolites of catechols formed from this reaction are then isolated by solvent extraction and separated by paper or thin-layer chromatography. Norepinephrine cannot be concentrated on alumina from CSF in this assay unless elaborate precautions are taken to eliminate aluminum ion, which inhibits the enzyme COMT. The assay is thus less sensitive and more variable than the PNMT-based assay, but it can provide a useful estimate of norepinephrine levels in CSF. The COMT assay is in use in a larger number of laboratories than the PNMT assay because the enzyme COMT is simpler to purify and the assay also provides an estimate of dopamine and epinephrine levels. The assay has the advantage of requiring a smaller volume of CSF, and in our experience, norepinephrine levels determined by this assay compare well with norepinephrine levels determined by the PNMT method when levels are relatively high. The assay method used to measure norepinephrine levels in CSF is critical, as revealed in several early studies using fluorometric techniques to measure norepinephrine. Tissue breakdown products can interfere with these fluorometric assays and give spuriously elevated norepinephrine levels. Recent studies with radioenzymatic techniques failed to support some of the earlier findings of very high norepinephrine levels in CSF after bleeding into the CSF.

The low levels of norepinephrine found in the CSF are not great enough to have much physiological importance. The significance of these levels depends on how well they reflect noradrenergic processes in the brain. Brain areas with the highest norepinephrine levels are located in close proximity to the CSF.[48] CSF norepinephrine levels are kept low by an avid reuptake mechanism; norepinephrine injected into the cerebral ventricles is actively and rapidly taken up into neurons lying adjacent to the cerebral ventricles.[18,19] Norepinephrine release should have a dramatic effect on CSF norepinephrine, since some brain areas have 1000 times as great a concentration of norepinephrine as the CSF.[76] CSF norepinephrine should reflect release and reuptake of norepinephrine from brain areas lying adjacent to the CSF. Since norepinephrine is a labile molecule subject to oxidation, some of it might be destroyed while in the CSF. Norepinephrine that is synthesized and destroyed intracellularly need not affect norepinephrine levels in CSF, since the CSF communicates with the extracellular space. Thus, CSF norepinephrine may not reflect the total turnover of brain norepinephrine, but instead it should more closely reflect release from areas lying adjacent to the CSF. On the other hand, metabolites of norepinephrine such as 3-methoxy-4-hydroxyphenylethylene glycol (MHPG) should be derived from brain norepinephrine turnover. Since turnover is loosely linked to norepinephrine release from nervous tissue, especially locus coeruleus activity,[1] one would expect the loose correlation between CSF norepinephrine levels and MHPG levels that has been found.[80] Norepinephrine in CSF probably also does not represent whole brain norepinephrine synthesis, since both the norepinephrine content[7] and the rate of norepinephrine synthesis[81] are highest in whole brain at the time of day when animals are least active and norepinephrine in CSF is highest at the time of day when animals are most active.[70] Unlike metabolites of the monoamine neurotransmitters, norepinephrine is both chemically labile and subject to rapid uptake by brain transport mechanisms. Because of this, norepinephrine in CSF sampled from one region should be greatly affected by brain areas immediately surrounding that region. Most samples of CSF taken from people are obtained by lumbar puncture, which drains CSF surrounding the spinal cord. Spinal cord norepinephrine is concentrated in the cord gray matter,[82] where noradrenergic nerve terminals derive from neurons with cell bodies in the A1 and A2 groups of brainstem nuclei and the locus coeruleus.[38] Thus, lumbar CSF norepinephrine levels may be strongly dependent on nerve terminals with cell bodies originating in these small brainstem nuclei. It seems possible that CSF norepinephrine represents neuronal release of norepinephrine in brain areas near the CSF. This would make CSF norepinephrine an important tool, since the functional activity of nerve cells depends on their rate of neurotransmitter release. Readers can conclude for themselves whether CSF norepinephrine actually reflects central noradrenergic nerve activity. Additional discussion of this topic is contained in Chapter 5.

2. CSF Gradient for Norepinephrine

Since the cell bodies of origin for central noradrenergic neurons occur in the brainstem and the greatest concentration of noradrenergic terminals is in the area of the hypothalamus, one might expect that CSF in the spinal canal would have less norepinephrine than CSF close to the brainstem. In seven patients with Huntington's chorea, CSF was removed by lumbar puncture with a total of 20 ml of fluid taken. There was a gradient with increasing levels of norepinephrine until the 12th milliliter, when there was no further increase in norepinephrine levels (Fig. 1).[76] The average volume of CSF in man is 140 ml.[70] Of this, 110 ml is contained within the skull and 30 ml in the spinal subarachnoid space. Since removal of only 20 ml of CSF by lumbar puncture presumably includes only CSF from the spinal areas, a further experiment was performed with CSF removed by lumbar drainage from patients undergoing neurosurgical procedures.[80] A total of 38 ml of CSF was removed from these patients, and still the gradient for norepinephrine appears to plateau after about 16 ml has been removed. Since there is 30 ml of CSF in the spinal subarachnoid space, the last sample presumably contains a mixture of CSF from the cisterna magna, basal cisterns, and the cerebral ventricles. The norepinephrine gradient in lumbar CSF appears to be of spinal origin and not dependent on brain areas containing norepinephrine. Spinal norepinephrine is localized mainly in the gray matter,[82] and it is unlikely that the lower lumbar area in the region of the cauda equina contributes significant quantities of norepinephrine to the CSF first obtained during lumbar puncture. The local spinal gradient for norepinephrine in CSF indicates that the level of norepinephrine in CSF obtained by lumbar puncture is probably strongly influenced by noradrenergic nerves that originate in the brainstem and terminate in the gray matter of the spinal cord. There appears to be a relatively small gradient for MHPG, the major metabolite of central norepinephrine, in CSF obtained by lumbar puncture.[80] Several earlier investigations did not find such an MHPG gradient,[6,21,59] probably because of its small size. The gradient, however, does parallel the gradient for norepinephrine in lumbar CSF, and this indicates that there is probably some correlation between norepinephrine and MHPG levels in CSF. In patients with various neurological or psychiatric disorders, the relationship between levels of norepinephrine and MHPG and CSF ranged from $r = 0.29$ to $r = 0.76$.[43,80] There is an avid neuronal reuptake mechanism for norepinephrine, but a similar mechanism is not present for MHPG. MHPG levels are about 40 times as great as those of its parent compound, norepinephrine, reflecting the rapid reuptake and metabolism of norepinephrine. MHPG should have a much slower turnover rate in CSF than does norepinephrine, and this might explain why the gradient for MHPG is less steep than for norepinephrine. MHPG levels are also much less variable than those of norepinephrine. The coefficient of variation between patients for MHPG is approximately 15%, while that for norepinephrine is about 40%.[80]

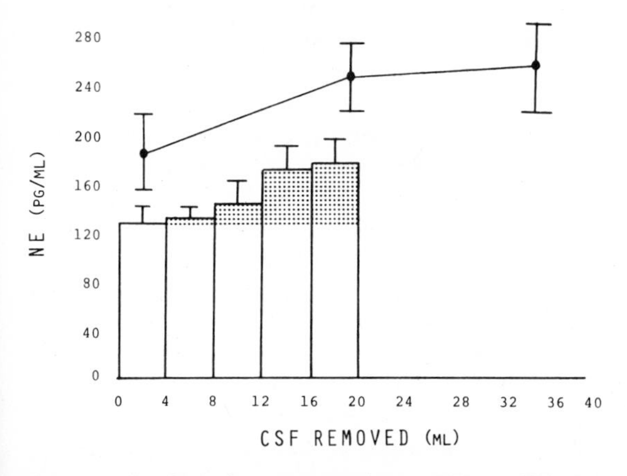

Figure 1. Gradient for norepinephrine (NE) in CSF. Successive samples of CSF were removed by lumbar puncture, and 4-ml samples were assayed with a PNMT-based radioenzymatic technique. Bars represent samples from patients with Huntington's chorea; (•) CSF from patients with other neurological disorders. Vertical brackets indicate S.E.M.

3. Circadian Rhythm of Norepinephrine in CSF

The overall rate of synthesis of norepinephrine[81] and brain content of norepinephrine[7] are highest during the time of day when animals are least active. However, norepinephrine content in areas of the brain containing predominantly noradrenergic nerve terminals is highest when animals are most active. The cervical spinal cord,[48] midbrain,[47] and hypothalamus[33] all have significant circadian variations in their norepinephrine content. Norepinephrine-releasing agents such as amphetamine increase activity, as does infusion of norepinephrine into the cerebral ventricles. When CSF is continuously collected from the lateral cerebral ventricle of monkeys, which are most active in the daytime, norepinephrine content in the CSF is highest in midafternoon and lower early in the morning (Fig. 2).[79] These monkeys were fed twice daily and stayed in a room that was very active Monday through Friday and had few visitors on weekends. However, the circadian rhythm for norepinephrine was the same on weekdays and weekends. One monkey was confined to a closed box that was illuminated only at night, but it continued to have elevated norepinephrine levels in its CSF during the day. This persistent diurnal rhythm of norepinephrine in CSF of monkeys is also present in man. When lumbar puncture is performed on patients at 03:00, 09:00, and 15:00 hours, the norepinephrine levels are higher in the daytime and highest at 15:00 hours, as in monkeys (Fig. 2). Noradrenergic nerve terminals adjacent to the CSF have their highest norepinephrine content in animals during periods of greatest activity.[12,16,31,33,47,48] Drugs that block norepinephrine synthesis or deplete brain norepinephrine decrease motor activity,[25] and many stimulant drugs such as amphetamine release norepinephrine.[13] Norepinephrine infusion into the cerebral ventricles increases motor activity. The diurnal rhythm of norepinephrine in CSF parallels the functional effects of norepinephrine on activity. The metabolites of norepinephrine, MHPG and 3-methoxy-4-hydroxymandelic acid [vanillymandelic acid (VMA)], are less rapidly altered by the body and do not show significant diurnal variation.[41]

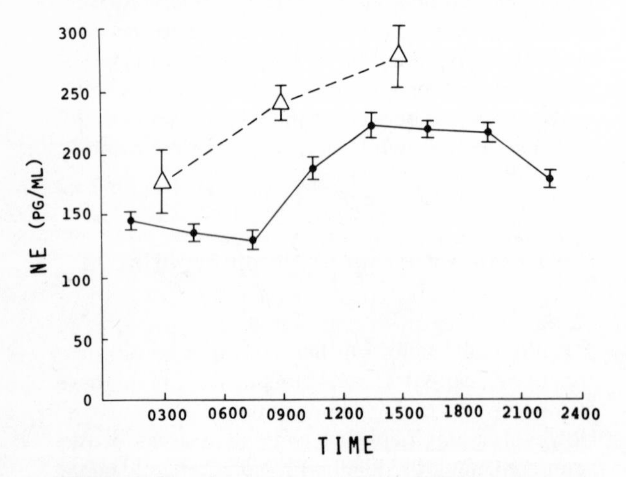

Figure 2. Circadian rhythm of norepinephrine (NE) in CSF. (●) CSF continuously collected from the lateral ventricles of monkeys; (Δ) lumbar CSF collected from human subjects with various neurological disorders at three different times of day. The vertical brackets indicate S.E.M.

4. Relationship between Blood and CSF Norepinephrine

There is a blood–brain barrier for catecholamines,[30,32,69] and this includes a blood–CSF barrier for norepinephrine. When radioactive norepinephrine was infused intravenously into a monkey and CSF samples collected for 2 days, the peak level of radioactive norepinephrine in CSF reached only 2% of peak plasma levels. VMA and MHPG, metabolites of norepinephrine, penetrated the CSF slightly better, but still reached only 5% of peak plasma levels.[78] One patient with high norepinephrine levels from a pheochromocytoma had a plasma norepinephrine level of 9680 pg/ml, while CSF obtained simultaneously had only 200 pg/ml of norepinephrine. There appears to be a very effective blood–CSF barrier to the passage of norepinephrine. Surprisingly, in 159 neurological patients from whom both blood and CSF were obtained, there is a strong relationship between norepinephrine in blood and CSF ($r = 0.78$, $p < 0.0001$).[78] This relationship between blood and CSF levels of norepinephrine clearly cannot be explained by penetration of norepinephrine from blood to the CSF. If norepinephrine in CSF were derived from peripheral sympathetic nerves, as is norepinephrine in blood, then this might give rise to the relationship. However, sympathetic innervation of central nervous system blood vessels is largely limited to a few major vessels,[50] and norepinephrine injected into the CSF is largely taken up by catecholaminergic neurons of central, not peripheral, origin.[18] It appears that norepinephrine in blood is derived from peripheral sympathetic nerves and norepinephrine in CSF from central noradrenergic neurons. The similarity between blood and CSF levels of norepinephrine can be explained in terms of functional similarities in the handling of norepinephrine by the central and peripheral nervous systems. Rates of reuptake, release, and metabolism of catecholamines may be genetically determined factors that are similar in both central and peripheral nervous systems in any individual. Central noradrenergic neurons appear to be involved in the control of peripheral noradrenergic nerves. The A1 and A2 groups of norepinephrine-containing cells send axons down the spinal cord with terminals close to preganglionic sympathetic neurons. Firing of these neurons might release norepinephrine in the spinal cord and simultaneously stimulate sympathetic neuronal activity. Since noradrenergic neurons are involved in the control of blood flow centrally[45] and peripherally, similarity of function can partly explain the similarity of blood and CSF norepinephrine levels.

5. Effects of Drugs on Norepinephrine in CSF

5.1. Amphetamine

Many of the important effects of amphetamine are mediated by norepinephrine. Lesions of the ventral noradrenergic bundle antagonize amphetamine-induced anorexia.[1a] Blockade of the α-receptors stimulated by norepinephrine suppresses the hyperexcitability, aggressiveness, and mortality associated with high doses of amphetamine. When norepinephrine is infused into the cerebral ventricles, it has an effect similar to an infusion of amphetamine.[55] On the other hand, the effects of amphetamine on behavior are greatly attenuated when endogenous stores of monoamines in the brain are depleted. Amphetamine is taken up by brain tissue, where it releases catecholamines[24] and prevents their reuptake.[66] This combined blockade of reuptake and stimulation of secretion should lead to high levels of norepinephrine in the synaptic cleft and increase norepinephrine levels in CSF. Amphetamine enhances release of radioactivity from noradrenergic nerve endings when they are previously incubated with radioactive tyrosine or radioactive norepinephrine. To study the effect of amphetamine on release of endogenous norepinephrine into CSF, monkeys were kept in a chair and CSF collected from the lateral ventricle into a refrigerated fraction collector set to advance one tube every 3 hr. Monkeys were given intravenous amphetamine sulfate at noon, when the circadian rhythm for norepinephrine in CSF has ordinarily reached its peak. A dose of 1.0 mg/kg *d*-amphetamine increases norepinephrine levels in CSF from 160 to over 400 pg/ml (Fig. 3), a striking response since this dose is only about one tenth that used in many other pharmacological experiments. Norepinephrine levels in CSF remained elevated for over 36 hr. Doses of amphetamine as small as 0.1 mg/kg gave smaller but consistent elevations in CSF norepinephrine that were of shorter duration than seen after larger doses of the drug. Both *d*- and *l*-amphetamine gave roughly comparable increases in CSF norepinephrine levels, so this study tends to support the *in vitro* studies of Baldessarini and Harris[2] and Thornburg and Moore,[67,68]—which find equal responses of *d*- and *l*-amphetamine on noradrenergic neurons—and fails to support several other investigations. In a single experiment, 0.5 mg/kg *d*-amphetamine increased norepinephrine in lumbar CSF of a monkey by 200 pg/ml with a roughly similar duration of effect (Fig. 4). This may reflect both local release of norepinephrine in the spinal cord and norepinephrine released from the brainstem. Although norepinephrine levels in CSF are quite labile in response to a number of stimuli and even undergo a circadian rhythm,[79] the increase in norepinephrine levels after amphetamine is as large a response as

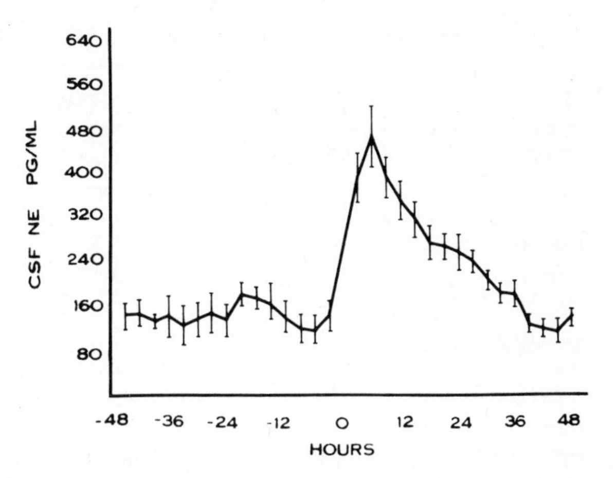

Figure 3. Level of norepinephrine (NE) in monkey ventricular CSF before and after 1.0 mg/kg *d*-amphetamine. Drug was given intravenously at zero hour (noon). The vertical brackets indicate S.E.M.

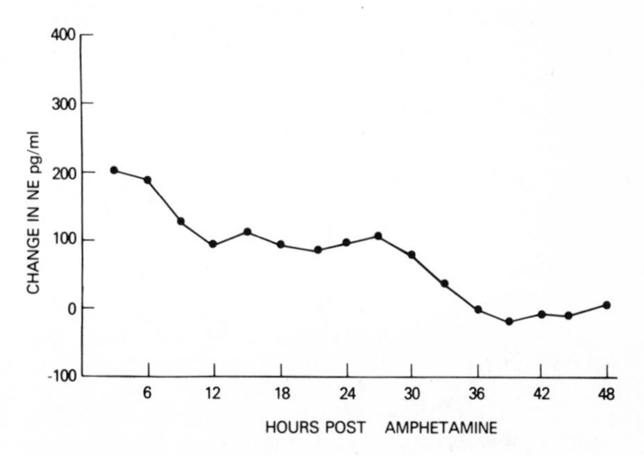

Figure 4. Level of norepinephrine (NE) in lumbar CSF collected from one monkey after 0.5 mg/kg *d*-amphetamine. Drug was given intravenously at zero hour (noon).

we have seen. Thus, doses of amphetamine in the range of those used therapeutically in man release endogenous norepinephrine in the monkey with an effect lasting up to 36 hr. Although amphetamine has been noted to inhibit the formation of endogenous norepinephrine,[26,61] there is no rebound depression of norepinephrine levels apparent after the effects of amphetamine have worn off (Figs. 3 and 4).

5.2. Bromocriptine

Bromocriptine, a drug that activates dopamine receptors, was recently released for clinical use in the United States. Like other dopamine agonist drugs, bromocriptine can cause hypotension,[14,23,29] and it has been used to treat hypertension.[27,63] Stumpe *et al.*[63] proposed that the drug lowers blood pressure through interaction with central dopaminergic mechanisms. Steinsland and Hieble[60] have shown that dopamine agonist drugs inhibit the peripheral release of norepinephrine through stimulation of presynaptic dopaminergic receptors. Bromocriptine inhibits the release of norepinephrine from peripheral sympathetic nerves.[77] The peripheral dopamine receptors are very similar to central dopamine receptors,[60] so it is possible that a dopamine agonist drug might also inhibit central norepinephrine release. To investigate this possibility, samples of CSF were obtained by lumbar puncture from six patients with parkinsonism before and after they received 60–70 mg bromocriptine daily. These patients had some improvement in their parkinsonian symptoms while taking bromocriptine and a small decrease in their blood pressure. Their norepinephrine level in CSF decreased from 201 to 99 pg/ml, a reduction of 102 ± 34 pg/ml ($p < 0.02$). This 50% decrease in CSF norepinephrine was comparable to the decrease in their plasma norepinephrine at the same time. Thus, a dopamine agonist can inhibit the release of central as well as peripheral norepinephrine, as manifested by a decrease in CSF norepinephrine. The large increase in CSF norepinephrine after amphetamine and decrease after bromocriptine demonstrate the sensitivity of CSF norepinephrine levels and suggest that they might be a very useful tool in investigation of drug effects in patients who are to undergo diagnostic lumbar puncture.

5.3. Probenecid

Probenecid is frequently used in the study of monoamine metabolism in man. It is a weak organic acid that interferes with the active transport of acidic metabolites out of the brain and into the blood. Homovanillic acid (HVA) (a metabolite of dopamine) and 5-hydroxyindoleacetic acid (a metabolite of serotonin) are cleared from the CSF by this acid-transport mechanism. When the mechanism is blocked by probenecid, these acidic monoamine metabolites accumulate in the CSF, and their rate of accumulation has been used to assess the rates of metabolism of dopamine and serotonin.[4,65] Since probenecid blocks the transport of only acidic metabolites, the neutral metabolite of norepinephrine, MHPG, should not accumulate significantly in CSF after probenecid. However, after large doses of probenecid, there has been a report of a 60% increase of MHPG levels in CSF[22] One possible explanation might be that probenecid affects norepinephrine itself and that this is reflected in increased levels of its metabolite MHPG. This hypothesis has been studied in 20 neurological patients.[28] A lumbar spinal tap was performed on the patients at 9:00 A.M. after an overnight fast, and then probenecid was given orally in four divided doses to a total of 100 mg/kg. At 18 hr after the first dose of probenecid, CSF was again collected by lumbar puncture. After probenecid administration, levels of norepinephrine in CSF were significantly elevated above levels found before the drug was administered. Twelve of the patients noted no adverse side effects from probenecid, and their CSF norepinephrine increased by 84 ± 34 pg/ml. Eight of the patients complained of anorexia, nausea, or vomiting from the probenecid, and their CSF norepinephrine increased by 96 ± 31 pg/ml. These are both large and significant increases in CSF norepinephrine, but the difference between patients who were ill and those who felt well is not significant. This increase in CSF norepinephrine levels is large enough to explain the previously reported 60% increase[22] in MHPG levels after probenecid. Since both norepinephrine and its metabolite MHPG are increased after probenecid, it appears that probenecid alters the turnover of norepinephrine, not the clearance of MHPG.

6. CSF Norepinephrine in Disease

The level of norepinephrine in CSF of patients with neurological diseases has a wide distribution ranging from 40 to 2000 pg/ml. Some of this wide range is due to diseases that decrease norepinephrine levels in CSF, such as Huntington's chorea, or increase levels, such as amyotrophic lateral scle-

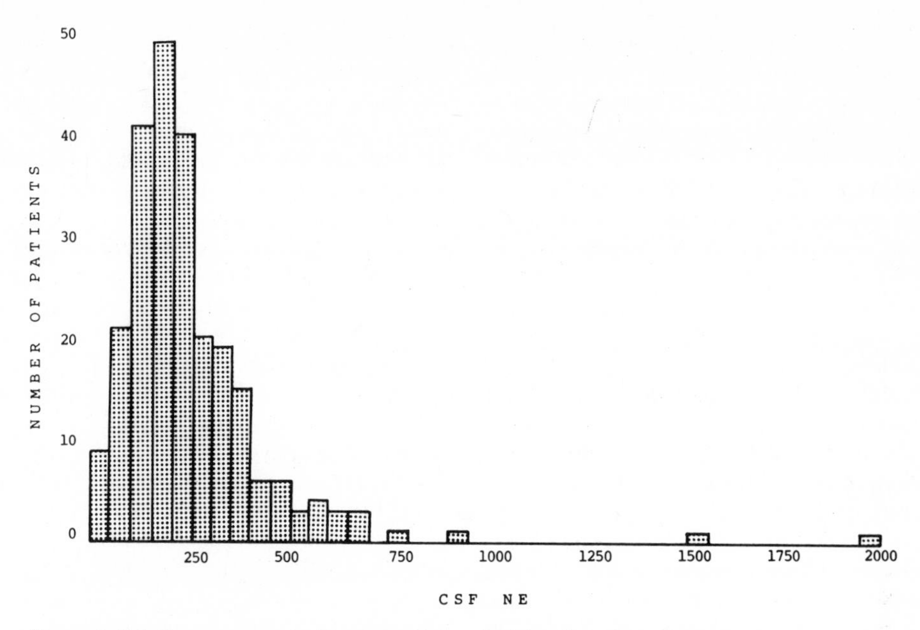

Figure 5. Relative occurrence of norepinephrine (NE) levels in CSF of 240 patients with various neurological disorders not known to affect norepinephrine.

rosis. When only patients with diseases not known to affect norepinephrine levels are included, there is a distribution of patients skewed toward those with higher levels (Fig. 5). This is similar to the distribution of levels of plasma norepinephrine and many other biological variables. This can be converted to a normal distribution by taking the natural logarithm of CSF norepinephrine (Fig. 6), which then permits use of linear methods for statistical evaluation of CSF norepinephrine levels. Because of the skewing toward higher values, the normal range for norepinephrine levels in CSF is considerably higher than that computed by the mean ± 2 standard deviations. Skewing of the levels is not so extreme that evaluation by t test or linear regression gives results far from true values, however.

6.1. Psychiatric Illness

Central norepinephrine appears to be involved in regulation of mood and may be altered in patients with manic–depressive illnesses.[5,52] Pharmacological studies implicate catecholamines in the regulation of mood, since drugs that deplete norepinephrine, such as reserpine, tend to cause depression, and drugs that increase norepinephrine, such as amphetamine, cause stimulation. The main metabolite of central norepinephrine in man is MHPG, and it can be measured in CSF or urine. Urinary

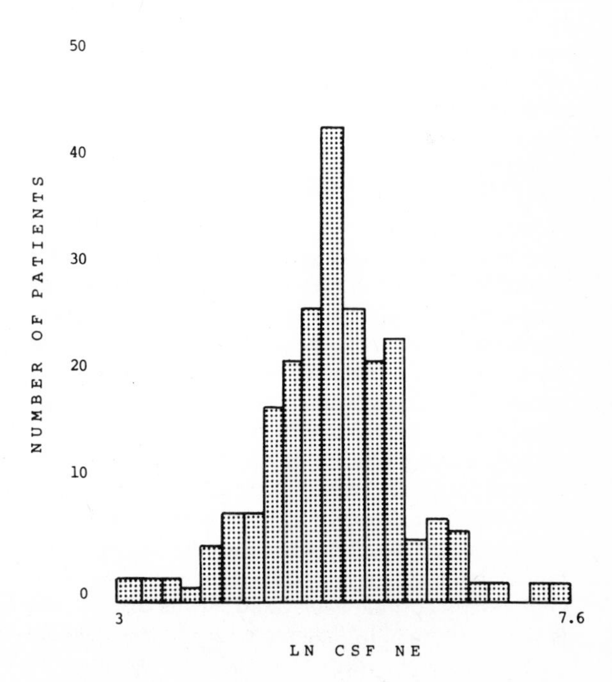

Figure 6. Relative occurrence of natural logarithm of CSF norepinephrine (ln CSF NE) in same patients shown in Fig. 5.

MHPG is lower in the depressed phase of manic–depressive illness than in the manic phase.[3,10,44,51] In two studies MHPG in CSF was lower in depressed patients than in controls[42,64]; however, other studies have found MHPG normal in depression, but elevated in mania.[56,57,71] Since these reports on norepinephrine metabolites do not fully agree, direct measurement of norepinephrine in CSF is of interest. In studies with Post *et al.*,[43] six manic patients had CSF norepinephrine levels more than twice as high as norepinephrine levels in depressed patients or in patients with neurological disease but without affective disorders ($p < 0.001$). Among a group of depressed patients, those with a high degree of anxiety had higher levels of norepinephrine in their CSF than patients who were depressed but not anxious ($p < 0.01$). When the manic patients were treated and had improved, their CSF norepinephrine levels tended to decrease toward normal. CSF norepinephrine is elevated in mania, but not clearly different from normal in depression. This parallels the findings of Shopsin *et al.*[57] that MHPG levels are elevated in mania, but not low in depression. These studies may indicate an increase in central noradrenergic activity during the manic phase of manic–depressive illness and are supported by findings that amphetamine, a drug that can cause a manic type of behavior, increases norepinephrine in CSF, and that slow infusions of norepinephrine directly into the ventricular CSF of animals produce an increase in motor activity.[20,54]

6.2. Amyotrophic Lateral Sclerosis

Amyotrophic lateral sclerosis (ALS) is a progressive paralyzing disease characterized by degeneration of pyramidal tract fibers and anterior horn cells in the spinal cord. Mendell *et al.*[34] had already shown that patients with ALS have abnormally low levels of HVA, a metabolite of dopamine. A study of 34 ALS patients with Dr. B. R. Brooks revealed other neurochemical abnormalities.[4a,75] The patients with ALS have CSF norepinephrine levels of 365 ± 44 pg/ml significantly higher than a control group level of 251 ± 14 pg/ml. Those most severely affected with the disease were bedridden and had CSF norepinephrine levels of 595 ± 65 pg/ml, so that CSF norepinephrine is elevated in this disease and is most abnormal in patients with the most severe disease. Patients with ALS also had significantly elevated plasma norepinephrine, and these plasma norepinephrine levels are highest in patients with most severe disease. Plasma and CSF norepinephrine correlate in patients with ALS ($r = 0.51$, $p < 0.05$) as they do in other subjects. These patients thus have both a central and a peripheral increase in noradrenergic discharge. It is of interest that these same patients have a depressed level of γ-aminobutyric acid (GABA) in their CSF, and there is a significant negative correlation ($r = -0.50$, $p < 0.05$) between GABA and norepinephrine levels in CSF. This correlation between GABA and norepinephrine in CSF is present only in patients with ALS, not in control subjects ($r = -0.03$). The correlations among plasma norepinephrine, CSF norepinephrine, and CSF GABA suggest that these neurochemical derangements might be caused by the same central defect. Electrical stimulation of the human cerebellum can also elevate lumbar CSF norepinephrine[73] and depress GABA levels.[72] Chemical neurostimulants can also lead to a decrease in synthesis of GABA by lowering glutamic acid decarboxylase, its synthetic enzyme.[39,40] GABA is inhibitory on noradrenergic neurons, and the decreased GABA levels can lead to increased norepinephrine turnover.[8] Thus, it is possible that a single agent can cause the neurochemical changes observed in these patients with ALS.

The increase in norepinephrine release in patients with ALS is not without consequence. These patients' blood pressure of 132 ± 3/87 ± 3 was significantly higher than a control group's blood pressure of 117 ± 3/74 ± 2 ($p < 0.01$). Other investigators have noted that patients with ALS show evidence of increased sympathetic nervous tone[15] and have a rigid monophasic pattern of blood flow through their arterioles[17] and a high incidence of nonatherosclerotic angiopathy in peripheral arteries.[62] These phenomena may all be the result of increased norepinephrine release.

6.3. Huntington's Chorea

Abnormalities in vasoregulatory activities have been reported in patients with Huntington's chorea, who tend to have low blood pressure and an increased incidence of postural hypotension. Plasma levels of norepinephrine are significantly lower in patients with Huntington's chorea than in an age-matched control group.[58] Nine untreated patients with Huntington's chorea had CSF norepinephrine levels of 171 ± 18 pg/ml significantly lower ($p < 0.02$) than the 239 ± 17 pg/ml norepinephrine measured in nine age- and sex-matched control patients.[74] Thus, in contrast to patients with ALS, patients with Huntington's chorea have low CSF norepinephrine, low plasma norepinephrine, and low blood pressure. This topic shall be discussed in detail in Chapters 11 and 12.

6.4. Hypertension

Alterations in CSF norepinephrine in Huntington's chorea and ALS are associated with changes in blood pressure. Central noradrenergic nerves are involved in control of blood pressure,[53] and destruction of central noradrenergic nerves can prevent hypertension.[46] These observations have led to studies of central noradrenergic abnormalities in human hypertension. Murray *et al.*[37] found no relationship between MHPG in CSF and blood pressure, but in a subsequent study, Saran *et al.*[49] reported a close correlation between CSF levels of MHPG and blood pressure in hypertensive patients. In a study of 128 hospitalized patients who underwent diagnostic lumbar puncture, there is a highly significant correlation of blood pressure with CSF norepinephrine ($r = 0.41$, $p < 0.0001$). This correlation does not appear to reflect the effects of drugs or neurological diseases. In 87 drug-free patients, the correlation is $r = 0.36$, and in 14 patients with no neurological disease, the correlation is $r = 0.45$. However, there is some effect of age on this relationship, since CSF norepinephrine increases with age ($r = 0.27$, $p < 0.01$) and blood pressure increases with age. Even correcting for this age effect, there is still a significant correlation between CSF norepinephrine and blood pressure ($r = 0.37$, $p < 0.0001$). In this same group of patients, there was no correlation between plasma norepinephrine and blood pressure ($r = 0.04$), so there appears to be a central noradrenergic abnormality associated with the elevated blood pressure that many of these patients exhibited, but there is no evidence for a peripheral noradrenergic abnormality in these patients.

Numerous studies on experimental animals have demonstrated a strong link between central norepinephrine and the regulation of blood pressure. In man, CSF norepinephrine is elevated in mania, agitated depression, and ALS, and these diseases are all associated with some elevation in blood pressure. In Huntington's chorea, there is a decrease in CSF norepinephrine, and blood pressure tends to be low.

Amphetamine increases norepinephrine in CSF and raises blood pressure; bromocriptine lowers norepinephrine in CSF and lowers blood pressure. Norepinephrine in CSF is highest at the time of day when blood pressure is highest. There is a relationship between central norepinephrine in man and blood pressure. Findings of both elevated norepinephrine and MHPG in CSF of hypertensive patients provides evidence for increased norepinephrine turnover in human hypertension. An appropriate central response to hypertension would be to decrease central noradrenergic activity to lower blood pressure toward normal. Instead, central noradrenergic activity is increased in hypertensive patients. This makes one wonder whether a central nervous system defect is causative in some cases of essential hypertension.

References

1. Adèr, J.-P., Aizenstein, M. L., Postema, F., Korf, J.: Origin of free 3-methoxy-4-hydroxyphenylethylene glycol in rat cerebrospinal fluid. *J. Neurol. Trans.* **46**:279–290, 1979.

1a. Ahlskog, J. E.: Food intake and amphetamine anorexia after selective forebrain norepinephrine loss. *Brain Res.* **82**:211–240, 1974.

2. Baldessarini, R. J., Harris, J. E.: Effects of amphetamines on the metabolism of catecholamines in the rat brain. *J. Psychiatr. Res.* **11**:41–43, 1974.

3. Bond, P. A., Jenner, F. A., Sampson, G. A.: Daily variations in the urine content of 3-methoxy-4-hydroxyphenyl-glycol in two manic–depressive patients. *Psychol. Med.* **2**:81–85, 1972.

4. Bowers, M. B.: Deficient transport mechanisms for the removal of acid monoamine metabolites from cerebrospinal fluid. *Brain Res.* **15**:522–524, 1969.

4a. Brooks, B. R., Ziegler, M. G., Lake, C. R., Wood, J. H., Enna, S. J., Engel, W. K.: Cerebrospinal fluid norepinephrine and free γ-aminobutyric acid in amyotrophic lateral sclerosis. *Brian Res. Bull.* (in press).

5. Bunney, W. E., Jr., David, J. M: Norepinephrine in depressive reactions. *Arch. Gen. Psychiatry* **27**:312–317, 1972.

6. Chase, T. N., Gordon, E. K., Ng, L. K. Y.: Norepinephrine metabolism in the central nervous system of man: Studies using 3-methoxy-4-hydroxyphenylethylene glycol levels in cerebrospinal fluid. *J. Neurochem.* **21**:581–587, 1973.

7. Collu, R., Jequier, J. C., Letarte, J., Leboeuf, G., Ducharne, J. R.: Diurnal variations of plasma growth hormone and brain monoamines in adult male rats. *Can. J. Physiol. Pharmacol.* **51**:890–892, 1973.

8. Cott, J., Engel, J.: Suppression by GABAergic drugs of the locomotor stimulation induced by morphine, amphetamine, and apomorphine: Evidence for both pre- and post-synaptic inhibition of catecholamine systems. *J. Neural Transm.* **40**:253–268, 1977.

9. Cummins, B. H., Lothian, D.: Amine levels in cerebrospinal fluid after subarachnoid hemorrhage. *Br. J. Surg.* **60**(2):910, 1973.

10. DeLeon-Jones, F., Maas, J. W., Dekirmenjian, J.,

Fawcett, J. A.: Urinary catecholamine metabolites during behavioral changes in a patient with manic–depressive cycles. *Science* **179**:300–302, 1973.

11. Denker, S. J., Häggendal, J., Ilves-Häggendal, M.: Presence of free and conjugated noradrenaline in human cerebrospinal fluid. *Acta Physiol. Scand.* **69**:140–146, 1967.
12. Eleftheriou, B. S.: Circadian rhythm in blood and brain biogenic amines and other biochemical changes in rabbits. *Brain Res.* **75**:145–152, 1974.
13. Evans, H. L., Ghiselli, W. B., Patton, R. A.: Diurnal rhythm in behavioral effects of methamphetamine, *p*-chloromethamphetamine and scopolamine. *J. Pharmacol. Exp. Ther.* **186**:10–17, 1973.
14. Finch, I., Hersom, A.: Studies on the centrally mediated cardiovascular effects of apomorphine in the anaesthetized rat. *Br. J. Pharmacol.* **56**:366, 1976.
15. Forrester, J. M.: Amyotrophic lateral sclerosis and bedsores. *Lancet* **1**:7966, 1976.
16. Friedman, A. H., Walker, C. A.: Circadian rhythms in rat midbrain and caudate nucleus biogenic amine levels. *J. Physiol. (London)* **197**:77–85, 1968.
17. Furakawa, T., Toyokura, Y.: Amyotrophic lateral sclerosis and bedsores: Plethysmographic analysis. *Lancet* **1**:8056, 1978.
18. Fuxe, K., Ungerstedt, U.: Localization of catecholamine uptake in rat brain after intraventricular injection. *Life Sci.* **5**:1817–1824, 1966.
19. Fuxe, K., Ungerstedt, U.: Histochemical studies on the effects of (+)-amphetamine, drugs of the imipramine groups, and tryptamine on central catecholamine and 5-hydroxytryptamine neurons after intraventricular injection of catecholamines and 5-hydroxytryptamine. *Eur. J. Pharmacol.* **4**:135–144, 1968.
20. Geyer, M. A., Segal, D. S., Mandell, A. J.: Effect of intraventricular infusion of dopamine and norepinephrine on motor activity. *Physiol. Behav.* **8**:653–658, 1972.
21. Gordon, E. K., Oliver, J.: 3-Methoxy-4-hydroxyphenylethlene glycol in human cerebrospinal fluid. *Clin. Chim. Acta* **35**:145–150, 1971.
22. Gordon, E. K., Oliver, J., Goodwin, F. K., Chase, T. N., Post, R. M.: Effect of probenecid on free 3-methoxy-4-hydroxyphenylethylene glycol (MHPG) and its sulfate in human cerebrospinal fluid. *Neuropharmacology* **12**:391–396, 1973.
23. Greenacre, J. K., Teychenne, P. F., Petrie, A., Calne, D. B., Leigh, P. N., Reid, J. L.: The cardiovascular effects of bromocriptine in parkinsonism. *Br. J. Cline. Pharmacol.* **35**:571–574, 1976.
24. Heikkila, R. E., Orlansky, H., Mytilineou, C., Cohen, G.: Amphetamine: Evaluation of *d*- and *l*-isomers as releasing agents and uptake inhibitors for ^{3}H-dopamine and ^{3}H-norepinephrine in slices of rat neostriatum and cerebral cortex. *J. Pharmacol. Exp. Ther.* **194**(1):47–56, 1975.
25. Hutchins, D. A., Rogers, K. H.: Some observations on the circadian rhythm of locomotor activity of mice after depletion of cerebral monoamines. *Psychopharmacologia* **31**:343–348, 1973.
26. Jacquot, C., Rapin, J., Renault, H., Cohen, Y.: Effect de l'amphétamine et de ses métabolites sur l'activité de la dopamine-β-hydroxylase. *Pharmacologie*, pp. 479–484, 1974.
27. Kaye, S. B., Shaw, K. M., Ross, E. J.: Bromocriptine and hypertension. *Lancet* **1**:1176, 1976.
28. Lake, C. R., Wood, J. H., Ziegler, M. G., Ebert, M. H., Kopin, I. J.: Probenecid-induced norepinephrine elevations in plasma and CSF. *Arch. Gen. Psychiatry* **35**:237–240, 1978.
29. Lamberts, S. W. J., Birkenhager, J. C.: Bromocriptine in Nelson's syndrome and Cushing's disease. *Lancet* **2**:811, 1976.
30. Leimdorfer, A., Arana, R., Hack, M. H.: Hyperglycemia induced by the action of adrenalin on the central nervous system. *Am. J. Physiol.* **150**:588–595, 1947.
31. Lew, G. M., Quay, W. B.: The mechanism of circadian rhythms in brain and organ contents of norepinephrine: Circadian changes in the effects of methyltyrosine and 6-hydroxydopamine. *Comp. Gen. Pharmacol.* **4**:375–381, 1973.
32. Liljedahl, S. O., Von, Euler, U. S.: Elimination of noradrenaline from the spinal fluid. *Acta Chir. Scand.* **108**:163–169, 1954.
33. Manshardt, J., Wurtman, R. J.: Daily rhythm in the noradrenaline content of rat hypothalamus. *Nature (London)* **217**:574–575, 1968.
34. Mendell, J. R., Chase, T. N., Engel, W. K.: Amyotrophic lateral sclerosis. *Arch. Neurol.* **25**:320–325, 1972.
35. Meyer, J. S., Stoica, E., Pascu, I., Shimazu, K., Hartman, A.: Catecholamine concentrations in CSF and plasma of patients with cerebral infarction and hemorrhage. *Brain* **96**:277–288, 1973.
36. Meyer, J. S., Welch, K. M. A., Okamoto, S., Shimazu, K.: Disordered neurotransmitter function. *Brain* **97**:655–664, 1974.
37. Murray, S., Jones, D. H., Davies, D. S., Dollery, C. T., Reid, J. L.: Free and total 3-methoxy-4-hydroxyphenylethylene glycol in human cerebrospinal fluid: Relationship to blood pressure in hypertensives and patients with neuropsychiatric disease. *Clin. Chim. Acta* **79**:63–68. 1977.
38. Nygren, L. G., Olson, L.: A new major projection from locus coeruleus: The main source of noradrenergic nerve terminals in the ventral and dorsal columns of the spinal cord. *Brain Res.* **132**:85–93, 1977.
39. Olney, J. W., Ho, O. L., Rhee, V.: Cytotoxic effects of acidic and sulphur-containing amino acids on the infant mouse central nervous system. *Exp. Brain Res.* **14**:51–76, 1972.
40. Olney, J. W., Rhee, V., Ho, O. L.: Kainic acid: A powerful neurotoxic analogue of glutamate. *Brain Res.* **77**:507–512, 1974.

41. PERLOW, M. EBERT, M. H., GORDON, E. K., ZIEGLER, M. G., LAKE, C. R., CHASE, T. N.: The circadian variation of catecholamine metabolism in the subhuman primate. *Brain Res.* **139:**101–113, 1978.
42. POST, R. M., GORDON, E. K., GOODWIN, F. K., BUNNEY, W. E., JR., Central norepinephrine metabolism in affective illness: MHPG in the cerebrospinal fluid. *Science* **179:**1002–1003, 1973.
43. POST, R. M., LAKE, C. R., JIMERSON, D. C., BUNNEY, W. E., JR., WOOD, J. H., ZIEGLER, M. G., GOODWIN, F. K.: Cerebrospinal fluid norepinephrine in affective illness. *Am. J. Psychiatry* **135**(8):907–912, 1978.
44. POST, R. M., STODDARD, F. J., GILLIN, J. C., BUCHSBAUM, M. S., RUNKLE, D. C., BLACK, K. E., BUNNEY, W. E., JR.: Alterations in motor activity, sleep, and biochemistry in a cycling manic–depressive patient. *Arch. Gen. Psychiatry* **34:**470–477, 1977.
45. RAICHLE, M. E., HARTMAN, B. K., EICHLING, J. O., SHARPE, L. G.: Central noradrenergic regulation of cerebral blood flow and vascular permeability. *Proc. Natl. Acad. Sci. U.S.A.* **72:**3726–3730, 1975.
46. REID, J. L., ZIVIN, J. A., KOPIN, I. J.: Central and peripheral adrenergic mechanisms in the development of deoxycorticosterone–saline hypertension in rats. *Circ. Res.* **37:**569–579, 1975.
47. REIS, D. J., WEINBREN, M., CORVELLI, A.: A circadian rhythm of norepinephrine regionally in cat brain: Its relationship to environmental lighting and to regional diurnal variations in brain serotonin. *J. Pharmacol. Exp. Ther.* **164:**135–145, 1968.
48. REIS, D. J., WURTMAN, R. J.: Diurnal changes in brain noradrenaline. *Life Sci.* **7:**91–98, 1968.
49. SARAN, R. K., SAHUJA, R. C., GUPTA, N. N., HASAN, M., BHARGAVA, K. P., SHANKER, K., KISHOR, K.: 3-Methoxy-4-hydroxyphenyl-ethylene glycol in cerebrospinal fluid and vanillylmandelic acid in urine of humans with hypertension. *Science* **200:**317–318, 1978.
50. SATO, S., SUZUKI, J.: Anatomical mapping of the cerebral nervi vasorum in the human brain. *J. Neurosurg.* **43:**559–568, 1975.
51. SCHILDKRAUT, J. J.: Norepinephrine metabolites as biochemical criteria for classifying depressive disorders and predicting responses to treatment: Preliminary findings. *Am. J. Psychiatry* **130:**695–699. 1973.
52. SCHILDKRAUT, J. J.: The catecholamine hypothesis of affective disorders: A review of supporting evidence. *Am. J. Psychiatry* **122:**509–522, 1974.
53. SCRIABINE, A., CLINESCHMIDT, B. V., SWEET, C. S.: Central noradrenergic control of blood pressure. *Annu. Rev. Pharmacol. Toxicol.* **16:**113–123, 1976.
54. SEGAL, D. S., MANDELL, A. J.: Behavioral activation of rats during intraventricular infusion of norepinephrine. *Proc. Natl. Acad. Sci. U.S.A.* **66:**289–293, 1970.
55. SEGAL, D. S., MCALLISTER, C., GEYER, M. A.: Ventricular infusion of norepinephrine and amphetamine: Direct versus indirect action. *Pharmacol. Biochem. Behav.* **2:**79–86, 1974.
56. SHAW, D. M., O'KEEFE, R., MACSWEENEY, D. A., BROOKSBANK, B. W. L., NOGUERA, R., COPPEN, A.: MHPG in depression. *Psychol. Med.* **3:**333–336, 1973.
57. SHOPSIN, B., WILK, S., GERSHON, S., DAVIS, K., SUHL, M.: An assessment of norepinephrine metabolism in affective disorders. *Arch. Gen. Psychiatry* **28:**230–233, 1973.
58. SHOULSON, I., ZIEGLER, M. G., LAKE, C. R.: Huntington's disease (HD): Determination of plasma norepinephrine (NE) and dopamine-beta-hydroxylase (DBH). *Neurosci. Abstr.* **2:**800, 1976.
59. SJÖSTRÖM, R., EKSTEDT, J., ÄNGGÅRD, E.: Concentration gradients of monoamine metabolites in human cerebrospinal fluid. *J. Neurol. Neurosurg. Psychiatry* **38:**666–668, 1975.
60. STEINSLAND, O. S., HIEBLE, J. P.: Dopaminergic inhibition of adrenergic neurotransmission as a model for studies on dopamine receptor mechanisms. *Science* **199:**443–445, 1978.
61. STOLK, J. M.: Evidence for reversible inhibition of brain dopamine-β-hydroxylase activity *in vivo* by amphetamine analogues. *J. Neurochem.* **24:**135–142, 1975.
62. STÖRTEBECKER, P., NORDSTRÖM, G., PAP DE PESTENY, M., SEEMAN, T., BJÖRKERUD, S.: Vascular and metabolic studies of amyotrophic lateral sclerosis. *Neurology* **20:**1157–1160, 1970.
63. STUMPE, K. O., KOLLOCH, R., HIGUCHI, M., DRUCK, F., VETTER, H.: Hyperprolactinaemia and antihypertensive effect of bromocriptine in essential hypertension. *Lancet* **2:**211–214, 1977.
64. SUBRAHMANYAM, S.: Role of biogenic amines in certain pathological conditions. *Brain Res.* **87:**355–362, 1975.
65. TAMARKIN, N. R., GOODWIN, F. K., AXELROD, J.: Rapid elevation of biogenic amine metabolites in human CSF following probenecid. *Life Sci.* **9:**1397–1408, 1970.
66. TAYLOR, K. M., SNYDER, S. H.: Amphetamine: Differentiation by *d*- and *l*-isomers of behavior involving brain norepinephrine or dopamine. *Science* **168:**1487–1489, 1970.
67. THORNBURG, J. E., MOORE, K. E.: Dopamine and norepinephrine uptake by rat brain synaptosomes: Relative inhibitory potencies of *l*- and *d*-amphetamine and amantadine. *Res. Commun. Chem. Pathol. Pharmacol.* **5**(1):81–89, 1973.
68. THORNBURG, J. E., MOORE, K. E.: The relative importance of dopaminergic and noradrenergic neuronal systems for the stimulation of locomotor activity induced by amphetamine and other drugs. *Neuropharmacology* **12:**853–866, 1973.
69. WEIL-MALHERBE, H., AXELROD, J., TOMCHICK, R.: Blood–brain barrier for adrenaline. *Science* **129:**1226–1227, 1959.
70. WESTON, P. G.: Sugar content of the blood and spinal fluid of insane subjects. *J. Med. Res.* **35:**199–208, 1916.
71. WILK, S., SHOPSIN, B., GERSHON, S., SUHL, .: Cerebrospinal fluid levels of MHPG in affective disorders. *Nature (London)* **235:**440–441, 1972.

72. WOOD, J. H., GLAESER, B. S., HARE, T. A., SODE, J., BROOKS, B. R., VAN BUREN, J. M.: Cerebrospinal fluid GABA reductions in seizure patients evoked by cerebellar surface stimulation. *J. Neurosurg.* **47**:582–589, 1977.
73. WOOD, J. H., LAKE, C. R., ZIEGLER, M. G., SODE, J., BROOKS, B. R., VAN BUREN, J. M.: Cerebrospinal fluid norepinephrine alterations during electrical stimulation of cerebellar, cerebral surfaces in epileptic patients. *Neurology* **27**:716–724, 1977.
74. WOOD, J. H., ZIEGLER, M. G., LAKE, C. R., SHOULSON, I., BROOKS, B. R., VAN BUREN, J. M.: Norepinephrine reductions in cerebrospinal fluid in man after electrical stimulation and degeneration of the caudate nucleus. *Trans. Am. Neurool. Assoc.* **101**:68–72, 1976.
75. ZIEGLER, M. G., BROOKS, B. R., LAKE, C. R., WOOD, J. H., ENNA, S. J.: Norepinephine and gamma-aminobutyric acid in amyotrophic lateral sclerosis. *Neurology* **30**(1): 98–101, 1980.
76. ZIEGLER, M. G., LAKE, C. R., KOPIN, I. J.: Norepinephrine in cerebrospinal fluid. *Brain Res.* **108**:436–440, 1976.
77. ZIEGLER, M. G., LAKE, C. R., WILLIAMS, A. C., TEYCHENNE, P. F., SHOULSON, I., STEINSLAND, O.: Bromocriptine inhibits norepinephrine release. *J. Clin. Pharmacol.* **25**:137–142, 1979.
78. ZIEGLER, M. G., LAKE, C. R., WOOD, J. H., BROOKS, B. R., EBERT, M. H.: Relationship between norepinephrine in blood and cerebrospinal fluid in the presence of a blood cerebrospinal fluid barrier for norepinephrine. *J. Neurochem.* **28**:677–679, 1976.
79. ZIEGLER, M. G., LAKE, C. R., WOOD, J. H., EBERT, M. H.: Circadian rhythm in cerebrospinal fluid noradrenaline of man and monkey. *Nature (London)* **264**:656–658, 1976.
80. ZIEGLER, M. G., WOOD, J. H., LAKE, C. R., KOPIN, I. J.: Norepinephrine and 3-methoxy-4-hydroxyphenylglycol gradients in human cerebrospinal fluid. *Am. J. Psychiatry* **134**(5):565–568, 1977.
81. ZIGMOND, M. H., WURTMAN, R. J.: Daily rhythm in the accumulation of brain catecholamines synthesized from circulating H^3-tyrosine. *J. Pharmacol. Exp. Ther.* **172**:416–422, 1970.
82. ZIVIN, J. A., REID, J. L., SAAVEDRA, J. M., KOPIN, I. J.: Quantitative localization of biogenic amines in the spinal cord. *Brain Res.* **99**:293–301, 1975.

CHAPTER 11

Huntington's Disease

Biogenic Amines in Cerebrospinal Fluid

Ian J. Butler, William E. Seifert, and S. J. Enna

1. Introduction

Patients with Huntington's disease, a dominantly inherited disorder characterized by choreiform movements and dementia, have a marked loss of neurons in the caudate nucleus and putamen.[5] Recent neurochemical studies indicate that some of these cells utilize γ-aminobutyric acid (GABA) and acetylcholine as transmitter agents.[13] Nevertheless, for many years, physicians have observed beneficial effects on the movement disability of patients with Huntington's disease after treatment with drugs that modify metabolism of dopamine, serotonin, and norepinephrine. In particular, the improvements noted following administration of dopamine-receptor-blocking agents such as phenothiazine and butyrophenone neuroleptics have been used as evidence to suggest that the dopaminergic system is overactive in this disorder. However, measurements of dopamine function in the brains of Huntington patients indicate that dopamine metabolism is normal or reduced, rather than increased.

These findings may be explained by the fact that within the degenerative brain areas, there are various neurons and terminals containing a number of different transmitters. Neurobiological studies have suggested that the interactions among these neurons in the basal ganglion region are of fundamental importance in motor control. Thus, the concept of an imbalance among neurotransmitter systems as a basis for the pathophysiology of disorders of the human nervous system may apply to Huntington's disease as well as Parkinson's disease, Gilles de la Tourette syndrome, and other movement disorders. Accordingly, studies on neurotransmitter metabolism and interactions in Huntington's disease have enhanced our understanding of the neurochemical mechanisms that underlie abnormal movements and cognitive behavior in neurological and psychiatric disorders. This information has been utilized to develop new, rational therapeutic approaches to Huntington's disease and has added impetus to the development of biochemical techniques to determine which at-risk patients will develop the disease in order to facilitate genetic counseling with appropriate individuals.

In this chapter, the evidence indicating disordered biogenic amine metabolism in Huntington's chorea is reviewed. The studies to be discussed include both direct and indirect experimental approaches that have been taken utilizing biological tissues and cerebrospinal fluid (CSF).

Ian J. Butler, M.B., F.R.A.C.P. • Departments of Neurology, Pediatrics, and Neurobiology and Anatomy, The University of Texas Medical School at Houston, Houston, Texas 77025. **William E. Seifert, Ph.D.** • Department of Biochemistry and Molecular Biology, The University of Texas Medical School at Houston, Houston, Texas 77025. **S. J. Enna, Ph.D.** • Departments of Pharmacology and Neurobiology and Anatomy, The University of Texas Medical School at Houston, Houston, Texas 77025.

2. Neurochemistry

2.1. Brain Tissue Analysis

In Huntington's disease, the major pathological features are seen in the corpus striatum and cerebral cortex. There is extensive astrocytic proliferation and marked loss of small and, to a lesser degree, large neurons in the caudate nucleus and diffuse cerebral atrophy with a severe reduction in number of neurons in cortical layers 3, 5, and 6. Although the movement disorder in adult-onset Huntington's disease is characteristically of the hyperkinetic, choreiform type, bradykinesia and rigidity may be prominent either in the terminal phases of the disorder or in the less common juvenile-onset variant.[6] In addition to the loss of small interneurons in caudate nucleus, patients with rigidity appear to have an even greater loss of large striatal neurons than those individuals experiencing less clinical rigidity.

Although it seems reasonable to assume that the movement disorder in Huntington's disease is related to the striatal-cell degeneration and the mental symptoms to the cortical pathology, the precise pathophysiology of the symptoms is still uncertain. In fact, the modulatory role of dopaminergic neurons in the rigid form may be different than in the classic form due to a differential rate or degree of neuronal degeneration in the striatum. Studies of dopamine metabolism in brain tissue, particularly striatum, from patients with Huntington's disease have increased our understanding of the pathophysiology of this condition as well as yielded significant new information about normal and developmental aspects of dopaminergic interactions at the level of the basal ganglia. The role of biogenic amines in Huntington's disease has been examined using brain samples obtained at autopsy or biopsy by analyzing biosynthetic enzymes, neurotransmitter and metabolite levels, and postsynaptic receptor binding. Furthermore, indirect approaches to study cerebral metabolism of the biogenic amines have included quantitation of these amines and their metabolites in CSF.

Despite the difficulty in controlling variables such as premortem clinical states, drug administration, postmortem delays, age at death, sex, and genetic factors, there are extensive data on quantitation of biogenic amine metabolism in various regions of the brain in tissue from patients with Huntington's disease. One concern encountered is interpretation of quantitative data based on wet weight of brain specimens taken from areas in which marked atrophy has occurred. For example, a 2-fold elevation in tyrosine hydroxylase activity and dopamine concentration per unit wet weight of substantia nigra from autopsied choreic brain is probably due to atrophy in this nucleus, since tyrosine hydroxylase levels in caudate nucleus and putamen are normal and dopamine content is decreased in the caudate and normal in the putamen and olfactory tubercle.[3] Since many of the parameters studied, such as biosynthetic enzymes, metabolites, and receptors, appear stable after death, most neurochemical studies have been performed on brain specimens obtained at autopsy. To overcome problems of stability of some enzymes and neurotransmitters, there have been recent studies using brain tissue removed from cortex and basal ganglia at biopsy.[19]

Brain tissue from 10 patients with Huntington's disease was studied by Hornykiewicz and his coworkers. Brains were obtained 4–18 hr postmortem, and dopamine, homovanillic acid (HVA), norepinephrine, and serotonin were analyzed in various regions.[2] Serotonin levels were significantly different only in the central gray region of the midbrain, where they were higher than in controls. Brain norepinephrine levels were normal in all brain regions studied. Dopamine and HVA levels were significantly reduced, by approximately 40%, in caudate nucleus, but not in the putamen, globus pallidus, or substantia nigra. Reasons suggested for the reduction in dopamine and its major metabolite in caudate nucleus included a preferential loss of dopamine neurons projecting to this nucleus or reduced dopamine turnover in these neurons. Furthermore, the differential decrease in dopamine in caudate nucleus compared to putamen was postulated to contribute to the hyperkinesis.

In a study by Bird and Iversen[4] of brain obtained at autopsy from choreic patients, there was no significant difference in dopamine levels in caudate nucleus and putamen from control brains, although in a later study, dopamine was found to be significantly decreased in caudate.[3] Normal levels of tyrosine hydroxylase were found in the putamen. However, in patients with the rigid form of the disease, mean dopamine levels in putamen were elevated compared to levels in nonrigid patients (185%) and controls (150%). One explanation suggested for the elevated dopamine levels observed in putamen was that the increase was due to a greater atrophy in basal ganglia in rigid Huntington patients.

Another approach to the study of neurotransmitter function in Huntington's disease has been utilization of new biochemical techniques to quantitate and characterize binding properties of specific receptors with which neurotransmitters interact. These procedures require the availability of specific radioactive ligands that are capable of binding specifically to membrane receptor sites. Using low concentrations of these labeled compounds and careful washing techniques, it is possible to discriminate specific neurotransmitter-receptor-binding from nonspecific binding sites. Kinetic analysis of the binding permits estimation of the number and affinity of the receptors in brain tissue. In a study of brain tissue obtained at autopsy from patients with Huntington's chorea, serotonin and cholinergic muscarinic receptor-binding sites were significantly reduced in globus pallidus, putamen, and caudate nucleus.[12] In another group of patients, dopamine-receptor-binding sites were shown to be reduced by 50% in caudate nucleus and putamen and by 65% in frontal cortex of brain at autopsy.[26] These studies are consistent with the known loss of neurons in striatum in patients with Huntington's disease, since this would result in a loss of the specialized neuronal receptor sites in caudate and putamen with which dopaminergic, cholinergic, and serotonergic neurons normally synapse.

2.2. CSF Analysis

As a result of variables associated with studies using brain obtained at postmortem and ethical constraints in using biopsied brain tissue for neurochemical estimations, many investigators have measured biogenic amines, and particularly their metabolite levels, in CSF.[29b] HVA, the major acidic metabolite of dopamine, can be measured in ventricular, cisternal, and lumbar CSF and is an indicator of brain dopamine metabolism, particularly in those areas adjacent to the ventricles such as the basal ganglia, which contain more than 80% of brain dopamine.[23] The acidic metabolite of serotonin, 5-hydroxyindoleacetic acid (5-HIAA), can also be measured in CSF, but since this transmitter is more widely distributed in brain and spinal cord, metabolite levels in CSF are not necessarily indicative of serotonin metabolism in specific brain regions.[14] Norepinephrine metabolism in brain has been studied by measurement in CSF of its major metabolite, 3-methoxy-4-hydroxy-phenylethylene glycol (MHPG),[9] and more recently norepinephrine itself has been quantitated in CSF.[31]

Initially, CSF studies in Huntington's chorea were performed under steady-state conditions, but in an attempt to study the dynamic aspects of biogenic amine metabolism, the probenecid technique used in animal studies to investigate neurotransmitter turnover was adapted to human investigations.[11a,24] This technique relies on complete blockage of reuptake of the acidic metabolites from CSF to blood, which results in a linear accumulation of metabolites in lumbar CSF.[28] After 9–18 hr of probenecid administration, the ratio of accumulated metabolite (HVA and 5-HIAA) to level of metabolite before probenecid is measured and compared to the ratio in control patients. Probenecid levels in CSF must be measured to evaluate the degree of blockade of metabolite transport.[11a,25,29a] The rate of accumulation of HVA and 5-HIAA reflects the end result of a number of processes including synthesis, storage, and metabolism of the neurotransmitters, and therefore, it does not isolate defects at specific steps in the biosynthetic pathways.

We have recently studied seven patients with the classic form of Huntington's chorea in which baseline and accumulated levels of HVA and 5-HIAA after oral probenecid administration were measured by gas chromatography–mass spectrometry (GC-MS) in lumbar CSF obtained under standard conditions (see Table 2 and Fig. 1). There were seven male patients with an average age of 57 years (range 47–69). Four patients were evaluated prior to administration of a neuroleptic drug such as haloperidol, and three patients were studied 2 weeks after discontinuation of neuroleptics. CSF was obtained by lumbar puncture in the lateral decubitus position at approximately 9:00 A.M. after a 12-hr period of bedrest and fasting with plentiful fluid intake.* The initial 4–5 ml of CSF was collected in glass tubes containing ascorbic acid, 2 mg/ml, as preservative, and rapidly transported on ice and stored at −80°C. Probenecid, 125–150 mg/kg body weight, was administered orally in three divided doses. At 18 hr after starting probenecid and 6 hr after the last dose, a further sample of lumbar CSF was obtained at approximately 9:00 A.M. HVA, 5-HIAA, and MHPG were assayed by a quantitative GC-MS method.[27] The ratio of accumulated (postprobenecid) to baseline (preprobenecid) levels of HVA and 5-HIAA were compared to ratios in published studies using similar techniques.[15,25,28] Probenecid levels in CSF were measured by a modified gas chromatographic method[29] and were sufficient (range 18.8–47.1 μg/ml) to block egress of metabolites from CSF to blood.[11a]

Brain norepinephrine metabolism in patients with Huntington's disease appears normal as indicated by levels of MHPG in CSF[8] (Table 1) and levels of norepinephrine in brain.[2] The baseline levels of MHPG in CSF in ten Huntington patients that we studied were comparable to those in control subjects

* This project was approved by an institutional committee for human studies, and informed consent for lumbar punctures and probenecid loading was obtained from the patients or from their spouses in appropriate cases.

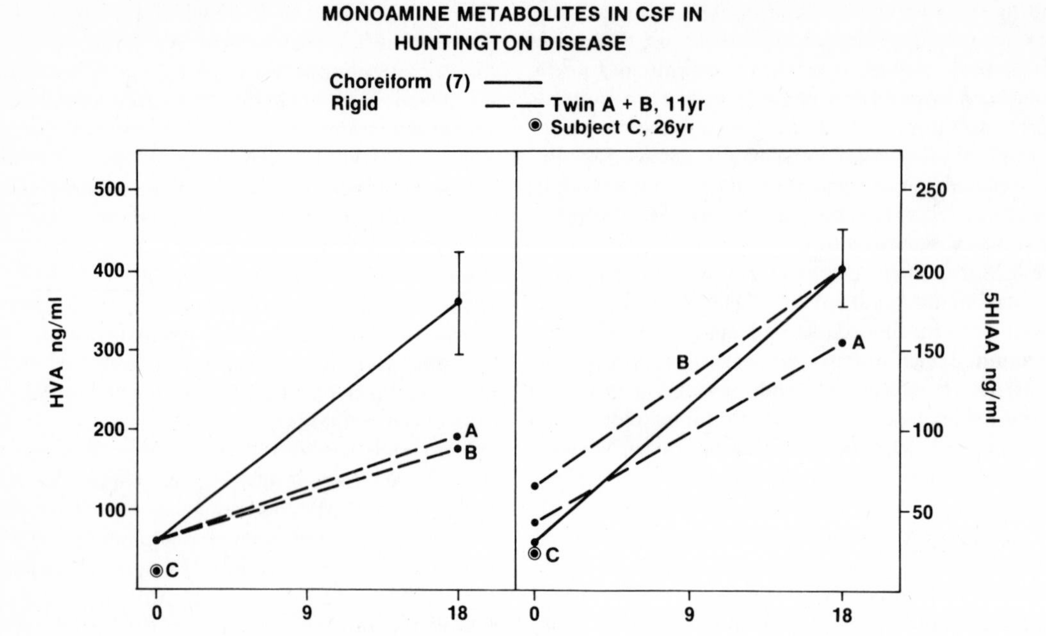

Figure 1. Levels of homovanillic acid (HVA) and 5-hydroxyindoleacetic acid (5-HIAA) in lumbar CSF before and after oral probenecid (125–150 mg/kg) was administered in three divided doses over 18 hr. Values for group of seven classic choreic patients are means in ng/ml ± S.E.M., and single values are given for three patients with rigid form of Huntington's disease.

Table 1. CSF Biogenic Amine Metabolites in Huntington's Disease[a]

Study	HVA		5-HIAA		MHPG		Ref. No.
Control subjects	36 ± 4	(25)	28 ± 2	(25)	15 ± 2	(11)	
Huntington's disease	14 ± 3	(11)	31 ± 4	(11)	18 ± 4	(6)	8
Control subjects	37 ± 3	(17)	23 ± 2	(17)	—	—	
Huntington's disease	23 ± 4	(16)	29 ± 2	(16)	—	—	11
Control subjects	47 ± 3	(12)	30 ± 3	(12)	—	—	
Huntington's disease	24 ± 5	(15)	17 ± 3	(15)	—	—	7
Control subjects	34 ± 4	—	20 ± 2	—	—	—	
Huntington's disease	29 ± 5	(9)	15 ± 2	(9)	—	—	21
Control subjects	41 ± 4	(30)	16 ± 1	(30)	—	—	
Huntington's disease	34 ± 7	(6)	24 ± 4	(6)	—	—	30
Control subjects	66	(7)	—	—	—	—	
Huntington's disease	64	(6)	—	—	—	—	17
Control subjects	36 ± 6	(34)	40 ± 8	(34)	—	—	
Huntington's disease	55	(3)	35	(3)	—	—	16

[a] HVA, 5-HIAA, and MHPG were measured in lumbar CSF. Values (ng/ml) are means ± S.E.M. and number of subjects is in parentheses.
Abbreviations: HVA, homovanillic acid; 5-HIAA, 5-hydroxyindoleacetic acid; MHPG, 3-methoxy-4-hydroxyphenylethylene glycol.

(Table 2 and unpublished data). Furthermore, the synthetic enzyme for norepinephrine, dopamine-β-hydroxylase, is normal in hypothalamus and locus ceruleus.[3] However, in another group of patients with Huntington's disease, CSF norepinephrine levels were significantly reduced compared to controls.[29c,31] and norepinephrine levels were reported to be low in hypothalamic tissue obtained at autopsy from choreic patients.[3]

Brain serotonin metabolism in Huntington's disease has been reported normal as studied by measurement of baseline levels of 5-HIAA and also accumulated levels after probenecid. Baseline levels of 5-HIAA in CSF were normal (Table 1), and after an oral probenecid load was administered to 8 patients for 9 hr, the rate of accumulation of 5-HIAA was comparable to the increase seen in 8 control subjects.[8] In the study reported herein, brain serotonin metabolism also appeared normal as indicated by baseline levels of 5-HIAA and a ratio of accumulated 5-HIAA in CSF of 7 compared to the normal 4- to 12-fold increase seen by other investigators.[15,25,28] We are unable to explain the results of a study of 15 patients in which a significantly low baseline level of 5-HIAA in CSF suggested impaired serotonin metabolism in brain.[7]

The evidence for a disturbance in dopamine metabolism in brain in Huntington's disease is difficult to interpret in that variability has been observed in brain levels of dopamine, with some investigators showing normal striatal dopamine and others demonstrating low levels in specific areas of striatum from brain at autopsy. Similar difficulties are encountered in interpretation of levels of the dopamine metabolite HVA in CSF as measured by different investigators (see Table 1). In three series of patients, the mean baseline level of HVA was about 50% of the control value, and this observation suggested an impaired turnover of dopamine in brain and in basal ganglia in particular.[7,8,11] These studies of HVA in CSF should be compared to the decrease in dopamine and HVA observed in caudate nuclei of autopsied brain.[2] However, in three smaller series, mean baseline HVA levels were normal,[17,21,30] and in a single series of three patients, the mean HVA level in CSF was elevated.[16] In our study of seven patients, the mean baseline HVA level was similar to that in control studies of other investigators (Tables 1 and 2). Thus, these CSF studies may be comparable to the previously reported normal levels of dopamine in the brains of patients with Huntington's disease.[4]

CSF neurotransmitter metabolite studies after probenecid are an indication of the dynamics of biogenic amine biosynthesis and may detect functional disturbances in dopamine metabolism not detected by steady-state levels of CSF HVA. In eight patients with Huntington's disease, there was impaired accumulation of HVA after probenecid loading for 9 hr as compared to eight control subjects.[8] In our seven patients investigated after 18 hr of probenecid loading (see Table 2 and Fig. 1), we observed a mean 5-fold increase in HVA after adequate blockade of uptake of metabolites by probenecid, which is appreciably lower than the 9- to 40-fold increase normally observed in control subjects.[15,25,28] Thus, these two probenecid studies are consistent with a defect in dopamine metabolism in brain in the classic form of Huntington's disease.

Intrigued by the marked clinical differences in the juvenile-onset rigid variant of Huntington's disease, we studied three young rigid Huntington subjects, including two after probenecid loading. Serotonin metabolism was normal as shown by normal base-

Table 2. CSF Biogenic Amines after Probenecid in Huntington's Disease[a]

	HVA				5-HIAA				MHPG	
	Pre-probenecid		Post-probenecid		Pre-probenecid		Post-probenecid		Pre-probenecid	
Huntington's disease	59 ± 9	(7)	273 ± 26	(7)	30 ± 4	(7)	202 ± 22	(7)	12 ± 1	(10)
Post/pre ratio										
Huntington's disease	—	—	5	—	—	—	7	—	—	—
Published controls[b]	—	—	9–40	—	—	—	4–12	—	—	—

[a] HVA, 5-HIAA, and MHPG were measured in lumbar CSF by a GC-MS method. The values are means ± S.E.M. in ng/ml of CSF (I. J. Butler and W. E. Seifert, unpublished data). The number of patients is given in parentheses, and the ratios are post-probenecid/pre-probenecid metabolite levels.

[b] Adapted from Goodwin *et al.*[15] Perel *et al.*,[25] and Tamarkin *et al.*[28]

Abbreviations: HVA, homovanillic acid; 5-HIAA, 5-hydroxyindoleacetic acid; MHPG, 3-methoxy-4-hydroxyphenylethylene glycol.

line and accumulated levels of 5-HIAA in lumbar CSF (Fig. 1). However, dopamine metabolism appeared even more defective than in the classic form in that one patient had a very low baseline HVA level in CSF and two patients showed defective accumulation of HVA after probenecid loading (Fig. 1). These findings are consistent with the beneficial response of rigidity, but not chorea, to administration of dopamine precursor levodopa[1] and the low levels of HVA in CSF reported in three patients with rigid Huntington's disease.[7,11]

As a result of technical advances in utilization of the sensitive and specific GC-MS methods of neurochemical analysis, a new method to study the dynamic aspects of biogenic amine metabolism *in vivo* has been developed using stable-isotope labeling and tracing techniques.[10] The rate-limiting enzyme in the biosynthesis of dopamine and norepinephrine is tyrosine hydroxylase and for serotonin is tryptophan hydroxylase. Since both these hydroxylases require molecular oxygen, a stable isotope of oxygen (^{18}O) can be introduced at this step. Oxygen in the form of stable isotope ($^{18}O_2$) can be administered to patients, and the rate of incorporation of the label into the biosynthetic pathway is quantified by GC-MS measurement of labeled metabolites in human biological tissues or fluids, particularly in CSF, blood, and urine. In a study of six patients with Huntington's disease compared to four with Parkinson's disease, dopamine turnover was much more rapid in the Parkinson patients than in the Huntington subjects, suggesting once again that dopamine turnover is reduced in Huntington's disease.[10]

3. Neuropharmacology

The clinical responses observed following administration of therapeutic agents having known actions on neurotransmitters in brain may be used as evidence to indicate a biochemical basis for a disorder. Clinical investigations of patients with Huntington's disease have involved both the acute and chronic administration of agents affecting brain biogenic amine metabolism.[18] A pharmacological approach to the study of Huntington's disease has been used since the observation that reserpine, which depletes brain dopamine, norepinephrine, and serotonin, improves hyperkinesia in these patients. Tetrabenazine, which has a similar mode of action, also improves hyperkinesia in choreic patients. Improvement in the choreiform movements appears as a result of depletion of dopamine rather than serotonin stores in neurons, since administration of levodopa, a precursor of dopamine but not of serotonin, will reverse the clinical effect of reserpine.

Agents that affect biogenic amine biosynthesis also modify the clinical manifestations of Huntington's disease. For example, α-methyltyrosine, an inhibitor of tyrosine hydroxylase, depletes catecholamines and has been used in the treatment of this disorder. Again, the defect in dopamine biosynthesis appears most important, since *p*-chlorophenylalanine, a potent inhibitor of serotonin but not catecholamine biosynthesis, has no effect on Huntington patients.[8] Like the dopamine-depleting drugs, pharmacological agents that block dopamine receptors in the central nervous system are effective in reducing the excessive movements in Huntington's disease. Phenothiazines, such as chlorpromazine and trifluroperazine, and a butyrophenone, haloperidol, have been used extensively to control chorea and other hyperkinesias in this disorder.

Aromatic amino acid precursors of dopamine and serotonin biosynthesis such as levodopa, tryptophan, and 5-hydroxytryptophan have also been administered to patients with Huntington's chorea.[8] Levodopa, while lessening the rigidity seen in some Huntington patients, exacerbates chorea in these individuals and may initiate chorea in subjects at risk for the disorder.[18] Oral 5-hydroxytryptophan also seems to increase chorea, although this may be due primarily to the displacement of dopamine by the serotonin formed from this precursor.[20] Large doses of tryptophan, a serotonin precursor, have no effect on chorea,[22] whereas imipramine, which is thought to increase synaptic levels of dopamine and serotonin by inhibiting neuronal reuptake of these transmitters, exacerbates the choreic symptoms.

Thus, pharmacological evidence strongly suggests a functional excess of brain dopamine in Huntington's disease, since agents that will increase or decrease the effects of dopamine in the central nervous system can exacerbate or ameliorate, respectively, the motor manifestations of this disease.

4. Conclusion

Although neurochemical and pathological studies indicate a loss of cholinergic and GABAergic neurons in the striatum, there also appears to be a disturbance in dopamine neurons terminating in this region. Currently, the evidence points to a decrease in function in dopaminergic neurons, as both the neurotransmitter and its metabolite HVA are decreased in specific brain regions. Similarly, most studies quantitating cerebral dopamine metabolism by measurement of the dopamine metabolite HVA in CSF under basal conditions, and after oral probenecid loading, also indicate impaired function of dopaminergic neurons in striatum. Furthermore, not only is there a decrease in striatal dopamine, but also the number of dopaminergic receptors is decreased in this region. Thus, the neurochemical data strongly suggest impaired function of the presynaptic dopamine neuron and of the postsynaptic dopamine receptor.

However, despite this neurochemical evidence for decreased dopamine function in brain, neuropharmacological studies indicate excessive brain dopamine activity. An explanation for this apparent discrepancy may be that there is a relative excess of dopamine in brain compared to the normal balanced state. As a result of the progressive loss of striatal GABAergic and cholinergic neurons, there is a decrease in function in dopaminergic neurons to compensate for the resultant neurotransmitter imbalance. This compensatory mechanism is apparently insufficient, and the relative dopamine excess results in chorea and hyperkinesia. This neurotransmitter imbalance is partially restored by neuroleptic drugs, alleviating the hyperkinetic symptoms.

ACKNOWLEDGMENTS. The research reported herein was supported in part by USPHS Grant NS-13803, by Research Career Development Award NS-00335 (S.J.E.), and by the Huntington's Chorea Foundation.

References

1. BARBEAU, A.: L-DOPA and juvenile Huntington's disease. *Lancet* **2:**1066, 1969.
2. BERNHEIMER, H., HORNYKIEWICZ, O.: Brain amines in Huntington's chorea. In Barbeau, A., Chase, T. N., Paulson, G. W. (eds.): *Advances in Neurology,* Vol. 1. New York, Raven Press, 1973, pp. 525–531.
3. BIRD, E. D.: Biochemical studies on γ-aminobutyric acid metabolism in Huntington's chorea. In Bradford, H. F., Marsden, C. D. (eds.): *Biochemistry and Neurology.* New York, Academic Press, 1976, pp. 83–92.
4. BIRD, E. D., IVERSON, L. L.: Huntington's chorea: Postmortem measurement of glutamic acid decarboxylase, choline acetyltransferase and dopamine in basal ganglia. *Brain* **97:**457–472, 1974.
5. BLACKWOOD, W., CORSELLIS, J. A. N.: *Greenfield's Neuropathology.* Chicago, Edward Arnold, 1976, pp. 822–827.
6. BYERS, R. K., GILLES, F. H., FUNG, C.: Huntington's disease in children: Neuropathologic study of four cases. *Neurology* **23:**561–569, 1973.
7. CARACENI, T., CALDERINI, G., CONSOLAZIONE, A., RIVA, E., ALGERI, S., GIROTTI, F., SPREAFICO, R., BRANCIFORTI, A., DALL'OLIO, A., MORSELLI, P. L.: Biochemical aspects of Huntington's chorea. *J. Neurol. Neurosurg. Psychiatry* **40:**581–587, 1977.
8. CHASE, T. N.: Biochemical and pharmacologic studies of monoamines in Huntington's chorea. In Barbeau, A., Chase, T. N., Paulson, G. W. (eds.): *Advances in Neurology,* Vol. I. New York, Raven Press, 1973, pp. 533–542.
9. CHASE, T. N., GORDON, E. K., NG, L. K. Y.: Norepinephrine metabolism in the central nervous system of man: Studies using 3-methoxy-4-hydroxyphenylethylene glycol levels in cerebrospinal fluid. *J. Neurochem.* **21:**581–587, 1973.
10. CHASE, T. N., NEOPHYTIDES, A., SAMUEL, D., SEDVALL, G., SWAHN, C.-G.: Oxygen-18 use for clinical studies of central monoamine metabolism. In Kopin, I. J. (ed.): Proceedings of the Fourth International Catecholamine Symposium, Pacific Grove, California, 1978, abstract 192.
11. CURZON, G.: Involuntary movements other than parkinsonism: Biochemical aspects. *Proc. R. Soc. Med.* **66:**873–876, 1973.

11a. EBERT, M. H., KARTZINEL, R., COWDRY, R. W., GOODWIN, F. K.: Cerebrospinal fluid amine metabolities and probenecid text. In Wood, J. H. (ed.): *Neurobiology of Cerebrospinal Fluid I.* New York, Plenum Press, 1980.

12. ENNA, S. J., BIRD, E. D., BENNETT, J. P., JR., BYLUND, D. B., YAMAMURA, H. I., IVERSEN, L. L., SNYDER, S. H.: Huntington's chorea: Changes in neurotransmitter receptors in the brain. *New Engl. J. Med.* **294:**1305–1309, 1976.
13. ENNA, S. J., STERN, L. Z., WASTEK, G. J., YAMAMURA, H. I.: Neurobiology and pharmacology of Huntington's disease. *Life Sci.* **20:**205–212, 1977.
14. GARELIS, E., YOUNG, S. N., LAL, S., SOURKES, T. L.: Monoamine metabolites in lumbar CSF: The question of their origin in relation to clinical studies. *Brain Res.* **79:**1–8, 1974.
15. GOODWIN, F. K., POST, R. M., DUNNER, D. L., GORDON, E. K.: Cerebrospinal fluid amine metabolites in affective illness: The probenecid technique. *Am. J. Psychiatry* **130:**73–79, 1973.
16. GUILLEMINAULT, C., THARP, B. R., COUSIN, D.: HVA and 5-HIAA CSF measurement and 5HTP trials in some patients with involuntary movements. *J. Neurol. Sci.* **18:**435–441, 1973.
17. KLAWANS, H. L.: Cerebrospinal fluid homovanillic acid in Huntington's chorea. *J. Neurol. Sci.* **13:**277–279, 1971.
18. KLAWANS, H. L., WEINER, W. J.: The pharmacology of choreatic movement disorders. *Prog. Neurobiol.* **6:**49–80, 1976.
19. KREMZNER, L. T., BERL, S., STELLAR, S., COTE, L. J.: GABA, homocarnosine, amino acids, polyamines, and

GAD activity in cortical biopsies from patients with Huntington's disease. In Chase, T. N., Wexler, N., Barbeau, A. (eds.): Proceedings of the Second International Huntington's Disease Symposium, San Diego, California, 1978, abstract 26.

20. LEE, D. K., MARKHAM, C. H., CLARK, W. G.: Serotonin (5-HT) metabolism in Huntington's chorea. *Life Sci.* **7**(1):707–712, 1968.
21. MCLELLAN, D. L., CHALMERS, R. J., JOHNSON, R. H.: A double-blind trial of tetrabenazine, thiopropazate, and placebo in patients with chorea. *Lancet* **1**:104–107, 1974.
22. MCLEOD, W. R., DE L. HORNE, D. J.: Huntington's chorea and tryptophan. *J. Neurol. Neurosurg. Psychiatry* **34**:510–513, 1972.
23. MOIR, A. T. B., ASHCROFT, G. W., CRAWFORD, T. B. B., ECCLESTON, D., GULDBERG, H. C.: Cerebral metabolites in cerebrospinal fluid as a biochemical approach to the brain. *Brain* **93**:357–368, 1970.
24. NEFF, N. H., TOZER, T. N.: *In vivo* measurement of brain serotonin turnover. In Garattini, S., Shore, P. A. (eds.): *Advances in Pharmacology*, Vol. 6A. New York, Academic Press, 1968, pp. 97–109.
25. PEREL, J. M., LEVITT, M., DUNNER, D. L.: Plasma and cerebrospinal fluid probenecid concentrations as related to accumulation of acidic biogenic amine metabolites in man. *Psychopharmacology* **35**:83–90, 1974.
26. REISINE, T. D., FIELDS, J. Z., STERN, L. Z., JOHNSON, P. C., BIRD, E. D., YAMAMURA, H. I.: Alterations in dopaminergic receptors in Huntington's disease. *Life Sci.* **21**:1123–1128, 1977.
27. SWAHN, C.-G., SANDGÄRDE, B., WIESEL, F. A., SEDVALL, G.: Simultaneous determination of the three major monoamine metabolites in brain tissue and body fluids by a mass fragmentographic method. *Psychopharmacology* **48**:147–152, 1976.
28. TAMARKIN, N. R., GOODWIN, F. K., AXELROD, J.: Rapid elevation of biogenic amine metabolites in human CSF following probenecid. *Life Sci.* **9**(1):1397–1408, 1970.
29. WATSON, E., WILK, S.: Determination of probenecid in small volumes of cerebrospinal fluid. *J. Neurochem.* **21**:1569–1571, 1973.

29a. WOOD, J. H.: Neurochemical analysis of cerebrospinal fluid. *Neurology* **30:** 645–651, 1980.

29b. WOOD, J. H.: Sites of origin and concentration gradients of neurotransmitters, their precursors and metabolites and cyclic nucleotides in cerebrospinal fluid. In Wood, J. H: *Neurobiology of Cerebrospinal Fluid I.* New York, Plenum Press, 1980.

29c. WOOD, J. H., ZIEGLER, M. G., LAKE, C. R., SHOULSON, I., BROOKS, B. R., VAN BUREN, J. M.: Cerebrospinal fluid norepinephrine reductions in man after degeneration and electrical stimulation of the caudate nucleus. *Ann. Neurol.* **1**:94–99, 1977.

30. YATES, C. M., MAGILL, B. E. A., DAVIDSON, D., MURRAY, L. G., WILSON, H., PULLAR, L. A.: Lysosomal enzymes, amino acids and acid metabolites of amines in Huntington's chorea. *Clin. Chem. Acta* **44**:139–145, 1973.
31. ZIEGLER, M. G., LAKE, C. R., FOPPEN, F. H., SHOULSON, I., KOPIN, I. J.: Norepinephrine in cerebrospinal fluid. *Brain Res.* **108**:436–440, 1976.

CHAPTER 12

Cerebrospinal Fluid Noradrenergic and Behavioral Alterations Associated with Stimulation and Atrophy of the Caudate Nucleus

James H. Wood, Michael G. Ziegler, C. Raymond Lake, Ira Shoulson, and Benjamin Rix Brooks

1. Introduction

The caudate nucleus, a portion of the striatum, is the basal ganglia structure that most lends itself to physiological, biochemical, and pathological evaluation. This chapter reviews both animal and human investigations involving behavioral and cerebrospinal fluid (CSF) norepinephrine alterations induced by caudate nucleus stimulation. In addition, the CSF noradrenergic alterations among patients with Huntington's disease and caudate nucleus degeneration are discussed with respect to observed defects in central blood pressure control.

James H. Wood, M.D. • Division of Neurosurgery, University of Pennsylvania School of Medicine, Philadelphia, Pennsylvania 19104. **Michael G. Ziegler, M.D.** • Department of Medicine, University of California, San Diego, California 92103. **C. Raymond Lake, M.D., Ph.D** • Departments of Psychiatry and Pharmacology, Uniformed Services University of Health Sciences, School of Medicine, Bethesda, Maryland 20014. **Ira Shoulson, M.D.** • Division of Neurology, University of Rochester School of Medicine, Rochester, New York 14642. **Benjamin Rix Brooks, M.D.** • Department of Neurology, The Johns Hopkins University School of Medicine, Baltimore, Maryland 21205.

2. Neurobiology of Caudate Nucleus Stimulation

Afferent axonal projections to the caudate nucleus have their cell bodies in the substantia nigra,[51] thalamus,[48] and cerebral cortex.[14,72] Histological studies of the synaptic organization of the caudate nucleus suggest that input fibers synapse *en passant* with many caudate neurons.[40,42] Stimulation of these monosynaptic afferent projections evokes biphasic responses of initial excitation followed by long-duration inhibition in caudate neurons.[12] These physiological observations suggest that the afferent fibers can excite a caudate neuron, and collateral excitation of caudate interneurons may then result in an inhibitory influence on this caudate neuron.[13,35] The short-armed inhibitory interneurons comprise approximately 95% of the caudate cell population. Low-frequency stimulation of the cat caudate nucleus evokes sequences of excitatory postsynaptic potentials followed by inhibitory postsynaptic potentials in ipsilateral dorsal thalamic neurons.[30] Thus, electrical stimulation of the caudate nucleus would be expected to produce initial excitation and subsequent inhibition of its efferent projections.

The caudate nucleus is innervated by axons projecting from dopaminergic neurons of the substantia nigra.[9,34] Unilateral surgical ablation of these dopamine-containing neurons results in head- and tail-turning to the unoperated side.[4] After dopaminergic receptors have had time to develop hypersensitivity, apomorphine induces turning of unilaterally striatomized rats to the operated side, and the action of this dopamine agonist is inhibited by drugs that block dopaminergic receptors.[5] Electrical stimulation of the substantia nigra increases dopamine concentrations in the caudate nucleus.[55] Nigral stimulation also evokes frequency- and intensity-related release of labeled dopamine into cerebroventricular perfusates.[70] Dopamine application to caudate neurons results in excitation[38,39]; however, an inhibitory mechanism operates within the caudate nucleus.[12,24,35]

Histochemical[25] and electrophysiological[26] investigations have detected excitatory monosynaptic nigrostriatal tracts. These neurons contain acetylcholine[53,62] and thus persist after the surgical and pharmacological destruction of striatal dopaminergic neurons.[37,61] Stimulation-induced release of caudate acetylcholine[50] augments the output of a minority of caudate neurons.[61] Buchwald *et al.*[12] noted that purely excitatory postsynaptic potentials are induced in caudate neurons only during thalamic stimulation. This excitation of caudate neurons may augment the inhibitory influence of efferent caudate projections to distant neuronal systems.[75]

Both the spatial prominence and the periventricular location of the caudate nucleus enable reliable intranuclear electrode placement. Electrical stimulation of the caudate nucleus has been reported to evoke somatic, visceral, facilitatory, and inhibitory responses.

Contraversive head-turning and circling in dogs during caudate stimulation was demonstrated by Ferrier[27] in 1873. In addition, Cools[18] and Chandler and Crosby[16] described contralateral forelimb movements in approximately 80% of stimulated animals. These contralateral body responses were demonstrated to have somatotopic localization within the head of the caudate nucleus.[28] These evoked motor responses were found not to be secondary to current spread to the internal capsule or stimulation of fibers of passage.[16,44,64] However, the vegetative responses induced by Laursen[44] were due to current spread from the caudate electrode to the septum. Multiple investigations[13,29,31,49] have demonstrated caudate-stimulation-induced inhibition of movement. Buchwald *et al.*[13] described an evoked interruption of spontaneous motor activity and slow tonic turning of the neck and trunk musculature with maintenance of this posture throughout the period of stimulation in cats. This response was simular to the "arrest" reaction observed by Hunter and Jasper[36] during cat thalamic stimulation. Spiegel and Szekely[63] demonstrated catatonia and loss of spontaneous movements in cats during prolonged alumina cream stimulation of the caudate nucleus and proved that this inhibition was not an artifact of current spread to the internal capsule.

Early clinical investigations employing electrical stimulation of the caudate nucleus in man demonstrated sleep induction,[33] disturbances of speech, and confusion.[65] Van Buren[66] observed that stimuli delivered to the caudate head are more prone to alter higher psychic processes than are those placed in adjacent posterofrontal white matter, which cause a simple inhibition of motor function.

Recently, Wood *et al.*[75,77] described a modification of the Van Buren depth-coagulating electrode[65-67] that allows selective stimulation at contact points along its shaft. Flexible depth electrodes were constructed of 7-mil wires (Med Wire Corporation, Mt. Vernon, New York), each insulated with Teflon and wound about a central 9-mil wire to make 11 stimulating contacts at 5.0-mm intervals along the shaft including the coagulating tip (Fig. 1). The terminal

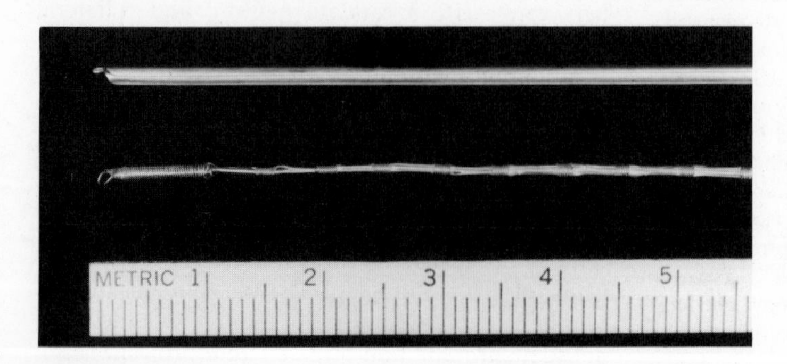

Figure 1. *Top:* Rigid pointed probe employed for stereotaxic insertion of electrodes into brain. *Bottom:* Flexible depth-coagulating electrode with stimulating contacts at 5.0-mm intervals along wire shaft. Terminal loop of electrode permits attachment of electrode to probe during insertion into brain. Reprinted with permission from Wood *et al.*[75]

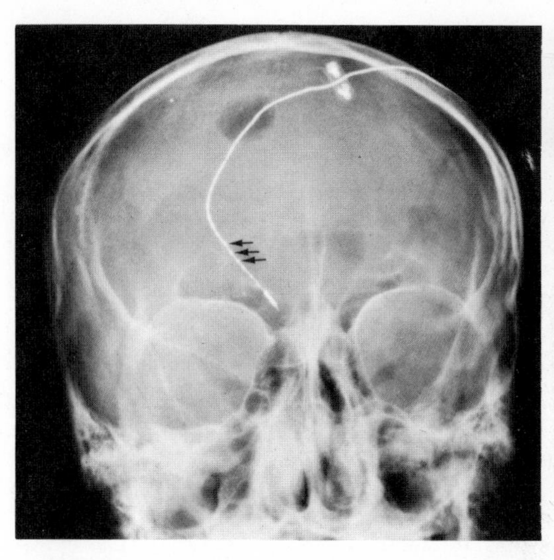

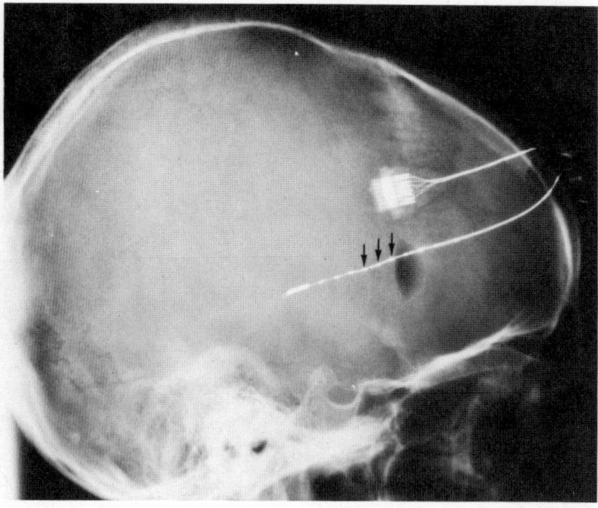

Figure 2. Patient A. Inclined posteroanterior (*left*) and lateral (*right*) skull radiographs. Coagulating tip of flexible depth electrode is located within right thalamus. Arrows indicate stimulating points along electrode shaft passing through right caudate nucleus. Reprinted with permission from Wood *et al.*[75]

loop of the central wire permits attachment of the electrode to the rigid, pointed probe during insertion into the brain. The exact anatomical location of each stimulating contact along the electrode shaft is then stereotaxically calculated using ventricular landmarks[68] (Fig. 2). In confirmation of previous studies, Wood *et al.*[75–77] induced slurred speech, immediate memory difficulty, confusion, inappropriate smiling, and pleasant feelings of intoxication employing intermittent biphasic 60/sec 10-mA peak-to-peak 2.5-msec square-wave stimulation of the human caudate nucleus (Table 1). In these studies (Fig. 3), CSF samples were collected preoperatively and 12 days after electrode installation. The CSF norepinephrine concentrations determined by radioenzymatic assay[79] in these two groups of CSF samples were not significantly different, indicating that the stereotaxic procedure itself did not alter CSF norepinephrine levels. Analysis of CSF samples from these patients obtained 12 hr following caudate stimulation[20] revealed a significant mean 29% reduction ($p < 0.02$, two-tailed paired Student's t test) in steady-state CSF norepinephrine concentrations following electrical stimulation of the caudate nucleus in the five patients listed in Table 1.

The striatum contains only scant quantities of nor-

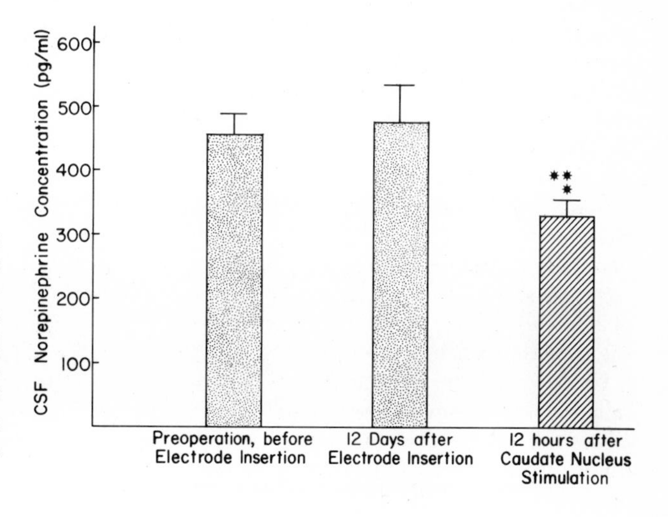

Figure 3. CSF noradrenergic alterations with respect to electrode insertion and human caudate nucleus stimulation. *Significant reductions ($p < 0.01$, two-tailed paired Student's t-test) in CSF norepinephrine levels after caudate stimulation when compared with preoperative CSF norepinephrine concentrations.[75] **Significant reductions ($p < 0.03$) in CSF norepinephrine levels after caudate stimulation when compared with postoperative, prestimulation CSF norepinephrine concentrations.[75]

Table 1. Neurophysiological and CSF Noradrenergic Responses to Electrical Caudate Nucleus Stimulation in Five Patients[a]

Patient	Age	Sex	Diagnosis	Surgical procedure	Neurophysiological response	Stimulation-evoked change of CSF norepinephrine (%)[b]
A	16	M	Dystonia musculorum deformans	Right thalamotomy	None	−33
B	26	M	Dystonia musculorum deformans	Right thalamotomy	Slurred speech, anxiety, pleasant feeling of intoxication	−16
C	45	M	Right hemiparkinsonism	Left thalamotomy	Immediate memory difficulty	−29
D	63	F	Torticollis, rectrocollis	Left thalamotomy	Dizziness, immediate memory difficulty	−41
E	38	F	Thalamic pain syndrome	Bilateral thalamotomy	Confusion, inappropriate smiling	−24

[a] From Wood *et al.*[75,76]
[b] CSF norepinephrine concentration after caudate stimulation compared with average of CSF norepinephrine levels before and after electrode insertion but prior to stimulation.

epinephrine.[11,60] Dopamine-β-hydroxylase, a necessary enzyme for the synthesis of norepinephrine, is not present in the caudate nucleus.[60] Incubation of caudate slices with labeled tyrosine produces labeled dopamine but not labeled norepinephrine.[47] Almost all caudate norepinephrine resides in the terminals of the vasa nervorum.[58] Thus, norepinephrine does not play a primary role in interneuronal transmission within the caudate nucleus. Caudate-stimulation-evoked CSF norepinephrine alterations in these patients must be a result of synaptic events occurring outside the caudate nucleus.[75–77] Thus, caudate efferent projections apparently decrease the activity of noradrengeric neurons distant to the caudate nucleus.

The report by Von Voigtlander and Moore[69] of exogenous norepinephrine release into ventricular perfusates during electrical stimulation of the cat caudate nucleus does not suggest that endogenous norepinephrine is released by caudate stimulation. Exogenous norepinephrine is taken up by both dopaminergic and noradrenergic neurons.[60] Thus, electrical stimulation would release the labeled norepinephrine located in dopaminergic terminals within the caudate nucleus.[55,70]

Chinese reports[1] indicate that pain thresholds of rabbits could be elevated by stimulation of the dorsal portion of the head of the caudate nucleus. In addition, destruction of this region of the striatum markedly reduced the analgesic effect of acupuncture stimulation, suggesting that the caudate nucleus may be involved in the mechanism of acupuncture analgesia. Later, these Chinese investigators[2] reported induced elevation of pain thresholds among patients with intractable pain syndromes during electrical caudate stimulation employing depth electrodes. Depletion of whole-brain catecholamines potentiates morphine-induced analgesia.[8] Alpha adrenergic blockers can cause naloxone-reversible analgesia and potentiation of morphine-induced analgesia.[17] Wood *et al.*[75–77] observed caudate-stimulation-evoked reductions in CSF norepinephrine levels, thus demonstrating that noradrenergic pathways in man are inhibited by direct caudate stimulation. Although the possibility exists that this caudate inhibition of noradrenergic activity may alter pain thresholds, no clinical confirmation of pain suppression was reported by Wood *et al.*[75,76] More recent communications[21,78] suggest that enkephalins or endogenous opiates may play a role in the caudate mediation of analgesia and behavior.

Avakyan and Arushanyan[6] suggest that the epileptogenic properties of the caudate nucleus can be modified by the administration of catecholaminergic drugs. Depletion[45,73] or reduction[45] in synthesis of norepinephrine facilitates experimental seizures. Endogenous norepinephrine is necessary for normal resistance to seizures.[74] Administration of norepinephrine via the CSF inhibits curare-induced convulsions in cats.[23] Since caudate stimulation reduces CSF norepinephrine in nonepileptic patients,[75-77] modulation of noradrenergic activity by the caudate nucleus may play a role in the maintenance of seizure thresholds.

Sympathetic responses such as vasoconstrictor reflexes[54,71] and galvanic skin reflexes[71] induced in the cat's paw by radial-nerve stimulation are reduced in amplitude by electrical stimulation of the caudate nucleus. The centrally acting hypotensive drug clonidine depresses brainstem norepinephrine levels[7] and sympathetic outflow.[56] Barbeau *et al.*[7] suggest that the caudate nuclei modulate the blood-pressure centers of the brainstem and hypothalamus and contribute to the postural regulation of blood pressure.

3. Central Noradrenergic Depression in Huntington's Disease

Degeneration of the caudate nucleus is a consistent autopsy finding among patients with Huntington's disease.[52] This degenerative brain disorder is inherited as an autosomal dominant trait with complete penetrance. Symptoms of dementia and chorea begin in early or middle adult life and progress to death over the subsequent 12–15 years.[52,59] Caudate flattening, cortical atrophy, and *ex vacuo* ventricular enlargement can be demonstrated employing computed axial tomography in advanced cases (Fig. 4). These neuroradiological features correlate with the atrophy and cell loss noted in the caudate nucleus, putamen, cerebral cortex,[59] cerebellum,[46] and ventrolateral thalamus[22] among autopsied patients with Huntington's disease.

Postmortem neurochemical analysis of brains from patients with Huntington's disease demonstrates decreased levels of dopamine and of its major metabolite, homovanillic acid (HVA), in the caudate nucleus.[10] γ-Aminobutryic acid (GABA) concentrations are also low in the substantia nigra, globus pallidus–putamen, and caudate nucleus in choreic patients.[43] The lumbar CSF steady-state concentrations of HVA[13a,19,41] and GABA[22a,32,32a] are reduced in patients with Huntington's disease. Therefore,

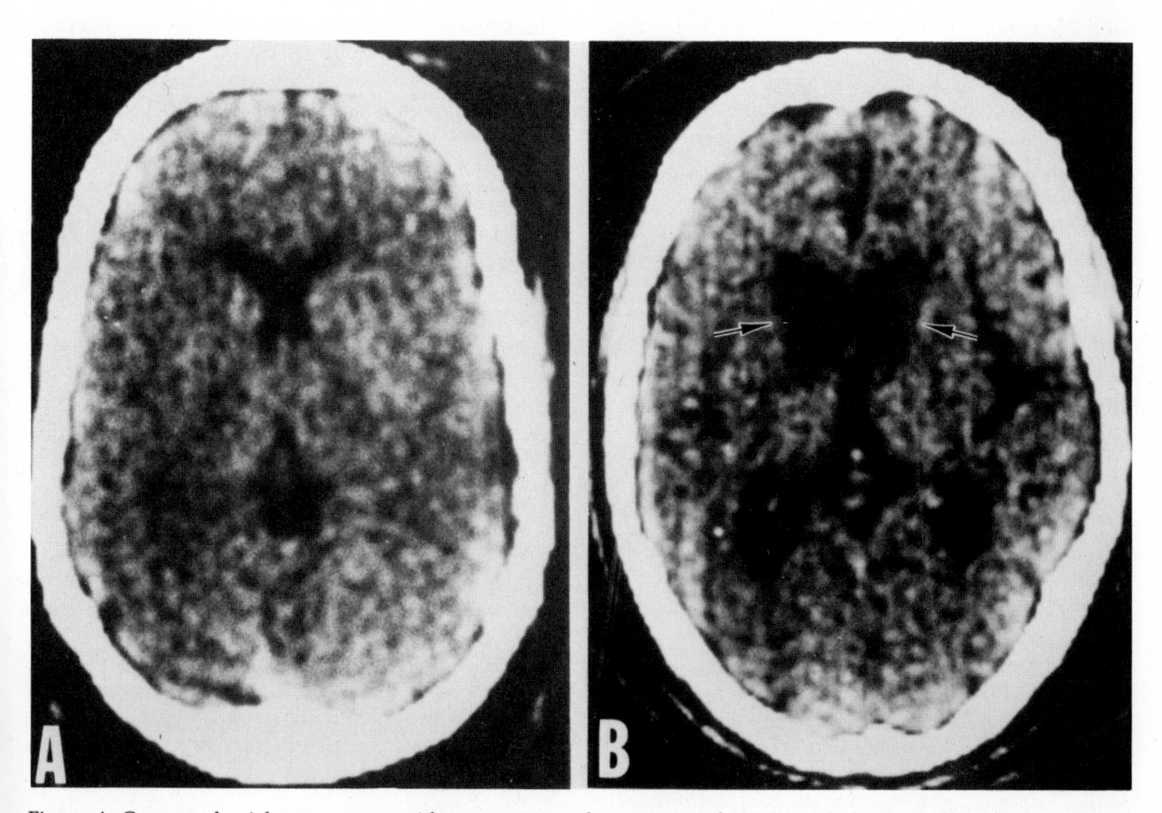

Figure 4. Computed axial tomograms, without contrast enhancement, of (A) control patient with intrasellar pituitary adenoma and (B) choreic patient with flattened caudate nucleus (*arrows*), cortical atrophy, and *ex vacuo* ventricular enlargement. Reprinted with permission from Wood *et al.*[76]

brain neurochemical alterations appear to be reflected by changes in lumbar CSF.

Increasing CSF norepinephrine levels have been noted in successive lumbar CSF fractions collected from patients with Huntington's disease.[79,80] This rise in CSF norepinephrine concentration plateaus after 12 ml of CSF has been removed and remains constant in later fractions.[82] The norepinephrine in lumbar CSF is probably derived from descending spinal noradrenergic tracts the cell bodies of which originate in a brainstem region such as the locus coeruleus.[2a,80,82]

Central noradrenergic neurons are involved in the control of blood pressure.[57] Ziegler *et al.*[80,81] demonstrated that CSF norepinephrine is more consistent than plasma norepinephrine in its elevation in association with increasing blood pressure. Experimental destruction of noradrenergic neurons by administration of 6-hydroxydopamine via the CSF prevents reflex elevations in blood pressure.[15] Thus, the noradrenergic neurons mediating blood pressure control are in reasonably close proximity to the CSF such that changes in their activity should be reflected in alterations in CSF norepinephrine concentrations.

Aminoff and Gross[3] implied central defects in the vasoregulatory mechanism of patients with Huntington's disease and caudate atrophy. Communications by Wood *et al.*[75–77] report lumbar CSF norepinephrine levels determined by radioenzymatic assay[79] in nine patients with caudate atrophy associated with Huntington's disease to be significantly lower ($p < 0.02$, two-tailed unpaired Student's t test) than those in nine age- and sex-matched control patients (Fig. 5). Choreic patients may also have lower resting blood pressures (unpublished results). Thus, the function of these noradrenergic pathways modulating reflex blood pressure alterations in man appear to be impaired in choreic patients. In addition, further depression of noradrenergic activity by extrinsic electric stimulation of the caudate nucleus[76,77] should not be expected to benefit patients with Huntington's disease. These same choreic patients who had decreased CSF norepinephrine also had reduced plasma norepinephrine levels. Thus, central control of peripheral sympathetic nerves may be defective in Huntington's disease, which might result in the reported vasoregulatory insufficiency[3] in these patients.

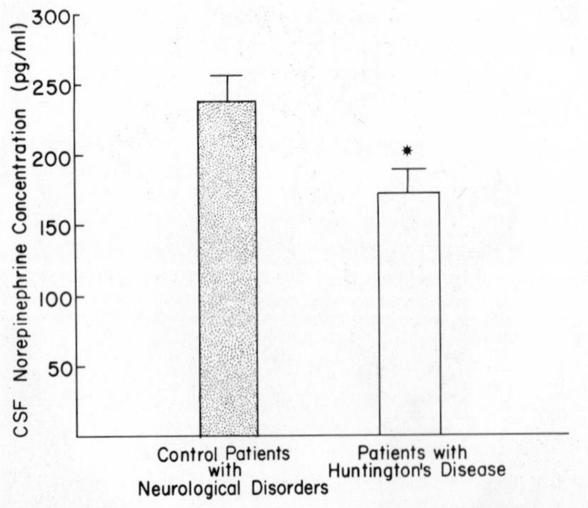

Figure 5. CSF norepinephrine concentrations in age- and sex-matched control patients with various neurological disorders and patients with Huntington's disease. *Significant reductions ($p < 0.02$, two-tailed unpaired Student's t test) in CSF norepinephrine levels in patients with Huntington's disease compared with control patients.[76]

References

1. Acupuncture Anesthesia Coordinating Group of Shanghai First Medical College (China): Observations on electrical stimulation of the caudate nucleus of human brain and acupuncture in treatment of intractable pain. *China Med. J.* **3**:117–124, 1977.
2. Acupuncture Anesthesia Group, 2nd Laboratory of Shanghai Institute of Physiology (China): Possible role of the caudate nucleus in acupuncture anesthesia. *Zhonghua Yixal Zazhi* **55**:33–38, 1974.

2a. Adèr, J.-P., Aizenstein, M. L., Postema, F., Korf, J.: Origin of free 3-methoxy-4-hydroxyphenylethylene glycol in rat cerebrospinal fluid. *J. Neural Trans.* **46**:279–290, 1979.

3. Aminoff, M. J., Gross, M.: Vasoregularity activity in patients with Huntington's chorea. *J. Neurol. Sci.* **21**:33–38, 1974.
4. Anden, N.-E., Dahlstrom, A., Fuxe, K.: Functional role of the nigroneostriatal dopamine neurons. *Acta Pharmacol. Toxicol.* **24**:263–274, 1966.
5. Anden, N.-E., Rubenson, A., Fuxe, K.: Evidence for dopamine receptor stimulation by apomorphine (letter to the editor). *J. Pharm. Pharmacol.* **19**:627–629, 1967.
6. Avakayan, R. M., Arushanyan, E. B.: Effect of catecholaminergic drugs on epileptogenic properties of the caudate nucleus. *Neurosci. Behav. Physiol.* **7**:13–16, 1976.
7. Barbeau, A. L., Gillo-Joffroy, L., Brossard, Y.: Renin, dopamine and Parkinson's disease. In Barbeau, A., McDowell, F. H. (eds.): *L-dopa and Parkinsonism.* Philadelphia, F. A. Davis, 1970, pp. 286–293.
8. Bauxbaum, D. M., Yarbrough, G. G., Carter, M. D.: Biogenic amines and narcotic effects. I. Modification of morphine-induced analgesia and motor activity

after alterations of cerebral amine levels. *J. Pharmacol. Exp. Ther.* **185**:317–326, 1973.
9. BEDARD, P., LAROCHELLE, L., PARENT, A., POIRIER, L. J.: The nigrostriatal pathway: A correlative study based on neuroanatomical and neurochemical criteria in the cat and the monkey. *Exp. Neurol.* **25**:365–377, 1969.
10. BERNHEIMER, H., BIRKMAYER, W., HORNYKIEWICZ, O., JELLINGER, K., SEITELBERGER, F.: Brain dopamine and the syndrome of Parkinson and Huntington. *J. Neurol. Sci.* **20**:415–455, 1973.
11. BERTHER, A., ROSENGREN, E.: Occurrence and distribution of dopamine in brain and other tissues. *Experimentia* **15**:10–11, 1959.
12. BUCHWALD, N. A., PRICE, D. D., VERNON, L., HULL, C. D.: Caudate intracellular response to thalamic and cortical inputs. *Exp. Neurol.* **38**:311–232, 1973.
13. BUCHWALD, N. A., WYERS, E. J., LAUPRECHT, B. A., HEUSER, G.: The "caudate-spindle." IV. A behavioral index of caudate-induced inhibition. *Electroencephalogr. Clin. Neurophysiol.* **13**:531–537, 1961.
13a. BUTLER, I. J., SEIFERT, W. E., ENNA, S. J.: Huntington's disease: Biogenic amines in cerebrospinal fluid. In Wood, J. H. (ed.): *Neurobiology of Cerebrospinal Fluid I.* New York, Plenum Press, 1980.
14. CARMAN, J. B., COWAN, W. M., POWELL, T. P. S., WEBSTER, K. E.: A bilateral corticostriate projection. *J. Neurol. Neurosurg. Psychiatry* **28**:71–77, 1965.
15. CHALMERS, J. P., REID, J. L.: Participation of central noradrenergic neurons in arterial baroreceptor reflexes in the rabbit: A study with intracisternally administered 6-hydroxydopamine. *Circ. Res.* **31**:789–804, 1972.
16. CHANDLER, W. F., CROSBY, E. C.: Motor effects of stimulation and ablation of the caudate nucleus in the monkey. *Neurology (Minneapolis)* **25**:1160–1163, 1975.
17. CICERO, T. S., MEYER, E. R., SMITHLOFF, B. R.: Alpha adrenergic blocking agents: Anti-nociceptive activity and enhancement of morphine-induced analgesia. *J. Pharmacol. Exp. Ther.* **189**:72–82, 1974.
18. COOLS, A. R.: Chemical and electrical stimulation of the caudate nucleus in freely moving cats: The role of dopamine. *Brain Res.* **58**:437–451, 1973.
19. CURZON, G., GUMPERT, J., SHARPE, D.: Amine metabolites in the cerebrospinal fluid in Huntington's chorea. *J. Neurol. Neurosurg. Psychiatry* **35**:514–519, 1972.
20. DICHIRO, G., HAMMOCK, M. K., BLEYER, W. A.: Spinal descent of cerebrospinal fluid in man. *Neurology (Minneapolis)* **26**:1–8, 1976.
21. DILL, R. E., COSTA, E.: Behavioral dissociation of the enkephalinergic systems of nucleus accumbens and nucleus caudatus. *Neuropharmacology* **16**:323–326, 1977.
22. DOM, R., MALFRAID, M., BARO, F.: Neuropathology of Huntington's chorea: Studies of the ventrobasal complex of the thalamus. *Neurology (Minneapolis)* **26**:64–68, 1976.
22a. ENNA, S. J., ZIEGLER, M. G., LAKE, C. R., WOOD, J. H., BROOKS, B. R., BUTLER, I. J.: Cerebrospinal fluid γ-aminobutyric acid: Correlation with cerebrospinal fluid and blood constituents and alterations in neurological disorders. In Wood, J. H. (ed.): *Neurobiology of Cerebrospinal Fluid I.* New York, Plenum Press, 1980.
23. FELDBERG, W., SHERWOOD, S. L.: Injections of drugs into the lateral ventricle of the cat. *J. Physiol. (London)* **123**:148–167, 1954.
24. FELZ, P., ALBE-FESSARD, D.: A study of ascending nigro-caudate pathway. *Electroencephalogr. Clin. Neurophysiol.* **33**:179–193, 1972.
25. FELZ, P., DECHAMPLAIN, J.: Persistence of caudate unitary responses to nigral stimulation after destruction and functional impairment of the striatal dopaminergic terminals. *Brain Res.* **43**:595–600, 1972.
26. FELTZ, P., MACKENZIE, J. S.: Properties of caudate unitary responses to repetitive nigral stimulation. *Brain Res.* **13**:612–626, 1969.
27. FERRIER, D.: Experimental researches in cerebral physiology and pathology. *West Riding Lunatic Asylum Med. Rep.* **3**:30–96, 1873.
28. FORMAN, D., WARD, J. W.: Responses to electrical stimulation of caudate nucleus in cats in chronic experiments. *J. Neurophysiol.* **20**:230–244, 1957.
29. FREEMAN, G. L., KRASNO, L.: Inhibitory function of the corpus striatum. *Arch. Neurol. Psychiatry* **44**:323–327, 1940.
30. FRIGYERI, T. L., MARCHEK, J.: Basal ganglia–diencephalon synaptic relations in the cat. I. An intracellular study of dorsal thalamic neurons during capsular and basal ganglia stimulation. *Brain Res.* **20**: 201–217, 1970.
31. GEREBTZOFF, M. A.: Contribution a la physiologie du corps strié. *Arch. Int. Physiol.* **51**:333–352, 1941.
32. GLAESER, B. S., VOGEL, D. B., OLEWEILER, D. B., HARE, T. A.: GABA level in cerebrospinal fluid of patients with Huntington's chorea: A preliminary report. *Biochem. Med.* **12**:380–385, 1975.
32a. HARE, T. A., MANYAM, N. V. B., GLAESER, B. S.: Evaluation of cerebrospinal fluid γ-aminobutyric acid content in neurologic and psychiatric disorders. In Wood, J. H. (ed.): *Neurobiology of Cerebrospinal Fluid I.* New York, Plenum Press, 1980.
33. HEATH, R. G., HODES, R.: Induction of sleep by stimulation of the caudate nucleus in macaques rhesus and man. *Trans. Am. Neurol. Assoc.* **77**:204–210, 1952.
34. HÖKFELT, T., UNGERSTEDT, U.: Electron and fluorescence microscopical studies on the nucleus caudatus putamen of the rat after unilateral lesions of ascending nigro-neostriatal dopamine neurons. *Acta Physiol. Scand.* **76**:415–426, 1969.
35. HULL, C. D., BERNARDI, G., PRICE, D. D., BUCHWALD, N. A.: Intracellular responses of caudate neurons to temporally and spatially combined stimuli. *Exp. Neurol.* **38**:324–336, 1973.
36. HUNTER, J., JASPER, H. H.: Effects of thalamic stimulation in unanesthetized cats: Arrest reactions and petit mal-like siezures, activation patterns and generalized

convulsions. *Electroencephalogr. Clin. Neurophysiol.* **1**:305–324, 1949.

37. JONES, B. E., GUYENET, P., CHERAMY, A., GAUCHY, C., GLOWINSKI, J.: The *in vivo* release of acetylcholine from cat caudate nucleus after pharmacological and surgical manipulations of dopaminergic nigrostriatal neurons. *Brain Res.* **64**:355–369, 1973.
38. KITAI, S. T., SUGIMORI, M., KOCSIA, J. D.: Excitatory nature of dopamine in the nigro-caudate pathway. Exp. Brain Res. **24**:351–363, 1976.
39. KITAI, S. T., WAGNER, A., PRECHT, W., OHNO, T.: Nigro–caudate and caudate–nigral relationship: An electrophysiological study. *Brain Res.* **85**:44–48, 1975.
40. KEMP, J. M., POWELL, T. P. S.: The site of termination of afferent fibers in the caudate nucleus. *Philos. Trans. R. Soc. London (Ser. B)* **262**:413–427, 1971.
41. KLAWANS, H. L.: Cerebrospinal fluid homovanillic acid in Huntington's chorea. *J. Neurol. Sci.* **13**:277–279, 1971.
42. KEMP, J. M., POWELL, T. P. S.: The synaptic organization of the caudate nucleus. *Philos. Trans. R. Soc. London (Ser. B)* **262**:403–412, 1971.
43. PERRY, T. L., HANSEN, S., KLOSTER, M.: Huntington's chorea: A deficiency of gamma-animobutyric acid in brain. *N. Engl. J. Med.* **288**:337–342, 1973.
44. LAURESEN, A. M.: Movements evoked from region of the caudate nucleus in cats. *Acta Physiol. Scand.* **54**:175–184, 1962.
45. LESSIN, A. W., PARKS, M. W.: The effects of reserpine and other agents on leptazol convulsions in mice. *Br. J. Pharmacol.* **14**:108–111, 1959.
46. LISS, L., PAULSON, G. W., SOMMER, A.: Rigid form of Huntington's chorea. In Barbeau, A., Chase, T., Paulson, G. (eds.): *Advances in Neurology*, Vol. 1. New York, Raven Press, 1973, pp. 405–424.
47. MATSUOKA, D. T., SCHOOT, H. F., PETRIELLO, L.: Formation of catecholamines by various areas of cat brain. *J. Pharmacol. Exp. Ther.* **139**:73–76, 1963.
48. MEHLER, W. R.: Further notes on the center median nucleus of Luys, In Pupura, D. P., Yahr, M. D. (eds.): *The Thalamus.* New York, Columbia University Press, 1966, pp. 109–127.
49. METTLER, F. A., ADES, H. W., LIPMAN, E.: The extrapyramidal system: An experimental demonstration of function. *Arch. Neurol. Psychiatry* **41**:984–995, 1939.
50. MITCHELL, J. F., SZERB, J. C.: The spontaneous and evoked release of acetycholine from the caudate nucleus. In *International Congress Series 48: 12th International Physiology Congress*, Vol. 2. Amsterdam, Excerpts Medica, 1962, p. 111 (abstract 819).
51. MOORE, R. Y., BHATNAGAR, R. K., HELLER, A.: Anatomical and chemical studies of a nigroneostriatal projection in the cat. *Brain Res.* **30**:119–135, 1971.
52. MYRIANTHOPOULOS, N. C.: Huntington's chorea. *J. Med. Genet.* **3**:298–314, 1966.
53. PORTIG, P. J., VOGT, M.: Release into the cerebral ventricles of substances with possible transmitter function in the caudate nucleus. *J. Physiol.* **204**:687–715, 1969.
54. PROUT, B. J.: Supraspinal control of direct and reflex actions of the autonomic nervous system. Ph.D. thesis, University of London, 1963.
55. RIDDELL, D., SZERB, J. C.: The release *in vivo* of dopamine synthesized from labelled precursors in the caudate nucleus of the cat. *J. Neurochem.* **18**:989–1006, 1971.
56. SCHMITT, H., SCHMITT, H., BOISSIER, J. R., GUIDICELLI, J. F., FICHELLE, J.: Cardiovascular effects of 2-(2,6-dichlorophenyl-amino)-2-imidazoline hydrochloride. II. Central sympathetic structures. *Eur. J. Pharmacol.* **2**:340–346, 1968.
57. SCRIABINE, A., CLINESCHMIDT, B. V., SWEET, C. S.: Central noradrenergic control of blood pressure. *Annu. Rev. Pharmacol. Toxicol.* **16**:113–123, 1976.
58. SERCOMBE, R., AUBINEAU, P., EDVINSSON, L., MAMO, H., OWMAN, C. H., PINARD, E., SEYLAZ, J.: Neurogenic influence of local cerebral blood flow: Effect of catecholamines or sympathetic stimulation as correlated with the sympathetic innervation. *Neurology (Minneapolis)* **25**:954–963, 1975.
59. SHOULSON, I., CHASE, T. N.: Huntington's disease. *Ann. Rev. Med.* **26**:419–426, 1975.
60. SNYDER, S. H.: Catecholamines and serotonin. In Albers, R. W., Siegel, G. J., Katzman, R., Agranoff, B. W. (eds.): *Basic Neurochemistry.* Boston, Little, Brown, 1972, pp. 89–104.
61. SPEHLMANN, R.: The effects of acetylcholine and dopamine on the caudate nucleus depleted of biogenic amines. *Brain* **98**:219–230, 1975.
62. SPEHLMANN, R.: The susceptibility to acetylcholine and dopamine in the caudate nucleus of cats with chronic nigrostriatal lesions. *Trans. Am. Neurol. Assoc.* **99**:255–257, 1974.
63. SPIEGEL, B. A., SZEKELY, E. G.: Prolonged stimulation of the head of the caudate nucleus. *Arch. Neurol.* **4**:55–65, 1961.
64. STEVENS, J. R., KIM, C., MACLEAN, P. D.: Stimulation of caudate nucleus: Behavioral effects of chemical and electrical excitation. *Arch. Neurol.* **4**:47–54, 1961.
65. VAN BUREN, J. M.: Confusion and disturbance of speech from stimulation in vicinity of the head of the caudate nucleus. *J. Neurosurg.* **20**:148–157, 1963.
66. VAN BUREN, J. M.: Evidence regarding a more precise localization of the posterior frontal–caudate arrest response in man. *J. Neurosurg.* **24**:416–417, 1966.
67. VAN BUREN, J. M.: Incremental coagulation in stereotaxic surgery. *J. Neurosurg.* **24**:458–459, 1966.
68. VAN BUREN, J. M., MACCUBBIN, D. A.: A standard method of plotting loci in human depth stimulation and electrography with an estimation of errors. *Confin. Neurol.* **22**:259–264, 1962.
69. VON VOIGTLANDER, P. F., MOORE, K. E.: *In vivo* electrically evoked release of H^3-noradrenaline from cat brain (letter to the editor). *J. Pharm. Pharmacol.* **23**:381–382, 1971.
70. VON VOIGTLANDER, P. F., MOORE, K. : The release of H^3-dopamine from cat brain following electrical stim-

ulation of the substantia nigra and caudate nucleus. *Neuropharmacology* **10**:733–741, 1971.

71. WANG, G. H., BROWN, V. W.: Suprasegmental inhibition of an automatic reflex. *J. Neurophysiol.* **19**:564–572, 1956.
72. WEBSTER, K. E.: The cortico-striatal projection in the cat. *J. Anat.* **99**:329–337, 1965.
73. WENGER, G. R., STITZEL, R. E., CRAIG, C. R.: The role of biogenic amines in the reserpine-induced alteration of minimal electroshock seizure thresholds in the mouse. *Neuropharmacology* **12**:693–703, 1973.
74. WOOD, J. H., LAKE, C. R., ZIEGLER, M. G., SODE, J., BROOKS, B. R., VAN BUREN, J. M.: Cerebrospinal fluid norepinephrine alterations during electrical stimulation of cerebellar and cerebral surfaces in epileptic patients. *Neurology (Minneapolis)* **27**:716–724, 1977.
75. WOOD, J. H., LAKE, C. R., ZIEGLER, M. G., VAN BUREN, J. M.: Neurophysiological and neurochemical alterations during electrical stimulation of human caudate nucleus. *J. Neurosurg.* **46**:716–724, 1977.
76. WOOD, J. H., ZIEGLER, M. G., LAKE, C. R., SHOULSN, I., BROOKS, B. R., VAN BUREN, J. M.: Cerebrospinal fluid norepinephrine reductions in man after degeneration and electrical stimulation of the caudate nucleus. *Ann. Neurol.* **1**:94–99, 1977.
77. WOOD, J. H., ZIEGLER, M. G., LAKE, C. R., SHOULSON, I., BROOKS, B. R., VAN BUREN, J. M.: Norepinephrine reductions in cerebrospinal fluid in man after electrical stimulation and degeneration of the caudate nucleus. *Trans. Am. Neurol. Assoc.* **101**:68–72, 1976.
78. YANG, H. Y., FRATTA, W., HONG, J. S., DIGIULIO, A. M., COSTA, E.: Detection of two endorphin-like peptides in nucleus caudatus. *Neuropharmacology* **17**:433–438, 1978.
79. ZIEGLER, M. G., LAKE, C. R., FOPPEN, F. H., SHOULSON, I., KOPIN, I. J.: Norepinephrine in cerebrospinal fluid. *Brain Res.* **108**:436–440, 1976.
80. ZIEGLER, M. G., LAKE, C. R., WOOD, J. H., EBERT, M. H.: Norepinephrine in cerebrospinal fluid: Basic studies, effect of drugs and disease. In Wood, J. H. (ed.): *Neurobiology of Cerebrospinal Fluid I.* New York, Plenum Press, 1980.
81. ZIEGLER, M. G., LAKE, C. R., WOOD, H. J., BROOKS, B. R.: Relationship between cerebrospinal fluid norepinephrine and blood pressure in neurologic patients. *Clin. Exp. Hypertension* (in press).
82. ZIEGLER, M. G., WOOD, J. H., LAKE, C. R., KOPIN, I. J.: Norepinephrine and 3-methoxy-4-hydroxyphenyl glycol gradients in human cerebrospinal fluid. *Am. J. Psychiatry* **134**:565–568, 1977.

CHAPTER 13

Evaluation of Cerebrospinal Fluid γ-Aminobutyric Acid Content in Neurological and Psychiatric Disorders

Theodore A. Hare, N. V. Bala Manyam, and Bruce S. Glaeser

1. Introduction

γ-Aminobutyric acid (GABA) is present at micromolar concentrations in the brain and spinal cord[13,31,46,48,52,68,69] and thus has been subject to widespread investigation of its role as a neurotransmitter. It functions primarily as an inhibitory neurotransmitter,[30,43,47,64] and alterations of its function have been associated with a variety of neurological and mental disorders, based either on attractive hypotheses or on implicating evidence, or on both. For example, reduced GABA level has been associated with epilepsy[41,66,72] on a hypothetical basis because epilepsy can easily be pictured as resulting from a loss of inhibition, and on experimental bases because many convulsant drugs act by interfering with vitamin B_6. Vitamin B_6 deficiency itself can produce convulsions, and glutamic acid decarboxylase (GAD), the primary GABA synthetic enzyme, is particularly susceptible to reduced availability of B_6. Reduction of GABA level has been implicated in Huntington's disease (HD) through observation of reduced levels of GABA[4,53,56,67] as well as GAD[3,4,39,40,62,67] in autopsied brain tissue. Reduction of GABA function has been linked hypothetically with various psychiatric disorders,[7,32,33,57–59] again because of its action as an inhibitory neurotransmitter and because of clinical associations between schizophrenia and HD. Reduced GABA levels have also been observed in autopsied brains of schizophrenic individuals[56] and in certain autopsied brain areas from patients with Parkinson's disease.[27]

Until recently, GABA had been thought to be present exclusively in central nervous system tissue, since available analytical methods were either not sufficiently specific or not sensitive enough to detect it outside the central nervous system. Those methods included amino acid analysis, which was not sufficiently sensitive because it is based on colorimetric detection, and enzymatic procedures, which are subject to interference[12] in samples in which GABA is not a major constituent.

The efficient glial-cell reuptake of GABA had led researchers to conclude that GABA was not present at measurable levels in cerebrospinal fluid (CSF).[8,50,54,70,71] This absence of measurable GABA in CSF precluded the useful possibility that CSF

Theodore A. Hare, Ph.D. • Department of Pharmacology, Jefferson Medical College of Thomas Jefferson University, Philadelphia, Pennsylvania 19107. **N. V. Bala Manyam, M.D.** • Department of Neurology, Jefferson Medical College of Thomas Jefferson University, Philadelphia, Pennsylvania 19107; Neurology Service, Veterans Administration Medical and Regional Office Center, Wilmington, Delaware 19805. **Bruce S. Glaeser, Ph.D.** • Department of Pharmacology, Jefferson Medical College of Thomas Jefferson University, Philadelphia, Pennsylvania 19107. Dr. Glaeser's present address is: Department of Nutrition and Food Science, Massachusetts Institute of Technology, Cambridge, Massachusetts 02139.

GABA measurements could be carried out in living patients. Such measurements would perhaps yield useful information in terms of understanding and monitoring the natural history of the various diseases as well as the responses to therapeutic manipulations. CSF measurements have the advantage of not requiring autopsy material and of bypassing the use of an animal model, thus minimizing the difficulties of extrapolation to the living human disease state.

Since CSF is in intimate contact with both the brain and the spinal cord, it seemed possible that there might be present in CSF a low steady-state level of GABA that could be exploited for human benefit, especially if it reflected anomalies of central GABA function. To begin investigating this possibility, it was first necessary to develop more specific and sensitive procedures for measuring low levels of GABA in the presence of high levels of possible interfering constituents.

2. Measurement of GABA in CSF

Recent advancements of analytical technique have resulted in methods for quantifying GABA to a sensitivity approximately 1000-fold greater than that of conventional amino acid analysis. These methods have been utilized to document not only that GABA is present in CSF at nanomolar concentrations but also that it is present at detectable levels in many fluids and tissues outside the central nervous system.

2.1. Ion-Exchange/Fluorometric Method

We described, in 1975,[18] a method for measuring GABA in CSF that, like amino acid analysis, was based on separation using a highly resolving ion-exchange column and sequential elution with two lithium citrate buffers. The procedure achieved sensitivity greater than that of amino acid analysis because it used a detection system based on the fluorescence resulting from the reaction between primary amines and the fluorogenic reagent orthophthalaldehyde.[60,61]

The method was compared to full amino acid analysis, utilizing fluorescent detection rather than the usual colorimetric detection, and shown to provide equivalent resolution. Originally, the ion-exchange/fluorometric (I-E/F) method was sensitive to about 50 pmol GABA, subsequently improved to 10 pmol,[73] and required 2 hr per analysis. Bohlen *et al.*[5] reported improvement of the I-E/F procedure resulting in further increase of sensitivity to 5 pmol and decrease of time required to 1 hr. Recently, we have described[26] improvements of the I-E/F method that have resulted in an automatic three-column analyzer capable of producing an analysis every 15 min using 100-μl samples to a sensitivity of 1 pmol. This high-performance liquid chromatographic (HPLC) procedure utilizes three stainless steel microbore columns containing spherical cation-exchange resin. GABA is eluted using a single lithium citrate buffer and then detected in the flow stream after reaction with the fluorogenic reagent *o*-phthalaldehyde. Each column utilizes a timer-controlled, pneumatically actuated two-way ten-port valve having a 100-μl loop for sample injection and a 1-ml loop for regeneration. Similar four-port valves alternately direct the output from one of the columns to the reaction manifold. A flow diagram of the apparatus and typical analytical results are shown in Fig. 1 and in Fig. 2, respectively.

In a study to evaluate results obtained by the I-E/F procedure, Perry and Hansen[55] in 1976 attempted to measure GABA in CSF concentrates using conventional amino acid analysis with colorimetric detection. They utilized concentrates from as much as 82 ml of CSF and noted a peak that cochromatographed with GABA, but concluded that it was not GABA following paper electrophoresis of the "GABA fraction." Following ninhydrin spray of the electrophoretogram, they noted a component that did not migrate like GABA, and they did not observe a GABA component even though as little as 2 nmol of authentic GABA could be detected when applied to the electrophoresis strip. This study cannot be regarded as definitive, at least in light of the description presented, because (1) the non-GABA component on the electrophoretogram could have originated from the eluting buffer rather than from the sample and (2) the reported lower limit of detection (2 nmol) was determined for a pure GABA standard rather than a "spiked" internal standard and was very close to the maximum amount (approximately 8 nmol) that might have been anticipated in their sample as judged from the peak size they reported.

2.2. Radioreceptor Method

Enna and Snyder[10] in 1976 described a radioreceptor method for measuring GABA in brain based on the principle that unlabeled GABA in a sample displaces radioactive GABA bound to rat brain synaptic membranes. Subsequently, the sensitivity of

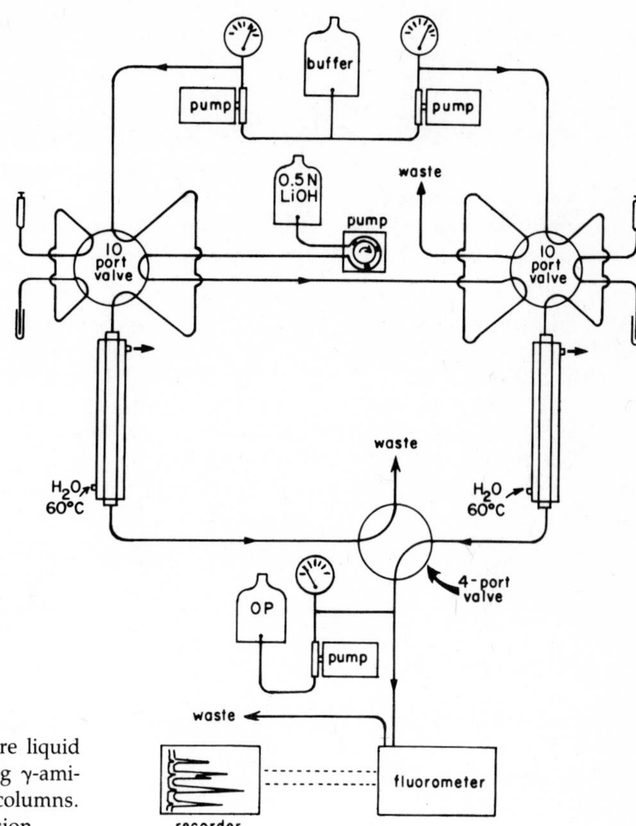

Figure 1. Flow diagram of automatic high-pressure liquid chromatographic (HPLC) procedure for measuring γ-aminobutyric acid (GABA) showing two of the three columns. Reprinted from Hare and Manyam,[26] with permission.

this procedure was improved to 10 pmol, and it was utilized to measure GABA in CSF.[12] In a collaborative study,[73] identical CSF specimens from 24 individuals were analyzed for their GABA content by both the I-E/F and the radioreceptor methods. The data from the two laboratories showed a high correlation coefficient ($r = 0.85$) when compared by linear regression analysis. This high degree of correlation between the two procedures provided strong evidence that each is specific for GABA in CSF in

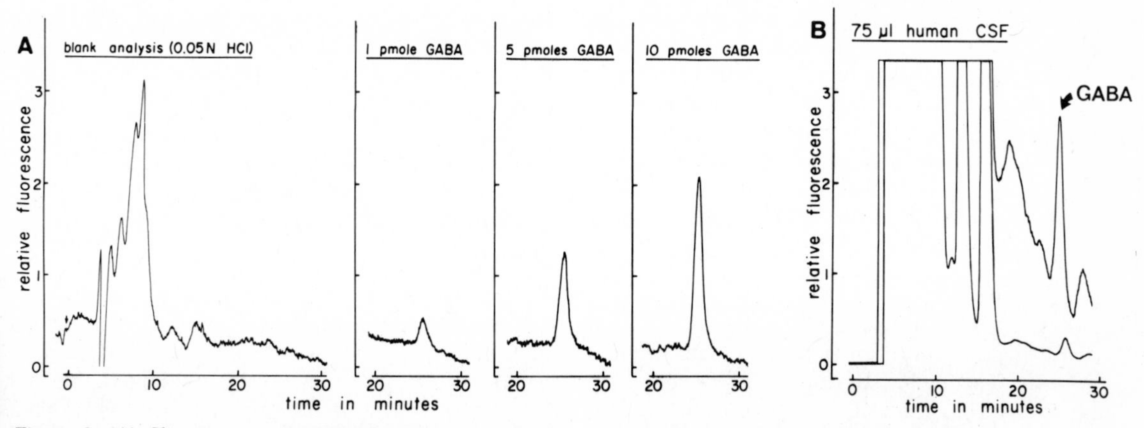

Figure 2. (A) Chromatograms of blank analysis and of 1, 5, and 10 pmol of γ-aminobutyric acid (GABA) standard by high-pressure liquid chromatographic ion-exchange fluorometric (HPLC I-E/F) method. (B) Chromatogram of 75 μl of human CSF by HPLC I-E/F method. Reprinted from Hare and Manyam,[26] with permission.

view of the facts that the underlying bases of the two procedures are fundamentally different and that each has been shown to be specific for GABA by other comparisons.

2.3. Gas Chromatographic–Mass Spectrometric Methods

In 1977, two methods for measuring GABA in CSF using gas chromatography–mass spectrometry (GC-MS) were described in the literature. Huizinga *et al.*[28,29] utilized a GC-MS method to show human CSF GABA levels over a range very similar to the range of the original study using the I-E/F procedure.[19] Enna *et al.*[11] showed direct correlation ($r = 0.82$) between results obtained by the radioreceptor method and by a GC-MS method. More recently, the presence of GABA in CSF was unequivocally confirmed through a report of complete mass-spectral data by Faull *et al.*[14]

Even though the I-E/F, radioreceptor, and GC-MS methods are fundamentally different analytical procedures, they have produced remarkably consistent data showing the presence of GABA in human CSF in the nanomolar range. These three procedures have also produced similar data when used to study patient CSF specimens. For example, the range of 145–913 pmol/ml for CSF from patients with various neurological disorders seen in our initial study[19] is very close to the range of 155–815 pmol/ml determined subsequently by GC-MS measurement for a similar group of patients.[28,29] Analyses in our laboratory have shown a mean GABA level of 239 ± 91 (S.D.) pmol/ml in CSF from 20 normal volunteers[74] and 239 ± 76 pmol/ml in CSF from 19 neurologically normal controls.[36,37] These values are similar to a mean control value of 220 ± 81 pmol/ml reported by Bohlen *et al.*[5] for a group of 38 patients with intervertebral disc disorders. Likewise, the I-E/F and radioreceptor procedures have produced consistent data showing reduced GABA levels in CSF from patients with Huntington's disease.[11,19,35,36]

2.4. Enzymatic Method

Achar *et al.*[1] in 1976 also reported studies of GABA in CSF using a modification of an enzymatic method previously described by Baxter[2] for measurement in brain tissue. They found CSF GABA levels ranging from undetectable to 1.4 nmol/ml in patients with various neurological disorders and were unable to detect GABA in any of 19 control individuals. The results obtained with CSF measured by this method have been criticized[5] as being difficult to interpret because the modifications of the original method were not described. In addition, even though the method itself was stated to be sensitive to 2 pmol when pure GABA standards were used, the actual limit of sensitivity for patient specimens cannot be determined from the presentation because the use of internal standards or "blank" analyses was not reported. It is difficult, therefore, to judge the significance of these data, especially in instances where no GABA was detected, in that the lower limit of detection is ambiguous. The lack of experimental detail is particularly unfortunate in view of the report[12] that a substance interfering with the enzymatic procedure is present in CSF.

A similar enzymatic procedure was used by another group of investigators[33] to measure CSF GABA levels in schizophrenic and control individuals. In contrast to the results of the previous investigation, this study found a mean (± S.D.) level of 301 ± 65 pmol/ml in the CSF from the nine control individuals. This study is also subject to criticism in that the use of internal standards or "blank" analyses was not described; in addition, the samples were subjected to acid conditions and elevated temperatures prior to the enzymatic assay. These conditions are similar to conditions that have been shown to produce decomposition of homocarnosine and elevation of CSF GABA levels,[22] a fact that could explain the difference of results between these two studies (see further discussion below).

3. Basic Considerations

Before the study of GABA in clinical specimens can be reliably carried out on a large scale, certain basic questions must be answered. One of these questions is: under what conditions of collection and storage is the CSF GABA level stable? Knowledge of stability during storage is essential before drawing conclusions about patient specimens and also when comparing experimental results from the various laboratories.

Furthermore, to assess the significance of measuring GABA in lumbar CSF of patients with various neurological and mental disorders, it is important to answer the question: to what extent do altered CSF levels reflect alterations of brain levels? One initial method of approaching this question is to determine whether a gradient exists between lumbar CSF and CSF closer to the brain. The existence of such a gradient would argue that at least part of the lumbar CSF GABA comes from the brain, since in that case glial reuptake[38,63] would be expected to produce a continually decreasing gradient as the CSF moves further from brain. Lack of a gradient would suggest that GABA in lumbar CSF comes from the spinal cord, since release and reuptake would be taking place simultaneously, resulting in a more steady-state situation. Another approach to evaluating the reflection of brain levels in CSF would be to directly measure levels in the two areas and to correlate the influence on CSF levels of drugs or surgical procedures known to specifically alter brain GABA levels.

3.1. Instability of GABA Concentration in CSF during Improper Collection and Storage

In a series of studies[20,22] to measure the stability of GABA during storage, GABA levels were measured in CSF specimens before and after storage under various conditions. The results are summarized in Table 1 for untreated samples and in Table 2 for samples deproteinized with sulfosalicylic acid prior to storage.

These studies show that in CSF stored under frozen conditions, the GABA level remained stable at −20° in untreated aliquots and at −70° in both the deproteinized and the untreated samples. On the other hand, storage of the deproteinized samples at −20°C resulted in a statistically significant ($P < 0.001$) 2-fold increase of GABA level during an 11-month period.

In the untreated samples maintained under nonfrozen conditions, the level of GABA approximately doubled during 2 hr at room temperature, although the increase was not significant during 10 min at

Table 1. Stability of GABA[a] Concentration in Human CSF during Storage: Untreated Specimens

Conditions	Time	Number	Fold increase ± S.E.
−70°C	25 months	12	1.03 ± 0.04
−70°C	40 months	8	1.09 ± 0.11
−20°C	16 months	7	1.04 ± 0.07
Room temp.	10 min	7	1.03 ± 0.07
Room temp.	120 min	7	2.32 ± 0.30[b]
2–4°C	120 min	Pool[c]	1.03

[a] GABA, γ-Aminobutyric acid.
[b] $p < 0.005$, paired Student's *t* test.
[c] Pool of ten specimens.

Table 2. Stability of GABA[a] Concentrations in Human CSF during Storage: Deproteinized Specimens

Conditions	Time	Number	Fold increase ± S.E.
−70°C	11 months	5	0.98 ± 0.02
−20°C	11 months	13	1.97 ± 0.22[b]
Room temp.	24 hr	5	1.00 ± 0.04
Room temp.	49 hr	5	1.04 ± 0.03
Room temp.	25 hr	Pool[c]	1.01
Room temp.	3 weeks	Pool[c]	1.99
2–4°C	3 weeks	Pool[c]	1.35

[a] GABA, γ-Aminobutyric acid.
[b] $p < 0.001$, paired Student's *t* test.
[c] Pool of ten specimens.

room temperature or during 2 hr at 2–4°C. In the deproteinized samples maintained under nonfrozen conditions, the GABA level was stable at room temperature for 2 days, but increased during 3 weeks both at room temperature and at 2–4°C.

Bohlen *et al.*[5] have also reported an increase of GABA in CSF as a function of time at room temperature. Their results are generally consistent with those summarized above, except that: (1) they concluded that the GABA level is stable in sulfosalicylic-acid-deproteinized samples for at least 3 weeks at 4°C; (2) they suggested that the increase of GABA level may result from enzymatic generation of GABA *de novo* and therefore recommended that samples be deproteinized immediately prior to storage; and (3) they suggested that storing untreated CSF at −20°C may result in formation of new GABA.

Our studies show stability of GABA levels in untreated CSF at −20°C but instability in deproteinized CSF at −20°C and demonstrate that a nonenzymatic mechanism must account, at least in part, for the increase of CSF GABA levels during certain conditions of storage. Thus, the increase may not be entirely dependent on the formation of new GABA, but alternatively may result from the instability of other constituents, perhaps peptides such as homocarnosine, which would produce GABA as a breakdown product. Homocarnosine, a dipeptide of histidine and GABA, is known to be present in human[51] and rat[6] CSF at a level of 2–3 nmol/ml, which is about 10-fold higher than that of GABA in CSF. The fact that homocarnosine spontaneously hydrolyzes during acid storage was demonstrated by subjecting homocarnosine standards to sensitive amino acid analysis following storage in 0.1 N HCl under various conditions.[22] Results of analyses of the "fresh" standard, "6 days at room temperature" standard, and "22 months at −20°C" standard are presented in Table 3. Homocarnosine was the major peak in each analysis, being present at concentrations of 100, 100, and 90.6 nmol/ml, respectively. The data revealed a 22-fold increase of GABA and a 19-fold increase of histidine during 22 months at −20°C in 0.1 N HCl. During 6 days at room temperature, the GABA and histidine both increased 1.7-fold. The results of this study lead to the conclusion that homocarnosine is hydrolyzed to its constituent amino acids during storage in dilute acid at room temperature and at −20°C, a fact that may account for the increase of GABA in CSF during storage, at least in the deproteinized specimens.

The studies of the stability of CSF GABA level are significant from several standpoints. The data demonstrate that CSF specimens must be frozen immediately after collection if reliable GABA data are to be obtained from CSF specimens. The cumulative tolerable time that a fresh CSF specimen can be exposed at room temperature is less than 10 min, although a total of 100 min could be tolerated if the specimen were chilled with ice (4°C). Deproteinization should be carried out quickly and at low temperature just prior to analysis and then the deproteinized sample should be stored at −70°C if subsequent analysis is anticipated. Storage of the fresh sample at −70°C is preferable to storage at higher temperatures, but if specimens are to be deproteinized immediately after collection, storage at −20°C or higher would result in an elevation of the GABA level. Except for those specimens of our initial report,[19] the standard procedure utilized in our laboratory has been to freeze specimens in dry ice immediately after collection and store them at −70°C (or lower). Prior to analysis, the samples are thawed and deproteinized within 3–5 min. In some cases, use of less stringent conditions could produce artifactually elevated GABA levels. As discussed in other sections of this chapter, the instability of CSF GABA levels under certain conditions appears to explain discrepancies of data among some of the laboratories carrying out CSF GABA measurements.

Table 3. Results of Amino Acid Analyses of 100 nmol/ml Homocarnosine Standards in 0.1 N HCl during Storage

Conditions	GABA[a] (nmol/ml)	Histidine (nmol/ml)	Homocarnosine (nmol/ml)
Fresh	0.65	0.67	100
6 days at room temperature	1.13	1.15	100
22 months at −20°C	14.6	12.9	90.6

[a] GABA, γ-Aminobutyric acid.

The observation that GABA levels are stable for several days at room temperature in deproteinized CSF is also significant in a more practical analytical sense because a change of level would not be expected during storage on an automatic sampler when these measurements are carried out on a large scale.

3.2. CSF GABA Concentration Gradients

It has been extensively established that GABA is present in mammalian brain and spinal cord at micromolar concentrations.[13,31,48,52] Presumably, the nanomolar concentrations of GABA seen in CSF arise either from the spinal cord or brain, or both, since CSF is in direct communication with both areas. Efficient reuptake of GABA by glial and neuronal cells has also been well established,[38,63] a fact that likely accounts for the relatively low level of GABA in CSF.

It seems reasonable to assume that during lumbar puncture, the first aliquots drawn would be lumbar CSF, whereas subsequent aliquots would be more nearly representative of CSF from higher levels in the spinal canal. Thus, analysis of sequential aliquots of CSF should provide evidence of whether or not a rostrocaudal concentration gradient of GABA level exists. The amount of CSF in the lumbar region below the level of the tap could mix with that from higher levels, but this would tend only to decrease the apparent magnitude of an existing gradient and would not be expected to artificially create a gradient if none existed physiologically.

Enna *et al.*[12] reported GABA levels in CSF from different paired fractions obtained from 13 patients. Their results showed the mean values from the 24th and 40th milliliters to be 1.20- and 1.48-fold higher, respectively, than that of the 8th milliliter, reflecting a rate of increase of about 1.5%/ml. Studies in our laboratory have also demonstrated the existence of a similar rostrocaudal CSF GABA gradient. In one of these studies,[20] lumbar CSF from 5 patients with normal-pressure hydrocephalus showed a mean (± S.D.) GABA level of 124 ± 56 pmol/ml, while ventricular CSF from 2 normal-pressure hydrocephalus patients showed higher values, i.e., 301 and

250 pmol/ml. In another more detailed study,[26] 1-ml aliquots were sequentially obtained from the first 20 ml during lumbar puncture from 7 patients. The results of analysis of these aliquots are summarized in Fig. 3, showing an increasing gradient in the sequential CSF aliquots. Analysis of the data by linear regression revealed a correlation coefficient of 0.94 and a mean rate of increase of 2%/ml. The magnitude of this rostrocaudal GABA concentration gradient appears to be more pronounced in females.[26a]

3.3. Is There a Relationship between Brain and CSF GABA Concentrations?

The existence of a gradient of GABA level in the CSF provides some indication that there is a relationship between brain and lumbar CSF GABA concentrations. Recently, more direct evidence of this relationship has been achieved through comparison of brain and ventricular CSF GABA concentrations in rats and cats before and after treatment with γ-vinyl GABA and γ-acetylenic GABA,[6] drugs that are known to produce elevation of brain GABA levels. Both drugs produced elevation of brain as well as ventricular CSF GABA levels; moreover, the more detailed study with γ-vinyl GABA revealed a dose-dependent increase of the CSF GABA concentration as well as a high correlation ($r = 0.92$) between the brain GABA levels and the log of the CSF GABA levels. In that same study, GABA was measured following acid hydrolysis of the CSF specimens, and it was shown that "conjugated" (presumably peptide) GABA responded similarly with a high linear correlation ($r = 0.84$) to the brain GABA concentration. In another study[34] using dogs, a dose-dependent increase of cisternal CSF GABA was noted following γ-acetylenic GABA treatment, and increases were also observed in cisternal CSF following treatment with γ-vinyl GABA and sodium valproate.

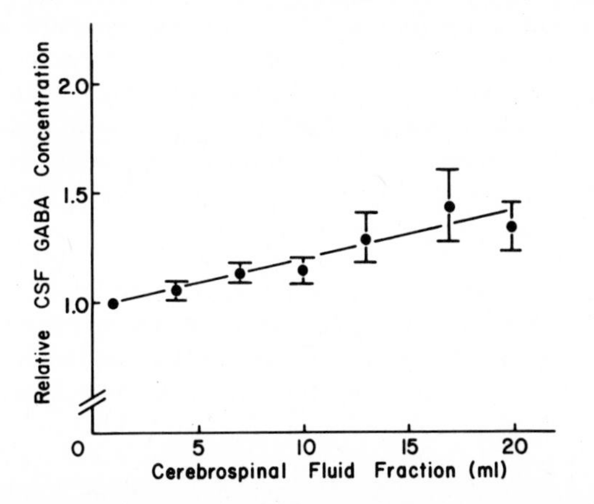

Figure 3. GABA levels in sequential aliquots of human CSF of seven patients showing an increasing gradient of GABA concentration. Data points show mean values ± S.E. Line is result of linear regression analysis, which revealed correlation coefficient of 0.94 and mean rate of increase of 2%/ml CSF. From Hare and Manyam.[26]

The fact that these brain-GABA-elevating drugs produce similar effects on brain and ventricular or cisternal CSF provides evidence that CSF levels are a reflection of brain levels. When viewed in light of the gradient of GABA in the CSF between the lumbar area and that closer to the brain, it seems likely that lumbar levels would also be a reflection of brain levels. However, in view of the efficient glial-cell reuptake of GABA, it is still possible that the drug-induced alterations observed in ventricular or cisternal CSF would not have been evident in lumbar CSF. In addition, the drugs in the above studies were administered peripherally and therefore would be expected to alter spinal cord as well as brain GABA level. If both spinal cord and brain levels were similarly altered, even measuring lumbar concentrations following drug treatment would not definitely establish whether or not the brain is a source of GABA in lumbar CSF. Recent studies[76] demonstrated reduced lumbar CSF GABA levels in patients with bulbar or spinal cord degenerative disorders. The spinal cord therefore appears to contribute to the GABA content of lumbar CSF. Thus, even though current data are consistent with the hypothesis that lumbar CSF GABA reflects brain GABA content, unequivocal evidence must await studies in which lumbar CSF GABA measurements are carried out following treatments such as stereotaxically placed lesions or drug applications that specifically influence brain levels without influencing levels in the spinal cord or areas of the periphery.

3.4. Diurnal or Circadian Variation of CSF GABA Level

Another basic question that must be approached has to do with whether or not there is diurnal or circadian variation of CSF GABA level. A recent report[49] suggested that CSF GABA concentrations are higher during the day than during the night, but the results are ambiguous by virtue of the fact that the values fluctuated widely and irregularly. In this report, the unsubstantiated statement was made that "CSF GABA appears to be stable at room temperature for at least 24 hr." This statement is in conflict with the data of Grossman *et al.*[20,22] and Bohlen *et al.*,[5] which show a substantial increase of GABA level in CSF maintained at room temperature. In the aforementioned study of daily variation,[49] CSF samples were obtained through catheters under conditions that resulted in their being exposed to room temperature for approximately 2 hr and then at 4°C for variable periods up to 24 hr prior to being frozen at −20°C. It is possible that exposure to these conditions accounts in large measure for the variations seen during this study.

4. CSF GABA Levels in Normal Individuals

To evaluate the use of lumbar CSF GABA measurements for study of various neurological disorders, it is first essential to establish the level in normal individuals. Several studies of GABA levels in CSF from control and normal individuals have been reported. In our original study of CSF GABA levels,[19] the diseased control population consisted of 9 individuals with various neurological disorders for whom GABA involvement was not suspected. The mean (± S.D.) CSF GABA level for this population was 534 ± 266 pmol/ml. It was later concluded that the GABA concentrations in these samples had probably become elevated during storage because these control samples, unlike those from the experimental individuals, were stored for 10 months at −20°C following deproteinization—a condition now known[22] to produce an approximate doubling of the GABA concentration. Enna *et al.*[11] reported a level of 230 ± 120 (S.D.) pmol/ml in CSF from 26 diseased control individuals (mean age 45 years, range 7–80 years) who were patients having a variety of neurological and psychiatric complaints. For their study, the aliquot utilized was from the 6th to the 10th milliliters drawn. The lumbar punctures were performed between 5 P.M. and 8 P.M. following 24 hr of bedrest, and the specimens were immediately placed in a −20°C freezer.

Two studies of GABA in CSF from normal controls using the enzymatic procedure have been reported. Achar *et al.*[1] were unable to detect GABA in CSF from 19 controls who had been found to have no organic neurological disease. Their analyses were carried out on the first milliliter obtained, which was frozen immediately and stored at −85°C. On the other hand, Lichtshtein *et al.*[33] found a mean (± S.D.) GABA level of 301 ± 65 pmol/ml in CSF from 9 normal individuals. Difficulties in the interpretation of data from the enzymatic assay have been discussed above, and in addition, during the latter study, samples were exposed to conditions that have been shown to produce elevation of the GABA level.

Bohlen *et al.*[5] reported a level of 220 ± 81 pmol/ml in lumbar CSF from 38 patients with intervertebral disc disorders without clinical, biochemical, or radiological evidence of central nervous system disease. The specimens were the first fractions drawn, which were deproteinized immediately with sulfosalicylic acid and stored for less than 3 days at 4°C prior to analysis.

We have analyzed CSF samples obtained from two different populations of normal individuals, collected and stored under conditions known to not alter CSF GABA concentrations. One of these studies utilized specimens from 19 persons (mean age 42 ± 15 years), of whom 3 were normal volunteers and 16 were patients who had no evidence of neurological or psychiatric disorders.[36,37] CSF used in the other study[74] was from 20 normal volunteers (mean age 32 ± 14 years). In the former study, specimens were from the first 12 ml and were immediately frozen in dry ice or at −20°C prior to storage at −80°C. In the latter study, subjects were placed on a low-monoamine diet 2 weeks prior to sampling, and oral intake and physical activity were avoided 18 hr preceding CSF collection. The lumbar punctures were performed at 9 A.M., and the specimens analyzed were from pools consisting of the 13–26 ml of CSF obtained. The CSF was immediately placed on ice and then frozen at −70°C within 30 min. Analysis of the specimens showed GABA to be present at 239 ± 76 and 239 ± 91 pmol/ml in the former and latter study, respectively. In the report[36] of one of these studies, it was noted that males showed a relatively higher level than females ($P < 0.05$), and there seemed to be a decrease of GABA level with advancing age. More definitive indication of these differences becomes apparent when the data from these two studies are combined.

4.1. CSF GABA Concentrations in Normal Individuals with Respect to Age and Sex

These data are summarized in Figs. 4 and 5. The mean CSF GABA level in females was 228 ± 84 pmol/ml ($N = 13$, age 37 ± 11 years) and that in males was 245 ± 84 pmol/ml ($N = 26$, age 37 ± 17 years), values which are not significantly different. When the data were analyzed by linear regression, there was a decrease with age for the entire population as well as for both the male and female subgroups; however, the correlation coefficient was

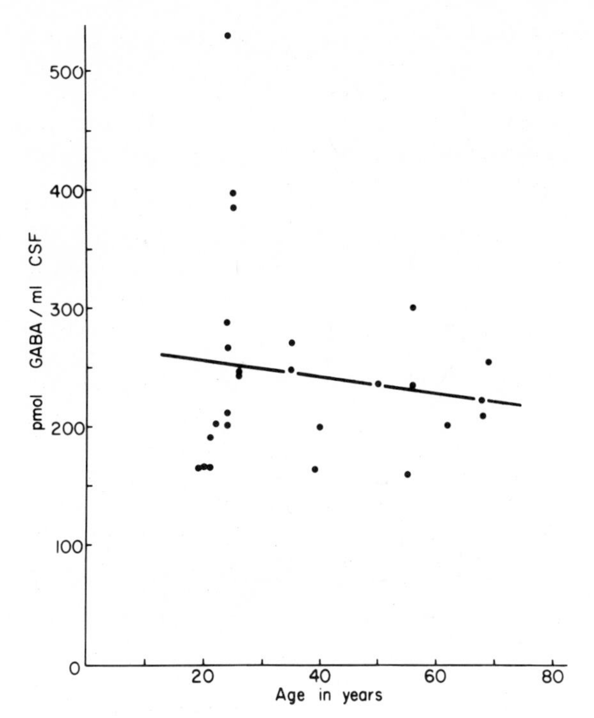

Figure 4. Combined data from two studies[37,74] showing γ-aminobutyric acid (GABA) levels in CSF from normal males as function of age. Line shows result of linear regression analysis, which did not reveal a significant correlation coefficient.

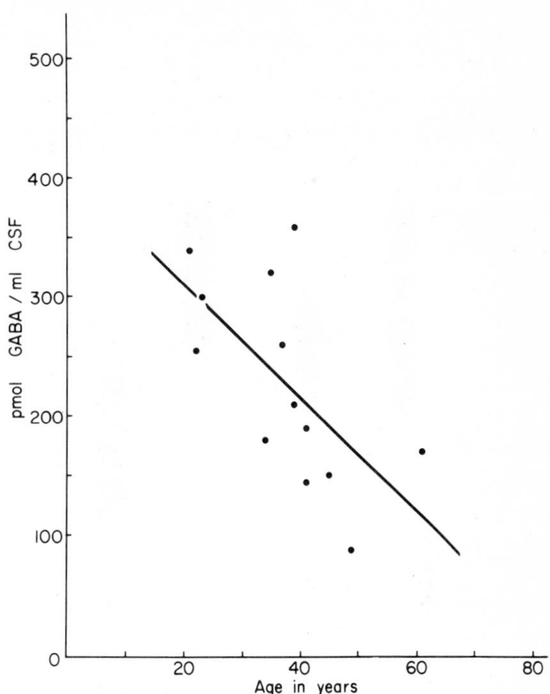

Figure 5. Combined data from two studies[37,74] showing γ-aminobutyric acid (GABA) levels in CSF from normal females as function of age. Line shows result of linear regression analysis, which revealed correlation coefficient of 0.63.

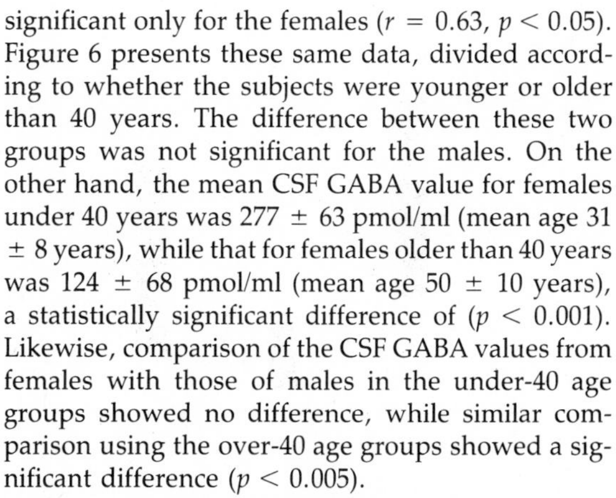

significant only for the females ($r = 0.63$, $p < 0.05$). Figure 6 presents these same data, divided according to whether the subjects were younger or older than 40 years. The difference between these two groups was not significant for the males. On the other hand, the mean CSF GABA value for females under 40 years was 277 ± 63 pmol/ml (mean age 31 ± 8 years), while that for females older than 40 years was 124 ± 68 pmol/ml (mean age 50 ± 10 years), a statistically significant difference of ($p < 0.001$). Likewise, comparison of the CSF GABA values from females with those of males in the under-40 age groups showed no difference, while similar comparison using the over-40 age groups showed a significant difference ($p < 0.005$).

Thus, it seems that normal adult males have a mean CSF GABA level of about 250 pmol, which may decrease somewhat with age. Females under age 40 seem to have the same general range of CSF GABA values as do males; however, this decreases substantially in older females. This observation suggests a possible hormonal association by virtue of the fact that menopause corresponds roughly to the age at which the decrease of CSF GABA was noted. In practical terms, the significance of this observation is that for any clinical research studies, an appropriate reference point must be established and differences of age and sex must be taken into consideration.[26a]

5. CSF GABA Levels in Patients with Huntington's Disease and Those "at Risk for" Huntington's Disease

The involvement of GABA in HD is of substantial interest because GABA has been shown to be reduced in the brains of patients who had died of HD.[4,53,56,67] In agreement with this observation is the fact that the enzyme primarily involved in the synthesis of GABA, glutamic acid decarboxylase, is also reduced in the brains of HD patients.[3,4,39,62,67] The deficiency of brain GABA could be more readily exploited for benefit if it were also reflected and could be detected in the CSF, since CSF can be obtained from living patients. The measurements could pro-

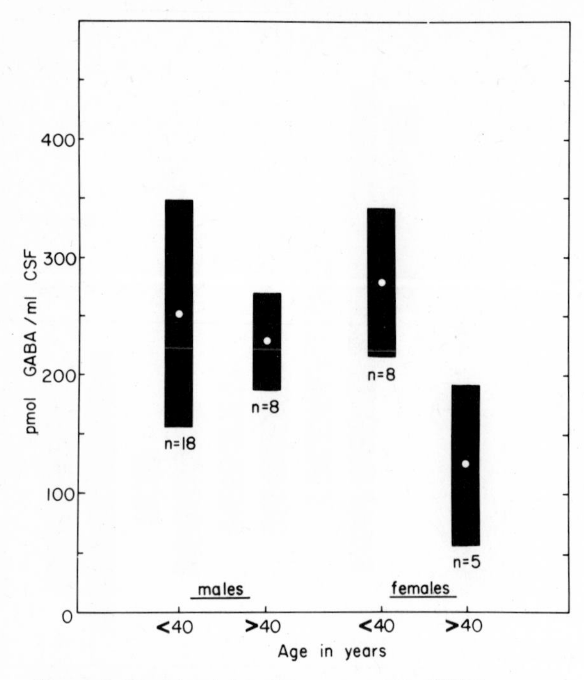

Figure 6. Combined data from two studies[37,74] showing γ-aminobutyric acid (GABA) levels in CSF from normal males and females under age 40 years compared with levels from those over age 40 years. Points show mean values; bars show ± S.D. Results of analysis by Student's *t* test showed GABA level in CSF from females older than 40 years to be significantly lower than that from females younger than 40 years ($p < 0.001$) and males older than 40 years ($p < 0.005$).

vide diagnostic benefit, as well as therapeutic usefulness in terms of monitoring response to drug treatment. In the case of HD, there is also potential for using such measurements as the basis for presymptomatic detection and perhaps even presymptomatic treatment if the reduced GABA level is present in the subjects and reflected in their CSF prior to the appearance of symptoms. Four studies have been reported in which GABA measurements were carried out on CSF from patients with HD and from control individuals.[11,19,35,36,42] All these studies concluded that the GABA level in CSF from the HD patients was significantly lower than that of the respective control group. The values from our studies[19,35,36] are shown in Fig. 7, along with that of normal control individuals. The difference between these two groups is significant ($p < 0.001$). Also included in Fig. 7 is the distribution of GABA levels in CSF from individuals "at risk for" HD. The level for some of the "at-risk" individuals was within the normal range, while for others it was below the normal range. This result is compatible with the hypothesis that CSF GABA levels are reduced in HD prior to the appearance of symptoms. Of course, definitive conclusions cannot be drawn until the appearance of symptoms can be correlated to the GABA level prior to onset. One other study of GABA in CSF from "at-risk" individuals has been reported.[42] In this study, an HPLC procedure was used to measure GABA, and the results were confirmed by GC-MS analysis. The results were reported as follows (in pmol/ml): normal subjects ($N = 24$), 753 ± 62; HD ($N = 14$), 244 ± 40; "at risk for" HD ($N = 37$), 235 ± 27. The HD and "at-risk" group values were both below that of the normal controls, but not different from each other. This result is not consistent with the statistical probability that half the "at-risk" population should fall within the normal range. A possible explanation of the fact that the reported CSF GABA level in the normals was more than 2-fold higher than that reported by several other investigators could be that the GABA level was elevated during collection or storage or both, although these conditions were not specified. We have measured GABA in CSF from another pop-

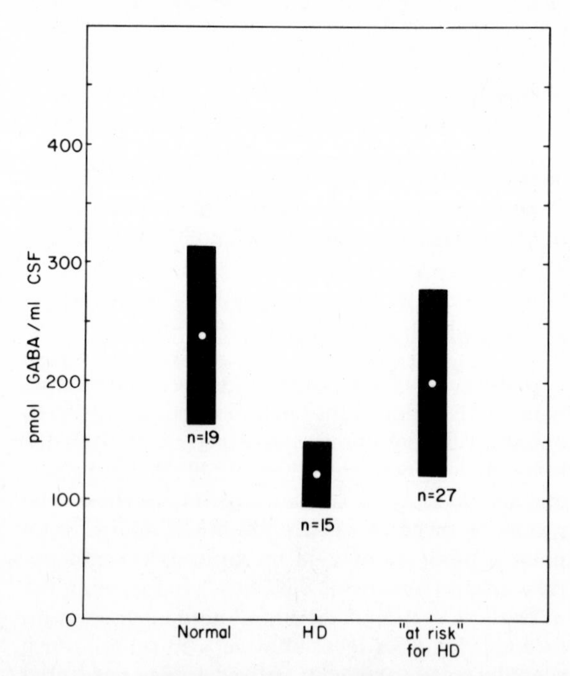

Figure 7. γ-Aminobutyric acid (GABA) levels in CSF of normal individuals, individuals with Huntington's disease (HD), and individuals at risk for HD.[19,35–37] Points show mean values; bars show ± S.D. Analysis by Student's *t* test showed values from HD patients to be significantly lower ($p < 0.001$) than that of normals.

ulation ($N = 12$) of normal volunteers. These specimens had been allowed to stand at room temperature for up to 24 hr prior to freezing, and the mean GABA level was 514 ± 159 pmol/ml.[20] The difference between this level and the lower levels seen for other populations of normal individuals is probably accounted for by the storage conditions in view of the elevation of GABA seen during storage at room temperature.[5,22]

6. CSF GABA Levels in Patients with Epilepsy

GABA levels have also been studied in CSF from patients with epilepsy. In one study,[12] lumbar CSF GABA concentrations were seen to be lower than that of controls, but the number of subjects was small and the differences were not statistically significant. In two other studies[37,74] involving epileptic patients and normal controls, the reduction was shown to be significant at the $p < 0.025$ and $p < 0.001$ levels, respectively. For these two studies,[37,74] there was virtually no difference between the manner of specimen collection and storage, and the analyses were carried out in the same laboratory. When data from these two studies are combined, the mean value (± S.D.) for the 29 epileptic patients is found to be 146 ± 60 pmol/ml (mean age 32 ± 13 years), that for the 39 normal controls is 239 ± 83 pmol/ml (mean age 37 ± 15 yr), and the difference is statistically significant ($p < 0.001$). Chapter 19 by Wood and Brooks contains further discussion of CSF GABA alterations in seizure disorders.

7. CSF GABA Levels in Patients with Other Neurological Disorders

GABA levels have been studied in CSF from patients with various other neurological disorders. In our study,[37] levels were measured in CSF from 109 individuals, of whom 19 were established to be normal and 90 had various neurological disorders. The results of the analyses are presented in Fig. 8. Of these patients, only the acute hypoxic encephalopathy group ($N = 6$) showed levels higher than the normals ($p < 0.05$). The central nervous system infections group ($N = 9$) showed a mean value essentially equal to normal, but with a larger standard deviation.

Individuals with stroke ($N = 16$) showed a level somewhat lower than, but similar to, that for the normal controls, and the difference was not statistically significant. In 5 of these patients, the lumbar

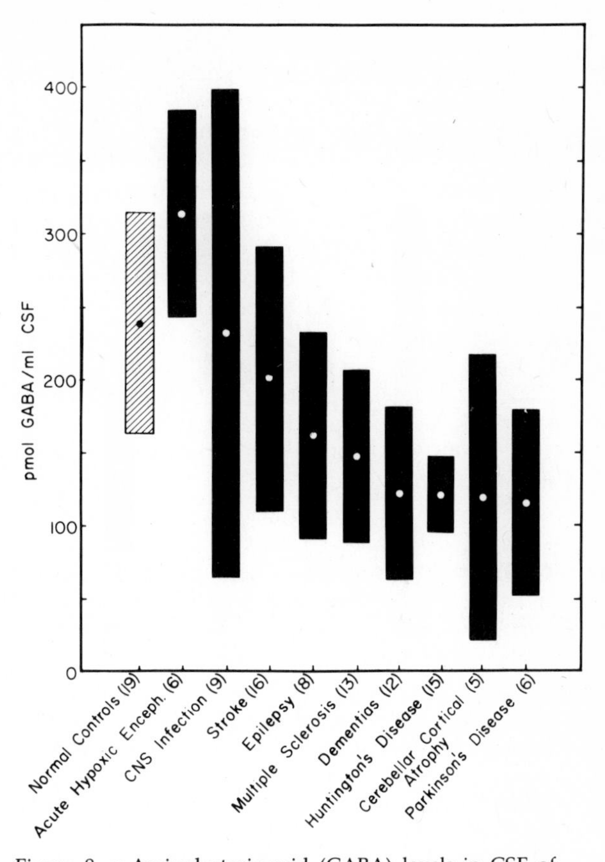

Figure 8. γ-Aminobutyric acid (GABA) levels in CSF of normal individuals and patients with various neurological disorders.[37] Points show mean values; bars show ± S.D.; numbers in parentheses are numbers of subjects.

puncture was performed within 4 weeks of the onset of symptoms, but the GABA level (199 ± 99 pmol/ml) was not different from that of the 11 patients for whom the interval was greater than 4 weeks (202 ± 92 pmol/ml). These data conflict with previous reports by investigators who used the enzymatic assay and observed an increased GABA level in CSF of about half of 41 patients following stroke.[1,71a]

As discussed in previous sections of this chapter, the CSF GABA values from patients with HD and epilepsy were significantly lower than those of the normal controls ($p < 0.001$ and 0.025, respectively). CSF from patients with multiple sclerosis ($N = 13$) showed a mean GABA level of 148 ± 58 pmol/ml, which was significantly lower ($p < 0.005$) than that of the normals. In the group of patients with various types of dementia ($N = 12$), the mean GABA level in CSF was 123 ± 59 pmol/ml (mean age 59 ± 7 yr) and was significantly lower than that of the controls ($p < 0.001$). Of these dementia cases, 5 were Alzheimer's disease, 1 senile dementia, 1 normal-pressure hydrocephalus, and the remaining 5 had dementia associated with chronic alcoholism. No statistically significant difference in CSF GABA levels was seen between dementia of chronic alcoholism (mean GABA 124 ± 47 pmol/ml) and Alzheimer's disease (mean GABA 144 ± 72 pmol/ml). These data from Alzheimer patients are in agreement with the report of Enna *et al.*[11] who used the radioreceptor assay and found a reduced level in 3 patients with Alzheimer's disease. The data from CSF of 5 patients with cerebellar cortical atrophy (120 ± 99 pmol/ml) showed a particularly large standard deviation, and the CSF GABA values were significantly different ($p < 0.01$) from those of the normal controls.

CSF from 6 patients with Parkinson's disease (PD) showed GABA values (115 ± 65 pmol/ml) significantly lower than those of the normals ($P < 0.005$). In this instance, there may be a drug effect influencing GABA levels because 3 of the patients were untreated and showed a mean GABA level of 68 ± 19 pmol/ml, whereas the other 3 patients had received L-DOPA plus carbidopa for at least 3 years prior to the lumbar puncture. The small number of samples in the two subgroups eliminates the possibility of drawing firm conclusions, but the data may be consistent with the report that GAD is deficient in certain areas of brain from untreated patients with PD but nearly normal in brains of PD patients who had received long-term treatment with L-DOPA.[27] The data from the treated PD patients are similar to those reported by Enna *et al.*[11] using the radioreceptor assay, who studied CSF from 5 PD patients receiving L-DOPA plus carbidopa treatment. These investigators found a mean (± S.D.) level of 190 ± 48 pmol/ml in the CSF of the PD patients, a value that was lower than but not significantly different from that of their controls (220 ± 120 pmol/ml). Further discussions of the CSF GABA concentrations in patients with PD are contained in Chapters 14 and 15.

8. CSF GABA Levels in Patients with Psychiatric Disorders

GABA has been hypothesized to be reduced in brains of patients with schizophrenia on the basis of its role as an inhibitory neurotransmitter[57–59] and because of the reported association between HD and schizophrenia.[32] A recent report[56] of GABA levels from autopsied brains has confirmed the reduced level of GABA in brains of HD patients and also showed an equally marked reduction in the brains of the schizophrenic individuals when compared to that of the normal controls. To consider the possibility that chronic treatment with antipsychotic drugs could account for the reduced brain GABA content in both these patient groups, rats were chronically treated with large doses of chlorpromazine or haloperidol in this same study, but these drugs did not cause a decrease of mesolimbic GABA in comparison with control animals.

A report[33] comparing normal and untreated schizophrenic CSF GABA levels as measured by an enzymatic procedure concluded that the levels were not significantly different, although the mean GABA level was lower for the schizophrenic group. In that study, CSF GABA levels were also measured in some of the patients after neuroleptic treatment, and it was concluded that the treatment produced a significant, although variable, decline in GABA levels. Difficulties with this study, as discussed earlier, include the fact that the reliability of the enzymatic GABA assay procedure is controversial. This assay procedure has been criticized as being subject to interfering constituents in CSF[12] and has produced data inconsistent with the I-E/F, radioreceptor, and GC-MS procedures in other population groups. An additional possible problem stems from the fact that the samples were deproteinized with trichloracetic acid, which was removed by chromatography on a Dowex-1–acetate column eluted with 0.5 N acetic acid. The eluates, still in the 0.5 N acetic acid, were then subjected to drying in a rotary evaporator. During this treatment, which preceded the enzymatic assay for GABA, the samples would have been exposed to acid conditions and elevated temperature, which could have resulted in artifactually elevated GABA levels in view of the observation[22] that homocarnosine decomposes in dilute acid even at reduced temperatures to release GABA and histidine. Finally, although it was not specified, it seems appropriate to surmise that the lower limit of sensitivity for the assay reflected about 200 pmol GABA/ml of CSF, since the 6 lowest of the 17 values for the schizophrenia were reported to be 200 pmol/ml.

9. Studies of GABA Levels in Blood, Amniotic Fluid, and Peripheral Tissues

There is also current interest in extending the use of the new GABA analytical procedure to studies of other areas where GABA had previously not been detected. For example, current studies have revealed the presence of GABA in blood[15,21,23] and various peripheral organs,[16,17,65] especially the pancreas.

The two studies of GABA in blood are in general agreement in concluding that GABA is present at nanomolar concentrations in mammalian blood and that the concentration in formed elements is higher than that in the plasma. One of these studies, using the radioreceptor GABA assay procedure,[15] showed GABA to be present in blood at levels ranging from 500 to 1200 pmol/ml in eight mammalian species with human values at a level of about 900 pmol/ml. Plasma values were one fourth to one half the whole-blood values. In that study, correlation ($r = 0.94$) was shown between results by the radioreceptor procedure and a GC-MS procedure. The other study,[21,23] using the I-E/F procedure, showed a level of 488 ± 166 (S.D.) pmol/ml in fasting blood from 18 normal individuals. In the latter study, the mean blood GABA level in the females ($N = 7$) was 401 ± 106 pmol/ml, while that in the males ($N = 11$) was 544 ± 177 pmol/ml, a difference which was statistically significant ($p < 0.05$). Analyses of subfractionated blood revealed that the red blood cell fraction contained 80% and the platelet-free plasma fraction 20% of the whole-blood GABA concentration. The GABA level in blood from 6 male patients with HD was 523 ± 146, which was not significantly different from that of the normal males. In the same study, GABA was measured in amniotic fluid ($N = 8$) and shown to be present at a mean (± S.D.) level of 538 ± 208 pmol/ml. The results seen with both blood and amniotic fluid from the I-E/F method were confirmed by comparison with results obtained from full amino acid analyses showing correlation coefficients of 0.91 and 0.83, respectively.

In contrast to the observed instability of GABA level in CSF at room temperature,[22] both the I-E/F[23] and radioreceptor[15] studies agree that blood GABA levels are stable at room temperature. Likewise, stability of GABA level at room temperature was shown for amniotic fluid.[23]

These observations are important to the study of GABAergic metabolism of the central nervous system. Xanthochromic or bloody CSF is often obtained from patients with subarachnoid hemorrhage or hemorrhagic cerebral infarctions and from those undergoing a traumatic lumbar puncture. The GABA content in CSF specimens from these patients would be expected to be in excess of that derived from central GABAergic activity. Unfortunately, removal of the erythrocytes by centrifugation would not eliminate contaminating peripheral GABA present in the supernatant which was contributed by the plasma during the hemorrhagic event. The same methodological problems exist in the study of amniotic fluid which is usually xanthochromic and infrequently bloody.

Several recent studies of GABA-elevating drugs have included measurements of blood, serum, or plasma GABA concentrations. Ferkany *et al.*[15] administered aminooxyacetic acid, an inhibitor of GABA-transaminase, to rats and noted a 90% increase of blood GABA level and a 74% increase of brain GABA level. Bohlen *et al.*[6] found that intraperitoneal administration of either γ-vinyl GABA or γ-acetylenic GABA to rats produced increases of serum GABA concentrations as well as of the brain and CSF GABA. These investigators also administered ethanolamine-*O*-sulfate, a peripheral GABA transaminase inhibitor, to rats intraperitoneally, at a dose that did not alter brain GABA levels but that increased serum and CSF GABA levels approximately 4-fold and 2-fold, respectively. As they pointed out, this action of ethanolamine-*O*-sulfate may suggest at least some contribution of GABA from the periphery to that found in the CSF. Loscher[34] administered γ-acetylenic GABA or γ-vinyl GABA intravenously to dogs at a dose half that used in the study of Bohlen *et al.*[6] and concluded that plasma GABA was increased by γ-acetylenic GABA but not by γ-vinyl GABA. In Loscher's study, sodium valproate also did not produce elevation of plasma GABA, although all three drugs produced elevation of CSF GABA levels. Taken together, these studies appear to support the hypothesis that CSF GABA is derived primarily from the CNS, although a minimal contribution from the periphery is also possible.

Unlike most amino acids, GABA had been generally thought to be absent from peripheral tissues. A number of investigators have recently presented data that are in disagreement with this conclusion.[9,16,17,24,25,44,45,65,75] In several reports[43,45,65] since 1975, a group of investigators in Japan using an enzymatic-cycling GABA assay procedure reported GABA in rat pancreatic islet of Langerhans tissue at a level of 18.75 mmol/kg dry weight corresponding to 4–5 nmol GABA/mg on a wet weight basis. Their values for whole rat pancreas ranged from 87 pmol/mg to 0.6 nmol/mg wet weight—the lower values resulting from the enzymatic procedure without NADPH cycling. They showed GABA to be localized in the beta cells because in a functional beta-cell tumor, GABA was increased about 10-fold, and because streptozotocin pretreatment of rats decreased islet GABA content 10-fold without effecting the GABA content of pancreatic exocrine (acinar) tissue.

Studies in our laboratory, using the I-E/F procedure for measuring GABA, are in general agreement with the reports discussed above, except that lower values were seen in pancreas and islet of Langerhans tissue, i.e., 40 and 190 pmol/mg, respectively.[17] Use of streptozotocin pretreatment produced a reduction of whole-pancreas GABA levels to 37% of control values. The utility of the I-E/F procedure in peripheral organs was confirmed by us during a peripheral screen for GABA. In 11 organs from male Sprague–Dawley rats, the I-E/F procedure correlated well with full amino acid analysis using fluorescence detection ($r = 0.95$). This peripheral screen[17] further revealed GABA to be present in kidney, liver, spleen, and testis, but again at levels 2–5 times lower than previously reported.[65]

10. Summary and Conclusion

Recent improvements of analytical techniques have demonstrated the presence of GABA in human CSF. Measurements using the ion-exchange/fluorometric, radioreceptor, and gas chromatographic–mass spectrometric methods are in agreement that it is present at nanomolar concentrations. Basic studies have provided evidence that CSF GABA measurements may be a reflection of brain GABA levels. Studies using GABA-elevating drugs have shown a relationship between brain and ventricular or cisternal CSF, and an increasing gradient of GABA concentration between CSF from the lumbar sac and that closer to the brain has been demonstrated. These observations supply the rationale for clinical investigations of central GABA activity employing GABA analysis of lumbar CSF specimens. In addition, it has been shown that GABA levels are unstable under certain conditions in that they increase in fresh samples maintained at room temperature and in acid-deproteinized samples during frozen storage at −20°C. This increase, at least in the deproteinized samples, apparently results from hydrolysis of GABA peptides such as homocarnosine and may explain discrepancies in the reports from various laboratories.

Normal adult males have a mean CSF GABA level of about 250 pmol/ml, which may decrease somewhat with age. Females under age 40 seem to have the same range of CSF GABA levels as do the males; however, this decreases substantially in older females. CSF GABA levels are reduced in various neurological diseases such as Huntington's disease (HD), epilepsy, dementias, and multiple sclerosis. Individuals "at risk for" HD showed a mean CSF GABA level between those of the normals and the HD patients, with levels in some being close to that seen in HD patients while levels in others were in the range of normals.

GABA has also been shown to be present in blood and many peripheral tissues. Cellular constituents of blood contain higher GABA levels than does plasma. Mean blood GABA levels in normal adults were seen to be 488 ± 166 pmol/ml and were not altered in six patients with HD. Peripheral studies have shown GABA to be present at low levels in various organs, with highest levels seen in pancreas and kidney.

Thus, recent studies have established several of the basic criteria necessary for the study of GABA in CSF as well as in peripheral tissue and fluids. It can be expected that future studies will expand and refine these criteria. Continued utilization of sensitive procedures for GABA measurement of patient specimens offers potential for diagnostic and therapeutic benefit and may provide valuable knowledge of the underlying bases of several disorders.

References

1. Achar, V. S., Welch, K. M. A., Chabi, E., Bartosh, K., Meyer, J. S.: Cerebrospinal fluid gamma-aminobutyric acid in neurologic disease. *Neurology* **26**:777–780, 1976.
2. Baxter, C. F.: Assay of gamma-aminobutyric acid and enzymes involved in its metabolism. In Fried, R. (ed.): *Methods of Neurochemistry*, Vol. 3. New York, Marcel Dekker, 1972, pp. 1–73.
3. Bird, E. D., MacKay, A. V. P., Rayner, C. N., Iversen, L. L.: Reduced glutamic acid decarboxylase activity of postmortem brain in Huntington's chorea. *Lancet* **1**:1090–1092, 1973.
4. Bird, E. D., Iversen, L. L.: Huntington's chorea: Postmortem measurement of glutamic acid decarboxylase, choline acetylase and dopamine in basal ganglia. *Brain* **97**:457–472, 1974.
5. Bohlen, P., Schechter, P. J., van Damme, W., Coquillat, G., Dosch, J.-C., Koch-Weser, J.: Automatic assay of gamma-aminobutyric acid in human cerebrospinal fluid, *Clin. Chem.* **24**:256–260, 1978.
6. Bohlen, P., Hout, S., Palfreyman, M. G.: The relationship between GABA concentrations in brain and cerebrospinal fluid. *Brain Res.* **167**:297–305, 1979.

7. COSTA, E., GUIDOTTI, A., MAO, C. C.: A GABA hypothesis for the action of benzodiazepines. In Roberts. E., Chase, T. N., Tower, D. B. (eds.): *GABA in Nervous System Function*. New York, Raven Press, 1976, pp. 413–426.
8. DICKINSON, J. C., HAMILTON, P. B.: The free amino acids of human spinal fluid determined by ion exchange chromatography. *J. Neurochem.* **13**:1179–1187, 1966.
9. DRUMMOND, R. J., PHILLIPS, A. T.: L-Glutamic acid decarboxylase in non-neural tissues of the mouse. *J. Neurochem.* **23**:1207–1213, 1974.
10. ENNA, S. J., SNYDER, S. H.: A simple and specific radioreceptor assay for endogenous GABA in brain tissues. *J. Neurochem.* **26**:221–224, 1976.
11. ENNA, S. J., STERN, L. Z., WASTEK, G. J., YAMAMURA, H. I.: Cerebrospinal fluid gamma-aminobutyric acid variations in neurological disorders. *Arch. Neurol.* **34**:683–685, 1977.
12. ENNA, S. J., WOOD, J. H., SNYDER, S. H.: Gamma-aminobutyric acid (GABA) in human cerebrospinal fluid: Radioreceptor assay. *J. Neurochem.* **28**:1121–1124, 1977.
13. FAHN, S.: Regional distribution studies of GABA and other putative neurotransmitters and their enzymes. In Roberts, E., Chase, T. N., Tower, D. B. (eds.): *GABA in Nervous System Function*. New York, Raven Press, 1976, pp. 169–186.
14. FAULL, K. P., DOAMARAL, J. R., BERGER, P. A., BARCHAS, J. D.: Mass spectrometric identification and selected ion monitoring quantitation of gamma-aminobutyric acid (GABA) in human lumbar cerebrospinal fluid. *J. Neurochem.* **31**:1119–1122, 1978.
15. FERKANY, J. W., SMITH, L. A., SEIFERT, W. E., JR., CAPRIOLI, R. M., ENNA, S. J.: Measurement of gamma-aminobutyric acid (GABA) in blood. *Life Sci.* **22**:2121–2128, 1978.
16. GERBER, J. C., III, HARE, T. A.: GABA in peripheral organs: Emphasis on endocrine pancreas of rat and catfish, *Fed. Proc. Fed. Am. Soc. Exp. Biol.* **38**:375, 1979 (abstract).
17. GERBER, J. C., III, HARE, T. A.: Gamma-aminobutyric acid in peripheral tissues with emphasis on the endocrine pancreas: Presence in two species and reduction by streptozotocin. *Diabetes* **28**:1073–1076, 1979.
18. GLAESER, B. S., HARE, T. A.: Measurement of GABA in human cerebrospinal fluid. *Biochem. Med.* **12**:274–282, 1975.
19. GLAESER, B. S., VOGEL, W. H., OLEWEILER, D. B., HARE, T. A.: GABA levels in cerebrospinal fluid of patients with Huntington's chorea. *Biochem. Med.* **12**:380–385, 1975.
20. GROSSMAN, M. H., HARE, T. A., TOURTELLOTTE, W. W., ALDERMAN, J. L., KATZ, L., MANYAM, N. V. B.: GABA measurement in human cerebrospinal fluid: Basic considerations. *Soc. Neurosci. Abstr.* **3**:407, 1977.
21. GROSSMAN, M. H., HARE, T. A., MANYAM, N. V. B.: Measurement of gamma-aminobutyric acid in human whole blood and amniotic fluid. *Fed. Proc. Fed. Am. Soc. Exp. Biol.* **38**:375, 1979 (abstract).
22. GROSSMAN, M. H., HARE, T. A., MANYAM, N. V. B., GLAESER, B. S., WOOD, J. H.: Stability of GABA levels in CSF under various conditions of storage. *Brain Res.* **182**:99–106, 1980.
23. GROSSMAN, M. H., HARE, T. A., MANYAM, N. V. B.: GABA in human blood and amniotic fluid (submitted).
24. GYLFE, E.: Changes of free amino acids in pancreatic B cells after starvation and substrate deprivation. *Acta Endocrinol.* **75**:105–118, 1974.
25. GYLFE, E., SEHLIN, J.: Interactions between the metabolism of L-leucine and D-glucose in the pancreatic B-cells. *Horm. Metab. Res.* **8**:7–11, 1976.
26. HARE, T. A., MANYAM, N. V. B.: Rapid and sensitive ion-exchange/fluorometric method for measuring GABA in physiological fluids, *Anal. Biochem.* **101**:349–355, 1980.
26a. HARE, T. A., WOOD, J. H., MANYAM, N. V. B., BALLENGER, J. C., POST, R. M., GERNER, R. H.: Selection of control populations for clinical cerebrospinal fluid GABA investigations based on comparison with normal volunteers. *Brain Res. Bull.* (in press).
27. HORNYKIEWICZ, O., LLOYD, K. G., DAVIDSON, L.: The GABA system, function of the basal ganglia, and Parkinson's disease. In Roberts, E., Chase, T. N., Tower, D. B. (eds.): *GABA in Nervous System Function*. New York, Raven Press, 1976, pp. 479–486.
28. HUIZINGA, J. D., TEELKEN, A. W., MUSKIET, F. A. J., MUELEN, J. V. D., WOLTHERS, B. G.: Identification of GABA in human CSF by gas liquid chromatography and mass spectrometry. *N. Engl. J. Med.* **296**:692, 1977.
29. HUIZINGA, J. D., TEELKEN, A. W., MUSKIET, F. A. J., JEURING, H. J., WOLTHERS, B. G.: Gamma-aminobutyric acid determination in human cerebrospinal fluid by mass-fragmentography. *J. Neurochem.* **30**:911–913, 1978.
30. KRNJEVIC, K.: Inhibitory action of GABA and GABA-mimetics on vertebrate neurons. In Roberts, E., Chase, T. N., Tower, D. B. (eds.): *GABA in Nervous System Function*. New York, Raven Press, 1976, pp. 269–282.
31. KURIYAMA, K.: Subcellular localization of the GABA system in brain. In Roberts, E., Chase, T. N., Tower, D. B. (eds.): *GABA in Nervous System Function*. New York, Raven Press, 1976, pp. 187–196.
32. LANGER, D. H., BROWN, G. L. V., BUNNEY, W. E., JR., VAN KAMMEN, D. P.: Gamma-aminobutyric acid in CSF in schizophrenia. *N. Engl. J. Med.* **293**:201, 1975.
33. LICHTSHTEIN, D., DOBKIN, J., EBSTEIN, R. P., BIEDERMAN, J., RIMON, R., BALMAKER, R. H.: Gamma-aminobutyric acid (GABA) in the CSF of schizophrenic patients before and after neuroleptic treatments. *Br. J. Psychiatry* **132**:145–148, 1978.
34. LOSCHER, W.: GABA in plasma and cerebrospinal fluid of different species: Effects of gamma-acetylenic GABA, gamma-vinyl GABA, and sodium valproate. *J. Neurochem.* **32**:1587–1591, 1979.
35. MANYAM, N. V. B., HARE, T. A., KATZ, L., GLAESER,

B. S.: Huntington's disease: Cerebrospinal fluid GABA levels in at-risk individuals. *Arch. Neurol.* **34:**728–730, 1978.

36. MANYAM, N. V. B., HARE, T. A., KATZ, L.: Cerebrospinal fluid GABA levels in Huntington's disease, at "risk for" Huntington's disease and normal controls. In Chase, T. N., Wexler, N. S., Barbeau, A. (eds.): *Advances in Neurology*, Vol. 23. New York, Raven Press, 1979, pp. 547–556.
37. MANYAM, N. V. B., KATZ, L., HARE, T. A., GERBER, J. C., III, GROSSMAN, M. H.: Cerebrospinal fluid GABA levels in various neurologic disorders. *Arch. Neurol.* **37:** 352–355, 1980.
38. MARTIN, D. L.: Carrier-mediated transport and removal of GABA from synaptic regions. In Roberts, E., Chase, T. N., Tower, D. B. (eds.): *GABA in Nervous System Function.* New York, Raven Press, 1976, pp. 347–386.
39. MCGEER, P. L., MCGEER, E. G., FIBIGER, H. C.: Choline acetylase and glutamic acid decarboxylase in Huntington's chorea. *Neurology* **23:**912–917, 1973.
40. MCGEER, P. L., MCGEER, E. G.: The GABA system and function of the basal ganglia: Huntington's disease. In Roberts, E., Chase, T. N., Tower, D. B. (eds.): *GABA in Nervous System Function.* New York, Raven Press, 1976, pp. 487–496.
41. MELDRUM, B. S.: Epilepsy and gamma-aminobutyric acid-mediated inhibition. In Pfeiffer, C. C., Smythies, J. R. (eds.): *International Review of Neurobiology*, Vol. 17. New York, Academic Press, 1975, pp. 1–36.
42. NEOPHYTIDES, A. N., SURIA, A., CHASE, T. N.: Cerebrospinal fluid GABA in neurologic disease. *Neurology* **28:**359, 1978 (abstract).
43. OBATA, K.: Excitatory effects of GABA. In Roberts, E., Chase, T. N., Tower, D. B. (eds.): *GABA in Nervous System Function.* New York, Raven Press, 1976, pp. 283–286.
44. OKADA, Y., TANIGUCHI, H., SHIMADA, C., KUROSAWA, F.: High concentration of gamma-aminobutyric acid (GABA) in the Langerhan's islets of the pancreas. *Proc. Jpn. Acad.* **51:**760–762, 1975.
45. OKADA, Y., TANIGUCHI, H., SHIMADA, C.: High concentration of GABA and high glutamate decarboxylase activity in rat pancreatic islets and human insulinoma. *Science* **194:**620–622, 1976.
46. OTSUKA, M., OBATA, K., MIYATA, T., TANAKA, Y.: Measurement of gamma-aminobutyric acid in isolated nerve cells of cat central nervous system. *J. Neurochem.* **18:**287–295, 1971.
47. OTSUKA, M.: GABA in the crustacean nervous system: A historical review. In Roberts, E., Chase, T. N., Tower, D. B. (eds.): *GABA in Nervous System Function.* New York, Raven Press, 1976, pp. 245–250.
48. OTSUKA, M., KONISHI, S.: GABA in the spinal cord. In Roberts, E., Chase, T. N., Tower, D. B. (eds.): *GABA in Nervous System Function.* New York, Raven Press, 1976, pp. 197–202.
49. PERLOW, M. J., ENNA, S. J., O'BRIEN, P. J., HOFFMAN, H. J., WYATT, R. J.: Cerebrospinal fluid gamma-aminobutyric acid: Daily pattern and response to haloperidol, *J. Neurochem.* **32:**265–268, 1979.
50. PERRY, T. L., JONES, R. T.: The amino acid content of human cerebrospinal fluid in normal individuals and in mental defectives. *J. Clin. Invest.* **40:**1363–1372, 1961.
51. PERRY, T. L., HANSEN, S., STEDMAN, D., LOVE, D.: Homocarnosine in human cerebrospinal fluid: An age-dependent phenomenon. *J. Neurochem.* **15:**1203–1206, 1968.
52. PERRY, T. L., HANSEN, S., BERRY, K., MOK, C., LESK, D.: Free amino acids and related compounds in biopsies of human brain. *J. Neurochem.* **18:**521–528, 1971.
53. PERRY, T. L., HANSEN, S., KLOSTER, M.: Huntington's chorea: Deficiency of gamma-aminobutyric acid in brain. *N. Engl. J. Med.* **288:**337–342, 1973.
54. PERRY, T. L., HANSEN, S., KENNEDY, J.: CSF amino acids and plasma CSF amino acid ratios in adults. *J. Neurochem.* **24:**587–589, 1975.
55. PERRY, T. L., HANSEN, S.: Is GABA detectable in human CSF? *J. Neurochem.* **27:**1537–1538, 1976.
56. PERRY, T. L., KISH, S. J., HANSEN, S., BUCHANAN, J. L.: Brain gamma-aminobutyric acid deficiency in schizophrenia and Huntington's chorea. *Program of the Second International Huntington's Disease Symposium,* San Diego, November 16–18, 1978, p. 42.
57. ROBERTS, E.: An hypothesis suggesting that there is a defect in the GABA system in schizophrenia. *Neurosci. Res. Program Bull.* **10:**468–480, 1972.
58. ROBERTS, E.: Disinhibition as an organizing principle in the nervous system—The role of the GABA system: Application to neurologic and psychiatric disorders. In Roberts, E., Chase, T. N., Tower, D. B. (eds.): *GABA in Nervous System Function.* New York, Raven Press, 1976, pp. 515–539.
59. ROBERTS, E.: The gamma-aminobutyric acid system and schizophrenia. In Usdin, E., Hamburg, D. A., Barchas, J. D. (eds.): *Neuroregulators and Psychiatric Disorders.* New York, Oxford University Press, 1976, pp. 347–357.
60. ROTH, M.: Fluorescence reaction for amino acids. *Anal. Chem.* **43:**880–882, 1971.
61. ROTH, M., HAMPAI, A.: Column chromatography of amino acids with fluorescence detection. *J. Chromatogr.* **83:**353–356, 1973.
62. STAHL, W. L., SWANSON, P. D.: Biochemical abnormalities in Huntington's chorea brains. *Neurology* **24:**813–819, 1974.
63. STORM-MATHIESEN, J., FONNUM, F., MALTHE-SORENSSEN, O.: GABA uptake in nerve terminals. In Roberts, E., Chase, T. N., Tower, D. B. (eds.): *GABA in Nervous System Function.* New York, Raven Press, 1976, pp. 387–394.
64. TAKEUCHI, A.: Studies of inhibitory effects of GABA in invertebrate nervous systems. In Roberts, E., Chase, T. N., Tower, D. B. (eds.): *GABA in Nervous System Function.* New York, Raven Press, 1976, pp. 255–268.
65. TANIGUCHI, H., OKADA, Y., SHIMADA, C., BABA, S.:

GABA in pancreatic islets. In Kobayashi, S., Chiba, T. (eds.): *Paraneurons: New Concepts on Neuroendocrine Relatives, Arch. Hist. Jpn.* **40** (Suppl.). Nigata, Japan, Japan Society of Histological Documentatia, 1977, pp. 87–97.

66. Tower, D. B.: GABA and seizures: Clinical correlates in man. In Roberts, E., Chase, T. N., Tower, D. B. (eds.): *GABA in Nervous System Function*. New York, Raven Press, 1976, pp. 461–478.
67. Urquhart, N., Perry, T. L., Hansen, S., Kennedy, J.: GABA content and glutamic acid decarboxylase activity in brain of Huntington's chorea patients and control subjects. *J. Neurochem.* **24**:1071–1075, 1975.
68. van Gelder, N. M.: Hydrazinopropionic acid: A new inhibitor of aminobutyrate transaminase and glutamate decarboxylase. *J. Neurochem.* **15**:747–757, 1968.
69. van Gelder, N. M.: The action *in vivo* of a structural analogue of GABA: Hydrazinopropionic acid. *J. Neurochem.* **16**:1355–1360, 1969.
70. van Sande, M., Mardens, Y., Adrianssens, K., Lowenthal, A.: The free amino acids in human cerebrospinal fluid. *J. Neurochem.* **17**:125–135, 1970.
71. Welch, K. M. A., Chabi, E., Bartosh, K., Achar, V. S., Meyer, J. S.: Cerebrospinal fluid GABA levels in migraine. *Br. Med. J.* **3**:516–517, 1975.

71a. Welch, K. M. A., Meyer, J. S.: Neurochemical alterations in cerebrospinal fluid in cerebral ischemia and stroke. In Wood, J. H. (ed.): *Neurobiology of Cerebrospinal Fluid I*. New York, Plenum Press, 1980.

72. Wood, J. D.: The role of gamma-aminobutyric acid in the mechanism of seizures. In Kerkut, G. A., Phillis, J. W. (eds.): *Progress in Neurobiology*, Vol. 15. New York, Pergamon Press, 1975, pp. 79–95.
73. Wood, J. H., Glaeser, B. S., Enna, S. J., Hare, T. A.: Verification and quantification of GABA in human cerebrospinal fluid. *J. Neurochem.* **30**:291–293, 1978.
74. Wood, J. H., Hare, T. A., Glaeser, B. S., Ballenger, J. C., Post, R. M.: Low cerebrospinal fluid γ-aminobutyric acid content in seizure patients. *Neurology* **29**:1203–1208, 1979.
75. Zachmann, M., Tocci, P., Nyhan, W. L.: The occurrence of gamma-aminobutyric acid in human tissues other than brain. *J. Biol. Chem.* **241**:1355–1358, 1966.
76. Ziegler, M. G., Brooks, B. R., Lake, C. R., Wood, J. H., Enna, S. J.: Norepinephrine and gamma-aminobutyric acid in amyotropic lateral sclerosis. *Neurology* **30**:98–101, 1980.

CHAPTER 14

Cerebrospinal Fluid γ-Aminobutyric Acid

Correlation with Cerebrospinal Fluid and Blood Constituents and Alterations in Neurological Disorders

S. J. Enna, Michael G. Ziegler, C. Raymond Lake, James H. Wood, Benjamin Rix Brooks, and Ian J. Butler

1. Introduction

Biochemical and electrophysiological studies indicate that γ-aminobutyric acid (GABA) functions as an inhibitory neurotransmitter and may be, quantitatively at least, the predominant transmitter in brain.[2,20] Recent findings have indicated that brain GABA may be involved in the central regulation of blood pressure and in neuroendocrine function.[1,22] Furthermore, numerous neurochemical and pharmacological studies have provided evidence that alterations in GABAergic function may be related to a variety of neuropsychiatric disorders. Abnormal GABAergic activity may underlie seizure activity in some epilepsies[23] and may account for some of the symptoms related to Huntington's disease,[6,7,10,27] Parkinson's disease,[18] and possibly schizophrenia.[26] For example, using autopsy material, it has been demonstrated that there is a severe loss of GABA-containing neurons in the basal ganglia regions of Huntington patients.[27] These losses appear to be related, not to drug treatment, but rather to the neuronal degeneration that is characteristic of this disorder. With regard to epilepsy, drugs that enhance GABAergic transmission, such as the barbituates, benzodiazepines, and GABA transaminase inhibitors, have antiepileptic properties, whereas agents that inhibit brain GABA synthesis, such as the hydrazides, or block GABA receptors, such as picrotoxin or bicuculline, produce seizures.[9,23]

Because of these findings, it would be of value to study brain GABAergic activity during the course of an illness and drug treatment in an attempt to determine whether brain GABA plays any role in the etiology of the disorder and in the effectiveness of therapy. However, as for other neurotransmitter systems, ethical constraints preclude the direct anal-

S. J. Enna, Ph.D. • Departments of Pharmacology and of Neurobiology and Anatomy, University of Texas Medical School at Houston, Houston, Texas 77025. **Michael G. Ziegler, M.D.** • Department of Medicine, University of California, San Diego, California 92103. **C. Raymond Lake, M.D., Ph.D.** • Departments of Psychiatry and Pharmacology, Uniformed Sciences, University of Health Sciences, School of Medicine, Bethesda, Maryland 20014. **James H. Wood, M.D.** • Division of Neurosurgery, University of Pennsylvania School of Medicine, Philadelphia, Pennsylvania 19104. **Benjamin Rix Brooks, M.D.** • Department of Neurology, The Johns Hopkins University School of Medicine, Baltimore, Maryland 21205. **Ian J. Butler, M. B., F.R.A.C.P.** • Departments of Neurology, of Pediatrics, and of Neurobiology and Anatomy, The University of Texas Medical School at Houston, Houston, Texas 77025.

ysis of brain GABA levels in patients, making it necessary to gain this information by less direct means, such as by studying the GABA content of cerebrospinal fluid (CSF).

It is only recently that analytical methods sensitive enough to measure the minute quantities of GABA in human CSF have become available. At present, there appear to be at least three methods that are capable of measuring CSF GABA: radioreceptor assay,[11,12,30,] gas chromatography–mass spectrometry,[11,19] and ion-exchange chromatography.[3,17] Using these procedures, results from different laboratories have obtained a similar range (200–400 pmol/ml) of values for human CSF GABA content. Also, reports [12,17a,29a] have indicated that there is a small, but significant, gradient for GABA in human CSF, indicating that the CSF levels of this amino acid probably reflect the brain content of GABA. Further support for this hypothesis has been accumulated in animal studies indicating that drug-induced increases or decreases in brain GABA levels result in alterations in CSF GABA content of a similar magnitude.[12] In addition, using these assay methods, studies with human CSF have indicated that the concentration of GABA in this fluid is significantly reduced in Huntington's disease,[11,176] myclonic epilepsy,[8] generalized and focal seizure disorders,[29b,30a] and possibly in Alzeheimer's disease[11] or schizophrenia.[27a]

These reports indicate that CSF GABA analysis may be a useful clinical tool to further investigate the relationship between GABA and neurological and psychiatric disorders and the biochemical effects of drugs used to treat these states. However, to interpret CSF GABA data appropriately and to compare results from different studies, it is necessary to know what factors, such as age of the patient and time of sample collection, significantly influence the CSF content of this amino acid. Furthermore, it would be useful to know whether CSF GABA levels are related to other chemical parameters that can be measured in CSF and blood. Knowledge of such relationships may shed light on the GABA function in brain and on dietary and environmental factors that may influence brain GABAergic activity.

This chapter describes a comprehensive study comparing human CSF GABA content to a number of variables such as physical signs and chemical constituents of blood and CSF. The results suggest that CSF GABA correlates with a select group of variables, indicating that measurement of this substance in CSF may be a useful indicator of various aspects of brain function.

2. CSF GABA Correlations

2.1. Subject Treatment and Sample Collection

This study was conducted using over 200 patients being treated for a variety of complaints including myopathies, amyotrophic lateral sclerosis, arteriovenous malformations, neoplasms, peripheral neuropathies, Huntington's disease, and epilepsy. Of the subjects, 39% were males, and the age range for all patients was 1–75 years, with a mean of 39 years.

Prior to obtaining CSF and blood specimens, all patients were at bedrest and fasting since the previous night, during which time oral intake was avoided. Lumbar punctures were usually performed at 9 A.M. in the standard fashion with the patient in the lateral decubitus position. In some cases, lumbar puncture was performed at 9 P.M. or 3 A.M. on patients who had also been at bedrest for the previous 10 hr. CSF was collected in plastic tubes and the 12th–16th or 16th–18th milliliters removed and placed on ice until stored in a −70°C freezer. In general, GABA analysis was performed on the 12th–16th milliliters of CSF, which contained 10 mg ascorbic acid. Other portions were sent to the clinical laboratory for routine analysis. Prior to lumbar puncture, a 15-ml blood sample was withdrawn into tubes containing acid–citrate–dextrose (NIH-ACD solution) for plasma norepinephrine assay, and in some cases, plasma was obtained and portions allocated for routine analysis and cyclic nucleotide assays. Whole-blood samples were used for glucose, urea nitrogen, and GABA analysis. Pulse, respiration, and blood pressure were also noted at the time of lumbar puncture.

The CSF and blood samples were obtained over a 13-month period beginning in January 1975. All samples were analyzed for GABA content during February 1977.

2.2. CSF and Blood Analysis

Portions of the CSF and blood were assayed for electrolyte, glucose, protein, urea nitrogen, uric acid, creatinine, albumin, bilirubin, and blood cell content using standard clinical laboratory procedures. The cyclic AMP (cAMP) and cyclic (cGMP) content of the fluids was analyzed using a ligand-binding procedure,[16] and the norepinephrine content was determined using a previously described radioenzymatic assay.[21,31] The CSF content of homovanillic acid (HVA) and 3-methoxy-4-hydroxyphenylethylene glycol (MHPG) was determined by gas chromatography–mass spectrometry,[4,28] whereas CSF 5-hydroxyindoleacetic acid (5-HIAA) was measured fluorometrically.[29] The GABA concentration in CSF and blood was determined using a radioreceptor assay.[11,15]

All data were analyzed by computer at the NIH central computer facility. Levels of significance were determined using a two-tailed Student's *t* test or from correlation coefficients derived by least-squares analysis.

2.3. Correlation with Time, Age, and Physical Signs

The CSF GABA content of each individual was compared to the time of day the sample was taken, the age of the patient, and the duration of the illness, and to selected physical signs (Table 1). No significant correlation was observed with any of the variables studied, though there was the suggestion of a positive correlation between the time of day and GABA content. Thus, GABA levels tended to be somewhat higher at midday and during the afternoon than at earlier hours. No significant correlations were found with regard to age, pulse or respiratory rate, systolic, diastolic, or mean blood pressure, or duration of the illness.

2.4. Correlation with CSF Constituents

The CSF GABA concentration was also compared to 16 other CSF constituents, including electrolytes, glucose, norepinephrine, neurotransmitter metabolites, cAMP, and cGMP (Table 2). Of these variables, 4 correlated in a significant fashion with CSF GABA.

Significant positive correlations were observed between levels of CSF GABA and CSF cAMP, 5-HIAA, and HVA (Table 2). In contrast, there was a significant negative correlation between the concentration of glucose in the CSF and the GABA content, indicating that GABA levels are generally higher when CSF glucose levels are depressed. No significant correlation was observed with respect to CSF norepinephrine, cGMP, MHPG, blood cell content, or electrolytes (Table 2).

2.5. Correlation with Blood Constituents

Of the 18 blood constituents measured, only 2 substances correlated in a significant fashion with CSF GABA content. As measured in 28 patients, there was a negative correlation between both blood uric acid and urea nitrogen and CSF GABA (Table 3). As opposed to results with CSF, neither blood cAMP nor glucose correlated significantly with CSF GABA. Similar to the CSF data, the GABA in CSF did not vary in a significant fashion with regard to blood norepinephrine, cGMP, protein, or electrolytes. In addition, blood GABA, creatinine, albumin, and bilirubin appeared to vary independently of CSF GABA content (Table 3).

2.6. Comparison of CSF GABA Content in Various Disorders

Using patients having myopathies without known neurological disorders as controls, there do not appear to be any significant differences in the CSF GABA content in individuals diagnosed as having arteriovenous malformations, brain tumors, or peripheral neuropathies (Table 4). It should be pointed

Table 1. Correlation of CSF GABA[a] Content with Time, Age, and Physical Signs

Variable	Correlation coefficient	Probability	Number of patients
Time[b]	0.30	0.07	38
Age	−0.08	0.33	131
Pulse	0.15	0.22	70
Respiration	0.06	0.69	53
Systolic blood pressure	0.13	0.29	70
Diastolic blood pressure	0.19	0.11	70
Mean blood pressure	0.17	0.15	70
Duration of illness[c]	0.02	0.90	33

[a] GABA, γ-Aminobutyric acid.
[b] Time of day sample was taken.
[c] Time since diagnosis of illness.

Table 2. Correlation of CSF GABA Content with Other CSF Constituents[a]

Constituent	Correlation coefficient	Probability	Number of patients
Norepinephrine	−0.03	0.70	139
Red blood cells	−0.06	0.65	64
White blood cells	0.06	0.66	64
Protein	0.21	0.08	68
Cyclic AMP	0.40	< 0.01	95
Cyclic GMP	−0.09	0.52	57
5-HIAA	0.49	< 0.01	29
HVA	0.40	0.03	30
MHPG	0.25	0.29	20
Glucose	−0.54	< 0.01	28
Chloride	−0.05	0.79	28
Magnesium	0.21	0.28	29
Calcium	0.02	0.94	27
Sodium	0.07	0.74	26
Potassium	−0.08	0.70	27
Phosphate	−0.07	0.75	26

[a] Abbreviations: GABA, γ-aminobutyric acid; AMP, adenosine monophosphate; GMP, guanosine monophosphate; 5-HIAA, 5-hydroxyindoleacetic acid; HVA, homovanillic acid; MHPG, 3-methoxy-4-hydroxyphenylethylene glycol.

Table 3. Correlation of CSF GABA Content with Blood Constituents[a]

Constituent[b]	Correlation coefficient	Probability	Number of patients
Norepinephrine	0.04	0.69	91
GABA	−0.20	0.10	63
Cyclic AMP	0.12	0.42	50
Cyclic GMP	0.06	0.82	17
Uric acid	−0.40	0.03	28
Blood urea nitrogen	−0.49	< 0.01	28
Glucose	−0.06	0.75	29
Creatinine	−0.16	0.43	27
Albumin	−0.11	0.63	21
Bilirubin	−0.13	0.57	20
Protein	−0.14	0.47	29
Magnesium	0.17	0.37	28
Sodium	0.08	0.67	28
Potassium	−0.11	0.56	29
Chloride	−0.04	0.82	29
Bicarbonate	−0.20	0.30	28
Phosphate	−0.19	0.32	29
Calcium	0.11	0.59	27

[a] Abbreviations: GABA, γ-aminobutyric acid; AMP, adenosine monophosphate; GMP, guanosine monophosphate.
[b] Values for blood urea nitrogen, glucose, and GABA were derived from analysis of whole blood, while all other values were determined from plasma only.

Table 4. CSF GABA[a] Content in Various Disorders

Disorder	CSF GABA (pmol/ml)[b]	
Myopathies	251 ± 51	(14)
Arteriovenous malformations	162 ± 41	(4)
Brain tumors	161 ± 46	(5)
Peripheral neuropathies	246 ± 42	(14)
Huntington's disease	118 ± 18[c,d]	(25)
Parkinson's disease	228 ± 30[d]	(5)
Alzheimer's disease	130 ± 22[c,d]	(3)
Myoclonic epilepsy	116 ± 15[c,e]	(10)
Tourette syndrome	231 ± 32[e]	(18)

[a] GABA, γ-aminobutyric acid.
[b] Measured using a radioreceptor assay. Each value is the mean ±S.E. of the number of patients in parentheses.
[c] $P < 0.05$ compared to myopathies.
[d] Adapted from Enna *et al.*[11]
[e] Adapted from Enna *et al.*[8]

out that it is unknown whether the neuropathy patients had any central involvement. In contrast, CSF GABA content was significantly reduced in patients with Huntington's disease, Alzheimer's disease, and myoclonic epilepsy relative to those with myopathies.

3. Significance of CSF GABA Correlations

The results of the investigation reported herein provide information with regard to the study of CSF GABA content as an indicator of brain GABAergic activity. The finding that heart rate, blood pressure, and respiratory rate do not correlate in a significant manner with CSF GABA content indicates that these variables probably need not be taken into consideration when studying this amino acid in CSF. Although this study of sexually mixed populations did not demonstrate CSF GABA correlations with age, Hare *et al.*[17b] have shown the existence of a tendency for CSF GABA concentrations to decrease with age in females over 40 years of age. This finding, if confirmed, would require future CSF GABA investigations to employ only age- and sex-matched control groups for comparison purposes. In contrast, the tendency for a positive correlation between time of sample collection and GABA level ($p = 0.07$) suggests that for comparison, CSF samples should be withdrawn at the same time of day. This finding hints at the possibility of a diurnal variation in CSF GABA content, as exists for other brain substances.[24] In fact, this suggestion of a diurnal variation corresponds to the findings of a recent study with monkey cisternal CSF in which it was found that GABA levels tended to be higher during the day than at night.[25]

The significant positive correlations observed comparing GABA to other CSF constituents are subject to a number of interpretations. Lacking further data, it is impossible to make definitive statements about the significance of the correlations found in this study. However, drawing on results from previous studies, it is possible to speculate about the biological significance of the present results. For example, the fact that GABA levels rise and fall in tandem with CSF cAMP content suggests that central nervous system GABAergic activity may regulate the intracellular production and release of this secondary messenger. While GABA itself cannot activate production of cyclic nucleotides,[5] it is known that GABAergic transmission can modify catecholamine activity, which in turn can stimulate production of cAMP. However, a problem with this interpretation is that in general, GABA appears to exert an inhibitory action on catecholamine systems, and thus an increase in GABA outflow would be expected to produce a decrease in cAMP formation.

Another possible explanation is that increases in CSF cAMP are primarily related to the leakage of this substance from damaged brain cellular elements. If this is the case, the rise in GABA levels associated with increasing cAMP concentrations may also reflect brain cell damage, since GABA is found in virtually all areas of the central nervous system.[13] Thus, CSF GABA and cAMP content may not be causally related at all, but rather may both be secondary to cell damage.

The majority of HVA, a dopamine metabolite, in the CSF probably derives from dopamine stores located in the corpus striatum, and since GABA modifies dopaminergic activity in this brain area, it is not surprising to find that there is a relationship between GABA activity and the CSF content of HVA. However, as with cAMP, it is curious that the relationship is positive, since it is generally felt that GABA is inhibitory on striatal dopamine activity, making it more reasonable to suspect that an increase in HVA, which suggests an increase in dopamine release and metabolism, would be accompanied by a decrease in GABA content in CSF.

A more parsimonious interpretation is that an increase in brain GABA activity sufficient to cause an increase in the CSF content of this amino acid occurs at a time when there is an apparent increase in dopaminergic firing, since it is probably unrealistic to believe that even a substantial increase in GABA activity in a single brain area could result in a significant increase in CSF GABA content. Thus, the concentrations of neurotransmitters and their metabolites in CSF reflect brain neurotransmitter activity in a global sense, and therefore the increase in HVA probably represents stimulation of the dopamine system in several brain areas, just as an increase in GABA content undoubtedly reflects an activation of a number of GABAergic pathways. This finding may be of fundamental importance, since it implies that there is an interplay between the brain GABAergic and dopaminergic systems in many central nervous system areas.

Similar speculation can be applied to the finding of a positive correlation between CSF GABA and CSF 5-HIAA, a serotonin metabolite. While numerous investigations have suggested that there is an interaction between the GABA and dopamine systems in brain, little is known about interactions between GABA and serotonin. Thus, the results of the study reported herein provide evidence that increased GABAergic activity is accompanied by a net increase in serotoninergic firing. As with HVA, it is impossible to determine which neurotransmitter system activates the other, or what effect GABA has on serotonin firing in any particular brain regions, such as the raphe nucleus. Nevertheless, the highly significant positive correlation between CSF 5-HIAA and GABA suggests the possibility that alterations in GABAergic function may lead to an alteration in activity of the serotonin system. Some caution should be reserved in these interpretations since this data does not correct for the differential blockade of the amine metabolites by probenecid.

The significant negative correlations between CSF GABA and CSF glucose may be a reflection of the fact that GABA is derived from glucose, or it may indicate that hypoglycemia induces a net increase in the release of central nervous system GABA. However, blood glucose content did not correlate with CSF GABA levels.

The significant negative correlations found between CSF GABA, blood urea nitrogen, and blood uric acid are difficult to interpret. While these correlations may be meaningless, it is interesting that of all the blood variables studied, only these two correlated in a significant fashion and the blood contents of these substances generally rise and fall in tandem, being indicators of kidney function and purine and protein metabolism. Although probenecid administration may have increased the blood levels of urea nitrogen and uric acid, previously reported data[12] suggest that probenecid-loading does not significantly elevate CSF GABA concentrations.

It is also of interest that CSF GABA does not correlate in a significant manner with blood levels of this amino acid. This finding suggests that blood GABA is not derived primarily from CSF[8] and is consistent with the premise that the GABA content of CSF is central in origin.[29a]

The finding that CSF GABA may be significantly reduced in certain neurological states indicates that measurement of this substance may have diagnostic value and may be useful for determining biochemical deficits present in certain disorders. Thus, it is not surprising that CSF GABA is reduced in patients diagnosed as having Huntington's disease, since, neuropathologically, this disorder is characterized by a profound loss of GABA-containing neurons in the corpus striatum.[10] However, the reduced GABA content may be of diagnostic value, since it is often difficult to differentiate Huntington's disease from other disorders such as alcoholism or dementia. Along these lines, it is noteworthy that CSF GABA appears to be significantly reduced in Alzheimer's disease, a form of presenile dementia, which symptomologically is often confused with Huntington's disease. Thus, CSF GABA determinations may have little value in the differential diagnosis of these two disorders.

The lack of a significant reduction in CSF GABA in Tourette syndrome and Parkinson's disease demonstrates that a decrease of GABA in CSF is not characteristic of all neurological disorders. Also, it provides evidence that the reduction found in Huntington's disease is not related to drug treatment, since, like Huntington patients, individuals with Tourette syndrome are maintained on neuroleptics. This finding confirms an earlier study with monkeys that indicated that neuroleptic treatment does not alter CSF GABA content.[24]

Our study demonstrated significantly lower CSF GABA levels in patients with myoclonic epilepsy than in those with myopathies. This finding is consistent with communications of Wood *et al.*[29b,30a] which report low CSF GABA concentrations in seizure patients who have associated cerebellar atrophy.

In conclusion, while correlations do not prove causality, the results of the investigation reported herein suggest that measurement of CSF GABA content may be a useful indicator of brain GABAergic function and may provide information with regard to neurotransmitter interactions, the biochemical action of therapeutic agents, and the biochemical abnormalities present in neuropsychiatric disorders. In addition, CSF GABA analysis may have diagnostic value. While, in most cases, CSF analysis of GABA, or of any other neurotransmitter, reveals little about specific neuronal pathways, such information may be valuable in obtaining some idea about the overall neurochemical state of the central nervous system.

ACKNOWLEDGMENTS. This research was supported in part by USPHS grants 5P5O-NS-07377, NS-13803, by a Research Career Development Award NS-00335 (S.J.E.), and by the Huntington's Chorea Foundation.

References

1. ANTONACCIO, M., TAYLOR, D.: Involvement of central GABA receptors in the regulation of blood pressure and heart rate of anesthetized cats. *Eur. J. Pharmacol.* **46:**283–287, 1977.
2. BLOOM, F. E., IVERSEN, L. L.: Localizing ^{3}H-GABA in nerve terminals of rat central cortex by electron microscopic autoradiography. *Nature (London)* **229:**629–630, 1971.
3. BOHLEN, P., SCHECHTER, O., DAMME, W., COQUILLAT, G., DOSCH, J., KOCH-WESER, J.: Automated assay of gamma-aminobutyric acid in human cerebrospinal fluid. *Clin. Chem.* **24:**256–260, 1978.
4. CHASE, T., GORDON, E., ING, L. K. Y.: Norepinephrine metabolism in central nervous system of man: Studies using 3-methoxy-4-hydroxyphenylethylene glycol levels in cerebrospinal fluid. *J. Neurochem.* **21:**581–587, 1973.
5. COSTA, E., GUIDOTTI, A., MAO, C. C.: A GABA hypothesis for the action of benzodiazepines. In Roberts, E., Chase, T. N., Tower, D. B. (eds.): *GABA in Nervous System Function.* New York, Raven Press, 1976, pp. 413–426.
6. ENNA, S. J., BENNETT, J., BYLUND, D., SNYDER, S. H., BIRD, E., IVERSEN, L. L.: Alterations of brain neurotransmitter receptor binding in Huntington's chorea. *Brain Res.* **116:**531–537, 1976.
7. ENNA, S. J., BIRD, E., BENNETT, J., BYLUND, D., YAMAMURA, H., IVERSEN, L., SNYDER, S. H.: Huntington's chorea: Changes in neurotransmitter receptors in the brain. *N. Engl. J. Med.* **294:**1305–1309, 1976.
8. ENNA, S. J., FERKANY, J. W., VAN WOERT, M., BUTLER, I. J.: Measurement of GABA in biological fluids: Effect of GABA transaminase inhibitors. In Chase, T. N., Wexler, N., Barbeau, A. (eds.): *Huntington's Chorea: 1972–1978.* New York, Raven Press, 1979, pp. 741–750.
9. ENNA, S. J., MAGGI, A.: Biochemical pharmacology of GABAergic agonists. *Life Sci.* **24:**1727–1738, 1979.
10. ENNA, S. J., STERN, L. Z., WASTEK, G., YAMAMURA, H.: Neurobiology and pharmacology of Huntington's disease. *Life Sci.* **20:**205–212, 1977.
11. ENNA, S. J., STERN, L. Z., WASTEK, G., YAMAMURA, H.: Cerebrospinal fluid GABA variations in neurological disorders. *Arch. Neurol.* **34:**683–685, 1977.
12. ENNA, S. J., WOOD, J. H., SNYDER, S. H.: Radioreceptor assay for γ-aminobutyric acid (GABA) in human cerebrospinal fluid. *J. Neurochem.* **28:**1121–1124, 1974.
13. FAHN, S.: Regional distribution studies of GABA and other putative neurotransmitters and their enzymes. In Roberts, E., Chase, T. N., Tower, D. B. (eds.): *GABA in Nervous System Function.* New York, Raven Press, 1976, pp. 169–188.
14. FERKANY, J. W., BUTLER, I. J., ENNA, S. J.: Effect of drugs on rat brain, cerebrospinal fluid and blood GABA content. *J. Neurochem.* **33:**29–33, 1979.
15. FERKANY, J. W., SMITH, L. A., SEIFERT, W., CAPRIOLI, R. M., ENNA, S. J.: Measurement of γ-aminobutyric acid (GABA) in blood. *Life Sci.* **22:**2121–2128, 1978.
16. GILMAN, A. G.: A protein binding assay for adenosine 3':5'-cyclic monophosphate. *Proc. Natl. Acad. Sci. U.S.A.* **67:**305–312, 1970.
17. GLAESER, B. S., HARE, T. A.: Measurement of GABA in human cerebrospinal fluid. *Biochem. Med.* **12:**274–282, 1975.
17a. HARE, T. A., MANYAM, N. V. B.: Rapid and sensitive ion-exchange/fluorometric measurement of GABA in physiological fluids. *Anal. Biochem.* **101:**349–355, 1980.
17b. HARE, T. A., MANYAM, N. V. B., GLAESER, B. S.: Evaluation of cerebrospinal fluid γ-aminobutyric acid content in neurologic and psychiatric disorders. In Wood, J. H. (ed.): *Neurobiology of Cerebrospinal Fluid I.* New York, Plenum Press, 1980.
18. HORNYKIEWICZ, O., LLOYD, K. G., DAVIDSON, L.: The GABA system and function of the basal ganglia—Parkinson's disease. In Roberts, E., Chase, T. N., Tower, D. B. (eds.): *GABA in Nervous System Function.* New York, Raven Press, 1976, pp. 479–486.
19. HUIZINGA, J. D., TEELKEN, A. W., MUSKIET, F., JEURING, H., WOLTHERS, B.: Gamma-aminobutyric acid determination in human cerebrospinal fluid by mass-fragmentography. *J. Neurochem.* **30:**911–913, 1978.
20. IVERSEN, L. L., BLOOM, F. E.: Studies of the uptake of ^{3}H-GABA and ^{3}H-glycine in slices and homogenates of rat brain and spinal cord by electron microscopic autoradiography. *Brain Res.* **41:**131–143, 1972.
21. LAKE, C. R., ZIEGLER, M. G., KOPIN, I. J.: Use of plasma norepinephrine for evaluation of sympathetic neuronal function in man. *Life Sci.* **18:**1315–1326, 1976.
22. LAMBERTS, S., MACLEOD, R.: Studies on the mechanism of the GABA-mediated inhibition of prolactin secretion. *Proc. Soc. Exp. Biol. Med.* **158:**10–13, 1978.
23. MELDRUM, B.: Epilepsy and gamma-aminobutyric acid mediated inhibition. *Int. Rev. Neurobiol.* **17:**1–36, 1975.
24. PERLOW, M., ENNA, S. J., O'BRIEN, P., HOFFMAN, H., WYATT, R.: Cerebrospinal fluid γ-aminobutyric acid:

Daily pattern and response to haloperidol. *J. Neurochem.* **32:**265–268, 1979.
25. PERLOW, M., FESTOFF, B., GORDON, E., EBERT, M., JOHNSON, D., CHASE, T. N.: Daily fluctuation in the concentration of cAMP in the conscious primate brain. *Brain Res.* **126:**341–396, 1977.
26. ROBERTS, E.: Gamma-aminobutyric acid and nervous system function—a perspective. *Biochem. Pharmacol.* **23:**2637–2649, 1974.
27. URQUHART, N., PERRY, T. L., HANSEN, S., KENNEDY, J.: GABA content and glutamic acid decarboxylase activity in brain of Huntington's chorea patients and control subjects. *J. Neurochem.* **24:**1071–1075, 1975.
27a. VAN KAMMEN, D. P., STERNBERG, D. E.: Cerebrospinal fluid studies in schizophrenia. In Wood, J. H. (ed.): *Neurobiology of Cerebrospinal Fluid I.* New York, Plenum Press, 1980.
28. WATSON, E., TRAVIS, B., WILK, S.: Simultaneous determination of 3,4-dihydroxyphenylacetic acid and homovanillic acid in milligram amounts of rat striatal tissue by gas–liquid chromatography. *Life Sci.* **15:**2167–2178, 1974.
29. WEIR, R. L., CHASE, T. N., ING, L. K. Y., KOPIN, I. J.: 5-Hydroxyindole acetic acid in spinal fluid: Relative contribution from brain and spinal cord. *Brain Res.* **52:**409–412, 1973.
29a. WOOD, J. H.: Sites of origin and concentration gradients of neurotransmitters, their precursors and metabolites, and cyclic nucleotides in cerebrospinal fluid. In Wood, J. H. (ed.): *Neurobiology of Cerebrospinal Fluid I.* New York, Plenum Press, 1980.
29b. WOOD, J. H., BROOKS, B. R.: Neurotransmitter, metabolite, and cyclic nucleotide alterations in cerebrospinal fluid of seizure patients. In Wood, J. H. (ed.): *Neurobiology of Cerebrospinal Fluid I.* New York, Plenum Press, 1980.
30. WOOD, J. H., GLAESER, B. S., ENNA, S. J., HARE, T. A.: Verification and quantification of GABA in human cerebrospinal fluid. *J. Neurochem.* **30:**291–293, 1978.
30a. WOOD, J. H., HARE, T. A., GLAESER, B. S., BALLENGER, J. C., POST, R. M.: Low cerebrospinal fluid γ-aminobutyric acid content in seizure patients. *Neurology (Minneapolis)* **29:**1203–1208, 1979.
31. ZIEGLER, M. G., LAKE, I. J., KOPIN, I. J.: Norepinephrine in cerebrospinal fluid. *Brain Res.* **108:**436–440, 1976.

CHAPTER 15

Cerebrospinal Fluid Studies in Parkinson's Disease

Norepinephrine and γ-Aminobutyric Acid Concentrations

Paul F. Teychenne, C. Raymond Lake, and Michael G. Ziegler

1. CSF and Plasma Concentrations of Norepinephrine in Parkinson's Disease

1.1. Introduction

Dopamine is the precursor of the neurotransmitter norepinephrine. The enzyme dopamine-β-hydroxylase catalyzes the conversion of dopamine to norepinephrine. Reserpine, a rauwolfia alkaloid, depletes the presynaptic stores of dopamine and norepinephrine. In the rodent, reserpine suppresses motor activity, inducing akinesia. In man, with large doses, it produces a Parkinson-like syndrome. Levodopa, the precursor of dopamine and subsequently norepinephrine, reverses the action of reserpine on rodents.[6]

While deficiencies in the concentrations of both dopamine and norepinephrine were at first considered responsible for the clinical features of parkinsonism, subsequent evidence established the deficiency of dopamine as the major factor in the pathogenesis of Parkinson's disease. The chemical and physiological definition of dopamine as a neurotransmitter, the recognition that the main pathology in parkinsonism involved the dopaminergic neurons in the substantia nigra, and the therapeutic effect of levodopa pointed to a deficiency of dopaminergic transmission in the nigrostriatal pathway as the principal deficit in Parkinson's disease.

In rodents with 6-hydroxydopamine-induced lesions of the nigrostriatal pathway, dopamine, dopamine-releasing agents, and dopamine agonists produce specific rotation. This model has been useful in predicting which drugs would be effective for treating parkinsonism[62] and has provided further evidence for the prime role of dopaminergic dysfunction in Parkinson's disease. However, only about 40% of patients exhibit a good response to levodopa therapy, while 30% show a moderate response and 30% a negligible response.[42] Similarly, levodopa is most effective in treating bradykinesia and rigidity and least effective in treating tremor.[46]

A deficiency of dopamine and norepinephrine was found during histochemical studies of postmortem brain tissue from parkinsonian patients.[12] The pathology in parkinsonism is not limited to areas of high dopamine content but also involves the locus coeruleus, the dorsal motor nucleus of the vagus, and neurons of the sympathetic ganglia. These areas contain mainly noradrenergic neurons.[14]

Paul F. Teychenne, M.D. • Department of Neurology, George Washington University Medical School, Washington, D.C. 20037. **C. Raymond Lake, M.D., Ph.D.** • Departments of Psychiatry and Pharmacology, Uniformed Services University of Health Sciences, School of Medicine, Bethesda, Maryland 20014. **Michael G. Ziegler, M.D.** • Department of Medicine, University of California, San Diego, California 92103.

Though no synaptic connections between the noradrenergic and dopaminergic neuronal systems have been found, norepinephrine terminals are found within the substantia nigra. A dense noradrenergic terminal network innervates the area containing the A8 dopaminergic neurons that participate in dopamine innervation of the striatum. A direct interaction between dopamine and norepinephrine may exist in this area.[18]

There is physiological evidence of a link between the dopaminergic and noradrenergic neuron systems. Inhibition of central noradrenergic activity reduces the amount of locomotor activity in rodents induced by dopaminergic drugs. This activity can be restored by direct-acting noradrenergic compounds.[41] Catalepsy, an akinetic syndrome in laboratory animals, is induced by dopaminergic antagonists and bilateral lesions of the nigrostriatal pathway.[32] Chemical and electrolytic lesions in the locus coeruleus potentiate haloperidol (a dopamine antagonist)-induced catalepsy.[49] The α-adrenergic blocker phentolamine enhances haloperidol-induced catalepsy.[27] While not playing a dominant role in the pathogenesis of parkinsonism, norepinephrine may be significantly involved in modulating the activity of dopamine neurons.

Sensitive and specific assays for norepinephrine have been developed.[26,66] We measured the concentration of CSF norepinephrine, comparing the values among untreated parkinsonian patients, treated parkinsonian subjects, and normal controls.

1.2. Patients and Methods

The CSF and plasma concentration of norepinephrine was measured in 44 subjects, who were divided into three groups: (A) patients suffering from parkinsonism not receiving any therapy; (B) subjects with Parkinson's disease receiving levodopa/carbidopa; (C) normal subjects. All participants were matched as closely as possible for age and sex (Table 1). Plasma and CSF were collected under standard conditions of time and rest to avoid errors due to the circadian rhythm and activity.[68] All samples were taken between midday and 3 P.M., plasma samples the first day, CSF the following day. Patients were on standard hospital diets, since diet does not alter norepinephrine concentration.[66]

1.2.1. Collection of Plasma Samples. With the subjects at complete bed rest, a 19-gauge butterfly cannula connected to a heparin lock was inserted into a large antecubital vein. After 20 min (when blood pressure and pulse were stable), a 12-ml sample of blood was taken through the cannula, transferred into a chilled tube containing acid–citrate–dextrose anticoagulant, and placed on ice. Plasma was separated by centrifuging the sample at 4°C within 30 min of collection. The plasma was stored at −70°C until assayed for norepinephrine 4 weeks later.

1.2.2. Collection of CSF samples. Lumbar punctures were performed with the patient supine lying on his left side. The subjects had been at complete bedrest for 2 hr and had not eaten for 3 hr prior to each tap. A 4-ml sample of CSF was obtained after 16 ml CSF had been removed for other tests. This sample was placed in a tube containing 10 mg ascorbic acid, put immediately on ice, then frozen to −70°C 30 min later for assay within 6 weeks.

1.2.3. Norepinephrine Assay. Levels of norepinephrine in CSF and plasma were determined by a modification of the radioenzymatic method of Henry *et al.*[26,66] In this reaction, catalyzed by the enzyme phenylethanolamine-*N*-methyl-transferase, norepinephrine is converted to [^{3}H]epinephrine. The [^{3}H]methyl group is enzymatically transferred fron [^{3}H]methyl-*S*-adenosylmethionine to the primary amine of norepinephrine, and [^{3}H]epinephrine

Table 1. CSF and Plasma Norepinephrine in Parkinsonian Patients and Controls[a]

Groups	Number and sex of patients	Mean age (yr) ± S.E.M.	Mean CSF NE (pg/ml) ± S.E.M.	Mean plasma NE (pg/ml) ± S.E.M.	Mean levodopa/ carbidopa dose (mg/day) ± S.E.M.
A. Parkinsonian, drug-free	16 (10 M, 6 F)	60 ± 2	205 ± 32	198 ± 23	—
B. Parkinsonian	12 (7 M, 5 F)	62 ± 2	250 ± 70[b]	370 ± 58[c]	1008/100 ± 196/ 20
C. Control	16 (9 M, 7 F)	53 ± 2	489 ± 39[d]	353 ± 45[d]	—

[a] The two-tailed nonpaired Student's *t* test was used for all statistical analyses. [b] $p < 0.05$ vs. Group A. [c] $p < 0.01$ vs. Group A. [d] $p < 0.001$ vs. Group A.

Table 2. CSF and Plasma Norepinephrine in Parkinsonian Patients with and without Significant Tremor[a]

Group	Number and sex of patients	Mean age (yr) ± S.E.M.	Mean score for tremor (%) ± S.E.M.	Mean CSF NE (pg/ml) ± S.E.M.	Mean plasma NE (pg/ml) ± S.E.M.	Mean levodopa/carbidopa dose (mg/day) ± S.E.M.
A.a. Parkinsonian: drug-free, tremor	8 (5 M, 3 F)	58 ± 3	40 ± 3	317 ± 48[b,c]	205 ± 41[c]	—
A.b. Parkinsonian, drug-free	8 (5 M, 3 F)	63 ± 3	13 ± 3	122 ± 20[e]	191 ± 23[d]	—
B.a. Parkinsonian, tremor	9 (5 M, 4 F)	61 ± 3	40 ± 2	407 ± 82	420 ± 86	922/92 ± 254/25
B.b. Parkinsonian	4 (3 M, 1 F)	64 ± 2	8 ± 3	181 ± 98[d]	268 ± 78	933/93 ± 377/37

[a] The two-tailed nonpaired Student's t test was used in all statistical analyses. [b] $p < 0.01$ A.a vs A.b. [c] $p < 0.02$ vs. normal controls (Group C, Table 1).
[d] $p < 0.01$ vs. normal controls (Group C, Table 1). [e] $p < 0.001$ vs. normal controls (Group C, Table 1).

is measured by liquid scintillation spectrometry. The assay can detect concentrations of norepinephrine as low as 20 pg/ml of norepinephrine in CSF or plasma.

1.2.4. Clinical Assessments. The parkinsonian patients were assessed by a physician unaware of their current therapy. Scores for tremor, rigidity, and bradykinesia were rated on a 0 (no deficit) to 4 (maximum deficit) scale. Time tests for walking, writing, and finger dexterity were also used in the clinical assessments.[59] The scores for tremor, rigidity, and bradykinesia were converted to percentage (%) scores, with 100% representing the maximum deficit.

1.3. Results

The mean concentrations of CSF and plasma norepinephrine (NE) in each group of subjects are shown in Table 1. The untreated parkinsonian patients had a significantly lower mean concentration of norepinephrine, in both CSF and plasma, than the treated parkinsonian subjects and the normal control individuals.

There was considerable variation in the individual values for CSF norepinephrine within the untreated parkinsonian group. The coefficient of variation was high, 64%. However, this group of patients could be divided into two subgroups on the basis of their individual scores for tremor. The mean percentage scores for tremor in each subgroup are shown in Table 2. Of the eight patients in Subgroup A.a, none had a lower score for tremor than 30%. In Subgroup A.b, the highest score for tremor was 20%. The mean scores for tremor in these two subgroups were significantly different ($p < 0.001$); however, the mean scores for rigidity, bradykinesia, and time tests were not different.

Those patients with the greater mean score for tremor (Subgroup A.a) had a significantly higher mean concentration of norepinephrine in the CSF than those patients with the lower mean score for tremor (Subgroup A.b). The mean CSF norepinephrine concentrations in both subgroups were significantly lower than the mean CSF norepinephrine concentration in the normal controls (Table 2). The mean plasma concentrations of norepinephrine in the two untreated parkinsonian subgroups were almost identical and significantly lower than the mean plasma norepinephrine concentration in the normal subjects (Table 2).

There was a correlation between the severity of tremor and the values for CSF norepinephrine in the untreated parkinsonian group ($r = 0.62$). But no correlation was found between the degree of tremor and the plasma concentration for norepinephrine in these subgroups.

Therapy with levodopa/carbidopa raised the concentration of norepinephrine in both plasma and CSF (see Table 1). The treated parkinsonian group could also be differentiated into two subgroups, using scores for tremor noted either during periods of placebo administration or prior to levodopa/carbidopa therapy. Patients with the greater mean score for tremor (Subgroup B.a) had a higher concentration of CSF norepinephrine, but the mean plasma norepinephrine concentrations and the mean doses of levodopa/carbidopa were not significantly different between these subgroups (Table 2). Those patients with the lower mean score for tremor (Subgroup B.b) had a significantly lower CSF norepinephrine concentration than the normal controls.

1.4. Discussion

There is a gradient in the concentration of norepinephrine obtained from lumbar CSF.[68] The concentration of norepinephrine increases until 12 ml of CSF have been taken, after which the concentration does not alter even after the spinal subarachnoid space has been drained.[69] Examination of CSF for concentrations of norepinephrine appears useful as a reflection of the tissue concentration of this chemical. In untreated parkinsonian patients, the decreased level of CSF norepinephrine conforms to the deficiency of tissue norepinephrine in the locus coeruleus, substantia nigra, and hypothalamus.

There has been debate on whether levodopa therapy increases the level of norepinephrine in the brain. Levodopa administration increased the level of brain norepinephrine in reserpinized rodents when a monoamine oxidase inhibitor was given.[22] In one postmortem study, the concentrations of norepinephrine in the substantia nigra, hypothalamus, and caudate nucleus of levodopa-treated parkinsonian subjects were similar to those of control subjects, but not significantly higher than the levels in untreated parkinsonian subjects.[51]

The time period that elapses between the last dose of levodopa, death, and postmortem examination of brain tissue is always variable. This unavoidable lack of control may be a source of error in postmortem studies of neurotransmitter concentrations in brain tissue. CSF taken from live patients can be timed appropriately to the last dose of levodopa.

Levodopa therapy increased the level of norepinephrine in both plasma and CSF. The increase in plasma norepinephrine in the presence of carbidopa, a peripheral decarboxylase inhibitor, suggests that at this dose carbidopa does not completely inhibit peripheral dopa-decarboxylase. We used substantial doses of carbidopa to block the peripheral hypertensive response to levodopa when given in the presence of a monoamine oxidase inhibitor.[57] Another possibility is that the plasma norepinephrine concentration is determined by central activity, possibly through the descending norepinephrine tracts that project to the sympathetic cell bodies in the intermediolateral columns of the spinal cord.

The derangement producing parkinsonian tremor may not be the same as that producing bradykinesia. Dopamine antagonists induce bradykinesia rather than tremor as their main parkinsonian feature.[31] An ergot derivative and putative dopamine agonist, CF 25–397,[33] exacerbated bradykinesia yet improved tremor.[58] Our findings suggest that the noradrenergic system is involved in the genesis of parkinsonian tremor. Muscle activity due to tremor cannot be the cause of the increased CSF concentration of norepinephrine, since both the patients with and those without significant tremor had almost equal mean concentrations of plasma norepinephrine (Table 2).

Parkinsonian tremor may be produced by a relative imbalance between the noradrenergic and the dopaminergic systems. Drugs that have the ability to inhibit central noradrenergic activity while increasing dopaminergic activity may be more effective in treating tremor than drugs that either raise the central concentration of both dopamine and norepinephrine or simultaneously increase the activity of these respective systems.

Therapy with propranolol (a β-adrenergic antagonist) suppresses the amplitude of essential tremor but is marginally effective against parkinsonian tremor.[43] Motor activity is more significantly altered by central α-adrenergic mechanisms.[27] Treatment with α-adrenergic antagonists may improve parkinsonian tremor, though bradykinesia may be increased.

1.5. Summary

Untreated parkinsonian patients had a lower mean concentration of norepinephrine, in both CSF and plasma, than levodopa-treated parkinsonian patients and normal subjects. There was no significant difference in either the CSF or plasma concentration of norepinephrine between the treated parkinsonian patients and the normal controls.

The untreated parkinsonian patients with marked tremor had a significantly higher mean concentration of CSF norepinephrine, but not plasma norepinephrine, than the untreated parkinsonian patients without significant tremor.

The reduced concentration of CSF norepinephrine probably reflects the reduced concentration of brain tissue norepinephrine noted in postmortem studies of untreated parkinsonian patients. Levodopa therapy raises the concentration of norepinephrine both in CSF and in plasma.

There may be a relative noradrenergic–dopaminergic imbalance producing tremor. Drugs that have the ability to inhibit central noradrenergic activity while increasing dopaminergic activity may be more effective in treating tremor.

2. Lergotrile Mesylate Therapy: Effect on CSF Norepinephrine Concentrations in Parkinson's Disease

2.1. Introduction

Lergotrile mesylate, a dihydrogenated ergot alkaloid, is a dopaminergic agonist.[8] It is effective in the treatment of Parkinson's disease,[39,59] and is structurally similar to another dopaminergic dihydrogenated ergot derivative, bromocriptine.[10]

Lergotrile and bromocriptine have a similar profile of therapeutic response in parkinsonism. Like levodopa, they alleviate bradykinesia; they are more effective than levodopa in treating tremor, yet less effective in relieving rigidity.[46,59] Biochemically and pharmacologically, these ergot derivatives differ from levodopa. They do not stimulate dopamine-sensitive striatal adenyl cyclase, and they inhibit dopamine activation of this receptor molecule.[35,54,61] A cross-tolerance exists between lergotrile and bromocriptine, but neither is cross-tolerant with levodopa.[60]

Bromocriptine decreases the concentration of norepinephrine in both CSF and plasma.[67] Other dopaminergic drugs also effect noradrenergic systems. Dopamine, the chemical precursor of norepinephrine, increases the concentration of CSF and brain tissue norepinephrine.[51] Piribedil, a dopaminergic agonist structurally distinct from the ergot derivatives,[11] decreases the concentration of whole-brain norepinephrine in rodents, but increases the concentration of its major metabolite, 3-methoxy-4-hydroxyphenylethylene glycol sulfate ($MHPG{\cdot}SO_4$).[9]

We measured the CSF concentration of norepinephrine in parkinsonian patients both before they received lergotrile and during lergotrile therapy.

2.2. Patients and Methods

Nine patients suffering from Parkinson's disease were studied. There were eight males and one female with a mean age of 57 ± 3 years (± S.E.M.). Five subjects were not receiving any therapy for parkinsonism; four were taking levodopa/carbidopa at a mean dose of 1294/129 ± 303/30 mg per day. The doses of levodopa/carbidopa were not altered during the study. Subjects were on standard hospital diets.

Each patient had two lumbar punctures. All lumbar punctures were performed between midday and 3 P.M. with the patient lying supine on the left side. The first lumbar puncture was performed just prior to starting lergotrile therapy (baseline), the second after the patient received 50 mg lergotrile/day for 10 days. All patients took gradually increasing doses of lergotrile for 3–4 weeks before reaching this dose.

During both lumbar punctures, 4 ml CSF was collected after 16 ml had been taken for other tests. The sample was placed in a chilled tube containing 10 mg ascorbic acid, put on ice, and 30 min later frozen to −70°C. Norepinephrine assay on the CSF was carried out 4 weeks later. For details of the norepinephrine assay, refer to Section 1.2.

2.3. Results

The mean (± S.E.M.) concentration of CSF norepinephrine during the baseline period was 238 (± 65) pg/ml. During lergotrile therapy, the mean concentration of CSF norepinephrine fell to 159 (± 56) pg/ml ($p < 0.03$, two-tailed paired Student's t test).

2.4. Discussion

Norepinephrine undergoes rapid turnover but maintains a steady level because of a variety of regulatory mechanisms.[2] High concentrations of norepinephrine inhibit tyrosine hydroxylase, the rate-limiting enzyme in the synthesis of norepinephrine. When the concentration of intraneuronal norepinephrine falls (with rapid neuronal firing), the activity of tyrosine hydroxylase increases.[63] Released norepinephrine stimulates presynaptic α-adrenoreceptors to inhibit further release of norepinephrine.[37] In the isolated guinea pig "nerve–atria" preparation, phentolamine (an α-adrenergic antagonist) significantly increases both norepinephrine release and atrial rate as a result of nerve stimulation,[38] suggesting that presynaptic α-adrenoreceptors have a physiological role.

The ergot derivatives are α-adrenergic agonists and antagonists. Antagonist activity is increased and agonist activity decreased with dihydrogenation of the lysergic acid nucleus.[23] Lergotrile, like bromocriptine,[20] is an α-adrenergic antagonist as well as a dopaminergic agonist (Eli Lilly & Co., personal communication). Actions of ergot derivatives on the central nervous system occur at low doses; side effects prevent administration at doses that could produce more than minimal α-adrenergic blockade in man.[23] As an α-adrenergic antagonist, lergotrile should increase norepinephrine release.

In *in vitro* preparations, dopaminergic agonists (dopamine, piribedil, and apomorphine) inhibit the release of norepinephrine induced by stimulation of sympathetic nerves.[28,44] Dopaminergic antagonists prevent dopaminergic agonists from inhibiting norepinephrine release, but cannot prevent norepinephrine from inhibiting its own release.[29] There may be two populations of presynaptic receptors on the adrenergic neuron, one activated by dopaminergic drugs and one by adrenergic drugs.[29] The CSF concentration of norepinephrine should be determined by the concentration of extracellular or released norepinephrine. Lergotrile, like bromocriptine, may inhibit norepinephrine release, possibility by activating presynaptic dopamine receptors on noradrenergic neurons.[29,67]

Lergotrile, bromocriptine, and piribedil are all significantly effective in alleviating parkinsonian tremor.[53,59] They could achieve some of this therapeutic response by inhibiting the release of norepinephrine to correct a putative relative noradrenergic–dopaminergic imbalance. This imbalance may be one factor producing parkinsonian tremor.

2.5. Summary

Lergotrile, an ergot alkaloid and dopaminergic agonist, is structurally similar to bromocriptine. Bromocriptine decreases the concentration of CSF norepinephrine in parkinsonian patients and inhibits the presynaptic release of sympathetic neuronal norepinephrine in rabbits.

In patients with Parkinson's disease, there was a significant fall in the concentration of CSF norepinephrine during lergotrile therapy. Lergotrile probably inhibits the presynaptic release of norepinephrine.

Untreated parkinsonian patients with marked tremor have a higher mean concentration of CSF norepinephrine than those without significant tremor. Lergotrile and bromocriptine may be more effective than levodopa in treating tremor because of their ability to lower the concentration of norepinephrine, while at the same time acting as dopaminergic agonists.

3. Levodopa Therapy: Effect on CSF GABA Concentrations in Parkinson's Disease

3.1. Introduction

The amino acid γ-aminobutyric acid (GABA) is predominantly an inhibitory neurotransmitter.[55] L-Glutamic acid decarboxylase (GAD) catalyzes the decarboxylation of glutamic acid to GABA. The major catabolic route for GABA is reversible transamination with α-ketoglutarate, catalyzed by an aminotransferase, GABA-transaminase. The synthesis and breakdown of GABA occur in different sites. GAD is present in presynaptic nerve endings, whereas GABA-transaminase occurs in the mitochondria of postsynaptic cell bodies, their dendrites, glia, and endothelial cells.[4] GAD regulates the steady-state concentration of GABA.[55]

An inhibitory GABAergic striatal–nigral neuronal pathway has been demonstrated by anatomical, pharmacological, and biochemical studies.[17,19,24,25,52,65] Inhibition of GABA synthesis results in a decline in the concentration of GABA in the substantia nigra proportional to the degree of activity in the striatal–nigral pathway.[56] Hemitransections at the subthalamic level or destruction of the striatum markedly lowers the concentration of GABA in the substantia nigra.[34,36] GABA injected into the globus pallidus is transported to the substantia nigra.[45] GABA- and GAD-containing terminal boutons impinge on the cell bodies and dendrites of the substantia nigra.[3] GABA suppresses the activity of cells in the substantia nigra, while GABA antagonists block this effect of GABA.[16,48]

The highest concentrations of GABA and GAD are in the substantia nigra and globus pallidus.[15] In Parkinson's disease, the concentrations of GAD in the substantia nigra and globus pallidus are lower than normal.[30,55] Drugs increasing GABA concentrations in the brain reduce dopamine turnover in the striatum, decrease locomotor activity, and enhance neuroleptic-induced catalepsy.[1,5,30,50,64] Prolonged administration of levodopa increases the activity of GAD in the striatum.[30,40] We measured the concentration of GABA in lumbar CSF to determine whether untreated parkinsonian patients varied from control subjects, or levodopa therapy altered the concentration of GABA in CSF, an indirect measure of GABAergic activity.[25a,63a]

3.2. Patients and Methods

Lumbar CSF for GABA assay was obtained from 26 patients. These subjects came from three groups: (A) untreated parkinsonian patients; (B) treated parkinsonian patients; (C) control subjects. The control group were patients without any neurological or systemic disease. The number of subjects, mean age, and sex distribution for each group are shown in Table 3. The mean dose of levodopa/carbidopa in the treated parkinsonian group is also documented in Table 3. Patients were matched as closely as possible in regard to age and sex. The CSF samples were collected under standard conditions of time and rest.

3.2.1. Collection of CSF Samples. Lumbar puncture were done with the patient supine on his left side. Subjects had been at bedrest for 3 hr prior to CSF sampling. Since the spinal cord contains considerable amounts of GABA, the sample of CSF for GABA assay was collected after 20 ml CSF had been taken for other tests. A 4-ml sample was put into a tube with no preservative, placed on ice, and then deep-frozen at −70°C within 30 min of collection.

3.2.2. GABA Assay. GABA was measured in CSF by the radioreceptor assay of Enna *et al.*,[13] within 6 weeks of collection. This assay can measure down to 10 pmol/ml GABA. The assay is based on the displacement of synaptic-membrane-bound radioactive GABA by the GABA in the CSF sample.[13] Crude synaptic membranes are prepared from rat brain, frozen, and stored. Just prior to assay, the membranes are homogenized and incubated with Triton X-100. Membranes containing 1 mg protein are then incubated in a 2-ml volume with 10 nM triated GABA and 250 μl CSF. The sample is centrifuged and the radioactivity in the pellet measured. Standard curves for the displacement of [^{3}H]-GABA binding by unlabeled GABA was determined from six concentrations of GABA in 250 μl water and compared with the mean result from four determinations on each CSF sample.

3.3. Results

The mean CSF GABA concentration was determined in each of the three groups. These values were not significantly different (Table 3).

3.4. Discussion

A concentration gradient for GABA in lumbar CSF has been previously reported.[13,25a,63a] In our control subjects, 40 ml CSF was serially collected via lumbar puncture. Using the same method for assaying GABA in CSF, the mean concentration of GABA in the 24th ml of CSF removed was 16% lower than the mean GABA concentration in the 40th ml,[13] but this difference was not significant in this study.

In patients with Huntington's chorea, there is a significant decrease in the mean CSF concentration of GABA,[25a] consistent with the low tissue levels of GABA in the striatum of these patients.[21,47] Contrary to the findings of Hare *et al.*,[25a] our data did not indicate any deficiency of CSF GABA in Parkinson's disease. Different methods were used to assay GABA, but the mean concentration of CSF GABA in our untreated parkinsonian patients was 70% higher than the mean CSF GABA level in subjects with Huntington's chorea.[21] Though the concentration of GAD is decreased in the striatum of parkinsonian patients, there may not be a decrease in the level of tissue or released GABA.

Neither dopaminergic agonists nor antagonists alter the tissue concentration of GABA in the substantia nigra of rodents.[7] Levodopa therapy did not significantly alter the CSF GABA concentration in

Table 3. CSF GABA[a] in Parkinsonian Patients and Controls

Groups	Number and sex of patients	Mean age (yr) ± S.E.M.	Mean GABA (pmol/ml) ± S.E.M.	Mean dose levodopa/ carbidopa (mg/day) ± S.E.M.
A. Parkinsonian, untreated	9 (8 M, 1 F)	56 ± 3	181 ± 34	—
B. Parkinsonian	9 (7 M, 2 F)	60 ± 2	184 ± 31	1150/115 ± 240/24
C. Control	8 (1 M, 7 F)	51 ± 4	203 ± 5	—

[a] GABA, γ-Aminobutyric acid.

our parkinsonian patients when compared to controls or to untreated patients. Dopaminergic neurons do not appear to modify GABA neurons in this study. It may be necessary to measure turnover of GABA in the striatum to demonstrate whether dopaminergic drugs modify GABAergic activity.

3.5. Summary

The concentration of GABA in lumbar CSF was measured in untreated parkinsonian patients, levodopa-treated parkinsonian patients, and control subjects. There was no significant difference in the mean concentrations of CSF GABA among these groups of subjects. Levodopa therapy does not alter the concentration of CSF GABA. While GABAergic neurons appear to modulate dopaminergic activity, there is no evidence, from our study, that dopaminergic neurons modulate GABAergic activity.

ACKNOWLEDGMENT. We thank Barbara Ann Johnson for her indispensable secretarial and administrative help in preparing this manuscript.

References

1. ANDEN, N. E.: Inhibition of the turnover of brain dopamine after treatment with gamma-aminobutyric: 2-Oxyglutarate transaminase inhibitor of aminooxyacetic acid. *Arch. Pharmacol.* **283**:419–424, 1974.
2. AXELROD, J.: Relationship between catecholamines and other hormones. *Recent Prog. Horm. Res.* **31**:1–35, 1975.
3. BAK, I. J., CHOI, W. B., HASSLER, R., USUNOFF, K. G., WAGNER, A.: Fine structural synaptic organization of the corpus striatum and substantia nigra in rat and cat. In Calne, D. B., Chase, T. N., Barbeau, A. (eds.): *Advances in Neurology,* New York, Raven Press, 1975, pp. 24–41.
4. BARBER, R. P., SAITO, K.: Light microscopic visualization of GAD and GABA-T in immunocytochemical preparations of rodent CNS. In Roberts, E., Chase, T. N., Tower, D. B. (eds.): *GABA in Nervous System Function.* New York, Raven Press, 1976, pp. 113–132.
5. BISWAS, B., CARLSSON, A.: Effect of intraperitoneally administered GABA on locomotor activity of mice. *Psychopharmacology* **59**:91–94, 1978.
6. CARLSSON, A., LINDQUIST, M., MAGNUSSON, T.: 3,4-Dihydroxyphenylalanine and 5-hydroxytryptophan as reserpine antagonists. *Nature (London)* **180**:1200, 1957.
7. CATTABENI, F., EROS, T., GALLI, C. L., TONON, C. G.: Mass fragmentographic assay of GABA in substantia nigra: Effects of agonists and antagonists of dopaminergic and cholinergic systems. *Pharmacol. Res. Commun.* **7**(5):421–427, 1975.
8. CLEMENS, J.A., SMALSTIG, E. G., SHAAR, C. J.: Inhibition of prolactin secretion by lergotrile mesylate: Mechanism of action. *Acta Endocrinol.* **79**:230–237, 1975.
9. CONSOLO, S., FANELLI, R., GARATTINI, S., GHEZZI, D., JORI, A., LADINSKY, H., MARC, V., SAMANIN, R.: Dopaminergic cholinergic interaction in the striatum: Studies with piribedil. In Calne, D. B., Chase, T. N., Barbeau, A. (eds.): *Advances in Neurology,* Vol. 9. New York, Raven Press, 1975, pp. 257–272.
10. CORRODI, H., FUXE, K., HÖKFELT, T.: Effect of ergot drugs on central catecholamine neurones: Evidence for stimulation of central dopamine neurons. *J. Pharm. Pharmacol.* **25**:409–412, 1973.
11. CORRODI, H., FUXE, K., UNGERSTEDT U.: Evidence for a new type of dopamine receptor stimulating agent. *J. Pharm. Pharmacol.* **23**:989–991, 1971.
12. EHRINGER, H., HORNYKIEWICZ, O.: Verteilung von Noradrenalin and Dopamin (3 Hydroxytyramin) im Gehirn des Menschen und ihr verhalten Systems. *Klin. Wochenschr.* **38**:1236–1239, 1960.
13. ENNA, S. J., WOOD, J. H., SNYDER, S. H.: Radioreceptor assay for γ-aminobutyric acid (GABA) in human cerebrospinal fluid. *J. Neurochem.* **28**:1121–1124, 1977.
14. ESCOUROLLE, R. POIRIER, J.: *Manual of Basic Neuropathology.* Philadelphia, W. B. Saunders, 1978.
15. FAHN, S., COTE, L. J.: Regional distribution of gamma-aminobutyric acid (GABA) in brains of the rhesus monkey. *J. Neurochem.* **15**:209–213, 1968.
16. FELTZ, P.: Gamma-aminobutyric acid and a caudate–nigral inhibition. *Can. J. Physiol. Pharmacol.* **49**:1113–1115, 1971.
17. FONNUM, F., GROFOVA, I., RINVIK, E., STORM-MATHISEN, J., WALBERG, F.: Origin and distribution of glutamate decarboxylase in substantia nigra of the cat. *Brain Res.* **71**:77–92, 1974.
18. FUXE, K., HÖKFELT, T., OLSON, L., UNGERSTEDT, U.: Central monoaminergic pathways with emphasis on their relation to the so called "extrapyramidal motor system." *Pharmacol. Ther. B* **3**:169–210, 1977.
19. GALE, K., GUIDOTTI, A.: GABA mediated control of rat striatal tyrosine hydroxylase revealed by use of intranigral muscimol. *Pharmocologist* **18**:131, 1976.
20. GIBSON, A., SAMINI, M.: Bromocriptine is a potent alpha-adrenoceptor antagonist in the perfused mesenteric blood vessels of the rat. *J. Pharm. Pharmacol.* **30**:314–315, 1978.
21. GLAESER, B. S., VOGEL, W. H., OLEWEILER, D. B., HARE, T. A.: GABA levels in cerebrospinal fluid of patients with Huntington's chorea: A preliminary report. *Biochem. Med.* **12**:380–385, 1975.
22. GLOWINSKI, J., IVERSEN, L. L., AXELROD, J.: Storage and synthesis of norepinephrine in the reserpine treated rat brain. *J. Pharmacol. Exp. Ther.* **151**:385–399, 1966.

23. GOODMAN, L. S., GILMAN, A. (eds.): *The Pharmacological Basis of Therapeutics*. New York, Macmillan, 1975.
24. GROFOVA, I., RINVICK, E.: An experimental electron microscopic study on the striatonigral projection in the cat. *Exp. Brain Res.* **11**:249–262, 1970.
25. HAJDU, F., HASSLER, R., BAK, I. J.: Electron microscopic study of the substantia nigra and striatal–nigral projection in the rat. *Z. Zellforsch.* **146**:207–221, 1973.
25a. HARE, T. A. MANYAM, N. V. B., GLAESER, B. S.: Evaluation of cerebrospinal fluid α-aminobutyric acid content in neurologic and psychiatric disorders. In Wood, J. H. (ed.): *Neurobiology of Cerebrospinal Fluid I.* New York, Plenum Press, 1980.
26. HENRY, D. P., STARMAN, B. J., JOHNSON, D. G., WILLIAMS, R. H.: A sensitive radioenzymatic assay for norepinephrine in tissues and plasma. *Life Sci.* **16**:375–384, 1975.
27. HONMA, T., FUKUSHIMA, H.: Role of brain norepinephrine in neuroleptic induced catalepsy in rats. *Pharmacol. Biochem. Behav.* **7**:501–506, 1977.
28. HOPE, W., LAW, M. MCCULLOCH, M. W., RAND, J. M., STORY, D. F.: Effects of some catecholamines on noradrenergic transmission in the rabbit ear artery. *Clin. Exp. Pharmacol. Physiol.* **3**:15, 1975.
29. HOPE, W., MCCULLOCH, M. W., STORY, D. F., RAND, M. J.: Effects of pimozide on noradrenergic transmission in rabbit isolated ear arteries. *Eur. J. Pharmacol.* **46**:101–111, 1977.
30. HORNYKIEWICZ, O.: Neurohumoral interactions and basal ganglia function and dysfunction. In Yahr, M. D. (ed.): *The Basal Ganglia.* New York, Raven Press, 1976, pp. 269–280.
31. HORNYKIEWICZ, O.: Parkinsonism induced by dopaminergic antagonists. In Calne, D. B., Chase, T. N., Barbeau A. (eds.): *Advances in Neurology,* Vol. 9. New York, Raven Press, 1975, pp. 155–164.
32. HORNYKIEWICZ, O.: Biochemical and pharmacological aspects of akinesia. In Siegfried, J. (ed.): *Parkinson's Disease.* Bern, Huber, 1972.
33. JATON, A. L., LOEW, D. M., VIGOURET, J. M.: CF 25–397 (9,10-di-dehydro-6-methyl-8-beta-[2-pyridylthiomethyl] erogoline), a new central dopamine receptor agonist. *Br. J. Pharmacol.* **56**:371P, 1976.
34. KATAOKA, K., BAK, I. J., HASSLER, R., KIM, J. S., WAGNER, A.: L-Glutamate decarboxylase and choline acetyltransferase activity in the substantia nigra and the striatum after surgical interruption of the strio-nigral fibres of the baboon. *Exp. Brain Res.* **19**:217–227, 1974.
35. KEBABIAN, J. W., CALNE, D. B., KEBABIAN, P. R.: Lergotrile mesylate: An *in vivo* dopamine agonist which blocks dopamine receptors *in vitro*. *Commun. Psycholpharmacol.* **1**:311–318, 1977.
36. KIM, J. S., BAK, I. J., HASSLER, R., AKADA, Y.: Role of gamma-aminobutyric acid (GABA) in the extrapyramidal motor system. 2. Some evidence for the existence of a type of GABA-rich strio-nigral neurons. *Exp. Brain Res.* **14**:95–104, 1971.
37. LANGER, S. Z.: Presynaptic regulation of catecholamine release. *Biochem. Pharmacol.* **23**(13): 1793–1800, 1974.
38. LANGER, S. Z., ADLER-GASCHINSKY, E., GIORGI, O.: Physiological significance of alpha-adrenoceptor mediated negative feedback mechanism regulating noradrenaline release during nerve stimulation. *Nature (London)* **265**:648–650, 1977.
39. LIEBERMAN, A. L., MIYAMOTO, T., BATTISTA, A. F., GOLDSTEIN, M.: The antiparkinsonian efficacy of lergotrile. *Neurology (Minneapolis)* **25**:459–462, 1975.
40. LLOYD, K. G., HORNYKIEWICZ, O.: L-Glutamic acid decarboxylase in Parkinson's disease: Effect of L-dopa therapy. *Nature (London)* **243**:521–523, 1973.
41. LLOYD, K. G., HORNYKIEWICZ, O.: Catecholamines in regulation of motor function. In Friedhoff, A. J. (ed.): *Catecholamines and Behavior,* Vol. 1. New York, Plenum Press, 1975, pp. 41–57.
42. MARSDEN, C. D., PARKES, J. D., REES, J. E.: A comparison of treatment of Parkinson's disease with levodopa and with Sinemet. In Yahr, M. D. (ed.): *Current Concepts in the Treatment of Parkinsonism.* New York, Raven Press, 1973, pp. 21–36.
43. MCALLISTER, R. G., MARKESBERY, W. R., WARE, R. W., HOWELL, S. M.: Suppression of essential tremor by propranolol: Correlation of effect with drug plasma levels and intensity of beta-adrenergic blockade. *Ann. Neurol.* **1**:160–166, 1977.
44. MCCULLOCH, M. W., RAND, J. M., STORY, D. F.: Evidence for a dopaminergic mechanism for modulation of adrenergic transmission in the rabbit ear artery. *Br. J. Pharmacol.* **49**:41P, 1973.
45. MCGEER, P. L, FIBIGER, H. C., HATTON, T., SINGH, V. K., MCGEER, E. G., MALER, L.: Biochemical neuroanatomy of the basal ganglia. In Myers, R. D., Drucker-Colin, R. R. (eds.): *Advances in Behavioral Biology,* Vol. 10. New York, Plenum Press, pp. 27–47. 1974.
46. OLANOW, W. C., SCHWARTZ, A. M.: A controlled, blind double observor study of MK-486 (carbidopa) in Parkinson's disease. In Yahr, M. D. (ed.): *Current Concepts in the Treatment of Parkinsonism.* New York, Raven Press, 1973, pp. 69–86.
47. PERRY, T. L., HANSEN, S., KLOSTER, M.: Huntington's chorea: Deficiency of gamma-aminobutyric acid in brain. *N. Engl. J. Med.* **288**:337–342, 1973.
48. PRECHT, W., YOSHIDA, M.: Blockage of caudate evoked inhibition of neurons in the substantia nigra by picrotoxin. *Brain Res.* **32**:229–233, 1971.
49. PYCOCK, C.: Noradrenergic involvement in dopamine-dependent stereotyped and cataleptic responses in the rat. *Arch. Pharmacol. (Weinheim)* **298**:15–22, 1977.
50. RACAGNI, G., BRUNO, F., CATTABENI, F., *et al.*: Functional interaction between rat substantia nigra and striatum: GABA and dopamine interrelation. *Brain Res.* **134**:353–358, 1977.
51. RINNE, V. K., SONNINEN, V., RIEKKINEN, P., LAAKSONEN, H.: Postmortum findings in parkinsonian pa-

tients treated with L-dopa: Biochemical considerations. In Yahr, M. D. (ed.): *Current Concepts in the Treatment of Parkinsonism.* New York, Raven Press, 1973, pp. 211–233.

52. ROBERTS, E., KURIYAMA, K.: Biochemical–physiological correlations in studies of the gamma-aminobutyric acid system. *Brain Res.* **8**:1, 1968.
53. RONDOT, P., BATHIEN, N., RIBORDEAU DUMAS, J. L.: Indications of piribedil in L-dopa treated parkinsonian patients: Physiopathologic implications. In Calne, D. B., Chase, T. N., Barbeau, A. (eds.): *Advances in Neurology,* Vol. 9. New York, Raven Press, 1975, pp. 373–381.
54. SCHMIDT, M. J., HILL, L. E.: Effects of ergots on adenylate cyclase activity in the corpus striatum and pituitary. *Life Sci.* **20**:798, 1977.
55. SIEGEL, G. J., ALBERS, W. R., KATZMAN, R., AGRANOFF, B. W. (eds.): *Basic Neurochemistry.* Boston, Little, Brown, 1976.
56. SWIFT, R. M., HOFFMAN, P. C., HELLER, A.: Activity dependent changes in substantia nigra gamma-aminobutyric acid. *Brain Res.* **156**:181–186, 1978.
57. TEYCHENNE, P. F., CALNE, D. B., LEWIS, P. J., FINDLEY, L. J.: Interactions of levodopa with inhibitors of monoamine oxidase and L-aromatic amino acid decarboxylase. *Clin. Pharmacol. Ther.* **18**:273–277, 1975.
58. TEYCHENNE, P. F., PFEIFFER, R., BERN, S. M., CALNE, D. B.: Experiences with a new ergoline (CF 25-397) in parkinsonism. *Neurology (Minneapolis)* **27**:1140–1143, 1977.
59. TEYCHENNE, P. F., PFEIFFER, R. F., BERN, S. M., MCINTURFF, D., CALNE, D. B.: Comparison between lergotrile and bromocriptine in parkinsonism. *Ann. Neurol.* **3**:319–324, 1978.
60. TEYCHENNE, P. F., PLOTKIN, C. N., ROSIN, A. J., CALNE, D. B.: Cross tolerance between bromocriptine and lergotrile. *Neurology (Minneapolis)* **27**:406–407, 1977.
61. TRABUCCHI, M., SPANO, P. F., TONON, G. C., FRATTOLA, L.: Effects of bromocriptine on central dopaminergic receptors. *Life Sci.* **19**:225–232, 1976.
62. UNGERSTEDT, U.: Striatal dopamine release after amphetamine or nerve degeneration revealed by rotational behavior. *Acta Physiol. Scand. (Suppl. 83)* **367**:49–68, 1971.
63. WEINER, N., RABADJIJA, M.: The regulation of norepinephrine synthesis: Effect of puromycin on the accelerated synthesis of norepinephrine associated with nerve stimulation. *J. Pharmacol. Exp. Ther.* **164**:103–114, 1968.

63a. WOOD, J. H.: Sites of origin and concentration gradients of neurotransmitters, their precursors and metabolites,, and cyclic nucleotides in cerebrospinal fluid. In Wood, J. H. (ed.): *Neurobiology of Cerebrospinal Fluid I.* New York, Plenum Press, 1980.

64. WORMS, P., WILLIGENS, M. T., LLOYD, K. G.: GABA involvement in neuroleptic induced catalepsy. *J. Pharm. Pharmacol.* **30**:716–718, 1978.
65. YOSHIDA, M., PRECHT, W.: Monosynaptic inhibition of neurones of the substantia nigra by caudatonigral fibers. *Brain Res.* **32**:225–228, 1971.
66. ZIEGLER, M. G., LAKE, C. R., KOPIN, I. J.: Norepinephrine in cerebrospinal fluid. *Brain Res.* **108**:436–440, 1976.
67. ZIEGLER, M. G., LAKE, C. R., WILLIAMS, A. C., TEYCHENNE, P. F., SHOULSON, I., STEINSLAND, O.: Bromocriptine inhibits norepinephrine release. *Clin. Pharmacol. Ther.* **25**(2):137–142, 1978.
68. ZIEGLER, M. G., LAKE, C. R., WOOD, J. H., EBERT, M. H.: Circadian rhythm in cerebrospinal fluid noradrenaline of man and monkey. *Nature (London)* **264**:656–658, 1976.
69. ZIEGLER, M. G., WOOD, J. H., LAKE, C. R., KOPIN, I. J.: Norepinephrine and 3-methoxy-4-hydroxy-phenylglycol gradients in human cerebrospinal fluid. *Am. J. Psychiatry* **134**(5):565–568, 1977.

CHAPTER 16

Neurochemical Alterations in Parkinson's Disease

Thomas N. Chase

1. Introduction

In the early 1960s, reports appeared describing characteristic reductions in cerebral monoamine levels in patients with Parkinson's disease (for references, see Hornykiewicz[58]). A few years earlier, the observation had been made that central monoamine depletion by reserpine produced a Parkinson-like syndrome in rodents that could be reversed when brain dopamine levels were restored through the systemic administration of the immediate precursor of dopamine, L-DOPA.[20] These key observations pointed the way to the successful development of dopamine-replacement therapy for the relief of parkinsonian signs.

During the same period, newly developed assay techniques revealed the presence of monoamine metabolites in cerebrospinal fluid (CSF) and indicated that CSF levels of these substances in various animal species tended to reflect, albeit imperfectly, the central metabolism of the parent amines.[3,8,14,40,54,79,86] The clinical application of this technology quickly led to the discovery that dopamine and serotonin metabolite levels were significantly reduced in CSF from parkinsonian patients,[2,11,53,61] thus demonstrating that transmitter-system abnormalities could be identified in life through CSF measurements. With the development of improved assay techniques, as well as the introduction of methods for estimating the turnover of central transmitter amines, CSF studies have continued to play an important role in the search for biochemical abnormalities that might definitively elucidate the pathophysiology of parkinsonism and assist in the rational development of improved symptomatic therapies.

Thomas N. Chase, M.D. • National Institute of Neurological and Communicative Disorders and Stroke, National Institutes of Health, Bethesda, Maryland 20205.

2. Parkinson's Disease

2.1. Clinical and Pathological Features

Tremor at rest, muscular rigidity, and akinesia constitute the cardinal clinical features of Parkinson's disease. Associated signs may include facial masking, speech disturbances, micrographia, stooped posture, and a festinating gait. Onset of symptoms is typically insidious during middle or late life, and the course is of slow, inexorable progression.

Degeneration of neurons in the pars compacta of substantia nigra and certain other pigmented brainstem nuclei occurs as the distinctive neuropathological finding of idiopathic or postencephalitic parkinsonism. Although there is some evidence of neuronal loss elsewhere in the central nervous system (e.g., in the striatum, pallidum, and hypothalamus), changes in these areas are inconsistently observed and of uncertain significance.

2.2. Etiology

The cause of the selective neuronal degeneration in Parkinson's disease remains obscure. Parkinsonian symptoms can arise during the acute phase of a variety of encephalitides, or appear as either an early or late sequel of Von Economo's type A encephalitis. With this in mind, the recent finding that certain unconventional viruses can produce chronic, noninflammatory degeneration of specific populations of central neurons has led to speculation that idiopathic Parkinson's disease may be a consequence of a similar process. To date, however, no convincing evidence linking any form of parkinsonism with atypical virus infection has been uncovered.

2.3. Pathophysiology

Our understanding of the pathophysiology of the parkinsonian syndrome has advanced considerably during the past decade. An absolute prerequisite for this disorder is the loss of striatal inputs from the pigmented, dopamine-containing neurons having their cell bodies in the substantia nigra. The degeneration of this system and possibly other systems of dopaminergic neurons is now known to account for the reductions in dopamine, and in certain of its catabolites, that are characteristic of Parkinson's disease.[12]

Biochemical evidence also suggests that abnormalities in other transmitter systems occur in parkinsonian patients. For example, there is a diffuse decrease in brain norepinephrine,[58,91] presumably due to destruction of noradrenergic neurons having their cell bodies in the locus ceruleus. In addition, direct assays of parkinsonian brain tissues have disclosed reductions in serotonin[58] and in the activity of glutamic acid decarboxylase,[74,77,92] the enzyme that mediates γ-aminobutyric acid (GABA) synthesis, and possibly in nigral choline acetyltransferase, which catalyzes acetylcholine formation.[74] Consistent histopathological changes have yet to be associated with any of these latter biochemical alterations, and the precise contribution of norepinephrine, serotonin, acetylcholine, or GABA system function to the parkinsonian syndrome remains uncertain.

2.4. CSF Studies

To date, most investigations carried out on parkinsonian CSF have involved attempts to ascertain the functional state of various transmitter systems, the relationship of these abnormalities to aspects of the clinical syndrome and response to therapy, and the effect of centrally active drugs on these systems. Nonempirical clinical studies using CSF to elucidate the etiology of Parkinson's disease have been relatively infrequent and generally unrevealing.

3. Dopamine

3.1. Homovanillic Acid Determinations

Most CSF studies of dopaminergic function in parkinsonian patients have relied on measurements of homovanillic acid (HVA), the major product of dopamine catabolism in the mammalian central nervous system. In view of the characteristic reductions in brain dopamine and HVA in Parkinson's disease,[58] and since structures principally affected by the disease process (such as the caudate nucleus) lie close to the ventricular system, it is not surprising that the most consistently abnormal CSF finding has been a decrease in HVA levels in ventricular,[53,84,85,107] cisternal,[59] and lumbar CSF.[11,13,30,45,48,61,90,94,111,113] This reduction is observed in patients with either idiopathic or postencephalitic parkinsonism,[11,38,94] as well as in a variety of other neurological disorders that have parkinsonian features, such as parkinsonism–dementia of Guam, striatonigral degeneration, progressive supranuclear palsy, and olivopontocerebellar degeneration.[23,88]

Numerous attempts have been made to relate steady-state HVA values in CSF with various clinical characteristics of the disorder. Despite reports that the basal CSF content of HVA correlates with the severity of parkinsonian signs or with the response to L-DOPA,[19,55,59,78,85] most studies document the general inconsistency of such associations.[21,30,32,38,49,70,90,94,96,111,113,114] Indeed, HVA values typically vary considerably from patient to patient, and levels in parkinsonian individuals not infrequently overlap with those found in normal subjects. Moreover, since low HVA levels may occur in a variety of disorders (e.g., amyotrophic lateral sclerosis, multiple sclerosis, Alzheimer's disease, and certain affective states) that are not associated with a primary abnormality in the dopamine system, it is generally conceded that basal CSF HVA measurements contribute little to the diagnosis or staging of Parkinson's disease or to the prediction of the eventual response to drug therapy.

3.2. Other Dopamine Metabolites

In addition to HVA, various other metabolites of dopamine have been measured in human CSF.[39,76,100,112] Most of these substances occur at relatively low levels in CSF and have received only scant attention in parkinsonian patients. A partial exception has been CSF measurements of 3,4-dihydroxyphenylacetic acid (DOPAC).[100,112] Levels of this deaminated derivative of dopamine may reflect more accurately than HVA the amount of dopamine released extraneuronally and thus be specially useful in future studies of Parkinson's disease.

3.3. Techniques to Estimate Turnover

Various factors in addition to the functional state of the dopamine system can influence CSF HVA measurements. These include the effects of patient physical activity and diet; differences in the sensitivity and specificity of available assay methods[60,104]; sampling errors introduced by the steep concentration gradient for HVA along the neuraxis[64,100,103,104,118]; the dynamics of CSF production, flow, and absorption; and the fact that only a small fraction of HVA produced in the central nervous system is excreted via the CSF.[65] Since CSF HVA levels reflect rates of efflux from the CSF compartment as well as the rate of synthesis and influx, drugs that inhibit efflux mechanisms, without substantially influencing dopamine metabolism, should provide an index to the rate of formation of this metabolite and thus to the turnover of the parent amine. An agent widely used to achieve this goal is probenecid.[52,117] Although no rigorous verification of the validity of probenecid test results is possible in man, studies in the experimental animal as well as those involving drugs that influence central monoamine metabolism in a known way are generally consistent with this view.[47,108]

Despite differing techniques for the administration of probenecid and for the timing of CSF collections, results using this approach have been reasonably consistent. Parkinson's disease is associated with a substantial reduction in the probenecid-induced accumulation of HVA in CSF.[29,30,70,82,96] Moreover, the overall severity of this disorder, as well as the severity of both akinesia and rigidity, appears to correlate inversely with dopamine turnover as estimated by the probenecid technique.[29,30,68,94] The notable lack of a relationship between apparent dopamine turnover and tremor suggests that this cardinal parkinsonian feature may reflect dysfunction in a nondopaminergic system. Although these observations have not been confirmed by every investigator,[15] they are consistent with direct biochemical measurements made at autopsy.[12]

Since the response to L-DOPA varies considerably, a pretherapeutic test to predict antiparkinsonian efficacy and toxicity would be of considerable value. Unfortunately, a number of studies have failed to find a consistent relationship between central dopamine metabolism as estimated by the probenecid test and the degree of improvement with L-DOPA.[15,30,96] On the other hand, biochemical investigations of postmortem tissues suggest that parkinsonian individuals with the most advanced symptoms and the most profound loss of striatal dopamine may have the most favorable response to L-DOPA.[12] Consistent with this observation, some investigators have reported a negative correlation between post-probenecid HVA values and improvement on L-DOPA in a subgroup of parkinsonian patients who have relatively normal pretreatment HVA elevations with probenecid.[68,70,71] Uncertainties concerning the validity of the probenecid technique for estimating central dopamine metabolism, as well as the variability of probenecid test results (due to such factors as interpatient differences in the pharmacokinetics of the drug, and the inability to obtain a total HVA blockade), have prompted the search for alternative approaches to the study of central monoamine turnover.

Some attention has been given to the use of radioisotopic labeling techniques. Although the systemic administration of L-[^{14}C]-DOPA can result in measurable amounts of labeled HVA in human lumbar CSF,[89] use of this approach has provided few useful data, largely because administration of maximum permissible radiation-dose levels yields HVA specific activities in parkinsonian CSF very close to background levels.[43]

Nonradioactive (stable) isotopes afford another potential means to study central monoamine metabolism. In the experimental animal, brain HVA labeling following exposure to a breathing mixture containing oxygen-18 appears to reflect the rate of central dopamine turnover.[99] Recent clinical studies indicate that inhalation of oxygen-18 can produce easily detectable labeling of monoamine metabolites in CSF.[28] Moreover, preliminary results suggest that despite the reduction in endogenous HVA levels, central dopamine turnover may be relatively rapid in parkinsonian patients.

3.4. Drug Effects

CSF dopamine metabolite studies have been widely applied to the clinical investigation of the pharmacological effects of drugs used to alleviate parkinsonism. The administration of L-DOPA is associated with a dose-dependent rise in CSF HVA,[32,36,38,45,57,90,96,111,113] but little or no change in

DOPAC.[100,112] As might be expected, L-DOPA, given alone or with a peripheral decarboxylase inhibitor, also increases CSF DOPA, dopamine, and iso-HVA levels;[39,90,100] these elevations seem, however, to vary widely in patients receiving the same L-DOPA dose and appear unrelated to the clinical effects observed.[83]

HVA elevations during treatment with L-DOPA in combination with a peripheral decarboxylase inhibitor are significantly less than those occurring when L-DOPA is given alone at therapeutically equivalent dose levels.[19,32,38,96] This difference suggests that a substantial portion of the L-DOPA-induced HVA rise in CSF reflects DOPA-decarboxylation in the walls of brain capillaries.[9] Moreover, it is probable that the metabolism of exogenous L-DOPA in nondopaminergic neurons and at other sites unrelated to its antiparkinsonian actions also contributes to dopamine metabolite levels in CSF.[81] Thus, despite reports that CSF HVA increments during L-DOPA treatment alone or with a peripheral carboxylase inhibitor correlate with the antiparkinsonian response or the appearance of centrally mediated adverse effects,[45,57,78] it is not surprising that a majority of studies have failed to find this relationship.[36,38,55,59,66,96,111,114] This view is in agreement with a report of direct biochemical measurements in postmortem tissues of parkinsonian patients.[95]

Numerous preclinical studies have demonstrated the ability of anti-psychotic drugs that block dopamine receptors to stimulate the synthesis and release of this transmitter amine. In man, the subacute administration of ordinary therapeutic dose levels of the antipsychotic agent haloperidol has been shown to increase both basal levels and probenecid-induced accumulations of CSF HVA.[24,26] Studies of the probenecid-induced accumulation of HVA during treatment with drugs that influence receptor mechanisms may be useful as a provocative test of dopaminergic function in man. For example, the relative ability of haloperidol to accelerate dopamine metabolism might serve as an index to the latent capacity of the presynaptic neuron to synthesize and release this amine as well as to the integrity of feedback mechanisms affecting dopamine formation.[24] Dopamine turnover values obtained prior to haloperidol administration, on the other hand, may provide information relative to the basal condition of the dopamine system. Analyses of these biochemical data together with clinical observations might enable certain inferences relative to the functional state of postsynaptic dopamine receptors as well as of the presynaptic dopamine neurons.

Preliminary attempts to apply this approach using CSF assays suggest that in disorders such as Huntington's disease, where the dopamine system probably remains intact, haloperidol substantially increases dopamine turnover as estimated by the probenecid technique.[26] On the other hand, haloperidol has no significant effect on the probenecid-induced accumulation of HVA in the CSF of patients with naturally occurring Parkinson's disease. These findings are not unexpected in view of the reduced number and suspected hyperfunctional state of surviving dopamine neurons.[1,12]

Studies of patients who manifest parkinsonian signs only during treatment with antipsychotic agents, such as haloperidol, show essentially normal rates of apparent dopamine turnover during periods when they are not receiving centrally active drugs.[26] When these individuals are given haloperidol, parkinsonian signs emerge, and dopamine turnover (as reflected by the probenecid-induced accumulation of CSF HVA) increases significantly. These results may indicate that no major functional impairment of dopamine-containing neurons need be present in patients with drug-induced parkinsonism and are thus consistent with the view that blockade of postsynaptic dopamine receptors accounts for the appearance of parkinsonian signs during antipsychotic drug therapy. Moreover, in individuals with drug-induced parkinsonism, receptor blockade rather than augmentation of the synthesis and release of dopamine seems to be the overriding functional effect of haloperidol.

Dopamine-receptor agonists, as well as antagonists, might be expected to serve as provocative pharmacological agents to test the functional state of the dopamine system. On the other hand, in Parkinson's disease, where dopaminergic function is already impaired, use of agonists may be of limited practical value, since their effect is to further depress dopamine turnover.[24,33,93] One study has, however, demonstrated a relationship between the ability of dopamine receptor agonists (piribedil or bromocriptine) to reduce the probenecid-induced accumulation of HVA and their antiparkinsonian efficacy, such that patients evidencing the greatest improvement had the largest reduction in apparent dopamine turnover.[93] Individuals with the best response to treatment also tended to be those with the smallest reduction in dopamine turnover prior to therapy. That the destruction of dopaminergic neurons occurring in parkinsonian patients diminishes the attenuating effect of dopamine-receptor agonists on the turnover of this amine is further suggested by the results of studies in which parkinsonian and nonparkinsonian patients were compared.[33]

4. Norepinephrine

It is now well established from preclinical studies that noradrenergic mechanisms contribute to the regulation of locomotor activity, and from postmortem assays that cerebral norepinephrine levels are decreased in patients with Parkinson's disease.[58,91] The norepinephrine reductions are present diffusely, presumably reflecting degeneration of noradrenergic terminals having their cell bodies in the locus ceruleus. Pharmacological observations also suggest that central noradrenergic function may be influenced by L-DOPA loading. For these reasons, CSF studies have attempted to explore the relationship of norepinephrine system function to the pathogenesis of parkinsonism and to the clinical effects of L-DOPA or dopamine-agonist therapy.

In the past, the principal approach to CSF studies of noradrenergic mechanisms in man has been the measurement of 3-methoxy-4-hydroxyphenylethylene glycol (MHPG), a major product of central norepinephrine metabolism. Although CSF MHPG levels appear to bear a close relationship to central norepinephrine metabolism,[27] no consistent alteration in MHPG values has been found in the CSF of parkinsonian individuals.[27,38,119] Moreover, MHPG values tend to vary widely from one parkinsonian patient to another, but no correlation has been observed between clinical severity and MHPG level.[27,38] Attempts to use probenecid to study the dynamics of central noradrenergic metabolism have also been unsuccessful, since probenecid has little effect on CSF concentrations of either free or conjugated MHPG.[27] The administration of therapeutic dose levels of L-DOPA to parkinsonian patients does not significantly alter MHPG values in CSF.[27,38,119]

Several factors appear to contribute to the lack of MHPG depression in the lumbar CSF of patients with Parkinson's disease and the unresponsiveness of MHPG to probenecid. In contrast to the situation with dopamine, considerable norepinephrine metabolism occurs within noradrenergic neurons in the spinal cord. Clearly, MHPG synthesis in the spinal cord is an important source of MHPG in lumbar CSF,[105] and noradrenergic dysfunction occurring only in brain might not produce any detectable change in norepinephrine metabolite levels in CSF. Preclinical studies also indicate that MHPG diffuses rapidly from CSF to blood[120] and may be cleared from the central nervous system largely by this means, rather than by a probenecid-sensitive transport mechanism. This situation undoubtedly contributes to the absence of any substantial response to probenecid.

Certainly, available CSF data bearing on such questions as the participation of noradrenergic mechanisms in the pathophysiology of parkinsonism or in modifying the response to dopaminergic agents remain inconclusive. On the other hand, in view of pharmacological evidence suggesting an influence of central noradrenergic mechanisms on the antiparkinsonian efficacy of L-DOPA,[101,102] the exploration of CSF markers of central norepinephrine metabolism other than MHPG would seem appropriate. In this regard, considerable investigative effort has recently been devoted to CSF studies of the norepinephrine metabolite vanillylmandelic acid,[60] to norepinephrine itself, [121] and to the norepinephrine-synthesizing enzyme dopamine-β-hydroxylase.[46] The potential for studying Parkinson's disease through CSF measurements of this type warrants further attention.

5. Serotonin

Most CSF studies of serotoninergic function in parkinsonian patients have measured 5-hydroxyindoleacetic acid (5-HIAA), the principal product of serotonin degradation. In parkinsonian patients, 5-HIAA levels in ventricular or lumbar CSF tend to be decreased, although not nearly as consistently as the reduction in HVA.[25,30,48,53,55,61,63,90,96,111] That some decline in 5-HIAA levels might occur is certainly not unexpected in view of the diffuse loss of serotonin in the cerebral tissues of parkinsonian patients studied at autopsy.[58]

A substantial portion of 5-HIAA in CSF originates from serotonin metabolism in the spinal cord, while only a fraction of 5-HIAA deriving from cerebral serotonin metabolism ever reaches the lumbar CSF.[17,18,37,105,115] These factors undoubtedly contribute to the variability of results of CSF 5-HIAA assays in parkinsonian patients. Supporting this view is the observation that 5-HIAA reductions occur more consistently in samples of ventricular CSF.[53] Moreover, attempts to correlate 5-HIAA values in lumbar CSF with the severity of parkinsonian signs or the subsequent therapeutic response to dopaminergic agents have generally been unsuccessful.[30,49,55,70,90,94,96] Two studies have, however, observed that the antiparkinsonian efficacy of L-DOPA may be better in patients with relatively high basal 5-HIAA levels.[38,55] These findings, although not observed consistently,[68] might indicate that the presence of an intact serotonin system contributes to the ability of L-DOPA to ameliorate parkinsonian signs.

Probenecid interferes with the active transport of 5-HIAA from CSF in much the same way as it effects HVA efflux. Studies of serotonin metabolism using this technique suggest that Parkinson's disease is attended by a significant reduction in serotonin turnover.[25,30,82,96] A negative association between the severity of parkinsonian signs and the decline in apparent serotonin turnover has been observed in some studies,[30,68] although not in others.[15,96] No relationship has been found between the probenecid-induced elevation in 5-HIAA in lumbar CSF and the response to L-DOPA.[70,96] Conceivably, biochemical evidence of serotoninergic dysfunction in Parkinson's disease may reflect a secondary or compensatory response to the primary changes affecting the dopamine system. Pharmacological observations in parkinsonian patients fail to support the contention that the serotonin system plays a significant role in the pathophysiology of this disorder.[10,22,25,31]

Treatment of parkinsonian patients with L-DOPA, alone or with a peripheral decarboxylase inhibitor, has been reported to diminish basal 5-HIAA levels in lumbar CSF in some studies,[96,111] although not in others.[19,32,38,45,62,90] Consistent with the former observations, some investigators have obtained evidence, using the probenecid technique, suggesting that a reduction in central serotonin turnover attends L-DOPA therapy.[62,111] Such a change is not entirely implausible, since L-DOPA loading can depress serotonergic function due to competition among the amine precursors for uptake and metabolism and among the amines themselves for intraneuronal storage in the central nervous system.

6. Amino Acids

6.1. γ-Aminobutyric Acid

Since the limited sensitivity of the ninhydrin reaction hindered GABA detection by conventional amino acid analysis, considerable controversy has surrounded attempts to measure GABA levels in human CSF. Recently, however, amino acid analysis using fluorometric, radioreceptor, or gas chromotographic-mass spectrometric procedures has begun to yield reasonably consistent results, thus allowing interest to turn to the possible diagnostic value of CSF GABA determinations.

Significant reduction in CSF GABA levels has been reported in several studies of parkinsonian patients,[51,72,80] but not in all.[42] Some alteration in CSF GABA levels might be expected on the basis of the reportedly diminished activity of glutamic acid decarboxylase (the enzyme mediating GABA synthesis) in the basal ganglia and cerebral cortex of parkinsonian brains.[74,77,92] (See Chapters 13–15.)

The reliability of CSF GABA measurements has yet to be fully accepted, and their biological meaning or diagnostic potential in patients with Parkinson's disease remains to be determined. It is not at all unlikely that CSF GABA may largely reflect metabolism in pools unrelated to GABA-mediated synaptic transmission. If such be true, even though CSF GABA values provide an accurate index to central nervous system concentrations of this amino acid, CSF GABA assays may be expected to contribute little to our understanding of GABA system function.

6.2. Other Amino Acids

Abnormalities of other free amino acids have been the subject of numerous empirical searches in patients with Parkinson's disease. Amino acids in CSF are essential constituents of metabolic processes in the central nervous system; in addition, some appear to serve as transmitters or precursors of transmitters. Levels of these substances are regulated within fairly close limits by carrier-mediated transport mechanisms, which often may be rate-limiting.[87]

Various abnormalities in free amino acid concentrations in the CSF of parkinsonian patients have been reported.[16,44,51,69,72,109] These investigations have usually found elevations in clusters of amino acids, although there is some variability from study to study in exactly which amino acids are affected. Most of these changes do not, however, appear specific to Parkinson's disease and are thus of uncertain significance. Studies of the effects of L-DOPA, alone or with a peripheral decarboxylase inhibitor, on amino acid concentrations in parkinsonian CSF have also tended to yield inconclusive results.[56,69]

7. Acetylcholine

Since some dopaminergic neurons originating in the substantia nigra make synaptic contact with striatal cholinergic interneurons, and in view of the well-known ability of cholinergic drugs to influence parkinsonian signs, biochemical evidence of altered cholinergic function might be expected in the CSF of parkinsonian patients. To date, however, no consistent alteration in CSF choline or acetylcholine has been found.[6,50,116] On the other hand, autopsy studies of parkinsonian brains report either essentially normal choline acetyltransferase activity in the major extrapyramidal nuclei[77] or diminished activity only in the substantia nigra.[74] Moreover, even in disorders (such as Huntington's disease) in which striatal choline acetyltransferase activity is known to be substantially decreased, CSF acetylcholine levels are reportedly normal,[116] while choline concentrations have been found to be normal[116] or reduced.[6]

The foregoing observations raise some question about the reliability of existing CSF approaches to the evaluation of central cholinergic function. Although it has been suggested that CSF choline may reflect acetylcholine turnover in the central nervous system,[7] more recent studies have cast doubt on the validity of this notion.[97] Similarly, the relationship between the CSF content of acetylcholine (which occurs at extremely low levels, in part due to the effect of specific and nonspecific cholinesterases) and the functional state of the central cholinergic system remains to be determined. In the future, assays of choline acetyltransferase activity in CSF may provide a more useful approach to the study of central cholinergic function.[5]

8. Miscellaneous Substances

8.1. Peptides

A variety of endogenous peptides in the mammalian central nervous system are now recognized as having neurotransmitter or neuromodulator activity. Some of these compounds, e.g., substance P, somatostatin, angiotensin, and certain peptides active at opiate receptors (endorphins and enkephalins), occur in high concentrations in the basal ganglia, and their pharmacological manipulation in the experimental animal influences motor function. Further knowledge of these peptidergic systems might be expected to contribute to our understanding of parkinsonism and especially to the development of improved pharmacotherapies for this disorder. Assay systems for many of these peptides in human CSF are now rapidly becoming available.

8.2. Proteins

Analysis of the protein content of CSF from parkinsonian patients has yielded essentially normal results using paper and agar electrophoresis;[73] isoelectric focusing of CSF proteins has revealed only a nonspecifically abnormal pattern in some individuals, which might reflect CSF-barrier alterations.[106]

8.3. Cyclic Nucleotides

An adenylate cyclase system is associated with the receptors for a number of central transmitters, and the cyclic nucleotides, adenosine 3′,5′-monophosphate (cyclic AMP) and guanosine 3′,5′-monophosphate (cyclic GMP), appear to function as second messengers in mediating the cellular effects of these transmitters. Monoamines and histamine have been shown to stimulate cyclic AMP production, while cholinergic receptor function has been linked with the accumulation of cyclic GMP. Although not all receptors (including some dopamine receptors) function with cyclic nucleotides, in many cases, changes in the activity of the presynaptic neurons or in the sensitivity of postsynaptic receptors might be expected to influence cyclic nucleotide levels.

Cyclic AMP is present in measurable quantities in human CSF, appears to be of central origin, and is eliminated by a probenecid-sensitive mechanism.[34,98] On the other hand, measurements of cyclic AMP in the CSF of parkinsonian patients thus far have failed to reveal any characteristic abnormalities. Moreover, neither the acute nor the chronic administration of L-DOPA significantly affects CSF cyclic AMP concentrations.[35] Nevertheless, future studies of this type may be of value, since the receptor origin of the cyclic nucleotides complements CSF studies of transmitters and their metabolites arising from presynaptic terminals.

8.4. Hydroxylase Cofactor

Assay of CSF levels of tetrahydrobiopterin (BH_4) may serve as another approach to the evaluation of central monoaminergic function. BH_4 is a cofactor for the enzyme that hydroxylates tyrosine and tryptophan, and thus contributes to the regulation of the synthesis of dopamine, norepinephrine, and serotonin. Alterations in CSF BH_4 levels might therefore reflect functionally important changes in the production of monoamines. Assays of BH_4 levels in the CSF of Parkinson patients show a substantial reduction in comparison to age-matched control subjects.[75] Moreover, there is a close correlation between BH_4 and HVA values, but not with 5-HIAA concentrations. These results support the view that dopaminergic neurons may be a major source of BH_4 in the lumbar CSF.

8.5. Antibodies

As an approach to the search for a possible viral agent in Parkinson's disease, CSF antibodies to arborviruses have recently been sought in patients with either the idiopathic or the postencephalitic form of this disorder.[41] Hemagglutination inhibition results were negative, however, with all 17 arborvirus antigens tested.

9. Conclusions

Assays of substances in CSF, especially those that appear to reflect the functional state of a chemically defined neuronal pathway, serve as a unique source of information for studies of central nervous system disease mechanisms and therapeutic interventions in the living patient. Once a linkage can be established between dysfunction of a specific neuronal system and the appearance of a characteristic clinical syndrome, drugs can be sought that selectively and potently manipulate the affected system or subsystem in the appropriate direction. Such was the basis for initial therapeutic trials with L-DOPA as well as current investigations of dopamine-receptor agonists in the treatment of Parkinson's disease.

Measurement of monoamine metabolites in parkinsonian CSF provided early confirmation of the presence of characteristic abnormalities in these neural systems and later afforded insight into the complex relationships between monoaminergic function and various aspects of the clinical syndrome. Data deriving from CSF studies have also supplied important information on the diverse pharmacological actions of drugs used in the treatment of Parkinson's disease. Considering the rapid discovery of additional substances that serve as transmitters or modulators within the mammalian central nervous system as well as the development of more sensitive and selective assay techniques, future CSF studies can be expected to continue to further our understanding of the pathogenesis and symptomatic treatment of Parkinson's disease.

Major uncertainties do, however, plague the collection and interpretation of CSF data: for many of the conditions studied and for many of the substances assayed, it has yet to be precisely established what is being measured biologically and, in some instances, even chemically. In order that future CSF studies attain their full potential, such questions must be satisfactorily resolved, especially in ways that allow assessment of the dynamic state of neuronal function. Of no less importance are steps to extend current CSF studies to include nonempirical evaluations of etiological mechanisms, for only such approaches offer hope for the eventual development of solutions to the problems of prevention and definitive intervention.

References

1. AGID, Y., JAVOY, F., GLOWINSKY, J.: Hyperactivity of remaining dopaminergic neurons after partial destruction of the nigrostriatal dopaminergic system in the rat. *Nature (London) New Biol.* **245**:150, 1973.
2. ANDEN, N., ROOS, B., WERDINIUS, B.: On the occurrence of homovanillic acid in brain and cerebrospinal fluid and its determination by a fluorometric method. *Life Sci.* **7**:448–458, 1963.
3. ANDERSSON, H., ROOS, B.: 5-Hydroxyindoleacetic acid in cerebrospinal fluid after administration of 5-hydroxytryptophan I. *Acta Pharmacol.* **26**:293–297, 1968.
4. ANDERSSON, H., VON ESSEN, C., ROOS, B.: 5-Hydroxyindole acetic acid and homovanillic acid in cerebrospinal fluid after intrathecal and intravenous administration of probenecid to normal and hydrocephalic dogs. *Acta. Pharmacol.* **32**:139–146, 1973.
5. AQUILONIUS, S.-M., ECKERNÁS, S.-A.: Choline acetyltransferase in human cerebrospinal fluid: Non enzymatically and enzymatically catalysed acetylcholine synthesis. *J. Neurochem.* **27**:317–318, 1976.
6. AQUILONIUS, S.-M., NYSTROM, B., SCHUBERTH, J., SUNDWALL, A.: Cerebrospinal fluid choline in extrapyramidal disorders. *J. Neurol. Neurosurg. Psychiatry* **35**:720–725, 1972.
7. AQUILONIUS, S.-M., SCHUBERTH, J., SUNDWALL, A.: Choline in the cerebrospinal fluid as a marker for the release of acetylcholine. In Heilbronn, E., Winter, A. (eds.): *Drugs and Cholinergic Mechanisms in the CNS.* Stockholm, Forsvarets Forskningsanstatt, 1970, pp. 399–410.
8. BARTHOLINI, G., PLETSCHER, A., TISSOR, R.: On the origin of homovanillic acid in the cerebrospinal fluid. *Experientia* **15**:609–610, 1966.
9. BARTHOLINI, G., TISSOT, R., PLETSCHER, A.: Brain capillaries as a source of homovanillic acid in cerebrospinal fluid. *Brain Res.* **27**:163–168, 1971.
10. BEASLEY, B. L., DAVENPORT, R. W., CHASE, T. N.: Fenfluramine treatment of parkinsonism. *Arch. Neurol.* **34**:255–256, 1977.
11. BERNHEIMER, H., BIRKMAYER, W., HORNYKIEWICZ,

O.: Homovanillic acid in the cerebrospinal fluid in Parkinson's syndrome and other diseases of the CNS. *Wien. Klin. Wochenschr.* **23**:417–419, 1966.

12. Bernheimer, H., Birkmayer, W., Hornykiewicz, O., Jellinger, K., Seitelberger, F.: Brain dopamine and the syndromes of Parkinson and Huntington: Clinical, morphological and neurochemical correlations. *J. Neurol. Sci.* **20**:415–455, 1973.
13. Bernheimer, H., Hornykiewicz, O.: Das Verhalten des Dopamin-Metaboliten Homovanillinsäure im Gehirn von normalen und Parkinson-kranken Menschen. *Arch. Exp. Pathol. Pharmakol.* **247**:305–306, 1964.
14. Bowers, M. B., Jr.: 5-Hydroxyindoleacetic acid in the brain and cerebrospinal fluid of the rabbit following administration of drugs affecting 5-hydroxytryptamine. *J. Neurochem.* **17**:827–828, 1970.
15. Bowers, M. B., Jr., Van Woert, M. H.: The probenecid test in Parkinson's disease. *Lancet* **2**:926–927, 1972.
16. Bruck, H., Gerstenbrand, F., Gnad, H., Grundig, E., Prosenz, T.: Vergleichende biochemische Untersuchungen bei organischen Parkinson Syndromen und medikamentösem Parkinsonoid. *Psychiatr. Neur. (Basel)* **151**:81–87, 1966.
17. Bulat, M., Živković, B.: Origin of 5-hydroxyindoleacetic acid in the spinal fluid. *Science* **173**:738–740, 1971.
18. Burns, D., London, J., Brunswick, D. J., Pring, M., Garfinkel, D., Rabinowitz, J. L., Mendels, J.: A kinetic analysis of 5-hydroxyindoleacetic acid excretion from rat brain and CSF. *Biol. Psychiatry* **11**:125–157, 1976.
19. Campanella, G., Algeri, S., Cerletti, C., Dolfini, E., Jori, A., Rinaldi, F.: Correlation of clinical symptoms, HVA and 5-HIAA in CSF and plasma L-dopa in parkinsonian patients treated with L-dopa and L-dopa + RO 4-4602. *Eur. J. Clin. Pharmacol.* **11**:255–261, 1977.
20. Carlsson, A., Lindqvist, M., Magnusson, T.: 3,4-Dihydroxyphenylalanine and 5-hydroxytryptophan as reserpine antagonists. *Nature (London)* **180**:1200, 1957.
21. Casati, C., Agnoli, A., Jori, A., Dolfini, E.: On the relationship between L-dopa therapy and CSF monoamine metabolites in Parkinson's disease. *Z. Neurol.* **204**:149–154, 1973.
22. Chase, T. N.: Serotonergic mechanisms in Parkinson's disease. *Arch. Neurol.* **27**:354–356, 1972.
23. Chase, T. N.: Catecholamine metabolism and neurologic disease. In Usdin, E., Snyder, S. (eds.): *Frontiers in Catecholamine Research.* New York, Pergamon Press, 1973, pp. 1127–1132.
24. Chase, T. N.: Central monoamine metabolism in man: Effect of putative dopamine receptor agonists and antagonists. *Arch. Neurol.* **29**:349–351, 1973.
25. Chase, T. N.: Serotonergic mechanisms and extrapyramidal function in man. In McDowell, F., Barbeau, A. (eds.): *Second Canadian American Conference on Parkinson's Disease.* New York, Raven Press, 1974, pp. 31–39.
26. Chase, T. N.: Antipsychotic drugs, dopaminergic mechanisms and extrapyramidal function in man. *Wenner-Gren Center International Symposium Series,* Vol. 25. Oxford, Pergamon Press, 1976, pp. 321–329.
27. Chase, T. N., Gordon, E. K., Ng, L. K. Y.: Norepinephrine metabolism in the central nervous system of man: Studies using 3-methoxy-4-hydroxyphenylethylene glycol levels in cerebrospinal fluid. *J. Neurochem.* **21**:581–587, 1973.
28. Chase, T. N., Neophytides, A., Samuel, D., Sedvall, G., Swahn, C.-G.: Oxygen-18 use for clinical studies of central monoamine metabolism. In Usden, E., Kopin, I. J., Barkus, J. (eds.): *Catecholamines: Basic and Clinical Frontiers,* Vol. 2. New York, Pergamon Press, 1979, pp. 1569–1571.
29. Chase, T. N., Ng, L. K. Y.: Probenecid test in Parkinson's disease. *Lancet* **2**:1265–1266, 1971.
30. Chase, T. N., Ng, L. K. Y.: Central monoamine metabolism in Parkinson's disease. *Arch. Neurol.* **27**:486–491, 1972.
31. Chase, T. N., Ng, L. K. Y., Watanabe, A. M.: Parkinson's disease: Modification by 5-hydroxytryptophan. *Neurology (Minneapolis)* **22**:479–484, 1972.
32. Chase, T. N., Watanabe, A. M.: Methyldopahydrazine as an adjunct to L-dopa therapy in parkinsonism. *Neurology (Minneapolis)* **22**:384-392, 1972.
33. Chase, T. N., Woods, A. C., Glaubiger, G. A.: Parkinson disease treated with a suspected dopamine receptor agonist. *Arch. Neurol.* **30**:383–386, 1974.
34. Cramer, H., Ng, L. K. Y., Chase, T. N.: Effect of probenecid on levels of cyclic AMP in human cerebrospinal fluid. *J. Neurochem.* **19**:1601–1602, 1972.
35. Cramer, H., Ng, L. K. Y., Chase, T. N.: Adenosine 3′,5′-monophosphate in cerebrospinal fluid: Effect of drugs and neurologic disease. *Arch. Neurol.* **29**:197–199, 1973.
36. Curzon, G., Godwin-Austen, R. B., Tomlinson, E. B., Kantamaneni, B. D.: The cerebrospinal fluid homovanillic acid concentration in patients with Parkinsonism treated with L-dopa. *J. Neurol. Neurosurg. Psychiatry* **33**:1–6, 1970.
37. Curzon, G., Gumpert, E. J. W., Sharpe, D. M.: Amine metabolites in lumbar cerebrospinal fluid of humans with restricted flow of cerebrospinal fluid. *Nature (London) New Biol.* **231**:189–191, 1971.
38. Davidson, D. L., Yates, C. M., Mawdsley, C., Pullar, I. A., Wilson, H.: CSF studies on the relationship between dopamine and 5-hydroxytryptamine in parkinsonism and other movement disorders. *J. Neurol. Neurosurg. Psychiatry* **40**:1136–1141, 1977.
39. Dziedzic, S. W., Gitlow, S. E.: Cerebrospinal fluid homovanillic acid and iso-homovanillic acid: A gas–liquid chromatographic method. *J. Neurochem.* **22**:333–335, 1974.
40. Eccleston, D., Ashcroft, G. W., Moir, A. T. B., Parker-Rhodes, A., Lutz, W., O'Mahoney, D. P.:

A comparison of 5-hydroxyindoles in various regions of dog brain and cerebrospinal fluid. *J. Neurochem.* **15**:947–957, 1968.
41. Elizan, T. S., Schwartz, J., Yahr, M. D., Casals, J.: Antibodies against arboviruses in postencephalitic and idiopathic Parkinson's disease. *Arch. Neurol.* **35**:257–260, 1978.
42. Enna, S. J., Stern, L. Z., Wastek, G. J., Yamamura, H. I.: Cerebrospinal fluid γ-aminobutyric acid variations in neurological disorders. *Arch. Neurol.* **34**:683–685, 1977.
43. Extein, I., Van Woert, M. H., Roth, R. H., Bowers, M. B., Jr.: ^{14}C-Homovanillic acid in the cerebrospinal fluid of parkinsonian patients after intravenous ^{14}C-L-dopa. *Biol. Psychiatry* **11**:227–232, 1976.
44. Gjessing, L. R., Gjesdahl, P., Dietrichson, P., Presthus, J.: Free amino acids in the cerebrospinal fluid in old age and in Parkinson's disease. *Eur. Neurol.* **12**:33–37, 1974.
45. Godwin-Austen, R. B., Kantamaneni, B. D., Curzon, G.: Comparison of benefit from L-dopa in parkinsonism with increase of amine metabolites in the CSF. *J. Neurol. Neurosurg. Psychiatry* **34**:219–223, 1971.
46. Goldstein, D. J., Cubeddu, X. L.: Dopamine-β-hydroxylase activity in human cerebrospinal fluid. *J. Neurochem.* **26**:193–195, 1976.
47. Goodwin, F. K., Post, R. M.: Studies of amine metabolites in affective illness and in schizophrenia: A comparative analysis in biology of the major psychoses. *Res. Publ. Assoc. Res. Nerv. Ment. Dis.* **54**:299–332, 1975.
48. Gottfries, C. G., Gottfries, I., Roos, B. E.: Homovanillic acid and 5-hydroxyindoleacetic acid in the cerebrospinal fluid of patients with senile dementia, presenile dementia and parkinsonism. *J. Neurochem.* **16**:1341–1345, 1969.
49. Granerus, A.-K., Magnusson, T., Roos, B.-E., Svanborg, A.: Relationship of age and mood to monoamine metabolites in cerebrospinal fluid in parkinsonism. *Eur. J. Clin. Pharmacol.* **7**:105–109, 1974.
50. Growdon, J. H., Cohen, E. L., Wurtman, R. J.: Effects of oral choline administration on serum and CSF choline levels in patients with Huntington's disease. *J. Neurochem.* **28**:229–231, 1977.
51. Gründig, E., Gerstenbrand, F., Bruck, J., Gnad, H., Prosenz, P., Teuflmayr, R.: Der Einfluss der Verabreichung von Aminosäuren, speziell von L-dopa und Methyldopa, auf die Zusammensetzung des Liquor cerebrospinalis bei extrapyramidalen Syndromen. I. Veränderungen der Liquorzusammensetzung nach L-dopa-Gaben bei Parkinson-Patienten und Gesunden. *Dtsch. Z. Nervenheilkd.* **196**:236–255, 1969.
52. Guldberg, H. C., Ashcroft, G. W., Crawford, T. B. B.: Concentrations of 5-hydroxy-indoleacetic acid and homovanillic acid in the cerebrospinal fluid of the dog before and during treatment with probenecid. *Life Sci.* **5**:1571–1575, 1966.
53. Guldberg, H. C., Turner, J. W., Hanich, A., Ashcroft, G. W., Crawford, T. B. B., Perry, W. L. M., Gillingham, F. J.: On the occurrence of homovanillic acid and 5-hydroxyindol-3-ylacetic acid in the ventricular CSF of patients suffering from parkinsonism. *Confin. Neurol.* **29**:73–77, 1967.
54. Guldberg, H. C., Yates, C. M.: Some studies of the effects of chlorpromazine, reserpine and dihydroxyphenylalanine on the concentrations of homovanillic acid, 3,4-dihydroxyphenylacetic acid and 5-hydroxyindol-3-ylacetic acid in ventricular cerebrospinal fluid of the dog using the technique of serial sampling of the cerebrospinal fluid. *Br. J. Pharmacol. Chemother.* **33**:457–471, 1968.
55. Gumpert, J., Sharpe, D., Curzon, G.: Amine metabolites in the cerebrospinal fluid in Parkinson's disease and the response to levodopa. *J. Neurol. Sci.* **19**:1–12, 1973.
56. Hare, T. A., Beasley, B. L., Chambers, R. A., Boehme, D. H., Vogel, W. H.: Dopa and amino acid levels in plasma and cerebrospinal fluid of patients with Parkinson's disease before and during treatment with L-dopa. *Clin. Chim. Acta* **45**:273–280, 1973.
57. Hinterberger, H., Andrews, C. J.: Catecholamine metabolism during oral administration of levodopa. *Arch. Neurol.* **26**:245–252, 1972.
58. Hornykiewicz, O.: Neurochemistry of parkinsonism. In Lajtha, A. (ed.): *Handbook of Neurochemistry*, Vol. 7. New York, Plenum Press, 1972, p. 465.
59. Jéquier, E., Dufresne, J.-J.: Biochemical investigations in patients with Parkinson's disease treated with L-dopa. *Neurology (Minneapolis)* **22**:15–21, 1972.
60. Jimerson, D. C., Gordon, E. K., Post, R. M., Goodwin, F. K.: Homovanillic acid in human CSF: Comparison of fluorimetry and gas chromatography–mass spectrometry. *Commun. Psychopharmacol.* **2**:343–350, 1978.
61. Johansson, B., Roos, B.-E.: 5-Hydroxyindoleacetic and homovanillic acid levels in the cerebrospinal fluid of healthy volunteers and patients with Parkinson's syndrome. *Life Sci.* **6**:1449–1454, 1967.
62. Johansson, B., Roos, B.-E.: 5-Hydroxyindoleacetic acid in cerebrospinal fluid of patients with Parkinson's syndrome treated with L-dopa. *Eur. J. Clin. Pharmacol.* **3**:232–235, 1971.
63. Johansson, B., Roos, B.-E.: 5-Hydroxyindoleacetic acid and homovanillic acid in cerebrospinal fluid of patients with neurological diseases. *Eur. Neurol.* **11**:37–45, 1974.
64. Johansson, B., Roos, B.-E.: Concentrations of monoamine metabolites in human lumbar and cisternal cerebrospinal fluid. *Acta. Neurol. Scand.* **52**:137–144, 1975.
65. Kessler, J. A., Fenstermacher, J. D., Patlak, C. S.: Homovanillic acid transport by the spinal cord. *Neurology (Minneapolis)* **26**:434–440, 1976.
66. Klawans, H. L., Jr.: Effect of vitamin B_6 on levodopa-

induced changes in spinal fluid homovanillic acid. *J. Neurol. Sci.* **14:**421–426, 1971.

67. KORF, J., VAN PRAAG, H. M.: Amine metabolism in the human brain: Further evaluation of the probenecid test. *Brain Res.* **35:**221–230, 1971.
68. KORF, J., VAN PRAAG, H. M., SCHUT, D., NIENHUIS, R. J., LAKKE, J. P. W. F.: Parkinson's disease and amine metabolites in cerebrospinal fluid: Implications for L-dopa therapy. *Eur. Neurol.* **12:**340–350, 1974.
69. KREMZNER, L. T., BERL, S., MENDOZA, M., YAHR, M. D.: Cerebrospinal fluid levels of dopa and 3-*O*-methyldopa in parkinsonism during treatment with L-dopa and MK-486. *Adv. Neurol.* **2:**79–89, 1973.
70. LAKKE, J. P. W. F., KORK, J., VAN PRAAG, H. M., SCHUT, T.: Predictive value of the probenecid test for the effect of L-dopa therapy in Parkinson's disease. *Nature (London) New Biol.* **236:**208–209, 1972.
71. LAKKE, J. P. W. F., KORF, J., HOORNTJE, D., SCHUT, T., VAN PRAAG, H. M.: The probenecid test in Parkinson's disease. *Psychiatr. Neurol. Neurochir.* **76:**139–146, 1973.
72. LAKKE, J. P. W. F., TEELKEN, A. W.: Amino acid abnormalities in cerebrospinal fluid of patients with parkinsonism and extrapyramidal disorders. *Neurology (Minneapolis)* **26:**489–493, 1976.
73. LATERRE, E. C., CALLEWAERT, A., HEREMANS, J. F., SFAELLO, Z.: Electrophoretic morphology of gamma globulins in cerebrospinal fluid of multiple sclerosis and other diseases of the nervous system. *Neurology (Minneapolis)* **20:**982–990, 1970.
74. LLOYD, K. G., MÖHLER, H., HEITZ, P., BARTHOLINI, G.: Distribution of choline acetyltransferase and glutamate decarboxylase within the substantia nigra and in other brain regions from control and parkinsonian patients. *J. Neurochem.* **25:**789–795, 1975.
75. LOVENBERG, W., LEVINE, R. A., ROBINSON, D. S., EBERT, M., WILLIAMS, A. C., CALNE, D. B.: Hydroxylase cofactor activity in cerebrospinal fluid of normal subjects and patients with Parkinson's disease. *Science* **204:**624–626, 1979.
76. MATHIEU, P., REVOL, L., TROUILLAS, P.: The 4-*O* methyl metabolites of catecholamines in man: Chromatographic identification of 3-hydroxy,4-methoxyphenylacetic acid (homo-iso-vanillic acid) in cerebrospinal fluid. *J. Neurochem.* **19:**81–86, 1972.
77. MCGEER, P. L., MCGEER, E. G.: Enzymes associated with the metabolism of catecholamines, acetylcholine and GABA in human controls and patients with Parkinson's disease and Huntington's chorea. *J. Neurochem.* **26:**65–76, 1976.
78. MIACHON, S., DALMAZ, Y., COTTET-EMARD, J. M., PEYRIN, L.: Cerebrospinal homovanillic acid and parkinsonism. *Biomedicine* **20:**303–308, 1974.
79. MOIR, S. T. B., ASHCROFT, G. W., CRAWFORD, T. B. B., ECCLESTON, D., GULDBERG, H. C.: Cerebral metabolites in cerebrospinal fluid as a biochemical approach to the brain. *Brain* **93:**357–368, 1970.
80. NEOPHYTIDES, A. N., SURIA, A., WANIEWSKI, R. D., CHASE, T. N.: Cerebrospinal fluid GABA in neurologic disease. *Neurology* **28:**359, 1978 (abstract).
81. NG, L. K. Y., CHASE, T. N., COLBURN, R. W., KOPIN, I. J.: L-Dopa in parkinsonism: A possible mechanism of action. *Neurology (Minneapolis)* **22:**688–696, 1972.
82. OLSSON, R., ROOS, B.-E.: Concentrations of 5-hydroxyindoleacetic acid and homovanillic acid in the cerebrospinal fluid after treatment with probenecid in patients with Parkinson's disease. *Nature (London)* **219:**502–503, 1968.
83. PAPAVASILIOU, P. S., COTZIAS, G. C., LAWRENCE, W. H.: Levodopa and dopamine in cerebrospinal fluid. *Neurology (Minneapolis)* **23:**756–759, 1973.
84. PAPESCHI, R., MOLINA-NEGRO, P., SOURKES, T. L., HARDY, J., BERTRAND, C.: Concentration of homovanillic acid in the ventricular fluid of patients with Parkinson's disease and other dyskinesias. *Neurology (Minneapolis)* **20:**991–995, 1970.
85. PAPESCHI, R., MOLINA-NEGRO, P., SOURKES, T. L., ERBA, G.: The concentration of homovanillic and 5-hydroxyindoleacetic acids in ventricular and lumbar CSF. *Neurology (Minneapolis)* **22:**1151–1159, 1972.
86. PAPESCHI, R., SOURKES, T. L., POIRIER, L. J., BOUCHER, R.: On the intracerebral origin of homovanillic acid of the cerebrospinal fluid of experimental animals. *Brain Res.* **28:**527–533, 1971.
87. PARTRIDGE, W. M.: Kinetics of competitive inhibition of neutral amino acid transport across the blood–brain barrier. *J. Neurochem.* **28:**103–108, 1977.
88. PELTON, E. W., CHASE, T. N.: L-Dopa and the treatment of extrapyramidal disease. *Adv. Pharmacol. Chemother.* **13:**253–304, 1975.
89. PLETSCHER, A., BARTHOLINI, G., TISSOT, R.: Metabolic fate of L[14C] DOPA in cerebrospinal fluid and blood plasma of humans. *Brain Res.* **4:**106–109, 1967.
90. PULLAR, I. A., WEDDELL, J. M., AHMED, R., GILLINGHAM, F. J.: Phenolic acid concentrations in the lumbar cerebrospinal fluid of parkinsonian patients treated with L-dopa. *J. Neurol. Neurosurg. Psychiatry* **33:**851–857, 1970.
91. RIEDERER, P., BIRKMAYER, W., SEEMANN, D., WUKETICH, S.: Brain-noradrenaline and 3-methoxy-4-hydroxyphenylglycol in Parkinson's syndrome. *J. Neural. Transm.* **41:**241–251, 1977.
92. RINNE, U. K., LAAKSONEN, H., RIEKKINEN, P.: Brain glutamic acid decarboxylase activity in Parkinson's disease. *Eur. Neurol.* **12:**13–19, 1974.
93. RINNE, U. K., MARTTILA, R., SONNINEN, V.: Brain dopamine turnover and the relief of parkinsonism. *Arch. Neurol.* **34:**626–629, 1977.
94. RINNE, U. K., SONNINEN, V.: Acid monoamine metabolites in the cerebrospinal fluid of patients with Parkinson's disease. *Neurology (Minneapolis)* **22:**62–67, 1972.
95. RINNE, U. K., SONNINEN, V.: Brain catecholamines and their metabolites in parkinsonian patients. *Arch. Neurol.* **28:**107–110, 1973.
96. RINNE, U. K., SONNINEN, V., SIIRTOLA, T.: Acid mon-

oamine metabolites in the cerebrospinal fluid of parkinsonian patients treated with levodopa alone or combined with a decarboxylase inhibitor. *Eur. Neurol.* **9**:349–362, 1973.
97. SCHUBERTH, J., JENDEN, D. J.: Transport of choline from plasma to cerebrospinal fluid in the rabbit with reference to the origins of choline and to acetylcholine metabolism in brain. *Brain. Res.* **84**:245–256, 1975.
98. SEBENS, J. B., KORF, J.: Cyclic AMP in cerebrospinal fluid: Accumulation following probenecid and biogenic amines. *Exp. Neurol.* **46**:333–344, 1975.
99. SEDVALL, G., MAYERSKY, A., FRI, C.-G., SJOQUIST, B., SAMUEL, D.: *Adv. Biochem. Pharmacol.* **7**:57–68, 1973.
100. SHARPLESS, N. S., ERICSSON, A. D., MCCANN, D. S.: Clinical and cerebrospinal fluid changes in parkinsonian patients treated with L-3,4-dihydroxyphenylalanine (L-dopa). *Neurology (Minneapolis)* **21**:540–549, 1971.
101. SHOULSON, I., CHASE, T. N.: Clonidine and the antiparkinsonian response to L-dopa or piribedil. *Neuropharmacology* **15**:25–27, 1976.
102. SHOULSON, I., GLAUBIGER, G. A., CHASE, T. N.: On–off response: Clinical and biochemical correlations during oral and intravenous levodopa administration in parkinsonian patients. *Neurology (Minneapolis)* **25**:1144–1148, 1975.
103. SIEVER, L, DRAEMER, H., SACK, R., ANGWIN, P., BERGER, P., ZARCONE, V., BARCHAS, J., BRODIE, H. K. H.: Gradients of biogenic amine metabolites in cerebrospinal fluid. *Dis. Nerv. Syst.* **36**:13–16, 1975.
104. SJOQUIST, B., JOHANSSON, B.: A comparison between fluorometric and mass fragmentagraphic determinations of homovanillic acid and 5-hydroxyindoleacetic acid in human cerebrospinal fluid. *J. Neurochem.* **31**:621–625, 1978.
105. SJÖSTRÖM, R., EKSTEDT, J., ANGGÅARD, E.: Concentration gradients of monoamine metabolites in human cerebrospinal fluid. *J. Neurol. Neurosurg. Psychiatry* **38**:666–668, 1975.
106. STIBLER, H., KJELLIN, K. G.: Isoelectric focusing and electrophoresis of the CSF proteins in tremor of different origins. *J. Neurol. Sci.* **30**:269–285, 1976.
107. TABADDOR, K., WOLFSON, L. I., SHARPLESS, N. S.: Ventricular fluid homovanillic acid and 5-hydroxyindoleacetic acid concentration in patients with movement disorders. *Neurology (Minneapolis)* **28**:1249–1253, 1978.
108. VAN PRAAG, H. M., FLENTGE, F., KORF, J., DOLS, L. C. W., SCHUT, T.: The influence of probenecid on the metabolism of serotonin, dopamine and their precursors in man. *Psychopharmacologia (Berlin)* **33**:141–151, 1973.
109. VAN SANDE, M., MARDENS, Y., ADRIAENSSENS, K., LOWENTHAL, A.: The free amino acids in human cerebrospinal fluid. *J. Neurochem.* **17**:125–135, 1970.
110. VAN WOERT, M. H., AMBANI, L., BOWERS, M. B., JR.: Levodopa and cholinergic hypersensitivity in Parkinson's disease. *Neurology Suppl.*, pp. 86–93, 1972.
111. VAN WOERT, M. H., BOWERS, M. B., JR.: The effect of L-dopa on monoamine metabolites in Parkinson's disease. *Experientia* **26**:161–163, 1970.
112. WATSON, E., WILK, S.: Assessment of cerebrospinal fluid levels of dopamine metabolites by gas chromatography. *Psychopharmacologia* **42**:57–62, 1975.
113. WEINER, W., HARRISON, W., KLAWANS, H.: L-Dopa and cerebrospinal fluid homovanillic acid in parkinsonism. *Life Sci.* **8**:971–976, 1969.
114. WEINER, W. J., KLAWANS, H. L., JR.: Failure of cerebrospinal fluid homovanillic acid to predict levodopa response in Parkinson's disease. *J. Neurol. Neurosurg. Psychiatry* **36**:747–752, 1973.
115. WEIR, R. L., CHASE, T. N., NG, L. K. Y., KOPIN, I. J.: 5-Hydroxyindoleacetic acid in spinal fluid: Relative contribution from brain and spinal cord. *Brain Res.* **52**:409–412, 1973.
116. WELCH, M. J., MARKHAM, C. H., JENDEN, D. J.: Acetylcholine and choline in cerebrospinal fluid of patients with Parkinson's disease and Huntington's chorea. *J. Neurol. Neurosurg. Psychiatry* **39**:367–374, 1976.
117. WERDINIUS, B.: Effect of probenecid on the levels of monoamine metabolites in the rat brain. *Acta. Pharmacol. Toxicol.* **25**:18–23, 1967.
118. WEST, K. A., ROOS, B.-E.: The ventricular fluid pressure and the concentration of homovanillic acid in different parts of the cerebrospinal fluid system in patients with Parkinson's disease. *Confin. Neurol.* **34**:136–142, 1972.
119. WILK, S., MONES, R.: Cerebrospinal fluid levels of 3-methoxy-4-hydroxyphenylethylene glycol in parkinsonism before and after treatment with L-dopa. *J. Neurochem.* **18**:1771–1773, 1971.
120. WOLFSON, L. I., ESCRIVA, A.: Clearance of 3-methoxy-4-hydroxyphenylglycol from the cerebrospinal fluid. *Neurology (Minneapolis)* **26**:781–784, 1976.
121. ZIEGLER, M. G., LAKE, C. R., FOPPEN, F. H., SHOULSON, I., KOPIN, I. J.: Norepinephrine in cerebrospinal fluid. *Brain Res.* **108**:436–440, 1976.

CHAPTER 17

Cerebrospinal Fluid Monoamine Metabolites in Neurological Disorders of Childhood

Bennett A. Shaywitz, Donald J. Cohen, and
Malcolm B. Bowers, Jr.

1. Introduction

Intensive investigations over the past two decades provide strong evidence that abnormalities in central monoamines may lie at the core of, and significantly influence the development and course of clinically relevant aberrations in central nervous system (CNS) functioning. Such a notion mandates a more detailed understanding of brain neurotransmitter mechanisms if we are to better comprehend and ultimately unravel and treat those complex and enigmatic neuropsychiatric disorders affecting children. However, the study of central monoamines in human disorders presents both theoretical and practical obstacles. Thus, while a number of different strategies have been employed to assess monoamine metabolism within the CNS of man, they all have in common the dependence on indirect estimates of CNS monoaminergic function. One approach relies on the determination of monoamines or their related enzymes in peripheral blood or urine. A considerable and rapidly expanding literature documents the peripheral concentrations of monoamine oxidase (MAO),[56] dopamine-β-hydroxylase (DBH),[33] and serotonin[77] in a variety of human disorders. However, there is no assurance that the activity or level of a monoaminergic enzyme or monoamine itself in the plasma reflects the activity of that enzyme in brain.

It is reasonable to assume that examination of concentrations of monoamines and their metabolites in cerebrospinal fluid (CSF) provides a more reliable index of CNS function than the determination of neurotransmitters or metabolites in urine, blood or their constituents. For the most part, the limitations in analytical technology have narrowed the scope of such investigations to the determination of the acid metabolites of the central monoamines; until recently, the concentrations of the amines themselves in CSF have been too low to measure by conventional techniques. While there exist a number of minor CSF metabolites, particularly of dopamine [e.g., 3,4-dihydroxyphenylacetic acid (DOPAC),[90] iso-homovanillic acid (iso-HVA),[8] 3,4-dihydroxyphenylethanol (DHPE),[90] and 3-methoxy-4-hydroxyphenylethanol (MHPE)[90]], their origin and fate in human CSF remain largely unexplored; virtually nothing is known of their concentrations in the CSF of children. Methodology for the determination of the principal metabolite of norepinephrine in human CSF, 3-methoxy-4-hydroxyphenylethylene (MHPG),[51,52] was developed substantially later than that for HVA and 5-hydroxyindoleacetic acid (5-HIAA), the principal metabolites of dopamine

Bennett A. Shaywitz, M.D. • Departments of Pediatrics and Neurology, Yale University School of Medicine, New Haven, Connecticut 06510. **Donald J. Cohen, M.D.** • Departments of Pediatrics, Psychiatry and Psychology, Child Study Center, Yale University School of Medicine, New Haven, Connecticut 06510. **Malcolm B. Bowers, Jr., M.D.** • Department of Psychiatry, Child Study Center, Yale University School of Medicine, New Haven, Connecticut 06510.

and serotonin, respectively. Thus, information regarding MHPG, particularly in children, has been limited in comparison to the data available for HVA and 5-HIAA.

In the following discussion, it should be recognized that the underlying assumption, and indeed the central theoretical justification on which all investigators have relied, is the belief that the concentrations of the amine metabolites in the CSF can provide information about the rate of metabolism of their respective parent amines within the CNS. Evidence to support such a notion was reviewed earlier in this volume (see Chapter 5) and is based on (1) ventriculospinal concentration gradients for the monoamine metabolites; (2) the presence of a blood–brain barrier for the amines and their metabolites that prevents peripheral contamination; and (3) animal studies demonstrating that drugs that alter *brain* metabolite levels alter *CSF* levels appropriately.

Several practical concerns must be considered in the interpretation of CSF monoamine metabolite concentrations. Variation in HVA and 5-HIAA over the course of 24 hr has been documented in ventricular CSF,[39] and presumably occurs in lumbar CSF as well. In clinical research, obtaining samples at the same time of day in all patients can reduce circadian variation. Similarly, metabolites in lumbar CSF are not homogeneously distributed,[40] and clinical investigators must be consistent in utilizing particular aliquots. If the first milliliter of CSF is compared with the 5th or 10th milliliter, apparent differences may reflect nonuniformity in distribution of the metabolites within the large lumbar CSF space. Unfortunately, no convention has been established among investigators.

A theoretically more powerful approach is that of probenecid loading, a method that is believed to provide an estimate of the combined processes of synthesis, utilization, and metabolism of the monoamines—processes grouped together as "turnover"[11,34] (see Chapter 8).

However, the use of probenecid presents a number of complications in human studies. Thus, while a dose of 200 mg/kg is used in animal experiments, we have generally observed that nausea and vomiting will occur if a dose of greater than 125–150 mg/kg is employed in children. Lower doses may not totally inhibit monoamine efflux from CSF.[19–22,68,100]

Potential variability in probenecid action may occur if differing amounts of probenecid reach the site of the transport system in different patients or patient populations or if the effects of probenecid act erratically on the transport system of different subjects. In theory, such a problem may be circumvented by determining the concentrations of the metabolites in relation to that of probenecid. Whether such a relationship exists, however, continues to remain controversial. We have repeatedly noted a significant positive correlation between probenecid and both HVA and 5-HIAA in CSF after oral probenecid (see Chapter 46). In the only other published reference using the probenecid method in children, Winsberg *et al.*[98] were unable to document any correlation between CSF probenecid and either HVA or 5-HIAA. However, the extremely high concentrations of HVA reported by this group in lumbar CSF are unusual and suggest that they might be employing different methodology than that used by other investigative groups. If their extreme values are omitted, a positive correlation between 5-HIAA and probenecid concentration appears to be present.

Probenecid loading may alter behavioral systems. Thus, probenecid in doses customarily used to inhibit egress of monoamine metabolites causes a marked reduction in spontaneous motor activity in rats.[50] This may be a sign of toxicity, since we have noted quieting in very disturbed children when nauseated by probenecid. However, spontaneous motor activity is related to the integrity of central catecholamines, and reduced activity may indicate that probenecid is influencing catecholaminergic mechanisms as well as efflux of acid metabolites from CSF. Such an effect would not be surprising, since Westerink and Korf[97] noted increased concentrations of HVA in striatum, nucleus accumbens, and olfactory tubercle 2 hr after administration of probenecid (200 mg/kg), and a slight but significant increase of DOPAC in the nucleus accumbens. The action of probenecid on HVA in human brain is, of course, not known; extrapolation from animal studies suggests that probenecid not only inhibits egress of monoamine metabolites from CSF but also may increase the concentration of HVA in brain.

Probenecid may affect the concentrations of monoamine metabolites differently in normal and disease states, and any CSF monoamine metabolite concentrations observed may reflect more than the disease entity in question (e.g., previous use of medication, secondary changes related to chronic adaptation). Finally, the effects of probenecid on the developing organism are not well defined, and any extrapolation must not only bridge the wide gulf between animal and human systems but also take into consideration the unknown responses of the immature CNS to agents such as probenecid. Limited evidence suggests that such responses mature soon after birth. Thus, Bass *et al.*[7] indicate that probenecid is effective in increasing concentrations of 5-HIAA in brain and CSF as early as 10 days of age in the rat pup.

Despite these limitations, examination of HVA and 5-HIAA in lumbar CSF is at present the most useful method for the examination of central monoaminergic mechanisms in children.

2. Monoamines and Epilepsy

That brain monoamines play an important role in epilepsy has become increasingly apparent. For example, pharmacological agents that decrease the concentrations of endogenous brain amines increase the susceptibility to experimental seizures, while agents that increase biogenic amine concentration decrease the susceptibility to convulsions. Thus, reserpine or tetrabenzamine, compounds known to deplete brain amines, decrease the seizure threshold to pentylenetetrazol,[18,43,48,61] hyperbaric oxygen,[29,89] audiogenic stimuli,[10,47] and electroshock.[4] MAO inhibitors and other drugs that increase brain amines appear to protect animals from experimental seizures. The tricyclic antidepressant imipramine increases brain amine concentrations by preventing their deactivation and reuptake and appears to be a useful anticonvulsant in some patients with petit mal or minor motor epilepsy.[30] Paradoxically, however, both the MAO inhibitors and imipramine may rarely precipitate convulsions.[41]

The relationship between monoamines and seizures has been difficult to establish in human investigations (Table 1). Barolin and Hornykiewicz[6] examined lumbar CSF in six patients with epilepsy and found reduced concentrations of HVA compared to normal controls. In two additional patients with seizures, somewhat higher values were noted, though still below normal. Dubowitz and Rogers[26] examined 5-HIAA in lumbar CSF. They included seizures of various types in a group of 21 children considered as "controls." One child had infantile spasms, another (age 2 months) had repeated convulsions, a 6-year-old had epilepsy, and a fourth child is listed as "convulsive." Although the mean CSF 5-HIAA concentrations in the four were slightly lower than in the control group, this difference was not significant.

Papeschi *et al.*[58] examined ventricular CSF HVA and 5-HIAA in four patients (mean age 25 years) with temporal lobe epilepsy. A marked reduction in HVA was noted compared to patients with obsessive–compulsive neurosis or pain syndromes; 5-HIAA concentrations were not markedly reduced.

Garelis and Sourkes[31] examined lumbar CSF in patients undergoing pneumoencephalography. Twelve patients, ages 18–62 (mean 32 years), with epilepsy were included. Their HVA concentrations were similar to those of ten controls, but 5-HIAA concentrations were significantly reduced.

Chadwick *et al.*[17] compared 15 untreated patients with epilepsy (mean age 24.1 years, range 13–53) with 22 controls (ages 18–59) suspected of having multiple sclerosis. Both 5-HIAA and HVA were similar in both groups. A third group composed of 27 treated epilepsy patients (receiving phenytoin, phenobarbital, and primidone) demonstrated slightly elevated concentrations of 5-HIAA. It is difficult to interpret such a finding, however, since probenecid loading was not employed and the concentrations of both monoamine metabolites were at the lower

Table 1. CSF Monoamine Metabolites in Seizure Disorders[a]

		5-HIAA (ng/ml)		HVA (ng/ml)	
Investigators	CSF	Seizure	Controls	Seizure	Controls
Barolin and Hornykiewicz[6]	L	—	—	21	71
Dubowitz and Rogers[26]	L	50	71	—	—
Papeschi *et al.*[58]	V	39	58	56	331
Garelis and Sourkes[31]	L	24.6	32.1	42.8	43.1
Chadwick *et al.*[17]	L	33.1	32.1	47.7	50.1
Shaywitz *et al.*[68]	L	63.6	117	89.1	172
Metabolite/probenecid ratio:		4.78	8.18	6.70	12.5

[a] (L) Lumbar CSF; (V) ventricular CSF; (5-HIAA) 5-hydroxyindoleacetic acid; (HVA) homovanillic acid.

limits of sensitivity for the fluorometric assay employed.

We have examined the concentrations of 5-HIAA and HVA in children with epilepsy using the technique of probenecid loading. We initially reported a series of 14 children with epilepsy compared to a contrast group of 17 children with a variety of psychiatric and neurological disorders exclusive of epilepsy. All children were admitted to our Children's Clinical Research Center from 1 day to 2 months after their most recent convulsion. Oral probenecid was administered over 10–12 hr in four divided doses for a total of 100–150 mg/kg. Lumbar puncture was performed 2 hr after the last dose, and the CSF was frozen and analyzed within 2 weeks. 5-HIAA and HVA were determined by the method of Gerbode and Bowers[32] and probenecid by the method of Korf and van Praag.[44] The children with epilepsy comprised a group of 8 boys and 6 girls ranging in age from 6 months to 17 years (mean 7.2 years). All had histories of grand mal convulsions; in addition, 6 had minor motor and 4 had partial complex (psychomotor) seizures. Three were studied before initiation of any medication, while the remaining 11 were taking various drugs or combinations: phenobarbital, phenytoin, phenytoin and phenobarbital; phenytoin and primidone or acetazolamide.

As shown in Figs. 1 and 2, the mean CSF concentrations of both 5-HIAA and HVA were significantly reduced in children with epilepsy compared with the control group; probenecid concentrations did not differ significantly. Both 5-HIAA and HVA varied with probenecid concentrations. The ratio of monoamine metabolite concentration to probenecid was significantly reduced for both 5-HIAA/probenecid and HVA/probenecid in the epilepsy group, compared to controls.

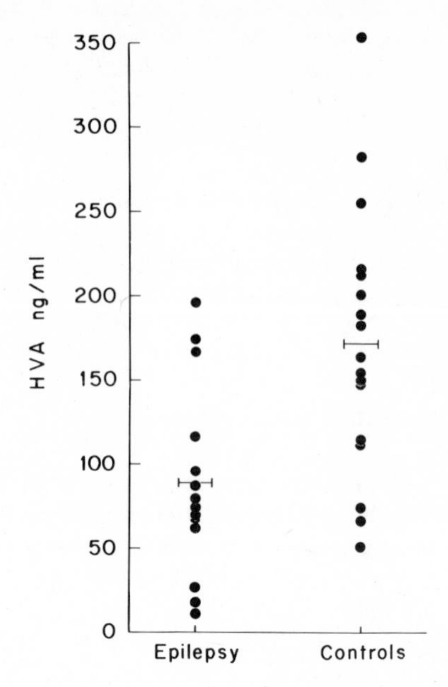

Figure 2. Comparison of CSF homovanillic acid (HVA) concentrations in children with epilepsy and in control patients. Each point represents one specific child. Horizontal brackets indicate mean concentrations. Reprinted from Shaywitz *et al.*,[68] with permission.

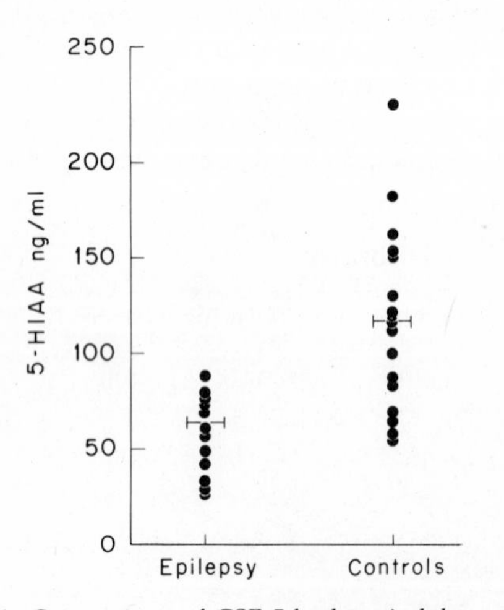

Figure 1. Comparison of CSF 5-hydroxyindoleacetic acid (5-HIAA) concentrations in children with epilepsy and in control patients. Each point represents one specific child. Horizontal brackets indicate mean concentrations. Reprinted from Shaywitz *et al.*,[68] with permission.

The effect of anticonvulsant medication presents a confounding factor. Chadwick *et al.*[17] found both 5-HIAA and HVA to increase in patients with therapeutic and toxic serum concentrations of phenytoin, phenobarbital, or primidone. No such phenomenon has been found by other groups of investigators, however. Thus, in our group of 14 children with epilepsy, 3 were investigated prior to initiating anticonvulsant treatment, and CSF amine metabolites in all 3 were reduced compared to nonepileptic controls. Four children were taking phenobarbital alone, and 3 more were taking phenobarbital and phenytoin. The remaining 4 were taking combinations of phenobarbital, phenytoin, acetazolamide, ethosuximide, and primidone. Serum anticonvulsant concentrations were either at therapeutic levels or slightly low. In all, and as shown in Fig. 3, we found no relationship between 5-HIAA or HVA and serum concentrations of phenobarbital or phenytoin. Thus, it is unlikely that reduced con-

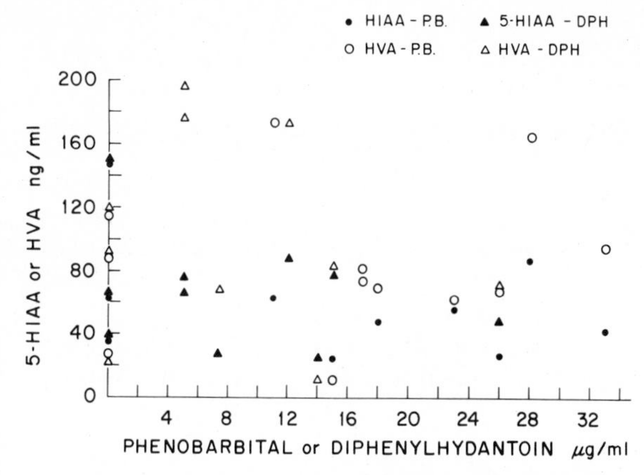

Figure 3. Relationship between CSF 5-hydroxyindoleacetic acid (5-HIAA) or homovanillic acid (HVA) and serum concentration of phenobarbital (P.B.) or diphenylhydantoin (DPH) in children with epilepsy. Each point represents CSF monoamine metabolite concentration corresponding with specific serum level of anticonvulsant drugs. CSF 5-HIAA and HVA for three untreated patients are also shown for comparison. Reprinted from Shaywitz *et al.*,[68] with permission.

centration of 5-HIAA and HVA in the CSF of children with epilepsy can be attributed to the effect of anticonvulsant medication.

Since our original report, we have expanded the number of children in our series with seizures, as well as our contrast population. We continue to find a significant correlation between monoamine metabolite and probenecid concentrations in the lumbar CSF; reduction in monoamine metabolite accumulations in CSF (expressed as the 5-HIAA/probenecid or HVA/probenecid ratio) continues to be noted.

The development of high-performance liquid chromatography has increased assay sensitivity so that we are able to analyze as little as 20 μl CSF. Such an approach makes possible the elucidation of monoamine metabolites in the CSF of neonates with seizures, a determination impossible with fluorometric methods that require 4 ml CSF.

2.1. Myoclonic Epilepsy and Myoclonus

Two groups of investigators have examined lumbar CSF in patients with postanoxic, intention myoclonus (Table 2). Van Woert and Sethy,[85] utilizing probenecid loading, noted reduction in both 5-HIAA and HVA. Chadwick *et al.*[16] reported on nine patients with myoclonus. While mean CSF concentrations of 5-HIAA and HVA in the entire myoclonic group did not differ from controls, both 5-HIAA and HVA were reduced in the CSF of three patients who exhibited a positive response to oral tryptophan.

We have examined one adolescent with postanoxic myoclonus. This 16-year-old was described in a report by Van Woert and Sethy[85] although no CSF values were available at that time. The 5-HIAA/probenecid and HVA/probenecid ratios are substantially below those obtained in our contrast population and add further support to the belief that disturbances in brain monoaminergic mechanisms may be observed in myoclonus.

3. Juvenile Parkinsonism

Juvenile parkinsonism is an uncommon disorder for which etiology and pathogenesis remain less well established than for classic Parkinson's disease.[13] Duvoisin and Yahr[27] obtained a history of onset of Parkinson-like symptoms between 10 and 19 years in 6 of 215 patients (2.8%) in a Parkinson's disease clinic, and Scott and Brody[67] found 1 patient in 21 (4.8%) with the onset of symptoms in his teens (14 years). Martin *et al.*[54] described two brothers with parkinsonism at 10 and 19 years of age, per-

Table 2. CSF Monoamine Metabolites in Myoclonus[a]

Investigators	5-HIAA (ng/ml)		HVA (ng/ml)	
	Myoclonus	Controls	Myoclonus	Controls
Van Woert and Sethy[85]	42	90.3	50	108.2
Chadwick *et al.*[16]	30.0	32.8	45.2	49
Shaywitz *et al.* (unpublished data)				
Metabolite/probenecid ratio	5.41	8.48	4.36	12.5

[a] (5-HIAA) 5-Hydroxyindoleacetic acid; (HVA) homovanillic acid.

haps resulting from a deficiency of tyrosine hydroxylase in brain. In most cases of juvenile parkinsonism, a postinfectious etiology has been suspected.

While it is well established that the concentration of HVA in the CSF of adults with parkinsonism is markedly reduced (see Chapter 16), reflecting a disruption in the nigrostriatal dopaminergic system, such determinations are rare in children with juvenile parkinsonism. Sachdev *et al.*[66] reported two sisters, ages 10 and 12, with juvenile parkinsonism, and found concentrations of 5-HIAA and HVA in lumbar CSF too low to measure. We have examined monoamine metabolite concentrations in three teenagers with symptoms and signs of parkinsonism. Patient J.K. was well until 9 years of age, when decreased activity, staring, and drooling were noted. Patient O.N. developed symptoms at 11 years of age. When seen at 14 years of age, both children had mask-like facies, staring, bradykinesia, tremor, and cogwheel rigidity. A third child, G.M., initially referred with parkinsonism, differed both clinically and biochemically from these two patients. In this child, tremor and drooling began at 18 months of age. When the child was seen at 14 years, cogwheel rigidity and tremor without bradykinesia were demonstrated.

Lumbar CSF concentrations of 5-HIAA, HVA, and probenecid are shown in Table 3, along with similar concentrations in a large contrast group without symptoms of parkinsonism. It is clear that in the first two patients, reductions in HVA are evident (shown by the reduced HVA/probenecid ratio). In patient O.N., reduction in 5-HIAA also may be considered. However, in patient G.M., no reduction in either HVA or 5-HIAA was detected. For this third patient, the Parkinson's symptoms may reflect a distinctly different pathological process. However, it must be remembered that available indices of dopaminergic functioning may be too insensitive to reflect superficially similar but functionally distinct CNS mechanisms. It is also possible that profound alterations in CSF metabolites are found in later stages of parkinsonism in adults and not in the progressive disease in childhood. An additional distinction between childhood and adult parkinsonism is the response to medication. None of our patients responded as favorably to oral L-DOPA as is observed in adult-onset parkinsonism. Clearly, studies of more children and longitudinal investigations will be required to understand the rare childhood variant and its relationship to the adult-onset disease.

3.1. Dystonia

Idiopathic torsion dystonia, or dystonia musculorum deformans (DMD), is an autosomal recessive (among Jews) or autosomal dominant (in non-Jewish populations) syndrome characterized by irregular, sustained, involuntary movements often resulting in abnormal postures. Onset of DMD almost always follows a normal perinatal and early developmental history, and there is generally no precipitating illness or exposure to drugs known to provoke dystonia. Cognitive, pyramidal, cerebellar, and sensory systems appear to be spared on clinical examination.[28]

Curzon[25] (Table 4) described normal lumbar CSF 5-HIAA and HVA in patients with DMD. Tabaddor *et al.*[82] (Table 4) compared ventricular concentrations of monoamine metabolites in 14 children and young adults exhibiting dystonia of onset during childhood to CSF metabolites in 6 adults with onset of dystonia in later life, most often the third or fourth decade. HVA was significantly lower in adult-onset dystonia than in the childhood variety, while 5-HIAA concentrations were similar. Whether the reduced CSF in the adult disorder reflects a developmental effect

Table 3. CSF Monoamine Metabolites in Juvenile Parkinsonism and Contrasting Patients[a]

Patient	5-HIAA (ng/ml)	HIAA/P	HVA (ng/ml)	HVA/P[a]	Probenecid (μg/ml)
J.K.	147	11.8	44	3.5	12.5
O.N.	52	3.5	107	7.1	15.0
G.M.	72	16.0	98	21.8	4.5
Controls (n = 27)	107 ± 9.5[b]	7.1	190 ± 16.6[b]	12.7	15 ± 1.6[b]

[a] (5-HIAA/P, HVA/P) Metabolite/probenecid ratios; (5-HIAA) 5-hydroxyindoleacetic acid; (HVA) homovanillic acid.
[b] Mean ± S.E.M.

Table 4. CSF Monoamine Metabolites in Dystonia[a]

Investigators	CSF	5-HIAA (ng/ml)		HVA (ng/ml)	
		Dystonia	Controls	Dystonia	Controls
Curzon[25]	L	21	23	31	37
Tabaddor *et al.*[82]	V				
Adult-onset		65	—	179	—
Childhood-onset		82	—	275	—
Ouvrier[57]	L	1.7, 4.6	30.5	0	20.8

[a] (L) Lumbar CSF; (V) ventricular CSF; (5-HIAA) 5-hydroxyindoleacetic acid; (HVA) homovanillic acid.

or a difference in pathology cannot be clarified with available data. However, in our studies, CSF HVA appears to decrease with age while 5-HIAA remains stable (see Chapter 46).

Within the last few years, a new entity termed "progressive dystonia" has been described. Ouvrier[57] (Table 4) studied an 8-year-old girl and a 12-year-old boy with dystonia appearing between the ages of 1 and 9 years and progressing insidiously to involve first one arm and leg and then the contralateral extremities. Lumbar CSF HVA was too low to measure in both children; 5-HIAA was low in one child and normal, on two occasions, in the other.

We have examined three boys with DMD utilizing probenecid loading (Table 5). The 5-HIAA/probenecid and HVA/probenecid ratios were lower than comparable values in contrast children. Such findings support the notion that serotonergic and dopaminergic mechanisms may be involved in the pathogenesis of this disorder.

4. CSF Monoamine Metabolites in Coma

4.1. Coma of Traumatic Origin

A number of investigators have examined the concentrations of monoamine metabolites after head injury in children and adults (Table 6). Vecht *et al.*[86–88] examined the concentrations of HVA and 5-HIAA in lumbar CSF in 98 patients between 13 and 64 years of age, before and after probenecid administration. Both metabolites were found to be reduced in patients who remained in a coma for 5 days or more, but concentrations did not correlate with the state of consciousness. The data suggested a decreased metabolism of both dopamine and serotonin after head injury. Van Woerkom *et al.*[84] demonstrated a modest reduction of 5-HIAA in lumbar CSF and a more pronounced decrease in HVA in 11 patients (average age 22 years) with frontal–temporal lobe contusion compared to patients with diffuse cerebral contusions and control patients.

Obviously, all these studies required examination of lumbar CSF, and lumbar punctures presumably were not done in the presence of clinical evidence of increased intracranial pressure (ICP). Thus, the lumbar CSF monoamine metabolite levels in these studies do not reflect overt increased ICP. Hyyppa *et al.*[38] examined the relationship between monoamine metabolites in ventricular CSF and ICP in eight patients (ages 3–48 years, mean 22.3 years) with posttraumatic coma. Patients with ICP pressure greater than 25 mm Hg exhibited concentrations of CSF tryptophan 80% greater than controls; concen-

Table 5. CSF Monoamine Metabolites in Dystonia Musculorum Deformans[a]

Patient	Age	HIAA (ng/ml)	5-HIAA/P	HVA (ng/ml)	HVA/P	Probenecid (μg/ml)
R.W.	15	82	4.55	152	8.40	18
J.B.	11	124	7.25	170	10.60	16
J.M.	15	127	4.30	230	7.80	29.6
Controls	—	—	8.48	—	12.5	—

[a] (5-HIAA/P, HVA/P) Metabolite/probenecid ratios; (5-HIAA) 5-hydroxyindoleacetic acid; (HVA) homovanillic acid.

Table 6. CSF Monoamine Metabolites in Coma[a]

		5-HIAA (ng/ml)		HVA (ng/ml)	
Investigators	CSF	Coma	Controls	Coma	Controls
Vecht *et al.*[86]	L	21.0	35.5	9.5	42.0
Van Woerkem *et al.*[84]	L	20.6	28.5	26.2	45.3
Hyyppa *et al.*[38]	V	76	88	136	264
Bareggi *et al.*[5]	L	31	30	21	47
Porta *et al.*[62]	V	249	109	339	210
Shaywitz *et al.* (unpublished data)	V	209	187	465	317

[a] (L) Lumbar CSF; (V) ventricular CSF; (5-HIAA) 5-hydroxyindoleacetic acid; (HVA) homovanillic acid.

trations of HVA and 5-HIAA were reduced compared to patients with normal CSF pressures.

Porta and co-workers[5,62,63] examined monoamine metabolites in CSF of patients after severe head injury. HVA in lumbar CSF in ten patients (ages 4–28, mean 14.1 years) was reduced, but no alteration in 5-HIAA was noted. Ventricular CSF was determined in six patients (ages 4–37, mean 21.5 years), and compared to that in four patients with pseudotumor cerebri. Both 5-HIAA and HVA were markedly elevated.

In our ongoing series of investigations of monoamine metabolite concentrations in neuropsychiatric disorders of childhood, we have had the opportunity to examine ventricular CSF serially in children in coma after cardiac arrest in whom a ventricular catheter had been placed to monitor ICP. A total of 32 samples were obtained in five children who ranged in age from 3 to 7 years. The concentration of 5-HIAA was slightly above that observed in hydrocephalic children (our contrast group) at the time of shunting, but the concentration of HVA was more clearly elevated, reflecting what we believe are the effects of cerebral edema (see discussion below). As was the case in Reye syndrome (see below), we found *no* relationship between the degree of elevation of ICP and elevation in CSF monoamine metabolites.

4.2. Coma in Metabolic Encephalopathy

Reye syndrome, the most common encephalopathy seen in children, is characterized by persistent vomiting and progressive obtundation. There is usually evidence of hepatic dysfunction with elevation of blood ammonia, serum glutamic oxalacetic transaminase (SGOT), and prothrombin time. Our treatment protocol includes the early placement of a ventricular catheter to monitor ICP immediately on admission. This has allowed serial sampling of ventricular CSF throughout the course of the illness, and we have described CSF monoamine metabolites in nine girls and six boys ranging in age from 2 years to 14 years 10 months (mean 6.3 years) with Reye syndrome.[71] Monoamine metabolite concentrations in ventricular CSF were compared to those in a contrast population (primarily children with communicating hydrocephalus) who had ventricular CSF sampled at the time of ventriculoperitoneal shunt installations.

The peak ventricular CSF concentrations of HVA and 5-HIAA in Reye syndrome patients and contrasts are shown in Fig. 4. The CSF 5-HIAA level averaged 187 ng/ml in the contrast population, with a median of 189 ng/ml. The concentration of this metabolite in Reye syndrome patients averaged 238 ng/ml, with a median concentration of 198 ng/ml. As is evident from Fig. 4, these CSF metabolite concentrations did not differ significantly between groups. The ventricular CSF concentration of HVA in the contrast population averaged 317 ng/ml, with a median concentration of 282 ng/ml. The mean concentration of HVA in Reye syndrome patients averaged 1296 ng/ml, with a median concentration of 887 ng/ml. Utilizing the Mann Whittney U Test, these CSF HVA concentrations are significantly different ($p < 0.001$).

Figure 5 represents the serial ventricular CSF HVA concentrations in two children whose intraventricular pressure was monitored for periods of 6 days. In both patients, initial HVA concentrations were markedly elevated in the 500 ng/ml range. C.D., a 9-year-old boy, demonstrated elevations of HVA up to 3000 ng/ml on day 4. This child's blood ammonia peaked on day 2 at 556 μg/dl. He made an uneventful recovery. Patient K.B., a 12-year-old

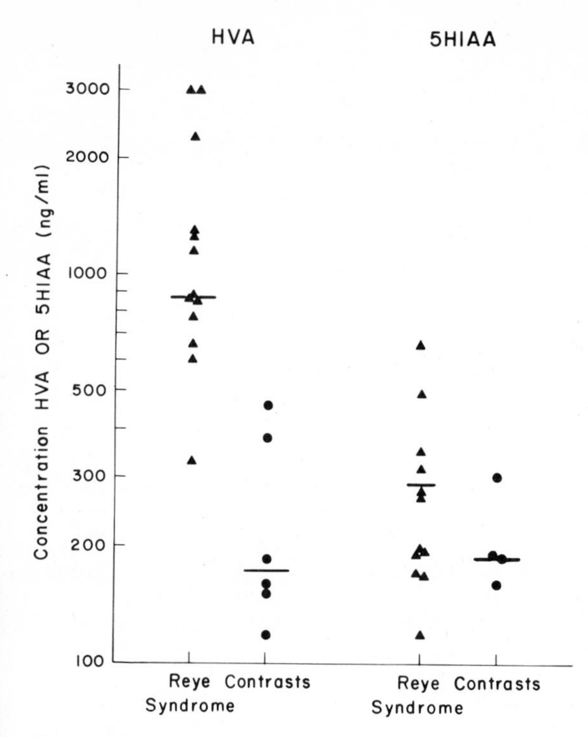

Figure 4. Peak concentrations of homovanillic acid (HVA) and 5-hydroxyindoleacetic acid (5-HIAA) in ventricular CSF of children with Reye syndrome compared to contrast pediatric population. Horizontal bars indicate median concentrations. Reprinted from Shaywitz *et al.*,[71] with permission.

girl, demonstrated ventricular HVA concentrations that ranged between 200 and 800 ng/ml. Her blood ammonia peaked on day 2, at 660 μg/dl, and she too made an uneventful recovery. Figure 6 represents serial ventricular CSF HVA concentrations in children who were monitored for longer periods of time. Patient S.W. had an initial HVA concentration of 600 ng/ml, and over the 11 days of monitoring, this concentration varied between 2200 and 250 ng/ml. This child's recovery was prolonged, but 1 year post–Reye syndrome, he appeared normal. Patient S.C. was a 4-year-old child whose blood ammonia was 1016 μg/ml. His early course was not marked by episodes of hypotension, hypoglycemia, or increased ICP sufficient to reduce cerebral perfusion pressure below 60 mm Hg. However, ICP became increasingly difficult to control, and management was confounded by the development of renal toxicity believed to be related to prolonged administration of mannitol. By the 7th hospital day, usual methods of controlling ICP proved futile, and a bifrontal decompressive craniectomy was performed. As shown in the Fig. 6, this child's initial HVA was 400 ng/ml and increased to 3000 ng/ml prior to his craniectomy. Immediately postcraniectomy, HVA concentration varied between 100 and 400 ng/ml; however, later, HVA increased to 2200 ng/ml at a time when ICP was normal. In contrast are the results on patient G.F., an 8-year-old boy whose ICP was monitored for 10 days after near drowning in a pool. ICP was never above 20 mm Hg; HVA was moderately elevated initially, fell rapidly, and remained relatively low. Serial 5-HIAA values showed no appreciable variation.

We found no relationship between the degree of difficulty in controlling ICP and elevation in monoamine metabolites. Our treatment regimen mandates aggressive therapeutic measures (mannitol, hyperventilation) when ICP rises above 20–30 mm Hg, and this would preclude observing any correlations between extreme ICP and CSF metabolites.

Our finding of markedly elevated concentrations of HVA in children with Reye syndrome adds further support to the notion that brain monoaminergic systems may be involved in the pathophysiology of this disorder. Our results suggest that alterations in

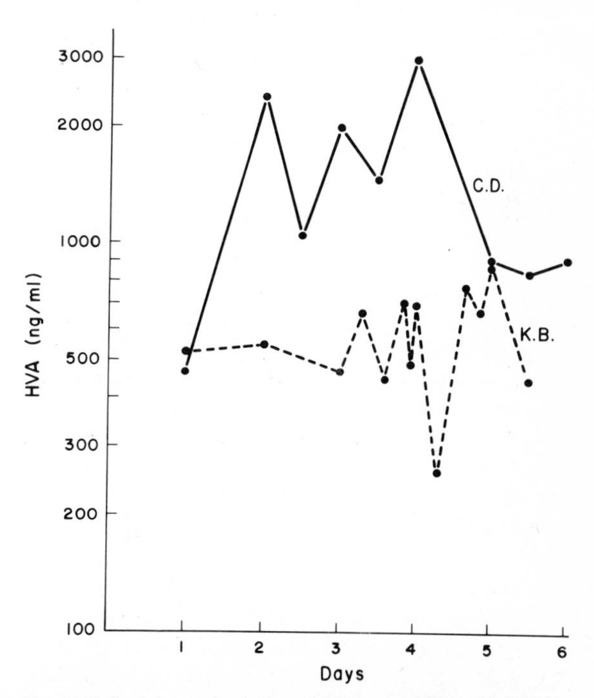

Figure 5. Serial concentrations of homovanillic acid (HVA) in ventricular CSF. Patient C.D. is a 9-year-old male, while patient K.B. is a 12-year-old female. Both patients have Reye syndrome. Reprinted from Shaywitz *et al.*,[71] with permission.

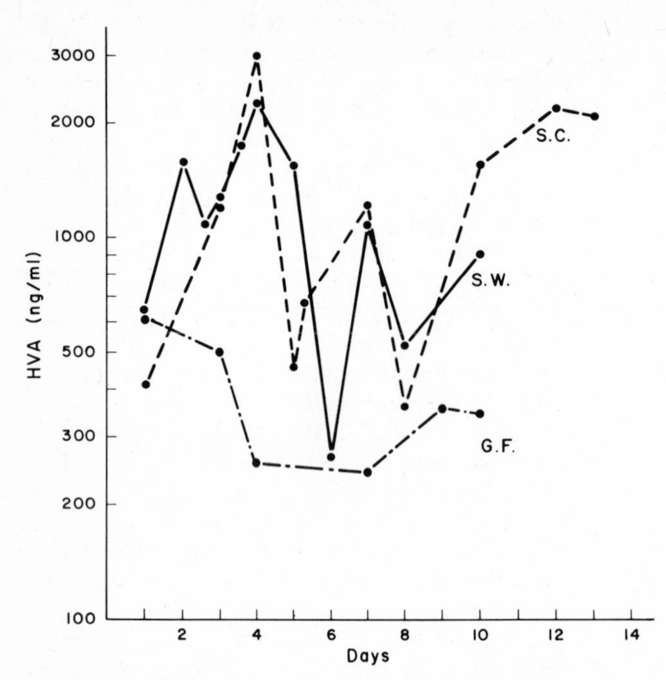

Figure 6. Serial concentration of homovanillic acid (HVA) in ventricular CSF. Patients S.W. and S.C. are males with Reye syndrome, ages 14 and 4 years, respectively. Patient G.F. was an 8-year-old boy who had nearly drowned. Reprinted from Shaywitz *et al.*,[71] with permission.

cerebral dopamine rather than serotonergic mechanisms are most important, since 5-HIAA appears to be unchanged. These monoamine results can be explained most parsimoniously by considering the relationship between brain catecholamines and cerebral ischemia. For example, Zervas *et al.*[101] have demonstrated a significant decline in brain dopamine within 3 hr after ligation of the common carotid artery in gerbils. They suggest that ischemia provokes the release of large amounts of vasoactive amines and their metabolites from effector neurons, and these may in turn extend the tissue injury.[46,99]

In Reye syndrome, an undefined, generalized metabolic insult, as suggested recently by Aprille,[3] appears to initiate cerebral ischemia by restriction of oxygen supply and leads to the development of brain edema. Vasoactive amines are released into the brain and act locally; cerebral vessels in compromised brain areas constrict, and the tissue damage is exacerbated. Therapeutic intervention that interrupts this sequence of events may ameliorate some of the consequences of Reye syndrome. Such measures might include the utilization of pharmacological agents to prevent release of amines or block their actions on receptors. Controlled trials of such medications remain the subject of future investigations.

The elevations of HVA observed by us in Reye syndrome are even greater than those reported in studies of coma of traumatic origin (see above). In contrast to Porta and associates,[5,62,63] we were not able to discern any abnormality in serotonergic mechanisms as reflected in the concentrations of 5-HIAA. Discrepancies between individual investigative groups may be biased on several factors, e.g., differences between lumbar and ventricular CSF and among populations studied. Another confounding variable might well be the problem of homogeneity of CSF, as noted earlier.

5. Monoamines in Hydrocephalus

The most extensive study of CSF monoamine metabolites in hydrocephalus remains that performed a decade ago by Andersson and Roos[2] (Table 7). They examined 5-HIAA in either lumber or ventricular CSF in a total of 191 children (122 boys and 69 girls), ranging from prematurely born newborns to children 10 years of age. Their population consisted of a "control" group (102 children) with diagnoses of subdural effusion, mental retardation, cerebral atrophy, convulsions, macrocephaly, meningitis, dysplasia, spina bifida cystica, and cerebral tumor (averaged together as "nonhydrocephalic") and 89 hydrocephalic children. Because the concentration of 5-HIAA tended to decrease with age, Andersson and Roos compared the hydrocephalic to

Table 7. CSF Monoamine Metabolites in Hydrocephalus[a]

Investigators	CSF	5-HIAA (ng/ml)		HVA (ng/ml)	
		Hydroceph.	Controls	Hydroceph.	Controls
Andersson and Roos[2]	L	154	87	—	—
Andersson and Roos[2]	V	315	130	—	—
Maira *et al.*[53]					
Obstructive	L	26.2	30.3	7.5	46.3
Nonobstructive	L	29.6	30.3	63	46.3
Tabaddor *et al.*[81]	V	100	74	341	233
Shaywitz *et al.*[71]	V	187	—	317	—

[a] (L) Lumbar CSF; (V) ventricular CSF; (5-HIAA) 5-hydroxyindoleacetic acid; (HVA) homovanillic acid.

nonhydrocephalic groups according to age and observed that in the "very young" hydrocephalics, 5-HIAA in both lumbar and ventricular CSF was elevated compared to the "very young" nonhydrocephalic group. Similar findings were observed in the older children, although the number of older children was very small and the authors show their results in graphic form only.

Maira *et al.*[53] (Table 7) examined lumbar and ventricular monoamine metabolites in 13 adults with normal-pressure hydrocephalus. The first group of 8 (ages 27–59, mean 43 years) manifested ventricular filling on isotope cisternography and a positive Katzman infusion test, indicating obstructive hydrocephalus. Lumbar CSF HVA was reduced compared to control values, while 5-HIAA was unchanged. The second group of 5 was of comparable age and severity but exhibited negative infusion tests and poor ventricular filling on isotope cisternography, indicating that CSF circulation was not qualitatively altered, but only slower than normal. In this group, lumbar CSF HVA was slightly elevated compared to controls, and 5-HIAA was unchanged. Ventricular CSF metabolite concentrations in 4 patients did not differ from the values reported by West *et al.*[96] in control patients. Lumbar CSF was examined in 6 patients after shunting. HVA and 5-HIAA were reduced in 4 and increased in 2, but there seemed to be little relationship with clinical course. Tabaddor *et al.*[82] (Table 7) found that concentrations of both 5-HIAA and HVA in the ventricular CSF of a 24-year-old adult with hydrocephalus and cerebral palsy were elevated in comparison to those reported for 7 patients with cerebral palsy.

We have examined ventricular CSF concentrations of monoamine metabolites in six children with hydrocephalus, ranging in age from 4 months to 12 years (mean 3.6 years); five had communicating hydrocephalus and one had hydrocephalus in association with a brainstem glioma. Concentrations of HVA and 5-HIAA are shown in Table 7. It is impossible to know normal values for children, since normal children do not undergo ventriculostomy and ventriculograms are now performed only rarely in an era of computed tomography.

Explanation for the increase in CSF monoamines in hydrocephalus is not at all clear. It may be simply a result of reduced absorption of the metabolites because of the disturbances in CSF production and absorptive mechanisms in hydrocephalus. An alternative hypothesis, however, is that the metabolism of monoamines themselves is disrupted by the cerebral damage resulting from hydrocephalus. However, one would expect reduced concentrations, rather than elevated levels, if this were the case.

6. Mental Retardation and Cerebral Palsy

Although these cases have generally been considered to represent "normal" CSF, rationale for such a distinction is difficult to document. Thus, Dubowitz and Rogers[26] reported 16 children with retardation, some of whom were considered to have cerebral palsy with hyptonia. Concentrations of CSF 5-HIAA in these 16 were reduced compared to those in other children without retardation or seizures.

We have examined lumbar CSF using probenecid loading in 12 children ranging in age from 8 months to 16 years and comprising a heterogeneous group with retardation. Two were siblings exhibiting congenital microcephaly and retardation, while the remaining 10 were handicapped as a result of perinatal insults. The 5-HIAA/probenecid ratio in CSF was comparable to that observed in children with headache, mononeuropathy, and conversion symptoms, but the HVA/probenecid ratio was slightly reduced in the children with retardation.

Recently, Brewster *et al.*[14] have described a low concentration of HVA in lumbar CSF after probenecid in an 8-month-old child with psychomotor retardation and dihydropteridine reductase deficiency. They suggest that absence of this cofactor resulted in a reduced synthesis of dopamine. However, probenecid levels were not assayed, and it is possible that the absence of an increase in HVA after probenecid was the result of poor ingestion or absorption of probenecid, rather than an indication of disturbances in monoaminergic mechanisms.

6.1. Down's Syndrome

A number of investigators have demonstrated reduced levels of serotonin in platelets of children with Down's syndrome (DS),[42,65,83] but to date, documentation of a disturbance in central serotonergic mechanisms in DS has been difficult to discern (Table 8). Dubowitz and Rogers[26] found comparable values for lumbar 5-HIAA in 10 children with DS (ages 7 months to 3 years) compared to 21 contrast children. Partington *et al.*[59] reported no difference in CSF concentrations of 5-HIAA in 5 children with DS compared to 5 other retarded individuals of comparable age. Lott *et al.*[49] examined an older group of 9 DS (18–25 years) and 12 non-mentally retarded control subjects. Basal levels of HVA and 5-HIAA in lumbar CSF were slightly higher in DS than in controls, but these disparities failed to achieve statistical significance. Oral probenecid administration resulted in an equal increase in CSF concentrations for HVA and 5-HIAA in both DS and controls. Airaksinen and Kauko[1] studied a group of older adults with DS using oral probenecid loading and measuring its concentration in CSF, as well. While they found a reduced concentration of 5-HIAA in DS compared to controls, concentrations of probenecid were significantly reduced in the DS group compared to controls. When this difference was corrected by calculating the 5-HIAA/probenecid ratio, comparable values were obtained in each group. Thus, to date, abnormalities in central monoamines have not been convincingly demonstrated in Down's syndrome.

7. Monoamines and Minimal Brain Dysfunction

Minimal brain dysfunction (MBD) or attention-deficit disorder with hyperactivity (ADD) is perhaps the most common problem in current pediatric practice, affecting an estimated 5–10% of the school-age population.[37,45,79,93] It is viewed most parsimoniously as a symptom complex that includes various combinations of impairment in perception, conceptualization, language, memory, and control of attention, impulse, and motor function. Abundant evidence from several lines of investigation supports the belief that the cardinal symptoms of this disorder may be related to disturbances in central monoaminergic systems.[73]

Epidemiological evidence provided the earliest suggestion that monoamines might play an important role in the genesis of MBD. Following the worldwide epidemic of von Economo's encephalitis in 1917–1918, many adults developed a clinical syndrome that was interpreted as postencephalitic parkinsonism. The pediatric literature of the 1920's reveals that many of the childhood victims of encephalitis developed a clinical picture of hyperactivity, impulsivity, short attention span, and school learning difficulties, which today would lead to the diagnosis of MBD.[36] This clinical observation

Table 8. CSF Monoamine Metabolites in Down's Syndrome[a]

Investigators	CSF	5-HIAA (ng/ml)		HVA (ng/ml)	
		Down's	Controls	Down's	Controls
Dubowitz and Rogers[26]	L	56.5	47.1	—	—
Partington *et al.*[59]	L	35.4	32.0	—	—
Lott *et al.*[49]	L	38	30	56	38
Airaksinen and Kauko[1]	L	81.4	121	—	—
5-HIAA/probenecid ratio:		6.1	7.9		

[a] (L) Lumbar CSF; (V) ventricular CSF; (5-HIAA) 5-hydroxyindoleacetic acid; (HVA) homovanillic acid.

led to the speculation that the virus of von Economo's encephalitis produced disturbances in central dopaminergic mechanisms resulting in symptoms of parkinsonism in adults and, in the less mature nervous system of childhood, in the symptoms of MBD.

Further evidence supporting an association between MBD and brain monoamines is derived primarily from the ameliorating effect of amphetamines on the hyperactivity of many children with MBD, as initially reported in 1937 by Bradley.[12] It is well established that the primary actions of amphetamine in the brain are mediated by monoaminergic, primarily catecholaminergic, mechanisms[35]. amphetamines stimulate the release of catecholamines from neuronal terminals, prevent reuptake, and inhibit catecholamine degradation to some extent by inhibiting MAO. The sum total of these actions appears to be an increase in the concentration of catecholamines at the synaptic cleft. Numerous investigations have documented that the administration of amphetamine or the clinically related agent, methylphenidate, to children with MBD results in a reduction in activity and distractability and an increase in attention span.[24,55,76,80,91,92] This therapeutic effect, along with the known pharmacological actions of amphetamine, prompted Wender[93] to suggest that since amphetamines ameliorate the symptoms of MBD, and since amphetamines act via central catecholaminergic mechanisms, the symptoms of MBD may be related to catecholaminergic systems.

However, direct documentation of an abnormality in central monoaminergic function in children with MBD has proven difficult to document. Urinary concentrations of HVA and 5-HIAA are normal in children with MBD,[94] but Shekim and Dekirmenjian[74] have demonstrated reductions in urinary concentrations of MHPG in hyperactive boys compared to age-matched controls. One group of investigators has reported reductions in platelet serotonin in hyperactive children,[23] but another group has not been able to confirm this finding. Rapoport *et al.*[64] have shown normal concentrations of plasma serotonin in children with MBD, as well as normal concentrations of DBH and MAO.

Such studies suffer serious limitation, since studies involving urine or blood reflect the combined biochemical activity of the peripheral, autonomic, and central nervous systems. With the exception of MHPG, the activity or level of a monoamine or metabolite or its related enzyme in plasma has not been related to concentrations of that compound in brain. Only two investigative groups have examined monoamine metabolite concentrations in the CSF of children with MBD. Shetty and Chase[75] (Table 9) measured concentrations of HVA and 5-HIAA in lumbar CSF in 23 hyperactive children (19 boys and 4 girls, ages 2–13 years) compared to 6 controls (4 boys and 2 girls, ages 5–15 years). Baseline CSF concentrations of HVA and 5-HIAA in hyperactives and controls were not significantly different. HVA concentration decreased, while 5-HIAA increased slightly, in 10 hyperactive children treated with 0.5 mg/kg amphetamine for 2–14 days (mean 12 days), but neither metabolite changed significantly in 5 children receiving placebo for similar time periods. The authors suggest that the physiological effects of amphetamine on catecholaminergic mechanisms were associated with therapeutic response. Clinical reduction in hyperactivity in children taking amphetamine was correlated with the reduction in concentration in lumbar CSF HVA. These changes in HVA and 5-HIAA with treatment may reflect feedback inhibition of dopaminergic activity and inhibitory serotonergic compensation.

Table 9. CSF Monoamine Metabolites in Minimal Brain Dysfunction[a]

Investigators	CSF	5-HIAA (ng/ml)		HVA (ng/ml)	
		MBD	Controls	MBD	Controls
Shetty and Chase[75]	L	37	37	78	62
Amphetamine		53[b]	41[c]	64[b]	97
Shaywitz *et al.*[69]	L	102	106	177	195
Metabolite/probenecid ratio:		6.4	8.0	9.8	16.5

[a] (L) Lumbar CSF; (5-HIAA) 5-hydroxyindoleacetic acid; (HVA) homovanillic acid.
[b] CSF values 2–14 days (±12 days) after 0.5 mg/kg *d*-amphetamine.
[c] Baseline CSF values in 10 hyperactive children prior to amphetamine.

We[69] (Table 9) have employed the technique of probenecid loading in a study of children with MBD. Our experimental group was selected to be homogeneous and representative of the major features of the syndrome: short attention span, impulsivity, hyperactive motor behavior, inability to adjust to a new environment, a preponderance of males (3:1 to 9:1, in various studies),[95] the presence of nonlocalizing neurological findings suggesting immaturity and difficulties in motor control,[60] nonspecific EEG abnormalities,[15,78] and globally normal intellectual abilities. Our patients were all male; three had abnormal EEGs, and all exhibited "soft" neurological findings, and normal IQ's on the Wechsler Intelligence Scale for Children (WISC). The control group consisted of 26 children, ages 2–15 years, evaluated for a variety of neurological difficulties, including headache, muscle weakness, back pain, and personality disorder. None of the children had symptomatic hyperactivity, and none was receiving stimulants or any other medication. We utilized the probenecid loading technique, and concentrations of monoamine metabolites are shown in Table 9. If only the absolute values of metabolites are compared, the average CSF HVA in MBD is only slightly below that found in controls. The children with MBD, however, attained a higher average probenecid concentration in CSF than did the control children. To correct for this difference in probenecid concentrations, metabolite concentrations per microgram of probenecid concentration, a ratio shown as 5-HIAA/probenecid and HVA/probenecid, was calculated. When represented in this manner, HVA/probenecid in CSF of children with MBD is significantly lower than that found in control subjects ($p < 0.05$, Student's t test).

Findings in a very select and clinically homogeneous population should not be generalized to all children characterized as MBD. Alterations in CSF HVA are, however, consistent with an experimental model of childhood hyperactivity produced by administration of 6-hydroxydopamine to neonatal rat pups.[70,72] This treatment produces a persistent, selective reduction of brain dopamine and a pattern of increased motor activity, which abates as the rat approaches maturity. In addition, dopamine-depleted rat pups have persistent cognitive difficulties. Stimulant medications yield a "paradoxical" or normalizing reduction in hyperactivity, an effect similar to that found in children with MBD.

Thus, pharmacological evidence, along with epidemiological studies of von Economo's encephalitis, reduced urinary MHPG, and our CSF findings, support the belief that brain catecholamines play a role in at least some MBD children.

8. Miscellaneous Disorders

8.1. Leukemia

Various neurological dysfunctions have been associated with CNS leukemia or its treatment. We studied CSF monoamine metabolites in 22 children (3–10 years of age) with acute lymphocytic leukemia who were examined to define CNS involvement while in remission and during the acute stages of illness. Probenecid loading was not employed. While concentrations of both metabolites were normal in most children, a small number exhibited significantly reduced levels. In general, the abnormalities appeared to correlate with the administration of intrathecal methotrexate; reductions may thus reflect abnormal CSF dynamics resulting from toxic effects of methotrexate on CSF absorption sites or toxic effects on specific monoaminergic mechanisms themselves. Assessment of CSF monoamines may help clarify the neurological sequelae of chemotherapeutic regimens in leukemia.

8.2. Periodic Hypersomia

Billiard *et al.*[9] reported CSF studies in a 13-year-old girl with periodic hypersomnia, a syndrome they relate to Kleine–Levin syndrome. While baseline concentrations of both 5-HIAA and HVA were normal, concentrations of both metabolites after probenecid loading were markedly increased. Thus, the 5-HIAA/probenecid ratio was 14.6 and the HVA/probenecid ratio 25.7, both considerably above our "normal" population.

8.3. Spinal Cord Disorders

A number of investigators have examined CSF monoamine metabolites in spinal-cord transections in adults but no reports are available for these same problems in children. In our large ongoing study of CSF in children, we have studied several children with diseases affecting the spinal cord. In one of our cases, CSF 5-HIAA appeared to be reduced in an 8-year-old with familial spastic paraplegia. This was not a consistent finding, and in other cases, e.g., a 15-year-old with transverse myelitis, concentrations were normal.

9. Summary

This review has focused on a number of investigations that have explored concentrations of monoamine metabolites in a variety of neurological disorders affecting children. We emphasized the inherent difficulties in such studies, but conclude that investigations of monoamines and their metabolites in CSF provide the most reasonable approach today to investigations of monoaminergic mechanisms in children. We noted that monoamine metabolites are reduced in a number of conditions, including epilepsy, myoclonus, juvenile parkinsonism and dystonia, and minimal brain dysfunction (MBD). Lumbar monoamine metabolites appear to be elevated in traumatic coma and in ventricular CSF of children with Reye syndrome and postanoxic coma, as well as hydrocephalus.

Clinical–pathological neurochemical research in childhood has already led to new hypotheses about etiology, pathophysiology, and possible interventions. However, this area is only in the first phases of development, and today's hypotheses must remain open to radical revision in the light of new data, including information about possible artifacts. In the future, more sensitive analytical methods (such as mass fragmentography and high-performance liquid chromatography), uniformity of sampling techniques, and greater sophistication in the classification of patients (by natural history, symptomatology, associated biological findings) will allow clinical investigators to measure more compounds in the CSF relevant to neurological integrity, to assay compounds in smaller volumes of CSF and at much lower concentrations, and to relate neurochemical findings to subtle differences in clinical disorders. Such studies offer the promise of increasing our basic understanding of the origin and course of profound, and often perplexing, neurological disorders of childhood.

ACKNOWLEDGMENTS. These studies were supported by NIH Grants NS 12384, HD 03008, and MH 24393, Children's Clinical Research Center Grant RR 00125, NIMH Clinical Research Center Grant No. 1 P50 MM3 0929, The Hood Foundation, the Nutrition Foundation, the William T. Grant Foundation, Mr. Leonard Berger, The Schall Family Trust, and The Solomon R. and Rebecca D. Baker Foundation, Inc. We are extremely grateful for the collaboration of Dr. J. Gerald Young, Ms. Barbara Caparulo, and Ms. Claudia Carbonari.

References

1. Airaksinen, E. M., Kauko, K.: Effect of probenecid on 5-hydroxyindoles in cerebrospinal fluid in Down's syndrome. *Ann. Clin. Res.* **5**:392–394, 1973.
2. Andersson, H., Roos, B.-E.: 5-Hydroxyindoleacetic acid in cerebrospinal fluid of hydrocephalic children. *Acta Paediatr. Scand.* **58**:601–608, 1969.
3. Aprille, J. R.: Reye's syndrome: Patient serum alters mitochondrial function and morphology *in vitro*. *Science* **197**:908–910, 1977.
4. Azarro, A. J., Wenger, G. R., Craig, C. R., Stitzel, R. E.: Reserpine induced alterations in brain amines and their relationship to changes in the incidence of minimal electroshock seizures in mice. *J. Pharmacol. Exp. Ther.* **180**:558–568, 1972.
5. Bareggi, S. R., Porta, M., Selenati, A., Assael, B. M., Calderini, G., Collice, M., Rossanda, M., Morselli, P. L.: Homovanillic acid and 5-hydroxyindole-acetic acid in the CSF of patients after a severe head injury. *Eur. Neurol.* **13**:528–544, 1975.
6. Barolin, G. S., Hornykiewicz, O.: Zur diagnostischen Wertigkeit der Homovanillinsäure im Liquor cerebrospinalis. *Wien. Klin. Wochenschr.* **79**:815–818, 1967.
7. Bass, N. H., Fallstrom, S. P., Lundborg, P.: Digoxin-induced arrest of the cerebrospinal fluid circulation in the infant rat: Implications for medical treatment of hydrocephalus during early postnatal life. *Pediatr. Res.* **13**:26–30, 1979.
8. Bertilsson, L., Palmer, L.: Determination of isomeric acid dopamine metabolites in human cerebrospinal fluid by mass fragmentography. *Life Sci.* **13**:859–866, 1973.
9. Billiard, M., Guilleminault, C., Dement, W. C.: A menstruation-linked periodic hypersomnia Kleine–Levin symdrome or new clinical entity? *Neurology* **25**:436–443, 1975.
10. Boggan, W. O., Seiden, L. S.: Dopa reversal of reserpine enhancement of audiogenic seizure susceptibility in mice. *Physiol. Behav.* **6**:215–217, 1971.
11. Bowers, M. B., Jr.: Clinical measurements of central dopamine and 5-hydroxytryptamine metabolism: Reliability and interpretation of cerebrospinal fluid acid monoamine metabolite measures. *Neuropharmacology* **11**:101–111, 1972.
12. Bradley, C.: The behavior of children receiving benzedrine. *Am. J. Psychiatry* **94**:577–585, 1937.
13. Brett, E. M.: Juvenile parkinsonism. *Dev. Med. Child Neurol.* **14**:391–402, 1972.
14. Brewster, T. G., Moskowitz, M. A., Kaufman, S., Breslow, J. L., Milstien, S., Abroms, I. F.: Dihydropteridine reductase deficiency associated with severe neurologic disease and mild hyperphenylalaninemia. *Pediatrics* **63**:94–99, 1979.
15. Capute, A. J., Niedermeyer, E. F. L., Richardson, F.: The electroencephalogram in children with mini-

mal cerebral dysfunction. *Pediatrics* **41**:1104–1114, 1968.

16. CHADWICK, D., HARRIS, R., JENNER, P., REYNOLDS, E. H., MARSDEN, C. D.: Manipulation of brain serotonin in the treatment of myoclonus. *Lancet* **2**:434–435, 1975.
17. CHADWICK, D., JENNER, P., REYNOLDS, E. H.: Amines, anticonvulsants, and epilepsy. *Lancet*, **1**:473–476, 1975.
18. CHEN, G., ENSOR, C. R., BOHNER, B.: A facilitation action of reserpine on the central nervous system. *Proc. Soc. Exp. Biol. Med.* **86**:507–510, 1954.
19. COHEN, D. J., YOUNG, J. G.: Neurochemistry and child psychiatry. *J. Am. Acad. Child Psychiatry* **16**:353–411, 1977.
20. COHEN, D. J., CAPARULO, B. K., SHAYWITZ, B. A.: Neurochemical and developmental models of childhood autism. In Serban, G. (ed.): *Cognitive Defects in the Development of Mental Illness*, New York, Brunner/Mazel, 1978, pp. 66–100.
21. COHEN, D. J., SHAYWITZ, B. A., CAPARULO, B., YOUNG, J. G., BOWERS, M. B., JR.: Chronic, multiple tics of Gilles de la Tourette's disease. *Arch. Gen. Psychiatry* **35**:245–250, 1978.
22. COHEN, D. J., SHAYWITZ, B. A., YOUNG, J. G., CARBONARI, C. M., NATHANSON, J. A., LIEBERMAN, D., BOWERS, M. B., JR., MAAS, J. W.: Central biogenic amine metabolism in children with the syndrome of chronic multiple tics of Gilles de la Tourette. *J. Am. Acad. Child. Psychiatry* **18**:320–328, 1979.
23. COLEMAN, M.: Serotonin level in children with "minimal brain dysfunction." *Lancet* **2**:1012, 1970.
24. CONNERS, C. K., EISENBERG, L., BARCAI, A.: Effect of dextroamphetamine on children: Studies on subjects with learning disabilities and school behavior problems. *Arch. Gen. Psychiatry* **17**:478–485, 1967.
25. CURZON, G., Involuntary movements other than parkinsonism: Biochemical aspects. *Proc. R. Soc. Med.* **55**:873–876, 1973.
26. DUBOWITZ, V., ROGERS, K. J.: 5-Hydroxyindoles in the cerebrospinal fluid of infants with Down's syndrome and muscle hypotonia. *Dev. Med. Child Neurol.* **11**:730–734, 1969.
27. DUVOISIN, R. C., YAHR, M. D.: Encephalitis and parkinsonism, *Arch. Neurol.* **12**:227–239, 1965.
28. ELDRIDGE, R.: The torsion dystonias: Literature review and genetic and clinical studies. *Neurology* **20**(2):1–78, 1972.
29. FAIMAN, M. D., MEHL, R. G., MYERS, M. B.: Brain norepinephrine and serotonin in central oxygen toxicity. *Life Sci.* **10**:21–34, 1971.
30. FROMM, G. H., AMORES, C. Y., THIES, W.: Imipramine in epilepsy. *Arch. Neurol.* **27**:198–204.
31. GARELIS, E., SOURKES, T. L.: Sites of origin in the central nervous system of monoamine metabolites measured in human cerebrospinal fluid. *J. Neurol. Psychiatry* **4**:625–629, 1973.
32. GERBODE, F. A., BOWERS, M. B., JR.: Measurement of acid monoamine metabolites in human and animal cerebrospinal fluid. *J. Neurochem.* **15**:1053–1055, 1968.
33. GOLDSTEIN, M., FREEDMAN, L. S., EBSTEIN, R. P., PARK, D. M., KASHIMOTO, T.: Human serum dopamine-β-hydroxylase: Relationship to sympathetic activity in physiological and pathological states. In Usdin, E. (ed.): *Neuropsychopharmacology of Monoamines and Their Regulatory Enzymes*. New York, Raven Press, 1974, pp. 105–119.
34. GOODWIN, F. K., POST, R. M., DUNNER, D. L., GORDON, E. K.: Cerebrospinal fluid amine metabolites in affective illness: The probenecid technique. *Am. J. Psychiatry* **130**:73–79, 1973.
35. HEIKKILA, R. E., ORLANSKY, H., MYTILINEOUS, C., COHEN, G.: Amphetamine: Evaluation of *d* and *l*-isomers as releasing agents and uptake inhibitors for ^{3}H-dopamine and ^{3}H-norepinephrine in slices of rat neostriatum and cerebral cortex, *J. Pharmacol. Exp. Ther.* **194**:47–56, 1975.
36. HOHMAN, L. B.: Post encephalitic behavior disorders in children. *Johns Hopkins Hosp. Bull.* **380**:372–374, 1922.
37. HUESSEY, H. R.: Study of the prevalence and therapy in the choreaiform syndrome or hyperkinesis in rural Vermont. *Acta Paedopsychiatr. Int. J. Child Psychiatry* **34**:130–135, 1967.
38. HYYPPA, M. T., LANGVIK, V. A., NIEMINEN, V., VAPALAHTI, M.: Tryptophan and monoamine metabolites in ventricular cerebrospinal fluid after severe cerebral trauma. *Lancet* **1**:1367–1368, 1977.
39. INGAWA, T., MORI, S., YOSHIMOTO, H., ISHIKAWA, S., UOZUMI, T., KAJIKAWA, H.: Parallel variation of homovanillic acid and 5-hydroxyindoleacetic acid in ventricular cerebrospinal fluid of man. *Hiroshima J. Med. Sci.* **25**:79–87, 1976.
40. JAKUPCEVIC, M., LACKOVIC, Z., STEFOSKI, D., BULAT, M.: Nonhomogeneous distribution of 5-hydroxyindoleacetic acid and homovanillic acid in the lumbar cerebrospinal fluid of man. *J. Neurol. Sci.* **31**:165–171, 1977.
41. JARVIK, M. E.: Drugs used in the treatment of psychiatric disorders. In Goodman, L. S., Gilman, A. (eds.): *The Pharmacological Basis of Therapeutics*. New York, Macmillan, 1970, pp. 151–203.
42. JEROME, H., LEJEUNE, J., TURPIN, R.: Étude de l'excretion urinaire de certains metabolites du tryptophane chez les enfants mongoliens. *C. R. Acad. Sci. Ser. D* **251**:474–476, 1970.
43. JONES, B. J., ROBERTS, D. J.: The effects of intracerebroventricularly administered noradrenaline and other sympathomimetic amines upon leptazol convulsions in mice. *Br. J. Pharmacol.* **34**:27–31, 1968.
44. KORF, J., VAN PRAAG, H. M.: Amine metabolism in the human brain: Further evaluation of the probenecid test. *Brain Res.* **35**:221–230, 1971.
45. LAUFER, M. W., DENHOFF, E.: Hyperkinetic behavior syndrome in children. *J. Pediatr.* **50**:463–474, 1957.
46. LAVYNE, M. D., MOSKOWITZ, M. A., LARIN, F., ZER-

VAS, N. T., WURTMAN, R. J.: Brain H^3-catecholamine metabolism in experimental cerebral ischemia. *Neurology* **25:**483–485, 1975.
47. LEHMAN, A.: Audiogenic seizures data in mice support new theories of biogenic amine mechanisms in the central nervous system. *Life Sci.* **6:**1423–1431, 1967.
48. LESSIN, A. W., PARKS, M. W.: The effects of reserpine and other agents upon leptazol convulsions in mice. *Br. J. Pharmacol.* **14:**108–111, 1959.
49. LOTT, I. T., MURPHY, D. L., CHASE, T. N.: Down's syndrome: Central monoamine turnover in patients with diminished platelet serotonin. *Neurology* **22:**967–972, 1972.
50. LYNESS, W. H., MYCEK, M. J.: The effect of probenecid on spontaneous motor activity in rats. *Neuropharmacology* **17:**211–213, 1978.
51. MAAS, J. W., HATTOX, S. E., LANDIS, D. H., ROTH, R. H.: The determination of a brain arteriovenous difference for 3-methoxy-4-hydroxyphenethyleneglycol (MHPG). *Brain Res.* **118:**167–173, 1976.
52. MAAS, J. W., HATTOX, S. E., LANDIS, D. H., ROTH, R. H.: A direct method for studying 3-methoxy-4-hydroxyphenethyleneglycol (MHPG) production by brain in awake animals. *Eur. J. Pharmacol.* **46:**221–228, 1977.
53. MAIRA, G., BAREGGI, S. R., DI ROCCO, C., CALDERINI, G., MORSELLI, P. L.: Monoamine acid metabolites and cerebrospinal fluid dynamics in normal pressure hydrocephalus: Preliminary results. *J. Neurol. Neurosurg. Psychiatry* **38:**123–128, 1975.
54. MARTIN, W. E., RESCH, J. A., BAKER, A. B., Juvenile parkinsonism. *Arch. Neurol.* **25:**495–500, 1971.
55. MILLICHAP, J. G., BOLDREN, E. E.: Studies in hyperkinetic behavior. II. Laboratory and clinical evaluations of drug treatments. *Neurology* **17:**467–471, 1967.
56. MURPHY, D. L., DONNELLY, C. H.: Monoamine oxidase in man: Enzyme characteristics in platelets, plasma and other human tissues. In Usdin, E. (ed.): *Neuropsychopharmacology of Monoamines and Their Regulatory Enzymes,* Vol. 12. New York, Raven Press, 1974, pp. 71–85.
57. OUVRIER, R. A.: Progressive dystonia with marked diurnal fluctuation. *Ann. Neurol.* **4:**412–417, 1978.
58. PAPESCHI, R., MOLINA-NEGRO, P., SOURKES, T. L., ERBA, G.: The concentration of homovanillic and 5-hydroxyindole-acetic acids in ventricular and lumbar CSF: Studies in patients with extrapyramidal disorders, epilepsy and other diseases. *Neurology (Minneapolis)* **22:**1151–1159, 1972.
59. PARTINGTON, M. W., MACDONALD, M. R. A., TU, J. B.: 5-Hydroxytryptophan (5-HTP) in Down's syndrome. *Dev. Med. Child Neurol.* **13:**362–372, 1971.
60. PETERS, J. E., ROMINE, J. S., DYKMAN, R. A.: A special neurological examination of children with learning disabilities. *Dev. Med. Child Neurol.* **17:**63–78, 1975.
61. PFIEFER, A. K., GALAMBOS, E.: Action of alpha methyldopa on the pharmacologic and biochemical effect of reserpine in rats and mice. *Biochem. Pharmacol.* **14:**37–40, 1965.
62. PORTA, M., BAREGGI, S. R., COLLICE, M., ASSAEL, B. M., SELENATI, A., CALDERINI, G., ROSSANDA, M., MORSELLI, P. L.: Homovanillic acid and 5-hydroxyindoleacetic acid in the CSF of patients after a severe head injury. *Eur. Neurol.* **13:**545–554, 1975.
63. PORTA, M., BAREGGI, S. R., SELENATI, A., ASSAEL, B. M., BEDUSCHI, A., MORSELLI, P. L.: Acid monoamine metabolites in ventricular and lumbar cerebrospinal fluids of patients in post-traumatic coma. *J. Neurosurg. Sci.* **17:**230–237, 1974.
64. RAPOPORT, J. L., QUINN, P., SCRIBANU, N., MURPHY, D. L.: Platelet serotonin of hyperactive school age boys. *Br. J. Psychiatry* **125:**138–140, 1974.
65. ROSNER, F., ONG, B. H., PAINE, R. S.: Blood serotonin activity in trisomic and translocation Down's syndrome. *Lancet* **1:**1191–1193, 1965.
66. SACHDEV, K. K., SINGH, N., KRISHNAMOORTHY, M. S.: Juvenile parkinsonism treated with levodopa. *Arch. Neurol.* **34:**244–245, 1977.
67. SCOTT, R. M., BRODY, J. A.: Benign early onset of Parkinson's disease: A syndrome distinct from classic postencephalitic parkinsonism. *Neurology* **21:**366–368, 1971.
68. SHAYWITZ, B. A., COHEN, D. J., BOWERS, M. B., JR.: Reduced cerebrospinal fluid 5-hydroxyindoleacetic acid and homovanillic acid in children with epilepsy. *Neurology* **25:**72–79, 1975.
69. SHAYWITZ, B. A., COHEN, D. J., BOWERS, M. B., JR.: CSF monoamine metabolites in children with minimal brain dysfunction—evidence for alteration of brain dopamine. *J. Pediatr.* **90:**67–71, 1977.
70. SHAYWITZ, B. A., KLOPPER, J. H., YAGER, R. D., GORDON, J. W.: A paradoxical response to amphetamine in developing rats treated with 6-hydroxydopamine. *Nature (London)* **261:**153–155, 1976.
71. SHAYWITZ, B. A., VENES, J., COHEN, D. J., BOWERS, M. B., JR.: Reye syndrome: Monoamine metabolites in ventricular fluid. *Neurology* **29:**467–472, 1979.
72. SHAYWITZ, B. A., YAGER, R. D., KLOPPER, J. H.: An experimental model of minimal brain dysfunction (MBD) in developing rats. *Science* **191:**305–308, 1976.
73. SHAYWITZ, S. E., COHEN, D. J., SHAYWITZ, B. A.: The biochemical basis of minimal brain dysfunction. *J. Pediatr.* **92:**179–187, 1978.
74. SHEKIM, W., DEKIRMENJIAN, J.: Catecholamine metabolites in nonhyperactive boys with arithmetic learning disability: A pilot study. *Am. J. Psychiatry* **135**(4):490–491, 1978.
75. SHETTY, T., CHASE, T. N.: Central monoamines and hyperkinesis of childhood. *Neurology* **26:**1000–1002, 1976.
76. SROUFE, L. A., STEWART, M. A.: Treating problem children with stimulant drugs. *New Engl. J. Med.* **289:**407–413, 1973.
77. STALL, S. M.: The human platelet. *Arch. Gen. Psychiatry* **34:**509–516, 1977.

78. Stevens, J., Sachdev, K., Milstein, V.: Behavior disorders of childhood and the electroencephalogram. *Arch. Neurol.* **18**:160–177, 1968.
79. Stewart, M. A., Pitts, F. N., Jr., Craig, A. G., Dieruf, W.: The hyperactive child syndrome. *Am. J. Orthopsychiatry* **36**:861–867, 1966.
80. Sykes, D. H., Douglas, V. I., Morgenstern, G.: The effect of methylphenidate (Ritalin) on sustained attention in hyperactive children. *Psychopharmacologia* **22**:282–294, 1972.
81. Tabaddor, K., Wolfson, L. I., Sharpless, N. S.: Ventricular fluid homovanillic acid and 5-hydroxyindoleacetic acid concentrations in patients with movement disorders. *Neurology* **28**:1249–1253, 1978.
82. Tabaddor, K., Wolfson, L. I., Sharpless, N. S.: Diminished ventricular fluid dopamine metabolites in adult-onset dystonia. *Neurology* **28**:1254–1258, 1978.
83. Tu, J. B., Zellweger, H.: Blood serotonin deficiency in Down's syndrome. *Lancet* **2**:715–716, 1965.
84. van Woerkom, T. C. A. M., Teelken, A. W., Minderhoud, J. M.: Difference in neurotransmitter metabolism in frontotemporal lobe contusion and diffuse cerebral contusion. *Lancet* **1**:812–813, 1977.
85. Van Woert, M. N., Sethy, V. H.: Therapy of intention myoclonus with L-5-hydroxytryptophan and a peripheral decarboxylase inhibitor MK 486. *Neurology* **25**:135–140, 1975.
86. Vecht, C. J., van Woerkom, T. C. A. M., Teelken, A. W., Minderhoud, J. M.: Homovanillic acid and 5-hydroxyindole-acetic acid cerebrospinal fluid levels. *Arch. Neurol.* **32**:792–797, 1975.
87. Vecht, C. J., van Woerkom, T. C. A. M., Teelken, A. W., Minderhoud, J. M.: 5-Hydroxyindoleacetic acid (5-HIAA) levels in the cerebrospinal fluid in consciousness and unconsciousness after head injury. *Life Sci.* **16**:1179–1186, 1975.
88. Vecht, C. J., van Woerkom, T. C. A. M., Teelken, A. W., Minderhoud, J. M.: On the nature of brain stem disorders in severe head injured patients. *Acta Neurochir.* **34**:11–21, 1976.
89. Wada, J. A., Terao, A., Scholtmeyer, H., Trapp, W. G.: Susceptibility to audiogenic stimuli induced by hyperbaric oxygenation and various neuroactive agents. *Exp. Neurol.* **33**:123–219, 1971.
90. Waterbury, L. D., Pearce, L. A.: Separation and identification of neutral and acidic metabolites in cerebrospinal fluid. *Clin. Chem.* **18**:258–262, 1972.
91. Weiss, G., Kruger, E., Danielson, V., Elmen, M.: Effect of long-term treatment of hyperactive children with methylphenidate. *Can. Med. Assoc. J.* **112**:159–165, 1975.
92. Weiss, G. Minde, K., Werry, J. S., Douglas, V., Nemth, E.: Studies on the hyperactive child. VIII. Five-year follow-up. *Arch. Gen. Psychiatry* **24**:409–414, 1971.
93. Wender, P. H.: *Minimal Brain Dysfunction in Children.* New York, John Wiley, 1971, pp. 12–30.
94. Wender, P. H., Epstein, R. S., Kopin, I. J., Gordon, E. K.: Urinary monoamine metabolites in children with minimal brain dysfunction. *Am. J. Psychiatry* **127**:1411–1415, 1971.
95. Werry, J. S.: Studies of the hyperactive child. IV. An empirical analysis of the minimal brain dysfunction syndrome. *Arch. Gen. Psychiatry* **19**:9–16, 1968.
96. West, K. A., Edvinsson, L., Nielsen, K. C., Roos, B.-E.: Concentration of acid monoamine metabolites in ventricular CSF of patients with posterior fossa tumors. In Brock, M., Dietz, H. (eds.): *Intracranial Pressure.* Berlin, Springer-Verlag, 1972, pp. 331–337.
97. Westerink, B. H. C., Korf, J.: Regional rat brain levels of 3,4-dihydroxyphenylacetic acid and homovanillic acid: Concurrent fluorometric measurement and influence of drugs. *Eur. J. Pharmacol.* **38**:281–291, 1976.
98. Winsberg, B. G., Hurwic, M. J., Sverd, J., Klutch, A.: Neurochemistry of withdrawal emergent symptoms in children. *Psychopharmacology* **56**:157–161, 1978.
99. Wurtman, R. J., Zervas, N. T.: Monoamine neurotransmitters and the pathophysiology of stroke and central nervous system trauma. *J. Neurosurg.* **40**:34–36, 1974.
100. Young, J. G., Cohen, D. J.: The molecular biology of development. In Nosphitz, J. (ed.): *Handbook of Child Psychiatry.* New York, Basic Books, 1979, pp. 22–69.
101. Zervas, N. T., Hori, H., Negora, M., Wurtman, R. J., Larin, F., Lavyne, M. H.: Reduction in brain dopamine following experimental cerebral ischemia. *Nature (London)* **247**:283–284, 1974.

CHAPTER 18

Seizure-Induced Metabolic Alterations in Human Cerebrospinal Fluid

Benjamin Rix Brooks

1. Introduction

The severe changes in systemic and cerebral metabolism associated with generalized tonic–clonic seizures have been a concern of physicians since the graphic description by Lucretius[56] portraying a seizure victim:

> Oft too some wretch, before our startled sight
> Struck as with lightning, by some keen disease
> Drops sudden: —by the dread attack o'erpowered
> He foams, he groans, he trembles, and he faints;
> Now rigid, now convulsed, his laboring lungs
> Heave quick, and quivers each exhausted limb,
> Spread through the frame, so deep the dire disease
> Perturbs his spirit: as the briny man
> Foams through each wave beneath the tempest's ire
>
> ....
>
> But when, at length, the morbid cause declines,
> And the fermenting humors from the heart
> Flow back—with staggering foot the man first treads
> Led gradual on to intellect and strength.

Attempts to define the changes in oxidative and anaerobic metabolism associated with generalized tonic–clonic seizures in man have concentrated on changes in systemic and cerebral metabolism associated with chemically or electrically induced seizures. Little attention has been paid, until recently, to the course of metabolic changes associated with spontaneous generalized tonic–clonic convulsions.[16,124]

We will define those investigations in man that serve to delineate the systemic and cerebral metabolic changes attendant upon (1) electrically induced seizures (EIS), (2) drug-induced seizures (DIS), and (3) spontaneous seizures (SS) in man. Where pertinent, crucial data available only in animal models of seizures will be compared with those from clinical seizures. The major questions to be answered are: (1) what are the metabolic consequences of a single seizure or of a series of seizures to the body as a whole and the brain in particular and (2) what diagnostic studies will permit evaluation of these consequences in a single patient?

2. Systemic Metabolic Changes in Seizures in Man

Experimentally defined conditions have permitted description of the systemic metabolic changes in man following electrically induced and drug induced seizures.[15,67] Under appropriate conditions, hyperventilation can induce in man spontaneous seizures that can be carefully studied.[67] Occasionally, serial studies of spontaneous seizures in the clinical setting can provide additional information.[7,16,124]

2.1. Electrically Induced Seizures

A severe systemic lactic acidosis occurs transiently after a single electroconvulsive seizure (EIS)

Benjamin Rix Brooks, M.D. • Department of Neurology, The Johns Hopkins University School of Medicine, Baltimore, Maryland 21205.

in paralyzed, passively ventilated man.[55,72] This lactic acidosis persists 30 min longer than the lesser lactic acidosis associated with intense muscular work in normal man.[55,85] Unparalyzed patients display a more profound systemic lactic acidosis.[72] Intraarterial pressure monitoring in such patients provides evidence for an abrupt rise in arterial pressure during the tonic phase of the EIS or even momentary cessation of cardiac pulsations followed by increased pulse pressure during the clonic phase of EIS.[40] Venous pressure is also increased. Occasionally, decreases in arterial pressure during the tonic phase are noted.[57] These effects on the systemic circulation are mitigated to a great degree by administration of muscle-paralyzing agents.[72,80]

Arterial oxygen saturation is significantly reduced and whole-body oxygen consumption significantly increased (600%) by 1 min post-EIS in unparalyzed patients compared with paralyzed patients artificially ventilated with 100% oxygen.[137] Unparalyzed patients show a 500% increase in CO_2 production compared with paralyzed patients at the same time following EIS. A delayed decrease in arterial blood pressure may occur in severely acidotic patients following EIS.[50] In such patients, the arterial pH fell as far as 7.1 with a 25% decrease in arterial CO_2-combining power. Normally, maximal exercise of short duration may severely decrease arterial pH and increase lactic acid without significant prolonged effect on arterial blood pressure,[85,134] although prolonged metabolic acidosis may cause hypotension.[92]

2.2. Drug-Induced Seizures

During the apneic phase of a drug-induced seizure (DIS) in unparalyzed, unventilated man, the arterial oxygen tension (Pa_{O_2}) and pH fall, while the carbon dioxide tension (Pa_{CO_2}) and potassium increase.[67] Heart rate, blood pressure, and cerebrospinal fluid (CSF) pressure increase 100, 115, and 600%, respectively, within 75 sec following DIS in paralyzed patients.[136] The rise in heart rate, blood pressure, and CSF pressure is reduced to 30, 8, and 25% by high spinal anesthesia. Similar results occur in dogs following DIS and are mitigated by spinal cord section.[92,95]

Bicuculline-induced seizures in anesthetized baboons result in metabolic acidosis due to lactic acid and initial hypertension similar to that observed in DIS in man.[63,65] Prolonged seizures lasting up to 5 hr may result in hyperpyrexia, hyperkalemia, and hypoglycemia. Death secondary to cardiovascular collapse supervenes, as has been noted clinically in man.[65,74]

2.3. Spontaneous Seizures

As in EIS and DIS, spontaneous seizures (SS) in unanesthetized, unparalyzed man are associated with a severe metabolic acidosis.[16,31,60,83,84,98,103,123,124] Two patterns of blood pressure responses to SS are noted: (1) hypertension and (2) hypotension.[57] The former is due to a central neurogenic response and is mitigated by spinal anesthesia.[136] The latter is thought to be due to a Valsalva maneuver during the tonic phase of the seizure, but completely paralyzed patients can also show this effect.[80]

Arterial P_{O_2} may decrease, as will pH. Arterial P_{CO_2} is increased within 1–4 min, but the CO_2 content is decreased for up to 30 min.[60,67] Hyperglycemia occurs with a slight fall in serum sodium and no changes in serum potassium.[73] The anion gap that is observed in the serum is not totally explained by the change in lactate, pyruvate, or β-hydroxybutyrate concentrations.[81,83]

3. Cerebral Metabolic Changes in Seizures in Man

A variety of metabolic balance studies in man and animals have served to define changes in central nervous system metabolism during and following induced and spontaneous seizures. Arteriovenous metabolic input–output studies as well as CSF analyses have been employed in man following EIS and DIS.[15,50,67,103] In man as well as animals, such studies have been correlated with measurement of cerebral blood flow by several techniques.[15,50,55,110,141] Studies in animals have allowed direct analysis of cortical surface gas tensions and pH changes as well as direct tissue sampling.[18] Attempts have been made to distinguish between the metabolic effects on the central nervous system of single seizures as compared to a series of recurrent or continous seizures (status epilepticus). Such studies necessarily have been accomplished only in experimental animals.[12]

3.1. Electrically Induced Seizures

The effects of EIS on cerebral metabolism have been studied in paralyzed patients who were anesthetized with rapidly acting hypnotics.[15,50] Artificial ventilation proceeded with either 21, 33, or

100% oxygen[15,50] to mitigate the effects on respiration of muscular activity in unparalyzed patients. In three studies, the jugular bulb oxygen tension ($Pj O_2$) rises during EIS and can remain elevated for 3–15 min following the onset of seizure.[15,50,103] The jugular bulb carbon dioxide tension ($Pj CO_2$) rises during the seizure, while the arterial ($Pa CO_2$) remains constant. The 50% decrease in the arteriovenous oxygen saturation difference indicates increased cerebral blood flow, which was confirmed by the Xenon (^{133}Xe) washout method.[15,50] Under these conditions, no apparent tissue hypoxia can be demonstrated, and the arteriovenous oxygen saturation difference returns toward normal as the cerebral blood flow stabilizes slightly below normal.[15]

The venous–arterial lactate concentration difference increased immediately following the EIS in three studies. The nonhypoxic production of lactic acid is reflected in the alteration in respiratory quotient (RQ), defined as the ratio of the arteriovenous carbon dioxide content difference to the arteriovenous oxygen content difference. During a single seizure, the cerebral RQ increases 30% consistent with production of fixed acid in the brain and conversion to carbonic acid. The cerebral RQ is decreased 39% below normal in the first 3 min postictally, but normalizes by 15 min after the onset of the seizure.[15] These changes occur during the development of an arterial hyperglycemia despite little change in the arteriovenous glucose difference (Fig. 1).

In mice with EIS, paralysis and artificial ventilation with 100% oxygen result in only a 20–50% rise in brain lactate compared to the 100% increase in nonparalyzed, spontaneously ventilated animals.[21] Cerebral high-energy metabolic substrates—phosphocreatine and ATP—decrease by 50% only in unparalyzed mice compared with paralyzed mice. Brain pH, calculated from the phosphocreatine/ATP ratio, is reduced by 90% in unparalyzed animals, but not, apparently, in paralyzed, ventilated animals. Direct measurement of brain pH was not accomplished.

Postictally, hyperventilation accompanied by increased cerebral blood flow raises brain oxygen tensions, allowing oxidative metabolism to resynthesize phosphocreatine and ATP. Brain glucose is normalized more slowly after an initial rapid decline, suggesting ongoing glucose consumption. The slow normalization of brain lactate may result from slow oxidation to pyruvate or the extrusion of lactate from the cell into the brain extravascular space. Once in the extracellular fluid, the lactate may be unable to return to the cell to be oxidized

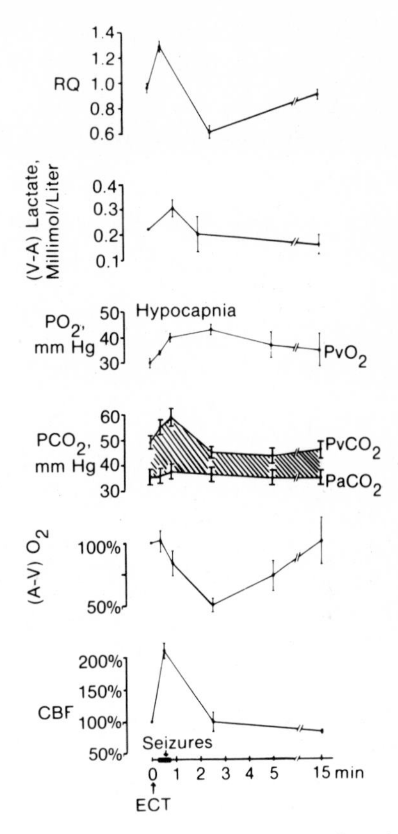

Figure 1. Metabolic changes after electrically-induced seizures (EIS) in man. Internal carotid artery (a) and jugular venous bulb (v) catheters measured PO_2, PCO_2, oxygen saturation, and lactate before and at various times following electroconvulsive therapy in 11 patients. The RQ was determined as rate of CO_2 production to O_2 consumption between internal carotid artery and jugular venous bulb. Cerebral blood flow (CBF) was determined by Xe^{133} washout method. Reproduced with permission from Brodersen *et al.*[15]

Such compartmental shifts in lactate may be important in determining the physiological changes that are due to the metabolic consequences of seizures.

Serial EIS in unparalyzed cats resulted in a systemic lactic acidosis that worsened with attenuation of blood pressure and death secondary to cardiac arrhythmias.[37,134] Bicarbonate administration mitigated seizure mortality. Serial EIS in paralyzed, oxygen-ventilated cats significantly increased arterial lactate by 100% and CSF lactate by 150% without any increase in CSF sodium or potassium.[134]

3.2. Drug-Induced Seizures

The effects of DIS on cerebral metabolism in man have not been studied as extensively as those of EIS. In nonparalyzed, nonventilated man, $Pj\text{O}_2$ decreased 60% and $Pj\text{CO}_2$ increased 30%, while pH in the jugular bulb decreased more than 90%.[67] Serum potassium in the jugular bulb also increased, as did the cerebral blood flow. Postictally, there was hyperoxia in the jugular-bulb effluent similar to that seen in EIS. Oxidative and anaerobic metabolism substrates were not studied (Fig. 2).

DIS in paralyzed, ventilated cats were studied by means of heat-clearance-probe determination of cerebral blood flow, arterial and venous acid–base balance, and chemical determination of energy metabolites, as well as by direct measurement of brain surface pH changes.[41] DIS resulted in systemic hypertension, increased cerebral blood flow, and hyperoxia of the jugular-bulb effluent. The cerebral RQ increased as brain ATP, phosphocreatine, bicarbonate, and calculated pH decreased during DIS. Cortical surface pH, however, correlated with brain surface $P\text{CO}_2$ changes. Brain lactate increased 500%, but the lactate/pyruvate ratio increased only 300%. The lactate increase reflects a shift in the cytoplasmic redox potential due to a reduction in NAD to NADH.[19] Using cortical fluorometry, however, there is evidence for a shift in the mitochondrial redox potential toward increased oxidation, suggesting adequate tissue oxygenation.[44] This compartmentalized metabolic response to increased energy needs is seen in the heart during increased contractile activity and in the electrically stimulated superior cervical ganglion.[44,117] One theory favors the view that this nonhypoxic lactate increase affects extracellular pH by decreasing extracellular bicarbonate at constant extracellular $P\text{CO}_2$.[41,92] The resultant decreased pH is felt to be instrumental in the change in cerebral vascular resistance that augments cerebral blood flow.[53] However, the reduced sensitivity of cerebral vessels to carbon dioxide postictally may not be compatible with the hypothesis of an extracellular pH–cerebral blood flow couple.[78,79]

Prolonged DIS in paralyzed, artifically ventilated, anesthetized rats resulted in an 80% decrease in cortical glycogen and a 70% decrease in cortical glucose.[19] Cortical lactate progressively increased 900% and pyruvate increased 200%, while the lactate/pyruvate ratio increased 400%. The calculated cortical pH decreased, and the calculated NADH/NAD ratio increased, then decreased. Phosphocreatine decreased 50%, but the calculated adenylate energy change decreased only 2% except in previously starved animals, in which it decreased 4%.[6] The cytoplasmic redox potential shifted toward reduction, while the mitochondrial redox potential shifted toward oxidation. Concomitantly, cerebral blood flow increased 900% with an 85% decrease in cerebrovascular resistance.[64] Oxygen consumption increased 300% initially and fell to 200% with recurrent seizures. Oxidative phosphorylation in the brain can increase only 300%. Glucose consumption increased 400% initially, but fell to 200% during recurrent seizures, matching the oxygen-consumption increase.[6] Glycolytic flux increased 200–600%, which is not explained by the decrease in brain glucose and glycogen. Increased glucose transport into brain could make up part of the difference. Alternative use of endogenous substrates in the citric acid cycle is also important.[117] High arterial oxygen and glucose content is imperative to meet the metabolic needs of the brain during the seizures. If arterial hypotension, systemic hypoxia, and hypoglycemia supervene, then these needs are not met.[61]

3.3. Spontaneous Seizures

No comprehensive study of cerebral metabolism during SS in man has been accomplished. A limited arteriovenous acid–base balance study in nonparalyzed, nonventilated man[67] has shown evidence for increased cerebral oxygen consumption and carbon dioxide production in generalized tonic–clonic convulsions. Arterial oxygen gas tension, saturation, and pH decreased during the apneic phase of the tonic seizure in conjunction with increased carbon dioxide gas tension. Arterial sodium, and to a more marked degree potassium, rise and are cleared in less than 5 min after a generalized tonic–clonic seizure. The oxygen gas tension rises in the jugular bulb and remains elevated for at least 15 min after spontaneous seizures, as has been noted in EIS and DIS. The carbon dioxide gas tension increases in the jugular bulb, but a component of this increase derives from the systemic increase in carbon dioxide gas tension. Potassium increases in the jugular bulb and is cleared in the same time period as arterial potassium (Fig. 3).

Recovery of the patient and the EEG during the postictal period (Fig. 4) is associated with a return to normal in the jugular-bulb pH and a decline in the $Pj\text{O}_2$ to normal.[67,70]

4. CSF Changes in Seizures in Man

The brain extracellular fluid is in continuity with the CSF.[133] Changes in acid–base balance and po-

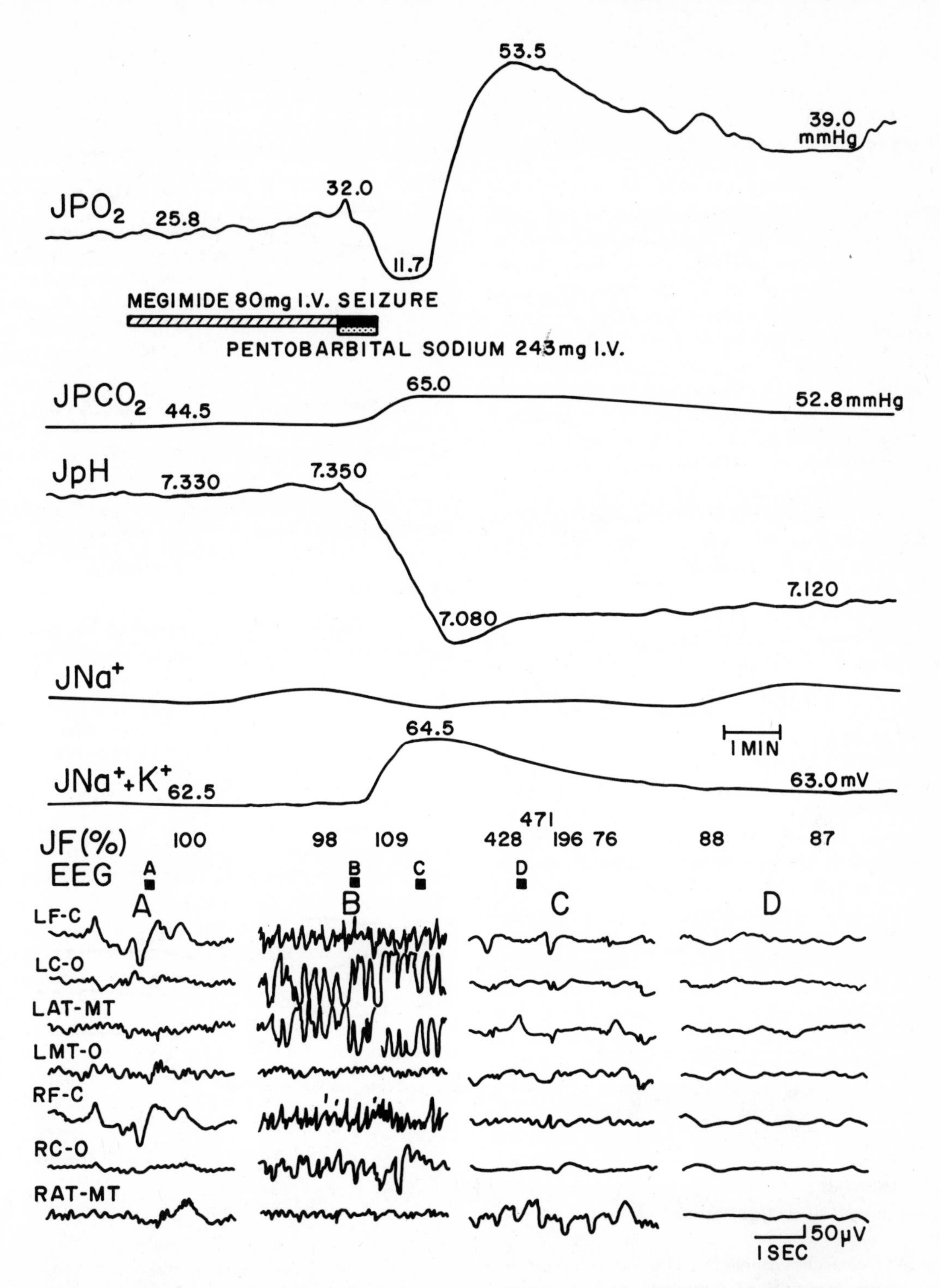

Figure 2. Metabolic changes after drug-induced seizures (DIS) in man. Jugular venous bulb (J) PO_2, PCO_2, pH, Na^+, and $Na^+ + K^+$ concentrations were measured in one patient before, during, and following a megimide-induced seizure. The EEG changes before (A), during (B), immediately following (C), and 6 min following (D) DIS were correlated with jugular venous bulb flow determined by indwelling thermistor-probe technique. Reproduced with permission from Meyer *et al.*[67]

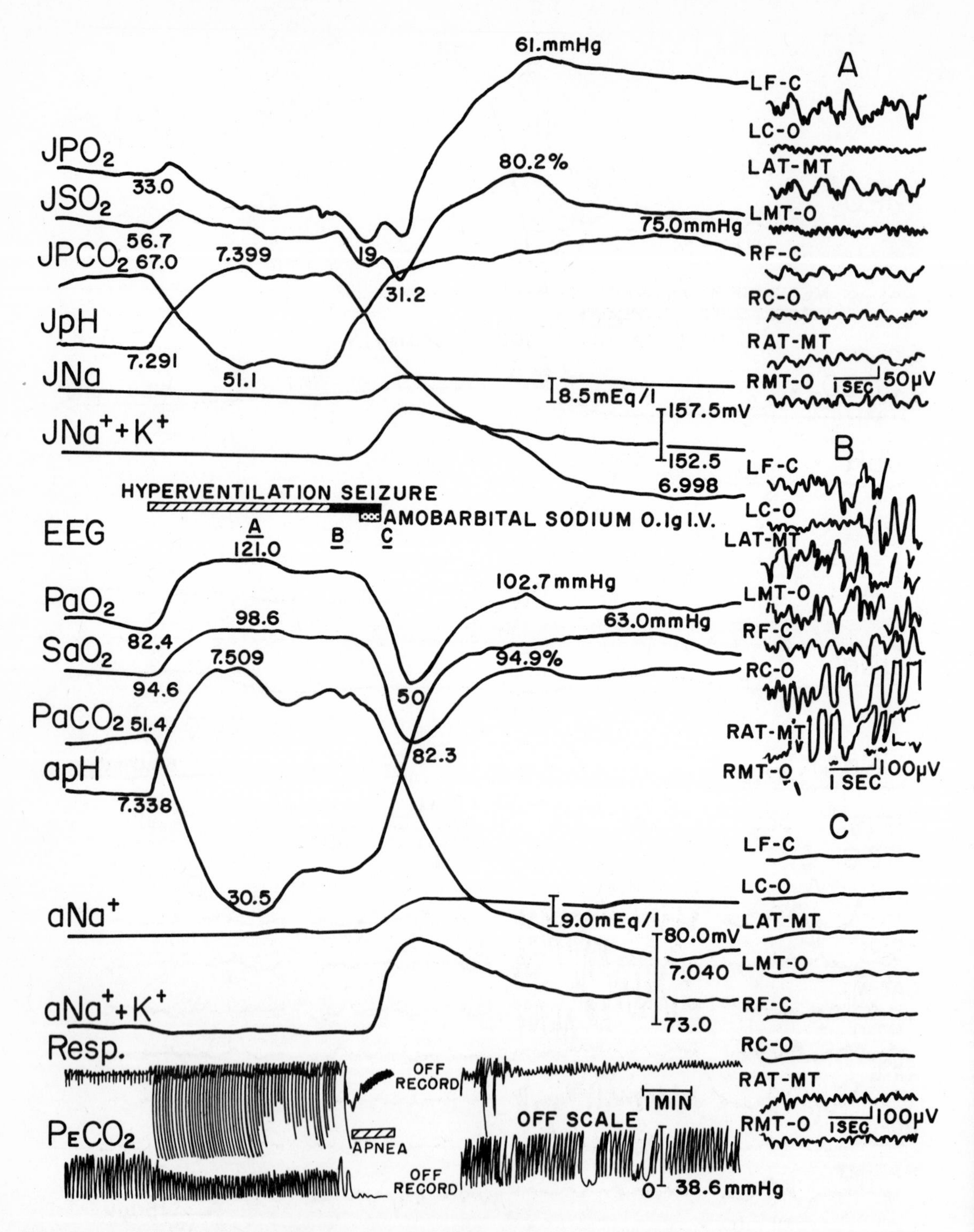

Figure 3. Metabolic changes after spontaneous seizures (SS) in man. Internal carotid artery (a) and jugular venous bulb (J) P_{O_2}, P_{CO_2}, pH, O_2 saturation, Na^+ and $Na^+ + K^+$ concentrations were measured in one patient before, during, and following a generalized tonic–clonic seizure induced by hyperventilation. The EEG changes before (A), during (B), and following (C) the SS were correlated with respiration rate and expired air $P_{E}CO_2$. Reproduced with permission from Meyer *et al.*[67]

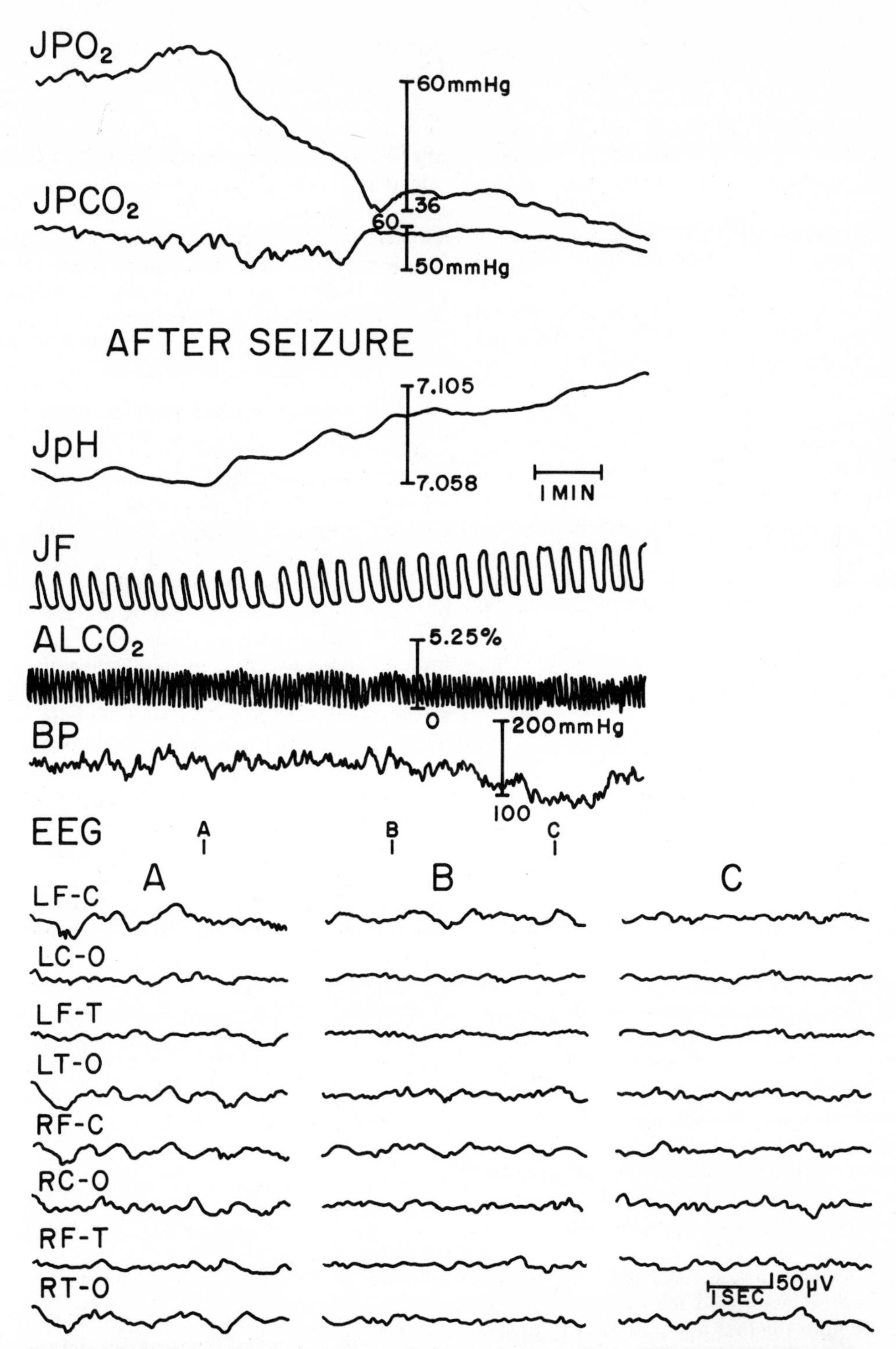

Figure 4. Metabolic changes during recovery from spontaneous seizures (SS) in man. Jugular venous bulb (J) PO_2, PCO_2, and pH were measured in one patient during the recovery from generalized spontaneous tonic–clonic seizure. EEG changes just before (A), immediately following normalization of jugular bulb oxygenation (B), and 3 min later (C) were correlated with jugular bulb venous flow (JF), arterial blood pressure (BP), and expired alveolar CO_2 ($ALCO_2$). Reproduced with permission from Meyer *et al.*[67]

tassium occur in the extracellular fluid in response to metabolic changes in the underlying cortical tissue.[10,22] Acutely, some of these changes may be reflected in the cisternal CSF and to a lesser extent in the lumbar CSF.[30,46,51,128] Chronically, in response to systemic changes in acid–base balance, alterations in CSF P_{CO_2}, bicarbonate, and pH are more pronounced in the lumbar CSF, although the pH is corrected toward normal in the cisternal CSF.[28,29,86,91,96,97,100–102,104,112,115,125]

Lactate and pyruvate can diffuse from central nervous system cells into the CSF.[117,118] Lactate and pyruvate are related to the cytoplasmic redox potential and may indirectly reflect the intracellular redox potential.[99,117–120] The lactate/pyruvate ratio may change in response to hypoxia, alkalosis, and intracellular acidosis.[34,111,120–122,126] In severe cerebral hypoxia, lumbar CSF acid–base and metabolite concentrations can reflect the severity of the metabolic insult to the brain, but can underestimate these changes relative to the cisternal CSF.[87] If spontaneous seizures in man are associated with intraictal systemic hypoxia, concomitant cerebral hypoxia may occur. The extent of such hypoxia may be revealed by analysis of the CSF in patients following generalized tonic–clonic seizures.[14,33,46]

4.1. CSF Acid–Base Balance in Seizures

Ten clinical investigations have evaluated CSF acid–base and metabolite changes in man interictally and following SS.[7,16,17,27,73,84,90,105,123,124,143] Direct measurements of oxygen and carbon dioxide gas tensions in the CSF of control patients and interictal seizure patients indicate a correlation between arterial and CSF gas tension (Fig. 5A). In seizure patients with repetitive seizures less than 1 hr apart and patients within 3 hr following a generalized tonic–clonic seizure, there is no alteration in this relationship (Fig. 5B). Although arterial hypocarbia is present in patients with alcohol-withdrawal seizures, there is no evidence for altered oxygen gas tension in the CSF (Fig. 5B–D). CSF oxygen gas tension, however, is not an adequate measure of cerebral tissue oxygenation.[108]

A single spontaneous generalized tonic–clonic seizure in man is associated with a severe fall in arterial pH within the first few minutes of the seizure.[83,98,107] This pH change resolves partially by 1 hr and completely by 6 hr postictally (Fig. 6). Intraictally, the CSF pH is normal and changes little postictally.[16,17] However, an occasional patient may have a markedly decreased pH.*

The blood carbon dioxide content (bicarbonate and dissolved carbon dioxide) decreases acutely after a seizure and resolves partially by 1 hr postictally (Fig. 7). The CSF bicarbonate is normal intraictally and decreases only slightly postictally.[16] The CSF–arterial blood bicarbonate difference, however, has been shown in three studies to shift from negative to positive postictally (Table 1), indicating a resolving metabolic acidosis that is corrected by 6 hr after the seizure.[13,16,71,91,96,97,102,104]

4.2. CSF Electrolyte Changes in Seizures

Although CSF potassium is elevated acutely following experimental anoxia[69] and bemegride-induced seizures[68] in primates, potassium is cleared from the cisternal CSF within minutes.[68] After a single seizure in man, blood and lumbar CSF potassium are not significantly altered (Fig. 8) despite the rapid development and resolution of the seizure-induced metabolic acidosis.[16] The role of the seizure-induced hyperglycemia in modulating the potassium changes has not been systematically evaluated.[73] Lumbar CSF sodium and chloride are also unchanged in man as in experimental primates (rhesus monkey) following generalized seizures.[68]

The status of divalent cations in the CSF after seizures in man is still controversial. Two studies measuring magnesium by atomic absorption spectrometry and Lang's complexometry suggest that the CSF magnesium is elevated for at least a week following a generalized seizure.[38,39] Several other studies, including our own investigation using atomic absorption spectrometry, have not confirmed this observation (Table 2). Calcium and phosphate in the CSF have not been shown to be acutely altered during seizures in man[8,9,38,39] although alterations in these CSF constituents have been observed among some medicated seizure patients (see Chapter 49).

4.3. CSF Metabolite Changes in Seizures

Glycolytic metabolic products change in the CSF following generalized tonic–clonic seizures in man. Arterial hyperglycemia occurs following EIS and SS

* One such patient studied 1 hr postictally had a decreased CSF pH to 7.16 and an elevated CSF P_{CO_2} to 67 mm Hg with normal CSF (HCO_3^-). He was somnolent and difficult to arouse for 12 hr postictally.[16]

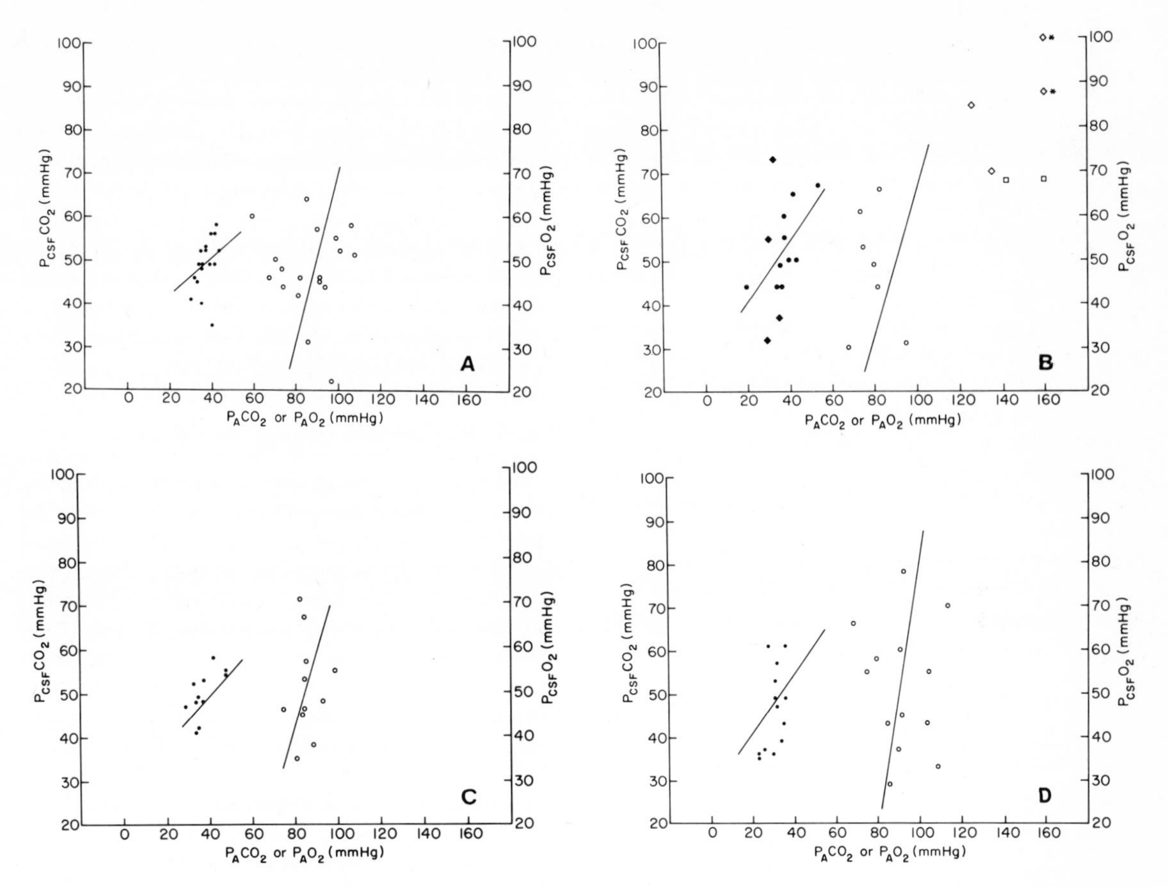

Figure 5. CSF and arterial PO_2 and PCO_2 in seizure patients. (A) PO_2 (○) and PCO_2 (●) in patients without seizure disorder. (B) PO_2 [(○) spontaneous ventilation; (□) artificial ventilation] and PCO_2 (●) in patients less than 3 hr following a single tonic–clonic seizure; PO_2 [(◇) spontaneous ventilation; (◇*) artificial ventilation] and PCO_2 (♦) following recurrent persistent tonic–clonic seizures. (C) PO_2 (○) and PCO_2 (●) 3–12 hr following a single tonic–clonic seizure. (D) PO_2 (○) and PCO_2 (●) 5–12 hr following single alcohol-withdrawal seizure. Data from Brooks and Adams.[16,17]

in man.[15,73,90] In the CSF, glucose has been shown to be increased by 8–9% postictally ($p < 0.05$) in SS in man even in the absence of a documented rise in arterial glucose in the same patient.[73] These changes may reflect seizure-induced alterations in the blood–brain barrier.[23,45,88]

Following a single generalized tonic–clonic seizure in unanesthetized, nonventilated, nonparalyzed adults, the blood lactate rises precipitously approximately 12-fold (Fig. 9). The raised lactate is decreased 30% by 1 hr and 90% between 1 and 3 hr postictally. During the 4- to 6-hr period postictally, blood lactate is actually decreased compared to the intraictal concentration ($p < 0.05$). CSF lactate, however, increased 30% by 1–3 hr postically and remained at this level for 4–6 hr postictally.[16] In other studies, only 40% of adult patients showed a lactate increase of 30–40% after a single tonic–clonic seizure.[73,90] The durations of these seizures were not described in detail. In a third study in children (Fig. 10), a single seizure less than 30 min in duration raised the CSF lactate in only 4 of 31 patients.[123,124] In both studies, the lactate/pyruvate ratio was increased consistent with an altered NAD/NADH ratio.[117,118]

Repeated seizures in adults and children will result in a rise in CSF lactate in 50–100% of patients, as will single seizures of longer than 30-min duration.[124] In three patients with status epilepticus (repeated seizures less than 1 hr apart), CSF lactate was

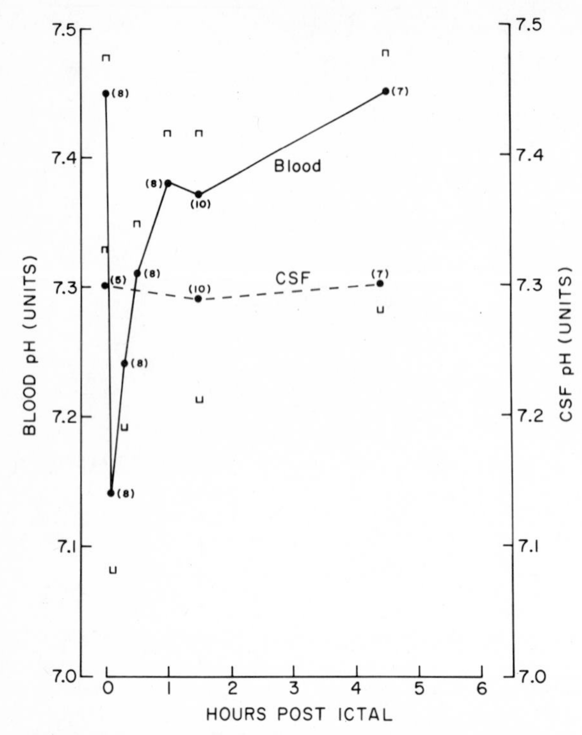

Figure 6. Arterial and CSF pH following spontaneous seizures (SS) in man. Mean pH (± S.E.M.) before and at various times following a single tonic–clonic seizure. Number of patients in parentheses. Data from Brooks and Adams[16] and Orringer *et al.*[83]

increased by more than 30% beyond that observed following prolonged passive ventilation.[16] Patients were stuporous for 24–36 hr following control of seizure activity. Clinical improvement paralleled the return to normal of the CSF lactate and not the normalization of the blood lactate (Fig. 11).

Following seizures in man, the exact source of the elevation in CSF lactate has not been defined. The elevated CSF lactate may result (1) directly from cerebral metabolic changes during the seizure; (2) indirectly from the muscular activity associated with the seizure; or (3) indirectly from cerebral metabolic changes due to a complex change in the ventilatory state secondary to the metabolic stress of a seizure leading to reactive hyperventilation.

Small amounts of lactate are measured in the cerebral-venous effluent after a seizure in man.[15,103] Brain lactate increases in animals during a seizure under conditions that parallel the spontaneous seizures in unparalyzed, unventilated man.[21] Small increases in CSF lactate are measured in animals immediately after DIS.[96] This increase is approximately that seen in man with SS or EIS.[16,72,90] In SS in man, the cerebral oxygen supply probably does not meet the cerebral oxygen demand of the seizure because of apnea and increased oxygen requirements of contracting muscles,[4,21] suggesting that brain is the source of increased CSF lactate in man. Careful attention to minimizing the peripheral demands for oxygen can evidently modify the amount of oxygen available to the brain and decrease the amount of lactate production by the brain during a seizure.[134]

A tonic–clonic seizure in nonparalyzed, nonventilated man causes an increased arterial lactate concentration acutely secondary to changes in muscle metabolism.[55,83] Since blood lactate falls in time after the seizure, one possibility is that arterial lactate diffuses into the brain and hence into the CSF.[82] The blood–brain barrier for some anions is altered during DIS.[11,23,45,116] In addition, animals with electrically induced muscular activity for 20 min had a rise in CSF lactate over the subsequent 20 min.[130] Infusion of lactic acid in the absence of muscular activity did not cause a rise in CSF lactate.[3,52,130] Muscular activity in these animals, however, led to reactive hyperventilation and CSF lactate accumulation on this basis.[26,129]

The mild increase in CSF lactate in man between 1 and 6 hr after an SS is not significantly different from that increase seen in passively hyperventilated man.[42] Hyperventilation in animals leads to in-

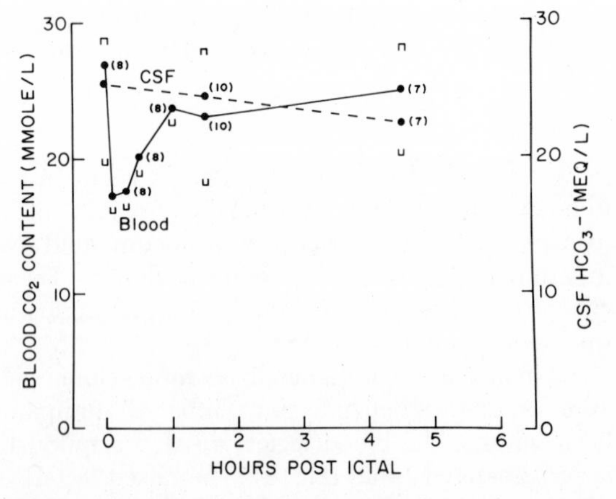

Figure 7. Arterial and CSF carbon dioxide content and bicarbonate concentrations following spontaneous seizures (SS) in man. Mean arterial carbon dioxide content (± S.E.M.) and mean CSF bicarbonate concentration (± S.E.M.) before and at various times following a single tonic–clonic seizure. Number of patients in parentheses. Data from Brooks and Adams[16] and Orringer *et al.*[83]

Table 1. CSF–Arterial Blood Acid–Base Difference after Seizures in Man

Patients	Number	CSF–blood difference P_{CO_2} (mm Hg)	HCO_3^- (meq/liter)
Controls			
No seizure disorder[16]	13	+12 ± 5	−1.8 ± 2.0
Interictal seizure patients			
Lumbar puncture 4–8 days after idiopathic seizure[16]	5	+16 ± 15	−1.6 ± 7.2
Recurrent persistent seizures			
Lumbar puncture within 1 hr after idiopathic seizure[16]	4	+18 ± 19	+1.5 ± 8.8
Idiopathic seizures			
Adult[16]			
Lumbar puncture less than 3 hr after single seizure	10	+15 ± 9	+1.5 ± 4.8[a]
Lumbar puncture 3–6 hr after single seizure	7	+12 ± 6	−2.3 ± 3.4
Children[123,124]			
Lumbar puncture 1–22 hr after febrile seizure[123]	29	+12 ± 4	+1.3 ± 2.1
Lumbar puncture 3–8 hr after prolonged seizure[124]	22	+11 ± 6	+1.5 ± 3.5
Alcohol-withdrawal seizures			
Lumbar puncture 5–12 hr after single seizure[17]	13	+17 ± 10	−0.4 ± 8.0
Delirium tremens			
Lumbar puncture on day 2 to day 7 of delirium[17]	7	+16 ± 16	−3.7 ± 9.4

[a] $p < 0.05$, two-tailed Student's t test vs. controls.

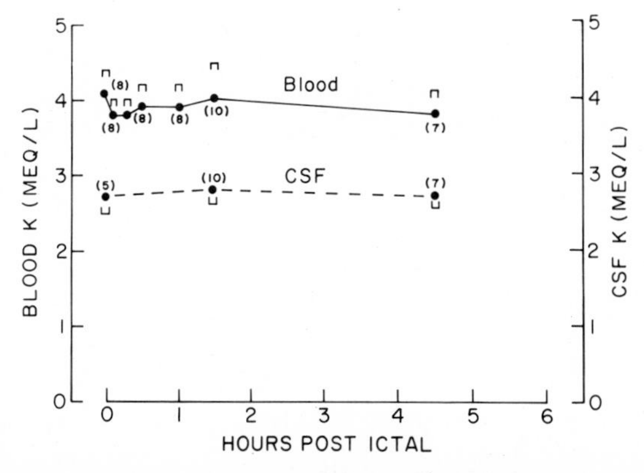

Figure 8. Arterial and CSF potassium concentrations following spontaneous seizures (SS) in man. Mean potassium concentration (± S.E.M.) before and at various times following single tonic–clonic seizure in man. Number of patients in parentheses. Data from Orringer *et al.*[83] and Brooks, B. R. (unpublished data).

creases in brain and CSF lactate[93] that are completely corrected by hyperbaric oxygenation during hyperventilation.[94] It is not clear whether the increased lactate signifies tissue hypoxia, since high-energy substrates—ATP and phosphocreatine—in brain remain unchanged with hyperventilation.[117,119,122] In man, hyperventilation leads to metabolic changes in the cerebral-venous effluent, suggesting increased fixed acid production,[2,32,142] and although cerebral blood flow decreases, there is no apparent alteration in the cerebral metabolic rate of oxygen utilization.[59]

The rise in CSF lactate in man after a single spontaneous tonic–clonic seizure may be related in part to hyperventilation, since hypocapnea is documented in a few of the patients studied. It is possible that a brief period of hyperventilation occurs immediately after the seizure, increasing the brain and CSF lactate. Because of the delayed clearance of lac-

Table 2. Arterial Blood and CSF Electrolyte Changes after Seizures in Man

		Blood			CSF		
Patients	Number	K^+ (meq/liter)	Ca^{2+} (meq/liter)	Mg^{2+} (meq/liter)	K^+ (meq/liter)	Ca^{2+} (meq/liter)	Mg^{2+} (meq/liter)
Controls							
No seizure disorder[16,17]	8	4.2 ± 0.4	9.7 ± 0.4	1.5 ± 0.2	2.8 ± 0.2	4.8 ± 0.5	2.1 ± 0.1
Idiopathic seizures							
Lumbar puncture 1–6 hr after idiopathic seizure[16]	11	4.0 ± 0.5	9.3 ± 0.6	1.6 ± 0.2	2.8 ± 0.2	5.1 ± 0.3	2.0 ± 0.4
Alcohol-withdrawal seizures							
Lumbar puncture 5–12 hr after alcohol-withdrawal seizure[17]	11	3.8 ± 0.5[a]	9.5 ± 0.9	1.2 ± 0.3[a]	2.8 ± 0.3	5.0 ± 0.4	1.7 ± 0.3
Delirium tremens without antecedent seizure							
Lumbar puncture on day 2 to day 7 of delirium[17]	6	4.1 ± 1.2	9.2 ± 0.8	1.6 ± 0.4	2.7 ± 0.3	5.0 ± 0.3	2.0 ± 0.2
Cerebrovascular disease							
Lumbar puncture 2–10 hr after carotid occlusion	7	4.1 ± 0.2	9.0 ± 0.3	1.7 ± 0.3	2.9 ± 0.4	4.8 ± 0.4	2.1 ± 0.3

[a] $p < 0.05$, two-tailed Student's t test vs. controls.[16,17]

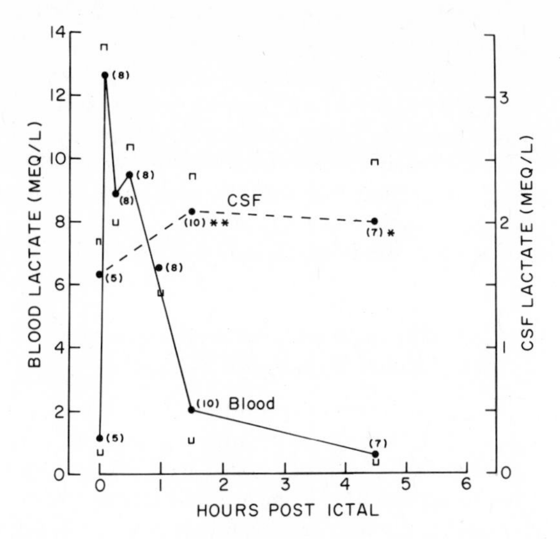

Figure 9. Arterial and CSF lactate concentrations following spontaneous seizures (SS) in man. Mean lactate concentration (± S.E.M.) before and at various times following single tonic–clonic seizure in man. Number of patients in parentheses. Statistical significance: $^{**}p < 0.005$, $^{*}p < 0.05$ compared with preseizure values (unpaired two-tailed Student's *t* test). Data from Brooks and Adams[16] and Orringer *et al.*[83]

tate from the CSF,[106,127] the measured CSF lactate represents partially the effect of reactive hyperventilation due to the demands of muscular metabolism during the seizure.[129,130] Against the possibility that the total measured CSF lactate increase after a seizure in man is due to reactive hyperventilation alone is the fact that at least 1 hr of steady hyperventilation is required to attain the level of CSF lactate measured in neurological patients,[42] and reactive hyperventilation after a single seizure lasts only a few minutes.[67] Considering the experimental determinations of brain lactate changes in nonparalyzed, nonventilated animals,[20] it is likely that the CSF lactate increase is related primarily to the early increase in brain lactate during the apneic stage of the seizure, when oxygen demand is in excess of oxygen supply, and secondarily to reactive hyperventilation.

The patients with recurrent persistent seizures demonstrated the extent to which reactive hyperventilation may lead to a respiratory alkalosis and further increases in CSF lactate. These patients had a series of seizures prior to intubation and increased oxygenation. CSF lactate was markedly increased above that seen in passive hyperventilation for 1 hr,[42] suggesting that the major portion of the increase is due to the persistent seizures. However, prolonged hyperventilation. In primates kept well continuing elevation of the CSF lactate despite a constant brain lactate level.[47–49] Data in man are, at present, unavailable to determine what proportion of the elevated CSF lactate in these patients is due to the systemic and cerebral metabolic demands of the seizures and what proportion is due to the prolonged hyperventilation. In primates kept well oxygenated during drug-induced persistent or recurrent seizures for several hours, there is an early doubling of the cerebral-venous lactate coupled with a systemic metabolic acidosis. As seizures progress and inspired oxygen is increased, there is normalization of the arterial and venous pH, but cerebral-venous lactate remains elevated until late in the course of the seizures.[61–63] These experiments parallel the situation in the patients with recurrent per-

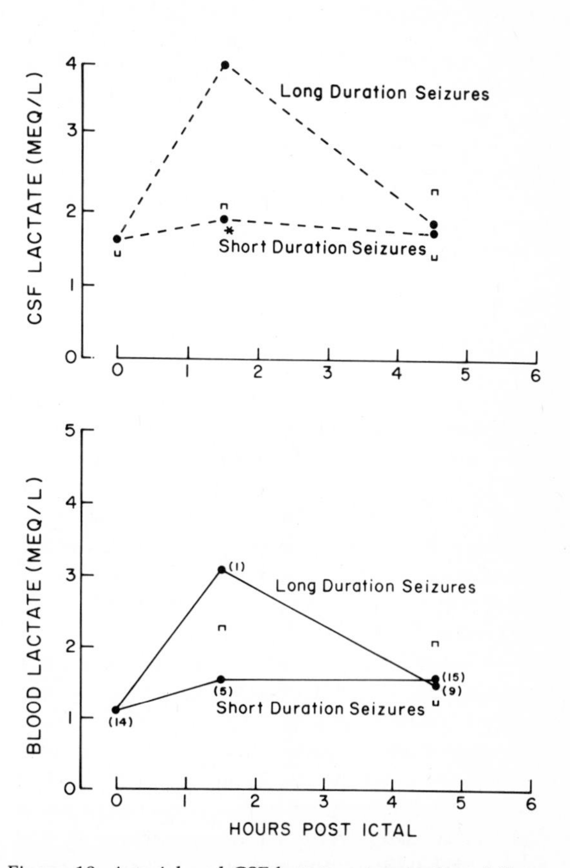

Figure 10. Arterial and CSF lactate concentrations following spontaneous seizures (SS) of varying durations in children. Mean lactate concentration (± S.E.M.) in controls and at various times following short-duration and long-duration seizures. Statistical significance: $^{*}p < 0.05$. Data from Simpson *et al.*[123,124]

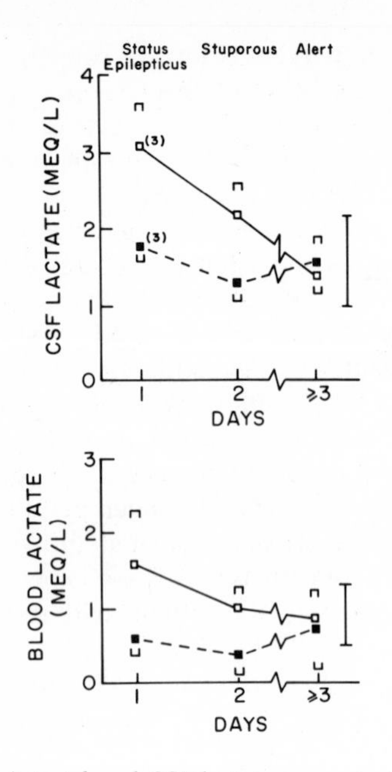

Figure 11. Arterial and CSF lactate concentrations following status epilepticus in man. Serial simultaneous lumbar CSF and arterial lactate concentrations (± S.D.) were measured in three patients with status epilepticus (□) and three alert patients with other neurological disease (■). Vertical brackets at right denote 95% confidence limit of normal concentrations based on 13 patients. CSF lactate concentration decreases in parallel with clinical improvement.

sistent seizures and suggest that hyperventilation alone does not explain the entire rise in CSF lactate. Thus, although lactate can enter brain via a carrier-mediated transport mechanism from the blood,[75,76,109,113] the primary source of the CSF lactate following seizures is production directly by the central nervous system.[92] Clearance of lactate from the CSF is primarily via diffusion.[106,127] With intense lactate production locally, clearance of lactate from the CSF is even further slowed. Delayed clearance of lactate from the CSF may also be seen in severe generalized systemic acidosis.[58]

Several studies provide evidence that a single tonic–clonic seizure in nonparalyzed, spontaneously ventilating man can lead to a small rise in CSF lactate with minimal derangement of CSF acid–base balance in the majority of patients. Those patients with normal CSF acid–base balance are usually alert in the postictal period of a single seizure. The mechanism of the rise in CSF lactate is probably production by brain tissue due to a complex interaction of the metabolic events during the acute seizure and the changes in ventilatory status that occur during the postictal period. Recurrent persistent seizures in man lead to a disproportionate rise in CSF lactate accompanied by a respiratory alkalosis in the postictal period. These patients have normal CSF acid–base balance but are not alert in the postictal period, indicating that the CSF acid–base balance does not correlate directly with the state of consciousness in the postictal period.[96,102]

4.4. CSF Acid–Base and Metabolite Changes after Alcohol-Withdrawal Seizures

Patients with alcohol-withdrawal seizures were evaluated 5–12 (7 ± 3) hr after a grand mal convulsion.[17,131,132,138–140] All patients were alkalotic and hypocapneic at a time when arterial bicarbonate was not significantly different from that in controls. The CSF acid–base balance was not significantly different from that in controls, and the CSF–arterial blood difference for bicarbonate was not significantly altered (see Table 1). Arterial lactate was significantly elevated compared to that in controls, but not significantly different from that seen in passively hyperventilated controls.[42] CSF lactate was disproportionately increased compared to that in controls, hyperventilated controls, and patients with idiopathic seizures evaluated from 3 to 6 hr after the seizure (Table 3). Eight patients who developed prolonged delirium tremens had higher CSF lactate concentrations (3.2 ± 0.5 meq/liter) than five patients who had milder withdrawal syndromes (2.3 ± 0.3 meq/liter) (Table 3).

Both mean arterial and mean CSF magnesium concentrations (see Table 2) were decreased compared to those in controls and patients with idiopathic seizures. The decreased arterial magnesium concentration is comparable to that seen in previous studies.[8,9,24,138–140]

Patients with delirium tremens occurring without antecedent withdrawal seizure were evaluated from day 2 to day 7 of the delirium. Patients had received chlordiazepoxide, thiamine, and magnesium chloride in pharmacological doses prior to evaluation. Marked arterial alkalosis was present, but there was no significant difference from controls in arterial carbon dioxide tension and arterial bicarbonate concentration.[17] The CSF was alkaline although there was no statistically significant change in CSF carbon dioxide tension or bicarbonate concentration, suggesting a mixed respiratory and metabolic alkalosis.[17,43,114] Arterial blood and CSF lactate were elevated compared to values in controls[16,17] and hyperventilation controls,[42] but were not signifi-

Table 3. Arterial Blood and CSF Lactate Changes after Seizures in Man

Patients	Number	Arterial lactate (meq/liter)	CSF lactate (meq/liter)
Controls			
No seizure disorder[16]	13	0.9 ± 0.2	1.6 ± 0.3
Hyperventilation[42]	14	1.5 ± 0.4[a]	1.9 ± 0.5[a]
Interictal seizure patients			
Lumbar puncture 4–8 days after idiopathic seizure[16]	5	1.1 ± 0.4	1.7 ± 0.3
Recurrent persistent seizures			
Lumbar puncture within 1 hr after idiopathic seizure[16]	4	1.4 ± 0.5	3.3 ± 0.7[a]
Idiopathic seizures			
Lumbar puncture less than 3 hr after single seizure[16]	10	2.0 ± 0.9[a]	2.1 ± 0.3[a]
Lumbar puncture 3–6 hr after single seizure[16]	7	0.6 ± 0.2[a]	2.0 ± 0.5[a]
Alcohol-withdrawal seizures			
Lumbar puncture 5–12 hr after single seizure[17]	13	1.2 ± 0.5	2.8 ± 0.6[a]
Patients who developed prolonged delirium	8	1.4 ± 0.5	3.2 ± 0.5[a]
Patients who had short withdrawal syndrome	5	1.0 ± 0.6	2.3 ± 0.3
Delirium tremens			
Lumbar puncture on day 2 to day 7 of delirium[17]	7	1.3 ± 0.3	2.5 ± 0.5[a]

[a] $p < 0.05$, two-tailed Student's t test vs. controls.[16,17]

cantly different from values in patients with alcohol-withdrawal seizures (Table 3). The important clinical differences between those patients who developed delirium after seizures and those who had delirium without a history of seizures are (1) the presence of a pure respiratory alkalosis in patients with alcohol-withdrawal seizures[138–140] and (2) the fact that delirium developed subsequent to the initial examination in those patients with alcohol-withdrawal seizures.

Arterial blood and CSF magnesium concentrations are markedly reduced in patients with alcohol-withdrawal seizures,[17] but these values in patients with delirium who had been treated are not significantly different from those in controls, despite persisting delirium (see Table 2).

4.5. Relationship of Length of Delirium Tremens to CSF Lactate

The CSF lactate is elevated in patients 5–12 hr after an alcohol-withdrawal seizure and in patients who have developed delirium tremens without antecedent seizures (Fig. 12). Of 13 patients, 8 developed prolonged delirium tremens within 24 hr after the withdrawal seizure, while 5 patients had a milder and shorter withdrawal syndrome (Table 3). When the value for the CSF lactate concentration was graphed according to the day of delirium on which it was determined (Fig. 12), it was apparent that the CSF lactate concentration was higher early in the course of delirium tremens, suggesting that the fall in the CSF lactate concentration preceded clinical improvement.

The association of alcohol-withdrawal seizures with a marked respiratory alkalosis is well established,[8,9,24,138–140] but no admission clinical data up to this time predicted the severity and length of the delirium that a patient might develop. There is a marked decrease in magnesium concentration on admission in arterial blood and CSF in these patients, but there is no correlation between the arterial or CSF magnesium concentration on admission and the length of the delirium.[17] In addition,

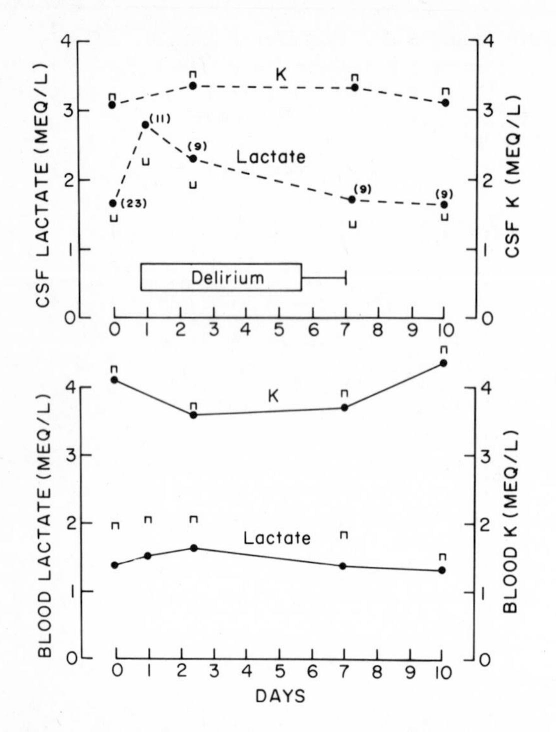

Figure 12. Arterial and CSF lactate and potassium concentrations during alcohol withdrawal with delirium tremens in man. Mean lactate and potassium concentrations (± S.E.M.) before and during course of delirium tremens. Mean duration of delirium (± S.E.M.) is correlated with metabolite changes. Number of patients in parentheses. Data from Brooks and Adams[17] and Schnaberth *et al.*[114]

treatment with exogenous magnesium chloride can normalize the magnesium concentration (see Table 2) in patients with delirium tremens, but this normalization is not associated with a rapid reversal of the delirium.[66,132]

In contrast to the nonpredictive value of either the arterial blood or the CSF magnesium concentration stands the CSF lactate on admission. There is little correlation between the arterial blood lactate after an alcohol-withdrawal seizure and the severity and length of the withdrawal syndrome that follows. However, patients can be divided clinically into two groups according to the severity and length of their withdrawal syndrome. The group of five patients with a mild withdrawal syndrome lasting at most 3 days after an alcohol-withdrawal seizure was characterized by a slightly elevated CSF lactate (2.3 ± 0.3 meq/liter), while the group of eight patients with prolonged delirium tremens had a disproportionately elevated CSF lactate (3.2 ± 0.5 meq/liter), (see Table 3). Within the restrictions placed by sample size, it is possible to state within 95% confidence limits that a CSF lactate greater than 3.0 meq/liter measured shortly after a single alcohol-withdrawal seizure may be associated with the development of severe and prolonged delirium tremens.

5. Conclusion

Years of clinical and experimental investigation have been directed at answering the question of what the metabolic consequences are of a single seizure or a series of seizures to the body as a whole and the brain in particular.[61,92,117,134,135] To date, several lines of evidence indicate that unrestricted seizure activity itself can be harmful to the brain even if the systemic consequences of unrestricted seizure activity—acidosis, hyperthermia, hypoxia, hypotension, and hypoglycemia—are controlled.[5,61,65,,117] In animals[5,134] and primates,[61,62] status epilepticus can lead to neuropathological alterations as seen in man.[1,77] This ischemic cell change is mitigated to some degree in experimental systems by reducing the amount of oxygen or glucose available to neurons during sustained seizure activity.[5,6] These data suggest that the combination of increased Pa_{O_2} and the increased metabolic rate of neurons during seizures may allow the rapid accumulation of free radicals that result in cell destruction.[5] Thus, the most important clinical goal is to stop seizure activity as soon as possible.[77]

Direct observation of the local neuronal environment in man during seizure activity indicates that the local cerebral temperature changes during seizure activity in synchrony with the increased cerebral blood flow resulting from the seizure.[25] In addition, while human neurons may alter their firing activity in response to changes in systemic factors such as Pa_{CO_2} and Pa_{O_2}, the postictal depression in these neurons is not related to systemic hypoxia.[35,36] Nevertheless, in the usual clinical situation, seizure activity is attended by hyperthermia and severe reductions in oxygen availability due to interruption of respiration and increased muscular activity.[67,92] In this context, neuronal damage results from failure of cellular energy metabolism.[61–65] Thus, in addition to terminating seizure activity, careful attention must be given to rectifying abnormalities in body temperature, blood pressure, oxygenation, acid–base balance, and blood glucose.[65,92,134]

Given the situation when a seizure patient presents postictally and the systemic factors noted above are corrected, it is occasionally necessary to determine the extent of the stress placed on cerebral metabolism by the seizure. In an individual patient, hypoxia, hypotension, acidosis, hypoglycemia, and hyperpyrexia may contribute to possible cerebral anoxia and resultant neuronal damage. An index of possible cerebral anoxia, albeit not entirely adequate, is the CSF lactate concentration.[16,17] The CSF lactate is increased in status epilepticus (see Fig. 11) and in patients with prolonged seizures (see Fig. 10) as well as in delirium tremens (see Fig. 12). Thus, it is now possible to assess the severity of the metabolic stress of single or serial seizures on the brain in clinical situations. Such an index will allow longitudinal studies correlating the severity of seizures with psychological and intellectual development in humans.[135] In addition, the course of certain clinical conditions may be predicted on the basis of the admission CSF lactate determination.[17] Further study will be required to fully exploit the possible role of CSF metabolite determinations in our understanding of the pathogenesis of seizure and other neurological disorders in man.

ACKNOWLEDGMENTS. We are extremely grateful for the encouragement and suggestions of Dr. R. D. Adams of the Department of Neurology, Dr. D. C. Shannon of the Department of Pediatrics, and Dr. H. Kazemi of the Department of Medicine at the Massachusetts General Hospital and Harvard Medical School. This review was prepared during the tenure of a Teacher-Investigator Development Award (1-KO7-NS-00385) from the National Institute of Neurological and Communicative Disorders and Stroke.

References

1. AICARDI, J., CHEURIE, J. J.: Convulsive status epilepticus in infants and children: A study of 239 cases. *Epilepsia* **11**:187–197, 1970.
2. ALEXANDER, S. C., SMITH, T. C., STROBEL, G.: Cerebral carbohydrate metabolism of man during respiratory and metabolic alkalosis. *J. Appl. Physiol.* **24**:66–72, 1968.
3. ALEXANDER, S. C., WORKMAN, R. D., LAMBERTSEN, C. J.: Hyperthermia, lactic acid infusion, and the composition of arterial blood and cerebrospinal fluid. *Am. J. Physiol.* **202**:1049–1054, 1962.
4. BERESFORD, H. R., POSNER, J. B., PLUM, F.: Changes in brain lactate during induced cerebral seizures. *Arch. Neurol.* **20**:243–248, 1969.
5. BLENNOW, G., BRIERLEY, J. B., MELDRUM, B. S., SIESJO, B. K.: Epileptic brain damage: The role of systemic factors that modify cerebral energy metabolism. *Brain* **101**:687–700, 1978.
6. BLENNOW, G., FOLBERGROVA, J., NILSSON, B., SIESJO, B. K.: Effects of bicuculline-induced seizures on cerebral metabolism and circulation of rats rendered hypoglycemic by starvation. *Ann. Neurol.* **5**:139–151, 1979.
7. BLENNOW, G., SVENNINGSEN, N. W.: Cerebrospinal fluid lactate/pyruvate ratio in children with febrile convulsions. *Neuropaediatrie* **5**:157–161, 1974.
8. BOGDEN, J. D., TROIANO, R. A.: Plasma calcium, copper, magnesium, and zinc concentrations in patients with the alcohol withdrawal syndrome. *Clin. Chem.* **24**:1553–1556, 1978.
9. BOGDEN, J. D., TROIANO, R. A., JOSELOW, M. M.: Copper, zinc, magnesium, and calcium in plasma and cerebrospinal fluid of patients with neurologic diseases. *Clin. Chem.* **23**:485–489, 1977.
10. BOLWIG, T. G., ASTRUP, J., CHRISTOFFERSON, G. R.: EEG and extracellular K^+ in rat brain during pentylenetetrazol seizures and during respiratory arrest. *Biomedicine* **27**:99–102, 1977.
11. BOLWIG, T. G., HERTZ, M. M., PAULSON, O. B., SPOTOFT, H., RAFAELSEN, O. J.: The permeability of the blood–brain barrier during electrically induced seizures in man. *Eur. J. Clin. Invest.* **7**:87–93, 1977.
12. BOLWIG, T. G., QUISTORFF, B.: *In vivo* concentration of lactate in the brain of conscious rats before and during seizures: New ultrarapid technique for the freeze sampling of brain tissue. *J. Neurochem.* **21**:1345–1348, 1973.
13. BRADLEY, R. D., SEMPLE, S. J. G.: A comparison of certain acid–base characteristics of arterial blood, jugular venous blood and cerebrospinal fluid in man, and the effect on them of some acute and chronic acid–base disturbances. *J. Physiol.* **160**:381–391, 1962.
14. BRODERSEN, P., JORGENSEN, E. O.: Cerebral blood flow and oxygen uptake, and cerebrospinal fluid biochemistry in severe coma. *J. Neurol. Neurosurg. Psychiatry* **37**:384–391, 1974.
15. BRODERSEN, P., PAULSON, O. B., BOLWIG, T. G., ROGON, Z. E., RAFAELSEN, O. J., LASSENS, N. A.: Cerebral hyperemia in electrically induced epileptic seizures. *Arch. Neurol.* **28**:334–338, 1973.
16. BROOKS, B. R., ADAMS, R. D.: Cerebrospinal fluid acid–base and lactate changes after seizures in unanesthesized man. I. Idiopathic seizures. *Neurology* **25**:935–942, 1975.
17. BROOKS, B. R., ADAMS, R. D.: Cerebrospinal fluid acid–base and lactate changes after seizures in unanesthetized man. II. Alcohol withdrawal seizures. *Neurology* **25**:943–948, 1975.
18. CASPERS, H., SPECKMAN, E.-J.: Cerebral Po_2, Pco_2 and pH: Changes during convulsive activity and their significance for spontaneous arrest of seizures. *Epilepsia* **13**:699–725, 1972.

19. CHAPMAN, A. G., MELDRUM, B. S., SIESJO, B. K.: Cerebral metabolic changes during prolonged epileptic seizures in rats. *J. Neurochem.* **28**:1025–1035, 1977.
20. COLLINS, R. C., KENNEDY, C., SOKOLOFF, L., PLUM, F.: Metabolic anatomy of focal motor seizures. *Arch. Neurol.* **33**:536–542, 1976.
21. COLLINS, R. C., POSNER, J. B., PLUM, F.: Cerebral energy metabolism during electroshock seizures in mice. *Am. J. Physiol.* **218**:943–950, 1970.
22. CORDINGLEY, G. E., SOMJEN, G. G.: The clearing of excess potassium from extracellular space in spinal cord and cerebral cortex. *Brain Res.* **151**:291–306, 1978.
23. CUTLER, R. W. P., LORENZO, A. V., BARLOW, C. F.: Changes in blood–brain permeability during pharmacologically induced convulsions. *Prog. Brain Res.* **29**:367–384, 1968.
24. DELANEY, R. L., LANKFORD, H. G., SULLIVAN, J. F.: Thiamine, magnesium and plasma lactate abnormalities in alcoholic patients. *Proc. Soc. Exp. Biol. Med.* **123**:675–679, 1966.
25. DYMOND, A. M., CRANDALL, P. H.: Intracerebral temperature changes in patients during spontaneous epileptic seizures. *Brain Res.* **60**:249–254, 1973.
26. ELDRIDGE, F., SALZER, J.: Effect of respiratory alkalosis on blood lactate and pyruvate in humans. *J. Appl. Physiol.* **22**:461–468, 1967.
27. EMPEY, L. W., PATTERSON, H. A., MCQUARRIE, I.: The pH and the CO_2 content of cerebrospinal fluid in epilepsy. *Proc. Soc. Exp. Biol. Med.* **29**:1003–1006, 1932.
28. FENCL, V., VALE, J. R., BROCH, J. A.: Respiration and cerebral blood flow in metabolic acidosis and alkalosis in humans. *J. Appl. Physiol.* **27**:67–76, 1969.
29. FISHER, V. J., CHRISTIANSON, L. C.: Cerebrospinal fluid acid–base balance during a changing ventilatory state in man. *J. Appl. Physiol.* **18**:712–716, 1963.
30. GERAUD, J., RASCOL, A., BES, A., GUIRAUD, B., GERAUD, G., CHARLET, J. P., CAUSSANEL, J. P., DAVID, J.: Acid–base equilibrium and partial gas pressures in blood and cerebrospinal fluid with acute cerebrovascular accidents. *Rev. Neurol.* **129**:153–172, 1973.
31. GIBBS, E. L., LENNOX, W. G., GIBBS, F. A.: Variations in the carbon dioxide content of the blood in epilepsy. *Arch. Neurol. Psychiatry* **43**:223–239, 1940.
32. GOTOH, F., MEYER, J. S., TAKAGI, Y.: Cerebral effects of hyperventilation in man. *Arch. Neurol.* **12**:410–423, 1965.
33. GRANHOLM, L., LUKJANOVA, L., SIESJO, B. K.: Evidence of cerebral hypoxia in pronounced hyperventilation. *Scand. J. Lab. Clin. Invest. Suppl.* **102**:IV–C, 1968.
34. GRANHOLM, L., SIESJO, B. K.: Signs of tissue hypoxia in infantile hydrocephalus. *Dev. Med. Child. Neurol.* **12**(Suppl. 22):73–77, 1970.
35. HALGREN, E., BABB, T. L., CRANDALL, P. H.: Responses of human limbic neurons to induced changes in blood gases. *Brain Res.* **132**:43–63, 1977.
36. HALGREN, E., BABB, T. L., CRANDALL, P. H.: Post-EEG seizure depression of human limbic neurons is not determined by their response to probable hypoxia. *Epilepsia* **18**:89–93, 1977.
37. HAMILTON, R. W., JR., SCHOOLMAN, A. S.: Acid–base changes in cats following experimental epileptic seizures. *Neurology* **15**:444–448, 1965.
38. HEIPERTZ, R. K., EICKHOFF, K., KARSTENS, K. H.: magnesium and inorganic phosphate content in CSF related to blood–brain barrier function in neurological disease. *J. Neurol. Sci.* **40**:87–95, 1979.
39. HEIPERTZ, R., EICKHOFF, K., KARSTENS, K. H.: Cerebrospinal fluid concentrations of magnesium and inorganic phosphate in epilepsy. *J. Neurol. Sci.* **41**:55–60, 1979.
40. HOLMBERG, G., THESLEFF, S., VON DARDEL, O., HARD, G., RAMQUIST, N., PETTERSSON, H.: Circulatory conditions in electroshock therapy with and without a muscle relaxant. *Arch. Neurol. Psychiatry* **72**:73–79, 1954.
41. HOWSE, D. C., CARONNA, J. J., DUFFY, T. E., PLUM, F.: Cerebral energy metabolism, pH, and blood flow during seizures in the cat. *Am. J. Physiol.* **227**:1444–1451, 1974.
42. HUNTER, A. R.: Lactate and pyruvate changes in cerebrospinal fluid during neurosurgical anesthesia. *Acta Anaesthesiol. Scand.* (*Suppl.*) **37**:149–151, 1970.
43. ILUCEV, D., KUKLADZIEV, B.: Acid–base and electrolyte correlations in blood and cerebrospinal fluid in alcoholic delirium. *Arch Psychiatr. Nervenkr.* **220**:393–403, 1975.
44. JOBSIS, F. F., O'CONNOR, M., VITALE, A., VREMAN, H.: Intracellular redox changes in functioning cerebral cortex. I. Metabolic effects of epileptiform activity. *J. Neurophysiol.* **34**:735–749, 1971.
45. JOHANSSON, B., NILSSON, B.: The pathophysiology of the blood–brain barrier dysfunction induced by severe hypercapnia and by epileptic brain activity. *Acta Neuropathol.* (*Berlin*) **38**:153–158, 1977.
46. KALIN, E. M., TWEED, W. A., LEE, J., MACKEEN, W. L.: Cerebrospinal fluid acid–base and electrolyte changes resulting from cerebral anoxia in man. *N. Engl. J. Med.* **293**:1013–1016, 1975.
47. KAZEMI, H., SHANNON, D. C., CARVALLO-GIL, E.: Brain CO_2 buffering capacity in respiratory acidosis and alkalosis. *J. Appl. Physiol.* **22**:241–246, 1967.
48. KAZEMI, H., SHORE, N. S., SHIH, V. E., SHANNON, D. C.: Brain organic buffers in respiratory acidosis and alkalosis. *J. Appl. Physiol.* **34**:478–482, 1973.
49. KAZEMI, H., VALENCA, L. M., SHANNON, D. C.: Brain and cerebrospinal fluid lactate concentration in respiratory acidosis and alkalosis. *Respir. Physiol.* **6**:178–186, 1969.
50. KETY, S. S., WOODFORD, R. B., HAMEL, M. H.: Cerebral blood flow and metabolism in schizoprenia: The effect of barbiturate semi-narcosis insulin coma and electroshock. *Am. J. Psychiatry* **104**:765–770, 1948.
51. KING, L. R., MCLAURIN, R. L., KNOWLES, H. C., JR.: Acid–base balance and arterial and CSF lactate levels following human head injury. *J. Neurosurg.* **40**:617–625, 1974.

52. KLEIN, J. R., OLSEN, N. S.: Distribution of intravenously injected glutamate, lactate, pyruvate, and succinate between blood and brain. *J. Biol. Chem.* **167:**1–5, 1947.
53. LASSEN, N. A., CHRISTENSEN, M. S.: Physiology of cerebral blood flow. *Br. J. Anaesth.* **48:**719–734, 1976.
54. LAVY, S., MELAMED, E., PORTNOY, Z., CARMON, A.: Interictal regional cerebral blood flow in patients with partial seizures. *Neurology* **26:**418–422, 1976.
55. LOWENBACH, H., GREENHILL, M.: The effect of oral administration of lactic acid upon the clinical course of depressive states. *J. Nerv. Ment. Dis.* **105:**343–458, 1947.
56. LUCRETIUS: Book III, lines 487 ff (Good's translation). In Penfield, W., Jasper, H.: *Epilepsy and the Functional Anatomy of the Human Brain.* Boston, Little-Brown, 1954, p. 8.
57. MAGNAES, B., NORNES, H.: Circulatory and respiratory changes in spontaneous epileptic seizures in man. *Eur. Neurol.* **12:**104–111, 1974.
58. MARKS, C. E., GOLDRING, R. M., VECCHIONE, J. J., GORDON, E. E.: Cerebrospinal fluid acid–base relationships in ketoacidosis and lactic acidosis. *J. Appl. Physiol.* **35:**813–819, 1973.
59. MCHENRY, L. C., SLOCUM, H. C., BIVENS, H. E.: Hyperventilation in awake and anesthetized man. *Arch. Neurol.* **12:**270–277, 1965.
60. MCLAUGHLIN, F. L., HURST, R. H.: Acid–base equilibrium of blood in epilepsy. *Q. J. Med.* **2:**419–429, 1933.
61. MELDRUM, B.: Physiological changes during prolonged seizures and epileptic brain damage. *Neuropaediatrie* **9:**203–212, 1978.
62. MELDRUM, B. S., BRIERLEY, J. B.: Prolonged epileptic seizures in primates: Ischemic cell change and its relation to ictal physiological events. *Arch. Neurol.* **28:**10–17, 1973.
63. MELDRUM, B. S., HORTON, R. W.: Physiology of status epilepticus in primates. *Arch. Neurol.* **28:**1–9, 1973.
64. MELDRUM, B. S., NILSSON, B.: Cerebral blood flow and metabolic rate early and late in prolonged epileptic seizures induced in rats by bicuculline. *Brain* **99:**523–542, 1976.
65. MELDRUM, B. S., VIGOUROUX, R. A., BRIERLEY, J. B.: Systemic factors and epileptic brain damage. *Arch. Neurol.* **29:**82–87, 1973.
66. MENDELSON, J. H., WEXLER, D., KUBZANSKY, P., LEIDERMAN, H., SOLOMON, P.: Serum magnesium in delirium tremens and alcoholic hallucinosis. *J. Nerv. Ment. Dis.* **128:**352–357, 1959.
67. MEYER, J. S., GOTOH, F., FAVALE, E.: Cerebral metabolism during epileptic seizures in man. *Electroencephalogr. Clin. Neurophysiol.* **21:**10–22, 1966.
68. MEYER, J. S., KANDA, T., SHINOHARA, Y., FUKUUCHI, Y.: Changes in cerebrospinal fluid sodium and potassium concentrations during seizure activity. *Neurology* **20:**1179–1184, 1970.
69. MEYER, J. S., KANDA, T., SHIOHARA, Y., FUKUUCHI, Y.: Effects of anoxia on cerebrospinal fluid sodium and potassium concentrations. *Neurology* **21:**889–895, 1971.
70. MEYER, J. S., PORTNOY, H. D.: Post-epileptic paralysis: A clinical and experimental study. *Brain* **82:**162–185, 1959.
71. MITCHELL, R. A., HERBERT, D. A., CARMAN, C. T.: Acid–base constants and temperature co-efficients for cerebrospinal fluid. *J. Appl. Physiol.* **20:**27–30, 1965.
72. MITIS, Z. K., HARRIS, T., NOWINSKI, W. W.: The levels of peripheral blood lactic acid in psychiatric patients treated by EST, with or without the use of a muscle paralysant. *Tex. Rep. Biol. Med.* **12:**305–312, 1954.
73. MOLNAR, L., KOVACS, M.: Effect of epileptic seizures on the composition of the cerebrospinal fluid. *Arch. Psychiatr. Nervenkr.* **219:**285–296, 1974.
74. MUNSON, E. S., WAGMAN, I. H.: Acid–base changes during lidocaine-induced seizures in *Macaca mulatta. Arch. Neurol.* **20:**406–412, 1969.
75. NEMOTO, E. M., HOFF, J. T., SEVERINGHAUS, J. W.: Lactate uptake and metabolism by brain during hyperlactatemia and hypoglycemia. *Stroke* **5:**48–53, 1974.
76. NEMOTO, E. M., SEVERINGHAUS, J. W.: Stereospecific permeability of rat blood–brain barrier to lactatic acid. *Stroke* **5:**81–84, 1974.
77. NICOL, C. F.: Status epilepticus. *J. Am. Med. Assoc.* **234:**419–420, 1975.
78. NILSSON, B., NORBEG, K., SIESJO, B. K.: Biochemical events in cerebral ischemia. *Br. J. Anaesth.* **47:**751–755, 1975.
79. NILSSON, B., RHENCRONA, S., SIESJO, B. K.: Coupling of cerebral metabolism and blood flow in epileptic seizures, hypoxia and hypoglycemia. In Purves, M. (ed.). *Cerebral Vascular Smooth Muscle and Its Control.* Ciba Foundation Symposium 56 (New Series), London, Ciba Foundation, 1977, pp. 199–218.
80. NOWILL, W. K., WILSON, W., BORDERS, R.: Succinylcholine chloride in electroshock therapy. II. Cardiovascular reactions. *Arch. Neurol. Psychiatry* **71:**189–197, 1954.
81. OH, M. S., CARROLL, H. J.: The anion gap. *N. Engl. J. Med.* **297:**814–817, 1977.
82. OLDENDORF, W. H.: Blood–brain barrier permeability to lactate. *Eur. Neurol.* **6:**49–55, 1971/1972.
83. ORRINGER, C. E., EUSTACE, J. C., WUNSCH, C. D., GARDNER, L. B.: Natural history of lactic acidosis after grand-mal seizures—A model for the study of an anion-gap acidosis not associated with hyperkalemia. *N. Engl. J. Med.* **297:**796–799, 1977.
84. OSNATO, M., KILLIAN, J. A., GARCIA, T.: Comparative chemical studies of the blood and spinal fluid in epilepsy. *Brain* **50:**581–200, 1927.
85. OSNES, J. B., HERMANSEN, L.: Acid–base balance after maximal exercise of short duration. *J. Appl. Physiol.* **32:**59–63, 1972.
86. PAULI, H. G., VORBURGER, C., REUBI, F.: Chronic derangements of cerebrospinal fluid acid–base components in man. *J. Appl. Physiol.* **17:**993–1001, 1962.
87. PAULSON, G. W., WISE, G., CONKLE, R.: Cerebrospi-

nal fluid lactic acid in death and in brain death. *Neurology* **22**:505–509, 1972.

88. PETITO, C. K., SCHAEFER, J. A., PLUM, F.: Ultrastructural characteristics of the brain and blood–brain barrier in experimental seizures. *Brain Res.* **127**:251–267, 1977.
89. PIERCE, N. F., FEDSON, D. S., BRIGHAM, K. L., PERMUTT, S., MONDAL, A.: Relation of ventilation during base deficit to acid–base values in blood and spinal fluid. *J. Appl. Physiol.* **30**:677–683, 1971.
90. PINTILIE, C., MISON-CRIGHEL, N., TUDOR, I.: Correlations between some biochemical constituents of the blood and cerebrospinal fluid, the clinical forms and the electroencephalographic patterns in epileptics. *Rev. Roum. Neurol.* **8**:197–207, 1971.
91. PLUM, F.: Ion homeostasis in cisternal and lumbar cerebrospinal fluid. *N. Engl. J. Med.* **293**:1041–1042, 1975.
92. PLUM, F., HOWSE, D. C., DUFFY, T. E.: Metabolic effects of seizures. *Res. Publ. Assoc. Res. Nerv. Ment. Dis.* **53**:141–157, 1974.
93. PLUM, F., POSNER, J. B.: Blood and cerebrospinal fluid lactate during hyperventilation. *Am. J. Physiol.* **212**:864–870, 1967.
94. PLUM, F., POSNER, J. B., SMITH, W. W.: Effect of hyperbaric–hyperoxic hyperventilation on blood, brain, and CSF lactate. *Am. J. Physiol.* **215**:1240–1244, 1968.
95. PLUM, F., POSNER, J. B., TROY, B.: Cerebral metabolic and circulatory responses to induced convulsions in animals. *Arch. Neurol.* **18**:1–13, 1968.
96. PLUM, F., PRICE, R. W.: Acid–base balance of cisternal and lumbar cerebrospinal fluid in hospital patients. *N. Engl. J. Med.* **289**:1346–1351, 1973.
97. PLUM, F., SIESJO, B. K.: Recent advances in CSF physiology. *Anesthesiology* **42**:708–730, 1975.
98. POLI, S., DE KALBERMATTEN, J. P., ENRICO, J. F.: Crise epileptique et pertubations acido–basiques. *Helv. Med. Acta (Suppl.)* **50**:120, 1970.
99. PONTEN, U., KJALLQUIST, A., SIESJO, B. K., SUNDBARG, G., SVENGAARD, N.: Relation of selective acidosis of CSF to increased lactate concentrations and a discussion of the lactate/pyruvate ratios. *Scand. J. Clin. Invest.* (*Suppl.*) **120**:1XD, 1968.
100. POSNER, J. B., PLUM, F.: Lack of rapid equilibrium between blood and CSF lactate. *Neurology* **16**:316, 1966.
101. POSNER, J. B., PLUM, F.: Independence of blood and cerebrospinal fluid lactate. *Arch. Neurol.* **16**:492–496, 1967.
102. POSNER, J. B., PLUM, F.: Spinal fluid pH and neurologic symptoms in systemic acidosis. *N. Engl. J. Med.* **277**:605–613, 1967.
103. POSNER, J. B., PLUM, F., VAN POZNAK, A.: Cerebral metabolism during electrical induced-seizures in man. *Arch. Neurol.* **20**:388–395, 1969.
104. POSNER, J. B., SWANSON, A. G., PLUM, F.: Acid–base balance in cerebrospinal fluid. *Arch. Neurol.* **12**:479–496, 1965.
105. PRIOR, G. P. U., EDWARDS, A. T.: Lumbar punctures and acidosis in epilepsy and allied convulsive disorders. *Med. J. Austr.* **13**:507–513, 1926.
106. PROCKOP, L. D.: Cerebrospinal fluid lactic acid-clearance and effect on facilitated diffusion of a glucose analogue. *Neurology* **18**:189–196, 1968.
107. ROGROVE, H. J., ALABASTER, S.: Lactic acidosis in seizures. *N. Engl. J. Med.* **297**:1352, 1977.
108. ROSSANDA, M., SGANZERLA, E. P.: Acid–base and gas tension measurements in cerebrospinal fluid. *Br. J. Anaesth.* **48**:753–560, 1976.
109. SACKS, W.: The cerebral metabolism of L- and D-lactate-C^{14} in humans *in vivo. Ann. N. Y. Acad. Sci.* **119**:1091–1108, 1965.
110. SAKAI, F., MEYER, J. S., NARITOMI, H., HSU, M. C.: Regional cerebral blood flow and EEG in patients with epilepsy. *Arch. Neurol.* **35**:648–657, 1978.
111. SALFORD, L. G., BRIERLEY, J. B., PLUM, F., SIESJO, B. K.: Energy metabolism and histology in the brain during combined hypoxemia and ischemia. *Eur. Neurol.* **6**:329–334, 1971/72.
112. SAMBROOK, M. A., HUTCHINSON, E. C., ABER, G. M.: Metabolic studies in subarachnoid hemorrhage and strokes. I. Serial changes in acid–base values in blood and cerebrospinal fluid. *Brain* **96**:171–190, 1973.
113. SCHEINBERG, P., BOURNE, B., REINMUTH, O. M.: Human cerebral lactate and pyruvate extraction. I. Control subjects. *Arch. Neurol.* **12**:246–250, 1965.
114. SCHNABERTH, G., GELL, G., JAKLITSCH, H.: Disturbances of the acid–base balance in the cerebrospinal fluid with delirium tremens. *Arch. Psychiatr. Nervenkr.* **215**:417–428, 1972.
115. SCHWAB, M., MOTEL, M. V. E.: The acid–base status in blood and cerebrospinal fluid during chronic renal insufficiency. *Klin. Wochenschr.* **40**:765–772, 1962.
116. SIEMES, H., SIEGERT, M., HANEFELD, F.: Febrile convulsions and blood–cerebrospinal fluid barrier. *Epilepsia* **19**:57–66, 1978.
117. SIESJO, B. K.: Epileptic seizures. In *Brain Energy Metabolism.* Chichester and New York, John Wiley, 1978, pp. 345–380.
118. SIESJO, B. K., GRANHOLM, L., KJALLQUIST, A.: Regulation of lactate and pyruvate levels in the CSF. *Scand. J. Lab. Clin. Invest.* (*Suppl.*) **102**:IF, 1968.
119. SIESJO, B. K., JOHANNSSON, H., LJUNGGREN, B.: Brain dysfunction in cerebral hypoxia and ischemia. *Res. Publ. Assoc. Res. Nerv. Ment. Dis.* **53**:75–123, 1974.
120. SIESJO, B. K., KJALLQUIST, A., ZWETNOW, N.: The CSF lactate/pyruvate ratio in cerebral hypoxia. *Life Sci.* **7**:45–52, 1968.
121. SIESJO, B. K., PLUM, F.: Cerebral energy metabolism in normoxia and in hypoxia. *Acta Anaesthesiol. Scand.* (*Suppl.*) **45**:81–101, 1971.
122. SIESJO, B. K., PLUM, F.: Pathophysiology of anoxic brain damage. In Gaull, G. E. (ed.): *Biology of Brain Dysfunction,* Vol. 1. New York, Plenum Press, 1973, pp. 319–372.
123. SIMPSON, H., HABEL, A. H., GEORGE, E. L.: Cerebro-

spinal fluid acid–base status and lactate and pyruvate concentrations after short first febrile convulsions in children. *Arch. Dis. Child.* **52:**836–843, 1977.

124. SIMPSON, H., HABEL, A. H., GEORGE, E. L.: Cerebrospinal fluid acid–base status and lactate and pyruvate concentrations after convulsions of varied durations and aetiology in children. *Arch. Dis. Child.* **52:**844–849, 1977.
125. SORENSEN, E., OLESEN, J., RASK-MADSEN, J., RASK-ANDERSEN, H.: The electrical potential difference and impedance between CSF and blood in unanesthetized man. *Scand. J. Clin. Lab. Invest.* **38:**203–207, 1978.
126. SVENNINGSEN, N. W., SIESJO, B. K.: Cerebrospinal fluid lactate/pyruvate ratio in normal and asphyxiated neonates. *Acta Paediatr. Scand.* **61:**117–124, 1972.
127. VALENCA, L. M., SHANNON, D. C., KAZEMI, H.: Clearance of lactate from the cerebrospinal fluid. *Neurology* **21:**615–620, 1971.
128. VAN HEIJST, A. N. P., MAAS, A. H. J., VISSER, B. F.: Comparison of the acid–base balance in cisternal and lumbar cerebrospinal fluid. *Pfluegers Arch.* **287:**242–247, 1966.
129. VAN VAERENBERGH, P. J. J., DEMEESTER, G., LEUSEN, I.: Lactate in cerebrospinal fluid during hyperventilation. *Arch. Int. Physiol. Biochim.* **73:**738–737, 1965.
130. VAN VAERENBERGH, P. J. J., LACROXI, E., DEMEESTER, G.: Lactate in cerebrospinal fluid during muscular exercise. *Arch. Int. Physiol. Biochim.* **73:**729–737, 1965.
131. VICTOR, M.: The role of hypomagnesemia and respiratory alkalosis in the genesis of alcohol withdrawal symptoms. *Ann. N. Y. Acad. Sci.* **215:**235–248, 1973.
132. VICTOR, M., BRAUSCH, C.: Role of abstinence in the genesis of alcoholic epilepsy. *Epilepsia* **8:**1–20, 1967.
133. WALD, A., HOCHWALD, G. M., GANDHI, M.: Evidence for the movement of fluid, macromolecules and ions from the brain extracellular space to the CSF. *Brain Res.* **151:**283–290, 1978.
134. WASTERLAIN, C. G.: Mortality and morbidity from serial seizures—an experimental study. *Epilepsia* **15:**155–176, 1974.
135. WASTERLAIN, C. G.: Neonatal seizures and brain growth. *Neuropaediatrie* **9:**213–228, 1978.
136. WHITE, P. T., GRANT, P., MOSIER, J., CRAIG, A.: Changes in cerebral dynamics with seizures. *Neurology* **11:**354–361, 1961.
137. WILSON, W. P., HICKAM, J. B., NOWILL, W. K., FRAYSER, R.: Succinylcholine chloride in electroshock therapy. III. Oxygen consumption and arterial oxygen saturation. *Arch. Neurol. Psychiatry* **72:**550–554, 1954.
138. WOLFE, S., MENDELSON, J., OGATA, M., VICTOR, M., MARSHALL, W., MELLO, N.: Respiratory alkalosis and alcohol withdrawal. *Trans. Assoc. Am. Physicians* **82:**344–352, 1969.
139. WOLFE, S., VICTOR, M.: The relationship of hypomagnesemia and alkalosis to alcohol withdrawal symptoms. *Ann. N. Y. Acad. Sci.* **162:**973–984, 1969.
140. WOLFE, S. M., VICTOR, M.: The physiological basis of the alcohol withdrawal syndrome. In Mello, N. K., Mendelson, J. H. (eds.): *Recent Advances in Studies of Alcoholism.* Washington, D.C., National Institute of Mental Health, 1971, pp. 188–199.
141. WOLLMAN, H., SMITH, A. L., NEIGH, J. L., HOFFMAN, J. C.: Cerebral blood flow and oxygen consumption in man during electro-encephalographic seizure patterns associated with ethrane anesthesia. In Brock, M. (ed.): *Cerebral Blood Flow: Clinical and Experimental Results.* Berlin, Springer-Verlag, 1969, pp. 246–248.
142. WOLLMAN, H., SMITH, T. C., STEPHEN, G. W., COLTON, E. T., GLEATON, H. E., ALEXANDER, S. C.: Effects of respiratory and metabolic alkalosis on cerebral blood flow in man. *J. Appl. Physiol.* **24:**60–65, 1968.
143. WORTIS, S. B., MARSH, F.: Lactic acid content of the blood and of the cerebrospinal fluid. *Arch. Neurol. Psychiatry* **35:**717–722, 1936.

CHAPTER 19

Neurotransmitter, Metabolite, and Cyclic Nucleotide Alterations in Cerebrospinal Fluid of Seizure Patients

James H. Wood and Benjamin Rix Brooks

1. Introduction

Recently, Maynert *et al.*[102] reviewed the role of neurotransmitters in the initiation, spread, and termination of seizures. Abnormalities in water, electrolyte, acid–base, vitamin, and hormone metabolism alter seizure thresholds. In addition, the administration of chemical agents and electrical stimulation of the brain have been shown to modify seizure susceptibility. These manipulations influence neuronal activity, possibly by influencing the synthesis, storage, and release of neurotransmitters. The metabolic activity of these neurotransmitters may in turn be reflected in the degree of accumulation of their respective metabolites.

Brain extracellular fluid is separated from the cerebrospinal fluid (CSF) to only a thin pial or ependymal membrane that does not appear to present a significant barrier to the diffusion of material between these two fluid compartments.[176] Thus, the CSF composition tends to approximate that of brain extracellular fluid. CSF circulation patterns have been assessed in man,[33,34] and sensitive assay techniques have been developed that are capable of determining neurotransmitter concentrations in CSF.

The purpose of this chapter is to review briefly the CSF neurochemistry of convulsive disorders and CSF alterations evoked by anticonvulsant medications. In addition, CSF alterations induced by electrical brain-surface stimulation in epileptic patients will be discussed. The significance of neurochemical analysis of CSF has recently been reviewed in a separate communication.[174]

2. CSF Alterations Associated with Anticonvulsant Medications in Seizure Patients

2.1. γ-Aminobutyric Acid

The exact role of γ-aminobutyric acid (GABA) in the determination of seizure thresholds is unknown. GABA, a putative neurotransmitter, is found in highest concentrations in the extrapyramidal areas, base of the pons, hypothalamus, and cerebellum.[40,44,96] GABA provides inhibition by inducing hyperpolarization and decreasing membrane resistance secondary to elevation of chloride ion permeability.[36,80] In accordance with the reviews of Meldrum,[105,106] GABA-containing interneurons narrow or sharpen the area of neuronal activity and

James H. Wood, M.D. • Division of Neurosurgery, University of Pennsylvania School of Medicine and Hospital, Philadelphia, Pennsylvania 19104. **Benjamin Rix Brooks, M.D.** • Department of Neurology, The Johns Hopkins University School of Medicine, Baltimore, Maryland 21205.

shorten or terminate the period of this discharge. Thus, these GABAergic interneurons would appear to provide a major defense against the spread or buildup of convulsive activity.

Experimental interruption of GABA synthesis or release and impediment of its postsynaptic action reduces seizure thresholds.[168] Cerebral GABA concentrations are decreased during cobalt-induced focal seizures in cats,[77,159] and degeneration of GABAergic inhibitory synapses has been observed at seizure foci, which may promote epileptic activity in cortical pyramidal neurons.[133] The early communications of Van Gelder *et al.*[160] reported diffusely low cerebral GABA levels in seizure patients; however, Perry *et al.*[123] noted no reduction in GABA concentrations in rapidly frozen epileptogenic foci excised from the temporal or frontal cortex of patients with focal convulsions. More recent communications[90] have associated pyridoxine-dependent epilepsy with deficient brain GABA formation in man. Preliminary studies noted only mildly lower CSF GABA levels in epileptic patients as compared to patients with various nonconvulsive neurological disorders[42,60]; however, the report of Wood *et al.*[171] documents significantly lower CSF GABA concentrations among medicated seizure patients as compared with unmedicated normal volunteers (Fig. 1). Reductions of CSF GABA concentrations have also been demonstrated in patients with degenerative disorders[41,97,113]; thus, low CSF GABA levels are not an exclusive characteristic of epilepsy.

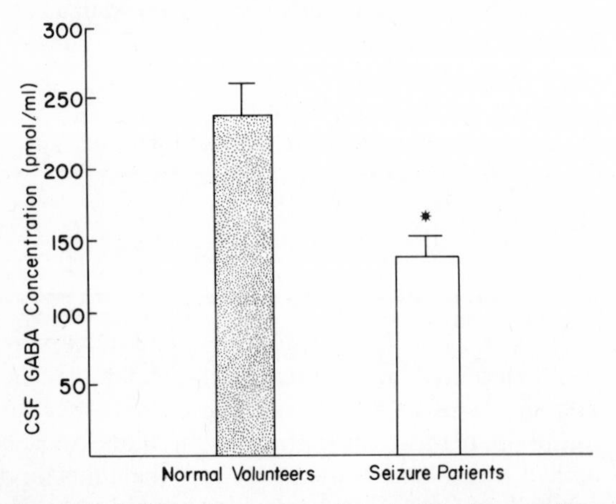

Figure 1. CSF GABA concentrations in 21 seizure patients maintained on phenytoin, phenobarbital, and primidone and in 20 normal volunteers.[171,172] Mean CSF GABA level of seizure group (139 ± 12 pmol/ml, S.E.M.) is significantly lower than that of normal volunteer group (239 ± 21 pmol/ml): $p < 0.001$, two-tailed Student's t test.

Convulsive activity has been associated with elevated cerebral energy requirements; however, the tonic phases of convulsions may relatively interfere with oxygen availability to the brain.[38,125] Purkinje cell degeneration in the cerebellum has been associated with periods of hypoxia[114,137] despite a lesser interictal energy debt,[103] with chronic severe seizure disorders of all types,[131,135] and with phenytoin therapy.[132] The exact mechanism of this cerebellar degeneration (Fig. 2) is unknown. The inhibitory neurotransmitter of Purkinje cells is thought to be GABA[115]; thus such Purkinje cell degeneration would be expected to decrease cerebellar GABA content. An update[172] of the investigations by Wood *et al.*[170,171] suggests that patients with chronic grand mal (generalized tonic–clinic) and psychomotor (complex partial) seizures may have lower CSF GABA concentrations than those with focal motor–sensory (simple partial) seizures (Fig. 3). This differential alteration of GABA metabolism may reflect the variance of the involved cell population with respect to seizure type or represent a greater destructive potential for grand mal or psychomotor epilepsy. The patient population in this study had severe intractable convulsions for most of their lives; thus, variation in the cumulative effects of the chronicity and frequency of the seizures did not detract from the relevance of their CSF findings.

Experimental elevations in brain GABA content by augmentation of its production or reduction of its catabolism have been associated with seizure suppression.[102,106,168] Grand mal and petit mal convulsions have been reported to be inhibited by orally administered GABA[155] despite poor penetration of GABA through the normal blood–brain barrier.[83] Topically applied, vascularly injected, or intraventricularly administered GABA has been demonstrated to elevate seizure thresholds.[57,136,155] Julien and Halpern[70] have shown that phenytoin increases the firing rate of GABA-containing neurons. However, toxic administration of phenytoin is associated with both a loss of epileptic control and progressive loss of GABA from cerebellar Purkinje cells.[63] Phenobarbital enhances GABA-mediated postsynaptic inhibition[94] and possibly alters GABA uptake at the synapse[166] without altering GABA release.[21] However, Wood *et al*[171] were not able to demonstrate significant correlations between lumbar CSF GABA levels and serum phenytoin (r = 0.56), phenobarbital (r = 0.22), or primidone (r = 0.28) concentrations in a small group of seizure patients.

Intraperitoneally injected sodium valproate (so-

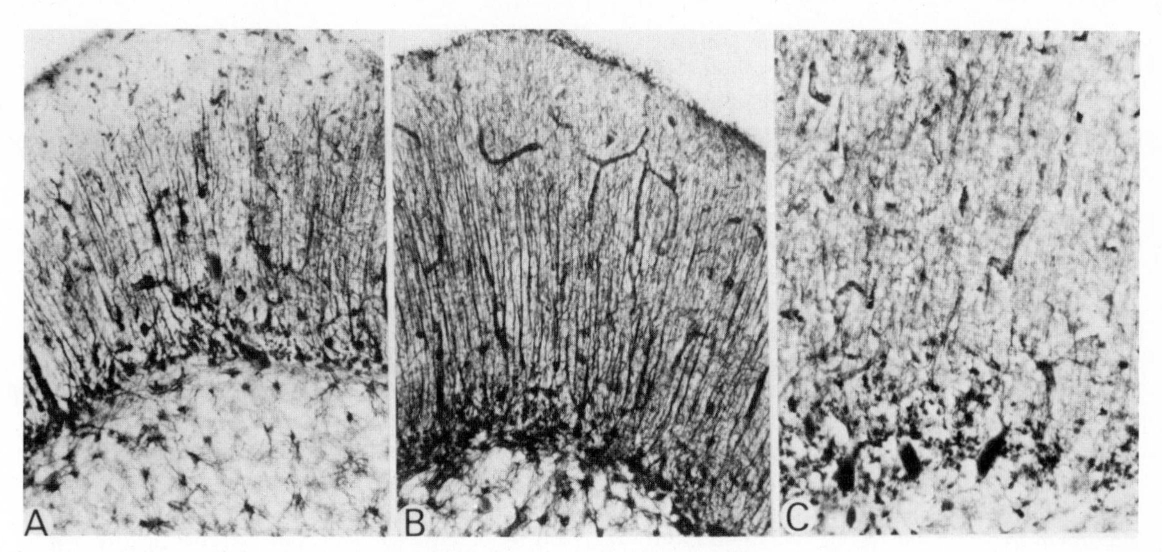

Figure 2. Cajal's gold sublimate stain of cerebellar sections. (A,B) Sections obtained from anticonvulsant-treated epileptic patients prior to cerebellar electrode implantation. Sections demonstrate Purkinje cell loss with isomorphic gliosis of cortices. (C) Section obtained from autopsied normal patient. Section contains plentiful Purkinje cells. All three photomicrographs are at equal magnification. Reprinted from Rajjoub *et al.*,[131] with permission.

dium *n*-dipropylacetate) increases whole-brain and cerebellar GABA content and suppresses experimental seizures.[78,122,144] This anticonvulsive action may be mediated by GABAergic postsynaptic inhibition of neurons.[95] At present, valproate is thought to be a competitive inhibitor of both GABA-transaminase and succinic semialdehyde dehydrogenase, the enzyme that follows GABA-transaminase in the GABA shunt pathway.[39,78] The determination of CSF GABA concentrations might be expected to be useful in the monitoring of patients undergoing valproate therapy,[122] since Gram *et al.*[55] demonstrated a lack of correlation between serum valproate levels and seizure-frequency reductions. However, Emson,[39] Stone,[149] and Kukino and Deguchi[82] suggest that the anticonvulsive action of valproate is not primarily mediated via the GABA system. In addition, Neophytides *et al.*[113] were unsuccessful in documenting oral-valproate-induced rises in CSF GABA concentrations. In accordance, Perry and Hansen[122] report that equivalent doses of oral valproate do not elevate brain GABA content, whereas similar intraperitoneal injections of this agent induce rat brain GABA accumulation. Inter-

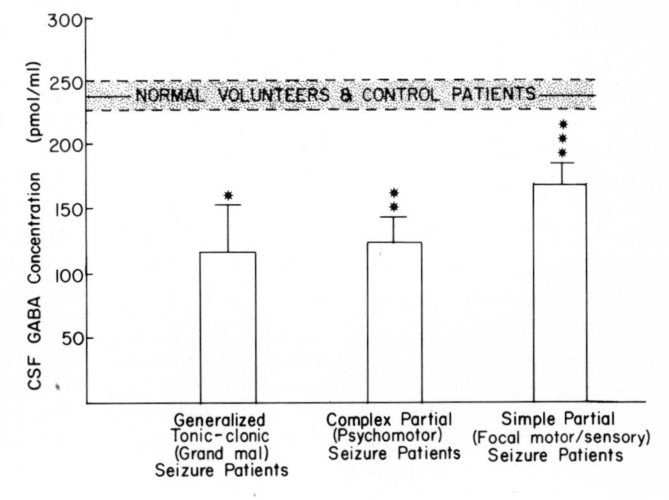

Figure 3. Mean CSF GABA concentrations of medicated patients with predominately generalized tonic–clonic (grand mal), complex partial (psychomotor), and simple partial (focal sensory–motor) seizures. Horizontal stippled area combines 20 previously reported normal volunteers[171,172] whose mean CSF GABA level was 239 ± 21 pmol/ml (S.E.M.) and 19 nonepileptic patients with normal neurological examinations[97] whose mean CSF GABA concentration was 239 ± 13 pmol/ml. Mean CSF GABA concentrations of *5 patients with grand mal seizures (118 ± 33 pmol/ml), **8 patients with psychomotor epilepsy (123 ± 20 pmol/ml), and ***8 patients with focal sensory–motor seizures (169 ± 16 pmol/ml) were significantly lower ($p < 0.005$, $p < 0.001$, and $p < 0.05$, respectively; two-tailed Student's t test) than that of the combined control group.

estingly, oral valproate is effective in reducing electroencephalographic convulsive activity.[1] Recently, close correlations have been demonstrated between pharmacologically induced elevations in brain GABA content and ventricular or cisternal CSF GABA concentrations.[10] Intravenously injected sodium valproate at dosages of 60 mg/kg elevates cisternal CSF GABA levels[89]; however, chronic intraperitoneal administration of 20 mg/kg of this drug does not significantly alter brain or CSF GABA content.[47] Thus, the use of CSF GABA monitoring in the evaluation of seizure patients receiving sodium valproate requires further study.[171,172,174]

Pinder *et al.*,[124] in their review, imply that valproate may be more efficacious in controlling grand mal and psychomotor seizures than in controlling focal motor–sensory convulsions. The preliminary reports of Wood *et al.*[171,172] demonstrate more dramatic CSF GABA reductions among patients with grand mal and psychomotor convulsions than among those with focal motor–sensory seizures (Fig. 3). The lack of availability of controlled evaluations of the differential effects of valproate on various types of convulsions precludes the formulation of correlations between valproate-evoked seizure inhibition and respective CSF GABA levels.

2.1.1. Cyclic GMP and Its Relationship to GABA. Gross and Ferrendelli[56] suggest that elevated brain levels of cyclic guanosine monophosphate (GMP) may have an epileptogenic effect in initiating or maintaining seizure activity. Mao *et al.*[98,99,100] demonstrated an inverse relationship between GABA and cyclic GMP in rat cerebellum following the administration of various anticonvulsant drugs. Experimental elevations of cerebral cortical and cerebellar cyclic GMP levels have been correlated with reductions in seizure thresholds.[91,93,148] Microiontophoretic application of cyclic GMP on pyramidal cells of hippocampal transplants *in oculo* evokes epileptic discharges.[64] Conversely, the anticonvulsant drugs diazepam, phenytoin, phenobarbital, and valproate reduce brain cyclic GMP levels[78,92,93,100] or inhibit the accumulation of cyclic GMP.[49] Thus, the role of GABA in convulsive disorders may be mediated in part by intracellular cyclic GMP. In accordance with this hypothesis, Seligmann *et al.*[141] have suggested that cyclic nucleotides may regulate GABA synthesis. Our preliminary data analyzed in collaboration with Drs. J. Sode, B. S. Glaeser, and T. A. Hare did not reveal either significant alterations in lumbar CSF levels of this cyclic nucleotide (Fig. 4) or correlations between CSF cyclic GMP and GABA levels among medicated seizure patients (unpublished data).

Additional information concerning the role of cyclic nucleotides in central nervous system function has been reviewed by Nathanson,[112] Daly,[29] Ferrendelli *et al.*,[48,49] and Brooks *et al.*[12]

2.2. Norepinephrine

Most neurochemical reviews[102] indicate that brain catecholamines reduce cerebral susceptibility to epileptogenic stimuli. More specifically, the effect of norepinephrine on cortical neurons is usually inhibitory.[81] Rodents have circadian rhythms in brain norepinephrine content and have lower seizure thresholds during periods when cerebral norepinephrine levels are depressed.[138] Ziegler *et al.*[179] have documented a circadian noradrenergic rhythm in man in which CSF concentrations are decreased at night and early morning, when convulsive thresholds are reduced.

Accordingly, α-adrenergic blocking agents increase the susceptibility of mice to epileptogenic stimuli.[74,134] Disulfiram inhibition of dopamine-β-hydroxylase, the enzyme that synthesizes norepinephrine,[2] induces a prompt depression of brain norepinephrine content[54] and augments evoked seizures in rats.[74] Reserpine depletes brain norepinephrine and lowers convulsive thresholds.[86,167]

Maynert *et al.*[102] in their review, suggest that endogenous norepinephrine is necessary for normal resistance to seizures. However, brain norepinephrine content cannot be raised by systemic administration of norepinephrine because this neurotransmitter cannot effectively penetrate the blood–CSF[178] or blood–brain[165] barrier. Oral administration of tyrosine, the dietary precursor of norepinephrine, is not efficacious because tyrosine hydroxylase is saturated at physiological levels of the amino acid.[102]

Tyramine promotes release of norepinephrine from the synapse and inhibits epileptiform afterdischarges in isolated strips of cat cerebral cortex.[79] Presynaptic terminal reuptake of norepinephrine is the primary mechanism for termination of its action.[3] Tricyclic antidepressant medications increase free synaptic norepinephrine[3] and suppress experimentally induced seizures.[147] Jobe *et al.*[69] correlated the anticonvulsant action of imipramine with increased brain norepinephrine content. However, the inhibition of norepinephrine uptake by carbamazepine, a compound structurally related to imipramine, is insufficient to account for its anticonvulsant action.[129] The anticonvulsant drugs phenytoin and pentobarbital also antagonize the presynaptic reuptake of norepinephrine.[59,166]

Unfortunately, the CSF norepinephrine alterations associated with seizure disorders have not yet been evaluated in a strictly controlled fashion. However, Wood and associates reported mean CSF norepinephrine levels of 238 ± 33 pg/ml (S.E.M.) among five electrode-implanted seizure patients without effective cerebellar stimulation who were maintained on phenytoin, phenobarbital, and primidone[173] and 239 ± 17 pg/ml among nine unmedicated nonepileptic patients.[175]

2.2.1. Cyclic AMP and Its Relationship to Norepinephrine. Brain cyclic adenosine monophosphate (AMP) metabolism has been implicated in epileptogenesis and the action of anticonvulsant drugs. Intracellular cyclic AMP influences depolarization of neuronal membranes, release and reuptake of neurotransmitters, and prolongation of negative after-potentials.[29,37,112,162] Walker *et al.*[164] demonstrated elevated cyclic AMP levels in epileptogenic cortical foci. Dibutyryl cyclic AMP evokes experimental seizures when instilled into ventricular CSF[51] or directly applied to the cerebral surface.[163] Dibutyryl cyclic AMP, phosphodiesterase inhibitors, and adenosine (which stimulates adenyl cyclase) increase the uptake of norepinephrine.[162] Since the noradrenergic action on cortical neurons is usually inhibitory,[81] this augmentation of the presynaptic uptake of norepinephrine by these drugs may explain their stimulant and epileptogenic properties.[162] However, Gross and Ferrendelli[56] suggest that norepinephrine and perhaps other biogenic amines have a regulatory effect on cyclic AMP levels in epileptic brain and that elevations in brain cyclic AMP content may have an antiepileptic effect leading to seizure termination.

The anticonvulsant drugs carbamazepine and phenytoin inhibit ouabain-induced elevations in cerebral cyclic AMP content, whereas carbamazepine and phenobarbital suppress norepinephrine-evoked accumulation of cyclic AMP.[87] Pretreatment of mice with anticonvulsant drugs—phenytoin, phenobarbital, carbamazepine, clonazepam, and diazepam—prevents pentylenetetrazol-induced elevations in brain cyclic AMP content.[119] Ferrendelli and Kinscherf[49] confirmed these findings with respect to all of these anticonvulsants except clonazepam in low dosage. In addition, phenytoin and phenobarbital impede the reuptake of norepinephrine at the synapse.[59,166] Phenytoin blocks the cyclic-AMP-mediated calcium influx that is associated with neurotransmitter release.[37] These mechanisms may be important to the anticonvulsant action of phenytoin. Interestingly, disordered calcium metabolism[130] and reduced CSF calcium concentrations[68] have been reported in seizure patients receiving long-term anticonvulsant therapy. Myllylä[110] has demonstrated that phenobarbital, phenytoin, and carbamazepine treatment decreased basal CSF cyclic AMP levels and partly inhibited the rise in CSF cyclic AMP after convulsions in rabbits. In addition, low CSF cyclic AMP concentrations have been observed in anticonvulsant-treated epileptic patients free from convulsive attacks.[109]

The presence of a blood–CSF barrier for cyclic AMP in man suggests that the CSF cyclic AMP concentration might be a reflection of central nervous system cyclic AMP metabolism.[11] However, lumbar CSF has been less useful than ventricular CSF in clinical investigations of brain cyclic nucleotide metabolism. Although Myllylä *et al.*[109] reported increases in lumbar CSF cyclic AMP concentrations within 3 days of an epileptic attack, more recent studies suggest that the brain need not be the major source of cyclic AMP in lumbar CSF.[156] Similarly, the acute administration of diazepam, which increases brain cyclic AMP,[139] has no effect on lumbar cyclic AMP levels.[156] Unilateral frontal lobectomy for intractable seizures in one of our patients studied twice before and twice after craniotomy did not alter the lumbar CSF cyclic AMP levels.[13]

Our own preliminary investigations in collaboration with Drs. J. Sode and M. H. Ebert revealed similar baseline lumbar CSF cyclic AMP concentrations in intractable-seizure patients treated with phenytoin and phenobarbital and unmedicated control patients with various nonepileptic, nonischemic disorders. The uncorrected CSF cyclic AMP accumulation after probenecid administration appeared to be higher in the treated seizure patients, but this difference was not statistically significant. The CSF accumulation of cyclic AMP (in pmol/ml) varied directly with the CSF concentration of probenecid (in μg/ml) for both the unmedicated control patients ($r = 0.7849$; < 0.05, two-tailed Student's t test) and the treated seizure patients ($r = 0.7997$; $p < 0.05$). In addition, the slopes of the plots of this CSF cyclic AMP–CSF probenecid relationship for the control and seizure patient groups were not significantly different, indicating similar probenecid blockage of cyclic AMP egress from the CSF in both patient populations. The CSF cyclic AMP accumulation when corrected for variation in CSF probenecid levels was significantly elevated among the anticonvulsant-treated seizure patients in comparison to that of the neurological control patients (Fig. 4). Our data imply that this elevation in CSF cylic AMP during probenecid blockade in the medicated seizure patients may be secondary to an augmentation in cyclic AMP synthesis or possibly transport into CSF.

2.3. Dopamine and Its Metabolite Homovanillic Acid

The primary function of dopaminergic pathways is the regulation of motor function, hypothalamic releasing factors, and arousal. The influence of dopamine on seizure thresholds is controversial because most abnormalities in dopamine metabolism evoke secondary alterations in noradrenergic activity. Dopamine is a precursor of norepinephrine, and the reuptake mechanism of dopaminergic neurons functions for either norepinephrine or dopamine.[146] Methods of selective destruction of dopaminergic neurons by blocking the uptake of 6-hydroxydopamine into adrenergic neurons with protriptyline[43] have not been applied to the study of seizures.[102] The influence of phenothiazines, which block dopaminergic receptors and appear to lower seizure thresholds, requires additional evaluation.[102]

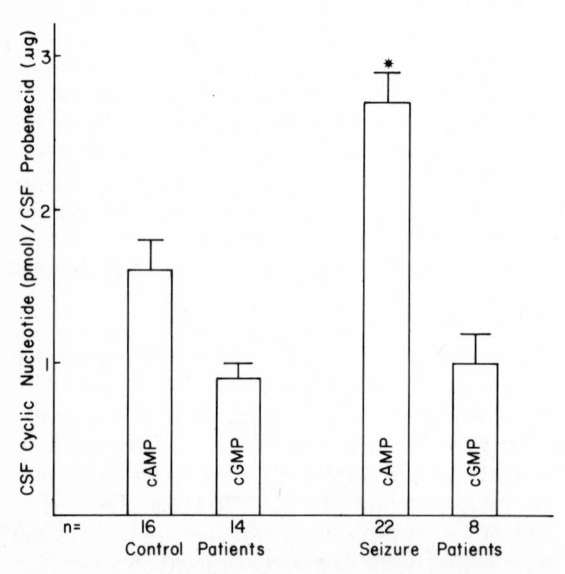

Figure 4. CSF cyclic adenosine 3′,5′-monophosphate (cAMP) and cyclic guanosine 3′,5′-monophosphate (cGMP) accumulations with respect to CSF probenecid concentrations after oral probenecid (100 mg/kg over 18 hr) loading in unmedicated, nonepileptic control patients and seizure patients chronically medicated with phenytoin and phenobarbital. *Mean CSF cAMP accumulation per microgram CSF probenecid of 2.7 ± 0.2 pmol/μg in medicated seizure patients was significantly higher ($p < 0.005$, two-tailed Student's t test) than that of 1.6 ± 0.2 pmol/μg calculated in control patients. Mean CSF cGMP accumulations in pmol/μg CSF probenecid were similar for seizure and control patients.

A metabolite of dopamine, homovanillic acid (HVA), is detectable in CSF. Preliminary reports[6,120] imply that the concentration of CSF HVA is reduced in epileptic patients; however, Garelis, Sourkes, Laxer, and co-workers[50,85] have not been able to confirm these findings. Shaywitz *et al.*[142] documented reduced lumbar CSF HVA levels in children with idiopathic seizure disorders after probenecid loading and did not demonstrate correlations between serum phenytoin or phenobarbital levels and CSF HVA concentrations. Chadwick *et al.*[17] and Laxer *et al.*[85] have noted no significant alterations in lumbar CSF HVA levels during anticonvulsant therapy.

Our own data analyzed in collaboration with Drs. R. M. Post and M. H. Ebert preliminarily suggest that anticonvulsant-medicated seizure patients experience a reduced lumbar CSF HVA accumulation after probenecid administration. However, correction of individual CSF HVA levels for the lowered CSF probenecid concentrations in these seizure patients has demonstrated no significant difference in probenecid-evoked accumulation of CSF HVA in unmedicated control and in seizure patients treated with phenytoin and phenobarbital (Fig. 5).

Dopamine administered to mice via ventricular CSF facilitates pentylenetetrazol-evoked convulsions at low doses and suppresses seizures at higher doses.[15] Similarly instilled dopamine in rats augments convulsions at higher doses. The administration of L-DOPA is usually ineffective in altering seizure thresholds[20]; however, some rodent studies demonstrated L-DOPA-evoked convulsion suppression, especially when combined with monoamine oxidase inhibitors.[104] Chadwick *et al.*[18] recently noted neither a significant lumbar CSF HVA rise nor clinical seizure suppression in a small group of epileptic patients during chronic L-DOPA therapy.

2.4. Serotonin and Its Metabolite 5-Hydroxyindoleacetic Acid

Serotonin-containing neurons located in the raphe nuclei of the pons and mesencephalon

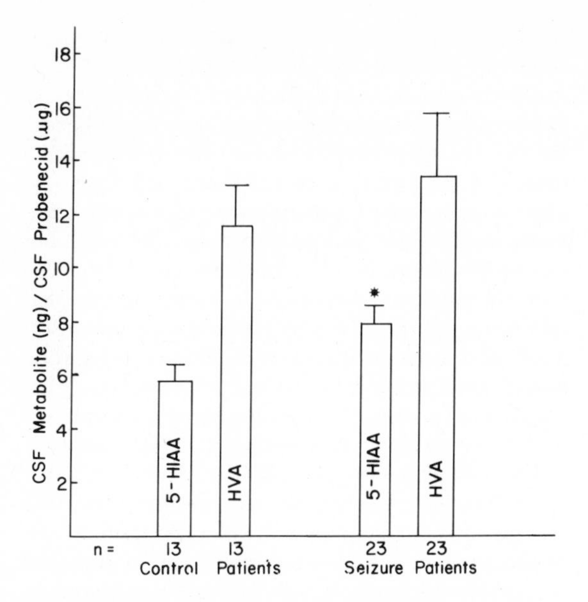

Figure 5. CSF serotonin and dopamine metabolite accumulations with respect to CSF probenecid concentrations after oral probenecid (100 mg/kg over 18 hr) loading in 13 unmedicated, nonepileptic control patients and 23 seizure patients chronically medicated with phenytoin and phenobarbital. *Mean CSF 5-hydroxyindoleacetic acid (5-HIAA) accumulation per microgram CSF probenecid of 7.9 ± 0.7 ng/µg in medicated seizure patients was significantly higher ($p < 0.05$, two-tailed Student's t test) than that of 5.9 ± 0.6 ng/µg calculated in control patients. Mean CSF homovanillic acid (HVA) accumulations in nanograms of CSF HVA per microgram of CSF probenecid were similar for seizure and control groups.

innervate all regions of the diencephalon and telencephalon. Serotonin tracts of the spinal cord originate in the medulla oblongata.[102] Chlorophenylalanine, an inhibitor of tryptophan hydroxylase that markedly reduces brain serotonin,[76] facilitates experimental seizure activity.[127]

Several groups of investigators[50,120,142] have reported low lumbar CSF concentrations of the serotonin metabolite 5-hydroxyindoleacetic acid (5-HIAA) in seizure patients. Shaywitz *et al.*[142] noted no apparent relationship between serum phenytoin and phenobarbital levels and CSF 5-HIAA levels after probenecid administration. Contrarily, Chadwick *et al.*[16,17] demonstrated normal lumbar CSF 5-HIAA levels among untreated epileptic patients and elevated CSF 5-HIAA concentrations among seizure patients treated with phenytoin and phenobarbital. Later studies by Chadwick's group[17] documented phenytoin- and phenobarbital-evoked elevations in CSF levels of tryptophan among medicated seizure patients and concluded that anticonvulsant drugs elevate brain serotonin. Chase *et al.*[19] have reached similar conclusions in rats after noting phenytoin-, diazepam-, and possibly phenobarbital-induced elevations in brain serotonin and 5-HIAA. Oral administration of sodium valproate also appears to elevate rat brain concentrations of tryptophan and 5-HIAA.[82] Recently, however, Laxer *et al.*[85] reported that untreated epileptic patients have lower concentrations of 5-HIAA in lumbar CSF than control patients, but the differences were not statistically significant. Furthermore, treatment of these seizure patients with phenytoin, phenobarbital, primidone, and/or carbamazepine did not evoke alterations in the CSF amine metabolite levels.

Our own data analyzed with the collaboration of Drs. R. M. Post and M. H. Ebert preliminarily revealed identical baseline lumbar CSF 5-HIAA levels in epileptic patients treated with phenytoin and phenobarbital and unmedicated control patients. The uncorrected CSF 5-HIAA accumulation after probenecid administration appeared to be lower among the treated seizure patients. The CSF accumulation of 5-HIAA (in ng/ml) varied directly with the CSF concentration of probenecid (in µg/ml) for both the unmedicated control patients ($r = 0.8025$; $p < 0.01$, two-tailed Student's t test) and the treated seizure patients ($r = 0.8001$; $p < 0.01$). In addition, the slopes of the plots of this CSF 5-HIAA–CSF probenecid relationship for the control and seizure patient groups were not significantly different, indicating similar probenecid blockage of 5-HIAA egress from the CSF in both patient groups. Thus, the CSF 5-HIAA accumulation when corrected for variation in CSF probenecid concentrations was actually significantly elevated among the seizure patients treated with phenytoin and phenobarbital when compared to that of the control patients (Fig. 5). Our data suggest that this increase in CSF 5-HIAA during probenecid administration in the medicated seizure patients may be secondary to an augmentation of 5-HIAA synthesis or possibly transport into CSF.

Recently, intention myoclonus secondary to anoxic brain insults or progressive myoclonic epilepsy has been associated with low CSF 5-HIAA levels. The CSF concentration of this metabolite markedly increases during treatment with 5-hydroxytryptophan, a precursor of serotonin, and a peripheral decarboxylase inhibitor.[161]

3. CSF Alterations Associated with Electrical Stimulation of Brain in Seizure Patients

3.1. Cerebellar Surface Stimulation

Cerebellar cortical stimulation has been reported to significantly reduce experimentally evoked cerebral epileptiform discharges[22,35,58,66,67,108,150] and to suppress strychnine-induced spinal cord convulsions.[154] Conversely, disruption of cerebellar cortical function by surgical ablation or cooling augments the severity of induced cerebral epileptic activity.[35] Efforts at experimental confirmation of these findings have not been totally supportive.[4,7,62,88a,117,154a]

Clinical trials of cerebellar stimulation have likewise been inconclusive. Cooper and associates[23–25] and later other investigators[32,46,61,107,121] reported clinical improvement in patients with epilepsy, cerebral palsy, aggressive behavior, and schizophrenia. However, Van Buren *et al.*[158] did not find significant differences in seizure frequency when comparing intervals of about 7 days of on-and-off stimulation in both the double-blind and unblinded conditions among epileptic patients.

3.1.1. Description of Surgical Procedure and Stimulation Apparatus. Exhaustive descriptions of our clinical experience with chronic cerebellar stimulation are contained in communications by Wood *et al.*[170,173,175] and Van Buren *et al.*[158] In brief, two Avery E333 electrode arrays (Avery Laboratories, Inc., Farmingdale, New York) were installed over the paravermian regions of the anterodorsal cerebellum (Fig. 6). These arrays (Fig. 7) were connected via subcutaneous wiring to two Avery I110 radiofrequency receivers placed subcutaneously in the chest (Fig. 8). External Avery 901 loop antennae connected to an Avery S227 transmitter were taped over the subcutaneous receiver implants (Fig. 8). These patients received 3–12 mA, capacitively coupled monophasic pulses with exponential decay and 1-msec pulse duration alternating between electrode arrays every 8 min. All patients were stimulated with 10/sec pulses, except the patient with atypical myoclonic epilepsy, who received 200/sec pulses.

3.1.2. GABA and Cyclic GMP. Wood *et al.*[170] reported evoked GABA reductions in lumbar CSF during periods of unilateral, alternating-followed-by-bilateral, continuous cerebellar stimulation and frequently during long-term unilateral, alternating cerebellar stimulation in seizure patients (Fig. 9). The evoked mean reductions in CSF GABA of 48 ± 9 and 37 ± 13% in our patients who underwent unilateral, alternating-followed-by-bilateral, continuous stimulation and those who received chronic unilateral cerebellar stimulation, respectively, were associated with mildly increased convulsive activity.[158] These findings are consistent with reviews[101] associating spontaneous generalized convulsions with experimental brain GABA reductions of approximately 40%. Such inhibition of GABA release,

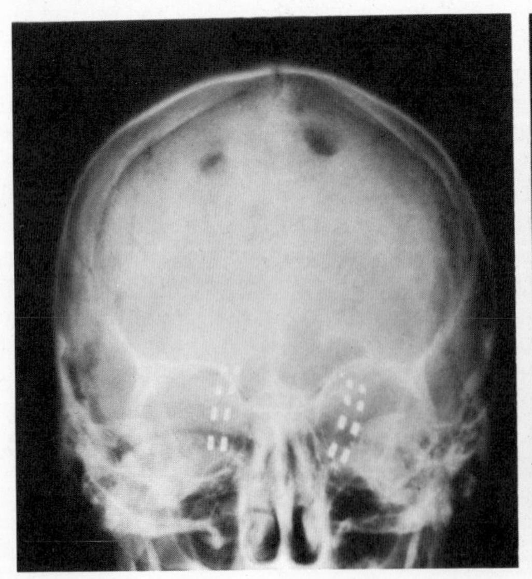
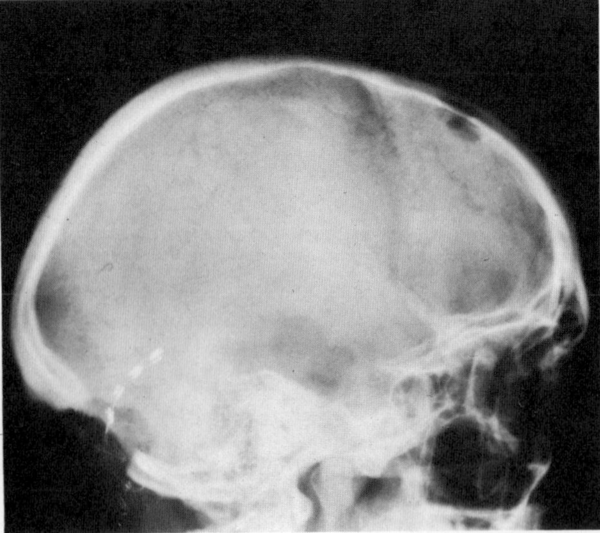

Figure 6. Caldwell posteroanterior and lateral skull radiographs of a patient after surgical implantation of two electrode arrays over paravermian regions of the anterodorsal cerebellum. Reprinted from Wood *et al.*,[173] with permission.

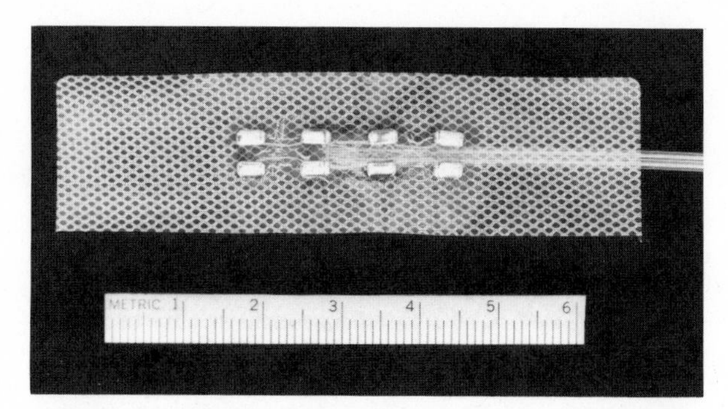

Figure 7. Cerebellar electrode array consisting of silicon-coated Dacron (E.I. duPont de Nemours & Co., Wilmington, Delaware) mesh pad with four pairs of platinum disc electrodes. Reprinted from Wood *et al.*,[173] with permission.

when evoked by penicillin,[27] depresses synaptic inhibitory mechanisms and may cause seizures.[126]

The specific site of this stimulation-evoked alteration of CSF GABA metabolism has not been identified. However, the induced CSF GABA depressions in our patients were probably cerebellar-mediated, since variation of cerebellar stimulation was the only manipulation in our investigation. Within the cerebellar system, basket-cell axons on Purkinje cell bodies and Purkinje cell terminals on deep cerebellar nuclei[115] or the lateral vestibular nucleus[26] contain GABA. Our findings are also consistent with the experimental investigations of Hablitz.[58] Accordingly, more recent neurophysiological studies in animals have demonstrated sequences of initial augmentation followed by more prolonged suppression of Purkinje cell discharge[30,31] and little effect on cortical evoked potentials[62] at low stimulation frequencies. Contrary to the experimental work of Obata and Takeda,[116] the percentage de-

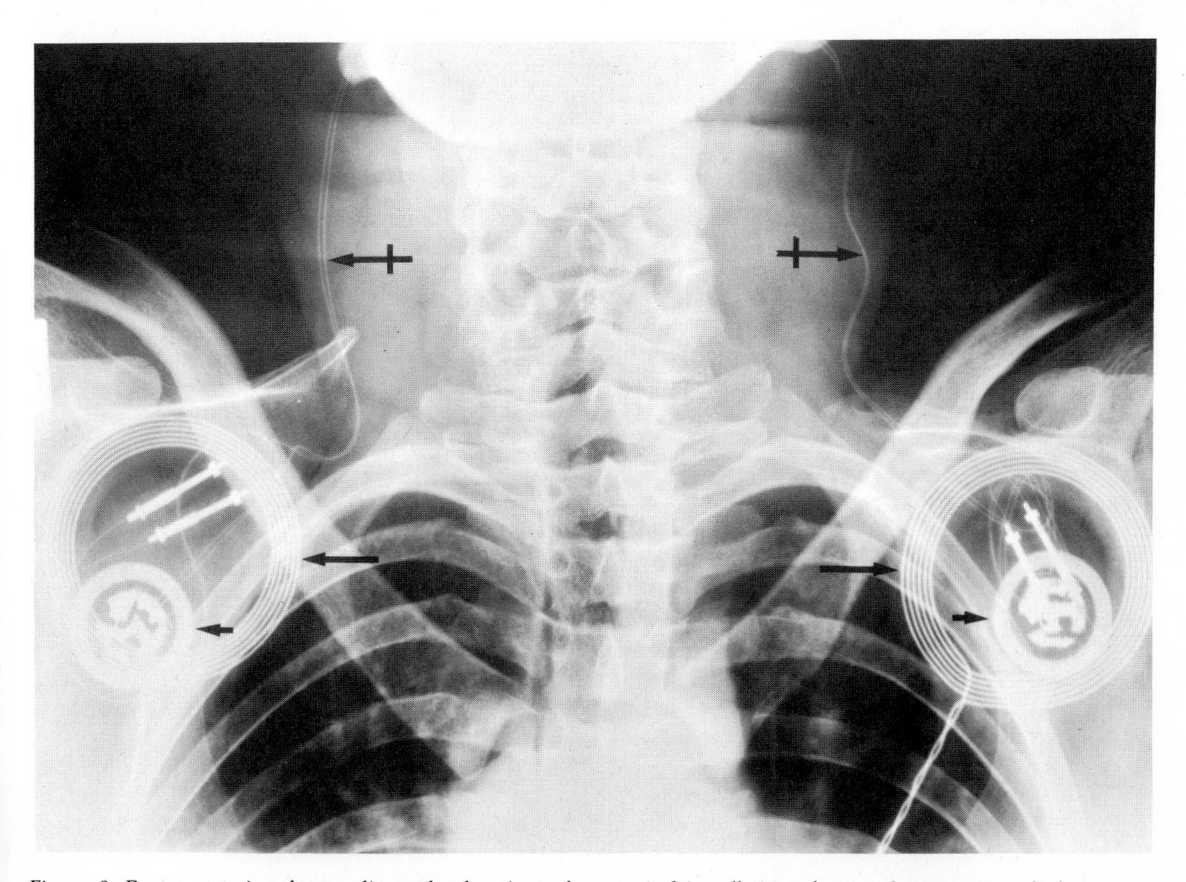

Figure 8. Posteroanterior chest radiograph of patient after surgical installation of two subcutaneous radiofrequency receivers (short arrows) connected to cerebellar electrodes via subcutaneous wiring (crossed arrows). Note external loop antennae (long arrows) taped on skin overlying subcutaneous receivers. Reprinted from Wood *et al.*,[173] with permission.

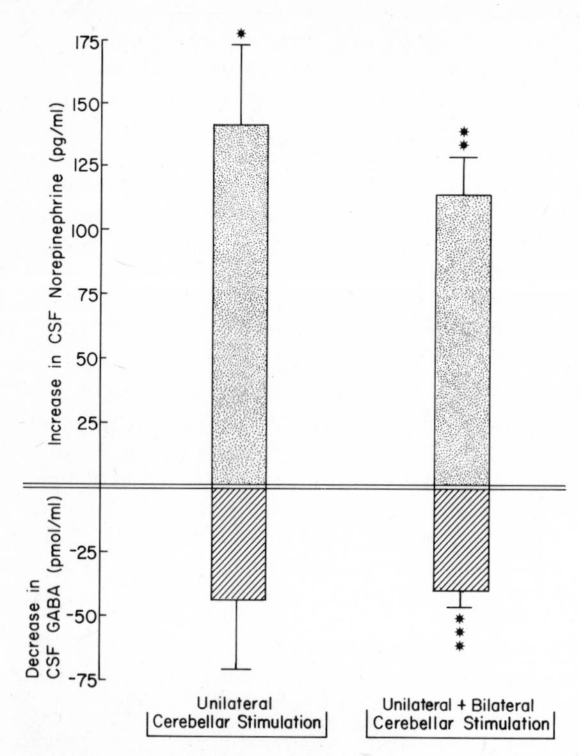

Figure 9. Elevations in CSF norepinephrine levels and reductions in CSF GABA concentrations during cerebellar stimulation in four electrode-implanted seizure patients. Zero change line represents mean CSF neurotransmitter "baseline" level after 1 week without cerebellar stimulation. *Mean CSF norepinephrine elevation of 141 ± 32 pg/ml (S.E.M.) after 5–7 days of unilateral cerebellar stimulation and **that of 114 ± 15 pg/ml after 5–7 days of unilateral stimulation followed by 16 hr of bilateral cerebellar stimulation were significant ($p < 0.03$ and $p < 0.005$, respectively; two-tailed paired Student's t test) when compared with mean baseline CSF norepinephrine level. ***Mean CSF GABA reduction of 42 ± 8 pmol/ml after 5–7 days of unilateral stimulation followed by 16 hr of bilateral cerebellar stimulation was significant ($p < 0.02$) when compared with mean baseline CSF GABA concentration.

crease in CSF GABA in our seizure patients appeared to be independent of cerebellar stimulation frequency.[170] CSF GABA levels in electrode-implanted seizure patients not undergoing electrical stimulation were similar to those in seizure patients without cerebellar electrodes; stimulation-evoked CSF GABA alterations returned to baseline levels within 1 week after the termination of stimulation. Therefore, neither the presence of cerebellar electrodes nor stimulation-induced cellular degeneration[153] contributed to the evoked CSF GABA reduction reported by Wood *et al.*[170]

Despite the plentiful quantities of cyclic GMP in Purkinje cells,[101] the inverse relationship between cyclic GMP and GABA in the cerebellum,[98,99] and documentation of cerebellar stimulation-induced CSF GABA depressions, no significant alterations in lumbar CSF cyclic GMP concentrations were demonstrated during periods of stimulation.[170] Evoked intracellular cerebellar cyclic GMP alterations, if present, may not be reflected in lumbar CSF or may be marked by contributions from the spinal cord.[75] In addition, Purkinje cell degeneration secondary to chronic exposure to anticonvulsants[132] may blunt the effect of electrical stimulation on intracellular cyclic GMP metabolism.

3.1.3. Norepinephrine and Cyclic AMP. Administration of 6-hydroxydopamine via the CSF in rats reduces brain norepinephrine and dopamine[157] and increases susceptibility to convulsive stimuli.[14] Intraventricularly injected norepinephrine is distributed throughout the catecholamine-rich regions of the brain[52] and suppresses seizures evoked by intraventricular injection of curare in cats.[45] Thus, most catecholaminergic neurons are accessible to agents delivered via the CSF, and elevation of the ventricular CSF norepinephrine content appears to have anticonvulsant activity. Wood *et al.*[175] demonstrated significant elevations in CSF norepinephrine levels during unilateral and unilateral plus bilateral cerebellar stimulation in epileptic patients (Fig. 9); however, their preliminary double-blind evaluations did not confirm increased clinical seizure control.[158]

Slow-frequency (10/sec) cerebellar stimulation has been reported to be more efficacious in suppressing seizures than high-frequency (200/sec) stimulation.[23,62] The percentage of increase in CSF norepinephrine evoked by 200/sec cerebellar stimulation appears to be 3 times higher than that induced during 10/sec stimulation.[175] Acceleration of stimulation frequencies may further increase neuronal discharge rates or recruit additional noradrenergic neurons that require high-frequency pulses for activation. High-frequency cerebellar stimulation may be required for the reported suppression of myoclonus and muscular tone[24,25,32] and reduction of somatosensory and dentate-evoked cortical potentials.[62] The application of norepinephrine to cerebral cortical neurons inhibits their activity.[81] This apparent frequency dependence of cerebral cortical-evoked response suppression during cerebellar stimulation may be mediated via noradrenergic neurons.

The exact mechanism of these CSF norepinephrine elevations induced by cerebellar stimulation is not known. The cerebellum has been shown to indirectly mediate cerebral catecholamine metabolism.[145] Noradrenergic cell bodies within the locus coeruleus send axons to the cerebellar cortex.[118] Stimulation of the rat cerebellar cortex evokes antidromic activation of the locus coeruleus.[111] Stimulation in the vicinity of the locus coeruleus suppresses epileptiform types of electrocortical activity in rats.[88] This augmentation of locus coeruleus activity may diffusely increase the noradrenergic inhibition of cerebral neurons and the input to the reticular formation.[81,140] Although this hypothesis may explain the reported ability of cerebellar stimulation to inhibit cerebral seizures,[22,66,67,108] our own clinical trials did not associate elevations in CSF norepinephrine with seizure control.[158] As yet, the previously reported cerebellar stimulation-induced activation of the reticular formation[5,8] and nonspecific thalamic nuclei[5] has not adequately explained this evoked increase in CSF noradrenergic activity. However, preliminary evidence suggests that the autonomic (possibly noradrenergic) hypothalamic outburst elicited by cerebellar stimulation in cats is mediated by the ascending reticular formation.[177]

The blood–CSF barrier for norepinephrine impedes diffusion of peripheral norepinephrine into the CSF,[178] and the blood–brain barrier is restored within 1 month during chronic cerebellar stimulation.[128] The stimulation-induced CSF norepinephrine elevations return to baseline within 1 week after termination of cerebellar stimulation.[173,175] Thus, tissue damage does not account for the CSF norepinephrine alterations observed during cerebellar stimulation. Wood *et al.*[173] also reported similar CSF norepinephrine elevations during both unblinded and double-blinded conditions; thus, patient stress did not contribute significantly to the evoked increases in CSF noradrenergic activity (Fig. 10). These evoked elevations in noradrenergic activity do not appear to be an epiphenomenon of brain-surface stimulation, but rather seem to be relatively singular to cerebellar stimulation.[173]

Noradrenergic afferent neurons to the cerebellum synapse primarily with the Purkinje cell dendrites.[9] Direct application of norepinephrine to Purkinje cells results in suppression of their spontaneous discharge.[65] The reported inhibition of Purkinje cell firing during cerebellar stimulation[30] may be mediated by activation of this noradrenergic system. This suppression of Purkinje output might account for the reported reduction in CSF GABA levels[170] and increased seizure frequencies[158] among our patients. Thus, augmentation of noradrenergic activity

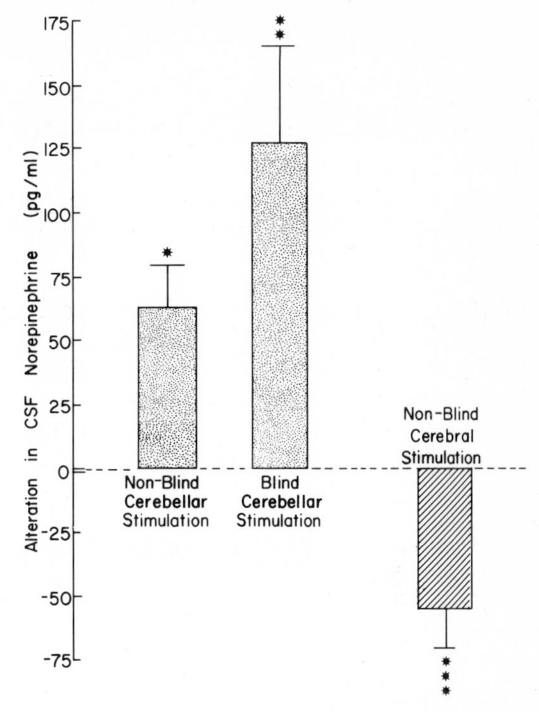

Figure 10. Alteration in CSF norepinephrine concentrations with respect to cerebellar stimulation in five electrode-implanted seizure patients and with respect to cerebral stimulation in three electrode-implanted seizure patients. Zero change line represents mean CSF norepinephrine "baseline" level after 1 week without cerebellar stimulation. *Mean CSF norepinephrine elevation of 63 ± 16 pg/ml (S.E.M.) after 5–7 days of nonblind cerebellar stimulation and **that of 127 ± 38 pg/mg after 5–7 days of double-bind cerebellar stimulation were significant ($p < 0.05$ and $p < 0.005$, respectively; two-tailed paired Student's t test) when compared with mean baseline CSF norepinephrine level. ***Mean CSF norepinephrine reduction of 55 ± 16 pg/ml determined 12 hr following cerebral-surface stimulation was significant ($p < 0.03$) when compared with mean baseline CSF norepinephrine level.

not only may diffusely elevate the seizure threshold of cerebral neurons, but also may promote convulsions by inhibiting Purkinje output. These functionally opposing noradrenergic and GABAergic alterations might increase the variance of clinical responses evoked by cerebellar stimulation.

The plentiful quantities of cyclic AMP present in the rat cerebellum[151] increase further after administration of catecholamines[72] and may mediate the inhibitory action of norepinephrine on rat Purkinje cells.[143] Electrical stimulation usually causes significant elevations in brain cyclic AMP content[53,71]; however, our studies did not demonstrate significant induced cyclic AMP alterations in lumbar CSF during cerebellar stimulation.[173,175]

3.1.4. HVA and 5-HIAA. Lumbar CSF levels of HVA and 5-HIAA, the metabolites of, respectively, dopamine and serotonin, were not significantly altered by cerebellar stimulation in seizure patients treated with anticonvulsant medications.[175] These findings suggest that dopaminergic and serotoninergic activity are not affected by electrical stimulation of the cerebellum. Recently, Tabaddor *et al.*[152] reported markedly increased ventricular CSF HVA and 5-HIAA levels in stimulated cerebral palsy patients; however, the significance of these findings may have been compromised by methodological difficulties.[169]

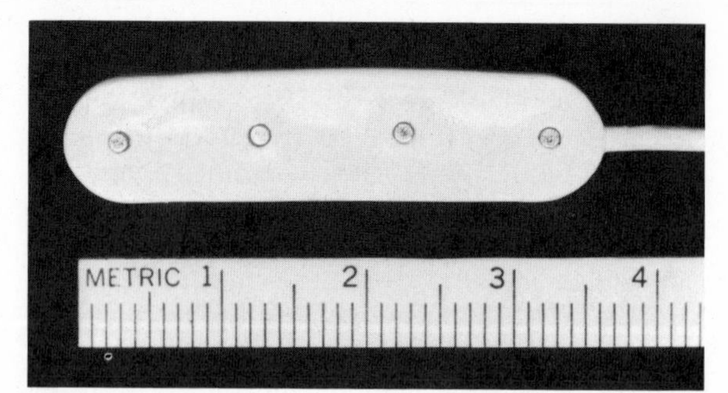

Figure 11. Cerebral electrode array consisting of segmented polyurethane flap with four 2-mm-diameter platinum–iridium disk electrodes set at 10-mm intervals. Reprinted from Wood *et al.*,[173] with permission.

3.2. Cerebral Surface Stimulation

Wood *et al.*[173] described the trephination and installation of subdural flap electrodes (Fig. 11) over the surface of the superior frontal gyrus (Fig. 12) and parasaggital cerebral cortex (Fig. 13) in seizure patients. The lumbar CSF norepinephrine levels were significantly depressed by volleys of 4–12 mA (peak-to-peak), 60/sec biphasic square pulses of 2.5-msec duration (see Fig. 10).

3.2.1. Norepinephrine and Cyclic AMP. Electrical stimulation of isolated cerebral cortical slices evokes a release of norepinephrine into the incubation medium,[73] which appears contrary to the response to such stimulation in our patients. Thus, the noradrenergic response of isolated cerebral cortical preparations to electrical stimulation may not reliably mimic those elicited by *in vivo* cerebral surface stimulation.

The direction of change in noradrenergic activity after cerebral surface stimulation appears to be opposite that occurring during cerebellar surface stimulation. This finding suggests that the CSF noradrenergic response to brain surface stimulation may have regional specificity and provides more evidence that the CSF norepinephrine elevations induced during cerebellar stimulation are not the result of tissue[128] or blood vessel[28] damage, changes in blood–CSF barrier,[178] or other nonspecific phenomena.

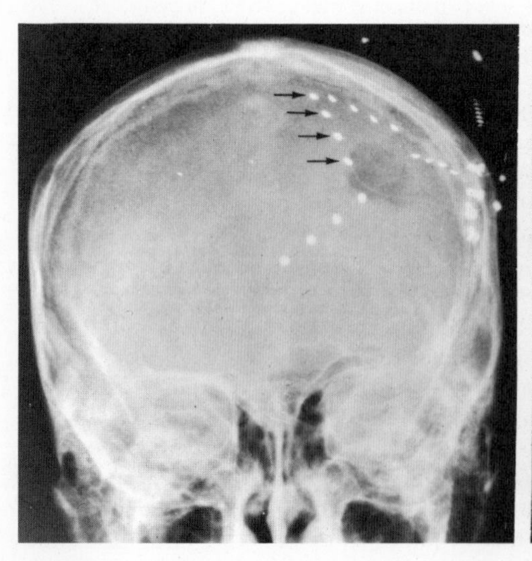

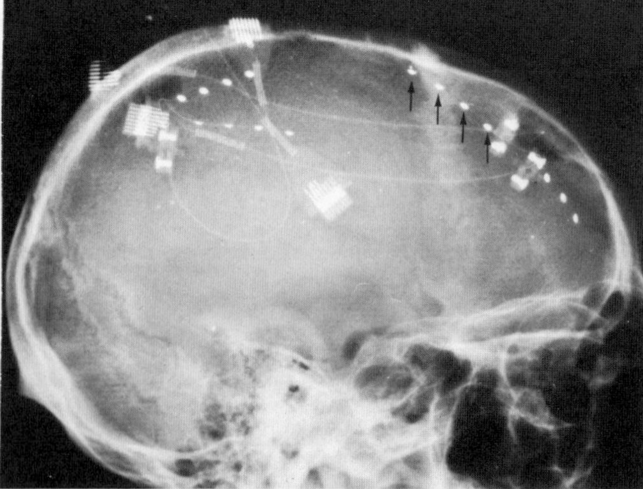

Figure 12. Caldwell posteroanterior and lateral skull radiographs of patient after surgical installation of subdural flap electrodes over left cerebral convexity. Arrows indicate electrode sites stimulated in this study. Reprinted from Wood *et al.*,[173] with permission.

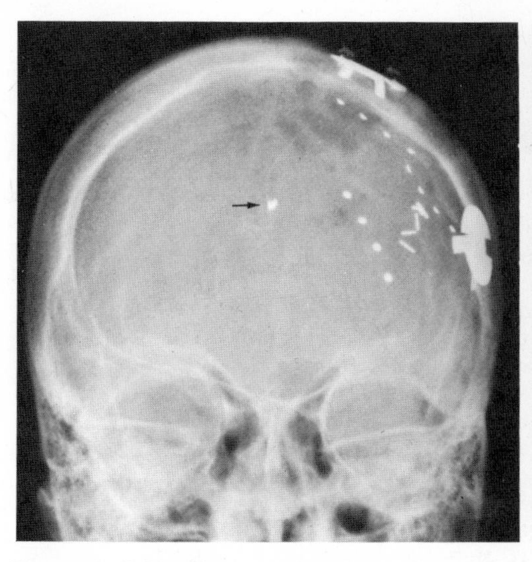
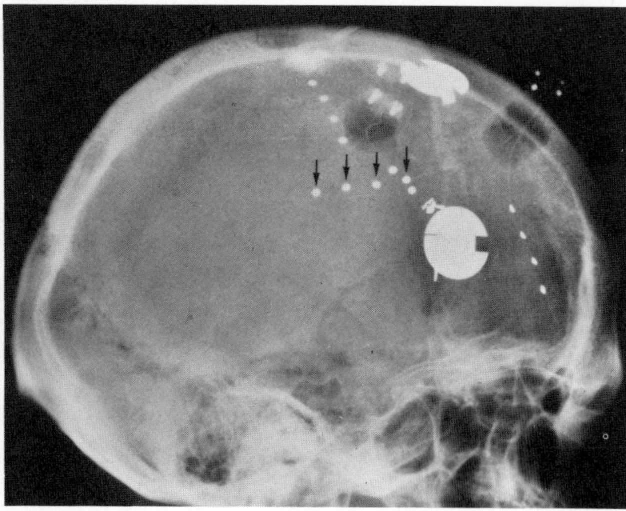

Figure 13. Caldwell posteroanterior and lateral skull radiographs after surgical installation of subdural flap electrodes over left cerebrum. Arrows indicate parasagittal-surface electrode sites stimulated. Note evidence of left frontotemporal craniotomy and cortical resection performed 20 years prior to cerebral electrode implantation. Reprinted from Wood *et al.*,[173] with permission.

Electrical stimulation of the cerebral cortex *in vitro*[71] and *in vivo*[53] elicits a prominent elevation in cyclic AMP synthesis, which has been shown not to be mediated by norepinephrine.[71] Contrary to its action on cerebellar Purkinje cells, the depressant influence of norepinephrine on cerebral cortical neurons may not be mediated by cyclic AMP.[84] The lumbar CSF cyclic AMP concentrations do not appear to be significantly altered by cerebral surface stimulation in man.[173]

4. Conclusions

Neurochemical alterations in lumbar CSF reflect central nervous system metabolism. As discussed in previous chapters, lumbar concentrations of GABA and HVA correlate with the brain content of GABA and dopamine. The norepinephrine in lumbar CSF reflects the activity of descending noradrenergic tracts originating in the brainstem. Both the brain and the spinal cord make contributions to lumbar CSF levels of 5-HIAA and the cyclic nucleotides. Thus, the neurochemical examination of lumbar CSF offers an experimental method of monitoring brain metabolic alterations in living patients with convulsive disorders who are being treated with anticonvulsant drugs and cerebellar stimulation.

Lumbar CSF GABA levels are significantly lower in anticonvulsant-treated seizure patients, especially those having psychomotor and grand mal convulsions. Cerebellar stimulation evokes a further reduction in GABAergic activity and has been associated with augmented seizure activity in some patients. This depression in GABA metabolism among seizure patients may correlate with the efficacy of those anticonvulsant drugs the mode of action of which is the stimulation of brain GABAergic activity.

According to unmatched preliminary comparisons, CSF norepinephrine concentrations do not appear to be altered in medicated seizure patients; however, cerebellar stimulation tends to augment noradrenergic metabolism in this patient population. The CSF norepinephrine elevations evoked by cerebellar stimulation appear to to be relatively specific for cerebellar stimulation and may be expected to raise convulsive thresholds. This increase in noradrenergic activity during cerebellar stimulation may have an inhibitory effect on the GABAergic Purkinje cell output, which would tend to provoke seizure activity.

The augmented accumulation of 5-HIAA, the major serotonin metabolite, and of cyclic AMP during probenecid blockade suggests either an increase in their synthesis or an augmentation of their release into the CSF of seizure patients. Contrarily, the CSF concentrations of HVA, the major dopamine metabolite, and of cyclic GMP in treated epileptic patients are not different from those in unmedicated control patients. Cerebellar stimulation did not alter CSF levels of these monoamine metabolites or cyclic nucleotides in our seizure patients.

Clinical data concerning the effects of anticonvulsant medications on brain and CSF neurochemical activity are both scant and controversial. Future investigations should evaluate central nervous system metabolism as reflected in CSF both prior to and following the initiation of anticonvulsant therapy and correlate these neurochemical alterations with clinical seizure activity. Although this review concerns mainly CSF neurotransmitters, their metabolites, and cyclic nucleotides, future CSF studies may also document a role of neurohormones and peptides in the neurochemistry of epilepsy.

ACKNOWLEDGMENTS. The author thanks Drs. Michael G. Ziegler, C. Raymond Lake, Robert M. Post, Theodore A. Hare, Bruce S. Glaeser, James C. Ballinger, Michael H. Ebert, and J. Sode for their valuable assistance in CSF neurochemical and data analysis. In addition, the author thanks Dr. Rodwan K. Rajjoub, who counted Purkinje cells in cerebellar biopsy and autopsy specimens; Dr. John M. Van Buren, who implanted cerebellar and cerebral surface electrodes; and Dr. Theodore A. Hare, who prepared the graphic illustrations.

References

1. ADAMS, D. J., LUDERS, H., PIPPENGER, C.: Sodium valproate in the treatment of intractable seizure disorders: A clinical and electroencephalographic study. *Neurology (Minneapolis)* **28:**152–157, 1978.
2. AXELROD, J.: Dopamine β-hydroxylase: Regulation of its synthesis and release from nerve terminals. *Pharmacol. Rev.* **24:**233–243, 1972.
3. AXELROD, J., WEINSHILBOUN, R.: Catecholamines. *N. Engl. J. Med.* **287:**237–242, 1972.
4. BANTLI, H., BLOEDEL, J. R., ANDERSON, G., MCROBERTS, R., SANDBERG, E.: Effects of stimulating the cerebellar surface on the activity in penicillin foci. *J. Neurosurg.* **48:**69–84, 1978.
5. BANTLI, H., BLOEDEL, J. R., TOLBERT, D.: Activation of neurons in the cerebellar nuclei and ascending reticular formation by stimulation of the cerebellar surface. *J. Neurosurg.* **45:**539–554, 1976.
6. BERNHEIMER, H., BIRKMAYER, W., HORNYKIEWICZ, O.: Homovanillinsaur im Liquor cerebrospinalis: Untersuchungen beim Parkinson-syndrom und anderen Erkrankungen des ZNS. *Wien. Klin. Wochenschr.* **78:**417–419, 1966.
7. BLACK, P., FISCHELL, R. E., MARKOWITZ, R. S., POWELL, W. R.: Cerebellar stimulation: Comparison of effect on electrically-induced and "spontaneous" alumina seizures. *Epilepsia* **18:**287, 1977 (abstract).
8. BLOEDEL, J. R., BURTON, J. E.: Electrophysiological evidence for a mossy fiber input to the cerebellar cortex activated indirectly by collaterals of spinocerebellar pathways. *J. Neurophysiol.* **33:**308–320, 1970.
9. BLOOM, F. E., HOFFER, B. J., SIGGINS, G. R.: Studies on norepinephrine-containing afferents to Purkinje cells of rat cerebellum. I. Localization of the fibers and their synapses. *Brain Res.* **25:**501–521, 1971.
10. BÖHLEN, P. S., HUOT, S., PALFREYMAN, M. G.: The relationships between GABA concentrations in brain and cerebrospinal fluid. *Brain Res.* **167:**297–305, 1979.
11. BROOKS, B. R., ENGEL, W. K., SODE, J.: Blood-to-cerebrospinal fluid barrier for cyclic adenosine monophosphate in man. *Arch. Neurol.* **34:**468–469, 1977.
12. BROOKS, B. R., WOOD, J. H., DIAZ, M., CZERWINSKI, C., GEORGES, L. P., SODE, J., EBERT, M. H., ENGEL, W. K.: Extracellular cyclic nucleotide metabolism in the human central nervous system. In Wood, J. H. (ed.): *Neurobiology of Cerebrospinal Fluid I.* New York, Plenum Press, 1980.
13. BROOKS, B. R., WOOD, J. H., SODE, J., ENGEL, W. K.: Cyclic nucleotide metabolism in neurological disease. *Trans. Am. Neurol. Assoc.* **101:**221–222, 1976 (abstract).
14. BROWNING, R. A., MAYNERT, E. W.: Increased seizure susceptibility in 6-hydroxydopamine-treated rats. *Fed. Proc. Fed. Am. Soc. Exp. Biol.* **29:**966, 1970.
15. BROWNING, R. A., MAYNERT, E. W.: Role of body temperature in the monoamine-induced facilitation of seizures. *Fed. Proc. Fed. Am. Soc. Exp. Biol.* **20:**589, 1971.
16. CHADWICK, D., JENNER, P., REYNOLDS, E. H.: Amines, anticonvulsants and epilepsy. *Lancet* **1:**474–476, 1975.
17. CHADWICK, D., JENNER, P., REYNOLDS, E. H.: Serotonin metabolism in human epilepsy; The influence of anticonvulsant drugs. *Ann. Neurol.* **1:**218–224, 1977.
18. CHADWICK, D., TRIMBLE, M., JENNER, P., DRIVER, M. V., REYNOLDS, E. H.: Manipulation of cerebral monoamines in the treatment of human epilepsy: A pilot study. *Epilepsia* **19:**3–10, 1978.
19. CHASE, T. N., KATZ, R. I., KOPIN, I. J.: Effect of anticonvulsants on brain serotonin. *Trans. Am. Neurol. Assoc.* **94:**236–238, 1969 (abstract).
20. CHEN, G., ENSOR, C. R., BOHNER, B.: Drug effects on the disposition of active biogenic amines in the CNS. *Life Sci.* **7:**1063–1074, 1968.
21. COLEMAN-RIESE, D., CUTLER, R. W. P.: Inhibition of γ-aminobutyric acid release from rat cerebral cortex slices by barbiturate anesthesia. *Neurochem. Res.* **3:**423–429, 1978.
22. COOKE, P. M., SNIDER, R. S.: Some cerebellar influences on electrically-induced cerebral seizures. *Epilepsia* **4:**19–28, 1955.
23. COOPER, I. S., RIKLAN, M., AMIN, I., CULLINAN, T.: A long-term follow-up study of cerebellar stimulation

for the control of epilepsy. In Cooper, I. S. (ed.): *Cerebellar Stimulation in Man.* New York, Raven Press, 1978, pp. 19–38.
24. COOPER, I. S., RIKLAN, M., TABADDOR, K., CULLINAN, T., AMIN, I., WATKINS, E. S.: A long-term follow-up study of chronic cerebellar stimulation for cerebral palsy. In Cooper, I. S. (ed.): *Cerebellar Stimulation in Man.* New York, Raven Press, 1978, pp. 59–99.
25. COOPER, I. S., UPTON, A. R. M.: Use of chronic cerebellar stimulation for disorders of disinhibition. *Lancet* **1**:595–600, 1978.
26. CURTIS, D. R., DUGGAN, A. W., FELIZ, D.: GABA and inhibition of Deiter's neurones. *Brain Res.* **23**:117–120, 1970.
27. CUTLER, R. W. P., YOUNG, J.: The effect of penicillin on the release of γ-aminobutyric acid from cerebral cortex slices. *Brain Res.* **170**:157–163, 1979.
28. DAHL, E., NELSON, E.: Electron microscope observation on human intracranial arteries. II. Innervation. *Arch. Neurol.* **10**:56–62, 1964.
29. DALY, J. W.: The formation, degradation and function of cyclic nucleotides in the nervous system. In Smythies, J. R., Bradley, R. J. (eds.): *International Review of Neurobiology,* Vol. 20. New York, Academic Press, 1977, pp. 105–168.
30. DAUTH, G. W., DELL, S., GILMAN, S.: Alterations of Purkinje cell activity from transfolial stimulation of the cerebellum in the cat. *Neurology (Minneapolis)* **28**:654–600, 1978.
31. DAUTH, G. W., YOKOYAMA, T., GILMAN, S.: The effects of transfolial stimulation on fastigial neurons. *Neurology (Minneapolis)* **29**:597, 1979 (abstract).
32. DAVIS, R. M., CULLEN, R. F., FLITTER, M., DUENAS, D., ENGLE, H., ENNIS, B.: Control of spasticity and involuntary movements. *Neurosurgery* **1**:205–207, 1977.
33. DAVSON, H.: *Physiology of the Cerebrospinal Fluid.* London, Churchill, 1967.
34. DICHIRO, G., HAMMOCK, M. K., BLEYER, W. A.: Spinal descent of cerebrospinal fluid in man. *Neurology (Minneapolis)* **26**:1–8, 1976.
35. DOW, R. S., FERNANDEZ-GUARDIOLA, A., MANNI, E.: The influences of the cerebellum on experimental epilepsy. *Electroencephalogr. Clin. Neurophysiol.* **14**:383–398, 1962.
36. DREIFUSS, J. J., KELLY, J. S., KRYJEVIC, K.: Cortical inhibition and γ-aminobutyric acid. *Exp. Brain Res.* **9**:137–154, 1969.
37. DRETCHEN, K. L., STANDAERT, F. G., RAINES, A.: Effects of phenytoin on the cyclic nucleotide system in the motor nerve terminal. *Epilepsia* **18**:337–348, 1977.
38. DUFFY, T. E., HOWSE, D. C., PLUM, F.: Cerebral energy metabolism during experimental status epilepticus. *J. Neurochem.* **24**:925–134, 1975.
39. EMSON, P. C.: Effects of chronic treatment with amino-oxyacetic acid or sodium *n*-dipropylacetate on brain GABA levels and the development and regression of cobalt epileptic foci in rats. *J. Neurochem.* **27**:1489–1494, 1976.
40. ENNA, S. J., BENNETT, J. P., BYLUND, D. B., CRESE, I., BURT, D. R., CHARNESS, M. E., YAMAMURA, H. I., SEMANTOV, R., SNYDER, S. H.: Neurotransmitter receptor binding: Regional distribution in human brain. *J. Neurochem.* **28**:233–236, 1977.
41. ENNA, S. J., STERN, L. Z., WASTEK, G. J., YAMAMURA, H. I.: Cerebrospinal fluid γ-aminobutyric acid variations in neurologic disorders. *Arch. Neurol.* **34**:683–685, 1977.
42. ENNA, S. J., WOOD, J. H., SNYDER, S. H.: γ-Aminobutyric acid in human cerebrospinal fluid: Radioreceptor assay. *J. Neurochem.* **28**:1121–1124, 1977.
43. EVETTS, K. D., IVERSEN, L. L.: Effects of protriptyline on the depletion of catecholamines induced by 6-hydroxydopamine in the brain of the rat. *J. Pharmacol.* **22**:540–542, 1970.
44. FAHN, S., COTE, L. J.: Regional distribution of γ-aminobutyric acid (GABA) in brain of the rhesus monkey. *J. Neurochem.* **15**:209–213, 1968.
45. FELDBERG, W., SHERWOOD, S. L.: Injections of drugs into the lateral ventricle of the cat. *J. Physiol. (London)* **123**:148–167, 1954.
46. FENTON, G. W., FENWICK, P. B. C., BRINDLEY, G. S., FALCONER, M. A., POLKEY, C. H. RUSHTON, D. M.: Chronic cerebellar stimulation in the treatment of epilepsy: A preliminary report. In Penry, J. K. (ed.): *Epilepsy: The Eighth International Symposium.* New York, Raven Press, 1977, pp. 333–340.
47. FERKANY, J. W., BUTLER, I. J., ENNA, S. J.: Effect of drugs on rat brain, cerebrospinal fluid and blood GABA content. *J. Neurochem.* **33**:29–33, 1979.
48. FERRENDELLI, J. A., GROSS, R. A., KINSCHERF, D. A., RUBIN, E. H.: Effects of seizures and anticonvulsant drugs on cyclic nucleotide regulation in the CNS. In Palmer, G. (ed.): *Neuropharmacology of Cyclic Nucleotides.* Baltimore, Urban and Schwarzenberg, 1979.
49. FERRENDELLI, J. A., KINSCHERF, D. A.: Inhibitory effects of anticonvulsant drugs on cyclic nucleotide accumulation in brain. *Ann. Neurol.* **5**:533–538, 1979.
50. GARELIS, E., SOURKES, T. L.: Use of cerebrospinal fluid drawn at pneumoencephalography in the study of monoamine metabolism in man. *J. Neurol. Neurosurg. Psychiatry* **37**:704–710, 1974.
51. GESSA, G. L., KRISHNA, G., FORN, J., TAGLIAMONTE, A., BRODIE, B. B.: Behavioral and vegetative effects produced by dibutyryl cyclic AMP injected into different areas of the brain. In Greengard, P., Costa, E. (eds): *Role of Cyclic AMP in Cell Function: Advances in Biochemical Pharmacology,* Vol. 3. New York, Raven Press, 1970, pp. 371–381.
52. GLOWINSKI, J., IVERSEN, L. L.: Regional studies of catecholamines in the rat brain. I. The deposition of H^3-norepinephrine, H^3-dopamine and H^3-DOPA in various regions of the brain. *J. Neurochem.* **13**:655–666, 1966.
53. GOLDBERG, N. D., LUST, W. D., O'DEA, R. F., WEI, S., O'TOOLE, A. G.: A role of cyclic nucleotides in brain metabolism. In Greengard, P., Costa, E. (eds):

Role of Cyclic AMP in Cell Function: Advances in Biochemical Psychopharmacology, Vol. 3. New York, Raven Press, 1970, pp. 67–87.

54. Goldstein, M.: Inhibition of norepinephrine biosynthesis at the dopamine-β-hydroxylase stage. *Pharmacol. Rev.* **18:**77–82, 1966.
55. Gram, L., Wulff, K., Rasmussen, K. E., Flachs, H., Wurtz-Jorgensen, H., Sommerbeck, K. W., Lohren, V.: Valproate sodium: A controlled clinical trial including monitoring of drug levels. *Epilepsia* **18:**141–148, 1977.
56. Gross, R. A., Ferrendelli, J. A.: Effects of reserpine, propranolol, and aminophylline on seizure activity and CNS cyclic nucleotides. *Ann. Neurol.* **6:**296–301, 1979.
57. Gulati, O. D., Stanton, H.C.: Some effects of the central nervous system of gamma-*N*-butyric acid (GABA) and certain related amino acids administered systemically and intracerebrally to mice. *J. Pharmacol. Exp. Ther.* **29:**178–185, 1960.
58. Hablitz, J. J.: Intramuscular penicillin epilepsy in the cat: Effect of chronic cerebellar stimulation. *Exp. Neurol.* **50:**505–514, 1976.
59. Hadfield, M. G.: Uptake and binding of catecholamines: Effect of diphenylhydantoin and a new mechanism of action. *Arch. Neurol.* **26:**78–84, 1972.
60. Hare, T. A., Wood, J. H., Manyam, N. V. B., Ballenger, J. C., Post, R. M., Gerner, R. H.: Selection of control populations for clinical cerebrospinal fluid GABA investigations based on comparison with normal volunteers. *Brain Res. Bull.* (in press).
61. Heath, R. G.: Modulation of emotion with a brain pacemaker: Treatment for intractable psychiatric illness. *J. Nerv. Ment. Dis.* **165:**300–317, 1977.
62. Hemmy, D. C., Larson, S. J., Sances, A., Millar, E. A.: The effect of cerebellar stimulation on focal seizure activity and spasticity in monkeys. *J. Neurosurg.* **46:**648–653, 1977.
63. Hitchcock, E., Gabra-Saunders, T.: Effect of diphenylhydantoin on gamma aminobutyric acid (GABA) and succinate activity in rat Purkinje cells. *J. Neurol. Neurosurg. Psychiatry* **40:**565–569, 1977.
64. Hoffer, B., Seiger, A., Freedman, R., Olson, L., Taylor, D.: Electrophysiology and cytology of hippocampal formation transplants in the anterior chamber of the eye. II. Cholinergic mechanisms. *Brain Res.* **119:**107–132, 1977.
65. Hoffer, B. J., Siggins, G. R., Bloom, F. E.: Studies on norepinephrine-containing afferents to Purkinje cells of rat cerebellum. II. Sensitivity of Purkinje cells to norepinephrine and related substances administered by microiontophoresis. *Brain Res.* **25:**523–534, 1971.
66. Hutton, J. T., Frost, J. D., Foster, J.: The influence of the cerebellum in cat penicillin epilepsy. *Epilepsia* **13:**401–408, 1972.
67. Iwata, K., Snider, R. S.: Cerebello-hippocampal influences on the electroencephalogram. *Electroencephalogr. Clin. Neurophysiol.* **11:**439–446, 1959.
68. Jimerson, D. C., Post, R. M., Carman, J. S., van Kammen, D. P., Wood, J. H., Goodwin, F. K., Bunney, W. E.: CSF calcium: Clinical correlates in affective illness and schizophrenia. *Biol. Psychiatry* **14:**37–51, 1979.
69. Jobe, P. C., Picchioni, A. L., Chin, L.: Role of brain norepinephrine in audiogenic seizures in the rat. *J. Pharmacol. Exp. Ther.* **184:**1–9, 1973.
70. Julien, R. M., Halpern, L. M.: Effects of diphenylhydantoin and other antiepileptic drugs on epileptic activity and Purkinje cells discharge rate. *Epilepsia* **12:**387–400, 1972.
71. Kakinchi, S., Rall, T. W., McIlwain, H.: The effect of electrical stimulation upon the accumulation of adenosine 3′,5′-phosphate in isolated cerebral tissue. *J. Neurochem.* **16:**485–491, 1969.
72. Kakinchi, S., Rall, T. W.: The influence of chemical agents on the accumulation of adenosine 3′,5′-phosphate in slices of rabbit cerebellum. *Mol. Pharmacol.* **4:**367–378, 1968.
73. Katz, R. I., Chase, T. N.: Neurohumoral mechanisms in the brain slice. *Adv. Pharmacol.* **8:**1–30, 1970.
74. Kilian, M., Frey, H. H.: Central monoamines and convulsive thresholds in mice and rats. *Neuropharmacology* **12:**681–692, 1973.
75. Kinscherf, D. A., Chang, M. M., Rubin, E. H., Schneider, D. R., Ferrendelli, J. A.: Comparison of the effects of depolarizing agents and neurotransmitters on regional CNS cyclic GMP levels in various animals. *J. Neurochem.* **26:**527–530, 1976.
76. Koe, K. B., Weissman, A.: *p*-Chlorophenylalanine: A specific depletor of brain serotonin. *J. Pharmacol. Exp. Ther.* **154:**499–516, 1966.
77. Koyama, I.: Amino acids in the cobalt-induced epileptogenic cat's cortex. *Can. J. Physiol. Pharm.* **50:**740–752, 1972.
78. Kupferberg, H. J., Lust, W. D., Penry, J. K.: Anticonvulsant activity of dipropylacetic acid (DPA) in relation to GABA and cGMP brain levels in mice. *Fed. Proc. Fed. Am. Soc. Exp. Biol.* **34:**283, 1975.
79. Krip, G., Vazquez, J.: Effects of some symptomimetic drugs and their antagonists on after discharges elicited in chronically isolated slabs of cerebral cortex. *Br. J. Pharmacol.* **43:**696–705, 1971.
80. Krnjevíc, K., Schwartz, S.: The action of γ-aminobutyric acid on cortical neurones. *Exp. Brain Res.* **3:**320–336, 1967.
81. Krnjevíc, K., Phillis, J. W.: Actions of certain amines on cerebral cortical neurons. *Br. J. Pharmacol.* **20:**471–490, 1963.
82. Kukino, K., Deguchi, T.: Effects of sodium dipropylacetate on γ-amimobutyric acid and biogenic amines in rat brain. *Chem. Pharm. Bull.* **25:**2257–2262, 1977.
83. Kuriyama, K., Sze, P.: Blood–brain barrier to H^3-γ-aminobutyric acid in normal and amino-oxyacetic acid-treated animals. *Neuropharmacology* **10:**103–108, 1971.
84. Lake, N., Jordan, L. M., Phillis, J. W.: Evidence against cyclic adenosine 3′,5′-monophosphate (AMP)

mediation of noradrenaline depression of cerebral cortical neurones. *Brain Res.* **60**:411–421, 1973.

85. LAXER, K. D., SOURKES, T. L., FANG, T. Y., YOUNG, S. N., GAUTHIER, S. G., MISSALA, K.: Monoamine metabolites in the CSF of epileptic patients. *Neurology (Minneapolis)* **29**:1157–1161, 1979.
86. LESSIN, A. Q., PARKES, M. W.: The effects of reserpine and other agents on leptazol convulsions in mice. *Br. J. Pharmacol.* **14**:108–111, 1959.
87. LEWIN, E., BLACK, V.: Cyclic AMP accumulation in cerebral cortical slices; Effect of carbamazepine, phenobarbital and phenytoin. *Epilepsia* **18**:237–242, 1977.
88. LIBET, B., GLEASON, C. A., WRIGHT, E. W., FEINSTEIN, B.: Suppression of an epileptiform type of electrocortical activity in the rat by stimulation in the vicinity of locus coeruleus. *Epilepsia* **18**:451–462, 1977.

88a. LOCKARD, J., OJEMANN, G. A., CONGDON, W., DUCHARME, L.: Cerebellar stimulation in alumina gel monkey model: Inverse relationship between clinical seizures and EEG interictal bursts. *Epilepsia* **20**:223–234, 1979.

89, LOSCHER, W.: GABA in plasma and cerebrospinal fluid of different species: Effects of γ-acetylenic GABA, γ-vinyl GABA and sodium valproate. *J. Neurochem.* **32**:1587–1591, 1979.

90. LOTT, I. T., COULOMBE, T., DIPAOLO, R. V., RICHARDSON, E. P., LEVY, H. L.: Vitamin B_6-dependent seizures: Pathology and chemical findings in brain. *Neurology (Minneapolis)* **28**:47–54, 1978.
91. LUST, W. D., GOLDBERG, N. D., PASSONNEAU, J. V.: Cyclic nucleotide in murine brains: The temporal relationship of changes in adenosine 3′,5′-monophosphate and guanosine 3′,5′-monophosphate following maximal electroshock or decapitation. *J. Neurochem.* **26**:5–10, 1976.
92. LUST, W. D., KUPFERBERG, H. J., PASSONNEAU, J. V., PENRY, J. K.: Brain cyclic nucleotides and gamma-aminobutyric acid; Effect of anticonvulsant agents. *Tran. Am. Soc. Neurochem.* **60**:170, 1975. (abstract).
93. LUST, W. D., KUPFERBERG, H. J., PASSONNEAU, J. V., PENRY, J. K.: On the mechanism of action of sodium valproate, the relationship of GABA and cyclic GMP levels to anticonvulsant activity. In Legg, N. J. (ed.): *Clinical and Pharmacological Aspects of Sodium Valproate (Epilim) in the Treatment of Epilepsy.* Tunbridge Wells, England, MCS Consultants, 1976, pp. 123–129.
94. MACDONALD, R. L., BARKER, J. L.: Anticonvulsant and anesthetic barbiturates: Different postsynaptic actions in cultured mammalian neurons. *Neurology (Minneapolis)* **29**:432–447, 1979.
95. MACDONALD, R. L., BERGEY, G. K.: Valproic acid augments GABA-mediated postsynaptic inhibition in cultured mammalian neurons. *Brain Res.* **170**:558–562, 1979.
96. MACKAY, A. V. P., DAVIES, P., DEWAR, A. J., YATES, C. M.: Regional distribution of enzymes associated with neurotransmission by monoamines, acetylcholine and GABA in the human brain. *J. Neurochem.* **30**:827–839, 1978.
97. MANYAM, N. V. B., KATZ, L., HARE, T. A., GERBER, J. C.,GROSSMAN, M. H.:Levels of γ-aminobutyric acid in cerebrospinal fluid in various neurologic disorders. *Arch. Neurol.* **37**:352–355, 1980.
98. MAO, C. C., GUIDOTTI, A., COSTA, E.: Evidence for an involvement of GABA in the mediation of the cerebellar cyclic GMP decrease and the anticonvulsant action of diazepam. *Naunyn-Schmiedeberg's Arch. Pharmacol.* **289**:369–378, 1975.
99. MAO, C. C., GUIDOTTI, A., COSTA, E.: Interactions between gamma-aminobutyric acid and guanosine 3′,5′-monophosphate in rat cerebellum. *Mol. Pharmacol.* **10**:736–745, 1974.
100. MAO, C. C., GUIDOTTI, A., COSTA, E.: The regulation of cyclic guanosine monophosphate in rat cerebellum: Possible involvement in putative amino acid neurotransmitters. *Brain Res.* **79**:510–514, 1974.
101. MAO, C. C., GUIDOTTI, A., LANDIS, S.: Cyclic GMP: Reduction of cerebellar concentrations in "nervous" mutant mice. *Brain Res.* **90**:335–339, 1975.
102. MAYNERT, E. W., MARCZYNSKI, T. J., BROWNING, R. A.: The role of the neurotransmitters in the epilepsies. In Friedlander, W. J. (ed.): *Advances in Neurology,* Vol. 13. New York, Raven Press, 1975, pp. 79–147.
103. MCCANDLESS, D. W., FEUSSNER, G. K., LUST, W. D., PASSONNEAU, J. V.: Sparing of metabolic stress in Purkinje cells after maximal electroshock. *Proc. Natl. Acad. Sci. U.S.A.* **76**:1482–1484, 1979.
104. MCKENZIE, G. M., SOROKO, F. E.: The effects of apomorphine, (+)-amphetamine and L-DOPA on maximal electroshock convulsions—a comparative study in the rat and mouse. *J. Pharm. Pharmacol.* **24**:696–701, 1972.
105. MELDRUM, B.: Convulsant drugs, anticonvulsants and GABA-mediated neuronal inhibition. In Kofod, H., Krogsgaard-Larsen, P., Scheel-Kruger, J. (eds.): *GABA-Neurotransmitters: Alfred Benzon Symposium 12.* Copenhagen, Munksgaard, 1978, pp. 380–405.
106. MELDRUM, B. S.: Epilepsy and γ-aminobutyric acid-mediated inhibition. In Pfeiffer, C. C., Smythies, R. (eds.): *International Review of Neurobiology,* Vol. 17. New York, Academic Press, 1975, pp. 1–36.
107. MIYASAKA, K., HOFFMAN, H. J., FROESE, A. B.: The influence of chronic cerebellar stimulation on respiratory muscle coordination in a patient with cerebral palsy. *Neurosurgery* **2**:262–265, 1978.
108. MUTANI, R., BERGAMINI, L., DORIGUSSI, T.: Experimental evidence of the existence of an extrarhinencephalic control of the activity of the cobalt rhinencephalic epileptogenic focus. Part 2. Effect of paleocerebellar stimulation. *Epilepsia* **10**:351–362, 1969.
109. MYLLYLÄ, V. V., HEIKKINEN, E. R., VAPAATALO, H., HOKKANEN, E.: Cyclic AMP concentration and enzyme activities of cerebrospinal fluid in patients with epilepsy on central nervous system damage. *Eur. Neurol.* **13**:123–130, 1975.
110. MYLLYLÄ, V. V.: Effect of convulsions and anticonvulsive drugs on cerebrospinal fluid cyclic AMP in rabbits. *Eur. Neurol.* **14**:97–107, 1976.

111. NAKAMURA, S., IWARMA, K.: Antidromic activation of the rat locus coeruleus neurons from hippocampus, cerebral and cerebellar cortices. *Brain Res.* **99**:372–376, 1975.
112. NATHANSON, J. A.: Cyclic nucleotides and nervous system function. *Physiol. Rev.* **57**:157–256, 1977.
113. NEOPHYTIDES, A. N., SURIA, A., CHASE, T. N.: Cerebrospinal fluid GABA in neurologic disease. *Neurology (Minneapolis)* **28**:359, 1978 (abstract).
114. NORMAN, R. N.: The neuropathology of status epilepticus. *Med. Sci. Law* **4**:46–51, 1964.
115. OBATA, K., ITO, M., OCHI, R., SATO, N.: Pharmacological properties of the postsynaptic inhibition by Purkinje cell axons and the action of γ-aminobutyric acid on Deiter's neurones. *Exp. Brain Res.* **4**:43–57, 1967.
116. OBATA, K., TAKEDA, K.: Release of γ-aminobutyric acid into the fourth ventricle induced by stimulation of the cat's cerebellum. *J. Neurochem.* **16**:1043–1047, 1969.
117. OJEMANN, G. A., OAKLEY, J. C.: Effect of chronic cerebellar stimulation on seizure frequency in the alumina monkey model of epilepsy. Presented to Annual Meeting of American Association of Neurological Surgeons, Toronto, Canada, April 24–28, 1977.
118. OLSON, L., FUXE, K.: On the projection from the locus coeruleus noradrenaline neurons: The cerebellar innervation. *Brain Res.* **28**:165–171, 1971.
119. PALMER, G. C., JONES, D. J., MEDINA, M.A., STAVINOHA, W. B.: Anticonvulsant drug actions on *in vitro* and *in vivo* levels of cyclic AMP in the mouse brain. *Epilepsia* **20**:95–104, 1979.
120. PAPESCHI, P., MOLINA-NEGRO, P., SOURKES, T. L., GIUSEPPE, E.: The concentration of homovanillic acid and 5-hydroxyindoleacetic acid in ventricular and lumbar CSF: Studies in patients with extrapyramidal disorders, epilepsy and other diseases. *Neurology (Minneapolis)* **22**:1151–1159, 1972.
121. PENN, R. D., GOTTLIEB, G. L., AGARWAL, G. C.: Cerebellar stimulation in man; Quantitative changes in spasticity. *J. Neurosurg.* **48**:779–786, 1978.
122. PERRY, T. L., HANSEN, S.: Biochemical effects in man and rat of three drugs which can increase brain GABA content. *J. Neurochem.* **30**:679–684, 1978.
123. PERRY, T. L., HANSEN, S., KENNEDY, J., WADA, J. A., THOMPSON, G. B.: Amino acids in human epileptogenic foci. *Arch. Neurol.* **32**:752–754, 1975.
124. PINDER, R. M., BROGDEN, R. N., AVERY, G. S.: Sodium valproate: A review of its pharmacological properties and therapeutic efficacy in epilepsy. *Drugs* **13**:81–123, 1977.
125. PLUM, F., HOWSE, D. C., DUFFY, T. E.: Metabolic effects of seizures. *Res. Publ. Assoc. Res. Nerv. Ment. Dis.* **53**:141–157, 1974.
126. PRINCE, D. A.: Topical convulsant drugs and metabolic antagonists. In Purpura, D. P., Penry, J. K., Tower, D., Woodbury, D. M., Walter, R. (eds.): *Experimental Models of Epilepsy—A Manual for the Laboratory Worker.* New York, Raven Press, 1972, pp. 51–83.
127. PRICHARD, J. W., GUROFF, G.: Increased cerebral excitability caused by *p*-chlorophenylalanine in young rats. *J. Neurochem.* **18**:153–160, 1971.
128. PUDENZ, R. H., BULLARA, L. A., JACQUES, S., HAMBRECHT, F. T.: Electrical stimulation of the brain. III. The neural damage model. *Surg. Neurol.* **4**:389–400, 1975.
129. PURDY, R. E., JULIEN, R. M., FAIRHURST, A. S., TERRY, M. D.: Effect of carbamazepine on the *in vitro* uptake and release of norepinephrine in adrenergic nerves of rabbit aorta and in whole brain synaptosomes. *Epilepsia* **18**:251–257, 1977.
130. PYLYPCHUK, G., OREOPOULOS, D. G., WILSON, D. R., HARRISON, J. E., MCNEILL, K. G., MEEMA, H. E., OGILVIE, R., STURTRIDGE, W. C., MURRAY, T. M.: Calcium metabolism in adult outpatients with epilepsy receiving long-term anticonvulsant therapy. *Can. Med. Assoc. J.* **118**:635–638, 1978.
131. RAJJOUB, R. K., WOOD, J. H., VANBUREN, J. M.: Significance of Purkinje cell density in seizure suppression by chronic cerebellar stimulation. *Neurology (Minneapolis)* **26**:645–650, 1976.
132. RAPPORT, R. L., SHAW, C.-M.: Phenytoin-related cerebellar degeneration without seizures. *Ann. Neurol.* **2**:437–439, 1977.
133. RIBAK, C. E., HARRIS, A. B., VAUGHN, J. E., ROBERTS, E.: Inhibitory, GABAergic nerve terminals decrease at sites of focal epilepsy. *Science* **205**:211–214, 1979.
134. RUDZIK, A. D., MENNEAR, J. H.: Antagonism of anticonvulsants by adrenergic blocking agents. *Proc. Soc. Exp. Biol. Med.* **122**:278–280, 1966.
135. SALCMAN, M., DEFENDINI, R., CORRELL, J., GILMAN, S.: Neuropathological changes in cerebellar biopsies of epileptic patients. *Ann. Neurol.* **3**:10–19, 1978.
136. SCHLESINGER, K., STAVNES, K. L., BOGGAN, W. O.: Modification of audiogenic and pentylenetetrazol seizures with gamma-aminobutyric acid, norepinephrine and serotonin. *Psycholpharmacologia (Berlin)* **15**:226–231, 1969.
137. SCHOLZ, W.: The contribution of pathoanatomical research to the problem of epilepsy. *Epilepsy* **1**:36–55, 1959.
138. SCHREIBER, R. A., SCHLESINGER, K.: Circadian rhythms and seizure susceptibility: Effects of manipulations of light cycles on susceptibility to audiogenic seizures and on levels of 5-hydroxytryptamine and norepinephrine in brain. *Physiol. Behav.* **8**:699–703, 1972.
139. SCHULTZ, J.: Adenosine 3',5'-monophosphate in guinea pig cerebral cortical slices: Effect of benzodiazepines. *J. Neurochem.* **22**:685–690, 1974.
140. SEGAL, M., PICKEL, V., BLOOM, F.: The projections of the nucleus locus coeruleus: An autoradiographic study. *Life Sci.* **13**:817–821, 1973.
141. SELIGMANN, B., MILLER, L. P., BROCKMAN, D. E., MARTIN, D. L.: Studies of the regulation of GABA synthesis: The interaction of adenosine nucleotides

and glutamate with brain glutamate decarboxylase. *J. Neurochem.* **30**:371–376, 1978.

142. SHAYWITZ, B. A., COHEN, D. J., BOWERS, M. B.: Reduced cerebrospinal fluid 5-hydroxyindoleacetic acid and homovanillic acid in children with epilepsy. *Neurology (Minneapolis)* **25**:72–79, 1975.
143. SIGGINS, G. R., HOFFER, B. J., BLOOM, F. R.: Studies on norepinephrine-containing afferents to Purkinje cells in rat cerebellum. III. Evidence for mediation of norepinephrine afferents by cyclic 3',5'-adenosine monophosphate. *Brain Res.* **25**:535–553, 1971.
144. SIMBER, S., CIESIELSKI, L., MAITRE, M., RANDRIANARISOA, H., MANDEL, P.: Effect of sodium *n*-dipropylacetate on audiogenic seizures and brain γ-aminobutyric acid level. *Biochem. Pharmacol.* **22**:1701–1708, 1973.
145. SNIDER, S. R., SNIDER, R. S.: Phenytoin and cerebellar lesions: Similar effect on cerebral catecholamine metabolism. *Arch. Neurol.* **34**:162–167, 1977.
146. SNYDER, S. H.: Catecholamines and serotonin. In Albers, R. W., Siegel, G. J., Katzman, R., Agranoff, B. W. (eds.): *Basic Neurochemistry.* Boston, Little Brown, 1972, pp. 89–104.
147. STILK, G., SAYERS, A.: The effect of antidepressant drugs on the convulsive excitability of brain structures. *Neuropharmacology* **3**:605–609, 1964.
148. STONE, T. W., TAYLOR, D. A., BLOOM, F. E.: Cyclic AMP and cyclic GMP may mediate opposite neuronal responses in the rat cerebral cortex. *Science* **187**:845–847, 1975.
149. STONE, W. E.: Effects of alterations in the metabolism of γ-aminobutyrate on convulsant potencies. *Epilepsia* **18**:507–515, 1977.
150. STRAIN, G. M., VAN METER, W. G., BROCKMAN, W. H.: Elevation of seizure thresholds: A comparison of cerebellar stimulation, phenobarbital and diphenylhydantoin. *Epilepsia* **19**:493–504, 1978.
151. SUTHERLAND, E. W., RALL, T. W., MENON, T.: Adenyl cyclase. I. Distribution, preparation and properties. *J. Biol. Chem.* **237**:1220–1227, 1962.
152. TABADDOR, K., WOLFSON, L. I., SHARPLESS, N. S.: Ventricular fluid homovanillic acid and 5-hydroxyindoleacetic acid concentrations in patients with movement disorders. *Neurology (Minneapolis)* **28**:1249–1253, 1978.
153. TENNYSON, V. M., KREMZNER, L. T., DAUTH, G. W., GILMAN, S.: Chronic cerebellar stimulation in the monkey; Electron microscopic and biochemical observations. *Neurology (Minneapolis)* **25**:650–654, 1975.
154. TERZUOLO, C.: Influences supraspinales sur le tetanos strychnique de la moelle elineire. *Arch. Intern. Physiol.* **62**:179–196, 1954.
154a. TESTA, G., PELLEGRINI, A., GIARETTA, D.: Effects of electrical stimulation and removal of cerebellar structures in an experimental model of generalized epilepsy. *Epilepsia* **20**:447–454, 1979.
155. TOWER, D. B.: Neurochemistry of convulsive states. In Folch, P.-J. (ed.): *Chemical Pathology of the Nervous System.* London, Pergamon Press, 1961, pp. 307–344.
156. TSANG, D., LAL, S., SOURKES, T. L., FORD, R. M., ARONOFF, A.: Studies on cyclic AMP in different compartments of cerebrospinal fluid. *J. Neurol. Neurosurg. Psychiatry* **39**:1186–1190, 1976.
157. URETSKY, N. J., IVERSEN, L. L.: Effects of 6-hydroxydopamine on catecholamine-containing neurons in the rat brain. *J. Neurochem.* **17**:269–278, 1970.
158. VAN BUREN, J. M., WOOD, J. H., OAKLEY, J., HAMBRECHT, F.: Preliminary evaluation of cerebellar stimulation by double-blind stimulation and biological criteria in the treatment of epilepsy. *J. Neurosurg.* **48**:407–416, 1978.
159. VAN GELDER, N. M., COURTOIS, A.: Close correlation between changing content of specific amino acids in epileptogenic cortex of cats and severity of epilepsy. *Brain Res.* **43**:477–484, 1972.
160. VAN GELDER, N. M., SHERWIN, A. L., RASMUSSEN, T.: Amino acid content of epileptogenic human brain: Focal versus surrounding regions. *Brain Res.* **40**:385–393, 1972.
161. VAN WOERT, M. H., ROSENBAUM, D., HOWIESON, J., BOWERS, M. B.: Long-term therapy of myoclonus and other neurologic disorders with L-5-hydroxytryptophan and carbidopa. *N. Engl. J. Med.* **296**:70–75, 1977.
162. WALKER, J. E., GOODMAN, P., JACOBS, D., LEWIN, E.: Uptake and release of norepinephrine by slices of rat cerebral cortex: Effect of agents that increase cyclic AMP levels. *Neurology (Minneapolis)* **28**:900–904, 1978.
163. WALKER, J. E., LEWIN, E., MOFFITT, B.: Production of epileptiform discharges by application of agents which increase cyclic AMP levels in rat cortex. In Harris, P., Mawdsley, C. (eds.): *Epilepsy: Proceedings of the Hans Berger Centenary Symposium.* New York, Churchill Livingstone, 1975, pp. 30–36.
164. WALKER, J. E., LEWIN, E., SHEPPARD, J. R., CROMWELL, R.: Enzymatic regulation of adenosine 3',5'-monophosphate (cyclic AMP) in the freezing epileptogenic lesion of rat brain and in homologous contralateral cortex. *J. Neurochem.* **21**:79–85, 1973.
165. WEIL-MALHERBE, H., AXELROD, J., TOMCHICK, R.: Blood–brain barrier for adrenaline. *Science* **129**:1226–1227, 1959.
166. WEINBERGER, J., NICKLAS, W. J., BERL, S.: Mechanism of action of anticonvulsants: Role of the differential effects on the active uptake of putative neurotransmitters. *Neurology (Minneapolis)* **26**:162–166, 1976.
167. WENGER, G. R., STITZEL, R. E., CRAIG, C. R.: The role of biogenic amines in the reserpine-induced alteration of minimal electroshock seizure thresholds in the mouse. *Neuropharmacology* **12**:693–703, 1973.
168. WOOD, J. D.: The role of γ-aminobutyric acid in the mechanism of seizures. In Kerkut, G. A., Phillis, J. W. (eds.): *Progress in Neurobiology,* Vol. 15, Part 1. New York, Pergamon Press, 1975, pp. 79–95.

169. WOOD, J. H.: CSF HVA and 5-HIAA. *Neurology (Minneapolis)* **29**:910–911, 1979.

170. WOOD, J. H., GASESER, B. S., HARE, T. A., SODE, J., BROOKS, B. R., VAN BUREN, J. M.: Cerebrospinal fluid GABA reductions in seizure patients evoked by cerebellar surface stimulation. *J. Neurosurg.* **47**:582–589, 1977.

171. WOOD, J. H., HARE, T. A., GLAESER, B. S., BALLENGER, J. C., POST, R. M.: Low cerebrospinal fluid γ-aminobutyric acid content in seizure patients. *Neurology (Minneapolis)* **29**:1203–1208, 1979.

172. WOOD, J. H., HARE, T. A., GLAESER, B. S., BROOKS, B. R., BALLENGER, J. C., POST, R. M.: Cerebrospinal fluid GABA variations with seizure type and cerebellar stimulation in man. *Brain Res. Bull.* (in press).

173. WOOD, J. H., LAKE, C. R., ZIEGLER, M. G., SODE, J., BROOKS, B. R., VAN BUREN, J. M.: Cerebrospinal fluid norepinephrine alterations during electrical stimulation of cerebellar and cerebral surfaces in epileptic patients. *Neurology (Minneapolis)* **27**:716–724, 1977.

174. WOOD, J. H.: Neurochemical analysis of cerebrospinal fluid. *Neurology* **30**:645–651, 1980.

175. WOOD, J. H., ZIEGLER, M. G., LAKE, C. R., SODE, J., BROOKS, B. R., VAN BUREN, J. M.: Elevations in cerebrospinal fluid norepinephrine during unilateral and bilateral cerebellar stimulation in man. *Neurosurgery* **1**:260–265, 1977.

176. WRIGHT, E. M.: Factors influencing the composition of the cerebrospinal fluid. In Loeschcke, H. H. (ed.): *Acid Base Homeostasis of the Brain Extracellular Fluid and the Respiratory Control system.* Pliezhausen, Federal Republic of Germany, Georg Thieme, 1976, pp. 2–8.

177. ZANCHETTI, A., ZOCCOLINI, A.: Autonomic hypothalamic outbursts elicited by cerebellar stimulation. *J. Neurophysiol.* **17**:475–483, 1954.

178. ZIEGLER, M. G., LAKE, C. R., WOOD, J. H., BROOKS, B. R., EBERT, M. H.: Relationship between norepinephrine in blood and cerebrospinal fluid in the presence of a blood–cerebrospinal fluid barrier for norepinephrine. *J. Neurochem.* **28**:677–679, 1977.

179. ZIEGLER, M. G., LAKE, C. R., WOOD, J. H., EBERT, M. H.: Circadian rhythm in cerebrospinal fluid noradrenaline of man and monkey. *Nature (London)* **264**:656–658, 1976.

CHAPTER 20

Subarachnoid Hemorrhage

John P. Laurent

1. Introduction

Studies of the cerebrospinal fluid (CSF) can be complicated by abnormal amounts of protein products gaining access to the fluid through subarachnoid hemorrhage. Since the CSF may serve as a major conduit in the transfer of protein for neural integrations and physiological function, it is important to review the products of erythrocytes that may change the characteristics of this conduit. Erythrocytes in the subarachnoid space have been associated with cerebral vasospasm,[52,52a] hydrocephalus,[22,23,76] and aseptic meningitis.[25,29] Whether these conditions are caused by the erythrocytes themselves or by other released products remains controversial. This chapter will review published information concerning the clearance of erythrocytes from the subarachnoid space.

2. Definition of Subarachnoid Hemorrhage

The entrance of a single red blood cell into the subarachnoid space is—by definition—a subarachnoid hemorrhage. An arbitrary number of 100 erythrocytes/mm^3 of CSF has been suggested as diagnostic of clinical subarachnoid hemorrhage. These cells enter through a tear in the vascular endothelium, pass through connective tissue, and then into the CSF of the subarachnoid space. Obviously, other protein products from the blood will follow the same pathway. The etiology of the hemorrhage depends on several factors. If the initial episode of subarachnoid hemorrhage occurred in a patient less than 20 years old, the bleeding is likely to be caused by an arteriovenous malformation. In those patients beyond 30 years, the majority will have congenital (berry) aneurysms or atherosclerotic dilations. Patients between 20 and 30 years old will have a variety of etiologies for subarachnoid hemorrhage, including tumors, aneurysms, and arteriovenous malformations. The cause of subarachnoid hemorrhage remained obscure in 20% of 5000 patients in 1969.[42] Computed tomographic (CT) scanning methods have reduced this idiopathic group to 10%. In a cooperative study of subarachnoid hemorrhage, Locksley[42] showed that aneurysms had a frequency of bleeding of 50% compared to 10% for arteriovenous malformation. Hypertension, atherosclerotic disease, and infarction presented with subarachnoid bleeding in fewer than 15% of cases. Miscellaneous causes of subarachnoid hemorrhage included major and minor head trauma, bleeding diathesis, primary or metastatic tumors, collagen diseases, and bacterial vasculitides. Supratentorial lesions as a source of bleeding exceeded the combined total of the other areas.

John P. Laurent, M.D. • Division of Neurosurgery, St. Luke's–Texas Children's Hospital, Baylor University School of Medicine, Houston, Texas 77030.

3. Clinical Presentation

The clinical symptomatology of blood in the subarachnoid space can be separated into general and focal effects. Fifty percent of the patients will present with an acute generalized headache. Nuchal pain or rigidity begins within 2–5 min of the headache. Sudden loss of consciousness will occur in 20% of these patients. If the total time course of subarachnoid hemorrhage is evaluated, loss of consciousness will occur in 80%. The Botterel neurological grading system was devised to determine the prognosis and the timing of possible surgical intervention, and is based primarily on the patient's level of consciousness.[11] Confusion, vomiting, hyperpyrexia, vertigo, and extremity dysfunction will be present in 2–5% of the patients.[70] Focal neurological deficits are limited to the vascular location or compressive effects of the bleeding area. Compressive effects may be due to the size of the aneurysm itself, such as the oculomotor palsy from a posterior-internal-carotid aneurysm. Vasospastic effects can be focal or general, resulting in ischemic areas if the blood flow is reduced to a critical value.

4. Analysis of CSF

Analysis of the CSF characteristics of subarachnoid hemorrhage will depend on the location, the amount, and the etiology of the bleeding. Examination of the CSF is done after thorough consideration of the necessity for traumatic intervention into the subarachnoid space.

In view of the 50% accuracy of CT scanning in noninvasive detection of the presence of acute subarachnoid hemorrhage, the number of diagnostic lumbar punctures has decreased. When the clinical presentation does not agree with the CT findings, a lumbar puncture should be considered. Standard methods to obtain CSF are discussed by Wood in Chapter 7. However, the number of erythrocytes in the lumbar CSF does not correlate with that contained in the cephalic subarachnoid CSF, since the lumbar area seems to clear earlier.[7,25,66] The fluid will be uniformly bloodstained or pink to red in the majority of the patients at the time of the initial puncture. Occasionally, blood from a ruptured intracranial aneurysm will not extend to the lumbar subarachnoid space for 6–12 hr. Middle cerebral artery aneurysms commonly rupture intraparenchymally and may not extravasate into the subarachnoid space. CSF should normally be crystal clear when viewed from the top over a white paper and compared to water viewed in a similar manner. The fluid may contain as many as 500 erythrocytes/mm^3 and appear clear.[53] Gross evidence of blood in the hemorrhage (erythrocytes > 4000/mm^3) may disappear within 24 hr, but generally persists for 7–14 days. If cells are obviously present, the supernatant should be removed by centrifugation and inspected for the presence of color.

4.1. Traumatic and Atraumatic Subarachnoid Hemorrhage

The differentiation between a bloody traumatic puncture of the subarachnoid space and true subarachnoid hemorrhage can be determined by a number of methods. Traumatic punctures will show a decline in the number of erythrocytes as more CSF is serially drained from the spinal needle, and the proportion of leukocytes to erythrocytes will be the same as that of peripheral blood. The presence of crenated erythrocytes does not suggest an earlier bleeding episode.[37] If the supernatant exhibits xanthochromia (yellow discoloration), true subarachnoid hemorrhage can be suspected to antedate the lumbar puncture, but caution should be used in further interpretation. Xanthochromia may be seen in three other conditions. When the protein content of CSF exceeds 150 mg%, the fluid will be significantly tinged. Second, if the erythrocyte contamination is high (>1,500,000 erythrocytes/mm^3) there may be a trace of color from the plasma component of the contaminating blood.[9,65] Third, necrotic tissue with high lipid levels in the CSF may cause xanthochromia.[14] If a clot forms in the collecting vial, this is strong evidence for a traumatic puncture, since the erythrocyte count usually exceeds 200,000 cells/mm^3. True subarachnoid hemorrhage is seldom this torrential. The supernatant should be analyzed for protein. Bleeding from any source will elevate the CSF protein 1 mg/1000 erythrocytes. If the protein level exceeds this expected value, primary subarachnoid hemorrhage has likely occurred. The presence of a CSF leukocytosis suggests that the subarachnoid bleeding preceded the lumbar puncture by at least 24 hr. Recently, the appearance of fluorescein in the CSF of patients undergoing intravenous fluorescein administration during lumbar puncture was found to be a sensitive indicator of CSF contamination with systemic blood during the traumatic lumbar puncture.[8a]

4.2. Intracerebral Hemorrhage and Cerebral Infarction

Aring and Merritt[8] presented, in considerable detail, the CSF cytology in cerebral hemorrhage and thrombosis. In recent studies, further delineation of CSF pathology was accomplished.[40,47] If lumbar puncture is performed within 1 week of the onset of neurological deficits from either cerebral hemorrhage or cerebral infarction, the measurements of CSF color and number of erythrocytes are more useful than pressure protein and leukocyte values. With intracerebral hemorrhage, 75% of the patients will present with bloody or xanthochromic fluid. The CSF in cerebral infarction is never grossly bloody, but the fluid can be xanthochromic if hemorrhagic infarction has occurred (Table 1). Spectrophotome-

Table I. CSF Alterations in Cerebrovascular Disorders[a]

Disorder	Appearance	WBCs/mm^3	Protein	Comments
Traumatic puncture	Streaked, clearing No xanthochromia	2 WBCs/1000 RBCs	1 mg/1000 RBCs	Crenated RBCs not significant.
Subarachnoid hemorrhage	Bloody, 90% Xanthochromia in less than 12 hr	Initially proportional to RBCs Leukocytosis in 24–48 hr	Initially proportional to RBCS Increased by 24–48 hr	Oxyhemoglobin and bilirubin content determined spectrophoto-metrically.
Intracerebral hemorrhage	Bloody, Xanthochromia	Initially proportional to RBCs Leukocytosis in 48–72 hr Confirmation with CT scan	Increased	None
Cerebral thrombosis	Normal	Normal, increased in 30% Leukocytosis in 48–72 hr	Normal, increased in 40%	Hemorrhagic infarction may produce bloody CSF.
Subdural hematoma	Normal; bloody if brain is lacerated	Normal	May increase	Methemoglobin may be present in CSF.

[a] References: 8, 40, 44, 46, 53, 57, 62, 64, 68, 73; abbreviations: (RBCs) erythrocytes; (WBCs) leucocytes; (CSF) cerebrospinal fluid; (CT) computed tomography.

tric analysis of the CSF approximately one week following the onset of neurologic symptoms increases the accuracy of detecting intracerebral hematomas or hemorrhagic infarctions to greater than 95%.[50a,61a]

4.3. Aseptic Meningitis

Erythrocytes initiate an irritative response of the meninges.[25,29] Normally, 2 leukocytes/1000 erythrocytes are added to the CSF with traumatic lumbar punctures. With subarachnoid hemorrhage, an inflammatory response is initiated within 24–48 hr, and the leukocyte count will rise above this expected value.[74] Hammes[25] described a leukocytosis within 2 hr of subarachnoid hemorrhage. Richardson and Hyland[55] showed a relative lymphocytosis on the third to fifth day following subarachnoid hemorrhage while the neutrophil count was falling. The decrease in neutrophils is theoretically ascribed to adsorption in clots in the fissures or to their involvement in phagocytosis. Jackson[29] presented an elegant demonstration that the supernatant of lysed erythrocytes caused an intense cellular reaction in the CSF exceeding 3000 white blood cells/mm^3. Jackson[29] further reported that both oxyhemoglobin and methemoglobulin caused this severe meningeal reaction. Hemolyzed erythrocytes (free hemoglobin) elicited a greater cellular response than fresh whole blood, but less than degenerated old blood or oxyhemoglobin. The marked leukocytosis may be confused with meningitis.[48]

4.4. Xanthochromia

Xanthochromia is the term used to describe the color of CSF when lysis of erythrocytes has occurred. The pigments give CSF a multicolored appearance seen in subarachnoid hemorrhage: yellow, bilirubin; red, oxyhemoglobin; brown, methemoglobin. The quantity can be determined spectrophotometrically.[9,28,38] Kjellin and Steiner[36] described their utilization of spectrophotometry to detect xanthochromia in CSF originally judged to be colorless. They divided the spectrophotometry into the H-pattern for subdural hematoma and the S-pattern for patients with subarachnoid hemorrhage. Spectrophotometric analysis of the CSF pigment profile is important in the diagnosis of subarachnoid hemorrhage or hemorrhagic infarction in patients whose ictus has occurred more than one week prior to lumbar puncture, whose CSF leukocytosis has cleared, and whose CT scan is negative.[50a] CT and CSF spectrophotometry have been compared in patients with cerebrovascular disorders in whom the presence or absence of brain parenchymal bleeding was correctly diagnosed in 97% by spectrophotometry but only in 65% by CT.[61a] Biochemical methods can also be used to determine the presence of each pigment: bilirubin (positive in the Van den Bergh test), oxyhemoglobin (positive in the benzidine test), and methemoglobin (positive in the potassium cyanide test).[3]

Xanthochromia appears in the supernatant within 2 hr in a few cases, within 6 hr in 70% of the cases, and within 12 hr in 90% of the cases after onset of clinical symptoms.[9,38] The initial pigment is oxyhemoglobin, and its red color becomes maximal within 2–3 days. This color diminishes over the next 7–10 days. Oxyhemoglobin converts to bilirubin by enzymatic destruction of the hemoglobin molecule. Bilirubin is the iron-free derivative of hemoglobin (Fig. 1). Enzymes necessary for oxyhemoglobin reduction are present in macrophages, the arachnoid, and the choroid plexus.[54,56] Although the initial subarachnoid hemorrhage releases both bound bilirubin and unbound bilirubin into the subarachnoid space, the new bilirubin (yellow color in the CSF) appears 2–4 days after the hemorrhage and may persist for 2–4 weeks. Methemoglobin, the ferric form of iron in hemoglobin, is found when the blood has been encapsulated (subdural hematoma), and is not usually found in subarachnoid hemorrhage.

5. Clearance of Erythrocytes from Subarachnoid Space

Clearance of erythrocytes from the subarachnoid space can be accomplished by three basic methods. Lysis of erythrocytes with subsequent phagocytosis

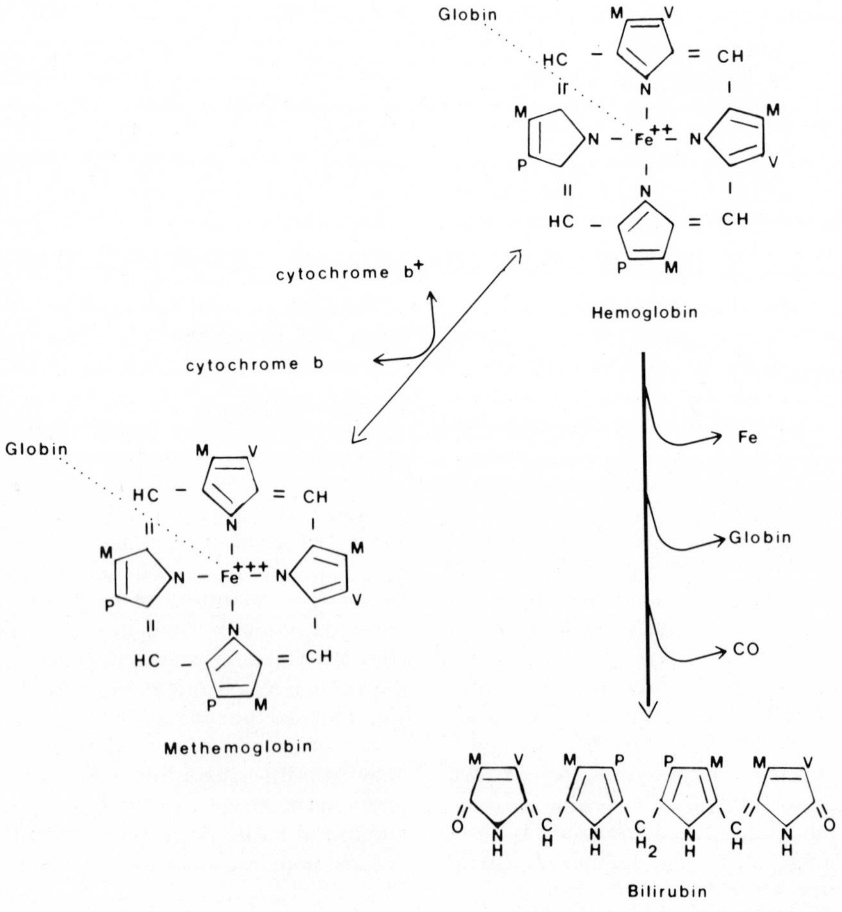

Figure 1. Hemoglobin degradation products in CSF. (M) Methyl; (P) propionyl; (V) vinyl.

accounts for the major pathway of removal of erythrocytes from the subarachnoid space.[1,10,15,67] The hemolyzed products (see above) have significant physiological effects (vasospasm, meningitis, hydrocephalus). Erythrocytes may remain intact for 16 hr after hemorrhage into the subarachnoid space.[7,25] However, most observers agree that hemolysis begins within 24 hr of hemorrhage. A small number of swollen and ghost forms of erythrocytes can be seen within a few hours of hemorrhage. Maximal hemolysis is seen on the 5th to 7th day posthemorrhage. Sprang[63] hypothesized that as early as 1 hr after ictus, the erythrocytes become entangled in a clot or the meshes of the arachnoid trabeculae and are evacuated by lysis and phagocytosis. These enmeshed, clumped erythrocytes are invaded by macrophages, engulfed, and released the hemolyzed hemoglobin pigments many days after the hemorrhage.[15] Approximately 75% of subarachnoid erythrocytes will become enmeshed in the arachnoid villi and disposed of in this manner.

From experiments with dogs and rabbits, lymphatic type "pathways" have been postulated as channels to remove erythrocytes from the subarachnoid space.[12,32,45,60] The olfactory route via the meningeal–arachnoid sheath shows accumulation of labeled erythrocytes injected into the subarachnoid space. However, these lymphatic channels and olfactory accumulations of labeled erythrocytes have not been seen in monkeys or man, and probably are insignificant.

Simmond[60,61] stated that whole erythrocytes were returned to the bloodstream through the arachnoid villi, with phagocytosis and lysis involved in a minor role. Courtice and Simmond[13] in 1951 further concluded that plasma proteins were eliminated from the subarachnoid space in much the same manner. They have demonstrated the transfer of 2–60% of whole erythrocytes from the subarachnoid space to the bloodstream. Whole tagged erythrocytes (chromium-51) have been found in the peripheral blood within 3 min after being deposited in the subarachnoid space.[17,20,33,69,75] It is estimated that 25% of whole erythrocytes in the CSF will be absorbed directly into the blood stream.[1] Welch and co-workers[71,72] demonstrated the presence of tubular spaces in the arachnoid villi and postulated the movement of erythrocytes through these channels. This theory, although supported by others,[24,27,30,34] has not been substantiated morphologically.[58,59]

Hydrostatic pressures and distention of arachnoid granulations can cause increased movement of CSF across the arachnoid membrane.[4–6] Low pressures decrease the transfer of cells into the bloodstream.[47] The movement of erythrocytes across this villus membrane propelled by hydrostatic pressure is countered by the blockage of these small passages by erythrocytes and subsequent inflammatory response. The blockage and entrapment of erythrocytes may explain the clinical syndrome of "normal-pressure" hydrocephalus.[2,8,18,23,26]

6. Hydrocephalus

Of patients with subarachnoid hemorrhage, 10% will develop chronic communicating hydrocephalus.[2,76] Obstruction to the normal flow of CSF is prevented by the blockage of the arachnoid structures by fibrosis of the leptomeninges and the irritative aseptic meningitis.[35,51] Since this pia-arachnoid reaction does not develop for 10 days after subarachnoid hemorrhage, the onset of acute communicating hydrocephalus is thought to be caused by obstruction of the basilar cisterns by the erythrocytes themselves.[39] Ventricular and spinal lavage have been attempted in both clinical and experimental situations to remove subarachnoid erythrocytes. These irrigations have been uniformly unsuccessful, with recovery of less than 10% of the subarachnoid erythrocytes.[33,43,63]

7. Computed Tomographic Scanning in Subarachnoid Hemorrhage

The ability of CT scans to accurately map the relative attenuation coefficients and predict the composition of different pathological or anatomical structures makes available a completely new approach to the evaluation of subarachnoid hemorrhage. Many times, the lumbar puncture will not be necessary (see above) to detect subarachnoid blood. High attenuation coefficients are seen with fresh hemorrhage or fresh clot formation and facilitate their identification. A linear relationship between the attenuation and increasing hemoglobin (24–82 Housfield units) suggests that CT is capable of determining the severity and intensity of subarachnoid hemorrhage, which until now were only inferential.[31,41,49,50] Attempts to correlate the protein of the CSF have been unreliable.

Patients presenting with signs and symptoms of subarachnoid hemorrhage should undergo immediate CT scanning. The presence of subarachnoid blood can be detected in 50% of proven cases.[21] Frequently, the site (86%) and type of lesion can be detected with the scan.[31,41] When the scan is not diagnostic, CSF must be obtained for cytologic, and, if possible, spectrophotometric analysis. If the patient with subarachnoid hemorrhage has a CT scan within 48 hr of the ictus, more than 50% will show blood in at least one cistern. If the scan is performed 3–4 days later, approximately 25% will show blood, while those scans completed at greater than 5 days will not show subarachnoid blood. It is possible to have bloody CSF without a positive cranial CT scan (i.e., cases of bleeding from spinal origin).[19] Bleeding sites in cases of multiple aneurysms can occasionally be identified with CT scanning. Subarachnoid blood will frequently be seen near the falx region following head injury.[16]

8. Summary

Erythrocytes are normally absent in the CSF, and their presence (erythrocytes greater than 500 cells/mm^3) suggests subarachnoid hemorrhage. Sophisticated knowledge of the time course of both the cellular and chemical alterations in CSF following an ictus in the central nervous system is essential in making a differential diagnosis and in formulating therapy such as anticoagulation or antifibrinolysis. The etiologies and complications of subarachnoid hemorrhage are numerous. CT scanning has contributed greatly to their identification; however, the resulting physiological and biochemical changes may not be readily explainable. Clearance of erythrocytes remains a major area of controversy in respect to the contribution of erythrocyte breakdown products to the development of vasospasm and hydrocephalus.

References

1. Adams, J. E., Prawirohardjo, S.: Fate of red blood cells injected into cerebrospinal fluid pathways. *Neurology* **9:**561–564, 1959.
2. Adams, R. D., Fisher, C. M., Hakin, S., Ojemann, R. G., Sweet, W. H.: Symptomatic occult hydrocephalus with normal cerebrospinal fluid pressure: A treatable syndrome. *N. Engl. J. Med.* **273:**117–126, 1956.
3. Alajouanine, T., Thurel, R., Durupt, T.: Cerebral hemorrhage and quantitation of the elements of the blood. *Rev. Neurol.* **78:**617–618, 1946.
4. Alksne, J. F., Lovings, E. T.: The role of arachnoid villus in the removal of red blood cells from the subarachnoid space: An electron microscope study in the dog. *J. Neurosurg.* **36:**192–200, 1972.
5. Alksne, J. F., Lovings, E. T.: Functional ultrastructure of arachnoid villus. *Arch. Neurol.* **27:**371–377, 1972.
6. Alksne, J. F., White, L. E.: Electron microscopic study of the effects of increased intracranial pressure on the arachnoid villus. *J. Neurosurg.* **22:**481–488, 1965.
7. Alpers, B. J., Forster, F. M.: The repairative processes in subarachnoid hemorrhage. *J. Neuropathol. Exp. Neurol.* **4:**262–268, 1945.
8. Aring, C. D., Merritt, H. H.: Differential diagnosis between cerebral hemorrhage and cerebral thrombosis. *Arch. Intern. Med.* **56:**435–456, 1935.

8a. Barnhart, B. J., Lace, J. K., Yount, J. E.: Diagnosis of intracranial hemorrhage: Technique using fluorescein. *J. Pediatr.* **95:**289–292, 1979.

9. Barrows, L. J., Hunter, F. T., Banker, B. O.: The nature and clinical significance of pigments in the cerebrospinal fluid. *Brain* **78:**59–80, 1955.
10. Bagley, C.: Blood in the cerebrospinal fluid: Resultant functional and organic alteration in the central nervous system. A. Experimental data. B. Clinical data. *Arch. Surg.* **17:**18–81, 1928.
11. Botterell, E. H., Lougheed, W. M., Scott, J. W., Vanderwater, S. L.: Hydrothermia and interruption of carotid or carotid and vertebral circulation in the surgical treatment of intracranial aneurysms. *J. Neurosurg.* **13:**1–42, 1956.
12. Bradford, F. K., Johnson, P. C.: Passage of intact iron-labelled erythrocytes from the subarachnoid space to systemic circulation in dogs. *J. Neurosurg.* **19:**332–336, 1962.
13. Courtice, F. C., Simmond, W. J.: Removal of protein from the subarachnoid space. *Aust. J. Exp. Biol. Med. Sci.* **29:**255–263, 1951.
14. Crosby, R. M. N., Weiland, G. L.: Xanthochromia of the cerebrospinal fluid. *Arch. Neurol. Psychiatry* **69:**732–736, 1953.
15. Crompton, M. R.: The pathogenesis of cerebral infarction following the rupture of cerebral berry aneursyms. *Brain* **87:**491–510, 1964.
16. Dolinskas, C. A., Zimmerman, R. A., Bilaniuk, L. T.: A sign of subarachnoid bleeding on cranial computed tomograms of pediatric head trauma patients. *Radiology* **126:**409–411, 1978.
17. Dupont, J. R., Van Wart, C. A., Kraintz, L.: The clearance of major components of whole blood from cerebrospinal fluid following simulated subarachnoid hemorrhage. *J. Neuropathol. Exp. Neurol.* **20:**450–455, 1961.
18. Ellington, E., Margolis, G.: Block of the arachnoid villus by subarachnoid hemorrhage. *J. Neurosurg.* **30:**651–657, 1969.
19. Fincher, E. F.: Spontaneous subarachnoid hemorrhage in intradural tumors of the lumbar sac: Clinical syndrome. *J. Neurosurg.* **8:**576–584, 1951.

20. FISHMAN, R. A.: Exchange of albumin between plasma and cerebrospinal fluid. *Am. J. Physiol.* **175:**96–105, 1953.
21. FLEISCHER, A. S., TINDALL, G. T.: Preoperative management of ruptured cerebral aneurysms. *Contemp. Neurosurg.* **1:**1–6, 1978.
22.. FOLTZ, E. L., WARD, A. A.: Communicating hydrocephalus from subarachnoid bleeding. *J. Neurosurg.* **13:**546–566, 1956.
23. GALERA, R., GREITZ, T.: Hydrocephalus in the adult secondary to rupture of intracranial arterial aneurysms. *J. Neurosurg.* **32:**634–641, 1970.
24. GOMEZ, D. G., POTTO, G., PEONARINE, V.: Arachnoid granulations of the sheep: Structural and ultrastructural changes with varying pressure differences. *Arch. Neurol.* **30:**169–175, 1974.
25. HAMMES, E. M.: Reaction of meninges to blood. *Arch. Neurol. Psychiatry* **52:**505–508, 1944.
26. HAKIM, S., ADAMS, R. D.: The special clinical problem of symptomatic hydrocephalus with normal cerebrospinal fluid pressure. *J. Neurol. Sci.* **7:**481–493, 1965.
27. HAYES, K. C., MCCOMBS, H. L., FAHERTY, T. P.: The fine structure of vitamin A deficiency. II. Arachnoid granulations and cerebrospinal fluid pressure. *Brain* **94:**213–224, 1971.
28. HELLSTRÖM, B., KJELLIN, K. G.: The diagnostic value of spectrophotometry of the CSF in the newborn period. *Dev. Med. Child Neurol.* **13:**789–797, 1971.
29. JACKSON, I. J.: Aseptic hemogenic meningitis: Experimental study of aseptic meningeal reactions due to blood and its breakdown products. *Arch. Neurol. Psychiatry* **62:**572–575, 1949.
30. JAYATILAKA, A. D. P.: An electron microscopic study of arachnoid granulations. *J. Anat.* **99:**635–649, 1965.
31. KENDALL, B. E., LEE, B. C. P., CLAVERIA, E.: Computerized tomography and angiography in subarachnoid hemorrhage. *Br. J. Radiol.* **49:**483–501, 1976.
32. KENNADY, J. C.: Investigations of the early fate and removal of subarachnoid blood. *Pac. Med. Surg.* **75:**163–168, 1967.
33. KENNADY, J. C.: Early fate of subarachnoid blood and removal by irrigation. *Trans. Am. Neurol. Assoc.* **91:**265–267, 1966.
34. KEY, A., RETZIUS, G.: *Studien in der Anatomie des Nerven Systems und des Bindegewebes,* Hälfte 2, Abt. 1. Stockholm, Samson and Wallin, 1876.
35. KIBLER, R. F., COUCH, R. S. C., CROMPTON, M. R.: Hydrocephalus in the adult following spontaneous subarachnoid hemorrhage. *Brain* **84:**45–61, 1961.
36. KJELLIN, K. G., STEINER, L.: Spectrophotometry of cerebrospinal fluid in subacute and chronic subdural haematomas. *J. Neurol. Neurosurg. Psychiatry* **37:**1121–1127, 1974.
37. KRIEG, A. F.: Cerebrospinal fluid and other body fluids and secretions. In Davidsohn, I., Henry, J. B. (eds.): *Clinical Diagnosis.* Philadelphia, W. B. Saunders, 1969, pp. 1161–1169.
38. KRONHOLM, V., LINTRUP, S.: Spectrophotometric investigations of the cerebrospinal fluid in near-ultraviolet region: A possible diagnostic aid in diseases of the central nervous system. *Acta Psychiat. Scand.* **35:**314–329, 1960.
39. KUSSKE, J. A., TURNER, P. T., OJEMANN, G. A., HARRIS, A. B.: Ventriculostomy for the treatment of acute hydrocephalus following subarachnoid hemorrhage. *J. Neurosurg.* **38:**591–595, 1973.
40. LEE, M. C., HEANEY, L. M., JACOBSON, R. L.: Cerebrospinal fluid in cerebral hemorrhage and infarction. *Stroke* **6:**638–641, 1975.
41. LILIEQUIST, B., LINDQUIST, M., VALDIMARSSON, E.: Computed tomography and subarachnoid hemorrhage. *Neuroradiology* **14:**21–26, 1977.
42. LOCKSLEY, H. B.: Natural history of subarachnoid hemorrhage, intracranial aneurysms, and arteriovenous malformation. Part II. In Sahs, A. L., Perret, G. E., Locksley, H. B., Nishioka, H. (eds.): *Intracranial Aneurysms and Subarachnoid Hemorrhage: A Cooperative Study.* Philadelphia, J. B. Lippincott, 1969, pp. 58–108.
43. MEREDITH, J. M.: The efficacy of lumbar puncture for removal of red blood cells from the cerebrospinal fluid. *Surgery* **9:**524–533, 1941.
44. MCMENEMEY, W. H.: The significance of subarachnoid bleeding. *Proc. R. Soc. Med.* **47:**701–704, 1954.
45. MCQUEEN, J. D., NORTHRUP, B. E., LEIBROCK, L. G.: Arachnoid clearance of red blood cells. *J. Neurol. Neurosurg. Psychiatry* **37:**1316–3121, 1974.
46. MOLLE, W. E.: Leukocytosis in the cerebrospinal fluid in cerebral hemorrhage. *Ohio State Med. J.* **38:**325–327, 1942.
47. MORTENSEN, O. A., WEED, L. H.: Absorption of isotonic fluids from the subarachnoid space. *Am. J. Physiol.* **108:**458–468, 1934.
48. MYOUNG, C. L., HEANEY, L. M., JACOBSON, R. L., KLASSER, A. C.: Cerebrospinal fluid in cerebral hemorrhage and infarction. *Stroke* **6:**638–641, 1975.
49. NEW, P. F. J., SCOTT, W. R., SEHNAR, J. A.: Computerized axial tomography with the EMI scanner. *Radiology* **110:**109–123, 1974.
50. NORMAN, D., PRICE, D., BOYD, D., FISHMAN, R., NEWTON, T. H.: Quantitative aspects of computed tomography of the blood and cerebrospinal fluid. *Radiology* **123:**335–338, 1972.
50a. NORRVING, B., OLOSSON, J.-E.: The diagnostic value of spectrophotometric analysis of the cerebrospinal fluid in cerebral hematomas. *J. Neurol. Sci.* **44:**105–114, 1979.
51. OJEMANN, R. G.: Normal pressure hydrocephalus. In Tindall, G. T., (ed.): *Clinical Neurosurgery,* Vol. 18. Baltimore, Williams and Wilkins, 1971, pp. 33–370.
52. OSAKA, K.: Prolonged vasospasm produced by the breakdown products of erythrocytes. *J. Neurosurg.* **47:**403–441, 1977.
52a. OSAKA, K., HANDA, H., OKAMOTO, S.: Subarachnoid erythrocytes and their contribution to cerebral vasos-

pasm. In Wood, J. H. (ed.): *Neurobiology of Cerebrospinal Fluid I.* New York, Plenum Press, 1980.
53. PATTEN, B. M: How much blood makes cerebrospinal fluid bloody? *J. Am. Med. Assoc.* **206**:378, 1968.
54. PIMSTONE, N. R., TENHUNEN, R., SEITZ, P. T., MARVER, H. S., SCHMID, R.: The enzymatic degradation of hemoglobin to bile pigments by macrophages. *J. Exp. Med.* **133**:1264–1281, 1971.
55. RICHARDSON, J. C., HYLAND, H. H.: Intracranial aneurysms: Clinical and pathological study of subarachnoid and intracerebral hemorrhage caused by berry aneurysms. *Medicine (Baltimore)* **20**:1–83, 1941.
56. ROOST, K. T., PIMSTONE, N. A., DIAMOND, I., SCHMID, R.: The formation of cerebrospinal fluid xanthochromia after subarachnoid hemorrhage: Enzymatic conversion of hemoglobin to bilirubin by the arachnoid and choroid plexus. *Neurology* **22**:973–977, 1972.
57. SCHAAFSMA, S.: The differential diagnosis between cerebral hemorrhage and infarction. *J. Neurol. Sci.* **7**:83–95, 1968.
58. SHABO, A. L., MAXWELL, D. S.: The morphology of the arachnoid villi: A light and electron microscopic study in the monkey. *J. Neurosurg.* **29**:451–463, 1968.
59. SHABO, A. L., MAXWELL, D. S.: Electron microscopic observations on the fate of particulate matter in the cerebrospinal fluid. *J. Neurosurg.* **29**:464–474, 1968.
60. SIMMOND, W. J.: The absorption of labelled erythrocytes from the subarachnoid space in rabbits. *Aust. J. Exp. Biol. Med. Sci.* **31**:77, 1953.
61. SIMMOND, W. J.: Absorption of blood from cerebrospinal fluid in animals. *Aust. J. Exp. Biol. Med. Sci.* **30**:261–270, 1953.
61a. SÖDERSTRÖM, C. E.: Diagnostic significance of CSF spectrophotometry and computer tomography in cerebrovascular disease. *Stroke* **8**:606–612, 1977.
62. SORNAS, R., OSTLUND, H., MULLEN, R.: Cerebrospinal fluid cytology after stroke. *Arch. Neurol.* **26**:489–501, 1972.
63. SPRANG, W.: Disappearance of blood from cerebrospinal fluid in traumatic subarachnoid hemorrhage: Ineffectiveness of repeated lumbar punctures. *Surg. Gynecol. Obstet.* **58**:705–708, 1934.
64. TOURTELLOTTE, W. W., SOMERS, J. F., PARKER, J. A.: Study of traumatic lumbar punctures. *Neurology* **8**:129–134, 1958.
65. TOURTELLOTTE, W. W., QUAN, K. C., HAERER, A. F., BRYAN, E. R.: Neoplastic cells in the cerebrospinal fluid. *Neurology* **13**:866–876, 1963.
66. TOURTELLOTTE, W. W., METZ, L. N., Bryan, E. R.: Spontaneous subarachnoid hemorrhage: Factors affecting rate of clearing of the cerebrospinal fluid. *Neurology* **14**:301–306, 1964.
67. TOURTELLOTTE, W. W., SIMPTON, J. F., METZ, L. N., BRYAN, E. R.: Intracranial hemorrhage and cerebrospinal fluid. In Fields, W. S., Sahs, A. L. (eds.): *Intracranial Aneurysms and Subarachnoid Hemorrhage.* Springfield, Illinois, Charles C. Thomas, 1965, pp. 85–95.
68. TOWSEND, S. R., CRAIG, R. L., BRAUNSTEIN, A. L.: Neutrophilic leukocytosis in spinal fluid associated with cerebral vascular accidents. *Arch. Intern. Med.* **63**:848–857, 1939.
69. VANWART, C. A., DUPONT, J. R., KRAINTZ, L.: Transfer of radioiodinated human serum albumin (RIHSA) from cerebrospinal fluid to blood plasma. *Proc. Soc. Exp. Biol. Med.* **103**:708, 1960.
70. WALTON, J. N.: *Subarachnoid Hemorrhage.* Edinburgh, Scotland, E & S Livingstone, 1956.
71. WELCH, K., POLLAY, M.: Perfusion of particles through arachnoid villi of the monkey. *Am. J. Physiol.* **201**:651–654, 1960.
72. WELCH, K., FRIEDMAN, V.: The cerebrospinal fluid values. *Brains* **83**:454–469, 1960.
73. WESTLAKE, P. T., MARKOVITS, C. T., STELLAR, S.: Cytologic evaluation of cerebrospinal fluid with clinical and histologic correlation. *Acta Cytol.* **16**:224–239, 1972.
74. WILKINSON, H. A., WILSON, R. B., PATEL, P. P., ESMALI, M.: Corticosteroid therapy of experimental hydrocephalus after intraventricular subarachnoid hemorrhage. *J. Neurol. Neurosurg. Psychiatry* **37**:224–229, 1974.
75. WOLLARD, H. H.: Vital staining of leptomeninges. *J. Anat.* **58**:87, 1923.
76. YASARGIL, M. G., YONEKAWA, Y., ZUMSTEIN, B., STAHL, H. J.: Hydrocephalus following spontaneous subarachnoid hemorrhage: Clinical features and treatment. *J. Neurosurg.* **39**:474–479, 1973.

CHAPTER 21

Effects of Subarachnoid Blood and Spasmodic Agents on Cerebral Vasculature

Ronald J. Cohen and George S. Allen

1. Introduction

Cerebral arterial spasm, which is an abnormally severe constriction of the larger cerebral arteries, remains a major source of morbidity and mortality in patients who suffer a subarachnoid hemorrhage from the rupture of a cerebral aneurysm, an abnormal outpouching due to a congenital weakness in the wall of one of the larger cerebral arteries at the base of the brain. Deprivation of blood flow, caused by more severe degrees of cerebral arterial spasm, can cause ischemia in a brain that has already been injured by the subarachnoid hemorrhage. In addition, the poor neurological condition of many patients with cerebral arterial spasm often causes delays in the definitive surgical treatment of the aneurysm. With these delays comes the risk of rebleeding before the aneurysm can be obliterated. The successful treatment of cerebral arterial spasm, then, is directed at both the prevention of ischemic neurological deficits and the earlier surgical therapy of the cerebral aneurysm.

The importance of constriction of the cerebral arteries following subarachnoid hemorrhage from intracranial aneurysm rupture was noted in a report by Robertson[33] in 1949. Ecker and Riemenschneider,[13] in 1951, described in detail the general angiographic characteristics of cerebral vasospasm in patients with subarachnoid hemorrhage from ruptured aneurysms, and since then it has become established that in the human, cerebral arterial spasm is often a diffuse, prolonged process. Most important, the close association of vasospasm with rupture of an intracranial aneurysm was confirmed.

In a retrospective study of patients with subarachnoid hemorrhage from ruptured aneurysm, Allcock and Drake[1] found that the incidence of angiographic cerebral arterial spasm in patients with ruptured intracranial aneurysms was 40%, and that it was the major determining factor in clinical outcome. In an analysis of the records of 50 consecutive patients with subarachnoid hemorrhage from an angiographically demonstrated intracranial aneurysm, Fisher *et al.*[15] found that 50% of the cases developed a delayed, new neurological deficit (strokelike syndrome) that in each case was ischemic in origin, and 50% of the patients remained free of any sign of a delayed, new neurological deficit. In each case in which a delayed ischemic deficit occurred, the deficit corresponded to the site of worst vasospasm on the angiogram, usually in the corresponding middle cerebral distribution. The delayed deficit occurred between the 4th and 16th days after the onset of the subarachnoid hemorrhage, with a peak incidence on day 8. All patients who developed ischemic deficits had severe vasospasm, and none of the patients with milder degrees of arterial spasm on angiography developed a neurological deficit. On the basis of these results, the authors concluded that severe vasospasm "appears to be the only

Ronald J. Cohen, M.D., and **George S. Allen, M.D., Ph.D.** • Department of Neurosurgery, The Johns Hopkins University School of Medicine and Hospital, Baltimore, Maryland 21205.

cause" of delayed ischemic deficits after a subarachnoid hemorrhage. In a comment following this paper, Peerless reported an incidence of "significant" angiographic vasospasm of 39% in a prospective group of patients with subarachnoid hemorrhage, but stated that in only 22% of the patients in the whole series was the arterial narrowing thought to be the major factor in the production of the neurological deficit.

The time of onset of the cerebral arterial spasm after aneurysm rupture can make early surgical intervention difficult. In the series of Fisher *et al.*,[15] the peak incidence of the onset of the delayed ischemic deficit occurred on day 8 after the subarachnoid hemorrhage, with a range from the 4th to the 16th day. Since it is impractical to subject patients to daily angiograms, it is impossible to state with certainty the exact course of the angiographic spasm, particularly since lesser degrees of arterial constriction were commonly asymptomatic. Therefore, one can be relatively certain only about the course of the more severe degrees of arterial narrowing, which Fisher and co-workers[15] found occurred most commonly about 1 week after hemorrhage. The Cooperative Aneurysm Study[26] reported that in 471 patients with intracranial aneurysms who were treated with antifibrinolytic medication, 12.7% rebled, with most of the rebleeds occurring between the 6th and 11th days following the initial hemorrhage, paralleling closely the highest incidence of symptoms from arterial spasm. Thus, surgical therapy sometimes must be delayed (because of spasm) just at the time when the probability of rebleeding from the aneurysm is greatest.

The actual cause of cerebral arterial spasm has been a source of controversy for many years. The larger cerebral arteries at the base of the brain normally communicate with each other in an efficient anastomotic circuit known as the circle of Willis. The distances separating the arteries from each other are often minimal, usually less than a few centimeters, and the common pool of cerebrospinal fluid (CSF) that bathes all the larger vessels offers a means by which chemical factors influencing one vessel can spread to influence other vessels, even across the midline, from one side of the brain to the other. It is the purpose of this chapter to examine some of the more important landmarks in our developing theories about the causes of cerebral arterial spasm and to suggest how these ideas might be used in the prevention and treatment of patients with cerebral arterial spasm.

2. *In Vivo* Model of Cerebral Arterial Spasm

An intrinsic controlling mechanism for the cerebral circulation was suggested by Roy and Sherrington[35] as early as 1890, and Bayliss[7] postulated such an internal controlling mechanism in 1902. An early attempt to establish a reproducible experimental model of artificially induced cerebrovascular constriction was made in 1925 by Florey,[16] who used both mechanical and electrical stimulation to produce marked constriction of the pial arteries of the exposed cerebral hemisphere of the cat. In 1942, Echlin[11] confirmed these observations in 30 cats, 1 dog, and 4 monkeys. Direct mechanical and electrical stimulation of the cerebral arteries produced vasoconstriction significant enough to cause histopathological changes of focal cerebral infarction. From the results of these experiments, Echlin drew three important conclusions: (1) There appear to be species differences in the susceptibility to vasospasm generated either by stretching the arteries or by electrical stimulation, with the cat most vulnerable, then the dog, and the monkey least susceptible. (2) The spasm produced by mechanical or electrical stimulation is short-lived (several seconds to 12 min in the cats) and never propogated; i.e., it was always a local phenomenon at the site of stimulation. (3) Focal ischemic necrosis of the brain could be produced in the cats if repeated stimuli were delivered to vessels supplying that area, and the areas of focal ischemia were often wedge-shaped corresponding to the arterial supply. The constriction induced by mechanical and electrical stimulation was independent of any neurovascular mechanism, since the changes were local, nonpropogated, and brief. In one cat the pial vasculature reacted the same before and during cervical sympathetic stimulation.

In a series of experiments on the cortical vessels of 34 cats, 8 dogs, 3 monkeys, and 1 guinea pig, Lende[23] was able to reproduce the brief (up to 30 min) arterial spasms described by Echlin, by either mechanical stimulation with a nerve hook or gauze sponge or electrical stimulation with a bipolar electrode (9-V current). In agreement with Echlin, Lende found that the monkey's vessels were least reactive of all the species tested, and that the constriction, when produced, was always maximal at first, then gradually dissipated. None of the constrictions was prolonged or propogated. In confirmation of Echlin's opinion that the constrictions were independent of neural influence, Lende showed that the local arterial response occurred even in the presence of local anesthetic and acute midcervical sympathectomy (unilateral and bilateral). Finally, Lende studied the effects of a number of stimulatory and inhibitory drugs on the constricted cerebral artery, and found that the α-adrenergic blocking agent phentolamine was the most effective drug tested in releasing the spasm induced by mechanical or electrical stimuli. Furthermore, he showed that while topical application of phentolamine to the adventitial surface of the artery was highly successful in relieving the spasm, intracarotid injection of the drug failed to relieve the spasm, thus focusing attention on the qualitative differences of these two routes of administration. Although Lende did not mention the blood–brain barrier, we now consider this differential effect of phentolamine to be due to its inability to pass from the circulation to the CSF in sufficient concentration to exert its effect on the smooth-muscle-cell receptors of the cerebral arteries. Similar differences in the effects of other drugs have been shown for these two routes of administration.

In an attempt to mimic as closely as possible the rupture of an intracranial aneurysm in humans, Simeone *et al.*[36] reported that prolonged vasospasm could be produced in monkeys by a 30-gauge needle puncture and brief bleeding of the middle cerebral artery near its origin. These authors found that while only brief periods (about 10 min) of localized vasospasm could be produced by injecting blood into the subarachnoid space of rhesus monkeys, prolonged and diffuse spasm followed the simultaneous puncture of the intracranial internal carotid artery or one of its branches with a 30-gauge needle and bleeding from that puncture site. Increased vasoconstriction was often noted on angiograms several hours after the puncture. Angiography 4 days later demonstrated even more intense and diffuse spasm. By 7 days later, the vessel caliber had returned almost to normal. Of 48 monkeys, 65% were shown to have developed prolonged vasospasm in a similar manner, thus providing an early reproducible animal model. In 1968, Echlin[12] used a series of monkeys to try to differentiate between mechanical factors and subarachnoid bleeding in the pathogenesis of cerebral arterial spasm. He showed that by bloodlessly transfixing the basilar artery with a 30-gauge needle, a brief local, nonpropogated spasm could be produced. If, however, 4 ml of fresh arterial blood was injected at the level of the foramen magnum or C_1 (3–4 cm from the arteries at the base of the brain) through a catheter placed in the high anterior cervical subarachnoid space of otherwise normal monkeys, marked diffuse spasm of the basilar, middle, and anterior cerebral arteries resulted, usually lasting 10–30 min, sometimes longer. If needle puncture of the artery was followed by diffuse bleeding from the puncture site, a "marked local and widespread constriction of the intracranial arteries" followed, which persisted in some cases for at least 2 hr after the subarachnoid hemorrhage. Unfortunately, all the animals expired during the night before delayed angiograms could be performed. Echlin concluded from these experiments that while the pathogenesis of local spasm around an aneurysmal rupture might be due, in part, to mechanical factors, the diffuse, prolonged spasm characteristic of the human chronic condition was primarily related to the presence in blood of some "vasoconstrictor substance" that, when in contact with the adventitial surface of the cerebral arteries, produced arterial constriction.

The existence of certain factors in blood that can cause vasoconstriction when applied to the pial vasculature is generally accepted, but the role of these substances in the pathogenesis of human vasospasm following subarachnoid hemorrhage remains controversial. In 1944, Marjorie Zucker[46] was able to isolate from the buffy coat of human blood a fraction that was heat-stable and dialyzable, and by demonstrating its activity in stimulating the smooth muscle of the rat intestine, virgin rat uterus, carotid artery, nictitating membrane of the cat, and vessels of the cat tail, but not the rabbit ear vessels, Zucker was able to postulate that it was a single substance. Because of the inability of this fraction to constrict the vessels in the perfused rabbit ear, even though serum had been shown to cause constriction of these vessels, Zucker concluded that more than one substance was responsible for the activity of serum on smooth muscle indicators. In 1947, Zucker[47] defined the role of platelets in the production of vasoconstriction at the site of hemorrhage in the rat mesoappendix, and showed that the presence of adequate platelet agglutination at the site of the incision in the mesoappendix was prerequisite both for the cessation of bleeding and for vasoconstriction, linking the two in the natural process of hemostasis. Not only did the platelet plugs at the site of bleeding induce vasoconstriction in the injured vessel, but also vessels adjacent to the injured vessel were regularly seen to constrict in the presence of these nearby collections of agglutinated platelets. In thrombocytopenic rats, these platelet plugs did not form and vasoconstriction did not occur. Zucker concluded that a vasoconstrictor substance was released by the platelets in the process of agglutination, thus aiding in the normal mechanism of hemostasis.

One year later, Rapport *et al.*[30] reported the purification from beef serum of an almost colorless solid that possessed 25–50% of the vasoactivity found in the original serum, when tested on the vessels of the perfused isolated rabbit ear. In the same preparation, this substance had more than twice the vasoconstrictive properties of epinephrine.[31] Because it was isolated from blood and because it was a potent vasoconstrictor, these authors named the compound "serotonin." Plasma separated from traumatized blood or blood that had been aged was rich in vasoconstrictor activity, but plasma separated from fresh blood was found to possess very little or no vasoconstrictor activity or serotonin.[22,29] By administering [^{14}C]tryptophan to rabbits, Udenfriend and Weissbach[41] found that platelets contained all the serotonin in blood, and no more could be detected in the plasma. Moreover, they found that the half-life of serotonin in rabbits was about 2.5 days, in close agreement with the reported half-life of the platelet. This finding suggested strongly that platelets are the source of serotonin, and indeed, serotonin was isolated from bovine platelets by several investigators.[18,49]

A biochemical basis was thus established for the hypothesis that the presence of bleeding, *alone,* into the subarachnoid space was sufficient to produce cerebral arterial spasm.

An observation was made by several investigators about cerebral arterial spasm produced in animals by experimental subarachnoid hemorrhage, namely, that the spasm was of a biphasic nature. In ten adult dogs, Brawley *et al.*[10] continuously measured the circumference of the intracranial carotid artery before, during, and after avulsion of the anterior cerebral artery, which caused subarachnoid hemorrhage. Except in one animal that succumbed immediately, there was an immediate constriction, within 5 min in all the carotid arteries, that showed a return to near-baseline size within 1 hr. The size of the internal carotid was found to parallel the degree of neurological deficits that the dog suffered. Recurrence of arterial narrowing was found to some degree in all the surviving dogs at 24 hr after rupture. The maximal degree of spasm was found in all animals on the 3rd day post hemorrhage.

In a series of mongrel dogs, Nagai *et al.*[25] demonstrated a biphasic time course for vasospasm with experimental subarachnoid hemorrhage from puncture of the posterior communicating artery. These authors performed both serial cerebral blood flow determinations and serial angiograms before and after the subarachnoid hemorrhage, which was caused by pulling on a suture connected to a needle (0.33 mm in diameter) that had been placed through one wall of the posterior communicating artery by craniotomy 2–3 days earlier. In 7 dogs followed by angiography, there was uniformly a decrease of 25–40% in the caliber of the larger arteries at the base of the brain within 30 min after rupture. By 2 hr, the vessels were practically back to original size. Recurrent spasm was noted in all angiograms 24 hr after rupture. In 20 dogs in which cerebral blood flow was monitored, the cerebral blood flow fell a few minutes after rupture and had returned to baseline about 2–3 hr later. In 8 dogs in which the cerebral blood flow was measured for 4 days, there was an initial fall in cerebral blood flow in the first 30 min, followed by a "restoration" of blood flow thereafter (although it is not clear to exactly what extent), then a gradual significant reduction in cerebral blood flow by 24 hr after rupture. If, instead of rupture of the posterior communicating artery, subarachnoid hemorrhage was caused by injection of 5 ml or more of fresh autogenous blood into the cisterna magna (7 dogs), diffuse constriction of the larger arteries of the base of the brain was seen within 1 hr. By 3 hr after injection, only 2 of the 7 dogs still had diffuse spasm; 4 of the 7 dogs were observed at 24 hr, and all 4 were found to have diffuse spasm. In an attempt to separate the factors leading to this biphasic response, Kuwayama *et al.*[21] induced spasm in the basilar artery of 9 dogs by slowly injecting 2 ml of autogenous fresh blood into the cisterna magna under atraumatic microsurgical control, thereby minimizing mechanical factors, and then subjected the animals to angiography at 15 min, 30 min, 60 min, 120 min, 2 days, and on later days. There was early constriction to 29–59% of the control diameter, with recovery by 2 hr. Most of the arteries reexpanded to control size or greater. By 48 hr, vasoconstriction had recurred (25–59% of control), and by 7 days, the arterial diameters had returned to normal. Once spasm had subsided, it did not recur.

While each of the three groups of investigators has clearly documented that in dogs, following either rupture of an intracranial vessel or injection of blood into the subarachnoid space, there appears to be an early severe arterial constriction, followed by partial recovery, followed again by more severe and diffuse vasoconstriction, the significance of this "biphasic" nature of the response remains obscure. Even in the dog, which is the laboratory animal most thoroughly studied, it seems reasonable to accept that the initial hemorrhage would produce an immediate vessel response, due to release of vasoconstrictor factors from the blood; it is reasonable to suppose that, thereafter, a spectrum of responses by the cerebral arteries would ensue, from intense vasoconstriction to milder degrees of spasm. This wide variation in the chronic phase of the spasm is illustrated well in all three of the reports discussed above. The biphasic nature of cerebral arterial spasm has never been demonstrated satisfactorily in man, but the acute phase, if it exists, may play a role in the cause of immediate death following the hemorrhage in a significant percentage of patients.

This careful documentation of the biphasic nature of spasm in the experimental model called for an explanation. In 1955, Zucker and Borrelli[48] reported that when the platelets participate in coagulation in their own native plasma, they liberate about half their stored serotonin. Raynor *et al.*[32] applied serotonin topically to the exposed cortical vessels of the cat and found that marked vasoconstriction resulted, with transient blanching of areas of the cortex. The spasm sometimes lasted over 3 hr, and in some animals, the spasm did not subside prior to the end of the experiment. These results, along with Zucker and Borrelli's findings of partial release of serotonin during coagulation, led Raynor and associates to conclude that the biphasic nature of vasospasm documented in the experimental animal was due directly to the timing of serotonin release. Initially, when the bleeding occurs or when the blood is injected into the subarachnoid space, platelets release the majority of their serotonin, and spasm occurs quickly. As the serotonin is cleared by the CSF, the spasm abates. Finally, as the platelets slowly release their remaining serotonin, the delayed phase of cerebral arterial spasm ensues.

Although there has been a concentration of interest in serotonin as the agent responsible for cerebral arterial spasm, other fractions isolated from blood have been shown to possess vasoconstrictor properties. The dialyzable, heat-stable fraction isolated by Zucker[46] in 1944 had some of the biochemical properties of the crystalline substance purified from beef serum by Rapport *et al.*,[30] but it did not cause the rabbit ear vessels to constrict, which was precisely the bioassay used by the latter authors for their "serotonin." Kapp *et al.*[19] isolated from cat platelets another fraction that was heat-stable and dialyzable, was not inactivated by trypsin or carboxypeptidase, but was inactivated by pronase, and produced marked vasoconstriction of the basilar artery when applied to its adventitial surface. However, these authors state that they were able to differentiate this substance from serotonin by chromatography. Unfortunately, neither the work of Zucker nor that of Kapp and associates has been duplicated, nor has further work been done on these two fractions to isolate or purify any substance. It is not possible to determine whether Zucker's and Kapp and co-workers' fractions were different from serotonin.

It is now well established that cerebral arterial spasm can be achieved by introducing blood into the subarachnoid space of a variety of animals, and that at present this is the best *in vivo* experimental model for the study of cerebral arterial spasm following the rupture of an intracranial aneurysm. This model, however, has major disadvantages. First, there are species differences, as documented by Echlin.[11] Second, in nearly all cases in which cerebral arterial spasm is produced in these animals, there is an absence of any neurological deficit. Unlike the case in the human counterpart, even marked degrees of cerebral arterial spasm in many animals are asymptomatic, and this might be related to the greater number of extracranial–intracranial anastomotic channels in these animals. A second possibility as to why the experimental animals get fewer deficits might be that the degree of arterial constriction produced in these experiments is insufficient to significantly alter the cerebral blood flow. Simeone *et al.*[37] monitored cerebral blood flow in adult rhesus monkeys subjected to experimental subarachnoid hemorrhage either by injecting 3 ml of fresh arterial blood into the subarachnoid space through a fine catheter placed by craniotomy so that its tip was in contact with the intracranial internal carotid artery or by 30-gauge needle puncture of the intracranial internal carotid artery performed by craniotomy. The results of these experiments were interpreted by the authors as showing that at least 50% reduction in the vessel diameter was necessary to decrease the cerebral blood flow to 60% or less of its control value. Simeone and co-workers concluded, therefore, that some published experimental angiograms indicating "spasm" actually show degrees of arterial constriction that might have little or no effect on cerebral blood flow. This explanation is possible, but even with arterial constriction of less than 50% of control diameter, we have documented in dogs apparently normal neurological function. Thus, the rich extracranial–intracranial anastomoses appear to be the more plausible reason for the lack of symptoms of cerebral ischemia in the experimental animal. Finally, the *in vivo* model of subarachnoid hemorrhage is not quantitatively reproducible, since it relies on the variables inherent in the methods of introducing blood into the subarachnoid space and since it depends on the imprecision of serial angiography. The development of *in vitro* techniques of measuring the contractions of the cerebral artery, although even further removed from the clinical situation of cerebral arterial spasm following subarachnoid hemorrhage, has provided a quantitative understanding of the biochemical events that begin with the rupture of an aneurysm and that can lead to symptoms of cerebral ischemia.

3. *In Vitro* Model of Cerebral Arterial Spasm

In 1961, Bohr *et al.*[9] reported experiments testing the reactivity of isolated small resistance vessels of 200–300 μm in diameter obtained by microdissection from the dog and the rabbit. The authors cut helical strips from these tiny cerebral arteries, which had been stored at 4°C for up to 24 hr postmortem, and measured their isometric contractions in the presence of different chemical stimuli. The arterial strips were oriented with the long axis of the strips parallel to the long axis of the muscle fibers. After different test solutions were introduced into the buffer surrounding the artery segment, isometric contractions of the segment were measured. The authors demonstrated that these small cerebral arterial segments were barely responsive to catecholamines, whereas they were consistently responsive to serotonin. Moreover, the amplitude of contraction was dose-dependent, with greater concentrations of serotonin causing greater contractions. On the other hand, mesenteric arteries tested in a similar manner showed much greater responsiveness to epinephrine than to serotonin, which barely caused any response. The same dose dependency was manifested by the mesenteric arteries for epinephrine. The initial response of the cerebral vessels to serotonin was at a lower concentration of the drug than the initial response of the mesenteric vessels to epinephrine; i.e., the cerebral vessels showed a greater sensitivity of response. The maximum response shown by any of these tiny cerebral vessels was to a dose of approximately 10 mg. Large, nonphysiological concentrations of potassium (50–100 mM) caused contractions to occur in small arterial segments from the cerebral, mesenteric, pulmonary, and renal vascular beds, whereas other vasoactive substances, such as vasopressin, angiotensin, and acetylcholine, failed to cause contractions in artery segments from some of these vascular beds, demonstrating that the same chemical stimulus often produces different responses in different regions of the circulation of the same organism.

Six years later Uchida *et al.*[40] described a method whereby acutal flow through a tiny cerebral "resistance" vessel of 50–250 μm in outside diameter could be recorded, and these authors used this method to study responses of small branches of rabbit middle cerebral artery and the artery of the rabbit mesojejunum to a variety of vasoactive agents. Their results were generally in agreement with those of the earlier work of Bohr and co-workers,[9] but they found that these small cerebral vessels had a higher threshold (minimum concentration of agent at which the vessel responded) for both the catecholamines and serotonin and a lower threshold for vasopressin than the mesenteric vessels. However, they also found that the threshold dose for serotonin of the large cerebral arteries from the area of the circle of Willis was much lower than the threshold dose for serotonin of the small cerebral arteries. This finding was especially significant, since vasospasm after subarachnoid hemorrhage is known to involve the larger cerebral arteries. Serotonin consistently potentiated the effects of the catecholamines, even after tachyphylaxis to serotonin was seen. Although further additions of serotonin themselves produced no further contractions, serotonin continued to potentiate the contractions caused by the catecholamines. This potentiation was found to last 20–30 min after the introduction of serotonin, and it was specific for the catecholamines. Although this finding is of unknown significance, it introduces the theoretical possibility that more than one agent could be responsible for the arterial constriction after subarachnoid hemorrhage. Finally, the degree of response of the perfused artery was found to be dependent on the perfusion pressure, the magnitude of the response to any agent increasing with increasing perfusion pressure to 25–30 mm Hg, remaining unchanged from this point to about 60 mm Hg, then decreasing with further increases in perfusion pressure. Later, our laboratory provided quantitative confirmation that there is a narrow range of intraluminal pressures at which a cerebral artery manifests a maximal response to a vasoactive stimulus, by showing that the isolated canine basilar artery segment responds maximally when placed under a tension of 2–6 g, in terms of both its sensitivity (K_{ED50}) and its amplitude of response (C_{max}), but at tensions both above and below these, there is a significant decrease in both the sensitivity and the magnitude of response.[5]

Both Bohr *et al.*[9] and Uchida *et al.*[40] demonstrated the differences in reactivity of different vascular beds to the same chemical stimulus. The pitfalls of using noncerebral vascular smooth muscle for an assay system to mimic the events observed in cerebral arterial spasm are apparent. In 1967, Wilkins *et al.*[45] described a method whereby spirally cut strips of rabbit aorta was suspended between metal hooks and their isometric responses to different vasoactive agents tested. By showing that normal blood levels of catecholamines, angiotensin, and serotonin are at or below the minimum sensitivity of the aortic-strip assay, while 67% of the plasma and 92% of the serum specimens tested showed "significant activity," these authors concluded that vasoactive compounds other than those tested must be responsible for the activity of the serum and plasma on the aortic strip. These authors imply that the mere demonstration of activity of a vasoactive compound is not necessarily of physiological importance unless the concentration of the compound used is in the range expected under actual clinical conditions. Although this principle is useful, the conclusions reached by the authors are valid only for the rabbit aorta segment, and do not necessarily extend to the cerebral vasculature.

In 1971, Nielsen and Owman[27] described a technique by which the vasoactivity of large cerebral arteries could be directly measured. These are the same size arteries that are seen on angiograms to develop spasm after subarachnoid hemorrhage. A 4-mm segment from the middle cerebral artery of a cat was mounted in an isometric contraction apparatus, and the acutal circumferential constriction of the arterial segment could be measured both before and after application of a variety of chemical agents. The entire segment was immersed in buffer solution, so that the vasoactive agents tested had access to both the luminal and the external surface of the artery. By the use of this technique, reproducible dose–response curves for different vasoactive agents could be generated, so that the relative sensitivity of the arterial segments for these agents could be compared. When comparing their results with the earlier results of Bohr and co-workers,[9] who used smaller, spirally cut arterial strips, Nielsen and Owman[27] were able to show about the same sensitivity of response to serotonin ($K_{ED50} = 5 \times 10^{-7}$ M), but maximal contractions were about 10–25 times as great as those of the former authors, for corresponding doses of serotonin. The techniques of Nielson and Owman proved more applicable to the testing of reactivity of larger cerebral arterial segments and have been adopted by several laboratories for *in vitro* work. Again, these investigators showed that the cerebral artery appeared more sensitive to serotonin than to any of the catecholamines or acetylcholine.

In 1973, Toda and Fujita,[39] using a similar *in vitro* technique, demonstrated once more that canine cerebral arteries responded both at a lower dose and to a greater magnitude with serotonin than with norepinephrine. In addition, human cerebral arteries removed postmortem were shown to possess the same characteristics of response, including reproducible dose–response curves for each vasoactive agent. Using the dose–response curves for comparison, these authors were able to develop the following hierarchy of response to serotonin: cerebral artery > internal carotid artery > external carotid artery > common carotid artery > distal mesenteric artery > proximal mesenteric artery. The magnitudes of the norepinephrine-induced contractions were in the reverse order. Phentolamine, which effectively blocked the norepinephrine-induced contractions, failed to affect the contractions caused by 30 mm KCl in five cerebral artery strips. In a similar manner, methysergide was shown to be effective in blocking the serotonin-induced contraction, but was not effective in blocking the KCl-induced contraction. Thus, the concept of a certain specificity of receptor sites for different vasoactive agents, with relatively specific antagonists operating at these sites, in the manner of enzyme–substrate interactions, was implied. The contraction caused by KCl remained exceptional in that it was unaffected by the usual blocking agents.

In 1974, we reported a series of experiments[5,6] using a similar *in vitro* technique, with a small volume chamber measuring isometric contractions of isolated canine basilar and middle cerebral artery segments. By mounting of a 3-mm arterial segment in the manner described by Nielsen and Owman,[27] and by careful control of temperature and pH, reproducible dose–response curves of individual arterial segments could be obtained. Our results showed 100 times the sensitivity (K_{ED50}) and 20 times the maximal contraction (C_{max}) reported previously.

These differences might be accounted for in the

case of Nielsen and Owman[27] by species differences between the dog and cat, and perhaps in the case of Toda and Fujita[39] by the differences in the preparation and mounting of the arterial strips. However, in addition, we have been careful to adjust the resting tension of the artery segments to approximately 3 g, at which they respond to vasoactive stimuli with both the best $K_{ED_{50}}$ and the best C_{max}. The effect of the resting tension on both the $K_{ED_{50}}$ and the C_{max} of the serotonin dose–response curve is shown in Fig. 1. The smallest $K_{ED_{50}}$ values (greatest sensitivity of response) and the highest C_{max} values occur between 2 and 8 g of tension. Second, the artery segment was not used for an experiment until successive cumulative serotonin dose–response curves were identical within experimental error. Such serotonin internal standard responses were obtained periodically on each artery throughout the experiment to demonstrate the reproducibility of responses. During the initial standardization, the artery usually manifests an increasing magnitude (C_{max}) as well as sensitivity ($K_{ED_{50}}$) to serotonin. During the standardization procedure, which usually takes about 3–4 hr, there is also a tendency for the

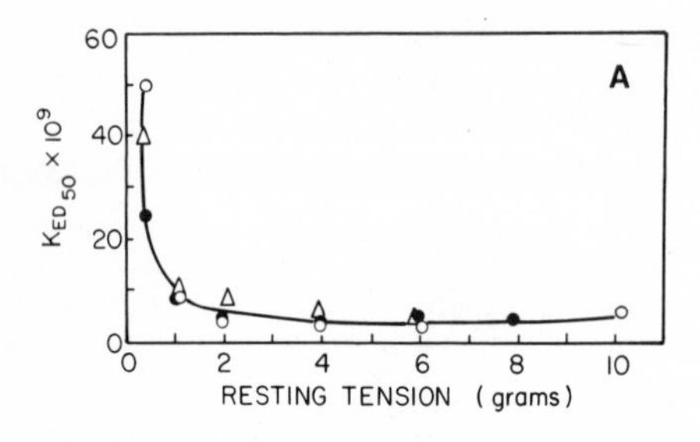

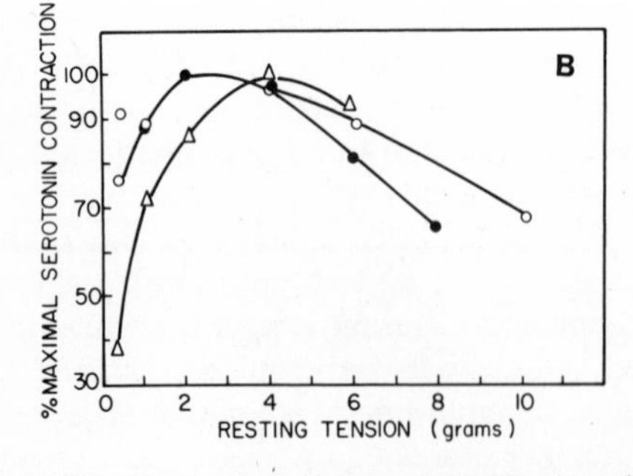

Figure 1. (A) Graph showing serotonin $K_{ED_{50}}$ values of three basilar arteries as a function of resting tension. Values for each artery are represented by the same symbol. (B) Graph showing percentage maximal serotonin contraction as function of resting tension. Values for each artery are represented by same symbol. Reproduced from Allen *et al.*,[5] by permission of publisher.

artery to continue to relax to below baseline between serotonin contractions. When it does so, the tension is gently increased to maintain a resting tension of 3 g. Although the precise reason for this improvement in response that the artery initially demonstrates is not known, it is seen in almost all arteries tested. It is possible that slight conformational changes in the receptors take place as tension is slowly increased on the artery, until an ideal conformation is reached.

Using this technique of careful standardization, four canine basilar and six human cerebral arteries were tested with a variety of vasoactive agents. Results of these dose–response curves are shown in Fig. 2. Notice the similarity in these results. Some agents gave contractions at concentrations that are not likely to be present in blood. Serotonin, however, was unique in that 90% of its maximal contraction was obtained with a concentration 10–30 times less than that present in clotted blood. Potassium ion, even at a concentration of 10 mM, did not produce contraction. At the potassium levels reported in the CSF following subarachnoid hemorrhage by Wilkins and Levitt,[44] none of the cerebral arterial spasm following such hemorrhage should be due to increased potassium. Thus, according to the results of this *in vitro* method, serotonin appears the most plausible candidate as the agent in blood that is responsible for arterial spasm following a subarachnoid hemorrhage.

It is interesting that prostaglandins A_1 and $F_2\alpha$ gave the highest contractions of all the agents tested (Fig. 2), but both at concentrations that are not likely to be present in the blood or in CSF. Holmes[17] has identified prostaglandins E_1, E_2, $F_1\alpha$, and $F_2\alpha$ from perfusates of the cerebral ventricles of anesthetized dogs. With control values of E_1 approximately 150 ng/hr, he found that by introducing serotonin into the perfusate, the release of prostaglandin E_1 increased 4-fold. This is still not in the range of concentrations at which one might expect any effect on the cerebral vasculature. Both Pennink *et al.*[28] and White *et al.*[43] have produced cerebral vasospasm in dogs by injecting $F_2\alpha$ and other prostaglandins into the chiasmatic cisterns and have followed the course of the vasoconstriction in serial angiograms. However, again, the dose of prostaglandin was far too high (1–20 mg/kg) to draw conclusions about its physiological significance. Ellis *et al.*[14] have shown that thromboxane A_2, a labile arachidonic acid metabolite generated by the action of prostaglandin cyclic endoperoxide H_2 on human platelet particles, causes a contraction equal to that produced by pros-

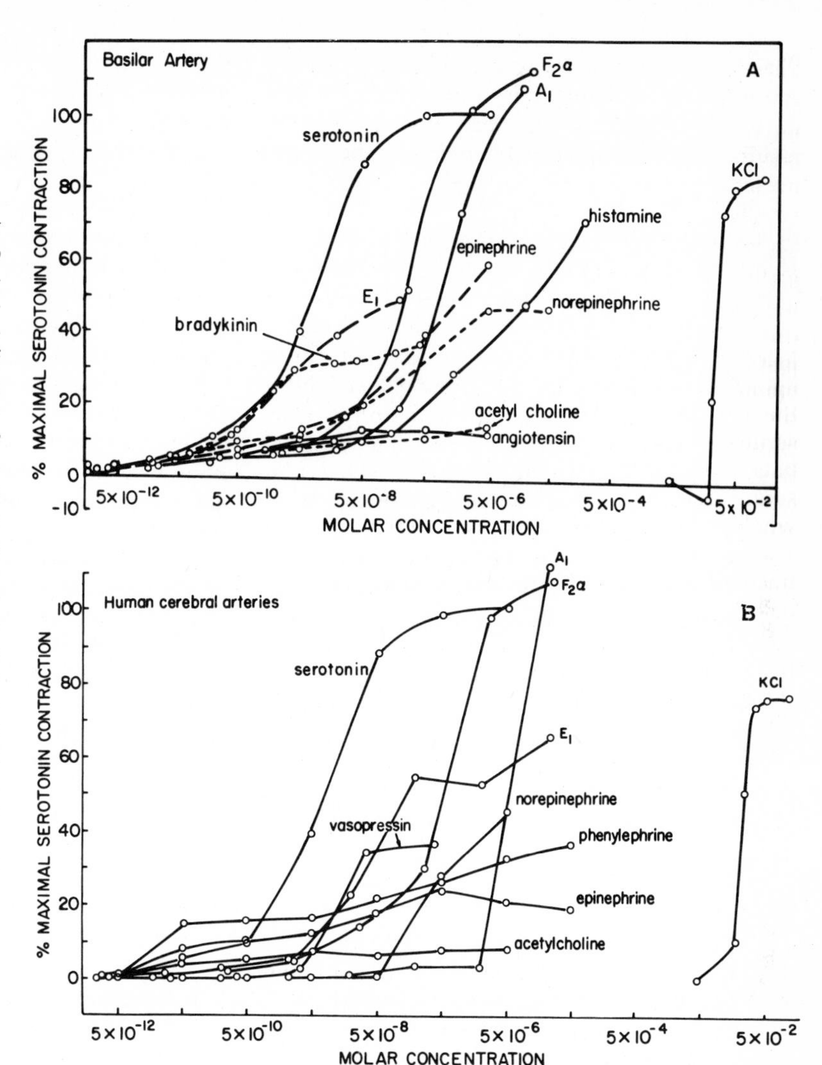

Figure 2. (A) Relative sensitivity of four canine basilar arteries to vasoactive agents shown as cumulative log-dose, isometric response curves. All data are presented in terms of percentage of maximal response to serotonin (100% response to serotonin = 7.4 g). (Serotonin C_{max} values for the four arteries were 5.0, 5.8, 6.9, and 11.9 g.) Each point is the mean value at that concentration of agent for the number of arteries that responded to agent. Reproduced from Allen *et al.*,[5] by permission of publisher. (B) Graph showing relative sensitivity of six human cerebral arteries to vasoactive agents shown as cumulative log-dose, isometric response curves. All data are presented in terms of percentage of maximum response to serotonin (100% response to serotonin = 12.2 g). (Serotonin C_{max} values for the six arteries were 14.2, 14.4, 13.2, 4.1, 12.7, and 14.6.) Each point is mean value at that concentration of agent for number of arteries tested that responded to agent. Reproduced from Allen *et al.*,[3a] by permission of publisher.

taglandin $F_2\alpha$, and greater than the serotonin response, but with a significantly greater K_{ED50} (lesser sensitivity). However, since thromboxane A_2 has a half-life of only 32 sec in aqueous medium, it is unlikely that it plays an important role in human cerebral arterial spasm. Rosenblum[34] has shown that the vasoconstrictor properties of certain prostaglandins were enhanced by serotonin as well as by norepinephrine, but states in his conclusion that the concentrations of prostaglandin observed in the CSF of man, even in disease states, are far lower than those found necessary to induce constriction after topical application in animals. The significance of prostaglandins in human cerebral arterial spasm remains to be demonstrated.

Further substantiation that serotonin is the agent responsible for arterial spasm after subarachnoid hemorrhage has been reported.[6] CSF samples obtained from patients who had suffered a subarachnoid hemorrhage from a ruptured intracranial aneurysm were shown to cause the canine basilar artery to contract in a dose-dependent fashion, whereas CSF samples from control patients who had not suffered a subarachnoid hemorrhage did not cause the artery segment to contract. The agent in the CSF responsible for this vasoactivity was isolated by methanol extraction and paper chromatography in two separate solvent systems, and was shown to be capable of producing contractions in the basilar artery segment lasting at least 2 hr, the length of the experiment.

Human serum has been shown to produce a typical dose–response curve when added in increasing amounts (Fig. 3) to the buffer surrounding the canine basilar artery in the chamber. By comparing the results of these serum experiments with the results of blocking studies, it was possible to exclude certain vasoactive agents as the responsible fractions of serum in the contraction of the basilar artery segment. First, it was observed that those arteries that did not respond to epinephrine responded to serum just as well as those arteries that were sensitive to epinephrine. Therefore, epinephrine could not be the agent responsible for the contractile activity of serum. Second, methysergide was found to block both the serotonin-induced contraction and the serum-induced contraction, but not the contraction when prostaglandin $F_2\alpha$ was added to the chamber. Therefore, $F_2\alpha$ could not be responsible for the contractile activity in serum. Phenoxybenzamine, usually regarded as an α-adrenergic blocking agent, was shown to irreversibly abolish the contraction of the artery to both serum and serotonin, but reduced the artery's response to $F_2\alpha$ by only 50%. The evidence from these studies therefore indicates that the contractile activity in serum is due to serotonin.

Using the same *in vitro* technique, we have been able to further characterize the serotonin receptor of the canine basilar artery. As is the case with other energy-requiring biological systems, the temperature at which the contraction of the basilar artery segment is maximal is close to 37°C (Fig. 4). The irreversible inactivation of arterial segments above 48°C is probably due to the denaturation of proteins in the system. Interestingly, it was possible to reduce the temperatures to 7°C without damaging the artery, and when the temperature was brought back to 37°C, the artery's response to serotonin was identical to that obtained at 37°C prior to the temperature reduction.

Large, nonphysiological doses of serotonin caused relaxation of artery segments that had been in a state of maximum serotonin contraction prior to the addition to the chamber of these large doses. The effects of these large doses of serotonin were reversible, and there was no demonstrable tachphylaxis to serotonin. The mechanism of the relaxation produced by these large doses is unknown.

Twelve serotonin analogues were tested in the same manner, and their dose–response curves are shown in Fig. 5. From these results and from a comparison of the chemical structures of those serotonin analogues, the portion of the serotonin molecule regarded as essential for its intrinsic activity is the ethylamine portion. By comparing the dose–response curves obtained in buffer solutions containing a variety of serotonin analogues, it is possible to conclude that the sites for binding and intrinsic activity in the serotonin receptor cannot be separated. 5-Hydroxyindoleacetic acid, the major metabolic product of serotonin, had no appreciable contractile activity itself, nor did it appear to competitively inhibit serotonin at the receptor site.

4. Role of Calcium in Cerebral Arterial Spasm

Extracellular calcium was found to be necessary for the serotonin-induced contraction to occur. If calcium is removed from the buffer solution bathing

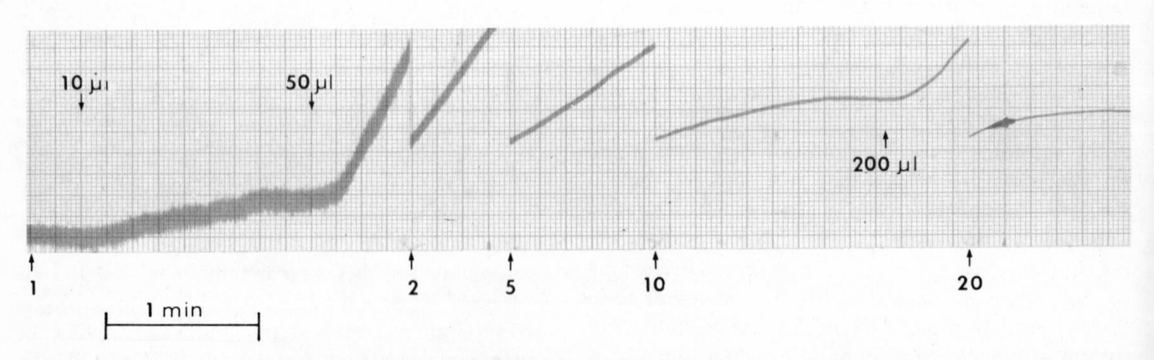

Figure 3. Polygraph tracing of isometric contraction of a basilar artery segment to cumulative additions of human serum. Numbers at base of tracing indicate grams of tension at full scale reading, and arrows at base of tracing show when this tension was changed. The arrows on tracing indicate when serum was added in volumes shown. Reproduced from Allen *et al.*,[6] by permission of publisher.

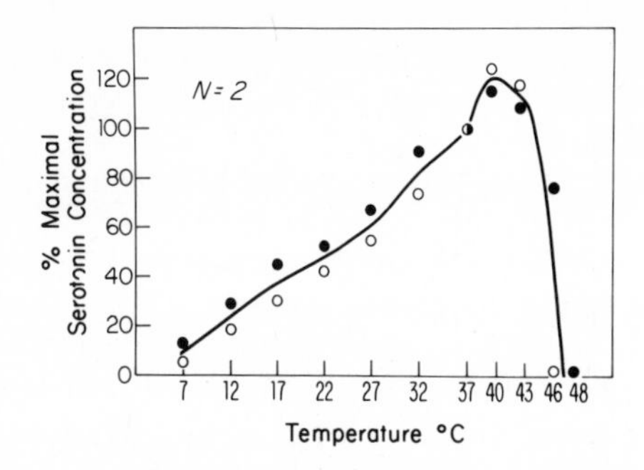

Figure 4. Graph showing effect of temperature on maximum contraction in response to serotonin of two canine basilar artery segments (100% = maximum serotonin contraction at 37°C). Response of one artery; (○) response of the second artery; (—) mean value. Maximum responses to serotonin at 37°C for the two arteries were 7.0 and 10.2 g. Reproduced from Allen *et al.*,[4] by permission of publisher.

the arterial segments, the usual contractile doses of serotonin produce only a minimal and extremely short-lived contraction. If, in addition to removing the calcium from the buffer, lanthanum ion is added in a 1–2 mM concentration, there is complete blockage of any contractile activity of the basilar artery to serotonin, prostaglandin $F_2\alpha$, and phenylephrine. Na_4EDTA in 1 mM concentration will likewise block the artery's response to serotonin as well as to other vasoactive agents.[4]

The dependence of the serotonin-induced contraction on the concentration of extracellular calcium is shown in Fig. 6. If the Krebs–Ringer buffer was made 1.15 mM in calcium (the concentration of free calcium in CSF and plasma), the response of the artery was maximal. If the calcium concentration was raised to 4 mM, there was progressive alteration of the serotonin dose–response curve. These results are to be compared with those in Fig. 7, in which progressively larger concentrations of calcium were added to the arterial segment in the absence of serotonin. Significant contractions resulted at very high concentrations of calcium. The time course of these high-calcium contractions was much slower than those resulting from serotonin at physiological calcium concentrations.

Bohr[8] has shown for vascular smooth muscle that calcium is the final activator of the contractile proteins. In the resting smooth muscle cell, the concentration of activator (free, ionized) calcium is less than 10^{-7} M, although the concentration of extracellular calcium is greater than 10^{-3} M; thus, for the smooth muscle cell to be in the relaxed state, the plasma membrane must generate a calcium concentration ratio of 10,000:1.

The intracellular activator calcium concentration can be increased to a level sufficient to activate the contractile proteins either by increasing the rate at which calcium enters the cell or by decreasing the rate at which intracellular free calcium is sequestered by intracellular binding sites or is extruded by the cell. The site at which intracellular calcium is sequestered in noncerebral arterial smooth muscle has not been conclusively demonstrated. Sarcoplasmic reticulum and mitochondria are two sites that have been proposed.[38] Nothing is known about calcium-binding sites in cerebral arterial smooth muscle cells. Van Breemen *et al.*[42] have demonstrated that the trivalent lanthanum ion effectively blocks entry of extracellular calcium into the cell,

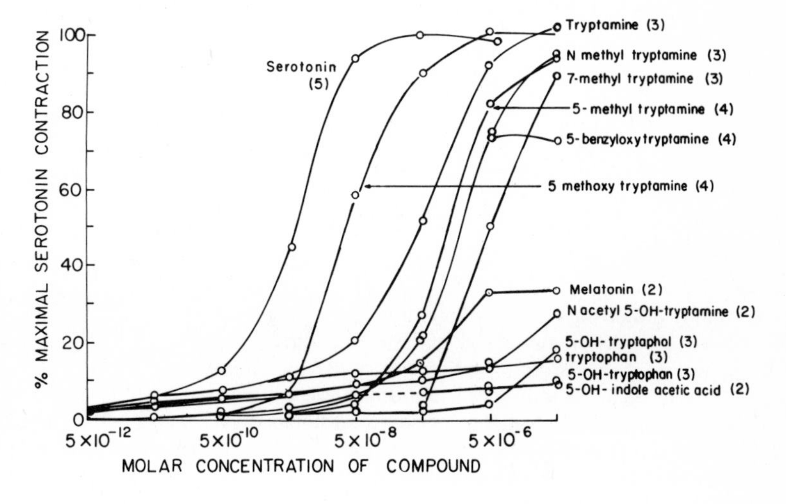

Figure 5. Graph showing relative sensitivity of canine basilar artery to various analogues of serotonin shown as cumulative log-dose, isometric response curves. Number in parentheses is number of arteries tested for each compound. Values are presented in terms of percentage of maximum response to serotonin. Each point is the mean value of the number of arteries tested at that concentration of agent. Reproduced from Allen *et al.*,[4] by permission of publisher.

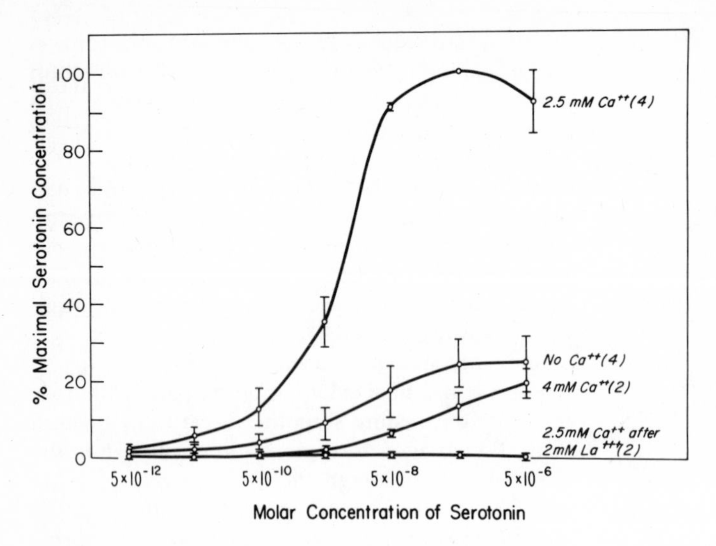

Figure 6. Graph showing effect of varying calcium concentrations and addition of lanthanum on the response of canine basilar arterial segment to serotonin. Number in parentheses is number of preparations tested, each from a different dog. Vertical brackets indicate standard error of mean (S.E.M.). The 100% response to serotonin is 11.2 ± 0.7 g S.E.M. Reproduced from Allen *et al.*,[4] by permission of publisher.

probably by binding very tightly to extracellular sites normally occupied by calcium. They have demonstrated that norepinephrine, angiotensin, and histamine will produce one contraction in the isolated rabbit aorta segment after the application of lanthanum, but no further contractions can be elicited, and they have interpreted this as evidence that these vasoactive agents are able to cause the release of bound intracellular calcium from a common pool that is ordinarly replenished by an influx of extracellular calcium. Lanthanum was thought to block the replenishment of the intracellular calcium but not to block the release of intracellular calcium. Our results with the canine basilar artery show that after the application of lanthanum, not even one contraction can be elicited by serotonin, phenylephrine, or prostaglandin $F_2\alpha$, indicating that these agents must rely primarily on extracellular calcium to activate the contractile apparatus of the cerebral artery smooth muscle cell.

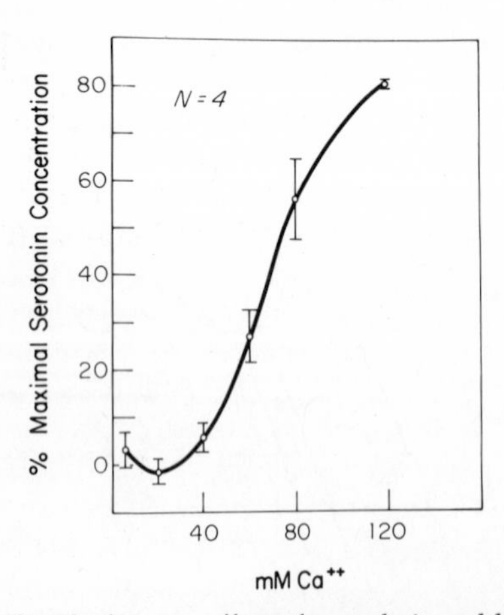

Figure 7. Graph showing effect of cumulative additions of calcium on state of contraction of canine basilar arterial segment. (*N* Number of preparations, each obtained from a different dog. Vertical brackets indicate standard error of mean (S.E.M.). The 100% response to serotonin is 9.2 ± 1.2 g S.E.M. Reproduced from Allen *et al.*,[4] by permission of publisher.

Removing the calcium from the buffer solution resulted in an inability of the smooth muscle cell to contract to the usual vasoactive compounds.[4] Likewise, 10^{-3} M Na_4EDTA produced essentially the same result. Although the finding has not been previously reported, we have also found that the contraction elicited by high KCl concentration in the cerebral artery is also abolished by omission of calcium from the buffer.

Using our small-volume-chamber method, we have been able to demonstrate a fundamental difference between the vascular smooth muscle of the cerebral artery and the rabbit aorta. The cerebral artery is absolutely dependent on extracellular calcium (free calcium in its environment of CSF), whereas the rabbit aorta can exhibit a sustained contraction in the absence of buffer calcium. Moreover, there appears to be an ideal calcium concentration at which the cerebral artery segment responds maximally, and this is the normal concentration of free calcium in the CSF, 1.15 mM see (Fig. 6). Calcium alone was able to cause the artery segment to contract, if the extracellular concentration was high enough to overcome the impermeability of the smooth muscle cell membrane to calcium (see Fig. 7).

The effects of calcium were not found to be entirely excitatory for contraction. As mentioned above, calcium was found to possess a definite modulating effect on the serotonin-induced contraction, as shown in Fig. 6. In addition, if small concentrations of calcium (0.1–1.0 mM) were added to a calcium-free buffer surrounding an arterial segment, sustained contractions occurred, whereas if higher concentrations of calcium (1.5–2.5 mM) were used, an initial short-lived contraction resulted, followed by relaxation (Fig. 8). Calcium seems, therefore, to possess an autoinhibitory effect on its own influx in contractions of the basilar artery segment. Thus, when small calcium concentrations are used (0.5 mM), the artery spontaneously contracts because of the normal permeability of the membrane to calcium, but as the calcium concentration is increased, the artery contracts, then rapidly relaxes. This relaxation is caused by the autoinhibitory effect calcium exerts on its own influx. As very high calcium concentrations are reached, the autoinhibitory effect is overcome by the diffusion gradient, and contraction once again occurs.

The absolute requirement of the cerebral arterial smooth muscle cell for extracellular calcium can be used to therapeutic advantage. Theoretically, contraction of the cerebral arterial smooth muscle cell could be blocked either by removing calcium from the CSF and extracellular fluid or by blocking its transmembrane flux from the extracellular fluid into the cerebral arterial smooth muscle cell. The former is impractical in patients, since there is no safe way

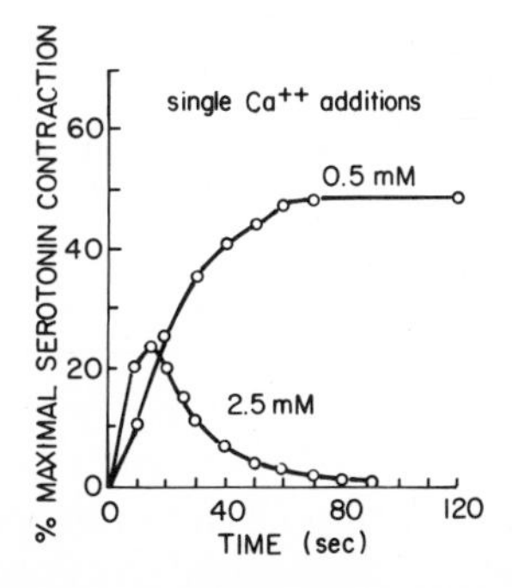

Figure 8. Graph showing time course of change in resting state of contraction of one canine basilar arterial segment following two different single additions of calcium to calcium-free bath. This response was found to be reproducible on three separate preparations. Reproduced from Allen *et al.*,[4] by permission of publisher.

to wash out or chelate extracellular calcium in the living patient. However, the arterial constriction could theoretically be blocked by any one of a variety of pharmacological agents that block the transmembrane flux of calcium into the arterial smooth muscle cell. We have demonstrated in our laboratory that nifedipine, a dihydropyridine derivative that has been used to treat coronary artery spasm (Prinzmetal's angina),[24] is successful in blocking the contractions of the canine basilar artery segment to all vasoactive chemical stimuli tested, including high-potassium depolarization.[3] Interestingly, the contractions elicited by serotonin and phenylephrine in the canine femoral artery—which resembles the rabbit aorta in its use of primarily an intracellular pool of calcium for contraction—were not significantly diminished by nifedipine, while the high-potassium-induced contraction was significantly diminished, although not totally abolished. This partial selectivity in response of the cerebral artery to nifedipine provides an unusual opportunity to treat patients with cerebral arterial spasm. Most therapies directed at the prevention or relief of cerebral arterial spasm have as their major side effect significant hypotension, since most vasoparalytic agents have a universal effect on all vascular beds. In patients with cerebral arterial spasm, hypotension is especially dangerous, since any decrease in perfusion pressure can lead to irreversible cerebral ischemia and infarction. From our *in vitro* data, it would seem that the calcium-blocking agent nifedipine, however, has a much more profound relaxant effect on the cerebral vasculature *in vitro* than on the systemic vaculature, and so might be useful to prevent and treat cerebral arterial spasm in patients who have suffered a subarachnoid hemorrhage.

Although nifedipine and certain other calcium-blocking agents have been shown to work well *in vitro* to block the contraction of the isolated cerebral artery segment, it is uncertain whether they will prove useful clinically, since in order to work, they must gain access to the cerebral artery smooth muscle cell. Nifedipine is highly lipid-soluble, and therefore it is rapidly absorbed after sublingual administration, producing measurable arterial blood levels in minutes.[20] It probably also traverses the blood–brain barrier without great difficulty, gaining access to the CSF side of the cerebral vasculature. This access is highly important, since even the most effective blocking agent *in vitro* can fail *in vivo* if it cannot reach its target tissue.

We have used nifedipine successfully both to treat

and to prevent acute cerebral arterial spasm produced experimentally in a series of anesthetized dogs by injecting fresh autogenous blood into the cisterna magna. Nifedipine was able to completely reverse the arterial spasm produced acutely. Dogs given nifedipine sublingually after the production of acute cerebral arterial spasm and studied angiographically 30 min later showed dramatic relief in their spasm. In addition, dogs in which chronic cerebral arterial spasm was documented angiographically 2 days after the intracisternal injection of blood were also given nifedipine sublingually. In these dogs, too, the arterial spasm was relieved within 30 min. Statistical analysis of the results of these *in vivo* experiments showed a significant difference between the control and experimental groups of dogs, with $p < 0.05$.[2]

We are hopeful that these results will provide the foundation for future therapy of cerebral arterial spasm in humans. If such therapy fails in its clinical trials, then we must reexamine and modify our current *in vivo* and *in vitro* models for cerebral arterial spasm.

References

1. Allcock, J., Drake, C.: Ruptured intracranial aneurysms—the role of arterial spasm. *J. Neurosurg.* **22**:21–29, 1965.
2. Allen, G., Bahr, A.: Cerebral arterial spasm. Part 10. The reversal of acute and chronic spasm in dogs with orally administered nifedipine. *Neurosurgery* **4**:43–47, 1979.
3. Allen, G., Banghart, S.: Cerebral arterial spasm. Part 9. *In vitro* effects of nifedipine on serotonin, phenylephrine, and potassium-induced contractions of canine basilar and femoral arteries. *Neurosurgery* **4**:37–42, 1979.

3a. Allen, G., Gross, C., French, L., Chou, S.: Cerebral arterial spasm. Part 5. *In vitro* contractile activity of vasoactive agents including human CSF on human basilar and anterior cerebral arteries. *J. Neurosurg.* **44**:594–600, 1976.

4. Allen, G., Gross, C., Henderson, L., Chou, S.: Cerebral arterial spasm. Part 4. *In vitro* effects of temperature, serotonin analogues, large nonphysiological concentrations of serotonin, and extracellular calcium and magnesium on serotonin-induced contractions of the canine basilar artery. *J. Neurosurg.* **44**:585–593, 1976.
5. Allen, G., Henderson, L., Chou, S., French, L.: Cerebral arterial spasm. Part 1. *In vitro* contractile activity of vasoactive agents on canine basilar and middle cerebral arteries. *J. Neurosurg.* **40**:433–441, 1974.
6. Allen, G., Henderson, L., Chou, S., French, L.: Cerebral arterial spasm, Part 2. *In vitro* contractile activity of serotonin in human serum and CSF on the canine basilar artery, and its blockage by methylsergide and phenoxybenzamine. *J. Neurosurg.* **40**:442–450, 1974.
7. Bayliss, W.: On the local reactions of the arterial wall to changes of internal pressure. *J. Physiol.* **28**:220–231, 1902.
8. Bohr, D.: Vascular smooth muscle updated. *Circ. Res.* **32**:665–672, 1973.
9. Bohr, D., Goulet, P., Taquini, A.: Direct tension recording from smooth muscle of resistance vessels from various organs. *Angiology* **12**:478–485, 1961.
10. Brawley, B., Strandness, D., Kelly, W.: The biphasic response of cerebral vasospasm in experimental subarachnoid hemorrhage. *J. Neurosurg.* **28**:1–8, 1968.
11. Echlin, F.: Vasospasm and focal cerebral ischemia. *Arch. Neurol. Psychiatry* **47**:77–96, 1942.
12. Echlin, F.: Current concepts in the etiology and treatment of vasospasm. *Clin. Neurosurg.* **15**:133–160, 1968.
13. Ecker, A., Riemenschneider, P.: Arteriographic demonstration of spasm of the intracranial arteries, with special reference to saccular arterial aneurysms. *J. Neurosurg.* **8**:660–667, 1951.
14. Ellis, E., Nies, A., Oates, J.: Cerebral arterial smooth muscle contraction by thromboxane A_2. *Stroke* **8**:480–483, 1977.
15. Fisher, C., Roberson, G., Ojemann, R.: Cerebral vasospasm with ruptured saccular aneurysm—the clinical manifestations. *Neurosurgy* **1**:245–248, 1977.
16. Florey, H.: Microscopical observations on the circulation of the blood in the cerebral cortex. *Brain* **48**:43–64, 1925.
17. Holmes, S.: The spontaneous release of prostaglandins into the cerebral ventricles of the dog and the effect of external factors on this release. *Br. J. Pharmacol.* **38**:653–658, 1970.
18. Humphrey, J., Jaques, R.: The histamine and serotonin content of the platelets and polymorphonuclear leukocytes of various species. *J. Physiol.* **124**:305–310, 1954.
19. Kapp, J., Mahaley, M., Odom, G.: Cerebral arterial spasm. Part 3. Partial purification and characterization of a spasmogenic substance in feline platelets. *J. Neurosurg.* **29**:350–356, 1968.
20. Kroneberg, G.: Pharmocology of nifedipine. In Lochner, W., Braasch, W., Kroneberg, G. (eds.): *2nd International Adalat Symposium.* Berlin, Springer-Verlage, 1975, pp. 12–19.
21. Kuwayama, A., Zervas, N., Belson, R., Shintani, A., Pickren, K.: A model for experimental cerebral arterial spasm. *Stroke* **3**:49–56, 1972.
22. Landis, E., Wood, S., Guerrant, J.: Effect of heparin on the vasoconstrictor action of shed blood tested by perfusion of the rabbit's ear. *Am. J. Physiol.* **139**:26–38, 1943.
23. Lende, R.: Local spasm in cerebral arteries. *J. Neurosurg.* **17**:90–103, 1960.

24. MULLER, J., GUNTHER, S.: Nifedipine therapy for Prinzmetal's angina. *Circulation* **57**:137–139, 1978.
25. NAGAI, H., SUZUKI, Y., SUGIURA, M., NODA, S., MABE, H.: Experimental cerebral vasospasm. Part 1. Factors contributing to early spasm. *J. Neurosurg.* **41**:285–292, 1974.
26. NIBBELINK, D., TORNER, J., HENDERSON, W.: Intracranial aneurysms and subarachnoid hemorrhage. A cooperative study. Antifibrinolytic therapy in recent onset subarachnoid hemorrhage. *Stroke* **6**:622–629, 1975.
27. NIELSEN, K., OWMAN, C.: Contractile responses and amine receptor mechanisms in isolated middle cerebral artery of the cat. *Brain Res.* **27**:33–42, 1971.
28. PENNINK, M., WHITE, R., CROCKARELL, J., ROBERTSON, J.: Role of prostaglandin $F_{2}\alpha$ in the genesis of experimental cerebral vasospasm. *J. Neurosurg.* **37**:398–406, 1972.
29. RAND, M., REID, G.: Source of serotonin in serum. *Nature (London)* **168**:385, 1951.
30. RAPPORT, M., GREEN, A., PAGE, I.: Partial purification of the vasoconstrictor in beef serum. *J. Biol. Chem.* **174**:735–741, 1948.
31. RAPPORT, M., GREEN, A., PAGE, I.: Crystalline serotonin. *Science* **108**:329–330, 1948.
32. RAYNOR, R., MCMURTRY, J., POOL, J.: Cerebrovascular effect of topically applied serotonin in the cat. *Neurology* **11**:190–195, 1961.
33. ROBERTSON, E.: Cerebral lesions due to intracranial aneurysms. *Brain* **72**:150–185, 1949.
34. ROSENBLUM, W.: Effects of prostaglandins on cerebral blood vessels: Interaction with vasoactive amines. *Neurology* **25**:1169–1171, 1975.
35. ROY, C., SHERRINGTON, C.: On the regulation of the blood-supply of the brain. *J. Physiol.* **11**:85–108, 1890.
36. SIMEONE, F., RYAN, K., COTTER, J.: Prolonged experimental cerebral vasospasm. *J. Neurosurg.* **29**:357–366, 1968.
37. SIMEONE, F., TREPPER, P., BROWN, D.: Cerebral blood flow evaluation of prolonged experimental vasospasm. *J. Neurosurg.* **37**:302–311, 1972.
38. Somlyo, A. P., SOMLYO, A. U., SHUMAN, H., GARFIELD, R.: Calcium compartments in vascular smooth muscle: Electron probe analysis. In Betz, E. (ed.): *Ionic Actions on Vascular Smooth Muscle.* Berlin and Heidelberg, Springer-Verlag, 1976, pp. 17–20.
39. TODA, N., FUJITA, Y.: Responsiveness of isolated cerebral and peripheral arteries to serotonin, norepinephrine, and transmural electrical stimulation. *Circ. Res.* **33**:98–104, 1973.
40. UCHIDA, E., BOHR, D., HOOBLER, S.: A method for studying isolated resistance vessels from rabbit mesentery and brain and their responses to drugs. *Circ. Res.* **21**:525–536, 1967.
41. UDENFRIEND, S., WEISSBACH, H.: Studies on serotonin (5-hydroxytryptamine) in platelets. *Fed. Proc. Fed. Am. Soc. Exp. Biol.* **13**:412–413, 1954.
42. VAN BREEMAN, C., FARINAS, B., GERBA, P., MCNAUGHTON, E.: Excitation–contraction coupling in rabbit aorta studied by the lanthanum method for measuring cellular calcium influx. *Circ. Res.* **30**:44–54, 1972.
43. WHITE, R., HAGEN, A., MORGAN, H., DAWSON, W., ROBERTSON, J.: Experimental study on the genesis of cerebral vasospasm. *Stroke* **6**:52–57, 1975.
44. WILKINS, R., LEVITT, P.: Potassium and the pathogenesis of cerebral arterial spasm in dog and man. *J. Neurosurg.* **35**:45–50, 1971.
45. WILKINS, R., WILKINS, G., GUNNELLS, J., ODOM, G.: Experimental studies of intracranial arterial spasm using aortic strip assays. *J. Neurosurg.* **27**:490–500, 1967.
46. ZUCKER, M.: A study of the substances in blood serum and platelets which stimulate smooth muscle. *Am. J. Physiol.* **142**:12–26, 1944.
47. ZUCKER, M.: Platelet agglutination and vasoconstriction as factors in spontaneous hemostasis in normal, thrombocytopenic, heparinized, and hypoprothrombinemic rats. *Am. J. Physiol.* **148**:275–288, 1947.
48. ZUCKER, M., BORRELLI, J., Quantity, assay, and release of serotonin in human platelets. *J. Appl. Physiol.* **7**:425–431, 1955.
49. ZUCKER, M., FRIEDMAN, B., RAPPORT, M.: Identification and quantitative determination of serotonin (5-hydroxytryptamine) in blood platelets. *Proc. Soc. Exp. Biol. Med.* **85**:282–285, 1954.

CHAPTER 22

Response of Extraparenchymal Cerebral Arteries to Biochemical Environment of Cerebrospinal Fluid

Frederick A. Simeone, Phillip E. Vinall, and John D. Pickard

1. Introduction

The contractile activity of cerebral arteries is governed by a continuously changing environment, both internally and externally, provided by the blood and cerebrospinal fluid (CSF), respectively. The influence of CSF makes this vascular system physiologically and pathologically unique. This environment supplies both the metabolic substrates and the ions necessary for vascular smooth muscle contraction. The purpose of this chapter is to assess the effect of the external metabolic and ionic environment on the function of large cerebral arteries. Smaller cerebral arteries lying within the substance of the brain are not bathed by CSF and thus will be excluded from this discussion, although they are more important than larger arteries in the regulation of cerebral blood flow.

Recent investigations[7,10,15,26,29] suggest that cerebral arteries respond differently from extracranial arteries. Cerebral arteries are more sensitive to serotonin than catecholamines in comparison to peripheral arteries.[4,20] Membrane calcium is more easily depleted in cerebral arterial smooth muscle cells.[15] The sodium content of cerebral arteries is greater than that of other arteries.[30] Reduction in extracellular potassium concentration in cerebral arterial smooth muscle causes a significant contraction, whereas peripheral arteries are relatively insensitive to a paucity of this ion.[29] Thus, cerebral arterial smooth muscle appears more sensitive overall to organic and inorganic ionic changes than noncerebrovascular smooth muscle. This suggests a different balance between the ionic mechanisms controlling cerebral and extracranial circulations.

We used an *in vitro* model of fresh large bovine middle cerebral arteries suspended in an artificial CSF-like medium to investigate some of the basic mechanisms of contraction. Arteries were cut helically, and changes in length were measured isotonically. The metabolic, ionic, and pH compositions of the bathing solution were manipulated to delineate some of the responses that may mimic those occurring *in vivo* during various physiological and pathological CSF alterations.

Frederick A. Simeone, M.D., and **Phillip E. Vinall, B.S.** • Division of Neurosurgery, Pennsylvania Hospital, Philadelphia, Pennsylvania 19107. **John D. Pickard, M.D.** • Division of Neurosurgery, Institute of Neurological Sciences, Southern General Hospital, GB Glasgow, G51 4TF, Scotland.

2. Energy Metabolism and Ions in CSF

Brain metabolism accounts for 20% of total basal oxygen consumption and is supported almost entirely by glucose utilization.[25] Consequently, the brain is very susceptible to both oxygen and glucose deprivation.[28] Furthermore, brain tissue has only small stores of oxygen and substrate, and its capacity for anaerobic respiration is limited.[24] The following experiments were undertaken to explore the effects of hypoxia and hypoglycemia, and of the interaction of these conditions with ions in the CSF, on the contractile activity and calcium permeability of the middle cerebral artery.

2.1. Effects of Energy Metabolism on Potassium-Induced Contractions

When large cerebral arteries are exposed to a high-potassium solution (K_o = 70 mM) in a bath environment containing a physiological concentration of calcium, a sustained contraction occurs (Fig. 1). If extracellular calcium is reduced, the duration of the contraction is decreased. Arteries challenged with a medium that is high in potassium content, but free of calcium, will demonstrate only a phasic contraction that promptly relaxes to baseline. The addition of calcium in increasing concentrations evokes progressively more tonic contractions (Fig. 2). If potent inhibitors of oxidation [2,4-dinitrophenol (DNP) and sodium azide (NaN_3)] or glycolytic inhibitors [iodoacetic acid (IAA)] are present, these calcium-dependent tonic contractions are depressed. The combination of DNP and IAA produces a complete inhibition of this calcium contractile curve. Similar, though not as dramatic, results were evident when oxygen and glucose were omitted from the artificial CSF medium. These results confirm that potassium stimulates calcium influx down its electrochemical gradient and that this gradient is dependent on a supply of energy for its maintenance. Blockade of both glycolytic and oxidative pathways is required for complete suppression.

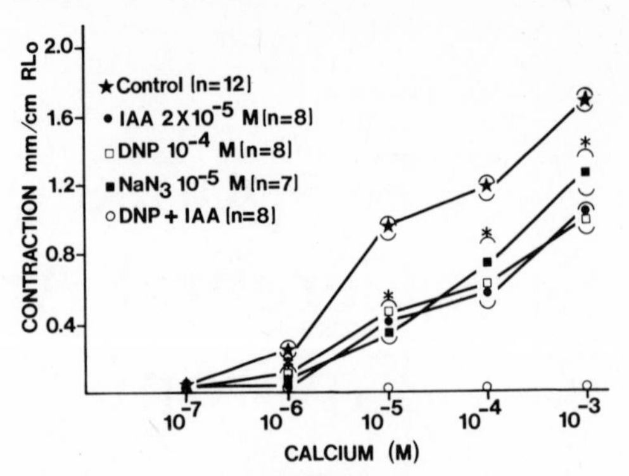

Figure 2. Cumulative calcium dose–response curve in 70 mM potassium solution before and after oxidative and glycolytic inhibition. (IAA) Iodoacetic acid; (DNP) 2,4-dinitrophenol); (NaN_3) Sodium azide. Each point is the mean ± S.D. All points below * are significantly different from control, $p < 0.01$ (two-tailed Student's t test). (RL_o) Optimal resting length of cerebral arterial strip; (n) number of strips.

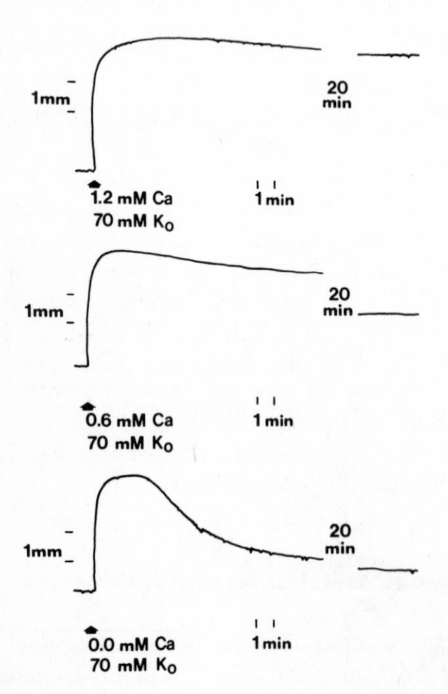

Figure 1. Effect of calcium reduction on tonicity of potassium (Ko) contraction in bovine middle cerebral artery.

2.2. Effects of Energy Metabolism on Biogenic-Amine-Induced Contractions

Cerebral arteries exposed to anoxia or a substrate-free environment for periods up to 4 hr demonstrate a modified ability to contract in response to serotonin (Fig. 3). When the bathing solution is free of both oxygen and glucose, serotonin responses are depressed significantly. Total glucose deprivation alone does not inhibit the serotonin dose–response curve, while the total lack of oxygen produces significant inhibition. Serotonin responses recovered in all preparations despite 4 hr of total deprivation of oxygen or glucose or both. A similar suppression of the serotonin dose–response curve was evident with the oxidative inhibitors DNP and NaN_3 and with the glycolytic inhibitor IAA (Fig. 4). Serotonin-evoked contractions, unlike those evoked by potassium, appear more dependent on the oxidative than on the glycolytic pathways for energy.

The tonicity of the contraction is also affected by hypoxia, but not by glucose deprivation. Serotonin

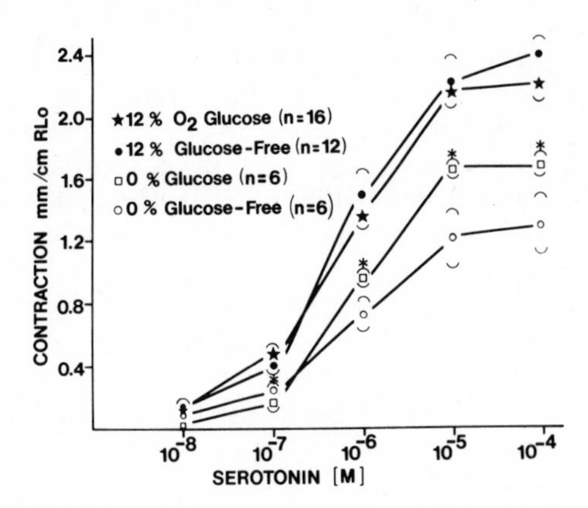

Figure 3. Effects of anoxia and glucose-free environments on serotonin cumulative dose–response curve. Each point is the mean ± S.D. All points below * are significantly different from control (12% O_2–glucose), $p < 0.01$ (two-tailed Student's t test). (RL_o) Optimal resting length of cerebral arterial strip; (n) number of strips.

induced tonic contractions can persist up to 8 hr following a single chemical stimulus. If the environment is anoxic, a phasic contraction of lesser amplitude develops, then quickly relaxes (Fig. 5). This phasic serotonin-evoked contraction in the presence of anoxia is similar to that seen with high-potassium, calcium-free solutions and may indicate impediment of calcium influx. That this may be the case is indicated by the effects of sodium nitroprusside on serotonin-evoked contractions. Sodium

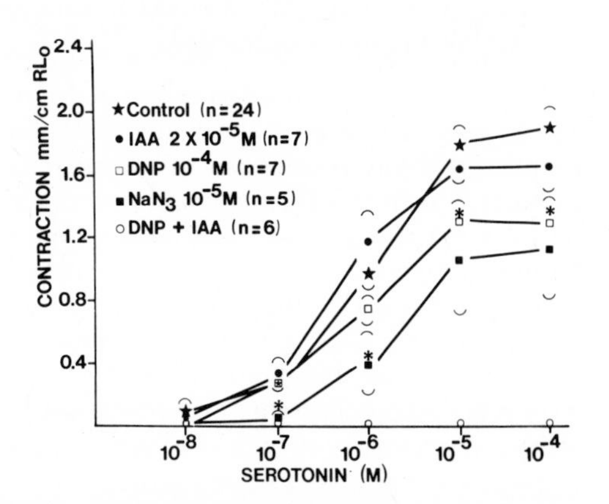

Figure 4. Comparison of the effect of glycolytic (IAA, iodoacetic acid) and oxidative (DNP, 2, 4-dinitrophenol; NaN_3, sodium azide) inhibition of serotonin cumulative dose–response curves. Each point is the mean ± S.D. All points below * are significantly different from controls, $p < 0.01$ (two-tailed Student's t test). (RL_o) Optimal resting length of cerebral arterial strip; (n) number of strips.

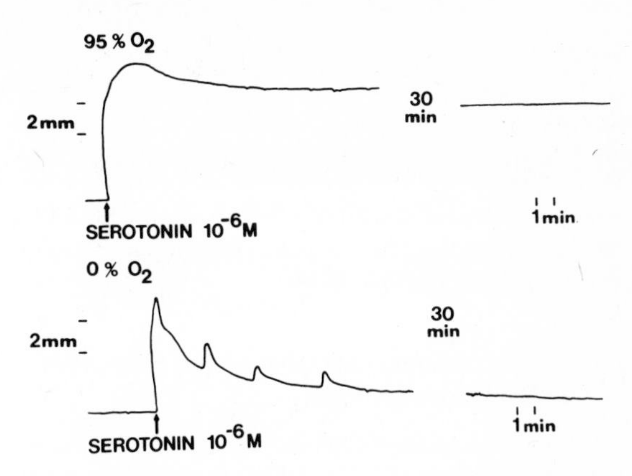

Figure 5. Effect of anoxia on tonicity of serotonin contraction. The tissue was made anoxic 30 min before the addition of serotonin. (%O_2) Oxygen content of gas bubbled into the bath.

nitroprusside, a calcium antagonist that inhibits the tonic portion of smooth muscle contractions,[3] is effective in masking the tonic part of the serotonin-induced contraction in the cerebral artery (Fig. 6), leaving the phasic portion intact.

2.3. Effects of Energy Metabolism on Calcium Movement

Labeled-calcium (^{45}Ca)-uptake studies demonstrated a direct relationship between the availability of oxygen and the amount of intracellular calcium (Fig. 7). In these investigations, intracellular calcium was measured by van Breemen's lanthanum technique.[32] This method is based on the principle that 10 mM lanthanum displaces extracellular calcium, blocks both calcium uptake and efflux, and does not enter the cell to alter intracellular calcium distribution. Intracellular ^{45}Ca decreases with reduction of bath oxygen and conversely increases with elevation of bath oxygen concentration. When oxygen is available, serotonin significantly increases ^{45}Ca uptake as compared to uncontracted tissue. This activation of ^{45}Ca uptake is dependent on the availability of

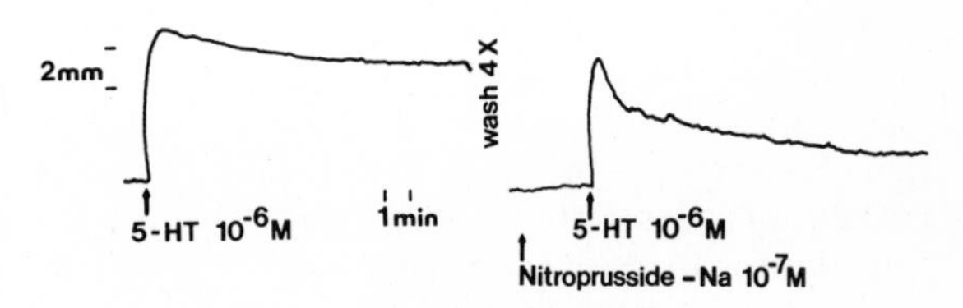

Figure 6. Effect of sodium nitroprusside on tonicity of serotonin (5-HT)-induced contraction.

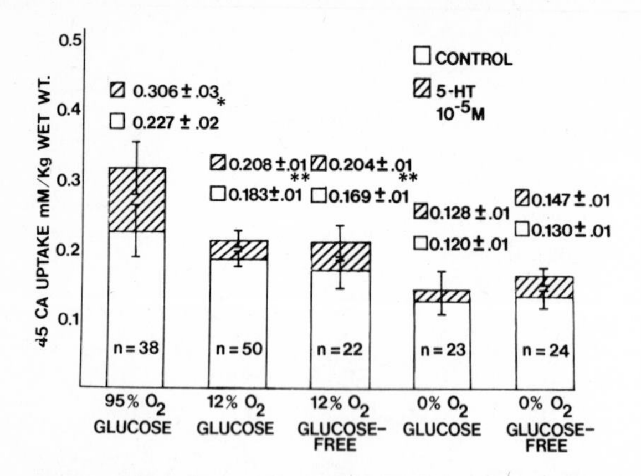

Figure 7. Effect of varying oxygen and glucose availability on labeled calcium (^{45}Ca) uptake in bovine middle cerebral artery. The uptake was determined in serotonin (5-HT)-stimulated and unstimulated (control) arterial strips in a physiological solution. The absolute values ± S.D. of ^{45}Ca uptake are above the bars; (n) number of strips. Significant differences between control and serotonin-stimulated strips are shown: $*p < 0.001$; $**p < 0.01$ (two-tailed Student's t test).

oxygen in the bathing solution. No significant difference in the amount of calcium uptake between serotonin-stimulated and control arteries was observed in an oxygen-free environment. Glucose deprivation did not alter calcium uptake significantly.

Although hypoxia can cause contraction of some isolated arterial smooth muscle preparations,[5] low oxygen tension usually depresses the reactivity of isolated arteries to catecholamines and other agents.[18,19] In our cerebral artery preparation, contractility and tonicity are suppressed by hypoxia. Energy for serotonin-evoked contractions appeared more dependent on oxygen than on glucose availability, whereas potassium-induced contractions may derive its energy from either source.

The physiology of calcium utilization during potassium-induced arterial contraction differs from that occurring during amine-evoked arterial contraction. Van Breemen and Deth[31] have demonstrated that a small intracellular pool of calcium is available for amine activation, and that this pool is sufficient to permit a brief contraction when stimulated by biogenic amines.[12] Prolonged excitation depletes this store and leads to relaxation. The secondary tonic contraction is dependent on the continued presence of the vasoactive substance and on an external calcium source.[9] This secondary contraction reflects a steady transmembrane influx of calcium ions, principally via the so-called T-activation system.[3] In the middle cerebral artery preparation, the tonic portion of the serotonin-evoked contraction was inhibited by oxygen deprivation. Similar effects were noted with the calcium antagonist sodium nitroprusside. Calcium-45 uptake was reduced significantly in a hypoxic environment. Thus, hypoxia in cerebral arteries appears to affect predominantly the T-activation system.

The exact mechanism by which hypoxia reduces smooth muscle calcium uptake is unknown, but interference with adenosine triphosphatase (ATP) mechanisms that provide energy for the transport of calcium across the membrane[6] seems to be a most reasonable explanation. Hypoxia probably interferes with the ability of the smooth muscle cell to regulate its cytoplasmic calcium level and with its ability to maintain ATP available for contraction. As in other tissues, the ATP content of rabbit *Taenia coli* is reduced by hypoxia.[27] Furthermore, exogenous ATP will partially reverse the depression in myocardial contractility produced by anoxia.[21] In addition, cerebrovascular smooth muscle appears to be more sensitive to oxygen changes than to glucose alterations of its environment.

Cerebral arterial spasm is a clinical phenomenon precipitated by the presence of blood products in CSF. In studies[27a] that mimic subarachnoid hemorrhage *in vitro*, cerebral arteries are first contracted with fresh whole blood, then rendered hypoxic. Under these conditions, relaxation is incomplete and permanent contractions follow. When hypoxia develops after a contractile stimulus, the vessel may stay contracted in a non-energy-dependent state for very long intervals. This state, perhaps analogous to *rigor mortis* of skeletal muscle, might explain the extremely long duration of arterial constrictions seen angiographically following a subarachnoid hemorrhage in patients, during which vasoconstrictors are poured into the CSF that bathes cerebral arteries.

3. Hydrogen Ion Alterations and Cerebral Arterial Activity

The smaller arteries of the cerebral circulation, cerebral arterioles, are known to be sensitive to slight changes in hydrogen ion concentration. In fact, one explanation for the sensitivity of cerebral circulation to hypercapnia is the change in extracellular fluid pH evoked by rapid diffusion of carbon dioxide (CO_2) across the blood–brain barrier.[2,14]

Metabolic acidosis and alkalosis have little effect on cerebral blood flow in acute experiments provided the arterial carbon dioxide tension is kept constant, because the normal blood–brain barrier is not rapidly permeable to hydrogen ions.[13] In the bovine *in vitro* model described above (which is based on large cerebral arteries and not necessarily the resistance vessels that control cerebral blood flow), arterial tone was found to be highly sensitive to pH changes.[22] Changes in tone were produced by varying pH between 6.8 and 7.7 by both bicarbonate and CO_2 manipulation. As the pH became more acidotic, progressive arterial relaxation was observed.

3.1. Effects of pH Alterations on Potassium-Induced Contractions

The potassium concentration (K_o) of the artificial CSF solution was varied to manipulate the membrane potential of the smooth muscle cell and to determine whether or not the magnitude or direction of pH changes was potassium-dependent. Consequently, at various concentrations of potassium, the pH was changed and the tone of the artery constantly monitored by isotonic transducers.

The arterial contraction produced by elevation of pH with bicarbonate at K_o = 25 mM increased over 3-fold when compared to that at K_o = 6 mM. However, the magnitude of contraction decreased progressively with potassium concentrations above 25 mM until, at K_o = 100 mM, the contraction was negligible (Fig. 8). Following the reduction of potassium concentrations from 6 to 0 mM, further addition of bicarbonate to raise the pH produced only a transient increase in tone. With prolonged exposure to K_o = 0 mM, the effect of bicarbonate and elevated pH was markedly diminished.

When pH was changed by varying the percentage of CO_2 bubbled into the bath, the effect on tone at each potassium concentration was less than when the pH was changed with bicarbonate variations (Fig. 8). However, the qualitative dependence on potassium concentration was similar. At potassium concentrations of 50 and 100 mM, the effects of pH elevation on arterial strip contraction were the same.

When the CO_2 concentration of the mock CSF was altered and the bicarbonate concentration compensatorily adjusted to keep the pH constant, changes in tone were produced. At 6 mM K_o, the arteries contracted transiently with increasing bath CO_2, and they underwent only slight relaxation with decreasing medium CO_2. At 25 mM K_o, the contraction with increasing CO_2 was maintained, whereas the relaxation that followed the reduction of bath CO_2 content was moderate, but incomplete.

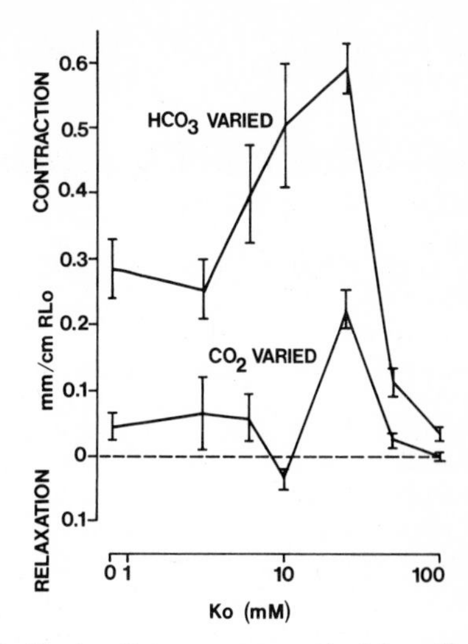

Figure 8. Contractile responses evoked by pH elevation from 6.8 to 7.7 at various potassium (K_o) concentrations. The pH was varied by either bicarbonate or carbon dioxide adjustment. Each point is the mean ± S.D. (vertical brackets) of eight strips.

As previously discussed,[23] these differential effects of carbon dioxide and pH may reflect the more rapid effects of carbon dioxide on intracellular pH. Extracellular pH, whether changed by CO_2 or HCO_3^- variation, will affect superficial membrane-controlled phenomena such as calcium influx, whereas intracellular pH, which only CO_2 will change quickly, might affect intracellular binding of calcium. To examine this possibility, we have studied the effects of pH alterations on calcium movement (see Section 3.3).

3.2. Effects of pH Alterations on Biogenic-Amine-Induced Contractions

Dose–response curves of biogenic-amine-evoked contractions are inhibited by acidity and potentiated by alkalinity of the bathing solution (Fig. 9). The effects of pH on serotonin-induced contractions when varied by bicarbonate alterations did not differ from those produced by CO_2 manipulation (Fig. 9). The contractile effects of norepinephrine, tyramine, isoproterenol, acetylcholine, and histamine were also sensitive to pH variations induced by changes in bicarbonate concentrations (Fig. 10). Both the phasic and the tonic components of the norepinephrine response were decreased by reduction in pH of the bathing solution.

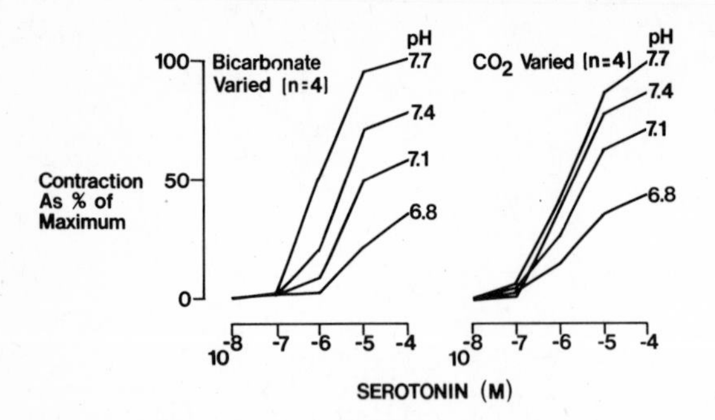

Figure 9. Effect of bicarbonate- and carbon dioxide-varied pH on cumulative dose–response curve of serotonin-induced arterial contractions. The magnitude of the contraction is expressed as the percentage of maximal contraction occurring with addition of serotonin at a pH of 7.7.

There is a modest difference between our results for bovine middle cerebral artery and those of other investigators for pial arteries studied with the perivascular micropuncture technique.[8,17] While acidity inhibits the arterial response to norepinephrine in both the middle cerebral artery and the pial arteriole, only the middle cerebral artery response is potentiated by alkalinity.

3.3. Effects of pH Alterations on Calcium Movement

The contractile effect of high potassium on the smooth muscle peripheral vascular bed has been shown to depend on extracellular calcium. In an effort to define the pH–calcium interaction, cerebral arteries were washed for an hour in a calcium-free solution. These arteries were subsequently stimulated with potassium at 25 mM concentration before the reintroduction of calcium. The pH was altered by adjusting the bicarbonate concentration (Fig. 11). Low pH (6.98) reduced the augmenting effect of calcium on potassium-evoked arterial contractions, whereas high pH (7.60) potentiated this effect. The additional elevation of bath calcium concentrations above 10^{-3} M did not progressively increase this contractile effect, but rather produced mild reductions in vascular tone.

By employing van Breemen's lanthanum technique,[32] we measured the calcium uptake by the cerebral arteries after incubation for 1 hr in three different potassium concentrations at two pH levels (altered by bicarbonate buffer variation). Atomic absorption spectrometry confirmed that calcium content after lanthanum was elevated as the bath potassium concentration increased (Table 1). Calcium-45 uptake was both potassium- and pH-dependent (Table 2). Calcium-45 uptake was not affected by varying the CO_2 content of the bath at a pH held constant by bicarbonate buffering. However, ^{45}Ca uptake was enhanced by increasing the bath potassium concentration, and in addition was reduced by lowering the pH from 7.7 to 6.8 at a bath postassium concentration of 25 mM.

Calcium is the most significant ion in the final common pathway of smooth muscle contraction. Its

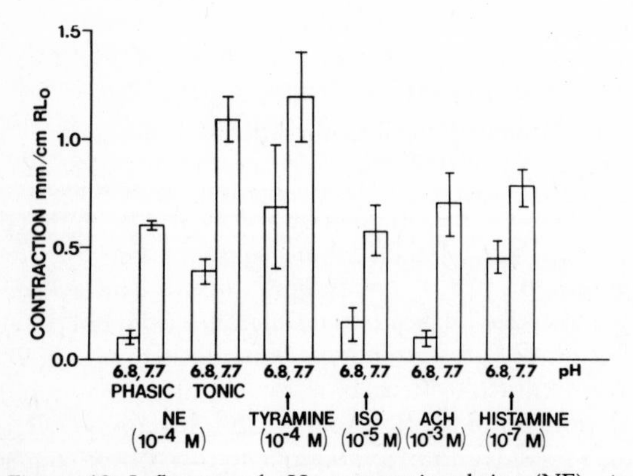

Figure 10. Influence of pH on norepinephrine (NE)-, tyramine-, isoproterenol (ISO)-, acetylcholine (ACH) and histamine-induced contractions of bovine cerebral arteries. (RL_o) Optimal resting length of arterial strip. Each bar is the mean ± S.D. (vertical brackets) of four strips.

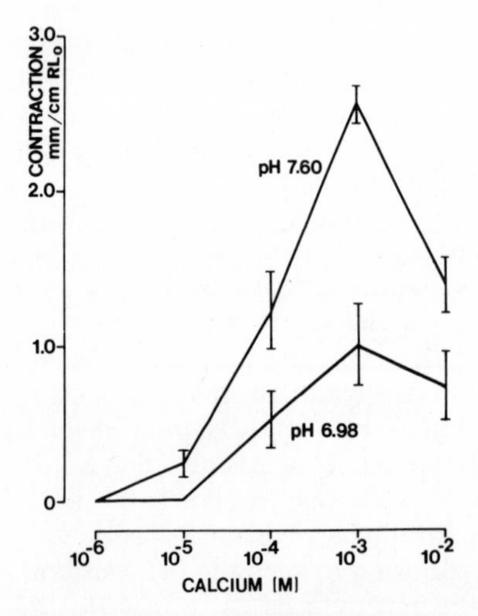

Figure 11. Effect of pH on cumulative calcium dose–response curve. The arterial contraction stimulus was elevation of the bath potassium concentration from 3 to 25 mM. (RL_o) Optimal resting length of arterial strip. Each point is the mean ± S.D. (vertical brackets) of eight strips.

Table 1. Effects of Potassium on Calcium Uptake as Measured by Atomic Absorption Spectrometry

K_o (mM)	Calcium (mM/kg wet wt. of tissue)	
6	0.32 ± 0.04	
		$p < 0.005$[a]
25	0.55 ± 0.05	
		$p < 0.05$[a]
100	0.69 ± 0.04	

[a] Level of significance found with two-tailed Student's *t* test.

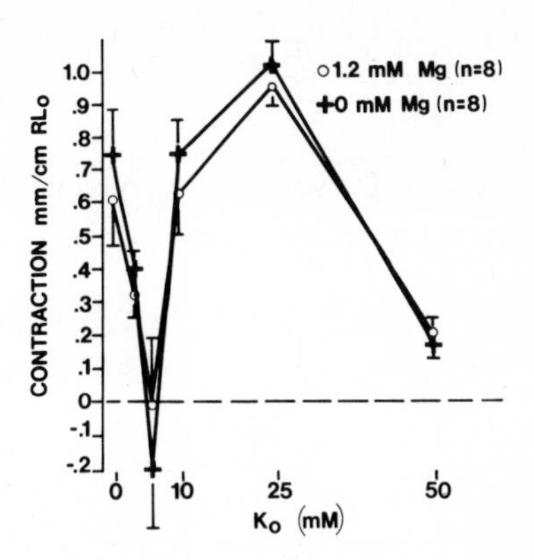

Figure 12. Effect of the presence or absence of magnesium (Mg) on cerebral arterial contractions induced by pH elevation from 6.8 to 7.7 with bicarbonate at various extracellular potassium (K_o) concentrations. Points below the dashed line represent relaxation. Each point is the mean ± S.D. (vertical brackets) of eight strips.

availability controls the tone of the cerebral vessel by its interaction with other ions and neurotransmitters. As seen above, potassium, pH, and serotonin effects were all dependent on the presence of calcium in the extracellular fluid.

3.4. Effects of Magnesium on pH-Induced Contractions

Magnesium ions are involved in the regulation of the permeability of the cell membrane to various ions and may modulate vascular reactivity.[1] Sensitivity and maximal response of smooth muscle to acetylcholine, potassium, and angiotensin are greater in the absence of magnesium.[11] The response to serotonin is also enhanced by magnesium depletion.[11]

In our experiments, pH elevations from 6.8 to 7.7 with bicarbonate manipulation were employed to stimulate arterial contractions in the presence of various potassium concentrations in the bath. The magnitude of these contractions in the presence of physiological levels of magnesium was compared to that of those occurring in the absence of extracellular magnesium (Fig. 12). No significant differences were noted at any potassium (K_o) concentration; thus, magnesium may not be involved in tone alterations produced by pH alterations of the CSF environment.

3.5. Effects of pH Alterations at Various CSF Osmolarities

The brain is constantly adjusting its osmolarity to maintain its steady state between the extracellular fluid, CSF and blood. Hyperosmolarity is a cerebral vasodilator *in vivo,* since hyperosmolar solutions dilate pial arteries on local application.[33] We investigated the interaction between pH and osmolarity because the changes in ionic equilibria provoked by the latter might help to elucidate the effects of the former. Sodium chloride was used to alter the osmolarity of the bathing solution. These experiments were performed at a bath potassium concentration of 25 mM to produce baseline tone in the bovine middle cerebral artery. Increasing osmolarity relaxed, and decreasing osmolarity contracted, the cerebral artery (Fig. 13). Unlike the effect of extra-

Table 2. Effects of Potassium and pH on ^{45}Ca Uptake (mM/kg wet wt. of Tissue) as Measured by the Lanthanum Method[a]

K_o (mM)	pH 7.7	(*N*)		pH 6.8	(*N*)	
6	0.221 ± 0.018	(19)		0.209 ± 0.019	(19)	
			$p < 0.005$[b] (6 vs 25, pH 7.7)			
25	0.314 ± 0.022	(8)	← $p < 0.005$[b] →	0.229 ± 0.012	(8)	
			$p < 0.05$[b] (25 vs 100, pH 7.7)			$p < 0.005$[b] (25 vs 100, pH 6.8)
100	0.414 ± 0.034	(16)		0.369 ± 0.033	(8)	

[a] Reproduced in part from Pickard *et al.*[23]
[b] Level of significance found with two-tailed Student's *t* test.

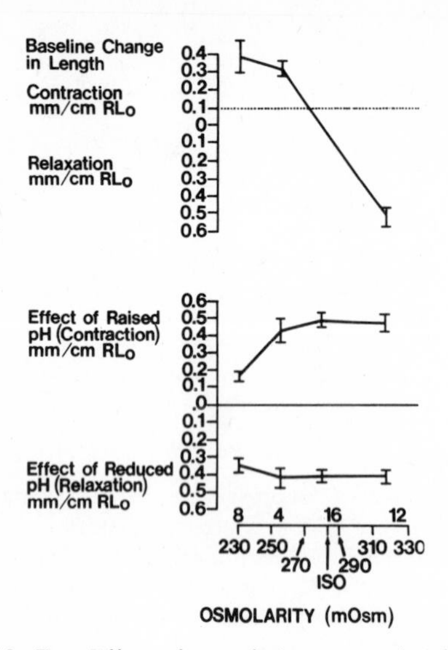

Figure 13. *Top:* Effect of osmolarity on resting length of middle cerebral artery in mock CSF. *Bottom:* Effect of increased pH (6.8–7.7) and decreased pH (7.7–6.8) on resting length of middle cerebral artery in relation to the osmolarity of the bathing solution. The numbers above the horizontal axis are the numbers of strips tested at each osmolarity. (RL_o) Optimal resting length of cerebral arterial strip; (ISO) isotonic osmolarity.

cellular potassium on contractions induced by pH elevations, the arterial response to pH was not altered by osmolarity except at the low value of 233 mosmol. Correlations between osmolarity and pH are not as yet adequately explained.

4. Summary

In vivo experiments on pial vessels have shown the influence of various CSF ion changes on vasomotor tone. It has been suggested (for a review, see Heuser[16]) that the CSF ions are information mediators from neuronal tissue to the cells of cerebrovascular smooth muscle by which blood flow adjusts to the metabolic demands of the brain tissue. We have demonstrated a dependence of the tone of large cerebral vessels on a steady supply of oxygen that is related to an ultimate dependence on a steady influx of transmembrane calcium. Cerebral arterial smooth muscle is pH-sensitive within the physiological range. Acidity causes a decrease in vascular tone, whereas alkalinity evokes an increase in tone. These effects can be modified by alterations of the potassium concentration in the external environment. Consistent differences between contractions stimulated by biogenic amines and potassium exist. At different potassium concentrations, the effects of pH on cerebral vessel contraction appear to be less when these pH changes are produced by CO_2, rather than by bicarbonate manipulation. These effects of extracellular pH on the vascular tone seem to be mediated by calcium uptake. Calcium uptake by the cerebral artery increases with alkalinity and decreases with acidity of the external environment. Small changes in osmolarity and the elimination of magnesium from the artificial CSF do not affect the pH response of the cerebral vessel. The net effects of metabolic and ionic changes in the CSF on large cerebral arterial contractions are consistent with those found in the smaller cerebral vessels, which have a greater role in the regulation of blood flow. These effects, as discussed in this chapter, suggest a role of CSF as a medium for the exchange of information involved in the regulation of cerebral blood flow and metabolism.

Because of the *in vitro* nature of these experiments, and the species differences, clinical inferences must be drawn cautiously from these data. However, the *in vitro* preparation provides a high degree of precise environmental control that is not attainable by *in vivo* techniques such as the pial window method.

References

1. ALTURA, B. M., ALTURA, B. T.: Influence of magnesium on drug-induced contractions and ion content in rabbit aorta. *Am. J. Physiol.* **220:**938–944, 1971.
2. BETZ, E., HEUSER, D.: Cerebral cortical blood flow during change of acid–base equilibrium of the brain. *J. Appl. Physiol.* **23:**726–733, 1967.
3. BOEV, K., GOLENHOFEN, K., LUKANOW, J.: Selective suppression of phasic and tonic activation mechanisms in stomach smooth muscle. *Pfluegers Arch.* **343:**56, 1973.
4. BOHR, D., GOULET, P., TAQUINI, A.: Direct tension recording from smooth muscle of resistance vessels from various organs. *Angiology* **12:**478–485, 1961.
5. DETAR, R., BOHR, D. F.: Contractile responses of isolated vascular smooth muscle during prolonged exposure to anoxia. *Am. J. Physiol.* **222:**1269–1277, 1972.
6. DHALLA, N. S.: Defect in calcium regulatory mechanisms in heart failure. In Dhalla, N. S. (ed.): *Myocardial Biology: Recent Advances in Studies on Cardiac Structures and Metabolism.* Baltimore, University Park Press, 1974, pp. 331–345.
7. EDVINSSON, L., HARDEBO, J. E., OWMAN, C.: Pharmacological analysis of 5-HT receptors in isolated intracranial and extracranial vessels of cat and man. *Circ. Res.* **42:**143–151, 1978.

8. EDVINSSON, L., SERCOMBE, R.: Influence of pH and pCO_2 on alpha-receptor mediated contraction in brain vessels. *Acta Physiol. Scand.* **97**:325–331, 1976.
9. FASTIER, F. N., PURVES, R. D., TAYLOR, K. M.: Observations of fade: A complication of the contractile response of smooth muscle to a large dose of an agonist. *Br. J. Pharmacol.* **49**:490–497, 1973.
10. FUJIWARA, M., MURAMATSU, I., SHIBATA, S.: Gamma aminobutyric acid receptors on vascular smooth muscle of dog cerebral arteries. *Br. J. Pharmacol.* **55**:561–562, 1975.
11. GOLDSTEIN, S., ZSOTER, T. T.: The effect of magnesium on the response of smooth muscle to 5-hydroxytryptamine. *Br. J. Pharmacol.* **62**:507–514, 1978.
12. HAEUSLER, G.: Cellular calcium stores and contraction in vascular smooth muscle. *Experientia* **29**:762–763, 1973.
13. HARPER, A. M., BELL, R. A.: The failure of intravenous urea to alter the blood flow through the cerebral cortex. *J. Neurol. Neurosurg. Psychiatry* **26**:69–70, 1963.
14. HARPER, A. M., GLASS, H. I.: Effect of alterations in arterial carbon dioxide tension on blood flow through cerebral cortex at normal and low arterial blood pressure. *J. Neurol. Neurosurg. Psychiatry* **28**:449–452, 1965.
15. HAYASHI, S., TODA, N.: Inhibition by Cd^{2+}, verapamil, and papaverine of Ca^{2+} induced contractions in isolated cerebral and peripheral arteries of the dog. *Br. J. Pharmacol.* **60**:35–43, 1977.
16. HEUSER, D.: The significance of cortical extracellular H^+, K^+, and Ca^{2+} activities for regulation of local cerebral blood flow under conditions of enhanced neuronal activity. In Purves, J. (ed.): *Ciba Foundation Symposium 56: Cerebral Vascular Smooth Muscle and Its Control.* Amsterdam, Elsevier/North-Holland, 1978, pp. 339–348.
17. KUSCHINSKY, W., WAHL, M., BOSSE, G., THURAU, W.: Perivascular potassium and pH as determinants of local pial arterial diameter in cats. *Circ. Res.* **16**:240–247, 1972.
18. LLOYD, T. C., JR.: Responses to hypoxia of pulmonary arterial strips in nonaqueous baths. *J. Appl. Physiol.* **28**:566–569, 1970.
19. NEEDLEMAN, P., BLEHN, D. J.: Effect of epinephrine and potassium chloride on contraction and energy intermediates in rabbit thoracic aorta strips. *Life Sci.* **9**:1181–1189, 1970.
20. NIELSON, K., OWMAN, C.: Contractile responses and amine receptor mechanisms in isolated middle cerebral artery of the cat. *Brain Res.* **27**:33–42, 1971.
21. PARRATT, J. R., MARSHALL, R. J.: The response of isolated cardiac muscle to acute anoxia: Protective effect of adenosine triphosphate and creatine phosphate. *J. Pharm. Pharmacol.* **26**:427–433, 1974.
22. PICKARD, J. D., SIMEONE, F. A., SPURWAY, N. C., VINALL, P. E., LANGFITT, T. W.: Mechanisms of the pH effect on tone of bovine middle cerebral arterial strips *in vitro.* In Harper, A. M., Jennett, B., Miller, D., Rowan, J. O., (eds.): *Blood Flow and Metabolism in the Brain.* Edinburgh, Churchill Livingstone, 1975, pp. 9.17–9.18.
23. PICKARD, J. D., SIMEONE, F. A., VINALL, P. E.: H^+, CO_2, Prostaglandins and cerebrovascular smooth muscle. In Betz, E. (ed.): *Ionic Actions on Vascular Smooth Muscle.* Berlin, Springer-Verlag, 1976, pp. 101–104.
24. PURVES, M. J.: Control of cerebral blood vessels: Present state of the art. *Ann. Neurol.* **3**:377–383, 1978.
25. RAPOPORT, S. I.: *Blood Brain Barrier in Physiology and Medicine.* New York, Raven Press, 1976.
26. ROSENBLUM, W. I., CHEN, M.: Comparison of nerves to cerebral and extracerebral blood vessels: A differential effect of alpha methyl tyrosine and norepinephrine content. *Stroke* **8**:391–392, 1977.
27. SIEGMAN, M. J., BUTLER, T. M., MOOERS, S. U., DAVIES, R. E.: Crossbridge attachment, resistance to stretch, and viscoelasticity in resting mammalian smooth muscle. *Science* **191**:383–385, 1976.
27a. SIMEONE, F. A., VINALL, P. E.: Effect of oxygen and glucose deprivation on vasoactivity and calcium movement in isolated middle cerebral arteries. In Wilkins, R. H. (ed.) : *Cerebral Arterial Spasm: Proceeding of the Second International Workshop.* Baltimore, Waverly Press, 1980.
28. SOKOLOFF, L.: Circulation and energy metabolism of the brain. In Albers, G. J., Siegal, G. J., Katzman, R., Agranoff, B. W. (eds.): *Basic Neurochemistry.* Boston, Little, Brown, 1972, pp. 299–325.
29. TODA, N.: Mechanical responses of isolated dog cerebral arteries to reduction of external K, Na and Cl. *Am. J. Physiol.* **234**:H404–411, 1978.
30. TODA, N.: Potassium-induced relaxation in isolated cerebral arteries contracted with prostaglandin F_2 alpha. *Pfluegers Arch.* **364**:235–242, 1976.
31. VAN BREEMAN, C., DETH, R.: La^{+++} and excitation contraction coupling in vascular smooth muscle. In Betz, E. (ed.): *Ionic Actions on Vascular Smooth Muscle.* Berlin, Springer-Verlag, 1976, pp. 26–33.
32. VAN BREEMAN, C., FARINAS, B. R., GERBA, P., MCNAUGHTON, E. D.: Excitation–contraction coupling in rabbit aorta studied by the lanthanum method for measuring cellular calcium influx. *Circ. Res.* **30**:44–53, 1972.
33. WAHL, M., KUSCHINSKY, W., BOSSE, O., THURAU, K.: Dependency of pial arterial and arteriolar diameter on perivascular osmolarity in the cat: A microapplication study. *Circ. Res.* **32**:162–169, 1973.

CHAPTER 23

Subarachnoid Erythrocytes and Their Contribution to Cerebral Vasospasm

Kunihiko Osaka, Hajime Handa, and Shinichiro Okamoto

1. Introduction

Erythrocytes are not present in normal cerebrospinal fluid (CSF), and their presence there implies subarachnoid hemorrhage. The differential diagnosis of subarachnoid hemorrhage includes ruptured aneurysms or arteriovenous malformations, hypertensive cerebral hemorrhage, head trauma, and meningitis. Although erythrocytes are commonly found in the CSF of patients with vascular pathology, their clinical importance may not be well appreciated.

In patients with ruptured aneurysms, large volumes of extravasated blood may form space-occupying hematomas within the brain or subdural spaces. Usually, however, the volume is small and the extravasated blood spreads only through the subarachnoid space. Although a "small-volume" hemorrhage produces no significant "mass effect," such subarachnoid bleeding may induce severe neurological deterioration. These deficits are apparently produced by the toxic effect of the subarachnoid blood itself. Among various constituents of the blood, erythrocytes seem to be the most noxious component. In this chapter, the role of the subarachnoid erythrocytes will be reviewed, mainly in relation to "cerebral vasospasm" and "erythrogenic meningitis."

Kunihiko Osaka, M.D., Hajime Handa, M.D., and **Shinichiro Okamoto, M.D.** • Department of Neurosurgery, Kyoto University School of Medicine, Kyoto, Japan.

2. Natural History of CSF Erythrocytes after Subarachnoid Hemorrhage

The extravasated erythrocytes are eventually cleared from the CSF. Their rate of clearing is highly variable, ranging from 6 to 30 days or more.[1,4,6,60,64] The clearing tends to be slower in patients of advanced age and with a past or family history of diabetes, hypertension, and arteriosclerotic cardiovascular and cerebrovascular disease.[64]

Most erythrocytes placed in the subarachnoid space become enmeshed and fixed in the arachnoid or form clumped masses.[1,4,6,24] Between 2 and 4 hr after hemorrhage, the release of the contents of some erythrocytes results in xanthochromia of the CSF.[40] Most extravasated erythrocytes, however, remain in an intact state for the first 12–16 hr following hemorrhage.[4,24] The histological evidence of significant breakdown of erythrocytes (hemosiderin deposits and brown pigment) does not appear until 12–32 hr. The breakdown of erythrocytes continues with time and reaches a peak at approximately the 7th day after hemorrhage. During this period, segmented, clumped erythrocytes forming small islands are gradually penetrated by thin bands arising from the pia and the arachnoid. These erythrocytes are then broken down and engulfed by macrophages, and blood escapes into the neighboring tissue. Such packed erythrocytes may be present up to 35 days or more after the initial hemorrhage.[4,13,24] The number of erythrocytes in lumbar CSF is not an accurate indicator of the clearance of subarachnoid erythrocytes, since lumbar CSF seems to be cleared earlier than the intracranial subarachnoid space.[4,24,64]

Lyzed erythrocytes release various substances into the CSF, including electrolytes, amino acids, and hemoglobin. Of these, the blood pigments seem noxious to the brain, as will be discussed later. The blood pigments released from erythrocytes include oxyhemoglobin, bilirubin, and methemoglobin,[7,36] and can be easily identified by spectrophotometry. Oxyhemoglobin is the initial pigment liberated into the CSF, appearing, though in very small amounts, 2 hr after subarachnoid hemorrhage. The amount of oxyhemoglobin increases to a maximum over several days, then gradually diminishes over a 7- to 9-day period.[7,36] Bilirubin is the iron-free derivative of hemoglobin, and its production from heme-compounds requires an appropriate reducing system. The enzymes necessary for the degradation of these heme-compounds are present in the arachnoid, macrophages, and choroid plexus.[40,53,55] Bilirubin appears 2–3 days after hemorrhage, increasing in amount as oxyhemoglobin decreases, and persists for 2–3 weeks. Methemoglobin, the ferric form of hemoglobin, is absent or present only in very small amounts in the CSF of patients with subarachnoid hemorrhage. This blood pigment is found mainly in subdural and intracerebral hematomas or in fluid near encapsulated blood.[7]

3. Role of Erythrocytes in Cerebral Vasospasm

The pathogenesis of cerebral vasospasm is not well understood, despite intensive investigative efforts. In 1965, Echlin[15] demonstrated that severe vasoconstriction was induced by topical application of fresh blood to the basilar artery of the monkey. Although this vasoconstriction was of short duration, many later investigators[9,16,19,42,63] succeeded in producing prolonged vasospasm by cisternal injection of blood or by sectioning small cerebral arteries. It is now accepted that the subarachnoid blood itself is the cause of vasospasm. Of the numerous substances contained in blood, vasoconstrictor substances released from the platelets were considered to be the primary cause of vasospasm.[3,29,30,37,51,54,62] Serotonin[3,54] and prostaglandins[37,51,52] from platelets were demonstrated to induce vasoconstriction in the cerebral arteries. Contrary to this general belief, our investigations have suggested that subarachnoid erythrocytes, not the platelets, release the substance causing delayed vasospasm.[46,47]

In our experiments, the basilar arteries of cats were exposed and subjected to the topical application of various blood fractions that were either fresh or incubated for 1–7 days. Incubation was performed to evaluate the stability of the vasoconstrictors. Severe vasospasm was induced by application of fresh and incubated fractions of lyzed erythrocytes. Fresh, intact erythrocytes had no vasoactivity, but during incubation, cell lysis released vasoconstrictive substances (Fig. 1 and Tables 1 and 2). Vasospasm induced by lyzed erythrocytes, either fresh or after prolonged incubation, did not relax but tended to increase in severity up to 24 hr (Fig. 2). These findings strongly indicate that lyzed erythrocytes are the cause of delayed clinical vasospasm.

Fresh serum and platelet-rich plasma fractions contain such vasoconstrictors as serotonin and prostaglandins. Application of these fractions resulted in vasoconstriction, as other investigators have re-

Table 1. Response of Basilar Artery to Topical Applications of Fresh Blood Fractions[a]

Fresh Blood Fractions	−60% −50% −40% −30% −20% −10% 0 +10% +20%	Average
heparinized whole blood		−22%
platelet-poor plasma		−2%
platelet-rich plasma		−18%
serum		−20%
intact erythrocytes		−2%
lysed erythrocytes		−23%

[a] Reproduced from Osaka,[47] with permission. Each dot represents data from one animal.

Table 2. Response of Basilar Artery to Topical Application of Incubated Blood Fractions[a]

Incubated Blood Fractions	−60% −50% −40% −30% −20% −10% 0 +10% +20%	Average
platelet-poor plasma	••• ••••	+6 %
platelet-rich plasma	••••• • ••	+10%
serum	•• •••••••	−1 %
intact erythrocytes	• ••• • •• •••• ••• • •	−29%
lysed erythrocytes	•• •• • • •	−30%

[a] Reproduced from Osaka,[47] with permission. Each dot represents data from one animal.

ported.[3,30,37,54] However, the vasoconstrictive activity of these fractions was lost after incubation for 24 hr. The vasoconstrictors in platelets appear to be unstable and cannot survive incubation for 24 hr. The relaxation within a few hours of the vasospasm due to fresh subarachnoid bleeding (Fig. 3) also suggests that platelet-evoked vasospasm is of short term. This vasospasm is presumed to be platelet-induced, since erythrocytes are not yet lyzed. Clinical cerebral vasospasm is a prolonged reduction of the diameter of the cerebral arteries, which usually persists for 1–2 weeks. The agent causing vaso-

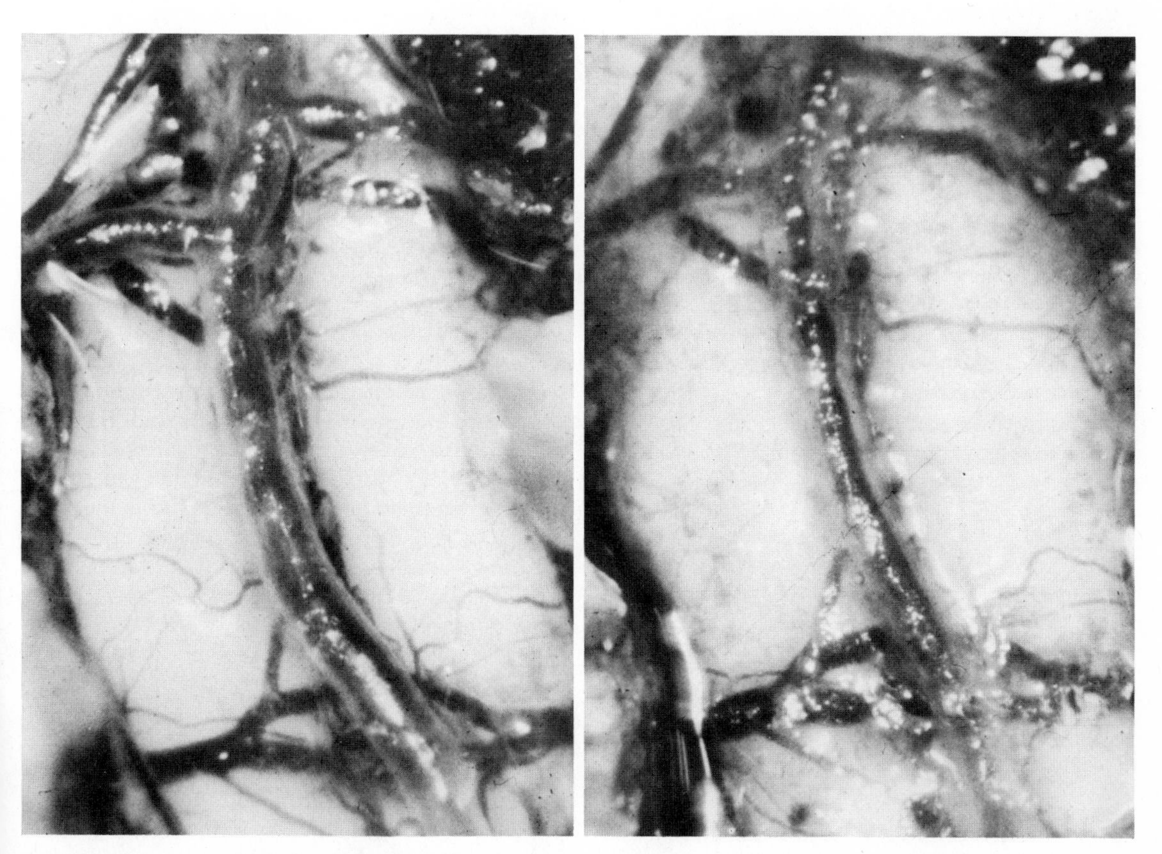

Figure 1. Vasospasm produced by topical application of incubated lyzed erythrocyte fraction to basilar artery of cat. *Left:* control. *Right:* 1 hr after application of incubated lyzed erythrocyte fraction. Reproduced from Osaka,[47] with permission.

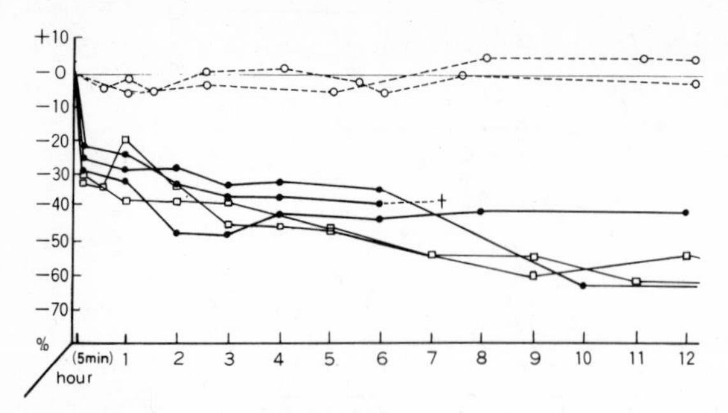

Figure 2. Sequential changes of vasospasm induced by topical application of lyzed erythrocyte fraction. Each line represents data from one animal. (○) Controls; (□) application of fresh lyzed erythrocytes; (●) application of incubated lyzed erythrocytes. Observation was extended to 24 hr in three cats to which fresh or incubated lyzed erythrocytes had been applied. No relaxation of vasospasm was observed.

spasm should therefore be stable at body temperature and should be able to induce vasoconstriction that persists for at least several days. Therefore, prolonged vasospasm is unlikely to be caused by the unstable vasoconstrictors in platelets. Even if vasoconstrictors in platelets do play some role, their duration of action would be limited to the period immediately after the episode of subarachnoid hemorrhage. Allen *et al.*[3] postulated that serotonin is constantly released by the lysis of agglutinated platelets in blood clots, thus keeping the affected cerebral arteries in spasm. But Tani and co-workers[62,67] reported that the platelets in the subarachnoid blood clot are devoid of serotonin. In addition, Buckell[10] found no correlation between the presence of vasospasm and the content of serotonin in CSF around the ruptured cerebral aneurysm.

Vasoconstrictor activity of lyzed erythrocytes has been reported by other investigators. The vasoconstrictive activity in lyzed erythrocytes observed by Zucker[71] was not initially felt to be clinically significant because the vasoactivity per unit volume of lyzed erythrocytes amounts to only 1/80th of that of platelets. Kapp *et al.*[29] found that lyzed erythrocytes diluted 10-fold produced significant vasoconstriction; however, this observation was attributed to vasospasm induced by an unidentified polypeptide contained in platelets.[30] Wilkins and co-workers[65,66] reported that hemolyzed plasma, xanthochromic CSF, and incubated purple-brown serum containing substances from lyzed erythrocytes had vasoconstrictive activity, whereas nonhemolyzed plasma, clear, colorless CSF, and incubated serum had little or no vasoactivity. From these findings, it was speculated that vasoconstrictive substances were liberated with hemolysis.

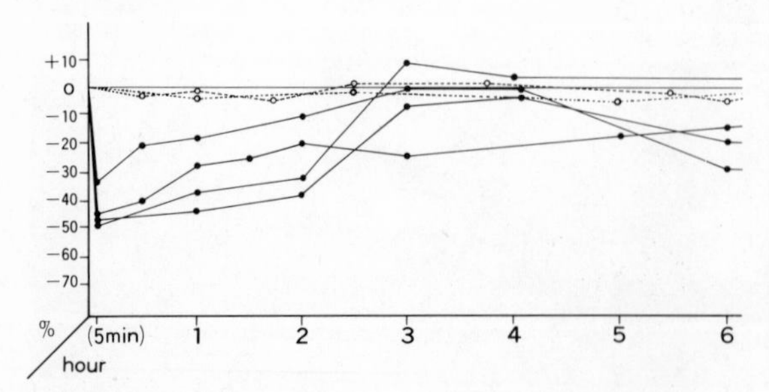

Figure 3. Sequential changes of spasm induced by fresh subarachnoid bleeding. Each line represents data from one animal. (○) Controls; (●) fresh subarachnoid bleeding.

Autopsy findings also suggest that lyzed erythrocytes, especially their iron pigments, cause the vasospasm seen in subarachnoid hemorrhage. Most of the blood extravasated from ruptured aneurysms remains in the basal cisterns or spreads along the major cerebral vessels, and blood pigments are more abundant near the arteries in vasospasm.[12,13] Prominent staining of the vasospastic arteries by iron pigments was found in experimental animals.[16,57]

Vasoconstrictor substances appear to be released with the lysis of erythrocytes. The lysis of the erythrocytes in the subarachnoid space proceeds gradually to reach the maximum level in 5–10 days after subarachnoid hemorrhage, and most of the erythrocytes are cleared of CSF in about 30 days.[1,5,6,64] This pattern of erythrocyte breakdown coincides well with the timing of the cerebral vasospasm. Clinically significant cerebral vasospasm is unusual shortly after aneurysmal rupture. Most frequently, this arterial constriction peaks at about the 2nd week, then subsides over the 3rd and 4th weeks after the episode.[8,38,56] Brawley [9] reported that vasospasm is a biphasic phenomenon consisting of a short initial phase and a prolonged late phase with relaxation of vasospasm between the two phases. Our experimental results indicate that the initial short phase (probably lasting for only a few hours) is of platelet origin, whereas the late phase of spasm appears with lysis of erythrocytes and persists for several weeks. Many investigators thought that vasospasm is caused by vasoconstrictors of platelet origin because cerebral arteries constrict by application of platelet products. The ability of a substance to induce severe vasoconstriction is not sufficient evidence to establish that substance as the etiology of clinical vasospasm. Vasospasm induced by that substance must be demonstrated to persist for at least several days, and thus mimic the time course of clinical vasospasm.

Our observations have been confirmed by several other groups in Japan[11,42,58] who reported that lyzed erythrocytes, fresh or aged for several days, can induce severe, long-lasting vasospasm, whereas serum rapidly loses its vasoconstrictive activity on incubation at 37°C. Identification of the vasoconstrictive substance or substances in erythrocytes has been accomplished by employing biochemical analysis, heat coagulation, ultrafiltration, Sephadex gel-chromatography, disk electrophoresis, or spectrophotometry. Chokyu[11] and Miyaoka[42] have suggested that the vasoconstrictive substance is a polypeptide closely allied to oxyhemoglobin. Sonobe and Suzuki[58,59] have concluded that the vasoconstrictor is oxyhemoglobin itself and that oxyhemoglobin loses its vasoconstrictor activity if it is converted to methemoglobin. However, Yoshida[69] found that both methemoglobin and oxyhemoglobin are potent vasoconstrictors. At present, oxyhemoglobin or an oxyhemoglobin-like protein has been shown to have vasoconstrictor activity, but the vasoconstrictor activity of methemoglobin has not yet been determined. The vasoconstrictor activity of bilirubin is much weaker than that of oxyhemoglobin, and is considered insufficient to produce clinical vasospasm.[11]

Although many investigators agree that lyzed erythrocytes are capable of producing prolonged vasospasm,[11,42,47] the exact mechanism is unknown. The toxic products released by the erythrocytes may stimulate the perivascular nerves or directly irritate the arterial smooth muscle. The function of the numerous perivascular nerve fibers is not fully understood, but vasoconstriction is generally considered to be induced by excitation of the perivascular sympathetic nerves.[44,48–50] Our studies employing Falck's fluorescence histochemical method[18] have identified a rich ground plexus of adrenergic fibers that appear as green fluorescent dots lying within the deeper layer of the adventitia (Fig. 4). This catecholamine fluorescence is rapidly lost after subarachnoid hem-

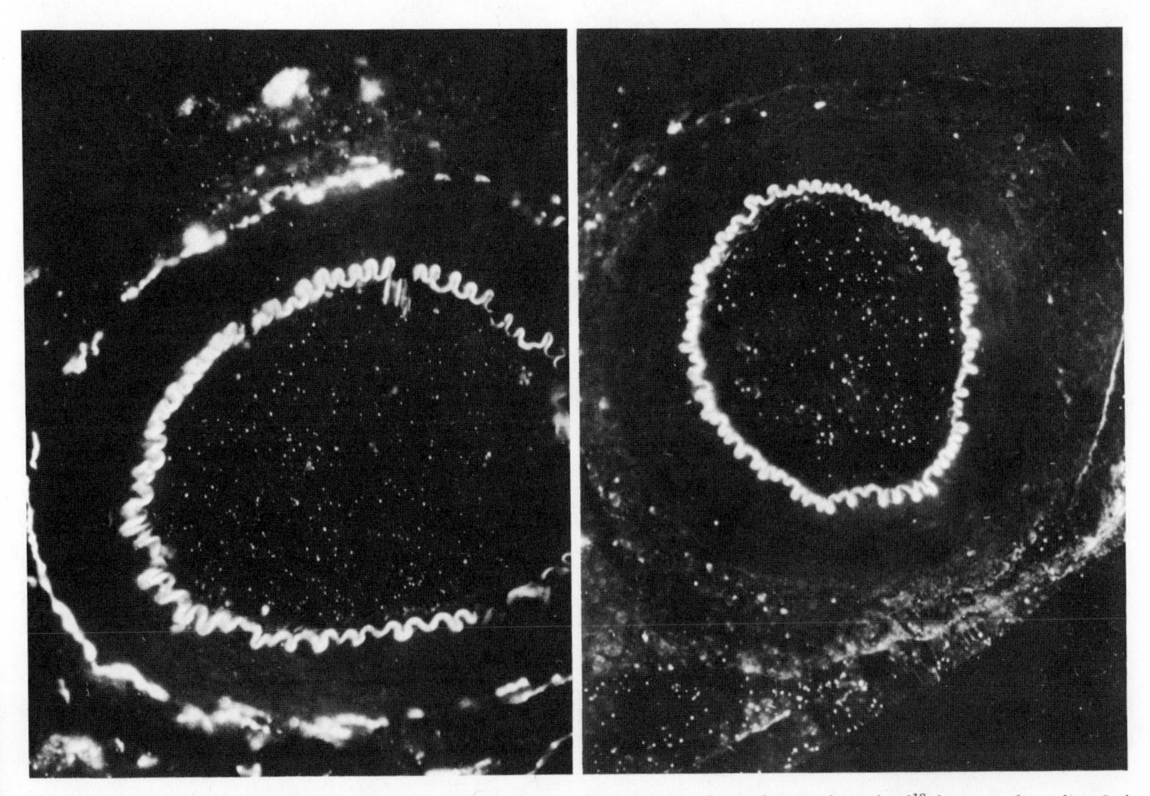

Figure 4. Section of basilar artery of cat treated by Falck's fluorescence histochemical method[18] for noradrenalin. *Left:* control. Many fluorescent dots of noradrenalin are seen lying in deeper layer of adventitia. *Right:* 1 week after extirpation of bilateral superior cervical sympathetic ganglia. The fluorescent dots of noradrenalin are not seen.

orrhage.[50] Endo *et al.*,[17] in their electron-microscopic study, similarly reported the gradual transformation, diminution, and disappearance of small cored vesicles of sympathetic nerve endings after subarachnoid bleeding. Sympathetic nerve fibers were also noted to be abundant in segments of arteries in severe vasospasm, but were scant in adjacent segments of the same arteries where vasospasm was milder. Endo *et al.*[17] concluded that adrenergic axons may play an important role in the genesis of erythrocyte-induced vasospasm.

However, erythrocyte-evoked vasospasm can develop under conditions in which the adrenergic fibers are inactive. The fluorescent dots of noradrenalin in the arterial wall disappear 1 week after removal of both superior cervical sympathetic ganglia[31,46] (Fig. 4). Despite perivascular sympathetic denervation, lyzed erythrocytes produce the same degree of vasospasm as seen prior to sympathectomy (Fig. 5). This finding indicates that augmented sympathetic activity is unlikely to play a major role. However, Peerless and Kendall[50] have suggested that vasospasm and its propagation to vessels remote from the portion bathed with subarachnoid blood may represent a form of denervation supersensitivity. Contrarily, Toda *et al.*[63] found that the response to various vasoconstrictors of cerebral arteries bathed with blood for 24 hr or 7 days was significantly less than that of controls, and concluded that vasoconstriction is not induced by arterial denervation supersensitivity.

4. Direct Irritation of Brain by Erythrocytes

Cerebral vasospasm has been associated with decreased cerebral blood flow and resultant neurological deterioration.[2,45,56] Fisher *et al.*[21] recently demonstrated that severe vasospasm is an important cause of the delayed neurological deficit that de-

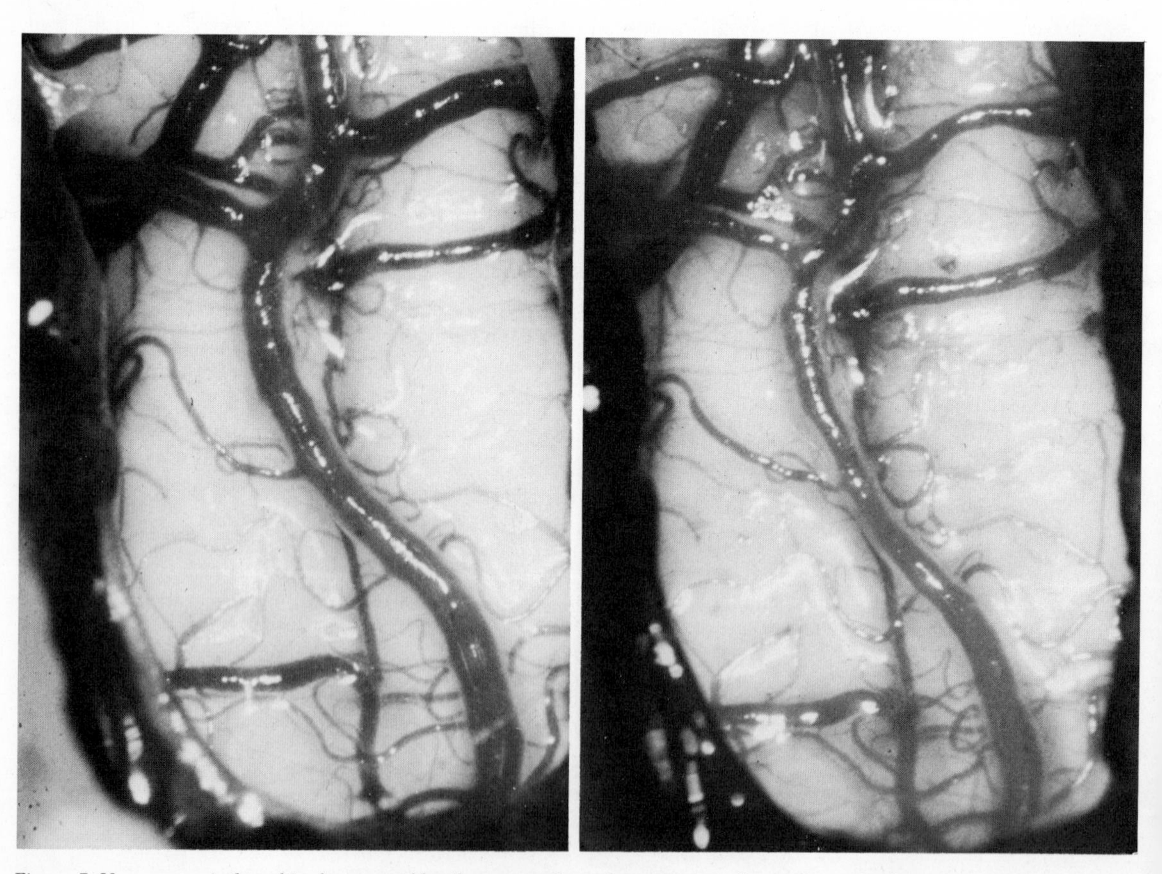

Figure 5. Vasospasm induced in denervated basilar artery. Experiment was performed 1 week after extirpation of bilateral superior cervical sympathetic ganglia. *Left:* control. *Right:* 5 min after topical application of lyzed erythrocyte fraction incubated for 24 hr.

velops more than 4 days after subarachnoid hemorrhage. However, the neurological status of some patients with severe vasospasm is unexpectedly good, and vasospasm is absent in many patients with severe neurological deficiency.[26,45] Some investigators[25,26,43,70] feel that cerebral vasospasm may not be the major factor in causing neurological deterioration. Direct brain damage by noxious agents in subarachnoid blood, especially by those released by lysis of the erythrocytes, may account for this occasional lack of association between angiographically demonstrated vasospasm and neurological deterioration.

The cause of aseptic meningitis developing after subarachnoid hemorrhage has been attributed to blood in the CSF[14,39] or to a breakdown of hematoma exposed to CSF.[20,27] In 1949, Jackson[27] clearly demonstrated that the substance in blood eliciting the greatest meningeal response and causing brain injury is present in erythrocytes. His observations of the reaction of dogs to the intracisternal injection of various blood fractions are summarized in Fig. 6.

Intracisternal injection of autogenous plasma or serum incubated at body temperature for several days does not induce meningeal signs, and the CSF undergoes only a mild cellular and protein response. This observation suggests that the noxious agent or agents are contained primarily in erythrocytes. When autogenous erythrocytes incubated for 3–8 days are injected, severe meningeal irritation results. In most dogs, malaise is severe, and the nuchal rigidity is so pronounced that slight flexion of the neck is impossible without eliciting pain. An intense cellular reaction occurs in the CSF, with leukocyte counts varying from 2300 to 6000 mm^3. Meningismus and the CSF reaction are prominent following intracisternal injection of the supernatant of the lyzed erythrocytes; however, the response to the residue portion is minimal. Therefore, the stroma of the erythrocytes can be discounted as a toxic

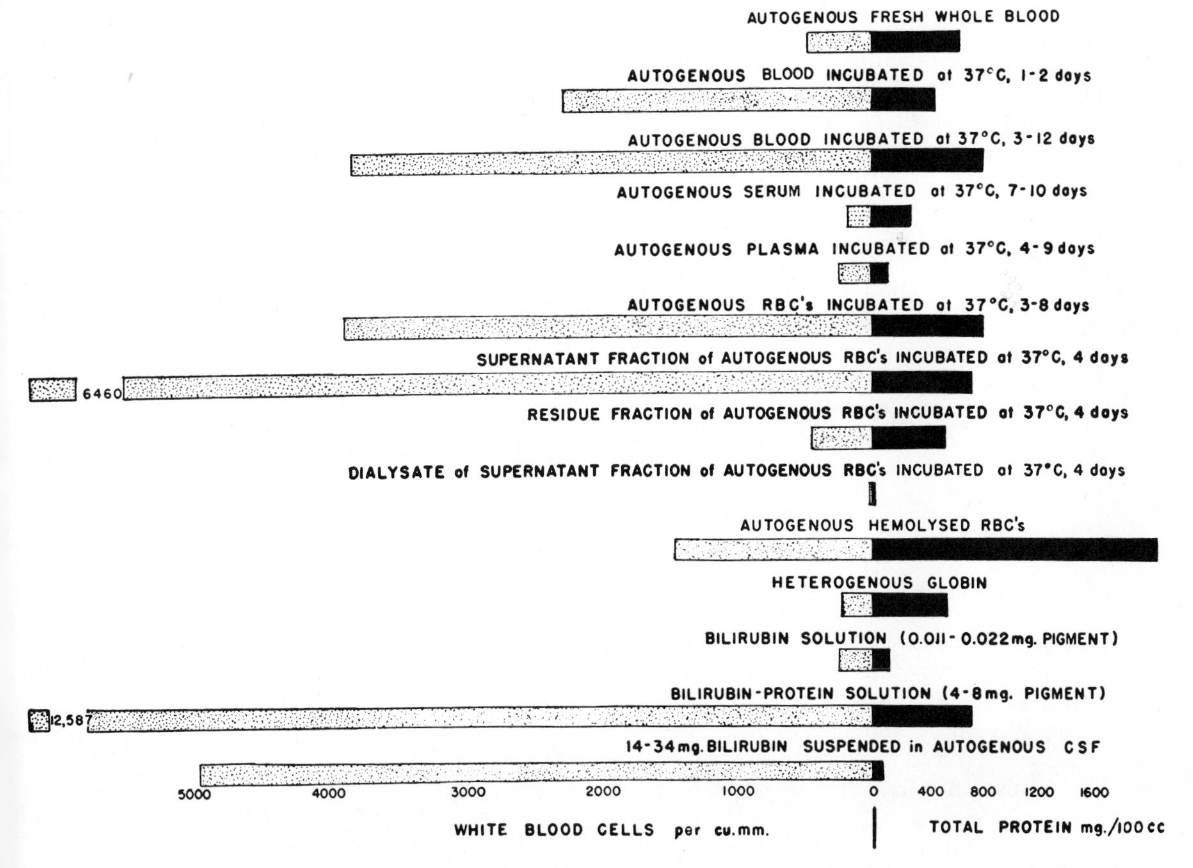

Figure 6. Summary of CSF and protein response 5 hr after intracisternal injection of various materials. Data from Jackson.[27]

agent. Since the dialysate of the supernatant fraction does not evoke this irritative response, such soluble substances as salts, urea, uric acid, amino acids, and lactic acid can also be discounted as toxic agents. Oxyhemoglobin, therefore, appears to be the probable major irritating factor produced by erythrocyte lysis. After administration of pure oxyhemoglobin solution, severe meningeal irritation develops and is associated with deterioration of the clinical status. Pure oxyhemoglobin solution can be prepared from pooled blood of dogs, and has been confirmed by spectrophotometric examination to contain no other blood products. Jackson[27] concludes that methemoglobin might also be toxic, since severe meningeal reactions have developed following injection of a chocolate-colored fluid, the color of which is presumably due to the presence of this protein.

5. Recent Trends in Treatment of Subarachnoid Hemorrhage

As discussed in the preceding sections, breakdown products of the subarachnoid erythrocytes are probably the major offending substances that induce neurological deterioration in patients with subarachnoid hemorrhage by the development of "cerebral vasospasm" and by the induction of "aseptic meningitis." The lyzed erythrocytes also contribute to the "post-subarachnoid hydrocephalus" that may complicate the "erythrogenic aseptic meningitis.[6,20,22,32,35]

To prevent these adverse and noxious effects of the lyzed erythrocytes on the brain, as many as possible of the erythrocytes in the subarachnoid space should be promptly removed, or the noxious component of the erythrocytes should be converted to a harmless substance if possible.

Efforts at removing subarachnoid erythrocytes prior to lysing have generally not been successful. According to Sprong[60] and Meredith,[41] the amount of erythrocytes recoverable, either by repeated lumbar puncture or even by continuous drainage of CSF, is insignificant. These investigators demonstrated that after the first 48 hr following cisternal blood injection in dogs, erythrocytes recovered by CSF drainage never amounted to more than 2–4.5% of the original number, and in most instances were considerably less (≤1%). Even after such a short interval as 90 min to 5 hr following injection, only 25–32% of the original erythrocytes were recoverable. In the clinical studies of these investigators involving patients with subarachnoid hemorrhage, the rate of disappearance of erythrocytes from the subarachnoid space was not appreciably affected by the withdrawal of CSF.

The failure of these attempts to remove subarachnoid erythrocytes implies that the erythrocytes freely suspended in the CSF represent only a fraction of those extravasated and that most of them are promptly bound in a clot or entangled in the meshes of the arachnoid trabeculae.

Saito *et al.*[56] recommended continuous CSF drainage from the prechiasmatic cistern for prevention of cerebral vasospasm in patients with ruptured aneurysms, but did not show how much of the erythrocyte substance was actually evacuated by such a method. Kennady[33,34] advocated irrigation of the subarachnoid space with artificial CSF to remove the subarachnoid erythrocytes after observing a 70–85% erythrocyte recovery. The addition of a fibrinolytic agent to the irrigation fluid appeared to increase the efficacy of the method. These agents, such as heparin and fibrinolysin, were felt to be necessary to lyze clots and to prevent further clotting of blood in the subarachnoid space. The practicality of this method is not yet known.

The converting of noxious agents into harmless substances, and thereby the relieving of vasospasm, is currently under investigation. Miyaoka *et al.*[42] suggest that the vasoconstrictive activity of oxyhemoglobin may be abolished through its binding to haptoglobin. Haptoglobin is a normal constituent of serum and is known to combine with hemoglobin to form a stable hemoglobin–haptoglobin compound.[23,28,68] These investigators have demonstrated that the hemoglobin-induced vasospasm decreases following topical application of haptoglobin to cerebral vessels in experimental animals, as well as in some patients with ruptured aneurysms. The results are encouraging, but further study is necessary to establish the practicality of the method.

Sonobe *et al.*[58,59] suggested that oxyhemoglobin-induced vasospasm could be relieved by oxidizing oxyhemoglobin to "inactive" methemoglobin with sodium nitrite ($NaNO_2$). However, several major problems in this method must be solved prior to its clinical application. First, methemoglobin may have some vasoconstrictor activity.[69] Second, methemoglobin, even if inert toward major blood vessels, probably evokes potent "direct" noxious stimulation to the brain.

6. Summary

Most erythrocytes released into the subarachnoid CSF are evacuated by gradual hemolysis and phagocytosis. Hemolysis of the extravasated erythrocytes begins shortly after hemorrhage and reaches a maximum around the end of the first week. The blood pigments released by lysis of erythrocytes are oxyhemoglobin, bilirubin, and methemoglobin. Oxyhemoglobin, the first pigment to appear in CSF, reaches its maximal level in a few days, then gradually diminishes over a 7- to 9-day period. Bilirubin appears after 2–3 days, increases in amount, and persists for 2–3 weeks. Little methemoglobin is found in CSF. Meningeal irritation accompanying subarachnoid hemorrhage is caused mainly by the blood pigments released by lysing erythrocytes. This irritation may cause pia–arachnoid adhesion leading to hydrocephalus.

Cerebral vasospasm may also be caused by breakdown products of erythrocytes. The currently popular hypothesis that vasospasm is induced by vasoconstrictors of platelet origin (e.g., serotonin, prostaglandins) has not been supported by our experimental findings. Vasoconstrictive activity of platelet origin is destroyed by incubation for 24 hr, implying that platelet-induced vasoconstriction is of short duration. Severe vasospasm is induced by application of lyzed erythrocytes, and this vasoconstrictive activity of lyzed erythrocytes does not diminish after 1–7 days of incubation. Vasospasm induced by lyzed erythrocytes did not relax, but tended to increase in severity. Prolonged vasospasm in patients with subarachnoid hemorrhage appears to be caused by breakdown products of erythrocytes.

The role of perivascular nerves in inducing vasospasm remains controversial. Various methods of removing the subarachnoid erythrocytes or reducing the noxious effects of the lyzed erythrocytes are currently under investigation.

References

1. Adams, J. E., Prawirohardjo, S.: Fate of red blood cells injected into cerebrospinal fluid pathways. *Neurology* **9**:561–564, 1959.
2. Allcock, J. M., Drake, C. G.: Ruptured intracranial aneurysms—The role of arterial spasm. *J. Neurosurg.* **22**:21–29, 1965.
3. Allen, G. S., Henderson, L. M., Chou, S. N., French, L. M.: Cerebral arterial spasm. Part 2. *In vitro* contractile activity of serotonin in human serum and CSF on the canine basilar artery, and its blockage by methylsergide and phenoxybenzamine. *J. Neurosurg.* **40**:442–450, 1974.
4. Alpers, B. J., Forster, F. M.: The reparative processes in subarachnoid hemorrhage. *J. Neuropathol. Exp. Neurol.* **4**:262–268, 1945.
5. Bagley,C., Jr.: Blood in the cerebrospinal fluid: Resultant functional and organic alterations in the central nervous system. A. Experimental data. *Arch. Surg.* **17**:18–38, 1928.
6. Bagley, C., Jr.,: Blood in the cerebrospinal fluid: Resultant functional and organic alterations in the central nervous system. B. Clinical data. *Arch. Surg.* **17**:39–81, 1928.
7. Barrows, L. J., Hunter, F. T., Banker, B. Q.: The nature and clinical significance of pigments in cerebrospinal fluid. *Brain* **78**:59–80, 1955.
8. Bergvall, U., Galera, R.: Time relationship between subarachnoid hemorrhage, arterial spasm, changes in cerebral circulation and posthaemorrhagic hydrocephalus. *Acta Radiol. Diagn.* **9**:229–237, 1969.
9. Brawley, B. E., Strandness, D. E., Jr., Kelly, W. A.: The biphasic response of cerebral vasospasm in experimental subarachnoid hemorrhage. *J. Neurosurg.* **28**:1–8, 1968.
10. Buckell, M.: Demonstration of substances capable of contracting smooth muscle in the haematoma fluid from certain cases of ruptured cerebral aneurysm. *J. Neurol. Neurosurg. Psychiatry* **27**:198–199, 1964.
11. Chokyu, M.: An experimental study of cerebral vasospasm, especially on spasmogenic factors in red blood cells. *Osaka City Med. J.* **24**:211–221, 1975.
12. Conway, L. W., McDonald, L. W.: Structural changes of the intradural arteries following subarachnoid hemorrhage. *J. Neurosurg.* **37**:715–723, 1972.
13. Cromptom, M. R.: The pathogenesis of cerebral infarction following the rupture of cerebral berry aneurysms. *Brain* **87**:491–510, 1964.
14. Cushing, H.: Experiences with the cerebellar astrocytomas. *Surg. Gynecol. Obstet.* **52**:129–204, 1931.
15. Echlin, F. A.: Spasm of basilar and vertebral arteries caused by experimental subarachnoid hemorrhage. *J. Neurosurg.* **23**:1–11, 1965.
16. Echlin, F. A.: Experimental vasospasm, acute and chronic, due to blood in the subarachnoid space. *J. Neurosurg.* **35**:646–656, 1971.
17. Endo, S., Hori, S., Suzuki, J.: Experimental cerebral vasospasm and sympathetic nerve. *Neurol. Med. Chir.* **17**(Part II): 313–326, 1977.
18. Falck, B.: Observations on the possibilities of the cellular localization of monoamines by a fluorescence method. *Acta Physiol. Scand.* **56**(Suppl. 197):1–25, 1962.
19. Fein, J. M., Flor, W. J., Cohan, S. L., Parkhurst, J.: Sequential changes of vascular ultrastructure in experimental cerebral vasospasm: Myonecrosis of subarachnoid arteries. *J. Neurosurg.* **41**:49–58, 1974.
20. Finalayson, A. I., Penfield, W.: Acute postoperative aseptic leptomeningitis. *Arch. Neurol. Psychiatry* **46**:250–276, 1941.
21. Fisher, C. M., Robertson, G. H., Ojemann, R. G.: Cerebral vasospasm with ruptured saccular aneurysm—The clinical manifestations. *Neurosurgery* **1**:245–248, 1977.
22. Foltz, E. L., Ward, A. A.: Communicating hydro-

cephalus from subarachnoid bleeding. *J. Neurosurg.* **13**:546–566, 1956.

23. GORDON, S., BEAN, A. G.: Hemoglobin binding capacity of isolated haptoglobin polypeptide chains. *Proc. Soc. Exp. Biol. Med.* **121**:846–850, 1966.
24. HAMMES, E. M., JR.: Reaction of the meninges to blood. *Arch. Neurol. Psychiatry* **52**:505–514, 1944.
25. HEILBRUN, M. P., OLESEN, J., LASSEN, N. A.: Regional cerebral blood flow studies in subarachnoid hemorrhage. *J. Neurosurg.* **37**:36–44, 1972.
26. HEILBRUN, M. P.: The relationship of neurological status and angiographical evidence of spasm to prognosis in patients with ruptured intracranial saccular aneurysms. *Stroke* **4**:973–979, 1973.
27. JACKSON, I. J.: Aseptic hemogenic meningitis—An experimental study of aseptic meningeal reactions due to blood and its breakdown products. *Arch. Neurol. Psychiatry* **62**:572–589, 1949.
28. JAENICKE, R., PAULICEK, Z.: On the nature of the hemoglobin–haptoglobin interaction. *Z. Naturforsch.* **256**:1272–1277, 1970.
29. KAPP, J. P., MAHALEY, M. S., JR., ODOM, G. L.: Cerebral arterial spasm. Part 2. Experimental evaluation of mechanical and humoral factors in pathogenesis. *J. Neurosurg.* **29**:339–349, 1968.
30. KAPP, J. P., MAHALEY, M. S., JR., ODOM, G. L.: Cerebral arterial spasm. Part 3. Partial purification and characterization of a spasmogenic substance in feline platelets. *J. Neurosurg.* **29**:350–356, 1968.
31. KAJIKAWA, H.: Mode of the sympathetic innervation of the cerebral vessels demonstrated by the fluorescent histochemical technique in rats and cats. *Arch. Jpn. Chir.* **38**:227–235, 1969.
32. KAUFMANN, H. H., CARMAL, P. W.: Aseptic meningitis and hydrocephalus after posterior fossa surgery. *Acta Neurol. Chir.* **44**:179–196, 1978.
33. KENNADY, J. C.: Early fate of subarachnoid blood and removal by irrigation. *Trans. Am. Neurol. Assoc.* **91**:265–267, 1966.
34. KENNADY, J. C.: Investigations of the early fate and removal of subarachnoid blood. *Pac. Med. Surg.* **75**:163–168, 1967.
35. KIBLER, R. F., COUCH, R. S. C., CROMPTON, M. R.: Hydrocephalus in the adult following spontaneous subarachnoid hemorrhage. *Brain* **84**:45–61, 1961.
36. KRONHOLM, V., LINTRUP, J.: Spectrophotometric investigations of the cerebrospinal fluid in near-ultraviolet region: A possible diagnostic aid in diseases of the central nervous system. *Acta Psychiatr. Scand.* **35**:314–329, 1960.
37. LATORRE, E., PATRONO, C., FORLUNA, A., GROSSI-BELLONI, D.: Role of prostaglandin F_2 in human cerebral vasospasm. *J. Neurosurg.* **41**:292–299, 1974.
38. MARSHALL, W. H., JR.: Delayed arterial spasm following subarachnoid hemorrhage. *Radiology* **106**:325–327, 1973.
39. MATSON, D. D.: *Neurosurgery of Infancy and Childhood*, 2nd ed. Springfield, Illinois, Charles C Thomas, 1969.
40. MATTHEWS, W. F., FROMMEYER, W. B., JR.: The *in vitro* behavior of erythrocytes in human cerebrospinal fluid. *J. Lab. Clin. Med.* **45**:508–515, 1955.
41. MEREDITH, J. M.: The inefficacy of lumbar puncture for the removal of red blood cells from the cerebrospinal fluid. *Surgery* **9**:524–533, 1941.
42. MIYAOKA, M., NONAKA, T., WATANABE, H., CHIGASAKI, H., ISHII, S.: Etiology and treatment of prolonged vasospasm: Experimental and clinical studies. *Neurol. Med. Chir. (Tokyo)* **16**:103–114, 1976.
43. MILLIKAN, C. H.: Cerebral vasospasm and ruptured intracranial aneurysm. *Arch. Neurol.* **32**:433–449, 1975.
44. NELSON, E., RENNELS, M.: Innervation of intracranial arteries. *Brain* **93**:475–490, 1970.
45. OHTA, T., KAWAMURA, J., OSAKA, K., KAJIKAWA, H., HANDA, H.: Angiographic classification of so-called cerebral vasospasm: Correlation between existence of vasospasm and postoperative prognosis in subarachnoid hemorrhage. *Brain Nerve (Tokyo)* **21**:1019–1027, 1969.
46. OSAKA, K.: Experimental studies on cerebrovascular spasm. *Arch. Jpn. Chir.* **38**:349–371, 1969.
47. OSAKA, K.: Prolonged vasospasm produced by the breakdown products of erythrocytes. *J. Neurosurg.* **47**:403–411, 1977.
48. OWMAN, C. H., EDVINSSON, L., NIELSEN, K. C.: Autonomic neuroreceptor mechanisms in brain vessels. *Blood Vessels* **11**:2–31, 1974.
49. PEERLESS, S. J., YASARGIL, M. G.: Adrenergic innervation of the cerebral blood vessels of the rabbit. *J. Neurosurg.* **35**:148–154, 1971.
50. PEERLESS, S. J., KENDALL, M. J.: The significance of the perivascular innervation of the brain. In Cervos-Nararro, J., Betz, E., Matakas, F., Wüllenweber, R. (eds.): *The Cerebral Vessel Wall*. New York, Raven Press, 1976, pp. 175–182.
51. PELOFSKY, S., JACOBSON, E. D., FISHER, R. G.: Effects of prostaglandin E on experimental vasospasm. *J. Neurosurg.* **36**:634–639, 1972.
52. PENNINK, M., WHITE, R. P., CROCKARELL, J. R.: Role of prostaglandin F_2 in the genesis of experimental cerebral vasospasm: Angiographic study in dogs. *J. Neurosurg.* **37**:398–406, 1972.
53. PIMSTONE, N. R., TENHUNEN, R., SEITZ, P. T., MARVER, H. S., SCHMID, R.: The enzymatic degradation of hemoglobin to bile pigments by macrophages. *J. Exp. Med.* **133**:1264–1281, 1971.
54. RAYNOR, R. B., MCMURRY, J. G., POOL, J. L.: Cerebrovascular effects of topically applied serotonin in the cat. *Neurology* **11**:190–195, 1961.
55. ROOST, K. T., PIMSTONE, N. A., DIAMOND, I., SCHMID, R.: The formation of cerebrospinal fluid xanthochromia after subarachnoid hemorrhage—Enzymatic conversion of hemoglobin to bilirubin by the arachnoid and choroid plexus. *Neurology* **22**:973–977, 1972.
56. SAITO, I., UEDA, Y., SANO, K.: Significance of vasospasm in the treatment of ruptured intracranial aneurysms. *J. Neurosurg.* **47**:412–429, 1977.

57. SIMEONE, F. A., RYAN, K. G., COTTER, J. R.: Prolonged experimental cerebral vasospasm. *J. Neurosurg.* **29**:357–366, 1968.
58. SONOBE, M., SUZUKI, J.: Vasospasmogenic substance produced following subarachnoid hemorrhage, and its fate. *Neurol. Med. Chir. (Tokyo)* **18**(Part II):29–37, 1978.
59. SONOBE, M., SUZUKI, J.: Vasospasmogenic substance produced following subarachnoid haemorrhage, and its fate. *Acta Neurochir.* **44**:97–106, 1978.
60. SPRONG, W.: The disappearance of blood from the cerebrospinal fluid in traumatic subarachnoid hemorrhage—The ineffectiveness of repeated lumbar punctures. *Surg. Gynecol. Obstet.* **58**:705–710, 1934.
61. TANABE, Y., SAKATA, K., YAMADA, H., ITO, T., TAKADA, M.: Cerebral vasospasm and ultrastructural changes in cerebral arterial wall: An experimental study. *J. Neurosurg.* **49**:229–238, 1978.
62. TANI, E., YAMAGATA, S., ITO, Y.: Morphological study of experimental cerebral vasospasm. *Med. Neuro-Chir.*, Part II, pp. 9–18, 1978.
63. TODA, N., OZAKI, T., OHTA, T.: Cerebrovascular sensitivity to vasoconstricting agents induced by subarachnoid hemorrhage and vasospasm in dogs. *J. Neurosurg.* **46**:296–303, 1977.
64. TOURTELLOTTE, W. W., METZ, L. N., BRYAN, E. R, DEJONG, R. N.: Spontaneous subarachnoid hemorrhage: Factors affecting the rate of clearing of the cerebrospinal fluid. *Neurology* **14**:301–306, 1964.
65. WILKINS, R. H., WILKINS, G. K., GUNNELIS, C., ODOM, G. L.: Experimental studies of intracranial arterial spasm using aortic strip assays. *J. Neurosurg.* **27**:490–500, 1967.
66. WILKINS, R. H., LEVITT, P.: Intracranial arterial spasm in the dog: A chronic experimental model. *J. Neurosurg.* **33**:260–269, 1971.
67. YAMAGATA, S., ITO, Y., TANI, E.: Contractile activities of spastic canine basilar arteries and its surrounding clot. *Neurol. Med. Chir. (Tokyo)* **18**(Part II):489–494, 1978.
68. YAMAOKA, K., YAMAGUCHI, M., NAITO, S.: Haptoglobin. *Int. Med.* **15**:1105–1114, 1965.
69. YOSHIDA, Y.: Experimental studies on fibrinolytic activity and angiospasm following subarachnoid hemorrhage in dogs. *Arch. Jpn. Chir.* **47**:537–562, 1978.
70. ZINGESSER, L. H., SCHECHTER, M. M., DEXTER, J.: Regional cerebral blood flow in patients with subarachnoid hemorrhage. *Acta Radiol. Diagn.* **9**:573–588, 1969.
71. ZUCKER, M. B.: A study of the substances in blood serum and platelets which stimulate smooth muscle. *Am. J. Physiol.* **142**:12–26, 1944.

CHAPTER 24

Neurochemical Alterations in Cerebrospinal Fluid in Cerebral Ischemia and Stroke

K. M. A. Welch and John Stirling Meyer

1. Introduction

This chapter is a review of cerebrospinal fluid (CSF) studies that support the hypothesis that the disrupting effect of cerebral ischemia on certain central neurotransmitter systems may be responsible for the pathophysiological changes and deficits in brain function that complicate cerebrovascular disease. Any investigator interested in the precise mechanisms of disease should find the study of ischemic effects on brain function a daunting one. Even when the pattern of ischemia can, to a degree, be controlled, as it is in experimental models, the effects of reduced blood flow are catastrophic for the physiological function of many brain systems. Interpretation of the results is therefore hazardous at the outset and even more so in man, in whom ischemia is more often than not variable in onset, site, degree, and despite the advent of computed tomographic (CT) scanning, frequently difficult to define. Nevertheless, if, by examining the CSF, which is the "sink" for many products of brain metabolism, it can be determined that there is deviation from normality, then this primarily phenomenological approach can initiate and justify more extensive mechanistic studies in experimental animals, and with the development of positron-emission tomography, eventually in man.

K. M. A. Welch, M.B.Ch.B, M.R.C.P. (U.K.) • Department of Neurology, Baylor College of Medicine, Neurosensory Center, Houston, Texas 77030. **John Stirling Meyer, M.D.** • Department of Neurology, Baylor College of Medicine; Cerebral Blood Flow Laboratory, Veterans Administration Medical Center, Houston, Texas 77211.

The reasons for altered levels of a substance in CSF after cerebral ischemia or infarction may be many (Fig. 1). Increased levels could result from breakdown of the blood–brain or blood–CSF barrier, with passage from the circulation into the CSF. The impairment of transport mechanisms that normally clear substances from the CSF or brain into cerebrovenous blood is another cause of increase. Brain tissue and extracellular fluid are alternative sources of CSF increase and the usual site of abnormality if neurochemical levels are decreased, reflecting decreased synthesis.

There are considerable problems in the interpretation of altered neurochemical levels in CSF even if the source can be identified. For example, even if it can be determined that the changes take place in brain, it is still not known at what cellular level they occur, and there still remains the uncertainty as to whether they are due to altered uptake, storage, release, or synthesis of the compound in question. Another question concerns the pathophysiological relevance of a measured increase of any substance in CSF if it simply reflects leakage of that substance from necrotic brain. Finally, there is the added complexity of whether measurements in the customarily obtained lumbar CSF accurately represent changes in ventricular CSF and contiguous brain.

The bulk of the work reported concerns the effects of ischemia on brain neurotransmitter metabolism as reflected by neurochemical changes in CSF. Because of the foregoing considerations, CSF studies are, whenever possible, correlated with other physiological measurements in the clinical patients and supported by measurements in brain tissue and CSF of experimental animal models.

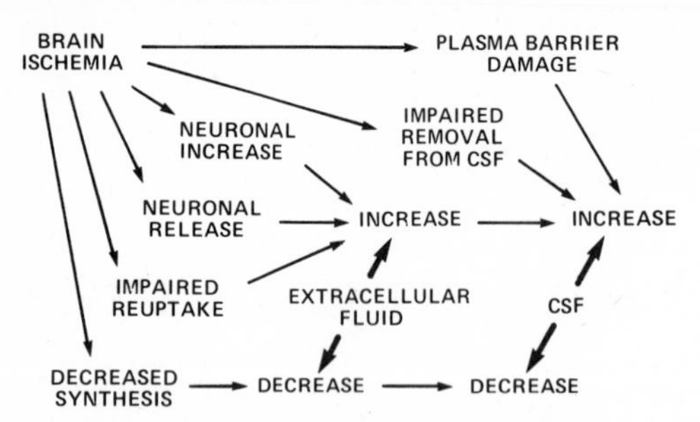

Figure 1. Schema of possible events that lead to neurochemical change in CSF following brain ischemia.

2. Studies in Stroke

2.1. Monoamines

The biogenic amines are among those substances popularly regarded as putative neurotransmitters in the central nervous system (CNS). They are localized within the cell bodies, axons, and synaptic terminals of neurons, which are considered specific for the individual monoamine, and from which they may be released by physiological, electrical, or pharmacological stimulation. These particular neurotransmitters were chosen for study because in the free state they had been shown to affect cerebral vasoactivity and brain energy metabolism and to promote edema in brain and other tissues.[26] Since altered cerebral hemodynamics, disordered cerebral energy metabolism, and cerebral edema are prominent features of ischemic cerebrovascular disease (CVD) we reasoned that abnormal monoamine metabolism might be implicated in the pathophysiology of the condition. A number of the questions that can be asked about the consequences of ischemia on a typical monoamine neuronal system are represented diagrammatically in Fig. 2.

Before 1973, there had been sporadic reports in the literature concerning elevated monoamine levels in the CSF of patients with CVD. These reports largely concerned CSF serotonin changes in small series of patients with ischemic and hemorrhagic states.[6,27] One of these studies had attempted to relate the severity of the clinical condition to the degree of serotonin elevation in plasma or CSF, with the general finding that higher levels correlated with greater neurological deficits.[6] In 1973 and 1974, we published two papers[25,26] in the journal *Brain* on the subject of catecholamine and serotonin levels in the CSF of a relatively large series of patients with ischemic and hemorrhagic CVD. In these papers, we attempted to establish the pathophysiological relevance of the neurochemical changes by relating the findings to the chemistry of plasma, to clinical functions such as duration of symptoms, hypertension, and severity of neurological deficit, and to cerebral blood flow and metabolism.

The earlier of the two studies concentrated on clinical and temporal profiles in relation to norepinephrine and epinephrine changes in CSF and plasma.[25] Highest catecholamine values were found in the CSF of patients with intracerebral hemorrhage (Table 1). Total norepinephrine and epinephrine levels were also elevated in patients with recent (<2 weeks) cerebral infarction, although the major significance was contributed by epinephrine (the mean value for CSF norepinephrine concentration closely approached but did not reach the level of statistical significance). The source of the epinephrine in CSF was probably plasma, since brain tissue levels are normally low and highly localized in distribution. Moreover, the same pattern of increase was found in plasma.

When the patients with recent cerebral infarction were further categorized into hypertensive and nor-

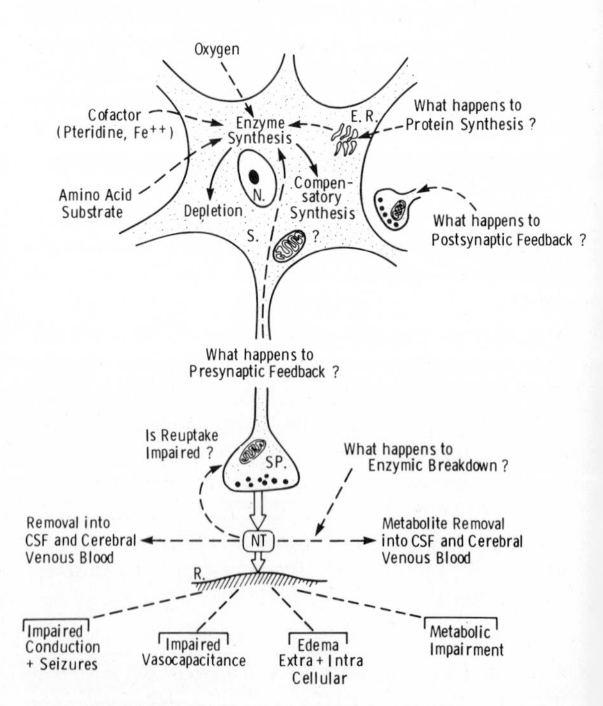

Figure 2. Possible effects of ischemia on monoaminergic neuronal function and its relationship to pathogenesis of cerebral infarction. (S.) Soma; (SP.) synaptic terminal; (R.) receptor; (E.R.) endoplasmic reticulum; (N.) nucleus; (NT) neurotransmitter.

Table 1. Catecholamine Concentrations in CSF[a]

Catecholamine	Control (N = 5)	Cerebral infarction: Acute (N = 15)	Cerebral infarction: Subacute and chronic (N = 8)	Cerebral hemorrhage and hemorrhagic infarction (N = 8)
Norepinephrine (μg/liter)	0.26 ± 0.11	0.47 ± 0.23	0.27 ± 0.15[c]	1.28 ± 0.83[b–d]
Epinephrine (μg/liter)	0.00 ± 0.00	0.14 ± 0.15[b]	0.07 ± 0.05[b]	1.39 ± 0.83[b,c,f]
Total (μg/liter)	0.26 ± 0.11	0.61 ± 0.33[b]	0.34 ± 0.17[c]	2.67 ± 1.47[b,e,g]

[a] Modified from Meyer *et al.*[25] Values are means ± S.D.; (*N*) number of measurements.
[b] Statistically significant difference as compared with control value.
[c] Statistically significant difference as compared with acute cerebral infarction, $p < 0.05$.
[d] Statistically significant difference as compared with subacute and chronic cerebral infarction, $p < 0.02$.
[e] Statistically significant difference as compared with acute cerebral infarction, $p < 0.01$.
[f] Statistically significant difference as compared with subacute and chronic cerebral infarction, $p < 0.05$.
[g] Statistically significant difference as compared with subacute and chronic cerebral infarction, $p < 0.005$.

motensive groups, there was significant elevation of both norepinephrine and epinephrine in the CSF of the former group (Table 2). Again, a similar pattern of increase was obtained in plasma. Although brain tissue as a source of catecholamine elevation in CSF could not be disproved, the most appropriate inference that could be made from the study was that the increase of catecholamines in the CSF of recent stroke victims occurred in part as a result of movement from plasma across a damaged blood–brain barrier.

Accordingly, another series of stroke patients was studied, and more extensive parameters of CNS function were correlated with changes in CSF catecholamines[26] (Fig. 3). Levels of norepinephrine, but not of epinephrine, correlated inversely with the duration of stroke, declining as patients showed recovery from neurological deficit over a 3-week period. Patients with more severe neurological deficits tended to show high CSF norepinephrine levels (Fig. 4). Finally, although CSF norepinephrine failed to correlate with cerebral hemispheric blood flow,

Table II. Catecholamine Concentrations Compared between Hypertensive and Normotensive Patients with Acute Cerebral Infarction[a]

Catecholamine	CSF: Hypertension (N = 8)	CSF: Normotension (N = 7)	Plasma: Hypertension (N = 6)	Plasma: Normotension (N = 6)
Norepinephrine (μg/liter)	0.61 ± 0.17[b]	0.30 ± 0.15	0.86 ± 0.24[c]	0.48 ± 0.17
Epinephrine (μg/liter)	0.20 ± 0.19	0.07 ± 0.06	0.39 ± 0.36	0.11 ± 0.05
Total (μg/liter)	0.81 ± 0.39[b]	0.38 ± 0.18	1.25 ± 0.55[d]	0.58 ± 0.18

[a] Modified from Meyer *et al.*[25] Values are means ± S.D.; (*N*) number of measurements.
[b] Statistically significant difference as compared with normotensive group, $p < 0.005$.
[c] Statistically significant difference as compared with normotensive group, $p < 0.02$.
[d] Statistically significant difference as compared with normotensive group, $p < 0.05$.

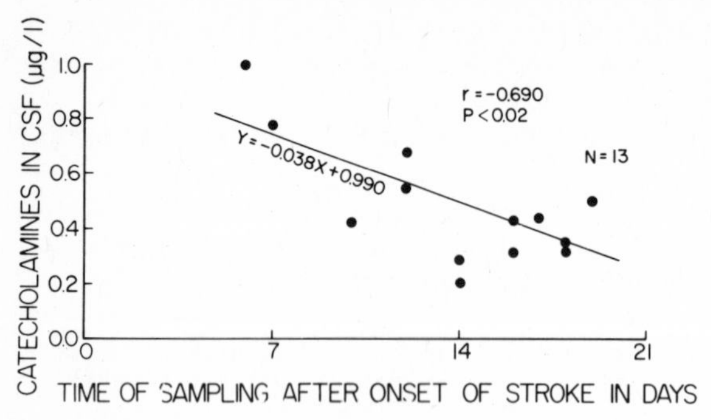

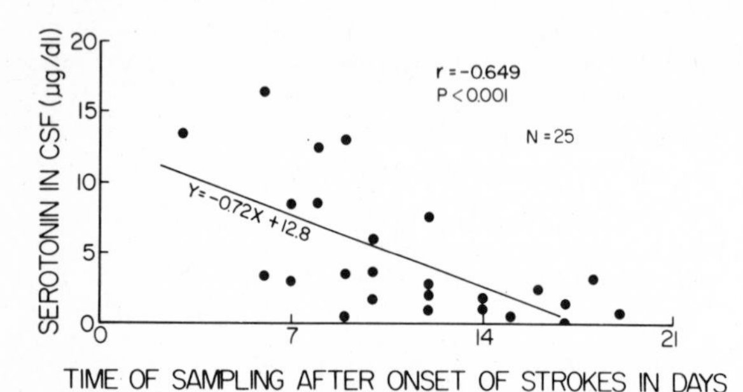

Figure 3. Concentrations of catecholamines and serotonin in CSF correlated with duration of stroke at time of sampling. Reproduced from Meyer *et al.*,[26] with permission.

there was a direct correlation with the oxygen consumption of the ischemic hemisphere (Fig. 5). The findings in this second series therefore more strongly suggested that the abnormal levels of norepinephrine in the CSF of patients with recent stroke originated in part from CNS tissue, indicating disturbance of central noradrenergic function. Breakdown products of norepinephrine were not measured in the study, so that there was no information as to whether norepinephrine was simply released from brain to diffuse passively into CSF or whether neuronal norepinephrine turnover was affected.

Serotonin levels were also found to be high at an early stage after stroke and, like those of norepinephrine, declined as patients recovered from neurological deficit over a 3-week period (see Fig. 3). Patients with more severe neurological deficits also had higher CSF serotonin levels (see Fig. 4). There was a strong inverse correlation between CSF serotonin levels and blood flow in the infarcted and, to a lesser extent, the noninfarcted hemisphere (Fig. 6), implying a possible role of serotonin in the bilateral reduction of hemispheric blood flow, a phenomenon known as diaschisis, which will be discussed in more detail in later paragraphs. However, there was no correlation between serotonin levels and cerebral oxygen consumption (see Fig. 5). The origin of the elevated serotonin in CSF could also be from the circulation as well as brain. However, total blood serotonin was measured in a number of the patients studied, and no significant changes were observed. Nevertheless, the potential for platelets to release serotonin into plasma after circulating through partially ischemic brain must be recognized and could contribute to the CSF pool.[49]

Measurement of 5-hydroxyindoleacetic acid (5-HIAA), the immediate breakdown product of serotonin, in CSF obtained from patients within 12 hr of an ischemic infarct shed no further light on the origin of CSF serotonin changes in this condition. The levels varied from nondetectable to high when compared to the expected norms (unpublished observations). Preliminary analysis has revealed no significant correlation with other parameters, e.g., clinical deficit.

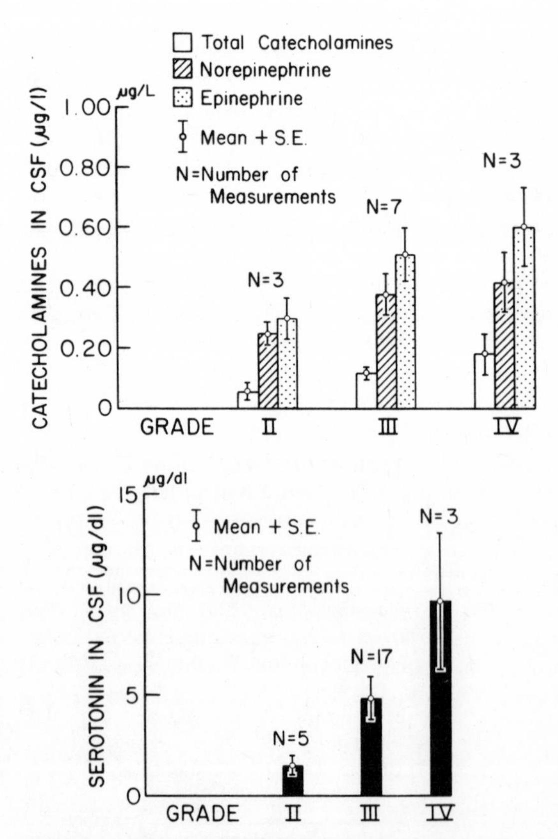

Figure 4. Concentrations of catecholamines and serotonin in CSF correlated with severity of neurological deficits. Reproduced from Meyer *et al.*,[26] with permission.

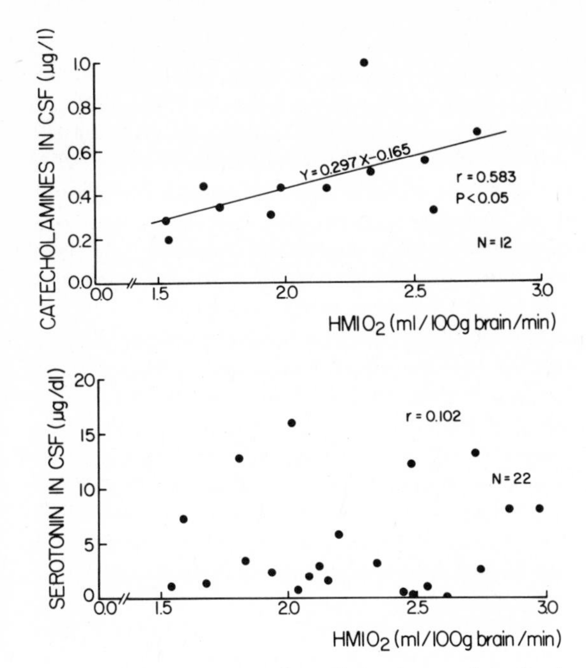

Figure 5. Concentrations of catecholamines and serotonin in CSF correlated with hemispheric oxygen consumption. Reproduced from Meyer *et al.*,[26] with permission.

Dopamine and its metabolite homovanillic acid (HVA) have also been measured in CSF from patients with CVD. In patients studied from 1 to 22 days after an ischemic infarct, there was wide variability in the levels of both substances.[23] HVA values measured in CSF obtained within 12 hr from patients with more acute ischemic stroke were equally inconsistent. In the latter study, HVA was either undetectable or increased (unpublished observations).

The problems of studying the clinical patients with stroke that were outlined in earlier paragraphs could probably explain the variability of CSF results in many instances. However, the accumulated data strongly indicated that ischemia caused abnormality in the neurochemistry of the brain biogenic amines. Animal experiments were therefore designed to test the clinical data and examine their relevance. In baboons that were subjected to cerebral hemispheric embolization by the method described by Molinari *et al.*,[28] 5-HIAA values became significantly increased in cisternal CSF after 1 hr of ischemia (Fig. 7). However, levels of serotonin were unaltered in areas of embolized cortex, basal ganglia, and brainstem selected for tissue measurements.[12] We interpreted these results as increased turnover of serotonin caused by release of serotonin from cortical neurons. The perikarya of the serotonergic neurons are located in the nonischemic vertebrobasilar territory and thus remained intact. The concept of serotonin release from ischemic neurons was supported by the finding of major increases of this monoamine in cerebral venous blood of baboons after surgical occlusion of all four extracranial arteries, but to a much greater extent when the brainstem was rendered ischemic by occlusion of the vertebral arteries.[49] Studies using the gerbil stroke model also confirmed that ischemia caused rapid release of serotonin[10] and that severe and prolonged ischemia resulted in persistent depletion of tissue serotonin content.[43] When ischemia was limited to periods of up to 1 hr, there was a slow though complete recovery of serotonin levels.[10] Both clinical and experimental studies, therefore, are compatible with disordered serotonin metabolism as a complication of cerebral ischemia.

Catecholamine changes have also been observed in experimental models of ischemia. In the baboon cerebral embolism model described above, dopamine levels were decreased in the caudate nucleus (the site of maximal ischemia) but increased in the gray matter of the cerebral cortex, particularly in

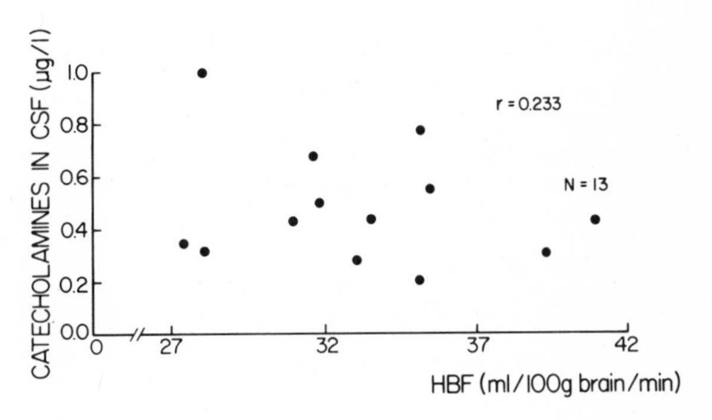

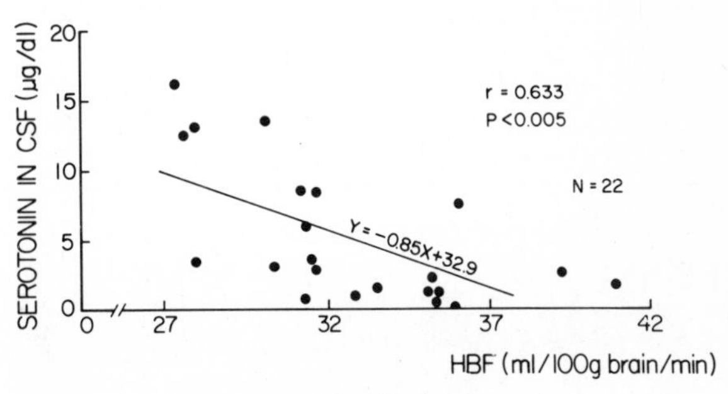

Figure 6. Concentrations of catecholamines and serotonin in CSF correlated with hemispheric blood flow in stroke. Reproduced from Meyer *et al.*,[26] with permission.

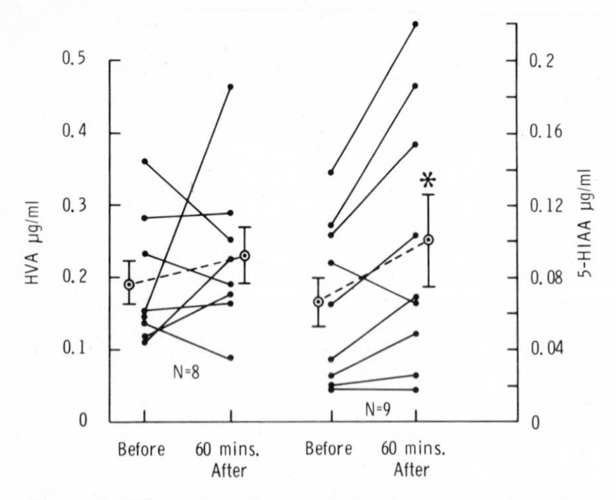

Figure 7. Effect of cerebral embolization on cisternal CSF homovanillic acid (HVA) and 5-hydroxyindoleacetic acid (5-HIAA) values before embolization and 60 min after embolization in the baboon. *Significant difference on Student's paired *t* test; ⨀ mean ± S.E. Reproduced from Ishihara *et al.*,[12] with permission.

regions of collateral flow.[12] No consistent HVA changes were measured in cisternal CSF (Fig. 7), probably because of the regional variation in tissue dopamine content described above. In the gerbil stroke model, both dopamine and norepinephrine were depleted in the cerebral cortex after severe, prolonged ischemia.[43,53] There was some evidence, however, that seizure activity (prominent after ischemia in this model) contributed to the cause of the catecholamine depletion.[51] After brief episodes (30–60 min) of cerebral hemispheric transient ischemia, catecholamine levels increased to twice normal levels in the recovery period.[10] The mechanism for this catecholamine rebound remains uncertain. Heterogeneity of regional flow is a common accompaniment of cerebral ischemia. Reactive hyperemia may also frequently develop in previously ischemic brain regions. Based on the observations of catecholamine depletion in gerbil brain followed by increase after transient ischemia, regional heterogeneity of catecholamine levels might also be expected after focal ischemia. This could account for the dopamine changes after cerebral embolism in the baboon and also explain some of the variability in the CSF changes in both the clinical and experimental situations. Whatever the explanation, just as with serotonin, ischemia seems to profoundly alter brain catecholamine metabolism.

The relevance of disorder of CNS catecholamine and serotonin function to the complications of cerebral ischemia has not yet been established in the clinical patient. Release of the neurotransmitters in increased amount into the synaptic cleft may impair synaptic transmission. In addition, diffusion onto cells not normally receiving input from these specific neurotransmitters may evoke an aberrant neurotransmitter effect. Such factors serve to make disruption of normal synaptic neurotransmission a likely event under ischemic conditions.

Release of abnormal amounts of cerebral neurotransmitters, such as norepinephrine and serotonin, that cause cerebral vasoconstriction when administered exogenously may also in part explain the areas of focal or spreading cortical pallor with associated vasoconstriction that are seen after cerebral artery occlusion.[21] Release and extracellular accumulation of such agents could impair the collateral circulation surrounding ischemic brain foci, thus contributing to the progression of infarction.[47] Cerebral edema of extracellular type may also transpire from diffusion of these monoamines.[20] On the other hand, intracellular edema may result from the influence of excessive neurotransmitter overflow on neuronal metabolism and neuronal membrane ionic exchange.[31]

Studies of blood flow and metabolism in patients with stroke have indicated a possible uncoupling of oxidative phosphorylation in areas of ischemic brain.[22] Relevant to this, the release of free fatty acids (FFA) into cerebrovenous blood in patients with stroke implies that they are also released into ischemic brain.[22] FFA are known to cause uncoupling of oxidative phosphorylation.[32] Norepinephrine has been found to promote FFA release from membrane polar lipids.[5] The studies reported here have shown greater oxygen consumption corresponding to higher levels of norepinephrine in CSF (see Fig. 5). Therefore, despite the overall reduction of oxygen consumption in infarcted brain of patients with stroke, it is possible that there are regions where uncoupling of oxidative phosphorylation is caused by extracellular accumulation of norepinephrine.

The remote effect of depressed cerebral blood flow and metabolism that follows focal ischemia, i.e., diaschisis,[11] could also be explained by disordered neurotransmission and transneuronally mediated inhibition of function. The increased CSF levels of serotonin measured during the early stages after acute cerebral hemispheric blood flow reintroduce a possible vascular component in the etiology of diaschisis.[24]

2.2. Cyclic AMP

After establishing abnormality of catecholamine and serotonin metabolism in ischemic brain, we sought evidence of related neurochemical changes. Because cyclic adenosine 3′,5′-monophosphate (AMP) participates in the molecular events of central and peripheral neurotransmission by acting as a postsynaptic intracellular second messenger for certain neurotransmitters, the biogenic amines included, this seemed important to study. Cyclic AMP has been measured in human CSF, although values show some variability among individual studies. In our laboratory, CSF values in control patients without evidence of neurological disease ranged from 10 to 19 pmol/ml. These values fall in much the same range as those reported by Robison *et al.*[35] of 14.2–16.2 pmol/ml in selected psychiatric and neurological patients. Myllylä *et al.*[29] reported a larger mean value of 24.3 ± 1.4 pmol/ml (range 18–67 pmol/ml) in CSF obtained from patients with unspecified neurological disease. (See Chapter 9.)

Cyclic AMP is apparently removed from the CSF via an active transport process that is sensitive to blockade by probenecid.[8] It has been suggested that the accumulation rates of cyclic AMP in CSF after administration of probenecid to clinical patients might qualitatively reflect central turnover of cyclic AMP and be useful in the assessment of those diseases that may involve disorder of cyclic AMP metabolism.[8] Sebens and Korf[37] have studied cisternal CSF samples in the rabbit and showed that cyclic AMP was increased by intraperitoneal probenecid, but that this increase was not influenced by tricyclic antidepressants, haloperidol, isoprenaline or L-DOPA. They did, however, show that intracisternally injected norepinephrine, isoprenaline, dopamine, and histamine, as well as intravenously injected isoprenaline, increased CSF cyclic AMP levels. These results substantiate the theory that cyclic AMP in the CSF is of central origin and that a close relationship exists between cyclic AMP in brain tissue and CSF.

Cyclic AMP levels were clearly elevated in the CSF of patients with recent cerebral infarction (Fig. 8).[48] Since plasma cyclic AMP levels were also elevated, the question again arose that this might be the source of the increase in CSF. Animal studies performed in this laboratory[50] and other laboratories[15,19] have proved that cyclic AMP is elevated in brain tissue both during ischemia and in the reflow period after episodes of transient ischemia. Further, in 18 clinical patients, cerebral venous blood

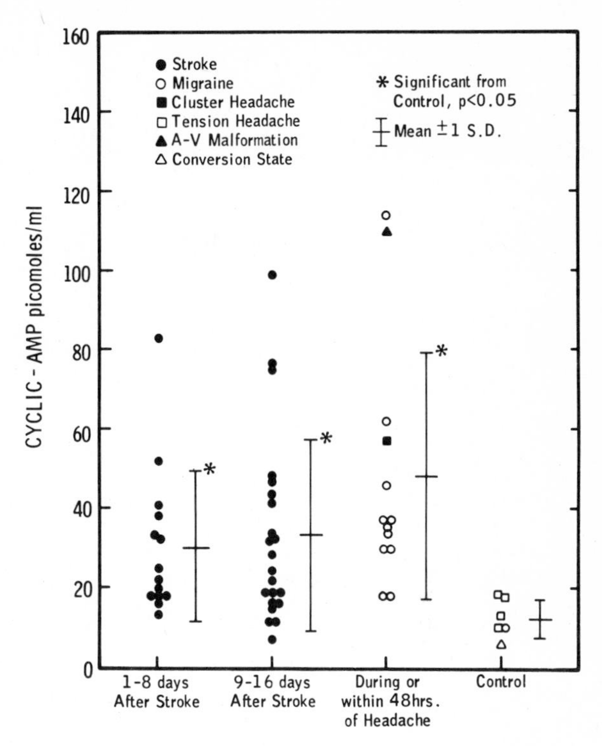

Figure 8. Increase of cyclic adenosine 3′,5′-monophosphate (AMP) and CSF with recent onset of thromboembolic cerebral hemispheric infarction. Levels are still increased 16 days after onset of stroke. Increase of cyclic AMP in CSF during or within 48 hr of an attack of migraine headache, and in one patient with cluster headache and arteriovenous malformation, is shown (symbols represent single measurements). Reproduced from Welch *et al.*,[45] with permission.

levels for cyclic AMP were significantly higher than arterial levels (Fig. 9). In 11 of the patients in whom both types of measurements were successful, there was a direct correlation between these cerebral arteriovenous differences and CSF levels of cyclic AMP ($r = 0.607$, $p < 0.05$). In both instances, this provided strong evidence that the abnormal CSF cyclic AMP levels had their origin in ischemic brain and were removed from CSF into cerebral venous blood, possibly by an active transport process or possibly by simple leakage from damaged brain.

The changes in brain cyclic nucleotide levels during and immediately after transient ischemia are striking. In our own gerbil studies, a 3- to 4-factor elevation in cyclic AMP during ischemia gave way to a 2000-fold increase immediately on reflow. The increased levels of cyclic AMP in ischemic brain could be due to release of a number of substances, including the catecholamines, e.g., adenosine and

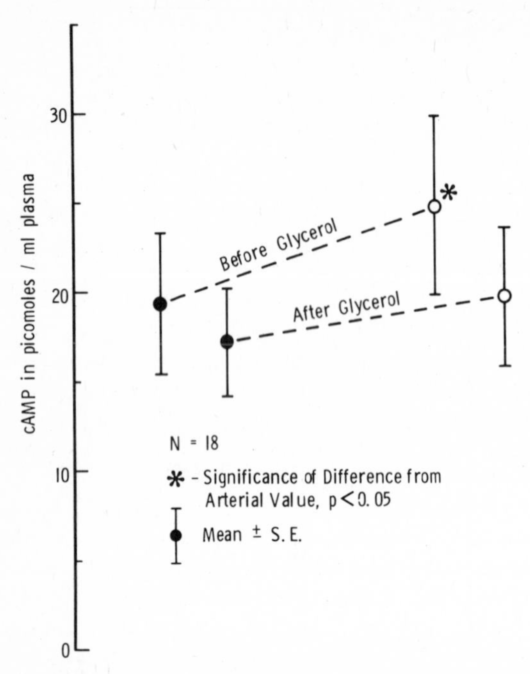

Figure 9. Movement of cyclic adenosine 3′,5′-monophosphate (AMP) from brain and CSF to cerebral venous blood of patients with recent cerebral infarction before and after intravenous 10% glycerol [carotid arterial (●) and cerebral venous blood (○)]. In this patient series, intravenous glycerol improved blood flow and cerebral metabolism. Decreased removal of cyclic AMP from central nervous system was probably a reflection of this reparative process. Reproduced from Welch *et al.*,[48] with permission.

potassium, all of which stimulate the enzyme adenyl cyclase. Presumably, these substances are still free to act when, in the reflow period after ischemia, adenosine triphosphate (ATP) is increasingly available for cyclic AMP synthesis. The fairly rapid recovery of cyclic AMP to near-normal levels after transient ischemia probably indicates neuronal reuptake of these substances as brain energy metabolism recovers. With the expected heterogeneity of regional cerebral blood flow in conditions such as occlusive stroke, there is likely to be marked regional variability in tissue cyclic AMP levels. This could explain the wide range of cyclic AMP elevation seen in the CSF of patients with cerebral ischemia.

The consequences of increased cyclic AMP levels in ischemic brain are unknown. Cyclic AMP can depress excitability of cortical neurons,[40,41] so that cyclic AMP might be responsible for the reduced neuronal excitability in foci of ischemia. Cyclic AMP has additional vasomotor effect, mediating cerebral vasodilatation possibly by an intracellular influence on calcium binding.[42] It remains to be seen, however, whether the extracellular accumulation of cyclic AMP is a significant influence, over and above that of pH change and other neurotransmitter effects, responsible for the fixed vasodilatation and vasomotor paralysis that occur in regions of brain ischemia. Cyclic AMP increase in ischemic brain might itself promote anaerobic glucose metabolism and lactacidosis via its regulatory role in brain glycolysis and glycogenolysis.[18] There are experiments, however, that call this into question, since, in the rat, graded hypoxia caused increased levels of brain lactate before cyclic AMP levels were altered.[36]

2.3. γ-Aminobutyric Acid

γ-Aminobutyric acid (GABA) is a putative inhibitory neurotransmitter in the CNS. Because of its involvement in cerebral energy metabolism, this seemed an important neurotransmitter to study in ischemia. GABA is widely distributed in brain at levels that exceed those of other amino acids.[34] Despite this, CSF GABA content is of the lowest order of all the amino acids,[7] probably because of high-affinity cellular-reuptake mechanisms.[9]

We used an enzymatic fluorometric assay to detect GABA in patients with stroke.[4] Despite being able to detect GABA in CSF at 10^{-12} mole per assay tube of CSF,[1] we detected no GABA in 19 control patients without neurological disease. When GABA was measured in CSF from patients 1–16 days after onset of cerebral hemispheric infarction, 14 of 28 patients showed elevated values (Fig. 10). Of 8 patients with transient vertebrobasilar ischemia, 5 patients, when studied within 48 hr of an attack, showed detectable GABA. Since these studies were performed, more sensitive assays that permit more quantitative analysis have become available (see Chapters 13 and 14). Hare *et al.*[10a] have observed significant elevations in CSF GABA levels in patients with acute hypoxic encephalopathy; however, they have not noted such elevations among stroke patients. Nevertheless, these early results strongly suggested disturbance of GABA metabolism in patients with ischemic CVD.

Detectable GABA in CSF probably results from an increased concentration in brain tissue extracellular fluid. This could be secondary to increased intracellular content of GABA, cellular leakage caused by impaired membrane integrity, increased synaptic release, or failure of reuptake mechanisms. In-

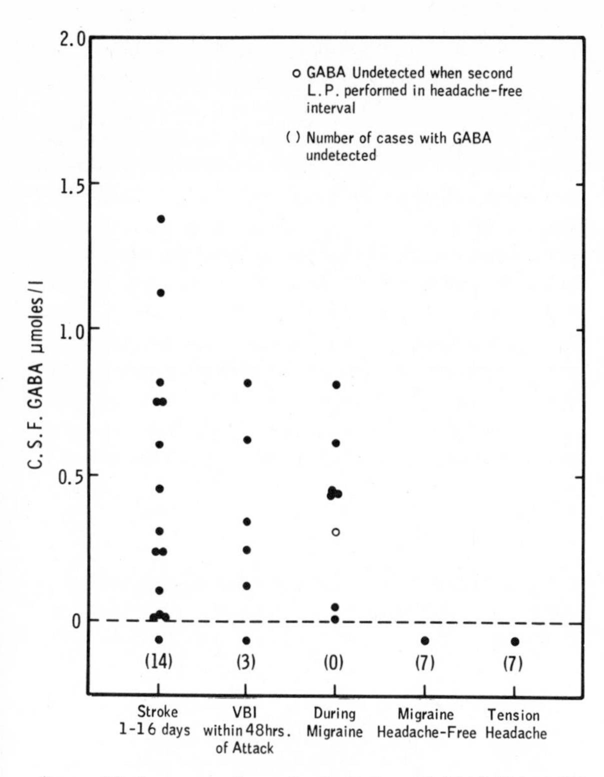

Figure 10. Increase of γ-aminobutyric acid (GABA) in CSF with recent onset of thromboembolic cerebral hemispheric infarction or vertebrobasilar ischemia (VBI). CSF GABA is also increased in migraine patients, but only during an attack of headache. No GABA was detected in CSF of controls with muscle-contraction (tension) headache. Symbols represent single measurements. Reproduced from Welch *et al.*,[45] with permission.

creased GABA concentrations have been reported in anoxic ischemic brain tissue of experimental animals.[17] The increases were profound and prolonged even after brief and transient cerebral ischemia.[14] The mechanisms of increase are as yet undefined, but probably result from disturbance of energy metabolism. An alteration in tricarboxylic acid cyclic intermediates because of impaired oxidative metabolism could result in an increase of GABA that is not utilized through the GABA shunt. Protein breakdown, ammonia detoxification, and altered amino acid transport also justify consideration.

GABA is a major inhibitory neurotransmitter in the CNS, producing hyperpolarization after neuronal depolarization.[16] Excess neuronal GABA levels or release and extracellular accumulation of GABA could therefore be another factor that promotes depression of neuronal activity associated with ischemic syndromes.

3. Studies in Migraine

Measurements of cerebral blood flow have, in some migraine patients, indicated that associated with the prodromal phase of an attack, there is cerebral ischemia that, although transient, is equivalent to that found in occlusive CVD.[30] There is also evidence that the headache phase of the attack is a residual of this ischemia.[46] The neurochemical findings in the CSF of patients with this condition seemed worthy of inclusion in this chapter, since the origin of some of the reported changes has been attributed to the ischemic phase of the syndrome.[45]

The number of CSF studies performed in this condition have in fact been few. Apart from the study of Skinhøj and Paulson,[39] who found elevated lactate levels in CSF obtained within 24–48 hr of an attack, most early studies concentrated on evidence for altered cerebral serotonin metabolism. Search for abnormality of this sytem was stimulated by the systemic serotonin changes reported from several laboratories.[3,38] Briefly, these results suggested that increased serotonin levels in the circulation caused the migraine prodrome, and that during the headache phase, there was release of serotonin from the platelets, consequent platelet serotonin depletion, and increased urinary excretion of 5-HIAA. However, findings from the few reported studies concerning the serotonin metabolite 5-HIAA in CSF from migraine patients have been too inconsistent to confirm altered CNS serotonin metabolism.[13,33]

This laboratory has studied cyclic AMP and GABA levels in the CSF of migraineurs and compared these results to those reported from stroke patients.[45] In 13 patients, the mean cyclic AMP value, calculated from levels in CSF sampled during or within 48 hr of an attack of vascular headache of migraine type, was significantly higher than that of controls (See Fig. 8). One of the highest individual values was recorded in a patient who presented with classic migraine but in whom cerebral arteriography demonstrated a large cerebral hemispheric arteriovenous malformation. In view of the similar findings in stroke cases, it seems reasonable to suggest that the cyclic AMP changes in the CSF of migraine patients are due to ischemia. The higher range of values in migraineurs compared to stroke patients is interesting in view of the animal experiments that showed more massive increase in brain tissue cyclic AMP concentration after transient ischemia. As in the case of stroke patients, the clinical relevance of these findings is possibly related to the role of cyclic AMP in depression of neuronal excitability, vasodilatation, and altered energy metabolism, since such pathophysiological changes may be associated with the migraine syndrome.

When lumbar punctures were performed in eight patients during a migraine attack, CSF GABA was measurable in all cases (see Fig. 10). One patient in the group had a second tap because of the detection of ventricular dilatation on CT scan. No CSF GABA was detected in the second sample, taken when the patient was free of headache. This patient was included in the group of seven patients in whom no GABA was detected in CSF obtained during a headache-free interval. No GABA was detected in CSF from series controls who had an eventual diagnosis of muscle-contraction headache.

It seems reasonable to suggest that ischemia is a possible cause of the GABA elevation and that the mechanisms are as those quoted for stroke patients. Any speculation concerning the relevance of the findings to the clinical features of migraine must include primarily the influence of GABA as an inhibitory neurotransmitter causing depression of neuronal excitability.

4. Therapeutic Implications of CSF Studies

Neurochemical investigations performed in human CSF have highlighted a possibly important relationship of disordered neurotransmitter function to the pathogenesis of neurological deficit resulting from cerebral ischemia and infarction. Obviously, further study is required to establish the specific stages of disordered neurotransmitter function, the clinical and functional consequences of such disorder, and the relationship of excessive neurotransmitter release, displacement, and accumulation in the cerebral parenchyma, extracellular space, and CSF to the pathogenesis of brain edema, changes in energy metabolism, paralysis, neurological symptoms, and progression of infarction. There is evidence that pursuing this course of study could benefit the patient with acute stroke, a disease state that as yet has yielded little to the therapeutic strivings of the clinician. Several phenomenological studies in the experimental animal have been aimed at prevention of abnormal neurotransmitter release and accumulation in ischemic brain and have shown favorable modification of the stroke process. Inhibition of catecholamine synthesis with α-methyltyrosine reduced morbidity and mortality in monkeys subjected to middle cerebral artery occlusion.[52] Pretreatment with *p*-chlorophenylalanine to reduce the brain serotonin levels reduced development of acute stroke in the gerbil.[43] Reserpine had a similar effect in the rat.[36] Receptor blockade to prevent the postsynaptic influence of neurotransmitters released by ischemia has also been effective in reducing stroke incidence.[36,44] To date, studies that have applied the same therapeutic principles in stroke management have been few. There was little evidence of improved cerebral energy metabolism in acute stroke patients treated with propranolol and phenoxybenzamine.[23] However, modification of serotonin function with the drug trazodone has also been attempted, with results that promised less morbidity and reduced hospital stay in treated acute stroke patients.[2] Further studies using these principles of approach now seem appropriate.

5. Summary

Neurochemical measurements in the CSF of patients with cerebral ischemia caused by cerebrovascular disease and migraine have indicated metabolic disorder involving a number of important neurotransmitter systems. The studies have been supported by data obtained from experimental animal models. As a result, pharmacological manipulation of neurotransmitter metabolism is being examined for possible therapeutic benefit.

ACKNOWLEDGMENT. This work was supported by USPHS grant NS 09287.

References

1. ACHAR, V. S., WELCH, K. M. A., CHABI, E., BARTOSH, K., MEYER, J. S.: Cerebrospinal fluid gamma-aminobutyric acid (GABA) in neurological disease. *Neurology* **26**:777–780, 1976.
2. ALLORI, L., CIOLI, V., SILVESTRINI, B.: Experimental and clinical data indicating a potential use of trazodone in acute stroke. *Curr. Ther. Res.* **18**:410–416, 1975.
4. ANTHONY, M., HINTERBERGER, H., LANCE, J. W.: Plasma serotonin in migraine and stress. *Arch. Neurol.* **16**:544–552, 1967.
4. BAXTER, C. F.: Assay of gamma aminobutyric acid. In Fried, R. (ed.): *Methods of Neurochemistry*, Vol. 3, New York, Marcell Dekker, 1972, pp. 1–73.
5. BAZÁN, N. G., JR.: Effects of ischemic and electroconvulsive shock on free fatty acid pool in the brain. *Biochim. Biophys. Acta* **218**:1–10, 1970.
6. BERZIN, Y. E., AUNA, Z. P., BREZHINSKY, G. Y.: The

significance of blood serotonin and CSF in the clinical pucture and pathogenesis of acute cerebral circulatory disturbance (in Russian), *Zh. Nevropathol. Psikhiatr. im. S. S. Korsakova* **69:**1011–1015, 1969.
7. BITO, L., DAVSON, H., LEVIN, E., MURRAY, M., SNIDER, N.: The concentrations of free amino acids and other electrolytes in cerebrospinal fluid, *in vivo* dialysate of brain, and blood plasma of the dog. *J. Neurochem.* **13:**1057–1067, 1966.
8. CRAMER, H., NG, N. K. Y., CHASE, T. N.: Effect of probenecid on levels of cyclic AMP in human cerebrospinal fluid. *J. Neurochem.* **19:**1601–1603, 1972.
9. ELLIOTT, K. A. C., VAN GELDER, N. M.: Occlusion and metabolism of γ-aminobutyric acid by brain tissue. *J. Neurochem.* **3:**28–40, 1958.
10. GAUDET, R., WELCH, K. M. A., CHABI, E., WANG, T.-P.: Effect of transient ischemia on monoamine levels in the cerebral cortex of gerbils. *J. Neurochem.* **30:**751–757, 1978.
10a. HARE, T. A., MANYAM, N. V., GLAESER, B. S.: Evaluation of cerebrospinal fluid γ-aminobutyric acid content in neurologic and psychiatric disorders. In Wood, J. H. (ed.): *Neurobiology of Cerebrospinal Fluid I.* New York, Plenum Press, 1980.
11. HØEDT-RASMUSSEN, K., SKINHØJ, E.: Transneural depression of the cerebral hemispheric metabolism in man. *Acta Neurol. Scand.* **40:**41–46, 1964.
12. ISHIHARA, N., WELCH, K. M. A., MEYER, J. S., CHABI, E., NARITOMI, H., WANG, T.-P. F., NELL, J. H., HSU, M.-C., MIYAKAWA, Y.: The influence of cerebral embolism on brain monoamines. *J. Neurol. Neurosurg. Psychiatry* **42:**847–853, 1979.
13. KANGASNIEMI, P., SONNINEN, V., RINNE, U. K.: Excretion of free and conjugated 5-HIAA and VMA in urine and concentration of 5-HIAA and HVA in CSF during migraine attacks and free intervals. *Headache* **12:**62–65, 1972.
14. KOBAYASHI, K., KAWAKAMI, S., HOSSMANN, K. A., KLEIHUES, P.: Free amino acids in the cat brain during cerebral ischemia and subsequent recirculation. In Harper, M., Jennett, B., Miller, D., Rowan, J. (eds.): *Blood Flow and Metabolism in the Brain.* Edinburgh, Churcmill Livingston, 1975, pp. 10.3–10.7.
15. KOBAYASHI, M., LUST, W. D., PASSONNEAU, J. V.: Concentrations of energy metabolites and cyclic nucleotides during and after bilateral ischemia in the gerbil cerebral cortex. *J. Neurochem.* **29:**53–59, 1977.
16. KRNJEVIĆ, K.: Chemical nature of synaptic transmission in vertebrates. *Physiol. Rev.* **54:** 418–540, 1974.
17. LOVELL, R. A., ELLIOTT, S. J., ELLIOTT, K. A. C.: The gamma aminobutyric acid and Factor I content of brain. *J. Neurochem.* **10:**479–488, 1963.
18. LOWRY, O. H., PASSONNEAU, J. V.: Kinetic evidence for multiple binding sites on phosphofructokinase. *J. Biochem.* **241:**2268–2279, 1966.
19. LUST, W. D., MRŚULJA, B. B., MRŚULJA, B. J., PASSONNEAU, J. V., KLATZO, I.: Putative neurotransmitters and cyclic nucleotides in prolonged ischemia of the cerebral cortex. *Brain Res.* **98:**394–399, 1975.
20. MAJNO, G., PALADE, G. E.: Studies on inflammation. Part 1. The effect of histamine and serotonon on vascular permeability: An electron microscopic study. *J. Biophys. Biochem. Cytol.* **11:**571–605, 1961.
21. MEYER, J. S.: Localized changes in properties of the blood and effects of anticoagulant drugs in experimental cerebral infarction. *N. Engl. J. Med.* **258:**151–159, 1958.
22. MEYER, J. S.: Localized changes in properties of the blood and effects of anticoagulant drugs in experimental cerebral infarction. *N. Engl. J. Med.* **258:**151–159, 1958.
22. MEYER, J. S., ITOH, Y., OKAMOTO, S., WELCH, K. M. A., MATHEW, N. T., OTT, E. O., SAKAKI, S., MIYAKAWA, Y., CHABI, E., ERICSSON, A. D.: Circulatory and metabolic effects of glycerol infusion in patients with recent cerebral infarction. *Circulation* **51:**701–712, 1975.
23. MEYER, J. S., MIYAKAWA, Y., WELCH, K. M. A., ITOH, Y., ISHIHARA, N., CHABI, E., NELL, J., BARTOSH, K., ERICSSON, A. D.: Influence of adrenergic receptor blockade on circulatory and metabolic effects of disordered neurotransmitter function in stroke patients. *Stroke* **7:**158–167, 1976.
24. MEYER, J. S., SHINOHARA, Y., KANADA, T., FUKUUCHI, Y., ERICSSON, A. D., KOK, N. K.: Diaschisis resulting from acute unilateral cerebral infarction: Quantitative evidence for man. *Arch. Neurol. (Chicago)* **23:**241–247, 1970.
25. MEYER, J. S., STOICA, E., PASCU, I., SHIMAZU, K., HARTMANN, A.: Catecholamine concentrations in CSF and plasma of patients with cerebral infarction and haemorrhage. *Brain* **96:**277–288, 1973.
26. MEYER, J. S., WELCH, K. M. A., OKAMOTO, S., SHIMAZU, K.: Disordered neurotransmitter function—Demonstration by measurement of norepinephrine and 5-hydroxytryptamine in CSF of patients with recent cerebral infarction. *Brain* **97:**655–665, 1974.
27. MISRA, S. S, SINGH, K. S., BHARGAVA, P.: Estimation of 5-hydroxytryptamine (5-HT) level in cerebrospinal fluid of patients with intracranial or spinal lesions. *J. Neurol. Neurosurg. Psychiatry* **30:**163–165, 1967.
28. MOLINARI, G. S., MOSELEY, J. I., LAURENT, J. P.: Segmental middle cerebral artery occlusion in primates: An experimental method requiring minimal surgery and anesthesia. *Stroke* **5:**334–339, 1974.
29. MYLLYLÄ, V. V., VAPAATALO, H., HOKKANEN, E., HEIKKINEN, E. R.: Cerebrospinal fluid concentration of cyclic adenosine-3′,5′-monophosphate and pneumoencephalography. *Eur. Neurol.* **12:**28–32, 1974.
30. O'BRIEN, M. D.: Cerebral blood flow changes in migraine. *Headache* **10:**139–143, 1971.
31. OSTERHOLM, J. L., BELL, J.: Experimental effects of free serotonin on the brain and its relation to brain injury. Part 1. The neurological consequences of intracerebral serotonin injections. *J. Neurosurg.* **31:**408–412, 1969.
32. OZAWA, K., ITADA, N., KUNO, S., SETA, K., HANDA, H., ARAKI, C.: Biochemical studies on brain swelling. Part 1. Changes in respiratory control 2,4-dinitrophenol-induced ATPase activity and phosphorylation:

Correlation between brain swelling and mitochondrial function. *Folia Psychiatr. Neurol. Jpn.* **20:**57–58, 1966.
33. POLONI, M., NAPPI, G., ARRIGO, A., SAVOLDI, F.: Cerebrospinal fluid 5-hydroxyindoleacetic acid level in migrainous patients during spontaneous attacks, during headache-free period and following treatment with L-tryptophan. *Experientia* **30:**640–641, 1974.
34. ROBINSON, N., WELLS, F.: Distribution and localization of sites of gamma-aminobutyric acid metabolism in the adult rat brain. *J. Anat.* **114:**365–378, 1973.
35. ROBISON, G. A., COPPEN, A. J., WHYBROW, P. C., PRANCE, A. J.: Cyclic AMP in affective disorders. *Lancet* **2:**1028–1029, 1970.
36. SCHEINBERG, P.: Correlation of brain monoamines and energy metabolism changes. In Scheinberg, P. (ed.): *Cerebrovascular Diseases: Tenth Princeton Conference.* New York, Raven Press, 1976, pp. 167–171.
37. SEBENS, J. B., KORF, J.: Cyclic AMP in cerebrospinal fluid: Accumulation following probenecid and biogenic amines. *Exp. Neurol.* **46:**333–344, 1975.
38. SICUTERI, F.: Vasoneuroactive substances and their implication in vascular pain. *Headache* **1:**6–45, 1967.
39. SKINHØJ, E., PAULSON, O. B.: Regional blood flow in internal carotid distribution during migraine attack. *Br. Med. J.* **3:**569–570, 1969.
40. SKINNER, J. E., WELCH, K. M. A., REED, J. C., NELL, J. H.: Psychological stress reduced cyclic 3′,5′-adenosine monophosphate levels in the cerebral cortex of conscious rats, as determined by a new cryogenic method of rapid tissue fixation. *J. Neurochem.* **39:**691–698, 1978.
41. STONE, T. W., TAYLOR, D. A., BLOOM, F. E.: Cyclic AMP and cyclic GMP may mediate opposite neuronal responses in the rat cerebral cortex. *Science* **187:**845–847, 1975.
42. TAGASHIRA, Y., MATSUDA, M., WELCH, K. M. A., CHABI, E., MEYER, J. S.: Effects of cyclic AMP and dibutyryl cyclic AMP on cerebral hemodynamics and metabolism in the baboon. *J. Neurosurg.* **46:**484–493, 1977.
43. WELCH, K. M. A., CHABI, E., BUCKINGHAM, J., BERGIN, B., ACHAR, V. S., MEYER, J. S.: Catecholamines and 5-hydroxytryptamine levels in ischemic brain: Influence of *p*-chlorophenylalanine. *Stroke* **8:**341–346, 1977.
44. WELCH, K. M. A., CHABI, E., DODSON, R. F., WANG, T.-P. F., NELL, J., BERGIN, B.: The role of biogenic amines in the progression of cerebral ischemia and edema: Modification by *p*-chlorophenylalanine, methysergide and pentoxyfilline. In Pappius, H. S., Feindel, W. (eds.): *Dynamics of Brain Edema.* Berlin and Heidelberg, Springer-Verlag, 1976, pp. 195–202.
45. WELCH, K. M. A., CHABI, E., NELL, J. H., BARTOSH, K., CHEE, A. N. C., MATHEW, N. T., ACHAR, V. S.: Biochemical comparison of migraine and stroke. Headache **16:**160–167, 1976.
46. WELCH, K. M. A., CHABI, E., NELL, J. H., BARTOSH, K., MEYER, J. S., MATHEW, N. T.: Similarities in biochemical effects of cerebral ischemia in patients with cerebrovascular disease and migraine. In Greene, R. (ed.): *Current Concepts in Migraine Research.* New York, Raven Press, 1978, pp. 1–9.
47. WELCH, K. M. A., HASHI, K., MEYER, J. S.: Cerebrovascular response to intracarotid injection of serotonin before and after middle cerebral artery occlusion. *J. Neurol. Neurosurg. Psychiatry* **26:**724–735, 1973.
48. WELCH, K. M. A., MEYER, J. S., CHEE, A. N. C.: Evidence for disordered cyclic AMP metabolism in patients with cerebral infarction. *Eur. Neurol.* **13:**144–154, 1975.
49. WELCH, K. M. A., MEYER, J. S., TERAURA, T., HASHI, K., SHINMARU, S.: Ischemic anoxia and cerebral serotonin levels. *J. Neurol. Sci.* **16:**85–92, 1972.
50. WELCH, K. M. A., NELL, J., CHABI, E.: The role of cyclic AMP in neurologic and affective disorders. In Volicer, L. (ed.): *Clinical Aspects of Cyclic Nucleotides.* New York, Spectrum, 1977, pp. 327–360.
51. WELCH, K. M. A., WANG, T.-P. F., CHABI, E.: Ischemia-induced seizures and cortical monoamine levels. *Ann. Neurol.* **3:**152–155, 1978.
52. ZERVAS, N. T., HORI, H.: Effect of alpha methyl tyrosine on cerebral infarction. *Stroke* **4:**331–339, 1973.
53. ZERVAS, N. T., HORI, H., NEGORA, M., WURTMAN, R. J., LARIN, F., LAVYNE, M. H.: Reduction in brain dopamine following experimental cerebral ischaemia. *Nature (London)* **247:**283–284, 1974.

CHAPTER 25

Cerebrospinal Fluid Cyclic Nucleotide Alterations in Traumatic Coma

Alan S. Fleischer and George T. Tindall

1. Introduction

Cyclic AMP [(cAMP) cyclic adenosine 3′,5′-monophosphate] was initially discovered by Sutherland and Rall[31] in 1958. Further investigations have led to the concept that cAMP may be involved as the intracellular mediator of the actions of other hormones and of various putative synaptic transmitters. According to this concept, first messengers, the hormones or transmitters themselves, move from their cells of origin and induce the synthesis of cAMP in their target cells. Thereby, cAMP, by activating an appropriate sequence of enzymes, can evoke the specific response of a target cell to the hormone.[33] During the course of research on cAMP, a second cyclic nucleotide, cyclic guanosine 3′,5′-monophosphate (cGMP), was discovered[1] and is now thought to act as a second messenger for the actions of various neurotransmitters, including acetylcholine.

cAMP is widespread in mammalian tissues along with the enzymes adenylate cyclase and phosphodiesterase, which control cAMP synthesis and degradation, respectively.[3,32] Particularly high concentrations are observed in brain and CSF.[5,16,20,24] The concentration of cAMP in the brain is uniformly high and is not localized in any particular area.

Synthesis of cAMP, specifically related to levels of adenylate cyclase in the brain, is hormonally controlled and is stimulated by catecholamines, histamine, serotonin, and melanotropic peptides.[4,10,19,27] The final role of cyclic nucleotides in brain and other nervous tissue is unclear, but increasing evidence suggests that cAMP is involved in the regulation of metabolism and function in these tissues.[14] These processes include neurotransmission and the accompanying ionic permeability changes, neurotransmitter synthesis, regulation of intraneural movements, neuronal metabolism, and even trophic and developmental processes. Of principal importance has been the implication of cyclic nucleotides in synaptic transmission, which has been confirmed by several investigators.[21,36]

Evidence suggests that measuring cerebrospinal fluid (CSF) cAMP and cGMP concentrations is more useful in studying alterations of these brain nucleotides than assaying plasma or urine levels.[17,18,38] A recent study in this laboratory[28] showed that normal human lumbar CSF contains 15–30 nM cAMP. Also, ventricular CSF from patients with normal intracranial pressure (ICP) and normal sensorium contained the same concentration of cAMP as normal lumbar CSF, although ventricular CSF cAMP was found to be higher than lumbar CSF cAMP in another series.[35] In previous reports,[13,26] we demonstrated a significant correlation between the level of consciousness and ventricular CSF cAMP levels in patients comatose following head trauma or intracranial hemorrhage. Improvement to normal sensorium was associated with cAMP levels returning toward normal values, whereas those patients who

Alan S. Fleischer, M.D., and **George T. Tindall, M.D.** • Division of Neurosurgery, Emory University School of Medicine, Atlanta, Georgia 30322.

remained comatose had persistent, markedly diminished ventricular CSF cAMP levels, emphasizing that prolonged traumatic coma is associated with a disturbance of cAMP metabolism within the central nervous system. This chapter will concentrate primarily on this observation.

2. Clinical Material and Methods

Over a three-year period, beginning in October 1975, 50 patients, 16–63 years of age, were studied. All were admitted to the Neurosurgery Service of Grady Memorial Hospital in a comatose state after severe craniocerebral trauma. The majority of these patients had blunt, closed head trauma with level of consciousness graded as follows:

Grade 0: Normal.
Grade I: Drowsy, lethargic, indifferent and uninterested, or belligerent and uncooperative; does not lapse into sleep when left undisturbed.
Grade III: Deep stupor, requires strong pain to evoke movement; may have focal neurological signs, but will respond appropriately to noxious stimuli.
Grade IV: Does not respond appropriately to any stimuli; may exhibit decerebrate or decorticate posturing; retains deep tendon reflexes; may have dilated pupils, absent corneal or oculocephalic reflexes.
Grade V: Does not respond appropriately to any stimuli; flaccid; no deep tendon reflexes; usually apneic.

All patients were studied with immediate computer-assisted tomographic (CAT) scan or cerebral angiography. Thirty-seven patients required emergency craniotomy for evacuation of intracranial hematomas, contusions, or debridement of missile wounds. Within 12 hr of admission, all 50 patients had a ventricular catheter inserted into a lateral ventricle and connected to a closed Rickham reservoir (Cordis Corp., Miami, Florida) seated in the burr hole and buried subcutaneously.[12] After wound closure, a No. 23 needle with a soft catheter (Abbott butterfly catheter, Abbott Hospital Supplies, Chicago, Illinois) attached was inserted into the reservoir and connected to a Statham pressure transducer (Statham Manufacturing Co., San Juan, Puerto Rico) for continuous recording of ICP (Fig. 1). During the ensuing 1–40 days, depending on the outcome in the individual patient, 1- to 2-ml samples of CSF were collected at intervals of 6–72 hr by percutaneous puncture of the subcutaneous reservoir. Grade of coma was recorded at the time of each sampling. Venous blood was obtained simultaneously with CSF in one half of the samples for measurement of plasma cAMP and cGMP levels.

Each sample of CSF and blood was centrifuged immediately after collection and then stored at −20°C. Within 1 week, the samples were analyzed for cAMP and cGMP by radioimmunoassay.[30] Experimental details of these analyses, and the verification of their specificities, have been described in previous reports.[25,27,28] CSF was also analyzed for cyclic nucleotide phosphodiesterase activity.[30]

For control purposes, cAMP was measured in plasma and in lumbar CSF of 20 patients aged 30–62 years, with no neurological disease, who were undergoing spinal anesthesia. Ventricular CSF from patients without head trauma was not analyzed in this study, but in a previous investigation,[28] it was found that in ventricular CSF from adult patients with normal intracranial pressure and mentation, cAMP concentration averaged 22 nM. This did not differ significantly ($P < 0.05$ by Student's t test) from that in lumbar CSF from patients without neurological disease.

3. Summary of Cases

3.1. Control Patients

In the 20 adults without neurological disease, plasma cAMP (mean ± S.D.) was 13.8 ± 2.4 nM, and lumbar CSF cAMP was 23 ± 4 nM.

3.2. Head-Trauma Patients

The concentration of ventricular CSF cAMP was determined in 228 samples from 50 patients admitted to the hospital in a comatose state following head trauma. When the data from all 50 patients were pooled (Fig. 2), the following values for CSF cAMP concentration (mean ± S.E.M.) corresponding to each degree of coma were obtained: Grade V, 1.2 ± 0.1 nM; Grade IV, 1.5 ± 0.3 nM; Grade III, 3.1 ± 0.7 nM; Grade II, 10.2 ± 3.3 nM; Grade I, 13.7 ± 1.6 nM. After the sensorium became normal (Grade 0), cAMP was 22.5 ± 1.9 nM. The correlation between the grade of coma and cAMP concentration was −0.80 ($p < 0.01$).

For each patient, a change in degree of coma was usually associated with a similar change in cAMP level. In Grade IV coma, cAMP was always less than

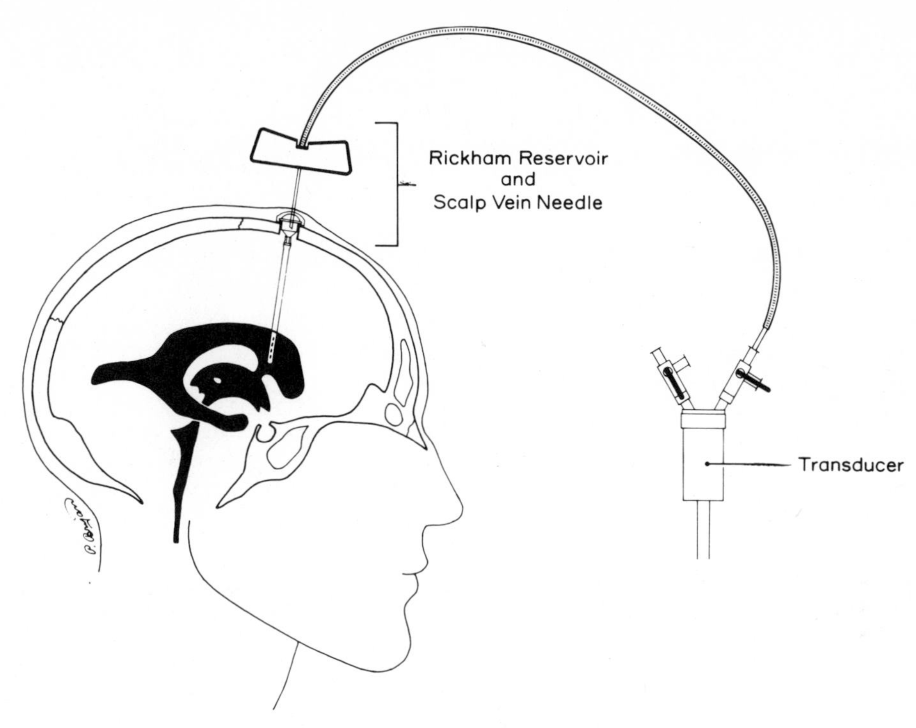

Figure 1. Drawing showing ventricular catheter in place in frontal horn of lateral ventricle connected to subcutaneously implanted Rickham reservoir. This setup may be used for continuous intracranial pressure (ICP) recording or intermittent sampling of ventricular CSF. Reproduced from Fleischer *et al.*,[12] with permission of publishers.

6 nM. A rise from less than 6 nM into the range of 6–12 nM in a patient with Grade IV coma was associated with a simultaneous or subsequent improvement of coma to Grade III or II. Yet, when cAMP remained less than 6 nM, coma persisted at the Grade IV level and was associated with either persistent vegetative survival or death. When cAMP in Grade III or II patients increased above 10 nM, this rise was accompanied or followed within 4 days by improvement in level of consciousness to Grade I or 0. Whenever cAMP rose above 15 nM, the patient made a full recovery.

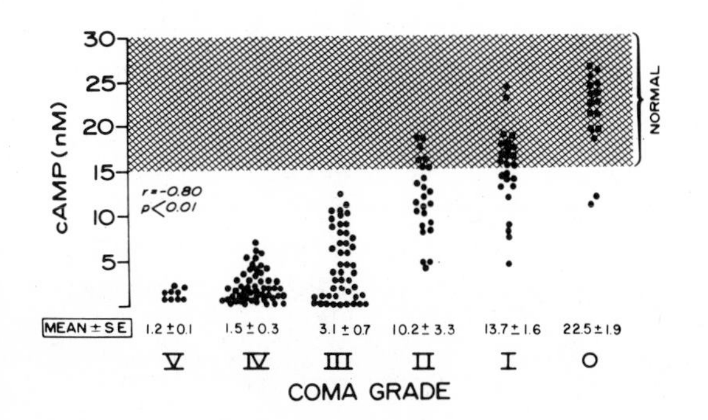

Figure 2. CSF cAMP levels obtained in 228 samples in 50 head-trauma patients in relation to coma grade.

Ventricular CSF cGMP did not correlate in any way with level of consciousness; however, there was a barely statistically significant correlation between elevated ICP and increased levels of CSF cGMP. ICP and CSF cAMP, however, had no apparent relationship.

3.2.1. Representative Cases

Case 1. This 24-year-old man was admitted in Grade IV coma following a wound from a large-caliber gun to the right frontal–temporal region (Fig. 3). Immediate CAT scan disclosed a right acute subdural hematoma of moderate size and evidence of an avascular intracerebral mass. The subdural and intracerebral hematomas were immediately evacuated, and a catheter was placed in the left lateral ventricle and connected to a Rickham reservoir. Transient, minimal neurological improvement ensued; however, the patient died on the 10th hospital

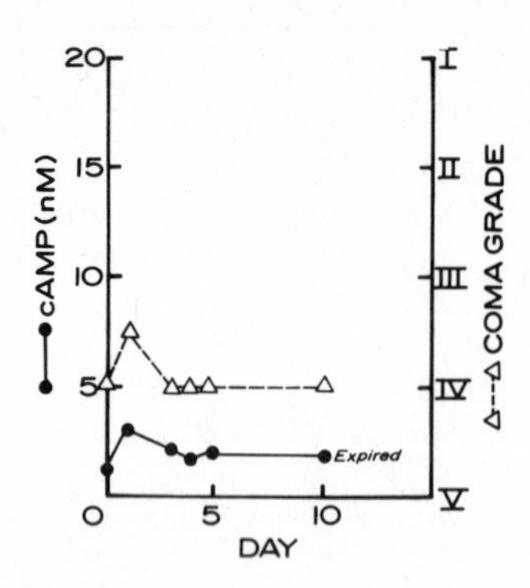

Figure 3. Clinical course in Case 1, showing relationship of ventricular CSF cAMP level to coma grade and time.

day after remaining in Grade IV coma. Preplateau and plateau waves were treated as they developed with intermittent mannitol infusion, hyperventilation, steroids, and CSF aspiration. The ventricular CSF cAMP never exceeded 3 nM.

Case 2. This 49-year-old woman suffered head trauma in an automobile accident and was admitted in Grade III coma (Fig. 4). Cerebral angiography disclosed an avascular left temporal mass lesion with a 3-mm shift of the midline structures. A Rickham reservoir was placed in the right lateral ventricle shortly after admission, and a CSF specimen obtained immediately disclosed a normal cAMP level, which fell precipitously shortly thereafter. The ICP was initially elevated in the 20–25 mm Hg range and responded to intermittent mannitol infusion, returning to normal. The patient deteriorated on the 3rd hospital day to Grade IV coma, associated with increasing ICP elevations. Repeat angiography disclosed an increase in the left temporal mass effect, and a temporal lobectomy was performed with some transient improvement to Grade III level postoperatively. Despite control of ICP with therapy including steroids, hyperventilation, and mannitol, the patient deteriorated to Grade IV and V coma and remained in a vegetative state throughout 60 days of hospitalization. After the initial normal cAMP level was obtained, CSF cAMP never exceeded 2 nM.

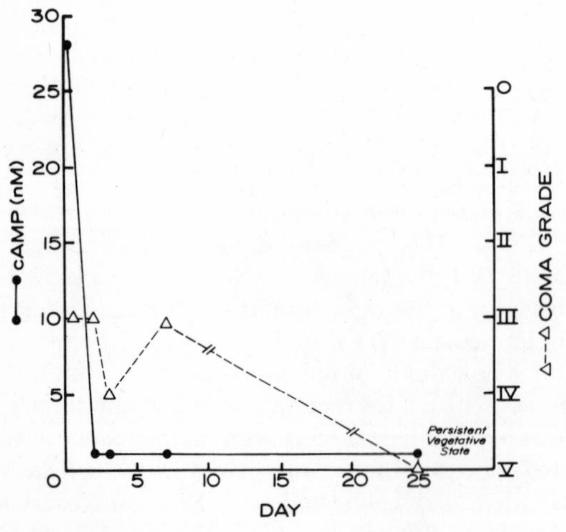

Figure 4. Clinical course in Case 2, showing relationship of ventricular CSF cAMP level to coma grade and time.

Case 3. This 26-year-old woman sustained multiple facial fractures in an automobile accident and was admitted in Grade IV coma (Fig. 5). Cerebral angiography was normal. A Rickham reservoir was inserted under local anesthesia in the right lateral ventricle. Continuous monitoring of ICP over a 1-week period did not reveal any pressure elevation, thus suggesting a clinical diagnosis of brainstem contusion. After a few days of transient improvement following a tracheostomy, she regressed to Grade IV coma and remained in a persistent vegetative state for several weeks after the last CSF cAMP sample, which was obtained on the 20th hospital day, before dying of pulmonary complications. Her CSF cAMP levels remained consistently below 6 nM throughout the sampling period.

Case 4. This 41-year-old man had a generalized seizure after a blow to the head and was admitted to Grady Memorial Hospital in Grade IV coma (Fig. 6). CAT scan revealed a small right subdural hematoma and a left temporal contusion without an associated shift of the midline structures. A Rickham reservoir was inserted into the right lateral ventricle under local anesthesia to monitor ICP, which was normal initially. On the 3rd day, ICP rose to 35 mm Hg in a sustained manner refractory to standard

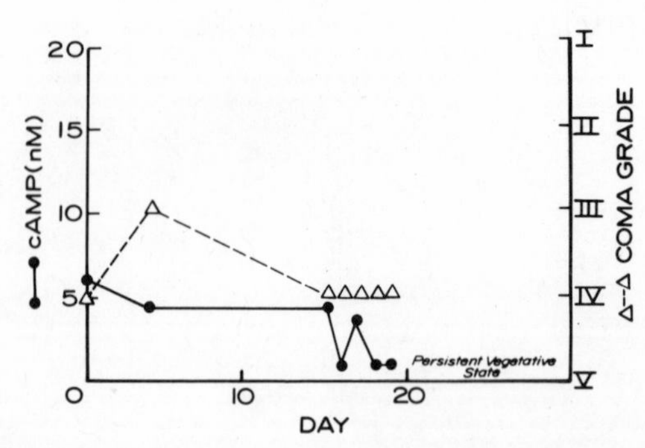

Figure 5. Clinical course in Case 3, showing relationship of ventricular CSF cAMP level to coma grade and time.

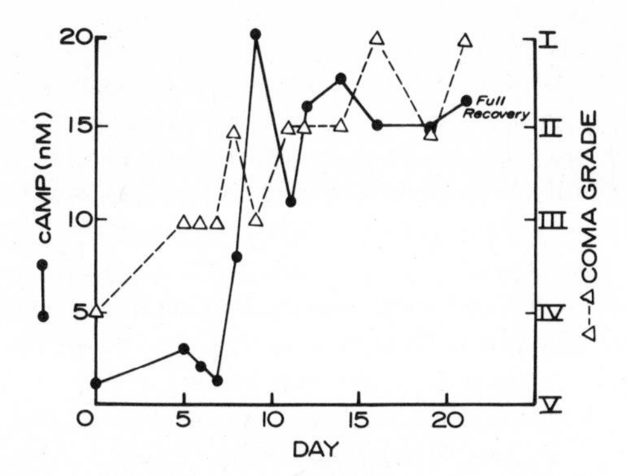

Figure 6. Clinical course in Case 4, showing relationship of ventricular CSF cAMP level to coma grade and time.

medical therapy. He remained unchanged clinically, however, with CSF cAMP levels below 4 nM. On the 4th day, bilateral subtemporal craniectomies were performed with evacuation of a right subacute subdural hematoma and a subtotal left temporal lobectomy for contusion. The patient's level of consciousness improved progressively from the 5th day, and his sensorium was normal 20 days after admission. The CSF cAMP levels returned to normal.

Case 5. This 49-year-old alcoholic man was found unresponsive in jail with evidence of facial contusions. On admission to Grady Memorial Hospital, the patient was in Grade III coma with an associated left hemiparesis (Fig. 7). CAT scan revealed a right acute subdural hematoma and a right temporal lobe hematoma. At surgery, the subdural hematoma was evacuated and a subtotal temporal lobectomy performed for hemorrhagic contusion. A Rickham reservoir was inserted in the left lateral ventricle. Postoperatively, he remained stable for the first 3 days with transient ICP elevations requiring medical therapy. By the 5th postoperative day, he had improved to Grade II coma, and by the 20th hospital day was fully recovered. Changes in ventricular CSF cAMP paralleled his clinical course.

Case 6. This 46-year-old man sustained blunt head trauma and was admitted to the hospital in Grade IV coma with a fixed, dilated right pupil and left-sided decerebration (Fig. 8). Cerebral angiography disclosed an acute right subdural hematoma, which was immediately evacuated. A Rickham reservoir was inserted into his left lateral ventricle. Postoperatively, he was initially unchanged, and ICP elevations to 60 mm Hg were successfully treated with standard medical measures. By the 7th day, his level of consciousness improved to Grade III with an associated elevation of CSF cAMP from 3 to 12 nM. He continued to improve slowly, and at the time of the last cAMP sample, 27 days after admission, he was in Grade I stupor with a CSF cAMP of 15 nM. Over the ensuing 40 days of hospitalization, he made a full recovery.

4. Conclusions

Previous reports have demonstrated changes in cyclic nucleotide levels in brain tissue and in CSF in various clinical conditions in humans and animal experimental models. Results, however, have been conflicting and difficult to interpret. For example, Flamm *et al.*[11] recently demonstrated markedly diminished cAMP levels in cat temporal lobe cortex more than 3 hr following middle cerebral artery occlusion. Within 3 hr after the onset of ischemia, the

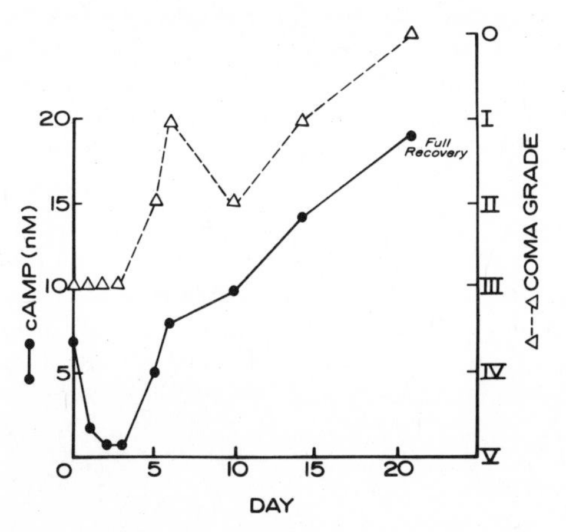

Figure 7. Clinical course in Case 5, showing relationship of ventricular CSF cAMP level to coma grade and time.

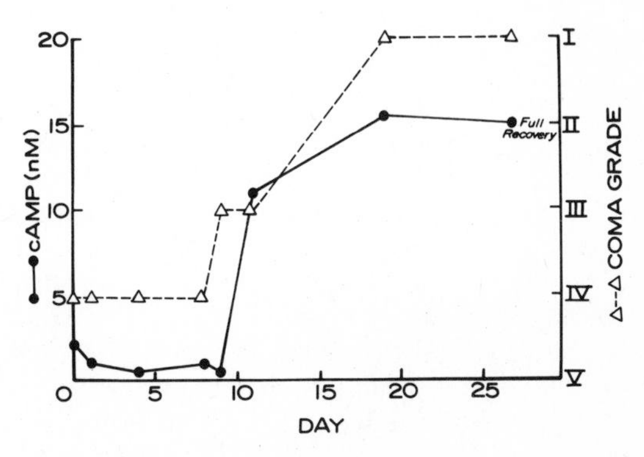

Figure 8. Clinical course in Case 6, showing relationship of ventricular CSF cAMP level to coma grade and time.

cAMP levels were normal and the lesion was considered reversible and thought to indicate that an initial step in the alteration of cellular metabolism following ischemia is possibly due to membrane perturbation and altered production of the second messenger. Although CSF cAMP levels have not been correlated with changes in brain tissue levels specifically, it is generally accepted that CSF and the metabolites in it are products of brain tissue and may therefore reflect brain metabolism.[2,9] However, in patients studied following cerebral infarction, elevated cAMP blood and CSF levels have been demonstrated,[15,16,39] rather than the expected diminished levels. (See Chapters 5, 9, 19, and 24.)

Similar conflicts occur in studying brain trauma.[37] At 1 min after stab trauma to the cerebrum of the mouse, brain cAMP concentration rose to 7 times normal.[32] Also, CSF cAMP in rabbits following experimental brain injury demonstrated a marked elevation for 2 weeks. From the 3rd week onward after the injury, however, the CSF cAMP concentration was significantly lower than the basal level before the injury.[23] The same investigator reported CSF cAMP concentrations significantly higher in patients with cerebral contusion than in those with cerebral concussion. Even in this latter group with minimal head trauma, however, the cAMP values were higher than in control patients.[23]

Rudman *et al.*[28] first attempted to relate CSF cAMP levels to sensorium, and demonstrated subnormal concentrations in children with diffuse brain disease causing severe psychomotor retardation. This same study demonstrated a direct correlation between ICP and CSF cGMP. Additional reports from our laboratory[13,26] revealed markedly reduced ventricular CSF cAMP levels in 31 patients who were comatose following either head trauma or spontaneous intracranial hemorrhage; in those cases, cAMP returned to normal levels with improvement in sensorium.

This chapter expands on the correlation of ventricular CSF cAMP and level of consciousness in a larger, homogeneous group of patients rendered comatose as a result of severe acute head trauma. These previous reports and our own observations suggest that the subnormal level of cAMP in the ventricular CSF of the present series of 50 patients does not represent a nonspecific effect of brain damage, but may reflect instead a depletion of brain cAMP that is uniquely associated with severe and prolonged coma after trauma or intracranial hemorrhage.

The plasma cAMP level of these comatose patients was normal. Therefore, their subnormal level of CSF cAMP was probably not caused by diminished transport of cAMP into CSF from plasma, but presumably resulted instead from a disorder of cAMP metabolism within the central nervous system. One possible cause of disorder might be decreased production of cAMP by brain adenylate cyclase, which could result from damage to the cyclase enzyme, from inadequate delivery of cyclase-activating hormones (catecholamines, histamine, serotonin, or melanotropic peptides), from deficiency of the enzyme's substrate, adenosine triphosphate (ATP), or from deficiency of the endogenous cyclase-activating metabolic intermediate adenosine,[29] which is derived from ATP. Another possible explanation is accelerated transport of cAMP out of CSF by the probenecid-sensitive active transport system.[8] The possibility of increased degradation of cAMP in CSF is unlikely because of the absence of the enzyme phosphodiesterase in the samples tested.

Therapeutic implications of these findings may be important. If the diminished CSF cAMP levels in comatose patients following head trauma do not represent an epiphenomenon, then it is possible that the low levels may be the cause of the diminished consciousness. There are several methods of increasing CSF cAMP that may be of potential therapeutic value. Probenecid has been demonstrated in the experimental animal[6] and in man[7] an ability to increase the CSF content of cAMP that may be of potential therapeutic value. In addition, any of the many adenylate-cyclase-stimulating agents or phosphodiesterase inhibitors may conceivably accomplish the same result. A synthetic cAMP, dibutyryl cAMP, may be instilled directly into the CSF. Indeed, Tagashira *et al.*[34] recently demonstrated increase in cerebral blood flow, reduction in intravascular resistance, and increased cerebral oxygen and glucose consumption in baboons following the administration of intracarotid cAMP and intracisternal dibutyryl cAMP.

Whether or not the correlation of CSF cAMP level with degree of coma is of therapeutic significance, it may serve as an excellent prognosticating factor in severe head trauma. No patient in this series whose CSF cAMP level remained below 6 nM for longer than 10 days recovered, whereas all patients whose CSF cAMP levels exceeded 15nM recovered.

References

1. Ashman, D. F., Lipton, R., Melicow, M. M., Price, T. D.: Isolation of adenosine 3′,5′-monophosphate and

guanosine 3′,5′-monophosphate from rat urine. *Biochem. Biophys. Res. Commun.* **11**:330–334, 1963.

2. Bering, E. J., Jr.: Cerebrospinal fluid production and its relationship to cerebral metabolism and cerebral blood flow. *Am. J. Physiol.* **197**:825–828, 1959.
3. Butcher, R. W., Sutherland, E. W.: Adenosine 3′,5′-phosphate in biological materials: Purification and properties of cyclic 3′,5′-nucleotide phosphodiesterase and use of this enzyme to characterize adenosine 3′,5′-phosphate in human urine. *J. Biol. Chem.* **237**:1244–1250, 1962.
4. Chasin, M., Rivkin, I., Mamrak, F., Samaniego, S. G., Hess, S. M.: α- and β-adrenergic receptors as mediators of accumulation of cyclic adenosine 3′,5′-monophosphate in specific areas of guinea pig brain. *J. Biol. Chem.* **246**:3037–3041, 1971.
5. Cramer, H., Goodwin, F. K., Post, R. M., Bunney W. E.: Effects of probenecid and exercise on cerebrospinal-fluid cyclic A.M.P. in affective illness (letter). *Lancet* **1**:1346–1347, 1972.
6. Cramer, H., Lindl, T.: Probenecid inhibits efflux of adenosine 3′,5′-monophosphate (cAMP) from cerebrospinal fluid (CSF) in the rat. *Psychopharmacologia* **26** (Suppl.):49, 1972 (abstract).
7. Cramer, H., Ng, L. K. Y., Chase, T. N.: Adenosine 3′,5′-monophosphate in cerebrospinal fluid: Effect of drugs and neurologic disease. *Arch. Neurol.* **29**:197–199, 1973.
8. Cramer, H., Ng, L. K. Y., Chase, T. N.: Effect of probenecid on levels of cyclic AMP in human cerebrospinal fluid. *J. Neurochem.* **19**:1601–1602, 1972.
9. Cserr, H. F.: Relationship between cerebrospinal fluid and interstitial fluid of brain. *Fed. Proc. Fed. Am. Soc. Exp. Biol.* **33**:2075–2078, 1974.
10. Daly, J. W., Huang, M., Shimizu, H.: Regulation of cyclic AMP levels in brain tissue. *Adv. Cyclic Nucleotide Res.* **1**:375–387, 1972.
11. Flamm, E. S., Schiffer, J., Viau, T. A., Naftchi, E. N.: Alterations of cyclic AMP in cerebral ischemia. *Stroke* **9**:400–402, 1978.
12. Fleischer, A. S., Patton, J. M., Tindall, G. T.: Monitoring intraventricular pressure using an implanted reservoir in head injured patients. *Surg. Neurol.* **3**:309–311, 1975.
13. Fleischer, A. S., Rudman, D. R., Fresh, C. B., Tindall, G. T.: Concentration of 3′,5′ cyclic adenosine monophosphate in ventricular CSF of patients following severe head trauma. *J. Neurosurg.* **47**:517–524, 1977.
14. Greengard, P., Costa, E. (eds): *Role of Cyclic AMP in Cell Function: Advances in Biochemistry and Psychopharmacology*, Vol. 3. New York, Raven Press, 1970.
15. Heikkinen, E. R., Myllylä, V. V., Hokkanen, E., Vapaatalo, H.: Cerebrospinal fluid concentration of cyclic AMP in cerebrovascular diseases. *Eur. Neurol.* **14**:129–137, 1976.
16. Heikkinen, E. R., Myllylä, V. V., Vapaatalo, H., Hokkanen, E.: Urinary excretion and cerebrospinal fluid concentration of cyclic adenosine-3′,5′-monophosphate in various neurological diseases. *Eur. Neurol.* **11**:270–280, 1974.
17. Heikkinen, E. R., Simila, S., Myllylä, V. V., Hokkanen, E.: Cyclic adenosine-3′,5′-monophosphate concentration and enzyme activities of cerebrospinal fluid in meningitis of children. *Z. Kinderheilkd.* **20**:243–250, 1975.
18. Heikkinen, E. R., Vapaatalo, H., Myllylä, V. V., Hokkanen E.: Role of cyclic AMP in neurological diseases. Presented at the Second International Conference on Cyclic AMP, Vancouver, B.C., 1974.
19. Kakiuchi, S., Rall, T. W.: The influence of chemical agents on the accumulation of adenosine 3′,5′-monophosphate in slices of rabbit cerebellum. *Mol. Pharmacol.* **4**:367–378, 1968.
20. Krishna, G., Forn, J., Voigt, K., Paul, M., Gessa, G. L.: Dynamic aspects of neuro-hormonal control of cyclic 3′,5′-AMP synthesis in brain. *Adv. Biochem. Psychopharmacol.* **3**:155–172, 1970.
21. Marx, J. L.: Cyclic AMP in brain: Role in synaptic transmission. *Science* **178**:1188–119–, 1972.
22. Myllylä, V. V., Heikkinen, E. R., Vapaatalo, H., Hokkanen, E..: Cyclic AMP concentration and enzyme activities of cerebrospinal fluid in patients with epilepsy or central nervous system damage. *Eur. Neurol.* **13**:123–130, 1975.
23. Myllylä, V. V.: Effect of cerebral injury on cerebrospinal fluid cyclic AMP concentration. *Eur. Neurol.* **14**:413–425, 1976.
24. Robison, G. A., Coppen, A. J., Whybrow, P. C., Prance, A. J.: Cyclic A.M.P. in affective disorders (letter). *Lancet* **2**:1028–1029, 1970.
25. Rudman, D. R.: Injection of melatonin into cisterna magna increases concentration of 3′,5′ cyclic guanosine monophosphate in cerebrospinal fluid. *Neuroendocrinology* **20**:235–242, 1976.
26. Rudman, D., Fleischer, A. S., Jutner, M. S.: Concentration of 3′,5′ cyclic adenosine monophosphate in ventricular cerebrospinal fluid of patients with prolonged coma after head trauma or intracranial hemorrhage. *N. Engl. J. Med.* **295**:635–638, 1976.
27. Rudman, D., Isaacs, J. W.: Effect of intrathecal injection of melantropic–lipolytic peptides on concentration of 3′,5′ cyclic adenosine monophosphate in cerebrospinal fluid. *Endocrinology* **97**:1476–1480, 1975.
28. Rudman, D., O'Brien, M. S., McKinney, A. S., Hoffman, J. C., Patterson, J. H.: Observations on the cyclic nucleotide concentrations in human cerebrospinal fluid. *J. Clin. Endocrinol. Metab.* **42**:1088–1097, 1976.
29. Sattin, A., Rall, T. W.: The effect of adenosine and adenine nucleotides on the cyclic adenosine 3′,5′-phosphate content of guinea pig cerebral cortex slices. *Mol. Pharmacol.* **6**:13–23, 1970.
30. Steiner, A. L., Parker, C. W., Kipnis, D. M.: The measurement of cyclic nucleotides by radioimmunoassay. *Adv. Biochem. Psychopharmacol.* **3**:89–111, 1970.
31. Sutherland, W. E., Rall, T. W.: Fractionation and characterization of a cyclic adenine ribonucleotide

formed by tissue particles. *J. Biol. Chem.* **232:**1077–1091, 1958.

32. SUTHERLAND, E. W., RALL, T. W., MENON, T.: Adenyl cyclase: Distribution, preparation, and properties. *J. Biol. Chem.* **237:**1220–1227, 1962.
33. SUTHERLAND, W. E., ROBISON, G. A., BUTCHER, R. W.: Some aspects of the biological role of adenosine 3′,5′-monophosphate (cyclic AMP). *Circulation* **37:**279–306, 1968.
34. TAGASHIRA, Y., MATSUDA, M., WELCH, K. M. A., CHABI, E., MEYER, J. S.: Effects of cyclic AMP and dibutyryl cyclic AMP on cerebral hemodynamics and metabolism in the baboon. *J. Neurosurg.* **46:**484–493, 1977.
35. TSANG, D., LAL, S., SOURKES, T. L., FORD, R. M., ARANOFF, A.: Studies on cyclic AMP in different compartments of cerebrospinal fluid. *J. Neurol. Neurosurg. Psychiatry* **39:**1186–1190, 1976.
36. VAPAATALO, H.: Role of cyclic nucleotides in the nervous system. *Med. Biol.* **52:**200–207, 1974.
37. WATANABE, H., PASSONNEAU, J. V.: Cyclic adenosine monophosphate in cerebral cortex: Alterations following trauma. *Arch. Neurol.* **32:**181–184, 1975.
38. WELCH, K. M. A., MEYER, J. S.: Effects of cerebral infarction in man on cyclic AMP levels in CSF and blood. Second International Conference on Cyclic AMP. Vancouver, B.C., 1974.
39. WELCH, K. M. A., MEYER, J. S., CHEE, A. N. C.: Evidence for disordered cyclic AMP metabolism in patients with cerebral infarction. *Eur. Neurol.* **13:**144–154, 1975.

CHAPTER 26

Acetylcholine Metabolism in Intracranial and Lumbar Cerebrospinal Fluid and in Blood

Bernard Haber and Robert G. Grossman

1. Introduction

Most studies of the biochemical composition of human cerebrospinal fluid (CSF) have been performed on fluid obtained from the lumbar sac, since lumbar puncture has been the common method of obtaining CSF for diagnostic study. However, there is reason to believe that CSF obtained from intracranial sites would more accurately reflect biochemical processes in specific brain areas. In the course of diagnosis and treatment of certain serious neurosurgical disorders, we have obtained CSF from intracranial sites and have compared the content of various neurotransmitters and related enzymes in intracranial CSF with their content in lumbar CSF and in the blood.

We here report on the values obtained for acetylcholine (ACh), its precursor and degradation product, choline (Ch), and the enzymes involved in the synthesis and degradation of ACh, choline acetyltransferase (ChAT) and cholinesterase (ChE), respectively. Values were obtained for cases with minimal areas of brain pathology as well as in cases with extensive areas of brain pathology due to brain tumor, ruptured aneurysm, and head injury.

Bernard Haber, P.D. • Departments of Neurology and Human Biological Chemistry and Genetics, The Marine Biomedical Institute, Galveston, Texas 77550. **Robert G. Grossman, M.D.** • Division of Neurosurgery, Baylor University School of Medicine, Houston, Texas 77030.

The concentration of ACh was found to be related to the point in the CSF circulation pathway at which the sample was obtained, both in cases in which there was minimal brain pathology and in cases with extensive pathology.

2. Methods

2.1. CSF Sampling Sites and Categorization

Intracranial CSF was obtained at surgery during fluid aspiration from the following sites: the subarachnoid space over the cerebral convexities, generally over the frontal lobes; the basilar cisterns, generally the chiasmatic cistern or cisterna magna; and from the ventricles via ventriculostomy. When a ventriculostomy was inserted for monitoring of the intracranial pressure, CSF was obtained at the time of insertion and in some cases when CSF was removed for control of increased pressure. Lumbar CSF was obtained intraoperatively when lumbar drainage was necessary to provide exposure of the circle of Willis or of the sella. With the exception of three stereotaxic cases, all the CSF obtained during surgery was obtained during the course of general anesthesia, usually with N_2O–O_2 mixtures and 0.2–0.5% fluether, and the majority of cases were also hyperventilated to a Pa_{CO_2} of about 25 mm Hg. In operative cases, arterial blood was drawn into heparinized tubes at intervals to monitor blood gas levels during anesthesia, and an aliquot was used for measurement of the transmitters under study. The CSF samples that were selected as being suitable for normative values were obtained from cases in which there was no gross pathology of the brain and in which CSF samples were obtained in the course of CSF removal as technically required during stereotaxic surgery for the relief of pain of central and peripheral origin, and during craniotomy for trigeminal neuralgia, for ligation of the supraclinoid carotid artery for trapping of intracavernous aneurysms or carotid–cavernous fistulas, and for excision of small pituitary and parasellar tumors. Values obtained from these cases were designated as coming from cases with minimal brain pathology.

These values were compared with those obtained in three types of intracranial pathology: brain tumor, ruptured intracranial aneurysms, and head injury. The brain tumors were either intrinsic tumors (gliomas, metastases) or extrinsic tumors (meningiomas). In most cases, there was shift of the midline structures of the brain and ventricular distortion.

CSF from cases with ruptured aneurysms was obtained at the time of craniotomy for obliteration of the aneurysm 10–18 days after the aneurysmal rupture. The patients were neurologically normal at the time of sampling (Grade I) or with minor deficits (third nerve palsy, Grade II). The CSF was grossly clear and colorless or slightly xanthochromic, and under normal pressure.

In aneurysm cases as well as in some cases with cavernous sinus or sellar pathology as described above, CSF was obtained from intracranial and lumbar sites. The technique for relaxation of the brain to facilitate elevation of the frontal lobe to expose basilar structures (sella, circle of Willis) utilized lumbar drainage of CSF, as well as hyperventilation. After induction of anesthesia, while the craniotomy was being made, but prior to opening of the dura, CSF was slowly removed via a lumbar catheter in 8 to 10-ml aliquots. A total of 50–60 ml of CSF was removed in a 20- to 25-min period, making five to eight serial samples. We estimate that the subarachnoid space around the spinal cord contains about 40 ml of CSF; therefore, there is reason to think that the final aliquots contain an admixture of CSF from the basilar cisterns and possibly from over the convexities and the ventricles as well. The data given herein for lumbar CSF are for the initial aliquot only.

In most cases, CSF drainage resulted in removal of sufficient CSF so that only a very small amount of CSF could subsequently be aspirated from over the convexities or the chiasmatic cistern. However, in a few cases, we were still able to obtain some CSF in the process of aspirating the cisterns to visualize basilar structures.

2.2. Neurochemical Methods

CSF samples were placed in iced tubes immediately after being obtained and were taken to the laboratory, where they were centrifuged at 4°C and blood cells were removed in less than 15 min after being aspirated.

Despite all care taken in sampling, small amounts of fresh blood were often present in the CSF samples, particularly those obtained by aspiration of CSF from over the convexity of the brain, and in some samples of ventricular CSF as well. In cases of head injury, hemorrhage into the CSF at the time of injury resulted in the presence of blood in many of the samples.

ACh and Ch were determined by the radiometric method of McCaman and Stetzler.[3] Total ChE and pseudo-ChE and true ChE and ChAT were determined as described by Fonnum.[1] CSF protein was measured with the method of Lowry *et al.*[2]

3. Results in Patients with Minimal Pathology, Tumor, Aneurysm, or Head Injury

The data obtained from different sampling sites along the pathway of CSF circulation are presented in Table 1–4 for cases with minimal pathology, tumors, aneurysms, and head injury, respectively. Table 5 summarizes the data obtained for each category of case at ventricular, convexity, and lumbar sampling sites, and also gives plasma values.

ACh was present in CSF in concentrations ranging from 100 to 1500 pmol/ml at different sites in cases with minimal pathology. Although only a small number of samples were obtained for cases with minimal pathology, the similarity of the values obtained from the same sites in cases having larger amounts of brain pathology, with two exceptions discussed below, suggests that the ranges of values given for the cases with minimal pathology are probably in the range of the true normative values for the physiological states existing at the time of sampling.

ACh levels were comparatively low in lateral ventricular CSF (297 ± 125 pmol/ml) in cases with minimal pathology. Proceeding in the direction of CSF flow, the level rose to 479 ± 142 pmol/ml at the

Table 1. Patients with Minimal Pathology[a]

Fluid	ACh[b] (pmol/ml)	Ch[b] (pmol/ml)	ChAT[b] (μmol prod./ml fl./hr)	ChE[b] (μmol prod./ml fl./hr) Total	True	Protein (mg/ml)
CSF						
Ventricular	297 ± 125 (3)	3,157 ± 626 (3)	0.88 ± 0.15 (3)	0.55 ± 0.20 (6)	0.08 ± 0.02 (6)	0.24 ± 0.03 (6)
Cisterna magna	479 ± 142 (5)	4,135 ± 738 (6)	1.15 ± 0.40 (2)	1.03 ± 0.20 (7)	0.28 ± 0.05 (7)	0.97 ± 0.22 (8)
Cerebral convexities	1350 ± 376 (8)	10,198 ± 3157 (8)	—	5.64 ± 1.54 (12)	0.48 ± 0.10 (11)	4.01 ± 0.95 (13)
Lumbar sac	187 ± 84 (2)	1,745 ± 709 (6)	0.48 ± 0.11 (8)	1.42 ± 0.19 (13)	0.32 ± 0.08 (12)	0.62 ± 0.19 (11)
Plasma	1689 ± 426 (16)	6,173 ± 672 (13)	—	—	—	—

[a] Each value is the mean ± S.E. with the number of patients in parentheses.
[b] Abbreviations: (ACh) acetylcholine; (Ch) choline; (ChAT) choline acetyltransferase; (ChE) cholinesterase; (fl.), fluid.

Table 2. Patients with Tumors[a]

Fluid	ACh[b] (pmol/ml)	Ch[b] (pmol/ml)	ChAT[b] (μmol prod./ml fl./hr)	ChE[b] (μmol prod./ml fl./hr) Total	True	Protein (mg/ml)
CSF						
Ventricular	258 ± 63 (7)	3048 ± 689 (7)	—	1.07 ± 0.42 (7)	0.33 ± 0.19 (6)	1.07 ± 0.42 (9)
Cisterna magna	150 ± 23 (3)	4616 ± 2158 (4)	—	3.41 ± 0.89 (5)	0.40 ± 0.13 (4)	1.47 ± 0.32 (4)
Cerebral convexities	1001 ± 297 (10)	8663 ± 2375 (8)	0.69 ± 0.24 (3)	4.56 ± 1.62 (9)	0.48 ± 0.08 (8)	4.65 ± 2.33 (10)
Lumbar sac	317 ± 222 (2)	2139 ± 1124 (4)	—	1.11 ± 0.66 (2)	—	0.43 ± 0.07 (2)
Plasma	1037 ± 228 (17)	7019 ± 943 (18)	—	—	—	—

[a,b] See Table 1 footnotes.

Table 3. Patients with Aneurysm[a]

Fluid	ACh[b] (pmol/ml)	Ch[b] (pmol/ml)	ChAT[b] (μmol prod./ml fl./hr)	ChE[b] (μmol prod./ml fl./hr) Total	True	Protein (mg/ml)
CSF						
Ventricular	252 ± 53 (4)	3,239 ± 339 (4)	—	0.33 ± 0.08 (4)	0.18 ± 0.10 (4)	0.52 ± 0.26 (4)
Cerebral convexities	4206 ± 2010 (9)	17,201 ± 3616 (11)	—	7.82 ± 2.59 (10)	0.57 ± 0.10 (10)	7.67 ± 2.83 (10)
Lumbar sac	173 ± 43 (15)	2,722 ± 310 (17)	—	1.36 ± 0.24 (19)	0.37 ± 0.11 (19)	1.75 ± 0.40 (22)
Plasma	1296 ± 251 (15)	10,442 ± 3338 (16)	—	—	—	—

[a,b] See Table 1 footnotes.

Table 4. Patients with Head Injuries[a]

Fluid	ACh[b] (pmol/ml)	Ch[b] (pmol/ml)	ChAT[b] (μmol prod./ ml fl./hr)	ChE[b] (μmol prod./ml fl./ hr) Total	True	Protein (mg/ml)
CSF						
Ventricular	736 ± 245 (24)	3577 ± 1005 (14)	0.43 ± 0.08 (12)	2.93 ± 1.17 (23)	0.67 ± 0.38 (22)	4.36 ± 2.56 (20)
Cerebral convexities	1152 ± 311 (12)	8728 ± 1912 (11)	—	9.99 ± 2.00 (12)	0.71 ± 0.17 (10)	12.01 ± 2.07 (8)
Plasma	1533 ± 641 (14)	6016 ± 833 (24)	—	—	—	—

[a,b] See Table 1 footnotes.

cisterna magna and to 1350 ± 376 pmol/ml over the cerebral convexities. ACh levels in lumbar CSF were in the same range as those in ventricular CSF, or a little lower. Ch levels at the sampling sites paralleled the ACh levels. Ch was 10 times higher than ACh in ventricular CSF and 8 times higher than CSF over the cerebral convexities.

The relative levels of ACh and Ch in CSF and in plasma are of considerable interest. ACh levels in plasma ranged from 1037 to 1689 pmol/ml in different conditions. Average ACh blood levels were higher than average levels in the CSF in cases with minimal pathology. Plasma Ch levels were 2–3 times higher than ventricular or lumbar CSF levels, but not as high as Ch levels in CSF over the convexities. These data bring up the question of the

Table 5. Summary of ACh and Ch Concentrations[a]

Category	ACh[b] (pmol/ml)		Ch[b] (pmol/ml)		Protein (mg/ml)	
Ventricular CSF						
Minimal	297 ± 125	(3)	3,157 ± 626	(3)	0.24 ± 0.03	(6)
Tumor	258 ± 63	(7)	3,048 ± 689	(7)	1.07 ± 0.42	(9)
Aneurysm	252 ± 53	(4)	3,239 ± 339	(4)	0.52 ± 0.26	(4)
Head injury	736 ± 245	(24)	3,577 ± 1005	(14)	4.36 ± 2.56	(20)
Convexity						
Minimal	1350 ± 376	(8)	10,198 ± 3157	(8)	4.01 ± 0.95	(13)
Tumor	1001 ± 297	(10)	8,663 ± 2375	(8)	4.65 ± 2.33	(10)
Aneurysm	4206 ± 2010	(9)	17,201 ± 3616	(11)	7.67 ± 2.83	(10)
Head injury	1152 ± 311	(12)	8,728 ± 1912	(11)	12.01 ± 2.07	(8)
Lumbar CSF						
Minimal	187 ± 84	(2)	1,745 ± 709	(6)	0.62 ± 0.19	(11)
Tumor	317 ± 222	(2)	2,139 ± 1124	(4)	0.43 ± 0.07	(2)
Aneurysm	173 ± 43	(15)	2,722 ± 310	(17)	1.75 ± 0.04	(22)
Head injury	—		—		—	
Plasma						
Minimal	1689 ± 426	(16)	6,173 ± 672	(13)	—	
Tumors	1037 ± 228	(17)	7,019 ± 943	(18)	—	
Aneurysm	1296 ± 251	(15)	10,442 ± 338	(16)	—	
Head injury	1533 ± 641	(14)	6,016 ± 833	(24)	—	

[a,b] See Table 1 footnotes.

blood–brain equilibrium of these substances and the question of possible contribution by blood contamination of CSF to the values obtained. The extent of blood contamination of all the convexity fluids can be appreciated by inspection of the protein values; head-injury ventricular CSF and lumbar-aneurysm CSF also contained blood. However, much higher levels of ACh were seen in ventricular CSF in head-injury patients, and in convexity CSF in aneurysm cases, than were seen in the other categories. The convexity aneurysm values were 3–4 times higher than plasma values and therefore are unlikely to be due to only a breakdown of the blood–brain barrier to plasma ACh. In addition, head-injury convexity CSF was as bloody as aneurysm CSF, but did not have the high ACh content. Since ventricular CSF in head-injured patients was high in ACh, it might be expected that convexity CSF would also be high in ACh. The failure to find this expected result may be an artifact of case selection and sampling. The head-injury cases from which convexity CSF was obtained were not equivalent to those from which ventricular CSF was obtained. The ventricular CSF cases were deeply comatose and had ventriculosomoties for intracranial pressure monitoring. The convexity CSF cases generally had depressed fractures and were usually conscious on admission to the hospital. Many of these cases had only a focal area of cortical damage, in contrast to the diffuse damage of many of the cases with ventriculostomies.

In several head-injury cases, serial ventricular samples were obtained over several days that showed changes in ACh levels that paralleled changes in the clinical status of the patient. In one case showing progressive improvement, ventricular ACh values were (in pmol/ml): day 1, 1159; day 2, 709; day 4, 492; day 8, 382. In a case with deterioration, serial values were: day 1, 110; day 4, 116, day 6, 512. Much further work, including blood levels, will have to be done to evaluate these preliminary findings.

ChAT levels ranged from 0.43 to 1.15 μmol product/ml CSF per hr at different sites in different conditions. Total ChE levels were directly related to blood contamination, but true ChE levels appeared to be less correlated with blood contamination and to be somewhat lower in ventricular than in convexity CSF.

4. Discussion

The question of normative values for CSF constituents is a difficult one because of the number of factors that influence the levels of CSF constituents. The study reported in this chapter indicates the effect of the site of sampling on the values obtained. Furthermore, there is reason to believe that certain constituents will be a function of the patient's age and level of brain activity and that some substances will have diurnal or other cyclical variations. Additional factors that might have affected the values obtained for neurotransmitter, precursor and enzyme substances in this study were general anesthesia and hyperventilation, which, by changing the level of neural activity, could possibly change the levels of these constituents in the CSF. It is also possible that even in those patients who were essentially normal neurologically—those with pituitary tumors, carotid–cavernous fistulas, or trigeminal neuralgia—treatment with anticonvulsants or pain-relieving drugs in the preoperative period may have affected the levels of substances studied. The state of the blood–brain barrier, and direct blood contamination of samples, which is technically impossible to avoid in some cases and is part of the pathological process in others, affect the values obtained. Finally, the adequacy of CSF formation and absorption will also affect CSF constituent levels. These considerations make it clear that in addition to accurate normative data, turnover studies are needed to interpret the nature of changes found in levels of various constituents in pathological states.

Despite the multiplicity of factors involved in regulation of CSF constituents, the values found for the substances in this study were reasonably tightly grouped, especially when it is considered that the values normally found for certain substances of clinical interest, such as CSF protein, can vary by a factor of 2–3 or more among normal individuals. The data presented herein suggest that head injury and aneurysmal rupture may produce modest increases in CSF ACh at certain intracranial sites. Such changes are probably related to changes in neuronal activity and in membrane integrity. We have the impression that in those cases with the most brain disruption and with ischemic–anoxic damage, the values may be the highest. Subcategorization of cases into different grades of brain injury and statistical analysis of more data will be required to determine the significance of these findings.

ACKNOWLEDGMENTS. This research was supported by USPHS Grant NS 07377-09. The technical assistance of Mr. George Karp and Ms. Cathy Jones is gratefully acknowledged.

References

1. FONNUM, F.: Radiochemical microassays for the determination of choline acetyltransferase and acetylcholinesterase activities. *Biochem. J.* **115**:465–472, 1969.
2. LOWRY, O., ROSENBROUGH, N. J., FARR, A. L., RANDALL, R. J.: Protein measurement with the Folin phenol reagent. *J. Biol. Chem.* **193**:265–275, 1951.
3. MCCAMAN, R. E., STETZLER, J.: Radiochemical assay for ACh: Modifications for sub-picomole measurements. *J. Neurochem.* **28**:669–671, 1977.

CHAPTER 27

Cerebrospinal Fluid Pressure Dynamics and Brain Metabolism

Derek A. Bruce

1. Introduction

The deleterious effects of elevated pressure in the head have been known since the time of the ancient Egyptians, and the poor outcome associated with a tight dura following open cranial injury is alluded to in the Smith papyrus. Elevated intracranial pressure (ICP) can occur in association with many pathological conditions in the brain (e.g., infection, trauma, neoplasm). The signs and symptoms of elevated ICP may be florid, with severe headache and vomiting followed by the onset of deep coma, or they may be minimal, with occasional morning headache. The height of the pressure in the head will be only one determinant of the pathological consequences of raised ICP. The rate of rise of pressure, the preexisting state of the brain, the presence or absence of brain edema, and the length of time during which pressure is elevated will all play a role. It is unusual for the pressure to be continuously high; more frequently, the pressure rises and falls episodically (pressure wave).[26] High static pressures have not been found to interfere with cell metabolism until levels of 2000–3000 lb/sq. in. are reached, and mitosis can still occur at pressures of 2000 lb/sq. in. Thus, we ask, why does elevated ICP interfere with cerebral metabolism? The purpose of this chapter is to explore and, it is hoped, to answer this question.

Abbreviations and terms used in this chapter: (CBF) cerebral blood flow (normal = 50 ml/100 g brain per min); (CBV) cerebral blood volume; ($CMRo_2$) cerebral metabolic rate for oxygen; (compliance) $\Delta V/\Delta P$ (reciprocal of *elastance*); (CPP) cerebral perfusion pressure (= *SAP* − *ICP*); (CSF) cerebrospinal fluid; (CT scan) computed tomographic scan; (elastance) $\Delta P/\Delta V$; (ICP) intracranial pressure; ($Paco_2$) arterial partial pressure of carbon dioxide (in mmHg); (R_o) outflow resistance (see Fig. 1 and the text); (SAP) systemic arterial pressure [= (systolic − diastolic pressures)/3 + diastolic pressure]; (torr) 1 torr = 13.7 mm H_2O; (PVI) pressure–volume index.

Derek A. Bruce, M.B., Ch.B. • Children's Hospital of Philadelphia, Philadelphia, Pennsylvania 19104; Division of Neurosurgery, University of Pennsylvania School of Medicine, Philadelphia, Pennsylvania 19104.

2. Homeostasis of Intracranial Pressure

Before examining the effects of raised pressure on cerebral metabolism, we must understand the interactions that occur among the various intracranial contents under normal circumstances and that maintain normal intracranial volume–pressure relationships. The ICP is not a static pressure; it changes with phase of respiration and with the cardiac cycle. Alterations in body position affect the ICP such that going from a flat recumbent position to the upright standing position will lower the ICP by 8–10 torr (1 torr = 13.7 mm H_2O) in a normal individual; yet no changes in cerebral function occur. Coughing, sneezing, or straining may raise the ICP from a resting value of 0–5 torr to 40 or 50 torr. Because the rise is transient, no changes in metabolism usually occur. Unconsciousness following a severe coughing spell, however, can occur, and is probably due to prolonged elevation of ICP with resultant cerebral ischemia.

The intracranial contents are: blood, which is mostly in the venous sinuses and pial veins; cerebrospinal fluid (CSF); and brain. The skull in adults and children over the age of 2–3 years is essentially unexpandable and may be considered as a fixed-volume container. The spinal dural sac, as a result of changes in volume in the spinal epidural venous plexus, does undergo volumetric change, and thus the intradural contents, spinal cord and brain, are considered to be enclosed in a partially expandable container as opposed to the classic Monroe–Kellie hypothesis, which envisioned a nonexpandable, fixed-volume container. Once the expansion of the dural sac is complete, the relationship between the intracranial components may be expressed as

$$V_{\text{blood}} + V_{\text{CSF}} + V_{\text{brain}} + V_{\text{other}} = \text{constant}$$

where V_{blood} is a volume of blood; V_{CSF} is the volume of CSF; and V_{brain} is the volume of brain. Approximately 80% of the intracranial volume is brain and extracellular fluid; 10%, CSF; 10%, blood.

2.1. CSF

CSF is produced at a rate of 0.35 ml/min.[2] Of this production, 80% or more occurs via the choroid plexus of the lateral third and fourth ventricles, while 10–20% may be produced from the brain extracellular space.[2,45] CSF production is relatively independent of changes in ICP until the pressure is high enough to decrease choroid plexus blood flow. After formation, the CSF circulates by bulk flow out of the third ventricle via the aqueduct of Sylvius, exits from the brain via the fourth ventricle foraminae (Luschka and Magendie), and passes via the perimesencephalic and basal cisterns into the supratentorial space, where it is then reabsorbed. Reabsorption occurs through the arachnoid villi, which are one-way flap valves that, in man, open at an approximate pressure difference of 5 torr.The villi or arachnoid granulations occur mainly along the posterior sagittal sinus and lateral sinuses. CSF flow across the valves increases with rising CSF-venous sinus pressure gradients.[53] Thus, the rate of clearance of CSF will depend on (1) the presence of an open subarachnoid space, (2) the pressure difference between the sagittal sinus and the subarachnoid space, and (3) the patency of the subarachnoid valves. The formation rate of CSF, while unaffected by mild alterations in ICP, can be decreased by a number of drugs (Diamox, furosemide, corticosteroids), but none has been shown to be valuable on a long-term basis.[39,52] (See Chapter 1.)

CSF is readily displaced from the skull to the spinal subarachnoid spaces, and this displacement constitutes a major portion, 30%[27] to 70%,[23] of the mechanism that compensates for an expanding intracranial mass. As the extra volume in the cranium increases, CSF is gradually displaced to the spinal subarachnoid space. This mechanism seems to be most important when sudden changes in ICP occur (e.g., coughing). Thus, in a patient with an intracranial mass who has fully expanded the spinal sac, simple physiological maneuvers (e.g., coughing, sneezing) that normally lead to small increases in intracranial volume may precipitate significant increases in ICP because the compensatory effect of expressing CSF into the spinal sac is lost.

The increased absorption of CSF that occurs with increasing ICP is a second compensatory mechanism and is dependent on free flow of CSF through the subarachnoid space. Once the ability to displace CSF into the spinal sac is lost, further increase in intracranial volume and ICP will be compensated for by increased absorption of CSF. For this to occur, the subarachnoid spaces and the subarachnoid villi must be patent. The measure of the combined resistance to CSF flow along the subarachnoid spaces and across the valves has been designated R_o as the outflow resistance and can be calculated by adding a small volume of CSF to the cranium and recording the rate of return of the ICP back to a stable baseline[28] (Fig. 1):

$$R_0 = \frac{t_2 P_0}{\text{PVI} \log\left[P_2/P_p \cdot (P_p - P_0)/(P_2 - P_0)\right]}$$

where PVI is the volume of CSF infusion required to raise ICP by a factor of 10. Any event that interferes with the outflow of CSF may lead to increased ICP (e.g., basal meningitis, subarachnoid hemorrhage, and meningeal spread of tumor). In a patient with an expanding mass lesion in the cranium, the CSF pathways may suddenly become occluded by tentorial herniation, which may cause occlusion of the cistern at the tentorial notch; diffuse brain swelling, which may occlude the cortical subarachnoid space; or subarachnoid hemorrhage or

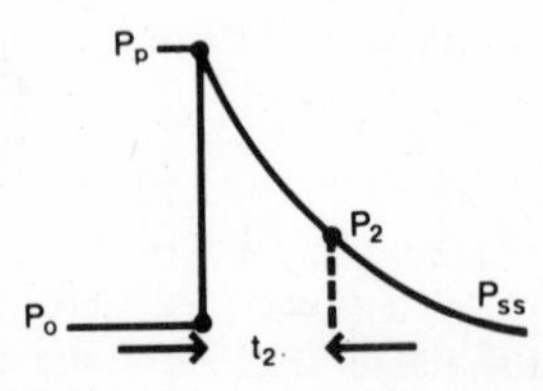

Figure 1. Calculation of CSF outflow resistance (R_o) following bolus injection into intracranial space (see text). (P_o) resting CSF pressure; (P_p) peak CSF pressure; (t_2) time following bolus injection; (P_2) CSF pressure at time t_2; (P_{ss}) steady-state CSF pressure.

infection, which occludes the basal cisterns or the arachnoid villi or both. A change in intracranial volume that might have been well tolerated when the outflow resistance (R_o) was low may now lead to a dangerous rise in ICP. This phenomenon has been well demonstrated experimentally.[3] Since changes in brain bulk will affect the size of the CSF subarachnoid space, changes in the arterial partial pressure of carbon dioxide ($PaCO_2$) would be expected to alter the R_o. Thus, changes in the ability of the CSF to circulate freely through the subarachnoid spaces or to cross the arachnoid villi may well lead to a sudden decrease in the ability of the intracranial contents to tolerate a sudden increase in volume (elastance $= \Delta P/\Delta V$) such that a change in volume that previously was well tolerated may now cause a sudden rise in ICP. Alterations in intracranial volume may be produced by coughing, sneezing, straining, seizures, change of position, change in respiratory state, sleep, subarachnoid hemorrhage, sudden tumor expansion, or other factors.

A further factor affecting CSF absorption is the pressure in the sagittal sinus. It is the pressure gradient across the arachnoid valves (i.e., ICP − sagittal sinus pressure) that controls CSF absorption. The sagittal sinus pressure varies with position, but to the same degree as the ICP. Thus, the relationship between the two is unchanged. Normally, the posterior sagittal sinus pressure is less than the ICP. Increased sinus pressure will interfere with reabsorption of CSF by decreasing the pressure difference across the arachnoid villi. High intrathoracic pressure due to venous obstruction or respiratory difficulties, major venous sinus occlusion, or high cerebral blood flow (CBF) will all lead to increased sinus pressure and therefore potentially decrease the intracranial compliance and eventually cause an increase in ICP.

In summary, the CSF production rate is relatively fixed, and therefore increased production can be generally ignored as a cause of increased intracranial volume. The ability to displace CSF into the spinal sac and to increase CSF absorption appear to be major mechanisms for compensation of the essentially fixed cranial space for an increase in volume or pressure. Sudden occlusion of the subarachnoid space may lead to an acute decrease of intracranial elastance, which in turn may result in dangerous increases of ICP resulting from small changes in intracranial volume. Finally, increases in sagittal sinus pressure, by decreasing CSF absorption, may also lead to some decrease in intracranial compliance. If increase in ICP occurs in concert with increase in sinus pressure, the difference between the two may remain unchanged, and thus no change in compensatory ability may occur as a result of this mechanism; yet the ICP may rise to a higher basal level. The CSF appears to be the major intracranial buffering component.

2.2. Blood

The second intracranial component is blood. The total intracranial blood volume is approximately 150 ml in an adult. However, most of this is in the major sinuses and cerebral veins. The actual intracerebral blood volume is 2–4 ml/100 g,[20a,21,40] with higher values in the gray matter and lower values in the white matter. The cerebral blood volume (CBV) is directly affected by changes in $PaCO_2$. The actual sensitivity of the adult brain to changes in CO_2 is approximately 0.049 ml/100 g brain per mm Hg $PaCO_2$.[21] Changes in cerebral perfusion pressure [cerebral perfusion pressure (CPP) = systemic arterial pressure (SAP) − ICP] will also affect blood volume via cerebrovascular autoregulation, which is the ability of the brain's blood vessels to maintain a normal flow despite changes in the CPP. This is controlled by an intrinsic myogenic response of the vessels to alterations in the transmural pressure such that when the arterial pressure drops, producing a decrease in perfusion pressure and a decrease in transmural pressure, the muscle tone relaxes and vasodilatation occurs. Thus, as the CPP is lowered, cerebrovascular dilatation occurs. This increase in volume will occur mainly in the muscular arterioles, and as progressive arteriolar dilatation occurs in response to a dropping CPP, there will be a gradual shift of blood from the venous to the arterial side of the circulation that may change the elastic properties of the brain. An increase in total CBV has been experimentally confirmed with increased ICP.[13,24] Grubb *et al.*[13] found in monkeys that decreasing CPP resulted in a maximum increase in CBV of 66%. This occurred with 60% decrease in CPP and an ICP of approximately 70 torr. Increases in CBV will also accompany increases in CBF.

The recent techniques of measuring CBV consist in measuring the intraparenchymal volume, and it is possible that reciprocal changes in CBV are occurring in the larger pial veins or in the venous sinuses such that no change or even a decrease of total intracranial blood volume is occurring. On the other hand, there is no specific evidence for this. Studies by Shapiro *et al.*[47] in monkeys suggested that significant compression of the intracranial venous sinuses can occur at high ICP. In man, however, change in size or shape of the venous sinuses seems to be an uncommon event even with significant intracranial hypertension.[34]

The profile of changes in intracranial blood volume that occur with volumetric changes in the other compartments of the intracranial space are not clear, and thus the exact role of CBV as a compensatory mechanism during changes in intracranial volume is unclear. It is likely that many of the rapid increases in ICP that occur physiologically (e.g., with coughing) are produced by sudden rise in sagittal sinus pressure and sudden engorgement of the brain with blood due to restricted outflow. It is unlikely in this case, then, that changes in blood volume represent a compensatory as well as a causative mechanism. Increases in mass lesions inside the head acutely (e.g., as with an epidural hematoma) may lead to displacement of blood from the cortical veins and possibly from the venous sinuses.[20a] However, as soon as the ICP rises, there will be an increase in the blood volume within the cerebrum itself as myogenic autoregulation becomes active and the blood vessels dilate to keep up blood flow. On a chronic basis, increased ICP leads to increases in blood volume due to cerebrovascular dilatation.[40] Arterial and capillary dilatation will occur with long-standing increases in CO_2 (e.g., in patients with chronic lung disease or in patients with chronically increased ICP). Increased sagittal sinus pressure may also produce cerebral venous congestion and an increase in CBV, although clinically, other possibly than in pseudotumor cerebri,[40] this does not appear to be a major problem. As the intracerebral blood volume increases, the brain will increase in bulk and may become less compliant. This increase in bulk may decrease the size of the CSF spaces and lead to an increase in outflow resistance for CSF. This increased outflow resistance plus a change in compliance of the brain substance may lead to a dramatic overall increase in intracranial elastance and a marked decrease in compensatory ability. The intracranial elastance may also change concomitantly with a decreasing CPP because the redistribution of blood associated with myogenic autoregulation may alter the brain's elastic properties. In this circumstance, however, the decrease in blood pressure or in perfusion pressure will tend to alter brain elastance in the opposite direction.[25] The result of these two factors has not yet been demonstrated. Clinically, increases in CBV have been suggested as important causes of increased ICP following head injury, particularly in children,[5] although the recently reported mean CBV in head-injured adults appears reduced.[20a] Increases in CBV have been demonstrated in patients with pseudotumor cerebri.[40] The mechanisms for these increases in blood volume are, at present, unclear. At present, there is no evidence that alteration in CBV plays a role as a compensatory mechanism for slow increases in ICP.

2.3. Brain

The final normal intracranial component is the brain substance itself. Although it is generally stated that the brain is incompressible, rapid and significant compression and displacement of the brain can occur. The rapid occurrence of an epidural hematoma is frequently associated with marked compression and displacement of cerebral tissue. This has been well demonstrated in patients by angiography and computed tomographic (CT) scan. Slowly growing tumors (e.g., meningiomas) may also lead to a major degree of cerebral displacement and compression.

The dynamic response of the brain to a sudden pressure load is quite different from that of the whole intracranial space. Schettini and Walsh[46,51] have shown that a linear relationship exists between brain displacement and changes in pressure. The elastic modulus of the normal brain at normal blood pressure is $2.8–4.1 \times 10^5$ dynes/cm^2. It is reasonable to expect that the brain elasticity will be altered in the face of cerebral edema or when brain tissue is invaded or replaced by tumor. Changes in brain elasticity will also occur secondary to changes in blood pressure[25] and blood volume.

Cerebral edema is an increase in the water content of the brain. In general, two types of cerebral edema are differentiated based on the location of the fluid and the status of the blood–brain barrier. Vasogenic edema results from a disruption of the blood–brain barrier and is manifest as an accumulation of protein-containing fluid in the extracellular space of the white matter.[11,41] This is the variety of edema most frequently encountered in clinical practice. Cytotoxic edema is an accumulation of excess water within cell membranes of neurons and glia and is produced by any substance that interferes with energy metabolism and membrane pumping (e.g., ouabain, anoxia). The importance of cerebral edema in the context of intacranial pressure–volume relationships is the potential effect of edema on the brain's elastic properties and the patency of the subarachnoid spaces. As yet, there is no information on the effect of edema on the cerebral elastic properties. The elastic properties of the brain tissue influence overall intracranial elastic responses in a number of ways. Changes in the bulk of the brain (e.g., by edema) have been shown to alter the relationship between CBF and CPP[35] and may alter R_o by changing the size of the subarachnoid space and by altering free communication within that space (e.g., secondary to herniation). Thus, changes in brain bulk and brain elastic properties will influence the overall intracranial elastance to an acute change in volume and will also influence the intracranial compensatory mechanisms, particularly by affecting CSF circulation and absorption. Finally, if a local change in brain elastic response occurs (e.g., due to tumor or hemorrhage), the pressure changes in the local area that may be produced by small changes in total intracranial volume may be larger than those produced in the rest of the brain and lead to displacement of cerebral tissue and herniation of brain substance.

3. Pressure–Volume Relationships

If we now consider the relationship expressed above ($V_{\text{blood}} + V_{\text{CSF}} + V_{\text{brain}} + V_{\text{other}} = \text{constant}$), it is clear that the only way in which a pathological mass lesion can occur in the brain without a rapid rise in ICP is if there is a reduction in volume of one of the normal intracranial constituents or if the container size increases in volume. This latter event occurs only in children with open sutures and fontanelle. A transient rise in pressure secondary to an acute increase in intracranial volume is a normal physiological response that comprises at least two phases. One is the instantaneous response of the ICP, which will be dependent on the volume added and the summation of the elastic properties of all the intracranial components, and the second is the return of the ICP to a normal level, which will be dependent on the ability of the various compensatory mechanisms to correct for the additional volume load. There is still considerable disagreement about which intracranial constituents are the most readily displaceable to accommodate for both an acute and a slow change in intracranial volume. Furthermore, it seems likely that the accommodation to a slowly increasing mass will be different from that which occurs with a rapid volume change. Displacement of CSF from the intracranial to the spinal sac is a significant compensatory mechanism and has been estimated to make up from 30 to 70% of the total intracranial compensatory ability. When an acute rise in ICP (ΔP) is precipitated by a sudden increase in volume (ΔV), as with coughing or sneezing, the rise in pressure occurs both intracranially and intraspinally and therefore cannot be modified by displacement of CSF from the cranium to the spinal sac or by displacement of venous blood into major sinuses, but possibly is associated with a sudden decrease in blood entering the brain as a result of the sudden change in ICP and drop in perfusion pressure. If the increase in pressure is severe enough or long enough, even in the normal brain, cerebral dysfunction can occur due to cerebral ischemia (e.g., as in cough syncope). If the increase in intracranial volume is short-lived, or if the compensatory mechanism of CSF, blood, and brain can compensate, then the system will return to its previously resting pressure. If the various intracranial volumes have also returned to their resting state, and if the brain's elastic properties have not been changed, both the acute ($\Delta P/\Delta V$ = elastance) and slower responses to another similar change in volume will be the same.

If, however, the intracranial volume does not return to its former resting level, or if the CSF communication is blocked, or if the brain elastic properties are changed, then the addition of a similar volume (ΔV) may now produce a greater change in pressure (ΔP), which may be much longer lasting. The response will depend on the immediate elastic properties of the intracranial contents and the residual compensatory reserve of blood, brain, and CSF. In general, these will reflect each other [e.g., if the immediate elastance ($\Delta P/\Delta V$) is high, it is likely that there has been some exhaustion of the residual compensatory mechanisms]. However, it is possible that the immediate elastance may be high, yet outflow resistance or residual compensatory reserve to a slower change in volume may still be quite good. Thus, conceptually, these two properties (immediate elastic response of the intracranial space, or elastance, and compensatory reserve) must be considered separately. The property ($\Delta P/\Delta V$) is called the elastance and its reciprocal ($\Delta V/\Delta P$), the compliance. It is clear that the changes in pressure that occur in the intracranial space will depend on the rate and amount of volume added to the system—since all the compensatory mechanisms can be overcome if sufficient volume is added with sufficient rapidity—and on the resting elastance of the system. It is also clear that a change in volume that may cause very little change in ICP on one occasion may produce very large increases at another time when the compensatory reserve has decreased. The major factor that influences compensatory reserve is the free communication of pressure via the CSF pathways. If this is suddenly lost, a relatively small volume change in one compartment of the brain may lead to a large increase in pressure within that compartment because the compensatory mechanisms are abolished. This most commonly occurs with tentorial herniation,[19] but may also occur suddenly as a result of subarachnoid hemorrhage. When a slow increase in intracranial volume occurs, as in a growing tumor or hydrocephalus, the major compensation is via an increase in CSF reabsorption and distortion or compression of brain. CSF is expressed to the spinal sac, but this will become distended quite early in the course of the expanding mass, with a decreased capacity for further compensation. Blood volume within the brain tends to increase rather than to decrease under these circumstances. In this situation of a chronic increase in intracranial volume, the pressure may still be within a normal range, but now the ability to compensate for an acute increase in volume will be poor. The elastance ($\Delta P/\Delta V$) is very high, and residual compensation may be lost. Since the major mechanism (displacement of CSF) is already exhausted, decreases in blood volume, compression of the brain, and increased CSF reabsorption are the only compensatory factors remaining. If there is a mass lesion that has produced brain displacement and herniation as well, the CSF pathways may be occluded, and an even greater rise in pressure in one compartment may occur. For these reasons, patients who have chronically increased intracranial volume even though the ICP may be within the normal range at any given time are at great risk of developing extreme intracranial hypertension induced by small changes in blood volume (e.g., during coughing, small change in CO_2, seizures), particularly if the basal increased volume is being produced by a mass that also causes cerebral shift and herniation. Marmarou *et al.*[28] have recently published a classic paper that treats the factors influencing the volume–pressure relationship within the cranium mathematically and have derived an equation that endeavors to predict static and dynamic responses of the intracranial space to volume alterations:

$$P(t) = \psi(t) / \left\{ 1/P_0 + (k/R)_0 \int^{t} \psi(\tau) d\tau \right\}$$

where

$$\psi(t) = e^{k}{}_{0} \int^{t} I(\tau) d\tau$$

and $P(t)$ is the time course of intracranial pressure, P_0 is the intracranial pressure, R_o is the outflow resistance, and $I(\tau)$ is the volume input.[28]

It is clear that knowing the ICP at one point in time does not allow an adequate definition of the intracranial volume–pressure balance or residual compensatory reserve. If the pressure is above normal, it is reasonable to assume that a disturbance of the intracranial volume–pressure relationship has occurred. However, if the ICP is within the normal range, it is not possible to state that there is no such disturbance. As we have already noted, the residual displaceable volume of the system may be very low at a normal ICP.

4. Cerebral Herniation

The result of an expanding lesion inside the cranium may be not only diffuse elevation of ICP, but also distortion or displacement of cerebral substance.[1] Displacement of cerebral substance from one compartment within the cranium to another is called cerebral herniation, and this is a well-recognized clinical event. Most frequently, herniation syndromes are produced by the expansion of a local mass lesion. However, diffuse elevations of ICP may also produce herniation syndromes at the tentorium or foramen magnum. Significant cerebral herniation occurs in the closed skull at three major sites; (1) herniation from one side to the other within the supratentorial space, or subfalcial or cingulate hernition; (2) herniation around the tentorial notch, or transtentorial herniation; (3) herniation around the foramen magnum, or tonsillar herniation.

Subfalcial or cingulate herniation occurs from one side of the supratentorial space to the other. The cingulate gyrus of one hemisphere herniates below the falx cerebri into the opposite side of the calvarium. The major pathological effect of this is occlusion of the anterior cerebral artery, usually ipsilaterally, but occasionally bilaterally. This type of herniation is not common, but occurs most frequently with relatively superiorly and anteriorly situated mass lesions within the cranium. Actual cerebral infarction may occur locally within the cingulate gyrus, due to local vascular compression, or more generally, within the distal distribution of the anterior cerebral artery territories.

Transtentorial herniation occurs when one or both parahippocampal gyrii and unci are displaced inferiorly through the incisural notch. This may occur unilaterally with an expanding mass lesion or bilaterally when diffuse cerebral swelling, edema, or hydrocephalus is present. Pathologically, the most frequent occurrence is grooving and infarction within the mesial temporal structures. However, infarction within the distribution of the posterior cerebral arteries may occur, and local infarction, either ischemic or hemorrhagic, may occur within the brainstem. The posterior cerebral infarction occurs as a result of compression of the posterior cerebral arteries as they cross the tentorial edge. This compression may occur transiently during a pressure wave and may be relieved as the ICP returns to normal. Thus, local distal posterior arterial infarction without focal uncal infarction and with a patent posterior cerebral artery may occur as a result of an episode of intracranial hypertension. Furthermore, occipital lobe infarction may be the only residual finding resulting from such an episode of intracranial hypertension (Fig. 2). The mechanism of progressive rostral-to-caudal deterioration[38] associated with transtentorial herniation is due to brainstem ischemia. Whether this ischemia is produced by direct compression and distortion of the brainstem and peduncles by the herniating temporal lobe or by caudal displacement of the brainstem and distortion of the small perforating arteries that feed the brainstem from the basilar artery is not clear.[16,19] Once compression has been present, relief of the elevated ICP or brain herniation or both may result in secondary hemorrhages into the brainstem and failure of the patient to recover despite removal of the lesion that resulted in the cerebral herniation. Occasionally, when a mass is present in the posterior fossa, upward herniation of the cerebellum may occur, producing a similar clinical picture of transtentorial herniation. Upward herniation is most frequently seen when a posterior fossa lesion has resulted in hydrocephalus. If decompression of the hydrocephalus is achieved too rapidly, the posterior fossa contents may decompress upward, producing the herniation syndrome.[6a]

The final area within the cranium at which herniation of brain tissue occurs is at the foramen magnum. In this region, the herniated tissue is usually one or both of the cerebellar tonsils. Here also, the deleterious effects of such herniation are believed to be due to compression of the underlying medulla and secondary ischemia.

All the signs, symptoms, and pathological changes that are seen following cerebral herniation can be best explained by secondary cerebral ischemia. The effects of simple distortion of cerebral tissue are not known, and at present, the importance of distortion and displacement of the brain tissue as a causual factor in neuronal or axonal dysfunction is purely speculative. Recent studies by Rowan and co-workers[42,43] have demonstrated that local distortion and compression within the posterior fossa produced by inflation of a posterior fossa balloon led to an increase in systemic arterial pressure before any changes in brainstem blood flow, as measured by hydrogen clearance, were apparent. These studies suggest that, in fact, distortion or compression may result in neuronal stimulation.

Cerebral herniation syndromes may occur at relatively normal baseline intracranial pressures be-

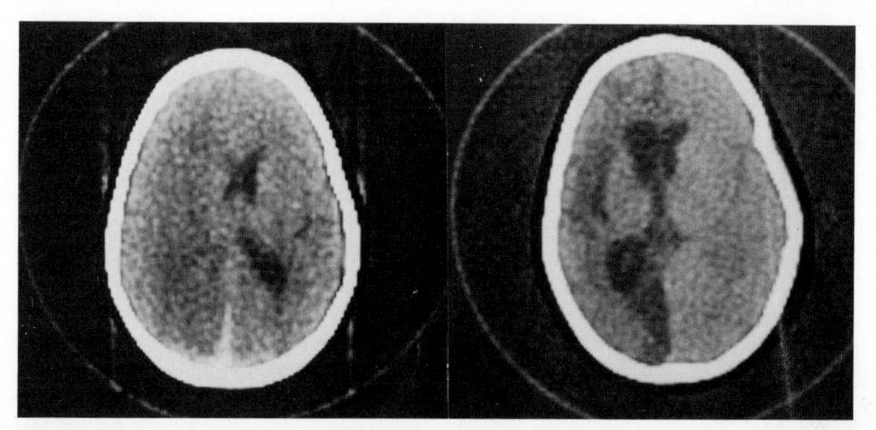

Figure 2. CT scans of battered child. *Left*: Unenhanced CT scan immediately following head trauma demonstrating small occipital–posterior interhemispheric subdural and left-sided brain swelling. *Right*: Unenhanced CT scan of same patient 4 months later demonstrating posterior cerebral arterial territory infarct with ventricular dilatation and scattered areas of encephalomalacia in left hemisphere.

cause of local distortion effect, or they may occur suddenly as a result of an acute rise in ICP. Clinically, it is important to recognize the signs and symptoms of herniation so that therapy can be begun early before distal infarction has occurred. The herniation must be relieved as early as possible to avoid the occurrence of secondary brainstem hemorrhages when transtentorial herniation is present. Finally, the sudden occurrence of cerebral herniation may result in obstruction of CSF pathways and a dramatic increase in outflow resistance (R_o). This, in turn, may result in dramatic changes in ICP and CBF due to small changes in volume that prior to the herniation were readily tolerated. The presence of cerebral distortion and herniation signifies a very unstable intracranial volume pressure state, and every effort should be made to remedy this situation at the earliest possible time.

5. Cerebral Blood Flow

Before examining the effects of generally elevated ICP on cerebral metabolism, it is necessary to understand the relationship of ICP to CBF. Poisseulle's equation, which describes flow through infinitely narrow-walled tubes, has been used as a close approximation to examine the factors governing blood flow through the brain:

$$F = \frac{\pi r^4 (P_1 - P_2)}{8 \eta l}$$

where F = CBF, $P_1 - P_2$ = SAP − ICP, r is the radius of the muscular arterioles, l is the length of the vascular bed, and η is the viscosity. For most practical purposes, the viscosity of the blood and the length of the vascular bed may be considered constants, and thus the major factors controlling CBF will be the radius of the blood vessels (primarily the muscular arteries and arterioles) and the CPP. Numerous experiments have demonstrated that CBF is stable over a wide range of CPP, which means that as the CPP changes, a reciprocal change in the vessel radius occurs (i.e., as the CPP decreases, the blood vessel diameter increases such that the resistance is lowered and CBF remains constant).[14,20] Whether the perfusion pressure is lowered by increasing ICP or by decreasing SAP, the CBF remains constant over perfusion pressures from 50–60 torr up to 150–200 torr. This phenomenon has been called pressure autoregulation. The CBF is preserved to slightly lower levels when CPP is lowered by increasing ICP rather than by decreasing SAP, and Reivich *et al.*[44] have shown that this may be related to a sympathetic catecholamine effect causing vasoconstriction of the larger delivery vessels (i.e., internal carotid and possibly middle and anterior cerebral trunks). The mechanism responsible for this type of autoregulation of CBF is believed to be a myogenic response within the cerebral vessels. Present evidence suggests that there are factors such as chemical and neurogenic influences that may modify the response of the muscular arterioles and reset their resting tone. However, at present, there is nothing to refute the hypothesis that the major factor involved in pressure autoregulation is the intrinsic myogenic response of the muscular arterioles. This mechanism may be interfered with as a result of trauma, neoplasia, infarction, anoxia, and alterations of the chemical composition of the CSF (see Chapter 22). As autoregulation becomes increasingly defective, CBF will vary directly with the CPP. Autoregulation may be defective focally or in a global fashion. When it is focally lost, an apparent adequate global CPP may result in local decreased CBF and ischemia. This situation is likely to be present in many patients. At the lower limits of autoregulation, the CBF becomes passively dependent on the CPP whether or not autoregulation is intact or defective. Functionally, this is believed to represent the point at which maximum cerebral vasodilatation has occurred. This lower limit begins around 50 torr. There is also an upper limit for autoregulation, and at mean pressures somewhere around 150–200 torr, an increase in CBF will occur. This is believed to represent forced dilatation of the cerebral vessels as a result of the increased arterial pressure.

A second type of autoregulation, metabolic autoregulation, exists in the cerebral circulation. This term describes a mechanism whereby increases in local neuronal activity result in local increases in CBF. This is probably the result of the accumulation of local metabolic end products and possibly increased requirements for substrate. When metabolism is decreased, a reduced CBF will occur. Thus, despite an adequate CPP, changes in CBF may occur as a result of primary changes in metabolism. If the lower limits of pressure autoregulation are passed such that a decrease in CBF has occurred, then local increase in metabolism will be unable to produce a further increase in CBF because maximal vasodilatation is already present. Under these circumstances, it is possible that functional tissue ischemia may result because the metabolic demand exceeds the ability of the cerebral vessels to deliver adequate substrate and to clear the products of metabolism. This situation may occur with no change in CPP and therefore may be completely unrecorded by present monitoring methods. Metabolic autoregulation may also be lost. This is seen following head injury in children where a high CBF with a low cerebral metabolism has been recorded[5] and in the luxury perfusion that occurs around an area of cerebral infarction.[17]

Loss of pressure autoregulation, since it may be local or global, can result in areas of ischemia within the brain despite an apparently adequate CPP. For this reason, recent reviews of the clinical need to treat ICP have defined an ICP of greater than 20 torr as being significantly elevated even though an apparently normal CPP is present.[4,36] This is in the hope of avoiding focal areas of ischemia. It can be seen that it is very difficult within the clinical setting to define an ideal or correct CPP or ICP.

In the presence of cerebral edema, CPP appears to be a poor reflection of CBF. Most studies of CBF in the presence of cerebral edema have shown a decrease in blood flow despite an apparently normal perfusion pressure, and the CBF appears to correlate better with the water content of the tissue than with the CPP.[31,35] Since edema increases total tissue volume, the flow per gram of dry weight of tissue may remain unchanged, yet the measured flow per gram of wet weight of tissue has decreased (i.e., the actual flow through the total tissue area is unchanged). A second possible cause for the decrease in CBF found in areas of edema is that there is an increase in tissue pressure in these areas such that the functional CPP is, in fact, lowered. Marmarou *et al.*[29] have demonstrated that tissue pressure is elevated, but only for a brief time, during the formation of edema. Thus, these authors feel that this mechanism is an unlikely one to explain the decreased CBF. Finally, the decrease in CBF may be due to decreased metabolism and intact metabolic autoregulation. This concept is supported by recent unpublished studies by Marmarou and associates.[30] Edema, produced by infusing mock CSF into the white matter, had no effect on CBF. The cortex was undamaged under these circumstances, and presumably metabolism was unchanged. When a cold lesion was produced on the cortex and the same increase in white matter water content occurred, a decrease in CBF in the white matter was seen. In this case, the cortex is injured, and it is likely that the local metabolism within the white matter is also decreased. Whichever of these mechanisms explains the altered CBF, it is important to realize than when edema is present, the CPP may be a poor indicator of tissue flow.

6. Effects of Pressure on Metabolism

In considering the potentially deleterious effects of raised CSF pressure on cerebral function, a number of factors have to be considered. One is the actual CPP. Studies[37] suggest that a decreased blood flow may occur at a higher CPP if the primary increase in pressure occurs in the posterior fossa rather than in the supratentorial space.[37,42] The status of cerebral pressure autoregulation will have an effect in defining the level of CBF at any given CPP. Whether the pathological process resulting in increased ICP is generalized or focal will certainly affect the degree of cerebral displacement and may affect focal autoregulation. If cerebral herniation is present, then even at relatively low intracranial pressures, significant disturbances of vital functions may occur as a result of focal compression and ischemia. Finally, the metabolic level of the brain will be important in defining the needs of the cerebral tissue for substrate. It is probable that higher cerebral function may be interfered with before any global changes in metabolism result. As we shall see later, these changes may reflect the inability of the cerebral vessels to deliver increased substrate when increased demand is made despite their ability to supply adequate substrate for a continuous lowered level of function.

Studies in normal man suggest that certain alterations in cognitive function (yawning, staring, confusion) occur when the CPP is decreased to levels between 30 and 50 torr.[10,48] In these studies, the CPP, which was decreased by reducing SAP, varied from 65 to 35 torr, and despite so-called evidence of clinical hypoxia, as outlined above, no changes in the cerebral metabolic rate for oxygen ($CMRO_2$) occurred. These studies support previous studies in man in which the CPP was reduced by increasing CSF pressure by lumbar infusion to levels of 100–150 torr.[9] Apart from some agitation, other signs of neuronal dysfunction were not observed. These combined observations suggest that increased ICP alone has not been shown to have a major effect on cerebral metabolism and that even low CPP can be well tolerated.

Hammer *et al.*[15] found in dogs that the effects of

decreased CPP on cerebral metabolism were similar whether the CPP was decreased by raising ICP or by lowering SAP.[18] They did find, however, that the CBF was better preserved at lower CPP's produced by elevating the ICP rather than by lowering SAP. This observation has been confirmed by Miller *et al.*[37] and Grubb *et al.*[13] Regardless of the ICP, there was no measurable change in oxygen metabolism until CBF was 50% of normal. These authors did observe a small increase in lactate production within the brain at a CPP of approximately 70 torr, although no change in CBF or oxygen or glucose metabolism had occurred. Recent studies by Eklof *et al.*[8] failed to show an increase in cerebral tissue lactate before frank signs of cerebral ischemia, such as an increase in adenosine monophosphate. This increased lactate production is not yet explained, but seems not to be due to ischemia or anoxia.

Bruce *et al.*[6] showed that in the dog, progressive increase in ICP, and therefore decrease in CPP, had no effect on $CMRO_2$ or CMR glucose until the CBF fell to levels of 50–60% of control. At this level of CBF, there was evidence of a general mild decrease of cerebral metabolism associated with some slowing of the electroencephalogram (EEG), but without evidence of increase in anaerobic glycolysis. As the CBF was lowered below 40% of normal, with a CPP of approximately 30 torr, frank evidence of diffuse cerebral ischemia occurred with an increase in glucose metabolism and a marked increase in anaerobic glycolysis. The EEG became flat at this level of CBF.

Grubb *et al.*[13] also found that despite a low CPP of 21 torr, no change in $CMRO_2$ occurred provided the CBF was greater than 45% of control. Marshall *et al.*[32,33] found normal brain energy metabolites providing the CBF was maintained above 40% of control. These findings are in agreement with those of Eklof and Siesjo,[7] who found that a sudden deterioration of energy metabolism occurred below a level of CBF of approximately 45% of normal.

Recently, studies by Sundt *et al.*[49] in patients undergoing carotid endarterectomy have also suggested that a CBF of over 50% of normal is adequate to support a normal EEG. As the CBF falls to 40–50% of control, some EEG slowing occurs, but only when the CBF falls below 40% does the EEG become flat.

Thus, with respect to global measurements of CBF, metabolism, and brain energetics, increased ICP appears to interfere with these parameters only when the CPP is decreased sufficiently to cause a fall in CBF below 40–45% of normal. There is no evidence to suggest that increased ICP has any effect on oxygen or glucose metabolism of the brain other than via a decrease in CBF. Further, there is no evidence that alterations in ICP affect the chemical modulation of synaptic activity. No studies have examined the effect of increased ICP alone on chemicals known to influence synaptic function (e.g., γ-aminobutyric acid, norepinephrine, serotonin), and this is a possible fruitful area for research, since minor alterations in chemical transmitter function may explain some of the clinical findings associated with an increased ICP.

How then might such symptoms as confusion, restlessness, and others be explained in the face of a normal CBF and metabolism? As discussed above, there are two types of cerebral vascular autoregulatory response: one to pressure changes and the other to local metabolic demands. Grossman *et al.*[12] have shown that the evoked potential response is abolished at a higher CPP than that at which the EEG decreases.[50] Local increases in metabolism are accompanied by local increases in CBF. This has been shown in experimental seizures[17] and in local cortical activity recorded on the EEG. Local increases in CBF have also been demonstrated in man in association with simple muscular movements and cognitive functions.[22] If cerebral pressure autoregulation is stressed to its limit by the increased ICP such that maximum cerebrovascular dilatation is present, the local metabolic products may be unable to increase blood flow further, and therefore a functionally inadequate tissue flow may occur. The resting flow may be quite adequate to maintain a normal metabolic state, but local increases in metabolic need cannot be met, and therefore higher cognitive functions are compromised. In areas of already stressed homeostasis (e.g., edema, spasm, diseased blood vessels), the changes in CBF may occur at higher perfusion pressures and to a greater degree than in normal brain. Thus, under pathological conditions, neither the CPP nor measurements of global CBF or metabolism are adequate to ensure that all areas of brain are being adequately perfused. Recent works by Welsh *et al.*[54] have clearly shown that even in small areas of brain, significant variation in local metabolic and redox state can occur during and after an ischemic insult. This marked inhomogeneity makes any estimate of the local ratio of flow to metabolism even more difficult in damaged brain.

7. Conclusions

In the healthy brain, there is no evidence that increased ICP produces any deleterious effects on metabolism until either cerebral herniation occurs or the CPP is lowered such that CBF falls below 40–45% of control values. The metabolic alterations that occur as a result of a lowered CPP appear to be similar whether the decreased perfusion is produced by increasing ICP or by decreasing SAP. The deleterious effects of herniation on cerebral function clinically are well described and occur as a result of secondary cerebral ischemia, rather than purely as the effect of distortion and displacement of brain tissue. However, the effects of brain distortion on the metabolism and function remain very poorly understood. At the limit of cerebral pressure autoregulation, the evoked cortical responses may become abnormal or higher intellectual tasks may be poorly performed before any evidence of changes in the resting metabolic state appears. This phenomenon may occur because the increase in flow required by augmented local metabolism can no longer be produced or because there is indeed inhomogeneous perfusion within the cerebral tissue such that although the global perfusion appears adequate, regional perfusion is decreased below a level adequate to sustain function. In conclusion, both global alterations of ICP and focal herniation and distortion of brain appear to produce deleterious effects on cerebral function and metabolism by secondary ischemia.

Unfortunately, there is no method readily available in the clinic for measuring the degree of pressure autoregulation, the adequacy of collateral circulation, or the degree of cerebral edema in patients with brain insults. Therefore no way exists to predict at what ICP any individual may develop signs of local or global cerebral ischemia. It is clear that many of the brain's homeostatic mechanisms can be disturbed by disease, and the assumptions about the state of autoregulation, regional blood flow, and metabolism, or the influence of small changes in ICP on CBF and metabolism, cannot be accurately made. It would appear that uncal herniation can occur at quite low ICP if diffuse brain swelling is present, and therefore even the probability of herniation cannot be easily predicted. The purpose of this chapter is to point out that the protective mechanisms which function in the healthy brain to protect against the deleterious effects of elevated ICP cannot be relied on once the brain has suffered an insult from trauma or disease. Thus, while as yet there is no evidence that elevated ICP alone interferes with cerebral metabolism, there is good evidence that within the damaged brain, the normal protective mechanisms may dysfunction. At present, therefore, we believe that controlling the ICP within the so-called normal range in patients with disturbed volume–pressure relationships and intracranial disease affords the best protection available.

References

1. ADAMS, J. H., GRAHAM, D. I.: The relationship between ventricular fluid pressure and the neuropathology of raised intracranial pressure. *J. Neuropathol. Appl. Neurobiol.* **2**:323–332, 1976.
2. BERING, E. A., JR., SATO, O.: Hydrocephalus: Changes in formation and absorption of cerebrospinal fluid within the cerebral ventricles. *J. Neurosurg.* **20**:1050–1063, 1963.
3. BROCK, M., FURUSE, M., HASUO, M., DIETZ, H.: Influence of CSF resorption pathways on intracranial capacitance. In Penzholz, H., Brock, M., Hamer, J., *et al.* (eds): *Advances in Neurosurgery,* Vol. 3. New York, Springer-Verlag, 1975, pp. 109–114.
4. BRUCE, D. A., BERMAN, W. A., SCHUT, L.: Cerebrospinal fluid pressure monitoring in children: Physiology, pathology and clinical usefulness. In Barness, L. A. (ed.): *Advances in Pediatrics,* Vol. 24. Chicago, Year Book Medical Publishers, 1977, pp. 233–290.
5. BRUCE, D. A., RAPHAELY, R. C., GOLDBERG, A. I., ZIMMERMAN, R. A., BILANIUK, L. T., SCHUT, L., KUHL, D. E.: The pathophysiology, treatment and outcome following severe head injury in children. *Child's Brain* **5**:174–191, 1979.
6. BRUCE, D. A., SCHUTZ, H., VAPALAHTI, M., GUNBY, N., LANGFITT, T. W.: An intrinsic mechanism to protect the brain during progressive cerebral ischemia. In Langfitt, T. W., McHenry, L. C., Jr., Reivich, M., Wollman, H. (eds): *Cerebral Circulation and Metabolism.* New York, Springer-Verlag, 1975, pp. 203–206.

6a. CUNEO, R. A., CARONNA, J. J., PITTS, L., TOWNSEND, J., WINESTOCK, D. P.: Upward transtentorial herniation. *Arch. Neurol.* **36**:618–623, 1979.

7. EKLOF, B., SIESJO, B. K.: The effect of bilateral carotid artery ligation upon the blood flow and the energy state of the rat brain. *Acta Physiol. Scand.* **86**:155–165, 1972.
8. EKLOF, B., MACMILLAN, V., SIESJO, B. K.: Cerebral energy state and cerebral venous pO_2 in experimental hypotension caused by bleeding. *Acta Physiol. Scand.* **86**:515–527, 1972.
9. EVANS, J. P., ESPEY, F. F., KRISTOFF, F. V., KIMBELL, F. D., RYDER, H. W.: Experimental and clinical observations on rising intracranial pressure. *Arch. Surg.* **63**:107–114, 1951.
10. FINNERTY, F. A., JR., WITKIN, L., FAYEKAS, J. F.: Cerebral hemodynamics during cerebral ischemia induced by acute hypotension. *J. Clin. Invest.* **33**:1227–1232, 1954.
11. FISHMAN, R. A.: Brain edema. *N. Engl. J. Med.* **293**:706–711, 1975.

12. Grossman, R. G., Turner, J. W., Miller, J. D., Rowan, J. O.: The relationship between cortical electrical activity, cerebral perfusion pressure and cerebral blood flow during increased intracranial pressure. In Langfitt, T. W., McHenry, L. C., Jr., Reivich, M., Wollman, H. (eds): *Cerebral Circulation and Metabolism*. New York, Springer-Verlag, 1975, pp. 231–234.
13. Grubb, R. L., Raichle, M. W., Phelps, M. E., Ratcheson, R. E.: Effects of increased intracranial pressure on cerebral blood volume, blood flows and oxygen utilization in monkeys. *J. Neurosurg.* **43:**385–398, 1975.
14. Haggendal, E., Lofgren, J., Nilsson, N. J., Zwetnow, N. N.: Effects of varied cerebrospinal fluid pressure on cerebral blood flow in dogs. *Acta Physiol. Scand.* **79:**262–271, 1970.
15. Hamer, J., Moyer, S., Stoeckel, H., Alberti, E., Weinhardt, F.: Cerebral blood flow and cerebral metabolism in acute increase of intracranial pressure. *Acta Neurochir.* **28:**95–100, 1973.
16. Hassler, O.: Arterial pattern of human brain stem: Normal appearance and deformation in expanding supratentorial conditions. *Neurology* **17:**368–375, 1967.
17. Heiss, W. D., Turnheim, M., Vollmer, R., Hoyer, J.: Neuronal activity and focal flow in experimental seizures. In Ingvar, D. H., Lassen, N. A. (eds): *Cerebral Function, Metabolism and Circulation*. Copenhagen, Munksgaard, 1977, pp. 230–231.
18. Hoyer, S., Hamer, J., Alberti, E., Stoeckel, N., Weinhardt, F.: The effect of stepwise arterial hypotension on blood flow and oxidative metabolism of the brain. *Pfluegers Arch.* **351:**161–172, 1974.
19. Jefferson, G.: The tentorial pressure cone. *Arch. Neurol. Psychiatry* **40:**857–876, 1938.
20. Johnson, I. H., Rowan, J. O., Harper, A. M., Jennett, W. B.: Raised intracranial pressure and cerebral blood flow. I. Cisterna magna infusion in primates. *J. Neurol. Neurosurg. Psychiatry* **35:**285–296, 1972.
20a. Kuhl, D. E., Alaui, A., Hoffman, E. J., Phelps, M. E., Zimmerman, R. A., Obrist, W. D., Bruce, D. A., Greenberg, J. H., Uzzell, B.: Local cerebral blood volume in head-injured patients: Determination by emission computed tomography of ^{99M}Tc-labeled red cells. *J. Neurosurg.* **52:**309–320, 1980.
21. Kuhl, D. E., Reivich, M., Alavi, A., Nyary, I., Staum, M. M.: Local cerebral blood volume determined by three-dimensional reconstruction of radionuclide scan data. *Circ. Res.* **36:**610–619, 1975.
22. Lassen, N. A., Ingvar, D. H., Skinhoj, E.: Brain function and blood flow. *Sci. Am.* **239:**62–71, 1978.
23. Lofgren, J., Zwetnow, N. N.: Cranial and spinal components of the cerebrospinal fluid pressure–volume curve. *Acta Neurol. Scand.* **49:**575–585, 1973.
24. Lofgren, J., Zwetnow, N. N.: Intracranial blood volume and its variation with changes in intracranial pressure. In Beks, J. W. F., Bosch, D. A., Brock, M. (eds.): *Intracranial Pressure III*. New York, Springer-Verlag, 1976, pp. 25–28.
25. Lofgren, J.: Effects of variations in arterial pressure and arterial carbon dioxide tension on the cerebrospinal fluid pressure–volume relationships. *Acta Neurol. Scand.* **49:**586–598, 1973.
26. Lundberg, N., Ponten, U., Brock, M. (eds.): *Intracranial Pressure II*. New York, Springer-Verlag, 1975.
27. Marmarou, A., Shulman, K.: Pressure–volume relationships—basic aspects. In McLaurin, R. L. (ed): *Head Injuries*. New York, Grune & Stratton, 1976, pp. 233–236.
28. Marmarou, S., Shulman, K., Rosende, R.: A nonlinear analysis of the cerebrospinal fluid system and intracranial pressure dynamics. *J. Neurosurg.* **48:**332–334, 1978.
29. Marmarou, A., Shulman, K., Shapiro, K., Poll, W.: The time course of brain tissue pressure and local CBF in vasogenic edema. In Pappius, H. M., Feindel, W. (eds.): *Dynamic Aspects of Cerebral Edema*. New York, Springer-Verlag, 1976, pp. 113–121.
30. Marmarou, A.: Personal communications.
31. Marshall, L. F., Bruce, D. A., Graham, D. I., Langfitt, T. W.: Alterations in behaviour, brain electrical activity, cerebral blood flow and intracranial pressure produced by triethyl tin sulfate-induced cerebral edema. *Stroke* **7:**21–25, 1976.
32. Marshall, L. F., Welsh, F., Durity, F., Lounsbury, R., Graham, D. I., Langfitt, T. W.: Experimental cerebral oligoaemia and ischemia produced by intracranial hypertension. Part 3. Brain energy metabolism. *J. Neurosurg.* **43:**323–328, 1975.
33. Marshall, L. F., Durity, F., Lounsbury, R., Graham, D. I., Welsh, F., Langfitt, T. W.: Experimental cerebral oligemia and ischemia produced by intracranial hypertension. Part 1. Pathophysiology, electroencephalography, cerebral blood flow, blood–brain barrier, and neurological function. *J. Neurosurg.* **43:**308–317, 1975.
34. Martin, A. N., Kobrine, A. I., Larssen, D. F.: Pressure in the sagittal sinus during intracranial hypertension in man. *J. Neurosurg.* **40:**603–608, 1974.
35. Mienig, G., Reulen, H. J., Magawly, C.: Regional cerebral blood flow and cerebral perfusion pressure in global brain edema induced by water intoxication. *Acta Neurochir.* **29:**1–13, 1973.
36. Miller, J. D., Becker, D. P., Ward, J. D., Sullivan, H. G., Adams, W. E., Rosner, M. J.: Significance of intracranial hypertension in severe head injury. *J. Neurosurg.* **47:**503–516, 1977.
37. Miller, J. D., Stanek, A., Langfitt, T. W.: Concepts of cerebral perfusion pressure and vascular compression during intracranial hypertension. *Prog. Brain Res.* **35:**411–432, 1972.
38. Plum, F., Posner, J. B.: *Diagnosis of Stupor and Coma*. Philadelphia, Davis, 1972.
39. Pollay, M.: Formation of CSF: Relation of studies of isolated choroid plexus to the standing gradient hypothesis. *J. Neurosurg.* **42:**665–673, 1975.
40. Raichle, M. E., Grubb, R. L., Jr., Phelps, M. E., Gado, M. H., Caronna, J. J.: Cerebral hemodynamics

and metabolism in pseudotumor cerebri. *Ann. Neurol.* **4**:104–111, 1978.

41. Reulen, H. J., Graham, R., Spatz, M., Klatzo, I: Role of pressure gradients and bulk flow in dynamics of vasogenic brain edema. *J. Neurosurg.* **46**:24–35, 1977.
42. Rowan, J. O., Johnston, I. H.: Blood pressure to raised CSF pressure. In Lundberg, N., Ponten, V., Brock, M. (eds.): *Intracranial Pressure II.* Berlin, Springer-Verlag, 1975, pp. 298–302.
43. Rowan, J. O., Teasdale, G.: Brain stem blood flow during raised intracranial pressure. In Ingvar, D. H., Lassen, N. A. (eds): *Cerebral Function, Metabolism and Circulation.* Copenhagen, Munksgaard, 1977, pp. 520–521.
44. Reivich, M., Kovách, A. G. B., Spitzer, J. J., Sandor, P.: Cerebral blood flow and metabolism in hemorrhagic shock in baboons. In Kovách, A. G. B. (ed.): *Neurohumoral and Metabolic Aspects of Injury.* New York, Plenum Press, 1973, pp. 19–26.
45. Sato, O., Bering, E. A.: Extraventricular formation of cerebrospinal fluid. *Brain Nerv.* **19**:883–885, 1967.
46. Schettini, A., Walsh, E. K.: Experimental identification of the subarachnoid and subpial compartments by intracranial pressure measurements. *J. Neurosurg.* **40**:609–616, 1974.
47. Shapiro, H. M., Langfitt, T. W., Weinstein, J. D.: Compression of cerebral vessels by intracranial hypertension. II. Morphological evidence for collapse of vessels. *Acta Neurochir.* **12**:223–233, 1966.
48. Stone, H. H., MacKrell, T. N., Wechsler, J.: The effect on cerebral circulation and metabolism in man of acute reduction in blood pressure by means of intravenous hexamethonium bromide and head up tilt. *Anesthesiology* **16**:168–176, 1955.
49. Sundt, T. M., Jr., Sharbrough, F. W., Anderson, R. E., Michenfelder, J. D.: Cerebral blood flow measurements and electroencephalogram during carotid end-arterectomy. *J. Neurosurg.* **41**:310–320, 1974.
50. Teasdale, G., Rowan, J. O., Turner, J., Grossman, R., Miller, J. D.: Cerebral perfusion failure and cortical electrical activity. In Ingvar, D. H., Lassen, N. A. (eds.): *Cerebral Function, Metabolism and Circulation.* Copenhagen, Munksgaard, 1977, pp. 430–431.
51. Walsh, E. K., Schettini, A.: Elastic behaviour of brain tissue *in vivo. Am. J. Physiol.* **230**:1058–1062, 1976.
52. Weiss, M. H., Nulsen, F. E.: The effect of glucocorticoids on CSF flow in dogs. *J. Neurosurg.* **32**:452–458, 1970.
53. Welch, K., Friedman, V.: The cerebrospinal fluid valves. *Brain* **83**:454–469, 1961.
54. Welsh, F. A., O'Connor, M. J., Marcy, V., Reider, W.: Two distinct types of inhomogeneous metabolic failure in cerebral oligaemia. In Ingvar, D. H., Lassen, N. A. (eds.): *Cerebral Function, Metabolism and Circulation.* Copenhagen, Munksgaard, 1977, pp. 118–119.

CHAPTER 28

Non-Steady-State Measurements of Cerebrospinal Fluid Dynamics

Laboratory and Clinical Applications

Frederick H. Sklar

1. Introduction

Numerous techniques have been used to quantify rates of cerebrospinal fluid (CSF) production and absorption in animals and man. A recent monograph by Welch[39] critically reviews the various approaches developed to study the CSF compartment. The ventriculocisternal perfusion technique described by Pappenheimer *et al.*[24] has proved to be a valuable laboratory tool with which to define the physiology of the CSF system. Such open-ended perfusion techniques have been applied not only to animal preparations, but also to patients with presumed normal CSF dynamics,[6,30] absorptive defects,[19,20] and choroid plexus tumors.[9] However, ventriculocisternal and ventriculospinal perfusions are restricted greatly by an absolute requirement for steady-state equilibrium. In this chapter, two methods that do not require long time periods for steady-state determinations are described. The feasibility and clinical pertinence of measuring CSF dynamics in patients will be emphasized.

2. Open Perfusion Technique

The infusion circuitry as described by Pappenheimer *et al.*[24] is shown in Fig. 1. The height of the outflow tubing determines the level of intracranial pressure (ICP). Tracer-clearance measurements provide an indication of the rate of CSF absorption (A); dilution calculations estimate the formation rate (F):

$$A = \frac{R_{in}C_{in} - R_{out}C_{out}}{C_{out}} \tag{1}$$

$$F = \frac{C_{in}C_{out}}{C_{out}}(R_{in}) \tag{2}$$

where R_{in} and R_{out} represent the rates of inflow and outflow, respectively, and C_{in} and C_{out} represent the concentration of nondiffusible tracer in the inflow and outflow fluids.

The time required to reach steady-state equilibrium depends on the perfusion volume as well as the infusion rate. Resistance across the aqueduct sets the upper limit as to the rate of infusion. Experience with dogs in our laboratory has indicated that an infusion rate of 0.2 ml/min will result in steady-state equilibrium after approximately 2 hr. The time required to measure rates of CSF production and absorption with the ventriculocisternal perfusion technique may approach 2–3 hr per determination under specific circumstances. Were the physiological conditions to change during an equilibration period, extended perfusions would be necessary. To define CSF dynamics in dogs at three or four intracranial pressures may require 9–12 hr of infusion. Certainly, patient studies that last 8 and 9 hr are of limited value when they provide only two or three data points. Thus, the great limitation

Frederick H. Sklar, M.D. • Division of Neurological Surgery, University of Texas Health Science Center at Dallas, Dallas, Texas 75235.

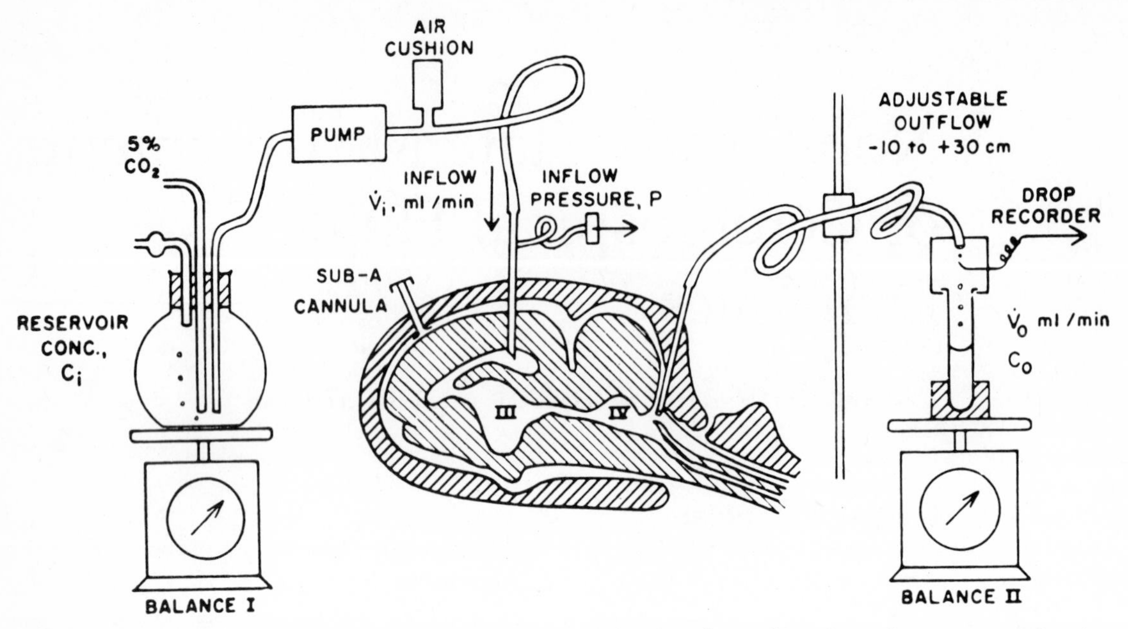

Figure 1. Circuitry for open ventriculocisternal perfusion. Height of outflow tubing determines pressure of system. Reproduced from Pappenheimer *et al.*,[24] with permission from publisher.

of open-ended ventriculocisternal and ventriculospinal perfusion techniques is that of time. The method does not lend itself to the study of labile experimental conditions. Application of this test to patients for routine clinical evaluations is not thought practical.

3. Closed Recirculatory Perfusion Technique

A recirculatory spinal perfusion technique has been described for use in animals.[38] This method utilizes tracer-clearance calculations to determine rates of CSF formation in addition to manometric calculations of concurrent CSF absorption. Measurements are made under non-steady-state conditions. Accordingly, the time requirements are greatly reduced—a distinct advantage over ventriculocisternal perfusion techniques.

The circuitry is outlined in Fig. 2. Mock CSF with nondiffusible fluorescent tracer is rapidly recirculated through the spinal subarachnoid space in a retrograde fashion. A peristaltic pump circulates the fluid at 4 ml/min from the cisterna magna through a fluorometric flow cell, the syringe of an infusion pump, and finally back into a lumbar subarachnoid catheter. The volume of the external circuit can be changed with the syringe pump. Rapid recirculation of the fluid ensures mixing of fluorescein-tagged albumin. Newly formed CSF dilutes the concentration of circulating tracer. Measurements of tracer concentration (C), the rate of change of concentration with time (dC/dt), the volume of the external circuitry (V_1), as well as the volume of that part of the subarachnoid space actually perfused in the recirculatory system (V_{20}), provide the basic parameters with which to calculate CSF formation at resting pressure once mixing oscillations disappear:

$$F = \frac{-dC/dt}{C}(V_1 + V_{20}) \tag{3}$$

At resting pressure, CSF formation equals absorption. The volume of the external circuitry (V_1) is measured directly. The sum of V_1 and the internal perfusion volume (V_{20}) is equivalent to the total perfusion volume, which can be determined from dilution calculations. Extrapolation of the concentration–time curve back to time zero provides an estimate of this total volume necessary to dilute a known amount of tracer.

Intracranial pressures other than resting pressure can be studied by controlling the external circuit volume (V_1) with the syringe infusion pump (Fig. 2). Fluid can be infused into the animal to raise ICP. The resulting change in the subarachnoid volume can be estimated from elasticity calculations relating pressure to volume as an exponential function[37]:

$$\ln P = bV + \alpha \quad (4)$$

$$\Delta V = [(1/b) \ln P + \alpha] - [(1/b) \ln P_0 + \alpha]$$

or

$$\Delta V = (1/b) \ln P/P_0 \quad (5)$$

where the elasticity slope b is determined empirically for each animal with rapid subarachnoid infusions.

With the syringe pump, ICP can be changed to any desired pressure. A simple feedback servocontrolled system can regulate the syringe pump to maintain ICP constant automatically. A pressure plateau can be held long enough for adequate mixing in the recirculatory spinal perfusion system, as indicated by a disappearance of oscillations in the concentration–time curve. Formation can then be calculated:

$$F = \frac{-dC/dt}{C} [V_1 + V_{20} + (1/b) \ln P/P_0] \quad (6)$$

The rate of CSF absorption (A) can also be calculated:

$$A = F + dV_1/dt \quad (7)$$

where dV_1/dt is the rate of the syringe pump for a given pressure plateau.

The recirculatory spinal perfusion technique can provide rapid quantitative assessment of the CSF compartment without steady-state requirements. Measurements of CSF dynamics can be completed after only a brief time for mixing. In dogs, this initial mixing usually requires 45 min. Subsequent determinations need only 5–10 min for mixing and an additional 10 min for data collection. The technique is therefore applicable to the study of labile pathological conditions. It is difficult to maintain all the various physiological parameters in a strict state of equilibrium for the many hours necessary to complete an experiment with classic ventriculocisternal perfusion techniques. Cserr[5] has suggested that ependymal permeabilities to high-molecular-weight tracers such as inulin change after only several hours of ventriculocisternal perfusion. In contrast, the recirculatory perfusion system is well suited to study labile experimental conditions. The effect of drugs on CSF dynamics can be assessed with this technique.

4. Manometric Determinations of CSF Dynamics

The preceding discussions summarize two tracer-clearance techniques to measure CSF formation. The major advantage of the recirculatory perfusion system is that steady-state equilibrium is not required; CSF dynamics can be measured relatively quickly at numerous constant-pressure plateaus. The laboratory value of the recirculatory perfusion technique has been emphasized. However, because of its invasive nature, application of this method to study patients is not recommended. The recirculatory perfusion technique includes a manometric determi-

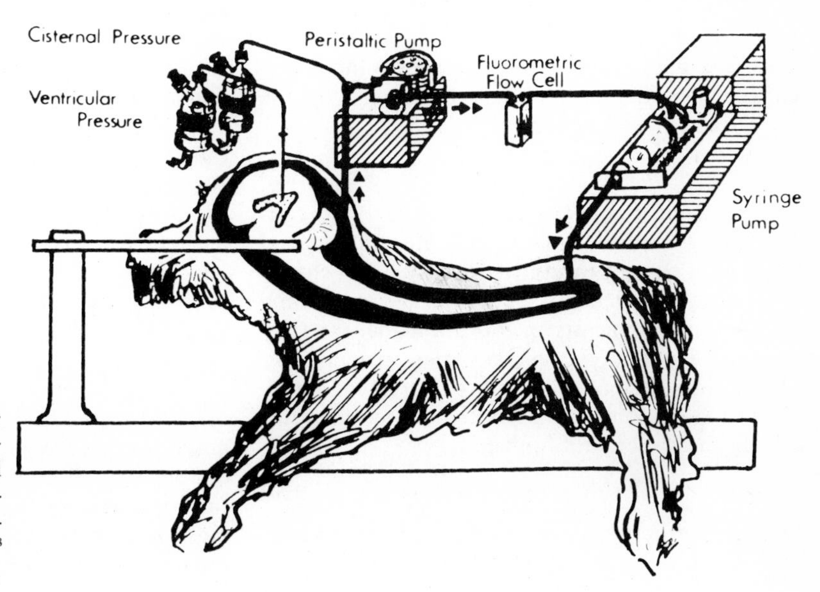

Figure 2. Circuitry of closed recirculatory perfusion system. Syringe infusion pump is connected in series to regulate ICP by changing external circuitry volume. Reproduced from Sklar and Long,[38] with permission from publisher.

nation of CSF absorption (equation 7). In patients, valuable clinical information can be ascertained from such manometric measurements alone, but the rates of CSF formation and absorption cannot be defined individually.

A simple feedback system can be used to hold intracranial pressure constant at multiple pressure plateaus. At constant pressure, the infusion rate approximates the arithmetic difference between the rates of CSF absorption and formation, and this parameter can be examined as a function of pressure. Unlike the recirculatory perfusion system, the variable-rate, constant-pressure infusion technique is less invasive and offers little risk to the patient. This manometric method will be referred to as the *servocontrolled lumbar infusion* technique.

In servocontrolled lumbar infusions, pump rates necessary to maintain pressure constant at multiple pressure plateaus are measured. A linear relationship exists between infusion rate and ICP,[8,10,15,36] and the slope of this relationship is an indication of net CSF absorptive capacity.[36] It will be shown that the infusion rate required to maintain pressure at a desired level is equivalent to the difference between CSF absorption and formation at that particular pressure.

In the infusion system, the rate of change of intracranial volume will reflect the rate at which fluid is infused or withdrawn as well as any concurrent volume changes related to CSF formation (F), absorption (A), and changes in cerebral blood volume (CBV) within a given period of time:

$$dV/dt = R + F - A + \Delta\mathrm{CBV}/\Delta t \qquad (8)$$

At constant intracranial pressure, dV/dt becomes zero. The pump rate (R) then measures the difference between CSF absorption and formation and the change in vascular volume:

$$R = (A - F) - \Delta\mathrm{CBV}/\Delta t \qquad (9)$$

If it is assumed that the cerebrovascular compartment reaches steady state quickly after intracranial pressure is changed and that the CBV is not changing during a particular pressure plateau at the time the manometric measurements are made, the CBV term in equation 9 then becomes insignificant. Under these circumstances, the pump rate is equivalent to the arithmetic difference between the rates of CSF absorption and formation:

$$R = (A - F) \qquad (10)$$

Multiple constant-pressure plateaus can be studied, and the resulting pump rates can be considered as a linear function of pressure. The slope of the infusion rate vs. ICP curve is determined with the least-squares technique. This slope value is equivalent to the net CSF absorptive capacity. It is important that this calculated slope value reflect changes in either or both CSF absorption and formation with respect to pressure. For instance, diminished CSF formation at high ICP may appear as increased net absorptive capacity with this manometric technique.

The x-axis intercept represents the resting pressure as determined by CSF dynamics. This is the pressure at which $(A - F)$ is equivalent to zero: absorption equals formation.

The details of the variable-rate, servocontrolled infusion technique have been reported previously.[36] Because of its potential value as a diagnostic tool to investigate difficult clinical problems involving the CSF compartment, the method will be reviewed briefly.

Patients are studied under light sedation. Infants occasionally require endotracheal anesthesia to eliminate excessive movement artifact, but this is relatively uncommon. The patients are positioned on their side with their head supported by a pillow. Small infants are studied prone with the head turned to one side. The infusion arrangement is illustrated in Fig. 3. A single large-gauge Touhy needle is introduced into the lumbar subarachnoid space. A small catheter is introduced through the needle into the subarachnoid space to a length of 4–6 cm. A Y-connector with a gasket arrangement allows for the fluid to be infused and withdrawn through the needle around the catheter while artifact-free pressure measurements are made through the catheter (Fig. 4). A pressure transducer is zeroed at the intracranial midline. Arterial pressure is monitored with a percutaneously placed arterial catheter.

A servocontrolled infusion pump with reciprocally arranged 12-cc syringes is used for the study. A solenoid-valve system switches the circuitry from the patient to a fluid reservoir when the syringes reach their limits as indicated by a position-detector, which monitors syringe-plunger movement. Accordingly, it is unnecessary to stop the infusion protocol to refill a syringe. Lactated Ringer's solution is used as the infusion fluid.

Intracranial pressure is recorded on a multichannel polygraph. The output voltage from the polygraph is compared to a regulated reference voltage of a servoamplifier, which activates the pump to infuse or withdraw fluid. The servo system continuously adjusts the pump rate until the voltage differential is reduced to zero. In this way, the pump

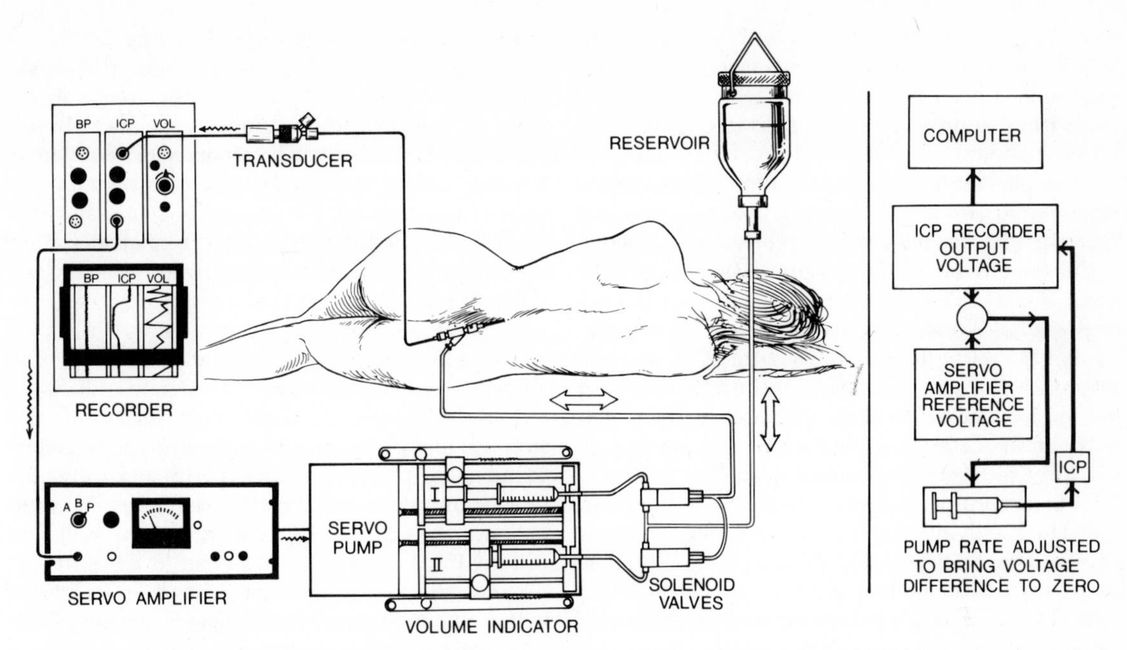

Figure 3. Servocontrolled lumbar infusion circuitry. ICP is maintained constant at desired level with servocontrolled infusion pump, which utilizes reciprocal syringes and solenoid-valve switching device. Reproduced from Sklar *et al.*,[36] with permission from publisher.

actively stabilizes ICP to a desired level. To minimize servooscillations, ICP measurements are converted electronically to mean values and are dampened further with an intermediate amplifier.

Intracranial pressure, mean arterial blood pressure, and syringe-plunger movement as indicated by the position-detector are sampled by a computer at 1-sec intervals. The computer also activates the solenoid-valve switching device when the syringes reach their limits. Pressure-vs.-time and volume-vs.-time data are displayed on a graphics computer terminal. Figure 5A shows the original computer

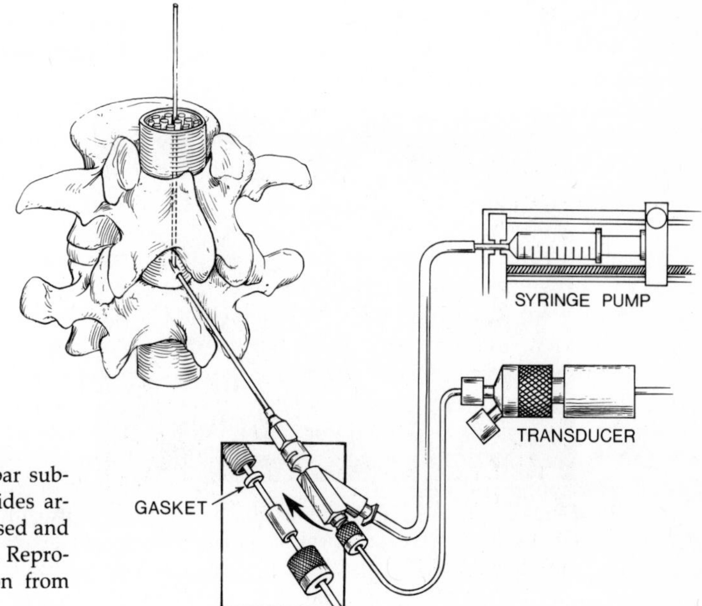

Figure 4. Touhy needle positioned in lumbar subarachnoid space. Indwelling catheter provides artifact-free pressure recordings. Fluid is infused and withdrawn through needle around catheter. Reproduced from Sklar *et al.*,[36] with permission from publisher.

display of a 63-year-old patient with enlarged ventricles, dementia, gait ataxia, and urinary incontinence after a head injury and surgery for subdural hematoma. Editing to exclude pressure artifacts related to coughing, talking, and movement is done, and the plateaus appropriate for data interpretation are selected (I–I to O–O in Fig. 5A). For a given pressure plateau, the computer calculates the slope of the volume–time data with the least-squares technique (DVDT in Fig. 5A). This infusion rate is equivalent to the difference between absorption and formation at that particular ICP. Figure 5B shows the relationship of $(A - F)$ vs. ICP for the patient shown in Fig. 5A. The slope of this function is calculated with linear-regression techniques and represents the net CSF absorptive capacity. In this example, the CSF absorptive capacity is 0.067 ml/min per mm Hg. The x-axis intercept is -1.2 mm Hg, and this corresponds to the resting pressure (P_0) as determined by CSF dynamics. Obviously, the absolute value for the resting pressure depends on the exact zero calibration of the ICP transducer. Errors in transducer placement probably account for the negative P_0.

The servo system allows for rapid and efficient study of the CSF compartment. Constant pressure is quickly achieved and maintained long enough to define the effective infusion rate. On-line computer analysis facilitates data interpretation; the study can be stopped when the results have reached acceptable levels of statistical significance. In general, a study can be completed in 45–60 min.

Simple manometric techniques to measure CSF absorptive capacity have been described previously in the literature. In 1948, Foldes and Arrowood[12] reported a constant-rate infusion technique in patients. With constant-rate lumbar infusions at 0.2–0.8 ml/min, CSF pressures were noted to equilibrate after 40 min. Similarly, Davson *et al.*[8] found that varied rates of intrathecal infusions in animals result in corresponding constant-pressure plateaus. They reported a linear relationship between pressure and infusion rate.

Katzman and Hussey[18] have described a simple constant-rate infusion technique to describe CSF absorptive capacity. At a constant rate of 0.76 ml/min, the CSF pressure normally shows a brief pressure rise that plateaus at a level only slightly greater than resting pressure. Patients with absorptive defects show plateaus at higher pressures. In some patients, the CSF pressure fails to plateau at all, and ICP continues to rise in response to the infusion. An abnormal pressure response has been suggested to indicate a CSF absorptive defect.[17] In essence, the constant-rate Katzman–Hussey infusion looks at only a single pressure plateau of the technique described by Davson *et al.*[8]

The constant-pressure infusion method of Davson *et al.*[8] was further modified by Ekstedt[10] for use in patients. Ekstedt has used a servo system that allows him to define the relationship between infusion rate and ICP in a relatively brief time period, and this technique is very similar to the servocontrolled lumbar infusion developed in our laboratory. Ekstedt has found that five or more manometric measurements can be made in the same time re-

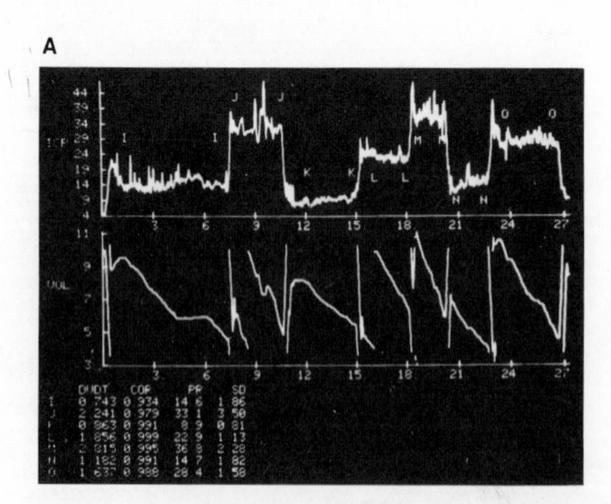

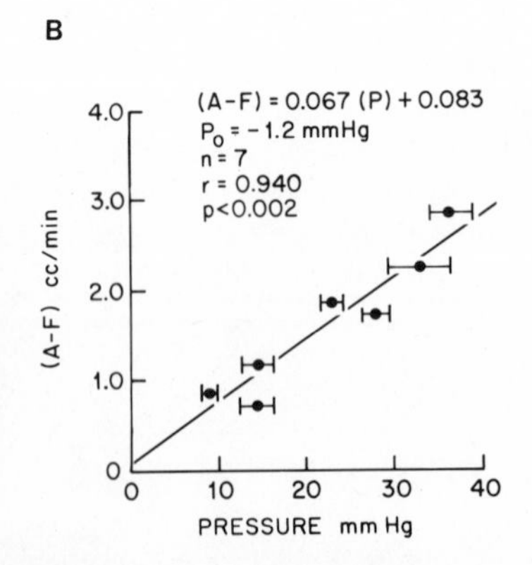

Figure 5. (A) Original computer graphics display for patient V. L. *Top graph:* pressure (mm Hg) vs. time (min); *bottom graph:* syringe volume (mm) vs. time (min). Slope of volume-vs.-time curve (DVDT) is calculated by computer along with coefficient of correlation (COR). Computer determines average pressure for given plateau (PR) and its standard deviation (SD). (B) $(A - F)$ vs. pressure from data of (A). These data represent one third of data points collected during 80 min of infusion time. Horizontal brackets indicate standard deviation.

quired for a single constant-rate infusion protocol as described by Katzman and Hussey. He has reported a linear relationship between pressure and infusion rate.[10] Portnoy and Croissant[25] have also used a constant-pressure infusion technique in patients. Similarly, their results suggest a linear relationship between infusion rate and pressure.

In a more recent report, Portnoy and Croissant[26] relate their experience with variable-rate, constant-pressure ventricular infusions in meningomyelocele children before and after a shunting procedure for hydrocephalus. However, the practice of infusing fluid into a lateral ventricle at rates up to 2 and 3 ml/min[26] requires critical review. The anatomy of the normal ventricular system dictates that the highest resistance to flow of infused fluid is across the cerebral aqueduct. As the infusion rate is gradually increased, exit of fluid from the third ventricle will eventually be limited not only by the absorptive capacity of the arachnoid granulations, but also by the resistance across the aqueduct. A given infusion rate would then result in a higher ventricular pressure than would be recorded if there were no aqueductal resistance to flow. The infusion rate-vs.-pressure curve would have a correspondingly lower slope, yet not necessarily be representative of an absorptive defect.

Moreover, rapid-rate ventricular infusions could be potentially dangerous. Ashcroft *et al.*[1] reported a recirculatory ventriculocisternal perfusion technique in dogs and found that the animals regulated their own ICP as long as the perfusion rate was less than 0.33 ml/min. Similarly, constant-rate ventriculospinal perfusions in dogs at rates greater than 0.4 ml/min have been reported to result frequently in herniation as indicated by rapidly increasing ICP, reflex blood pressure changes, and a significant lateral ventricle–cisterna magna pressure gradient, presumably related to resistance to flow across the aqueduct.[38] It is concluded that the value of variable-rate, constant-pressure ventricular infusions in patients in whom the aqueduct may be stenotic or occluded is diminished by these theoretical considerations and potential dangers.

5. Clinical Studies

5.1. Definition of Normal

Ranges of normal CSF absorptive capacity have been established based on a cumulative experience with the servocontrolled variable-rate lumbar infusion in adults and children with active hydrocephalus, arrested hydrocephalus, brain atrophy, and pseudotumor cerebri. In this laboratory, CSF slope measurements greater than 0.2 ml/min per mm Hg are considered normal with a high degree of certainty. Below 0.1 ml/min per mm Hg, a significant absorptive defect is suggested. Values intermediate between 0.1 and 0.2 ml/min per mm Hg are more difficult to classify and probably represent the lower range of normal. Although it may not be possible to define an absolute value of the lower limits of normal, a slope of less than 0.13 ml/min per mm Hg is considered to be probably abnormal.

To determine an absolute value of the lower limits of normal would require an extensive experience with normal volunteers, but such measurements are not available. Based on the results of over 700 infusion studies in patients with a variety of clinical conditions, Ekstedt[10] has suggested a slope value of 0.1 ml/min per mm Hg to represent the lower limits of normal with his experimental protocol. More recently, he has reported "normative data" based on measurements in 58 patients who proved to be free of progressive neurological disorders on long-term follow-up.[11] However, at the time of their infusion studies, these patients were suspected clinically of having disorders of the CSF system and therefore cannot be considered to represent accurately the normal population. Nonetheless, his results show that 95% of these patients had slope values greater than 0.09 ml/min per mm Hg; 84% had slopes of 0.12 ml/min per mm Hg or greater. Ekstedt's experimental protocol involves increasing ICP in a step-like fashion. Directional ordering of pressure plateaus is avoided in our laboratory, because it is thought that random selection of pressures may define the system more accurately. Experience in animals[15,36] and patients[36] has indicated that the infusion rate required to maintain pressure constant at a given ICP may be different when the order of testing is changed (Fig. 6). It is therefore concluded that an ordered selection of pressures may introduce a systematic error. Random pressure selection may require more points for the data to reach statistical significance, but such an approach appears to be a more physiological evaluation. These procedural variances probably account for the small differences in what is considered normal by Ekstedt and this laboratory.

5.2. Sources of Artifact

Manometric determinations of CSF dynamics assume that there are no concurrent changes in intracranial volume other than those related to CSF absorption and formation. It is hypothesized that with

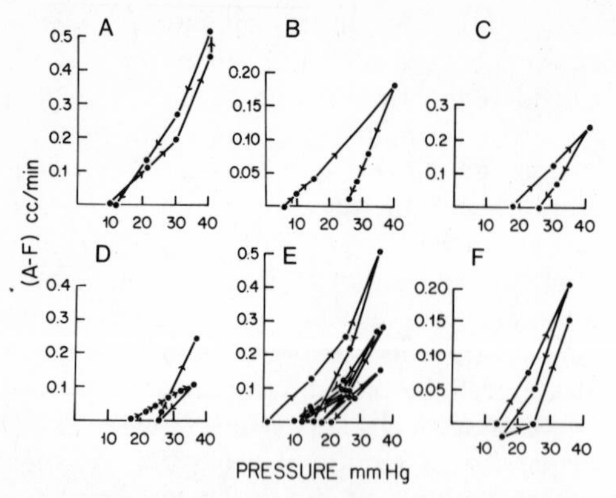

Figure 6. ($A - F$)-vs.-pressure data from six dogs studied with servocontrolled lumbar infusions. Data points were plotted according to sequence studied. In first animal (A) there was no significant difference between results determined as pressure was sequentially increased and those measured as pressure was returned to baseline levels. In (B–F), pressure sequence seemed to influence measurements of CSF dynamics. These results suggest that measurements of CSF dynamics may not be constant for given intracranial pressure. Such observations indicate the necessity to choose pressures at random. Reproduced from Sklar *et al.*,[36] with permission of publisher.

each change in ICP, the cerebrovascular compartment reaches steady state quickly; cerebral blood volume (CBV) cannot be changing during a constant-pressure plateau (equation 9). However, cerebral vascular engorgement may be associated with significant intracranial hypertension. Sagittal sinus venous pressure (SSVP) is normally lower than CSF pressure,[33] and may actually decrease with raised ICP.[2,16] A varied SSVP response to graded intracranial hypertension was noted by Rowen *et al.*[29] They found that some animals showed SSVP not to change, while others showed modest pressure increases, but to levels less than the CSF pressure. Still other animals showed delayed SSVP responses to levels approaching ICP when the latter exceeded 54 mm Hg, and this was attributed to sagittal sinus collapse. CSF pressures to 68 mm Hg in patients do not cause collapse of the cortical venous bed in angiographic studies.[14] Nagai *et al.*[23] have demonstrated increased CBV associated with decreasing SSVP in the face of reductions in CBF in response to graded increases of ICP. These observations suggest venous engorgement with pressure. Risberg *et al.*[28] have documented increased regional CBV with spontaneous plateau waves. Morphological studies show compression and collapse of the straight and sagittal sinuses with intracranial hypertension[32]; distal subarachnoid veins develop "cuff constrictions" as ICP is increased.[40] Thus, collapse of the distal venous system with engorgement of the more proximal venous bed may actually occur with intracranial hypertension.

Moreover, the measurements of Risberg *et al.*[28] indicate that the CBV changes with intracranial hypertension continue to occur over 10–13 min (Fig. 7). In the servocontrolled lumbar infusion protocol, ICP is rapidly changed to a desired level and held constant by the variable-rate infusion pump. In the early minutes of each pressure plateau, the pump rate necessary to maintain ICP constant may reflect not only the difference between the rates of CSF absorption and formation but also this rate of change of CBV:

$$R = (A - F) - \Delta CBV/\Delta t \qquad (9)$$

With increasing CBV secondary to venous congestion, a lower pump rate would be required to main-

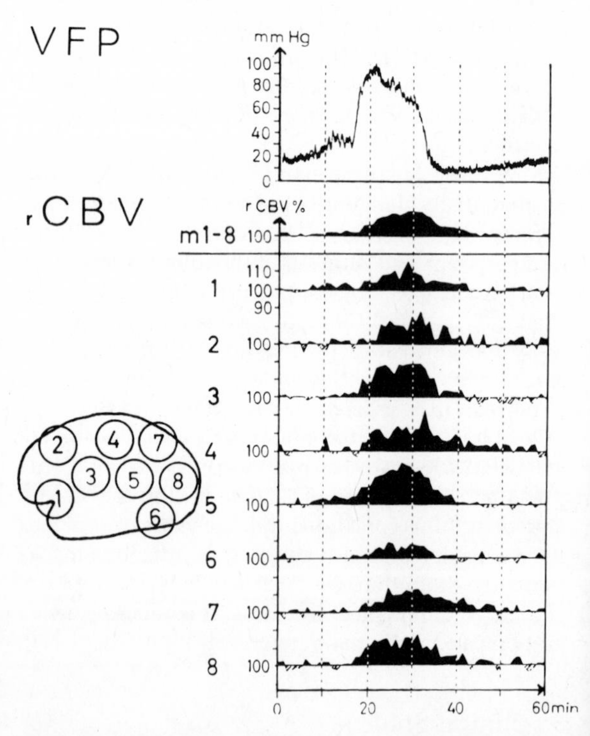

Figure 7. Regional cerebral blood volume (rCBV) measurements during spontaneous pressure plateau wave. These data illustrate that blood volume changes with intracranial hypertension may occur gradually, with maximum at 10–13 min. (VFP) Ventricular fluid pressure. Reproduced from Risberg *et al.*,[28] with permission from publisher.

tain pressure constant. However, even if CBV were a sensitive function of ICP, artifactual lowering of measured CSF absorptive capacity would probably be present in patients with diverse clinical conditions. In other words, the artifact would be "built in" to what is considered normal.

The magnitude by which CBV changes affect measurements of CSF dynamics was examined in a single patient with pseudotumor cerebri.[35] This patient was studied at numerous pressure plateaus maintained for 30 min rather than the usual 3–5 min. The effective pump rate necessary to maintain ICP constant was determined both in the early (first 15 min) and the late (last 15 min) phase of each plateau. It would be predicted that the calculated values of absorptive capacity from the early minutes of the 30-min plateaus would be less steep than the curve defined by the late minutes if significant CBV artifact were present. CSF absorptive capacities as calculated from the "early" and "late" data were not significantly different, and these results are summarized in Table 1. It is concluded that in this single patient, vascular artifact contributed little to the determination of CSF dynamics.

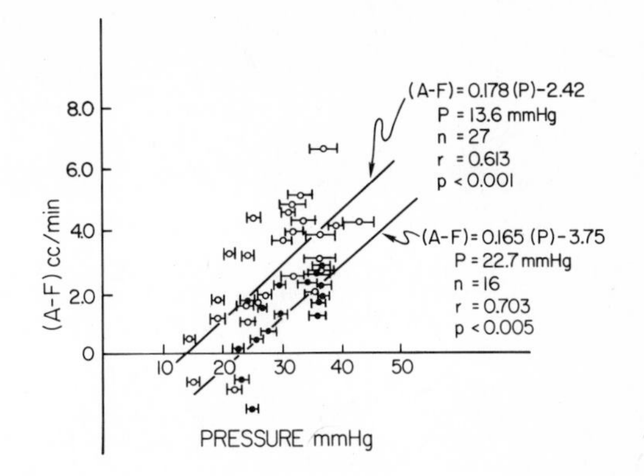

Figure 8. CSF dynamics as measured with servocontrolled lumbar infusion technique in patient symptomatic of pseudotumor cerebri on oral furosemide, 160 mg/day (●). Measurements were repeated 1 week later after daily medication dose was increased to 320 mg (○). Patient's symptoms had disappeared on this drug regimen. CSF absorptive capacity as indicated by regression slope was not significantly changed. However, resting pressure had become lower on new drug dosage. Horizontal brackets indicate standard deviation.

It is necessary to emphasize what the servocontrolled variable-rate lumbar infusion technique actually measures. The test determines the quantitative relationship between $(A - F)$ and ICP. The results give only indirect information about either CSF absorption or formation individually. The slope of the $(A - F)$-vs.-ICP curve represents the net absorptive capacity and is equivalent to the actual CSF absorptive capacity only under conditions of constant formation, independent of ICP. For instance, if CSF production is halved by various medications, no change in slope would be expected, but the curve would be shifted to the left on the x-axis. This would result in a lower resting pressure P_0. Figure 8 shows the CSF dynamics of a patient symptomatic with pseudotumor cerebri on oral furosemide, 160 mg/day. The medication dose was increased to 320 mg, and the infusion test was repeated 1 week later. The patient had become asymptomatic. Although the slope value remained unchanged, the resting pressure fell from 22.7 to 13.6 mm Hg. On the other hand, parenteral administration of furosemide at 1 mg/kg in another patient resulted in an increased net CSF absorptive capacity measured 60 min after intravenous administration of the drug (Fig. 9). With the assumption that the drug functions to inhibit CSF production, the slope increase is attributed to a more pronounced inhibition of CSF formation at higher pressures. Nonetheless, it is important that the results of the infusion technique reflect changes in either absorption or formation, or both.

Table 1. CSF Dynamics as a Function of ICP in a Patient Symptomatic of Pseudotumor Cerebri[a]

Data used for calculations	CSF absorptive capacity (ml/min per mm Hg)	P_0 (mm Hg)[b]	Probability (p)
All points	0.0612	12.2	<0.002
Early points	0.0638	13.2	<0.05
Late points	0.0563	10.2	0.057

[a] The CSF dynamics were studied with 30-min pressure plateaus.
[b] Measured P_0 = 13 mm Hg.

Table 2. CSF Dynamics in Eight Adults with Communicating Hydrocephalus

Patient									
No.	Age	Sex	Symptoms	History of etiological importance	Isotope cisternography	CSF absorptive capacity (ml/min per mm Hg)	P_0 (mm Hg)	Probability (p)	Effect of shunt on symptoms
1	47 yr	F	Dementia, gait ataxia, incontinence, headaches	Subarachnoid hemorrhage	Not done	0.032	16.7	0.05	Good
2	25 yr	M	Headaches	Undefined granulomatous meningitis	Not done	0.087	11.5	<0.001	Good
3	63 yr	M	Dementia, gait ataxia, incontinence	Head trauma	Positive	0.105	−1.0	<0.001	Good
4	43 yr	F	Dementia, gait ataxia, incontinence	Subarachnoid hemorrhage	Negative	0.068	0.8	<0.001	Good
5	61 yr	M	Dementia, gait ataxia, incontinence	None	Positive	0.033	−2.4	<0.02	Not shunted
6	70 yr	M	Dementia, gait ataxia	Minor head injury	Positive	0.131	1.7	<0.001	Not shunted
7	61 yr	M	Dementia, gait ataxia, incontinence	Sarcoid	Positive	0.080	6.2	<0.005	Not shunted
8	58 yr	M	Dementia, gait ataxia	None	Positive	0.133	6.3	<0.001	Not shunted

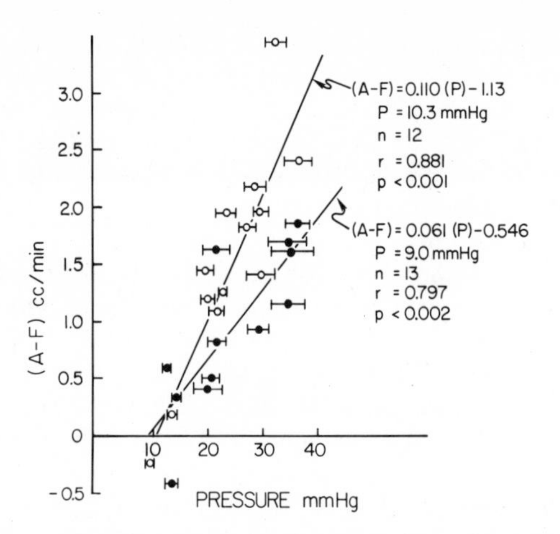

Figure 9. CSF dynamics as determined by servocontrolled lumbar infusion technique in patient symptomatic with pseudotumor cerebri. Measurements were made before (●) and after (○) parenteral administration of furosemide. No significant change in resting pressure is noted, but there is significant increase in CSF absorptive capacity as indicated by increase in regression slope ($p < 0.05$). Horizontal brackets indicate standard deviation.

6. Disease Entities

6.1. Adult Communicating Hydrocephalus

CSF dynamics have been measured with the servocontrolled lumbar infusion technique in patients with a variety of clinical disorders. The test may prove to be particularly useful in the diagnostic evaluation of adults with dementia and enlarged ventricles. Table 2 summarizes the clinical data and CSF dynamics in eight adults with communicating hydrocephalus. Six patients showed absorptive defects, and two were at the lower limits of normal (0.13 ml/min per mm Hg). Four of the patients have been shunted at the time of this writing. Two had a history of subarachnoid hemorrhage, one had had surgery for subdural hematoma after head trauma, and one had a granulomatous meningitis that could not be clearly defined. Each of these four patients with histories of etiological significance has shown impressive clinical improvement with shunting. Isotope cisternal scans were done in six patients and were abnormal in all but one. Obviously, an experience with many more patients is required to define the diagnostic accuracy of the servocontrolled infusion test in distinguishing active hydrocephalus from brain atrophy in adults.

6.2. Arrested Hydrocephalus

Foltz and Shurtleff[13] have suggested that the term "arrested hydrocephalus" implies resolution of the underlying pathological process responsible for the ventricular dilatation. "Compensated hydrocephalus" suggests that other mechanisms function to prevent further ventricular enlargement, i.e., transependymal absorption or shunts. The servocontrolled lumbar infusion test has been used in 11 children with presumed disorders of the CSF system.[34] In each child, there was clinical confusion as to disease activity and whether surgical intervention was indicated. Of the 11 children, 4 had normal CSF dynamics, ventricular enlargement, and a static neurological picture without clinical settings suggestive of an atrophic process. These children

A

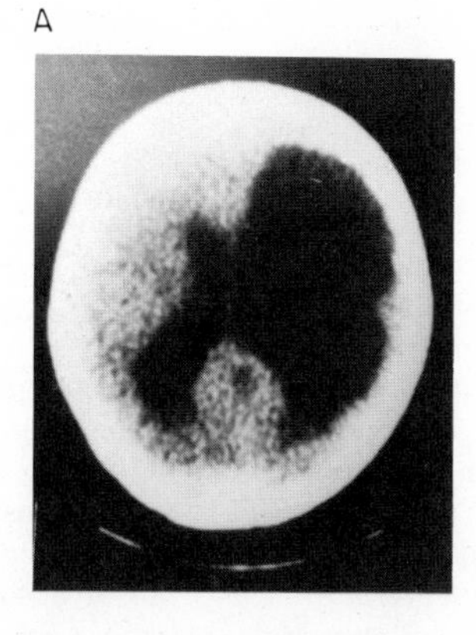

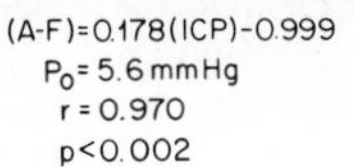

B

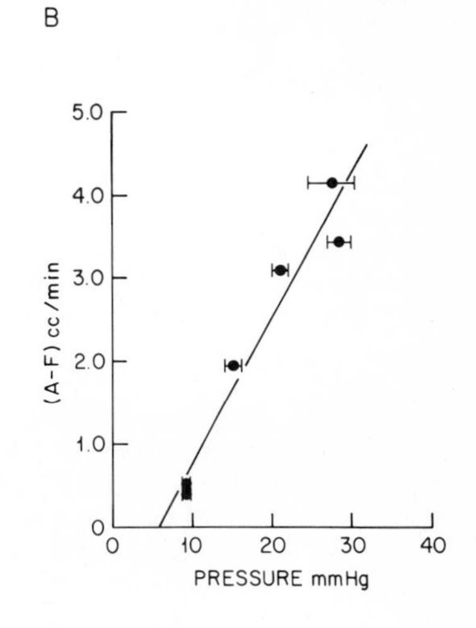

Figure 10. (A) CT scan of 9-year-old Caucasian male who presented with seizures, behavior disorder, right frontal proencephaly, and panventricular enlargement. Ventricles were not enlarged on air study done in neonatal period, which did demonstrate the porencephalic cyst. (b) CSF dynamics as determined by servocontrolled lumbar infusion technique. Results indicate normal CSF absorptive capacity (0.178 ml/min per mm Hg) with resting pressure of 5.6 mm Hg. Actual resting pressure was 5 mm Hg. On basis of these measurements, child was thought to meet criteria of having arrested hydrocephalus. One-year follow-up has shown no evidence for disease progression. Symptoms have improved with anticonvulsant therapy. Horizontal brackets indicate standard deviation. Reproduced from Sklar *et al.*[34]

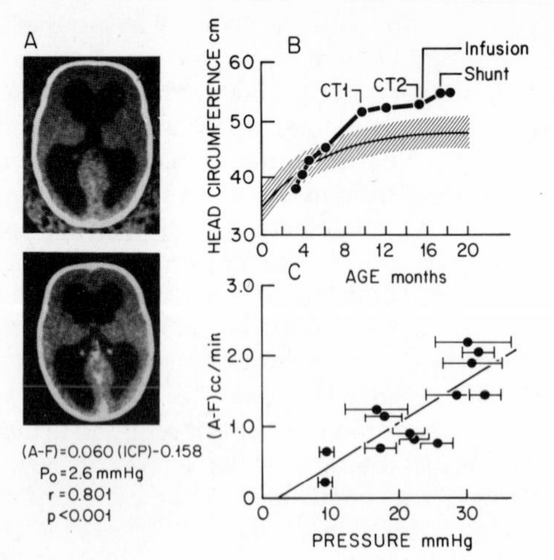

Figure 11. Serial CT scans (A) and head-circumference measurements (B) in Caucasian male child with severe communicating hydrocephalus. When child came to neurosurgical attention, it appeared that head growth had plateaued at level parallel to 98th percentile. No progression was apparent on follow-up CT scan. Servocontrolled lumbar infusion was done at 15 months of age, and results indicate significant CSF absorptive defect at 0.06 ml/min per mm Hg. This case represents example of active hydrocephalus as determined by measurements of CSF dynamics. Horizontal brackets indicate standard deviation. Reproduced from Sklar *et al.*[34]

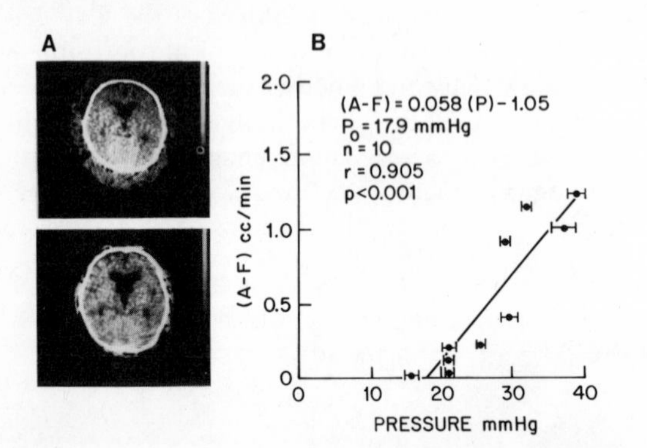

Figure 12. (A) *Top*: CT scan of 5-month-old Caucasian male with achondroplasia who presented with bulging fontanelle and enlarging head. Scan shows only mild ventricular enlargement with generous subarachnoid spaces. *Bottom*: CT scan was repeated 7 months later and revealed progressive ventricular enlargement. (B) CSF dynamics are shown as function of ICP, measured with servocontrolled lumbar infusion technique at time of first scan. Results indicate severe absorptive defect at 0.058 ml/min per mm Hg and resting pressure of 17.9 mm Hg. Shunt procedure was postponed because of recurrent problems with upper respiratory infections and uncertainty whether surgical intervention was definitely indicated. Horizontal brackets indicate standard deviation.

therefore meet the criteria of arrested hydrocephalus. One child was thought to have compensated hydrocephalus; 3, active hydrocephalus; and 3, brain atrophy. Case examples of arrested and active hydrocephalus are shown in Figs. 10 and 11, respectively.

Retrospective management decisions based on clinical presentations, physical findings, and traditional diagnostic tests disagreed with management as indicated by the CSF measurements in 8 of 11 cases. In 6 of these 8 instances, a nonsurgical approach was suggested by the CSF studies. Whether physiological measurements of CSF dynamics should take priority over clinical intuition remains to be determined. Several of the patients have been followed for over a year without surgery and have shown no progression of signs or symptoms. An-

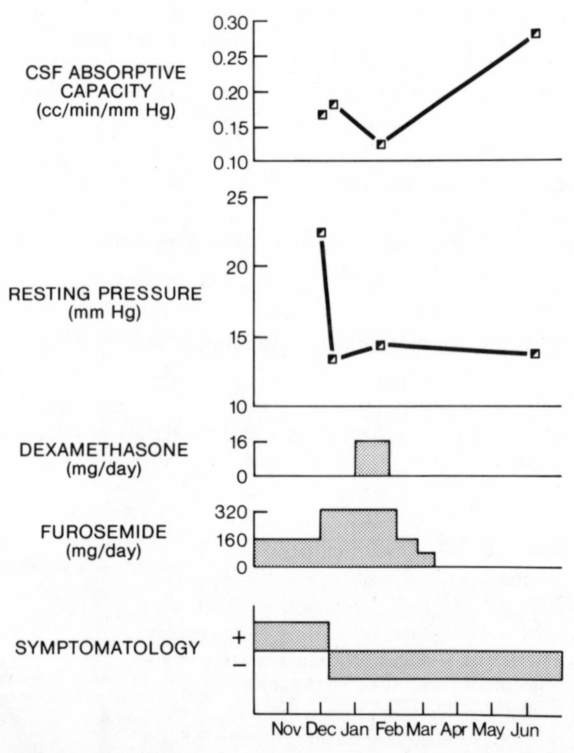

Figure 13. Serial measurements of CSF dynamics, medication regimen, and clinical symptomatology in 21-year-old Negro female with pseudotumor cerebri. Increased medications were associated with fall in resting pressure and disappearance of symptoms. However, measurements of CSF absorptive capacity did not change. On the other hand, CSF absorptive capacity was noted to increase significantly when patient was restudied months later off all medications. Reproduced from Sklar *et al.*[35] with permission of publisher.

other child with mild ventricular enlargement and a severe absorptive defect was followed without surgery and showed computed tomographic (CT) scan evidence of progression of the hydrocephalus over a 7-month period (Fig. 12). Nevertheless, the value of this test as a management indicator must be more clearly defined.

6.3. Pseudotumor Cerebri

Ten patients with the clinical diagnosis of benign intracranial hypertension were studied with the servocontrolled lumbar infusion technique.[35] Serial measurements were made in five patients. All patients had papilledema, headache, and/or visual symptoms. All patients had elevated pressures (> 200 mm H_2O) on lumbar puncture at some time in the course of their clinical follow-up. The CSF dynamics are summarized in Table 3.

Nearly all the patients had abnormally low CSF absorptive capacities. On the other hand, marked elevations of resting pressure were not a constant feature of the disease. Resting pressure P_0 as determined by CSF dynamics was not elevated in half the patients studied. Calculated values of P_0 agreed closely with actual measured resting pressures, recorded when the patients were left undisturbed once the infusion study had been completed.

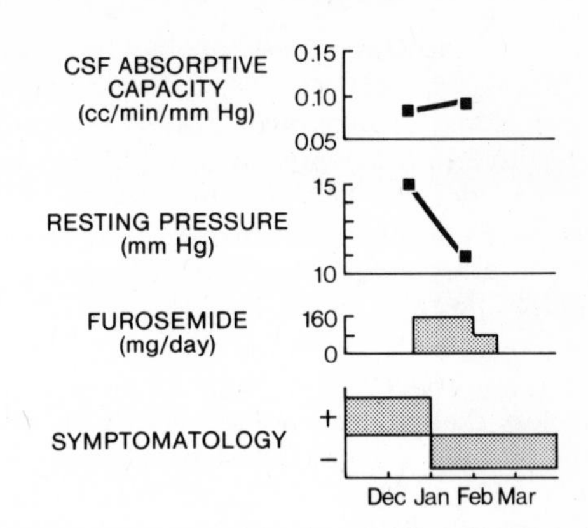

Figure 14. Serial measurements of CSF dynamics, medication regimen, and clinical symptomatology in 16-year-old Negro male with pseudotumor cerebri. Medications appeared to induce reduction in resting pressure and resolution of symptomatology but had no effect on CSF absorptive capacity. Unfortunately, measurements were not repeated once the medications had been discontinued. Reproduced from Sklar *et al.*[35] with permission of publisher.

The physiological correlate of the calculated P_0 is not fully understood. This parameter represents the *x*-axis intercept and is that pressure at which the rate of CSF absorption should equal formation. Perhaps in some patients or in various "types" of pseu-

Table 3. CSF Dynamics in Ten Patients with Pseudotumor Cerebri

Patient No.	Age	Sex	Study	Medications (daily dose)	CSF absorptive capacity (ml/min per mm Hg)	P_0 (mm Hg)	Probability (p)
1	42 yr	F	1	None	0.0187	6.0	<0.02
			2	None	0.0590	16.2	<0.001
2	11 yr	F	1	Prednisone (4 mg/q.o.d.)	0.146	7.3	<0.001
			2	Prednisone (4 mg/q.o.d.)	0.195	8.7	<0.001
3	19 yr	F	1	None	0.109	8.2	<0.001
4	45 yr	F	1	None	0.0992	9.5	<0.001
			2	None	0.0610	9.0	<0.002
5	17 yr	F	1	Dexamethasone (12 mg) Chlorothiazide (1 g)	0.0609	7.9	<0.001
6	16 yr	M	1	None	0.0778	14.9	<0.001
			2	Furosemide (160 mg)	0.0882	11.0	<0.001
7	29 yr	F	1	None	0.110	21.0	<0.001
8	25 yr	F	1	Dexamethasone (4 mg) Furosemide (40 mg)	0.0928	11.0	<0.002
9	21 yr	F	1	Furosemide (160 mg)	0.165	22.7	<0.005
			2	Furosemide (320 mg)	0.178	13.6	<0.001
			3	Furosemide (320 mg) Dexamethasone (16 mg)	0.125	14.4	<0.001
			4	None	0.272	13.4	<0.001
10	32 yr	F	1	None	0.0612	12.2	<0.002

dotumor, the disease process involves episodic intracranial hypertension. Perhaps the mechanism for these pressure increases involves physiological parameters that have no effect on the calculations of P_0. The relationship between the pressure at rest (which seems to be approximated by P_0) and average daily pressure requires further study.

Multiple infusions were done in five patients. Three showed resting pressure changes to correlate with changes in clinical conditions. However, symptomatology changes were not necessarily associated with impressive changes in CSF absorptive capacities (Figs. 13 and 14).

The pathophysiology of pseudotumor cerebri remains controversial. In the literature, vascular engorgement,[7,22,27] brain swelling,[31] and CSF absorptive defects[3,4,21] have been considered as important factors. The servocontrolled-infusion data suggest that impaired CSF absorptive capacity may be of etiological importance.

7. Conclusions

In this chapter, two methods for determining CSF dynamics without steady-state requirements have been discussed. From tracer-clearance and manometric measurements, the recirculatory spinal perfusion technique provides individual values of the rates of CSF formation and absorption. The relatively invasive nature of the procedure limits its application to the animal laboratory. Non-steady-state measurements make this technique valuable to study labile pathological conditions and the effects of medications on the CSF compartment.

On the other hand, manometric determinations of CSF absorptive capacity with the servocontrolled lumbar infusion method have definite clinical application. However, the technique does not provide individual assessment of the rates of absorption or formation, but instead defines the arithmetic difference between these parameters as a linear function of ICP. Measurements of absorptive capacity are sensitive to changes in either CSF absorption or formation. Studies in patients may be indicated to help sort out difficult clinical problems such as distinguishing communicating hydrocephalus from hydrocephalus ex vacuo or arrested hydrocephalus from a slowly progressive active process. Obviously, the diagnostic value of this test will be defined only with extensive clinical experience.

References

1. Ashcroft, G. W., Dow, R. C., Moir, A. T. B.: The active transport of 5-hydroxyindol-3-ylacetic acid and 3-methoxy-4-hydroxy-phenylacetic acid from the recirculatory perfusion system of the cerebral ventricles of the unanesthetized dog. *J. Physiol. (London)* **199**:397–425, 1968.
2. Bedford, T. H. B.: The effect of variations in the subarachnoid pressure on venous pressure in the superior longitudinal sinus and in the torcular of the dog. *J. Physiol.* **101**:362–368, 1942.
3. Bercaw, B. K., Greer, M.: Transport of intrathecal ^{133}I RISA in benign intracranial hypertension. *Neurology (Minneapolis)* **20**:787–790, 1970.
4. Calabrese, V. P., Selhorst, J. B., Harbison, J. W.: Cerebrospinal fluid infusion test in pseudotumor cerebri. *Ann. Neurol.* **4**:173, 1978.
5. Cserr, H.: Potassium exchange between cerebrospinal fluid, plasma, and brain. *Am. J. Physiol.* **209**:1219–1226, 1965.
6. Cutler, R. W. P., Page, L., Galichich, J., Watters, G. V.: Formation and absorption of cerebrospinal fluid in man. *Brain* **91**:707–720, 1968.
7. Dandy, W. E.: Intracranial pressure without brain tumor. *Ann. Surg.* **106**:492–513, 1937.
8. Davson, H., Hollingsworth, G., Spegal, M. B.: The mechanism of drainage of the cerebrospinal fluid. *Brain* **93**:665–678, 1970.
9. Eisenberg, H. M., McComb, J. G., Lorenzo, A. V.: Cerebrospinal fluid over-production: Hydrocephalus associated with choroid plexus papilloma. *J. Neurosurg.* **40**:381–385, 1974.
10. Ekstedt, J.: CSF hydrodynamic studies in man. I. Method of constant pressure CSF infusion. *J. Neurol. Neurosurg. Psychiatry* **40**:105–119, 1977.
11. Ekstedt, J.: CSF hydrodynamic studies in man. II. Normal hydrodynamic variables related to CSF pressure and flow. *J. Neurol. Neurosurg. Psychiatry* **41**:345–353, 1978.
12. Foldes, F. F., Arrowood, J. G.: Changes in cerebrospinal fluid pressure under the influence of continuous subarachnoid infusion of normal saline. *J. Clin. Invest.* **27**:346–351, 1948.
13. Foltz, E. L., Shurtleff, D. B.: Five-year comparative study of hydrocephalus in children with and without operation (113 cases). *J. Neurosurg.* **20**:1064–1079, 1963.
14. Greenfield, J. C., Jr., Tindall, G. T.: Effect of acute increase in intracranial pressure on blood flow in the internal carotid artery of man. *J. Clin. Invest.* **44**:1343–1351, 1965.
15. Guinane, J. E.: An equivalent circuit analysis of cerebrospinal fluid hydrodynamics. *Am. J. Physiol.* **223**:425–430, 1972.
16. Hedges, T. R., Weinstein, J. D., Kassell, N., Stein, S.: Cerebrovascular responses to increased intracranial pressure. *J. Neurosurg.* **21**:292–297, 1964.

17. HUSSEY, F., SCHANZER, B., KATZMAN, R.: A simple constant-infusion manometric test for measurement of CSF absorption. II. Clinical studies. *Neurology (Minneapolis)* **20**:665–680, 1970.
18. KATZMAN, R., HUSSEY, F.: A simple constant-infusion manometric test for measurement of CSF absorption. I. Rationale and method. *Neurology (Minneapolis)* **20**:534–544, 1970.
19. LORENZO, A. V., BREFNAU, M. J., BARLOW, C. F.: Cerebrospinal fluid absorption deficit in normal pressure hydrocephalus. *Arch. Neurol.* **30**:387–393, 1974.
20. LORENZO, A. V., PAGE, L. K., WATTERS, G. V.: Relationship between cerebrospinal fluid formation, absorption, and pressure in human hydrocephalus. *Brain* **93**:679–692, 1970.
21. MARTINS, A. N.: Resistance to drainage of cerebrospinal fluid: Clinical measurement and significance. *J. Neurol. Neurosurg. Psychiatry* **36**:313–318, 1973.
22. MATHEW, N. T., MEYER, J. S., OTT, E. O.: Increased cerebral blood volume in benign intracranial hypertension. *Neurology (Minneapolis)* **25**:646–649, 1975.
23. NAGAI, H., IKEYAMA, A., FURUSE, M., MAEDA, S., BANNO, K., HASUO, M., KUCHIWAKI, H.: Effect of increased intracranial pressure on cerebral hemodynamics; observations of CBF, CBV, and pO2. *Eur. Neurol.* **8**:52–56, 1972.
24. PAPPENHEIMER, J. R., HEISEY, S. R., JORDAN, E. F., DOWNER, J. C.: Perfusion of the cerebral ventricular system in unanesthetized goats. *Am. J. Physiol.* **203**:763–774, 1962.
25. PORTNOY, H. D., CROISSANT, P. D.: A practical method for measuring hydrodynamics of cerebrospinal fluid. *Surg. Neurol.* **5**:273–277, 1976.
26. PORTNOY, H. D., CROISSANT, P. D.: Pre- and postoperative cerebrospinal fluid absorption studies in patients with meningomyelocele shunted for hydrocephalus. *Child's Brain* **4**:47–64, 1978.
27. RAICHLE, M. E., GRUBB, R. L., JR., PHELPS, M. E., GADO, M. H., CARONNA, J. J.: Cerebral hemodynamics and metabolism in pseudutumor cerebri. *Ann. Neurol.* **4**:104–111, 1978.
28. RISBERG, J., LUNDBERG, N., INGVAR, D. H.: Regional cerebral blood volume during acute transient rises of intracranial pressure (plateau waves). *J. Neurosurg.* **31**:303–310, 1969.
29. ROWAN, J. O., JOHNSTON, I. H., HARPER, A. M., JENNETT, W. B.: Perfusion pressure in intracranial hypertension. In Brock, M., Dietz, H., (eds.): *Intracranial Pressure.* New York/Heidelberg/Berlin, Springer-Verlag, 1972, pp. 165–170.
30. RUBEN, R. C., HENDERSON, E. S., OMMAYA, A. K., WALKER, M. D., RAL, D. P.: The production of cerebrospinal fluid in man and its modification by acetazolamide. *J. Neurosurg.* **25**:430–436, 1966.
31. SAHS, A. L., JOYNT, R. J.: Brain swelling of unknown cause. *Neurology (Minneapolis)* **6**:791–803, 1956.
32. SHAPIRO, H. M., LANGFITT, T. W., WEINSTEIN, J. D.: Compression of cerebral vessels by intracranial hypertension. II. Morphologic evidence for collapse of vessels. *Acta Neurochir.* **15**:223–233, 1966.
33. SHULMAN, K., YARNELL, P., RANSOHOFF, J.: Dural sinus pressure in normal and hydrocephalic dogs. *Arch. Neurol.* **10**:575–580, 1964.
34. SKLAR, F. H., BEYER, C. W., JR., RAMANATHAN, M., CLARK, W. K.: Servo-controlled lumbar infusions in children: A quantitative approach to the problem of arrested hydrocephalus. *J. Neurosurg.* **52**:87–98, 1980.
35. SKLAR, F. H., BEYER, C. W., JR., RAMANATHAN, M., COOPER, P. R., CLARK, W. K.: Cerebrospinal fluid dynamics in patients with pseudotumor cerebri. *Neurosurgery* **5**:208–216, 1979.
36. SKLAR, F. H., BEYER, C. W., JR., RAMANATHAN, M., ELASHVILI, I., COOPER, P. R., CLARK, W. K.: Servo-controlled lumbar infusions: A clinical tool for the determination of CSF dynamics as a function of pressure. *Neurosurgery* **3**:170–175, 1978.
37. SKLAR, F. H., ELASHVILI, I.: The pressure–volume function of brain elasticity: Physiological considerations and clinical applications, *J. Neurosurg.* **47**:670–679, 1977.
38. SKLAR, F. H., LONG, D. M.: Recirculatory spinal subarachnoid perfusions in dogs: A method for determining CSF dynamics under non-steady state conditions. *Neurosurgery* **1**:48–56, 1977.
39. WELCH, K.: The principles of physiology of the cerebrospinal fluid in relation to hydrocephalus including normal pressure hydrocephalus. *Adv. Neurol.* **13**:247–332, 1975.
40. WRIGHT, R. D.: Experimental observations on increased intracranial pressure. *Aust. N. Z. J. Surg.* **7**:215–235, 1938.

CHAPTER 29

Experimental Studies Relating to Diagnostic Imaging in Disorders of Cerebrospinal Fluid Circulation

A. Everette James, Jr., Gary R. Novak, Ernst-Peter Strecker, Barry B. Burns, Francisco Correa-Paz, and William J. Flor

1. Introduction

Much is known about the mechanisms of production and absorption of cerebrospinal fluid (CSF), but there remains little general agreement regarding the specifics of these physiological phenomena.[10] In this chapter, the authors will discuss CSF physiology from the bias of the diagnostic imaging studies employed in evaluating hydrocephalus and from a series of experiments utilizing animal models developed by our laboratory.[29,33,38,39]

In a simplistic consideration of a very complex subject, hydrocephalus results from an imbalance of CSF production and absorption.[21,26–28,30] Attendant to a continued circumstance of relative volume overproduction, the CSF spaces enlarge at the expense of brain substance. The progression of these processes results in cerebral ventricular dilatation and loss of brain substance or neuropil. As hydrocephalus develops, physiological and pathological changes are produced; attempts at compensation for this imbalance in CSF absorption and production occur. Utilizing a model that we developed for chronic communicating hydrocephalus, we have attempted to elucidate a number of these pathophysiological changes, document whether or not certian compensatory mechanisms are operative, and in so doing gain understanding of normal and abnormal CSF physiology.

Certain diagnostic imaging studies such as angiography and pneumoencephalography have provided excellent and useful anatomical information regarding the size of the intracranial CSF spaces. Radionuclide cisternography not only offered an opportunity to observe the anatomical features, but also, in a gross, semiquantitative manner, allowed simultaneous evaluation of CSF flow and absorption.[27,30,35,37] It was found that following injection of an appropriately labeled molecule, the movement of the radionuclide could be observed by serial imaging studies (usually with a scintillation camera) and the transfer from the CSF space to the intravascular compartment documented.[30,31,66,67] If the radiopharmaceutical was injected into the intrathecal space of the spine or in the cisterna magna, movement would be observed to the arachnoid area

A. Everette James, Jr., Sc.M., J.D., M.D., Gary R. Novak, L.A.T., and **Francisco Correa-Paz, M.D.** • Department of Radiology and Radiological Sciences, Vanderbilt University School of Medicine, Nashville, Tennessee 37232. **Ernst-Peter Strecker, M.D.** • Institut für Röntgendiagnostik der Universität Freiburg, Freiburg, West Germany. **Barry B. Burns, Ph.D.** • Department of Environmental Medicine, The Johns Hopkins University School of Hygiene and Public Health, Baltimore, Maryland 21205. **William J. Flor, Ph.D.** • Department of Experimental Pathology, Armed Forces Radiobiological Research Institute, National Naval Medical Center, Bethesda, Maryland 20014.

in the parasagittal region without apparent ventricular entry in the normal patient. In patients with significant ventricular enlargement, the radiopharmaceutical injected would enter the enlarged ventricular system. Additionally, delayed radiopharmaceutical movement over the cerebral convexity and delayed transfer to the intravascular compartment could be observed.

Computerized axial tomography (CT) has greatly changed the evaluation of patients with suspected CSF abnormalities.[45] The exquisite anatomical detail that can be achieved by this modality allows clear identification of early ventricular enlargement. Combined with the use of an appropriate water-soluble contrast medium injected into the CSF space, serial studies and blood-level determinations will provide significant clinical information regarding CSF dynamics of flow and transfer. Because of the inherent advantages of CT, this method has practically replaced pneumoencephalography and radionuclide cisternography as the initial diagnostic procedure in patients suspected of hydrocephalus or other abnormalities of CSF physiology.

Despite our ability to accurately identify patients with enlarged CSF spaces, pathological conditions impressing upon and obstructing CSF pathways, and abnormalities of flow and transfer into the intravascular compartment, we have no diagnostic test to properly select those patients who will benefit from a CSF diversionary shunt procedure. Greater understanding of the dynamics of CSF may allow development of such a test, but more importantly might permit development of a more physiological and less invasive procedure to treat CSF abnormalities.

We will attempt to relate the experimental findings presented in this chapter to the manifestations noted on diagnostic-imaging studies and to certain clinical and physiological phenomena observed in patients with chronic communicating hydrocephalus.

2. Experimental Model for Chronic Communicating Hydrocephalus

We have utilized dogs, cats, and nonhuman primates as animal models in producing chronic communicating hydrocephalus.[29] The animals are placed in the left lateral recumbent position under intravenous sodium pentothal or light inhalation anesthesia, with tracheal intubation. By use of the external occipital protuberance and wings of the atlas as anatomical landmarks, the middle of the cisterna magna can be identified and the overlying skin shaved, washed with an antiseptic solution, and draped. A longitudinal skin incision of 3–5 mm is made, and the midline is punctured with a 17-gauge needle the inside diameter of which will accept a 19-gauge polyethylene catheter. The needle is then angled laterally and superiorly to avoid the medulla and facilitate passage of the polyethylene catheter into the area of the basal cisterns (Fig. 1). This catheter manipulation may be performed under fluoroscopic control if the catheter tip has been rendered radiopaque with tantalum powder (Fansteel Metals Corp., Baltimore, Maryland). After clear CSF appears at the needle hub, the catheter is advanced into the area of the pontine and interpeduncular cisterns. Proper subarachnoid placement of the catheter can be monitored by radiography. The flow of CSF through the catheter and the CSF pressure are recorded by a pressure transducer and multichannel recording system (Hewlett Packard, Waltham, Massachusetts).

When the tip of the catheter is in the appropriate position, 1–2 ml of a silicone rubber material (Dow Corning Corp., Midland, Michigan) is injected into the basal subarachnoid space. This mixture contains 3 ml polysiloxane polymer, 3.5 ml dimethylpolysiloxane, 2 drops of the catalyst stannous octoate, and 1 ml powdered tantalum. The animal is then placed supine with the head dependent for 20–30 min to aid flow of the mixture anteriorly and superiorly.

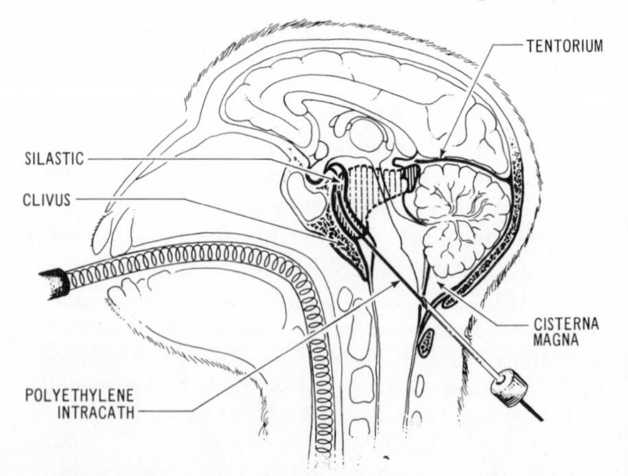

Figure 1. Drawing of animal model (rhesus monkey) for communicating hydrocephalus. Hatched area shows location of silicone rubber mixture at necropsy. Reproduced from James *et al.*,[29] with permission of the *Journal of Neurosurgery*, American Association of Neurological Surgeons, publishers.

In dogs, this flow to the parasagittal region is easily facilitated, but in nonhuman primates, location of the mixture on hardening in the communicating pathways (such as the ambient cisterns) is more common.

Serial radionuclide cisternograms computed tomographic images and measurements of transfer of the injected radiopharmaceutical from the subarachnoid space into the blood are made to monitor the development of communicating hydrocephalus, which requires 14–30 days in dogs and cats and 45–60 days in nonhuman primates.[26,29,33,48,62] Cisternograms are performed by injection of an appropriate radiopharmaceutical {150 μCi [^{131}I]- or 1 mCi [^{99m}Tc]serum albumin or 500 μCi [^{111}In]- or [^{169}Yb]diethylenetriaminepentaacetic acid (DTPA)} into the subarachnoid space of the cisterna magna. Lateral and vertex views are subsequently made at 30 min, 4 hr, and 24 hr using a scintillation camera.

The animals tolerate this procedure well. The silastic or silicone rubber mixture does not appear to cause any significant meningeal irritation if the puncture of the cisterna magna is atraumatic. Rarely, proximal localization of the injection is followed by rapid development of hydrocephalus with clinical signs of increased intracranial pressure. These symptoms will abate after several days of supportive therapy. The physical signs of the procedure will disappear in 7–10 days, and the neurological signs of developing hydrocephalus often become evident. These signs and symptoms include depression of mental alertness, and muscle incoordination, which are more dramatic in dogs than in nonhuman primates.

As hydrocephalus develops, the cisternograms reveal entry of the radiopharmaceutical into the ventricles. At first, there is evidence of distal subarachnoid radioactivity and ventricular "clearing" in the delayed views. However, as the ventricles continue to enlarge, they retain the radioactivity on the CSF images at 24 hr. Following the initial ventricular enlargement, the hydrocephalus progresses for several months and then appears to become "arrested." With the development of the chronic communicating hydrocephalus, measurement of transfer of radiopharmaceutical from the subarachnoid space into the blood reveals that this process is delayed throughout each time period.

Reconstruction of the clinical circumstance of chronic communicating hydrocephalus is important in any related experimental study. The most common pathological findings have been obliteration of the subarachnoid space in the basal cisterns, the communicating CSF pathways (such as the ambient cisterns), the tentorial region, or over the cerebral convexities and around the arachnoid villi.[70] These findings correlate well with reported pneumoencephalographic manifestations and appearances on CSF images (cisternograms). Although the clinical history in many patients with chronic communicating hydrocephalus is often incomplete, antecedent episodes of subarachnoid hemorrhage, meningitis, or trauma can frequently be elicited. It appears that the most common etiology for chronic communicating hydrocephalus is obliteration of the peripheral subarachnoid pathways from whatever cause. Silastic, introduced by an atraumatic procedure, produces this obstruction with minimal morbidity or mortality in the experimental animal. By employing the proper mixture of silicone rubber, diluting fluid, and catalyst, the required properties of flow and localization can be obtained. When a mixture is used that is viscous when injected but becomes solid in a predictable time period, obstruction of the ventricular outlets can be avoided. This, we believe, is a very desirable feature of this particular animal model.

Several animal models have been proposed to produce hydrocephalus.[9,18,33] Some cause such an inflammatory meningeal response that many animals die as a result of the procedure. In the animals that survive, histological and ultrastructural changes secondary to the inflammatory process are difficult to separate from those due to the communicating hydrocephalus. Other preparations have been reported to produce communicating hydrocephalus rarely; thus, the yield is too low to be experimentally useful. The ventricular outlets are obstructed in other preparations, and noncommunicating hydrocephalus results. None of these models duplicates the anatomical substrate of chronic communicating hydrocephalus as it is encountered clinically in humans.

The ventricles in our animal model progressively enlarge for 4–6 weeks in the dog and 6–12 weeks in the nonhuman primate before some type of compensatory level is reached. The observed temporal difference probably represents a relationship of the amount of injected silastic to the size of a peripheral subarachnoid space or to a species variation. This chronic model lends itself especially well to repeated study of physiological parameters in animals with developing communicating hydrocephalus. Correlation of cisternograms, transfer of radioactive labels or water-soluble contrast media from the subarachnoid space into the blood, appearance on serial CT studies, and CSF pressure measurements at different times are possible with this model.

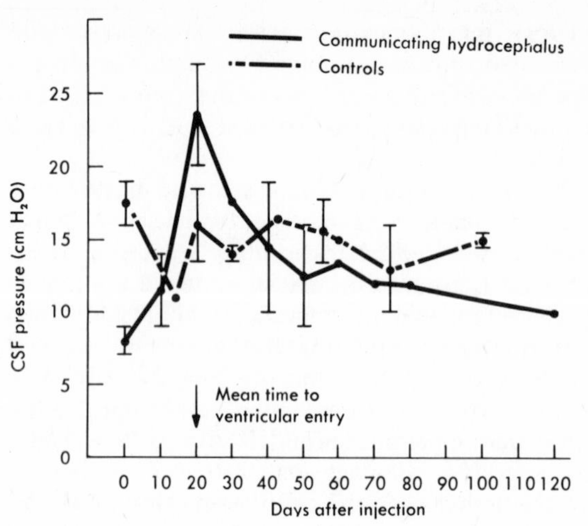

Figure 3. Graph comparing CSF pressure measurements in normal and hydrocephalic animals. *Abscissa:* days after injection of silastic into subarachnoid space to produce communicating hydrocephalus. Vertical brackets represent 95% confidence limit of mean. In animals with chronic communicating hydrocephalus, it is noted that prior to ventricular enlargement sufficient to cause radiopharmaceutical entry from lumbar intrathecal injection, pressure is increased. Following dilatation of ventricles to extent that radiopharmaceutical will enter ventricular system freely and remain there for protracted time periods, CSF pressure decreases, and after approximately 30 days, in dogs, enters normal range, where it remains. The final result is markedly enlarged ventricles with chronically normal CSF pressure. Reproduced from James *et al.*,[40] with permission of *Neurology*, Harcourt Brace Jovanovich Publications.

3. CSF Pressure Measurements

In normal animals, the CSF pressure measurements will be in the range of 6–21 cm H_2O on serial cisterna magna punctures.[48] Recordings made with animals intubated, positioned lying on the left side, and anesthetized with sodium pentothal revealed the findings in Fig. 2. The cisterna magna was punctured and correct placement of the needle assured by the presence of clear CSF, radiographic monitoring, and pressure tracings. Without loss of CSF, the needle is connected to a pressure transducer and recording system (Sanborn 350-1100 C carrier preamplifier and a Sanborn 7700 series recorder, Hewlett Packard, Waltham, Massachusetts). Validity of CSF pressure measurements is confirmed by the characteristic "arterial-like" waveforms that are believed to be induced by the pulsations of the choroid plexus and cerebral vessels. In the animals in which communicating hydrocephalus is induced, elevated CSF pressures were observed initially. Later, the CSF pressure measurements decreased and eventually returned to the normal range (Fig. 3).

Comparison of cisternograms with pressure measurements in animals with communicating hydrocephalus shows that the pressure begins to increase before ventricular radiopharmaceutical entry is detected on the cisternograms and continues to increase during the phase of ventricular entry and "clearing."[40] Pressures are significantly elevated ($P < 0.05$) and begin to decrease from the elevated state about the time that ventricular entry and "stasis" appear on the cisternograms. Eventually these pressures decrease into the normal range. The decline of CSF pressure could be correlated with the increased localization of radiopharmaceutical in the ventricles and the prolonged amount of time the radiopharmaceutical remained there. Utilizing indwelling ventricular catheters and subdural pressure transducers to monitor CSF pressure both intraventricularly and over the cerebral cortex, the same general findings are obtained.

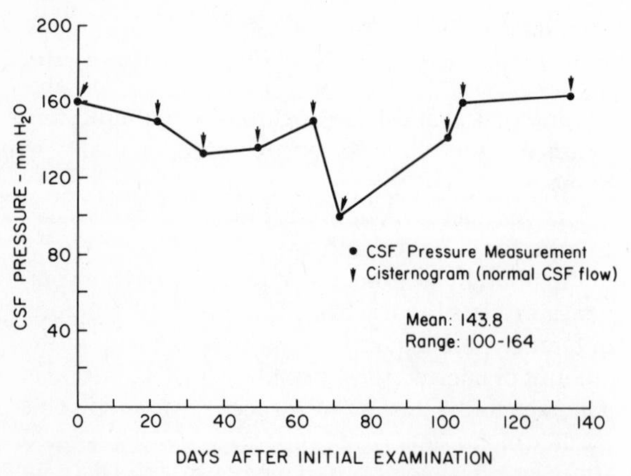

Figure 2. Graph demonstrating normal CSF pressure measurements at time of cisterna magna puncture. Reproduced from Novak *et al.*,[48] with permission of *Laboratory Animals*, Laboratory Animal Science Association, publishers.

A number of theories and mathematical expressions attempt to explain the findings of normal CSF pressure measurements and large cerebral ventricles. Although the CSF pressures are in the normal range, their effective force on the brain may be greater than the same pressure would exert with ventricles of normal size.[22,24] The forces act on the ependymal surface of the ventricles in a lateral or outward direction. Since the ventricular surface area is enlarged in hydrocephalus, the pressure is distributed over a large surface area and the effect of the outward force is increased. While these forces may be reflected by increase in CSF pressure peripherally around the cerebral cortex, there is purportedly a transient imbalance when the internal CSF pressure exceeds that in the peripheral subarachnoid space, and (at this time) the outward force is opposed by the rigid bony calvarium. During these periods of imbalance, which may be initiated by a reduced drainage capacity despite a normal production of CSF, the ventricles enlarge. This enlargement will produce an increase in cranial vault size if the sutures have not fused. If the cranial vault is rigid and resists expansion of intracranial contents, then the more compliant brain is compressed between the outward force of the CSF in the ventricles and the inner table of the bony calvarium. During this time period, the CSF pressure measurements are elevated. At some point, the ventricles enlarge enough that radiopharmaceutical injected into the subarachnoid space in the lumbar spine or cisterna magna will enter the ventricles and be retained there. The pressures then decrease and after some time enter the normal range. From these experimental observations, their pathological substrate, and clinical human diagnostic correlations, we have developed a concept of ventricular radiopharmaceutical entry.[27,30,37,41]

4. Response of CSF Pressure to Alterations in Arterial CO_2 Tension

The original description of "normal-pressure" hydrocephalus assumed that CSF pressure was within normal limits.[2,22,23] However, in these patients, CSF pressure measurements were performed for only a short duration of time. Recently, measurements performed over a duration of 48 hr revealed a nocturnal plateau in CSF pressure.[68] These intermittent CSF pressure elevations have been related to an improved prognosis after CSF diversionary shunt operations. Serial CSF monitoring is technically difficult, has a definite morbidity, and is subject to artifacts. For these reasons, we attempted to develop a rapid CSF-pressure-monitoring test that could provide accurate information about CSF pressure waves.

For these experiments, communicating hydrocephalus was induced in seven dogs (15–23 kg body weight), with a control group of five dogs of equal weight.

To monitor CSF-pressure response to alterations in CO_2 tension, general anesthesia was induced with intravenous injection of sodium pentothal. An endotracheal tube was inserted and respiration controlled by a mechanical respirator. Anesthesia was maintained with a mixture of O_2, NO, and alloferin (Hoffman–LaRoche, Inc., Nutley, New Jersey). A polyethylene catheter (2.0 mm I.D.) was inserted percutaneously through a femoral artery into the aorta to monitor arterial blood pressure and to obtain samples for blood gas determinations. A second polyethylene catheter (19-gauge) was inserted into the cisterna magna to measure CSF pressure. After regulation of the respirator to maintain normal blood gas values, arterial and CSF pressures were measured and recorded on a multichannel recorder. Arterial CO_2 tension ($Paco_2$) was then increased by disconnecting the CO_2 absorber of the closed anesthetic system, and CSF and arterial pressures were recorded simultaneously. Multiple blood samples for blood gas analysis were obtained. Alterations of the CSF pressure in response to changes in $Paco_2$ were compared in the control group and animals with communicating hydrocephalus.

The CSF pressures in all dogs ($Paco_2$ 40 torr with normal pH values) were recorded. CSF pressures were increased in hydrocephalic animals 1–2 weeks after silicone injection, but have been shown by other studies to return to normal limits when communicating hydrocephalus develops in a chronic time frame.[40] The response of CSF pressure to increasing $Paco_2$ in a control animal and an animal with hydrocephalus is demonstrated in Fig. 4. The control animals (Fig. 4, N) had a moderate increase in CSF pressure due to elevated $Paco_2$. The response of CSF pressure to elevated $Paco_2$ in hydrocephalic animals (Fig. 4, H) was significantly different since the CSF pressure shows a more rapid increase to a higher level than in the control animals. These CSF pressure elevations can be decreased to the normal range by lowering $Paco_2$ to 40 torr. By comparing the response of CSF pressure to increasing $Paco_2$, one can separate the control group and the group with hydrocephalus (Fig. 5). Animals with hydrocephalus respond with a dramatic pressure increase (Fig. 5, H). Between 30 and 70 torr $Paco_2$ there is a significant increase in CSF pressure with only slight changes in $Paco_2$. Analysis of the CSF pressure response reveals that the rate

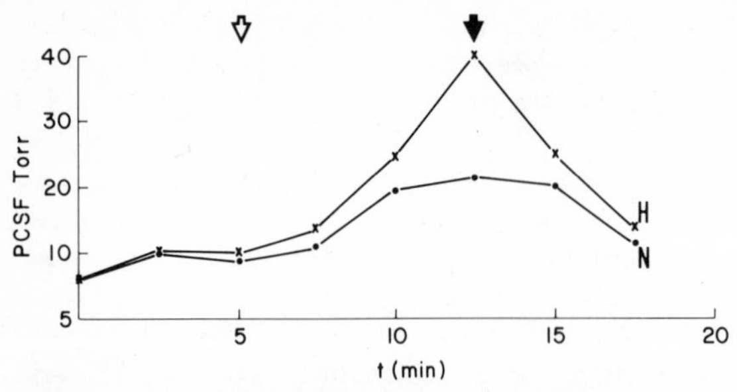

Figure 4. Graph demonstrating response of CSF pressure (PCSF) to increasing levels of arterial carbon dioxide tension in control animal (N) and animal with communicating hydrocephalus (H). CO_2 absorber was disconnected (⇩) when $Paco_2$ was 40 torr and response of CSF vs. time plotted. CO_2 absorber was reconnected (⬇) when $Paco_2$ was 64 torr. CSF pressure returned to normal range when $Paco_2$ was lowered to 40 torr.

of increase of CSF pressure per unit increase in $Paco_2$ is increased by a factor of approximately 3 in the hydrocephalic animals relative to the controls.

There was only moderate CSF pressure elevation in normal animals due to an increase of $Paco_2$ because the CSF volume can be redistributed under normal conditions. CSF can transfer into the intravascular space through the arachnoid villi under relatively low CSF pressure increase when there is an increase of cerebral blood volume.[16,52,57,71] In communicating hydrocephalus, the CSF transfer to the blood compartment is limited by obliteration of the normal transit and exit CSF pathways.[64,65] When cerebral blood volume is enlarged by vasodilatation, CSF pressure elevation probably occurs due to CSF redistribution through alternative pathways that very likely have a higher resistance than the arachnoid villi.

In patients with an abnormal CSF pressure elevation, response to increased $Paco_2$ elevation, and secondary enlargement of cerebral blood volume, there will be significant changes of cerebral blood flow and progressive loss of brain tissue. In these patients, a CSF diversionary shunt operation should be indicated and theoretically may alleviate the physiological process causing significant neurological symptoms. A disturbance of cerebral blood flow autoregulation is inferred when an increase in cerebral blood flow is required by the physiological state and no vasodilatation occurs. These circumstances have been described in patients with encephalopathies and chronic hydrocephalus.[49] In these patients, normal CSF pressure waves do not appear to occur and there will be no expected clinical improvement after a CSF shunt operation because of irreversible brain tissue damage. A diagnostic test *in vivo* to determine which patients have such profound changes may result in a more rational approach to those patients for definitive surgical procedures.

5. Pathological Studies

The gross pathology of these animals with chronic communicating hydrocephalus is quite remarkable.[33] There is marked enlargement of the lateral ventricles in all animals that have ventricular entry and stasis of radiopharmaceutical at cisternography (Fig. 6). These findings correlate well with the image manifestations using data from CT. Varied enlargement of the third and fourth ventricles and the aqueduct of Sylvius is noted on gross inspection. The silastic (silicone rubber) is usually localized in the anterior basal cisterns and rostrally. The outlets of

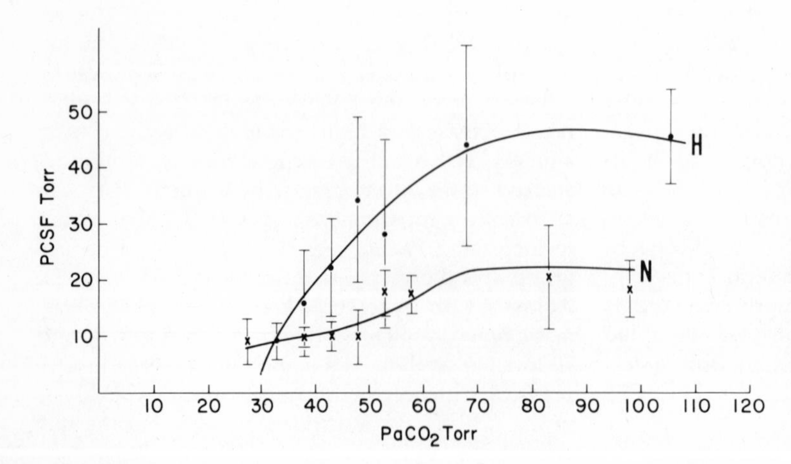

Figure 5. Graph demonstrating response of CSF pressure (PCSF) to increasing levels of arterial carbon dioxide tension ($Paco_2$) in control group (N) and group with communicating hydrocephalus (H). Two groups can be clearly separated by analyzing response to increasing $Paco_2$ and by characteristic shape of curves. Vertical brackets represent 95% confidence limits of mean.

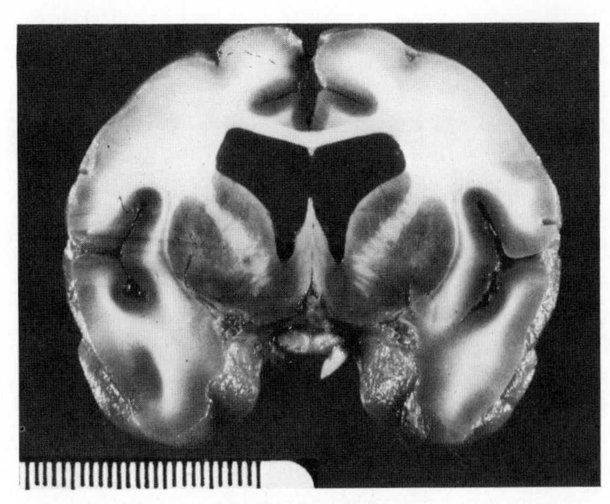

Figure 6. Gross coronal brain section from monkey with chronic experimental communicating hydrocephalus 990 days after silastic implantation. Note great enlargement of lateral ventricles and obliteration of septum in this animals. Scale is graduated in millimeters.

the fourth ventricle are patent and often enlarged to the extent that they can be readily identified.

We have studied a number of animals at varying times after implantation and in various phases of the disease for histopathological and ultrastructural changes. The primates we have studied to date can be divided into two general groups, one population at about 100 days postimplantation of silastic and one at approximately 1000 days postimplantation, both with ventricular radiopharmaceutical entry. From these animals taken as two distinct population groups, a number of observations can be made. The earliest gross pathological change is the rounding of the angles of the lateral ventricles, which can be noted on pneumoencephalography and CT and has been described as an early manifestation of hydrocephalus. As ventricular enlargement increases, the contours become even more rounded. Histologically, the ependyma near the angles becomes stretched, flattened, and ultimately denuded with what appears to be a loss of the integrity of the ependymal lining at the most severely affected vertex of the dorsolateral angles.[8,9,20,27,31]

In the 100-day animals, the ependymal cells are lost in the region immediately adjacent to the dorsolateral angles, while normal ependyma remains further from the angles (Fig. 7). There is an increase in extracellular fluid space in the adjacent periventricular tissue, especially in the periventricular white matter, and this fluid is in communication with the CSF of the ventricular lumen (Fig. 8). An increase in the perivascular space of small vessels in the periventricular region is also observed. This appearance correlates well with the decreased attenuation noted on CT. The excess fluid appears to dissect along white matter tracts in the parenchyma and along vessels by following and/or splitting the basement membrane. In general, astrocyte endfeet along basement membranes retain their close apposition despite the presence of significant quantities of fluid in the subjacent parenchyma. Vascular endothelial cells also appear generally to maintain tightly apposed junctions.[7,56]

Evidence of periventricular white matter degeneration changes is present to some extent in the 100-day animals. In addition to the generally edematous nature of the tissue, axonal destruction is observed and reactive astrocytes are present. Degenerating

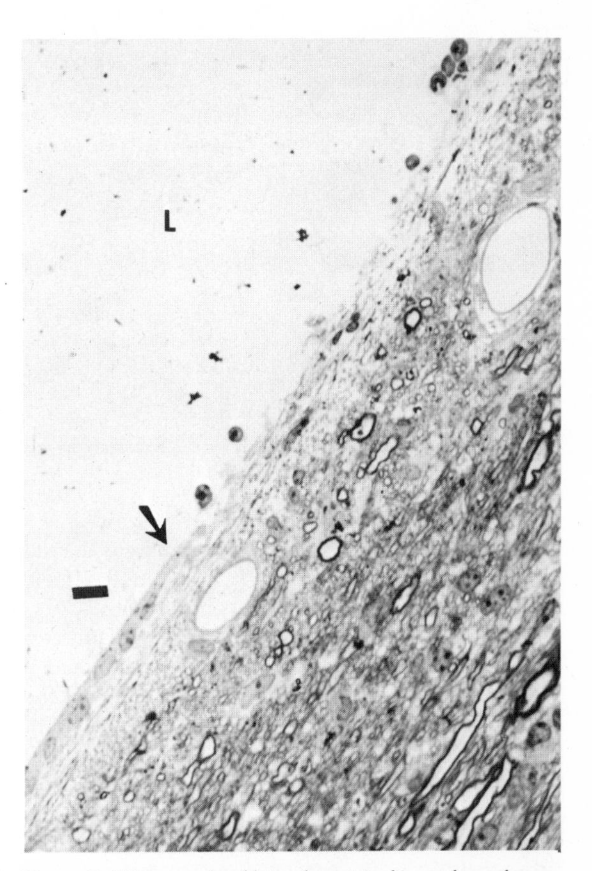

Figure 7. Micrograph of lateral ventricular surface of monkey 83 days after silastic implantation. This is transition region between normal ciliated ependymal lining (below arrow) and denuded surface (above arrow). Increased parenchymal extracellular space is evidence by lightened subependymal zone. Dorsolateral angle is off figure to top. (L) Ventricular lumen. Scale bar: 10 μm.

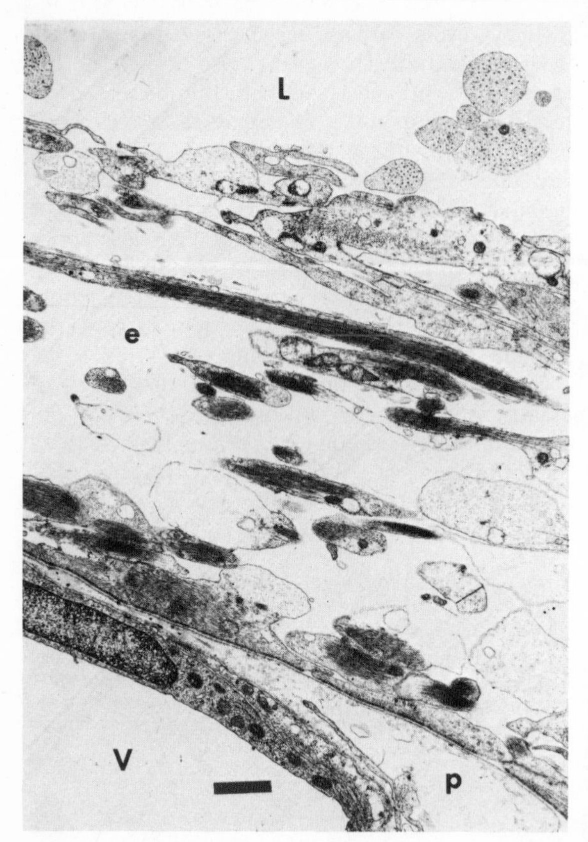

Figure 8. Electron micrograph of denuded zone of ventricular surface of animal in Fig. 7. (L) Ventricular lumen; (e) increased parenchymal extracellular space; (V) vascular lumen; (p) increased perivascular fluid space. Loose organization of surface elements and abundant parenchymal extracellular fluid are evident. Scale bar: 1 μm.

profiles of corticospinal tract fibers can be observed in spinal cord sections (Fig. 9). These motor fibers originate in cerebral cortical neurons and must pass in close proximity to the expanding lateral ventricles as they course in the direction of the spinal cord.[73] Small vacuolar changes have been observed in the cells of the choroid plexus, but the significance, if any, of these alterations is not fully understood at this time. The possible implications of these histological and ultrastructural changes with respect to choroid plexus function will be considered subsequently.

The 1000-day specimens exhibit more extensive changes of the types described above. In these animals, the ventricular system is grossly enlarged, primarily at the expense of periventricular white matter. Large areas of ventricular lining on either side of the dorsolateral angles of the lateral ventricles are devoid of normal ciliated ependyma. However, processes of ependymal or astrocytic origin, or both, now cover much of this surface (Fig. 10). Many areas of increased extracellular space that appear to communicate with the ventricular lumen are still observable in this region. Prominent fiber loss is evident in edematous parenchyma. The motor tract fiber loss is clearly present in both the lateral (crossed) and ventral (uncrossed) corticospinal tracts (Fig. 11) and is traceable throughout the spinal cord. Many spinal motor neurons show increased pigmentation and clumping of Nissl material consistent with postchromatolytic changes.[54] Collateral tracts to the cerebellum also contain evidence of fiber degeneration.

We have not observed enlargement of the central canal of the spinal cord in any animals (Fig. 12).[32] Since the outlets of the fourth ventricle are patent and enlarged in this model, there seems to be no reason for development of central canal dilatation.

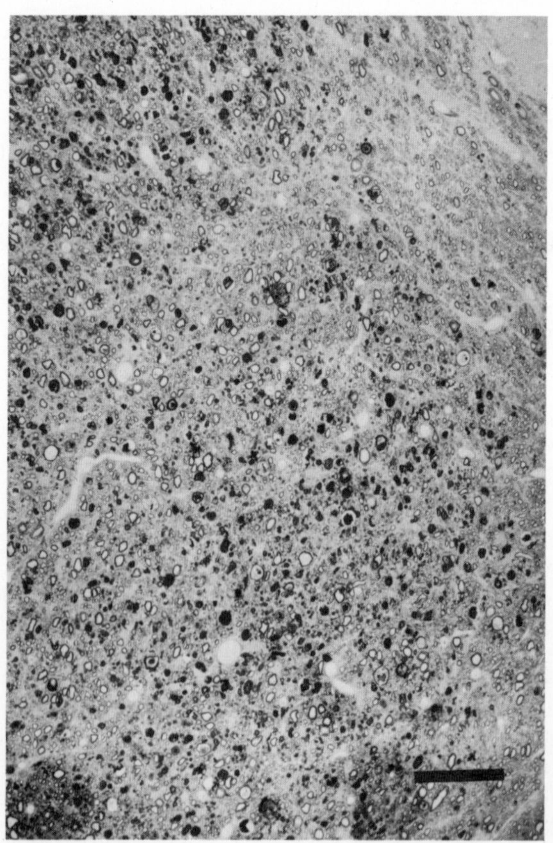

Figure 9. Light micrograph of lateral corticospinal tract in animal of Fig. 7, showing degenerating axonal profiles (dark dots). Scale bar: 100 μm.

Figure 10. Electron micrograph of ventricular surface of animal in Fig. 6. (L) Ventricular lumen; (e) increased parenchymal extracellular space. Filamentous processes of astrocytic or ependymal origin, or both, have proliferated to "scar" over tissue surface, but increased fluid in extracellular space is still evident. Scale bar: 1 μm.

This apparently occurs in kaolin-induced hydrocephalus,[18] in which these outlets are obstructed, leaving the central canal connected to a closed and enlarging ventricular system. The changes observed in our animal model clearly reflect the kind of underlying neuroanatomical alterations that would be responsible for the constellation of clinical signs observed in patients with communicating hydrocephalus.[29]

6. Molecular-Transfer Analysis

As we have discussed, diagnostic images obtained after injection of radioactively labeled macromolecules into the subarachnoid space have been used to characterize the various forms of hydrocephalus.[16,17,30,35,53] One might well increase the accuracy and sensitivity of these studies by quantitative measurement of the movement of these radiopharmaceuticals into the intravascular compartment as well as by monitoring the rate of transfer of these labeled molecules from the subarachnoid space.[1,63,66,67]

For a more accurate assessment of these phenomena, we have applied mathematical analysis of the CSF movement and transfer following subarachnoid injection of radiopharamceuticals.[66] This methodology should reflect the dynamics of molecular movement across the CSF–brain–blood interface. The mongrel dogs used in this study were divided into two groups: eight animals with chronic communicating hydrocephalus and an equal population of normal controls. Hydrocephalus was induced by the method previously described[29,39] (see Section 2).

The development of communicating hydrocephalus in these animals was monitored by serial cisternograms utilizing ^{99m}Tc-labeled serum albumin. CSF pressure was measured serially at separate time intervals in the control animals as well as in those after development of chronic communicating hydrocephalus. The CSF pressure was found to be within normal limits at the time of measurement.

To assess transfer from the subarachnoid space into the blood, each dog was given an intracisternal injection of 1–2 mCi in 1 ml ^{99m}Tc-labeled serum albumin (10 mg serum albumin/ml radiopharmaceutical) and blood samples obtained at varying time intervals up to 26 hr after intrathecal injection of the radiopharmaceutical. Radioactivity of all collected blood plasma samples and a standard solution was measured in a scintillation well counter. Counts of 1 ml of each plasma sample were compared with those of 1 ml of a 1 : 1000 dilution of a ^{99m}Tc standard. By determination of the difference in radioactivity before and after injection, the amount of radiophar-

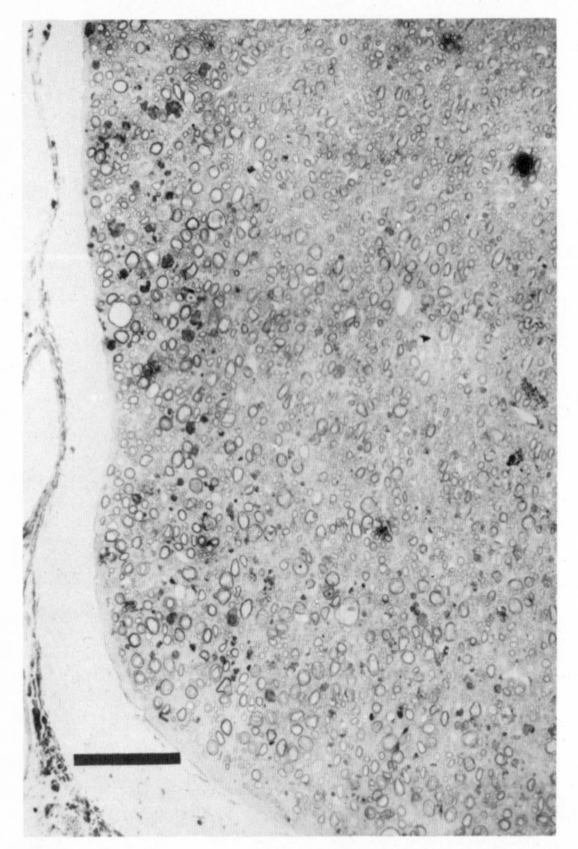

Figure 11. Light micrograph of the ventral corticospinal tract of an animal from the 1000-day group, showing degenerating axonal processes (dark profiles in the left half of the figure). Scale bar: 100 μm.

maceutical injected intrathecally and into the dilution bottle was calculated. At the time of counting, the samples were normalized taking the standard as 100%. This served to correct for radioactive decay during the time that elapsed between intracisternal injection and sample counting. Plasma radioactivity was defined as the amount of radioactivity that would have been present in the total circulating plasma if the radiopharmaceutical had been administered intravenously. Total plasma volume (TPV) was calculated by a method previously described.[66]

To accurately calculate the actual radiopharmaceutical transport from subarachnoid space to blood from the true measured radiopharmaceutical blood concentrations, the rate of disappearance of the radioactive labeled albumin from the circulatory system into other compartments (such as liver, kidneys, lymphatic system, and extracellular space) was measured. The measured blood concentration (which represents transfer from the CSF) was corrected for the disappearance of the radiopharmaceutical from the blood to other compartments at 2-hr intervals for both groups. Sequentially, each concentration value for total transferred radiopharmaceutical was considered to be the sum of the corrected value of the preceding time interval and the amount of radiopharmaceutical that transferred from the CSF to the plasma during the following interval. These calculations were performed from the time of injection for 26–28 hr afterward.

The sum of the values of radiopharmaceutical that transferred from the subarachnoid space into the intravascular compartment from time zero (injection) until 26–28 hr after cisterna magna injection was plotted for all animal groups on linear paper and arithmetic probability paper (National 12-083). These derived functions were used to calculate the mean transit time ($\bar{t}$) of the labeled albumin from

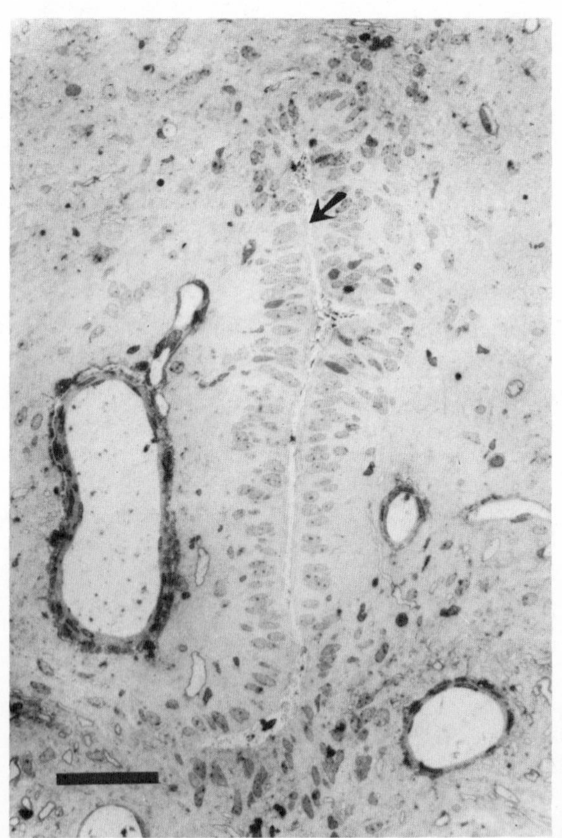

Figure 12. Light micrograph of the central canal (arrow) of the cervical cord of a dog with communicating hydrocephalus 37 days after silastic implantation, showing no dilatation of the canal. Scale bar: 50 μm. Reproduced from James *et al.*,[32] with permission of the *Journal of Neurosurgery*, American Association of Neurological Surgeons, publishers.

the site of injection to the exit from the CSF system. By employment of the method described by Zierler,[74] the mean transit time was calculated. The mean transit time is determined by the fraction of the area under the reciprocal of the blood accumulation function and the zero time radioactivity in the CSF (Fig. 13):

$$\bar{t} = \frac{\text{area under curve}}{\text{peak (or zero line) concentration}} \qquad (1)$$

The mean transit time through the system in this context is considered to be the ratio of the volume in which the indicator is distributed to the flow from the system. From this value, one obtains by equation

$$\bar{t} = \frac{V}{F} \text{ or } \frac{F}{V} = \frac{I}{\bar{t}} \qquad (2)$$

a flow-to-volume ratio of radiopharmaceutical leaving the CSF space to the volume of the radiopharmaceutical distribution space.

In the circumstance of equal dispersion of labeled albumin within the CSF, these calculations provide information regarding the relationship of CSF flow entering the circulation of the CSF distribution space:

$$\frac{F}{V} = \frac{I}{\bar{t}} \qquad (3)$$

In normal dogs, the ventricular CSF volume was not included in the radiopharmaceutical space because there was no evidence of radiopharmaceutical entry into the ventricular system. In animals with communicating hydrocephalus, the ventricular volume was considered as part of the radiopharmaceutical distribution space because the ventricles are clearly delineated on the cisternographic images because of the radiopharmaceutical they contained. Multiple scintillation camera images from different views have provided sufficient anatomical resolution to identify all regions of the ventricular system. Therefore, the entire ventricular volume was included.

The observed circulating blood levels of [^{99m}Tc]serum albumin were plotted against sampling time for all groups and the blood levels recorded as percentages of injected activity:

$$\frac{\text{Total activity circulating in blood}}{\text{Total activity injected into CSF}} \times 100 \qquad (4)$$

Certain features of these curves indicate that the animals with hydrocephalus may be distinguished from the controls on the basis of their transfer kinetics. Normal animals reach maximum concentration at a different time than those with chronic communicating hydrocephalus. The apparent transfer of radiopharmaceutical, not corrected for disappearance from the vascular circulation after intracisternal injection, is shown by curves of the mean plasma concentration (as a percentage of the injected dose) of the radiopharmaceutical plotted against elapsed time after injection (in hours) (Fig. 14).

The blood may be regarded as the end organ of CSF blood transfer at ^{99m}Tc-labeled albumin if the measured radiopharmaceutical blood concentration is corrected for passage of the radiopharmaceutical from the circulation into other compartments. After this correction, one can obtain the amount of labeled albumin transferred. In the control group, the maximum rate of transfer occurred 10 hr after subarachnoid injection and was 9.7%/2 hr. The radiopharmaceutical accumulation curve demonstrates that after 13.6 hr, 50%, and after 24 hr, 85%, of the injected radiopharmaceutical exited from the CSF compartment in the control animals. In dogs with chronic communicating hydrocephalus, transfer of labeled albumin from the CSF to the blood was slower than in normals. The accumulation curve in the first 3 hr showed only a slight upward slope; the maximum slope occurred between 4 and 8 hr. The

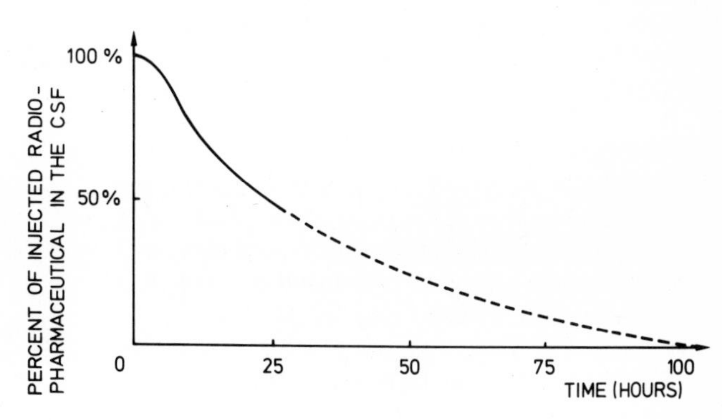

Figure 13. Graph of reversed or reciprocal radiopharmaceutical blood accumulation function representing amount remaining in CSF after intracisternal injection in dogs with chronic communicating hydrocephalus. Area under this curve divided by peak concentration (100%) gives mean transit time ($\bar{t}$) of radiopharmaceutical through CSF system. Reproduced from Strecker *et al.*,[66] with permission of *European Neurology*, S. Karger, publisher.

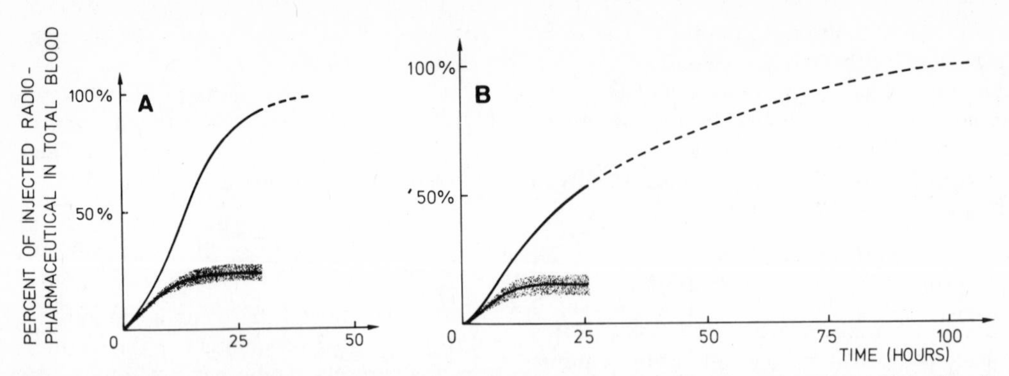

Figure 14. Graph of measured radiopharmaceutical blood concentration after intracisternal injection in (A) control dogs and (B) dogs with chronic communicating hydrocephalus, represented by lower shaded (± S.D.) curves. Corrected transfer is represented by upper curves. Reproduced from Strecker *et al.*,[66] with permission of *European Neurology*, S. Karger, publisher.

maximum amount of radiopharmaceutical that was transferred during any 2-hr time interval was only 6.8%; this peak of transfer occurred 8 hr after injection. The time required for 50% of the labeled albumin injected into the CSF to be transferred into the circulation was 24 hr.

The mean transit time ($\bar{t}$) of the radiopharmaceutical through the CSF spaces and the ratios of CSF flow from CSF space per CSF distribution space showed the characteristics of transfer in these two different physiological circumstances. The $\bar{t}$ in dogs with communicating hydrocephalus was more than twice that of controls. This would suggest that radiopharmaceutical flow from the CSF space (ml/hr) to intravascular space is less than one half of that found in control dogs or that the volume of the CSF compartment is twice that of normals (Table 1).

The radiopharmaceutical accumulation functions can be regarded as cumulative relative frequency distributions. When the values of these accumulation functions are plotted on standard probability graph paper, the degree to which all points lie on a straight line on this type of graph paper determines the closeness of fit of a given distribution to a normal distribution.[61] The data of the control group form a straight line on standard probability paper during the period of 12–26 hr post-intracisternal radiopharmaceutical injection (Fig. 15). This period was preceded by a curve that ascended in a convex fashion. In dogs with communicating hydrocephalus, a curve of similar contour was obtained that formed a straight line between 16 and 26 hr (Fig. 15). A line with shallow slope was preceded by a curved portion between 0 and 16 hr after radiopharmaceutical injection. The ascending curved part of these two functions represents the time from injection until the radiopharmaceutical is homogeneously distributed in the CSF space. This process appears to require approximately 12 hr in normal dogs and 16 hr in hydrocephalic animals. The following straight lines in both functions reflect equal distribution of the radiopharmaceutical within the total CSF distribution space. Analysis of these data allowed characterization of CSF-to-blood transfer in normals and animals with communicating hydrocephalus.

Table 1. Mean Transit Time of Radiopharmaceutical through CSF Space and CSF Flow/Volume Ratios in Control Animals and Dogs with Chronic Communicating Hydrocephalus[a]

Animals	$\bar{t}$ (hr)	$\frac{I}{\bar{t}} = \frac{\text{ml/hr}}{\text{ml}} \left(\frac{\text{flow from CSF space}}{\text{CSF space}}\right)$
Control animals	14.6	0.69
Chronic communicating hydrocephalus	32.4	0.31

[a] Reproduced from Strecker *et al.*,[66] with permission of *European Neurology*, S. Karger, publisher, Switzerland.

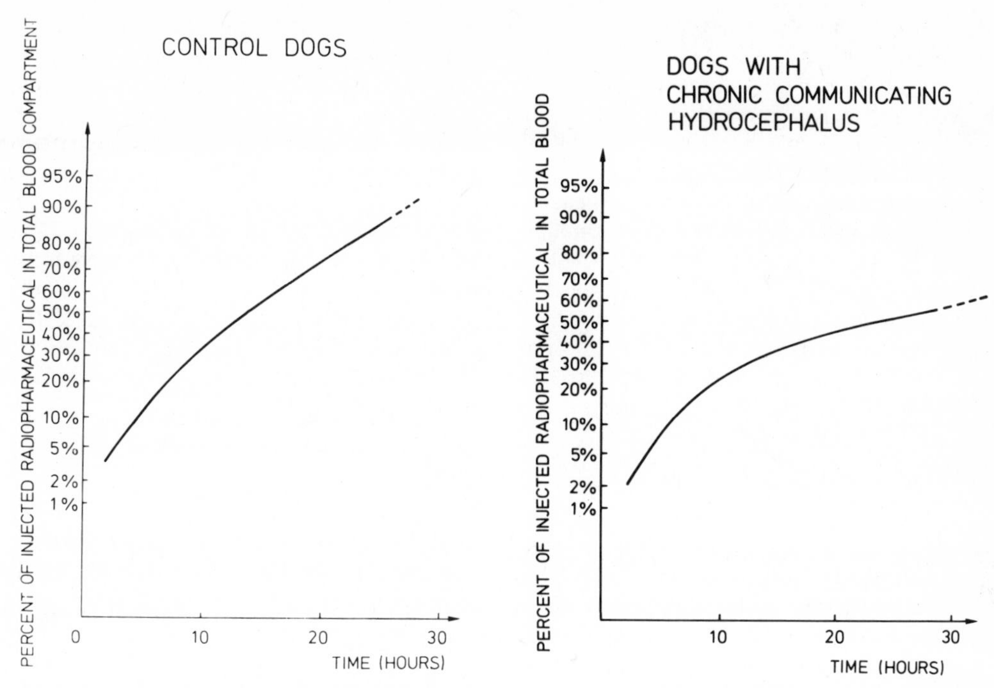

Figure 15. Graph of radiopharmaceutical accumulation function plotted on probability paper. Data partially fit straight line with characteristic slope for control dogs and dogs with chronic communicating hydrocephalus. Reproduced from Strecker *et al.*,[66] with permission of *European Neurology*, S. Karger, publisher.

The degree of slope of the accumulation frequency functions, the slope of the straight lines on the probability paper, and the radiopharmaceutical-transfer-vs.-time curves in our studies demonstrated that there are individual frequency functions of transit time for each cumulative hour. The acute slope of the radiopharmaceutical accumulation curve and the short mean transit time reflect a small radiopharmaceutical distribution space in normal animals. In hydrocephalic animals, the transfer time is more than twice as long as in the control group, which probably reflects the fact that the total radiopharmaceutical space is more than doubled. This space represents several anatomical compartments in animals with communicating hydrocephalus:

1. The extraventricular subarachnoid space of the spine and cranium (some subarachnoid cisterns are obliterated by the injected silastic).
2. The considerably enlarged ventricular space.
3. The altered enlarged extracellular space of the brain, especially the periventricular tissue.

The data from these experiments suggest that it is possible to determine clearance functions, mean transit times of the radiopharmaceutical through the CSF system, and CSF absorption/CSF distribution space ratios in normals and animals with chronic communicating hydrocephalus. It also suggests that this method may be used as an adjunct to cisternographic image interpretation in the diagnosis of communicating hydrocephalus, since an analysis of this type may prove sensitive in the early detection of CSF absorptive abnormalities. Since this method is easily performed in combination with cisternography and provides data regarding the CSF and transfer dynamics, its use in difficult clinical situations should be considered.

7. CSF Exit-Pathway Size

Studies have resulted in conflicting data regarding the mechanism by which macromolecules move from the CSF into the intravascular compartment.[67,69,72] Structurally, the arachnoid villi appear to represent a circuitous pathway that provides direct channels from the CSF space into the intravascular compartment.[60,69,72] Certain electron-microscopic studies have revealed large tortuous channels 4–12 μm in size,[72] while others have failed to demonstrate spaces of this magnitude.[60]

Ultrastructural studies employing histochemical markers have led to two concepts by which this movement of macromolecules across the "CSF–blood barrier" occurs. Tripathi[69] has shown vacuoles in direct communication with the CSF of the subarachnoid space that may be responsible for formation of a dynamic system of transcellular pores. The formation of micropinocytotic transport vesicles is believed by others to be the mechanism by which macromolecules are transferred from the CSF into the blood.[6,7,42] These observations would correlate with the persistence of radioactivity in the parasagittal area on delayed views of the cisternogram especially when labeled albumin (mol. wt. 69,000) is employed as the radiopharmaceutical (Fig. 16).

To study the dynamics of macromolecular transfer from the CSF to blood *in vivo,* we employed radioactive labels of different molecular size in an animal model. We believed this method could accurately determine effective CSF pathway size *in vivo,* which might imply the mechanism of transfer and would provide important information regarding CSF clearance.

New Zealand white rabbits (1.7–3.1 kg) were chosen because of availability and familiarity with studies on the passage of certain substances out of the CSF in this animal.[11–13] A total of 16 animals were utilized, 4 animals in each group, the groups being divided according to the size of the labeled molecule. The average for each group provided the data shown in the graphs. Rabbits were anesthetized with urethane supplemented when necessary by pentobarbital.

The animal's skull, anterior neck, and a small area on each side of the abdomen were shaved. Arterial catheterization and a tracheostomy were performed. The kidneys were exteriorized and the renal artery, vein, and ureter ligated to prevent radiopharmaceutical removal by this route (Fig. 17). Potassium perchlorate (200 mg) was given intravenously and infused with a standard artificial CSF to diminish choroid plexus and glandular uptake of the technetium ion. Bilateral ventricular infusion was accomplished by the method of Pollay and Davson.[51] The infusion apparatus was that described by Davson and Purvis[14] (Fig. 17). By use of a simultaneous reference marker (^{99m}Tc- or ^{111}In-labeled DTPA), transfer of labeled macromolecules from CSF to blood was measured for the following: (1) [^{14}C]dextran (5 and 8 nm); (2) [^{3}H] dextran (5 and 8 nm); (3) [^{99m}Tc]serum albumin "mini" microspheres (approximately 1.0 μm) (3M Corp., St. Paul, Minnesota); (4) [^{99m}Tc]- or [^{113}In]serum albumin microspheres (15–20 μm) (3M Corp.).

The infusion rate (0.17 ml/min) was regulated to maintain CSF pressure within normal range (10–180 mm H_2O) at approximately 140 mm H_2O. CSF pressure measurements were obtained every 5 min until they became stable (±10 mm H_2O on consecutive measurements).[14] Serial blood samples were collected from a peripheral vein commencing 15 min after initiation of infusion and continued for 1–4 hr, depending on known radiopharmaceutical stability tested *in vitro* and *in vivo* by other techniques.

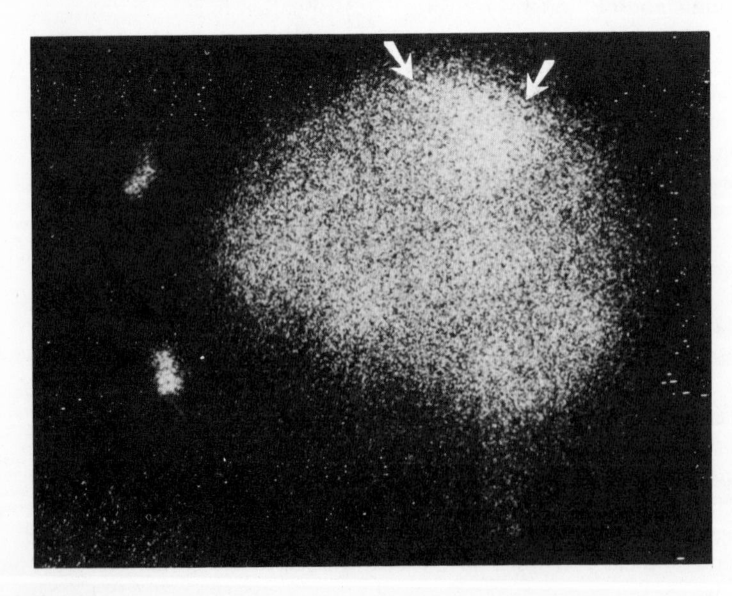

Figure 16. Left lateral cisternogram in human following intrathecal injection of 100 μCi [^{131}I]human serum albumin. Twenty-four-hour view shows persistence of radioactivity in the parasagittal area (arrows).

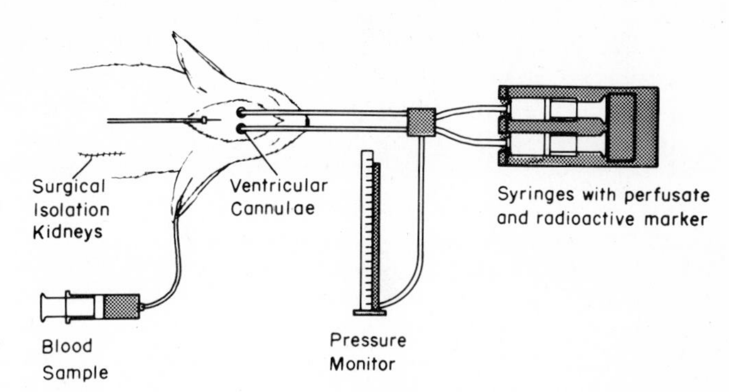

Figure 17. Drawing of experimental model for measurement of CSF drainage pathways.

The animals were euthanized and tissue specimens obtained to determine radiopharmaceutical location and distribution. Tissue specimens also served as an internal radiopharmaceutical quality control. Intravenous injections of radiopharmaceuticals were made in control animals and the disappearance rate and tissue distribution studies obtained.

Samples were prepared and counted in a well scintillation counter or an autoanalyzer. With the use of a standard prepared from a count of that radioactivity present in the CSF infusate, the plasma samples were measured against the standard. The values were expressed as percentage transfer of each time period in which a blood sample was obtained. Tissue samples were similarly counted and recorded as a single measurement.

The two dextran molecules transferred readily across the CSF–blood barrier, in the time period studied, at a rate and in an amount not significantly different from those of the reference marker (Fig. 18). The small albumin microspheres (1.0-μm "mini" microspheres) did not transfer significantly into the blood for the time period measured (Fig. 19). Tissue samples showed no increased accumulation in organs containing a large number of reticuloendothelial system (RES) cells. When the radiopharmaceutical was injected intravenously, the organ localization demonstrated marked accumulation in the RES cells. The large albumin microspheres (15–20 μm) did not transfer into the blood (Fig. 20), nor was entrapment in the lung capillaries present. With intravenous injection of the 15–20 μm albumin microspheres, localization to the lung capillaries was seen.

The use of labeled particles to measure effective size or pathways or intercellular channels has been employed in electron microscopy and by many investigators to measure arteriovenous communications. One objection to this method is that the particles, due to either surface configuration, charge, or other properties, may form aggregates larger than their individual effective diameters. With the labeled molecules and particles employed, these properties had been determined or were evaluated by us. Labeled dextran has been used extensively and found to be a stable radiopharmaceutical for long periods of time *in vivo* and *in vitro*.[13]

Serial blood samples were taken for only that period of time for which the radiopharmaceutical was known to be biologically stable. Certain radiopharmaceuticals were repeatedly washed of "free" radioactive label. Microscopic studies, pore filtration, and column chromatography were used to document radiopharmaceutical size and stability. With

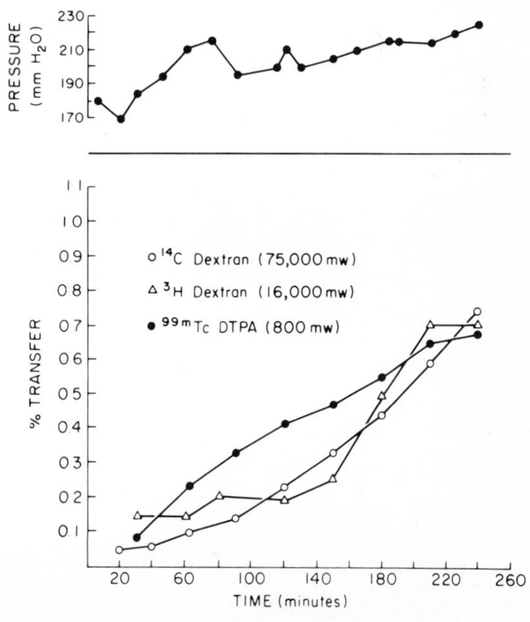

Figure 18. Graph of measured transfer of radioactive label from subarachnoid space into blood for 4 hr after start of ventricular perfusion. Pressure recording during same time period is illustrated.

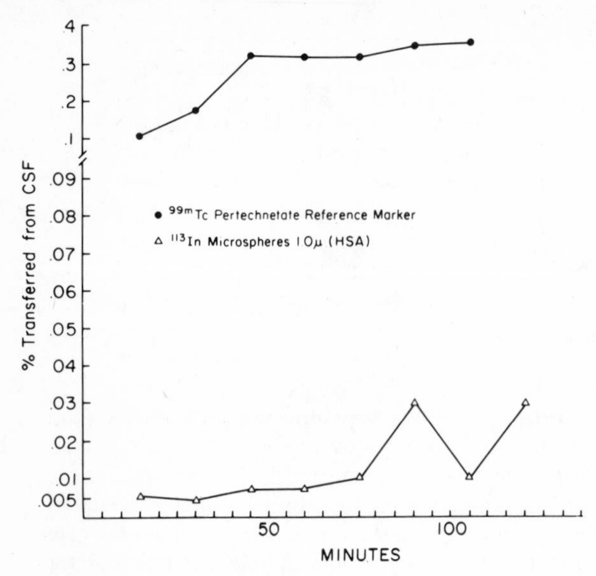

Figure 19. Graph of measured transfer of 1.0-μm labeled albumin particles from CSF to blood.

the larger particles, the infusion syringes were agitated at frequent time intervals to prevent settling of the larger labeled particles.

Separation of the radioactive label from the molecule or microspheres would result in spuriously high radioactivity values in the blood. If significant separation of the label from the molecule or microspheres was known to occur, repeated washing, counting of the radioactivity remaining, and appropriate timing of the study were utilized to minimize this potential source of error. However, our results

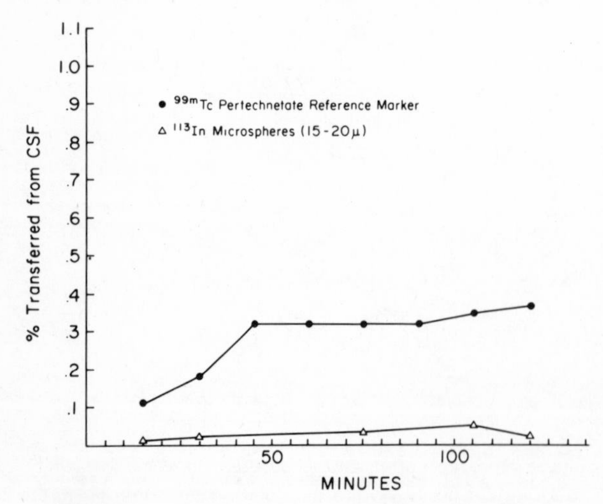

Figure 20. Graph of measured transfer of 15–20 μm albumin particles from CSF to blood.

suggested less transfer than expected, reducing the probability that radiopharmaceutical instability was an important factor.

A labeled molecule intraventricularly injected as a bolus will, in a matter of minutes, be homogeneously distributed throughout the ventricle. Movement from the ventricle to the remainder of the subarachnoid space also occurs rapidly.[19,26,55,64] Ventricular infusion was constant to prevent artifactually low values due to uneven radiopharmaceutical distribution or failure to reach CSF drainage sites. Settling of particles in a dependent portion of the CSF compartment was a potential source of error that was not entirely excluded by this technique. In animals with larger CSF compartments, the error this may cause would be minimized. The infusion rate was regulated to keep the CSF pressure in the normal range.[34] In recognition that the CSF–transfer rate between normal animals was likely to vary, a reference marker was used.[12,13] This rate was found to be relatively constant, so that comparison among animals of the relative transfer was possible.

Although certain ultrastructural studies show the CSF drainage pathways to be 4–12 μm in size, our studies in the rabbit do not support this. These data would suggest that the limiting size in the rabbit is less than 1.0 μm. This experimental model can be utilized in a more appropriate animal such as the nonhuman primate. Additionally, more specific documentation can be made by employing different-sized labeled molecules and particles.

A number of compensatory mechanisms relating CSF overproduction and abnormalities of arachnoid transfer and subsequent CSF absorption have been proposed. One mechanism may well be development or utilization of pathways of CSF drainage or absorption in hydrocephalus that are not operative in the normal state. A number of alternative pathways of drainage have been considered, including (1) increased movement down the central canal and the subarachnoid space of the spinal cord and absorption in the arachnoid granulations[17,18] and (2) transependymal movement of CSF into the periventricular space and eventual removal from that site.[20,27,30,41,47,58,63] Neither of these mechanisms has been established as the proper explanation, and other possibilities certainly exist.

8. Autoradiography

Using the radiopharmaceuticals injected for CSF imaging studies, we were able to correlate autoradiographic findings (gross, histological, and semiquantitative) with the histological and ultrastructural appearance to consider these two potential alternative pathways.[25,26,31,32,58] Investigations using normal animals or animals with noncommunicating hydrocephalus have demonstrated transependymal movement of high-molecular-weight substances consistent with diffusion through the periventricular extracellular space of the brain.[19,25,26,55,58]

We have demonstrated transependymal bulk flow of these large molecules in animals with chronic communicating hydrocephalus. Autoradiography was employed to trace pathways of altered CSF movement and to compare them with normal pathways. Iodine-131-labeled serum albumin (150–300 μCi) was initially used for the autoradiographic studies. Before intracisternal or intraventricular injection, the unbound ^{131}I was determined by electrophoresis. Radiopharmaceutical preparations with more than 2% "free" iodine were excluded because the movement pattern reflected by autoradiographs may be altered by the unbound ^{131}I.[67] Proposed sites of CSF absorption were studied by both general and specific area analysis for distribution and gradients of radioactivity. Areas for microscopic study were selected from those showing radioactivity on gross autoradiographs.

Following intracisternal injection of the radiopharmaceutical in animals with severe communicating hydrocephalus, silver grains representing the labeled albumin were found adjacent to the free border of the ventricular ependymal wall, over the ependymal cell layer, and in the choroid plexus both on the surface and in the interstitial tissue. Labeled albumin was detected in the walls of the veins and intraluminally in vessels of the choroid plexus. Increased radioactivity was present in the periventricular brain tissue in the white matter as well as the gray.

The greatest concentration of silver granules was present in the first 2–3 mm below the ventricular ependymal surface, and there was a definite gradient from the ependyma to the cerebral cortex. Radioactivity appeared to localize around the walls of capillaries and venules, especially in the subependymal veins (Fig. 21).

There was no significant difference in distribution of the grains in animals with communicating hy-

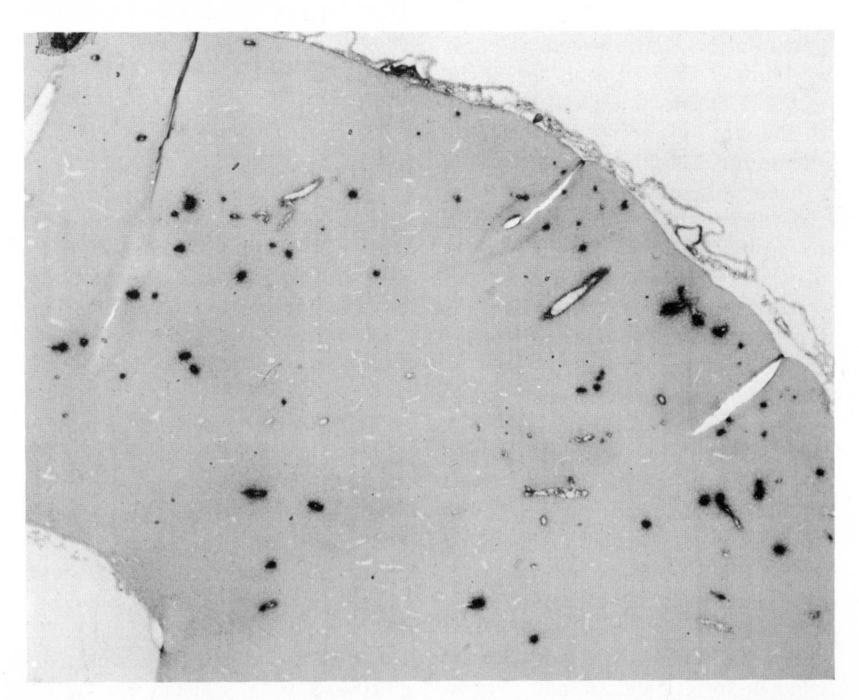

Figure 21. Low-power autoradiograph of nonhuman primate with communicating hydrocephalus showing penetration of radioactivity into brain substance and orientation of labeled albumin around venous structures. Ventricle is in lower left corner; cortex, is in upper right corner. Reproduced from James *et al.*,[33] with permission of *The Ocular Cerebrospinal Fluids*, Academic Press, Inc. publishers.

drocephalus regardless of whether the injection of radiopharmaceutical was intraventricular or intracisternal. In those animals that had communicating hydrocephalus and radiopharmaceutical entry and "stasis" in cisternograms, labeled albumin was observed in the periventricular area. Radioactivity was found in the same distribution in the animals with ventricular radiopharmaceutical entry and "clearing." In addition, silver grains were detected in the arachnoid and pia mater over the cerebral convexities, in arachnoid vessels, in the subpial area especially around the subpial capillaries, and in the first 2 mm of cerebral cortex below the pia.

In control animals with cisterna magna injections of radiopharmaceutical, no labeled albumin was present in the ventricular system or adjacent tissue. Most of the labeled albumin was detected in the subarachnoid space surrounding the cervical spine, the basal cisterns, the CSF space over the cerebral hemispheres, and in the area of the arachnoid villi. In control dogs following ventricular injection of the labeled albumin, some penetration of the labeled albumin into the periventricular tissue and choroid plexus was noted. The greatest movement of radioactivity appeared to be in the lateral ventricles in the angles between the corpus callosum and caudate nucleus. Less penetration was observed through other ventricular borders. The number of silver granules beneath the dorsolateral angles of the lateral ventricles in the control animals was significantly less than the number observed in the hydrocephalic animals (even with intracisternal injection), especially if the comparison was made at distances greater than 2 mm from the ventricular ependyma. The greater the distance from the ependymal surface (toward the cerebral cortex) the measurements were made, the greater became the differences between hydrocephalic and normal animals (Fig. 22).

The findings of the gross autoradiographs parallel those of the microscopic autoradiography. In the control animals with intracisternal injection, no radioactivity could be detected in the lining of the ventricular system. In animals with communicating hydrocephalus, the greatest amount of radioactivity was related to the ventricular system, with little in the subarachnoid space over the cerebral hemispheres. After intraventricular injection, radioactivity was present in the periventricular tissue in both the control and the hydrocephalic animals. However, the density of radioactivity was much greater in animals with communicating hydrocephalus, and the label penetrated into the periventricular tissue to a greater distance from the ependyma in the animals with communicating hydrocephalus than in the controls. These findings would correlate with the periventricular "edema" seen in CT scans (Fig. 23).

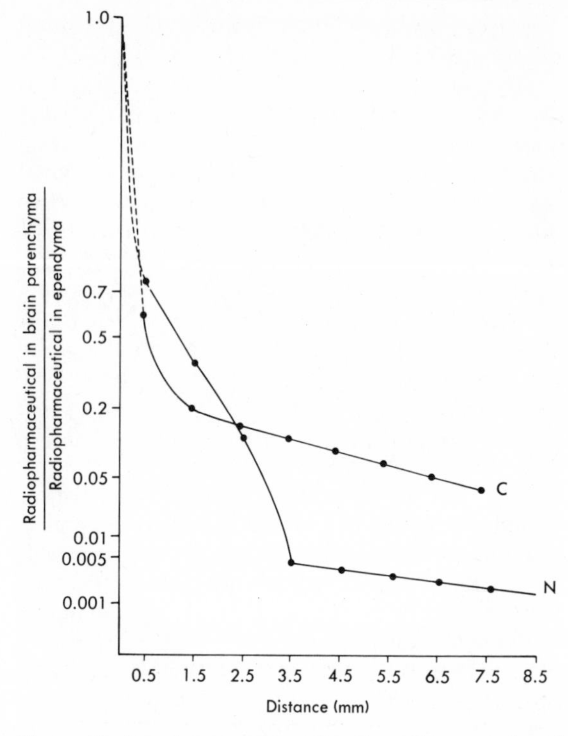

Figure 22. Graph comparing parenchymal to ependymal radioactivity vs. distance from ventricular surface plotted on probability paper for control animal (N) and animal with chronic communicating hydrocephalus with ventricular radiopharmaceutical stasis (C). Reproduced from Strecker *et al.*,[63] with permission of *Radiology*, Radiological Society of North America, Inc., publishers.

9. CSF Production

Another compensatory mechanism in a relative imbalance of CSF production and adsorption might be a decrease in CSF production.[4,25,36] To test this hypothesis, ventriculocisternal perfusion was carried out in the manner described by Pappenheimer *et al.*[50] and Fenstermacher *et al.*[19] The animals were first weighed and then perfused under pentothal anesthesia with assisted respiration. A 19-gauge cannula was introduced into the lateral ventricle through a parietal burr hole 16–20 mm posterior to the coronal suture and 10 mm lateral to the midline. Perfusion pressure and flow were constantly mon-

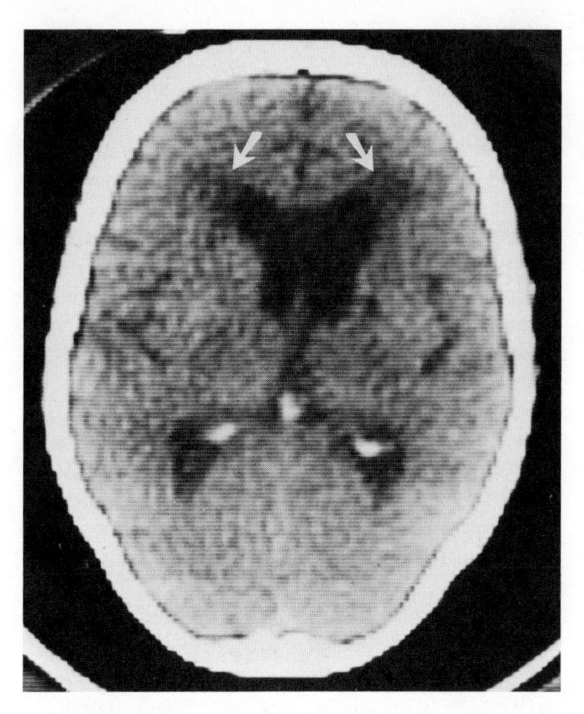

Figure 22. CT image of human with communicating hydrocephalus. Area of decreased attenuation (arrows) in periventricular region of frontal horns may represent penetration of CSF (periventricular "edema").

itored. Initial pressure was recorded prior to loss of CSF. The perfusion pressure was maintained between 10 and 15 cm H_2O with reference to the interaural line by adjusting the vertical position of the cisternal outflow tubing. Initial pressure measurements were correlated with the findings on cisternograms, the time period following injection of the silicone rubber, and appearance of the ventricles at necropsy. Cannulas were placed in the lateral ventricle and the cisterna magna (Fig. 24). Free flow in the system was documented prior to beginning the perfusion with artifical CSF.

Artificial CSF containing trace quantities of either [^{14}C]inulin (400 mg/100 ml) or [^{3}H]polyethylene glycol (mol. wt. 900) was used as the perfusion fluid. The inflow rate of 0.12 ml/min did not elevate CSF pressure. Following establishment of steady-state conditions, the outflow from the cisternal cannula was collected in glass test tubes at fixed time intervals for subsequent measurements of volume and concentration of the nondiffusible marker. Outflow was assumed to be at a steady state when the concentration of the reference marker was constant within $\pm 2\%$ measured on successive 10-min samples. Perfusion was continued in the manner described for 2–4 hr.

Following collection, 0.5-ml samples from the cisternal outflow were counted directly after solubilization in 1.0 M NCS reagent. Animals were euthanized at the end of the ventriculocisternal perfusion. The brain was removed and examined. The volume of CSF production was calculated by the indicator dilution method, which is expressed mathematically as follows[36]:

$$Q_{\text{in}}(C_i) = Q_{\text{out}}(C_o) \qquad (5)$$

where Q_{in} is flow in, Q_{out} is flow out, and C_i and C_o are the concentrations of indicator.

The term for flow out (Q_{out}), however, is actually composed of two terms: flow in, plus the flow due to CSF production (Q_{CSF}). Therefore, $Q_{\text{out}} = Q_{\text{in}} + Q_{\text{CSF}}$, and equation (5) can be expressed as

$$Q_{\text{in}}(C_i) = C_o\,(Q_{\text{in}} + Q_{\text{CSF}}) \qquad (6)$$

and, on solving for CSF production

$$Q_{\text{CSF}} = \frac{C_i - C_o}{C_o}(Q_{\text{in}}) \qquad (7)$$

We found that this calculation for CSF production rate agreed closely with that given volumetrically on the same sample, in the steady state. The dilution

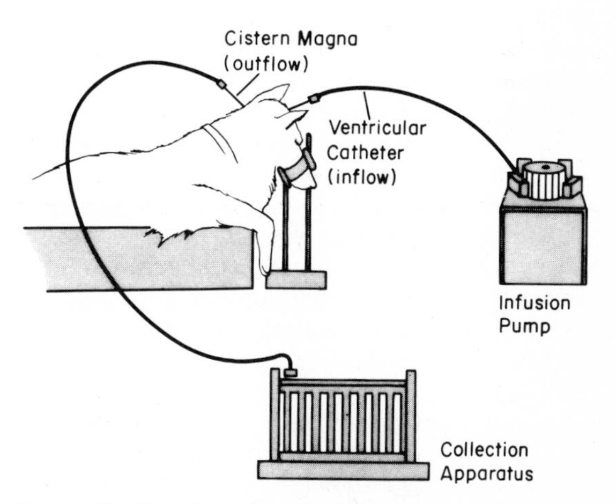

Figure 24. Drawing of ventriculocisternal perfusion apparatus. Ventricular cannula is inserted in stereotaxic manner and system filled with perfusate. Cisternal cannula is then placed by suboccipital puncture, and free flow in system is established. Perfusate is introduced into lateral ventricle at constant rate, and outflow from cisternal cannula is collected once steady state has been established. Reproduced from James *et al.*,[36] with permission of *Experimental Brain Research*, Springer-Verlag, publishers.

method was less subject to error from transient changes in blood pressure. Changes in blood pressure may tend to change the volume of the ventricular system being perfused. This effect is diminished using the methodology described because the production calculation is based on the radioactivity per milliliter of CSF on the outflow. A sudden change in ventricular volume would thus not change the concentration-per-milliliter of the labeled nondiffusible indicator; it would change only the total volume collected for that time period. Both methods would be in error during transient changes, however, if the CSF volume added originated from the subarachnoid compartment, which was not being perfused. This error would affect only transient changes, and the steady-state measurements in either instance would be valid. For these reasons, we believed that this methodology would yield accurate results in our animal model.[36]

Ventriculocisternal perfusion was established in both normal animals and those with communicating hydrocephalus. Continuous and uniform flow from the cisternal catheter was achieved. CSF production in normals varied from 4.21 to 5.10 μl/min per kg body weight, with a mean of 4.42 μl/min per kg body weight (Table 2). All CSF pressure measurements were in the normal range of 100–180 mm H_2O. On the cisternographic images following injection of 0.5–1.0 mCi [^{111}In]-DTPA into the cisterna magna, no ventricular entry was seen in normal controls, and there was generalized distribution of radiopharmaceutical over the cerebral cortex by 24 hr.[48]

In the animals studied within 2 weeks after silastic injection, CSF pressure measurements ranged from 250 to 400 mm H_2O. On the early cisternographic scans at 1–2 hr after injection, ventricular entry or "penetration" was present. Radiopharmaceutical appeared to move from the ventricle ("clearing") by the 6- to 24-hr study. In this group of animals, CSF production was 55% of normal, which was significantly lower than that of controls (Table 2).

In the animals studied 2 weeks to several months after silastic injection, CSF pressure measurements were all in the normal range. The CSF production was lower in this test group (65% of normal), and the difference was significant (Table 2). At cisternography, radiopharmaceutical concentrated in the ventricles on the 1- to 2-hr images and persisted there for 6–24 hr. When the animals with communicating hydrocephalus (documented by cisternography *in vivo* and at necropsy) were considered as a group, the mean CSF production was again significantly different ($P < 0.005$) from that of controls.

Because the animals with elevated CSF pressures tended to have lower CSF production values, a regression analysis of the measured pressure vs. CSF production was obtained. CSF production and CSF pressure were positively correlated in our experiments ($r = 0.682$).

Thus, from these studies, CSF volume production is significantly decreased during the early (elevated-pressure) phase of development of communicating hydrocephalus and remains decreased when the ventricles enlarge markedly and the CSF pressure returns to normal. In our experiments, three distinct groups of animals were included: (1) normal; (2) those with moderate changes in the ventricular ependyma with elevated CSF pressures; and (3) animals with marked histological changes in the ventricular ependyma and periventricular area with normal CSF pressure.

There is a significant difference in CSF production between the normals and the latter groups (animals with chronic communicating hydrocephalus).[36] The tendency toward lower CSF production in animals with chronic communicating hydrocephalus could be accounted for by increased movement into the extracellular space in the periventricular region.[63,65] From the cisternographic and CT images, we believe

Table 2. Effect of Experimentally Induced Communicating Hydrocephalus on CSF Production[a]

Condition	N	CSF pressure (mm H_2O)	Q_{csf} (μl/min/kg)[b]	P
Normal control	5	100–180	4.42 ± 0.20	
Normal-pressure hydrocephalus	5	100–180	2.87 ± 0.49	< 0.02
Increased-pressure hydrocephalus	6	250–400	2.42 ± 0.45	< 0.005
All hydrocephalus	11	100–400	2.63 ± 0.32	< 0.005

[a] Reproduced from James *et al.*,[36] with permission of *Experimental Brain Research*, Springer-Verlag, publishers.
[b] Mean values of CFS production ± SEM of number of experiments (N).

this to be the case. The trend to lower CSF production values in the animals with increased CSF pressure and in animals with enlarged ventricles supports the findings of others.[43,44] Although statistically significant, this may not be sufficient to represent an effective compensatory mechanism, since the relative CSF production may be a multiple of the absorptive capacity.

10. Discussion

Considerable data regarding the physiology of CSF have been omitted from this chapter. We selected the experimental studies that seemed to provide information that directly or indirectly assists in our understanding of the diagnostic images we produce in patients with communicating hydrocephalus. We were especially interested in the relationship of the periventricular edema and increased extracellular space in this region to the findings at cisternography and CT.

Partial blockage of the normal CSF absorptive pathways seems to initiate a sequence of events that most probably includes a period of normal CSF production but diminished absorption, an episode and prolonged periods of elevated CSF pressure, and at least some but not sufficient compensation in the dynamic system by the diminution of CSF production. Ventricular enlargement follows, which we can readily depict by angiography, pneumoencephalography, and CT scans, but probably does not occur until the force, as a result of pressure on the ventricular surface, exceeds the resistance provided by the brain.[21,22,31,33] The ventricular angles first become rounded on pneumoencephalograms and computed tomograms. As the ventricles enlarge, the ependyma becomes flattened and stretched and loses its property as a partial barrier to the movement of large molecules from the CSF space into the extracellular periventricular space.[5,9,15,70] This is manifest on CT as decreased attenuation in the periventricular area and loss of distinct anatomical margins of the ventricles.

Although CSF is produced in somewhat reduced volume by the choroid plexus, the drainage or absorption remains inadequate, and the net movement of CSF from the ventricles to the arachnoid villi is diminished. An alternative route for the absorption of CSF appears to be established across the compromised ventricular ependyma.

Identification of detailed structural anatomy present on cisternograms is now of secondary clinical importance due to widespread use of computerized axial tomography (CT) devices. This CT instrumentation elegantly resolves small changes in anatomical configuration and size and will adequately document the presence or absence of ventricular enlargement as well as structural changes of the cerebral sulci. The general direction and time course of CSF movement, however, remain of significant importance, and serial radionuclide studies continue to provide accurate and useful information. CT with water-soluble contrast media, serial imaging, and blood determinations for transfer dynamics may obviate the use of radiopharmaceuticals.

The alterations of the ependymal surface of the CSF–brain barrier described may not directly influence our pattern recognition of diagnostic studies. However, structural alterations we see are expressions of basic pathophysiological changes, a number of which are incompletely understood at present. The silastic model that we have described provides a system in which we can continue to investigate the changes in the systems involved in CSF production and absorption.

Treatment of certain patients with chronic communicating hydrocephalus by CSF diversionary shunts has been associated with dramatic improvement.[3,46,59] However, more specific preoperative selection of patients is recommended not only to avoid a number of the complications due to surgery, but also to improve overall results. A number of modifications in shunt design have diminished complications, but at the very least, surgical revision at certain time intervals is predictable. Surgical establishment of an alternative drainage pathway is a traumatic procedure and is to be undertaken only in those patients in whom benefit is most likely. It has also been recommended that alternative methods of treatment be attempted, but to date no other effective form of therapy is available. It would appear from data regarding the basic mechanisms associated with the development of communicating hydrocephalus and from fundamental understanding of CSF production and absorption that more appropriate methods of patient management can be developed.

References

1. Abbott, M., Alksne, J. F.: Transport of intrathecal I^{125} RISA to circulating plama: A test for communicating hydrocephalus. *Neurology (Minneapolis)* **18**:870–874, 1968.

2. ADAMS, R. D., FISHER, C. M., HAKIM, S., OJEMANN, R. G., SWEET, W. H.: Symptomatic occult hydrocephalus with normal cerebrospinal fluid pressure. *N. Engl. J. Med.* **23**:117–126, 1965.
3. BENSON, D. F., LEMAY, M., PATTEN, D. H., RUBENS, A. B.: Diagnosis of normal pressure hydrocephalus. *N. Engl. J. Med.* **283**:609–615, 1970.
4. BERING, E. A., SATO, O.: Hydrocephalus: Changes in formation and absorption of cerebrospinal fluid within the cerebral ventricles. *J. Neurosurg.* **20**:1050–1063, 1963.
5. BLAKEMORE, W. F., JOLLY, R. D.: The subependymal plate and associated ependyma in the dog: An ultrastructural study. *J. Neurocytol.* **1**:69–89, 1972.
6. BRIGHTMAN, M. W.: The distribution within the brain of ferritin injected into the CSF compartments. *J. Cell Biol.* **26**:99–123, 1965.
7. BRIGHTMAN, M. W., KLATZO, I., OLSSON, Y., REESE, T. S.: The blood brain barrier to proteins under normal and pathological conditions. *J. Neurol. Sci.* **10**:215–239, 1970.
8. BRIGHTMAN, M. W., PALAY, S. L.: The fine structure of ependyma in the brain of the rat. *J. Cell Biol.* **19**:415–439, 1963.
9. CLARK, R. G., MILHORAT, T. H.: Experimental hydrocephalus. Part III. Light microscopic findings in acute and subacute obstructive hydrocephalus in the monkey. *J. Neurosurg.* **32**:400–413, 1970.
10. DAVSON, H.: *The Physiology of CSF.* London, Churchill, 1967.
11. DAVSON, H.: The rates of disappearance of substances injected into the subarachnoid space of the rabbit. *J. Physiol.* **129**:52–53, 1955.
12. DAVSON, H., HOLLINGSWORTH, S., SEGAL, M. B.: The mechanism of drainage of CSF. *Brain* **93**:665–678, 1970.
13. DAVSON, H., KLEEMAN, C. R., LEVIN, E.: Quantitative studies of the passage of different substances out of the CSF. *J. Physiol. (London)* **161**:126–142, 1962.
14. DAVSON, H., PURVIS, C.: An apparatus for controlled injection over long periods of time. *J. Physiol.* **149**:135–143, 1952.
15. DELAND, F. H., JAMES, A. E., LADD, D. J., KONINGSMARK, B. W.: Normal pressure hydrocephalus: A histological study. *Am. J. Clin. Pathol.* **58**:58–63, 1972.
16. DICHIRO, G.: Movement of cerebrospinal fluid in human beings. *Nature (London)* **204**:290–291, 1964.
17. DICHIRO, G., LARSON, S. M., HERRINGTON, T., JOHNSTON, G. S., GREEN, M. W., SWAN, S. J.: Descent of CSF to spinal subarachnoid space. *Acta Radiol.* **14**:379–384, 1973.
18. EISENBERG, H. M., MCLENNAN, J. E., WELCH, K.: Ventricular perfusion in cats and kaolin induced hydrocephalus. *J. Neurosurg.* **41**:20–28, 1974.
19. FENSTERMACHER, J. D., RALL, D. P., PATLAK, C. S., LEVIN, V. A.: Ventriculocisternal perfusion as a technique for analysis of brain capillary permeability and extracellular transport. In Crone, C., Lassen, N. A. (eds.): *Alfred Benson Symposium II: Capillary Permeability.* Copenhagen, Munksgaard, Academic Press, 1970, pp. 483–490.
20. FLOR, W. J., JAMES, A. E., JR., RIBAS, J. L., PARKER, J. L., SICKEL, W. L.: Brain ventricular surface changes in experimental communicating hydrocephalus. In: Johari, O. M., Becker, R. P. (eds.): *Scanning Electron Microscopy 1979/I–III (12th) Annual Symposium, April 1979, Washington, D.C.* AMF O'Hare, Illinois, SEM Inc., pp. 47–54, 1979.
21. GESCHWIND, N.: The mechanism of normal pressure hydrocephalus. *J. Neurol. Sci.* **7**:481–493, 1968.
22. HAKIM, S.: Biomechanics of hydrocephalus. In Harbert, J. C. (ed.): *Cisternography and Hydrocephalus.* Springfield, Illinois, Charles C. Thomas, 1972, pp. 25–55.
23. HAKIM, S., ADAMS, R. D.: A special clinical problem of symptomatic hydrocephalus with normal cerebrospinal fluid pressure. *J. Neurol. Sci.* **2**:307–327, 1965.
24. HAKIM, S., VENEGAS, J. G., BURTON, J. D.: The physics of the cranial cavity in hydrocephalus and normal pressure hydrocephalus: Mechanical interpretation and mathematical model. *Surg. Neurol.* **5**:187–210, 1976.
25. HOCHWALD, G. M., LUX, W. E., SAHAR, A., RANSOHOFF, J.: Experimental hydrocephalus: Changes in cerebral fluid dynamics as a function of time. *Arch. Neurol.* **26**:120, 1972.
26. JAMES, A. E., BURNS, B., FLOR, W. J., STRECKER, E.-P., MERTZ, T., BUSH, M., PRICE, D. L.: Pathophysiology of chronic communicating hydrocephalus in dogs (*Canis familiaris*): Experimental studies. *J. Neurol. Sci.* **24**:151–178, 1975.
27. JAMES, A. E., DAVSON, H.: Basic pathophysiological considerations of cerebrospinal fluid and the relation to imaging. In Potchen, E. J. (ed.): *Current Concepts in Radiology.* St. Louis, C. V. Mosby, 1977, pp. 348, 390.
28. JAMES, A. E., FLOR, W. J., BURNS, B., STRECKER, E.-P.: Pathophysiological changes in chronic communicating hydrocephalus, with particular reference to children. Gomez-Lopez, J., Bonmati, J. (eds.): *Proceedings of the XIII International Congress of Radiology,* Vol. 2. Amsterdam, Excerpta Medica, pp. 232–237.
29. JAMES, A. E., FLOR, W., BUSH, M., MERZ, T., RISH, B.: An experimental model for chronic communicating hydrocephalus. *J. Neurosurg.* **41**:32–37, 1974.
30. JAMES, A. E., FLOR, W. J., MERZ, T., STRECKER, E.-P., BURNS, B. B.: A pathophysiologic mechanism for ventricular entry of radiopharmaceutical and possible relation to chronic communicating hydrocephalus. *Am. J. Roentgenol. Radium Ther. Nucl. Med.* **122**:38–43, 1974.
31. JAMES, A. E., HOSAIN, F., SOM, P., REBA, R. C., WAGNER, H. N., DELAND, F. H.: Ytterbium 169 diethylenetriaminepentaacetic acid (DTPA): with special reference to cisternography. In *Proceedings of the International Congress of Radiology,* Vol. 2. Amsterdam, Excerpta Medica, 1971.
32. JAMES, A. E., FLOR, W. J., NOVAK, G. R., STRECKER,

E.-P., Burns, B.: Evaluation of the central canal of the spinal cord in experimentally induced hydrocephalus. *J. Neurosurg.* **48:**970–974, 1978.
33. James, A. E., Flor, W. J., Novak, G. R., Strecker, E.-P., Burns, B., Epstein, M.: Experimental hydrocephalus. In Bito, L. Z., Davson, H., Fenstermacher, J. D. (eds.): *The Ocular Cerebrospinal Fluids.* London, Academic Press, 1977, pp. 435–459.
34. James, A. E., McComb, S. G., Christian, J., Davson, H.: The effect of cerebrospinal fluid pressure on the size of drainage pathways. *Neurology* **26:**659–663, 1976.
35. James, A. E., New, P. F. J., Heinz, E. R., Hodges, F. J., DeLand, F. H.: A cisternographic classification of hydrocephalus. *Am. J. Roentgenol. Radium Ther. Nucl. Med.* **115:**39–49, 1972.
36. James, A. E., Jr., Novak, G., Bahr, A. L., Burns, B.: The production of cerebrospinal fluid in experimental communicating hydrocephalus. *Exp. Brain Res.* **27:**553–557, 1977.
37. James, A. E., Sperber, E., Strecker, E.-P., Diegel, C., Novak, G., Bush, M.: Use of serial cisternograms to document dynamic changes in the development of communicating hydrocephalus: A clinical and experimental study. *Acta Neurol. Scand.* **50:**153–170, 1974.
38. James, A. E., Strecker, E.-P., Bush, R. M.: A catheter technique for the production of communicating hydrocephalus. *Radiology* **106:**437–439, 1973.
39. James, A. E., Strecker, E.-P., Bush, R. M., Merz, T.: Use of silastic to produce chronic communicating hydrocephalus. *Invest. Radiol.* **8:**105–110, 1973.
40. James, A. E., Strecker, E.-P., Novak, G. R., Burns, B.: Correlation of serial cisternograms and cerebrospinal fluid pressure measurements in experimental communicating hydrocephalus. *Neurology* **23:**1226–1233, 1973.
41. James, A. E., Strecker, E.-P., Sperber, E., Flor, W. J., Merz, T., Burns, B.: An alternative pathway of cerebrospinal fluid absorption in communicating hydrocephalus: Transependymal movement. *Radiology* **111**(1):143–146, 1974.
42. Karnovsky, M.: The ultrastructural basis of capillary permeability studies with paraoxidase as a tracer. *J. Cell Biol.* **35:**213–236, 1967.
43. Lorenzo, A. V., Bresnan, M. S., Barlow, C. F.: Cerebrospinal fluid absorption deficit in normal pressure hydrocephalus. *Arch. Neurol.* **30:**387–393, 1974.
44. Lorenzo, A. V., Page, L. K., Watters, G. V.: Relationship between CSF formation, absorption and pressure in human hydrocephalus. *Brain* **93:**6790–692, 1970.
45. Lowry, J., Bahr, A. L., Allen, J. H., Meacham, W. F., James, A. E., Jr.: Radiological techniques in diagnostic evaluation of dementia. In Wells, C. E. (ed.): *Dementia,* 2nd ed. Philadelphia, Davis, 1977, pp. 223–245.
46. McCullough, D. C., Harbert, J. C., DiChiro, G., Ommaya, A. E.: Prognostic criteria for CSF shunting from cisternography in communicating hydrocephalus. *Neurology* **20:**594–598, 1970.
47. Milhorat, T. H., Clark, R. G.: Some observations on the circulation of PSP in the CSF: Normal flow and flow in hydrocephalus. *J. Neurosurg.* **32:**522, 1970.
48. Novak, G., Digel, C., Burns, B., James, A. E.: Cerebrospinal fluid pressure measurements and radioisotope cisternography in dogs. *Lab. Anim.* **8:**85–91, 1974.
49. Olesen, J.: Cerebral blood flow: Methods for measurement, regulation, effects of drugs in disease. *Acta Neurol. Scand. Suppl. 57,* **50,** 1974.
50. Pappenheimer, J. R., Heisey, S. R., Jordan, E. F., Downer, D. E. C.: Perfusion of cerebral ventricular system in unanesthetized goats. *Am. J. Physiol.* **203:**763–774, 1962.
51. Pollay, M., Davson, H.: The passage of certain substances out of the CSF. *Brain* **86:**137–150, 1963.
52. Potts, B. G., Reilly, K. F., Deornarine, V.: Morphology of the arachnoid villi in granulations. *Radiology* **105:**333–341, 1972.
53. Price, D. L., James, A. E., Sperber, E. P.: Communicating hydrocephalus: Cisternographic and neuropathologic studies. *Arch. Neurol.* **33:**15–20, 1976.
54. Price, D. l., Porter, K. R.: The response of amphibian ventral horn neutrons to axonal transection. *J. Cell Biol.* **53:**24–37, 1972.
55. Rall, D. P., Oppelt, W. W., Patlak, C. S.: Extracellular space of brain as determined by diffusion of inulin from the ventricular system. *Life Sci.* **2:**43–48, 1962.
56. Reese, T. S., Karnovksy, M. J.: Fine structural localization of a blood brain barrier to exogenous peroxidase. *J. Cell Biol.* **34:**207–217, 1967.
57. Reivich, M., Isaacs, G., Evarts, E., Kety, S.: Regional cerebral blood flow during REM and slow wave sleep. *Trans. Am. Neurol. Assoc.* **92:**70–74, 1967.
58. Sahar, A., Hochwald, G. M., Ransohoff, J.: Alternate pathway of cerebrospinal fluid absorption in animals with experimental obstructive hydrocephalus. *Exp. Neurol.* **25:**200–205, 1969.
59. Salmon, J. H.: Adult hydrocephalus: Evaluation of shunt therapy in 80 cases. *J. Neurosurg.* **37:**423–428, 1972.
60. Shabo, A. L., Maxwell, D. S.: Electron microscopic observations on the fate of particulate matter in CSF. *J. Neurosurg.* **29:**464, 1968.
61. Spiegel, M. R.: *Theory of Problems of Statistics.* Schaums Outline Series, New York, McGraw-Hill, 1961.
62. Strecker, E.-P., Bush, R. M., James, A. E.: Cerebrospinal fluid imaging as a method to evaluate communicating hydrocephalus in dogs. *Am. J. Vet. Res.* **34:**101–104, 1973.
63. Strecker, E.-P., James, A. E., Kelly, J. E., Merz, T.: Semi-quantitative studies of transependymal albumin movement in communicating hydrocephalus. *Radiology* **111:**341–346, 1974.
64. Strecker, E.-P., James, A. E., Koningsmark, B.,

MERZ, T.: Autoradiographic observations in experimental communicating hydrocephalus. *Neurology* **24:**192–197, 1974.

65. STRECKER, E.-P., KELLY, J. E. T., MERZ, T., JAMES, A. E.: Transventricular albumin absorption in communicating hydrocephalus. *Arch. Psychiatr. Nervenkr.* **218:**369–377, 1974.
66. STRECKER, E.-P., NOVAK, G. R., JAMES, A. E.: Compartmental analysis of cerebrospinal fluid–blood albumin transfer: Consideration of kinetics in normal animals and animals with chronic communicating hydrocephalus. *Eur. Neurol.* **16:**203–212, 1977.
67. STRECKER, E.-P., SCHEFFEL, E., KELLY, J., JAMES, A. E.: CSF absorption in communicating hydrocephalus: Evaluation of transfer of radioactive albumin from subarachnoid space to plasma. *Neurology (Minneapolis)* **23:**854–864, 1973.
68. SYMON, L., DORSCH, W. C.: Use of long term intracranial pressure measurement to assess hydrocephalic patients prior to shunt surgery. *J. Neurosurg.* **42:**258–273, 1975.
69. TRIPATHI, R.: Tracing of the bulk outflow route of CSF by transmission and scanning electron microscopy. *Brain Res.* **80:**503–506, 1974.
70. VESSAL, K., SPERBER, E., JAMES, A. E.: Chronic communicating hydrocephalus with normal CSF pressures: A cisternographic pathologic correlation. *Ann. Radiol.* **17:**785–793, 1974.
71. WEED, L. H.: Studies on the CSF: The pathway of escape with particular reference to the arachnoid villi. *J. Med. Res.* **26:**51, 1914.
72. WELCH, K., FRIEDMAN, V.: The cerebrospinal fluid valves. *Brain* **83:**554–564, 1960.
73. YAKOVLEV, P. I.: Paraplegias of hydrocephalus. *Am. J. Ment. Defic.* **51:**561–576, 1947.
74. ZIERLER, K. L.: Equations for measuring blood flow by external monitoring of radioisotopes. *Circ. Res.* **6:**309–321, 1965.

CHAPTER 30

Dynamics of Cerebrospinal Fluid System as Defined by Cranial Computed Tomography

Burton P. Drayer and Arthur E. Rosenbaum

1. Introduction

With the development of cranial computed tomography (CT), the neuroradiological evaluation of the cerebrospinal fluid (CSF) system has been markedly altered. A definitive representation of the ventricles and subarachnoid spaces (SASs) is now noninvasively obtained within a matter of minutes. Associated pathological processes are also readily demonstrated.

It must be remembered, however, that nonenhanced CT is primarily a morphological modality from which physiologic information may only be surmised. To obtain further dynamic information, a marker must be introduced intrathecally and serial imaging performed (positive-contrast CT cisternography).

2. Techniques and Rationale

The physics and design of CT scanning (transmission tomography) equipment have been amply reviewed and are beyond the scope of this survey. The excellent spatial resolution provided by CT permits a routine, fairly detailed visualization of the SASs and ventricles. Thus, accurate and consistent information concerning the size, configuration, and position of the CSF-containing spaces is obtained from even a *nonenhanced* CT scan.

The *intravenous infusion* of iodinated contrast medium (30 or 60% meglumine diatrizoate) is most useful in delineating alterations in the blood–brain barrier that occur with neoplasms, abscesses, and other inflammatory processes. In unusual circumstances, the nonenhanced CT scan shows no evidence of any process causing ventricular (obstructive) hydrocephalus, while the intravenously enhanced scan shows an obvious enhancing mass lesion. When a paraventricular area of diminished density is associated with ventricular enlargement, the infusion of intravenous contrast medium permits the distinction of cyst (nonenhancing) from neoplasm (enhancing).

If direct information concerning the CSF circulation is required, *intrathecal contrast enhancement* is necessary.[6,7,22,27,28] The agent of choice at present is metrizamide (Amipaque, Winthrop Laboratories, New York), a nonionic, water-soluble, intrathecal contrast medium that is similar in osmolality to CSF when used at a concentration of 160–190 mg I/ml. Metrizamide CT cisternography (CTC) is performed by injecting 6 ml of 160–190 mg I/ml of contrast medium through a 22-gauge lumbar puncture needle. The needle is removed, and either of two techniques may then be used. The patient may be turned supine

Abbreviations used in this chapter: (CNS) central nervous system; (CSF) cerebrospinal fluid; (CT) computed tomography; (CTC) *CT* cisternography; (CTV) *CT* ventriculography; (NPH) normal-pressure hydrocephalus; (PFEAC) posterior fossa extra-axial cyst; (SAS) subarachnoid space.

Burton P. Drayer, M.D. • Division of Neuroradiology, Department of Radiology, Duke University Medical Center, Durham, North Carolina 27710. **Arthur E. Rosenbaum, M.D.** • Division of Neuroradiology, Department of Radiology, University of Pittsburgh Health Center, Pittsburgh, Pennsylvania 15261.

on the myelographic tilt table and placed in a 20–30° Trendelenberg position for 2 min or no positional maneuvers may be performed with the patient returned to his or her room to move about freely. In addition to a baseline CT scan, serial scans are performed at 6 (optional), 12, 24, and 48 (optional) hr following the lumbar introduction of metrizamide to follow its intracranial course. The interpretative criteria used are similar to those applied to radionuclide cisternographic studies.[34,47] The water-soluble contrast medium may also be introduced into the lateral ventricles via a shunt tube in a similar dose [metrizamide CT ventriculography (CTV)].[7,44]

3. Normal CSF System

3.1. Nonenhanced CT

Although various authors[23,26,29] have provided numerical data concerning the "normal" size of the ventricles and SASs and related them to patient age, most individuals experienced in CT interpretation use a gross general impression technique ("eyeballing it"). The range of "normal," particularly in the elderly,[26] has great latitude. In addition, the varying resolution afforded by different scanners may affect a strictly numerical interpretation of subtle differences (e.g., size of sulci).

The Evans index and cella media index are used to evaluate frontal horn and body size.[23] The anterolateral borders of the frontal horns are normally angular as opposed to rounded, and an adjacent area of intermediate density between CSF and white matter (periventricular interstitial edema) is not present. The third ventricle is normally less than 0.7 mm in width, and the temporal horns of the lateral ventricles are barely visualized. An evaluation of the fourth ventricle has not proven helpful, since the size may vary with slight changes in scanning angle. In younger individuals, some visualization of the cortical sulci is acceptable, particularly on the high vertex sections, while in older individuals, the sulci may be quite prominent without corresponding symptomatology.

3.2. Intravenously Enhanced CT

The primary reason for infusing iodinated contrast medium intravenously in the evaluation of CSF circulation disorders is to exclude a neoplastic or inflammatory process that might obstruct the ventricles or the SASs. When these abnormalities are not present and the blood–brain barrier is otherwise intact, no pathological enhancement occurs. Symmetrical enhancement of the circle of Willis, surface vasculature, cortical capillary bed, deep venous system, and choroid plexus is consistently seen.

3.3. Intrathecally Enhanced CT

In the first few hours following the introduction of metrizamide into the lumbar SAS (CTC), the contrast medium accurately defines the basal subarachnoid cisterns (Fig. 1) as well as the cerebellar, Sylvian fissure, and cortical CSF spaces.[6,22] Within 6–12 hr, the superficial brain substance adjacent to the SASs shows a definite staining (increased attenuation coefficient) that occurs symmetrically in both hemispheres including the parasagittal region[6,8] (Figs. 2 and 3). Due to difficulties in the radionuclide tagging of metrizamide, it is impossible to state whether this stain occurs in the intracellular or the extracellular–Virchow Robin space.[8,18] By 24 hr in the individual with a normal CSF circulation pattern, the blush is decreased, and at 48 hr it is no longer visualized. Metrizamide may normally, though unusually, reflux into the lateral ventricles (Fig. 4), particularly when the positional (Trendelenberg) technique is used. However, metrizamide is never normally present (stasis) in the lateral ventricles on the 24-hr serial scan.[6,7,22,28,30,37,45,46]

When the contrast medium is injected via a lateral ventricular shunt device (CTV), free flow into the third and fourth ventricles as well as the basal subarachnoid cisterns occurs within 15–30 min. The evaluation of ventricular stasis in this situation is far more complex, since the ventricular volume plays an important role in calculating the clearance of metrizamide.[41,42] With newer CT scanners, the ventricular volume may be estimated to a degree of accuracy that permits consistent and reproducible calculations of the CSF production rate.[41,42] A measurement of clearance using CTC requires ventricular entry of the contrast medium which is not often achieved in the normal individual.[6]

4. Abnormal CSF System

CT permits a simple classification of CSF system abnormalities based on the site of structural involvement (Fig. 5). The general groups *ventricular, subarachnoid,* and *brain* correspond to the previous categories of noncommunicating obstructive, communicating obstructive, and ex vacuo. A fourth group,

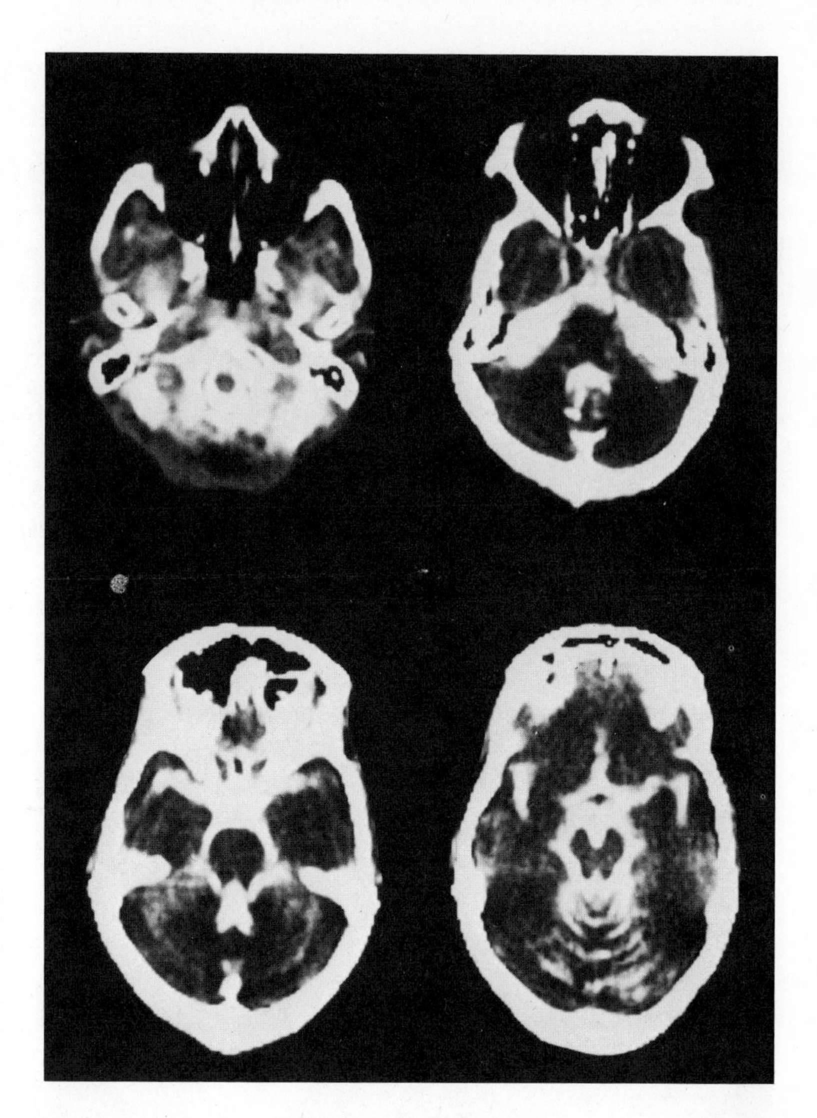

Figure 1. Normal subarachnoid cisterns: metrizamide CTC. Sequential CT scans performed at successively higher (10-mm) levels to show the detailed visualization of the basal subarachnoid cisterns, fourth ventricles, brainstem, and upper cervical spinal cord consistently obtained using CTC.

external hydrocephalus, seems to occur as a concomitant of chronic subdural hygroma in young children. Each group will be evaluated in terms of the different enhancement techniques and the various associated pathological processes.

4.1. Ventricular Hydrocephalus

4.1.1. Neoplastic. *a. Nonenhanced CT.* An area of diminished density (which usually enhances following intravenous infusion of iodinated contrast medium) within or compressing the ventricular system may be present with any type of neoplasm. Even subtle calcification associated with a tumor is readily detected using CT. A careful evaluation of the ventricles and SASs often proves quite helpful in classifying the type of hydrocephalus (Table 1), the acuteness, and the need for further radiological evaluation. The features most consistent with ventricular hydrocephalus of any etiology include marked enlargement of the temporal horns, prominent periventricular interstitial edema[5,9] (particularly if the obstructing process is acute or subacute), and normal cortical sulci and basal SASs.

b. Intravenously Enhanced CT. Neoplasms adjacent to or within the third or lateral ventricles may *obstruct the foramen of Monro* (Fig. 6) *or the adqueduct of Syvlius.* Significant enhancement occurs with higher-grade glial neoplasms, meningiomas, and colloid cysts, while lower-grade glial neoplasms, craniopharyngiomas, and germinomas (e.g., teratoma, pinealoma) enhance less prominently or not at all.

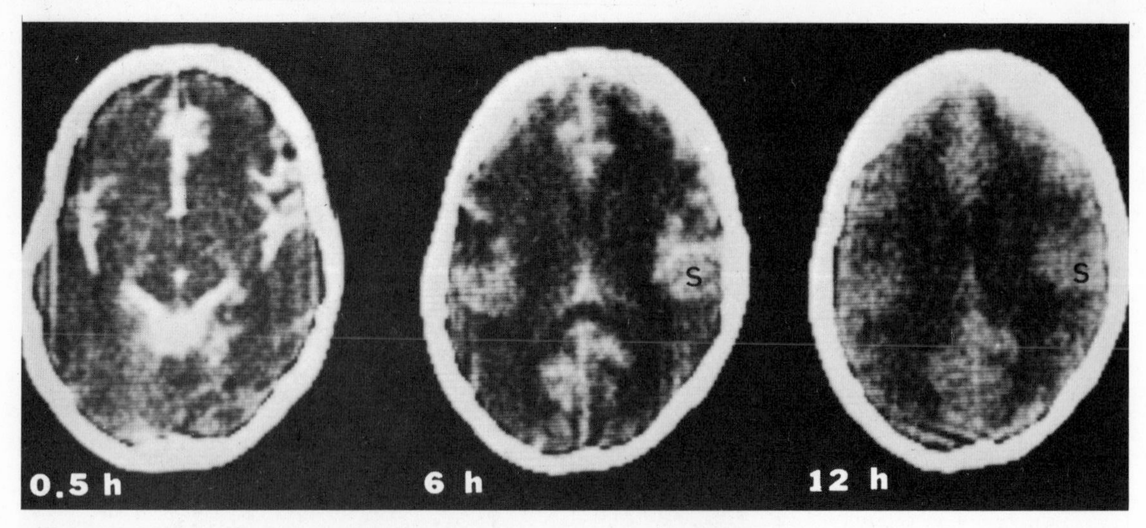

Figure 2. Normal brain staining: metrizamide CTC. Prominent staining (S) of the more superficial brain substance adjacent to the convexity SASs is most prominent from 6 to 12 hr after gravitational movement of metrizamide into the basal subarachnoid cisterns.

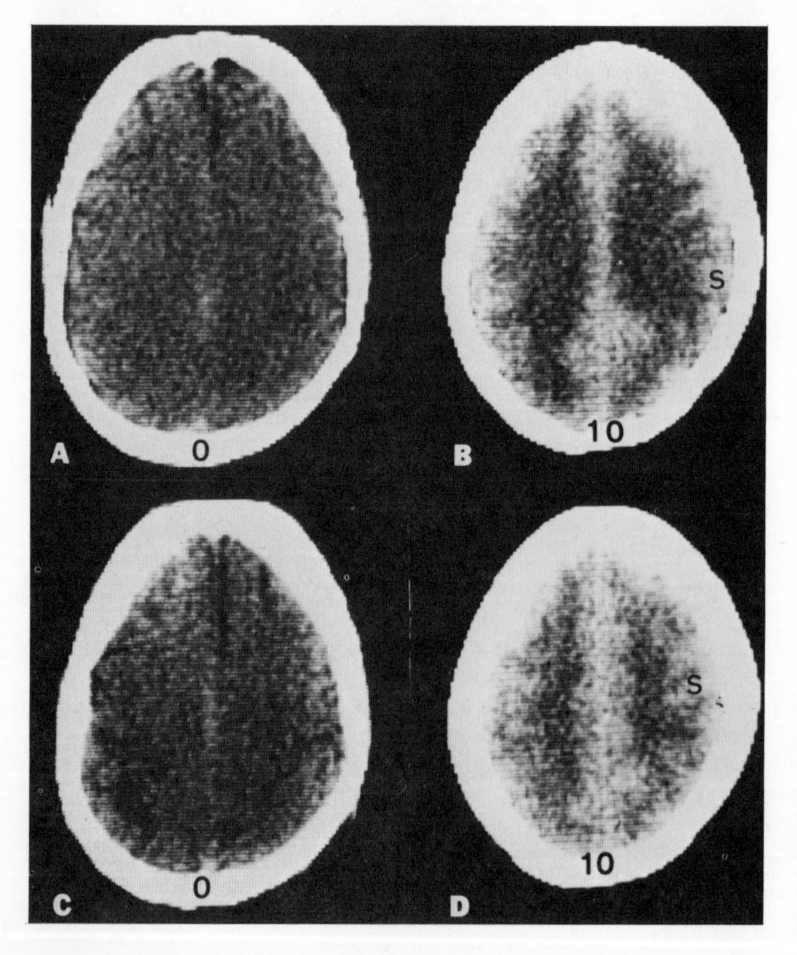

Figure 3. Normal brain staining: metrizamide CTC. (A,C) Nonenhanced CT. (B,D) Intrathecal (metrizamide) enhancement. A high convexity and parasagittal "blush" (S) are prominent 10 hr after lumbosacral instillation and initial positional movement of contrast medium into the posterior fossa SASs.

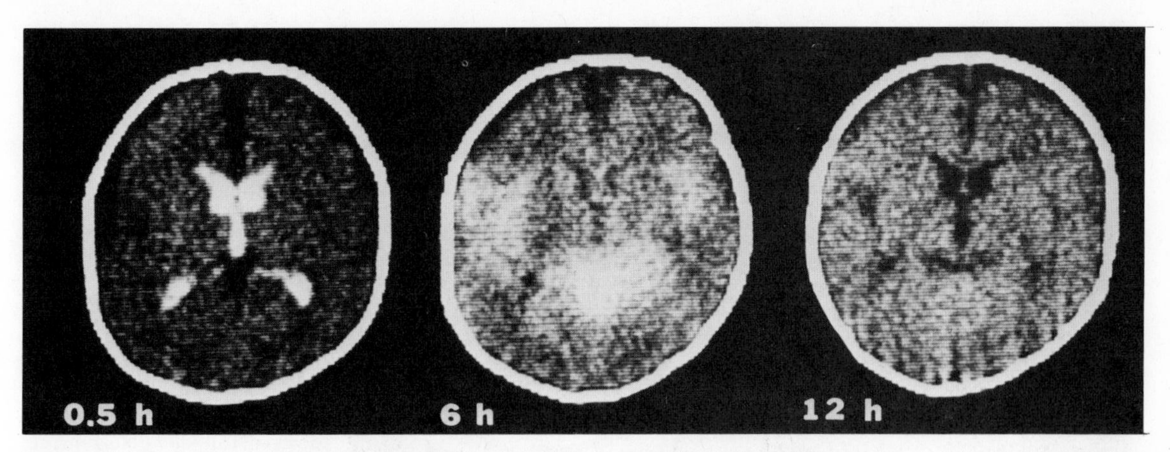

Figure 4. Normal ventricular entry: metrizamide CTC. CT scans performed at 30 min, 6 hr, and 12 hr after lumbar instillation of metrizamide in a child using the Trendelenberg supine positioning technique. After initial Amipaque entry, metrizamide clears from the lateral ventricles within 12 hr and a symmetrical cerebral "blush" is evident.

Subacute ventricular hydrocephalus (Fig. 7) is an extremely common concomitant of *posterior fossa neoplasms* (medulloblastoma, ependymoma, cerebellar astrocytoma, hemangioblastoma, metastasis, neurilemmoma, and meningioma). In contradistinction, brainstem neoplasms often produce no hydrocephalus and may not enhance.

c. Intrathecally Enhanced CT. Following the introduction of metrizamide into the lumbosacral SAS (CTC), contrast medium does not enter the ventricular system that is proximal to the site of obstruction. If an adjacent mass is present (e.g., craniopharyngioma, germinoma), a filling defect is obvious in the appropriate basal subarachnoid cistern[10] (Fig. 8). The ascent of metrizamide into the SASs overlying the cerebral convexities and the staining of the adjacent brain occur normally and symmetrically.

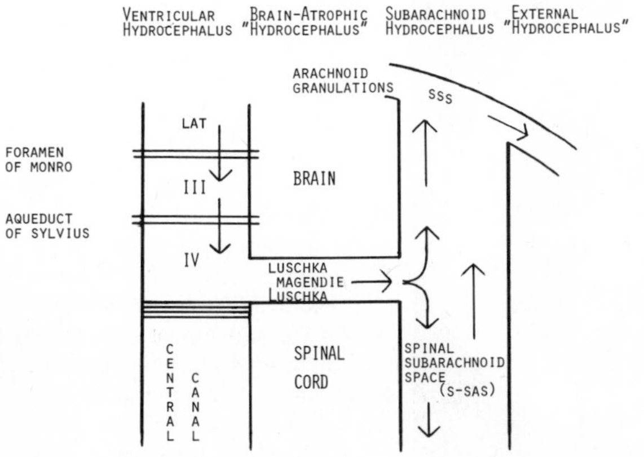

VENTRICULAR HYDROCEPHALUS
"BRAIN-ATROPHIC HYDROCEPHALUS"
SUBARACHNOID HYDROCEPHALUS
"EXTERNAL HYDROCEPHALUS"
ARACHNOID GRANULATIONS
SSS
LAT
FORAMEN OF MONRO
III
BRAIN
AQUEDUCT OF SYLVIUS
IV
LUSCHKA
MAGENDIE
LUSCHKA
CENTRAL CANAL
SPINAL CORD
SPINAL SUBARACHNOID SPACE (S-SAS)

Figure 5. Classification of CSF circulation abnormalities.

The site of obstruction may be further characterized by injecting metrizamide through a shunt tube in a lateral ventricle (CTV).[7,44] It should be noted that in a situation in which a large intravenously enhancing mass is present, further characterization with subarachnoid contrast medium is usually not necessary.

4.1.2. Congenital. *a. Nonenhanced CT.* Congenital disorders are often associated with complete or incomplete obstruction of the ventricular system. The nonenhanced CT scan is the most useful diagnostic tool for defining the location and extent of the morphological abnormality. In *aqueductal stenosis,* the third and lateral ventricles are prominently enlarged. Drawing diagnostic conclusions from fourth ventricular size is treacherous, since the fourth ventricle often appears small with CT scanning when it is actually normal in size or even enlarged. With the use of CT, both intraaxial and extraaxial *intracranial cysts*[35] are sharply defined as low-density, well-circumscribed abnormalities.[33a] With a Dandy–Walker cyst, a distinct fourth ventricle cannot be defined and is incorporated into the posterior fossa cyst, creating a cystic anterior peak near the midline (sombrero sign) (Fig. 9). Posterior fossa extraaxial cysts (PFEACs) are often unilateral and may displace the adjacent brainstem and fourth ventricle. However, the dynamic significance and detailed communications of these cysts cannot be delineated without the use of intrathecal contrast media.[7,11] Agenesis of the corpus callosum, absence of the septum pellucidum, and other associated midline anomalies are well visualized using both axial and coronal scanning. The ventricular enlargement and

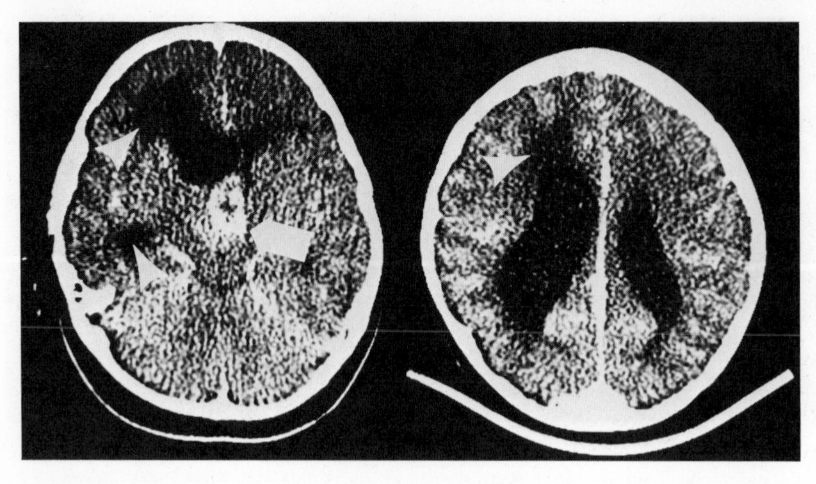

Figure 6. Ventricular hydrocephalus with periventricular edema. Large enhancing mass (arrows) in the hypothalamic region (pathology: astrocytoma, Grade 2–3) obstructing the left lateral ventricle more than the right. Prominent ipsilateral periventricular interstitial edema (arrowheads) and temporal horn enlargement (arrowheads).

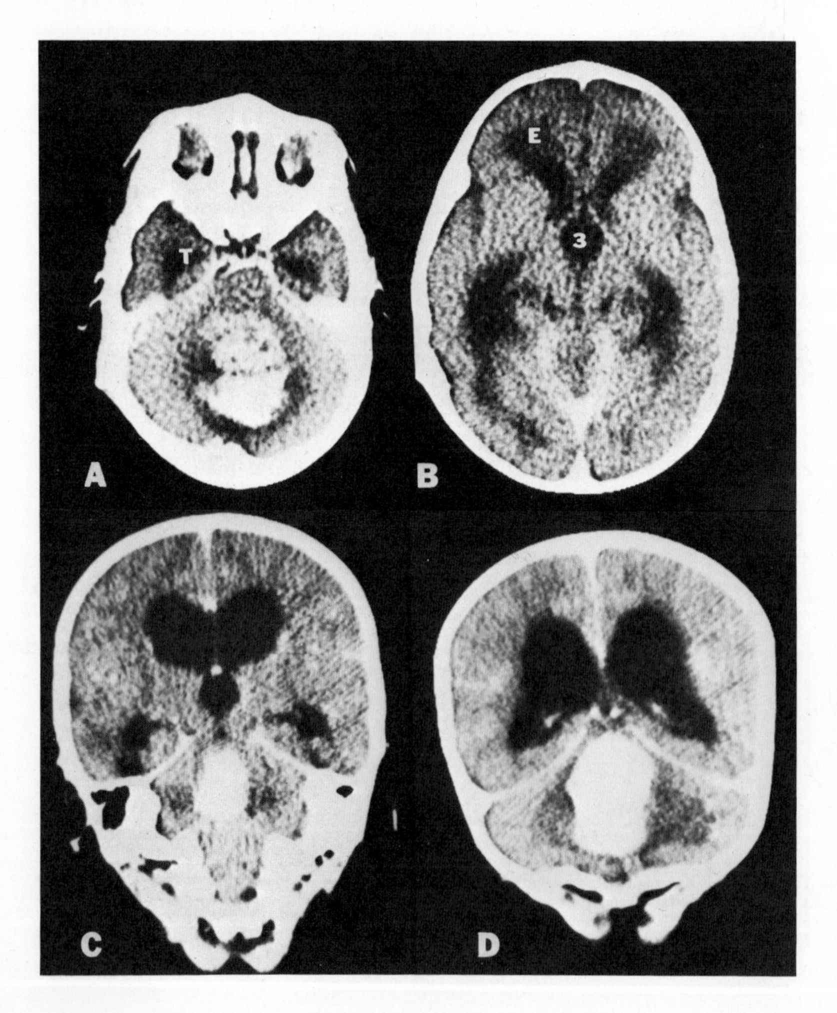

Figure 7. Ventricular hydrocephalus with periventricular edema. (A,B) Axial CT. Large enhancing midline cerebellar mass with surrounding diminished density (pathology: astrocytoma, partially cystic) resulting in dilatation of the third ventricle (3), enlargement of the temporal horns (T) of the lateral ventricles, and extensive periventricular interstitial edema (E). (C,D) Coronal CT. The relationship of the midline cerebellar mass to the tentorium is better defined, but the periventricular edema is not as well visualized due to the decreased contrast resolution on coronal CT.

Table 1. Hydrocephalus: Observations from Nonenhanced CT

CSF circulation abnormality	Enlarged third ventricle[a]	Round anterolateral frontal horns[a]	Enlarged temporal horns[a]	Periventricular interstitial edema[a]	Enlarged basal SASs and sulci[a]
Normal	0	0	0	0	0
Ventricular	2–3	2–3	2–3	2–3	0–1
Subarachnoid	2–3	2–3	1–2	1–2	0–1
Brain	1–2	1–2	0–1	0–1	2–3

[a] Severity: (0) none; (1) mild; (2) moderate; (3) marked.

CSF circulation abnormalities associated with *Chiari II malformations* and *basilar impression* are discussed below.

b. Intravenously Enhanced CT. All the congenital abnormalities are characterized by an absence of abnormal enhancement following the infusion of iodinated intravenous contrast medium (Figs. 10 and 11).

c. Intrathecally Enhanced CT. The use of CTC and CTV has obviated the need for traditional pneumoencephalography, ventriculography, and angiography in the evaluation of congenital hydrocephalus. The diagnosis of *aqueductal stenosis* is made with ease and certainty by injecting metrizamide into a ventricular shunt and obtaining a CT scan (Fig. 12). The presence of contrast medium in the lateral and third ventricles, an irregular configuration of the contrast-filled cephalad aqueduct and posterior third ventricle, and the absence of metrizamide in the fourth ventricle and basal subarachnoid cisterns are definitive diagnostic criteria. If the aqueductal stenosis occurs with an *Arnold–Chiari II malformation*[4,20,35a,38] (Fig. 13), other associated features such as a hypoplastic falx cerebri (flat callosal angle of the lateral ventricular roofs), inferomedial pointing of the inferior border of the lateral ventricles (coronal CT), an absent septum pellucidum, a large massa intermedia, and an anteroinferiorly directed outpouching of the third ventricle may be visualized. Following CTV, CTC may be performed to define the fourth ventricle as well as other Chiari II abnormalities (elongated, inferiorly displaced

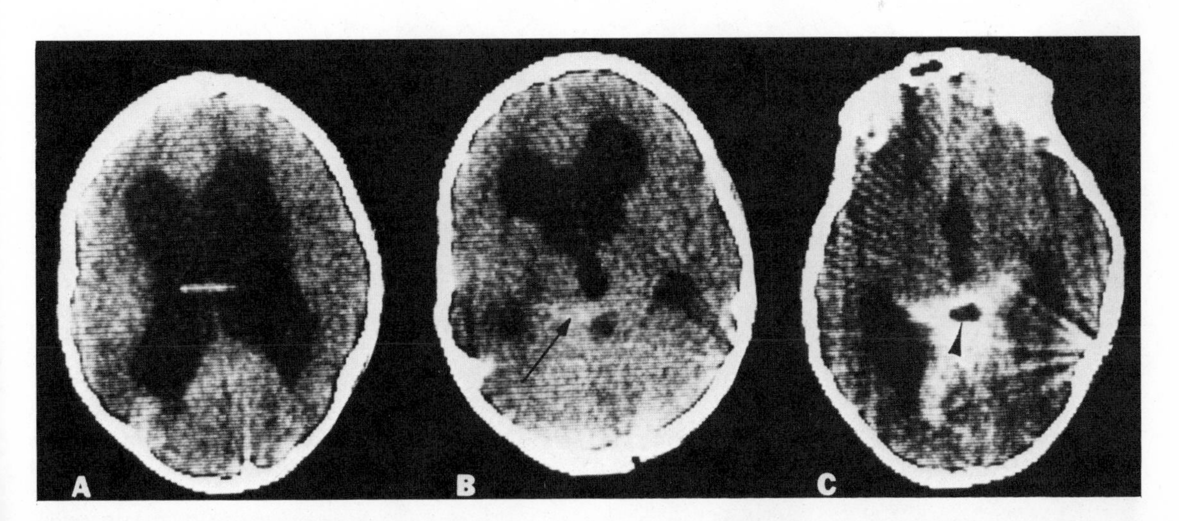

Figure 8. Posterior third ventricular region mass with ventricular hydrocephalus: metrizamide CTC. (A,B) CT: intravenous enhancement. A shunt tube is present in the markedly dilated lateral ventricles. Irregular speckles of increased density are seen in the posterior third ventricular region (arrow), but are quite subtle. (C) CT: intrathecal enhancement. An obvious irregular filling defect (arrowhead) at the confluence of the quadrigeminal, superior cerebellar, and vein of Galen cisterns represents a germinoma. No metrizamide has entered the dilated third ventricle (ventricular obstructive pattern).

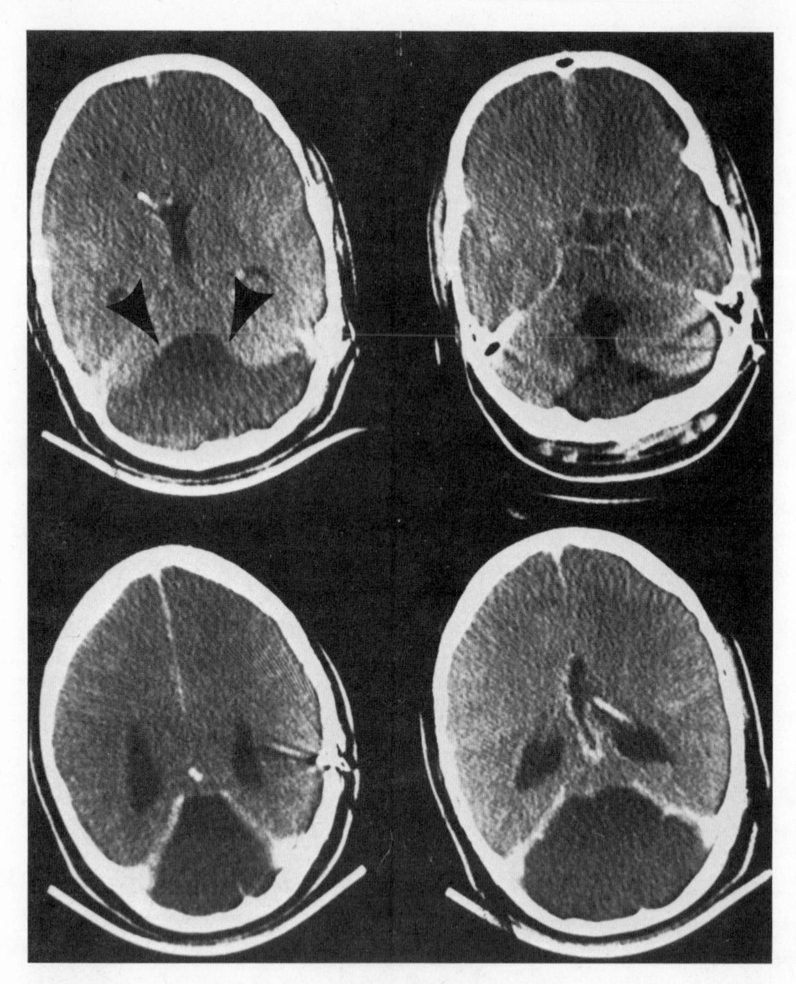

Figure 9. Dandy–Walker cyst: nonenhanced CT. A large, well-circumscribed area of diminished density in the posterior fossa is in continuation with the fourth ventricle [sombrero sign (arrowheads)]. Hydrocephalus and agenesis of the corpus callosum accompanied the posterior fossa cyst.

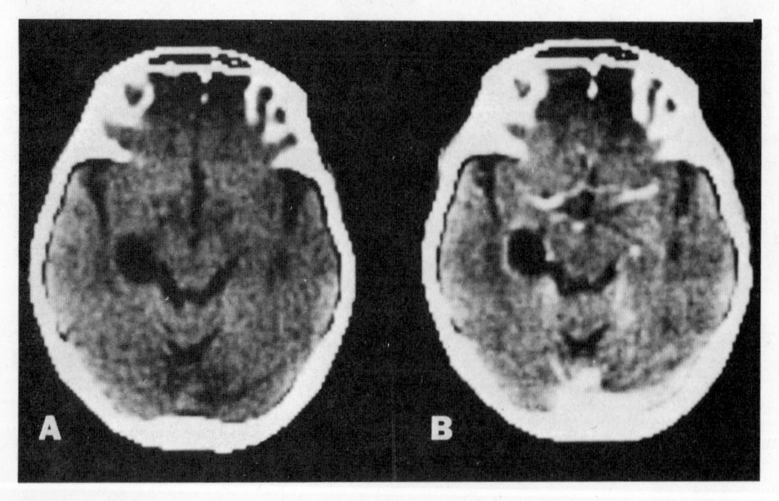

Figure 10. Perimesencephalic extraaxial cyst. (A) Nonenhanced. Area of diminished density with subtle distortion of the adjacent midbrain and ambient cistern. (B) Intravenous enhancement. Vessels on the surface of the low-density abnormality enhance, but no enhancement occurs within the cyst. The subtle midbrain and ambient cistern compression are again seen.

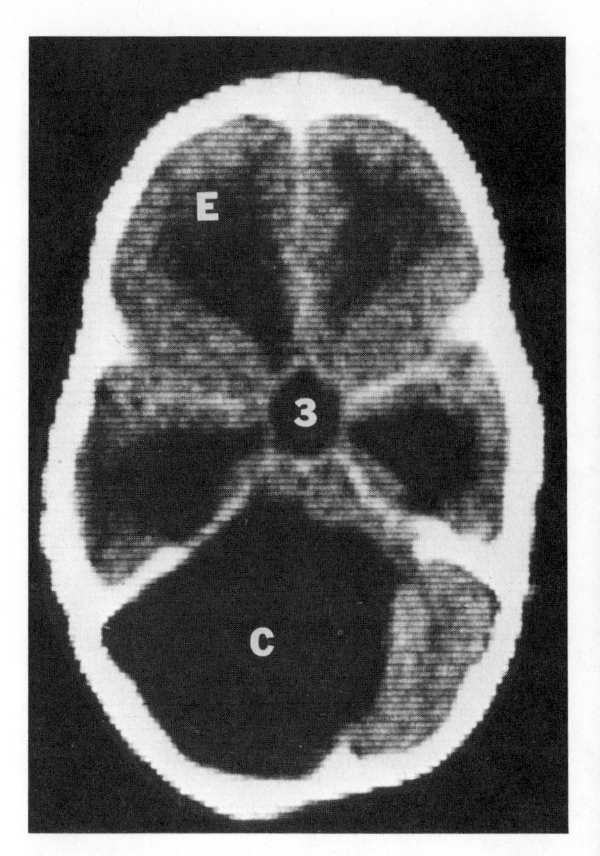

Figure 11. Posterior fossa extra–axial cyst (PFEAC) with ventricular hydrocephalus. A large, nonenhancing, sharply circumscribed, diminished-density abnormality (C) fills the left posterior fossa, compressing the midbrain and cerebellum. The characteristic features of ventricular obstructive hydrocephalus are present, including extensive third ventricular (3) enlargement, prominently dilated temporal horns (T), and periventricular interstitial edema (E).

fourth ventricle; hydromyelia; beaking deformity of the tectum[2]; anterior compression of the quadrigeminal cistern).

An accurate definition of *intracranial cysts* and their dynamic significance is permitted by using positive-contrast CTC, CTV, or both.[7,11] A simplified classification of cystic abnormalities permitting therapeutic decisions has been developed (Table 2). Thus, the previously complicated distinction of a large cisterna magna vs. a noncommunicating (or delayed communicating) PFEAC vs. a Dandy–Walker cyst is simplified. Metrizamide introduced by lumbar puncture will rapidly fill a *large cisterna magna* (Fig. 14). The *PFEAC* will not fill with contrast medium, and distortion of the adjacent brainstem and basal SAS is often evident. On delayed scans (6–12 hr after metrizamide introduciton), a small amount of contrast medium may increase the attenuation coefficient of a PFEAC, suggesting slow diffusion into the cyst from the contiguous SAS and distin-

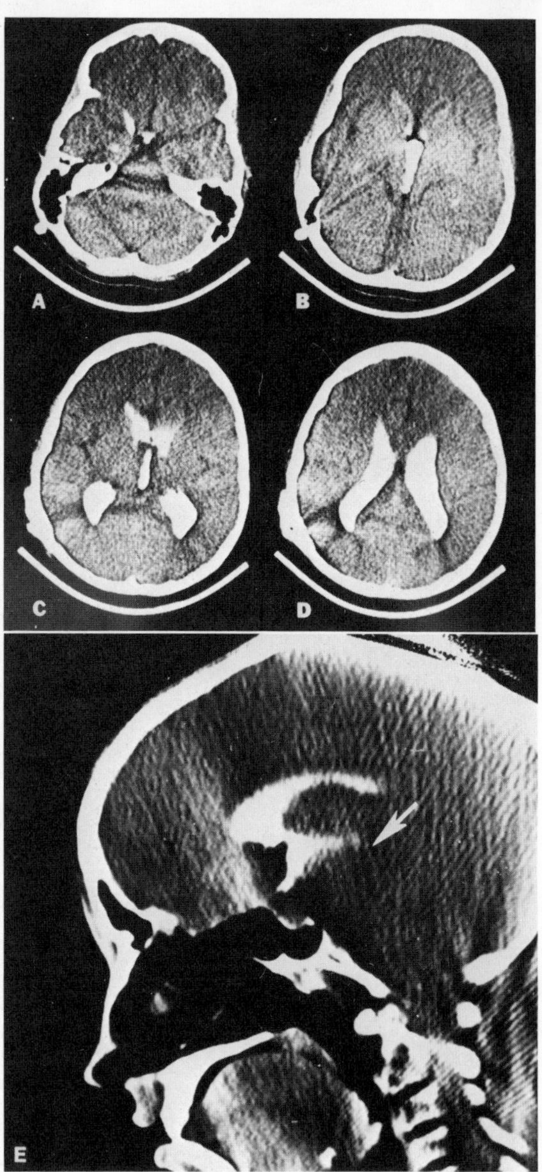

Figure 12. Aqueductal stenoisis: metrizamide CTV. (A–D) Following the instillation of 5 ml of 190 mg I/ml metrizamide via a ventricular shunt, the contrast medium fills the third and lateral ventricles but does not enter the fourth ventricle or basal subarachnoid cisterns, thereby definitively establishing the presence of aqueductal stenosis. The irregularity of the contrast-filled posterior third ventricle is commonly seen. (E) Saggital CTV further establishes the presence of an occluded aqueduct (arrow).

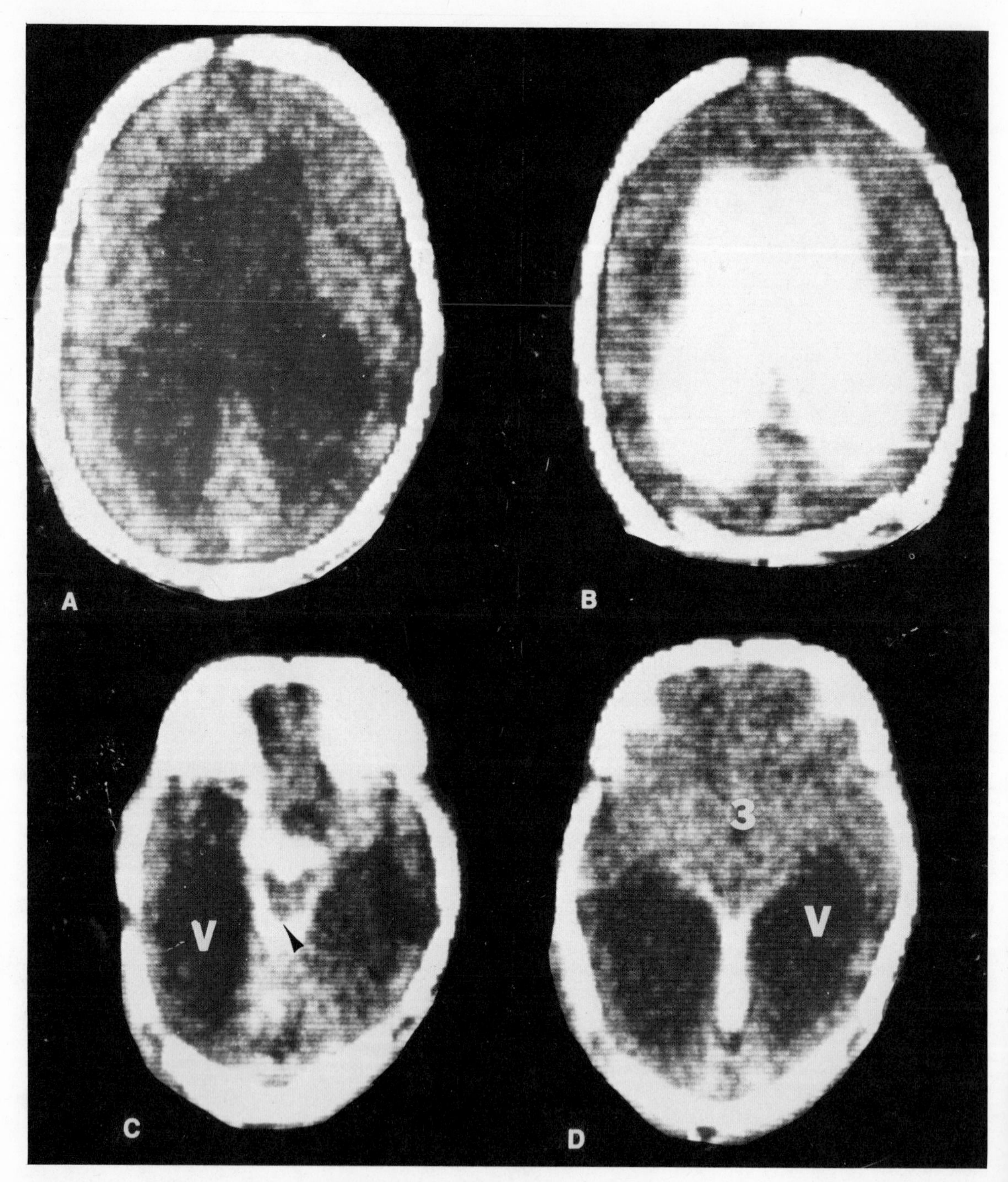

Figure 13. Chiari II with aqueductal stenosis. (A) Nonenhanced CT. Dilated lateral ventricles with an unusual acuteness of the anterior margin. (B) Metrizamide CTV. Metrizamide injected into the lateral ventricles is mixed evenly but does not proceed caudal to the aqueduct of Sylvius. (C,D) Metrizamide CTC. Metrizamide introduced by lumbar puncture does not enter the third (3) or lateral (V) ventricles. A posterior beaking (arrowhead) of the midbrain is demarcated by the contrast medium in the surrounding cisterns.

Table 2. CT Classification of Intracranial Cysts[a]

				CTC		CTV	
Type of cyst	Hydrocephalus	Ventricle/SAS displacement	Intravenous enhancement	Cyst filling	Adjacent SAS filling	Cyst filling	Adjacent SAS filling
Large cisterna magna	0	0	0	+	+	+	+
Extraaxial							
Communicating							
Immediate	±	0	0	+	+	+	+
Delayed	+	+	0	±	+	±	+
Noncommunicating	+	+	0	0	+	0	+
Dandy–Walker	+	±	0	0	+	+	0
Neuroepithelial	+	+	0	0	+	0	±
Intraaxial							
Porencephalic	+	0	0	±	+	+	+
Cerebral/cerebellar	±	+	0	0	+	0	+

[a] (SAS) Subarachnoid space; (CTC) CT cisternography; (CTV) CT ventriculography.

guishing the PFEAC from an *intraaxial cerebellar cyst* into which no contrast material enters. The *Dandy–Walker cyst* is best delineated using CTV, which shows filling of a large posterior fossa cystic region with no (or minimal) filling of the basal SASs (Fig. 15). In addition, a supratentorial *porencephalic cyst* may be distinguished from an intra- or extraaxial cyst by the free entry of metrizamide from the lateral ventricle into the previously low-density abnormality, as opposed to nonentry (e.g., colloid neuroepithelial cyst, Sylvian fissure extraaxial cyst). A subtle area of *encephalomalacia* is often defined by the absence of cerebral staining on delayed scans.[8,12,13]

4.1.3. Degenerative: Nonenhanced CT, Intravenously Enhanced CT, and Intrathecally Enhanced CT. Ventricular hydrocephalus does not occur with degenerative abnormalties (see Sections 4.2 and 4.3).

4.1.4. Vascular. *a. Nonenhanced CT.* Although vascular abnormalities are rare causes of ventricular hydrocephalus, two processes may lead to obstruction of the ventricular system at the aqueduct or posterior third ventricular level. On the nonenhanced CT scans, prominent third and lateral ventricular enlargement and an irregular increased density in the vein of Galen cistern region are seen with *vein of Galen vascular malformations.* Variable obstruction of the cephalad aqueduct or posterior third ventricle has been noted with *basilar artery ectasia* or *aneurysm.*[14,15]

b. Intravenously Enhanced CT. The marked dilatation of the vein of Galen and associated enlarged draining veins are well demarcated following the injection of iodinated contrast medium. Intravenous contrast enhancement may also assist in delineating an ectatic or aneurysmal basilar artery.

c. Intrathecally Enhanced CT. Although the extent of and cisternal distortion resulting from basilar artery abnormalities may be better defined using intrathecal metrizamide, CTC is usually not necessary when evaluating vascular abnormalities.

4.2. Subarachnoid Hydrocephalus

4.2.1. Neoplastic, Infectious and Toxic. *a. Nonenhanced CT.* The cortical sulci and subarachnoid cisterns are usually normal in size or mildly dilated when the meninges are violated by neoplasm, infection, or blood. Enlargement of the lateral ventricles is usually present. Bulbous dilatation of the temporal horns of the lateral ventricles and the third ventricle may be visualized, and the extent of periventricular edema relates to the acuteness of the arachnoid insult.

b. Intravenously Enhanced CT. Abnormal enhancement in a subarachnoid location reflects the abnormal porosity of the vasculature in the *meninges infiltrated* by neoplasm [e.g., central nervous system (CNS) neoplasm seeding, leukemia, lymphoma, melanoma] or infection (e.g., bacteria, fungus, tuberculosis). With long-standing asymptomatic ventricular enlargement of undetermined origin (often unrecognized birth or childhood insult), enhanced CT excludes a neoplastic mass.

c. Intrathecally Enhanced CT. Irregular mixing with multiple small filling defects may be noted in the basal subarachnoid cisterns. Reflux of cisternal metrizamide (CTC) and delayed clearance of ventric-

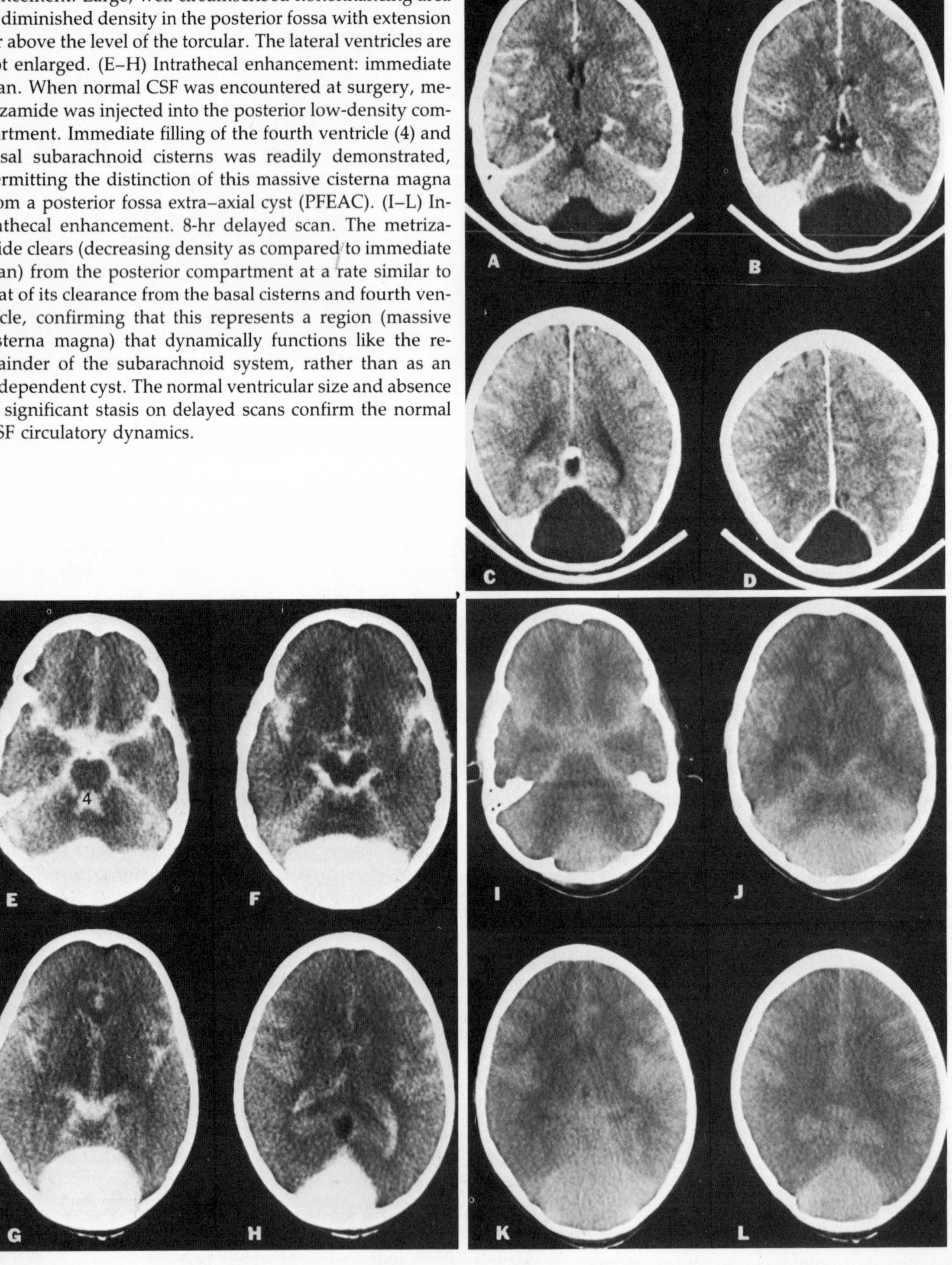

Figure 14. Massive cisterna magna. (A–D) Intravenous enhancement. Large, well-circumscribed nonenhancing area of diminished density in the posterior fossa with extension far above the level of the torcular. The lateral ventricles are not enlarged. (E–H) Intrathecal enhancement: immediate scan. When normal CSF was encountered at surgery, metrizamide was injected into the posterior low-density compartment. Immediate filling of the fourth ventricle (4) and basal subarachnoid cisterns was readily demonstrated, permitting the distinction of this massive cisterna magna from a posterior fossa extra–axial cyst (PFEAC). (I–L) Intrathecal enhancement. 8-hr delayed scan. The metrizamide clears (decreasing density as compared to immediate scan) from the posterior compartment at a rate similar to that of its clearance from the basal cisterns and fourth ventricle, confirming that this represents a region (massive cisterna magna) that dynamically functions like the remainder of the subarachnoid system, rather than as an independent cyst. The normal ventricular size and absence of significant stasis on delayed scans confirm the normal CSF circulatory dynamics.

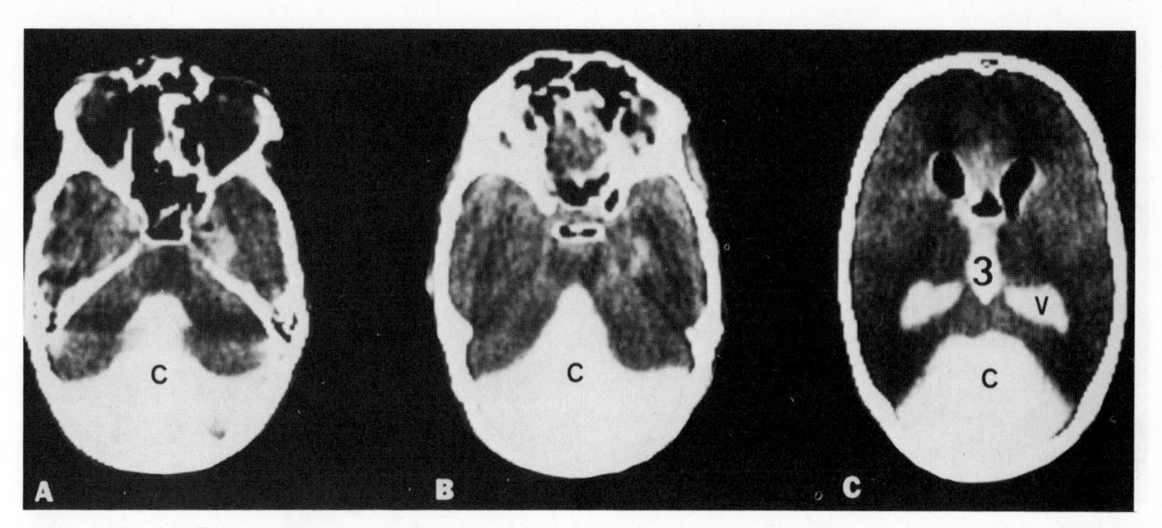

Figure 15. Dandy–Walker cyst: metrizamide CTV. (A–C) Filling of the third (3) and lateral (V) ventricles as well as a large posterior fossa cyst (C) that encompasses the region of the fourth ventricle with metrizamide. Only a minimal amount of contrast medium enters the basal subarachnoid cisterns, confirming the presence of almost total fourth ventricular outlet obstruction as seen with a Dandy–Walker cyst.

ular metrizamide (CTV) are associated with the delayed appearance of the contrast medium over the cerebral convexities and in the parasagittal region. When *iophendylate (Pantopaque) is encysted* in the SASs from a previous examination, incomplete filling of the basal SASs and asymmetrical staining of the brain have been recognized.[8]

4.2.2. Congenital. *a. Nonenhanced CT.* Multiple anomalies of the brain substance and SASs (see Section 4.1.2) are often associated with the *Arnold–Chiari II malformation*.[2,4,20,38] Hydrocephalus most prominently involves the frontal horns of the lateral ventricles in infants with this disorder. The shunted child often has persistent enlargement of occipital horns even when the frontal horns are collapsed, as well as a peculiar dilatation of the superior cerebellar, quadrigeminal, and other posterior subarachnoid cisterns. In some children, the cortical sulci and perimesencephalic SASs may be dilated.

b. Intravenously Enhanced CT. There is no evidence for abnormal leakage of the blood–brain barrier. An abnormal position of the torcula and tentorium cerebelli and associated anomalous vascular structures are sometimes noted with congential defects, as has been reported angiographically.

c. Intrathecally Enhanced CT. Metrizamide may fill SASs that are unusual in size and location. In general, the positive-contrast medium ascends symmetrically and at the proper time to the convexity and pasasagittal CSF spaces; however, delayed ascent and ventricular reflux and stasis may be seen. Although we have not studied any achondroplastics or other patients with severe basilar invagination using CTC, a basal or incisural block would be expected in some individuals.

4.2.3. Normal-Pressure Hydrocephalus. *a. Nonenhanced CT.* The combination of gait apraxia, dementia, incontinence, normal or mildly elevated CSF opening pressure at lumbar puncture, and a positve CSF infusion study is considered consistent with the nonspecific diagnosis of normal-pressure hydrocephalus (NPH).[1,24,34,47] Although often considered iodiopathic, this abnormality may reflect a previous insult to the SASs (e.g., hemorrhage, infection, other toxin). The usual CT picture (Fig. 16) consists of mild to moderate periventricular interstitial edema, prominent ventricular enlargement, and normal SASs,[8,17,31] However, the basal and cortical SASs may be mild to moderately dilated. Particularly prominent enlargement of the third ventricle and temporal horns of the lateral ventricles is common. It seems that in some individuals with Alzheimer's disease, an accompanying CSF circulation abnormality of varying degree is present.

b. Intravenously Enhanced CT. No enhancing abnormality is present. Symmetrical enhancement occurs normally in the choroid plexi of the lateral ventricles.

c. Intrathecally Enhanced CT. Using CTC, the characteristic picture (Table 3) consists of ventricular reflux and stasis (24+ hr) of positive-contrast medium[6,7,22,23,30,36,45,46] (Figs. 17 and 18), delayed or

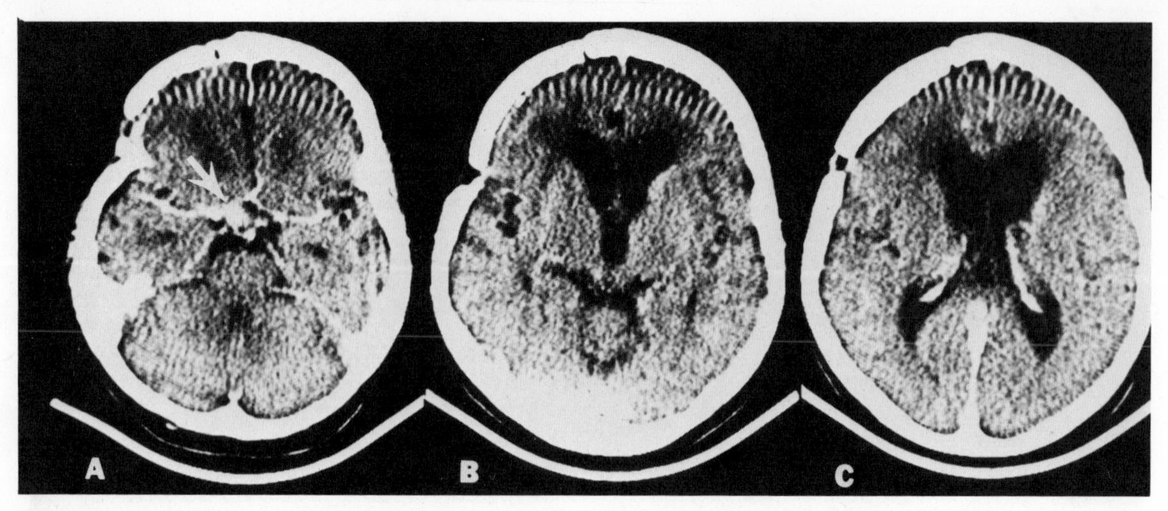

Figure 16. Subarachnoid (communicating) hydrocephalus: intravenously enhanced CT. (A) Visualization of the circle of Willis during iodinated contrast medium infusion suggests an internal carotid–posterior communicating artery junction aneurysm (arrow). (B,C) Moderate enlargement of the third and lateral ventricles after previous subarachnoid hemorrhage and surgery. Periventricular interstitial edema is well visualized, suggesting the acute or subacute nature of the hydrocephalus.

absent appearance of the convexity and parasagittal blush,[6,8,45] and a periventricular halo of diminished density that variably increases in attenuation coefficient on delayed scans[6,7,30] (Figs. 17–19). If the contrast medium is directly placed (CTV) or gravitationally positioned (CTC) in the lateral ventricles, the ventricular volume becomes an important factor in defining the abnormal clearance rate.[41,42] Further clinical experience is needed before deciding on the utility of these and other observations[39] in choosing the individuals who will successfully respond to a ventricular diversionary shunt.

4.3. Brain Hydrocephalus

4.3.1. Neoplastic, Infectious, and Metabolic: Nonenhanced CT, Intravenously Enhanced CT, and Intrathecally Enhanced CT. Mild ventricular enlargement and even prominent SAS enlargement may accompany neoplastic processes, particularly when the patient is nutritionally deficient or receiving chemotherapy. Chronic CNS infections (e.g., fungus) and metabolic disorders (e.g., vitamin B_{12} deficiency, hypothryoidism) may exhibit enlarged cortical sulci that return to normal following successful therapy.[29] No specific findings are present with either intravenous or intrathecal enhancement.

4.3.2. Congenital: Nonenhanced CT, Intravenously Enhanced CT, and Intrathecally Enhanced CT. An associated loss of brain substance (dilated SASs) commonly accompanies many of the complex congenital anomalies that afflict the CNS.

4.3.3. Degenerative. *a. Nonenhanced CT.* The CT findings in most of the degenerative disorders (e.g., Alzheimer's, Pick's, Creutzfeld–Jakob, Parkinson's, Huntington's) are nonspecific, consisting of mild to marked enlargement of the cortical sulci

Table 3. CSF Circulation Patterns: Metrizamide CT Cisternography

Condition	Delayed reflux[a]	24-hr stasis[a]	Periventricular low density[a]	Convexity stain[a]		
				Absent	Delayed	Assymetrical[a]
Normal circulation	−	−	−	−	−	−
Ventricular hydrocephalus	−	− (+ + + on CTV)	−	−	−	−
Subarachnoid hydrocephalus						
Abnormal	+ + +	+ + +	+ +	+	+ +	+
Intermediate	+	+	+	−	+ +	−
Brain hydrocephalus	−	−	−	−	+	−

[a] Frequency: (−) never; (+) sometimes; (+ +) often; (+ + +) always.

(Fig. 20) and basal subarachnoid cisterns with less severely involved lateral ventricles.[17,29,31] The temporal horns of the lateral ventricles and the third ventricle are normal in size or only mildly enlarged. Periventricular interstitial edema is rarely seen. The spinocerebellar degenerations (e.g., olivopontocerebellar degeneration) may be distinguished by the more extensive involvement of the cerebellar and perimesencephalic SASs (Fig. 21).

b. Intravenously Enhanced CT. No enhancing abnormalities are present. However, in the evaluation of dementia, the enhanced CT scan has a definite role in excluding primary and metastatic malignancy.

c. Intrathecally Enhanced CT. Most commonly, CTC is normal, with symmetrical ascent of metrizamide to the convexity and parasagittal region, symmetrical cerebral staining, and no significant ventricular reflux or stasis (Table 3). However, an intermediate CSF circulation pattern[6,7] is sometimes noted consisting of a delayed ascent of the contrast medium, some degree of ventricular reflux, and a small amount of ventricular stasis at 24 hr (Fig. 22). This delayed (intermediate) pattern may be the result of either an increase in the general CSF volume or a CSF circulation abnormality occurring in association with Alzheimer's disease.

4.3.4. Vascular: Nonenhanced CT, Intravenously Enhanced CT, and Intrathecally Enhanced CT. The lateral ventricle ipsilateral to encephalomalacia (secondary to infarction,trauma, or other condition) is dilated (Fig. 23). Research in a nonhuman primate model suggests that mild generalized ventricular enlargement may occur in the acute stages of cerebral infarction.[13]

4.4. External Hydrocephalus

4.4.1. Nonenhanced CT. A well-circumscribed, extracerebral region of diminished density is present overlying both hemispheres, predominantly over the anterior frontal convexities.[40] Usually, no ventricular shift is present, and the ventricles may or may not be enlarged. This abnormal CT finding is often associated with a large head in infants and young children, and its incidence is surprisingly high. When CT scans are repeated 6 months to 1 year later, the extracerebral low-density collection may have disappeared or it may still be present with associated ventricular enlargement. In the latter situation, the persistence or increase in hydrocephalus suggests CSF circulatory significance.

4.4.2. Intravenously Enhanced CT. Neither the anterior extracerebral collection of decreased density nor its inner margin enhances with iodinated contrast medium.

4.4.3. Intrathecally Enhanced CT. When metrizamide is introduced by lumbar puncture (CTC) and rapidly moved by gravity into the intracranial SASs, the extracerebral area of diminished attenuation coefficient does not change in density. The cortical SASs underlying the extracerebral collection fill with metrizamide (Fig. 24). We therefore believe that in almost all instances, this common pediatric abnormality (external hydrocephalus) actually represents a *chronic subdural hygroma.* The hygroma may compress or previous occult bleeding may have damaged the arachnoid granulations, producing a subarachnoid (communicating) hydrocephalus (Fig. 25), or the process may not affect the SASs and therefore not produce hydrocephalus.

5. Therapy for Hydrocephalus

One of the most important applications of CT scanning has been in the noninvasive evaluation of ventricular diversionary shunts. Both ventricular size and shunt position are readily defined, as are the various complications (Table 4) that may accompany this therapeutic modality.

Many of the complications of ventricular shunting are demonstrated by nonenhanced CT scanning. The prominent increased density of *blood within the ventricles* (Fig. 26) following shunt insertion is readily apparent. A diminished density in the brain substance adjacent to a present or previous diversionary catheter tract represents iatrogenic *encephalomalacia. Extracerebral hematomas* (high-density) and *hygromas* (low-density) are common complications of shunt-

Table 4. Diversionary Ventricular Shunt Complications

Extracranial malfunction
Blockage, migration, disconnection
Peritonitis, adhesion, perforation
Intracranial malfunction
Insertion
Ventricular hemorrhage
Encephalomalacia (along shunt tract)
Excess function
Shunt dependency
Extracerebral hematoma/hygroma
Premature craniostenosis
Inflammation
Ventriculitis
Meningitis
Abscess
Aqueductal stenosis

Figure 17. Subarachnoid (communicating) hydrocephalus: metrizamide CTC. (A) Intravenously enhanced CT. Moderate to marked ventricular enlargement with normal cortical sulci. (B–E) Metrizamide CTC. Serial scans over 48 hr map the *stasis* of the contrast medium in the lateral ventricles (metrizamide injected into the lumbosacral SAS). In addition, the *cerebral stain is delayed* in appearance and a *periventricular halo* of low density at 6 and 12 hr becomes isodense with the lateral ventricles on the 24- and 48-hr scans, suggesting transependymal migration of metrizamide. (F–I) Radionuclide (^{113}In) cisternography (lateral view). Serial scans over 48 hr confirm the prominent stasis of radionuclide in the lateral ventricles in this same patient with the clinical characteristics of normal-pressure hydrocephalus (NPH).

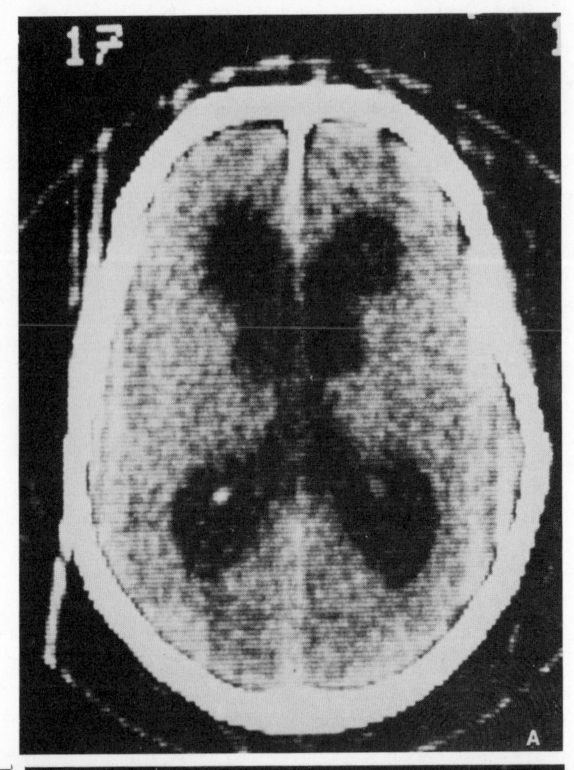

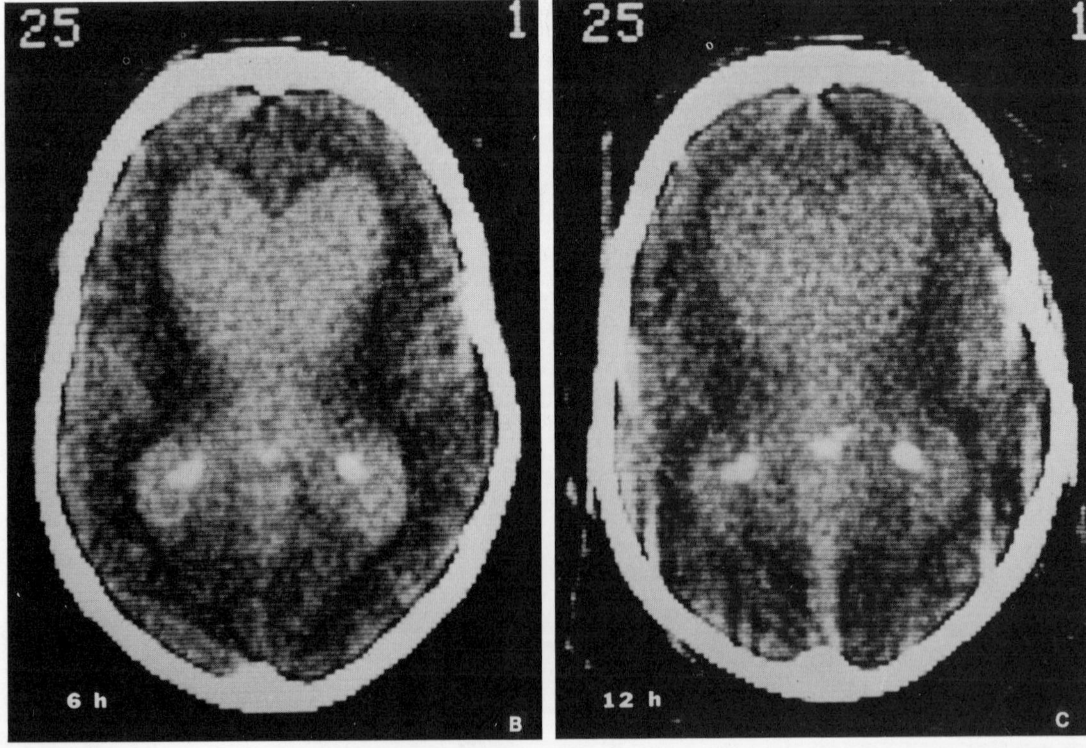

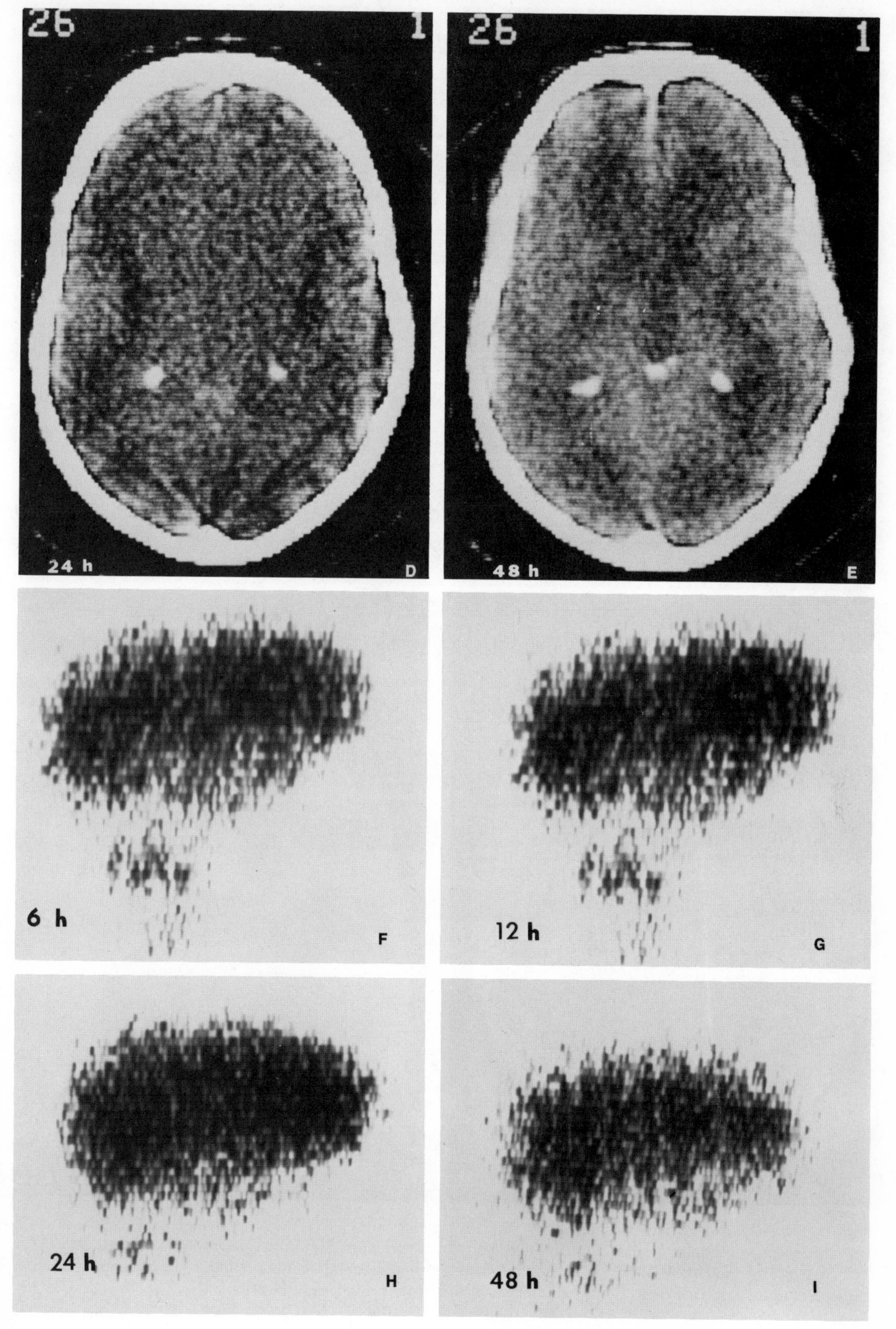

Figure 17. (*Continued*)

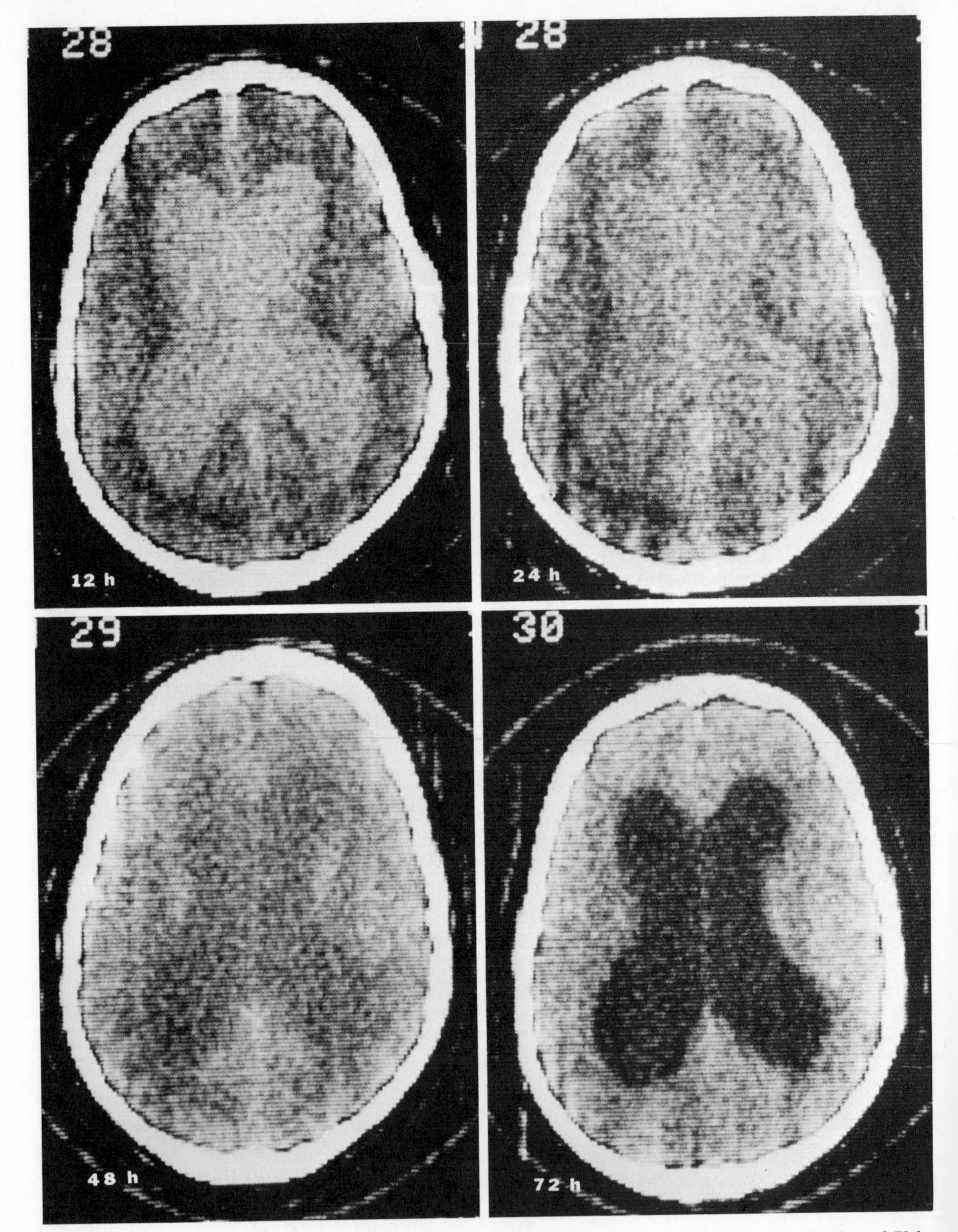

Figure 18. Subarachnoid (communicating) hydrocephalus: metrizamide CTC. Serial CT scans at 12, 24, 48, and 72 hr after lumbosacral injection of metrizamide define the typical abnormal pattern of subarachnoid hydrocephalus: (1) ventricular stasis of greater than 24 hr; (2) delayed and diminished cerebral staining; and (3) a periventricular halo of decreased attentuation coefficient that mildly increases in density on subsequent scans.

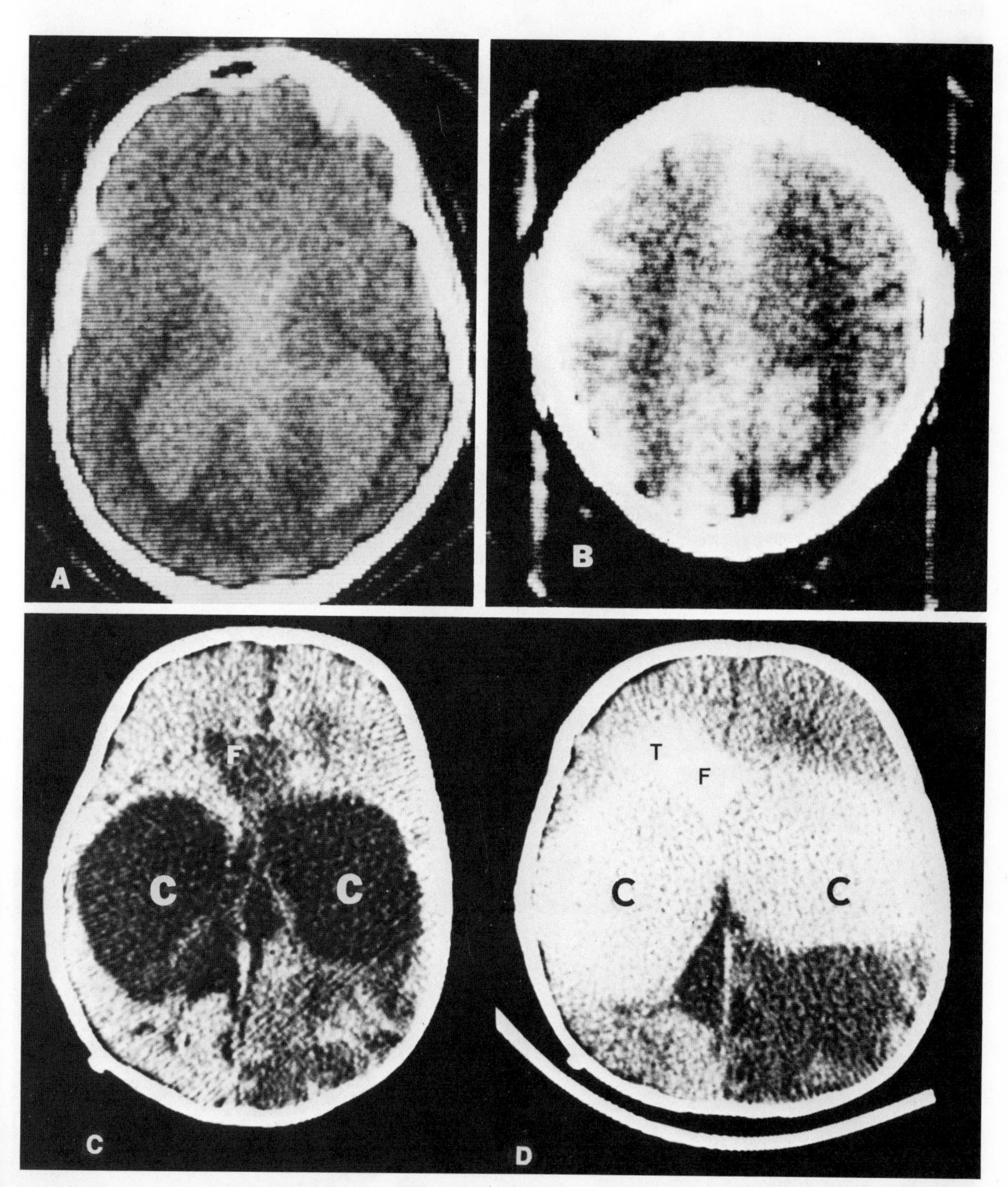

Figure 19. CSF circulation abnormalities: ancillary findings with metrizamide CTC and CTV. (A) CTC. Perventricular diminished density. (B) CTC. Asymmetrical cerebral stain (decreased on the right). (C) Nonenhanced CT. Multiple large inflammatory (status postshunt infections) cysts (C) obstructing and isolating the frontal horns of the lateral ventricles (F). (D) CTV. Scan performed 6 hr after the direct introduction of metrizamide into two of the large cysts (C) and the frontal horns (F). The extensive transependymal migration (T) of metrizamide into the adjacent white matter suggests acute ventricular (obstructive) hydrocephalus.

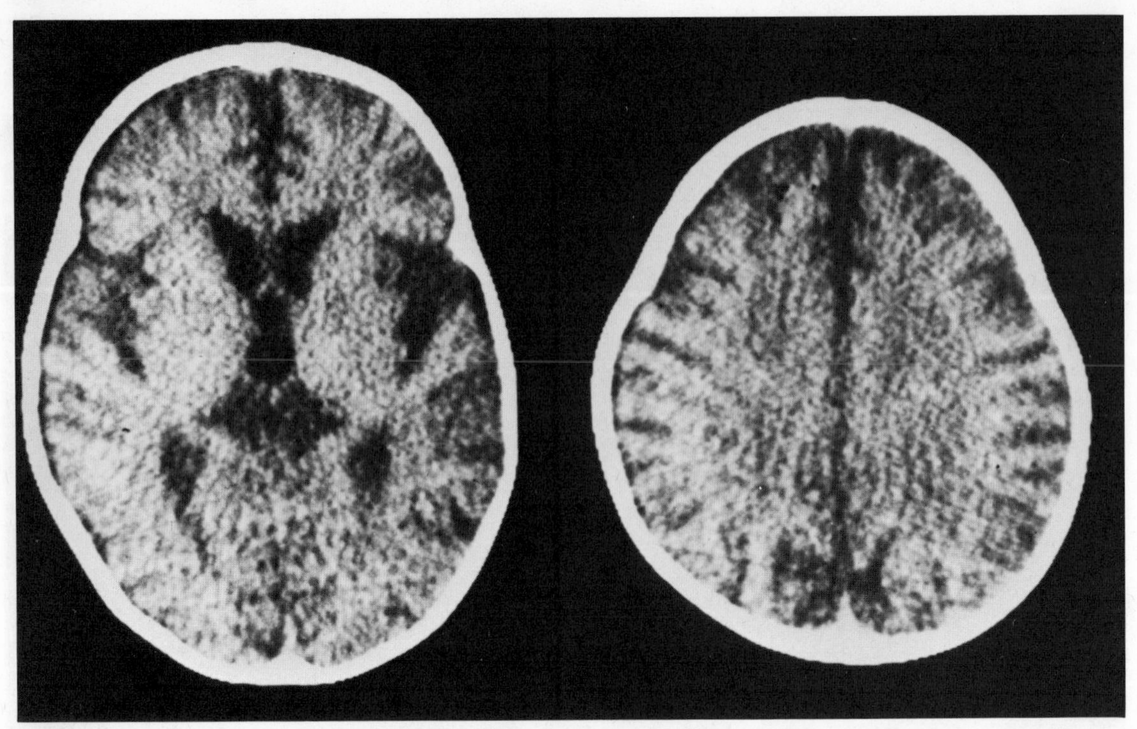

Figure 20. Brain (ex vacuo) hydrocephalus with atrophic degenerative disease. Mild to moderate ventricular dilatation as compared to prominent enlargement of the cortical sulci. No periventricular interstitial edema is present, and the temporal horns of the lateral ventricle were not visualized at more caudal scan levels.

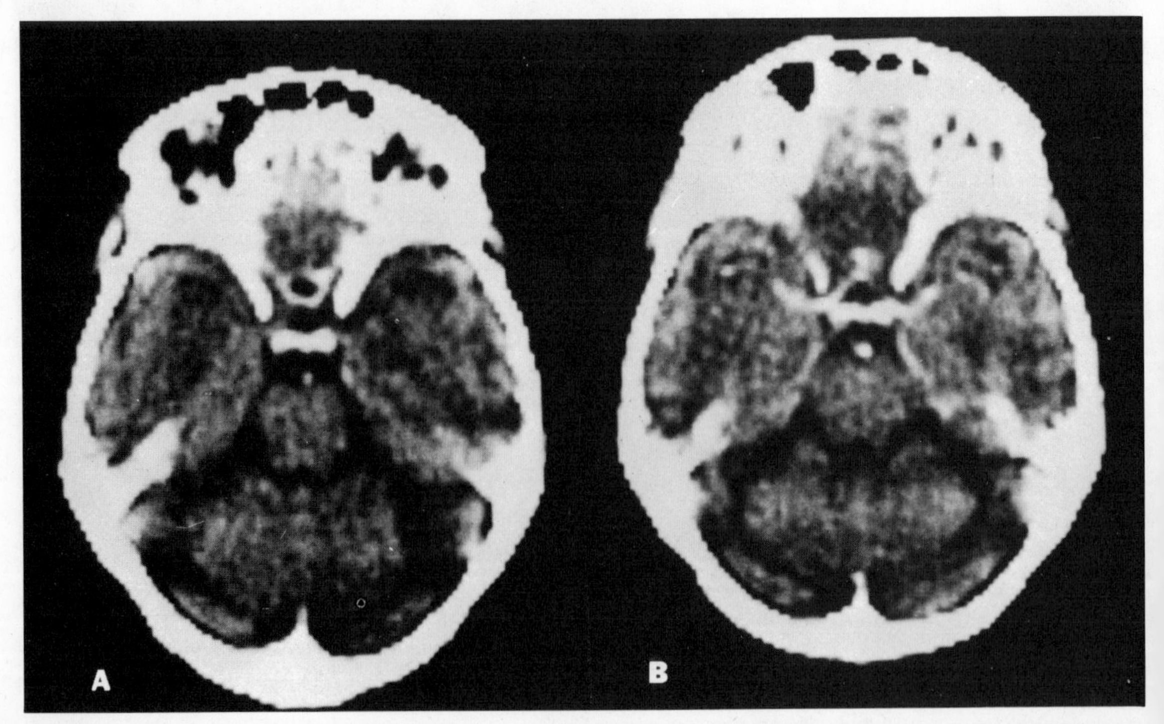

Figure 21. Spinocerebellar degeneration. (A,B) Prominent enlargement of the cerebellar and perimesencephalic SASs.

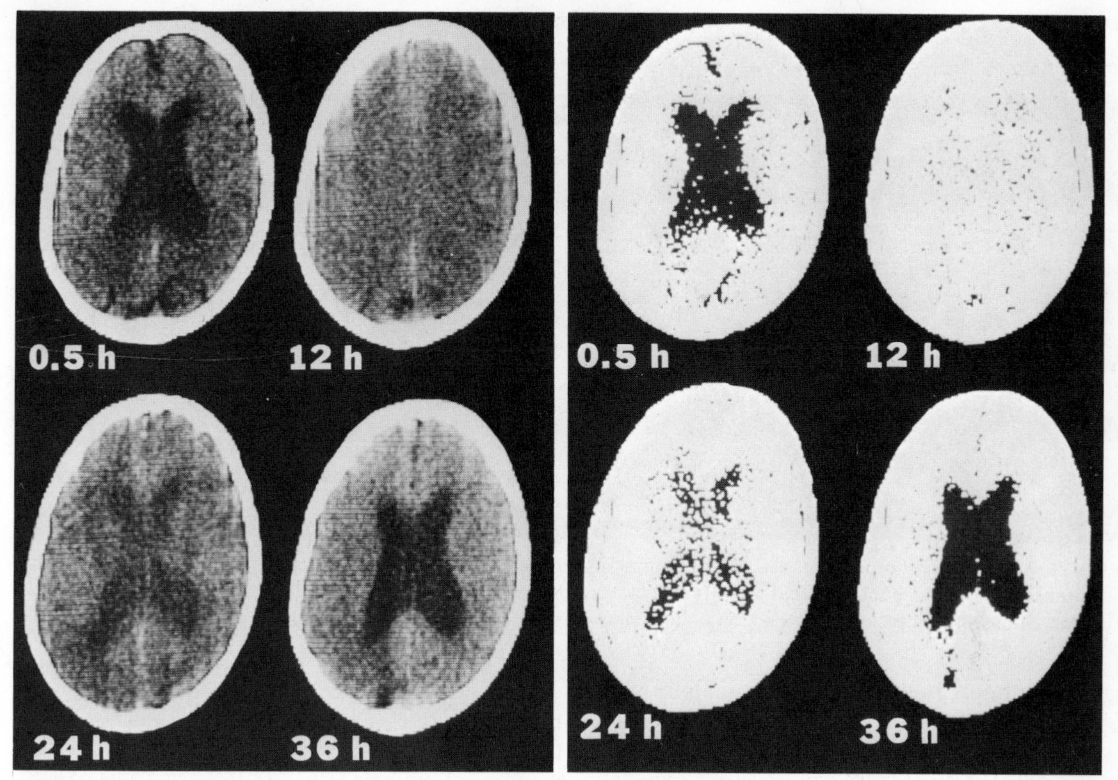

Figure 22. Intermediate CSF circulatory pattern. *Left:* Serial CT scan performed 0.5, 12, 24, and 36 hr following the lumbar instillation of metrizamide shows ventricular reflux, mild ventricular stasis at 24 hr with clearance at 36 hr, and a delayed cerebral stain. *Right:* Same CT scans as at left using measure-mode (window width 2, window level 16) photography to further display the density changes that occur in the CSF within the lateral ventricles.

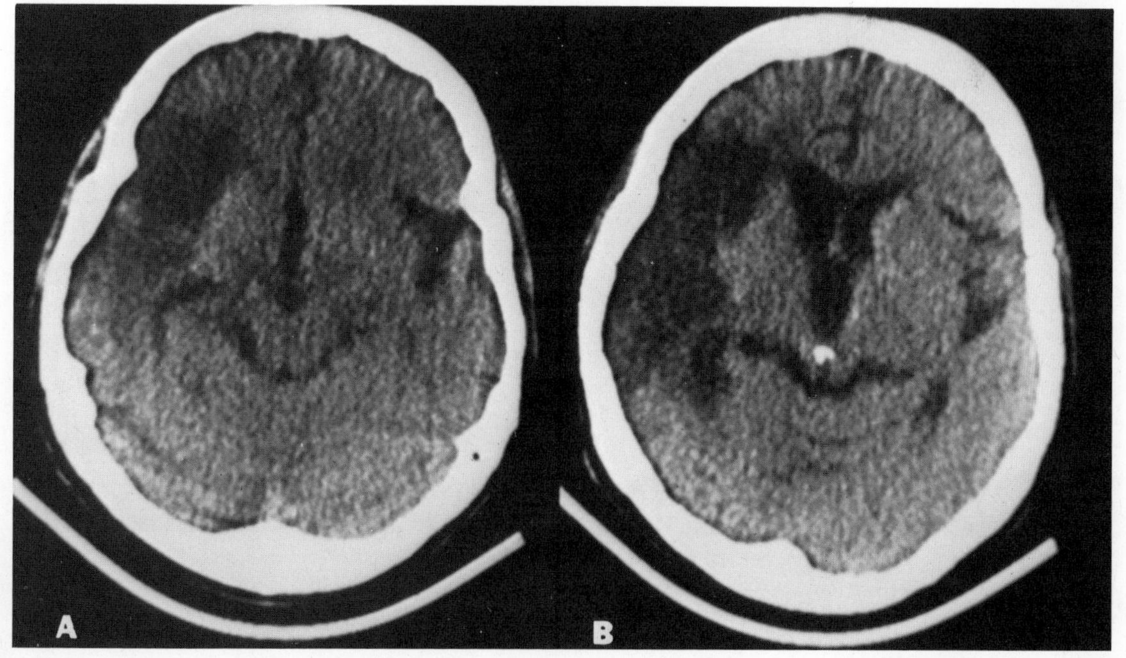

Figure 23. Brain hydrocephalus following cerebral infarction. (A,B) A large area of diminished density is present in the distribution of the middle cerebral artery. The ipsilateral lateral ventricle is enlarged, suggesting the periventricular loss of brain substance and the chronicity of the abnormality. The third ventricle and SASs are also dilated.

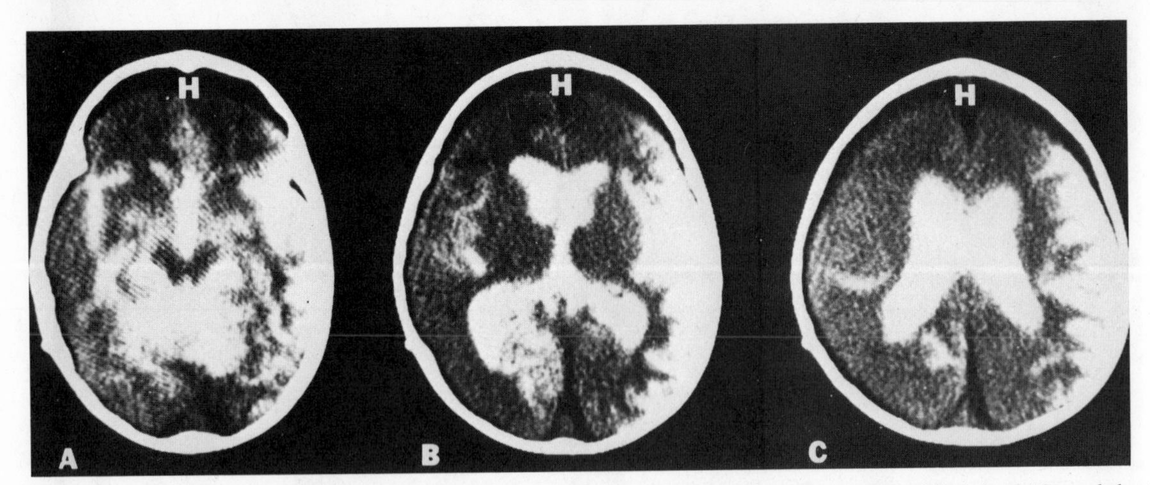

Figure 24. External hydrocephalus (subdural hygroma): metrizamide CTC. (A–C) Metrizamide filled the SASs and the entire ventricular system but did not fill the extracerebral region of decreased density (H), confirming that this represents a noncommunicating extracerebral collection (subdural hygroma).

ing procedures and are often bilateral, thereby causing no midline shift (Fig. 27). With newer-generation CT scanners and the use of intravenous contrast enhancement (if necessary), almost all extracerebral collections will be visualized. *Premature craniostenosis,*[33] another relatively common association with diversionary shunting, may be seen using CT, but is more easily observed with conventional skull radiographs.

The continued presence of large or enlarging ventricles or periventricular interstitial edema on serial CT scanning reflects probable *shunt malfunction* or insufficient function (Fig. 28). If the ventricles begin to enlarge following a period of apparent quiescence, the possibility of superimposed *aqueductal stenosis*[16] as a shunt complication should be considered. Some have suggested that collapsed frontal horns with dilated occipital horns are a sign of *shunt de-*

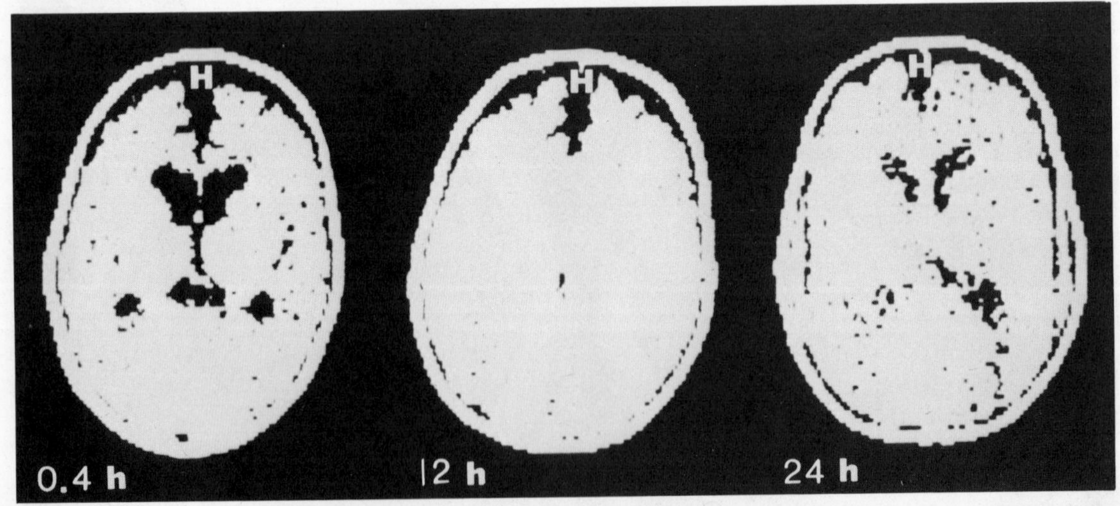

Figure 25. External hydrocephalus (subdural hygroma): metrizamide CTC. Serial CT scans (measure-mode: window width 2, window level 14) performed 0.4, 12, and 24 hr after metrizamide was introduced into the lumbosacral SAS. Even on delayed scans, no contrast medium enters the anterior extracerebral region of diminished density (H). This finding was consistently seen in a large group of children and appears to represent subdural hygroma, which may resolve without therapy. In addition, ventricular stasis of metrizamide at 12 and 24 hr suggests an accompanying disorder of CSF circulation.

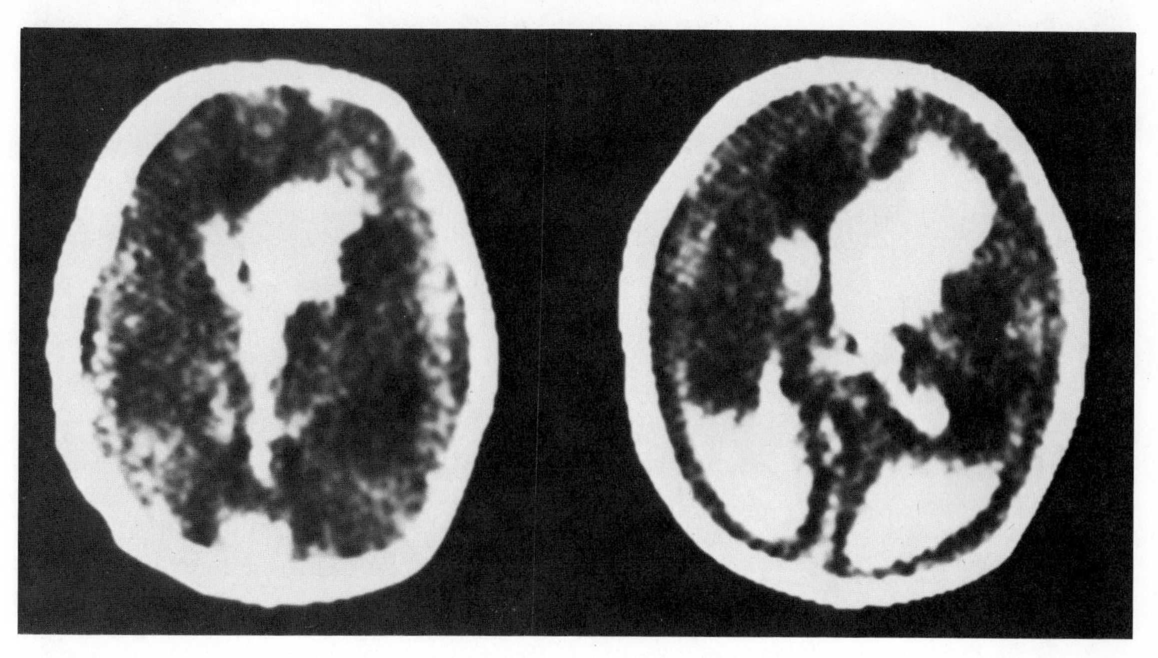

Figure 26. Shunt complication: *Intraventricular hemorrhage* following insertion of diversionary shunt in a child with a clotting disorder.

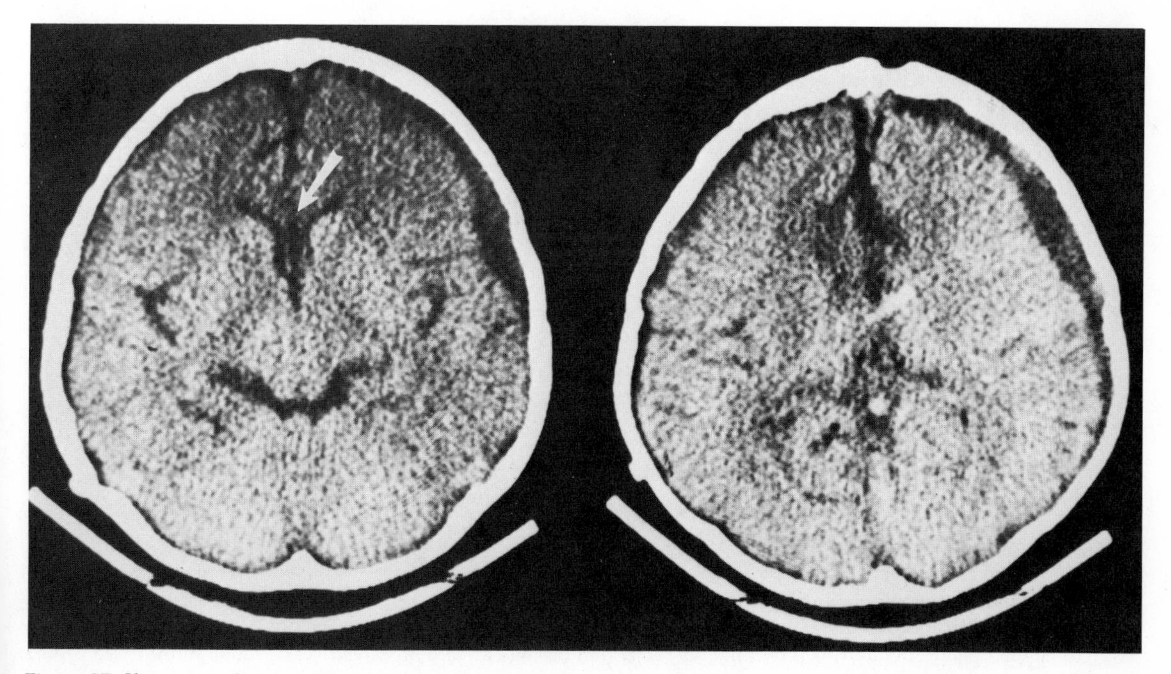

Figure 27. Shunt complication: *Chronic subdural hematomas* in a child with small ventricles following diversionary shunting. An incidental cavum septum pellucidum (arrow) is present.

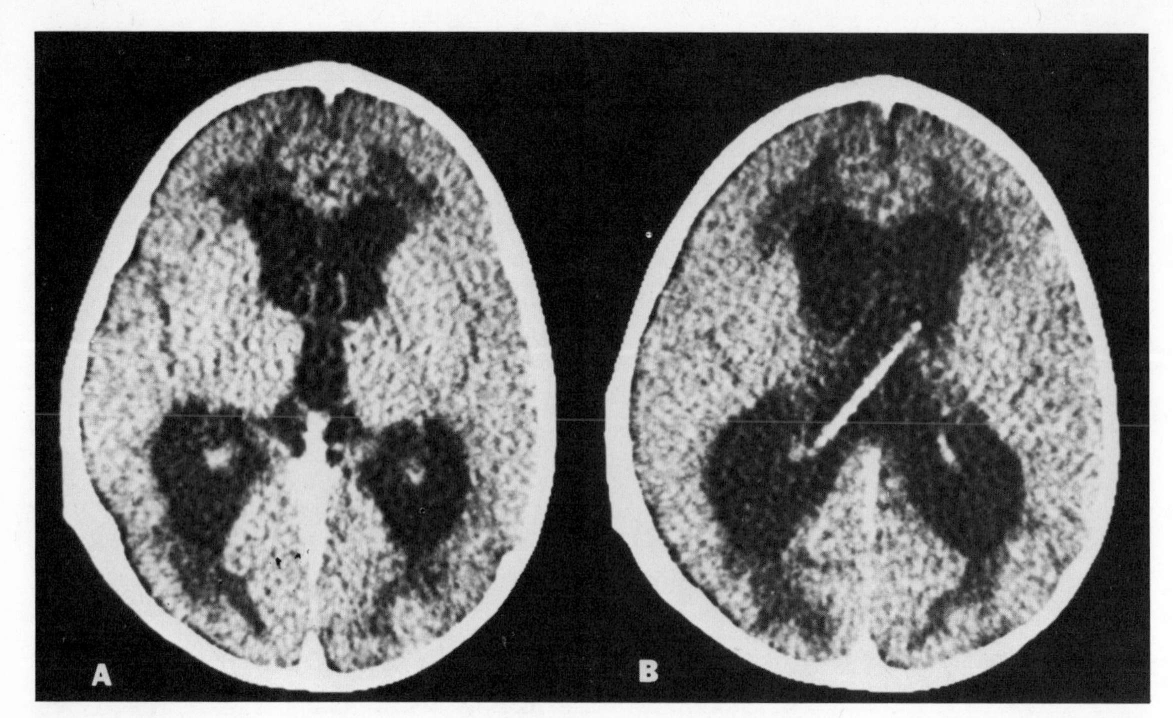

Figure 28. Shunt malfunction. (A) Prominent enlargement of the third and lateral ventricles with periventricular interstitial edema prior to shunting. (B) Shunt malfunction with no improvement in ventricular size and the continued presence (3 weeks after shunting) of periventricular edema.

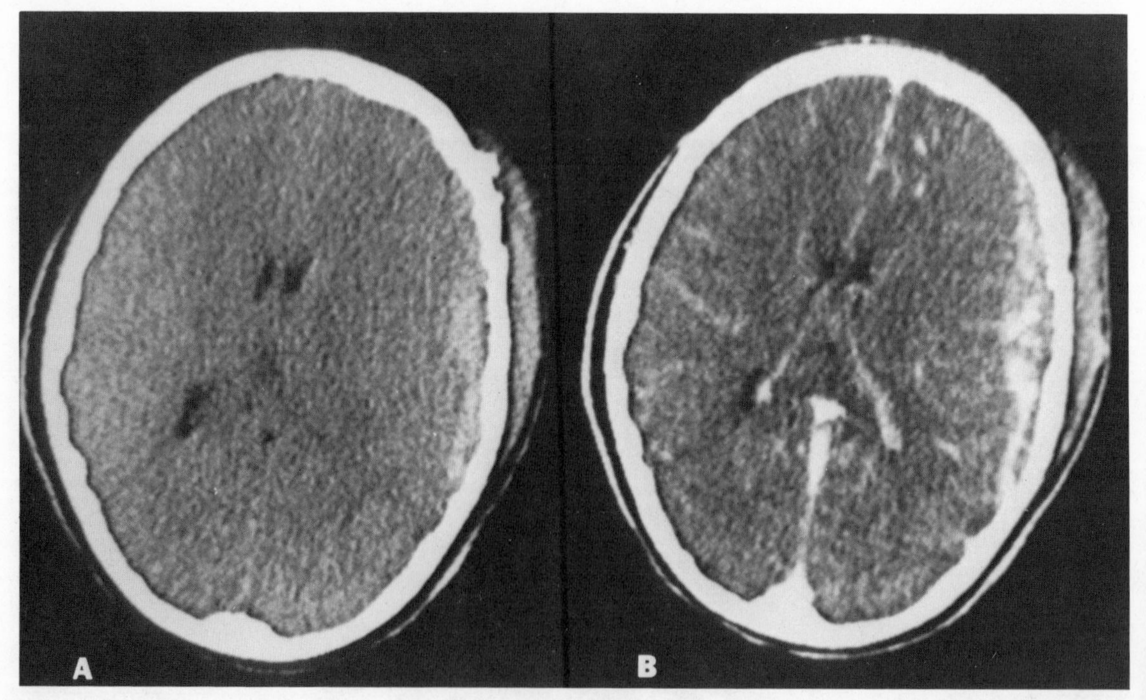

Figure 29. Subdural hematoma: intravenous enhancement. (A) Nonenhanced CT. Subtle crescent of extracerebral increased density adjacent to skull fracture and overlying soft tissue hematoma (acute trauma). (B) Intravenously enhanced CT. Prominent enhancement in the inner membrane region permits greatly improved visualization of the extracerebral hematoma.

pendence,[25] though this finding is also commonly seen with shunting in a child with an Arnold–Chiari malformation and a myelomeningocele.

Intravenous contrast enhancement is usually not necessary in evaluating diversionary shunts. However, prominent enhancement of the ependymal borders of the lateral ventricles occurs with *ventriculitis*. In addition, a prominent enhancing mass with associated vasogenic edema may represent an *abscess*. In certain subtle, *isodense extracerebral hematomas*, intravenous enhancement may prove diagnostically useful due to enhancement in the inner membrane region (Fig. 29).

Preliminary work has been performed concerning the evaluation of ventricular shunt function following the infusion of metrizamide into the lateral ventricles via the shunt.[7] *Distal (e.g., abdominal) blockage or disconnection* of the diversionary shunt is suggested by the absence of peritoneal visualization of metrizamide (body CT) following infusion into the shunt or reservoir. The rate of CSF formation and the clearance of a marker (e.g., metrizamide) from the lateral ventricles may be calculated if first-order kinetics and ventricular bulk flow clearance are assumed.[41,42] The technique is dependent on achieving uniform mixing of the infused contrast medium and ascertaining an accurate measurement of ventricular volume. A markedly delayed clearance of metrizamide from the lateral ventricles is consistent with shunt malfunction.

6. Discussion

Cranial computed tomography (CT) now dominates the radiographic evaluation of abnormalities of the CSF system. Straightforward criteria permit a reasonably organized classification of these disorders. A nonenhanced and an intravenously enhanced scan are usually adequate for diagnostic purposes. However, if more detailed dynamic information is needed, intrathecal enhancement (CTC or CTV) may be performed using a water-soluble contrast medium.

For the purposes of this review, *metrizamide* was the only intrathecal contrast medium used. It is to be hoped that even less toxic contrast media will be developed in the near future. Metrizamide is a nonionic, water-soluble iodinated glucosamide of metrizoic acid with a molecular weight of 789, an osmolality of 300 mmol/1000 g in a 170 mg I/ml solution (isotonic), and an estimated octanol–water partition coefficient of -0.19.[32] The improved safety of metrizamide as compared to other intrathecal contrast agents is related to its lower osmolality and lipid solubility.[3,32] When it is used in appropriate volumes and concentrations in both laboratory animals and man, its epileptogenic potential is mild, particularly when compared to that of meglumine iothalamate (Conray) and methylgucamine iocarnate (Dimer X).[43] Toxic side effects with cisternal metrizamide were found to be directly dose-dependent in the baboon.[48] It is suggested that neuroleptic drugs may potentiate the epileptogenic effects of metrizamide.[19,21]

Adverse reactions in man (headache, nausea, vomiting, hallucinations, EEG changes) are of great interest in that they are delayed in appearance. If they were related to maximal subarachnoid metrizamide concentrations, one would expect them to occur within the initial 30 min. However, the delayed presence of adverse reactions suggests that they are instead related to the cerebral staining, which is maximal in the 6- to 18-hr range.[6,8] For even greater safety, we advise the use of low doses (<1.2 g iodine) and good general hydration. It is of interest that no alterations in the serial EEG appearance occurred in children less than 1 year of age unless they had prior EEG abnormalities or a history of seizures, while delayed EEG changes such as projected rhythms (e.g., frontal intermittent rhythmic delta activity) were far more common in older children and adults.

An ideal contrast medium for performing cisternography would be of low toxicity, chemically inert, of adequate atomic number and *k*-edge for radiographic imaging without excessive dosage, nonpyrogenic, noninflammatory to the arachnoid membranes, stable in a solution with a pH of 7.4, of osmolality similar to that of CSF, poorly lipid-soluble, and cleared from the CSF predominantly via the arachnoid granulations. Metrizamide appears to meet many of these criteria, and CT scanning provides both a visual and a numerical means of mapping its circulation through the ventricular–subarachnoid system. The excellent spatial resolution of CT as compared to radionuclide studies permits an analysis of the initial intracranial ascent of metrizamide, which occurs via both the anterior and posterior subarachnoid cisterns. Symmetrical ascent over the convexities to the parasagittal region is the consistent finding in the normal individual. The molecular weight of 789([^{169}Yb] diethlenetriaminepentaacetic acid, mol. wt. 600) is sufficient to prevent excessive local absorption of the iodinated contrast medium. The staining of the adjacent cerebrum and cerebellum as compared to the brainstem has been confirmed in dogs[8] and rabbits[18] autoradiographically, albeit a very small quantity of impurity marred the ^{131}I labeling.

In summary, a detailed analysis of abnormalities of the CSF circulation as well as intracranial cyst dynamics, CSF leakage, and diversionary shunt function may be obtained using a combination of careful clinical examination, biochemical testing, and CT scanning. The need for further review of diagnostic criteria continues, particularly in the evaluation of subarachnoid (communicating) hydrocephalus for shunt therapy. CT scanning with various contrast media provides a fertile source for future *in vivo* research into the mechanics of CSF production, circulation, and absorption.

References

1. Adams, R. D., Fisher, C. M., Hakim, S., Ojemann, R. G., Sweet, W. H.: Symptomatic occult hydrocephalus with "normal" cerebrospinal fluid pressure: A treatable syndrome. *N. Engl. J. Med.* **273**:117–126, 1965.
2. Adeloye, A.: Mesencephalic spur (beaking deformity of the tectum) in Arnold–Chiari malformation. *J. Neurosurg.* **45**:315–320, 1976.
3. Almen, T.: Contrast agent design: Some aspects of the synthesis of water-soluble contrast agents of low osmolality. *J. Theor. Biol.* **24**:216–223, 1969.

3a. Bilaniuk, L. T., Zimmerman, R. A., Littman, P., Gallo, E., Rorke, L. B., Bruce, D. A., Schut, L.: Computed tomography of brain stem gliomas in children. *Radiology* **134**: 89–95, 1980.

4. Cameron, A. H.: The Arnold Chiari and other neuroanatomical malformations associated with spina bifida. *J. Pathol. Bacteriol.* **73**:195–211, 1957.
5. DiChiro, G., Arimitsu, T., Brooks, R. A., Morgenthaler, D. G., Johnston, G. S., Jones, A. E., Keller, M. R.: Computer tomographic profiles of periventricular hypodensity in hydrocephalus and leukoencephalopathy. *Radiology* **130**:661–666, 1979.
6. Drayer, B. P., Rosenbaum, A. E., Higman, H. B.: Cerebrospinal fluid imaging using serial metrizamide CT cisternography. *Neuroradiology* **13**:7–17, 1977.
7. Drayer, B. P., Rosenbaum, A. E.: Studies of the third circulation: Amipaque CT cisternography and ventriculography. *J. Neurosurg.* **48**:946–956, 1978.
8. Drayer, B. P., Rosenbaum, A. E.: Metrizamide brain penetrance. *Acta Radiol. (Suppl.)* **355**:280–292, 1977.
9. Drayer, B. P., Rosenbaum, A. E.: Brain edema defined by cranial computed tomography. *J. Comput. Asist. Tomogr.* **3**:317–323, 1979.
10. Drayer, B. P., Rosenbaum, A. E., Kennerdell, J. S., Robinson, A. G., Bank, W. O., Deeb, Z. D.: Computed tomographic diagnosis of suprasellar masses by intrathecal enhancement. *Radiology* **123**:339–344, 1977.
11. Drayer, B. P., Rosenbaum, A. E., Maroon, J. C., Bank, W. O., Woodford, J. E.: Posterior fossa extraaxial cyst: Diagnosis with metrizamide CT cisternography. *Am. J. Roentgenol.* **128**:431–436, 1977.
12. Drayer, B. P., Rosenbaum, A. E., Reigel, D. B., Bank, W. O., Deeb, Z. L.: Metrizamide computed tomography cisternography: Pediatric applications. *Radiology* **124**:349–357, 1977.
13. Drayer, B. P., Dujovny, M., Wolfson, S. K., Boehnke, M., Cook, E. E., Rosenbaum, A. E.: Comparative cranial CT enhancement in a primate model of cerebral infarction. *Ann. Neurol.* **5**:48–58, 1979.
14. Ekbom, K., Greitz, T., Kugelberg, E.: Hydrocephalus due to ectasia of the basilar artery. *J. Neurol. Sci.* **8**:465–477, 1969.
15. Ekbom, K., Greitz, T.: Syndrome of hydrocephalus caused by saccular aneurysm of the basilar artery. *Acta Neurochir.* **24**:71–77, 1971.
16. Foltz, E. L., Shurtleff, D. B.: Conversion of communicating hydrocephalus to stenosis or occlusion of the aqueduct during ventricular shunt. *J. Neurosurg.* **24**:520–529, 1966.
17. Gado, M. H., Coleman, R. E., Lee, K. S., Mikhael, M. A., Alderson, P. O., Archer, C. R.: Correlation between computerized transaxial tomography and radionuclide cisternography in dementia. *Neurology* **26**:555–560, 1976.
18. Galmon, K.: Distribution and retention of ^{125}I-labelled metrizamide after intravenous and suboccipital injection in rabbit, rat and cat. *Acta Radiol. (Suppl.)* **335**:300–309, 1973.
19. Gonsette, R. C., Brucher, J. M.: Potentiation of Amipaque epileptogenic activity by neuroleptics. *Neuroradiology* **14**:27–30, 1977.
20. Gooding, C. A., Carter, A., Hoare, R. D.: New ventriculographic aspects of the Arnold–Chiari malformation. *Radiology* **89**:626–632, 1967.
21. Grepe, A., Widen, L: Effects of cisternal applications of metrizamide. *Acta Radiol. Diagn. (Suppl.)* **335**:119–124, 1973.
22. Greitz, T., Hindmarsh, T.: Computer assisted tomography of intracranial CSF circulation using a water-soluble contrast medium. *Acta Radiol. Diagn.* **15**:497–507, 1974.
23. Gyldensted, C.: Measurements of the normal ventricular system and hemispheric sulci of 100 adults with computed tomography. *Neuroradiology* **14**:183–192, 1977.
24. Hakim, S., Adams, R. D.: The special clinical problems of symptomatic hydrocephalus with normal cerebrospinal fluid pressure: Observations on cerebrospinal fluid hydrodynamics. *J. Neurol. Sci.* **2**:307–327, 1965.
25. Harwood-Nash, D. C., Fitz, C. R.: *Neuroradiology in Infants and Children.* St. Louis, C. V. Mosby, 1976.
26. Haug, G.: Age and sex dependence of the size of normal ventricles on computed tomography. *Neuroradiology* **14**:201–204, 1977.

27. HINDMARSH, T.: Elimination of water soluble contrast media from the subarachnoid space: Investigation with computed tomography. *Acta Radiol. (Suppl.)* **346:**45–49, 1975.
28. HINDMARSH, T., GREITZ, T.: Computed cisternography in the diagnosis of communicating hydrocephalus. *Acta Radiol. (Suppl.)* **346:**91–97, 1975.
29. HUCKMAN, M. S., FOX, J., TOPEL, J.: The validity of criteria for the evaluation of cerebral atrophy by computed tomography. *Radiology* **116:**85–92, 1975.
30. INABA, Y., HIRATSUKA, M., TSUYUMU, M., SUGANUMA, Y., OKADA, K., FUJIWARA, K., TAKASATO, Y.: Diagnostic value of CT cisternography with intrathecal metrizamide enhancement: Comparison with isotope cisternography. *Neuroradiology* **16:**214–215, 1978.
31. JACOBS, L., KINKEL, W.: Computerized axial transverse tomography in normal pressure hydrocephalus. *Neurology* **26:**501–507, 1976.
32. LEVITAN, H., RAPOPORT, S. L.: Contrast media: Quantitative criteria for designing compounds with low toxicity. *Acta Radiol. Diagn.* **17:**81–92, 1976.
33. LOOP, J. W., FOLTZ, E. L.: Craniostenosis and diploic lamination following operation for hydrocephalus. *Acta Radiol. Diagn.* **13:**8–13, 1972.
33a. MENEZES, A. H., BELL, W. E., PERRET, G. E.: Arachnoid cysts in children. *Arch. Neurol.* **37:**168–172, 1980.
34. MESSERT, B., WANNAMAKER, B. B.: Reappraisal of the adult occult hydrocephalus syndrome. *Neurology* **24:**224–231, 1974.
35. NAIDICH, T. P., EPSTEIN, F., LIN, J. P., KRICHEFF, I. I., HOCHWALD, G. M.: Evaluation of pediatric hydrocephalus by computed tomography. *Radiology* **119:**337–353, 1976.
35a. NAIDICH, T. P., PUDLOWSKI, R. M., NAIDICH, J. B., GORNISH, M., RODRIGUEZ, F. J.: Computed tomographic signs of the Chiari II malformation, part 1, skull and dural partitions. *Radiology* **134:**65–71, 1980.
36. NORMAN, D., ENZMANN, D. R., PRICE, D., NEWTON, T. H.: Metrizamide computed tomographic cisternography. *Neuroradiology* **15:**135, 1978 (abstract).
37. OSTERTAG, C. B., MUNDINGER, F.: Diagnosis of normal pressure hydrocephalus using CT with CSF enhancement. *Neuroradiology* **16:**216–219, 1978.
38. PEACH, B.: Arnold–Chiari malformation: Anatomic features of 20 cases. *Arch. Neurol.* **12:**613–621, 1965.
39. PARTAIN, C. L., SCATLIFF, J. H., STAAB, E. V., WU, H. P.: Quantitative multiregional CSF kinetics using serial metrizamide enhanced computed tomography. *J. Comput. Assist. Tomogr.* **2:**467–470, 1978.
40. ROBERTSON, W. C., GOMEZ, M. R.: External hydrocephalus. *Arch. Neurol.* **35:**541–544, 1978.
41. ROTTENBERG, D. A., HOWIESON, J., DECK, F.: The rate of CSF formation in man: Preliminary observations on metrizamide washout as a measure of CSF bulk flow. *Ann. Neurol.* **2:**503–510, 1977.
42. ROTTENBERG, D. A., DECK, F., ALLEN, J. C.: Metrizamide washout as a measure of CSF bulk flow. *Neuroradiology* **16:**203–206, 1978.
43. SAWHNEY, B. B., OFTEDAL, S. I.: Reactions to suboccipital injection of water-soluble contrast media in rabbits. *Acta Radiol. Diagn. (Suppl.)* **335:**67–83, 1973.
44. STRAND, R. D., BAKER, R. A., ORDIA, I. J., ARKINS, T. J.: Metrizamide ventriculography and computed tomography in lesions about the third ventricle. *Radiology* **128:**405–410, 1978.
45. TAKAHASHI, A. H., TAMAKAWA, T.: Comparison of metrizamide CT cisternography with radionuclide cisternography: Abnormal cerebrospinal fluid dynamics. *Neuroradiology* **16:**199–202, 1978.
46. TAMAKI, N., KANAZAWA, Y., ASADA, M., KUSUNOKI, T., MATSUMOTO, S.: Comparison of cerebrospinal fluid dynamics studied by computed tomography (CT) and radioisotope (RI) cisternography. *Neuroradiology* **16:**193–198, 1978.
47. WOOD, J. H., BARTLET, D., JAMES, A. E., UDVARHELYI, G. B.: Normal pressure hydrocephalus: Diagnosis and patient selection for shunt surgery. *Neurology* **24:**517–526, 1974.
48. WYLIE, I. G., AFSHAR, F., KOEZE, T. H.: Results of the use of a new water-soluble contrast medium (metrizamide) in the posterior fossa of the baboon. *Br. J. Radiol.* **48:**1007–1012, 1975.

CHAPTER 31

Cerebrospinal Fluid Alterations Associated with Central Nervous System Infections

Enrique L. Labadie

1. Introduction

This chapter represents an extensive literature review of the cellular and chemical alterations in cerebrospinal fluid (CSF) associated with bacterial, fungal, mycobacterial, viral, and parasitic infections of the central nervous system. In addition, CSF reactions secondary to noninfectious meningoencephalitic disorders are discussed. The data herein have been summarized in tabular form to facilitate clinical differential diagnosis.

2. CSF Cellular Reactions

The CSF obtained from a normal adult contains less than 5 white blood cells/mm^3, of which 2 or less may be polymorphonuclear leukocytes. Normally, no erythrocytes are present in CSF. The CSF cell count in young children is normally less than 10 lymphocytes/mm^3.[60,132] The clinical data concerning pathological CSF cellular reactions are presented in Table 1. CSF leukocyte counts per cubic milliliter associated with each disorder are divided into three categories: Group I, from 1000 to more than 10,000 leukocytes; Group II, from 250 to 1000 leukocytes; Group III, from 6 to 250 leukocytes. The CSF cellularity associated with each disorder is reported as the percentage of cases within each category. Potentially treatable diseases are arbitrarily placed at the beginning of each list rather than according to their statistical incidence in relation to each other. Disorders that induce the appearance of unusual cell types in CSF are listed in Table 2.

The widespread belief that the central nervous system is an "immunologically privileged" site should be considered as a "relative" concept.[39,175] Under normal conditions, the subarachnoid space is a tightly sealed compartment subserved by the blood–brain barrier, which impedes the entry of most low-molecular-weight amino acids, other normal blood constituents, and microorganisms.[162]

The three types of immunological reactions that may take place within the central nervous system are similar to those that occur systemically and include cell-mediated immunity, humoral antibody production, and macrophage activation. Whenever immunologically active agents penetrate the central nervous system, reactions occur by recruitment of local and systemic defense elements. At the incipient stages of meningeal infection, reproducing microorganisms enter the bloodstream via the CSF drainage into the venous sinuses at the arachnoid granulations.[177] Similarly, chemotactic substances elaborated within the subarachnoid space pass with the CSF into the systemic circulation. Both these phenomena elicit systemic recognition by blood-borne immunological elements.[122,213] Experimental brain irradiation in animals induces enhanced susceptibility to bacterial growth after cerebral inoculation, thus providing evidence that the infectious

Enrique L. Labadie, M.D. • Department of Neurology, Tucson Veterans Administration Medical Center, Tucson, Arizona 85723; University of Arizona School of Medicine, Tucson, Arizona 85724.

Table 1. CSF Cellular Reactions

Category	Ref. Nos.
First category: Leukocyte count from 1000 to over 5000 cells/mm^3 and polymorphonuclears predominate. Glucose below 55 mg/dl in practically all except viral infections.	
Over 5000 cells	
Bacterial meningitis (all types)	29, 42, 43, 54, 56, 57, 60, 62, 77, 125, 132, 133, 152, 172, 176, 188, 197, 205
Subdural abscess in infants	53
Amoebic meningitis	4, 25, 168
1000–5000 cells	
Bacterial	
Early bacterial meningitis; partially treated bacterial meningitis; unusual bacteria such as *Clostridia*, anthrax, and *Nocardia*	32, 33, 36, 70, 83, 86, 91, 107, 132, 133, 142, 158, 176, 188, 200, 205, 207 (additional references in third category)
Fungal	
Mainly histoplasmosis, coccidioidomycosis, blastomycosis, and *Cryptococcus*	44, 58, 75, 126, 196 (additional references in second category)
Parasitic	
Amoebic meningitis (some)	4, 25, 168 (additional references in third category)
Viral	
Herpes simplex, Echo type 9, mumps, measles, varicella, and lymphocytic choriomeningitis	1, 2, 26, 28, 94, 132, 135, 137, 186, 204, 208 (additional references in third category)
Chemical – inflammatory	
Usually instilled dyes, e.g., pantopaque, radioactive albumin; subarachnoid hemorrhage and rarely ruptured CNS tumors, glioblastoma, craniopharyngioma, etc.	69, 90, 96, 102, 103, 131, 132, 136, 192 (additional references in second category)
Acute hemorrhagic leukoencephalitis	7, 10a
Unknown etiology	
Mollaret's meningitis	(References in third category)
Second category: Leukocyte count from 250 to 1000 cells with mixed cellularity, but lymphocytes predominate. Glucose commonly below 55 mg/dl with exceptions.	
Tuberculosis	151, 195
Leptospirosis	93, 132, 133
Meningovascular syphilis	
Abscesses (up to 40% of them)	
Parameningeal, brain, subdural, epidural, and spinal abscesses	8, 48, 51, 53, 54, 79, 95, 115, 138, 139, 143, 153, 158, 167, 169, 178, 198, 200, 201, 206, 207, 210
Fungal	
Cryptococcus, cocci, *Candida*, *Histoplasma*, blastomycosis, cladosporiosis, actinomyces, *Aspergillus*, mucormycosis	17, 20, 27, 30, 48, 50, 58, 60, 67, 75, 79, 105, 109, 133, 140, 155, 173, 185, 196, 201, 203, 214 (additional references in first category)
Viral	(References in first and third categories)
Lymphocytic choriomeningitis, herpes simplex and zoster, Eastern equine, mumps, Echo 9, California virus, St. Louis	
Chemical – inflammatory	
Subarachnoid hemorrhage	69, 96 (additional references in first category)
Acute hemorrhagic leukoencephalitis	10a
Malignancies	
Meningeal carcinomatosis, metastases from intracranial tumors, ruptured necrotic CNS tumors	10, 13, 63, 69, 81, 102, 103, 106, 123, 132, 133, 136, 150, 182–184
Unknown etiology	(References in third category)
Mollaret's meningitis, Behcet's disease	

Table 1. *(Continued)*

Category	Ref. Nos.
Third category: Cell count from 6 leukocytes to 250 lymphocytes.	
Abscesses	
Up to 60% of them (see second category)	
Bacterial	(Additional references in first category)
Syphilis-tabes, general paresis, and some meningovascular; brucellosis, *Nocardia*	
Subacute bacterial endocarditis	3, 132, 138
Fungal	
Up to 20% of them (see first and second categories)	
Viral	
Most common are: Echo, Coxsackie, influenza, rubella, vaccinia, cytomegalovirus, California virus, mononucleosis, rabies, poliomyelitis, Western equine, progressive multifocal leukoencephalopathy, and SSPE[a]	12, 14, 21, 41, 45, 60, 64, 85, 93, 94, 97, 119, 121, 132, 133, 135, 179, 181, 193, 199, 202, 208 (additional references in first category)
Inflammatory or vascular	
Postvaccinal encephalomyelitis, strokes, subarachnoid hemorrhage, hypertensive encephalopathy, chronic adhesive arachnoiditis, and all arteritides, e.g., lupus, temporal arteritis	7, 10,10a,31, 34, 47, 49, 50, 55, 59, 68, 80, 82, 90, 96, 98, 111, 116, 130, 132–134, 137, 145, 146, 180, 192
Parasitic	
Toxoplasma; trichinosis, cysticercosis, malaria, schistosomiasis, trypanosomiasis, etc.	5, 6, 9, 15, 16, 66, 78, 101, 104, 110, 117, 133, 159, 170, 185, 191, 194, 211 (additional references in first category)
Malignancies	(References in second category)
Carcinomatous meningitis, brain and cord tumors	
Unknown etiology	
Multiple sclerosis, Behcet's disease, Vogt–Harada–Koyanagi, chronic benign lymphocytic meningitis, lymphomatoid granulomatosis, sarcoidosis	50, 52, 87, 89, 112, 120, 124, 133, 147, 156, 165, 166, 209

[a] (SSPE) Subacute sclerosing panencephalitis.

resistance of central nervous system structures also relies heavily on local macrophage activity.[88] Cell transformation of blast forms into mature antibody-producing plasma cells has been well documented in CSF.[72,148] Similarly, basophils, the prototype blood elements involved in cell-mediated immunity, have been found in the CSF of patients harboring malignancies.[71,148]

The CSF cellular data presented in Table 1 indicate that central nervous system defense reactions are very similar to those observed systemically. The existing system-specific differences in cellular reactions to inflammation tend to be only minor variations of the intensity of the reactions and thus are not so qualitatively divergent as previously thought.[175]

Reproducing microorganisms and their antigens or tumor antigens when present in CSF exit from the subarachnoid space via the arachnoid granulations. Systemic recognition of their presence should occur promptly following entry into the bloodstream. However, in some central nervous system disorders, direct antigenic access into blood may be immunologically counterproductive. Animal experiments have demonstrated that the route of immunization is critical for induction or inhibition of cellular-mediated immunity. Injection of progressively increasing antigenic doses directly into the bloodstream impedes development of cell-mediated immunity entirely, or greatly attenuates its functions.[130] Chronic viral, fungal, mycobacterial, parasitic, and carcinomatous meningitides are quite resistant to humoral-antibody destruction. Their progressive antigen release into the general circulation may lead to suppression of cellular-mediated immunity, concomitant with limited CSF cellular response and systemic anergy of the host.[13,18,20,23,27,29,30,63,84,100,108,118,123,128,136,138,141,154,155,165,182-184,195,196,200,202,213]

Clinical examples of these complex relationships

Table 2. Unusual Cells in CSF

Cell type	Etiology	Ref. Nos.[a]
Eosinophils	Cysticercosis	6, 15, 66, 78, 185, 191, 195
	Trichinosis	104, 110, 133
	Echinococcosis	5, 101
	Toxoplasmosis	194
	Schisosomiasis	16
	Coccidioidomycosis	20, 58
	Chemical arachnoiditis	60
	Meningeal lymphomatosis	106
	Filariasis	46
	Angiostrongylus cantonensis (Pacific Islands)	170
	Paragonimiasis (Korea)	149
	Gnathostoma spinigerum (Asia)	22
	Foreign bodies (e.g., rubber catheters)	60
Plasma cells (morular cells of Mott)	Trypanosomiasis	72, 132, 211
	Viruses (?)	71, 72
Basophils	Meningeal lymphomatosis	71, 81
Many endothelial cells	Mollaret's meningitis	87
Malignant cells	Tumors in CNS or meninges	13, 63, 69, 81, 103, 106, 123, 136, 150, 182–184, 187

[a] For reviews, see Ref. Nos. 50, 58, 60, 132, 133, 148, 185, and 196.

can be observed during meningeal coccidioidomycosis. Patients initially show negative skin reactivity to coccidiodin (cell-mediated immunity), and yet demonstrate high complement-fixation titers in both blood and CSF (humoral-mediated immunity). Delayed skin hypersensitivity usually reappears after prolonged intrathecal therapy and once CSF complement-fixation titers have decreased below 1:4 or 1:2.[20,27] In this context, CSF defense reactions observed during some chronic central nervous system infections are comparable to those systemic reactions elicited by indolent subacute bacterial endocarditis.

Detailed discussion of the immunoinflammatory reactions in CSF is beyond the scope of this chapter. In brief, intricate interrelationships exist among humoral immunity, the kinins system, the sequential activation of complement, hemostatic–fibrinolytic reactions, and cellular-mediated immunity.[18,37,100,122,129,154,157,163,171,189] Variations in the net balance of the aforenamed participating mechanisms catalyze and control the type and intensity of CSF cellular reactions occurring during central nervous system infections.

During antibiotic treatment of meningitis, serial reevaluation of the drug efficacy is clinically desirable. Such an analysis depends on the rapidity of the recovery of the CSF alterations following complete sterilization of meningeal tissues. However, inflammatory reactions do not cease instantly, but rather diminish progressively at a varying rate. Three clinical forms of noninfectious meningitis may serve as guideline examples of CSF recovery following sterile meningeal inflammation.

Pneumoencephalography frequently causes an aseptic meningeal leukocytosis in CSF. This cellular response diminishes rapidly during the following 48 hr, although on occasion, a minimally elevated leukocyte count may persist for up to 1 week.[132] A similar CSF leukocytosis has been observed following radionuclide-labeled serum albumin (RISA) cisternography.[134] After a single hemorrhagic event within the subarachnoid space or after infusion of blood into the CSF of animals, an aseptic hemogenic meningitis usually follows. Erythrocyte lysis releases irritative hemoglobin degradation byproducts that evoke a chemical meningitis usually peaking in 3–4 days. Fever, CSF leukocytosis, and decreased CSF glucose content may be observed during this reaction.[96] Spontaneous recovery gradually occurs during the subsequent week, and thereafter CSF leukocytosis and xanthochromia usually subside within a 14- to 20-day period.[96,192] This last example of the CSF recovery from aseptic meningitis may more closely resemble the time course that may be observed following optimal antibiotic treatment of bacterial meningitis, because, despite microorganism eradication, bacterial antigens may persist in CSF for some time. CSF outflow drainage during experimental bacterial meningitis is markedly diminished to about one fifth the normal rate, and this reduction persists even after antibiotic eradication of the bacteria.[35] Similarly, a pronounced delay of radionuclide movement in CSF, communicating hydrocephalus, and accumulation of antigens in the lumbar sac have been demonstrated in patients with bacterial meningitis.[57,174] During the treatment of most bacterial meningitis, worsening or unduly prolonged CSF recovery should be alarming. However, some infectious disorders demonstrate unusual CSF responses to appropriate therapy and follow a protracted recovery course (see Section 5).

3. CSF Glucose Alterations

The content of glucose in CSF is diagnostically important, since a relatively select number of diseases alter its concentration in CSF. Bacterial, mycobacterial, and fungal infections of the central nervous system are treatable; however, viral disorders and neoplastic inflitrations defy curent modes of therapy.

Absolute CSF glucose levels are directly related to blood glucose concentrations. Glucose enters the CSF via passive diffusion, and also by a more rapid non-energy-dependent, facilitating mechanism.[61] Although the net influx of glucose into the CSF remains in close parallel to the serum glucose content, a relatively slow saturation effect may decrease the CSF/blood glucose ratio in the presence of high serum glucose levels.[60,132] Net balance of this important CSF component demonstrates a strong tendency toward the maintainance of stable levels. When serum glucose concentrations are rapidly reduced, equilibrium in CSF is reached slowly in about 2 hr.[60] Similarly, acute elevations of blood glucose levels are reflected in CSF within 60 to 120 min.[22a] Ventricular and cisternal levels are usually higher than those measured in lumbar CSF samples, in contrast to the protein content, which increases rostrocaudally.[40,132] Unfortunately, little is known about glucose utilization by the central nervous system structures bordering the subarachnoid space.

Most lumbar punctures are not performed after a prolonged fasting state, but rather after regular meals or concomitant with intravenous glucose administration. Thereby, relatively higher CSF glucose content may be encountered in these patients, especially if the patients have diabetes mellitus. The recommended method of obtaining a simultaneous blood glucose determination is therefore clinically invaluable. Unfortunately, when one reviews the available literature, including individual case reports, clinical reviews, or textbooks, it is apparent that most authors reporting CSF glucose values seldom include the simultaneous blood glucose or the CSF/blood glucose ratios.

Reduction in CSF glucose levels associated with meningoencephalitis was believed to be caused exclusively by enhanced glycolytic leukocyte utilization. However, more recent evidence suggests that hypoglycorrhachia does not strictly parallel the leukocytic response, since this finding may be observed in viral illnesses and carcinomatous meningitis with low cellularity. On occasion, decreased CSF glucose may persist during recovery from treated bacterial meningitis, even when bacterial cultures are sterile and the CSF protein levels have returned toward normal.[60] In addition, the acuteness of the clinical illness correlates with rapidly lowered CSF glucose, while in chronic disorders, glucose content decreases slowly and later in the clinical course.[8,17,20,27,44,50,51,54,58,60,67,81,87,89,123,124,138,150,185,195,196]

These observations led investigators to postulate that infectious agents or even leukocytes may release substances that may block the CSF glucose transport, either by a competitive mechanism or by noncompetitive "toxic" inhibition. Although experimental verification of this concept is lacking, mercuric chloride ($HgCL_2$) has been used frequently in experimental animal models to block glucose CSF entry, while phloretin and glycoside phlorezin have both been shown to reversibly block central nervous system capillary and erythrocyte glucose transport.[162] However, inflammatory substances may be the most common causative agents. Diminished CSF glucose occurs frequently following asceptic subarachnoid hemorrhages; this effect has been reproduced in experimental animals.[24,69,96]

Many clinical disorders, such as parameningeal abscesses, osteomyelitis of cranial or spinal bones, brucellosis, toxoplasmosis, fungal disorders, meningovascular syphilis, and meningeal carcinomatosis, present clinically with a modest CSF cellular response. Some of these disorders are associated with a slight CSF glucose decrement, frequently reported in the range of 45–55 mg/dl. As would be expected, in the majority of these disorders, CSF cultures may be repeatedly negative.

In many hospitals, accurate glucose measurements are performed in an automated fashion by the Hexokinase® (Dupont Corporation), glucose-oxidase® (Beckman Company), or *o*-toluidine® (Coulter Company) method, all of which which measure exclusively glucose. These automated methods show normal fasting blood glucose levels to be 70–110 mg/dl, while previously employed systems were unable to differentiate other reducing substances from glucose, thereby providing higher normal fasting values of 80–120 mg/dl.[132,160] Since reducing substances in serum may not readily penetrate the blood–brain barrier, CSF glucose values reported in older literature may still be valid.

With the aid of Drs. P. R. Finley and F. Griffith at the University of Arizona, the CSF data from 770 selected cases stored in the Pathology Laboratory computer were analyzed. CSF samples obtained from children under 3 years of age were analyzed separately because systemic hypoglycemia and hence CSF hypoglycorrachia is encountered frequently in that age group. For inclusion in this CSF study, cases were selected only when their leukocyte counts demonstrated less than 5 mononuclear cells and were free of overt subarachnoid hemorrhage. Diabetic patients were not excluded from this analysis. Of the 770 cases reviewed, 9.5% had a CSF glucose concentration of less than 55 mg/dl and only 4.5% had a level of less than 50 mg/dl. Interestingly, of the cases that had a CSF glucose of less than 50 mg/dl, 70% were female.

Merritt and Fremont-Smith[132] reviewed 842 patients with brain tumors, epilepsy, polyneuritis, neuroses, and psychoses, but all without evidence of meningeal inflammation, and noted that 96.8% had CSF glucose levels above 50 mg/dl. Only 3.2% of the cases in their study had CSF glucose concentrations below 50 mg/dl (Table 3). These findings are strikingly similar to the data reported in our patient analysis.

These two studies suggest that in clinical situations where the CSF analysis demonstrates even a modest leukocytosis and minimal elevation in protein content, a glucose level less than 50 mg/dl could

Table 3. CSF/Blood Glucose Ratios Normogram for "Normal Range" (mg/dl)

Glucose content																								
Blood:	120	115	110	105	103	101	99	97	95	93	91	89	87	85	83	81	79	77	75	73	71	69	67	65
CSF 66%:	79	76	73	69	68	67	65	64	63	61	60	59	57	56	55	53	52	51	50	48	47	46	44	43
CSF 60%:	72	69	66	63	62	61	59	58	57	56	55	53	52	51	50	49	47	46	45	44	43	41	40	39

	CSF glucose (mg/dl)	
Cases	Below 50	Over 50
Statistical Probability: Normal CSF Series		
CSF glucose in cases of central neurological disease without inflammatory meningeal reaction (From Merritt and Fremont-Smith[132]): 842	3.2%	96.8%
Author series of CSF glucose in selected cases, excluding children under age 3: 770	4.5% (70% women)	95.5%
Both series above combined and averaged, total: 1612	3.85% (4%)	96.15% (96%)
CSF Glucose in Parameningeal–Meningeal Infections and Subarachnoid Hemorrhage at Initial Lumbar Puncture[132]		
Parameningeal infections: 177	11%	89%
Subarachnoid hemorrhage: 336	15%	85%
Neurosyphilis, mostly meningovascular: 142	21%	79%
Bacterial, mycobacterial, and fungal meningitis: 1263	85%	15%
Total cases analyzed in series above: 3530		

Table 4. Hypoglycorrachia[a] with respect to CSF Leukocyte Count

Disorders	Approximate percentage	Leukocyte count		
		10,000–1000	1000–50	50–6
All bacterial meningitis	>98%	>95%	—	—
Amoebic meningitis	>98%	90%	10%	—
Subdural abscess in infants	>98%	50%	50%	—
Tuberculosis	>90%	—	90%	10%
Brucellosis	>90%	—	10%	90%
Parameningeal abscesses	31%	—	60%	40%
Meningovascular syphilis	50%	—	90%	10%
Toxoplasmosis	67%	—	25%	75%
Cryptococcosis	90%	Unusual	70%	30%
Coccidiodomycosis	90%	Unusual	70%	30%
Blastomycosis	90%	Unusual	Most	Some
Actinomycosis	Some	Unusual	Most	Some
Histoplasmosis	60%	28%	72%	Few
Nocardiosis	90%	Unusual	70%	Some
Candidiasis	90%	Unusual	70%	Some
Cladosporiosis	90%	Unusual	70%	Some
Lymphocytic choriomeningitis	Up to 20%	Some	Most	Few
Mumps	Up to 50%	10%	50%	40%
Echo type 9	Up to 35%	16%	37%	47%
Herpes simplex	Up to 35%	10%	80%	10%
Herpes zoster	10–20%	Rare	Most	Some
Eastern equine	10–20%	Rare	Most	Some
Measles	10–20%	Rare	Most	Some
Poliomyelitis	10–20%	Rare	Most	Some
SSPE[b]–rubella	>50%	—	Few	>90%
Meningeal carcinomatosis	40–75%	—	34%	66%
Meningeal metastases from intracranial tumor	40–75%	—	34%	66%
Chemical meningitis	>98%	90%	10%	—
Mollaret's meningitis	>98%	64%	36%	—
Chronic benign lymphocytic meningitis	80%	Rare	66%	34%
Sarcoidosis	80%	—	27%	73%
Lymphomatoid granulomatosis	2 known cases	—	Some	Most
Acute hemorrhagic leukoencephalitis	2 known cases	30%	60%	10%
Aseptic hemogenic meningitis	Frequent	Some	Some	Some
Systemic hypoglycemia	Frequent	—	—	Most

[a] Hypoglycorrachia defined as CSF glucose concentration less than 55 mg/dl.
[b] (SSPE) Subacute sclerosing panencephalitis.

be termed "statistically probably abnormal" even in the presence of a "borderline" CSF/blood glucose ratio (between 0.60 and 0.66) as noted in Table 3.

In the case of male patients, a slightly higher CSF glucose content between 50 and 55 mg/dl would be more suspicious, since our series of "normal CSF" only 4.5% of all cases fell below 50 mg/dl, but males comprised less than 1.5% (Table 3) and below 55 mg/dl, only 3%.

From our extensive literature review of CNS infections and other disorders which may depress the CSF glucose content levels, we compiled a list presented in Table 4. Since the literature often did not provide CSF/blood glucose ratios and since both large series of "normal CSF glucose" values show a confidence level of less than 5% (Table 3), some disorders were then selected with the assumption that reported CSF glucose contents below 55 mg/dl were "suspicious" and those less than 50 mg/dl were "statistically abnormally low."

Table 4 provides a clinically valuable list of comprehensive differential diagnosis, but tabulated percentages should be considered only approximate, since those percentages may be modified by future prospective clinical investigation or case reports that do include CSF/blood glucose ratios.

4. Nonimmunoglobulin-Protein Alterations

The total protein concentration in the ventricles is 6–15 mg/dl, whereas the total concentrations in the cisterna magna and lumbar sac are 15–25 and 20–50 mg/dl, respectively.[60] Although the exact physiological origin of these intriguing rostrocaudal gradients is not known, the physical characteristics of the lumbar sac and CSF circulation within it, coupled with local transudation of the relatively heavier protein molecules, could account for their accumulation at the lumbar level. Indirect evidence lends favor to this notion, since bacterial meningitis evokes a large accumulation of bacterial antigens preferentially in the lumbar sac.[57]

The combination of a marked elevation in CSF protein—commonly above 200 mg/dl—with a low leukocyte count (usually less than 100 white cells), with complete or partial CSF blockage, is known as Froin's syndrome. This syndrome is characterized by deeply xanthochromic, viscous CSF that may or may not clot spontaneously. Some patients with drastically elevated CSF protein do not develop clinical symptoms of increased intracranial pressure, papilledema, or hydrocephalus, while others demonstrate such symptons in the presence of only moderate CSF protein elevations.[10,60,65,68,69,127,132,158,164]

Fibrinogen, a protein of high molecular weight as compared to immunoglobulins and albumin, does not normally penetrate into the CSF. Plasminogen, the inactive precursor of plasmin (fibrinolysin), and plasmin, the active fibrinolytic enzyme, also do not cross the intact blood–brain–CSF barrier (Fig. 1). However, during bacterial and mycobacterial meningitis, fibrinogen or fibrin degradation products, or both, are frequently observed in the CSF.[19,92,132,212] This transudation of coagulable proteins into the subarachnoid space during central nervous system infections may explain CSF clotting *in vitro*. Similarly, the chronic presence of significant quantities of fibrinogen within the circulating CSF may lead

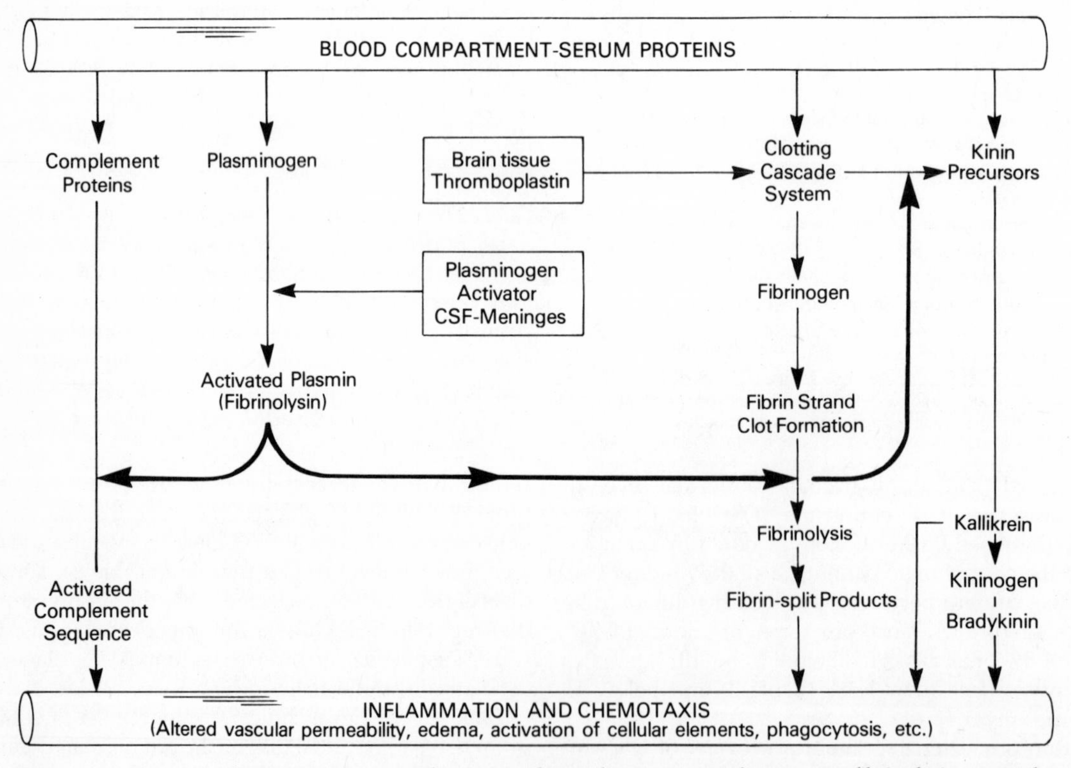

Figure 1. Diagram demonstrating complex interrelationships among hemostatic–fibrinolytic, complement, kinin, and inflammatory mechanisms. The role of brain, meninges, and CSF has been included.[37,38,76,113,114,129,154,157,161,163,171,190]

to fibrin deposition throughout the CSF pathways. This deposition of fibrin may produce obstruction of CSF circulation, increased intracranial pressure, and communicating hydrocephalus. In support of this notion, injection of plasma into the subarachnoid space of animals has precipitated elevations in CSF outflow resistance and evoked obstruction of the arachnoid villi by clotted proteinaceous material.[98]

Of those clinical disorders that cause Froin's syndrome with a markedly elevated CSF protein content and a minimal CSF cellular response, most are either parameningeal inflammatory disorders or tumors (Table 5). Approximately 25 cases of spinal tumors have been reported to be associated with a minimal cellular response and the triad of elevated CSF protein content, hydrocephalus, and dementia. One of these patients whose CSF was shown to contain fibrinogen without evidence of bleeding underwent improvement of his hydrocephalus and dementia following excision of a cauda equina

Table 5. Disorders That Cause Froin's Syndrome[a]

Spinal epidural abscess includes all bacterial, all fungal, tuberculosis, *Toxoplasma*, Brucella, cysticercosis		Most
Cord tumors		60%
Most intradural:	70%	
Most lumbar–thoracic		
Cervical tumors only:	17%	
In children, repeated bleeding ependymoma:	Likely	
Neurofibromas of the nerve roots		
Cauda equina		Most
Cervical region		Some
Meningiomas		
Cord		Most
Tentorial		Some
Foramen magnum		Some
Cerebellopontine angle tumors		54%
Meningeal metastases from intracranial tumors		40%
Meningeal carcinomatosis		40%
Cerebral gliomas		13%
Guillain-Barré		
First puncture		17%
Punctures after 2nd week yield a higher percentage		
Some neuropathies (e.g., Refsum's)		Unusual
Severe viral encephalitis (e.g., herpes)		Unusual
Chronic adhesive arachnoiditis		Most

[a] Froin's syndrome is the combination of CSF protein between 250 and less than 1000 mg/dl with low cellularity (<100), with or without CSF block by Queckenstedt's maneuver. Ref. Nos.: 8, 10, 58, 60, 63, 65, 69, 81, 123, 127, 132, 133, 136, 150, 158, 164, 182, 191, 194, 207, 214.

neurilemmoma.[10] Several other studies have shown evidence that central nervous system tumors may indeed "leak" excessive amounts of plasma proteins including plasminogen without associated bleeding.[13,84,141]

The brain itself contains procoagulant moieties that, when in direct contact with blood, rapidly induce coagulation.[76] Contrarily, the meninges, cerebral blood vessels, and CSF contain an activator of the fibrinolytic system.[161,190] However, this proteolytic activator requires the presence of plasminogen, which is not normally available in the subarachnoid space (see Fig. 1).

This alluded balance of hemostatic–fibrinolytic mechanisms within the central nervous system is probably dynamic and may be altered in favor of either increased coagulation or accelerated fibrinolysis. Within chronic reforming subdural hematomas, an acceleration of coagulation, fibrinolysis, liquefaction, and accumulation of breakdown products occurs to produce a net anticoagulant effect.[114] This observed phenomenon in man has been reproduced subcutaneously in rats by the injection of large autologous blood clots.[113] In other instances, however, coagulation occurring within the subarachnoid space may not be followed by fibrinolysis. Therefore, the development of hydrocephalus may follow mechanical blockage of CSF pathways by fibrinogen–fibrin complexes (e.g., Froin's syndrome) or active inflammatory reactions producing CSF obstruction (e.g., bacterial or fungal meningitis). The simultaneous measurement of fibrinogen and fibrin-split products in CSF may be useful in differentiating those cases in which clotting proteins are suspected to be the cause of increased intracranial pressure or hydrocephalus. To prevent proteolytic degradation of fibrinogen *in vitro* after CSF sampling, ε-aminocaproic acid (Amicar®, Lederle Company) should be added to one of the CSF samples.[113,114,157]

The knowledge that fibrinolytic activator is a normal constituent of CSF has led to the clinical use of ε-aminocaproic acid and tranexamic acid in cases of subarachnoid hemorrhage secondary to arterial aneurysm rupture. Both these drugs are potent inhibitors of the plasminogen activator in CSF and thereby impede fibrinolysis of the recently formed clot surrounding the aneurysm.[144] The possibility exists that these drugs impede fibrinogen degradation and may contribute to the generation of communicating hydrocephalus.

Corticosteroids such as dexamethasone may be efficacious in disorders in which fibrinolysis and inflammation participate simultaneously. Anecdotal cases of bacterial, mycobacterial, and parasitic meningitis that have been successfully treated with both antibiotic and corticosteroids have been reported in the literature. However, controlled studies will be required to verify that concomitant short-term corticosteroid administration is beneficial in decreasing the incidence of postmeningitic hydrocephalus and other sequelae of infectious meningitis.[42,117,151,153a] The beneficial effects of corticosteroid therapy have also been suggested in other noninfectious inflammatory disorders of the meninges such as postoperative asceptic hemogenic meningitis and chronic subdural hematoma.[11,24,73] Theoretically, increasing fibrinolysis with urokinase-streptokinase combined with the simultaneous administration of antiinflammatory agents might be advantageous in some disorders, whereas in others concomitant inhibition of both fibrinolysis and inflammation may be more appropriate. Additional research in this area may provide more rational therapeutic approaches for the prevention of postmeningitic hydrocephalus or possibly chronic adhesive arachnoiditis.[90,131]

5. Conclusions

CSF glucose and leukocyte alterations indicating the beneficial effects of appropriate antibiotic therapy should be apparent within 48–72 hr after initiation of medication, and complete resolution should be expected in 2 or 3 weeks thereafter. If improvement in these CSF abnormalities does not follow this time course during antibiotic therapy, several possibilities may be inferred. Either the diagnosis is incorrect, inappropriate drugs or routes of administration have been initiated, the offending organisms are drug-resistant, the microorganisms have become loculated elsewhere (e.g., parameningeal abscesses or ventricular septations), or the host is immunologically compromised. These assertions apply to the vast majority of bacterial infections. However, there are some noteworthy exceptions which may lead to clinical misinterpretations; patients with tuberculous meningitis frequently demonstrate worsening of clinical and CSF parameters as an initial response to appropriate systemic and intrathecal chemotherapy. The CSF leukocyte count may fluctuate markedly with a preponderant rise of polymorphonuclear cells, mycobacteria and CSF protein. With continued antibiotic therapy the CSF profile recovers gradually several months later.[151,153a] A similar paradoxical worsening may be encountered during initial treatment of some patients with meningovascular syphilis and general paresis. This untoward therapeutic response to appropriate antibiotics may be part of the Jarisch–Herxheimer reaction, phenomena believed to be caused by a massive antigenic release of dying microorganisms.[132,132a] In addition, present-day therapeutic agents for use in many nonbacterial infections (e.g., fungal infections) are toxic and only partially successful, leading to very prolonged recovery periods that may span months or even years.

As previously discussed, knowledge of the physiology of CSF is necessary for valid interpretation of CSF data. Significant differences exist in the leukocyte, glucose, and protein contents of CSF samples obtained from ventricular, cisternal, and lumbar levels.[132] Therefore, during drug efficacy reevaluations, laboratory data obtained from ventricular or cisternal CSF may be less abnormal than those from CSF obtained from a simultaneous lumbar puncture.[74] In addition, the glucose content of CSF may be modified by the blood glucose level; thus, the analysis of a blood sample drawn at the time of CSF sampling is imperative for the meaningful interpretation of CSF glucose levels.

This review has implicated nonimmunological proteins such as fibrinogen and fibrin in the pathophysiology of postmeningitic CSF outflow obstruction, hydrocephalus, ventricular septations, and possibly chronic adhesive arachnoiditis. Additional research in this area may provide prophylactic measures that would prevent these complications of meningitis.

References

1. Adair, C. V., Gauld, R. L., Smadel, J.: Aseptic meningitis—clinical and etiologic studies on 854 cases. *Ann. Intern. Med.* **39**:675–704, 1953.
2. Adams, R. D., Cammermeyer, J.: Acute necrotizing hemorrhagic encephalopathy. *J. Neuropathol. Exp. Neurol.* **8**:1–30, 1949.
3. Aita, J. A.: Neurologic manifestations of infectious endocarditis. *Nebr. Med. J.* **58**:200–202, 1973.
4. Apley, J., Clarke, S. K. R., Roome, A. P. C.: Primary amoebic meningoencephalitis in Britain. *Br. Med. J.* **1**:596–599, 1970.
5. Arana-Iniquez, R., Fernandez-Lopez, J. R.: Parasitosis of the nervous system with special reference to echinococcosis. *Clin. Neurosurg.* **14**:123–144, 1967.
6. Arseni, C.: Cysticercosis of the brain. *Br. Med. J.* **2**:494–497, 1957.
7. Babinski, J., Gendron, A.: Leucocytose du liquide cephalorachidien au cors du ramollissement de l'ecorce cerebrale. *Bull. Soc. Med. Hosp. Paris* **33**:370, 1912.

8. Baker, A. S., Ojemann, R. G., Swartz, M. N., Richardson, E. P.: Spinal epidural abscess. *N. Engl. J. Med.* **293**:463–468, 1975.
9. Bamford, C. R.: Toxoplasmosis mimicking a brain abscess in an adult with treated scleroderma. *Neurology* **25**:343–345, 1975.
10. Bamford, C. R., Labadie, E. L.: Reversal of dementia in normotensive hydrocephalus after removal of locally asymptomatic cauda equina tumor. *J. Neurosurg.* **45**:104–107, 1976.
10a. Behan, P. O., Moore, M. J., Lamarche, J. B.: Acute necrotizing hemorrhagic encephalopathy. *Postgrad. Med.* **54**:154–160, 1973.
11. Bender, M. B., Christhoff, N.: Nonsurgical treatment of subdural hematomas. *Arch. Neurol.* **31**:73–79, 1974.
12. Bennett, D. R., Peters, H. A.: Acute cerebellar syndrome secondary to infectious mononucleosis in a 52-year-old man. *Ann. Intern. Med.* **55**:47–48, 1961.
13. Bernat, J. L.: Glioblastoma multiforme and the meningeal syndrome. *Neurology* **26**:1071–1074, 1976.
14. Bhatt, D. R., Hattwick, A. W., Gerdsen, R.: Human rabies: Diagnosis, complications and management. *Am. J. Dis. Child.* **127**:862–869, 1974.
15. Bickerstaff, E. R.: Cerebral cysticercosis. *Br. Med. J.* **7**:1055–1058, 1955.
16. Blankfein, R. J., Chirico, A. C.: Cerebral schistosomiasis. *Neurology* **15**:957–967, 1965.
17. Bolten, C. F., Ashenhurst, E. M.: Actinomycosis of the brain: Case report and review of the literature. *Can. Med. Assoc. J.* **90**:922–928, 1964.
18. Broder, S., Waldmann, T. A.: The suppressor-cell network in cancer. Part I. *N. Engl. J. Med.* **299**:1281–1284, 1978.
19. Brueton, M. J., Tugwell, P., Whittle, A. C.: Fibrin degradation products in the serum and cerebrospinal fluid of patients with Group A meningococcal meningitis. *J. Clin. Pathol.* **27**:402–404, 1974.
20. Buchsbaum, W. H.: Clinical management of coccidiodal meningitis. In Ajello, L. (ed.): *Coccidioidomycosis—Current Clinical and Diagnostic Status.* Miami, Symposia Specialists, 1977, pp. 191–199.
21. Buescher, E. L., Artenstein, M. S., Olson, L. C.: CNS infections of viral etiology: The changing pattern. In Zimmerman, H. M. (ed.): *Infections of the Nervous System.* Baltimore, Williams & Wilkins, 1978.
22. Bunnang, T., Comer, D. S., Punyagupta, S.: Eosinophilic myeloencephalitis caused by *Gnathostoma spinigerum*. *J. Neurol. Sci.* **10**:419–434, 1970.
22a. Brooks, B. R., Wood, J. H., Diaz, M., Czerwinski, C., Georges, L. P., Sode, J., Ebert, M. H., Engel, W. K.: Extracellular cyclic nucleotide metabolism in human central nervous system. In Wood, J. H. (ed.): *Neurobiology of Cerebrospinal Fluid I.* New York, Plenum Press, 1980.
23. Brooks, W. H., Netsky, M. G., Normansell, D. E., Horwitz, D. A.: Depressed cell-mediated immunity in patients with primary intracranial tumors. *J. Exp. Med.* **136**:1631–1647, 1972.
24. Carmel, P. W., Fraser, R. A. R., Stein, B. M.: Aseptic meningitis following posterior fossa surgery in children. *J. Neurosurg.* **41**:44–48, 1974.
25. Carter, R. F.: Primary amoebic meningoencephalitis: An appraisal of present knowledge. *Trans. R. Soc. Trop. Med. Hyg.* **66**:193–213, 1972.
26. Castleman, R.: Herpes simplex encephalitis: Case records. *N. Engl. J. Med.* **284**:1023–1031, 1971.
27. Caudill, R. G., Smith, C. E., Reinarz, J. A.: Coccidiocal meningitis, a diagnostic challenge. *Am. J. Med.* **49**:360–364, 1970.
28. Ch'ien, L. T., Cannon, N. J., Charamella, L. J.: Effect of adenine arabinoside on severe herpes virus hominis infections in man. *J. Infect. Dis.* **128**:658–662, 1973.
29. Chilcote, R. R., Baehner, R. L., Hammond, D.: Septicemia and meningitis in children splenectomized for Hodgkin's disease. *N. Engl. J. Med.* **295**:798–800, 1976.
30. Chmmel, H., Grieco, M. H.: Cerebral mucormycosis and renal aspergillosis in heroin addicts without endocarditis. *Am. J. Med. Sci.* **266**:225–231, 1973.
31. Cone, W., Barrera, S. E.: The brain and the cerebrospinal fluid in acute aseptic cerebral embolism. *Arch. Neurol. Psychiatry* **25**:523–547, 1931.
32. Conomy, J. P., Dalton, J. W.: *Clostridium perfringens* meningitis. *Arch. Neurol.* **21**:44–50, 1969.
33. Corwin, N. D.: *Listeria monocytogenes*—report of a case and review. *J. Med. Soc. N. Y.* **70**:105–109, 1973.
34. Cravioto, H., Feigin, I.: Non-infectious granulomatous angiitis with predilection for nervous system. *Neurology* **9**:599–609, 1959.
35. Dacey, R. G., Welsh, J. E., Scheld, W. M., Winn, M. A., Jane, J. A.: Alterations of cerebrospinal fluid outflow resistance in experimental meningitis. *Ann. Neurol.* **4**:173, 1978 (abstract).
36. Dalton, H. P., Allison, M. J.: Modification of laboratory results by partial treatment of bacterial meningitis. *Am. J. Clin. Pathol.* **49**:410–413, 1968.
37. David, J. R.: Lymphocyte mediators and cellular hypersensitivity. *N. Engl. J. Med.* **288**:143–148, 1973.
38. Dannenberg, A. M.: Macrophages in inflammation and infection. *N. Engl. J. Med.* **293**:489–493, 1975.
39. Dixit, S. P., Coppola, E. D.: The fate of intracerebral adrenal grafts in dogs. *Arch. Surg.* **99**:352–355, 1969.
40. Davson, H.: *Physiology of the Cerebrospinal Fluid.* Boston, Little, Brown, 1967.
41. Dayan, A. D., Stokes, M. I.: Rapid diagnosis of viral meningoencephalitis by immunofluorescent exam of cerebrospinal cells. *Lancet* **1**:177–179, 1973.
42. de Lemos, R. A., Haggerty, R. J.: Corticosteroids as adjunct to treatment in bacterial meningitis. *Pediatrics* **44**:30–34, 1969.
43. Destaing, F.: Recurrent purulent meningitis. *Sem. Hosp. Paris* **49**:1759–1792, 1973.
44. Diamond, R. D., Bennett, J.: Prognostic factors in cryptococcal meningitis. *Ann. Intern. Med.* **80**:176–181, 1974.

45. Di Benedetto, R. J., Jurgensen, P. F.: Infectious mononucleosis meningoencephalitis. *South. Med. J.* **67**:736–738, 1974.
46. Dobson, C., Welch, J. S.: Dirofilariasis as a cause of eosinophilic meningitis in man, diagnosed by immunofluroescene and arthus hypersentisitivity. *Trans. R. Soc. Trop. Med. Hyg.* **68**:223–228, 1974.
47. Dodge, P. R.: Headache and weakness of the limbs following spinal anesthesia. Case records. *N. Engl. J. Med.* **255**:138–145, 1956.
48. Duque, O.: Meningoencephalitis and brain abscesses caused by *Cladosporium*. *Am. J. Clin. Pathol.* **36**:505–517, 1961.
49. Edmonds, M. S.: The role of proteolytic enzymes in demyelination in experimental allergic encephalomyelitis. *Neurochem. Res.* **2**:233–246, 1977.
50. Ellner, J. J., Bennett, J. E.: Chronic meningitis (review). *Medicine (Baltimore)* **55**:341–369, 1976.
51. Faden, A.: Neurological sequelae of malignant external otitis. *Arch. Neurol.* **32**:204–205, 1975.
52. Fadli, M. E., Youssef, M. M.: Neuro-Behcet's syndrome in the United Arab Republic. *Eur. Neurol.* **9**:76–89, 1973.
53. Farmer, T. W., Wise, G. R.: Subdural empyema in infants, children and adults. *Neurology* **23**:254–261, 1973.
54. Feigin, R. D., Shackelford, P. G.: Value of repeat lumbar puncture in differential diagnosis of meningitis. *N. Engl. J. Med.* **289**:571–574, 1973.
55. Feldman, R. G., Southgate, T. M.: Progressive personality changes in a middle-aged man, *J. Am. Med. Assoc.* **225**:143–153, 1973.
56. Feldman, W. E.: *Bacteroides fragilis* ventriculitis and meningitis: Report of two cases. *Am. J. Dis. Child.* **130**:880–883, 1976.
57. Feldman, W. E.: Relation of concentrations of bacterial antigen in CSF to prognosis in patients with bacterial meningitis. *N. Engl. J. Med.* **296**:433–435, 1977.
58. Fetter, B. F., Klintworth, G. K., Hendry, W. S.: *Mycoses of the Central Nervous System*. Baltimore, Williams & Wilkins, 1967.
59. Fisher, C. M.: Ocular palsy in temporal arteritis. *Minn. Med.* **142**:1258–1268, 1430–1437, 1617–1630, 1959.
60. Fishman, R. A.: Cerebrospinal fluid. In Baker, A. B., Baker, L. H. (eds.): *Clinical Neurology*, Chapt. 5. New York, Harper & Row, 1974, pp. 1–40.
61. Fishman, R. A.: Carrier transport of glucose between blood and cerebrospinal fluid. *Am. J. Physiol.* **206**:836–844, 1964.
62. Fox, H. A., Hagen, P. A., Turner, D. J.: Immunofluorescence in the diagnosis of acute bacterial meningitis. *Pediatrics* **43**:44–49, 1969.
63. Fragoyannis, S., Yalcin, S.: Ependymomas with distant metastases: Review of the literature. *Cancer* **19**:246–256, 1966.
64. Freeman, J. M.: The clinical spectrum and early diagnosis of Dawson's encephalitis. *J. Pediatr.* **75**:590–603, 1969.
65. Gautier-Smith, P. C.: Clinical aspects of spinal neurofibromas. *Brain* **90**:359–394, 1967.
66. Gelfand, M., Jeffrey, C.: Cerebral cysticercosis in Rhodesia. *J. Trop. Med. Hyg.* **76**:87–89, 1973.
67. Gerber, H. J., Schoonmaker, F. W., Vazquez, M. D.: Chronic meningitis associated with *Histoplasma endocarditis*. *N. Engl. J. Med.* **275**:74–76, 1966.
68. Geschwind, N., Richardson, E. P.: A 77-year-old woman with a gait disorder and confusion. Case records. *N. Engl. J. Med.* **292**:852–857, 1975.
69. Gibberd, F. B., Ngan, H., Swann, G. F.: Hydrocephalus, subarachnoid hemorrhage and ependymomas of the cauda equina. *Clin. Radiol.* **23**:422–426, 1972.
70. Gilligan, B. S., Williams, I., Perceval, A. K.: Norcardial meningitis: Report of a case with bacteriological studies. *Med. J. Aust.* **2**:747–752, 1962.
71. Glasser, L., Corrigan, J. J., Payne, C.: Basophilic meningitis. *Neurology* **26**:899–902, 1976.
72. Glasser, L., Payne, C., Corrigan, J. J.: The *in vivo* development of plasma cells: A morphologic study of human CSF. *Neurology* **27**:448–459, 1977.
73. Glover, D., Labadie, E. L.: Physiopathogenesis of subdural hematomas. Part 2. Inhibition of growth of experimental hematomas with dexamethasone. *J. Neurosurg.* **45**:393–397, 1976.
74. Goldstein, E., Winship, M. J., Pappagianis, D.: Ventricular fluid and the management of coccidiodal meningitis. *Ann. Intern. Med.* **77**:243–246, 1972.
75. Goodman, J. W., Kaufman, L., Koenig, M. G.: Diagnosis of cryptococcal meningitis. *N. Engl. J. Med.* **285**:434–436, 1971.
76. Goodnight, S. H., Kenoyer, G., Rapaport, S. I., Patch, M. J., Lee, J. A., Kurze, T.: Defibrination after brain-tissue destruction. *N. Engl. J. Med.* **190**:1043–1047, 1974.
77. Gorbach, S. L., Bartlett, J. G.: Anaerobic infections (three parts). *N. Engl. J. Med.* **290**:1237–1245; **290**:1289–1294, 1974.
78. Greenspan, G., Stevens, L.: Infection with cysticercosis cellulosae. *N. Engl. J. Med.* **264**:751–753, 1961.
79. Greer, H. D., Geraci, J. E., Corbin, K. B.: Disseminated histoplasmosis presenting as a brain tumor and treated with amphotericin B. *Proc. Mayo Clin.* **39**:490–494, 1964.
80. Griffin, J.: Granulomatous angiitis of the CNS with aneurysms on multiple cerebral arteries. *Trans. Am. Neurol. Assoc.* **98**:145–148, 1973.
81. Griffin, J. W., Thompson, R. W., Mitchinson, M. J.: Lymphomatous leptomeningitis. *Am. J. Med.* **51**:200–208, 1971.
82. Griffith, J. F.: Nervous system manifestations of cytomegalovirus infection. *Dev. Med. Child. Neurol.* **13**:520–522, 1971.
83. Haight, T. H.: Anthrax meningitis: Review of the literature. *Am. J. Med. Sci.* **224**:57–69, 1952.

84. HASS, W. K.: Soluble tissue antigens in human brain tumor and cerebrospinal fluid. *Arch. Neurol.* **14**:443–447, 1966.
85. HAYNES, R. E., HILTY, M. D., AZIMI, P. H.: California encephalitis in children. *Am J. Dis. Child.* **124**:530–533, 1972.
86. HEATH, C. W., ALEXANDER, A. D., GALTON, M. M.: Leptospirosis in the U.S.A. Part II. Analysis of 483 cases. *N. Engl. J. Med.* **273**:915–922, 1965.
87. HERMANS, P. E., GOLDSTEIN, N. P., WELLMAN, W. E.: Mollaret's meningitis and differential diagnosis of recurrent meningitis. *Am. J. Med.* **52**:128–140, 1972.
88. HOPEWELL, J. W., ADAMS, G. A.: Modification of subsequent intracerebral infection with *Bordetella pertussis* by previous local brain irradiation. *Exp. Neurol.* **26**:173–182, 1970.
89. HOPKINS, A. P., HARVEY, P. K. P.: Chronic benign lymphocytic meningitis. *J. Neurol. Sci.* **18**:443–453, 1973.
90. HOWLAND, W. J., CURRY, J. L., BUTLER, A. K.: Pantopaque arachnoiditis: Experimental study with blood as potentiating agent. *Radiology* **80**:489–491, 1963.
91. HUBLENT, W. T., HUMPHREY, G. L.: Leptospirosis: A cause of aseptic meningitis. *Calif. Med.* **108**:113–117, 1968.
92. HUNTER, R., THOMSON, T., REYNOLDS, C. M.: Fibrin–fibrinogen degradation products in CSF of patients admitted to a psychiatric unit. *J. Neurol. Neurosurg. Psychiatry* **37**:249–251, 1974.
93. HURST, E. W., PAWAN, J. L.: Rabies paralytic form—a further account of the Trinidad outbreak of acute rabic myelitis. *J. Pathol. Bacteriol.* **35**:301, 1932; *Brain* **59**:1, 1936.
94. ILLIS, L. W., GOSTLING, J. V. T.: *Viral Diseases of the CNS.* London, Baillere Tindall, 1975, pp. 56–75.
95. INGHAM, H. R., SELKON, J. B., ROXBY, C. M.: Bacteriological study of otogenic cerebral abscesses: Chemotherapeutic role of metronidazole. *Br. Med. J.* **2**:991–993, 1977.
96. JACKSON, I. J.: Aseptic hemogenic meningitis. *Arch. Neurol. Psychiatry* **62**:572–589, 1949.
97. JAMIESON, W. M., KERR, M., SOMMERVILLE, R. G.: Echo type 9 meningitis in East Scotland. *Lancet* **1**:581, 1958.
98. JOHNSON, R. N., MAFFEO, C. J., DACEY, R. W., BUTLER, A. B., BASS, N. H.: Mechanism for intracranial hypertension during experimental subarachnoid hemorrhage: Acute malfunction of arachnoid villi by components of plasma. *Ann. Neurol.* **4**:193, 1978 (abstract).
100. KANTOR, F. S.: Infection, anergy and cell mediated immunity. *N. Engl. J. Med.* **292**:629–634, 1975.
101. KATZ, A. M., PAN, C. T.: *Echincococcus disease in the U.S.A. Am. J. Med.* **25**:759, 1958.
102. KELLY, R.: Colloid cysts of the third ventricle, 29 cases. *Brain* **74**:23–65, 1951.
103. KEPES, J. J., MAXWELL, J. A., HEDEMAN, L.: Primary diffuse malignant lymphoma of the leptomeninges presenting as pseudo-tumor cerebri. *Neurochirurgia* **14**:188–196, 1971.
104. KERSHAW, W. E., ST. HILL, C. A., SEMPLE, A. B.: *Trichinella spirallis:* Distribution of the larva in muscle, viscera, CNS. *Ann. Trop. Med. Parasitol.* **50**:355, 1956.
105. KHOO, T. K., SUGAI, K., LEONG, T. K.: Disseminated aspergillosis: Case report and review of the world literature. *Am. J. Clin. Pathol.* **45**:697–703, 1966.
106. KING, D. K., LOH, K. K., AYALA, G. G., GAMBLE, J. F.: Eosinophilic meningitis and lymphomatous meningitis. *Ann. Intern. Med.* **82**:228, 1975.
107. KING, R. B., STOOPS, W. L., FITZGIBBONS, J.: *Nocardia asteroides* meningitis: A case successfully treated with large doses of sulfadiazine and urea. *J. Neurosurg.* **24**:749–751, 1966.
108. KORNBLITH, P. L., DOHAN, F. C., WOOD, W. C., WHITMAN, B. O.: Human astrocytoma: Serum mediated immunologic response. *Cancer* **33**:1512–1519, 1974.
109. KOZINN, P. J., TASCHDJIAN, C. L., PISHUAZADEH, P.: *Candida* meningitis successfully treated with amphotericin B. *N. Engl. J. Med.* **268**:881–884, 1963.
110. KRAMER, M. D.: Trichinosis with CNS involvement: A case report and review of the literature. *Neurology* **22**:485–491, 1972.
111. KRAYENBÜHL, H. A.: Cerebral and sinus thrombosis. *Clin. Neurosurg.* **14**:1–24, 1966.
112. LABADIE, E. L., VAN ANTWERP, J., BAMFORD, C. R.: Abnormal lumbar isotope cisternography in an unusual case of spontaneous hypoliquorreic headache. *Neurology* **26**:135–139, 1976.
113. LABADIE, E. L., GLOVER, D.: Physiopathogenesis of subdural hematomas. Part 1. Histological and biochemical comparisons of subcutaneous hematoma in rats with subdural hematoma in man. *J. Neurosurg.* **45**:382–392, 1976.
114. LABADIE, E. L., GLOVER, D.: Local alterations of hemostatic–fibrinolytic mechanisms in reforming subdural hematomas. *Neurology* **25**:669–675, 1975.
115. LANDAU, J. M., NEWCOMER, V. D.: Acute cerebral mucormycosis (phycomicosis). *J. Pediatr.* **61**:363–385, 1962.
116. LEE, M. C., HEANEY, L. M., JACOBSON, R. L.: Cerebrospinal fluid in cerebral hemorrhage and infarction. *Stroke* **6**:638–641, 1975.
117. LEITNER, M. J., GRYNKEWICH, S.: Encephalopathy associated with trichinosis—treatment with ACTH and cortisone: Report of two cases. *Am. J. Med. Sci.* **236**:546, 1958.
118. LEVY, N. L., MAHALEY, M. S., DAY, E. D.: *In vitro* demonstration of cell-mediated immunity to human brain tumors. *Cancer Res.* **32**:477–482, 1972.
119. LIBRACH, I. M.: Acute encephalitis in infectious mononucleosis. *Br. J. Clin. Pathol.* **26**:379–380, 1972.
120. LIEBOW, A. A., CARRINGTON, C. R. B., FRIEDMAN, P. J.: Lymphomatoid granulomatosis. *Human Pathol.* **3**:457–558, 1972.
121. LIKOSKY, W. H.: United States cases in 1968: Epide-

miology of Echo virus aseptic meningitis. *Health Serv. Rep.* **87**:638–642, 1972.

122. LISAK, R. P., ZWEIMAN, B.: Reactivity of CSF lymphocytes to basic protein. *N. Engl. J. Med.* **297**:850–854, 1977.
123. LITTLE, J. R., DALE, A. J. D., OKAZAK, H.: Meningeal carcinomatosis. *Arch. Neurol.* **30**:138–143, 1974.
124. LIVANAINEN, M.: Benign recurrent aseptic meningitis of unknown etiology (Mollaret's). *Acta Neurol. Scand.* **49**:133–138, 1973.
125. LOGUE, V., TILL, K.: Posterior fossa dermoid cysts with special reference to intracranial infection. *J. Neurol. Neurosurg. Psychiatry* **15**:1–12, 1952.
126. LOUDON, R. G., LAWSON, R. A.: Systemic blastomycosis: Recurrent neurological relapse in a case treated with amphotericin B. *Ann. Intern. Med.* **55**:139–147, 1961.
127. LUZECHY, M., SIEGEL, B. A., COXE, W. S.: Papilledema and communicating hydrocephalus associated with lumbar neurofibroma. *Arch. Neurol.* **30**:487–489, 1974.
128. MAHALEY, M. S.: Experiences with antibody production from human glioma tissue. *Prog. Exp. Tumor Res.* **17**:31–39, 1972.
129. MCKAY, D. G.: Participation of components of blood coagulation system in the inflammatory response. *Am. J. Pathol.* **67**:181–210, 1972.
130. MACKANESS, G. B., LAGRANGE, P. H., MILLER, T. E., ISHIBASHI, T.: Feedback inhibition of specifically sensitized lymphocytes. *J. Exp. Med.* **139**:528–559, 1974.
131. MASON, M. S., RAAF, J.: Complications of pantopaque myelography—case and review. *J. Neurosurg.* **19**:302–311, 1962.
132. MERRITT, H. H., FREMONT-SMITH, F.: *The Cerebrospinal Fluid*. Philadelphia, W. B. Saunders, 1938.

132a. MERRITT, H. H., ADAMS, R. D., SOLOMON, H. C.: *Neurosyphilis*. New York, Oxford, Oxford University Press, 1946.

134. MESSERT, B., RIEDER, M.: RISA cisternography–study of CSF changes associated with RISA injection. *Neurology* **22**:789–792, 1972.
135. MEYER, H. M., JOHNSON, R. T., CRAWFORD, I. P.: Central nervous system syndromes of viral etiology: A study of 713 cases. *Am. J. Med.* **29**:334–347, 1960.
136. MILLER, A. A., RAMSDEN, F.: Malignant meningioma with extracranial metastases and seeding of subarachnoid space and ventricles. *Pathol. Eur.* **7**:167–175, 1972.
137. MILLER, J. A., HARTER, D. H.: Acute viral encephalitis. *Med. Clin. North Am.* **56**:1393–1404, 1972.
138. MOORE, C. M.: Acute bacterial meningitis with absent or minimal cerebrospinal fluid findings. *Clin. Pediatr.* **12**:117–118, 1973.
139. MORGAN, H., WOOD, M. H., MURPHEY, F.: Experience with 88 consecutive cases of brain abscess. *J. Neurosurg.* **38**:698–704, 1973.
140. MUKOYAMA, M., GIMPLE, K., POSER, C. M.: Aspergillosis of the central nervous system. *Neurology* **19**:967–974, 1969.
141. MURRAY, K. J., AUSMAN, J. I., CHOU, S. N., DOUGLAS, S. D.: Immuno-proteins in human brain tumor cyst fluids. *J. Neurosurg.* **46**:314–319, 1977.
142. NACHUM, R., LIPSEY, A., SIEGEL, S. E.: Rapid detection of gram negative bacterial meningitis by the Limulus lysate test. *N. Engl. J. Med.* **289**:931–934, 1973.
143. NADER, G. T.: Mastoid and paranasal sinus infection and their relation to the CNS. *Clin. Neurosurg.* **14**:288–313, 1966.
144. NIBBELINK, D. W., TURNER, J. C., HENDERSON, W. E.: Intracranial aneurysms and subarachnoid hemorrhage: A cooperative study of antifibrinolytic therapy in recent onset subarachnoid hemorrhage. *Stroke* **6**:622–629, 1975.
145. NICKEL, S. N., FRAME, B.: Neurologic manifestations of myxedema. *Neurology* **8**:511–517, 1958.
146. O'CONNOR, J. F., MUSKER, D. M.: Central nervous system involvement in systemic lupus erythematosus. *Arch. Neurol.* **14**:157–164, 1966.
147. O'DUFFY, J. D., GOLDSTEIN, N. P.: Neurologic involvement in seven patients with Behcet's disease. *Am. J. Med.* **61**:170–178, 1976.
148. OELMICHEN, M.: *Cerebrospinal Fluid Cytology: An Introduction and Atlas*. Philadelphia, W. B. Saunders, 1976.
149. OH, S. J.: Cerebral and spinal paragonimiasis. *J. Neurol. Sci.* **9**:205–235, 1969.
150. OLSON, M. E., CHERNIK, N. L., POSNER, J. B.: Infiltration of the leptomeninges by systemic cancer. *Arch. Neurol.* **30**:122–137, 1974.
151. O'TOOLE, R. D., THORNTON, G. F., MUKHERJEE, M. K.: Dexamethasone in tuberculous meningitis. *Ann. Intern. Med.* **70**:39–48, 1969.
152. O'TOOLE, R. D., THORNTON, G. F., MUKHERJEE, M. K.: Cerebrospinal fluid immunoglobulins in bacterial meningitis. *Arch. Neurol.* **25**:218–224, 1971.
153. ORRISON, W. W., LABADIE, E. L., RAMGOPAL, V.: Fatal meningitis secondary to undetected bacterial psoas abscess: Report of three cases. *J. Neurosurg.* **47**:755–760, 1977.

153a. PARSONS, M.: *Tuberculous Meningitis—A Handbook for Clinicians*. New York, Toronto, Oxford University Press, 1979, pp. 9, 16, 48.

154. PARKER, C. W.: Control of lymphocyte function. *N. Engl. J. Med.* **295**:1180–1186, 1976.
155. PARRILLO, J., MEILBERGER, M., ELSTON, H.: *Candida* meningitis complicating Hodgkin's disease: Apparent recovery with amphotericin therapy. *J. Am. Med. Assoc.* **182**:189–191, 1962.
156. PATTISON, E. M.: Uveomeningoencephalitic syndrome (Vogt–Harada–Koyanagi). *Arch. Neurol.* **12**:197–205, 1965.
157. PECHET, L.: Fibrinolysis. *N. Engl. J. Med.* **273**:1024–1034, 1965.
158. PEDRO-PONS, A., FOZ, M., CODINA, A.: Neurobru-

cellosis: Study of 41 cases. *Minerva Med.* **64**:846–854, 1973.

159. Perot, P., Lloyd-Smith, D., Libman, I., Gloor, P.: Trichinosis encephalitis: A study of electroencephalographic and neuropsychiatric abnormalities. *Neurology* **13**:477–485, 1963.
160. Pilleggi, V. J., Szustkiewicz, C. P.: *Carbohydrates in Clinical Chemistry: Principles and Technics*, 2nd ed. Hagerstown, Maryland, Harper & Row, 1974.
161. Porter, J. M., Acinapura, A. J., Kapp, J.: Fibrinolytic activity of the spinal fluid and meninges. *Surg. Forum* **17**:425–427, 1966.
162. Rapoport, S. I.: *Blood–Brain Barrier in Physiology and Medicine.* New York, Raven Press, 1976.
163. Ratnoff, O. D.: Some relationships among hemostasis, fibrinolytic phenomena and the inflammatory response. *Adv. Immunol.* **10**:145–227, 1969.
164. Richardson, E. P.: Marked disorientation with pleocytosis and elevated protein in CSF: Case records. *N. Engl. J. Med.* **262**:623–627, 1960.
165. Richardson, E. P.: Progressive multifocal leukoencephalopathy. *N. Engl. J. Med.* **265**:815–823, 1961.
166. Riehl, J. L., Andrews, J. M.: Uveomeningoencephalitic syndrome. *Neurology* **16**:603–609, 1966.
167. Riley, O., Mann, S. H.: Brain abscess caused by *Cladosporium trichoides*—review of three cases and case report. *Am. J. Clin. Pathol.* **33**:525–531, 1960.
168. Rober, V. B., Rorke, L. B.: Primary amoebic encephalitis. *Ann. Intern. Med.* **79**:174–179, 1973.
169. Rose, A. S.: Case record of the Massachusetts General Hospital: Acute meningitis with proptosis. *N. Engl. J. Med.* **240**:267–270, 1949.
170. Rosen, L., Loison, G., Leignet, J.: Studies on eosinophilic meningitis. III. Epidemiologic and clinical observations of Pacific Islands and the possible etiologic role of *Angiostrongylus cantonensis. Am. J. Epidemiol.* **85**:17–44, 1967.
171. Ruddy, S., Gigli, I., Austen, F.: The complement system of man (four-part review). *N. Engl. J. Med.* **287**:489–494, 545–549, 592–596, 642–646, 1972.
172. Salmon, J. H., Berger, T. S.: *Salmonella* meningitis. *Surg. Neurol.* **3**:75–78, 1975.
173. Sarosi, G. A., Parker, J. D., Doto, I. L.: Amphotericin B in cryptococcic meningitis: long term results. *Ann. Intern. Med.* **71**:1079–1089, 1969.
174. Strecker, E.-P., James, A. E.: Evaluation of the changes of CSF movement associated with meningitis: A cisternographic analysis. *Am. J. Roentgenol. Radiumther. Nucl. Med.* **118**:147–154, 1973.
175. Scheinberg, L. E., Kotsilimbas, D. J., Karpf, R.: Is the brain an immunologically privileged site? III. Studies based on homologous skin grafts to the brain tissue. *Arch. Neurol.* **15**:62–67, 1966.
176. Schlesinger, J. J., Ross, A. L.: *Propionibacterium acnes* meningitis in a previously normal adult. *Arch. Intern. Med.* **127**:921–923, 1977.
177. Scheld, M., Park, T. S., Dacey, R. G., Winn, H., Sande, H., Jane, J. A.: Clearance of bacteria from CSF to blood in experimental meningitis. Paper No. 80 (abstract), American Association of Neurological Surgeons Meeting, Los Angeles, California, April 26, 1979.
178. Schulhof, L. A., Russell, J. R.: Intracerebral toxoplasmosis presenting as a mass lesion. *Surg. Neurol.* **4**:9–11, 1975.
179. Sells, C. J., Carpenter, R. L., Ray, C. G.: Sequelae of central nervous system enterovirus infections. *N. Engl. J. Med.* **293**:1–4, 1975.
180. Shiraki, H., Otani, S.: Rabies post-vaccinial allergic encephalomyelitis. In Kies, M. W., Alvord, E. C.(eds.): *Allergic Encephalomyelitis.* Springfield, Illinois, Charles C. Thomas, 1959, pp. 59–122.
181. Silverstein, A., Steinberg, G., Nathanson, M.: Nervous system involvement in infectious mononucleosis. *Arch. Neurol.* **26**:353–358, 1972.
182. Smith, D. R., Hardman, J. M., Earle, K. M.: Metastasizing neuroectodermal tumors of the CNS. *J. Neurosurg.* **31**:50–58, 1969.
183. Snitzer, L. S., McKinney, E. C., Tejada, F.: Cerebral metastases and carcinoembryonic antigen in CSF. *N. Engl. J. Med.* **293**:1101, 1975.
184. Spatako, J., Sachs, O.: Oligodendroglioma with remote metastases. *J. Neurosurg.* **38**:373–379, 1968.
185. Spillane, J. D.: *Tropical Neurology.* London, Oxford Press, 1973.
186. Stalder, H., Oxman, M. N., Dawson, D. M.: Herpes simplex meningitis: Isolation of herpes simplex virus type 2 from CSF. *N. Engl. J. Med.* **289**:1296–1298, 1973.
187. Stokes, H. B., O'Hara, C. M., Buchanan, R. D.: An improved method for examination of cerebrospinal fluid cells. *Neurology* **25**:901–906, 1975.
188. Swartz, M. N., Dodge, P. R.: Bacterial meningitis. *N. Engl. J. Med.* **272**:725–730, 1965; **272**:954–960, 1965 (Part II); **272**:1003–1009, 1965 (Part III).
189. Tabira, T., Webster, H. DeF., Wray, S. H.: Multiple sclerosis CSF produces myelin lesions in tadpole optic nerves. *N. Engl. J. Med.* **295**:644–649, 1976.
190. Takashima, S., Koga, M., Tanaka, K.: Fibrinolytic activity of human brain and CSF. *Br. J. Exp. Pathol.* **50**:533–539, 1969.
191. Thomas, J. A., Kothane, S. N., Baptist, S. J.: *Cysticercus cellulosae. J. Trop. Med. Hyg.* **76**:106–110, 1973.
192. Tourtellote, W. W., Mitz, L., Brian, E. R., De Jong, R.: Spontaneous subarachnoid hemorrhage—factors affecting the clearance of cerebrospinal fluid. *Neurology* **14**:301–306, 1964.
193. Townsend, J. J., Baringer, J. R., Wolinsky, J. S.: Progressive rubella panencephalitis: Late onset after congenital rubella. *N. Engl. J. Med.* **292**:990–994, 1975.
194. Towsend, J. J., Wolinsky, J. W., Baringer, J. R., Johnson, P. C.: Acquired toxoplasmosis: A neglected cause of treatable CNS disease. *Arch. Neurol.* **32**:335–343, 1975.

195. Udani, P. M., Dastur, D. K.: Tuberculous encephalopathy with and without meningitis: Clinical features and pathological correlations. *J. Neurol. Sci.* **10**:541–561, 1970.
196. Utz, J. P.: Fungal infections of the central nervous system. *Clin. Neurosurg.* **74**:86–100, 1966.
197. Vaze, S.: *Salmonella* meningitis in infants. *J. Indian Med. Assoc.* **100**:298, 1973.
198. Verdura, J., White, R. J., Resnikoff, R.: Interhemispheric subdural empyema: Angiographic diagnosis and surgical treatment. *Surg. Neurol.* **3**:89–92, 1975.
199. Victor, M.: Convulsions, bilateral neurologic signs and pleocytosis. Case records. *N. Engl. J. Med.* **252**:633–641, 1955.
200. Viroslav, J., Williams, T. W.: Nocardial infection of the pulmonary and central nervous system: Successful medical treatment. *South. Med. J.* **64**:1382–1385, 1971.
201. Waisbren, B. A., Ullrich, D.: An isolated blastomycetoma of the posterior fossa treated successfully with surgery and amphotericin B. *Am. J. Med.* **32**:621–624, 1962.
202. Weil, M. L., Itabashi, H. H., Cremer, N. E.: Chronic progressive panencephalitis due to rubella virus simulating SSPE. *N. Engl. J. Med.* **292**:994–998, 1975.
203. White, B. E.: Cerebral candidiasis (letter). *N. Engl. J. Med.* **286**:321, 1972.
204. Wilfert, C. M.: Mumps meningoencephalitis with low cerebrospinal fluid glucose, prolonged pleocytosis and protein elevation. *N. Engl. J. Med.* **280**:855–859, 1969.
205. Willis, A. T., Jacobs, S. I.: A case of meningitis due to *Clostridium welchi. J. Pathol. Bacteriol.* **88**:312–314, 1964.
206. Weintraub, M. I., Glaser, G. H.: Nocardial brain abscess and pure motor hemiplegia. *N. Y. State J. Med.* **70**:2717–2121, 1970.
207. Welsh, J. D., Rhoades, E. R., Jaques, W.: Disseminated nocardiosis involving spinal cord. *Arch. Intern. Med.* **108**:141–147, 1961.
208. Wolf, S. M.: Decreased cerebrospinal fluid glucose in herpes zoster meningitis. *Arch. Neurol.* **30**:109, 1974.
209. Wolf, S. M., Schotland, D. L., Phillips, L. L.: Involvement of nervous system in Behcet's syndrome. *Arch. Neurol.* **12**:315–325, 1965.
210. Wolfowitz, B. L.: Otogenic intracranial complications. *Arch. Otolaryngol.* **96**:220–222, 1972.
211. Woody, N. C., Woody, H. B.: American trypanosomiasis: Clinical and epidemiological background of Chagas disease in U.S.A. *J. Pediatr.* **58**:568, 1961.
212. Wu, K. K., Jacobsen, C. D., Hoak, J. C.: Plasminogen in normal and abnormal human CSF. *Arch. Neurol.* **28**:64–66, 1973.
213. Wyler, D. J., Wasserman, S. I., Karchmer, A. W.: Substances which modulate leukocyte migration are present in CSF during meningitis. *Ann. Neurol.* **5**:322–325, 1979.
214. Young, W. B.: Actinomycosis with involvement of vertebral column: Case report and review of the literature. *Clin. Radiol.* **11**:175–182, 1960.

CHAPTER 32

Penetration of Antimicrobial Agents into Cerebrospinal Fluid

Pharmacokinetic and Clinical Aspects

Ragnar Norrby

1. Introduction

Most often, bacterial and fungal infections in the central nervous system are life-threatening infections that require acute antimicrobial treatment. These disorders are rarely self-limited, secondary to the limited ability of the immune reactions of the cerebrospinal fluid (CSF) to eliminate pathogenic organisms. Even in the presence of meningeal inflammation, the CSF contains less than 0.1% of the number of circulating immunocompetent leukocytes found in peripheral blood, and scant quantities of immunoglobulins. Thus, the rapid achievement of therapeutic CSF concentrations of appropriate antibiotics is imperative for the successful treatment of meningitis.

Despite the existence of numerous publications concerning the theoretical and clinical aspects of antimicrobial treatment of meningitis, only limited information is available concerning the CSF pharmacokinetics of antibiotics and chemotherapeutic agents in man. The reasons for the lack of data are obvious: ethical limitations prohibit repeated sampling in the same patient, and the considerable variability of the factors that influence drug penetration into CSF make comparisons among patients difficult. Some drugs have well-documented clinical effects; however, these impediments make the evaluation of the efficacy of new products difficult.

Ragnar Norrby, M.D., Ph.D. • Department of Infectious Diseases, East Hospital, University of Gothenburg, S-41685 Gothenburg, Sweden.

The purpose of this chapter is to discuss the theoretical factors that affect the penetration of antibiotics into CSF and to review available clinical literature in order to evaluate which antibiotics are most appropriate in the treatment of various microbial infections within the CSF space.

2. Theoretical Aspects

2.1. Blood–CSF Barrier

The barriers restricting penetration of drugs into CSF or brain tissue are often referred to as the "blood–brain barrier." The presence or absence of significant differences between the blood–CSF barrier and the blood–brain barrier is controversial.[104] Obvious and important differences appear to exist for many antimicrobial drugs in their capacity to penetrate the CSF and into brain tissue. In addition, histological differences may be noted on light and electron microscopy. The blood–brain barrier consists of the walls of the brain capillaries and the surrounding layers of glia cells. The blood–CSF barrier, on the other hand, consists of the apical tight junctions between the epithelial cells of the choroid plexus.[128]

2.2. Factors that Affect Penetration of Antimicrobial Agents into CSF

As a pharmacokinetic compartment, the CSF space differs from all other compartments in the body. This fluid compartment possesses some unique properties that compromise the achievement of therapeutic drug concentrations. In addition, several factors are known to influence the CSF penetration of an antimicrobial drug. With the exception of meningeal inflammation, the importance of these factors, dealt with below, is not clearly defined for all drugs used for treatment of infections.

2.2.1. Lipid Solubility, Ionization, and pH Gradients. All organic compounds entering CSF must traverse a lipid membrane in the nonfenestrated brain capillary or the epithelial layer of the choroid plexus.[7] Consequently, the lipid solubility of an antimicrobial agent will influence its ability to penetrate the blood–CSF barrier. The importance of lipid solubility for some antibiotics has been demonstrated. Within the tetracycline group, the lipophilic drugs doxycycline and minocycline penetrate considerably better into CSF than the less lipid-soluble oxytetracycline and tetracycline.[1,5] However, highly lipophilic fusidic acid does not penetrate the blood–CSF barrier to any significant degree.[40] This discrepancy can be explained by variations in ionization that influence the penetration of antimicrobial drugs into CSF.[4,20,73] Only the nonionized molecules of an antimicrobial agent are lipid-soluble, whereas the ionized molecules lack lipophilicity.[4] Thus, the lower the ionization of a drug at the pH of the plasma and CSF, the better is its penetration on the basis of lipid solubility. Conversely, the combination of a low or high pK_a, resulting in a high degree of ionization at physiological pH, and a low lipid solubility reduces the ability of an acidic or basic antibiotic, respectively, to penetrate into CSF. Examples of antibiotics possessing these properties that impede penetration of the blood–CSF barrier in the absence of inflammatory changes are benzyl penicillin,[18,25,127] ampicillin,[118,120] carbenicillin,[7,18] cephaloridine,[66] cephalothin,[65,121] and cephacetrile.[82] Of these, cephalothin has a lower penetrative capacity, probably due to the high degree of deacetylation of this cephalosporin. Antibiotics that have a high degree of lipophilicity and a low degree of ionization at physiological pH and that have a good CSF penetration include trimethoprim,[36,127a] chloramphenicol,[58] and the antimycotic drug flucytosin.[28]

Another factor that primarily influences drug ionization, and secondarily its lipophilicity, is the pH gradient that normally exists between CSF and plasma. The pH of CSF and plasma is normally about 7.3 and 7.4, respectively. In purulent meningitis, this gradient is increased by the presence of acid metabolites in the CSF, which lowers CSF pH. Accordingly, Rall *et al.*[95] induced acidosis in dogs by infusing hydrochloric acid and increased the serum/CSF ratio of sulfadiazine from 0.8 to 1.25, which considerably decreased the CSF penetration of the sulfonamide. Similarly, in man, metabolic acidosis will decrease the pH gradient and may decrease CSF penetration of antibiotics. Contrarily, the metabolic alkalosis evoked by protracted vomiting is likely to increase the gradient and thus may increase the penetration of some drugs, especially weak acids, through the blood–CSF barrier.

2.2.2. Protein Binding. The effect of protein binding of antibiotics on their penetration into peripheral compartments is much debated. Experimental models simulating deep compartments (e.g., subcutaneous silicon cylinders and fibrin clots) have been used for studies of the problem.[6,48] From these experiments, protein binding of an antibiotic below 80% appears to be of little consequence for the penetration into the peripheral compartments, whereas protein binding above 80% constitutes a hindrance for the antibiotic to leave the central compartment. Several other studies indicate that the degree of protein binding of an antibiotic is of greater importance for its penetration into CSF than into other peripheral compartments.[31,73] The review of sulfonamides by Garrod and O'Grady[37] states that sulfanilamide, which has a protein binding of 5–20%, has a CSF/plasma ratio of 100%. This ratio is considerably lower for those sulfonamides with higher protein binding such as sulfadiazine and sulfamethoxazol, which have CSF/plasma ratios of 40–80% and protein bindings of 20–60% and 60–70%, respectively.

Kunin[62] studied the influence of protein binding on the CSF penetration of penicillins and found that highly protein-bound penicillins had a poorer CSF penetration than those with low protein binding. Moreover, he could demonstrate that orthocresotinic acid, which prevents binding of penicillin to rabbit albumin, increased the penetration of penicillin into CSF. Experimentally induced uremia, which reduces the protein binding of benzyl penicillin, has been found to increase the CSF concentrations of benzyl penicillin in rabbits.[114] Of the cephalosporins, cefazoline, which is protein-bound to about 80%, appears to penetrate the blood–CSF barrier to a slightly lesser extent than the cephalosporins, which have lower protein binding.[9]

2.2.3. Molecular Size and Structure. High molecular weight or complex molecular structure, or both, seem to constitute hindrances for penetration of antimicrobial drugs into CSF. This might be an explanation of the poor penetration into CSF of the aminoglycosides despite their low protein binding.[14,41,83] Polymyxin B, which combines a complex molecular structure with a high molecular weight, has a poor penetration into CSF even in the presence of meningeal inflammation.[67]

2.2.4. Active-Transport Mechanisms. The possibility of an active system that transports antimicrobial agents across the blood–CSF barrier has been postulated by Fishman.[31] Stereospecific carrier systems have been shown to be partly responsible for the transportation of sugars into CSF.[30] Accordingly, animal studies by Lithander and Lithander[70] demonstrate that increasing administered dosages of benzyl penicillin does not cause a corresponding increase of the CSF concentration. If these findings are the result of an active-transport mechanism, then antibiotic administration schedules may need to be modified to enhance CSF penetration. Rather than administering large doses with long intervals, lower antibiotic doses would have to be given with shorter intervals to utilize the transport mechanism. At present, conclusive evidence for the existence of an active transport of antibiotics into CSF is insufficient to motivate such a change of the currently used modes of administration.

2.2.5. Inflammation of Meninges. Without doubt, absence of inflammation of the meninges is the most important factor limiting the penetration of antimicrobial agents over the blood–CSF barrier. Several studies, both in animals and man, have convincingly demonstrated that even a moderate inflammatory change of the meninges results in a markedly increased penetration of antibiotics into the CSF. Geering and Just[38] noted an increase of more than 100% in CSF ampicillin concentrations among patients with serous meningitis as compared to CSF ampicillin concentrations among patients without inflamed meninges. In patients with acute purulent meningitis, Thrupp *et al.*[119] found CSF ampicillin concentrations 5 times higher than could be detected in the same patients in the recovery phase. This high correlation of the degree of meningeal inflammation and antibiotic penetration into CSF has been demonstrated in dogs by Oppenheimer *et al.*[89] for cephalothin, cephaloridine, and methicillin.

2.3. Kinetics of Penetration of Antibiotics into CSF

For ethical reasons, it is normally not possible to follow the kinetics of CSF antibiotic penetration in volunteers or patients. Kinetic studies must therefore be based on comparisons among groups of patients sampled at various times after systemic administration of the drug. Taking into account the aforementioned variability of meningeal inflammation and other factors that affect the penetrative capacity, it is difficult to give exact data concerning the kinetics of antibiotic penetration into CSF. However, most studies report peak CSF concentrations approximately 2–4 hr after intravenous administration, and slow antibiotic elimination. This slow elimination from CSF may account for the accumulation of drugs such as cefoxitin during the first three doses administered to patients with purulent meningitis.[2a] This literature review appears to suggest that it is important to assay CSF concentrations of antimicrobial drugs after repeated doses and that such analysis is not meaningful earlier than 2 hr after parenteral administration.

2.4. Transportation of Antimicrobial Agents within CSF Space

The direction of flow of CSF as described by Ingraham *et al.*[50] in 1948 is from the cisterna magna, then progressively through the basal cisterns, around the brainstem in the ambient cisterns, into the corpus callosum cistern, and finally, reaching the subarachnoid spaces. Comprehension of the pharmacokinetics of drugs in CSF requires knowledge of the direction of the CSF flow as well as of the fact that in the normal adult, CSF is constantly produced in an amount of 500–600 ml/24 hr and eliminated at the same rate. One important consequence of the flow of CSF in a distal direction is that the site of sampling of CSF may influence the drug concentration. Dumoff-Stanley *et al.*[25] demonstrated that in infants, considerably higher concentrations of benzyl penicillin could be found in ventricular than in lumbar CSF. Thus, the lumbar CSF does not always reflect the concentrations of drug in all parts of the CSF space. Moreover, a considerable time may be required before an antibiotic reaches the lumbar space, and this may partially explain the relatively late peak concentration of antibiotics in lumbar CSF relative to the serum concentration.

The concentration of an antimicrobial agent at a specific site in the CSF space frequently depends on the route of administration, especially when drugs are instilled directly into the CSF space. The frequently employed intrathecal administration of antibiotics into the lumbar space appears to have considerable limitations. Moellering and Fisher[83] have demonstrated that direct gentamicin into the lumbar subarachnoid space yields high lumbar CSF concentrations but very low ventricular CSF concentrations. Contrarily, intraventricular antibiotic administration resulted in high concentrations in both ventricular and lumbar CSF. Similar results have been reported by Kaiser and McGee[55] in children with gram-negative meningitis. In monkeys, Rieselbach *et al.*[98] demonstrated that a foreign compound injected into lumbar CSF will give measurable concentrations in the basal cisterns if the volume used for injection is 10% or more of the total CSF volume. Achievement of therapeutic concentrations of antibiotics in ventricular CSF requires the lumbar instillation of volumes of 25% of the total CSF volume.

Moellering and Fisher[83] have suggested that the diffusion of an aminoglycoside in CSF can be facilitated by repeated lumbar punctures following the intraventricular injection. Obstruction of the foramina of Monroe, the aqueduct of Sylvius, or the outlets of the fourth ventricle has been demonstrated by Schultz and Leeds[105] to produce hydrocephalus in patients with ventriculitis. Kalsbeck *et al.*[55a] has also reported compartmentalization of the cerebral ventricles as a sequela of neonatal meningitis. In these cases, instillation of antibiotic both above and below the obstruction may be indicated. External CSF drainage or shunting followed by intraventricular antibiotic administration achieves therapeutic effects in patients with ventriculitis and obstructive hydrocephalus in whom lumbar drainage should be avoided.

2.5. Elimination of Antibiotics from CSF

In vitro and *in vivo* experiments have demonstrated that weak organic acids, including benzyl penicillin, are transported actively from CSF by the choroid plexus.[20,31,73,93,113] The elimination of ^{14}C-labeled benzyl penicillin from cisternal CSF in dogs was found to be 2.5–3%/min, suggesting a half-life of approximately 35 min. This active transport of benzyl penicillin will proceed even against a concentration gradient. This system appears to have a limited capacity, since the half-life of benzyl penicillin in CSF is lower when the initial concentrations are high than when concentrations are low. Probenicid can inhibit the transport of benzyl penicillin over the choroid plexus both *in vitro* and *in vivo*.[31,113] Probenicid has also been shown to increase the CSF levels of benzyl penicillin in man.[19] This elevation in CSF concentrations was larger than could be explained by the increased serum penicillin levels produced by the effect of probenicid on the renal tubules.

The active transport of penicillin via the choroid plexus can be inhibited by factors other than probenicid administration. Spector and Lorenzo[111] demonstrated an inhibition of this transport in the presence of meningeal inflammation by bacterial infections. Similar observations have been made in animals with experimentally induced uremia.[114] Salicylates and probably other acid compounds seem capable of causing a competitive inhibition of the elimination of penicillin from CSF.[112]

Bass and Lundborg[8] suggest that the maturity of the brain may influence the elimination of antibiotics from CSF via active-transport mechanisms. Acid metabolites formed in the brain of immature rats are eliminated via secretion into the CSF followed by transport across the choroid plexus. In the mature rat, these metabolites are removed across the blood–brain barrier. Since acid metabolites (e.g., 5-hydroxyindoleacetic acid) utilize the same transport system as benzyl penicillin for their elimination from CSF, it is possible that the excretion of benzyl penicillin from the CSF of neonates and prematures is slower than that of older children and adults. Although the presence of the active transport of benzyl penicillin was known in 1966, the importance of this transport mechanism for the elimination of antibiotics from CSF in man is still obscure. The elevation of CSF concentrations of benzyl penicillin evoked by probenecid administration does suggest a clinically important role for this active-transport mechanism.[19] However, the elimination of antibiotics from CSF by these mechanisms might explain the surprisingly low concentrations of drugs found in persons with noninflamed meninges. In these cases, the low CSF antibiotic concentrations could be a result of not only rapid elimination from CSF, but also difficulties in drug penetrance of the blood–CSF barrier. With the exception of benzyl penicillin, few antibiotics have been studied with respect to their possible elimination from CSF via active transport. The probenicid effect has also been demonstrated experimentally with ampicillin, carbenicillin, cephacetrile, cefazoline and nafcillin.[18] *In vitro,* Spector[113] has demonstrated that gentamicin is taken up by the choroid plexus and that this uptake mechanism can be inhibited by other aminoglycosides, as well as by decreased oxygen tension, lowered temperature, and decreased glucose concentration.

Studies of the elimination of cephalothin and cefuroxime from the CSF space of hydrocephalic children after intraventricular administration have demonstrated extended half-lives (4 hr or more), indicating the absence of major elimination by active transport over the choroid plexus in these children (Rylander *et al.*, to be published).

3. Concentration of Antibiotics in CSF

Table 1 represents an effort to list the penetration of various antimicrobial agents into the CSF of man in the absence and presence of meningeal inflammation. For reasons mentioned below, steady-state data have been used, when available. The degree of penetration has been expressed in general terms—poor, fair, and good—to describe the great variations (and sometimes confusion) in data available in the literature. The available data are based on sampling of CSF at various times after the antibiotic doses were systematically administered. Moreover, the authors tend to express the penetrance of the agents in nonuniform terms; CSF/serum ratios, serum/CSF ratios, ratios between peak serum concentrations and peak CSF concentrations, and so on. Another difficulty in the interpretation of the results of CSF penetration studies is the lack of an accepted and reliable technique to determine and quantify the degree of meningeal inflammation. Although the CSF protein concentration tends to parallel the CSF penetration of an antibiotic, data concerning CSF protein concentrations are often lacking. All these factors result in large variations among different reports that *per se* can be of high quality.

It is obvious from the data presented in Table 1 that very few antimicrobial agents penetrate well over an intact blood–CSF or blood–brain barrier. Only chloramphenicol, isoniazid, flucytosine, metronidazol, trimethoprim, and some of the sulfonamides enter the CSF in concentrations approximating those present in the serum. Doxycycline may have a good penetrance if repeated doses are administered; however, single doses given to healthy volunteers yield low CSF concentrations.[1,52] Notably, all β-lactam antibiotics have a poor ability to enter CSF in patients with noninflamed meninges. This drawback can be partially compensated by the use of very high systemic doses, which are permissible with drugs with low toxicity. However, with other drugs (e.g., the aminoglycosides and amphotericin B), increasing the dosage to achieve better CSF concentrations is contraindicated secondary to the high toxicity of these agents. Thus, with these drugs, therapeutic concentrations cannot be achieved after systemic administration even in the presence of marked meningeal inflammation.

4. Clinical Aspects

The most common etiological agents of purulent meningitis and suggested therapy for patients with meningitis are listed in Table 2. When choosing this therapy, several factors have to be considered.

4.1. Factors in the Choice of Therapy

4.1.1. Antimicrobial Sensitivity. For some bacterial species (e.g., *Streptococcus pneumoniae, Staphylococcus aureus,* and *Neisseria meningitidis*), the antibiotic susceptibility can be predicted with a high degree of certainty. For other species, especially gram-negative enteric bacilli and *Hemophilus influenzae,* is it not always possible to predict accurately which single drug alternative is likely to be effective. In these cases, combination therapy is often used to ensure antibacterial activity.

4.1.2. Penetration across the Blood–CSF Barrier. Several antibiotics will have to be excluded from the therapeutic arsenal for treatment of meningitis due to their inability to cross the blood–CSF barrier even in the presence of profound inflammatory reactions. Examples of antibiotics that cannot be used systemically in meningitis are cephalothin, the aminoglycosides, amphotericin B, oxytetracycline, and polymyxin B. For some antibiotics, limited, if any, information is available in the literature on therapeutic efficacy in patients with meningitis.

4.1.3. Route of Administration. Meningitis is normally treated with intravenously (or intramuscularly) administered antimicrobial agents. These routes are acceptable if the drug has the ability to penetrate the blood–CSF barrier and to yield therapeutic CSF concentrations. When this is not the case, administration directly into the CSF space must be considered. Since the flow of CSF is directed out of the ventricles,[128] lumbar subarachnoid administration is unlikely to be effective in cases of ventriculitis, and thus antibiotics must be administered directly into the ventricles. Access to the ventricular system is relatively easy in neonates and infants whose cranial sutures are open; however, subependymal vein puncture may be complicated by intracranial hemorrhage. In older children and adults, however, intraventricular administration can be made only after an access has been neuro-

Table 1. Penetration of Antimicrobial Agents into Human CSF after Systemic Administration

Antimicrobial agent	CSF penetration[a]		Ref. Nos.
	Normal meninges	Inflamed meninges	
Amikacin	NDA	Poor	14
Amphotericin B	Poor	Fair[b]	10
Ampicillin	Poor	Fair–good	38, 56, 118, 120
Benzyl penicillin	Poor	Fair–good	2, 19, 25
Carbenicillin	Poor	NDA	89
Cefamandole	Poor	Fair–good	60, 71
Cefazoline	Poor	Fair–good	6, 59
Cefoxitin	Poor	Fair–good	2a, 71, 84a
Cefuroxime	Poor	Fair–good	35a, 87, 96
Cephacetrile	Poor	Fair	82
Cephaloridine	Poor	Fair–good	9, 66, 74, 82, 125
Cephalothin	Poor	Poor–fair	59, 121
Chloramphenicol	Good	Good	58, 79, 99
Chlortetracycline	NDA	Poor	3, 130
Clindamycin	Poor	Fair	68
Demethylchlortetracycline	Poor	NDA	12
Doxycycline	Fair–good	NDA	1, 52
Erythromycin	Poor	Fair	43, 52
Ethambutol	Poor	Fair	11, 90, 91
Ethionamide	Fair–good	Good	49
Flucytosin	NDA	Good	28
Fosfomycin	Poor–fair	Fair	23, 72,108
Fusidic acid	Poor	NDA	40
Gentamicin	Poor	Fair[b]	55, 77, 83, 91
Kanamycin	NDA	Fair[b]	27, 47, 80
Isoniazid	Fair–good	Good	26, 32
Isoxazoyl penicillins	Poor	Poor	52
Lincomycin	Poor	Poor–fair	65, 81
Methicillin	Poor	NDA	22, 52
Metronidazol	Good	Good	53
Minocycline	Fair–good	NDA	3
Nafcillin	NDA	Fair	34, 100
Oxytetracycline	NDA	Poor	130
p-aminosalicyclic acid	Poor	Poor–fair	24
Polymyxin B	Poor	Poor	67
Rifampicin	Fair	Good	21, 57
Sulfonamides	Fair–good	Fair–good	13, 37, 89, 127a
Tinidazole	Good	Good	53
Trimethoprim	Good	Good	36, 127a
Vancomycin	Poor	Good	39, 89, 110

[a] Poor, fair, and good indicate CSF concentrations in relation to concurrent plasma concentrations of <10%, 10–30%, and >30%. (NDA) No data available.
[b] Therapeutic levels cannot be achieved in CSF when the drug is used in nontoxic doses.

surgically gained through the skull bone. Methods of chronic cannulation of the cerebral ventricles are discussed in detail by Wood[129] in Chapter 7.

4.1.4. Dosage. The size and frequency of the doses of antimicrobial agents used in treatment of meningitis have not been discussed to any large extent in the literature. Looking on the CSF as an extremely deep compartment, it seems justified to suggest that the doses used should be as high as possible without inducing severe dose-dependent adverse reactions. Normally, one tends to administer the doses three or four times daily in severe

Table 2. Etiology and Recommended Treatment of Purulent Meningitis in Various Categories of Patients

Patient category	Etiology	Antibiotics	Route of administration[a]
Neonates	Gram-negative aerobic enteric bacilli	Ampicillin	P
		Aminoglycosides	I + P
		Cerfuroxime[b]	P
		Cefamandole[b]	P
		Cephalothin	I + P
	β-Hemolytic streptococci, Group B	Ampicillin	P
		Benzyl penicillin	P
Infants	*Hemophilus influenzae*	Chloramphenicol	P
		Ampicillin	P
		Cefamandole[b]	P
		Cefuroxime[b]	P
	Neisseria meningitidis	Benzyl penicillin	P
		Ampicillin	P
		Chloramphenicol	P
Adolescents and adults	*Streptococcus pneumoniae*	Benzyl penicillin	P
		Ampicillin	P
		Chloramphenicol	P
	Gram-negative aerobic enteric bacilli	Chloramphenicol	P
		Aminoglycosides	I + P
		Ampicillin	P
		Cefamandole[b]	P
		Cefuroxime[b]	P
		Cefoxitin[b]	P
		Doxycycline[b]	P
		Co-trimoxazole[b]	P
	Staphyloccocus aureus	Nafcillin	P
		Vancomycin	P
		Chloramphenicol	P
		Cefamandole[b]	P
		Cefuroxime[b]	P
		Co-trimoxazole[b]	P
Compromised patients	*Listeria monocytogenes*	Ampicillin	P
	Pseudomonas aeruginosa	Carbenicillin	I + P
		Aminoglycosides	I + P
		Fosfomycin[b]	P
	Candida spp.	Amphotericin B	I + P
		Flucytosin	P
	Cryptococcus neoformans	Amphotericin B	I + P
		Flucytosin	P
	Enterococci	Ampicillin	P
		Aminoglycosides	I + P
Hydrocephalic patients	*Staphylococcus epidermidis*	As for *Staphylococcus aureus* above	

[a] (P) Parenteral administration (intravenously or intramuscularly); (I) intraventricular administration.
[b] Limited, if any, clinical information available.

infections. Taking into account the possibility of active transport of antibiotics into the CSF, it might be justified to split the total daily dose into six or even eight doses.

4.1.5. Treatment in the Convalescent Phase. Nobody disputes the use of high parenteral doses in the acute phase of a meningitis. The size of the doses and the route of administration become more controversial when the patient is improving. In many cases, therapy is changed to lower doses and sometimes oral administration. This type of alteration should probably be avoided and can be of benefit only to the patient having an infectious focus outside the CSF space from which a recurrence of the meningitis may originate. In most cases, only two alternatives appear rational when the patient has improved: either the parenteral dose is increased to compensate for the augmented difficulties of the drug in crossing the healing blood–CSF barrier or the antibiotic is discontinued entirely.

4.1.6. Treatment Time. The treatment time must be adjusted with respect to the etiological agent causing the meningitis. Meningococcal meningitis requires rather short treatment, since the meningococci are extremely susceptible to antibiotics like benzyl penicillin. In meningitis caused by pneumococci or *Hemophilus influenzae,* a considerably longer time is required for complete eradication of the bacteria. The type of meningitis requiring the longest treatment times is the fungal infections, in which several months of treatment must be used to prevent relapses. The possible usefulness of the most common antimicrobial agents in treating meningitis will now be discussed.

4.2. β-Lactam Antibiotics

4.2.1. Benzyl Penicillin. With its bactericidal activity in low concentrations on virtually all strains of *Streptococcus pneumoniae* and *Neisseria meningitidis,* benzyl penicillin remains one of the prime choices for treatment of bacterial meningitis. Its relatively poor passage across the blood–CSF barrier is compensated by its low toxicity, enabling the use of extensively high doses without adverse reactions. The possibility of neurotoxic reactions, mainly convulsions, is probably overemphasized, since they have been reported only after direct application of the drug into brain tissue[124] or after administration of very high doses to patients with severe renal impairment.[33,86]

4.2.2. Ampicillin. The CSF penetration of ampicillin is comparable to that of benzyl penicillin. Ampicillin also has a low toxicity, enabling the use of high doses. Its spectrum adds *Hemophilus influenzae* and *Listeria monocytogenes* to that of benzyl penicillin. In addition, ampicillin is active against many gram-negative enteric bacilli, but the frequency of resistant strains is too high to allow the use of ampicillin in purulent meningitis caused by gram-negative bacteria other than *Hemophilus influenzae* before antibiotic-susceptibility testing has been performed. The increasing frequency of β-lactamase-producing strains of *Hemophilus influenzae* that are ampicillin-resistant limits the efficacy of this drug.[54,118,120] In areas with a significant risk of infections caused by β-lactamase-producing strains of *Hemophilus influenzae,* treatment of gram-negative meningitis in infants should therefore be initiated with another antibiotic. Several reports indicate that chloramphenicol and ampicillin are equally effective in treatment of *Hemophilus* meningitis.[69,106] The possibility of an interaction between chloramphenicol and ampicillin when used in combination has been suggested by Wallace *et al.*[126] on the basis of *in vitro* studies. In one clinical survey,[69] a tendency toward higher rates of sequelae was found in patients treated with the combination as compared to those treated with either of the antibiotics alone.

Ampicillin treatment of meningitis caused by *Listeria monocytogenes* has proved efficacious.[42,51,64,75] This disease is most common in patients compromised by malignancies,[16] but is also found in previously healthy individuals.[51]

4.2.3. Carbenicillin. The antibacterial spectrum of carbenicillin is similar to that of ampicillin, but in addition, carbenicillin is active against many strains of *Pseudomonas aeruginosa* and indole-positive *Proteus.* Carbenicillin has a low toxicity, can be used in high dosages without adverse reactions, and penetrates the blood–CSF barrier in a degree similar to that of benzyl penicillin.[90] Unfortunately, therapeutic concentrations can rarely be achieved in CSF in cases of *Pseudomonas aeruginosa* meningitis because the minimal inhibitory concentrations for this species are quite high. However, intraventricular carbenicillin administration has been used with success in meningitis caused by *Pseudomonas aeruginosa.*[97]

4.2.4. Penicillinase-Resistant Penicillins. This group consists of the isoxazolyl penicillins (oxacillin, cloxacillin, dicloxacillin, and flucloxacillin), methicillin, and nafcillin. Very limited information is available about the isoxazolyl penicillins. The existing data[52] in combination with the fact that these antibiotics are highly protein-bound (93% or more) makes it unlikely that they will be useful for treatment of staphylococcal meningitis. Also, methicillin appears unable to penetrate the blood–CSF barrier well enough to be used clinically.[22] Nafcillin appears to be the only member of the group of penicillinase-resistant penicillins that might have a place in therapy of staphylococcal meningitis. Several studies indicate that although low concentrations are found in patients with normal meninges, the penetration of nafcillin over the blood–CSF barrier should be sufficient to protect against and, when meningeal inflammation is present, to treat staphylococcal meningitis.[34,100]

4.2.5. Cephalosporins. The use of cephalosporins for treatment of purulent meningitis is controversial. Theoretically, this group of antibiotics offers an alternative to benzyl penicillin in penicillin-allergic patients, although the risk of cross-allergy is a significant 8%. The wider antibacterial spectrum of the cephalosporins should offer an alternative in treatment of gram-negative meningitis. In addition, cephalosporins are active against *Staphylococcus aureus* irrespective of penicillinase production. The clinical results reported with cephalosporins in treatment of meningitis are, however, discouraging. As reviewed by Fisher *et al.*,[29] cephalothin should not be used for treatment of meningitis, since therapeutic failures have been reported in a considerable number of patients. In addition, development of meningitis caused by cephalothin-susceptible organisms during systemic treatment with cephalothin has been reported.[35,76] However, cephalothin has been used successfully in treatment of meningitis when administered intraventricularly and is one of the few β-lactam antibiotics with documented safety and efficacy when administered by that route.[109]

Cephaloridine gives higher CSF concentrations than cephalothin, secondary to lower protein binding and less catabolism. However, several failures have been reported when cephaloridine was employed in the treatment of pneumococcal meningitis.[29,74] Japanese clinicians have reported cefazoline to be effective in the treatment of meningitis.[59] The limited available information and the high protein binding of cefazoline preclude its recommendation for meningitis until further data are reported.

Cephacetrile has been found to have a fairly good penetration over inflamed meninges.[82] However, its pharmacokinetic characteristics make cephacetrile less likely to be superior to cephaloridine.

Cefamandole has been found to be effective, both experimentally and clinically, in the treatment of meningitis.[9a,60,71,103,107,116] However, failures have been reported when cefamandole was used to treat *Hemophilus influenzae* meningitis.[115] The possible place of cefamandole in treatment of gram-negative meningitis cannot yet be defined.

Cefuroxime, which like cefamandole is a new cephalosporin with improved β-lactamase stability and broader spectrum than the previous cephalosporins, has also been evaluated for penetrative capacity into the CSF and efficacy in treatment of meningitis.[35a,88,97] The limited data available, in combination with the excellent activity of this antibiotic on some bacterial species (e.g., *Hemophilus influenzae*), make cefuroxime an interesting alternative to other β-lactam antibiotics and chloramphenicol in some types of meningitis.

4.2.6. Cefoxitin. Cefoxitin is a cefamycin with an unusually wide antibacterial spectrum and high resistance to degradation by β-lactamases. Its penetration over an intact blood–CSF barrier seems to be approximately the same as for ampicillin.[71] Cefoxitin has a considerably better penetration in patients with meningitis and has been used successfully in two cases.[2a,71,84a] The future place of cefoxitin in treatment of meningitis remains to be defined.

4.3. Chloramphenicol

Chloramphenicol remains one of the drugs of choice for treatment of purulent meningitis. Chloramphenicol has a wide antibacterial spectrum, and resistant strains are rare. Its penetration over the blood–CSF barrier is probably superior to that of most other antibacterial agents. The toxicity of chloramphenicol has been an argument against its use. However, comparing the mortality of purulent meningitis with that of adverse reactions to chloramphenicol, the disproportion is obvious. Thus, in meningitis caused by *Hemophilus influenzae,* chloramphenicol should be the antibiotic of choice, especially in areas with a high frequency of β-lactamase-producing strains of this species.

4.4. Erythromycin

Theoretically, erythromycin should be an alternative to penicillin and ampicillin in treatment of meningococcal and pneumococcal meningitis in penicillin-hypersensitive patients. Unfortunately, very little information is available concerning its clinical efficacy in these diseases. The concentrations attained in CSF are low,[43,52] and chloramphenicol is a better alternative than erythromycin.

4.5. Lincomycin and Clindamycin

Lincomycin and clindamycin are active against anaerobes and gram-positive aerobes, but lack activity against gram-negative aerobes. Their CSF penetration is less than would be expected from their general pharmacokinetics.[65,68,81] No data are currently available concerning their clinical efficacy in treatment of meningitis.

4.6. Vancomycin

Although vancomycin seems unable to cross an intact blood–CSF barrier,[39,110] it has been reported to moderately penetrate inflamed meninges[90] and to be effective in treatment of staphylococcal meningitis.[46] Thus, vancomycin might offer an alternative for treatment of this type of meningitis.

4.7. Sulfonamides and Sulfonamide–Trimethoprim Combinations

As discussed previously, the low-protein-bound sulfonamides have an excellent penetration into CSF even when the meninges are not inflamed.[37,90] Their usefulness in treatment of meningitis is limited by their narrow antibacterial spectrum and by the increasing occurrence of sulfonamide-resistant strains.

Co-trimoxazole has an antibacterial spectrum that covers most aerobic pathogens causing bacterial meningitis. Trimethoprim easily crosses the intact blood–CSF barrier,[36,127a] and sulfamethoxazole has also been found to have an adequate CSF penetration.[13,127a] Despite reports of clinical success of co-trimoxazole in the treatment of meningitis,[15,63,84,102] this drug has not been employed in meningitis to an extent justified by its pharmacokinetics and antibacterial spectrum.

4.8. Metronidazole and Tinidazole

Systemically administered metronidazole yields high concentrations in CSF, as does tinidazole.[53] Both these drugs are active exclusively against anaerobic bacteria and thus have a limited place in the treatment of meningitis.

4.9. Aminoglycosides

The systemic toxicity of the aminoglycosides relative to their penetration across the blood–CSF barrier invalidates their parenteral use in treatment of purulent meningitis. Although the clinical efficacy of aminoglycosides administered in the lumbar subarachnoid space has been reported,[26,44,101] there are also a considerable number of reports of failures and subsequent ventriculitis when aminoglycosides were given by this route.[44,55,78,123] The large controlled study of McCracken and Mize[78] demonstrated no significant increase of the survival rate when intrathecal administration of gentamicin was compared to systemic maintenance. Intraventricular administration of aminoglycosides has been well documented to give therapeutic effect in gram-negative meningitis.[43,55,61]

4.10. Tetracyclines

Repeated doses of tetracyclines yield therapeutic CSF concentrations.[1,45,52,130] Taking into account the lipophilicity of various tetracyclines, doxycycline and minocycline are likely to give the highest concentrations.[3] However, data on clinical results of the use of tetracyclines in purulent meningitis are essentially lacking.

4.11. Antibiotics for Treatment of Tuberculous Meningitis

Meningitis caused by *Mycobacterium tuberculosis* is almost invariably treated with a combined therapy to avoid development of resistance. Therefore, evaluations of the therapeutic efficacy of the various drugs used for treatment of this serious condition are difficult to make. The pharmacokinetics of drugs used for treatment of tuberculous meningitis have been reviewed by Barling and Selkon.[3] Streptomycin appears to follow the pattern described above for the aminoglycoside group, having poor penetration over the blood–CSF barrier and not yielding therapeutic CSF concentrations. Isoniazid has been found to penetrate readily into CSF in a degree comparable to that of chloramphenicol.[32] Rifampicin, which is highly protein-bound, penetrates poorly into CSF.[21,57] However, it still appears to be effective in treating tuberculous meningitis. p-Aminosaliscylic acid appears to be less efficacious than isoniazid and rifampicin for the treatment of tuberculous meningitis, since the concentrations in CSF are low.[24]

Ethambutol enters the CSF in adequate levels in patients with tuberculous meningitis[91,92] and has been shown to be therapeutically effective.[11] Ethionamide appears to have a CSF penetration equal to that of isoniazid and is moderately independent of the degree of meningeal inflammation.[49] The drug may be useful both for treatment and for prophylaxis of tuberculous meningitis. Pyrazinamide is a drug that requires streptomycin for optimal antituberculous activity and thus has a limited usefulness in treatment of tuberculous meningitis despite its adequate penetration into CSF.

4.12. Agents for Treatment of Fungal Meningitis

The increasing use of immunosuppressive drugs and prolonged survival times in diseases complicated by immune deficiencies have led to an increased frequency of fungal meningitis caused by *Cryptococcus neoformans* and *Candida* spp. Until a few years ago, amphortericin B was the only drug available for treatment of these diseases. Systemically administered amphotericin B is highly toxic and does not yield therapeutic CSF concentrations even in the presence of inflamed meninges.[10] A considerably better penetration across the blood–CSF barrier has been found with flucytosin (5-fluorocytosine), which yields CSF concentrations almost equal to those found in the serum.[28] A limited number of clinical studies have indicated that flucytosin may be efficacious for the treatment of fungal meningitis.[17,28,87]

5. Conclusions

Treatment of infections within the CSF space cannot be chosen only on the basis of the activity of an antimicrobial drug against the agent causing the infection. In addition to antimicrobial activity, pharmacokinetic properties enabling the drug to cross the blood–CSF barrier will determine the efficacy of the drug. The CSF penetration is determined by several properties of the drug and the infected patient. Most important is the degree of meningeal inflammation. Noninflamed meninges are resistant to penetration by most antimicrobial drugs, while marked inflammation, such as that caused by bacterial meningitis, often allows passage of a large number of drugs to an extent sufficient to give therapeutic CSF concentrations. Other host factors that might influence the CSF concentrations of antimicrobial drugs include active transport into the CSF and active transport out of the CSF at the choroid plexus. Physiochemical characteristics that influence the penetrability over the blood–CSF barrier for an antibiotic are lipophilicity, ionization at physiological pH, protein binding, and molecular size and structure.

Taking into consideration these numerous factors, it is not surprising that only a few antimicrobial drugs have both an adequate spectrum and high capacity to penetrate the blood–CSF barrier, irrespective of the degree of inflammation. Examples of such drugs are chloramphenicol, trimethoprim, some of the sulfonamides, and the antifungal drug flucytosin (5-fluorocytosine). Since systemic administration of aminoglycosides, isoxazolyl penicillins, and amphotericin B does not yield therapeutic CSF concentrations, systemic administration of these drugs is not efficacious in the treatment of meningitis; thus, their intraventricular administration should be considered optimal. This review suggests that intrathecal administration into lumbar CSF may yield insufficient and uneven concentrations when the entire CSF space is infected. Thus, the lumbar route seems to have limited, if any, place in therapy of meningitis.

Finally, the available data concerning the pharmacokinetics of antimicrobial agents in CSF are limited, and more information is needed. In addition, new drugs are required for treatment of some types of meningitis, such as gram-negative meningitis in the neonate and the compromised adult patient.

References

1. ANDERSSON, H., ALESTIG, K.: Penetration of doxycycline into CSF. *Scand. J. Infect. Dis. Suppl.* **9**:17–19, 1976.
2. AUWARTER, W., MAURER, H., FORTNER, H.: Nachweis von Penicillin-G und Ampicillin in Blut und Liquor bei eitriger Meningitis im Verlauf einer hochdosierten Therapie. *Infection* **1**:98–105, 1973.

2a. AYROSA GALVAO, P. A., LOMAR, A. V., FRANCISCO, W., DEGODOY, C. V., NORRBY, R.: Cetoxitin penetration to cerebrospinal fluid in patients with purulent meningitis. *Antimicrob. Agents Chemother.* **17**:526–529, 1980.

3. BARLING, R. W. A., SELKON, J. B.: The penetration of antibiotics into cerebrospinal fluid and brain tissue. *J. Antimicrob. Chemother.* **4**:203–227, 1978.
4. BARLOW, C. F.: Clinical aspects of the blood–brain barrier. *Annu. Rev. Med.* **15**:187–202, 1964.
5. BARZA, M., BROWN, R. B., SHANKS, C., GAMLE, C., WEINSTEIN, L.: Relation between lipophilicity and pharmacological behavior of minocycline, doxycycline, tetracycline and oxytetracycline in dogs. *Antimicrob. Agents Chemother.* **8**:713–720, 1975.
6. BARZA, M.: The effect of protein-binding on the distribution of antibiotics and the problem of continuous versus intermittent infusion: Including a review of some controversial aspects. *Infection* **4**(*Suppl. 2)*:144–155, 1976.

7. BARZA, M., WEINSTEIN, L.: Pharmacokinetics of the penicillins in man. *Clin. Pharmacol. Ther.* **1**:297–308, 1976.
8. BASS, N. H., LUNDBORG, P.: Transport mechanisms in the cerebrospinal fluid system for removal of acid metabolites from developing brain. *Adv. Exp. Med. Biol.* **69**:31–40, 1976.
9. BASSARIS, H. P., QUINTILIANI, R., MODERAZO, E. G., TILTON, R. G., NIGHTINGALE, C. H.: Pharmacokinetics and penetration characteristics of cefazoline into human spinal fluid. *Curr. Ther. Res.* **19**:110–120, 1976.

9a. BEATY, H. N., WALTERS, E.: Pharmacokinetics of cefamandole and ampicillin in experimental meningitis. *Antimicrob. Agents Chemother.* **16**:584–588, 1979.

10. BINDSCHADLER, D. D., BENNET, J. E.: A pharmacologic guide to the clinical use of amphotericin B. *J. Infect. Dis.* **120**:427–433, 1969.
11. BOBROWITZ, I. D.: Ethambutol in tuberculous meningitis. *Chest* **61**:629–632, 1973.
12. BOGER, W. P., GAVID, J. J.: Demethylchlorotetracycline: Serum concentration studies and cerebrospinal fluid diffusion. In Marti-Ibanez, F. (ed.): *Antibiotics Annual 1959–1969,* New York, Antibiotics, Inc., 1969, pp. 393–408.
13. BOGER, W. P., GAVIN, J. J.: Sulfamethoxazole: Comparison with sulfisoxazole and sulfaethidole and cerebrospinal fluid diffusion. *Antibiot. Chemother.* **10**:527–580, 1960.
14. BRIEDIS, D. J., ROBSON, H. G.: Cerebrospinal fluid penetration of amikacin. *Antimicrob. Agents Chemother.* **13**:1042–1043, 1978.
15. CALOUGHI, G. F., IELASI, G.: Intravenous therapy with a combination of sulfamethoxazole–trimethoprim in purulent meningitis. *G. Mal. Infect. Parassit.* **24**:9–12, 1972.
16. CHERNIC, N. L., ARMSTRONG, D., PORNER, J. B.: Central nervous system infections in patients with cancer. *Medicine (Baltimore)* **52**:563–568, 1973.
17. CHESNEY, P. J., TEETS, K. C., MULVIHILL, J. J., SALIT, I. E., MARKS, M. I.: Successful treatment of *Candida* meningitis with amphotericin B and 5-fluorocytosin in combination. *J. Pediatr.* **89**:1017–1019, 1976.
18. DACEY, R. G., SANDE, M. D.: Effect of probenecid on cerebrospinal fluid concentrations of penicillin and cephalosporin derivatives. *Antimicrob. Agents Chemother.* **6**:437–441, 1974.
19. DEWHURST, K.: The use of probenecid for increasing penicillin concentrations in cerebrospinal fluid. *Acta Neurol. Scand.* **45**:253–256, 1969.
20. DIXON, R. L., OWENS, E. S., RALL, D. P.: Evidence of active transport of benzyl-[^{14}C]-penicillin from cerebrospinal fluid to blood. *J. Pharm. Sci.* **58**:1106–1109, 1969.
21. D'OLIVEIRA, J. J. G.: Cerebrospinal fluid concentrations of rifampicin in meningeal tuberculosis. *Am. Rev. Respir. Dis.* **106**:432–437, 1972.
22. DOUTHWAITE, A. H., TRAFFORD, J. A. P., MCGILL, D. A. F., EVANS, I. E.: Methicillin. *Br. Med. J.* **2**:6–8, 1961.
23. DROBNIC, L., QUILES, M., RODRIQUES, A.: A study on the levels of fosfomycin in the cerebrospinal fluid in adult meningitis. *Chemotherapy* **23**(*Suppl. 1*):180–188, 1977.
24. DUBOIS, R.: The treatment of acute tuberculosis in childhood. In *The Treatment of Tuberculosis in Childhood.* Paris, Centre International de L'Enfrance, Travaux et Documents—X, 1956, pp. 155–193.
25. DUMOFF-STANLEY, E., DOWLING, H. F., SWEET, L. K.: The absorption into and distribution of penicillin in the cerebrospinal fluid. *J. Clin. Invest.* **24**:87–93, 1946.
26. Editorial: Intrathecal antibiotics in purulent meningitis. *Lancet* **2**:1068, 1976.
27. EICHENWALD, H. F.: Some observations on dosage and toxicity of kanamycin in premature and full term infants. *Ann. N. Y. Acad. Sci.* **132**:984–991, 1966.
28. FASS, R. J., PERKINS, R. L.: 5-Fluorocytosine in the treatment of cryptococcal and *Candida* mycoses. *Ann. Intern. Med.* **74**:535–539, 1971.
29. FISHER, L. S., CHOW, A. W., YOSHIKAWA, T. T., GUZE, L. B.: Cephalothin and cephaloridin therapy for bacterial meningitis. *Ann. Intern Med.* **83**:689–693, 1975.
30. FISHMAN, R. A.: Carrier transport of glucose between blood and cerebrospinal fluid. *Am. J. Physiol.* **206**:836–844, 1964.
31. FISHMAN, R. A.: Blood–brain and CSF barriers to penicillin and related organic acids. *Arch. Neurol.* **15**:113–124, 1966.
32. FLETCHER, A. P.: Cerebrospinal fluid isoniazid levels in tuberculous meningitis. *Lancet* **2**:694–696, 1953.
33. FOSSIECH, B., JR., PARKER, R. H.: Neurotoxicity during intravenous infusion of penicillin. *J. Clin. Pharmacol.* **14**:504–508, 1974.
34. FOSSIECK, B., JR., KANE, J. G., DIAZ, C. R., PARKER, R. H.: Nafcillin entry into human cerebrospinal fluid. *Antimicrob. Agents Chemother.* **11**:965–967, 1977.
35. FREIJ, L., HEBELKA, M., SEEBERG, S.: Meningitis developing during cephalothin therapy of septicaemia. *Scand. J. Infect. Dis.* **7**:153–155, 1975.

35a. FRIEDRICH, H., HAENSEL-FRIEDRICH, G., LANGMAAK, H., DASCHNER, F. D.: Investigations of cefuroxime levels in the cerebrospinal fluid of patients with and without meningitis. *Chemotherapy* **2**:91–97, 1980.

36. FRIES, N., KEUTH, U., BRAUN, J. S.: Untersuchungen zur Liquorgängigkeit von Trimethoprim in Kindesalter. *Fortschr. Med.* **93**:1178–1183, 1975.
37. GARROD, L. P., O'GRADY, F.: *Antibiotics and Chemotherapy.* Edinburgh, Churchill-Livingston, 1971, pp. 17–22.
38. GEERING, J. M., JUST, M.: Blut–Liquor Verteilung von Ampicillin bei Kindern mit und ohne Meningitis. *Monatschr. Kinderheilkd.* **123**:545–547, 1975.
39. GERACI, J. E., HEILMAN, F. R., NICHOLS, D. R., WELLMAN, W. E., ROSS, G. T.: Some laboratory and clinical experience with a new antibiotic, vancomycin. In Marti-Ibanez, F. (ed.): *Antibiotics Annual 1956–1957.* New York, Antibiotics, Inc., 1957, pp. 90–105.

40. GODTFREDSEN, W., ROHOLT, K., TYBRING, L.: Fucidin: A new orally active antibiotic. *Lancet* **1**:928–931, 1962.
41. GOITEN, K., MICHEL, J., SACHS, T.: Penetration of parenterally administered gentamicin into the cerebrospinal fluid in experimental meningitis. *Chemotherapy* **21**:181–188, 1975.
42. GORDON, R. G., BARRET, F., YOW, M. D.: Ampicillin treatment in listeriosis. *J. Pediatr.* **77**:1067–1070, 1970.
43. GRIFFITH, R. S., BLACK, H. R.: Erythromycin. *Med. Clin. North Am.* **54**:1199–1215, 1970.
44. HAMORY, B., IGNOTIADIS, P., SANDE, M. A.: Intrathecal amikacin administration: Use in the treatment of gentamicin-resistant *Klebsiella pneumoniae* meningitis. *J. Am. Med. Assoc.* **236**:1973–1974, 1976.
45. HANSON, H., KOCH, R.: Tetracycline serum and cerebrospinal fluid levels in acute purulent meningitis. In Marti-Ibanez, F. (ed.): *Antibiotics Annual 1956–1957*. New York, Antibiotics, Inc., 1957, pp. 319–325.
46. HAWLEY, H. B., GUMP, D. W.: Vancomycin therapy of bacterial meningitis. *Am. J. Dis. Child.* **126**:261–264, 1973.
47. HOWARD, J. B., MCCRACKEN, G. H., JR.: Reappraisal of kanamycin usage in neonates. *Pediatrics* **86**:949–956, 1975.
48. HOWELL, A., SUTHERLAND, R., ROLINSON, G. N.: Effect of protein binding on levels of ampicillin and cloxacillin. *Clin. Pharmacol. Ther.* **13**:724–732, 1972.
49. HUGHES, I. E., SMITH, H., KANE, P. O.: Ethionamide: Its passage into the cerebrospinal fluid in man. *Lancet* **1**:616–617, 1962.
50. INGRAHAM, F. D., MATSON, D. D., ALEXANDER, F., WOODS, P. P.: Studies on the treatment of experimental hydrocephalus. *J. Neuropathol. Exp. Neurol.* **7**:123–132, 1948.
51. IWARSON, S., LIDIN-JANSON, G., SVENSSON, R.: Listeric meningitis in the non-compromised host. *Infection* **5**:204–206, 1977.
52. IWISKENETSKAJA, V. F.: Blood–CSF permeability for antibiotics in neurosurgical patients. In Daikos, G. (ed.): *Progress of Chemotherapy*, Proceedings of the 8th International Congress of Chemotherapy, Vol. 1, Athens, 1974, pp. 467–479.
53. JOKIPII, A. M. M., MYLLYLÄ, V. V., HOKKANEN, E., JOKIPII, L.: Penetration of the blood brain barrier by metronidazole and tinidazole. *J. Antimicrob. Chemother.* **3**:239–245, 1977.
54. KAHN, W., ROSS, S., RODRIQUEZ, W., CONTRONI, G., SAZ, H.: *Haemophilus influenzae* type B resistant to ampicillin: A report of two cases. *J. Am. Med. Assoc.* **229**:298–301, 1974.
55. KAISER, A. B., MCGEE, Z. A.: Aminoglycoside therapy of gram-negative bacillary meningitis. *N. Engl. J. Med.* **293**:1215–1220, 1975.
55a. KALSBECK, J. E., DESOUSA, A. L., KLEIMAN, M. B., GOODMAN, J. M., FRANKEN, E. A.: Compartmentalization of the cerebral ventricles as a sequela of neotonal meningitis. *J. Neurosurg.* **52**:547–552, 1980.
56. KAPLAN, J. M., MCCRACKEN, G. H., HORTON, J. L.: Pharmacological studies in neonate given large doses of ampicillin. *J. Pediatr.* **84**:571–578, 1974.
57. KEBERLE, H., SCHMID, K., MEYER-BRUNOT, H. G.: The metabolic fate of rimactane in the animal and in man. *A Symposium on Rimactane*. Basle, Ciba, 1968, pp. 20–27.
58. KELLY, R. S., HUNG, A. D., JR., TASHMAN, S. C.: Studies on the absorption and distribution of chloramphenicol. *Pediatrics* **8**:362–367, 1951.
59. KOBAYASHI, Y.: Chemotherapy of purulent meningitis in children. *Jpn. J. Antibiot.* **28**:567–580, 1975.
60. KORZENIOWSKI, O. M., CARVALHO, E. M., JR., ROCHA, H., SANDE, M. A.: Evaluation of cefamandole therapy of patients with bacterial meningitis. *J. Infect. Dis.* **137**:S169–179, 1978.
61. KOURTOPOULOS, H., HOLM, S. E.: Intraventricular treatment of *Serratia marcescens* meningitis with gentamicin. *Scand. J. Infect. Dis.* **8**:57–60, 1970.
62. KUNIN, C. M.: Effect of serum binding on the distribution of penicillin in the rabbit. *J. Lab Clin. Med.* **65**:406–415, 1965.
63. LAFAIX, C., CAMERLYNCK, P., DIOP MAR, I., GUERIN, M., REY, M.: Therapy of bacterial meningitis with sulfamethoxazole–trimethoprim. *Presse Med.* **78**: 1375–1377, 1970.
64. LAVETTER, A., LEEDOM, J. M., MATHIES, A. W., JR., IVLER, D., WEHRLE, P. F.: Meningitis due to *Listeria monocytogenes:* A review of 25 cases. *N. Engl. J. Med.* **285**:593–600, 1971.
65. LERNER, P. I.: Penetration of cephalothin and lincomycin into the cerebrospinal fluid. *Am. J. Med. Sci.* **257**:125–131, 1969.
66. LERNER, P. & I.: Penetration of cephaloridine into cerebrospinal fluid. *Am. J. Med. Sci.* **261**:321–326, 1971.
67. LERNER, P. & I.: Selection of antimicrobial agents in bacterial infections of the nervous system. *Adv. Neurol.* **6**:169–203, 1974.
68. LERNER, P. & I.: Antimicrobial considerations in anaerobic infections. *Med. Clin. North Am.* **58**:533–544, 1974.
69. LINDBERG, J., ROSENHAL, U., NYLEN, O., RINGNER, A.: Long-term outcome of *Hemophilus influenzae* meningitis related to antibiotic treatment. *Pediatrics* **60**:1–6, 1977.
70. LITHANDER, A., LITHANDER, B.: The passage of penicillin into cerebrospinal fluid after parenteral administration in staphylococcal meningitis. *Acta Pathol. Microbiol. Scand.* **56**:534–450, 1962.
71. LIU, C., HINTHORN, D. R., HODGES, G. R., HARMS, J. L., COUCHONNAL, G., DWORZACK, D. L.: Penetration of cefoxitin into human cerebrospinal fluid: Comparison with cefamandole, ampicillin and penicillin. *Rev. Infect. Dis.* **1**:127–131, 1979.
72. LLORENS, J., LOBATO, A., OLAY, T.: The passage of fosfomycin into the cerebrospinal fluid in childhood meningitis. *Chemotherapy* **23**(*Suppl. 1):*185–195, 1977.
73. LORENZO, A. V., SPECTOR, R.: The distribution of

drugs in the central nervous system. *Adv. Exp. Med. Biol.* **69**:447–461, 1976.
74. LOVE, W. C., MCKENZIE, P., LAWSON, J. H., PINKERTON, I. W., JAMIESON, W. M., STEVENSON, J., ROBERTS, W., CHRISTIE, A. B.: Treatment of pneumococcal meningitis with cephaloridine. *Postgrad. Med. J.* **46**(*Suppl.*):155–159, 1970.
75. MACNAIR, D. R., WHITE, J. E., GRAHAM, J. M.: Ampicillin in the treatment of *Listeria monocytogenes* meningitis. *Lancet* **1**:16, 1968.
76. MANGI, R. J., KINDARGI, R. S., QUINTILIANI, R., ANDRIOLE, V. T.: Development of meningitis during cephalothin therapy. *Ann. Intern. Med.* **78**:347–351, 1973.
77. MCCRACKEN, G. H., JR., CHRANE, D. F., THOMAS, M. L.: Pharmacologic evaluation of gentamicin in newborn infants. *J. Infect. Dis.* **124**(*Suppl.*):S214–223, 1973.
78. MCCRACKEN, G. H., MIZE, S. G.: A controlled study on intrathecal antibiotic therapy in gram-negative enteric meningitis in infancy. *J. Pediatr.* **89**:66–72, 1976.
79. MCCRUMB, F. R., JR., SNYDER, M. J., HICKEN, W. J.: The use of chloramphenicol acid succinate in the treatment of acute infections. In Marti-Ibanez, F. (ed.): *Antibiotics Annual 1957–1958.* New York, Antibiotics, Inc., 1958, pp. 837–841.
80. MCDONALD, L. L., ST. GEME, J. W., JR.: Cerebrospinal fluid diffusion of kanamycin in newborn infants. *Antimicrob. Agents Chemother.* **2**:41–45, 1972.
81. MEDINA, A., FISKE, N., HJELT-HARVEY, I., BROWN, C. D., PRIGOT, A.: Absorption, diffusion and excretion of a new antibiotic, lincomycin. In Sylvester, J. C. (ed.): *Antimicrobial Agents and Chemotherapy.* Ann Arbor, American Society for Microbiology, 1963, pp. 179–203.
82. MEYER-BRUNOT, H. G., SCHENK, C., SCHMID, K., SPRING, P., THEOBALD, W., WAGNER, J.: Einige Aspekte zur Pharmakokinetik von Celospor insbesondere Liquorgängigkeit. *Wien. Med. Wochenschr.* **125**(*Suppl.*):23–31, 1975.
83. MOELLERING, R. C., FISHER, E. G.: Relationship of intraventricular gentamicin levels to cure of meningitis. *J. Pediatr.* **81**:532–537, 1972.
84. MORZARIA, R. N., WALTON, I. G., PICKERING, D.: Neonatal meningitis treated with trimethoprim and sulfamethoxazol. *Br. Med. J.* **2**:511, 1969.
85. NAIR, S., CHERUBIN, C. E., WEINSTEIN, M.: Penetration of cefoxitin into cerebrospinal fluid and treatment of meningitis caused by gram-negative bacteria. *Rev. Infect. Dis.* **1**:134–141, 1979.
86. NEW, P. S., WELLS, C. E.: Cerebral toxicity associated with massive intravenous penicillin therapy. *Neurology (Minneapolis)* **15**:1053–1058, 1965.
87. NORDSTROM, L., OISTAMO, S., OLMEBRING, F.: *Candida* meningoencephalitis treated with 5-fluorocytosine. *Scand. J. Infect. Dis.* **9**:63–65, 1977.
88. NORRBY, R., FOORD, R. D., PRICE, D., HEDLUND, P.: Pharmacokinetic and clinical studies on cefuroxime. *Proc. R. Soc. Med.* **70**(*Suppl. 9*):25–33, 1977.
89. OPPENHEIMER, S., BEATY, H. N., PETERSDORF, R. G.: Pathogenesis of meningitis. VIII. Cerebrospinal fluid and blood concentrations of methicillin, cephalothin, and cephaloridine in experimental pneumococcal meningitis. *J. Lab. Clin. Med.* **73**:535–543, 1969.
90. OTTEN, H., PLEMPEL, M.: Antibiotika und Chemotherapeutika in Einzeldarstellung. In Otten, H., Plempel, M., Zimmerman, K. G. (eds.): *Antibiotika-Fibel.* Stuttgart, Georg Thieme Verlag, 1975, pp. 116, 234, 522.
91. PILHEU, J. A., MAGLIO, F., CETRANGOLO, R., PLEUS, A. D.: Concentrations of ethambutol in the cerebrospinal fluid after oral administration. *Tubercle* **52**:117–122, 1971.
92. PLACE, V. A., PYLE, M. M., DE LA JUERGA, J.: Ethambutol in tuberculous meningitis. *Am. J. Respir. Dis.* **99**:783–785, 1968.
93. POLLAY, M.: Transport mechanisms in the choroid plexus. *Fed. Proc. Fed. Am. Soc. Exp. Biol.* **33**:2064–2069, 1975.
94. RAHAL, J. J., JR., HYAMS, P. J., SIMBERKOFF, M. S., RUBINSTEIN, E.: Combined intrathecal and intramuscular gentamicin for gram-negative meningitis: Pharmacologic study of 21 patients. *N. Engl. J. Med.* **290**:1394–1398, 1974.
95. RALL, D. P., STABENAU, J. R., ZUBRUD, C. G.: Distribution of drugs between blood and cerebrospinal fluid: General methodology and effects of pH gradients. *J. Pharmacol. Exp. Ther.* **125**:185–193, 1959.
96. RENLUND, M., PETTAY, O.: Pharmacokinetics and clinical efficacy of cefuroxime in the newborn period. *J. R. Soc. Med.* **70**(*Suppl. 9*):179–182, 1977.
97. RICHARDSON, A. E., SPITTLE, C. R., JAMES, K. W., ROBINSON, O. P. W.: Experience with carbenicillin in the treatment of septicaemia and meningitis. *Postgrad. Med. J.* **44**:844–847, 1968.
98. RIESELBACH, R. E., DICHIRO, G., FREIRICH, E. J., RALL, D. D.: Subarachnoidal distribution of drugs after lumbar injections. *N. Engl. J. Med.* **267**:1273–1276, 1962.
99. ROSS, S., PUIG, J. R., ZAREMBA, E. A.: Chloramphenicol acid succinate: Some preliminary clinical and laboratory observations in infants and children. In Marti-Ibanez, F. (ed.): *Antibiotics Annual 1957–1958.* New York, Antibiotics, Inc., 1958, pp. 803–820.
100. RUIZ, D. E., WARNER, J. F.: Nafcillin treatment of *Staphylococcus aureus* meningitis. *Antimicrob. Agents Chemother.* **8**:554–555, 1976.
101. SAAD, A. F., FARRAR, W. E.: Intracisternal and intrathecal injections of gentamicin in *Enterobacter* meningitis. *Arch. Intern. Med.* **134**:738–740, 1974.
102. SABEL, K. G.: The treatment of meningitis in infants with co-trimoxazole administered parenterally. *Scand. J. Infect. Dis. Suppl.* **8**:86–89, 1974.
103. SANDE, M. A., SHERERTZ, R. J., ZAK, O., STRAUSBAUGH, L. J.: Cephalosporin antibiotics in therapy of experimental *Streptococcus pneumoniae* and *Haemophi-*

lus influenzae meningitis in rabbits. *J. Infect. Dis.* **137**:S161–168, 1978.

104. SCHANKER, L. S.: Passage of drugs across body membranes. *Pharmacol. Rev.* **14**:501–525, 1962.
105. SCHULTZ, P., LEEDS, N. E.: Intraventricular septation complicating neonatal meningitis. *J. Neurosurg.* **38**:620–627, 1973.
106. SCHACKELFORD, P. J., BOBINSKI, J. E., FEIGIN, R. D., CHERRY, J. D.: Therapy of *Haemophilus influenzae* meningitis reconsidered. *N. Engl. J. Med.* **287**:634–636, 1972.
107. SHERERTZ, R. J., DACEY, R., SANDE, M. A.: Cefamandole in the therapy of experimental pneumococcal meningitis. *J. Antimicrob. Chemother.* **2**:159–165, 1976.
108. SILICA, T., FADON, A., RODRIQUES, A., SOTO, J.: Fosfomycin in pneumococcal meningitis. *Chemotherapy* **23**(*Suppl. 1*):429–434, 1977.
109. SIMPSON, P. B., JR., WARREN, G.-G., SMITH, R. R.: Intraventricular cephalothin in childhood meningitis. *Surg. Neurol.* **4**:279–280, 1975.
110. SPEARS, R. L., KOCH, R.: The use of vancomycin in pediatrics. In Marti-Ibanez, F. (ed.): *Antibiotics Annual 1959–1960.* New York, Antibiotics, Inc., 1960, pp. 798–803.
111. SPECTOR, R., LORENZO, A. V.: Inhibition of penicillin transport from the cerebrospinal fluid after intracisternal inoculation of bacteria. *J. Clin. Invest.* **54**:316–325, 1974.
112. SPECTOR, R., LORENZO, A. V.: The effect of salicylate and probenicid on the cerebrospinal fluid transport of penicillin, aminosalylic acid and iodide. *J. Pharmacol. Exp. Ther.* **188**:55–65, 1974.
113. SPECTOR, R.: The transport of gentamicin in the choroid plexus and cerebrospinal fluid. *J. Pharmacol. Exp. Ther.* **194**:82–88, 1975.
114. SPECTOR, R., SNODGRASS, S. R., LEVY, P.: The effect of uremia on penicillin flux between blood and cerebrospinal fluid. *J. Lab. Clin. Med.* **87**:749–759, 1976.
115. STEINBERG, E. A., OVERTURF, G. D., WILKONS, J., BARAFF, L. J., STRENG, J. M., LEEDOM, J. M.: Failure of cefamandole in treatment of meningitis due to *Haemophilus influenzae* type B. *J. Infect. Dis.* **137**:S180–186, 1978.
116, STRAUSBAUGH, L. J., MANDELERIS, C. D., SANDE, M. A.: Cefamandole and ampicillin therapy in experimental *Haemophilus influenzae* meningitis. *J. Infect. Dis.* **135**:210–216, 1977.
117. TABER, L. H., YOW, M. D., NIEBERG, F. G.: The penetration of broad-spectrum antibiotics into the cerebrospinal fluid. *Ann. N. Y. Acad. Sci.* **145**:437–481, 1967.
118. THOMAS, W. J., MCREYNOLDS, J. W., MOCK, C. R., BAILEY, D. W.: Ampicillin-resistant *Haemophilus influenzae* meningitis. *Lancet* **1**:313, 1974.
119. THRUPP, L. D., LEEDOM, J. M., IVLER, D., WEHRLE, P. F., PORTNOY, B., MATHIES, A. W.: Ampicillin levels in cerebrospinal fluid during treatment of bacterial meningitis. In Sylvester, J. C. (ed.): *Antimicrobial Agents and Chemotherapy.* Ann Arbor, American Society for Microbiology, 1966, pp. 206–213.
120. TOMEH, M. O., STARR, S. E., MCGOWAN, J. E., JR., TERRY, P. M., NAHMIAS, A. J.: Ampicillin-resistant *Haemophilus influenzae* type B infection. *J. Am. Med. Soc.* **229**:295–297, 1974.
121. VIANNA, N. J., KAYE, D.: Penetration of cephalothin into the spinal fluid. *Am. J. Med. Sci.* **254**:216–220, 1967.
122. VISUDHIPHAN, P., CHIEMCHANYA, S.: Evaluation of rifampicin in the treatment of tuberculous meningitis in children. *Pediatrics* **87**:983–986, 1975.
123. YEUNG, C. Y.: Intrathecal antibiotic therapy for neonatal meningitis. *Arch. Dis. Child.* **51**:686–690, 1976.
124. WALKER, A. E., JOHNSON, H. C., KOLLROS, J. J.: Penicillin convulsions: The convulsive effect of penicillin applied directly to the cerebral cortex of monkeys and man. *Surg. Gynecol. Obstet.* **81**:692–701, 1945.
125. WALKER, S. H., COLLINS, C. G., JR.,: Failure of cephaloridine in *Haemophilus influenzae* meningitis. *Am. J. Dis. Child.* **116**:285–291, 1968.
126. WALLACE, J. F., SMITH, R. H., GARCIA, M., PETERSDORF, R. G.: studies on the pathogenesis of meningitis. VI. Antagonism between penicillin and chloramphenicol in experimental pneumococcal meningitis. *J. Lab. Clin. Med.* **70**:408–412, 1967.
127. WELLMAN, W. E., DODGE, H. W., JR., HEILMAN, F. R., PETERSEN, M. C.: Concentrations of antibiotics in the brain. *J. Lab. Clin. Med.* **43**:275–279, 1954.

127a. WILKINSON, P. J., REEVES, D. S.: Tissue penetration of trimetoprim and sulphonamides. *J. Antimicrob. Chemother.* **5**(Supp. 13):159–168, 1979.

128. WOOD, J. H.: Physiology, pharmacology, and dynamics of cerebrospinal fluid. In Wood, J. H. (ed.): *Neurobiology of Cerebrospinal Fluid I.* New York, Plenum Press, 1980.
129. WOOD, J. H.: Technical aspects of clinical and experimental cerebrospinal fluid investigations. In Wood, J. H. (ed.): *Neurobiology of Cerebrospinal Fluid I.* New York, Plenum Press, 1980.
130. WOOD, W. S., KIPNIS, G. P.: The concentration of tetracycline, chlortetracycline and oxytetracycline in the cerebrospinal fluid after intravenous administration. In Marti-Ibanez, F. (ed.): *Antibiotics Annual 1953–1954.* New York, Antibiotics, Inc., 1953, pp. 98–101.

CHAPTER 33

Pathophysiology of Cerebrospinal Fluid Immunoglobulins

John L. Trotter and Benjamin Rix Brooks

1. Introduction

The presence of elevated cerebrospinal fluid (CSF) immunoglobulin G (IgG) in central nervous system (CNS) syphilis with a resultant abnormal colloidal gold curve was the first use of the immune response to an infectious agent in the nervous system to diagnose that condition. Today, our laboratory tests are more specific, and evidence is overwhelming for CNS synthesis of immunoglobulins in a number of conditions. The presence of this abnormality is not pathognomonic of any particular disease, nor by itself sufficient to define an immunological or infectious etiology, but by statistical inference has provided clues to such an etiology in degenerative diseases of the nervous system with unknown pathophysiology. The following review of the literature may well contain some obsolete theoretical views by the time it reaches the reader's hands, but in that case it will at least provide a historical record of this rapidly moving field. The disproportionate emphasis on multiple sclerosis (MS) reflects the volume of literature as well as the authors' special interests.

Abbreviations used in this chapter: (AGE) agarose gel electrophoresis; (Alb) albumin; (CNS) central nervous system; (CSF) cerebrospinal fluid; (EID) electroimmunodiffusion; (IEF) isoelectric focusing; (Ig) immunoglobulin; (κ chain) kappa chain (class of the *Ig* molecule light chain); (λ chain) lambda chain (class of the *Ig* molecule light chain); (MS) multiple sclerosis; (ON) optic neuritis; (RIA) radioimmunoassay; (SSPE) subacute sclerosing panencephalitis.

John L. Trotter, M.D. • Department of Neurology and Neurosurgery (Neurology), Washington University School of Medicine, St. Louis, Missouri 63110. **Benjamin Rix Brooks, M.D.** • Department of Neurology, The Johns Hopkins University School of Medicine, Baltimore, Maryland 21205.

2. Quantitative Abnormalities of IgG in CSF of Patients with Multiple Sclerosis

Lumsden[89] and Tourtellotte[152,153] have exhaustively reviewed the evidence for, and methods of determining the presence of, an increased content of gamma globulin relative to that of other proteins in the CSF of the majority of MS patients. In particular, this phenomenon was established by Kabat *et al.*[57] and confirmed by Yahr *et al.*[176] Although other investigators had noted an absolute increase in gamma globulin content, it was only when the gamma globulin was related to total protein that a significant difference between MS patients and controls was found. The reason for this empirical finding is that a breakdown of the blood–brain barrier will raise the CSF total protein, which includes serum gamma globulins; thus, an elevated absolute gamma globulin may be found in many neurological conditions. However, in MS patients, there is good evidence for CNS synthesis of the relative increased

IgG that is seen in the CSF (see below). When the CSF gamma globulin (or more specifically IgG) is divided by the total protein (or albumin), changes due to a blood–brain barrier breakdown are partially compensated, and this value served to better identify the abnormalities seen in the CSF and MS patients.

The technical advances of IgG quantitation using the immunochemical methods of electroimmunodiffusion (EID)[48,68,161] or radial immunodiffusion[7,91] have allowed more precise evaluation of the low concentration of proteins in CSF. Although both methods give comparable results,[4] the EID method may have less variance.[121] Since the numbers of interest for diagnostic considerations are ratios, the ability to measure both albumin and IgG on the same EID plate[161] allows many technical variations to cancel and improves the reproducibility of data.

To optimally express the presence of CNS IgG synthesis, the serum content of IgG (IgG_s) and albumin (Alb_s), as well as the CSF IgG (IgG_{csf}) and albumin (Alb_{csf}), must be taken into account. Tourtellotte[155] has devised an equation in an attempt to utilize these four measured values, to approximate the CNS synthesis of IgG (IgG_{syn}) per day:

$$12IgG_{syn} = \left[\left(IgG_{csf} - \frac{IgG_s}{369}\right) - \left(Alb_{csf} - \frac{Alb_s}{230}\right)\left(\frac{IgG_s}{Alb_s}\right)(0.43)\right] \times 5$$

This equation has been validated by employing [^{131}I]-IgG injected intravenously into 11 chronic MS patients. Serial simultaneous CSF and serum samples were obtained and the dynamic equilibrium of [^{131}I]-IgG graphed. There was good concordance of the equation when compared to the empirically determined estimates.[156] The IgG index, also designed to incorporate the four measurements, has been used by others[45,81,115,151] and also identifies the MS population more accurately than the IgG/Alb ratio of CSF alone:

$$\text{CSF IgG index} = \left(\frac{\text{CSF IgG/CSF Alb}}{\text{Serum IgG/serum Alb}}\right)$$

Our laboratory uses commercial EID plates (AEID plates®, Antibodies, Inc. Davis, California). These plates contain anti-human IgG and anti-human Alb in agar–agarose. A 2-μl aliquot of human CSF is applied to the wells in the center of the plate, and an electric field is placed across the plate. The Alb moves toward the anode and the IgG toward the cathode. Precipitin cones ("rockets") are formed; the distance to the tip of the cone is related to the quantity of antigen applied. A standard curve derived from three standard samples is plotted and the concentrations of the unknown determined graphically. Since CSF and (diluted) serum may be run on the same plate, technical variation due to a particular run generally cancels out when the ratios are calculated. Using the CSF IgG, CSF Alb, serum IgG, and serum Alb, the CSF IgG index or CNS IgG synthesis may be calculated. The results of analysis of 89 neurological controls are shown in Table 1 and in scattergram form in Fig. 1. It is readily apparent that the CSF IgG/Alb ratio has a higher rate of "false positives" than the two methods of analysis that correct for serum proteins. A similar analysis (Fig. 2, Table 2) was performed on 46 MS patients who fulfilled the requirements for clinically definite MS patients as defined by Rose *et al.*[130]

Thus, IgG index and IgG synthesis calculations identify a larger percentage of "definite" MS patients than the IgG/Alb ratio. In this series, the IgG index appears to correlate better with the disease diagnosis than the IgG synthesis. However, a larger series would be required to prove this point. A summary of a retrospective hospital chart review of

Table 1. Comparison of Quantitative CSF IgG Abnormalities in 89 Control Patients

Neurological controls	IgG/Alb ratio	IgG index	IgG synthesis (mg/day)
Mean ± S.D.	0.14 ± 0.07	0.44 ± 0.11	−4.28 ± 4.38
Mean + 2 S.D.	0.28	0.66	+4.47
Mean − 2 S.D.	0.005	0.23	−13.04
Number of patients > 2 S.D. above mean	5	1	1
Number of patients > 2 S.D. below mean	0	2	2

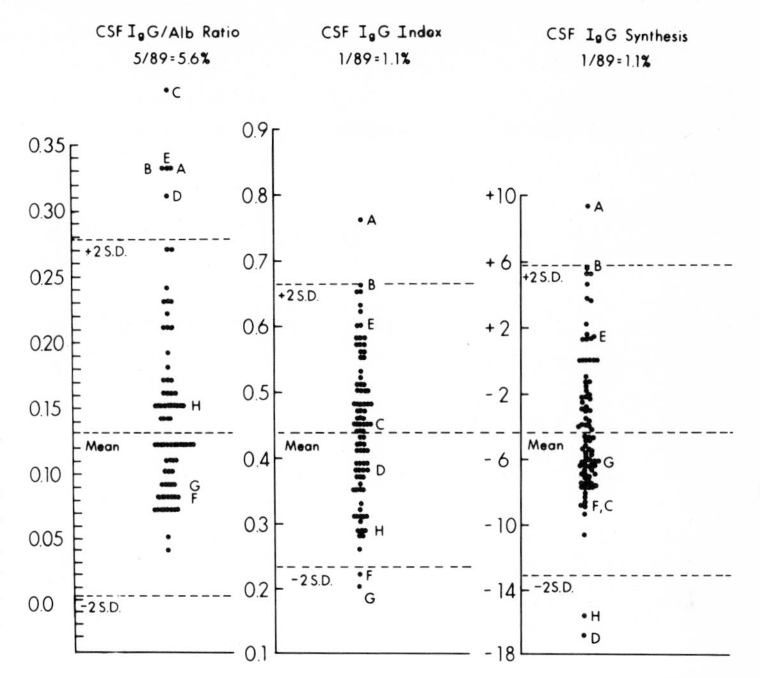

Figure 1. Scattergrams of quantitative data on 89 neurological control patients. Each dot represents one patient, and the same 89 patients were used for all three analyses. Values falling outside of ±2 S.D. in one or more of three scattergrams are marked with letters. Each letter represents same patient on all three analyses.

quantitative and qualitative IgG studies on CSF from patients with MS and other neurological diseases is presented in Table 3 and discussed below.

3. Oligoclonal IgG in CSF of Patients with Multiple Sclerosis

Since Lowenthal *et al.*[85] first applied agar gel electrophoresis (AGE) to the analysis of CSF proteins, several authors have described and confirmed the presence of a qualitatively abnormal gamma globulin in the CSF of patients with MS as compared to controls.[66,75,85] This pattern, consisting of striations or discrete bands in the gamma globulin region (representing restricted electrophoretic heterogeneity), has been termed the "oligoclonal aspect."[66] This subject has recently been reviewed.[74,56] The importance of using a constant amount of CSF IgG, the coelectrophoresis of serum in parallel (to detect serum bands that might leak into the CSF),[73] and the recognition of mid-cathodal and end-cathodal "normal" bands in many controls[73] have increased the specificity of the test. Although "oligoclonal bands" have been found in 75–94% of CSF samples from clinically definite MS patients[56,66,73,75,77,171] (81% in our laboratory), they have also been described in 40% of CNS infections and up to 5% of other neurological diseases[77] (39 and 3% in our laboratory). The development of purified agarose and use of a barbital buffer containing calcium lactate[67] have made the electrophoresis of CSF reliable for most laboratories. The availability of a commercial plate and buffer (Panagel®, Worthington Diagnostics,

Table 2. Comparison of Quantitative CSF IgG Abnormalities in 46 Definite MS[a] Patients

Definite MS	IgG/Alb ratio	IgG index	IgG synthesis (mg/day)
Mean ± S.D.	0.38 ± 0.20	1.35 ± 0.68	20.15 ± 15.96
Number of patients with normal values	27	4	10

[a] (MS) Multiple sclerosis.

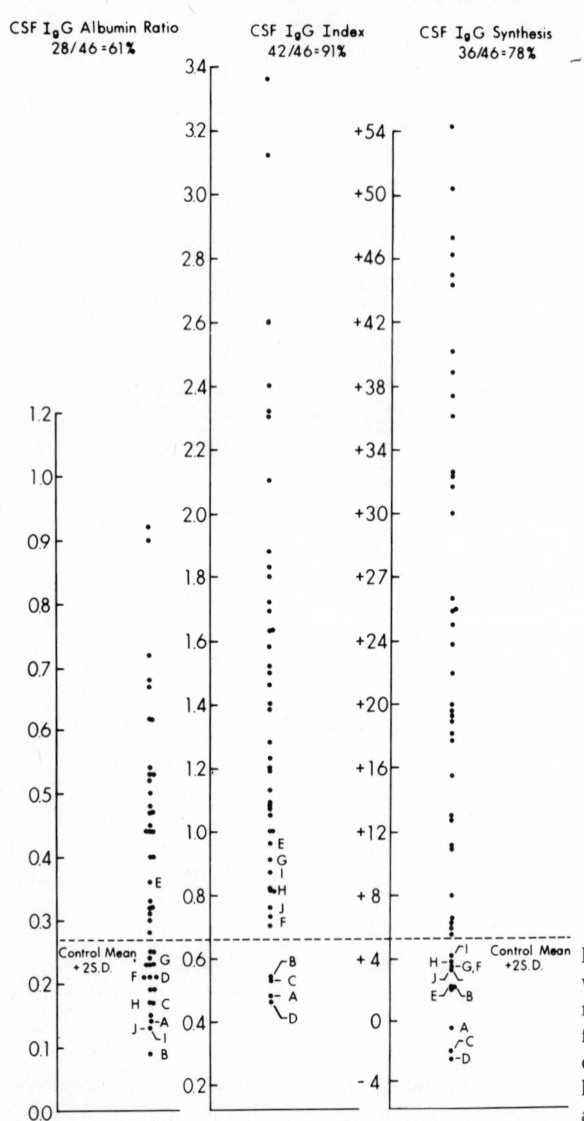

Figure 2. Scattergrams of quantitative data on 46 patients with clinically definite multiple sclerosis (MS). Each dot represents one patient, and same 46 patients were used for all three analyses. Values falling outside of ±2 S.D. in one or more of the three scattergrams are marked with letters. Each letter represents the same patient on all three analyses.

Freehold, New Jersey) has made these techniques available to the clinical laboratory[55] (Fig. 3). Our laboratory finds oligoclonal IgG bands in the CSF of 82% of clinically definite MS, 61% of probable MS, and 25% of possible MS patients as defined by Rose *et al.*[130] (Table 3). Despite the precautions of applying an equal amount of IgG, discounting bands that were also present in serum, and discounting CSF bands found in many controls, there remained a "false-positive" incidence of 3% among patients with diseases other than MS and CNS immunological/infectious disease. The patients with "false-positive" CSF oligoclonal bands had diseases such as stroke, cervical disc disease, and transient ischemic attacks (Table 4). Unfortunately, data are not available for large numbers of "normal" (nondisease) controls. Such data would be of interest in interpreting the significance of these bands.

As mentioned above, oligoclonal bands may be seen in up to 40% of infectious CNS diseases[66,77] (see Table 3). The infectious diseases include neurosyphilis, subacute sclerosing panencephalitis (SSPE), acute bacterial and aseptic meningitis,[66] acute measles encephalitis,[142] cryptococcal meningitis,[125] and progressive rubella panencephalitis.[175] In our experience, the oligoclonal bands in acute encephalitides tend to disappear in the convalescent period, accompanied by a drop in the previously

Table 3. Comparison of Methods for the Analysis of CSF IgG

Patient group	Elevated IgG/Alb ratio	Elevated IgG index	Oligoclonal bands By AGE[a]	Oligoclonal bands By IEF[a]
Definite MS	41/68 (60%)	42/46 (91%)	55/68 (81%)	37/44 (84%)
Probable MS	17/31 (55%)	15/20 (75%)	19/31 (61%)	11/12 (92%)
Possible MS	9/42 (21%)	12/30 (40%)	6/11 (55%)	5/9 (56%)
Neurological controls	9/167 (5.4%)	1/89 (1.1%)	5/157 (3.2%)	6/88 (6.8%)
Immunological/ infectious disease controls	11/30 (37%)	8/17 (47%)	11/28 (39%)	6/13 (46%)
Definite MS (on steroids)	5/11 (45%)	5/9 (56%)	6/11 (55%)	5/9 (56%)

[a] (AGE) Agarose gel electrophoresis; (IEF) isoelectric focusing; (MS) multiple sclerosis.

elevated CSF IgG index. This phenomenon has been described previously by others.[142] The oligoclonal bands may also be seen in diseases that may not be related or may be related only indirectly to infection or immune responses, e.g., Guillain-Barré,[66] myasthenia gravis,[3] as a remote effect of carcinoma (unpublished results), and in some diseases for which an immunological–infectious etiology would be surprising.[16,51]

Isoelectric focusing (IEF) in polyacrylamide gels may also be used to separate CSF proteins and identify oligoclonal bands.[35,36,62,63,69,70,105,162] Delmotte and Gonsette[36] and our early study[162] suggested that this method might reveal a higher incidence of oligoclonal bands than the AGE method. Furthermore, our method could utilize commercial kits (Pagplate®, LKB Instruments, Rockville, Maryland) for routine clinical analysis[162] (Fig. 4). Further experience with this technique has revealed several findings. The IEF method is somewhat more time-consuming than the AGE method, but is technically more consistent in that the Panagel plates frequently have a smeared pattern. The IEF technique also requires (as does the AGE method) parallel focusing of corresponding serum samples, since some alkaline bands in serum are transferred to CSF. Furthermore, the IEF method occasionally shows "extra" bands in the alkaline region, some of which may not be IgG.[105] The ammonium sulfate precipitation step used in our method[162] lessens this problem. A comparison of the AGE and IEF methods is shown in Table 3. A similar comparison was published while

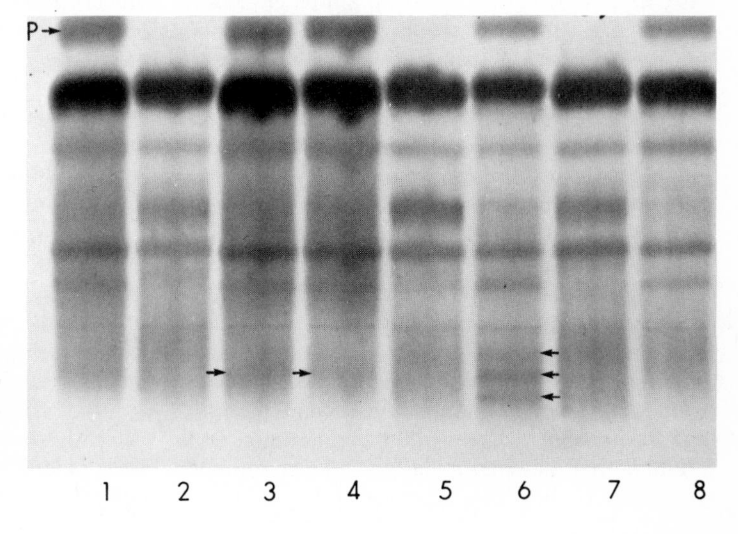

Figure 3. Agarose gel electrophoresis (AGE) of concentrated CSF (500 mg% protein) and serum (1:4 dilution) from multiple sclerosis (MS) and control patients. The CSF slots are recognizable (1, 3, 4, 6, 8) by presence of prealbumin (P). Arrows directed to right (→) pont to midcathodal band found in many CSF samples from controls. Arrows directed to left (←) (slot 6) point to oligoclonal IgG bands from patient with MS.

Table 4. False Positives: Neurological Controls

Disease	Elevated I/A[a]	Elevated C/S[a]	AGE[a]	IEF[a]	Abnormal C/S[a] + bands
Pseudotumor cerebri	1/8	0/8	0/6	0/6	0/8
Psychiatric	1/18	0/8	0/17	0/9	0/8
Migraine	0/9	0/4	0/9	0/4	0/4
Miscellaneous non-CNS damage	0/14	0/9	0/14	2/8	0/9
Stroke	1/14	0/12	1/15	1/12	0/12
CNS tumor	0/4	0/3	0/4	0/3	0/3
Parkinson's	0/3	0/2	0/3	0/1	0/2
Dementia	0/12	0/8	0/10	0/9	0/8
Huntington's chorea	0/1	—	0/1	0/1	—
Epilepsy/seizures	2/15	0/8	1/15	1/5	0/8
CNS trauma	0/3	0/1	0/2	0/1	0/1
Amyotrophic lateral sclerosis	0/5	0/2	1/4	0/3	0/2
Ischemic optic neuropathy	0/3	—	0/2	0/1	—
Alcoholism complications	0/2	—	0/2	—	—
Cervical spondylosis	1/4	1/1	1/3	1/4	1/1
Congenital disorders	0/2	0/2	0/2	—	—
Myasthenia gravis	0/3	—	0/3	0/2	—
Radiation myelopathy	0/1	0/2	0/1	—	0/2
Polyneuropathy	1/6	0/3	0/6	1/4	0/3
Transient ischemic attacks	0/7	—	1/7	0/1	—
Cerebellar degeneration	2/9	0/2	0/9	0/4	0/2
Miscellaneous CNS damage	1/24	0/4	0/22	0/10	0/4
Total	9/167	1/89	5/157	6/88	1/89
Percentage	5.4%	1.1%	3.2%	6.6%	1.1%

[a] (I/A) IgG/Alb ratio; (C/S) IgG index; (AGE) agarose gel electrophoresis; (IEF) isoelectric focusing.

this chapter was in press.[114a] IEF, in our series, defined oligoclonal bands in a higher proportion of probable and possible MS patients, but had double the rate of "false positives" that AGE had. Laurenzi and Link[69] have also compared these two methods. As more experience, and immunological techniques, e.g., immunofixation,[70] reveal the location of "nonpathological" bands, this false-positive rate may be decreased. The use of IEF with ampholines selected to give an alkaline pH has recently been proposed to give an increased resolution of oligoclonal bands in CSF from MS patients.[37a]

In MS, the pattern of the oligoclonal bands is unique for each patient and stable over time.[113] We have followed patients at 6-month intervals over 2 years, during which the oligoclonal pattern of the CSF remained unchanged. Using the technique of immunofixation, Cawley *et al.*[25] have shown that the oligoclonal bands demonstrated by AGE of CSF from MS patients are entirely IgG. Vandvik *et al.*[168] and Palmer *et al.*[116] have shown by immunoelectrophoresis that the oligoclonal IgG belongs primarily to the IgG_1 subclass. Link *et al.*[76a] used immunofixation to identify κ (heavy) and λ (light) chains. Although IgG bands may contain κ chains, λ chains, or both κ and λ chains, there appeared to be a preponderance of κ chains. The method also identified free λ chains, although free κ chains were not to be found. Another approach to this same problem has been the use of crossed immunoelectrofocusing, which has yielded similar results.[140a]

4. CSF IgG Abnormalities in Multiple Sclerosis in Relation to Clinical Parameters

Some investigators have found an increased relative concentration of gamma globulin in MS patients during exacerbations as opposed to remissions,[6,15] while this finding has not been confirmed in other studies.[77,176] Olsson and Link[113] criticized the early studies because they did not use each patient as his own control, which, due to the great

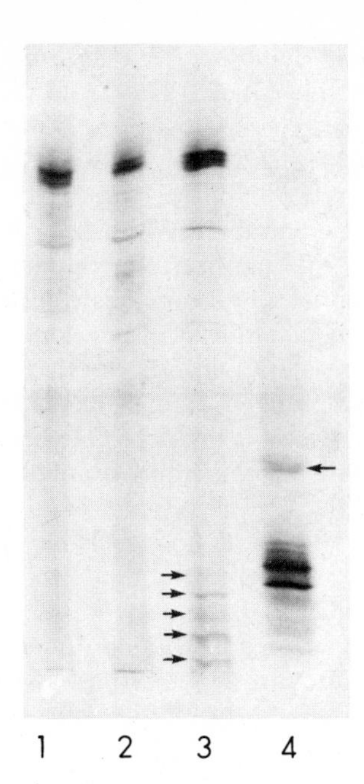

Figure 4. Isoelectric focusing (IEF) of CSF from control (slots 1, 2) and multiple sclerosis (MS) (slot 3) patients. Slot 4 contained the hemoglobin references. Arrows directed to right (→) represent oligoclonal IgG. Arrow directed to left (←) point to minor hemoglobin band cathodal to the location where most oligoclonal IgG is found.

overlap of IgG values, could obscure significant differences. Olsson and Link[113] obtained serial CSF samples from each patient during exacerbations and remissions. There was a statistically significant mean increase of the relative concentration of CSF IgG during exacerbations when the paired data were compared. However, this was not necessarily true for each individual, since in some patients, a decreased relative CSF IgG was observed during exacerbation. Using the CSF IgG index, which is a more reliable indicator of CNS IgG synthesis, Christiansen *et al.*[26] were unable to demonstrate significant differences between a group of patients during an exacerbation and another group during remission. There does not appear to be a variation in the presence or absence of oligoclonal bands in the CSF during exacerbations and remissions,[26,113] and although the κ/γ light-chain ratios vary over time in the CSF, there is no consistent pattern.[113]

Schmidt *et al.*[138] matched 40 MS patients whose CSF contained extremely high relative gamma globulin values with 40 MS patients who had normal values. They found a much higher incidence of oligoclonal IgG in the CSF samples from the former group, but the patients themselves did not differ significantly in age, type of progression, duration of illness, presence of remission, or disease severity. Christiansen *et al.*[26] were also unable to correlate the IgG index, presence of CSF oligoclonal bands, and their clinical data. On the other hand, Olsson *et al.*[114] found the highest CSF IgG index among the most disabled patients. They concluded that CSF IgG abnormalities were encountered more often and were more pronounced in the patients with the most malignant course of the disease. MS patients with a late age of onset (>35 years) showed a lesser degree of abnormalities than the rest of the group.[114]

5. Corticosteroids Inhibit CNS IgG Synthesis

We[30] and others[22,122] have shown that prednisone therapy induces a decrease in serum IgG levels. This is most likely due to decreased IgG synthesis, which may persist several days after a single dose of prednisone.[22,163] The IgG content of the CSF is also decreased by steroid therapy.[19,40,159] Using Tourtellotte's calculation of CNS IgG synthesis, which compensates for changes in serum IgG,[155] one group found that ACTH or exogenous steroids would decrease this synthesis, but not abolish the oligoclonal pattern of IgG.[40,159] We have studied 11 patients before and at several time points following "pulse" methylprednisolone therapy.[163] After 1 week of laboratory tests, each patient received three consecutive daily intravenous infusions of 1 g methylprednisolone. Of the 11 patients, 7 showed a marked (20 mg/day) mean decrease in CNS IgG synthesis as determined by the Tourtellotte equation. The remaining 4 patients had minimal decreases or a slight (laboratory variation?) increase in IgG synthesis. These 4 patients could be distinguished by an initial "normal" level of IgG synthesis (4/4). Of these 4 "nonresponder" patients, 2 were the only patients in the study taking chronic daily oral steroids. Unfortunately, there is no definite connection between a decrease in CNS IgG synthesis while taking steroids and clinical improvement.[159]

6. CSF Immunoglobulin Abnormalities in Optic Neuritis

Optic neuritis (ON) is one of the most common presenting lesions of MS, and a large percentage of

ON patients subsequently fulfill the clinical criteria for the diagnosis of MS. The percentage of ON patients who later develop MS varies widely in different series, and these data have been reviewed.[133] Many investigators (see below) have analyzed laboratory parameters in patients with ON to further elucidate this relationship. Significant correlations with CSF IgG abnormalities would not only be clinically useful for diagnosis and treatment evaluations, but also might suggest pathogenic factors that influence the severity and degree of disease involvement in MS.

Elevated relative CSF IgG concentrations occur in 16–50% of ON patients.[79,103,133] An oligoclonal pattern of the IgG is seen on AGE in 24–40% and a mononuclear leucocytosis in 33–60%.[79,103,133] Link *et al.*[79] compared other laboratory features between patients with CSF oligoclonal bands and those without. A mononuclear pleocytosis was found in 15 of 21 patients with oligoclonal IgG, but in only 3 of the 20 patients without the bands. The oligoclonal group had significantly increased total protein, relative IgG concentration, and frequency of occurrence of abnormal light-chain ratios. Measurable measles-virus hemolysis-inhibiting antibody titers were found in CSF from 20 of 21 ON patients with oligoclonal bands. Reduced serum/CSF antibody titers were found in 12 of 21 oligoclonal patients and only 1 of 20 of the nonpathological CSF group.[79]

Elevated measles-antibody titers occur in 62% of ON patients and 21% of controls[104]; there is also an increased CSF titer to other viruses and a decreased serum/CSF measles titer ratio. Histocompatibility type HL-A3 was found to correlate with elevated measles antibody titers irrespective of the diagnosis of MS, ON, or controls in a series of patients described by Arnason *et al.*[5] They found an increase in this antigen among MS patients, but not in isolated ON. Contrary results were reported by Sandberg-Wollheim *et al.*,[136] who found increased frequencies of the HL-A3, HL-A7, and LD-7a (DrW2) determinants of approximately the same magnitude in patients with MS and ON. There was a significant increase in the frequency of the HL-A3 determinant in patients with oligoclonal IgG in their CSF at the onset of illness. Stendahl *et al.*[143] studied similar parameters and found that ON patients with oligoclonal bands in their CSF had an incidence of the LD-7a antigen similar to that of MS patients. This association was absent in those ON patients who lacked CSF oligoclonal IgG. They found the antigens HL-A3 and HL-A7 to occur at similar frequencies in ON and controls. Only those patients with oligoclonal IgG had demonstrable decreased serum/CSF measles-antibody ratios.

The evaluation of laboratory parameters that might predict which ON patients will subsequently develop symptoms indicative of MS requires a prospective study. Sandberg-Wollheim[135] has reported such a study, although the follow-up for most of the patients was brief. Of 61 previously healthy patients who developed acute ON, 11 developed symptoms and signs consistent with MS during the follow-up period, which ranged from 7 months to 6 years (mean: 2.6 years). Of these 11 patients, 5 had had oligoclonal IgG at their initial examination of ON. In 5 others, the CSF developed oligoclonal IgG at a later date. There was no correlation between the onset of symptoms and the time of appearance of oligoclonal IgG. Six patients who had normal CSF at initial evaluation subsequently developed oligoclonal IgG without new symptoms. The author concluded that the CSF was often normal in ON patients who later developed MS. The series was not large enough to give specific risk factors. Similar prospective studies in larger groups of patients with longer follow-up might prove to be of great interest.

7. CSF Immunoglobulin Abnormalities in Chronic Myelopathy

Among the patients presenting with a chronic progressive myelopathy, the etiological agent frequently remains unknown after myelography has excluded mass lesions and there is no evidence of vitamin deficiency or diagnosable degenerative disorders. The Rose *et al.*[130] classification of MS would include many of these patients in the category of "possible" MS. An early study[73] found 12 of 19 patients with chronic myelopathy to have oligoclonal IgG in their CSF. Subsequent studies have confirmed the presence of CSF oligoclonal IgG in 48/73,[78] 16/25,[23] and 31/72[118] patients with this syndrome.

In addition to CSF oligoclonal IgG, these patients have other laboratory similarities to the MS population.[23,78,118] Of patients with CSF oligoclonal IgG, 81% had measles antibodies in the CSF and 65% had reduced serum/CSF measles-virus-antibody ratios. Of chronic myelopathy patients, 76% had delayed latencies on visual-evoked-response testing,[23,118] suggesting subclinical disease, and another brain lesion. Occasional chronic myelopathy patients have evidence of abnormalities on CAT scan,[118] supporting the concept of subclinical disseminated lesions. These findings strengthen the diagnosis of MS for these patients.

8. Quantitative and Qualitative Abnormalities of Light Chains in CSF

Radial immunodiffusion with antisera against human light chains was used by Link and Zettervall[82] to quantitate the light-chain ratios in the CSF of normal controls and neurological patients. They arbitrarily defined the kappa/lambda (κ/λ) light-chain ratio in normal serum as 1.0. The κ/λ ratios of CSF and serum from normal controls showed no statistically significant variation. Among 11 MS patients with oligoclonal bands in the CSF, 6 had significant elevations of the κ/λ ratio. Using preparative electrophoresis in barbital-buffered agar, eluates of fractions of CSF and sera from MS patients were analyzed for light-chain determinants.[178] Although some variation in κ/λ ratios (1.6–2.8) occurred in eluates from serum samples, the CSF eluates varied from 1.5 to 17.0. The protein fractions with abnormal κ/λ ratios did not contain IgA or IgM. The high and low peaks for IgG did not correlate with the κ/λ ratio peaks, but the regions with restricted IgG heterogeneity had the highest κ/λ ratios.

In a study of 64 MS patients, 39 patients with CNS infections, 81 patients with other neurological disorders, and 30 controls, increased κ/λ ratios were found in 53% of the MS CSF samples, 0% of the infectious disease group, and 3% (1 patient) with other neurological diseases.[77] The κ/λ ratio was abnormally high in all MS CSF samples in which immunoelectrophoresis of CSF showed a doubling of the IgG precipitation time. In 11 cases in which only a cathodic extra arc was obtained, the κ/λ ratio was abnormally high in 4.

An elevated κ/λ ratio was first reported as being found more often when there was a relative increase of CSF IgG, and only when there was an oligoclonal pattern of the IgG electrophoresis.[77] MS patients without oligoclonal banding had abnormally low (related to controls) κ/λ ratios. There was a higher incidence (64%) of abnormal κ/λ ratios among CSF samples from patients who had duration of MS less than 10 years than among those patients who had had the disease more than 10 years (39%). There was no relationship between the incidence of abnormal ratios and the interval since previous exacerbations. The κ/λ ratios of sera from MS patients were all normal. The quantitative abnormalities of κ/λ ratios in MS CSF have been confirmed in other laboratories[9,11,64,165] and have also been described in SSPE.[11] Furthermore, not all investigators agree as to the presence of abnormal κ/λ ratios and the correlation with oligoclonal banding patterns.[9,11,77]

Free κ and λ light chains have been described in the CSF of patients with SSPE and MS using specific free-light-chain antisera, radial immunodiffusion, immunoelectrophoresis, and crossed immunoelectrophoresis.[9,11,165] Free light chains have also been reported in some patients with cervical myelopathy, polyneuritis, meningitis, viral myelitis, encephalitis, and neurosyphilis.[11,165]

In addition to an abnormal κ/λ ratio and the presence of free light chains in the CSF from MS patients, there is also a phenomenon of "double-ring formation" that occurs in single radial immunodiffusion for light chains of type κ.[11] Iwashita *et al.*[52] who first described this phenomenon, found the double-ring formation in 24/42 patients with MS, 1/3 with neurosyphilis, 1/1 patient with cerebral cysticercosis, and 0/73 patients with other neurological diseases. Subsequent studies[11] have confirmed these findings but also found double-ring precipitation of κ light chains in other inflammatory or infectious diseases, especially SSPE.

The significance of the quantitative and qualitative abnormalities of CSF light chains, like the significance of elevated IgG synthesis or oligoclonal IgG, is not known. Antibodies of the IgG class with abnormal light-chain composition may account for the abnormal light-chain ratios. This phenomenon has also been reported with antigenic stimulation with polysaccharides in man[42,92] and with haptens in the guinea pig.[100] Free light chains may also affect the κ/λ ratio. The presence of free light chains has been suggested to represent proteolysis of oligoclonal immunoglobulins synthesized within the CNS, or to imply a regulatory defect, such as desynchronization between heavy- and light-chain assembly in antibody-producing cells under intense immunogenic stimulation.[129,165] Alternatively, the free light chains may be part of a defective immunoglobulin synthesis with an overproduction of light chains.[11]

The double-ring precipitation phenomenon has been hypothesized by Bollengier *et al.*[11] to indicate that immunoglobulins carry incomplete antigenic determinants that are specific for the disease. The hypothesis of Lietze *et al.*[72] is that the double precipitation indicates blocking of antigenic determinants by other bound proteins, e.g., immune complexes.

It is of interest that light-chain analysis of brain extracts from some MS patients also demonstrates abnormal ratios,[10] with κ-chain predominance in all cases. The MS-brain-extract ratios correlated well with the CSF ratios. However, control brain extracts, which also contained oligoclonal IgG, were λ-predominant and were different from the CSF ratios. The authors theorized that in controls, the CSF IgG is derived from serum, but the MS CSF IgG is derived from CNS synthesis.[10]

9. CSF Immunoglobulins in Subacute Sclerosing Panencephalitis

SSPE is probably due to an infection with measles or a measles-related virus. There are high measles-antibody titers in serum and CSF, demonstrable measles antigen within nerve cells, and isolatable virus from brain.[28,49] The CSF has a relative increased gamma globulin[49] and relative increased IgG concentrations.[32] The cathodal region of CSF electrophoresed on agar or agarose contains discrete "oligoclonal" bands.[84] These oligoclonal bands may also be seen in serum.[84] The κ/γ ratios are frequently elevated in CSF from SSPE patients.[80]

Link *et al.*[80] examined the relationship between the elevated CSF IgG concentrations and the measles-antibody titers as determined by several techniques. There was no correlation between the IgG concentration and measles titers; however, preparative electrophoresis showed that the measles-antibody peak activities corresponded to the IgG bands. They found a low serum/CSF measles-antibody titer ratio, which confirmed the observations of Connolly *et al.*[28] and was consistent with the evidence of Cutler *et al.*[32] for CNS synthesis of IgG. Mattson *et al.*[93a] have suggested that, in SSPE, the IEF oligoclonal IgG pattern is superimposed on a low background IgG. In MS, the peaks were lower (less dense bands), and there was a higher background IgG. Futhermore, in MS, but not in SSPE, IEF band number correlated with CSF IgG and IgG/albumin. The number of CSF IgG oligoclonal bands did not correlate with disease severity or measles titer. The authors suggested differences in genesis of bands in MS and SSPE.[93a]

Vandvik *et al.*[170] absorbed CSF and serum from SSPE patients with measles-infected cells or purified virus, resulting in a significant reduction of measles-antibody titers as measured by several serological techniques. Serial elutions at progressively lower pH allowed the isolation of specific measles antibodies. The dominant class of immunoglobulin isolated from SSPE sera and CSF was IgG; IgA and IgM were barely detected in eluates from CSF precipitates. Absorption of the CSF from SSPE patients by different measles antigens resulted in disappearance of some, but not all, of the oligoclonal IgG bands as demonstrated by electrophoresis.[170] Homogeneous IgG bands corresponding to the CSF oligoclonal bands occurred in eluates of absorbed CSF samples. Cell-associated-virus preparations and purified viral antigens differed in the ability to absorb the oligoclonal IgG. Some bands were absorbed by two measles antigens, whereas other bands were absorbed preferentially by one or the other.[170] Eluates of precipitates from control sera and CSF contained electrophoretically heterogeneous IgG only. The IgG absorbed by measles antigen from control sera and CSF was mainly eluted at pH 4, whereas the majority of IgG from SSPE fluids was eluted at pH 2 or 3.

The oligoclonal pattern of IgG that is detectable so frequently in the CSF from SSPE patients may be uncovered in the sera if first absorbed with measles antigen and subsequently eluted.[166,170] The eluate of antibody absorbed by cell-associated measles virus contains mostly antibody to the virus nucleocapsids. The quantity of IgG antibodies against measles and the number of elutable oligoclonal bands in the serum vary over the course of the disease, with a tendency toward increasing quantities of both.[166] In all cases, the antibodies obtainable from sera reflected those in the CSF, although the converse was not always true.[166] The serum antibodies may all have derived from CNS synthesis and subsequently been transported via CSF, or, less likely, they may represent synthesis outside the CNS. The presence of measles antibody in SSPE oligoclonal bands has been confirmed in another laboratory.[95]

Idiotypes are antigenic determinants located in the hypervariable region of the antigen-combining sites of immunoglobulins.[24,65] Nordal *et al.*[106] have succeeded in preparing an antiidiotypic antiserum against one of the oligoclonal bands representing measles-virus-nucleocapsid-specific IgG in the serum of a patient with SSPE. The antiserum was specific to only one band in the serum and CSF of this patient and did not cross-react with bands from other patients with SSPE or MS as determined by immunodiffusion. This had previously been achieved by others.[38] The future use of antiidiotypic antisera may be to elucidate the nature of, and reasons for, the CNS-synthesized oligoclonal IgG in SSPE and MS.[38b,106] Furthermore, radioimmunoassays (RIAs) for idiotypic determinants could be developed to detect and quantitate the oligoclonal bands in CSF and possibly serum over prolonged observation periods.

In addition to IgG, measles-specific IgM has been

detected in the CSF and serum from patients with SSPE using immunofluorescence,[29] RIA,[59] or absorption and immunodiffusion.[95] Measles-specific IgM is present in serum in only half the patients with SSPE.[59,95] Whether the measles-specific IgM is present during only certain phases of the course of SSPE is not known. Recently, virus-specific IgD has also been described in SSPE patients.[90]

10. Source of Abnormal CSF IgG

Tourtellotte[155] has reviewed the source of CSF IgG and albumin in normal and pathological conditions. The CSF is an expansion of the brain extracellular space and acts as a "sink," draining into the serum through the arachnoid villi. Most of the CSF proteins originate in the serum, with the choroid plexus serving as a "filter" that lowers the protein concentration to 1/200 that of serum. Under certain pathological conditions, the blood–brain barrier is damaged and the total CSF protein is increased, with the lower-molecular-weight proteins more likely to leak across in cases of minor damage. In the conditions, such as MS, that are of interest in this discussion, there is an increase of IgG that is proportionately greater than the increase in total protein. Tourtellotte and co-workers have presented data suggesting that the relative increase in IgG is due to *de novo* synthesis of IgG in the CNS.[155]

Frick and Scheid-Seydel[44] injected radioactive (iodinated) IgG or Alb intravenously into patients with varying neurological conditions. In cases with a normal blood–brain barrier, a normal CSF total protein, and a normal IgG concentration, the IgG penetrated into the CSF at a slower rate than Alb, but eventually equilibrated at a steady-state ratio of unity, indicating that the CSF IgG in these cases is entirely derived from the serum. In cases where there was breakdown of the blood–brain barrier, the labeled IgG equilibrated in the CSF more rapidly, but all the CSF IgG was again derived from serum. However, in patients with MS who had a normal blood–brain barrier and a normal CSF total protein, the data were not compatible with the CSF IgG being totally derived from serum. In these patients, 16–92% with the CSF IgG being totally derived from some other source, which these workers suggested to be the CNS itself. Similar studies by other workers[34,83] confirmed the presence of IgG in the CSF that was derived from a nonserum (presumed CNS) source. There is also considerable evidence that the excess IgG in the CSF of patients with SSPE is due to CNS synthesis.[33,160]

The site of the CNS synthesis of IgG has been of interest for some time. Using a direct immunofluorescence technique, Simpson *et al.*[141] localized most of the IgG in MS brains at the edge of active plaques. Several other workers have utilized immunofluorescence or immunoperoxidase methods to demonstrate the presence of IgG^{-}, IgA^{-}, and, more rarely, IgM-bearing cells in the CNS of MS patients.[37,41,87,99,126,149] The immunoglobulin-bearing cells have included lymphocytes,[37,41,99,126,149] plasma cells,[126] and reactive astrocytes.[37,99,126] Prineas and Raine[126] consider the reactive-astrocyte staining to represent nonspecific absorption of macromolecules; Mussini *et al.*[99] also believe this phenomenon to have little pathological significance.

The pathology of MS, especially concerning lymphocytes, has recently been reviewed.[127] Perivascular cuffs or inflammatory cells have long been believed to be present in early lesions.[1] Tanaka *et al.*[148] described two types of cuffs in MS brains. The acute type of perivascular cuffs consist of predominantly lymphocytes, lymphoid cells, and a few lipid-laden macrophages that are seen in nondemyelinated white matter and the margins of plaques. The chronic type of cuffs consist mainly of macrophages and plasma cells, and are always seen within plaques. Prineas and Wright[127] have confirmed these findings and especially emphasized the persistence of plasma cells in chronic plaques. They believe that this large permanent population of plasma cells is the source of the elevated IgG in MS CSF and that it explains the persistence of elevated CSF IgG and lack of significant correlation with disease activity. They also hypothesize that this population of cells may indicate the continued expression of antigens in chronic MS lesions. Tourtellotte *et al.*[158] have proposed that the elevated IgG in the CSF of MS patients derives from both B-lymphocytes—the synthesis of which is decreased by steroids and that synthesize the elevated viral-antibody titers—and the aforementioned plasma cells, which are steroid-resistant.

Biochemical studies have supported the kinetic and morphological studies indicating that the CNS is the site of synthesis for the elevated and abnormal IgG in the CSF from MS patients. Tourtellotte *et al.*,[157] in early studies, found an elevated IgG content in MS brains, both in plaques and in apparently normal white and gray matter. Link[76] confirmed this finding, and found that the IgG eluted from MS brains was oligoclonal [However, IgG from control brains may also be oligoclonal (see below)]. Excess and oligoclonal IgG are also present in brains from SSPE patients.[112] The abnormal IgG found in brain extracts is of the IgG_1 subclass, as is that in CSF.[168]

Cohen and Bannister[27] demonstrated that CSF leukocytes from an MS patient synthesized IgG and IgA, but not IgM. Sandberg-Wollheim *et al.*[137] confirmed this phenomenon and later showed that the IgG synthesized by these cells was oligoclonal.[134] However, Tourtellotte[155] has performed calculations that suggest that less than 0.1% of the CSF IgG in MS patients could be produced by the apparent number of CSF lymphocytes in an average patient. This suggests that the major synthesis of IgG must come from cells located in the CNS tissue itself.

11. Viral Antibodies in CSF of Patients with Multiple Sclerosis

Adams and Imagawa[2] published the first laboratory data supporting the epidemiological evidence that MS might be a viral disease. They found increased measles-antibody titers in sera from MS patients, a phenomenon that has since been confirmed in numerous laboratories using a variety of serological techniques.[17,18,47,131] Subsequently, elevated measles-antibody titers were also found in the CSF from patients with MS.[18,21,31,47,110,111,131,132,167] The methods used in these studies included neutralizing, complement-fixing, hemagglutination, hemagglutination inhibition, hemolysis inhibition, immunodiffusion, and immunofluorescent antibody techniques. Antibodies were found to react with measles ribonucleoproteins, envelope structures, nucleocapsid, and whole virus. The ratio of serum/CSF measles-virus titers was found to be greatly reduced in 8 of 15 cases of MS as compared to the corresponding ratio of an unrelated adenovirus-antibody titer,[131] suggesting local CNS synthesis of antibodies. This finding was confirmed by others.[31,110,111] An RIA method has demonstrated that different MS patients have antibodies against different protein components of the measles virus in their CSF.[31]

The emerging concept of slow virus disease, and the knowledge that SSPE was (1) a modified viral disease, (2) had apparent CNS synthesis of antimeasles antibodies, and (3) had CNS synthesis of oligoclonal IgG that mainly represented antimeasles activity (see above), all contributed to the prominent theory that MS was a viral (presumably measles?)-related disease. However, several lines of evidence have tempered enthusiasm for this particular mechanism. First, antibody titers to vaccinia virus were also reported to be elevated in the CSF of MS patients,[58] as were antibodies to rubella[96,111] and other viruses.[47,96] Furthermore, there was a suggestion of CNS synthesis of these other viral antibodies in some cases.[96] However, the elevated titers, and presumed CNS synthesis of measles antibodies, have been found much more consistently in differing geographical areas than the other viruses, and the titers to measles have been in general higher than to other viruses, even in groups of patients who have apparent concomitant CNS synthesis of antibodies to these other viruses.[46]

There are several lines of evidence that the elevated titers of measles antibody in serum and perhaps even the apparent CNS synthesis of virus may be an epiphenomenon reflecting genetic or environmentally induced influences that are only remotely linked to the pathogenesis of disease. The titers to virus do not correlate with the course of the disease.[128] Furthermore, Brody *et al.*[18] found that although serum measles-antibody titers were elevated in MS patients when compared to randomly selected controls, the viral titers were not significantly different than those of siblings of the same sex born within 3 years of the patient. Arnason *et al.*[5] studied the histocompatibility typing and serum measles-antibody titers of MS, ON, and control patients, and concluded that the presence of elevated virus titers reflected the HLA-A3 histocompatibility type more than the presence of demyelinating disease. Jersild *et al.*[53] reported increased measles-antibody titers in MS patients carrying the HLA-A7 and HLA-A3 antigens as compared to controls. Bertrams *et al.*,[8] on the contrary, found no correlation between viral titers and histocompatibility types. Whitaker *et al.*[173] found a more complex relationship; they found no relationship between viral titers and HLA antigens among MS patients, but a definite relationship among controls. These confusing data were analyzed in an editorial.[94] More recent studies again indicate a relationship between serum viral titers and HLA antigens.[71,94,119,120] Unfortunately, correlation studies between elevated CSF titers to viruses and histocompatibility antigens in siblings, patients, and controls have not been reported.

The significance of elevated viral-antibody titers in the CSF of MS patients may also be explained by phenomena observed in other situations. Lucas *et al.*[86] found elevated titers to measles virus in the sera of patients with systemic lupus erythematosus, a recognized "autoimmune" disease, as well as in sera from patients with various autoantibodies. Phillips and Christian[123] found that the serum total IgG concentration influenced the titer of antibodies to measles virus so that an elevation of IgG due to any cause would result in an elevated measles titer without necessarily a reinfection. Since NZB mice show a hyperactive humoral response to some antigens,[124] the serological findings in MS may merely reflect a hyperreactive humoral immune response in patients with autoimmune disease. (Symington *et al.*[144] have demonstrated an increased humoral immune response to flagellin among MS patients.)

Whereas the elevated titers to measles antibodies in SSPE sera and CSF correlate well with the bands of oligoclonal IgG (see above), and present a unified theory as to the pathogenesis of the disease, this clear relationship is not seen in MS. This lack of correlation has led to the statement that most of the IgG produced in the CNS of MS patients is not against measles or rubella, especially considering the lack of titer variation in the course of the disease.[128]

Haire *et al.*[47] used indirect immunofluorescence methods to identify IgM antibodies specific for measles in the sera of a significant number (22/56) of MS patients. Such a finding would be of great significance, since IgM elevation implies a persistent viral infection, as is found in SSPE.[29,49] However, another laboratory was unable to confirm this finding,[93] and Fraser[43] later concluded that the measles-specific IgM was a serological artifact. Furthermore, although elevated IgM levels occur in a significant number of MS CSF samples, this IgM is not measles-specific as determined by absorption.[174] Norrby[109] has recently reviewed the subject of viral antibodies in MS.

12. Significance of Abnormal CSF IgG in Multiple Sclerosis

The significance of the oligoclonal pattern of IgG in MS CSF is unknown. In some diseases, such as collagen vascular disease, hepatic cirrhosis, and chronic infection, there is a diffuse, heterogeneous, polyclonal elevation in serum IgG.[172] Oligoclonal IgG has been reported to occur in serum early in infections, possibly indicating the early recruitment of only a few clones of lymphocytes,[67] and in experimental hyperimmunization, especially with carbohydrate-bearing antigens.[61,97] Homogeneous antibodies are produced in lymphoproliferative disease such as multiple myeloma, and in Waldenström's macroglobulinemia, but also in primary immunodeficiency, autoimmune disease, prolonged antigen stimulation, and, occasionally, old age.[177] Evidence suggests that the tendency to synthesize the homogeneous immunoglobulin patterns, in some cases, may relate to histocompatibility type,[39,147,164] although the synthesis of oligoclonal CSF IgG in MS may not relate to histocompatibility antigens.[150]

Bornstein and Appel[13,14] pioneered the tissue-culture studies that described the demyelinative effects of sera from animals with experimental allergic encephalomyelitis (induced with whole white matter) and patients with MS on tissue-culture explants of rat cerebellum. Reviews of the large number of subsequent studies by these and other authors have recently been published.[88,140] We will not attempt to review this subject except to emphasize the presence of "cytotoxicity" in variable numbers of normal and neurological control sera[50] that have prevented a clear assessment of this entire area of research. The presence of the abnormal IgG in the CSF of MS patients has led to the theory that this IgG represents antibodies directed against CNS antigens and may somehow be involved in the pathogenesis of the disease. The most vocal proponent of this theory has been Tourtellotte.[154] Although unconcentrated CSF does not have apparent demyelinating activity in the *in vitro* preparations used by Lumsden,[88] concentrated CSF, especially those samples with elevated IgG, will demyelinate explanted cultures according to some workers.[60,88]

The most recent work on CSF toxicity has used whole mounts of tadpole optic nerves as the target tissue.[145,146] Unconcentrated CSF was injected into the right eye of 8–10 tadpoles. After 48 hr, the optic nerves were examined morphologically, using the left nerve as control. Tadpoles injected with CSF from MS patients frequently had loss of myelin, with lesions usually not involving axons. The toxic activity was inactivated by heating to 56°C for 30 min, but was not reactivated by the addition of fresh guinea pig serum. This phenomenon was present in only 1 of 20 control CSF samples. The toxic CSF tended to be from MS patients with a longer and more severe course of disease, possibly suggesting a nonspecific effect, but also correlated with acute attacks, which may support a role of this "evil humor" in the pathogenesis of MS. We feel that more work should be done in this area, especially with more "neurological" controls. We are unaware of experiments using MS CSF to search for the antibody-dependent cell-mediated demyelination that has been described to occur with sera from CNS-immunized animals and lymphocyte products.[20]

In our laboratory, we have absorbed five MS CSF specimens containing oligoclonal IgG with various brain antigens. We compared the IgG content by EID and the IgG pattern by AGE, prior to and after absorption. Whole human brain homogenate, purified bovine and human myelin, the nonmyelin pellet (at the bottom of the gradient) found on human myelin isolation, and cerebroside micelles were used in these experiments. There were no qualitative changes in the CSF IgG following absorption with these materials. These negative results with a crude method do not exclude the possibility of "hidden" antigenic determinants in human brain, however, and further experiments with purified brain antigens are planned. Panitch *et al.*[117] have reported that there are antibodies to myelin basic protein in the CSF from some patients with MS, and their incidence correlates with the presence of oligoclonal bands. However, further studies must be done with isolated bands to demonstrate a more significant relationship. In a later abstract, using electroimmunofixation, this group[117a] suggested that oligoclonal IgG with antimeasles activity (obtained from SSPE patients) also showed cross-reaction with myelin basic protein. We eagerly await confirmation of these results.

The abnormal relative increase in oligoclonal CSF IgG that occurs in cryptococcal meningitis[125] and SSPE[169] may be absorbed by the infectious agents involved in these diseases. Johnson and Nelson,[56] however, have found at least one band in SSPE CSF that is not directed against measles virus, and Booe *et al.*[12] have suggested that the bands in SSPE that remain after absorption with measles virus are antibodies directed against brain antigens. The CSF oligoclonal bands in MS patients do not correspond to peaks of antimeasles activity.[112] Vandvik *et al.*[170] absorbed CSF from MS and SSPE patients with measles antigens and subsequently eluted any bound antibodies in the antigen–antibody precipitates with acidic buffer. Electrophoretically homogenous (oligoclonal) measles antibodies were found in five of five patients with SSPE and in five of seven MS patients. In SSPE, most of the oligoclonal IgG was represented by antimeasles activity, whereas in MS, none of the oligoclonal IgG found in native CSF was found to be removed by absorption.[170]

Nordal *et al.*[108] have used an imprint electroimmunofixation technique to study antibodies to measles, rubella, mumps, and herpes simplex virus in CSF and sera from MS patients. An oligoclonal-type pattern could be demonstrated in nine of ten patients using this method. There were bands against measles virus in five sera and seven CSF, against rubella in no sera and four CSF, against mumps virus in five sera and seven CSF, and against herpes simplex virus in nine sera and nine CSF. Polyclonal patterns of antibody to all four viruses occurred in both the serum and CSF of only one patient. *However, there was no direct correspondence between the pattern of virus-antibody bands and the pattern of IgG bands in the CSF by ordinary agarose gel electrophoresis.* Thus, bands occurred in the CSF, but not in the serum (presumed indication of CNS synthesis), against one or more of the four viruses in nine of the ten patients. Previous studies[111] have shown that local synthesis of antibodies to two or more viruses occurred in 16% of 150 patients with MS.

Oligoclonal antibodies as demonstrated by imprint electroimmunofixation may be seen in the sera from healthy adults and neurological controls[107] However, in these cases, the patterns in CSF and serum are similar, implying a simple transfer to the CSF. Furthermore, extracts from human and rat "normal" control brain have "oligoclonal" IgG.[10] We would like to put forth the theory that synthesis of "oligoclonal" bands may be common when some individual antigens are examined. Futhermore, scattered immunological elements in normal brain may synthesize IgG that appears oligoclonal. In ongoing antigenic stimulation, as occurs in a persistent infection, the CSF concentration of this "oligoclonal" IgG becomes sufficiently elevated to be visible with ordinary AGE. In MS, the antigen or antigens are unknown, but are not likely ordinary measles virus, which does not absorb out the "macro"-oligoclonal bands seen in the CSF of this disease. The presence of these bands in MS, along with evidence of increased synthesis of IgG to known antigens, i.e., virus, suggest the possibility that there is a defect in the regulation of antibody synthesis in the CNS of MS patients. Whether all patients with MS have "macro"-oligoclonal antibodies to the same antigen or whether each patient has antibodies to only one antigen remains to be discovered. This knowledge may be expedited by the discovery of *the* MS antigen or the development of one or more antiidiotypic determinant antisera.[38a]

13. CSF IgM and IgA

Although CSF IgG may be measured reliably using radial immunodiffusion or EID, these meth-

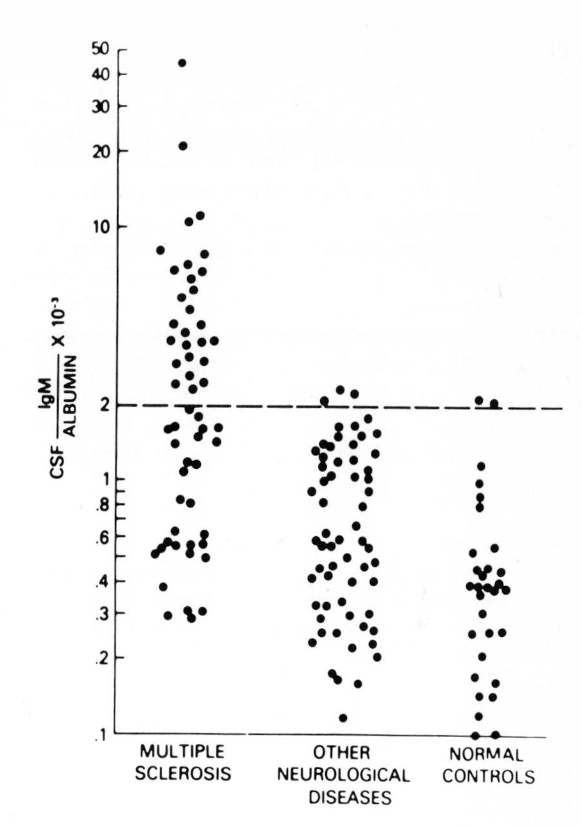

Figure 5. CSF IgM. CSF IgM measured by radioimmunoassay (RIA) is increased in 50% of patients with multiple sclerosis (MS) as compared with patients with other neurological diseases and normal controls. Reproduced with permission from Williams *et al.*[174]

ods are not of sufficient sensitivity to measure the low concentrations of other immunoglobulins in unconcentrated CSF. The conflicting findings in early studies[139] are most likely due to a lack of assay sensitivity. The powerful and sensitive technique of RIA has recently been applied to this problem.[98,101,108,174]

Using an RIA for IgM, Williams *et al.*[174] found an increased reactive CSF IgM in 50% of MS patients (Fig. 5); an even higher incidence was found in the study of Nerenberg *et al.*[102] There was no correlation between serum and CSF IgM concentrations or CSF IgG and IgM concentrations, results compatible with but not proving *de novo* CNS synthesis.[174] Of MS patients with normal CSF IgG levels, 40% have abnormal IgM levels. However, elevation of either immunoglobulin is not specific for MS, being present in other inflammatory or infectious diseases.[77,102,174] The presence of elevated serum IgM occurs with thymus-independent antigen stimulation, early in the course of antigen stimulation, and in persistent infections.[174] This may provide a clue to the significance of elevated CSF IgM. CSF IgM may be decreased by steroid therapy.[19]

CSF IgA has been found by several authors to be elevated in some patients with MS[102,139] and is also seen in various inflammatory or infectious diseases. Nerenberg and co-workers found a *decreased* content of IgD and IgE in MS CSF. The investigators involved in these studies are now searching for specific immunoglobulin patterns in disease states and for clues to disease pathogenesis from such patterns.[102,174]

14. Conclusion

In conclusion, modern immunological techniques to identify qualitative and quantitative abnormalities in the CSF have strengthened our diagnostic armamentarium and suggested clues to the pathogenesis of diseases of unknown etiology. In diseases of known infectious etiology such as rubella encephalitis and subacute sclerosing panencephalitis, the presence of elevated viral titers correlating with elevated specific immunoglobulins and an oligoclonal pattern of IgG on electrophoresis is consistent with our understanding of the pathogenesis of the disease. In MS, the oligoclonal IgG is directed against unknown antigens, and although compatible with an infectious etiology, could be an epiphenomenon reflecting a defect in immunoregulation.

ACKNOWLEDGMENTS. The authors thank Patti Vessell and Jane McGinnis for their excellent secretarial and editing assistance. The authors are recipients of Teacher-Investigator Development Awards (J.L.T.: 5-K07-NS-00221; B.R.B.: 1-K07-NS-00385) from the National Institute of Neurological and Communicative Disorders and Stroke. This work was supported in part by the Kroc Foundation and a Washington University Clinical Research Center Grant (PHS-RR-36) (J.L.T.) and by the Amyotrophic Lateral Sclerosis Society of America (77-04) (B.R.B.).

References

1. ADAMS, C. W. M.: Pathology of multiple sclerosis: Progression of the lesion. *Br. Med. Bull.* **33**:15–20, 1977.
2. ADAMS, J. M., IMAGAWA, D. T.: Measles antibodies in multiple sclerosis. *Proc. Soc. Exp. Biol. Med.* **111**:562–566, 1962.
3. ADORNATO, B. T., HOUFF, S. A., ENGEL, W. K., DALAKAS, M., MADDEN, D. L., SEVER, J. L.: Abnormal immunoglobulin bands in cerebrospinal fluid in myasthenia gravis. *Lancet* **2**:367–368, 1978.
4. ANSARI, K. A, WELLS, B. S., VATASSERY, G. T.: Quantitative estimation of cerebrospinal fluid globulins in multiple sclerosis. *Neurology* **25**:688–692, 1975.
5. ARNASON, B. G. W., FULLER, T. C., LEHRICH, J. R., WRAY, S. H.: Histocompatibility types and measles antibodies in multiple sclerosis and optic neuritis. *J. Neurol. Sci.* **22**:419–428, 1974.
6. BAUER, H., GOTTLESLEBEN, A.: Quantitative immunochemical studies of cerebrospinal fluid proteins in relation to clinical activity of multiple sclerosis. *Int. Arch. Allergy* **36**:643–648, 1969.
7. BERNER, J. J., CLEMINS, V. A., SCHROEDER, E. F.: Radial immunodiffusion of cerebrospinal fluid. *Am. J. Clin. Pathol.* **58**:145–152, 1972.
8. BERTRAMS, J., VON FISENNE, E., HOHNER, P. G., KUWERT, E.: Lack of association between HL-A antigens and measles antibody in multiple sclerosis. *Lancet* **2**:441, 1973.
9. BOLLENGIER, F., DELMOTTE, P., LOWENTHAL, A.: Biochemical findings in multiple sclerosis. III. Immunoglobulins of restricted heterogeneity and light chain distribution in cerebrospinal fluid of patients with multiple sclerosis. *J. Neurol.* **212**:151–158, 1976.
10. BOLLENGIER, F., MAHLER, A., CLINET, G., LOWENTHAL, A.: Multiple sclerosis: Oligoclonal IgG, κ–γ light chain distribution and measles antibodies in brain extracts. *Brain Res.* **152**:133–144, 1978.
11. BOLLENGIER, F., RABINOVITCH, N., LOWENTHAL, A.: Oligoclonal immunoglobulins, light chain ratios and free light chains in cerebrospinal fluid and serum from patients affected with various neurological diseases. *J. Clin. Chem. Biochem.* **16**:165–173, 1978.
12. BOOE, I., TOURTELLOTTE, W. W., BRANDES, D. W.: Brain and measles-specific immunoglobulin G in cerebrospinal fluid of patients with subacute sclerosing panencephalitis. *Neurology* **26**:377, 1976.
13. BORNSTEIN, M. B., APPEL, S. H.: Tissue culture studies of demyelination. *Ann. N. Y. Acad. Sci.* **122**:280–287, 1965.
14. BORNSTEIN, M. B., APPEL, S. H.: The application of tissue culture to the study of experimental allergic encephalomyelitis. I. Patterns of demyelination. *J. Neuropathol. Exp. Neurol.* **20**:141–157, 1961.
15. BRADSHAW, P.: The relation between clinical activity and the level of gamma globulin in the cerebrospinal fluid in patients with multiple sclerosis. *J. Neurol. Sci.* **1**:374–379, 1964.
16. BRITTON, D. E., HOUFF, S. A., EIBEN, R. M., MADDEN, D. L., SEVER, J. L.: Studies of viral antibodies, oligoclonal IgG, *in situ* central nervous system IgG production, and lymphocyte rosetting in sex-linked recessive adrenoleukodystrophy. *Neurology* **27**:396, 1977.
17. BRODY, J. A., SEVER, J. L., EDGAR, A., MCNEW, J.: Measles antibody titers in multiple sclerosis patients and their siblings. *Neurology* **22**:492–499, 1972.
18. BRODY, J. A., SEVER, J. L., HENSON, T. W.: Virus antibody titers in multiple sclerosis patients, siblings, and controls. *J. Am. Med. Assoc.* **216**:1441–1446, 1971.
19. BROOKS, B. R., COOK, J. D., TROTTER, J. L., MCFARLIN, D. E., ENGEL, W. K.: Effect of prednisone therapy on cerebrospinal fluid immunoglobulins and lymphocytes in chronic idiopathic relapsing polyneuropathy and other neuromuscular disorders. *Ann. Neurol.* **1**:510, 1977.
20. BROSNAN, C. F., STONER, G. L., BLOOM, B. R., WISNIEWSKI, H. M.: Studies on demyelination by activated lymphocytes in the rabbit eye. II. Antibody dependent cell-mediated demyelination. *J. Immunol.* **118**:2103–2110, 1977.
21. BROWN, P., CATHALA, F., GAJDUSEK, D. C., GIBBS, C. J.: Measles antibodies in the cerebrospinal fluid of patients with multiple sclerosis. *Proc. Soc. Exp. Biol. Med.* **137**:956–961, 1971.
22. BUTLER, W. T., ROSEN, R. D.: Effects of corticosteroids on immunity in man. II. Alteration in serum protein components after methylprednisolone. *Transplant. Proc.* **5**:1215–1219, 1973.
23. BYNKE, H., OLSSON, J. E., ROSEN, I.: Diagnostic value of visual evoked response, clinical eye examination and CSF analysis in chronic myelopathy. *Acta Neurol. Scand.* **56**:55–69, 1977.
24. CAPRA, J. D., KEHOE, J. M.: Hypervariable regions, idiotypy, and antibody-combining site. *Adv. Immunol.* **20**:1–40, 1975.
25. CAWLEY, L. P., MINARD, B. J., TOURTELLOTTE, W. W., MA, B. I., CHELLE, C.: Immunofixation electrophoretic

techniques applied to identification of proteins in serum and cerebrospinal fluid. *Clin. Chem.* **22**:1262–1268, 1976.

26. Christiansen, O., Clausen, J., Fog, T.: Relationship between abnormal IgG index, oligoclonal bands, acute phase reactants and some clinical data in multiple sclerosis. *J. Neurol.* **218**:237–244, 1978.
27. Cohen, S., Bannister, R.: Immunoglobulin synthesis within the central nervous system in disseminated sclerosis. *Lancet* **1**:366–367, 1967.
28. Connolly, J. H., Allen, I. V., Hurwitz, L. J., Millar, J. H. D.: Measles virus antibody and antigen in subacute sclerosing panencephalitis. *Lancet* **1**:542–544, 1967.
29. Connolly, J. H., Haire, M., Hadden, D. S. M.: Measles immunoglobulins in subacute sclerosing panencephalitis. *Br. Med. J.* **1**:23–26, 1971.
30. Cook, J. D., Trotter, J. L., Engel, W. K., Sciabbarrasi, J. S.: The effects of single dose alternate-day prednisone therapy on the immunological status of patients with neuromuscular disease. *Ann. Neurol.* **3**:166–176, 1978.
31. Cunningham-Rundles, C., Jersild, C., Dupont, B., Posner, J. B., Good, R. A.: Detection of measles antibodies in cerebrospinal fluid and serum by a radioimmunoassay. *Scand. J. Immunol.* **4**:785–790, 1975.
32. Cutler, R. W. P., Merler, E., Hammerstad, J. P.: Production of antibody by the central nervous system in subacute sclerosing panencephalitis. *Neurology* **18**(2):129–132, 1968.
33. Cutler, R. W. P., Watters, G. V., Hammerstad, J. P., Merler, E.: Origin of cerebrospinal fluid gammaglobulin in subacute sclerosing leukoencephalitis. *Arch. Neurol.* **17**:620–628, 1967.
34. Cutler, R. W. P., Watters, G. V., Hammerstad, J. P.: The origin and turnover rates of cerebrospinal fluid albumin and gamma globulin in man. *J. Neurol. Sci.* **10**:259–268, 1970.
35. Delmotte, P.: Comparative results of agar electrophoresis and isoelectric focusing examination of the gamma globulins of the cerebrospinal fluid. *Acta Neurol. Belg.* **72**:226–234, 1972.
36. Delmotte, P., Gonsette, G.: Biochemical findings in multiple sclerosis. IV. Isoelectric focusing of the CSF gamma globulins in multiple sclerosis and other neurological diseases. *J. Neurol.* **215**:27–37, 1977.
37. Dubois-Dalq, M., Schumacher, G., Worthington, E. K.: Immunoperoxidase studies on multiple sclerosis brain. *Neurology* **25**:496, 1975.

37a. Ebers, G. C., Paty, D. W.: Alkaline isoelectric focusing in polyacrylamide gel of cerebrospinal fluid in multiple sclerosis. *Ann. Neurol.* **6**:163, 1979.

38. Ebers, G. C., Fraser, B., van de Rign, I., Zabriskie, J. B., Kunkle, H. G.: Idiotypic determinants in SSPE. *Clin. Res.* **25**:482a, 1977.

38a. Ebers, G. C., Zabriskie, J. B., Kunkle, H. G.: Oligoclonal immunoglobins in subacute sclerosing panencephalitis and multiple sclerosis: A study of idiotypic determinants. *Clin. Exp. Immunol.* **35**:67–75, 1979.

39. Eichmann, K., Lachland, H., Hood, L., Krause, R. M.: Induction of rabbit antibodies with molecular uniformity after immunization with group C streptococci. *J. Exp. Med.* **131**:207–221, 1970.
40. Elsner, W., Tourtellotte, W. W., Murthy, K. N., Booe, I., Potvin, A. R., Syndulko, K.: Multiple sclerosis: Effect of dexamethasone on *in situ* central nervous system IgG synthesis. *Neurology* **28**:403–404, 1978.
41. Esiri, M. M.: Immunoglobulin-containing cells in multiple sclerosis plaques. *Lancet* **2**:478–480, 1977.
42. Franklin, E. C., Fudenberg, H. H.: Antigenic heterogeneity of human Rh antibodies, rheumatoid factors, and cold agglutinins. *Arch. Biochem.* **104**:433–437, 1964.
43. Fraser, K. B.: False-positive measles-specific IgM in multiple sclerosis. *Lancet* **1**:91–92, 1978.
44. Frick, E., Scheid-Seydel, L.: Untersuchungen mit I^{131}-markiertem globulin zur Frage der Abstammung der Liquoreweiss-Körper, *Klin. Wochenschr.* **36**:857–863, 1958.
45. Ganrot, K., Laurell, C. B.: Measurement of IgG and albumin content of cerebrospinal fluid and its interpretation. *Clin. Chem.* **20**:571–573, 1976.
46. Haire, M.: Significance of virus antibodies in multiple sclerosis. *Br. Med. Bull.* **33**:40–44, 1977.
47. Haire, M., Fraser, K. B., Miller, J. H. D.: Virus-specific immunoglobulins in multiple sclerosis. *Clin. Exp. Immunol.* **14**:409–416, 1973.
48. Harley, T. F., Merrill, D. A., Claman, H. N.: Quantitation of immunoglobulins in cerebrospinal fluid. *Arch. Neurol.* **15**:472–479, 1966.
49. Horta-Barbosa, L., Fucillo, D. A., Sever, J. L.: Subacute sclerosing panencephalitis: Isolation of measles virus from a brain biopsy. *Nature (London)* **221**:974, 1969.
50. Hughes, D., Field, E. J.: Myelinotoxicity of serum and spinal fluid in multiple sclerosis: A critical assessment. *Clin. Exp. Immunol.* **2**:205–309, 1967.
51. Iivanainen, M., Leinikki, P., Taskinen, E., Shekarchi, I., Madden, D., Sever, J.: Oligoclonal IgG and virus antibodies in cerebrospinal fluid in progressive myoclonus epilepsy. *Ann. Neurol.* **4**:176, 1978.
52. Iwashita, H., Grunwald, F., Bauer, H.: Double ring formation in single radial immunodiffiusion for kappa chains in multiple sclerosis cerebrospinal fluid. *J. Neurol.* **207**:45–52, 1974.
53. Jersild, C., Svejgaard, A., Fog, T., Amnitzboll, T.: HL-A antigens and disease. I. Multiple sclerosis. *Tissue Antigens* **3**:243–250, 1973.
54. Johansson, B. G.: Agarose gel electrophoresis. *Scand. J. Clin. Lab. Invest.* **124**:7–19, 1972.
55. Johnson, K. P., Arrigo, S. C., Nelson, B. J., Ginsberg, S.: Agarose electrophoresis of cerebrospinal fluid in multiple sclerosis. *Neurology* **27**:273–277, 1977.
56. Johnson, K. P., Nelson, B. J.: Multiple sclerosis:

Diagnostic usefulness of cerebrospinal fluid. *Ann. Neurol.* **2**:425–431, 1977.

57. KABAT, E. A., GLUSMAN, M., KNAUB, V.: Quantitative estimation of the albumin and gamma globulin in normal and pathological cerebrospinal fluid by immunochemical methods. *Am. J. Med.* **4**:653–662, 1948.
58. KEMPE, C. H., TAKABAYASHI, K., MIYAMOTO, H., MCINTOSH, K., TOURTELLOTTE, W., ADAMS, J. M.: Elevated cerebrospinal fluid vaccinia antibodies in multiple sclerosis. *Arch. Neurol.* **28**:278–279, 1973.
59. KIESSLING, W. R., HALL, W. W., YUNG, L. L. TER MEULEN, V.: Measles-virus-specific immunoglobulin-M response in subacute sclerosing panencephalitis. *Lancet* **1**:324–327, 1977.
60. KIM, S. U., MURRAY, M. R., TOURTELLOTTE, W. W., PARKER, J. A.: Demonstration in tissue culture of myelinotoxicity in cerebrospinal fluid and brain extracts from multiple sclerosis patients. *J. Neuropathol. Exp. Neurol.* **29**:420–431, 1970.
61. KIMBALL, J. W., PAPPENHEIMER, A. M., JATON, J. C.: The response in rabbits to prolonged immunization with type III pneumococci. *J. Immunol.* **106**:1177–1184, 1971.
62. KJELLIN, K. G., SIDÉN, A.: Abberant CSF protein fractions found by electrofocusing in multiple sclerosis. *Eur. Neurol.* **15**:40–50, 1977.
63. KJELLIN, K. G., VESTERBERG, O.: Isoelectric focusing of CSF proteins in neurological disease. *J. Neurol. Sci.* **23**:199–213, 1974.
64. KOLAR, O. J.: Light chains in cerebrospinal fluid in multiple sclerosis. *Lancet* **2**;1030, 1977.
65. KUNKEL, H. G., MANNIK, M., WILLIAMS, R. C.: Individual antigenic specificity of isolated antibodies. *Science* **140**:1218–1219, 1963.
66. LATERRE, E. C., CALLEWAERT, A., HEREMANS, J. F., SFAELLO, Z.: Electrophoretic morphology of gamma globulins in cerebrospinal fluid of multiple sclerosis and other diseases of the nervous system. *Neurology* **20**:982–990, 1970.
67. LAURELL, C. B.: Composition and variation of the gel electrophoretic fractions of plasma, cerebrospinal fluid, and urine. *Scand. J. Clin. Lab. Invest.* **29**(124):71–82, 1972.
68. LAURELL, C. B.: Quantitative estimation of proteins by electrophoresis in agarose gel containing antibodies. *Anal. Biochem.* **15**:45–52, 1966.
69. LAURENZI, M. A., LINK, H.: Comparison between agarose gel electrophoresis and isoelectric focusing of CSF for demonstration of oligoclonal immunoglobulin bands in neurological disorders. *Acta Neurol. Scand.* **58**:148–156, 1978.
70. LAURENZI, M. A., LINK, H.: Localization of the immunoglobulins G, A, M, β-trace protein and γ-trace protein on isoelectric focusing of serum and cerebrospinal fluid by immunofixation. *Acta Neurol. Scand.* **58**:141–147, 1978.
71. LEHRICH, J. R., ARNASON, B. G. W.: Histocompatibility types and viral antibodies. *Arch. Neurol.* **33**:404–405, 1976.
72. LIETZE, A., COLIN, S., NOWE, H. A.: An intrinsic inaccuracy of radial immunodiffusion measurements of incomplete antigens. *Clin. Biochem.* **3**:335–338, 1970.
73. LINK, H.: Comparison of electrophoresis on agar gel and agarose in the evaluation of gamma globulin abnormalities in cerebrospinal fluid and serum in multiple sclerosis. *Clin. Chem. Acta* **46**:383–389, 1973.
74. LINK, H.: Immunoglobulin abnormalities in multiple sclerosis. *Ann. Clin. Res.* **5**:330–336, 1973.
75. LINK, H.: Immunoglobulin G and low molecular weight proteins in human cerebrospinal fluid. *Acta Neurol. Scand.* **43**(28):1–136, 1967.
76. LINK, H.: Oligoclonal immunoglobulin G in multiple sclerosis brains. *J. Neurol. Sci.* **16**:103–114, 1972.

76a. LINK, H., LAURENZI, M. A.: Immunoglobin class and light chain type of oligoclonal bands in CSF in multiple sclerosis determined by agarose gel electrophoresis and immunofixation. *Ann. Neurol.* **6**:107–110, 1979.

77. LINK, H.: MÜLLER, R.: Immunoglobulins in multiple sclerosis and infections of the nervous system. *Arch. Neurol.* **25**:326–344, 1971.
78. LINK, H., NORRBY, E., OLSSON, J. E.: Immunoglobulin abnormalities and measles antibody response in chronic myelopathy. *Arch. Neurol.* **33**:26–32, 1976.
79. LINK, H., NORRBY, E., OLSSON, J. E.: Immunoglobulins and measles antibodies in optic neuritis. *N. Engl. J. Med.* **289**:1103–1107, 1973.
80. LINK, H., PANELIUS, M., SALMI, A. A.: Immunoglobulins and measles antibodies in subacute sclerosing panencephalitis. *Arch. Neurol.* **28**:23–30, 1973.
81. LINK, H., TIBBLING, G.: Principles of albumin and IgG analysis in neurological disorders. II. Relation of the concentration of the proteins in serum and cerebrospinal fluid. *Scand. J. Lab. Invest.* **37**:391–396, 1977.
82. LINK, H., ZETTERVALL, O.: Multiple sclerosis: Disturbed kappa:lambda chain ratio of immunoglobulin G in cerebrospinal fluid. *Clin. Exp. Immunol.* **6**:435–438, 1970.
83. LIPPINCOTT, S. W., KORMAN, S., LAX, L. C., CORCORAN, A. B.: Transfer rates of gamma globulin between cerebrospinal fluid and blood plasma. *J. Nucl. Med.* **6**:632–644, 1965.
84. LOWENTHAL, W.: *Agar Gel Electrophoresis.* Amsterdam, Elsevier Press, 1965.
85. LOWENTHAL, W., VAN SANDE, M., KARCHER, D.: The differential diagnosis of neurological diseases by fractionating electrophoretically the CSF-globulins. *J. Neurochem.* **6**:51–56, 1960.
86. LUCAS, C. J., BROUWER, R., FELTKAMP, T. E. W., TENVEEN, J. H., VAN LOGHEM, J. J.: Measles antibodies in sera from patients with autoimmune diseases. *Lancet* **1**:115–116, 1972.
87. LUMSDEN, C. E.: Immunogenesis of the MS plaque. *Brain Res.* **28**:365–390, 1971.

88. LUMSDEN, C. E.: The clinical immunology of multiple sclerosis. In MaAlpine, D., Lumsden, C. D., Acheson, E. D., (eds.): *Multiple Sclerosis: A Reappraisal,* 2nd ed. Baltimore, Williams and Wilkins, 1972, pp. 559–568.
89. LUMSDEN, C. E.: The proteins of cerebrospinal fluid—findings in multiple sclerosis. In McAlpine, D., Lumsden, C. E., Acheson, E. D. (eds.): *Multiple Sclerosis: A Reappraisal.* Baltimore, Williams and Wilkins, 1972, pp. 368–432.
90. LUSTER, M. I., ARMEN, R. D., HALLUM, J. V., LESLIE, G. A.: Measles-virus-specific IgD antibodies in patients with subacute sclerosing panencephalitis. *Proc. Natl. Acad. Sci. U.S.A.* **73**:1297–1299, 1976.
91. MANCINI, G., CARBONARA, A. O., HEREMAN, J. F.: Immunochemical quantitation of antigens by single radial immunodiffusion. *Immunochemistry* **2**:235–254, 1965.
92. MANIK, M., KUNKEL, H. G.: Localization of antibodies in group I and group II γ-globulins. *J. Exp. Med.* **118**:817–826, 1963.
93. MASSARO, A. R., AGLIANO, A. M., GRILLO, R.: Immunoglobulin M specific for measles and cerebrospinal fluid of patients with multiple sclerosis and other neurological diseases. *J. Neurol.* **217**:191–194, 1978.
93a. MATTSON, D. H., ROOS, R. P., ARNASON, B. G. W.: Comparison of agar gel electrophoresis (AGE) and isoelectric focusing (IEF) in subacute sclerosing panencephalitis (SSPE), multiple sclerosis (MS), and other demyelinating diseases. *Neurology* **29**:549, 1979.
94. MCFARLIN, D. E., MCFARLAND, H. F.: Histocompatibility studies and multiple sclerosis. *Arch. Neurol.* **33**:395–398, 1976.
95. MEHTA, P. D., KANE, A., THORMAR, H.: Quantitation of measles virus-specific immunoglobulins in serum, CSF, and brain extract from patients with subacute sclerosing panencephalitis. *J. Immunol.* **118**:2254–2261, 1977.
96. MEURMAN, O. H., ARSTILA, P. P., PANELIUS, M., REUNANEN, M. I., VILJANEN, M. K., HALONEN, P. E.: Solid-phase radioimmunoassay detection of rubella virus IgG antibody in serum and CSF of patients with multiple sclerosis. *Acta Pathol. Microbiol. Scand.* **85**:113–116, 1977.
97. MILLER, E. J., OSTERLAND, C. K., DAVIE, J. M., KRAUSE, R. M.: Electrophoretic analysis of polypeptide chains isolated from antibodies in the serum of immunized rabbits. *J. Immunol.* **98**:710–715, 1967.
98. MINGIOLI, E. S., STROBER, W., TOURTELLOTTE, W. W., WHITAKER, J. H., MCFARLIN, D. E.: Quantitations of IgG, IgA, and IgM in the CSF by radioimmunoassay. *Neurology* **28**:991–995, 1978.
99. MUSSINI, J. M., HAUW, J. J., ESCOUROLLE, R.: Immunofluorescence studies of intracytoplasmic immunoglobulin binding lymphoid cells in the central nervous system. *Acta Neuropathol.* **40**:227–232, 1977.
100. NASSENZWERG, V., LAMM, M. E., BENCCERRA, B.: Presence of two types of L polypeptide chains in guinea pig 7 S immunoglobulins. *J. Exp. Med.* **124**:787–803, 1966.
101. NERENBERG, S. T., PRASAD, R.: Radioimmunoassays for Ig classes G, A, M, D, and E in spinal fluids: Normal values of different age groups. *J. Lab. Clin. Med.* **86**:887–898, 1975.
102. NERENBERG, S. T., PRASAD, R., ROTHMAN, M. E.: Cerebrospinal fluid IgG, IgA, IgM, IgD, and IgE levels in central nervous system disorders. *Neurology* **28**:988–990, 1978.
103. NIKOSKELAINEN, E., IRJALA, K. SALMI, A.: Cerebrospinal fluid findings in patients with optic neuritis. *Acta Ophthalmol.* **53**:105–119, 1975.
104. NIKOSKELAINEN, E., NIKOSKELAINEN, J., SALMI, A., HALONEN, P. E.: Virus antibody levels in the cerebrospinal fluid from patients with optic neuritis. *Acta Neurol. Scand.* **51**:347–364, 1975.
105. NILSSON, K., OLSSON, J. E.: Analysis for cerebrospinal fluid proteins by isoelectric focusing on polyacrylamide gel. *Clin. Chem.* **24**:1134–1139, 1978.
106. NORDAL, H. J., VANDVIK, B., NATVIG, J. B.: Idiotypy of measles virus nucleocapsid-specific IgG K antibody in serum and cerebrospinal fluid in subacute sclerosing panencephalitis. *Scand. J. Immunol.* **6**:1351–1356, 1977.
107. NORDAL, H., VANDVIK, B., NORRBY, E.: Demonstration of oligoclonal virus-specific antibodies in serum and cerebrospinal fluid by imprint electroimmunofixation. *Scand. J. Immunol.* (in press).
108. NORDAL, H. J., VANDVIK, B., NORRBY, E.: Multiple sclerosis: Local synthesis of electrophoretically restricted measles, rubella, mumps, and herpes simplex virus antibodies in the central nervous system. *Scand. J. Immunol.* **7**:473–479, 1978.
109. NORRBY, E.: Viral antibodies in MS. *Med. Virol.* **24**:1–39, 1978.
110. NORRBY, E., LINK, H., OLSSON, J. E.: Measles virus antibodies in multiple sclerosis—comparison of antibody titers in cerebrospinal fluid and serum. *Arch. Neurol.* **39**:285–292, 1974.
111. NORRBY, E., LINK, H., OLSSON, J. E., PANELIUS, M., SALMI, A., VANDVIK, B.: Comparison of antibodies against different viruses in cerebrospinal fluid and serum samples from patients with multiple sclerosis. *Infect. Immunol.* **10**:668–694, 1974.
112. NORRBY, E., VANDVIK, B.: Relationship between measles virus–specific antibody activities and oligoclonal IgG in the central nervous system of patients with subacute sclerosing panencephalitis and multiple sclerosis. *Med. Microbiol. Immunol.* **162**:63–72, 1975.
113. OLSSON, J. E., LINK, H.: Immunoglobulin abnormalities in multiple sclerosis. *Arch. Neurol.* **28**:392–399, 1973.
114. OLSSON, J. E., LINK, H., MÜLLER, R.: Immunoglobulin abnormalities in multiple sclerosis. *J. Neurol. Sci.* **27**:233–245, 1976.

114a. OLSSON, J. E., NILSSON, K.: Gamma globulins of CSF and serum in multiple sclerosis: Isoelectric focusing on polyacrylamide gel and agar gel electrophoresis. *Neurology* **29**:1383–1391, 1979.
115. OLSSON, J. E., PETTERSON, B.: A comparison between agar gel electrophoresis and CSF serum quotients of IgG and albumin in neurological diseases. *Acta Neurol. Scand.* **53**:308–322, 1976.
116. PALMER, D. L., MINARD, B. J., CAWLEY, L. P.: IgG subgroups in cerebrospinal fluid in multiple sclerosis. *N. Engl. J. Med.* **294**:447–448, 1976.
117. PANITCH, H. S., HAFLER, D. A., JOHNSON, K. P.: Antibodies to myelin basic protein in multiple sclerosis: Clinical correlations. *Neurology* **28**:394, 1978.
117a. PANITCH, H. S., SWOVELAND, P., JOHNSON, K. P.: Antibodies to measles virus react with myelin basic protein. *Neurology* **29**:548–549, 1979.
118. PATY, D. W., BLUME, W. T., BROWN, W. F., JAATOUL, N., KERTESZ, A., MCINNIS, W.: Chronic progressive myelopathy: Investigation with CSF electrophoresis, evoked potentials, and CAT scan. *Ann. Neurol.* **6**:419–424, 1979.
119. PATY, D. W., DOSSETOR, J. B., STILLER, C. R., COUSIN, H. K., MARCHUK, L., FURESZ, J., BOUCHER, D. W.: HLA in multiple sclerosis: Relationship to measles antibody, mitogen responsiveness and clinical course. *J. Neurol. Sci.* **32**:371–379, 1977.
120. PATY, D. W., FURESZ, J., BOUCHER, D. W., RAND, C. G., STILLER, C. R.: Measles antibodies as related to HL-A types in multiple sclerosis. *Neurology* **26**:651–655, 1976.
121. PERRY, J. J., BRAY, P. F., HACKETT, T. M.: Comparison of electroimmunodiffusion and radial immunodiffusion for measurement of IgG in the laboratory diagnosis of multiple sclerosis. *Clin. Chem.* **20**:1441–1443, 1974.
122. PETRANYI, G., BENIZUN, M., ALFODY, D.: The effect of single large dose hydrocortisone treatment on IgM and IgG production, morphological distribution of antibody producing cells and immunological memory. *Immunology* **21**:157–158, 1971.
123. PHILLIPS, P. E., CHRISTIAN, C. L.: The influence of serum immunoglobulin concentration on measles antibody level. *Proc. Soc. Exp. Biol. Med.* **140**:1340–1343, 1972.
124. PLAYFAIR, J. H. L.: Strain differences in the immune response of mice. *Immunology* **15**:35–50, 1968.
125. PORTER, K. G., SINNAMON, D. G., GILLIES, R. R.: *Cryptococcus neoformans*-specific oligoclonal immunoglobulins in cerebrospinal fluid in cryptococcal meningitis. *Lancet* **1**:1262, 1977.
126. PRINEAS, J. W., RAINE, C. S.: Electron microscopy and immunoperoxidase studies of early multiple sclerosis lesions. *Neurology* **16**(Part 2):29–32, 1976.
127. PRINEAS, J. W., WRIGHT, R. G.: Macrophages, lymphocytes, and plasma cells in the perivascular compartment in chronic multiple sclerosis. *Lab. Invest.* **38**:409–421, 1978.
128. REUNANEN, M., ARSTILA, P., HAKKARAINEN, H., NIKOSKELAINEN, J., SALMI, A., PANELIUS, M.: A longitudinal study on antibodies to measles and rubella viruses in patients with multiple sclerosis. *Acta Neurol. Scand.* **54**:366–370, 1976.
129. RIBERI, M., BERNHARD, D., DEPIEDS, R.: Evidence for the presence of γ chain dimers in cerebrospinal fluid of patients suffering from subacute sclerosing panencephalitis. *Clin. Exp. Immunol.* **19**:45–53, 1975.
130. ROSE, A. S., ELLISON, G. W., MYERS, L. W., TOURTELLOTTE, W. W.: Criteria for the clinical diagnosis of multiple sclerosis. *Neurology* **26**(Part 2):20–22, 1976.
131. SALMI, A., NORRBY, E., PANELIUS, M.: Identification of different measles virus-specific antibodies in the serum and cerebrospinal fluid from patients with subacute sclerosing panencephalitis and multiple sclerosis. *Infect. Immun.* **6**:248–254, 1972.
132. SALMI, A., PANELIUS, M., VAINIONPAA, R.: Antibodies against different viral antigens in cerebrospinal fluid of patients with multiple sclerosis and other neurological diseases. *Acta Neurol. Scand.* **50**:183–193, 1974.
133. SANDBERG, M., BYNKE, H.: Cerebrospinal fluid in 25 cases of optic neuritis. *Acta Neurol. Scand.* **49**:443–452, 1973.
134. SANDBERG-WOLLHEIM, M.: Immunoglobulin synthesis *in vitro* by cerebrospinal fluid cells in patients with multiple sclerosis. *Scand. J. Immunol.* **3**:717–730, 1974.
135. SANDBERG-WOLLHEIM, M.: Optic neuritis: Studies on the cerebrospinal fluid in relation to clinical course in 61 patients. *Acta Neurol. Scand.* **52**:167–178, 1975.
136. SANDBERG-WOLLHEIM, M., PLATZ, P., RYDER, L. P., NIELSEN, L. S., THOMSEN, M.: HL-A histocompatibility antigens in optic neuritis. *Acta Neurol. Scand.* **53**:161–166, 1975.
137. SANDBERG-WOLLHEIM, M., ZETTERVALL, O., MÜLLER, R.: *In vitro* synthesis of IgG by cells from the cerebrospinal fluid in a patient with multiple sclerosis. *Clin. Exp. Immunol.* **4**:401–405, 1969.
138. SCHMIDT, R., RIEDER, H. P., WÜTHRICH, R. P.: The course of multiple sclerosis cases with extremely high gamma-globulin values in the cerebrospinal fluid. *Eur. Neurol.* **15**:241–248, 1977.
139. SCHNECK, S. A., CLAMAN, H. N.: CSF immunoglobulins in multiple sclerosis and other neurologic diseases: measurement by electroimmunodiffusion. *Arch. Neurol.* **20**:132–139, 1969.
140. SEIL, F. J.: Tissue culture studies of demyelinating disease: A critical review. *Ann. Neurol.* **2**:345–355, 1977.
140a. SIDEN, A.: Isoelectric focusing and crossed immunoelectrofocusing of CSF immunoglobulins in MS. *J. Neurol.* **221**:39–51, 1979.
141. SIMPSON, J. F., TOURTELLOTTE, W. W, KOKMEN, E., PARKER, J. A., ITABASHI, H. H.: Fluorescent protein tracing in multiple sclerosis brain tissue. *Arch. Neurol.* **20**:373–377, 1969.
142. SKOLDENBERG, B., CARLSTROM, A., FORSGREN, M.,

NORRBY, E.: Transient appearance of oligoclonal immunoglobulins and measles virus antibodies in the cerebrospinal fluid in a case of acute measles encephalitis. *Clin. Exp. Immunol.* **23**:451–455, 1976.

143. STENDAHL, L., LINK, H., MOLLER, E., NORRBY, E.: Relation between genetic markers and oligoclonal IgG in CSF in optic neuritis. *J. Neurol. Sci.* **27**:93–98, 1976.
144. SYMINGTON, G. R., MACKAY, I. R., WHITTINGHAM, S., WHITE, J., BUCKLEY, J. D.: A "profile" of immune responsiveness in multiple sclerosis. *Clin. Exp. Immunol.* **31**:141–149, 1978.
145. TABIRA, T., WEBSTER, H. DE F., WRAY, S. H.: *In vivo* test for myelinotoxicity of cerebrospinal fluid. *Brain Res.* **120**:103–112, 1977.
146. TABIRA, T., WEBSTER, H. DE F., WRAY, S. H.: Multiple sclerosis cerebrospinal fluid produces myelin lesions in tadpole optic nerves. *N. Engl. J. Med.* **295**:644–649, 1976.
147. TAKUGUCHI, T., ADLER, W. H., SMITH, R. T.: Strain specificity of monodisperse gammaglobulin appearance after immunization of inbred mice. *Proc. Soc. Exp. Biol. Med.* **145**:868–873, 1974.
148. TANAKA, R., IWASAKI, Y., KOPROWSKI, H.: Ultrastructural studies of perivascular cuffing cells in multiple sclerosis brain. *Am. J. Pathol.* **81**:467–478, 1975.
149. TAVOLATO, B. G.: Immunoglobulin-G distribution in multiple sclerosis brain: An immunofluorescence study. *J. Neurol. Sci.* **24**:1–11, 1975.
150. THORSBY, E., HELGESEN, A., SOLHEIM, B. G., VANDRICK, B.: HLA antigens in multiple sclerosis. *J. Neurol. Sci.* **32**:187–193, 1977.
151. TIBBLING, G., LINK, H., OHMAN, S.: Principles of albumin and IgG analysis in neurological disorders. I. Establishment of reference values. *Scand. J. Clin. Lab. Invest.* **37**:385–390, 1977.
152. TOURTELLOTTE, W. W.: Cerebrospinal fluid immunoglobulins and the central nervous system as an immunological organ particularly in multiple sclerosis and subacute sclerosing panencephalitis. *Res. Publ. Assoc. Res. Nerv. Ment. Dis.* **49**:112–115, 1971.
153. TOURTELLOTTE, W. W.: Cerebrospinal fluid in multiple sclerosis. In Vinken, P. J., Bruyn, G. W. (eds.): *Handbook of Clinical Neurology*. Amsterdam, North-Holland, 1970, pp. 324–382.
154. TOURTELLOTTE, W. W.: Interaction of local central nervous system immunity and systemic immunity in the spread of multiple sclerosis demyelination. In Ellison, G. W., Stevens, J. G., Andrews, J. M., (eds.): *Multiple Sclerosis—Immunology, Virology, and Ultrastructure*. New York, Academic Press, 1972, pp. 385–332.
155. TOURTELLOTTE, W. W.: On cerebrospinal fluid immunoglobulin—quotients in multiple sclerosis and other diseases: A review and a new formula to estimate the amount of IgG synthesized per day by the central nervous system. *J. Neurol. Sci.* **10**:279–304, 1970.
156. TOURTELLOTTE, W. W.: What is multiple sclerosis? Laboratory criteria for diagnosis. In Davison, A. N., Humphrey, J. H., Liversedge, A. L. (eds.): *Multiple Sclerosis Research*. New York, Elsevier, 1975, pp. 9–26.
157. TOURTELLOTTE, W. W., ITABASHI, H. H., PARKER, J. A.: Multifocal areas of synthesis of immunoglobulin-G in multiple sclerosis brain tissue and the sink action of the cerebrospinal fluid. *Trans. Am. Neurol. Assoc.* **92**:288–290, 1967.
158. TOURTELLOTTE, W. W., MA, B. I.: Multiple sclerosis, the blood brain barrier and the measurement of *de novo* central nervous system IgG synthesis. *Neurology* **28**(Part 2):78–83, 1978.
159. TOURTELLOTTE, W. W., MURTHY, K., BRANDES, D. W., SAJBEN, B., COMISO, P., POTVIN, A., COSTANZA, A., KORETLITZ, J.: Schemes to eradicate the multiple sclerosis central nervous system immune reaction. *Neurology* **26**:59–61, 1976.
160. TOURTELLOTTE, W. W., PARKER, J. A., HERNDON, R. M., CUADROS, C. V.: Subacute sclerosing panencephalitis, brain immunoglobulin G, measles antibody, and albumin. *Neurology* **18**(Part 2):117–121, 1968.
161. TOURTELLOTTE, W. W., TAVALATO, B., PARKER, J. A., COMISO, P.: Cerebrospinal fluid electroimmunodiffusion. *Arch. Neurol.* **25**:345–350, 1971.
162. TROTTER, J. L., BANKS, G., WANG, P.: Isoelectric focusing of gamma globulins in cerebrospinal fluid from patients with multiple sclerosis. *Clin. Chem.* **23**:2213–2215, 1977.
163. TROTTER, J. L., GARVEY, W. F.: Prolonged decrease in CSF IgG synthesis after pulse methylprednisolone therapy. *Neurology* (in press).
164. VAN CAMP, B. G. K., COLE, J., PEETERMAN, M. E.: HLA antigens and homogeneous immunoglobulins. *Clin. Immunol. Immunopathol.* **7**:315–318, 1977.
165. VANDVIK, B.: Oligoclonal IgG and free light chains in the cerebrospinal fluid of patients with multiple sclerosis and infectious diseases of the central nervous system. *Scand. J. Immunol.* **6**:913–922, 1977.
166. VANDVIK, B.: Oligoclonal measles virus-specific IgG antibodies isolated from sera of patients with subacute sclerosing panencephalitis. *Scand. J. Immunol.* **6**:641–649, 1977.
167. VANDVIK, B., DEGRE, M.: Measles virus antibodies in serum and cerebrospinal fluid in patients with multiple sclerosis and other neurological disorders, with special reference to measles antibody synthesis within the central nervous system. *J. Neurol. Sci.* **24**:201–219, 1975.
168. VANDVIK, B., NATVIG, J. B., WIGER, D.: IgG_1 subclass restriction of oligoclonal IgG from cerebrospinal fluids and brain extracts in patients with multiple sclerosis and subacute encephalitides. *Scand. J. Immunol.* **5**:427–436, 1976.
169. VANDVIK, B., NORRBY, E.: Oligoclonal IgG antibody response in the central nervous system to different measles virus antigens in subacute sclerosing pan-

encephalitis. *Proc. Natl. Acad. Sci. U.S.A.* **70**:1060–1063, 1973.

170. VANDVIK, B., NORRBY, E., NORDAL, J., DEGRE, M.: Oligoclonal measles virus–specific IgG antibodies isolated from cerebrospinal fluids, brain extracts, and sera from patients with subacute sclerosing panencephalitis and multiple sclerosis. *Scand. J. Immunol.* **5**:979–992, 1976.
171. VANDVIK, B., SKREDE, S.: Electrophoretic examination of cerebrospinal fluid proteins in multiple sclerosis and other neurological diseases. *Eur. Neurol.* **9**:224–241, 1973.
172. WALDMANN, T. A., STROBER, W.: Metabolism of immunoglobulins. *Prog. Allergy* **13**:1–110, 1969.
173. WHITAKER, J. N., HERRMANN, K. L., ROGENTINE, G. N. STEIN, S. F., KILLINS, L. L.: Immunogenetic analysis and serum viral antibody titers in multiple sclerosis. *Arch. Neurol.* **33**:399–403, 1976.
174. WILLIAMS, A. C., MINGIOLI, E. S., MCFARLAND, H. F., TOURTELLOTTE, W. W., MCFARLIN, D. E.: Increased CSF IgM in multiple sclerosis. *Neurology* **28**:996–998, 1978.
175. WOLINSKY, J. S., BERG, B. O., MAITLAND, C. J.: Progressive rubella panencephalitis. *Arch. Neurol.* **33**:722–723, 1976.
176. YAHR, M. D., GOLDENSOHN, S. S., KABAT, E. A.: Further studies on the gamma globulin content of cerebrospinal fluid in multiple sclerosis and other neurological diseases. *Ann. N.Y. Acad. Sci.* **58**:613–624, 1954.
177. ZAWADZKI, Z. A., EDWARDS, G. A.: Nonmyelomatous monoclonal immunoglobulinemia. In Schwartz, R. S. (ed.): *Progress in Clinical Immunology.* New York, Grune and Stratton, 1972, p. 105.
178. ZETTERVALL, O., LINK, H.: Electrophoretic distribution of kappa and lambda immunoglobulin light chain determinants in serum and cerebrospinal fluid in multiple sclerosis. *Clin. Exp. Immunol.* **7**:365–372, 1970.

CHAPTER 34

Myelin Basic Protein in Cerebrospinal Fluid

Index of Active Demyelination

Steven R. Cohen, Benjamin Rix Brooks,
Burk Jubelt, Robert M. Herndon, and Guy M. McKhann

1. Introduction

Medical diagnosis is greatly aided by laboratory tests that indicate damage or malfunction in specific tissues. Since most tissues have direct access to the blood, changes in the physiological or pathological state of specific organs are often reflected by changes in blood composition or the appearance of specific tissue markers. For example, specific enzymes are elevated in the serum after damage to heart, liver, or muscle. Development of diagnostic tests for disorders of the nervous system, however, has been hindered by lack of specific neural markers. Moreover, the central nervous system (CNS) usually lacks direct exchange with blood because of the blood–brain barrier. The fluid compartment in direct contact with the CNS is the cerebrospinal fluid (CSF), and one might expect that pathological changes in the CNS would be reflected in the CSF.

Among the most common disorders of the nervous system are the demyelinating disorders, in particular multiple sclerosis. Since the demyelination often occurs acutely, it is reasonable to expect that products of this demyelination may be released into the CSF during the active phases of the disease. Indeed, in 1970, Herndon and Johnson,[9] using electron microscopy, identified myelin fragments in the 100,000 *g* sediment from the CSF of two patients with multiple sclerosis. Several laboratories then began to search for myelin components in the CSF of multiple sclerosis patients.

Myelin contains about 70% lipid and 30% protein.[12] Although the lipids, especially sulfatide, are characteristic of myelin, they are not specific to the myelin sheath. The major myelin proteins, however, do appear to be specific to myelin. The most highly characterized of the myelin proteins is the basic protein.[3] Strongly basic, as its name implies, the protein has a pI of 12 and a molecular weight of 18,600, and the complete amino acid sequence is known for several species. It is readily purified and its antigenic properties are known.[10] It is this protein that is the antigen responsible for the induction of experimental allergic encephalomyelitis when whole brain, myelin, or purified basic protein is injected into susceptible species. For these reasons, we developed a radioimmunoassay for myelin basic protein with a view to determining whether this specific protein appeared in the CSF of multiple sclerosis patients undergoing an acute exacerbation.

Our original aim was to provide a means for objective evaluation of demyelinating activity in multiple sclerosis. With the radioimmunoassay,[5] we were able to show that the basic protein was indeed detectable in the CSF of multiple sclerosis patients undergoing exacerbation.[6,7] Moreover, the levels of myelin basis protein correlated with the clinical

Steven R. Cohen, Ph.D., Benjamin Rix Brooks, M.D., Burk Jubelt, M.D., and **Guy M. McKhann, M.D.** • Department of Neurology, The Johns Hopkins University School of Medicine, Baltimore, Maryland 21205. **Robert M. Herndon, M.D.** • Center for Brain Research and Department of Neurology, University of Rochester School of Medicine and Dentistry, Rochester, New York 14642.

course of multiple sclerosis. In contrast, we could not detect any myelin basic protein in CSF from inactive multiple sclerosis patients or patients with nondemyelinative neurological disease. Thus, radioimmunoassay of myelin basic protein in CSF is a useful adjunct to the diagnosis and management of multiple sclerosis. In this chapter, based on over 800 CSF samples from patients with a wide variety of neurological diseases, we show that the usefulness of this test has clearly been extended beyond the original aim of evaluating activity in multiple sclerosis. The test is helpful in the clinical assessment of many neurological diseases in which myelin breaks down acutely. These include leukodystrophies, severe anoxia, and the myelopathies and encephalopathies due to radiation or chemotherapy.

2. Materials and Methods

2.1. Patients

All patients except some of those with optic neuritis were from Johns Hopkins Hospital. Approximately half the optic neuritis CSF samples came from Dr. Shirley Wray of the Massachusetts General Hospital.

We have divided our patients into the following categories based on clinical symptoms:

1. Classic multiple sclerosis: At least two attacks occurring in different parts of the nervous system more than 1 month apart and not explainable on the basis of other disease processes. Age between and 10 and 50 years.
   a. Active: Within 1 week from the onset of new neurological symptoms.
   b. Inactive (remission): More than 2 weeks from the onset of any new neurological symptoms or change in existing symptoms.
2. Chronic multiple sclerosis: A progressive disease of more than 6 months' duration affecting more than one area of the CNS (this may occur in the absence or presence of previous exacerbations and remissions). For purposes of this study, typical remitting disease that has become progressive is included in this category.
3. Optic neuritis: Without evidence of other nervous system involvement.
4. Myelinopathy: Diseases other than multiple sclerosis that affect myelin.
5. Nondemyelinating neurological disease.

Patients with suspected disease have not been included in this study. However, we are following these patients to determine the outcome of their disease and its relationship to CSF basic protein values.

2.2. Assay of CSF

All samples are obtained by lumbar punctures and are stored at $-10°C$ until assay. The CSF basic protein appears to be stable, since identical values for this protein were obtained on a sample before and after incubation for 1 week at room temperature. However, because of the possibility of elevated protease activity in an occasional sample, we suggest freezing the sample until assay.

For the assay, 0.05 ml of a 10-fold concentrated assay buffer (2 M tris-acetate, pH 7.5, containing 10 mg histone/ml) and antiserum at the appropriate concentration were added directly to 0.5 ml CSF. This mixture was incubated for 1 hr at 37°C, 15,000 cpm of ^{125}I-labeled basic protein (specific activity, 10–20 μCi/μg) was added, and the mixture was incubated for an additional 10–24 hr at 4°C. The antibody–basic protein complex was then precipitated with cold ethanol, the pellet and supernatant fraction were separated by centrifugation, and each was assayed for radioactivity. The percentage of [^{125}I]basic protein bound (i.e., in the pellet) was then determined.[5] Results are reported as either negative (<4 ng/ml), weakly positive (5–8 ng/ml), or positive (>9 ng/ml).

These studies were done using two similar batches of basic protein antisera. Our experience has demonstrated that not all antisera react equally well with the CSF basic protein. However, in studies with different sera, elevated basic protein is consistently found in CSF from patients with active demyelinating diseases.

3. Results of Clinical Investigations

Samples have been assayed from 846 patients with known neurological disease (Fig. 1). These include 187 samples from multiple sclerosis patients, 11 from patients with other active demyelinating diseases, 626 from patients with nondemyelinating neurological diseases, and 22 with optic neuritis.

3.1. Multiple Sclerosis Patients

Of 60 multiple sclerosis patients in acute exacerbation, 56 had CSF basic protein levels greater than 8 ng; the remaining 4 had levels between 5 and 8 ng/ml. The lumbar punctures on these 4 patients

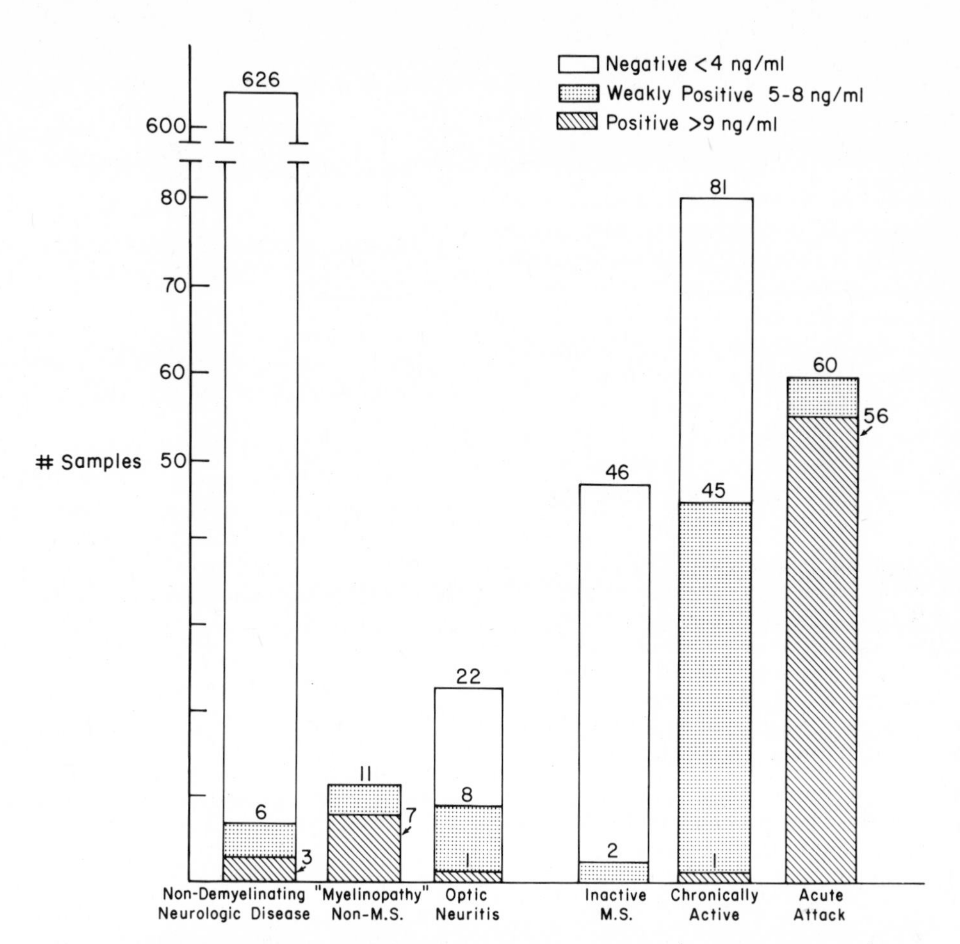

Figure 1. Presence of myelin basic protein in CSF. Nondemyelinating neurological diseases with no detectable CSF myelin basic protein included at least one sample from each of the following: arteriovenous malformation, seizures, neuroblastoma, arthrogryposis, temporal arteritis, cerebellar degeneration, microcephaly, peripheral neuropathy, presenile dementia, subacute sclerosing panencephalitis, stroke, subarachnoid meningitis, hydrocephalus, migraine, labyrinthitis, Moya-Moya syndrome, vascular headache, progressive supranuclear palsy, lacunar infarct, Guillain-Barré, spastic paraplegia, pseudotumor cerebri, senile dementia, progressive multifocal leukoencephalopathy, meningoencephalitis, sixth nerve palsy, fifth nerve palsy, trigeminal anesthesia, vasculitis, postanoxic encephalopathy, third nerve sarcoid, progressive spastic paraparesis, progressive external ophthalmoplegia, striatonigral degeneration, cervical degenerative arthritis and radiculopathy, polymyositis, trauma, motor neuron disease, Meniere's disease, vertebral basilar ischemia, herpes simplex encephalomyelitis, neurosyphilis. Non-multiple-sclerosis patients with positive CSF basic protein are listed in Table 1.

were obtained between 5 and 15 days after the onset of the acute exacerbation. Of 81 patients with the slowly progressive form of the disease, 45 had basic protein levels between 5 and 8 ng/ml. All but 2 of the 46 patients with inactive disease had levels below 4 ng/ml (Fig. 1). These 2 patients were recovering from acute exacerbations that occurred 14–21 days prior to the lumbar puncture. In addition, 5 of the patients in this group presented with the first clinical evidence of multiple sclerosis, and their CSF revealed elevated myelin basic protein. These patients all had second acute attacks, confirming the diagnosis of multiple sclerosis.

Serial samples were obtained from patients before, during, and after attacks (Figs. 2–4). The CSF basic protein levels rose and fell with the exacerbation. In one patient (Fig. 2) who subsequently had a very severe attack, the CSF basic protein was

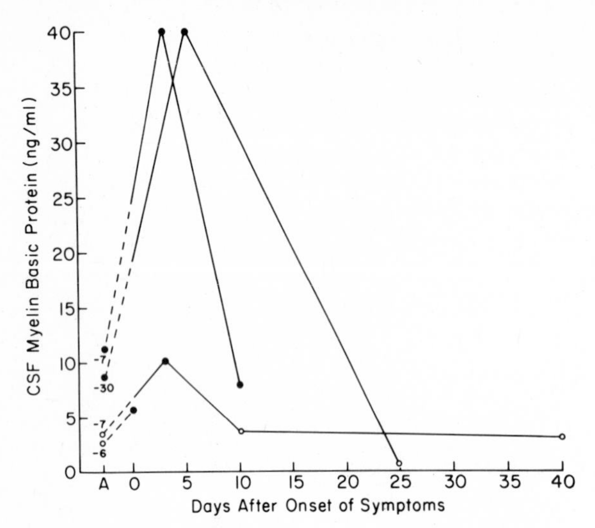

Figure 2. Myelin basic protein levels in multiple sclerosis patients before, during and after acute exacerbations. This figure represents CSF myelin basic protein determination from four patients, prior to acute attacks of multiple sclerosis. Samples were obtained 6 days before attack for one patient, 7 days prior for two patients, and 30 days before acute attack for fourth patient. (●) Positive myelin basic protein value; (○) negative myelin basic protein value.

slightly elevated (11 ng/ml) 7 days before the actual clinical attack. At this time, the patient was showing chronic progression of the disease. At the height of the exacerbation, the CSF basic protein value was 40 ng/ml. In a second patient in whom a sample was obtained 7 days prior to a multiple sclerosis attack, the CSF basic protein was negative, and the value rose to 10 ng/ml at the height of the attack. Those with significant CSF basic protein 7–16 days after the exacerbation were still demonstrating some progression of their disease (Fig. 3, patients A and D), while those whose basic protein had returned to normal levels were recovering from the exacerbation (Fig. 3, patients C and G). Thus, as the patient improved, the CSF myelin basic protein returned to normal levels. Those patients who went from acute attacks to slow progression usually showed low, but significant, levels of CSF basic protein months after the attack (Fig. 3, patients H, I, J, and K). Thus, of the slowly progressive group of patients, those with CSF myelin basic protein (approximately half) are frequently the ones with exacerbations superimposed on their chronic progression.

Generally, the CSF basic protein levels decrease rapidly after an acute attack (Fig. 4). Within 2 weeks following an exacerbation, most patients have nor-

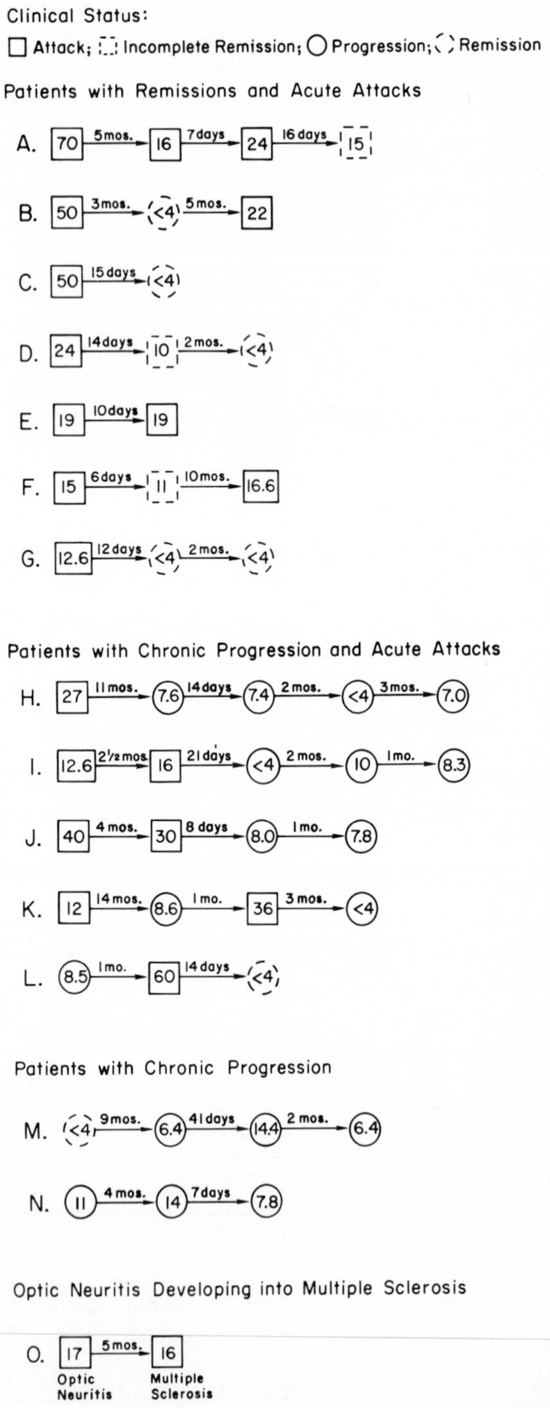

Clinical Status:

□ Attack; ⬚ Incomplete Remission; ○ Progression; ◌ Remission

Patients with Remissions and Acute Attacks

A. 70 —5 mos.→ 16 —7 days→ 24 —16 days→ 15

B. 50 —3 mos.→ <4 —5 mos.→ 22

C. 50 —15 days→ <4

D. 24 —14 days→ 10 —2 mos.→ <4

E. 19 —10 days→ 19

F. 15 —6 days→ 11 —10 mos.→ 16.6

G. 12.6 —12 days→ <4 —2 mos.→ <4

Patients with Chronic Progression and Acute Attacks

H. 27 —11 mos.→ 7.6 —14 days→ 7.4 —2 mos.→ <4 —3 mos.→ 7.0

I. 12.6 —2½ mos.→ 16 —21 days→ <4 —2 mos.→ 10 —1 mo.→ 8.3

J. 40 —4 mos.→ 30 —8 days→ 8.0 —1 mo.→ 7.8

K. 12 —14 mos.→ 8.6 —1 mo.→ 36 —3 mos.→ <4

L. 8.5 —1 mo.→ 60 —14 days→ <4

Patients with Chronic Progression

M. <4 —9 mos.→ 6.4 —41 days→ 14.4 —2 mos.→ 6.4

N. 11 —4 mos.→ 14 —7 days→ 7.8

Optic Neuritis Developing into Multiple Sclerosis

O. 17 —5 mos.→ 16

Optic Neuritis — Multiple Sclerosis

Figure 3. Relationship of myelin basis protein to clinical course of multiple sclerosis. Each symbol represents clinical status of patient at time of lumbar puncture. Number in each symbol is CSF myelin basic protein value (ng/ml).

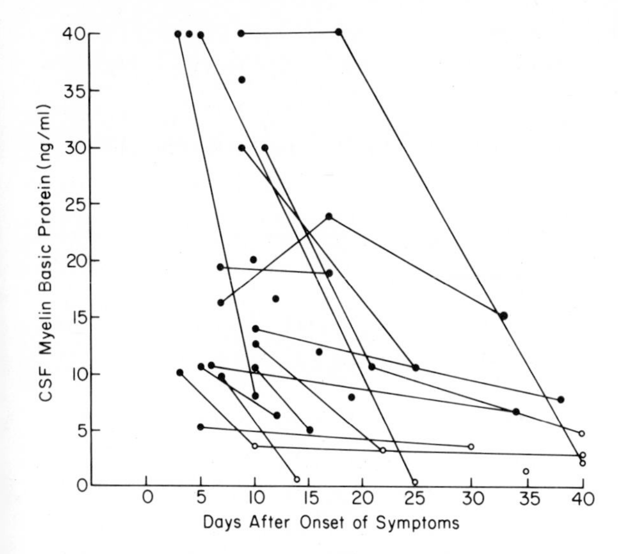

Figure 4. Decrease in CSF myelin basic protein following acute attacks. CSF samples were obtained from multiple sclerosis patients at various times after onset of acute exacerbation, and samples from same patient are connected by solid lines. (●) Positive myelin basic protein value; (○) negative myelin basic protein value.

mal to low values of CSF basic protein. In a few patients, usually those with continually exacerbating symptoms, the levels remain elevated for as long as 20–35 days after the onset of symptoms.

Of 22 patients with optic neuritis, 9 have basic protein in the CSF. All these patients are being followed to determine the outcome of this first attack. One of these patients has subsequently developed multiple sclerosis (Fig. 3, patient O).

3.2. Other Diseases of Myelin

The category of "myelinopathies" included patients with systemic lupus erythematosus involving the CNS, metachromatic leukodystrophy, central pontine myelinolysis, adrenal leukodystrophy, an undescribed hereditary leukodystrophy, methotrexate myelopathy, and Pelizaeus–Merzbacher disease. These patients also had elevated CSF basic protein (Table 1). The additional patients with Pelizaeus–Merzbacher disease had no detectable CSF basic protein.

The patient with methotrexate myelopathy received a bone marrow transplant, whole-body irradiation, and prolonged treatment with intrathecal methotrexate. He began to show photophobia, and 2 weeks later developed a transverse myelitis. The patient died 2 weeks after this, and the spinal cord pathology revealed severe diffuse microvacuolization of the long tracts of the spinal cord. During this time, he also had elevated CSF basic protein. We are currently completing a study in conjunction with the Pediatric Oncology Branch of the National Cancer Institute on patients who have received injections of intrathecal methotrexate. Of 100 patients, 4 had clinically detectable encephalopathies, and these 4 also had elevated CSF myelin basic protein.

3.3. Controls

All but 6 of the 626 controls had less than 4 ng basic protein/ml CSF. These 6 patients are listed in Table 1. They include two severe strokes on the surface of the brain, Wernicke's disease, a microglioma, encephalitis, and severe anoxia following surgery. With respect to this last patient, it is interesting that Kohlschütter[11] has recently reported the application of the basic protein radioimmunoassay to the CSF from children. He found that 8 of 41 CSF samples

Table 1. Levels of CSF Myelin Basic Protein in Patients without Multiple Sclerosis

Diseases	Myelin basic protein (ng/ml)
Nondemyelinative neurological disease	
Lateral medullary infarction	60
Cerebellar infarction	56
Wernicke's disease	18
Anoxic encephalopathy	8
Microglioma	8
Encephalitis	13
Myelinopathies other than multiple sclerosis	
Leukodystrophies	
Hereditary leukodystrophy	23
Metachromatic leukodystrophy	12
Adrenal leukodystrophy, patient 1	8
Adrenal leukodystrophy, patient 2	8
Pelizaeus–Merzbacher, patient 1	8
Pelizaeus–Merzbacher, patient 2	8
Other demyelinating disorders	
Transverse myelitis and systemic lupus erythematosus	100
Systemic lupus erythematosus with CNS involvement	20
Systemic lupus erythematosus	24
Central pontine myelinolysis	50
Methotrexate myelopathy	17

were positive for myelin basic protein. These 8 samples were from 6 children with severe hypoxia and 1 case of encephalitis with cardiac arrest. In 5 of 6 cases, the brain damage led to death. Thus, this test may be useful in assessing brain tissue destruction in patients experiencing hypoxia.

4. Discussion

These studies indicate that myelin basic protein or fragments thereof are released into the CSF as part of the myelin-breakdown process, thus confirming and extending our previously reported results.[7] Myelin basic protein is present not only in the CSF of patients with active multiple sclerosis, but also in that of patients with other active demyelinating conditions such as metachromatic leukodystrophy and central pontine myelinolysis. The few positive tests in other neurological conditions clearly indicate that myelin-breakdown products may also be released into the CSF as a result of damage by different processes such as hypoxia and necrosis stemming from radiation or chemotherapy.

Similar results have been obtained in the laboratories of Whitaker and co-workers[14,15] and Trotter *et al.*[13] on patients undergoing acute demyelinative episodes, while Kohlschütter[11] has used a CSF basic protein assay to assess brain tissue destruction in hypoxic children. Carson *et al.*[4] have recently reported the presence of a large component (mol. wt. 50,000 daltons) in the CSF of patients with multiple sclerosis that cross-reacts with antibody to myelin basic protein.

In a given pathological condition in which myelin breaks down, several factors may determine how much myelin basic protein will be released and whether it will appear in the lumbar CSF. These include (1) location of myelin breakdown and direction of CSF flow and (2) the nature of the pathological process.

The location of myelin breakdown is important, since material released from superficial lesions has good access to CSF pathways. Thus, the predilection of multiple sclerosis plaques for the periventricular white matter and the superficial white matter of the brainstem and spinal cord would be likely to result in the release of substantial amounts of basic protein into the CSF. Much of this myelin basic protein would then appear in the lumbar CSF.

On the other hand, in optic neuritis, where the lesion is in the optic nerve or chiasm, myelin basic protein might not appear in the lumbar region, since the CSF bathing the chiasm would tend to flow up over the hemispheres to be absorbed by the arachnoid granulations. In this situation, the myelin basic protein could reach the lumbar sac only through mixing and back-diffusion against the main direction of CSF flow. This may account for the low percentage of positive results for CNS basic protein seen in the optic neuritis cases, even though the lesion is fairly superficial and substantial myelin basic protein may be released into the CSF. CSF basic protein is rarely elevated after cerebrovascular thrombosis, probably because in most cerebrovascular occlusions the area of necrosis is fairly deep, and there is a considerable thickness of surviving tissue between the area of necrosis and the ventricular system. If the infarct is in the cortex, there is a substantial amount of gray matter, which contains very little myelin, superficial to the infarcts. Only with infarcts of the brainstem or the rare instance in spinal cord would superficial myelin destruction be likely to cause release of myelin basic protein that would reach the lumbar CSF; indeed, one of our seven positive results in nondemyelinating disorders is a case of lateral medullary infarction.

The nature of the pathological process may also determine the properties of the basic protein that appears in CSF during an attack of multiple sclerosis. The CSF basic protein may be present as intact protein or peptide fragments. It may be whole myelin, myelin fragments, or cells with ingested myelin. Alternatively, this CSF basic protein may be in a lipid, protein, or nucleic acid complex. For example, the finding of peptide fragments in the CSF would suggest extracellular enzymatic degradation as the mechanism of demyelination, whereas the presence of whole myelin would indicate that the entire sheath was being removed by cellular attack. In this regard, it is interesting that Cammer *et al.*[2] have reported the degradation of myelin basic protein by neutral proteases secreted by stimulated macrophages. Whitaker and co-workers[1,14] have reported the presence of peptide fragments in the CSF of multiple sclerosis patients undergoing exacerbation. Our own data (unpublished) indicate that antigenic sites spread over the entire basic protein molecule are present in the CSF basic protein. However, it remains to be determined whether these antigenic sites represent peptide fragments or intact myelin basic protein. In the experimental demyelinating disease allergic encephalomyelitis, which is clearly autoimmune, the basic protein released into the CSF is bound to its antibody.[8] The presence of antibody-bound basic protein has not been demonstrated in multiple sclerosis, although very low levels of antibody may be present. We are continuing our investigation of the properties of the CSF basic protein in multiple sclerosis.

It is important to distinguish between a test that is diagnostic and one that measures activity of the disease. CSF myelin basic protein levels rise and fall with the exacerbations and remissions that are typical of multiple sclerosis. This is consistently observed, although in individual patients there are considerable variations in the amount of CSF basic protein at the height of the attack, probably reflecting the amount of tissue undergoing demyelination. In patients with inactive disease, no CSF basic protein is found, and thus the radioimmunoassay for myelin basic protein should not be regarded as a definitive diagnostic test for multiple sclerosis. However, it is clearly a useful indicator of active myelin degradation and as such, when used in conjunction with the clinical data, is a very useful adjunct in the diagnosis and management of multiple sclerosis. In using the test for diagnostic purposes, it should be noted that CSF basic protein levels return to normal rapidly with recovery from acute attacks. This must be taken into account in interpreting the data clinically. Currently, the test is being used in a prospective fashion to assess the efficacy of steroid treatment. We anticipate that the test will be used in a similar fashion as newer methods of therapy for multiple sclerosis become available.

There are other situations in which the physician wishes to know whether a patient is experiencing an acute demyelinating process. Thus, in the case of leukodystrophies, anoxia, and encephalopathy or myelopathy due to cancer, radiation, or chemotherapy, the presence of myelin basic protein in the CSF may be an indicator that myelin breakdown is actually occurring. Radioimmunoassay of myelin basic protein is the first test to indicate damage to a specific component of neural tissue, myelin.

As knowledge of neurochemistry increases, one can visualize specific tests for each of the cellular components of the nervous system. With immunochemical methods, it may eventually be possible to identify breakdown products of highly specific cell types and transmitter systems in the CSF.

5. Summary

Specific laboratory tests for neurological diseases are not available at present for routine use. We have developed and used such a test for measurement of a specific neurological component, the basic protein of myelin, in the CSF. The test can be used for objective evaluation of disease activity in multiple sclerosis patients, since the levels of basic protein in the CSF correlate precisely with the clinical activity of the disease. Moreover, it is a useful adjunct to diagnosis and potential therapy. In addition, the test may be helpful in the diagnosis of many other diseases in which there is an acute breakdown of myelin, including severe anoxia, necrosis due to radiation or chemotherapy, and the leukodystrophies.

ACKNOWLEDGMENTS. The authors are grateful to Dr. Pamela Talalay for assistance in preparation of the manuscript. This research was supported by grants (1052-A-2, 1052-B-3, and 1052-C-5) from the National Multiple Sclerosis Society and by grants (NS10920 and NS14167) from the United States Public Health Service. Dr. Cohen is the recipient of a Research Career Development Award (NS00315) from the United States Public Health Service. Dr. Brooks is a recipient of Teacher-Investigator Award 1 K07 NS00385-01 from the United States Public Health Service.

References

1. BASHIR, R. M., WHITAKER, J. N.: Molecular features of myelin basic protein fragments in cerebrospinal fluid of persons with multiple sclerosis. *Ann. Neurol.* **4**:175, 1978.
2. CAMMER, W., BLOOM, B. R., NORTON, W. T., CTORDON, S.: Degradation of basic protein in myelin by neutral proteases secreted by stimulated macrophages: A possible mechanism of inflammatory demyelination. *Proc. Natl. Acad. Sci. U.S.A.* **75**:1554–1558, 1978.
3. CARNEGIE, P. R., DUNKLEY, P. R.: Basic protein of central and peripheral nervous system myelin. In Agranoff, B. W., Aprison, M. H. (eds.): *Advances in Neurochemistry, Vol. 1.* New York, Plenum Press, 1975, pp. 95–135.
4. CARSON, J. H., BARBARESE, E., BRAUN, P. E., MCPHERSON, T. A.: Components in multiple sclerosis cerebrospinal fluid that are detected by radio-immunoassay for myelin basic protein. *Proc. Natl. Acad. Sci. U.S.A.* **75**:1976–1978, 1978.
5. COHEN, S. R., MCKHANN, G. M., GUARNIERI, M: A radioimmunoassay for myelin basic protein and its use for quantitative measurements. *J. Neurochem.* **25**:371–376, 1975.
6. COHEN, S. R., BRUNE, M. J., HERNDON, R. M., MCKHANN, G. M.: Cerebrospinal fluid myelin basic

protein and multiple sclerosis. In Palo, S. (ed.): *Myelination and Demyelination*. New York, Plenum Press, 1978, pp. 513–519.

7. COHEN, S. R., HERNDON, R. M., MCKHANN, G. M.: Radioimmunoassay of myelin basic protein in spinal fluid: An index of active demyelination. *N. Engl. J. Med.* **295**:1455–1457, 1976.
8. GUTSTEIN, H. S., COHEN, S. R.: Spinal fluid differences in experimental allergic encephalomyelitis and multiple sclerosis. *Science* **199**:301–303, 1978.
9. HERNDON, R. M., JOHNSON, M.: A method for the electron microscopic study of cerebrospinal fluid sediment. *J. Neuropathol. Exp. Neurol.* **29**:320–330, 1970.
10. KIES, M. W.: Immunology of myelin basic proteins. In Tower, D. B. (ed.): *The Nervous System*, Vol. 1, *The Basic Neurosciences*. New York, Raven Press, 1975, pp. 637–646.
11. KOHLSCHÜTTER, A.: Myelin basic protein in cerebrospinal fluid from children. *Eur. J. Pediatr.* **127**:155–161, 1978.
12. NORTON, W. T.: Isolation and characterization of myelin. In Morell, P. (ed.): *Myelin*. New York, Plenum Press, 1977, pp. 161–199.
13. TROTTER, J. L., HUSS, B., BLANK, W. P., O'CONNELL, K., HAGAN, S., SHEARER, W. T., AGRAWAL, H. C.: Myelin basic protein in cerebrospinal fluid and normal and pathological brains. *Trans. Am. Soc. Neurochem.* **9**:59, 1978.
14. WHITAKER, J. N.: Myelin encephalitogenic protein fragments in cerebrospinal fluid of persons with multiple sclerosis. *Neurology* **27**:911–920, 1977.
15. WHITAKER, J. N., LISAK, R. P., BASHIR, R. M., KRANCE, R., LAWRENCE, J. A., CH'IEN, L. T., O'SULLIVAN, P.: Immunoreactive myelin basic protein in the cerebrospinal fluid in neurological disorders. *Ann. Neurol.* **4**:178, 1978.

CHAPTER 35

Electron-Microscopic Studies on Cerebrospinal Fluid Sediment

Robert M. Herndon

1. Introduction

In a brilliant series of reports beginning with his introduction of lumbar puncture in 1891, Quincke[12] described the normal cerebrospinal fluid (CSF) and most of the tests that we routinely perform today. He described the increased cell count in serous meningitis and the increased cell count, low sugar, and bacteria in bacterial meningitis. In addition, he described the changes in sugar, protein, and cell count and the presence of tubercle bacilli in the pellicle in tuberculous meningitis. Early cytological studies were carried out by Quincke and a number of other investigators, but the field was largely neglected until exfoliative cytology, in general, expanded with the introduction of the Papanicolaou stain in 1942.[11] Since then, the cytological examination of CSF has been carried out in an increasing number of centers. Results have, in general, been rather disappointing and have, with a few exceptions, been limited to the study of malignancies. More recently, these studies have included special stains for ribonucleic acid, studies of thymidine uptake by CSF cells, and studies of cultured cells.[9,14,17] In addition, immunofluorescent stains have been used to detect infectious agents in CSF.[16]

The introduction of methods for the electron-microscopic study of CSF sediment[3,4,7,8,15,20] has added new dimension to the field, and the finding of free myelin fragments in CSF[4,6] has led to the development of tests for the detection of myelin-breakdown products in the CSF.[1,2,22] (See Chapter 34.)

There are three preparative techniques applicable to the electron-microscopic study of CSF constituents. These are (1) Millipore filter techniques[7]; (2) low-speed centrifugation either with repeat centrifugation during processing[3] or with the use of serum to hold the cells[7]; and (3) high-speed centrifugation.[4]

The Millipore filter, which is in common usage in exfoliative cytology including CSF cytology, is also used for the concentration of suspended cells for electron microscopy and has the advantage that it causes less cell distortion than centrifugation techniques. Its main disadvantages from the electron-microscopic standpoint are the presence of the filter material in the pellet and, more important, the fact that most of the noncellular constituents of interest are likely to pass through the filter and be lost.

Low-speed centrifugation has been used successfully by Duffy *et al.*[3] and produces excellent morphology. While the lighter, non-cellular constituents are not spun down by this technique and are therefore lost, the cells are not distorted as much as they are with high-speed centrifugation, and the danger of losing the material during processing is minimal. Similar excellent cytology has been achieved using serum to hold the cells together,[7] but this necessitates ignoring extracellular elements that may have been present in the serum.

High-speed centrifugation, which was used in this study, as detailed below, has one major advantage and two distinct disadvantages vis-à-vis the other techniques. The main advantage is that minute constituents, including cell fragments, viruses, collagen fibers, and other noncellular elements, will

Robert M. Herndon, M.D. • Center for Brain Research and Department of Neurology, University of Rochester School of Medicine and Dentistry, Rochester, New York 14642.

be included in the pellet. The disadvantages are: (1) unless there is a significant pleocytosis, the pellets are extremely small, often barely visible, and easily lost during processing; and (2) the high *g* forces can produce considerable cellular distortion so that in some instances, the cellular organelles settle to the side of the cell nearest the bottom of the pellet. Despite occasional distortion of this type, the morphology in most pellets has been surprisingly good.

2. Materials and Methods

CSF for electron-microscopic study was allowed to drip directly from the lumbar puncture needle into a sterile 10-ml polycarbonate ultracentrifuge tube. The tube was immediately capped, chilled, and transported to the laboratory. It was then placed in a Spinco Model L ultracentrifuge and spun at 100,000 *g* for 30 min. The supernatant fluid was then removed and replaced with 3–5 ml 4.5% glutaraldehyde in 0.1 N cacodylate buffer. After fixation for 2–15 hr, the fixative was replaced by 10% buffered sucrose, and the pellet was kept in this solution for periods of up to 1 week before processing. The pellets were then postfixed for 1–2 hr in 1% osmium tetroxide and dehydrated in a graded series of alcohols to the 95% alcohol level.

At this point, one of two alternative procedures was used:

1. If the pellet was large enough to be readily visible, it was removed with a slim, pointed stainless steel weighing spatula and transferred to a glass bottle containing 100% ethanol. The pellet was then stained, in block, for 1 hr with 1% uranyl acetate in ethanol, transferred to propylene oxide, and embedded in Epon[10] or, in the latter part of the study, in Spurr low-viscosity embedding medium.
2. If the pellet was so small as to be barely visible, in order to eliminate the danger of losing the pellet, it was left in the tube, and the 95% ethanol was replaced briefly by 100% ethanol, which was then removed and replaced by Epon. The pellet was allowed to infiltrate overnight in a vacuum, the Epon was replaced with fresh Epon, and the block was polymerized in the ultracentrifuge tube. The ultracentrifuge tube was then cut away and the tissue block cut out of the end of the large block formed in the tube.

The blocks resulting from these alternative procedures were then trimmed and sectioned for light and electron microscopy in the usual manner. The sections for electron microscopy were stained with uranyl acetate[19] and lead citrate[18,21] to enhance contrast. The electron micrographs used to illustrate this study were taken with a Siemens Elmiskop 1A or a Philips 200 electron microscope.

Attention to several specific details is necessary to obtain satisfactory pellets. The following avoidable problems have been encountered: (1) Gross contamination of the pellet with talc or starch granules. This occurs when the hub of the lumbar puncture needle or the attached three-way stopcock is brushed by the glove of the person performing the lumbar puncture. A surprising amount of starch or talc (depending on the type of glove powder used) can get into the CSF in this way and, in some instances, cause the pellet to break up during processing. (2) Bacterial contamination, contamination with *Mycoplasma*, and contamination by extraneous bits of paper and dirt occurred on a number of occasions early in the study before it became apparent that strict sterility and cleanliness must be maintained up through the ultracentrifugation stage if such contamination is to be avoided. (3) Loss of pellets during processing has been a problem only with very small pellets and is usually avoided by use of the modified procedure described above. In a few instances where the pellets had a high lipid content, the pellet appeared to dissolve in the ethanol, and on at least one occasion, this resulted in the loss of a pellet.

3. Results with CNS Demyelination and Cancer

All the pellets encountered had a layered structure in which erythrocytes, when present, were on the bottom, with a mixed layer of leukocytes next, and finally a layer of finely filamentous debris and noncellular elements on top. Pellets from normal CSF in which the lumbar puncture was atraumatic consisted entirely of a thin, usually one-cell-thick, discontinuous layer of lymphocytes overlaid by a thin layer of finely filamentous material. In other specimens, erythrocytes, neutrophilic leukocytes, monocytes, eosinophils, macrophages, transitional lymphocytes, and rarely plasma cells were encountered.

3.1. Demyelinating Disease

The results of CSF examination in 32 cases of known or suspected demyelinating disease have been reported elsewhere.[5] Of particular interest in this group was the presence of myelin fragments intermingled with the filamentous material overlying the pellet (Fig. 1). These fragments had a recognizable alternation of major dense lines and intraperiod lines and were seen in 8 of 12 cases of multiple sclerosis (MS) where the specimen was taken within 2 weeks of the onset of new neurological symptoms. They were rarely seen in cases of MS in remission. Similar myelin fragments were seen in CSF from 1 to 2 cases of progressive multifocal leukoencephalopathy. Identifiable myelin fragments were not found in any of the more than 200 CSF specimens examined from patients with diseases not in the demyelinating category.

The less specific finding of laminated lipid without recognizable alternating major dense lines and intraperiod lines was found in a large number of cases in a variety of disorders. This finding is of little diagnostic value, since it appears to be essentially nonspecific, though it has not been seen in normal CSF specimens.

The presence of lipid-laden macrophages in 27 of the 32 cases of demyelinating disease, including many inactive cases, is of considerable interest.[6] While lipid-laden macrophages were occasionally seen in other disease processes, they were much more common in the demyelinating diseases. They were not seen in normal CSF specimens and clearly indicate the presence of tissue breakdown, though they are in no way specific for demyelinating diseases. Lymphocytes, transitional cells, and occasional plasma cells were regular components of the CSF in patients with demyelinating disease. Mitotic cells were rarely encountered.

3.2. Malignancies Involving the CNS

The findings in six cases of malignancy involving the central nervous system (CNS) are shown in

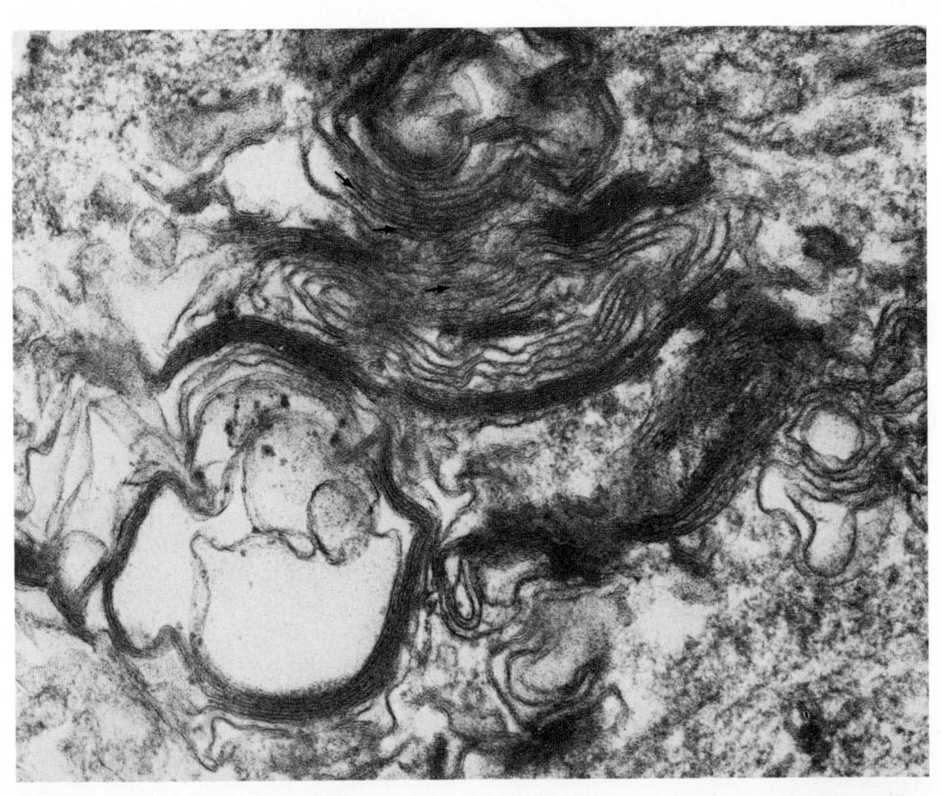

Figure 1. Electron micrograph illustrating typical appearance of myelin fragments embedded in filamentous material overlying pellet. The alternation of major dense lines with intraperiod lines (arrows) can be clearly seen in several areas. ×73,000.

Table 1. Electron-Microscopic Findings in Six Cases of Malignancy Involving CNS

Malignancy	Total cases	Macrophages	Malignant cells	Myelin figures	Collagen fibers
Cancer of the meninges	2	1	2	1	1
Meningeal leukemia	2	1	2	—	1
Metastatic melanoma	1	—	1	1	—
Reticulum cell sarcoma	1	—	1	1	—
TOTALS:	6	2	6	3	2

Table 1. The number of positive findings as regards malignant cells obviously reflects a sampling bias, and such a high percentage of positives would be unlikely if primary brain tumors and early cases of cerebral metastasis were included. Tumor cells could be identified by a variety of characteristics including size, nucleocytoplasmic ratio, cytoplasmic inclusions, and, in some cases, desmosomes (Fig. 2).

The finding of most interest in these cases was the presence of free collagen fibers in two of the specimens (Fig. 3). Collagen fibers were also seen in a case of herpes simplex encephalitis and in a case of MS, indicating that this finding is not limited to malignancies. Laminated lipid bodies were also found in a few cases.

Malignant cells apparently in the lymphocyte series were seen in a case of generalized lymphomatosis and in a case of meningeal leukemia. The cells were readily recognized by their very dense ap-

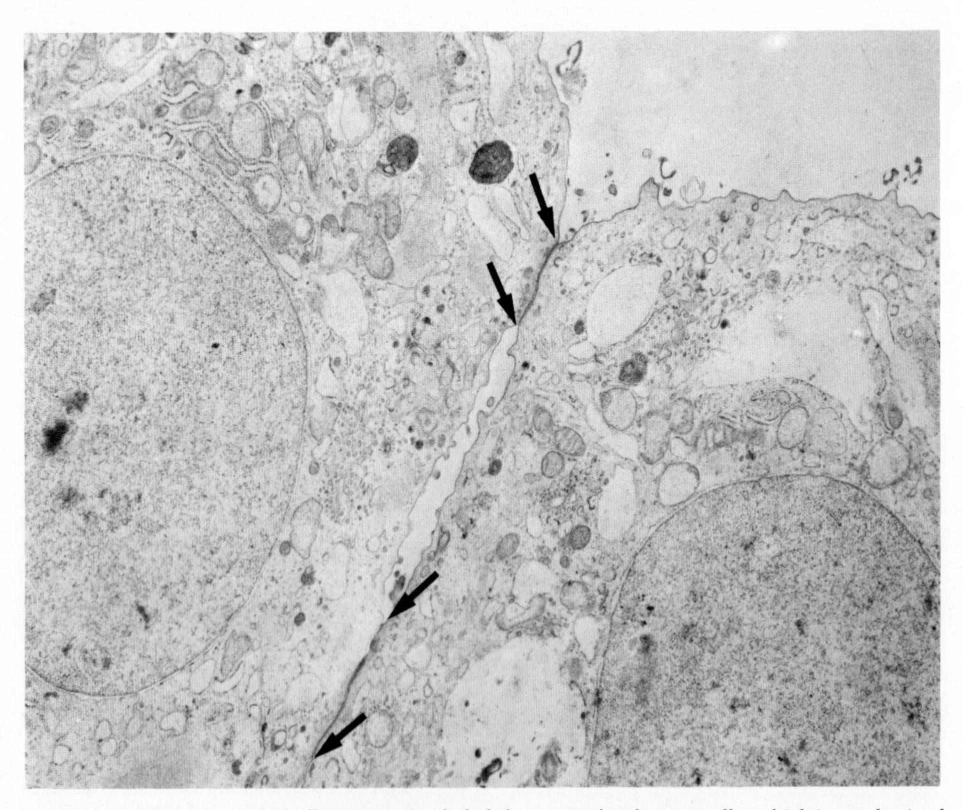

Figure 2. Electron micrograph illustrating epithelial character of malignant cells, which is emphasized by prominent desmosomes joining them. This specimen is from patient with carcinomatosis of meninges, metastatic from adenocarcinoma of breast. ×9500.

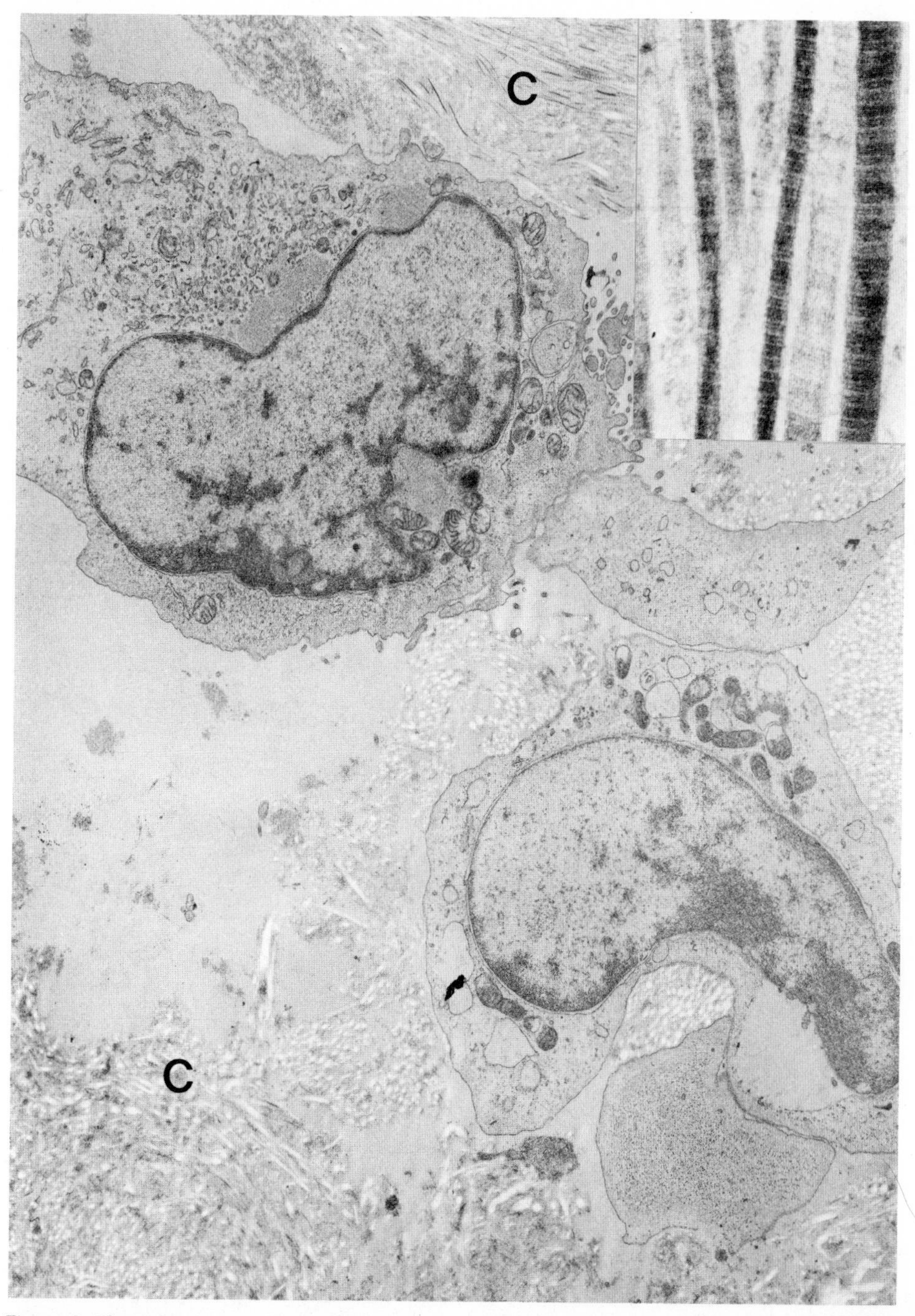

Figure 3. Electron micrograph showing a macrophage (upper left) and lymphocyte (lower right) surrounded by large numbers of free collagen fibers (C), some of which have pulled out of section during cutting process. Characteristic collagen banding can be seen at higher magnification (inset) from same case as Fig. 2. ×7500; inset: ×83,000.

pearance, which resulted from the presence of large numbers of free ribosomes.

4. Miscellaneous Clinical Findings

4.1. Mollaret's Meningitis

CSF was obtained from a single case of benign recurrent lymphocytic meningitis during the patient's fourth attack. The pellet was made up almost entirely of lymphocytes and neutrophil leukocytes (Fig. 4). These cells were remarkable in that they were packed with small, extremely regular, approximately 30-nm electron-dense granules. Histochemically, these granules stained with the methenamine silver technique, the electron-microscopic equivalent of the periodic acid–Schiff stain[13] (Fig. 5), indicating that these granules are probably glycogen.

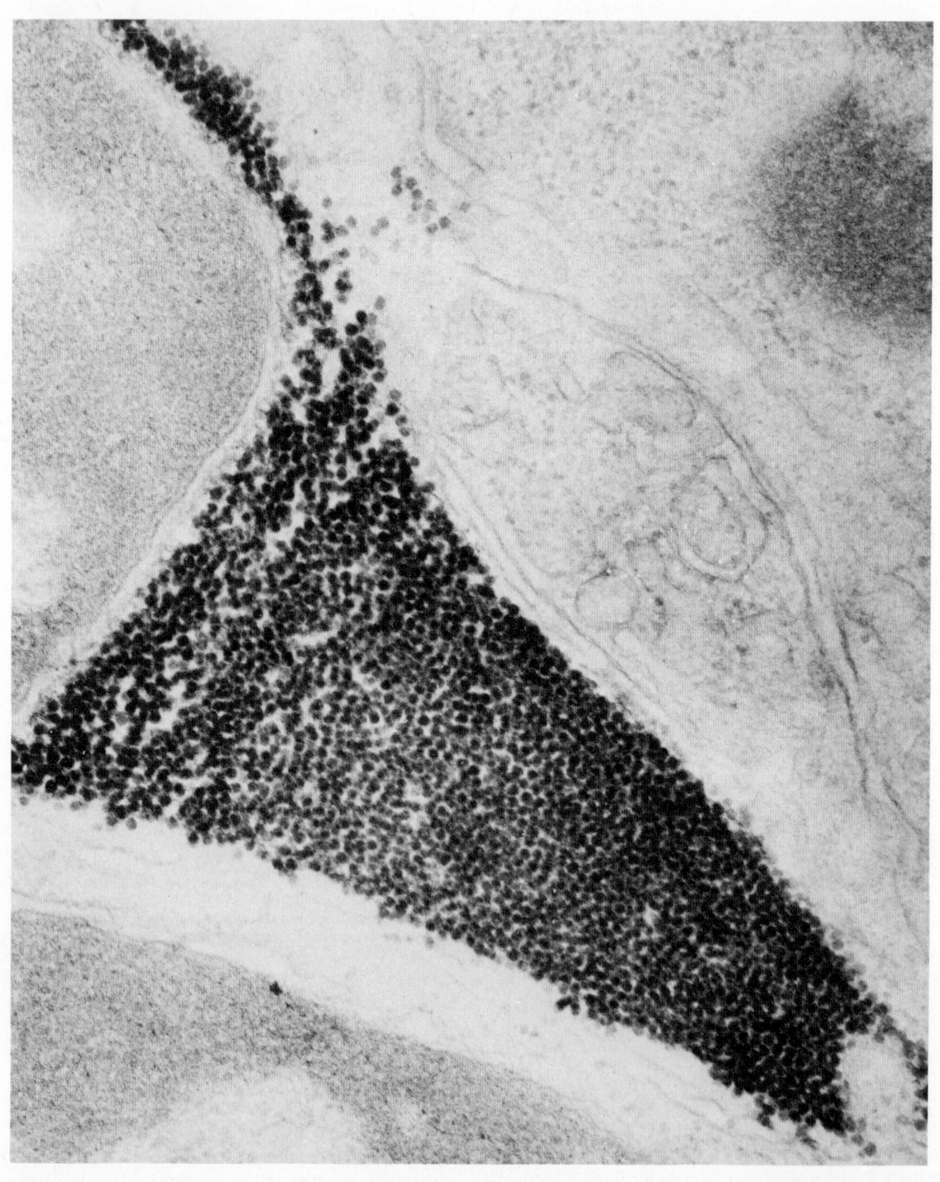

Figure 4. Electron micrograph showing densely stained, relatively uniform 20- to 25-nm β-glycogen particles, which could be easily mistaken for viral particles. This heavy concentration was seen in very high proportion of lymphocytes and neutrophilic leukocytes in this pellet from patient with benign recurrent meningitis. ×56,000.

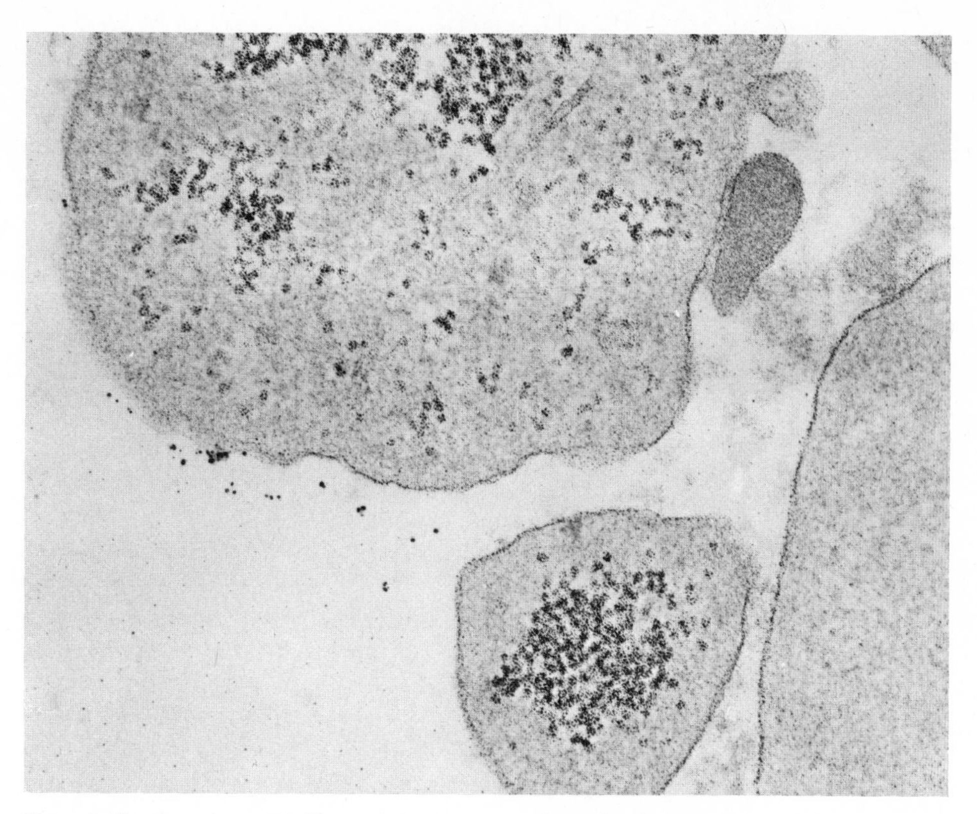

Figure 5. Electron micrograph illustrating appearance of particles shown in Fig. 4 after methenamine silver staining, supporting view that they are indeed glycogen. ×59,000.

Attempts at digestion of the epoxy-embedded material with saliva gave equivocal results with evidence of partial digestion of the granules. There was insufficient material to repeat the studies. Aside from the excessive amount, the glycogen did not differ in appearance from that seen in other leukocytes.

4.2. Herpes Encephalitis

Two specimens from patients with *Herpesvirus hominis* (Herpes simplex) encephalitis were examined. These two pellets contained lymphocytes and neutrophilic leukocytes, and one of the two contained free collagen fibers. No viral particles were seen. One specimen from a patient with herpes zoster encephalitis was examined. The specimen contained lymphocytes, neutrophil leukocytes, and an occasional macrophage. In one cell, intranuclear herpes virus nucleocapsids were seen. In this regard, it should be pointed out that viruses can be cultured from CSF in many cases of zoster encephalitis, whereas herpes simplex can rarely be cultured from CSF.

4.3. Mumps Meningitis

Mumps virus nucleocapsids (Fig. 6) were seen in six cases of benign mumps meningitis.[5] These characteristic 18-nm coated tubules were easily identified in lymphocytes and in ependymal cells (Fig. 7) and choroid plexus cells that were shed as a result of the infection.[5] In addition, large collections of membranous whorls were found in several cases. Although these whorls may have been derived from myelin, the characteristic alternation of major dense and intraperiod lines was never seen, so positive identification as myelin fragments was not possible. The prominence of the CSF changes and the occasional occurrence of obstructive hydrocephalus following mumps suggests that "benign mumps encephalitis" is not always so benign.

4.4. Lipidol

Lipiodol, which consisted of small micelles embedded in a filamentous matrix, was regularly seen in specimens from patients who had had previous myelography (Fig. 8). These amorphous glob-

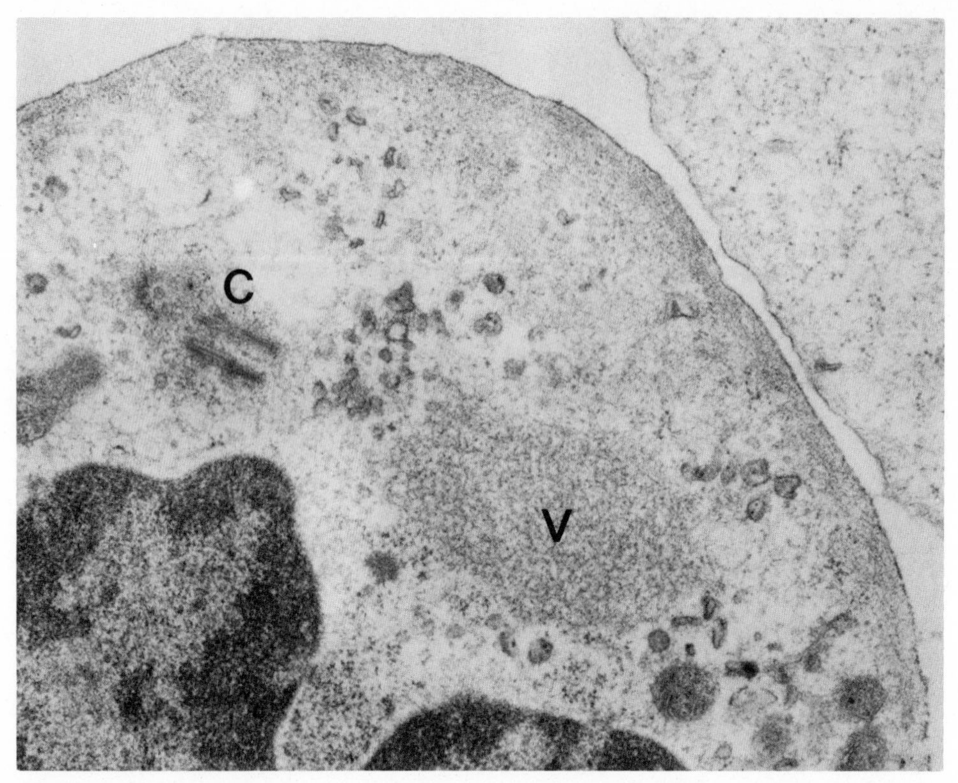

Figure 6. Electron micrograph showing aggregate of 18-nm fuzzy tubules of mumps virus nucleocapsid (V) in lymphocyte from patient with mumps meningitis centriole (C). ×30,000.

ules varied in diameter and could be seen in specimens taken many years after the myelogram was done.

5. Discussion

A number of methods for electron microscopy of CSF cells have been reported. Quite good cytology has been achieved using low-speed centrifugation with[7] or without serum[4] to protect the cells during fixation, and similarly good results have been reported using Millipore filters.[8] These methods preserve the overall shape of the suspended cells much better than the method used in this study. Nevertheless, ultracentrifugation has the distinct advantage of including a substantial number of noncellular elements of interest, including collagen fibers, cell fragments, myelin fragments, extracellular myelin figures, and droplets of myelogram dye. The technique was designed in the hope of finding extracellular viral elements, but the only viral elements seen thus far have been intracellular. The relatively good cytological preservation was unexpected.

The high *g* forces produced by ultracentrifugation do not appear to destroy or fragment many of the cells found in most CSF specimens. However, in a number of cases of malignancy, a substantial portion of the cells were fragmented. Whether this was due to spontaneous degeneration of the cells or the effect of ultracentrifugation on unusually fragile cells was not determined in this study. A combination of the two probably accounts for the finding.

The finding of myelin fragments (see Fig. 1) in the "demyelinating diseases" is of considerable interest. The presence of these large fragments in a substantial proportion of cases in which active demyelination is occurring suggests that the myelin is breaking down extracellularly. This is not totally unexpected, since the myelin in cultures exposed to MS serum appears to flow off the axon and ball up into large fragments (Bornstein, M., personal communication). The presence of these fragments in the CSF indicates that a portion of the myelin

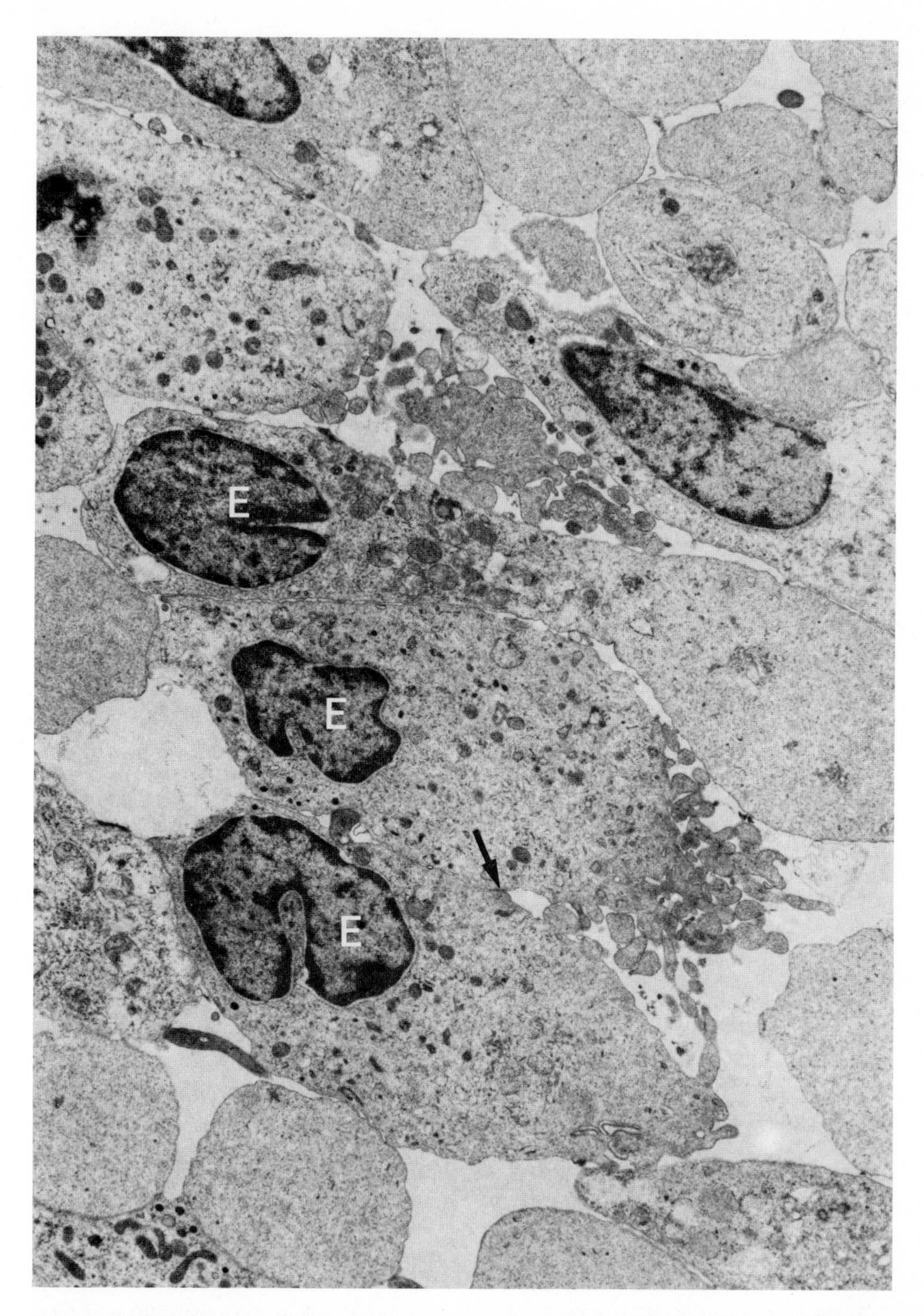

Figure 7. Electron micrograph illustrating short rows of ependymal cells (E) such as were regularly seen in cases of mumps meningitis. No viral inclusions are evident in this particular group of cells. Junction complex joining two of the cells is barely visible at this magnification (arrow). ×7500.

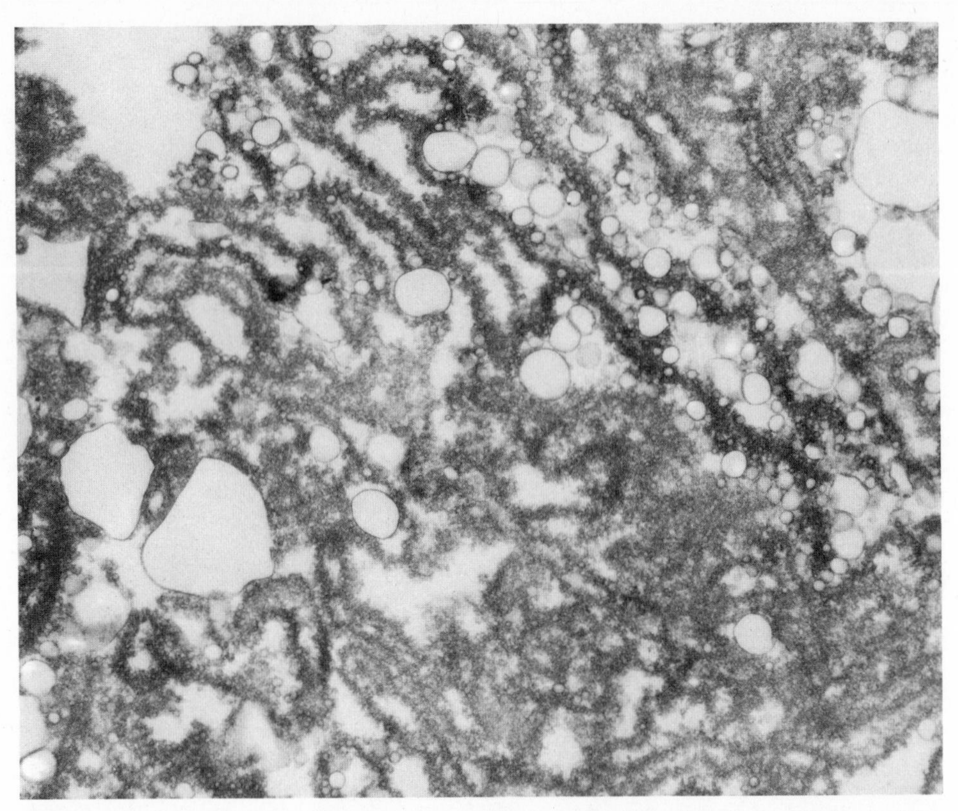

Figure 8. Electron micrograph illustrating characteristic appearance of lipiodol micelles embedded in filamentous material. Lipiodol is dissolved out in processing, leaving empty spheres in proteinaceous matrix. This is a very frequent finding in specimens from patients who have undergone past myelography. ×14,000.

breakdown in MS occurs extracellularly and thus suggests that stripping of myelin from the axons by macrophages may not be the only mechanism of myelin destruction, since we would expect any myelin fragments removed by the stripping process to be immediately phagocytized.

The question of how the myelin fragments, once they have separated from the axon, get into the subarachnoid space remains to be answered. Since they must (1) break through the ependymal lining or (2) break through the pia mater, one or both of these mechanisms must occur. Of interest with regard to the first is the fact that in one patient suffering from an acute medial longitudinal fasciculus syndrome, ependymal cells as well as myelin fragments were seen. This suggested that the myelin fragments broke through the ependymal floor of the fourth ventricle. The fact that the largest myelin fragments we have found occurred in a case of progressive spinal cord MS suggests that the fragments are capable of passing through the pia into the subarachnoid space.

The unexpected finding of free collagen fibers in four cases—two of carcinomatosis of the meninges, one of herpes encephalitis, and one of leukemic meningitis—raises the question of how such material occurs in the CSF. There are two major possibilities: (1) the collagen is newly formed by fibroblasts as part of the meningeal reaction; or (2) the free collagen results from the breakdown of preexisting collagen in the meninges. Both mechanisms are possible. Neutrophilic leukocytes are capable of elaborating a collagenase that could partially break down collagen in the meninges, releasing it into the subarachnoid space. On the other hand, fibrosis with formation of new collagen is a common response to injury and certainly occurs in the meninges. This could also result in the release of collagen fibers into the CSF.

Since adequate study of a single CSF specimen

requires from 5 to 30 man-hours of work, the routine electron-microscopic study of CSF sediment is clearly not feasible. The technique is of considerable interest for research purposes and may eventually prove to be useful in the diagnosis of some disease states. Since we discovered the presence of myelin fragments in the CSF in demyelinating diseases, we have devised more efficient techniques for their detection. Attempts to utilize antimyelin antibodies tagged with fluorescein isothiocyanate to stain CSF sediment after concentration by centrifugation proved unsuccessful. Subsequently, a radioimmunoassay for myelin basic protein has been used with considerable success.[1,2] (See Chapter 34.)

Because of its characteristic amino acid composition, collagen could probably be detected by chemical means, and if the finding is sufficiently common or sufficiently specific, such a test might be clinically useful.

The marked accumulation of glycogen in the cells in the case of Mollaret's meningitis raises the possibility that glycogen accumulation in leukocytes could contribute to the lowered CSF sugar in specimens from cases of bacterial meningitis. Since few specimens from patients with bacterial meningitis have been examined by this technique, such conjecture remains highly speculative.

As is the case with most new research techniques, these studies have raised more questions than they have answered. Whether new findings as unexpected as some of those presented here will result from further studies remains to be seen. Thus far, we have only scratched the surface, and the potential of the technique remains to be determined.

ACKNOWLEDGMENTS. The author gratefully acknowledges the assistance of Mrs. Lilliana Descalzi, who is responsible for Figs. 6 and 8, and of Mr. John Kasckow in carrying out this work.

References

1. COHEN, S., BRUNE, M. J., HERNDON, R. M., MCKHANN, J. M.: Cerebrospinal fluid myelin basic protein and multiple sclerosis. In J. Palo (ed.): *Myelination and Demyelination. Adv. Exp. Med. Biol.* **100**:513–520, 1978.
2. COHEN, S. R., BRUNE, M. J., HERNDON, R. M., MCKHANN, G. M.: Diagnostic value of myelin basic protein in cerebrospinal fluid. *Ann. Neurol.* (in press).
3. DUFFY, P. E., SIMON, J., DEFENDINI, R., KARALIAN, S.: The study of cells in cerebrospinal fluid by electron microscopy. *Arch. Neurol.* **21**:358–362, 1969.
4. HERNDON, R. M., JOHNSON, M.: A method for the electron microscopic study of cerebrospinal fluid sediment. *J. Neuropathol. Exp. Neurol.* **29**:320–330, 1970.
5. HERNDON, R. M., JOHNSON, R. T., DAVIS, L. E., DESCALZI, L. R. Ependymitis in mumps virus meningitis: Electron microscopical studies of cerebrospinal fluid. *Arch. Neurol.* **30**:475–479, 1974.
6. HERNDON, R. M., KASCKOW, J.: Electron microscopic studies of cerebrospinal fluid sediment in demyelinating disease. *Ann. Neurol.* **4**:515–523, 1978.
7. ITO, U., INABA, Y.: Electron microscopic observation of cerebrospinal fluid cells—A new method for embedding of CSF cells. *J. Electron Microsc.* **19**:265–270, 1970.
8. JOHNSTON, W. W., GINN, F. L., AMATULLI, J. M.: Light and electron microscopic observations on malignant cells in cerebrospinal fluid from metastatic alveolar cell carcinoma. *Acta Cytol.* **15**:365–371, 1971.
9. KOLMEL, H. W., CHONE, B. K. F.: Combined application of cytodiagnostic technics to the cerebrospinal fluid. *J. Neurol.* **216**:1–8, 1977.
10. LUFT, J. H.: Improvements in epoxy resin embedding methods. *J. Biophys. Biochem. Cytol.* **9**:409–414, 1961.
11. PAPANICOLAOU, G. N.: A new procedure for staining vaginal smears. *Science* **95**:438–439, 1942.
12. QUINCKE, H. I.: Ueber Hydrocephalus. *Verh. Kong. Innere Med. Wiesb.*, pp. 321–331, 1891.
13. RAMBOURG, A.: An improved silver methenamine technique for the detection of periodic acid reactive complex carbohydrates with the electron microscope. *J. Histochem. Cytochem.* **15**:409–412, 1967.
14. SCHMIDT, R. M., SCHMIDT, T., SEIFERT, B.: Beitrag zur Fluorescenz und Elektronen Mikroskopie der Liquorzellen. *Psychiatr. Neurol. Med. Psychol.* **18**:201–203, 1966.
15. SCHMIDT, R. M., SEIFERT, B.: Beitrag zur ultrastrukturellen Darstellung der Liquorzelle. *Dtsch. Z. Nervenheilkd.* **192**:209–225, 1967.
16. SOMMERVILLE, R. G.: Rapid identification of neurotropic viruses by an immunofluorescent technique applied to cerebrospinal fluid cellular deposits. *Arch. Gesamte Virusforsch.* **19**:63–69, 1966.
17. TOURTELLOTTE, W. W.: A selected review of reactions of the cerebrospinal fluid to disease. In Fields, W. S. (ed.): *Neurological Diagnostic Techniques.* Springfield, Illinois, Charles C. Thomas, 1966, pp. 25–49.
18. VENABLE, J. H., COGGESHALL, R.: A simplified lead citrate stain for use in electron microscopy. *J. Cell Biol.* **25**:407–408, 1965.
19. WATSON, M.: Staining of tissue sections for electron microscopy with heavy metals. *J. Biophys. Biochem. Cytol.* **4**:475–478, 1958.
20. WENDER, M., SNIATALA,: Zur Feinstructur der Liquorzellen. *Wien. Z. Nervenheilkd.* **27**:38–44, 1969.
21. WESTRUM, L. E.: A combination staining technique for electron microscopy. *J. Microsc.* **4**:275–278, 1965.
22. WHITAKER, J. N.: Myelin encephalitogenic protein fragments in cerebrospinal fluid of persons with multiple sclerosis. *Neurology* **27**:911–920, 1977.

CHAPTER 36

Lymphocyte Subpopulations in Human Cerebrospinal Fluid

Effects of Various Disease States and Immunosuppressive Drugs

Jay D. Cook and Benjamin Rix Brooks

1. Introduction

Normal cerebrospinal fluid (CSF) obtained from the lumbar subarachnoid space contains 0–10 nucleated cells/mm^3.[57] The majority of these cells are lymphocytes (approximately 60%) and histiocytes (approximately 30%). The origin of these cells is still controversial. What functional role they play is only now beginning to be understood. We will review methods for determining the cellular composition of the CSF, and subpopulations of these cells in specific disease states, proposed functions of these cells in specific disease states, and the response of these subpopulations to immunosuppressive agents.

2. CSF Cytology

With the introduction of the lumbar puncture in 1891,[75] interest in the pathophysiology of cell populations flourished. It was not long before centrifugation was used to collect the cells in the CSF.[103] A method to enumerate the number of cells in the unconcentrated CSF and to perform differential cell counts was then developed.[27] The next major advance was the introduction of the spontaneous sedimentation technique.[101] This technique greatly preserved cell morphology; however, the CSF was lost for other studies. A membrane filtration technique[44] was then developed that seemingly maximized the benefits of each of the two previous techniques: (1) centrifugation (use of CSF for other studies) and (2) sedimentation (preservation of cell morphology). The usefulness of CSF cytological examination has been substantiated clinically[36] for diagnosing some primary and metastatic central nervous system (CNS) tumors.

Many morphological studies have attempted to provide evidence for function of the cells in CSF. Several techniques for studying the subtypes of lymphocytes in CSF have been introduced.[1,3,9,11,15,25,29,31,35,54,55,61,63,67,69,79,85] A major advance in understanding cell function in the CSF has been the development of techniques for studying the transformation of CSF lymphocytes to mitogens and specific antigens in an attempt to characterize the immunological competence of CSF.[51–53]

2.1. Techniques for Studying CSF Cytology

2.1.1. Collection. No matter which technique is used, there are some general axioms that, if not followed, will negate the potential benefit of any of the techniques. These involve the *length of time* from collection to cell fixation, the *temperature* to which

Jay D. Cook, M.D. • Department of Neurology, University of Texas Southwestern Medical School and Veterans Administration Hospital, Dallas, Texas 75216. **Benjamin Rix Brooks, M.D.** • Department of Neurology, The Johns Hopkins University School of Medicine, Baltimore, Maryland 21205.

unfixed cells are exposed, and the *material* of which the vials used to collect CSF for cytology are made.

CSF is not a suitable transport medium for cells.[36] Within 2 hr, there are changes in the cytomorphology and autolysis. The most sensitive are granulocytes and eosinophils. Tumor cells, heavily laden phagocytes–macrophages, and plasma cells are the next most sensitive. Lymphocytes are the least sensitive.[70] The sensitivity to morphological change apparently has a direct relationship to the baseline metabolic activity of the cell.

The degree of cytolysis is proportional not only to the time of exposure of CSF but also to the temperature.[70] The cytolysis can be retarded if the cells are maintained at 1–4°C for a much longer period of time (up to 7 days).[88] Thus, if cells are not going to be immediately processed for morphological studies, they must be maintained at 1–4°C. This is done with the realization that the longer they are refrigerated, the greater the alteration of the cell population will be.

Last, when doing qualitative and quantitative cell counts, a glass tube is unsuitable for transferring specimens unless it has been silicon-coated to prevent adhesion of mononuclear phagocytes, which have a strong affinity for glass.[70]

Thus, the CSF should be collected in either siliconized glass tubes or plastic tubes and rapidly transferred to the laboratory, and if there is any time delay, it should be stored at 1–4°C.

There are basically three approaches to study CSF cytology: centrifugation, sedimentation, and membrane filtration. Each has some practical advantages and disadvantages (see Table 1).

2.1.2. Centrifugation.[20,57,70,94] This method is the oldest of all techniques used to study the cell population of CSF. The CSF sample is centrifuged at 400–800*g* for 10–15 min at 4°C. The supernatant is removed and can be used for other studies. The cell pellet is resuspended in 5 μl of a 20% serum albumin and 0.9% sodium chloride solution. The albumin prevents drying artifacts. A 1-μl aliquot of this concentrated CSF sample is then smeared on a glass slide. After drying, the smear may be stained with any number of stains. Wright's stain is usually the most convenient and helpful. However, various stains, including histochemical stains, may be used[70] as well as electron microscopy.[41,42]

The efficiency of recovery (50–80%) of cells by this technique has been documented in only one study.[70] There have been two extensive studies of the cellular composition in the CSF of large normal populations. Marks and Marrack[57] found that 10–30% of the cells were lymphocytes. "M" cells (superficially resembling lymphocytes but larger, being 10–20 μm in diameter) constituted 70% of these cells, and "G" cells (resembling blood monocytes) comprised the balance, with no evidence of granulocytes. Tourtellotte,[97] on the other hand, studying 135 normal university students, found them to have 63 ± 18% large lymphocytes, 16 ± 10% monocytes, 2% "ghost" cells, and less than 1% granulocytes. The discrepancy between these studies may be resolved by the observation that the morphology of cells is distorted by both centrifugation and fixation.[70]

2.1.3. Sedimentation.[2,4,6,21,36,37,48,61,80,83,90–93] In 1949, Schönberg[83] demonstrated that the morphological features of CSF cells could be enhanced by collecting them by sedimentation rather than centrifugation. An apparatus to collect cells that enhanced the rate of sedimentation by "controlled diffusion" of the liquid was developed by Sayk.[80] Various techniques are currently in use.

The technique of Sornas[90] uses a glass cylinder 1 cm in diameter and 4 cm in height. One edge of the cylinder is coated with heated liquid Vaseline and placed on the slide. As the Vaseline cools, it fixes the cylinder to the slide. When 0.7 ml of fresh CSF is placed in the cylinder and allowed to settle for 1 hr, lymphocytes settle at 1 mm/10 min.[23] The supernatant is then removed and the sample examined by phase microscopy with a water-immersion lens. The sample can be stained with May–Grünwald/Giemsa stain. The recoveries with this technique range between 52 and 74% for the wet and 10 and 34% for the fixed and stained material. The differential counts consist of 87% lymphocytes and 12% monocytoids (nonmacrophage cells). Normal CSF counted in a Janssen chamber (10 mm^3) contained 1.45 cells/mm^3.

The simplest device for sedimentation using controlled diffusion is that described by Oehmichen.[70] Materials needed are a cylinder with a rim, a metal plate spring, filter paper, and a paper punch that makes holes the same size as the internal diameter of the cylinder. The volume of CSF (3–5 cc) used can be much greater than that for the Sornas technique. The cylinder is placed directly over the hole in the filter paper, and the metal plate spring holds the cylinder tightly to the paper. The CSF is placed in the cylinder and allowed to be absorbed by the paper. This can be accomplished in 1–3 hr at room temperature or overnight at 4°C. This process can be enhanced by using pressure from above.[6,23,48]

Cell recoveries are extremely low (10–30%) and variable.[65] But some authors[6,50] claim up to 90% recovery with their technique. Normal CSF contains by this technique 2.5 cells/mm^3 consisting of 19% small lymphocytes, 45% large lymphocytes, 6% monocytes, 30% reticulomonocytes, and 1% granulocytes. The cell concentration in the smear will be different at different areas of the filter. Cells are far more concentrated around the edge of the filter paper than in the center.[82]

2.1.4. Cytocentrifugation.[7,39,43,92,96,100,104] A compromise between the centrifugation and sedimentation technique was introduced by Watson.[100] CSF is concentrated onto a small area of a glass slide by means of a column holding the CSF and a piece of filter paper that absorbs the CSF. Pressure is applied to the CSF column by centrifugation (Shandon cytocentrifuge). The added benefits of this technique are a decrease in time of slide preparation, the small area on which the cells are concentrated, and the decreased centrifugal force required to prepare the cells (<200*g*). The results from various controlled studies by this method show recoveries of only 10–30% of the cells.[39] The cellular differential of "normal" CSF samples is in the range of 54–76% lymphocytes, 21–30% monocytes, and 26% granulocytes.[7,28,43,92] The cell concentration was found to vary from 0.03 to 1.17 cells/mm^3.

2.1.5. Membrane Filtration.[5,34,45,46,102] While there are several techniques used routinely by cytopathologists, the method described by Barringer[5] is the simplest and most practical method yet described. Materials needed are a 5-mm syringe barrel, a 10-inch plastic tube, a 13-mm diameter Millipore filter (5-μm pore diameter for cells, 0.45-μm pore diameter for bacteria). The principle employed in collecting the filter fluid is gravitational pull on a column of liquid. The plastic tubing is attached to the adapter. Both are filled with 0.9% sodium chloride and clamped. The adapter is then placed on the syringe barrel. The CSF sample is placed in the barrel and the tube is unclamped, allowing the CSF to flow through the adapter and cells to be collected on the filter. If the CSF is to be collected for other studies, it can be saved by discarding the volume of the adapter and what the tube holds, saving the rest of the liquid. After the CSF is filtered, the cellulose acetate filter is fixed, stained, cleared with xylene, and mounted. Since the recovery is greater than 90%,[16,70] quantitative cell concentrations and differentials can be determined. Only one study has examined CSF from patients without disease.[46] These workers found that normal CSF contained 0.885 cell/mm^3 consisting of 45% lymphocytes, 30% monocytes, 7% granulocytes, 15% reticulohistocytes, and 3% meningeal cells.

2.1.6. Comparison of Techniques (Table 1). There are good qualitative and quantitative studies comparing the various techniques.[33,39,50,58,59,70,81] From these studies, it is quite clear that the sedimentation and membrane filter techniques are clearly much better than the cytocentrifugation or the centrifugation method. For optimal morphological results, the sedimentation technique is clearly superior.[70] However, the cellulose acetate filter technique clearly collects the largest percentage of cells, and some investigators feel that the morphological abnormalities can be overcome with experience.[32,33,40,58,59] Yet yields as high as 90% have been recorded by some investigators using the sedimentation method.[6,50] It would thus seem that the sedimentation methodology can be perfected to allow both a fine morphological result and high recovery.

2.2. Cellular Composition of CSF

There is considerable disagreement in the literature concerning the nomenclature of cellular elements found in CSF.[38] The same cell type is given different names depending on the observer's biases as to what the cell's origin and function are. Even when there is a consensus on nomenclature, it has been our experience using the membrane filter technique that often there is difficulty in classification of a particular cell by different observers. A current classification system is based on the similarity between blood and CSF cellular elements and current knowledge of cell function.[70] The hypothesis is that CSF and blood cellular elements are similar in pathological states.[68–72] A complete review of the hematological origin of CSF cells is available.[70] Those cells that may be found in the CSF include immunocompetent cells, mononuclear phagocytes, polymorphonuclear granulocytes, erythrocytes, cells lining the CSF space, abnormal cellular elements, and tumor cells[8,69] (Table 2).

The cells found in cytological examination of CSF do not necessarily reflect the cellular elements normally found in CSF. Erythrocytes found in the CSF may be the result of a pathological state or of the process of obtaining the sample. Also, because a specific cell is found on examination, it does not necessarily follow that it was collected from the subarachnoid space. The importance of an unusual cell

Table 1. Comparison of the Different Techniques for CSF Cytological Examination

Technique	Centrifugation	Cytocentrifugation	Membrane filter	Sedimentation
Materials required	20% Albumin in 0.9% saline Pasteur pipettes Slides	Filter paper Pasteur pipette Slide	Filter Pasteur pipette Slide, xylene, alcohol Peramont Syringe, tubing	Slide Pasteur pipette
Equipment needed	Refrigerated centrifuge	Cytocentrifuge	Filter holder	Sediment apparatus (not commercially available).
Time to process	<30 min	< 30 min	<30 min	90–120 min
Time between tap and processing	< 2 hr	< 2 hr	< 2 hr (if fixed, 7 days)	< 2 hr
Technical skill	Minimal	Moderate	Moderate	Moderate
Optimum stain(s)	Wright's or Methyl Green pyronine	Wright's Papanicolaou	Papanicolaou	May–Grünwald Giemsa
Histochemistry	Yes	Yes	Yes (if not fixed)	Yes
Can enumerate cells?	No	No	Yes	No
CSF supernatant available?	Yes	Yes	Yes (if not fixed)	No
Results				
Cell recovery	50–80%	10–30%	>90%	10–90%
Cell morphology	Poor	Fair–good	Fair–excellent	Excellent
Normal values (%)				
Round cells (lymphocytes)	70–90	54–76	45–67	63–90
Mononuclear cells (phagocytes)	10–16	21–30	30	0–37
Granulocytes	<1	0–26	<1	<1
Other	0–2	0–25	4–28	0–30
Total cells (cells/mm^3)	—	0.03–1.17	0.855–1.06	1.45–2.60
Ref. Nos.	57, 70, 97	7, 28, 43, 94	5, 12, 16, 46, 57	21, 57, 91

type must be based on the number of such cells, the other cellular elements found in the sample, and the techniques of collecting the sample, (e.g., time interval, transport-tube materials, traumatic lumbar puncture).

Cellular differentials of CSF cells in 139 patients indicated the following constitution (means ± S.D.): lymphocytes, 59 ± 17%; histocytes (monocytes), 32 ± 17%; polymorphonuclear leukocytes, 3 ± 7%. Other cell types constituted the remaining few percent. The total cell count was 1.064 ± 1.543 cells/mm^3. Of the 139 patients, 11 were felt to have no findings of any systemic or neurological disease; their cellular differentials comprised 70 ± 15% lymphocytes, 23 ± 12% histocytes, and 7 ± 6% polymorphonuclear leukocytes. The total cell count was 0.826 ± 0.753 cells/mm^3 (Figs. 1 and 2). Comparing the eight other diagnostic categories to these "controls," significantly fewer lymphocytes were found in several groups (peripheral neuropathy, cerebellar degeneration, and amyotrophic lateral sclerosis), with a concomitant increase in the percentage of macrophages. While these findings were statistically significant, it is not clear at present what role such cells play in the pathogenesis of these various disorders. The greater number of histocytes seems correlated with a continuous degenerative disorder of the CNS. This observation may reflect their role of "clearing" material found in the tissue and subarachnoid space.

Table 2. Classification of CSF Cellular Elements

Normal
1. Immunocompetent cells (66%):
   Small lymphocytes, large lymphocytes, plasma cells
2. Mononuclear phagocytes (33%):
   Monocytoid cells, activated monocytoid cells, macrophages (lipo-, erythro-, sidero-, and leukophages), giant cells
3. Polymorphonuclear granulocytes (1%):
   Neutrophilic, eosinophilic, and basophilic granulocytes

Abnormal

4. Erythrocytes
5. Cells lining the CSF space:
   Choroid plexus cells, ependymal cells, arachnoid cover cells, naked nuclei
6. Abnormal cellular elements:
   Cartilage, bone marrow, glia (?), muscle
7. Tumor cells:
   Primary, metastatic

3. Immunocompetent Cells in CSF

3.1. Classification of Immunocompetent Cells

It is generally accepted that the blood immunocompetent cell, the lymphocyte, is derived from the bone marrow. Its functional role is dependent on humoral and/or other influences. There are three major subgroups of immunocompetent lymphocytes: thymus-dependent (T cell), null cell (N or K cell), and bursa equivalent (B cell).[13] The latter group of lymphocytes (B cells) are the precursor cells of antibody production. The role that T cells play is more complicated; they are involved with antibody production (helper cells), delayed-type hypersensitivity (cytotoxic cells), and modulation of immune response (suppressor cells). The null cell (K cell) mediates antibody-dependent, but not complement-dependent, cellular cytotoxicity. These cells can be identified by specific membrane determinants: T cells, by sheep red blood cell receptors (E_S^+); null cells, by complement receptors; and B

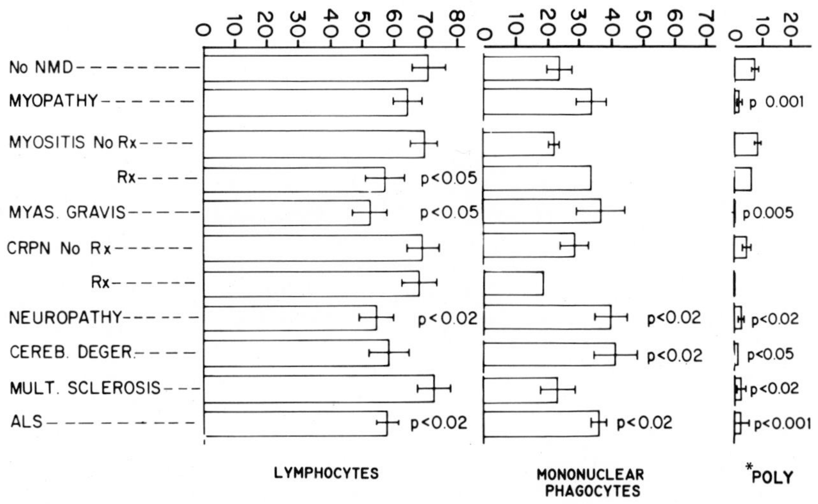

Figure 1. Cellular differential of human CSF. Known amount (1–2 ml) of CSF was passed through cellulose acetate filter. Filter was then stained using Papanicolaou stain. Total number of cells counted was 50–100 per sample. Results are presented as mean percentages ± S.E.M. Statistical analysis employed two-tailed Student's t test. Patient groups and numbers of patients: No NMD (no neurological or systemic disease), 8; MYOPATHY, 17; MYOSITIS (dermato/polymyositis complex): No Rx (not on therapy), 9; Rx (on immunosuppressive therapy), 9; MYAS. GRAVIS (myasthenia gravis), 8; CRPN (chronic relapsing polyneuropathy); No Rx (not on therapy), 9; Rx (on immunosuppressive therapy), 7; NEUROPATHY, 19; CEREB. DEGENER. (cerebellar degeneration), 5; MULT. SCLEROSIS (multiple sclerosis), 10; ALS (amyotrophic lateral sclerosis), 36. *POLY: Polymorphonuclear granulocytes.

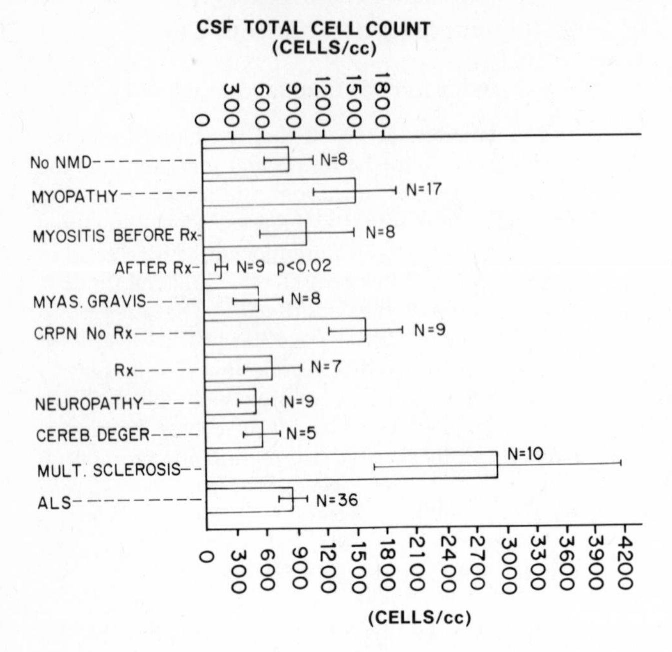

Figure 2. Total cellular content of human CSF. Number of cells in 20 random fields of cellulose acetate filter was multiplied by number of fields (which varied with microscope used) divided by 20 and total volume (ml) of CSF filtered. Results are presented as mean total cell count per cubic centimeter ± S.E.M. Statistical analysis employed two-tailed Student's *t* test. Abbreviations and patient population same as in Figure 1.

cells, by surface immunoglobulin (SIg^+) (Table 3). The percentages of these three separate groups found in blood are: 60–70% T cells, 20–30% null cells, and 10–20% B cells. With the use of these markers, it has been possible to enumerate these subpopulations of lymphocytes in other body fluids.[55]

The value of enumerating the subpopulations of lymphocytes in the peripheral blood derives from the fact that these surface markers may define the *in vivo* function of these cells.[13] The evidence for this statement comes from studies of congenital immunological deficiencies. Patients with thymic aplasia (no evidence of delayed hypersensitivity) have very few E_S^+ cells (T lymphocytes) but normal numbers of SIg^+ cells (B lymphocytes), whereas infantile X-linked agammaglobulinemia patients have normal numbers of E_S^+ cells but no SIg^+ cells. Using these markers, it has been possible to obtain pure populations of either T (E_S^+) or B (SIg^+) lymphocytes. With the use of these techniques, a functional definition of disease states has been possible. While there are relative normal numbers of E_S^+ cells (T lymphocytes) and SIg^+ cells (B lymphocytes) in common variable hypogammaglobulinemia, the de-

Table 3. Immunocompetent Cell Surface Markers[a]

	Marker[b]							
Cell type	E_S	SIg	Comp	Fc	Th	Ia	IaR	Present in CSF
T cell								
Helper	+	−	−	+ (IgM)	+	−	−	+
Suppressor	+	−	−	+ (IgG)	−	−	(−)?	+
Cytotoxic	+	−	−	−	−	−	+	+
B cell	−	+	+	+	−	+	−	+
Null cell	−	−	−	+	−	−	−	+
Macrophage	−	−	+	+	−	+	(+)?	+

[a] Data from Chess and Schlossman[13] and Lisak and Zweiman.[52]

[b] Markers: (E_S) receptor to bind sheep red cells; (SIg) surface immunoglobulin; (Comp) receptor for complement; (Fc) receptor for the Fc (constant fragment) of IgG (or IgM); (Th) lysed by antisera to purified T cells in the presence of complement; (Ia) alloantigens on surface—stimulate mixed-lymphoctye reactions; (IaR) receptor for alloantigens—respond to mixed-lymphocyte reactions.

fect was shown to be the result of a hyperactive suppressor-T-cell population.[22] Immunosuppression by corticosteroids has been correlated with decreased E_S^+ and SIg^+ cell numbers.[99] Thus, the role of the immunocompetent cells in the CSF may be inferred by enumerating such subpopulations in the CSF.

3.2. Origin of CSF Immunocompetent Cells and Exchange with Peripheral Blood

The immunocompetent cells found in the CSF must now be considered as being of hematogenic origin. Evidence for this hypothesis was provided by Schwarze[86] and Oehmichen and co-workers.[68–72] Blood leukocytes labeled with [^{3}H]thymidine were transfused into animals with experimentally induced meningoencephalitis. Labeled cells were found in the subarachnoid space. Therefore, in the pathological state, the cells in the CSF arise from the peripheral blood. Presumably, the same is true for the healthy state.

Schwalbe[85] first demonstrated that there was a direct physiological, anatomical connection between the subarachnoid space and the peripheral lymphatic system. Studies by Brierley and Field[10] using India ink (0.4–1.5 μm) infused into the subarachnoid space showed that within 4 hr the cervical nodes and within 6 hr the sacral nodes were labeled. While the exact mechanism for this transport is not known, particulate material of the size, 0.4–1.5 μm, found in the subarachnoid space has access to the peripheral lymphoid system. Thus, there is a mechanism for the systemic sensitization to material found in the subarachnoid space. Whether CSF lymphocytes use this pathway is unknown.

Thin-walled channels in the perivascular space have recently been demonstrated by electron microscopy.[74] These channels contain small lymphocytes and macrophages. Collections of lymphoid tissue not unlike the antibody-producing medullary cords of peripheral lymph nodes were found in the perivascular space of plaques in the brains of two of the multiple sclerosis (MS) patients of Prineas.[74] While the anatomical connections of these channels were not determined, they presumably drain into the subarachnoid space. Lymphocytes infused into the subarachnoid space of rabbits disappear quickly ($t_{1/2}$ = 4.5 days).[64] The site of their exit was suggested to be the superior sagittal sinus; however, no direct proof of this pathway was demonstrated. Thus, there are lymphatic channels in the perivascular spaces of the CNS, and in a disease state presumably immunologically mediated, lymph-node-like tissue can be found organized in the perivascular space of the CNS.

3.3. Quantitation of Immunocompetent Cells in CSF

All our CSF samples were processed by cellulose acetate filters for cytological examination. Direct enumeration indicated that only two thirds of the cells were lymphocytes. The next largest category of cellular elements was mononuclear phagocytes, approximating a third of the cells.

Initial attempts to identify T lymphocytes in the CSF did not involve a conversion factor for the percentage of lymphocytes in each sample, We, like many others, assumed that the percentage of lymphocytes determined by sedimentation was greater than 90% so that one need only count the number of mononuclear cells binding sheep red blood cells and the total number of mononuclear leukocytes. We accepted this assumption, and our measured percent of T lymphocytes in the CSF was 39 ± 20% for 111 patients with neuromuscular disease. This value was significantly lower than the percent of T cells normally found in the blood. This value was similar to the value found in three studies[1,3,85] for patients with chronic or stable MS, but was much lower than the value for patients with neuropsychiatric disease.[54]

The number of B lymphocytes was enumerated using the technique of fluorescein-conjugated goat anti-human immunoglobulin. The percentage of total cells that were B lymphocytes is small (6 ± 5%).[15] Seemingly, the largest subpopulation (≈55%) of lymphocytes in the CSF was the null-lymphocyte. This would have meant that the lymphocyte-subpopulation composition of the CSF compartment was markedly different from the peripheral-blood-monocyte compartment.

By taking the value for the percentage of total cells constituted by lymphocytes, the value of the "correct" percentages of total cells comprised by T, B, and null cells can be calculated. With this correction, the percentages of T, B, and null lymphocytes showed less variability, and the values were similar to those found in the peripheral-blood-monocyte compartment. In addition, the values agreed exactly with those values in a study that counted only the lymphocytes.[54] Thus, any current method for identifying and enumerating immunocompetent-cell subpopulations in the CSF from patients with a variety of neurological disorders requires a determination not only of the percentages of E^+ and Ig^+

cells but also of the percentage of lymphocytes in the CSF sample.

3.3.1. Percentages of Subgroups of Immunocompetent Cells in CSF. The percentage of T, B, or null lymphocytes found in the CSF (Fig. 3A) is approximately the same as found in the peripheral blood (Fig. 4A). There were no significant differences between the "control" group and the other neurological disease groups in the percentage of T, B, or null lymphocytes found in either the CSF or the blood. The percentage of T, B, or null lymphocytes in the CSF is not merely a reflection of the peripheral-blood-lymphocyte subpopulations because the correlation coefficient between the blood and CSF for each group was very small.

3.3.2. Total Numbers of Immunocompetent Cells in CSF. The total numbers of CSF (Fig. 3B) and blood (Fig. 4B) T, B, and null lymphocytes were cal-

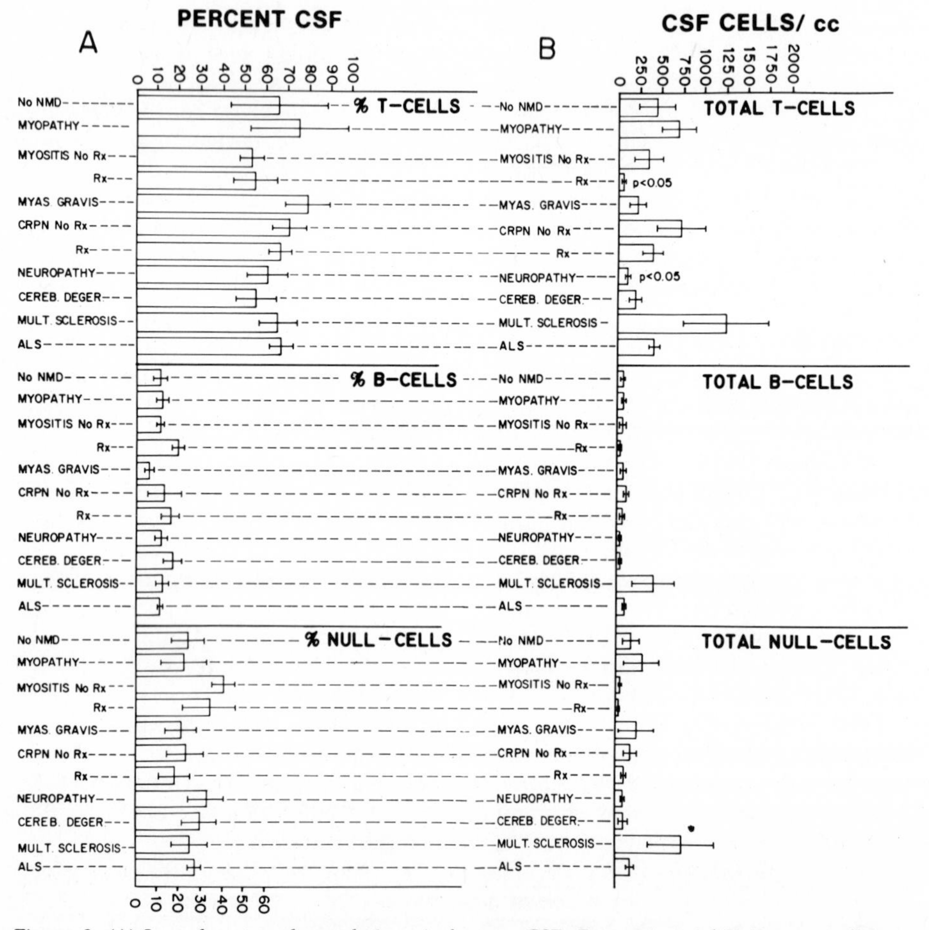

Figure 3. (A) Lymphocyte subpopulations in human CSF. Proportions of T, B, and null lymphocytes in human CSF were determined by synthesis of sedimentation and membrane filtration techniques. Percentage of T cells is defined as number of cells that bind sheep red blood cells (ES^+) ($\times$ 100) divided by total number of cells then divided by fraction of cells that are lymphocytes. Percentage of B cells is defined as number of cells that have immunoglobulin on their surface (SIg^+) ($\times$ 100) divided by total number of cells and then divided by fraction of cells that are lymphocytes. Percentage of null cells is defined as fraction that is not ES^+ or SIg^+. Patient categories are defined as in Fig. 1. (B) Lymphocyte-subpopulation cell concentrations in human CSF. Total numbers of T, B, and null lymphocytes were calculated by multiplying proportion of each type of lymphocyte times total number of lymphocytes (total CSF cells $\times$ fraction that are lymphocytes) divided by 100. Statistical analysis was made using two-tailed Student's t test. Abbreviations and patient population same as in Figure 1.

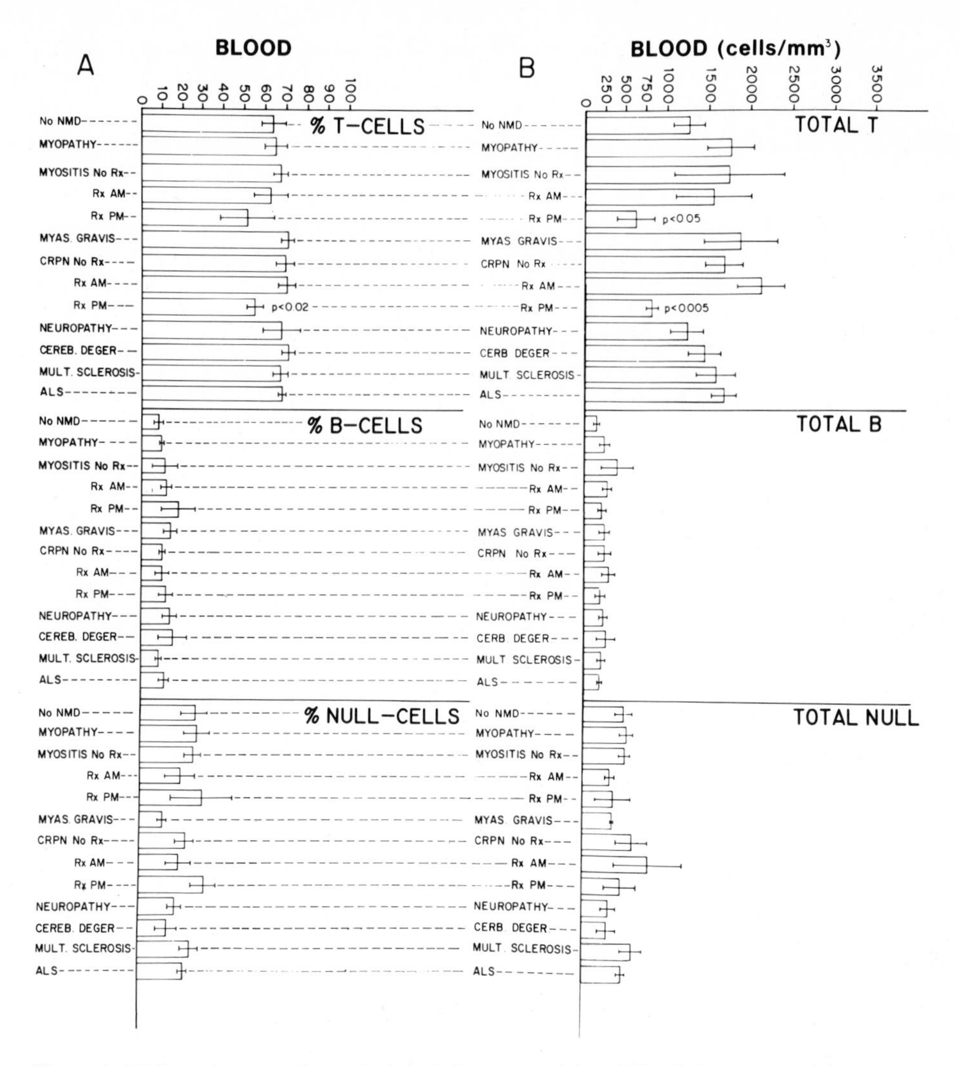

Figure 4. (A) Lymphocyte subpopulations in human peripheral blood. Proportions of T, B, and null lymphocytes in human peripheral blood were determined simultaneously in patients who had lymphocyte subpopulations determined in CSF.[16] Patient categories are defined as in Fig. 1 with exception that Rx AM and Rx PM refer to those groups of patients on immunosuppressive therapy in morning and 6 hr later, respectively. (B) Lymphocyte-subpopulation cell concentrations in human peripheral blood. Lymphocyte-subpopulation concentrations were determined by multiplying proportion of each type of lymphocyte by total number of lymphocytes present.[16] Abbreviations and patient population same as in Figure 1.

culated for each of the 11 diagnostic categories. While the numbers of the total CSF T, B, and null lymphocytes for the MS group are obviously much greater than in the CSF of the other diagnostic categories, they were not significantly different because of the large group variation (range 156–8000 cells/ml). These MS patients were not in an acute exacerbation at the time of evaluation.

3.4. Reproducibility

Of the patients studied, there were six with amyotrophic lateral sclerosis (ALS) who underwent repeated lumbar punctures. Using each patient as his own control, the coefficient of variation for the cellular differentials was 12%. However, the total cell count coefficient of variation was 60% (Fig. 5). This

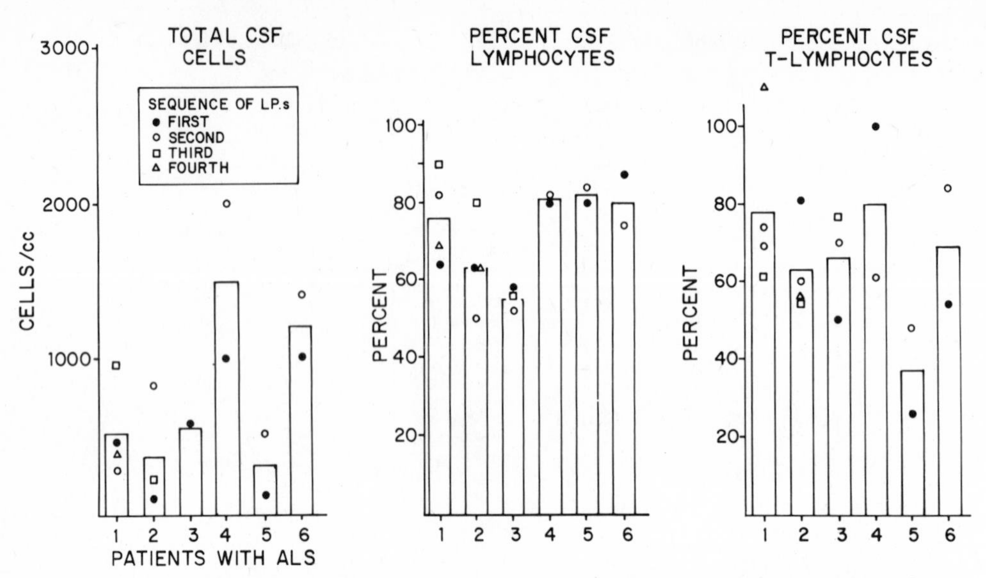

Figure 5. Reproducibility of repetitive CSF values for six amyotrophic sclerosis (ALS) patients. Six patients who underwent repeated lumbar punctures had subpopulations of immunocompetent cells determined. Greatest variation is seen in total cell count, with coefficient of variation of 60%.

variation is probably biological, since the methodology shows less than a 10% variation on duplicate samples. Thus, despite careful control of the conditions for obtaining the sample, there can be a vast variability in the cell count numbers.

3.5. Comparison of Determinations of CSF Immunocompetent Cells

The percentages of CSF T, B, and null lymphocytes found by other investigators are compared with the results of this study in Table 4. Because of the different clinical conditions, it is difficult to make group comparisons among all these different studies. However, those studies investigating MS can be compared. Initial studies[1,85] suggested that in chronic inactive MS, the percentage of CSF T lymphocytes was much lower (30–35%) than in the acute state (55–67%). We, like most of the other investigators,[47,57,63,67] have found that the corrected percentage of CSF T lymphocytes was 65 ± 10%, or about the same as found in blood. This apparent difference can be explained by methodological differences. The initial studies[1,85] did not take into account the CSF cytological differentials. Since it is known that the percentage of lymphocytes increases during an acute MS attack,[83] the initial differences in percentage of CSF T lymphocytes most likely reflected this difference. Those studies taking differentials into account have found, as presented here, that the percentages of T, B, and null lymphocytes are similar but not identical to those found in the blood. While there were no significant differences in various disease states in the percentages of T, B, and null cells, total cell numbers did demonstrate significant differences. Thus, in those studies purporting to enumerate the various cell populations, techniques that allow for cellular enumeration and differentiation must be used.

3.6. Immunological Functions of CSF Immunocompetent Cells

Functional activity of CSF cells has been demonstrated *in vitro* in several studies (Table 5).

3.6.1. Antibody Synthesis. The B lymphocyte is the precursor cell for the plasma cell that synthesizes immunoglobulins. Evidence that the B cell functions in the CSF compartment has been suggested by several different studies.[14,19,30,66,76–79,84,88,89] A patient with MS was studied who had elevated levels of IgG and IgA in the CSF, and the lambda/kappa ratio of CSF immunoglobulins was different from the ratio found in the blood.[14] *In vitro* incubation of CSF lymphocytes provided evidence that these cells produced IgG and IgA. Specific antibody synthesis in the CNS has been corroborated by evidence of oligoclonal immunoglobulin formation in MS,[76] rubella panencephalitis,[89] optic neuritis,[66] and lymphoreticular neoplasms.[88] The initial appearance of antibodies determined by complement fixation for lym-

Table 4. Lymphocyte Subpopulations in Human CSF

Ref. No.	Disease		Number	CSF[a] T	CSF[a] B	CSF[a] Null	Blood[a] T	Blood[a] B	Blood[a] Null	Enumeration[b]	Differentiation[b]
87	Control		1	23	16	(61)	—	—	—	–	–
	Viral	1 wk	12	8	16	(76)	—	—	—	–	–
	Encephalomyelitis	3 wk	12	29	13	(58)	—	—	—	–	–
	Post infectious	1 wk	12	66	6	(28)	—	—	—	–	–
	encephalomyelitis	3 wk	12	51	17	(22)	—	—	—	–	–
		6 mon	12	2	0	(98)	—	—	—	–	–
	MS	1 wk	12	67	10	(23)	—	—	—	–	–
		2 wk	12	51	9	(40)	—	—	—	–	–
		3 wk	12	31	27	(42)	—	—	—	–	–
		4 wk	12	19	35	(46)	—	—	—	–	–
		26 wk	12	21	37	(42)	—	—	—	–	–
		52 wk	12	10	1	(89)	—	—	—	–	–
	Chronic stable MS		12	37	3	(60)	—	—	—	–	–
31	Acute viral meningitis		3	(60)	—	—	(30)	—	—	–	–
	Acute viral meningitis treated		1	(42)	—	—	(15)	—	—	–	–
	Acute tubercular meningitis		1	(100)	—	—	(10)	—	—	–	–
	Untreated MS		1	(0)	—	—	(9)	—	—	–	–
	Treated MS		1	(0)	—	—	(9)	—	—	–	–
	After treatment for MS		1	(0)	—	—	(45)	—	—	–	–
79	Acute MS		6	74	5	(21)	62	11	(27)	–	–
1	Stable MS			30	7	(63)	—	—	—	–	–
	Relapsing MS		10	55	13	(32)	54	26	(20)	–	–
60	Neurological disorders		52	73	16	(11)	64	15	—	–	+
62	Syncope		1	59	—	—	—	—	—	–	–
	Encephalopathy		7	72	—	—	69				
	Myelopathy		5	60	—	—	67	—	—	–	–
	Viral meningitis		2	44	—	—	52	—	—	–	–
	MS		4	95	—	—	64	—	—	–	–
54	Neuropsychiatric disorders		8	93	2	(5)	42	8	(50)	–	+
3	Acute lymphocytic leukemia			15	1	(84)	55	13	(32)	–	–
29	Not stated		9	—	13	—	—	24	—	+	+
15	Neuromuscular disease		78	72	16	22	—	—	—	+	+
9	Mumps meningocephalitis		2	(50)	(4)	(46)	28	27	(45)	–	–
35	Gambian sleeping sickness		40	11	61	(28)	—	—	—	–	+
55	Neuropsychiatric disorders		25	95	2	(3)	52	21	(27)	–	+
63	Active MS		7	94	—	—	47	—	—	+	+
	Stable MS		8	71	—	—	68	—	—	+	+
11	Myositis untreated		8	52	11	40					
	treated		7	54	19	34	68	11	26	+	+
	CRPN untreated		5	70	13	23	70	10	22	+	+
	treated		5	66	16	18	56	12	32	+	+

[a] Values in parentheses are calculated from data.
[b] (NA) Not available. (MS) Multiple sclerosis; (CRPN) chronic relapsing polyneuropathy.

(Continued)

Table 4. (*Continued*)

Ref. No.	Disease	Number	CSF[a] T	CSF[a] B	CSF[a] Null	Blood[a] T	Blood[a] B	Blood[a] Null	Enumeration[b]	Differentiation[b]
25	Normal	10	—	—	—	72	10	(18)	NA	NA
	Mumps	17	81	4	(15)	76	12	(12)	−	+
	Aseptic meningitis	15	82	4	(14)	74	10	(16)	−	+
	Neurological disease	8	82	3	(15)	75	8	(17)	−	+
62	Controls	30	77	—	—	66	25	(9)	+	+
	Idiopathic postinfectious polyneuropathy		85	—	—	48	37	(15)	+	+
	Normal	39	—	—	—	72	10	(18)	NA	NA
49	Mumps meningitis	17	81	4	(15)	76	12	(12)	+	+
	Optic neuritis	5	80	4	10	72	5	24	+	+
	MS	18	88	4	9	70	5	24	+	+
98	Bacterial meningitis	5	72	23	5	66	26	8	+	−
	Aseptic meningitis	7	79	17	4	74	23	3	+	−
	Optic neuritis	9	69	22	9	78	20	2	+	−
	MS	15	76	21	3	72	27	1	+	−

phocyte choriomeningitis antigen in the CSF prior to the blood also indicated a compartmentalized B-cell response.[19] Correlative morphological studies in inflammatory CNS conditions support this conclusion.[30]

3.6.2. Cell-Proliferation Reactions. Recent investigations have suggested that there is a direct correlation between *in vivo* lymphocyte replication to antigen and *in vivo* immune responses.[13] If the cellular replication is blocked, then there is no immune response. CSF cells from tubercular meningitis patients responded to purified protein derivative (PPD), as evidenced by autoradiographic techniques using tritiated thymidine, whereas the cells of patients with other diseases did not.[53] The response of these cells to phytohemagglutinin (PHA) was decreased.[73] Thus, this response to PPD in this disease appeared to be specific. However, nonspecific increase in cellular replication was demonstrated in patients with MS and infections of the CNS compared to patients with stable MS when their CSF lymphocytes were incubated with PHA, a nonspecific mitogen.[51] With the use of similar techniques, CSF lymphocytes from 25 patients with MS or acute disseminated encephalomyelitis responded to basic myelin protein, whereas lymphocytes from a group of 9 patients with other neurological disorders did not.[52] CSF lymphocytes from patients with mumps meningitis responded to mumps antigen in 4 of 5 cases, whereas peripheral blood lymphocytes responded in only 2 of 5 cases. CSF lymphocytes from 7 patients with other viral meningitis did not respond at all.[26]

Other cell-mediated immune functions may be differentially expressed by CSF lymphocytes compared to blood lymphocytes. In a patient with histoplasmosis meningitis who had a low CSF immu-

Table 5. Evidence for Specific CSF Compartment Immunologic Function

I. Immunoglobulin synthesis
   A. Morphological studies showing plasma cells in the CSF[30,84]
   B. Specific antibody titers appearing in the CSF before blood[19,84]
   C. Oligoclonal bands in CSF[14,88,89]
   D. Immunoglobulin synthesis *in vitro* by CSF lymphocytes[54,76–78]
II. Cellular-proliferation response
   A. CSF lymphocyte response to a nonspecific mitogen (PHA)[51,52,73]
   B. CSF lymphocyte response to a specific antigen in a disease state
      1. Tuberculosis (PPD)[53,73]
      2. Multiple sclerosis (basic myelin protein)[52]
      3. Mumps encephalitis (mumps viral antigen)[26]
      4. Herpes encephalitis (herpes simplex antigen)[73]
      5. *S. schenckii* meningitis (*S. schenckii* antigen)[73]
   C. Modified immunological response by hyperactive suppressive population[18]

noglobulin level, isolated CSF T lymphocytes suppressed the immunoglobulin synthesis of both the CSF and peripheral-blood B lymphocytes.[18] When the CSF B lymphocytes were incubated with T lymphocytes from the peripheral blood, there was a normal synthesis of immunoglobulin. This study was the first demonstration of hyperactive suppressor-cell function confined to the CSF. This observation might suggest that an altered population of lymphocytes may be isolated in the CSF compartment and result in a susceptibility to a disease state. Further corrobrative studies will have to be made before the hypothesis that hyperactive suppressor-cell function in the CSF results in localized

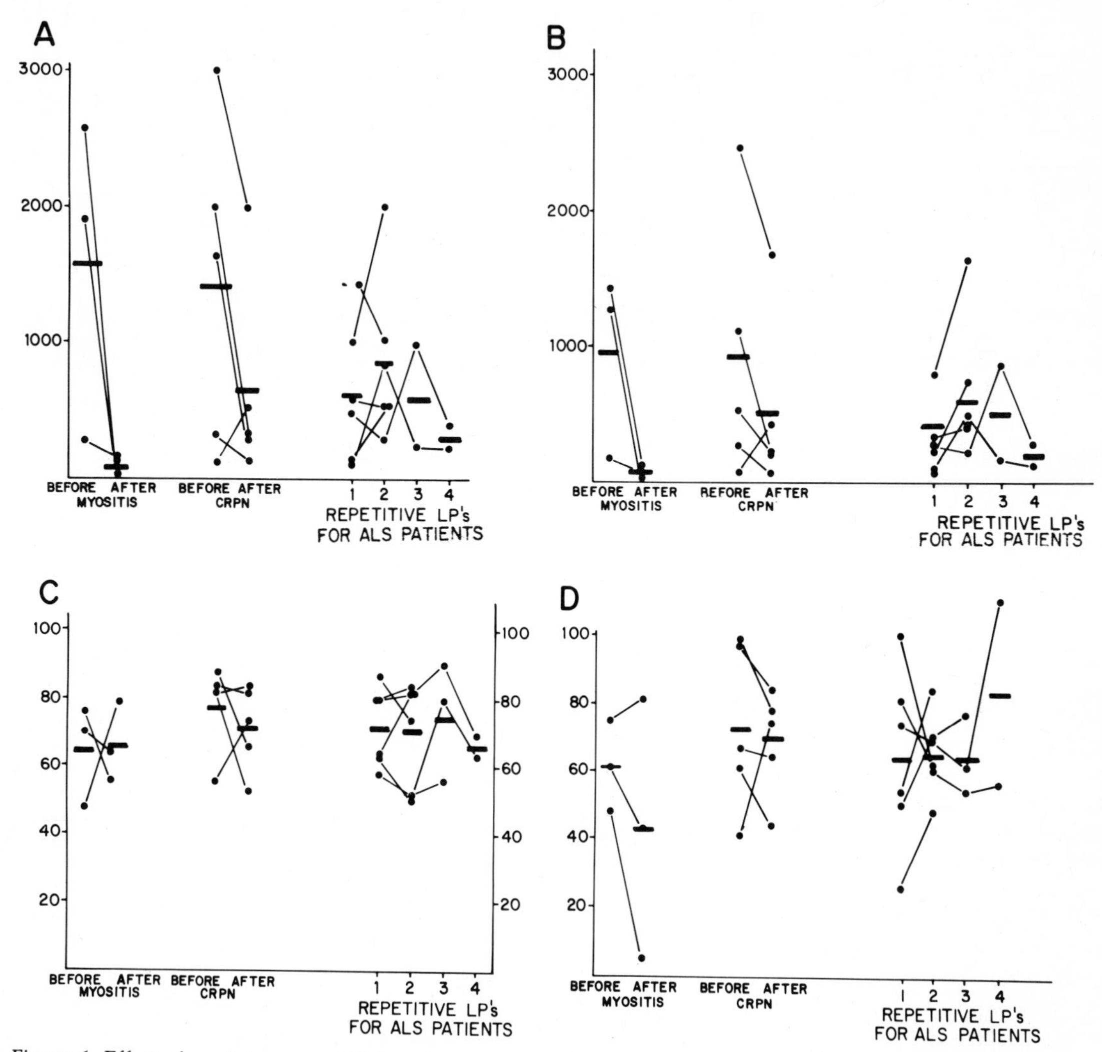

Figure 6. Effect of prednisone on CSF lymphocytes. (A) Total cells. Three patients with myositis and five patients with chronic relapsing polyneuropathy (CRPN) had CSF cellular determinations before therapy and after clinical response to immunosuppressive therapy. Six patients with amyotrophic lateral sclerosis (ALS) underwent at least two lumbar punctures. Individual values are represented by dots and connected by straight lines. Mean values for each group are represented by horizontal bars. (B) Total lymphocytes. Mean total lymphocyte count decreased significantly ($p < 0.05$) in myositis group. Total lymphocyte number decreased in three of three observations for myositis patients, in five of six observations for CRPN patients, and in only four of ten observations for ALS patients. (C) Lymphocyte proportion. While total lymphocytes decreased, percentage of lymphocytes did not vary significantly. (D) T-lymphocyte proportion. Percentage of T lymphocytes decreased significantly in some neuromuscular disease patients on prednisone, but also varied in untreated ALS patients.

infection of only the CNS can be generalized to explain specific CNS infections.

3.7. Effect of Immunosuppressive Drugs on Blood and CSF Lymphocytes

The effect of corticosteroids on peripheral-blood immunocompetent cells has been studied.[17] It has been known that intrathecal corticosteroids reduce the pleocytosis associated with adhesive arachnoiditis.[24] We have had the opportunity to study three patients with myositis (a non-CNS immunological disease) and five patients with chronic relapsing polyneuropathy (CRPN) with sequential blood and CSF samples obtained before initiating therapy and after a clinical response to prednisone (100 mg q.o.d.).[11] The percentages of T, B, and null lymphocytes were unchanged. The total number of CSF cells and likewise CSF lymphocytes was significantly reduced after steroid treatments in all patients (Fig. 6). There was a reduction in the total blood T and B but not null lymphocytes at 6 hr after prednisone, as previously described.[17] Associated with this reduction was a decrease in CSF immunoglobulins and clinical improvement in most patients.[11] This effect was seen in both the non-CNS and CNS immunological diseases.

Because of the large variation in cell concentrations in the CSF, we compared the results of sequential samples of CSF in six patients with ALS to those for neuromuscular disease patients before and after treatment with prednisone. Total CSF lymphocytes were decreased in seven of the eight patients on prednisone, but the lymphocyte subpopulation proportions did not change. In the ALS patients, the total cell count and lymphocyte subpopulation proportions varied within a smaller range than those parameters in patients treated with prednisone.

One patient with myositis had sequential blood and CSF sampling at various times after systemic methotrexate treatment. There was depression of the total lymphocyte count and all subpopulations for the first 5 days. The blood values were not similarly depressed, nor was there any correlation between blood and CSF for the total cell count or lymphocyte-subpopulation proportion.

4. Conclusion

The normal cellular elements found in the CSF consist primarily of lymphocytes (67%), monocyte–histocytes (30%), and other cellular elements making up the remaining few percent. The origin of these cellular elements is primarily hematogenic. There are "lymphatics" associated with the perivascular space in the CNS. A direct anatomical communication between the CSF compartment and the lymphatic system that can accommodate particles up to several microns in diameter exists. The exact trafficking of lymphocytes between blood and CSF and vice versa has not yet been determined. The percentages of T, B, and null lymphocytes found in the CSF are similar but not identical to those found in the blood. Significant differences between normal and disease states are found only when the total numbers of cells, as opposed to lymphocyte-subpopulation proportions, are compared. Thus, comparison of total numbers rather than proportions is the more valid means to compare different disease states. There is now substantial evidence for the local immunological function of lymphocytes found in the CSF, i.e., local immunoglobulin synthesis, as well as increased nonspecific and specific cellular responses to antigens. Corticosteroids, which suppress the total number of peripheral-blood lymphocytes, primarily T lymphocytes, also have a similar effect in the CSF compartment. This cellular reduction is associated with a reduction in CSF immunoglobulin level and clinical improvement in some patients with chronic relapsing polyneuropathy.

ACKNOWLEDGMENTS. We would like to acknowledge the support and encouragement to carry out these studies given by Dr. W. K. Engel, the technical expertise of Ms. Janet Harner and Mr. Joseph Sciabarrassi, and the typing skills of Ms. Phyllis Darnel, Ms. Priscilla Leventhal, and Ms. Linda Kelly. B. R. B. is the recipient of a Teacher-Investigator Development Award (1-KO7-NS-00385) from the National Institute of Neurological and Communicative Disorders and Stroke. J. D. C. was supported by a grant from the Muscular Dystrophy Association of America and a Research Advisory Group Grant from the Veterans Administration, Department of Medicine and Surgery.

References

1. ALLEN, J., SHEREMATA, W., COSGROVE, J., OSTERLAND, L., SHEA, M.: Cerebrospinal fluid T and B lymphocyte kinetics related to exacerbations of multiple sclerosis. *Neurology* **26**:579–583, 1976.
2. APPEL, E.: Preliminary data concerning the presence of alkaline phosphatase on the CSF cellular sediment. *Rev. Roum. Neurol.* **9**:287–289, 1972.

3. Astaldi, A., Pasino, M., Rosanda, C., Massimo, L.: T and B cells in cerebrospinal fluid in acute lymphocytic leukemias. *N. Engl. J. Med.* **294**:550–551, 1976.
4. Balhizen, J. C., Bots, G., Schaberg, A., Bosman, F.: Value of cerebrospinal fluid cytology for the diagnosis of malignancies in the central nervous system. *J. Neurosurg.* **48**:747–753, 1978.
5. Baringer, J.: A simplified procedure for spinal fluid cytology. *Arch. Neurol.* **22**:305–308, 1970.
6. Battifora, H., Hiduelgi, D.: Improved apparatus for spinal fluid cytomorphology. *Acta Cytol.* **22**:170–171, 1978.
7. Bennett, J., Ruberg, J., Dixon, S.: A new semiautomated method for the concentration of cerebrospinal fluid for cytologic examination. *Am. J. Clin. Pathol.* **50**:533–536, 1968.
8. Bosch, I., Oehmichen, M.: Eosinophilic granulocytes in cerebrospinal fluid: Analysis of 94 cerebrospinal fluid specimens and review of the literature. *J. Neurol.* **219**:93–105, 1978.
9. Braito, A., Giacchino, R., Viscoli, C.: Immunological studies in cerebrospinal fluid or acute viral and tubercular meningitis. *Boll. Ist. Sieroter. Milan.* **55**:491–494, 1976.
10. Brierley, J., Field, E.: The connections of the spinal subarachnoid space with the lymphatic system. *J. Anat.* **68**:153–166, 1948.
11. Brooks, B., Cook, J., Trotter, J., McFarland, D., Engel, W.: Effect of prednisone therapy on cerebrospinal fluid immunoglobulins and lymphocytes in chronic idiopathic relapsing polyneuropathy (CIRPN) and other neuromuscular disorders. *Trans. Am. Neurol. Assoc.* **102**:180–183, 1977.
12. Burechailo, F., Cunningham, T.: Counting cells in cerebrospinal fluid collected directly on membrane filters. *J. Clin. Pathol.* **27**:101–105, 1974.
13. Chess, L., Schlossman, S. E.: Human lymphocyte subpopulations. In Kunkel, H., Dixon, F. (eds.): *Advances in Immunology*. New York, Academic Press, **25**:213–241, 1977.
14. Cohen, S., Bannister, R.: Immunoglobulin synthesis within the central nervous system in disseminated sclerosis. *Lancet* **1**:366–367, 1967.
15. Cook, J., Trotter, J., Brooks, B., Engel, W.: Identifying T and B lymphocytes in human cerebrospinal fluid. *Neurology* **26**:376, 1976.
16. Cook, J., Trotter, J., Brooks, B., Sciabarrassi, J., Engel, W. K.: A method to quantitatively identify subpopulations of lymphocytes in cerebrospinal fluid *Neurology* (submitted).
17. Cook, J., Trotter, J. L., Engel, W. K., Siabbarrassi, J. S.: The effects of single-dose alternate-day prednisone therapy on the immunological status of patients with neuromuscular diseases. *Ann. Neurol.* **3**:166–176, 1978.
18. Couch, J., Abou, N., Sagawa, A.: *Histoplasma* meningitis with hyperactive suppressor T cells in cerebrospinal fluid. *Neurology* **28**:119–123, 1978.
19. Deibel, R., Schryver, G.: Viral antibody in the cerebrospinal fluid of patients with acute central nervous system infections. *J. Clin. Microbiol.* **3**:397–401, 1976.
20. Duffy, P., Simon, J., Defendini, R., Karalian, S.: The study of cells in cerebrospinal fluid by electron microscopy. *Arch. Neurol.* **21**:358–362, 1969.
21. Dyken, P.: Cerebrospinal fluid cytology: Practical clinical usefulness. *Neurology* **25**:210–217, 1975.
22. Dwyer, J. M.: Identifying and enumerating human T and B lymphocytes. *Prog. Allergy* **21**:178–260, 1976.
23. Enestrom, S.: Die Saugkammermethode. *Sammlung zwangloser Abhandlungen aus dem Gebiet der Psychiatrie und Neurologie* **31**:23, 1966.
24. Feldman, S., Behar, A.: Effect of intrathecal hydrocortisone on advanced adhesive arachnoiditis and cerebrospinal fluid pleocytosis. *Neurology* **9**:251–258, 1959.
25. Fryden, A.: B and T lymphocytes in blood and cerebrospinal fluid in acute aseptic meningitis. *Scand. J. Immunol.* **6**:1283–2388, 1977.
26. Fryden, A., Link, H., Moller, E.: Demonstration of cerebrospinal fluid lymphocytes sensitized against virus antigens in mumps meningitis. *Acta Neurol. Scand.* **57**:396–404, 1978.
27. Fuchs, A., Rosenthal, R.: Physikalischchemishe, zytologische und anderweitige Untersuchungen der Zerebrospinalflüssigkeit. *Wien. Med. Presse* **45**:2081, 1904.
28. Gadson, D., Emery, J.: Fatty changes in the brain in perinatal and unexpected death. *Arch. Dis. Child.* **51**:42–48, 1976.
29. Gangi, D., Collard-Ronge, E., Balleriaux-Waha, D., Hildebrandt, J., Stryckmans, P.: T lymphocytes in cerebrospinal fluid. *N. Engl. J. Med.* **294**:902, 1976.
30. Glasser, L., Payne, C., Corrigan, J.: The *in vivo* development of the plasma cell: A morphologic study of human cerebrospinal fluid. *Neurology* **27**:448–459, 1977.
31. Goasgen, J., Sabouraud, O.: Rosettes mouton sur lymphocytes de liquide cephalo-rachidien. *Nouv. Presse Med.* **3**:2266, 1974.
32. Gondos, B.: Cytology of cerebrospinal fluid: Technical and diagnosis considerations. *Ann. Clin. Lab. Sci.* **6**:152–157, 1976.
33. Gondos, B., King, E.: Cerebrospinal fluid cytology: Diagnostic accuracy and comparison of different techniques. *Acta Cytol.* **20**:542–547, 1976.
34. Green, G., Wagstaff, J.: Nucleopore membrane filter techniques for diagnostic cytology of urine and other body fluids. *Med. Lab. Technol.* **30**:265–271, 1973.
35. Greenwood, B., Whittle, H., Oduloju, K., Dourmashikin, R.: Lymphocytic infiltration of the brain in sleeping sickness. *Br. Med. J.* **2**:1291–1292, 1976.
36. Guseo, A.: Morphological signs as indications of functions of cells in the cerebrospinal fluid. *J. Neurol.* **212**:159–170, 1976.
37. Guseo, A., Lechner, G., Bierleutgeb, F.: A simple method for demonstrating cells in the cerebrospinal

fluid by scanning electron microscopy. *Acta Cytol.* **21**:352–355, 1977.
38. Guseo, A.: Classification of cells in the cerebrospinal fluid. *Eur. Neurol.* **15**:169–176, 1977.
39. Hansen, H., Bender, R., Shelton, B.: The cyto-centrifuge and cerebrospinal fluid cytology. *Acta Cytol.* **18**:259–262, 1974.
40. Harms, D.: Zum diagnostischen Wert der Liquorzytologie. *Kin. Paediatr.* **187**:142–150, 1975.
41. Herndon, R. M.: The electron microscopic study of cerebrospinal fluid sediment. *J. Neuropathol. Exp. Neurol.* **28**:128, 1969.
42. Herndon, R. M.: Electron-microscopic studies on cerebrospinal fluid sediment. In Wood, J. H. (ed.): *Neurobiology of Cerebrospinal Fluid I.* New York, Plenum Press, 1980.
43. Hoectge, G., Furlan, A., Hoffman, G.: The differential cytology of cerebrospinal fluids prepared by cyto-centrifugation. *Cleveland Clin. Q.* **43**:237–246, 1976.
44. Hutton, W. E.: A survey of the application of the "molecular" membrane filter to the study of cerebrospinal fluid cytology. *Am. J. Clin. Pathol.* **30**:407–412, 1958.
45. Iivanainen, M., Taskinen, E.: Cytological examination of the cerebrospinal fluid. *Ann. Clin. Res.* **5**:80–86, 1973.
46. Iivanainen, M., Taskinen, E.: Differential cellular increase in cerebrospinal fluid after encephalography in mentally retarded patients. *J. Neurol. Neurosurg. Psychiatry* **37**:1252–1258, 1974.
47. Kam-Hansen, S., Fryden, A., Link, H.: B and T lymphocytes in cerebrospinal fluid and blood in multiple sclerosis, optic neuritis, and mumps meningitis. *Acta Neurol. Scand.* **58**:95–103, 1978.
48. Kolar, O., Zeman, W.: Spinal fluid cytology. *Arch. Neurol.* **18**:44–51, 1968.
49. Kolmel, H.: A method for concentrating cerebrospinal fluid cells. *Acta Cytol.* **21**:154–157, 1977.
50. Krentz, M., Dyken, P.: Cerebrospinal fluid cytomorphology: Sedimentation vs. filtration. *Arch Neurol.* **26**:253–257, 1972.
51. Levinson, A., Lisak, R., Zweiman, B.: Immunologic characterization of cerebrospinal fluid lymphocytes: Preliminary report, *Neurology* **26**:693–695, 1975.
52. Lisak, R., Zweiman, B.: *In vitro* cell-mediated immunity of cerebrospinal fluid lymphocytes to myelin basic protein in primary demyelinating diseases. *N. Engl. J. Med.* **297**:850–853, 1977.
53. Malashkhia, Y., Geladze, M.: Autoradiographic studies of cultures of cerebrospinal fluid lymphocytes in nonsuppurative meningitis. *Neurology* **26**:1081–1099, 1976.
54. Manconi, P., Zaccheo, D., Bugiani, O., Fadda, M., Grifoni, V., Mantovani, G., Giacco, G., Tognella, S.: T and B lymphocytes in normal cerebrospinal fluid. *N. Engl. J. Med.* **294**:49, 1976.
55. Manconi, P., Fadda, M., Cadoni, A., Cornaglia, P., Zaccheo, D., Grifoni, V.: Subpopulations of T lymphocytes in human extravascular fluids. *Int. Arch. Allergy Appl. Immunol.* **56**:385–390, 1978.
56. Manconi, P., Zaccheo, D., Bugiani, O., Fadda, M., Cadoni, A., Marrosu, M., Cianchetti, C., Grifoni, V.: Surface markers on lymphocytes from human cerebrospinal fluid. *Eur. Neurol.* **17**:87–91, 1978.
57. Marks, V., Marrack, D.: A technique for staining of cells from the CSF. *Confin. Neurol.* **20**:310–315, 1960.
58. Mathios, A., Neilsen, S., Barrett, D., King, E.: Cerebrospinal fluid cytomorphology: Identification of benign cells originating in the central nervous system. *Acta Cytol.* **21**:403–412, 1977.
59. McGarry, P., Holmquist, N., Carmel, A.: A postmortem study of cerebrospinal fluid with histological correlation. *Acta Cytol.* **13**:48–52.
60. Moser, R., Robinson, J., Prostko, E.: Lymphocyte subpopulations in human cerebrospinal fluid. *Neurology* **26**:726–238, 1976.
61. Mussini, J. M., Hauw, J. J., Escourolle, R.: Combined Nomarski interference contrast and immunofluorescent study of neuropathological specimens: CSF sediments and paraffin embedded brain tissues. *Biomedicine* **27**:280–282, 1977.
62. Naess, A.: Demonstration of T-lymphocytes in cerebrospinal fluid. *Scand. J. Immunol.* **5**:165–168, 1976.
63. Naess, A., Nyland, H.: Multiple sclerosis: T-lymphocytes in cerebrospinal fluid and blood. *Eur. Neurol.* **17**:61–65, 1978.
64. Neuwelt, E., Doherty, D.: Toxicity kinetics and clinical potential of subarachnoid lymphocyte infusions. *J. Neurosurg.* **47**:205–217, 1977.
65. Neuwelt, E. A., Hill, S. A.: Intrathecal lymphocyte infusions: Clinical and animal toxicity studies. In Wood, J. H. (ed.): *Neurobiology of Cerebrospinal Fluid I.* New York, Plenum Press, 1980.
66. Nikoskelainen, E., Irjala, K., Salmi, T. T.: Cerebrospinal fluid findings in patients with optic neuritis. *Acta Ophthalmol.* **53**:105–119, 1974.
67. Nyland, H., Naess, A.: Lymphocyte subpopulations in blood and cerebrospinal fluid from patients with acute Guillain-Barré syndrome. *Eur. Neurol.* **17**:247–252, 1978.
68. Oehmichen, M.: Characterization of mononuclear phagocytes in human CSF using membrane markers. *Acta Cytol.* **20**:548–552, 1976.
69. Oehmichen, M.: *Mononuclear Phagocytes in the Central Nervous System—Origin, Mode of Distribution, and Function of Progressive Microglia, Perivascular Cells of Intracerebral Vells, Free Subarachnoid Cells, and Epiplexus Cells,* Berlin, Springer-Verlag, 1978.
70. Oehmichen, M.: *Cerebrospinal Fluid Cytology: An Introduction and Atlas.* Philadelphia, W. B. Saunders, 1976.
71. Oehmichen, M., Genicic, M.: Experimental studies on kinetics and functions of mononuclear phagocytes

of the central nervous system. *Acta Neuropathol. Suppl.* **VI:**285–290, 1975.
72. OEHMICHEN, M., GRUNINGER, H.: Cytokinetic studies on the origin of cells of the cerebrospinal fluid. *J. Neurol.* **22:**165–176, 1976.
73. PLOUFFE, J. F., SILVA, J., FEKETY, R., BAIRD, I.: Cerebrospinal fluid lymphocyte transformations in meningitis. *Arch. Intern. Med.* **139:**191–194, 1979.
74. PRINEAS, J. W.: Multiple sclerosis: Presence of lymphatic capillaries and lymphoid tissue in the brain and spinal cord. *Science* **203:**1123–1125, 1979.
75. QUINCKE, H.: Die Lumbarpunktion des Hydrocephalus. *Berl. Klin. Wochenschr.* **28:**929–933, 1891.
76. SANDBERG-WOLLHEIM, M.: Immunoglobulin synthesis *in vitro* by cerebrospinal fluid cells in patients with multiple sclerosis. *Scand. J. Immunol.* **3:**717–730, 1974.
77. SANDBERG-WOLLHEIM, M.: Immunoglobulin synthesis *in vitro* by cerebrospinal fluid cells in patients with meningoencephalitis of presumed viral origin. *Scand. J. Immunol.* **4:**617–622, 1975.
78. SANDBERG-WOLLHEIM, M., ZETTERUALL, O., MÜLLER, R.: *In vitro* synthesis of IgG by cells from the cerebrospinal fluid in a patient with multiple sclerosis. *Clin. Immunol.* **4:**401–405, 1969.
79. SANDBERG-WOLLHEIM, M., TURESSON, I.: Lymphocyte subpopulations in the cerebrospinal fluid and peripheral blood in patients with multiple sclerosis. *Scand. J. Immunol.* **4:**831–836, 1975.
80. SAYK, J.: Ergebnisse neuer liquor-cytologischer Untersuchungen mit dem Sedimentierkammer-Verfahren. *Arzt Wochenschr.* **9:**1042–1045, 1954.
81. SCELSI, R., BASSI, P., POLONI, M., NAPPI, G.: Cytomorphologic changes in the cerebrospinal fluid sediment: A comparative study by cytocentrifugation and cytosedimentation techniques. *Boll. Soc. It. Biol. Sper.* **111:**1229–1233, 1976.
82. SCHEDIFKA, V., BLUME, R.: Zur Auswertung von Liquorzellpräparaten: Die Zusammensetzung des Zellbildes am Rand und im Zentrum von Sedimentkammer Präparaten. *Z. Aerztl. Fortbild.* **72:**435–436, 1978.
83. SCHÖNBERG, H.: Eine einfache Methode zur Herstellung gut differenzierbaren Liquorzellpräparate. *Dsch. Med. Wochenschr.* **74:**881, 1949.
84. SCHULLER, E., DELASNERIE, N., DELOCHE, G., LORIDAN, M.: Multiple sclerosis: A two phase disease? *Acta Neurol. Scand.* **49:**453–460, 1973.
85. SCHWALBE, G.: Die Arachnoidalraum, ein Lymphraum und sein Zusammenhang mit dem Perichoriordalraum. *Zentralbl. Med. Wiss.* **7:**465–467, 1869.
86. SCHWARZE, E.: Zytomorphologischer Vergleich der kleinen und "grossen Rundzelle" des Liquor cerebrospinalis mit Lymphozyten des peripheren Blutes. *Verh. Dtsch. Ges. Pathol.* **57:**262–268, 1973.
87. SHEREMATA, W., ALLEN, J., SAZANT, A., COSGROVE, J., OSTERLAND, K.: Cerebrospinal fluid T and B lymphocyte responses in exacerbations of multiple sclerosis. *Trans. Am. Neurol. Assoc.* **101:**40–45, 1976.
88. SIDEN, A., KJELLIN, K.: Isoelectric focusing of CSF and serum proteins in neurological disorders combined with benign and malignant proliferations of reticulocytes, lymphocytes, and plasmocytes. *J. Neurol.* **216:**251–264, 1977.
89. SKOLDENBERG, B., CARLSTRON, A., FORESGEN, M., NORRBY, E.: Transient appearance of oligoclonal immunoglobulins and measles virus antibodies in the cerebrospinal fluid in a case of acute measles encephalitis. *J. Exp. Immunol.* **23:**451–455, 1976.
90. SORNAS, R.: A new method for the cytological examination of the cerebrospinal fluid. *J. Neurol. Neurosurg. Psychiatry* **30:**568–577, 1967.
91. SORNAS, R.: Transformation of mononuclear cells in cerebrospinal fluid. *Acta Cytol.* **15:**545–552, 1971.
92. SORNAS, R.: The cytology of the normal cerebrospinal fluid. *Acta Neurol. Scand.* **48:**313–320, 1972.
93. SORNAS, R., OSTLUND, H., MÜLLER, R.: Cerebrospinal fluid cytology after stroke. *Arch. Neurol.* **26:** 489–501, 1972..
94. STOKES, H., O'HARA, C., BUCHANAN, R., OLSON, W.: An improved method for examination of cerebrospinal fluid cells. *Neurology* **25:**901–906, 1975.
95. STOLZE, H.: Zur Frage des "Zerfalls" und "Alterns" der Liquorzellen. *Arch. Psychiatr. Nervenkr.* **116:**263–267, 1943.
96. TANNENBERG, W., JEHN, U.: The life cycle of antibody-forming cells. *Immunology* **22:**589–600, 1972.
97. TOURTELLOTTE, W.: A selected review of reactions of the cerebrospinal fluid to disease. In Fields, W. S. (ed.): *Neurological Diagnostic Techniques.* Springfield, Illinois, Charles C. Thomas, 1966, pp. 25–50.
98. TRAUGOTT, U.: T and B lymphocytes in the cerebrospinal fluid of various neurological diseases. *J. Neurol.* **219:**185–197, 1978.
99. WALDMAN, T. A., DURM, M., BRODER, S., BLACKMAN, M.,BLAESE, R. M., STROBER, W.: Role of suppressor cells in the pathogenesis of common variable hypogammaglobulinemia. *Lancet* **2:**609–613, 1974.
100. WATSON, P.: A slide centrifuge: An apparatus for concentrating cells in suspension onto a microscope slide. *J. Lab. Clin. Med.* **68:**494–501, 1966.
101. WEDEMEYER, H. E.: Ueber die Zellen im Liquor cerebrospinalis bei der Meningitis tuberculosa. *Klin. Wochenschr.* **14:**858, 1935.
102. WERTLAKE, P., MARKOVITS, B., STELLAR, S.: Cytologic evaluation of cerebrospinal fluid with clinical and histologic correlation. *Acta Cytol.* **16:**224–239, 1972.
103. WIDAL, J., SICARD, L., RAVAUT, G.: Cytologie du liquide cephalorachidienne au cours de quelques processus meninges chroniques. *Bull. Soc. Med. Hop. Paris* **18:**31, 1901.
104. WOODRUFF, J. H.: Cerebrospinal fluid cytomorphology using cytocentrifugation. *Am. J. Clin. Pathol.* **60:**621–627, 1973.

CHAPTER 37

Intrathecal Lymphocyte Infusions

Clinical and Animal Toxicity Studies

Edward A. Neuwelt and Suellen A. Hill

1. Introduction

Malignant gliomas are universally fatal. Untreated patients with this disease survive about 6 weeks from onset of symptoms. Surgery, which may vary from needle biopsy to radical resection, is the most widely accepted therapeutic modality, but results in a median survival of only 6 months.[8] Radiologists have demonstrated that megavoltage irradiation after surgical decompression can extend the median survival to 9 months.[36] Wilson[40] has shown that the nitrosoureas, 1,3-*bis*(2-chlorethyl)-1-nitrosourea (BCNU) and 1-(2-chlorethyl)-3-cyclohexyl-1-nitrosourea (CCNU), may have efficacy in therapy of malignant astrocytomas. They are not so effective as radiation, but the combination of radiation and BCNU results in an additional 6 weeks of median survival.[39] Therefore, today, the combination of surgery, radiation, and chemotherapy is the best available therapy for malignant gliomas. Nonetheless, it results in a median survival of only 10 months, which is not very impressive.

Several factors should be considered as responsible for these dismal therapeutic results. First, the infiltrative nature of these tumors makes primary total resection virtually impossible, and if it is attempted, it results in a devastating deficit to the patient. Second, since the brain is contained in a rigid and closed cavity, even minor changes in tumor size may result in an elevation of intracranial pressure with brain herniation, leading to death. Finally, the blood–brain barrier, a function of the tight junctions present between endothelial cells in central nervous system (CNS) capillaries and the parenchyma, is impermeable to most chemotherapeutic agents available at present.

The blood–brain barrier is also important in malignant gliomas because it partitions the CNS from the body's immune surveillance system. In fact, the blood–brain barrier plus the absence of CNS lymphatics are the main reasons that the CNS is designated as an immunologically privileged site.[15] Yet, glioma patients have been shown to have circulating tumor-specific killer lymphoid cells.[4,6,9,14,38] The most recent, largest, and best-controlled study of this tumor-specific cytotoxicity was reported by Levy.[12] He reported specific lymphocytotoxicity in 85% of 41 glioma patients. However, he also observed significant levels of serum blocking factors in 80% of his patients.[13]

As reported by Ridley and Cavanaugh,[21–27] lymphoid cells, although present in 50% of malignant gliomas, are rarely able to penetrate the tumor parenchyma beyond the perivascular Virchow–Robin spaces. Nonetheless, these lymphoid cells seem to play a role in the pathogenesis of at least some glioma patients. Brooks *et al.*[3] found a statistically significant correlation between the presence of lymphocytic infiltrates in gliomas and prognosis. Palma *et al.*[20] came to the same conclusion in a series of 228 glioma patients studied. Takakura and co-workers[33–35] demonstrated enhanced survival of glioma patients in a small series when autogenous white blood cells were placed in the tumor cavity. Young

Edward A. Neuwelt, M.D., and **Suellen A. Hill, R.N.** • Division of Neurosurgery, The University of Texas Health Science Center at Dallas, Dallas, Texas 75235.

et al.[41] have obtained similar results. It is possible, therefore, that patients with malignant gliomas may derive significant benefit from direct exposure of the tumor bed to white cells free of serum blocking factor.

The particular type of white cell that should have efficacy, if any does, is the lymphoid cell, which not only is responsible for immunological memory, but also is the effector of cellular immunity. In addition, because of the paucity of cytoplasmic organelles, as compared to other white cells, a lymphoid cell may have less toxicity when infused into the subarachnoid space, or tumor bed.

This chapter presents a series of animal and clinical studies investigating the feasibility of intrathecal lymphocyte infusions and the toxicity associated with xenogeneic-and syngeneic-lymphocyte infusions in animals and autologous-lymphocyte infusions in man. No attempt is made to draw conclusions regarding the therapeutic efficacy of such a practice in patients; rather, the intent is to present a particularly interesting and innovative avenue for further investigation.

2. Feasibility Studies: Animals

2.1. Acquisition of Cells

Xenogeneic lymphocytes used in our animal investigations were obtained from human volunteers via a Haemonetics Model 30 Cell Separator (Haemonetics Corp., Natick, Massachusetts) according to the method of Aisner *et al.*[1] A Ficoll–Hypaque gradient was used to harvest a highly purified lymphocyte concentration,[5] and the cells were resuspended in synthetic cerebrospinal fluid (CSF) (Elliot's B Solution, obtained from the National Cancer Institute, Bethesda, Maryland).

Cell counts were determined on a ZB-I Coulter Counter (Coulter Electronics, Hialeah, Florida), and cellular composition was evaluated by examining Wright's-stained smears. Lymphocytes comprised 98% or more of the leukocytes present, with less than 1 erythrocyte/white blood cell (WBC), and 5–10 platelets/WBC. Viability was determined by dye exclusion (using trypan blue), and all suspensions were cultured for bacterial and fungal contaminants. [With the Haemonetics Model 30 Cell Separator, it was possible to obtain 6–8 $\times$ 10^9 lymphocytes in only 2–3 hr in a volume of 200–300 ml with negligible red blood cell (RBC) loss to the donor. Further purification and resuspension of the cells yielded a concentration of 10^8–10^9 cells/ml.]

Syngeneic lymphocytes for each experiment were obtained by cardiac puncture from nine highly inbred, B-strain female Dutch rabbits (obtained from Jackson Laboratory, Bar Harbor, Maine). The blood was pooled and the lymphocytes purified as described above, noting that the purified rabbit lymphocyte suspension contained up to 10 RBCs/WBC.

EL-4 lymphoma cells were passaged weekly in C-57 mice by injection of 10^6 cells intraperitoneally. Cells were harvested 7–10 days after passage by peritoneal lavage after the animal was sacrificed by cord transsection, and purified as described above.

2.2. CSF Studies and Analysis of Cells

CSF cell counts were determined using both a hemacytometer and a ZB-I Coulter Counter on CSF that contained less than 500 cells/mm^3 but the ZB-I Coulter Counter alone if the CSF contained more than 500 cells/mm^3. Cellular composition and cell viability were determined as described above. The rabbit peripheral blood lymphocytes and postinfusion CSF cells were further analyzed on a Model 6301 Cytograf (Bio-Physics Instrument Co., Mahpac, New York). The instrument assigns each cell to one of 100 different channels, according to its absorbance of a laser beam.[28] Since cell size bears a direct relation to laser-beam absorbance, the instrument provides an accurate profile of the size of various cell suspensions. The CSF glucose was determined by the glucose oxidase method[37] and the CSF protein by the Biuret method.[11]

2.3. Intrathecal Infusion of Xenogeneic Lymphocytes into Rabbit Subarachnoid Space

Xenogeneic lymphocytes were infused into anesthetized New Zealand white rabbits via a spinal needle inserted into the cisterna magna percutaneously. If the tap was atraumatic and CSF easily aspirated, 1–2 ml of CSF was obtained for baseline studies and 2 ml of the lymphocyte suspension injected. (In a total of 32 initial xenogeneic lymphocyte infusions, the average number of cells infused was 8 $\times$ 10^8, of which 92% were viable and 98% were lymphocytes.) A larger inoculum was not practical, since the entire volume of the rabbit subarachnoid space is only 10–12 ml. A second cisternal puncture was performed either 12 or 36 hr later, then a third at 4–10 days, and finally a fourth 4 weeks after infusion. In most cases, the second tap was performed at 36 hr to avoid anesthetic complications. The results of the diagnostic studies done on CSF obtained from successive cisternal taps are summarized in Table 1 and Fig. 1.

Preinfusion rabbit CSF was found to be very similar to normal human CSF, containing 0–3 lymphocytes/mm^3. At 12 hr postinfusion, the CSF cell counts ranged from 1150 to 70,000 cells/mm^3. The variability of this figure was due to the different numbers of cells infused, the quality of the cisternal tap, and possibly the amount of leakage of CSF and cells back through the needle tract. Differential cell counts and viability studies on the 12-hr-postinfusion CSF revealed that the cells were all viable lymphocytes. By 36 hr postinfusion, the total cell count had dropped dramatically (Fig. 1), and no polymorphonuclear leukocytes were observed. At 10 days postinfusion, only a few hundred cells, again all viable lymphocytes, remained, and by 4 weeks postinfusion, the CSF cell count had returned to normal.

The high cell viabilities following intrathecal lymphocyte infusion suggest that cell destruction was not the cause of the rapid and marked drop in total cell count observed following intrathecal cell infusion. Furthermore, if a host response had occurred, resulting in the destruction of the infused lymphocytes, systemic morbidity, hyperthermia, and a leukocytosis associated with hypoglycorrhachia and infiltration of the CSF with polymorphonuclear leukocytes might have resulted. None of these events was observed (Table 1). Indeed, the only parameter that seemed to change following lymphocyte infusion was a transient rise in the total CSF protein (Fig. 1). Therefore, it appeared that the rapid drop in CSF cell count following lymphocyte infusion was due to either sequestration in the CNS or escape from the CNS.

Only 4 of 43 rabbits showed any evidence of toxicity to a single infusion of xenogeneic lymphocytes (Table 2). Two rabbits developed a transient mono-

Table 1. Intrathecal Infusion of Xenogeneic and Syngeneic Lymphocytes into New Zealand White Rabbits

	CSF glucose (mg%)			CSF protein (mg%)			Rectal temperature (°C)		
Time of measurement	*N*	Mean	Range	*N*	Mean	Range	*N*	Mean	Range
Xenogeneic lymphocytes									
First administration									
Preinfusion	23	75	62–98	23	70	23–129	22	38.34	36.5–39.9
Postinfusion									
12 hr	2	56.6	53–64	2	142	52–233	3	38.4	—
36 hr	24	79.4	60–112	24	192.8	40–618	23	38.78	36.5–40.5
4 weeks	14	69	52–80	13	64	39–124	20	38.5	37.5–39.5
Second administration									
Preinfusion	4	64	53–75	4	11	74–280	6	38	37–39.5
Postinfusion									
36 hr	—	—	—	—	—	—	5	38.5	38–39.5
4 weeks	1	73	—	1	58	—	1	38	—
Syngeneic lymphocytes									
First administration									
Preinfusion	3	83	82–87	—	—	—	4	38	37.5–39
Postinfusion									
36 hr	2	75	73–77	4	68	29–119	4	38.1	38–38.75
4 weeks	1	95	—	1	58	—	—	—	—
Second and third administrations									
Preinfusion	3	83	91–86	3	45	7–76	3	38	37–38.5
Postinfusion									
36 hr	3	103	82–142	14	57	44–82	3	38	37–39
4 weeks	2	38.3	38–38.7	2	38.5	38.5	2	38	38

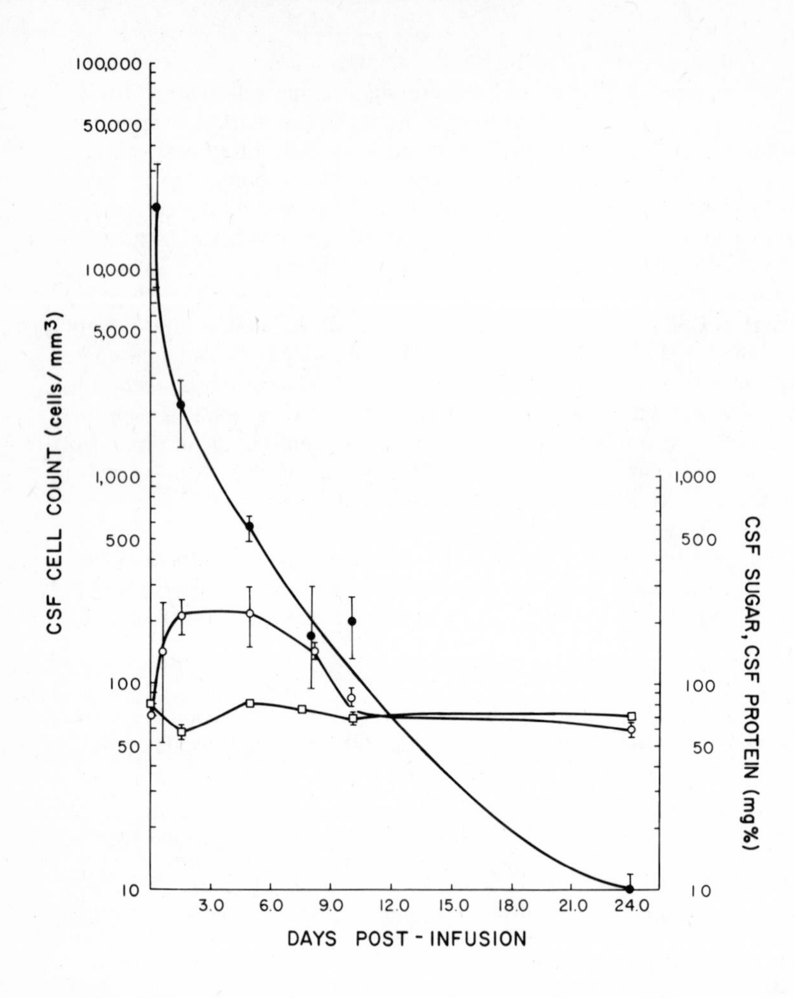

Figure 1. Graph showing changes in total CSF cell count, CSF protein, and CSF glucose following intrathecal infusion of human lymphocytes into New Zealand white rabbits. Vertical brackets indicate ± S.E.M. (●) Total CSF cell count; (□) CSF glucose; (○) CSF protein. Reproduced from Neuwelt and Doherty,[19] with permission of publisher.

Table 2. Toxicity of Intrathecal Infusion of Lymphocytes in 43 Rabbits Followed for an Average of 75 Days (Range: 2–197 Days)

Toxicity	Number of rabbits
Mortality	1
Neurological status	
Transient monoparesis	2
Transient torticollis	1
Allergic reactions[a]	2

[a] Both allergic reactions occurred in rabbits receiving xenogeneic lymphocytes for the second time. Two other rabbits received the same lymphocyte preparation without developing allergic symptoms, but had not previously been exposed to human lymphocytes. Reproduced from Neuwelt and Doherty,[19] with permission of publisher.

paresis, one rabbit developed torticollis that resolved over 2 weeks, and one rabbit died after infusion. It is worth noting that the rabbits could have developed these complications from the cisternal taps rather than as a direct result of the lymphocyte infusions. The single mortality was in a rabbit that died unnoticed 24–72 hr after infusion, over a weekend; an autopsy was not possible. Because of the possible need for multiple intrathecal infusions in clinical studies, the toxicity of a second infusion of xenogeneic lymphocytes was evaluated. Although there was no histological evidence of inflammation in the cervical spinal cord and leptomeninges, it was often difficult to obtain a free flow of CSF after the initial cisternal punctures. It was possible in a few rabbits, nonetheless, to freely aspirate CSF and infuse cells a second time.

In one such study, there were two rabbits that

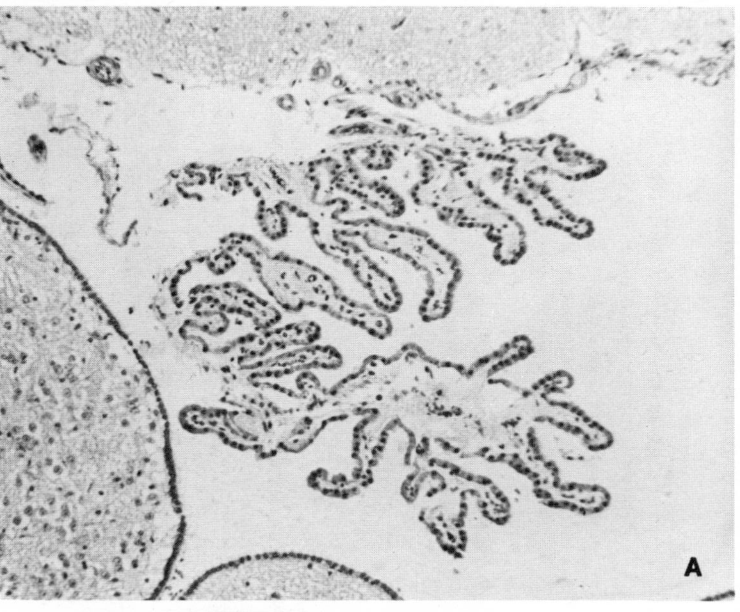

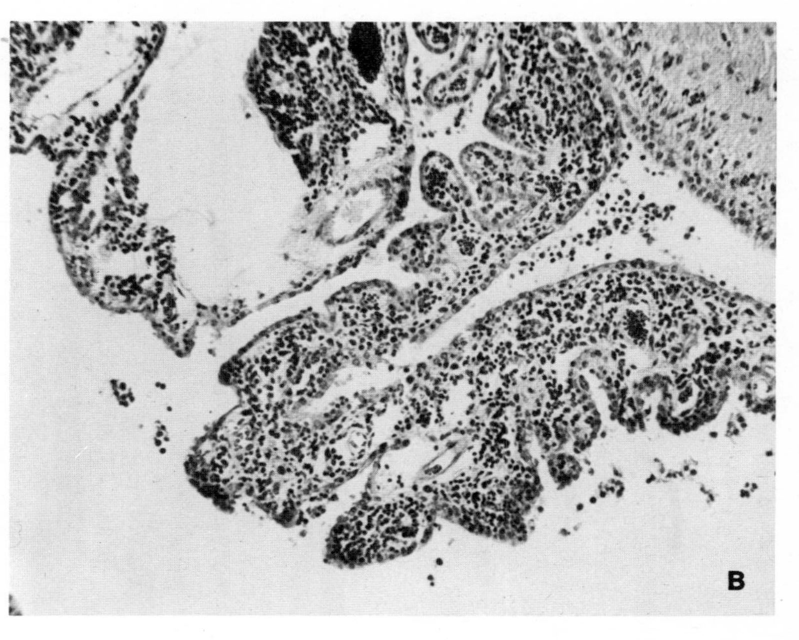

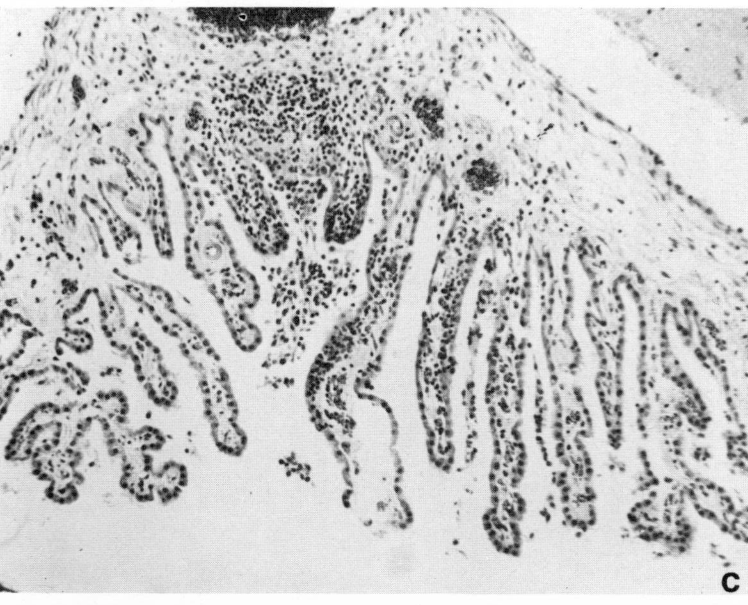

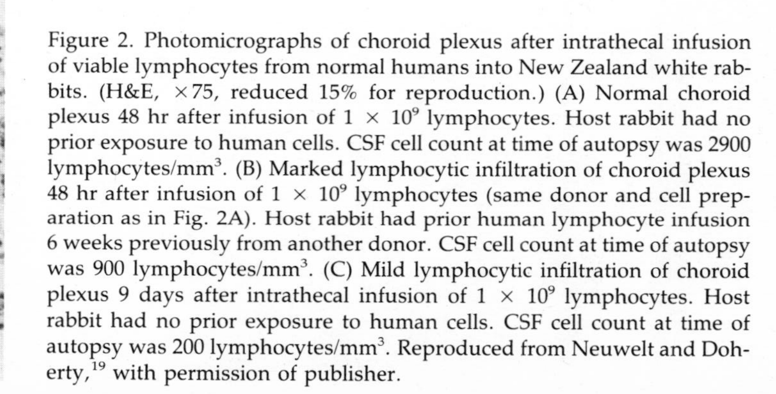

Figure 2. Photomicrographs of choroid plexus after intrathecal infusion of viable lymphocytes from normal humans into New Zealand white rabbits. (H&E, ×75, reduced 15% for reproduction.) (A) Normal choroid plexus 48 hr after infusion of 1×10^9 lymphocytes. Host rabbit had no prior exposure to human cells. CSF cell count at time of autopsy was 2900 lymphocytes/mm^3. (B) Marked lymphocytic infiltration of choroid plexus 48 hr after infusion of 1×10^9 lymphocytes (same donor and cell preparation as in Fig. 2A). Host rabbit had prior human lymphocyte infusion 6 weeks previously from another donor. CSF cell count at time of autopsy was 900 lymphocytes/mm^3. (C) Mild lymphocytic infiltration of choroid plexus 9 days after intrathecal infusion of 1×10^9 lymphocytes. Host rabbit had no prior exposure to human cells. CSF cell count at time of autopsy was 200 lymphocytes/mm^3. Reproduced from Neuwelt and Doherty,[19] with permission of publisher.

had previously received intrathecal xenogeneic lymphocytes and two that had not; the latter two were given intrathecal xenogeneic lymphocytes from a donor who had not been used previously. Over the next 24–48 hr, the two rabbits that had received xenogeneic cells for a second time became noticeably tachypneic and appeared ill (Table 2). One of the ill and one of the asymptomatic rabbits were selected for autopsy 48 hr after infusion. The other ill rabbit recovered after 72–96 hr. The CSF of all four rabbits 36 hr after infusion was of similar composition. No polymorphonuclear leukocytes were observed in any of the CSF samples, nor were there any changes in the CSF glucose. Similarly, gross and histological examinations of the CNS were unremarkable with one important exception: there was a marked mononuclear infiltration of choroid plexus stroma in the rabbit that received cells for a second time (Fig. 2B). This infiltration was not seen in the choroid plexus of the rabbit that received a single intrathecal lymphocyte infusion (Fig. 2A). Indeed, a mononuclear infiltration of the choroid plexus was never seen 24–48 hr after infusion, but was seen to a mild degree in one rabbit 9 days after infusion when the total CSF cell count was returning to normal (Fig. 2C). Therefore, infiltration of the choroid plexus by the rabbit's own lymphocytes (choroid plexitis) was a more likely explanation for the origin of these cells than infiltration of choroid plexus cells by infused lymphocytes from the ventricular space. Besides, few infused cells appeared to migrate into the ventricles on histological sections, making infiltration from the ventricle an even less tenable explanation.

To investigate the origin of the tachypnea in the symptomatic rabbits, a careful postmortem examination of the lungs of the two autopsied rabbits was performed. Grossly, the symptomatic rabbit's lung was congested in contrast to the asymptomatic rabbit's lung. Microscopic examination revealed pulmonary edema in the symptomatic rabbit in contrast to the asymptomatic rabbit (Fig. 3). Thus, the two rabbits that had received xenogeneic cells a second time developed what appeared to be a prolonged systemic allergic reaction associated with marked pulmonary edema and choroid plexitis. The prolonged course of the reaction was consistent with continual escape of the xenogeneic cells into the systemic circulation over 2–3 days. The rapid drop in CSF lymphocyte count over the first 2–3 days after infusion in the 32 rabbits studied supports such a pathogenesis (see Fig. 2).

2.4. Intrathecal Infusion of Syngeneic Lymphocytes into Rabbit Subarachnoid Space

The use of autologous rabbit lymphocytes permitted the infusion of only 10^6 cells, since the lymphocytes had to be purified from whole blood. This problem was partially alleviated by infusing pooled syngeneic lymphocytes as previously mentioned. Even so, it was feasible to infuse only 6×10^7 lymphocyte into the inbred rabbits; this resulted in a maximum CSF cell count at 36 hr of 625 cells/mm^3. As with the infusion of the xenogeneic lymphocytes, the cell viability after infusion was always nearly 100% (see Table 1).

There was no change in CSF glucose (Table 1), mean systemic white cell count, or mean body temperature. No morbidity or complications were observed even after a second or a third syngeneic lymphocyte infusion, and pathological examination 48 hr after the second and third infusions revealed neither choroid plexitis nor pulmonary edema. Thus, the allergic reaction seen following a second xenogeneic lymphocyte infusion was not seen when syngeneic cells were used.

2.5. Intrathecal Infusion of EL-4 Lymphoma Cells

Since it was impossible to differentiate donor and recipient lymphocytes morphologically, it was possible that there was a migration of host lymphocytes into the CSF in response to the intrathecal lymphocyte infusions. In an attempt to explore this possibility, it was first necessary to distinguish between infused and host cells. The EL-4 lymphoma cells, which are larger than normal rabbit lymphocytes, were examined, as described, on the Cytograf, an instrument that can distinguish two cell populations of different sizes using laser-beam absorbance.[28] In *in vitro* experiments (Fig. 4), the Cytograf could detect 5% rabbit lymphocytes in a cell suspension containing 95% EL-4 cells. The EL-4 cells were then infused intrathecally, and the Cytograf was used to detect any infiltration of host lymphocytes into the CSF. On this basis, 10^9 EL-4 cells, purified from mouse ascites fluid on Ficoll–Hypaque gradients and resuspended in synthetic CSF, were infused into the rabbit subarachnoid space. At 24 hr after infusion, the CSF cell count was 30,000 cells/mm^3, and over 80% of the cells were viable. At 72 hr, the CSF cell count was 500, and only 20–30% of the cells

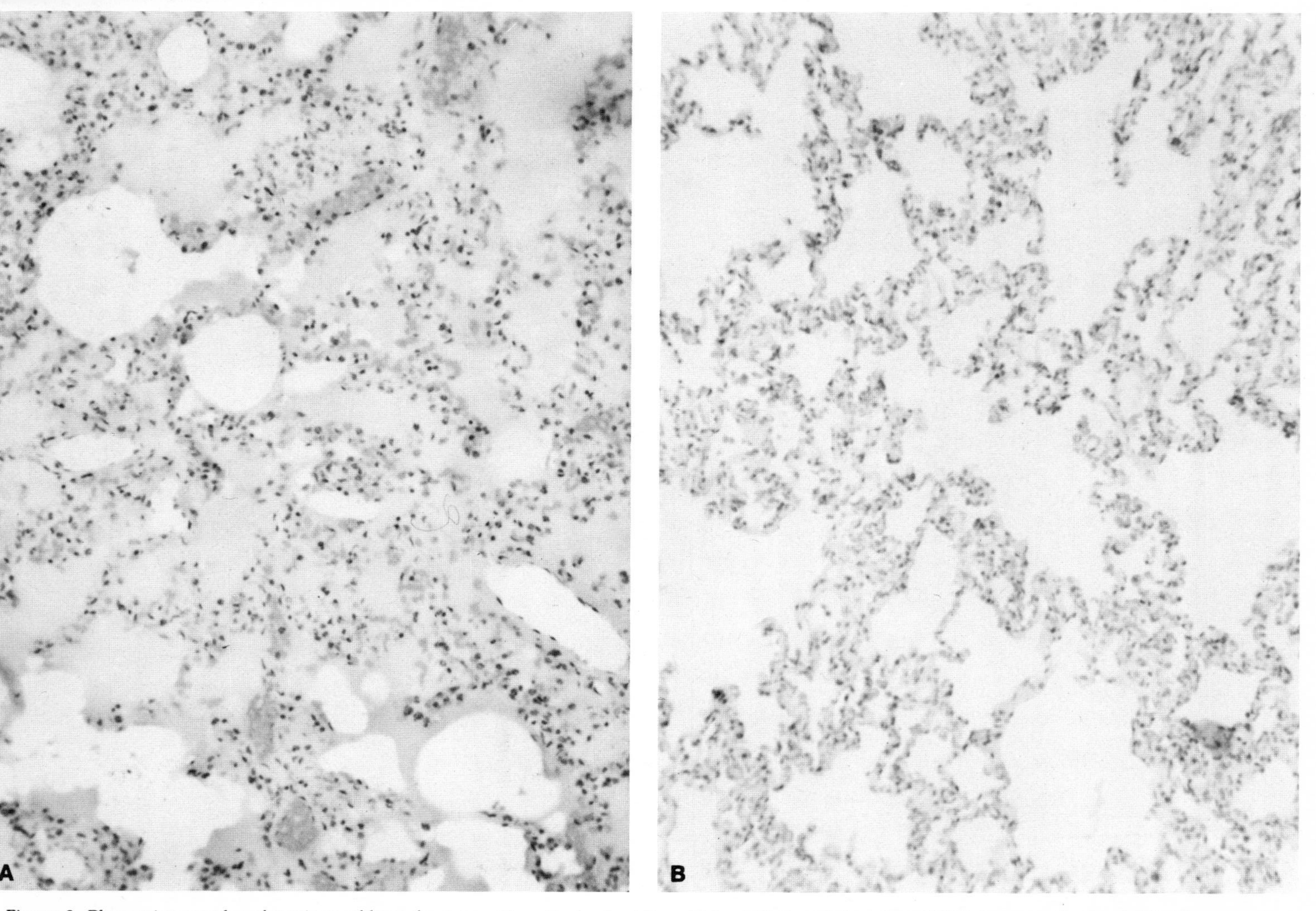

Figure 3. Photomicrographs of sections of lung from two New Zealand white rabbits. (H&E, $\times 100$, reduced 15% for reproduction.) (A) Absence of pulmonary edema 48 hr after initial intrathecal infusion of xenogeneic lymphocytes (same rabbit whose normal choroid plexus is illustrated in Fig. 2A). (B) Pulmonary edema 48 hr after second intrathecal infusion of xenogeneic lymphocytes (same rabbit whose choroid plexus is illustrated in Fig. 2B). Reproduced from Neuwelt and Doherty,[19] with permission of publisher.

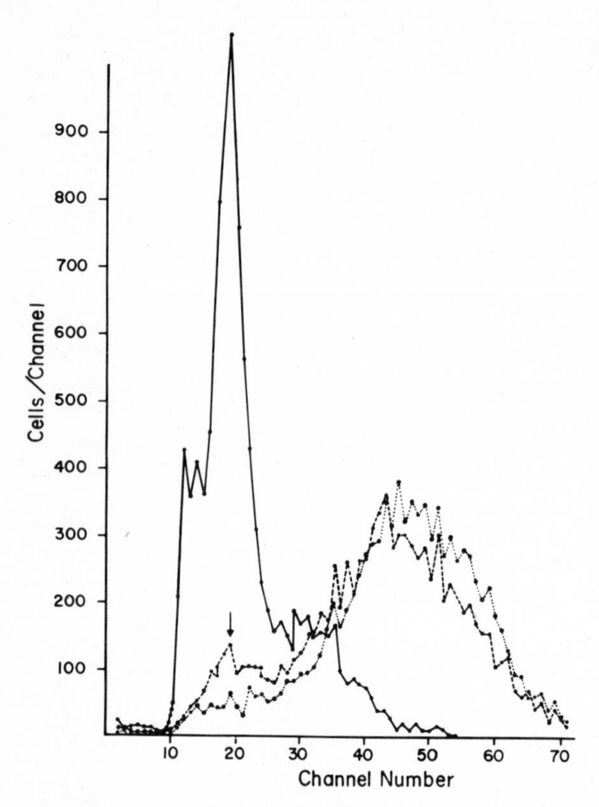

Figure 4. Graph illustrating use of Cytograf to detect rabbit lymphocytes in suspension of EL-4 lymphoma cells. The rabbit lymphocyte suspension (——) and EL-4 lymphoma cell suspension (. . .) were prepared as described and cell concentration of each adjusted to 1×10^6 cells/ml. Aliquot of each suspension and aliquot of mixture containing 95% EL-4 cells and 5% rabbit lymphocytes (- - - -) were added to Cytograf as described. Each curve represents 10,000 cells, and Cytograf assigned each of these cells to one of 100 channels according to cell diameter. Channel numbers increase in proportion to cell diameter. (↓) Peak due to presence of 5% lymphocytes in cell suspension containing 95% EL-4 lymphoma cells. Reproduced from Neuwelt and Doherty,[19] with permission of publisher.

were viable, as determined by dye exclusion, and the CSF contained a large amount of cell debris. To examine viable EL-4 cells, CSF was obtained 24 hr after intrathecal EL-4 cell infusion and an aliquot examined on the Cytograf (Fig. 5). In addition, the host rabbit's blood was obtained by cardiac puncture 24 hr after infusion, the lymphocytes purified, and an aliquot of this cell suspension examined on the Cytograf. These experiments revealed that at 24 hr after infusion, no infiltration of the CSF by host lymphocytes was detectable and no EL-4 cells were detectable in the rabbit's systemic circulation (Fig. 5).

Although these experiments provide evidence that host cells do not appear to infiltrate the CSF in response to lymphoid-cell infusion, the absence of detectable EL-4 cells in the systemic circulation could be explained in a variety of ways. Because of their large size or their rapidly decreasing viability, EL-4 cells might not have been able to escape from the CNS and either the Cytograf was not able to detect them or they did not remain in the systemic circulation long enough to be detected. As a result, experiments are now under way to follow the migration of infused normal lymphocytes using radioisotopes.

2.6. Necropsy Studies

Animals were sacrified at different times following intrathecal syngeneic and xenogeneic lymphocyte infusions in an attempt to find evidence of CNS

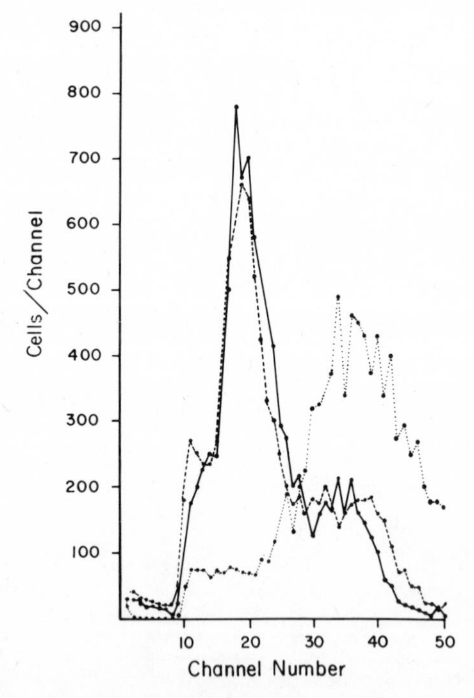

Figure 5. Graph showing absence of host lymphocytes in CSF and absence of EL-4 lymphoma cells in host blood 24 hr after intrathecal infusion of EL-4 lymphoma cells into New Zealand white rabbit. This figure is analogous to Fig. 4 except: (——) normal, purified, New Zealand white rabbit lymphocytes; (- - - -) host lymphocytes, purified from blood of New Zealand white rabbit 24 hr after intrathecal infusion of EL-4 lymphoma cells; (· · ·) CSF cells from same host rabbit 24 hr after intrathecal infusion of EL-4 lymphoma cells. Reproduced from Neuwelt and Doherty,[19] with permission of publisher.

parenchymal damage and to study the migration of the infused cells. On gross examination, after coronal slices were made of formalin-fixed brain, brainstem, and spinal cord, no abnormalities were seen in any of the specimens. There was no evidence of hydrocephalus or areas of pallor suggestive of demyelination. On microscopic examination, except for the findings in the choroid plexus discussed above, no abnormalities of the CNS parenchyma were noted. The subarachnoid space, but not the ventricles, was diffusely infiltrated at 12–36 hr after infusion with lymphocytes (Fig. 6), but the majority of cells had disappeared by 10 days. This is strong evidence that the rapid drop in total cell count following intrathecal lymphocyte infusion was not the result of sequestration of lymphocytes in the CNS. Although adequate sections of the superior sagittal sinus were difficult to obtain, cells did appear to accumulate around this structure transiently, suggesting a possible route of escape (Fig. 7).

Autopsy studies also revealed that the distribution of EL-4 lymphoma cells in the subarachnoid space and ventricles (Fig. 8) seemed to be identical to the pattern seen with normal xenogeneic and syngeneic lymphocytes. However, the EL-4 cell architecture was hard to discern, and the cells fragmented, suggesting cell death, at 72 hr after infusion. This finding is consistent with the dye exclusion studies of EL-4 cells at 72 hr noted earlier. It is not clear exactly why these cells did not remain viable, in contrast to normal lymphocytes, which did.

Microscopically, the only cells seen in the subarachnoid space following intrathecal EL-4 cell infusion were EL-4 lymphoma cells. No normal host lymphocytes were observed (Fig. 8). Thus, neither the Cytograf (see Fig. 5) nor histological studies (see Fig. 8) demonstrated any host-cell infiltration of the CSF following lymphoma-cell infusions. Up to 72 hr after intrathecal EL-4-lymphoma-cell infusion, no infiltrates of the choroid plexus were seen.

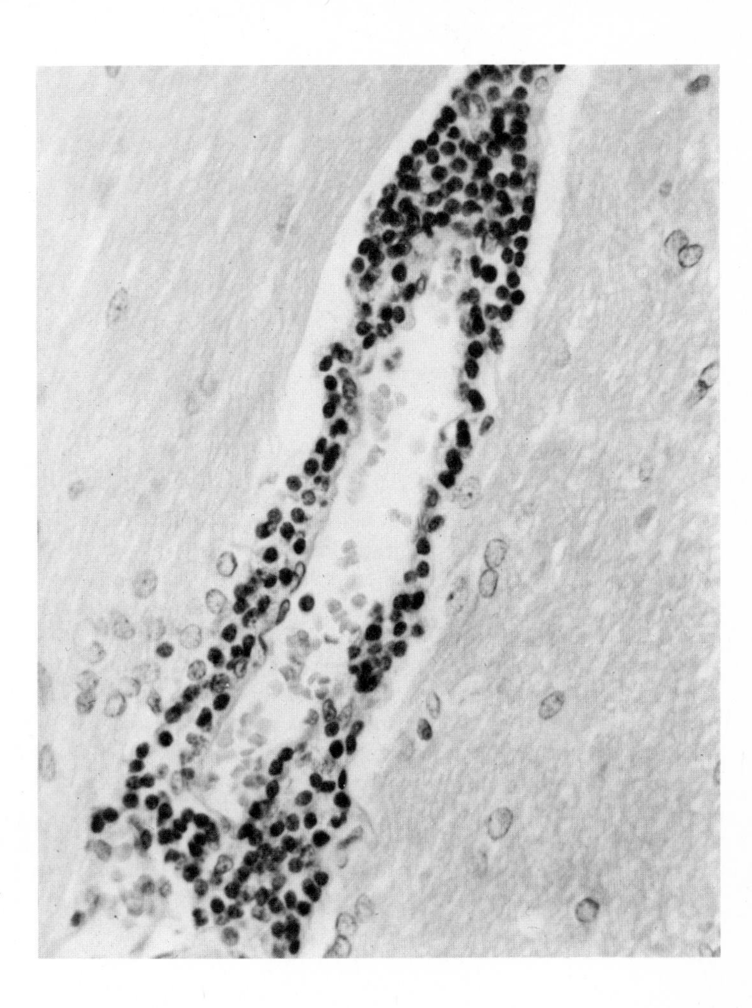

Figure 6. Photomicrograph showing lymphocytic infiltration of subarachnoid space of New Zealand white rabbit 36 hr after intrathecal infusion of 2×10^9 viable human lymphocytes. CSF cell count just before autopsy was 3000 cells/mm^3, all of which were viable lymphocytes. (H&E, ×350.) Reproduced from Neuwelt and Doherty,[19] with permission of publisher.

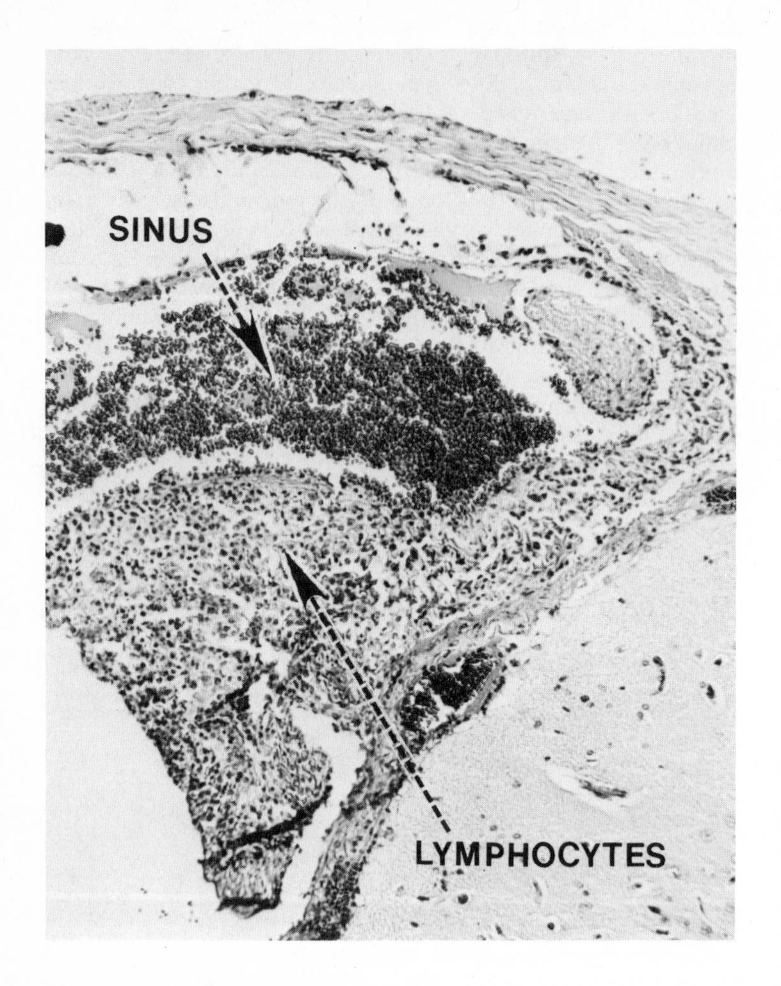

Figure 7. Photomicrograph showing accumulation of lymphocytes around superior sagittal sinus 48 hr after intrathecal infusion of xenogeneic lymphocytes into New Zealand white rabbit. Mononuclear cells in mass inferior to sinus were all lymphocytes. (H&E, ×100.) Reproduced from Neuwelt and Doherty,[19] with permission of publisher.

2.7. Discussion

The question asked in our animal studies was whether rabbits could tolerate intrathecal infusions of normal syngeneic or xenogeneic lymphocytes. Since preliminary studies using moderate numbers of autologous lymphocytes and later studies using larger numbers of syngeneic lymphocytes revealed no toxicity, we wanted to determine whether the animals would also tolerate the infusion of massive numbers of xenogeneic lymphocytes. To investigate that question, we used the Haemonetics Model 30 Cell Separator, which could furnish enormous numbers of normal lymphocytes from healthy human volunteers without a significant RBC loss to the donor.[1] With this method, even though 100 times the number of cells was used and species barriers crossed, toxicity after a single infusion was again minimal.

However, two rabbits did develop a systemic reaction to a second xenogeneic lymphocyte infusion. No such reactions were seen with a second or even a third administration of syngeneic lymphocytes. The systemic reaction might have been an allergic response to cells that escaped from the subarachnoid space over a 48- to 72-hr period into the systemic circulation of an already sensitized host. This could have resulted in the subacute pulmonary edema and the choroid plexitis we observed. The findings in the choroid plexus of several electively autopsied animals certainly support an immune mechanism, since after an initial xenogeneic lymphocyte infusion, a mononuclear infiltrate in the choroid plexus was not seen until 9 days after infusion, which is consistent with the time sequence of primary immune sensitization. However, in the two rabbits with pulmonary edema that had a prior exposure to human lymphocytes, there was a

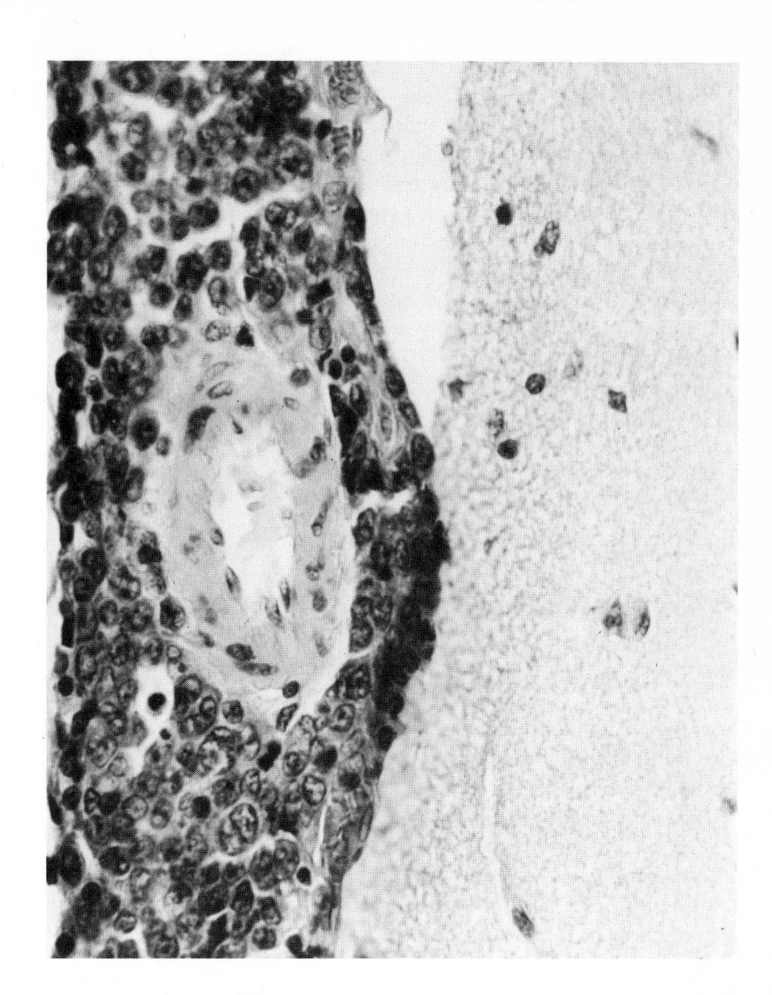

Figure 8. Photomicrograph showing infiltration of subarachnoid space of New Zealand white rabbit with EL-4 lymphoma cells 24 hr after intrathecal infusion of 1×10^9 EL-4 lymphoma cells. Notice that lymphoma cells in subarachnoid space are larger than lymphocytes illustrated in Fig. 6. CSF cell count just before autopsy was 30,000 cells/mm^3, of which 85% were viable. (H&E, ×350.) Reproduced from Neuwelt and Doherty,[19] with permission of publisher.

marked mononuclear infiltration of the choroid plexus at 48 hr. Two other rabbits that received the same human-lymphocyte preparation but did not have a prior exposure to human cells developed neither pulmonary edema nor choroid plexitis. Thus, xenogeneic cells transplanted into the CNS were able to activate both limbs of the immune mechanism.

Scheinberg and Taylor[31] reported that skin homografts transplanted into brain could accelerate rejection of subsequent systemic skin grafts from the same donor; this finding indicates stimulation of the afferent limb of the immune response. However, the actual mechanism of activation was not clear in their report.[29,32] In the study reported herein, escape of the infused lymphocytes from the subarachnoid space into the systemic circulation might account for sensitization of the immune system. The evidence that the infused lymphocytes were indeed escaping from the CNS is as follows: The infused lymphocytes disappeared very rapidly from the subarachnoid space; the total CSF cell count was as high as 70,000 cells/mm^3 12 hr after infusion and dropped to only 200–300 cells/mm^3 4–5 days later. It did not appear that the precipitous fall in total CSF cell count was due to cell death when normal lymphocytes were infused, since cell viability remained near 100%. In contrast, when EL-4 lymphoma cells were infused, viability dropped dramatically between 24 and 72 hr after infusion. If the infused normal lymphocytes were not dying, then they must have been either sequestered in or escaping from the CNS. Histological studies of autopsied animals essentially eliminated the former possibility. That is, although the subarachnoid space was filled with lymphocytes on histological sections 12 hr after infusion (see Fig. 6), these cells had all but disappeared from histological sections obtained several days later. The only other remaining possibility, then, was that the lymphocytes were escaping from the subarachnoid space systemically. The accumulation of lymphocytes around the superior sagittal sinus (see Fig. 7) suggested that this structure might be the avenue of escape. More detailed studies are at present under way to investigate further the migration of lymphocytes infused into the subarachnoid space and the possible role that the superior sagittal sinus may play in the escape of these cells into the systemic circulation. Regardless of the role of the superior sagittal sinus, however, the evidence presented above does strongly suggest that the infused lymphocytes were escaping from the CNS.

3. Phase I Studies: Man

3.1. Delineation of Population

Our initial studies[18a] dealt with the toxicity associated with infusions of autologous lymphocytes in patients with malignant gliomas. Patients were considered for admission to this study on the basis of the following criteria: all patients had histologically proven malignant astrocytomas, i.e., astrocytoma Grade 3 or 4; all had undergone standard surgical extirpation and postoperative whole-head biplane megavoltage cranial irradiation; all were alert and able to communicate to varying degrees, and the general physical condition of the patients was reasonably good. This research was carried out with the approval of the Human Research Committee of the University of Texas Southwestern Medical School and the Dallas Veterans Administration Hospital. Informed consent was obtained from four patients as well as from their immediate families.

3.2. Acquisition of Cells

Lymphocytes were obtained by placing the patients on the Haemonetics Model 30 Cell Separator as described by Aisner *et al.*[1] The lymphoid-cell-rich blood obtained from the cell separator was purified on Ficoll–Hypaque gradients[19] and resuspended in either Elliot's B solution or physiological normal saline. Cell counts, cellular composition, cell viability, and cell cultures were carried out as described earlier (Section 2.2). The only positive culture in either our animal or our human studies was a purified autologous lymphoid cell preparation from patient A.V.B., which grew *Staphylococcus epidermidis* in the thioglycolate broth but not on the agar plate, and was thought to be a contaminant rather than a break in sterile technique.

3.3. Intrathecal Infusions of Autologous Lymphocytes

The purified autologous lymphoid cells were infused into the patient's CSF by lumbar puncture within 6 hr of procurement. The total volume infused by lumbar puncture was in the range of 5 ml, and it was administered by barbotage. A total of 18 infusions were carried out in these four patients, with initial infusions of 1×10^6 viable lymphoid cells injected. Generally, the infusion was repeated every 2 weeks with 1×10^7, 1×10^8, 1×10^9, and finally 5×10^9 viable lymphoid cells being given sequentially. The viability of the cells infused was generally in the range of 90–95%. The amount of RBC contamination was variable in the range of 1 RBC per 2–6 WBCs. There were, at times, considerable numbers of platelets in the purified lymphocyte preparation. Despite multiple infusions, the systemic hemoglobin of all the patients remained stable.

Case Study 1. A 56-year-old white male (S.C.L.) presented with a history of diplopia, intermittent behavioral changes, and staggering gait. He underwent a subtotal resection of a glioblastoma in the pineal region through an occipital transtentorial approach. He completed a course of whole-head irradiation, and a 6-month postop (CT) scan showed evidence of residual tumor in both occipital lobes. Six months following his surgery, the patient entered this study.

A total of five intrathecal autologous-lymphocyte infusions were administered over the course of 2 months, with the number of lymphoid cells infused increased by one log value with each infusion from 1×10^6 to a final dose of 5×10^9 viable cells. Though the initial infusion produced little change, a transient rise was noted in the total CSF cell count with each successive infusion with a return to normal within several days. With the infusion of 1×10^9 and 5×10^9 lymphoid cells, the CSF white count rose to as high as 1500 white cells/mm^3, of which 90% were mononuclear cells with a viability of 95% (Fig. 9).

No significant change in the patient's CSF glucose content (Fig. 10) accompanied these increased CSF cell counts. This patient was diabetic with erratic CSF glucose values that may have masked a consistent pattern of hypoglycorrhacia as seen in two of the other glioma patients. A gradual increase of total CSF protein over the 8 weeks of study occurred

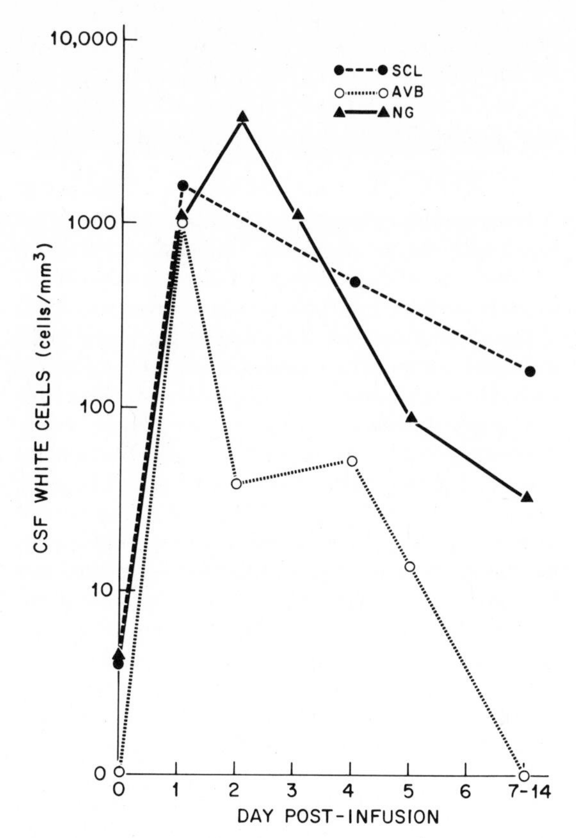

Figure 9. Serial CSF white cell counts following intrathecal infusions of large numbers of lymphoid cells in three glioma patients (S.C.L., A.V.B., and N.G.). Number of lymphoid cells infused varied from 1 to 5 × 10^9 viable cells. Reproduced from Neuwelt *et al.*,[18a] with permission of publisher.

(Fig. 11), which is probably most consistent with a recurrence of tumor. The general autopsy revealed lobar pneumonia and acute pyelonephritis. Pathological examination of the CNS revealed glioblastoma present in both occipital lobes, epithalamus, and thalamus, with seeding of the tumor into the spinal subarachnoid space. No evidence of any inflammatory reaction or demyelination was seen within the CNS. There was no evidence of obstructive hydrocephalus, and evaluation of the arachnoid granulations was normal. There was little evidence of comunication between the tumor and the subarachnoid space.

Case Study 2. Patient J.G., a 56-year-old black male with a history of severe headaches, a left hemiparesis, and a left homonymous hemianopsia, underwent a right parietal craniotomy with resection of a malignant astrocytoma. A course of whole-head irradiation was completed, and a cisternogram showed no flow of the radioisotope over the right cerebral hemisphere. He tolerated his initial infusion of 1 × 10^6 cells with only a mild frontal headache. Grand mal seizures 2 weeks postinfusion were controlled by anticonvulsants.

As with patient S.C.L. (Case Study 1), a mild rise in the CSF cell count was observed, but there were no significant trends in the CSF sugar and protein, and the systemic WBC remained stable throughout the protocol.

The patient's course was continued deterioration consistent with tumor recurrence, and he expired 11 days following his second infusion.

The general autopsy revealed acute bronchopneumonia and multiple organizing and recanalizing thrombotic pulmonary emboli. The pathological findings in the CNS revealed the presence of recurrent tumor in the right parietal area, but no ev-

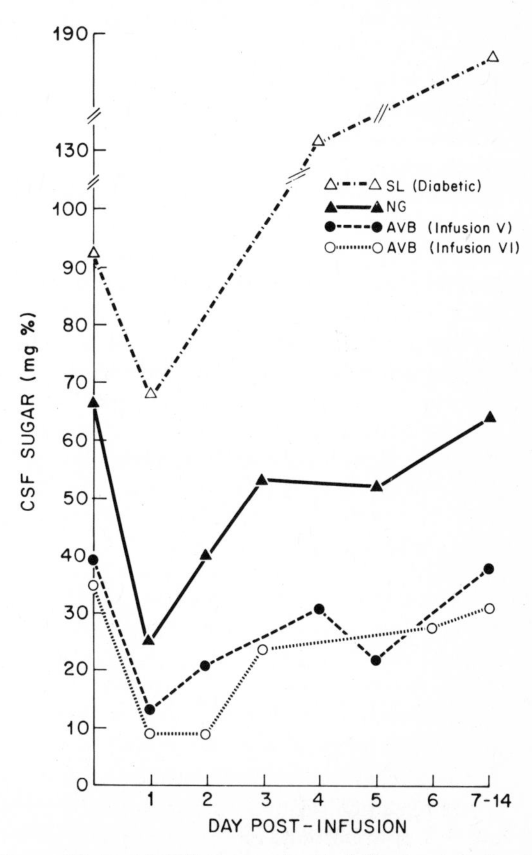

Figure 10. Serial CSF glucose levels following intrathecal infusion of 1–5 × 10^9 viable lymphoid cells in three glioma patients (S.C.L., N.G., and A.V.B.). Reproduced from Neuwelt *et al.*,[18a] with permission of publisher.

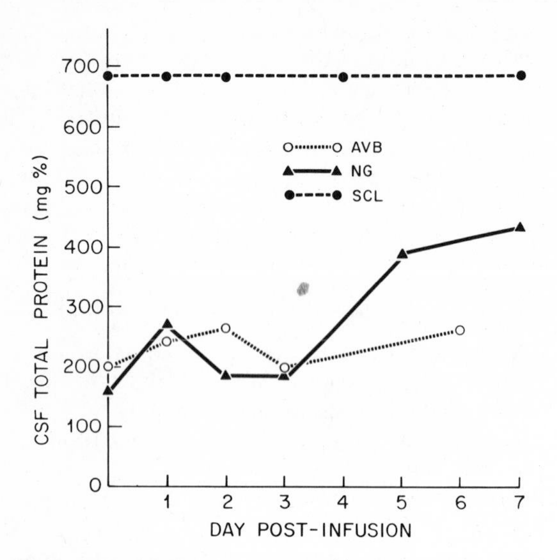

Figure 11. Serial CSF total protein levels following intrathecal infusion of 1 to 5 × 10^9 viable lymphoid cells in three glioma patients (A.V.B., N.G., and S.C.L.). Reproduced from Neuwelt *et al.*,[18a] with permission of publisher.

idence of inflammation or demyelination. The entire CNS was examined, including the spinal cord and overlying leptomeninges.

Case Study 3. A 52-year-old white male (A.V.B.) presented with 3-month history of behavioral changes followed by a 3-week history of dysarthria and progressive difficulty with gait, and subsequently underwent a left temporal lobectomy for a malignant astrocytoma. Postoperatively, he received 6000 rads of whole-head irradiation and was entered into the protocol 2 months following surgery.

Again, the initial infusion of 1 × 10^6 viable autologous lymphocytes via a lumbar puncture was tolerated without any evidence of untoward reaction. The patient underwent five additional infusions, with the dose of lymphoid cells infused increased as previously outlined until a maximum of 5 × 10^9 lymphoid cells were infused. An aliquot of the purified lymphoid cells used for the final two intrathecal infusions was evaluated for the percentage of T cells present. Using the sheep red cell E rosette method,[7] 72 and 74% of the lymphoid cells were found to be T cells.

The patient's CSF pressure and systemic white count remained relatively stable throughout the six courses of immunotherapy (Fig. 12).

Over the course of the first 8 weeks of the protocol, the patient seemed to show some improvement in his speech and began to sing spontaneously. Similarly, a CT scan after five intrathecal lymphoid-cell infusions over a 13-week period demonstrated a decreased shift of the midline, a decreased deformity of the lateral ventricles, and less tumor enhancement in the tumor bed (Fig. 13A,B).

A CT scan 20 weeks after beginning immunotherapy showed enlargement of both lateral ventricles due to tumor involvement around the foramen of Monro, but no evidence of third or fourth ventricular enlargement (Fig. 13C). The patient subsequently expired 7 months postoperatively.

The general autopsy revealed that the patient had pulmonary edema and cirrhosis. Gross neuropathological examination revealed that there was hernition across the midline of at least 2.5 cm. When the brain was cut, the tumor was noticed to extend down to the level of the mid pons. There was edema of the centrum semiovale on the side of the tumor. There was no gross evidence of any significant demyelination. Microscopic evaluation revealed a marked perivascular lymphocytic infiltrate in the area of the tumor that was almost completely necrotic. The degree of perivascular lymphocytic infil-

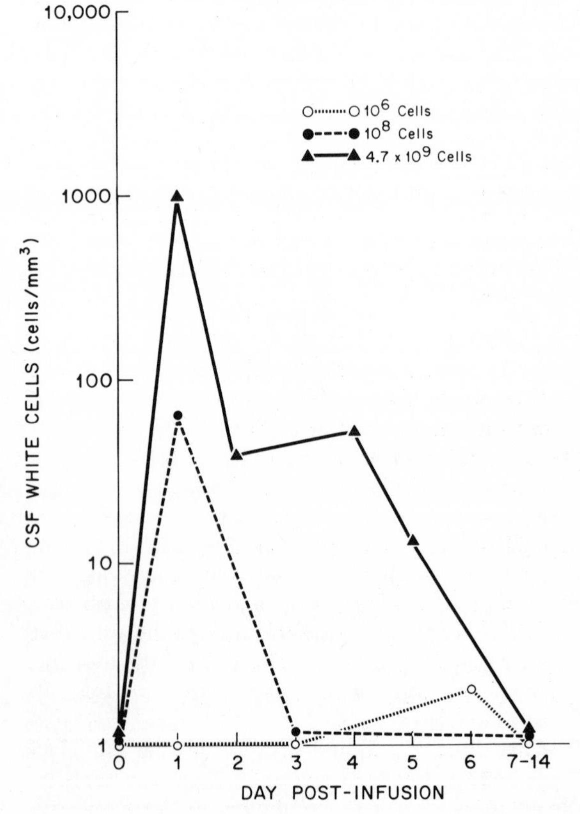

Figure 12. Serial CSF cell counts following intrathecal infusions of varying numbers of lymphoid cells in glioma patient (A.V.B.).

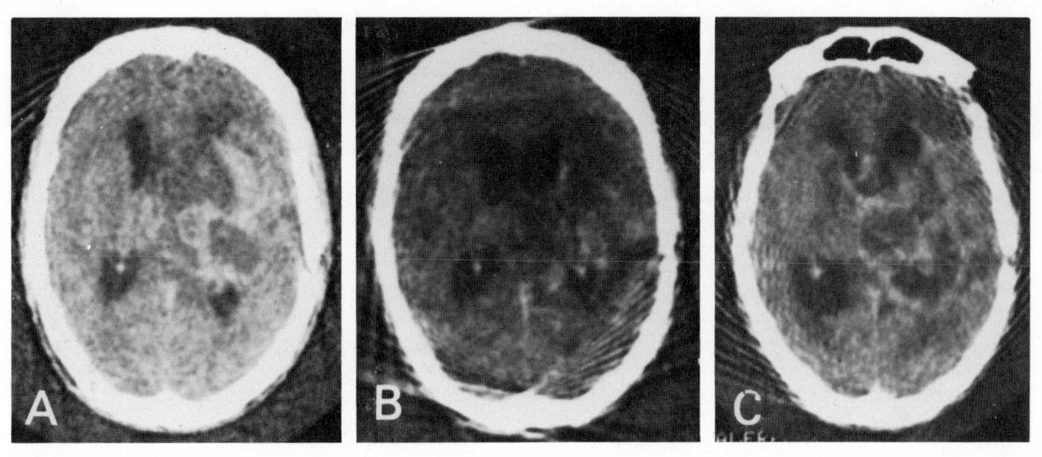

Figure 13. CT scan (A) before immunotherapy but following surgery and right after completion of radiotherapy; (B) following five intrathecal lymphoid cell infusions over 16 weeks; (C) 6 weeks after CT scan illustrated in Fig. 13B, just prior to time patient expired (patient A.V.B.). Reproduced from Neuwelt and Clark,[18] with permission of publisher.

tration in the tumor bed was greater than is normally seen following surgery and radiation (Fig. 14A). These lymphocytic infiltrates not only were perivascular but also seemed to extend into the infiltrating margins of the tumor that had seeded the subarachnoid cisterns (Fig. 14D). Away from the area of the tumor, there was no lymphocytic infiltration in the CNS parenchyma, but there were lymphocytes in the subarachnoid space (Fig. 14E) and in the perivascular spaces (Fig. 14F). With the exception of one small perivascular area that may have been adjacent to the tumor, there was no evidence of demyelination anywhere in the CNS. The choroid plexus was also normal and without evidence of choroid plexitis (Fig. 14G).

Case Study 4. A 58-year-old right-handed white male (N.G.) presented with a 3-month history of a progressive memory loss and difficulty in expressing himself. A computerized coaxial tomogram and arteriography revealed a vascular mass in the left temporal area. He underwent a subtotal excision of a Grade 3 astrocytoma and received 6000 rads of whole-head irradiation postoperatively. He was begun on the study 6 months after tumor resection.

The number of lymphocytes infused was increased over the course of five treatments from 1×10^6 to 1×10^9 cells, as has been described. The infusions were tolerated well, and his neurological exam remained essentially unchanged.

After both infusions of 1×10^9 and 9.9×10^8 viable white cells, the CSF cell count rose to over 1000 cells/mm^3, of which 75–78% were mononuclear cells (see Fig. 9), but following the final infusion, the CSF white cell count at 24 hr was less than that at 48 hr and 72 hr (Table 3). In addition, the percentage of white cells that were mononuclear was really rather small. The exact significance of this atypical CSF-cell-count pattern at 48–72 hr is unclear. However, by the 5th day postinfusion, the total white count had fallen back down to 88 white cells/mm^3, of which the vast majority were lymphocytes.

Consistent with the previous three case reports, the CSF total protein showed little consistent change with the lymphocyte infusions (see Fig. 11). As in the third patient (A.V.B.), with lymphoid-cell infusions in the range of 1×10^9 cells, the CSF glucose did seem to fall significantly (see Fig. 10), returning to normal within 2–3 days.

The patient's level of consciousness began to decrease 3 weeks after his fifth infusion, and a CT scan revealed an increased mass effect and increased tumor enhancement suggestive of tumor recurrence (Fig. 15C). The family requested that no further infusions be administered.

In the toxicity studies discussed above, the lymphoid cells were always infused via the lumbar subarachnoid space even though the radio-iodinated serum albumin cisternograms in these patients often did not document communication with the tumor bed. The toxicity studies were carried out this way for three reasons: (1) the animal studies demonstrated that cells infused into the cisterna

Table 3. CSF Cell Counts Following Intrathecal Infusion of Lymphoid Cells

		CSF cell count (% mononuclear/total WBCs/RBCs)					
Patient	Day post-infusion	Infusion I	Infusion II	Infusion III	Infusion IV	Infusion V	Infusion VI
		(2/16/77)	(3/2/77)	(3/16/77)	(3/30/77)	(4/14/77)	—
S.C.L.	0	47/236/2,310,000	100/4/0	100/6/1	100/25/31	100/4/35	
	1	– /0/2573	10/15/678	23/186/3554	98/1000/154	98/1590/5246	
	2	—	—	—	—	—	
	3	80/38/17,502	84/6/3,348	66/45/960	96/172/127	—	
	4	—	—	—	—	100/490/7250	
	5	97/312/3,888	40/9/65	75/4/21	100/311/137	—	
	8–13	—	—	—	—	100/161/579	
		(4/6/77)	(5/4/77)	(5/26/77)	(6/9/77)	(6/22/77)	(7/20/77)
A.V.B.	0	– /0/1	– /9/4	– /0/15	100/2/2	– /0/0	100/2/2
	1	– /5/6	– /0/0	73/63/0	—	91/1015/682	91/1440/158
	2	—	—	—	90/10/0	100/38/14	31/80/57
	3	– /0/0	– /0/2	– /0/1	—	—	100/5/37
	4	—	—	—	—	100/51/12	—
	5	—	—	—	– /0/1000	100/13/358	—
	6	100/2/2	—	—	—	—	96/25/0
	7	—	—	—	—	– /0/22	—
	8–13	—	– /0/9	—	—	—	88/16/61
		(6/16/77)	(6/29/77)	(7/13/77)	(7/27/77)	(8/10/77)	—
N.G.	0	100/1/2	100/2/1	Lost	– /0/0	75/4/5700	
	1	16/8/1005	0/4/29	Lost	78/1400/ –	75/1124/2816	
	2	—	—	—	79/470/80	13/4030/25,320	
	3	—	– /0/19,666	—	82/214/1011	37/1184/4013	
	4	—	—	—	—	—	
	5	– /0/90	—	—	95/52/466	80/88/460	
	6	—	—	– /0/12	—	—	
	7	—	—	—	—	—	
	8–13	—	– /0/648	—	100/9/9	80/34/15	
		(5/11/77)	(6/2/77)	—	—	—	—
J.G.	0	80/12/5	– /0/0				
	1	88/25/1299	40/22/1				
	2	—	—				
	3	85/8/343	—				
	4	—	—				
	5	90/17/34	—				
	6	—	100/2/0				
	7	—	—				
	8–13	—	—				

magna migrated throughout the subarachnoid space over the cerebral and cerebellar hemispheres; (2) there are theoretical advantages to infusions into the subarachnoid space as opposed to infusions into the tumor bed, as enumerated in Section 3.5 below; and (3) the use of the Ommaya reservoir is associated with a significant rate of complications, particularly infectious complications. It was also felt that the major potential complication of the immunotherapy was infection. Therefore, the use of Ommaya reservoirs could have complicated the interpretation of the toxicity studies.

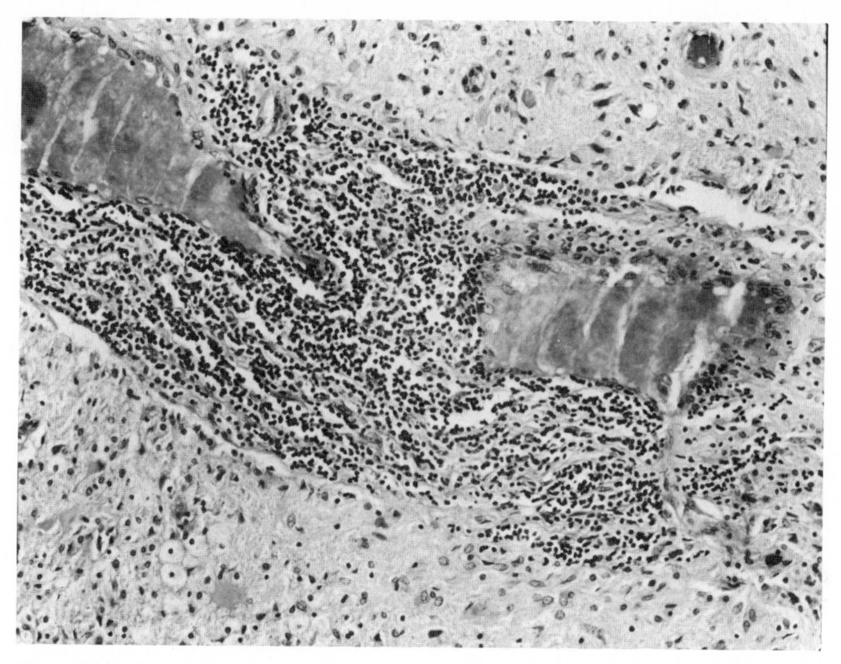

Figure 14A. Histopathology in patient A.V.B. Low-power illustration: Vessel within tumor is surrounded by dense infiltrate of lymphocytes and other mononuclear cells. (H&E, ×40, reduced 10% for reproduction.) Reproduced from Neuwelt *et al.*,[18a] with permission of publisher.

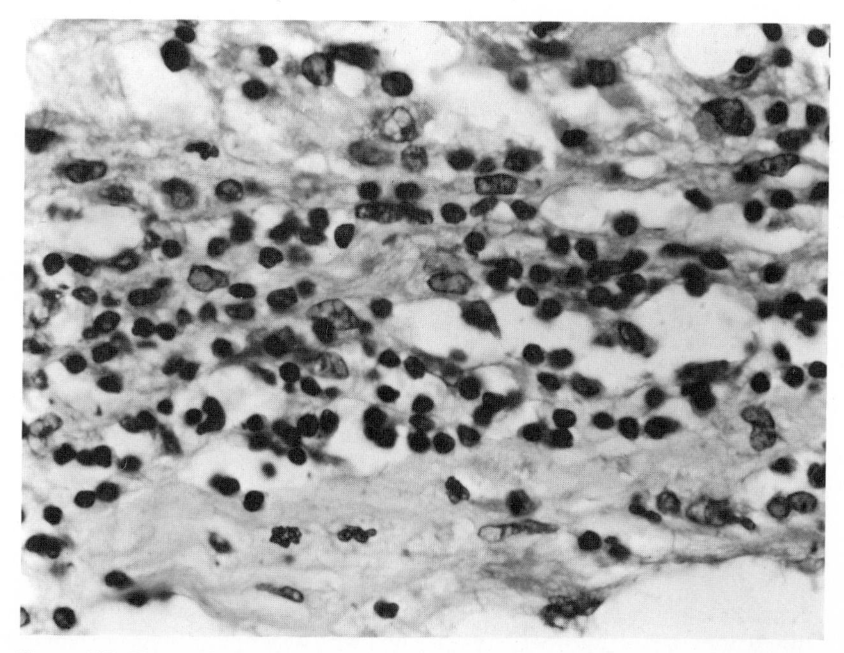

Figure 14B. Histopathology in patient A.V.B. Medium-power illustration: Perivascular infiltrate adjoining tumor. Margin of vessel is at bottom of photograph with tumor below. Lymphocytes and mononuclear cells from perivascular space are extending into tumor. (H&E, ×160, reduced 10% for reproduction.) Reproduced from Neuwelt *et al.*,[18a] with permission of publisher.

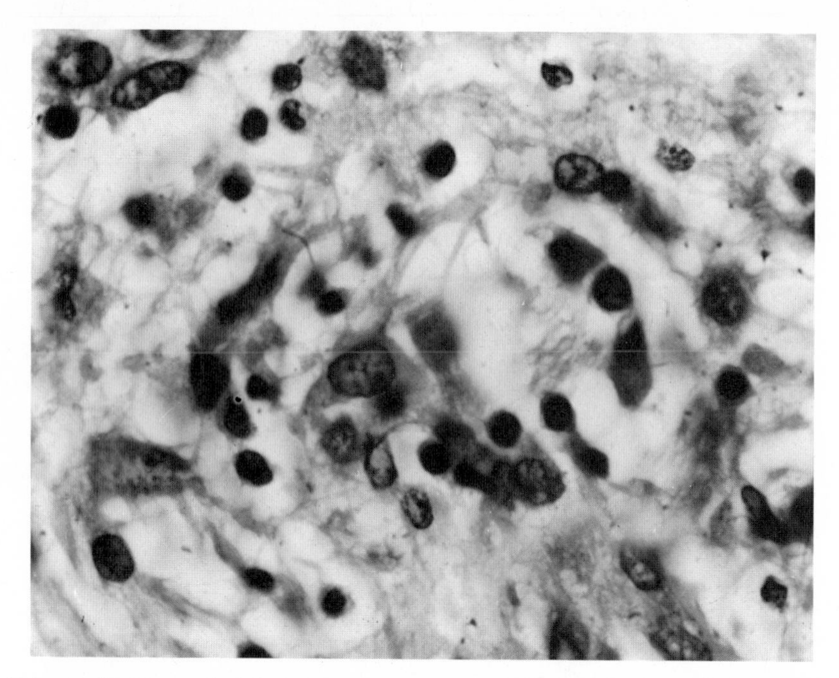

Figure 14C. Histopathology in patient A.V.B. High-power magnification: Lymphocytes in tumor can be identified by dark round nucleus and scant cytoplasm. (H&E, ×250, reduced 10% for reproduction.) Reproduced from Neuwelt *et al.*,[18] with permission of publisher.

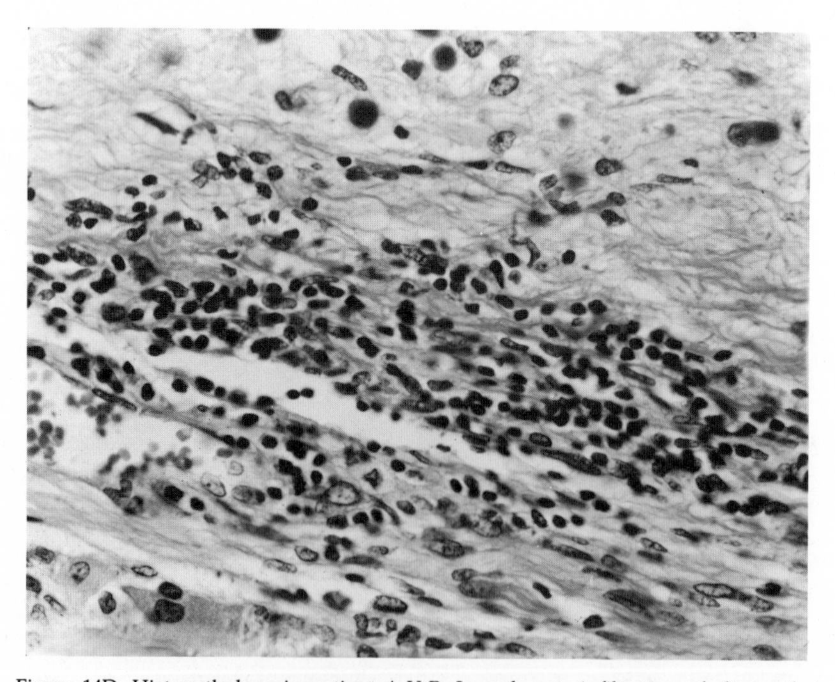

Figure 14D. Histopathology in patient A.V.B. Lymphocyte infiltration of glioma that has seeded into pontine cistern. Uninvolved pons can be seen at top of photomicrograph. (H&E, ×40, reduced 10% for reproduction.) Reproduced from Neuwelt *et al.*,[18a] with permission of publisher.

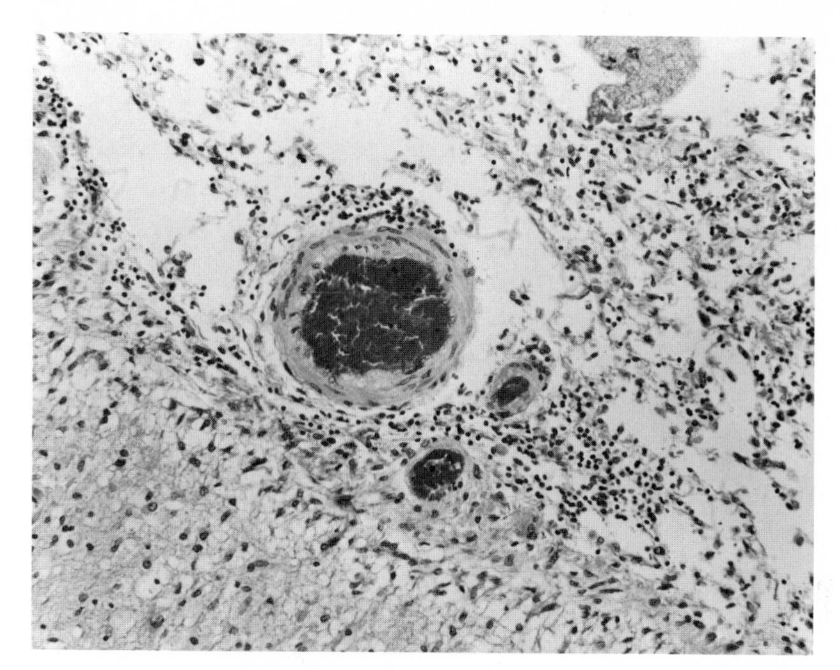

Figure 14E. Histopathology in patient A.V.B. Subarachnoid space at time of autopsy, which was 4 weeks following last lymphoid-cell infusion. Note persistent marked lymphocyte infiltration of subarachnoid space (top of the figure). (H&E, ×40, reduced 10% for reproduction.) Reproduced from Neuwelt *et al.*,[18a] with permission of publisher.

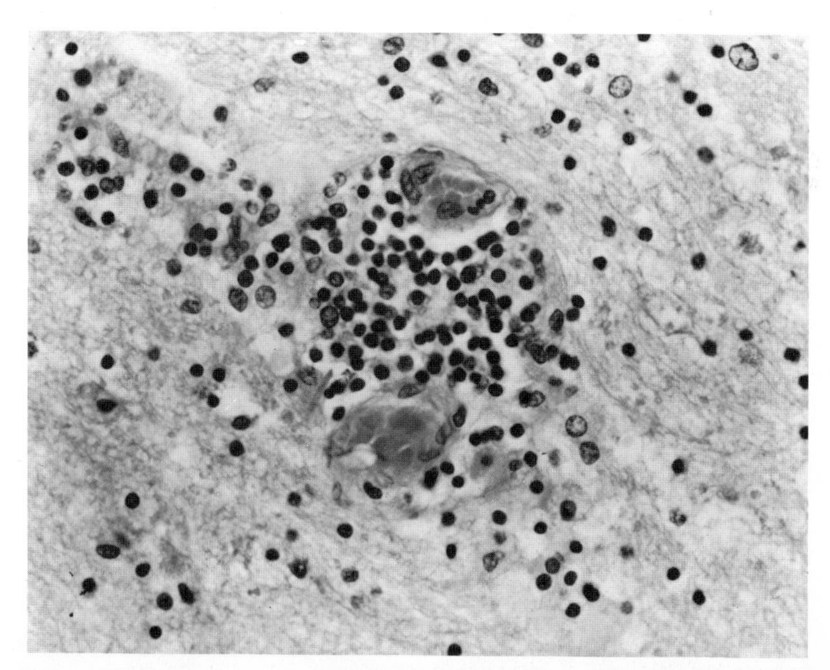

Figure 14F. Histopathology in patient A.V.B. Perivascular lymphocyte infiltration at time of autopsy in brain uninvolved with tumor. There was no evidence of perivascular demyelination in association with these lymphocyte infiltrates. (H&E, ×100, reduced 10% for reproduction.) Reproduced from Neuwelt *et al.*,[18a] with permission of publisher.

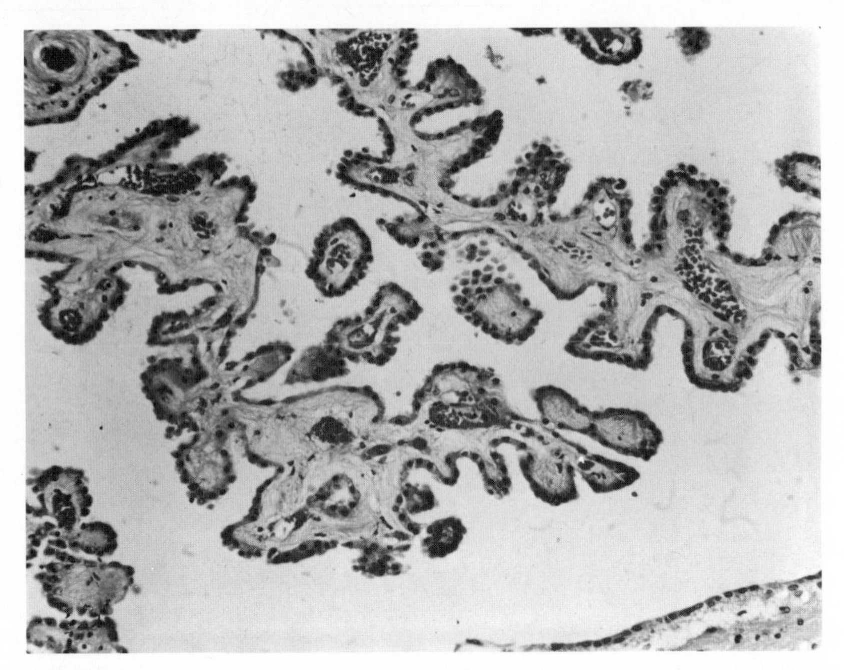

Figure 14G. Histopathology in patient A.V.B. Normal choroid plexus at time of autopsy. (H&E, ×40, reduced 10% for reproduction.) Reproduced from Neuwelt *et al.*,[18a] with permission of publisher.

Intrathecal infusions may not permit migration of lymphoid cells to the tumor bed because of leptomeningeal scarring in the postoperative period. Yet, intrathecal autologous-lymphoid-cell infusions in a patient whose tumor bed was been adequately exposed to the subarachnoid space may be better than direct intratumoral infusion for two reasons: (1) The infused lymphocytes, after exposure of the tumor bed and subsequent return to the systemic circulation, may further activiate the afferent limb of the immune response (Table 4). Thus, the proportion of sensitized lymphoid cells may progressively increase after each infusion. (2) There is less risk of creating a localized mass effect.

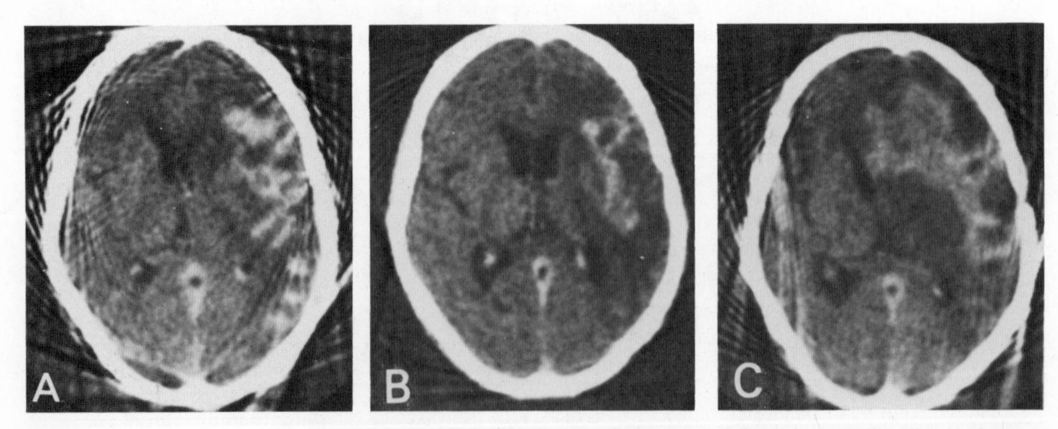

Figure 15. CT scan (A) prior to start of immunotherapy but following surgery and at completion of radiotherapy (reproduced from Neuwelt and Clark,[18] with permission of publisher); (B) following two intrathecal lymphocyte infusions (first with 1×10^6 and second with 1×10^7 lymphoid cells); (C) 2 months following CT scan illustrated in Fig. 15A and following five intrathecal lymphoid cell infusions (patient N.G.).

Table 4. Comparison of Cell Count, Glucose, and Protein between CSF and Ommaya Reservoir Fluid before Initiation of Immunotherapy

	CSF			Ommaya reservoir fluid		
Patient	% Monocytes/ WBC[a]	Glucose[b] (ng%)	Protein (ng%)	% Monocytes/ WBC[a]	Glucose (ng%)	Protein (ng%)
J.B.	–/0	67	51	40/1100	—	—
J.M.	–/0	79	150	38/7100	—	—
J.R.	63/8	55	88	96/1870	51	2.4 g/dl

[a] Total WBCs/mm^3.

The decision as to whether intrathecal or intratumoral infusion is the most efficacious route of administration is still under study.[18b]

3.4. Current Studies

The toxicity studies are now complete, so that more recently we have begun Phase II studies. Surgery was designed to give maximum exposure of the tumor bed to the subarachnoid space and cisterns as well as to the ventricular system. Ommaya reservoirs were placed in the tumor beds of four patients, and three of the reservoirs subsequently became infected; therefore, their use has been discontinued. No other infections have occurred. Postoperatively, we have been evaluating the communication of the subarachnoid and ventricular CSF with the tumor bed in two ways: (1) by comparing the CSF sugar, protein, and cell count with those of the tumor-bed fluid obtained (Table 4); and (2) by injecting metrizamide into the lumbar subarachnoid space and following the flow of this contrast agent using the CT scanner (i.e., performing a metrizamide cisternogram). This technique is illustrated in Fig. 16.

In the seven patients studied to date by either or both of these two techniques, communication of the subarachnoid space and ventricular CSF and the tumor bed has been documented in only two patients (29%).[18b] This high degree of noncommunication between the CSF and the tumor bed may comprise the possible efficacy of such lymphocyte infusions delivered via the CSF, thus necessitating direct injection into the tumor bed.[186]

3.5. Discussion

It should be noted that glioma cells shed from the main tumor mass in a glioma patient may escape to the systemic circulation by the same pathway as proposed in our animal studies, i.e., escape of the lymphocytes from the superior sagittal sinus. The presence of intravascular tumor cells in glioma patients has been reported previously.[10,16] The escape of glioma cells into the systemic circulation may be a mechanism by which glioma patients develop immunologically specific, cytotoxic lymphocytes.[4,6,9,14,15,38]

On the basis of previous *in vitro*[4,6,9,14,15,38] evidence demonstrating that glioma patients have circulating immunologically specific, cytotoxic lymphocytes,

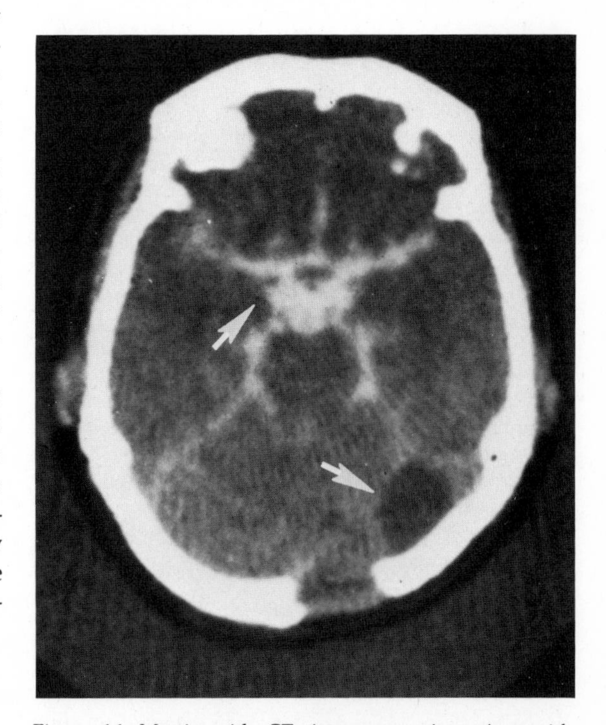

Figure 16. Metrizamide CT cisternogram in patient with recurrent cerebellar glioma. Note absence of metrizamide in tumor bed (lower arrow), but excellent filling of basilar cisterns (upper arrow).

and on the basis of our animal studies, we proceeded to conduct a Phase I human trial to evaluate the toxicity of intrathecal infusions of autologous lymphocytes in glioma patients. Despite multiple lymphoid-cell infusions, none of the patients developed choroid plexitis. In previous animal studies, choroid plexitis was observed following repeated xenogeneic-lymphoid-cell infusions, but not after multiple syngeneic-lymphoid-cell infusions.[19]

On a short-term follow-up, our Phase I human trial studies revealed no toxicity. Indeed, the infusions were tolerated very well. On the other hand, no conclusions can be made from these studies either in support of or against the efficacy of intrathecal autologous-lymphocyte infusions in patients with primary CNS tumors.

References

1. AISNER, J., SCHIFFER, C. A., WOLFF, J. H., WIERNIK, P. H.: A standardized technique for efficient platelet and leukocyte collection using the Model 30 Blood Processor. *Transfusion* (5):437–445, 1976.
2. BROOKS, W. H., CALDWELL, H. D., MORTARA, R. H.: Immune responses in patients with gliomas. *Surg. Neurol.* **2**:419–423, 1974.
3. BROOKS, W. H., MARKESBERY, W. R., GUPTA, G. D., ROSZMAN, T. L.: Relationship of lymphocyte and survival of brain tumor patients. *Ann. Neurol.* **1**:219–224, 1978.
4. BROOKS, W. H., NETSKY, M. G., NORMANSELL, D. E., HORWITZ, D. A.: Depressed cell-mediated immunity in patients with primary intracranial tumors: Characterization of a humoral immunosuppressive factor. *J. Exp. Med.* **136**:1631–1647, 1972.
5. CUTTS, H. J.: *Cell Separation: Methods in Hematology.* New York and London, Academic Press, 1970.
6. EGGERS, A. E.: Auto-radiographic and fluorescence antibody studies of the human host immune response to gliomas. *Neurology* **22**:246–250, 1972.
7. EVANS, J., SMITH, M. A., STEEL, C. M.: Rosetting test. *Lancet* **1**:96–97, 1974.
8. JELSMA, R., BUCY, P. C.: Glioblastoma multiforme: Its treatment and some factors affecting survival. *Arch. Neurol.* **20**:161–171, 1969.
9. KUMAR, S., TAYLOR, G., STEWARD, J. K., WAGHE, M. A., MORRIS-JONES, P.: Cell-mediated immunity and blocking factors in patients with tumors of the central nervous system. *Int. J. Cancer* **12**:194–205, 1973.
10. KUNG, P. C., LEE, J. C., BAKAY, L.: Vascular invasion by glioma cells in man: An electron microscopic study. *J. Neurosurg.* **31**:339–345, 1969.
11. LAYNE, E.: *Biuret Procedure: Methods in Enzymology,* Vol. 3. New York and London, Academic Press, 1957.
12. LEVY, N. L.: Specificity of lymphocyte-mediated cytotoxicity in patients with primary intracranial tumors. *J. Immunol.* **121**(3):903–915, 1978.
13. LEVY, N. L.: Cell-mediated cytotoxicity and serum-mediated blocking: Evidence that their associated determinants on human tumor cells are different. *J. Immunol.* **121**(3):916–922, 1978.
14. LEVY, N. L., MAHALEY, M. S., JR., DAY, E. D.: *In vitro* demonstration of cell-mediated immunity to human brain tumors. *Cancer Res.* **32**:447–482, 1972.
15. MEDAWAR, P. B.: Immunity to homologous grafted skin. III. Fate of skin homografts transplanted to the brain, to subcutaneous tissue, and to the anterior chambers of the eye. *Br. J. Exp. Pathol.* **299**:58–69, 1948.
16. MORLEY, T. P.: The recovery of tumor cells from venous blood draining cerebral gliomas: A preliminary report. *Can. J. Surg.* **2**:363–365, 1959.
17. MURPHY, J. B., STURM, E.: Conditions determining the transplantation of tissue in the brain. *J. Exp. Med.* **38**:183, 1923.
18. NEUWELT, E. A., CLARK, W. K.: *Clinical Aspects of Neuroimmunology.* Baltimore, Williams & Wilkins, 1978.

18a. NEUWELT, E. A., CLARK, W. K., KIRKPATRICK, J., TOBEN, H.: Clinical studies of intrathecal autologous lymphocyte infusions in patients with malignant glioma: A toxicity study. *Ann. Neurol.* **4**:307–312, 1978.

18b. NEUWELT, E. A., DIEHL, J. T., HILL, S. A., MARAVILLA, K. R.: Use of metrizamide computerized tomographic cisternography in evaluation of patients with malignant glioma for immunotherapy. *Neurosurgery* **5**:576–582, 1979.

19. NEUWELT, E. A., DOHERTY, D.: Toxicity kinetics and clinical potential of subarachnoid lymphocyte infusions. *J. Neurosurg.* **47**:205–217, 1977.
20. PALMA, L., DI LORENZO, N., GUIDETTI, B.: Lymphocytic infiltrates in primary glioblastomas and recidivous gliomas. *J. Neurosurg.* **49**:854–861, 1978.
21. RIDLEY, A., CAVANAUGH, J. B.: The cellular reactions to heterologous, homologous, and autologous skin implanted into brain. *J. Pathol.* **99**:193, 1969.
22. RIDLEY, A.: Survival of guinea pig skin grafts in the brains of rats under treatment with antilymphocytic serum. *Transplantation* **10**:86, 1970.
23. RIDLEY, A.: The effect of cyclophosphamide and antilymphocytic serum on the cellular reactions to homologous skin implanted into the brain. *Acta Neuropathol.* **15**:351–358, 1970.
24. RIDLEY, A.: Anti-lymphocytic serum and tumors transplanted in the brain. *Acta Neuropathol.* **19**:307–317, 1971.
25. RIDLEY, A.: Clinical significance of immunopathological mechamisms in diseases of the nervous system. *Clin. Allergy* **1**:311–342, 1971.
26. RIDLEY, A., CAVANAUGH, B.: Lymphocytic infiltration in gliomas: Evidence of possible host resistance. *Brain* **94**:117–124, 1971.
27. RIDLEY, A.: Immunological reactions to gliomas. In Williams, D. (ed.): *Modern Trends in Neurology.* London and Boston, Butterworths, 1975, pp. 249–266.

28. RUCKDESCHEL, J. C., DOUKAS, J., MARDINEY, M. R., JR.: Direct, quantitative analysis of *in vitro* lymphocyte responsiveness (IVLR) to antigen, mitogen and allogenic cells using a new laser based cell analyzer. *Fed. Proc. Fed. Am. Soc. Exp. Biol.* **34**:995, 1975 (abstract).
29. SCHEINBERG, L. C., EDELMAN, F. L., LEVY, W. A.: Is the brain an immunologically privileged site? I. Studies based on intracerebral tumor homotransplantation and isotransplantation to sensitized hosts. *Arch. Neurol.* **11**:248–264, 1964.
30. SCHEINBERG, L. C., LEVY, A., EFELMAN, F.: Is the brain an immunologically privileged site? II. Studies in induced host resistance to transplantable mouse glioma following irradiation of prior implants. *Arch. Neurol.* **13**:283–286, 1965.
31. SCHEINBERG, L. C., TAYLOR, J. M.: Is the brain an immunologically privileged site? III. Studies based on homologous skin grafts to the brain and subcutaneous tissue. *Arch. Neurol.* **15**:62–67, 1966.
32. SCHEINBERG, L. C., TAYLOR, J. M.: Immunological aspects of brain tumors. *Prog. Neurol. Surg.* **2**:267–291, 1968.
33. TAKAKURA, K., MIKI, Y., KUBO, O., OGAWA, N., MATSUTANI, M., SANO, K.: Adjuvant immunotherapy for malignant brain tumors. *Jpn. J. Clin. Oncol.* **2**:109–120, 1972.
34. TAKAKURA, K., SING, O. K.: Inhibitory role of immunologically compentent cells on glioblastoma and its clinical application: *In vitro* studies. In Fusek, I., Kunc, A. (eds.): *Present Limits of Neurosurgery*. Amsterdam; Excerpta Medica, 1972, pp. 47–51.
35. TAKAKURA, K., MIKI, Y., KUBO, O.: Adjuvant immunotherapy for malignant brain tumors in infants and children. *Child's Brain* **1**:141–147, 1975.
36. TAVERAS, J. M., HARTWELL, G. T., JR., POOL, J. L.: Should we treat glioblastoma multiforme: A study of survival in 425 cases. *Am. J. Roentgenol. Radium Ther. Nucl. Med.* **87**:473–479, 1962.
37. TRINDER, P.: Determination of glucose in blood using glucose oxidase with an alternative oxygen acceptor. *Ann. Clin. Biochem.* **6**:24–27, 1969.
38. WAHLSTROM, T., SAKSELA, E., TROUPP, H. L.: Cell-bound antiglial immunity in patients with malignant tumors of the brain. *Cell. Immunol.* **6**:161–170, 1973.
39. WALKER, M. D., WEISS, H. D.: Chemotherapy in the treatment of malignant brain tumors. In Friedlander, W. J. (ed.): *Advances in Neurology*, Vol. 13: *Current Reviews*. New York, Raven Press, 1975, pp. 149–192.
40. WILSON, C. B.: Chemotherapy of brain tumors. In Thompson, R. A., Green, J. R. (eds.): *Advances in Neurology*, Vol. 16: *Neoplasia in the Central Nervous System*. New York, Raven Press, 1976, pp. 361–367.
41. YOUNG, H., KAPLAN, A., REGELSON, W.: Immunotherapy with autologous white cell infusions ("lymphocytes") in the treatment of recurrent glioblastoma—A preliminary report. *Cancer* **40**(3):1037–1044, 1977.

CHAPTER 38

Cerebrospinal Fluid Analysis in Central Nervous System Cancer

S. Clifford Schold and Dennis E. Bullard

1. Introduction

Traditional cerebrospinal fluid (CSF) analysis has been of limited value in the diagnosis of central nervous system (CNS) cancer.[92] Abnormalities of biochemical and cellular composition have, with certain exceptions, been nonspecific and of limited diagnostic, prognostic, and therapeutic significance. The development of more sophisticated methods of CSF analysis offers the hope that more useful information can be gained by judicious application of these methods to the CSF of patients with brain cancer.

The purposes of CSF analysis in patients with CNS tumors include: (1) early diagnosis, allowing early and appropriate therapy; (2) to provide prognostic information; (3) as an index of activity of neoplastic disease, especially during and following the institution of therapy; (4) as an index of the extent of CNS damage; and (5) to rule out secondary, nonneoplastic complications to which these patients are susceptible.

2. Types of CNS Cancer

2.1. Primary Brain Tumors

Primary brain tumors are subdivided by presumed cell of origin and degree of histological malignancy, both of which have prognostic significance.[13] The most common CSF abnormaility in this group of diseases is an elevated total protein concentration,[61] thought to represent interruption of the blood–brain barrier. The degree of protein elevation varies with the location and origin of the primary tumor and possibly with the degree of histological malignancy. However, this correlation has never been sufficiently consistent to be of value in monitoring the course of the disease in these patients.

S. Clifford Schold, M.D. • Division of Neurology, Duke University Medical Center, Durham, North Carolina 27710 **Dennis E. Bullard, M.D.** • Department of Microbiology and Immunology and Division of Neurosurgery, Duke University Medical Center, Durham, North Carolina 27710.

2.2. Parenchymal Metastatic Brain Tumors

Metastatic tumors constitute the majority of parenchymal brain tumors.[81] Intraparenchymal metastases occur in approximately 15% of all patients dying from systemic cancer, a figure that varies with the type of primary tumor and that appears to be increasing as chemotherapy of the underlying neoplasms becomes more effective.[80] Certain patients, such as those with metastatic melanoma or oat-cell carcinoma of the lung, are at particularly high risk for the development of brain metastases.

2.3. Leptomeningeal Carcinomatosis

Leptomeningeal carcinomatosis is diffuse or multifocal involvement of the cerebral and spinal meninges by metastatic systemic cancer. The importance of this complication in metastatic tumors was first recognized in the childhood leukemias.[27] Currently, the most common sources of leptomeningeal metastases are lung and breast cancer and the lymphomas, and the incidence of this complication in solid tumors appears to be increasing with the advent of successful systemic chemotherapy.[79] The diagnosis rests on clinical evidence of multifocal involvement of the nervous system and a typical CSF profile, including a variably elevated protein concentration, a depressed glucose concentration, and mild pleocytosis. The diagnosis is established with certainty, however, only when neoplastic cells can be demonstrated by cytological examination of the CSF. While this eventually occurs in the majority of patients with leptomeningeal carcinomatosis, early in the course the cytological examination is frequently negative. Olson *et al.*[74] have shown that the CSF cytology is positive in only about 50% of such patients on the first lumbar puncture, and in about 20% it is repeatedly negative.

Delay in diagnosis of all forms of brain cancer results in more severe residual neurological difficulties after the institution of treatment and in less overall success in extending survival. Hence, improved methods of early diagnosis of CNS cancer and consequent early institution of aggressive therapy should result in better therapeutic results. Also, the development of effective forms of therapy in primary and secondary brain cancers emphasizes the importance of establishing reliable indicators of disease activity. If preclinical recurrence can be detected, alteration of therapy is more likely to be successful than if one relies on clinical evidence of disease activity alone.

3. Biological Tumor Markers

The discovery of tumor-associated markers in recent years has led to significant theoretical and clinical advances in oncological medicine.[68] Oncofetal antigens, neoplasm-associated enzymes, hormones, and morphological abnormalities such as marker chromosomes have been valuable both in the diagnosis of systemic tumors and in following the course of the disease during therapy. Tumor markers have been found in urine, pleural and peritoneal fluid, and tumor cyst fluid, as well as blood of cancer patients. In principle, similar analysis of CSF should provide useful information about the nature, extent, and activity of cancer affecting the CNS, although these investigations are complicated by the necessity of a simultaneous consideration of blood–brain barrier phenomena and CSF dynamics. Ideally, one would like to find abnormalities, or markers, that are specific for particular histological forms of brain cancer and that correlate with tumor bulk and activity.

4. CSF Tumor Markers in CNS Cancer

4.1. Human Chorionic Gonadotropin

A number of biochemical tumor markers have been analyzed in the CSF of patients with known or suspected CNS cancer (see Table 1). The exhaustive analysis by Bagshawe and Harland[6] of human chorionic gonadotropin (HCG) has established it as the prototype of CSF tumor markers. This 45,000-dalton glycoprotein, secreted normally by the human placenta, is structurally related to the anterior pituitary glycoprotein hormones.[104] Its more immunologically specific β chain (βHCG) has been quantified recently in both blood and CSF.[103,105] It is elevated in the serum of patients with uterine choriocarcinoma and in some patients with testicular carcinoma.[93] In the latter disease, an elevated serum βHCG level appears to be specific for syncytiotrophoblastic elements in the tumor and is probably a poor prognostic sign. These patients also appear to be at increased risk for the development of brain metastases.[106] A small percentage of patients with tumors of breast, female genital tract, and a number of other organs also have elevated serum βHCG levels.[12] In all these conditions, evidence suggests that serum levels parallel disease activity.

In the absence of CNS involvement by tumor, the CSF level of βHCG is approximately 1–2% of the blood level. A higher ratio is a reliable indicator of CNS metastasis.[6,95] In addition, the level of the marker in the CSF appears to accurately reflect the activity of CNS disease in that treated metastatic neoplasms are accompanied by a return of the ratio to normal levels, and recurrence is paralleled by

Table 1. Clinically Useful Tumor Markers in CSF

Marker	Primary brain tumors	Parenchymal metastases	Leptomeningeal metastases
Human chorionic gonadotropin		X	
Carcinoembryonic antigen			X
α-Fetoprotein	X	X	
Polyamines	X		
Desmosterol	X		
Lactate dehydrogenase isoenzymes			X
β-Glucuronidase			X

pathological levels of CSF βHCG.[6] One must be cautious in interpreting this ratio in the presence of a rapidly changing serum level, however, since the CSF tends to lag behind serum, and CSF analysis at the time of a rapidly falling blood level after the institution of chemotherapy will result in a falsely elevated ratio. The relative sensitivities of the CSF βHCG determination, computerized tomography (CT), and the clinical state of the patient are as yet unknown; however, in recent reports, elevated levels of βHCG in the CSF of patients with metastatic choriocarcinoma preceded clinical evidence of neurological disease.[5,95]

4.2. Carcinoembryonic Antigen

Carcinoembryonic antigen (CEA) is a protein with a molecular weight of 200,000 that is found in fetal entodermal cells and is reexpressed by certain neoplastic cells.[37] This fetal antigen was originally thought to be specific for carcinoma of the colon, and its greatest utility remains in following the course of that disease.[54] Subsequent experience indicated that it was also elevated in the plasma of patients with a variety of other neoplasms and that there was a correlation between plasma levels and disease activity.[38] The "physiological" ratio of CSF CEA to plasma CEA has not been established with certainty, but preliminary data suggest that it is on the order of 1%.[94] Elevated CSF CEA levels were first reported in leptomeningeal carcinomatosis secondary to breast cancer[102]; subsequently, elevated values have also been reported in leptomeningeal spread of lung cancer, carcinoma of the bladder, and even malignant melanoma.[94] The presence of CEA in this latter disease is somewhat surprising, since it is one of the neoplasms not known to be associated with elevated plasma CEA. CSF elevations of CEA in leptomeningeal malignant melanoma have been modest, but their presence suggests that low levels of this antigen are produced by many neoplasia and that the relatively low volume of the CSF system allows detection of this protein in amounts that would not normally be detected in plasma. In addition to being of diagnostic value, CSF CEA levels appear to parallel the activity of the leptomeningeal neoplasm. In a number of reported patients, levels have fallen with successful treatment of the leptomeningeal cancer and have again become elevated with clinical recurrence.[102,113] The sensitivity of CSF CEA determination relative to more conventional methods of CSF analysis, especially cytological examination, has not been established, nor has its relative accuracy in predicting recurrent disease, but this remains an area of active investigation.

In contrast, the presence of CEA in the CSF of patients with intracerebral metastases from systemic tumors known to be secreting CEA is irregular.[94] Proximity to the leptomeningeal surface appears to be critical to the appearance of CSF CEA. This distinguishes CEA from βHCG, and this distinction remains unexplained, though it may be related to the relative molecular weights of the substances.

Patients with primary brain tumors generally do not have detectable CSF CEA.[45] While others have found measurable levels of CEA in the CSF of patients with "CNS malignancies," the lack of a description of the nature and location of the tumor makes the data difficult to interpret.[42] One patient with a craniopharyngioma who had a CSF CEA of 2.2 ng/ml has been reported.[101]

4.3. α-Fetoprotein

α-Fetoprotein (AFP) is a glycoprotein produced normally by yolk-sac elements and fetal liver. It is elevated in the serum of patients with carcinoma of the testis and primary hepatoma.[82] It has a molecular weight of 70,000 and is considered an α^1-globulin by electrophoretic mobility.[97] Elevated amniotic fluid levels of AFP have been found in a number of fetal abnormalities.[97]

Limited published data are available on the value of CSF measurement of AFP.[95,100,103] While hepatoma rarely metastasizes to the CNS, the institution of successful chemotherapy in the management of carcinoma of the testis has been accompanied by an increasing attention to the importance of CNS metastases in that disease.[95] Since serum AFP has been found to be a reliable indicator of disease activity in carcinoma of the testis, in principle CSF analysis of this protein should be valuable as an indicator of the presence of CNS metastases. The published data on this subject are ambiguous, however, and detailed studies have not been performed relating the CSF AFP level to the serum level, the incidence of detectable CSF AFP in the presence of CNS metastases, and the sensitivity of this measurement relative to established methods of diagnosis such as CT.[103] CSF AFP may be more valuable as a marker for primary intracranial germ-cell tumors, in which elevated CSF levels may be diagnostic, sparing the patient a difficult operation and allowing accurate assessment of the state of the disease.[2]

4.4. Polyamines

The polyamines are small, naturally occurring cationic compounds that are associated with nucleic acid metabolism and function.[4,15,89] The polyamines spermidine and spermine and their precursor putrescine are formed in mammalian cells by four enzymes: ornithine decarboxylase, *S*-adenosyl-L-methionine decarboxylase, spermidine synthase, and spermine synthase. Russell *et al.*[90] in 1971 made the observation that these compounds were elevated in the urine of cancer patients. Subsequent studies demonstrated that these compounds were also elevated in the serum[70] and the CSF[56] of patients with various types of cancer. However, it soon became evident that these compounds were not tumor-specific, and instead were related more to cellular proliferation.[16,88] Nevertheless, the polyamines are potentially valuable as tumor-associated markers for monitoring of tumor burden and response to therapy.

Marton *et al.*[56] demonstrated in 1976 that polyamine concentrations were elevated in the CSF of patients with untreated malignant CNS tumors. Utilizing a high-pressure liquid chromatographic technique, these investigators showed that putrescine levels were significantly elevated in patients with glioblastomas, medulloblastomas, and astrocytomas, although there was some overlap between patients with astrocytomas and control patients. Patients with pituitary adenomas had putrescine levels generally within normal limits. With spermidine, similar results were seen with the exception that patients with meningiomas also tended to have elevated levels. In this study, spermine levels were sporadically elevated and had no diagnostic pattern. The polyamine concentrations in the CSF had no relationship to the protein concentrations, and the authors concluded that elevation in polyamine concentrations were not the direct result of the disruption of the blood–brain barrier. In this initial study, polyamine concentrations in patients with malignant gliomas and medulloblastomas tended to return to normal after treatment. Moreover, when short-term serial CSF polyamine analyses were performed, increasing putrescine concentrations frequently preceded tumor regrowth. However, in certain instances, increased polyamine levels were seen with infection, degenerative disease, and cerebrovascular disease.[56]

Longer serial evaluation of the 16 patients with medulloblastomas demonstrated that in 15 of the 16 patients, CSF polyamine levels decreased with response to therapy. In 7 of these 15 patients, increased polyamine levels in the CSF were noted prior to clinical recurrence. No patient with increased CSF polyamine levels failed to develop recurrent tumor, and only one false-negative value was obtained. Increased polyamine levels were again seen with nonneoplastic conditions such as infection or vascular occlusion. However, the increases in putrescine noted with recurrence of tumor usually significantly exceeded the increases seen with these other conditions.[57]

Elevated CSF polyamine levels have also been demonstrated in children with acute lymphoblastic leukemia.[86] Three of three patients with positive CSF cytologies had elevated CSF spermidine and spermine levels. A number of other patients with active systemic disease but no evidence on CNS involvement also had elevated spermidine or spermine levels, but no correlation was made with simultaneous blood levels.

4.5. Sterols

It has been proposed that the presence of desmosterol (24-dehydrocholesterol) in the CSF may be a useful biochemical test for the presence of brain tumors.[31,75,76,84] Desmosterol is the immediate precursor of cholesterol in the CNS and is a major constituent of sterol synthesis.[32,75,107] Desmosterol levels in normal adult human brain are extremely low,[35] but it is detectable in larger quantities in fetal brain,[33,34] in several experimental animal tumors of the nervous system,[77,108] and in certain human brain tumors.[31,34] However, the detection of desmosterol in the CSF of patients with brain tumors requires the administration of triparanol (MER-29), an inhibitor of the enzymatic conversion of desmosterol to cholesterol, prior to the collection of CSF. This agent has had toxic effects following long-term usage.[1,50] In addition, the technique for measuring the levels of CSF desmosterol requires a combination of thin-layer chromatography and gas–liquid chromatography, both of which are time-consuming and have technical shortcomings.[55]

Moreover, attempts to utilize the levels of this sterol in the CSF as a marker for the presence of brain tumors have yielded conflicting results. There are several reports of elevated levels of CSF desmosterol in patients with CNS tumors.[31,34,75] Others have not successfully duplicated this finding.[55]

Attempts have also been made to utilize desmosterol levels in the CSF in the evaluation of the response of glial tumors to chemotherapy and as a monitor for tumor recurrence.[75,76,84,107] Although the data at present support a relationship between desmosterol levels in the CSF and tumor burden, the disadvantages associated with this method of analysis have prevented its measurement from being widely accepted.

4.6. Lactate Dehydrogenase

Data and opinion vary about the value of lactate dehydrogenase (LDH) analysis in the CSF of patients with brain cancer. While a number of investigators have noted consistently elevated LDH levels in CSF of patients with both primary and secondary brain tumors, others have found elevations only in the presence of metastatic lesions or not at all.[21,43,112] In most reports, no statements were made regarding concomitant LDH blood levels, the CSF total protein concentration, or the location and number of metastatic lesions. Recently, total CSF LDH activity was found not to be useful in detecting CNS relapse in acute lymphocytic leukemia in children.[85]

Histochemical analysis of LDH isoenzyme patterns from biopsied brain tumors has revealed interesting and consistent patterns that correlate with histological change and prognosis.[59,83] The LDH four and five fractions, not normally found in brain, are increased in the tissue of malignant brain tumors and in the CSF of patients with bacterial meningitis.[8,59] However, analysis of the CSF of patients with malignant primary brain tumors revealed only a "slight shift" toward the LDH four and five fractions in some.[73,83] In a recent report, Fleisher *et al.*[29] have found a significant shift toward the LDH four and five fractions in the CSF of patients with leptomeningeal carcinomatosis. This isoenzyme pattern shifted toward normal with successful institution of intrathecal chemotherapy. It may be that this marker of neoplastic growth, like many others, will be of use principally in those tumors that directly invade the subarachnoid pathways.

4.7. β-Glucuronidase

β-Glucuronidase is a lysosomal enzyme known to be associated with neoplastic tissue of epithelial origin.[62] Its activity is measured colorimetrically by the liberation of phenolphthalein from phenolphthalien mono-β-glucuronic acid.[78] It has been analyzed in CSF in a number of neurological conditions. Shuttleworth and Allen[98] in 1968 first noted the striking elevations of glucuronidase in meningeal carcinomatosis, also noting stabilization of enzyme activity in two patients undergoing treatment. Subsequent experience has confirmed the value of this enzyme level as a marker for leptomeningeal spread of cancer.[94,99] Elevated levels have been found in lung, breast, bladder, parotid, and even glial cancer metastatic to the meninges. Like other CSF alterations in leptomeningeal tumor, ventricular CSF shows less marked abnormalities than lumbar CSF.[94] However, with treatment, both lumbar and ventricular abnormalities tend to return toward normal, paralleling the clinical course of the patient. The value of this determination as an indicator of recurrence and hence as an indication for altering therapy remains to be determined.

CSF β-glucuronidase activity is usually normal in cases of parenchymal metastatic brain tumors and malignant primary brain tumors without leptomeningeal spread.[3,94] It is elevated in the presence of polymorphonuclear leukocyte infiltration, but usually normal in nonneoplastic mononuclear inflammation.

4.8. Miscellaneous Tumor-Associated Substances

Lysozyme is a low-molecular-weight anionic protein that is found in various human tissues but is absent in normal CSF.[30] Elevated lysozyme levels in the CSF have been found with both primary and metastatic CNS tumors.[22,58,69] However, elevations have also been found in patients with CNS infections and CNS sarcoid,[58,69] while some patients with CNS tumors and normal CSF cellular and chemical parameters have normal CSF lysozyme levels.[22]

Since the original extraction and purification of glial fibrillary acidic protein (GFAP) by Eng *et al.*[26] in 1971, several apparently similar proteins have been extracted by different techniques from normal brain[18] and human glioma tissue.[63] The relationships among these proteins and their specificity have not been defined.[51] Using a direct immunoperoxidase staining method, Deck *et al.*[19] were able to demonstrate GFAP in astrocytic tumor cells both in brain and in the subarachnoid space. No attempts to directly analyze CSF cells for GFAP have been reported. The only studies of CSF levels of glial-derived proteins have utilized a human-glioma-derived antiserum,[39b,63,64] the specificity of which has been questioned.[51]

A number of other enzymes,[96] proteins,[49] and lipids[14,28,41] have been evaluated as potential CNS tumor markers, but have not been definitively investigated in CSF. Various potential brain-tumor antigens have not as yet been analyzed in CSF.[9,10,23]

5. CSF Cytology

Identification of neoplastic cells in the CSF is the most reliable indication of the presence of CNS malignancy.[36] However, the yield of cytological examination of the CSF is quite variable, depending on the amount of CSF available, the technique of CSF processing and analysis, the location of the tumor in relation to the ventricular and subarachnoid pathways, and the total tumor burden, as well as the experience and judgment of the cytologist.[17,20,25,39,48,53,67,87] Recently, Glass *et al.*[36] reported that CSF cytological examinations were positive in only 26% of 117 patients with primary or metastatic involvement of the CNS with tumor. Malignant cells in the CSF were observed in only 59% of 51 patients with autopsy-proven leptomeningeal tumor. Fortunately, false-positive cytologies were rare. The morphological features of the cellular components of CSF have been dealt with in detail by Kolmel[47] and Oehmichen.[71]

5.1. Primary Brain Tumors

Positive CSF cytological examination has been found in a variable percentage of patients with primary brain tumors[7,20,39,46,67,71,109] (Fig. 1). This percentage varies with the type of tumor and probably with the degree of histological malignancy.[7,11,20,46,60] Malignant cells are most often identified in the CSF of patients with tumors that arise near the ventricular surface. For example, positive cytology has been reported in up to 91% of cases of medulloblastomas.[7] Ependymomas and pineal tumors are also commonly associated with positive CSF cytology.[7,20,71] Longer survival in gliomas and other primary brain tumors may be associated with a higher incidence of subarachnoid spread of tumor and consequent positive CSF cytology.[114] In a series of 122 patients with gliomas,[7] the initial 14% incidence of tumor cells observed in CSF increased to 40% following surgical intervention.

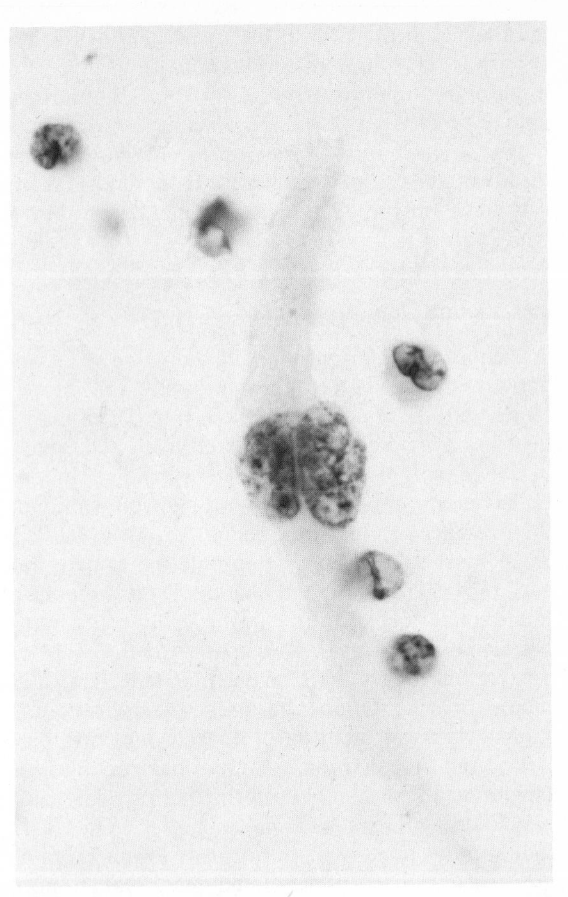

Figure 1. Large binucleate malignant cell from CSF of patient with meningeal spread of pontine glioblastoma multiforme (Millipore filter, Papanicolaou stain; ×1000, oil immersion). From Duke Medical Center Cerebrospinal Fluid Cytology Workshop.

5.2. Metastatic Brain Tumors

The presence of malignant exfoliative cells in the CSF is more commonly found with metastatic brain tumors than with primary tumors.[46,60,87,111] (Fig. 2). Unfortunately, the distinction between purely parenchymal metastatic lesions and leptomeningeal metastasis is not usually made. Balhuizen *et al.*[7] made this distinction on clinical grounds and still found positive preoperative CSF cytologies in 20% of their patients with single or multiple secondary CNS tumors. On the other hand, in an autopsy study, Glass *et al.*[36] found that positive cytological examination of the CSF virtually always indicated some degree of leptomeningeal involvement. Subclinical leptomeningeal metastasis may be more common than has been appreciated, and a positive CSF cytology may be an accurate marker for meningeal invasion.

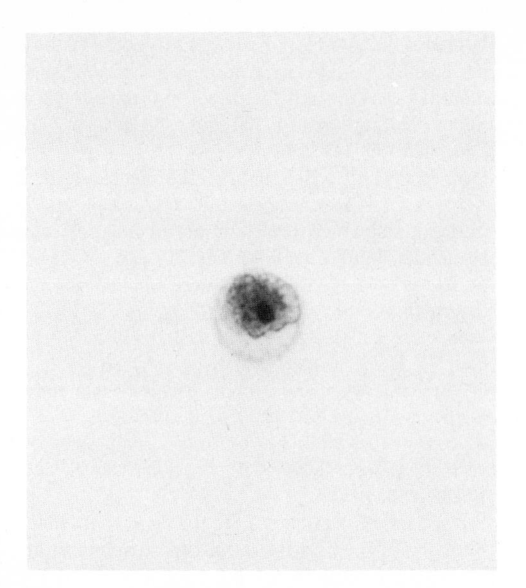

Figure 2. Representative malignant cell from CSF of patient with adenocarcinoma of stomach and meningeal carcinomatosis. There is extensive nuclear folding and prominent nucleolus (Millipore filter, Papanicolaou stain; ×1000, oil immersion). From the Duke Medical Center Cerebrospinal Fluid Cytology Workshop.

5.3. Other Techniques

In addition to standard cytological techniques, a number of new diagnostic procedures have been introduced in recent years. Some of the more promising methods include the immunological characterization of the lymphocytes in the CSF by their surface markers,[39a] rosette formation and phytohemagglutinin responsiveness,[52,65] scanning and routine electron microscopy,[24,40] tritiated thymidine autoradiography,[40,72] and CSF cell culture.[44,66,91]

6. Summary and Future Directions

It is likely that the full potential of CSF analysis in patients with CNS cancer has not been approached. Many tumor-related substances have not yet been measured in CSF. Others have not been adequately correlated with other CSF parameters, extent of CNS disease, other diagnostic methods, and corresponding blood levels. No biochemical CSF abnormality rivals in accuracy the identification of malignant cells in the CSF. Reliable tumor-specific markers have not as yet been identified in primary tumors of the brain.

On the other hand, a number of abnormalities have been found in the CSF of patients suffering from various forms of brain cancer. Some, especially βHCG and CEA, are of established diagnostic value. Many, notably β-glucuronidase, CEA, βHCG, LDH isoenzyme pattern, and polyamines, appear to reflect the acitivity of the CNS neoplasm. Until tumor-specific markers can be reliably identified and quantified, consistent, though nonspecific, alterations of CSF content can be of considerable value in following the course of neoplastic disease. These abnormalities of CSF content vary both with the type of CNS cancer and with the extent of disease, but primarily with the location of the neoplastic cells, in particular their proximity to subarachnoid pathways.

As more effective forms of treatment become available, the importance of reliable objective measurements of disease activity increases. Recent characterization of tumor markers[97] and development of more specific brain-tumor markers[23,110] suggest that these measurements will become available and will have an impact on treatment decisions in neurooncology. In the immediate future, multiple marker analysis may provide more accurate information about tumor activity than individual analyses alone.

ACKNOWLEDGMENTS. The authors wish to thank Dr. Jerome B. Posner for guidance in preparation of the manuscript and Dr. Sandra H. Bigner for preparation of the cytological material.

References

1. ACHOR, R. W. P., WINKELMAN, R. K., PERRY, H. O.: Cutaneous side effects from use of triparanol (MER-29): Preliminary data on ichthyosis and loss of hair. *Mayo Clin. Proc.* **36**:217–228, 1961.
2. ALLEN, J. C., NISSELBAUM, J., EPSTEIN, F., ROSEN, G., SCHWARTZ, M. K.: Alphafetoprotein and human chorionic gonadotropin: An aid in the diagnosis and management of intracranial germ cell tumors. *J. Neurosurg.* **S1**:368–374, 1979.
3. ALLEN, N.: Acid hydrolytic enzymes in brain tumors. *Prog. Exp. Tumor Res.* **17**:291–307, 1972.
4. BACHRACH, U.: *Function of Naturally Occurring Polyamines.* New York, Academic Press, 1973.
5. BAGSHAWE, K. D., HARLAND, S.: Detection of intracranial tumours with special reference to immunodiagnosis. *Proc. R. Soc. Med.* **69**:51–53, 1976.
6. BAGSHAWE, K. D., HARLAND, S.: Immunodiagnosis and monitoring of gonadrotropin-producing metastases in the central nervous system. *Cancer* **38**:112–118, 1976.

7. BALHUIZEN, J. C., BOTS, G. T. A. M., SCHABERG, A., BOSMAN, F. T.: Value of cerebrospinal fluid cytology for the diagnosis of malignancies in the central nervous system. *J. Neurosurg.* **48**:747–753, 1978.
8. BEATY, H. N., OPPENHEIMER, S.: Cerebrospinal fluid lactic dehydrogenase and its isoenzymes in infections of the central nervous system. *N. Engl. J. Med.* **279**:1197–1202, 1968.
9. BOGOCH, S.: Astrocytin and malignin: Two polypeptide fragments (recognins) related to brain tumors. *Natl. Cancer Inst. Monogr.* **46**:133–137, 1977.
10. BOGOCH, S., BOGOCH, E. S., FAGER, C. A., GOLDENSOHN, E. S., HARRIS, J. H., HICKOLS, D. F., LOWDEN, J. A., LUX, W. E., RANSOHOFF, J., WALKER, M. D.: Elevated serum anti-malignin antibody in glioma and other cancer patients: A seven-hospitial blood study. *Neurology* **29**:584, 1979.
11. BOTS, G. T. A. M., WENT, L. N., SCHABERG, A.: Results of a sedimentation technique for cytology of cerebrospinal fluid. *Acta Cytol.* **8**:234–241, 1964.
12. BRAUNSTEIN, G. D., VAITUKAITIS, J. L., CARBONE, P. P., ROSS, G. T.: Ectopic production of human chorionic gonadotropin by neoplasms. *Ann. Intern. Med.* **78**:39–45, 1973.
13. BURGER, P. C., VOGEL, F. S.: *Surgical Pathology of the Central Nervous System and Its Coverings.* New York, John Wiley, 1976.
14. CHRISTENSEN, L. H., CLAUSEN, J., BIERRING, F.: Phospholipids and glycolipids of tumors in the central nervous system. *J. Neurochem.* **12**:619–627, 1965.
15. COHEN, S. S.: *Introduction to the Polyamines.* Englewood Cliffs, New Jersey, Prentice-Hall, 1971.
16. COHEN, S. S.: Conference on polyamines in cancer. *Cancer Res.* **37**:939–942, 1977.
17. COOK, J. D., BROOKS, B. R.: Lymphocyte subpopulations in human cerebrospinal fluid: Effects of various disease states and immunosuppresive drugs. In Wood J. H. (ed.): *Neurobiology of Cerebrospinal Fluid I.* New York, Plenum Press, 1980.
18. DAHL, D., BIGNAMI, A.: Immunochemical and immunofluoresence studies of the glial fibrillary acidic protein in vertebrates. *Brain Res.* **57**:343–360, 1973.
19. DECK, J. H. N., ENG, L. F., BIGBEE, J., WOODCOCK, S. M.: The role of glial fibrillary acidic protein in the diagnosis of central nervous system tumors. *Acta Neuropathol. (Berlin)* **42**:183–190, 1978.
20. DEN HARTOG JAGER, W. A.: Cytopathology of the cerebrospinal fluid examined with the sedimentation technique after Sayk. *J. Neurol. Sci.* **9**:155–177, 1969.
21. DHARKER, S. R., DHARKER, R. S., CHAURASIN, B. D.: Lactate dehydrogenase and aspartate transaminase of the cerebrospinal fluid in patients with brain tumours, congenital hydrocephalus, and brain abscess. *J. Neurol. Neurosurg. Psychiatry* **39**:1081–1085, 1976.
22. DI LORENZO, N., PALMA, L.: Spinal-fluid lysozyme in diagnosis of central-nervous-system tumours. *Lancet* **1**:1077, 1976.
23. DITTMAN, L., AXELSEN, N. H., NORGAART-PEDERSON, B., BOCK, E.: Antigens in human glioblastomas and meningiomas: Search for tumor and onco-foetal antigens. Estimation of S-100 and GFA protein. *Br. J. Cancer* **35**:135–141, 1977.
24. DOMAGALA, W., EMESON, E. E., GREENWOLD, E., KOSS, L. G.: A scanning electron microscopic and immunologic study of B-cell lymphosarcoma cells in cerebrospinal fluid. *Cancer* **40**:716–720, 1977.
25. EL-BATATA, M.: Cytology of cerebrospinal fluid in the diagnosis of malignancy. *J. Neurosurg.* **28**:317–326, 1968.
26. ENG, L. F., VANDERHAAGHEN, J. H., BIGNAMI, A., GERSTL, B.: An acidic protein isolated from fibrous astrocytes. *Brain Res.* **28**:351–354, 1971.
27. EVANS, A. E., CRAIG, M.: Central nervous system involvement in children with acute leukemia. *Cancer* **17**:256–258, 1964.
28. FLEISHER, J. H., MARTON, L. J., BACHUR, N. R., MANN-KAPLAN, R. S.: Cholesterol in cerebrospinal fluid of brain tumor patients. *Life Sci.* **13**:1517–1526, 1973.
29. FLEISHER, M., SCHOLD, S. C., SCHWARTZ, M. K., POSNER, J. B.: Tumor markers in cerebrospinal fluid for the differential diagnosis of central nervous system metastasis. *Clin. Chem.* **24**:1002, 1978.
30. FLEMING, A.: On a remarkable bacteriolytic element found in tissues and secretions. *Proc. R. Soc. London Ser. B* **93**:306–317, 1922.
31. FUMAGALLI, R., PAOLETTI, P.: Sterol test for human brain tumors: Relationship with different oncocytes. *Neurology* **21**:1149–1156, 1971.
32. FUMAGALLI, R., PEZZOTTA, S., PAOLETTI, P.: Sterols in cerebrospinal fluid during nitrosourea chemotherapy of brain tumors. *Pharmacol. Res. Commun.* **8**:127–141, 1976.
33. FUMAGALLI, R., PAOLETTI, R.: The identification and significance of desmosterol in the developing human and animal brain. *Life Sci.* **2**:291–295, 1963.
34. FUMAGALLI, R., GROSSI, E., PAOLETTI, P., PAOLETTI, R.: Studies on lipids in brain tumours. I. Occurrence and significance of sterol precursors of cholesterol in human brain tumors. *J. Neurochem.* **11**:561–565, 1964.
35. GALLI, G., GROSSI-PAOLETTI, E., WEISS, J. F.: Sterol precursors of cholesterol in adult human brain. *Science* **162**:1495–1496, 1968.
36. GLASS, J. P., MELAMED, M., CHERNIK, N. L., POSNER, J. B.: Malignant cells in cerebrospinal fluid (CSF): The meaning of a positive CSF cytology. *Neurology* **29**:1369–1375, 1979.
37. GOLD, P., FREEDMAN, S. O., SHUSTER, J.: Carcinoembryonic antigen: Historical perspective, experimental data. In Herberman, R. B., McIntire, K. R. (eds.): *Immunodiagnosis of Cancer.* New York, Marcel Dekker, 1979, pp. 147–164.
38. GOLDENBERG, D. M., PAVIA, R. A., SHARKEY, M., BENNETT, S.: Biology of carcinoembryonic antigen: An overview. In Griffiths, K., Neville, A. M., Pierrepoint,

C. G. (eds.): *Tumor Markers: Determination and Clinical Role.* Baltimore, University Park Press, 1977, pp. 29–40.

39. GONDOS, B., KING, E. B.: Cerebrospinal fluid cytology: Diagnostic accuracy and comparison of different techniques. *Acta Cytol.* **20:**542–547, 1976.

39a. GOODSON, J. D., STRAUSS, G. M.: Diagnosis of lymphomatous leptomeningitis by cerebrospinal fluid lymphocyte cell surface markers. *Am. J. Med.* **66:**1057–1059, 1979.

39b. HAYAKAWA, T., MORIMOTO, K., USHIO, Y., MORI, T., YOSHIMINE, T., MYOGA, A., MOGAMI, H.: Levels of astroprotein (an astrocyte-specific cerebroprotein) in cerebrospinal fluid of patients with brain tumors: An attempt at immunochemical diagnosis of gliomas. *J. Neurosurg.* **52:**229–233, 1980.

40. HERNDON, R. M.: Electron-microscopic studies on cerebrospinal fluid sediment. In Wood, J. H. (ed.): *Neurobiology of Cerebrospinal Fluid I.* New York, Plenum Press, 1980.

41. HILDEBRAND, J.: Early diagnosis of brain metastases in an unselected population of cancerous patients. *Eur. J. Cancer* **9:**621–626, 1973.

42. HILL, S. A., MARTIN, E. C., HUNT, W. E.: *Carcinoembryonic Antigen in CNS Neoplasia,* Proceedings of the 47th AANA, Los Angeles, April 22–26, 1975, p. 142.

43. JAKOBY, R. K., JAKOBY, W. B.: Lactic dehydrogenase of cerebrospinal fluid in the differential diagnosis of cerebrovascular disease and brain tumor. *J. Neurosurg.* **15:**45–51, 1958.

44. KAJIKAWA, H., OHTA, T., OHSHIRO, H., HARADA, K., ISHIKAWA, S., UOZUMI, T., KODAMA, M., OKADA, T.: Cerebrospinal fluid cytology in patients with brain tumors: A simple method using the cell culture technique. *Acta Cytol.* **21:**162–167, 1977.

45. KIDO, D. K., DYCE, B. J., HAVERBACK, B. J., RUMBAUGH, C. L.: Carcinoembryonic antigen in patients with untreated central nervous system tumors. *Bull. Los Angeles Neurol. Soc.* **41:**47–54, 1976.

46. KLINE, T., SPEIGEL, I. J., TINSLEY, M.: Tumor cells in the cerebrospinal fluid. *J. Neurosurg.* **19:**679–684, 1962.

47. KOLMEL, H. W.: *Atlas of Cerebrospinal Fluid Cells.* New York, Springer-Verlag, 1977.

48. KOLMEL, H. W., CHONE, B. K.: Combined application of cytodiagnostic technics to the cerebrospinal fluid. *J. Neurol.* **216:**1–8, 1977.

49. KUUSELA, P., VAHERI, A., PALO, J., RUOSLAHTI, E.: Demonstration of fibronectin in human cerebrospinal fluid. *J. Lab. Clin. Med.* **92:**595–601, 1978.

50. LAUGHLIN, R. C., CAREY, T. F.: Cataracts in patients treated with triparanol. *J. Am. Med. Assoc.* **181:**339–340, 1962.

51. LIEM, R. K. H., SHELANSKI, M. L.: Identity of the major protein in "native" glial fibrillary acidic protein preparation with tubulin. *Brain Res.* **145:**196–201, 1978.

52. LEVINSON, A. I., LISAK, R. P., ZWEIMAN, B.: Immunologic characterization of cerebrospinal fluid lymphocytes: Preliminary report. *Neurology* **26:**693–695, 1976.

53. MARKS, V., MARRACK, D.: Tumor cells in the cerebrospinal fluid. *J. Neurol. Neurosurg. Psychiatry* **23:**194–201, 1960.

54. MARTIN, E. W., JR., KIBBEY, W. E., DI VECCHIA, L., ANDERSON, G., CATALANO, P., MINTON, J. P.: Carcinoembryonic antigen: Clinical and historical aspects. *Cancer* **37:**62–81, 1976.

55. MARTON, L. J., GORDON, G. S., BARKER, M., WILSON, C. B., LUBICH, W.: Failure to demonstrate desmosterol in spinal fluid of brain tumor patients. *Arch. Neurol.* **28:**137–138, 1973.

56. MARTON, L. J., HEBY, O., LEVIN, V. A., LUBICH, W. P., CRAFTS, D. C., WILSON, C. B.: The relationship of polyamines in cerebrospinal fluid to the presence of central nervous system tumors. *Cancer Res.* **36:**973–977, 1976.

57. MARTON, L. J., EDWARDS, M. S., LEVIN, V. A., LUBICH, W. P., WILSON, C. B.: Predictive value of cerebrospinal fluid polyamines in medulloblastoma. *Cancer Res.* **39:**993–997, 1979.

58. MASON, D. Y., ROBERTS-THOMSON, P.: Spinal-fluid lysozyme in diagnosis of central-nervous-system tumours. *Lancet* **2:**952–953, 1974.

59. MCCORMICK, D., ALLEN, I. V.: The value of LDH isoenzymes in the rapid diagnosis of brain tumors. *Neuropathol. App. Neurobiol.* **2:**269–278, 1976.

60. MCCORMICK, W. E., COLEMAN, S. A.: A membrane filter technic for cytology of spinal fluid. *Am. J. Clin. Pathol.* **38:**191–197, 1962.

61. MERRITT, H. H., FREMONT-SMITH, F.: *The Cerebrospinal Fluid.* Philadelphia, W. B. Saunders, 1937.

62. MONIS, B., BANKS, B. M., RUTENBURG, A. M.: β-Glucuronidase activity in malignant neoplasms of man: A histochemical study. *Cancer* **13:**386–393, 1960.

63. MORI, R., MORIMOTO, K., USHIO, Y., HAYAKAWA, T., MOGAMI, H.: Radioimmunoassay of astroprotein (an astrocyte-specific cerebroprotein) in cerebrospinal fluid from patients with glioma: A preliminary study. *Neurol. Med.-Chir. (Tokyo)* **15:**23–25, 1975.

64. MORI, T., MORIMOTO, K., HAYAKAWA, T., USHIO, V., MOGAMI, H., SEKIGUCHI, K.: Radioimmunoassay of astroprotein (an astrocyte-specific cerebroprotein) in cerebrospinal fluid and its clinical significance. *Neurol. Med.-Chir. (Tokyo)* **18:**25–31, 1978.

65. MOSER, R. P., ROBINSON, J. A., PROSTKO, E. R.: Lymphocyte subpopulations in human cerebrospinal fluid. *Neurology* **26:**726–728, 1976.

66. NAGAI, M.: Diagnosis of brain tumors by cell culture of the cerebrospinal fluid. *Adv. Neurol. Sci. (Tokyo)* **13:**573–589, 1969.

67. NAYLOR, B.: The cytologic diagnosis of cerebrospinal fluid. *Acta Cytol.* **8:**141–149, 1964.

68. NEVILLE, A. M.: Tumor markers—an overview. In Griffiths, K., Neville, A. M., Pierrepoint, C. G. (eds.):

Tumor Markers: Determination and Clinical Role. Baltimore, University Park Press, 1977, pp. 1–14.

69. NEWMAN, J., CACATAIN, A., JOSEPHSON, A. S., TSANG, A.: Spinal-fluid lysozyme in the diagnosis of central nervous system tumours. *Lancet* **2:**756–757, 1974.
70. NISHIOKA, K., ROMSDAHL, M. M.: Elevation of putrescine and spermidine in sera of patients with solid tumors. *Clin. Chim. Acta* **57:**155–161, 1974.
71. OEHMICHEN, M.: *Cerebrospinal Fluid Cytology: An Introduction and Atlas.* Philadelphia, W. B. Saunders, 1976.
72. OHTA, H., FUKUI, M., YAMAKAWA, Y., MATSUNO, H., KITAMURA, K.: A study of CSF cells by ^{3}H-thymidine autoradiography and cytology regarding the subarachnoid disseminations of brain tumor. *J. Neurol.* **217:**243–251, 1978.
73. OLDENKOTT, P., HELLER, W., BLANKENHORN, H., ELIES, W.: The enzyme-pathobiochemistry of intracranial processes. *Acta Neurochir.* **25:**57–68, 1971.
74. OLSON, M. E., CHERNIK, N. C., POSNER, J. B.: Infiltration of the leptomeninges by systemic cancer: A clinical and pathologic study. *Arch. Neurol.* **30:**122–137, 1974.
75. PAOLETTI, P., VANDENHEUVEL, F. A., FUMAGALLI, R., PAOLETTI, R.: The sterol test for the diagnosis of human brain tumors. *Neurology* **19:**190–197, 1969.
76. PAOLETTI, P., FUMAGALLI, R., WEISS, J. F., PEZZOTTA, S.: Desmosterol: A biochemical marker of glioma growth. *Surg. Neurol.* **8:**399–405, 1977.
77. PAOLETTI, R., FUMAGALLI, R., GROSSI, E., PAOLETTI, P.: Studies on brain sterols in normal and pathological conditions. *J. Am. Oil Chem. Soc.* **42:**400–404, 1965.
78. PLAICE, C. H. J.: A note on the determination of beta-glucuronidase activity. *J. Clin. Pathol.* **14:**661–665, 1961.
79. POSNER, J. B.: Management of central nervous system metastases. *Semin. Oncol.* **4:**81–91, 1972.
80. POSNER, J. B., CHERNIK, N. L.: Intracranial metastases from systemic cancer. *Adv. Neurol.* **19:**575–587, 1978.
81. POSNER, J. B., SHAPIRO, W. R.: Brain tumor: Current status of treatment and its complications. *Arch. Neurol.* **32:**781–784, 1975.
82. PURVES, L. R.: Alpha-feotoprotein: Its chemistry and biology. In Griffiths, K., Neville, A. M., Pierrepoint, C. G. (eds.): *Tumor Markers: Determination and Clinical Role.* Baltimore, University Park Press, 1977, pp. 53–64.
83. RABOW, L., KRISTENSSON, K.: Changes in lactate dehydrogenase isoenzyme patterns in patients with tumors of the central nervous system. *Acta Neurochir.* **36:**71–81, 1977.
84. RANSOHOFF, J., WEISS, J. F.: Cerebrospinal fluid sterols in the evaluation of patients with gliomas. *Natl. Cancer Inst. Monogr.* **46:**119–124, 1977.
85. REEVES, J. D., HUTTER, J. J., JR., FAVARA, B. E.: Spinal fluid LDH activity in children with acute leukemia. *Am. J. Dis. Child.* **132:**634–635, 1978.
86. RENNERT, O. M., LAWSON, D. L., SHUKLA, J. B., MIATE, T. D.: Cerebrospinal fluid polyamine monitoring in central nervous system leukemia. *Clin. Chim. Acta* **75:**365–369, 1977.
87. RICH, J. R.: A survey of cerebrospinal fluid cytology. *Bull. Los Angeles Neurol. Soc.* **34:**115–131, 1969.
88. RUSSELL, D. H. (ed.) *Polyamines in Normal and Neoplastic Growth.* New York, Raven Press, 1973.
89. RUSSELL, D. H., DURIE, B. G. M.: *Polyamines as Biochemical Markers of Normal and Malignant Growth: Progress in Cancer Research and Therapy,* Vol. 8. New York Raven Press, 1978.
90. RUSSELL, D. H., LEVY, C. C., SCHIMPF, S. C., HAWK, I. A.: Urinary polyamines in cancer patients. *Cancer Res.* **31:**1555–1558, 1971.
91. SANO, K., NAGAI, M., TSUCHIDA, T.: New diagnostic method for brain tumors by cell culture of the cerebrospinal fluid: Millipore filter-cell culture method. *Neurol. Med.-Chir. (Tokyo)* **8:**17–27, 1966.
92. SAYK, J.: The cerebrospinal fluid in brain tumors. In Vinken, P. J., Bruyn, G. W. (eds.): *Handbook of Clinical Neurology,* Vol. 16, Chapt. 12. Amsterdam, North-Holland, 1974, pp. 360–417.
93. SCARDINO, P. T., COX, H. D., WALDMAN, T. A., MCINTIRE, K. R., MITTEMEYER, B., JAVADPOUR, N.: The value of serum tumor markers in the staging and prognosis of germ cell tumors of the testis. *J. Urol.* **118:**994–999, 1977.
94. SCHOLD, S. C., FLEISHER, M., SCHWARTZ, M., POSNER, J. B.: CSF biochemical "markers" of CNS metastasis. *Trans. Am. Neurol. Assoc.* **103:**179–181, 1978.
95. SCHOLD, S. C., VURGRIN, D., GOLBEY, R. B., POSNER, J. B.: Central nervous system metastases from germ cell carcinoma of testis. *Semin. Oncol.* **6:**102–108, 1979.
96. SEIDENFELD, J., MARTON, L. J.: Biochemical markers of central nervous system tumors in cerebrospinal fluid. *Ann. Clin. Lab. Sci.* **8:**459–466, 1978.
97. SHUSTER, J., FREEDMAN, S. O., GOLD, P.: Oncofetal antigens. *Am. J. Clin. Pathol.* **68:**679–687, 1977.
98. SHUTTLEWORTH, E. C., ALLEN, N.: Early differentiation of chronic meningitis by enzyme assay. *Neurology* **18:**534–542, 1968.
99. SHUTTLEWORTH, E., ALLEN, N.: Cerebrospinal fluid β-glucuronidase assay as an aid to the diagnosis of diffuse meningeal dissemination of neoplasm. *Ann. Neurol.* **4:**175, 1978.
100. SMITH, J. A., FRANCIS, T. I., EDINGTON, G. M., WILLIAMS, A. O.: Human alpha fetoprotein in body fluids. *Br. J. Cancer* **25:**337–342, 1971.
101. SNITZER, L. S., MCKINNEY, E. C.: Carcinoembryonic antigen in cerebrospinal fluid. *Proc. Am. Soc. Clin. Oncol.* **17:**249, 1976.
102. SNITZER, L. S., MCKINNEY, E. C., TEJADA, F., SIGEL, M. M., ROSOMOFF, H. L., ZUBROD, C. G.: Cerebral metastases and carcinoembryonic antigen in CSF. *N. Engl. J. Med.* **293:**1101, 1975.
103. SUNDARESAN, N., VUGRIN, D., NISSELBAUM, J., GAL-

ICICH, J. H., CVITKOVIC, E., SCHWARTZ, M. K.: Cerebrospinal fluid markers in central nervous system metastases from testicular carcinoma. *Neurosurgery* **4**:292–295, 1979.

104. VAITUKAITIS, J. L.: Human chorionic gonadotropin: Chemical and biological characterization. In Herberman, A. B., McIntire, K. R. (eds.): *Immunodiagnosis of Cancer*. New York, Marcel Dekker, 1979, pp. 369–383.
105. VAITUKAITIS, J. L., BRAUNSTEIN, G. D., ROSS, G. T.: A radioimmunoassay which specifically measures human chorionic gonadotropin in the presence of human luteininzing hormone. *Am. J. Obstet. Gynecol.* **113**:751–758, 1972.
106. VUGRIN, D., CVITKOVIC, E., POSNER, J. B.: Biology of brain metastases of testicular germ cell carcinomas. *Proc. Am. Soc. Clin. Oncol.* **19**:197, 1978.
107. WEISS, J. F., RANSOHOFF, J., KAYDEN, H. J.: Cerebrospinal fluid sterols in patients undergoing treatment for gliomas. *Neurology* **92**:187–193, 1972.
108. WEISS, J. F., GROSSI-PAOLETTI, E., PAOLETTI, P., SCHIFFER, D., FABIANI, A.: Occurrence of desmosterol in tumors of the nervous system induced in the rat by nitrosourea derivatives. *Cancer Res.* **30**:2107–2109, 1970.
109. WERTLAKE, P. T., MARKOVITS, B. A., STELLER, S.: Cytologic evaluation of cerebrospinal fluid with clinical correlation. *Acta Cytol.* **16**:224–239, 1972.
110. WIKSTRAND, C. J., MAHALEY, M. S., BIGNER, D. D.: Surface antigenic characteristics of human glial brain tumor cells. *Cancer Res.* **37**:4267–4275, 1977.
111. WILKINS, R. H., ODOM, G. L.: Cytologic changes in cerebrospinal fluid associated with resections of intracranial neoplasms. *J. Neurosurg.* **25**:24–34, 1966.
112. WROBLEWSKI, F., DECKER, B., WROBLEWSKI, R.: The clinical implications of spinal fluid lactic dehydrogenase activity. *N. Engl. J. Med.* **258**:635–639, 1958.
113. YAP, B., YAP, H., BENJAMIN, R. S., BODEY, G. P., FREIREICH, E. J.: Cerebrospinal fluid (CSF) carcinoembryonic antigen (CEA) in breast cancer patients with meningeal carcinomatosis. *Proc. Am. Soc. Clin. Oncol.* **19**:98, 1978.
114. YUNG, W., HORTEN, B., SHAPIRO, W. R.: Meningeal gliomatosis: A review of 13 cases. *Ann. Neurol.* **6**:150–151, 1979.

CHAPTER 39

Pharmacology of Antineoplastic Agents in Cerebrospinal Fluid

David G. Poplack, W. Archie Bleyer, and Marc E. Horowitz

1. Introduction

The treatment of central nervous system malignancy represents one of the single greatest challenges in the field of cancer medicine. The great strides in the treatment of extracerebral malignancy achieved in the past two decades have not been accompanied by an appreciable improvement in the treatment of intracerebral malignancies, either primary or metastatic. Now that effective therapy is available for acute lymphocytic leukemia, brain tumors are emerging as the most common cause of nonaccidental death in children over 6 months of age. In adults, malignant glioma, the most common brain tumor, remains a lethal disease in almost all of affected individuals. Also, the number of patients with central nervous system metastases from systemic cancer continues to increase (Table 1), mainly because the risk of central nervous system infiltration in such patients appears to be directly proportional to the duration of survival. As survival has increased, so has the incidence of central nervous system metastases. Ironically, success in the treatment of systemic malignancy has created a new problem, one that underscores the need for more effective therapy for central nervous system neoplasia.

It is hoped that better understanding of the pharmacological principles that govern the chemotherapy of primary and secondary central nervous system malignancy will lead to improvements in therapy. This chapter will summarize the most important pharmacological principles of central nervous system cancer therapy, discuss the current pharmacological approaches to central nervous system malignancy with emphasis on the important role of cerebrospinal fluid (CSF) physiology, and offer certain perspectives for future directions in this field.

David G. Poplack, M.D. and **Marc E. Horowitz, M.D.** • Pediatric Oncology Branch, National Cancer Institute, National Institutes of Health, Bethesda, Maryland 20205. **W. Archie Bleyer, M.D.** • Division of Hematology–Oncology, Department of Pediatrics, Children's Orthopedic Hospital Medical Center, Seattle, Washington 98105.

2. Important Considerations in Pharmacological Approach to Central Nervous System Malignancy

Rational chemotherapy of central nervous system malignancy requires an appreciation of a number of important factors regarding both the biology of the tumor being treated and the pharmacology of the agents employed. Some of the more important factors include: (1) the anatomical location of the tumor, (2) tumor-cell kinetics, (3) the route of drug administration, (4) the dose and schedule of drug administration, (5) drug distribution, (6) the physicochemical characteristics of the drug, and (7) drug toxicity.

2.1. Location of Tumor

The location and anatomical presentation of the tumor may significantly influence the pharmacological approach to its treatment. For example, the optimal route of administration may be different depending on whether one is dealing primarily with meningeal involvement vs. an intraparenchymal

Table 1. Incidence of Central Nervous System Metastases

Malignancy	Incidence	Ref. Nos.
Acute leukemia[a,b]		
Childhood ALL	35/42 (83%)	6
Adult ALL	17/23 (74%)	74
	12/24 (50%)	48
Childhood AML	9/19 (47%)	32
	21/118 (18%)	44
Adult AML	7/41 (17%)	29
	4/20 (20%)	48
Lymphoma	15/140 (11%)	39
Burkitt's lymphoma	37/77 (48%)	77
Diffuse histiocytic	15/52 (29%)	22
Hodgkin's disease	22/210 (10%)	28
Multiple myeloma	43/125 (34%)	28
Carcinoma	5/70 (7%)	65
Lung		
Small-cell	5/21 (24%)	27
Squamous	11/98 (11%)	27
Breast	25/500 (5%)	75
Prostate	4/91 (4%)	24
Sarcoma	5/14 (36%)	35
Rhabdomyosarcoma	20/409 (5%)	68

[a] (ALL) Acute lymphocytic leukemia; (AML) acute myelocytic leukemia.
[b] Figures for acute leukemia represent incidence in studies in which CNS prophylaxis was not administered.

mass lesion. Although it is not always possible to delineate central nervous system involvement along these lines, it may be important. Meningeal disease, which may occur in leukemia, lymphoma, and the central nervous system spread of a variety of carcinomas, is usually treated utilizing intrathecal chemotherapy. The direct administration of drugs into the CSF, bypassing the blood–brain barrier, allows one to achieve high concentrations of active agents in close proximity to the affected meninges. Although deep intraparenchymal penetration of drugs may be less feasible by the intrathecal approach, this route of administration does have certain advantages (see Section 3). In contrast, a central nervous system tumor which presents as a mass lesion may be more optimally treated with systemic agents, not only because of the technical contraindications to performing a lumbar puncture (see Section 2.3), but also because transport of drug to the tumor mass is more effectively accomplished via the vascular channels that supply the neoplasm.

2.2. Tumor-Cell Kinetics

An understanding of the growth kinetics of brain-tumor cells can provide information useful in choosing appropriate chemotherapy. Although there is a paucity of data available on human brain tumors, recent studies have helped define the growth characteristics of some of the more common central nervous system tumors.[41,42,47] Knowledge of kinetic parameters such as the cell-cycle time, the duration of DNA synthesis, the labeling index, and the growth fraction may be helpful both in the choice of the drug(s) to be employed and in the scheduling of drug administration. For example, antineoplastic agents differ in regard to the time during the cell cycle when they are effective. Agents such as vincristine, cytosine arabinoside, and methotrexate act at a specific point in the cell cycle and kill only those cells in a particular, sensitive phase of the growth cycle at the time of drug administration. Because of its cell-cycle specificity, a single dose of such an agent can be expected to kill only a small proportion of growing tumor cells. It has been estimated that the highest cell-kill which a single course of a drug of this type could effect against a glioblastoma is 30%.[41] Although such kinetic considerations may reduce the attractiveness of cell-cycle-specific agents, appropriate methods of scheduling (see Section 2.4) and use in combination can overcome some of the inherent limitations of these types of drugs.

In contrast to cell-cycle-specific drugs, another class of chemotherapeutic agents is capable of killing tumor cells throughout their cell cycle. These "cell-cycle-nonspecific" agents, such as the nitrosoureas and alkylating agents, act against both proliferating and nonproliferating cells and have been the agents most active against brain gliomas.

Treatment with one or another class of antineoplastic agents itself alters the growth kinetics of a tumor. Recognition of this fact may allow more optimal use of sequential chemotherapy. For example, initial use of a cell-cycle-nonspecific agent may be followed by an increase in the tumor growth fraction as the remaining viable tumor begins to regrow. This may result in an increase in the number of cells going through the synthetic (S) phase of the cell cycle. Exposure to an S-phase-cell-cycle-specific agent at this time would theoretically result in a greater total cell-kill.

The availability of newer, more rapid methods of estimating kinetic parameters, such as flow microfluorometry,[70] is likely to lead to better understanding of the kinetics of intracranial neoplasms. This information should lead to the development of improved chemotherapy treatment schedules.

2.3. Route of Drug Administration

Several routes of administration are available to dispense antineoplastic agents to the central nervous system: intrathecal, systemic, intraarterial, and intratumor. The two most commonly employed are the intrathecal and systemic approaches. The choice of a particular route of administration depends on the anatomical location of the tumor, certain physical characteristics of the drug being utilized (relative to its action, distribution, and metabolism), and the clinical situation in question. As discussed above, the location and type of malignant process may dictate the more appropriate route of drug therapy; e.g., while meningeal disease may be more approachable from the intrathecal route, intraparenchymal solid tumors may be more advantageously treated via systemic therapy. Certain drugs that are appropriate for intrathecal use are either impractical to use or less effective via the systemic route. For example, cytosine arabinoside is rapidly deaminated when given intravenously, and although drug concentrations equivalent to 40% of plasma levels can be achieved in the CSF, maintenance of constant plasma levels requires a constant intravenous infusion. In contrast, when given intrathecally, this drug disappears more slowly from CSF (initial half-time of 2 hr as compared to the initial plasma half-time of 11 min after intravenous administration).[40] The route of administration may also be dictated by certain physical characteristics of the particular agent under consideration. Thus, drugs that are primarily bound to protein would be a poor choice for systemic administration, since only that small fraction of unbound drug is likely to cross the blood–brain barrier.[8] In addition, the systemic toxicity of a particular agent may preclude its use intravenously, while the same drug may be given intrathecally with relative safety. Classically, methotrexate has been considered such a drug. However, newer techniques of administration and improved methods of prevention of systemic toxicity have made the intravenous administration of this agent to treat central nervous system malignancy more feasible (see Section 4.2).

Physiological factors may influence the route of drug administration. Intrathecal chemotherapy has traditionally been administered to the lumbar subarachnoid space. Recent information has confirmed the rapidity of caudally directed CSF bulk flow.[31] For an intralumbar-administered agent to reach the cerebral ventricles, it must ascend the spinal subarachnoid space against the movement of CSF, risk absorption along the spinal pathway, and finally enter the ventricles by overcoming the unidirectional flow of CSF in the opposite direction through the aqueduct of Sylvius and the ventricular outlet foramina.[31] Recent studies have confirmed that there is limited entry of drug into the ventricular compartment following injection into spinal CSF.[18,66] In an effort to circumvent this problem, investigators have turned to the use of intraventricular drug administration, via subcutaneously implanted Ommaya reservoirs[57] (see Section 3.3).

In addition to the aforementioned considerations, preference for a particular route of administration is often dictated by the practical realities of a given clinical situation. For instance, as alluded to previously, the presence of an intracerebral mass lesion may preclude the possibility of the intralumbar route. Similarly, the neurosurgeon confronted with a partially resectable tumor may choose to optimize local drug administration to the residual lesion by implanting a subcutaneous reservoir which is attached to a catheter, the tip of which is inserted into the tumor. This intratumor route of administration has been reported to be useful in the treatment of certain intracerebral malignancies.[56] As these examples illustrate, a number of factors must be considered in selecting the appropriate route of drug administration.

2.4. Dose and Schedule of Drug Administration

The importance of drug dose and schedule of administration to the planning of a rational chemotherapeutic approach to the treatment of central nervous system malignancy cannot be overemphasized. The most obvious concern is that regardless of the route of administration, therapeutic levels of drug must be achieved at the tumor site. Unfortunately, from a practical point of view, measurement of the actual drug concentrations achieved at the level of the tumor cell is not feasible. Consequently, assessment of drug concentrations within the CSF offers the nearest approximation. The ability to monitor CSF drug concentrations is of considerable value in helping to estimate and assess therapeutic efficacy. For example, determination of the

CSF/plasma ratio of a systemically administered compound offers an indication of how well it penetrated the blood–brain barrier, as well as providing necessary information regarding optimal dosage. However, the extent to which the measured level of a drug within CSF reflects drug concentration in the tumor depends on the site of the malignancy; e.g., CSF levels of drug probably more accurately represent the concentrations achieved in a meningeal neoplasm than in an intraparenchymal mass lesion.[13] As discussed in Section 2.7, measurement of CSF drug levels may also be helpful in preventing drug-related neurotoxicity.

Of equal importance to the identification of a therapeutic, nontoxic dose is the recognition of the role of an appropriate schedule of drug administration. Rational drug scheduling requires knowledge of tumor-growth kinetics as well as the mechanism of a drug's action, metabolism, distribution, and pharmacokinetics. The illustration of the preferability of administering systemic cytosine arabinoside by continuous intravenous infusion because of its rapid deamination (Section 2.3) is a case in point. A good example of the influence of schedule on a drug's effectiveness is offered by methotrexate. Over 100-hundred fold more methotrexate is required to produce a given lethal effect in rats when administered as a single dose compared to a low-dose protracted infusion.[76] Recently, an intraventricular administration schedule designed to take advantage of this "concentration × time" approach has been used successfully to treat patients with meningeal disease[16] (see Section 3.3).

Blasberg[10] has emphasized the importance of considering the clearance half-time of a compound in the planning of an appropriate administration schedule. The concept of clearance half-time refers to the rate of clearance of a drug from a particular physiological compartment. Drugs with a rapid clearance half-time may not be present at their target site for a sufficient length of time to obtain an optimal therapeutic effect. This type of agent may be more advantageously administered by a constant infusion. The alkylating agent N,N^1,N^{11}-triethylenethiophosphoramide (thio-TEPA) is an example of a drug with a rapid clearance half-time that might be more optimally given systematically by an infusion technique. In contrast, a drug with a relatively long clearance half-time might be more appropriately administered by bolus administration. Such is the case for intrathecal methotrexate.

Consideration of clearance half-time is particularly important in optimizing intrathecal therapy. In a series of elegant experiments, Blasberg and his colleagues evaluated the penetration of methotrexate into the brain during ventriculocisternal and ventriculolumbar perfusions in subhuman primates.[9,13] They observed that decreasing the rate of methotrexate clearance from CSF resulted in the maintenance of higher methotrexate concentrations in brain tissue and that higher levels were achieved deeper within brain tissue. These observations have stimulated the development of new approaches to intrathecal methotrexate therapy (see Section 3.4).

2.5. Drug Distribution

In addition to dose and schedule of administration, a number of complex factors and processes are involved in the delivery of a chemotherapeutic agent to its target site. The parameters that affect drug distribution, recently reviewed in depth,[10,26] differ somewhat depending on the route of administration. Some of the most important of these parameters are listed in Table 2.

To be effective, a given drug obviously must be transported to the site of the tumor. Initially, it is carried via bulk flow by either CSF (in the case of intrathecal administration) or blood (systemic administration). The extent to which the tumor is exposed to the drug depends on its proximity to CSF, in the former circumstance, and the adequacy and rate of perfusion with blood, in the latter.

Distribution of drug also depends on the presence of a gradient that favors transport to the interior of the cell.[26] Interference with such a gradient by any process that reduces the amount of available active drug, such as protein binding, may interfere with the process of drug diffusion toward tumor cell.

The transport of most systemically administered

Table 2. Factors That Affect Distribution of Drugs Used to Treat Central Nervous System Neoplasms[a]

Perfusion of tumor—bulk flow
Activity gradients
Blood–brain barrier—capillary permeability
Diffusion through extracellular fluid
Diffusion in CSF and other tissue fluids
Cellular uptake

[a] Adapted from Blasberg[10] and Cowles and Fenstermacher.[26]

agents into brain tissue is restricted by the "blood–brain barrier." This term is used to describe the unusual characteristics of the membranes of the brain capillaries, which readily permit nonpolar, lipid-soluble compounds to cross from the vascular space into brain extracellular fluid but restrict the transcapillary movement of polar, lipid-insoluble, and water-soluble materials.[30,51] The permeability of brain capillaries to a given drug can be expressed in terms of a transfer constant that expresses the amount of drug that crosses into brain per unit time.[10] Knowledge regarding the transcapillary movement of an agent into brain is an important factor in the selection of agents for systemic administration.

Once a drug has permeated the cerebral capillary, it diffuses through brain extracellular fluid at a rate that is dependent on the physical characteristics of the molecule. The aqueous diffusion constants for most drugs are between 5 and 7×10^{-6} cm^2/sec at 37°C.[10] The actual concentration of a drug that exists in the extracellular fluid not only is a function of the transcapillary transfer constant, but also is influenced by biotransformation, cell uptake, diffusion back into the capillaries, and movement by bulk flow into the CSF. Drug may also move through brain tissue itself by diffusion. This process is believed to be more rapid for nonpolar, lipid-soluble compounds.

Movement of drug into tumor cells takes place by passive diffusion and in some situations by active transport. Detailed understanding of this process is lacking for most drugs and tumor types. Goldman[36] has demonstrated that intracellular accumulation of methotrexate is limited to one twentieth of the existing extracellular drug concentration, possibly because of an energy-dependent pump that extrudes the drug from the cell. Vincristine may augment the intracellular accumulation of methotrexate by blocking this energy-dependent efflux.[34] Our knowledge of the intracellular accumulation of other chemotherapeutic agents is limited.

2.6. Physicochemical Characteristics of Antineoplastic Agents

The physical and chemical properties of antineoplastic agents greatly affect their pharmacological behavior and influence their activity against central nervous system neoplasms. Several factors are now appreciated as being important determinants of a drug's ability to cross the blood–brain barrier, thereby influencing its delivery to the tumor site. These factors include the degree of protein binding, lipid solubility, ionization characteristics, and molecular weight. The importance of protein binding, alluded to in the previous section, has been well studied.[5,46] Because albumin does not penetrate into the central nervous system, the presence of a significant degree of drug–protein binding reduces the amount of unbound drug available for entrance into the central nervous system. The influence of drug–protein binding on the actual CSF/plasma ratio obtained with a given systemically administered drug may be more important than the impermeability of the brain capillaries.[10]

The degree to which a drug is soluble in lipid is, as mentioned previously, a crucial determinant of a drug's capacity to cross the blood–brain barrier. Log-P values have been used to rank drugs in terms of their lipid solubility. A high positive log-P value (+1.0 to +3.0, indicating 10- to 1000-fold preference for partitioning in octanol vs. water) identifies a drug with high lipid solubility, one that is likely to penetrate the central nervous system well.[53] Table 3 presents data demonstrating how the penetration of the central nervous system correlates to some degree with the lipid solubility of a drug.

The ionization characteristics of a drug may also be important. Information regarding the pK_a value for a drug allows an estimate of the degree of ionization that occurs at physiological pH. The portion of a drug that is in an unionized form will generally penetrate more readily.

The molecular size of a drug is also important to consider, since compounds with a high molecular

Table 3. Relationship of Lipid Solubility to Drug Penetration of Central Nervous System[a]

Agent[b]	Dose (mg/kg) and route	Log-P value	Ratio of brain/serum levels
DTIC	317 i.p.	−0.24	0.10
5-FU	50 i.v.	−0.95	0.33
BCNU	20 i.p.	+1.53	0.5–1.5
CCNU	—	+2.83	4.0–5.0

[a] Adapted from Mellet.[53] Data presented refer to studies performed in mice with intracerebrally implanted L1210.

[b] (DTIC) Dimethyl triazeno imidozole carboxamide; (5-FU) 5-fluorouracil; (BCNU) 1,3-*bis*(2-chloroethyl)-1-nitrosourea; (CCNU) 1-(2-chloroethyl)-3-cyclohexyl-1-nitrosourea; (i.p.) intraperitoneal; (i.v.) intravenous.

weight (> 500 daltons) may not penetrate the central nervous system to a significant degree.[53]

2.7. Drug Toxicity

A large number of agents currently being utilized to treat central nervous system malignancy are potentially toxic both to the central nervous system and systemically. A major task of the chemotherapist is to achieve maximal tumor-cell-kill with a minimum of drug-related toxicity. Systemic toxicity has been associated with both systemic and intrathecal administration. For many of the most effective systemically administered drugs, systemic toxicity, such as drug-induced bone-marrow hypoplasia, may limit the amount of drug that can be given safely. Similarly, systemic toxicity may also occur following intrathecal drug administration. The disappearance half-time of methotrexate in the CSF is longer than in the plasma. Following intrathecal methotrexate administration, the CSF may act as a reservoir, with drug gradually being returned to the systemic circulation over a prolonged period of time. The attendant increased risk of bone-marrow toxicity occasionally necessitates concomitant systemic administration of the rescue agent citrovorum factor.

The neurotoxicity associated with the use of antineoplastic agents has recently been reviewed.[59,73] Acute, subacute, and delayed types of neurotoxicity can occur. In a given clinical situation, the occurrence of a neurotoxic reaction may not be predictable. However, for certain drugs, such as methotrexate, the relationship between CSF drug levels and the occurrence of toxicity is reasonably well determined. As shown in Fig. 1, acute and subacute neurotoxic reactions to intralumbar-administered methotrexate are associated with higher than normal CSF methotrexate concentrations. This information has provided the rationale for adjusting an individual's intrathecal methotrexate dosage according to the concentration of drug in his CSF, increasing the drug dosage in patients with low levels and reducing it in those with elevated levels. Such a policy of pharmacological monitoring and individual dosage modification can dramatically reduce the amount of methotrexate-related neurotoxicity (Table 4). Thus, an understanding of the pharmacokinetic behavior of the drug in both normal and pathological states may help avert toxicity.

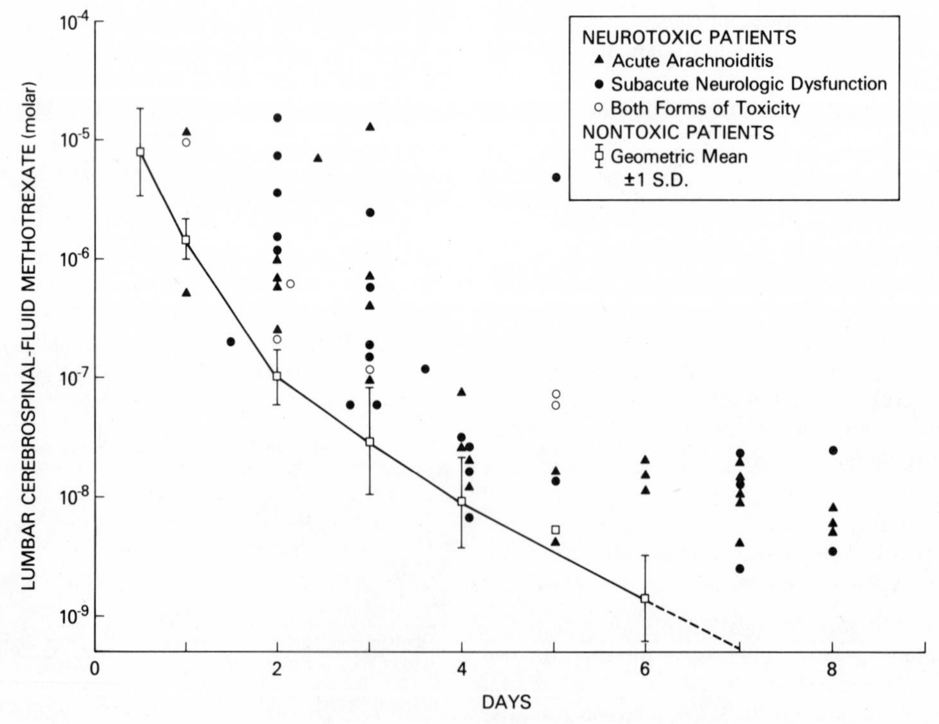

Figure 1. Lumbar CSF methotrexate concentrations after lumbar methotrexate 12 mg/m^2. Curve connects geometric mean (± 1 S.D.) values in 76 patients with no evidence for toxicity (□). (▲, ●, ○) Patients who had neurotoxic reactions.

Table 4. Incidence of Intrathecal Methotrexate Neurotoxicity[a]

Patients	1971–1972	1973–1974
Number	56	36
Number with neurotoxicity	21 (38%)	1 (2.8%)
Number with severe or life-threatening neurotoxicity	6 (11%)	0

[a] This table presents a comparison of the incidence of acute and subacute neurotoxic reactions to intrathecal methotrexate in the year 1971–1972 vs. the year 1973–1974 on the Pediatric Oncology Branch, National Cancer Institute. The marked decrease in the year 1973–1974 was associated with the introduction of a policy by which an individual's CSF methotrexate level was used to guide the dosage of subsequent intrathecal methotrexate. Adjustment of dosage in this manner provided CSF methotrexate levels in the therapeutic but nontoxic range.

In addition to a knowledge of pharmacokinetic parameters, appreciation of the molecular pharmacology of different compounds may provide methods of averting the toxicities of antineoplastic therapy. The identification of the "rescue" agent citrovorum factor has enabled prolonged, high-dose intravenous methotrexate infusions to be explored as a possible therapeutic avenue.[23]

Perhaps the most rational method of averting the neurotoxicity of antineoplastic agents is through identification of equally effective but less neurotoxic structural analogues of a given compound. Although not directly applicable to the treatment of central nervous system neoplasia, an example of this type of phenomenon has recently been identified for the drug cis-diamminedichloroplatinum (DDP). This agent, which is known to cross the blood–brain barrier, is used to treat a variety of non-central-nervous-system malignancies.[21] Its use has been associated with seizures and other forms of neurotoxicity. Recent studies in subhuman primates have demonstrated that a DDP analogue, sulfato 1,2-diaminocyclohexane platinum, does not penetrate the central nervous system, suggesting that use of this latter analogue may not be associated with neurotoxicity.[37] A similar search for less toxic analogues of those agents currently being used to treat central nervous system malignancy would appear warranted.

In the preceding sections, we have outlined some of the more important principles to be considered in the pharmacological approach to central nervous system malignancy. In the following sections, we will discuss the intrathecal and systemic chemotherapy of central nervous system malignancy, focusing attention on the important role of the CSF in these pharmacological approaches. Where possible, examples drawn from the experience with methotrexate will be emphasized, since more is understood about this drug than about the other agents currently being used to treat central nervous system cancer.

3. Intrathecal Chemotherapy

3.1. Rationale of Intrathecal Approach

The major rationale for the use of intrathecal chemotherapy is that this route of administration allows one to bypass the blood–brain barrier and introduce drug, in high concentration, at the site where it is most needed. Higher concentrations of drug can be achieved with the intrathecal route than by systemic administration, in large part because the initial volume of distribution following intrathecal injection is considerably smaller (140 ml for CSF vs. 3500 ml for plasma). In addition, the clearance half-life of a drug is longer in the CSF than in the systemic circulation. For example, the initial half-life of methotrexate in the CSF is 4.5 hr vs. 45 min in plasma.[43] Following injection into the CSF, the distribution of a drug is dictated by a number of processes. These include: (1) bulk flow of entrained drug via the normal pathways of CSF circulation; (2) diffusion into and through the extracellular spaces of the brain parenchyma and spinal cord; (3) back flow from the extracellular fluid of brain parenchyma into the CSF; (4) removal from the central nervous system by the normal pathways of CSF absorption; and (5) removal by diffusion from the extracellular fluid into the capillaries of the brain and spinal cord. Other processes that may be involved are diffusion into the arachnoid and dural membranes and removal via the circulating blood of the meninges, absorption from ventricular CSF by an active transport mechanism located in the choroid plexus, and uptake by certain cells in the brain and spinal cord.[14] For a high drug concentration to be achieved at the site of a tumor that lies within the brain parenchyma, there must be little loss of drug as it moves along the CSF pathways and diffuses within the parenchyma toward the tumor. The ideal intrathecal agent should possess the following characteristics: it should (1) be slowly cleared from the CSF; (2) rapidly diffuse through the extracellular space; (3) have a long time course of action; and (4) undergo little uptake by the capillaries or cells within the central nervous system.[10,14] In addition, the drug should be tumoricidal and without significant neurotoxicity. However, it has been pointed out that the conditions for optimal drug delivery are different if one is treating meningeal disease rather than an intraparenchymal mass lesion. In the former situation, factors such as the rate of mixing and the distribution of drug within the CSF compartments must be considered together with the rapidity of CSF clearance. For example, a drug such as thio-TEPA may be cleared too rapidly from CSF to allow adequate therapeutic concentrations to be achieved throughout the various CSF compartments.[10]

A crucial question regarding intrathecal chemotherapy is the extent to which an intrathecally administered drug can penetrate deep into brain tissue to reach an intraparenchymal tumor that may reside several millimeters or even centimeters from the CSF–brain interface. Although the concept of treating solid tumors that reside on the external or ventricular surfaces of brain with intrathecally administered chemotherapy has been well accepted, the issue of drug delivery to deeper intraparenchymal lesions has been controversial. Blasberg and his colleagues have shed considerable light on this subject through studies in which ventriculocisternal perfusions were performed with different chemotherapeutic agents in subhuman primates.[9,12] Following a perfusion, the drug concentrations were measured in brain tissue and evaluated in terms of the distance from the perfused ependymal surface. The greatest degree of penetration was observed with methotrexate, since there is negligible uptake of this drug by brain cells and limited diffusion into brain capillaries. The distance into the brain at which drug concentrations of 1% of the CSF concentration were obtained was 3.2 mm for methotrexate, 1.8 mm for cytosine arabinoside, 1.5 mm for thio-TEPA, and 1.4 mm for 1,3-*bis*(2-chloroethyl)-1-nitrosourea (BCNU). Cytosine arabinoside diffuses to a lesser depth than methotrexate primarily because it is accumulated to a greater degree by brain cells. Thio-TEPA and BCNU, relatively nonpolar, lipid-soluble compounds, penetrate smaller distances into the brain because they readily diffuse across brain capillaries and are removed by the systemic circulation before they can penetrate deeply into the brain parenchyma.

An important result of these studies was the observation made regarding the relationship between depth of penetration and perfusion time. The longer the perfusion time, the deeper methotrexate penetrated into the brain. Since the depth of penetration increases with the square root of the perfusion time, it took 26 min for a drug level of 10% perfusate concentration to be reached at a depth of 0.1 cm and 43 hr of perfusion to attain that concentration at 1.0 cm below the ependymal surface.[12] The major point of these findings is that decreasing the rate of methotrexate clearance from CSF can result in deeper penetration into brain tissue. In addition, because slowing CSF methotrexate clearance also decreases the back flux of this compound into the CSF, more drug remains in the brain. This type of data has raised the possibility that through manipulation and reduction of CSF drug clearance, intrathecal chemotherapy may be rendered more effective. Some experimental techniques designed to improve intrathecal therapy by reducing CSF drug clearance are discussed in Section 3.4.

3.2. Limitations of Intrathecal Approach

In addition to the concerns regarding the degree of drug penetration from CSF into brain tissue following intrathecal administration, there are a number of other factors that limit the effectiveness of intrathecal therapy. Most intrathecal therapy is administered by lumbar puncture, with the drug being injected directly into lumbar CSF. Lumbar administration has a number of potential drawbacks. Inadvertent injection into the subdural or epidural space, rather than the arachnoid space, is a not infrequent accompaniment of the lumbar approach.[7,45] Likewise, the unintentional creation of a meningeal CSF leak may change the pressure, volume, and flow of CSF and thus alter the distribution of administered drug. In addition, the requirement for an intralumbar-administered drug to ascend the spinal subarachnoid space and enter the cerebral ventricles against the unidirectional flow of CSF out of the ventricles is known to compromise the drug concentrations that can ultimately be achieved in ventricular CSF. We studied this latter phenomenon in 18 acute lymphoblastic leukemia patients with indwelling Ommaya reservoirs who were receiving intrathecal methotrexate by lumbar puncture. Following the intralumbar administration of methotrexate (12 mg/m^2 methotrexate in 12 ml/m^2 Elliott's B solution), mean ventricular methotrexate concentrations were 2.1×10^{-6}, 4.1×10^{-7}, and 1.2×10^{-7} M at 6, 12, and 24 hr after injection, respectively. These values for ventricular CSF antifolate concentrations were approximately 1 log lower than values obtained for lumbar CSF, confirming that limited entry of the drug occurs in the ventricular compartment following lumbar injection.[14,66] The results for a representative patient are shown in Fig. 2.

Another potential problem with intrathecal therapy is that the tumor being treated may interfere

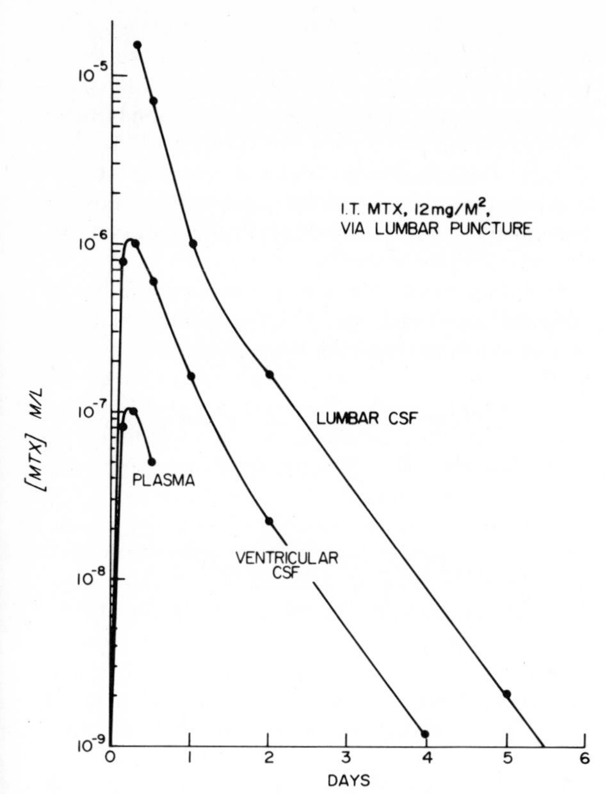

Figure 2. Methotrexate (MTX) concentrations in plasma, lumbar CSF, and ventricular CSF after intralumbar injection of 12 mg/m^2. Ventricular CSF was obtained via indwelling Ommaya reservoir. Mean MTX concentrations were 1 log lower in ventricular CSF than in lumbar CSF, confirming limited entry of drug into ventricular compartment after intralumbar administration. Reprinted, with permission, from Bleyer and Poplack.[14]

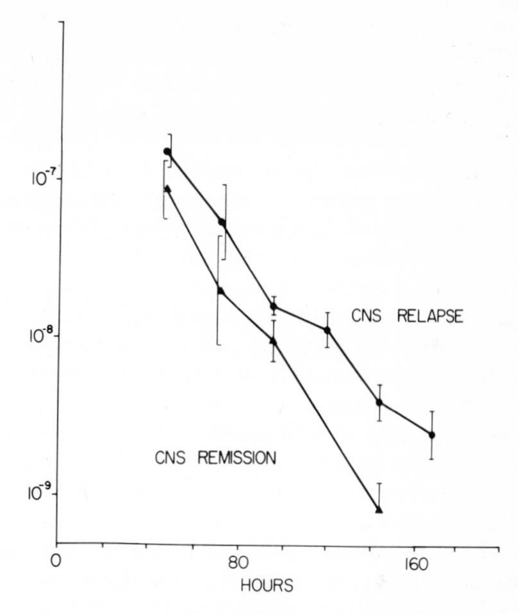

Figure 3. Elevated lumbar CSF methotrexate concentrations after intralumbar methotrexate (12 mg/m^2) in presence of overt meningeal leukemia. (CNS) Central nervous system. Reprinted, with permission, from Bleyer and Poplack.[14]

with CSF flow by obstruction of the subarachnoid or ventricular pathways of CSF bulk flow. Interference with the pattern of CSF flow may have profound effects on the subsequent distribution and pharmacokinetic behavior of an intrathecally administered agent. This situation has been documented in patients with overt meningeal leukemia being treated with intrathecal chemotherapy. As shown in Fig. 3, the presence of meningeal leukemia results in higher CSF levels of methotrexate following lumbar administration than occur in patients without central nervous system disease. This phenomenon has also been observed in individuals with communicating hydrocephalus (Fig. 4).

CSF protein content may affect intrathecal chemotherapy. In situations of high CSF protein, drug–protein binding may reduce the amount of

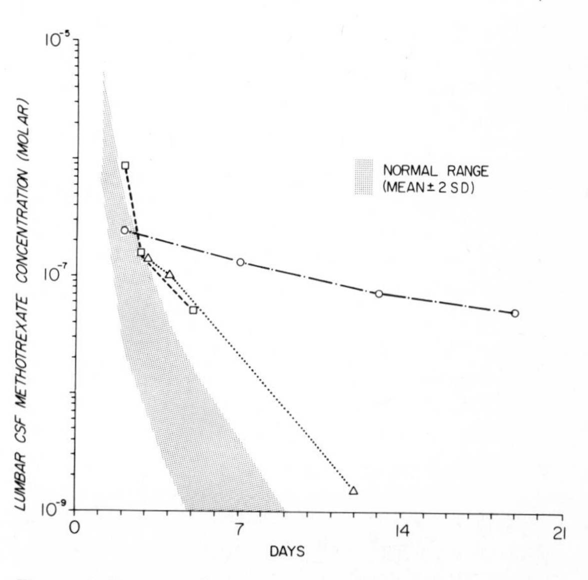

Figure 4. Decreased clearance of methotrexate from CSF in three patients (□, ○, △) with communicating hydrocephalus. Reprinted, with permission, from Bleyer and Poplack.[14]

free drug available for transport into tumor cells. In addition, markedly elevated CSF protein levels have been associated with reductions in CSF outflow through the arachnoid villi.

As mentioned earlier (Section 2.7), intrathecal chemotherapy may be associated with significant toxicity. We have shown that acute and subacute neurotoxic reactions following intrathecal methotrexate are related to the CSF methotrexate concentration (see Fig. 1). CSF pharmacokinetic considerations also play a role in the systemic toxicity (i.e., bone-marrow hypoplasia) that may be associated with intrathecal drug administration. For example, intrathecal methotrexate administration results in more prolonged toxic plasma levels of methotrexate than occur with either intravenous or oral administration of the same dose. Presumably, this is due to the "reservoir" effect of the CSF. Intrathecally administered drug is released slowly into the systemic circulation, resulting in more sustained plasma levels at toxic concentrations. This situation is graphically illustrated in Fig. 5. It should be emphasized, however, that this type of phenomenon does not occur with every intrathecally administered agent. As mentioned earlier, systemic toxicity following intrathecal administration of cytosine arabinoside is not associated with significant systemic toxicity, probably because the rate of deamination of this compound in the systemic circulation is more rapid than the rate at which it leaves the CSF. Although a detailed account of the various systemic and neurological toxicities of intrathecal therapy is beyond the scope of this chapter, it is necessary to emphasize their importance.

Finally, a serious limitation of lumbar intrathecal chemotherapy mentioned previously is that in many patients with central nervous system mass lesions, lumbar puncture is contraindicated because of the risk of brain herniation. This fact, together with the other limitations of lumbar administration discussed above, has stimulated interest in the intraventricular route of administration.

3.3. Intraventricular Chemotherapy

The availability of indwelling subcutaneously implanted Ommaya reservoirs has provided the opportunity to administer chemotherapy directly into ventricular CSF. The intraventricular approach has a number of advantages over intralumbar therapy.

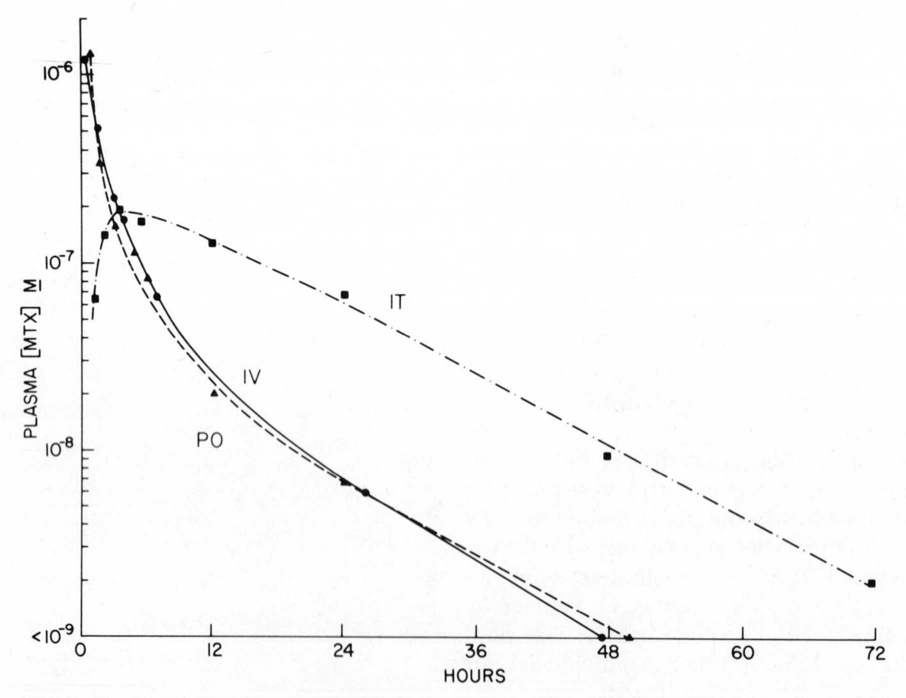

Figure 5. Plasma methotrexate (MTX) concentrations after oral (PO), intravenous (IV), and intrathecal (IT) administration of 12 mg/m^2 in same patient. Plasma half-lives are considerably greater after IT than after PO or IV administration. Reprinted, with permission, from Bleyer and Poplack.[14]

First, this approach assures the delivery of drug into the CSF, avoiding some of the technical limitations (e.g., CSF leakage, inadvertent epidural administration) encountered with intralumbar therapy. Second, intraventricular injection results in a more complete distribution of drug throughout the CSF. Studies have shown that intraventricular injection provides consistently higher drug levels in ventricular CSF than intralumbar injection.[18,66] Distribution of drug via the intraventricular route is believed to be maximized, since the ventricles are the site of CSF production and drug injected within ventricular CSF is carried by CSF bulk flow throughout the various CSF compartments.[12] Third, use of the reservoir eliminates the need for frequent lumbar punctures and thus is associated with considerably less patient discomfort. Fourth, the Ommaya reservoir system allows repetitive intraventricular administration, enabling novel therapeutic approaches such as the "concentration × time" (C × T) technique outlined below. The increased efficacy of intraventricular injection vs. intralumbar chemotherapy of meningeal neoplasms has been confirmed in the literature.[15,65]

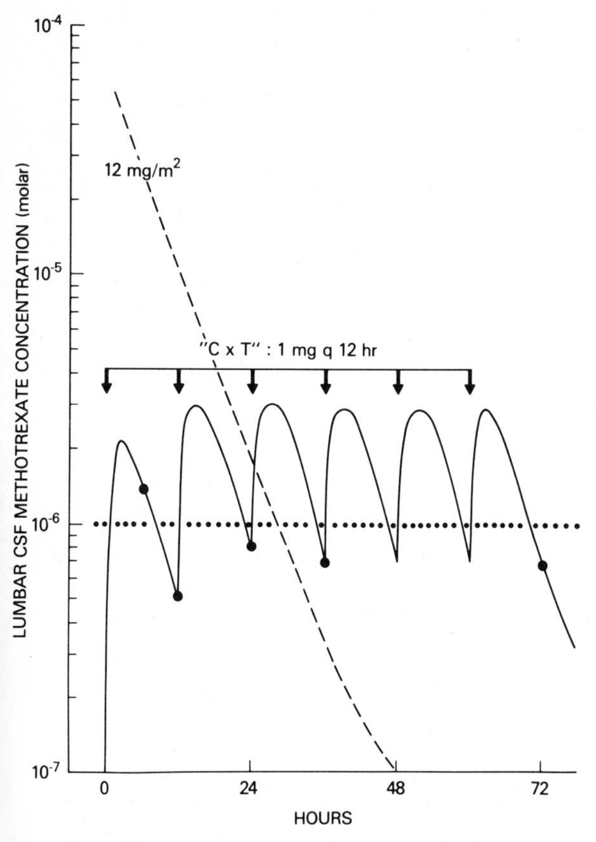

Figure 6. CSF methotrexate concentrations achieved with a concentration (C) × time (T) dosage schedule. (●) Values measured in single patient undergoing "C × T" course during which 1 mg methotrexate is given via an Ommaya reservoir; (- - - -) mean levels achieved with conventional dosage regimen; (· · · · ·) line at 1×10^{-6} M estimating maximum cytocidal concentration for most leukemic cells. With "C × T," cytocidal drug level can be maintained for 72 hr (vs. less than 32 hr by conventional regimen) using one half the total methotrexate dose. Reprinted, with permission, from Bleyer and Poplack.[14]

Based on a knowledge of the CSF pharmacokinetics of methotrexate, we devised a "concentration × time" (C × T) type of regimen that is designed to maintain a minimal cytocidal extracellular methotrexate concentration in the CSF for 72 hr. The rationale for this approach is provided in part by the observations that drug penetration from the CSF into the brain is greater the longer CSF levels remain constant (as discussed in Section 3.1) and also by tumor-cell kinetic considerations that dictate that the maintenance of prolonged cytocidal levels of a cell-cycle-specific drug such as methotrexate exposes more tumor cells to the drug as they pass through the S phase of their cell cycle. As shown in Fig. 6, the intraventricular injection via the reservoir of 1 mg methotrexate every 12 hr for six doses (a total of 6 mg) maintains a therapeutic methotrexate level in the CSF for a total of 72 hr in comparison to a single 12 mg/m² intraventricular injection, which maintains this level for less than 32 hr. This C × T approach has been demonstrated to be equally effective against meningeal leukemia with less associated neurotoxicity than the single-injection intraventricular approach.[15,16] In current studies, a more prolonged C × T approach (up to 120 hr) is being investigated. This regimen may also be of value in treating nonmeningeal central nervous system neoplasms.

3.4. Improving Intrathecal Chemotherapy by Altering Drug CSF Disappearance Kinetics

The observation that slowing the rate of methotrexate CSF clearance from ventricular CSF results in increased penetration of this agent into brain tissue[9] has prompted a search for clinically feasible methods of altering CSF clearance in an attempt to improve intrathecal chemotherapy.

We have studied various methods of decreasing the CSF clearance of methotrexate. Disappearance of methotrexate from CSF is believed to occur by two processes: (1) CSF bulk flow and (2) active transport, which presumably takes place at the level of the choroid plexus:[9,19,60] Probenecid, a weak inorganic acid, is known to inhibit the active transport of methotrexate (another weak acid) from the kidney. We evaluated the ability of this compound to decrease methotrexate CSF clearance by inhibiting its active transport via the choroid plexus.

Initially, subhuman primates with subcutaneously implanted Ommaya reservoirs were given intraventricular methotrexate and its CSF clearance was determined. On a different occasion, these same animals (acting as their own controls) were given a second intrathecal dose of methotrexate, this time shortly after receiving parenteral probenecid treatment in a dose sufficient to obtain meaningful levels of probenecid in the CSF. As shown in Fig. 7, the CSF clearance of methotrexate was dramatically reduced by probenecid pretreatment.[19,60] Probenecid-induced prolongation of the CSF methotrexate disappearance half-time has been observed in a patient treated similarly.[19] These observations suggest the possibility that probenecid may be an effective adjunct to intrathecal methotrexate therapy by decreasing methotrexate CSF disappearance.

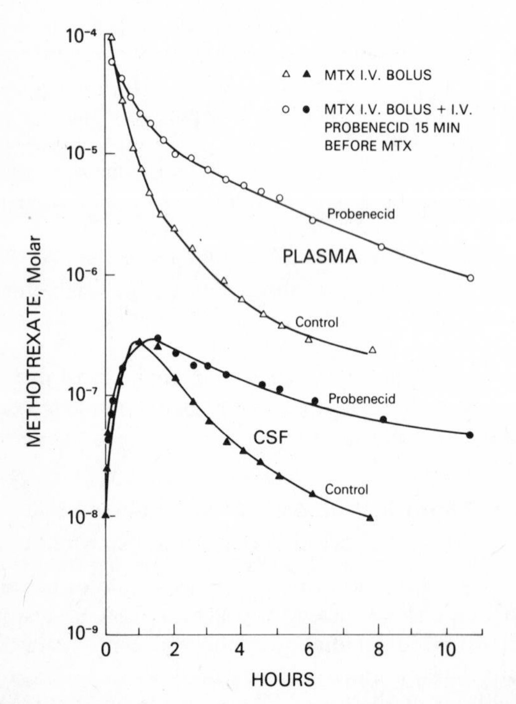

Figure 7. Retention of methotrexate (MTX) in CSF after pretreatment with probenecid. Subhuman primate was given intravenous (IV) MTX (10 mg/kg) on two separate occasions, once without probenecid (control) and once with probenecid pretreatment. Increased MTX levels seen with probenecid reflect inhibition of active-transport component of MTX efflux from CSF.

We have also evaluated the effect on methotrexate disappearance from CSF of reducing the bulk-flow turnover of CSF by pretreatment with an inhibitor of CSF production, acetazolamide. Use of this compound results in a significant increase in the disappearance half-time of methotrexate from CSF, again suggesting that inhibition of CSF methotrexate clearance may be a clinically applicable technique.[19]

An alternative approach to inhibiting the processes by which drugs are cleared form the CSF is to find new agents or analogues of currently utilized drugs that have longer CSF clearance half-times. We have investigated the CSF clearance of the methotrexate analogue aminopterin. Although this folate antagonist was the first drug to be used intrathecally to treat neoplasia, it was replaced by methotrexate when the latter became the drug of choice for systemic therapy.[63] However, new evidence suggests that there may be a wider margin of safety and greater therapeutic index with intrathecal aminopterin than with intrathecal methotrexate. Not only does the drug appear to be less neurotoxic than methotrexate, but also, as illustrated in Fig. 8, our pharmacological studies in monkeys indicate that amimopterin is cleared from the CSF more slowly than methotrexate. Clinical trials have substantiated these findings in man.[17]

4. Systemic Chemotherapy

4.1. Rationale and Limitations of Systemic Approach

The systemic approach to the treatment of central nervous system malignancy has at least two distinct advantages. First, it is far more convenient and less invasive than intrathecal therapy. Second, from a theoretical viewpoint, administration of drug by this route is more "physiological," in that drug is carried to the tumor by the very vascular channels that support its growth. The amount of drug that leaves the vascular system, crosses the extracellular fluid, and eventually accumulates within the tumor cell is influenced by a variety of factors, including (1) the vascularity of the target area, (2) the capillary permeability for the drug, (3) the concentration of active drug and the length of time the drug is exposed to the target area, (4) the diffusion rate through extracellular fluid, and (5) the cellular uptake of the drug.

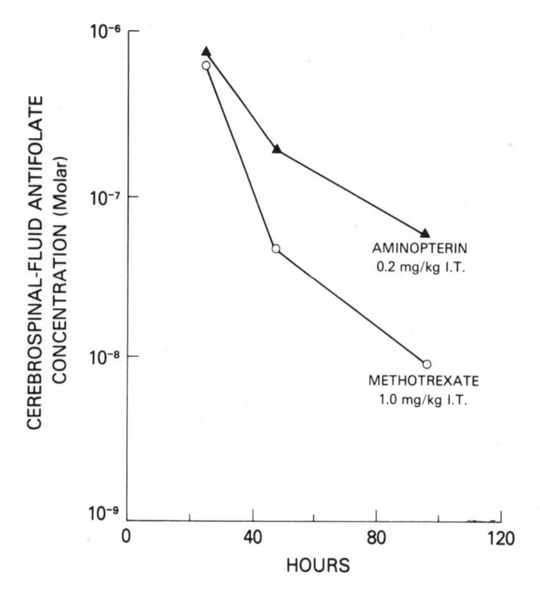

Figure 8. Disappearance of aminopterin and methotrexate from CSF of rhesus monkey following intrathecal (I.T.) injection via indwelling Ommaya reservoir. One fifth methotrexate dose was used as aminopterin dose, since equivalent therapeutic effect is believed to be obtained with this dose. Rate of disappearance from CSF is significantly slower for aminopterin than for methotrexate, providing concentrations in excess of 10^{-7} M for 75 hr with aminopterin and 40 hr with methotrexate, despite administration of 5 times as much methotrexate.

Although it is currently the most widely utilized route of administration for treating primary central nervous system tumors, the systemic route has serious limitations, some of which have been discussed earlier in this chapter. First, delivery of a drug to a central nervous system tumor requires that it penetrate the "blood–brain" barrier. As mentioned previously, the particular physicochemical requirements necessary for an agent to cross the brain capillaries are stringent; consequently, the number of potentially therapeutic agents available is small. Second, even in the presence of a suitable compound, the blood supply to certain areas of a tumor may be insufficient to deliver therapeutic concentrations of the drug.[11,49] Third, the use of the systemic route of administration carries the attendant risk of systemic toxicity, usually in the form of drug-induced marrow hypoplasia. Fourth, few cell-cycle-specific agents are available for systemic use, in large part because few fulfill the physicochemical requirements for penetration of the blood–brain barrier. Thus, the identification of combinations of cell-cycle-specific and cell-cycle-nonspecific agents for systemic use has been difficult. Finally, the extracellular concentration of a drug which slowly crosses from the blood into the brain extracellular fluid is rapidly diluted by the "sink" effect of the relatively large volume of CSF.

Attempts to overcome these limitations of systemically administered chemotherapy have taken different approaches. There is an ongoing effort aimed at development of new tumoricidal agents with characteristics that facilitate their penetration into the central nervous system. In addition, there is intense investigation aimed at identifying new techniques of drug administration that may maximize the efficacy of existing agents.

In the following sections, using methotrexate as an example, we will discuss this latter avenue of investigation, emphasizing how an understanding and appreciation of CSF pharmacokinetics has led to new systemic approaches to central nervous system tumors.

4.2. Increasing Central Nervous System Penetration with High-Dose Systemic Chemotherapy

Most systemically administered chemotherapeutic agents have a defined CSF/plasma ratio. As discussed previously, this is dictated by the physicochemical characteristics of the particular drug. Theoretically, for many agents, if a high enough systemic dose of drug could be given, meaningful therapeutic levels could be achieved in the CSF. The obvious constraint on this hypothesis is that the use of high-dose systemic chemotherapy is limited by increased systemic toxicity. Although methotrexate therapy for central nervous system disease has traditionally been limited to the intrathecal route, the availability of citrovorum-factor rescue to prevent the excessive systemic toxicity of this compound has created the possibility of utilizing high doses of this agent to treat central nervous system disease. We have explored the feasibility of replacing intrathecal methotrexate chemotherapy with effective systemic therapy. Our approach has been to infuse large amounts of methotrexate intravenously at a dose and rate sufficient to overcome those processes that ordinarily limit methotrexate penetration into the central nervous system using delayed citrovorum-factor rescue to prevent systemic toxicity. In initial studies, we utilized a technique that provided relatively constant plasma drug levels for 42 hr. A 1-hr priming intravenous infusion, administered at 1.5×10^{-7} times the desired plasma methotrexate concentration, was followed immediately by a 41-hr maintenance infusion given at an hourly rate of one fifth the priming dose. For instance, a plasma methotrexate level of 5×10^{-6} M required a dose of "75/15" mg/m^2, the "75" referring to the priming dose and the "15" to the hourly maintenance dose. Figure 9 gives the results obtained in ten patients who received such an infusion. As can be seen in this figure, after the first 8 hr of the infusion, both CSF and plasma methotrexate were relatively constant, with a mean CSF/plasma drug ratio of 1:33. In subsequent patients, treated with successively higher priming and maintenance doses (up to 600/120 mg/m^2), the mean CSF/plasma methotrexate ratio remained relatively constant.[14]

These results confirmed the ability of high-dose intravenous methotrexate administration to achieve therapeutic levels of methotrexate in the CSF. The results also suggested that a specified CSF antifolate concentration could be achieved with a 33-fold greater plasma drug level. In addition, a comparison of CSF samples obtained from ventricular CSF (via indwelling Ommaya reservoirs) and lumbar CSF revealed that the mean CSF/plasma drug ratios from the two sampling sites were similar. This suggests that systemic therapy has the advantage of providing consistent drug concentrations throughout the CSF axis. Whether higher central nervous system parenchymal levels of drug can be achieved by the intravenous than by the intrathecal route requires additional study. Another interesting facet of the high-dose intravenous methotrexate infusion approach is that a continuous infusion of the type discussed above fulfills the criteria of a "concentration × time" approach. As mentioned previously (Section 3.3), this type of approach is particularly important for cell-cycle-specific agents such as methotrexate. Its apparent feasibility via the intravenous route is appealing.

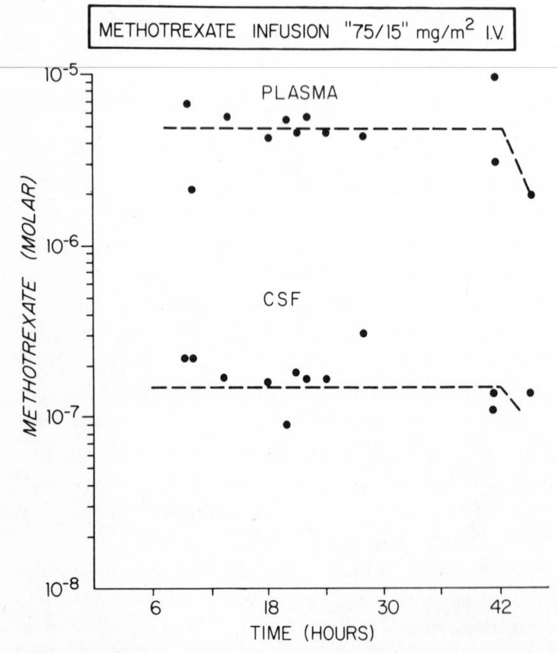

Figure 9. Plasma and CSF methotrexate concentrations in ten patients treated with "75/15" mg/m^2 high-dose intravenous methotrexate regimen. Mean CSF/plasma drug ratio was 1:33. Reprinted, with permission, from Bleyer and Poplack.[14]

4.3. Improving Systemic Chemotherapy by Altering CSF Drug Disappearance Kinetics

Some of the experimental techniques that are capable of altering the disappearance kinetics of drugs from the CSF and are of potential value as adjuncts to intrathecal methotrexate therapy may also be applied to the systemic use of this agent. For example, reduction of CSF production by acetazolamide and inhibition of the active transport component of methotrexate efflux by probenecid conceivably could be utilized to improve systemic methotrexate therapy (see Section 3.4). In addition to these agents, two other drugs have been noted to prolong the clearance of methotrexate from the CSF following intravenous methotrexate administration. Tejada and Zubrod[69] have reported that concomitant vincristine therapy slows the exit of systemically administered methotrexate from the CSF. The mechanism may be similar to that shown for vincristine-induced augmentation of intracellular methotrexate: blockage of an energy-dependent efflux system.[34]

A unique and exciting approach has been explored by Abelson *et al.*,[1] who have utilized the enzyme carboxypeptidase G_1 immediately following intravenous methotrexate infusions. Intravenous administration of this enzyme, which does not penetrate the central nervous system, rapidly cleaves methotrexate into nontoxic metabolites. Carboxypeptidase G_1 may allow very prolonged systemic methotrexate infusions to be used to treat central nervous system tumors without significant risk of systemic toxicity. In addition, this agent, through a mechanism yet to be determined, appears to decrease the efflux of methotrexate from the CSF.[2] Clinical studies of the combined intravenous methotrexate and carboxypeptidase approach are currently under way.

4.4. Improving Systemic Chemotherapy with Agents That Perturb Blood–Brain Barrier

A more experimental area currently being investigated involves efforts to improve the central nervous system penetration of chemotherapeutic agents by altering the blood–brain barrier. Various pharmacological and physical manipulations have been shown to increase the permeability of the brain capillaries to a variety of indicator compounds. Amphotericin, an antifungal polyene antibiotic that has been demonstrated to alter membrane lipids, increases the permeability of galactitol across the blood–brain barrier of the rat.[50] Another experimental approach has been to employ carotid infusions of hypertonic solutions of water-soluble nonelectrolytes. This has been shown to shrink capillary endothelial cells and widen the interendothelial tight junctions.[61] Using a hypertonic infusion of 1.6–1.8 molal arabinose to osmotically disrupt the blood–brain barrier, Rapoport *et al.*[62] demonstrated a 20-fold increase in the permeability of the rat cerebrovasculature to both [^{14}C]sucrose and Evans blue albumin. The ability of the hyperosmolar-infusion technique to improve chemotherapeutic penetration was examined by Allen *et al.*[3] They found that intracarotid hyperosmolar mannitol augmented ipsilateral brain methotrexate concentrations in normal rats. Although experimental at present, the hypertonic-infusion method may eventually prove to be of clinical value.

The effects of ionizing irradiation on the blood–brain barrier are a subject of obvious interest to those involved in the chemotherapy of brain tumors, since radiation is employed in most brain-tumor treatment protocols. A number of studies have shown that radiation can perturb the blood–brain barrier. Clemente and Holst[25] showed that a dose of 6000 rads resulted in significant uptake of trypan blue in monkey brains. This substance is normally excluded from the central nervous system following intravenous administration. Using much higher doses, other investigators have demonstrated increased penetration of sulfate-35 into rat brains.[54] More recently, Griffin *et al.*[38] demonstrated detectable brain methotrexate levels in rats injected with 100 mg/kg of the drug intraperitoneally 24 hr following treatment with a single 2000-rad fraction of cranial irradiation. In these experiments, both rats treated with less than 2000 rads and unirradiated controls had undetectable brain methotrexate levels at the same methotrexate dose.[38]

The possibility of selective alteration of the blood–brain barrier in a particular area of the brain has also been explored employing the use of ultrasound. In initial studies, Shealy and Crafts[67] demonstrated selective uptake of trypan blue and mercury-203 into isolated ultrasound-produced lesions of the cat cerebral cortex following arterial injection of either 50% Hypaque, 15% ethanol, or distilled water. No increase in the uptake of these substances was noted in normal brain tissue.[67] Similar results were obtained in selectively focused ultrasound-produced lesions in cat brains following 2000–3000 rads of X-irradiation.[64]

Alteration of the blood–brain barrier has also been reported with the use of microwave radiation. In early experiments, increases in the permeability of brain tissue of rats to intravenously injected flourescein was observed when the brains of these animals were exposed to low-power, pulsed, or continuous-wave microwave energy at 1.2 GHz.[33] Subsequent work has confirmed these observations and has demonstrated that the permeability of low-molecular-weight saccharides is increased by either pulse or continuous-wave microwave energy. Notably, the increased permeability was a temporary phenomenon that was observed immediately and 4 hr after microwave exposure, but not 24 hr later.[58]

Whether or not any of the aforementioned methods for increasing the permeability of the blood–brain barrier will have clinical applicability is not yet clear. This is not only because these techniques are still preliminary and experimental, but also because the exact status of the blood–brain barrier within a tumor and in the area of brain adjacent to tumor is the subject of much controversy. For example, several studies suggest that the microvasculature within a tumor and brain adjacent to tumor is distinctly abnormal, being characterized by fenestrated and discontinuous endothelium, in contrast to normal brain parenchymal endothelial cells, which have relatively tight endothelial-cell junctions.[20,55,71] Using the extracellular protein tracer horseradish peroxidase (mol.wt. 44,000) in a virally induced glioma system, Vick and co-workers[20,71,72] demonstrated the *in vivo* permeability of fenestrated and discontinuous tumor endothelium to constituents of the blood vascular compartment. The presence of peroxidase in brain adjacent to tumor, indicating leakage from abnormal capillaries at the tumor–brain border, was also observed. These findings have been used to argue that the blood–brain barrier may not be as important a factor as previously appreciated.[72] However, others have found reduced capillary permeability in brain and brain adjacent to tumor.[49,50] Levin *et al.*,[49] using an implanted 9L sarcoma or Walker 256 carcinoma model system, found that the exchange between blood and brain adjacent to tumor for [^{14}C]urea and ^{22}Na was 53% less than for comparable regions of normal brain. These investigators reasoned that drug delivery of smaller lipid-soluble agents to the area of brain adjacent to tumor would not be hampered, but that water-soluble drugs administered by single intravenous injection, resulting in a short peak concentration, may not achieve adequate therapeutic concentrations in this "marginal zone" of tumor growth. Whether the techniques described above may be of value in increasing the concentration of such water-soluble drugs within the tumor or in brain adjacent to tumor awaits definitive confirmation.

Recent results of studies performed by Allen *et al.*[4] are of interest in this regard. These investigators studied the influence of intracarotid mannitol on intracerebral methotrexate concentrations achieved within and surrounding intracranially implanted Walker 256 carcinomas in the rat brain. They noted that without mannitol, the methotrexate concentration ratio (tissue/serum) was highest in tumor, next highest in brain adjacent to tumor, and lowest in areas of brain distant from tumor. However, after intracarotid mannitol, no change was noted in the methotrexate concentration ratio in tumor or brain adjacent to tumor, although this ratio did increase dramatically in areas of brain distant from tumor. Their results confirmed the absence of a significant blood–brain barrier within tumor and implied that the blood–brain barrier in areas of brain adjacent to tumor was at least partially defective. These data suggest that the therapeutic value of any technique designed to increase the permeability of the blood–brain barrier will depend on the degree of brain-barrier function that remains preserved within and surrounding the tumor.[4]

5. Conclusion

Brain-tumor treatment has traditionally relied most heavily on the modalities of surgery and radiation therapy. It is only in the past decade that the addition of chemotherapy has made the "combined-modality approach" to central nervous system malignancy a reality. However, the development of rational chemotherapeutic protocols to treat central nervous system neoplasia requires a thorough understanding of the unique pharmacological problems encountered in this area. In particular, the chemotherapist must have an appreciation for the pharmacological dynamics of antineoplastic agents within the CSF. As emphasized in this chapter, better understanding of the pharmacology of antineoplastic agents in the CSF should lead to improved treatment of central nervous system cancer.

References

1. Abelson, H. T., Ensminger, W. D., Rosowsky, A.: Serum and cerebrospinal fluid pharmacokinetic studies of high dose methotrexate–carboxypeptidase G. *Proc. Am. Assoc. Cancer Res.* **20**:213, 1979.
2. Abelson, H. T., Ensminger, W., Rosowsky, A., Uren, J.: Comparative effects of citrovorum factor and carboxypeptidase G on cerebrospinal fluid–methotrexate pharmacokinetics. *Cancer Treatment Rep.* **62**:1549–1552, 1978.
3. Allen, J. C., Hasegawa, H., Mehta, B., Shapiro, W. R., Posner, J. B.: CNS penetration of methotrexate is enhanced by hyperosmolar intracarotid mannitol and meningeal carconimatosis. *Neurology* **28**:351, 1978.
4. Allen, J. C., Hasegawa, H., Mehta, B. M., Shapiro, W. R., Posner, J. B.: Influence of intracarotid mannitol on intracerebral methotrexate concentrations surrounding experimental brain tumors. *Proc. Am. Assoc. Cancer Res.* **20**:286, 1979.
5. Anton, A. H., Solomon, H. M.: Drug–protein binding. *Ann. N. Y. Acad. Sci.* **226**:1–362, 1973.
6. Aur, R. J. A., Simone, J., Hustur, H. P., Walters, T., Borella, L., Pratt, C., Pinkel, D.: Central nervous system therapy and combination chemotherapy of childhood lymphocytic leukemia. *Blood* **37**:272–281, 1971.
7. Benson, D. F., LeMay, M., Patten, D. H., Rubeno, A. B.: Diagnosis of normal pressure hydrocephalus. *N. Engl. J. Med.* **283**:609–615, 1970.
8. Bering, E. A., Rall, D. P., Walker, M., Leventhal, C., Ommaya, A.: Intrathecal chemotherapy of gliomas: Rationale and current status. *Ann. N. Y. Acad. Sci.* **159**:599–602, 1969.
9. Blasberg, R. G.: Methotrexate, cytosine arabinoside and BCNU concentration in brain after ventriculocisternal perfusion. *Cancer Treatment Rep.* **61**:625–631, 1977.
10. Blasberg, R. G.: Pharmacodynamics and the blood–brain barrier. In *Modern Concepts in Brain Tumor Therapy: Laboratory and Clinical Investigations. Natl. Cancer Inst. Monogr.* **46**:19–27, 1977.
11. Blasberg, R. G.: Pharmacokinetics and metastatic brain tumor chemotherapy. In Weiss, L., Gilbert, H., and Posner, J. (eds.): *Brain Metastases.* Boston, G. K. Hall and Company, 1980.
12. Blasberg, R. G., Patlak, C., Fenstermacher, J. D.: Intrathecal chemotherapy: Brain tissue profiles after ventricular perfusion. *J. Pharmacol. Exp. Ther.* **195**:73–83, 1975.

13. Blasberg, R. G., Patlak, C. S., Shapiro, W. R.: Distribution of methotrexate in the cerebrospinal fluid and brain after intraventricular administration. *Cancer Treatment Rep.* **61**:633–641, 1977.
14. Bleyer, W. A., Poplack, D. G.: Clinical studies on the central-nervous-system pharmacology of methotrexate. In Pinedo, H. M. (ed.): *Clinical Pharmacology of Antineoplastic Drugs.* Amsterdam, Elsevier/North-Holland, 1978, pp. 115–131.
15. Bleyer, W. A., Poplack, D. G.: Intraventricular vs. intralumbar methotrexate for central nervous system leukemia: Prolonged remission with the Ommaya reservoir. *Med. Pediatr. Oncol.* **6**:207–213, 1979.
16. Bleyer, W. A., Poplack, D. G., Simon, R. M.: "Concentration × time" methotrexate via a subcutaneous reservoir: A less toxic regimen for intraventricular chemotherapy of central nervous system neoplasms. *Blood* **51**:835–842, 1978.
17. Bleyer, W. A., Savitch, J. L., Hartmann, J. R.: Is aminopterin a safer drug for intrathecal injection than methotrexate? *Proc. Am. Assoc. Cancer Res.* **20**:179, 1979.
18. Bleyer, W. A., Savitch, J., Poplack, D. G.: Methotrexate in cerebrospinal fluid. *N. Engl. J. Med.* **293**:1152, 1975.
19. Bode, U., Magrath, I., Bleyer, W., Poplack, D., Glaubiger, D.: Mechanism for methotrexate efflux from the cerebrospinal fluid in man. *Proc. Am. Assoc. Clin. Oncol.* **20**:375, 1979.
20. Brightman, M. W., Reese, T. S., Vick, N. A., Bigner, D. D.: A mechanism underlying the lack of a blood–brain barrier to peroxidase in virally induced brain tumors. *J. Neuropathol. Exp. Neurol.* **30**:139–140, 1971.
21. Bruckner, H. W., Cohen, C. J., Kabakow, B., Wallach, R. C., Greenspan, E. M., Gusberg, S. B., Holland, J. F.: Combination chemotherapy of ovarian carcinoma with platinum: Improved therapeutic index. *Proc. Am. Assoc. Cancer Res.* **18**:339, 1977.
22. Bunn, P., Schein, P.: Meningeal involvement in lymphoma. *N. Engl. J. Med.* **290**:517, 1974.
23. Capizzi, R. L., DeConti, R. C., Marsh, J. C. Bertino, J. R.: Methotrexate therapy of head and neck cancer: Improvement in therapeutic index by use of leucovorin "rescue." *Cancer Res.* **30**:1782–1788, 1970.
24. Catane, R., Kaufman, J., West, C., Merrin, C., Tsukada, Y., Murphy, G. P.: Brain metastasis from prostatic carcinoma. *Cancer* **38**:2583–2587, 1976.
25. Clemente, C. D., Holst, E. A.: Pathological changes in neurons, neuroglia and blood–brain barrier induced by X-irradiation of heads of monkeys. *Arch. Neurol. Psychiatry* **71**:66–79, 1954.
26. Cowles, A. L., Fenstermacher, J. D.: Theoretical considerations in the chemotherapy of brain tumors. In Sartorelli, A. C., Johns, D. G. (eds.): *Handbook of Experimental Pharmacology,* Vol. 38. Berlin and New York, Springer-Verlag, 1979, pp. 319–329.
27. Cox, J. D., Petrovich, Z., Paig, C., Stanley, K.: Prophylactic cranial irradiation in patients with inoperable carcinoma of the lung. *Cancer* **42**:1135–1140, 1978.
28. Currie, S., Henson, R. A.: Neurological syndromes in the reticuloses. *Brain* **94**:307–320, 1971.
29. Dawson, D. M., Rosenthal, D. S., Moloney, W. C.: Neurological complications of acute leukemia in adults: Changing rate. *Ann. Intern Med.* **79**:541–544, 1973.
30. Dawson, H.: A comparative study of the aqueous humor and cerebrospinal fluid in the rabbit. *J. physiol.* **129**:111–133, 1955.
31. DiChiro, G., Hammock, M. K., Bleyer, W. A.: Spinal descent of cerebrospinal fluid in man. *Neurology* **26**:1–8, 1976.
32. Fleming, I., Simone, J., Jackson, R., Johnson, W., Walters, T., Mason, C.: Splenectomy and chemotherapy in acute myelocytic leukemia of childhood. *Cancer* **33**:427–434, 1974.
33. Frey, A. H., Feld, S. R., Frey, S.: Neural function and behavior: Defining the relationship. *Ann. N. Y. Acad. Sci.* **247**:433–439, 1975.
34. Fyfe, M. J., Goldman, I. D.: Characteristics of vincristine induced augmentation of methotrexate uptake in Ehrlich ascites tumor cells. *J. Biol. Chem.* **248**:5067–5073, 1973.
35. Gerovich, F. G., Luna, M. A., Gottlieb, J. A.: Increased indicence of cerebral metastases in sarcoma patients with prolonged survival from chemotherapy. *Cancer* **36**:1843–1851, 1975.
36. Goldman, I. D.: Effects of methotrexate on cellular metabolism: Some critical elements in the drug–cell interaction. *Cancer Treatment Rep.* **61**:549–558, 1977.
37. Gormley, P., Poplack, D., Pizzo, P.: The cerebrospinal fluid pharmacokinetics of *cis*-diamminedichloroplatinum (II) and several platinum analogues. *Proc. Am. Assoc. Cancer Res.* **20**:1131, 1979.
38. Griffin, T. W., Rasey, J. S., Bleyer, W. A.: The effect of photon irradiation in blood–brain barrier permeability to methotrexate in mice. *Cancer* **40**:1109–1111, 1977.
39. Griffin, J. W., Thompson, R. W., Mitchinson, M. J.: Lymphomatous leptomeningitis. *Am. J. Med.* **51**:200–208, 1971.
40. Ho, D. H. W., Frei, E., III: Clinical pharmacology of 1-β-D-arabinofuranosylcytosine. *Clin. Pharmacol. Ther.* **12**:944–954, 1971.
41. Hoshino, T.: Therapeutic implications of brain tumor cell kinetics. In *Modern Concepts in Brain Tumor Therapy: Laboratory and Clinical Investigations. Natl. Cancer Inst. Monogr.* **46**:171–178, 1977.
42. Hoshino, T., Barker, M., Wilson, C. B., Boldrey, E. B., Fewer, D.: Cell kinetics of human gliomas. *J. Neurosurg.* **37**:15–26, 1972.
43. Huffman, D. H., Wan, S. H., Azarnoff, D. L., Hoogstraten, B.: Pharmacokinetics of methotrexate. *Clin. Pharmacol. Ther.* **14**:572–579, 1973.
44. Kay, H. E. M.: Development of CNS leukemia in acute myeloid leukemia in childhood. *Arch. Dis. Child.* **51**:73–74, 1976.
45. Kieffer, S. A., Wolff, J. M., Prentice, W. B., Loken, M. K.: Scinticisternography in individuals without

known neurological disease. *Am. J. Roentgenol.* **112**:225–239, 1971.
46. Koch-Wesser, J., Sellers, E. M.: Drug therapy: Binding of drugs to serum albumin. *N. Engl. J. Med.* **294**:311–316, 1976.
47. Kury, G., Carter, H. D.: Autoradiographic study of human nervous system tumors. *Arch. Pathol.* **80**:38–42, 1965.
48. Law, I. P., Blom, J.: Adult acute leukemia: Frequency of central nervous system involvement in long term survivors. *Cancer* **40**:1304–1306, 1977.
49. Levin, V. A., Freeman-Dove, M., Dandahl, H. D.: Permeability characteristics of brain adjacent to tumors in rats. *Arch. Neurol.* **32**:785–791, 1975.
50. Levin, V. A., Landahl, H. D., Freeman-Dove, M. A.: The application of brain capillary permeability coefficient measurements to pathological conditions and the selection of agents which cross the blood–brain barrier. *J. Pharmacokinet. Biopharm.* **4**:499–519, 1976.
51. Mayer, S., Maickel, R. P., Brodie, B. B.: Kinetics of penetration of drugs and other foreign compounds into cerebrospinal fluid and brain. *J. Pharmacol.* **127**:205–211, 1959.
52. Mehta, Y., Hendrickson, F. R.: CNS involvement in Ewing's sarcoma. *Cancer* **33**:859–862, 1974.
53. Mellet, L. B.: Physiochemical considerations and pharmacokinetic behavior in delivery of drugs to the central nervous system. *Cancer Treatment Rep.* **61**:527–531, 1977.
54. Nair, V., Roth, L. J.: Effect of x-irradiation and certain other treatments in blood–brain barrier permeability. *Radiat. Res.* **23**:249–264, 1964.
55. Nystrom, S.: Pathological changes in blood vessels of human glioblastoma multiforme: Comparative studies using plastic casings, angiography, light microscopy, and with references to some other brain tumors. *Acta Pathol. Microbiol. Scand.*, Suppl. 137, 1960.
56. Ommaya, A. K.: Immunotherapy of gliomas: a review. *Adv. Neurol.* **15**:337–359, 1976.
57. Ommaya, A. K.: Subcutaneous reservoir and pump for sterile access to ventricular cerebrospinal fluid. *Lancet* **2**:983–984, 1963.
58. Oscar, K. J., Hawkins, T. D.: Microwave alteration of the blood–brain barrier system of rats. *Brain Res.* **126**:281–293, 1977.
59. Pizzo, P. A., Poplack, D. G., Bleyer, W. A.: The neurotoxicities of current leukemia therapy. *Am. J. Pediat. Hematol. Oncol.* **1**:127–140, 1979.
60. Poplack, D. G., Bleyer, W. A., Wood, J. H.: A primate model for the study of the CNS pharmacokinetics of antineoplastic agents. In Whitehouse, J. M. (ed.): *CNS Complications of Malignant Disease.* Hampshire, England, Macmillan Press, 1979, pp. 395–404.
61. Rapoport, S. I., Hori, M., Klatzo, I.: Testing of a hypothesis for osmotic opening of the blood–brain barrier. *Am. J. Physiol.* **223**:323–331, 1972.
62. Rapoport, S. I., Ohno, K., Fredericks, W. R., Pettigrew, K. D.: Regional cerebrovascular permeability to ^{14}C sucrose after osmotic opening of the blood–brain barrier. *Brain Res.* **150**:653–657, 1978.
63. Sansone, G.: Pathomorphosis of acute infantile leukemia treated with modern therapeutic agents; "meningoleukemia" and Fröhlich obesity. *Ann. Paediatr.* **183**:33, 1954.
64. Schettler, T., Shealy, C. N.: Experimental selective alteration of blood–brain barrier by X-irradiation. *J. Neurosurg.* **32**:89–94, 1970.
65. Shapiro, W. R., Posner, J. B., Ushio, Y., Chernik, N. L., Young, D. F.: Treatment of meningeal neoplasms. *Cancer Treatment Rep.* **61**:733–743, 1977.
66. Shapiro, W. R., Young, D. F., Mehta, B. M.: Methotrexate: Distribution in cerebrospinal fluid after intravenous ventricular and lumbar injections. *N. Engl. J. Med.* **293**:161–166, 1975.
67. Shealy, C. N., Crafts, D.: Selective alteration of the blood–brain barrier. *J. Neurosurg.* **23**:484–487, 1955.
68. Tefft, M., Fernandez, C., Donaldson, M., Newton, W., Moon, T. E.: Incidence of meningeal involvement by rhabdomyosarcoma of the head and neck in children. *Cancer* **42**:253–258, 1978.
69. Tejada, F., Zubrod, C. G.: Vincristine effect on methotrexate cerebrospinal fluid concentration. *Cancer Treatment Rep.* **63**:143–145, 1979.
70. Van Dilla, M. A., Trujillo, T. T., Mullaney, P. E., Coulter, J. R.: Cell microfluorometry: A method for rapid fluorescence measurement. *Science* **163**:1213–1214, 1969.
71. Vick, N. A., Bigner, D. D.: Microvascular abnormalities in virally-induced canine brain tumors: Structural basis for altered blood–brain barrier function. *J. Neurol. Sci.* **17**:29–39, 1972.
72. Vick, N. A., Khandekar, J. D., Bigner, D. D.: Chemotherapy of brain tumors: The "blood–brain barrier" is not a factor. *Arch. Neurol.* **34**:523–526, 1977.
73. Weiss, H. D., Walker, M. D., Wiernik, P. H.: Neurotoxicity of commonly used antineoplastic agents. *N. Engl. J. Med.* **291**:75–81, 127–233, 1974.
74. Wolk, R. W., Masse, S. R., Conklin, R., Freireich, E. J.: The incidence of central nervous system leukemia in adults with acute leukemia. *Cancer* **33**:863–869, 1974.
75. Yap, H. Y., Yap, B. S., Tashiwa, C. K., DiStefano, A., Blumenschein, G. R.: Meningeal carcinomatosis in breast cancer. *Cancer* **42**:283–286, 1978.
76. Zaharko, D. S.: Pharmacokinetics and drug effect. *Biochem. Pharmacol.* **23**(suppl. 2):1–8, 1974.
77. Ziegler, J. L., Bluming, A. Z., Morrow, R. M., Fass, L., Carbone, P. P.: Central nervous system involvement in Burkitt's lymphoma. *Blood* **36**:718–728, 1970.

CHAPTER 40

Cerebrospinal Fluid Melatonin

Steven M. Reppert, Mark J. Perlow, and David C. Klein

1. Introduction

The history of melatonin starts in 1917, with the discovery that mammalian pineal gland extracts have the ability to lighten amphibian skin.[14] About 40 years later, the skin-lightening constituent of the pineal gland, melatonin, was isolated and chemically identified as *N*-acetyl-5-methoxytryptamine by Lerner *et al.*[13]

Melatonin is a small (mol.wt. 232), highly lipophilic molecule derived from serotonin (Fig. 1). The lipophilic nature of melatonin appears to explain why it readily passes through the blood–brain barrier,[22,28] the placenta,[20] and most cell membranes.

Various physiological effects have been ascribed to melatonin.[15] The most consistent effect in mammals is on reproductive function.[25] Experimental evidence from studies on the Syrian hamster strongly supports the notion that melatonin is the pineal hormone involved in mediating pineal-dependent seasonal changes in reproductive function.[31]

2. Melatonin in CSF

In 1976, 17 years after the chemical identification of melatonin, it was reported to be present in human cerebrospinal fluid (CSF).[29] Subsequently, melatonin has been identified in CSF obtained from calves,[4] sheep,[28] and rhesus monkeys.[22] The methods used to measure this compound in CSF are radioimmunoassay (RIA)[1,4,22,28,30,32] and gas chromatography—mass spectrometry (GC-MS).[22,29,33]

The reported concentrations of melatonin in CSF vary substantially among and within species from below 2 pg/ml (8.2 pM) to above 800 pg/ml (3.4 nM) (see Table 1). This variation may be due to true species differences. However, it may also have arisen from differences in sampling procedures, in the age or state of the animals and subjects from which the samples were obtained, and in the precise location from which CSF was removed. It should also be considered that as of this writing, the science of measuring melatonin in CSF is at an early stage of development. Thus, it remains to be seen whether all the reported values of melatonin will be confirmed as the ability to measure melatonin improves in the various laboratories throughout the world interested in this problem.

3. Entry of Melatonin into CSF

This topic has been thoroughly reviewed from an anatomical standpoint by Reiter *et al.*,[26] and only a brief summary will be presented here. Three possible modes of entry of pineal melatonin into CSF exist in mammals. First, melatonin could be directly secreted into CSF in the subarachnoid space around the pineal gland.

Second, the pineal gland of some mammals, including primates and artiodactyla, forms part of the roof of the third ventricle, and thus melatonin could be secreted directly into ventricular CSF. In other

Steven M. Reppert, M.D., and **David C. Klein, Ph.D.** • Section on Neuroendocrinology, Laboratory of Developmental Neurobiology, National Institute of Child Health and Human Development, National Institutes of Health, Bethesda, Maryland 20205. **Mark J. Perlow, M.D.** • Laboratory of Clinical Psychopharmacology, National Institute of Mental Health, National Institutes of Health, Washington, D.C. 20032. Dr. Reppert's present address is: Children's Service, Massachusetts General Hospital, Boston, Massachusetts 02114.

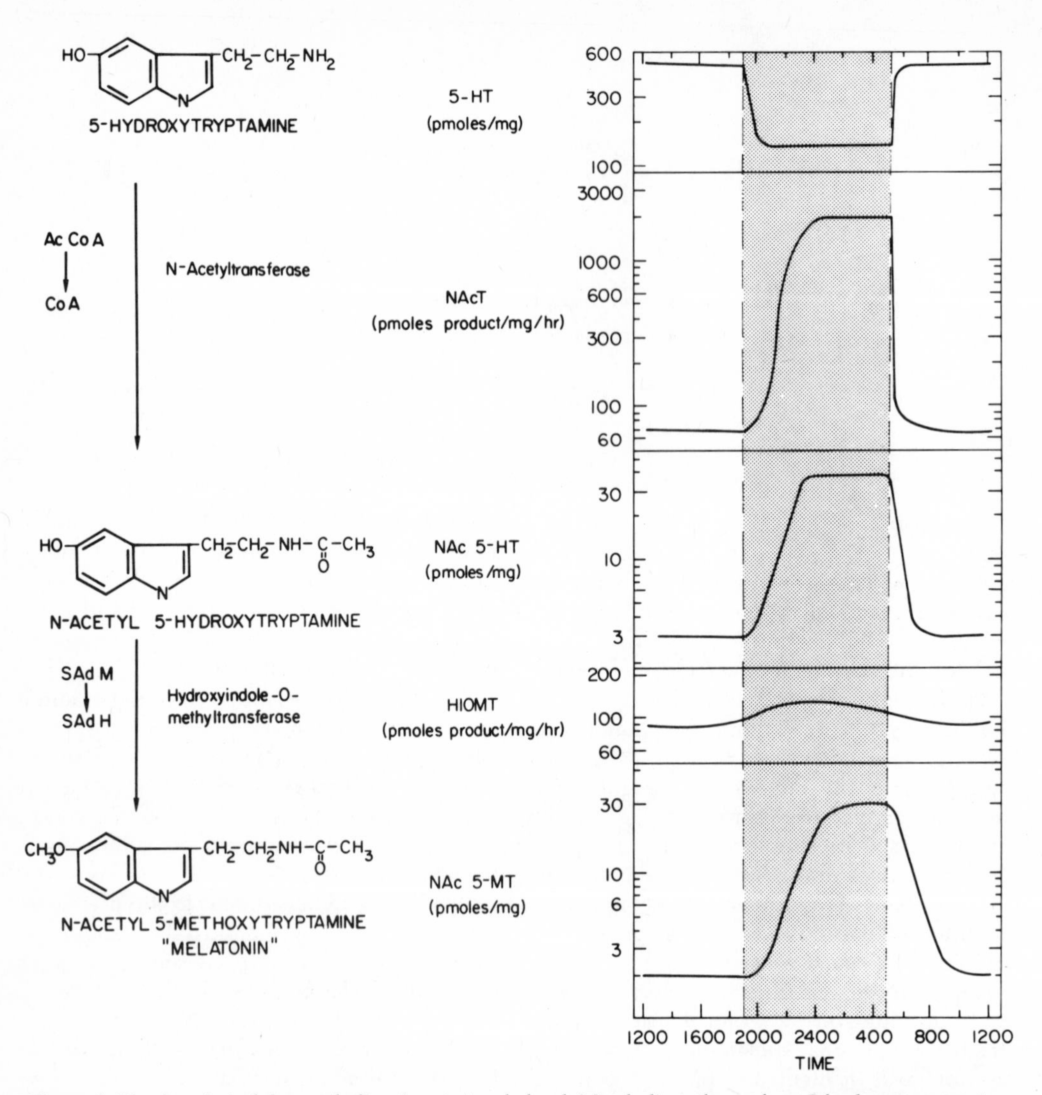

Figure 1. Rhythms in indole metabolism in rat pineal gland. Metabolic pathway from 5-hydroxytryptamine to melatonin is on left. Daily variations in concentrations of metabolites and activities of enzymes are on right. Shaded portion indicates dark period of light/dark cycle. Data have been abstracted from reports in literature. (Ac CoA) Acetyl coenzyme A; (CoA) coenzyme A; (SAd M) *S*-adenosyl methionine; (SAd H) *S*-adenosyl homocysteine; (5-HT) 5-hydroxytryptamine (serotonin); (NAcT) *N*-acetyl-transferase; (NAc 5-HT) *N*-acetyl 5-hydroxytryptamine; (HIOMT) hydroxyindole-*O*-methyltransferase; (NAc 5-MT) *N*-acetyl-5-methoxytryptamine (melatonin). Reproduced from Klein,[7] with permission.

mammals, including rodents such as the rat, the major portion of the pineal gland migrates away from the roof of the third ventricle during embryonic development. This results in the bulk of the gland becoming completely separated from the third ventricle. Thus, direct secretion of melatonin into ventricular CSF in these animals seems quite unlikely.

The third potential mode of entry of melatonin in all mammals is from the circulation. The pineal gland is well vascularized, and melatonin could be released from pineal cells into the vasculature and then enter CSF from vascularized areas contacting CSF, with the choroid plexus being an obvious potential site of transfer.[16a] The rapid entry of melatonin from blood into CSF has been demonstrated in the sheep[28] and in the rhesus monkey (Fig. 2).[22]

The relative contribution to total CSF melatonin of melatonin entering either by direct secretion into

Table 1. CSF and Circulating Melatonin Concentrations in Mammals

Order	Age	Method	CSF sample collection		Melatonin (pg/ml)		Comments
			Site	Time	CSF	Serum or plasma[a]	
Primates							
Human	2–6 years	RIA	Lumbar	Day	< 25–240	< 10–25	Range of values for 7 children undergoing therapy for leukemia[30]
	Adult	GC-MS	Lumbar	Day	60,80	—	Values for 2 individuals[33]
	Adult	RIA	Lumbar	Day	< 29–118	< 28–147	Range of values for 15 individuals[1]
	Adult	RIA	Lumbar	Day	< 10–30	21–91	Lumbar values: ranges from 15 individuals with intracranial disease; ventricular values: from 2 individuals.[32]
			Lateral ventricle	Day	13,19		
Rhesus monkey	Adult	RIA + GC-MS	Cervical	Day	<2–6	7–18	Ranges of day and night values for 2 animals from Fig. 4.
				Night	9–32	8–32	
	Adult	RIA + GC-MS	Cervical	Day	<2–10	—	Ranges of day and night values for 13 animals[22]
				Night	10–50	—	
Artiodactyla							
Cattle	9 months	RIA	Lateral ventricle	Day	10–100	3–40	Ranges of mean day and night values for 6 animals[4]
				Night	540–880	60–200	
Sheep	Adult	RIA	Cisterna magna	Day	10–100	10–400	Ranges of day and night values for 4 animals[28]
				Night	10–300	10–2000	

[a] Serum or plasma samples were collected at the same time(s) as CSF samples. (RIA) Radioimmunoassay; (GC–MS) gas chromatography–mass spectrometry.

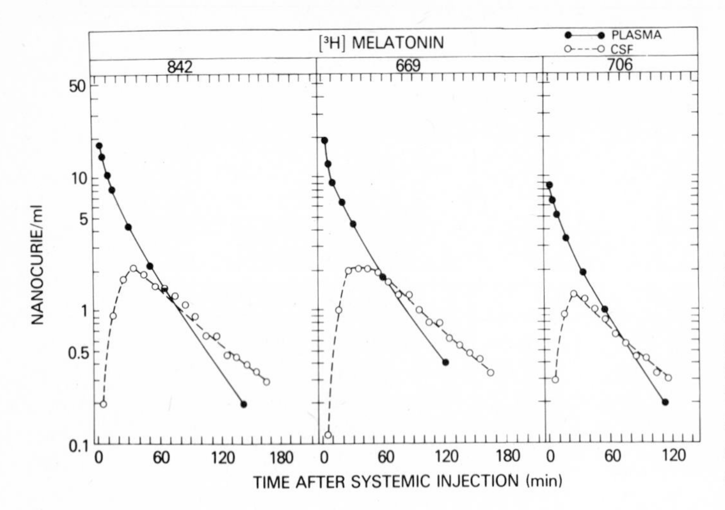

Figure 2. [^{3}H]Melatonin in blood and CSF. Two animals (842, 669) were given an intravenous bolus injection of 90 μCi (0.6 μg) [^{3}H]melatonin and one animal (706), 70 μCi (0.5 μg). CSF was collected as 10-min fractions, and data are plotted at midpoint of each collection period. Plasma was obtained at times indicated. Reproduced from Reppert *et al.*,[22] with permission.

CSF or from the circulation has been experimentally examined only in the sheep.[28] From these studies, it appears that most pineal melatonin is secreted into the blood before entering CSF, and that direct secretion into the ventricles accounts for no more than 1% of the total melatonin entering CSF.

Even though it appears that entry of melatonin from blood is far more important in maintaining CSF melatonin concentrations than is direct secretion, it seems possible that direct secretion into the ventricular space could be of special physiological significance. In those animals in which the pineal gland forms part of the roof of the third ventricle, direct secretion of melatonin into ventricular CSF might elevate concentrations of melatonin in the ventricular space immediately surrounding the pineal gland. Interestingly, this includes the hypothalamus, which has been speculated to be a site of action of melatonin.[15]

It should be noted that even though the pineal gland is thought to be the major source of melatonin, it has been reported to be present in the blood and urine of pinealectomized rats and sheep.[6,17] The extent of the contribution of circulating melatonin from extrapineal sources to the CSF melatonin levels in intact animals is not known.

4. Disappearance and Fate of CSF Melatonin

Melatonin has a relatively short biological life in the vasculature.[11,12,20–22,28] It disappears in a multiphasic fashion, with a rapid component of disappearance (halving time of about 3 min) and a slower component (having time of about 30 min).

On the basis of studies in the rhesus monkey, it appears that melatonin disappears from CSF slightly slower than from blood.[22] Following a bolus injection of [^{3}H]melatonin into the vasculature, [^{3}H]melatonin levels in blood rapidly decrease in a biphasic manner as predicted (Fig. 2). Disappearance from CSF appears monophasic, with a halving time of 30–40 min. Similar rates of disappearance of melatonin from blood and CSF are seen following a 2-hr infusion of radiolabeled melatonin into the circulation (Fig. 3).

The half-life of melatonin in CSF have also been investigated using sheep,[28] and the results of these studies are in general agreement with those involving monkeys.

There appear to be two possible fates of melatonin in CSF: transfer to blood and metabolism by brain tissue. The rapid transfer of CSF melatonin into blood has been documented in the sheep.[28] After being transferred to blood, melatonin is transported to the liver, where it is converted to 6-hydroxymelatonin—which is conjugated primarily to sulfate—and also to glucuronide.[11] More than 85% of injected melatonin is excreted into urine as 6-hydroxymelatonin or conjugates of this metabolite.[11,12,21] The remainder (< 1%) is excreted unchanged and, as a metabolite, suspected to be formed in the brain.[5]

The metabolism of melatonin by brain tissue has recently been studied, and it appears that *N*-acetyl-5-methoxykynurenamine may be a metabolic product.[5] It is formed *in vitro* by a two-step enzymatic route that opens the five-membered ring and removes the carbon adjacent to the ring nitrogen.

None of the metabolites mentioned above has been reported to have biological activity.

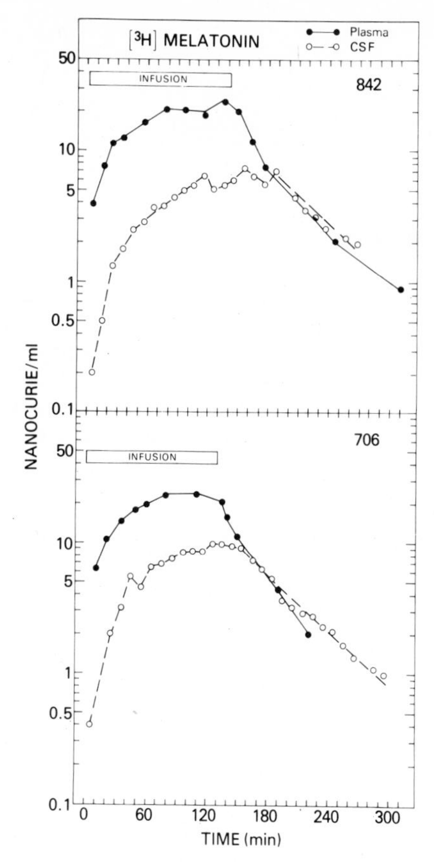

Figure 3. [^{3}H]Melatonin in blood and CSF during and following intravenous [^{3}H]melatonin infusion. [^{3}H]Melatonin was infused at constant rate of 7 μCi/min (47 ng/min) for 132 min (706) and 140 min (842). CSF was collected as 10-min fractions, and data are plotted at midpoint of each collection period. Plasma was obtained at times indicated. Reproduced from Reppert *et al.*,[22] with permission.

5. Relationship between Blood and CSF Melatonin

One would predict from the rapid transfer of melatonin from blood to CSF and from CSF to blood that melatonin concentrations in blood and CSF might be similar. However, this is not the case (Table 1). In the calf[4] and in children undergoing therapy for acute lymphocytic leukemia,[30] CSF values are reported higher than paired circulating levels. In contrast, CSF melatonin levels are lower than corresponding circulating levels in adult humans,[1,32] rhesus monkeys,[22] and sheep.[28]

One factor that might influence the relative amounts of melatonin in blood and in CSF is binding to blood proteins. It has been demonstrated that melatonin binds to blood proteins, including albumin.[2] In view of this, it seems reasonable to suspect that the lower CSF values in adult humans, rhesus monkeys, and sheep may in part be a reflection of the binding of melatonin to serum albumin. As a result, only a fraction of the total melatonin in blood would be available to equilibrate with CSF.

The higher concentrations of melatonin in the CSF of calves and in children could be age-associated. Perhaps there is closer anatomical association of the pinealocytes to ventricular CSF early in development, or perhaps there is more melatonin-binding protein in the CSF early in development, as compared to blood. Another factor that should be considered in the child study is that the high melatonin values may be associated primarily with leukemia or antileukemic therapy, and may not be a true reflection of CSF melatonin in normal children.

As mentioned above, CSF melatonin is higher than blood melatonin in the calf; the opposite is true in the monkey. Despite this contrasting quantitative relationship, however, there is a consistent dynamic relationship; in both species, similar daily rhythms in both blood and CSF melatonin occur (Fig. 4).[4,22]

6. Daily Rhythm in CSF Melatonin

The daily rhythm in CSF melatonin in the rhesus monkey was studied using animals that were partially restrained; this made it possible to continually withdraw CSF from an individual animal for periods up to 10 days (Fig. 5)[18] The values of melatonin in monkey CSF, as measured by RIA,[27] were quantitatively confirmed by GC-MS analysis of the same samples.[22]

All 15 monkeys studied exhibit a daily rhythm in CSF melatonin under conditions of diurnal lighting (light/dark, 12:12). There is apparent variation from individual to individual in both the nadir and the zenith of rhythmic changes (Fig. 6). In contrast, there is little day-to-day variation in the rhythm when an individual is examined for periods of 3 days (Fig. 7) and 6 days (Fig. 8). Whereas these observations suggest that individual differences in the melatonin rhythm exist, the possibility that variations in the precise location of the sampling catheter (from cervical to lumbar subarachnoid space) cause apparent individual differences has not been eliminated.

In the study that demonstrated the CSF melatonin

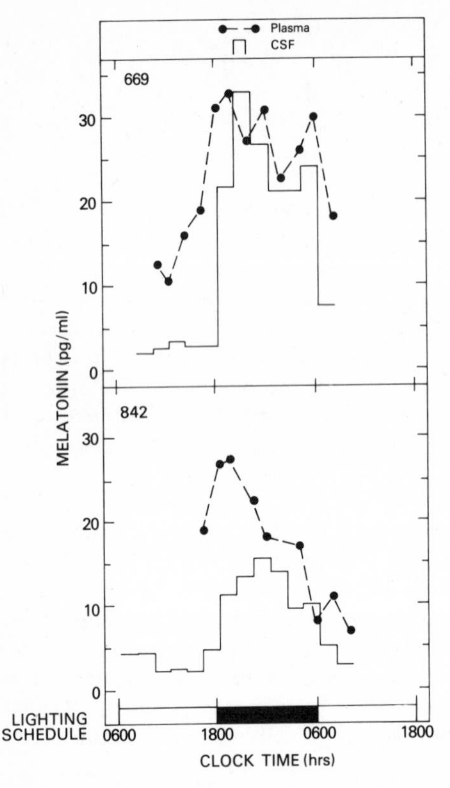

Figure 4. Diurnal pattern of melatonin in plasma and CSF. Plasma and CSF were collected over same 24-hr time period for each animal. CSF was collected as 120-min fractions. Lower limit of detection was 2 pg/ml. Reproduced from Reppert *et al.*,[22] with permission.

rhythm in the calf, six animals were sampled for a 24-hr period; the data were presented as mean values. Thus, it is not possible to comment on individual-to-individual or day-to-day variation in this species. Diurnal rhythms of CSF and serum melatonin level have recently been confirmed in green sea turtles.[16a]

7. Regulation of Melatonin Rhythm

7.1. Regulation of Melatonin Production in Rat

According to studies in the rat, the daily rhythm of plasma and urinary melatonin is a direct result of daily changes in the production of melatonin by the pineal gland.[7,8] Neural signals drive the daily oscillations in pineal melatonin production; these signals appear to originate in the suprachiasmatic nuclei (SCN) of the anterior hypothalamus (Fig. 9). A neural circuit, which pases through both central and peripheral neural structures, connects the SCN to the pineal gland and transmits signals that increase the activity of pineal *N*-acetyltransferase. This enzyme is involved in melatonin synthesis (see Fig. 1). The increase in *N*-acetyltransferase activity causes a decrease in pineal serotonin and an increase in pineal *N*-acetyl-serotonin and melatonin.

The daily rhythm in melatonin production can be modified by environmental lighting acting through the eye and a retinohypothalamic projection to the SCN.[9,16] In diurnal lighting, the endogenous oscillations generated in the SCN are entrained to the 24-hr day by the environmental lighting cues. As a result, the daily rhythm in melatonin production is coordinated with the daily lighting cycle, with increases production occurring only at night.

When animals are deprived of photic input, by blinding or by housing them in darkness, the daily rhythm in pineal melatonin production persists,

Figure 5. Drawing of experimental equipment used for continuous collection of CSF from partially restrained rhesus monkey. Male animals are adapted to chronic restraint in primate-restraining chairs for period of 4–5 weeks. Following adaptation, polyethylene catheter is inserted into lumbar subarachnoid space and advanced cephalad so that tip terminates at cisternal subarachnoid region. CSF is continuously withdrawn by peristaltic pump at constant rate of approximately 1 ml/hr, and collected as either 90- or 120-min fractions by automated fraction collector.

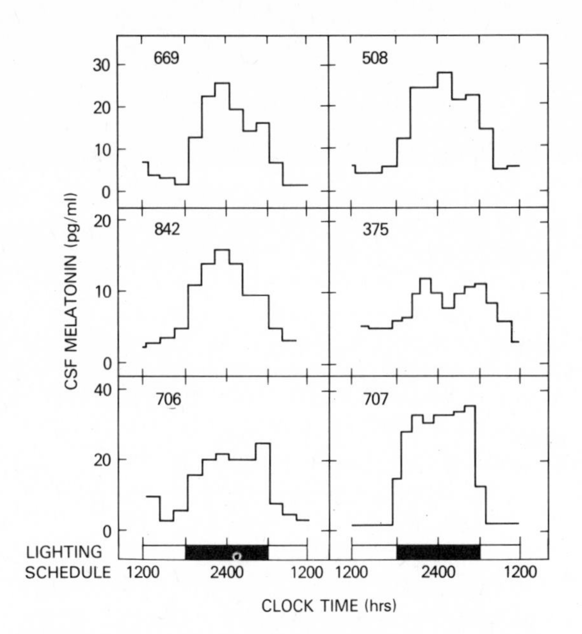

Figure 6. Diurnal pattern of CSF melatonin for six animals. CSF was collected as 120-min fractions for four animals (669, 508, 842, 706) and as 90-min fractions for two animals (375, 707). Lower limit of detection for each assay was 2 pg/ml. Reproduced from Reppert *et al.*,[22] with permission.

which is a characteristic of a true endogenous circadian rhythm.[7,8] The endogenous nature of this rhythm is further supported by studies showing that when animals are exposed to darkness during the normal light period of the day, there is no increase in melatonin production. Presumably, this is because the SCN are not in the proper phase during the normal light period of the day to stimulate melatonin synthesis.

In contrast to the lack of an effect of darkness on melatonin production during the light period, there is a marked effect of light during the dark period. After melatonin production has increased at night in the dark, exposure to light acutely suppresses melatonin synthesis.[7,8] Similarly, if animals are exposed to constant light for several days, the daily rhythm is abolished for the duration of light exposure.

7.2. Regulation of CSF Melatonin in Rhesus Monkey

The effects of environmental lighting on the CSF melatonin rhythm in the rhesus monkey[19] have been found to be similar to the effects of environmental lighting on melatonin synthesis in the rat described above.

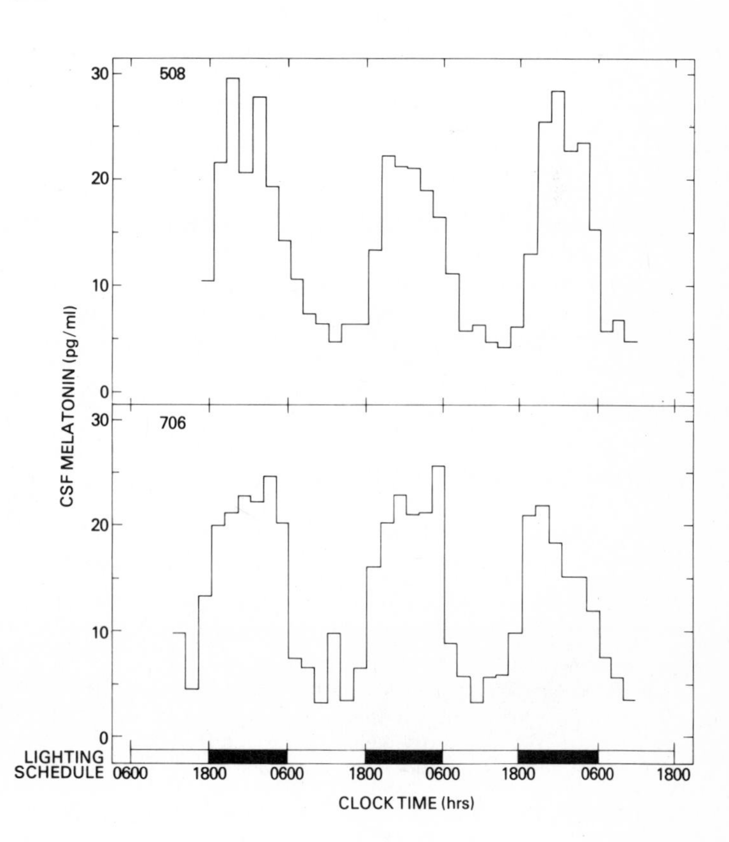

Figure 7. Diurnal pattern of CSF melatonin for individual animals studied for 3 consecutive days. CSF was collected as 120-min fractions. Lower limit of detection was 2 pg/ml. Reproduced from Reppert *et al.*,[22] with permission.

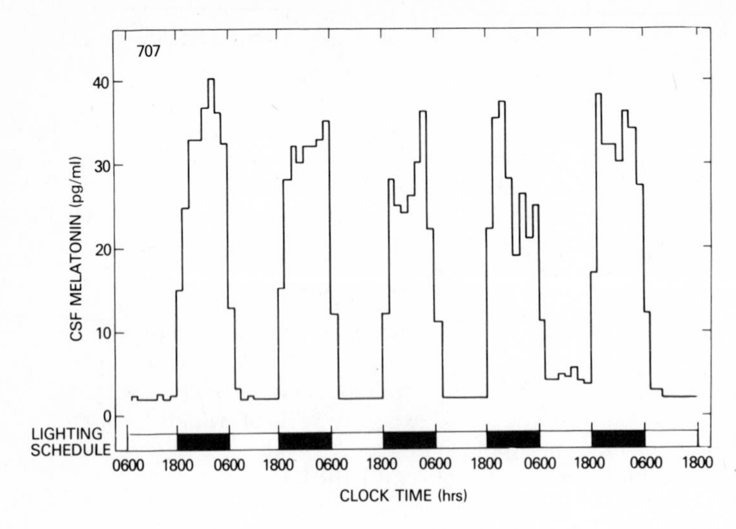

Figure 8. Diurnal pattern of CSF melatonin for animal studied for 6 consecutive days. CSF was collected as 90-min fractions. Lower limit of detection was 2 pg/ml. Reproduced from Reppert *et al.*,[22] with permission.

As previously mentioned, a large daily rhythm in CSF melatonin occurs in diurnal lighting (see Figs. 4 and 6–8). This rhythm appears to be strongly coupled to diurnal lighting, with elevated levels occurring only during the dark period.

The daily CSF melatonin rhythm is modified by environmental lighting as follows: (1) Constant light completely suppresses the rhythm (Fig. 10). (2) Exposure to light during the dark period of diurnal lighting acutely suppresses melatonin values to low

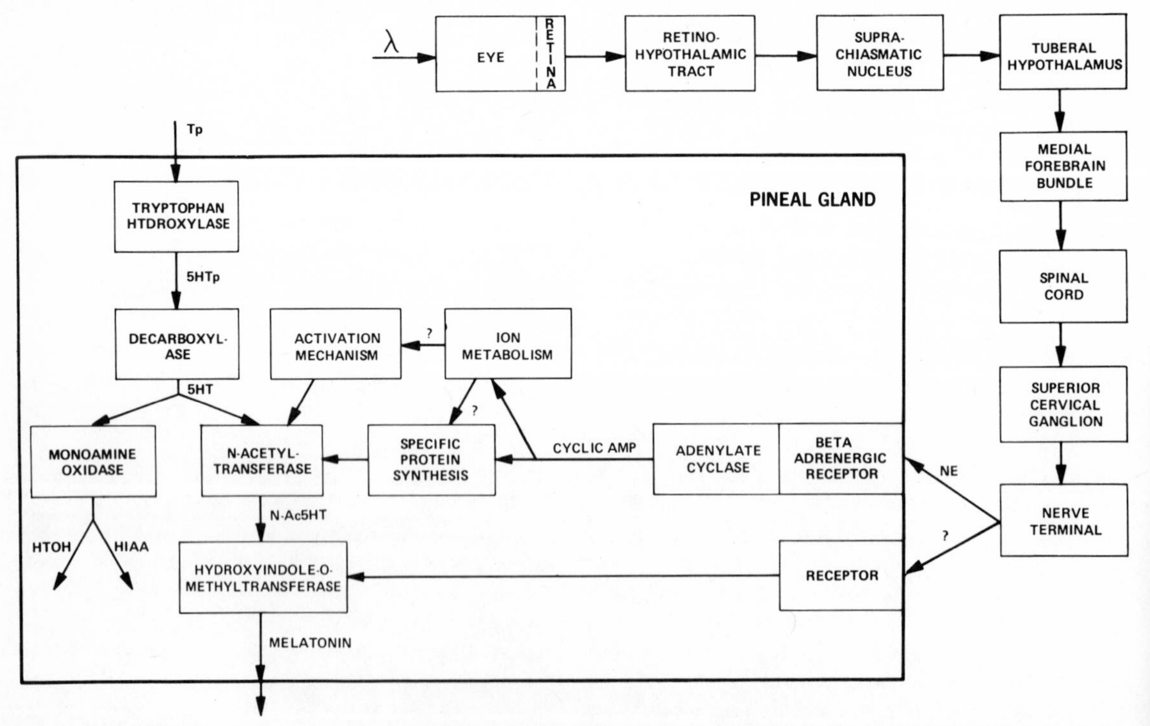

Figure 9. Schematic representation of the neural control of the pineal gland. (NE) Norepinephrine; (CYCLIC AMP) adenosine 3′,5′-cyclic monophosphate; (Tp) tryptophan; (5HTp) 5-hydroxytryptophan; (5-HT) 5-hydroxytryptamine (serotonin); (HIAA) 5-hydroxyindoleacetic acid; (HTOH) 5-hydroxytryptophol; (N-Ac5HT) *N*-acetyl-serotonin; (?) unproven hypotheses. Reproduced from Klein and Moore,[9] with permission.

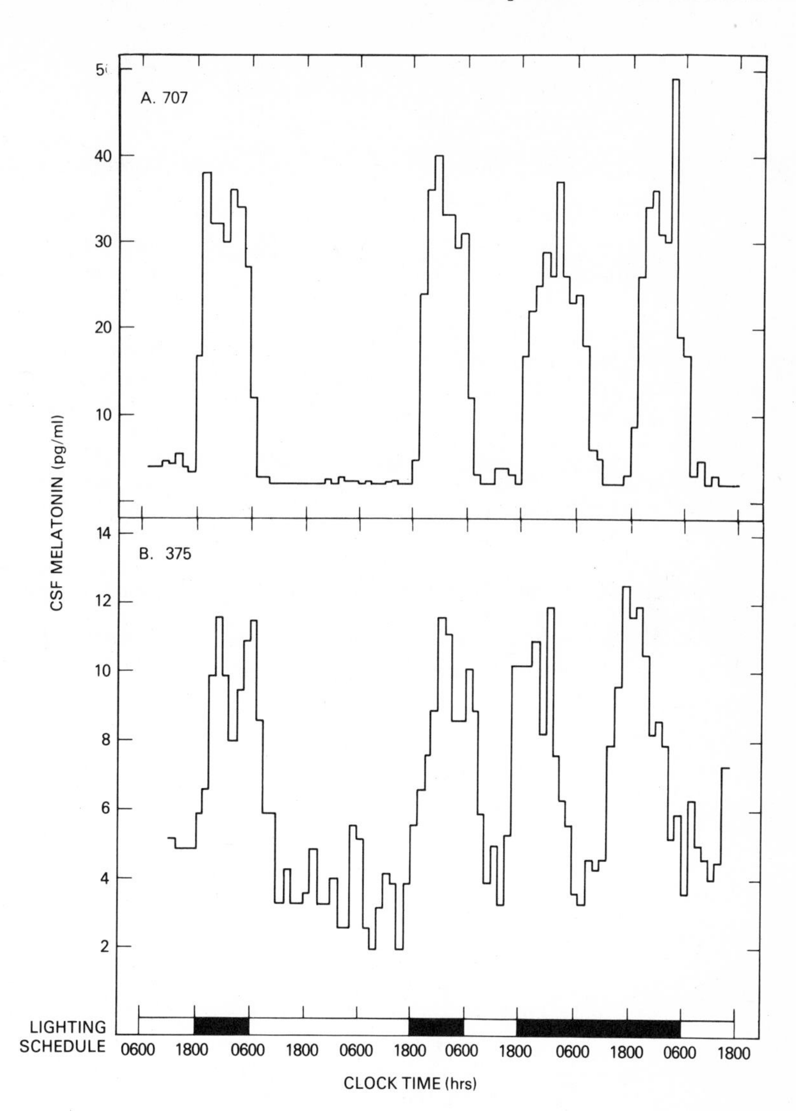

Figure 10. Pattern of CSF melatonin concentration of two rhesus monkeys during periods of diurnal lighting (light/dark, 12:12), constant light (24:0), and constant darkness (0:24). CSF was collected in 90-min fractions. Lower limit of sensitivity was 2 pg/ml. Reproduced from Perlow *et al.*,[19] with permission.

daytime levels.[24] In addition, the daily rhythm persists in constant darkness (Fig. 10), and exposure to darkness during the normal light period of diurnal lighting does not increase CSF melatonin levels (Reppert, S. M., Perlow, M. J., Klein, D. C., unpublished results). These latter two findings indicate that the CSF melatonin rhythm is endogenously generated in the monkey, as in the rat.

Only limited studies on the neural circuit involved in the regulation of the CSF melatonin rhythm in the rhesus monkey have been performed. Interest has focused on the role of the SCN in the generation of the CSF melatonin rhythm, because destruction of this nucleus in the rat results in disappearance of the rhythm in *N*-acetyltransferase activity.[9,16]

Complete destruction of the SCN area in the monkey markedly alters the regulation of the CSF melatonin rhythm.[23] First, the daily rhythm is either absent or abnormal in diurnal lighting. Second, constant light does not consistently suppress CSF melatonin levels. The third and most surprising finding is that there is some evidence that a daily rhythm of reduced amplitude persists in constant darkness. Thus, it appears that the SCN are important for the effects of environmental lighting that are normally seen on the CSF rhythm and may also be important in the generation or coordination of this rhythm. However, in the monkey, in contrast to the situation in the rat, it appears possible that an oscillator other than the SCN may also be involved in the generation of the melatonin rhythm.

8. Final Comments

According to the evidence cited in this review, there is little doubt that melatonin normally occurs in CSF, that it exhibits the same dynamic changes that occur in blood and urine, and that its diurnal rhythm is regulated by the same mechanism that regulates melatonin production. However, the large variation in reported CSF melatonin concentrations raises the possibility that other factors, independent of this regulatory mechanism, may also control the amount of melatonin in CSF. Further investigations will be required to settle this question.

The physiological significance of melatonin in CSF has not been established. Perhaps some areas of the brain bathed in CSF, such as the hypothalamus, contain melatonin receptors[3] or a melatonin-concentrating system[10]; these may be involved in mediating the effects of melatonin on reproductive function.

References

1. ARENDT, J., WETTERBURG, L., HEYDEN, T., SIZONENKO, P. C., PAUNIER, L.: Radioimmunoassay of melatonin: Human serum and cerebrospinal fluid. *Horm. Res.* **8:**65–75, 1977.
2. CARDINALI, A. P., LYNCH, H. J., WURTMAN, R. J.: Binding of melatonin to human and rat plasma proteins. *Endocrinology* **91:**1213–1218, 1972.
3. CARDINALI, A. P., HYYPPÄ, M. T., WURTMAN, R. J.: Fate of intracisternally injected melatonin in the rat brain. *Neuroendocrinology* **12:**30–40, 1973.
4. HEDLUND, L., LISCHKO, M. M., ROLLAG, M. D., NISWENDER, G. D.: Melatonin: Daily cycle in plasma and cerebrospinal fluid of calves. *Science* **195:**686–687, 1977.
5. HIRATA, F., HAYAISHI, O. Y., TOKUYAMA, T., SENOH, S.: *In vitro* and *in vivo* formation of two new metabolites of melatonin. *J. Biol. Chem.* **249:**1311–1313, 1974.
6. KENNAWAY, D. J., FIRTH, R. G., PHILLIPOU, G., MATTHEWS, C. D., SEAMARK, R. F.: A specific radioimmunoassay for melatonin in biological tissue and fluids and its validation by gas chromatography–mass spectrometry. *Endocrinology* **101:**119–127, 1977.
7. KLEIN, D. C.: Circadian rhythms in indole metabolism in the rat pineal gland. In Schmidt, F. O. (ed.): *The Neurosciences: Third Study Programme.* Cambridge, Massachusetts, MIT Press, 1974, pp. 509–515.
8. KLEIN, D. C.: The pineal gland: A model of neuroendocrine regulation. In Reichlin, S., Baldessarini, R. J., Martin, M. B. (eds.): *The Hypothalamus.* New York, Raven Press, 1978, pp. 303–327.
9. KLEIN, D. C., MOORE, R. Y.: Pineal *N*-acetyltransferase and hydroxyindole-*O*-methyltransferase: Control by the retinohypothalamic tract and the suprachiasmatic nucleus. *Brain Res.* **174:**245–262, 1979.
10. KNIGHT, B. K., HAYES, M. M. M., SYMINGTON, R. B.: Anatomical evidence to support the role of the cerebrospinal fluid as a transmitter medium in the pineal–hypothalamic axis. *S. Afr. J. Med. Sci.* **39:**25–31, 1974.
11. KOPIN, I. J., PARE, C. M. B., AXELROD, J., WEISSBACH, H.: The fate of melatonin in animals. *J. Biol. Chem.* **236:**3072–3075, 1961.
12. KVEDER, S., MCISAAC, M. W.: The metabolism of melatonin (*N*-acetyl-5-methoxytryptamine) and 5-methoxytryptamine. *J. Biol. Chem.* **236:**3214–3220, 1961.
13. LERNER, A. B., CASE, J. D., HEINZELMAN, R. V.: Structure of melatonin. *J. Am. Chem. Soc.* **81:**6084–6087, 1959.
14. MCCORD, C. P., ALLEN, F. P.: Evidence associating the pineal gland function with alterations in pigmentation. *J. Exp. Zool.* **23:**207–224, 1917.
15. MINNEMAN, K. P., WURTMAN, R. J.: The pharmacology of the pineal gland. *Annu. Rev. Pharmacol. Toxicol.* **16:**33–51, 1976.
16. MOORE, R. Y., KLEIN, D. C.: Visual pathways and the central neural control of a circadian rhythm in pineal *N*-acetyltransferase activity. *Brain Res.* **71:**17–33, 1974.

16a. OWENS, D. W., GERN, W. A., RALPH, C. L.: Melatonin in the blood and cerebrospinal fluid of the green sea turtle (*Chelonia mydas*). *Gen. Comp. Endocrinol.* **40:**180–187, 1980.

17. OZAKI, Y., LYNCH, H. J.: Presence of melatonin in plasma and urine of pinealectomized rats. *Endocrinology* **99:**641–644, 1976.
18. PERLOW, M. J., FESTOFF, B., GORDON, E. K., EBERT, M. H., JOHNSON, D. K., CHASE, T. N.: Daily fluctuation in the concentration of cAMP in the conscious primate brain. *Brain Res.* **126:**391–396, 1977.
19. PERLOW, M. J., REPPERT, S. M., TAMARKIN, L., WYATT, R. J., KLEIN, D. C.: Photic regulation of the melatonin rhythm: Monkey and man are not the same. *Brain Res.* **182:**211–216, 1980.
20. REPPERT, S. M., CHEZ, R. A., ANDERSON, A., KLEIN, D. C.: Maternal–fetal transfer of melatonin in the non-human primate. *Pediatr. Res.* **113:**788–791, 1979.
21. REPPERT, S. M., KLEIN, D. C.: Transport of maternal [^{3}H]melatonin to suckling rats and the fate of [^{3}H]melatonin in the neonatal rat. *Endocrinology* **102:**582–588, 1978.
22. REPPERT, S. M., PERLOW, M. J., TAMARKIN, L., KLEIN, D. C.: A diurnal melatonin rhythm in primate cerebrospinal fluid. *Endocrinology* **104:**295–301, 1979.

23. REPPERT, S. M., PERLOW, M. J., MISHKIN, M., TAMARKIN, L., KLEIN, D. C.: Effects of damage to the suprachiasmatic area of the anterior hyphothalamus on the daily melatonin rhythm in the rhesus monkey. 61st Annual Endocrine Society Meeting, 1979.
24. REPPERT, S. M., PERLOW, M. J., TAMARKIN, L., KLEIN, D. C.: Photic regulation of the melatonin rhythm: A distinct difference between man and monkey. *Pediatr. Res.* **13**:362, 1979.
25. REITER, R. J.: Comparative physiology: Pineal gland. *Annu. Rev. Physiol.* **35**:305–328, 1973.
26. REITER, R. J., VAUGHAN, M. K., BLASK, D. E.: Possible role of cerebrospinal fluid in the transport of pineal hormones in mammals. In Knigge, K. M., Scott, D. E., Kobayashi, H., Ishii, J. (eds.): *Brain Endocrine Interaction II.* Basel, S. Karger, 1975, pp. 337–354.
27. ROLLAG, M., NISWENDER, G.: Roadioimmunoassay of serum concentrations of melatonin in sheep exposed to different lighting regimens. *Endocrinology* **98**:482, 1976.
28. ROLLAG, M. D., MORGAN, R. J., NISWENDER, G. D.: Route of melatonin secretion in sheep. *Endocrinology* **102**:1–8, 1978.
29. SMITH, I., MULLEN, P. E., SILMAN, R. E., SNEDDEN, W., WILSON, B. W.: Absolute identification of melatonin in human plasma and cerebrospinal fluid. *Nature (London)* **260**:718–719, 1976.
30. SMITH, J. A., MEE, T. J., BARNES, N. A., THORBURN, R. J., BARNES, J. L. C.: Melatonin in serum and cerebrospinal fluid. *Lancet* **2**:425, 1976.
31. TAMARKIN, L., HOLLISTER, C., LEFEBVRE, N. G., GOLDMAN, B. D,: Melatonin induction of gonadal quiescence in pinealectomized Syrian hamsters. *Science* **198**:953–955, 1977.
32. VAUGHAN, G. M., McDONALD, S. A., JORDON, R. M., ALLEN, J. P., BOHMFALK, G. L., ABOU-SAMRA, M., STORY, J. L.: Melatonin concentration in human blood and cerebrospinal fluid: Relationship to stress. *J. Clin. Endocrinol. Metab.* **47**:220–223, 1978.
33. WILSON, B. W., SNEDDEN, W., SILMAN, R. E., SMITH, I., MULLEN, P.: A gas chromatography–mass spectrometry method for the quantitative analysis of melatonin in plasma and cerebrospinal fluid. *Anal. Biochem.* **81**:283–291, 1977.

CHAPTER 41

Cerebrospinal Fluid Pituitary Hormone Concentrations in Patients with Pituitary Tumors

Kalmon D. Post, Bruce J. Biller, and Ivor M. D. Jackson

1. Introduction

Adenohypophyseal hormones are frequently detectable in the cerebrospinal fluid (CSF) of patients with pituitary adenomas,[1,3,26,33,35,37,40,59,65] especially those showing suprasellar extension (SSE).[33,37,40] However, the issue of their presence in the CSF of normal subjects is controversial. Some workers readily find anterior pituitary hormones in CSF,[2,62] while others do not.[1,33,37,40,65] It has been proposed that in the normal individual, the blood-brain barrier (BBB) is relatively impermeable to the anterior pituitary hormones and that measurable levels of these substances in the CSF of patients with pituitary tumors indicate a breakdown in the blood–CSF barrier.[33,35,59] However, the possibility that adenohypophyseal hormones are normal constituents of CSF, though present at the limits of sensitivity of available assays, is not excluded. Evidence in favor of this view is provided by studies from Linfoot *et al.*,[40] who reported that growth hormone (GH) levels in the CSF of acromegalic patients without SSE correlated with the weight of the plasma concentration of GH. Similar findings have also been reported for prolactin (PRL).[5]

In this chapter, we will review previously published evidence concerning the role of anterior pituitary hormones in the CSF and discuss the physiological and clinical significance of their presence in this biological fluid. In addition, we will present a summary of our own studies in this area.

Kalmon D. Post, M.D. • Department of Neurosurgery, Tufts–New England Medical Center Hospital, Boston, Massachusetts 02111. **Bruce J. Biller, M.D.** • Department of Medicine, Massachusetts Institute of Technology, Cambridge, Massachusetts 02139; Division of Endocrinology, Department of Medicine, Tufts–New England Medical Center Hospital, Boston, Massachusetts 02111. **Ivor M. D. Jackson, M.D.** • Division of Endocrinology, Department of Medicine, Tufts–New England Medical Center Hospital, Boston, Massachusetts 02111.

2. Factors That Regulate Presence of Anterior Pituitary Hormones in CSF

2.1. Molecular Weight

Although the process underlying the formation of CSF is complex,[16] filtration of plasma at the choroid plexus appears to be the most important factor. The CSF content of serum proteins in man is a reflection of the hydrodynamic volume of the protein molecule. For proteins with an overall globular conformation, the CSF concentration correlates with the molecular weight.[5] In keeping with this view, serum/CSF concentration ratios of 237 for albumin (mol. wt. 69,000) and up to 6332 for β-lipoprotein (mol. wt. $> 2 \times 10^6$) have been reported.[20] Since the molecular weight of the adenohypophyseal hormones is less than that of albumin, they should be present in CSF, though detection may be a function of assay sensitivity. In the original studies reported by Assies *et al.*,[3] these workers were unable to detect PRL in the CSF of normoprolactinemic individuals without pituitary disease. In all cases, the hormone concentration was below the detection limit of the assay used at that time (< 2.5 ng/ml). However, with a 5-fold increase in assay sensitivity, these workers[5] and others[58] have shown that the PRL level of CSF is related to the plasma level in both normoprolactinemic and hyperprolactinemic patients with or without pituitary tumors. GH, which has a molecular weight similar to that of PRL, also appears in the CSF in concentrations related to its level in the peripheral circulation.[57] Indeed, in patients with nonendocrine disease, the PRL plasma/CSF ratio was 6,[5] and was identical to that reported by Schaub *et al.*[57] for GH in 43 normal subjects. Although rat PRL injected intravenously into rabbits entered the CSF,[42] infusions of ACTH[2] and [^{125}I]-ACTH[33] in man have been reported not to cross the BBB.

Evidence favoring the integrity of the pituitary–CSF barrier for adenohypophyseal hormones, at least for a pituitary gland not affected by tumor, is provided by studies reported in 12 cases of empty sella syndrome.[31,32] All levels of ACTH, thyroid-stimulating hormone (TSH), luteinizing hormone (LH), follicle-stimulating hormone (FSH), PRL, and GH were at the limits of detectability, and were low despite elevated plasma levels induced by stress, menopause, TSH-releasing hormone (TRH), or LH-releasing factor (LRF).

2.2. Retrograde Transport from Pituitary to Brain

Assies *et al.*[4] studied simultaneous plasma and CSF levels of PRL and human chorionic somatomammotropin (HCS) in six pregnant women without pituitary disease. Plasma and CSF levels were closely correlated for each. Although it was felt that the CSF concentration of a protein hormone depended on the plasma concentration and on its molecular size, the plasma/CSF concentration ratio for HCS (24.6) was significantly different from the PRL ratio (7.2). Since the CSF PRL concentration[6] is higher than would be expected from the concentration in peripheral blood, an additional, but not necessarily alternate, process must be considered. Were pituitary PRL to reach CSF by filtration from blood having a much higher PRL content than peripheral blood, the relative increase in CSF prolactin concentration compared with HCS, which is derived solely from the placenta, could be explained.

It has generally been accepted that anterior pituitary secretions drain directly into the cavernous sinuses from lateral hypophyseal veins, and thence into the systemic circulation.[68] Support for this view was provided by the studies of Ganong and Hume,[22] who found high levels of ACTH within the cavernous sinuses of dogs. However, there have been few reports of anatomical studies demonstrating significant veins draining from the anterior pituitary.[7,8]

Many years ago, Szentagothai *et al.*[64] and Torok[66] questioned this concept, and suggested the possibility of retrograde venous transport from pituitary to brain. More recently, elegant anatomical studies by Bergland and colleagues[7,8,51,52] have reawakened interest in the concept of retrograde vascular transport from the pituitary. Utilizing vascular casts subjected to scanning electron microscopy, Bergland and Page[8] failed to demonstrate lateral hypophyseal veins to the cavernous sinus in the rhesus monkey, but did see Y-shaped confluent pituitary veins joining the pars distalis and the infundibular process to the cavernous sinuses through a common trunk. However, few of these veins were encountered. An arterial supply was found only to the neurohypophysis, and none to the anterior lobe. Numerous anastomatic connections existed between these arteries and a continuous neurohypophyseal capillary bed that connected the infundibulum, infundibular stem, and infundibular process (neurohypophysis). The entire afferent blood supply to the anterior gland was via portal vessels. Many short portal vessels were interposed between the adenohypophysis and the infundibular stem and process, with arrangements that implied alternate efferent routes from the adenohypophysis. This anatomy suggested some circular flow within the gland, i.e., from pars distalis to infundibular process, thence to the infundibulum, and back again to the pars distalis. Efferent routes for the pituitary blood would then be: (1) from adenohypophysis via short portal vessels to neurohypophysis; (2) from adenohypophysis and neurohypophysis to systemic circulation via lateral confluent veins joining the cavernous sinus; (3) from adenohypophysis and infundibular process to hypothalamus via capillaries and portal vessels; (4) to the cerebral arterial system via flow reversal in hypophyseal arteries; (5) to hypothalamus via retrograde axonal transport[12]; and (6) to CSF in the third ventricle and subarachnoid space by tanycyte transport or fenestrations in the portal vessels. Thus, from the neurohypophysis, blood may drain to the anterior gland, the systemic circulation, or toward the brain, with the most significant vessels suggesting flow toward the brain[7,8] (see Fig. 1). This latter concept was also suggested by the report of Torok[66] of flow within the neurohypophysis going toward the infundibulum and thence to the hypothalamus. Moreover, Oliver *et al.*[49] demonstrated high concentrations of hormones from the pars dis-

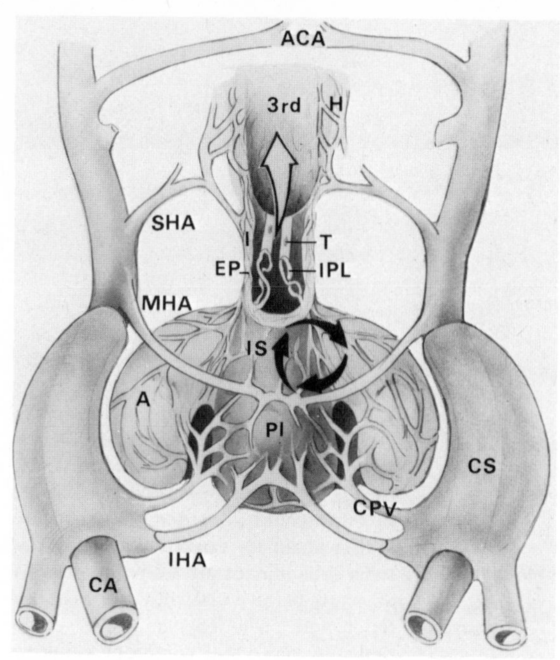

Figure 1. Abbreviations: (A) adenohypophysis; (ACA) anterior communicating artery; (CA) carotid artery; (CPV) confluent pituitary vein; (CS) cavernous sinus; (EP) external plexus; (H) hypothalamus; (I) infundibulum; (IHA) inferior hypophyseal artery; (IPL) internal plexus; (IS) infundibular stem; (MHA) middle hypophyseal artery; (PI) pars intermedia; (SHA) superior hypophyseal artery; (T) tanycyte; (3rd) third ventricle. Figure demonstrates findings from vascular cast studies visualized posteriorly:

1. Three pairs of arteries serve neurohypophysis: SHA, MHA, IHA. They form extensive anastomatic links, but do not pass through adenohypophysis.
2. Confluent pituitary veins to cavernous sinus are scant.
3. Numerous short vessels connect the adenohypophysis to infundibular process, forming common capillary bed.
4. Within infundibulum is thin external plexus and coiled complex internal plexus.
5. All arterial supply is to external plexus.
6. Tanycytes are stretched between internal plexus and third ventricle.
7. Phenomenon of circular blood flow (solid arrows) within pituitary could provide tanycytes high concentrations of adenohypophyseal hormones.
8. Tanycyte transport may be toward third ventricle (outline arrow), delivering hormones to brain.

Reproduced with the kind permission of Dr. Richard M. Bergland.

talis and pars intermedia in the neurohypophysis, lending support to the circular-flow concept.

In a study in sheep,[7] increased concentrations of PRL and GH were found in blood sampled from the internal carotid artery above the level of the pituitary gland and also from the sagittal sinus. It was concluded that the neurohypophyseal capillary bed not only received trophic hormones produced in the adenohypophysis, but also, under certain physiological circumstances, delivered those same hormones directly to the brain. Studies by Nakai and Naito[47] in the frog after systemic injection of horseradish peroxidase have shown that the endothelium of the capillaries of the hypophyseal portal vessels is permeable to this substance. It was taken up by pinocytosis of the ependymal tanycytes in the median eminence (ME) and carried by an ascending transport mechanism to be secreted into the third ventricle. The concepts of Bergland and Page[5] applied to the anterior pituitary hormones in mammalian species might similarly allow the ependymal tanycytes to take up pituitary peptides that have reached the ME via retrograde vascular transport for transport to the third ventricle CSF. Alternately, or in addition, the absence of the BBB at the ME[38,55] could allow pituitary hormones to reach the ME from the systemic circulation and subsequently be transported by the ependymal tanycytes to the third ventricle (see Fig. 2A).

2.3. Pituitary Tumors and CSF Secretion

Normal anterior pituitary cells extrude hormone-containing granules primarily at the interface between the cell membrane and the capillaries of the pituitary capillary plexus.[19] But a neoplastic pituitary cell can have multiple sites of granule exocytosis remote from the pericapillary space, termed "misplaced exocytosis" by Horvath and Kovacs.[28] Such misplaced exocytosis could occur because a tumor outgrows its vascular supply, resulting in fewer available pericapillary exocytosis sites, or as a consequence of "dedifferentiation" of the neoplastic cells, resulting in an abnormal secretory process.

Pituitary tumors have no capsules as such, but rather are contained by the aponeurotic sheath of the sella.[25] This is bounded superiorly by reflections of the dura that form the diaphragma sellae, with the arachnoid surrounding the aperture of the diaphragm through which the stalk passes.[36] Accordingly, when a tumor extends beyond the diaphragma, only arachnoid separates it from CSF, and via misplaced exocytosis, it might release hormone directly into the subarachnoid space. With the distortion of the normal anatomy, either the tumor or the normal gland would be adjacent to the CSF space, and therefore either the tumor hormone or normal gland hormones might be seen in the CSF in elevated concentrations (see Figure 2B). Such a hypothesis has been proposed by Jordan and co-

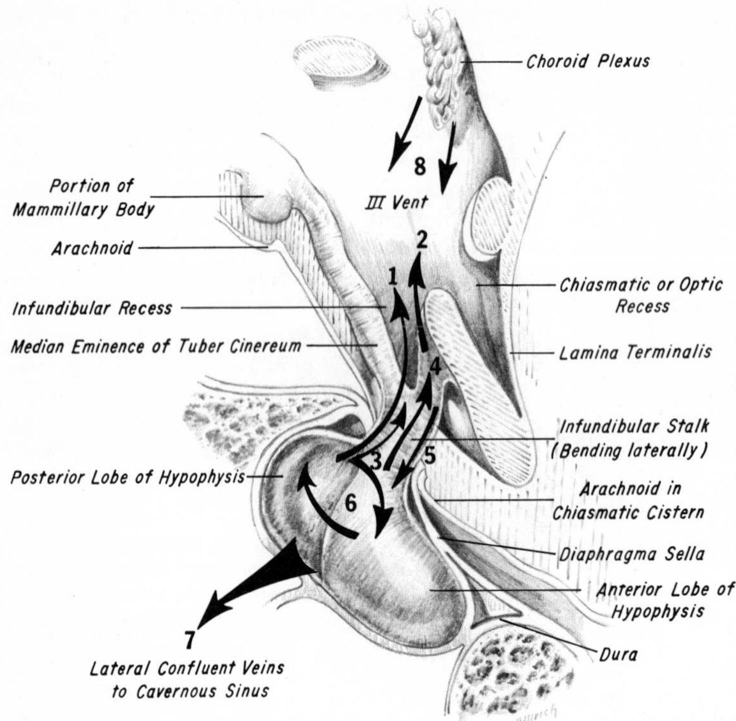

Figure 2A. Representation of potential pathways for anterior pituitary hormones as discussed in text: (1) flow in long portal vessels toward third ventricle; (2) flow from tuber cinereum toward third ventricle; (3) flow in long portal vessels toward tuber cinereum; (4) retrograde axonal flow within stalk; (5) flow down stalk from tuber cinereum toward adenohypophysis; (6) short portal vessel flow between anterior and posterior gland; (7) venous drainage via lateral confluent veins toward the cavernous sinus; (8) filtration or secretion of pituitary hormones through choroid plexus.

workers[30,33] to explain their studies. These workers reported increased CSF levels of pituitary hormones only in pituitary tumor patients with SSE. In most instances, but not all, the hormone secreted by the tumor into the peripheral blood appeared in the CSF. Occasionally, more than one hormone was present in the CSF: the abnormally elevated hormone was secreted by the tumor along with another, e.g., GH with PRL or a gonadotropin, all of which could be derived from the tumor or adjacent normal tissue. Occasionally, presumably nonfunctional pituitary adenomas were associated with elevated CSF adenohypophyseal hormones supposedly derived from normal tissue, but secretion by the tumor only into the CSF and not into the systemic circulation might have occurred.[43] Surprisingly, the larger-molecular species of PRL ("big" PRL) has been identified in the CSF at concentrations far below that in the plasma, even in patients with tumors showing SSE.[5,30] Such findings are in conflict with the concept of misplaced exocytosis.

2.4. Other Factors That Might Affect Anterior Pituitary Hormones in CSF

Large size and lipid insolubility of a substance, factors that will tend to limit the uptake of adenohypophyseal hormones from the systemic circulation, also favor exclusion from transfer across the BBB to the CSF.[48] Although protein binding is not thought to be an important means by which the adenohypophyseal hormones are transported in the blood (cf. thyroxine, cortisol), any such binding by plasma proteins, either specific or nonspecific, would limit transfer across the BBB.[48] Regardless of the relative contribution of the choroid plexus, the transependymal route from brain parenchyma, or the pial blood vessels,[16] the polypeptide transport system may have a finite capacity. This was illustrated for PRL by Login and MacLeod[42] when they showed that there was a plateau of CSF concentration in rats with PRL-secreting pituitary tumors, despite the presence of up to 38,000 ng/ml in serum.

3. Functional Role of Anterior Pituitary Hormones in CSF

There is increased recognition of the potential importance of the CSF as a link in neuroendocrine control mechanisms.[38] Hormones in the CSF do possess functional activity, but their physiological significance is uncertain. Kendall *et al.*[34] found that when a systemically ineffective dose of cortisol was introduced into the ventricular CSF system, the typical stress-related elevation of serum ACTH was prevented or suppressed. Similarly, administration of thyroxine caused TSH suppression, perhaps suggesting feedback inhibition. Following injection into the lateral ventricle, TRH, ACTH, thyroxine, and cortisol are concentrated in the ME.[34] Ondo *et al.*[50] also showed that certain compounds in the CSF (such as labeled corticosterone, LH, PRL, and hemoglobin), after injection into the third ventricle, can pass through the ME and enter the hypophyseal blood, while only trace amounts are found in simultaneously obtained arterial blood.

This type of CSF hormonal circulation leads to several interesting hypotheses for short-loop feedback inhibitions. For instance, when PRL is secreted by the pituitary into the systemic circulation, it may then enter the CSF at the choroid plexus or it might be secreted directly into the CSF. PRL might then circulate in the CSF to a localized receptor area at the infundibular recess of the third ventricle, where it is then transported by tanycyte ependyma to PRL receptors on the dopamine-containing cells of the ME. Here the dopamine [PRL-inhibiting factor (PIF)] system is activated to complete the loop and regulate the anterior pituitary secretion of PRL.[44] It is possible that the other pituitary hormones may similarly regulate their own secretion through interaction at the hypothalamic level.[46]

There is also much evidence that pituitary hormones may have a role in brain function.[18] PRL[21] and ACTH[17] have been reported to influence behavior in mammals and submammalian vertebrates, and transport from the CSF may be an important route of delivery to specific brain areas.

4. Specific Hormones and CSF

4.1. Growth Hormone

Wright *et al.*[71] reported 84 cases of acromegaly and showed a significant correlation between pituitary

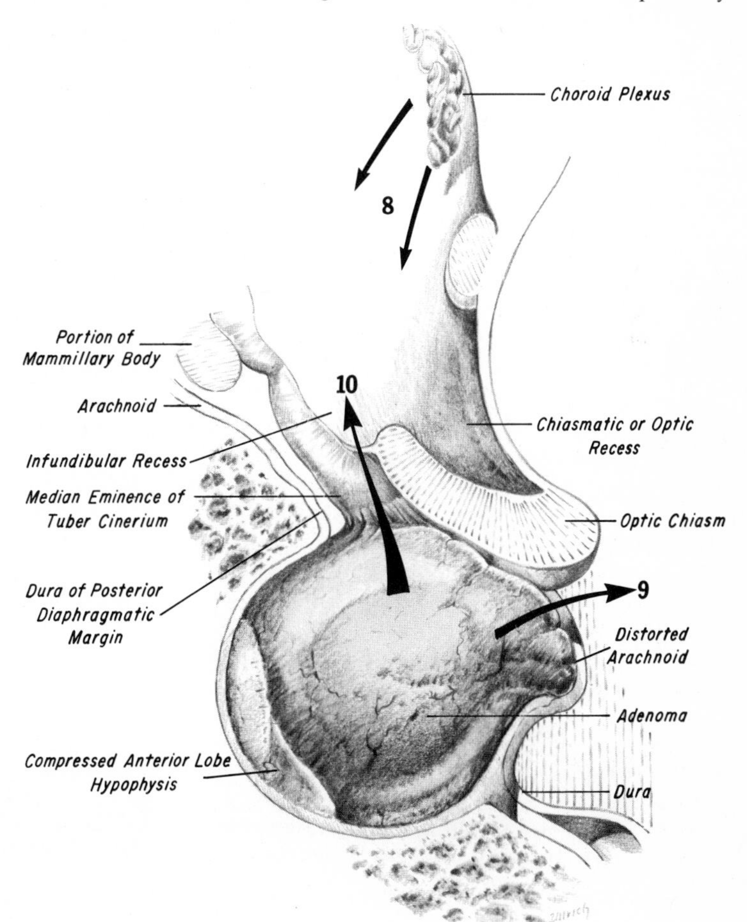

Figure 2B. Representation of potential pathways for anterior pituitary hormones as discussed in text: (9) direct secretion from tumor through distorted arachnoid, which is in direct contact with tumor; (10) direct secretion from tumor into third ventricle or stalk.

tumor size and serum GH levels. Linfoot *et al.*,[40] in 1970, however, published the first report of human GH levels in CSF studied by radioimmunoassay (RIA). They compared the results of 8 normals with those of 12 acromegalics, 3 of whom had SSE and 1 of whom also had a coexistent partially empty sella. The control GH values were extremely low (mean < 0.09 ± 0.01 ng/ml, with a range of < 0.05 to 0.12 ng/ml), while the acromegalic patients without SSE had higher levels (mean 0.49 ± 0.11 ng/ml). Simultaneous mean blood GH levels were 0.89 ± 0.30 ng/ml (range 0.16–1.14 ng/ml) for the normals and 14.1 ± 2.24 ng/ml (range 3.0–26.3 ng/ml) for the acromegalic patients. The acromegalic patients had a mean CSF GH level 5 times higher and a mean plasma GH level 16 times higher than the controls, but no significant correlation was appreciated between the CSF and plasma concentrations. However, the 3 patients with SSE of their tumors had levels of GH in their CSF 140–800 times greater than acromegalics whose tumors were confined to the sella. The plasma GH was 16 times higher in the former group, but again, no correlation was appreciated between the CSF and plasma levels. This may have reflected merely the frequent fluctuations seen in circulating HGH in both normals and acromegalics. After treatment of the tumor with heavy-particle therapy, there was a decrease in both the CSF and plasma GH, but again without clear correlation between them. In a patient with acromegaly and a partially empty sella, the CSF GH was not higher than in the other patients without SSE. Based on these data, Linfoot and colleagues suggested that a breakdown in the blood–CSF barrier accompanying SSE might have been a causative mechanism for increased CSF GH.[40]

In 1972, Thomas *et al.*[65] reported the results of RIA for GH in the CSF of acromegalics using an activated-charcoal technique. No GH was detected in the CSF of patients without acromegaly (CSF GH < 0.4 ng/ml). GH was measurable in patients with acromegaly (range 0.6–1.2 ng/ml) and in one patient with SSE, the CSF levels were considerably elevated (12.6 ng/ml—while plasma GH was 7.7 ng/ml). They suggested that in conjunction with pneumoencephalography (PEG), CSF GH might be a useful test to diagnose SSE. Thomas and co-workers'[65] series of 18 cases of acromegaly, in contrast to the report of Wright *et al.*,[71] failed to show any correlation between tumor size and serum GH concentration.

Allen *et al.*,[1] in 1974, reported on 26 patients, 3 of whom were acromegalic, who had simultaneous plasma and CSF determinations of ACTH and GH on samples obtained during PEG. There were 17 normal controls in whom a rise in both plasma ACTH and GH was seen during PEG, but without change in CSF levels. The 3 acromegalic patients had elevations in both CSF GH and plasma GH, but the patients with SSE had much higher basal CSF GH and CSF ACTH levels than those without SSE. In contrast, 5 patients with nonfunctioning chromophobe adenomas or craniopharyngiomas that extended into the basilar cisterns did not have significant elevation of GH or ACTH in the CSF.

In 1976, Hanson *et al.*[24] reported a case of acromegaly in which baseline plasma GH levels were minimally elevated (11.3 ng/ml) and suppressed partially during a glucose tolerance test (11.3 → 6.6 ng/ml). However, a markedly elevated CSF GH (23.7 ng/ml, with normal < 1) was appreciated. PEG and subsequent craniotomy clearly demonstrated significant suprasellar and parasellar extension. While in the previously cited reports, some question could be raised as to the correlation between plasma and CSF levels (because plasma GH was also elevated), this case was unique and suggested disruption of the normal pituitary–CSF barrier.

Jordan *et al.*,[33] in 1976, found that only when a pituitary tumor extended beyond the diaphragma sellae were elevated levels of pituitary hormones seen in the CSF. Moreover, successful treatment resulted in a fall in CSF hormone levels. A total of 83 patients were studied at PEG: 28 with neurological disease, but no evidence of endocrine disease; 49 suspected of harboring pituitary tumors (22 with SSE and 27 without SSE); and 6 with craniopharyngiomas. The control group had a CSF GH of less than 2 ng/ml, and the 27 tumor patients without SSE did not differ significantly from them. One patient without SSE had a coexistent partially empty sella and did not have significant elevation of CSF GH despite this distorted anatomy. All but 1 of the 22 patients with SSE had elevations of one or more CSF hormones. The hormone actually produced by the tumor was found in elevated CSF concentrations in all cases of SSE except 1 acromegalic who had increased FSH and LH only. Furthermore, no correlation was found between CSF hormone and plasma hormone levels in patients with SSE, suggesting that a breakdown of a blood–CSF barrier was unlikely. Rather, it was felt more likely that direct secretion by the tumor into CSF existed, i.e., a disturbed pituitary–CSF barrier. Last, since 21 of 22 patients with SSE had elevated adenohypophyseal hormone concentrations in the CSF, Jordan *et al.*[33] felt this had important clinical relevance for predicting tumor expansion.

Schaub *et al.*[57] also studied GH concentrations in blood and CSF in controls and acromegalic patients. In 43 patients without pituitary disease undergoing PEG, CSF GH was 0.35 ± 0.03 ng/ml, while plasma GH was 1.95 ± 0.20 ng/ml, giving a plasma/CSF ratio of 6. A group of 27 patients with diabetic retinopathy had similar values. Ten acromegalic patients without SSE (on lipiodol ventriculography) were studied as well. In 7 acromegalic patients, the CSF GH was consistently more elevated than in the normal subjects (mean 3.35 ± 1.4 ng/ml, with a plasma/CSF mean ratio of 11). In 1 patient, the CSF GH was extremely elevated (3850 and 250 ng/ml on two samples) and, indeed, was higher than the plasma levels taken simultaneously (78 and 27 ng/ml), i.e., a plasma/CSF ratio of less than one. In 2 acromegalics with SSE, repeated CSF measurements showed markedly increased CSF GH concentration, but lower than basal plasma GH levels (ratio > 2). Since elevated GH was seen in the CSF secondary to hypersecreting tumors both with and without SSE, these workers concluded that a breakdown of the BBB from SSE was not necessary. They suggested that an active-transport process existed for the secretion of pituitary hormones into the CSF. However, the nature of this mechanism was not defined.

4.2. ACTH

Kleerekoper *et al.*[37] were the first to report elevated CSF ACTH levels in patients with Nelson's syndrome. If GH (with a molecular weight of around 21,000) could cross the BBB, then ACTH (with a molecular weight of 4500) should be expected to cross more easily if molecular size were the sole or major determinant of pituitary hormone concentrations in the CSF. By RIA, they measured both plasma and CSF ACTH in six patients without endocrine disease (CSF ACTH 0–85 pg/ml). In three adrenalectomized patients with high plasma ACTH (113–4500 pg/ml), ACTH was also high in the CSF. The highest value was in a patient with SSE of an ACTH-producing tumor. Kleerekoper and associates suggested that ACTH-secreting pituitary tumors did not give rise to very high levels of the hormone in the CSF unless there was SSE of the tumor.[37]

Allen *et al.*[2] reported studies of CSF ACTH in 22 patients with nonendocrine disease. The mean immunoreactive CSF ACTH concentration was 98 pg/ml, which was slightly higher than the mean concentration in simultaneously obtained plasma (74 pg/ml). However, in a subsequent report[33] from the same laboratory, 27 subjects with nonendocrine disease had CSF ACTH levels that were low, a finding that raises questions regarding the specificity of the measurements reported.[2] In the latter study, there was, however, no correlation between the concentrations of the individual pairs of ACTH values obtained from plasma and CSF, suggesting that ACTH did not readily cross the blood–CSF barrier. Studies were also undertaken during a 48-hr ACTH infusion to see whether increasing plasma ACTH would cause an increase in CSF ACTH. Despite achievement of plasma ACTH levels of greater than 10,000 pg/ml, the CSF ACTH concentration remained low. In two normal subjects, a single bolus of ACTH labeled with ^{125}I failed to cause a significant increase in CSF ACTH at 60 min or 6 hr, suggesting that the blood–CSF barrier was relatively impermeable to ACTH. During infusion of labeled ACTH in cats, a steady-state plasma/CSF radioactivity ratio was reached after 4–6 hr, at which time the concentration of the labeled ACTH was 100 times higher in plasma than in CSF. These findings supported the view that the blood–CSF barrier is relatively impermeable to ACTH in both man and cat. The relatively high CSF ACTH concentration found in patients with nonendocrine disease suggested that some or most of the ACTH present entered by a mechanism that bypassed the blood–CSF barrier, such as a direct leak from the pituitary surface into the adjacent basilar cistern or by retrograde transport through the stalk and basal hypothalamus to the third ventricle. Supporting the concept of retrograde transport through either or both the stalk and the basal hypothalamus to CSF was the finding that in an anencephalic infant who lacked both hypothalamus and pituitary stalk, ACTH was undetectable in CSF, while a concentration of 60 pg/ml was seen in plasma. It should be noted that significant levels of immunoreactive ACTH have been detected in the hypothalamus and other brain locations even in hypophysectomized rats,[39] so that CSF ACTH could be derived from extrapituitary brain sites wherein ACTH may be synthesized *in situ*. Another possibility was that the immunoassay being employed might be measuring fragments of ACTH that possibly entered the CSF more readily than the whole molecule. Supporting the direct-leak theory was one patient with SSE of an ACTH-secreting tumor who showed a CSF ACTH of 3220 pg/ml and a plasma ACTH of 1070 pg/ml. The comparatively higher CSF levels of ACTH as opposed to GH might be due to cell position within the gland.

In a further report from the same group,[1] a series of 26 normal patients and 10 patients with pituitary–hypothalamic disease were described in whom ACTH and GH levels in both CSF and plasma were studied during PEG. Of 3 patients with acromegaly, 2 were seen to have minimal elevation of CSF ACTH, 113–116 pg/ml (normal < 100), while the plasma ACTH was normal. Two patients with Nelson's disease were studied. One patient with SSE had marked plasma ACTH elevation, 1069 pg/ml, but even higher CSF ACTH, 3217 pg/ml. The other patient, without SSE, had elevation of plasma ACTH to 261 pg/ml, while CSF ACTH was 139 pg/ml. Five other patients, 3 with chromophobe adenomas and 2 with craniopharyngiomas, all had normal plasma and CSF ACTH levels, even though all 3 patients with chromophobe adenomas had SSE. It was noted that although CSF and plasma ACTH concentrations were about equal, only the plasma ACTH rose substantially during the stress of PEG. However, there may not have been enough time for equilibration between plasma and CSF.

In the 1976 report of Jordan *et al.*,[33] three patients with Nelson's syndrome, all having SSE, had increased ACTH in the CSF. One of these patients also had elevated FSH, LH, and PRL in the CSF, but no correlation between CSF and plasma hormone levels was seen in any patient with SSE. One patient with Nelson's syndrome and SSE had a CSF ACTH level of 3100 pg/ml, and was studied further using [^{125}I]-ACTH. At 60 min after intravenous injection, there was no radioactivity in the CSF, suggesting to the authors that the blood–CSF barrier was intact even with SSE, and supporting direct secretion by the tumor into CSF. However, it could be argued that 60 min was not long enough for equilibration after intravenous injection.

Hoffman *et al.*,[27] in 1974, described a patient with Nelson's syndrome whose serum and CSF ACTH levels were elevated. They administered intravenous hydrocortisone and observed that only the serum ACTH was partially suppressed. The CSF ACTH level increased and remained elevated. While explanation was not clear, the suggestion was made that serum and CSF ACTH are not in equilibrium and that the ACTH might enter the CSF via direct transport from the hypothalamic–pituitary unit.

Thus far, studies have not been done to show a diurnal variation in CSF ACTH, as has been shown by Shambaugh *et al.*[61] for CSF cortisol (A.M. level: 0.68 ± 0.08 mg/100 ml mean; P.M.: 0.38 ± 0.02 mg/100 ml mean).

4.3. Prolactin

Clemens and Sawyer[15] clearly demonstrated in their studies with rat PRL in the CSF of rabbits that PRL can pass the blood–CSF barrier. They also concluded that in rats under normal physiological conditions, PRL was present in the CSF.

Jimerson *et al.*,[29] Sedvall *et al.*,[60] and Wode-Helgodt *et al.*[69] also confirmed PRL levels in the CSF of humans and found that psychoactive drugs given to psychiatric patients produced predictable changes in CSF PRL. Moreover, the changes were in the same direction as the plasma PRL response to these compounds. The plasma/CSF PRL concentration ratios varied from 5 to 10 in these patients. Since it had been shown for ACTH[2] and insulin[70] that CSF hormone levels respond slowly to an increase in the plasma level and do not mirror the moment-to-moment changes in hormone secretion, this slow equilibration between blood and CSF might account for the low levels of CSF PRL (< 1 ng/ml) found in several patients in whom a lumbar puncture was performed 90 min after administration (I.M.) of 25 mg chlorpromazine, even though plasma PRL increased 3- to 5-fold.[5] It is possible that sustained and not circadian hormone secretion in hyperprolactinemic tumor patients could be a contributory factor in the increased CSF PRL concentrations reported in such subjects.

In 1974, Assies *et al.*[3] reported that PRL was not detectable in the CSF of patients without pituitary disease who had normal plasma PRL. It was considered that either the hormone was absent or the concentration was below the detection limit of the assay at that time, which was less than 2.5 ng/ml. PRL was measurable, however, in the CSF of several patients with SSE of chromophobe adenomas, and the authors concluded that PRL gained access to the CSF by direct secretional leakage into the third ventricle or basilar cisterns, or both. Assies and coworkers speculated that CSF measurements might be a worthwhile adjunct to PEG in patients with SSE of tumors. Subsequent studies[5] by these workers using a PRL RIA with a sensitivity of 0.5 ng/ml enabled them to demonstrate that PRL was indeed present in the CSF of non-pituitary-tumor patients and that the level was a function of the plasma concentration. They suggested that the entry of the hormones into the CSF occurred via filtration at the choroid plexus with a concentration correlated to molecular size.[5] In their tumor patients, equal plasma and CSF levels of PRL were found in one sixth, while almost half of them had a plasma/CSF ratio of less than 3. In comparison, none of the nontumor patients had a CSF level equal to the plasma level, and almost 90% had ratios greater than 3. In their explanation for these abnormally low ratios, the authors considered the possibility that in the tumor patients with SSE, there was direct secretion or leakage from the tumor into the CSF because of misplaced exocytosis, since there was no true tumor capsule,[33] as suggested by Jordan *et al.*[33] However, since an adenoma is partially bounded by the aponeurotic sheath of the sella and dura, even when it expands above the sella,[25,26] this explanation may be incorrect. Additionally, if there was direct leakage from the pituitary, both "big" and "little" PRL would be expected in the CSF, similar to that found in the plasma, pituitary extracts, and pituitary culture media. However, neither Assies *et al.*[5] nor Jordan and Kendall[30] were able to detect substantial quantities of "big" PRL in the CSF, even in patients with SSE, whereas "big" PRL constituted 11–25% of the total plasma concentration. As already discussed, pituitary hormones could reach the CSF by a more direct route, such as retrograde transport via the vessels in the pituitary stalk and ME.[52,56,64]

Assies *et al.*[5] proposed that PRL and presumably other large protein hormones are filtered from the blood into the CSF, the filtration rate being directly proportional to the plasma concentration. They suggested that the filtration process is located in the choroid plexus, but might occur elsewhere in the vicinity of the pituitary gland, the pituitary stalk, or the ME. They concluded that in most patients with PRL-producing tumors with or without SSE, the blood–CSF barrier seemed to be intact and that detectable or high CSF hormone levels *per se* should not be construed as evidence of SSE of the tumor.

Schroeder *et al.*[58] measured PRL in the serum and CSF of control subjects, pregnant women, and patients with pituitary disease. In the 30 control subjects, the mean serum PRL concentration was 7.0 ng/ml and mean CSF PRL concentration 1.2 ng/ml. A statistically significant relationship was demonstrated between the serum and CSF PRL levels of 12 hyperprolactinemic patients with pituitary tumors. Additionally, 3 pregnant women were found to have elevated CSF PRL levels, whereas in 15 patients with primary empty sella syndrome and a defective diaphragma sellae, low CSF PRL concentrations were found. This latter finding argued against the postulate that disruption of the diaphragma sellae by a pituitary tumor significantly influences pituitary hormone concentrations in CSF. Schroeder and colleagues concluded that the CSF PRL concentration was influenced by the serum PRL level, but admitted that mechanisms other than passive diffusions (i.e., direct secretion of PRL from tumor into CSF) may have determined the CSF PRL levels. Of their tumor patients with SSE, 2 had a higher PRL level in the CSF than in the serum, and in 3 normoprolactinemic tumor patients, 2 had SSE with elevated CSF PRL levels. The posibility that decreased clearance of PRL from the CSF of these patients may have contributed to the elevated CSF levels was also suggested.

Jordan *et al.*[33] reported 13 patients with elevated CSF PRL levels, 12 of whom had chromophobe adenomas and 1 of whom had an embryonal cell carcinoma. There was not felt to be a correlation between CSF and plasma hormone levels in patients with SSE. All had normal CSF proteins. Patients with markedly elevated plasma PRL levels but no SSE had low CSF PRL levels. The conclusion of these authors was that the finding of an elevated adenohypophyseal hormone concentration in the CSF was a sensitive indicator of SSE of a pituitary tumor.

Login and MacLeod[42] reported that rats treated with haloperidol or with implants of PRL-secreting tumors had elevated CSF PRL levels compared to control rats. These levels were commensurate with the serum level of PRL, although there appeared to be an upper limit to the CSF PRL concentration (200 ng/ml). These authors also reported that patients with PRL-secreting pituitary adenomas had elevated CSF hormone levels compared to patients with nonendocrine neurological disease. This CSF elevation, associated with elevated serum PRL, occurred regardless of whether the tumor was intra- or extrasellar, since four of the five patients with PRL-secreting adenomas had totally intrasellar tumors. The authors concluded that an elevation in CSF PRL should not be used to diagnose SSE, and that the CSF PRL reflected the serum PRL regardless of the cause for the increase. They suggested that entry into the CSF occurred at the choroid plexus via a peptide-hormone-carrier mechanism with a finite capacity. They raised the possibility that a retarded CSF clearance because of binding, CSF bulk flow, or a less efficient enzyme system might contribute to elevated levels of CSF hormones. The difference in observations between Login and MacLeod[42] and Jordan *et al.*[33] was emphasized again by Login[41] in a letter to the editor of the *Annals of Internal Medicine*.

In studies from our own group, Biller *et al.*[9] have reported on 21 patients with PRL-secreting adenomas (19 female and 2 male), 5 of whom had SSE. These patients were compared with 10 endocrinologically and neurologically normal patients. All patients with hyperprolactinemia had elevated CSF PRL compared to normals (mean 53.7 ng/ml, *r* 2.6–315.1; normal: < 2.5 ng/ml). CSF PRL was markedly elevated in 4 of the 5 patients with SSE (mean 211.1 ng/ml, *r* 124.3–315.1), but in the 16 patients without SSE, mean CSF PRL was 15.3 ng/ml (*r* 2.6–40.5).

CSF levels of TSH were elevated compared to normal in 10 of 18 patients with prolactinomas (mean 4.9 μU/ml), but there was no significant difference between tumor patients with and without SSE. Likewise, CSF GH levels were elevated compared to normal in 7 of 18 patients (mean 1.2 ng/ml, *r* 0.4–3.7), but there was no level that distinguished patients with SSE.

Of 18 patients, 14 had elevations of CSF LH (mean 3.3 mIU/ml, *r* 1.8–6.0) and 8 had elevations of CSF FSH (mean 3.8 mIU/ml, *r* 3.3–4.3) compared to normals, but again there was no difference between those with and without SSE. Overall, the frequency of CSF pituitary hormone elevations in this group of 21 patients with PRL-secreting tumors was: PRL, 100%; LH, 77%; FSH, 44%; TSH, 55%; GH, 38%. Marked elevation of CSF PRL (> 100 ng/ml) was always associated with SSE, although SSE could occur without such marked elevation. Biller and colleagues therefore proposed that CSF PRL levels could be used as an adjunct to other studies to diagnose SSE in patients with prolactinomas.

Supporting Biller and colleagues was the study of Matsumura *et al.*,[45] who found a similar relationship between the CSF/plasma ratio of PRL levels and SSE in prolactin-secreting tumors.

4.4. Thyroid-Stimulating Hormone

Since subcutaneous injection of human CSF produced a hyperplasia of the thyroid epithelium of young rabbits, Caulaert *et al.*,[14] in 1931, postulated that TSH was present in the CSF. Borell,[11] in 1945, studying a variety of mammals, concluded that TSH from the pars distalis reaches the CSF, from where it is absorbed by the choroid plexus. In 1974, Seaich *et al.*[59] reported on a patient with a chromophobe adenoma with SSE who had elevation of CSF TSH by RIA.

Schaub *et al.*,[57] in 1977, reported on TSH levels in CSF, finding a normal mean of 2.65 ± 0.2 ng/ml while plasma levels were 5.95 ± 0.3 μU/ml. There appeared to be a good correlation between CSF and plasma levels for TSH as well as for GH, but the regression curves for both were distinctly different and appeared specific for each polypeptide hormone.

4.5. Luteinizing Hormone and Follicle-Stimulating Hormone

The first report of the presence of gonadotropins in the CSF was provided in 1932 by Pighini,[53] who described a substance in the third ventricle CSF with gonadal-stimulating properties. Subsequently, in a study of 22 patients harboring pituitary tumors with SSE, 16 had elevation of CSF LH or CSF FSH or both.[33] When the levels of these CSF gonadotropins were elevated, both hormones were usually seen. This might be expected, since production is probably from a common cell.

Luboshitzky and Barzilai[43] reported on a male patient with a pituitary adenoma, partial hypopituitarism, and normal basal serum FSH and LH, but blunted FSH rise following the administration of LH-releasing hormone. Subsequent assays for CSF LH and FSH showed elevated levels, correctly predicting SSE.

4.6. β-Lipotropin and "β-Melanocyte-Stimulating Hormone"

Until recently, human pituitary β-melanocyte-stimulating hormone (β-MSH) was thought to be a peptide comprising 22 amino acids. However, it now seems likely that human "β-MSH" is an extraction artifact, being a fragment of human β-lipotropin (β-LPH), a 91-amino-acid structure, present not only in the anterior pituitary, but also in other parts of the brain including the hypothalamus, wherein it may be synthesized.[67] It appears likely that assays purported to measure β-MSH in man are measuring either the whole molecule or a fragment of β-LPH (see Rees[54] for a review). More recently, evidence has been gathered to suggest that ACTH and β-LPH may be derived from a common precursor molecule and be secreted from the same cell in the anterior pituitary.[23]

With this in mind, Smith and Schuster[63] have reported immunoreactive β-MSH in human CSF. The mean CSF β-MSH was 60.1 ± 8.9 ng/ml (*r* 0–188) in adult subjects who had CSF removed during routine lumbar puncture, 24 of whom had varying neurological diseases and 6 of whom had no specific diagnosis. Although simultaneous blood determinations were not performed in these subjects, plasma β-MSH was found to be 16.1 ± 1.1 ng/ml (*r* 0–33) in a group of normals. There was no obvious association between CSF protein and CSF β-MSH levels, and the significance and source of these levels are unknown.

In a later report,[62] the same group studied plasma β-MSH levels in 19 patients with hypopituitarism, and 5 of these patients had CSF measurements as well. In neither plasma nor CSF were the β-MSH concentrations significantly different from those in normals, whereas ACTH levels were apparently low. (CSF β-MSH was 78.5 ± 15.8 ng/ml and plasma levels were 23.6 ± 5.6 ng/ml.) This suggests a dissociation between the secretion of β-MSH (or more probably its precursor β-LPH) and the secretion of ACTH. Moreover, the possibility that the β-MSH in CSF may be derived from extrapituitary sites is suggested by reports of considerable amounts of immunoreactive β-MSH in various regions of the brain, though total brain β-MSH is only about 5% of that of the pituitary.[13] Clearly, the nature of the immunoreactive β-MSH needs to be reexamined with specific measurements of β-LPH.

Data available at present on CSF β-LPH and endorphins are discussed by Bloom and Segal[10] in Chapter 45.

5. Extrapituitary Sources of Pituitary Hormones

Evidence now exists demonstrating that anterior pituitary hormones may be synthesized in extrapituitary locations within the brain. Significant amounts of GH and ACTH have been demonstrated in the rat amygdaloid nucleus, and immunoperoxidase staining has demonstrated dense accumulations of the peroxidase antibody to GH complex in amygdala cells.[72] These workers reported that the immunoreactive GH content within the amygdaloid nucleus showed a dramatic increase 30 days following hypophysectomy. ACTH has been demonstrated to be widely distributed throughout the limbic system of normal and hypophysectomized animals.[73] TSH has been demonstrated throughout the brain of rats and in the hypothalamic region of the human brain.[74] Finally, δ-MSH ($ACTH^{1-13}$) has been demonstrated in appreciable quantities in various regions in the rat brain.[75] PRL has also been demonstrated in brain neurons.[76] With the use of antibodies against rat PRL, fluorescent nerve terminals in many hypothalamic nuclei were seen. These fibers did not disappear after hypophysectomy. These data seem to imply that anterior pituitary hormones can be produced *in situ* in extrapituitary regions of the brain. Whether the hormones produced in these sites contribute to the CSF levels is thus far unknown.

6. Conclusions

The significance and role of anterior pituitary hormones in CSF have been reviewed. Although some reports suggest that anterior pituitary hormones are not present in the normal CSF, the evidence from a number of sources indicates that with increased radioimmunoassay sensitivity, these peptides can be detected in normal subjects. Since the anterior pituitary hormones are polar and poorly lipophilic, these substances do not readily cross the blood–brain barrier. It does appear, however, that their presence and concentration in the CSF are a reflection primarily of their molecular weight as well as their concentration in the systemic circulation in patients with a normal hypothalamic–pituitary axis as well as in some patients with pituitary tumors. Failure to see changes in the CSF after infusion of ACTH or after acute elevation of plasma PRL may reflect the slow time of equilibration from blood to CSF, with subsequent retarded clearance. The exclusion of the nontumorous pituitary gland from the brain is also shown by the low levels of CSF pituitary hormones in the "empty sella syndrome."

In some reports, especially in patients with SSE of pituitary tumors, the CSF/plasma ratio is higher than might be expected solely by filtration through the choroid plexus. Retrograde vascular transport from pituitary to brain has been discussed as a possible mechanism that might account for the higher CSF levels in such patients. Some reports suggest that the presence of anterior pituitary hormones in the CSF indicates SSE of a pituitary tumor with breakdown of the normal pituitary–CSF barrier due to the process of misplaced exocytosis. An argument against this view is the failure to find the higher-molecular-weight species of PRL ("big" PRL) in CSF in concentrations approaching that found in the peripheral blood.

Nonetheless, in our own studies of CSF pituitary peptide levels in patients with prolactinomas, we found that a CSF PRL level of 100 ng/ml always indicates SSE of a pituitary tumor. Such findings suggest that CSF measurements of pituitary hormones may be helpful diagnostically in certain clinical situations.

Finally, increased recognition of the CSF as a means of regulating neuroendocrine function suggests that the CSF may serve as a conduit in the transfer of anterior pituitary hormones to the hypothalamus for the purpose of regulating secretions through "short-loop feedback" mechanisms. It is also possible that secretion into the CSF may allow anterior pituitary hormones to reach distant parts of the nervous system and thereby regulate aspects of brain function, including behavior.

References

1. Allen, J. P., Kendall, J. W., McGilvra, R., Lamorena, T. L., Castro, A.: Adrenocorticotrophic and growth hormone secretion: Studies during pneumoencephalography. *Arch. Neurol.* **31**:325–328, 1974.
2. Allen, J. P., Kendall, J. W., McGilvra, R., Vancura, C.: Immunoreactive ACTH in cerebrospinal fluid. *J. Clin. Endocrinol. Metab.* **38**:586, 1974.
3. Assies, J., Schellekens, A. P. M., Touber, J. L.: Prolactin secretion in patients with chromophobe adenoma of the pituitary gland. *Neth. J. Med.* **17**:163, 1974.
4. Assies, J., Schellekens, A. P. M., Touber, J. L.: Protein hormones in cerebrospinal fluid: Evidence for retrograde transport of prolactin from the pituitary to the brain in man. *Clin. Endocrinol.* **8**:487–491, 1978.
5. Assies, J., Schellekens, A. P. M., Touber, J. L.: Prolactin in human cerebrospinal fluid. *J. Clin. Endocrinol. Metab.* **46**(4):576, 1978.
6. Bagshawe, K. D., Harland, S.: Immunodiagnosis and monitoring of gonadotrophin-producing metastases in the central nervous system. *Cancer* **38**:112–118, 1976.
7. Bergland, R. M., Davis, S. L., Page, R. B.: Pituitary secretes to brain. *Lancet* **2**:276–278, 1977.
8. Bergland, R. M., Page, R. B.: Can the pituitary secrete directly to the brain? (affirmative anatomical evidence). *Endocrinology* **102**(5):1325–1338, 1978.
9. Biller, B., Post, K. D., Molitch, M., Reichlin, S., Jackson, I.: CSF pituitary hormone concentrations in patients with pituitary tumors. Abstract presented at the American Association of Neurological Surgeons Meeting, New Orleans, Louisiana, April 25, 1977 (Paper No. 28).
10. Bloom, F. E., Segal, D. S.: Endorphins in cerebrospinal fluid. In Wood, J. H. (ed.): *Neurobiology of Cerebrospinal Fluid I.* New York, Plenum Press, 1980.
11. Borell, U.: On the transport route of the thyrotropic hormone, the occurrence of the latter in different parts of the brain and its effect on the thyroidea. *Acta Med. Scand.* **161**(Supp.):1–227, 1945.
12. Broadwell, R. D., Brightman, M. W.: Entry of peroxidase into neurons of the central and peripheral nervous system from extracerebral and cerebral blood. *J. Comp. Neurol.* **166**:257, 1976.
13. Carter, R. J., *et al.*: Unpublished observations.
14. Caulaert, V. C., Aron, M., Stahl, J.: Sur la presence de l'hormone prehypophysaire excito-secretoire e la thyroide dans le sang et le liquide cephalorachidien et sur sa repartition dans ces milieux et dans l'urine. *C. R. Seances Soc. Biol. Fil.* **106**:607–609, 1931.
15. Clemens, J. A., Sawyer, B. D.: Identification of prolactin in cerebrospinal fluid. *Exp. Brain Res.* **21**:399–402, 1974.
16. Davson, H. (ed.): *Physiology of the CSF.* Boston, Little, Brown, Company, 1967.
17. de Wied, D.: Pituitary adrenal system hormones and behavior. Symposium on Developments in Endocrinology. In honour of Dr. G. A. Overbeek, Organon International, Oss, The Netherlands, October 1976.
18. de Wied, D.: Peptides and behavior. *Life Sci.* **20**:195–204, 1977.
19. Farquhar, M. G.: Origin and fate of secretory granules in cells of the anterior pituitary gland. *Trans. N. Y. Acad. Sci.* **1960**:347–351.
20. Felgenhauer, K.: Protein size and cerebrospinal fluid composition. *Klin. Wochenschr.* **52**:1158–1164, 1974.
21. Frantz, A. G.: Prolactin. *N. Engl. J. Med.* **298**(4):201–207, 1978.
22. Ganong, W. F., Hume, D. M.: Concentration of ACTH in cavernous sinus and peripheral arterial blood in the dog. *Proc. Soc. Exp. Biol. Med.* **92**:721, 1956.
23. Guillemin, R.: Beta-lipotropin and endorphins: Implications of current knowledge. *Hosp. Prac.* **13**:53–60, 1978.
24. Hanson, E. J., Jr., Miller, R. H., Randall, R. V.: Suprasellar extension of tumor associated with increased cerebrospinal fluid activity of growth hormone. *Mayo Clin. Proc.* **51**:412–416, 1976.
25. Hardy, J.: Transsphenoidal microsurgery of the normal and pathological pituitary. *Clin. Neurosurg.* **16**:185–217, 1969.
26. Hardy, J.: Trans-sphenoidal microsurgical removal of pituitary microadenoma. *Prog. Neurol. Surg.* **6**:200, 1975.
27. Hoffman, J. D., Baumgartner, J., Gold, E. M.: Dissociation of plasma and spinal fluid ACTH in Nelson's syndrome. *J. Am. Med. Assoc.* **228**:491, 1974.
28. Horvath, E., Kovacs, K.: Misplaced exocytosis. *Arch. Pathol.* **97**:221–224, 1974.
29. Jimerson, D. C., Post, R. M., Skyler, J., Bunney, W. E.: Prolactin in cerebrospinal fluid and dopamine function in man. *J. Pharmacol.* **28**:845–847, 1976.
30. Jordan, R. M., Kendall, J. W.: Dissociation of plasma and CSF prolactin heterogeneity. *Clin. Res.* **24**:273, 1976.
31. Jordan, R. M., Kendall, J. W., Kerber, C. W.: The

primary empty sella syndrome. *Am. J. Med.* **62:**569–580, 1977.

32. JORDAN, R. M., KENDALL, J. W., SEAICH, J. L.: CSF hormone studies in pituitary diseases. *Clin. Res.* **24:**143A, 1976.
33. JORDAN, R. M., KENDALL, J. W., SEAICH, J. L., ALLEN, J. P., PAULSEN, C. A., KERBER, C. W., VANDERLAAN, W. P.: Cerebrospinal fluid hormone concentration in the evaluation of pituitary tumors. *Ann. Intern. Med.* **85:**49–55, 1976.
34. KENDALL, J. W., JACOBS, J. J., KRAMER, R. M.: Studies on the transport of hormones from the cerebrospinal fluid to the hypothalamus and pituitary. In Knigge, K. M., Scott, D. E., Weidl, A. (eds.): *Brain–Endocrine Interaction. Median Eminence: Structure and Function.* Basel, S. Karger, 1971, pp. 342–349.
35. KENDALL, J. W., SEAICH, J. L., ALLEN, J. P., VANDERLAAN, W. P.: Pituitary–CSF relationships in man. In Knigge, K. M., Scott, D. E., Kobayashi, J., Ishu, S. (eds.): *Brain–Endocrine Interaction II. The Ventricular System.* Basel, S. Karger, 1975, pp. 313–323.
36. KIRGIS, H. D., LOCKE, W.: Anatomy and embryology. In Locke, W., Schally, A. V. (eds.): *The Hypothalamus and Pituitary in Health and Disease.* Springfield, Illinois, Charles C. Thomas, 1972, pp.57–58.
37. KLEEREKOPER, M., DONALD, R. A., POSEN, S.: Corticotrophin in cerebrospinal fluid of patients with Nelson's syndrome. *Lancet* **1:**74–76, 1972.
38. KNIGGE, K. M., SCOTT, D. E.: Structure and function of the median eminence. *Am. J. Anat.* **129:**223, 1970.
39. KRIEGER, D. T., LIOTTA, A., BROWNSTEIN, M. J.: Presence of corticotropin in limbic system of normal and hypophysectomized rats. *Brain Res.* **128:**575–579, 1977.
40. LINFOOT, J. A., GARCIA, J. F., WEI, W., FINK, R., SARIN, R., BORN, J. L., LAWRENCE, J. H.: Human growth hormone levels in cerebrospinal fluid. *J. Clin. Endocrinol. Metab.* **31:**230–232, 1970.
41. LOGIN, I. S.: Spinal fluid prolactin (letter to the editor). *Ann. Intern. Med.* **86:**119, 1977.
42. LOGIN, I. S., MACLEOD, R. M.: Prolactin in human and rat serum and cerebrospinal fluid. *Brain Res.* **132:**477–483, 1977.
43. LUBOSHITZKY, R., BARZILAI, D.: Suprasellar extension of tumor associated with increased cerebrospinal fluid activity of LH and FSH. *Acta Endocrinol. (Copenhagen)* **87**(4)**:**673–684, 1978.
44. MACLEOD, R. M., LOGIN, I.: Control of prolactin secretion by the hypothalamic catecholamines. *Adv. Sex Horm. Res.* **2:**211–231, 1976.
45. MATSUMURA, S., MORI, S., YOSHIMOTO, H., OHTA, M., UOZUMI, T.: Endocrinological evaluation of sellar and suprasellar tumor cases (the ninth report)—on the PRL levels in the CSF (author's translation). *No. Shinkei Geka* **5**(10)**:**1057–1063, 1977.
46. MOTTA, M., FRASCHINI, F., MARTINI, L.: Short feedback mechanisms in the control of anterior pituitary function. In Ganong, W. F., Martini, L. (eds.): *Frontiers in Neuroendocrinology.* New York, Oxford, 1969, pp. 211–254.
47. NAKAI, U., NAITO, N.: Uptake and bidirectional transport of peroxidase injected into the blood and cerebrospinal fluid by ependymal cells of the median eminence. In Knigge, K. M., Scott, D. E., Kobayashi, H., Ishi, S. (eds.): *Brain–Endocrine Interaction II. The Ventricular System.* Basel, S. Karger, 1975, p. 94.
48. OLDENDORF, W. H.: Blood–brain barrier permeability to drugs. *Am. Rev. Pharmacol.* **14:**239–248, 1974.
49. OLIVER, D., MICAL, R. S., PORTER, J. C.: Hypothalamic–pituitary vasculature: Evidence of retrograde blood flow in the pituitary stalk. *Endocrinology* **101:**598, 1977.
50. ONDO, J. G., MICAL, R. S., PORTER, J. C.: Passage of radioactive substances from CSF to hypophysial portal blood. *Endocrinology* **91**(5)**:**1239–1246, 1971.
51. PAGE, R. B., BERGLAND, R. M.: The neurohypophyseal capillary bed: Anatomy and arterial supply. *Am. J. Anat.* **148:**345, 1977.
52. PAGE, R. B., MUNGER, B. L., BERGLAND, R. M.: Scanning microscopy of pituitary vascular casts: The rabbit pituitary portal system revisited. *Am. J. Anat.* **146:**273, 1976.
53. PIGHINI, G.: Sulla presenza dell'ormone anteipofisario nel "tuber cinereum" nel "liquor" ventricolare dell'uomo. *Rev. Sper. Greniatria Med. Leg. Alienazioni Mentali* **56:**575–622, 1932.
54. REES, L. H.: ACTH, lipotrophin, and MSH in health and disease. *Clin. Endocrinol. Metab.* **6:**137–153, 1977.
55. RODRIGUEZ, E. M.: Comparative and functional morphology of the median eminence. In Knigge, K. M., Scott, D. E., Weindl, A. (eds.): *Brain–Endocrine Interaction. Median Eminence: Structure and Function.* Basel, S. Karger, 1972, p. 319.
56. RODRIGUEZ, E. M.: The cerebrospinal fluid as a pathway in neuroendocrine integration. *J. Endocrinol.* **71:**407–443, 1976.
57. SCHAUB, C., BLUET-PAJOT, M. T., SZIKLA, G., LORNET, C., TALAIRACH, J.: Distribution of growth hormone and thyroid-stimulating hormone in cerebrospinal fluid and pathological compartments of the central nervous system. *J. Neurol. Sci.* **13:**123, 1977.
58. SCHROEDER, L. L., JOHNSON, J. D., MALARKEY, W. B.: Cerebrospinal fluid prolactin: A reflection of abnormal prolactin secretion in patients with pituitary tumors. *J. Clin. Endocrinol. Metab.* **43:**1255–1260, 1976.
59. SEAICH, J. L., ALLEN, J. P., KENDALL, J. W.: Diagnostic values of CSF hormone determinations. Abstract presented at the Annual Meeting of the Endocrine Society, June 1974 (Abstract No. 126).
60. SEDVALL, G., ALFREDDSON, G., JERKENSTEDT, L. B., ENEROTH, P., FYRO, G., HARNRYD, C., SWAHN, C. G., WEISEL, F. A., WODE-HELGODT, B.: Selective effects of psychoactive drugs on levels of monoamine metabolites and prolactin in cerebrospinal fluid of psychiatric patients. *Proc. Sixth Int. Congr. Pharmacol.* **3:**255, 1975.
61. SHAMBAUGH, G. E., III, WILBER, J. F., MONTOYA, E.,

REIDER, H., BLONSKY, E. R.: Thyrotrophin-releasing hormone (TRH): Measurements in human spinal fluid. *J. Clin. Endocrinol. Metab.* **41**(1):131–134, 1975.

62. SHUSTER, S., SMITH, A., PLUMMER, N., THODY, A., CLARK, F.: Immunoreactive beta-melanocyte-stimulating hormone in cerebrospinal fluid and plasma in hypopituitarism: Evidence for an extrapituitary origin. *Br. Med. J.* **1**:1318–1319, 1977.
63. SMITH, A. G., SHUSTER, S.: Immunoreactive beta-melanocyte-stimuating hormone in cerebrospinal fluid. *Lancet* **1**:1321, 1976.
64. SZENTAGOTHAI, J., GLERKO, B., MESS, B., HALASZ, B.: Hypothalamic control of anterior pituitary. *Budapest Akademiai Kiado* **1968**:90.
65. THOMAS, F. J., LLOYD, J. M., THOMAS, M. J.: Radioimmunoassay of human growth hormone: Technique and application to plasma, cerebrospinal fluid and pituitary extracts. *J. Clin. Pathol.* **25**:774–782, 1972.
66. TOROK, B.: Structure of the vascular connections of the hypothalamohypophyseal region. *Acta Anat.* **59**:84, 1964.
67. WATSON, S. J., BARCHAS, J. D., LI, C. H.: Beta-lipotropin: Localization of cells and axons in rat brain by immunocytochemistry. *Proc. Natl. Acad. Sci. U.S.A.* **74**(11):5155–5158, 1977.
68. WISLOCKI, G. B.: The vascular supply of the hypophysis cerebri of the rhesus monkey and man. *Proc. Assoc. Nerv. Ment. Dis.* **17**:48, 1938.
69. WODE-HELGODT, B., ENEROTH, P., FYRO, B., GULLBERG, B., SEDVALL, G.: Effect of chlorpromazine treatment on prolactin levels in cerebrospinal fluid and plasma of psychotic patients. *Acta Psychiatr. Scand.* **56**(4):280–293, 1977.
70. WOODS, S. C., PORTE, D., JR.: Insulin and the set-point regulation of bodyweight. In Novin, D., Wyrwicka, W., Bray, G. (eds.): *Hunger: Basic Mechanisms and Clinical Implication.* New York, Raven Press, 1976, p. 273.
71. WRIGHT, A. D., MCLACHLAN, M. S., DOYLE, F. H., FRASER, T.: Serum GH levels and size of pituitary tumor in untreated acromegaly. *Br. Med. J.* **4**:582–584, 1968.
72. DAVIS, W. J., GILLETTE, R.: Biologically active pituitary hormones in the rat brain amygdaloid nucleus. *Science* **199**:804–805, 1978.
73. KRIEGER, D. T., LIOTTA, A., BROWNSTEIN, M. J.: Presence of corticotropin in limbic system of normal and hypophysectomized rats. *Br. Res.* **128**:575–579, 1977.
74. MOLDOW, R. L., YALOW, R. S.: Extrahypophysial distribution of thyrotropin as a function of brain size. *Life Sci.* **22**:1859–1864, 1978.
75. OLIVER, C., PORTER, J. C.: Distribution and characterization of alpha-melanocyte-stimulating hormone in the rat brain. *Endocrinology* **102**(3):697–705, 1978.
76. FUXE, K., HÖKFELT, T., ENEROTH, P., GUSTAFSSON, J. A., SKETT, P.: Prolactin: Localization in nerve terminals of the rat hypothalamus. *Science* **196**:899–900, 1977.

CHAPTER 42

Cerebrospinal Fluid Steroid Hormones

Samuel P. Marynick, James H. Wood, and D. Lynn Loriaux

1. Introduction

Sex steroid secretion by the gonads is regulated by luteinizing hormone and follicle-stimulating hormone secreted from the pituitary gland. The secretion of these gonadotropins is, in turn, regulated by the circulating levels of the sex steroid hormones. Thus, the ability of these sex steroid hormones to enter the central nervous system from the plasma is likely to be important in the feedback regulation of gonadotropin secretion. This short chapter reviews this aspect of steroid physiology. The discussion centers on the sex steroids of primary importance in man: testosterone, dihydrotestosterone, estradiol, progesterone, and 17-hydroxyprogesterone. Cortisol, the major glucocorticoid in man, is also examined for comparison. The movement of the sex steroid hormones from plasma into the cerebrospinal fluid (CSF), the concentrations of the sex steroid hormones in CSF, and the movement of sex steroid hormones from plasma into various areas of the brain, including the hypothalamus and the pituitary gland, are discussed. Data generated in our own laboratories relevant to these subjects are first presented, and then discussed in relation to data available from the literature.

Samuel P. Marynick, M.D. • Division of Endocrinology, Department of Internal Medicine, Baylor University Medical Center, Dallas, Texas 75246. **James H. Wood, M.D.** • Division of Neurosurgery, University of Pennsylvania School of Medicine and Hospital, Philadelphia, Pennsylvania 19104. **D. Lynn Loriaux, M.D., Ph.D.** • Developmental Endocrinology Section, Endocrinology and Reproduction Research Branch, National Institute of Child Health and Human Development, National Institutes of Health, Bethesda, Maryland 20205. This chapter is dedicated to the memory of Robert Boyar, M.D.

2. Movement of Steroid Hormones from Plasma to CSF

It is generally believed that the primary factors governing the movement of small molecules from plasma into CSF are molecular size, charge, and lipid solubility.[21] The steroid hormones under consideration are all of approximately the same size (mol. wt. ≈300), and without charge at neutral pH. Hence, it would be predicted that the major determinant of steroid penetration into CSF would be their relative lipid solubility. This hypothesis was examined using six adult male rhesus monkeys (*Macaca mulata*), each having an indwelling catheter in a lateral ventricle of the brain, placed according to a method previously described.[15] This technique allows the continuous collection of CSF. The study animals were confined to chairs for the duration of the study and received no medication before or during the studies, which began at 09:00 hours and were completed at 15:00 hours. An indwelling venous catheter was placed at the start of the study to allow the serial collection of blood. The animals studied were separated into three groups of two each and given a bolus of 0.5 mCi of the ^{3}H-labeled steroid followed by a constant infusion of 0.5 mCi of the same steroid over a 6-hr period via a Harvard infusion pump. Isotope infusion was given in an extremity different from that used for serial blood collections.

Ventricular CSF was collected continuously in 15-min fractions, beginning 1 hr before and extending throughout the 6-hr steroid infusion. CSF was collected at 4°C at a rate of 300–400 μl/hr and stored at −20°C. Venous blood was collected in heparinized tubes immediately before and every 15 or 30 min during the steroid infusion. The blood was immediately centrifuged, and the plasma was separated and frozen at −20°C. CSF and plasma, 50 and 10 μl, respectively, from each collection period were counted. All scintillation counting was carried out to ±2% error. One pool of CSF and another of plasma composed of 50 μl from each 30-min collection of CSF and 50 μl from each 30-min sample of plasma, respectively, were then extracted with 5 volumes of diethyl ether. The diethyl ether extracts were concentrated to 1 ml, and 50 μl of each sample was counted. This allowed the determination of the percentage of unconjugated isotopic activity present in the pooled samples of CSF and plasma. To the diethyl ether extracts were then added 10 mg authentic cold steroid and 1000 dpm of the ^{14}C-labeled steroid, and the extracts were recrystallized to constant specific isotope ratio using an acetone–water mixture.[2] From the known volume of CSF and plasma in each specimen pool, the quantity of labeled unconjugated authentic steroid per volume of CSF or plasma was determined, and calculations of the ratio of unconjugated authentic CSF to plasma steroids were made. The results are shown in Table 1.

It is seen that at equilibrium, the CSF/plasma ratio of tritiated steroid varies from a low of 0% for progesterone (no recrystallizable progesterone counts were found in the CSF) to a high of 22% for cortisol. If lipid solubility plays a major role in the regulation of steroid entry from plasma into the CSF, it would be expected that progesterone would be found to be the least lipid-soluble of the steroids examined, and cortisol the most.

To determine the relative lipid–aqueous partitioning of estradiol, testosterone, dihydrotestosterone, 17-hydroxyprogesterone, cortisol, and progesterone, 2 ml of a 6 nM concentration of the [^{14}C]nuclide of each steroid in petroleum ether was placed in a centrifuge tube, and 2 ml distilled water was added. The tubes were agitated for 2 min and allowed to stand for 30 min before the aqueous and petroleum ether phases were separated. The separate phases were then dried under a constant air flow and counted as described above. The lipid–aqueous partition ratios for these steroids were, in fact, just the reverse of what was expected, progesterone being the most lipid-soluble and cortisol being the least (Table 1). These findings suggest that some mechanism other than lipid solubility must be influencing steroid plasma-to-CSF transfer.

3. Plasma Protein Binding of Steroid Hormones

There are specific high-affinity plasma binding proteins for two groups of steroids, the glucocorticoids (and progestins) and the sex steroids (estradiol and testosterone).[24] Albumin, in addition, binds many steroid hormones with low affinity. The plasma steroids examined in these studies are transported, in part, tightly bound to a specific binding protein. They are also transported loosely bound to albumin and plasma-protein-unbound (plasma free steroid). It is believed that free steroid in plasma is the fraction that is able to cross membrane barriers, while protein-bound steroid is not.[16] Hence, a reasonable hypothesis might propose that the fraction of radiolabeled steroid entering the CSF compartment from the plasma is a reflection of the plasma protein-unbound steroid.

Table 1. Relationships among CSF/Plasma Radioactive Steroid Concentrations, Lipid–Aqueous Partition Coefficients, and Plasma Free-Hormone Fractions

Steroid	CSF/plasma ratio	Lipid–aqueous partition	Plasma free-hormone fraction
Progesterone	0	6.1	2%
Dihydrotestosterone	0.001	4.4	3%
Testosterone	0.016	3.8	4%
Estradiol	0.035	2.4	3%
17-Hydroxyprogesterone	0.103	2.7	6%
Cortisol	0.225	0.8	22%

The plasma binding of the steroids studied in the rhesus monkeys was assessed using equilibrium dialysis. Two milliliters of prestudy plasma from each monkey used in the study of plasma to CSF movement of steroid hormones was placed into Union Carbide $\frac{7}{16}$-inch dialysis tubing containing 9×10^{-15} mol of the ^{14}C-labeled steroid being examined. The sealed dialysis bag was placed in 50 ml Ringer's solution and incubated for 12 hr at 37°C in a shaker. Total radioactivity was then determined in 1 ml of dialysand and 10 ml of dialyzate. All counting was carried out to 10^4 observations ($\pm 1\%$ coefficient of variation). Bound and unbound (free) plasma steroid was calculated as outlined by Slaunwhite.[22] The results are shown in Table 1. A correlation with the CSF/plasma ratios of radioactive steroids is shown in Fig. 1. In contrast to the negative correlation found between steroid lipid solubility and CSF penetrance, a rough positive correlation was found to exist between plasma free steroid and the ability of the steroid to gain entry into the CSF from the plasma compartment.

4. Radioimmunoassay of CSF Steroid Hormones

The finding that the CSF radioactive steroid concentrations correlated with free plasma steroid suggested that the true CSF concentrations of these hormones might similarly reflect plasma free steroid. The correlations between plasma protein-unbound steroid and CSF steroid levels as measured by radioimmunoassay are shown in Table 2. The table contains data from the studies of Backstrom *et al.*[3] and Carrol *et al.*[4] correlating the CSF concentrations of estradiol, testosterone, progesterone, and cortisol with the plasma total and unbound concentrations of these steroids. The table also contains previously unpublished data from our laboratory in which correlations were made between CSF levels of testosterone, 17-hydroxyprogesterone, and cortisol and the plasma free fractions of these steroids in men and women undergoing diagnostic pneumoencephalography. The CSF levels of cortisol, estradiol, and testosterone correlated strongly with the concentrations of plasma-protein-unbound steroids. Only 17-hydroxyprogesterone failed to show a significant correlation.

5. Discussion and Literature Review

The concept that the CSF concentrations of steroid hormones are a reflection of the concentrations of plasma-protein-unbound steroids is supported by a number of other studies. Abelson *et al.*[1] showed that CSF cortisol levels in man were of the order of

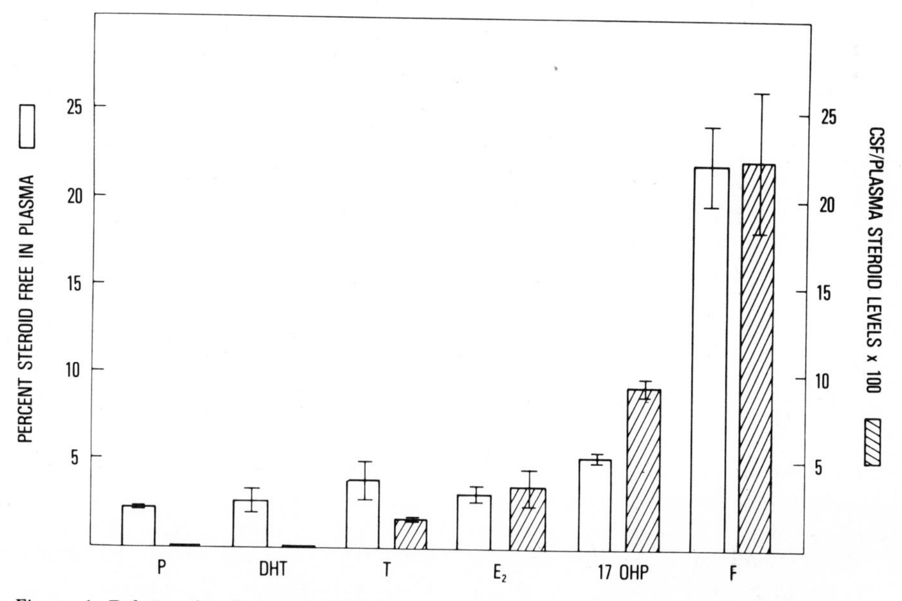

Figure 1. Relationship between CSF/plasma steroid ratio and fraction of circulating steroid that is not plasma-protein-bound. (P) Progesterone; (DHT) dihydrotestosterone; (T) testosterone; (E_2) estradiol; (17 OHP) 17-hydroxyprogesterone; (F) cortisol.

Table 2. Relationship between CSF Steroid Concentrations and Plasma Protein-Unbound Steroid Concentrations

Steroid	Patient group	CSF steroid concentration	Percentage of plasma steroid concentration	Correlation between CSF and plasma free steroid concentration	Ref. No.
Cortisol	Depressed	0.8 μg%	6.1%	$p < 0.001$	5
	Normal	1.6 μg%	6.4%	$p < 0.01$	Previously unpublished
Estradiol	Normal	2.3 pg/ml	4.0%	$p < 0.01$	4
Progesterone	Normal	4.3 ng/ml	10.8%	Not done	4
17-Hydroxyprogesterone	Normal	135 pg/ml	27.2%	Not significant	Previously unpublished
Testosterone	Normal	5.1 μg%	2.5%	$p < 0.01$	4
	Normal	11.0 μg%	1.7%	$p < 0.01$	Previously unpublished

0.2–0.4 μg/dl, values commensurate with the known free fraction of cortisol in the plasma of man. Fishman and Christy[9] also concluded that CSF cortisol was about 20% of the circulating concentration of plasma cortisol in man. Murphy *et al.*[20] found in 1967 that CSF cortisol was about 7% of that of concurrently measured plasma cortisol, and that this ratio was stable over a wide range of plasma cortisol concentrations, the result of estrogen-induced increases in cortisol-binding globulin. CSF cortisol concentrations ranged between 0.6 and 1.3 μg%. Carrol *et al.*[4] have also presented data showing that plasma-protein-free cortisol correlates well with CSF cortisol levels. Many other workers have reported CSF cortisol concentrations to be in a range that would agree well with expected concentrations of plasma-protein-unbound cortisol.[6,11,18]

Progesterone has been found by several workers to be of low concentration in the CSF. Lurie and Weiss[14] found CSF progesterone concentrations varying between 0.005 and 0.55 μg% in pregnant women having plasma progesterone levels of 8.7–11 μg%. Backstrom *et al.*[3] found somewhat higher CSF progesterone concentrations than our studies would predict, but did show a good correlation with plasma free progesterone.

Estradiol and testosterone have not been extensively studied, but the available data[3,15,16] tend to support the hypothesis that CSF steroid concentrations reflect plasma-protein-unbound hormone concentrations. 17-Hydroxyprogesterone, interestingly, does not fit neatly into the hypothesized mechanism. Although the CSF/plasma ratio of 17-hydroxyprogesterone (10%) compared reasonably well with the plasma-protein-unbound fraction of 17-hydroxyprogesterone (6.3%) in subhuman primate studies, comparison of human CSF 17-hydroxyprogesterone concentrations with concurrently measured plasma-free 17-hydroxyprogesterone levels failed to show a significant correlation. Why this particular steroid should behave differently from all others studied is difficult to understand. David and Anand-Kumar[7] were the first to show the movement of this steroid from plasma to CSF in the rhesus monkey. They found that the CSF radioactive 17-hydroxyprogesterone concentrations were higher than plasma concentrations measured several hours after an intravenous injection of labeled 17-hydroxyprogesterone, suggesting that 17-hydroxyprogesterone was concentrated in the CSF. We found in the rhesus monkey undergoing a continuous infusion of labeled 17-hydroxyprogesterone that the CSF 17-hydroxyprogesterone concentration is about 10% of the concurrent plasma level of labeled 17-hydroxyprogesterone. This does not confirm the observation of David and Anand-Kumar. One mechanism that may explain the divergent findings is related to the rapid plasma metabolic clearance of 17-hydroxyprogesterone compared to that of other steroids studied.[23] If CSF clearance is markedly slower than the clearance of 17-hydroxyprogesterone from plasma, the chance of achieving a close correlation between CSF and concurrent plasma levels would be much diminished, especially if 17-hydroxyprogesterone was secreted episodically. Most of the 17-hydroxyprogesterone in men, however, is of gonadal origin, implying that episodic secretion may not be prominent. The lack of a significant CSF–plasma free 17-hydroxyprogesterone correlation remains unexplained.

6. Movement of Steroid Hormones from Plasma to Brain in Rhesus Monkey

The interesting differences found among steroids regarding their ability to enter the CSF from the plasma led to a series of studies designed to examine transfer of steroid hormones from plasma into brain tissue itself. Dihydrotestosterone and estradiol were chosen for study because of the differences they exhibited in their ability to gain entry into CSF, dihydrotestosterone essentially being blocked from entry into the CSF and estradiol entering with apparent ease. Each steroid was studied in four rhesus monkeys. The monkeys were anesthetized with ketamine, and 1 mCi of tritiated steroid (about 3 μg) was administered intravenously. In two monkeys of each group studied with a given steroid, a 5000-fold molar excess of cold steroid (about 15 mg) was given with the tritiated-steroid injection. At the same time, 500 μCi of chromium-tagged erythrocytes was given intravenously. The animals were sacrificed 60 min following the steroid injection, a CSF specimen obtained, and the carotid arteries isolated. The brain was perfused via the carotid arteries with iced Ringer's–lactate buffer until the perfusate was free of radioactivity. This excluded plasma contamination as a source of radioactivity. The brains were removed and divided into pituitary gland, hypothalamus, cerebral cortex, and cerebellum. The tissues were homogenized and separated into nuclear and cytosol fractions. The fractions were extracted with ether–ethanol (3:1, vol./vol.) and the steroids recrystallized to a constant isotopic ratio.[2] The data from these studies are shown graphically in Figs. 2 and 3. It can be seen that in each case, the specific (displaceable by 5000-fold excess steroid) steroid binding in the adenohypophysis compared to the plasma radioactive steroid concentrations (adenohypophysis/plasma ratio) was greater than the CSF/plasma steroid ratio. Additionally, the adenohypophysis ratio was greater than the hypothalamus/plasma ratio, which was greater than either the cerebrum/plasma or the cerebellum/plasma ratio. Estradiol was more actively concentrated than dihydrotestosterone. The ability of the 5000-fold molar excess of nonradioactive steroid to block nuclear translocation of tritiated dihydrotestosterone is shown in Table 3. It can be seen that in both adenohypophysis and hypothalamus, the presence of a 5000-fold molar excess of nonradioactive steroid blocked the nuclear accumulation of radioactivity. This was also seen with the estradiol infusion, and suggests that the steroid is binding to a saturable receptor protein. This is further suggested by the demonstration that some of the activity extracted

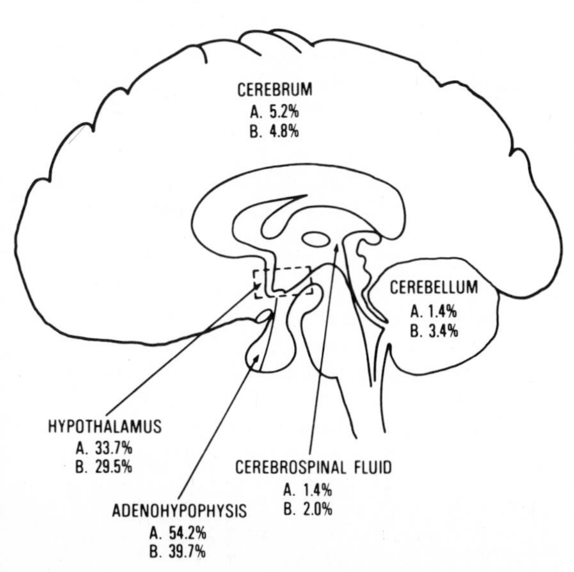

Figure 2. [^{3}H]Estradiol as percentage of plasma estradiol 60 min after infusion of 1 mCi [^{3}H]estradiol (2.2×10^{-8} M). Data for two experiments (A, B) are shown.

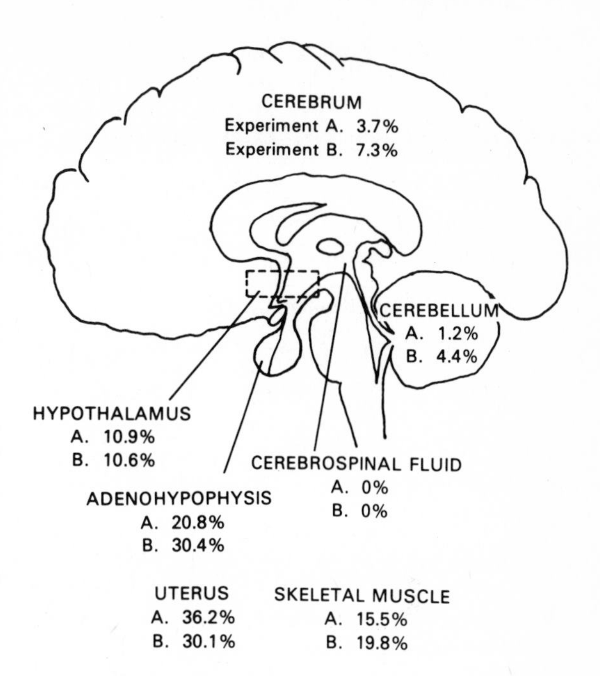

Figure 3. [^{3}H]Dihydroxytestosterone (DHT) as percentage of plasma DHT 60 min after infusion of 2 mCi [^{3}H]-DHT (5×10^{-8} M). Data for two experiments (A, B) are shown.

Table 3. Distribution of [^{3}H]Dihydrotestosterone (DHT) in Cytoplasm and Nuclei of Hypothalamus and Adenohypophysis of Rhesus Monkey with and without Competition with Excess Steroid

	Subcellular [^{3}H]-DHT (dpm/g tissue)	
Fraction	[^{3}H]-DHT (1.2×10^{-8} mol)	[^{3}H]-DHT (1.2×10^{-8} mol) plus Cold DHT (5.9×10^{-5} mol)
Adenohypophysis		
Nuclei	33,196	0
Cytosol	52,804	35,584
Hypothalamus		
Nuclei	19,890	205
Cytosol	25,110	5,099

from the nuclei with 0.4 M KCl is associated with high-molecular-weight proteins that elute in the void volume of a Sephadex G-100 column and by the presence of high-affinity, low-capacity binding for at least one of these steroids, dihydrotestosterone, in cerebral cortex ($K_d = 2.7 \times 10^{-9}$ M).

Although the data for plasma-to-brain steroid hormone movement are limited, they suggest that some factor other than lipid solubility or plasma protein binding is influencing the tissue accumulations of these steroids. The several lines of evidence mentioned above point toward specific steroid receptors as playing an important role in this process. Many investigators have demonstrated the presence of cytosol receptors in the rodent adenohypophysis and hypothalamus for androgens[12,13] and estrogens,[5,8,19] and also for glucocorticoids.[10,17]

7. Conclusions

From the data available, a few tentative conclusions may be drawn. First, the movement of steroid hormones from plasma into CSF and the CSF concentration of the steroid hormones examined appear to be a reflection of the concentration of plasma-protein-unbound steroid. Second, steroid hormones can be concentrated in various areas of the brain, in some cases without measurably entering the CSF (e.g., dihydrotestosterone). The concentration mechanism may well involve cytoplasmic steroid receptor proteins.

References

1. Abelson, D., Baron, D. N., Toakley, J. G.: Studies of cerebrospinal fluid following oral administration of cortisone acetate or hydrocortisone. *Endocrinology* **12**:87–92, 1955.
2. Axelrod, L. R., McHigssen, C., Galdzreber, J. W., Pulliam, J. E.: Definitive identification of microquantities of radioactive steroids by recrystallization to constant specific activity. *Acta Endocrinol. Suppl.* **99**:1–77, 1965.
3. Backstrom, T., Cartensen, H., Sudergard, R.: Concentration of estradiol, testosterone, and progesterone in cerebrospinal fluid compared to plasma unbound and total concentrations. *J. Steroid Biochem.* **7**:469–472, 1976.
4. Carroll, B. J., Cuites, G. C., Mendels, J.: Cerebrospinal fluid and plasma free cortisol concentrations in depression. *Psychol. Med.* **6**:235–244, 1976.
5. Chader, G., Vilbe, C. A.: Uptake of oestradiol by the rabbit hypothalamus, *Biochem. J.* **118**:93–97, 1970.
6. Coppen, A., Brooksback, R., Noguera, R., Wilson, D.: Cortisol in cerebrospinal fluid in patients suffering from affective disorders. *J. Neurol. Neurosurg. Psychiatry* **34**:432–455, 1971.
7. David, G. F. X., Anand-Kumar, T. C.: Transfer of steroid hormones from blood to cerebrospinal fluid in rhesus monkey. *Neuroendocrinology* **14**:114–120, 1974.
8. Eisenfeld, A. J.: ^{3}H-estradiol: *In vitro* binding to macromolecules from the rat hypothalamus, anterior pituitary, and uterus. *Endocrinology* **86**:1313–1318, 1970.
9. Fishman, R. A., Christy, N. P.: Fate of adrenal cortical steroids following intrathecal injection. *Neurology* **15**:1–6, 1965.
10. Grosser, B. E., Steven, W., Bruenger, F. W., Reed, D. J.: Corticosterone binding by rat brain cytosol. *J. Neurochem.* **18**:1725–1732, 1971.
11. Haberfellner, H., Auer, B., Gleispach, H.: Preliminary studies in plasma and cerebrospinal fluid cortisol in children with different diseases. *Clin. Chim. Acta* **54**:5–10, 1974.
12. Joan, P., Samperez, S., Thieulant, M. L., Mercer, L.: Etude du recepteur cytoplasmique de la (1,2-^{3}H)testosterone dans l'hypophyse anterienne et

l'hypothalamus du rat. *J. Steroid Biochem.* **2**:223–236, 1971.
13. KATO, J., ONOUCHIS, T.: 5α-Dihydrotestosterone "receptor" in the rat hypothalamus. *Endocrinol Jpn.* **20**:429–432, 1973.
14. LURIE, A. O., WEISS, J. B.: Progesterone in cerebrospinal fluid during human pregnancy. *Nature (London)* **215**:1178, 1967.
15. MARYNICK, S. P., HAVENS, W. W., EBERT, M. H., LORIAUX, D. L.: Studies on the transfer of steroid hormones across the blood–cerebrospinal fluid barrier in the rhesus monkey. *Endocrinology* **99**:400–404, 1976.
16. MARYNICK, S. P., SMITH, G. B., EBERT, M. H., LORIAUX, D. L.: Studies on the transfer of steroid hormones across the blood–cerebrospinal fluid barrier in the rhesus monkey II. *Endocrinology* **101**:562–567, 1977.
17. MCEWAN, B. S., MAGNUS, C., WALLACH, G.: Soluble corticosterone-binding macromolecules extracted from rat brain. *Endocrinology* **90**:217–226, 1972.
18. MCCLURE, D. J., CLEGHORN, R. A.: Hormone deficiency in depression. In Masserman, J. H. (ed.): *Science and Psychoanalysis*, Vol. 17. New York, Grune and Stratton, 1970, pp. 12–20.
19. MOWLES, T. F., ASHKANAZY, T. B., MIX, B., SHEPPARD, J. H.: Hypothalamus and hypophyseal estradiol-binding complexes. *Endocrinology* **89**:484–491, 1971.
20. MURPHY, B. E. P., COSGROVE, J. B., MCELQUHAM, M. C., PATTIE, C. J.: Adrenal corticoid levels in human cerebrospinal fluid. *Can. Med. Assoc. J.* **97**:13–17, 1967.
21. RALL, D. P.: Transport through the ependymal linings. In Lagthen, A., Ford, D. M. (eds.): *Progress in Brain Research.* Amsterdam, Elsevier, 1967, pp. 159–172.
22. SLAUNWHITE, R.: The binding of estrogens, androgens and progesterone by plasma proteins *in vitro.* In Antoniades, H. N. (ed.): *Hormones in Human Plasma.* Boston, Little Brown, 1960, pp. 478–494.
23. STROTT, C. A., YOSHIMI, T., BARDIN, C. W., LIPSETT, M. B.: Blood progesterone and 17-hydroxyprogesterone levels and production rates in a boy with virilizing congenital adrenal hyperplasia. *J. Clin. Endocrinol. Metab.* **28**:1085–1088, 1968.
24. WESPHAL, U. (ed.): *Steroid Protein Interactions.* New York, Springer-Verlag, 1971.

CHAPTER 43

Cerebrospinal Fluid Vasopressin and Vasotocin in Health and Disease

Thomas G. Luerssen and Gary L. Robertson

1. Introduction

It is now recognized that integrative central nervous system function involves not only electrophysiological phenomena and chemical neurotransmission, but also interaction between brain and endocrine hormones. With the recent upsurge in interest in "peptidergic" neurons, it has become evident that many hypothalamic and pituitary peptides occur in brain outside the boundaries of the hypothalamo-pituitary axis. Anterior and posterior pituitary peptides, hitherto characterized as "peripheral hormones," have now been identified in cerebrospinal fluid (CSF). This chapter reviews our current knowledge about the occurrence of the antidiuretic hormone, arginine vasopressin (AVP), and the closely related nonapeptide hormone, arginine vasotocin (AVT), in human CSF.

The suggestion was first advanced by Herring[28] in 1908 that hyaline material observed in the neurohypophysis represented neurosecretory material that could ultimately find its way into third ventricular CSF. Citing his work, Cushing and Goetsch[13] reported that the intravenous injection of a concentrate of human CSF into an experimental animal resulted in a marked pressor response similar to that produced by an intravenous injection of pituitary extracts. Following this report, most attempts to measure posterior pituitary peptides in CSF with antidiuretic or pressor bioassays resulted in negative findings.[26] It was not until 1968 that Vorherr *et al.*[73] and Heller *et al.*,[27] working in separate laboratories with a refined bioassay technique, reported antidiuretic activity in the CSF of anesthetized, but not of conscious, animals. Using a similar bioassay, Gupta[25] found low levels of antidiuretic activity in only 1 of 12 samples of CSF obtained from conscious humans. Subsequently, Pavel[45] and co-workers reported finding high levels of antidiuretic activity in the CSF of anesthetized, but not of conscious, adults. On the basis of the other chemical and biological properties of this activity, these workers tentatively proposed that this antidiuretic activity was due largely to AVT rather than AVP.

These findings set the stage for recent studies using more sensitive and specific radioimmunoassays. One such method, which was originally developed for the measurement of AVP in plasma,[54] has now been used without modification to measure the hormone in CSF.[36] This technique has made it possible for the first time to study AVP at physiological concentrations in the plasma and CSF of adults without interference from AVT or other antidiuretic substances. In the following sections, we will review recent information obtained with this method and other methods and attempt to summarize what it reveals about the nature, origin, and control of the antidiuretic activity in human CSF. We will also review briefly some recent studies that may shed some light on possible functions of the hormones in this compartment.

Thomas G. Luerssen, M.D. • Department of Neurosurgery, Indiana University School of Medicine, Indianapolis, Indiana 46223. **Gary L. Robertson, M.D.** • Division of Endocrinology, Department of Medicine, University of Chicago, Pritzker School of Medicine, Chicago, Illinois 60637.

2. Chemistry

2.1. Identity

Chemical identification of the antidiuretic activity in CSF previously has been limited to inactivation studies with thioglycollate and trypsin. This approach established that the active principle was a sulfhydryl- and basic-amino-acid-containing peptide like AVP, but did not exclude other possibilities such as AVT (Fig. 1). Immunoassay studies have now established for the first time that human CSF does indeed contain appreciable amounts of AVP (Fig. 2). Thus, in an assay system that effectively discriminates among AVP, AVT, and oxytocin, serial dilutions of an acetone extract of human CSF produce a displacement curve that parallels completely that of purified synthetic AVP (Fig. 2). The foregoing results do not totally exclude the possibility that AVT is present as well. However, the lack of any flattening in the slope of the displacement curve generated by the CSF extract indicates that if AVT was present, its concentration must have been low compared to that of AVP. This observation appears to be inconsistent with the tentative identification of high concentrations of AVT-like bioactivity in the CSF of other patients with diabetes insipidus.[12] Further studies using more specific chemical or immunological tests, or both, for AVT will be needed to fully resolve the question about the presence of this hormone in the CSF of man under normal as well as abnormal conditions.

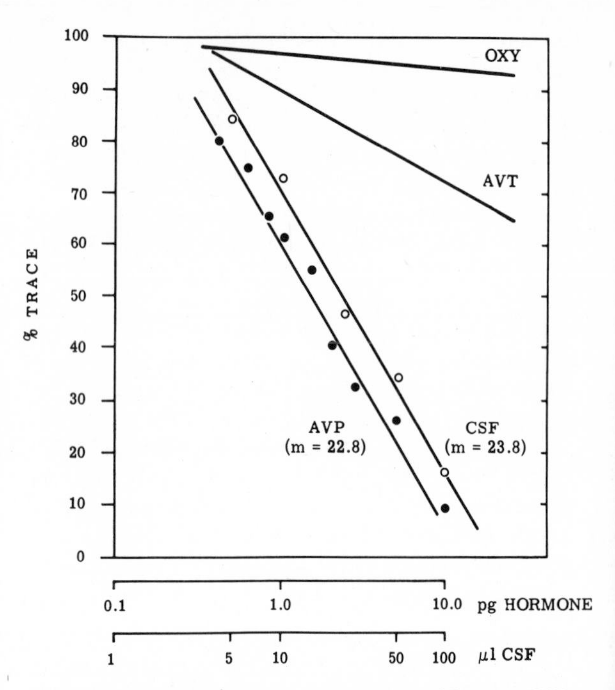

Figure 2. Immunological cross-reactivity of human CSF, relative to standard arginine vasopressin (AVP), arginine vasotocin (AVT), and oxytocin (OXY). Sample was obtained from patient with hydrocephalus and diabetes insipidus.

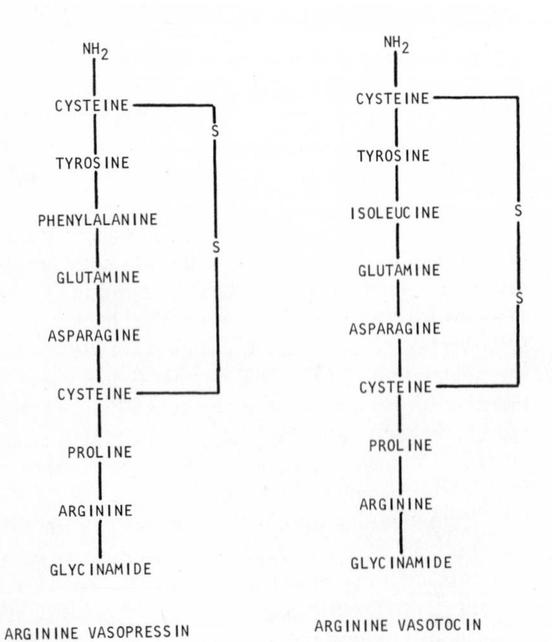

Figure 1. Amino acid sequence of arginine vasopressin and arginine vasotocin.

2.2. Stability

The AVP content of CSF is not altered by several hours of incubation at room temperature or by several weeks of storage at −20°C.[37] This stability greatly simplifies clinical studies of AVP in CSF and indicates that CSF does not contain the vasopressinase-like activity that has been observed in frozen plasma.[14,55] Because of this difference in stability in the two fluids, the AVP concentration of plasma and CSF cannot be meaningfully compared unless the plasma is extracted shortly after collection.

3. Concentration

3.1. Normal Neurohypophyseal Function

In normally hydrated patients without clinical evidence of abnormal pituitary function, the concentration of AVP in CSF obtained by lumbar puncture in the lateral decubitus position ranges from less than 0.5 to 1.8 pg/ml (Fig. 3). Using a different immunoassay method, Dogterom *et al.*[21] have reported 5-fold higher concentrations of AVP in the CSF of conscious adults. The reason for the discrepancy in the immunoassay values is not immediately apparent. It should be noted, however, that the higher values greatly exceed those found previously by the most sensitive bioassay method[25] and the activity was not tested for identity with AVP in any immunological, biological, or chemical system. Thus, these values could reflect at least in part some degree of nonhormonal interference in the assay.[54]

The levels of AVP in CSF are slightly but significantly lower than those found in plasma collected simultaneously and extracted immediately. This gradient between plasma and CSF is similar to that observed previously using bioassay methods in both experimental animals[27,73] and man.[25] A plasma–CSF gradient in the opposite direction has been reported by Dogterom *et al.*[21] using another immunoassay system. It should be noted, however, that the latter study is also atypical in other respects. Besides finding singularly high levels of AVP in human CSF, it is unique in reporting the hormone to be consistently undetectable (< 0.5 pg/ml) in plasma from normally hydrated adults. The cause of these anomalous results is uncertain, but probably involves several factors including a lack of appropriate action to prevent degradation of plasma AVP during storage (see Section 2.2). The previously noted possibility of interference by nonhormonal immunoactivity in human CSF could also have contributed to a spurious reversal of the concentration gradient between plasma and CSF. Recently, Jenkins *et al.*[28a] have reported values for CSF and plasma AVP obtained with another radioimmunoassay. In general, they report absolute concentrations and gradients similar to those found by Luerssen and Robertson (Fig. 4).

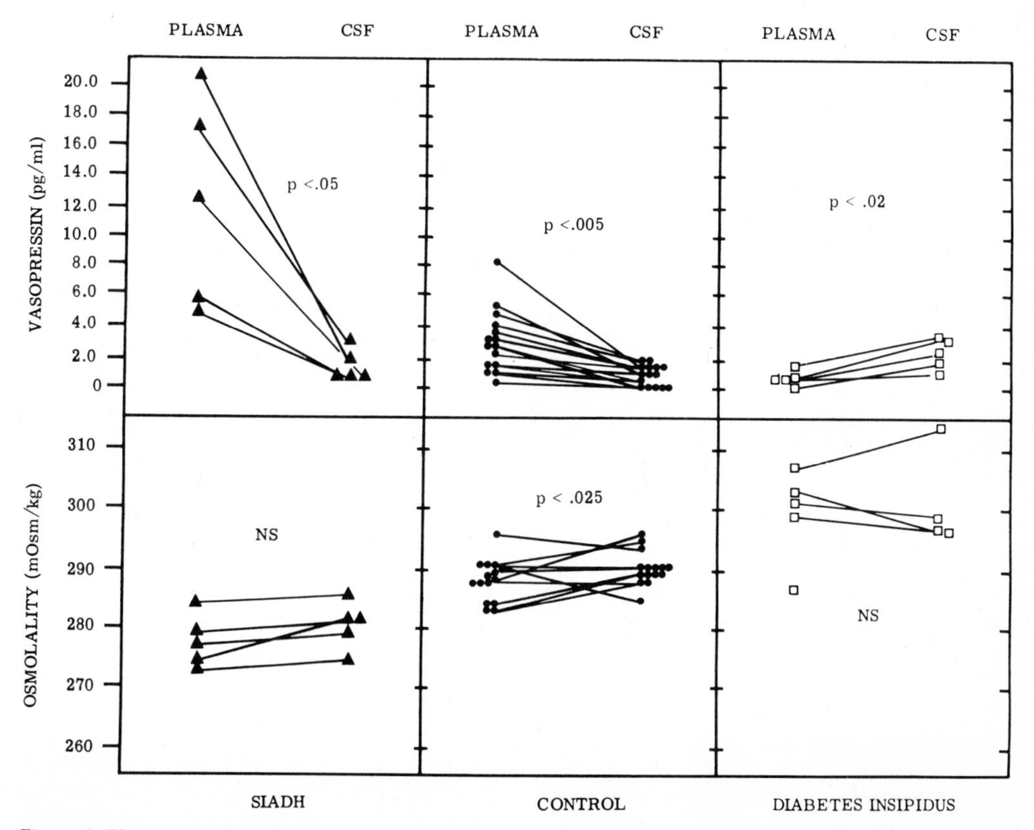

Figure 3. Plasma and CSF osmolality and arginine vasopressin (AVP) concentration in patients with normal neurohypophyseal function and patients with diabetes insipidus and syndrome of inappropriate antidiuresis (SIADH). Note reversed relationship of AVP levels in plasma and CSF in patients with diabetes insipidus.[36]

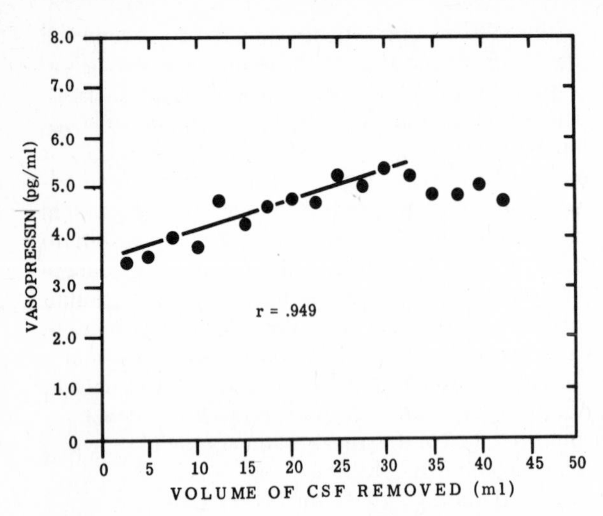

Figure 4. Arginine vasopressin (AVP) concentration in serial 2.5-ml fractions of CSF removed during therapeutic spinal drainage. In first 12 samples, there was gradual progressive rise in AVP concentration that correlated significantly ($r = 0.949$) with volume of CSF removed. In last 5 samples, AVP concentration was constant.

In some patients with various brain injuries, CSF levels were increased.

The concentration of AVP in CSF obtained from the lumbar cistern closely approximates that obtained from more rostral portions of the neuraxis.[37] However, serial assays on relatively large volumes of CSF rapidly drained from conscious patients undergoing therapeutic lumbar drainage (Fig. 4) indicate that a rostrocaudal gradient in AVP concentration may exist. This gradient is not of sufficient magnitude that comparisons between AVP concentrations in plasma and CSF are affected by the level within the neuraxis at which CSF is obtained. In other studies, lumbar cisternal AVP concentrations measured before and immediately after complete pneumoencephalography are similar. Moreover, the AVP content of CSF obtained directly from an intraventricular catheter is in the same range as that found in the lumbar cistern. These findings suggest that the concentration of AVP is relatively uniform throughout the CSF compartment.

The AVT content of CSF has not been studied with immunoassay methods. By the relatively insensitive rat antidiuretic and frog bladder hydroosmotic assays, the hormone is undetectable (< 5 μU antidiuretic activity/ml) in CSF obtained by lumbar or suboccipital puncture in conscious healthy adults.[45] The concentration, if any, of AVT in plasma or other extracranial body fluids has not been determined.

The neurophysins, a class of binding proteins that are found complexed to the hormones in the neurohypophysis, have also been measured in the CSF of humans and monkeys.[56] Using an immunoassay that does not distinguish among the different neurophysins associated with oxytocin and vasopressin, Robinson and Zimmerman[56] found a mean level of about 5 ng/ml in the CSF of conscious adults. In molar terms, this concentration is about 500 times greater than that of AVP. The concentration of neurophysins in plasma averaged 0.7 ng/ml and, when paired samples were collected simultaneously, was consistently lower than that in CSF. Thus, the plasma–CSF gradient for the neurophysins is the reverse of that found by most laboratories for AVP and antidiuretic activity. This dissociation as well as the large molar excess of the neurophysins in CSF may be due to the inability of the immunoassays then in use to distinguish among the different neurophysins associated with oxytocin and AVP. It could also reflect different rates of clearance of the hormone and its binding protein from CSF or plasma or both.

3.2. Abnormal Neurohypophyseal Function

In patients with the clinical syndrome of inappropriate antidiuresis (SIADH),[5] the concentration of AVP in CSF is slightly higher than in control patients (see Fig. 3). However, the normal gradient between CSF and plasma is retained since, in SIADH, plasma AVP is also increased. These observations indicate that the plasma–CSF barrier to AVP is not disrupted by various types of intracranial pathology or by the cerebral swelling that accompanies hyponatremia.

In patients with diabetes insipidus, the concentration of AVP in CSF is significantly increased even though plasma AVP is subnormal (Fig. 3). Hence, the plasma–CSF gradient for AVP is the reverse of that found in control patients and those with SIADH. These findings indicate that the plasma–CSF barrier to AVP is bidirectional. As discussed later, these findings may also have important implications for our understanding of both the origin and function of the AVP in CSF.

The CSF of patients with diabetes insipidus is also reported to contain increased concentrations of AVT-like activity.[12] On the basis of the reported bioassay levels and assigned antidiuretic and hydroosmotic potencies of 250 and 34,000 IU/mg, respectively, the concentration of AVT in the CSF of these patients ought to be in the range of 84–111 pg/ml, or almost 40 times the average AVP level as determined by immunoassay. AVT concentrations of this magnitude might be expected to cross-react very slightly in the AVP immunoassay. In limited studies, however, such interference cannot be detected by immunological criteria (see Fig. 2). Further studies using more refined immunological and chromatographic techniques for differentiating AVT from AVP need to be performed in these patients.

4. Origin

Several lines of evidence now indicate that little, if any, AVP enters the CSF by diffusion or transport from peripheral plasma. Intravenous administration of the hormone to humans[37] as well as experimental animals[73,77] results in little or no rise in the AVP content of CSF even though plasma AVP is markedly increased. Heller *et al.*[27] initially reported significant increases in antidiuretic activity in CSF following the administration of very large amounts of AVP to rabbits. However, in a subsequent discussion[77] of these results, the same group suggested that the normal blood–brain barrier may have been disrupted by the hypertensive effect of the very high plasma AVP levels produced. Moreover, the concentration of AVP in the CSF of patients with diabetes insipidus does not seem to be further increased when peripheral hormone levels are chronically increased by treatment with pitressin tannate in oil.[36] Finally, the reversed plasma–CSF gradient observed in patients with diabetes insipidus (see Fig. 3) is also inconsistent with passive diffusion of hormone from plasma. These results do not totally exclude the presence of a high-affinity, low-capacity, active-transport system between peripheral plasma and CSF, but they do indicate that if such a system exists, it cannot be the only or even the major source of AVP in the CSF of man. Exogenous neurophysin also fails to penetrate the blood–brain barrier,[56] indicating that it too must enter the CSF by some mechanism other than diffusion from peripheral plasma.

Diffusion or transport from plasma containing a locally high concentration of AVP must also be considered as a possible source of the hormone in CSF. Such a condition might occur if there were retrograde flow from the capillary plexi of the pars nervosa or infundibulum to the hypothalamus and third ventricle.[6] However, the finding of elevated levels of AVP in the CSF of patients with diabetes insipidus (Fig. 3) would seem to indicate that AVP is not derived directly or indirectly from hormone-enriched plasma leaving the neurohypophysis.

By exclusion, therefore, it seems likely that AVP is secreted directly into CSF. From what is known about the direction of flow of CSF[19] and the similarity of AVP concentration in ventricular and subarachnoid space (see above), it is likely that most, if not all, of the secretion occurs in the lateral or third ventricles, or in both. And since AVP concentration is not depressed in the CSF of patients with diabetes insipidus (Fig. 3), most of the hormone in CSF must be secreted by neurons that are anatomically separate from those that secrete into blood (Fig. 5). Otherwise, the neurosecretory pathway(s) to the CSF would undergo the same irreversible degeneration that occurs in the supraopticohypophyseal tract of patients with surgical and other causes of diabetes insipidus.[38,53]

The location of the neurosecretory cells that release AVP into the brain ventricles has not been defined with certainty. However, several possibilities have been suggested by the histochemical studies. In experimental animals, the ependyma of the third ventricle has been shown to be innervated by magnocellular neurosecretory fibers that, in some cases, have been shown histochemically to contain AVP or its associated neurophysin or both.[8,30,56,57,64,66,68,74,75] Although uptake from CSF cannot be excluded, the presence of these peptides strongly suggests that these neurons are engaged in some type of secretory activity. The pathway to the third ventricle is probably not the only source, or even the major source, of AVP in CSF. The apparent preservation of activity in patients with diabetes insipidus and other evidence of significant hypothalamic damage (see Fig. 3) suggests that most secretion into CSF is performed by neurons that arise and terminate at some distance from the supraopticoneurohypophyseal tract. It has been shown recently that AVP is present in a number of other hypothalamic nuclei[22,23] and that there are also extrahypothalamic neurosecretory pathways that supply the lateral ventricles.[33] The possibility that one or more of these other pathways is the major source of AVP in CSF warrants further study.

The source of AVT in CSF has not been specifically studied. However, since this hormone is not known to occur in plasma or other extracranial fluids, it is probably also produced locally by one or more parts of the brain. Of these, the pineal is the most likely source, since this organ has been found to contain substantial amounts of AVT in man[35,48,65] as well as animals.[9,10,40,48,49,59,65] Located in the roof of the third

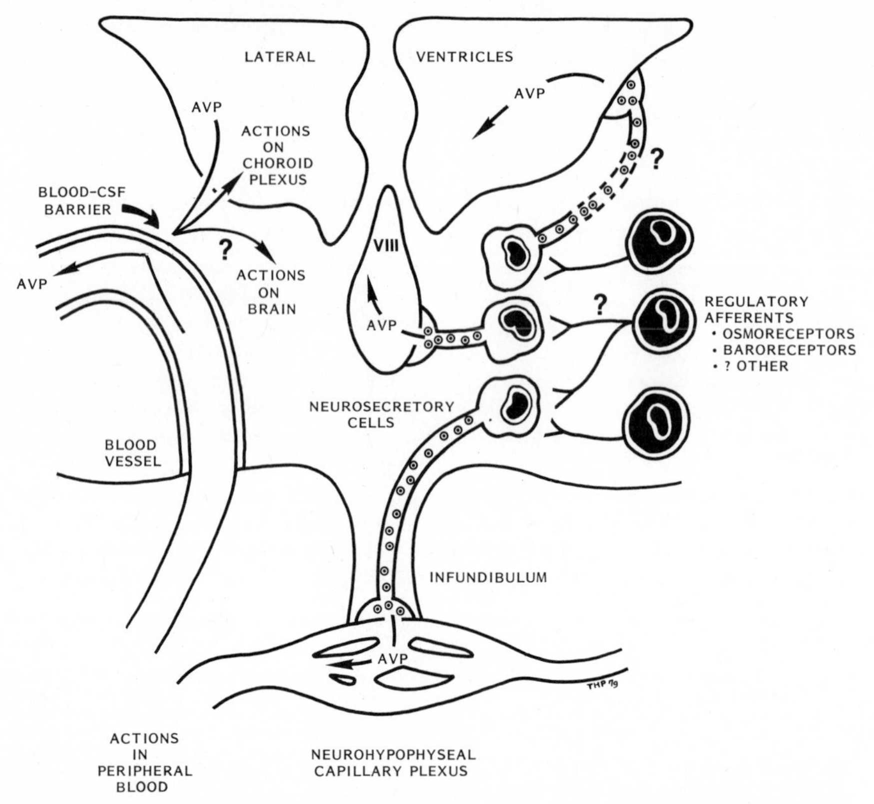

Figure 5. Schematic representation of proposed neurosecretory pathway to CSF. Note that blood and CSF compartments are separated by barrier impervious to arginine vasopressin (AVP) and are innervated by separate neurosecretory pathways. It is not known whether these pathways receive regulatory afferents from the same receptors or whether they are under independent control.

ventricle, the pineal is also well situated to release hormones into the CSF. On the other hand, the pineal of man is a comparatively rudimentary structure that often begins to involute and calcify during adolescence. Thus, it is possible that the AVT reported to be present in the CSF of adult patients may have come from some other area of brain. Recently, Pavel[51a] reported finding much higher AVT concentrations in the CSF of newborns and infants than in adults. These differences would be consistent with an origin in the pineal.

5. Regulation

Although originating in different neurons, the secretion of AVP into CSF appears to be influenced by the same stimuli that cause AVP to be released into plasma. Thus, in patients with normal neurohypophyseal function, there is a significant positive correlation between the concentration of AVP in plasma and that in CSF (Fig. 6). An identical relationship is observed in patients with SIADH, indicating that even pathological stimuli exert a similar action on the two neurosecretory pathways. As noted previously, the abnormal plasma–CSF relationship observed in patients with diabetes insipidus probably reflects selective or disproportionate destruction of the pars nervosa or its major regulatory afferents or both.

The relative importance of different stimuli in regulating AVP secretion into CSF has not been defined. In healthy recumbent adults, the osmolality of body water appears to be the major determinant of plasma AVP concentration.[54] The same is probably true for AVP secreted into CSF, since, when multiple paired samples are obtained from a single individual in different states of water balance, the AVP content and osmolality of plasma and CSF cross-correlate completely.[37] This conclusion is also

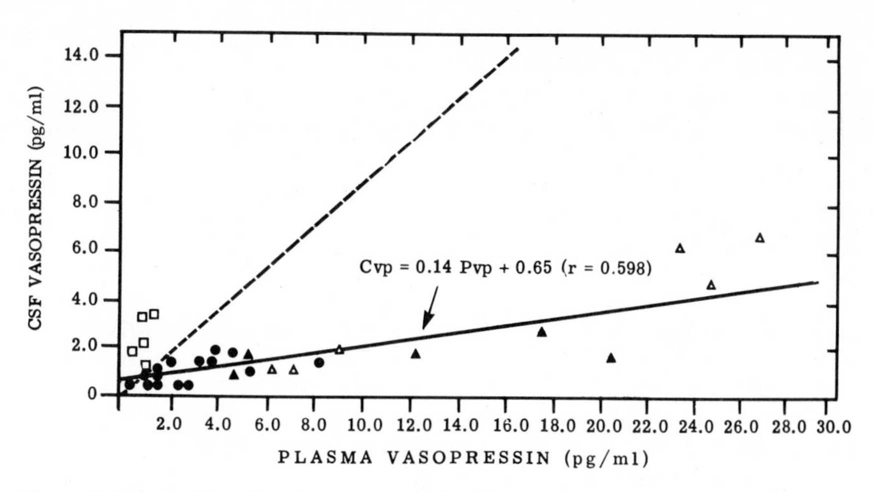

Figure 6. Relationship of arginine vasopressin (AVP) in CSF to AVP in plasma in normal subjects (●) and patients with syndrome of antidiuresis (SIADH) (▲, △) or diabetes insipidus (□). The regression line (—) was calculated from the values obtained in normal subjects. Note proximity of data obtained from patients with SIADH to this line. All patients with diabetes insipidus exhibited CSF/plasma ratios greater than unity.[36]

consistent with the finding of elevated hormone levels in the CSF of patients who are mildly dehydrated due to diabetes insipidus (see Fig. 3). It is not known whether the osmoreceptors that regulate secretion into plasma and CSF are different functionally or anatomically or both.

Nonosmotic variables also appear to have a significant influence on the secretion of AVP into both CSF and plasma. This may be inferred from the correlation observed in patients with SIADH (see Fig. 5) as well as from previous studies in animals showing that hemorrhage[73] or stimulation of the vagus nerve[27] increases the antidiuretic activity of CSF as well as plasma. Anesthesia with pentobarbital also causes an increase in the antidiuretic activity of CSF in animals.[27,73] A similar increase has been observed in man, although the antidiuretic activity was attributed to AVT rather than AVP.[45] It is not known whether other nonosmotic, nonhemodynamic stimuli such as nausea, glucopenia, and angiotensin or inhibitors such as alcohol and opiates have the same effect on AVP in CSF as they do on AVP in plasma.[61]

The factors that influence the AVT concentration in CSF have been studied even less well, but may be similar in some respects to those that affect AVP. In addition to the stimulation by anesthesia noted above, the AVT content also appears to be markedly increased in patients with diabetes insipidus.[12] Whether this increase reflects true control by an osmoreceptor or is due to some nonosmotic consequence of neurohypophyseal disease will have to be determined by further studies in patients with normal pituitary function. The release of AVT from pineal into CSF has also been reported to be stimulated by intracarotid injection of melanocyte-stimulating-hormone-release-inhibiting factor,[50] luteinizing-hormone-releasing hormone, thyrotropin-releasing hormone, and growth-hormone-release-inhibiting hormone.[24] It is not known whether the effects of these various peptides are direct or indirect, since changes in other possible stimuli such as blood osmolality and pressure were not determined.

6. Function

The function, if any, of the AVP or AVT in CSF is unknown. It is clear that the AVP does not contribute significantly to the regulation of water excretion, because patients with pituitary diabetes insipidus continue to exhibit polydipsia, polyuria, and a deficiency of plasma AVP despite elevated levels of this hormone in their CSF.

Sterba[67] has hypothesized that the CSF acts as a transport medium in a system of "ascending neurosecretory pathways" in which neurohormones act on brain at sites distant from the hypothalamus. AVP and its related peptides in CSF may function in this way, since they have been shown to exert a direct effect on the electrical activity in certain populations of neurons. Although AVP has been proposed as a neurotransmitter,[76] it apparently regulates the electrical activity of neurons in a different manner from conventional neurotransmitters. Furthermore, it may have opposing effects in two different cell populations. Thus, when AVP is applied iontophoretically to supraoptic neurosecretory cells, inhibition of extracellularly recorded electrical activity occurs, but when it is applied to cortical neurons, the responses obtained are excitatory in nature.[42] The molecular mechanisms of these effects are not known, but AVP may directly affect membrane conductance or may cause changes in membrane affinity for extracellular Ca^{2+}.[2-4]

Several proposals concerning the specific function of AVP in the central nervous system have been advanced. Studying the effects of parenteral vasopressin on water intake in humans with diabetes insipidus, Pasqualini and Codevilla[43,44] suggested that AVP may influence the cortical appreciation of thirst. However, the relationship of these studies to the function of AVP in CSF is uncertain, since it is now clear that the blood–CSF barrier is impervious to exogenously administered hormone.[73,77] It also should be noted that thirst appears to be normal in Brattleboro rats, which lack the ability to synthesize AVP.

Knowles and Vollrath,[31,32] studying lower vertebrates, have suggested that neurosecretory material, which they presume to be antidiuretic hormone, may be discharged into the CSF and serve as a direct-feedback regulator of the supraoptic nucleus. There are no conclusive studies to support or refute this hypothesis. It has also been proposed that AVP secreted into CSF may serve as a releasing factor for anterior pituitary hormones, particularly ACTH.[34,78] However, this suggestion is not consistent with the finding that Brattleboro rats exhibit a relatively normal pituitary–adrenal response to stressful stimuli.[1,39] Similarly, it has been suggested that AVT may serve as a releasing factor for prolactin[29,47,71,72] and may inhibit the release of gonadotropins[7,10] and ACTH.[46,51] However, other laboratories have reported that AVT inhibits the release of prolactin[11] and has no effect on the release of gonadotropins.[15] Furthermore, chemical analysis of the "pineal antigonadotropic factor" has indicated that AVT is not the active principle.[60] These conflicting results indicate that the relationship, if any, among AVP, AVT, and anterior pituitary function will be defined only by further investigation.

Much recent investigative activity has centered around the effects of AVP on learning and memory. De Wied and co-workers,[16-18,69] studying rats, have suggested that AVP functions at the level of the dorsal hippocampus and rostral septal areas to facilitate memory processes. In their studies, AVP administered into the cerebral ventricles induces resistance to extinction of a conditioned avoidance response. Interestingly, the Brattleboro rat has been shown to exhibit impairment in long-term memory that is apparently corrected by the administration of AVP.[70] Like the thirst studies, however, the impermeability of the blood–CSF barrier to AVP suggests that an extracerebral site of action may have been responsible for the observed effect on memory.

An entirely different function for the AVP in CSF has been proposed by Rodriguez and Heller,[57,58] who have studied the effects of AVP on choroid plexus epithelium. They have suggested that AVP may act at the choroid plexus to modify the rate of formation or chemical composition of CSF. Such an action would be analogous to its well-established effects on water transport across renal tubular epithelium. Antidiuretic material has been demonstrated in mammalian choroid plexus,[62] and recently, AVP has been shown to cause ultrastructural changes in choroid plexus epithelium.[63] Remarkably, the cellular responses obtained following intravenous infusion of AVP indicated that CSF was being absorbed by the choroid plexus epithelium against the hydroosmotic pressure gradient. The concept of transchoroidal absorption of CSF is not new,[20,41] but the possibility that CSF formation or absorption may be hormonally regulated is interesting and requires further investigation.

Finally, a recent report has linked centrally released AVP to the regulation of brain water permeability.[52] In this experiment, AVP administered into the lateral ventricles of the rhesus monkey induced a transient increase in cerebral capillary permeability without affecting cerebral blood flow. Although further studies are necessary, these findings have important implications in regard to the understanding and clinical management of cerebral edema.

7. Summary

This report reviews current information regarding the occurrence of the antidiuretic peptides AVP and AVT in human CSF. Clinical investigations utilizing radioimmunoassay techniques have provided clarification and confirmation of previous studies utilizing bioassay and have added new data about the origin and control of antidiuretic peptides in CSF. Thus, it is now apparent that AVP contributes to most, if not all, of the antidiuretic activity measured in human CSF by bioassay. Furthermore, the AVP in CSF is functionally isolated from peripheral blood by a bidirectionally impervious blood–CSF barrier.

Studies performed in patients with normal and abnormal neurohypophyseal function have indicated that AVP is secreted directly into the CSF by a group of neurosecretory cells that are functionally separate from the supraopticohypophyseal tract. Moreover, the neurosecretory pathway or pathways to the CSF are spared by disease processes that effectively ablate normal neurohypophyseal secretion into peripheral blood, suggesting that the two pathways are also anatomically distinct. Although the neurosecretory cells that secrete AVP into CSF are separate from those that secrete into peripheral blood, the two cell populations appear to respond similarly to physiological and pathophysiological stimuli.

The functional significance of these antidiuretic peptides in human CSF is not known. However, several hypotheses have been advanced and are at present undergoing further investigation. Most interesting are the proposals that AVP may influence memory and learning or regulate the formation of CSF.

References

1. Arimura, A., Saito, T., Bowers, C. Y., Schally, A. V.: Pituitary–adrenal activation in rats with hereditary hypothalamic diabetes insipidus. *Acta Endocrinol.* **54:**155–165, 1967.
2. Barker, J. L.: Peptide regulation of neuronal excitability: Evidence for a neurohormonal role. In Buckley, J. P., Ferrario, C. M. (eds.): *Central Actions of Angiotensin and Related Hormones*. New York, Pergamon Press, 1977, pp. 29–51.
3. Barker, J. L., Gainer, H.: Peptide regulation of bursting pacemaker activity in a molluscan neurosecretory cell. *Science* **184:**1371–1373, 1974.
4. Barker, J. L., Smith, T. G.: Peptides as neurohormones. In Cowan, W. M., Ferrendelli, J. A. (eds.): *Approaches to Cell Biology of Neurons*. Bethesda, Maryland, Society for Neuroscience, 1977, pp. 340–373.
5. Bartter, F. C.: The syndrome of inappropriate secretion of antidiuretic hormone (SIADH). *Dis. Mon.* **11:**1–47, 1973.
6. Bergland, R. M., Davis, S. L., Page, R. B.: Pituitary secretes to brain. *Lancet* **2:**275–278, 1977.
7. Blask, D. E., Vaughan, M. K., Reiter, R. J., Johnson, L. Y.: Influence of arginine vasotocin in the estrogen-induced surge of LH and FSH in adult ovarectomezed rats. *Life Sci.* **23:**1035–1040, 1978.
8. Brightman, M. W., Polay, S. L.: The fine structure of ependyma of the brain of the rat. *J. Cell Biol.* **19:**415–439, 1967.
9. Calb, M., Goldstein, R., Pavel, S.: Diurnal rhythm of vasotocin in the pineal of the male rat. *Acta Endocrinol.* **84:**523–526, 1977.
10. Cheesman, D. W.: Structural elucidation of a gonadotropin-inhibiting substance from the bovine pineal gland. *Biochim. Biophys. Acta* **207:**247–253, 1970.
11. Cheesman, D. W., Osland, R. B., Forsham, P. H.: Effects of 8-arginine vasotocin on plasma prolactin and follicle-stimulating hormone surges in the proestrous rat. *Proc. Soc. Exp. Biol. Med.* **156:**369–372, 1977.
12. Coculescu, M., Pavel, S.: Arginine vasotocin-like activity of cerebrospinal fluid in diabetes insipidus. *J. Clin. Endocrinol. Metab.* **36:**1031–1032, 1973.
13. Cushing, H., Goetsch, E.: Concerning the secretion of the infundibular lobe of the pituitary body and its presence in the cerebrospinal fluid. *Am. J. Physiol.* **27:**60–86, 1910.
14. Czaczkes, J. W., Kleeman, C. R., Koenig, M.: Physiologic studies of antidiuretic hormone by its direct measurement in human plasma. *J. Clin. Invest.* **43:**1625–1640, 1964.
15. Demoulin, A., Hudson, B., Franchimont, P., Legros, J. J.: Arginine-vasotocin does not affect gonadotrophin secretion *in vitro*. *J. Endocrinol.* **72:**105–106, 1977.
16. de Wied, D.: Effects of peptide hormones on behavior. In Ganong, W. F., Luciano, M. (eds.): *Frontiers in Neuroendocrinology 1969*. New York, Oxford University Press, 1969, pp. 97–140.
17. de Wied, D.: Behavioral effects of intraventricularly administered vasopressin and vasopressin fragments. *Life Sci.* **19:**685–690, 1976.
18. de Wied, D., Bohus, B.: Long and short term effects on retention of a conditioned avoidance response in rats with long acting pitressin and alpha MSH. *Nature (London)* **212:**1484–1486, 1966.
19. DiChiro, G., Hammock, M. K., Bleyer, W. A.: Spinal descent of cerebrospinal fluid in man. *Neurology* **26:**1–8, 1976.
20. Dodge, P. R., Fishman, M. A.: The choroid plexus—two way traffic? *N. Engl. J. Med.* **283:**316–317, 1970.
21. Dogterom, J., van Wimersma Greidanus, T. J. B., deWied, D.: Vasopressin in cerebrospinal fluid and plasma of man, dog, and rat. *Am. J. Physiol.* **234:**E463–E467, 1978.
22. George, J. M.: Immunoreactive vasopressin and oxytocin: Concentrations in individual human hypothalamic nuclei. *Science* **200:**342–343, 1978.
23. George, J. M., Jacobowitz, D. M.: Localization of vasopressin in discrete areas of the rat hypothalamus. *Brain Res.* **93:**363–366, 1975.
24. Goldstein, R., Pavel, S.: Vasotocin release into the cerebrospinal fluid of cats induced by luteinizing hormone releasing hormone, thyrotrophin releasing hormone and growth hormone release-inhibiting hormone. *J. Endocrinol.* **75:**175–176, 1977.

25. GUPTA, K. K.: Antidiuretic hormone in cerebrospinal fluid. *Lancet* **1**:581, 1969.
26. HELLER, H.: Neurohypophyseal hormones in the cerebrospinal fluid. In Sterba, G. (ed.): *Zirkumventrikulare Organe und Liquor*. Jena, Gustav Fischer Verlag, 1969, pp. 235–242.
27. HELLER, H., HASAN, S. H., SAIFI, A. Q.: Antidiuretic activity in the cerebrospinal fluid. *J. Endocrinol.* **41**:273–280, 1968.
28. HERRING, P. T.: The histological appearances of the mammalian pituitary body. *Q. J. Exp. Physiol.* **1**:121–159, 1908.
28a. JENKINS, J. S., MATHER, H. M., ANG, V.: Vasopressin in human cerebrospinal fluid. *J. Clin. Endocrinol. Metab.* **50**:364–367, 1980.
29. JOHNSON, L. Y.: The effects of arginine vasotocin, a pineal peptide, on prolactin secretion in the female rat. *Anat. Rec.* **2**:143, 1978.
30. KNIGGE, C. M., SCOTT, D. E.: Structure and function of the median eminence. *Am. J. Anat.* **129**:223–244, 1970.
31. KNOWLES, F., VOLLRATH, L.: A functional relationship between neurosecretory fibres and pituicytes in the eel. *Nature (London)* **208**:1343, 1965.
32. KNOWLES, F., VOLLRATH, L.: Neurosecretory innervation of the pituitary of the eels *Anguilla* and *Conger*. I. The structure and function of the neuro-intermediate lobe under normal and experimental conditions. *Philos. Trans. R. Soc. Lond. Ser. B* **1**:121–159, 1966.
33. KOZLOWSKI, G. P., BROWNFIELD, M. S., SCHULTZ, W. J.: Neurosecretory pathways to the choroid plexus. *IRCS Med. Sci.* **4**:299, 1976.
34. KWAAN, H. C., BARTELSTONE, J. H.: Corticotropin release following injections of minute doses of arginine vasopressin in the third ventricle of the dog. *Endocrinology* **65**:982–985, 1959.
35. LEGROS, J. J., LOUIS, F., DEMOULIN, A., FRANCHIMONT, P.: Immunoreactive neurophysins and vasotocin in human foetal pineal glands. *J. Endocrinol.* **69**:289–290, 1976.
36. LUERSSEN, T. G., SHELTON, R. L., ROBERTSON, G. L.: Evidence for separate origin of plasma and cerebrospinal fluid vasopressin. *Clin. Res.* **25**:14A, 1977.
37. LUERSSEN, T. G., ROBERTSON, G. L.: Arginine vasopressin in human cerebrospinal fluid. *J. Lab. Clin. Med.* (in press).
38. MACCUBBIN, D. A., VAN BUREN, J. M.: A quantitative evaluation of hypothalamic degeneration and its relation to diabetes insipidus following interruption of the human hypophyseal stalk. *Brain* **86**:443–468, 1963.
39. MCCANN, S. M., ANTUNES-RODRIGUEZ, J., NALLAR, R., VALTIN, H.: Pituitary–adrenal function in the absence of vasopressin. *Endocrinology* **79**:1058–1064, 1966.
40. MILCU, S. M., PAVEL, S., NEACSU, C.: Biological and chromatographic characterization of a polypeptide with pressor and oxytocic activities isolated from bovine pineal gland. *Endocrinology* **72**:563–566, 1963.
41. MILHORAT, T. H., MOSHER, M. B., HAMMOCK, M. K., MURPHY, C. F.: Evidence for choroid plexus absorbtion in hydrocephalus. *N. Engl. J. Med.* **283**:286–289, 1970.
42. NICOLL, R. A., BARKER, J. L.: The pharmacology of recurrent inhibition in the supraoptic neurosecretory system. *Brain Res.* **35**:501–511, 1971.
43. PASQUALINI, R. Q.: Diabetes insipidus without thirst. *Lancet* **1**:889–890, 1959.
44. PASQUALINI, R. Q., CODEVILLA, A.: Thirst suppressing ("antidipsetic") effect of pitressin in diabetes insipidus. *Acta Endocrinol.* **30**:37–41, 1959.
45. PAVEL, S.: Tentative identification of arginine vasotocin in human cerebrospinal fluid. *J. Clin. Endocrinol. Metab.* **31**:369–371, 1970.
46. PAVEL, S.: Opposite effects of vasotocin injected intrapituitarily and intraventricularly on corticotropin release in mice. *Experientia* **31**:1469–1470, 1975.
47. PAVEL, S., CALB, M., GEORGESCU, M.: Reversal of the effects of pinealectomy on the pituitary prolactin content in mice by very low concentrations of vasotocin injected into the third cerebral ventricle. *J. Endocrinol.* **66**:289–290, 1975.
48. PAVEL, S., DORCESCU, M., PETRESCU-HOLBAN, R., GHINEA, E.: Biosynthesis of a vasotocin-like peptide in cell cultures from pineal glands of human fetuses. *Science* **181**:1252–1253, 1973.
49. PAVEL, S., GOLDSTEIN, R., CALB, M.: Vasotocin content in the pineal gland of foetal, newborn and adult male rats. *J. Endocrinol.* **66**:283–284, 1975.
50. PAVEL, S., GOLDSTEIN, R., GHEORGHIU, C., CALB, M.: Pineal vasotocin: Release into cat cerebrospinal fluid by melanocyte-stimulating hormone release–inhibiting factor. *Science* **197**:179–180, 1977.
51. PAVEL, S., MATRESCU, L., PETRESCU, M.: Central corticotropin inhibition by arginine vasotocin in the mouse. *Neuroendocrinology* **12**:371–375, 1973.
51a. PAVEL, S.: Presence of relatively high concentrations of arginine vasotocin in the cerebrospinal fluid of newborns and infants. *J. Clin. Endocrinol. Metab.* **50**:271–273, 1980.
52. RAICHLE, M. E., GRUBB, R. L.: Regulation of brain water permeability by centrally released vasopressin. *Brain Res.* **143**:191–194, 1978.
53. RASMUSSEN, A. T., GARDNER, W. J.: Effects of hypophyseal stalk section on the hypophysis and hypothalamus of man. *Endocrinology* **27**:219–226, 1940.
54. ROBERTSON, G. L.: The regulation of vasopressin function in health and disease. *Recent Prog. Horm. Res.* **33**:333–385, 1977.
55. ROBERTSON, G. L., MAHR, E. A., ATHAR, S., SINHA, T.: Development and clinical application of a new method for the radioimmunoassay of arginine vasopressin in human plasma. *J. Clin. Invest.* **52**:2340–2352, 1973.
56. ROBINSON, A. G., ZIMMERMAN, E. A.: Cerebrospinal fluid and ependymal neurophysin. *J. Clin. Invest.* **52**:1260–1267, 1973.
57. RODRIGUEZ, E. M.: Morphological and functional relationship between the hypothalamo–neurohypo-

physeal system and cerebrospinal fluid. In Bargmann, W., Scharrer, E. (eds.): *Aspects of Neuroendocrinology*. New York, Springer-Verlag, 1969, pp. 354–365.
58. Rodriguez, E. M., Heller, H.: Antidiuretic activity and ultrastructure of the toad choroid plexus. *J. Endocrinol.* **46**:83–91, 1970.
59. Rosenbloom, A. A., Fisher, D. A.: Radioimmunoassayable AVT and AVP in adult mammalian brain tissue: Comparison of normal and Brattleboro rats. *Neuroendocrinology* **17**:354–361, 1975.
60. Rosenblum, I. Y., Benson, B., Hruby, V. J.: Chemical differences between bovine pineal antigonadotropin and arginine vasotocin. *Life Sci.* **18**:1367–1374, 1976.
61. Rowe, J. W., Baylis, P. H., Robertson, G. L.: Stimulation of vasopressin secretion by pitressin in man—influence of sodium depletion and age. *Clin. Res.* **26**:494A, 1978.
62. Rudman, D., Chawla, R. K.: Antidiuretic peptide in mammalian choroid plexus. *Am. J. Physiol.* **230**:50–55, 1976.
63. Schultz, W. J., Brownfield, M. S., Kozlowski, G. P.: The hypothalamo-choroidal tract. II. Ultrastructural response of the choroid plexus to vasopressin. *Cell Tussue Res.* **178**:129–141, 1977.
64. Scott, D. E., Dudley, G. K., Gibbs, F. P., Brown, G. M.: The mammalian median eminence. In Knigge, K. M., Scott, D. E., Weindl, A. (eds.): *Brain–Endocrine Interation. Median Eminence: Structure and Function*. Basel, S. Karger, 1972, pp. 35–49.
65. Skowsky, W. R., Fisher, D. A.: Fetal neurohypophyseal arginine vasopressin and arginine vasotocin in man and sheep. *Pediatr. Res.* **11**:627–630, 1977.
66. Smoller, C. G.: Neurosecretory processes extending into the third ventricle: Secretory or sensory? *Science* **47**:882–884, 1965.
67. Sterba, G.: Ascending neurosecretory pathways of the peptidergic type. In Knowles, F., Vollrath, L. (eds.): *Neurosecretion—The Final Neuroendocrine Pathway*. New York, Springer-Verlag, 1974, pp. 39–42.
68. Sterba, G.: Cerebrospinal fluid and hormones. In Mitro, A. (ed.): *Ependyma and Neurohormonal Regulation*. Bratislava, Veda, 1974, pp. 143–179.
69. van Wimersma Greidanus, T. B., Bohus, B., de Wied, D.: The role of vasopressin in memory consolidation. *J. Endocrinol.* **64**:30p, 1975.
70. van Wimersma Greidanus, T. B., Bohus, B., de Wied, D.: The role of vasopressin in memory processes. *Prog. Brain Res.* **42**:135–142, 1975.
71. Vaughan, M. K., Blask, D. E., Johnson, L. Y., Reiter, R. J.: Prolactin releasing activity of arginine vasotocin *in vitro*. *Horm. Res.* **6**:342–350, 1975.
72. Vaughan, M. K., Little, J. C., Johnson, L. Y., Blask, D. E., Vaughan, G., Reiter, R. J.: Effects of malatonin and natural and synthetic analogues of arginine vasotocin on plasma prolactin levels in adult male rats. *Horm. Res.* **9**:236–246, 1978.
73. Vorherr, H., Bradbury, M. W. B., Hoghoughi, M., Kleeman, C. R.: Antidiuretic hormone in cerebrospinal fluid during endogenous and exogenous changes in its blood level. *Endocrinology* **83**:246–250, 1968.
74. Vigh, B., Vigh-Teichmann, I.: A comparison between ependymosecretion and neurosecretion. In Sterba, G. (ed.): *Zirkumventrikulare Organe und Liquor*. Jena, Gustav Fischer Verlag, 1969, pp. 227–230.
75. Vigh-Teichmann, I.: Fiber connections of the hypothalamic CSF contracting neurosecretory cells. In Knowles, F., Vollrath, L. (eds.): *Neurosecretion—The Final Neuroendocrine Pathway*. New York, Springer-Verlag, 1974, pp. 324–325.
76. Vincent, J. D., Arnauld, E.: Vasopressin as neurotransmitter in the CNS: Some evidence from the supraoptic neurosecretory system. *Prog. Brain Res.* **42**:57–66, 1975.
77. Zaidi, S. M., Heller, H.: Can neurohypophysial hormones cross the blood–cerebrospinal fluid barrier? *J. Endocrinol.* **60**:195–196, 1974.
78. Zimmerman, E. A., Kozlowski, G. P., Scott, D. E.: Axonal and ependymal pathways for the secretion of biologically active peptides into hypophyseal portal blood. In Knigge, K. M., Scott, D. E., Kobayashi, H., Ishi, S. (eds.): *Brain–Endocrine Interaction II. The Ventricular System*. Basel, S. Karger, 1975, pp. 123–134.

CHAPTER 44

Significance and Function of Neuropeptides in Cerebrospinal Fluid

Ivor M. D. Jackson

1. Introduction

Although the mammalian anterior pituitary gland lacks a direct nerve supply from the brain, it is now well established that the central nervous system (CNS) exercises control over the secretion of each adenohypophyseal hormone through "releasing factors" synthesized and secreted by peptidergic neurons in the hypothalamus.[13,85] As postulated over 30 years ago ("portal vessel chemotransmitter hypothesis"), the hypothalamus secretes into the portal capillaries of the median eminence (ME) specific pituitary regulatory substances that are transported to the anterior pituitary by the portal vessels of the pituitary stalk[47] (see Fig. 1). The isolation and synthesis of three of these hypothalamic hypophysiotropic hormones, thyrotropin-releasing hormone (TRH), luteinizing-hormone-releasing hormone (LH-RH), and growth-hormone-release-inhibiting hormone (somatostatin) (Fig. 2), have provided powerful tools for the investigation of hypothalamic–pituitary function and have permitted the development of specific radioimmunoassays for the measurement of these substances at low concentrations.[120,153] An unanticipated outcome of these methodological developments was the finding that most of neural TRH and somatostatin are located in regions of the CNS outside the hypothalamus.[59,60] Even more surprising was the discovery that both TRH[60,86] and somatostatin are present in anatomical locations outside the nervous system altogether. Further, these hypothalamic hormones are present in primitive species wherein no pituitary occurs.[61] The functional significance of the widespread phylogenetic distribution is not known for sure, but it has been postulated that these substances may initially have evolved as primitive neurotransmitters or modulators of neurotransmission and that only late in evolution did they acquire the function of regulating anterior pituitary hormone secretion.[60]

It is believed that the hypothalamic peptidergic neurons that produce the hypophysiotropic hormones are in turn regulated by neurotransmitters, especially of the monoaminergic variety, and that the peptidergic neuron acts as a "neuroendocrine transducer," converting neuronal information from the brain into chemical information.[158] In addition to the classical hypothalamic releasing hormones, a number of other neural peptides, including substance P, neurotensin, and the endorphins, have recently been recognized to be present in the hypothalamus as well as in extrahypothalamic brain locations and extraneural sites. Some of these peptides have the ability to influence pituitary function, but their physiological roles in the hypothalamus and elsewhere have not been fully elucidated at this time.

The possible role of the cerebrospinal fluid (CSF) in the regulation of pituitary function was first put forward by Vesalius, the father of modern anatomy, in his *De Humani Corporis Fabrica,* which was published in 1543. Vesalius described the drainage of

Ivor M. D. Jackson, M.D. • Division of Endocrinology, Department of Medicine, Tufts–New England Medical Center Hospital, Boston, Massachusetts 02111.

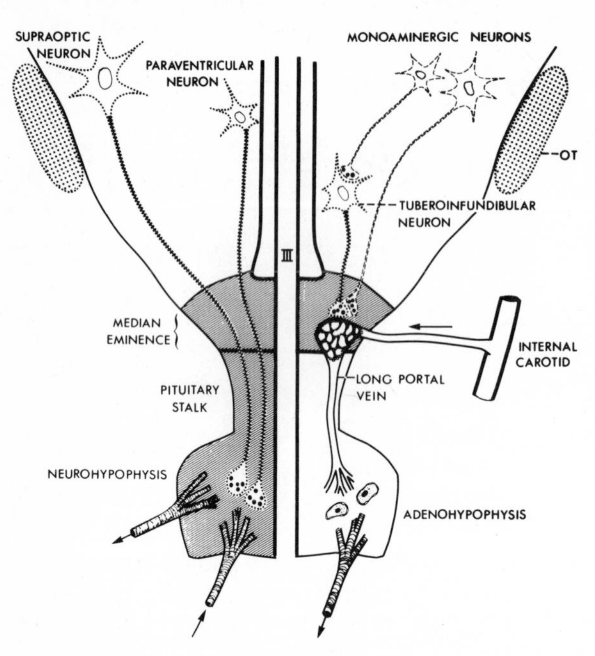

Figure 1. Diagram of hypothalamic–pituitary axis in coronal section. *Left:* Hypothalamic–neurohypophyseal system. Supraoptic and paraventricular axons terminate on blood vessels in posterior pituitary (neurohypophysis). *Right:* Hypothalamic–adenohypophyseal system. Tuberoinfundibular neurons, believed to be source of hypothalamic regulatory hormones, terminate on capillary plexus in median eminence (ME). Pituitary portal system is derived from branches of internal carotid, which forms a primary capillary bed in ME. Long portal veins drain the capillary plexus into sinusoids of anterior pituitary (adenohypophysis). Supraoptic, paraventricular, and tuberoinfundibular neurons are all classed as neurosecretory cells. Activity of tuberoinfundibular neurons is influenced by monoaminergic cells. Reproduced from Martin *et al.*,[85] with permission.

CSF through the floor of the third ventricle [named *infundibulum* (Lat.) because of its resemblance to a funnel] into the pituitary, and thence into the nose to form mucus [*pituita* (Lat.), from which the modern term pituitary is derived]. In more recent years, the possible role of the CSF in neuroendocrine function has received much attention.[76] The ventromedial hypothalamus contains specialized cells (ependymal tanycytes) that extend from the floor of the third ventricle through the interstitial space of the ME and come into intimate contact with the capillaries of the primary portal plexus. It has been postulated that the hypophysiotropic hormones are secreted into the ventricular system and taken up by the luminal processes of the tanycytes for release in apposition to the primary capillaries of the hypophyseal portal vessels.[76] The presence of the hypophysiotropic hormones in extrahypothalamic brain locations provides the CSF with even more potential significance, since this medium may be a conduit by which peptides produced in the hypothalamus may reach extrahypothalamic brain sites and/or pep-

TRH

Pyro-GLU-HIS-PRO-NH_2

LH-RH

Pyro-GLU-HIS-TRP-SER-TYR-GLY-LEU-ARG-PRO-GLY-NH_2
1 2 3 4 5 6 7 8 9 10

SOMATOSTATIN

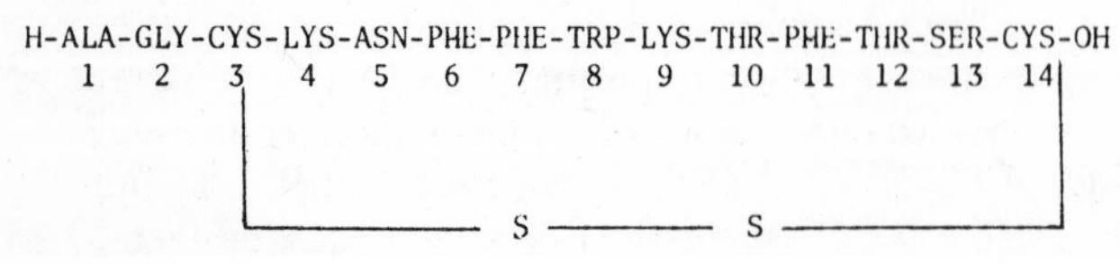

Figure 2. Chemical structure of hypothalamic releasing hormones thyrotropin-releasing hormone (TRH), luteinizing-hormone-releasing hormone (LH-RH), and somatostatin.

tides produced in such brain areas may reach other locations in the CNS or pituitary or both to affect function.[57a]

In this review, I will describe the anatomical distribution of the hypothalamic hormones and other neural peptides throughout the hypothalamus and extrahypothalamic brain and discuss what is known about their function with respect to the regulation of pituitary hormone secretion and the modulation of neuronal function. The mechanisms by which neural peptides might reach the CSF will be examined, and the potential physiological role of this biological fluid, as a means through which their effects are mediated, will be discussed. The physiological effects of neural peptides placed in the CSF will be described, and finally, the data on measurements of endogenous peptides in CSF and their significance will be considered.

2. Distribution of Neural Peptides in CNS

2.1. Hypothalamus

2.1.1. Hypophysiotropic Hormones. *a. TRH.* Immunoreactive (IR) TRH is readily detectable in extracts of hypothalami from mammalian and submammalian vertebrates, although in species lower than birds, the tripeptide does not stimulate pituitary thyroid function.[60] More recent studies have also demonstrated substantial quantities of TRH throughout the hypothalamus and stalk–median eminence (SME) of humans.[63] TRH has been reported to show immunofluorescence and staining of nerve terminals in the medial part of the external layer of the ME.[50] With the use of immunofluorescence, TRH-positive nerve cell bodies have been detected in the dorsomedial nucleus of the hypothalamus as well as the perifornical region.[53] No immunopositive tanycytes have been reported.

Table 1. Effect of Lesion of "Thyrotropic Area" of Hypothalamus on Brain Distribution of Thyrotropin-Releasing Hormone (TRH) in Rat[a]

	TRH (ng/organ) (mean ± S.E.M.)		
	Lesion	Control	Significance
Hypothalamus	3.6 ± 0.3	9.3 ± 0.5	$P < 0.001$
Extrahypothalamic brain	17.0 ± 9.0	18.9 ± 0.7	NS

[a] From Jackson and Reichlin.[62]

TRH is of physiological importance in the regulation of thyroid-stimulating hormone (TSH) secretion from the anterior pituitary. Evidence for this is demonstrated by the effects of ablation of the "thyrotrophic area" of the hypothalamus[62]—a region that involves the paraventricular nucleus. Such a lesion induces hypothyroidism in the rat and reduces the hypothalamic content of TRH by two thirds (Table 1). Utilizing a microdissection technique that allows discrete nuclei to be dissected from neural tissue, Brownstein *et al.*[18] showed that the highest concentration of TRH in the hypothalamus was found in the "thyrotrophic area." These findings are consistent with our report of a gradient of TRH from dorsal hypothalamus (0.05 ng/mg tissue) to SME (3.6 ng/mg tissue.)[61]

b. LH-RH. The luteinizing-hormone (LH)-releasing factor or hormone (LH-RH) was isolated from porcine[133] and ovine[24] hypothalamic tissue and its structure shown to be that of a simple decapeptide (Fig. 2). Following the development of radioimmunoassays, extracts of SME of rat and man,[4] as well as sheep, monkey, and pig,[93] were shown to cross-react equally well with antisera to LH-RH. This peptide releases LH as well as follicle-stimulating hormone (FSH) in a wide variety of mammalian species,[134] suggesting that it is the FSH-RH as well, so that some call this decapeptide gonadotropin-releasing hormone (GnRH).

The preparation of specific antibodies to LH-RH has allowed the utilization of immunohistochemical techniques to study the neural distribution of LH-RH (Fig. 3). Specific fluorescence has been demonstrated in the ME of a number of mammalian species,[9] and under electron microscopy, the nerve endings of the external zone of the ME were shown to contain secretory granules with immunoperoxidase staining.[108] LH-RH reactive perikarya have been reported in many mammals including man, monkey, rabbit, rat, cat, dog, and guinea pig (see reference 8 for full references).

Using the immunoperoxidase system, Zimmerman *et al.*[161] have reported remarkable amounts of IR LH-RH throughout the course of numerous tanycytes and in some arcuate perikarya of the mouse hypothalamus (these workers also reported vasopressin and neurophysin in tanycytes of the ME). Zimmerman *et al.*[161] have suggested that the presence IR LH-RH in tanycytes provides evidence that the ventricular system provides a pathway that is of importance in the neural regulation of gonadal function. Studies by Naik,[90] who reported that the intensity of LH-RH immunofluorescence changes in

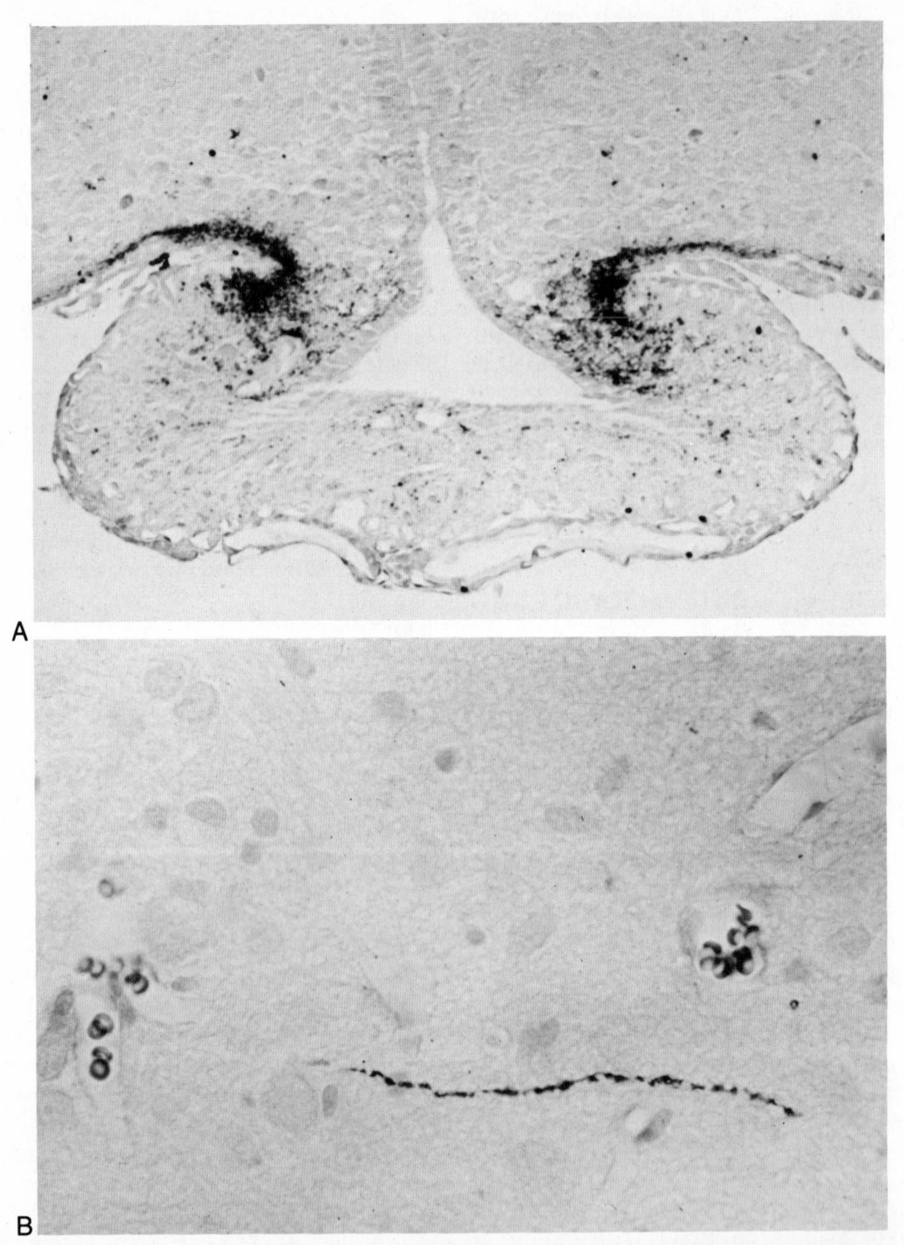

Figure 3. Distribution of luteinizing-hormone-releasing hormone (LH-RH) in rodent hypothalamus as demonstrated by immunohistochemical staining (peroxidase technique). (A) Frontal section at level of median eminence (ME) in mouse. LH-RH-containing cell processes are seen to sweep around lateral walls to enter contact zone of nerves and portal vessels in ME. (B) Beaded LH-RH-containing nerve process in anterior hypothalamus of rat. It is believed that localized swellings in neuron are accumulations of secretory product; these are analogous to Herring body of supraoptic and paraventricular–hypophyseal pathway.[85] Courtesy of L. Alpert.

tanycyte cells during the estrus cycle of the rat, provide support for this view, but no staining of the tanycytes was found by Setalo *et al.*,[136a] Gross,[44] or Lechan *et al.*[80] in the rat or mouse hypothalamus.

c. Somatostatin. Following the development of radioimmunoassays for this tetradecapeptide (see Fig. 2), it was shown that high concentrations are present in the rat hypothalamus, especially in the ME and arcuate nucleus regions,[19] providing anatomical evidence that this peptide is of physiological importance in its inhibition of growth hormone (GH) and TSH secretion from the adenohypophysis.[40] Immunohistochemical studies, including electron microscopy, have shown somatostatin located in membrane-bound secretory granules within nerve endings of the external layer of the ME near the portal capillaries and in the ventromedial nucleus (VMN).[109] The cellular source of the somatostatin in the medial basal hypothalamus (MBH) appears to be somatostatin-containing perikarya in the preoptic and anterior periventricular nuclei as shown by immunohistochemical staining.[1]

d. Other Hypothalamic Releasing Factors (Table 2). Corticotropin-releasing factor (CRF) activity is present in extracts of hypothalami,[130] but the chemical nature of a specific physiological CRF that regulates adrenocorticotropic hormone (ACTH) secretion from the adenohypohysis has not been reported. Vasopressin, which is known to effect the release of both ACTH and GH from the mammalian pituitary,[13] is synthesized in the magnocellular neurons of the supraoptic nuclei (SON). Although the terminals of the SON end around systemic capillaries of the posterior pituitary, some axons appear to abut the portal capillary blood. Since vasopressin was occasionally noted in the tanycytes of rodent hypothalami, it remains possible that vasopressin is normally secreted in CSF, which may serve as a source for tanycyte absorption.[162] The physiological significance of vasopressin in relation to ACTH regulation is not known.

Table 2. Hypothalamic Releasing Factors Postulated But Not Yet Characterized

Corticotropin-releasing factor (CRF)
Growth-hormone-releasing factor (GH-RF)
Prolactin-inhibiting factor (PIF)
Prolactin-releasing factor (PRF)
Melanocyte-stimulating-hormone-releasing and -inhibiting factors (MSH-RF and MSH-IF)

GH-releasing factor (GH-RF) activity is present in hypothalamic extracts and has been localized to the VMN[84] and identified by bioassay in pituitary portal blood.[154] In addition to vasopressin, a number of amino acids and peptides have been reported to stimulate GH secretion in man,[84] but the site and mechanisms of action of each have not been fully established. Substance P (SP), neurotensin, and the endorphins can also stimulate the release of GH and prolactin (PRL)[124,125] (see also later).

There is some evidence that a specific peptide with PRL-inhibiting factor (PIF) activity may exist in the hypothalamus,[14] though it has been suggested that hypothalamic PIF activity can be accounted for solely by dopamine.[133] Although TRH can stimulate the release of PRL directly from pituitary tissue, its physiological importance in PRL regulation is uncertain; however, a specific hypothalamic PRL-releasing factor (PRF) may exist.[14] It should be noted that arginine vasopressin from the neurohypophysis and arginine vasotocin, the related peptide present in the mammalian pineal gland, can also stimulate PRL release in the rat.[14,148] Their possible physiological importance as PRFs requires further evaluation.

2.1.2. Other Neural Peptides. SP is found in high concentration in mammalian hypothalamus, but is present only to a very small extent in the rat ME, especially in the external layer.[52] However, in this species, numerous hypothalamic SP-positive cell bodies have been observed by Hökfelt *et al.*[52] in the VMN and dorsomedial nucleus (DMN). In contrast to these findings in the rat, Hökfelt *et al.*[54] observed a dense plexus of SP-positive nerve terminals in the external layer of the primate ME close to blood vessels supplying the anterior pituitary gland. These findings suggest that in both monkey and man, SP may be directly involved in regulating the secretion of anterior pituitary hormones. Neurotensin, as well as opioid peptides, have also been found in high concentration by immunohistochemistry in the hypothalamus and ME.[53]

Vasoactive intestinal peptide (VIP) has been found in hypophyseal portal blood of rats in concentrations approximately 19 times those in peripheral blood, suggesting that this peptide may function as a hypophysiotropic hormone. High concentrations of IR VIP have been found in human ME, contrasting with the low content of this peptide in the same area of rat brain.[131]

A number of other neural peptides have also been localized to the hypothalamus by immunoassay of tissue extracts or immunohistochemistry. These include angiotensin, gastrin, cholecystokinin (CCK),[53] and bombesin.[17] However, their physiological role in the regulation of anterior pituitary function is unknown.

Although immunohistochemical staining of the heptadecapeptide (gastrin$_{17}$) has been observed in the hypothalamus, periventricular area, and medial external layer of the ME,[53] it now appears that the sole location of a brain peptide reacting with antibody *specific* to gastrin-17 or the tetratriaconta peptide, gastrin-34, is limited to the pituitary gland—especially the neurohypophysis.[119] Although CCK may have a functional role in appetite regulation, it is not known whether this peptide directly affects pituitary hormone secretion.[33]

Immunoreactive angiotensin II (AII)-like material has been demonstrated in high concentration in the external layer of the ME.[53] Cell bodies have been detected in the DMN and VMN. AII may be of physiological importance in the regulation of vasopressin and ACTH secretion.[121]

Bombesin (a tetradecepeptide, first isolated from amphibian skin, and present in the mammalian gastrointestinal tract) immunoreactivity has been demonstrated in the rat hypothalamus,[17] and may be a physiological regulator of PRL and TSH secretion.

The opioid peptides and neurohypophyseal peptides are also present in the hypothalamus, but they will be covered elsewhere in this volume (Chapters 45 and 43, respectively). A list of neural peptides found in the CNS is given in Table 3.

2.2. Extrahypothalamic Brain and Spinal Cord

2.2.1. Hypophysiotropic Hormones

a. TRH. Significant concentrations of TRH are found in the rat extrahypothalamic brain.[61] Although such concentrations are small when compared with the levels in the hypothalamus, quantitatively, over 70% of total brain TRH is found outside this region.[97,155] To determine whether this extrahypothalamic brain TRH might be derived from the hypothalamus, possibly via the CSF, we studied the effects of classic hypothalamic "thyrotrophic area" lesions, which brought about a reduction in hypothalamic TRH by two thirds.[62] The extrahypothalamic brain TRH content in these animals was unaffected, suggesting that synthesis in such areas occurs *in situ* (see Table 1). Following hypothalamic deafferentation, not only were the levels of TRH in the extrahypothalamic brain unaltered, but also there was a marked reduction in hypothalamic content.[20] These findings suggest that much of hypothalamic TRH may be synthesized by cells outside this area or dependent on extrahypothalamic neural connection. In addition, they provide evidence *against* the view that TRH located in the hypothalamus is transported from extrahypothalamic brain sites *via the CSF*.

Table 3. Neural Peptides Found in Hypothalamus and Extrahypothalamic Brain

Hypothalamic releasing hormones (see Fig. 2)
Neurohypophyseal hormones (Chapter 43)
Opioid peptides (Chapter 45)
Adenohypophyseal hormones (Chapter 41)
Substance P (SP)
Neurotensin
Vasoactive intestinal peptide (VIP)
Angiotensin II (AII)
Gastrin/cholecystokinin (CCK)
Bombesin

Substantial quantities of TRH have been detected in the spinal cord,[68] and immunohistochemical studies have localized TRH around the motoneurons of the spinal cord[50] so that TRH present in spinal cord, as well as in brain, may contribute to levels of the endogenous peptide found in the CSF, especially if drawn from the lumbar region. Thus the concentration of TRH in third ventricular CSF may have different implications from cisternal or lumbar determinations. Similar caveats operate with respect to studies of other neural peptides in CSF.

b. LH-RH. Initial reports of large quantities of LH-RH in extracts of rat extrahypothalamic brain[152] have not been confirmed by others in the rat or mouse.[2,80] In this laboratory, we have found that the rat hypothalamic content of LH-RH was 7 ng and the content in cortex, brainstem, and cerebellum combined amounted to 1.6 ng (about 17% of total brain content). The spinal cord did not contain LH-RH. As with TRH, deafferentation of the rat hypothalamus causes a marked reduction in LH-RH content of the MBH,[21] suggesting that such LH-RH arises from, or is controlled by, cells elsewhere in the brain. These findings suggest that such brain regulation involves neural connections rather than CSF transport, which should be unaffected by deafferentation.

c. Somatostatin. Somatostatin is widely distributed throught the mammalian extrahypothalamic brain.[19,105] As for the other hypophysiotropic hormones, hypothalamic deafferentation caused no reduction in the extrahypothalamic content.[23] Like

TRH, somatostatin is present in the spinal cord but localized to the sensory side, as shown by immunofluorescence in some neuronal cell bodies of dorsal spinal root ganglia as well as in fibers in the substantia gelatinosa.[51]

d. Other Hypophysiotropic Factors. Extrahypothalamic brain tissue of stressed rats has been reported to contain CRF activity.[82] PIF activity has been reported to be widely distributed throughout extrahypothalamic rat brain,[147] but whether this reflects peptide or catecholamine activity has not been fully clarified.

2.2.2. Other Neural Peptides. SP is widely but unevenly distributed throughout the CNS. High concentrations of IR SP have been reported in the mesencephalon, preoptic region, and substantia nigra of the rat brain.[22] Immunohistochemical staining has localized SP to the spinal cord,[52] where it is found in higher concentration in the dorsal roots than in the ventral roots. Neurotensin is distributed extensively throughout the extrahypothalamic brain, and immunohistochemical staining has revealed cell bodies in the brainstem region.[53] Specific immunofluorescence has been reported by the aforementioned workers in the dorsal horn of the spinal cord, probably in interneurons.

VIP is widely distributed throughout the human brain[131] as well as in the ventral horn of the spinal cord.[55]

All the components of the renin–angiotensin system have been demonstrated in the brain.[111] Renin-like activity has been demonstrated in the brain of nephrectomized animals and has been shown to be similar to renal renin, with a molecular weight of 40,000–60,000 daltons. Angiotensinogen (renin substrate) is present in brain tissue. Converting enzyme, which is required to convert angiotensin I (AI) to AII, is found in brain at concentrations higher than those found in any other tissues except lung. Recently, there have been reports of immunocytochemical demonstration of AII in brain tissue unaffected by nephrectomy. Cell bodies have been demonstrated in the perifornical area. Outside the hypothalamus, AII-like immunoreactivity has been found in the brainstem, mesencephalon, and amygdaloid complex, as well as in the dorsal and ventral horns of the spinal cord.[55]

CCK has been identified immunohistochemically in neuronal perikarya of the rabbit cerebral cortex.[142] It now seems likely that the gastrin-like immunoreactivity previously reported throughout the brain represents CCK.[53] Gastrin and CCK belong to the group of hormonal peptides with a dual localization to endocrine cells or nerves or both in the gut on the one hand, and to neurons in the CNS on the other. These peptides include SP, somatostatin, neurotensin, enkephalin, and VIP. Compared to these substances, CCK shows a unique regional distribution.[119] Rehfeld and Kruse-Larsen[119] report concentrations of CCK on the order of 1 nmol/g tissue (wet weight) in the telencephalic gray matter, which they state are 10 times or more above the amounts so far reported for other hormonal peptides, releasing factors, or release-inhibiting factors.

3. Significance and Function of Peptides Located Outside Hypothalamus

3.1. Amine Precursor Uptake and Decarboxylation Hypothesis

The observation that neurons and some endocrine cells that produce peptide hormones share a set of common cytochemical and ultrastructural characteristics was first made by Pearse[107] over a decade ago. This is the amine precursor uptake and decarboxylation (APUD) hypthesis. It was at first believed that these endocrine cells were derived from a common neuroectodermal ancester, the neural crest, but it is now postulated that all peptide-hormone-producing cells are derived from the neural ectoderm, as are all neurons.[45]

Accordingly, the diffuse anatomical distribution of neural peptides throughout the hypothalamus, brain neurons, and endocrine cells of the gastrointestinal tract and their presence in the skin of amphibia[63] can be explained on the basis of a similar ontogenetic history. Pearse[107] has postulated that these peptides are part of a diffuse neuroendocrine system.

3.2. Function of Neural Peptides in Extrahypothalamic Brain and Spinal Cord

There is now much evidence supporting a role for these peptides in brain function quite apart from their effects on the regulation of pituitary function. This evidence will be summarized below.

3.2.1. Hypophysiotropic Hormones. *a. TRH.* The location of TRH in several cranial nerve nuclei of the brainstem and in the anterior horn of the spinal cord[50] suggests a transmitter function for this peptide, especially in the motor system. This hypothesis is supported by pharmacological studies. Hypophysectomized mice pretreated by pargyline, a monoamine oxidase inhibitor, show enhancement of motor activity induced by L-DOPA when TRH is concomitantly administered,[112] thus indicating that the TRH effect is independent of pituitary–thyroid function. Other CNS effects of TRH include potentiation of the excitatory action of acetylcholine (ACh) on cerebral cortical neurons,[160] enhancement of cerebral norepinephrine (NE) turnover,[71] potentiation of behavioral changes following increased 5-hydroxytryptamine (serotonin) accumulation in rats,[42] and alteration of rotational activity in the rat [believed to be a dopamine (DA)-mediated action].[29] Thus, TRH may directly or indirectly influence the action of the commonly accepted neurotransmitters, DA, NE, serotonin, and ACh. TRH has a depressed effect on the electrical activity of single neurons[123] and its presence in the synaptic vesicles and synaptosomes of the extrahypothalamic brain of the rat[156] support a neurotransmitter role for this substance. The psychobiological effects in man, including its reversal of human depression,[115] may reflect its action in modulating neurotransmitter function.

b. LH-RH. Like TRH, LH-RH has been shown to have a depressant action on the excitability of neurons in several areas of the CNS.[123] This decapeptide also been shown to have a role in sex behavior unrelated to gonadal function, for it potentiates lordotic behavior in hypophysectomized, ovariectomized female rats.[110] These findings bespeak a role for LH-RH in neuronal function.

c. Somatostatin. Similar to the findings for TRH and LH-RH, neurophysiological studies have demonstrated the somatostatin affects the excitability of single neurons.[123] Its subcellular location in brain synaptosomes[35] is consistent with a role in neurotransmission. The finding that a certain population of primary sensory nerves contain somatostatin suggests that this substance may act as a depressant neurotransmitter in sensory neurons.[51] It is noteworthy that somatostatin administration induces prolongation of barbiturate anesthesia and shortening of strychnine seizure activity in the rat, effects that are opposite those produced by TRH.[16] These behavioral effects (stimulation of motor activity by TRH but enhancement of sensory depression by somatostatin) are consistent with the anatomical location of these peptides—TRH being located in the motor and somatostatin in the sensory side of the nervous system.

3.2.2. Other Neural Peptides. The presence of SP-like immunoreactivity in high concentration in the dorsal root ganglion of the spinal cord is consistent with its postulated role as a sensory excitatory neurotransmitter.[52] There are electrophysiological studies demonstrating an excitatory action of SP at the supraspinal level and reports that it stimulates adenylate cyclase in a number of brain regions.[52]

There is now much evidence linking a number of neural peptides with specific functional roles in the CNS. Neurotensin is involved in temperature regulation,[92] CCK may affect appetite control,[132,143] and AII appears to influence thirst and blood-pressure regulation.[111] These last two effects are discussed further in Section 4.2.2. Vasopressin, oxytocin, and enkephalin-like immunoreactivity are widely distributed throughout the brain and spinal cord, and there is much evidence to indicate that these neural peptides also have important roles in the function of the nervous system.[53]

4. Mechanisms by Which Neural Peptides Might Reach CSF

4.1. Trans-Median-Eminence Transport

4.1.1. Ependymal Tanycyte Theory. The hypothesis that the specialized ependymal cells (tanycytes) located at the floor of the third ventricle play a role in endocrine regulation was originally proposed on a purely morphological basis by Löfgren.[81] Cytochemical and peroxidase experiments have demonstrated that tanycytes of the median eminence (ME) absorb substances from the ventricular CSF. These cells are most abundant in those regions of the ME that correspond to the "hypophysiotrophic area," and their structural features are compatible with the suggestion that they take up substances from the ventricular CSF and transport them for local release into the adjacent hypothalamic neuropil (Fig. 4). The ME can accumulate amino acids and thyroxine *in vitro* through an energy-dependent mechanism.[138] It has been shown that the cells responsible for this process are probably not neurons, since ME tissue grown in organ culture continues to concentrate certain substances after the neuronal elements have degenerated.[138]

The ependymal tanycytes are particularly suited for transporting substances from CSF to portal blood,[146] since their apices form the floor of the third ventricle and their opposite ends terminate in proximity to the primary capillaries of the long hypophyseal portal vessels.[157] Although the evidence demonstrating a truly secretory function for these cells is tenuous, located in the neck processes of

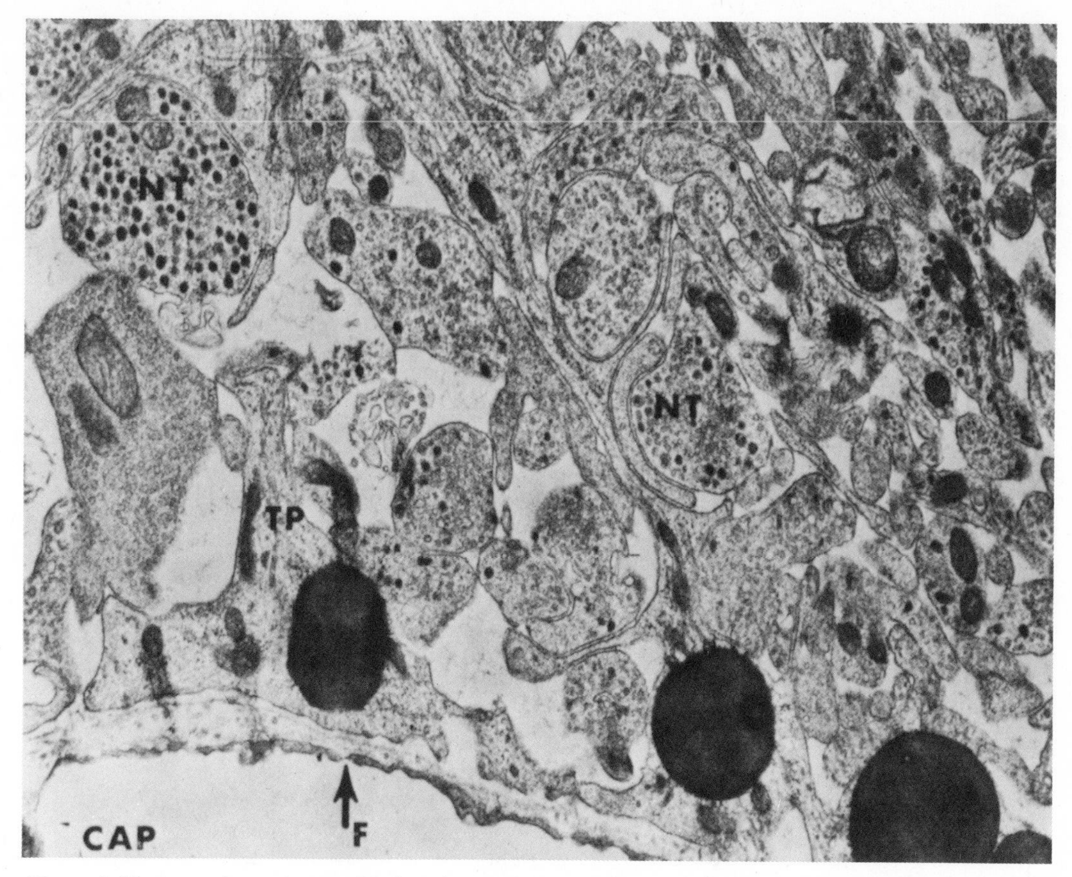

Figure 4. Electron photomicrograph of median eminence (ME) of rat. Capillary lumen (CAP) is evident in lower left corner of figure. Fenestrations (F) of capillary endothelium can be identified. Lipid droplets are contained in tanycyte processes (TP), which together with nerve terminals (NT) abut on capillary. Dense core and "synaptic" vesicles are evident in nerve terminals of neurosecretory neurons. These terminals are believed to contain hypophysiotropic hormones and catecholamines norepinephrine and dopamine.[85] ×18,000. Courtesy of J. Brawer.

ventrally located tanycytes are concentric layers of endoplasmic reticulum that may indicate the elaboration of unique substances.[87]

A large number of hormones, including thyroxine, ACTH, corticosterone, and estradiol, and radiolabeled materials such as NaI, ovine LH, corticosterone, and PRL are transferred from rat ventricular CSF to the hypophyseal portal system, though the time of crossing the ME varies among these substances.[78] The role of tanycytes in the transfer of hormones from CSF to pituitary portal capillaries has been reviewed by Joseph and Knigge.[66] Tight junctions at their apical (ventricular) boundaries suggest that movement of substances from CSF occurs by passage through tanycytes rather than between them. Tanycytes, which line the wall of the ventricular recess, course through the bed of the arcuate nucleus, and it is of note that an unusual system of gap junctions exists between cell bodies of the arcuate nucleus and tanycyte processes. These junctions allow the passage of low-molecular-weight substances between the junctional partners; the implications of this with respect to possible transfer of releasing hormones to tanycytes are apparent.[66] A physiological role for the ependymal cells is supported by evidence of structural variations during the estrus cycle.[15,77] (Fig. 5).

4.1.2. Retrograde Portal-Vessel Transport. The initial suggestion by Popa and Fielding[113] that blood flowed in the portal vessels from pituitary to hypothalamus has recently gained support from the studies of Bergland *et al.*,[12] who have provided ev-

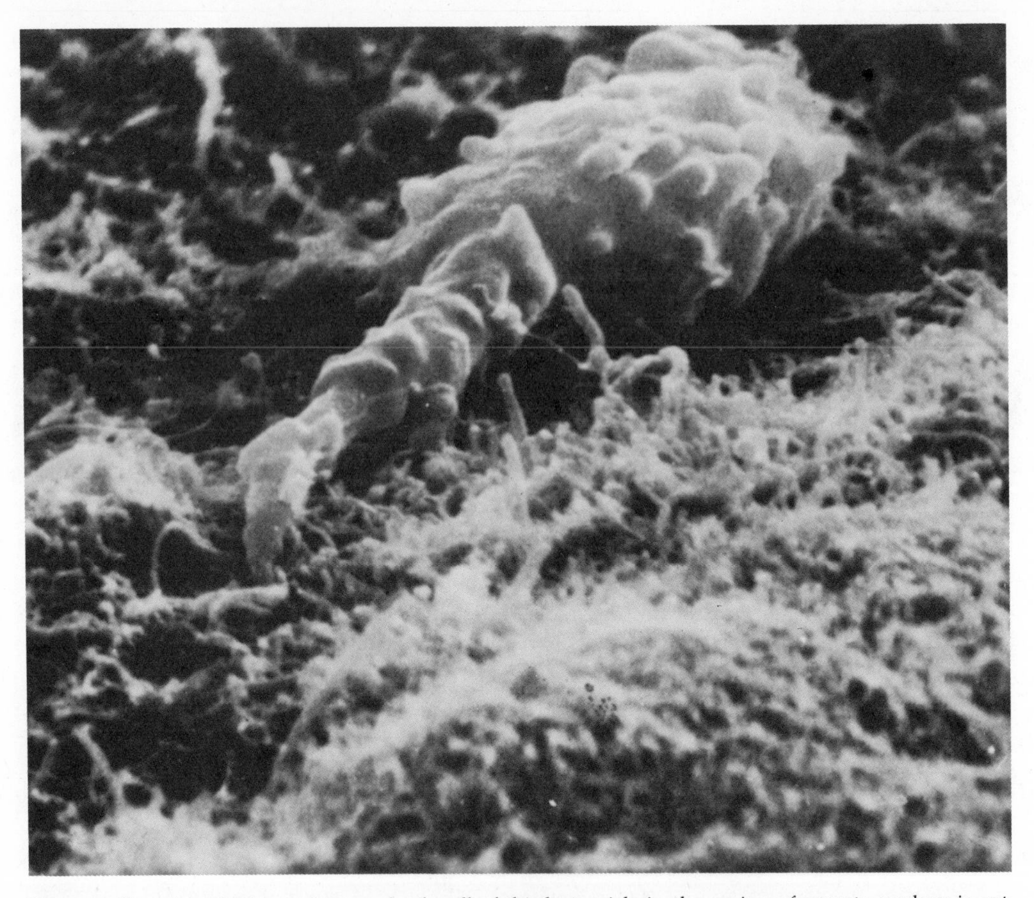

Figure 5. Scanning electron micrograph of wall of third ventricle in the region of arcuate nucleus in rat. This surface is comprised of tanycyte apices that exhibit protrusions of microvilli and other surface irregularities. Large mace-shaped body lying on tanycyte surface is supraependymal cell characteristic of variety commonly seen within third ventricle.[85] This micrograph is of tissue taken from sexually mature, normally cycling female rat in proestrus. ×5000. Courtesy of J. Brawer.

idence that the pituitary secretes to the brain. Studies favoring retrograde transport back up the portal vessels have been further provided by Oliver *et al.*,[100] raising the possibility that by this means, pituitary hormones might feed back on the hypothalamic releasing hormones ("short feedback loop").

There is evidence supporting the concept of bidirectional transport of substances from the ME to the CSF,[157] for Nakai and Naito[91] have demonstrated that ependymal cells in the frog ME have intracellular bidirectional (ascending and descending) transport activities of substances coming from both hypophyseal blood and the CSF. Such ascending transport could allow pituitary hormones to reach hypothalamic nuclei (short feedback loop) and hypothalamic hormones to either feed back on their own ("ultrashort feedback loop") or other hypothalamic nuclei. In addition, this mechanism would permit the distribution of hypothalamic peptides and pituitary hormones to distant parts of the brain.

4.2. Circumventricular Organs

4.2.1. Structure. The circumventricular organs are a group of specialized midline structures that differ from typical brain tissue with regard to the ultrastructure of their vascular, ependymal, glial, and neuronal components.[150] They comprise not only the well-recognized ME and the neural lobe of the neurohypophysis, but also the organum vasculosum of the lamina terminalis (OVLT), the subfornical organ (SFO), the subcommissural organ (SCO), and the area postrema (AP) (Fig. 6). With the exception of the SCO, these structures, along with the pineal gland, are areas wherein the blood–brain barrier (BBB) (see also Section 4.4) is absent and might be termed the six "windows of the brain."[74] Increased interest in the circumventricular organs has recently been generated by the finding that TRH, LH-RH, and somatostatin have been located in the OVLT, SFO, SCO, and AP in high concentration.[73,102] The anatomical characteristics of these tissues have been summarized by Kizer and associates[73,102]: The organs are composed of a specialized ependyma capable of active pinocytosis, specialized ependymal cells (tanycytes), small unmyelinated axon terminals not synapsing with any effector cells but terminating on capillary walls, small nerve cells containing neurosecretory granules and giving rise to the small unmyelinated axon terminals showing synaptoid contacts with the small nerve cells, ependymal cells, and a fenestrated capillary bed. Histologically, the structure of the OVLT bears a striking similarity to that of the ME—that is, a brain region wherein neurosecretory neurons terminate on the perivascular space of fenestrated capillaries and wherein a modified nonciliated ependymal lining may functionally link the CSF and the peripheral vasculature.[66]

4.2.2. Neural Peptide Distribution. Immunohistochemical studies in the rodent have localized LH-RH to the OVLT,[10,162] where 80% of the LH-RH present in the preoptic region occurs.[73] The concentration of LH-RH in the OVLT is over 50% of that found in the ME, with concentrations of LH-RH in the SFO, SCO, and AP being somewhat lower.[73] There is no evidence to suggest that LH-RH in the OVLT is present in cell bodies that have axons passing to the ME, for a "portal" relationship between the OVLT capillary vasculature and sinusoids of the adenohypophysis does not occur. No LH-RH-staining perikarya occur in the OVLT, and terminal de-

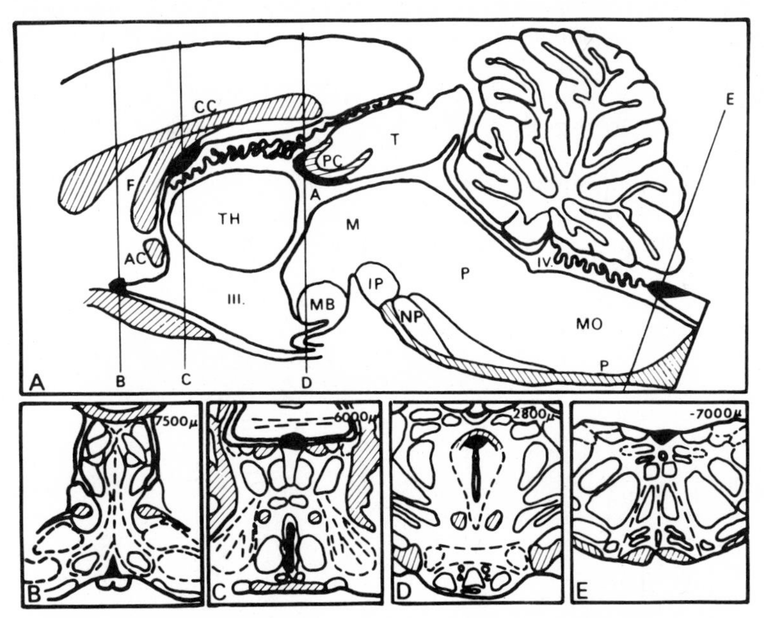

Figure 6. Schematic representation of single midline parasagittal section (A) and four coronal sections (B–E) of rat brain, demonstrating location of four circumventricular organs: (B) organum vasculosum of lamina terminalis (OVLT); (C) subfornical organ (SFO); (D) subcomissural organ (SCO); (E) area postrema (AP). Apart from the SCO, the blood–brain barrier (BBB) is absent at these structures as well as the median eminence (ME), neural lobe of neurohypophysis, and pineal gland (anatomic areas not outlined here). In (B–E), figures in the upper right-hand corner indicate distances (in μm) from interauricular line. Abbreviations: (F) fornix; (AC) anterior commissure; (CC) corpus callosum; (TH) thalamus; (A) adqueduct of Sylvius; (PC) posterior commissure; (T) tectum; (MB) mamillary body; (IP) interpenducular nucleus; (FB) fibrae pyramidalae; (M) midbrain; (P) pons; (MO) medulla oblongata; (III) third ventricle; (NP) nucleus pontis. Reproduced from Kizer *et al.*,[73] with permission.

generation has not been observed in the ME after rostral knife cuts.[66,73] Hormones such as LH-RH released into the capillaries of the OVLT would reach the systemic circulation. The architecture of the OVLT (and of other circumventricular organs) may allow LH-RH (and other neuropeptides present there) to be released into the CSF; it is also possible that peptides may be taken up from the CSF by the circumventricular organs.

TRH was located in the OVLT, SFO, SCO, and AP, but at concentrations lower than that of LH-RH. It is noteworthy that the concentrations of LH-RH and TRH in the tissue immediately adjacent to the OVLT were nearly 50 times and 4 times less, respectively, than the concentrations of these two releasing factors within the OVLT itself.[73] Significant quantities of somatostatin have also been located in these circumventricular organs at concentrations one fourth to one fifth of that measured in the arcuate nucleus of the hypothalamus.[102] Significant concentrations of LH-RH have also been reported in the circumventricular organs of the human,[95] the levels being especially high in the OVLT. Significant concentrations of TRH were also reported, but the levels in the OVLT were much lower than those of LH-RH.

The pineal gland contains a number of neural peptides, including arginine vasotocin,[106] TRH,[64] and somatostatin,[104] that may be of importance in neuroendocrine regulation. However, Okon and Koch[94] did not find significant concentrations of TRH and LH-RH in the human pineal.

Substances given intravenously are deposited in the circumventricular organs in concentrations equal to those in the blood. However, they are not taken up by nerve cells, though small amounts may be taken up by glial processes. On the other hand, substances are taken up from ventricular CSF and transported to adjacent capillaries from which they reach the systemic circulation, thereby providing a means of communication between the CSF and the systemic circulation.[73] The function of TRH, LH-RH, and somatostatin in the circumventricular organs other than the ME is unknown. It is possible that neural peptides concentrated there (either from CSF or by neuronal connections via axoplasmic flow) may be released into the CSF to affect brain or hypothalamic–pituitary function. The LH-RH released into the systemic circulation from the OVLT could constitute an additional effect on gonadotropin secretion from the adenohypophysis.

Although the role of the classic hypophysiotrophic hormones in the circumventricular organs is unknown, there is evidence suggesting their importance in mediating the central effects of AII.[111] The SFO is a thirst center of the brain and contains AII receptors. AII is the most potent dipsogenic substance known, and as little as 0.1 pg applied to the SFO can induce a dipsogenic response, and removal of the SFO abolishes drinking behavior caused by systemic AII in the rat.[139] Simpson *et al.*[139] suggest that the regulation of drinking behavior by the SFO is effected by systemic and *not* centrally produced AII. The OVLT, which, like the SFO, is located in the anterior wall of the third ventricle (Fig. 6), is also a dipsogenic receptor site for AII.[111]

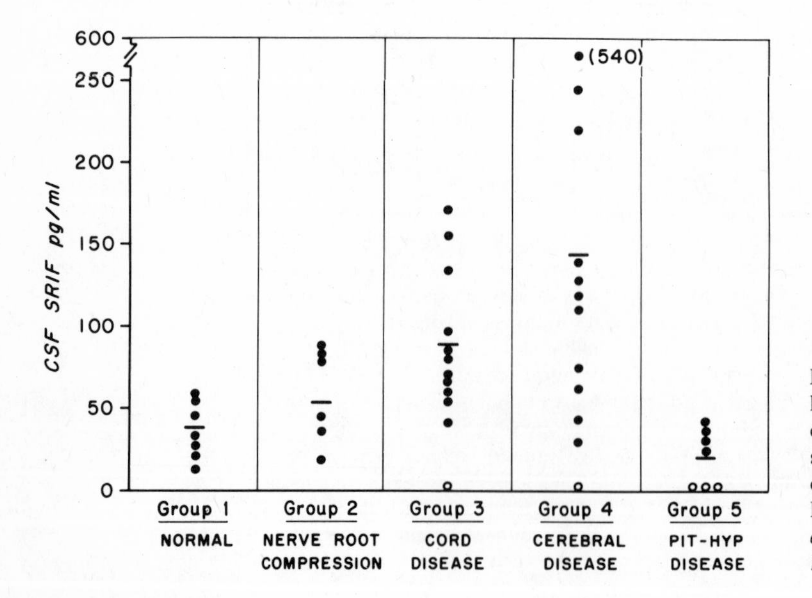

Figure 7. CSF somatotropin-release-inhibiting factor (SRIF), or somatostatin, concentrations in normal (Group 1) subjects and patients with neurological or pituitary–hypothalamic (PIT-HYP) disease. Reproduced from Patel *et al.*,[104] with permission from *The New England Journal of Medicine*.

The AP appears to be an important area in blood-pressure regulation, and a pressure response is produced by the application of AII to this structure. However, whether the physiological regulation of blood pressure might be influenced by central or peripheral AII is unclear. Species differences appear to exist. Whereas systemic AII is more effective in the dog, intraventricular AII is more effective in the rat in producing a central pressure effect.[111]

The SCO is involved in the secretion of a mucopolysaccharide substance, termed Reissner's fiber, the function of which is unknown. However, it has been proposed that the SCO might be involved in regulating the composition of the CSF particularly with respect to its catecholamine concentration.[49] Since biogenic amines and related synthesizing enzymes[129] as well as hypophysiotropic hormones abound in circumventricular organs, it is possible that the neural peptides might regulate the release of catecholamines from the SCO into the CSF.

4.3. Diffusion from Nerve Cells

A "blood–CSF barrier" is distinguished at times from a "blood–brain barrier" to expain why intravascular substances enter CSF and brain at different rates. However, these kinetic differences can be interpreted by taking into consideration gross anatomical relationships among CSF, brain, and blood. Since the ependymal surfaces of the cerebral ventricles do *not* limit exchange between CSF and brain, they do not constitute a subbarrier[116]; thus, a "blood–CSF" and a "blood–brain barrier" should not be distinguished. Accordingly, measurement of neural peptides in CSF may reflect diffusion from nerve cells. The levels recorded may be influenced by the location sampled, rates of clearance and degradation within the CSF, and accessibility of particular brain regions to the ventricular system.

4.4. Transport across Blood–Brain Barrier

The ability of a molecule to penetrate the blood–brain barrier (BBB) is primarily a reflection of its lipid solubility.[31] Water-soluble molecules released into the extracellular space of the brain can readily pass into the CSF. To cross from blood to brain extracellular space, a water-soluble molecule must pass through the cerebral capillaries—site of the BBB—which, unlike the capillaries in other parts of the body, are joined by tight junctions. To cross from blood to CSF, the molecule must traverse either the choroid plexus, the fronds of which float in the CSF, or the arachnoid membrane. The epithelial cells of these "blood–CSF" barriers are also joined by tight junctions that impede the entry of water-soluble molecules.[141]

Since the neural peptides are in general low-molecular-weight, water-soluble substances, their ability to traverse the BBB would appear severely restricted on that account alone. Such restriction is bidirectional. However, "leaky" areas between blood and brain do occur at the circumventricular organs, which are discussed in Section 4.2. Such tissues may therefore permit communications between the brain and CSF on the one hand and the systemic circulation on the other.[150]

Although the BBB is impermeable to many polar small molecules, it is permeable to some metabolic substrates, such as glucose, despite extremely lipophobic qualities. As reviewed by Oldendorf,[96] several specific BBB systems have been demonstrated with sufficient affinity to compete successfully against even strong hydrogen bonding to water. Specific transport systems exist not only for glucose but also for various classes of amino accids that may act as precursors of neurotransmitters in the brain. No evidence has been provided to date with regard to the existence of specific transport systems for the neural peptides. Nevertheless, it has been reported that dipeptide transport occurs in mouse brain slices *in vitro*,[159] and Kastin and colleagues[43,69] have suggested that peptides and enkephalins readily penetrate the BBB *in vivo* following intracarotid injection. The studies reported by these workers were not confirmed by Cornford *et al.*,[31] who reported a brain uptake index (BUI) of 2–3% for Met-enkephalin compared with 15% by Kastin and his colleagues, whose methodology was strongly criticized by the former group. The BUI of TRH was reported to be 1%, as was that of carnosine β-alanyl-histidine,[31] a putative neurotransmitter.

The findings cited above are consistent with the absence of significant accumulation of radioactivity following the intravenous injection of radiolabeled TRH[118] and LH-RH.[88] As discussed earlier, many of the behavioral effects of neural peptides that can be readily elicited following intraventricular administration cannot be obtained after injection into the systemic circulation, studies that provide evidence against important transfer for neural peptides across the BBB. This issue is of importance, since many of the brain neural peptides are also found in the gastrointestinal tract (e.g., SP, neurotensin, somatostatin) and peripheral nervous system, so that these substances might reach the systemic circulation from locations other than the brain. Somatostatin,[48] SP,[114] and AII[111] have all been detected in the systemic circulation, but it seems unlikely that the peptides, in the concentrations reported, will reach CNS locations with the exception of the circumventricular organs. As previously discussed, some workers believe that the dipsogenic effect of AII is mediated at the level of the SFO by peripheral and *not* central AII.[139]

Conversely, it seems unlikely, though not wholly excluded, that endogenous neural peptides present in the CSF can cross the BBB and reach the systemic circulation in concentrations sufficient to significantly influence pituitary function. However, in some mammalian species, the BBB is not fully developed at birth, so that neural peptides might "leak" from brain or hypothalamus into the peripheral blood in the neonatal period.[136] Indeed, it has been reported that TRH, which is present in only meager quantities in the blood of adult rats,[63] is significantly raised in blood taken from neonatal rats.[145] The significance of the functional absence of the BBB in the neonatal period, and whether this absence affects neural peptides, requires study.*

5. Physiological Effects of Peptides Placed into CSF

5.1. Hypophysiotropic Effects

5.1.1. TRH. To examine the possible physiological role of TRH absorption from the ventricular system in the regulation of pituitary–thyroid function, Kendall *et al.*,[72] using thyroidal iodide release as a measure of TSH secretion, showed that 300 ng synthetic TRH was equally effective, at least at 2 hr, whether injected into the lateral ventricle or intravenously. In subsequent studies reported from this laboratory, using somewhat smaller doses of TRH, Gordon *et al.*[41] reported that the plasma TSH response to TRH was much less when the TRH was given directly into the third ventricle compared with that produced when the same dose was given intravenously or directly into the median eminence (ME) or pituitary gland. These authors interpreted their findings to suggest that TRH transport from the CSF was not of major importance in the regulation of pituitary TSH secretion. In agreement with these two reports, Oliver *et al.*[99] showed that the administration of TRH into a lateral ventricle results in a significant release of TSH. There was a delay in reaching maximal TSH concentration following intravenous injection, although the TSH levels following ventricular injection were maintained for a longer time than were those induced by intravenous injection of TRH. There was a good correlation between the timed course of TSH release and the appearance of radioactivity in hypophyseal portal blood following either intravenous or intraventricular injection of [^{3}H]-TRH, and these workers reported that 289 pg [^{3}H]-TRH was recovered in portal blood after intraventricular injection of 50 ng [^{3}H]-TRH, whereas only 19 pg was recovered in portal blood following intravenous injection of the same quantity of [^{3}H]-TRH. Overall, the data reported support the view that TRH is able to cross the ME from CSF into hypophyseal portal blood and that it is capable of stimulating the pituitary gland to release TSH.[99]

When rat ME was incubated *in vitro* with [^{3}H]-TRH, or when radiolabeled hormone was infused intraventricularly, autoradiographic evidence indicated that the hormone was present in the tanycytes of the ME. In lower vertebrates also, the ME appears capable of concentrating TRH from the CSF. In the duck, following intraventricular injection of [^{3}H]-TRH, there is an intense radioautographic reaction in all cellular elements of the ME, viz., ependyma, neurons, glia, and endothelial vascular cells.[26] The labeling of the tanycytes was also intense after administration of [^{3}H]proline—findings that suggested to the authors that tanycytes may have a selective function in both the transport and the synthesis of TRH.

5.1.2. LH-RH. Following the injection of ^{125}I-labeled LH-RH into the third ventricle of the rat, autoradiographs demonstrated [^{125}I]-LH-RH in the ependymal perikarya and their processes as well as in portal capillaries and adenohypophysis.[78] Uptake of [^{3}H]-LH-RH by cells lining the floor of the third ventricle has also been demonstrated.[39] Light- and electron-microscopic autoradiographic analysis of the mediobasal hypothalamus of rat brains incubated for 30–60 min in [^{3}H]-LH-RH revealed selective uptake in the form of silver grains over the ependymal tanycytes and perivascular glial cells in the palisade-contrast zone of the ME.[135] Scott *et al.*[135] also observed that some arcuate neurons were labeled and suggested that this may offer morphological support for the concept that cells of this nucleus may either function in a receptor role or,

* Engler and Jackson (Program, 62nd Annual Meeting Endocrine Society, Washington, D.C., 1980, abstract #126) have recently shown that circulating TRH in the neonatal rat is derived from the pancreas and gastrointestinal tract and that encephalectomy does not alter plasma TRH levels.

alternatively, be responsible for the synthesis of LH-RH. (However, studies by Lechan *et al.*[80] in this laboratory suggest that the arcuate nucleus of the mouse is *not* a site of LH-RH synthesis.)

A number of studies have been reported on the effect of intraventricular administration of LH-RH on the release of LH. Ondo *et al.*[101] injected LH-RH into the third ventricle and cisterna magna and demonstrated a rise in plasma LH. Weiner *et al.*[151] also found a rise in LH, but concluded from their studies in ovariectomized, estrogen-primed rats that intraventricularly administered LH-RH was less potent in releasing LH than intravenously injected LH-RH. In further studies by Ben-Jonathan *et al.*,[11] LH-RH was found in the portal blood after injection of 125 ng LH-RH intraventricularly and in some rats injected with a little as 5 ng of the decapeptide—the recovery in portal blood being approximately 5% of the quantity injected into the ventricular system. During the first 20–30 min following injection, intravenously administered LH-RH was slightly more effective in stimulating LH release than was intraventricularly administered LH-RH. After this time, the plasma LH levels in rats given LH-RH intravenously fell steadily, while the plasma LH levels in rats given LH-RH intraventricularly remained elevated. These workers concluded that LH-RH can be transported from CSF to hypophyseal portal blood in significant quantity and that the decapeptide given intraventricularly is more effective on a prolonged basis in stimulating LH release than when given intravenously.

The effect of estrogen on the uptake of exogenous LH-RH from the third ventricle was studied by Recabarren and Wheaton.[117] These workers showed that treatment of ovariectomized rats with estradiol resulted in greater release of LH when compared with control animals following the intraventricular administration of LH-RH, but not after its systemic injection. Estrogen also faciliated a greater uptake of [^{125}I]-LH-RH in the ME from the third ventricle.

5.1.3. Other Neural Peptides. SP, an undecapeptide, and neurotensin, a tridecapeptide, are widely distributed throughout the CNS, including the hypothalamus, and the gastrointestinal tract. These peptides stimulate the release of GH and PRL in the rat when administered by systemic injection *in vivo.*[124] Similar effects are also shown by certain peptide fragments of β-lipotropin (β-LPH), including β-LPH_{61-65} (methionine-enkephalin) and β-LPH_{61-91} (β-endorphin)[125]; however, in common with SP and neurotensin, these opioid peptides do not increase GH and PRL secretion from pituitary fragments *in vitro.*[124,125] Placement of SP directly into the lateral ventricle did not increase serum PRL, and in fact lowered serum GH.[27] However, simulataneous injection of β-endorphin and SP caused greater GH and PRL responses than did β-endorphin alone. The findings suggested to Chihara *et al.*,[27] first, that intraventricular administration of SP may stimulate hypothalamic somatostatin release into the portal vessels, thereby decreasing GH secretion, and second, that SP may potentiate the action of opiate-receptor stimulators on GH and PRL release at the level of the CNS. VIP, a 28-amino-acid polypeptide originally isolated from the porcine duodenum, has been reported to cause a significant and dose-related increase in plasma PRL levels in urethane-anesthetized rats when administered by either the intraventricular or the intravenous route.[70] VIP does not stimulate the release of PRL from cultured pituitary cells, but attenuates the inhibitory effect of DA.[70]

Bombesin, a tetradecapeptide, has been reported to induce a rise in serum PRL and to inhibit the cold-induced rise in pituitary TSH secretion following intraventricular injection.[17]

AII induces a release of vasopressin and ACTH after intraventricular but not consistently after intravenous administration.[121]

5.2. Behavioral Effects of Neural Peptides apart from Pituitary Regulation

5.2.1. TRH. TRH interacts with other neural peptides with respect to temperature regulation and other behavioral effects. Some of these actions have been demonstrated following intraventricular injection and may not be seen with systemic administration (see Section 4.4). Following intraventricular injection in the rabbit, TRH produces hyperthermia and behavioral excitation[57] and antagonizes the hypothermic effect of morphine in the rabbit[57] and of β-endorphin in the rat.[56] Intracisternal administration of TRH produces hyperthermia, and the elevation of temperature is antagonized by intracisternal injection of bombesin.[17]

Other effects of TRH that can be recognized following intraventricular administration, but not intravenous administration, include suppression of feeding and drinking activity in the rat[149] and stimulation of colonic activity, probably due to stimulation of central cholinergic receptors.[140]

5.2.2. Somatostatin. Although evidence has been presented that neural peptides are measurable in the CSF, the physiological significance of these substances present in the CSF is not yet determined. Recent studies to clarify the physiological role of endogenous somatostatin in the CSF have been reported.[27a] Sheep anti-somatostatin γ-globulin was administered intraventricularly to the rat and was found to decrease significantly the duration of strychnine-induced seizures and increase the pentobarbital LD_{50} as compared to controls. These findings support the view that endogenous somotostatin in the CSF or periventricular tissue, or both, modulates the response of the CNS to strychnine and pentobarbital in rats and that the CSF may be a conduit important in its physiological regulation of brain function.

5.2.3. SP, Neurotensin, and VIP. Intraventricular, but not intravenous, administration of eledoisin, a peptide related to SP, at as little as 10^{-11} mol stimulated drinking behavior in the pigeon.[37] SP itself was also shown to stimulate dipsogenic behavior. The central administration of neurotensin produces a marked hypothermic effect in rats and mice[92] unaffected by passive immunization intraventricularly.[17] Decreases in locomotor activity in rats and a marked dose-related enhancement in pentobarbital-induced mortality, sedation, and hypothermia also occur. None of these effects was observed after peripheral administration of neurotensin. In contrast to neurotensin, the injection of 10–40 μg VIP into the third ventricle of cats was reported to elicit shivering and a hyperthermic response.[28]

5.2.4. AII. A great deal of information has been reported regarding the CNS effects of angiotensin.[111] In addition to its peripheral vasoconstrictor effects, intraventricular administration results in stimulation of drinking, blood-pressure increase, and vasopressin release. Since the intraventricular route is about 1000 times more potent than the intravenous route in stimulating drinking behavior,[36] the dipsogenic action of AII appears to be under CNS control (see also Section 4.2). All the components of the renin–angiotensin system are present in the CNS,[111] and when exogenous renin is injected into the third ventricle, there is a marked stimulation of drinking, a prolonged increase in arterial blood pressure, and a stimulation of vasopressin secretion—responses that are abolished by intraventricular saralasin acetate (an antagonist of AII), suggesting that these effects are mediated via the formation of AII.[121]

5.3. Evidence Favoring Specificity of Behavioral Responses

Evidence favoring the specificity of the responses is supported by the studies of Malthe-Sørenssen *et al.*[83] Following intraventicular injection of TRH, somatostatin, neurotensin, and AII into rats, evidence was obtained that these peptides modulated the turnover rate of acetylcholine (TR-ACh) in the brain. However, whereas TRH increased the TR-ACh in the parietal cortex, the other peptides were ineffective. Somatostatin and neurotensin increased the TR-ACh in diencephalon, whereas AII and TRH did not, and only somatostatin affected the TR-ACh in the brainstem. The selective changes in the TR-ACh produced by each individual neural peptide in specific brain regions, as well as reports of high-affinity receptors for these peptides,[25] support their postulated role as specific agents in brain function.

It is probable that peptides placed in the CSF reach a specific neuronal region by diffusion from the CSF, since there is no brain–CSF barrier. The presence of high-affinity receptors for these peptides in brain tissue not only supports their postulated role as specific agents in brain function, but also may aid their uptake from the CSF.

6. Endogenous Neural Peptides in CSF

The importance of the CSF in relation to the action of neural peptides at the hypothalamus, pituitary, and brain locations has been discussed. Clearly, however, direct evidence that endogenous neural peptides are present in the CSF, and that the levels of these substances show appropriate changes in response to physiological stimuli, is necessary to support the role of the CSF as a conduit for neuroendocrine regulation. However, since the neural peptides occur not only in the hypothalamus and brain but also in the spinal cord, caution must be exercised in the interpretation of peptide levels measured in lumbar CSF. It is also possible that significant changes could occur in the concentration of a peptide in the third ventricle without producing alterations in the levels elsewhere in the ventricular system. Further, even if lumbar CSF peptide concentrations reflect central levels, it is not clear whether the peptide comes from the hypothalamus, brain, or both. This issue is of particular importance, since it remains to be determined whether hypothalamic hormones located in extrahypothalamic sites are subject to the same feedback regulation as those present in the hypothalamus. Despite these caveats, such data as exist on the measurement of endogenous neural peptides in the CSF will be presented. A list of peptides reported present in the CSF is provided in Table 4.

Table 4. Peptides Detected in CSF

Thyrotropin-releasing hormone (TRH)
Luteinizing-hormone-releasing hormone (LH-RH)
Somatostatin
Corticotropin-releasing factor (CRF)
Growth-hormone-releasing factor (GH-RF)
Gastrin
Cholecystokinin (CCK)
Vasoactive intestinal peptide (VIP)
Angiotensin II (AII)
Substance P (SP)
Sleep factor
Antimelanotropic protein
Adenohypophyseal hormones
Neurohypophyseal hormones
Opioid peptides

6.1. Thyrotropin-Releasing Hormone

The presence of TSH-releasing-factor activity in the CSF was first reported in the rat by Averill and Kennedy.[5] TRH activity has also been detected in human CSF drawn from the third ventricle of cadavers 2–5 hr after death. This material, when concentrated, showed TSH-releasing activity when incubated with human anterior pituitary tissue *in vitro*.[58] In subsequent studies by Shambaugh and Wilber,[137] TRH was quantitated in human lumbar CSF by radioimmunoassay. TRH was stable in CSF stored at 4°C, and TRH levels in A.M. or P.M. samples obtained from 15 women and 12 men were determined. No sex difference was evident, and TRH concentrations were around 40 pg/ml with no diurnal rhythm. By contrast, CSF cortisol levels obtained concurrently were 2-fold higher in A.M. than in P.M. samples. Immunoreactive (IR) TRH has also been reported in CSF obtained by cisternal puncture from patients undergoing neuroradiological examination.[98] The levels reported ranged from 60 to 290 pg/ml. TRH biological activity equivalent to 18.5 ng/ml has been reported in rat third ventricular CSF,[75] but the authenticity of the material, especially because of the extraordinarily high values reported, needs confirmation. The latter laboratory also reported that the TRH biological activity rose with acute cold as well as after treatment with methimazole (to cause hypothyroidism) and after hypophysectomy.

6.2. Luteinizing-Hormone-Releasing Hormone

Apart from vasopressin, LH-RH is the only neural peptide reported in ependymal tanycytes of the median eminence (ME) by immunohistochemistry.[162] However, controversy exists regarding the presence of LH-RH in the CSF. In third ventricular CSF taken from rats, Cramer and Barraclough[32] reported an LH-RH concentration of 0.33 pg/ml. However, following electrical stimulation of the medial preoptic area, which caused an elevation of serum LH, there was no rise in CSF levels of LH-RH. These authors concluded that CSF does not serve as a vehicle for transport of LH-RH to the ME under physiological conditions. Furthermore, Coppings *et al.*,[30] reported that IR LH-RH was not detectable (< 100 pg/ml) in ovine CSF collected from the third ventricle. These reports contradict the astronomical quantities of LH-RH (141–344 ng/ml) reported from Knigge's laboratory in ovariectomized female rats[67] and male rats[89] after death. The significance of the rise in CSF LH-RH reported by this group to occur following ovariectomy and deafferentation remains unclear at this time.

In human CSF, Gunn *et al.*[46] reported that LH-RH was absent (< 1 pg/ml) in 23 of 26 samples obtained for the purpose of venereal disease serology. In the other 3 samples, levels of 22, 25, and 120 pg/ml were obtained, but dilutions showed nonparallelism, thus creating doubts about identity. Rolandi *et al.*,[126] however, reported LH-RH in the third ventricle of 5 hydrocephalic patients, the level ranging from 50 to 150 pg/ml.

6.3. Somatostatin

IR somatostatin has been reported in the CSF taken from seven neurologically normal persons to range from 15 to 55 pg/ml.[104] To determine whether brain disease might lead to abnormal CSF somatostatin, these workers studied 30 patients with neurological disease, of whom 20 of 24 with cord or cerebral disease had concentrations above the highest normal level (Fig. 7). The authors suggested that the wide variety of neurological disease with CSF somatostatin elevation may indicate nonspecific leakage from damaged brain tissue. Nevertheless, CSF somatostatin measurements may provide a marker of brain insult or injury. Kronheim *et al.*[79] also reported IR somatostatin in normal CSF, with levels of 55 ± 28 pg/ml, and the immunoreactivity found showed mobility on various chromatographic procedures identical to synthetic somatostatin standard. Urosa and Reichlin (personal communication) have carried out additional studies of human immunoassayable somatostatin. On chromatographic analysis, most somatostatin coincides with the region of the cyclic compound on both Sepha dex G-25 and Biogel P2 chromatography, A small portion of the total material appears as "big" somatostatin, a form that is demonstrable in tissue extracts. Mean somatostatin concentration for a group of 16 apparently neurologically normal men and women was 42.9 ± 6.5 (S.D.) pg/ml. The material is relatively stable in normal CSF, less than 20% degradation occurring after 24 hr at 4°C. The presence of small amounts of blood in the CSF brings about rapid degradation, presumably due to the proteolytic activity (plasmin-like) of normal plasma.

6.4. Other Hypothalamic Releasing Factors

A CRF has been reported in dog CSF,[34] and in the CSF of acromegalic patients, a GH-RF has been reported.[6] The latter material was apparently capable of stimulating pituitary protein synthesis and GH release *in vitro*.

6.5. Other Neural Peptides

6.5.1. Gastrin and CCK. Both gastrin and CCK have been reported in human CSF by Rehfeld and Kruse-Larsen.[119] These workers, utilizing specific radioimmunoassays, reported the concentration of gastrin in CSF from ten neurologically normal persons to range from 1.5 to 3.0 pM, whereas the concentration of CCK ranged from 4 to 55 pM. There was molecular heterogeneity of these peptides, for chromatography revealed that gastrin was present in molecular forms corresponding to gastrin-34 ("big gastrin") and gastrin-17. CCK was present in molecular forms corresponding to the COOH-terminal octapeptide amide of CCK-33 and a fragment corresponding to sequence 25–29 of CCK-33. Also, a peptide corresponding to a COOH-terminal tetrapeptide common to both gastrin and CCK was found. The authors concluded that true gastrin as well as CCK are present in CSF, and both peptides display a molecular heterogeneity similar to that found in extracts of brain tissue.

6.5.2. VIP. VIP, which is present in neurons of both the central and the peripheral nervous system, has also been shown to be richly represented immunohistochemically in fibers innervating brain arteries.[38] Such findings led Fahrenkrug *et al.*[38] to examine the CSF for VIP in patients undergoing myelography or pneumoencephalograpy. Mean values obtained were 50 pmol/liter in "normal" CSF. Significantly lower levels were found in patients with cerebral atrophy.

6.5.3. AII. AII levels in the rat cisterna magna have been reported on the order of 100 pg/ml,[139] and significant levels have also been reported in dog CSF.[121] High levels of angiotensin have also been reported in the CSF of hypertensive patients and the CSF of hypertensive rats.[111] However, other studies have been unable to confirm the presence of IR AII in the CSF.[111] Reid and Moffat[122] injected a purified preparation of renin substrate (angiotensinogen) into the third ventricle of anesthetized dogs. The injection increased CSF angiotensinogen concentrations 3-fold, but CSF AII concentration, which was undetectable (< 6.25 fmol/ml) before injection, did not change. In contrast, after intraventricular injection of renin, AII levels rose to 2900 fmol/ml. Since these workers were able to demonstrate AII formation centrally after administration of exogenous renin but not substrate, they concluded that renin activity may not be present in the brain *in vivo*.

6.5.4. SP. SP was not detected by Powell *et al.*[114] in human CSF (< 7 pg/ml) taken from six patients under investigation for various neurological studies. However, Nutt *et al.*[93a] found SP-like immunoreactivity in lumbar CSF of 18 neurologically normal adults that was in the range of 2.9–11.1 fmol/ml. This lumbar CSF SP-like immunoreactivity was significantly reduced in patients with neuropathy and seemed to arise largely from the spinal cord. There have been no reports of neurotensin levels in the CSF.

6.5.5. Sleep Factor. Pappenheimer *et al.*[103] found that the CSF from goats deprived of sleep for 72 hr produces profound sleep when injected into cats and rats by the intraventricular but not the intravascular route. This material is a peptide of low molecular weight (350–700 daltons) and has also been found by these workers in human CSF. In addition, these workers also report the presence of an excitatory peptide in CSF. The sleep (or S) peptide factor shows a gradual increase in the CSF on sleep deprivation and may be of physiological importance in the induction of natural sleep. The excitatory (E) peptide factor could operate as a physiological antagonist of the S-factor.

6.5.6. Antimelanotropic Protein. Human and simian but not bovine CSF contains an antimelanotropic protein (20,000–40,000 daltons) that inhibits melanotropic activity *in vitro* utilizing the frog skin preparation.[128] Melatonin, which is an indoleamine derived from the pineal, was apparently excluded as a cause of this effect. The significance of this protein is unknown.

Peptides with both melanotropic and lipolytic effects have been isolated from the CSF of both monkeys and man.[127] Whether these peptides are related to, or identical with, β-LPH, a peptide (91 amino acids) with lipolytic and pigmentatory effects that is now known to occur in the CSF,[65] is unclear at this time.

6.5.7. Other Peptides. Anterior pituitary hormones (see Chapter 41), vasopressin and the related neurophypophyseal peptides oxytocin and vasotocin[106] (see Chapter 43) and opioid peptides[65] (see Chapter 45) are other peptides that have been detected in the CSF. Their significance is discussed in the chapters indicated.

7. Summary and Conclusions

It has been clearly established, following the characterization of TRH, LH-RH, and somatostatin, that the hypothalamus regulates the secretion of adenohypophyseal hormones. Two major hypotheses have been proposed regarding the means by which hypothalamic hormones reach the portal-vessel circulation of the pituitary stalk for transport to the anterior pituitary. The first view, and the most widely accepted one, is that the hypothalamic peptides are transported along a neural pathway (tubero-infundibular tract) for release in apposition to the portal-vessel capillaries. The second hypothesis (ependymal tanycyte theory) proposes that the hypothalamic neurons release the hypophysiotropic hormones into the CSF, and from there the peptides are taken up by specialized ependymal cells and carried by their processes to the median eminence. Much evidence, both anatomical and autoradiographic, has been put forward demonstrating the functions of the ependyma and their ability to concentrate peptides from the CSF. Although TRH and LH-RH can both effect the release of anterior pituitary hormones following intraventricular administration, the pituitary response appears either less than or slower than that obtained following systemic administration. Further, only LH-RH, of the hypothalamic hormones, has been localized in the tanycytes, and a number of investigators have not confirmed this finding. Some of the releasing hormones have been detected in the CSF by some but not all workers, so that the physiological significance of the CSF in the regulation of pituitary function remains controversial. It is possible that the ependymal tanycytes provide an alternate or supplementary means by which the hypothalamus is able to regulate pituitary function.

Following the development of specific radioimmunoassays, it was found that the hypothalamic hormones were widely distributed throughout the extrahypothalamic brain, and could be found in the spinal cord and indeed outside the CNS altogether. It is now clear that the hypothalamic hormones are part of a family of neural peptides that includes substance P (SP), neurotensin, angiotensin II (AII), vasocative intestinal peptide (VIP), cholecystokinin (CCK), opioid peptides, and the neurohypophyseal hormones. These substances are also widely distributed throughout the CNS and gastrointestinal tract and may also have a role in pituitary regulation. It seems likely that these peptides function as neuromodulators having a role similar to neurotransmitters but with longer-lasting effects.[7] Evidence supporting a significant role for these substances in neuronal functions is provided by their anatomical, subcellular, and phylogenetic distribution and the presence of specific brain receptors and active neuronal synthesizing and degrading systems, as well as by neurophysiological and behavioral studies. As postulated by Pearse,[107] the neural peptides are located in cells of the APUD series derived from neuroectoderm. This neuroendocrine system of neurons and cells is held to constitute a third division of the nervous system in addition to the autonomic and somatic branches[107] (see Section 3.1).

The function of these peptides located in the brain tissues is not known for sure, though specific roles have been assigned to certain peptides. AII appears to have a specific action in thirst regulation acting through the subfornical organ, and LH-RH may have specific effects on sexual behavior unrelated to pituitary function. Although many behavioral effects have been achieved by neural peptides following intraventricular administration, but not after systemic injection, the physiological as opposed to the pharmacological significance of these responses is not established.

The widespread distribution of these peptides throughout the CNS necessitates a reappraisal of the significance of the CSF as a conduit for these peptides in the modulation of neuronal function in different brain regions. Some of the endogenous brain peptides have been detected in CSF, but their physiological significance has not been determined. Hypothalamic deafferentation in the rat leads to a reduction of the hypothalamic content of TRH and LH-RH with concomitant hypothyroidism and hypogonadism without affecting the extrahypothalamic brain content of these peptides. Such findings would suggest that neuronal rather than CSF connections are important in the brain regulation of hypothalamic hypophysiotropic hormone secretion and function. However, the physiological significance of the CSF in neuronal functions is supported by the effect of passive immunization studies with antisomatostatin injected into the CSF, which leads to changes in response to the subsequent administration of strychnine or barbiturate. Further studies with antiserum to the neural peptides injected into the CSF are required to evaluate the roles of the brain peptides and the significance of their presence in the CSF.

The evidence strongly suggests that the neural peptides do not readily cross the blood–brain barrier (BBB) except possibly in the neonatal period. There are areas (circumventricular organs) wherein the BBB is not operative and where, interestingly, hypothalamic hormones have been detected in high concentration.

Physiological inferences from measurement of neural peptides in the CSF are fraught with danger, since it is unclear whether the peptides derive from hypothalamus, brain, or spinal cord, or whether lumbar CSF peptide concentration is representative only of changes in the spinal cord or is indicative of central function. Further studies are required to clarify and characterize changes in peptide hormone concentration in different parts of the ventricular system to determine whether the CSF regulates brain function through its distribution of neural peptides or whether the CSF is merely a biological fluid that provides the CNS with a means of disposing of its waste products. There is, however, some evidence, for both somatostatin and VIP, that CSF levels of these peptides are markedly altered in response to CNS pathology. Further evlauation is necessary, but the unique distribution of each of the neural peptides raises the exciting possibility that measurements of brain peptides may provide a sensitive "marker" for the anatomical diagnosis of specific areas of CNS damage.

ACKNOWLEDGMENT. This research was supported by NIH Grant AM 21863 to the author.

References

1. ALPERT, L. C., BRAWER, J. R., PATEL, Y. C., REICHLIN, S.: Somatostatinergic neurones in anterior hypothalamus: Immunohistochemical localization. *Endocrinology* **98:**255–258, 1976.
2. ARAKI, S., FERIN, M., ZIMMERMAN, E. A., VANDE WIELE, R. L.: Ovarian modulation of immunoreactive gonadotropins-releasing hormone (Gn-GH) in the rat brain: Evidence for a differential effect on the anterior and mid-hypothalamus. *Endocrinology* **96:**644–650, 1975.
3. ARIMURA, A., SATO, H., DUPONT, A., NISHI, N., SCHALLY, A. V.: Somatostatin: Abundance of immunoreactive hormone in rat stomach and pancreas. *Science* **189:**1007–1009, 1975.
4. ARIMURA, A., SATO, H., KUMASAKA, T., WOROBEC, R. B., DEBELJUK, L., DUNN, J., SCHALLY, A. V.: Production of antiserum of LH-releasing hormone (LH-RH) associated with gonadal atrophy in rabbits: Development of radioimmunoassay for LH-RH. *Endocrinology* **93:**1092–1103, 1973.
5. AVERILL, R. L. W., KENNEDY, T. H.: Elevation of thyrotopin release by intrapituitary infusion of crude hypothalamic extracts. *Endocrinology* **81:**113–120, 1967.
6. BARBATO, T., LAWRENCE, A. M. KIRSTEINS, L.: Cerebrospinal fluid stimulation of pituitary protein synthesis and growth hormone release *in vitro*. *Lancet* **1:**599–600, 1974.
7. BARKER, J. L.: Peptides: Roles in neuronal excitability. *Physiol. Rev.* **56:**435–452, 1976.
8. BARRY, J.: Immunofluorescence study of LRF neurons in man. *Cell Tissue Res.* **181:**1–16, 1977.
9. BARRY, J., DUBOIS, M. P., POULOIN, P., LEONARDELLI, J.: Caracterisation et topographie des neurones hypothalamiques immunoreactifs avec des anticorps anti-LRF synthese. *C. R. Acad. Sci.* **276:**3191–3193. 1973.
10. BARRY, J., DUBOIS, M. P., CARETTE, B.: Immunofluorescence study of the preoptico–infundibular LRF neuroscretory pathway in the normal, castrated or testosterone treated male guinea pig. *Endocrinology* **95:**1416–1423, 1974.
11. BEN-JONATHAN, N., MICAL, R. S., PORTER, J. C.: Transport of LRF from CSF to hypophyseal portal and systemic blood and release of LH. *Endocrinology* **95:**18–25, 1974.
12. BERGLAND, R. M., DAVIS, S. L., PAGE, R. B.: Pituitary secretes to brain. *Lancet* **2:**276–277, 1977.
13. BLACKWELL, R. E., GUILLEMIN, R.: Hypothalamic control of adnenohypophysial secretions. *Annu. Rev. Physiol.* **35:**357–390, 1973.
14. BOYD, A. E., SANCHEZ-FRANCO, F., SPENCER, E., PATEL, Y. C., JACKSON, I. M. D., REICHLIN, S.: Characterization of hypophysiotropic hormones in porcine hypothalamic extracts. *Endocrinology* **103:**1075–1083, 1978.
15. BRAWER, J., SUN LIN, P., SONNENSCHEIN, C.: Morphological plasticity in the wall of the third ventricle during the estrous cycle of the rat: A scanning electron microscopic study. *Anat. Rec.* **179:**481–490, 1974.
16. BROWN, M., VALE, W.: Central nervous system effects

of hypothalamic peptides. *Endocrinology* **96:**1333–1336, 1975.

17. BROWN, M., RIVER, J., KOBAYASHI, R., VALE, W.: Neurotensin-like and bombesin like peptides: CNS distribution and actions. In Bloom, S.R. (ed.): *Gut Hormones.* Endinburgh, Churchill-Livingstone, 1978, pp. 550–558.
18. BROWNSTEIN, M. J., PALKOVITS, M., SAAVEDRA, J. M., BASSIRI, R. M., UTIGER, R. D.: Thyrotropin-releasing hormone in specific nuclei of rat brain. *Science* **185:**267–269, 1974.
19. BROWNSTEIN, M., ARIMURA, A., SATO, H., SCHALLY, A. V., KIZER, J. S.: The regional distribution of somatostatin in the rat brain. *Endocrinology* **96:**1456–1461, 1975.
20. BROWNSTEIN, M. J., UTIGER, R. D., PALKOVITS, M., KIZER, J. S.: Effect of hypothalamic deafferentation on thyrotopin releasing hormone levels in rat brain. *Proc. Natl. Acad. Sci. U.S.A.* **72:**4177–4179, 1975.
21. BROWNSTEIN, M. J., ARIMURA, A., SCHALLY, A. V., PALKOVITS, M., KIZER, J. S.: The effect of surgical isolation of the hypothalamus on its luteinizing hormone-releasing hormone content. *Endocrinology* **98:**662–665, 1976.
22. BROWNSTEIN, M. J., MROZ, E. A., KIZER, J. S., PALKOVITS, M., LEEMAN, S. E.: Regional distribution of substance P in the brain of the rat. *Brain Res.* **116:**299–305, 1976.
23. BROWNSTEIN, M. J., ARIMURA, A., FERNANDEZ-DURANGO, R., SCHALLY, A. V., PALKOVITS, M., KIZER, J. S.: The effect of hypothalamic deafferentation on somatostatin-like activity in the rat brain. *Endocrinology* **100:**246–249, 1977.
24. BURGUS, R., BUTCHER, M., AMOSS, M., LING, N., MONAHAN, M., RIVER, J., FELLOUN, R., BLACKWELL, R., VALE, W., GUILLEMIN, R.: Primary structure of the ovine hypothalamic luteinizing hormone-releasing factor (LRF). *Proc. Natl. Acad. Sci. U.S.A.* **69:**278–282, 1972.
25. BURT, D. R., SNYDER, S. H., Thyrotropin releasing hormone (TRH): Apparent receptor binding in rat brain membranes. *Brain Res.* **93:**309–328, 1975.
26. CALAS, A.: The avian median eminence as a model for diversified neuroendocrine routes. In Knigge, K. M., Scott, D. E., Kobayashi, H., Ishii, S. (eds.): *Brain–Endocrine Interaction II. The Ventricular System.* Basel, S. Karger, 1975, pp. 54–69.
27. CHIHARA, K., ARIMURA, A., COY, D. H., SCHALLY, A. V.: Studies on the interaction of endorphine, substance P and endogenous somatostatin in growth hormone and prolactin release in rats. *Endocrinology* **102:**281–290, 1978.

27a. CHIHARA, K., ARIMURA, A., CHIHARA, M., SCHALLY, A. V.: Effects of intra-ventricular administration of anti-somatostatin γ-globulin on the lethal dose-50 of strychnine and pentobarbital in rats. *Endocrinology* **103:**912–916, 1978.

28. CLARK, W. G., LIPTON, J. M., SAID, S. I.: Hyperthermic responses to vasoactive intestinal polypeptide (VIP) injected into the third cerebral ventricle of cats. *Neuropharmacology* **17:**883–885, 1978.
29. COHN, M. L., COHN, M., TAYLOR, H.: Thyrotopin releasing factor (TRF) regulation of rotation in the non-lesioned rat. *Brain Res.* **96:**134–137, 1975.
30. COPPINGS, R. J., MALVEN, P. V., RAMIREZ, V. D.: Absence of immunoreactive luteinizing hormone releasing hormone in ovine cerebrospinal fluid collected from the third ventricle. *Proc. Soc. Exp. Biol. Med.* **154:**219–223, 1977.
31. CORNFORD, E. M., BRAUN, L. D., CRANE, P. D., OLDENDORF, W. H.: Blood–brain barrier restriction of peptides and the low uptake of enkephalins. *Endocrinology* **103:**1297–1303, 1978.
32. CRAMER, O. M., BARRACLOUGH, C. A.: Failure to detect luteinizing hormone-releasing hormone in third ventricle cerebral spinal fluid under a variety of experimental conditions. *Endocrinology* **96:**913–921, 1975.
33. DOCKRAY, G. J.: Immunochemical evidence of cholecystokinin-like peptides in brain. *Nature (London)* **264:**568–570, 1976.
34. EIK-NES, K. B., BROWN, D. M., BRIZZEE, D. R., SMITH, E. L.: Partial purification and properties of a corticotropin influencing factor (CIF) from human spinal fluid: An assay method for CIF in the trained dog. *Endocrinology* **69:**411–421, 1961.
35. EPELBAUM, J., BRAZEAU, P., TSANG, D., BRAWER, J., MARTIN, J. B.: Subcellular distribution of radioimmunoassayable somatostatin in rat brain. *Brain Res.* **126:**309–323, 1977.
36. EPSTEIN, A. N., FITZSIMMONS, J. T., ROLLS, B. J.: Drinking induced by injection of angiotensin in the brain of the rat. *J. Physiol.* **210:**457–474, 1970.
37. EVERED, M. D., FITZSIMMONS, J. T., DECARO, G.: Drinking behaviour induced by intracranial injections of eledoisin and substance P in the pigeon. *Nature (London)* **268:**332–333, 1977.
38. FAHRENKRUG, J., SCHAFFALITZKY DE MUCKADELL, O. B., FAHRENKRUG, A.: Vasoactive intestinal polypeptide (VIP) in human cerebrospinal fluid. *Brain Res.* **124:**581–584, 1977.
39. GOLDGEFTER, L.: Non-diffusional distribution of radioactivity in the rat median eminence after intraventricular injection of ^{3}H-LH-RH. *Cell Tissue Res.* **168:**411–418, 1976.
40. GOMEZ-PAN, A., HALL, R.: Somatostatin (growth hormone-release inhibiting hormone). *Clin. Endocrinol. Metab.* **6:**181–200, 1977.
41. GORDON, J., BOLLINGER, J., REICHLIN, S.: Plasma thyrotropin responses to thyrotropin releasing hormone after injection into the third ventricle, systemic circulation, median eminence and anterior pituitary. *Endocrinology* **91:**696–701, 1972.
42. GREEN, A. R., GRAHAME-SMITH, D. G.: TRH potentiates behavioural changes following increased brain 5-hydroxytryptamine accumulation in rats. *Nature (London)* **251:**524–526, 1974.

43. GREENBERG, R., WHALLEY, C. E., JOURDIKIAN, R., MENDELSON, I. S., WALTER, R., NICOLICS, K., COY, D. H., SCHALLY, A. V., KASTIN, A. J.: Peptides readily penetrate the blood brain barrier: Uptake of peptides by synaptosomes is passive. *Pharmacol. Biochem. Behav. (Suppl. 1)* **5**:151, 1976.
44. GROSS, D. S.: Distribution of gonadotropin-releasing hormone in the mouse brain as revealed by immunohistochemistry. *Endocrinology* **98**:1408–1417, 1976.
45. GUILLEMIN, R.: Peptides in the brain: The new endocrinology of the neuron. *Science* **202**:390–402, 1978.
46. GUNN, A., FRASER, H. M., JEFFCOATE, S. L., HOLLAND, D. T., JEFFCOATE, W. J.: CSF and release of pituitary hormones. *Lancet* **1**:1057, 1974.
47. HARRIS, G. W.: Neural control of the pituitary gland. *Physiol. Rev.* **28**:139–179, 1948.
48. HARRIS, V., CONLON, J. M., SRIKANT, C. B., MCCORKLE, K., SCHUSDZIARRA, V., IPPE, E., UNGER, R. H.: Measurements of somatostatin-like immunoreactivity in plasma. *Clin. Chim. Acta* **87**:275–283, 1978.
49. HESS, J., DIEDEREN, J. H. B., VULLINGS, H. G. B.: Influence of changes in composition of the cerebrospinal fluid on the secretory activity of the subcommissural organ in *Rana esculenta*. *Cell Tissue Res.* **185**:505–514, 1977.
50. HÖKFELT, T., FUXE, K., JOHANSSON, O., JEFFCOATE, S., WHITE, N.: Thyrotropin releasing hormone (TRH)-containing nerve terminals in certain brain stem nuclei and in the spinal cord. *Neurosci. Lett.* **1**:133–139. 1975.
51. HÖKFELT, T., ELDE, R., JOHANSSON, O., LUFT, R., ARIMURA, A.: Immunohistochemical evidence for the presence of somatostatin a powerful inhibitory peptide in some primary sensory neurons. *Neurosci. Lett.* **1**:231–235, 1975.
52. HÖKFELT, T., JOHANSSON, O., KELLERTH, J. O., LJUNGDAHL, A., NILSSON, G., NYGORDS, A., PERNOW, B.: Immunohistochemical distribution of substance P. In VON EULER, V. S., Pernow, B. (eds.): *Substance P*. New York, Raven Press, 1977, pp. 117–145.
53. HÖKFELT, T., ELDE, R., FUXE, K., JOHANSSON, O., Ljungclahl, H., Goldstein, M., Luft, R., Efendic, S., Nilsson, G., Terenius, L., Ganten, D., Jeffcoate, S. L., Rehfeld, J., Said, S., Perez de la Mola, M., Passani, L., Tapia, R., Teran, L., Palacios, R.: Aminergic and peptidergic pathways in the nervous system with special reference to the hypothalamus. In Reichlin, S., Baldessarini, R. J., Martin, J. B. (eds.): *The Hypothalamus*. New York, Raven Press, 1978, pp. 69–135.
54. HÖKFELT, T., PERNOW, B., NILSSON, G., WETTERBERG, L., GOLDSTEIN, M., JEFFCOATE, S. L.: Dense plexus of substance P immunoreactive nerve terminals in eminentia medialis of the primate hypothalamus. *Proc. Natl. Acad. Sci. U.S.A.* **75**:1013–1015, 1978.
55. HÖKFELT, T., SCHULTZBERG, M., JOHANSSON, O., LJUNGDAHL, A., Elfvin, L., Elde, R., Terenius, L., Nilsson, G., Said, S., Goldstein, M.: Central and peripheral peptide producing neurons. In Bloom, S. R. (ed.): *Gut Hormones*. Churchill-Livingstone, Edinburgh, 1978, pp. 423–433.
56. HOLADAY, J. W., TSENG, L. F., LOH, H. H., LI, C. H.: Thyrotropin releasing hormone antagonizes β-endorphin hypothermia and catalepsy. *Life Sci.* **22**:1537–1544, 1978.
57. HORITA, A., CARINO, M. A.: Thyrotropin-releasing hormone (TRH)-induced hyperthermia and behaviorial excitation in rabbits. *Psychopharmacol. Commun.* **1**:403–414, 1975.
57a. HYYPPA, M. T., LIRA, J., SKETT, P., GUSTAFSSON, J. A.: Neuropeptide hormones in cerebrospinal fluid: Experimental and clinical aspects. *Med. Biol.* **57**:367–373, 1979.
58. ISHIKAWA, H.: Study of the existence of TRH in the cerebrospinal fluid in humans. *Biochem. Biophys. Res. Commun.* **54**:1203–1209, 1973.
59. JACKSON, I. M. D.: Extrahypothalamic and phylogenetic distribution of hypothalamic peptides. In Reichlin, S., Baldessarini, R. J., Martin, J. B. (eds.): *The Hypothalamus*. New York, Raven Press, 1978, pp. 217–231.
60. JACKSON, I. M. D.: Phylogenetic distribution and function of the hypophysiotropic hormones of the hypothalamus, *Am. Zool.* **18**:385–399, 1978.
61. JACKSON, I. M. D., REICHLIN, S.: Thyrotropin-releasing hormone (TRH): Distribution in hypothalamic and extrahypothalamic brain tissues of mammalian and submammalian chordates. *Endocrinology* **95**:854–862, 1974.
62. JACKSON, I. M. D., REICHLIN, S.: Brain thyrotrophin-releasing hormone is independent of the hypothalamus. *Nature (London)* **267**:853–854, 1977.
63. JACKSON, I. M. D., REICHLIN, S.: Distribution and biosynthesis of TRH in the nervous system. In Collu, R., Barbeau, A., Ducharme, J. R., Rochefort, J. G. (eds.): *Central Nervous System Effects of Hypothalamic Hormones and Other Peptides*. New York, Raven Press, 1979, pp. 3–54.
64. JACKSON, I. M. D., SAPERSTEIN, R., REICHLIN, S.: Thyrotropin releasing hormone (TRH) in pineal and hypothalamus of the frog: Effect of season and illumination. *Endocrinology* **100**:97–100, 1977.
65. JEFFCOATE, W. J., MCLOUGHLIN, L., HOPE, J., REES, L. H., RATTER, S. H., LOWRY, P. J., BESSER, G. M.: β-Endorphin in human cerebrospinal fluid. *Lancet* **2**:119–121, 1978.
66. JOSEPH, S. A., KNIGGE, K. M.: The endocrine hypothalamus: Recent anatomical studies. In Reichlin, S., Baldessarini, R. J., Martin, J. B. (eds.): *The Hypothalamus*. New York, Raven Press, 1978, pp. 15–47.
67. JOSEPH, S. A. SORRENTINO, S., SUNDBERG, D. K.: Releasing hormones, LRF and TRF, in the cerebrospinal fluid of the third ventricle. In KNIGGE, K. M., SCOTT, D. E., KOBAYASHI, H., ISHII, S. (eds.): *Brain–Endocrine Interaction II. The ventricular system*. Basel, S. Karger, 1975, pp. 306–321.
68. KARDON, F. C., WINOKUR, A., UTIGER, R. D.: Thy-

rotropin-releasing hormone (TRH) in rat spinal cord. *Brain Res.* **122**:578–581, 1977.

69. KASTIN, A. J., NISSEN, C., SCHALLY, A. V., COY, D. H.: Blood barrier half time disappearance and brain distribution of labeled enkephalin and potent analog. *Brain Res. Bull.* **1**:583, 1976.
70. KATO, Y., IWASAKI, Y., IWASAKI, J., ABE, H., YANAIHARA, N., IMURA, H.: Prolactin release by vasoactive intestinal polypeptide in rats. *Endocrinology* **103**:554–558, 1978.
71. KELLER, H. H., BARTHOLINI, G., PLETSCHER, A.: Enhancement of cerebral noradrenaline turnover by thyrotropin-releasing hormone. *Nature (London)* **248**:528–529, 1974.
72. KENDALL, J. W., REES, L. H., KRAMER, R.: Thyrotropin-releasing hormone (TRH) stimulation of thyroidal radioiodine release in the rat: Comparison between intravenous and intraventricular administration. *Endocrinology* **88**:1503–1506, 1971.
73. KIZER, J. S., PALKOVITS, M., BROWNSTEIN, M. J.: Releasing factors in the circumventricular organs of the rat brain. *Endocrinology* **98**:311–317, 1976.
74. KNIGGE, K. M.: Opening remarks. In Knigge, K. M., Scott, D. E., Kobayashi, H., Ishii, S. (eds.): *Brain–Endocrine Interaction II. The Ventricular System.* Basel, S. Karger, 1975, pp. 1–2.
75. KNIGGE, K. M., JOSEPH, S. A.: Thyrotrophin releasing factor (TRF) in cerebrospinal fluid of the 3rd ventricle of rat. *Acta Endocrinol. (Copenhagen)* **76**:209–213, 1974.
76. KNIGGE, K. M., SILVERMAN, A. J.: Transport capacity of the median eminence. In Knigge, K. M., Scott, D. E., Weindl, A. (eds.): *Brain–Endocrine Interactions I. Median Eminence: Structure and Function.* Basel, S. Karger, 1972, pp. 350–363.
77. KNOWLES, F. W. G., ANAND KUMAR, T. C.: Structural changes related to the hypothalamus and pars tuberalis of the monkey. I. The hypothalamus; II. The pars tuberalis. *Philos. Trans. R. Soc. London Ser. B* **256**:357–375, 1969.
78. KOBAYASHI, H.: Absorption of cerebrospinal fluid by ependymal cells of the median eminence. In Knigge, K. M., Scott, D. E., Kobayashi, H., Ishii, S. (eds.): *Brain–Endocrine Interaction II. The Ventricular System.* Basel, S. Karger, 1975, pp. 109–122.
79. KRONHEIM, S., BERELOWITZ, M., PIMSTONE, B. L.: The presence of immunoreactive growth hormone release-inhibiting hormone in normal cerebrospinal fluid. *Clin. Endocrinol.* **6**:411–415, 1977.
80. LECHAN, R. M., ALPERT, L. C., JACKSON, I. M. D.: Synthesis of luteinizing hormone releasing factor and thyrotropin releasing factor in glutamate lesioned mice. *Nature (London)* **264**:463–465, 1976.
81. LÖFGREN, F.: New aspects of the hypothalamic control of the adenohypophysis. *Acta Morphol. Neerl.-Scand.* **2**:220–229, 1959.
82. LYMANGROVER, J. R., BRODISH, A.: Tissue CRF: An extra-hypothalamic Corticotrophin releasing factor (CRF) in the peripheral blood of stressed rats. *Neuroendocrinology* **12**:225–235, 1973.
83. MALTHE-SØRENSSEN, D., WOOD, P. L., CHENEY, D. L., COSTA, E.: Modulation of the turnover rate of acetylcholine in rat brain by intraventricular injections of thyrotropin-releasing hormone, somatostatin, neurotensin and angiotensin II. *J. Neurochem.* **31**:685–691, 1978.
84. MARTIN, J. B.: Brain regulation of growth hormone secretion. In Martini, L., Ganong, W. F. (eds.): *Frontiers in Neuroendocrinology.* New York, Raven Press, 1976, pp. 129–168.
85. MARTIN, J. B., REICHLIN, S., BROWN, G. M.: *Clinical Neurendocrinology.* Philadelphia, F. A. Davis, 1977.
86. MARTINO, E., LERNMARK, A., SEO, H., STEINER, D. F., REFETOFF, S.: High concentration of thyrotropin-releasing hormone in pancreatic islets. *Proc. Natl. Acad. Sci. U.S.A.* **75**:4265–4267, 1978.
87. MILLHOUSE, O. E.: Lining of the third ventricle in the rat. In Knigge, K. M., Scott, D. E., Kobayashi, H., Ishii, S. (eds.): *Brain–Endocrine Interaction II. The Ventricular System.* Basel, S. Karger, 1975, pp. 3–18.
88. MIYACHI, Y., MECKLENBURG, R. S., HANSEN, J. W., LIPSETT, M. B.: Metabolism of ^{125}I-luteinizing hormone-releasing hormone. *J. Clin. Endocrinol.* **37**:63–67, 1973.
89. MORRIS, M., TANDY, B., SUNDBERG, D. K., KNIGGE, K. M.: Modification of brain and CSF LH-RH following deafferentation. *Neuroendocrinology* **18**:131–135, 1975.
90. NAIK, D. V.: Immunohistochemical localization of LH-RH during different phases of estrus cycle of rat with reference to the preoptic and arcuate neurons and the ependymal cells. *Cell Tissue Res.* **173**:143–166, 1976.
91. NAKAI, Y., NAITO, N.: Endocytic uptake and transport of intravascularly injected peroxidase by ependymal cells of the frog median eminence. *J. Electron Microsc.* **23**:19–32, 1974.
92. NEMEROFF, C. B., BISSETTE, G., PRANGE, A. J., JR. LOOSEN, P. T., BARLOW, T. S., LIPTON, M. A.: Neurotensin: Central nervous system effects of a hypothalamic peptide. *Brain Res.* **128**:485–496, 1977.
93. NETT, T. M., AKBAR, A. M., NISWENDER, G. D., HEDLUND, M. T., WHITE, W. F.: A radioimmunoassay for gonadotropin releasing hormone (GN-RH) in serum. *J. Clin. Endocrinol. Metab.* **36**:880–885, 1973.

93a. NUTT, J. G., MROZ, E. A., LEEMAN, S. E., WILLIAMS, A. C., ENGEL, W. K., CHASE, T. N.: Substance P in human cerebrospinal fluid: Reductions in neuropathy and autonomic dysfunction. *Neurology (Minneapolis)* (in press).

94. OKON, E., KOCH, Y.: Localization of gonadotropin releasing and thyrotropin releasing hormones in human brain by radioimmunoassay. *Nature (London)* **263**:345–347, 1976.
95. OKON, E., KOCH, Y.: Localization of gonadotropin

releasing hormone in the circumventricular organs of human brain. *Nature (London)* **268**:455–447, 1977.

96. OLDENDORF, W. H.: Blood–brain barrier permeability to drugs. *Annu. Rev. Pharmacol.* **14**:239–248, 1974.
97. OLIVER, C., ESKAY, R. L., BEN-JONATHAN, N., PORTER, J. C.: Distribution and concentration of TRH in the rat brain. *Endocrinology* **95**:540–546, 1974.
98. OLIVER, C., CHARVET, J. P., CODACCIONI, J. L., VAGUE, J., PORTER, J. C.: TRH in human CSF. *Lancet* **1**:873, 1974.
99. OLIVER, C., BEN-JONATHAN, N., MICAL, R. S., PORTER, J. C.: Transport of thyrotropin-releasing hormone from cerebrospinal fluid to hypophysial portal blood and the release of thyrotropin. *Endocrinology* **97**:1138–1143, 1975.
100. OLIVER, C., MICAL, R. S., PORTER, J. C.: Hypothalamic–pituitary vasculature: Evidence for retrograde blood flow in the pituitary stalk. *Endocrinology* **101**:598–604, 1977.
101. ONDO, J. G., ESKAY, R. L., MICAL, R. S., PORTER, J. C.: Release of LH by LRF injected into the CSF: A transport role for the median eminence. *Endocrinology* **93**:231–237, 1973.
102. PALKOVITS, M., BROWNSTEIN, M. J., ARIMURA, A., SATO, H., SCHALLY, A. V., KIZER, J. S.: Somatostatin content of the hypothalamic, ventromedial and arcuate nuclei and the circumventricular organs in the rat. *Brain Res.* **109**:430–434, 1976.
103. PAPPENHEIMER, J. R., FENCL, V., KARNOVSKY, M. L., KOSTI, G.: Peptides in cerebrospinal fluid and their relation to sleep and activity. In Plum, F. (ed.): *Brain Dysfunction in Metabolic Disorders.* New York, Raven Press, 1974, pp. 201–208.
104. PATEL, Y. C., RAO, K., REICHLIN, S.: Somatostatin in human cerebrospinal fluid. *N. Engl. J. Med.* **269**:529–533, 1977.
105. PATEL, Y. C., REICHLIN, S.: Somatostatin in hypothalamus, extrahypothalamic brain, and peripheral tissues of the rat. *Endocrinology* **102**:523–530, 1978.
106. PAVEL, S., GOLDSTEIN, R., GHEORGHIA, C., CALB, M.: Pineal vasotocin: Release into cat cerebrospinal fluid by melanocyte-stimulating hormone release-inhibiting factor. *Science* **197**:179–180, 1977.
107. PEARSE, A. G. E.: The diffuse neuroendocrine system and the "common peptides." In MacIntyre and Szelke (eds.): *Molecular Endocrinology.* Amsterdam, Elsevier/North-Holland, 1977, pp. 309–323.
108. PELLETIER, J., LABRIE, F., PUVIANI, R., ARIMURA, A., SCHALLY, A. V.: Immunohistochemical localization of luteinizing hormone releasing hormone in the rat median eminence. *Endocrinology* **95**:314–317, 1974.
109. PELLETIER, G., LECLERC, R., DUBE, D., LABRIE, F., PUVIANI, R., ARIMURA, A., SCHALLY, A. V.: Localization of growth hormone release-inhibiting hormone (somatostatin) in the rat brain. *Am. J. Anat.* **142**:397–401, 1975.
110. PFAFF, D. W.: Luteinizing hormone-releasing factor potentiates lordosis behavior in hypophysectomized ovarectomized female rats. *Science* **182**:1148–1149, 1973.
111. PHILLIPS, M. I.: Angiotensin in the brain. *Neuroendocrinology* **25**:354–377, 1978.
112. PLOTNIKOFF, N. P. PRANGE, A. J., JR., BREESE, G. R., ANDERSON, M. S., WILSON, I. C.: Thyrotropin-releasing hormone: Enhancement of dopa activity by a hypothalamic hormone. *Science* **178**:417–418, 1972.
113. POPA, G., FIELDING, U.: A portal circulation from the pituitary to the hypothalamic region, *J. Anat.* **65**:88–91, 1930.
114. POWELL, D., SKRABANEK, P., CANNON, D.: Substance P: Radioimmunoassay studies. in von Euler, U. S., Pernow, B. (eds.): *Substance P.* New York, Raven Press, 1977, pp. 35–40.
115. PRANGE, A. J., JR., LARA, P. P., WILSON, I. C., ALLTOP, L. B., BREESE, G. R.: Effects of thyrotropin-releasing hormone in depression. *Lancet* **2**:999–1002, 1972.
116. RAPAPORT, S. L.: *Blood–brain Barrier in Physiology and Medicine.* New York, Raven Press, 1976.
117. RECABARREN, S. E., WHEATON, J. E.: Estradiol potentiation of hypothalamic uptake of LH-RH from the CSF. *Neuroendocrinology* **27**:1–8, 1978.
118. REDDING, T. W., SCHALLY, A. V.: On the half life of thyrotropin-releasing hormone in rats. *Neuroendocrinology* **9**:250–256, 1972.
119. REHFELD, J. F., KRUSE-LARSEN, C.: Gastrin and cholecystokinin in human cerebrospinal fluid: Immunochemical determination of concentrations and molecular heterogeneity. *Brain Res.* **155**:19–26, 1978.
120. REICHLIN, S., SAPERSTEIN, R., JACKSON, I. M. D., BOYD, A. E., PATEL, Y.: Hypothalamic hormones. *Annu. Rev. Physiol.* **38**:389–424, 1976.
121. REID, I. A.: The brain renin–angiotensin system: New observations. In Onesti, G., Fernandes, M., Kim, K. E. (eds.): *The Regulation of Blood Pressure by the Central Nervous System.* New York, Grune and Stratton, 1976, pp. 149–202.
122. REID, I. A., MOFFAT, B.: Angiotensin II concentration in cerebrospinal fluid after intraventricular injection of angiotensinogen or renin, *Endocrinology* **103**:1494–1498, 1978.
123. RENAUD, L. P., MARTIN, J. B., BRAZEAU, P.: Depressant action of TRH, LH-RH and somatostatin on activity of central neurons. *Nature (London)* **255**:233–235, 1975.
124. RIVIER, C., BROWN, M., VALE, W.: Effect of neurotensin, substance P and morphine sulfate on the secretion of prolactin and growth hormone in the rat. *Endocrinology* **100**:751–754, 1977.
125. RIVIER, C., VALE, W., LING, N., GUILLEMIN, R.: Stimulation of prolactin and growth hormone by β-bendorphin. *Endocrinology* **100**:238, 241, 1977.
126. ROLANDI, E., BARRECA, T., MASTUIZO, P., GIANROSSI, R., PALLERI, A., PERRIA, C.: CSF and release of pituitary hormones. *Lancet* **1**:1080, 1976.
127. RUDMAN, D., DEL RIO, A. E., HOLLINS, B. M.,

HOUSER, D. H., KEELING, M. E., SUTIN, J., SCOTT, J. W., SEARS, R. A., ROSENBERG, M. Z.: Melanotropic–lipolytic peptides in various regions of bovine, simian and human brains and in simian and human cerebrospinal fluids. *Endocrinology* **92**:372–379, 1973.

128. RUDMAN, D., DEL RIO, A. E., CHAWLA, R. K., HOUSER, D. H., SHEEN, S.: An antimelanotropic protein in human cerebrospinal fluid. *Am. J. Physiol.* **226**:693–697, 1974.
129. SAAVEDRA, J. M., BROWNSTEIN, M. J., KIZER, J. S., PALKOVITS, M.: Biogenic amines and related enzymes in the circumventricular organs of the rat. *Brain Res.* **107**:412–417, 1976.
130. SAFFRAN, M., SCHALLY, A. V., BENFEY, B. G.: Stimulation of the release of corticotropin from the adenohypophysis by a neurohypophysial factor. *Endocrinology* **57**:439–444, 1955.
131. SAMSON, W. K., SAID, S. I., GRAHAM, J. W., MCCANN, S. M.: Vasocative-intestinal-polypeptide concentrations in median eminence of hypothalamus. *Lancet* **2**:901–902, 1978.
132. SCHANZER, M. C., JACOBSON, E. D., DAFNY, N.: Endocrine control of appetite: Gastrointestinal hormonal effects on CNS appetitive structures. *Neuroendocrinology* **25**:329–342, 1978.
133. SCHALLY, A. W., ARIMURA, A., KASTIN, A. J., MATSUO, H., BABA, Y., REDDING, T. W., NAIR, R. M. G., DEBELJUK, L., WHITE, W. F.: Gonadotropin-releasing hormone: One polypeptide regulates secretion of luteinizing and follicle stimulating hormones. *Science* **173**:1036–1038, 1971.
134. SCHALLY, A. V., ARIMURA, A., KASTIN, A. J.: Hypothalamic regulatory hormones. *Science* **119**:341–350, 1973.
135. SCOTT, D. E., DUDLEY, G. K., KNIGGE, K. M., KOZLOWSKI, G. P.: *In vitro* analysis of the cellular localization of luteinizing hormone releasing factor (LRF) in the basal hypothalamus of the rat. *Cell Tissue Res.* **149**:371–378, 1974.
136. SESSA, G., PEREZ, M. M.: Biochemical changes in rat brain associated with the development of the blood–brain barrier. *J. Neurochem.* **25**:779–782, 1975.

136a. SETALO, G., VIGH, S., SCHALLY, A. V., ARIMURA, A., FLERKO, B.: Immunohistological study of the origin of LH-RH containing nerve fibers of the rat hypothalamus. *Brain Res.* **103**:597–602, 1976.

137. SHAMBAUGH, G. E., III, WILBER, J. F.: Thyrotropin-releasing hormone: Measurements in human spinal fluid by radioimmunoassy. *Clin. Res.* **22**:634A, 1974.
138. SILVERMAN, A. J., KNIGGE, K. M., RIBAS, J. L., SHERIDAN, M. N.: Transport capacity of the median eminence. III. Amino acid and thyroxine transport of the organ cultured median eminence. *Neuroendocrinology* **11**:107–118, 1973.
139. SIMPSON, J. B., SAAD, W. A., EPSTEIN, A. N.: The subfornical organ, the cerebrospinal fluid, and the dipsogenic action of angiotensin. In Onesti, G., Fernandes, M., Kim, K. E. (eds.): *Regulation of Blood Pressure by the Central Nervous System*. New York, Grune and Stratton, 1976, pp. 191–202.
140. SMITH, J. R., LAHANN, T. R., CHESNUT, R. M., CARINO, M. A., HORITA, A.: Thyrotropin-releasing hormone: Stimulation of colonic activity following intracerebroventricular administration. *Science* **196**:660–662, 1976.
141. SPECTOR, R.: Vitamin homeostasis in the central nervous system. *N. Engl. J. Med.* **296**:1393–1398, 1977.
142. STRAUS, E., MULLER, J. E., CHOI, H. S., PARONETTO, F., YALOW, R. S.: Immunohistochemical localization of a peptide resembling the COOH-terminal octapeptide of cholecystokinin. *Proc. Natl. Acad. Sci. U.S.A.* **75**:5711–5714, 1977.
143. STRAUS, E., YALOW, R. S.: Cholecystokinin in the brains of obese and non-obese mice. *Science* **203**:68–69, 1979.
144. TAKAHARA, J., ARIMURA, A., SCHALLY, A. V.: Suppression of prolactin release by a purified porcine PIF preparation and catecholamines infused into a rat hypophysial portal vessel. *Endocrinology* **95**:462–465, 1974.
145. THEODOROPOULOS, T., FANG, S. L., HINERFELD, L., BRAVERMAN, L. E., VAGENAKIS, A. G.: Lack of a physiologic role of TRH in the regulation of TSH secretion in the fetal and neonatal rat. *Program of the 54th Meeting of the American Thyroid Association, Portland* p. T-13, 1978.
146. UREMURA, H., ASAI, T., NOZAKI, M., KOBAYASHI, H.: Ependymal absorption of luteinizing hormone releasing hormone injected into third ventricle of rats. *Cell Tissue Res.* **160**:443–452, 1975.
147. VALE, W., RIVIER, C., PALKOVITS, M., SAAVEDRA, J. M., BROWNSTEIN, M.: Ubiquitous brain distribution of inhibition of adenohypophysial secretion. Program of the 56th Annual Meeting of the Endocrine Society, Atlanta (Paper No. A-128), 1974 (abstract).
148. VAUGHAN, M. K., BLASK, D. E., VAUGHAN, G. M., REITER, R. J.: Dose-dependent prolactin releasing activity of arginine vasotocin in intact and pinealectomized estrogen–progesterone treated adult male rats. *Endocrinology* **99**:1319–1322, 1976.
149. VIJAYAN, E., MCCANN, S. M.: Suppression of feeding and drinking activity in rats following intraventricular injection of thyrotropin releasing hormone (TRH). *Endocrinology* **100**:1727–1730, 1977.
150. WEINDL, A.: Neuroendocrine aspects of circumventricular organs. In Ganong, W. F., Martini, L. (eds.): *Frontiers in Neuroendocrinology*. London, Oxford University Press, 1973, pp. 1–32.
151. WEINER, R. I., TERKEL, J., BLAKE, C. A., SCHALLY, A. V., SAWYER, C. H.: Changes in serum luteinizing hormone following intraventricular and intravenous injections of luteinizing hormone-releasing hormone in the rat. *Neuroendocrinology* **10**:261–272, 1972.
152. WHITE, W. F., HEDLUND, M. T., WEBER, G. F., RIPPEL, R. H., JOHNSON, E. S., WILBER, J. F.: The pineal gland:

A supplemental source of hypothalamic releasing hormones. *Endocrinology* **94**:1422–1426, 1974.

153. WILBER, J. F., MONTOYA, E., PLOTNIKOFF, N. P., WHITE, W. F., GENDRICH, R., RENAUD, L., MARTIN, J. B.: Gonadotropin-releasing hormone and thyrotropin releasing hormone: Distribution and effects in the central nervous system. *Recent Prog. Horm. Res.* **32**:117–159, 1976.
154. WILBER, J. F., PORTER, J. C.: Thyrotropin and growth hormone releasing activity in hypophysial portal blood. *Endocrinology* **87**:807–811, 1970.
155. WINOKUR, A., UTIGER, R. D.: Thyrotropin releasing hormone: Regional distribution in rat brain. *Science* **185**:265–267, 1974.
156. WINOKUR, A., DAVIS, R., UTIGER, R. D.: Subcellular distribution of thyrotropin releasing hormone (TRH) in rat brain and hypothalamus. *Brain Res.* **120**:423–432, 1977.
157. WITTKOWSKI, W.: On the functional morphology of ependymal and extraependymal glia within the framework of neurosecretion: Electron microscopical studies on the neurohypophysis of the rat. *Z. Zellforsch.* **86**:111–128, 1968.
158. WURTMAN, R. T.: Brain monomamines and endocrine function. *Neurosci. Res. Program Bull.* **9**:172–297, 1971.
159. YAMAGUCHI, M., LAJTHA, A.: Inhibition of dipeptide transport in mouse brain slices. *J. Neurol. Sci.* **10**:323, 1970.
160. YARBROUGH, G. G.: TRH potentiates excitatory actions of acetylcholine on cerebral cortical neurones. *Nature (London)* **263**:523–524, 1976.
161. ZIMMERMAN, E. A., HSU, K. C., FERRIS, M., KOZLOWSKI, G. P.: Localization of gonadotropin-releasing hormone (Gn-RH) in the hypothalamus of the mouse by immunoperoxidase technique. *Endocrinology* **95**:1–8, 1974.
162. ZIMMERMAN, E. A., KOZLOWSKI, G. P., SCOTT, D. E.: Axonal and ependymal pathways for the secretion of biological active peptides into hypophysial portal blood. In Knigge, K. M., Scott, D. E., Kobayashi, H., Ishii, S. (eds.): *Brain–Endocrine Interaction II. The Ventricular System*. Basel, S. Karger, 1975, pp. 123–134.

CHAPTER 45

Endorphins in Cerebrospinal Fluid

Floyd E. Bloom and David S. Segal

1. Introduction

Endorphins are endogenous brain peptides (see Goldstein[43]) that are opiate-like in their pharmacology. Despite their relatively brief recorded history, they have already become household words in neuropharmacology. At the time of the original molecular identification of Met^5-enkephalin (M-e) and Leu^5-enkephalin (L-e),[59] the possibility of one or more other endorphins of pituitary origin had already been suggested.[23,43,44,103,120] When sequencing studies of the purified M-e revealed it to be the N-terminal pentapeptide[58] of the erstwhile pituitary hormone β-lipotropin (β-LPH)[77–79] (see the sequence in Fig. 1), the possibility that β-LPH was the prohormone of pituitary M-e was temporarily viable. That possible relationship appeared to be strengthened by the subsequent isolation, purification, sequencing, and synthesis of α-endorphin (A-E)[47,82,83]; β-endorphin (B-E), called also C-fragment[14–16,29,30,44,45,79–81,84]; γ-endorphin (G-E)[82–84]; and δ-endorphin (D-E), called also C′-fragment.[44,116] All these fragments of β-LPH were found in extracts of brain and pituitary, exhibited some action as specific opioid agonists, and contained M-e as their N-terminal pentapeptide (see Fig. 1 for structures). When subsequent tests *in vitro*[14,29,75] and *in vivo*[13,14,16,45,86] revealed that B-E was by far the most potent and longest-acting of the natural peptides, some workers concluded that the transient opioid actions of M-e and L-e indicated that these substances were merely weakly active breakdown products of the naturally active hormone, B-E.[16,35,40,43,116] Others interpreted the same data to mean that the natural "neurotransmitter" opioid peptides were the succinctly acting M-e and L-e,[125,129] while B-E was regarded exclusively as a pituitary product the longer duration of action of which arose from proteolytic protection afforded by the greater length of its peptide chain. Curiously, this greater length did not improve the potency or duration of action of A-E, D-E, or G-E.[12] Nevertheless, all workers seemed agreeable to the idea that opiate receptors in innervated tissues really represented the natural receptors for the endorphins and enkephalins.

Better definition of the functional roles of and relationships among these peptides required the development of perfected methods for the optimal preservation and extraction[48,69,72,104,105,114] of the individual peptides and the development of specific antisera for radioimmunoassay (RIA)[47] and immunocytochemical localization of their storage sites in brain and pituitary.[10,11,120] The results of such studies have just become known and force the realization that B-E-containing cells exist independently from enkephalins and pituitary.[4,10,11,120] Moreover, while B-E in brain and pituitary may be derived from β-LPH[44,75] and may also be the source of A-E and G-E,[131] B-E and β-LPH obviously cannot be the metabolic source of L-e. Finally, comparison of the actions of B-E with those of the enkephalin on central

Abbreviations of brain opioid peptides used throughout this chapter: (A–E) α-endorphin (β-LPH_{61-76}); (B–E) β-endorphin (β-LPH_{61-91}); (D–E) δ-endorphin (β-LPH_{61-87}); (G–E) γ-endorphin (β-LPH_{61-77}); (L-e) Leu^5-enkephalin; (β-LPH) β-lipotropin; (M-e) Met^5-enkephalin (β-LPH_{61-65}).

Floyd E. Bloom, M.D. • Arthur Vining Davis Center for Behavioral Neurobiology, The Salk Institute, San Diego, California 92112. **David S. Segal, Ph.D.** • Department of Psychiatry, University of California School of Medicine, San Diego, California 92112.

H·Glu-LeuAla-Gly-Ala-Pro-Pro-Glu-Pro-Ala-Arg-Asp-Pro-Glu-Ala-
5 10 15
Pro-Ala-Glu-Gly-Ala-Ala-Ala-Arg-Ala-Glu-Leu-Glu-Tyr-Gly-Leu-
20 25 30
Val-Ala-Glu-Ala-Gln-Ala-Ala-Glu-Lys-Lys-Asp-Glu-Gly-Pro-Tyr-
35 40 45
Lys-Met-Glu-His-Phe-Arg-Trp-Gly-Ser-Pro-Pro-Lys-AspLys Arg-
50 55 60
|Tyr-Gly-Gly-Phe-Met|Thr-Ser-Glu-Lys-Ser-Gln-Thr-Pro-Leu Val-
61 65 e 70 75
Thr|Leu|Phe-Lys-Asn-Ala-Ile-Val-Lys-Asn-Ala-His-Lys-Lys Gly-
α γ 80 85 90
porcine Gln OH
β

Figure 1. β-Lipotropin and its neurotropic subunits.

and peripheral receptors has led to the postulation that there may not be simply a single monolithic class of endorphin receptors that is acceptable to all the peptides, but rather that some receptors may be peptide-specific.[52,87]

With these rapidly evolving views in mind, we have structured this review chapter with the intention of surveying the various reports pertaining to two main categories of information: (1) the cellular and behavioral effects of injecting endorphins into the cerebrospinal fluid (CSF) and (2) the measurements of endorphins in samples of normal or pathological CSF.

2. Actions of Endorphins via CSF

The endorphins—particularly B-E—exhibit several very potent actions that emulate various elements of the pharmacological profile of opiates. As an overview, these actions after intracerebroventricular injection consist of akinesia and analgesia,[6,9,13,20,34,42,50–52,56,76,86,109,123] hypothermia,[13,54] and hyperglycemia.[34,46] At threshold doses for analgesia, respiratory depression is more pronounced in primates than it is in rodents[94] (also Bloom, Foote, Henriksen, and Ommaya, unpublished observations). Cats also show cardiovascular responses, especially elevation of blood pressure.[36]

In hippocampus, where there are relatively few opiate receptors, few enkephalin fibers, and no endorphin innervation at all, intracisternal or intraventricular injection of very low doses of B-E produced a highly reproducible, naloxone-reversible behavioral response[18,39,50–52,110,111] that is accompanied by EEG signs of electroconvulsive activity[34,57,121,125] without overt motor seizures. After doses of 1–5 nmol B-E, these electrographic convulsive actions can persist for 3 hr or more.[50,87]

One perplexing action that is shared by all the endorphin peptides and enkephalins regardless of their relative potency in eliciting these other effects is that all can produce "wet-dog shakes" on initial intracerebroventricular injection in opiate-naïve rats.[13,76] Morphine does not produce wet-dog shaking after intraventricular administration, but both morphine and B-E induced extensive wet-dog shaking behavior when injected into the medial amygdala.[17] It is conceivable, therefore, that the appearance of wet-dog shaking during withdrawal in morphine-tolerant animals might reflect the unmasking of a morphine action at specific brain sites. Although wet-dog shakes are ordinarily attributed to withdrawal from the opiate-dependent state,[133] this behavior when induced by opioid peptides can be prevented by prior injection of an opiate antagonist.[13] It should be noted, however, that wet-dog shaking is not uniquely associated with administration of opioid peptides or with the opiate-withdrawal syndrome, having been observed after systemic or intracerebral administration of a variety of substances.[22,73,74,89,99,100,134]

With this brief overview, we can now examine the actions of CSF endorphins in somewhat greater detail.

2.1. Analgesia

Many workers have examined the ability of the natural and synthetic endorphins and enkephalins (see Bradbury *et al.*,[16] Lord *et al.*,[87] and McGregor *et al.*[91]) to produce analgesia (i.e. an antinocisponsive effect.* Those data sufficiently detailed to permit threshold and duration relationships are summarized in Tables 1 and 2. Although it is clear that analgesia is a complicated pharmacological action that may be the result of many independent receptor events at various levels of the nervous system, the results support the general contention that B-E is the most potent and longest-acting natural endorphin, and that it is 30 or more times more potent than morphine on a molar basis. However, the transient nature of the enkephalin effects in these tests makes the comparisons complicated.

Two related pharmacological issues concern the importance of route of administration and tolerance development.[42,123,132] Tseng *et al.*[122] have observed that B-E can produce modest analgesia in the mouse tail-flick after intravenous administration in the dose

* The term "nocisponsive" was suggested by Cline-schmidt *et al.*[21] to reflect more accurately what is measured in such tests, namely, responsiveness to noxious stimuli.

Table 1. Analgesic Effect by Intracerebral Administration in Various Species

Species	Test used	ED_{50} (nmol/animal)			Length of analgesia after B-E[a]		Reference
		B-E[a]	Morphine	M-e[a]	Dose (nmol)	Time (min)	
Mouse	Tail-flick	0.09	2.7	119	—	—	Roemer *et al.* (1977)[102]
Mouse	Tail-flick	0.04	1.26	—	0.29	120	Loh *et al.* (1976)[86]
	Hot plate	0.06	1.02	—	0.29	60	Loh *et al.* (1976)[86]
	Righting	0.02	0.45	—	0.15	30	Loh *et al.* (1976)[86]
Rat	Tail-flick	0.6	3.2	—	—	—	Szekeley *et al.* (1977)[117]
Rat	Tail-flick	—	—	—	0.8	60–120	Graf *et al.* (1976)[44]
Rat	Hot plate	—	—	—	1.5	27	Bradbury *et al.* (1976)[15]
	Tail-pinch	—	—	—	3.0	100	Bradbury *et al.* (1976)[15]
Rat	Tail-pinch	—	—	—	9.0	100	Bloom *et al.* (1976)[13]
Cat	Jaw-opening reflex	—	—	—	7.3	120	Meglio *et al.* (1977)[88]
Cat	Tail-pinch	—	—	—	5.8	120	Feldberg and Smyth (1976)[34]

[a] (B-E) β-Endorphin; (M-e) Met^5-enkephalin.

range of 8–20 mg/kg, a dose level at which some peptide may be presumed to have entered certain portions of the central nervous system (CNS), although this remains to be determined. When expressed on a molar basis, even these seemingly larger amounts of B-E still indicate it to be 3–4 times more potent in producing analgesia than morphine.[13,122,124] In cats (tail-pinch test), Feldberg and Smyth[35] have also reported that B-E (250 μg/kg) administered intravenously produced some analgesia that lasted about 20 min.

In the rat, subcortical EEG recordings have been found to be a highly sensitive index of central endorphin actions, with the limbic electroconvulsive activity being detectable earlier and at lower dosages than analgesia to tail-pinch or loss of corneal reflexes.[39,50–52,125] With this index, in the rat, B-E produced no detectable actions by the intravenous route at cumulative doses up to 20 mg/kg (Henriksen and Bloom, unpublished results).

In Swiss–Webster mice, intravenous injection of the metabolically resistant enkephalin analogue D-Met^2-Pro^5-enkephalinamide (D-Met^2-Pro^5-NH_2), as with opiates, resulted in a dose-related increase in stereotyped locomotion; D-Met^2-Pro^5-NH_2 also elicited tail elevations.[110,111] In contrast to the effects of opiates and D-Met^2-Pro^5-NH_2, behavioral activation was not observed after intravenous injection of B-E. Mice treated with 5 or 10 mg/kg of B-E were indistinguishable from saline controls with respect to both locomotor activity and general appearance. With the highest dose tested (20 mg/kg), locomotion was significantly reduced. Therefore, after systemic administration, B-E may not accumulate in the brain in amounts sufficient to produce opiate-like changes in locomotion. However, even in the rat, intravenous B-E and enkephalin will cause the release of vasopressin,[46] prolactin,[31,101] and growth hormone,[31,101] although these central effects could result from activating peripheral autonomic receptors.

Table 2. Analgesic Properties of Opioid Peptides in Mouse (Tail-Flick Test)

Administration route	ED_{50}			Reference
	B-E[a]	Morphine	M-e[a]	
Intracerebrally	0.09	2.7	119	Roemer *et al.* (1977)[102]
(nmol/mouse)	0.04	1.26	—	Loh *et al.* (1976)[86]
Intravenously	2.7	11.4	—	Tseng *et al.* (1976)[123]
(μmol/kg)	—	4.8	267	Roemer *et al.* (1977)[102]

[a] (B-E) β-Endorphin; (M-e) Met^5-enkephalin.

Tolerance to B-E administered by continuous intraventricular infusion was reported early in the course of events when analgesia and locomotor depression were used as the primary indices.[42,117,123,124,133] As with morphine, no dependence is seen if 24 hr of drug-free state elapses between test doses.[13] If the EEG activity is employed as the index, tolerance is not seen with intertest intervals of 8–12 hr.[51] Since the electroconvulsive activity of even a single low dose of B-E may last for as long as 4 hr,[50,51] it may be difficult to distinguish between "tolerance" and reduced effectiveness due to refractoriness from a recent response. Other studies employing analgesia and withdrawal symptoms as indices have been interpreted as indicating that rats made dependent on morphine by continuous subcutaneous release show cross-tolerance to the effects of centrally administered B-E.[42,117,123,124,133] However, as mentioned earlier, the meaning of withdrawal signs such as the wet-dog shaking behavior may need some reevaluation, since this response is produced in opiate-naïve animals on the first central injection of B-E and M-e.[13,76,109]

In accord with their hypothesis that endorphins are involved in reward functions, Belluzzi and Stein[7] have shown that rats developed high rates of intraventricular self-administration for the pentapeptides L-e and M-e and for morphine. Although more rapid learning of self-administration behavior resulted with a pentapeptide dose of 1 μg per injection, a more sustained performance level was generated with a 10-μg dose. These results were interpreted to indicate that tolerance develops to the reinforcing action of the enkephalins as it does with morphine. Thus, the 10-μg dose produces a progressively more optimal effect as tolerance develops.

Although little pharmacological work has been directed toward modifying the rate of synthesis or degradation of B-E by brain peptidases,[3,4,70,116] available evidence suggests that B-E is relatively resistant to degradation in both plasma and CSF.[104] Other biochemical manipulations such as stabilizing the molecule or improving its penetrance into specific CNS target areas may be anticipated to yield improved peptides capable of producing analgesia without addiciton.

2.2. Effects on Other Central Neurotransmitters

Opiates have been shown to inhibit release of transmitter from electrically activated autonomic nerve fibers.[71,98] Similar actions of M-e have been demonstrated on acetylcholine,[67] noradrenaline,[118] and substance P[66] release from brain slices, and B-E shows a similar presynaptic effect on dopamine release.[8,85] In addition, intracerebroventricular injection of B-E in doses that produce akinesia, analgesia, and muscular rigidity[13,61] (see below) has been reported to produce slight elevations of midbrain serotonin levels and to produce a slight delay in the "turnover" of midbrain dopamine.[8,61,126,127] After injections of small analgesic doses of B-E, a decrease in the turnover of acetylcholine has also been observed. This effect is especially pronounced in cerebral cortex, hippocampus, globus pallidus, and nucleus accumbens.[92,93] Most of these neurochemical effects are compatible with the commonly observed inhibitions of discharge rate produced by iontophoretic B-E[96,137] in all brain areas except the atypical response of hippocampus.[38,96,104] Present research is unclear as to which—if any—of the peptidergic or monoaminergic central systems are able to mediate or modify the intensity of the opiate-tolerance and -withdrawal syndrome.

2.3. Behavioral Pharmacology

Our group has found that intraventricular administration of B-E, but not of other naturally occurring opioid peptides, induces in rats a profound state of immobilization, characterized by the absence of spontaneous movement, loss of the righting response, and extreme generalized muscular rigidity.[13,17–19,109–111] Similar results have been obtained by others.[60,95,124] Furthermore, although Jacquet and Marks[61] reported that injections of B-E into the periaqueductal gray (PAG) elicited a neuroleptic-like behavioral state, our studies show that the immobility induced by B-E can be differentiated from that associated with neuroleptics such as haloperidol.[19]

Within 30 min after injection into the cisterna or lateral ventricle, B-E produced a dose-related increase in muscular rigidity (Table 3). During the rigidity phase, which persisted for at least 6 hr at the highest dose tested (300 μg), animals remained motionless if left undisturbed or handled gently; however, they could be provoked into moving with the presentation of relatively mild auditory, visual, or tactile stimulation. This effect was particularly apparent before and subsequent to the period of peak rigidity and occurred at a time when the animals were not responsive to noxious stimuli. Furthermore, the intensity of the rigidity appeared to be reduced during tests at night, a time when rats are normally active. These results indicate that during the rigidity phase, animals were capable of coordi-

Table 3. Opioid-Peptide-Induced Immobility[a]

Treatment[b]	Dose (μg/10 μl)	Number	Righting response[c] (10 sec)	Rigidity[d] (0–4)
Saline	10 μl	13	+	0
βp-Endorphin	2.5	4	−	0
	5.0	19	−	2.0 ± 0.2
	10.0	12	−	2.8 ± 0.3
	25.0	4	−	3.5 ± 0.5
	50.0	43	−	3.7 ± 0.1
	100.0	4	−	4
	150.0	3	−	4
	300.0	5	−	4
βh-Endorphin	50.0	3	−	3.6 ± 0.3
D-Met^2-Pro^5·NH_2	25.0	3	−	4
D-Ala^2-Met^5·NH_2	50.0	5	−	1.6 ± 0.2

[a] From Segal *et al.*,[110] with permission of the Elsevier Publishing Co.
[b] Synthetic porcine β-endorphin; (βh) synthetic human β-endorphin; (D-Met^2-Pro^5·NH_2) D-Met^2-Pro^5-enkephalinamide; (D-Ala^2-Met^5·NH_2) D-Ala^2-Met^5-enkephalinamide.
[c] A minus righting response was designated when the rat stayed in a supine position for 10 sec.
[d] The rigidity score represents a composite measure derived from three tests: (1) stiffness, assessed during handling (scored 0–3); (2) trunk rigidity, based on the time (up to 4 sec) that the animal remained in an upright posture when held above the knee joints of the hind limbs (scored 0–4); and (3) bridge test, a positive score being assigned when the animal remained self-supporting for 10 sec after being placed across metal bookends. Values (expressed as means ± S.E.M.) indicate peak effects after intraventricular injection.

nated motor activity, and that the behavioral immobility may have been partially due to an impaired ability to initiate voluntary movement.

All doses of B-E tested produced wet-dog shaking behavior within 15 min after infusion. Initially, these shaking episodes were followed by brief periods of activity; at lower doses, the hyperactivity was eventually interrupted by recurrent episodes of a "trance-like" state, during which the animals remained motionless in a standing or rearing position for up to 15 min.[110] A 2.5-μg dose of B-E that did produce detectable rigidity did induce wet-dog shaking behavior as well as periods of the "trance-like" behavioral state. Rigid immobility was also induced by synthetic human B-E (50 μg) and by the metabolically resistant enkephalin analogues D-Met^2-Pro^5-enkephalinamide (D-Met^2-Pro^5-NH_2) (25 μg) and D-Ala^2-Met^5-enkephalinamide (D-Ala^2-Met^5-NH_2) (50 μg) (Table 3).

Immobility has frequently been assessed using horizontal-bar and vertical-grid tests.[37] Most neuroleptics produce positive responses on both these measures.[37,109] In contrast, a positive response was observed only on the horizontal-bar test after B-E administration.[17,19] Thus, animals injected with B-E would quickly climb off the vertical grid before and after the period of rigidity and typically would slide or fall off the grid during the rigidity phase. Animals that received haloperidol grasped the grid tightly and remained stationary for relatively long periods of time (Table 4). Furthermore, doses as high as 12 mg/kg of haloperidol did not produce rigidity or loss of the righting response. Instead, animals injected with haloperidol (0.5–12 mg/kg, s.c.) typically displayed a hunched posture and abducted limbs, and vocalized when handled.

Rigidity resulting from injection of B-E could be rapidly reversed by naloxone administered subcutaneously (0.1 mg/kg) or intraventricularly (0.5 μg/10 μl).[109] In contrast, the effects of haloperidol were unaltered by doses of naloxone as high as 2 mg/kg. Furthermore, animals made rigid with B-E and later injected with haloperidol (2 mg/kg) became flaccid. In rats rendered flaccid by this combined treatment, naloxone (2 mg/kg) administration resulted in the emergence of the typical haloperidol effects. Thus, naloxone, by selectively antagonizing the effect of B-E, unmasked the behavioral pattern induced by haloperidol.

Like B-E, opiates such as morphine, methadone, and etonitazene produced a dose-dependent increase in rigidity and loss of righting response (Table

Table 4. Characteristics of Immobility Induced by Haloperidol[a]

Treatment	Dose (mg/kg)	Number	Righting response[b] (10 sec)	Rigidity[c] (0–4)	Vertical Grid[d] (0–3)
Saline	1 ml/kg	10	+	0	0
Haloperidol	0.5	12	+	0	1.8 ± 0.3
	1.0	12	+	0	2.9 ± 0.1
	2.0	15	+	0	2.6 ± 0.1
	4.0	10	+	0	2.5 ± 0.2
	8.0	15	+	0	2.3 ± 0.2
	12.0	6	+	0	2.6 ± 0.2

[a] From Segal *et al.*,[110] with permission of the Elsevier Publishing Co.
[b] See Table 3, footnote *c*.
[c] See Table 3, footnote *d*.
[d] The vertical grid test was scored on a 0–3 scale, based on the time (up to 60 sec) that the rat remained immobile on the grid.

5). Furthermore, opiate-treated rats are also selectively nonresponsive to noxious stimuli. Intraventricular administration of morphine or etonitazene produced a similar response profile (Table 6). In addition, naloxone (0.05–10 μg) injected intraventricularly or into the caudate, globus pallidus, amygdala, or PAG rapidly reversed all these opiate-induced actions.[17,19]

A prolonged period of hyperactivity follows the opiate-induced immobility in rats.[5,25,28,97,110,111] We have recently found that a similar biphasic pattern results from intraventricular administration of B-E and the enkephalin analogues D-Met^2-Pro^5-NH_2 and D-Ala^2-Met^5-NH_2.[18,110,111] Oral stereotypy is a prominent feature of the behavioral activation produced by higher doses of the opiates or opioid peptides. Both the stereotypy and the locomotion produced by 7.5 mg/kg of methadone (s.c.) or by 50 μg B-E (intraventricular) were antagonized by administration of either naloxone (0.05 mg/kg, s.c.) or halo-

Table 5. Opiate-Induced Immobility: Subcutaneous Administration[a]

Treatment	Dose	Number	Righting response[b] (10 sec)	Rigidity[c] (0–4)
Saline	1 ml/kg	13	+	0
Morphine (mg/kg)	5.0	18	+	0
	7.5	11	±	0.7 ± 0.2
	10.0	18	−	1.9 ± 0.3
	20.0	10	−	3.5 ± 0.2
Methadone (mg/kg)	1.0	10	+	0
	2.5	17	+	0
	5.0	21	−	2.5 ± 0.6
	7.5	20	−	3.6 ± 0.2
	10.0	11	−	3.9 ± 0.1
Etonitazene (μg/kg)	1.0	8	+	0
	5.0	11	±	2.1 ± 0.4
	10.0	11	−	3.7 ± 0.2
	20.0	4	−	4

[a] From Segal *et al.*,[110] with permission of the Elsevier Publishing Co. Subcutaneous injection of morphine, methadone, and etonitazene resulted in rigidity and accompanying loss of righting.
[b] See Table 3, footnote *c*; (±) rats remained supine for at least 5 sec, but were capable of self-righting within 10 sec.
[c] See Table 3, footnote *d*.

Table 6. Opiate-Induced Immobility: Intraventricular Administration[a]

Treatment	Dose (μg/10 μl)	Number	Righting response[b] (10 sec)	Rigidity[c] (0–4)
Saline	10 μl	13	+	0
Morphine	25.0	3	±	0
	50.0	4	−	1.2 ± 0.6
	100.0	5	−	2.4 ± 0.4
Etonitazene	0.5	2	+	0
	1.0	10	±	1.5 ± 0.6
	2.0	2	−	3.5 ± 0.5
	5.0	10	−	3.2 ± 0.3
	10.0	6	−	4

[a] From Segal *et al.*,[110] with permission of the Elsevier Publishing Co.
[b] See Table 5, footnote *b*.
[c] See Table 3, footnote *d*.

peridol (0.5 mg/kg, s.c.). Furthermore, naloxone, when administered during the hyperactivity phase, was also effective in reducing locomotion and stereotypy. This finding suggests that the hyperactivity phase may also be mediated through opiate-receptor activation.

In summary, B-E and other opioid peptides produce a broad spectrum of dose- and time-related behaviors, ranging from locomotor excitation to extreme muscular rigidity. These effects are reversed by naloxone and closely resemble the behavior actions of opiates, and thus appear to be mediated through the activation of opiate receptors in the brain.

Although we have shown that the behavioral profile of the opioid peptides differs from that elicited by neuroleptics such as haloperidol, de Wied *et al.*[26,27] have suggested that fragmentation of B-E may result in compounds with neuroleptic-like properties. These workers reported that the effects of haloperidol and (Des-Tyr^1)-γ-endorphin [DTγE (β-LPH_{62-77})] are comparable with respect to passive and active avoidance and performance on various "grip" tests. In accord with these observations in animals, DTγE was also found by this group[128] to have antipsychotic activity in schizophrenic patients resistant to conventional neuroleptic drugs. However, we have recently found that subcutaneous administration of this peptide (0.21–1.23 mg/kg) did not produce haloperidol-like immobility in either the horizontal-bar or the vertical-grid test.[135] Furthermore, in contrast to most neuroleptics, DTγE did not antagonize amphetamine- or apomorphine-induced locomotion stereotypy.

3. Endorphins in CSF

The pharmacological effects of the various natural and synthetic endorphins produced after injection into the CSF provide a strong basis for examining the physiological role of the natural substances in the CSF and their possible involvement in pathological states. However, before proceeding to examine the available measurements of endorphins in human CSF, some consideration must be given to what such measurements may mean. As with the other substances covered in this compendium of CSF constituents, the detection of a substance in the CSF may have at least one of two major interpretations: (1) The substance "escapes" or is "secreted" into CSF only during periods of extreme activity (such as the extremes to which some experiments or disease states perturb the system). (2) Alternatively, the substance may normally be secreted into CSF by nerve fibers adjacent to or penetrating into the ventricular system, as one means of reaching targets to which intrinsic central fibers have not been extended. In the case of the maps of the endorphins[11,12,122] and enkephalins,[32,53,69,106,113,115,130,136] no fibers have been observed to extend into CSF, but in every case, fibers have been seen close to the ependyma. The possibility that the first explanation of CSF-measured materials may hold does not necessarily exclude the second explanation, and vice versa. Nevertheless, the tenability of the "escape" view also raises the possibility that materials not detected within CSF [such as the enkephalins, on the basis of present reports[1,2,58] (but also see Sarne *et al.*[107])] may also still perform functional roles within the substance of the brain. We previously considered that the profound behavioral changes induced by B-E in rats might indicate that these peptides have etiological significance in some mental illnesses[8] and therefore that opiate antagonists might be efficacious in their treatment. Although one group[49] has reported that 0.4 mg naloxone reduced auditory hallucination intensity in 4 of 6 chronic schizophrenics, others[24,129] as well as ourselves,[62,63] using 2- to 3-fold higher doses and double-blind protocols, have been unable to repeat these positive results with naloxone. However, on the basis of our behavioral, EEG, and unit recording observations, these doses of naloxone may have been too low to reverse the central effects of endorphins. Furthermore, in the subclassification of opiate receptors proposed by Martin and associates,[41,90] some opiate-agonist actions may require 3–6 times more naloxone for reversal. More recently, Emrich *et al.*,[33] Watson *et al.*,[130] and Schenk *et al.*[108] have observed clinical improvement by increasing the dose of naloxone. In addition, Judd *et al.*[68] have reported that 4 of 12 manic patients exhibited a marked reduction in manic behavior after administration of relatively high doses of naloxone (20 mg/kg i.v.).

An entire separate set of "endorphin-like" substances, which are presumably not peptides,[112] have also been detected in the CSF of patients with chronic pain syndromes[119] and in psychotic patients.[49] These levels have been found to normalize after treatment. However, Terenius[119] and his colleagues justify calling these substances "endorphins" only because of their activity against labeled opiate agonists and antagonists in receptor displacement assays *in vitro*, and not from any direct demonstration of chemical resemblance to any of the known endorphin peptides already described above. Although this arbitrary chemical distinction has excluded them from extended discussion here, these undetermined "endorphin factors," and others (see Shorr *et al.*[112]) that may also represent "opioid" activity on *in vitro* assays, may well be relevant clinically as pathogenic agents. We must therefore await their chemical characterization with eager anticipation.

There are, however, good reasons to believe that among the various materials in human CSF with opiate-like pharmacological actions are the fully characterized endorphin peptides, especially B-E. Jeffcoate *et al.*[63,64] have combined N-terminal and C-terminal RIAs for B-E immunoreactive substances in the lumbar CSF of normal humans to demonstrate that B-E (rather than β-LPH, which is read by all C-terminally directed anti-B-E sera) is indeed present under basal conditions along with lesser amounts of larger fragments of β-LPH. However, the amounts present are in the 50–150 pM range. In their subjects, Jeffcoate and associates observed more endorphin-like material in CSF than in plasma. That observation, together with substantial evidence from direct assay of normal and hypophysectomized animal brains, strongly suggests the conclusion that CSF B-E arises from brain.[105]

Akil *et al.*[1] also reported bioassay data indicating that in patients with chronic pain syndromes, electrical activation of the periaqueductal gray (PAG) region, which is rich in both enkephalins and endorphins, releases into the third-ventricular CSF an opiate-like material that was originally thought to be immunoreactive enkephalin. However, repetition of these studies by the same authors employing a B-E RIA system revealed that the material released was in fact B-E, which rose from below their limit of detectability (25 pM) to amounts in the range of hundreds of femtomoles per milliliter.[2] This reinterpretation was more in keeping with the work of Hosobuchi and associates,[55,56,58] who have observed that in some humans, PAG stimulation for intractable pain is naloxone-sensitive;[55] analgesia can be produced in these patients with third-ventricular injections of B-E, but not with enkephalin.[56]

These reports on B-E electrically released into human CSF serve to illustrate the general interpretive caveats with which this section began. The experimental perturbation of the PAG system in these patients releases into CSF an endorphin-like immunoreactive substance. This observation could mean that the electrical activation was excessive, and that as a result of the extraordinary release of B-E, material could be detected in CSF. Alternatively, this release as well as the detectable basal levels could indicate that the PAG normally releases endorphin into CSF, either as a means of excretion or to propagate the peptide to other brain regions via the CSF. Neither interpretation can exclude the possibilities that the same pain-suppressing systems of the PAG may also employ enkephalins as their transmitter and that this substance cannot be detected in CSF (see Sarne *et al.*[107]) for other reasons that are not yet clear.

In more recent studies, Hosobuchi *et al.*[58] selected patients to verify the direct involvement of the PAG endorphin fibers in the pain relief produced by electrical stimulation. Two groups of patients were studied. One group had pain relieved by PAG stimulation that was naloxone-sensitive.[55] A second group of patients had a chronic deafferentation pain syndrome that is sensitive neither to opiates nor to PAG stimulation, but is sensitive to electrical stimulation of the internal capsule.

In both groups of patients, electrical stimulation induced pain relief, but elevated CSF levels of radioimmunoassayed B-E were found only in those patients in whom PAG-stimulating electrodes were located in the PAG. In this study, basal levels were well within the detectable range of the immunoassay, and varied between 140 and 210 pg/ml (46–70 fm/ml, or 46–70 pM), or slightly higher than baseline sensitivity of the study of Akil *et al.*[2] In the patients in whom PAG electrodes produced pain relief, and only in those patients, the electrical stimulation gave a 2- to 7-fold increase in CSF B-E, with the maximal value (approximately equal to that observed by Akil *et al.*[2]) being detected between 5 and 15 min after onset of stimulation.

4. Conclusions

The major pharmacologically active endorphin peptides are comprised of enkephalin pentapeptides and β-endorphin (B-E). These peptides can produce significant effects on neuronal activity, as detected chemically or electrophysiologically, and on whole-animal behaviors, such as responsiveness to noxious stimuli, locomotor activity, posture, and possibly on memory, learning, and reward systems.

Although these pharmacological actions of naturally occurring peptides have generated considerable interest, the physiological functions of these substances in the CSF, and in regions of the brain adjacent to the cerebral ventricles, remain unknown. Studies on the stimulation-induced release of endorphins into CSF suggest that B-E may be involved in the pain suppression mediated by electrical activation of the periaqueductal gray region of the midbrain. Further detailed studies of immunoassayable materials are needed to determine the involvement of endorphins in psychological processes.

References

1. AKIL, H. RICHARDSON, D. E., HUGHES, J., BARCHAS, J. D.: Enkephalin-like material elevated in ventricular cerebrospinal fluid of pain patients after analgetic focal stimulation. *Science* **201**:463–465, 1978.
2. AKIL, H., RICHARDSON, D. E., BARCHAS, J. D., LI, C. H.: Appearance of β-endorphin-like immunoreactivity in human ventricular cerebrospinal fluid upon analgesic electrical stimulation. *Proc. Natl. Acad. Sci. U.S.A.* **75**:5170–5172, 1978.
3. AUSTEN, B. M., SMYTH, D. G.: The NH_2-terminus of C-fragment is resistant to the actions of aminopeptidases. *Biochem. Biophys. Res. Commun.* **76**:477–482, 1977.
4. AUSTEN, B. M., SMYTH, D. G.: Specific cleavage of lipotropin C-fragment by endopeptidases; evidence for a preferred conformation. *Biochem. Biophys. Res. Commun.* **77**:86–94, 1977.
5. BABBINI, M., DAVIS, W. M.: Time–dose relationships for locomotor activity effects of morphine after acute and repeated treatment. *Br. J. Pharmacol.* **46**:213–224, 1972.
6. BELLUZZI, J. D., GARNT, N., GARSKY, V., SAFRANTAKIS, D., WISE, C. D., STEIN, L.: Analgesia induced *in vivo* by central administration of enkephalin in rat. *Nature (London)* **260**:625–626, 1976.
7. BELLUZZI, J. D., STEIN, L.: Enkephalin may mediate euphoria and drive-reduction reward. *Nature (London)* **266**:556–558, 1977.
8. BERNEY, S., HORNYKIEWICZ, O.: The effect of β-endorphin and met-enkephalin on striatal dopamine metabolism and catalepsy: Comparison with morphine. *Commun. Psychopharmacol.* **1**:597–604, 1977.
9. BHARGAVA, H. N.: New *in vivo* evidence for narcotic agonistic property of leucine enkephalin. *J. Pharm. Sci.* **67**:136–137, 1978.
10. BLOOM, F., BATTENBERG, E., ROSSIER, J., LING, N., LEPPALUOTO, J., VARGO, T. M., GUILLEMIN, R.: Endorphins are located in the intermediate and anterior lobes of the pituitary gland, not in the neurophypophysis. *Life Sci.* **20**:43–48, 1977.
11. BLOOM, F. E., ROSSIER, J., BATTENBERG, E. L. F., BAYON, A., FRENCH, E., HENRIKSEN, S., SIGGINS, G. R., SEGAL, D., BROWNE, R., LING, N., GUILLEMIN, R.: β-Endorphin: Cellular localization, electrophysiological and behavioral effects. In Costa, E. (ed.): *Endorphins.* New York, Raven Press, 1979, pp. 89–110.
12. BLOOM, F., ROSSIER, J., BATTENBERG, E., VARGO, T., MINICK, S., LING, N., GUILLEMIN, R.: Regional distribution of β-endorphin and enkephalin in rat brain: A biochemical and cytochemical study. *Soc. Neurosci. Abstr.* **3**:286, 1977.
13. BLOOM, F. E., SEGAL, D., LING, N., GUILLEMIN, R.: Endorphins: Profound behavioral effects in rats suggest new etiological factors in mental illness. *Science* **194**:630–632, 1976.
14. BRADBURY, A. F., FELDBERG, W. F., SMYTH, D. G., SNELL, C. R.: Lipotropin C-fragment: An endogenous peptide with potent analgesic activity. In Kosterlitz, H. W. (ed.): *Opiates and Endogenous Opioid Peptides.* Amsterdam, North-Holland/Elsevier, 1976, pp. 9–17.
15. BRADBURY, A. F., SMYTH, D. G., SNELL, C. R., BIRDSALL, N. J. M., HOLME, E. C.: C-fragment of lipotro-

pin has a high affinity for brain opiate receptors. *Nature (London)* **260**:793–795, 1976.
16. BRADBURY, A. F., SMYTH, D. G., SNELL, C. R., DEAKIN, J. F. W., NENDLANDT, S.: Comparison of the analgesic properties of lipotropin C-fragment and stabilized enkephalins in the rat. *Biochem. Biophys. Res. Commun.* **74**:748–754, 1977.
17. BROWNE, R. G., DERRINGTON, D. C., SEGAL, D. S.: Comparison of opiate- and opioid peptide-induced immobility. *Life Sci.* **24**:933–942, 1979.
18. BROWNE, R. G., SEGAL, D. S.: β-Endorphin and opiate-induced immobility: Behavioral characterization and tolerance development. In Van Ree, J. M., Terenius, L. (eds.): *Characteristics and Function of Opioids: Developments in Neuroscience*, Vol. 4, Amsterdam, Elsevier, 1978, pp. 413–414.
19. BROWNE, R. G., SEGAL, D. S.: Behavioral activating effects of opiates and opioid peptides. *Biol. Psychiatry* **15**:77–86, 1980.
20. BÜSCHER, H. H., HILL, R. C., ROMER, D., CARDINAUX, F., CLOSSE, A., HAUSER, D., PLESS, J.: Evidence for analgesic activity of enkephalin in the mouse. *Nature (London)* **261**:423–425, 1976.
21. CLINESCHMIDT, B. V., MCGUFFIN, J., BUNTING, P. B.: Neurotensin: Antinocisponsive action in rodents. *Eur. J. Pharmacol.* **54**:129–140, 1979.
22. COLASANTI, B. K., KOSA, J. E., CRAIG, C. R.: Appearance of wet dog shake behavior during cobalt experimental epilepsy in the rat and its suppression by reserpine. *Psychopharmacologia* **44**:33–36, 1975.
23. COX, B. M., OPHEIM, K. E., TESCHEMACHER, H., GOLDSTEIN, N. A.: A peptide-like substance from pituitary that acts like morphine. *Life Sci.* **16**:1777–1782, 1975.
24. DAVIS, G. C., BUNNEY, W. E., JR., DE FRAITES, E. G., KLEINMAN, J. E., VAN KAMMEN, D. P., POST, R. M., WYATT, R. J.: Intravenous naloxone administration in schizophrenia and affective illness. *Science* **197**:74–77, 1977.
25. DAVIS, W. M., BRISTER, C. C.: Acute effects of narcotic analgesics on behavioral arousal in the rat. *J. Pharm. Sci.* **62**:974–979, 1973.
26. DE WIED, D., BOHUS, B., VAN REE, J. M., URBAN, I.: Behavioral and electrophysiological effects of peptides related to lipotropin (β-LPH). *J. Pharmacol. exp. Ther.* **204**:570–580, 1978.
27. DE WIED, D., KOVÁCS, G. L., BOHUS, B., VAN REE, J. M., GREVEN, H. M.: Neuroleptic activity of the neuropeptide β-LPH_{62-77} ([Des-Tyr[1]]-γ-endorphin; DTγE). *Eur. J. Pharmacol.* **49**:427–436, 1978.
28. DOMINO, E. F., VASKO, M. R., WILSON, A.: Mixed depressant and stimulant actions of morphine and their relationship to brain acetylcholine. *Life Sci.* **18**:361–376, 1976.
29. DONEEN, B. A., CHUNG, D., YAMASHIRO, D., LAW, P. Y., LOH, H. H., LI, C. H.: β-endorphin structure activity relationships in the guinea pig ileum and opiate receptor binding assays. *Biochem. Biophys. Res. Commun.* **74**:656–662, 1977.
30. DRAGON, N., SEIDAH, N. G., LIS, M., ROUTHIER, R., CHRETIEN, M.: Primary structure and morphine-like activity of human β-endorphin. *Can. J. Biochem.* **55**:666–670, 1977.
31. DUPONT, A., CUSAN, L., GARON, M., LABRIE, F., LI, C. H.: β-Endorphin: Stimulation of growth hormone release *in vivo*. *Proc. Natl. Acad. Sci. U.S.A.* **74**:358–359, 1977.
32. ELDE, R., HÖKFELT, T., JOHANNSON, O., TERENIUS, L.: Immunohistochemical studies using antibodies to leucine enkephalin: Initial observations on the nervous system of the rat. *Neuroscience* **1**:349–351, 1976.
33. EMRICH, H. M., CORDING, C., PIREE, S., KOLLING, A., ZERSSEN, D., HERZ, A.: Indication of an antipsychotic action of the opiate antagonist naloxone. *Pharmakopsychiatr./Neuro-Psychopharmakol.* **10**:265–270, 1977.
34. FELDBERG, W., SMYTH, D. G.: The C-fragment of lipotropin: A potent analgesic in cat. *J. Physiol. (London)* **260**:30–31P, 1976.
35. FELDBERG, W., SMYTH, D. G.: Analgesia produced in cats by the C-fragment of lipotropin and by a synthetic pentapeptide. *J. Physiol. (London)* **265**:25–27P, 1977.
36. FELDBERG, W., WEI, E.: Central cardiovascular effects of enkephalins and C-fragment of lipotropin. *J. Physiol. (London)* **280**:18P, 1978.
37. FOG, R.: On stereotypy and catalepsy: Studies on the effects of amphetamines and neuroleptics in rats. *Acta Neurol. Scand.* **48**(Suppl.50):10–66, 1972.
38. FRENCH, E. D., SIGGINS, G. R., HENRIKSEN, S. J., LING, N.: Iontophoresis of opiate alkaloids and endorphins accelerates hippocampal unit firing by a non-cholinergic mechanism: Correlation with EEG seizures. *Soc. Neurosci. Abstr.* **3**:291, 1977.
39. FRENK, H., URCA, G., LIEBESKIND, J. C.: Epileptic properties of leucine and methionine enkephalin: Comparison with morphine and reversibility by naloxone. *Brain Res.* **147**:327–337, 1978.
40. GEISOW, M. J., DEAKIN, J. F. W., DOSTROVSKY, J. O., SMYTH, D. G.: Analgesic activity of lipotropin C-fragment depends on carboxyl terminal tetrapeptide. *Nature (London)* **269**:167–168, 1977.
41. GILBERT, P. E., MARTIN, W. R.: The effect of morphine- and nalorphine-like drugs in the nondependent, morphine-dependent and cyclazocine-dependent chronic spinal dog. *J. Pharmacol. Exp. Ther.* **198**:66–82, 1976.
42. GISPEN, W. H., WIEGENT, V. M., BRADBURY, A. F., HULME, E. C., SMYTH, D. G., SNELL, C. R., DE WIED, D.: Induction of tolerance to the analgesic action of lipotropin C-fragment. *Nature (London)* **264**:792–794, 1976.
43. GOLDSTEIN, A.: Opioid peptides (endorphins) in pituitary and brain. *Science* **193**:1081–1086, 1976.
44. GRAF, L., RONAI, A. Z., BAJUSZ, S., CSEH, G., SZEKELY, J. I.: Opioid agonist activity of β-lipotropin fragments: A possible biological source of morphine-like substances in the pituitary. *FEBS Lett.* **64**:181–185, 1976.

45. GRAF, L., SZEKELY, J. I., RONAI, A. Z., DUNAI-KOVACS, Z., BAJUSZ, S.: Comparative study on analgesic effect of Met[5]-enkephalin and related lipotropin fragments. *Nature (London)* **263**:240–242, 1976.

46. GUILLEMIN, R., BLOOM, F. E., ROSSIER, J., MINICK, S., HENRIKSEN, S., BURGUS, R., LING, N.: Current physiological studies with the endorphins. In McIntyre, I. (ed.): *Sixth International Conference on Endocrinology*. Amsterdam, Elsevier/North-Holland, 1977, pp. 221–235.

47. GUILLEMIN, R., LING, N., BURGUS, R.: Endorphins, peptides d'origine hypothalamique et neurohypophysaire a activite morphinomimetrique: Isolement et structure moleculaire de l'α-endorphin. *C. R. Acad. Sci. Ser. D* **283**:783–785, 1976.

48. GUILLEMIN, R., LING, N., VARGO, T. M.: Radioimmunoassays for α- and β-endorphin. *Biochem. Biophys. Res. Commun.* **77**:361–366, 1977.

49. GUNNE, L. M., LINDSTROM, L., TERENIUS, L.: Naloxone-induced reversal of schizophrenic hallucinations. *J. Neural Transm.* **40**:13–19, 1977.

50. HENRIKSEN, S. J., BLOOM, F. E., LING, N., GUILLEMIN, R.: Induction of limbic seizures by endorphins and opiate alkaloids: Electrophysiological and behavioral correlates. *Soc. Neurosci. Abstr.* **3**:293, 1977.

51. HENRIKSEN, S. J., BLOOM, F. E., MCCOY, F., LING, N., GUILLEMIN, R.: β-Endorphin induces nonconvulsive limbic seizures. *Proc. Natl. Acad. Sci. U.S.A.* **75**:5221–5225, 1978.

52. HENRIKSEN, S. J., MCCOY, F., FRENCH, E., BLOOM, F. E.: β-Endorphin induced epileptiform activity: Effects of lesions and specific opiate receptor agonists. *Soc. Neurosci. Abstr.* **4**:409, 1978.

53. HÖKFELT, T., ELDE, R., JOHANNSON, O., TERENIUS, L., STEIN, L.: The distribution of enkephalin-immunoreactive cell bodies in the rat central nervous system. *Neurosci. Lett.* **5**:25–31, 1977.

54. HOLADAY, J. W., LAW, P.-Y., TSENG, L.-F., LOH, H. H., LI, C. H.: β-Endorphin: Pituitary and adrenal glands modulate its action. *Proc. Natl. Acad. Sci. U.S.A.* **74**:4628–4632, 1977.

55. HOSOBUCHI, Y., ADAMS, J. E., LINCHITZ, R.: Pain relief by electrical stimulation of the central grey matter in humans and its reversal by naloxone. *Science* **197**:183–186, 1977.

56. HOSOBUCHI, Y., LI, C. H.: The analgesic activity of human β-endorphin in man. *Commun. Psychopharmacol.* **2**:33–37, 1978.

57. HOSOBUCHI, Y., MEGLIO, M., ADAMS, J. E., LI, C. H.: β-Endorphin: Development of tolerance and its reversal by 5-hydroxytryptophane in cats. *Proc. Natl. Acad. Sci. U.S.A.* **74**:4017–4019, 1977.

58. HOSOBUCHI, Y., ROSSIER, J., BLOOM, F. E., GUILLEMIN, R.: Electrical stimulation of periaqueductal grey for pain relief in humans is accompanied by elevation of immunoreactive β-endorphin in ventricular fluid. *Science* **203**:279–281, 1979.

59. HUGHES, J., SMITH, T. W., KOSTERLITZ, H. W., FOTHERGILL, L. A., MORGAN, B. A., MORRIS, H. R.: Identification of two related pentapeptides from the brain with potent opiate agonist activity. *Nature (London)* **258**:577–580, 1975.

60. IZUMI, K., MOTOMATSU, T., CHRETIEN, M., BUTTERWORTH, R. F., LIS, M., SEIDA, N., BARBEAU, A.: β-Endorphin akinesia in rats: Effect of apomorphine and α-methyl-*p*-tyrosine and related modifications of dopamine turnover in the basal ganglia. *Life Sci.* **20**:1149–1156, 1977.

61. JACQUET, Y. F., MARKS, N.: The C-fragment of β-lipotropin: An endogenous neuroleptic or antipsychotogen? *Science* **194**:632–635, 1976.

62. JANOWSKY, D. S., SEGAL, D. S., ABRAMS, A., BLOOM, F., GUILLEMIN, R.: Negative naloxone effects in schizophrenic patients. *Psychopharmacology* **53**:295–297, 1977.

63. JANOWSKY, D. S., SEGAL, D. S., BLOOM, F., ABRAMS, A., GUILLEMIN, R.: Lack of effect of naloxone on schizophrenic symptoms. *Am. J. Psychiatry* **134**:926–927, 1977.

64. JEFFCOATE, W. J., REESE, L. H., LOWRY, P. J., HOPE, J., BESSER, G. M.: β-Lipotropin in human plasma and cerebrospinal fluid: Radio-immunoassay evidence for γ-lipotropin and β-endorphin. *Proc. Soc. Endocrinol. (U.K.)* **1978**:27P–28P.

65. JEFFCOATE, W. J., REESE, L. H., MCLOUGHLIN, L., HOPE, J., RETTER, S. J., LOWRY, P. J., BESSER, G. M.: β-Endorphin in human cerebrospinal fluid. *Lancet* **2**:119–121, 1978.

66. JESSEL, T. M., IVERSEN, L. L.: Opiate analgesics inhibit substance P release from rat trigeminal nucleus. *Nature (London)* **268**:549–551, 1977.

67. JHAMANDAS, K., SAWYNOK, J., SUTAK, M.: Enkephalin effects on brain acetylcholine. *Nature (London)* **260**:433–434, 1977.

68. JUDD, L. L., JANOWSKY, D. S., SEGAL, D. S., HUEY, L.: Naloxone-induced behavioral and physiological effects in normal and manic subjects. *Arch. Gen. Psychiatry* (in press).

69. KOBAYASHI, R. M., PALKOVITS, M., MILLER, R. J., CHANG, K.-J., CUATRECASAS, P.: Brain enkephalin distribution is unaltered by hypophysectomy. *Life Sci.* **22**:527–530, 1978.

70. KOSTERLITZ, H. W., HUGHES, J., LORD, J. A. H., WATERFIELD, A. A.: Enkephalins, endorphins, and opiate receptors. In Cowan, W. M., Ferrendelli, J. A. (eds.): *Society for Neuroscience Symposia*, Vol. II, 1977, pp. 291–301.

71. KOSTERLITZ, H. W., LORD, J. A. H., WATT, A. J.: Morphine receptor in the myenteric plexus of the guinea-pig ileum. In Kosterlitz, H. W., Collier, H. O. J., Villareal, J. E. (eds.): *Agonist and Antagonist Actions of Narcotic Analgesic Drugs*. Baltimore, University Park Press, 1973, pp. 45–61.

72. KRIEGER, D. T., LIOTTA, A., SUDA, T., PALKOVITS, M., BROWNSTEIN, M. J.: Presence of immunoassayable β-lipotropin in bovine brain and spinal cords: Lack of concordance with ACTH concentrations. *Biochem. Biophys. Res. Commun.* **26**:930–936, 1977.

73. La Bella, F. A., Havlicek, V., Pinsky, C., Leybin, L.: Opiate-like naloxone-reversible effects of androsterone sulfate in rats. *Soc. Neurosci. Abstr.* **3**:295, 1977.
74. Lanthorn, T., Isaacson, R. L.: Studies of kainate-induced wet dog shakes in the rat. *Life Sci.* **22**:171–178, 1978.
75. Lazarus, L. H., Ling, N., Guillemin, R.: β-Lipotropin as a prohormone for the morphinomimetic peptides, endorphins and enkephalins. *Proc. Natl. Acad. Sci. U.S.A.* **73**:2156–2159, 1976.
76. Leybin, L., Pinsky, C., La Bella, F. S., Havlicek, V., Rezek, M.: Intraventricular Met[5]-enkephalin causes unexpected lowering of pain threshold and narcotic withdrawal signs in rats. *Nature (London)* **264**:458–459, 1976.
77. Li, C. H.: Lipotropin, a new active peptide from pituitary glands. *Nature (London)* **201**:924, 1964.
78. Li, C. H., Barnafi, L., Chretien, M., Chung, D.: Isolation and amino-acid sequence of β-LPH from sheep pituitary gland. *Nature (London)* **208**:1093–1094, 1965.
79. Li, C. H., Chung, D.: Isolation and structure of an untriakontapeptide with opiate activity from camel pituitary glands. *Proc. Natl. Acad. Sci. U.S.A.* **73**:1145–1148.
80. Li, C. H., Lemaire, S., Yamashiro, D., Doneen, B. A.: The synthesis and opiate activity of β-endorphin. *Biochem. Biophys. Res. Commun.* **71**:19–25, 1976.
81. Li, C. H., Yamashiro, D., Tseng, L. F., Loh, H. H.: Synthesis and analgesic activity of human β-endorphin. *J. Med. Chem.* **20**:325–328, 1977.
82. Ling, N.: Solid phase synthesis of porcine α-endorphin and β-endorphin, two hypothalamic–pituitary peptides with opiate activity. *Biochem. Biophys. Res. Commun.* **74**:248–256, 1977.
83. Ling, N., Burgus, R., Guillemin, R.: Isolation, primary structures and synthesis of α-endorphin and β-endorphin, two peptides of hypothalamic–hypophysial origin with morphinomimetic activity. *Proc. Natl. Acad. Sci. U.S.A.* **73**:3942–3946, 1976.
84. Ling, N., Guillemin, R.: Morphinomimetic activity of synthetic fragments of β-lipotropin and analogs. *Proc. Natl. Acad. Sci. U.S.A.* **73**:3308–3310, 1976.
85. Loh, H. H., Brase, D. A., Sampath-Khanna, S., Mar, J. B., Way, E. L., Li, C. H.: β-Endorphin *in vitro* inhibition of striatal dopamine release. *Nature (London)* **264**:567–568, 1976.
86. Loh, H. H., Tseng, L. F., Wei, E., Li, C. H.: β-Endorphin is a potent analgesic agent. *Proc. Natl. Acad. Sci. U.S.A.* **73**:2895–2898, 1976.
87. Lord, J. A. H., Waterfield, A. A., Hughes, J., Kosterlitz, H. W.: Endogeneous opioid peptides: Multiple agonists and receptors. *Nature (London)* **267**:495–499, 1977.
88. Meglio, M., Hosobuchi, Y., Loh, H. H., Adams, J. E., Li, C. H.: β-Endorphin: Behavioral and analgesic activity in rats. *Proc. Natl. Acad. Sci. U.S.A.* **74**:774–776, 1977.
89. Martin, B. R., Dewey, W. L., Chau-Pham, T., Prange, A. J., Jr.: Interactions of thyrotropin releasing hormone and morphine sulfate in rats. *Life Sci.* **20**:715–722, 1977.
90. Martin, W. R., Eades, C. G., Thompson, J. A., Huppler, R. E., Gilbert, P. E.: The effects of morphine- and nalorphine-like drugs in the nondependent and morphine-dependent chronic spinal dog. *J. Pharmacol. Exp. Ther.* **197**:517–532, 1976.
91. McGregor, W. H., Stein, L., Belluzi, J. D.: Potent analgesic activity of the enkephalin-like tetrapeptide H-TYR-D-ALA-GLY-PHE-NH. *Life Sci.* **23**:1371–1378, 1978.
92. Moroni, F., Cheney, D. L., Costa, E.: β-Endorphin inhibits ACH turnover in nuclei of rat brain. *Nature (London)* **267**:267–269, 1977.
93. Moroni, F., Cheney, D. L., Costa, E.: The turnover rate of acetylcholine in brain nuclei of rats injected intraventricularly and intraseptally with alpha and beta endorphin. *Neuropharmacology* **171**:191–198, 1978.
94. Moss, I. R., Friedman, E.: β-Endorphin: Effects of respiratory regulation. *Life Sci.* **23**:1271–1276, 1978.
95. Motomatsu, T., Lis, M., Seidah, N., Chretien, N.: Cataleptic effect of 61–91 beta-lipotropic hormone in rat. *Can. J. Neurol. Sci.* **4**:49–52, 1977.
96. Nicoll, R. A., Siggins, G. R., Ling, N., Bloom, F. E., Guillemin, R.: Neuronal actions of endorphins and enkephalins among brain regions: A comparative microiontophoretic study. *Proc. Natl. Acad. Sci. U.S.A.* **74**:2584–2588, 1977.
97. Oka, T., Hosoya, E.: Effects of humoral modulators and naloxone on morphine-induced changes in the spontaneous activity of the rat. *Psychopharmacology* **47**:243–248, 1976.
98. Paton, W. D. M.: The action of morphine and related substances on contraction and on acetylcholine output of electrically stimulated guinea-pig ileum. *Br. J. Pharmacol. Chemother.* **12**:119–127, 1957.
99. Prange, A. J., Jr., Breese, G. R., Cott, J. M., Martin, B. R., Cooper, B. R., Wilson, I. C., Plotnikoff, N. P.: Thyrotropin releasing hormone: Antagonism of pentobarbital in rodents. *Life Sci.* **14**:447–455, 1974.
100. Rezek, M., Havlicek, V., Leybin, L., La Bella, F. S., Friesen, H.: Opiate-like naloxone reversible actions of somatostatin given intracerebrally. *Can. J. Physiol. Pharmacol.* **56**:227–231, 1978.
101. Rivier, C., Vale, W., Ling, N., Brown, M., Guillemin, R.: Stimulation *in vivo* of the secretion of prolactin and growth hormone by β-endorphin. *Endocrinology* **100**:238–241, 1977.
102. Roemer, D., Büscher, H. H., Hill, R. C., Pless, J., Bauer, W., Cardinaux, F., Closse, A., Hauser, D., Hugenin, R.: A synthetic enkephalin analogue with prolonged parenteral and oral analgesic activity. *Nature (London)* **268**:547–549, 1977.
103. Ross, M., Dingledine, R., Cox, B. M., Goldstein, A.: Distribution of endorphins (peptides with mor-

phine-like pharmacological activities) in pituitary. *Brain Res.* **124:**523–532, 1977.
104. Rossier, J., Bayon, A., Vargo, T., Ling, N., Guillemin, R., Bloom, F.: Radioimmunoassay of brain peptides: Evaluation of a methodology fo the assay of β-endorphin and enkephalin. *Life Sci.* **21:**847–852.
105. Rossier, J., Vargo, T. M., Minick, S., Ling, N., Bloom, F., Guillemin, R.: Independent variation of β-endorphin and enkephalin levels in rat brain regions. *Proc. Natl. Acad. Sci. U.S.A.* **74:**5162–5164, 1977.
106. Sar, M., Stumpf, W. E., Miller, R. J., Chang, K. J., Cuatrecasas, P.: Immunohistochemical localization of enkephalin in rat brain and spinal cord. *J. Comp. Neurol.* **182:**17–38, 1978.
107. Sarne, Y., Azov, R., Weissman, B. A.: A stable enkephalin-like immunoreactive substance in human CSF. *Brain Res.* **151:**399–403, 1978.
108. Schenk, G. K., Enders, P., Engelmeier, M. P., Ewert, T., Herdemerten, S., Kohler, K.-H., Lodemann, E., Matz, D., Pach, J.: Application of the morphine antagonist naloxone in psychic disorders. *Drug Res.* **28:**1274–1277, 1978.
109. Segal, D. S., Browne, R. G., Bloom, F., Ling, N., Guillemin, R.: β-Endorphin endogenous opiates or neuroleptics. *Science* **198:**411–414, 1977.
110. Segal, D. S., Browne, R. G., Arnsten, A., Derrington, D. C.: Behavioral effects of β-endorphin. In van Ree, J. M., Terenius, L. (eds.): *Characteristics and Function of Opioids: Developments in Neuroscience,* Vol. 4. Amsterdam, Elsevier, 1978, pp. 377–388.
111. Segal, D. S., Browne, R. G., Arnstein, A., Derrington, D. C.: Characteristics of β-endorphin-induced behavioral activation and immobilization. In Usdin, E., Bunney, B. E., Kline, N. (eds.): *Endorphins in Mental Health Research,* London, Macmillan, 1979, pp. 307–324.
112. Shorr, J., Foley, K., Spector, S.: Presence of a nonpeptide morphine-like compound in human cerebrospinal fluid. *Life Sci.* **23:**2057–2062, 1978.
113. Simantov, R., Kuhar, M. J., Uhl, G. R., Snyder, S. H.: Opioid peptide enkephalin: Immunohistochemical mapping in rat central nervous system. *Proc. Natl. Acad. Sci. U.S.A.* **74:**2167–2171, 1977.
114. Simantov, R., Snyder, S. H.: Morphine-like peptides in mammalian brain: Isolation, structure elucidation, and interactions with the opiate receptors. *Proc. Natl. Acad. Sci. U.S.A.* **73:**2515–2519, 1976.
115. Simantov, R., Snyder, S. H.: Brain pituitary opiate mechanisms: Pituitary opiate receptor binding radioimmunoassays for methionine enkephalin and leucine enkephalin and H^3-enkephalin interactions with the opiate receptor. In Kosterlitz, H. W. (ed.): *Opiates and Endogeneous Peptides,* Amsterdam, Elsevier/North-Holland, 1976, pp. 41–48.
116. Smyth, D. G., Snell, C. R.: Metabolism of the analgesic peptide lipotropin C-fragment in rat striatal slices. *FEBS Lett.* **78:**225–228, 1977.
117. Szekeley, J. I., Ronai, A. Z., Dunai-Kovacs, Z., Miglecz, E., Bajusz, S., Graf, L.: Cross tolerance between morphine and β-endorphin *in vivo. Life Sci.* **20:**1259–1264, 1977.
118. Taube, H. D., Borowski, E., Endo, T., Starke, K.: Enkephalin: A potential modulator of noradrenaline release in rat brain. *Eur. J. Pharmacol.* **38:**377–380, 1976.
119. Terenius, L.: Significance of endorphins in endogenous antinociception. *Adv. Biochem. Psychopharmacol.* **18:**321–332, 1977.
120. Teschemacher, H., Opheim, K. E., Cox, B. M., Goldstein, A.: A peptide-like substance from pituitary that acts like morphine. I. Isolation. *Life Sci.* **16:**1771–1775, 1976.
121. Tortella, F. C., Moreton, J. E., Khazan, N.: Electroencephalographic and behavioral effects of *d*-ala-methionine enkephalin amide and morphine on the rat. *J. Pharmacol. Exp. Ther.* **206:**636–642, 1978.
122. Tseng, L.-F., Loh, H. H., Li, C. H.: β-Endorphin as a potent analgesic by intravenous injection. *Nature (London)* **263:**239–240, 1976.
123. Tseng, L.-F., Loh, H. H., Li, C. H.: β-Endorphin: Cross tolerance to and cross physical dependence on morphine. *Proc. Natl. Acad. Sci. U.S.A.* **73:**4187–4189, 1976.
124. Tseng, L. F., Loh, H. H., Li, C. H.: Human β-endorphin development of tolerance and behavioral activity in rats. *Biochem. Biophys. Res. Commun.* **74:**390–396, 1977.
125. Urca, G., Frenk, H., Liebeskind, J. C., Taylor, A. N.: Morphine and enkephalin: Analgesic and epileptic properties. *Science* **197:**83–86, 1977.
126. Van Loon, G. R., De Souza, E. B.: Effects of β-endorphin on brain serotonin and metbolism. *Life Sci.* **23:**971–978, 1978.
127. Van Loon, G. R., Kim, C.: β-Endorphin induced increase in striatal dopamine turnover. *Life Sci.* **23:**961–970, 1978.
128. Verhoeven, W. M. A., van Praag, H. M., Botter, P. A., Sunier, A., van Ree, J. M., de Wied, D.: [Des-Tyr[1]]-γ-endorphin in schizophrenia. *Lancet* **1**(8072):1076–1077, 1978.
129. Volavka, J., Marya, A., Balg, S., Perez-Cruet, J.: Naloxone in chronic schizophrenia. *Science* **106:**1227–1228, 1977.
130. Watson, S. J., Akil, H., Richards, C. W., III, Barchas, J. D.: Evidence for two separate opiate peptide neuronal systems. *Nature (London)* **275:**225–227, 1978.
131. Watson, S. J., Akil, H., Sullivan, S., Barchas, J. D.: Immunocytochemical localization of methionine enkephalin: Preliminary observations. *Life Sci.* **21:**733–738, 1977.
132. Wei, E., Loh, H. H.: Physical dependence on opiate-like peptides. *Science* **193:**1262–1264, 1976.
133. Wei, E., Loh, H. H., Way, E. L.: Brain sites of precipitated abstinence in morphine-dependent rats. *J. Pharmacol. Exp. Ther.* **185:**108–115, 1973.

134. WEI, E., SIGEL, S., LOH, H., WAY, E. L.: Regional sensitivity of the rat brain to the inhibitory effects of morphine on wet shake behavior. *Nature (London)* **253:**739–740, 1975.
135. WEINBERGER, S. B., ARNSTEIN, A., SEGAL, D. A.: Des-tyrosine[1]-γ-endorphin and haloperidol: Behavioral and biochemical differentiation. *Life Sci.* **24:**1637–1644, 1979.
136. YANG, H. Y., HONG, J. S., COSTA, E.: Regional distribution of Leu and Met enkephalin in rat brain. *Neuropharmacology* **16:**303–307, 1977.
137. ZEIGLGÄNSBERGER, W., FRY, J. P., HERZ, A., MORODER, L., WUNSCH, E.: Enkephalin-induced inhibition of cortical neurons and the lack of this effect in morphine tolerant dependent rats. *Brain Res.* **115:**160–164, 1976.

CHAPTER 46

Cerebrospinal Fluid Monoamine Metabolites in Neuropsychiatric Disorders of Childhood

Donald J. Cohen, Bennett A. Shaywitz, J. Gerald Young, and Malcolm B. Bowers, Jr.

1. Introduction

The severe neuropsychiatric disorders of childhood, phenotypically and genotypically diverse behavioral syndromes apparent during the first years of life, affect up to 3% of all children. Organically definable causes include structural malformations of the brain, chromosomal abnormalities, inborn errors of metabolism, anoxia, infection, head trauma, and numerous other types of inborn and acquired pathological influences on central nervous system (CNS) maturation. When such medically explicable disorders are excluded, the neuropsychiatric disorders consist of a broad range of syndromes of unknown etiology involving the normal unfolding of linguistic, cognitive, and social competence. In their most severe forms, such as primary childhood autism, all spheres of development may be disturbed pervasively; in less severe syndromes, one or another facet of development is more prominently involved (such as expressive and receptive language in the developmental aphasias or the regulation of attention and motor activity in attention-deficit syndromes).[13,15,18,58,59]

Various types of neurological and metabolic abnormalities have been postulated as the basis for the disorganization, desynchronization, and retardation in maturation that characterize such severe disorders as autism and schizophrenia. The viewpoint that these disorders rest on an organic substrate is supported by a range of suggestive evidence. The disorders sometimes are manifest during the first weeks of life, as when the parents of autistic children recognize that the child is not normally attentive to their faces or responsive to their care. The syndromes occur throughout the world, with fairly predictable incidence (e.g., 1:3000 live births for autism) and with a characteristic clinical presentation and natural history. Children with neuropsychiatric syndromes often have minor atypicalities on neurological examination or on electroencephalography and have a high incidence of minor physical anomalies. The response of the disorders to medication may sometimes be dramatic (as when children with the attention-deficit syndromes respond therapeutically to amphetamine or those with multiple tics respond to haloperidol) and pharmacologically explicable in relation to theories of neurotransmitter function.[21,28] The families of some children with neuropsychiatric disorders may have an increased incidence of developmental or psychiatric pathology, e.g., language difficulties in families of children with autism or character disturbances in families of hyperactive, attentionally disabled children. Finally, the neuropsychiatric disorders for which no organic basis is now known often bear striking similarities

Donald J. Cohen, M.D., Bennett A. Shaywitz, M.D., J. Gerald Young, M.D., and **Malcolm B. Bowers, Jr., M.D.** • Child Study Center and Departments of Pediatrics, Neurology, and Psychiatry, Yale University School of Medicine, New Haven, Connecticut 06510.

to disorders of known organic causation, such as lead encephalopathy, homocystinuria, and the rubella syndrome.

Although organic factors appear to play the central role in the origin of the severe neuropsychiatric disorders of childhood, experiential factors occasionally precipitate or markedly exacerbate a child's constitutional difficulties. For most children with these disorders, eventual competence represents the complementary relationships between endowment and the quality of the child's care, the degree of family stress, the opportunities for adequate stimulation, and luck. Interactions between environmental influences and CNS metabolism, and their mutual impact on natural history, remain to be investigated.[14,77]

In studies of the CNS metabolism of children with neuropsychiatric disorders, major attention has been focused on the acid metabolites of dopamine [homovanillic acid (HVA)] and serotonin [5-hydroxyindoleacetic acid (5-HIAA)]. Much less is known about cerebrospinal fluid (CSF) metabolites of norepinephrine [e.g., 3-methoxy-4-hydroxyphenylethylene glycol (MHPG)] in childhood. This chapter reviews the systematic research on CSF metabolites conducted by the Yale neuropsychiatric research group[20,21,23,25–28,61,63,76] (see Ritvo *et al.*[54] for a review).

2. Diagnostic Issues: Autism and Related Disorders

Biological research in child psychiatry requires operationally defined, reliable, and valid diagnostic criteria. The Yale research diagnoses were consistent with the proposed criteria of the *Diagnostic and Statistical Manual* (DSM-III) of the American Psychiatric Association and the National Society for Autistic Children.[4,53] Interobserver reliability studies using these criteria on the research cohort have been conducted.[17] The diagnostic process for each child included evaluation, over the course of numerous sessions, of the child's developmental history, current functioning, and neurological status. Standard interviewing techniques, free-field observation, neurological testing, and psychological testing were used. Children were observed at school, and reports from schools, clinics, and other professionals were reviewed. Biomedical evaluation included a range of blood screening tests, urine for genetic screening, electroencephalography, skull and wrist X-rays, computed brain tomograms, and brain scans. Children were evaluated by specialists in endocrinology, human genetics, and neurology.

Diagnostic issues concerning autism and childhood schizophrenia have been discussed repeatedly.[31,33,35,38,41,51,52,75] Disagreements involve (1) whether the concepts should be used narrowly, for a small, clinically homogeneous population, or broadly, for a larger, diverse group; and (2) whether the psychiatric diagnosis should imply etiology (specificity of organic or environmental genotype) or only development and current behavior (phenotype). Associated with these issues are concerns about whether autism lies on a spectrum of disturbances linked with other types of schizophrenia, mental retardation, or atypical development, or is a discrete entity; whether globally defective, retarded children, with or without other signs of organic brain syndromes such as seizures or inborn errors of metabolism, should be grouped as "autistic" along with bright but socially and linguistically deviant youngsters; whether developmental course or current status together, or separately, should carry most weight in diagnosis; or whether the changing picture of disturbance as children develop requires new criteria or a change of diagnosis. Despite conflict, consensus exists about areas of disturbance (social, language, and emotional), early age of onset, and the presence of some, as yet undefined, congenital dysfunction in psychological processing.

In the Yale studies, the diagnosis of *primary childhood autism* required age of onset before 2 years, profound disturbance in social attachment (aloofness, aloneness, unconcern for others); disturbance in the acquisition and use of social language and nonverbal communication (ranging from muteness through specific language peculiarities, such as pronoun reversal, echolalia, and answering questions with repetition); unusual use of the physical environment (manneristic use of objects, hyperactivity or hypoactivity, fascinations); marked disturbance in the expression or appreciation of emotions and in the regulation of anxiety and arousal (lack of normal experience of personal happiness or sadness in various contexts and appreciation of these affects in others, with sudden rushes of terror and anxiety or silly excitement); adaptive skills, special competencies, or behavioral acts, sometimes negativistic, demonstrating the presence of intellectual abilities that may be far above a basal level of globally retarded functioning; and motor development within the normal range or only mildly slow, with no gross dysfunction in motor skills (walking, running, climbing stairs, fine motor coordination). The diagnosis also required absence of any evidence from history, examination, or laboratory studies indicating a specifiable organic brain syndrome (e.g., inborn error of metabolism, brain malformation, seizure disorder) or highly suggestive that the behavioral disturbance was associated with such a disorder.

If all the other criteria were satisfied for the diagnosis of primary childhood autism but there was evidence for a specific or associated organic brain syndrome (markedly unusual appearance with various stigmata, slow motor development, abnormal neurological evaluation), the child was diagnosed as suffering from *secondary childhood autism* or *autism associated with the specific disorder*.

Primary childhood or developmental aphasia was diagnosed on the basis of criteria emphasizing communicative intent and social motivation in the presence of profound expressive (and less serious receptive) language dysfunction. As reviewed in detail elsewhere,[11,18] children with disturbances in the acquisition of expressive language may become overly attached to their mothers and develop hyperactivity, short attention, and extreme negativism. Their relatively more intact inner language, use of gesture and mime, attachment to others, concern for the feelings of other individuals, and more rapid response to educational intervention separate them from autistic children.

Early or atypical (nonautistic) childhood psychosis was diagnosed on the basis of the presence of a definable disorder in cognitive or thought processes and content (loose associations, bizarre preoccupations, concrete thinking, tangentiality); profoundly disturbed social relationships; some capacity for normal emotional responses to situations and people, but idiosyncrasies and inappropriate reactions; abnormal concerns about self (physical worries or preoccupations, sensitivity to criticism); and adaptive or other skills indicative of intellectual abilities within the normal range. The criteria for the diagnosis of early childhood psychosis in the proposed DSM III classification were satisfied; these children could also be considered as suffering from childhood schizophrenia.[33,36]

3. Probenecid Method in Childhood

The most direct method for studying biogenic amine metabolism in children involves sampling of CSF for the major metabolites of dopamine (i.e., HVA), serotonin (i.e., 5-HIAA), and norepinephrine (i.e., MHPG). HVA and 5-HIAA are actively excreted into the CSF. When membrane transport is inhibited by administration of probenecid, the egress of these acid metabolites from the CSF is inhibited, and their concentrations increasingly reflect parent amine turnover in the brain during a specified period of time (10–18 hr).[6,70,71] (See Chapters 7 and 8.)

We have used the probenecid method in studies of autism, aphasia, severe atypical development (early onset, nonautistic, schizophrenic-like psychosis), multiple-tic syndrome, and minimal brain dysfunction.[19,20,25–27,61] Due to ethical and practical restrictions, CSF could not be obtained from normal children for study. Instead, it has been necessary to contrast one diagnostic group with another and to utilize children suffering from various other conditions in which a lumbar puncture and examination of CSF were indicated (e.g., headache and disc disease). A standard methodology was employed. Probenecid was administered orally in four divided doses over 10–12 hr, for a total dose of 125–150 mg/kg. A lumbar puncture was performed using the usual technique between 8:00 A.M. and 9:30 A.M. Approximately 10–15 ml of CSF was obtained, and the second or third aliquot of 5 ml was immediately frozen for assays of metabolites and probenecid, performed within 2 weeks by described methods.[6,43,46,47] CSF was also examined for cells, protein, glucose, immunoglobulins, and folate.

In investigations with adults in which the probenecid method is used, it has generally been accepted that membrane blockade becomes effective when a substantial amount of probenecid has been ingested (usually 100 mg/kg orally over 18–20 hr), or when CSF probenecid levels above 20 μg/ml have been achieved.[69] Interpretation of metabolite data without actual measurement of CSF probenecid is hazardous because of the relationship between metabolite concentrations and CSF probenecid levels at lower CSF probenecid concentrations.[65] In a study of 43 children, the levels of metabolites were highly related to the levels of probenecid achieved (HVA–probenecid: $r = 0.54$, $p = 0.001$; 5-HIAA–probenecid: $r = 0.54$, $p = 0.001$).[19] For statistical analysis, the entire population was subdivided into three groups, based on the amount of probenecid in the CSF. The relationship between probenecid and amine metabolites was nonsignificant within the two groups of low (3–9.9 μg/ml, $N = 14$) and middle (10–19.9 μg/ml, $N = 18$) CSF probenecid levels. However, there was a highly significant HVA–probenecid correlation ($r = 0.74$, $p = 0.009$) in the group ($N = 11$) with CSF probenecid levels above 20 μg/ml. The lack of statistical correlation in the lower probenecid groups appeared to result from the restriction of the range of the independent variable.

Different methods of data transformation may be used to take into account the metabolite–probenecid relationship. The amine metabolites may be statistically adjusted by analysis of covariance,[26] or the metabolites may be expressed as a ratio to probenecid (nanograms of metabolite per micrograms of probenecid), which we refer to as the HVA/P or 5-HIAA/P ratio. Theoretically, higher levels of probenecid may be expected to yield more effective membrane blockade, analogous to the log dose-vs.-response curves for drug effects. Thus, another method of data transformation involves expression of the metabolite concentration as a ratio to the natural logarithm of probenecid (nanograms of metabolite/log of micrograms of probenecid), HVA/log P or 5-HIAA/log P. It is not clear whether the simple ratio or the log ratio model of metabolite–probenecid relationship is more valid.[19,20,26,28]

In addition to the methodological complications resulting from the metabolite–probenecid relationship, interpretation of data must consider the stomach irritation, nausea, and emotional upset that are experienced by many children. Autistic children, in particular, resist ingesting the medication.

4. Childhood Psychosis

Without probenecid loading, the concentrations of the major metabolites of dopamine and serotonin (HVA and 5-HIAA) in autistic children were low, closely clustered, and within the range roughly defined for adults (Table 1). Determined in six autistic boys (ages 6–15), HVA ranged from 45 to 100 ng/ml (mean ± S.E. = 65.0 ± 7.7 ng/ml), and 5-HIAA ranged from 36 to 60 ng/ml (41.2 ± 4.2 ng/ml). Without probenecid loading, there appeared to be a negative relationship between HVA and 5-HIAA in these autistic children ($r = -0.52$), most apparent in the values of the two boys with the most extreme metabolite values. Their HVA and 5-HIAA concentrations were as follows: (Patient 4) 100 ng/ml HVA,

Table 1. CSF Monoamine Metabolites in 6 Autistic Boys without Probenecid and in 12 Autistic Children Following Probenecid[a]

Patient[b]	Age	HVA (ng/ml CSF)	HVA/log probenecid	5-HIAA (ng/ml CSF)	5-HIAA/log probenecid	Probenecid (μg/ml CSF)	5-HIAA/HVA ratio
1	15	67	—	38	—	0	0.57
2	10	59	—	31	—	0	0.53
3	8	45	—	60	—	0	1.33
4	6	100	—	37	—	0	0.37
5	7	55	—	45	—	0	0.82
6	7	64	—	36	—	0	0.56
			—		—		
Mean ± S.E.:		65.0 ± 7.7	—	41.2 ± 4.2	—	0	0.70 ± 0.14
Range:		45–100		31–60			0.37–1.33
7	3	299	238.2	99	78.9	18.0	0.34
8[b]	8	87	89.0	64	65.6	9.5	0.74
9	4	180	257.5	65	93.0	5.0	0.37
10	8	235	187.2	86	68.5	18.0	0.37
11	9	133	102.2	162	124.5	20.0	1.22
12	9	173	176.9	54	55.2	9.5	0.31
13	9	414	343.8	182	151.1	16.0	0.44
14	7	134	140.4	56	58.7	9.0	0.42
15	9	243	201.8	103	85.5	16.0	0.42
16	21	104	76.9	88	65.1	22.5	0.85
17	7	60	125.8	69	144.6	3.0	1.15
18	5	149	213.2	58	83.0	5.0	0.39
Mean ± S.E.:		184.2 ± 28.8	179.4 ± 22.6	90.5 ± 12.0	89.5 ± 9.5	12.6 ± 1.9	0.58 ± 0.09
Range:		60–414	76.9–343.8	54–182	55.2–151.1	3.0–22.5	0.31–1.22

[a] (HVA) Homovanillic acid; (5-HIAA) 5-hydroxyindoleacetic acid. [b] Patient No. 8 is female; all others are male.

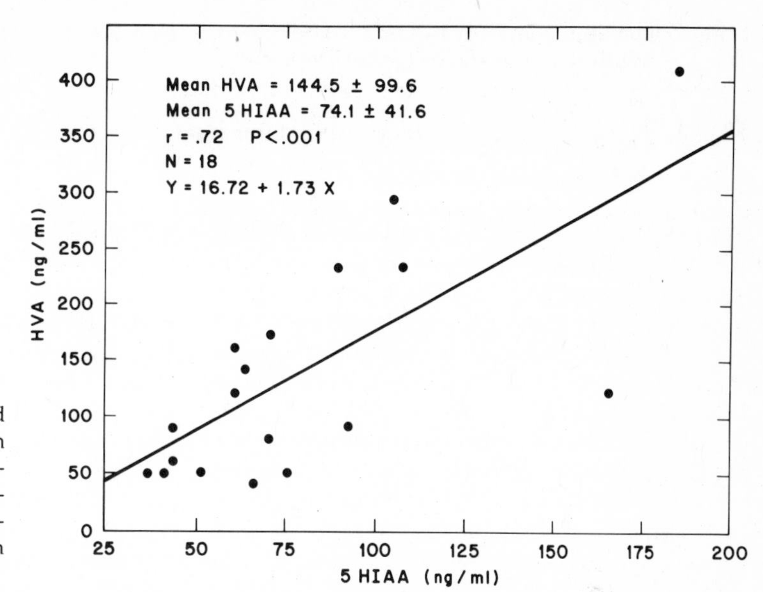

Figure 1. Homovanillic acid (HVA) and 5-hydroxyindoleacetic acid (5-HIAA) in autistic children. Means, standard deviations, regression equation, and best-fit regression line are shown. Reproduced from Cohen *et al.*,[21] with permission.

37 ng/ml 5-HIAA; (Patient 3) 45 ng/ml HVA, 60 ng/ml 5-HIAA.

Following 10–12 hr of oral probenecid administration, the accumulations of the amine metabolites increased significantly (Table 1). For 12 children with primary autism, HVA following probenecid ranged from 60 to 414 ng/ml (mean ± S.E. = 184.2 ± 28.8 ng/ml), and 5-HIAA ranged from 54 to 182 ng/ml (90.5 ± 12.0 ng/ml). The two metabolites were clearly correlated in a larger group of autistic children ($r = 0.72$, $p < 0.001$) (Fig. 1). The acid metabolites were also correlated with the levels of probenecid achieved in the CSF: 0.47 for HVA (N.S.) and 0.72 ($p < 0.05$) for 5-HIAA. In a study of 34 other neuropsychiatrically impaired children using probenecid loading, the correlation between HVA and 5-HIAA was 0.57 ($p < 0.001$). The range for this relationship, in different diagnostic subgroups, was from 0.41 to 0.80 (Fig. 2). Mean values for the metabolites are given in Table 2. The relationship of each metabolite to CSF probenecid was statistically significant: probenecid and HVA, $r = 0.46$ ($p <$

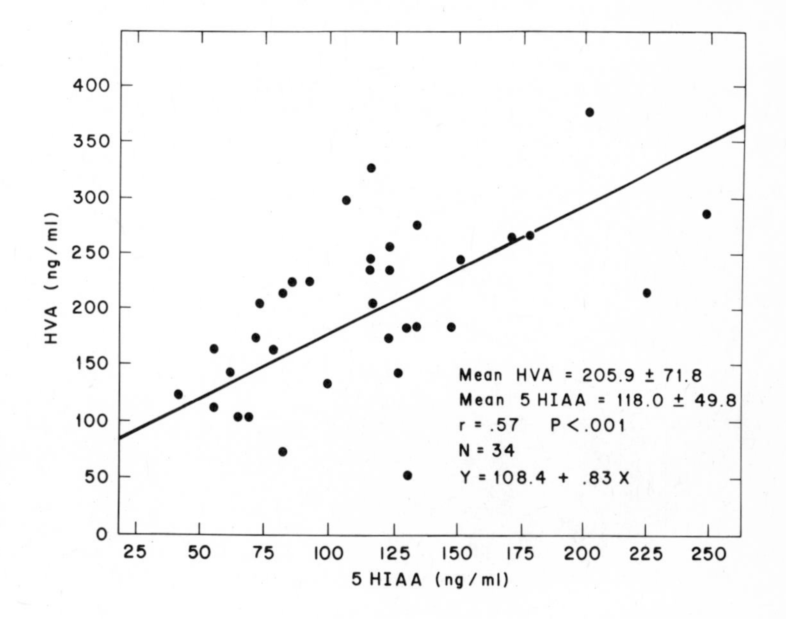

Figure 2. Homovanillic acid (HVA) and 5-hydroxyindoleacetic acid (5-HIAA) in nonautistic neuropsychiatrically impaired and medical contrast children. Reproduced from Cohen *et al.*,[21] with permission.

Table 2. Dopamine and Serotonin Metabolites in Nonautistic Neuropsychiatrically Impaired and Medical Contrast Groups of Children[a]

Compound[b]	Nonautistic early-onset psychosis (*N* = 11)	Central processing disturbance (*N* = 7)	Aphasia (*N* = 7)	Medical contrast (*N* = 9)	Pooled (*N* = 34)
HVA (ng/ml)	215.1 ± 24.7 (59–384)	191.4 ± 29.7 (105–292)	218.7 ± 17.0 (167–298)	196.1 ± 26.2 (78–332)	205.9 ± 12.3 (59–384)
HVA/log probenecid	167.4 ± 16.2 (39.2–235.1)	171.7 ± 19.9 (106.5–254.7)	207.0 ± 21.5 (141.9–313.7)	176.4 ± 20.1 (98.4–289.7)	178.8 ± 9.5 (39.2–313.7)
5-HIAA (ng/ml)	138.5 ± 9.8 (96–235)	119.7 ± 25.9 (58–256)	97.9 ± 9.8 (79–136)	107.2 ± 20.5 (42–230)	118.0 ± 8.5 (42–256)
5-HIAA/log probenecid	107.5 ± 6.5 (81.6–150.0)	108.2 ± 19.1 (46.2–198.4)	93.3 ± 12.6 (58.1–156.2)	95.0 ± 13.2 (30.8–162.5)	101.5 ± 6.0 (30.8–198.4)
Probenecid (μg/ml)	21.3 ± 2.9 (10–43)	16.5 ± 5.0 (4.5–44.0)	14.8 ± 3.3 (5.5–19.5)	15.6 ± 3.1 (5.0–31.0)	17.4 ± 1.7 (4.5–44.0)
5-HIAA/HVA ratio	0.77 ± 0.15	0.62 ± 0.07	0.45 ± 0.05	0.57 ± 0.10	0.62 ± 0.06

[a] Adapted from Cohen *et al.*[19,21] Means, standard errors, and ranges are given.
[b] (HVA) Homovanillic acid; (5-HIAA) 5-hydroxyindoleacetic acid.

0.001), and probenecid and 5-HIAA, $r = 0.50$ ($p < 0.001$) (Figs. 3 and 4). For this population of autistic and other neuropsychiatrically disabled children, the major difference between diagnostic groups was a reduced level of CSF 5-HIAA accumulation in the autistic children as compared to the age- and sex-comparable, nonautistic, early-onset psychotic children.[19] This finding may be related to differences in the severity of the disorders, since the autistic children were more pervasively and severely afflicted. This hypothesis will gain additional force in light of other findings, to be discussed later.

The functional relationships between HVA and 5-HIAA may differ among diagnostic groups, reflecting differences in the balance between the parent neurotransmitter systems. To assess this relationship, regression curves may be compared or a 5-HIAA/HVA ratio may be constructed both for individuals (within a group) and for diagnostic groups (Fig. 5). This ratio may be especially important in

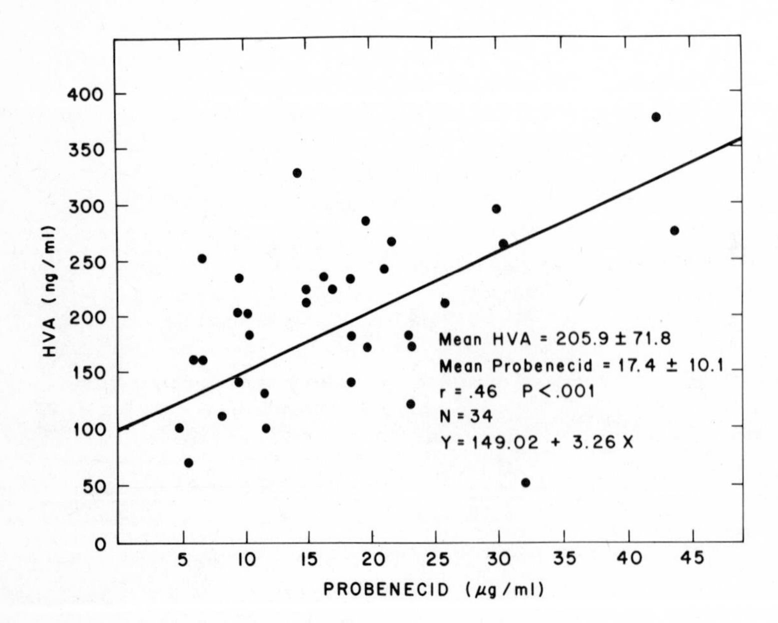

Figure 3. Homovanillic acid (HVA) and probenecid in nonautistic neuropsychiatrically impaired and medical contrast children. Reproduced from Cohen *et al.*,[21] with permission.

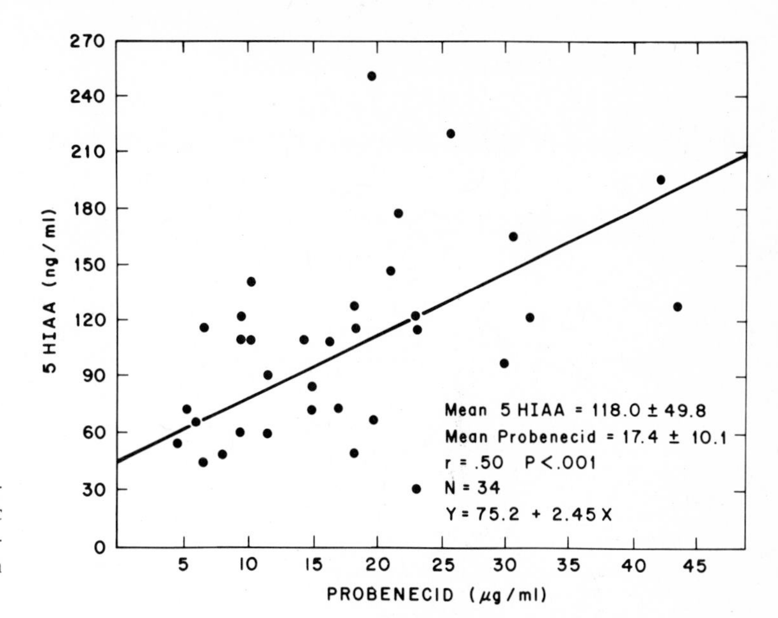

Figure 4. 5-Hydroxyindoleacetic acid (5-HIAA) and probenecid in nonautistic neuropsychiatrically impaired and medical contrast children. Reproduced from Cohen *et al.*,[21] with permission.

light of the different roles played by serotoninergic and dopaminergic mechanisms in the organization of behavior. For example, the serotonergic midbrain raphe system may subserve sensory modulating or gating functions, while the dopaminergic system appears to be involved in motor activation and arousal.[28] As will be discussed, the ratio of the metabolites of these two systems may be related to aspects of behavioral disorganization. Most studies of adult schizophrenia and depressed patients using the probenecid method have reported a 5-HIAA/HVA ratio between 0.5 and 0.7, although several reports (noted later) report values on both sides of this range. For autistic children, the 5-HIAA/HVA ratio is at the lower end of the adult range. This may reflect, in part, the relatively higher HVA levels observed in child patients and a negative relationship between CSF dopamine metabolites and age.

Since CSF metabolite concentrations span such a considerable range within the autistic and nonautistic early-childhood psychosis groups, the detection of between-group differences is quite difficult. Thus, we have utilized alternative strategies involving (1) delineation of subpopulations within diagnostic groups and (2) correlation of metabolites with explicit dimensions of behavior, within and across diagnostic groups. Similar approaches have been applied in adult psychiatry, e.g., in adult depression.[72,73] In studies of childhood psychosis, we have scored dimensions such as language comprehension and expression, activity, movement abnormalities, and social relatedness, using rating scales completed by clinicians, parents, and teachers.[17] One subgroup within the autistic population was found to have especially elevated levels of CSF HVA, both absolutely and in comparison with 5-HIAA (as reflected in 5-HIAA/HVA ratios). This subgroup was behaviorally distinguished by the greatest degree of stereotypic, repetitive behavior (flapping, twirling, finger-flicking, and the like) and locomotor activity, and was overall the most severely afflicted group.

Correlations between CSF metabolites and ratings of behavioral dimensions, done by clinicians familiar with all aspects of the child's history and condition and by clinicians "blind" to everything but current behavior, have suggested hypotheses about serotonin and dopamine metabolites and the organization of behavior. For example, for ten autistic children, HVA/P was negatively correlated with behavioral ratings of social responsiveness and attention; the 5-HIAA/HVA ratio was also very highly correlated ($r = 0.97$, $p < 0.001$) with these behavioral ratings. Autistic children with higher functional competence had higher levels of 5-HIAA and lower levels of CSF HVA. For 33 neuropsychiatrically handicapped children who were intensively rated by clinically involved and also independent observers, social responsiveness and attention was positively related to 5-HIAA ($r = 0.31$, $p = 0.08$) and 5-HIAA/log P ($r = 0.38$, $p = 0.03$). Thus, increased serotonin turnover, as assessed by measurement of metabolite concentration following probenecid, was associated with less impairment in social and atten-

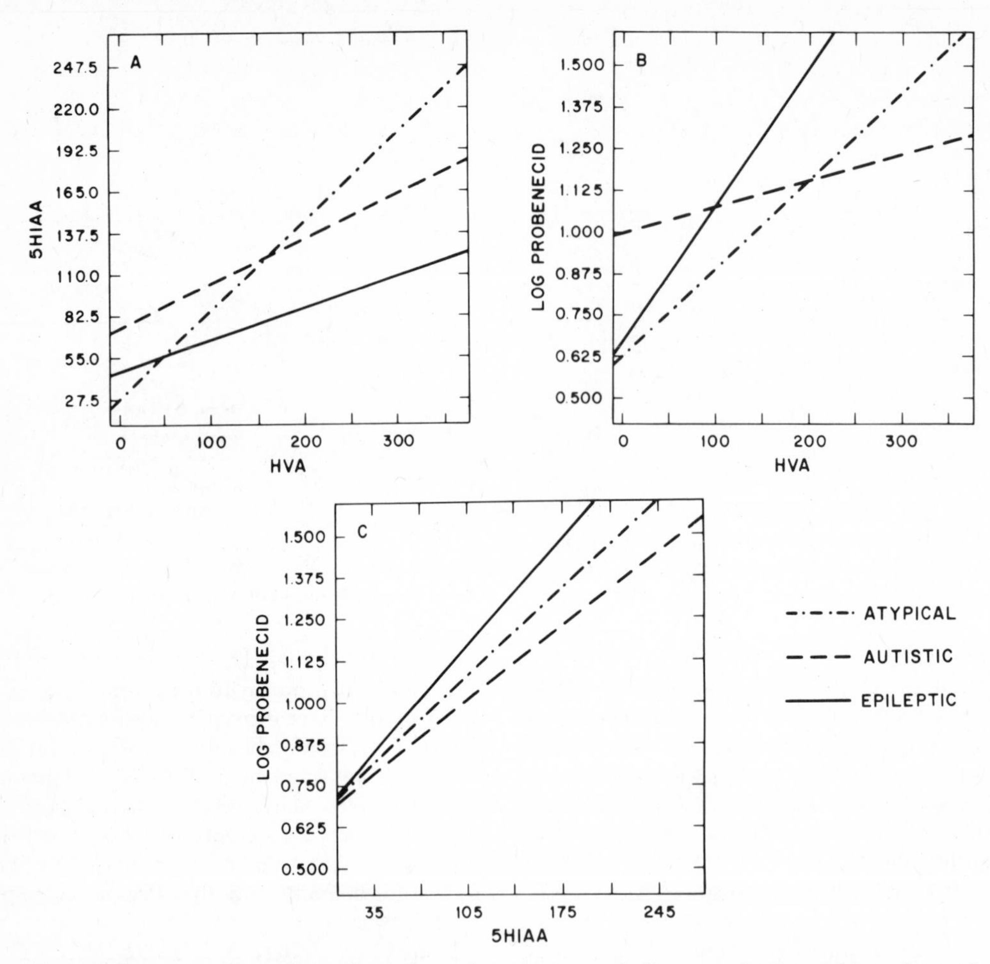

Figure 5. Regression lines relating dopamine and serotonin metabolites to each other and to probenecid for three groups of developmentally disabled children: autistic, nonautistic or atypical childhood psychosis, and epileptic patients. (HVA) Homovanillic acid; (5-HIAA) 5-hydroxyindoleacetic acid.

tional functioning in autistic and neuropsychiatrically disturbed children.

CSF amine metabolites in a monozygotic twinship concordant for childhood autism were consistent with this generalization, on the basis of statistical correlations between behavior and metabolite concentrations. In this twinship, one twin was more hyperactive and showed considerably more motoric stereotypy. Both with and without probenecid administration, the CSF HVA concentrations were higher in this active twin than in the less active twin. Both children were treated with haloperidol. The more stereotypic, hyperactive child tolerated a larger dose before displaying toxicity, a pharmacological effect that can be interpreted as consistent with relatively greater dopaminergic activity and the finding of increased CSF HVA.

In summary, findings concerning CSF metabolites in the serious, early-onset neuropsychiatric disorders of childhood are similar in type to those found in adult psychiatric disorders, such as schizophrenia. There is great variability within each diagnostic syndrome, and there are no very major differences among the subgroups. However, there are suggestions of relationships between CSF metabolites and aspects of behavioral impairment (such as attention): following probenecid administration, more disorganized children appear to have relatively lower concentrations of CSF 5-HIAA and higher concentrations of CSF HVA.

5. Chronic Multiple Tics

The syndrome of chronic multiple tics of Gilles de la Tourette (TS) consists of multiform motor, phonic, and psychological symptoms, usually starting between ages 4 and 12 years and often persisting throughout life.[1,16,49,60] Motor symptoms include rapid tics, such as blinking, facial twitches, shoulder- and trunk-jerking, and contraction of other muscle groups; organized compulsive actions, such as touching, clapping, arm-extending, kicking, and jumping; and dystonic-like writhing and stretching. Phonic symptoms include repetitive noises (hissing, grunting, barking, coughing), echolalia (repeating others' words), and palilalia (repeating oneself). Only 40% of patients develop coprolalia (explosive swearing), and fewer develop echopraxia (obscene gesturing). Initiation of motor and speech performances may be impaired by profound blocking and the release of paroxysms of tics; ongoing performance may be interrupted by sudden and frequent (80/min) rushes of motor and phonic symptoms. Disturbances in attention, academic achievement, and personality may represent basic manifestations of the underlying psychobiological disorder and quite frequently follow the isolation and terrible embarrassment produced by the disease.

A biological substrate for TS has been supported by neurological and electroencephalographic abnormalities and by positive family histories (for tics, compulsive personality, or full-blown TS) in many cases.[60,68] Catecholamine metabolism has been implicated by the therapeutic response to medication that inhibits dopaminergic activity (particularly haloperidol) and by exacerbations often experienced with dopaminergic agonists (such as dextroamphetamine).

In an initial study, CSF monoamine metabolites with probenecid were studied in six children (five boys, one girl), aged 6–15 years (mean ± S.E. = 10.5 ± 3.5 years), and one man (28 years) with characteristic histories and current symptoms of the chronic multiple-tic syndrome.[25] All patients had grossly normal neurological examinations without localized findings; two had coordination difficulties, and two others had nonspecifically, borderline abnormal EEGs. All had difficulties in academic performance, although intelligence quotients (IQ's) were within normal limits except for one girl with an IQ in the mid-60's. Contrast patients for this study consisted of children with central processing disturbances ($N = 6$), medical conditions (such as headache, $N = 8$), and serious developmental disturbances ($N = 13$) without stereotypic behavior. The TS group and contrasting groups were comparable in sex distribution (mostly males), age (8–10 years typically), socioeconomic position (middle class), and race (almost exclusively white). None of the patients was receiving medication, and most were drug-free for at least 1 year.[25]

For the total of 34 child patients in this study, the amine metabolites and probenecid were significantly related, as follows: for HVA, $r = 0.38$, $p = 0.03$; for 5-HIAA, $r = 0.31$, $p = 0.08$. The amine metabolites for the individual children and adult patients with TS and for the four child groups (TS and contrast groups) are in Tables 3 and 4. The TS patients had significantly lower amine metabolites. Compared with the children ($N = 27$) in the contrast groups, children with TS had lower serotonin turnover, as assessed by 5-HIAA/log P, $p = 0.003$ ($t = 3.3$, df 26), and 5-HIAA, $p = 0.09$ ($t = 1.7$, df 26). Also, the children with TS had lower levels of dopamine turnover, as assessed by HVA/log P, $p =$

Table 3. CSF Amine Metabolites in Tourette's Syndrome[a]

Patient	HVA (ng/ml CSF)	5-HIAA (ng/ml CSF)	Probenecid (μg/ml CSF)	5-HIAA/HVA	HVA/P	5-HIAA/P	HVA/log P	5-HIAA/log P
1	187	79	22	0.42	8.5	3.6	139.3	58.9
2	246	108	34	0.44	7.2	3.2	160.7	70.5
3	133	82	8.5	0.62	15.6	9.6	143.3	88.4
4	96	119	22	1.24	4.4	5.4	71.5	88.7
5	206	75	14	0.36	14.7	5.4	178.8	65.4
6	95	111	31	1.06	3.1	3.6	63.7	74.4
7A[b]	238	86	18.5	0.36	12.9	4.6	187.5	67.8
7B[b]	68	131	13.5	1.93	5.0	9.7	60.2	115.9

[a] (HVA) Homovanillic acid; (5-HIAA) 5-hydroxyindoleacetic acid; (P) probenecid.
[b] CSF values at baseline (A) and after a 48-hr course of dextroamphetamine (B).

Table 4. Mean CSF Amine Metabolites: Childhood Tourette's Syndrome and Pediatric Contrast Groups[a]

Group	Age (years)	HVA (ng/ml CSF)	5-HIAA (ng/ml CSF)	Probenecid (μg/ml CSF)	5-HIAA/HVA	HVA/P	5-HIAA/P	HVA/log P	5-HIAA/log P
Childhood Tourette's syndrome (N = 6)	10.5 ± 3.5	160.5 ± 25.4	95.7 ± 7.8	21.9 ± 4.0	0.71 ± 0.16	8.9 ± 2.1	5.1 ± 1.0	126.4 ± 19.5	74.4 ± 4.9
Processing disturbances (N = 6)	7.5 ± 2.7	204.7 ± 31.5	127.8 ± 29.0	17.3 ± 5.8	0.62 ± 0.08	16.3 ± 3.3	10.4 ± 2.3	183.1 ± 20.0	115.4 ± 21.4
Medical contrast group (N = 8)	10.5 ± 3.0	205.6 ± 27.7	113.5 ± 22.2	16.5 ± 3.4	0.59 ± 0.11	15.7 ± 2.9	8.4 ± 1.6	181.9 ± 21.9	99.0 ± 14.3
Serious developmental disturbances (N = 13)	8.2 ± 3.5	232.5 ± 15.7	119.0 ± 14.8	16.8 ± 3.0	0.52 ± 0.05	17.8 ± 2.4	9.2 ± 1.5	205.5 ± 12.7	105.7 ± 11.8

[a] All values are means ± S.E., except ages, which are means ± S.D. (HVA) Homovanillic acid; (5-HIAA) 5-hydroxyindoleacetic acid; (P) probenecid.

Table 5. Mean CSF Amine Metabolites with Probenecid: Tourette's Syndrome and Contrast Groups[a]

Group	Age (years)	HVA (ng/ml CSF)	5-HIAA (ng/ml CSF)	Probenecid (μg/ml CSF)	HVA/P	5-HIAA/P	HVA/log P	5-HIAA/log P
Tourette's syndrome (N = 9)	10.7 ± 2.5	174.3 ± 20.6	104.3 ± 13.1	22.9 ± 3.3	8.9 ± 1.5	5.0 ± 0.6	133.5 ± 14.6	78.3 ± 6.9
Contrast group (N = 51)	8.8 ± 3.2	205.0 ± 10.7	110.6 ± 6.7	15.8 ± 1.2	16.3 ± 1.2	8.6 ± 0.6	184.2 ± 8.3	98.8 ± 4.7
t Value	N.S.	N.S.	N.S.	N.S.	3.88	3.97	3.02	2.46
Degrees of freedom	—	—	—	—	20.4	27.5	13.7	16.7
Probability	—	—	—	—	0.001	0.001	0.009	0.025

[a] All values are means ± S.E., except ages, which are means ± S.D. (HVA) Homovanillic acid; (5-HIAA) 5-hydroxyindoleacetic acid; (P) probenecid.

0.006 ($t = 2.9$, df 29), and HVA, $p = 0.06$ ($t = 1.9$, df 31). Singer *et al.*[64] reported a replication of the findings of reduced CSF HVA, with and without probenecid, in TS; they also reported that several of their patients had reduced levels of 5-HIAA, although the group mean was not reduced. Their brief report provides no details about patient characteristics or probenecid levels.

To study the metabolic and behavioral response to a dopamine agonist, the 28-year-old patient noted above with a many-year history of TS was given a 2-day course of dextroamphetamine[25] (Table 3). With treatment, the patient experienced a marked exacerbation of motor and phonic symptoms. Concomitantly, there was a profound change in CSF metabolites, determined on two occasions (before and after treatment) with probenecid administration: CSF HVA decreased from 238 to 68 ng/ml, and 5-HIAA increased from 86 to 131 ng/ml. Correcting for the higher level of CSF probenecid achieved in the second, postmedication study, HVA/P was reduced from 12.9 to 5.0 and 5-HIAA/P increased from 4.6 to 9.7.

The reduced acid monoamine metabolites in the children with TS were interpreted as suggesting the operation of two complementary processes: dopaminergic overactivity with feedback inhibition, accounting for reduced CSF HVA; and serotoninergic underactivity, accounting for the reduced CSF 5-HIAA and the pervasive difficulty in inhibition of motor activity experienced by patients with TS. The results of the amphetamine treatment study in the adult patient were thought to be consistent with this line of explanation. For the adult patient, increased dopaminergic activity elicited feedback inhibition (and reduced CSF HVA) and the mobilization of serotonergic inhibitory mechanisms (and increased CSF 5-HIAA). Despite this serotonergic facilitation, the patient felt overwhelmed by his exacerbated symptomatology.

The findings of the initial study of TS were extended with a larger population of children with TS ($N = 9$) and children (comparable in age, sex, and socioeconomic position) with other neuropsychiatric disorders ($N = 51$). The findings of the initial study were reproduced with the larger populations (Table 5). Compared to the contrast group of children, the patients with TS had significantly reduced turnover of both brain dopamine and serotonin, as assessed by several measures of CSF accumulation (Table 5). For example, HVA/P was $8.9 + 1.5$ (mean $\pm$ S.E.) for TS and $16.3 + 1.2$ for the contrast group. For 5-HIAA, the ratios were 5.0 ± 0.6 and 8.6 ± 0.6, for the TS and contrast children, respectively.

CSF monoamine metabolites have been measured without probenecid administration in two boys with TS, with results consistent with those found with probenecid loading.[27] One boy had CSF HVA of 26 ng/ml and 5-HIAA of 32 ng/ml, both quite low. For example, without probenecid, in autistic children CSF HVA averages 65 ng/ml (range 45–100), and CSF 5-HIAA averages 41.2 ng/ml (range 31–60) (see Table 1). The second boy also had a low concentration of 5-HIAA (33 ng/ml), but a higher level of HVA (109 ng/ml). With probenecid administration leading to CSF probenecid of 14 μg/ml, this child's HVA increased only 25% (to 136 ng/ml), as compared with a 3-fold increase observed in most children (see Table 1). His 5-HIAA remained in the low range (72 ng/ml). Thus, with and without probenecid loading, CSF HVA and 5-HIAA appear to be relatively reduced in TS.

CSF MHPG in three children with childhood autism and three children with TS was well within the range reported for adult patients and "normal" controls (8.9 and 9.8 ng/ml for autism and TS, respectively). One boy with extremely severe TS had low CSF HVA and 5-HIAA levels consistent with other patients; his MHPG of 15.9 ng/ml was elevated more than 3 standard deviations above the other patients studied by us or others.[64] A single dose of clonidine, a centrally active imidazoline-derivative antihypertensive medication, resulted in marked reductions in urinary MHPG excretion.[67] An extended therapeutic trial of clonidine elicited profound, sustained amelioration of this patient's symptomatology. A therapeutic trial of clonidine led to improvement for seven other patients with TS whose disorders were not responsive to haloperidol.[27]

Reduced serotonin and dopamine turnover in TS do not appear to be related to major alterations in the activities of two important enzymes involved in amine metabolism. Dopamine-β-hydroxylase (DBH) in 9 patients with TS was 40.3 ± 8.2 μmol/min per liter serum, within the very broad range found in normal individuals but higher than similar-aged controls (15.7 ± 14.2, $p = 0.02$). Platelet monoamine oxidase (MAO) in 11 patients with TS was 22.7 ± 10.6 nmol/mg protein per hr, within normal limits.[27,29,57,80] Normal levels of plasma DBH, plasma norepinephrine, and platelet MAO have been reported in TS by other investigators.[32,44] Although extrapolation from peripheral measures of enzyme activity to central enzyme regulation must be cautious, these findings suggest that the reduced CSF levels of the amine metabolites in TS are probably not related to unavailability of these synthesizing or degradative enzymes.

The underlying phenomenological dysfunction in TS appears to be disturbance in inhibition that affects attention and the modulation of strong aggressive and sexual impulses, motor activity, thoughts, and complex actions. There are natural similarities in some aspects of this syndrome and the more pervasive neuropsychiatric disorders, such as childhood autism.[16] In TS, motor discharges and sensory occurrences that remain subliminal in normal children and adults lead to direct expressions in behavior or consciousness; in autism, there appears to be a failure in the initial development of inhibitory mechanisms. In both syndromes, serotonergic mechanisms appear to be related to behavioral dysfunction.

In TS, the monoamine metabolite findings, in conjunction with the pharmacological evidence concerning dopamine agonists and antagonists, suggest several alternative hypotheses. The reduced CSF HVA concentrations are consistent with receptor supersensitivity and feedback inhibition, such as has been suggested for other neurological and behavioral disorders.[42] The rapidity and degree of such potential feedback inhibition were displayed by the adult patient with TS treated with amphetamine; it is also demonstrable in children with attention-deficit disorders who are treated with stimulants (as described in Chapter 17). The reduced CSF concentrations of 5-HIAA may be understood as the result of two quite different processes. In the initial report,[25] a reduction in serotonergic inhibitory activity was suggested, presumably reflecting a dysfunction of midbrain raphe organization.[2] However, it is possible that the reduction of CSF 5-HIAA reflects an adaptation to receptor supersensitivity, analogous to the explanation for the reduced CSF HVA. If this second hypothesis is accurate, aspects of TS may be similar to the behavioral syndrome (myoclonus) produced in rats by augmentation of serotonergic activity. This syndrome is characterized by repetitive behaviors, such as head-nodding and forepaw-treading.[37,39,40,66] McCall and Aghajanian[48] have demonstrated that the mechanism of serotonin-induced myoclonus may relate to the ability of serotonin to reduce the threshold of motor neurons to direct catecholaminergic and neuronal excitement. This observation helps integrate the CSF findings and a range of pharmacological observations that suggest that augmentation of either catecholamine or indoleamine metabolism may lead to exacerbation of TS, while reduction in the functioning of these systems may be ameliorative.[27] Clonidine, which had therapeutic benefit for some patients with TS, may be useful because of its effect on central serotonergic activity, or because noradrenergic pathways mediate or are primarily involved in TS.

The observations on CSF metabolites in TS, as in the other neuropsychiatric disorders of childhood, are quite preliminary. Of special importance for the clinical researcher is the way in which studies of monoamine metabolites, in the context of clinical description and other types of studies, can help organize observations about separate syndromes (such as autism and Tourette's syndrome) and suggest new avenues for investigation and treatment.

6. Developmental Processes

The ontogeny of brain monoamine systems has been studied intensively in animals, particularly in developing rats. In contrast, very little is known about the maturation of dopaminergic and serotonergic synthesizing enzymes, degradative enzymes, or neurotransmitter metabolism in children, especially very young children. Andersson and Roos[5] found a decrease in CSF 5-HIAA with increasing age, without probenecid administration, in pediatric neurological patients under the age of 10 years. Rogers and Dubowitz[56] found similarly elevated baseline CSF values in very young children compared to older children and adults. Reports on age effects in adults are skimpy and conflicting. An initial report suggested that patients below the age of 34 years and above 55 years had higher CSF 5-HIAA, and those over 55 years, higher HVA, as well.[9] These findings have not been consistently observed.[6] In 34 neuropsychiatrically impaired children, we found a significant negative relationship ($r = -0.36$, $p = 0.04$) between age and one measure of dopamine turnover (HVA/log probenecid) and a trend in the same direction for another measure of dopamine turnover (HVA/probenecid: $r = -0.24$, $p = 0.06$).[19]

Concentrations of CSF HVA (and HVA/P and HVA/log P) in children have appeared to be higher than those found in adults. There are major methodological difficulties in assessing differences reported in the literature because of differences in diagnoses, method of probenecid administration, levels of probenecid achieved in the CSF, medication histories, and other procedural differences. Three adult patient groups had CSF probenecid concentrations within the same range as that of our childhood population (schizophrenic and affective patients of Bowers[7] and various psychiatric patients reported by Van Praag *et al.*[71]); the mean HVA levels in these adult groups were lower (186, 167, 117 ng/ml) than those found in children (mean 208.4 ng/ml for 43 children). The adult means of 5-HIAA were, if anything, somewhat higher than found in the child patients. One way of comparing adult and child values for these two metabolites is to compute the 5-HIAA/HVA ratio. In adult populations, this ratio varies considerably, from 0.46 for depressed patients[8] to 1.1 for schizophrenic patients.[7] The ratios in childhood diagnostic subgroups range from 0.45 (in childhood aphasia) to 0.62 (early-onset, nonautistic psychosis and central cognitive processing disturbances). These childhood ratios are within the range reported in adulthood.

In collaboration with us, James Leckman[45] has intensively analyzed the developmental course of HVA and 5-HIAA in a large group of child and adult psychiatric patients, who have received probenecid administration. CSF HVA, 5-HIAA, and probenecid were measured in a total of 163 patients, ranging in age from 2 to 67 years (69 males and 94 females). For 102 adult patients, the diagnoses included major affective disorders, schizoaffective disorder, and schizophrenia. For 61 child patients, the diagnoses included autism, nonautistic psychosis, cognitive processing disturbances, aphasia, Tourette's syndrome, and medical controls. The previously noted negative correlation between age and HVA was strikingly confirmed. Within the entire population, age at the time of lumbar puncture was negatively correlated with CSF concentration of HVA ($r = -0.25$, $p < 0.001$), HVA/probenecid ($r = -0.43$, $p < 0.00001$) (Fig. 6), and HVA/log probenecid ($r = -0.40$, $p < 0.00001$). Age was positively correlated with the CSF probenecid achieved ($r = 0.31$, $p < 0.001$); thus, when the effect of probenecid concentration was statistically eliminated by the method of partial correlation, the negative relationship between age and HVA was enhanced ($r = -0.29$, $p < 0.001$). In confirmation of previous impressions, there was no relationship between age and CSF 5-HIAA concentration (Fig. 7); there was, however, a relationship between age and 5-HIAA/probenecid ($r = -0.23$, $p < 0.01$). The 5-HIAA/HVA ratio was not significantly correlated with age.

Because of diagnostic heterogeneity, within and across child and adult populations, the interpretation of age trends in amine metabolite concentration must be cautious. However, the major age effects for HVA occur within childhood, with the sharpest decline in CSF HVA occurring roughly during the second decade (8–20 years). This may reflect changes in behavioral organization (e.g., decreased activity or increased cognitive competence) coincident with brain maturation.

Developmental trends may be approached most fruitfully by analysis of comparable diagnostic groups: adults with schizophrenia and children with disorders that have been conceptualized as childhood schizophrenia (primary and secondary autism and

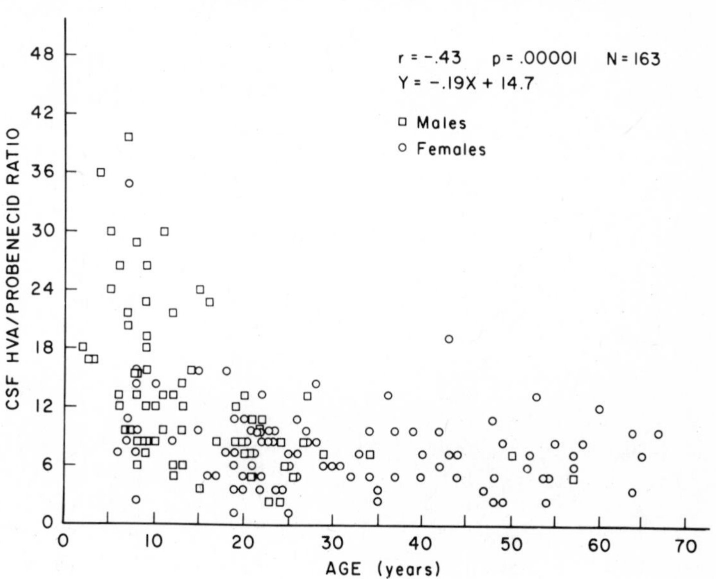

Figure 6. Relationship of CSF HVA/probenecid ratio to age for 163 pediatric and adult psychiatric patients. (HVA) Homovanillic acid.

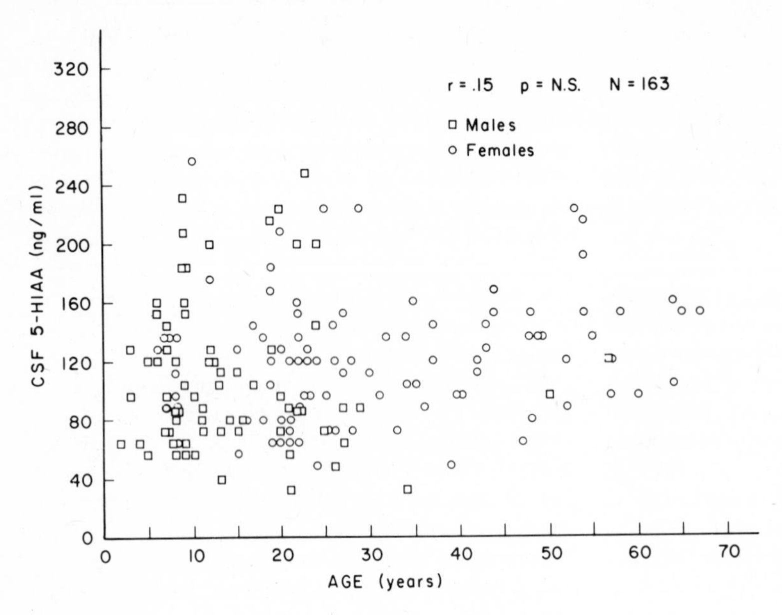

Figure 7. CSF 5-hydroxyindoleacetic acid (5-HIAA) accumulation following probenecid. CSF 5-HIAA appears not to be associated with age of neuropsychiatric patients.

early-onset, nonautistic psychosis).[34,35] This analysis may also cast light on the interesting theoretical question of the continuity or discontinuity between the child-and adult-onset schizophrenias. In our population, there were 21 children with early-onset psychosis and 26 adults with late-adolescent- and adult-onset schizophrenia. The psychotic children had a much higher mean CSF HVA accumulation (192.0 ± 20.2 ng/ml) than the adult schizophrenics (131.8 ± 10.6 ng/ml; $t = 2.64$, $p < 0.01$). Whether this significant difference represents a developmental or diagnosis-related effect cannot be answered with available data. The increased dopaminergic functioning in the children may be related to differences in the behavioral characteristics of the two syndromes (e.g., increased activity and more impoverished social relations in the childhood disorder).

The age-related changes in CSF HVA are consistent with an autopsy report of a decrease in brain MAO activity between the ages of 15 and 25 years.[55] Also, we have found a 25% reduction in platelet MAO activity among male autistic children from 4 to 20 years of age,[29,80] and Murphy *et al.*[50] have reported a 30% reduction in plasma MAO activity in normal men from 11 to 20 years of age. The age-related effects in CSF HVA, however, are not easily reconciled with considerable animal data on the increasing synthesis of brain dopamine during development and the achievement of an adult plateau at full brain maturation.[3,10,30,62] Changes in receptor sensitivity, metabolism, or other factors may account for an increasing brain concentration and the decreasing CSF HVA accumulation observed in our patients.

7. Sex Effects

As with age trends, there has been little systematic investigation of differences between CSF monoamine metabolites in males and females. Such studies may be of importance in understanding the difference between boys and girls in the incidence of certain types of disorders. For example, attention-deficit syndromes and childhood autism are 3- or 4-fold more common among boys. In neuropsychiatrically impaired children, we found a considerably higher dopamine turnover rate in boys ($N = 34$, HVA/probenecid = 17.4 ± 1.5, mean ± S.E.) than in girls ($N = 9$, HVA/P = 11.6 + 1.1, $t = 3.1$, $p = 0.004$).[19] This difference did not appear to be related to diagnostic clustering of the boys and girls in different neuropsychiatric subgroups. However, the small size of the sample prevented detailed analysis.

Leckman, in collaboration with us,[45] has analyzed CSF sex effects in a larger population of patients ages 3–20 years (drawn from the group described in the previous section). In this child and adolescent population, the mean CSF HVA, following probenecid, for males ($N = 49$) was 204 ng/ml; for females ($N = 22$), HVA was 164 ng/ml. The HVA/probenecid ratio was strikingly higher in the boys (15.7) than in the girls (9.5, $p < 0.01$) (Table 6). Diagnostic differences between the boys and the girls must be considered in evaluating the meaning of the increased HVA and HVA/probenecid measures in males. Also, since the females tended to be older, the age effects noted in the previous section may account, at least in part, for observed differences between the sexes.

Clearly, considerably more research will be required to assess the relationships of age and sex to CSF amine metabolites and implications of any CSF trends for psychopathology. Age effects may clarify vulnerable periods for the emergence of psychiatric disorders (e.g., late adolescence); also, there may be important interactions among neurochemical development (seen in CSF metabolites), endocrinological maturation, and the onset of exacerbation or behavioral disturbances at periods of biological upheaval (such as the normal moodiness at the transition into adolescence or the marked increase in aggression among autistic children at this phase of development).[22]

8. Associated Neurobiological Variables

The early-onset, pervasive behavioral disturbance and the high incidence of seizures and other neurological findings in children with autistic-like syndromes suggested the hypothesis that this syndrome may represent a congenital or early viral syndrome. Further support for this hypothesis is the high incidence of autistic-like syndromes in children with congenital rubella.[12] CSF protein, immunoglobulins, and colloidal gold were measured in autistic children without neurological findings and were found to be normal.[76] Since the report of this study, similar measures were obtained in many other children with primary and secondary autistic syndromes with negative results. The absence of abnormal CSF immunoglobulins does not rule out an etiological role for a slow-acting virus. However, the progressive improvement observed in some children with autism and the failure to develop other findings of CNS degeneration are inconsistent with other disorders presumably caused by slow virus infections.

CSF folate was measured in more than 60 children with childhood neuropsychiatric disorders, including 20 children with primary childhood autism. In most cases, CSF folate was above 25 ng/ml, a level considered within the normal range. In our series, the abnormal CSF folate levels occurred in children with dietary insufficiency. In most children with normal CSF folate, simultaneous measurements of plasma and red blood cell folate were within normal limits, but there were exceptions that are being explored.

Neurobiological studies of CSF must consider the relationship between structural variations and monoamine metabolites (e.g., the presence of abnormal ventricular morphology or occult hydrocephalus). Computed tomography of the brain was performed in a series of more than 75 neuropsychiatrically impaired children, including approximately 25 children with primary autism and 20 children with de-

Table 6. Mean of CSF HVA and 5-HIAA Concentrations and Related CSF Variables Classified by Age (5-Year Intervals) and Sex

Age group	Mean age (years) ± S.E.M.	N	HVA (ng/ml CSF)	HVA (ng/ ml CSF) / P (μg/ml CSF)	HVA/log P	5-HIAA (ng/ml CSF)	5-HIAA/P	5-HIAA/ log P	5-HIAA/ HVA ratio
				A. Male subjects					
0–4	3.0 ± 0.4	4	225 ± 35	21.9 ± 4.7	221 ± 21	90 ± 16	9.3 ± 2.6	91 ± 17	0.45 ± 0.14
5–9	7.7 ± 0.3	25	223 ± 17	17.2 ± 1.7	198 ± 12	118 ± 10	9.1 ± 1.0	104 ± 7	0.56 ± 0.05
10–14	11.9 ± 0.3	14	180 ± 17	12.6 ± 1.8	154 ± 13	97 ± 10	7.1 ± 1.2	84 ± 8	0.59 ± 0.08
15–19	16.8 ± 0.8	6	168 ± 22	12.9 ± 3.4	149 ± 23	119 ± 21	7.7 ± 1.0	99 ± 11	0.74 ± 0.12
				B. Female subjects					
5–9	7.4 ± 0.2	9	175 ± 21	12.2 ± 3.1	149 ± 20	108 ± 9	7.0 ± 1.4	89 ± 7	0.76 ± 0.19
10–14	11.0 ± 1.0	2	279 ± 13	11.8 ± 3.2	202 ± 24	215 ± 41	9.4 ± 3.8	158 ± 41	0.77 ± 0.11
15–19	17.6 ± 0.5	11	135 ± 16	7.8 ± 1.3	108 ± 14	115 ± 13	5.9 ± 0.4	89 ± 7	1.15 ± 0.34

[a] (HVA) Homovanillic acid; (P) probenecid; (5-HIAA) 5-hydroxyindoleacetic acid. All values are means ± S.E.

velopmental aphasia. In this population, approximately 25% of the computed tomograms in each neuropsychiatric diagnostic group revealed minor abnormalities, usually involving the temporal lobes. These abnormalities, however, were not specific for any diagnostic group. Radioactive brain scans in 25 autistic children were normal. Finally, while the EEG profiles of autistic and other neuropsychiatrically impaired children could be statistically discriminated, they did not reveal specific, diagnostic abnormalities.[74] Thus, alterations in CSF metabolites in childhood autism and similar disorders can be pursued without major concern that they reflect gross structural abnormalities, as observed in degenerative brain disorders or hydrocephalus.

CSF metabolites may become more meaningful if related to other indices of CNS metabolism. In two pilot studies, urinary MHPG and free catecholamines were measured in boys with childhood autism and normal, age- and sex-matched controls. Reduced levels of MHPG and catecholamines were found in the autistic children (538 μg/day of MHPG excretion compared to 766 μg/day for normal boys). These findings were contrary to the hypothesis that very active, anxious autistic boys would have elevated levels of urinary catecholamine metabolites.[78,79]

Reduced urinary MHPG and catecholamines in autism may reflect chronic adaptation rather than a primary dysfunction. Autistic children suffer from acute and repeated episodes of increased arousal and anxiety; this may result from an inborn dysfunction in the regulation of arousal and attention or from the emotional response to confusion about internal and external events because of impaired cognitive development.[11,21] In either case, to mute their overexcitement, autistic children appear to turn their attention inward, away from the external world, or toward the manipulation of easily controlled physical objects. We have speculated that these psychologically adaptive or defensive processes may affect the urinary MHPG and catecholamines in the direction we have observed. The absence of an elevation in autistic children's urinary MHPG and catecholamines is particularly striking in relation to the observation that their pulse rate and blood pressure tend to be elevated.[24]

The evaluation of biological findings in neuropsychiatric disorders must therefore consider state, as well as trait, variables. Especially for chronic conditions, underlying neurobiological alterations may be obscured or altered by many years of psychobiological adaptation. The concentration of monoamine metabolites at any one moment reflects the operation of diverse historical and current determinants; a specific finding may be associated with the "cause" or the "effects" of a disorder or, more likely, may represent a composite of various interacting forces.

Studies of CSF metabolites in childhood neuropsychiatric disorders are only in their infancy. Very few metabolites have been studied, and basic questions about their changes with state, development, and pharmacological intervention remain virtually unexplored. Important, measurable neurotransmitters, such as the peptides, have not been investigated yet in children. However, the basic methodology and practical possibility of studying metabolites in childhood neuropsychiatric disorders have been demonstrated; suggestive leads, associating dimensions of dysfunction and metabolite concentration, have emerged; and developmental trends have been outlined. Perhaps most important, the study of CSF monoamine metabolites has helped to focus sophisticated biological methods and concepts on the most enigmatic, disabling behavioral disorders of childhood.

ACKNOWLEDGMENTS. We are extremely grateful for the collaboration of Ms. Barbara Caparulo and Ms. Claudia Carbonari in all aspects of the clinical research. Dr. James Leckman collaborated in studies of developmental changes in monoamine metabolites and sex effects. Drs. James Maas and Susan Hattox collaborated in studies of norepinephrine metabolism. Clinical studies were conducted in the Yale Children's Clinical Research Center, directed by Dr. Myron Genel. These studies were supported by NIMH Clinical Research Center Grant 1 P50 MH 30929, Children's Clinical Research Center Grant RR00125, NIH Grants HD-03008 and MH 24393, the William T. Grant Foundation, Mr. Leonard Berger, the Schall Family Trust, and The Solomon R. and Rebecca D. Baker Foundation, Inc.

References

1. Abuzzahab, F. S., Anderson, F. O. (eds.): *Gilles de la Tourette's Syndrome*. St. Paul, Minnesota, Mason Publishing, 1976.
2. Aghajanian, G. K., Haigler, H. J., Bennett, J. L.: Amine receptors in CNS. III. 5-Hydroxytryptamine in brain. In Iverson, L. L., Iverson, S. D., Snyder, S. H. (eds.): *Handbook of Psychopharmacology*, Vol. 6. New York, Plenum Press, 1975, pp. 63–96.
3. Agrawal, H. C., Glisson, S. N., Himwich, W. A.:

Changes in monoamines of rat brain during postnatal ontogeny. *Biochim. Biophys. Acta* **130**:511–513, 1966.
4. AMERICAN PSYCHIATRIC ASSOCIATION: *Diagnostic and Statistical Manual of Mental Disorders* (DSM) III (draft version). Task Force on Nomenclature and Statistics, 1979.
5. ANDERSSON, H., ROOS, B. E.: 5-Hydroxyindoleacetic acid in cerebrospinal fluid of hydrocephalic children. *Acta Paediatr. Scand.* **58**:601–608, 1969.
6. BOWERS, M. B., JR.: Clinical measurements of central dopamine and 5-hydroxytryptamine metabolism: Reliability and interpretation of cerebrospinal fluid acid monoamine metabolite measures. *Neuropharmacology* **11**:101–111, 1972.
7. BOWERS, M. B., JR.: Central dopamine turnover in schizophrenic syndromes. *Arch. Gen. Psychiatry* **31**:50–54, 1974.
8. BOWERS, M. B., JR.: Lumbar 5-hydroxyindoleacetic acid and homovanillic acid in affective syndromes. *J. Nerv. Ment. Dis.* **158**:325–330, 1974.
9. BOWERS, M. B., GERBODE, F. A.: Relationship of monoamine metabolites in human cerebrospinal fluid with age. *Nature (London)* **219**:1256–1257, 1968.
10. BREESE, G. R., TRAYLOR, T. D.: Developmental characteristics of brain catecholamines and tyrosine hydroxylase in the rat: Effects of 6-OHDA. *Br. J. Pharmacol.* **44**:210–222, 1972.
11. CAPARULO, B. K., COHEN, D. J.: Cognitive structures, language, and emerging social competence in autistic and aphasic children. *J. Am. Acad. Child Psychiatry* **16**:620–645, 1977.
12. CHESS, S.: Followup report on autism in congenital rubella. *J. Autism Child. Schizophr.* **7**:69–81, 1977.
13. CHURCHILL, D. W., ALPERN, G. D., DEMYER, M. K. (eds.): *Infantile Autism*. Springfield, Illinois, Charles C. Thomas, 1971.
14. COHEN, D. J.: Competence and biology. In Anthony, E. J., Koupernik, C. (eds.): *The Child in His Family*. New York, John Wiley, 1974, pp. 361–394.
15. COHEN, D. J.: The diagnostic process in child psychiatry. *Psychiatr. Ann.* **6**:404–416, 1976.
16. COHEN, D. J.: The pathology of the self in two neuropsychiatric disorders of childhood. In Robson, K. (ed.): *Narcissism and the Psychopathology of the Self in Childhood*. New York, Jason Aronson (in press).
17. COHEN, D. J., CAPARULO, B. K., GOLD, J. R., WALDO, M. C., SHAYWITZ, B. A., RUTTENBERG, B. A., RIMLAND, B.: Agreement in diagnosis. *J. Am. Acad. Child Psychiatry* **17**:589–603, 1978.
18. COHEN, D. J., CAPARULO, B. K., SHAYWITZ, B. A.: Primary childhood aphasia and childhood autism. *J. Am. Acad. Child Psychiatry* **15**:604–645, 1976.
19. COHEN, D. J., CAPARULO, B. K., SHAYWITZ, B. A., BOWERS, M. B., JR.: Dopamine and serotonin metabolism in neuropsychiatrically disturbed children. *Arch. Gen. Psychiatry* **34**:545–550, 1977.
20. COHEN, D. J., CAPARULO, B. K., SHAYWITZ, B. A., BOWERS, M. B., JR.: Assessment of cerebrospinal monoamine metabolites in children using the probenecid method. *Isr. Ann. Psychiatry Relat. Discip.* **15**:47–57, 1977.
21. COHEN, D. J., CAPARULO, B. K., SHAYWITZ, B. A.: Neurochemical and developmental models of childhood autism. In Serban, G. (ed.): *Cognitive Defects in the Development of Mental Illness*. New York, Brunner/Mazel, 1978, pp.66–100.
22. COHEN, D. J., FRANK, R.: Preadolescence: A critical phase of biological and psychological development. In Sankar, D. V. S. (ed.): *Mental Health in Children*, Vol. 1. Westbury, New York, PJD Publications, 1975, pp. 129–165.
23. COHEN, D. J., JOHNSON, W., CAPARULO, B. K., YOUNG, J. G.: Creatine phosphokinase levels in children with severe developmental disturbances. *Arch. Gen. Psychiatry* **33**:683–686, 1976.
24. COHEN, D. J., JOHNSON, W. T.: Cardiovascular correlates of attention in normal and psychiatrically disturbed children. *Arch. Gen. Psychiatry* **34**:561–567, 1977.
25. COHEN, D. J., SHAYWITZ, B. A., CAPARULO, B., YOUNG, J. G., BOWERS, M. B, JR.: Chronic, multiple tics of Gilles de la Tourette's disease. *Arch. Gen. Psychiatry* **35**:245–250, 1978.
26. COHEN, D. J., SHAYWITZ, B. A., JOHNSON, W. T., BOWERS, M. B., JR.: Biogenic amines in autistic and atypical children: Cerebrospinal fluid measures of homovanillic acid and 5-hydroxyindoleacetic acid. *Arch. Gen. Psychiatry* **31**:845–853, 1974.
27. COHEN, D. J., SHAYWITZ, B. A., YOUNG, J. G., CARBONARI, C. M., NATHANSON, J. A., LIEBERMAN, D., BOWERS, M. B., JR., MAAS, J. W.: Central biogenic amine metabolism in children with the syndrome of chronic multiple tics of Gilles de la Tourette: Norepinephrine, serotonin, and dopamine. *J. Am. Acad. Child Psychiatry* **18**:320–341, 1979.
28. COHEN, D. J., YOUNG, J. G.: Neurochemistry and child psychiatry. *J. Am. Acad. Child Psychiatry* **16**:353–411, 1977.
29. COHEN, D. J., YOUNG, J. G., ROTH, J. A.: Platelet monoamine oxidase in early childhood autism. *Arch. Gen. Psychiatry* **34**:534–537, 1977.
30. COYLE, J. T., HENRY, D.: Catecholamines in fetal and newborn rat brain. *J. Neurochem.* **21**:61–67, 1973.
31. CREAK, M.: Schizophrenia syndrome in childhood. *Cerebral Palsy Bull.* **3**:501–504, 1961.
32. ELDRIDGE, R., SWEET, R., LAKE, C. R., ZIEGLER, M., SHAPIRO, A. K.: Gilles de la Tourette's syndrome. *Neurology* **27**:115–124, 1977.
33. FISH, B.: Biologic antecedents of psychosis in children. In Freedman, D. X. (ed.): *Biology of the Major Psychoses: Res. Publ. Assoc. Res. Nerv. Ment. Dis.*, Vol. 54. New York, Raven Press, 1975, pp. 49–83.
34. FISH, B.: Neurologic antecedents of schizophrenia in children: Evidence for an inherited, congenital neurointegrative defect. *Arch. Gen. Psychiatry* **34**:1297–1313, 1977.
35. FISH, B., RITVO, E.: Psychoses of childhood. In Nosh-

pitz, J. (ed.): *Basic Handbook of Child Psychiatry II: Disturbances of Development.* New York, Basic Books, pp. 249–304.

36. GOLDFARB, W.: *Growth and Change of Schizophrenic Children.* New York, John Wiley, 1975.
37. GRAHAME-SMITH, D. G.: Studies *in vivo* on the relationship between brain tryptophan, brain 5-HT synthesis and hyperactivity in rats treated with a monoamine oxidase inhibitor and L-tryptophan. *J. Neurochem.* **18:**1053–1066, 1971.
38. HINGTGEN, J., BRYSON, C.: Recent developments in the study of early childhood psychoses. *Schizophr. Bull.* **5:**8–55, 1972.
39. JACOBS, B. L.: Evidence for the functional interaction of two central neurotransmitters. *Psychopharmacology (Berlin)* **39:**81–86, 1974.
40. JACOBS, B. L.: An animal behavior model for studying central serotonergic synapses. *Life Sci.* **19:**777–786, 1976.
41. KANNER, L.: *Childhood Psychosis.* Washington, D.C., Winston & Sons, 1973.
42. KLAWANS, H., MARGOLIN, D.: Amphetamine-induced dopamine hypersensitivity in guinea pigs. *Arch. Gen. Psychiatry* **32:**725–732, 1975.
43. KORF, J., VAN PRAAG, H. M.: Amine metabolism in the human brain: Further evaluation of the probenecid test. *Brain Res.* **35:**221–230, 1971.
44. LAKE, C. R., ZIEGLER, M. G., ELDRIDGE, R., MURPHY, D. L.: Catecholamine metabolism in Gilles de la Tourette's syndrome. *Am. J. Pshchiatry* **134:**257–260, 1977.
45. LECKMAN, J. F., COHEN, D. J., SHAYWITZ, B. A., CAPARULO, B. K., HENINGER, G. R., BOWERS, M. B., JR.: CSF monoamine metabolites in child and adult psychiatric patients: A developmental perspective. *Arch. Gen. Psychiatry* **37:**677–684, 1980.
46. MAAS, J. W., GREENE, N. M., HATTOX, S. E., LANDIS, D. H.: Neurotransmitter metabolite production by human brain. In Usdin, E., Kopin, I. J., Barchas, J. (eds.): *Catecholamines.* New York, Pergamon Press, 1979, pp. 1878–1880.
47. MAAS, J. W., HATTOX, S. E., LANDIS, D. H., ROTH, R. H.: A direct method for studying 3-methoxy-4-hydroxyphenethyleneglycol (MHPG) production by brain in awake animals. *Eur. J. Pharmacol.* **46:**221–228, 1977.
48. MCCALL, R. B., AGHAJANIAN, G. K.: Serotonergic facilitation of facial motoneuron excitation. *Brain Res.* **169:**11–27, 1979.
49. MOLDOFSKY, H., TULLIS, C., LAMON, R.: Multiple tic syndrome (Gilles de la Tourette's syndrome). *J. Nerv. Ment. Dis.* **159:**282–292, 1974.
50. MURPHY, D. L., WRIGHT, C., BUCHSBAUM, M., NICHOLS, A., COSTA, J. L., WYATT, R. J.: Platelet and plasma amine oxidase activity in 680 normals: Sex and age differences and stability over time. *Biochem. Med.* **16:**254–265, 1976.
51. ORNITZ, E. M., RITVO, E. R.: The syndrome of autism. *Am. J. Psychiatry* **133:**609–621, 1976.
52. RITVO, E. R. (ed.): *Autism.* New York, Spectrum, 1976.
53. RITVO, E. R., FREEMAN, B. J.: The National Society for Autistic Children's definition of autism. *J. Am. Acad. Child Psychiatry* **17:**565–575, 1978.
54. RITVO, E. R., RABIN, K., YUWILER, A., FREEMAN, B. J., GELLER, E.: Biochemical and hematologic studies: A critical review. In Rutter, M., Schopler, E. (eds.): *Autism: A Reappraisal of Concepts and Treatment.* New York, Plenum Press, 1978, pp. 163–183.
55. ROBINSON, D. S., SOURKES, T. L., NIES, A., HARRIS, L. S., SPECTOR, S., BARTLETT, D. L., KAYE, I. S.: Monoamine metabolism in human brain. *Arch. Gen. Psychiatry* **34:**89–92, 1977.
56. ROGERS, J. J., DUBOWITZ, V.: 5-Hydroxyindoles in hydrocephalus: A comparative study of cerebrospinal fluid and blood levels. *Dev. Med. Child Neurol.* **12:**461–466, 1970.
57. ROTH, J. A., YOUNG, J. G., COHEN, D. J.: Platelet monoamine oxidase activity in children and adolescents. *Life Sci.* **18:**919–924, 1976.
58. RUTTER, M.: Brain damage syndromes in childhood: Concepts and findings. *J. Child Psychol. Psychiatry* **18:**1–21, 1977.
59. RUTTER, M., SCHOPLER, E. (eds.): *Autism: A Reappraisal of Concepts and Treatment.* New York, Plenum Press, 1978.
60. SHAPIRO, A. K., SHAPIRO, E., BRUNN, R. D., SWEET, R. D.: *Gilles de la Tourette Syndrome.* New York, Raven Press, 1978.
61. SHAYWITZ, B. A., COHEN, D. J., BOWERS, M. B., JR.: Reduced cerebrospinal fluid 5-hydroxyindoleacetic acid and homovanillic acid in children with epilepsy. *Neurology* **25:**74–79, 1975.
62. SHAYWITZ, B. A., GORDON, J. W., KLOPPER, J. H., ZELTERMAN, D. A., IRVINE, J.: Ontogenesis of spontaneous activity and habituation of activity in the rat pup. *Dev. Psychobiol.* **12:**359–367, 1979.
63. SHAYWITZ, B. A., VENES, J., COHEN, D. J., BOWERS, M. B., JR.: Reye syndrome: Monoamine concentrations in ventricular fluid. *Neurology* **29:**467–472, 1979.
64. SINGER, H. S., BUTLER, I. J., SEIFERT, W. E., CAPRIOLI, R., KOSLOW, S. H.: Tourette syndrome: A neurotransmitter disorder? *Ann. Neurol.* **4:**189–190, 1978.
65. SJÖSTRÖM, R.: Steady-state levels of probenecid and their relationship to acid monoamine metabolites in human cerebrospinal fluid. *Psychopharmacologia* **25:**96–100, 1972.
66. STEWART, R. M., GROWDON, J. H., CANCIAN, D., BALDESSARINI, R. J.: 5-Hydroxytryptophan-induced myoclonus. *Neuropharmacology* **15:**449–455, 1976.
67. SVENSSON, T. H., BUNNEY, B. S., AGHAJANIAN, G. K.: Inhibition of both noradrenergic and serotonergic neurons in brain by the α-adrenergic agonist clonidine. *Brain Res.* **92:**291–306, 1975.
68. SWEET, R. D., SOLOMON, G. E., WAYNE, H., SHAPIRO, E., SHAPIRO, A. K.: Neurological features of Gilles de la Tourette's syndrome. *J. Neurol. Neurosurg. Psychiatry* **36:**1–9, 1973.
69. TAMARKIN, N. R., GOODWIN, F. K., AXELROD, J.: Rapid

elevation of biogenic amine metabolites in human CSF following probenecid. *Life Sci.* **9**:1397–1408, 1970.
70. VAN PRAAG, H. M.: *Depression and Schizophrenia.* New York, Spectrum, 1977.
71. VAN PRAAG, H. H., FLENTGE, F., KORF, J., DOLS, L. C. W., SCHUT, T.: The influence of probenecid on the metabolism of serotonin, dopamine, and their precursors in man. *Psychopharmacologia* **33**:141–151, 1973.
72. VAN PRAAG, H. M., KORF, J., LAKE, J. P. W. T., SCHUT, T.: Dopamine metabolism in depressions, psychoses, and Parkinson's disease: The problem of the specificity of biological variables in behavior disorders. *Psychol. Med.* **5**:138–146, 1975.
73. VAN PRAAG, H. M., KORF, J., SCHUT, D.: Cerebral monoamines and depression. *Arch. Gen. Psychiatry* **28**:827–831, 1973.
74. WALDO, M. C., COHEN, D. J., CAPARULO, B. K., YOUNG, J. G., PRICHARD, J. W., SHAYWITZ, B. A.: EEG profiles of neuropsychiatrically disturbed children. *J. Am. Acad. Child Psychiatry* **17**:656–670, 1978.
75. WING, J. K.: Diagnosis, epidemiology, aetiology. In Wing, J. K. (ed.): *Early Childhood Autism.* Oxford, Pergamon Press, 1966, pp. 3–49.
76. YOUNG, J. G., CAPARULO, B. K., SHAYWITZ, B. A., JOHNSON, W. T., COHEN, D. J.: Childhood autism: Cerebrospinal fluid examination and immunoglobulin levels. *J. Am. Acad. Child Psychiatry* **16**:174–179, 1977.
77. YOUNG, J. G., COHEN, D. J.: The molecular biology of development. In Noshpitz, J. (ed.): *Handbook of Child Psychiatry.* New York, Basic Books, 1979, pp. 22–62.
78. YOUNG, J. G., COHEN, D. J., BROWN, S.-L., CAPARULO, B. K.: Decreased urinary free catecholamines in childhood autism. *J. Am. Acad. Child Psychiatry* **17**:671–678, 1978.
79. YOUNG, J. G., COHEN, D. J., CAPARULO, B. K., BROWN, S.-L., MAAS, J. W.: Decreased 24-hour urinary MHPG in childhood autism. *Am. J. Psychiatry* **136**:1055–1057, 1979.
80. YOUNG, J. G., COHEN, D. J., ROTH, J. A.: Association between platelet monoamine oxidase activity and hematocrit in childhood autism. *Life Sci.* **23**:797–806, 1978.

CHAPTER 47

Cerebrospinal Fluid Studies of Neurotransmitter Function in Manic and Depressive Illness

Robert M. Post, James C. Ballenger, and Frederick K. Goodwin

1. Introduction

This chapter reviews reported alterations in cerebrospinal fluid (CSF)-derived measures of a variety of neurotransmitter and modulator substances in patients with affective disorders. These studies are based on the assumption that measures in the CSF will provide an indirect reflection of central nervous system (CNS) processes. The data supporting this assumption have been reviewed elsewhere in this volume, as well as by Goodwin and Post,[60,61] Post and Goodwin,[112] and Moir *et al.*[96] The methodologies that involve the use of probenecid to block acid transport out of CSF in order to provide a more dynamic measure of neurotransmitter amine turnover have similarly been reviewed in Chapter 8. Both the baseline measures and those following probenecid pretreatment require considerable caution in interpretation, since for many of the neurotransmitter substances and metabolites mentioned, a contribution from the spinal cord is likely and critical transport characteristics appear to affect amine metabolite levels. However, within the context of these qualifications, many studies have documented that alterations in CSF levels can closely reflect changes occurring in CNS function. Thus, in this chapter, we will summarize data on a variety of the chemicals measured in the CSF, assuming from the outset that they are, at best, indirect and often global estimates for the more discrete changes that are being sought and investigated.

Increasing evidence suggests that the CSF pathway may participate in an active neuroregulatory process for endocrinological and behavioral modulation[78,122] (see also Chapters 41 and 44). To this extent, study of these substances in CSF may be of particular importance in elucidating particular chemical regulatory pathways even if CSF levels are not always directly reflective of occurrences within brain *per se*.

While we will not focus, in this chapter, on laboratory methodology or the assumptions underlying the CSF techniques, clinical–methodological aspects of the CSF techniques will be discussed and reviewed in some detail. As summarized in Table 1, a variety of factors can critically alter CSF neurotransmitter measurements and should therefore be taken into account in any CSF study of neuropsychiatric patients. Potential alterations in neurotransmitters and their metabolites in the affective disorders compared to control populations will then be reviewed. Pharmacologically induced alterations in CSF neurotransmitter function are also briefly discussed as they pertain to the biochemical findings in affective illness. Additionally, we will discuss recent findings of peptide and related neurohormones

Robert M. Post, M.D. • Section on Psychobiology, Biological Psychiatry Branch, National Institute of Mental Health, National Institutes of Health, Bethesda, Maryland 20205. **James C. Ballenger, M.D.** • Department of Behavioral Medicine and Psychiatry, University of Virginia School of Medicine, Charlottesville, Virginia 22908. **Frederick K. Goodwin, M.D.** • Clinical Psychobiology Branch, National Institute of Mental Health, National Institutes of Health, Bethesda, Maryland 20205.

Table 1. Clinical–Methodological Factors That Affect Amine Metabolite Measurements

Age	Disease state, including:
Sex	Diagnostic type and subtype
Menstrual status and hormonal state	Severity
Diurnal variation	Duration, acute–chronic, and other temporal relationships
Seasonal variation	Symptom vs. syndrome link of variable to behavior
Bedrest	Specificity
Activity	Nature of biological–behavioral linkage:
Acute stress	State-independent
Diet	Genetic; environmental; and experiential-based predispositions
Weight	State-dependent
Drugs (specific and nonspecific effects)	Etiological, correlative, secondary, tertiary
	Time lag for onset or offset of biological change in relation to behavior

in the CSF of affectively ill patients. These data will be presented preliminarily, with the assumption that further developments in peptide methodology may lead to important breakthroughs in this field in the near future. Finally, several of the theoretical implications of these findings in relation to heterogeneity of affective illness and prediction of clinical and pharmacological response will be discussed in an integrative overview.

2. Clinical–Methodological Issues

2.1. Diagnosis and Clinical Status

Two critical variables for the CSF methodologies are obviously the patient population studied and the diagnostic criteria employed. The criteria of Spitzer *et al.*,[138] recently incorporated into DSM III,[1a] appear to provide a working definition of the range of affective illnesses that are studied. These criteria are outlined in Table 2. Most investigators would agree that the diagnosis of bipolar manic–depressive illness is usually made with more certainty than that of unipolar depressive illness, which appears to be a highly heterogeneous group of diseases and perhaps involves multiple etiologies and underlying biochemical substrates. Thus, issues of clinical diagnosis and heterogeneity become critical in any CSF study of affective illness.

If one is dealing with acute, state-related changes in CSF biochemistry, alterations may be substantially different depending on duration and severity of illness. Changes early in a depressive episode may be highly different from those later in the illness. The findings in parkinsonism and a variety of other neuropsychiatric illnesses are relevant here. For example, early in Parkinson's disease, there may be little evidence of defects in CSF homovanillic acid (HVA), since remaining dopaminergic neurons provide compensatory increases in dopamine synthesis in response to early dopaminergic neuronal loss. As the illness progresses, HVA depletions in brain and CSF become more severe. Also, with pharmacological strategies, HVA increases early in neuroleptic administration may be substantially greater than those following chronic treatment with the same dose of neuroleptic[53,110,128,143] (see Fig. 12). Similarly, early and late phases of an affective episode may be associated with substantial differences in amine metabolite alterations. Additionally, the clinical presentation of a given syndrome may be important. For example, agitated and retarded depressed patients may have substantially different measures in some neurotransmitter systems.[98,147,148]

2.2. Psychomotor Activity

Alterations in motor activity are often intimately associated with the affective illnesses. Van Praag *et al.*[149,152] have argued that the alterations in motor activity may be associated with concomitant alterations in CSF neurotransmitters and that these changes are closely linked to this particular symptom cluster. However, it is also possible that alterations in activity may in themselves produce alterations in CSF measures. In studies in our laboratory, depressed patients were asked to be active for 4 hr

Table 2. Research Diagnostic Criteria for Primary Affective Disorders

I. *Depression (A through E required)*
- A. Dysphoric mood (sad, blue, depressed, low, irritable) that is prominent and persistent (i.e., does not shift rapidly from one dysphoric mood to another)
- B. At least five of the following symptoms:
  1. Poor appetite or weight loss or increased appetite or weight gain
  2. Sleep difficulty or sleeping too much
  3. Loss of energy, fatigue
  4. Psychomotor agitation or retardation
  5. Loss of interest in usual activities (e.g., social activities, sex)
  6. Feelings of self-reproach, excessive guilt
  7. Diminished ability to think or concentrate, slowed thinking, indecisiveness
  8. Recurrent thoughts of suicide or death
- C. Dysphoric features of illness lasting at least 1 week
- D. Help sought during dysphoric period, took medication, or was impaired socially at work or at home
- E. None of the following features suggesting schizophrenia present:
  1. Delusions of thought control, broadcasting, insertion, or withdrawal
  2. Nonaffective hallucinations throughout the day for several days or intermittently for 1 week
  3. Auditory hallucinations (in which voice carries on commentary or two voices converse)
  4. Period with delusions or hallucinations without depressive symptoms
  5. Preoccupation with a delusion or hallucination to exclusion of other symptoms (except delusions of guilt, sin, nihilism)
  6. Definite instances of formal thought disorder (as defined)

II. *Mania (A through E required)*
- A. Period of predominantly elevated or irritable mood
- B. At least three of the following symptoms if mood elevated (four if mood only irritable):
  1. More active than usual—socially, at work, sexually, or physically
  2. More talkative than usual or felt pressure to keep talking
  3. Flight of ideas (as defined) or subjective experience that thoughts are racing
  4. Inflated self-esteem (grandiosity)
  5. Decreased need for sleep
  6. Distractibility (i.e., attention too easily drawn to unimportant stimuli)
  7. Excessive involvement in activities without recognizing potential for painful consequences (e.g., buying sprees, sexual indiscretions)
- C. Overall disturbance so severe that at least one of the following is present:
  1. Meaningful conversation impossible
  2. Serious impairment socially (at home, work, with family)
  3. Hospitalization
- D. Duration of manic features at least 1 week
- E. None of symptoms of schizophrenia present (see item I.E above)

prior to lumbar puncture. This direct manipulation of motor activity produced significant alterations in 5-hydroxyindoleacetic acid (5-HIAA) and HVA[115] (Fig. 1) and in 3-methoxy-4-hydroxyphenylethylene glycol (MHPG)[111] (Fig. 2). The suggestion that activity is producing greater increases in CSF mixing does not appear adequate, since changes were observed in MHPG and there does not appear to be a marked rostal–caudal CSF gradient for MHPG as there is for 5-HIAA and HVA. It is likely that the motor activity itself or the stress associated with the procedure increased the CSF amine metabolite levels. On the basis of urinary MHPG studies, Goode *et al.*[58] and Sweeney *et al.*[140] suggest that alterations in activity may not be as important as alterations in the associated stress of a procedure in normal volunteers. Ebert *et al.*[48] and Wehr *et al.*[157] reported increased urinary MHPG in depressed patients, but not in volunteers during experimental periods of elevated activity.

Banki[8] found positive correlations between psychomotor activity and CSF HVA in both unipolar ($r = 0.62$, $p < 0.001$) and bipolar depressed ($r = 0.86$, $p < 0.001$) patients, but not in normal controls ($r = 0.09$, $p =$ N.S.). Van Praag *et al.*[152] also reported increases in HVA in a schizophrenic patient population that demonstrated increased motor agitation. Weiss *et al.*[158] and Kirstein *et al.*[77] did not find associations between monitor-measured activity and CSF HVA in depressed patients, but did report that lower 5-HIAA was correlated with increased motor agitation.[77]

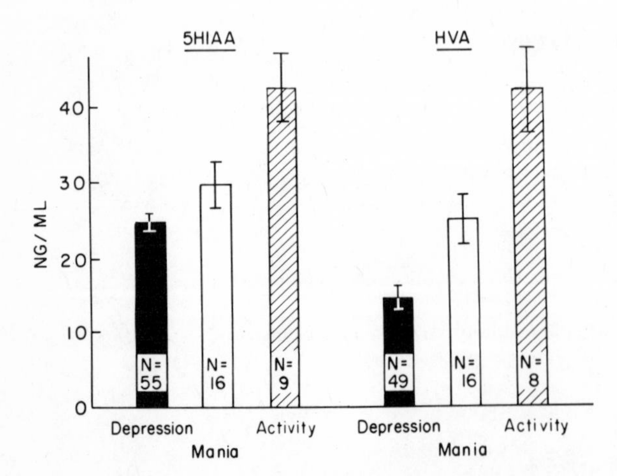

Figure 1. CSF amine metabolites in natural and experimentally induced behavioral states. Significant increases in CSF 5-hydroxyindoleacetic acid (5-HIAA) and homovanillic acid (HVA) were observed in depressed patients following 4 hr of moderately intense motor activity prior to lumbar puncture. Levels exceeded those in manic patients studied at bed rest.[115]

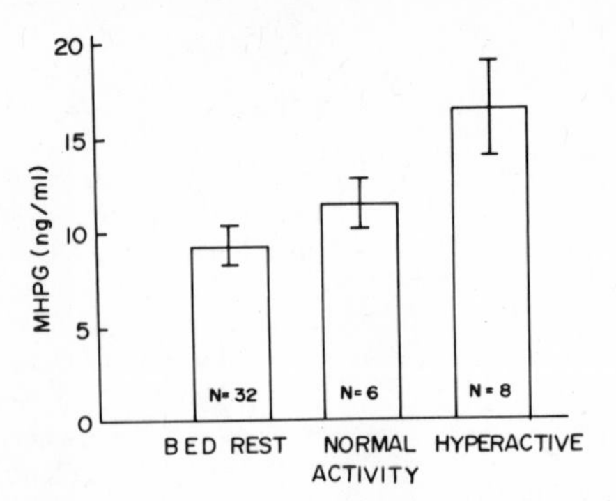

Figure 2. Effect of activity on CSF 3-methoxy-4-hydroxyphenylethylene glycol (MHPG) concentrations in depressed patients. CSF MHPG increased significantly following 4 hr of increased activity prior to lumbar puncture.[111] In one depressed patient, CSF norepinephrine doubled after 4 hr of activity compared to values obtained after bed rest.

Taken together, both naturalistic studies and those in which activity was directly manipulated support the notion that level of motor activity can be a partial determinant of amine metabolite levels in CSF.

2.3. Anxiety and Stress

Similar problems of interpretation arise in relation to variables such as anxiety or stress that could possibly be associated with nonspecific alterations in neurochemistry or with variables more intimately related with the illness studied. For example, as illustrated in Fig. 3, depressed patients with higher-rated anxiety had higher levels of CSF norepinephrine compared to those with lower anxiety. These preliminary findings in man are consistent with those of Redmond[119] implicating the noradrenergic locus coeruleus in the modulation of anxiety behaviors in the nonhuman primate.

2.4. Diet

Increasing evidence suggests that alterations in dietary intake can affect CNS transmitter function. Precursor loading strategies have been employed with compounds such as L-tryptophan and levodopa in affectively ill patients and are associated with appropriate alterations in respective CNS neurotransmitter systems (see Figs. 4 and 5). Animal studies suggest that similar findings may occur with a variety of amino acid compounds and in the cholinergic system as well. In a preliminary study, Lake

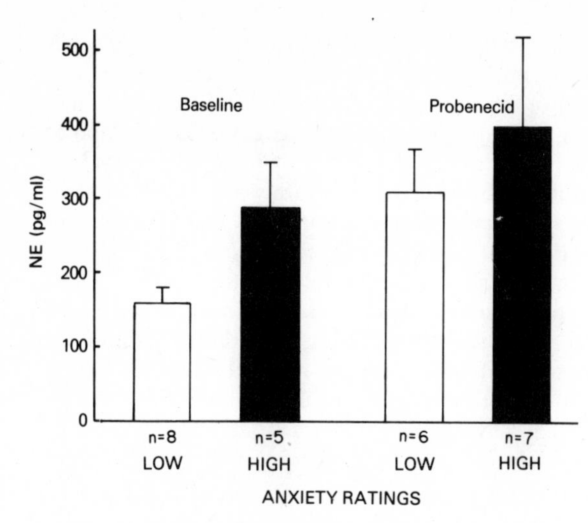

Figure 3. Elevated baseline CSF norepinephrine (NE) in high-anxiety depressed patients. Patients with higher anxiety ratings showed higher baseline CSF NE than less anxious depressed patients.[116]

et al.[81] found that a low- or high-monoamine diet, at least in patients with Huntington's chorea, did not affect levels of CSF norepinephrine. It is not clear whether this might be the case in patients with other illnesses, and data of Muscettola *et al.*[97] suggest that a low-monoamine diet may have differential effects on normal volunteers and depressed patients. In their study, they found that restricted intake of monoamines was associated with lower values of urinary MHPG in depressed patients, but not in normal volunteers.

2.5. Diurnal and Annual Variations in Neurotransmitter Amine Function

Diurnal variations in a variety of CNS neurotransmitters and their metabolites have been documented in the CSF of man and monkey. For example, Ziegler *et al.*[162] found that CSF norepinephrine was higher during the day than evening hours in both human and rhesus primate. Perlow *et al.*[100] have reported similar alterations in CSF HVA, adenosine-3′,5′-cyclic monophosphate (cAMP), and norepinephrine in the rhesus primate. Thus, not only is time of day of study important, but also the conceptual issue of whether illness-associated alterations are merely reflective of phase shifts in a given variable also requires careful attention.

Even more perplexing and confounding for the neuroscientist is the possible circannual variation of some neurotransmitter effects. Studies of Wirz-Justice *et al.*[160,161] reported seasonal alterations in melatonin, plasma dopamine-β-hydroxylase (DBH), and platelet monoamine oxidase (MAO). Carlsson and collaborators (personal communication, 1979) have documented seasonal and diurnal variations in brain amines measured in autopsy specimens. Seasonal alterations in CSF metabolites are currently

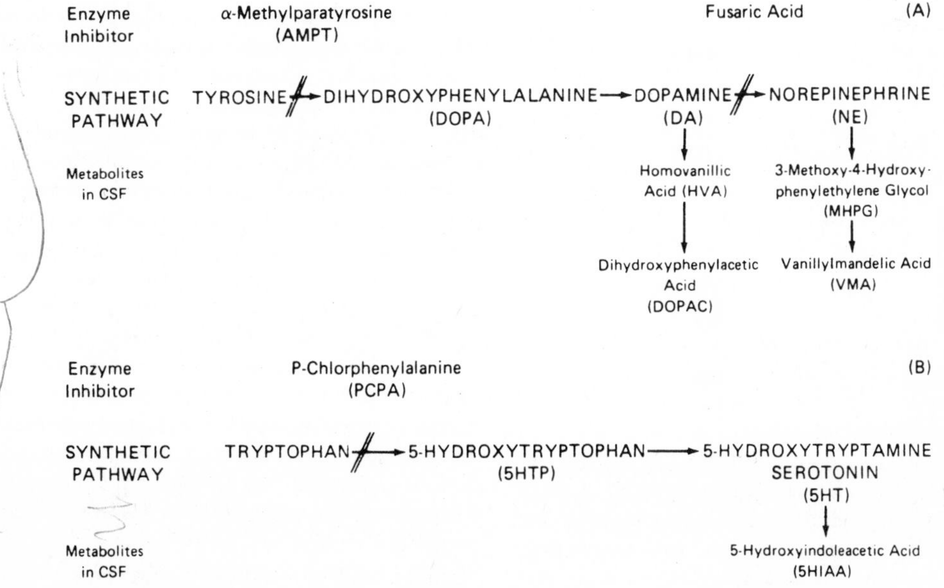

Figure 4. Illustration of simplified catecholamine (dopamine and norepinephrine) and indoleamine (serotonin) pathways.

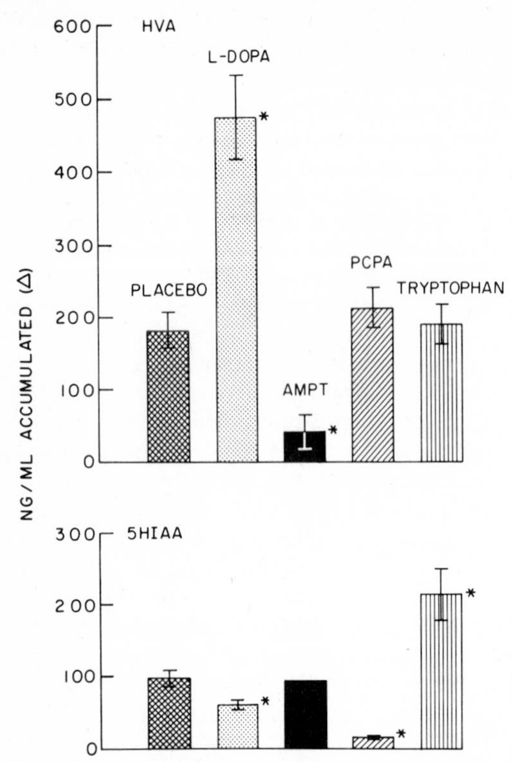

Figure 5. Probenecid-induced increase in amine metabolites: effects of precursors and inhibitors of enzyme synthesis. For each drug, average dose per day at time of lumbar puncture and duration of drug trial are as follows: L-DOPA, 812 mg, plus MK485, 656 mg, 16 days, $N = 8$ (HVA increase: $p < 0.01$; 5-HIAA decrease: $p < 0.05$; paired Student's t test); L-tryptophan, 8.7 g, 12.5 days, $N = 6$ (5-HIAA increase: $p < 0.01$; paired Student's t test); α-methyl-*p*-tyrosine (AMPT), 3.7 g, 21.2 days, $N = 4$ (HVA decrease $p < 0.001$; group t test); *p*-chlorphenylalanine (PCPA), 3.8 g, 20 days, $N = 2$ (5-HIAA decrease, $p < 0.001$; group t test).[64] (HVA) Homovanilllic acid; (5-HIAA) 5-hydroxyindoleacetic acid.

being studied in collaboration with Wirz-Justice and Wehr.

In recent studies conducted by Cools *et al.*[37] the findings of seasonal changes and their possible implications were extended. They found annual alterations not only in neurotransmitter levels, but also in pharmacological responsivity to a given drug challenge. Both dopamine and norepinephrine injections into the caudate nucleus produced different effects on turning behavior depending on the time of year. These seasonal changes were consistent over a 3-year period of testing.

Not only are these data intriguing in their own right as well as in possible relationship to the seasonal alterations in the occurrence of affective illness, but also they provide a possible critical and confounding variable for the neuroscientist. Particularly in human clinical investigations, collection of biological material such as CSF samples may be a prolonged endeavor, and rarely are these CSF samples collected from experimental and control populations at precisely the same time of year. Thus, considerable caution in relation to possible neurotransmitter alterations is required when experimental and control populations are studied at quite independent time periods, not only because of the usual methodological problems, such as storage time, assay reliability, and drift, but also because of possible real seasonal variations in these substances.

2.6. Age

Originally, Bowers and Gerbode,[20] Gottfries *et al.*,[68] and Asberg *et al.*[2] reported increases in either 5-HIAA or HVA in patients in the older decades. Their subsequent studies[15–18] and ours over a more restricted time span indicate only weak correlations between age and the CSF measures. In children, HVA may decrease with age.[35] In our depressed patients, age did not significantly correlate with baseline measures of CSF 5-HIAA ($r = 0.13$, $N = 70$), HVA ($r = 0.05$, $N = 62$), MHPG ($r = 0.30$, $N = 34$), vanillylmandelic acid (VMA) ($r = 0.12$, $N = 31$), or norepinephrine ($r = 0.16$, $N = 24$). Accumulations of HVA following probenecid were negatively correlated with age in depressed patients ($r = -0.34$, $N = 43$, $p < 0.02$).

Recent data of Carlsson[30] on measures of neurotransmitter levels and their metabolites in autopsy specimens suggest that many of these substances (dopamine, HVA, 3-methoxytyramine) decrease with age in normals and are decreased in those with presenile dementia. He raises the interesting possibility that some of the psychological deficits associated with depression may be intimately associated with a "panamine" deficit syndrome similar to that observed with premature aging.

Where correlations between CSF levels of a given substance with age of a patient population are substantial, obviously appropriate age-matched controls are essential.

2.7. Male–Female Differences

Similar caution is indicated in interpreting experimental and control data in relation to possible differences in sex distribution between the sample populations. In our studies, women have significantly higher probenecid-induced accumulations of both 5-HIAA and HVA, but not of MHPG (Table 3). These alterations may interact with activity of ovarian function in women and in the menstrual phase of the cycle. For example, postmenopausal amenorrheic depressed females had higher 5-HIAA than males or menstruating females (P. Gold, D. C. Jimerson, R. M. Post, F. K. Goodwin, unpublished data).

Again, these data are of importance not only in regard to potential problems of experimental design and methodological control, but also in their own right as they relate to possible biological differences regarding predisposition to psychiatric illness. The incidence of depression, particularly unipolar depression, is substantially higher among females than among males. Are amine metabolite alterations, possibly reflecting increased turnover of dopamine and serotonin, related to this differential incidence?

3. CSF Amine and Metabolite Alterations in Affective Illness

3.1. Norepinephrine

Reasoning from indirect pharmacological data and using relatively nonspecific drugs, investigators postulated that catecholamines, particularly norepinephrine and its metabolites, might be decreased in depression and increased in mania.[29,127] Studies of 3-methoxy-4-hydroxyphenylethylene glycol (MHPG) in CSF have been complicated by factors of assay unreliability at the low nanogram ranges and possible contributions from the spinal cord and periphery, as well as by the clinical–methodological factors reviewed above. As reviewed in Table 4, there is no consistent evidence for a deficit of either MHPG or vanillylmandelic acid (VMA) in depression or an excess of these compounds in mania, although several studies consistent with such a formulation are noted. In essentially every study, values in manic patients were higher than in depressed patients. The completeness of bedrest in manic patients is difficult to verify, but in our studies in which careful attention to bedrest was maintained, we did not find average CSF MHPG or VMA elevated in manic patients compared to controls. However, in our series of patients studied in collaboration with E. Gordon, the severity of mania correlated positively with CSF MHPG (r = 0.59, N = 13, $p < 0.03$). Ashcroft *et al.*[6] found a doubling of MHPG in agitated as compared to retarded depressed patients.

Our recent studies utilizing a sensitive radioenzymatic assay to measure norepinephrine directly in CSF in collaboration with C. R. Lake have not revealed any evidence of decreased norepinephrine during the depressed phase compared with either neurological or normal controls.[116] As indicated above, however, patients with higher levels of rated anxiety did have higher levels of norepinephrine (see Fig. 3), and the depressed patient population actually had significantly higher CSF norepinephrine than our current series of normal volunteers.

Our findings of higher levels of CSF norepinephrine in mania (Figs. 6 and 7) need to be viewed with particular caution because of increased motor activity and possibly stress during mania. However, in manic patients in whom complete bedrest prior to lumbar puncture was possible, norepinephrine values were still substantially elevated. This suggests the possibility that norepinephrine elevations may be more closely associated with the manic syndrome

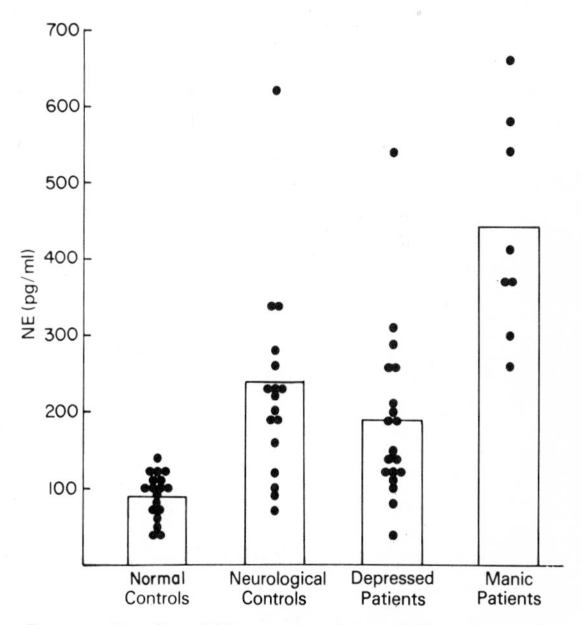

Figure 6. Baseline CSF norepinephrine (NE) in depression and mania. Manic patients had significantly higher baseline levels of CSF NE than depressive, neurological, or normal control patients studied under similar conditions in our laboratory.[116] Recent unpublished data (C. R. Lake, J. C. Ballenger) suggest that absolute values of CSF NE may be 200–400% higher in some individuals if CSF is collected directly into ascorbic acid and frozen immediately at bedside.

Table 3. Differences in Serotonin and Dopamine but Not Norepinephrine in Females Compared to Males

Patient population	Female values	Male values	Level of significance
Serotonin metabolite 5-HIAA (ng/ml)			
Baseline			
Depressed	23.7 ± 1.6 (*N* = 42)	20.9 ± 1.6 (*N* = 27)	$p < 0.10$
Manic	32.5 ± 3.9 (*N* = 12)	24.2 ± 3.4 (*N* = 11)	
Probenecid			
Depressed	161.8 ± 9.7 (*N* = 38)	123.4 ± 7.0 (*N* = 27)	$p < 0.001$
Manic	184.3 ± 13.1 (*N* = 17)	104.4 ± 14.1 (*N* = 5)	
Schizophrenic	163 ± 14 (*N* = 20)	115 ± 10 (*N* = 11)	$p < 0.05$
Dopamine metabolite HVA$_a$ (ng/ml)			
Baseline			
Depressed	32.6 ± 4.7 (*N* = 17)	26.6 ± 2.1 (*N* = 19)	N.S.
Manic	51.4 ± 10.3 (*N* = 9)	40.9 ± 10.6 (*N* = 7)	
Probenecid			
Depressed	230.5 ± 15.1 (*N* = 21)	197.7 ± 14.6 (*N* = 22)	$p < 0.10$
Manic	250.8 ± 26.0 (*N* = 11)	214.8 ± 37.5 (*N* = 5)	
Schizophrenic	249 ± 18 (*N* = 21)	182 ± 20 (*N* = 9)	$p < 0.05$
Norepinephrine metabolite MHPG$_a$ (ng/ml)			
Baseline			
Depressed	11.6 ± 1.6 (*N* = 16)	11.5 ± 1.1 (*N* = 12)	N.S.
Manic	11.4 ± 2.7 (*N* = 9)	9.8 ± 1.1 (*N* = 4)	
Probenecid			
Depressed	12.7 ± 1.0 (*N* = 18)	15.3 ± 1.0 (*N* = 22)	N.S.
Manic	17.8 ± 3.6 (*N* = 11)	15.8 ± 2.8 (*N* = 5)	
Schizophrenic	22.8 ± 2.0 (*N* = 16)	21.4 ± 3.1 (*N* = 8)	N.S.
Norepinephrine (pg/ml)			
Baseline			
Depressed	205.6 ± 45.7 (*N* = 10)	144.4 ± 17.2 (*N* = 14)	N.S.
Manic	279.2 ± 76.8 (*N* = 6)	512.7 ± 107.0 (*N* = 3)	
Probenecid			
Depressed	373.7 ± 126.2 (*N* = 7)	250.9 ± 27.0 (*N* = 14)	N.S.
Manic	660.7 ± 62.7 (*N* = 3)	856.3 ± 200.2 (*N* = 8)	

[a] Analyzed by mass fragmentographic technique. (5-HIAA) 5-Hydroxyindoleacetic acid; (HVA) homovanillic acid; (MHPG) 3-methoxy-4-hydroxyphenylethylene glycol.

Table 4. Noradrenergic Mechanisms in Affective Illness: MHPG and VMA in CSF

Investigators	Controls	Depressed	Depressed recovered	Manic
Baseline CSF MHPG[a]				
Wilk *et al.* (1972)[159]	16 ± 4.2 (*N* = 24)	17.6 ± 1.2 (*N* = 8)	—	31.6 ± 5.8[b] (*N* = 11)
Post *et al.* (1973)[113]	15.1 ± 3.6 (*N* = 44)	10.2 ± 2.4[d] (*N* = 55)	—	15.4 ± 5.5 (*N* = 26)
Shaw *et al.* (1973)[129]	10.8 ± 2.8 (*N* = 13)	11.9 ± 2.6 (*N* = 22)	10.2 ± 0.67 (*N* = 7)	—
Bertilsson (1973)[13]	—	13.4 ± 3.5[d] (*N* = 14)	—	—
Subrahmanyam (1975)[139]	20.6 ± 2.4 (*N* = 12)	14.2 ± 2.6[c] (*N* = 24)	16.6 ± 3.4 (*N* = 24)	—
Ashcroft *et al.* (1976)[6]	13 ± 8[d] (*N* = 11)	12 ± 6[e] (*N* = 7)	—	29 ± 18[b,d] (*N* = 5)
	—	24 ± 13[b,d] (*N* = 8)	—	—
Van Praag (1977)[144]	—	10.2 ± 1.2 (*N* = 8)	—	—
Vestergaard *et al.* (1978)[154]	10.4 ± 4 (*N* = 21)	12 ± 4 (*N* = 27)	11 ± 4 (*N* = 13)	15 ± 8 (*N* = 4)
F. K. Goodwin, R. M. Post, J. C. Ballenger, E. Gordon, unpublished data (1978)[g]		11.2 ± 0.9 (*N* = 34)	—	10.9 ± 1.9 (*N* = 13)
Baseline CSF VMA				
Jimerson *et al.* (1975)[74]	1.94 ± 0.25 (*N* = 16)	1.46 ± 1.51 (*N* = 32)	—	1.264 ± 0.65 (*N* = 12)
Probenecid CSF MHPG				
F. K. Goodwin, R. M. Post, J. C. Ballenger, E. Gordon, unpublished data (1978)[g]	—	14.1 ± 0.7 (*N* = 36)	—	17.2 ± 2.6 (*N* = 16)
Probenecid CSF VMA				
Jimerson *et al.* (1975)[74,g]	2.55 ± 0.46 (*N* = 13)	2.00 ± 1.16 (*N* = 35)	—	2.75 ± 1.56 (*N* = 15)

[a] (MHPG) 3-Methoxy-4-hydroxyphenylethylene glycol; (VMA) vanillylmandelic acid.
[b] Significantly higher than controls, $p < 0.05$. [c] Significantly lower than controls, $p < 0.05$. [d] Standard deviation. [e] Retarded. [f] Agitated.
[g] Determinations by mass fragmentographic techniques in collaboration with E. Gordon.

and not just an artifact of increased motor activity. The evidence of increased norepinephrine in CSF without concomitant elevations of CSF MHPG in our manic patients compared to controls deserves further study and may indicate differential release, reuptake, or transport characteristics of these two substances that could be of physiological significance.

Another possible marker for norepinephrine metabolism is the activity of its biosynthetic enzyme dopamine-β-hydroxylase (DBH) as measured in CSF. DBH catalyzes the conversion of dopamine to norepinephrine in noradrenergic neurons and has been measured in both serum and CSF. When norepinephrine is released from storage vesicles in the peripheral nervous system and adrenal medulla by exocytosis, DBH is also released and therefore may reflect norepinephrine release.[135,155] Although CSF DBH increases with peripheral nerve (sciatic) stimulation, there is little evidence available on the effect of central noradrenergic stimulation on DBH release into CSF.[44] However, brain levels of DBH decrease with ablation of locus coeruleus noradrenergic neurons with 6-OH-dopamine in rat.[40]

Several treatments that decrease or increase unit firing in the locus coeruleus alter CSF DBH in the same direction. Clonidine decreases firing and CSF DBH, and phenoxybenzamine increases firing and CSF DBH, suggesting a possible association of firing of central noradrenergic neurons and CSF DBH.[83] CSF DBH is correlated with CSF norepinephrine in

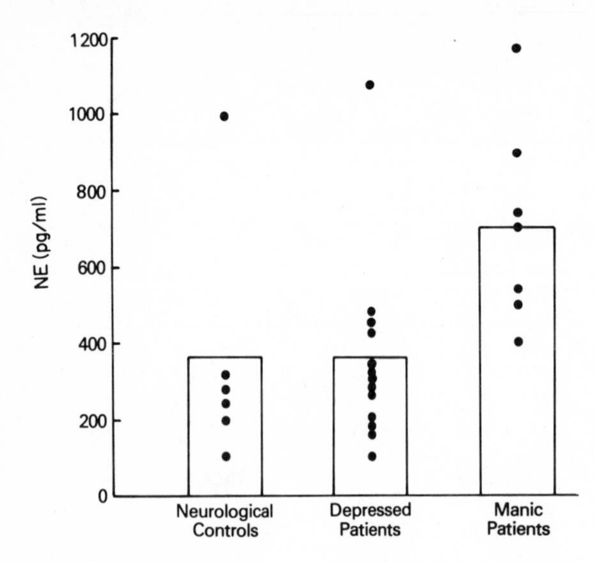

Figure 7. CSF norepinephrine (NE) following probenecid loading. CSF NE accumulations following probenecid correlated highly with baseline values in depressed patients ($r = 0.89$, $p < 0.001$), further supporting stability and reliability of these measures of noradrenergic function in separate assays and in lumbar puncture studies. Values after probenecid were again higher in manic patients than in depressive or control patients.[116]

depressed patients ($r = 0.80$, $N = 12$, $p < 0.005$) (P. Lerner *et al.*, unpublished data).

Serum DBH has been reported to be normal in affectively ill patients,[82] although psychotically depressed unipolar patients have been reported to have low serum DBH.[94] Grote *et al.*[70] found no DBH differences from controls in various brain areas in autopsy material from five depressed suicide victims.

We have measured CSF DBH in unipolar and bipolar affectively ill patients and normal controls[84] (see Table 5). Bipolar depressed patients tended to have higher CSF DBH than normal volunteers ($p < 0.05$). Manic patients had significantly lower DBH activity than bipolar depressed ($p < 0.002$) and euthymic patients ($p < 0.01$). DBH activity in manics was also nonsignificantly lower ($p < 0.10$) than in normals. Thus, paradoxically in relation to current theories and observed CSF norepinephrine elevations, DBH tended to be low in the CSF of manic patients. It is not clear whether the low values of DBH in mania are related to release and depletion of DBH, feedback decreases in synthesis, or altered CSF transport characteristics in relation to norepinephrine. Further exploration of the correlation of DBH and norepinephrine and its metabolites appears indicated.

Major *et al.*[90] reported an association between antidepressant response to monoamine oxidase inhibitors (MAOIs) and alterations in CSF norepinephrine in depressed patients. They found a high negative correlation between changes in CSF norepinephrine during MAOI treatment and global depression rating changes ($r = -0.95$, $p < 0.001$), such that increases in CSF norepinephrine were associated with the greatest improvement in depression.

In summary, definitive evidence is not yet available for alterations in noradrenergic tone in affective illness. However, in most studies reported to date, CSF MHPG and VMA values during mania are slightly higher than those observed during depression (see Table 4), and CSF norepinephrine is substantially higher in manics than in depressives or controls. Findings such as those of Major *et al.*[90] associating degree of clinical improvement with alterations in CSF norepinephrine are also highly suggestive of a catecholaminergic involvement in the control of affective tone even if baseline values are not necessarily different from controls. Depending on syndrome presentation and severity (Fig. 8), a subgroup of affectively ill patients may have significantly altered noradrenergic function in relation to controls. Alterations within a normal functional range, whatever the initial starting values, may also be of behavioral relevance (see Table 9).

3.2. Dopaminergic Function as Reflected in CSF

3.2.1. Medication-free Evaluation of Depressed and Manic Patients. Levels of homovanillic acid (HVA), the major dopamine metabolite in CSF, have been measured in depressed and manic patients,

Table 5. CSF DBH in Affectively Ill and Normal Populations[a]

Patient population	Number of patients	DBH[b] value[c] (nmol/ml/hr)
Depressed	23	0.74 ± 0.07
Unipolar	12	0.63 ± 0.08
Bipolar	11	0.86 ± 0.10
Recovered (bipolar)	6	0.76 ± 0.14
Manic	13	0.43 ± 0.07
Normal	28	0.60 ± 0.05

[a] P. Lerner, J. C. Ballenger, R. M. Post, L. F. Major, F. K. Goodwin, unpublished data (1979). DBH in manic patients was lower than in bipolar patients ($p < 0.002$), than in recovered patients ($p < 0.05$), than in depressed patients ($p < 0.005$), than in normal volunteers ($p < 0.10$); DBH in bipolar patients was higher than in normal volunteers ($p < 0.05$), than in manic patients ($p < 0.10$).
[b] (DBH) Dopamine-B-hydroxylase.
[c] Values are means ± S.E.M.

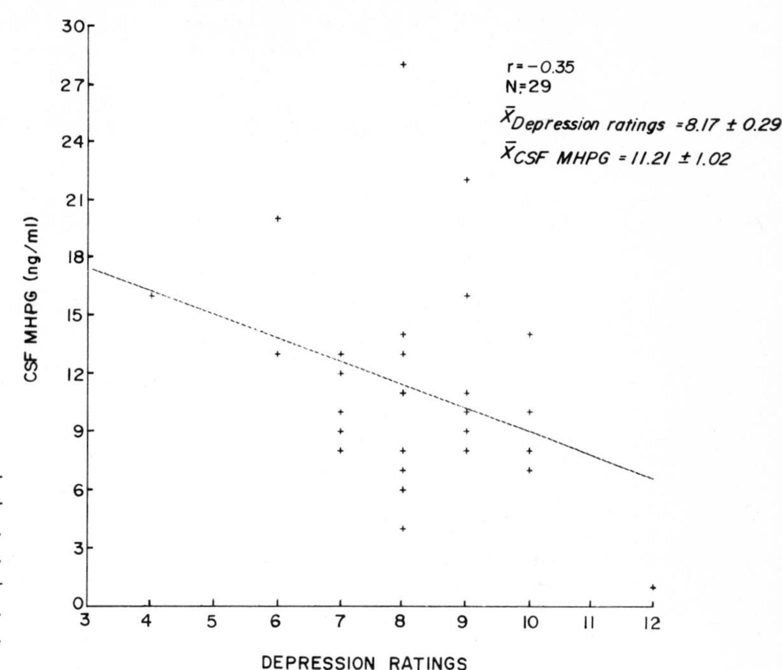

Figure 8. Correlation between depression ratings and CSF 3-methoxy-4-hydroxyphenylethylene glycol (MHPG) in depressed patients. Weak, but statistically significant ($p < 0.05$), negative correlation of CSF MHPG with severity of global depression rating is illustrated. MHPG was measured in collaboration with E. Gordon (unpublished data, 1978).

with and without probenecid, and are summarized in Tables 6A and 6B. Baseline levels of HVA are not consistently lower in undifferentiated groups of depressed patients compared to controls. However, Papeschi and McClure[98] and van Praag and Korf[147] reported low HVA in retarded depressed patients, and Banki[9] found lower values in bipolar, but not unipolar, depressives. In our studies, baseline HVA levels showed a trend toward an inverse relationship with severity of depression in all patients ($r = -0.28$, $N = 36$, $p < 0.10$, two-tailed Student's t test) and was significant in females ($r = -0.50$, $N = 17$, $p < 0.04$). In most studies (Table 6A), HVA levels were nonsignificantly higher in manics compared to depressed patients or controls.

Table 6B illustrates values of HVA following probenecid administration. Several studies report decreased probenecid accumulation of HVA in depressed patients compared to controls. Only one study reports significant elevations in probenecid-induced accumulations of HVA in manic patients compared to controls, although again most values in mania were nonsignificantly higher than those in depressed patients. HVA accumulations also did not significantly decrease when manics were restudied in a depressed or euthymic medication-free state, as illustrated in Fig. 9. However, in response to treatment with neuroleptic agents, HVA in-

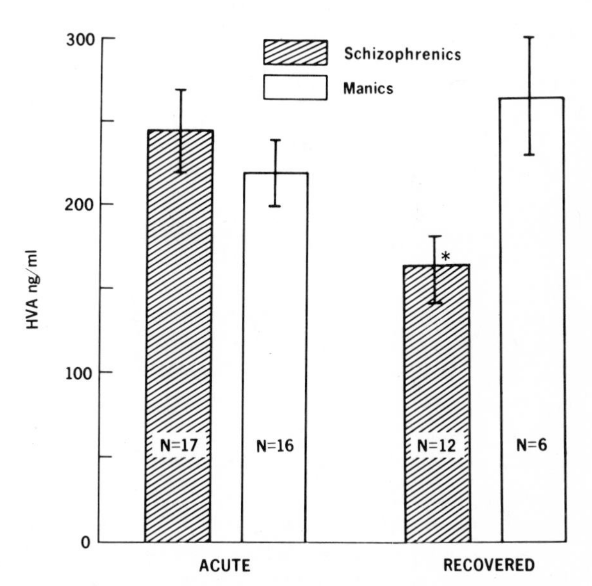

Figure 9. Differences in homovanillic acid (HVA) accumulation in recovered schizophrenic and manic patients. Probenecid-induced accumulations of dopamine metabolite HVA were not different in medication-free schizophrenic and manic patients. When patients were restudied during medication-free, improved states, schizophrenic patients showed significantly lower HVA compared to their ill states ($p < 0.05$), while recovered manic patients showed no change.[107]

Table 6A. Dopamine Metabolism in Affective Illness—HVA[a] in CSF: Baseline Values

Investigators	Controls[b]	Depressed[b]	Manic[b]
Roos and Sjöström (1969)[124]	44 ± 31 (7)	29 ± 7 (6)	41 ± 23 (7)
Bowers *et al.* (1969)[22]	—	22.7 ± 14.1[c] (8)	22.2 ± 16.3[c] (7)
Papeschi and McClure (1971)[98]	50 ± 6 (18)	19 ± 4[d] (17)	—
Van Praag and Korf (1971)[147]	42 ± 16 (12)	39 ± 16 (20)	—
	—	32 ± 8[e] (8, retarded)	—
Sjöström and Roos (1972)[134]	36 ± 4 (23)	34 ± 4 (27)	59 ± 6[f] (36)
Mendels *et al.* (1972)[95]	—	27.1 ± 22 (12)	44.2 ± 8.5 (2)
Goodwin *et al.* (1973)[64]	22.4 ± 2.4 (28)	15.2 ± 2.1 (49)	25.7 ± 4.3 (16)
Sjöström (1973)[132]	34 ± 3 (38)	24 ± 3 (16)	59 ± 10[d] (11)
Brodie *et al.* (1973)[25]	—	Low[e]	—
Ashcroft and Glen (1974)[7]	41 ± 23[c] (31)	34 ± 16[c] (9)	35 ± 15[c] (11)
Takahashi *et al.* (1974)[141]	37.5 ± 3.7 (30)	33.7 ± 3.7 (30)	—
Subrahmanyam (1975)[139]	40.2 ± 4 (12)	38.4 ± 3.4 (24)	—
Banki (1977)[9]	33.4 ± 1.0 (32)	24 ± 1.5 (55, unipolar)	44.4 ± 3.8[f] (10)
	—	15.9 ± 2.0[e] (16, bipolar)	—
Vestergaard *et al.* (1978)[154]	45 ± 26 (23)	83 ± 37[d] (29)	115 ± 66[f] (4)
F. K. Goodwin, R. M. Post, R. H. Gerner, E. Gordon, unpublished data (1978)	28.2 ± 3.89 (16)	29.4 ± 2.5 (36)	48.8 ± 7.3[f] (16)

[a] (HVA) Homovanillic acid.
[b] Values are means ± S.E.M. (in ng/ml), with the numbers of patients in parentheses.
[c] Standard deviation. [d] Significantly lower in retarded depressed patients only. [e] Significantly lower. [f] Significantly higher.

creases were larger in manic compared to nonmanic patients (Fig. 10).

Subgroup issues may be of particular importance in relation to these data. For example, Korf and van Praag[79] reported decreased HVA accumulations in retarded but not agitated depressed patients. Earlier, they reported that patients with low HVA also responded better to antidepressant treatment with levodopa. These data are interesting in light of two subsequent reports of possible associations of low HVA and pharmacological response. Van Scheyen *et al.*[153] reported that depressed patients with lower baseline HVA measured in cisternal CSF responded better to the dopamine active agent nomifensine. Post *et al.*[108] reported that low HVA [or 5-hydroxyindoleacetic acid (5-HIAA)] correlated with antidepressant responses to the dopamine agonist piribedil. Gerner *et al.*[55] found an association between low baseline (but not probenecid) HVA in CSF and subsequent sleep-deprivation response. They reported that those with lower HVA values prior to sleep deprivation showed the most positive mood change following this experimental treatment strategy (Fig. 11).

Thus, while HVA levels and probenecid-induced accumulations in affectively ill patients in relation

Table 6B. Dopamine Metabolism in Affective Illness—HVA[a] in CSF: Probenecid-Induced Accumulations

Investigators	Controls[b]	Depressed[b]	Manic[b]
Ross and Sjöström (1969)[124]	86 ± 43[c] (7)	48 ± 19[c,d] (6)	94 ± 57[c] (7)
Korf and van Prrag (1971)[79]	91 ± 26 (12)	106 ± 36 (12, nonretarded)	—
	—	53 ± 32 (8, retarded)[d]	—
Sjöström and Roos (1972)[134]	(240% increase)	(83%[d] increase)	(73%[d] increase)
Sjöström (1973)[132]	76 ± 14 (12)	44 ± 8[d] (7)	75 ± 15 (9)
Goodwin *et al.* (1973)[64]	—	204 ± 17 (26)	226 ± 35 (8)
Van Praag *et al.* (1973)[151]	170 ± 47.6 (12)	127 ± 68.4 (28)	—
Bowers (1974),[16] (1976)[19]	—	116 ± 11 (12)	169 ± 13 (10)
Berger *et al.* (1979)[12]	—	Decreased[d]	—
F. K. Goodwin, R. M. Post, R. H. Gerner, E. Gordon, unpublished data (1978)	164.8 ± 13.2 (14)	213.7 ± 47 (43)	239.6 ± 21.1[e] (16)

[a] (HVA) Homovanillic acid.
[b] Values are means ± S.E.M. (in ng/ml), with the numbers of patients in parentheses.
[c] Standard deviation. [d] Significantly lower. [e] Significantly higher.

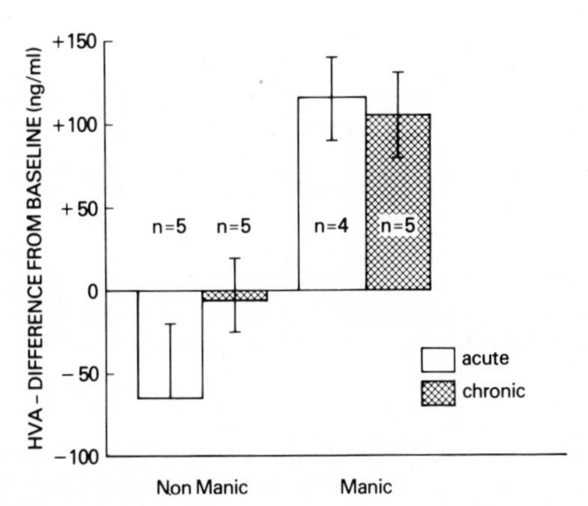

Figure 10. Differential homovanillic acid (HVA) response to pimozide in patients with and without mania. Patients rated as manic by blind nurse raters showed significantly greater probenecid-induced accumulations of HVA on pimozide than nonmanic patients (depressed and schizophrenic). Values during acute (4–10 days) and chronic (20–30 days) pimozide were subtracted from those obtained in the same patients during medication-free probenecid studies. A similar pattern of increased HVA accumulations in manic compared to nonmanic patients on neuroleptics was observed during treatment with thioridazine or chlorpromazine.[114]

to controls are not definitively altered, there are suggestive data across finer subgroupings and in relation to pharmacological responsivity that dopaminergic tone may be important in some aspects of the manic and depressive syndrome.

In collaboration with E. Gordon, we have found substantially lower levels of 3,4-dihydroxyphenylacetic acid (DOPAC) than of HVA in the CSF of affectively ill patients. Baseline levels of DOPAC were 1.54 ± 1.1 ng/ml in 27 depressed patients, and values increased markedly to 8.80 ± 4.5 ng/ml ($N = 23$) following probenecid. These concentrations are in a range similar to that observed in schizophrenic patients (E. Gordon, D. P. van Kammen, unpublished data), although definitive comparison with normals is not yet available. In 20 depressed patients, probenecid-induced accumulations of DOPAC were positively correlated with those of HVA ($r = 0.47$, $N = 20$, $p < 0.04$).

The recent development of assays sensitive enough to measure levels of dopamine in CSF[126] should also help clarify the role of dopaminergic mechanisms in the regulation and modulation of affective tone and psychomotor activity in both the depressive and the manic syndromes.

Particularly in regard to mania, there is substantial pharmacological evidence of a linkage of dopamine

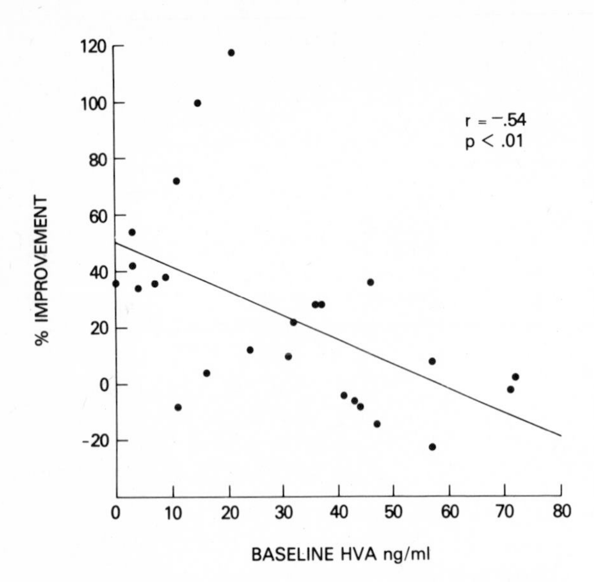

Figure 11. Low pretreatment homovanillic acid (HVA) in CSF and antidepressant response to sleep deprivation. Depressed patients with lowest levels of HVA in CSF show greatest mood improvement (nurse and self ratings) following sleep deprivation. Moderately to severely depressed patients show paradoxical antidepressant response to loss of one night's sleep. Responses, while dramatic in some patients, are usually transient, and patients often relapse following one night's recovery sleep, again in contrast to normal volunteers, who feel better after recovery sleep. Low HVA in CSF may be biological marker of those responding best to sleep deprivation. Low HVA could reflect low dopamine function, decreased metabolism or transport, or even altered circadian rhythms in these patients.[54]

in mania. A variety of neuroleptic compounds including pimozide appear to possess antimanic efficacy.[54,114] Lithium carbonate, the treatment of choice for manic illness, has recently been shown to have a wide variety of effects on dopaminergic neurotransmitter function and pre- and postsynaptic receptors.[52,101,102] Certain alterations in dopaminergic-receptor function that might be effectively treated by neuroleptics or lithium may not be adequately reflected in levels of HVA in CSF. Thus, failure to show consistent alterations in CSF HVA in depression and mania is not strong evidence against possible dopaminergic involvement in some phases of affective illness. A further complication of interpretation is that rather marked alterations could be occurring locally in limbic or cortical areas and not be reflected in CSF because of the larger contribution of HVA from the caudate nucleus,[137] which lines the lateral ventricles. Adaptive and "normalizing" changes could also occur over time (see below and Fig. 12).

3.2.2. Pharmacological Studies HVA alterations during drug treatment with several dopamine active compounds have been studied in affectively ill patients. As might be expected, levodopa, the precursor of dopamine, is associated with markedly increased accumulations of HVA in CSF, while the inhibitor of tyrosine hydroxylase α-methyl-*p*-tyrosine (AMPT) substantially reduces HVA levels in CSF (see Fig. 5).

As illustrated in Fig. 12, the neuroleptics, presumably secondary to their dopamine-receptor-blocking action, are initially associated with increased accumulations of HVA, consistent with animal studies reporting increased dopamine turnover following acute neuroleptic administration. However, with more chronic neuroleptic treatment, HVA accumulations tend to fall back toward pretreatment levels.[53,110,128,143] Bowers *et al.*[24] recently reported that HVA values on chronic neuroleptics remained elevated over baseline, but higher doses of drug were associated with the lowest HVAs, consistent with a tolerance phenomenon.

In contrast to the dopamine blockers, the dopamine agonist piribedil is associated with decreased HVA accumulations.[108] It is of interest that carbamazepine, an agent recently shown to be of use in treatment of both the manic and the depressive phase of the illness, also reduces HVA in CSF.[103]

3.3. γ-Aminobutyric Acid

Several investigators have suggested that a deficiency in γ-aminobutyruc acid (GABA) could account for the postulated dopaminergic hyperactivity in mania or schizophrenia. Our data in affective illness are inconsistent with this formulation, however.[116a] Values in depressed (213.6 ± 13.7, $n = 16$) and manic (211.4 ± 24.8, $n = 8$) patients were not significantly different from 41 normal subjects (231.8 ± 12.5 pmol/ml). In addition, in one rapidly cycling patient, GABA was significantly higher in the manic phase of illness (252.0 ± 3.7) than during the patient's depressions (195.6 ± 13.6).

GABA in CSF decreased with age in normal and depressed females ($r = -0.56$, $n = 25$, $p < 0.01$). Cerebrospinal fluid GABA was positively correlated with CSF MHPG ($r = 0.45$, $p < 0.02$) and CSF norepinephrine ($r = 0.41$, $p < 0.02$) and negatively correlated with CSF calcium and magnesium in the normal subjects. Carbamazepine, an anticonvulsant

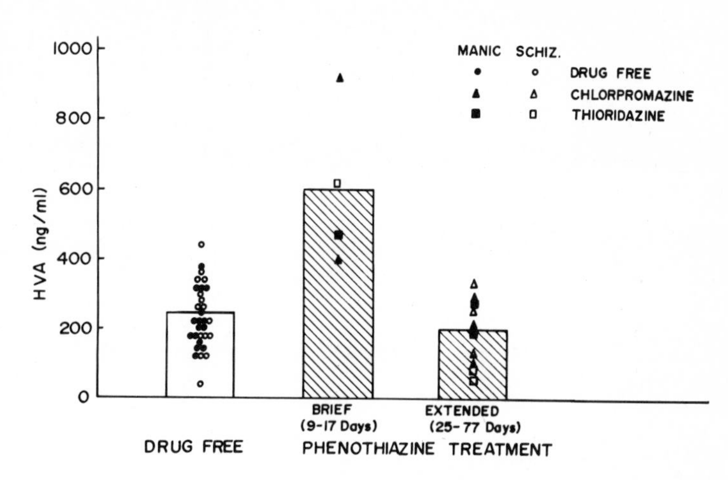

Figure 12. Effect of duration of phenothiazine treatment on homovanillic acid (HVA) accumulation in CSF. Patients treated with neuroleptic drugs acutely, but not chronically, have substantial HVA elevations after probenecid compared to medication-free values.[110] These findings are consistent with data in animals indicating that partial to complete tolerance develops to neuroleptic-induced increases in striatal dopamine turnover. Hypothalamic, mesolimbic, and mesocortical dopaminergic systems apparently develop tolerance less rapidly and completely than nigroneostriatal systems, which make the major contribution to CSF HVA. (SCHIZ.) Schizophrenic.

ness,[7a,b] did not significantly alter CSF GABA in nine affectively ill patients.

Our findings differ from those of Gold *et al.*[55a] who reported significantly lower GABA levels in depressed patients compared to neurological controls. They did not find GABA decreased in schizophrenia, as reported by van Kammen *et al.*[141a] Faull *et al.*[49] reported CSF GABA levels in six schizophrenics and one manic–depressive in a range similar to that in three Huntington's chorea patients (0.25 – 0.59 μmol/liter).

3.4. 5-Hydroxyindoleacetic Acid in Depression and Mania

3.4.1. Serotoninergic Continuum: Relationship to Clinical Phenomenology and Pharmacological Response. A number of investigators have postulated decreases in serotoninergic tone during depression. As illustrated in Tables 7A and 7B, baseline 5-HIAA levels and accumulations following probenecid tend to be significantly lower in depressed patients compared to controls in about half the studies reported.

The distribution of 5-HIAA in depressed patients and its relationship to psychiatric symptomatology and subsequent pharmacological response provides further data to support the suggestion of an alteration in serotoninergic tone in depression. For example, Asberg *et al.*[3,4] reported a bimodal distribution of baseline 5-HIAA in CSF. Those with low 5-HIAA had a higher incidence of serious suicide attempts, although a greater proportion of males in this low subgroup could partially explain these findings, since 5-HIAA is lower in males. Our own data following probenecid are illustrated in Fig. 13 and suggest a bimodal distribution in males but not in females. It is of some interest that Coppen *et al.*[38] reported in depressed and manic patients low 5-HIAA that did not increase following improvement. As illustrated in Fig. 14, 5-HIAA accumulations in our depressed patients appear to be a relatively stable variable over time and across clinical state. Mean values in our depressed patients were not significantly altered when they were restudied several months later when clinically improved during a 2-week medication-free period.

Values of these 11 patients studied during the depressed and improved state were positively correlated ($r = 0.74$, $p < 0.01$). These data are consistent with the view that CSF 5-HIAA may reflect a relatively stable measure of serotoninergic function in the same individual. Prange *et al.*[118] formulated a "permissive hypothesis," suggesting that a low level of serotoninergic function during both depressed and manic phases might predispose to mood swings in either direction. Also consistent with this formulation are the recent data of van Praag.[145] He noted that depressed patients with low 5-HIAA (who were clinically indistinguishable from those with higher 5-HIAA) showed better acute antidepressant responses to 5-hydroxytryptophan (5-HTP) or clomipramine alone or in combination. In subsequent work, he also found a higher incidence of 5-HTP long-term prophylaxis of depression in those with lower pretreatment CSF 5-HIAA. Thus, although 5-HIAA may not be low in all depressed patients (Table 7), it may be that those patients at the low end of the continuum have metabolically and behaviorally relevant alterations in the serotoninergic system that may relate either to a predisposition to depressive illness or to subsequent pharmacological responsivity, or to both.

Table 7A. Serotonin Metabolism in Affective Illness—5-HIAA[a] in CSF: Baseline Values

Investigators	Controls[b]	Depressed[b]	Manic[b]
Fotherby *et al.* (1963)[50]	11.5 ± 4.1[c] (11)	12.2 ± 8.2[c] (11)	—
Dencker *et al.* (1966)[43]	—	—	Low (6)
Ashcroft *et al.* (1966)[5]	19.1 ± 4.4[c] (21)	11.1 ± 3.9[c,d] (24)	18.7 ± 5.4[c] (4)
Bowers *et al.* (1969)[22]	39.5 ± 13.1[c] (8)	34.0 ± 11.5[c] (8)	32.0 ± 10.3[c] (8)
Van Praag and Korf (1971)[147]	40 ± 24 (11)	17 ± 17[d] (14)	—
Papeschi and McClure (1971)[98]	28 ± 3 (10)	22 ± 2 (12)	—
Coppen *et al.* (1972)[38]	42.3 ± 14[c] (20)	19.8 ± 8.5[c,d] (31)	19.7 ± 6.8[c] (18)
	—	19.9 ± 7.2[d,e] (8)	—
Roos (1972)[123]	29 ± 7[c] (26)	31 ± 8[c] (17)	36 ± 9[c] (19)
Mendels *et al.* (1972)[95]	—	12.9 ± 6.0 (2)	17.1 ± 14.6 (4)
McLeod and McLeod (1972)[93]	32.6 ± 11.4 (12)	20.5 ± 12.1[d] (25)	—
Goodwin *et al.* (1973)[64]	27.3 ± 1.6 (29)	25.5 ± 1.3 (58)	28.7 ± 2.5 (16)
Sjöström (1973)[133]	29 ± 1 (39)	30 ± 1 (23)	33 ± 2 (15)
Brodie *et al.* (1973)[25]	—	Nonsignificant	—
Ashcroft and Glen (1974)[7]	16 ± 8[c] (30)	18 ± 8[c] (9, bipolar)	15 ± 6[c] (11)
	—	10 ± 4[d] (11, unipolar)	—
Takahashi *et al.* (1974)[141]	30.4 ± 2.1 (30)	20.1 ± 1.8[d] (30)	—
Subrahmanyam (1975)[139]	40.6 ± 4.2 (12)	26.2 ± 4.2 (24)	—
	—	34.6 ± 4.8[e] (24)	—
Banki (1977)[9]	27.5 ± 1.2 (32)	14.6 ± 1.7[d] (16)	13.9 ± 2.9 (10)
Vestergaard *et al.* (1978)[154]	28 ± 28 (22)	29 ± 12 (28)	40 ± 16 (4)
F. K. Goodwin, R. M. Post, unpublished data (1978)	—	22.6 ± 1.1 (70)	28.5 ± 2.7 (23)

[a] (5-HIAA) 5-Hydroxyindoleacetic acid.
[b] Values are means ± S.E.M. (in ng/ml), with the numbers of patients in parentheses.
[c] Standard deviation. [d] Significantly lower. [e] Values for improved depressed patients.

This formulation is also supported by the findings of Asberg *et al.*[2] reporting decreased responsiveness to the predominantly noradrenergic tricyclic nortriptyline in patients with low 5-HIAA. Goodwin *et al.*[63,65] reported similar findings of decreased imipramine response in those with low 5-HIAA. At the same time, antidepressant response to piribedil[108] has also been associated with low 5-HIAA.

Low 5-HIAA in CSF has also been related to alterations in clinical profiles of pharmacological response to other agents. For example, Jimerson *et al.*[75] reported greater long-term dysphoric response

Table 7B. Serotonin Metabolism in Affective Illness—5-HIAA[a] in CSF: Probenecid-Induced Accumulations

Investigators	Controls[b]	Depressed[b]	Manic[b]
Roos and Sjöström (1969)[124]	46 ± 13[c] (11)	38 ± 11[c] (17)	43 ± 13[c] (19)
Van Praag *et al.* (1970)[150]	74 ± 26 (11)	39 ± 25[d] (14)	—
Korf and van Praag (1971)[79]	67 ± 25 (15)	40 ± 27[d] (15)	—
Sjöström and Roos (1972)[134]	(66% increase) (12)	(27%[d] increase) (24)	(20% increase) (21)
Sjöström (1973)[133]	52 ± 4 (21)	37 ± 4[d] (11)	40 ± 4 (10)
Goodwin *et al.* (1973)[64]	—	132 ± 7 (26)	133 ± 12 (8)
Van Praag *et al.* (1973)[151]	110 ± 46.7 (12)	76 ± 43.2[d] (28)	—
Bowers (1976)[19]	—	106 ± 14 (12)	120 ± 11 (10)
Banki (1977)[9]	67 ± 3.0 (30)	40 ± 2.1[d] (30)	—
Berger *et al.* (1979)[12]	—	Nonsignificant	—
F. K. Goodwin, R. M. Post, unpublished data (1978)	—	143 ± 54.0[c] (70)	164.6 ± 57.7[c] (24)

[a] (5-HIAA) 5-Hydroxyindoleacetic acid.
[b] Values are means ± S.E.M. (in ng/ml), with the numbers of patients in parentheses.
[c] Standard deviation. [d] Significantly lower.

to amphetamine in those depressed patients with low 5-HIAA. These data are consistent with a wide variety of animal studies suggesting that dopaminergic–serotoninergic balance is important for the modulation of psychomotor-stimulant and other dopamine-active behavioral effects. Of possible relevance to the low-serotonin permissive hypothesis of mania, Brown *et al.*[26] (Fig. 15) have reported increased aggressive acts, outbursts, and difficulties with the law in 26 patients with personality disorders who had low 5-HIAA in CSF ($r = -0.77$, $p < 0.001$) and high MHPG ($r = 0.65$, $p < 0.05$). Kirstein *et al.*[77] also reported that low 5-HIAA correlated with increased motor agitation in schizophrenic patients, and Banki[9] found lower 5-HIAA correlated with increased insomnia. Perhaps some components of the manic syndrome can be viewed as a "release" phenomenon in which low inhibitory serotoninergic tone insufficiently modulates other neurochemical systems mediating psychomotor activation, aggressivity, and other symptoms associated with affective dysregulation.

These data, potentially linking CSF measures of serotoninergic and other neurotransmitter function to a variety of clinical parameters and pharmacological responses, are not only of particular theoretical importance, but also may be of increasing practical clinical importance. The development of biological markers in psychiatric patients for drug responsivity would provide a major breakthrough for the drug treatment of these disorders. Since many of the antidepressant drugs require 3–6 weeks of administration before an adequate response can be observed, the choice of the proper drug from the outset becomes an extremely important clinical treatment variable. It is hoped that just as in the pneumonia syndromes, where the appropriate biochemical test for a bacterial, viral, or tubercular process leads to the proper choice of treatment modality, markers in the CSF of depressed and manic patients could lead to the appropriate choice of the treatment agent.

3.4.2. Pharmacological Studies. Treatment with the serotonin precursor L-tryptophan increases CSF 5-HIAA, while use of the tryptophan hydroxylase inhibitor *p*-chlorphenylalanine (PCPA) decreases 5-HIAA (see Fig. 5). A variety of tricyclic antidepressant treatments have been reported to consistently decrease 5-HIAA in CSF (Table 8). Treatment with MAOIs, which would decrease the catabolism of

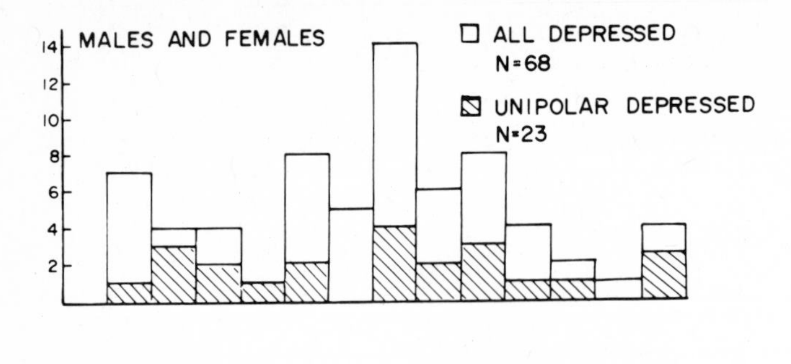

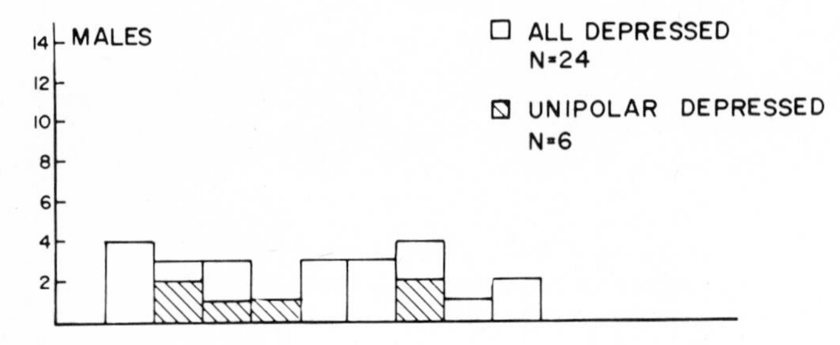

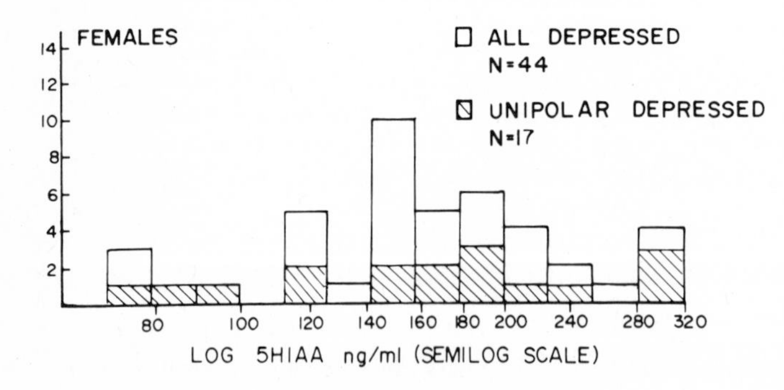

Figure 13. Frequency distribution of 5-hydroxyindoleacetic acid (5-HIAA) in CSF of depressed patients. A trend toward bimodal distribution of probenecid-induced accumulations of CSF 5-HIAA is observed in depressed patients. Lower 5-HIAA values in males compared to females should be noted. Probenecid (100 mg/kg) was administered orally in four divided doses beginning 18 hr prior to lumbar puncture, as previously described.[60,64]

intraneuronal serotonin to 5-HIAA, is also associated with decreases in CSF 5-HIAA.[23,80,91] Reports of the effects of other antidepressant treatments such as lithium and electroconvulsive therapy on 5-HIAA are inconsistent.[21,51,62,66,76,125] The decreases in 5-HIAA in CSF following tricyclic antidepressant treatment have been attributed to several mechanisms including blockade of reuptake and subsequent decreased availability of serotonin to intraneuronal MAO or presynaptic compensations in relation to increased serotonin in the synaptic cleft. Recently, de Montigny and Aghajanian[42] have described the development of serotoninergic-receptor supersensitivity following tricyclic administration in animals, suggesting the possibility that the serotonin-receptor supersensitivity may be involved in the appearance of decreased 5-HIAA in CSF in patients chronically treated with tricyclics. Militating against this explanation are findings that the 5-HIAA decrease is not time-dependent and appears to be as substantial in the first week of tricyclic administration as it is during the third and fourth weeks (Fig. 16). Since the serotonin supersensitivity developed only after several weeks in the animal studies, the immediate appearance of a decrease in 5-HIAA in depressed patients is not likely to be secondary to a change in receptor supersensitivity, and other mechanisms such as decreased reuptake and availability of serotonin to MAO should be considered.

Our observations of a strong positive correlation between CSF 5-HIAA and CSF norepinephrine in depressed patients[116] are particularly interesting in light of recent findings of a direct relationship between noradrenergic function and firing of the serotonin raphe system.[10] As illustrated in Fig. 17, both baseline and probenecid-induced accumulations of 5-HIAA and norepinephrine were signifi-

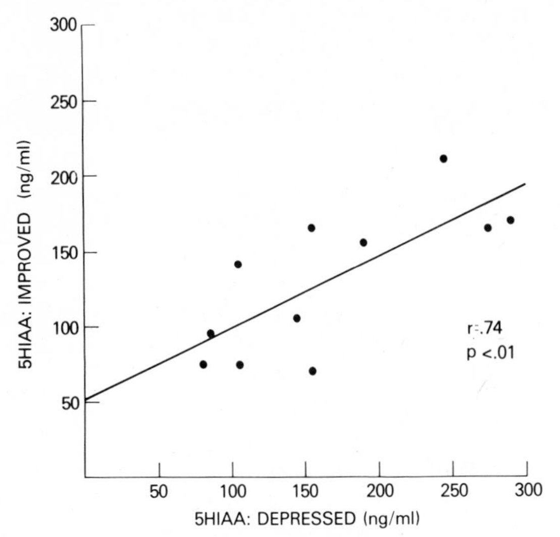

Figure 14. Consistency of CSF 5-hydroxyindoleacetic acid (5-HIAA) accumulations in depressed and improved clinical states. 5-HIAA values in 11 acutely depressed medication-free patients were significantly correlated with those observed when these patients were restudied during period of clinical improvement, again while medication-free. These data, showing that initially low or high 5-HIAA values during depression persist in recovered state, suggest that CSF 5-HIAA alterations in depressed patients may be a state-independent biological marker of serotonin metabolism.

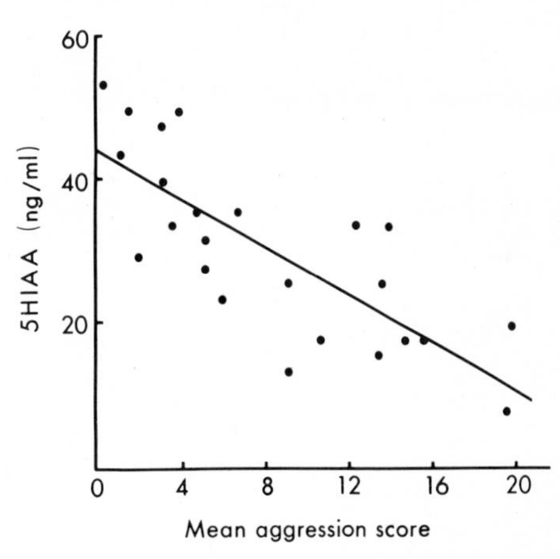

Figure 15. CSF 5-hydroxyindoleacetic acid (5-HIAA) in patients with personality disorders. Patients with higher incidence of aggressive acts, violent outbursts, and difficulties with law had lower CSF 5-HIAA than those with less aggressive difficulties. These patients also had significantly higher CSF 3-methoxy-4-hydroxyphenylethylene glycol (MHPG) levels.[26] Those who were subsequently given administrative discharges from military by totally independent board of inquiry after CSF samples were studied also had significantly lower 5-HIAA than those not discharged. (G. L. Brown, J. C. Ballenger, L. F. Major, P. Goyer, F. K. Goodwin, unpublished data.)

Table 8. Tricyclic Antidepressants That Decrease Baseline and Probenecid 5-HIAA[a] in CSF of Treated Patients

Investigators	Drug	5-HIAA (ng/ml) Pre	During	Decrease (%)
Baseline				
Ashcroft *et al.* (1966)[5]	Imipramine	11.1 ± 3.9	8.8 ± 24.0	21
Bowers *et al.* (1969)[22]	Amitriptyline	34.0 ± 11.5	21.3 ± 7.3	37
Asberg *et al.* (1973)[2]	Nortriptyline	19.2 ± 8.3	14.4 ± 6.8	25
Post and Goodwin (1974)[109]	Amitriptyline	22.8 ± 5.7	18.0 ± 2.4	21
	Imipramine	27.3 ± 3.0	21.2 ± 2.5	22
Bertilsson *et al.* (1974)[14]	Chlormipramine	20.2 ± 7.1	11.2 ± 4.9	45
	Nortriptyline	14.8 ± 5.7	13.3 ± 5.4	10
Walinder *et al.* (1976)[156]	Chlormipramine	23.1 ± 2.56	12.9 ± 1.22	44
Siwers *et al.* (1977)[131]	Zimelidine	18.1 ± 7.5	13.1 ± 4.5	20
Probenecid				
Bowers (1972)[15]	Amitriptyline	62.0 ± 6.0	39.0 ± 4.0	37
Post and Goodwin (1974)[109]	Amitriptyline	136.8 ± 1.7	84.2 ± 15.0	38
Bowers (1974)[17]	Amitriptyline and tryptophan	202.0 ± 55.0	104.0 ± 44.0	49

[a] (5-HIAA) 5-Hydroxyindoleacetic acid.

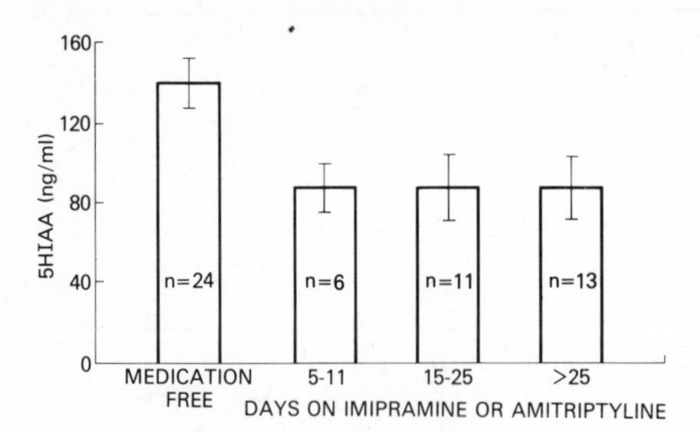

Figure 16. Time-independent decrease in CSF 5-hydroxyindoleacetic acid (5-HIAA) during tricyclic antidepressant treatment. Imipramine and amitriptyline decreased probenecid-induced accumulations of 5-HIAA in CSF of depressed patients. Decreased 5-HIAA was observed following both acute and chronic tricyclic administration.

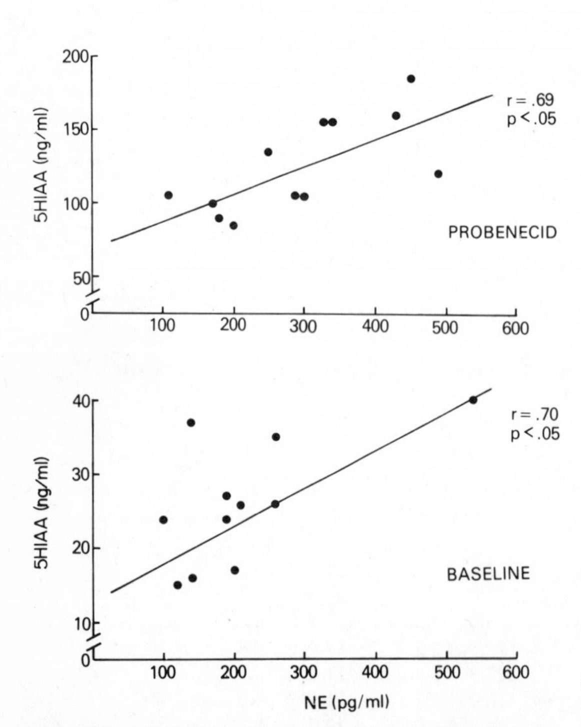

Figure 17. Correlation of norepinephrine and 5-hydroxyindoleacetic acid (5-HIAA) in CSF of depressed patients. Baseline levels and probenecid-induced accumulations of CSF 5-HIAA and norepinephrine were significantly correlated.[116] These data are consistent with recent studies in animals indicating a direct relationship between firing of serotoninergic raphe nucleus and noradrenergic locus ceruleus.[10]

cantly correlated. Although recent studies have emphasized tricyclic antidepressant effects on a variety of neurotransmitter receptor systems, earlier formulations of the antidepressant effects related to blockade of serotonin or norepinephrine reuptake may be considered in light of these data. That is, tricyclic antidepressants that block serotonin reuptake or those that block norepinephrine reuptake may, by either mechanism, decrease raphe unit firing. Whether or not this view is valid, the CSF findings do provide a direct confirmation, at least in depressed patients, of a positive relationship between noradrenergic and serotoninergic systems.

4. Biological Correlates of Clinical and Pharmacological Response

As discussed above and summarized in Table 9, altered amine metabolite values in CSF have been associated with differential clinical and pharmacological responses[59,106,146] in a variety of psychiatric patients. These data across several laboratories have a certain internal consistency, but should be regarded as preliminary until replications are undertaken. However, the wealth of data suggesting correlations with clinical or pharmacological response is noteworthy in relation to the findings in Parkinson's disease. As noted in Chapters 15 and 16, even in Parkinson's disease, a syndrome with known decreases in brain homovanillic acid (HVA), pretreatment values of HVA in CSF do not always correlate significantly with subsequent response to levodopa or a dopamine agonist.

It is intriguing that several of the studies in Table 9 report correlations of 5-hydroxyindoleacetic acid (5-HIAA), HVA, or 3-methoxy-4-hydroxyphenylethylene glycol (MHPG) not only with pharmacological responsivity, but also with degree of aggressivity, personality profile disturbance, or impairment in senile or Korsakoff's dementia. Thus, these findings indirectly support the importance of the continued effort to identify possible biochemical correlates of behavioral and cognitive disorders even in the absence of histopathological changes on gross examination of the brain. Recent reports of Crow *et al.*[41] and Mackey *et al.*[88] suggesting altered dopamine-receptor binding in autopsy specimens of schizophrenic patients also emphasize the importance of exploration of possible markers in CSF of pre- and postsynaptic receptor function in a variety of neuropsychiatric illnesses.

Table 9. Biological Correlates of Clinical or Pharmacological Response

Patient diagnosis (grouped by CSF finding)	Finding	Authors
Lower 5-HIAA[a]		
Depression	↓ Nortriptyline response	Asberg *et al.* (1973)[2]
Depression	↑ Suicide	Asberg *et al.* (1976)[4]
Depression	↑ Lithium response	Goodwin *et al.* (1973),[62] (1976)[66]
Depression	↓ Imipramine response	Goodwin *et al.*, unpublished data (1977)
Depression	↑ Dysphoric amphetamine response (30 hr)	Jimerson *et al.* (1977)[75]
Depression	↓ Amitriptyline response ↑ Anxiety ↑ Insomnia	Banki (1977)[9]
Depression	↑ 5-HTP response ↑ Chlormipramine (225 mg) response ↑ Acute response to 5-HTP and peripheral decarboxylase inhibitor, chlormipramine, or combination	Van Praag (1978)[145]
Depression	↑ Piribedil response	Post *et al.* (1978)[108]
Schizophrenia	↑ Hallucinations	Post *et al.* (1975)[107]
Mixed psychiatric	↑ Average evoked response/augmentation	Gottfries *et al.* (1976)[69]
Personality disorder	↑ Aggressive acts and trouble with the law	Brown *et al.* (1978)[26]
Gilles de la Tourette	↑ Tics	Cohen *et al.* (1978)[36]
Lower HVA		
Senile dementia	↑ Mental, not motor, impairment	Gottfries *et al.* (1970)[67]
Depression	↑ Levodopa response	Van Praag (1974)[142]
Depression	↑ Nomifensine response	Van Scheyen (1977)[153]
Depression	↑ Piribedil response	Post *et al.* (1978)[108]
Depression	↑ Sleep-deprivation response	Gerner *et al.* (1979)[55]
Schizophrenia	↑ Chronicity	Bowers (1974)[18]
Mixed psychiatric	↑ Average evoked response/augmentation	Gottfries *et al.* (1976)[69]
Lower DBH		
Alcoholics	↑ Disulfiram psychosis Abnormal MMPI profile	Major *et al.* (1980)[89]
Lower MHPG		
Schizophrenia	↑ Subjective distress in schizophrenia	Post *et al.* (1975)[107]
Depression	↑ Severity of depression	Goodwin *et al.*, unpublished data (1978)
Korsakoff's syndrome	↑ Dementia	McEntee and Mair (1978)[92]
Lower Norepinephrine		
Depression	↓ Amphetamine dysphoria ↑ Cognitive improvement	Reus *et al.* (1979)[121]

[a] (5-HIAA) 5-Hydroxyindoleacetic acid; (HVA) homovanillic acid; (MHPG) 3-methoxy-4-hydroxyphenylethylene glycol; (5-HTP) 5-hydroxytryptophan; (DBH) dopamine-β-hydroxylase; (MMPI) Minnesota Multiphasic Personality Inventory.

5. Cyclic Nucleotides in CSF of Affectively Ill Patients

One attempt to more directly approach measures of receptor function in the CSF has involved the study of the cyclic nucleotides adenosine and guanosine-3′, 5′-cyclic monophosphate (cAMP and cGMP) in CSF. As reviewed elsewhere,[104] the findings of altered urinary cAMP in depressed and manic states have been inconsistent. Although relatively little is known about the precise nature of the origin and transport characteristics of cAMP in CSF (see Chapter 9), data in depressed and manic patients do not support the notion that there are major alterations in CSF levels of either cAMP or cGMP[39,105,136] (Fig. 18).

Baseline levels of the cyclic nucleotides or those that accumulate following blockade of transport with probenecid may not, however, be the best approach to assessing cyclic nucleotide function in neuropsychiatric patients. In a variety of animal studies, basal levels during a given experimental manipulation often show little change, while a provocative test of stimulation-induced changes in cyclic nucleotides shows marked alterations. Thus, although it might be difficult in psychiatric patients, particularly in relation to the time course of drug effect, appropriate challenge strategies to study cyclic nucleotide function may be in order. Clearly, much more detailed knowledge of CSF cyclic nucleotide dynamics is required for accurate interpretation of the data. Other CSF approaches to more direct measurement of receptor function, such as measures of receptor protein, would be of great value.

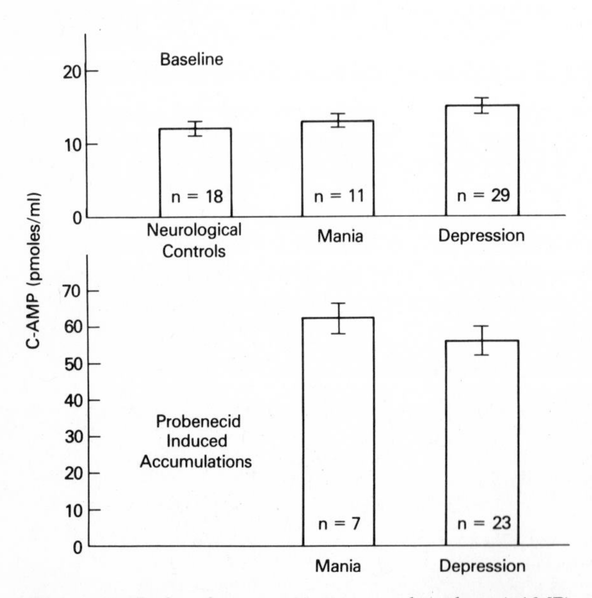

Figure 18. Cyclic adenosine-3′, 5′-monophosphate (cAMP) in CSF of manic and depressed patients. Baseline and probenecid-induced accumulations of cAMP were not significantly different in manic compared to depressed patients, although depressed patients had higher baseline cAMP than neurological control patients ($p < 0.05$).[104]

6. Neuroendocrine and Peptide Approaches to Depression and Mania

Various peptide hormones with well-described neuroendocrine actions [e.g., vasopressin, thyrotropin-releasing hormone (TRH), growth hormone (GH), adrenocorticotropic hormone (ACTH)] at the level of the pituitary and at peripheral target organs have also been demonstrated throughout the brain.[27,28] The action of these peptides in the CNS is thought to be distinct from their neuroendocrine roles. For example, in addition to their peripheral actions in vasoconstriction, renin and angiotensin I and II rapidly induce drinking behavior when infused centrally.[130] The hypothalamus releasing factors TRH and luteinizing-hormone-releasing hormone (LH-RH), in addition to their actions at the pituitary to release thyroid-stimulating hormone (TSH) and LH and follicle-stimulating hormone (FSH), respectively, have direct depressant effects on central neurons.[120] The bulk of current evidence suggests that these peptides as a class function as long-term regulators of neuronal systems,[11] perhaps modulating the action of neuronal systems at lower concentrations and for longer time periods than the traditional neurotransmitters.

Study of neuroendocrine and peptide mechanisms through the use of CSF appears to be a particularly important approach to affective illness for a variety of reasons. First, clinicians have long been aware that a variety of neuroendocrine disorders often present with psychiatric symptomatology and, conversely, the administration of a variety of peptide substances, such as ACTH or TRH, to animals and man produces important effects on behavior, mood, and cognition (for reviews, see Carroll,[33] de Weid *et al.*,[47] and Prange *et al.*[117]). Second, a variety of evidence indicates that depressed patients in particular may have increased glucocorticoid function, as measured in blood and urine and also using dexamethasone suppression tests (for a review, see Carroll[33]). Third, recent studies from a variety of disciplines have indicated that the pituitary peptide hormones act not only at peripheral end-organ sites, but also directly in brain[45,78,122] (see also Chapter 41). Fourth, as reviewed in Chapter 45, peptides such as endorphins and enkephalins may act as CNS neurotransmitter and neuromodulatory substances and be measurable in CSF as well. Finally, an encouraging possibility is the notion that the CSF pathway itself may participate in an active neuroregulatory feedback link in relation to a variety of transmitter substances including amines and peptides.[73a] If this were the case, assessment of CSF levels of these substances not only may provide closer assessment of CNS funtioning, but also may be a direct and measurable link of a regulatory pathway.

6.1. Adrenocorticotropic Hormone and Opiate-Like Substances in CSF

To our knowledge, relatively few studies of peptide substances have been carried out in patients with manic or depressive illness. Gunne *et al.*[72] and Lindstrom *et al.*[85] reported preliminary evidence of elevated opioid-like compounds in the CSF of four manic patients studied longitudinally compared to some control populations, but not to others. These provocative findings deserve further study, but existing methodologies utilizing receptor or radioimmunoassays appear to be only marginally reliable for measuring the low concentrations found in CSF. Akil *et al.*[1] and Hosobuchi *et al.*[73] have recently reported notable elevations in ventricular CSF levels of enkephalin-like or β-endorphin activity following stimulation of the periventricular area in patients with chronic pain.

Several groups have reported values of CSF cortisol in depressed patients compared to controls. Carroll and colleagues[32–34] found in depressed patients increased CSF cortisol that was highly correlated with values in plasma. Jimerson *et al.*,[75a] comparing depressed patients, schizophrenics, and neurological patients, failed to replicate the initial findings of higher levels of CSF cortisol in depression. In collaboration with Kendall, we have measured ACTH in the CSF of depressed, manic, and normal control populations. Preliminary data suggest that ACTH is measurable in the low picogram range in most normals and patients, and striking differences in relation to affective state were not observed. The anatomy and physiology of ACTH and opiate-like compounds in the CNS are discussed in detail in Chapters 44 and 45, respectively.

6.2. CSF Vasopressin

Vasopressin produced in the hypothalamus is stored in the posterior pituitary before release into the peripheral vascular system to promote retention of water at the renal distal tubule. However, it also appears to act as a "neurohormone" within the CNS in that it inhibits extinction of active-avoidance learning.[46] It is distributed throughout the brain either by the CSF or axonal pathways or via the intercellular space. There appears to be a bidirectionally impermeable barrier between plasma and CSF vasopressin, suggesting that vasopressin in the CSF is probably secreted by cells separate from those secreting into blood and therefore presumably subserves different functions than the regulation of plasma water balance.[86]

Vasopressin influences memory, pain sensitivity, and REM sleep, all phenomena that are disordered in affective illness.[57] We have begun to study the role of vasopressin in the affective illnesses, utilizing various techniques including the infusion of hypertonic saline, administration of synthetic analogues of vasopressin, and the measurement of CSF vasopressin across various affective mood changes.

In collaboration with P. Gold and G. L. Robertson,[56] we have measured CSF vasopressin by radioimmunoassay in 16 affectively ill patients in a preliminary study (see Table 10). Mean (± S.E.M.) vasopressin (1.0 ± 0.2 pg/ml) in 6 unipolar depressed patients was not significantly different from mean levels of a comparison group of 15 patients with various neurological disorders (1.0 ± 0.1 pg/ml). There was a nonsignificant trend for vasopressin to be lower in 6 bipolar depressed patients (0.8 ± 0.04 pg/ml). We are currently extending this study to include more patients during both the ill and the well state, as well as developing a normal control group to follow up these preliminary results.

6.3. Somatostatin

The tetradecapeptide somatostatin is distributed widely in both endocrine-like cells and nervous tissues. It has been demonstrated to inhibit secretion of a variety of hormones including the GH response to most stimuli, the TSH response to TRH, and the secretion of glucagon and insulin.[71] Cells containing somatostatin are widely distributed throughout the CNS, particularly in the hypothalamus and some cortical areas, as well as in peripheral nervous tissue (e.g., spinal ganglia, substantia gelatinosa of the spinal cord). In nervous tissue, it is generally a depressant and has multiple behavioral effects, e.g.,

Table 10. CSF Vasopressin in Affective Illness[a]

Diagnosis	Number of subjects	CSF vasopressin (pg/ml) (mean ± S.E.M.)	CSF osmolality
Depression			
Unipolar	6	1.0 ± 0.2	286.7 ± 1.2
Bipolar	6	0.8 ± 0.04	286.2 ± 1.6 (N = 5)
Neurological comparison group	15	1.0 ± 0.1	289 ± 1.5

[a] P. Gold, G. L. Robertson, J. C. Ballenger, R. M. Post, and F. K. Goodwin, unpublished data.

decreased spontaneous motor activity, potentiation of behavioral effects of L-DOPA, unusual rotational behavior in the rat, and a variety of behavioral, motor, and electrophysiological changes when injected into the hippocampus (for a review, see Luft *et al.*[87]). Current evidence suggests that somatostatin acts in nervous tissue as an inhibitory transmitter or modulator in a manner similar to other neuroendocrine peptides.

We have begun to study CSF somatostatin in affective illness in collaboration with Drs. P. Gold and S. Reichlin. In preliminary results with eight depressed patients, somatostatin levels were within the normal range,[99] but one patient had a markedly increased level. There was no observed probenecid effect. In one patient studied twice during depression and twice during mania, somatostatin was 133% higher during mania.

7. Summary and Conclusions

Throughout this essay, we have attempted to stress a variety of clinical and laboratory methodological variables that might interfere with accurate assessment of alterations in CSF neurotransmitter and modulator substances in manic and depressive patients. Motor activity and stress, as well as variables such as age, sex, and diurnal and seasonal differences in CSF substances, not only may affect comparison of patients to other populations, but also, as they relate to altered amine and peptide function, may be of interest in their own right. Despite major problems of clinical or laboratory methodology in many studies, in Table 11 we have tried to present an overview of probable or possible alterations in relation to affective state, with the appropriate reservations. The affective illnesses are het-

Table 11. Overview of the Literature on Potential CSF Neurotransmitter/ Modulator Alterations in Affective Illness

Neurotransmitter/ modulator	Mania[a]	Depression[a]
Norepinephrine		
Norepinephrine	High	Normal to elevated
MHPG[b]	Normal to elevated	Low to normal
VMA	Normal	Low
DBH	Low	Normal
Serotonin		
5-HIAA	Normal or low	Low or low subgroup
Dopamine		
HVA	Normal to high	Normal to low
GABA	(Higher than in depression)	Normal to low
Calcium	Lower than in depression	Normal to high
Calcitonin	(Low)	—
Cortisol	—	High or normal

[a] Parentheses indicate preliminary findings.
[b] (MHPG) 3-Methoxy-4-hydroxyphenylethylene glycol; (VMA) vanillylmandelic acid; (DBH) dopamine-β-hydroxylase; (5-HIAA) 5-hydroxyindoleacetic acid; (HVA) homovanillic acid; (GABA) γ-aminobutyric acid.

erogeneous, and no single alteration has been definitely documented. It is likely that deficits in indoleamines or catecholamines may occur at some time in the course of severe depressive illness, and preliminary evidence indicates that pharmacological treatment of these alterations may be associated with an increased clinical response. On the other hand, in mania, CSF norepinephrine appears elevated, while activity of its biosynthetic enzyme dopamine-β-hydroxylase (DBH) is low. 5-Hydroxyindoleacetic acid (5-HIAA) and homovanillic acid (HVA) alterations in mania require comparison to further study in normal controls and appropriate controls for motor activity. We are currently examining a variety of behavioral and personality measures in normals as they may relate to normal variations in CSF neurotransmitter and peptide substances (see Table 12). The very preliminary evidence of CSF peptide alterations in affective illness deserves careful follow-up in both patients and volunteers.

While the direct data from measurements in man

Table 12. Summary of NIMH Studies of Neurotransmitter Alterations in Mania Compared to Depression

Neurotransmitter	Manic[a]	Depressed[a]	Level of significance
Norepinephrine			
Baseline norepinephrine[b] (pg/ml)	357.0 ± 70.2 (9)	169.9 ± 21.8 (24)	$p < 0.03$
Probenecid norepinephrine[b] (pg/ml)	758.5 ± 103.5 (6)	291.9 ± 45.5 (21)	$p < 0.0001$
Baseline MHPG[c,f,h] (ng/ml)	10.9 ± 1.9 (13)	11.2 ± 0.9 (34)	N.S.
Probenecid MHPG[c,f] (ng/ml)	17.2 ± 2.6 (16)	14.1 ± 0.7 (36)	N.S.
Baseline VMA[c,f] (ng/ml)	1.2 ± 0.2 (12)	1.4 ± 0.3 (31)	N.S.
Probenecid VMA[c,f] (ng/ml)	2.7 ± 0.4 (15)	2.1 ± 0.18 (37)	N.S.
Dopamine-β-hydroxylase			
Baseline DBH[d] (nmol/ml)	0.43 ± 0.07 (13)	0.74 ± 0.07 (23)	$p < 0.01$
Dopamine			
Baseline HVA[c,f] (ng/ml)	48.8 ± 7.3 (16)	29.4 ± 2.5 (36)	$p < 0.04$
Probenecid HVA[c,f] (ng/ml)	239.6 ± 21.1 (16)	213.7 ± 47 (43)	N.S.
Baseline DOPAC[c,f] (ng/ml)	1.6 ± 0.8 (6)	1.4 ± 1.03 (21)	N.S.
Probenecid DOPAC[c,f] (ng/ml)	19.5 ± 15.1 (4)	8 ± 1.0 (22)	N.S.
Serotonin			
Baseline 5-HIAA[c,g] (ng/ml)	28.5 ± 2.7 (23)	22.6 ± 1.1 (70)	$p < 0.03$
Probenecid 5-HIAA[c,g] (ng/ml)	166.1 ± 12.2 (23)	145.9 ± 6.7 (65)	N.S.
γ-Aminobutyric acid			
Baseline GABA[e] (pmol/ml)	211.4 ± 24.8 (8)	231.6 ± 13.7 (16)	N.S.

[a] Values are means ± S.E.M., with the numbers of patients in parentheses.
[b] Unpublished data with R. Lake.
[c] Unpublished data, F. K. Goodwin, R. M. Post, E. K. Gordon, D. C. Jimerson, R. Cowdry.
[d] Unpublished data with P. Lerner.
[e] Unpublished data, R. M. Post, T. Hare, D. C. Jimerson, J. C. Ballenger, W. E. Bunney, Jr.
[f] Analyzed by mass fragmentographic technique.
[g] Analyzed by fluorometric technique.
[h] (MHPG) 3-Methoxy-4-hydroxyphenylethylene glycol; (VMA) vanillymandelic acid; (HVA) homovanillic acid; (DOPA) 3,4-dihydroxyphenylacetic acid; (5-HIAA) 5-hydroxyindoleacetic acid.

remain only suggestive, the indirect pharmacological data suggesting possible roles for several neurotransmitter systems in mania or depression are moderately strong. In addition, in most cases (see Table 13), the pharmacological agents, in addition to producing therapeutic effects in mania[105] and depression,[58,145] also produce consistent and expected neurotransmitter effects as reflected by alterations in CSF metabolites. To the extent that pharmacological treatments are producing large tonic and global neurotransmitter changes in the CNS, the CSF methodologies appear adequate to detect these changes at a suitable level of statistical significance. If similar global neurotransmitter changes are important in affective illness and its modulation, then existing CSF technologies ought to be sufficient to detect differences in relation to illness variables.

However, there may be occurring in areas of brain relatively distant from CSF discrete neurotransmitter changes that are not reflected in this accessible body fluid. We have also indicated that other variables, such as time-limited compensatory amine metabolite changes, may account in part for the failure to demonstrate consistent alterations in one or another CSF substance in affective illness. Clearly, the critical issues of CSF flow and transport in relation to brain concentrations (reviewed in Chapters 1, 5, and 7) deserve particular attention in relation to both positive and negative findings in affective illness. Recently, for example, Carlsson[31] has speculated that a global defect in amino acid transport or in enzyme function could account for the findings of deficient levels of several neurotransmitter metabolites in affectively ill patients. It will be increasingly important to measure a variety of substances in CSF simultaneously so that their alterations in relation to each other, as well as their possible relationships to personality variables and illness subtypes, can be assessed. The balance or ratio of several neurotransmitter systems may be important for behavioral modulation and therapeutic strategies, just as dopaminergic–acetylcholinergic balance is a critical variable in parkinsonism.

The continuum or subtype concept, highlighting possible biological heterogeneity in the affective illnesses (see Table 9), deserves a final reemphasis. Subtyping patients and their pharmacological res-

Table 13. Pharmacological Spectrum of Effects in Affective Illness: Implications for Etiology and Pathogenesis

	Mania[a]		Depression[a]		
Agents	Therap. effects	Precip. or Exacer.	Therap. effects	Precip. or Exacer.	Effects in CSF
Antimanic agents only					
Neuroleptics	++			√	↑ HVA
Reserpine	++			√√	Not studied
α-Methyl-*p*-tyrosine	++			√	↓ HVA, MHPG
Physostigmine	+			√	↑ HVA
Antidepressant agents only					
Tricyclics		√	++		↓ 5-HIAA
MAOIs[b]		√	++		↓ HVA, 5-HIAA,
Piribedil		(√)	(+)		↓ HVA
Combined acute effectiveness					
Electroconvulsive therapy (ECT)	++		++		± Effects on 5-HIAA, HVA
Tryptophan	+		+		↑ 5-HIAA
Combined acute and prophylactic effectiveness					
Lithium	++		(++)		± Effects on 5-HIAA, HVA
Carbamazepine	++		(++)		↓ HVA

[a] (Therap.) Therapeutic; (Precip.) precipitate; (Exacer.) exacerbate.
[b] (MAOI) Mono amine oxidase inhibitor; (HVA) homovanilliic acid; (5-HIAA) 5-hydroxyindoleacetic acid.

ponsivity on the basis of biological measures in CSF holds the promise for major clinical and theoretical advances in the understanding and treatment of the affective disorders.

ACKNOWLEDGMENTS. The expert technical assistance of Deborah Runkle, Marion Webster, Hester Bledsoe, and the nursing staffs of 3-West and 4-West, Clinical Center, National Institute of Mental Health, is gratefully acknowledged.

References

1. AKIL, H., RICHARDSON, D. E., HUGHES, J., BARCHAS, J.: Enkephalin-like material elevated in ventricular cerebrospinal fluid of pain patients after analgetic focal stimulation. *Science* **201**:463–465, 1978.

1a. AMERICAN PSYCHIATRIC ASSOCIATION: *Diagnostic and Statistical Manual of Mental Disorders* (DSM) III (draft version). Task Force on Nomenclature and Statistics, 1979.

2. ASBERG, M., BERTILSSON, L., TUCK, D., CRONHOLM, B., SJÖQVIST, F.: Indoleamine metabolites in the cerebrospinal fluid of depressed patients before and during treatment with nortriptyline. *Clin. Pharmacol. Ther.* **14**:277–286, 1973.

3. ASBERG, M., RINGBERGER, V. A., SJÖQVIST, F., THOREN, P., TRASKMAN, L., TUCK, J. R.: Monoamine metabolites in cerebrospinal fluid and serotonin uptake inhibition during treatment with chlormipramine. *Clin. Pharmacol. Ther.* **21**:201–207, 1977.

4. ASBERG, M., TRASKMAN, L., THOREN, P.: 5-HIAA in the cerebrospinal fluid. *Arch. Gen. Psychiatry* **33**:1193–1197, 1976.

5. ASHCROFT, G. W., CRAWFORD, T. B. B., ECCLESTON, D.: 5-Hydroxyindole compounds in the cerebrospinal fluid of patients with psychiatric or neurological diseases. *Lancet* **2**:1049–1050, 1966.

6. ASHCROFT, G. W., DOW, R. C., YATES, C. M., PULLAR, I. A.: Significance of lumbar CSF metabolite measurements in affective illness. In Tuomisto, J., Paasonen, M. K. (eds.): *CNS and Behavioral Pharmacology*, Vol. 3. Finland, University of Helsinki, 1976, pp. 277–284.

7. ASHCROFT, G. W., GLEN, A. I. M.: Mood and neuronal function: A modified amine hypothesis for the etiology of affective illness. *Adv. Biochem. Pharmacol.* **11**:335–339, 1974.

7a. BALLENGER, J. C., POST, R. M.: Therapeutic effects of carbamazepine in affective illness: A preliminary report. *Commun. Psychopharmacol.* **2**:159–175, 1978.

7b. BALLENGER, J. C., POST, R. M.: Carbamazepine (Tegretol) in manic-depressive illness: A new treatment. *Am. J. Psychiatry* **137**:782–790, 1980.

8. BANKI, C. M.: Correlation between cerebrospinal fluid amine metabolites and psychomotor activity in affective disorders. *J. Neurochem.* **28**:255–257, 1977.

9. BANKI, C. M.: Correlation of anxiety and related symptoms with cerebrospinal fluid 5-hydroxyindoleacetic acid in depressed women. *J. Neural Transm.* **41**:135–143, 1977.

10. BARABAN, J. M., WANG, R. Y., AGHAJANIAN, G. K.: Reserpine suppression of dorsal raphe neuronal firing: Mediation by adrenergic system. *Eur. J. Pharmacol.* **52**:27–36, 1978.

11. BARKER, J. L.: Peptides: Roles in neuronal excitability. *Physiol. Rev.* **56**:435–452, 1976.

12. BERGER, P. A., FAULL, K., DAVIS, K. L., BARCHAS, J. D.: Monoamine metabolites in CSF in psychiatric disorders. In Usdin, E., Kopin, I. J., Barchas, J. D. (eds.): *Catecholamines: Basic and Clinical Frontiers*. New York, Pergamon Press, 1979, pp. 1827–1829.

13. BERTILSSON, L.: Quantitative determination of 4-hydroxy-3-methoxyphenyl glycol and its conjugates in cerebrospinal fluid by mass fragmentography. *J. Chromatogr.* **87**:147–153, 1973.

14. BERTILSSON, L., ASBERG, M., THOREN, P.: Differential effect of chlormipramine and nortriptyline on cerebrospinal fluid metabolites of serotonin and noradrenaline in depression. *Eur. J. Clin. Pharmacol.* **7**:365–367, 1974.

15. BOWERS, M. B., JR.: Cerebrospinal fluid 5-hydroxyindoleacetic acid (5-HIAA) and homovanillic acid (HVA) following probenecid in unipolar depressives treated with amitriptyline. *Psychopharmacologia (Berlin)* **23**:26–33, 1972.

16. BOWERS, M. B., JR.: Lumbar CSF 5-hydroxyindoleacetic acid and homovanillic acid in affective syndromes. *J. Nerv. Ment. Dis.* **158**:325–330, 1974.

17. BOWERS, M. B., JR.: Amitriptyline in man: Decreased formation of central 5-hydroxyindoleacetic acid. *Clin. Pharmacol. Ther.* **15**:167–170, 1974.

18. BOWERS, M. B., JR.: Central dopamine turnover in schizophrenic syndromes. *Arch. Gen. Psychiatry* **31**:50–54, 1974.

19. BOWERS, M. B., JR.: CSF acid monoamine metabolites as a possible reflection of central MAO activity in chronic schizophrenia. *Biol. Psychiatry* **11**:245–249, 1976.

20. BOWERS, M. B., JR., GERBODE, F. A.: Reltaionship of monoamine metabolites in human cerebrospinal fluid to age. *Nature (London)* **219**:1256–1257, 1968.

21. BOWERS, M. B., JR., HENINGER, G. R.: Lithium: Clinical effects and cerebrospinal fluid acid monoamine metabolites. *Commun. Psychopharmacol.* **1**:135–145, 1977.

22. BOWERS, M. B., JR., HENINGER, G. R., GERBODE, F. A.: Cerebrospinal fluid 5-hydroxyindoleacetic acid and homovanillic acid in psychiatric patients. *Int. J. Neuropharmacol.* **8**:255–262, 1969.

23. BOWERS, M. B., JR., KUPFER, D. J.: Central monoamine oxidase inhibition and REM sleep. *Brain Res.* **35**:561–564, 1971.

24. Bowers, M. B., Jr., Meltzer, H. Y., Heninger, G. A.: Clinical indices of dopaminergic function. In Program and Abstracts of the Fourth International Catecholamine Symposium on Catecholamines: Basic and Clinical Frontiers. Presented in Pacific Grove, California, September 1978, p. 138 (Abstract No. 535).
25. Brodie, H. K. H., Sack, R., Siever, L.: Clinical studies of L-5-hydroxytryptophan in depression. In Barchas, J., Usdin, E. (eds.): *Serotonin and Behavior*. New York and London, Academic Press, 1973, pp. 549–627.
26. Brown, G. L., Goodwin, F. K., Ballenger, J. C., Goyer, P. F., Major, L. F.: CSF amine metabolites in human aggression. *Sci. Proc. Am. Psychiatr. Assoc.* **131:**88, 1978 (Abstract No. 180).
27. Brownstein, M. J.: Minireview: The pineal gland. *Life Sci.* **16:**1363–1374, 1975.
28. Brownstein, M. J.: Biologically active peptides in the mammalian central nervous system. In Gainer, H. (ed.): *Peptides in Neurobiology*. New York, Plenum Press, 1977, pp. 145–170.
29. Bunney, W. E., Jr., Davis, J. M.: Norepinephrine in depressive reactions. *Arch. Gen. Psychiatry* **13:**483–494, 1965.
30. Carlsson, A.: Antipsychotic drugs, neurotransmitters, and schizophrenia. *Am. J. Psychiatry* **135:**164–173, 1978.
31. Carlsson, A.: The impact of catecholamine research on medical science and practice. Presented at the Fourth International Catecholamine Symposium on Catecholamines: Basic and Clinical Frontiers. Pacific Grove, California, September 1978.
32. Carroll, B. J.: Limbic system–adrenal cortex regulation in depression and schizophrenia. *Psychosom. Med.* **38:**106–121, 1976.
33. Carroll, B. J.: Neuroendocrine function in psychiatric disorders. In Lipton, M. A., DiMascio, A., Killam, K. F. (eds.): *Psychopharmacology: A Generation of Progress*. New York, Raven Press, 1978, pp. 487–497.
34. Carroll, B. J., Curtis, G. C., Mendels, J.: Cerebrospinal fluid and plasma free cortisol concentrations in depression. *Psychol. Med.* **6:**235–244, 1976.
35. Cohen, D. J., Caparulo, B. K., Shaywitz, B. A., Bowers, M. B., Jr.: Dopamine and serotonin metabolism in neuropsychiatrically disturbed children. *Arch. Gen. Psychiatry* **34:**545–550, 1977.
36. Cohen, D. J., Shaywitz, B. A., Caparulo, B., Young, J. G., Bowers, M. B., Jr.: Chronic, multiple tics of Gilles de la Tourette's disease: CSF acid monoamine metabolites after probenecid administration. *Arch. Gen. Psychiatry* **35:**245–250, 1978.
37. Cools, A. R., van Dongen, P. A. M., Janssen, H.-J., Megens, A. A. P. H.: Functional antagonism between dopamine and noradrenaline within the caudate nucleus of cats: A phenomenon of rhythmically changing susceptibility. *Psychopharmacology* **59:**231–242, 1978.
38. Coppen, A. J., Prange, A. J., Jr., Whybrow, P. C., Noguera, R.: Abnormalities of indoleamines in affective disorders. *Arch. Gen. Psychiatry* **26:**474–478, 1972.
39. Cramer, H., Goodwin, F. K., Post, R. M., Bunney, W. E., Jr.: Effects of probenecid and exercise on cerebrospinal fluid cyclic-AMP in affective illness. *Lancet* **2:**1346–1347, 1972.
40. Cross, A. J., Crow, T. J., Killpack, W. S., Longden, A., Owen, F., Riley, G. J.: The activities of brain dopamine β-hydroxylase and catechol-*O*-methyl transferase in schizophrenics and controls. *Psychopharmacology* **59:**117–121, 1978.
41. Crow, T. J., Johnstone, E. C., Longden, A. J., Owen, F.: Dopaminergic mechanisms in schizophrenia: The antipsychotic effect and the disease process. *Life Sci.* **23:**563–568, 1978.
42. de Montigny, C., Aghajanian, G. K.: Tricyclic antidepressants: Long-term treatment increases responsivity of rat forebrain neurons to serotonin. *Science* **202:**1303–1306, 1978.
43. Dencker, S. J., Malm, V., Roos, B. E., Werdinius, B.: Acid monoamine metabolites of cerebrospinal fluid in mental depression and mania. *J. Neurochem.* **13:**1545–1548, 1966.
44. DePotter, W. P.: The presence of dopamine β-hydroxylase activity in rabbit cerebrospinal fluid and its increase after sciatic nerve stimulation. *J. Physiol.* **258:**26P–27P, 1976.
45. de Weid, D.: Peptides and behavior. *Life Sci.* **20:**195–204, 1977.
46. de Weid, D., Bohus, B., van Wimersma Greidanum, T.-J. B.: Memory deficit in rats with hereditary diabetes insipidus. *Brain Res.* **85:**152–156, 1975.
47. de Weid, D., Bohus, B., van Ree, J. M., Urban, I.: Behavioral and electrophysiological effects of peptides related to lipotropin (beta-LPH). *J. Pharmacol. Exp. Ther.* **204:**570–580, 1978.
48. Ebert, M. H., Post, R. M., Goodwin, F. K.: Effect of physical activity on urinary MHPG excretion in depressed patients. *Lancet* **2:**766, 1972.
49. Faull, K. F., DoAmaral, J. R., Berger, P. A., Barchas, J. D.: Mass spectrometric identification and selected ion monitoring quantitation of γ-amino-butyric acid (GABA) in human lumbar cerebrospinal fluid. *J. Neurochem.* **31:**1119–1122, 1978.
50. Fotherby, K., Ashcroft, G. W., Affleck, J. W., Forrest, A. D.: Studies on sodium transfer and 5-hydroxyindoles in depressive illness. *J. Neurol. Neurosurg. Psychiatry* **26:**71–73, 1963.
51. Fyro, B., Petterson, U., Sedvall, G.: The effect of lithium treatment on manic symptoms and levels of monoamine metabolites in cerebrospinal fluid of manic and depressive patients. *Psychopharmacologia (Berlin)* **44:**99–103, 1975.
52. Gallager, D. W., Pert, A., Bunney, W. E., Jr.: Haloperidol-induced presynaptic dopamine supersensitivity is blocked by chronic lithium. *Nature (London)* **273:**309–312, 1978.
53. Gerlach, J., Thorsen, K., Fog, R.: Extrapyramidal

reactions and amine metabolites in cerebrospinal fluid during haloperidol and clozapine treatment of schizophrenic patients. *Psychopharmacologia (Berlin)* **40**:341–350, 1975.

54. GERNER, R. H., POST, R. M., BUNNEY, W. E., JR.: A dopaminergic mechanism in mania. *Am. J. Psychiatry* **133**:1177–1180, 1976.
55. GERNER, R. H., POST, R. M., GILLIN, J. C., BUNNEY, W. E., JR.: Biological and behavioral effects of one night's sleep deprivation in depressed patients and normals. *J. Psychiatr. Res.* **15**:21–40, 1979.

55a. GOLD, B. I., BOWERS, M. B., ROTH, R. H., SWEENEY, D. W.: GABA levels in CSF of patients with psychiatric disorders. *Am. J. Psychiatry* **137**:362–364, 1980.

56. GOLD, P. W., BALLENGER, J. C., ZIS, A. P., ROBERTSON, G., POST, R. M., GOODWIN, F. K.: A vasopressin hypothesis of affective illness: Preliminary findings. In Usdin, E., Kopin, I. J., Barchas, J. D. (eds.): *Catecholamines: Basic and Clinical Frontiers*. New York, Pergamon Press, 1979, pp. 1807–1819.
57. GOLD, P., GOODWIN, F. K., REUS, V. I.: Vasopressin in affective illness—Hypothesis. *Lancet* **1**:1233–1236, 1978.
58. GOODE, D. J., DEKIRMENJIAN, H., MELTZER, H., MAAS, J. W.: Relation of exercise to MHPG excretion in normal subjects. *Arch. Gen. Psychiatry* **29**:391–396, 1973.
59. GOODWIN, F. K.: Drug treatment of affective disorders: General principles. In Janvik, M. E. (ed.): *Psychopharmacology in the Practice of Medicine*. New York, Appleton-Century-Crofts, 1977, pp. 241–252.
60. GOODWIN, F. K., POST, R. M.: Studies of amine metabolites in affective illness and schizophrenia: A comparative analysis. In Freedman, D. X. (ed.): *The Biology of the Major Psychoses*. New York, Raven Press, 1975, pp.299–332.
61. GOODWIN, F. K., POST, R. M.: Catecholamine metabolite studies in the affective disorders: Issues of specificity and significance. In Usdin, E., Hamburg, D. A., Barchas, J. D. (eds.): *Neuroregulators and Psychiatric Disorders*. New York, Oxford University Press, 1977, pp. 135–145.
62. GOODWIN, F. K., POST, R. M., MURPHY, D. L.: Cerebrospinal fluid amine metabolites and therapies for depression. *Sci. Proc. Am. Psychiatr. Assoc.* **126**:24–25, 1973 (Abstract No. 20).
63. GOODWIN, F. K., POST, R. M., WEHR, T.: Clinical approaches to the evaluation of brain amine function in mental illness: Some conceptual issues. In Yondim, M. B. H., Lovenberg, W., Sharman, D. F., Lagnado, J. R. (eds.): *Essays in Neurochemistry and Neuropharmacology*, Vol. II. London, John Wiley, 1977, pp. 71–104.
64. GOODWIN, F. K., POST, R. M., DUNNER, D. L., GORDON, E. K.: Cerebrospinal fluid amine metabolites in affective illness: The probenecid technique. *Am. J. Psychiatry* **130**:73–79, 1973.
65. GOODWIN, F. K., RUBOVITS, R., JIMERSON, D. C., POST, R. M.: Serotonin and norepinephrine subgroups in depression. *Sci. Proc. Am. Psychiatr. Assoc.* **130**:108–109, 1977 (Abstract No. 180).
66. GOODWIN, F. K., WEHR, T., SACK, R. L.: Studies on the mechanism of action of lithium in man: A contribution to neurobiological theories of affective illness. In Villeneuve, A. (ed.): *Lithium in Psychiatry: A Synopsis*. Quebec, Les Presses de L'Universite Laval, 1976, pp. 23–48.
67. GOTTFRIES, C. G., GOTTFRIES, I., ROOS, B. E.: Homovanillic acid and 5-hydroxyindoleacetic acid in cerebrospinal fluid related to mental and motor impairment in senile and presenile dementia. *Acta Psychiatr. Scand.* **46**:99–105, 1970.
68. GOTTFRIES, C. G., GOTTFRIES, I., JOHANSSON, B., OLSSON, R., PERSSON, T., ROOS, B. E., SJÖSTRÖM, R.: Acid monoamine metabolites in human cerebrospinal fluid and their relation to age and sex. *Neuropharmacology* **10**:665–672, 1971.
69. GOTTFRIES, C. G., VON KNORRING, L., PERRIS, C.: Neurophysiological measures related to levels of 5-hydroxyindoleacetic acid, homovanillic acid and tryptophan in cerebrospinal fluid of psychiatric patients. *Neuropsychobiology* **2**:1–8, 1976.
70. GROTE, S. S., MOSES, S. G., ROBINS, E., HUDGENS, R. W., CRONINGER, A. B.: A study of selected catecholamine metabolizing enzymes: A comparison of depressive suicides and alcoholic suicides with controls. *J. Neurochem.* **23**:791–802, 1974.
71. GUILLEMIN, R., GERICH, J. E.: Somatostatin: Physiological and clinical significance. *Annu. Rev. Med.* **27**:379–388, 1976.
72. GUNNE, L. M., LINDSTROM, L., TERENIUS, L.: Naloxone-induced reversal of schizophrenic hallucinations. *J. Neural Transm.* **40**:13–19, 1977.
73. HOSOBUCHI, Y., ROSSIER, J., BLOOM, F. E., GUILLEMIN, R.: Electrical stimulation of periaqueductal gray for pain relief increases immunoreactive β-endorphin in ventricular fluid. *Science* **203**:279–281, 1979.

73a. HYYPPA, M. T., LIRA, J., SKETT, P., GUSTAFSSON, J. A.: Neuropeptide hormones in cerebrospinal fluid: Experimental and clinical aspects. *Med. Biol.* **57**:367–373, 1979.

74. JIMERSON, D. C., GORDON, E., POST, R. M., GOODWIN, F. K.: Central noradrenergic function in man: Vanillylmandelic acid in CSF. *Brain Res.* **99**:434–439, 1975.
75. JIMERSON, D. C., POST, R. M., REUS, V. I., VAN KAMMEN, D. P., DOCHERTY, J., GILLIN, J. C., BUCHSBAUM, M., EBERT, M. H., BUNNEY, W. E., JR.: Predictors of amphetamine response in depression. *Sci. Proc. Am. Psychiatr. Assoc.* **130**:100–101, 1977 (Abstract No. 170).
76. JORI, A., DOLFINI, E., CASATI, C., ARGINTA, G.: Effect of ECT and imipramine treatment on the concentration of 5-hydroxyindoleacetic acid (5-HIAA) and homovanillic acid (HVA) in the cerebrospinal fluid of depressed patients. *Psychopharmacologia* **44**:87–90, 1975.
77. KIRSTEIN, L., BOWERS, M. B., JR., HENINGER, G.: CSF

amine metabolites, clinical symptoms, and body movement in psychiatric patients. *Biol. Psychiatry* **2**:421–434, 1976.

78. KNIGGE, K. M., MORRIS, M., SCOTT, D. E., JOSEPH, S. A., NOTTER, M., SCHOCK, D., KROBISIH-DUDLEY, G.: Distribution of hormones by cerebrospinal fluid. In Cserr, H. F., Fenstermacher, J. D., Fencl, V. (eds.): *Fluid Environment of the Brain*. New York, Academic Press, 1975, pp. 237–254.
79. KORF, J., VAN PRAAG, H. M.: Amine metabolism in human brain: Further evaluation of the probenecid test. *Brain Res.* **35**:221–230, 1971.
80. KUPFER, D. J., BOWERS, M. B., JR.: REM sleep and central monoamine oxidase inhibition. *Psychopharmacologia (Berlin)* **27**:183–190, 1973.
81. LAKE, C. R., ZIEGLER, M. G., SHOULSON, I., KOPIN, I. J.: No difference in cerebrospinal fluid levels of norepinephrine in patients when on high versus low monoamine diets. *Soc. Neurosci. Abstr.* **2**:252, 1977 (Abstract No. 794).
82. LAMPRECHT, F., EBERT, M. H., TURCK, I., KOPIN, I. J.: Serum dopamine β-hydroxylase in depressed patients and the effect of electroconvulsive shock treatment. *Psychopharmacologia* **40**:241–248, 1974.
83. LERNER, P., DENDEL, P. S., MAJOR, L. F.: Dopamine β-hydroxylase in cat cerebrospinal fluid and plasma. *Brain Res.* (in press).
84. LERNER, P., GOODWIN, F. K., VAN KAMMEN, D. P., POST, R. M., MAJOR, L. F., BALLENGER, J. C., LOVENBERG, W.: Dopamine β-hydroxylase in the cerebrospinal fluid of psychiatric patients. *Biol. Psychiatry* **13**:685–694, 1978.
85. LINDSTROM, L. H., WIDERLOV, E., GUNNE, L. M., WAHLSTROM, A., TERENIUS, L.: Endorphins in human cerebrospinal fluid: Clinical correlations to some psychotic states. *Acta Psychiatr. Scand.* **57**:153–164, 1978.
86. LUERSSEN, T. G., SHELTON, R. L., ROBERTSON, G. L.: Evidence for separate origin of plasma and CSF vasopressin. *Clin. Res.* **1**:14, 1977.
87. LUFT, R., EFENDIC, S., HÖKFELT, T.: Somatostatin—Both hormone and neurotransmitter? *Diabetologia* **14**:1–13, 1978.
88. MACKEY, A. V. P., DOBLE, A., BIRD, E. D., SPOKES, E. G., QUIK, M., IVERSEN, L. L.: ^{3}H-spiperone binding in normal and schizophrenic post-mortem human brain. *Life Sci.* **23**:527–532, 1978.
89. MAJOR, L. F., LERNER, P., GOODWIN, F. K., BALLENGER, J. C., BROWN, G. L., LOVENBERG, W.: Dopamine β-hydroxylase in cerebrospinal fluid: Relationship to personality measures. *Arch. Gen. Psychiatry* **37**:308–310, 1980.
90. MAJOR, L. F., LAKE, C. R., LIPPER, S., LERNER, P., MURPHY, D. L.: The central noradrenergic system and affective response to MAO inhibitors. *Neuropsychopharmacology* **3**:5–6, 1979.
91. MAJOR, L. F., MURPHY, D. L., LIPPER, S., GORDON, E. K.: Effects of clorgyline and pargyline on deaminated metabolites of norepinephrine, serotonin, and dopamine in human cerebrospinal fluid. *J. Neurochem.* **32**:229–231, 1979.
92. MCENTEE, W. J., MAIR, R. G.: Memory impairment in Korsakoff's psychosis: A correlation with brain noradrenergic activity. *Science* **202**:905–907, 1978.
93. MCLEOD, W. E., MCLEOD, M.: Indoleamines and the cerebrospinal fluid. In Davies, B. M., Carroll, B. J., Mowbray, R. M. (eds.): *Depressive Illness: Some Research Studies*. Springfield, Illinois, Charles C. Thomas, 1972, pp. 209–225.
94. MELTZER, H., CHO, H. W., CARROLL, B. J., RUSSO, P.: Serum dopamine β-hydroxylase activity in the affective psychoses and schizophrenia. *Arch. Gen. Psychiatry* **33**:585–591, 1976.
95. MENDELS, J., FRAZER, A., FITZGERALD, R. G., RAMSEY, T. A., STOKES, J. W.: Biogenic amine metabolites in cerebrospinal fluid of depressed and manic patients. *Science* **175**:1380–1381, 1972.
96. MOIR, A. T. B., ASHCROFT, G. W., CRAWFORD, T. B. B.: Central metabolites in cerebrospinal fluid as a biochemical approach to the brain. *Brain* **93**:357–368, 1970.
97. MUSCETTOLA, G., WEHR, T., GOODWIN, F. K.: Effect of diet on urinary MHPG excretion in depressed patients and normal control subjects. *Am. J. Psychiatry* **134**:914–916, 1977.
98. PAPESCHI, R., MCCLURE, D. J.: Homovanillic acid and 5-hydroxyindoleacetic acid in cerebrospinal fluid of depressed patients. *Arch. Gen. Psychiatry* **25**:354–358, 1971.
99. PATEL, Y., RAO, K., REICHLIN, S.: Somatostatin in human cerebrospinal fluid. *N. Engl. J. Med.* **296**:529–533, 1977.
100. PERLOW, M. J., GORDON, E. K., EBERT, M. H., HOFFMAN, H. J., CHASE, T. N.: The circadian variation in dopamine metabolism in the subhuman primate. *J. Neurochem.* **28**:1381–1383, 1977.
101. PERT, A., ROSENBLATT, J. E., SIVIT, C., PERT, C. B., BUNNEY, W. E., JR.: Long-term treatment lithium prevents the development of dopamine receptor supersensitivity. *Science* **201**:171–173, 1978.
102. PERT, C. B., ROSENBLATT, J. E., TALLMAN, J. F., PERT, A., BUNNEY, W. E., JR.: Lithium blocks dopamine receptor supersensitivity. In Program and Abstracts of the Fourth International Catecholamine Symposium on Catecholamines: Basic and Clinical Frontiers. Presented at Pacific Grove, California, September 1978, p. 148 (Abstract No. 574).
103. POST, R. M., BALLENGER, J. C., REUS, V. I., LAKE, C. R., LERNER, P., BUNNEY, W. E., JR.: Effects of carbamazepine in mania and depression. Presented at the 131st Annual Meeting of the American Psychiatric Association, Atlanta, May 1978. *New Res. Abstr.* No. 7.
104. POST, R. M., CRAMER, H., GOODWIN, F. K.: Cyclic AMP in cerebrospinal fluid in patients with affective

illness: Effects of probenecid, activity, and psychotropic medications. In Usdin, E., Hamburg, D. A., Barchas, J. D. (eds.): *Neuroregulators and Psychiatric Disorders*. New York, Oxford University Press, 1977, pp. 464–469.
105. POST, R. M., CRAMER, H., GOODWIN, F. K.: Cyclic AMP in cerebrospinal fluid of manic and depressive patients. *Psychol. Med.* 7:599–607, 1977.
106. POST, R. M., CUTLER, N. R.: The pharmacology of acute mania. In Klawans, H. L. (ed.): *Clinical Neuropharmacology*, Vol. IV, New York, Raven Press, 1979, pp. 39–81.
107. POST, R. M., FINK, E., CARPENTER, W., GOODWIN, F. K.: CSF amine metabolites in acute schizophrenia. *Arch. Gen. Psychiatry* **32**:1063–1069, 1975.
108. POST, R. M., GERNER, R. H., CARMAN, J. S., GILLIN, J. C., JIMERSON, D. C., GOODWIN, F. K., BUNNEY, W. E., JR.: Effects of a dopamine agonist piribedil in depressed patients: Relationship of pretreatment HVA to antidepressant response. *Arch. Gen. Psychiatry* **35**:609–615, 1978.
109. POST, R. M., GOODWIN, F. K.: Effects of amitriptyline and imipramine on amine metabolites in the cerebrospinal fluid of depressed patients. *Arch. Gen. Psychiatry* **30**:234–239, 1974.
110. POST, R. M., GOODWIN, F. K.: Time-dependent effects of phenothiazines on dopamine turnover in psychiatric patients. *Science* **190**:488–489, 1975.
111. POST, R. M., GOODWIN, F. K.: Studies of cerebrospinal fluid amine metabolites in depressed patients: Conceptual problems and theoretical implications. In Mendels, J. (ed.): *Biological Aspects of Depression*. New York, Spectrum, 1975, pp. 47–67.
112. POST, R. M., GOODWIN, F. K.: Approaches to brain amines in psychiatric patients: A re-evaluation of cerebrospinal fluid studies. In Iverson, L. L., Iverson, S. D., Snyder, S. H. (eds.): *Handbook of Psychopharmacology*, Vol. 13. New York, Plenum Press, 1978, pp. 147–185.
113. POST, R. M., GORDON, E. K., GOODWIN, F. K., BUNNEY, W. E., JR.: Central norepinephrine metabolism in affective illness: 3-Methoxy-4-hydroxyphenylethylene glycol in the cerebrospinal fluid. *Science* **179**:1002–1003, 1973.
114. POST, R. M., JIMERSON, D. C., BUNNEY, W. E., JR., GOODWIN, F. K.: Dopamine and mania: Behavioral and biochemical effects of the dopamine receptor blocker pimozide. *Psychopharmacology* **67**:297–305, 1980.
115. POST, R. M., KOTIN, J., GOODWIN, F. K., GORDON, E. K.: Psychomotor activity and cerebrospinal fluid amine metabolites in affective illness. *Am. J. Psychiatry* **130**:67–72, 1973.
116. POST, R. M., LAKE, C. R., JIMERSON, D. C., BUNNEY, W. E., JR., WOOD, J. H., ZIEGLER, M. G., GOODWIN, F. K.: Cerebrospinal fluid norepinephrine in affective illness. *Am. J. Psychiatry* **135**:907–917, 1978.
116a. POST, R. M., BALLENGER, J. C., HARE, T. A., GOODWIN, F. K., LAKE, C. R., JIMERSON, D. C., BUNNEY, W. E., JR.: Cerebrospinal fluid GABA in normals and patients with affective disorders. *Brain Res. Bull.* (in press).
117. PRANGE, A. J., NEMEROFF, C. B., LIPTON, M. A.: Behavioral effects of peptides: Basic and clinical studies. In Lipton, M. A., DiMascio, A., Killam, K. F. (eds.): *Psychopharmacology: A Generation of Progress*. New York, Raven Press, 1978, pp. 441–458.
118. PRANGE, A. J., WILSON, I. C., LYNN, C. W., ALLTOP, L. B., STIKELEATHER, R. A.: L-Tryptophan in mania: Contribution to a permissive hypothesis of affective disorders. *Arch. Gen. Psychiatry* **30**:56–62, 1974.
119 REDMOND, D. E.: Alterations in the function of the nucleus locus coeruleus: A possible model for studies of anxiety. In Hanin, I., Usdin, E. (eds.): *Animal Models in Psychiatry and Neurology*. New York, Pergamon Press, 1977, pp. 293–304.
120. RENAUD, L. P., MARTIN, J. B.: Thyrotropin releasing hormone: Depressant action on central neuronal activity. *Brain Res.* **86**:150–154, 1975.
121. REUS, V. I., SILBERMAN, E. K., POST, R. M., WEINGARTNER, H.: D-Amphetamine: Effects on memory in a depressed population. *Biol. Psychiatry* **14**:345–356, 1979.
122. RODRIGUEZ, E. M.: The cerebrospinal fluid as a pathway in neuroendocrine integration. *J. Endocrinol.* **71**:407–443, 1976.
123. ROOS, B. E.: CSF metabolites and psychopathology. Paper read at the Annual Meeting of the American College of Neuropsychopharmacology, Las Vegas, Nevada, January 18–22, 1972.
124. ROOS, B. E., SJÖSTRÖM, R.: 5-Hydroxyindoleacetic acid (and homovanillic acid) levels in the CSF after probenecid application in patients with manic–depressive psychosis. *Pharmacol. Clin.* **1**:153–155, 1969.
125. RUBOVITZ, R., GOODWIN, F. K., POST, R. M.: Effects of lithium on brain amine metabolism. *Sci. Proc. Am. Psychiatr. Assoc.* **129**:248–249, 1976 (Abstract No. 248).
126. SAAVEDRA, J. M.: Measurement of biogenic amines at the picogram level. *Prog. Anal. Chem.* **7**:33–44, 1974.
127. SCHILDKRAUT, J. J.: The catecholamine hypothesis of affective disorders: A review of supporting evidence. *Am. J. Psychiatry* **122**:509–522, 1965.
128. SEDVALL, G., ALFREDSSON, G., BJERKENSTEDT, L., ENEROTH, P., FYRO, B., HARNRYD, C., SWAHN, C. G., WIESEL, F. A., WODE-HELGODT, B.: Selective effects of psychoactive drugs on levels of monoamine metabolites and prolactin in cerebrospinal fluid of depressed patients. In Airaksinen, M. (ed.): *Proceedings of the Sixth International Congress of Pharmacology*, Vol. 3, *Central Nervous System and Behavioral Pharmacology*. New York, Pergamon Press, 1976, pp. 255–267.
129. SHAW, D. M., O'KEEFE, R. O., MACSWEENEY, D. A., BROOKSBANK, B. W. L., NOGUERA, R., COPPEN, A.:

3-Methoxy-4-hydroxyphenylethylene glycol in depression. *Psychol. Med.* **3**:333–336, 1973.
130. SIMPSON, J. B., ROUTTENBERG, A.: Subfornical organ: Site of drinking elicitation by angiotensin II. *Science* **181**:1172–1175, 1973.
131. SIWERS, B., RINGBERGER, V. A., TUCK, J. R., SJÖQVIST, F.: Initial clinical trial based on biochemical methodology of zimelidine (a serotonin uptake inhibitor) in depressed patients. *Clin. Pharmacol. Ther.* **21**:194–200, 1977.
132. SJÖSTRÖM, R.: Cerebrospinal fluid content of 5-hydroxyindoleacetic acid and homovanillic acid in manic–depressive psychosis. *Acta Univ. Ups.* **154**:5–35, 1973.
133. SJÖSTRÖM, R.: 5-Hydroxyindoleacetic acid and homovanillic acid in cerebrospinal fluid in manic–depressive psychosis and the effect of probenecid treatment. *Eur. J. Clin. Pharmacol.* **6**:75–80, 1973.
134. SJÖSTRÖM, R., ROOS, B. E.: 5-Hydroxyindoleacetic acid and homovanillic acid in cerebrospinal fluid in manic–depressive psychosis. *Eur. J. Clin. Pharmacol.* **4**:170–176, 1972.
135. SMITH, A. D., WINKLER, H.: Fundamental mechanisms in the release of catecholamines. In Blaschko, H., Muscholl, E. (eds.): *Handbook of Experimental Pharmacology,* Vol. 33, *Catecholamines.* Berlin, Springer-Verlag, 1972, pp. 538–617.
136. SMITH, C. C., TALLMAN, J. F., POST, R. M., VAN KAMMEN, D. P., JIMERSON, D. C., BROWN, G. L.: An examination of baseline and drug-induced levels of cyclic nucleotides in the cerebrospinal fluid of control and psychiatric patients. *Life Sci.* **19**:131–136, 1976.
137. SOURKES, T. L.: On the origin of homovanillic acid (HVA) in the cerebrospinal fluid. *J. Neural Transm.* **34**:153–157, 1973.
138. SPITZER, R., ENDICOTT, J., ROBINS, E.: *Research Diagnostic Criteria.* New York, New York State Psychiatric Institute, Biometrics Research, 1975.
139. SUBRAHMANYAM, S.: Role of biogenic amines in certain pathological conditions. *Brain Res.* **87**:355–362, 1975.
140. SWEENEY, D. R., MAAS, J. W., HENINGER, G. R.: Anxiety, physical activity, and urinary MHPG. *Sci. Proc. Am. Psychiatr. Assoc.* **130**:173, 1977 (Abstract No. 289).
141. TAKAHASHI, S., YAMANE, H., KONDO, H., TANI, N., KATO, N.: CSF monoamine metabolites in alcoholism: A comparative study with depression. *Folia Psychiatr. Neurol. Jpn.* **28**:347–354, 1974.
141a. VAN KAMMEN, D. P., STERNBERG, D. E., HARE, T., BALLENGER, J. C., POST, R. M.: GABA CSF levels in schizophrenia. *Brain Res. Bull.* (in press).
142. VAN PRAAG, H. M.: Towards a biochemical typology of depression? *Pharmakopsychiatr./Neuro-Psychopharmakol.* **7**:281–292, 1974.
143. VAN PRAAG, H. M.: The significance of dopamine for the mode of action of neuroleptics and the pathogenesis of schizophrenia. *Br. J. Psychiatry* **130**:463–474, 1977.
144. VAN PRAAG, H. M.: Significance of biochemical parameters in the diagnosis, treatment, and prevention of depressive disorders. *Biol. Psychiatry* **12**:101–131, 1977.
145. VAN PRAAG, H. M.: Central serotonin: Its relation to depression vulnerability and depression prophylaxis. In Obiols, J., Ballús, C., and González-Monclús, E. (eds.): *Biological Psychiatry Today.* Elsevier, North-Holland Biomedical Press, 1979, pp. 485–498.
146. VAN PRAAG, H. M.: *Psychotropic Drugs: A Guide for the Medical Practitioner.* New York, Brunner/Mazel, 1978.
147. VAN PRAAG, H. M., KORF, J.: Retarded depression and the dopamine metabolism. *Psychopharmacologia (Berlin)* **19**:199–203, 1971.
148. VAN PRAAG, H. M., KORF, J.: Serotonin metabolism in depression: Clinical application of the probenecid test. *Int. Pharmacopsychiatry* **9**:35–51, 1974.
149. VAN PRAAG, H. M., KORF, J.: Neuroleptics, catecholamines, and psychoses: A study of their interrelations. *Am. J. Psychiatry* **132**:593–597, 1975.
150. VAN PRAAG, H. M., KORF, J., PUITE, J.: 5-Hydroxyindoleacetic acid levels in the cerebrospinal fluid of depressive patients treated with probenecid. *Nature (London)* **255**:1259–1260, 1970.
151. VAN PRAAG, H. M., KORF, J., SCHUTT, D.: Cerebral monoamines and depression. *Arch. Gen. Psychiatry* **28**:827–831, 1973.
152. VAN PRAAG, H. M., KORF, J., LAKKE, J. P. W. F., SCHUT, T.: Dopamine metabolism in depressions, psychoses, and Parkinson's disease: The problem of the specificity of biological variables in behavior disorders. *Psychol. Med.* **5**:138–146, 1975.
153. VAN SCHEYEN, J. D., VAN PRAAG, H. M., KORF, J.: Controlled study comparing nomifensine and chlormipramine in unipolar depression using the probenecid technique. *Br. J. Clin. Pharmacol.* **4**:179–184, 1977.
154. VESTERGAARD, P., SØRENSEN, T., HOPPE, E., RAFAELSEN, O. J., YATES, C. M., NICOLAOU, N.: Biogenic amine metabolites in cerebrospinal fluid of patients with affective disorders. *Acta Psychiatr. Scand.* **58**:88–96, 1978.
155. VIVEROS, O. H., ARGUEROS, L., KIRSHNER, N.: Release of catecholamines and dopamine β-hydroxylase from the adrenal medulla. *Life Sci.* **7**:609–618, 1968.
156. WALINDER, J., SKOTT, A., CARLSSON, A., NAGY, A. Z., ROOS, B. E.: Potentiation of the antidepressant action of chlormipramine by tryptophan. *Arch. Gen. Psychiatry* **33**:1384–1389, 1976.
157. WEHR, T. A., MUSCETTOLA, G., GOODWIN, F. K.: Urinary 3-methoxy-4-hydroxyphenylglycol circadian rhythym: Early timing (phase-advance) in manic–depressives compared with normal subjects. *Arch. Gen. Psychiatry* **37**:257–263, 1980.
158. WEISS, B. L., KUPFER, D. J., FOSTER, F. G., DELGADO, J.: Psychomotor activity, sleep and biogenic amine metabolites in depression. *Biol. Psychiatry* **9**:45–54, 1974.

159. WILK, S., SHOPSIN, B., GERSHON, S., SUHL, M.: Cerebrospinal fluid levels of MHPG in affective disorders. *Nature (London)* **235**:440–441, 1972.
160. WIRZ-JUSTICE, A., FEER, H., RICHTER, R.: Circannual rhythm in human plasma free and total tryptophan, platelet serotonin, monoamine oxidase activity and protein. *Chronobiologia* **4**:165–166, 1977.
161. WIRZ-JUSTICE, A., LICHSTEINER, M., FEER, H.: Diurnal and seasonal variations in human platelet serotonin. *J. Neural Transm.* **41**:7–15, 1977.
162. ZIEGLER, M. G., LAKE, C. R., WOOD, J. H., EBERT, M. H.: Circadian rhythm in cerebrospinal fluid noradrenaline of man and monkey. *Nature (London)* **264**:656–658, 1976.
163. AMERICAN PSYCHIATRIC ASSOCIATION: *Diagnostic and Statistical Manual of Mental Disorders* (DSM) III (draft version). Task Force on Nomenclature and Statistics, 1979.

CHAPTER 48

Cerebrospinal Fluid Studies in Schizophrenia

Daniël P. van Kammen and David E. Sternberg

1. Introduction

The group of disorders known collectively as schizophrenia[15] continues to be a critical problem for modern psychiatry and accounts for large expenditures for the community (estimated at $19.6 billion annually)[66] and tragedy for many families. The chance that a person will be treated for schizophrenia in his lifetime has been reported to be about 2% (4 million people in the United States). Schizophrenia accounts for about 50% of the available mental hospital beds, or about 25% of all available hospital beds in the United States.[86] The disorder is equally common in males and females.

Many different patterns of behavior, connected by some commonly shared features, have come to be recognized and diagnosed as schizophrenia. It remains uncertain whether these are varieties of one illness or are, in fact, several different illnesses. This group of disorders is characterized by the disorganization of a previous level of functioning, by the presence of specific psychotic features during the "active" phase of the illness, by the absence of a full affective disorder, and by a tendency toward chronicity.[46] A "prodromal" and a postpsychotic "residual" phase occur frequently. The syndrome usually begins in young adulthood, although both earlier and later onsets occur and the onset can be acute or gradual. Symptoms include: *perceptual distortions*, involving hallucinations that can arise from any sensory modality, although auditory hallucinations are most common; *attentional deficits* and *disturbed thinking* that lead to impaired concept-formation and information-processing, loose associations of thought, incoherent speech and delusions, such as thinking that one is persecuted or that external forces are controlling one's thoughts and behavior, or that unrelated events (e.g., items on television) have special personal significance (i.e., ideas of reference); *disturbances of affective expression* that range from a severe reduction in the intensity of expressing affect to an inappropriateness of expressing emotions—the emotion expressed is discordant with the content of the patient's speech or ideas; and *altered motor behavior* that ranges from catatonic immobilization to frenetic non-goal-directed behavior and a number of peculiar mannerisms. Such profound disturbances obviously produce severe difficulties in accomplishing the tasks of everyday life.

Traditionally, symptom presentation established the different subtypes (e.g., catatonic, paranoid, hebephrenic, undifferentiated). Other subtypes have been defined by chronicity of the illness (e.g., acute vs. chronic, reactive vs. process schizophrenia) as well as by response to antipsychotic drugs (e.g., responders vs. nonresponders).

Although the work of Rosenthal and Kety[141] clearly indicated a major genetic contribution to the etiology of these disruptions of cognition, perception, and, presumably, brain function, nevertheless, the pathophysiology has remained obscure. The heterogeneity of symptomatology and clinical course makes the goal of finding a single biochemical etiological factor in this group of disorders suspect and suggests, rather, an underlying biochemical heterogeneity. Kety has pioneered a multifactorial view

Daniël P. van Kammen, M.D., Ph.D., and **David E. Sternberg, M.D.** • Section on Neuropsychopharmacology, Biological Psychiatry Branch, National Institute of Mental Health, National Institutes of Health, Bethesda, Maryland 20205.

of the etiology of schizophrenia, interpreting various biochemical factors as contributing to this syndrome by increasing vulnerability to the disorder. We think that such a "vulnerability hypothesis" may prove a better strategy for analyzing biochemical factors involved in schizophrenia. A vulnerability factor is a necessary but not sufficient element that contributes to the illness. Such a vulnerability factor also can be present, then, in other nonschizophrenic populations (e.g., relatives).

While remarkable progress has been made during the past thirty years in the pharmacotherapy of schizophrenia, these treatments are not ideal, for, while antipsychotic medications do improve schizophrenic symptoms, they do not seem to "cure" the illness; moreover, many patients do not respond to these drugs. In addition, treatment with these drugs may lead to permanent side effects that are at present untreatable (e.g., persistent tardive dyskinesia). However, these psychopharmacological agents have provided an important area for the interchange between psychiatry and modern biological research that offers some promise for understanding the pathophysiology of this calamitous mental illness.[11]

These antipsychotic medications and other psychiatric drugs have also prompted psychiatrists to become more rigorous in diagnosing mental disorders. While the limits of the concept of schizophrenia remain unclear,[46] the inconsistency in making this diagnosis and the lack of cross-validation of diagnostic systems among different researchers has been a major limitation to the interpretation of research data. There is a critical need to reduce diagnostic confusion through the use of reliable and valid diagnostic criteria that are based on specific objective phenomenology. The recent development of the *Research Diagnostic Criteria* (RDC)[161] and the *Diagnostic and Statistical Manual III* (DSM III)[46] should satisfy this need. Moreover, these new diagnostic systems require that to diagnose schizophrenia, the illness must persist for a minimum of at least 6 months because increasing evidence exists that "schizophrenia-like" illnesses of shorter duration have different features.

While uncertainty of classification and inconsistency of diagnosis have been major obstacles to clinical research in this area, the greatest barrier to research advancement, probably, is the inaccessibility of the living human brain to direct examination. The multiple biochemical and physiological changes that occur in the depths of various areas in the brain and that may produce the characteristics of this illness cannot be observed directly. While psychopharmacological studies using laboratory animals have led to an increase in our understanding of how drugs affect these disorders, the lack of a good animal model for schizophrenia is another serious limitation. Thus, most investigators have relied on the relatively crude approach of examining human body fluids including blood, urine, and cerebrospinal fluid (CSF), hoping to find alterations that may be clues to pathogenesis. Before we review our data and the published CSF studies in schizophrenia with an emphasis on the biogenic amines, we will discuss a number of questions that have been raised concerning the interpretation of such data.

2. Interpretation of CSF Studies in Schizophrenia

A variety of difficulties limit the interpretation of "indirect" approaches, such as CSF studies, to understanding the biology of schizophrenia. These difficulties include questions concerning the origins of the substances being measured, the limits of existing chemical assay methods, differences in the time frame of clinical vs. neurochemical effects of drugs, length of illness, biological rhythms, and difficulties in controlling for the variables that may contribute significantly to observed differences. Such intrinsic and extrinsic variations affecting the interpretation of CSF data have recently been reviewed by Wood.[187b]

2.1. Origins, Gradients, and Transport Mechanisms

An important methodological question in interpreting human CSF data concerns the extent to which lumbar CSF reflects the ventricular CSF: specifically, the question of what proportion of the substance under study is contributed by the spinal cord.[59] Concentration gradients existing for many CSF constituents has been discussed in Chapter 5. Moreover, CSF substances may largely reflect brain metabolism of those areas lying closest to CSF compartments, while those areas that are at some distance from CSF channels contribute less to substances in lumbar CSF. Another important concern in evaluating central nervous system (CNS) substances through CSF measurements is the proportion of the substances removed from the brain by the CSF system compared with the amount removed by the blood. Finally, because many substances are actively removed from CSF, the data interpretation is complicated further. Differences in transport-mechanism characteristics among individuals and psychiatric syndromes have been reported.[36] Such complications suggest that while we may be able to use CSF studies to evaluate a global, tonic state of the brain's biochemical function, we cannot use them to study the very discretely localized and rapid changes in specific brain biochemical systems.

2.2. Assay Methods

Limits in the sensitivity and specificity of existing chemical assay methods must be recognized. Lack of correlation between different assay methods (e.g., fluorometric vs. mass spectrometry, gas chromatography vs. mass spectrometry) for the same substances has been reported.[185] Clearly, this difficulty necessitates especially intensive communication and collaboration between basic scientists and clinical investigators if useful data are to be obtained.

2.3. Time Course

In studies of the effects of drugs on various substances in the CSF, the variable of time course is critical and should be specified in each study. This is emphasized by the repeated findings suggesting that a phenomenon of neurochemical adaptation occurs after the initial acute effects of drugs. Thus, studies of acute drug effects may differ significantly from investigations after chronic medication. Similar biological changes may occur during the course of the illness itself, with early stages differing greatly from later periods. The length of illness, therefore, should be stated. The need for longitudinal studies, with repeated measurements in the same patients, is obvious.

2.4. Biological Rhythms and States

Similarly, because lumbar punctures provide measurement of substances at a certain time of day, and a number of substances are known to have circadian rhythms,[192] the time of day when the tap is performed should be specified. Moreover, a number of studies have suggested that changes in biological rhythms may occur in various mental illnesses[185] and thus may alter the levels of the substances under investigation, even though the time of day or season may be controlled. Finally, the process of schizophrenic psychosis has been shown to occur in a characteristic sequence consisting of distinguishable and recognizable psychological states.[47] Each of these states may have different biological concomitants.

2.5. Other Sources of Variance

A review of the CSF literature on schizophrenia demonstrates the wide interindividual variability in the levels of various substances. The multiple potential sources of such variance and the difficulty in knowing whether these sources are or are not aspects of the disease itself are discussed in Chapter 47.

None of the studies reviewed in this chapter controlled for all the sources of variance listed. Even factors such as age, sex, age of onset, and time of day of CSF sampling, are rarely all controlled in any study. Another problem with CSF studies, similar to the problem of the lack of cross-validation of diagnostic systems among different investigators, has been the wide variety of "control" groups studied in comparison to schizophrenic patients. The nature of the control group usually is not well defined, and it is rarely a truly "normal" group. Controls are often neurological patients who may, in fact, have abnormalities of the substances under study.

3. Biogenic Amines

Many hypotheses about the etiology of schizophrenia have focused on changes in central catecholamine and indoleamine metabolism. The psychotomimetic effect of LSD, with its marked effect on serotonergic neurons, and the study of methylated serotonin metabolites have suggested a role for serotonin in psychosis.[55,188,189] Deficits in schizophrenia of norepinephrine (NE) and of dopamine-β-hydroxylase (DBH), the enzyme that converts dopamine (DA) to NE, have also been postulated.[69,163,187] Most evidence, however, has implicated a role for DA in schizophrenia.[109,177,179] Evidence supporting the "dopamine hypothesis of schizophrenia," which suggests that there is an increase in DA activity in specific CNS sites in schizophrenic patients, includes the following pharmacological observations: the antipsychotic neuroleptics block DA receptors[5] and produce a feedback increase in DA turnover[39]; their ability to block DA receptors correlates significantly with their clinical antipsychotic potency[36a,152,160]; drugs like amphetamine and methylphenidate in high and chronic dosages can produce a schizophrenic-like syndrome in normals[159]; and drugs that increase central DA function exacerbate schizophrenia.[6,76,77,177,179,180] Many investigators have concluded from such data that the neuroleptic medications act by decreasing central DA transmission.

3.1. Dopamine

If schizophrenia is associated with a generalized increase in brain DA activity, then the breakdown product of DA, homovanillic acid (HVA), might be elevated in the CSF. For example, the increase in brain DA turnover that occurs when the DA receptors are blocked by neuroleptic drugs is reflected in increased HVA levels in CSF.[56,126] However, measurement of HVA levels in the CSF of schizophrenic patients has failed to demonstrate differences in levels from controls (see Table 1). While Rimon *et al.*[136] found increased HVA in patients with paranoid symptoms, other investigators have not replicated this finding. Only Bowers[23,24] and Post *et al.*,[129] using the probenecid technique, have found significant CSF HVA differences within the schizophrenic group. Bowers reported that patients with Schneiderian symptoms or a poor prognosis tended to have lower HVA accumulations; Post and associates also found lower HVA accumulations in those patients with more Schneiderian symptoms within a population of good-prognosis patients. In addition, they found that after recovery from psychosis, the schizophrenic patients had lower probenecid-induced accumulations of HVA than during the active stage of psychosis. Post and co-workers[129] suggested that DA turnover may be reduced in the nonpsychotic phase and, possibly, is an indication of a trait-related receptor supersensitivity in schizophrenic patients, thereby causing the lower DA turnover that is observed during remission with increased DA turnover occurring during the periods of active psychosis.

Van Praag and Korf[182] emphasized the importance of DA metabolism for the antipsychotic effects of neuroleptics when they found a significant correlation, after 1 week's treatment with neuroleptics, between the percentage increase of HVA and the degree of clinical improvement. In rats, Stawarz *et al.*,[162] however, found no correlation between the antipsychotic potency of neuroleptics and their ability to produce an acute increase in HVA in the striatum or limbic forebrain.

The full antipsychotic response to neuroleptic drugs generally requires a number of weeks longer than their immediate DA receptor blockade, as evidenced by their causing an immediate rise in plasma prolactin levels.[108] While DA-receptor blockade seems necessary for the therapeutic effect of neuroleptics, the blockade may allow other slower "secondary processes" to take place, and they are probably responsible for the therapeutic change. Therefore, CSF studies during the period of most pronounced clinical effect—i.e., after chronic neu-

Table 1. CSF Homovanillic Acid in Schizophrenic Patients Compared to Controls

Ref. No.	HVA[a]	Diagnosis[b]	Controls[c]	Comments
Baseline studies				
126	N.S. (4)	CS	PSY, NORM	Patients on neuroleptics.
27	N.S. (6)	?	PSY, NORM	—
136	N.S. (30)	AS	PSY	Higher in paranoids compared to controls or nonparanoids.
129	N.S. (17)	AS	PSY, NORM, NEUR	—
Studies using probenecid accumulations				
22	N.S. (12)	AS	NORM	Low-dose probenecid.
23	N.S. (18)	AS, CS	PSY, NORM (inmates)	Low in Schneiderian-positive.
124	Low (17)	AS, CS	PSY	Lowest in poor prognosis.
151	N.S. (34)	AS, CS	PSY	Higher in women and manics.
129	N.S. (20)	AS	PSY	Good-prognosis patients only; no correlation with prognosis. Low in Schneiderian-positive. Not higher in paranoids. Lower on recovery compared to before or to manics.

[a] (N.S.) Not statistically significant; number of subjects in parentheses.
[b] (CS) Chronic schizophrenia; (AS) acute schizophrenia.
[c] (PSY) Psychiatric nonschizophrenic controls; (NORM) normal controls; (NEUR) neurological controls.

roleptic treatment—could prove especially elucidating. Although acute studies have shown that neuroleptics increase DA turnover and, thus, CSF HVA, presumably via the feedback response to their blockade of DA receptors, reduction in the feedback response over time has been noted in the striatal and, possibly, the limbic system.[29,116] Thus, neuroleptic-induced increases in CSF HVA have been found to decrease markedly after chronic treatment (3 weeks).[130,144] Indeed, Bunney and Grace[33] studied nigrostriatal DA neuronal activity in rats using extracellular single-neuron recording and microiontophoretic techniques; they found the expected increased firing of DA cells during acute neuroleptic treatment. However, chronic treatment produced an almost total absence of firing of DA cells that, they suggested, was secondary to the neuroleptic-induced increase in activity proceeding to a point of chronic depolarization block. Bowers *et al.*[28] recently reported that when they compared acute and chronic neuroleptic treatment of schizophrenic patients, they found that while some individual patients showed a decrease in CSF HVA during the chronic period, the mean CSF HVA was not significantly lower and thus no tolerance had developed in the group as a whole. Similarly, after chronic neuroleptic treatment (4–6 weeks), we have noted a trend for neuroleptic-responsive patients to continue to have increased CSF HVA ($\bar{x}$ change = $+102.29 \pm 133.22$ nmol/ml, $n = 7$) while nonresponders have the same or decreased ($\bar{x}$ change = -79.00 ± 87.50 nmol/ml, $n = 3$) levels.[103] We found, in addition, that after chronic neuroleptic treatment, the decrease in the patients' psychoses correlated with the increase in their CSF HVA ($r = 0.65$, $p < 0.05$). Additional longitudinal studies obviously are needed.

While van Praag and Korf[181] suggested that the level of CSF HVA reflected the patient's motor activity and Post *et al.*[131] showed that experimentally induced increases in psychomotor activity significantly elevated the levels of CSF HVA, Kirstein *et al.*[88] and Sedvall *et al.*[151] found no significant relationships between CSF HVA and measured motor activity.

The major contributor to CSF HVA seems to be the corpus striatum and, specifically, the caudate nucleus that borders on the lateral ventricles and provides its metabolites with direct access to CSF. Thus, many investigators criticize the CSF HVA findings and feel that CSF HVA may not reflect the DA metabolism of limbic and cortical areas, areas that are more relevant to emotional and cognitive functioning and that, while possibly critical to the pathophysiology of schizophrenia, may not contribute much to CSF HVA. However, there is evidence that the corpus striatum may be involved in schizophrenic behavior[110] and that the limbic and cortical DA systems are functionally and anatomically linked to the corpus striatum.

Another measurement of dopaminergic activity might be provided by future studies of CSF 3,4-dihydroxyphenylacetic acid (DOPAC), an intermediate metabolite of DA that is considered to reflect the amount of DA released and recaptured by nerve endings and then oxidized.[142] It may be a better index of the DA neuron firing rate than HVA.[143] In addition, we found no reports of direct measurement of CSF DA.

Another factor complicating these CSF studies is the concept of receptor supersensitivity: the variations of receptor response in relation to the availability of its neurotransmitter. Thus, with decreased neurotransmitter release, the receptor may become supersensitive. Such changes in the sensitivity of DA receptors have been demonstrated following a number of drugs, including neuroleptics.[42] Both Bowers[24] and Curzon[38] have pointed out that if there are supersensitive brain DA receptors in schizophrenia, then increased functional DA activity could exist in the face of low or normal HVA levels. Indeed, several groups of investigators have measured DA-receptor binding in autopsied brains of schizophrenic patients and controls.[92,98,119] Increased binding in the nucleus accumbens, putamen, or caudate of schizophrenic patients as compared to controls was reported using [^{3}H]spiroperidol and [^{3}H]haloperidol binding.[92,119] While the possibility that these findings are due to previous treatment with neuroleptics has not been sufficiently proven yet, such studies point to the importance of investigations directed at postsynaptic, rather than only presynaptic, phenomena.

3.2. Norepinephrine

Stein and Wise[163] proposed that 6-hydroxydopamine (6-OH-DA) could be formed endogenously from DA in the synaptic cleft by oxidation and suggested that the lack of goal-directed behavior and anhedonia noted in schizophrenic patients could be secondary to the destruction of the CNS noradrenergic "reward system" by 6-OH-DA. They also suggested that the formation of 6-OH-DA might be due to low activity of the enzyme DBH that converts DA into NE in NE neurons. Hartmann[69] suggested that low brain DBH, which could produce an excess of DA and a relative deficit of NE, may be the pathophysiology of schizophrenia. Further evidence for the support of noradrenergic mechanisms in schizophrenia comes from the many studies showing that the antipsychotic drugs interfere in many ways with central adrenergic transmission.[168] Recent evidence that the NE-receptor blocker propranolol may be effective chemically as an antipsychotic drug for schizophrenia further stresses the importance of noradrenergic systems for antipsychotic drug action.[138]

DBH has been measured in the postmortem brains of schizophrenic patients, as well as in their plasma and CSF. While Wise and Stein[187] reported significant reductions in brain DBH activity in schizophrenic patients as compared to controls, this was not replicated.[191] Moreover, there is no evidence that DBH content, a trait that is stable and genetically determined, rather than state-variable, is altered in the plasma or CSF of schizophrenic patients.[93,107] Using a highly sensitive radioenzymatic assay, our study found no significant difference in CSF DBH activity of schizophrenic patients ($\bar{x} = 0.0228 \pm 0.0095$ nmol/ml per hr per mg protein) and normal controls ($\bar{x} = 0.0232 \pm 0.0091$ nmol/ml per hr per mg protein).[164] However, using the Phillips prognostic scale, we did find that patients with reactive schizophrenia had lower DBH activity in their CSF than patients with process schizophrenia, that the DBH of these two groups differed significantly ($p < 0.02$, Wilcoxin rank sum test), and that DBH activity was a significant predictor of the process-reactive dichotomy ($X^2 = 8.881$, $p < 0.005$). Thus, while perhaps not of primary etiological significance, a relationship between DBH and schizophrenia may exist in that the effects of low DBH activity in a subgroup of patients may increase the vulnerability to psychosis in individuals made susceptible to schizophrenia by other factors. Support for such a vulnerability hypothesis comes from reports that: (1) those people who received disulfiram, a potent DBH inhibitor, during treatment of alcoholism and who subsequently became psychotic were those with genetically low DBH activity[50,100]; (2) psychiatrically normal controls with low serum DBH tended to have attentional deficits[32]; and (3) alcoholics with lower CSF DBH displayed significantly more personality pathology on psychological testing than did alcoholics with higher CSF DBH.[99] Finally, Böök and Wetterberg[21] recently reported that in a group of genetically isolated persons in Sweden, schizophrenia is significantly associated with both low serum DBH and low platelet monoamine oxidase (MAO).

Further attempts to investigate NE activity have included the measurement of NE metabolites in CSF (Table 2). A study of CSF vanillylmandelic acid (VMA), the acid metabolite of NE, showed no difference in schizophrenic patients from controls.[81] Two studies found no difference in 3-methoxy-4-hydroxyphenylethylene glycol (MHPG), the major metabolite of CNS NE, in the CSF of schizophrenic patients as compared to controls, nor any change during recovery.[129,155] Post *et al.*,[129] however, did find significantly lower MHPG levels in those acute

Table 2. CSF Norepinephrine and Its Metabolites in Schizophrenic Patients Compared to Controls

Ref. No.	MHPG[a]	VMA[a]	NE[a]	Diagnosis[b]	Controls[c]	Comments
Baseline studies						
155	N.S. (26)	—	—	AS	PSY, NORM	—
129	N.S. (18)	—	—	AS	PSY, NORM, NEUR	Lower in patients with higher anxiety.
81	—	N.S. (12)	—	AS	PSY, NEUR	No change on recovery.
165[90a]	—	—	High (26)	AS, CS	NORM	Possible high paranoid subgroup. Neuroleptic-induced decreased NE correlated with improvement.
Probenecid study						
129	N.S. (17)	—	—	AS	PSY	No change on recovery.

[a] (N.S.) Not statistically significant; number of subjects in parentheses. (NE) Norepinephrine; (MHPG) 3-methoxy-4-hydroxyphenylethylene glycol; (VMA) vanillylmandelic acid.
[b] (AS) Acute schizophrenia; (CS) chronic schizophrenia.
[c] (PSY) Psychiatric nonschizophrenic controls; (NORM) normal controls; (NEUR) neurological controls.

schizophrenic patients rated as experiencing high subjective stress. However, we have found a significant positive relationship between CSF MHPG and the acute change in psychosis when the patient was given amphetamine ($N = 28$, $r = 0.51$, $p < 0.005$).[178] The finding of Sedvall *et al.*[150] of a significant positive correlation between the decrease in CSF MHPG of schizophrenic patients treated with chlorpromazine and their decrease in psychosis emphasizes the importance of interaction with noradrenergic mechanisms for the antipsychotic effects of neuroleptic drugs.

Recently, a preliminary report found above-normal NE levels in the limbic forebrain by postmortem examinations of four patients with chronic paranoid schizophrenia.[51] We measured the CSF NE levels of drug-free schizophrenic patients and normal controls, finding a significantly increased level of CSF NE in schizophrenic patients [schizophrenics ($N = 35$): 125 ± 11 pg/ml; controls ($N = 29$): 91 ± 6 pg/ml; $p < 0.01$], as well as a possible subgroup of paranoid patients with markedly elevated CSF NE.[90a,165] However, other paranoid patients had NE levels within the range of the normal controls. Moreover, like Sedvall and associates' study of CSF MHPG,[150] we found that pimozide, a neuroleptic, produced a decrease in NE levels in all our patients (decrease $\bar{x}$ ± SDM NE = 74.83 ± 35.13 pg/ml) and that this pimozide-induced decrease in NE showed a significant positive relationship to the decrease in psychosis ($r = 0.71$, $p < 0.02$) (see Fig. 1). Since no significant correlation was found between CSF MHPG or VMA and NE (possibly reflecting differences in half-life and transport characteristics between NE and its metabolites), further study of NE in CSF may prove especially useful in elucidating the role of the noradrenergic system in schizophrenic patients and in antipsychotic drug action.

3.3. Serotonin

Two areas of investigation have led to the hypothesis that the serotoninergic system may play some role in schizophrenia. First is the evidence that the powerful psychotomimetic drug LSD profoundly affects serotonin, inasmuch as minute doses of LSD markedly inhibit serotoninergic cell-firing.[1] Because serotonin is an inhibitory neurotransmitter that seems to be involved in the control of sleep, sensation, attention, and mood, LSD seems to act as a psychotomimetic by disinhibiting these symptoms. The second indicator pointing to a relationship between psychosis and serotonin came after Osmond and Smythies[118] noticed the similarity in the chemical structure of the hallucinogen mescaline and the catecholamines. They then hypothesized that aberrant methylation of a number of biogenic amines (including serotonin) leads to the

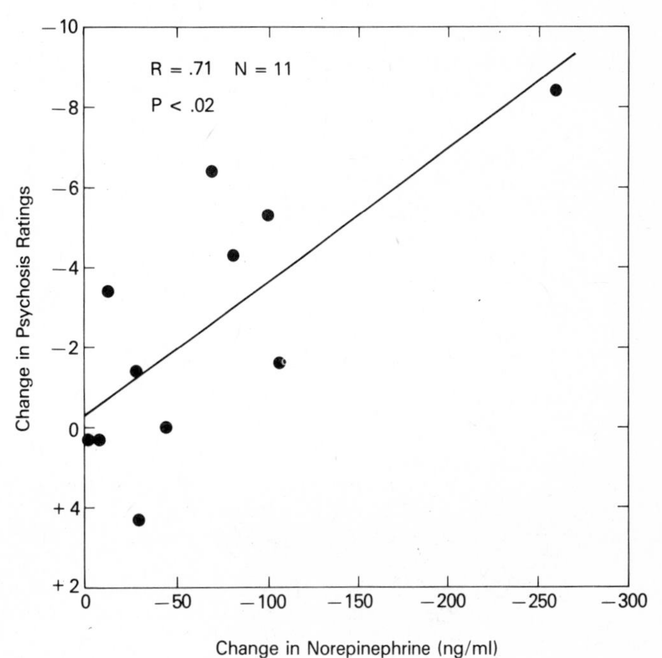

Figure 1. Graph showing the decrease in CSF NE relative to the decrease in psychosis ratings during pimozide treatment of 11 schizophrenic patients.

accumulation of endogenous psychomimetics such as psilocybin, bufotenine, and dimethyltryptamine and may thus play an etiological role in schizophrenia.

Therefore, a number of investigators have studied possible differences in serotonin and its principal metabolite, 5-hydroxyindoleacetic acid (5-HIAA), in the CSF of schizophrenic patients compared to controls. No consistent differences have been found (see Table 3). In the baseline studies of Ashcroft *et al.*,[8] lower CSF 5-HIAA in drug-free acute schizophrenic patients was found; Bowers *et al.*,[27] in their baseline study, found lower CSF 5-HIAA in drug-free psychotic patients, but most authors have found no difference. Although Bowers[23] found that probenecid-induced accumulations of CSF 5-HIAA tended to be high in schizophrenic patients having Schneiderian symptoms, Post *et al.*[129] found them to be lower. Similarly, while Bowers[23] and our group[103] found that neuroleptics decreased CSF 5-HIAA in the acute and chronic states, Goodwin and Post[64] and Ruther *et al.*[144] found no effects of neuroleptics on CSF 5-HIAA.

It appears unclear whether CSF 5-HIAA levels are a trait- or state-related phenomenon. While, in general, the higher the state of "arousal" the lower the serotonin turnover,[25] a negative correlation between motor activity of agitation and CSF 5-HIAA was found in schizophrenic patients.[88] Bowers[26] suggested that some acute psychoses, therefore, may have a primary inhibitory serotonin deficit or that they may have an endogenous psychotomimetic compound that acts on serotonin receptors (a trait-related phenomenon). He further suggested that the higher CSF 5-HIAA levels in the "unaroused" chronic schizophrenic patients may be secondary to their genetic deficiency of the type B form of the MAO enzyme.[26] Although there is no clear evidence pointing to a consistent abnormality of the serotonin system in schizophrenia, one way of exploring its role may be to investigate the relationship between serotonin and other neurotransmitters or enzymes and to identify subgroups with markedly low or high serotonin turnover.

3.4. Pteridine Cofactor

Another approach in evaluating catecholamine and indoleamine activity in the brain is to evaluate CSF tetrahydrobiopterin (BH_4) levels. There is considerable evidence that biopterin plays a major role in the regulation of catecholamine synthesis. BH_4 is a cofactor for the enzymes that hydroxylate tyrosine, phenylalanine, and tryptophan.[83,102,186] These enzymes are the rate-limiting steps in the synthesis of DA and NE, phenylethylamine (PEA), and serotonin, respectively. Mandell[101] concluded from his data that BH_4 could be particularly important in the regulation of DA synthesis.

Recently, mental retardation syndromes in children with low cofactor activity have been observed;[83,90] this association would preclude a role for low cofactor in schizophrenia, which usually begins in adolescence and early adulthood. However, McGeer and McGeer[104] reported that tyrosine hydroxylase activity decreases substantially in animals during puberty. This decrease in enzyme activity may be related to a change in the level of cofactor. Thus, it may be hypothesized that the decrease in

Table 3. CSF 5-Hydroxyindoleacetic Acid in Schizophrenic Patients Compared to Controls

Ref. No.	5-HIAA[a]	Diagnosis[b]	Controls[c]	Comments
Baseline studies				
53	N.S. (11)	AS	PSY	—
8	Low (7)	AS, CS	PSY, NEUR	Low only in AS.
126	N.S. (40)	CS	PSY, NORM	Patients on neuroleptics.
27	Low (7)	?	PSY, NORM	—
136	N.S. (22)	AS	PSY	—
129	N.S. (18)	AS	PSY, NORM, NEUR	—
Probenecid studies				
23	N.S. (18)	AS, CS	PSY, NORM (inmates)	Higher in Schneiderian-positive
24	N.S. (17)	AS, CS	PSY	Not different in poor prognosis.
129	N.S. (18)	AS	PSY	Lower in Schneiderian-positive. No change on recovery.

[a] (N.S.) Not statistically significant; number of subjects in parentheses. (5-HIAA) 5-Hydroxyindoleacetic acid.
[b] (AS) Acute schizophrenia; (CS) chronic schizophrenia.
[c] (PSY) Psychiatric nonschizophrenic controls; (NORM) normal controls; (NEUR) neurological controls.

BH_4 that may develop during puberty, once basic intellectual functions have been established, could play a role in the development of schizophrenia, which starts frequently in late adolescence or adulthood. Alterations in BH_4 levels could lead to functionally important changes in the production of DA, NE, serotonin, or PEA. Preliminary data on cofactor concentrations in the CSF of schizophrenic patients and normal comparison groups showed that there is no significant difference between the means in these two groups (18 schizophrenics: 17.5 ± 1.7 pmol/ml; 31 controls: 18.5 ± 1.3 pmol/ml). Further analysis of the data is needed to determine whether other CSF components and diagnostic subgroup divisions would relate to cofactor activity in these patients.[180a] Furthermore, psychotropic drugs might affect cofactor activity, since the affinity of the enzyme for the cofactor is regulated by substrate concentration, calcium, and firing rate of the neuron—factors known to be affected by neuroleptic medication. On the other hand, localized abnormalities in cofactor regulation cannot be excluded by measuring CSF levels. Moreover, a recent study in rats using a specific inhibitor of BH_4 synthesis reduced cerebral pools of BH_4 by half and yet produced no effect on cerebral levels of DA, NE, or serotonin.[154] Thus, the absolute level of BH_4 may not regulate amine biosynthesis *in vivo*. Studies evaluating the relationship between the monoamine metabolites and cofactor values, as well as whether cofactor concentration has effects on trait-related variables, need to be explored further.

3.5. Cyclic Nucleotides

There is increasing evidence that the many receptors with which various hormones and neurotransmitters bind have an adenylate cyclase system connected with the receptor[139] and that the cyclic nucleotides, adenosine-3′,5′-cyclic monophosphate (cAMP) and, possibly, guanosine-3′,5′-cyclic monophosphate (cGMP) function as second messengers in mediating the receptor action of numerous neurotransmitters. Dopamine, NE, serotonin, histamine, and perhaps substance P have been shown to stimulate cAMP production in neuronal tissue.[35,85] A number of studies have linked cholinergic-receptor function with cGMP accumulation in different tissues including brain.[52,120] Adenylate-cyclase-mediated receptors have been found for DA in the caudate nucleus, substantia nigra, and limbic forebrain[85,112] and for NE in the limbic forebrain.[18]

In the brain, this system has been described best for DA in areas wherein both DA and DA receptors have been identified.[85] Recently, Kebabian[84] classified brain DA receptors into those associated with and those independent of DA-sensitive adenylate cyclase. The postsynaptic action of DA includes the activation of a specific DA-sensitive adenylate cyclase that converts adenosine triphosphate to cAMP.

Logically, drugs that would interfere with the monoamine-induced formation of cAMP presumably should block the flow of cAMP-mediated postsynaptic responses. Thus, it is of interest that nearly all neuroleptics have been shown to potently inhibit DA-sensitive adenylate cyclase stimulation in striatal and mesolimbic areas and that within its chemical class, the potency of a drug that blocks DA-sensitive adenylate cyclase correlates well with its clinical antipsychotic potency.[96,168] Antipsychotic drugs may also inhibit the cAMP response to NE.[17,168]

The value of measuring CSF cyclic nucleotides for understanding schizophrenia lies in their alleged receptor origin, thereby complementing CSF studies of neurotransmitters with their presynaptic origin. Thus, changes in neurotransmitter levels or in the sensitivity of neurons to these neurotransmitters might be reflected by shifts in cyclic nucleotide levels in CSF and may reflect the activity of postsynaptic receptors. However, studies of cyclic nucleotides in CSF are limited because many receptors, including some DA receptors, function without adenyl cyclase;[84] at the same time, so many compounds do affect cyclic nucleotide synthesis that the meaning of a given level in CSF is ambiguous. Furthermore, it is unclear how levels of CSF cyclic nucleotides reflect levels in brain. Tsang *et al.*[174] suggested the spinal cord as the origin for human lumbar CSF cAMP, while Sebens and Korf,[149] in their study with rabbits, found that intraventricular injection of NE or DA raised CSF cAMP. However, in their attempt to identify the specific neurotransmitter system of origin for human CSF cAMP and cGMP, Belmaker *et al.*[10] found no effect of the DA precursor L-DOPA nor of the NE β-receptor blocker propranolol on the CSF cyclic nucleotides. (See Chapter 9.)

Recently, two studies found no significant difference between CSF cAMP in unmedicated schizophrenic patients and in controls[13,158] (see Table 4). However, one of the studies found significantly greater variance of CSF cAMP in the schizophrenic group: the two patients who had markedly high cAMP also had a very poor treatment response to neuroleptics and had signs of poor prognosis.[13] In a follow-up study, this same group reported that

Table 4. CSF Cyclic Nucleotides in Schizophrenic Patients

Ref. No.	cAMP[a]	cGMP[a]	Comments
Schizophrenic patients vs. controls			
158	N.S.	—	No change with neuroleptic.
13	N.S.	—	Neuroleptic → decrease in responders, no change in nonresponders. Greater variance in schizophrenic group.
158	—	Lower	Chronic neuroleptic → nonsignificant increase.
48	—	N.S.	Lower but not quite significantly. Chronic neuroleptic → significant increase.
29a	Lower	—	Compared to depressed patients chlorpromazine → significant decrease.
Among schizophrenic patients			
196	—	—	Higher cAMP in poor prognosis.
28	—	—	cAMP positively correlates with psychosis ratings. cAMP negatively correlates with change in psychosis on neuroleptics.

[a] (N.S.) Nonsignificant. (cAMP) Cyclic adenosine-3′,5′-monophosphate; (cGMP) cyclic guanosine-3′,5′-monophosphate.

factors related to poor prognosis (i.e., diagnosis of simple schizophrenia, slow first remission, poor intermorbid adjustment, young age of onset, absence of hallucinations) were significantly associated with higher levels of CSF cAMP.[196] They suggested that if CSF cAMP does reflect the activity of catecholamine receptors in the brain, the high levels in the CSF of poor-prognosis patients might be evidence of a catecholamine-receptor hypersensitivity in that subgroup. A postmortem measurement, however, of DA-sensitive adenylate cyclase activity in caudate from brains of schizophrenic patients and controls found no difference in either the baseline state or the response to stimulation by DA.[34] Although Biederman *et al.*[14] found a decrease of CSF cAMP in schizophrenic patients who improved with neuroleptic and no change in two nonresponders, Smith *et al.*[158] found no effect of neuroleptics. Recently, Bowers *et al.*[28,29a] reported a significant negative correlation between CSF cAMP and the change in psychosis during both acute and chronic neuroleptic treatment, as well as significant positive correlation with psychosis ratings.

Interestingly, Kafka *et al.*[82a] and Rotrosen[143a] observed that cAMP response to stimulation with prostaglandin E_1 in platelets of schizophrenic patients was decreased. This less responsive (i.e., subsensitive) adenylcyclase is linked to an α-receptor. Whether this subsensitivity is also present in the CNS remains a question for future study. (See also Section 6.3.)

While the origin of cGMP in human lumbar CSF is not known, some investigators have suggested it may be an index of brain cholinergic activity.[52,120] This could then elucidate the theory of a DA–acetylcholine imbalance in schizophrenia. Both Ebstein *et al.*[48] and Smith *et al.*[158] found lower levels of CSF cGMP in schizophrenic patients, achieving significance in the latter study (see Table 4). Both groups also found that chronic neuroleptic administration produced an increase in cGMP, but it was significant only in the former study. The studies disagree on the presence of a correlation between CSF cAMP and cGMP.

3.6. Endogenous Psychotogens

3.6.1. Transmethylation Hypothesis of Schizophrenia. Investigators of the hypothesis first suggested by Osmond and Smythies[118]—that aberrant methylation of the catecholamines and indolamines leads to an accumulation of an endogenous psychotogen—have focused on: (1) pharmacological interventions to alter the production of these substances; (2) studies to identify enzymes that might be involved in the synthesis of these compounds; and (3) attempts to detect these methylated derivatives in biological fluids. Corbett *et al.*[35a] did not find statistically different CSF levels of hallucinogenic *N*-methylated indolealkylamines in acute schizophrenia; however, several investigations of different methylated derivatives have found no consistent differences either quantitatively or qualitatively in blood and urine of schizophrenic patients as compared to controls.[60] However, because the number of candidates proposed as "schizotoxins" is large and not all have been studied, failure to identify the abnormally methylated "schizotoxin" does not rule out its existence, especially in a subgroup. Moreover, inasmuch as these substances generally are metabolized rapidly and may in fact be normal body constituents, longitudinal studies investigating changes in accumulation, excretion, or sensitivity may be far more valuable than measuring absolute concentrations one time. Again, urine, blood, and even CSF may be too far removed from the changes occurring in specific brain nuclei to be informative.

3.6.2. Phenylethylamine. Wyatt *et al.*[190] and Sandler and Reynolds[145] have suggested a role in schizophrenia for PEA, the compound formed endogenously in mammals by decarboxylation of phenylalanine. They base their proposal partly on the structural and pharmacological similarities of PEA and amphetamine. PEA readily passes through the blood–brain barrier, can produce in rats amphetamine-like hyperactivity and stereotypy that can be diminished by neuroleptics, and is metabolized rapidly to phenylacetic acid by MAO. Thus, the PEA hypothesis suggests that increased PEA activity could produce a syndrome similar to amphetamine psychosis. One mechanism for increased PEA activity would be decreased metabolism, which could be caused by the decreased MAO activity observed in some schizophrenic patients. A specific assay for urine PEA found no increase of PEA in schizophrenic patients compared to controls.[190] Recently, however, Sandler *et al.*,[146] using a sensitive and specific assay, reported a significant increase in the concentration of phenylacetic acid, the major metabolite of PEA, in the CSF of schizophrenic patients compared with controls. Potkin *et al.*[131a] reported increased urinary PEA excretion in paranoid schizophrenic patients. Further well-controlled (i.e., drug-free and with a controlled diet) acute and longitudinal studies of both PEA and phenylacetic acid in CSF are required to determine whether altered PEA metabolism plays a physiological role in the onset and course of schizophrenia.

3.6.3. LSD-Displacing Factor. Recently, Mehl *et al.*[106] reported on unidentified substances in human CSF that were capable of reversibly displacing LSD from its high-affinity binding sites, presumably by binding to specific LSD receptors. The "LSD-displacing factors" (LDFs) were different from serotonin, tryptamine, and 5-methoxytryptamine. Moreover, significantly higher levels of LDF were found in a group of unmedicated acutely psychotic patients, most of whom had been diagnosed schizophrenic. The highest levels of LDF were found in the subgroup of patients who responded best to neuroleptic treatment.[106] Stolzki *et al.*[167] found that a number of fractions of LDF could be separated from rat brain extract, and they confirmed the existence of multiple types of LSD-binding sites. Separating several of these fractions is required so that an LDF assay can investigate CSF from schizophrenic patients. Clearly, longitudinal studies using CSF are needed to determine whether such LSD-displacing factors are high only during psychosis and then return to normal during remission.

4. γ-Aminobutyric Acid

Although γ-aminobutyric acid (GABA) has long been known to be a constituent of brain tissue by playing a role in normal brain functioning as part of carbohydrate metabolism through the Krebs cycle, this amino acid has more recently been found to act also as an inhibitory neurotransmitter in the mammalian CNS.[37] While it is a widely distributed neurotransmitter, it appears to have a discrete distribution, often in the areas of CNS with high dopamine (DA) concentration,[124] and it seems to modulate the activity of dopaminergic neurons.[166] An inhibitory striatonigral GABA pathway has been described that presumably functions as a long-loop feedback from postsynaptic to presynaptic DA neurons.[117] Indeed, local applications of GABA on nigral DA neurons inhibit their spontaneous firing[2] but increase DA activity when applied in the zona reticulata. This and other evidence suggests that the GABA system may have partial control over the DA system or vice versa.

Roberts[137] proposed that the basic biochemical fault in schizophrenia may arise from a defective GABA system. As a variation of the DA hypothesis of schizophrenia, van Kammen[178] suggested that an underactive GABA system in schizophrenia would decrease the inhibitory influence of the GABA neurons on DA systems.[176] Then, DA activities that should be suppressed are not, thus producing inappropriate behavior. Suggestive evidence for this hypothesis was provided by Stevens *et al.*,[166] who found that microinjection of a GABA antagonist into a DA-rich area of a cat led to psychotic-like behavior. Further understanding of the feedback inhibition of DA neurons by GABA was reported, giving greater credence to a possible role for GABA in the pathophysiology of schizophrenia.[176] Pharmacological tests of this hypothesis in schizophrenic patients using drugs that presumably manipulate the GABA system have produced conflicting results, including a significant worsening of psychotic symptoms.[148]

Methods for measuring GABA in bodily fluids have been developed[49a,61,68a] and verified.[187a] Thus, measuring GABA in the CSF of drug-free schizophrenic patients and comparison groups might reveal a dysfunction of GABA neurons in schizophrenia.[91] While CSF GABA levels may probably reflect CNS GABA activity, Glaeser *et al.*[62] found reduced CSF GABA levels in patients with Huntington's chorea, a disease in which reduced brain GABA levels and glutamic acid decarboxylase (GAD) activity have been demonstrated. Two groups found that drug-free schizophrenic patients, in general, do not have lower GABA activity in CSF,[94,193] although six of the seven lower levels were found to belong to schizophrenic patients.[94] Yet van Kammen *et al.*[178] found that GABA levels were significantly lower in 17 schizophrenic patients compared to levels in 40 normal controls (191 ± 9 vs. 233 ± 13 pmol/ml, $p < 0.01$) (Fig. 2), which is consistent with the study by Perry *et al.*,[125] who reported a postmortem study with a decrease in GABA levels in specific brain areas of schizophrenic patients—a decrease as marked as in patients with Huntington's chorea. We noted a moderate negative correlation with CSF glucose ($r = -0.61$, $p < 0.02$), which suggests a relationship between GABA levels and metabolic activity and may actually reflect GABA transport in and out of the CSF. We also observed a positive correlation with duration of illness ($r = 0.54$, $p < 0.02$) and with the number of hospitalizations ($r = 0.50$, $p < 0.05$), which indicates that recently ill schizophrenic patients have lower levels than more chronically ill patients. In an analysis of variance with normal controls and affectively disordered patients, significantly lower levels were found for the schizophrenic patients ($p < 0.05$). The two highest levels in our patients belonged to schizoaffective schizophrenic patients who showed manic behavior (see Chapter 47).

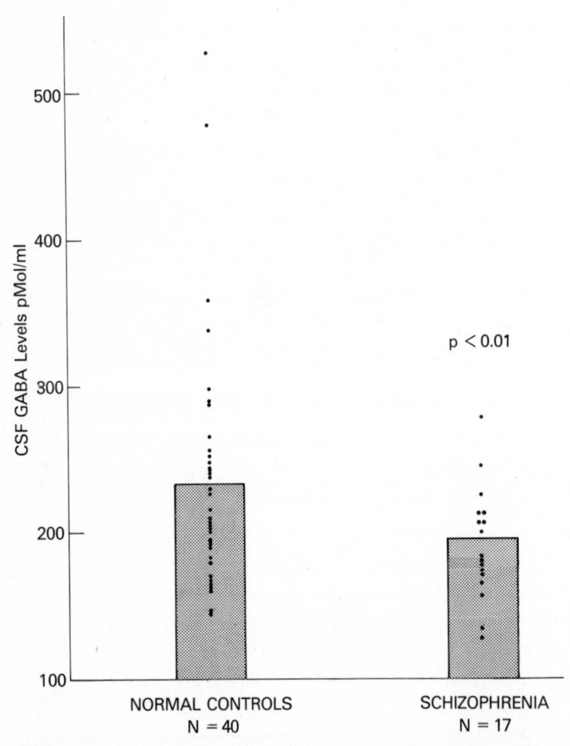

Figure 2. Low CSF γ-aminobutyric acid (GABA) levels in schizophrenia. Mean lumbar CSF GABA concentration among 17 drug-free schizophrenic patients were significantly lower than that found in 40 normal control individuals. CSF samples were collected on ice and stored immediately at −70°C. Thereafter, these specimens were assayed on blind basis by Dr. Theodore A. Hare employing ion-exchange/fluorometric methodology.[68a,187a] Mean CSF GABA level in these schizophrenic patients was not significantly different from that in affectively disordered patients (see Chapter 47).

Discrepant findings in the literature may be explained by methodological differences. All our samples were collected under identical conditions on two metabolic research units. Zimmer *et al.*[193] stored their samples at −40°C. Recently, it has been shown that CSF needs to be stored at temperatures lower than −40°C (i.e., −70°C), otherwise GABA levels will increase.[64a] In general, GABA levels were found to be higher in the studies by Lichtshtein *et al.*[94] and Zimmer *et al.*[193] than in our study. We found a nonsignificant increase in GABA levels following chronic pimozide treatment levels in six schizophrenic patients. Zimmer and colleagues attributed the low levels in the study of Lichtshtein and colleagues to that group's use of anticholinergic, antiparkinsonian drugs. However, the normal levels in our more chronic patients may have increased over time as a result of previous chronic neuroleptic treatment or as a compensation to the initially lower levels. So, our data suggest that brain autopsy studies that include only brains of chronic schizophrenic patients for reasons of certainty of retrospective diagnosis will not reveal a potential GABA deficit.

Increasing CNS GABA activity may not be helpful in schizophrenia, for even patients with normal or high CSF GABA levels may have such levels secondary to subsensitive GABA receptors. Drugs such as muscimol, a GABA-receptor agonist, or lioresal and γ-hydroxybutyrate that have inhibitory effects similar to GABA may actually induce worsening of psychotic symptoms in some schizophrenic patients.[148] Yet this drug-induced increased schizophrenic symptomatology still may imply dysfunctional (supersensitive) GABA and DA systems in schizophrenia when our understanding of these systems increases. Finally, persistent tardive dyskinesia, a movement disorder that probably results from neuroleptic-induced DA-receptor supersensitivity, may be a more appropriate target for raising GABA activity in the brain through low-dose treatment with these drugs.

5. Peptides

Two kinds of experimental studies have suggested that naturally occurring peptides may play a role in regulating CNS functions and, consequently, in influencing behavior[44,45]: (1) peptides that had been isolated from brain and pituitary were found to produce behavioral changes independent of their classic endocrine effects in animals[44,45]; and (2) immunohistochemical studies of peptides have revealed the distribution of some peptides that affect neurotransmission in discrete neuronal pathways in the brain.[111] Neuropharmacological animal and human studies using endogenous peptides, such as hypothalamic-releasing hormones, anterior and posterior pituitary hormones, or their fragments and oligopeptides such as substance P, have shown that many different peptides affect behavior.[44] We will discuss the opiate-like peptides—the endorphins—and other peptides with possible relevance to the pathophysiology of schizophrenia.

CSF may be an important link in neuroendocrine control mechanisms by acting as a route for the delivery of peptides to specific brain areas.[128,140] Thus, peptides, such as those from the pituitary, may regulate neuronal function in the CNS after getting to the brain via passage through the CSF as well as through peripheral blood. An example of the importance of such retrograde transport of pituitary hormones to the modulation of brain activity is the finding that the pituitary hormone prolactin in the CSF of animals selectively increases the activity of tuberoinfundibular DA neurons.[7] Furthermore, peptidergic neuronal pathways in the CNS have been found to release directly into the CSF as well.

Prange *et al.*[133] suggested the usefulness of categorizing the many brain peptides into two groups: those that are hypothalamic, hypophysiotropic hormones (HHHs) and those that are not. All HHH and many non-HHH brain peptides have been shown to exert behavioral effects completely independent of any endocrine effect. Moreover, some peptides previously considered only hormones now seem to be candidates for the definition of neurotransmitters.[65]

5.1. Thyrotropin-Releasing Hormone

Thyrotropin-releasing hormone (TRH), the first HHH to be isolated, has been shown to alter neuronal discharge rates and to affect behavior in animals; it reverses the CNS depression induced by ethanol and barbiturates,[132] while it facilitates L-DOPA-induced excitement.[127] Yet, the behavioral actions of TRH in humans remain controversial.[133] Moreover, a reliable measurement of TRH in human CSF still has not been achieved. Nevertheless, the similarity of the pharmacological profiles of TRH and amphetamine[97] warrants further study of this peptide in schizophrenia.

5.2. Oxytocin and Vasopressin

Oxytocin and vasopressin [antidiuretic hormone (ADH)] are synthesized in the supraoptic and paraventricular nuclei released into CSF, stored in the neurohypophysis, and then released into the blood. In addition to their well-known endocrine effects, they both powerfully inhibit those neurons that project to the neurohypophysis.[115] Moreover, de Wied[45] and his colleagues have pointed out that both peptides affect learning and memory processes. ADH exerts a long-term effect on the acquisition and maintenance of acquired behavior, stops retrograde amnesia, and seems to improve attention and memory in elderly patients. Oxytocin has an effect opposite to that of ADH inasmuch as it attenuates passive-avoidance behavior. Thus, these peptides may play a role in the cognitive and attentional difficulties in schizophrenia. Vasopressin release is inhibited by DA and NE and may also modulate NE activity.[184a] It is also involved in ACTH release. Finally, Raskind *et al.*[134] recently reported that plasma ADH by radioimmunoassay was higher in psychotics (mostly schizophrenics) than in anxious nonpsychotic patients or normal controls. Moreover, the ADH level showed a significant positive correlation with the degree of psychosis. Both oxytocin and vasopressin can be measured in human CSF.[63] Studies of schizophrenic patients are under way.

5.3. Substance P

Substance P, present in small neuronal systems in many parts of the CNS with high concentrations in the substantia nigra (SN), caudate putamen, amygdala, and cerebral cortex,[30] may be a neurotransmitter in the CNS and can be measured in human CSF.[63] Specifically, several investigations indicate that substance P may function as a neurotransmitter in the SN and participate with γ-aminobutyric acid (GABA) in regulating the dopamine (DA) neurons of the SN.[57] Hong *et al.*[71] reported that chronic administration of the neuroleptic haloperidol to rats significantly decreased the substance P content of the SN and that longer periods of treatment caused greater decreases. They speculated that substance P neurons are under tonic–inhibitory influence by DA neurons. Similar substance P regulation of DA neurons in other pathways may play a role in schizophrenia. Studies of CSF substance P in schizophrenic patients are under way.

5.4. Somatostatin

When injected intracerebrally into rats, somatostatin produced patterns of stereotyped behavioral changes in the sleep–waking cycle with loss of slow-wave REM sleep, increased appetite, and frequent dissociations of the EEG from behavior.[135] In man, no behavioral changes have been reported during clinical somatostatin studies.[156] Somatostatin levels have been measured in the CSF of neurologically ill patients and normals.[122] The discovery that patients with a wide variety of neurological diseases had levels of somatostatin above the highest normal level suggested that CSF somatostatin elevation may be a result of nonspecific leakage from damaged brain tissue and may provide an index of brain damage. We are at present analyzing our data on CSF somatostatin in schizophrenic and control patients.

5.5. Angiotensin I and II

A brain isorenin–angiotensin system has been described that apparently functions separately from the peripheral system[58] inasmuch as CSF and plasma angiotensin levels do not correlate. Numerous pieces of evidence indicate that angiotensin activates central sympathetic activity[194] and increases the biosynthesis of norepinephrine (NE).[19] In peripheral neurons, angiotensin has been shown both to blockade NE reuptake[121] and to facilitate NE release.[195] Finally, there is evidence that angiotensin increases the release and synthesis of serotonin in brain.[113] The marked relationship between angiotensin and these important neurotransmitters emphasizes the importance for future investigations of angiotensin in schizophrenia. Evaluating angiotensin-I-converting enzyme may yield positive results if angiotensin II appears to be decreased.

5.6. Prolactin and ACTH

The pituitary hormones, prolactin[54] and ACTH,[45] have been reported to influence behavior in animals. Moreover, prolactin can increase the activity of specific DA neurons.[7] De Wied[45] pointed out that ACTH and its related peptides are involved in motivational processes and have important effects on learning and remembering behaviors. Both hormones have been measured in CSF,[4,82] but to date, studies of how they relate to clinical phenomena in schizophrenia have not been reported.

5.7. Endorphins

The recent discoveries in brain and pituitary of endogenous polypeptides (endorphins) that bind to opiate receptors in the brain[72,171] have stimulated a search for the physiological role of these opioid substances. As discussed in Chapter 45, the effects of the opiate alkaloids include altering mood, cognition, sleep, pain perception, and release of pituitary hormones. At present, these functions are the best areas in which to investigate the physiological role of endorphin activity.

Brain enkephalins are found in especially high concentration in limbic structures and in the hypothalamus, areas involved in emotional and stress responses.[157] Furthermore, the literature contains many studies of opiate treatment of individuals with psychotic symptoms in which opiate treatment led to their behavioral improvement.[87] Indeed, Jaffe and Martin[75] suggest that some opiate addicts are "self-medicating" themselves against a psychiatric disorder that may be due to a metabolic deficiency of some type of endogenous opiate. Reports of psychotomimetic actions of both narcotics and certain narcotic antagonists have led to the suggestion that some psychosis may be associated with either an excess or a deficiency of endorphins in the CNS.[183] Finally, Bloom *et al.*[16] and Jacquette and Marks[74] reported the ability of β-endorphin to produce a catatonic state when injected intraventricularly into rats; this catatonic state could be reversed by the narcotic antagonist naloxone.[16] Bloom *et al.*[16] likened the drug-induced state to catatonic schizophrenia.

A variety of strategies are being utilized to study endorphins in the major psychiatric diseases, including schizophrenia. The direct administration of endorphins[89] and endorphin-related compounds[184] to schizophrenic patients is one such approach. An opposite strategy involves giving the patients narcotic antagonists, such as the short-acting naloxone or the longer-acting naltrexone, to block the effect of endorphins. Yet, the antagonists may produce increased endorphin release. While the first report was provocative in suggesting that naloxone temporarily reduced or eliminated the auditory hallucinations of chronic schizophrenics,[67] most subsequent studies were unable to replicate these findings.[40,78] Two groups, however, have found higher doses temporarily effective as an antipsychotic.[12,49] Clearly, the major liability of such studies consists of the uncertainty as to what dose to use as well as the problem of the short duration of naloxone action. Recently, in a preliminary study, we have found some antipsychotic effects of naltrexone.

Measuring endogenous opioid substances in urine, plasma, and CSF of schizophrenic patients and controls by both radioimmunoassays and radioreceptor assays is a third strategy of investigation. Terenius and Wahlstrom[169] and Lindstrom *et al.*[95] found opioid activity in human CSF as measured by a receptor-binding assay. They separated two substances from CSF samples and suggested that fraction 1 was not an enkephalin or a fragment of β-lipotropin and that fraction 2 behaved chemically like the endorphin methionine-enkephalin. Terenius *et al.*[170] reported that the levels of fraction 2 in CSF of four chronic schizophrenic patients varied with the severity of the disease. Yet, recently, these authors reported elevated levels of fraction 1 in nine drug-free schizophrenic patients as compared to controls, significantly greater variance in the schizophrenic group's fraction 1 level, and decreasing levels paralleling clinical improvement in four of six neuroleptic responders. In this later report, however, while the fraction 2 levels were nonsignificantly elevated in the patients, treatment response did not produce any consistent trend in fraction 2. The chemical characterization of these CSF endorphins and their anatomical sites of synthesis remain obscure. Another assay has been reported for an enkephalin-like substance in human CSF. But, while the material resembles methionine-enkephalin, it cannot be identified definitively as such.[3] Indeed, a nonpeptide opiate-like substance has been identified in human CSF.[147] Finally, Jeffcoate *et al.*[79] described a method for measuring β-endorphin in human CSF by radioimmunoassay. Following such technical advances, clinical investigations should focus on the possible role that altered endorphin homeostasis plays in schizophrenic psychosis.

Our own preliminary β-endorphin CSF data (radioreceptor assay) suggest that drug-free patients display some low normal levels. Recently, Burbach *et al.*[33a] reported that if levels of β-endorphin and met-enkephalin are altered in CSF of schizophrenic patients compared to normals, this alteration cannot be explained by a differential degradation in the spinal fluid of schizophrenic patients. Reports of β-endorphin metabolites that are physiologically active, such as α- and γ-endorphins and their des-tyrosine analogues measured in schizophrenic patients and normals, have not yet appeared in the literature. Relationships between β-endorphin and cortisol or vasopressin in the CSF may be informative.

6. Miscellaneous

6.1. Immunological and Viral Investigations

For many years, investigators have reported abnormalities of various protein fractions, including immunoglobulins, in the CSF and serum of schizophrenic patients. In 1942, Bruetsch *et al.*[31] reported that in a group of 1281 schizophrenic patients, 5.0% of the male and 3.2% of the female patients had increased CSF protein without evidence of syphilis. Autry,[9] Pearson,[123] and Bock and Rafaelson[20] have reviewed this literature.

The relationship between schizophrenia and immunological phenomena continues to be a focus of investigation. Moreover, clinical, epidemiological, and laboratory evidence suggests the hypothesis that some schizophrenic patients may have a chronic brain disease caused by a viral agent.[172]

Increased amounts of globulin protein in the CSF of schizophrenic patients involving α-globulins,[80] β-globulins,[68,153] and γ-globulins[43,73] have been reported by a number of groups. Torrey *et al.*[173] investigated the CSF immunoglobulins and reported that schizophrenic patients with multiple admissions showed the most significant abnormalities and found individual patients with markedly abnormal levels of specific fractions. Further immunological analysis of the CSF in schizophrenic patients, including our current attempt to demonstrate antibodies to specific neuronal proteins, may contribute to understanding the pathogenesis for a possible subgroup of this disorder.

6.2. Cholinergic Systems

Davis *et al.*[41] have reviewed the pharmacological evidence for a role of cholinergic mechanisms in relation to the pathogenesis of schizophrenia. Moreover, cholinergic–dopaminergic interactions have been well described in dopamine pathways.[109] McGeer and McGeer,[105] in a postmorten study, found elevated choline acetyltransferase activity, the enzyme responsible for synthesizing acetylcholine, in a number of striatal and limbic regions from the brains of 11 chronic schizophrenics compared to controls. They hypothesized that the high levels are a compensatory reaction to defective cholinergic receptors in the affected areas. CSF studies of cholinergic systems have proven difficult, however, because of the rapid enzymatic destruction of acetylcholine. Davis *et al.*[42a] recently reported no significant differences in lumbar CSF acetylcholinesterase activity in schizophrenic patients when compared to control patients.

6.3. Prostaglandins

It has been hypothesized (without evidence) that both increases and deficiencies of prostaglandins are present in schizophrenia.[51a,71a] Recently, we have reported that adenylcyclase linked to the α-adrenergic receptor in platelets of schizophrenic patients was subsensitive in response to prostaglandin E_1 stimulation (i.e., decreased cAMP production,[82a] which was independently observed by Rotrosen *et al.*[143a]). Since then, Mathé *et al.*[103a] observed increased CSF levels of prostaglandin E in drug-free schizophrenic males ($N = 8$, 2894 ± 486 pg/ml) compared to levels in normal controls ($N = 9$, 862 ± 89 pg/ml, $p < 0.01$), neurotic psychiatric patients (918 ± 130 pg/ml, $p < 0.01$), and neurological comparison group (985 ± 169 pg/ml, $p < 0.02$). CSF levels of thromboxane B_2, however, were not different. After replication, more extensive studies are needed to determine the specificity of this finding for schizophrenia.

7. Conclusion

Because of the clinical heterogeneity of people diagnosed as schizophrenic and the complex relationships among different neurobiological systems, we suggest that rather than attempting to find a single "cause" for schizophrenia as an approach to dissecting the behavioral–biochemical relationships in these disease states, CSF studies should concentrate on two more modest goals.

First, a finer delineation of diagnostic and biological heterogeneity would be produced by indentifying the following: (1) clinical (paranoid vs. catatonic, early vs. late onset); (2) pharmacological (neuroleptic responders vs. partial responders vs. nonresponders); and (3) biochemical (high vs. low CSF substance levels) subgroups in large populations of schizophrenic patients. Then, making links among these parameters could prove to be informative. The presumed heterogeneity of the disorder poses special problems for the clinical investigator. Statistically significant findings in a large group of patients are very likely to be secondary to the previously discussed nonspecific factors and to artifacts such as drug treatment (past or present). On the other hand, studies with a small patient sample are not likely to recognize an abnormality that may occur in only a small proportion of patients diagnosed as schizophrenic. Thus, an abnormality occurring in only 10% of patients will nevertheless occur in a sizable number of patients in the total population; yet, they will not be noticed in a study population of only 10–20 patients because only one or two patients would be expected to display this abnormality. A more fruitful approach to finding this subgroup would be to focus on those patients with extremely aberrant values, even though they may not affect the statistical significance of the entire study population. In this way, the biological value can be used as the independent variable to identify a subgroup of schizophrenic patients with consequences for etiology, course and treatment response.

The second goal should be to attempt to relate biological factors to specific component behaviors that make up the schizophrenic disorders: one can divide the behavioral components of a psychotic illness such as schizophrenia into separate groups to examine whether specific biological variables relate more to one of these component groups than to the variety of behavioral disorders grouped under the diagnosis schizophrenia. One such division that we think especially useful in conceptualizing schizophrenia is that of "state components" and "trait components." State components refer to aspects of the psychotic state itself, such as behavioral disorganization, hallucinations, and delusions. Specific "state-related" biological concomitants may relate primarily to the psychotic state and would be less evident during periods of remission. "Trait components" would be those aspects evident in the prepsychotic or postpsychotic period, such as social isolation, affective blunting, impaired role functioning, impaired eye tracking,[70] CAT scan abnormalities, or other as yet unknown behaviors. "Trait-related" biological concomitants would relate to "behaviors" of the nonpsychotic state, would not change over time, and thus could reflect a genetic vulnerability to psychotic decompensation. Some measured biological parameters may be of a mixed trait–state quality (i.e., they occur only during a specific state and only in those patients with a specific trait). Nevertheless, further delineation of biological measures that are state-related or are trait-related would provide an approach to understanding those aspects of the illness that are present in a range of people, including nonschizophrenics, as well as to understanding those aspects that are illness-specific.

References

1. Aghajanian, G. K.: LSD and CNS transmission. *Annu. Rev. Pharmacol.* **12**:157–168, 1962.
2. Aghajanian, G. K., Bunney, B. S.: Dopamine autoreceptors: Pharmacological characterization by microiontophoretic single cell recording studies. *Naunyn-Schmiedeberg's Arch. Pharmacol.* **297**:1–7, 1977.
3. Akil, H., Watson, S. J., Sullivan, S., Barchas, J. D.: Encephalin-like material in normal human CSF: Measurement and levels. *Life Sci.* **23**:121–126, 1978.
4. Allen, J. P., Kendall, J. W., McGilvra, R., Vancura, C.: Immunoreactive ACTH in cerebrospinal fluid. *J. Clin. Endocrinol. Metab.* **38**:586, 1974.
5. Andén, N. W., Butcher, S. G., Corrodi, H.: Receptor activity and turnover of dopamine and noredrenaline after neuroleptics. *Eur. J. Pharmacol.* **11**:303–314, 1970.
6. Angrist, B. M., Sathananthan, G. S., Gershon, S.: Behavioral effects of L-DOPA in schizophrenic patients. *Psychopharmacologia* **31**:1–12, 1973.
7. Annunziato, L., Moore, K. E.: Prolactin in CSF selectively increases dopamine turnover in the median eminence. *Life Sci.* **22**:2037–2042, 1978.
8. Ashcroft, G. W., Crawford, T. B. B., Eccleston, D.: 5-Hydroxyindole compounds in the cerebrospinal fluid of patients with psychiatric or neurological diseases. *Lancet* **2**:1049–1050, 1966.
9. Autry, J. H.: An overview of immunological research in schizophrenia. Presented at the Annual Meeting of the American Psychiatric Association, Honolulu, Hawaii, 1973.
10. Belmaker, R. H., Ebstein, R. P., Biederman, J., Stern, R., Berman, M., van Praag, H. M.: The effect of L-DOPA and propranolol on human CSF cyclic nucleotides. *Psychopharmacology* **58**:307–310, 1978.
11. Berger, P. A.: Medical treatment of mental illness. *Science* **200**:974–981, 1978.
12. Berger, P., Watson, S. J., Akil, H.: Some effects of opiate antagonists on severe psychiatric illness. In Usdin, E., Bunney, W. E. (eds.): *Endorphins in Mental Illness*. London, McMillan Press, 1978.
13. Biederman, J., Rimon, R., Ebstein, R., Belmaker, R. H., Davidson, J. P.: Cyclic AMP in the CSF of patients with schizophrenia. *Br. J. Psychiatry* **130**:64–67, 1977.
14. Biederman, J., Rimon, R., Ebstein, R., Zohar, J., Belmaker, R.: Neuroleptics reduce spinal fluid cyclic AMP in schizophrenic patients. *Neuropsychobiology* **2**:324–327, 1976.
15. Bleuler, E.: *Dementia Praecox or the Group of Schizophrenias*. New York, International Universities Press, 1950.
16. Bloom, F., Segal, D., Ling, N., Guillemin, R.: Endorphines: Profound behavioral effects in rats suggest new etiological factors in mental illness. *Science* **194**:630–632, 1976.
17. Blumberg, J. B., Taylor, R. E., Sulser, F.: Blockade by pimozide of a noradrenaline sensitive adenylate cyclase in the limbic forebrain: Possible role of limbic noradrenergic mechanisms in the mode of action of antipsychotics. *J. Pharm. Pharmacol.* **27**:125–128, 1975.
18. Blumberg, J. B., Vetulani, J., Stawarz, R. J., Sulser, F.: The noradrenergic cyclic AMP generating system in the limbic forebrain: Pharmacological characterization *in vitro* and possible role of limbic noradrenergic mechanisms in the mode of action of antipsychotics. *Eur. J. Pharmacol.* **37**:357–366, 1976.
19. Boadel, M. C., Hughes, J., Roth, R. H.: Angiotensin accelerates catecholamine biosynthesis in sympathetically innervated tissues. *Nature (London)* **222**:987–988, 1969.
20. Bock, E., Rafaelson, O. J.: Schizophrenia: Proteins in blood and cerebrospinal fluid. *Dan. Med. Bull.* **21**:93–105, 1974.
21. Böök, J. A., Wetterberg, L.: Genetics and biochemistry of schizophrenia in a defined population. Presented at the Fourth International Catecholamine Symposium, Asilomar, California, September 1978.
22. Bowers, M. B.: Acute psychosis induced by psychotomimetic drug abuse. II. Neurochemical findings. *Arch. Gen. Psychiatry* **27**:440–442, 1972.
23. Bowers, M. B.: 5-Hydroxyindoleacetic acid (5-HIAA) and homovanillic acid (HVA) following probenecid in acute psychotic patients treated with phenothiazines. *Psychopharmacologia* **28**:309–318, 1973.
24. Bowers, M. B.: Central dopamine turnover in schizophrenic syndromes. *Arch. Gen. Psychiatry* **31**:50–54, 1974.

25. BOWERS, M. B.: Serotonin (5HT) systems in psychotic states. *Psychopharmacol. Commun.* **1**:655–662, 1975.
26. BOWERS, M. B.: CSF acid monoamine metabolites in psychotic syndromes: What might they signify? *Biol. Psychiatry* **13**:375–383, 1978.
27. BOWERS, M. B., HENINGER, G. R., GERBODE, F. A.: Cerebrospinal fluid, 5-hydroxyindoleacetic acid and homovanillic acid in psychiatric patients. *Int. J. Neuropharmacol.* **8**:255–262, 1969.
28. BOWERS, M. B., HENINGER, G. R., MELTZER, H. Y.: Cerebrospinal fluid (CSF), homovanillic acid (HVA), cyclic adenosine mono-phosphate (cAMP), prolactin and serum prolactin in acute psychotic patients at two points during early chlorpromazine (CPZ) treatment. In Usdin, E., Kopin, I. J., Barchas, J. (eds.): *Catecholamines: Basic and Clinical Frontiers,* Vol. II. New York, Pergamon Press, 1979, pp. 1893–1895.
29. BOWERS, M. B., ROSITIS, A.: Regional differences in homovanillic acid concentrations after acute and chronic administration of antipsychotic drugs. *J. Pharm. Pharmacol.* **26**:743–745, 1974.
29a. BOWERS, M. B., STUDY, R. E.: Cerebrospinal fluid cyclic AMP and acid metabolites following probenecid: Studies in psychiatric patients (in prep.).
30. BROWNSTEIN, M., MROZ, E. A., KAIZER, J. S., PALKOVITS, M., LEEMAN, S.: Regional distribution of substance P in the brain of the rat. *Brain Res.* **116**:299–305, 1976.
31. BRUETSCH, W. L., BAHR, M. A., SKOBRA, J. S.: The group of dementia praecox patients with an increase of the protein content of the cerebrospinal fluid. *J. Nerv. Ment. Dis.* **95**:669–679, 1942.
32. BUCHSBAUM, M. S., MURPHY, D. L., COURSEY, R. D., LAKE, C. R., ZIEGLER, M. G.: Platelet monoamine oxidase, plasma dopamine-beta-hydroxylase and attention in a "biochemical high-risk" sample. *J. Psychiatr. Res.* **14**:215–234, 1978.
33. BUNNEY, B. S., GRACE, A. A.: Acute and chronic haloperidol treatment: Comparison of effects on nigral dopaminergic cell activity. *Life Sci.* **23**:1715–1728, 1978.
33a. BURBACH, J. P. H., LOEBER, J. G., BERHOEF, J., DE KLOET, E. R., VAN REE, J. M., DE WIED, D.: Schizophrenia and degradation of endorphins in cerebrospinal fluid. *Lancet* **II**:480–481, 1979.
34. CARENZI, A., GILLIN, J. C., GUIDOTTI, A., SCHWARTZ, M. A., TRABUCCHI, M., WYATT, R. J.: Dopamine-sensitive adenylyl cyclase in human caudate nucleus: A study in control subjects and schizophrenic patients. *Arch. Gen. Psychiatry* **32**:1056–1059, 1975.
35. CHOU, W. S., HO, A. K. S., LOH, H. H.: Neurohormones in brain adenyl cyclase activity *in vivo. Nature (London) New Biol.* **233**:280–281, 1971.
35a. CORBETT, L., CHRISTIAN, S. T., MORIN, R. D., BENINGTON, F., SMYTHES, J. R.: Hallucinogenic *N*-methylated indolealkylamines in the cerebrospinal fluid of psychiatric and control populations. *Br. J. Psychiatry* **132**:139–144, 1978.
36. COWDRY, R. W., EBERT, M. H., POST, R. M., GOODWIN, F. K.: CSF probenecid studies: A reinterpretation (unpublished manuscript), 1978.
36a. CREESE, I., BURT, D. R., SNYDER, S. H.: Dopamine receptor binding predicts clinical potential and pharmacological potencies of antischizophrenic drugs. *Science* **192**:481–483, 1976.
37. CURTIS, D. R., CRAWFORD, J. M.: Central synaptic transmission: Microelectrophorectic studies. *Annu. Rev. Pharmacol.* **9**:209–240, 1969.
38. CURZON, G.: CSF homovanillic acid: An index of dopaminergic hyperactivity. In Calne, D. B., Chase, T. N., Barbeau, A. (eds.): *Advances in Neurology,* Vol. 9. New York, Raven Press, 1975, pp. 349–357.
39. DAPRADA, N., FLETCHER, A.: Acceleration of the cerebro-dopamine turnover by chlorpromazine. *Experienta* **22**:465–466, 1966.
40. DAVIS, G. C., BUNNEY, W. E., DEFRAITES, E. G., KLEINMAN, J. E., VAN KAMMEN, D. P., POST, R. M., WYATT R. J.: Intravenous naloxone administration in schizophrenia and affective illness. *Science* **197**:74–77, 1977.
41. DAVIS, K. L., BERGER, P. A., HOLLISTER, L. E., BARCHAS, J. D. Cholinergic involvement in mental disorders. *Life Sci.* **22**:1865–1872, 1978.
42. DAVIS, K. L., HOLLISTER, L. E., FRITZ, W. C.: Induction of dopaminergic meso-limbic receptor supersensitivity by haloperidol. *Life Sci.* **23**:1543–1548, 1978.
42a. DAVIS, K. L., HOLLISTER, L. E., LIVESEY, J., BERGER, P. A.: Cerebrospinal fluid acetylcholinesterase in neuropsychiatric disorders. *Psychopharmacology* **63**:155–159, 1979.
43. DENCKER, S. J., MALM, U.: Protein pattern of cerebrospinal fluid in mental disease. *Acta Psychiatr. Scand. Suppl.* **203**:105–109, 1968.
44. DE WIED, D.: Pituitary–adrenal system hormones in behavior. In Schmitt, F. A., Worder, F. G. (eds.): *The Neurosciences: Third Study Program.* Cambridge, MIT Press, 1974, pp. 653–662.
45. DE WIED, D.: Peptides and behavior. *Life Sci.* **20**:195–204, 1977.
46. *Diagnostic and Statistical Manual of Mental Disorders,* 3rd Edition. Task Force on Nomenclature and Statistics, Robert L. Spitzer, Chairman. Washington, D.C., American Psychiatric Association, 1979.
47. DOCHERTY, J. P., VAN KAMMEN, D. P., SIRIS, S. G., MARDER, S. R.: Stages of onset of schizophrenic psychosis. *Am. J. Psychiatry* **135**:420–426, 1978.
48. EBSTEIN, R. P., BIEDERMAN, J., RIMON, R., ZOHAR, J., BELMAKER, R. H.: Cyclic GMP in the CSF of patients with schizophrenia before and after neuroleptic treatment. *Psychopharmacology* **51**:71–74, 1976.
49. EMRICH, H. M., CORDING, C., PIREE, S., KOLLING, A., VON ZERSSEN, D., HERZ, A.: Indication of an antipsychotic action of the opiate antagonist, nalox-

one. *Pharmakopsychiatr. Neuropsychopharmakol.* **10**:265–270, 1977.

49a. ENNA, S. J., WOOD, J. H., SNYDER, S. H.: γ-Aminobutyric (GABA) in human cerebrospinal fluid. *J. Neurochem.* **28**:1121–1124, 1977.

50. EWING, J. A., MULLER, R. A., BEATRICE, A., ROUSE, M., SILVER, D.: Low levels of dopamine-beta-hydroxylase and psychosis. *Am. J. Psychiatry* **134**:8–9, 1977.

51. FARLEY, I. F., PRICE, K. S., MCCULLOUGH, E., DECK, J. K. N., KORDYNSKI, W., HORNYKIEWICZ, O.: Norepinephrine in chronic paranoid schizophrenia: Above normal levels in limbic forebrain. *Science* **200**:456–458, 1978.

51a. FELDBERG, W.: Possible association of schizophrenia with a disturbance in prostaglandin metabolism: A possible physiological hypothesis. *Psychol. Med.* **6**:359–369, 1976.

52. FERRENDELLI, J. A., STEINER, A. L., MCDOUGAL, D. B., KIPNIS, D. M.: The effect of oxotremorine and atropine on cGMP and cAMP levels in mouse cerebrocortex and cerebellum. *Biochem. Biophys. Res. Commun.* **41**:1061–1067, 1970.

53. FOTHERBY, K., ASHCROFT, G. W., AFFLECK, J. W.: Studies on sodium transfer and 5-hydroxyindoles in depressive illness. *J. Neurol. Neurosurg. Psychiatry* **26**:71–73, 1963.

54. FRANTZ, A. G.: Prolactin. *Physiol. Med.* **298**:201–207, 1978.

55. FREEDMAN, D. X.: Effects of LSD-25 on brain serotonin. *J. Pharmacol. Exp. Ther.* **135**:160–166, 1961.

56. FRYO, B., WODE-HELGODT, B., BORG, S., SEDVALL, G.: The effect of chlorpromazine on homovanillic acid levels in cerebrospinal fluid of schizophrenic patients. *Psychopharmacologia* **35**:287–294, 1974.

57. GALE, K., GUIDOTTI, A., COSTA, E.: Dopamine sensitive adenylate cyclase: Location in substantia nigra. *Science* **195**:503–505, 1977.

58. GANTEN, D., FUXE, K., PHILLIPS, M. I., MANN, J. F. E., GANTEN, U.: The brain isorenin–angiotensin system: Biochemistry, localization, and possible role in drinking and blood pressure regulation. In Ganong, W. F., Martini, L. (eds.): *Frontiers in Neuroendocrinology*, Vol. 5. New York, Raven Press, 1978, pp. 61–99.

59. GARELIS, E., YOUNG, S. N., HAL, S., SOURKES, T. L.: Monoamine metabolites in lumbar CSF: The question of their origin in relation to clinical studies. *Brain Res.* **79**:1–8, 1974.

60. GILLIN, J. C., STOFF, D. M., WYATT, R. J.: Transmethylation hypothesis: A review of progress. In Lipton, M. A., DiMascio, A., Killam, K. F. (eds.): *Psychopharmacology: A Generation of Progress.* New York, Raven Press, 1978, pp. 1097–1112.

61. GLAESER, B. S., HARE, T. A.: Measurement of GABA in human cerebrospinal fluid. *Biochem. Med.* **12**:174–182, 1975.

62. GLAESER, B. S., VOGEL, W. H., OLEWEILER, D. B., HARE, T. A.: GABA levels in cerebrospinal fluid of patients with Huntington's chorea: A preliminary report. *Biochem. Med.* **12**:380–385, 1975.

63. GOLD, P.: Personal communication.

64. GOODWIN, F. K., POST, R. M.: Cerebrospinal fluid amine metabolites in affective illness and schizophrenia: Clinical and pharmacological studies. *Psychopharmacol. Commun.* **1**:641–653, 1975.

64a. GROSSMAN, M. H., HARE, T. A., MANYAM, N. V. B., GLAESER, B. S., WOOD, J. H.: Stability of GABA levels in CSF under various conditions of storage. *Brain Res.* **182**:99–106, 1980.

65. GUILLEMIN, R.: Peptides in the brain: The new endocrinology of the neuron. *Science* **202**:390–402, 1978.

66. GUNDERSON, J. G., MOSHER, L. R.: The cost of schizophrenia. *Am. J. Psychiatry* **132**:901–905, 1975.

67. GUNNE, L. M., LINDSTROM, L., TERENIUS, L.: Naloxone-induced reversal of schizophrenic hallucinations. *J. Neural. Transm.* **40**:13–19, 1977.

68. HABECK, V.: Electrophorese der Liquoreiweisskörper bei schizophrenen Erkrankungen. *Nervenarzt* **9**:396–400, 1959.

68a. HARE, T. A., MANYAM, N. V. B.: Rapid and sensitive ion-exchange fluorometric measurement of γ-aminobutyric acid in physiological fluids. *Anal. Biochem.* **101**:349–355, 1980.

69. HARTMANN, E.: Schizophrenia: A theory. *Psychopharmacology* **49**:1–15, 1976.

70. HOLZMAN, P. S., PROCTOR, L. R., LEVY, D. L., YASILLO, N. J., MELTZER, H. Y., HURT, S. W.: Eye-tracking dysfunctions in schizophrenic patients and their relatives. *Arch. Gen. Psychiatry* **31**:143–151, 1974. 1974.

71. HONG, J. S., YANG, H. Y. T., COSTA, E.: Substance P content of substantia nigra after chronic treatment with antischizophrenic drugs. *Neuropharmacology* **17**:83–85, 1978.

71a. HORROBIN, D. F., ALLY, A. L., KARMAZYN, M., MANKU, M. S., MORGAN, R. O.: Prostaglandins and schizophrenia. Further discussion of the evidence. *Psychol. Med.* **8**:43–48, 1978.

72. HUGHES, J.: Isolation of an endogenous compound from the brain with pharmacological properties similar to morphine. *Brain Res.* **88**:295–308, 1975.

73. HUNTER, R., JONES, M., MALLESON, A.: Abnormal cerebrospinal fluid total protein and gamma globulin levels in 256 patients admitted to a psychiatric unit. *J. Neurol. Sci.* **9**:11–38, 1969.

74. JACQUETTE, Y. F., MARKS, N.: C-fragment of beta-lipotropin: An endogenous neuroleptic or antipsychotogen. *Science* **194**:632–635, 1976.

75. JAFFE, J. H., MARTIN, W. R.: Narcotic analgesics and antagonists. In Goodman, L. S., Gilman, A. (eds.): *The Pharmacological Basis of Therapeutics.* New York, Macmillan, 1975, pp. 245–283.

76. JANOWSKY, D. S., DAVIS, J. M.: Methylphenidate, dextro-amphetamine and lev-amphetamine: Effects

on schizophrenic symptoms. *Arch. Gen. Psychiatry* **33**:304–308, 1976.

77. JANOWSKY, D. S., EL-YOUSEF, M. K., DAVIS, J. M., SEKERKE, H. J.: Provocation of schizophrenia symptoms by intravenous administration of methylphenidate. *Arch. Gen. Psychiatry* **28**:185–191, 1973.
78. JANOWSKY, D. S., SEGAL, D. S., BLOOM, F., ABRAMS, A., GUILLEMIN, R.: Lack of effect of naloxone on schizophrenic symptoms. *Am. J. Psychiatry* **134**:926–927, 1977.
79. JEFFCOATE, W. J., MCLOUGHLIN, L., HOPE, J., REES, L. H., RATTER, S. J., LOWRY, P. J., BESSER, G. M.: Beta-endorphin in human cerebrospinal fluid. *Lancet* **1**:119–121, 1978.
80. JENSEN, K., CLAUSEN, J., OSTERMAN, E.: Serum and cerebrospinal fluid proteins in schizophrenia. *Acta Psychiatr. Scand.* **40**:280–286, 1964.
81. JIMERSON, D. C., GORDON, E. K., POST, R. M., GOODWIN, F. K.: Central noradrinergic function in man: Vanillylmandelic acid in CSF. *Brain Res.* **99**:434–439, 1975.
82. JIMERSON, D. C., POST, R. M., SKYLER, J., BUNNEY, W. E.: Prolactin in cerebrospinal fluid and dopamine function in man. *J. Pharmacol.* **28**:845–847, 1976.

82a. KAFKA, M. S., VAN KAMMEN, D. P., BUNNEY, JR., W. E.: Reduced cyclic AMP production in blood platelets from schizophrenic patients. *Am. J. Psychiatry.* **136**:685–687, 1979.

83. KAUFMAN, S.: Establishment of tetrahydrobiopterin as a hydroxylase cofactor and review of some recent studies in man. *Psychopharmacol. Bull.* **14**:38–40, 1978.
84. KEBABIAN, J. W.: Multiple classes of dopamine receptors in mammalian central nervous system: The involvement of dopamine-sensitive adenylyl cyclase. *Life Sci.* **23**:479–484, 1978.
85. KEBABIAN, J. W., PETZOLD, G. L., GREENGARD, P.: Dopamine-sensitive adenylate cyclase in caudate nucleus of the rat brain and its similarity to the dopamine receptor. *Proc. Natl. Acad. Sci. U.S.A.* **69**:2145–2149, 1972.
86. KEITH, S. J., GUNDERSON, J. G., REIFMAN, A., BUCHSBAUM, S., MOSHER, L. R.: *Special Report: Schizophrenia 1976.* U.S. Dept. of Health, Education and Welfare, 1976.
87. KHANTZIAN, E. J., MACK, J. E., SCHATZERG, A. F.: Heroin use as an attempt to cope: Clinical observations. *Am. J. Psychiatry* **131**:160–164, 1974.
88. KIRSTEIN, L., BOWERS, M. B., HENINGER, G.: CSF amine metabolites, clinical symptoms, and body movement in psychiatric patients. *Biol. Psychiatry* **11**:421–434, 1976.
89. KLINE, N. S., LI, C. H., LEHMANN, H. E., LAJTHA, A., LASKI, E., COOPER, T.: Beta-endorphin-induced changes in schizophrenic and depressed patients. *Arch. Gen. Psychiatry* **34**:111–113, 1977.
90. KOSLOW, S. H., BUTLER, I. A.: Biogenic amine synthesis defect in dihydropterine reductase deficiency. *Science* **198**:522–523, 1977.

90a. LAKE, C. R., STERNBERG, D. E., VAN KAMMEN, D. P., BALLENGER, J. C., ZIEGLER, M. G., POST, R. M., KOPIN, I. J., BUNNEY, JR., W. E.: Schizophrenia: Elevated cerebrospinal fluid norepinephrine. *Science* **207**:331–333, 1980.

91. LANGER, D. H., BROWN, G. L., BUNNEY, W. E., VAN KAMMEN, D. P.: Gamma-aminobutyric acid in CSF in schizophrenia. *N. Engl. J. Med.* **293**:201, 1975.
92. LEE, T., SEEMAN, P., TOURTELLOTTE, W. W., FARLEY, I. J., HORNYKIEWICZ, O.: Binding of ^{3}H-neuroleptics and ^{3}H-apomorphine in schizophrenic brains. *Nature (London)* **274**:897–900, 1978.
93. LERNER, P., GOODWIN, F. K., VAN KAMMEN, D. P., POST, R. M., MAJOR, L. F., BALLENGER, J. C., LOVENBERG, W.: Dopamine-beta-hydroxylase in the cerebrospinal fluid of psychiatric patients. *Biol. Psychiatry* **13**:685–694, 1978.
94. LICHTSHTEIN, D., DOBKIN, J., EBSTEIN, R. P., BIEDERMAN, J., RIMON, R., BELMAKER, R. H.: Gamma-aminobutyric acid (GABA) in the CSF of schizophrenic patients before and after neuroleptic treatment. *Br. J. Psychiatry* **132**:145–148, 1978.
95. LINDSTROM, L. H., WIDERLOV, E., GUNNE, L. M., WAHLSTROM, A., TERENIUS, L.: Endorphins in human cerebrospinal fluid: Clinical correlations to some psychotic states. *Acta Psychiatr. Scand.* **57**:135–164, 1978.
96. LIPPMANN, W., PUGSLEY, T., MERKER, J.: Effect of Butaclamol and its enantiomers upon striatal homovanillic acid and adenyl cyclase of olfactory tubercle in rats. *Life Sci.* **16**:213–224, 1975.
97. LIPTON, M. A., PRANGE, A. J., NEMEROFF, C. B.: Thyrotropin-releasing hormone: Central effects in man and animals. In Usdin, E., Hamburg, D. A., Barchas, J. D. (eds.): *Neuroregulators and Psychiatric Disorders.* New York, Oxford University Press, 1977, pp. 258–266.
98. MACKAY, A. V. P., DOBLE, A., BIRD, E. D., SPOKES, E. G., QUIK, M., IVERSEN, L. L.: ^{3}H-Spiperone binding in normal and schizophrenic postmortem human brain. *Life Sci.* **23**:527–532, 1978.
99. MAJOR, L. F., LERNER, P., GOODWIN, F. K., BROWN, G. L., BALLENGER, J. C., LOVENBERG, W.: Dopamine-beta-hydroxylase in cerebrospinal fluid: Relationship to personality measures. *Arch. Gen. Psychiatry* **37**:308–313, 1980.
100. MAJOR, L. F., LERNER, P., BALLENGER, J. C., BROWN, G. L., GOODWIN, F. K., LOVENBERG, W.: Dopamine-beta-hydroxylase in the cerebrospinal fluid: Relation ship to disulfiram-induced psychosis. *Biol. Psychiatry* **14**:337–344, 1979.
101. MANDELL, A. J.: Redundant mechanisms regulating brain tyrosine and trytophane hydroxylases. *Annu. Rev. Pharmacol. Toxicol.* **18**:461–493, 1978.
102. MANDELL, A. J., BULLARD, W. P.: Regional and subcellular distribution and factors in the regulation of

reduced pterin in rat brain. *Psychopharmacol. Bull.* **14**:46–49, 1978.

103. MARDER, S. R., VAN KAMMEN, D. P., STERNBERG, D. E.: Pimozide effects on cerebrospinal fluid amine metabolites and psychosis in schizophrenia (in prep.).

103a. MATHÉ, A. A., SEDVALL, G., WIESEL, F. A., NYBACK, H.: Increased content of immunoreactive prostaglandin E in cerebrospinal fluid of patients with schizophrenia. *Lancet* **II**:16–18, 1980.

104. MCGEER, E. G., MCGEER, P. L.: Some characteristics of brain tyrosine hydroxylase. In Mandell, A. J. (ed.): *New Concepts in Neurotransmitter Regulation.* London, Plenum Press, 1973, pp. 53–68.

105. MCGEER, P. L., MCGEER, E. G.: Possible changes in striatal and limbic cholinergic systems in schizophrenia. *Arch. Gen. Psychiatry* **34**:1319–1323, 1977.

106. MEHL, E., RUTHER, E., REDEMANN, J.: Endogenous ligands of a putative LSD–serotonin receptor in the cerebrospinal fluid: Higher level of LSD-displacing factors (LDF) in unmedicated psychotic patients. *Psychopharmacology* **54**:9–16, 1977.

107. MELTZER, H. Y., CHO, H. W., CARROLL, B. J., RUSSO, P.: Serum dopamine-beta-hydroxylase activity in the affective psychoses and schizophrenia. *Arch. Gen. Psychiatry* **33**:585–591, 1976.

108. MELTZER, H. Y., GOODE, D. J., FANG, Y. S.: The effect of psychotropic drugs on endocrine function. I. Neuroleptics, precursors and agonists. In Lipton, M. A., Di Mascio, A., Killam, K. F. (eds.): *Psychopharmacology: A Generation of Progress.* New York, Raven Press, 1978, pp. 509–529.

109. MELTZER, H. Y., STAHL, S. M.: The dopamine hypothesis of schizophrenia: A review. *Schizophr. Bull.* **2**:19–76, 1976.

110. METTLER, F. A.: Perceptual capacity, functions of the corpus striatum and schizophrenia. *Psychiatr. Q.* **29**:89–111, 1955.

111. MILLER, L. H., SANDMAN, C. A., KASTIN, A. J. (eds.): *Neuropeptide Influences on the Brain.* New York, Raven Press, 1977.

112. MILLER, R. J., HORN, A. S., IVERSEN, L. L.: The action of neuroleptic drugs on dopamine-stimulated adenosine cyclic 3′-5′-monophosphate production in rat neostriatum and limbic forebrain. *Mol. Pharmacol.* **10**:759–766, 1974.

113. NAHMOD, V. E., FINKILMAN, S., BENARROCH, E. E., PIROLLA, C. J.: Angiotensin regulates release and synthesis of serotonin in brain. *Science* **202**:1091–1093, 1978.

114. NEMEROFF, C. B., PRANGE, A. J.: Peptides and psychoneuroendocrinology: A perspective. *Arch. Gen. Psychiatry* **35**:999–1010, 1978.

115. NICOLL, R. A., BARKER, J. L.: The pharmacology of recurrent inhibition in the supraoptic neurosecretory system. *Brain Res.* **35**:501, 1971.

116. OHMAN, R., LARSSON, M., NILSSON, I. N., ENGEL, J., CARLSSON, A.: Neurometabolic and behavioral effects of haloperidol in relation to drug levels in serum and brain. *Naunyn-Schmiedeberg's Arch. Pharmacol.* **299**:105–114, 1977.

117. OKADA, Y.: Role of GABA in the substantia nigra. In Roberts, E., Chase, T. N., Tower, D. B. (eds.): *GABA in Nervous System Function.* New York, Raven Press, 1976, pp. 235–243.

118. OSMOND, H., SMYTHIES, J.: Schizophrenia: A new approach. *J. Ment. Sci.* **98**:309–315, 1952.

119. OWEN, F., CROW, T. J., POULTER, M., CROSS, A. J., LONGDEN, A., RILEY, G. J.: Increased dopamine-receptor sensitivity in schizophrenia. *Lancet* **2**:223–226, 1978.

120. PALMER, G. C., DUSZYNSKI, C. R.: Regional cyclic GMP content in incubated tissue of rat brain. *Eur. J. Pharmacol.* **32**:375–379, 1975.

121. PANISSETT, J. C., BOURDOIS, P.: Effect of angiotensin on the response to noradrenaline and sympathetic nerve stimulation and on ^{3}H-noradrenaline uptake in cat mesenteric blood vessels. *Can. J. Physiol. Pharmacol.* **46**:125–131, 1968.

122. PATEL, Y. C., RAO, K., REICHLIN, S.: Somatostatin in cerebrospinal fluid. *N. Engl. J. Med.* **296**:529–533, 1977.

123. PEARSON, E. K.: Study of cerebrospinal fluid in schizophrenics: A review of the literature. *Psychopharmacol. Bull.* **9**:59–62, 1973.

124. PERRY, T. L., BERRY, K., HANSEN, S., DIAMOND, S., MOK, C.: Regional distribution of amino acids in human brain obtained at autopsy. *J. Neurochem.* **18**:513–519, 1971.

125. PERRY, T. L., KISH, S. J., HANSEN, S., BUCHANAN, G.: γ-Aminobutyric acid deficiency in schizophrenia and Huntington's chorea. Presented at the Second International Huntington's Disease Symposium, San Diego, California, November 1978.

126. PERSSON, T., ROOS, B. E.: Acid metabolites from monoamines in cerebrospinal fluid of chronic schizophrenics. *Br. J. Psychiatry* **115**:95–98, 1969.

127. PLOTNIKOFF, N. P., PRANGE, A. J., BREESE, G. R.: Thyrotropin-releasing hormone: Enhancement of DOPA activity by a hypothalamic hormone. *Science* **178**:417–418, 1972.

128. POST, K. D., BILLER, B. J., JACKSON, I. M. D.: Cerebrospinal fluid pituitary hormone concentrations in patients with pituitary tumors. In Wood, J. H. (ed.): *Neurobiology of Cerebrospinal Fluid. I.* New York, Plenum Press, 1980.

129. POST, R. M., FINK, E., CARPENTER, W. T., GOODWIN, F. K.: Cerebrospinal fluid amine metabolites in acute schizophrenia. *Arch. Gen. Psychiatry* **32**:1063–1069, 1975.

130. POST, R. M., GOODWIN, F. K.: Time-dependent effects of phenothiazines on dopamine turnover in psychiatric patients. *Science* **190**:488–489, 1975.

131. POST, R. M., KOTIN, J., GOODWIN, F. K., GORDON, E. K.: Psychomotor activity and cerebrospinal fluid

amine metabolites in affective illness. *Am. J. Psychiatry* **130**:67–72, 1973.

131a. Potkin, S. G., Karoum, F., Chuang, L.-W., Cannon-Spoor, H. E., Phillips, I., Wyatt, R. J.: Phenylethylamine in paranoid schizophrenia. *Science* **206**:470–471, 1979.

132. Prange, A. J., Breese, G. R., Catt, J. M.: Thyrotropin-releasing hormone: Antagonism of pentobarbital in rodents. *Life Sci.* **14**:447–455, 1974.

133. Prange, A. J., Nemeroff, C. B., Loosen, P. T., Wilson I. C.: Pharmacobehavioral effects of oligopeptides endogenous to brain. *Psychopharmacol. Bull.* **14**:20–23, 1978.

134. Raskind, M. A., Weitzman, R. E., Orenstein, H., Fisher, D. A., Courtney, N.: Is antidiuretic hormone elevated in psychosis? A pilot study. *Biol. Psychiatry* **13**:385–390, 1978.

135. Resek, M., Havlicek, V., Hughes, K. R., Friesen, H.: Central site of action of somatostatin (SRIF): Role of hippocampus. *Neuropharmacology* **15**:499–504, 1976.

136. Rimon, R., Roos, B. E., Rakkolainen, V.: The con tent of 5-HIAA and HVA in the CSF of patients with acute schizophrenia. *J. Psychosom. Res.* **15**:375–378, 1971.

137. Roberts, E.: A hypothesis suggesting that there is a defect in the GABA system in schizophrenia. *Neurosci. Res. Program Bull.* **10**:468–481, 1972.

138. Roberts, E., Amacher, T. (eds.): *Propranolol and Schizophrenia*. New York, Alan I. Liss, 1978.

139. Robison, G. A., Butcher, R. W., Sutherland, E. W. (eds.): *Cyclic AMP*. New York, Academic Press, 1971.

140. Rodriguez, E. M.: The cerebrospinal fluid as a pathway in neuroendocrine integration. *J. Endocrinol.* **71**:407–443, 1976.

141. Rosenthal, D., Kety, S. S. (eds.): *The Transmission of Schizophrenia*. London, Pergamon Press, 1968.

142. Roth, R. H., Murrin, L. C., Walters, J. R.: Central dopaminergic neurons: Effects of alterations in impulse flow on the accumulation of dihydroxyphenylacetic acid. *Eur. J. Pharmacol.* **36**:163–171, 1976.

143. Roth, R. H., Walters, J. R., Aghajanian, G. K.: Effect of impulse flow on the release and synthesis of dopamine in the rat striatum. In Usdin, E., Snyder, S. H. (eds.): *Frontiers in Catecholamine Research*. London, Pergamon Press, 1973, pp. 567–574.

143a. Rotrosen, J., Miller, A. D., Traficante, L. Y., Gershon, S.: Reduced PGE_1 stimulated ^{3}H-cAMP accumulation in platelets from schizophrenics. *Life Sci.* **23**:1989–1996, 1978.

144. Ruther, E., Schilkraut, R., Ackenheil, N., Eben, E., Hippius, H.: Clinical and biochemical parameters during neuroleptic treatment: Investigations with haloperidol. *Pharmakopsychiatr. Neuropsychopharmakol.* **9**:33–36, 1976.

145. Sandler, M., Reynolds, G. P.: Does phenylethylamine cause schizophrenia? *Lancet* **1**:70–71, 1976.

146. Sandler, M., Ruthvan, C. R. J., Goodwin, B. L., King, G. S., Pettit, B. R., Reynolds, G. P., Tyre, S. P., Weller, M. P., Hirsch, S. R.: Raised cerebrospinal fluid phenylacetic acid concentration: Preliminary support for the phenylethylamine hypothesis of schizophrenia? *Commun. Pharmacol.* **2**:199–202, 1978.

147. Schor, R. J., Foley, K., Spector, S.: Presence of a non-peptide morphine-like compound in human cerebrospinal fluid. *Life Sci.* **23**:2057–2062, 1978.

148. Schulz, S. C., van Kammen, D. P., Buchsbaum, M., Roth, R., Alexander, P. M., Bunney, W. E., Jr.: Gamma-hydroxybutyrate in the treatment of schizophrenia (submitted).

149. Sebens, J. B., Korf, J.: Cyclic AMP in cerebrospinal fluid: Accumulation following probenecid and biogenic amines. *Exp. Neurol.* **46**:333–344, 1975.

150. Sedvall, G., Alfredsson, G., Bjerkensteet, L., Eneroth, P., Fyro, B., Harnryd, C., Wode-Helgodt, B. W.: Central biochemical correlates to antipsychotic drug action in man. In Gershon, E. S., Belmaker, R. H., Kety, S. S., Rosenbaum, M. (eds.): *The Impact of Biology on Modern Psychiatry*. New York, Plenum Press, 1976, pp. 41–54.

151. Sedvall, G., Fyro, B., Nyback, H., Weisel, F., Wode-Helgodt, B.: Mass fragmentometric determination of homovanillic acid in lumbar cerebrospinal fluid of schizophrenic patients during treatment with antipsychotic drugs. *J. Psychiatr. Res.* **11**:75–80, 1974.

152. Seeman, P., Lee, T.: Antipsychotic drugs: Direct correlation between clinical potency and presynaptic actions on dopamine neurons. *Science* **188**:1217–1219, 1975.

153. Shanmugam, A.: A study of cerebrospinal fluid proteins in schizophrenia. *J. Indian Med. Assoc.* **57**:206–208, 1971.

154. Sherman, A. D., Gal, E. N.: Lack of dependence of amine or prostaglandin biosynthesis on absolute cerebral levels of pteridine cofactor. *Life Sci.* **23**:1675–1680, 1978.

155. Shopsin, B., Wilk, S., Gershon, S.: Collaborative psychopharmacologic studies exploring catecholamine metabolism in psychiatric disorders. In Usdin, E., Snyder, S. (eds.): *Frontiers in Catecholamine Research*. New York, Pergamon Press, 1973, pp. 1173–1179.

156. Siler, T. M., Vanden Berg, G., Yenn, S. S. C.: Inhibition of growth hormone release in humans by somatostatin. *J. Clin. Endocrinol. Metab.* **37**:632–638, 1973.

157. Simantov, R., Kuhar, M. J., Pasternak, G. W.: The regional distribution of a morphine-like factor, encephalin, in monkey brain. *Brain Res.* **106**:189–197, 1976.

158. Smith, C. C., Tallman, J. F., Post, R. M., van Kammen, D. P., Jimerson, D. C., Brown, G. L., Brooks, B. R., Bunney, W. E.: An examination of baseline and drug-induced levels of cyclic nucleo-

tides in the cerebrospinal fluid of control and psychiatric patients. *Life Sci.* **19**:131–136, 1976.

159. Snyder, S. H.: Catecholamines in the brain as mediators of amphetamine psychosis. *Arch. Gen. Psychiatry* **27**:169–179, 1972.
160. Snyder, S. H., Banerjee, S. P., Yamamura, H. I.: Drugs, neurotransmitters and schizophrenia. *Science* **184**:1234–1253, 1974.
161. Spitzer, R. L., Endicott, J., Robins, E.: Research Diagnostic Criteria for a Selected Group of Functional Disorders (RDC), 2nd ed. New York, New York State Psychiatric Institute, Biometrics Research, 1975.
162. Stawarz, R. J., Hill, H., Robinson, S. E., Setler, P., Dingell, J. B., Sulser, F.: On the significance of the increase in homovanillic acid (HVA) caused by antipsychotic drugs in corpus striatum and limbic forebrain. *Psychopharmacologia* **43**:125–130, 1975.
163. Stein, L., Wise, C. D.: Possible etiology of schizophrenia: Progressive damage to the noradrenergic reward system by 6-hydroxy-dopamine. *Science* **171**:1032–1036, 1971.
164. Sternberg, D. E., van Kammen, D. P., Ballenger, J. C., Lerner, P., Marder, S. R., Post, R. M.: Cerebrospinal fluid dopamine-beta-hydroxylase and schizophrenia. Presented at the Annual Meeting of the American Psychiatric Association, 1978. *New Res. Abstr., p. 32.*
165. Sternberg, D. E., van Kammen, D. P., Lake, C. R., Ballenger, J. C., Bunney, W. E., Jr.: Cerebrospinal fluid norepinephrine in schizophrenia. In Scientific Proceedings of the Annual Meeting, American Psychiatric Association, 1979.
166. Stevens, J., Wilson, K., Foote, W.: GABA blockade, dopamine and schizophrenia: Experimental studies in the cat. *Psychopharmacologia* **39**:105–119, 1974.
167. Stolzki, B., Kaiser, H. O., Mehl, E. L.: Heterogeneity of LSD-displacing factors and multiple types of high affinity LSD-binding sites. *Life Sci.* **23**:593–598, 1978.
168. Sulser, F., Robinson, S. E.: Clinical implications of pharmacological differences among antipsychotic drugs (with particular emphasis on biochemical central synaptic adrenergic mechanisms). In Lipton, M. A., DiMascio, A., Killam, K. F. (eds.): *Psychopharmacology: A Generation of Progress*. New York, Raven Press, 1978, pp. 943–954.
169. Terenius, L., Wahlstrom, A.: Morphine-like ligand for opiate receptors in human CSF. *Life Sci.* **16**:1759–1764, 1975.
170. Terenius, L., Wahlstrom, A., Lindstrom, L., Widerlov, E.: Increased CSF levels of endorphins in chronic psychosis. *Neurosci. Lett.* **3**:157–162, 1976.
171. Teschemacher, H., Opheim, K. E., Cox, B. M., Goldstein, A.: A peptide-like substance from pituitary that acts like morphine. I. Isolation. *Life Sci.* **16**:1771–1776, 1975.
172. Torrey, E. F., Peterson, M. R.: The viral hypoth esis of schizophrenia. *Schizophr. Bull.* **2**:136–146, 1976.
173. Torrey, E. F., Peterson, M. R., Brannon, W. L., Carpenter, W. T., Post, R. M., van Kammen, D. P.: Immunoglobulins and viral antibodies in psychiatric patients. *Br. J. Psychiatry* **132**:342–348, 1978.
174. Tsang, D., Lal, S., Sourkes, T. L.: Studies of cyclic AMP in different compartments of cerebrospinal fluid. *J. Neurol. Neurosurg. Psychiatry* **39**:1186–1190, 1976.
175. Vallors, M. B., Heninger, G. R., Gerbode, F. A.: Cerebrospinal fluid 5-hydroxyindoleacetic acid and homovanillic acid in psychiatric patients. *Int. J. Neuropharmacol.* **8**:255–262, 1969.
176. van Kammen, D. P.: Gamma-aminobutyric acid (GABA) and the dopamine hypothesis of schizophrenia. *Am. J. Psychiatry* **134**:138–143, 1977.
177. van Kammen, D. P.: The dopamine hypothesis of schizophrenia revisited. *Psychoneuroendocrinology* **4**:37–46, 1979.
178. van Kammen, D. P., Sternberg, D. E., Hare, T. A., Ballenger, J., Marder, S. R., Post, R., Bunney, W. E., Jr.: Cerebrospinal fluid GABA levels in schizophrenia. *Brain Res. Bull.* (in press).
179. van Kammen, D. P., Bunney, W. E., Jr.: Heterogeneity in response to amphetamine in schizophrenia: Effects of placebo, chronic pimozide and pimozide withdrawal. In Usdin, E., Kopin, I. J., Barchas, J. (eds.): *Catecholamines: Basic and Clinical Frontiers*. Oxford, Pergamon Press, 1979, pp. 1896–1898.
180. van Kammen, D. P., Bunney, W. E., Jr., Docherty, J. P., Jimerson, D. C., Post, R. M., Siris, S., Ebert, M., Gillin, J. C.: Amphetamine-induced catecholamine activation in schizophrenia and depression: Behavioral and physiological effects. *Adv. Biochem. Psychopharmacol.* **16**:655–659, 1977.

180a. van Kammen, D. P., Levine, R., Sternberg, D. E., Ballenger, J., Marder, S. R., Post, R., Bunney, W. E., Jr.: Preliminary evaluation of hydroxylase cofactor in human spinal fluid: Potential biochemical and clinical relevance in the study of psychiatric disease. *Psychopharm. Bull.* **14**(4):51–52, 1978.

181. van Praag, H. M., Korf, J.: Neuroleptics, catecholamines, and psychoses: A study of their interrelations. *Am. J. Psychiatry* **132**:593–597, 1975.
182. van Praag, H. M., Korf, J.: Importance of dopamine metabolism for clinical effects and side-effects of neuroleptics. *Am. J. Psychiatry* **133**:1171–1176, 1976.
183. Verebey, K., Volavka, J., Clouet, D.: Endorphins in Psychiatry. *Arch. Gen. Psychiatry* **35**:877–888, 1978.
184. Verhoeven, W. M. A., van Praag, H. M., Botter, P. A., Sunier, A., van Ree, J. M., de Wied, D.: (DES-TYR)-gamma-endorphin in schizophrenia. *Lancet* **II**:1046–1047, 1978.

184a. Versteegh, D. H. G., DeKloet, E. R., van Wim-

ERSMA GREIDANUS, T., DEWIED, D.: Vasopressin modulates the activity of catecholamine containing neurons in specific brain regions. *Neurosci. Lett.* **11**:69–73, 1979.

185. WEHR, T., GOODWIN, F. K.: Catecholamines in depression. In Burrows, K. (ed.): *Handbook of Studies on Depression*. New York, Excerpta Medica, 1977, pp. 283–301.

186. WEINER, N.: The role of pterin cofactor in the regulation of catecholamine synthesis. *Psychopharmacol. Bull.* **14**:40–44, 1978.

187. WISE, C. D., STEIN, L.: Dopamine-beta-hydroxylase deficits in the brains of schizophrenic patients. *Science* **181**:344–347, 1973.

187a. WOOD, J. H., GLAESER, B. S., ENNA, S. J., HARE, T. A.: Verification and quantification of GABA in human cerebrospinal fluid. *J. Neurochem.* **30**:291–293, 1978.

187b. WOOD, J. H.: Neurochemical analysis of cerebrospinal fluid. *Neurology* (in press).

188. WOOLEY, D. W., SHAW, E.: A biochemical and pharmacological suggestion about certain mental disorders. *Proc. Natl. Acad. Sci. U.S.A.* **40**:228–231, 1954.

189. WYATT, R. J., GILLIN, J. C., KAPLAN, J.: *N,n*-Dimethyltryptamine—a possible relationship to schizophrenia? In Costa, E., Gessa, G. L., Sandler, M. (eds.): *Serotonin: New Vistas*. New York, Raven Press, 1974, pp. 299–313.

190. WYATT, R. J., GILLIN, J. C., STOFF, D. M., MOJA, E. A., TINKLENBERG, J. R.: β-Phenylethylamine and the neuropsychiatric disturbances. In Usdin, E., Hamburg, D., Barchas, J. (eds.): *Neuroregulators and Psychiatric Disorders*. New York, Oxford University Press, 1977, pp. 31–45.

191. WYATT, R. J., SCHWARTZ, M. A., ERDELYI, E., BARCHAS, J. D.: Dopamine-beta-hydroxylase activity in brains of chronic schizophrenic patients. *Science* **187**:368–369, 1975.

192. ZIEGLER, M. G., LAKE, C. R., WOOD, J. H.: Circadian rhythm in cerebrospinal fluid noradrenalin of man and monkey. *Nature (London)* **264**:656–658, 1976.

193. ZIMMER, R., TEELKEN, A. W., MEIR, K. D., ACKENHEIL, M., ZANDER, K. J., FISCHER, H., HERTING, N.: Preliminary studies on gamma-aminobutyric acid (GABA) levels in schizophrenic patients before and during treatment with different psychopharmaca (in prep.).

194. ZIMMERMAN, B. L., GOMEZ, J.: Increased response to sympathetic stimulation in presence of angiotensin. *Int. J. Neuropharmacol.* **4**:185–193, 1965.

195. ZIMMERMAN, B. L., WHITMORE, L.: Effect of angiotensin and phenoxybenzamine on release of norepinephrine in vessels during sympathetic nerve stimulation. *Int. J. Neuropharmacol.* **6**:27–38, 1967.

196. ZOHAR, J., BIEDERMAN, J., RIMON, R., EBSTEIN, R., BELMAKER, R. H.: Clinical correlates of CSF cyclic nucleotides in schizophrenia. *Am. J. Psychiatry* **135**:253–255, 1978.

CHAPTER 49

Cerebrospinal Fluid Calcium

Clinical Correlates in Psychiatric and Seizure Disorders

David C. Jimerson, James H. Wood, and Robert M. Post

1. Introduction

Hypotheses linking affective illness to alterations in calcium function rest on indirect pharmacological and behavioral observations in laboratory animals and man. In this chapter, we will briefly review both indirect evidence supporting these hypotheses and data from cerebrospinal fluid (CSF) studies in psychiatric and neurological patients. Studies of serum and urinary calcium in affective illness have been reviewed elsewhere.[5,6,11]

Alterations in mood have been commonly recognized in patients with moderate alterations in calcium metabolism. These mood changes are qualitatively different from the organic brain syndromes of severe elevations or depletions of serum calcium.[34] Hyperparathyroidism and other chronic hypercalcemias are often associated with depressed mood.[18,28,53,57] Correction of the metabolic alteration, e.g., by parathyroidectomy, tends to produce improved mood and energy level.[2] In contrast, destabilization of affect and increased psychomotor activity have been associated with chronic hypocalcemias.[16,53]

Some drugs that alter mood in man alter brain calcium concentrations when administered to animals. Both alcohol and morphine, for example, lower regional brain calcium in the rat.[48] Naloxone administration selectively reverses both the behavioral effects and the brain calcium depletion produced by morphine.[9,49] Calcium has in common with many psychoactive drugs the capacity to alter biogenic amine function, influencing synthesis,[37,41,50] transport,[26] release of neurotransmitters,[51] and cyclic-AMP-dependent response to receptor activation.[54]

Measurement of calcium levels in CSF appears to offer advantages over measurement in serum for assessing the possible behavioral correlates of variations in extracellular brain calcium in psychiatric patients. Concentration of calcium in CSF is similar to that in brain extracellular fluid[29] and, like that in brain, is relatively stable in the face of acute alterations in serum concentration.[15,30] Thus, central nervous system (CNS) function tends to remain stable despite acute electrolyte changes that produce symptomatic dysfunction of peripheral nerves, cardiac muscle, smooth muscle, and skeletal muscle. Conversely, small alterations in CSF calcium produce marked behavioral effects in laboratory animals (see below). Calcium concentration in CSF is not in passive equilibrium with that in the plasma.[34] It has been proposed that calcium homeostasis in the CSF is a result of secretion by the choroid plexuses of a fluid that is near constant in composition in the presence of a blood–brain barrier relatively impermeable to calcium diffusion.[7] In man, calcium concentration in lumbar CSF reflects ventricular concentrations.[31]

David C. Jimerson, M.D., and **Robert M. Post, M.D.** • Biological Psychiatry Branch, National Institute of Mental Health, National Institutes of Health, Bethesda, Maryland 20205. **James H. Wood, M.D.** • Division of Neurosurgery, University of Pennsylvania School of Medicine and Hospital, Philadelphia, Pennsylvania 19104.

While behavioral and other clinical studies in CSF have to date relied on measurement of total (i.e., protein-bound plus free) calcium because of the technical difficulties in measuring levels of the free ion, the latter may provide a more sensitive index of altered calcium metabolism.[39] While approximately 50% of serum calcium exists in an ionized state, this fraction is increased to 73% in the CSF because of the lower protein concentration.[20] The constancy of the ionized : total ratio for calcium in CSF has not been determined. At least one study suggests that the proportion of ionized calcium in the CSF may be relatively stable in the face of physiological alterations—e.g., hyperventilation—known to alter the ionized fraction in the serum.[20]

2. CSF Calcium in Affective Illness

CSF calcium measurements in depressed patients have failed to demonstrate consistent differences from control groups (Table 1), in agreement with similar studies of serum calcium.[5,11] Because CSF samples from healthy volunteers are not usually available, most of the studies cited have used nondepressed psychiatric patients (e.g., character disorder) or neurological patients as control groups. Preliminary analysis of ionized-calcium concentration in CSF from depressed patients[25] and a healthy control group[3] has failed to demonstrate significant differences.

Improved methodology in the more recent CSF studies has included measurement of calcium by atomic absorption spectrometry, diagnosis of patients by more specific research criteria, study of patients in a drug-free state, and control of activity prior to study. Control of dietary calcium has not been attempted in these studies. The possibility that CSF calcium may have a diurnal rhythm like that of serum calcium[58] necessitates the collection of patient and control group samples at a consistent time of day. We have previously observed that CSF calcium measured at 3 P.M. following oral probenecid administration (in conjunction with amine metabolite studies[23]) gave higher values than a 9 A.M. drug-free sample in the same patients,[32] although possible effects of probenecid itself on calcium levels remain to be clarified.

Table 1. CSF Calcium in Depression Compared to Control Groups

No difference	Higher in depression
Jimerson *et al.* (1979)[32,a]	Hakin *et al.* (1975)[24]
Bech *et al.* (1978)[4]	Harris and Beauchemin (1956)[27]
Bjorum *et al.* (1972)[6]	Weston and Howard (1922)[59]
Ueno *et al.* (1961)[55]	
Katzenelbogen (1953)[35]	
Scholberg and Goodall (1926)[52]	

[a] In this study, CSF calcium concentration was positively correlated with severity of depression.

To improve the signal-to-noise ratio in CSF studies in affective illness, our group and others have examined longitudinal changes in individual patients across changes in mood and following therapeutic interventions. We have, in collaboration with Drs. J. S. Carman, D. P. van Kammen, F. K. Goodwin, and W. E. Bunney, Jr., studied three bipolar and one schizoaffective patient during drug-free depressed and manic episodes observed during periods of rapid mood cycles in the hospital (Fig. 1).[32] As a group, these patients had significantly higher CSF calcium levels during depression than during mania. Carman and Wyatt[12] found a decrease in CSF calcium following a switch from retarded to manic–excited states in patients with periodic psychoses. Evidence for alterations in plasma calcium accompanying such switches has been recently reviewed.[11] In a preliminary study of CSF calcitonin measured by radioimmunoassay, manic patients appear to have lower levels of the peptide than do bipolar depressed patients and healthy controls.[10]

In several studies of the effects of antidepressant treatment on CSF calcium levels in depressed patients, we have seen a relationship between clinical improvement and decreased calcium levels. These changes were observed following electroconvulsive therapy (ECT) in 6 patients responding to this treatment.[13] In 20 depressed patients, studied in collaboration with Drs. R. H. Gerner, J. C. Gillin, and W. E. Bunney, Jr., the direction of CSF calcium change on the day following one night's sleep deprivation showed an interaction with the direction of clinical change; those who improved following sleep deprivation had decreases in CSF calcium, while those who showed no change or became worse had increases in CSF calcium (Fig. 2).[21] In a group of depressed patients treated with lithium, there was a correlation between decrease in global depression ratings and decrease in CSF calcium in comparison to values dating from the initial drug-free depressed lumbar puncture.[32] Despite these observations, it is unclear whether there is a persistent decrease in CSF calcium accompanying clinical recovery in de-

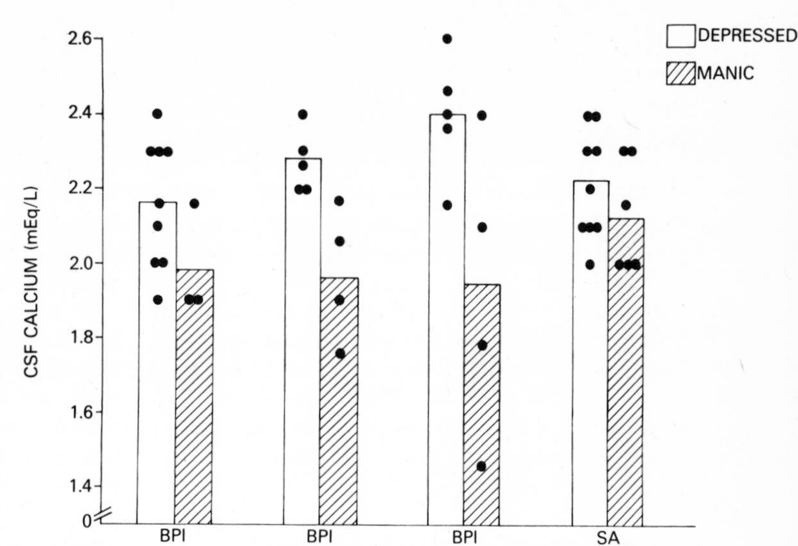

Figure 1. CSF calcium concentrations compared during periods of mania and depression in four rapidly cycling patients. CSF calcium levels were measured in bipolar I (BPI) and schizoaffective (SA) patients during alternating manic and depressed periods. For this patient group, CSF calcium was significantly higher during depression than during mania.[32]

pressed patients. We did observe a decrease in calcium in 3 P.M. post-probenecid CSF samples in 7 of 8 depressed patients restudied during a clinically improved, drug-free state.[32] In contrast, Bech *et al.*[4] found that CSF calcium values in depressed patients treated with ECT, imipramine, or lithium showed no apparent difference between initial depressed and subsequent improved lumbar punctures.

3. CSF Calcium in Schizophrenia

Studies of CSF calcium in schizophrenia, while less extensive than those in affective illness, have also failed to show significant differences from available control groups.[32,43,55] Similar observations have been made on serum calcium in schizophrenia.[1] Some association between psychosis and CSF calcium in schizophrenia was suggested by the observation that eight of nine patients had increased values when restudied in a state of clinical improvement (Fig. 3).[32] Possible contributions of nonspecific effects such as activity level[55] on this state-dependent increase in calcium need to be carefully evaluated.

4. CSF Calcium in Seizure Disorders

The possibility that altered brain calcium function may play a role in idiopathic seizure disorders is suggested by the regular occurrence of seizures in hypoparathyroid patients with severe hypocal-

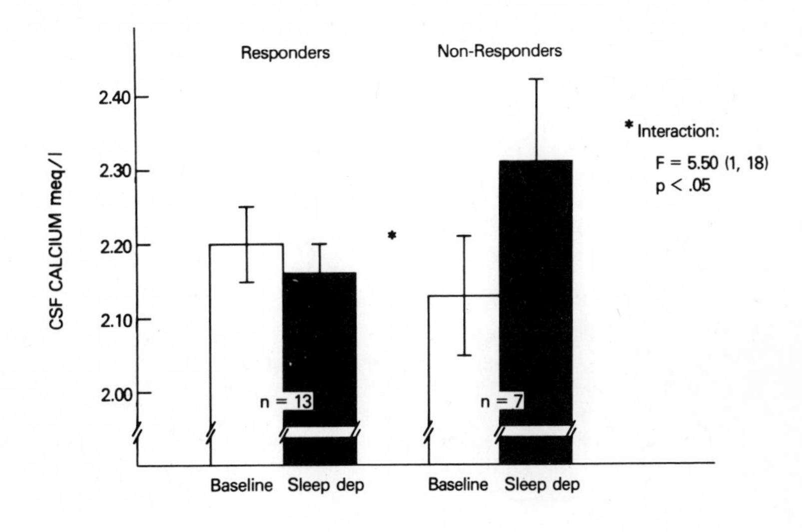

Figure 2. Interaction of sleep-deprivation response and calcium in CSF. Lumbar punctures performed following one night's sleep deprivation in depressed patients showed significant interactions between CSF calcium levels and mood response. Patients who felt better following night awake (Responders) had lower CSF calcium values than did unimproved group (Non-Responders).[21]

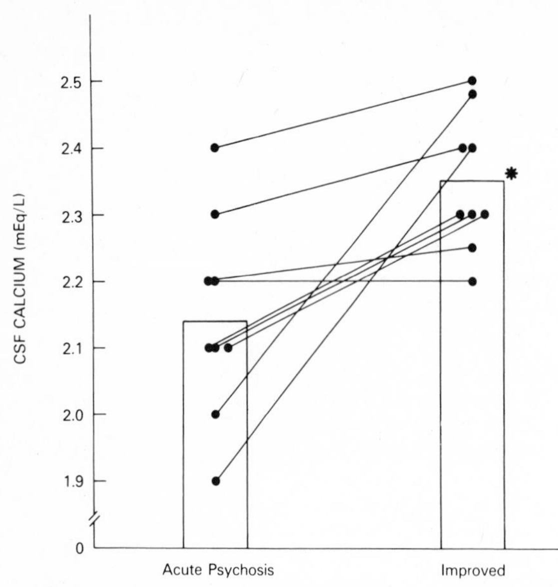

Figure 3. CSF calcium with respect to clinical course of schizophrenia and schizoaffective patients. In eight of nine such patients, CSF calcium showed significant increases (*$p < 0.01$) on clinical improvement.[32]

cemia.[22] We have observed decreased calcium in 3 P.M. post-probenecid CSF samples in a group of 12 patients with refractory generalized or partial complex seizure disorders who were taking combinations of anticonvulsant drugs including phenytoin, phenobarbital, primidone, diazepam, carbamazepine, or ethosuximide (Fig. 4).[32] The comparison

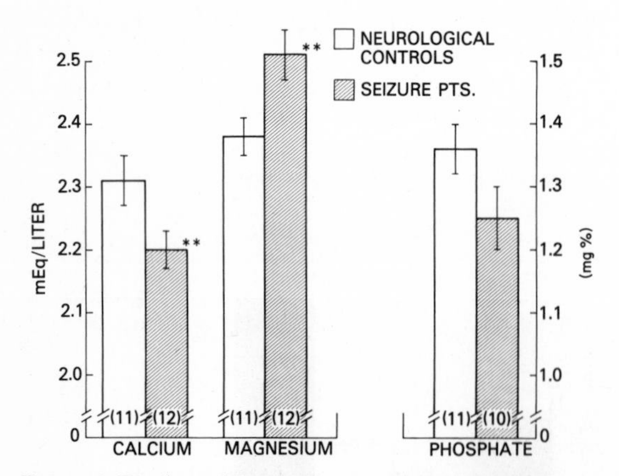

Figure 4. Divalent electrolyte concentrations in CSF in seizure disorders. In comparison to drug-free hospitalized nonepileptic neurological patients, anticonvulsant-treated seizure patients with generalized tonic–clonic (grand mal) and complex partial (psychomotor) convulsions showed significantly lower CSF calcium levels and higher CSF magnesium concentrations. **$p < 0.05$.

group for the study consisted of 11 drug-free neurological patients with spinal cord arteriovenous malformation, pituitary tumor, upper motor neuron disease, or peripheral neuralgia. Drug treatment of the seizure patients limits the interpretation of the CSF data, since hypocalcemia results from long-term anticonvulsant therapy *per se*.[45] The relationship between low CSF calcium and seizure disorders merits further study in view of laboratory evidence that decreased extracellular calcium enhances neuronal excitability,[19] that cortical seizures result in a measurable reduction in extracellular brain calcium,[29] and that some anticonvulsants reduce the hyperirritability of nerves placed in low-calcium solution.[47] In cats, increased CSF calcium acts as an anticonvulsant.[60]

5. CSF Calcium and Behavioral Activation

Laboratory studies have shown extracellular calcium concentration to have two principal, and potentially antagonistic, effects on neuronal function. Low extracellular calcium results in increased neuronal excitability through effects on the cell membrane.[19] Thus, direct iontophoretic application of calcium decreases neuronal excitability,[33,36] while extracellular application of calcium chelators produces excitation in spinal neurons.[14] It should be noted that increased neuronal activity may itself produce large decreases in regional extracellular calcium.[42] On the other hand, transmitter release at the presynaptic nerve ending is calcium-dependent and is inhibited by low extracellular calcium.[51] Other effects on monoaminergic neurons have been noted above.

The cumulative effect on animal behavior of decreased brain extracellular calcium, at least as reflected in CSF levels, appears to be psychomotor activation. Thus, artificially lowered CSF calcium produces arousal in goats,[44] while increased CSF calcium has sedative effects in cats.[17,56] Respiratory activity is similarly influenced in laboratory studies of altered CSF calcium.[40] Consistent with these observations in animals is the increased-amplitude visual average evoked response in patients with chronic hypocalcemia.[8] Thus, a variety of indirect data support a possible role of calcium alterations in mood and psychomotor activation. As reviewed above, the calcium-related endocrinopathies often present with psychiatric symptomatology that improves as calcium regulation is normalized. Several classes of psychotropic drugs alter regional calcium metabolism in a fashion closely linked to their behavioral effects. Neuronal activity on cellular level or level of activation or arousal on organismic level also tends to be inversely associated with alterations in calcium.

Many methodological difficulties need to be clarified before definite conclusions can be drawn regarding the indirect data on psychiatric patients implicating calcium alterations in mood or cognitive function. For example, the clinical importance of measuring ionized vs. total calcium requires further exploration. CSF calcium (ionized or total) would appear to be a poor measure of localized regional brain alterations that might be of clinical relevance, but may be a better measure of global calcium fluctuations that may also be of behavioral significance. Increasing evidence suggests that the CSF, in addition to its excretory and cushioning functions, may participate as a direct, active neuroregulatory pathway with its own time constants for feedback regulation compared to neuronal or blood pathways.[38,46] To the extent that CSF calcium is a physiologically and clinically relevant participant in this more tonic, global regulation, further study of its possible alterations in neuropsychiatric patients would appear indicated.

The preliminary data suggest in a limited way that CSF calcium may be altered in association with some behavioral-state changes. Whether calcium changes are related to selective changes in mood, cognition, and behavior that characterize depressed, manic, or schizophrenic syndromes remains for further study. CSF calcium could relate, as in animal studies, to the general level of physiological activation. Psychomotor activation tends to accompany recovery in depressed patients, as well as the switches from depression to mania, where CSF calcium decreases were observed at least in some studies. Similarly, decreased psychomotor activity is commonly observed with remission in schizophrenic psychoses, where increases in CSF calcium were noted. More specific measurement of motor and psychological activity in relation to potential calcium changes would appear indicated. While preliminary, the evidence for state-dependent alterations in CSF calcium in several clinical studies is consistent with the experimental data in animals linking decreases in CSF calcium and increases in psychomotor activation.

References

1. Alexander, P. E., van Kammen, D. P., Bunney, W. E., Jr.,: Serum calcium and magnesium in schizophrenia: Effects of neuroleptics and onset of EPS. *Br. J. Psychiatry* **133:**143–149, 1978.
2. Anderson, J.: Psychiatric aspects of primary hyperparathyroidism. *Proc. R. Soc. Med.* **61:**1123, 1968.
3. Ballenger, J. C., Hamilton, W. G., Jr., Jimerson, D. C., Hickey, T., Post, R. M., Bunney, W. E., Jr.: Unpublished observations.
4. Bech, P., Kirkegaard, C., Bock, E., Johannesen, M., Rafaelsen, O. J.: Hormones, electrolytes, and cerebrospinal fluid proteins in manic–melancholic patients. *Neuropsychobiology* **4:**99, 1978.
5. Bjorum, N.: Electrolytes in blood in endogenous depression. *Acta Psychiatr. Scand.* **48:**59, 1972.
6. Bjorum, N., Plenge, P., Rafaelsen, O. J.: Electrolytes in cerebrospinal fluid in endogenous depression. *Acta Psychiatr. Scand.* **48:**533, 1972.
7. Bradbury, M. W. B., Sarna, G. S.: Homeostasis of the ionic composition of the cerebrospinal fluid. *Exp. Eye Res. (Suppl.)* **25:**249–257, 1977.
8. Buchsbaum, M., Henkin, R. I.: Serum calcium concentration and the average evoked response. *Electroencephalogr. Clin. Neurophysiol.* **30:**10, 1971.
9. Cardenas, H. L., Ross, D. H.: Morphine induced calcium depletion in discrete regions of rat brain. *J. Neurochem.* **24:**487, 1975.
10. Carman, J. S.: Personal communication.
11. Carman, J., Jimerson, D. C., Post, R. M.: Electrolyte changes associated with shifts in affective states. In Alexander, P. (ed.): *Electrolyte and Neuropsychiatric Disorders.* New York, Spectrum (in press).
12. Carman, J. S., Wyatt, R. J.: Alterations in cerebrospinal fluid and serum total calcium with changes in psychiatric state. In Usdin, E., Hamburg, D. A., Barchas, J. D. (eds.): *Neuroregulators and Psychiatric Disorders.* New York, Oxford University Press, 1977, pp. 488–494.
13. Carman, J. S., Post, R. M., Goodwin, F. K., Bunney, W. E., Jr.: Calcium and electroconvulsive therapy of severe depressive illness. *Biol. Psychiatry* **12:**5, 1977.
14. Curtis, D. R., Perrin, D. D., Watkins, J. C.: The excitation of spinal neurons by iontophoretic application of agents which chelate calcium. *J. Neurochem.* **6:**1, 1960.
15. Davson H.: *Physiology of the Cerebrospinal Fluid.* London, Churchill, 1967, pp. 143–144.
16. Denko, J. D., Kaebling, R.: The psychiatric aspects of hypoparathyroidism. *Acta Psychiatr. Scand. Suppl.* **164:**5, 1962.
17. Feldberg, W., Sherwood, S. L.: Effects of calcium and potassium injected into the cerebral ventricles of the cat. *J. Physiol.* **139:**408, 1957.
18. Flanagan, T. A,. Goodwin, D. W., Alderson, P.: Psychiatric illness in a large family with familial hyperparathyroidism. *Br. J. Psychiatry* **117:**693, 1970.
19. Frankenhaeuser, B., Hodgkin, A. L.: The action of calcium in the electrical properties of squid axons. *J. Physiol. (London)* **137:**218–244, 1957.
20. Fuchs, C.: Ion selective electrodes in clinical medicine.

In Kessler, M., Clark, L. C., Lubbers, D. W., Silver, I. A., Simon, W. (eds.): *Ion and Enzyme Electrodes in Biology and Medicine*. Baltimore, University Park Press, 1976.

21. GERNER, R. H., POST, R. M., GILLIN, J. C. BUNNEY, W. E., JR.: Biological and behavioral effects of one night's sleep deprivation in depressed patients and normals. *J. Psychiatr. Res.* **15**:21–40, 1979.
22. GLASER, G. H., LEVY, L. L.: Seizures and ideopathic hypoparathyroidism: A clinical electroencephalographic study. *Epilepsia* **1**:454, 1959–1960.
23. GOODWIN, F. K., POST, R. M., DUNNER, D. L., GORDON, E. K.: Cerebrospinal fluid amine metabolites in affective illness: The probenecid technique. *Am. J. Psychiatry* **130**:73, 1973.
24. HAKIN, A. H., BOMB, B. S., PANDEY, S. K., SINGH, S. V.: A study of cerebrospinal fluid calcium and magnesium in depression. *J. Assoc. Physicians India* **23**:311, 1975.
25. HAMILTON, W. G., JR., BALLENGER, J. C., JIMERSON, D. C., HICKEY, T., POST, R. M., BUNNEY, W. E., JR.: Unpublished observations.
26. HAMMERSCHLAG, R., DRAVID, A. R., CHIU, A. Y.: Mechanism of axonal transport: A proposed role for calcium ions. *Science* **188**:273, 1975.
27. HARRIS, W. H., BEAUCHEMIN, J. A.: Cerebrospinal fluid calcium, magnesium, and their ratio in psychoses of organic and functional origin. *Yale J. Biol. Med.* **29**:117, 1956.
28. HECHT, A., GERSHBERG, H.: Primary hyperparathyroidism: Laboratory and clinical data in 73 cases. *J. Am. Med. Assoc.* **233**:519, 1975.
29. HEINEMANN, U., LUX, H. D., GUTNICK, M. J.: Extracellular free calcium and potassium during paroxysmal activity in the cerebral cortex of the cat. *Exp. Brain Res.* **27**:237–243, 1977.
30. HERBERT, F. K.: The total and diffusible calcium of serum and the calcium of cerebrospinal fluid in human cases of hypocalcemia and hypercalcemia. *Biochem. J.* **27**:1978–1991, 1933.
31. HUNTER, G., SMITH, H. V.: Calcium and magnesium in human cerebrospinal fluid. *Nature (London)* **186**:161, 1960.
32. JIMERSON, D. C., POST, R. M., CARMAN, J. S., VAN KAMMEN, D. P., WOOD, J. H., GOODWIN, F. K., BUNNEY, W. E., JR.: CSF calcium: Clinical correlates in affective illness and schizophrenia. *Biol. Psychiatry* **14**:37–51, 1979.
33. KATO, G., SOMJEN, G. G.: Effects of microionotophoretic administration of magnesium and calcium on neurons in the central nervous system of cats. *J. Neurobiol.* **2**:181, 1969.
34. KATZMAN, R., PAPPIUS, H. M.: *Brain Electrolyte and Fluid Metabolism*. Baltimore, Williams and Wilkins, 1973, pp. 246–264.
35. KATZENELBOGEN, S.: *The Cerebrospinal Fluid and the Blood*. Baltimore, Johns Hopkins Press, 1935, pp. 178–198.
36. KILLEY, J. S., KMJEVIC, K., SOMJEN, G.: Divalent cations and electrical properties of cortical cells. *J. Neurobiol.* **2**:197, 1969.
37. KNAPP, S., MANDELL, A. J., BULLARD, W. P.: Calcium activation of brain tryptophan hydroxylase. *Life Sci.* **16**:1583, 1975.
38. KNIGGE, K. M., MORRIS, M., SCOTT, D. E., JOSEPH, S. A., NOTTER, M., SCHOCK, D., KROBISIH-DUDLEY, G.: Distribution of hormones by cerebrospinal fluid. In Cserr, H. F., Fenstermacher, J. D., Fencl, V. (eds.): *Fluid Environment of the Brain*. New York, Academic Press, 1975, pp. 237–254.
39. LOW, J. C., SCHAAF, M., EARLL, J. M., PICCHOCKI, J. T., TING-KAI, L.: Ionic calcium determinations in primary hyperparathyroidism. *J. Am. Med. Assoc.* **223**:152, 1973.
40. LEUSEN, I.: Regulation of cerebrospinal fluid composition with reference to breathing. *Physiol. Rev.* **52**:1–56, 1972.
41. LOVENBERG, W., AMES, M. W., LERNER, P.: Mechanisms of short term regulation of tyrosine hydroxylase. In Lipton, M. A., DiMascio, A., Killam, K. F. (eds.): *Psychopharmacology: A Generation of Progress*. New York, Raven Press, 1978, pp. 247–259.
42. NICHOLSON, C., TEN BRUGGENCATE, G., STEINBERG, R., STACKLE, H.: Calcium modulation in brain extracellular microenvironment demonstrated with ion-selective micropipette. *Proc. Natl. Acad. Sci. U.S.A.* **74**:1287–1290, 1977.
43. PANDEY, S. K., DEVPURA, J. C., BEDI, H. K., BABEL, C. S.: An estimation of magnesium and calcium in serum and CSF in schizophrenia. *J. Assoc. Physicians India* **21**:203, 1973.
44. PAPPENHEIMER, J. R., HEISEY, S. R., JORDAN, E. F., DOWNER, J. C.: Perfusion of the cerebral ventricular system in unanesthetized goats. *Am. J. Physiol.* **203**:763, 1962.
45. RICHENS, A., ROWE, D. J. F.: Disturbance of calcium metabolism by anticonvulsant drugs. *Br. Med. J.* **4**:73–76, 1970.
46. RODRIGUEZ, E. M.: The cerebrospinal fluid as a pathway in neuroendocrine integration. *J. Endocrinol.* **71**:407–443, 1976.
47. ROSENBERG, P., BARTELS, E.: Drug effects on spontaneous electrical activity of the squid giant axon. *J. Pharmacol. Exp. Ther.* **155**:532, 1967.
48. ROSS, D. H., MEDINA, M. A., CARDENAS, H. L.: Morphine and ethanol: Selective depletion of regional brain calcium. *Science* **186**:63, 1974.
49. ROSS, D. H., LYNN, S. C., JR., CARDENAS, H. L.: Selective control of calcium levels by naloxone. *Life Sci.* **18**:789, 1976.
50. ROTH, R. H., SALZMAN, P. M., NOWYCKY, M. C.: Impulse flow and short-term regulation of transmitter biosynthesis in central catecholaminergic neurons. In Lipton, M. A., DiMascio, A., Killam, K. F. (eds.): *Psychopharmacology: A Generation of Progress*. New York, Raven Press, 1978, pp. 185–198.

51. RUBIN, R. P.: *Calcium and the Secretory Process.* New York, Plenum Press, 1974, pp. 25–66.
52. SCHOLBERG, H. A., GOODALL, E.: The phosphorus and calcium content of the blood plasma and cerebrospinal fluid in the psychoses. *J. Ment. Sci.* **72**:51, 1926.
53. SMITH, C. C., BARISH, J., CORREA, J., WILLIAMS, R. H.: Psychiatric disturbance in endocrinologic disease. *Psychosomat. Med.* **34**:69, 1972.
54. STEER, M. L., ATLAS, D., LEVITZKI, A.: Interrelations between β-adrenergic receptors, adenylate cyclase and calcium. *N. Engl. J. Med.* **292**:409, 1975.
55. UENO, Y., AUKI, N., YABUKI, T., KURAISHI, F.: Electrolyte metabolism in blood and cerebrospinal fluid in psychoses. *Folia Psychiatr. Neurol. Jpn.* **15**:304, 1961.
56. VEALE, W. L., MYERS, R. D.: Emotional behavior, arousal and sleep produced by sodium and calcium ions perfused within the hypothalamus of the cat. *Physiol. Behav.* **7**:601, 1971.
57. WATSON, L.: Clinical aspects of hyperparathyroidism. *Proc. R. Soc. Med.* **61**:1123, 1968.
58. WEHR, T. A., LEWY, A. J.: Circadian rhythms and affective illness. In Alexander, P. (ed.): *Electrolytes and Neuropsychiatric Disorders.* New York, Spectrum (in press).
59. WESTON, P. G., HOWARD, M. Q.: The determination of sodium, potassium, calcium, and magnesium in the blood and spinal fluid of patients suffering from manic–depressive insanity. *Arch. Neurol. Psychiatry* **8**:179, 1922.
60. ZUCKERMAN, E. C., GLASER, G. H.: Anticonvulsive action of increased calcium concentration in cerebrospinal fluid. *Arch. Neurol.* **29**:245, 1973.

Index